RELATIVE ATOMIC WEIGHTS (1965)

Based on the nuclidic mass $C^{12} = 12$

Name	*Symbol*	*Atomic Number*	*Atomic Weight*
Actinium	Ac	89	—
Aluminium	Al	13	26.9815
Americium	Am	95	—
Antimony	Sb	51	121.75
Argon	Ar	18	39.948
Arsenic	As	33	74.9216
Astatine	At	85	—
Barium	Ba	56	137.34
Berkelium	Bk	97	—
Beryllium	Be	4	9.0122
Bismuth	Bi	83	208.980
Boron	B	5	10.811 ±0.003[a]
Bromine	Br	35	79.904[b]
Cadmium	Cd	48	112.40
Calcium	Ca	20	40.08
Californium	Cf	98	—
Carbon	C	6	12.01115 ±0.00005[a]
Cerium	Ce	58	140.12
Cesium	Cs	55	132.905
Chlorine	Cl	17	35.453[b]
Chromium	Cr	24	51.996[b]
Cobalt	Co	27	58.9332
Copper	Cu	29	63.546[b]
Curium	Cm	96	—
Dysprosium	Dy	66	162.50
Einsteinium	Es	99	—
Erbium	Er	68	167.26
Europium	Eu	63	151.96
Fermium	Fm	100	—
Fluorine	F	9	18.9984
Francium	Fr	87	—
Gadolinium	Gd	64	157.25
Gallium	Ga	31	69.72
Germanium	Ge	32	72.59
Gold	Au	79	196.967
Hafnium	Hf	72	178.49
Helium	He	2	4.0026
Holmium	Ho	67	164.930
Hydrogen	H	1	1.00797 ±0.00001[a]
Indium	In	49	114.82
Iodine	I	53	126.9044
Iridium	Ir	77	192.2
Iron	Fe	26	55.847[b]
Krypton	Kr	36	83.80
Lanthanum	La	57	138.91
Lead	Pb	82	207.19
Lithium	Li	3	6.939
Lutetium	Lu	71	174.97
Magnesium	Mg	12	24.312
Manganese	Mn	25	54.9381
Mendelevium	Md	101	—
Mercury	Hg	80	200.59
Molybdenum	Mo	42	95.94
Neodymium	Nd	60	144.24
Neon	Ne	10	20.183
Neptunium	Np	93	—
Nickel	Ni	28	58.71
Niobium	Nb	41	92.906
Nitrogen	N	7	14.0067
Nobelium	No	102	—
Osmium	Os	76	190.2
Oxygen	O	8	15.9994 ±0.0001[a]
Palladium	Pd	46	106.4
Phosphorus	P	15	30.9738
Platinum	Pt	78	195.09
Plutonium	Pu	94	—
Polonium	Po	84	—
Potassium	K	19	39.102
Praseodymium	Pr	59	140.907
Promethium	Pm	61	—
Protactinium	Pa	91	—
Radium	Ra	88	—
Radon	Rn	86	—
Rhenium	Re	75	186.2
Rhodium	Rh	45	102.905
Rubidium	Rb	37	85.47
Ruthenium	Ru	44	101.07
Samarium	Sm	62	150.35
Scandium	Sc	21	44.956
Selenium	Se	34	78.96
Silicon	Si	14	28.086 ±0.001[a]
Silver	Ag	47	107.880[b]
Sodium	Na	11	22.9898
Strontium	Sr	38	87.62
Sulfur	S	16	32.064 ±0.003[a]
Tantalum	Ta	73	180.948
Technetium	Tc	43	—
Tellurium	Te	52	127.60
Terbium	Tb	65	158.924
Thallium	Tl	81	204.37
Thorium	Th	90	232.038
Thulium	Tm	69	168.934
Tin	Sn	50	118.69
Titanium	Ti	22	47.90
Tungsten	W	74	183.85
Uranium	U	92	238.03
Vanadium	V	23	50.942
Xenon	Xe	54	131.30
Ytterbium	Yb	70	173.04
Yttrium	Y	39	88.905
Zinc	Zn	30	65.37
Zirconium	Zr	40	91.22

[a] Atomic weights so designated are known to be variable because of natural variations in isotopic composition. The observed ranges are:

Hydrogen	±0.00001	Oxygen	±0.0001
Boron	±0.003	Silicon	±0.001
Carbon	±0.0005	Sulfur	±0.003

[b] Atomic weights so designated are believed to have the following experimental uncertainties:

Chlorine	±0.001	Iron	±0.003
Chromium	±0.001	Bromine	±0.001
Copper	±0.001	Silver	±0.001

For other elements the last digit is believed to be reliable to ±0.5.

WITHDRAWN
University of
Illinois Library
at Urbana-Champaign

Quantitative Chemical Analysis

QUANTITATIVE CHEMICAL ANALYSIS

FOURTH EDITION

I. M. KOLTHOFF
Professor Emeritus of Analytical Chemistry
Formerly Professor and Chief
Division of Analytical Chemistry
University of Minnesota

E. B. SANDELL
Professor of Analytical Chemistry
University of Minnesota

E. J. MEEHAN
Professor of Analytical Chemistry
University of Minnesota

STANLEY BRUCKENSTEIN
Professor of Chemistry
State University of New York at Buffalo

The Macmillan Company
Collier–Macmillan Limited *London*

First Printing

Earlier editions, entitled *Textbook of Quantitative Inorganic Analysis*, by I. M. Kolthoff and E. B. Sandell, copyright 1936, 1943, and 1952 by The Macmillan Company.

Library of Congress catalog card number: 69-10291

The Macmillan Company
Collier–Macmillan Canada, Ltd., *Toronto, Ontario*

Printed in the United States of America

Preface to the Fourth Edition

It is a very hard undertaking to seek to please everybody.—
PUBLILIUS SYRUS, *Maxim 675.*

The objectives and the organization of the fourth edition of this text remain much the same as in the earlier editions. Because of the introduction of some material that strictly does not fall under the heading of inorganic analysis, the title of the book has been changed. But most of its contents still deal with inorganic analysis or principles underlying both inorganic and organic analysis.

Among the additions that have been made in the revision, the following may be mentioned. The treatment of acid–base equilibria now includes a chapter dealing with these equilibria on a more advanced level; also, an introduction to acid–base equilibria in nonaqueous media is offered. New chapters have been needed to cover complex equilibria and compleximetric titrations adequately, with particular reference to metal–chelate and metal ion–indicator equilibria. Analytical separations, formerly discussed in one chapter, now require four. No one will be surprised to see that the section of the text dealing with physicochemical methods of analysis has been greatly expanded; familiarity with such methods is today required of every student majoring in chemistry. A discussion of the principles of electroanalysis and optical methods of analysis is followed by an experimental section in which a limited number of determinations are presented.

It seems to us that the scope of this textbook does not (and could not) call for a detailed exposé of all physicochemical and physical methods of analysis, but a number of such methods are surveyed. No doubt it is praiseworthy for a student to be familiar with nuclear magnetic resonance, for example, but there are many more mundane topics with which he should be acquainted before beginning the ascent to the more rarefied levels. We would be remiss if we slighted the fundamentals of quantitative analysis. We have encountered *graduate* students majoring in analytical chemistry who have not been able to calculate the calcium content of a sample from the titration of calcium oxalate with cerium(IV) or to detect chromium in chromite.

Some chapters have been set in reduced type to save space. To a large extent, these chapters cover material of a more or less advanced nature or that which is informational. Some of this material will need to be included in an elementary course, but since the scope of such a course will vary from one school to another, we have not attempted to indicate by type size the material that should or should not be included, which, after all, is a matter to be decided by the instructor. Proper attention to laboratory practice is always important and we have not neglected this aspect of the subject.

In the section on gravimetric analysis, as well as in other sections, more material is included than will be covered in any one course. This has been done in the belief that superior students, with a genuine interest in analysis, are likely to read chapters not formally assigned, and thus widen their knowledge. Moreover, when the treatment is not too parsimonious, the instructor can choose determinations most suitable for his course, and may, if he wishes, change determinations from one year to another. Need we say, in passing, that it is still essential for any but the most narrow of modern analysts to have some knowledge of *chemical* methods of analysis, either alone or in conjunction with physicochemical or physical methods? For example, chemical separations must often be applied in neutron activation analysis; indeed, gravimetric analysis may be a necessary part of such an analysis.

Analytical chemistry remains an essential part of a good chemistry curriculum. It is indispensable not only to chemistry but to other sciences and their applications as well. Believing this, we have thought it worth while once again to revise this book. We hope that the apportionment of space between theory and practice, and between the old and the new in analysis, represents a sound balance for instruction today. As much as anyone we regret the length of the revised edition, but the scope of modern chemical analysis and the need, as we see it, to treat the fundamentals with some thoroughness for the benefit of the serious student have not permitted a shorter treatment.

We are grateful to the many authors and editors of books, papers, and journals who have allowed us to use material in figures and tables. The source of the material has been indicated in each case. We are also indebted to the Arthur H. Thomas Company for the loan of cuts to illustrate certain laboratory apparatus.

Over the years, many users of the previous editions of this book have been very helpful and kind in offering comments and suggestions. We express our thanks for this interest and aid and welcome comments on this new edition.

The Authors

Minneapolis

From the Preface of the First Edition

An elementary textbook in quantitative analysis is necessarily limited in its scope, and many subjects of importance in this branch of chemistry cannot be treated. On the other hand, the advanced works in the field usually deal with highly specialized branches of the subject or selected topics, and there is little opportunity for the discussion of the broad fundamentals in a satisfactory way, at least not without repetition of much elementary matter which belongs in an introductory course. It seemed desirable to the authors to have available a book that could be used as an introductory text and which in addition would have the more or less comprehensive character of an advanced textbook, so that it would be suitable for use in beginning and advanced courses in analytical chemistry. With this aim in mind the authors have written the present work, in which there is contained material designed to make the student familiar with the fundamental theories of the subject, the practical methods of working, and the most important classical procedures. Highly specialized methods of analysis based on the measurement of physicochemical properties (conductometric and potentiometric titrations, gas analysis, nephelometry, spectrography, etc.) could naturally not be treated at length, but enough is said concerning these to enable the student to appreciate the advantages of such methods, and a sufficient number of references is given to serve as a starting point for him who is interested.

In writing this text, it has been our aim to offer a balanced outline of the theoretical and practical aspects of inorganic quantitative analysis. There appears to be a tendency to exaggerate the significance of "theory" at the expense of practical work in chemical analysis.

Theory and practice should go hand in hand. Many chemical students are faced in later life with analytical problems, of which the solution is not to be found in reference works, but which can be solved by a rational and intelligent application of the principles of the subject.

Taking into account the difficulties that the beginning student must encounter on being introduced to a subject which is entirely new to him, it seems to the authors that from the didactic point of view, gravimetric analysis should precede volumetric analysis. The theory of the former is much simpler than that of the latter. Therefore in beginning with gravimetric analysis the instructor can gradually develop the theoretical fundamentals of quantitative analysis without overburdening the student at the start. In general it is inadvisable to have the student make determinations before the theory underlying them has been offered.

No training in analytical chemistry can be considered satisfactory which does not give the student some opportunity to familiarize himself with the journal literature and standard reference books. For this reason rather frequent literature citations are given in this text. In advanced courses especially, the student should be encouraged to refer frequently to the original sources.

It does not seem to be generally realized that analytical chemistry is one of the fundamental branches of the science. The authors express the hope that books of the nature of the present one will aid in the recognition of the importance of quantitative analysis.

I. M. Kolthoff
E. B. Sandell

Minneapolis

Contents

Part Three
GRAVIMETRIC ANALYSIS 563

Part Four
THEORY OF VOLUMETRIC ANALYSIS 679

Part Five
VOLUMETRIC DETERMINATIONS 775

Part Six
THEORY OF ELECTROANALYTICAL CHEMISTRY 865

APPENDIX 1141

PART ONE

Theoretical Fundamentals

CHAPTER 1 *Introduction*

Analytical chemistry, or the art of recognizing different substances and determining their constituents, takes a prominent position among the applications of the science, since the questions which it enables us to answer arise wherever chemical processes are employed for scientific or technical purposes. Its supreme importance has caused it to be assiduously cultivated from a very early period in the history of chemistry, and its records comprise a large part of the quantitative work which is spread over the whole domain of the science.

—Wilhelm Ostwald (1894)

Qualitative and quantitative analysis

Qualitative chemical analysis deals with the detection and identification of the constituents of a sample, quantitative chemical analysis with the determination of their amounts (usually relative amounts expressed as percentages). By *sample* is meant that portion of the material in question which is examined. The constituents to be detected or determined may be elements with or without regard to their state of combination or oxidation, radicals, functional groups (as in organic substances), compounds, or even phases which may or may not be compounds.

The greater part by far of chemical analysis is quantitative. More often than not, the qualitative composition of a sample is known from its source and a quantitative analysis can be carried out directly. Occasionally the qualitative composition of a sample may be unknown or imperfectly known, and it must be analyzed qualitatively before a quantitative analysis can be made. Qualitative tests may entail identification of the form of an element, as whether iron is present as Fe(II) or Fe(III), or sulfur is present as sulfide or sulfate. All qualitative tests are in some degree quantitative and give an indication of the amount of reacting constituent, whether by the quantity of a precipitate, the intensity of a color, the density of a spectral line on a photographic plate, etc. A qualitative analysis providing a rough indication of the amounts of the constituents present may be of great value in the choice of methods for their determination. Qualitative tests may also be of importance in revealing the presence of unexpected elements in a sample, which could lead to gross errors if disregarded in the quantitative procedure. The familiar "wet" scheme of inorganic qualitative analysis, modernized with organic reagent spot reactions, can be of good service in occasional qualitative analyses, which usually need not be complete. A spectrographic analysis provides a quick and sensitive means of recognizing many elements (particularly the metals) and allows an estimate to be made of their relative amounts. Qualitative organic analysis is likely to require the identification of *functional groups*, which may be roughly defined as reactive atoms or groups of atoms, such as carbonyl $\left(\rangle CO\right)$, carboxyl (—COOH), azo (—N=N—), and one or two hundred others. Sometimes, the detection of an element, such as S or Cl, without regard to its form may suffice.

Qualitative tests of a nondestructive nature (not altering the physical or chemical identity of the sample or its amount) can be of great value. It may be of vital importance to learn whether or not a solid sample consists of more than one component (phase) and to obtain the identity of the components. A spectrographic or wet analysis might provide the gross composition of the sample but still leave the analyst ignorant of the actual composition, i.e., the number of components and their composition. Physical properties must, in general, be made the basis of such analyses. Microscopic examination (especially with a polarizing microscope) can show whether a finely divided solid is homogeneous or heterogeneous, and often permits identification of the components by their optical properties or from the measurement of their refractive indices. X-ray diffraction is of great value in the identification of solid phases. Simple tests such as determination of freezing point, boiling point, and refractive index can be of aid in identifying liquids.

The scope of quantitative chemical analysis

Science is essentially the study of quantitative relationships, and when these relationships involve the composition of substances, quantitative analysis must be brought in. It has played, and continues to play, a vital role in the development of chemistry and other sciences making use of chemistry. The formulation of the fundamental laws of chemistry has been based largely on the results of quantitative analyses. Our knowledge of the composition of the earth and all it contains and even of extraterrestrial matter has come from chemical analysis in one form or another. It has been said: "The sceptical chemist, and there is no other variety worthy of the name, draws conclusions regarding chemical materials under his scrutiny chiefly on the basis of quantitative chemical analysis, which in its proximate and ultimate forms, is the touchstone of all chemical hypothesis" (G. T. Morgan).

Chemical analysis is indispensable in many branches of science—in the earth sciences, in the biological sciences, even in archeology. Moreover, it finds constant application in technology and industry—in the analysis of raw materials and finished products, in the control of manufacturing processes—as well as in clinical medicine, agriculture, and other fields.

Quantitative chemical analysis is customarily divided into inorganic and organic. Both branches may be said to be based on essentially the same, or at least similar, fundamentals. However, just as inorganic analysis presupposes a knowledge of inorganic or general chemistry, so quantitative organic analysis presupposes some knowledge of organic chemistry and preferably of qualitative organic analysis. We do not on the whole[1] presuppose a knowledge of organic chemistry on the part of the student. Accordingly, this text deals almost wholly with inorganic or general quantitative analysis. The general principles of analysis can be illustrated very well by inorganic examples. Our treatment is in no way intended to belittle organic

[1] Certain sections of this book, as those on organic reagents and liquid–liquid extraction, can be read more meaningfully with an understanding of organic chemistry.

analysis, which in volume and perhaps in the importance of its applications over-shadows purely inorganic analysis. But organic quantitative analysis is best reserved for a later course, where it can be given full justice. A fair familiarity with inorganic analysis is a prerequisite for organic analysis.

The extent of modern inorganic analysis is so great that the whole subject cannot be treated, or even surveyed, in an introductory text such as this. Nor should it be. We shall stress the theoretical principles, especially of gravimetric and titrimetric analysis (which are still of fundamental importance), and illustrate them with a limited number of laboratory exercises. Some of these relate to samples of more or less complex composition, so that the aura of unreality tending to envelope academic analysis does not become all-pervasive. A number of physicochemical methods of analysis will be presented to demonstrate the usefulness of this type of analysis.

The general analytical process

The principal steps in an inorganic determination[2] may be outlined as follows:

1. Obtaining a representative sample of the material whose composition is desired. This may be no simple task when it is realized that a few tenths of a gram of material taken for analysis may be required to represent tons of parent material.
2. Bringing the weighed sample into solution or otherwise preparing it for analysis. In determinations by chemical methods the sample must in general be brought into complete solution. Dissolving the sample may not be necessary in determinations by physical methods, but some of these require other preparation of the sample.
3. Separating constituents that interfere in the determination of the desired constituent. This is an operation that is frequently required in applied analysis by chemical, and also by some physicochemical, and even physical, methods. Few if any methods are truly specific. Whenever possible, interfering substances are made innocuous by complexing or, less often, by a change in oxidation state. When such measures fail, separations must be made.
4. Carrying out the determination proper. The classification of methods of determination is considered in the next section.
5. Calculating the results. This step may include an evaluation of the precision and accuracy of the numerical result.

The methods of quantitative analysis

In quantitative chemical analysis we are almost always concerned with the relative amounts of constituents in a sample. The relative amounts, usually expressed as weight percentages, are obtained by finding the absolute amounts in a known amount of sample. A great variety of methods are available for this purpose. All methods of quantitative analysis are physical in the sense that the final step in a determination must involve the measurement of some physical property, the simplest being mass. However, in a more significant sense a distinction

[2] A point of terminology (and semantics): *Determination* is the process by which the amount of a given constituent is found in a sample. We say determination of calcium, not analysis *of* calcium. The latter means the determination of the constituents (there might be many) in a calcium sample. Determination of calcium is analysis *for* calcium. The term "determination" has also acquired, in some contexts, the connotation of the final step in an analysis, which has been preceded by separations or other treatment, and may then be used pejoratively (p. 10).

can be made between physical (including physicochemical) and chemical methods to give, in general, two classes of methods[3]:

Chemical methods. As the name implies, the methods of this class depend on the application of a chemical reaction involving the constituent being determined. They are distinguished from others by being ideally based on a stoichiometric reaction:

$$\underset{x}{mC} + \underset{v}{nR} \rightarrow \underset{w}{C_mR_n}$$

where C is the constituent and R is the chemical reagent. We can find the amount, x, of C from the amount, w, of the product C_mR_n formed, or from the amount, v, of R required to react with the constituent. When the amount of the reaction product is found by weighing we speak of *gravimetric analysis*. Usually the reaction product is isolated in the form of a very slightly soluble compound which is separated from the solution by filtration, washed free of soluble substances, and weighed after drying or conversion into another compound of definite composition. Since the atomic weights of the elements involved are known, the weight of C is readily calculated.

When the amount of constituent is found from the amount of the reagent required to react with it, we speak of *titrimetric analysis*. The process (titration) of finding the amount of reagent chemically equivalent to the constituent consists of incremental addition of the former as a solution of known concentration (standard solution) until an end point is reached (or found graphically). Because the quantity of reagent required is usually obtained by measuring the volume of the standard solution, this type of analysis is often called *volumetric analysis*.

Gravimetric and volumetric methods were the first to be used in quantitative analysis and they are sometimes referred to as *classical methods*.

If the reaction product is a gas, its amount can be found from the volume of the gas at known temperature and pressure. This is *gasometric analysis*.

Physical and physicochemical methods. If no chemical reaction is involved in a determination, the method is plainly physical. There are many physical properties which are functions either of the mass of a substance or of its concentration. For example, the color intensity of a solution of a permanganate is proportional to the concentration of MnO_4^-. Therefore, by measuring the intensity in a suitable way we can determine the concentration of manganese and therefrom its quantity. Usually the manganese would be present originally in the divalent state and it would be necessary to oxidize it to permanganate. This is a physicochemical method. A chemical reaction is an essential part of the method.

With few exceptions, physical and physicochemical methods are instrumental methods in the sense that they require the use of instruments other than the balance and buret. If a chemical reaction is involved it need not be stoichiometric, although this is always desirable. Physical methods are concerned largely with energy, in contrast to chemical methods, which are concerned with mass. The most important physical and physicochemical methods are those based on emission and absorption of radiation.

[3] There is a third class, the biological (or more accurately the microbiological) in which trace elements are usually determined by their effect on the growth rate of microorganisms. Compared to the other two classes, it is of minor importance.

The middle decades of the present century have been a period of spectacular development and application of instrumental methods in chemical analysis. The advantages of these methods lie partially in the possibility of carrying out analyses that earlier could not be made readily, if at all, and partially in their greater speed. Besides, what human being does not like to perform a near-miracle by not doing much more than pushing a button or turning a knob? It should not be concluded, however that the rise of instrumental methods has made classical analysis passé. The latter still play an important, even indispensable, role in modern analysis, although they have given much ground to instrumental methods. For most elements their accuracy is unexcelled as long as the constituent is not present in very small amounts. Moreover, they are absolute methods and do not as a rule require the use of standards, except a pure substance for standardization in titrimetric analysis. Many, if not most, instrumental methods require a series of standards, often with a gross composition resembling that of the sample, for calibration. If the standards cannot easily be prepared synthetically, they must be analyzed accurately, more likely than not by classical methods. Furthermore, some instrumental methods require the sample to be subjected to considerable chemical treatment, in which the operations of classical "wet analysis" are applied. For example, neutron-activation determination of trace elements normally involves the equivalent of a gravimetric determination (in fact, two, including the standard), with preceding separations.[4] It lies within the bounds of possibility that at some uncertain time in the future, analyses will be made almost entirely by purely physical methods, but today other methods are also necessary and cannot be ignored.[5]

The scale of analysis

An important consideration in chemical analysis is the size of sample needed. Quite often the amount of sample is limited (criminological, archeological investigations; examination of works of art; analysis of rare minerals, scarce bio material, corrosion products, and many others). Methods requiring decigram amounts of materials are of no use when only milligrams are available. One of the most significant advances in chemical analysis in the last half century has been the steady development of methods allowing the analysis of smaller and smaller samples and the determination of lower and lower percentages of constituents.

A macro sample may be defined as one larger than 0.1 g (the usual range is 0.1–1 g) and the method (or procedure) applied in the analysis of such a sample may be described as a macro method. A meso (semimicro) sample may be taken

[4] On the other hand, in other instrumental methods, such as optical and X-ray spectrography, chemical separations are rarely made, except sometimes for a preliminary concentration of trace constituents. Such methods will of course be favored for their rapidity and simplicity, if for no other reasons. In industrial laboratories, where many analyses must be made in a short time, instrumental methods of this type are strongly favored. The saving in labor costs soon pays for expensive instruments.

[5] Occasionally one hears the opinion expressed that students should not be exposed to classical methods of analysis because these methods are tedious and lack excitement. It is not wholly unexpected that in our age sybaritic considerations should be invoked in the choice of a method. Others may feel that the selection of an analytical method for a particular purpose should be determined primarily by its pertinence and utility, and that its pleasure-producing capabilities are of secondary or tertiary importance.

as one lying in the range 0.01–0.1 g, and a micro sample as one generally smaller than 0.01 g. A finer classification of the latter is

Micro: 0.001–0.01 g
Submicro: <0.001 g (sometimes called ultramicro; the terminology is not yet fixed)

Macro, meso, and micro methods and procedures may accordingly be roughly described as those calling for, respectively, decigram, centigram, and milligram (or smaller) samples.

The relative amount (concentration in the sample) of the constituent is also of importance in the classification of methods. Major, minor, and trace constituents may be assigned the respective ranges 100–1, 1–0.01, and <0.01 per cent. A trace analysis is any analysis in which less than ~0.01 per cent of constituent is determined; the size of the sample is immaterial. A systematic classification of analytical methods requires specification of both the size of sample and the relative amount of constituent (Fig. 1-1). The first term in the bipartite designations in the figure can be replaced by the corresponding sample-weight range. Thus a decigram-trace (macrotrace) method is one in which a sample in the approximate range 0.1–1 g can be used to determine less than 0.01 per cent of a constituent. The product of

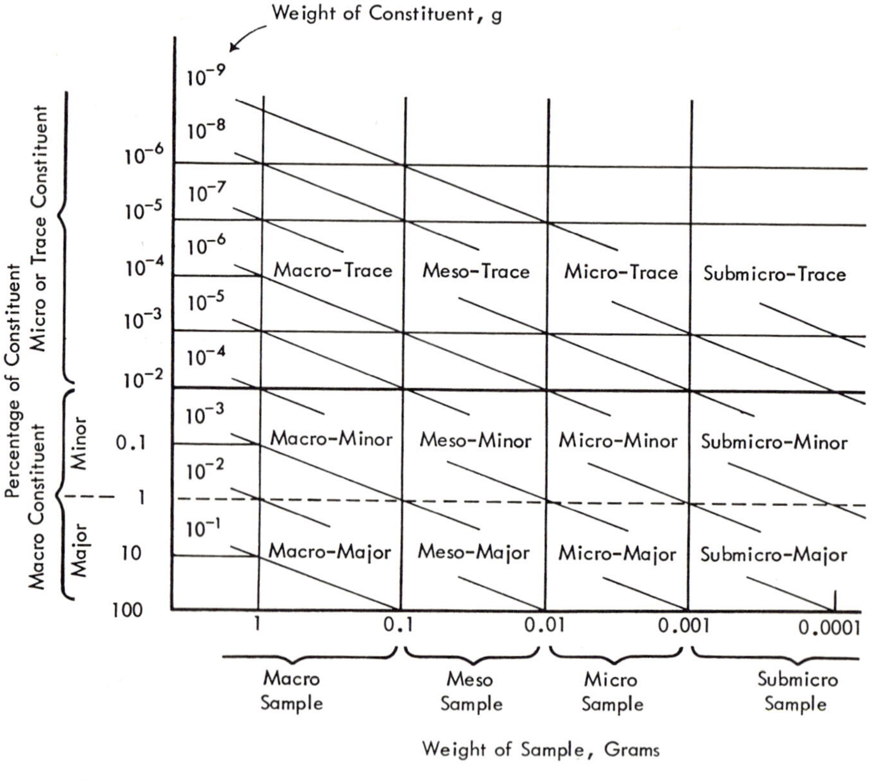

Fig. 1.1 Classification of analytical methods (or procedures) on the basis of the size of sample and relative amount (percentage) of constituent (From E. B. Sandell, *Colorimetric Determination of Traces of Metals*, 3rd ed., Wiley-Interscience, New York, 1959.)

the two numbers furnishes the weight of the constituent being determined. The weight will be the same in a series of diagonal classes but the required accuracy will vary, being relatively greater when a macro constituent must be determined.

Chemical analysis and analytical chemistry

Chemical analysis in its broader aspects is known as analytical chemistry. Some general observations on these topics and their relations follow.

There are various levels of analysis and concomitantly required abilities of the analyst. At the lowest, in carrying out a simple determination according to detailed directions, the operator at the worst is hardly more than an automaton and may be replaceable by a suitable machine. The minimum requirements here are manipulative skill and care, in short, good technique, which can be acquired in a reasonable length of time by many persons through practice and experience. Even at this level, however, something more than an ability to follow directions is desirable or perhaps necessary. Analysis by a technician or meter reader who does not understand the chemistry and physics of the analytical process is a risky affair. He requires supervision or, perhaps we should say, watching. Disturbing factors may enter even a simple determination and not be recognized by an operator deficient in basic knowledge or experience.

At a higher level we have the analyst who can decide what methods to apply in the analysis of a submitted sample whose composition is qualitatively known, and who also has sufficient skill and understanding to make the analysis, which may not always be a simple one. The choice of a method may or may not be easy. Some analyses are very well standardized. For example, analyses of water, agricultural materials, most ferrous and nonferrous alloys, ores, and many other materials of well-characterized composition have been thoroughly worked out. The analyst will usually find the information he needs for the analysis of such common samples in collections of standard methods, treatises, and monographs. At times he will need to consult the analytical journals, probably prefacing this with a search of the abstract journals, for newer methods, for methods for new materials, or for supplementary information. In general, he must choose a method on the basis of required accuracy and allowable time, within the limits of available facilities. Economy of time and effort is almost always an important factor in analysis. In general, the time expended on an analysis rises with the increase in accuracy, often in an unfavorable ratio, and the most accurate methods are not likely to be chosen unless they are really needed. It is highly desirable that the analyst have some first-hand experience with the methods among which a choice is to be made. Otherwise, he may not be able to say what accuracy can rightfully be expected or the length of time an analysis will require. Besides, to attempt to carry out an analysis by a method with which the analyst has had no previous experience is a hazardous undertaking, not to be embarked upon lightly.

The preceding problem grades into one of greater difficulty, that of modifying published procedures and devising new ones. New manmade products appear in which the less common elements occur with increasing frequency and for which available methods are inadequate; natural materials, such as minerals, may present complex combinations not explicitly resolvable by published methods. Moreover, it may be necessary to push the determination of many constituents to much lower limits than present methods permit. A wide familiarity with analytical methods in general and the chemistry of individual elements in particular is a prerequisite to the solution of problems of this kind. The analyst must also be well acquainted with separation methods. Tabulated data and constants, such as complexity constants, extraction equilibrium constants, etc., may be of great aid in planning new procedures or modifying old ones. This is only a first step, however. Procedures worked out on paper must always be tested experimentally. The theoretical approach indicates promising possibilities, and minimizes laboratory work.

Analysis versus determinations. There is a difference between making a simple determination and carrying out a real analysis in which one or more constituents must be determined in a sample of more or less complex composition.[6] This vital fact is not always made evident to students in an elementary course or even to those in so-called advanced courses, who may be left with the impression that analysis is after all rather simple, requiring nothing more than a routine precipitation or titration or perhaps turning a knob or twirling a dial or two.

Partly because of lack of time and partly because of current teaching styles and philosophies, few fresh Ph.D.'s with majors in analytical chemistry have come into contact with analytical realities. To a large extent their laboratory work and thinking has been limited to academic exercises and a specialized research which may have a peripheral relation to analysis. This state of affairs is perhaps to a large extent, but by no means wholly, unavoidable. Somewhere along the line, a student should be made aware that there is a difference between analyzing a pure solution or a synthetic mixture of salts handed to him by a laboratory instructor, on the one hand, and a real, nonfictitious sample, on the other. Attempts to simplify complexities by ignoring them are likely to result in unsound solutions. Even in an elementary course, such steps as preparing and decomposing a sample and making a separation or two can be incorporated to make a determination, in some measure, an analysis. In advanced courses much more should be done to simulate real analysis. The aim is not to turn out finished analysts. There would be little likelihood of that, even if it were attempted.

Erroneous concepts concerning analytical chemistry are not uncommon, and, it may be added, are not confined to nonchemists:

> *Today, very few analyses are made by those who need them, and many who need analyses have but little knowledge of the difficulties that may be involved, the labor that must be expended, or the results that can rightfully be expected.*
>
> *In the industrial field, a contributing factor to the present-day misconception of chemical analysis can be traced to the courses in so-called chemical analysis that must be taken by engineering students in many of our colleges and universities. Many of these students later employ analysts. The courses necessarily deal with chemical determinations that are so simple that they do not deserve the label "chemical analysis" at all, and leave the student with an entirely erroneous conception of what chemical analysis, particularly applied chemical analysis, is like. . . . The operations of chemical analysis may be divided into those that make the determination possible, and those that attend the determination itself. What is not generally understood is that in applied analysis the former are often by far the most important. The relation of the analysis to the determination might be compared with that between the art of fishing and the measurement of the fish. The catching of the fish may be quite difficult and*

[6] Serious students should read a notable article on the analysis of things as they are by the master analytical chemist G. E. F. Lundell, *Ind. Eng. Chem., Anal. Ed.* **5**, 221 (1933).

require considerable planning and patience. On the other hand, almost anyone can measure or weigh the fish after it has been caught.

In the schools, a contributing factor in the divorce of chemical determinations from chemical analysis was a swing from the applied toward the theoretical aspects of analytical chemistry which took place toward the close of the last century. In this, the cart was put before the horse, the determination was elected to pull the analysis instead of the analysis to pull the determination. This left such illustrious analysts as Hillebrand in the category of "cook-book" chemists, and attempted to substitute a strange hybrid disporting in a field of his own making, a field that was chiefly physical chemistry, and that ignored inorganic chemistry, which is the proving ground of analytical chemistry. The chemist who was working in the field of applied analysis was told he was using primitive methods and uncalibrated tools, and was given a map of the promised land free from the weeds of ignorance and ripe for tilling. Some have never come back to the old homestead, some have returned with a sigh of relief, and all agree that it was an "experience."[7]

Analytical chemical research. There are various types of analytical research, but the ultimate purpose in all is to improve existing methods of analysis or to devise new ones. We have hinted at problems arising in applied analysis which require specialized research. On the other hand, some investigations may be quite general in nature and may be undertaken to lay the groundwork for subsequent development of new methods. In other words, this type of research can take the form of studies of systems or processes on which analysis can be based. Thus the equilibria between species in a solution, the distribution of a substance between two immiscible solvents, the nature and properties of the reaction product of an organic reagent with a cation, the light-scattering properties of a suspension of a slightly soluble substance, and rates of reaction are all examples of subjects that the analytical chemist may wish to study because they are involved in actual or potential analytical methods. Such studies may supply information of interest to chemists other than the analytical, but they are carried out by the analytical chemist with the thought of analytical utilization, as means toward an end, and with an attack appropriate to that end.

Obviously, this type of research is, or may be, important or even essential. But what is necessary is not necessarily sufficient. It is only the first step in the development of a method. Studies of this type are a favorite form of academic research. They almost always yield results of some kind. They generally are not problems in the sense that a particular solution must be obtained. This being so, there is the danger that this kind of research can degenerate into studies of no particular importance from the analytical standpoint.

The analytical chemist of broad interests attempts to keep abreast of discoveries and developments in the other branches of chemistry and other sciences such as physics.

Sometimes the central idea on which an investigation hinges is provided by the application or transfer of a new principle or technique which has been discovered in another field. . . . This is probably the most fruitful and the easiest method in research, and the one most employed in applied research. It is, however, not to be in any way despised. Scientific advances are so difficult to achieve that every useful stratagem must be used.[8]

[7] G. E. F. Lundell, *Proc. ASTM*, **42**, 34 (1942).
[8] W. I. B. Beveridge, *The Art of Scientific Investigation,* Norton, New York, 1957.

It is curious to note how slow analytical chemists have often been to utilize discoveries or observations made in other branches of chemistry and even in their own. For example, the synthesis of diphenylthiocarbazone (dithizone) and its reaction with heavy metals were described in 1878, but no analytical use was made of this valuable colorimetric reagent until the 1920s. The differential adsorption of solutes by cellulose (filter paper) was observed at least as early as the 1850s, and although some attempts were made to apply this effect analytically in the next 50–60 years, it was not until the 1940s that paper chromatography began to be developed and used as an important separation method. The history of the use of ion-exchange materials is similar. The value of the combustion of organic samples in a flask filled with oxygen for elementary analysis (determination of Cl, S, P, even C) was not appreciated until the 1950s, although the same method had been applied to determine S in coal in 1892. It appears that it is not unusual to have an induction period of 50 years following a significant discovery or a pioneering essay before its use by thousands of analysts, while analytical chemists in the meantime have been belaboring topics of lesser importance.

Current trends. Finally, to conclude this introductory survey of analytical chemistry, it may be of interest to list the fields of activity of analytical chemists in the United States in the mid-1960s, as sampled by the American Chemical Society in a study based on information gathered by the National Science Foundation in 1966 (Table 1-1).

TABLE 1-1. **Fields of Activity of U.S. Analytical Chemists in 1966**[a]

Specialty	*Per cent*	*Estimated Number of Chemists*
Absorption spectroscopy	17.5	3,600
Biochemical analysis	4.4	900
Chemical microscopy	0.7	100
Chromatographic analysis	15.9	3,300
Distillation analysis	0.4	100
Electrochemical analysis	5.0	1,000
Electron microscopy	0.7	100
Emission spectroscopy	5.7	1,200
Extraction analysis	1.2	200
Fluorimetry, phosphorimetry, and Raman spectroscopy	0.6	100
Gas analysis	2.2	500
Gravimetry	8.3	1,700
Magnetic resonance spectroscopy	2.6	500
Mass spectroscopy	2.9	600
Microchemical analysis	3.7	800
Nucleonics and radiochemistry	4.7	1,000
Qualitative analysis	5.8	1,200
Titrimetry	6.8	1,400
X-ray and electron diffraction	2.8	600
Other	7.9	1,600
Total		20,500

[a] From *Chem. Eng. News*, June 19, 1967, p. 56.

REFERENCE WORKS IN QUANTITATIVE ANALYSIS
BOOKS

The following list is not intended to be complete. Books which are primarily textbooks are not included. Also see references in various chapters. With a few exceptions, books published before 1940 are not included.

I. General works on quantitative analysis

A. Theory of quantitative analysis

G. CHARLOT and R. GAUGIN, *Les Méthodes d'analyse des réactions en solution*, Masson et Cie, Paris, 1951.

H. A. LAITINEN, *Chemical Analysis*, McGraw-Hill, New York, 1960.

T. B. SMITH, *Analytical Processes*, E. Arnold, London, 2nd ed., 1952.

H. F. WALTON, *Principles and Methods of Chemical Analysis*, Prentice-Hall, Englewood Cliffs, N.J., 2nd ed., 1964.

B. Handbooks

L. MEITES (ed.), *Handbook of Analytical Chemistry*, McGraw-Hill, New York, 1963. The scope of this book is rather wider than the title indicates. In addition to extensive tables of analytical data, methods of analysis, methods of separation, and other useful information, there is considerable expository material on compleximetric titrations, ion exchange, and other topics.

The analyst will still need to have at hand more general handbooks, such as *Handbook of Chemistry and Physics* or *Handbook of Chemistry*, for general physical and chemical constants. Solubility data may be found in A. SEIDELL (W. F. Linke, ed.), *Solubilities of Inorganic and Metal Organic Compounds*, 2 vols., Am. Chem. Soc., Washington, 4th ed., 1958, 1965.

The best compilation of stability constants is L. G. SILLÉN and A. E. MARTELL, *Stability Constants of Metal-Ion Complexes*, Chemical Society, London, 1964. See also K. B. YATSIMIRSKII and V. P. VASIL'EV, *Instability Constants of Complex Compounds*, Pergamon, New York, 1961. Ionization constants of acids and bases may be found in G. KORTÜM, W. VOGEL, and K. ANDRUSSOW, *Dissociation Constants of Organic Acids in Aqueous Solution*, Butterworth, London, 1961, and D. D. PERRIN, *Dissociation Constants of Organic Bases in Aqueous Solution*, Butterworth, London, 1965.

C. Treatises[9]

A number of multivolume works aiming to cover the larger part of the field of chemical analysis in greater or less detail are available. Most of them were not complete in 1968.

R. FRESENIUS and G. JANDER (eds.), *Handbuch der analytischen Chemie*, part III, *Quantitative Bestimmungs- und Trennungsmethoden*, Springer, Berlin, 1940. Part II deals with qualitative analysis.

I. M. KOLTHOFF and P. J. ELVING, with the assistance of E. B. SANDELL (eds.), *Treatise on Analytical Chemistry*, Wiley-Interscience, New York, 1959. In three parts: I, Theory and Practice of Analytical Chemistry, to consist of 11 vols.; II, Analytical Chemistry of the Elements, to consist of 13 vols.; III, Analytical Chemistry in Industry, to consist of 5 vols.

C. L. WILSON and D. W. WILSON (eds.), *Comprehensive Analytical Chemistry*, Elsevier, Amsterdam (and Van Nostrand, Princeton, N.J.), 1959. Vols. IA, IB, and IC deal with classical analysis; vol. II with electrical methods and physical separation methods; vol. III with optical methods; vol. IV with industrial and other special applications; vol. V with miscellaneous methods.

Standard Methods of Chemical Analysis, Van Nostrand, Princeton, N.J., 6th ed., 1962–1966. Vol. I (N. H. Furman, ed.) deals with the individual elements; vols. IIA

[9] Although not strictly falling in this category, the following series of volumes may be mentioned here: F. Hecht, R. Kaiser, and H. Kriegsmann (eds.), *Methoden der Analyse in der Chemie*, Akademische Verlagsgesellschaft, Frankfurt, vols. 1, 2, and 3, 1964.

and IIB (F. J. Welcher, ed.) with industrial and natural products (detailed procedures are given) and noninstrumental methods; vols. IIIA and IIIB (F. J. Welcher, ed.) with instrumental analysis.

D. Chiefly gravimetric and titrimetric or gravimetric[10]

C. DUVAL (transl. by R. E. Oesper), *Inorganic Thermogravimetric Analysis*, Elsevier, Amsterdam, 2nd ed., 1963.

L. ERDEY (transl. by G. Svehla), *Gravimetric Analysis*, 3 vols., Pergamon, New York, 1964–65. A German translation (vols. 1 and 2, 1964) is also available (Akademiai Kiadó, Budapest).

W. F. HILLEBRAND, G. E. F. LUNDELL, H. A. BRIGHT, and J. I. HOFFMAN, *Applied Inorganic Analysis*, Wiley, New York, 2nd ed., 1953. The first edition (by Hillebrand and Lundell) was a classic work, based largely on the personal experience of the authors and their coworkers.

K. KODAMA, *Methods of Quantitative Inorganic Analysis*, Wiley-Interscience, New York, 1963. A compendium of almost all inorganic gravimetric, titrimetric, and colorimetric methods published up to 1957.

G. E. F. LUNDELL and J. I. HOFFMAN, *Outlines of Methods of Chemical Analysis*, Wiley, New York, 1938. Useful for tabular surveys of the behavior of almost all the elements with common reagents.

C. R. N. STROUTS, H. N. WILSON, R. T. PARRY-JONES, and J. H. GILFILLAN, *Chemical Analysis*, Oxford, New York, 2nd ed., 1962. Subtitled *The Working Tools*, this work describes most of the physical and chemical techniques used in analysis.

H. VAN TONGEREN, *Gravimetric Analysis*, Centen, Amsterdam, 1937. Emphasis on silicate analysis.

F. P. TREADWELL and W. T. HALL, *Analytical Chemistry*, vol. II (Quantitative), Wiley, New York, 9th ed., 1942. Obsolescent.

W. D. TREADWELL, *Tabellen und Vorschriften zur quantitativen Analyse*, Deuticke, Leipzig, 1938.

W. W. WENDLANDT, *Thermal Methods of Analysis*, Wiley-Interscience, New York, 1964.

H. H. WILLARD and H. DIEHL, *Advanced Quantitative Analysis*, Van Nostrand, Princeton, N.J., 1943. Surveys classical methods, element by element; now rather out of date.

E. Titrimetric analysis (see chapters in Part Four and Five)

II. Works on specialized methods of analysis

A. Separations

For some references on organic precipitants see Chapter 12; liquid–liquid extraction, Chapter 14; chromatography (including ion exchange), Chapter 13. A general treatment is given by E. W. Berg in *Physical and Chemical Methods of Separation*, McGraw-Hill, New York, 1963.

B. Electroanalytical methods (see Chapters 46–53)

C. Spectrophotometric and related methods (see Chapters 54–58)

D. Physical methods (see Chapter 59)

E. Gas analysis

V. J. ALTIERI, *Gas Analysis and Testing of Gaseous Materials*, American Gas Association, New York, 1945.

F. BAYER and G. WAGNER, *Gasanalyse*, Enke, Stuttgart, 1960.

J. S. HALDANE and J. I. GRAHAM, *Methods of Air Analysis*, Griffin, London, 4th ed., 1935.

[10] Two books dealing with recent developments in gravimetric, titrimetric, and other methods of analysis are R. Belcher and C. L. Wilson, *New Methods of Analytical Chemistry*, Reinhold, New York, 1964; A. G. Jones, *Analytical Chemistry—Some New Techniques*, Butterworth, London, 1959. Also see the paperback, H. N. Wilson, *An Approach to Chemical Analysis*, Pergamon, New York, 1966.

P. G. Jeffery and P. J. Kipping, *Gas Analysis by Gas Chromatography*, Pergamon, New York, 1964.

G. Lunge and H. R. Ambler, *Technical Gas Analysis*, Gurney and Jackson, London, 1934.

P. W. Mullen, *Modern Gas Analysis*, Wiley-Interscience, New York, 1955.

G. Tine, *Gas Sampling and Chemical Analysis in Combustion Processes*, Pergamon, New York, 1961.

Z. M. Turovtseva and L. L. Kunin, *Analysis of Gases in Metals*, transl. by Consultants Bureau, New York, 1961.

F. Microanalysis

I. P. Alimarin and M. N. Petrikova (transl. by M. G. Hell), *Inorganic Ultramicroanalysis*, Macmillan, New York, 1964.

R. Belcher, *Submicro Methods of Organic Analysis*, Elsevier, Amsterdam, 1966.

R. Belcher and A. L. Godbert, *Semi-micro Quantitative Organic Analysis*, Longmans, London, 2nd ed., 1954.

A. A. Benedetti-Pichler, *Introduction to the Microtechnique of Inorganic Analysis*, Wiley, New York, 1942.

A. A. Benedetti-Pichler and F. Hecht, *Handbuch der mikrochemischen Methoden*, vol. I, part II, *Waagen and Wägung, Geräte zur anorganischen Mikrogewichtsanalyse*, Springer, Berlin, 1959.

A. A. Benedetti-Pichler, *Identification of Materials*, Academic Press, New York, 1964. Qualitative, via physical properties, chemical tests, and microscopy.

H. V. A. Briscoe and P. F. Holt, *Inorganic Microanalysis*, E. Arnold, London, 1950.

E. M. Chamot and C. W. Mason, *Handbook of Chemical Microscopy*, vol. I, Wiley, New York, 3rd ed., 1958. (Chapter 14 is devoted to quantitative microscopical analysis.)

N. D. Cheronis, *Submicrogram Experimentation: Proceedings of the Symposium on Experimentation below the Microgram Range, Arlington, Va., 1960*, Wiley-Interscience, New York, 1961.

N. D. Cheronis, *Microchemical Techniques: Proceedings of the International Symposium on Microchemical Techniques, Pennsylvania State University, 1961*, Wiley-Interscience, New York, 1962.

N. D. Cheronis and T. S. Ma, *Organic Functional Group Analysis by Micro and Semimicro Methods*, Wiley-Interscience, New York, 1964.

S. J. Clark, *Quantitative Methods of Organic Microanalysis*, Butterworth, London, 1956.

E. J. Conway, *Microdiffusion Analysis and Volumetric Error*, Crosby Lockwood, London, 1957.

C. Duval, *Traité de micro-analyse minérale*, Presses Scientifiques Internationales, Paris, 1954–1956.

H. M. El-Badry, *Micromanipulators and Micromanipulation*, Academic Press, New York, 1963.

F. Emich, *Lehrbuch der Mikrochemie*, Bergmann, Munich, 2nd ed., 1926. Mostly of historical interest.

A. Friedrich, *Die Praxis der quantitativen organischen Mikroanalyse*, Deuticke, Leipzig, 1933.

J. Grant, *Quantitative Organic Microanalyses Based on the Methods of F. Pregl*, Churchill, London, 1951.

F. Hecht and J. Donau, *Anorganische Mikrogewichtsanalyse*, Springer, Berlin, 1940.

F. Hecht and M. K. Zacherl (eds.), *Handbuch der mikrochemischen Methoden*, Springer, Berlin, vol. I/1, 1954; vol. I/2, 1959; vol. II, 1955; vol. III, 1961; vol. IV (H. Malissa, *Elektronenstrahl-Mikroanalyse*), 1966.

G. Ingram, *Methods of Organic Elemental Microanalysis*, Chapman & Hall, London, 1962.

P. L. Kirk, *Quantitative Ultramicroanalysis*, Wiley, New York, 1950.

I. M. Korenman, *Introduction to Quantitative Ultramicroanalysis*, Academic Press, New York, 1965.

J. A. Kuch (ed.) (transl. by P. L. Bolton and K. Gingold), *Methods in Microanalysis*, vol. I, *Simultaneous Rapid Combustion*, Gordon & Breach, New York, 1964.

J. Mika, *Die exakten Methoden der Mikromassanalyse*, Enke, Stuttgart, 2nd ed., 1958.

R. F. Milton and W. A. Waters (eds.), *Methods of Quantitative Microanalysis*, Arnold, London, 2nd ed., 1955.

S. Natelson, *Microtechniques of Clinical Chemistry*, Thomas, Springfield, Ill., 2nd ed., 1961.

J. B. Niederl and V. Niederl, *Micromethods of Quantitative Organic Elementary Analysis*, Wiley, New York, 2nd ed., 1942.

H. Roth, *Pregl-Roth quantitative organische Mikroanalyse*, Springer, Berlin, 7th ed., 1958.

A. Steyermark, *Quantitative Organic Microanalysis*, Academic Press, New York, 2nd ed., 1961.

T. E. Wallis, *Analytical Microscopy*, Little, Brown, Boston, 2nd ed., 1957.

I. D. P. Wootton, *Micro-Analysis in Medical Biochemistry*, Churchill, London, 1964.

G. Trace analysis

N. L. Allport and J. W. Keyser, *Colorimetric Analysis*, vol. I, Chapman & Hall, London, 1957; N. L. Allport and J. E. Brocksopp, *Colorimetric Analysis*, vol. II, Chapman & Hall, London, 1963. (Vol. I deals with clinical and biochemical determinations, vol. II with determination of metals, acid radicals, and organic substances.)

Analytical Methods Committee, Society of Analytical Chemistry, *Determination of Trace Elements with Special Reference to Fertilisers and Feeding Stuffs*, Heffer, Cambridge, 1963.

D. F. Boltz (ed.), *Colorimetric Determination of Nonmetals*, Wiley-Interscience, New York, 1958.

M. S. Brooks and J. K. Kennedy (eds.), *Ultrapurification of Semiconductor Materials*, Macmillan, New York, 1962. (Conference proceedings, in part devoted to analysis.)

J. P. Cali (ed.), *Trace Analysis of Semiconductor Materials*, Macmillan, New York, 1964.

O. G. Koch and G. A. Koch-Dedic, *Handbuch der Spurenanalyse*, Springer, Berlin, 1964.

W. W. Meinke and B. F. Scribner, *Trace Characterization*, U. S. Government Printing Office, Washington, D.C., 1967.

O. I. Milner, *Analysis of Petroleum for Trace Elements*, Macmillan, New York, 1963.

G. H. Morrison (ed.), *Trace Analysis: Physical Methods*, Wiley-Interscience, New York, 1965.

M. Pinta, *Recherche et dosage des éléments traces*, Dunod, Paris, 1962. Chiefly absorptimetric, emission spectrographic and polarographic methods.

E. B. Sandell, *Colorimetric Determination of Traces of Metals*, Wiley-Interscience, New York, 3rd ed., 1959.

F. D. Snell and C. T. Snell, *Colorimetric Methods of Analysis* (vols. I, II, III, IV, IIA, IIIA; the two latter with C. A. Snell), Van Nostrand, Princeton, N.J., 3rd ed., 1948–1954. Both inorganic and organic.

J. H. Yoe and H. J. Koch (eds.), *Trace Analysis*, Wiley, New York, 1957. Miscellaneous symposium papers.

III. Works on analysis of particular materials

A. General technical analysis

A. H. Allen, *Commercial Organic Analysis*, Blakiston, Philadelphia, 5th ed., 1923–1933. Now rather old.

E. Berl and G. Lunge, *Chemisch-technische Untersuchungsmethoden*, Springer, Berlin, 8th ed., 1931–1934; 3 supplementary vols., 1939–1940.

Book of Standards, American Society for Testing Materials, Philadelphia.

S. C. JOLLY (Society for Analytical Chemistry), *Official, Standardized and Recommended Methods of Analysis*, Heffer, Cambridge, 1963.

F. D. SNELL and F. M. BIFFEN, *Commercial Methods of Analysis*, Chemical Publishing, New York, 1964.

F. D. SNELL and C. L. HILTON (eds.), *Encyclopedia of Industrial Chemical Analysis*, Wiley-Interscience, New York, 1966.

F. SPECHT, *Quantitative anorganische Analyse in der Technik*, Verlag Chemie, Weinheim, 1953.

R. S. YOUNG, *Industrial Inorganic Analysis*, Wiley, New York, 1953.

Standard Methods of Chemical Analysis (see IC)

B. Organic and related samples (see also IIF and IIIA)

American Association of Clinical Chemists, *Standard Methods of Clinical Chemistry*, Academic Press, New York, 1958 (vol 4 in 1963).

R. B. BRADSTREET, *The Kjeldahl Method for Organic Nitrogen*, Academic Press, New York, 1965.

F. E. CRITCHFIELD, *Organic Functional Group Analysis*, Macmillan, New York, 1962.

N. EVERS and W. SMITH, *Analysis of Drugs and Chemicals*, Griffin, London, 1955.

J. S. FRITZ and G. S. HAMMOND, *Quantitative Organic Analysis*, Wiley, New York, 1957.

D. C. GARRAT and others, *The Quantitative Analysis of Drugs*, Chapman & Hall, London, 3rd ed., 1964.

D. GLICK (ed.), *Methods of Biochemical Analysis*, Wiley, New York, 1954 (vol. 13 in 1965).

T. HIGUCHI and E. BROCKMANN-HANSSEN (eds.), *Pharmaceutical Analysis*, Wiley, New York, 1961.

B. KAKÁČ and Z. VEJDELÊK (transl. by E. Hachová), *Kolorimetrie in der Pharmazie* (vol. I of *Handbuch der Kolorimetrie*), Fischer, Jena, 1962.

S. MEITES (ed.), *Standard Methods of Clinical Chemistry*, vol. 5, Academic Press, New York, 1965.

J. MITCHELL, I. M. KOLTHOFF, E. S. PROSKAUER, and A. WEISSBERGER (eds.), *Organic Analysis*, Wiley-Interscience, New York, 1953. A series of collective vols. (vol. IV in 1960).

S. SIGGIA, *Quantitative Organic Analysis via Functional Groups*, Wiley, New York, 3rd ed., 1963.

S. SIGGIA and H. J. STOLTEN, *An Introduction to Modern Organic Analysis*, Wiley-Interscience, New York, 1956.

K. G. STONE, *Determination of Organic Compounds*, McGraw-Hill, New York, 1956.

F. WILD, *Estimation of Organic Compounds*, Cambridge, New York, 1953.

C. Agricultural materials, foods, etc.

American Association of Cereal Chemists, *Cereal Laboratory Methods*, St. Paul, Minn., 7th ed., 1964.

American Oil Chemists Society, *Official and Tentative Methods of AOCS*, AOCS, Chicago, 1962.

H. A. BOEKENOOGEN (ed.), *Analysis and Characterization of Oils, Fats and Fat Products*, vol. 1, Wiley-Interscience, New York, 1964.

A. BÖMER, A. JUCKENACK, and J. TILLMANS, *Handbuch der Lebensmittelchemie* Hirschwaldsche Buchhandlung, Berlin, 1933.

C. A. BROWNE and F. W. ZERBAN, *Physical and Chemical Methods of Sugar Analysis*, Wiley, New York, 3rd ed., 1941.

L. V. COCKS and C. VAN REDE, *Laboratory Handbook of Oil and Fat Analysis*, Academic Press, New York, 1966.

H. E. COX, *Chemical Analysis of Foods*, Churchill, London, 2nd ed., 1938.

H. E. COX and D. PEARSON, *Chemical Analysis of Foods*, Tudor, New York, 5th ed., 1962.

J. G. Davis, *Milk Testing*, Dairy Industries, London, 2nd ed., 1959.

H. C. S. DeWhalley (ed.), *ICUMSA Methods of Sugar Analysis*, Elsevier, Amsterdam, 1964.

S. Gericke, *Analytische Chemie der Düngmittel*, Enke, Stuttgart, 1949.

F. A. Gunther (ed.), *Residue Reviews*, vol. 5, Springer, Berlin, 1964. Some of the earlier and later volumes also contain material of analytical interest.

F. A. Gunther and R. C. Blinn, *Analysis of Insecticides and Acaricides*, Wiley-Interscience, New York, 1955.

W. Horwitz (ed.), *Official and Tentative Methods of Analysis*, Association of Official Agricultural Chemists, Washington, 10th ed., 1965.

IUPAC, *Standard Methods of the Oils and Fats Division of the IUPAC*, Butterworth, London, 5th ed., 1964.

M. B. Jacobs, *Chemical Analysis of Foods and Food Products*, Van Nostrand, Princeton, N.J., 3rd ed., 1958.

M. A. Joslyn, *Methods in Food Analysis*, Academic Press, New York, 1950.

A. E. Leach and A. L. Winton, *Food Inspection and Analysis*, Wiley, New York, 4th ed., 1924. Now rather old.

J. Lewkowitsch, *Chemical Technology and Analysis of Oils, Fats, and Waxes*, Macmillan, New York, 6th ed., 1921–1923. Good in its day.

V. C. Mehlenbacher, *Analysis of Fats and Oils*, Garrard, Champaign, Ill., 1960.

K. Paech and M. V. Tracey (eds.) (continued by H. F. Linskens and M. V. Tracey), *Modern Methods of Plant Analysis*, vol. 6, Springer, Berlin, 1963.

C. S. Piper, *Soil and Plant Analysis*, Wiley-Interscience, New York, 1947.

J. Schormüller (ed.), *Analytik der Lebensmittel. Physikalische und physikalisch-chemische Untersuchungsmethoden*, vol. II, part I of *Handbuch der Lebensmittelchemie*, Springer, Berlin, 1965.

H. O. Triebold and L. W. Aurand, *Food Composition and Analysis*, Van Nostrand, Princeton, N.J., 1963.

R. L. Whistler, M. L. Wolfrom, J. N. BeMiller, and F. Shafizadeh (eds.), *Methods in Carbohydrate Chemistry*, vol. I, *Analysis and Preparation of Sugars*, Academic Press, New York, 1962.

R. L. Whistler, M. L. Wolfrom, and J. N. BeMiller (eds.), *Methods in Carbohydrate Chemistry*, vol. II, *Reactions of Carbohydrates*, Academic Press, New York, 1963.

K. A. Williams, *Oils, Fats and Fatty Foods. Their Practical Examination*, Churchill, London, 1966.

A. L. Winton and K. B. Winton, *The Analysis of Foods*, Wiley, New York, 1945.

A. G. Woodman, *Food Analysis*, McGraw-Hill, New York, 4th ed., 1941.

C. H. Wright, *Agricultural Analysis*, T. Murby, London, 1938. (See also IIIA.)

H. Wüstenfeld and G. Haeseler, *Trinkbranntweine und Liköre*, Parey, Berlin, 4th ed., 1964. Partly analytical.

G. Zweig (ed.), *Analytical Methods for Pesticides, Plant Growth Regulators and Food Additives*, Academic Press, New York, 1963–1964.

D. Alloys

1. *Ferrous*

American Society for Testing and Materials, *ASTM Methods for Chemical Analysis of Metals*, ASTM, Philadelphia, 1960.

British Cast Iron Research Association, *Selected Methods of Analysis of Foundry Materials, part I. Pig Iron and Cast Iron*, Birmingham, 1960.

British Standards Institution, *Methods for the Sampling and Analysis of Iron and Steel*, BSI, London (as issued).

Chemists' Committee of the Subsidiaries of the U.S. Steel Corporation, *Sampling and Analysis of Carbon and Alloy Steels*, Reinhold, New York, 1938.

Iron and Steel Institute, Special Report 68: *Determination of Gases in Metals*, ISI, London, 1960.

C. M. Johnson, *Rapid Methods for the Chemical Analysis of Special Steels*, Wiley, New York, 4th ed., 1932.

W. Koch, *Metallkundliche Analyse*, Verlag Stahleisen, Düsseldorf, 1965.

N. W. Lord and D. J. Demorest, *Metallurgical Analysis*, McGraw-Hill, New York, 6th ed., 1952.

G. E. F. Lundell, J. I. Hoffman, and H. A. Bright, *Chemical Analysis of Iron and Steel*, Wiley, New York, 1931.

J. Mika, *Metallurgische Analysen*, Akadémiai Kiadó, Budapest, 1964. Also includes nonferrous alloy analysis and analysis of ores.

E. C. Piggot, *Ferrous Analysis*, Wiley, New York, 2nd ed., 1953.

United Steel Companies, *Standard Methods of Analysis of Iron, Steel and Ferro-Alloys*, Sheffield, 1959.

R. Weihrich and A. Winkel, *Die chemische Analyse in der Stahlindustrie*, Enke, Stuttgart, 4th ed., 1954.

W. Westwood and A. Mayer, *Chemical Analysis of Cast Iron and Foundry Materials*, G. Allen, London, 2nd ed., 1960.

2. *Nonferrous*

Aluminum Research Institute, *Analytical Methods for Aluminum Alloys*, Chicago, 1948.

American Society for Testing and Materials, *ASTM Methods for Chemical Analysis of Metals*, ASTM, Philadelphia, 1960.

British Aluminium Co., *Analysis of Aluminium and Its Alloys: Spectrochemical and Polarographic Methods*, 1946; *Analysis of Aluminium and Its Alloys: Chemical, Colorimetric and Photometric Methods*, London, 1949.

BNF Analytical Conference, Malvern, 1961, British Non-Ferrous Metals Research Association, London, 1962. Instrumental methods.

E. R. Caley, *Analysis of Ancient Metals*, Macmillan, New York, 1964.

Chemikerausschuss der Gesselschaft Deutscher Metallhütten- und Bergleute e.V., *Analyse der Metalle*, vol. 1, *Schiedsanalysen*, 3rd ed., 1966; vol. 2 (in two parts), *Betriebsanalysen*, Springer, Berlin, 2nd ed., 1961. Includes analysis of ores and industrial products other than alloys.

Chemikerausschuss der Gesellschaft Deutscher Metallhütten- und Bergleute e.V., *Edelmetall-Analyse: Probierkunde und Nassanalytische Verfahren*, Springer, Berlin, 1964.

H. V. Churchill and R. W. Bridges, *Chemical Analysis of Aluminum*, Aluminum Research Laboratories, New Kensington, Pa., 3rd ed., 1950.

M. Codell, *Analytical Chemistry of Titanium Metals and Compounds*, Wiley-Interscience, New York, 1959.

A. Cohen, *Rationelle Metallanalyse*, Birkhäuser, Basel, 1948.

C. M. Dozinel, *Modern Methods of Analysis of Copper and Its Alloys*, Thone, Liége, 1960.

W. T. Elwell and I. R. Scholes, *Analysis of Copper and Its Alloys*, Pergamon, New York, 1967.

W. T. Elwell and D. F. Wood, *The Analysis of Titanium, Zirconium and Their Alloys*, Wiley, New York, 1961.

W. T. Elwell and D. F. Wood, *Analysis of the New Metals—Titanium, Zirconium, Hafnium, Niobium, Tantalum, Tungsten and Their Alloys*, Pergamon, New York, 1966

H. Ginsberg (revised by H. Beusch, M. Köster, and H. Pfundt), *Leichtmetallanalyse*, de Gruyter, Berlin, 4th ed., 1965.

Imperial Chemical Industries, *Analysis of Titanium and Its Alloys*, London, 1959.

C. L. Lewis, W. L. Ott, and N. M. Sine, *Analysis of Nickel*, Pergamon, New York, 1966.

A. Mayer and W. J. Price, *Chemical and Spectrographic Analysis of Magnesium and Its Alloys*, Magnesium Elektron, London, 1955.

W. A. Naish, J. E. Clennell, and V. S. Kingswood, *Select Methods of Metallurgical Analysis*, Chapman & Hall, London, 2nd ed., 1953.

G. Norwitz, *Rapid Analysis of Nonferrous Metals and Alloys*, Chemical Publishing, New York, 1957.

G. H. Osborn and W. Stross, *Analysis of Aluminum Alloys*, Chemical Publishing, New York, 1953.

O. Proske and others, *Analyse der Metalle*, Springer, Berlin, 1949–1956.

F. J. Versagi, *Routine Analysis of Copper Base Alloys*, Tudor, New York, 1960.

E. Rocks, minerals and ores, and soils

L. H. Ahrens, *Quantitative Spectrochemical Analysis of Silicates*, Addison-Wesley, Reading, Mass., 1954.

H. Bennett and W. G. Hawley, *Methods of Silicate Analysis*, Academic Press, New York, 1964.

C. A. Black (ed.), *Methods of Soil Analysis, part 1: Physical and Mineralogical Properties, Including Statistics of Measurement and Sampling; part 2: Chemical and Microbiological Properties*, American Society of Agronomy, Madison, Wis., 1965.

Chemikerausschuss des Vereins Deutscher Eisenhüttenleute, *Handbuch für das Eisenhütten Laboratorium*, vol. I, *Die Untersuchung der Nichtmetallischen Stoffe*, Verlag Stahleisen, Düsseldorf, 1960.

A. W. Groves, *Silicate Analysis*, G. Allen, London, 2nd ed., 1951.

R. Hedin, *Colorimetric Methods for Rapid Analysis of Silicate Materials*, Swedish Cement and Concrete Institute, Royal Institute of Technology, Stockholm, 1947.

F. G. Hills, *The Technical Analysis of Ores and Metallurgical Products*, Tudor, New York, 1937.

M. L. Jackson, *Soil Chemical Analysis*, Prentice-Hall, Englewood Cliffs, 1958. (Also see Piper, IIID.)

J. Jakob, *Chemische Analyse der Gesteine und Silikatischen Mineralien*, Birkhäuser, Basel, 1951.

J. A. Maxwell, *The Chemical Analysis of Rocks and Minerals*, Wiley-Interscience, New York (to be published).

L. C. Peck, *Systematic Analysis of Silicates*, U.S. Geological Survey Bulletin 1170, U.S. Government Printing Office, Washington, 1964.

L. Shapiro and W. W. Brannock, *The Rapid Analysis of Silicate Rocks*, U.S. Geological Survey Circular 165, 1952, and 1036-C, 1956, U.S. Government Printing Office, Washington.

W. R. Schoeller and A. R. Powell, *The Analysis of Minerals and Ores of the Rarer Elements*, Hafner, New York, 3rd ed., 1956.

A. A. Smales and L. R. Wager (eds.), *Methods in Geochemistry*, Wiley-Interscience, New York, 1960.

Society of Chemical Industry, *Analysis of Calcareous Materials*, SCI Monograph 18, SCI, London, 1964.

A. P. Vinogradov and D. I. Ryabchikov, *Detection and Analysis of Rare Elements*, NSF, Washington, 1962 (transl. from Russian).

I. A. Voinovitch, J. Debras-Guédon, and J. Louvrier, *L'Analyse des silicates*, Hermann, Paris, 1962.

F. N. Ward, H. W. Lakin, F. C. Canney, and others, *Analytical Methods Used in Geochemical Exploration by the U.S. Geological Survey*, U.S. Geological Survey Bulletin 1152, U.S. Government Printing Office, Washington, 1963.

A. J. Weinig and W. P. Schoder, *Technical Methods of Ore Analysis for Chemists and Colleges*, Wiley, New York, 1939.

F. Reagents (See Chapter 17)

G. Water

American Public Health Association, *Standard Methods for the Examination of Water Sewage, and Industrial Wastes*, New York, 10th ed., 1955.

F. J. H. Mackereth, *Some Methods of Water Analysis for Limnologists*, Freshwater Biological Association, Westmorland, England, 1963.

F. H. Rainwater and L. L. Thatcher, *Methods for Collection and Analysis of Water Samples*, U.S. Geological Survey Paper 1454, U.S. Government Printing Office, Washington, 1960.

H. Miscellaneous

Commissariat à L'Energie Atomique, *Methodes d'analyse, Cetema—1964. Methodes 101 à 150*, Presses Universitaires, Paris, 1964. (First 100 methods published in 1962.)

M. B. Jacobs, *The Analytical Chemistry of Industrial Poisons, Hazards and Solvents*, Wiley-Interscience, New York, 2nd ed., 1949.

M. B. Jacobs, *The Chemical Analysis of Air Pollutants*, Wiley-Interscience, New York, 1960.

M. B. Jacobs and L. Scheflan, *Chemical Analysis of Industrial Solvents*, Wiley-Interscience, New York, 1960.

C. J. Rodden (ed.), *Analysis of Essential Nuclear Reactor Materials*, U.S. Government Printing Office, Washington, 1964.

C. J. Rodden (ed.), *Analytical Chemistry of the Manhattan Project*, McGraw-Hill, New York, 1950.

IV. Works on analytical chemistry of individual elements (see also IC)

Gmelin's *Handbuch der anorganischen Chemie* contains sections on the analytical chemistry of the elements.

Academy of Sciences of the USSR, *Analytical Chemistry of Uranium*, translated by Israel Program for Scientific Translations, Jerusalem, 1963. One of a series of some 50 volumes on the analytical chemistry of individual elements.

T. D. Avtokratova, *Analytical Chemistry of Ruthenium*, transl. by Israel Program for Scientific Translations, Jerusalem, 1963.

R. Z. Bachman and C. V. Banks, *Critical Review of the Analytical Methods of the Titanium, Vanadium and Chromium Transition Element Groups*, part 4 of vol. 3 of *Analytical Chemistry, Progress in Nuclear Energy*, series IX (C. E. Crouthamel, ed.), Pergamon, New York, 1963.

F. E. Beamish, *The Analytical Chemistry of the Noble Metals*, Pergamon, New York, 1966.

A. I. Busev, *The Analytical Chemistry of Indium*, Macmillan, New York, 1962.

A. I. Busev, *The Analytical Chemistry of Molybdenum*, Israel Program for Scientific Translations, Jerusalem, 1964.

S. V. Elinson and K. I. Petrov, *Analytical Chemistry of Zirconium and Hafnium*, Davey, New York, 1966.

I. M. Korenman, *Analytical Chemistry of Potassium*, Davey, New York, 1965.

I. M. Korenman, *Analytical Chemistry of Thallium*, Israel Program for Scientific Translations, Jerusalem, 1963.

A. Kurtenacher, *Analytische Chemie der Sauerstoffsäuren des Schwefels*, Enke, Stuttgart, 1938.

A. J. Moses, *Analytical Chemistry of the Actinide Elements*, Macmillan, New York, 1964.

R. W. Moshier, *Analytical Chemistry of Niobium and Tantalum*, Pergamon, New York, 1964.

D. I. Ryabchikov and E. K. Gol'braikh, *The Analytical Chemistry of Thorium*, Macmillan, New York, 1964.

W. R. Schoeller, *The Analytical Chemistry of Tantalum and Niobium*, Chapman & Hall, London, 1937.

L. E. Smythe and T. M. Florence, *Recent Advances in the Analytical Chemistry of Beryllium*, part 6 of vol. 3 of *Analytical Chemistry, Progress in Nuclear Energy*, series IX (C. E. Crouthamel, ed.), Pergamon, New York, 1963.

H. Tollert, *Analytik des Kaliums*, Enke, Stuttgart, 1962.

R. C. Vickery, *Analytical Chemistry of the Rare Earths*, Macmillan, New York, 1960.

A. WOGRINZ, *Analytische Chemie der Edelmetalle*, Enke, Stuttgart, 1936.
R. S. YOUNG, *Analytical Chemistry of Cobalt*, Pergamon, New York, 1966.

V. Miscellaneous

Government publications (as of the U.S. Atomic Energy Commission, U.S. Geological Survey, Bureau of Mines, and others) constitute a source of information on analytical methods. Certain institutions and industrial concerns publish their methods of analysis.

Yearbooks, such as the *Annual Reports on the Progress of Chemistry* (Chemical Society, London), may be referred to for summaries of recent progress in the analytical field. The series *Advances in Analytical Chemistry and Instrumentation* (C. N. Reilley, ed.), Wiley-Interscience, New York, 1960, presents new developments and evaluates current analytical methods.

A history of analytical chemistry is available in English translation: F. SZABADVARY, *History of Analytical Chemistry*, Pergamon, New York, 1965.

JOURNALS

The following journals are devoted entirely or largely to general or specialized analytical chemistry:

The Analyst, 1876–
Analytica Chimica Acta, 1947–
Analytical Chemistry, 1929– (until 1947, *Industrial and Engineering Chemistry, Analytical Edition*)
Analytical Letters, 1968–
Annales des Falsifications et des Fraudes, 1908–1940
Chemist-Analyst, 1911–
Chimie Analytique, 1896– (this title since 1942, earlier variant titles)
Japan Analyst (*Bunseki Kagaku*), 1957–
Journal of the Association of Official Agricultural Chemists, 1915– . Name changed in 1967 to *Journal of the Association of Official Analytical Chemists*
Journal of Chromatography, 1958–
Journal of Electroanalytical Chemistry, 1959–
Journal of Radioanalytical Chemistry, 1968–
Microchemical Journal, 1957–
Mikrochimica Acta, 1923– (earlier variant titles)
Spectrochimica Acta, 1947–
Talanta, 1958–
Zavodskaya Laboratoriya, 1935–
Zeitschrift für analytische Chemie, 1862–
Zeitschift für Lebensmittel-Untersuchung und -Forschung, 1882–
Zhurnal Analiticheskoi Khimii, 1946– (available in English translation as *Journal of Analytical Chemistry of the USSR*)

In addition, papers on chemical analysis appear in some of the general chemical journals (less often in science journals). Reference may also be made to the reports of the International Union of Pure and Applied Chemistry (earlier, International Congress of Applied Chemistry), some of which are of analytical interest.

The abstracting journals are of great value in literature searches:

Chemical Abstracts, 1907–
Chemisches Zentralblatt, 1897–
Analytical Abstracts, 1954–

Abstracts of analytical chemical work have appeared, and continue to appear, in some of the analytical or general chemical journals. For example, abstracts, of generous length, are included in *Z. Anal. Chem.*

Reviews of pure and applied analytical chemistry are published annually in *Anal. Chem.* and reviews on selected topics appear from time to time in the *Analyst* and other journals. The task of the analytical chemist to keep abreast of progress in his field is lightened by such review articles.

CHAPTER 2 *Stoichiometry. Expression of Results*

If from any art you take away that which concerns weighing, measuring and arithmetic, how little is left of that art!

—PLATO, *Philebus*

Atomic weights

Chemical compounds result from the combination, in fixed ratios, of atoms of elements, each of which has a definite mass, taking into account the existence of isotopes. The relative weights of the atoms, broadly termed atomic weights, are obviously of fundamental importance in the weight relations of elements and compounds (stoichiometry), which form the basis of quantitative chemical analysis.

Before the discovery of isotopes, the atomic weights of the elements were considered to be fundamental constants. Now it is known that only 21 of the elements consist of a single isotope, the remainder being mixtures of two or more isotopes. The natural proportions of the isotopes of an element can vary, although usually only slightly, so that the average atomic weights of such elements are subject to variation.

The *nuclidic* (or *atomic*) *mass* of a nuclidic species is constant and can be expressed in absolute mass units. However, it is advantageous to express it as a ratio with respect to some nuclide whose mass is represented by an integral number. Different scales of relative nuclidic masses are thus possible, depending on the reference nuclidic mass. The carbon-12 isotope (^{12}C) has now been adopted as the reference species by both chemists and physicists, following the suggestion of A. O. Nier and J. H. Mattauch, and has been assigned a relative mass equal to its mass number, **12**. The *atomic weight* of a mononuclidic element is identical with its mass value on this scale. The atomic weight of a polynuclidic element is the average nuclidic mass of the isotope mixture on the $^{12}C = 12$ basis.[1] Without qualification, *atomic weight* refers to the average for the isotopes of an element as it occurs in nature (see the table, inside front cover). The natural isotopic composition of some elements, especially those of lower atomic masses, may show very slight variations, almost always too small to be of analytical significance. Elements whose isotopic composition is affected by radioactive transformations can show appreciable

[1] A. E. Cameron and E. Wichers, Report of the International Commission on Atomic Weights (1961), in *J. Am. Chem. Soc.*, **84**, 4175 (1962). This report summarizes the considerations leading to the adoption of $^{12}C = 12$ as the atomic weight reference by the Commission at the Twenty-first Conference of the IUPAC at Montreal, 1961, and critically reviews the physical and chemical determinations of the atomic weights of elements 1–92.

differences in their atomic weights. Thus strontium in some lepidolites, resulting from decay of ^{87}Rb, consists of as much as 99 per cent ^{87}Sr, whereas normal strontium (at. wt. 87.62) contains $\sim$7 per cent ^{87}Sr.

In the half-century preceding 1961, the atomic weights published by the International Commission on Atomic Weights were based on 16 as the atomic weight of natural oxygen, which contains approximately 0.04 per cent ^{17}O and 0.20 per cent ^{18}O. The composition is not entirely constant—the ^{18}O content of freshwaters may be 3 per cent less than in atmospheric oxygen—so that natural oxygen is not an ideal reference. On the current scale, the atomic weight of oxygen becomes 15.9994 and the atomic weights of other elements are reduced proportionately. The reduction amounts to 4 parts per 100,000. Chemical factors, which are ratios of atomic and molecular weights, are not affected at all by the use of the new scale, but values of gram molecular (mole) weights, Avogadro's number, and the Faraday are affected, of course, but since the change is only $\sim$0.004 per cent it is often negligible from the practical standpoint.

Methods and their accuracy. Until the early 1940s, atomic weights were determined by chemical methods. For example, the combining weight of aluminum with oxygen can be found by converting a known weight of pure aluminum into Al_2O_3 and weighing this. Since the combining (or equivalent) weight of aluminum is one third its atomic weight, with the atomic weight of oxygen taken as 16 and its equivalent weight as 8, the atomic weight of aluminum is readily calculated. By such stoichiometric methods, the atomic weights of all the elements were found, except those of the inert gases, which were determined by gas-density measurements. The masters of experimental stoichiometry, T. W. Richards, G. P. Baxter, O. Hönigschmidt, and their coworkers, carried out most of these atomic-weight determinations. Many atomic weights were determined by finding the weight of silver equivalent to a known weight of the chloride or bromide of the element. The accuracy of such chemically determined atomic weights can be assessed by comparing them with the atomic weights of mononuclidic elements derived from physical determinations which are very accurate. In Table 2-1 this comparison is made for six elements whose atomic weights were determined chemically in the last 30 years. For some older determinations the comparison is less favorable, the worst case being that of scandium (1919), for which the physical value is 44.956, the chemical 45.095. The large error in the chemical value is no doubt due to impurity in the scandium bromide used.[2]

Since mass-spectrometric measurements were first used in the late 1930s to obtain atomic weights, they have steadily increased in importance and now eclipse the chemical stoichiometric method. Nuclidic masses can be determined very accurately with the mass spectrometer (the error is less than 1 part in 100,000),[3] so that the atomic weights of mononuclidic elements (Al, As, Be, Bi, Cs, Co, F, Au, Ho, I, Mn, Nb, P, Pr, Rh, Sc, Na, Tb, Tu, Th, Y) are known with the same high accuracy. The state of affairs is less favorable for the polynuclidic elements, because for them the isotope abundances must be known, and these cannot be determined

[2] It may be pointed out that the purity of the material in mass spectrometry is not of importance as long as elements forming isobaric ions are absent.

[3] For a critical examination of mass spectrometry in atomic-weight determinations, see A. E. Cameron, *Anal. Chem.*, **35**, 23A (1963).

TABLE 2-1. **Comparison of Atomic Weights of Some Mononuclidic Elements Determined by Physical and Chemical Methods**[a]

	Physical Value	*Chemical Value*	*Year*
As	74.9216	74.904	1933
Nb	92.906	92.904	1934
Ho	164.930	164.928	1940
P	30.9738	30.969	1940
Cs	132.905	132.901	1940
Be	9.0122	9.0123	1947

[a] From a more extensive table in A. E. Cameron and E. Wichers, *J. Am. Chem. Soc.*, **84**, 4180 (1962).

as accurately as the nuclidic masses with the mass spectrometer.[4] If only minor amounts of one or more isotopes are present (H, He, C, N, O, V, Ta), the atomic weight can be determined almost as accurately as that of a mononuclidic element. For about two thirds of the elements, an accurate mass-spectrometric determination of atomic weight requires calibration of the spectrometer with a synthetic mixture of the separated isotopes approximating the natural composition. Such calibrations have been carried out for isotope abundance determinations of B, N, Cl, Ar, Cr, Ag, and U. The atomic weights of these elements and of the seven previously mentioned, together with Br, can be considered known to within 1–3 parts in 100,000 according to Wichers.[5] This leaves 48 elements, of which 15 (Sb, Ba, Ca, Ce, Cu, Ga, Fe, La, Lu, Hg, Rb, Sr, S, Tl, and Zr) show such good agreement between physical and chemical values "that the last digit of the officially adopted value can be accepted with some confidence." The uncertainties in the atomic weights of the remaining 33 elements probably do not exceed 0.05 per cent, and some may be no more than 0.01 per cent.

It may be concluded that the atomic weights of all the elements are now known sufficiently accurately for all but the most unlikely analytical purposes. Undoubtedly the values of the more uncertain atomic weights will be refined, but it does not seem likely that anyone will have the courage—or foolhardiness—to do this by chemical, stoichiometric methods. However, stoichiometry will still play an important role in the future work, because the preparation of synthetic calibration mixtures of isotopes will require accurate knowledge of the composition of the weighing forms.

Stoichiometry

A chemical equation is a symbolic statement of the quantitative relation between the reactants and the reaction products. Thus the chemical equation

$$AgNO_3 + NaCl \rightarrow AgCl + NaNO_3$$

[4] Small mass differences between the initial and final nuclides in a series of nuclear reactions can be determined very accurately from energy differences.

[5] E. Wichers, *Anal. Chem.*, **35**, 23A (1963).

indicates that 1 mole[6] of silver nitrate reacts with 1 mole of sodium chloride to form 1 mole of silver chloride and 1 mole of sodium nitrate. If on mixing a solution of silver nitrate with a solution of sodium chloride the above reaction proceeds quantitatively to the right, the equation also indicates that 169.88 g of silver nitrate reacts with 58.44 g of sodium chloride to give 143.32 g of (insoluble) silver chloride and 84.99 g of sodium nitrate. Moreover, it indicates that 107.87 g of silver ion reacts with 35.45 g of chloride ion to yield 143.32 g of silver chloride. Thus it is an easy matter to calculate how many grams of silver chloride are formed from *a* grams of silver or *a′* grams of silver nitrate, or *b* grams of chloride or *b′* grams of sodium chloride. It may be stated also that the 143.32 g of silver chloride formed is equivalent to 107.87 g of silver and 35.45 g of chloride, respectively, or more generally *w* grams of silver chloride formed correspond to

$$\frac{107.87}{143.32}\,w \text{ grams of silver}$$

and

$$\frac{35.45}{143.32}\,w \text{ grams of chloride}$$

If gases react or are reaction products, it is also possible to derive weight–volume relationships from equations. Avogadro's law (1811) states: *Under the same conditions of temperature and pressure, equal volumes of all gases contain the same number of molecules.* One mole of an ideal gas occupies 22.41 liters under standard conditions of temperature and pressure (0°C, 760 mm Hg), and contains 6.02×10^{23} molecules (Avogadro's number).

The chemical reaction between calcium carbonate and hydrochloric acid takes place quantitatively under the proper conditions according to the equation

$$CaCO_3 + 2\,HCl \rightarrow CaCl_2 + CO_2 + H_2O$$

This means that 1 mole or 100.08 g of calcium carbonate yields 44.00 g of carbon dioxide, the latter occupying a volume of 22.41 liters at 0°C under the pressure of a normal atmosphere. The volume V occupied by the dry gas at a temperature t°C and a pressure p is found from the gas law:

$$\frac{pV}{T} = \text{constant} = R$$

in which $T = (273 + t\text{°C})$. This relation holds exactly only for an ideal gas, and actual gases show greater or smaller deviations.[7]

[6] "A mole is an amount of substance, of specified chemical formula, containing the same number of formula units (atoms, molecules, ions, electrons, quanta, or other entities) as there are in 12 grams (exactly) of the pure nuclide ^{12}C" (IUPAC definition, 1965).

[7] Regarding the calculations of quantitative analysis see, for example, the following: L. F. Hamilton and S. G. Simpson, *Calculations of Analytical Chemistry*, McGraw-Hill, New York, 6th ed., 1954; E. H. Miller, *The Calculations of Analytical Chemistry*, Macmillan, New York, 1921; J. A. Wilkinson, *Calculations in Quantitative Chemical Analysis*, McGraw-Hill, New York, 2nd ed., 1938; C. J. Engelder, *Calculations of Quantitative Analysis*, Wiley, New York, 1939; S. B. Arenson, *How to Solve Problems in Quantitative Analysis*, Crowell, New York, 1942; P. W. West, *Calculations in Quantitative Analysis*, Macmillan, New York, 1948.

Nonstoichiometric compounds. A number of inorganic compounds (solids) do not have constant, integral ratios of atoms. For example, ferrous oxide has the composition $FeO_{1.06-1.19}$. The deficiency in iron arises from cation vacancies in the lattice; electrical neutrality is maintained by replacement of some Fe^{++} ions by Fe^{3+}. Other solids may have an excess of the cation. This can result from the presence of cations in the interstices between the ions of the normal lattice, or from anion vacancies, and again charge balance is maintained by substitution of one ion by another of the same sign but different charge. The term nonstoichiometric compound is somewhat misleading because it does not refer to a pure compound. Nonstoichiometric ferrous oxide is actually a mixed crystal (see Chap. 11) of ferrous and ferric oxide.

Nonstoichiometric compounds are occasionally encountered in chemical analysis. Some weighing forms in gravimetric analysis tend to be nonstoichiometric (e.g., U_3O_8) and special care is needed to minimize error from variability in composition.

Expression of analytical results

Chemical expression of results. The amount of a certain constituent in a given sample is calculated from the analytical results by methods that will be considered later. The question that concerns us at this point is: In what form shall the results of an analysis be reported? For example, calcium has been determined in a limestone. Shall it be reported as the element, the carbonate, or the oxide?

Whenever possible, the element determined should be reported in the form in which it is present in the sample analyzed, unless there is a good reason for the contrary. Thus in reporting the result of a determination of nitrogen, it should be stated as nitrate (NO_3^- or N_2O_5), nitrite (NO_2^- or N_2O_3), or ammonia (NH_3), depending upon the state or states in which the element is present. The analysis of a simple salt may be reported in the ionic form (e.g., K_2SO_4 as K^+ and SO_4^{--}). If the composition of the sample is unknown and cannot readily be determined, the results are usually given in the form of the elements or their oxides. Often the purpose of the analysis decides the form in which the constituents are reported. Thus the calcium content of a limestone is conveniently given in terms of calcium oxide if the material is to be used for the manufacture of lime, the iron content of an ore is preferably stated in terms of metallic iron, etc.

In the complete analysis of a complex inorganic sample containing oxygen it is generally preferable to report the elements in the form of their oxides (K as K_2O, Ca as CaO, Fe(II) as FeO, Fe(III) as Fe_2O_3, S as SO_3, P as P_2O_5, etc.). There are a number of reasons for this. Oxygen is never determined directly and must therefore be apportioned among the other constituents with which it exists in combination. The state of oxidation of the metals is readily shown in the statement of results. The sum of the acid and basic oxides (including water), expressed in per cent, must equal 100 in a perfect analysis. There is no objection to this mode of expression as long as all acid-forming elements of the sample may be considered to be present as acid oxides, but difficulties arise when the acid elements have no corresponding oxides (halides, sulfides, etc.). However, in such cases it is possible to report the nonconforming acid elements in the elemental form (Cl, S(II), etc.) and then deduct the oxygen equivalent of these elements from the summation to obtain 100 per cent.

The results of the analysis of a solution of electrolytes are generally expressed in terms of the ions present.

Numerical expression of results. It is the relative amount of a constituent in a sample that is of importance. Accordingly, in expressing the quantity of a constituent present in the sample analyzed, the following general expression is used:

$$\frac{q}{Q} \times c$$

where q = the quantity of constituent present in the sample
Q = the quantity of the sample
c = a factor

The factor c is introduced in order that the result may be expressed in units of convenient magnitude.

If q and Q are both given in the same units of weight and c is put equal to 100, the result is the *percentage by weight* of constituent in the sample; if c is set equal to 1,000,000 and q and Q are expressed in the same weight units, the answer is the parts per million by weight of the constituent in the sample, and so on. The quantities q and Q may also be expressed in volume units, as we shall see.

Solids. When the sample is a solid, it is most usual to express the amount of the constituent in question in terms of percentage by weight:

$$\frac{\text{weight of constituent}}{\text{weight of sample}} \times 100$$

Sometimes when the relative amount of the constituent is very small, the result is stated in parts per million (ppm) by weight. For example, if a 1-g sample of potassium chloride is found to contain 0.002 mg of lead, it is more convenient to express the lead content of the potassium chloride as 2 ppm than as 0.0002 per cent.

Liquids. With liquid samples various modes of expression are possible:

1. *Percentage by weight.* The number of parts of constituent in 100 parts of sample, both in the same weight units, is the weight percentage of the constituent in the sample. This expression may be designated P^w_w.

2. *Percentage by volume.* The number of parts by volume of a constituent in 100 volumes of sample at a specified temperature gives the volume percentage of that constituent (P^v_v). This mode of expression is often used when the constituent is a liquid or gas.

3a. *Parts by weight* of constituent in 100 parts by volume of sample.
3b. *Parts by volume* of constituent in 100 parts by weight of sample.

Expressions 3a and 3b are designated by P^w_v and P^v_w. The temperature must be specified in each case. The former expression may be used to designate the concentration of a solid or liquid in a liquid and the latter the volume of liquid (or gas) in a given weight of another liquid.

In general, the expression P^w_w is preferable to the others because it is independent of the temperature. Evidently the following relations exist among these modes of expression if w is expressed in grams, v in milliliters, and density, d, in grams per milliliter:

$$P^w_v = d_l P^w_w \qquad (d_l = \text{density of liquid sample})$$

$$P^v_w = \frac{P^w_w}{d_c} \qquad (d_c = \text{density of liquid constituent})$$

and

$$P_v^v = \frac{d_s}{d_c} \times P_w^w$$

There are other ways of designating the concentration of solutions which need not be considered here (see Chap. 33).

When the amount of constituent present in a solution is small, it is advantageous to express it in parts per million by weight or volume. Thus in the analysis of natural waters the ions are reported in parts per million. For example, a water found to contain 0.030 g of magnesium in 1 liter is said to have 30 ppm of Mg^{++}, i.e., 30 parts by weight of Mg^{++} in 1,000,000 parts by volume of solution. Generally it makes little difference in the case of a *dilute* solution whether the quantity of the latter is expressed in weight or volume; i.e., 997 g of water at room temperature occupies a volume of about 1000 ml.

Gases. The composition of a gaseous mixture is commonly expressed in percentage by volume (P_v^v).

PROBLEMS

Most of the problems in this set are intended as a review (or a slight extension) of calculations of the type the student has already encountered in general chemistry courses. He is assumed to know the meaning of such terms as mole, molarity, equivalent, normality, and the like. In working these problems, and others that follow, he should obtain the answers through his own train of reasoning and should avoid the mechanical use of formulas and equations.

1. A solution of 100 ml volume contains 0.2083 g of $BaCl_2$.
   (a) How many moles of $BaCl_2$, Ba^{++}, and Cl^- are present?
   (b) How many equivalents of each of these are present?
   (c) What are the molarity and normality of the solution?
   (d) What weights of Ba and Cl are present?
   (e) Find the weight of $AgNO_3$ required to precipitate the chloride in the solution and the weight of AgCl obtained.
   (f) What volume of 0.100 *M* silver nitrate solution is needed for the precipitation of the chloride?
   (g) How many milliliters of silver nitrate solution containing 10.0 g of $AgNO_3$ per liter is needed to precipitate the chloride?
   (h) With what weight of $Al_2(SO_4)_3$ will the solution react and how much $BaSO_4$ will be obtained?

   Ans. (a) $0.2083/BaCl_2 = 0.2083/208.3 = 0.001000$; $0.2083/BaCl_2$; $(2 \times 0.2083)/BaCl_2$; (b) $(2 \times 0.2083)/BaCl_2$, etc.; (c) $1000/100 \times 0.2083/BaCl_2$ *M*; $2 \times (1000/100) \times 0.2083/BaCl_2$ *N*; (d) $(Ba/BaCl_2) \times 0.2083$ g; $(2Cl/BaCl_2) \times 0.2083$ g; (e) $(2AgNO_3/BaCl_2) \times 0.2083$ g; $(2AgCl/BaCl_2) \times 0.2083$ g; (f) $(2 \times 0.2083)/(BaCl_2 \times 0.100) \times 1000$ ml; (g) $(2 \times AgNO_3 \times 0.2083)/(BaCl_2 \times 10) \times 1000$ ml; (h) $(Al_2(SO_4)_3/3BaCl_2) \times 0.2083$ g; $(BaSO_4/BaCl_2) \times 0.2083$ g.
2. What weight of $H_2C_2O_4 \cdot 2\,H_2O$ is needed to prepare 1000 ml of a solution of such strength that 1.00 ml will just precipitate the cerium(III) in 1.00 ml of 0.100 *M* $CeCl_3$ solution?
3. What weight of NaCl contains the same amount of Cl as 1 g of $BaCl_2 \cdot 2H_2O$?
4. Assuming the availability of the necessary reactions, what weight of $K_4Fe(CN)_6$ can be obtained from 1.00 g of $K_3Fe(CN)_6$?
5. If K_2PtCl_6 sells for \$4.50 per gram and Pt for \$8.76 per gram, which is the cheaper source of platinum and by what percentage?

6. How many molecules of $CHCl_3$ are contained in 0.0010 ml of chloroform vapor at 25°C and a pressure of 0.0010 mm Hg?
7. 2 liters of dry air (0°C, 760 mm Hg) is passed over sodium hydroxide to absorb the carbon dioxide present. If a gain in weight of 1.2 mg is observed, what is the percentage by volume of carbon dioxide in the air? *Ans.* 0.03 per cent.
8. Water at room temperature in equilibrium with the atmosphere is 1.5×10^{-5} *M* in carbon dioxide. What are the weight and the volume (0°C, 760 mm Hg) of carbon dioxide in 1 liter of such water? *Ans.* 0.66 mg, 0.34 ml.
9. Concentrated phosphoric acid containing 85 per cent H_3PO_4 has a density of 1.69. Calculate the number of grams and moles of H_3PO_4 in 1 liter of acid. What is the molarity of the acid?
   Ans. $1000 \times (85/100) \times 1.69 = 1437$ g; $1437/H_3PO_4 = 1437/98.04 = 14.7$ moles.
10. The density of 16 *N* nitric acid is 1.40. How many grams of water are contained in 100 ml of the acid?
11. What volume of concentrated hydrochloric acid having a density of 1.200 and containing 39.11 per cent HCl is required to prepare (a) 1 liter of acid containing 20.01 per cent HCl (density 1.100); (b) 1 liter of 6 *N* acid?
    Ans. (a) 469.0 ml; (b) 466.2 ml.
12. If the density of anhydrous ethyl alcohol is 0.7939, what is the per cent by weight of alcohol in a 95 per cent by volume alcohol solution of density 0.8160?
    Ans. 92.43 per cent.
13. In a colorimetric determination of boron a sulfuric acid solution containing 98.5 per cent H_2SO_4 is required. It is to be prepared by adding fuming sulfuric acid containing 20 per cent free SO_3 to ordinary concentrated sulfuric acid containing 95 per cent H_2SO_4. How many grams of each are required to make 100 g of the desired mixture of 98.5 per cent H_2SO_4?
14. 1 mole of propanol ($d = 0.7854$ at 20°C, mol. wt. = 60.1) dissolves 4.6×10^{-5} mole of NaCl at 20°C. What is the solubility of NaCl in milligrams per liter of propanol?
15. 100 g of a saturated solution of lithium chloride in water which is in equilibrium with $LiCl \cdot H_2O$ as the solid phase contains 45.85 g of LiCl at 25°C. The solution density is 1.296. What is the solubility of $LiCl \cdot H_2O$ in (a) 1 kg of water; (b) 1 liter of the solution? What is the mole fraction of LiCl in the saturated solution?
16. It is desired to prepare a sample containing 50.0 per cent Cl by mixing NaCl and KCl. In what ratio must the salts be mixed?
17. In the determination of ethyl alcohol in a liquid, a 25.00-ml sample was taken at 20°C (specific gravity of liquid = 1.030, 20°C/4°C), diluted to 150 ml, and the mixture distilled. One hundred milliliters of distillate (containing all the alcohol in the original sample) was collected and its specific gravity determined; it was found to be 0.97953 (20°C/4°C), corresponding to 14.50 per cent alcohol by volume at 20°C and 11.68 per cent by weight. Calculate (a) the percentage by weight of alcohol in the original sample; (b) the percentage by volume at 20°C; (c) the number of grams of alcohol in 100 ml of sample at 20°C.
18. A hydrated cadmium salt contains 43.82 per cent Cd, 37.45 per cent SO_4, and 18.73 per cent H_2O. What is its formula?
19. What is the formula of a mineral containing 64.5 per cent SiO_2, 27.4 per cent Al_2O_3, and 8.4 per cent Li_2O? *Ans.* $LiAl(SiO_3)_2$ or $Li_2O \cdot Al_2O_3 \cdot 4SiO_2$.
20. A cesium iodobismuthite was found to contain 20.66 per cent Cs, 21.28 per cent Bi, and 58.15 per cent I. What is its formula? What would be the deviations (in absolute percentages) from the theoretical composition if the sample contained 5 per cent Cs_3BiI_6?
21. A compound composed only of C, H, and O was found to contain 65.5 per cent C and 5.5 per cent H and to have a molecular weight of 110. What is its empirical formula? *Ans.* $C_6H_6O_2$.
22. The chlorosity of seawater is defined as the chlorinity expressed in grams per liter at 20°C, and the chlorinity as the total amount of Cl, Br, and I in 1 kg of seawater,

assuming that bromine and iodine have been replaced by chlorine. Obtain the chlorosity of a sample of seawater from the following data:

	mg/kg
Cl	18,980
Br	65
I	0.05

$d_{20} = 1.0242$.

23. An excess of oxygen is added to a mixture of CO, CH_4, and N_2 having a volume of 25.0 ml and combustion is carried out. After cooling to the original (room) temperature, absorption analysis shows that 12.5 ml of O_2 has been consumed and 10.0 ml of CO_2 has been produced. What was the volume percentage composition of the sample mixture? *Ans.* 60 per cent N_2, 20 per cent CO, 20 per cent CH_4.
24. A suitable volume of a solution containing 102 g of RCl (molecular weight 110) and 11.1 g of the dimer $(RCl)_2$ per liter is titrated with standard $AgNO_3$ solution, whereby all the chloride in the solution is transformed into AgCl. The titration volume represents the *analytical concentration* of RCl, i.e., as if it were all present as RCl. What is the analytical concentration in moles per liter? What is the actual concentration in moles per liter?
25. The abundances of ^{17}O and ^{18}O in atmospheric oxygen are 0.0374 and 0.2039 per cent; the nuclidic masses, based on $^{12}C = 12$, are ^{16}O, 15.994915; ^{17}O, 16.999134; and ^{18}O, 17.999160. Calculate the atomic weight of oxygen.
26. Before 1960, physicists used a scale of nuclidic masses in which the reference species was $^{16}O = 16$. On this scale $^{12}C = 12.00386$. Find the atomic weight of oxygen when $^{16}O = 16$ (see Problem 25). What is an advantage in taking $^{12}C = 12$ instead of $^{16}O = 16$ as the reference species for atomic weights?
27. In 1937, the accepted atomic weight of carbon was 12.00, in 1938, 12.011. If the assay value of a sample of benzoic acid (C_6H_5COOH, mol. wt. = 122.1) based on the 1937 atomic weight was 99.90 per cent what was it in 1938 with the new atomic weight of carbon?
28. If the terrestrial extremes of the $^{18}O/^{16}O$ ratio are represented by polar ice and atmospheric carbon dioxide, which can differ by 10 per cent, what is the difference in the atomic weight of oxygen in the two sources? *Ans.* 0.0004.
29. Berzelius in 1843 determined the atomic weight of calcium by converting calcium oxide into calcium sulfate by treatment with sulfuric acid, the excess of the latter being removed by strong heating. In a typical experiment 2.5040 g of CaO yielded 6.0750 g of $CaSO_4$. Assuming S = 32.06 and O = 16.00, calculate the atomic weight of Ca, neglecting air buoyancy. *Ans.* 40.14.
30. Hönigschmid and Sachtleben in 1929 determined the atomic weight of silver and barium by converting anhydrous barium perchlorate into barium chloride by heating and then finding the weight of silver required to precipitate the chloride in the latter. In one analysis 11.69950 g of $Ba(ClO_4)_2$ gave 7.24609 g of $BaCl_2$, which was found to be equivalent to 7.50692 g of Ag (all weights are vacuo weights). Calculate the atomic weight of silver and barium from these data, taking O = 15.9994 and Cl = 35.453.
31. In a comparison of primary volumetric standards at the University of Minnesota in 1929, a solution of hydrochloric acid was standardized against borax, $Na_2B_4O_7 \cdot 10H_2O$, and found to be 0.18081 *N* (equivalents per kilogram of solution). A solution of sodium hydroxide standardized against U.S. Bureau of Standards and Kahlbaum's benzoic acid, C_6H_5COOH, gave the respective normalities 0.189255 and 0.189215. The weight ratio HCl/NaOH was 1.0463. (Normalities were calculated on the basis of 1929 atomic weights and vacuum corrections

were applied to air weights.) From these results calculate the atomic weight of carbon on the current scale, using 1962 atomic weights for H, B, O, and Na.

	At. Wt., 1929	*At. Wt., 1962*
H	1.008	1.00797
B	10.82	10.811
O	16	15.9994
Na	22.997	22.9898
C	12.00	(12.01115)

Ans. 12.007 (B.S.); 12.003 (Kahlbaum).

32. The unit cell of sodium chloride (cubic) consists of 4 Na^+ and 4 Cl^-. Taking the cell edge as 5.627 Å and the density of sodium chloride as 2.1637 (20°C), find the value of Avogadro's number.

CHAPTER 3 *The Mass-Action Law. Elementary Discussion of Acid–Base Equilibria*

Equilibrium in a reversible reaction. The mass-action law

When two substances, A and B, react with each other and there is no additional change in the relative masses of the two substances, *equilibrium* has been reached. When the reaction products, C and D, can react with each other to form A and B, the reaction is said to be *reversible* and is generally written in the form

$$A + B \rightleftharpoons C + D \tag{1}$$

Reaction (1) applies to all types of equilibria. For example, it describes the reaction of an acid with water, the solubility of an ionic or molecular substance, the reaction of an oxidizing agent and a reducing agent, and the esterification of acetic acid by ethanol.

In analytical chemistry we usually are concerned with reversible reactions in which homogeneous equilibrium is rapidly established. The two arrows separating the left and right sides of Eq. (1) indicate that the equilibrium is dynamic, not static. At equilibrium, A and B are reacting with each other to form C and D at the same rate that C and D are reacting to form A and B. Thus the equilibrium concentrations of A, B, C, and D remain unchanged, even though a particular ion or molecule upon which we focus our attention may appear to have only a transient existence. In other words, when equilibrium has been reached, the velocity of the reaction between A and B is equal to that between C and D. The speed of the reactions is expressed by the *law of mass action*, first clearly stated by Guldberg and Waage.[1] *The rate of reaction is proportional to the active concentrations of the reacting substances, each raised to the power equal to the number of molecules appearing in the balanced equation.*

According to the mass-action law the velocity, v_1, of the reaction between A and B in Eq. (1) is

$$v_1 = k_1[\mathrm{A}][\mathrm{B}]$$

and that between C and D is

$$v_2 = k_2[\mathrm{C}][\mathrm{D}]$$

where k_1 and k_2 are constants and the terms in brackets represent the concentration of the species in gram formula weights per liter, the molar concentration. When equilibrium is reached, $v_1 = v_2$, and

$$k_1[\mathrm{A}][\mathrm{B}] = k_2[\mathrm{C}][\mathrm{D}]$$

or

$$\frac{[\mathrm{C}][\mathrm{D}]}{[\mathrm{A}][\mathrm{B}]} = \frac{k_1}{k_2} = K \qquad (t = \text{const.})$$

If we consider the reaction between 2A and B:

$$2\mathrm{A} + \mathrm{B} \rightleftharpoons \mathrm{C} + \mathrm{D}$$

or

$$\mathrm{A} + \mathrm{A} + \mathrm{B} \rightleftharpoons \mathrm{C} + \mathrm{D}$$

it is easily shown that

$$\frac{[\mathrm{C}][\mathrm{D}]}{[\mathrm{A}]^2[\mathrm{B}]} = K \qquad (t = \text{const.})$$

The equilibrium constant for the general reversible reaction,

$$a\mathrm{A} + b\mathrm{B} + c\mathrm{C} + \cdots \rightleftharpoons p\mathrm{P} + q\mathrm{Q} + r\mathrm{R} + \cdots$$

is

$$K = \frac{[\mathrm{P}]^p[\mathrm{Q}]^q[\mathrm{R}]^r \cdots}{[\mathrm{A}]^a[\mathrm{B}]^b[\mathrm{C}]^c \cdots} \tag{2}$$

K is called the *equilibrium constant* (constant at a *given temperature*).

Equation (2) has been derived in an approximate manner and applies only to a particular set of experimental conditions. Using the methods of chemical thermodynamics it can be shown that the rigorously correct expression corresponding to Eq. (2) is

$$K^0 = \frac{(\mathrm{P})^p(\mathrm{Q})^q(\mathrm{R})^r \cdots}{(\mathrm{A})^a(\mathrm{B})^b(\mathrm{C})^c \cdots} \tag{3}$$

[1] C. M. Guldberg and P. Waage, *J. Prakt. Chem.* [2]**19** (1879).

where the parentheses denote **activities** and K^0 is called the *thermodynamic*, or *activity constant*. Generally, we shall use units of moles per liter for activities in this text, as in the case of concentrations. In dilute solution, it is often permissible to approximate the activity of a species with its molar concentration.

The equilibrium constant, K, in Eq. (2) is called the *concentration constant*.

When a chemical species has an activity exactly equal to unity it is said to be in its standard state. By convention all solids have an activity of unity at 25°C and 1 atm pressure. This convention is of importance in discussing some types of equilibria encountered in analytical chemistry.

Example: Reduction of silver ion by zinc metal, excess of zinc metal:

$$2\,Ag^+ + Zn \rightleftharpoons 2\,Ag + Zn^{++}$$

Since (Ag) = 1 and (Zn) = 1,

$$K^0 = (Zn^{++})/(Ag^+)^2$$

Example: Solution of silver chloride, excess of silver chloride:

$$AgCl \rightleftharpoons Ag^+ + Cl^-$$

Since (AgCl) = 1,

$$K^0_{AgCl} = (Ag^+)(Cl^-)$$

The equilibrium constant for the last reaction has a special name; it is the *solubility product* of silver chloride.

Other equilibrium constants have particular names associated with them. When a weak acid, such as nitrous acid, dissociates in water,

$$HNO_2 + H_2O \rightleftharpoons H_3O^+ + NO_2^-$$

and the equilibrium constant for this reaction is

$$K = \frac{(H_3O^+)(NO_2^-)}{(H_2O)(HNO_2)}$$

In dilute solutions the activity of water is constant (about 55 M), and

$$K^0_{HNO_2} = \frac{(H_3O^+)(NO_2^-)}{(HNO_2)}$$

where $K^0_{HNO_2}$ [$= K(H_2O)$] is called the *dissociation* (or *ionization*) *constant of the acid*,[2] and includes the activity of water.

The equilibrium constant for the reaction of the base, B, with water,

$$B + H_2O \rightleftharpoons BH^+ + OH^-$$

$$K^0_B = \frac{(BH^+)(OH^-)}{(B)}$$

is called the *dissociation constant of the base*.

[2] Protons are solvated in water and conventionally we indicate this solvation by writing H_3O^+. Actually the principal species appears to be $H_9O_4^+$. For simplicity we shall use H_3O^+ when it is necessary to indicate the solvation of the proton and H^+ when it is unnecessary. Hydroxyl ions are also solvated.

The very important equilibrium constant for

$$2\,H_2O \rightleftharpoons H_3O^+ + OH^-$$
$$K_w^0 = (H_3O^+)(OH^-)$$

is known as the *self-dissociation* (or *autoprotolysis*) *constant* of water.

Equilibrium constants for the formation of complex ions, e.g.,

$$Cd^{++} + Cl^- \rightleftharpoons CdCl^+$$

$$K^0 = \frac{(CdCl^+)}{(Cd^{++})(Cl^-)}$$

are known as *stability constants.* The reciprocal of the stability constant is called the *instability constant,* and corresponds to the reverse reaction.

It should be understood that there is no relation between the rate at which a reaction reaches equilibrium and the magnitude of its equilibrium constant. The speed of most reactions increases rapidly with increasing temperature (usually more than twofold for every increase of 10°C) and is also affected by the presence of certain substances that have a specific effect. Such substances are called *positive catalysts* if they *increase* the rate of reaction.

Positive catalysts are used in analysis to increase the speed of slow quantitative reactions. Some oxidation–reduction reactions employed in volumetric analysis proceed rather slowly. This, for example, is the case with the reaction between hydrogen peroxide and iodide in acid medium, or bromate with iodide in acid medium:

$$H_2O_2 + 2\,H^+ + 2\,I^- \rightarrow 2\,H_2O + I_2$$
$$BrO_3^- + 6\,H^+ + 6\,I^- \rightarrow Br^- + 3\,I_2 + 3\,H_2O$$

The addition of a trace of molybdate catalyzes the reactions so strongly that the titration of iodine can be made without waiting.

Factors affecting the position of equilibrium

Theorem of LeChâtelier–van't Hoff. This theorem may be stated as follows: *Any alteration in the factors that determine chemical equilibrium causes the equilibrium to become displaced in such a way as to oppose as far as possible the effect of the alteration.* An increase in the temperature of a system in equilibrium displaces the equilibrium in such a way that heat is absorbed. For example, the heat of neutralization of a strong acid with a strong base is determined by the reaction

$$H^+ + OH^- \rightleftharpoons H_2O + 13{,}700 \text{ cal}$$

The formation of water takes place with the evolution of much heat. According to the above rule, the equilibrium will be displaced from the right to the left by a rise in temperature, or, in other words, *the dissociation of water into hydrogen and hydroxyl ions increases with increasing temperature.* Changing the temperature from 25 to 100°C increases K_w^0 by a factor of 100.

Pressure changes have a relatively small influence on the position of solution equilibria, because volume changes for solution reactions are usually very small. On the other hand, gas-phase reactions may involve large volume changes and the position of equilibrium can be quite pressure-sensitive.

Strong and weak electrolytes in water

Electrolytes are substances that dissociate more or less completely into ions when dissolved in water. According to modern electrochemical theories a distinction may be made between *strong* and *weak electrolytes*. Strong electrolytes when dissolved in water are virtually completely dissociated into ions, whereas in solutions of weak electrolytes, ions and undissociated molecules of the electrolyte are both present.

To the group of strong electrolytes the following belong:

Various inorganic acids (in water): perchloric, nitric, hydrochloric, hydrobromic, and hydriodic acids.

The hydroxides of the alkalies, of silver, the alkaline earths, and quaternary ammonium ions. Almost all salts (exceptions: the halides, cyanides, and thiocyanates of mercury, cadmium, zinc, and a few others).

The alkali salts of acids: perchlorates, nitrates, chlorides, bromides, iodides, thiocyanates, acetates (and other monocarboxylates), and sulfonates, and many other acids.

To the group of weak electrolytes belong:

Some inorganic acids: e.g., phosphoric, sulfurous, boric, and carbonic acids.

Some inorganic hydroxides, such as ammonium hydroxide, most divalent and trivalent metal hydroxides, hydrazine, and hydroxylamine.

Virtually all the organic acids (with the exception of sulfonic acids).

Virtually all the organic bases.

A table of thermodynamic equilibrium constants for inorganic and organic acids and bases is given in the Appendix (Table A-2).

A very simple relation exists for estimating the dissociation constants of inorganic oxyacids. If we represent the formula of the acid as $XO_m(OH)_n$, where X is the central atom, the strengths of these acids fall in groups that depend only upon m, n, and the charge on the acid. Table 3-1 presents such a classification and it is clear that acid strength increases with the value of m but decreases about a factor of 10^{-5} for each negative charge placed on the acid.

The term pK in Table 3-1 represents the negative logarithm to the base 10 of the dissociation constant

$$pK = -\log_{10} K = -\log K$$

It is convenient to use this notation interchangeably with exponential notation.

Example: The dissociation constant of an acid is 2.00×10^{-5}. What is its pK?

$$\begin{aligned} pK &= -\log(2.00 \times 10^{-5}) = -\log 10^{-5} - \log 2.00 \\ &= 5.00 - 0.30 \qquad = 4.70 \end{aligned}$$

Example: The pK of an acid is 8.30. What is its equilibrium constant?

$$\begin{aligned} -\log K &= 8.30 = 9.00 - 0.70 \\ \log K &= -9.00 + 0.70 \\ K &= 10^{-9.00} \times 10^{+0.70} = 5.0 \times 10^{-9} \end{aligned}$$

TABLE 3-1. **Classification of the Strengths of Oxyacids**[a]

Example	*Structure*	pK_1	pK_2	pK_3
$m = 0$				
H_4GeO_4	$Ge(OH)_4$	8.68	12.7	
H_3AsO_3	$As(OH)_3$	9.22		
H_6TeO_6	$Te(OH)_6$	6.2, 8.8		
HClO	ClOH	7.53		
HBrO	BrOH	8.7		
HIO	IOH	11.0		
H_3BO_3	$B(OH)_3$	9.23		
H_3AlO_3	$Al(OH)_3$	11.2		
$m = 1$				
H_3PO_4	$PO(OH)_3$	2.23	7.2	12.33
H_3PO_3	$HPO(OH)_2$	2.0	6.58	
H_3PO_2	$H_2PO(OH)$	2.0		
H_3AsO_4	$AsO(OH)_3$	2.3	9.0	13.0
H_2SO_3	$SO(OH)_2$	1.9	7.0	
H_2SeO_3	$SeO(OH)_2$	2.6	8.3	
H_2TeO_3	$TeO(OH)_2$	2.7	8.0	
$HClO_2$	ClO(OH)	2.0		
H_5IO_6	$IO(OH)_5$	1.55	8.27	
HNO_2	NO(OH)	3.29		
$m = 2$				
H_2SO_4	$SO_2(OH)_2$	−3	1.9	
H_2SeO_4	$SeO_2(OH)_2$	Strong	2.0	
HIO_3	$IO_2(OH)$	0.8		
HNO_3	$NO_2(OH)$	−1.3		
$HClO_3$	$ClO_2(OH)$	−1		
$m = 3$				
$HClO_4$	$ClO_3(OH)$	Very strong		

[a] From J. E. Ricci, *J. Am. Chem. Soc.* **70**, 109 (1948).

Ion concentrations in solutions of strong electrolytes

Since strong electrolytes are completely dissociated in aqueous solutions, the ion concentrations are directly found from the molecular concentration of the solution.

Examples:

$$HCl \rightarrow H^+ + Cl^-$$

In a 0.1 *M* solution of hydrochloric acid the hydrogen-ion concentration, $[H^+]$, is equal to the chloride-ion concentration, $[Cl^-]$, and both are equal to 0.1 or 10^{-1}.

$$NaOH \rightarrow Na^+ + OH^-$$

In 0.01 *M* sodium hydroxide solution:

$$[Na^+] = [OH^-] = 0.01 = 10^{-2}$$

$$BaCl_2 \rightarrow Ba^{++} + 2\ Cl^-$$

In 0.01 *M* barium chloride solution:

$$[Ba^{++}] = 0.01 = 10^{-2}$$
$$[Cl^-] = 0.02 = 2 \times 10^{-2}$$

$$Ag_2SO_4 \rightarrow 2\ Ag^+ + SO_4^{--}$$

In 0.001 *M* silver sulfate solution:

$$[Ag^+] = 0.002 = 2 \times 10^{-3}$$
$$[SO_4^{--}] = 0.001 = 10^{-3}$$

$$M_aX_b \rightarrow aM^{b+} + bX^{a-}$$

In general, if the strong electrolyte has the formula M_aX_b, and C is the concentration,

$$[M^{b+}] = aC$$
$$[X^{a-}] = bC$$

Ion concentrations in solutions containing weak electrolytes

Weak electrolytes are incompletely dissociated and it is not possible to calculate the individual ion concentrations from the concentration of substance. However, simple relations exist between molecular (analytical) concentrations and equilibrium concentrations.

Examples:

Acetic Acid:

$$HAc \rightleftharpoons H^+ + Ac^-$$

If we denote molecular acetic acid by HAc, and acetate ion by Ac^-, the sum of acetic acid and acetate ion in a 0.01 *M* solution of acetic acid must be 0.01:

$$0.01 = [HAc] + [Ac^-]$$

Such expressions are sometimes called *material-balance equations.*

Phosphoric acid:

$$H_3PO_4 \rightleftharpoons H^+ + H_2PO_4^-$$
$$H_2PO_4^- \rightleftharpoons H^+ + HPO_4^{--}$$
$$HPO_4^{--} \rightleftharpoons H^+ + PO_4^{3-}$$

In 0.1 *M* phosphoric acid,

$$[H_3PO_4] + [H_2PO_4^-] + [HPO_4^{--}] + [PO_4^{3-}] = 0.1$$

Sodium sulfite:

$$Na_2SO_3 \rightarrow 2\ Na^+ + SO_3^{--}$$
$$SO_3^{--} + H_2O \rightleftharpoons HSO_3^- + OH^-$$
$$HSO_3^- + H_2O \rightleftharpoons H_2SO_3 + OH^-$$

In 0.05 *M* sodium sulfite,

$$[Na^+] = 2 \times 0.05 = 0.1$$
$$[H_2SO_3] + [HSO_3^-] + [SO_3^{--}] = 0.05$$

Cupric nitrate plus ammonium hydroxide[3]: In this mixture copper exists as a complex ion containing from zero to four ammonia molecules. If the concentration of cupric nitrate is $10^{-3}\ M$ and that of ammonia is $10^{-1}\ M$, we can write three material-balance equations:

$$[NO_3^-] = 2 \times (0.001) = 2 \times 10^{-3}$$

$$[Cu^{++}] + [CuNH_3^{++}] + [Cu(NH_3)_2^{++}] + [Cu(NH_3)_3^{++}] + [Cu(NH_3)_4^{++}] = 10^{-3}$$

$$[NH_3] + [CuNH_3^{++}] + 2[Cu(NH_3)_2^{++}] + 3[Cu(NH_3)_3^{++}] + 4[Cu(NH_3)_4^{++}] = 10^{-1}$$

Rule of electroneutrality

When a compound is dissolved in water, positive and negative ions may be produced. In addition, these ions may also undergo reaction with other solutes or the solvent to form other ions. The relationship between the concentration of positive and negative ions is given by the rule of electroneutrality, which states that the total concentration of positive charge exactly equals the total concentration of negative charge.

This is easily seen since every compound is electrically neutral (has no net charge) in the solid state, and therefore a solution of the compound must also be electrically neutral. In a solution a single salt, M_mX_n, the concentration of the cation, $[M^{n+}]$, multiplied by the charge on the cation, n, equals the concentration of the anion, $[X^{m-}]$, multiplied by the charge on the anion, m:

$$n[M^{n+}] = m[X^{m-}]$$

If several salts are present in the solution, all the terms of the form $n[M^{n+}]$ are summed and set equal to the sum of all terms of the form $m[X^{m-}]$.[4]

For example, if 1 gram formula weight of sodium perchlorate $NaClO_4$ is dissolved in 1 liter of water, the concentration of sodium ion is 1 M, as is that of perchlorate ion:

$$[Na^+] = [ClO_4^-]$$

In a 1 M solution of calcium perchlorate, $Ca(ClO_4)_2$, the concentration of calcium ion equals 1 M and that of perchlorate 2 M, or

$$2[Ca^{++}] = [ClO_4^-]$$

If the solution contains 1 M sodium perchlorate and 1 M calcium perchlorate, the total concentration of perchlorate ion is 3 M, and we see that

$$2[Ca^{++}] + [Na^+] = [ClO_4^-]$$

When cobaltous chloride is dissolved in aqueous sodium chloride, the solution

[3] We shall use the symbol NH_3 to represent the species present in an aqueous solution of ammonia (ammonium hydroxide). Evidence indicates that ammonia is coordinated to three water molecules in aqueous solution.

[4] Mathematically, the rule of electroneutrality is

$$\sum n_i[M^{n_i+}] = \sum m_i[X^{m_i-}]$$

where $[M^{n_i+}]$ represents the concentration of the ith cation of charge n_i, and $[X^{m_i-}]$ represents the concentration of the ith anion of charge m_i.

contains the following species: Na^+, Co^{++}, $CoCl^+$, $CoCl_2$, $CoCl_3^-$, and $CoCl_4^{--}$. The electroneutrality relation for such a solution is

$$[Na^+] + 2[Co^{++}] + [CoCl^+] = [Cl^-] + [CoCl_3^-] + 2[CoCl_4^{--}]$$

Note that the uncharged $CoCl_2$ species does not appear in the equation.

In the above discussion, a simplification has been made by omitting $[H^+]$ and $[OH^-]$, which arise from the dissociation of water, from the electroneutrality rule. This simplification is permissible as long as no acids or bases appear in the equation. Thus the exact electroneutrality rule for an aqueous solution of hydrochloric acid is

$$[H^+] = [OH^-] + [Cl^-]$$

In a solution of Na_2SO_3, the electroneutrality relation is

$$[Na^+] + [H^+] = [OH^-] + [HSO_3^-] + 2[SO_3^{--}]$$

since the reactions

$$SO_3^{--} + H_2O \rightleftharpoons HSO_3^-$$
$$HSO_3^- + H_2O \rightleftharpoons H_2SO_3$$

produce HSO_3^- and H_2SO_3.

Activity coefficients

In an *ideal solution* the activity and concentration of a species are identical. Most solutions are *nonideal*, particularly solutions containing strong electrolytes, and a difference exists between the activity and concentration of species. The activity of A is related to its concentration by

$$(A) = f_a[A]$$

where f_a is called the activity coefficient of A. f_A, as defined, becomes unity at *infinite dilution* (*when the concentrations of all solutes tend toward zero*). Therefore, K for reaction (2) is actually not a true constant and may vary considerably, depending upon the nature of the other solutes present in solutions. Such constants are known as concentration constants and are readily related to thermodynamic equilibrium constants. Dividing Eq. (3) by (2) we obtain the general relationship between concentration and activity constants:

$$\frac{K^0}{K} = \frac{f_P^p \cdot f_Q^q \cdot f_R^r \cdots}{f_A^a \cdot f_B^b \cdot f_C^c \cdots} \tag{4}$$

One of the great problems of solution equilibria is the evaluation of activity coefficients.

If the species is uncharged, a relationship of the form

$$\log f = b\mu$$

is applicable. b is a constant characteristic of the neutral molecule and also of the electrolyte present in solution. No adequate theory exists for predicting b, and it must be determined by experiment. μ is the ionic strength of the solution and is defined by

$$\mu = \tfrac{1}{2} \sum C_i Z_i^2$$

where C_i is the concentration of the ith ion, which has a charge Z_i.

Example: 0.01 M NaCl:

$$\mu = \tfrac{1}{2}\{C_{Na^+}(Z_{Na^+})^2 + C_{Cl^-}(Z_{Cl^-})^2\}$$
$$= \tfrac{1}{2}\{0.01(1)^2 + 0.01(-1)^2\} = 0.01$$

Example: 0.01 *M* K_2SO_4:

$$\mu = \tfrac{1}{2}\{C_{K^+}(Z_{K^+})^2 + C_{SO_4^{=}}(Z_{SO_4^{=}})^2\}$$
$$= \tfrac{1}{2}\{(2 \times 0.01)(1)^2 + 0.01(-2)^2\} = 0.03$$

Example: 0.01 *M* NaCl + 0.01 *M* K_2SO_4:

$$\mu = \tfrac{1}{2}\{C_{Na^+}(Z_{Na^+})^2 + C_{Cl^-}(Z_{Cl^-})^2 + C_{K^+}(Z_{K^+})^2 + C_{SO_4^{=}}(Z_{SO_4^{=}})^2\}$$
$$= \tfrac{1}{2}\{0.01(1)^2 + 0.01(-1)^2 + 0.02(1)^2 + 0.01(-2)^2\}$$
$$= 0.04$$

Hence if μ and b are known, the activity coefficient of the neutral molecule can be determined. Figure 3-1 shows some activity coefficient data for acetic acid in various strong electrolytes.

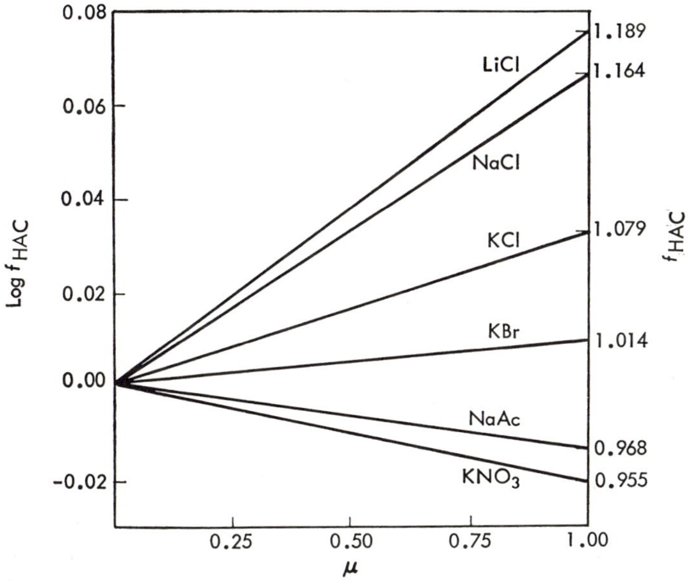

Fig. 3-1. The effect of various salts upon the activity coefficient of acetic acid.

In all cases, f_{HAC} is barely different from unity if μ is less than 0.1, and we shall assume ordinarily, lacking any other specific information, that the activity coefficient of a neutral molecule is unity.

The activity coefficients for ions for $\mu < 0.1$ may be evaluated using the Debye–Hückel expression:

$$-\log f_i = \frac{AZ_i^2\mu^{1/2}}{1 + \text{B}\mathring{a}\mu^{1/2}} \quad (5)$$

where f_i is the activity coefficient of the ith ion with charge Z_i in a solution of ionic strength μ. $\mathring{a}$ is the so called *ionic radius*, in angstrom units, B = 0.33, and A = 0.509 in water at 25°C.

For analytical purposes it is justifiable to use, up to an ionic strength of 0.1, averaged values of activity coefficients for ions of the same charge having different $\mathring{a}$ values. These data are given in Table A-1.

A comprehensive table of $\mathring{a}$ values with the corresponding activity coefficients calculated according to Eq. (5) has been compiled by Kielland.[5] These values of the

[5] J. Kielland, *J. Am. Chem. Soc.*, **59**, 1677 (1937).

single-ion activity coefficients are good estimates of f_i in 1 : 1, 1 : 2, 1 : 3, 3 : 1, and 2 : 1 electrolytes. Values of f_i for 2 : 2 electrolytes are not always reliable, because there is much evidence that species such as $CuSO_4$ and $ZnSO_4$ are not strong electrolytes. *It is always necessary to take incomplete dissociation into account before calculating the ionic strength of a solution.*

When $\mu < 0.01$, the Debye–Hückel limiting law may be used without making a serious error:

$$-\log f_i = 0.509 Z_i^2 \mu^{1/2} \tag{6}$$

Example: What is the activity of hydrogen ion in a solution 0.01 *M* in HCl and 0.09 *M* in KCl?

We calculate $\mu = 0.1$, and find from Table A-1 that f_{H^+} is 0.77. Therefore,

$$(H^+) = f_{H^+}[H^+] = 0.77 \times 0.01 = 7.7 \times 10^{-3}$$

Example: What is the concentration constant for uncharged weak acid HB, $K_{HB}^0 = 1.00 \times 10^{-7}$, if the ionic strength is 0.005?

$$K_{HB} = \frac{K^0 f_{HB}}{f_{H^+} f_{A^-}}$$

At such a low ionic strength $f_{HB} = 1$. f_{H^+} and f_{A^-} are equal and can be calculated from the Debye–Hückel limiting law, Eq. (6). $f_{H^+} = f_{B^-} = 0.92$ and

$$K_{HB} = \frac{1.00 \times 10^{-7}}{0.92 \times 0.92} = 1.18 \times 10^{-7}$$

Generally, we shall not distinguish between concentration constants and thermodynamic constants. In nearly all future calculation we shall write expressions in terms of equilibrium concentrations rather than activities, because the necessary information to estimate activity coefficients is available in relatively few situations of analytical interest.

Theories of acids and bases. Brønsted concept

Arrhenius defined an acid as a substance that in aqueous solution yields hydrogen ions and a base as a substance that yields hydroxyl ions. This theory is not correct. In the first place, free protons, H^+, do not exist to a measurable extent in any solvent. In aqueous solutions all the protons are associated with the solvent water, as already mentioned:[2] $H^+ + H_2O \rightarrow H_3O^+$. In anhydrous organic solvents (with basic character) the protons again are associated with the solvent $H^+ + S \rightarrow HS^+$; e.g., $H^+ + CH_3OH \rightarrow CH_3OH_2^+$. Thus "hydrogen ions" in aqueous solution are a different species than those in another solvent.

Second, when we consider a base such as ammonia it appears impossible to get a dissociation into hydroxyl ions in water without interaction with the solvent:

$$NH_3 + H_2O \rightarrow NH_3 \cdot H_2O \rightleftharpoons NH_4^+ + OH^-$$

Similarly, in a solvent such as methanol we have

$$NH_3 + CH_3OH \rightarrow NH_3 \cdot CH_3OH \rightleftharpoons NH_4^+ + CH_3O^-$$

While the dissociation of a base such as ammonia in water yields the OH^- ion, the corresponding CH_3O^- ion is obtained in methanol.

In this chapter we are concerned with the "hydrogen ion" concentration in aqueous solutions. Hydrogen ion concentrations in nonaqueous solutions are considered in Chapter 5. Formally, the use of the Arrhenius theory in aqueous solutions accounts quantitatively for all aqueous acid–base equilibria.

A more general theory of acidity and basicity, which is valid in all solvents, has been proposed by Brønsted.[6] Brønsted simply defines an acid as a substance that can split off protons; a base is a substance that can combine with protons. According to this definition, the substance HB is an acid if it can split off protons, and B is the corresponding, or *conjugate*, base of HB.

$$\underset{(acid)}{HB} \rightleftharpoons H^+ + \underset{(base)}{B}$$

According to Brønsted HB can have any charge and B has one less positive charge than HB; e.g.,

$$HCO_3^- \rightleftharpoons H^+ + CO_3^{--}$$

$$H_2CO_3 \rightleftharpoons H^+ + HCO_3^-$$

$$\underset{(acid)}{NH_4^+} \rightleftharpoons H^+ + \underset{(conjugate\ base)}{NH_3}$$

It has been mentioned that free protons practically do not exist in solution and that they are present in the solvated form. The reaction between protons and a solvent is an acid–base reaction according to the Brønsted definition:

$$H^+ + \underset{(solvent)}{H_2O} \rightarrow \underset{(acid)}{H_3O^+}$$

Thus, no "dissociation" of an acid can occur in a solvent such as ligroin, which has no basic properties. It is now clear that the "dissociation" of an acid is not similar to that of a salt (e.g., KCl). "Dissociation" of an acid occurs only by interaction between the base water and the acid HB:

$$\underset{(acid)}{HB} + \underset{(base)}{H_2O} \rightleftharpoons \underset{(acid)}{H_3O^+} + \underset{(base)}{B} \qquad (7)$$

Thus water can behave like a base (combines with protons) and also like an acid (splits off protons):

$$\underset{(acid)}{H_2O} \rightleftharpoons H^+ + \underset{(conjugate\ base)}{OH^-}$$

$$H^+ + \underset{(base)}{H_2O} \rightleftharpoons \underset{(conjugate\ acid)}{H_3O^+}$$

$$2\,H_2O \rightleftharpoons H_3O^+ + OH^- \qquad (8)$$

The dissociation of water according to Eq. (8) is an acid–base reaction. Solvents, such as water, which have both acidic and basic properties are called *amphiprotic*.

Dividing the equilibrium constant for reaction (7) into that of reaction (8) we obtain

$$\frac{K_w^0}{K_{HB}^0} = \frac{(H_3O^+)(OH^-)}{(H_3O^+)(B)/(HB)} = \frac{(HB)(OH^-)}{(B)}$$

The ratio K_w^0/K_{HB}^0 is the equilibrium constant for reaction (9):

$$B + H_2O \rightleftharpoons HB + OH^- \qquad (9)$$

which corresponds to the basic dissociation constant of B, the conjugate base of

[6] J. N. Brønsted, *Chem. Rev.*, 5, 231 (1923).

HB. Thus we have the very important relationship between the acid and basic dissociation constants of a conjugate acid and base:

$$K^0_{HB}K^0_B = K^0_w \tag{10}$$

$$pK^0_{HB} + pK^0_B = pK^0_w \tag{11}$$

From the table of acid dissociation constants given in the Appendix (Table A-2), pK°_B can be readily calculated for a number of inorganic and organic compounds using Eq. (11).

G. N. Lewis called a base a donor of a pair of electrons that is shared by an acid. In other words, an acid is an acceptor of a pair of electrons. All substances that are bases in the Brønsted sense are also bases in the Lewis sense. However, in the Brønsted sense acids are confined to proton donors, whereas a great number of substances that cannot split off protons can be acids in the Lewis theory (e.g., BCl_3 and Al^{3+}). In aqueous solutions all Lewis acids interact with the solvent to form a Brønsted acid; e.g.,

$$Al^{3+} + 6\ H_2O \rightleftharpoons Al(H_2O)_6^{3+}$$

The aquo aluminum ion is a Brønsted acid:

$$Al(H_2O)_6^{3+} \rightleftharpoons Al(H_2O)_5OH^{++} + H^+$$
$$Al(H_2O)_5OH^{++} \rightleftharpoons Al(H_2O)_4(OH)_2^+ + H^+$$
$$Al(H_2O)_4(OH)_2^+ \rightleftharpoons Al(H_2O)_3(OH)_3 + H^+$$

Evidently the $Al(H_2O)_5OH^{++}$ and $Al(H_2O)_4(OH)_2^+$ ions behave like acids and like bases. They are called amphoteric or, better, *amphiprotic.*

From an analytical viewpoint the Lewis theory of acids and bases is of no consequence in protonic solvents.

The dissociation of water. The hydrogen-ion exponent, pH

Water is an extremely weak electrolyte and is only very slightly dissociated into its ions:

$$H_2O \rightleftharpoons H^+ + OH^- \tag{12}$$

Application of the mass-action law yields, as previously shown,

$$K_w = [H^+][OH^-] \tag{13}$$

$pK^0_w = 13.996$ at 25.00°C. In the following discussion we shall assume that $K_w = 10^{-14}$ at room temperature.

The value of K_w increases approximately 8 per cent per 1 degree rise in temperature. With the aid of Eq. (13) it is a simple matter to calculate the hydrogen-ion concentration of pure water.

From a practical viewpoint it is important to note that it is very hard to prepare entirely pure water and to keep it unchanged. The atmosphere contains 0.03 volume per cent of carbon dioxide, which is fairly readily dissolved by water, dividing itself between water and air in a ratio of 1 : 1. In other words, water in equilibrium with the air contains 0.03 volume per cent of carbon dioxide. Such a solution has a hydrogen-ion concentration about 20 times larger than that of pure water (see Problem 9). Pure water is usually prepared by distillation from alkaline medium. If the original water contains a trace of ammonia, some of the ammonia will be carried into the distillate and will decrease the hydrogen-ion concentration of the latter (see Problem 8). If pure water is kept in ordinary glass containers, alkali from the glass will be dissolved, and the hydrogen-ion concentration will decrease. When exposed to the laboratory atmosphere, water quickly absorbs acid and alkaline fumes. Hence pure water must be handled with great care and must never be exposed to the atmosphere.

From Eq. (12) it follows that in the dissociation of water as many hydrogen as hydroxyl ions are formed. Therefore, in pure water it is found that $[H^+] = [OH^-]$, and from Eq. (13)

$$[H^+]^2 = [OH^-]^2 = K_w = 10^{-14} \text{ (room temperature)}$$

$$[H^+] = [OH^-] = 10^{-7}$$

Pure water contains 1 ten-millionth of a gram of hydrogen ions and 17 ten-millionths of a gram of hydroxyl ions per liter.

Any aqueous solution contains both hydrogen and hydroxyl ions. If by the addition of hydrogen ions (an acid) $[H^+]$ is made larger than 10^{-7}, $[OH^-]$ will become smaller; in basic solutions $[OH^-]$ will be greater than 10^{-7} and greater than $[H^+]$. The product of $[H^+]$ and $[OH^-]$ always remains constant.

$$[H^+] = \frac{K_w}{[OH^-]}$$

and

$$[OH^-] = \frac{K_w}{[H^+]}$$

Example: What is $[H^+]$ in 0.01 M sodium hydroxide?

$$[OH^-] = 10^{-2}; \quad \text{therefore,} \quad [H^+] = \frac{10^{-14}}{10^{-2}} = 10^{-12}$$

For various reasons it is often advantageous to express ion concentrations in terms of their negative logarithms to the base 10. This negative logarithm is usually called the *ion exponent* and is designated by the symbol p.

Thus pH designates the *hydrogen-ion exponent* and is equal to the *negative* logarithm of the hydrogen-ion concentration:

$$\text{pH} = -\log [H^+]$$

Similarly,

$$\text{pOH} = -\log [OH^-]$$

pOH is the hydroxyl-ion exponent. The relation between the ion concentration and the ion exponent is simply expressed algebraically:

$$\text{pH} = \log \frac{1}{[H^+]} = -\log [H^+]$$

$$[H^+] = 10^{-\text{pH}}$$

Example 1: $[H^+] = 5 \times 10^{-3}$. What is the pH?

$$\text{pH} = -\log (5 \times 10^{-3}) = (-\log 5) + 3 = -0.7 + 3 = 2.3$$

Example 2: pH = 4.4. What is $[H^+]$?

$$[H^+] = 10^{-4.40} = 10^{+0.60} \times 10^{-5} = x \times 10^{-5}$$
$$x = 10^{0.60}$$
$$\log x = 0.60$$
$$x = 3.98$$
$$[H^+] = 3.98 \times 10^{-5}$$

Now we can write Eq. (13) in terms of negative logarithms:

$$-\log [H^+] + (-\log [OH^-]) = -\log K_w$$

or

$$pH + pOH = pK_w$$

At room temperature we find

$$pH + pOH = -\log 10^{-14} = 14$$
$$pH = 14 - pOH$$
$$pOH = 14 - pH$$

If $[H^+] = [OH^-] = 10^{-7}$ and pH = pOH = 7, the solution has a *neutral reaction*. If $[H^+] > 10^{-7} > [OH^-]$, or pH < 7 < pOH, the solution has an *acid reaction*. If $[H^+] < 10^{-7} < [OH^+]$, or pH > 7 > pOH, the solution has an *alkaline reaction*. It should be realized that *decreasing* pH *means increasing hydrogen-ion concentration*.

In Fig. 3-2 the pH of various solutions is given.

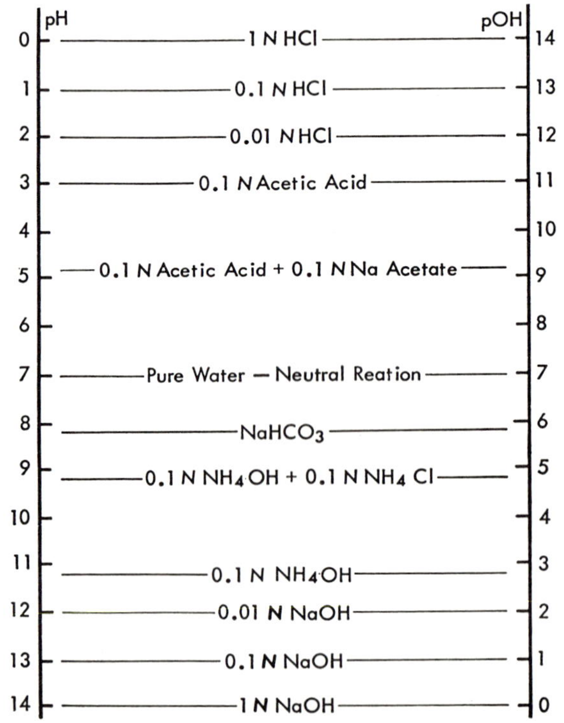

Fig. 3-2. Comparison of pH and pOH scales at 25°C. $K_w = 10^{-14}$.

In certain physiochemical measurements it is convenient to refer to the negative logarithm of the hydrogen-ion *activity*. This quantity is known as p_aH; i.e.,

$$p_aH = -\log (H^+)$$

The dissociation of weak acids and bases

In aqueous solutions a weak acid, HB, is only partially dissociated into hydrogen ions and its conjugate base, and the remaining part of the acid is present in the undissociated state:

$$HB \rightleftharpoons H^+ + B \tag{14}$$

From the Brønsted viewpoint the charge on an acid or base is not significant in describing the equilibrium properties of the solutions of an acid or base, and usually we shall not specify the charge on either the acid or its conjugate base. As is the case for most weak acids, suppose that the reaction is reversible, and equilibrium is attained instantaneously. Therefore, in any solution of a weak electrolyte the mass-action law may be applied:

$$\frac{[H^+][B]}{[HB]} = K_{HB} \qquad \text{(at constant temperature)} \tag{15}$$

In dissociating, the acid gives an equal number of hydrogen ions and B species therefore $[H^+] = [B]$, and instead of Eq. (15) we may write

$$\frac{[H^+]^2}{[HB]} = K_{HB} \tag{16}$$

If the molar concentration of the acid is c, then the concentration of the undissociated part, [HB], is equal to the total concentration less the concentration of the dissociated part:

$$[HB] = c - [H^+] = c - [B]$$

Therefore,

$$\frac{[H^+]^2}{c - [H^+]} = K_{HB} \tag{17}$$

and[7]

$$[H^+] = -\frac{K_{HB}}{2} + \left(\frac{K_{HB}^2}{4} + K_{HB}c\right)^{1/2} \tag{18}$$

In many cases it will be found that $[H^+]$ is so small compared with c that $c - [H^+]$ becomes virtually equal to c. In an approximate form we may write

$$[H^+]^2 = K_{HB}c$$

$$[H^+] = (K_{HB}c)^{1/2} \tag{19}$$

One can often use the approximate equation, Eq. (19), because if $[H^+]$ is less than about 5 per cent of c, the result obtained is sufficiently accurate; if $[H^+]$ is greater than 5 per cent, the exact equation, Eq. (18), should be employed.

Example: What is the hydrogen-ion concentration of a 0.1 M solution of an uncharged acid having an ionization constant of 10^{-5}? (For acetic acid $K_{HB} = 1.8 \times 10^{-5}$.)

[7] Do not memorize these expressions but derive them whenever needed.

According to Eq. (19),

$$[H^+] = (10^{-5} \times 10^{-1})^{1/2} = 10^{-3}$$

Since $c = 10^{-1}$, $c - H^+$ is virtually equal to 10^{-1}.

[Show that it is no longer permissible to calculate the hydrogen-ion concentration with the aid of Eq. (19) in a 0.0001 M solution of acetic acid.]

Example: What is $[H^+]$ in 0.1 M ammonium chloride solution at 25°C?

Ammonium chloride is a strong electrolyte that dissociates in water into Cl^- and NH_4^+. Cl^- has no basic properties in water. NH_4^+ is a cationic Brønsted acid whose acid dissociation constant is 5.8×10^{-9} (Table A-2). According to Eq. (19),

$$[H^+] = (5.8 \times 10^{-9} \times 10^{-1})^{1/2} = 2.4 \times 10^{-5}$$

Since $c = 10^{-1}$, $c - [H^+] \simeq c$.

In the older literature, the Arrhenius theory was used to treat solutions of this sort, equimolar mixtures of strong acid and uncharged weak base, and the term *hydrolysis* was applied to the reaction between a positively charged acid and water. This distinction between uncharged and charged acids is unnecessary using the Brønsted concept.

Dibasic acids, H_2A, have two ionization constants:

$$H_2A \rightleftharpoons H^+ + HA \tag{20}$$

$$HA \rightleftharpoons H^+ + A \tag{21}$$

$$\frac{[H^+][HA]}{[H_2A]} = K_1$$

$$\frac{[H^+][A]}{[HA]} = K_2$$

K_1 is called the first, and K_2 the second, ionization constant of the acid. In the calculation of the hydrogen-ion concentration in a solution of a dibasic acid, one can usually neglect the second dissociation [Eq. (21)] because the latter is repressed by the first dissociation according to Eq. (20). Therefore, Eq. (18) or (19) can be applied, in which K_{HB} is put equal to K_1. In certain special situations discussed in Chapter 4, both K_1 and K_2 must be used to calculate the hydrogen-ion concentration.

The calculation of the hydroxyl-ion concentration in a solution of a weak base B is simple:

$$B + H_2O \rightarrow BH + OH^-$$

$$\frac{[BH][OH^-]}{[B]} = K_B \tag{22}$$

Thus

$$[OH^-] = -\frac{K_B}{2} + \left(\frac{K_B^2}{4} + K_B c\right)^{1/2} \tag{23}$$

or, in an approximate form,

$$[OH^-] = (K_B c)^{1/2} \tag{24}$$

Example: What is $[OH^-]$ and $[H^+]$ in a 0.1 M solution of ammonia?

From Table A-2 we find that $pK_{NH_4^+} = 9.24$.

Using Eq. (11),

$$pK_{NH_3} = pK_w - pK_{NH_4^+} = 14.00 - 9.24 = 4.76$$

or

$$K_{NH_3} = 1.7 \times 10^{-5}$$

According to Eq. (24), $[OH^-] = 1.3 \times 10^{-3}$, and according to Eq. (13),

$$[H^+] = 7.7 \times 10^{-12}$$

Since $c = 10^{-1}$, $c - [OH^-] \simeq c$.

Example: What is $[OH^-]$ in 0.1 *M* sodium acetate?

Sodium acetate is a strong electrolyte. Na^+ has no acid properties, while acetate ion is a weak base,

$$K_{Ac} = \frac{K_w}{K_{HAc}} = \frac{10^{-14}}{1.8 \times 10^{-5}} = 5.5 \times 10^{-10}$$

Hence, according to Eq. (24), $[OH^-] = 7.4 \times 10^{-5}$.
Since $c = 10^{-1}$, $c - [OH^-] \simeq c$.

$[H^+]$ in a mixture of a weak acid and its conjugate base. Buffer action

In any solution containing a weak acid, HB, the hydrogen-ion concentration is determined by the magnitude of the ionization constant:

$$\frac{[H^+][B]}{[HB]} = K_{HB}$$

and

$$[H^+] = \frac{[HB]}{[B]} K_{HB} \tag{25}$$

Considering, in particular, a mixture of a weak acid and its salt, we must realize that the salt, KB, is a strong electrolyte. Thus the concentration of B is equal to the concentration c_b of the salt. The salt represses the dissociation of the acid, HB, because it furnishes the common species B.

In most cases this repression of the dissociation is so great that we can consider all the acid present to be in the undissociated form. If the analytical concentration of the acid is c_a, we find that $[HB] = c_a$ and $[B] = c_b$. Introduction of these values in Eq. (25) gives the hydrogen-ion concentration of a mixture of a weak acid and its salt (or conjugate base):

$$[H^+] = \frac{c_a}{c_b} K_{HB} \tag{26}$$

Actually this is only an approximate expression; it will, however, be used in this text for nearly all calculations. In a mixture of a weak acid and its salt having an *acid* reaction, the hydrogen ions are furnished by dissociation of the acid, and an equal amount of conjugate base B is formed. Therefore, $[HB] = c_a - [H^+]$, and $[B] = c_b + [H^+]$. Introducing these expressions in Eq. (25), it is found that

$$[H^+] = \frac{c_a - [H^+]}{c_b + [H^+]} K_{HB} \tag{27}$$

This is a quadratic equation which can be easily solved for $[H^+]$.

A mixture of a very weak acid containing an excess of the salt may have an *alkaline* reaction, owing to the protolysis reaction of the conjugate base with water:

$$B + H_2O \rightleftharpoons HB + OH^-$$

This protolysis reaction is repressed by the presence of HB in the mixture. However, as a result of the protolysis it is easily seen that

$$[\text{HB}] = c_a + [\text{OH}^-] = c_a + \frac{K_w}{[\text{H}^+]}$$

and
$$[\text{B}] = c_b - [\text{OH}^-] = c_b - \frac{K_w}{[\text{H}^+]}$$

Introducing these results in Eq. (25) it is found that

$$[\text{H}^+] = \frac{c_a + (K_w/[\text{H}^+])}{c_b - (K_w/[\text{H}^+])} K_{\text{HB}} \tag{28}$$

This again is a simple quadratic equation that can be solved easily for $[\text{H}^+]$.

In general, if c_a is the analytical concentration of the acid and c_b is the analytical concentration of the conjugate base of the acid, we can calculate the hydrogen-ion concentration from Eq. (26), (27), or (28), regardless of the charge types involved. For example, from the Brønsted viewpoint, there is no difference in the way in which one calculates the pH of a mixture of acetic acid and sodium acetate or of a mixture of ammonium chloride and ammonia; both are simple mixtures of an acid and its conjugate base. The distinction that has been made classically between mixtures of a weak acid and its salt with a strong base, and of a weak base and its salt with a strong acid, is an unnecessary complication, avoided by use of the Brønsted theory.

Mixtures of weak acids and their conjugate bases are called *buffer solutions, because they resist a change in hydrogen-ion concentration upon addition of slight amounts of acids or bases.* The pH of such buffer mixtures hardly changes on dilution.

Suppose, for example, that we have a mixture of acetic acid and sodium acetate in which $c_a = c_b = 0.1$. The hydrogen-ion concentration of this mixture is

$$[\text{H}^+] = \frac{c_a}{c_b} K_a = K_a = 1.8 \times 10^{-5}$$

$$\text{pH} = 4.74$$

Now, let an amount of base equivalent to 1 ml of 0.1 N sodium hydroxide be added to 100 ml of this mixture. The result is that an equivalent amount of acetic acid is neutralized and sodium acetate is formed. Therefore, [HB] becomes $0.1 - 0.001 = 0.099$, $[\text{B}^-] = 0.1 + 0.001 = 0.101$, and

$$[\text{H}^+] = \frac{0.099}{0.101} K_{\text{HB}} = 0.98 K_{\text{HB}} = 1.76 \times 10^{-5}$$

$$\text{pH} = 4.75$$

Therefore, the addition of this amount of base hardly changes the pH of the solution.

Buffer solutions play a very important role in the colorimetric determination of pH with the aid of indicators. Use is often made in analytical chemistry of the buffer actions of mixtures of weak acids and their salts, or of weak bases and their salts. Precipitations must often be made in a certain range of pH values. The pH is then adjusted by the addition of a suitable buffer mixture. In the precipitation of aluminum with ammonia, one is confronted with the difficulty that hydrous aluminum oxide dissolves partially in an excess of ammonia, owing to the formation of aluminate. The solubility of the oxide increases linearly with the hydroxyl-ion concentration. Addition of a suitable amount of ammonium chloride decreases the hydroxyl-ion concentration to such an extent that the solubility becomes negligibly small. Another advantageous application of the buffer action of the ammonia–ammonium salt system is made in separations in which magnesium hydroxide must be kept in solution. This will be discussed in more detail in Chapter 12, where other examples are also given.

In many cases the acidity of a strongly acid solution must be decreased before a certain operation can be carried out. This is often done by the addition of an adequate quantity of sodium acetate or sometimes by adding borax or sodium bicarbonate.

In the latter case we have the buffer system $H_2CO_3 + NaHCO_3$:

$$[H^+] = \frac{[H_2CO_3]}{[HCO_3^-]} K_1 = \frac{[H_2CO_3]}{[HCO_3^-]} \times 4.5 \times 10^{-7}$$

Carbonic acid is relatively slightly soluble in water. Its saturated solution (which loses carbon dioxide fairly rapidly on exposure to the air) has a concentration of about $5 \times 10^{-2}\ M$. In solutions saturated with carbon dioxide it is found that

$$[H^+] = \frac{1}{[HCO_3^-]} \times 5 \times 10^{-2} \times 4.5 \times 10^{-7} = \frac{1}{[HCO_3^-]} \times 2.3 \times 10^{-8}$$

If the sodium bicarbonate concentration were 0.23 M, we would find

$$[H^+] = \frac{2.3 \times 10^{-8}}{2.3 \times 10^{-1}} = 10^{-7} \text{ and a pH of 7}$$

Therefore, mixtures of sodium bicarbonate and acid will have a reaction in the neighborhood of the neutral point. A very useful application of this system is made in the titration of arsenious oxide with iodine:

$$As_2O_3 + 2I_2 + 2\,H_2O \rightleftharpoons As_2O_5 + 4\,H^+ + 4\,I^-$$

This reaction is reversible; in neutral solution it runs quantitatively from left to right, in a strongly acid medium quantitatively from right to left. Therefore, in the titration of the arsenic trioxide an excess of sodium bicarbonate is added to the acid solution before titration with iodine.

Tables A-15 and A-16 give the composition of a series of buffers from $p_aH = 1$ to $p_aH = 13$.

Solutions of amphiprotic substances

We denote an amphiprotic substance by A_2B_1, in which A_2 is the acid constituent with the conjugate base B_2 and B_1 is the basic constituent with the conjugate acid A_1.

A_2B_1 may be an uncharged compound like an amino acid or 8-hydroxyquinoline denoted by HOx. Here A_2 is identical with B_1:

$$HOx(A_2) \rightleftharpoons H^+ + Ox^-$$

$$HOx(B_1) + H^+ \rightleftharpoons H_2Ox^+$$

A_2B_1 may be a salt which dissociates in aqueous solution, e.g., ammonium acetate:

$$NH_4Ac \rightarrow \underset{(A_2)}{NH_4^+} + \underset{(B_1)}{Ac^-}$$

$$NH_4^+ \rightleftharpoons NH_3 + H^+$$

$$Ac^- + H^+ \rightleftharpoons HAc$$

Evidently $[A_2] = [B_1]$.

A_2B_1 may be an anion, such as HSO_3^-, $H_2PO_4^-$, HPO_4^-, or HCO_3^-.

$$H_2PO_4^-(A_2) \rightleftharpoons H^+ + HPO_4^{--}$$

$$H_2PO_4^-(B_1) + H^+ \rightleftharpoons H_3PO_4$$

Here A_2 is identical with B_1 and we have again: $[A_2] = [B_1]$.

A_2B_1 may be a cation, such as $Al(H_2O)_5OH^{++}$, and again $[A_2] = [B_1]$. Our problem is to find the pH in a solution of A_2B_1.

$$A_2 \rightleftharpoons H^+ + B_2 \tag{29}$$

$$B_1 + H^+ \rightleftharpoons A_1 \tag{30}$$

$$\frac{[H^+][B_2]}{[A_2]} = K_{A_2} \tag{31}$$

and

$$\frac{[H^+][B_1]}{[A_1]} = K_{A_1} \tag{32}$$

Multiplying K_{A_2} by K_{A_1} we find

$$\frac{[H^+]^2[B_2][B_1]}{[A_2][A_1]} = K_{A_2}K_{A_1} \tag{33}$$

It has been stated that $[A_2] = [B_1]$. We shall also make the approximation that $[B_2]$ formed according to Eq. (29) is equal to $[A_1]$ formed according to Eq. (30). This approximation is usually permissible unless the concentration of A_2B_1 is very small and the pH deviates much from 7.

With the above approximation, Eq. (33) becomes

$$[H^+]^2 = K_{A_2}K_{A_1}$$

and

$$[H^+] = (K_{A_2}K_{A_1})^{1/2} \tag{34}$$

Examples: A solution of KH_2PO_4 dissociates quantitatively into K^+ and $H_2PO_4^-$. From Eqs. (30) and (32) it is seen that K_{A_1} represents the first dissociation constant K_1 of phosphoric acid and K_{A_2} the second dissociation constant K_2. Hence $[H^+] = (K_1K_2)^{1/2} = (5.9 \times 10^{-3} \times 6.2 \times 10^{-8})^{1/2}$. pH = 4.71.

Solution of ammonium formate:

$$[A_2] = [NH_4^+] \quad \text{and} \quad [B_1] = [HCOO^-]$$

$$[H^+] = (K_{NH_4^+} \times K_{HCOOH})^{1/2}$$

When to an equimolar mixture of two monobasic weak acids A_1 and A_2 sodium hydroxide in an amount equivalent to the concentration of one of the two acids is added, it can be calculated again that in this mixture $[H^+] = (K_{A_2}K_{A_1})^{1/2}$.

Chapter 4 presents a rigorous, systematic treatment of aqueous acid–base equilibria. Methods are given for solving problems which occur less frequently than those presented in this chapter and which are ordinarily not discussed in an introductory course.

PROBLEMS

In the following six problems express the concentration constant at the stated electrolyte concentrations, or ionic strengths, in terms of the thermodynamic constants.

1. $BA \rightleftharpoons B^+ + A^-$; $K^\circ = \dfrac{(B^+)(A^-)}{(BA)}$: (a) 0.025 *M* NaCl; (b) 0.025 *M* Na_2SO_4.
2. $(Ag^+)(Cl^-) = K^\circ_{sp}$; $\mu = 0.1$.
3. $(Ba^{++})(SO_4^{--}) = K^\circ_{sp}$; $\mu = 0.1$.
4. $(Pb^{++})^2(Fe(CN)_6^{4-}) = K^\circ_{sp}$; $\mu = 0.1$.
5. $AgCl + Cl^- \rightleftharpoons AgCl_2^-$:

$$\frac{(AgCl_2^-)}{(AgCl)(Cl^-)} = K^\circ;\ \mu = 0.1$$

6. $NH_4^+ + H_2O \rightleftharpoons H_3O^+ + NH_3$:

$$\frac{(H_3O^+)(NH_3)}{(NH_4^+)} = K^\circ;\ \mu = 0.1$$

Unless stated otherwise in the problems given below, assume all activity coefficients are unity and that $K^0 = K$.

7. Calculate the hydrogen-ion concentration and the pH of the following solutions: (a) 0.01 *M* hydrochloric acid; (b) 0.1 *M* acetic acid; (c) 0.001 *M* acetic acid; (d) 1 *M* ammonia; (e) 0.01 *M* ammonia; (f) 0.1 *M* sodium hydroxide. A list of dissociation constants is given in Table A2. $pK_w = 14.00$.
Ans. (a) $[H^+] = 1.0 \times 10^{-2}$; pH = 2.00; (b) $[H^+] = 1.3 \times 10^{-3}$; pH = 2.88; (c) $[H^+] = 1.2 \times 10^{-4}$; pH = 3.91; (d) $[H^+] = 2.4 \times 10^{-12}$; pH = 11.62; (e) $[H^+] = 2.5 \times 10^{-11}$; pH = 10.61; (f) $[H^+] = 1.0 \times 10^{-13}$; pH = 13.00.
8. A sample of water contains 0.85 mg of ammonia per liter. Calculate the pH if ammonia is the only contaminant. $K_B = 1.8 \times 10^{-5}$. *Ans.* pH = 9.35.
9. Distilled water is in equilibrium with the air and contains carbon dioxide corresponding to a molar concentration of 1.35×10^{-5}. Calculate the pH of this water, given K_1 of $H_2CO_3 = 4.5 \times 10^{-7}$. *Ans.* pH = 5.65.
10. Calculate $[OH^-]$ and pH of a solution of 0.1 *M* ammonia (a) at 25°C; (b) at 100°C. $K_w = 10^{-14}$ at 25°C and 10^{-12} at 100°C. $K_b = 1.8 \times 10^{-5}$ at 25°C and at 100°C.
Ans. (a) $[OH^-] = 1.3 \times 10^{-3}$, pH = 11.13; (b) $[OH^-] = 1.3 \times 10^{-3}$; pH = 9.13.
11. Calculate the pH of the following solutions: (a) 1 *M* sodium orthoborate (NaH_2BO_3); (b) 0.01 *M* sodium orthoborate; (c) 0.1 *M* anilinium perchlorate. $K_{H_3BO_3} = 5.9 \times 10^{-10}$; $K_{aniline} = 4.2 \times 10^{-10}$.
Ans. (a) pH = 11.11; (b) pH = 10.10; (c) pH = 2.80.
12. Calculate the pH of the following solutions containing: (a) 0.05 mole acetic acid and 0.05 mole sodium acetate per liter; (b) 0.05 mole acetic acid and 0.005 mole sodium acetate per liter; (c) 0.005 mole acetic acid and 0.05 mole sodium acetate per liter; (d) 0.005 mole acetic acid and 0.5 mole sodium acetate per liter. $pK_{HAc} = 4.76$.
Ans. (a) pH = 4.76; (b) pH = 3.76; (c) pH = 5.76; (d) pH = 6.76.
13. Calculate the p_aH of the solution in Problem 12(a), (b), and (c). Use the averaged ionic activity coefficients in Table A-1. *Ans.* (a) 4.85; (b) 3.79; (c) 5.85.

14. Calculate the pH of the following solutions containing: (a) 0.1 mole NH_3 and 0.1 mole NH_4Cl per liter; (b) 0.01 mole NH_3 and 0.1 mole NH_4Cl per liter; (c) 0.1 mole NH_3 and 0.01 mole NH_4Cl per liter. $pK_{NH_4^+} = 9.24$.
Ans. (a) pH = 9.24; (b) pH = 8.24; (c) pH = 10.24.

15. To 100 ml of 0.5 *M* sodium hydroxide is added 5.35 g of ammonium chloride. Calculate the pH, assuming that the volume is not changed by the addition of the salt.
Ans. pH = 9.26.

16. It is desired to change the pH of 100 ml of 0.1 *M* hydrochloric acid solution from 1 to 4.4 by the addition of (a) solid sodium acetate trihydrate ($CH_3COONa \cdot 3\,H_2O$); (b) solid sodium hydroxide. How much solid compound must be added?
Ans. (a) 1.98 g; (b) 0.400 g.

17. Which of the solutions of pH = 4.4 in Problem 16 is the buffer? *Ans.* (a).

18. Calculate the pH of the following solutions containing: (a) 0.05 mole KH_2PO_4 and 0.05 mole Na_2HPO_4 per liter; (b) 0.05 mole Na_2HPO_4 per liter (derive equation): $K_1 = 5.9 \times 10^{-3}$, $K_2 = 6.2 \times 10^{-8}$, $K_3 = 4.8 \times 10^{-13}$. *Ans.* (a) 7.21; (b) 9.76.

19. To a mixture which is 0.1 *M* in acetic acid and 0.1 *M* in boric acid is added an amount of sodium hydroxide equivalent to one of the two acids. Calculate the pH after addition of the alkali. $pK_{HAc} = 4.76$, $pK_{H_3BO_3} = 9.23$. *Ans.* pH = 7.00.

20. Calculate the change in pH upon tenfold dilution of the following solutions: (a) 0.1 *M* HCl; (b) 0.1 *M* acetic acid; (c) 0.1 *M* ammonium chloride; (d) Mixture of 0.1 *M* acetic acid and 0.1 *M* sodium acetate.
Ans. (a) 1.00; (b) 0.50; (c) 0.50; (d) negligible.

21. Repeat Problem 20, calculating the change in p_aH.

22. The saturation concentration of uncharged 8-hydroxyquinoline (HOx) in water is 3.5×10^{-3} *M*. Calculate the solubility in buffers of pH 3 and 8.

$$HOxH^+ \rightleftharpoons H^+ + HOx \qquad K_1 = 1.0 \times 10^{-5}$$
$$HOx \rightleftharpoons H^+ + Ox \qquad K_2 = 1.4 \times 10^{-10}$$

Ans. 0.35 *M* and 3.5×10^{-3} *M*.

23. Calculate the pH of 0.1 *M* sodium carbonate solution. K_2 of $H_2CO_2 = 10^{-10.33}$.
Ans. pH = 11.65.

24. Given $pK_1^0 = 2.23$ and $pK_2^0 = 7.21$ for phosphoric acid, in 0.1 *M* KH_2PO_4 calculate (a) K_1 and K_2, using Table A-1 to estimate the single-ion activity coefficient; (b) pH (derive the equation); (c) p_aH using Table A-1 to estimate f_{H^+}; (d) p_aH using $\mathring{a} = 9$ and Eq. (5) to estimate f_{H^+}. (The solvated proton has an extraordinary large value of $\mathring{a}$. Most univalent ions have $\mathring{a} = 3.5 \pm 1.0$.)
Ans. (a) $K_1 = 9.80 \times 10^{-3}$, $K_2 = 1.62 \times 10^{-7}$; (b) 4.40; (c) 4.51; (d) 4.48.

CHAPTER 4 *Advanced Treatment of Aqueous Acid–Base Equilibria*

Relations between analytical and equilibrium concentrations

The analytical concentration of a solute possessing acid or base properties is defined solely in terms of the weight of solute, ignoring any protolytic (acid–base) equilibria. Thus if c_a gram formula weights of a monoprotic weak acid, HB, are dissolved in 1 liter of water, the analytical concentration of HB is said to be c_a M. Actually because of the equilibrium

$$\mathrm{HB} \rightleftharpoons \mathrm{H}^+ + \mathrm{B} \qquad K_{\mathrm{HB}} = \frac{[\mathrm{H}^+][\mathrm{B}]}{[\mathrm{HB}]}$$

the acid is present in two forms HB and B. The sum of the equilibrium concentration of HB and B is equal to the analytical concentration

$$c_a = [\mathrm{HB}] + [\mathrm{B}] \tag{1}$$

Note that in this chapter the symbol, c_a, will always represent the sum of all equilibrium concentrations arising from a particular acid as a result of protolytic reactions.

The equilibrium concentration of either HB or B is readily expressed as a function of $[\mathrm{H}^+]$, K_{HB}, and c_a. Eliminating [HB] from Eq. (1) by substituting from K_{HB} yields

$$c_a = \frac{[\mathrm{H}^+][\mathrm{B}]}{K_{\mathrm{HB}}} + [\mathrm{B}] \tag{2}$$

$$c_a = \left\{\frac{[\mathrm{H}^+]}{K_{\mathrm{HB}}} + 1\right\}[\mathrm{B}] \tag{3}$$

or, on solving for [B] and rearranging terms,

$$[\mathrm{B}] = c_a \frac{K_{\mathrm{HB}}}{[\mathrm{H}^+] + K_{\mathrm{HB}}} \tag{4}$$

Subtracting Eq. (4) from (1) yields, on rearranging terms,

$$[\mathrm{HB}] = c_a \frac{[\mathrm{H}^+]}{[\mathrm{H}^+] + K_{\mathrm{HB}}} \tag{5}$$

Thus, if $c_a = 0.0100\ M$, $K_{\mathrm{HB}} = 1.00 \times 10^{-4}$ and $[\mathrm{H}^+] = 3.00 \times 10^{-4}\ M$,

$$[\mathrm{B}] = 0.0100 \frac{1.00 \times 10^{-4}}{1.00 \times 10^{-4} + 3.00 \times 10^{-4}}$$

$$= 2.50 \times 10^{-3}\ M$$

A plot of the fraction of B, $\alpha_{\mathrm{B}} = [\mathrm{B}]/c_a$, and the fraction of HB, $\alpha_{\mathrm{HB}} = [\mathrm{HB}]/c_a$, is given in Fig. 4-1*a*. Equal amounts of B and HB are present when $[\mathrm{H}^+] = K_{\mathrm{HB}}$.

An alternative, logarithmic presentation of Eqs. (4) and (5) is shown in Fig. 4-1*b*, for an acid, $\mathrm{p}K_{\mathrm{HB}} = 5.00$, at an assumed analytical concentration of 1.00×10^{-2}. The

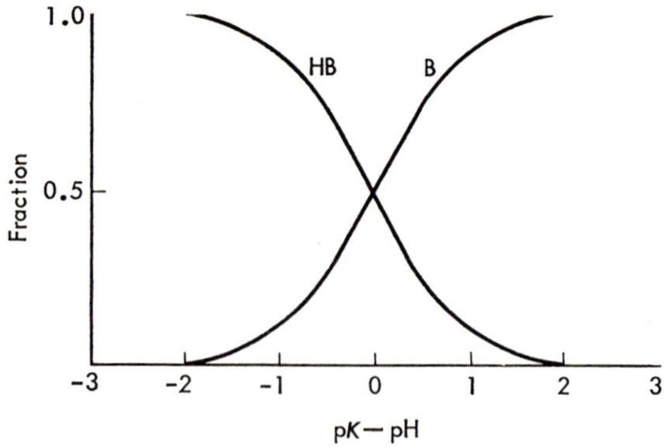

Fig. 4-1*a*. Fraction of the two forms of a monoprotic weak acid as a function of pH.

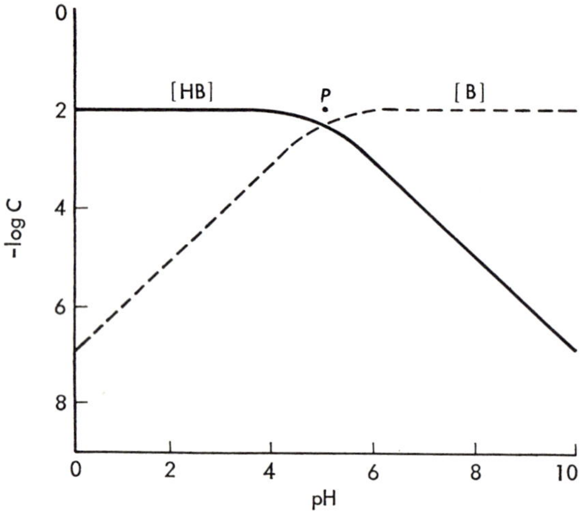

Fig. 4-1*b*. Logarithmic plot of monoprotic acid system. $pK_{HB} = 5.00$.

solid line represents [HB] and the dashed line [B]. Referring to Eqs. (4) and (5) it is seen that if $[H^+] \gg K_{HB}$,

$$\log [B] = \log c_a K_{HB} + pH$$
$$\log [HB] = \log c_a$$

and if $[H^+] \ll K_{HB}$,

$$\log [B] = \log c_a$$
$$\log [HB] = \log \frac{c_a}{K_{HB}} - pH$$

The log plot for HB consists of two straight-line segments, the first of zero slope at low pH and the second of -1 slope at high pH. The point of intersection, P, is called the system point and occurs at $pH = pK_{HB}$. In the region of $pH = pK_{HB}$, the plot of [HB] is a curve. At $pH = pK_{HB}$, the curve is exactly 0.30 log C units below point P (log 2). At $pH = pK \pm 0.5$, the curve is 0.12 unit under the extrapolated straight lines. The construction of the [HB] curve is quite simple; all that need be done is to draw straight lines of the appropriate slopes through the system point, and sketch the curved region through the points at pH $= pK_{HB} \pm 0.5$. The [HB] curve is a mirror image of the [B] curve about a vertical line pH $= pK_{HB}$. Logarithmic plots of this kind are quite valuable in the graphical solution of protolytic equilibrium problems.

The presence of other weak acids and weak bases in no way alters the above calculations, because we have assumed an *a priori* knowledge of c_a, K_{HB}, and $[H^+]$.

Expressions analogous to α_{HB} and α_B are readily obtained for a dibasic acid, H_2C. In any solution to which H_2C has been added the analytical concentration of H_2C, c_a, is given by

$$c_a = [H_2C] + [HC] + [C] \tag{6}$$

Writing the definition of the equilibrium constants in terms of chemical reactions we have

$$H_2C \rightleftharpoons H^+ + HC \qquad K_1 = \frac{[H^+][HC]}{[H_2C]} \tag{7}$$

$$HC \rightleftharpoons H^+ + C \qquad K_2 = \frac{[H^+][C]}{[HC]} \tag{8}$$

or

$$H_2C \rightleftharpoons 2\,H^+ + C \qquad K_1K_2 = \frac{[H^+]^2[C]}{[H_2C]} \tag{9}$$

Substituting Eqs. (8) and (9) into (6) to eliminate $[H_2C]$ and [HC],

$$c_a = \frac{[H^+]^2}{K_1K_2}[C] + \frac{[H^+]}{K_2}[C] + [C] \tag{10}$$

which yields, on factoring [C] and rearranging terms,

$$[C] = c_a\left\{\frac{K_1K_2}{[H^+]^2 + [H^+]K_1 + K_1K_2}\right\} \tag{11}$$

Similar expressions for [HC] and $[H_2C]$ are readily obtained:

$$[HC] = c_a\left\{\frac{[H^+]K_1}{[H^+]^2 + [H^+]K_1 + K_1K_2}\right\} \tag{12}$$

$$[H_2C] = c_a\left\{\frac{[H^+]^2}{[H^+]^2 + [H^+]K_1 + K_1K_2}\right\} \tag{13}$$

It can be shown by differentiation of Eq. (12) with respect to $[H^+]$ that the maximum fraction of HC, α_{HC}, is present when $[H^+] = (K_1K_2)^{1/2}$, and is equal to

$$\alpha_{HC_{max}} = \frac{1}{1 + 2(K_2/K_1)^{1/2}} \tag{14}$$

Thus, if H_2C were a long-chain, dicarboxylic acid for which K_2/K_1 approached the limiting statistical ratio of $\frac{1}{4}$, α_{HC} would be 0.5. The ratio K_2/K_1 is 3.46×10^{-3} for *o*-phthalic acid, so $(\alpha_{HC})_{max} = 0.90$, while for carbonic acid, $K_2/K_1 = 10^{-5}$ and $(\alpha_{HC})_{max} = 0.994$.

The fractions α_{H_2C}, α_{HC}, and α_C are shown in Fig. 4-2*a* as a function of pH for $K_2/K_1 = 10^{-6}$. For practical purposes, with such a large difference in strength between the first and second dissociation constants, α_{HC} and α_{H_2C} are accurately described by Eqs. (4) and (5), where $K_{HB} = K_1$ provided $[H^+] > (K_1K_2)^{1/2}$. Similarly, if $[H^+] < (K_1K_2)^{1/2}$, α_C and α_{HC} are accurately described by Eqs. (4) and (5), where $K_2 = K_{HB}$.

The logarithmic plot corresponding to Fig. 4-2*a* is shown in Fig. 4-2*b*, assuming $c_a = 1.00 \times 10^{-2}\ M$, $pK_1 = 1.00 \times 10^{-4}$ and $pK_2 = 1.00 \times 10^{-10}$. There are two

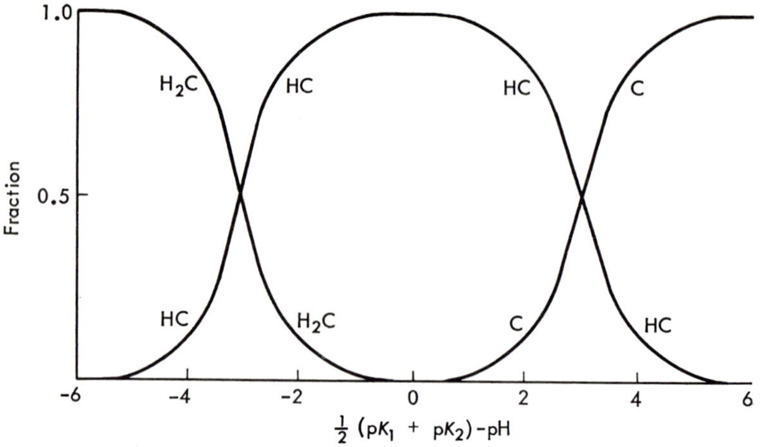

Fig. 4-2*a*. Fraction of the various forms of a diprotic acid as a function of pH. $pK_2 - pK_1 = 6$.

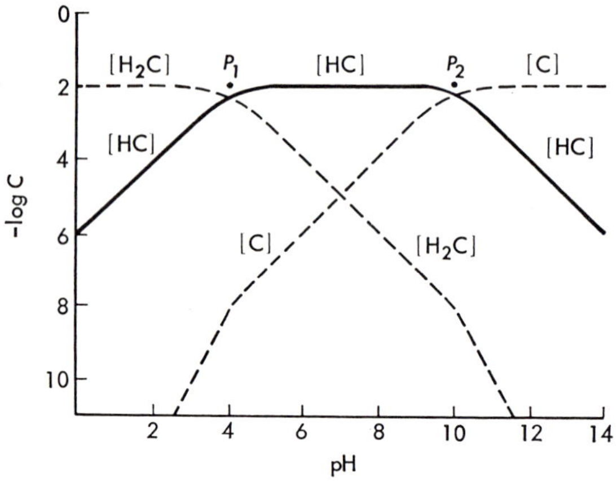

Fig. 4-2*b*. Logarithmic plot of a diprotic acid system. $pK_1 = 4.00$, $pK_2 = 10.00$.

system points, P_1 at pH $= pK_1$ and for P_2 at pH $= pK_2$. The $[H_2C]$ curve consists of three straight-line segments of slopes 0, -1, and -2 with curved regions in the vicinity of pK_1 and pK_2. The [C] curve is a mirror image of the $[H_2C]$ curve about pH $= \frac{1}{2}(pK_1 + pK_2)$. The [HC] curve has three limiting straight-line regions of slopes 1, 0, and -1. The construction of plots such as given in Fig. 4-2*b*, is done using the system points P_1 and P_2 in an analogous manner to that used for the simpler system, HB, shown in Fig. 4-1*b*.

The fraction of any form of a polybasic acid, H_nZ, is calculated as follows. Defining the analytical concentration, c_a,

$$c_a = [H_nZ] + [H_{n-1}Z] + \cdots + [HZ] + [Z] \tag{15}$$

$$c_a = \sum_{i=0}^{i=n} [H_iZ] \tag{16}$$

and the equilibrium constants for the reactions

$$H_nZ \rightleftharpoons H^+ + H_{n-1}Z \qquad K_1 \tag{17}$$

$$H_nZ \rightleftharpoons 2\,H^+ + H_{n-2}Z \qquad K_1K_2 \tag{18}$$

$$\vdots$$

$$H_nZ \rightleftharpoons (n-1)H^+ + HZ \qquad K_1K_2\cdots K_{n-2}K_{n-1} \tag{19}$$

$$H_nZ \rightleftharpoons nH^+ + Z \qquad K_1K_2\cdots K_{n-1}K_n \tag{20}$$

or, in general,

$$H_nZ \rightleftharpoons (n-m)H^+ + H_mZ \qquad \prod_{j=1}^{j=n-m} K_j \tag{21}$$

Substituting Eqs. (17)–(20) into Eq. (15) to obtain an expression in terms of $[H_nZ]$, c_a, $[H^+]$, and the various equilibrium constants, we obtain

$$c_a = [H_nZ]\left\{1 + \frac{K_1}{[H^+]} + \frac{K_1K_2}{[H^+]^2} + \cdots + \frac{K_1K_2\cdots K_{n-2}K_{n-1}}{[H^+]^{n-1}} + \frac{K_1K_2\cdots K_{n-1}K_n}{[H^+]^n}\right\} \tag{22}$$

$$c_a = [H_nZ]\left\{1 + \sum_{i=1}^{i=n} \frac{\prod_{j=1}^{j=i} K_j}{[H^+]^i}\right\} \tag{23}$$

or, on rearranging terms,

$$[H_nZ] = c_a\left\{\frac{[H^+]^n}{D(n)}\right\} \tag{24}$$

where

$$D(n) = [H^+]^n + [H^+]^{n-1}K_1 + [H^+]^{n-2}K_1K_2 + \cdots + [H^+]K_1K_2\cdots K_{n-1} + K_1K_2\cdots K_{n-1}K_n \tag{25}$$

Alternatively,

$$[H_nZ] = c_a\left\{\frac{1}{1 + \sum_{i=1}^{i=n} \left(\prod_{j=1}^{j=i} K_j\right)/[H^+]^i}\right\} \tag{26}$$

Equation (26) is a convenient general representation of the concentration of H_nZ and is commonly seen in the literature. Equation (24) is also used frequently, and has a convenient property which makes it easy to remember the expression for $[H_{n-1}Z]$, $[H_{n-2}Z]$, etc. Substituting Eq. (17) into (24), one obtains

$$[H_{n-1}Z] = c_a\left\{\frac{[H^+]^{n-1}K_1}{D(n)}\right\} \tag{27}$$

or, in general,

$$[H_{n-m}Z] = c_a\left\{\frac{[H^+]^{n-m}K_1K_2\cdots K_m}{D(n)}\right\} \tag{28}$$

As a mnemonic aid, note that the expressions for the concentration of species arising from the same acid by successive loss of protons have the same denominator. The denominator starts with $[H^+]$ raised to the nth power, ends with the products of all the stepwise equilibrium constants, and varies term by term in the simple way shown by Eq. (25); the numerator is the term in the denominator that has the hydrogen-ion concentration raised to the same power as the number of protons attached to Z.

For example, the expression for the amount of HPO_4^{--} ion is given by

$$[HPO_4^{--}] = c_a\left\{\frac{[H^+]K_1K_2}{D(3)}\right\} \tag{29}$$

where

$$D(3) = [H^+]^3 + [H^+]^2K_1 + [H^+]K_1K_2 + K_1K_2K_3 \tag{30}$$

The rule of electroneutrality

In the solution of certain equilibrium problems, a convenient approach is to make use of the rule of electroneutrality. The electroneutrality rule is a statement of total charge balance for a solution, and was discussed in Chapter 3 on page 41.

At this point it is possible to illustrate the ease with which it is possible to obtain an explicit algebraic expression for a complicated protolytic solution equilibrium problem. Consider the hypothetical titration of oxalic acid with ammonia. The rule of electroneutrality for this solution is

$$[NH_4^+] + [H^+] = [OH^-] + [HC_2O_4^-] + 2[C_2O_4^{--}] \tag{31}$$

If we let c_b equal the analytical concentration of ammonia, and c_a the analytical concentration of oxalic acid, we may immediately write

$$\frac{c_b[H^+]}{[H^+] + K_{NH_4^+}} + [H^+] = \frac{K_w}{[H^+]} + \frac{c_a\{[H^+]K_1 + 2K_1K_2\}}{[H^+]^2 + [H^+]K_1 + K_1K_2} \tag{32}$$

where

$$K_{NH_4^+} = \frac{[H^+][NH_3]}{[NH_4^+]}$$

and the other equilibrium constants have their customary significance. We have taken advantage of the expressions derived in the previous section in substituting for NH_4^+, $HC_2O_4^-$, and $C_2O_4^{--}$ in Eq. (31). Equation (32) is a fifth-power equation in $[H^+]$ but is of little *chemical* interest. It can be used to obtain a numerical value of $[H^+]$ only by someone familiar with numerical analysis methods or with access to a modern digital computer.

The object of the discussion of aqueous acid–base equilibria given below is to develop a single, logical approach to solution equilibria that requires a minimum of prior chemical experience. It is based on the use of the proton condition in its exact form, which is readily learned and leads directly to general, exact expressions for the pH of aqueous solutions in terms of easily estimated equilibrium concentrations. Approximations are made only after numerical substitution makes the magnitude of each term obvious.

The proton condition

Acid–base equilibria are most conveniently treated by considering all protolytic reactions that occur between the solvent (or its ions) and dissolved acids and bases. One mathematical statement of the proton condition is

$$[H^+] = a + \sum n_i[\text{base}]_i - b - \sum m_j[\text{acid}]_j, \tag{33}$$

where a and b represent the concentrations of strong acid and strong base, respectively. They are calculated from the amount originally added to the solution, while $[\text{base}]_i$ represents the equilibrium concentration of the ith base produced by assuming the reaction

$$H_{n_i}\text{base}_i \rightarrow n_iH^+ + \text{base}_i \tag{34}$$

for the species $H_{n_i}base_i$, and $[acid]_j$ represents the equilibrium concentration of the jth acid produced by assuming the reaction

$$\text{base}_j + m_j\text{H}^+ \rightarrow \text{H}_{m_j}\text{base}_j \text{ (i.e., [acid]}_j) \tag{35}$$

between the species $base_j$ and solvated protons.

For example, the proton condition for pure water is found by considering the reaction

$$H_2O \rightleftharpoons H^+ + OH^-, \tag{36}$$

from which it is apparent that

$$[H^+] = [OH^-] \tag{37}$$

while, if strong acid is present at a normal concentration, a,

$$[H^+] = [OH^-] + a \tag{38}$$

If a solution contains both a concentration a N strong acid, and b N strong base,

$$[H^+] = [OH^-] + a - b \tag{39}$$

The proton condition for a solution containing a weak monoprotic acid HB is found by considering the equilibria

$$H_2O \rightleftharpoons H^+ + OH^-$$

$$HB \rightleftharpoons H^+ + B$$

or

$$[H^+] = [OH^-] + [B] \tag{40}$$

If we add strong acid to the above solution, the proton condition becomes

$$[H^+] = [OH^-] + [B] + a \tag{41}$$

while if we add strong base, it becomes

$$[H^+] = [OH] + [B] - b \tag{42}$$

and is

$$[H^+] = [OH^-] + [B] + a - b$$

in a solution containing weak acid, strong acid, and strong base.

Obviously, if a solution contains a "weak" base G, in addition to HB, strong acid, and strong base, it is necessary to consider the reaction

$$G + H^+ \rightleftharpoons HG$$

to obtain the proton condition

$$[H^+] = [OH^-] + [B] - [HG] + a - b$$

As a final example of a proton condition, consider a solution of $NaNH_4HPO_4$. The relevant protolytic equilibria are as follows.

Case I:	*Reaction*	*Equilibrium Constant*
	$H_2O \rightleftharpoons H^+ + OH^-$	$K_w = 1.00 \times 10^{-14}$
	$NH_4^+ \rightleftharpoons H^+ + NH_3$	$K_{NH_4^+} = 5.8 \times 10^{-10}$
	$HPO_4^{--} \rightleftharpoons H^+ + PO_4^{3-}$	$K_3 = 4.8 \times 10^{-13}$
	$HPO_4^{--} + H^+ \rightleftharpoons H_2PO_4^-$	$1/K_2 = 1.6 \times 10^7$
	$HPO_4^{--} + 2H^+ \rightleftharpoons H_3PO_4$	$1/K_1K_2 = 2.8 \times 10^9$

to yield

$$[H^+] = [OH^-] + [NH_3] + [PO_4^{3-}] - [H_2PO_4^-] - 2[H_3PO_4]$$

The coefficient 2 preceding $[H_3PO_4]$ arises because two protons are consumed in converting the HPO_4^{--} ion to molecular phosphoric acid.

All species on the left side of the equilibrium expressions are considered to be the reference proton level or the principal species.

As a matter of practical convenience in solving protolytic equilibrium problems, *the species chosen as reference level should be that species of each Brønsted acid–base couple present in maximum concentration.*

For example, in the previous example, the solution of $NaNH_4H_2PO_4$ could have been considered an equimolar mixture of sodium hydroxide, *a M*, ammonium hydroxide, and phosphoric acid. Using these compounds as reference levels, we have

Case II:	*Reaction*	*Equilibrium Constant*
	$H_2O \rightleftharpoons H^+ + OH^-$	$K_w = 1.00 \times 10^{-14}$
	$NH_3 + H^+ \rightleftharpoons NH_4^+$	$1/K_{NH_4^+} = 1.7 \times 10^9$
	$H_3PO_4 \rightleftharpoons H^+ + H_2PO_4^-$	$K_1 = 5.9 \times 10^{-3}$
	$H_3PO_4 \rightleftharpoons 2H^+ + HPO_4^{--}$	$K_1K_2 = 3.6 \times 10^{-10}$
	$H_3PO_4 \rightleftharpoons 3H^+ + PO_4^{3-}$	$K_1K_2K_3 = 1.7 \times 10^{-22}$
	$(NaOH) + H^+ \rightleftharpoons (Na^+) + H_2O$	$1/K_w = 1.00 \times 10^{14}$

The parentheses in the last reaction indicate that the OH^- and the Na^+ originated from the *a M* sodium hydroxide. Hence the proton condition is

$$[H^+] = [OH^-] + [H_2PO_4^-] + 2[HPO_4^{--}] + 3[PO_4^{3-}] - [NH_4^+] - [Na^+]$$

Using these reference levels, the proton condition and the rule of electroneutrality are identical.

All possible proton conditions which can be written for a given solution by assuming different reference levels are related to each other and the rule of electroneutrality via material balance expressions. For example in the $NaNH_4HPO_4$ example just considered

$$[Na^+] = [NH_4^+] + [NH_3] \quad (43)$$

$$[Na^+] = [H_3PO_4] + [H_2PO_4^-] + [HPO_4^{--}] + [PO_4^{3-}] \quad (44)$$

and appropriate substitution of these latter two equations into the electroneutrality rule (or the proton condition obtained for case II) yields the proton condition given in case I.

The choice of reference levels used to obtain the proton condition for case II is not a natural one, because "chemical intuition" clearly indicates that reaction will occur between the sodium hydroxide and phosphoric acid to form $H_2PO_4^-$; also $H_2PO_4^-$ will react with NH_3 to form NH_4^+ and HPO_4^{--} at the analytical concentrations normally encountered in the laboratory. The natural choice of reference level is ordinarily obvious; however, situations do arise in which this is not the case, even for the experienced chemist. The novice may not have a well-developed chemical intuition, and a logical method for proceeding is given below for those lacking in prior experience.

It is always possible to decide whether or not a given choice of reference levels is the most appropriate one. This decision may be made by examination of the equilibrium constants for the protolytic reactions corresponding to the assumed reference level. In practice, questions arise only in fairly complicated equilibrium situations, and we shall immediately consider the $NaNH_4HPO_4$ solution.

An error in the choice of reference level will have been made if it is possible to write an acid–base reaction which will convert more than *half* of the assumed reference species into some other form(s). The most favorable equilibria for converting the assumed reference species will be formed from the sum(s) of the appropriate individual protolytic reactions considered in obtaining the particular proton condition. As a rule, it is not necessary to consider reactions that produce species more than one proton removed from the reference level. Thus the choice between case I and case II can be made using the following formal reasoning.

For case I, the sum of second and fourth reactions provides the principal path for the consumption of the assumed principal species, NH_4^+ and HPO_4^{--}, and is

$$NH_4^+ + HPO_4^{--} \rightleftharpoons NH_3 + H_2PO_4^-$$

$$K = \frac{K_{NH_4^+}}{K_2} = 9.3 \times 10^{-3}$$

The fraction of NH_3 present in solution which results from the reaction of NH_4^+ with HPO_4^{--} can be shown to be

$$\text{fraction} = \frac{[NH_3]}{[NH_4^+] + [NH_3]} = \frac{K^{1/2}}{1 + K^{1/2}} = \frac{0.096}{1.096} = 0.09$$

Hence the reaction between NH_4^+ and HPO_4^{--} converts less than 10 per cent of the NH_4 into NH_3, and the reference levels chosen for case I are the appropriate ones.

If the reference levels chosen initially corresponded to case II, we would proceed as follows. For case II, the sum of the third and last reactions represents the obvious way in which H_3PO_4 can be converted into $H_2PO_4^-$. Omitting Na^+ ion,

$$OH^- + H_3PO_4 \rightleftharpoons H_2PO_4^{--} + H_2O$$

$$K' = \frac{K_1}{K_w} = 5.9 \times 10^{11}$$

The magnitude of this equilibrium constant clearly demonstrates that all the H_3PO_4 is converted into $H_2PO_4^-$.

The question now arises as to whether NH_3 will react with $H_2PO_4^-$. The equilibrium constant for this reaction,

$$NH_3 + H_2PO_4^- \rightleftharpoons NH_4^+ + HPO_4^{--}$$

is

$$K'' = \frac{K_2}{K_{NH_4^+}} = \frac{6.2 \times 10^{-8}}{5.8 \times 10^{-10}} = 1.07 \times 10^2$$

and the fraction of NH_3 which does *not* react is

$$\text{fraction} = \frac{1}{1 + (K'')^{1/2}} = \frac{1}{1 + 10.3} = 0.09$$

Hence the principal species are NH_4^+ and HPO_4^{--}.

A problem of practical significance is the choice of reference level in a solution containing $2a$ M sodium hydroxide and a M H_2B, a dibasic weak acid with constants K_1 and K_2.

The usual assumption, that such a solution contains Na_2B, leads to case III.

Case III:	*Reaction*	*Equilibrium Constant*
	$H_2O \rightleftharpoons H^+ + OH^-$	K_w
	$B^{--} + H^+ \rightleftharpoons HB^-$	$1/K_2$
	$B^{--} + 2H^+ \rightleftharpoons H_2B$	$1/K_1K_2$

$$[H^+] = [OH^-] - [HB^-] - 2[H_2B] \qquad (45)$$

However, there are situations where it is more logical to assume that the solution consists of a M sodium hydroxide and a M NaHB, leading to

Case IV:	*Reaction*	*Equilibrium Constant*
	$H_2O \rightleftharpoons H^+ + OH^-$	K_w
	$HB^- \rightleftharpoons H^+ + B^{--}$	K_2
	$H^+ + HB^- \rightleftharpoons H_2B$	$1/K_1$
	$(NaOH) + H^+ \rightleftharpoons (Na^+) + H_2O$	$1/K_w$

$$[H^+] = [OH] + [B^{--}] - [H_2B] - [Na^+] \tag{46}$$

The choice of reference levels depends on the numerical values of K_1, K_2, and a, and is made as follows. In case III, the only possible acid with which the assumed principal species, B^{--}, can react is water; therefore, we add the first two reactions to obtain

$$B^{--} + H_2O \rightleftharpoons HB^- + OH^- \qquad \frac{K_w}{K_2} \tag{47}$$

In case IV we add the second and fourth reactions to describe the way in which the two assumed principal species, HB^- ion and OH^- ion, may react, to obtain

$$HB^- + OH^- \rightleftharpoons B^{--} + H_2O \qquad \frac{K_2}{K_w} \tag{48}$$

Thus we have demonstrated that the same equilibrium governs no matter what the initial assumption. In the example under consideration, the ratio $[B^{--}]/[HB^-]$ is dependent upon the values of a and K_2, and it can be shown that more $[HB^-]$ than $[B^{--}]$ is present when $aK_2/2K_w$ is less than unity.

Hence case III represents the reference levels we would choose if H_2B were carbonic acid, $K_1 = 4.5 \times 10^{-7}$, $K_2 = 4.7 \times 10^{-11}$, and $a > 4.3 \times 10^{-4}$ M, while case IV represents the reference levels we would choose if H_2B were hydrogen sulfide, $K_1 = 1.0 \times 10^{-7}$, $K_2 = 1.2 \times 10^{-13}$, and $a < 0.17$ M.

Very great emphasis has deliberately been placed upon the proper choice of reference level in applying the proton condition. The proton condition obtained using the reference level based upon the principal acid–base species in solution is the only one which can readily be used to obtain equations suitable for numerical substitution. The reason for this situation is easily explained.

Proton conditions obtained in the manner described *never* contain the equilibrium concentration of the assumed principal species, and can only contain equilibrium concentrations of acid–base species related to the principal species by an equilibrium constant. Hence, using the appropriate equilibrium constants, it is possible to substitute for all equilibrium concentrations appearing in the proton condition in terms of $[H^+]$, various K's, and principal species. The equilibrium concentration of all principal species can usually be stated (or estimated closely) in terms of analytical concentrations, thereby permitting the evaluation of $[H^+]$. Occasionally, this estimate of $[H^+]$ is not sufficiently precise and an iterative procedure is required to obtain a better value of $[H^+]$.

However, if the proton condition is obtained using the "wrong" level, the application of the equilibrium-constant expressions will produce a final expression containing equilibrium concentrations that cannot be readily estimated. Thus, although this final expression is completely correct, it will be of no value in obtaining a numerical result for the problem using the algebraic method described in the next section.

Graphical method of solution of protolytic equilibrium problems, originally introduced by Scandanavian chemists, have become quite popular in recent years. The choice of proton level is just as critical using these methods as when an algebraic method of solution is used.

CALCULATION OF [H^+]

Two principal methods exist for the calculation of hydrogen-ion concentration in an aqueous solution, and have been designated the "algebraic" and "graphical" methods. Both methods have their vigorous supporters, some in each group claiming their method is clearly superior to the other. We shall avoid this polemic and merely point out that both methods use exactly the same starting point, the proton condition, and diverge to the extent that one group treats the mass balance equations and equilibrium constants algebraically, whereas the other relies upon graphical methods. In an especially complicated problem the graphical methods can be of considerable help in getting a "feel" for the problem, while if especially accurate calculations are required, an algebraic method must be used to supplement the graphical method. Some of the problems discussed in this section will be treated both ways to familiarize the reader with both approaches.

Solutions of strong acids and bases

The proton condition for a mixture of a strong acid, a M, and strong base, b M, according to Eq. (39), is

$$[H^+] = [OH] + a - b \tag{49}$$

Since $[OH] = K_w/[H^+]$, we obtain

$$[H^+] = \frac{K_w}{[H^+]} + a - b \tag{50}$$

as the exact expression which may be used to solve for [H^+]. Equation (50), a quadratic expression in [H^+], ordinarily is unduly complicated for the typical problem and need be used only when $(a - b) \approx (K_w)^{1/2} = 10^{-7}$. Normally, either $a \gg b$ (the solution may be considered a mixture of strong acid $(a - b)$ M with b M neutral salt present), and

$$[H^+] = a - b \tag{51}$$

or $b \gg a$ (the solution contains $(b - a)$ M strong base and a M neutral salt) and

$$[H^+] = \frac{K_w}{b - a} \tag{52}$$

Example 1: 2.00×10^{-2} mole of hydrochloric acid is placed in three volumetric flasks of 1 liter capacity. The first flask is filled to the mark with distilled water; 2.00×10^{-2} mole of sodium hydroxide is added to the second flask and 3.00×10^{-2} mole of sodium hydroxide added to the third flask before the flasks are filled to the mark. What are the equilibrium concentrations of hydrogen ion in the three flasks?

In the first flask $a = 0.0200$ M and $b = 0$; therefore, according to Eq. (51),

$$[H^+] = 0.0200 - 0 = 2.00 \times 10^{-2}\ M$$

In the second flask $a = b = 0.0200$ M; therefore, according to Eq. (50),

$$[H^+] = \frac{K_w}{[H^+]} + 0.02 - 0.02 = \frac{K_w}{[H^+]}$$

or

$$[H^+] = \sqrt{K_w} = \sqrt{1.00 \times 10^{-14}} = 10^{-7}$$

In the third flask, $a = 0.0200$ M and $b = 0.0300$ M; therefore, according to Eq. (52),

$$[H^+] = \frac{1.00 \times 10^{-14}}{0.0300 - 0.0200} = 1.00 \times 10^{-12}$$

Neither Eq. (50), (51), nor (52) is sufficiently complicated to warrant using a graphical solution method under ordinary circumstances. However, in order to introduce the

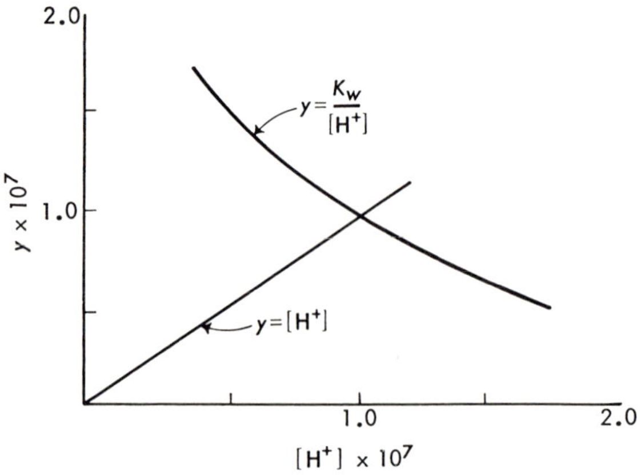

Fig. 4-3. Conventional graphical method for solving for $[H^+]$ in pure water at 25°C.

graphical method, we shall graphically solve the problem in the second part of the example just discussed. According to Eq. (49), the rigorous solution to this problem is given by

$$[H^+] = \frac{K_w}{[H^+]}$$

Using a conventional graphical procedure, we set

$$y = [H^+] \tag{53}$$

$$y = \frac{K_w}{[H^+]} \tag{54}$$

and proceed to plot the Eqs. (53) and (54) on the same graph. The result is shown in Fig. 4.3.

The intersection occurs at $[H^+] = 1.00 \times 10^{-7}$ and is the solution to the problem. This method is tedious because one of the equations is a hyperbola, and it is necessary to evaluate it for many values of $[H^+]$ over a wide range of numerical values in order

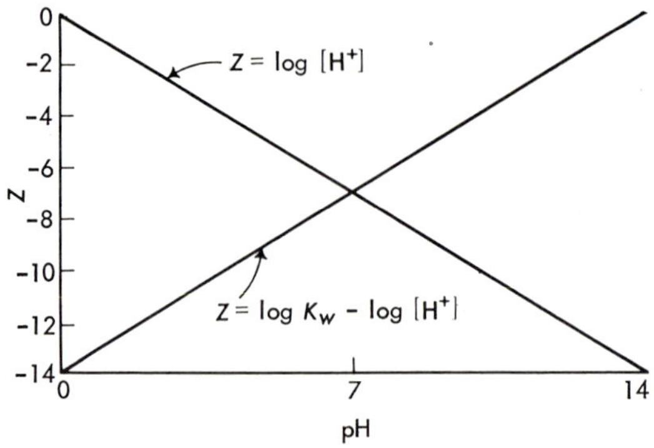

Fig. 4-4. Logarithmic diagram for finding the pH of pure water at 25°C.

to accurately construct the curve and find the region of the intersection. A far more convenient approach is to define a new variable Z, such that

$$Z = \log [H^+] \tag{55}$$

$$Z = \log \frac{K_w}{[H^+]}$$

$$Z = \log K_w - \log [H^+] \tag{56}$$

and proceed to plot Eqs. (55) and (56) as functions of $-\log [H^+]$ (pH) on the same graph. This result is shown in Fig. 4.4. The answer is pH $= 7.00$, and corresponds to the intersection of the two straight lines, the first having a slope of -1 and passing through $Z = 0$, and the second having a slope of $+1$ and passing through $Z = -14$.

The construction of Fig. 4.4 is far easier than Fig. 4.3 and has the additional advantage of displaying the entire pH range usually of interest in water in a compact form. The representation used in Fig. 4.4 is the basis of the graphical solution of protolytic equilibrium problems.

Solutions containing a weak monoprotic acid: Pure acid

The hydrogen-ion concentration of an c_a M solution of a weak monoprotic acid, HB, is readily calculated starting with the proton condition, Eq. (40), for this solution

$$[H^+] = [OH^-] + [B]$$

or

$$[B] = [H^+] - [OH^-] \tag{57}$$

Since

$$c_a = [HB] + [B] \tag{58}$$

we obtain

$$[HB] = c_a - [H^+] + [OH] \tag{59}$$

Substituting Eqs. (57) and (59) into Eq. (60),

$$K_{HB} = \frac{[H^+][B]}{[HB]} \tag{60}$$

yields

$$K_{HB} = \frac{[H^+]\{[H^+] - [OH^-]\}}{c_a - [H^+] + [OH^-]} \tag{61}$$

At this point the possibility of simplifying approximations becomes apparent. The first approximation that one recognizes is that in solutions of a weak acid $[H^+]$ is ordinarily much greater than $[OH^-]$, so that Eq. (61) can be written

$$K_{HB} = \frac{[H^+]^2}{c_a - [H^+]} \tag{62}$$

Equation (62) is a quadratic equation in $[H^+]$,

$$[H^+]^2 + [H^+]K_{HB} - c_a K_{HB} = 0$$

and can always be solved in the conventional manner.

$$[H^+] = \frac{-K_{HB} + (K_{HB}^2 + 4c_a K_{HB})^{1/2}}{2} \tag{63}$$

However, the term in the denominator of Eq. (62) suggests that when $c_a \gg [H^+]$ the very much simpler expression

$$[H^+] = (c_a K_{HB})^{1/2} \tag{64}$$

is sufficiently precise for practical purposes.

Example 2: Calculate the hydrogen-ion concentration in a solution of a weak acid, HB, $K_{HB} = 4.00 \times 10^{-5}$ when (a) $c_a = 0.100\ M$, (b) $c_a = 1.00 \times 10^{-3}\ M$.

Usually the most efficient way of solving any solution equilibrium problem is to substitute into the most approximate form of the solution, obtain a numerical result for $[H^+]$, and verify that the assumption originally used in obtaining the approximate equation is correct. If this is not the case, we then begin over using the equation which was used to produce the approximate term. Hence, for part (a), substituting into Eq. (64), we obtain

$$[H^+] \simeq [0.100(4.00 \times 10^{-5})]^{1/2} = 2.00 \times 10^{-3}$$

The approximation used in going from Eqs. (62) to (64) was that $c_a \gg [H^+]$; for these numerical values c_a is 50 times greater than $[H^+]$, and the approximation is valid to within 2 per cent, the usual accuracy required. Attempting part (b), we substitute into Eq. (64),

$$[H^+] \simeq [(1.00 \times 10^{-3})(4.00 \times 10^{-5})]^{1/2} = 2.00 \times 10^{-4}$$

to obtain the result that $[H^+]$ is 20 per cent of c_a. *Hence Eq.* (64) *may not be used,* because the approximation used in going from Eq. (62) to (64) is not valid. We then use Eq. (63),

$$[H^+]^2 + (4.00 \times 10^{-5})[H^+] - (1.00 \times 10^{-3}) \times (4.00 \times 10^{-5}) = 0$$

which has the numerical solution

$$[H^+] = 1.81 \times 10^{-4}$$

Problems requiring the use of Eq. (61) will be extremely rare and are of academic interest only. This is easily seen, because $[H^+] \approx [OH]$ only when $[H^+] \approx 10^{-6}$ to 10^{-7}; therefore, as long as $c_a > 10^{-5}$ to 10^{-6}, Eq. (62) suffices.

The graphical method may also be used to calculate the hydrogen-ion concentration of a weak acid solution. The basis of the method is easily seen if we substitute the expressions $[B] = c_aK_{HB}/[H^+] + K_{HB}$ and $[OH] = K_w/[H^+]$ into Eq. (40), the proton condition, to obtain

$$[H^+] = \frac{K_w}{[H^+]} + \frac{c_aK_{HB}}{[H^+] + K_{HB}} \tag{65}$$

Following the procedure used to solve the case of a mixture of a strong acid and strong base [Eqs. (55) and (56)], we write

$$Z = \log [H^+] \tag{66}$$

$$Z = \log \left\{\frac{K_w}{[H^+]} + \frac{c_aK_{HB}}{[H^+] + K_{HB}}\right\} \tag{67}$$

and immediately realize that, while Eq. (66) is easily plotted versus pH, in order to plot Eq. (67) versus pH, a lengthy series of calculations would be required to evaluate the terms in the braces exactly. However, in the usual case, this is unnecessary, because the term

$$\frac{c_aK_{HB}}{[H^+] + K_{HB}} \gg \frac{K_w}{[H^+]}$$

and it is sufficient to plot

$$Z = \log \left\{\frac{c_aK_{HB}}{[H^+] + K_{HB}}\right\} \tag{68}$$

in place of Eq. (67).

Figure 4-5 is a plot of Eqs. (66) and (68) for part (a) of the problem given in Example 2. Also included is a plot of Eq. (56). The answer corresponds to the intersection of the $[H^+]$ line and the [B] line (point C), which occurs at pH = 2.70 ($[H^+] = 2.00 \times 10^{-3}$).

For clarity in plotting graphs such as shown in Fig. 4.5, all species appearing on the right side of a proton condition will appear as a dashed curve. The point of intersection

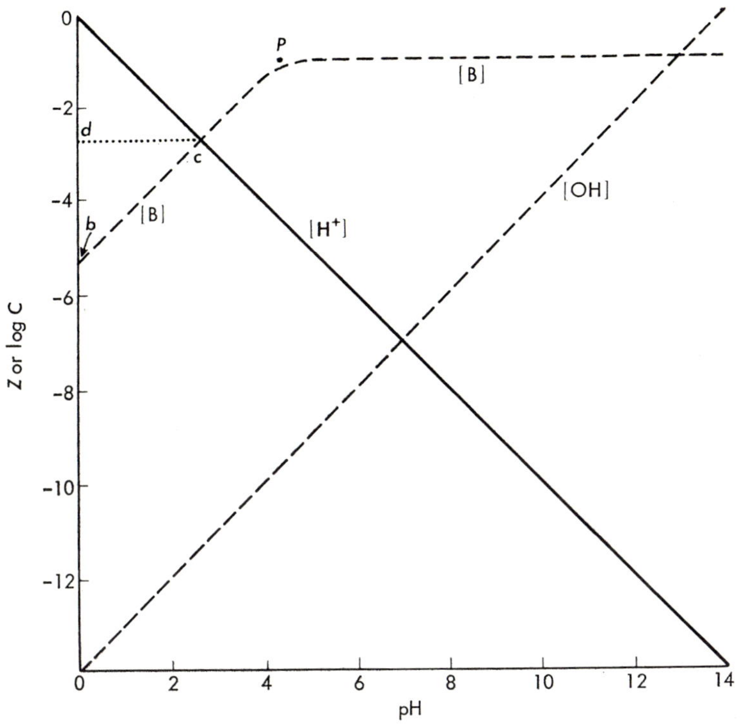

Fig. 4-5. Logarithmic diagram for finding the pH of a 0.1 M weak acid solution, $pK_{HB} = 4.40$.

can be determined by accurate plotting of Eqs. (66) and (68) and reading the result off the graph. Alternatively, the result may be calculated using elementary geometry.

When $pK_{HB} \gg pH$ (~ 3.5 in this case), Eq. (68) becomes

$$Z = \log \{c_a K_{HB}\} + pH$$

i.e., a straight line of slope = 1.00 intersecting the Z axis at $\log(0.100 \times 4 \times 10^{-5}) = -5.40$ (line cb).

The $[H^+]$ line has a slope = -1.00 and intersects the Z axis at 0.

The triangle formed by the points $0bc$ is an isosceles triangle, with side $0c$ = side bc. The dotted line, dc, is drawn perpendicular to $0b$, and bisects $0b$; therefore, $0d = db = 0b/2 = 5.40/2 = 2.70$. But $dc = db$; therefore, $dc = pH = 2.70$.

As mentioned earlier, the curve connecting the two straight-line portions of the [B] curve may be sketched in making use of the fact that it is 0.30 unit lower than P at pH = pK, and 0.12 unit lower than the limiting straight lines at pH = p$K \pm 0.5$. As long as the intersection of the $[H^+]$ and [B] lines does not occur near this curved region, the geometric method described above is quite straightforward, and corresponds to the solution of Eq. (64).

The graphical situation that would exist in part (b) of Example II corresponds to moving the [B] curve downward two units, since $c_a = 0.00100$, not 0.100 as in part (a). The intersection of the $[H^+]$ and [B] curves would now occur in the curved region and the answer could be obtained only by careful plotting of the curved region of the [B] curve, and reading the answer off the graph. This procedure corresponds to the solution of the quadratic equation (63).

The relative positions of the [B] and [OH^-] lines in Fig. 4-5 demonstrate that over nearly the entire pH range, [OH^-] $\ll$ [B], as assumed when we chose the correct answer to be point C. Only by making the analytical concentration of this acid less than 10^{-6} M will it be possible to move the intersection of the [B] and [H^+] lines close to [OH^-] line. In such a situation, the graphical method fails and an algebraic technique must be used. [It would be necessary to solve Eq. (61).]

The ordinate, Z, represents the logarithm of the concentration of the species each curve designates. For this reason, the variable Z will be replaced by $\log C$ in all future discussions.

Mixtures of weak acid and strong base

One proton condition for a solution containing a mixture of HB, analytical concentration, c_a, and strong base, analytical concentration b, has been derived earlier [Eq. (42)], using HB as the reference level:

$$[H^+] = [OH^-] + [B] - b$$

Solving for [B],

$$[B] = b + [H^+] - [OH^-] \tag{69}$$

Since $c_a = [HB] + [B]$ it follows that[1]

$$[HB] = c_a - b - [H^+] + [OH^-] \tag{70}$$

Substituting Eqs. (69) and (70) into Eq. (60), and rearranging, Eq. (71) is obtained:

$$[H^+] = K_{HB} \frac{c_a - b - [H^+] + [OH^-]}{b + [H^+] - [OH^-]} \tag{71}$$

Equation (71) has a number of more approximate forms, but from the computational viewpoint, only the most approximate need be considered. If $c_a - b \gg -[H^+] + [OH^-]$ and $b \gg [H^+] - [OH^-]$, the simple expression,

$$[H^+] = K_{HB} \frac{c_a - b}{b} \tag{72}$$

is obtained.[2] The most direct approach to an actual problem involves using Eq. (72) to obtain an estimate of [H^+] and [OH^-], and comparing these provisional values of [H^+] and [OH] with the magnitudes of $c_a - b$ and b. Equation (71) can then be simplified as far as is compatible with the numerical quantities in the problem being considered.

Example 3: 10.0 ml of 0.100 M sodium hydroxide is added to 40.0 ml of 0.100 M acetic acid ($K_{HB} = 1.74 \times 10^{-5}$). What is the hydrogen-ion concentration?

[1] Equation (70) can be obtained directly from the proton condition using B as the reference level; i.e., the solution is considered to be a mixture containing c_a M strong acid, c_a M weak base B, and b M strong base. The equilibria involved are

$$\begin{aligned} H_2O &\rightleftharpoons H^+ + OH^- \\ B + H^+ &\rightleftharpoons HB, \\ c_a\ M \text{ strong acid} &\rightarrow H^+ \\ b\ M \text{ strong base} + H^+ &\rightarrow H_2O \end{aligned}$$

and we write

$$[H^+] = [OH] - [HB] + c_a - b$$

which is equivalent to Eq. (70). Each term on the right side corresponds to a source or sink of protons.

[2] If we imagine that solution had been prepared by mixing the weak acid and its sodium salt (conjugate base), the student will recognize that the concentration term b corresponds to analytical concentration of salt (conjugate base), and the term $c_a - b$ corresponds to the analytical concentration of acid. Compare Eq. (72) with Eq. (26) on p. 51.

After mixing, $c_a = 0.0800\ M$ and $b = 0.0200\ M$. Substituting into Eq. (72),

$$[H^+] \approx (1.74 \times 10^{-5}) \frac{0.0800 - 0.0200}{0.0200} = 5.22 \times 10^{-5}$$

and

$$[OH^-] \approx \frac{1.00 \times 10^{-14}}{5.22 \times 10^{-5}} = 1.9 \times 10^{-10}$$

Since $(c_a - b) \gg -[H^+] + [OH^-]$ and $b \gg [H^+] - [OH^-]$, the use of Eq. (72) was appropriate and $[H^+] = 5.22 \times 10^{-5}$.

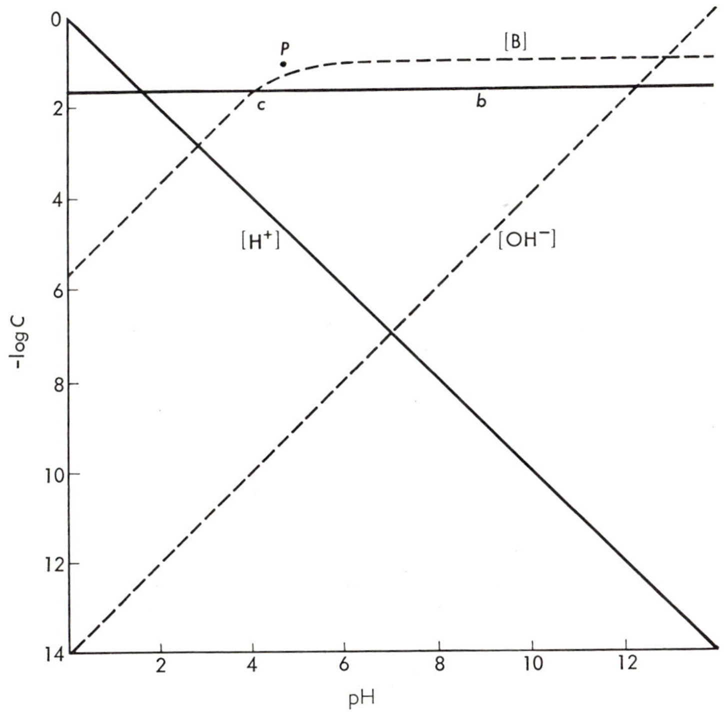

Fig. 4-6. Logarithmic diagram for finding the pH of a solution containing 0.0200 M strong base and 0.0800 M weak acid, $pK_{HB} = 4.76$.

Figure 4.6 represents the plot of the proton condition, Eq. (42), corresponding to Example 3, i.e.,

$$[H^+] + b = [OH^-] + [B]$$

$$[H^+] + 0.02 = [OH^-] + \frac{0.08\ K_{HB}}{[H^+] + K_{HB}}$$

The point of intersection of the uppermost solid line, $b = 0.020$, and the uppermost dashed line, $[B] = 0.08K_{HB}/([H^+] + K_{HB})$, occurs at point C, and is the graphical solution to the problem. If the curves are carefully constructed on a scale such that 2 pH units corresponds to 1 in., the correct answer, pH = 4.28, can be estimated to within 0.03 pH unit. Taking into consideration the effort required to carefully plot the curved region of the [B] curve as required in problems involving mixtures of an acid and its conjugate base, the graphical method has little to recommend it in this case.

Example 4: 0.200 millimole of solid sodium hydroxide is added to 100 ml. of 0.1 *M* formic acid ($K_{HB} = 1.78 \times 10^{-4}$). What is the hydrogen-ion concentration in this solution?

In this problem $c_a = 0.100$ and $b = 0.002$. Using Eq. (72), we have

$$[H^+] \simeq (1.78 \times 10^{-4}) \frac{0.100 - 0.002}{0.002} = 8.7 \times 10^{-3}$$

$[OH^-]$ is negligibly small compared to $[H^+]$ and can be neglected. However, using Eq. (72), we obtain the result that the first estimate of the value of $[H^+]$ is four times b and 10 per cent of $c_a - b$. We conclude that Eq. (71) can be simplified only to the form

$$[H^+] = K_{HB} \frac{c_a - b - [H^+]}{b + [H^+]} \tag{73}$$

which is a quadratic equation in $[H^+]$, with the explicit solution

$$[H^+] = -\frac{(b + K_{HB}) + [(b + K_{HB})^2 + 4K_{HB}(c_a - b)]^{1/2}}{2} \tag{74}$$

Numerical substitution yields

$$[H^+] = 3.2_2 \times 10^{-4}$$

for the correct solution to Example 4.

Example 5: A weak acid (HB) has a dissociation constant of 1.00×10^{-10}. What is the pH of a 1-liter solution to which 100 millimoles of solid HB and 99 millimoles of strong base has been added?

In this case, $c_a = 0.100$, $b = 0.099$, and $K_{HB} = 1.00 \times 10^{-10}$. Substituting into Eq. (72) yields

$$[H^+] \approx (1.00 \times 10^{-10}) \frac{0.100 - 0.099}{0.099} = 1.0 \times 10^{-12}$$

or $[OH] \approx 1.0 \times 10^{-2}$. Thus, the numerator of the fraction on the right side of Eq. (71) can be simplified to $c_a - b + [OH^-]$, and the denominator can be reduced to $\{b - [OH^-]\}$. The equation which must be solved in this case is

$$[H^+] = K_{HB}\left\{\frac{c_a - b + [OH^-]}{b - [OH^-]}\right\}$$

The solution is a quadratic expression in $[OH^-]$, and on substituting for b and c_a, we obtain

$$[OH^-] = 2.6 \times 10^{-3} \qquad \text{or} \qquad [H^+] = 3.8 \times 10^{-12}$$

If b is set equal to zero in Eq. (71), the resultant expression is identical to Eq. (61), the expression for the hydrogen-ion concentration of a pure acid. On the other hand, if $b = c_a$ in Eq. (71), we obtain the expression for the hydrogen-ion concentration in a solution of NaB,

$$[H^+] = K_{HB} \frac{[OH^-] - [H^+]}{c_a + [H^+] - [OH^-]} \tag{75}$$

Substituting $[H^+] = K_w/[OH^-]$ into Eq. (75) and rearranging, we obtain

$$\frac{K_w}{K_{HB}} = K_B = \frac{[OH^-]\{[OH^-] - [H^+]\}}{c_a + [H^+] - [OH^-]} \tag{76}$$

Equation (76) is similar in form to Eq. (61) and the approximate expressions corresponding to Eqs. (62) and (64) are

$$K_B = \frac{[OH^-]^2}{c_a - [OH^-]} \tag{77}$$

and

$$[OH^-] = (c_a K_B)^{1/2} \tag{78}$$

The choice between Eqs. (77) and (78) is made in an analogous manner to the choice between Eqs. (62) and (64).

Equation (76) can be derived directly using the proton condition (assuming B as the reference level) in a similar fashion to the derivation of Eq. (61).

Using Eq. (71) in its various forms it is possible to calculate the titration curve of a weak acid (dissociation constant $= K_{HB}$) with strong base in the range 0 to 100 per cent titrated. The identical equation also describes the titration curve of a weak base (basic dissociation constant $K_B = K_w/K_{HB}$) in the same titration range. This is readily seen because the term $c_a - b$ corresponds to the analytical concentration of the HB species while the term b corresponds to the analytical concentration of the B species.

Example 6: 100.0 ml of 0.100 M ammonia and 90.0 ml of 0.100 hydrochloric acid are diluted to a total volume of 200.0 ml. The dissociation constant of ammonia as a base, K_B, is 1.74×10^{-5}. What is the hydrogen-ion concentration of the resultant solution?[3]

After diluting to volume, the analytical concentration of ammonia species, c_b, is 0.0500 M, and the analytical concentration of hydrochloric acid, a, is 0.04500 M. The solution corresponds to a mixture of 0.0450 M ammonium chloride (acid) and 0.00500 M ammonia (conjugate base).

Before we can substitute into Eq. (72), the acid dissociation constant of ammonium ion, K_{HB}, must be calculated from

$$K_{HB} = \frac{K_w}{K_B} = \frac{1.00 \times 10^{-14}}{1.74 \times 10^{-5}} = 5.75 \times 10^{-10}$$

Substituting into Eq. (72), we obtain

$$[H^+] \simeq (5.75 \times 10^{-10}) \frac{0.0450}{0.00500} = 5.18 \times 10^{-9}$$

and $[OH] \simeq 2.3 \times 10^{-6}$. Making the usual checks of the approximations, we conclude that Eq. (72) holds, and $[H^+] = 5.18 \times 10^{-9}$.

Mixture of a weak and a strong acid

If a solution contains a mixture of a weak acid, HB, of molar analytical concentration $= c_a$ and a strong acid of molar concentration a, the appropriate proton condition is obtained by considering the following equilibria:

$$H_2O \rightleftharpoons H^+ + OH^-$$

$$HB \rightleftharpoons H^+ + B$$

$$\text{strong acid, } a\ M \rightarrow H^+$$

yielding

$$[H^+] = [OH^-] + [B] + a \tag{79}$$

[3] If we imagine the solution in Example 6 as having originated from analytical concentrations (after mixing) of c_a M ammonium chloride ($c_a = 0.0500$) and b M strong base ($b = 0.00500$), the equivalence between a weak acid–strong base titration with a weak base–strong acid titration is apparent.

In all practical problems $[OH^-]$ is negligibly small as compared to [B] and a. Substituting $[B] = c_a\{K_{HB}/[H^+] + K_{HB}\}$ into Eq. (79), we obtain

$$[H^+] = \frac{c_a K_{HB}}{[H^+] + K_{HB}} + a \tag{80}$$

The explicit solution to Eq. (80) for $[H^+]$ is

$$[H^+] = \frac{a - K_{HB} + [(a - K_{HB})^2 + 4K_{HB}(a + c_a)]^{1/2}}{2} \tag{81}$$

Even though Eq. (81) is readily evaluated numerically, it should be remembered that frequently $a \gg [OH^-] + [B]$ in Eq. (79), and that the approximate result

$$[H^+] \simeq a \tag{82}$$

very often yields a satisfactory answer. The error introduced by using Eq. (82) is easily checked, by comparing the magnitude of a and the term $c_a K_{HB}/(a + K_{HB})$. Unless the latter term is small compared to a, Eq. (81) rather than (82) must be used.

Example 7: A solution of 0.100 M in acetic acid, $K_{HB} = 1.74 \times 10^{-5}$, is made (a) 0.0100 M in strong acid, and (b) 0.00100 M in strong acid. What are the hydrogen-ion concentrations in these solutions?

For part (a), we use the approximate relationship, Eq. (82), to obtain a first approximation, $[H^+] \simeq 0.01$. The term

$$\frac{c_a K_{HB}}{[H^+] + K_{HB}} = \frac{0.1 \times 1.74 \times 10^{-5}}{0.01 + 1.74 \times 10^{-5}} = 1.74 \times 10^{-4}$$

which is negligibly small compared to a. Hence the correct answer is $[H^+] = 0.0100$, with an uncertainty of 1 or 2 in the last significant figure. The hydrogen ion from the strong acid has completely repressed the dissociation of the weak acid, HB.

Attempting part (b) in the same way, we estimate

$$[H^+] \simeq 0.001, \qquad [B] \simeq \frac{0.1 \times 1.74 \times 10^{-5}}{100.0 \times 10^{-5} + 1.74 \times 10^{-5}} = 1.7 \times 10^{-3}$$

a result which makes the use of Eq. (82) absurd. Using Eq. (81) we obtain

$$[H^+] = 1.91 \times 10^{-3}$$

as the correct answer. In this case only partial repression of the dissociation of acetic acid by the strong acid has occurred.

Mixture of a weak and a strong base

It is left as an exercise for the student to demonstrate that the hydroxyl-ion concentration of a solution with an analytical concentration c_b M of weak base and b molar in strong base is given by

$$[OH^-] = b \tag{83}$$

when repression of the dissociation of the weak base is almost complete, and by

$$[OH^-] = \frac{b - K_B + [(b - K_B)^2 + 4K_B(b - c_b)]^{1/2}}{2} \tag{84}$$

in the more general case, where K_B is the dissociation constant of the base.

Graphical solutions of the proton conditions that lead to Eqs. (81) and (84) are not practical. If, for example, Eq. (80) is plotted, it will be found that the answer corresponds to a pH where the [B] and a curves are so close together that the basic assumption concerning the graphical method no longer holds.

Equimolar mixture of a weak acid (HA) and weak base (B)

The hydrogen-ion concentration of an equimolar mixture of a weak acid and a weak base is of some interest. There are two common ways in which such a solution may occur. The first situation corresponds to the titration of the acid, HA, with the base, B, according to

$$\mathrm{HA} + \mathrm{B} \rightleftharpoons \mathrm{HB} + \mathrm{A} \qquad K_{eq} = \frac{K_{\mathrm{HA}}}{K_{\mathrm{HB}}} = \frac{K_{\mathrm{B}}K_{\mathrm{HA}}}{K_w} \tag{85}$$

When $K_{\mathrm{HA}} \gg K_{\mathrm{HB}}$, the titration reaction goes appreciably to the right.

This titration represents an undesirable analytical situation and is ordinarily avoided by choosing a strong base titrant. One would never elect to titrate acetic acid (HA) with ammonia (B) in preference to using sodium hydroxide.

However, another practical situation frequently occurs which cannot be avoided, i.e., the titration of a mixture of two different acids, HB and HA. If $K_{\mathrm{HA}} > 10^5 K_{\mathrm{HB}}$, it is possible to locate the first endpoint, corresponding to a mixture of A and HB, with considerable accuracy using potentiometric methods. An example would be the titration of a mixture of acetic acid ($K_{\mathrm{HA}} = 1.74 \times 10^{-5}$) and boric acid ($K_{\mathrm{HB}} = 5.85 \times 10^{-10}$) with sodium hydroxide, to the first end point.

The proton condition for such a solution is obtained by choosing as reference level the species A (acetate ion) and HB (boric acid). The reactions and equilibrium constants of interest are

$$\mathrm{H_2O} \rightleftharpoons \mathrm{H^+} + \mathrm{OH^-} \qquad K_w = [\mathrm{H^+}][\mathrm{OH^-}]$$

$$\mathrm{HB} \rightleftharpoons \mathrm{H^+} + \mathrm{B} \qquad K_{\mathrm{HB}} = \frac{[\mathrm{H^+}][\mathrm{B}]}{[\mathrm{HB}]}$$

$$\mathrm{A} + \mathrm{H^+} \rightleftharpoons \mathrm{HA} \qquad \frac{1}{K_{\mathrm{HA}}} = \frac{[\mathrm{HA}]}{[\mathrm{H^+}][\mathrm{A}]}$$

and the proton condition is

$$[\mathrm{H^+}] + [\mathrm{HA}] = [\mathrm{OH^-}] + [\mathrm{B}] \tag{86}$$

It is convenient to write species that correspond to the consumption of protons on the left side of the proton condition to simplify the algebra that must be performed.

Substituting for $[\mathrm{OH^-}]$, $[\mathrm{B}]$, and $[\mathrm{HA}]$ from the appropriate equilibrium constants we obtain

$$[\mathrm{H^+}] + \frac{[\mathrm{H^+}][\mathrm{A}]}{K_{\mathrm{HA}}} = \frac{K_w}{[\mathrm{H^+}]} + \frac{K_{\mathrm{HB}}[\mathrm{HB}]}{[\mathrm{H^+}]} \tag{87}$$

Equation (87) can then be solved for $[\mathrm{H^+}]$ in terms of the *equilibrium* concentrations of A and HB to yield

$$[\mathrm{H^+}] = \left(\frac{K_w + K_{\mathrm{HB}}[\mathrm{HB}]}{1 + [\mathrm{A}]/K_{\mathrm{HA}}}\right)^{1/2} \tag{88}$$

In obtaining Eq. (88) we have not considered the analytical concentrations of B and HA, which are related to the equilibrium concentrations of HB and A by the relations

$$[\mathrm{HB}] = \frac{c_{\mathrm{HB}}[\mathrm{H^+}]}{[\mathrm{H^+}] + K_{\mathrm{HB}}} \tag{89}$$

$$[\mathrm{A}] = \frac{c_{\mathrm{HA}}K_{\mathrm{HA}}}{[\mathrm{H^+}] + K_{\mathrm{HA}}} \tag{90}$$

where $c_{\mathrm{HB}} = [\mathrm{HB}] + [\mathrm{B}]$ and $c_{\mathrm{HA}} = [\mathrm{HA}] + [\mathrm{A}]$.

Obviously, substitution of Eqs. (89) and (90) into Eq. (88) would yield a very complicated algebraic expression and, in fact, is unnecessary. This is apparent because the choice of reference level implies that [HB] $\simeq c_{HB}$ and [A] $\simeq c_{HA}$. With these assumptions, Eq. (88) simplifies to

$$[H^+] = \left(\frac{K_w + K_{HB}c_{HB}}{1 + c_{HA}/K_{HA}}\right)^{1/2} \tag{91}$$

which can then be solved for $[H^+]$. Using the value of $[H^+]$ obtained from Eq. (91) it is possible to check the approximations concerning the equality of analytical and equilibrium concentrations of A and HB by substituting $[H^+]$ into Eqs. (89) and (90). These assumptions are nearly always verified. In certain rare cases, these assumptions are not sufficiently precise. In such instances, the estimates of [A] and [HB] obtained from Eqs. (89) and (90) may be substituted into Eq. (88) to obtain a better value of $[H^+]$. This new value of $[H^+]$ can be used again in Eqs. (89) and (90) to obtain better values of [A] and [HB], for use again in Eq. (88). This entire procedure is repeated until no change in $[H^+]$ occurs. (The iterative procedure has converged to the answer.)

For many practical problems, numerical substitution into Eq. (91) leads to the result that $1 \ll c_{HA}/K_{HA}$ and that $K_w \ll K_{HB}c_{HB}$ permitting simplification of Eq. (91) to

$$[H^+] = \left(\frac{c_{HB}}{c_{HA}} K_{HA}K_{HB}\right)^{1/2} \tag{92}$$

In the special case where $c_{HB} = c_{HA}$, the familiar formula

$$[H] = (K_{HA}K_{HB})^{1/2} = \left(\frac{K_wK_{HA}}{K_B}\right)^{1/2} \tag{93}$$

for an equimolar mixture of a weak acid and weak base (e.g., ammonium acetate) is obtained. The student should be able to derive Eq. (91), and make the permissible simplifications on the basis of pertinent numerical quantities.

Example 8: What is the hydrogen-ion concentration of a solution containing equimolar analytical concentrations (0.1 *M*) of formic acid ($K_{HA} = 1.78 \times 10^{-4}$) and ammonia ($K_B = 1.74 \times 10^{-5}$)?

The equilibrium constant for reaction (85) is $\sim 3 \times 10^5$; therefore the reaction of formic acid and ammonia to form ammonium and formate ions is virtually quantitative. Hence the reference proton levels are formate ion (A) and ammonium ion (HB). The numerical values in the problem are such that Eq. (93) seems appropriate.

$$[H^+] \cong \left[\frac{(1.00 \times 10^{-14})(1.78 \times 10^{-4})}{1.74 \times 10^{-5}}\right]^{1/2} = 3.22 \times 10^{-7}$$

Substitution of $[H^+]$ into Eq. (89),

$$[NH_4^+] = 0.100 \frac{3.22 \times 10^{-7}}{3.22 \times 10^{-7} + 10^{-14}/1.74 \times 10^{-5}} = 0.100$$

and into Eq. (90),

$$[\text{formate}] = 0.100 \frac{1.78 \times 10^{-4}}{1.78 \times 10^{-4} + 3.22 \times 10^{-7}} = 0.100$$

verifies the assumption that [HB] $= c_{HB}$ and [A] $= c_{HA}$. Thus $[H^+] = 3.22 \times 10^{-7}$.

Example 9: A solution of acetic acid and boric acid are titrated to the first end point using sodium hydroxide. The analytical concentrations of sodium acetate and boric acid are 0.200 *M* and 0.0500 *M*, respectively. What is the hydrogen-ion concentration in this solution?

The acid dissociation of acetic acid is 1.74×10^{-5} and that of boric acid is 5.85×10^{-10}. The numerical values suggest the use of Eq. (92), with $c_{HB} = 0.05$ and $c_{HA} = 0.200$:

$$[H^+] = \left[\frac{0.0500}{0.200}(1.74 \times 10^{-5})(5.85 \times 10^{-10})\right]^{1/2}$$

$$= 5.04 \times 10^{-8}$$

Substitution of $[H^+] = 5.04 \times 10^{-8}$ into Eqs. (89) and (90) verifies the appropriateness of using Eq. (92); therefore, $[H^+] = 5.04 \times 10^{-8}$.

Example 10: What is the hydrogen-ion concentration of a 0.0500 M solution of ammonium dichloroacetate?

The acid dissociation constant of ammonium ion is 5.85×10^{-10} and of dichloroacetic acid 5.00×10^{-2}. The numerical values require the use of the most general equation, Eq. (91).

$$[H^+] = \left[\frac{1.00 \times 10^{-14} + (0.0500 \times 5.85 \times 10^{-10})}{1 + 0.0500/0.0500}\right]^{1/2}$$

$$= \left(\frac{29.3 \times 10^{-12}}{2.00}\right)^{1/2}$$

$$= 3.8 \times 10^{-6}$$

On verifying that $[NH_4^+]$ = [dichloroacetate] = 0.05 using Eqs. (89) and (90), we conclude that $[H^+] = 3.8 \times 10^{-6}$. If we had used the approximate formula, Eq. (93), the calculated value of $[H^+]$ would have been 5.4×10^{-6}, in error by 0.15 pH unit.

Example 11: Solve the problem in Example 8 graphically.

The graphical solution requires logarithmic plotting of the individual terms in the proton condition, Eq. (86). The plot is shown in Fig. 4.7, and point C, the intersection of the formic acid curve and the $[NH_3]$ curve, occurs at pH = 6.50, in good agreement with the numerical answer obtained above using Eq. (93).

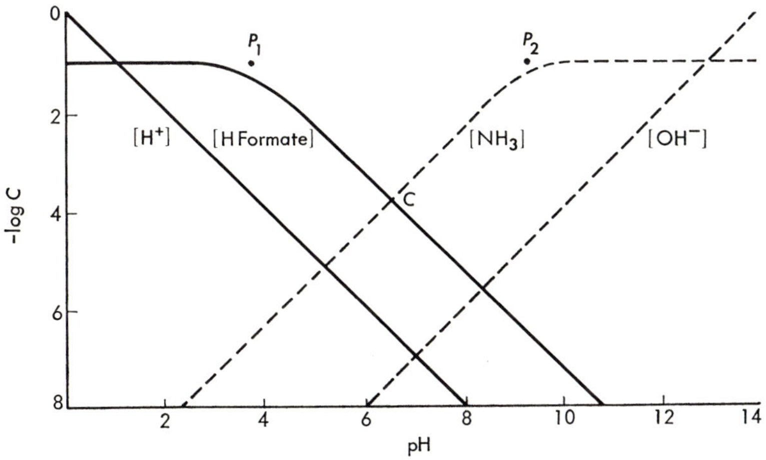

Fig. 4-7. Logarithmic diagram for finding the pH of a 0.100 M ammonium formate solution.

MULTIPLE ACID–BASE EQUILIBRIA

Consider a solution that contains either two or more monoprotic acids or a single polybasic acid. In the presence and absence of strong base, the pH of such a solution is given by a complicated algebraic expression, which can be solved only by a numerical method. Such numerical difficulties cannot be entirely avoided, but they can be minimized using the procedures described below. Whenever possible, a problem will be treated as if it were a perturbation upon a simpler problem, and an iterative numerical method developed. This procedure does not necessarily produce a numerical result with the least computational effort, but it does have the distinct advantage of providing a unified approach that helps develop chemical "intuition" and yields numerical results without excessive labor.

Mixture of two monobasic acids, HA and HB

Consider a solution of two monobasic acids, HA and HB, with equilibrium constants K_{HA} and K_{HB}, $K_{HA} > K_{HB}$, and analytical concentrations c_{HA} and c_{HB}.

The proton condition for such a solution, based upon the reference states, HA and HB, is

$$[H^+] = [OH^-] + [A] + [B] \tag{94}$$

Substituting for $[OH^-]$, [A], and [B] from K_w, K_{HA}, K_{HB} produces

$$[H^+] = \frac{K_w}{[H^+]} + \frac{K_{HA}[HA]}{[H^+]} + \frac{K_{HB}[HB]}{[H^+]} \tag{95}$$

which can be solved for $[H^+]$.

$$[H^+] = (K_w + K_{HA}[HA] + K_{HB}[HB])^{1/2} \tag{96}$$

where

$$[HA] = \frac{c_{HA}[H^+]}{[H^+] + K_{HA}} \tag{97}$$

and

$$[HB] = \frac{c_{HB}[H^+]}{[H^+] + K_{HB}} \tag{98}$$

As in the problem described in the previous section, the elimination of the principal species (the species used as the reference proton level) using Eqs. (97) and (98) would lead to an algebraic equation of high (fourth) power. It is more convenient in this problem to assume, initially, that $c_{HA} \simeq [HA]$ and $c_{HB} \simeq [HB]$ to obtain

$$[H^+] \simeq (K_w + K_{HA}c_{HA} + K_{HB}c_{HB})^{1/2} \tag{99}$$

After the approximate value of $[H^+]$ is obtained, Eqs. (97) and (98) are used to obtain better values of [HA] and [HB] for use in Eq. (96). This iterative cycle is continued until successive values of $[H^+]$ agree to the required accuracy.

Example 12: What is the concentration of hydrogen ion in a solution if $c_{HA} = 0.100\ M$, $K_{HA} = 1.00 \times 10^{-5}$, $c_{HB} = 0.100\ M$ and $K_{HB} = 1.00 \times 10^{-9}$?

Substituting into Eq. (99),

$$\begin{aligned}[H^+] &\cong [(1.00 \times 10^{-14}) + (1.00 \times 10^{-5})(0.100) + (1.00 \times 10^{-9}(0.100)]^{1/2} \\ &= (1.00 \times 10^{-6})^{1/2} = 1.00 \times 10^{-3}\end{aligned}$$

and $[H^+] \simeq 1.00 \times 10^{-3}$. Using Eqs. (97) and (98), we find that $[HA] = c_{HA}$ to within 1 per cent, and $[HB] = c_{HB}$ to within 10^{-4} per cent. Hence $[H^+] = 1.00 \times 10^{-3}$ with an error of $\frac{1}{2}$ per cent.

Example 13: If $c_{HA} = c_{HB} = 0.100$, $K_{HA} = 1.00 \times 10^{-3}$ and $K_{HB} = 1.00 \times 10^{-4}$, what is $[H^+]$?

Using Eq. (99),

$$[(1.00 \times 10^{-14}) + (1.00 \times 10^{-3})(0.100) + (1.00 \times 10^{-4})(0.100)]$$
$$= (1.10 \times 10^{-4})^{1/2}$$

or $[H^+] \simeq 1.05 \times 10^{-2}$.

Using Eq. (97) we calculate $[HA] \simeq 0.91_3$ and, using Eq. (98), $[HB] = 0.99_0$.

Substituting these new estimates of [HA] and [HB] into Eq. (96) yields (we neglect K_w as immeasurably small)

$$[H^+] \cong [(1.00 \times 10^{-3})(0.091_3) + (1.00 \times 10^{-4})(0.099_0)]^{1/2}$$
$$= (1.01_2 \times 10^{-4})^{1/2} = 1.01 \times 10^{-2}$$

or $[H^+] \simeq 1.01 \times 10^{-2}\ M$. Using Eq. (97), we calculate $[HA] = 0.909\ c_{HA}$, a change of 0.2 per cent in [HA], from the last estimate, while [HB] remains unchanged. Substitution of this new estimate of [HA] into Eq. (96) cannot produce a significant effect in $[H^+]$; hence $[H^+] = 1.01 \times 10^{-2}\ M$.

Mixture of two monobasic acids and a strong base

Consider a solution that contains two weak acids HA and HB, $K_{HA} > K_{HB}$, and a strong base. The relevant analytical concentrations are c_{HA}, c_{HB}, and b. We shall consider first the situation when $b < c_{HA}$. Proceeding in the same manner as we did for a solution containing one monoprotic acid, we write a proton condition based on the undissociated acids as reference proton levels.

$$[H^+] + b = [OH^-] + [A] + [B]$$

or

$$[A] = b + [H^+] - [OH^-] - [B] \tag{100}$$

Also,

$$c_{HA} = [HA] + [A] \tag{101}$$

Therefore,

$$[HA] = c_{HA} - b - [H^+] + [OH^-] + [B] \tag{102}$$

Substituting Eqs. (100) and (102) into the equilibrium constant expression for K_{HA} yields

$$[H^+] = K_{HA} \frac{c_{HA} - b - [H^+] + [OH^-] + [B]}{b + [H^+] - [OH^-] - [B]} \tag{103}$$

where

$$[B] = \frac{c_{HB}K_{HB}}{[H^+] + K_{HB}} \tag{104}$$

Comparison of Eq. (103) with (71), the exact expression for a mixture of a weak acid and a strong base, shows that the only difference is presence of the equilibrium concentration of the conjugate base of the weaker acid.

It is not correct to assume that it is frequently possible to neglect [B] in Eq. (103) and use Eq. (71). This procedure is equivalent to assuming that the dissociation of the weaker acid is completely repressed and neglects the fact that under certain conditions the [B] term is of the same order of magnitude, or more important than the $\{[H^+] - [OH]\}$ term. The simplest numerical procedure is to estimate $[H^+]$ using

$$[H^+] \approx K_{HA} \frac{c_{HA} - b}{b} \tag{105}$$

and then simplify Eq. (103) as much as possible, using an iterative numerical process whenever direct substitution of Eq. (104) into the simplified form of Eq. (103) would produce more than a quadratic equation.

Example 14: Calculate the hydrogen-ion concentration of solutions containing a mixture of weak acids, HA, $K_{HA} = 1.00 \times 10^{-5}$, $c_{HA} = 0.100\ M$, and $K_{HB} = 1.00 \times 10^{-7}$, $c_{HB} = 0.100\ M$ with the following amounts of strong base: (a) $b = 0.0500\ M$, (b) $b = 0.0990\ M$.

Solving part (a) we estimate $[H^+] \cong 1.00 \times 10^{-5}$ from Eq. (105); therefore, $[OH^-] \approx 10^{-9}$, and $[B] \simeq 10^{-5}$. Thus $[OH^-]$ is neglibly small as compared to all the other terms in Eq. (103). $[H^+]$ and $[B]$ exactly cancel each other, leaving Eq. (105) as the expression suitable for calculating $[H^+]$. Thus $[H^+] = 1.00 \times 10^{-5}\ M$.

Solving part (b) of Example 14, we estimate $[H^+] \simeq 1.0 \times 10^{-7}$ from Eq. (105). Therefore, $[OH] \simeq 1.0 \times 10^{-7}$, and $[B] \cong 1.0 \times 10^{-4}$. Since $c_{HA} - b = 0.00100\ M$, the simplest, accurate form of Eq. (103) becomes

$$[H^+] = K_{HA} \frac{c_{HA} - b + [B]}{b} \tag{106}$$

Substituting [B] for Eq. (104), we obtain the quadratic expression

$$[H^+]^2 + \left\{K_{HB} - \frac{K_{HA}(c_{HA} - b)}{b}\right\}[H^+] - \left\{\frac{c_{HA} - b + c_{HB}}{b}\right\}K_{HA}K_{HB} = 0 \tag{107}$$

Substituting the numerical quantities for the various terms in Eq. (107), we obtain $[H^+] = 1.62 \times 10^{-7}$.

When $b = c_{HA}$, we have a mixture corresponding to the first end point of the mixture of HA and HB. This problem has been solved in an earlier section, and the general solution is given by Eq. (88).

If $b > c_{HA}$, it is best to derive a result based upon A and HB as reference proton levels. The proton condition is

$$b' + [H^+] + [HA] = [OH^-] + [B] \tag{108}$$

or

$$[B] = b' + [H^+] - [OH^-] + [HA] \tag{109}$$

where $b' = b - c_{HA}$ = strong base added in excess of that required to titrate HA. Also,

$$c_{HB} = [HB] + [B] \tag{110}$$

Therefore,

$$[HB] = c_{HB} - b' - [H^+] + [OH^-] - [HA] \tag{111}$$

Substituting into the expression for K_{HB} we obtain

$$[H^+] = K_{HB} \frac{c_{HB} - b' - [H^+] + [OH^-] - [HA]}{b' + [H^+] - [OH^-] + [HA]} \tag{112}$$

where

$$[HA] = \frac{c_{HA}[H^+]}{[H^+] + K_{HA}} \tag{113}$$

Equation (112) may be used in much the same way as Eq. (103) to solve specific problems. The starting assumption is

$$[H^+] = K_{HB} \frac{c_{HB} - b'}{b'} \tag{114}$$

Example 15: Calculate the hydrogen-ion concentration for the acid mixture in Example 14, when (a) $b = 0.1500\ M$ ($b' = 0.0500$), (b) $b = 0.10100\ M$ ($b' = 0.00100$).

Solving part (a), we estimate $[H^+] \simeq 1.00 \times 10^{-9}$ from Eq. (114). Hence $[OH^-] \simeq 10^{-5}$ and, from Eq. (113), $[HA] \simeq 10^{-5}$. Numerical substitution of these approximate equilibrium concentrations demonstrates that Eq. (112) is adequately represented by Eq. (114), and $[H^+] = 1.00 \times 10^{-9}$.

Solving part (b), from Eq. (114) we estimate $[H^+] = \simeq 1.0 \times 10^{-7}$, $[OH^-] \approx 10^{-7}$, and $[HA] \approx 10^{-3}$ from Eq. (113). The simplest usable form of Eq. (112) becomes

$$[H^+] = K_{HB}\frac{c_{HB} - b'}{b' + [HA]} \tag{115}$$

We proceed to substitute Eq. (113) into Eq. (115), eliminating [HA], obtaining a quadratic expression for $[H^+]$:

$$[H^+]^2 + \left\{\frac{b'K_{HA} - (c_{HB} - b')K_{HB}}{(c_{HB} + b')}\right\}[H^+] - \frac{(c_{HB} - b')K_{HA}K_{HB}}{(c_{HB} + b')} = 0 \tag{116}$$

Substituting for b', c_{HB}, K_{HB}, and K_{HA}, we obtain $[H^+] = 6.15 \times 10^{-8}$.

Diprotic acid solutions

A general approach to obtaining numerical values for the hydrogen-ion concentration of a polyprotic acid solution will be illustrated using the diprotic acid, H_2C, as the specific example. Assume that the two equilibrium constants are K_1 and K_2 and that c_a is the analytical concentration of H_2C. Choosing H_2C as the reference proton level, the proton condition for this solution is

$$[H^+] = [OH^-] + [HC] + 2[C] \tag{117}$$

Eliminating $[OH^-]$, [HC], and [C] via the appropriate equilibrium constants, we obtain

$$[H^+] = \frac{K_w}{[H^+]} + \frac{[H_2C]}{[H^+]}K_1\left\{1 + \frac{2K_2}{[H^+]}\right\} \tag{118}$$

This equation is a cubic in $[H^+]$, and would be inconvenient to solve directly even if we knew the numerical value of $[H_2C]$. However, $[H_2C]$ is given by

$$[H_2C] = \frac{c_a[H^+]^2}{[H^+]^2 + [H^+]K_1 + K_1K_2} \tag{119}$$

which shows that the exact solution for $[H^+]$ in terms of c_a is a quartic equation.

If we consider the problem to be perturbation of the problem of finding $[H^+]$ for a solution of a weak, monoprotic acid, with a dissociation constant equal to K_1, we recognize that the term $2K_2/[H^+]$ corresponds to the perturbing term. Let us assume that this term is small (but, perhaps, not negligibly small) and write

$$[H^+] = \left(K_w + K_1[H_2C]\left\{1 + \frac{2K_2}{[H^+]}\right\}\right)^{1/2} \tag{120}$$

Equations (119) and (120) can be used to obtain a numerical result for $[H^+]$ by the conventional iterative procedure, provided we have a starting value for $[H^+]$. This initial value can be obtained by considering the limiting case, when $K_1 \gg K_2$ and c_a is large, $[H^+] \approx (c_aK_1)^{1/2}$. If c_a is of the order of K_1, Eq. (63) yields a better starting value of $[H^+]$.

If $1 \gg 2K_2/(c_aK_1)^{1/2}$, the second dissociation has been completely repressed by the first dissociation step, and the problem may be successfully treated as a monoprotic acid. However, if $1 \sim 2K_2/(c_aK_1)^{1/2}$, this simplification is not correct.

Example 16: What is the hydrogen-ion concentration of a 0.1 *M* oxalic acid solution? $K_1 = 5.37 \times 10^{-2}$ and $K_2 = 5.37 \times 10^{-5}$.

The value of $2K_2/(c_aK_1)^{1/2}$ for this solution is 0.0015. Therefore, the second stage of dissociation can be neglected, and this oxalic acid solution treated as if the acid were a monoprotic acid of $K = 5.37 \times 10^{-2}$. Hence, using Eq. (63), $[H^+] = 5.12 \times 10^{-2}$. Direct substitution into the right side of Eq. (120) using Eq. (119) and $[H^+] = 5.12 \times 10^{-2}$ verifies the latter numerical result.

Example 17: What is the hydrogen-ion concentration of 1.00×10^{-3} *M* tartaric acid solution? $K_1 = 9.1 \times 10^{-4}$ and $K_2 = 4.25 \times 10^{-5}$.

The numerical value of $2K_2/(c_aK_1)^{1/2}$ is 0.09; therefore Eq. (120) should be used, together with Eq. (119). Since c_a is of the order of K_1, we estimate $[H^+]$ from Eq. (63) and obtain $[H^+] \simeq 6.0_2 \times 10^{-4}$. Substituting into Eq. (119), we obtain a better estimate of $[H_2C] = 3.84 \times 10^{-4}$. Substituting these estimates for $[H^+]$ and $[H_2C]$ into Eq. (120) yields (K_w is negligible)

$$[H^+] \simeq \left[(9.1 \times 10^{-4})(3.84 \times 10^{-4})\left\{1 + \frac{2(4.25 \times 10^{-5})}{6.0 \times 10^{-4}}\right\}\right]^{1/2}$$

$$\simeq [(34.0 \times 10^{-6})1.14]^{1/2} = 6.3 \times 10^{-4}$$

Another cycle of iteration yields $[H_2C] \simeq 3.94 \times 10^{-4}$ and $[H^+] = 6.4 \times 10^{-4}$. Considering the accuracy required in a problem of this sort, the last iteration was unnecessary.

Solutions of a dibasic acid and a strong base

Three situations are of interest, $b < c_a$, $b = c_a$, and $b > c_a$. The first and last can be treated in a manner quite similar to that used to obtain the expressions for a mixture of two monoprotic acids. Thus it is left as an exercise for the student to show that when $b < c_a$,

$$[H^+] = K_1 \frac{c_a - b - [H^+] + [OH^-] + [C]}{b + [H^+] - [OH] - 2[C]} \tag{121}$$

and

$$[C] = \frac{c_aK_1K_2}{[H^+]^2 + [H^+]K_1 + K_1K_2} \tag{122}$$

Equations (121) and (122) should be compared to Eqs. (103) and (104). The numerical procedure in an actual problem is completely analogous to that demonstrated for Eqs. (103) and (104).

If $b > c_a$, we obtain

$$[H^+] = K_2 \frac{c_a - b' - [H^+] + [OH^-] - 2[H_2C]}{b' + [H^+] - [OH] + [H_2C]} \tag{123}$$

where $b' = b - c_a$ and

$$[H_2C] = \frac{c_a[H^+]^2}{[H^+]^2 + [H^+]K_1 + K_1K_2} \tag{124}$$

Equations (123) and (124) should be compared to Eqs. (112) and (113); they are used, after simplifying as much as is possible for the particular problem at hand, similarly to Eqs. (112) and (113).

Half-neutralized diprotic acids

The classic problem of the acid salt of a dibasic acid, e.g., NaHC, is readily solved by writing the proton condition, assuming HC as the reference proton level; i.e.,

$$[H^+] + [H_2C] = [OH^-] + [C] \tag{125}$$

Since $[H_2C] = [H^+][HC]/K_1$, $[C] = K_2[HC]/[H^+]$ and $[OH^-] = K_w/[H^+]$, these quantities can be substituted into Eq. (125) to obtain

$$[H^+] + \frac{[H^+][HC]}{K_1} = \frac{K_w}{[H^+]} + \frac{K_2[HC]}{[H^+]} \tag{126}$$

which can be rearranged to the form

$$[H^+] = \left(\frac{K_w + K_2[HC]}{1 + [HC]/K_1}\right)^{1/2} \tag{127}$$

The special case, which occurs when $K_w \ll K_2[\mathrm{HC}]$ and $1 \ll [\mathrm{HC}]/K_1$,

$$[\mathrm{H}^+] \approx (K_1K_2)^{1/2} \tag{128}$$

should not be used without verifying the approximations by numerical substitution.

For numerical purposes, one may as well use

$$[\mathrm{H}^+] \simeq \left(\frac{K_w + K_2c_a}{1 + c_a/K_1}\right)^{1/2} \tag{129}$$

obtained from Eq. (127) by the approximation $c_a \approx [\mathrm{HC}]$, since the numerical labor is virtually the same as using Eq. (128) and verifying the assumptions leading to it. Only in extraordinary situations is it necessary to get a better value of [HC] employing

$$[\mathrm{HC}] = \frac{c_a[\mathrm{H}^+]K_1}{[\mathrm{H}^+]^2 + [\mathrm{H}^+]K_1 + K_1K_2} \tag{130}$$

with the value of $[\mathrm{H}^+]$ from Eq. (129), and then substituting this value of [HC] into Eq. (127). This numerical situation exists since the hydrogen-ion concentration in a half-neutralized dibasic acid is not very sensitive to the equilibrium concentration of [HC]. [Note Eq. (128), a limiting situation in which there is no dependence on concentration.]

Example 18: What is the pH of a 0.05 M solution of potassium acid phthalate? $K_1 = 1.12 \times 10^{-3}$, $K_2 = 4.06 \times 10^{-6}$.

Substituting into Eq. (125) we obtain

$$[\mathrm{H}^+] \cong \left[\frac{10^{-14} + 0.0500(4.06 \times 10^{-6})}{1 + 0.0500/1.12 \times 10^{-3}}\right]^{1/2} = \left[\frac{10^{-14} + (2.03 \times 10^{-7})}{1 + 44.6}\right]^{1/2}$$

$$\simeq 6.7 \times 10^{-5}$$

It is apparent that applying Eq. (128) in this problem would lead to an error of $\sim 1\frac{1}{2}$ per cent. Therefore the actual equilibrium concentration of biphthalate ion is not important, and there is no need to use Eq. (130). Thus pH = 4.17.

Example 19: What is the hydrogen-ion concentration of 0.001 M solution of sodium bicarbonate? $K_1 = 4.46 \times 10^{-7}$ and $K_2 = 4.67 \times 10^{-11}$. Using Eq. (129),

$$[\mathrm{H}^+] \simeq \left[\frac{(1.00 \times 10^{-14}) + (1.00 \times 10^{-3})(4.67 \times 10^{-11})}{1 + (1.00 \times 10^{-3})/4.46 \times 10^{-7}}\right]^{1/2}$$

$$\simeq \left[\frac{(1.00 + 4.67) \times 10^{-14}}{1 + (2.14 \times 10^3)}\right]^{1/2}$$

$$\simeq 5.1 \times 10^{-9}$$

Substituting $[\mathrm{H}^+] = 5.1 \times 10^{-9}$ into Eq. (130) we obtain $[\mathrm{HCO_3^-}] \simeq 9.83 \times 10^{-4}$, verifying that the use of Eq. (129) is valid, and $[\mathrm{H}^+] = 5.1 \times 10^{-9}$. If Eq. (128) had been used, the value of $[\mathrm{H}^+]$ obtained would have been 4.7×10^{-9}, 8 per cent too low.

Solution of polyprotic acids

The general techniques for treating polyprotic acid equilibria were developed in the previous section. No attempt will be made to derive expressions for the multitude of possible situations. Instead, a few of the more important cases will be discussed, indicating where approximations are possible.

Consider a solution of a triprotic acid, H_3D. Proceeding as we did for the diprotic acid in the previous section, it is easily demonstrated that

$$[\mathrm{H}^+] = \left(K_w + K_1[\mathrm{H_3D}]\left\{1 + \frac{2K_2}{[\mathrm{H}^+]} + \frac{3K_2K_3}{[\mathrm{H}^+]^2}\right\}\right)^{1/2} \tag{131}$$

where

$$[H_3D] = \frac{c_a[H^+]^3}{[H^+]^3 + [H^+]^2K_1 + [H^+]K_1K_2 + K_1K_2K_3} \tag{132}$$

One could undertake to use Eqs. (131) and (132) in an iterative procedure, analogous to the method used for the dibasic acid, H_2C, which employs Eqs. (119) and (120). Although such an approach is rigorously correct, it is overly complicated and the student should realize that the third stage of dissociation cannot contribute significantly to the hydrogen-ion concentration of the solution in any practical situation. (We do not consider the unlikely situation of an extremely dilute triprotic acid where c_a approaches K_3.) This is mathematically equivalent to stating that $2K_2/[H^+] \gg 3K_2K_3/[H^+]^2$. This inequality immediately reduces Eq. (131) to Eq. (120), Eq. (132) to Eq. (119), and yields the practical result that no more than the first two dissociation constants of a polyprotic acid need be used to calculate the hydrogen-ion concentration of a solution of a polyprotic acid.

The calculation of the titration curve of a polyprotic acid is not a particularly difficult problem if the pH is not close to an end point; it is in the vicinity of an end point that expressions analogous to Eqs. (121) and (123) must be used. To illustrate the general approach let us obtain an expression for a solution that is a mixture of NaH_2D and Na_2HD. Such a solution arises in the titration of c_a *M* H_3D with NaOH. First, choosing H_2D as the reference proton level we write, where $b = [Na^+]$,

$$[H_3D] + b + [H^+] = [OH^-] + c_a + [HD] + 2[D]$$

which yields

$$[HD] = b - c_a + [H^+] - [OH^-] + [H_3D] - 2[D] \tag{133}$$

Second, choosing HD as the reference proton level,

$$[H_2D] = 2c_a - b - [H^+] + [OH^-] - 2[H_3D] + [D] \tag{133}$$

Defining $b' = b - c_a$, and substituting into the definition of K_2, we obtain

$$[H^+] = K_2 \frac{c_a - b' - [H^+] + [OH^-] + 2[H_3D] + [D]}{b' + [H^+] - [OH^-] + [H_3D] - 2[D]} \tag{134}$$

It is possible to proceed by making use of the expressions

$$[H_3D] = \frac{c_a[H^+]^3}{D(3)} \tag{135}$$

and

$$[D] = \frac{c_aK_1K_2K_3}{D(3)} \tag{136}$$

where

$$D(3) = [H^+]^3 + [H^+]^2K_1 + [H^+]K_1K_2 + K_1K_2K_3$$

by using

$$[H^+] \simeq K_2 \frac{c_a - b'}{b'} \tag{137}$$

as the starting estimate of $[H^+]$, to decide which terms must be included on the right side of Eq. (134). Usually, for a wide range of mixtures of NaH_2D and Na_2HD, Eq. (137) proves to be satisfactory. It is only when the solution contains much more NaH_2D than Na_2HD, or vice versa; that it becomes necessary to use an iterative numerical technique. Even then, in most practical cases, ordinarily it is not necessary to keep both $[H_3D]$ and $[D]$ in Eq. (134). For example, if b' is only slightly greater than zero (only a trace of Na_2HD is present) $[H_3D] \gg [D]$. In addition, Eq. (135) simplifies to

$$[H_3D] = \frac{c_a[H^+]^2}{[H^+]^2 + [H^+]K_1 + K_1K_2}$$

since $[H^+] \gg K_3$ in this solution.

Conversely, if b' is nearly equal to c_a (nearly all the NaH_2D is converted to Na_2HD), $[D] \gg [H_3D]$, and $[H_3D]$ may be eliminated from Eq. (134). In this case, Eq. (136) simplifies to

$$[D] = \frac{c_aK_2K_3}{[H^+]^2 + [H^+]K_2 + K_2K_3}$$

since $K_1 \gg [H^+]$ in this solution.

Although it is necessary to consider all three equilibrium constants in obtaining the general solution, in practice no practical calculation will require the use of more than two constants for a given pair of c_a and b values, and, in the region of $[H^+] = K_2$, Eq. (137) will usually suffice.

It should be apparent to the student that if the acid had been H_4D (or even H_nD) instead of H_3D, no change in expression actually used to calculate $[H^+]$ would occur. Thus the titration curve of a polyprotic acid can be calculated taking into account three successive equilibrium constants, K_{m+1}, K_m, and K_{m-1}, where the species being titrated in a particular pH range is assumed to be H_mY.

The student should realize that facility in deriving expressions such as Eqs. (131) *and* (134) *is no substitute for understanding the chemical situation. Such expressions are only a means of arriving at the simpler, commonly used approximations, familiar to experienced chemists. Mechanical derivation of complicated expressions, without the attendant ability to simplify the derived expressions to a useful form, is as pointless as the memorization of approximate formulas without understanding their origin or limitations.*

It is particularly instructive to consider the problem of calculating the hydrogen-ion concentration in a polyprotic acid solution containing an equivalent amount of base. One can demonstrate several features common to such mixtures by examining the equilibrium situation that exists in a solution containing equal analytical concentrations of H_3D, NH_3, and strong base, i.e., a solution containing c_a *M* $NaNH_4HPO_4$. It will be assumed that the reaction proceeds in solution to produce HD, NH_4^+, and the cation of the strong base. The proton condition is then

$$[H^+] + [H_2D] + 2[H_3D] = [OH^-] + [NH_3] + [D] \tag{138}$$

Substituting from K_w, K_1, K_2, K_3, and $K_{NH_4^+}(= [H^+][NH_3]/[NH_4^+])$, we obtain

$$[H^+] + \frac{[H^+][HD]}{K_2} + \frac{2[H^+]^2[HD]}{K_1K_2} = \frac{K_w}{[H^+]} + \frac{K_{NH_4^+}[NH_4^+]}{[H^+]} + \frac{K_3[HD]}{[H^+]}$$

which is written most conveniently in the form

$$[H^+] = \left(\frac{K_w + K_{NH_4^+}[NH_4^+] + K_3[HD]}{1 + ([HD]/K_2)\{1 + (2[H^+]/K_1)\}}\right)^{1/2} \tag{139}$$

Equation (139) is readily simplified by considering the significance of the various terms. The quantity $2[H^+]/K_1$ in the denominator is a measure of perturbation caused by the fact that H_2D can react with a proton. Thus it represents only a minor contribution, and usually $1 \gg 2[H^+]/K_1$. (It is always possible to use an iterative method, repeating the calculation including the latter term to correct for this approximation.) In addition, it is a good approximation that $[NH_4^+] \simeq [HD] \approx c_a$, when $K_{NH_4^+}$ is of the same magnitude as K_2 and $K_1 \gg K_2 \gg K_3$, leading to the expression

$$[H^+] \simeq \left[\frac{K_w + (K_{NH_4^+} + K_3)c_a}{1 + c_a/K_2}\right]^{1/2} \tag{140}$$

If $K_w \ll (K_{NH_4^+} + K_3)c_a$ and $1 \ll c_a/K_2$, the simple expression

$$[H^+] = [K_2(K_{NH_4^+} + K_3)]^{1/2} \tag{141}$$

is obtained.

Example 20: What is $[H^+]$ in 0.1 *M* $NaNH_4HPO_4$? $K_1 = 5.90 \times 10^{-3}$, $K_2 = 6.17 \times 10^{-8}$, $K_3 = 4.80 \times 10^{-13}$, and $K_{NH_4^+} = 5.75 \times 10^{-10}$.

The numerical quantities are such that Eq. (141) appears to hold. Substitution into Eq. (141) yields

$$[H^+] \simeq [(6.17 \times 10^{-8})(5.75 \times 10^{-10} + 4.80 \times 10^{-13})]^{1/2}$$
$$= 6.0 \times 10^{-9}$$

Note that $K_{NH_4^+} \gg K_3$; therefore, the very simple expression $[H^+] = [K_2[NH_4^+]]^{1/2}$ would also have yielded $[H^+] = 6.0 \times 10^{-9}$. We must now test the assumption concerning the equilibrium and analytical concentrations of $HPO_4^=$ and NH_4^+. Using $[H^+] = 6.0 \times 10^{-9}$, $[HPO_4^=] = [NH_4^+] = 0.91 \times 10^{-2}$. However, even though the equilibrium concentrations differ from c_a by 9 per cent since $K_w \ll (K_{NH_4^+}[NH_4^+] + K_3[HPO_4^=])$, $1 \ll [HPO_4^=]/K_2$ and $[HPO_4^=] = [NH_4^+]$, Eq. (141) may still be obtained from Eq. (139), and $[H^+] = 6.0 \times 10^{-9}$.

The advantage of the graphical method of solution becomes apparent when solving the problem in Example 20. A logarithmic plot for the proton condition, Eq. (138), is shown in Fig. 4-8. The answer corresponds to the intersection, *C*, of the uppermost

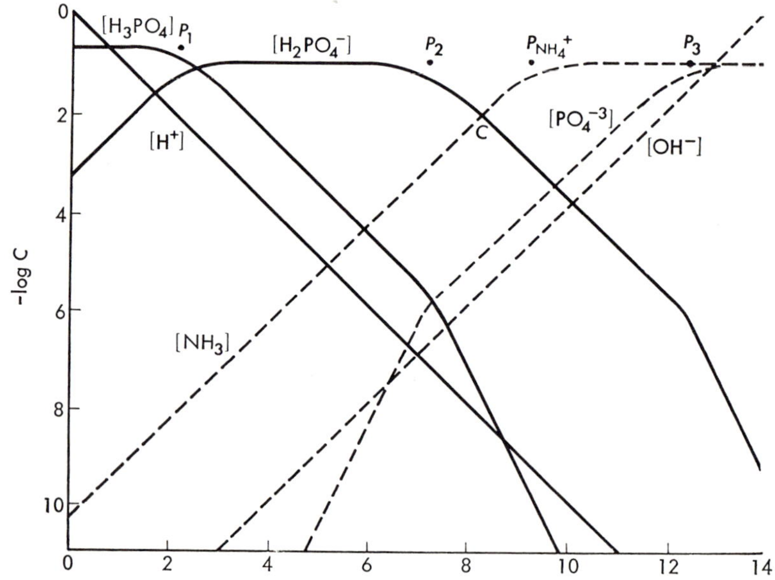

Fig. 4-8. Logarithmic diagram for finding the pH of a 0.100 *M* $NaNH_4HPO_4$ solution.

solid and dashed curves, pH = 8.23. The point of intersection clearly demonstrates that the concentrations of $H_2PO_4^-$ and NH_3 are the only important ones in the proton condition. Changing the analytical concentration (c_a) has the effect of vertically shifting all curves, except $[H^+]$ and $[OH^-]$, by a constant amount. As long as the point of intersection of the $[H_2PO_4^-]$ and $[NH_3]$ curves is distant from the $[OH^-]$ line, *C* must occur at $\frac{1}{2}(pK_2 + pK_{NH_4^+})$, independent of c_a. The concentrations of $H_2PO_4^-$ and NH_3 at point *C* are easily seen to be one tenth of the analytical concentration, c_a.

The graphical method has the distinct advantage of forcefully presenting the relative concentration of the species present in the proton condition, thus orienting the student in complicated situations, and suggesting starting approximations when employing an algebraic method using successive approximations.

Buffers

A solution that resists changes in pH on addition of strong acid or strong base is termed a *buffer*. Some common examples are solutions of strong acids or strong bases, and solutions of weak acids or bases for which pH $\sim$ pK.

The mechanism for the buffering action in the latter case is easily seen if we write

$$\mathrm{HB} \rightleftharpoons \mathrm{H}^+ + \mathrm{B}$$

and assume that [HB] $\approx$ [B]. If we add strong acid, the equilibrium is shifted to the left by reaction of H^+ and B. Conversely, if we add strong base, more HB dissociates to help maintain [H^+].

Van Slyke[4] proposed the rate of change of pH on the addition of strong base as a quantitative measure of the buffering capacity of a solution.

The quantity, $db/d\mathrm{pH}(=\beta)$, is known as the *buffer index* (or *buffer capacity*) and is readily calculated from the proton condition for the solution.

Consider a solution containing a M strong acid, b M strong base, c_{HB} M weak monoprotic acid HB, and $c_{\mathrm{H_2C}}$ M weak diprotic acid H_2C. The proton condition for this solution, using HB and H_2C as reference proton levels, is

$$b + [\mathrm{H}^+] = [\mathrm{OH}^-] + [\mathrm{B}] + [\mathrm{HC}] + 2[\mathrm{C}] + a$$

Therefore, the buffer index is

$$\frac{db}{d\mathrm{pH}} = -\frac{d[\mathrm{H}^+]}{d\mathrm{pH}} + \frac{d[\mathrm{OH}^-]}{d\mathrm{pH}} + \frac{d[\mathrm{B}]}{d\mathrm{pH}} + \frac{d[\mathrm{HC}] + 2[\mathrm{C}]}{d\mathrm{pH}} \tag{142}$$

$$\beta = \beta_{\mathrm{H}^+} + \beta_{\mathrm{OH}^-} + \beta_{\mathrm{HB}} + \beta_{\mathrm{H_2C}}$$

since $da/d\mathrm{pH} = 0$. (If the strong acid were hydrochloric acid, a would equal [Cl^-], which is constant.)

Equation (142) demonstrates that the total buffer capacity of a solution is the sum of individual buffer indices associated with H^+, OH^-, and the bases produced by dissociation of protons from HB and H_2C.

The explicit expressions for the individual buffer indices are evaluated in the following manner. Since

$$\mathrm{pH} = -\log[\mathrm{H}^+] = -\frac{1}{2.303}\ln[\mathrm{H}]$$

$$d\mathrm{pH} = -\frac{d[\mathrm{H}^+]}{2.303[\mathrm{H}^+]}$$

then

$$\frac{db}{d\mathrm{pH}} = -2.303[\mathrm{H}^+]\frac{db}{d[\mathrm{H}^+]}$$

To evaluate the buffer index in terms of the individual buffer indices, we write

$$\beta = 2.303[\mathrm{H}^+]\frac{d}{d[\mathrm{H}^+]}\left\{[\mathrm{H}^+] - \frac{K_w}{[\mathrm{H}^+]} - \frac{c_{\mathrm{HB}}K_{\mathrm{HB}}}{[\mathrm{H}^+] + K} - c_{\mathrm{H_2C}}\frac{[\mathrm{H}^+]K_1 + 2K_1K_2}{[\mathrm{H}^+]^2 + [\mathrm{H}^+]K_1 + K_1K_2}\right\}$$

[4] D. D. van Slyke, *J. Biol. Chem.*, **52**, 525 (1922).

and obtain as a final result

$$\beta = 2.303[H^+] + 2.303[OH^-] + \frac{2.303c_{HB}K_{HB}[H^+]}{([H^+] + K_{HB})^2}$$

$$\beta = \quad \beta_{H^+} \quad + \quad \beta_{OH^-} \quad + \quad \beta_{HB}$$

$$+ \left.\frac{2.303c_{H_2C}[H^+]K_1([H^+]^2 + 4[H^+]K_2 + K_1K_2)}{([H^+]^2 + [H^+]K_1 + K_1K_2)^2}\right\} \qquad (143)$$

$$+ \qquad \beta_{H_2C}$$

Figure 4.9 is a plot of β versus pH according to Eq. (143) assuming that $a = 0.0500\ M$, $c_{HB} = 0.100\ M$, $pK_{HB} = 7.00$, $c_{H_2C} = 0.1$, $pK_1 = 4.00$, $pK_2 = 10.00$, and that strong base was added until $[OH^-] = 0.05\ M$. The solid curve represents β; the dashed lines indicate the shapes of β_{H^+}, β_{HB}, β_{H_2C}, and β_{OH^-}.

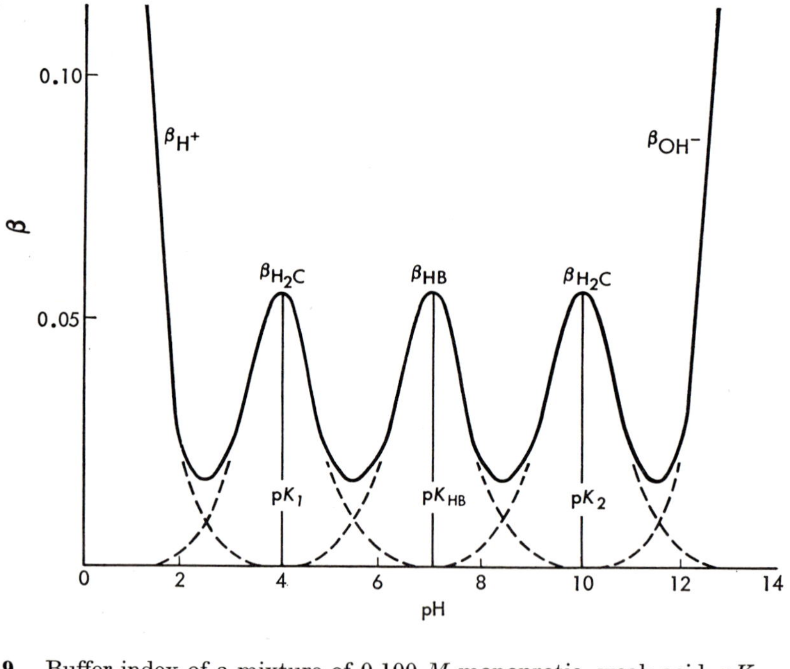

Fig. 4-9. Buffer index of a mixture of 0.100 M monoprotic, weak acid, $pK_{HB} = 7.00$ and 0.100 M diprotic weak acid, $pK_1 = 4.00$ and $pK_2 = 10.00$.

The maxima in the curves occur successively at pH values of pK_1, pK_{HB}, and pK_2, and correspond to $\beta = 0.0576$; $\beta_{HB} = 0.019$ when $pH = pK_{HB} \pm 1$. Despite the more complicated appearances of the equation for β_{H_2C}, when K_1 and K_2 differ by six units, the plot of β_{H_2C} is indistinguishable from that of a weak monobasic acid in the region of pK_1 or pK_2.

It is apparent that the pH range for which a conjugate acid–base pair can serve as a buffer depends upon the total analytical concentration, c_a, and pH.

Assume, for example, that the $pH = pK_{HB}$ for the solution in Fig. 4-9. Suppose that

1 ml of 0.1 *M* sodium hydroxide is added to 100 ml of this solution. This corresponds to a change in b of 0.001 *M*. Assuming

$$\frac{\Delta b}{\Delta \mathrm{pH}} \approx \frac{db}{d\mathrm{pH}} = \beta_{\max} = 0.576c_a$$

we calculate $\Delta\mathrm{pH} = 0.001/0.0576 = 0.017$ unit, a negligible change for most purposes.[5]

It should be noted that the approximation $\Delta\mathrm{pH} = \Delta b_s/\beta_{\max}$ gives results accurate to within 0.03 pH unit until $\Delta b/c_a > 0.2$.

Strong acids and bases make excellent buffers at either extreme of the pH scale.

Table A-16 gives the composition of standard buffers and mixtures.

The minima in Fig. 4-9 correspond to the various inflection points that are observed as strong base is added to the solution. The inflection points all have the same (relatively high) values of $\beta(\simeq 0.0075)$ and therefore have the same slope at the various end points when pH is plotted against milliliters of strong base added. *The lower the value of* β *at the end point* ($\beta_{\min}$), *the better the end point.*

Smith[6] suggested a quantity he terms the sharpness index, η,

$$\eta = \frac{d\mathrm{pH}}{dT} = \frac{c_a}{\beta} \tag{144}$$

as a quantitative measure of the sharpness of a titration. In Eq. (144), T represents the fraction titrated, and c_a the analytical concentration of the weak acid being titrated, i.e., $T = b/c_a$. By definition, when $\beta = \beta_{\min}$, $\eta = \eta_{\max}$.

Estimation of the error in performing a given titration is a problem of considerable importance. A very general, and powerful, method of treating this problem depends upon the fact that it is possible to describe the shape of the titration curve in the immediate vicinity of an end point in terms of $\eta_{\max}$, T, and pH. It can be demonstrated that if T is expanded in a Taylor series about the inflection point of a titration, that

$$2.303\eta_{\max}\,\Delta T = \sinh\,(2.30\Delta\mathrm{pH}) \tag{145}$$

where ΔpH is the pH measured with respect to the inflection point, and ΔT is the difference in the fraction actually titrated and the inflection point. Since there is only a negligible difference between the inflection point and the equivalence point, one can use Eq. (145) to calculate the titration error. For example, if in a series of titrations, it is found that the end point pH can be located with an uncertainty (reproducibility) ΔpH, the uncertainty (reproducibility) in the fraction titrated is ΔT. Equation (145) assumes that the titration curve is symmetrical about the end point, and is a good approximation for the region one or two pH units on either side of the end point. A much simpler expression, obtained from Eq. (145),

$$\Delta\mathrm{pH} = \eta_{\max}\,\Delta T \tag{146}$$

can be used with 0.01 pH unit accuracy if ΔpH is less than 0.22 unit.

[5] This same calculation can be performed using the expressions

$$[\mathrm{H}^+]_1 = K_{\mathrm{HB}}\frac{c_a - b_1}{b_1} \quad \text{and} \quad [\mathrm{H}^+]_2 = K_{\mathrm{B}}\frac{c_a - b_2}{b_2}$$

to yield $$|\Delta\mathrm{pH}| = \log\left\{\frac{c_a - b_2}{b_2}\,\frac{b_1}{c_a - b_1}\right\}$$

At $\mathrm{pH}_1 = \mathrm{p}K$, $b_1 = c_a/2 = 0.050$; at pH_2, $b_2 = 0.050 + 0.001$, or

$$|\Delta\mathrm{pH}| = \log\left\{\frac{0.049}{0.051}\,\frac{0.050}{0.050}\right\} = 0.017 \text{ unit}$$

[6] T. B. Smith, *Analytical Processes*, E. Arnold, London, 1940.

A discussion of the accuracy with which acid–base titrations can be performed is to be found elsewhere.[7]

Example 21: A 0.10 M solution of base $pK_{HB} = 7.00$ must be titrated acidimetrically with an accuracy of 0.3 per cent. What uncertainty in locating the end point pH is permissible?

We must evaluate $\beta(= \beta_{H^+} + \beta_{OH^-} + \beta_{HB})$ at the infection point. Using Eq. (64), at the equivalence point

$$[H^+] = (0.10 \times 10^{-7})^{1/2} = 10^{-4}$$

Therefore, $\beta_{H^+} = 2.3 \times 10^{-4}$, $\beta_{OH^-} = 2.3 \times 10^{-10}$, $\beta_{HB} = 2.3 \times 10^{-4}$, and

$$\beta_{min} = 2.3 \times 10^{-4} + 2.3 \times 10^{-10} + 2.3 \times 10^{-4} = 4.6 \times 10^{-4}$$

Hence $\eta_{max} = c_a/\beta_{min} = 0.01/4.6 \times 10^{-4} = 21.7$. From Eq. (146), $\Delta pH = \eta_{max}\,\Delta T = 21.7(0.003) = 0.07$ pH unit.

Example 22: Solve Example 14(b), page 82, and Example 15(b), page 83, using Eq. (146).

The problem consists of finding $[H^+]$ 1 per cent before and after the first equivalence point ($\Delta T = 0.01$) of the titration of 0.1 M HA ($K_{HA} = 1.00 \times 10^{-5}$) and 0.1 M HB ($K_{HB} = 1.00 \times 10^{-9}\ M$). According to Eq. (93), the equivalence-point hydrogen-ion concentration is

$$[H^+]_{eq.\,pt.} = (K_{HA}K_{HB})^{1/2} = [(1.00 \times 10^{-5})(1.00 \times 10^{-9})]^{1/2}$$
$$= 1.00 \times 10^{-7}$$

Also,

$$\beta = \beta_{H^+} + \beta_{OH^-} + \beta_{HA} + \beta_{HB}$$

$$\beta = 2.3[H^+] + \frac{2.3K_w}{[H^+]} + \frac{2.3c_{HA}[H^+]K_{HA}}{\{[H^+] + K_{HA}\}^2} + \frac{2.3c_{HB}[H^+]K_{HB}}{\{[H^+] + K_{HB}\}^2}$$

At the inflection point,

$$\beta = \beta_{min} = 2.3 \times 10^{-7} + 2.3 \times 10^{-7} + 2.3 \times 10^{-3} + 2.3 \times 10^{-3}$$
$$= 4.6 \times 10^{-3}$$

Since $\eta_{max} = c_{HA}/\beta_{min}$,

$$\eta_{max} = \frac{10^{-1}}{4.6 \times 10^{-3}} = 21.7$$

Substituting $\Delta T = 0.01$ and $\eta_{max} = 21.7$ into Eq. (146) yields

$$\Delta pH = 21.7(0.01) = 0.217$$

One per cent before the end point,

$$[H^+] = 10^{0.217}[H^+]_{eq.\,pt.} = 1.65 \times 10^{-7}$$

One per cent after the equivalence point,

$$[H^+] = 10^{-0.217}[H^+]_{eq.\,pt.} = 6.07 \times 10^{-8}$$

Both these results agree to within 2 per cent with the results obtained using Eqs. (106) and (115).

The use of the sharpness index, in conjunction with Eqs. (145) and (146), is the simplest way to calculate the pH change in the vicinity of the equivalence point for solutions containing mixtures of weak acids, or for solutions of polyprotic acids.

[7] I. M. Kolthoff and S. Bruckenstein, *Acids and Bases in Analytical Chemistry*, Wiley-Interscience, New York, 1964, p. 465.

PROBLEMS

(Use the equilibrium constants given in Table A-2.)

1. The pH of a solution containing 0.1 *M* acetic acid, 0.01 *M* oxalic acid, and 0.01 *M* phosphoric acid is adjusted to 7.00. Calculate the equilibrium concentrations of HAc, Ac^-, H_2Ox, HOx^-, Ox^{--}, H_3PO_4, $H_2PO_4^-$, HPO_4^{--}, and PO_4^{3-}.
   Ans. [HAc] = 5.7×10^{-4}, $[Ac^-] = 9.94 \times 10^{-2}$, $[H_2Ox] = 3.4 \times 10^{-11}$, $[HOx^-] = 2.04 \times 10^{-5}$, $[Ox^{--}] = 9.98 \times 10^{-3}$, $[H_3PO_4] = 1.05 \times 10^{-7}$, $[H_2PO_4^-] = 6.18 \times 10^{-3}$, $[HPO_4^{--}] = 3.82 \times 10^{-3}$, and $[PO_4^{3-}] = 1.83 \times 10^{-6}$.
2. Calculate the pH of the following concentrations of hydrochloric acid: (a) 1.00×10^{-3} *M*; (b) 1.00×10^{-5} *M*; (c) 1.00×10^{-6} *M*; (d) 5.00×10^{-7} *M*; (e) 2.00×10^{-7} *M*; (f) 1.00×10^{-7} *M*; (g) 1.00×10^{-8} *M*.
   Ans. (a) 3.00; (b) 5.00; (c) 6.00; (d) 6.29; (e) 6.62; (f) 6.79; (g) 6.98.
3. Calculate $[H^+]$ for 0.100 *M*, 0.0100 *M*, and 0.00100 *M* solutions of a weak acid HB if pK_{HB} = (a) 3.00; (b) 5.00; (c) 7.00.
   Ans. (a) 9.51×10^{-3}, 2.70×10^{-3}, 6.18×10^{-4}; (b) 9.95×10^{-4}, 3.11×10^{-4}, 9.51×10^{-5}; (c) 1.00×10^{-4}, 3.16×10^{-5}, 9.95×10^{-6}.
4. NaB is the salt of the weak acid HB, $pK_{HB} = 9.00$. What is the hydrogen-ion concentration in solutions containing 0.01 *M* NaB and (a) 10^{-3} *M* HB; (b) 10^{-4} *M* HB; (c) zero HB. *Ans.* (a) 1.09×10^{-10}; (b) 3.7×10^{-11}; (c) 3.2×10^{-11}.
5. A solution of 0.1 *M* amine, $K_B = 1.00 \times 10^{-5}$, is made (a) 0.00100 *M* in NaOH; (b) 0.00500 *M* in NaOH; (c) 0.0100 *M* in NaOH. Calculate $[OH^-]$.
   Ans. (a) 1.62×10^{-3}; (b) 5.21×10^{-3}; (c) 1.01×10^{-2}.
6. A solution contains 0.100 *M* formic acid and (a) 0.100 *M* acetic acid; (b) 0.100 *M* phenol. What is the pH? *Ans.* (a) 2.37; (b) 2.39.
7. What is the pH of a 0.01 *M* solution of KHtartrate? *Ans.* 3.72
8. What is the buffer index of (a) 0.1 *M* HCl; (b) 0.01 *M* HCl; (c) 0.001 *M* HCl; (d) 0.1 *M* NaOH; (e) 0.01 *M* NaOH; (f) 0.001 *M* NaOH; (g) 0.1 *M* acetic acid. (h) 0.05 *M* acetic acid + 0.05 *M* sodium acetate; (i) 0.1 *M* sodium acetate; (j) 0.05 *M* potassium acid phthalate.
   Ans. (a) 0.23; (b) 0.023; (c) 0.0023; (d) 0.23; (e) 0.023; (f) 0.0023; (g) 0.0061; (h) 0.0576; (i) 3.5×10^{-5}; (j) 0.0081.
9. Which is a better buffer at pH = 7.00, a mixture of weak acid, 0.01 *M*, and its salt 0.10 *M* ($K_{HB} = 1.00 \times 10^{-6}$), or a mixture of a weak base, 0.01 *M*, and its conjugate acid 0.10 *M* ($K_B = 1.00 \times 10^{-6}$)? Temperature 25°C.
   Ans. Both solutions have identical buffer indices.
10. How much does the pH change in going from 1 per cent to 1 per cent after equivalence point in the titration of 0.1 *M* Na_2CO_3 with strong acid to $NaHCO_3$?
    Ans. 0.44 pH unit.
11. Prove that the hydrogen-ion concentration of a 1 *M* NaH_2PO_4 is given by $[H^+] = \sqrt{K_1K_2}$. What assumptions have been made?
    Ans. $1 \gg 2K_3/[H^+]$, $1 \ll [H_2PO_4^-]/K_1$, $K_w \ll K_2[H_2PO_4^-]$.
12. Prove that the hydrogen-ion concentration of a 0.1 *M* solution of Na_2HPO_4 is given by $[H^+] = (K_2K_3)^{1/2}$. What assumptions have been made?
    Ans. $1 \gg 2[H^+]/K_1$, $1 \ll [HPO_4^{--}]/K_2$, $K_w \ll K_3[HPO_4^{--}]$.
13. Calculate $[H^+]$ in a 0.0100 *M* solution of NaH_2PO_4.
    Ans. Using $[H^+] = (K_2[H_2PO_4^-]/\{1 + ([H_2PO_4^-]/K_1)\})^{1/2}$, $[H^+] = 4.7 \times 10^{-5}$.

CHAPTER 5 *Acid–Base Equilibria in Nonaqueous Systems*

Classification of solvents

In a consideration of the qualitative and quantitative aspects of acid–base equilibria in nonaqueous solvents it is desirable to classify solvents in terms of their acid–base properties. If a solvent behaves both as a Brønsted acid and a base, it is termed an *amphiprotic* solvent. Amphiprotic solvents that are much stronger acids than water are designated protogenic solvents. Included in this class are, e.g., sulfuric acid, hydrofluoric acid, formic acid, acetic acid, and trifluoroacetic acid. Protophilic solvents are amphitrotic solvents which are much stronger bases than water, e.g., liquid ammonia, butylamine, and ethylenediamine. The hydroxylic solvents, water, methanol, ethanol, and other alcohols are classed together because their acid–base properties lie between the extremes of protophilic and protogenic solvents.

Solvents such as the ketones, ethers, nitriles, esters, dimethyl sulfoxide and pyridine, which have weak basic, but negligible acid properties as compared to water, do not fit in the previous classification. In addition, there are solvents whose acid and base strengths are so small that trace impurities, such as water, determine their effective acid–base properties under ordinary laboratory conditions. Benzene and its halogenated derivatives, hexane and similar hydrocarbons, chloroform, and carbon tetrachloride are examples of the latter solvents. We include all these compounds in the classification of aprotic solvents. Aprotic solvents have the common property that their autoprotolysis constants (see below) are so small that they cannot be determined using existing experimental methods.

Intrinsic acidity and basicity

It is convenient to imagine that we can describe the acid–base properties of a compound in terms of its *intrinsic acidity* or *intrinsic basicity*, represented by

$$\mathrm{HB} \rightleftharpoons \mathrm{H}^+ + \mathrm{B} \qquad K_a^{\mathrm{HB}} = \frac{(\mathrm{H}^+)(\mathrm{B})}{\mathrm{HB}} \tag{1}$$

$$\mathrm{B} + \mathrm{H}^+ \rightleftharpoons \mathrm{HB} \qquad K_b^{\mathrm{B}} = \frac{(\mathrm{HB})}{(\mathrm{H}^+)(\mathrm{B})} \tag{2}$$

which represent the tendency to split a proton off the acid, HB, or add a proton to the conjugate base B. (H^+) refers to the activity of the *unsolvated* proton and is not known.

Note that $K_a^{\mathrm{HB}} = 1/K_b^{\mathrm{B}}$.

For an amphiprotic solvent, HS, we can write

$$\mathrm{H}S \rightleftharpoons \mathrm{H}^+ + S^- \qquad K_a^{\mathrm{H}S} = \frac{(\mathrm{H}^+)(S^-)}{(\mathrm{H}S)} \tag{3}$$

$$\mathrm{H}S + \mathrm{H}^+ \rightleftharpoons \mathrm{H}_2S^+ \qquad K_b^{\mathrm{H}S} = \frac{(\mathrm{H}_2S^+)}{(\mathrm{H}^+)(\mathrm{H}S)} \tag{4}$$

where reactions (3) and (4) are measures of the intrinsic acidity and basicity of the solvent. The sum of reactions (3) and (4) is

$$2\,\mathrm{H}S \rightleftharpoons \mathrm{H}_2S^+ + S^- \qquad K_S^{\mathrm{H}S} = \frac{(\mathrm{H}_2S^+)(S^-)}{(\mathrm{H}S)^2} = K_a^{\mathrm{H}S}\cdot K_b^{\mathrm{H}S} \tag{5}$$

Taking the activity of the solvent equal to unity we obtain the *autoprotolysis constant*

$$K_S^{\mathrm{H}S} = (\mathrm{H}_2S^+)(S^-) = K_a^{\mathrm{H}S}K_b^{\mathrm{H}S} \tag{6}$$

Thus if HS were H_2O, we have the expression for the autoprotolysis constant of water

$$K_S^{\mathrm{H}S} = K_w = (\mathrm{H_3O^+})(\mathrm{OH^-})$$

For ethanol, C_2H_5OH, (EtOH) we get

$$K_S^{\mathrm{EtOH}} = (\mathrm{EtOH_2^+})(\mathrm{EtO^-})$$

while for liquid ammonia, we have

$$K_S^{\mathrm{NH_3}} = (\mathrm{NH_4^+})(\mathrm{NH_2^-})$$

The solvated proton is often denoted as the lyonium ion (H_3O^+, $EtOH_2^+$, or NH_4^+), while the anion produced by removing one proton from the solvent is denoted as the lyate ion (OH^-, EtO^-, or NH_2^-). The extent of solvation of the proton and the lyate ion is generally not known. For convenience the lyonium ion is written as a proton with one molecule of solvent and the lyate ion in the unsolvated form.

When an acid is dissolved in a high dielectric constant solvent, the net chemical reaction corresponds to the sum of reactions (1) and (4):

$$\mathrm{HB} + \mathrm{H}S \rightleftharpoons \mathrm{H}_2S^+ + \mathrm{B} \qquad K_{\mathrm{HB}} = \frac{(\mathrm{H}_2S^+)(\mathrm{B})}{(\mathrm{HB})} = K_a^{\mathrm{HB}}K_b^{\mathrm{H}S} \tag{7}$$

while the net chemical reaction for a base is obtained from the sum of reactions (2) and (3):

$$\mathrm{B} + \mathrm{H}S \rightleftharpoons \mathrm{HB} + S^- \qquad K_{\mathrm{B}} = \frac{(\mathrm{HB})(S^-)}{(\mathrm{B})} = K_b^{\mathrm{B}}K_a^{\mathrm{H}S} \tag{8}$$

Equations (7) and (8) describe the equilibria in solution, and illustrate the extent of dissociation of an acid (or a base) depends upon the intrinsic acidity (or basicity) of the solute and the intrinsic basicity (or acidity) of the solvent. [We assume (HS) = 1 in (7) and (8).]

Consider two solvents, HS and HS*, of identical dielectric constant, with $K_b^{\mathrm{H}S} = 1$, and $K_b^{\mathrm{H}S*} = 10^5$. (H$S$* is one hundred thousand times stronger a base than HS.) An acid, HB, with an intrinsic acidity, $K_a^{\mathrm{HB}} = 10^{-3}$, would appear to be a weak electrolyte in HS, since its dissociation constant, K_{HB}, would be 10^{-3}, while in HS*, the acid would appear to be a strong electrolyte, since K_{HB} would be 10^2.

Now consider an acid, HA, with an intrinsic acidity 10,000 times that of HB, i.e., $K_a^{\mathrm{HA}} = 10^2$. Its dissociation constant in HS would be 10^2 and in HS*, 10^7, and it would be a strong electrolyte in both solvents.

In solvent HS it would be a relatively simple matter to distinguish between the acid strengths of HA and HB, but it would not be possible to do so readily in solvent HS*. Solvent HS* has *leveled* HA and HB. This *leveling effect* also operates for bases on going to solvents with high values of intrinsic acidity. As an example we compare the dissociation of acid and bases in water and in formic acid. Water is a much stronger base and a much weaker acid than formic acid. Hydrochloric and nitric acids are strong acids in water but weak acids in formic acid. As a result of the leveling effect of water these acids are strong electrolytes in this solvent. Ammonia and amines are weak bases in water, but as a result of the much stronger acid character of formic acid as compared to that of water these bases are leveled in formic acid and become strong electrolytes.

Equation (7) makes it clear that the strongest acid that can exist in a solvent is the lyonium ion, H_2S^+, while the lyate ion, S^-, is the strongest base that can exist in a solvent [Eq. (8)].

Summing reactions (7) and (8) gives the relationship that exists between the dissociation constants of an acid and its conjugate base in any amphiprotic solvent:

$$K_S^{HS} = K_{HB}K_B \tag{9}$$

A compilation of autoprotolysis constants is given in Table 5-1.

TABLE 5-1. **Solvents for Acid–Base Reactions**[a]

Substance	pK_S^{HS}	*Dielectric Constant*
Amphiprotic, comparable to water		
Ethanol	19.1	24.3
Methanol	16.7	32.6
Water	14.0	78.5
Amphiprotic, protogenic		
Acetic acid	14.45	6.13
Formic acid	6.2	58.5
Hydrogen fluoride	>11.7 (0°C)	84 (0°C)
Sulfuric acid	3.6	100
Amphiprotic, protophilic		
Ammonia	27, 37.7 (−50°C)	16.9, 27 (−50°C)
Ethylenediamine	~13 ± 1	12.9
Aprotic		
Dimethyl sulfoxide	<20	46
Acetonitrile	>28	36.7
Methyl isobutyl ketone	>30	13.1

[a] Data refer to 25°C unless stated otherwise.

As will be discussed in Chapter 34, the choice of solvent for the titration of an acid is determined by reaction (10) and [(1) minus (3)]:

$$HB + S^- \rightleftharpoons HS + B \qquad \frac{K_a^{HB}}{K_a^{HS}} = \frac{(B)}{(HB)(S^-)} \tag{10}$$

while in selecting a solvent for the titration of a base, reaction (11) and [(2) minus (4)] must be considered:

$$B + H_2S^+ \rightleftharpoons HS + HB \qquad \frac{K_b^{B}}{K_b^{HS}} = \frac{(HS)(HB)}{(B)} \tag{11}$$

The titration of a weak acid, Eq. (10), is driven to completion by choosing a solvent with the *weakest acid* properties. The intrinsic basic strength.of the solvent plays no role. A weak base should be titrated in a solvent with a low intrinsic basicity.

Electrostatic and solvation effects on dissociation

The dissociation of an ionic compound MX in a solvent requires the separation of two ions of opposite charge,

$$MX \rightarrow M^+ + X^- \tag{12}$$

The work required to separate this charge is inversely proportional to the dielectric constant of the solvent, and the necessary energy for dissociation is supplied by solvation of the ions. In dissociating an uncharged acid,

$$HB + HS \rightleftharpoons H_2S^+ + B^- \qquad (13)$$

additional energy is supplied by the proton transfer from the acid to the solvent molecule. If the dielectric constant of a solvent is decreased, more energy will be required to separate the anion and cation, and the extent of dissociation of the acid will be decreased. The calculation of the magnitude of this decrease is beyond the scope of this text and the reader is referred elsewhere[1] for a more complete discussion.

For example, in going from water to ethanol as a solvent, the pK of benzoic acid increases about five units, in agreement with the qualitative prediction based upon Eq. (13). The reaction of ammonium ion with ethanol as the solvent,

$$NH_4^+ + EtOH \rightleftharpoons NH_3 + EtOH_2^+$$

does not involve the separation of oppositely charged ions, and therefore should be insensitive to dielectric constant. Actually, the dissociation constant of ammonium ion in ethanol is 10 per cent of that in water. Table 5-2 presents a comparison of some

TABLE 5-2. **Dissociation Constants in Water, Methanol, and Ethanol**[a]

Acid	pK_{H_2O}	pK_{CH_3OH}	$pK_{C_2H_5OH}$
Benzoic	4.18	10.4	9.7
o-Nitrobenzoic	2.21	8.4	7.4
m-Nitrobenzoic	3.47	9.2	8.3
p-Nitrobenzoic	3.40	8.9	8.1
Salicylic	2.97	8.7	7.9
Picric	0.80	3.8	3.8

[a] From I. M. Kolthoff, J. J. Lingane, and W. O. Larson, *J. Am. Chem. Soc.* **60**, 2512 (1938).

dissociation constants in water, methanol, and ethanol. For low-dielectric-constant solvents, < 13, the electrostatic forces have become so large that even compounds that exist as ionic species in a crystal, e.g., the alkali metal halides, are only slightly dissociated into ions. Bjerrum[2] proposed that such ionic compounds in solution exist as undissociated ion pairs, M^+X^-, in equilibrium with free ions, M^+ and X^-, and that an equilbrium

$$M^+X^- \rightleftharpoons M^+ + X^- \qquad K^\circ_{MX} = \frac{(M^+)(X^-)}{(MX)} \qquad (14)$$

existed that could be described by an ion-pair dissociation constant K°_{MX}. Ion pairing plays a very important role in describing equilibria in solvents such as acetic acid, ethylenediamine, and pyridine. In extremely low-dielectric-constant solvents, < 5, ion aggregates such as $X^-M^+X^-$ or $M^+X^-M^+$ (triple ions), $\begin{matrix} M^+X^- \\ X^-M^+ \end{matrix}$ (ion quadrupole or dimer), and more complicated species must be taken into account.

Table 5-1 lists the dielectric constant of some solvents.

[1] *Treatise on Analytical Chemistry* (I. M. Kolthoff and P. J. Elving, eds.), Wiley-Interscience, New York, 1959, chap. 13.

[2] N. Bjerrum, *Kgl. Danske Videnskab. Selskab* **7**, No. 9 (1926).

As stated above, the solvation of ions helps overcome coulombic forces and promotes dissociation. In the case of anions dissolved in an amphiprotic solvent, hydrogen bonding between the solvent molecule and the anion seems to be a major factor promoting a simple dissociation equilibrium such as Eq. (13). If the solvent has negligible acid properties (aprotic), it will be a poor hydrogen-bond former, and the anion will hydrogen-bond to any other acid present in solution. An example of such a reaction is the dissociation of hydrogen chloride in acetonitrile as a solvent.[3]

$$2HCl \rightleftharpoons H^+ + HCl_2^-$$

The ion HCl_2^- is called a (homo) conjugate ion. Conjugate-ion formation and titration curves of weak acids and a strong base are discussed in Chapter 34.

Acid–base equilibria in glacial acetic acid

The solvent glacial acetic acid is one of the most thoroughly investigated and best understood nonaqueous solvents. Conant and co-workers[4] demonstrated the great difference that exists between aqueous and acetous acid–base equilibria and applied the solvent acetic acid to titration of weak bases. A quantitative interpretation of acid–base equilibria in glacial acetic acid was presented by Kolthoff and Bruckenstein[5] and is presented in an abbreviated form below.

Acetic acid has a dielectric constant of only 6.13 at 25°C. Thus ion pairing is extensive and even the strongest electrolytes have dissociation constants of the order of only 10^{-5}. An acid (or base) must first react with the solvent before it can dissociate, and in acetic acid we must consider the equilibria

$$\begin{aligned} HX + HAc &\rightleftharpoons H_2Ac^+X^- \rightleftharpoons H_2Ac^+ + X^- \\ B + HAc &\underset{\text{(ionization)}}{\rightleftharpoons} BH^+Ac^- \underset{\text{(dissociation)}}{\rightleftharpoons} BH^+ + Ac^- \end{aligned}$$

where $H_2Ac^+X^-$ and BH^+Ac^- are the ion-paired forms of the product formed by the reaction between the acid, HX (or the base, B), and the solvent, HAc. This proton transfer reaction is an ionization reaction, and the equilibrium constants are

Ionization constant K_i:

$$K_i^{HX} = \frac{(H^+X^-)}{(HX)} = \frac{[H^+X^-]}{[HX]} \tag{15}$$

$$K_i^{B} = \frac{[BH^+Ac^-)}{(B)} = \frac{[BH^+Ac^-]}{[B]} \tag{16}$$

Dissociation constant K_d:

$$K_d^{HX} = \frac{(H^+)(X^-)}{(H^+X^-)} = \frac{[H^+][X^-]}{[H^+X^-]} \tag{17}$$

$$K_d^{B} = \frac{(BH^+)(Ac^-)}{(BH^+Ac^-)} = \frac{[BH^+][Ac^-]}{[BH^+Ac^-]} \tag{18}$$

[3] I. M. Kolthoff, S. Bruckenstein, and M. K. Chantooni, Jr., *J. Am. Chem. Soc.*, **83**, 3927 (1961).

[4] J. B. Conant and N. F. Hall, *J. Am. Chem. Soc.* **49**, 3047 (1927); N. F. Hall and J. B. Conant, *ibid.* **49**, 3062 (1927); N. F. Hall, *ibid.* **52**, 5115 (1930); N. F. Hall and T. H. Werner, *ibid.* **50**, 2367 (1928); J. P. Conant and T. H. Werner, *ibid.* **52**, 4436 (1930).

[5] (a) I. M. Kolthoff and S. Bruckenstein, *J. Am. Chem. Soc.* **78**, 1 (1956); (b) S. Bruckenstein and I. M. Kolthoff, *ibid.* **78**, 10 (1956); (c) S. Bruckenstein and I. M. Kolthoff, *ibid.* **78**, 2974 (1956); (d) I. M. Kolthoff and S. Bruckenstein, *ibid.* **79**, 1 (1957); (e) S. Bruckenstein and I. M. Kolthoff, *ibid.* **79**, 5915 (1957).

It is a fair approximation to equate activities to equilibrium concentrations, because the ionic strength of solutions in acetic acid is quite low.

The hydrogen-ion concentration of acetic acid solutions is determined by the *overall dissociation constant*, K_{HX} or K_B:

$$K_{HX} = \frac{(H^+)(X^-)}{\{(HX) + (H^+X^-)\}} = \frac{K_i^{HX}K_d^{HX}}{(1 + K_i^{HX})} \tag{19}$$

$$K_B = \frac{(BH^+)(Ac^-)}{\{(B) + (BH^+Ac^-)\}} = \frac{K_i^{B}K_d^{B}}{(1 + K_i^{B})} \tag{20}$$

The total concentration of undissociated species,

$$C_{HX} = [HX] + [H^+X^-] \tag{21}$$

$$C_B = [B] + [BH^+Ac^-] \tag{22}$$

must always be taken into account when two undissociated species are in equilibrium.

$[H^+]$ in a pure solution of acid or base

Since even the strongest electrolytes have dissociation constants of only $\sim 10^{-5}$, the simple expression for pure aqueous solutions of weak acids or bases obtained in Chapter 3 is adequate. For an acid

$$[H^+] = (K_{HX}C_{HX})^{1/2} \tag{23}$$

and for a base

$$[Ac^-] = (K_B C_B)^{1/2} \tag{24}$$

where C_{HX} and C_B refer to the analytical concentration of the species.

A list of overall dissociation constants for acids, bases and salts in acetic acid is given in Table 5-3.

TABLE 5-3. **Overall Dissociation Constants in Glacial Acetic Acid at 25°C**

	pK
Base	
Tribenzylamine	5.4
Pyridine	6.10
Dimethylaminoazobenzene	6.32
Ammonia	6.40
Potassium acetate	6.14
Sodium acetate	6.68
Lithium acetate	6.79
2,5-Dichloroaniline	9.48
Urea	10.24
Water	12.53
Acid	
Perchloric	4.87
Sulfuric	7.24
p-Toluenesulfonic	8.44
Hydrochloric	8.55
Salt	
Sodium perchlorate	5.48
Potassium chloride	6.88
Urea hydrochloride	6.96
Lithium chloride	7.08
$pK_S = 14.45$	

$[H^+]$ in mixtures of acids and bases

The electroneutrality rule in a solution containing both HX and B is

$$[H^+] + [BH^+] = [Ac^-] + [X^-] \tag{25}$$

The principal species remaining in solution at equilibrium will be either the undissociated acid or the base, depending on which is present in excess, and the salt, BH^+X^-.

The titration reaction is

$$B + HX \rightleftharpoons BH^+X^- \qquad K_f^{BHX} = \frac{C_{BHX}}{C_B C_{HX}} \tag{26}$$

where K_f^{BHX} is called the *formation constant* and is related to other equilibrium constants:

$$HAc \rightleftharpoons H^+ + Ac^- \qquad K_S^{HAc} = K_S = [H^+][Ac^-] \tag{27}$$

$$BHX \rightleftharpoons BH^+ + X^- \qquad K_{BHX} = \frac{[BH^+][Ac^-]}{C_{BHX}} \tag{28}$$

K_B and K_{HX}. We get

$$K_f^{BHX} = \frac{K_{HX}K_B}{K_S K_{BHX}} \tag{29}$$

Equations (19), (20), (26), (27), (28), and (29) may be substituted into Eq. (25) to obtain expressions useful for calculating titration curves in acetic acid. When excess base is present we solve for an expression in terms of the total equilibrium concentration of base, C_B, and salt, C_{BHX}, and obtain

$$[H^+] = \left[\frac{K_S + K_S K_{BHX} C_{BHX}/K_B C_B}{1 + K_B C_B/K_S}\right]^{1/2} \tag{30}$$

If excess acid is present, we solve for an expression in terms of the total equilibrium concentration of acid and salt and obtain

$$[H^+] = \left[\frac{K_S + K_{HX} C_{HX}}{1 + K_{BHX} C_{BHX}/K_{HX} C_{HX}}\right]^{1/2} \tag{31}$$

Since dissociation is very slight, analytical concentrations may be used in Eqs. (30) and (31).

In a solution with only pure salt, it is seen from Eq. (26) that $C_B = C_{HX} = (C_{BHX}/K_f^{BHX})^{1/2}$. Substitution of this result in either Eq. (30) or (31) yields

$$[H^+] = (K_S K_{HX}/K_B)^{1/2} \tag{32}$$

provided $1 \ll K_{BHX}C_{BHX}/K_B C_B$ and $1 \ll K_B C_B/K_S$ in Eq. (30) or $K_S \ll K_{HX}C_{HX}$ and $1 \ll K_{BHX}C_{BHX}/K_{HX}C_{HX}$ in Eq. (31). Various special cases of Eqs. (30) and (31) have been discussed.[5d] Equation (32) is simple and corresponds exactly to that for a salt of a weak acid and a weak base in water.

Equilibria in benzene

The dielectric constant of the aprotic solvent benzene is about 2.2 at 25°C. Ion association is very much more extensive than in acetic acid, and ion triplets, quadrupoles (dimers), etc., exist. As in acetic acid, the analytical concentration of a solute is closely approximated by the sum of all undissociated species. In the case of a strong electrolyte, these species are the monomer, dimer, trimer, etc., rather than the charged ionic species. Carboxylic acids are extensively dimerized (hydrogen-bonded) in benzene, whereas most amine bases are monomeric. Thus a good description of the reaction between a carboxylic acid and a base in benzene is given by the equilibria

$$2\ \mathrm{RCOOH} \rightleftharpoons (\mathrm{ROOH})_2$$

$$\mathrm{RCOOH} + \mathrm{B} \rightleftharpoons \frac{1}{m}\,(\mathrm{BH^{+}\ {}^{-}OOC{-}R})_m$$

In addition, homoconjugate ion formation occurs in benzene, so that the reaction

$$\frac{1}{m}\,(\mathrm{BH^{+}\ {}^{-}OOC{-}R})_m + \mathrm{RCOOH} = \frac{1}{n}$$

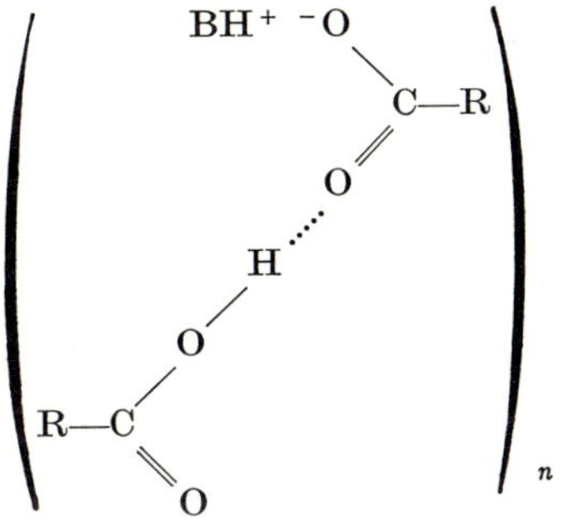

leading to the formation of an associated homoconjugated ion pair must be considered. The equilibria are involved, and are discussed by Bruckenstein and Saito.[6] A comprehensive discussion of acid-base chemistry in aprotic solvents is given by Davis.[7]

PROBLEMS

1. Acetonitrile and dimethyl sulfoxide have dielectric constants of the same order of magnitude. HCl and HNO_3 are weak acids in acetonitrile, but strong acids in dimethyl sulfoxide. Is dimethyl sulfoxide a weaker or stronger base than acetonitrile?
2. Benzoic acid reacts quantitatively in the solvent sulfuric acid,

$$C_6H_5COOH + H_2SO_4 \rightleftharpoons C_6H_5COOH_2^{+} + HSO_4^{-}$$

Interpret the significance of the reaction.

3. $K_S^{HS} = 10^{-14}$ for water and $10^{-19.1}$ for ethanol. (a) What is the pH of the pure solvents? (b) What is the pH and pS^- of 0.01 M HCl in both solvents? (c) What is the pH and pS^- of 0.001 M sodium lyate in both solvents?
Ans. (a) 7.00 in water, 9.55 in ethanol; (b) 2.00 and 12.00 in water, 2.00 and 17.10 in ethanol; (c) 11.00 and 3.00 in water, 16.10 and 3.00 in ethanol.

[6] S. Bruckenstein and A. Saito, *J. Am. Chem. Soc.* **87**, 698 (1965).

[7] M. M. Davis, *Acid-Base Behavior in Aprotic Organic Solvents*, N.B.S. Monograph **105**, Washington, D.C., 1968.

CHAPTER 6 *Complexation Equilibria*

This chapter is devoted to a discussion of the principles of equilibria involving complexes. A complex may be defined as an association of a constituent—either molecule or ion—with another species. In this sense most simple ions in water are complexes, e.g., the copper(II) ion, which is present as $Cu(H_2O)_4^{++}$, comparable to the complex $Cu(NH_3)_4^{++}$. It is common practice not to call hydrated ions complex ions. The question can be raised whether the association of ions to give an uncharged molecule is a complexation reaction. Actually such neutral molecules may indeed be called complexes, for the same reason the other species below are complexes:

$$Hg^{++} + Cl^- \rightleftharpoons HgCl^+ \quad (1)$$

$$Hg^{++} + 2\,Cl^- \rightleftharpoons HgCl_2 \quad (2)$$

$$Hg^{++} + 3\,Cl^- \rightleftharpoons HgCl_3^- \quad (3)$$

Other examples of complexation reactions are

$$Fe^{3+} + SCN^- \rightleftharpoons FeSCN^{++} \quad (4)$$

$$Ag^+ + 2\,NH_3 \rightleftharpoons Ag(NH_3)_2^+ \quad (5)$$

$$Al^{3+} + 3\,F^- \rightleftharpoons AlF_3 \quad (6)$$

Complex formation also occurs in nonaqueous solutions. For example, zinc amide is insoluble in liquid ammonia but dissolves upon the addition of sodium amide. This occurs because of the formation of a stable complex amide, and the process is analogous to the dissolution of zinc hydroxide, which is insoluble in water, upon the addition of sodium hydroxide to form a stable zincate.

Equations (1)–(6) as written represent the addition of one or more *ligands* (Cl^-, SCN^-, NH_3, F^-) to a metal ion. In many complexation reactions the solvent also acts as a ligand, and complexation of cations in water amounts to a displacement of a water ligand by another ligand. Thus reaction (4) actually should be written.

$$Fe(H_2O)_6^{3+} + SCN^- \rightleftharpoons Fe(H_2O)_5SCN^{++} + H_2O \quad (7)$$

This is comparable to the displacement reaction

$$Cu(NH_3)_4^{++} + 4\,CN^- \rightleftharpoons Cu(CN)_4^{--} + 4\,NH_3 \quad (8)$$

For simplicity, water or solvent molecules usually are not shown as ligands in complexation reactions, and this practice will be followed in this text.

For a substance to function as a ligand it must be an electron donor; i.e., it must have at least one pair of electrons to donate to a metal ion, thereby forming a shared electron-pair bond. The metal ion acts as a Lewis acid (see Chapter 3) and the ligand as a base. The ligand shares a pair of electrons with the acid to yield the complex, which is a neutralization product in the Lewis terminology. The addition of one or more protons to a base (see Brønsted definition of a base, p. 45) might be considered as a complexation reaction, for example:

$$PO_4^{3-} + H^+ \rightleftharpoons HPO_4^{--} \quad HPO_4^{--} + H^+ \rightleftharpoons H_2PO_4^- \quad H_2PO_4^- + H^+ \rightleftharpoons H_3PO_4$$

However, such reactions generally are denoted as acid–base rather than complexation reactions. Ligands such as F^- can form one such bond and are *unidentate* (literally, one-toothed). Substances of more complicated structure may contain two or more donors and may be able to form two or more bonds. Such substances are *bidentate*, . . . , or, in general, *multidentate*. Ethylenediamine, $NH_2 \cdot CH_2 \cdot CH_2 \cdot NH_2$, which contains two amino groups, is bidentate. One of the most important complexing agents is ethylenediamine tetraacetic acid, which has the formula $(HOOC \cdot CH_2)_2N \cdot CH_2 \cdot CH_2 \cdot N(CH_2COOH)_2$. This substance is commonly known as EDTA and the formula is abbreviated H_4Y. The anion of EDTA can be hexadentate, because it can be bonded through the two nitrogens and the four oxygens of the carboxyl groups. A multidentate ligand may be bonded to a metal ion in such a way that a ring, or *chelate*, is formed, and the ligand then is called a *chelating agent*. The word chelate is derived from Greek chele, claw, and chelation may be likened to the closing of a pincer upon the metal ion. The fact that in a great majority of cases a complex that includes a ring is more stable than a similar complex without a ring was noted by Schwarzenbach and termed by him the "*chelate effect*." For example, the complex of nickel with one ethylenediamine molecule is considerably more stable than the complex with two ammonia molecules. Calcium ion has a coordination number of six and can be bonded to six unidentate ligands, but all six bonds can be formed with the anion of EDTA. The resulting stable chelate has four five-membered rings, and has the structure represented in Fig. 6-1.

Not all bidentate or multidentate ligands form chelates. Instead, the ligand may bond two or more metal ions. A well-known example is the phosphate anion, which

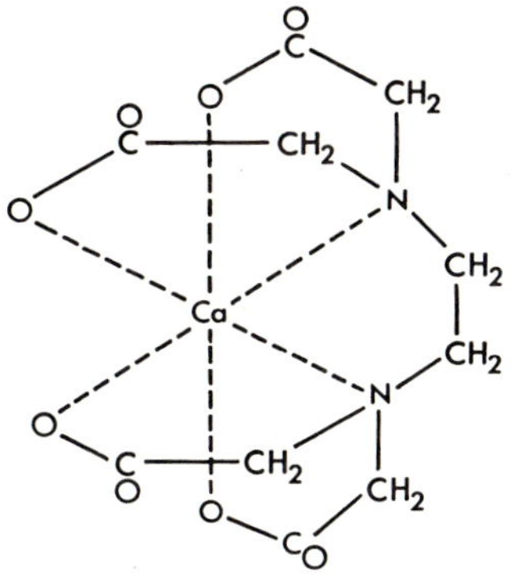

Fig. 6-1. Schematic representation of EDTA complex of calcium(II). The six octahedral bonds to ligands are shown as dotted lines.

can link two iron(III) ions; each of the latter in turn may be joined to two phosphates, leading to formation of a chain containing several iron and phosphate ions.

EDTA has become one of the most widely used reagents for titrations of many metal ions. Schwarzenbach[1] introduced it as an analytical reagent. Thereafter, many compounds were investigated which were similar to EDTA but formed metal complexes of stability different from EDTA. Typical examples include nitrilotriacetic acid, $N(CH_2COOH)_3$, and diethylenetriaminepentaacetic acid,

$$(HOOCCH_2)_2NCH_2CH_2\underset{\displaystyle CH_2COOH}{\underset{|}{N}}CH_2CH_2N(CH_2COOH)_2$$

The latter can be octadentate and is useful for titrations of lanthanides and actinides which have coordination numbers of eight. Also, by the introduction of chromophoric (color-forming) groups into chelating agents such as EDTA, it was possible to prepare strongly colored indicators for many metal ions. Such substances are used in titrations (Chapters 36 and 41). Higher amines such as diethylenetriamine, $H_2NCH_2CH_2NHCH_2 \cdot CH_2NH_2$; triaminotriethylamine, $N(CH_2CH_2NH_2)_3$ ("tren"); and triethylenetetramine, $H_2NCH_2CH_2NHCH_2CH_2NHCH_2CH_2NH_2$ ("trien") are more selective than EDTA and related compounds. They do not form complexes with alkali and alkaline earth ions.

The addition of more than one ligand generally occurs for unidentate ligands and may occur for other ligands as well. All such additions occur in consecutive steps. Thus reaction (5) is the sum of two step reactions:

$$Ag^+ + NH_3 \rightleftharpoons AgNH_3^+$$
$$AgNH_3^+ + NH_3 \rightleftharpoons Ag(NH_3)_2^+$$

Reaction between Fe^{3+} or Al^{3+} and F^- occurs in six steps:

$$Fe^{3+} + F^- \rightleftharpoons FeF^{++} \quad (9)$$

$$FeF^{++} + F^- \rightleftharpoons FeF_2^+ \quad (10)$$

$$FeF_2^+ + F^- \rightleftharpoons FeF_3 \quad (11)$$

$$FeF_3 + F^- \rightleftharpoons FeF_4^- \quad (12)$$

$$FeF_4^- + F^- \rightleftharpoons FeF_5^{--} \quad (13)$$

$$FeF_5^{--} + F^- \rightleftharpoons FeF_6^{3-} \quad (14)$$

Many equilibrium constants for such individual step reactions are known. Several numerical examples of the use of such constants are given later.

In many but not all cases of analytical interest the complex contains only one metal ion or *central atom* per molecule and is called *mononuclear*. Many *polynuclear* complexes exist, such as $Ag_2I_3^-$ and $Fe_2(OH)_2^{4+}$ and the Fe(III) phosphate mentioned earlier; polynuclear hydroxocomplexes are particularly common. The formation of mononuclear rather than polynuclear complexes is favored upon lowering the concentration of metal. *Mixed complexes* are formed when a metal ion

[1] G. Schwarzenbach, *Helv. Chim. Acta* **29**, 1338 (1946); G. Schwarzenbach and H. Flaschka, *Die Komplexometrische Titration*, Enke, Stuttgart, 1965.

[2] L. G. Sillen and A. E. Martell, *Stability Constants of Metal-Ion Complexes*, Chemical Society, London, 1964.

reacts with more than one kind of ligand. Such complexes, containing hydroxyl ion as one of the ligands, are relatively common, for example, $FeYOH^{--}$. Both trivalent bismuth and antimony form mixed complexes with bromide and chloride, for example, $BiBrCl_4^{--}$. The slightly soluble substance PbFCl can be regarded as a mixed complex.

Complexing trends among the metals

Some generalizations can be made concerning the stability of coordination compounds formed by different types of ligands with metals, or semimetals, on the basis of the electronic structure of these. This is not the place to go into this topic in any detail, but a few observations may be of interest.[3] Elements can be divided into three classes as shown (for metals and metalloids) in Table 6-1. In class I, the alkali metals are almost noncomplexable; Li and Na form a few weak complexes with very strong

TABLE **6-1. Classification of Metals on the Basis of Electronic Structure**[a]

I. Metal ions resembling inert gases							
Li	Be	B					
Na	Mg	Al					
K	Ca	Sc		Zr			
Rb	Sr	Y		Hf			
Cs	Ba	La, Ce, Lu		Th			
II. Metal ions having partially filled *d* or *f* orbitals (transition metals)							
Ti	V	Cr	Mn	Fe	Co	Ni	Cu(II)
Nb	Mo	Tc	Ru	Rh	Pd	Ag(II)	
Ta	W	Re	Os	Ir	Pt	Au(III)	
Rare earths and actinides							
III. Metal ions having filled *d* orbitals							
Cu(I)	Zn	Ga	Ge	As			
Ag(I)	Cd	In	Sn	Sb	Te		
Au(I)	Hg	Tl	Pb	Bi	Po		

[a] After D. D. Perrin, *Organic Complexing Reagents*, Wiley-Interscience, New York, 1964.

complexers, Rb and Cs rarely, if ever. The remaining metals of class I, together with those in the first column of class II, show a varying tendency to combine with ligands in organic compounds having donor O atoms, and with OH^- and F^-. Ca, Sr, and Ba are the most difficult metals to complex in this group. Although their fluorides are sparingly soluble, they are highly ionized in aqueous medium. Their hydroxides are almost completely ionized (some MOH^+ is formed). However, some chelating reagents (EDTA), with a donor O atom, form moderately strong complexes with the alkaline earths. The rare earth elements are similar to the alkaline earths but show somewhat stronger complexing tendencies. On the other hand, some metals of this group (class I + Ti, Nb, Ta, and rare earths) such as Al, are strongly complexed by both OH^- and

[3] See, e.g., G. Schwarzenbach, *Experentia Suppl.* **5**, 162 (1956); *Advances in Inorganic Chemistry and Radiochemistry*, vol. III, Academic Press, New York, 1961; S. Ahrland, *Acta Chem. Scand.* **10**, 723 (1956); S. Ahrland, J. Chatt, and N. R. Davies, *Quart. Rev.* (*London*), **12**, 265 (1958); D. D. Perrin, *Organic Complexing Reagents*, Wiley-Interscience, New York, 1964.

F^-, forming anionic species; others (Zr) are complexed by F^-. The elements of this group can perhaps best be characterized as oxyphilic, in that they show little tendency to coordinate with any ligand other than one containing a bonding O atom (OH^- or O^- in an organic group), or F^-, which, because of its small size and slight polarizability, can compete with OH^-. Complexing with Cl^-, Br^-, and I^- is weak (Zr, Th, and rare earths slightly). The cations of these metals hardly combine at all (Mg^{++}, Ca^{++}, and Sc^{++} slightly) with NH_3, which cannot successfully displace H_2O from the hydrated cation.

For many of the metals in class II and class III the strength of the bonding to atoms of O, N, and S in the ligand increases in the order O < N < S. This is especially true for the deformable and strongly covalent bonding atoms Cu(I), Ag(I), and Au(I). The more noble the metal and the less electronegative the donor atom of the ligand, the more stable the complex. Zn, Cd, and Hg behave similarly. Most of the metals in classes II and III can be broadly described as sulfophilic (Table 6-2). They combine

TABLE **6-2. Oxyphilic and Sulfophilic Elements**[a]

Li	Be*	B*					
Na	Mg	Al*					
K	Ca	Sc	Zr*				
Rb	Sr	Y	Hf*				
Cs	Ba	La, Ce, Lu	Th				
Ti*	V	Cr	Mn	Fe	Co	Ni	Cu(II)
Nb*	Mo	Tc	Ru	Rh	Pd	Ag(II)	
Ta*	W	Re	Os	Ir	Pt	Au(III)	
Rare earths, actinides							
Cu(I)	Zn	Ga	Ge	As			
Ag(I)	Cd	In	Sn	Sb	Te		
Au(I)	Hg	Tl	Pb	Bi	Po		

[a] The solid line encloses elements forming slightly soluble fluorides or slightly dissociated anionic fluoride complexes (M*) but having little tendency to complex with I^- or even Cl^-. These elements show a strong tendency to react with oxygen-containing anionic ligands. Most of them form sparingly soluble or anionic complex oxalates, citrates, etc., and combine with chromogenic organic reagents containing —OH groups (e.g., alizarin, eriochrome black T, morin).

The dashed line encloses elements capable of bonding more strongly to S than to O, thus forming slightly soluble sulfides or anionic sulfide complexes in aqueous solution, and combining with $—S^-$ groups in organic chelating reagents such as dithizone (p. 351). Many of these elements, of course, can also bond to O (for example, Bi cupferrate) and even F^- [Ga, In, Sn(IV), Ge, for example].

with S atoms (S^{--}, HS^-, and $—S^-$ in organic chelating reagents). With the exception of V, W, and Cr (and Ti, Nb, Ta, the rare earths, and actinides which we have already grouped with the elements of class I), they are capable of forming slightly soluble sulfides in aqueous solution and some yield anionic complex sulfides such as HgS_2^{--}.

The pronounced tendency for bonding to C, N, and the heavy halogens among many of the elements of classes II and III is illustrated in Table 6-3 and 6-4. For examples

TABLE **6-3. Metals Forming Ammine (NH_3) Complexes**[a]

Li	Be	B					
Na	Mg	Al					
K	Ca	Sc	Zr				
Rb	Sr	Y	Hf				
Cs	Ba	La, Ce, Lu	Th				
Ti	V	Cr	Mn	Fe	Co	Ni	Cu(II)
Nb	Mo	Tc	Ru	Rh	Pd	Ag(II)	
Ta	W	Re	Os	Ir	Pt	Au(III)	
Rare earths, actinides							
Cu(I)	Zn	Ga	Ge	As			
Ag(I)	Cd	In	Sn	Sb	Te		
Au(I)	Hg	Tl(III)	Pb	Bi	Po		

[a] Solid line encloses elements forming ammine complexes. Dashed line encloses elements forming weak ammine complexes.

TABLE **6-4. Metals Forming Slightly Dissociated Cyanide and Iodide Complexes**[a]

Li	Be	B					
Na	Mg	Al					
K	Ca	Sc	Zr				
Rb	Sr	Y	Hf				
Cs	Ba	La, Ce, Lu	Th				
Ti	V	Cr	Mn	Fe	Co	Ni	Cu(II)
Nb	Mo	Tc	Ru	Rh	Pd	Ag(II)	
Ta	W	Re	Os	Ir	Pt	Au(III)	
Rare earths, actinides							
Cu(I)	Zn	Ga	Ge	As			
Ag(I)	Cd	In	Sn	Sb	Te		
Au(I)	Hg	Tl(III)	(Pb)	(Bi)	Po		

[a] Solid line encloses elements forming complex cyanides. Pb and Bi complexes are weak, Tl(I) very weak, Tl(III) strong, Cu(II) unstable [→ Cu(I)]. Dashed line encloses elements forming more stable complexes with I^- than with F^-. (Uncharged or anionic complexes of In, Sn, Sb, and Te are sufficiently stable to have analytical utility.)

of chelate formation involving various groupings of S, N, and O see Chapter 12. These elements are, of course, not devoid of oxyphilic properties. Fe(III) is similar to Al in combining with O-containing ligands and also F^-. The metals in class III having a high valence (Ga, In, Ge, and Sn, for example) resemble the elements of class I. Tl(I) is, of course, similar to the metals having the inert-gas structure, even similar to the alkali metals. Pb, Bi, and Sb are more like elements of class I than of class II in some respects. Thus they show little tendency to combine with NH_3, and the cyanide complexes are weak.

Organic reagents with suitable chelating groups can form stable complexes with metal ions in all three classes. The outstanding example is EDTA and its analogues. It has both N and O donor atoms and so can complex Ca, Mg, and Al as well as Cu, Zn, and almost all other metals; the high charge of the Y^{4-} ion is also conducive to strong complexing. 8-Hydroxyquinoline (p. 207) is another near-universal chelating reagent, forming slightly soluble products; again O and N atoms bond the metal.

Factors affecting stability of chelate compounds

Obviously the most important factor in the stability of any chelate is the strength of the bonds between the metal ion and the donors contained in the chelating agent, discussed above. Superimposed on this main factor are other factors of more or less importance, which are discussed briefly in the following.[4]

As mentioned earlier, the chelate effect refers to the enhanced stability of a ring-containing complex compared to a similar complex without a ring. [Exceptions occur with ions of small coordination number such as Ag(I).] As might be anticipated the chelate effect is increased as the number of such rings per molecule is increased. Qualitatively this can be interpreted as follows: The more points of attachment or bonds there are between ligand and metal, the less likely it is for the metal to break all these bonds and separate from the ligand before the bonds are reformed.

Also the size of the ring is important. Usually the chelate effect is most pronounced for five- to six-membered rings. The lower stability of smaller rings is well known in organic chemistry and is accounted for by increased distortion of bond angles from their normal values. Specifically for chelate complexes it must be remembered that the most favorable arrangement of ligands about the metal ion depends upon the ion and its coordination number. Thus Mg(II) and Ca(II), with coordination numbers 4 and 6, form tetrahedral and octahedral complexes, respectively. (See Fig. 6-1.) Distortion from these structures will lower the stability of the complex; generally the distortions are a minimum with the five- and six-membered rings. On the other hand, Ag(I) with a coordination number of 2 forms linear complexes; this structure is more nearly approached, the larger the ring. Accordingly, Ag(I) complexes with seven-membered rings are more stable than those with five-membered rings.

The importance of the relation of the structure of the chelating agent to the preferred orientation of ligands around the metal ion is illustrated by the following comparison. Copper(II) and zinc(II) both have coordination numbers of 4 and form square planar and tetrahedral complexes, respectively.[5] Each metal ion forms complexes with trien and tren (see p. 104 for formulae of these chelating agents). From the formulae it may be anticipated that trien should be better adapted to the formation of planar than tetrahedral complexes, and that the reverse should be true for tren. The relative stabilities of the copper and zinc complexes are in accord with this anticipation (see the Appendix).

Steric effects may be of major importance. (See also Chapter 12.) These may be manifested in two different ways. If the ligands are too large, steric hindrance may

[4] For a detailed treatment see the books listed in Ref. 3, especially D. D. Perrin, chap. 3; also A. E. Martell and M. Calvin, *Chemistry of the Metal Chelate Compounds*, Prentice-Hall, Englewood Cliffs, N.J., 1952.

[5] Zinc(II) also has coordination number 6, forming octahedral complexes.

prevent the appropriate number of ligands from fitting into the coordinate sphere of the metal ion. Sometimes relatively minor changes in ligand structure cause such effects. Thus 1,10-phenanthroline forms a highly stable complex with iron(II), containing three phenanthroline molecules (i.e., a tris complex). The substituted 2,9-dimethyl-1,10-phenanthroline does not form a tris complex. Another striking example is afforded by 8-hydroxyquinoline, which forms a stable insoluble compound with aluminum(III), as compared to 2-methyl-8-hydroxyquinoline, which does not. A steric effect of the reverse sort is provided by the chelating agent known as calcichrome of the structure:

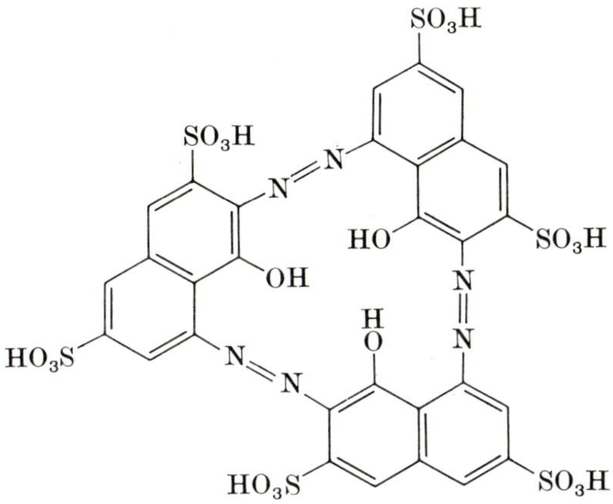

In this compound the azo and phenol groups which can bond to metal ions are arranged to form a "cage" of such a size that calcium ion can enter, with the formation of a colored chelate, but barium, strontium, and magnesium do not react.

To understand the principles of the analytical applications it is necessary to understand the equilibria involved, and therefore the method of making equilibrium calculations is the main subject of this chapter. Anticipating the numerical results, it may be stated that in general the complexes of analytical interest are so stable that metal ions can be converted completely into the complex form under suitable conditions. Almost all the reactions are practically instantaneous. In only relatively few cases the reactions are slow. This is the case with the reaction between chromium(III) and EDTA, which requires days to reach equilibrium at room temperature. Also, some replacement reactions are slow. For example, bismuth forms a more stable complex with EDTA than nickel does. However, when bismuth is added to a solution containing the nickel complex, the nickel is displaced very slowly. Most other metals, such as cobalt, zinc, and others, are displaced practically instantaneously. Also, some reactions of metal indicators are slow. These matters are referred to again in Chapters 36 and 41. The following discussion is restricted to equilibrium situations.[6]

Importance of complexation in analytical chemistry

Applications in gravimetric precipitations, in separations by ion exchange, precipitation, and extraction, and in spectrophotometric determinations are

[6] Studies of the rates of complexation reactions have been made by Margerum and coworkers: see, for example, D. W. Margerum, T. J. Bydalek, and J. T. Bishop, *J. Am. Chem. Soc.* **83**, 1791 (1961); D. W. Margerum and B. A. Zabin, *J. Phys. Chem.* **66**, 2214 (1962).

described elsewhere in the text. In addition to these, the reactions are made use of in two distinct ways. One is for the volumetric determination of a metal by titration with a standard solution of a complexing agent, or vice versa. An example is the Liebig titration of cyanide with a standard solution of silver nitrate. When silver ions are added to a solution of an alkali cyanide, a stable complex is formed in the reaction:

$$Ag^+ + 2\,CN^- \rightleftharpoons Ag(CN)_2^- \qquad (15)$$

When more silver than corresponds to Eq. (15) is added, the excess reacts according to Eq. (16) to form a precipitate of silver cyanide:

$$Ag^+ + Ag(CN)_2^- \rightleftharpoons 2\,AgCN(s) \qquad (16)$$

The equivalence point is marked by the appearance of turbidity. Other reactions, such as the reaction of mercuric ion with thiocyanate or with chloride, have been used for a long time in volumetric analysis. The most important kind of titration now is based upon the use of EDTA and related compounds, for example, in the determination of calcium. Numerous procedures have been developed which allow the determination of metal ions in mixtures by titrating at various values of pH and in the absence or presence of suitable *masking agents*. Use as a masking agent is another important application. The term "masking" refers to the addition of a complexing agent to form a metal complex of such stability that it is practically unreactive with the titrant.[7] Two examples will illustrate. Both copper and ferric ions oxidize iodide. By the addition of an excess of fluoride to a mixture of copper and ferric ions, the latter is transformed into a stable fluoride complex that does not react with iodide ion. In this case fluoride is a masking agent for ferric ion. As another example, both magnesium and nickel ions react with the anion of EDTA. Upon the addition of excess of cyanide at the proper pH, the nickel is transformed into a complex cyanide and no reaction of nickel with EDTA occurs (see p. 121). Such masking agents permit development of selective titrations and improve the selectivity of various methods of separation.

When a metal such as iron or cobalt can exist in two oxidation states which form complexes with a given agent of much different stability, the oxidation–reduction potential (see Chapter 8) is altered greatly by addition of the complexing agent. Two examples will illustrate. Both ferrous and ferric iron react with EDTA, but the ferric complex is much more stable than the ferrous. The difference in stability is so great that ferrous versenate is easily oxidized by air. Trivalent cobalt in the form of Co^{3+} is so strong an oxidizing agent that it oxidizes water, and solutions containing Co^{3+} are unstable. As in the case of iron, the complex of EDTA with Co(III) is much more stable than that with Co(II). In the presence of an excess of EDTA, divalent cobalt may be oxidized quantitatively by ferricyanide.

Calculations of complexation equilibria

The entire discussion is restricted to mononuclear complexes. The addition of ligand L to metal M occurs in the series of steps:

[7] Masking literally means converting to an unreactive form in any way. It can be brought about also by oxidation or reduction or adjustment of pH.

$$M + L \rightleftharpoons ML \tag{17}$$

$$ML + L \rightleftharpoons ML_2 \tag{18}$$

$$ML_2 + L \rightleftharpoons ML_3 \tag{19}$$

$$\vdots$$

$$ML_{n-1} + L \rightleftharpoons ML_n \tag{20}$$

Following general custom the electrical charges are omitted. n represents the maximum number of L ligands that can be attached to M. When L is unidentate, n is equal to the *coordination number* of M. The corresponding thermodynamic *stepwise-formation constants* are

$$K_1^0 = \frac{(ML)}{(M)(L)} \tag{21}$$

$$K_2^0 = \frac{(ML_2)}{(ML)(L)} \tag{22}$$

$$K_3^0 = \frac{(ML_3)}{(ML_2)(L)} \tag{23}$$

$$\vdots$$

$$K_n^0 = \frac{(ML_n)}{(ML_{n-1})(L)} \tag{24}$$

where the parentheses represent activities.

In all practical cases complexation reactions are carried out in electrolyte solutions, and the differences between activities and molar concentrations may be appreciable. For analytical purposes concentrations are wanted rather than activities. In some instances approximate corrections of activities to concentrations may be made (Chapter 3). However, it is preferable in general to use equilibrium "constants," expressed in molar concentration units, which apply in a specified medium. For example, the molar concentrations of Fe^{3+}, F^-, and FeF^{++} may be measured in 0.5 M potassium nitrate. The quantity $[FeF^{++}]/[Fe^{3+}][F^-]$, where brackets represent molar concentrations, is a constant *in this medium*. In any other medium, for example, 0.1 M sodium perchlorate, a different "constant" would apply. In this text the symbol K is used for *concentration constant*, applicable at a given ionic strength. When any confusion between concentration constant and thermodynamic constant [Eqs. (21)–(24)] may arise, the thermodynamic constant will be denoted K^0.

$$K_1 = \frac{[ML]}{[M][L]} \tag{25}$$

$$K_2 = \frac{[ML_2]}{[ML][L]} \tag{26}$$

$$K_3 = \frac{[ML_3]}{[ML_2][L]} \tag{27}$$

$$\vdots$$

$$K_n = \frac{[ML_n]}{[ML_{n-1}][L]} \tag{28}$$

Overall reactions can be written instead of step reactions:

$$M + L \rightleftharpoons ML \tag{29}$$

$$M + 2\,L \rightleftharpoons ML_2 \tag{30}$$

$$M + 3\,L \rightleftharpoons ML_3 \tag{31}$$

$$\vdots$$

$$M + nL \rightleftharpoons ML_n \tag{32}$$

The overall equilibrium constants are β_n:

$$\beta_1 = \frac{[ML]}{[M][L]} \tag{33}$$

$$\beta_2 = \frac{[ML_2]}{[M][L]^2} \tag{34}$$

$$\beta_3 = \frac{[ML_3]}{[M][L]^3} \tag{35}$$

$$\vdots$$

$$\beta_n = \frac{[ML_n]}{[M][L]^n} \tag{36}$$

The values of step and over-all constants are related:

$$\beta_1 = K_1 \tag{37}$$

$$\beta_2 = K_1K_2 \tag{38}$$

$$\beta_3 = K_1K_2K_3 \tag{39}$$

$$\vdots$$

$$\beta_n = K_1K_2K_3 \cdots K_n \tag{40}$$

Analogous relations of course exist for the thermodynamic constants. Some numerical values are given in the Appendix.

In the great majority of cases $K_1 > K_2 > \cdots > K_n$. Out of the thousands of formation constants now known, there are only a few exceptional cases in which this systematic trend is not observed. Exceptions appear in the systems

Ag^+–NH_3: $K_2 > K_1$

Zn^{++}–NH_3: $K_3 > K_2 > K_1$

Cd^{++}–CN^-: $K_4 > K_3$ (but $K_1 > K_2 > K_3$, as normal)

and a few others.

Addition of one ligand. First consider the simplest situation, in which Eq. (17) represents the only reaction that needs to be considered. The reaction of calcium ion with EDTA at pH a little above 11 is a practical example. At this pH practically all the EDTA is present as Y^{-4}, whereas at lower pH this is not true (see pp. 119–121). At higher pH the reaction of calcium ion with hydroxyl ion would have to be taken into account. The value of K_1 is 5.0×10^{10} at 20°C and an ionic strength of 0.1. Consider a titration of 10^{-3} *M* Ca^{++} with a solution of EDTA and for simplicity assume the volume remains practically constant. The *equivalent amount* of EDTA corresponds to 1 mole per mole of calcium. When this amount is added, the calcium is transformed quantitatively into the complex, so $[CaY^{--}] = 10^{-3}\,M$. Evidently the values of $[Ca^{++}]$ and $[Y^{4-}]$ must be exactly equal. Substitution in the expression for the equilibrium constant yields

$$[Ca^{++}] = ([CaY^{--}]/K_1)^{1/2} = 1.4 \times 10^{-7}\,M \qquad pCa = 6.85$$

Note that $[Ca^{++}]$ is negligibly small compared to $[CaY^{--}]$, confirming that the reaction is quantitative. If 99 per cent of the equivalent amount of EDTA is added, 1 percent of the calcium remains as Ca^{++}. Hence $[CaY^{--}] = 9.9 \times 10^{-4}\,M$, $[Ca^{++}] = 10^{-5}\,M$, i.e., pCa = 5.00, and

$$[Y^{4-}] = \frac{[CaY^{--}]}{[Ca^{++}]K_1} = 1.98 \times 10^{-9}\,M$$

On the other hand, if 101 per cent of the equivalent amount of EDTA is added, the excess over 100 per cent is present as Y^{4-}; hence $[Y^{4-}] = 10^{-5}\,M$, $[CaY^{--}] = 10^{-3}\,M$, and

$$[Ca^{++}] = \frac{[CaY^{--}]}{[Y^{4-}]K_1} = 2.00 \times 10^{-9}\,M \qquad pCa = 8.70$$

Thus $[Ca^{++}]$ changes 5000-fold, or pCa changes by 3.70 units, when the amount of EDTA is changed from 99 to 101 per cent of the equivalent amount. It is possible in various ways to detect such changes and thus to locate the equivalence point. The details are given in Chapter 36.

The calculations above were very simple because K_1 was so large that the reaction went practically to completion. This would not be the case if K_1 were not so large, or, for the same K_1, if the solutions were much more dilute. Fortunately, the exact and general solution of such problems is easy. Consider a solution in which are mixed C_M moles of M and C_L moles of L, per liter. There are three unknown concentrations, [M], [L], and [ML], and three independent algebraic relations are required to determine them. Equation (25) is one of the necessary relations and the other two are provided by material balance:

$$C_M = [M] + [ML] \tag{41}$$

$$C_L = [L] + [ML] \tag{42}$$

Equations (25), (41), and (42) are combined to yield

$$[ML] = K_1\{C_M - [ML]\}\{C_L - [ML]\} \tag{43}$$

from which the exact value of [ML] is found:

$$[ML] = \frac{1}{2}\left\{C_M + C_L + \frac{1}{K_1} - \left[\left(C_M + C_L + \frac{1}{K_1}\right)^2 - 4\,C_M C_L\right]^{1/2}\right\} \tag{44}$$

Analogous exact expressions can be written for [M] and [L]. It is left as an exercise for the student to prove that the exact algebraic expressions reduce to the simple expressions when C_M and C_L are not too small and K_1 is large (see Problem 1). Also, exact expressions may be derived easily for the general case in which arbitrary amounts of M, of L, and of ML are mixed.

Some illustrative values of the extent of reaction to form ML upon mixing equivalent amounts of M and L for various values of K_1 are given in Table 6-5.

TABLE 6-5. Extent of Reaction
$M + L \rightleftharpoons ML$
(Initial C_M = initial $C_L = C$)

C, M	K_1	Reaction, %	pM
10^{-1}	10^5	98.60	3.15
10^{-3}	10^5	86.82	4.18
10^{-3}	10^{10}	99.96	6.65
10^{-5}	10^5	26.8	5.44
10^{-5}	10^{10}	99.55	7.65
10^{-5}	10^{15}	99.999	10.15
10^{-7}	10^{15}	99.99	11.15

Also, the minimum value of K_1 that corresponds to at least 99 per cent reaction upon mixing equimolar amounts of M and L, each solution being of concentration c, is easily calculated:

$$K_1 = \frac{2 \times 0.99c}{(0.01c)^2} = 1.98 \times \frac{10^4}{c}$$

Thus a value of K_1 of about 2×10^{11} corresponds to 99 per cent reaction even for $c = 10^{-7}\,M$. Inspection of the formation constants of typical EDTA complexes (see the Appendix) shows that quantitative formation of the complexes may be expected even in very dilute solution. How these equilibria are affected by pH is considered on pages 119–121.

Addition of more than one ligand. When more than one complex is formed the calculations are more involved but still straightforward. If n different complexes are formed, there are $n + 2$ unknown concentrations: [M], [L], [ML], . . . , [ML_n]. Therefore, $n + 2$ independent algebraic relations are required. The equilibrium constants $K_1, \ldots, K_n$ provide n of these relations. The other two, just as in the preceding example, are obtained from material balances.

$$C_M = [M] + [ML] + [ML_2] + \cdots + [ML_n] \quad (45)$$

$$C_L = [L] + [ML] + 2[ML_2] + \cdots + n[ML_n] \quad (46)$$

In these systems it is advantageous to introduce the *average ligand number*, $\bar{n}$. The meaning is exactly what the name indicates; the formula $ML_{\bar{n}}$ represents the average composition of all species containing M. The value of $\bar{n}$ usually is determinable experimentally. Thus in a system containing 0.1 mole of M, 2.0 moles of ML, 0.2 mole of ML_2, and 0.05 mole of ML_3, the total amount of M is 2.35 moles, while the total combined L is 2.55 moles. Hence $\bar{n} = 2.55/2.35 = 1.085$. As another example; 0.100 mole of M and 0.250 mole of L are mixed, and the resultant solution contains 0.025 mole of free (uncombined) L. Evidently the combined L is $0.250 - 0.025 = 0.225$; hence the average L bound per M is $0.225/0.100 = 2.25$. Mathematically $\bar{n}$ is defined by

$$\bar{n} = \frac{C_L - [L]}{C_M} \quad (47)$$

from which Eqs. (48) and (49) are obtained:

$$\bar{n} = \frac{[ML] + 2[ML_2] + \cdots + n[ML_n]}{[M] + [ML] + [ML_2] + \cdots + [ML_n]} \tag{48}$$

$$\bar{n} = \frac{\beta_1[L] + 2\beta_2[L]^2 + \cdots + n\beta_n[L]^n}{1 + \beta_1[L] + \beta_2[L]^2 + \cdots + \beta_n[L]^n} \tag{49}$$

Equation (49) shows that $\bar{n}$ depends only upon [L] and the various equilibrium constants and is independent of C_M. This is true only for mononuclear systems, to which our discussion is restricted. In polynuclear systems $\bar{n}$ depends on C_M.

Given the values of the equilibrium constants $\bar{n}$ can be calculated *exactly* as a function of [L]. A numerical example is given for the system Me^{3+}–F^-, in which six complexes may be formed. The logarithms of $K_1, K_2, \ldots, K_6$ (concentration constants) are respectively 6.2, 5.0, 3.9, 2.7, 1.6, and 0.5. A pH is assumed such that HF is completely dissociated and no $MeOH^{++}$ is present.[8] The corresponding values of log β, K, and β are:[9] log $\beta_1, \ldots,$ log β_6: 6.2, 11.2, 15.1, 17.8, 19.4, and 19.9
$K_1, \ldots, K_6$: 1.585×10^6, 1.000×10^5, 7.94×10^3, 5.01×10^2, 39.8, and 3.16
$\beta_1, \ldots, \beta_6$: 1.585×10^6. 1.585×10^{11}, 1.259×10^{15}, 6.31×10^{17}, 2.51×10^{19}, and 7.94×10^{19}
When $[F^-] = 0.1\ M$ (note that this is *free* fluoride, not total fluoride), $\bar{n}$ is found from Eq. (49) to be

$$\bar{n} = \frac{1.585 \times 10^6 \times 10^{-1} + 2 \times 1.585 \times 10^{11} \times 10^{-2} + \cdots + 6 \times 7.94 \times 10^{19} \times 10^{-6}}{1 + 1.585 \times 10^6 \times 10^{-1} + 1.585 \times 10^{11} \times 10^{-2} + \cdots + 7.94 \times 10^{19} \times 10^{-6}}$$

$$= 1.987 \times 10^{15}/3.95 \times 10^{14} = 5.03$$

The student should write out the arithmetic in detail and verify that the main term corresponds to the species MeF_5^{--} with smaller terms corresponding to MeF_6^{3-} and MeF_4^-. The concentrations of $Me^{3+}, \ldots, MeF_3$ are negligibly small compared to those of the other species.

Repeating the calculation for $[F^-] = 10^{-4}\ M$, $\bar{n}$ is found to be $7.36 \times 10^3/3.07 \times 10^3 = 2.40$, with the main contributions from MeF_2^+ and MeF_3. In this manner $\bar{n}$ is calculated for any assigned $[F^-]$ with the results shown in Fig. 6-2.

As emphasized above, the concentration of free fluoride must be used in the calculation. The total concentration includes that present in the various complexes and depends upon C_{Me}. Evidently

$$C_L = [F^-] + \bar{n}C_{Me} \tag{50}$$

Thus in three different solutions, each having $[F^-] = 0.100\ M$ but with C_{Me} equal to 10^{-3}, 10^{-2}, or $10^{-1}\ M$, the total fluoride concentration is 0.1050, 0.1503, or 0.603 M, respectively.

[8] The values of $K_1, \ldots, K_6$ are approximately those for the Al^{3+}–F^- system. For the latter system the calculations would be more complex because, at any pH, the formation of hydroxo complexes of Al^{3+} or of HF would have to be taken into account.

[9] A uniform precision of about 0.1 per cent has been kept in the values of K and β for the illustrative calculation, although in fact the values of K usually are not known with such precision.

Using Fig. 6-2 it is possible to solve easily problems that at first glance appear difficult. For example, if a solution is 0.02 *M* in Me(III), how much fluoride must be added so that the main species is MeF_3? When MeF_3 is the main species $\bar{n}$ is practically equal to 3 (see Problem 2). From Figure 6-2 we see that at $\bar{n} = 3$, $\log [F^-] = -3.2$; hence $[F^-] = 6.3 \times 10^{-4}$ *M* and total fluoride $= 6.3 \times 10^{-4} + 3 \times 0.02 = 0.0606$ *M*. Thus a very slight excess over 3 moles of fluoride per mole of Me(III) is needed. However, if it is desired to have MeF_5^{--} as the main species, $\bar{n} = 5$, and $\log [F^-] = -1.0$. Therefore, $[F^-] = 0.100$ *M* and total fluoride = 0.200 *M*, which is 10 times the total concentration of Me(III).

A harder problem is the following: What is $\bar{n}$ in a mixture containing 0.01 mole of Me(III) and 0.08 mole of total fluoride per liter? This is readily solved by trial. Evidently, $\bar{n}$ can not be 6, since this is only attained at $[F^-]$ much larger than 1 *M*.

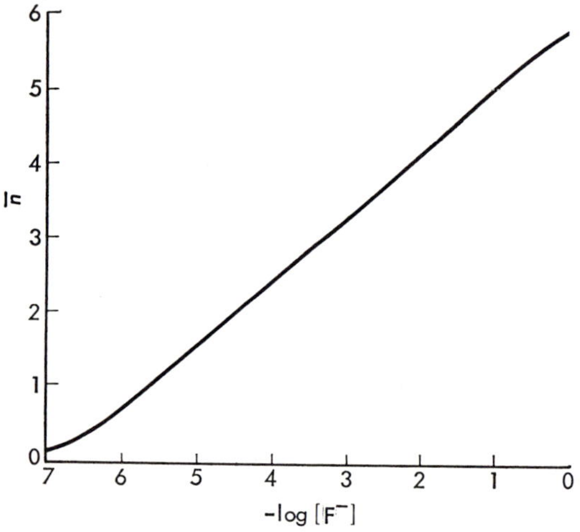

Fig. 6-2. Dependence of average ligand number $\bar{n}$ upon $[F^-]$ in the system $Me^{3+} - F^-$.

Assume various values of $\bar{n}$ smaller than 6, e.g., 5.5, 5.4, ... Then $[F^-]$ is calculated to be $0.08 - 0.01\,\bar{n} = 0.025, 0.026, \ldots$. These values are compared with the values of $[F^-]$ read from Fig. 6-2. In this way it is found that $\bar{n} = 4.62$, $[F^-] = 0.0338$ *M*.

The calculation of the fraction of total metal present in the various forms is as follows. We define f_j as the fraction of total metal present as ML_j; f_0 represents the fraction present as M.

$$f_j = \frac{\beta_j[L]^j}{1 + \beta_1[L] + \cdots + \beta_n[L]^n} \tag{51}$$

$$f_0 = \frac{1}{1 + \beta_1[L] + \cdots + \beta_n[L]^n} \tag{52}$$

Quite generally as [L] is increased from zero, [M] decreases; first [ML] reaches a maximum, then decreases as $[ML_2]$ comes to a maximum, and so on. At sufficiently high values of [L], the fraction present as the highest complexes approaches unity. This is illustrated for the Me(III)–fluoride system in Fig. 6-3.

A more detailed quantitative discussion of Figs. 6-2 and 6-3 is beyond the scope of this text, but some qualitative generalizations are in order. First, Fig. 6-2 is identical with the titration curve of a polyprotic acid with six replaceable hydrogens with a strong base (except for the H^+ or OH^- contributed by dissociation of water); pL corresponds to pH and the quantity of moles of base added per mole of acid is given by $\bar{n} + ([L]/C_M)$. Second, by straightforward but lengthy algebra it may be shown that the maximum value of f_j occurs approximately at the value $\bar{n} = j$; the accuracy of the approximation is better the more widely different the successive values of K are. The approximation is reasonably good even when the values differ only by a factor of about 10, as in the above example (see Problems 4 and 5). Next, it may be proved that $\bar{n}$ always increases with [L]; that is, the slope of the curve

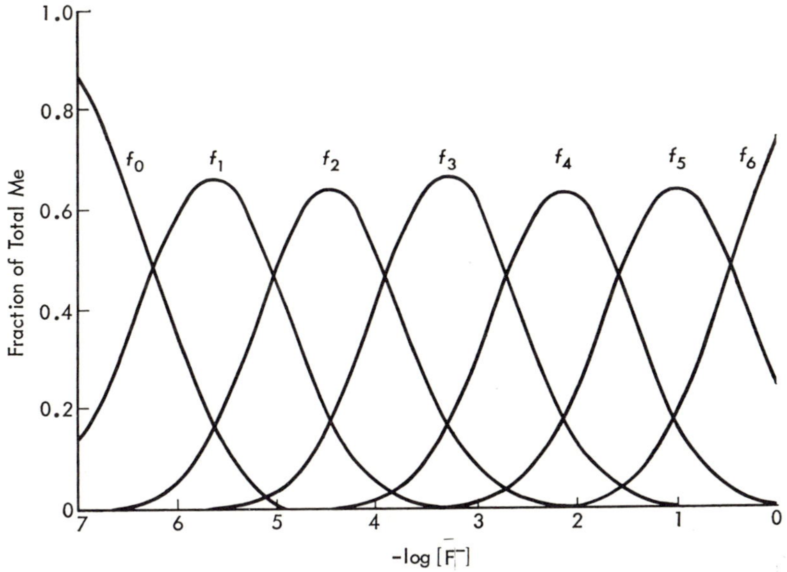

Fig. 6-3. Relative amounts of Me(III)-fluoride complexes at various values of $[F^-]$.

$\bar{n} - \log [L]$ cannot be negative. However, the exact form of the curve depends upon the relative values of the formation constants. When the successive K's do not differ much, as in the above example, $\bar{n}$ rises relatively smoothly to its limiting value. Also, the successive maxima of f_j occur at values of [L] that do not differ much, and no one complex comes close to being the only species present (Fig. 6-3), except, of course, ML_n at the highest values of [L]. Considering that the successive values of K for typical metal–ammonia systems also change by about a factor of 10, the student may judge the correctness of the statement frequently made that when ammonia is added to nickel solutions, the main species is $Ni(NH_3)_6^{++}$ with lesser amounts of $Ni(NH_3)_4^{++}$ (see also Problem 3). When the successive formation constants differ relatively more, the character of the n–log [L] curve changes, and the smooth curve of Fig. 6-2 assumes a step-like form. This is illustrated in Fig. 6-4 for the hypothetical situation: $K_1 = 10^9$, $K_2 = 10^6$; $K_3 = 10^3$, and $K_4 = 1$. The stepwise character is quite pronounced for those polyprotic acids, which have successive ionization constants differing by a factor of about 10^5 (Chap. 34). Also,

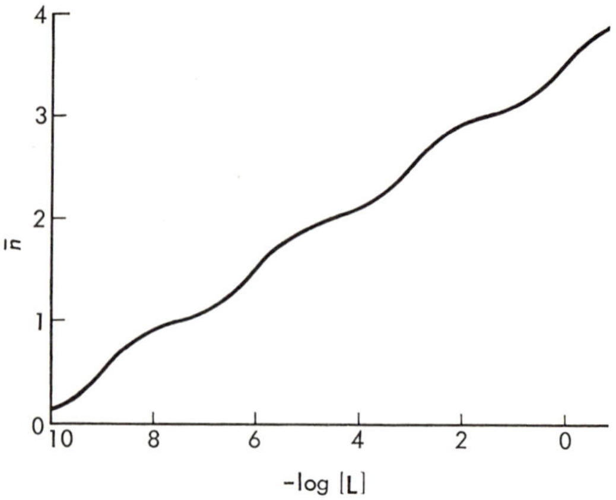

Fig. 6-4. Average ligand number $\bar{n}$ in the system $M - L$; $K_1 = 10^9$, $K_2 = 10^6$, $K_3 = 10^3$, $K_4 = 1$.

for the same values of $K_1, \ldots, K_4$, Fig. 6-5 shows that the maxima of f_j are more widely spaced than in Fig. 6-3; each in turn reaches the value of 0.941.

It is important to note that although the calculation of $\bar{n}$ as a function of [L] is exact and easy, the reverse calculation of [L] as a function of n is not simple

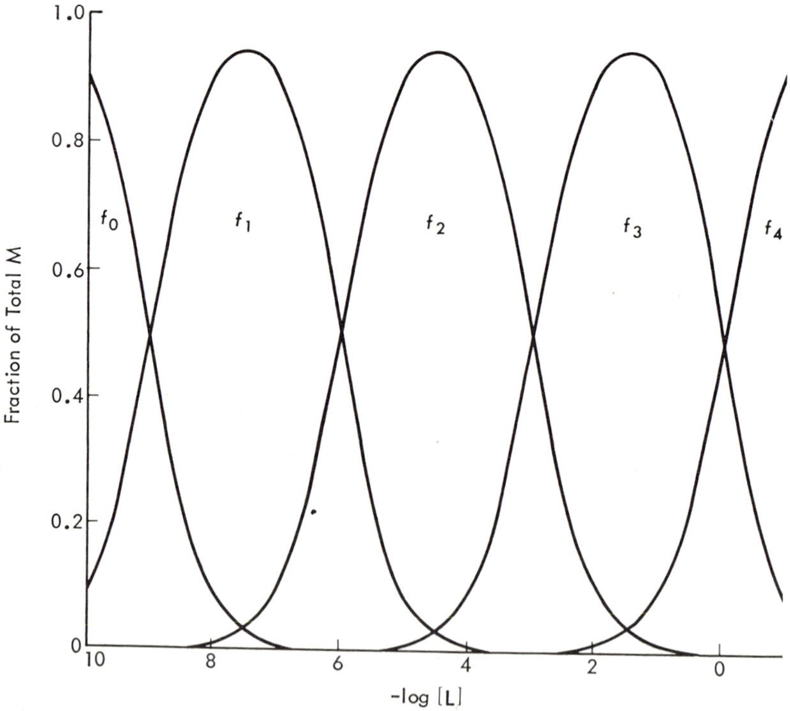

Fig. 6-5. Relative amounts of complexes in the system $M - L$; $K_1 = 10^9$, $K_2 = 10^6$, $K_3 = 10^3$, $K_4 = 1$.

when more than one complex is formed. In the M–L system when $n_{max} = 6$, if it were required to calculate [L] for a specified $\bar{n}$, it would be necessary in principle to solve a sixth-degree equation [Eq. (49)]. Naturally such equations always are solved by successive approximations. Most students learn by experience which terms in the numerator and denominator of Eq. (49) are significant and which may be neglected, but most students also do not realize that the reverse calculation is easy and requires no a priori simplification or approximation. Precisely the same remarks apply to the calculation of titration curves of polyprotic acids.[10]

Calculations in pH-dependent equilibria

In all practical cases complexation equilibria are dependent upon pH, either because the metal ion reacts with hydroxyl ion or because the ligand L is derived from a weak acid or weak base, or both. If the appropriate equilibrium constants are known, it is a simple matter to take the effects into account. Two examples will illustrate.

First, consider the reaction of aluminum ion with fluoride ion. Six fluoride complexes may be formed. At low pH the reaction of aluminum ion with hydroxyl ion may be neglected but the formation of hydrofluoric acid must be considered. The thermodynamic dissociation constant of hydrofluoric acid is 6.8×10^{-4}, $pK_a^0 = 3.17$; in this example an ionic strength of 0.53 will be assumed, at which

$$K_a = \frac{[H^+][F^-]}{[HF]} = 1.15 \times 10^{-3} \qquad pK_a = 2.94 \tag{53}$$

At low pH, therefore, Eq. (46) must be modified:

$$C_F = [F^-] + [HF] + [AlF^{++}] + 2[AlF_2^+] + \cdots + 6[AlF_6^{3-}] \tag{54}$$

Making use of Eq. (53) this becomes

$$C_F = [F^-]\left\{1 + \frac{[H^+]}{K_a}\right\} + [AlF^{++}] + 2[AlF_2^-] + \cdots + 6[AlF_5^{3-}] \tag{55}$$

Thus allowing for the difference between $[F^-]$ and $\{[F^-] + [HF]\}$, the calculations are identical with those for the Me(III)–fluoride system. At the given ionic strength and 20°C, $\log K_1 = 6.13$, $\log K_2 = 5.02$, $\log K_3 = 3.85$, $\log K_4 = 2.74$, $\log K_5 = 1.63$, and $\log K_6 = 0.47$. At a pH high enough so that virtually no HF is present, and $[F^-] = 0.01\ M$, $\bar{n}$ is calculated to be 4.06.[11] If the value of C_{Al} is $10^{-3}\ M$, $C_F = 0.01 + 0.00406 = 0.0141\ M$. But at pH 2.94, the concentrations of F^- and HF are equal. Hence to have $\bar{n} = 4.06$, $C_F = 0.01 + 0.01 + 0.00406 = 0.0241\ M$. Similarly, at pH = 1.94, $C_F = 0.01 + 0.1 + 0.00406 = 0.114\ M$. In each of the three solutions, $[F^-] = 0.01\ M$ and $\bar{n} = 4.06$.

The calculations for reactions involving EDTA are somewhat lengthier because there are four replaceable hydrogens. There is an added complication in practice

[10] Of course, the calculation of equilibrium constants from experimental data is not so simple. Available methods are described by F. J. C. Rossotti and H. Rossotti, *The Determination of Stability Constants*, McGraw-Hill, New York, 1961.

[11] Of course, at such a pH the aluminum not complexed with fluoride would be present as hydroxocomplexes. However, since $[Al^{3+}]$ would be negligibly small at $[F^-] = 0.01\ M$, practically no error is introduced into the calculation of $\bar{n}$.

in that more than one kind of complex—e.g., $FeOHY^{4-}$, $FeHY^{2-}$—may be formed; consideration of these is beyond the scope of this text. In practical applications, moreover, they are usually of minor importance. Therefore, this possibility will not be considered and only the effect arising from the acid properties of EDTA will be discussed.

The acid dissociation of EDTA is represented by the equations:

$$H_4Y \rightleftharpoons H^+ + H_3Y^-$$
$$H_3Y^- \rightleftharpoons H^+ + H_2Y^{--}$$
$$H_2Y^{--} \rightleftharpoons H^+ + HY^{3-}$$
$$HY^{3-} \rightleftharpoons H^+ + Y^{4-}$$

At 20°C and an ionic strength of 0.1, the respective dissociation constants are given by $pK_{a1} = 1.99$, $pK_{a2} = 2.67$, $pK_{a3} = 6.16$, and $pK_{a4} = 10.22$. Now consider the reaction of calcium ion with EDTA. Assuming that only one complex is formed;

$$C_{Ca} = [Ca^{++}] + [CaY^{--}] \tag{56}$$

$$C_L = [Y^{4-}] + [HY^{3-}] + [H_2Y^{--}] + [H_3Y^-] + [H_4Y] + [CaY_2^{--}] \tag{57}$$

Equation (56) thus is identical with Eq. (41), and Eq. (57) is a generalization of Eq. (42). Making use of the dissociation constants of EDTA, Eq. (57) becomes

$$C_L = [Y^{4-}]\left\{1 + \frac{[H^+]}{K_{a4}} + \frac{[H^+]^2}{K_{a4}K_{a3}} + \frac{[H^+]^3}{K_{a4}K_{a3}K_{a2}} + \frac{[H^+]^4}{K_{a4}K_{a3}K_{a2}K_{a1}}\right\} + [CaY^{--}] \tag{58}$$

Inserting the numerical values, there is obtained

$$C_L = [Y^{4-}]\{1 + 10^{10.22}[H^+] + 10^{16.38}[H^+]^2 + 10^{19.05}[H^+]^3 + 10^{21.04}[H^+]^4\} + [CaY^{--}] \tag{59}$$

The various terms in the braces are of more or less importance depending on the pH.

Thus at pH 11.22, the terms in the braces are $\{1 + 10^{-1.00} + 10^{-6.06} + \cdots\} = 1.10$; therefore 1/1.1, or 91 per cent of the EDTA is present as Y^{4-} and 9 per cent is present as HY^{3-}. At pH 7.00, the numerical values are $\{1 + 10^{3.22} + 10^{2.38} + 10^{-1.95} + \cdots\} = 1901$; at pH 7.00, only 1 part in 1901, or 0.053 per cent of the EDTA is present as Y^{4-}. Similarly, it is found that at pH 3.00, only 2.76×10^{-9} per cent of the total is present as Y^{4-}.

Evidently Eq. (58) can be written in the concise form

$$C_L = \alpha_L[Y^{4-}] + [CaY^{2-}] \tag{60}$$

in which α_L depends only upon $[H^+]$ and the acid dissociation constants of EDTA. Combining Eqs. (60), (56), and (25), there is obtained

$$[CaY^{--}] = \frac{K_1}{\alpha_L}\{C_{Ca} - [CaY^{--}]\}\{C_L - [CaY^{--}]\} \tag{61}$$

which is a generalization of Eq. (43). Equation (61) shows clearly how the pH dependence enters through the term α, which essentially acts to decrease the effective value of the formation constant. At pH much above 11, $\alpha = 1$ and Eq. (61) reduces exactly to Eq. (43). If Eq. (61) is written in the form

$$[CaY^{--}] = K_1'\{C_{Ca} - [CaY^{--}]\}\{C_L - [CaY^{--}]\} \qquad (62)$$

K_1' is a "constant" at a given pH. The values of K_1' at pH 11.22, 7.00, and 3.00 are 4.5×10^{10}, 2.6×10^7, and 1.5, respectively. From these values it may be calculated that if to a solution 10^{-4} *M* in calcium there is added a 100-fold excess of EDTA, the formation of CaY^{--} is quantitative at high pH and pH 7, but occurs only to the extent of 1.5 per cent at pH 3.00.[12]

Equilibria in systems containing two ligands

When two ligands are present there is competition for the metal ion, and which complex is formed depends upon the ratio of the formation constants and the ratio of the ligand concentrations.[13] A numerical example is given for the system nickel–cyanide–EDTA. At the concentrations of cyanide usually met with (0.1 *M* or more) the main reaction with nickel ion is

$$Ni^{++} + 4\,CN^- \rightleftharpoons Ni(CN)_4^{--}$$

the value of β_4 being $10^{31.3}$ at an ionic strength of 0.1. At the same ionic strength the equilibrium constant of the reaction

$$Ni^{++} + Y^{4-} = NiY^{--}$$

is $10^{18.6}$. Hence

$$\frac{[Ni(CN^-)_4^{--}]}{[NiY^{--}]} = 10^{12.7}\frac{[CN^-]^4}{[Y^{4-}]}$$

Evidently in the presence of (for example) 0.1 *M* CN^-, all the nickel is present as the cyanide complex.[14]

This is a simple example of masking of nickel by cyanide. It permits the titration, for example, of magnesium with EDTA without interference by nickel in the presence of cyanide. In the absence of cyanide the sum of magnesium and nickel is obtained. Other metals that may be masked similarly by cyanide include copper, cobalt, and zinc.

Graphical methods

Graphical methods based on the "proton condition" for solving acid–base equilibria have been presented in Chapter 4. Similar methods may be used for complexation calculations, and these are discussed briefly using the calcium–EDTA reaction for illustration. A pH of about 11 is assumed, so that all the EDTA is present as Y^{4-}. The student may easily adapt the method about to be described to a lower pH by replacing K_1 by K_1/α [see Eq. (61)]. A double logarithmic graph

[12] Many numerical values of α for complexing agents other than EDTA are given by A. E. Ringbom, *Complexation in Analytical Chemistry*, Wiley-Interscience, New York, 1963; also see A. E. Ringbom in I. M. Kolthoff and P. J. Elving (eds.), *Treatise on Analytical Chemistry*, part 1, vol. 1, Wiley-Interscience, New York, 1959, chap. 14.

[13] In some cases there may be a difference in the rates of reaction. Only equilibrium situations are considered here.

[14] In practice the solution may be buffered with ammonia. The student may show that the formation of zinc ammonia complexes is neglible in the presence of cyanide.

is constructed in which the concentrations of various substances present are plotted on the ordinate scale against the concentration of the substance chosen as the master variable. In this instance Y^{4-} is the proper choice, and accordingly $[Y^{4-}]$ is plotted on the abscissa (logarithmic scale). Evidently when $[Y^{4-}]$ is plotted on the ordinate (also on the logarithmic scale) the relation is exactly linear with 45° slope. The calcium is present as Ca^{++} and as CaY^{--}. When $[Y^{4-}] = 1/K_1$, $[Ca^{++}] = [CaY^{--}] = \frac{1}{2}$ [total calcium]. When $[Y^{4-}]$ is much larger than $1/K_1$, practically all the calcium is present as $[CaY^{--}]$. Accordingly,

$$[Ca^{++}] = \frac{[CaY^{--}]}{K_1[Y^{4-}]}$$

or

$$\log [Ca^{++}] = \log [\text{total calcium}] - \log K_1 - \log [Y^{4-}] \tag{63}$$

Thus in this region of $[Y^{4-}]$ a logarithmic plot of $[Ca^{++}]$ versus $[Y^{4-}]$ yields a straight line of unit slope, which passes exactly through the point at which $[Ca^{++}]$ = [total calcium] and $[Y^{4-}] = 1/K_1$. Exactly the same kind of considerations define $[CaY^{--}]$ and $[Ca^{++}]$ for $[Y^{4-}]$ much smaller than $1/K_1$. The resulting relations are in Fig. 6-6. This figure also includes C_Y. Various problems

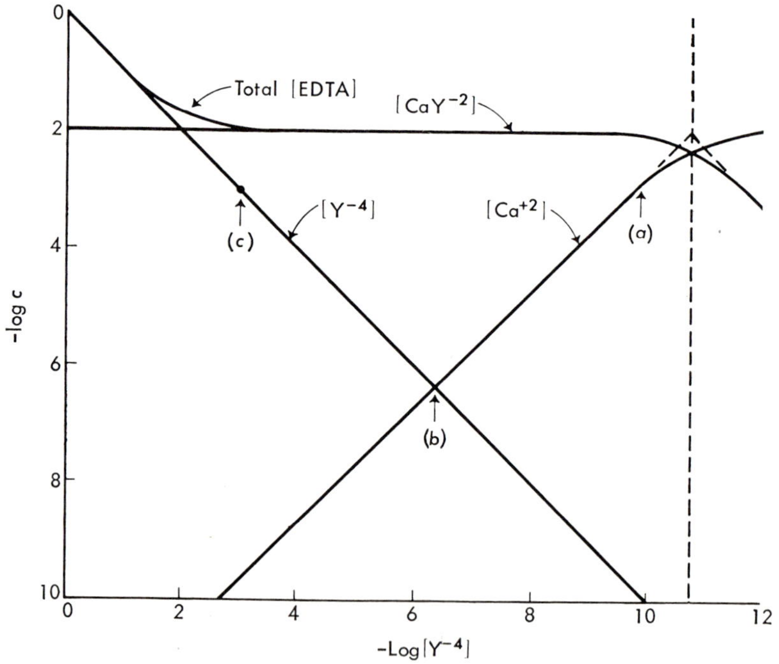

Fig. 6-6. Logarithmic diagram for the system Ca(II) – EDTA at pH around 11; $K_1 = 10^{10.7}$.

can be solved by inspection of the figure. For example, solutions made by adding 90, 100, and 110 per cent of the equivalent amount of EDTA (no change in volume) to a solution 0.01 M in calcium correspond to points (a), (b), and (c), respectively, and all the concentrations can be read from the graph. Note that at point (b), the graphical construction shows clearly that $[Ca^{++}] = \sqrt{C_{Ca}/K_1}$, which was derived in a different way on page 113. Naturally, the parts of the graph of interest in a given problem may be plotted on an enlarged scale to improve the accuracy, if desired.

Graphs of this sort provide a very convenient way to estimate the change in pM around the equivalence point in a titration. The present graph has been presented merely to explain the principle, because with such a simple system there is nothing gained by the use of the graphical method. If the concentration of calcium is much less than 10^{-2} *M*, however, or if the pH is much less than 11, thus lowering the effective K_1, the utility of the method becomes obvious (see Problem 8).

Conditional equilibria constants

The discussion of the calcium–EDTA system has illustrated the practical value of a "constant" which applies not only in a given medium, such as a concentration constant, but at a given pH. On page 121 the value of K_1/α was described as an "effective constant." This is a special case of a more general "*conditional constant.*" Such constants are becoming increasingly more widely used. The meaning is as follows. Suppose it is wished to determine the extent of the reaction $M + L \rightleftharpoons ML$. Since L in general is derived from a weak acid, the pH dependence of L must be considered. However, M may form various hydroxy complexes; if the reaction is carried out in ammoniacal solution, the formation of ammonia complexes must be considered, and this in turn is pH-dependent. Moreover, most metals form complexes of greater or less stability with acid anions that are ordinarily not considered as strong complexing agents, for example, chloride and sulfate. Suppose to take a specific and by no means unusually complicated example, it were desired to calculate the extent of reaction between ferric ions and EDTA in an ammonia–ammonium chloride buffer of given pH. A calculation based only on the thermodynamic formation constant $(FeY^-)/(Fe^{3+})(Y^{4-})$ would be entirely meaningless. In the specified medium, iron is present to a negligibly small extent as Fe^{3+} but is present to a greater or lesser extent as $FeOH^{++}$, $Fe(OH)_2^+$, $FeCl^{++}$ and higher chloro complexes, depending on the concentration of chloride in the buffer, polynuclear iron species if the iron concentration is much above 0.001 *M*, and several complexes with EDTA. The forms in which EDTA may be present have been dealt with already. It is evident that an exact calculation of the composition of the solution might be a formidable task, even if all the necessary concentration constants were available.

Fortunately such a detailed description of the system is rarely needed or desired. In particular analytical problems, the question usually is: To what extent does a given overall reaction occur? Or, more specifically, if reagent L is added to M, what fraction of M does *not* react with L, in some form or other, and thus escapes detection, for example by precipitation, titration, or extraction?

From this point of view the *conditional constant* for the reaction $M + L \rightleftharpoons ML$ is simply defined as

$$K_1' = \frac{[ML]'}{[M]'[L]'}$$

where [M]′ represents the total concentration of all metal in solution *not* combined with L. In the above example, [M]′ includes (the very small] concentration of

Fe^{3+}, the hydroxy complexes, and chloride complexes, both mononuclear and polynuclear. In other situations (e.g., if M were nickel ion) it would include the various ammonia complexes. Similarly, [L]′ represents the concentration of all L in solution *not* with M, and [ML]′ represents the total concentration of all species containing M and L, including those with hydrogen or hydroxyl ion. Thus we may write, in analogy to Eq. (60),

$$[L]' = \alpha_L[L]$$
$$[M]' = \alpha_M[M] \qquad (63)$$

Given such a conditional constant, all the calculations are carried in the same way, and just as simply, as the first example discussed in this chapter; all the complications have been absorbed into the value of K_1'.

A discussion at greater length lies beyond the scope of this text.[15] For convenience a few important pH-dependent conditional constants are collected in Table A-7.

PROBLEMS

(Consult the Appendix for any required equilibrium constants.)

1. Prove that under suitable conditions the exact equation, Eq. (44), for [ML] reduces to a simple form. Derive exact expressions for [M] and [L] and show how they similarly reduce to simple form.
2. Calculate $\bar{n}$ in the M–L system, given $K_1 = 10^6$, $K_2 = 10^3$, $K_3 = 1$, for $[L] = 10^{-5}$, 10^{-4}, 10^{-3}, 10^{-2}, 10^{-1}, and 1 *M*. At what [L] is the maximum concentration of HL obtained? Of ML_2? Note the connection with the values of *K*.
3. From the formation constants of the Cu(II)–ammonia system (see the Appendix) construct a $\bar{n}$–log[NH_3] graph. Under what conditions is it correct to speak of the main species as $Cu(NH_3)_4^{++}$?
4. In a system in which the only reactions are

$$M + L \rightleftharpoons ML \qquad ML + L \rightleftharpoons ML_2$$

prove that the maximum concentration of ML occurs *exactly* at the value $[L] = (K_1K_2)^{-1/2}$.
5. In the more complicated system M, ML, ML_2, ML_3, . . ., find the conditions under which the maximum concentration of ML occurs *approximately* at $[L] = (K_1K_2)^{-1/2}$.
6. The formation constants for the system Cd^{++}–Cl^- are approximately $K_1 = 20$, $K_2 = 8$, $K_3 = 1.2$, and $K_4 = 0.35$. Calculate the relative amounts of Cd^{++}. . ., $CdCl_4^{--}$ in solutions in which $[Cl^-]$ is 0.01 and 0.1 *M*. (Assume that the pH is low enough so that hydroxy complexes are neglible.) Is $CdCl_2$ a strong electrolyte?
7. Bonus Problem. Calculate the conditional constant of the nickel–EDTA complex in a solution 0.1 *M* in ammonia and 0.1 *M* in ammonium chloride. Ignore chloro complexes. *Ans.* $\log K' = 13.4$.
8. Construct a double logarithmic graph for the system Ca^{++}–EDTA at pH 9.0, $C_{Ca} = 10^{-3}$ *M*. What is $[Ca^{++}]$ at the equivalence point in a titration (at constant volume) of 10^{-3} *M* Ca^{++} with EDTA?

[15] For a thorough discussion, see Ringbom's book, Ref. 12.

CHAPTER 7 *Solubility of Precipitates*

This chapter deals mostly with the solubility of substances in aqueous solutions. Water is a strongly associated liquid whose molecules are hydrogen-bonded to each other. A substance cannot dissolve in water unless it is capable of disrupting the water structure, a process that involves bonding of water molecules to a solute or its component ions. If the solute is a crystalline solid, its tendency to enter the solution and become solvated is opposed by the forces binding the ions, atoms, or molecules in the lattice. The solubility of a solid is determined by the resultant of these opposing effects. The aqueous solubility of solids varies over an enormous range from kilograms down to less than 1 μg in 1 liter of water. These great differences are exploited in chemical analysis for separations by precipitation, in gravimetric determinations, and to some extent in volumetric and other branches of analysis.

Useful as it is, water is not the only solvent used in chemical analysis. The solubility of some inorganic substances and most metal chelate complexes in organic solvents not miscible with water allows separation of substances by liquid–liquid extraction (Chapter 14). Nonaqueous solutions are used in titrimetry. Fused salts or other fused substances are valuable solvents at times in analytical processes.

Solubility and solubility product

Consider an aqueous solution of a slightly soluble substance of the type BA. In a saturated solution the following equilibrium exists:

$$\underset{\text{solid}}{\mathrm{BA}} \rightleftharpoons \underset{\text{soln.}}{\mathrm{BA}} \rightleftharpoons \mathrm{B}^+ + \mathrm{A}^-$$

If no other species of A and B than those represented are present in solution, the molar solubility of BA is given by

$$S = [\mathrm{BA}] + [\mathrm{B}^+] = [\mathrm{BA}] + [\mathrm{A}^-] \tag{1}$$

[BA] is the molecular or intrinsic solubility of BA, which may be negligibly small compared to $[\mathrm{B}^+]$ and $[\mathrm{A}^-]$ for compounds that are highly ionized. It cannot always be ignored, however. Metal chelate compounds such as nickel dimethylglyoximate and ferric 8-hydroxyquinolate have intrinsic solubilities usually lying in the approximate range 10^{-6}–10^{-9} *M*. Even salts such as silver chloride and calcium sulfate have determinable molecular solubilities, the undissociated species of the latter being interpreted as an ion pair ($\mathrm{Ca}^{++}\mathrm{SO}_4^{--}$).

S is, of course, a function of the temperature. Its value depends upon the crystalline state of BA (for example, the solubility of calcite, rhombohedral

$CaCO_3$, is not the same as that of aragonite, orthorhombic $CaCO_3$) and to some extent on the particle size (see p. 210).

From the law of mass action, setting concentrations equal to activities,

$$\frac{[B^+][A^-]}{[BA]} = K \tag{2}$$

where K is the ionization constant of BA. In a saturated solution at a specified temperature

$$[B^+][A^-] = K[BA]_{sat} = \text{constant} = K_{sp} \tag{3}$$

This constant is known as the solubility-product constant of BA. For a precipitate B_mA_n,

$$K_{sp} = [B^{b+}]^m[A^{a-}]^n \tag{4}$$

If the intrinsic solubility of the precipitate is so small compared to the concentration of the dissociated species that it can be neglected, and no other forms of B and A (such as complex species BA_2^-, etc.) are present in the saturated solution, the molar solubility of BA is given by

$$S = [B^+] = [A^-] = K_{sp}^{1/2} \tag{5}$$

and that of B_mA_n by

$$S = \frac{[B^{b+}]}{m} = \frac{[A^{a-}]}{n} = \left(\frac{K_{sp}}{m^m n^n}\right)^{1/m+n} \tag{6}$$

Activity effects are ignored for the time being. It should be realized that there is no *necessary* relation between the solubility of a compound and its solubility-product constant formulated in terms of its simple component species. Thus the solubility product of mercuric chloride,

$$[Hg^{++}][Cl^-]^2$$

at room temperature has the value 2×10^{-14}, whereas the solubility of mercuric chloride in water is $\sim 0.25\ M$. Most of the mercury is present as un-ionized $HgCl_2$.

Common-ion effect. The molar solubility of a precipitate B_mA_n is represented by $[B^{b+}]/m$ when an excess of the precipitating ion A^{a-} is present (if there are no complicating effects) and conversely. Increasing the concentration of one of the common ions decreases the solubility of a precipitate in accordance with Eq. (4).[1] The solubility in the presence of a specified excess of A^{a-} or B^{b+} can be calculated if K_{sp} is known and if the concentrations of B_mA_n and other species of A and B formed by possible reaction with each other, with water, or with foreign species in the solution are negligibly small. Thus if A^{a-} is added in excess to a solution of B^{b+}, the (ionic) solubility of B_mA_n is given by

$$S = \frac{[B^{b+}]}{m} = \left\{\frac{K_{sp}}{[A^{a-}]^n}\right\}^{1/m} \Big/ m \tag{7}$$

The repression of the solubility of a sparingly soluble ionic substance by addition of an excess of one of its ions is important in gravimetric determinations and separations by precipitation and in some titrations. The effect of an excess of iodate ion on the solubility of $AgIO_3$ and $Pb(IO_3)_2$ is illustrated in the logarithmic

[1] The principle was first formulated (and demonstrated with silver acetate as the sparingly soluble salt) by W. Nernst, *Z. Physik. Chem.* (*Leipzig*) **4**, 372 (1889).

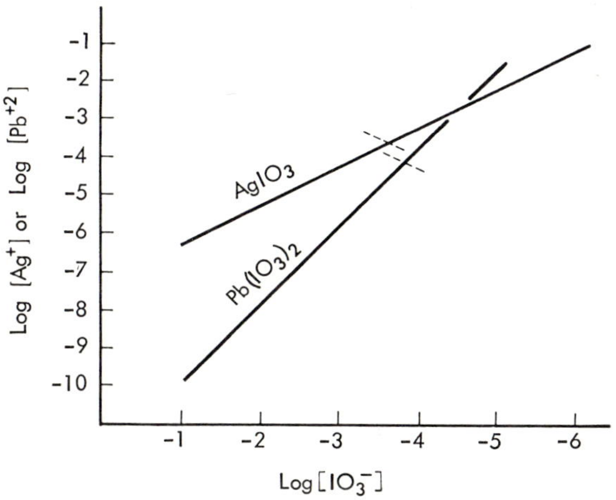

Fig. 7-1. Concentrations of Ag^+ and IO_3^- in saturated $AgIO_3$ solutions, and of Pb^{++} and IO_3^- in saturated $Pb(IO_3)_2$ solutions at 25° and an ionic strength of 0.1 *M*. $(K_{sp})_{AgIO_3} = 10^{-7.3}$, $(K_{sp})_{Pb(IO_3)_2} = 10^{-11.9}$.

concentration diagram of Fig. 7-1. The solid lines give the concentration of Ag^+ and Pb^{++} in a saturated solution of the iodates as a function of the IO_3^- concentration (or vice versa). The molar solubility of each salt in pure water is given by the ordinate of the intersection of these lines with the dashed segments representing $\log [IO_3^-] = \log [Ag^+]$ and $\log [IO_3^-] = 0.3 + \log [Pb^{++}]$. The repression of the solubility of $Pb(IO_3)_2$ by iodate is much greater than that of $AgIO_3$. In pure water $Pb(IO_3)_2$ is $\frac{1}{3}$ as soluble as $AgIO_3$; in 0.01 *M* IO_3^-, it is $\frac{1}{400}$ as soluble. Note also that with a sufficient identical excess of the respective metal ions, the solubility of $AgIO_3$ becomes less than that of $Pb(IO_3)_2$. Except when the pre-equilibrium concentration of iodate is comparable to the concentration in a saturated water solution, the solubility of the iodate in a solution of specified iodate (or metal) concentration may be found from the diagram (see Example 2).[2]

A precipitate will not necessarily separate from a solution in which the product of the ionic concentrations exceeds the solubility-product constant. The solution may remain supersaturated with the sparingly soluble substance for a long time. The tendency to form supersaturated solutions varies greatly from one substance to another, and depends also on various factors such as the temperature and the presence of nucleating materials. It is not uncommon to find that precipitation does not occur spontaneously until the ion product exceeds K_{sp} a hundred or a thousand times. Even after precipitation begins, some time may be needed for

[2] The formation of iodate–metal complexes has been ignored above. From available constants, it may be calculated that in a 0.1 *M* IO_3^- solution saturated with $AgIO_3$, less than 30 per cent of the total silver is present as un-ionized $AgIO_3$; in a saturated water solution the concentration of un-ionized $AgIO_3$ is negligible. The concentration of $Ag(IO_3)_2^-$ is less than that of $AgIO_3$. Figure 7-1 is correct as it stands, whether or not metal complexes are formed, but the solubility in the presence of excess iodate is a little greater than $[Ag^+]$ as read from the ordinate of a point on the $AgIO_3$ line. The extent of formation of lead–iodate complexes does not seem to be known. HIO_3 is quite a strong acid, so that the hydrolysis of IO_3^- can be neglected.

attainment of equilibrium. It should also be recognized that the precipitate formed at first may be an unstable (metastable) modification or hydrate whose solubility is not the same as that of the stable modification or hydrate. The particle size of the precipitate as well as its purity are also of importance. In short, calculations based on the solubility-product principle assume a solution in equilibrium with a pure solid phase of known identity and stable particle size. (See also Chapter 14.)

Example 1: The solubility of $BaSO_4$ in water at 25°C is 2.46 mg/liter or 1.06×10^{-5} *M*. Neglecting activity coefficients and taking the molecular solubility of $BaSO_4$ as negligible, find the weight of SO_4^{--} in 250 ml of saturated (neutral) barium sulfate solution which is 0.010 *M* in Ba^{++}.

Ans. $K_{sp} = (1.06 \times 10^{-5})^2 = 1.1 \times 10^{-10}$

$$S = [SO_4^{--}] = \frac{1.1 \times 10^{-10}}{[Ba^{++}]} = \frac{1.1 \times 10^{-10}}{1.0 \times 10^{-2}} = 1.1 \times 10^{-8}$$

$$\text{Weight of } SO_4^{--} = \frac{250}{1000} \times 1.1 \times 10^{-8} \times 96 \times 1000 = 0.0003 \text{ mg}$$

Example 2: Solid silver chloride is shaken with 1.00×10^{-4} *M* silver nitrate solution until equilibrium is attained. What is the concentration of chloride in the solution? K_{sp} of AgCl $= 1.00 \times 10^{-10}$.

Ans. $[Ag^+][Cl^-] = 1.00 \times 10^{-10}$

$$\text{Approximately, } [Cl^-] = \frac{1.00 \times 10^{-10}}{1.00 \times 10^{-4}} = 1 \times 10^{-6}$$

More exactly, since some Ag^+ is brought into the solution by dissolved AgCl,

$$[Cl^-] = \frac{1.00 \times 10^{-10}}{1.01 \times 10^{-4}} = 9.9 \times 10^{-7}$$

The same result is obtained by solving the quadratic

$$(1.00 \times 10^{-4} + [Cl^-])[Cl^-] = 1.00 \times 10^{-10}$$

Example 3: What is the molar solubility of Ag_2CrO_4 in a solution which is 0.010 *M* in CrO_4^{--}? $K_{sp} = 1.0 \times 10^{-12}$.

Ans. $[Ag^+] = \left(\frac{1 \times 10^{-12}}{1 \times 10^{-2}}\right)^{1/2} = 1 \times 10^{-5}, \qquad S = 5 \times 10^{-6}$

Example 4: Find the molar solubility of nickel dimethylglyoximate, $Ni(HDx)_2$, in a solution having a dimethylglyoximate ion concentration of 1×10^{-8}.

$$K_{sp} = [Ni^{++}][HDx^-]^2 = 4 \times 10^{-24}$$

$$[Ni(HDx)_2]_{satd} = 1.0 \times 10^{-6} \text{ (intrinsic solubility)}$$

Ans. $S = [Ni(HDx)_2]_{satd} + [Ni^{++}] = 1.0 \times 10^{-6} + \frac{4 \times 10^{-24}}{1 \times 10^{-16}}$

$$= 1.04 \times 10^{-6} \quad \text{or} \quad 1 \times 10^{-6}$$

In other words, the ionic solubility is so much smaller than the intrinsic solubility that it can be neglected.

Effect of indifferent electrolytes on solubility. The accurate formulation of all mass-action relationships requires the use of activities in place of concentrations, so that the expression for the solubility product of BA is

$$(B^+)(A^-) = K^0_{sp} \quad \text{(at a specified temperature in a specified solvent)} \tag{8}$$

where parentheses denote activities and K^0_{sp} is the activity or thermodynamic solubility-product constant, i.e., the constant at zero ionic strength. In precipitation processes the analytical chemist is interested in concentrations, not activities, so he writes Eq. (8) in terms of concentrations and activity coefficients:

$$f_B[B^+]f_A[A^-] = K^0_{sp} \tag{9}$$

from which

$$[B^+][A^-] = \frac{K^0_{sp}}{f_B f_A} = K_{sp} \tag{10}$$

K_{sp} is the concentration solubility-product constant, whose value varies with the ionic strength of the solution.

Evidently an increase in the concentration of an indifferent electrolyte (one not reacting chemically with the precipitate nor having an ion in common with it) will increase the solubility product and therewith the ionic solubility of the precipitate by decreasing the value of the activity coefficients of the component ions.[3] The solubility of a precipitate in a solution containing an indifferent electrolyte at a specified concentration can be calculated if the activity coefficients are known or can be estimated (p. 43) and K^0_{sp} is known. If values of K_{sp}, based on concentrations, are available, they can be used for calculations at the same, or almost the same, ionic strength without introducing activity coefficients.

The increase in solubility produced by an indifferent electrolyte at a given ionic strength depends upon the charges of the ions composing the precipitate. The relative effect is greater for a didivalent precipitate such as barium sulfate or calcium oxalate than for a diunivalent precipitate such as silver chloride (compare Tables 7-1, 7-2, and 7-3 and Fig. 7-2). At $\mu = 0.1$, K_{sp} for barium sulfate and other

TABLE 7-1. **Solubility of Silver Chloride[a] and Barium Sulfate[b] in Dilute Potassium Nitrate Solutions at 25°C**

(S_0 = molar solubility in water; S = molar solubility in electrolyte solution)

Concentration of KNO_3, M	AgCl, $S \times 10^5$	S/S_0	*Concentration of KNO_3, M*	$BaSO_4$, $S \times 10^5$	S/S_0
0	1.278 (S_0)	1.00	0	0.96 (S_0)	1.00
0.001	1.325	1.04	0.001	1.16	1.21
0.005	1.385	1.08	0.005	1.42	1.48
0.01	1.427	1.12	0.01	1.63	1.70
			0.036	2.35	2.45

[a] S. Popoff and E. W. Neuman, *J. Phys. Chem.* **34,** 1853 (1930).
[b] E. W. Neuman, *J. Am. Chem. Soc.* **55,** 879 (1933).

[3] At least this is true up to an electrolyte concentration of 1 or 2 *M* (or higher), above which f's for ions may have a value exceeding unity.

TABLE 7-2. Solubility of Calcium Oxalate Monohydrate in Sodium Chloride Solutions at 37°C[a]

Concentration of NaCl, M	0	0.005	0.01	0.04	0.1	0.4	0.6
Solubility of $CaC_2O_4 \cdot H_2O$, $M \times 10^3$	0.055	0.082	0.091	0.12	0.16	0.26	0.29

[a] G. Hammarsten, *Compt. Rend. Lab. Carlsberg* **17**, No. 11 (1929).

TABLE 7-3. Solubility of Lead Sulfate in Sodium Sulfate Solutions[a]

Na_2SO_4, M	0	0.001	0.01	0.02	0.04	0.100	0.200	0.500
Solubility of $PbSO_4$, moles $\times 10^5$/liter	15.2	2.4	1.6	1.4	1.3	1.6	2.3	2.3

[a] I. M. Kolthoff, R. W. Perlich, and D. Weiblen, *J. Phys. Chem.* **46**, 561 (1942).

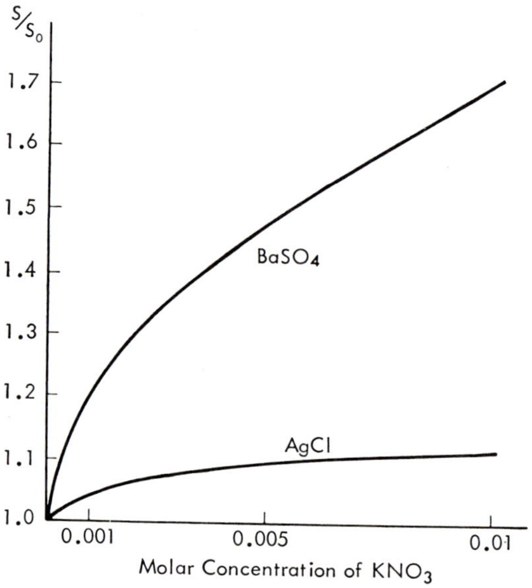

Fig. 7-2. Relative increase in solubility of silver chloride and barium sulfate in potassium nitrate solution.

didivalent precipitates is 7–8 times K^0_{sp}. This difference may be too large to ignore in analytical calculations for practical purposes. At the same ionic strength, $\mu = 0.1$, the K_{sp} of silver chloride and other diunivalent substances is roughly double K^0_{sp}, a difference small enough to neglect for most purposes. Above $\mu = 0.1$, the activity coefficients for a given ionic strength depend upon the particular ions, but the decrease with increasing electrolyte concentration is relatively small. For a singly charged ion we shall not be far wrong if we take $f = 0.7$ at $\mu = 0.5$ (compared to $f = 0.75$–0.8 at $\mu = 0.1$) as an average value. For an ion of charge 2, f may be taken as approximately 0.25 at $\mu = 0.5$ (compared to ~ 0.4 at $\mu = 0.1$). The approximation becomes quite rough for an ion of charge 3, but $f \sim 0.1$ at $\mu = 0.5$

for a 3+ cation will perhaps not be grossly in error ($f \sim 0.2$ for trivalent cations and ~ 0.1 for trivalent anions at $\mu = 0.1$).

Tables of K^0_{sp} values are not directly useful to the analyst because he usually requires a knowledge of K_{sp} at $\mu = 0.1$–0.5, but they have the virtue of furnishing approximate K_{sp} values with the aid of estimated activity coefficients.[4]

The reader will realize that the need to apply activity coefficients in analytical calculations varies from one practical problem to another. A change in a calculated concentration by a factor of $\frac{1}{2}$ or 2 is usually inconsequential. Even a tenfold change may be unimportant. Whether a solubility is 0.01 or 0.1 mg in 1 liter rarely matters—but the difference between 0.1 and 1 mg/liter may be analytically significant. All K_{sp} calculations should be looked upon as approximate because of uncertainties in the constants (p. 139) and subject to experimental check. The equilibria in some analytical systems (e.g., heavy metal sulfides) are so poorly known that meaningful solubility calculations cannot always be made.

Example 5: Assuming that the intrinsic solubility can be neglected, apply activity coefficients and find the solubility of $BaSO_4$ in the following solutions: (a) 0.01 *M* Ba^{++}, (b) 0.01 *M* Ba^{++} and 0.5 *M* NaCl.

Ans. (a) $\mu = 0.03$; from Table A-1 we estimate $f_{Ba}f_{SO_4}$ as 0.3:

$$[SO_4^{--}] = \frac{1.1 \times 10^{-10}}{0.3 \times 0.01} \sim 4 \times 10^{-8}\ M$$

The solubility is about three times greater than found in Example 1, where activity coefficients were taken as unity.

(b) $\mu = 0.53$ or ~ 0.5 and $f_{Ba}f_{SO_4} \sim 0.06$:

$$[SO_4^{--}] = \frac{1.4 \times 10^{-10}}{0.06 \times 0.01} = 2 \times 10^{-7}\ M$$

The solubility is still negligibly small from the analytical viewpoint (see Example 6).

Example 6: What is the approximate maximum allowable solubility of a precipitate for a gravimetric determination?

Ans. Certain variables must first be specified. A reasonable total volume for precipitation and washing is 250 ml. The precipitation may be considered complete for most analytical purposes if the solubility loss does not amount to more than ~ 0.1 per cent absolute in the percentage of a major constituent. The amount of sample taken in a macro gravimetric determination may be approximated as 0.01 mole (samples outside the range 0.005–0.02 mole are uncommon).

Accordingly, the amount of the slightly soluble substance which may remain in 250 ml of solution is $\sim \frac{1}{1000} \times 0.01 \sim 10^{-5}$ mole or $\frac{1000}{250} \times 10^{-5} \sim 4 \times 10^{-5}$ mole/liter. This limit barely (or not quite) meets the requirements, so it is best to round off the allowable solubility to 1×10^{-5} *M*. This solubility limit is a somewhat arbitrary but nevertheless useful approximation that allows a quick decision on the probable usefulness of a precipitate for a gravimetric determination from an inspection of its K_{sp} ($\mu \sim 0.1$) value. In such an inspection, an excess of precipitant equivalent to an ionic concentration of ~ 0.01 *M* may be taken as a

[4] See the Appendix for K^0_{sp} values. A useful compilation of solubility-product constants (and other equilibrium constants) for $\mu = 0.1$, as well as $\mu = 0$, is given by A. Ringbom in his book, *Complexation in Analytical Chemistry*, Wiley-Interscience, New York, 1963.

reasonable excess. The conclusion, if favorable, may be invalidated by strong complexing with a slight excess of reagent (including intrinsic solubility of the precipitate).

The increase in ionic strength resulting from the addition of a large excess of an electrolyte having an ion in common with a precipitate counteracts the repression of the solubility. The combined effects are illustrated in Fig. 7-3 for a 1:1 precipitate

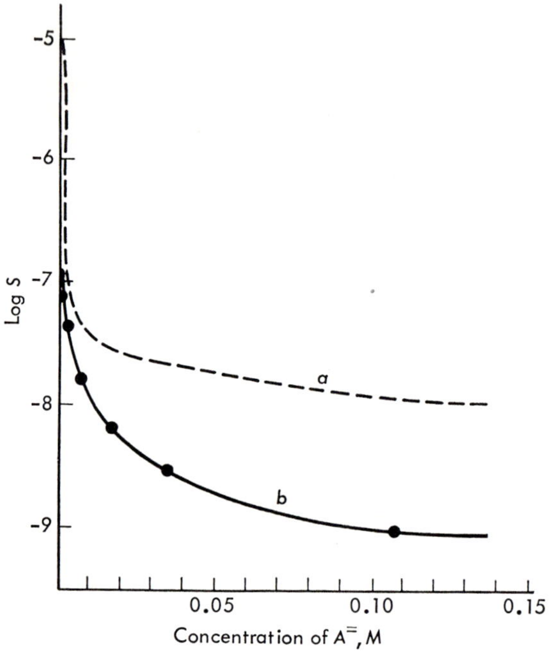

Fig. 7-3. Illustrating the molar solubility of $B^{++}A^{--}$ ($K^0_{sp} = 1 \times 10^{-10}$) in the presence A^{--} (as Na_2A) or B^{++} (as BCl_2) calculated (*a*) by using activity coefficients, (*b*) by taking all activity coefficients as 1.

of doubly charged ions whose K^0_{sp} is 1×10^{-10} (much the same as $BaSO_4$). The common-ion effect alone is also shown. It is seen that, for this precipitate, increasing the common-ion concentration above 0.02–0.03 *M* decreases the solubility but little more. An unnecessarily large excess of precipitant should be avoided not only because it may do little good, but also, and especially, because it may do serious harm by soluble complex formation. The actual effect of an excess of a precipitant is likely to be the resultant of a number of effects. The solubility of lead sulfate as a function of the sodium sulfate concentration is shown in Table 7-3. In the range 0.01–0.1 *M* Na_2SO_4 the solubility changes hardly at all. There is little tendency toward formation of soluble complexes such as $Pb(SO_4)_2^{--}$. There appears to be some formation of undissociated $PbSO_4$.

When a slightly soluble substance is incompletely dissociated in an aqueous solution, its solubility in the presence of a sufficient excess of the common ion may be less at high indifferent electrolyte concentrations than at low. This effect may occur when the ionic solubility is less than the molecular solubility, because the molecular solubility generally decreases as the electrolyte concentration is increased (salting-out effect, Chap. 14). For example, in a solution saturated with dimethylglyoxime above pH 4.5, the solubility of nickel dimethylglyoximate at a sodium chloride concentration of 0.05 *M* corresponds to 57 μg of Ni/liter, whereas at 2.0 *M* sodium chloride it corresponds to 30 μg of Ni.[5] Under these conditions, the concentration of Ni^{++} in the solution is negligibly small compared to that of $Ni(HDx)_2$ and the solubility of the latter falls with an increase in the ionic strength.

Complex formation with excess precipitant

It is often found, in spite of the common-ion effect, that an excess of reagent increases the solubility of a precipitate. The increase is due to the formation of soluble complex species by reaction of the precipitate with the precipitant; e.g., BA forms BA_2^- and possibly higher complexes with the precipitating ion A^-. Precipitation of silver ion by chloride ion provides an illustration. The complex ions $AgCl_2^-$, $AgCl_3^{--}$ and $AgCl_4^{3-}$ can be formed and the predominant species depend upon the chloride concentration. The ionization constants of the first two of these, as well as of AgCl, at 18°C are accurately known[6]:

$$\frac{(Ag^+)(Cl^-)}{(AgCl)} = 3.9 \times 10^{-4} \qquad \text{(parentheses denote activities)}$$

$$\frac{(Ag^+)(Cl^-)^2}{(AgCl_2^-)} = 5.1 \times 10^{-6}$$

$$\frac{(Ag^+)(Cl^-)^3}{(AgCl_3^{--})} = 5.7 \times 10^{-6}$$

With the aid of these constants and the known intrinsic solubility of AgCl in water (2.25×10^{-7} *M* at 18°C), the solubility of AgCl can be calculated as a function of the chloride ion activity (or concentration). For example, in 0.010 *M* HCl solution the solubility is given by

$$S = [AgCl] + [Ag^+] + [AgCl_2^-] \qquad (AgCl_3^{--} \text{ can be neglected})$$

$$= [AgCl] + \frac{3.9 \times 10^{-4}[AgCl]}{0.8[Cl^-]} + \frac{3.9 \times 10^{-4}[AgCl][Cl^-]}{5.1 \times 10^{-6}}$$

$$(f \text{ of } Ag^+ \text{ and } AgCl_2^- \sim 0.9 \text{ at } \mu = 0.01)$$

$$= [AgCl]\left\{1 + \frac{4.9 \times 10^{-4}}{[Cl^-]} + 76[Cl^-]\right\}$$

$$= 2.3 \times 10^{-7}(1 + 4.9 \times 10^{-2} + 0.76)$$

$$= 4 \times 10^{-7}\ M \text{ or } \sim 0.04 \text{ mg of } Ag^+/\text{liter}$$

[5] H. Christopherson and E. B. Sandell, *Anal. Chim. Acta* **10**, 4 (1954).

[6] K. H. Lieser, *Z. Anorg. Allgem. Chem.* **292**, 97 (1957): see also E. Klein, *Phot. Korr.* **92**, 139 (1956).

Note that the concentration of $AgCl_2^-$ is about 15 times that of Ag^+ and a little less than the intrinsic solubility of AgCl. As the concentration of Cl^- is increased the solubility of AgCl rises rapidly (Fig. 7-4).

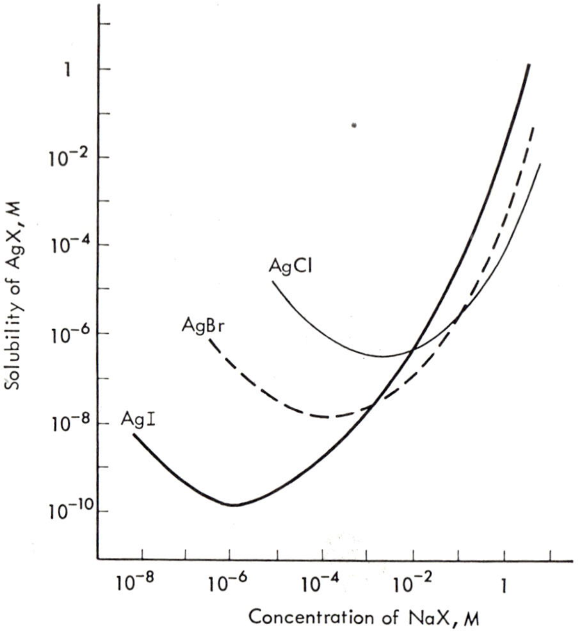

Fig. 7-4. Solubilities of silver halides in sodium halide solutions at 18°C. Curves begin with solubility in water. After Lieser, *Z. Anorg. Allgem. Chem.* **292**, 97 (1957).

Evidently addition of an unregulated excess of chloride in the precipitation of silver in its gravimetric determination could result in appreciable errors.

The chloride ion concentration at which the solubility of silver chloride is a minimum can be found by setting $dS/d[Cl^-] = 0$. Thus

$$\frac{dS}{d[Cl^-]} = -\frac{8.8 \times 10^{-11}}{f^2[Cl^-]^2} + 1.73 \times 10^{-5} = 0$$

or

$$[Cl^-] = 2.26 \times 10^{-3} f^{-1}$$

For the effect of excess Ag^+ on the solubility of silver chloride, see Chapter 24.

Formation of complexes with foreign substances

The solubility of a precipitate may be greatly increased by species in the sample solution that form soluble complexes with the substance being precipitated or with the precipitant. For example, the presence of Hg(II) can prevent quantitative precipitation of chloride with silver by forming slightly dissociated Hg–Cl complexes; the precipitation of silver chloride fails in the presence of much ammonia, because silver ammine complexes are formed to such an extent that the solubility product of silver chloride is never exceeded. A common type of calculation required in connection with gravimetric analysis involves finding the solubility of a precipitate in the presence of a foreign complexer of specified concentration when the excess of precipitant is specified.

Example 7: 50 ml of 0.02 M Br^- in 2.0 M ammonia is treated with 50 ml of 0.03 M Ag^+ in water. Find the concentration of Br^- in solution at equilibrium, neglecting activity coefficients and such species as $AgNH_3Br$.

$$\frac{[NH^+][OH^-]}{[NH_4OH]} = 2 \times 10^{-5}$$

$$[Ag^+][OH^-] = 1 \times 10^{-8} = K_{sp}$$

$$\frac{[Ag(NH_3)^+]}{[Ag^+][NH_3]} = 2.5 \times 10^3 \qquad \frac{[Ag(NH_3)_2^+]}{[Ag^+][NH_3]^2} = 2.5 \times 10^7$$

$$[Ag^+][Br^-] = 5 \times 10^{-13} = K_{sp}$$

Ans. We note that the species $Ag(NH_3)^+$ can be neglected because

$$\frac{[Ag(NH_3)^+]}{[Ag(NH_3)_2^+]} = \frac{2.5 \times 10^3}{2.5 \times 10^7} \times [NH_3] \sim 10^{-4} \times 1 = 10^{-4} \qquad ([NH_3] \text{ is taken equal to } [NH_4OH])$$

AgOH will not be precipitated, because the maximum concentration of Ag^+ will not exceed

$$[Ag^+] = \frac{[Ag(NH_3)_2^+]}{2.5 \times 10^7 \times 1} = \frac{1.5 \times 10^{-2}}{2.5 \times 10^7} = 6 \times 10^{-10}$$

which means that $[OH^-]$ would have to be ~ 20 M to exceed the solubility product of AgOH.

The following relations exist among the concentrations of the species $Ag(NH_3)_2^+$, Ag^+, and Br^-:

$$[Ag(NH_3)_2^+] = 0.005 + [Br^-]$$ (= analytical concentration of silver in solution, because $[Ag^+]$ is very low)

Therefore,

$$[Ag^+] = \frac{0.005 + [Br^-]}{2.5 \times 10^7}$$

Also, $$[Ag^+] = \frac{5 \times 10^{-13}}{[Br^-]}$$ (from the solubility-product relation)

Combination of the last two equations gives

$$[Br^-]^2 + 0.005[Br^-] - 1.25 \times 10^{-5} = 0$$

Therefore,

$$[Br^-] = 1.83 \times 10^{-3}$$

Approximately 18 per cent of the bromide is not precipitated.

A general expression for the solubility of a slightly soluble substance BA in the presence of excess precipitant A and foreign complexing agent at specified concentrations is readily derived. Let X be an anion or uncharged species forming the consecutive complex species $BX \cdots BX_n$ with B (charges of the species will be omitted here), having the formation constants

$$\frac{[BX]}{[B][X]} = \beta_1 \tag{11}$$

$$\vdots$$

$$\frac{[BX_n]}{[B][X]^n} = \beta_n \tag{11a}$$

The solubility of BA is given by

$$S = [B] + [BX] + \cdots + [BX_n]$$
$$= [B]\{1 + \beta_1[X] + \cdots + \beta_n[X]^n\} \tag{12}$$

$$[B] = \frac{K_{sp}}{[A]} = \frac{K_{sp}}{C_A + S} \tag{13}$$

where C_A is the concentration of the stoichiometric excess of A in ionic form (= [A] if the precipitate is entirely insoluble). Substitution of Eq. (13) into (12) gives

$$S^2 + C_A S - K_{sp}\{1 + \beta_1[X] + \cdots + \beta_n[X]^n\} = 0 \tag{14}$$

If S is small compared to C_A in Eq. (13), the denominator is $\sim C_A$ and a linear equation is obtained. Frequently the fraction of X bound in the complexes of B is small and the value of [X] corresponds to the amount added. If not, [X] must be found by successive approximations.

For precipitates of types other than BA, equations of higher degree are obtained, which can be solved by successive approximations.

Effect of hydrogen-ion concentration on solubility

The solubility of many sparingly soluble substances is strongly dependent on the pH of the solution in consequence of the alteration in concentration of one or both component ions by reaction with H^+ or OH^-. A slightly soluble metal hydroxide $M(OH)_n$ dissolves in acids by forming M^{n+} as well as intermediate charged hydroxo complexes $MOH^{(n-1)}$, $M(OH)_2^{(n-2)}$, . . ., and sometimes polymers of these. Salts of weak acids show increased solubility in acidic medium. Thus carbonates, phosphates, arsenates, chromates, sulfates (HSO_4^- is incompletely dissociated), and many other inorganic salts, as well as most metal chelates, dissolve in acidic solutions to an extent that is determined by the hydrogen-ion concentration, the solubility product, and the ionization constant of the weak acid that is formed. Organic substances containing a basic nitrogen atom combine with protons to give cations; e.g., 8-hydroxyquinoline gives the hydroxyquinolinium ion:

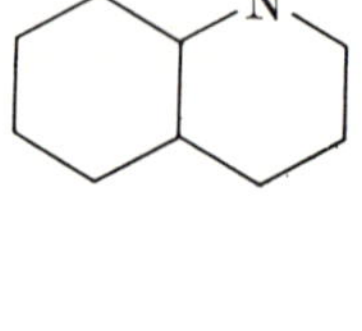

Substances showing increased solubility in alkaline solutions include acids (most metal-chelating reagents are weak acids) and metal hydroxides or salts that are amphoteric and form hydroxo complexes such as $M(OH)_{n+1}^{-}$.

Salts of quite weak acids or bases react with H^+ or OH^- furnished by water to give hydrolysis products. Thus calcium carbonate reacts with water to give OH^- and an equivalent amount of HCO_3^-. A saturated solution of calcium carbonate in water at 25°C is 3.9×10^{-5} *M* in CO_3^{--}, 8.8×10^{-5} *M* in HCO_3^- (and in OH^-), and 1.3×10^{-4} *M* in Ca^{++}. The hydrolysis of heavy metal salts can give a series of complex species: $MOH^{(n-1)}$, $M(OH)_2^{(n-2)}, \ldots, M(OH)_n$. The heavy metal salt may be entirely converted into $M(OH)_n$ or a basic salt, and a saturated solution of the salt will never be obtained.

Example 8: (a) Find the solubility of CaF_2 in 0.050 *M* HCl.

$$K_{sp} = 1.1 \times 10^{-10} \quad (\mu = 0.05) \quad K_a = \frac{[H^+][F^-]}{[HF]} = 1.0 \times 10^{-3} \quad (\mu = 0.05)$$

Ans. From the reactions

$$CaF_2 \rightleftharpoons Ca^{++} + 2F^-$$

$$F^- + H^+ \rightleftharpoons HF$$

it is seen that

$$S = [Ca^{++}] = \tfrac{1}{2}[\{F^-] + [HF]\}$$

From the ionization-constant expression for HF:

$$\frac{[F^-]}{[HF]} \cong \frac{1 \times 10^{-3}}{5 \times 10^{-2}} = 0.02$$

(the value is approximate because some H^+ is converted into HF by reaction with CaF_2)

or $$\frac{[F^-]}{[HF] + [F^-]} \cong \frac{0.02}{1 + 0.02} \cong 0.02$$

and $$[F^-] \cong 0.02\{[HF] + [F^-]\}$$

From the expression for the solubility product,

$$[Ca^{++}][F^-]^2 \cong S(2 \times 0.02S)^2 \cong 16 \times 10^{-4}S^3 \cong 1.1 \times 10^{-10}$$

$$S \cong 4.1 \times 10^{-3}\ M$$

This (approximate) value shows that the hydrogen-ion concentration has been reduced to $0.05 - (2 \times 4.1 \times 10^{-3}) = 0.042$ *M* by formation of HF. Carrying the calculation through another cycle, using $[H^+] = 0.042$, we find

$$S = 3.6 \times 10^{-3}\ M$$

The correct value lies between 3.6×10^{-3} and 4.1×10^{-3} *M* and for all practical purposes we can accept the former figure. A third cycle of calculation gives $S = 3.7 \times 10^{-3}$ *M*.

(b) Find the solubility of CaF_2 in a solution that has the analytical composition 0.050 *M* HCl and 0.010 *M* NaF. Use the same constants as in (a).

Ans. $$[Ca^{++}] = S$$

$$[F^-] + [HF] = 0.01 + 2S$$

$[H^+]$ in the mixture is approximately $0.05 - 0.01 = 0.04\ M$. Therefore,

$$[F^-] = \frac{1.0 \times 10^{-3}}{4 \times 10^{-2}}(0.01 + 2S) = 2.5 \times 10^{-2}(0.01 + 2S)$$

$$S\{2.5 \times 10^{-2}(0.01 + 2S)\}^2 = 1.1 \times 10^{-10}$$

$$S^3 + 1.0 \times 10^{-2}S^2 + 2.5 \times 10^{-5}S - 4.4 \times 10^{-8} = 0$$

$$S = 1.16 \times 10^{-3} \qquad \text{(see Fig. 7-5)}$$

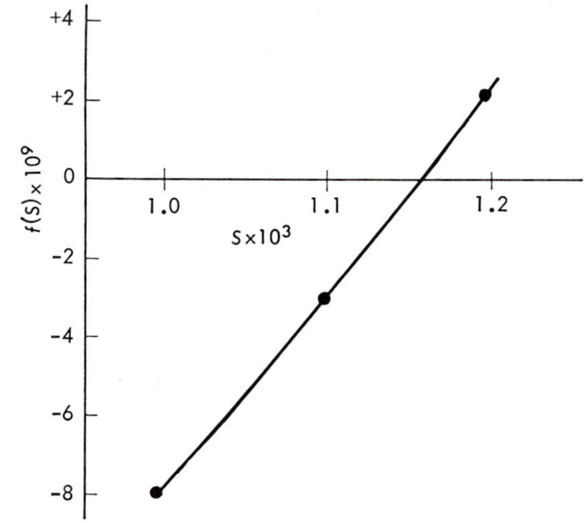

Fig. 7-5. Finding the positive real root of $S^3 + 1.0 \times 10^{-2}S^2 + 2.5 \times 10^{-5}S - 4.4 \times 10^{-8} = 0$. Evidently

$$S < \frac{4.4 \times 10^{-8}}{2.5 \times 10^{-5}} < 1.8 \times 10^{-3}.$$

Choose $S = 1 \times 10^{-3}$ (for ease of calculation) and calculate $f(S) = S^3 \ldots -4.4 \times 10^{-8} = 10^{-9} + 10^{-8} + 2.5 \times 10^{-8} - 4.4 \times 10^{-8} = -8 \times 10^{-9}$. Plot the point $S = 1 \times 10^{-3}, f(S) = -8 \times 10^{-9}$. Since S must be greater than 1×10^{-3} (but not much greater), calculate $f(S)$ for $S = 1.1$ and 1.2. Plot the values obtained, -3.1×10^{-9} and 2.1×10^{-9}, against S as in the figure. Draw a smooth curve (it is almost a straight line) through the three points. The curve crosses the S axis $[f(S) = 0]$ at $S = 1.16 \times 10^{-3}$, which is the required root.

This value shows that $[H^+] = 0.05 - 0.012 = 0.038$. Computation with the corrected hydrogen-ion concentration gives $S = 1.1 \times 10^{-3}$.

(c) Find the solubility of CaF_2 in 0.05 M HCl which is 0.01 M in $CaCl_2$.

Ans. $[Ca^{++}][F^-]^2 = (0.01 + S)(0.02 \times 2S)^2 = 1.1 \times 10^{-10}$

$$S^3 + 0.01S^2 + 6.9 \times 10^{-8} = 0$$

$$S = 2.4 \times 10^{-3}\ M$$

This value is slightly high because the hydrogen-ion concentration is actually $0.05 - 0.005 = 0.045$, from which $S = 2.25 \times 10^{-3}$.

Solubility of heavy metal hydroxides. At least two difficulties are encountered in considering the solubility of slightly soluble hydroxides. First, the solution may contain, at and near the precipitation pH, a number of metal–OH^- species, including polynuclear complexes, whose equilibria may be inadequately known; second, the solubility of the hydroxide may not be well defined under the conditions of precipitation. Many of the heavy metal hydroxides formed by more or less rapid precipitation under analytical conditions are amorphous and do not have a definite composition. They can be represented as hydrous oxides, $Fe_2O_3 \cdot xH_2O$ for example, with a variable amount of water, which changes with the aging of the precipitate, especially on heating. Anions such as Cl^- and SO_4^{--} may contaminate the precipitate.[7] The solubility of an unaged hydroxide, sometimes called "active," may be quite different from that of an aged or "inactive" hydroxide. The active form, not well characterized, will have a more or less variable solubility which is greater than that of the aged inactive form. Sometimes aging, promoted by heating above room temperature, leads to the formation of a definite oxide (e.g., Fe_2O_3) or hydroxide (or oxide–hydroxide), which has a fixed composition, crystal structure, and solubility. But equilibrium may be slowly attained and mixtures may result (aluminum, Chap. 26). The value of the solubility product of a given metal hydroxide that one finds listed in a table may not be the accurate value for the hydroxide obtained in an analytical procedure. In general, solubility calculations involving heavy metal hydroxides should be looked upon as quite approximate.

Except at very low concentrations, most heavy metals form charged polynuclear hydroxo species,[8] such as $Al_6(OH)_{15}^{3+}$, $Cu_2(OH)_2^{++}$, and $Fe_2(OH)_2^{4+}$. At pH values where hydroxide precipitates have minimum solubility and the concentration of M^{+n} is very small, the concentration of such polynuclear species may be negligible from the analytical standpoint. Cadmium, cobalt(II), iron(II), mercury(II), nickel, and zinc have little tendency to form polynuclear hydroxo complexes.

Logarithmic concentration–pH diagrams provide a convenient representation of metal hydroxide systems. Figure 7-6 is an example.

Precipitation of metal hydroxides from the standpoint of separations is considered on p. 251.

Example 9: (a) Find the concentrations of Al^{3+}, $AlOH^{++}$, $Al(OH)_4^-$, and $Al_6(OH)_{15}^{3+}$ at pH 7 in a solution in equilibrium with hydrous aluminum oxide precipitate. Is the precipitation of aluminum quantitative at this pH?

Take the equilibria in aluminum solutions to be represented by the following constants:

$$(K_{sp})_{Al(OH)_3} = 1 \times 10^{-32} \qquad \text{(variable with precipitate)}$$

$$\frac{[Al^{3+}][OH^-]}{[AlOH^{++}]} = 1.4 \times 10^{-9} \qquad \frac{[Al^{3+}]^6[OH^-]^{15}}{[Al_6(OH)_{15}^{3+}]} = 1 \times 10^{-163}$$

$$\frac{[Al^{3+}][OH^-]^4}{[Al(OH)_4^-]} = 5 \times 10^{-34}$$

[7] Some heavy metals can give precipitates of basic salts having almost stoichiometric composition.

[8] For a review, see L. G. Sillén, *Quart. Rev.* (*London*) **13,** 146 (1959).

Ans. $S = [Al^{3+}] + [AlOH^{++}] + [Al(OH)_4^-] + [Al_6(OH)_{15}^{3+}]$

$$= \frac{10^{-32}}{10^{-21}} + \frac{10^{-11} \times 10^{-7}}{1.4 \times 10^{-9}} + \frac{10^{-11} \times 10^{-28}}{5 \times 10^{-34}} + 6\left(\frac{10^{-66} \times 10^{-105}}{10^{-163}}\right)$$

$$= 10^{-11} + 7 \times 10^{-10} + 2 \times 10^{-6} + 6 \times 10^{-8} = 2.1 \times 10^{-6}\ M$$

This solubility represents 0.05 mg of Al per liter and the precipitation is therefore quantitative. If K_{sp} is actually 10^{-31}, the solubility would be 0.5 mg of Al per liter or 0.1 mg of Al for a reasonable analytical volume of 200 ml, and the precipitation would be barely quantitative. The contribution of $[Al(OH)_3]$ toward the

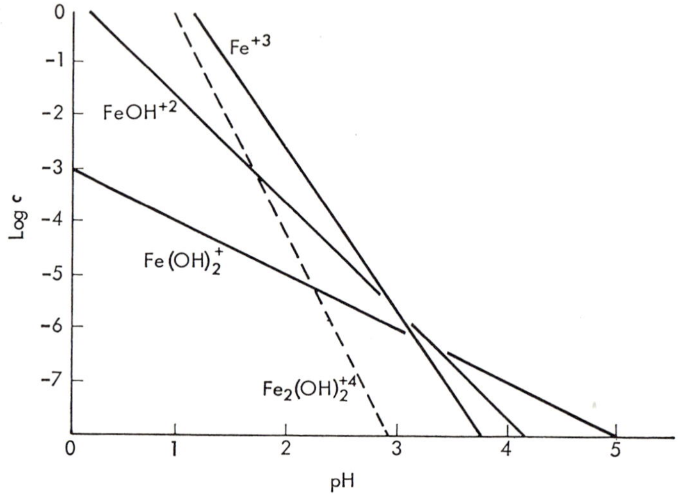

Fig. 7-6. Molar concentrations of iron species in a saturated ferric hydroxide solution as a function of pH at room temperature. The diagram is based on the following constants:

$$[Fe^{3+}][OH^-]^3 = K_{sp} = 2 \times 10^{-39}$$
$$[FeOH^{++}][H^+]/[Fe^{3+}] = 9 \times 10^{-4}$$
$$[Fe(OH)_2^+][H^+]^2/[Fe^{3+}] = 5 \times 10^{-7}$$
$$[Fe_2(OH)_2^{4+}][H^+]^2/[Fe^{3+}]^2 = 1.1 \times 10^{-3}$$

The molecular solubility of $Fe(OH)_3$ is so small ($< 2 \times 10^{-9}\ M$) that the line for $Fe(OH)_3$ does not appear in the field. Ferric hydroxide is only weakly amphoteric ($[Fe(OH)_4^-]/[OH^-] = 10^{-4} - 10^{-5}$ in a solution saturated with $Fe(OH)_3$), so that a significant concentration of $Fe(OH)_4^-$ will not be present below pH 11–12.

solubility can almost certainly be neglected. The constant for $Al(OH)_2^+$ has not been determined and the concentration of this species cannot be calculated. It would not be expected to exceed $[Al(OH)_4^-]$. Because of the uncertainties in the constants, and even in the species present, calculated solubilities merit little confidence and experimental values are called for (Chapter 26).

(b) Will it be satisfactory to precipitate aluminum hydroxide at pH 9 when quantitative recovery is desired?

Ans. Definitely not. $[Al(OH)_4^-] = 2 \times 10^{-4}$.

Solubility of heavy metal sulfides. The effect of acidity on the solubility of slightly soluble metal sulfides is of special analytical importance. Hydrogen sulfide is a very weak acid dissociating in two steps:

$$H_2S \rightleftharpoons H^+ + HS^-$$

$$HS^- \rightleftharpoons H^+ + S^{--}$$

At an ionic strength of 1.0 M (KCl) and 20°C the ionization constants are[9]

$$\frac{[H^+][HS^-]}{[H_2S]} = K_1 = 1.32 \times 10^{-7} \qquad pK_1 = 6.88 \tag{15}$$

$$\frac{[H^+][S^{--}]}{[HS^-]} = K_2 = 7.1 \times 10^{-15} \qquad pK_2 = 14.15 \tag{16}$$

$$\frac{[H^+]^2[S^{--}]}{[H_2S]} = K_1K_2 = 9.4 \times 10^{-22} \qquad \text{[From Eqs. (15) and (16)]} \tag{17}$$

In the precipitation of metal ions with hydrogen sulfide in acid medium, the sulfide–ion concentration is extremely small. Still, under equilibrium conditions, the following relation holds in the suspension of any (divalent) metal sulfide:

$$[M^{++}][S^{--}] = K_{sp} \tag{18}$$

The concentration of M^{++} is given by

$$[M^{++}] = \frac{K_{sp}}{[S^{--}]} = 1.1 \times 10^{21} \frac{K_{sp}[H^+]^2}{[H_2S]} \qquad \text{[from Eqs. (17) and (18); } \mu = 1.0\text{]} \tag{19}$$

If M^{++} is the only metal species present in solution in significant concentration compared to other possible metal species, the solubility of the sulfide increases as the square of the hydrogen-ion concentration. When the solution is saturated with hydrogen sulfide at room temperature (20–25°C), $[H_2S]$ = constant = 0.10 M, and

$$[M^{++}] = 1.1 \times 10^{22} K_{sp}[H^+]^2 \tag{20}$$

If the metal is univalent (Ag^+, Tl^+):

$$[M^+] = \left(1.1 \times 10^{21} \frac{K_{sp}[H^+]^2}{[H_2S]}\right)^{1/2} \tag{21}$$

The solubility products of various metal sulfides and their negative logarithms are listed in Table A-11. The column headed $[M^{n+}]_{[H^+]=1}$ gives the calculated concentration of simple metal ion in a solution of $[H^+] = 1$, saturated with hydrogen sulfide at room temperature, which is in equilibrium with the solid sulfide. As with the metal hydroxides, the solubility products of the sulfides depend on the state of the precipitate, including the particle size. The particle size will depend on the conditions of precipitation and the age of the precipitate. Some

[9] M. Widmer and G. Schwarzenbach, *Helv. Chim. Acta* **47**, 266 (1964). In 1.0 M $NaClO_4$, Schwarzenbach and Widmer, *ibid.* **49**, 111 (1966), found the values $pK_1 = 6.68$ and $pK_2 = 14.0$ at 20°C. According to A. J. Ellis and R. M. Golding, *J. Chem. Soc.* **111**, 127 (1959), the activity constants are $K_1 = 9.5 \times 10^{-8}$ and $K_2 = 1.0 \times 10^{-14}$ (25°C).

sulfides may exist in several crystal modifications. It should also be noted that the speed of precipitation of many sulfides is very slow. For example a 0.1 M zinc sulfate solution in 1 M hydrochloric acid remains clear when saturated with hydrogen sulfide. Nevertheless, the solution is supersaturated with zinc sulfide, because the equilibrium concentration corresponds to about 0.001 M Zn. Zinc sulfide will finally precipitate if the solution is kept saturated with hydrogen sulfide for 1 month or so. Precipitation once started is catalyzed by the solid sulfide formed (see also postprecipitation, p. 245).

The calculated value of $[M^{n+}]$ in a solution in equilibrium with the solid sulfide in the presence of excess hydrogen sulfide does not necessarily represent the solubility of the metal under the conditions. The solution may contain significant concentrations of molecular metal sulfide, metal bisulfide, and charged sulfide and bisulfide complexes of the metal. The concentrations of these may greatly exceed the concentration of the simple metal ion when this is very small. The equilibrium constants of the sulfo complexes of most metals are unknown and realistic solubility values for the sulfides cannot be calculated when the concentration of M^{n+} is very low. The intrinsic solubilities of heavy metal sulfides have not been determined. Only maximum values can be given. For example, the intrinsic (molecular) solubility of lead sulfide cannot exceed 4 μg/liter (2×10^{-8} M), which is the minimum total solubility which has been found for this sulfide. The intrinsic solubility of Ag_2S (or AgHS) at $\mu = 1.0$ is not greater than 2×10^{-9} M. It is of interest that the intrinsic solubility of lead dithizonate,

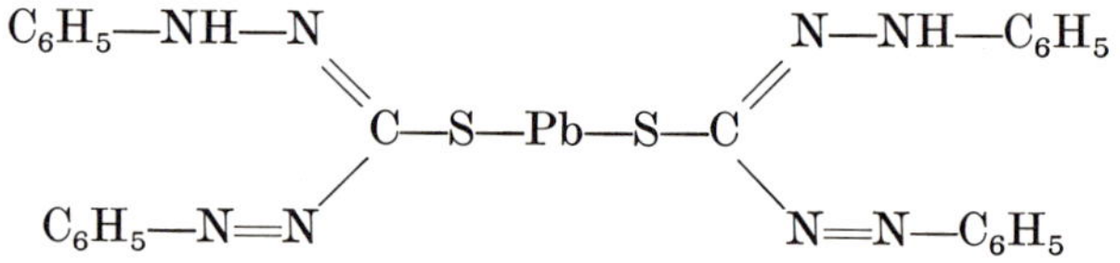

in water is 3×10^{-9} M at 25°C.[10]

The solubility of black HgS (metacinnabar) in the presence of HS^- and S^{--} has been studied carefully.[11] The complex species $Hg(HS)_2$ (or H_2HgS_2), $HgHS_2^-$ (or $HHgS_2^-$), and HgS_2^{--} are formed (Fig. 7-7). The equilibrium constants have the following values at 20°C and $\mu = 1.0$ (KCl):

$$HgS(s) \rightleftharpoons Hg^{++} + S^{--} \qquad \log K_{sp} = -50.96$$

$$HgS(s) + H_2S \rightleftharpoons Hg(HS)_2 \qquad \log K = -5.97$$

$$HgS(s) + HS^- \rightleftharpoons HgHS_2^- \qquad \log K = -5.28$$

$$HgS(s) + S^{--} \rightleftharpoons HgS_2^{--} \qquad \log K = 0.57$$

$$HgS_2^{--} + H^+ \rightleftharpoons HgHS_2^- \qquad pK = 8.30$$

$$HgHS_2^- + H^+ \rightleftharpoons Hg(HS)_2 \qquad pK = 6.19$$

Below pH 5, the principal dissolved species in the system is $Hg(HS)_2$; between pH 6 and 8, $HS{-}Hg{-}S^-$; and above pH 9, HgS_2^{--}. In the acid range (pH 0–5,

[10] O. B. Mathre and E. B. Sandell, *Talanta* **11**, 295 (1964).

[11] G. Schwarzenbach and M. Widmer, *Helv. Chim. Acta* **46**, 2613 (1963); see also W. D. Treadwell and F. Schaufelberger, *ibid.* **29**, 1936 (1946). For the solubility of Ag_2S in hydrogen sulfide and sulfide solutions, see Schwarzenbach and Widmer, *ibid.* **49**, 111 (1966). Below pH 4 the solubility is due to AgHS (or AgHS + Ag_2S).

$\mu = 1$) in a solution saturated with H_2S at 20°C the solubility of HgS is approximately

$$[Hg(HS)_2] = 10^{-1} \times 10^{-5.97} = 10^{-7} \quad \text{or} \quad \sim 20\ \mu g \text{ of Hg per liter}$$

A decrease in the H_2S concentration will decrease the solubility of HgS, e.g., to $\sim 0.2\ \mu g$ of Hg per liter in a 10^{-3} *M* H_2S solution, since the concentration of Hg^{++} is ridiculously small.[12]

The intrinsic solubility of HgS has not been determined, but (from some data of Schwarzenbach and Widmer) it is not likely to exceed 1×10^{-9} *M*. We may conclude that by restricting the excess of H_2S, the precipitation of HgS can be made so complete that less than 1 μg of Hg will remain in 1 liter of acid solution.

A brief discussion of separations of metals by sulfide precipitation appears on page 258.

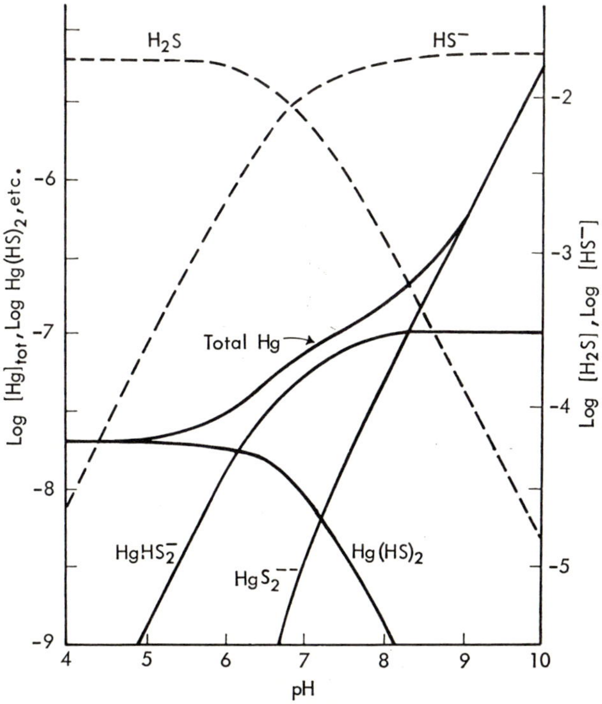

Fig. 7-7. Solubility of HgS (black) at 20°C as a function of the pH in a solution having a total sulfide ($H_2S + HS^-$) concentration of 0.019 *M* ($\mu = 1$). After Schwarzenbach and Widmer, *Helv. Chim. Acta* **46**, 2613 (1963).

Influence of temperature upon solubility

When heat is absorbed in the solution of a salt, the solid phase of which is in equilibrium with the solution,

$$BA + \text{heat} \rightleftharpoons B^+ + A^-$$

[12] The molar concentration of Hg^{++} in a 0.001 *M* H_2S solution at pH 0 is $10^{-51} \times 10^{21} \times 10^3 = 10^{-27}$. One ton (1000 liters) of such a solution in equilibrium with solid HgS would theoretically contain approximately one Hg^{++} ion.

raising the temperature of the solution will increase the solubility of the salt according to the theorem of Le Châtelier–van't Hoff. Most of the precipitates dealt with in quantitative analysis are more soluble in hot than in cold solution. The temperature effect is very pronounced for silver chloride (Fig. 7-8); for barium sulfate it is less so but still distinct.

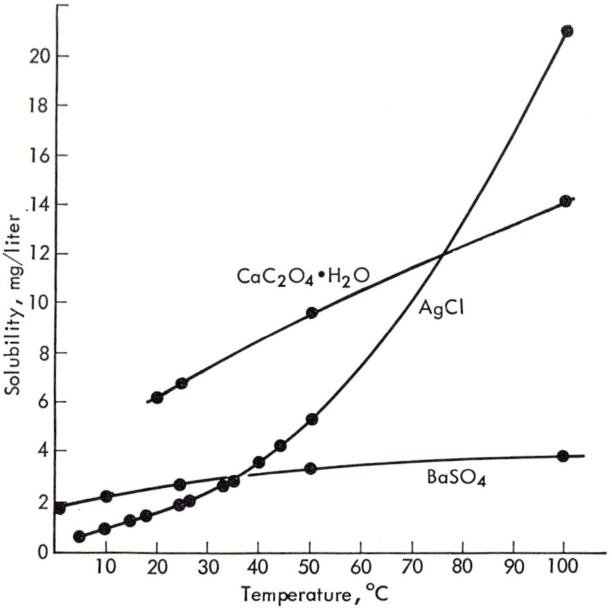

Fig. 7-8. Solubility in water of silver chloride, barium sulfate and calcium oxalate monohydrate as a function of the temperature.

Most precipitations are carried out in hot solution in order that an easily filterable precipitate may be obtained. Frequently the precipitate is digested near the boiling point for 1 hour or more. If the precipitate is very insoluble or if its temperature coefficient of solubility is small, the solution may be filtered hot (barium sulfate, ferric hydroxide, and other hydroxides). Otherwise, the solution is allowed to cool to room temperature before filtration; occasional stirring during cooling may be advisable to ensure attainment of equilibrium if the solubility factor is more or less critical. A few precipitation forms used in gravimetric analysis are relatively soluble and the solution, whose volume should be restricted, is best cooled in ice before filtration, especially if a double precipitation is required (magnesium ammonium phosphate and analogous double phosphates are examples).

Influence of the solvent

Most inorganic salts[13] are less soluble in organic solvents than in water. Their solubility in inert nonpolar or weakly polar organic liquids (such as chloroform) would be expected to be, and is, extremely small because of the negligible tendency

[13] Nonpolar inorganic substances such as iodine, stannic iodide, and germanium tetrachloride are often quite soluble in various organic solvents.

for solvation of their ions in such liquids. But the solubility of inorganic salts is also usually much less in oxonium-type solvents such as the alcohols, esters, ethers, and the like than in water. The solubility of sodium and potassium chloride in ethyl alcohol is roughly $\frac{1}{1000}$ the solubility in water at room temperature, and still less in the higher alcohols (Table 7-4). This behavior is typical of most inorganic salts. The decrease in solubility in organic solvents results from the generally lesser solvation of the ions in these solvents and the lower dielectric constant ($\varepsilon = 78.5$ for water and 24 for ethyl alcohol at 25°C), which tends to decrease ionization by increasing the attraction between positive and negative ions.

Some salts are quite soluble in oxonium-type solvents, albeit less soluble than in water. Lithium chloride is an example (Table 7-4). Its solubility in ethyl alcohol is roughly one third that in water. A useful rule states that deliquescent substances tend to be soluble in ethyl alcohol and similar solvents. Deliquescence of a salt signifies a strong tendency for its cation to hydrate, a tendency that extends to solvation by a water-like oxonium solvent. Strongly solvated ions may be said to prefer the solution to the crystal lattice and the salt dissolves. A solvated substance becomes similar to the solvent and mixes with it; like dissolves like.

TABLE 7-4. **Solubilities of Alkali Metal Chlorides in Water, Anhydrous Alcohols, and Acetone**[a]

	Solubility, g/100 g Solvent at 25°C		
	LiCl	NaCl	KCl
Water	84.7	36.0	35.9
Methanol	43.8 (20°C)	1.38	0.53
Ethanol	24.3 (20°C)	0.065	0.029
n-Propanol	16.2	0.012	0.0061
n-Butanol	10.6	0.005	0.0030
n-Pentanol	9.0	0.002	0.0022
n-Hexanol	7.1	0.001	5×10^{-5}
2-Ethylhexanol	3.6	0.0001	$<1 \times 10^{-5}$
Acetone	3.9	4×10^{-5} (18°C)	

[a] Values selected from A. Seidell, *Solubilities*, Van Nostrand, New York, vol. I, 3rd ed., 1940, except those for *n*-hexanol and 2-ethylhexanol, which are from E. R. Caley and H. D. Axilrod, *Ind. Eng. Chem., Anal. Ed.* **14**, 242 (1942). The value given by M. H. Brown and J. H. Reedy, *ibid.* **2**, 304 (1930), for the solubility of LiCl in acetone has been accepted in preference to others.

A few salts are more soluble in ethyl alcohol and similar solvents than in water. Lithium iodide is an example (251 g of LiI in 100 g of C_2H_5OH at 25°C, 165 g of LiI as the trihydrate, in 100 g of H_2O at 20°C). Its behavior is not unexpected in the light of the small size of the lithium ion and the polarizability of the iodide ion. Lithium perchlorate is another example (Table 7-5). It is almost twice as soluble in ethyl alcohol as in water, despite the nonpolarizability of the perchlorate anion. The perchlorates of the alkaline earths are almost as soluble in alcohol as in water. The ionic sizes of the alkaline earth metals are greater than that of lithium, but their higher charge counterbalances this factor and the ions have a strong tendency

TABLE 7-5. **Solubilities of Alkali and Alkaline Earth Metal Perchlorates in Water, Alcohol, and Other Organic Solvents**[a]
(grams in 100 ml of solution at 25°C)

	Water	*Methanol*	*Ethanol*	*n-Propanol*	*n-Butanol*	*Ethyl acetate*	*Acetone*	*Ethyl Ether*
$LiClO_4$	47.42	89.44	79.41	61.49	49.25	63.40	76.38	64.47
$NaClO_4$	113.9	35.83	11.13	3.87	1.50	8.43	36.60	Insol.
$KClO_4$	2.04	0.083	0.0094	0.008	0.0036	0.0013	0.118	Insol.
$RbClO_4$	1.33	0.047	0.0071	0.005	0.0016	0.0014	0.075	Insol.
$CsClO_4$	1.96	0.073	0.0086	0.0045	0.0048	Insol.	0.118	Insol.
NH_4ClO_4	21.91	5.27	1.488	0.309	0.014	0.029	1.77	Insol.
$Mg(ClO_4)_2$	73.45	37.75	18.40	50.48	44.64	54.17	32.41	0.206
$Ca(ClO_4)_2$	112.3	113.7	89.55	81.69	68.42	57.38	43.81	0.185
$Sr(ClO_4)_2$	157.5	113.95	100.0	83.31	71.21	76.67	89.92	Insol.
$Ba(ClO_4)_2$	129.0	119.85	78.54	52.31	41.72	80.81	81.05	Insol.

[a] From H. W. Willard and G. F. Smith, *J. Am. Chem. Soc.* **45**, 293 (1923). Some values have been rounded.

to be solvated. The combination of a large singly charged cation with the large perchlorate ion ($KClO_4$) results in relatively low solubility in water and still lower solubility in ethyl and higher alcohols.

Analytical chemists have not been slow to take advantage of the difference in solubility of salts in oxygen–solvents for effecting separations. Lithium chloride can be separated from sodium and the other alkali metal chlorides by extracting the dry mixture with a suitable solvent, e.g., acetone, or better (to avoid incomplete recovery of lithium from its occlusion in the other chlorides) by adding a mixture of ethyl alcohol and ether to a very small volume of aqueous solution to precipitate sodium and potassium chlorides. Another precipitation method entails evaporating the aqueous solution with 2-ethylhexanol and heating the alcohol at about 130°C until the water has been driven off and only traces of sodium and potassium chloride remain in solution (see Table 7-4).

The classical gravimetric method for the determination of potassium rests upon the insolubility of potassium chloroplatinate and the solubility of sodium and lithium chloroplatinates in strong ethyl alcohol solutions. Potassium can also be determined in the presence of sodium and lithium by extracting the latter two metals as perchlorates from a dry mixture with a suitable organic solvent in which potassium perchlorate is very slightly soluble. Ethyl acetate is the preferred solvent.

Repressing solubility with organic solvents. Occasionally an organic solvent miscible with water is added to an aqueous solution to diminish the solubility of a precipitate that otherwise might be too soluble for quantitative separations. Ethyl alcohol is the solvent commonly used for this purpose. Figure 7-9 shows the solubility of a number of water-soluble and slightly soluble salts as a function of the alcohol concentration. With a little more than 20 per cent by volume of ethyl alcohol the solubility of lead sulfate is decreased tenfold. The solubility will of course be reduced still more if a common ion is added also, but it should be realized when making calculations that the activity coefficients will not be the same in

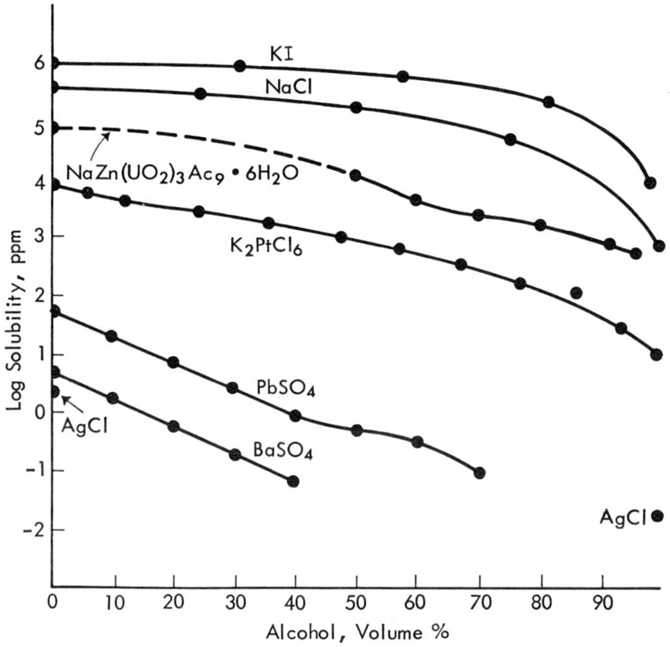

Fig. 7-9. Solubilities of some soluble and sparingly soluble salts in water-ethyl alcohol solutions. Alcohol concentration is expressed as volume per cent in the solvent at 20°C. Salt concentrations as follows: KI, mg/kg solution, 19.9°C; NaCl, mg/kg solvent, 25°C; $NaZn(UO_2)_3(C_2H_3O_2)_9 \cdot 6H_2O$, mg/l solution, 25°C; K_2PtCl_6, mg/kg solution, 20°C; $PbSO_4$, $BaSO_4$ and AgCl, mg/l solution, 25°C.

alcoholic solutions as in water. Repression of the solubility of a precipitate by addition of a miscible organic solvent must be applied with caution because the solubility of substances other than the desired one may also be repressed. This procedure is most likely to be of use when the precipitate is quite insoluble in water so that relatively little alcohol need be added, and there is little danger of precipitating soluble salts that are present. Moreover, when much alcohol is needed, the increase in volume may largely counteract the beneficial effect of the alcohol per se.

Soluble salts are sometimes crystallized from aqueous solution for purification, or for other reasons, by adding alcohol. For example, potassium bicarbonate, an acidimetric standard, can be precipitated in pure form at room temperature from a near-saturated solution in water by adding alcohol. Crystallization from a hot aqueous solution is open to the objection that some decomposition of bicarbonate (loss of CO_2) can take place under these conditions.

Organic solvent solutions of organic reagents. The use of chelating organic reagents as precipitants in gravimetric analysis rests on the difference in solubilities of the reagent and its chelate precipitate. The solubility of the reagent must be sufficiently large to permit the requisite excess to remain in solution under the conditions of quantitative precipitation of the metal chelate. The solubility of the acid form, HR, of the reagent may be orders of magnitude greater than that of the metal chelate, but the absolute solubility in water may be so small as to present problems. The solubility of HR in water-miscible organic solvents will almost always be greater than in water. The precipitant is often dissolved in ethyl alcohol. At the end of the precipitation, the solution may contain 10–20 per cent by volume of alcohol. If the pH is such that the reagent is present mostly as HR, there is danger that the excess will partially precipitate and contaminate the precipitate. Another danger is that the solubility of the metal chelate in the dilute alcohol solution may be significant. If an alcoholic solution of 8-hydroxyquinoline is used to precipitate aluminum, slightly low results are obtained because aluminum hydroxyquinolate is somewhat soluble in dilute alcohol solutions. It may be possible to avoid the use of alcohol or other water-miscible organic solvent by employing an aqueous solution, acidic or basic, to dissolve the reagent. Thus nitrosophenylhydroxylamine can be used as its ammonium salt, cupferron, 8-hydroxyquinoline can be dissolved in acetic acid or dilute hydrochloric acid, dimethylglyoxime can be dissolved in sodium hydroxide, etc. However, the problem of the possible precipitation of the excess of reagent remains. In many, if not most, precipitations with organic reagents the excess of reagent must be controlled to avoid this possibility. Sometimes the precipitate cannot be weighed as such after drying because of reagent contamination, and must be converted into another weighing form such as the metal oxide by ignition or other treatment.

Data on the solubility of chelating reagents in alcohol–water mixtures are scarce. Figure 7-10 shows the solubility of dithizone in such mixtures. It is evident that dithizone cannot be used as a precipitant because of its low solubility in alcohol–water.

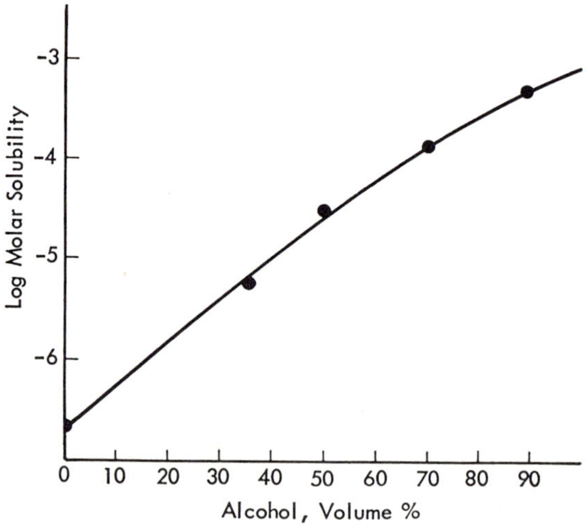

Fig. 7-10. Solubility of dithizone in ethyl alcohol-water mixtures at 25°. After Geiger and Sandell, *Anal. Chim. Acta* **8**, 200 (1953).

Differential precipitation

If two anions form slightly soluble salts with the same cation or if two cations form slightly soluble compounds with the same anion, the less soluble compound will precipitate first on the addition of a precipitating agent to a solution containing both. It is a matter of practical importance to know how completely one ion can be precipitated before the second ion reacts with the reagent.

Consider a solution of iodide and chloride which is treated with silver nitrate. Silver iodide, being less soluble than silver chloride, precipitates first, and after most of the iodide has reacted the chloride will precipitate. When silver chloride begins to separate, the supernatant liquid is in equilibrium with the two solid phases, AgI and AgCl. This means that the following relations are satisfied:

$$[Ag^+][I^-] = (K_{sp})_{AgI} = 10^{-16} \tag{22}$$

$$[Ag^+][Cl^-] = (K_{sp})_{AgCl} = 10^{-10} \tag{23}$$

From Eqs. (22) and (23) it is found that

$$[Ag^+] = \frac{(K_{sp})_{AgI}}{[I^-]} = \frac{(K_{sp})_{AgCl}}{[Cl^-]} \tag{24}$$

or, when both AgI and AgCl are present as solid phases,

$$\frac{[Cl^-]}{[I^-]} = \frac{(K_{sp})_{AgCl}}{(K_{sp})_{AgI}} = \frac{10^{-10}}{10^{-16}} = 10^6 \tag{25}$$

When $[Cl^-]/[I^-]$ is smaller than 10^6, the solubility product of silver chloride is not exceeded, and on addition of silver nitrate to the mixture silver iodide will precipitate until the above ratio has become 10^6. From then on, further addition of silver nitrate will result in a precipitation of the chloride.

If we start with a solution that is 0.01 *M* in potassium iodide and 0.1 *M* in potassium chloride, silver iodide alone will be precipitated on addition of silver nitrate until the iodide concentration has become equal to $[Cl^-]/10^6 = 10^{-7}$, or only 0.001 per cent of the original iodide remains in the solution. The separation of iodide and chloride is quantitative and could be applied for quantitative purposes if the point at which the iodide precipitation is complete could be indicated in some way or other. In volumetric analysis this can be done by using iodine–starch as an indicator. This remains blue until all the iodide has reacted and then changes with more silver nitrate to colorless before silver chloride precipitates. In the use of a silver electrode in potentiometric titrations an abrupt change of the potential of the electrode indicates the completeness of the precipitation of the iodide. In Chapter 12 (p. 254) and in the discussion of volumetric precipitation methods (Mohr's method, Chap. 35), we shall meet other cases where use is made of differential precipitation.

The preceding discussion has assumed that the two precipitates do not form mixed crystals with each other; otherwise the conclusions are invalidated (p. 234).

PROBLEMS

1. (a) Find the solubility of $La(IO_3)_3$ in water, $pK_{sp} = 11.20$; (b) Find the solubility of CaF_2 in 0.0010 *M* $CaCl_2$, $pK_{sp} = 8.5$.
2. At 100°C, the solubility of AgCl in water is 21.1 mg/liter. Calculate (a) K_{sp}; (b) the molar solubility in water; (c) the molar solubility in 0.001 *M* Ag_2SO_4 at 100° C, neglecting the intrinsic solubility and complex formation and taking the activity coefficients to be unity.

   Ans. (a) 2.2×10^{-8}; (b) 1.5×10^{-4}; (c) 1.1×10^{-5}.

3. If pK_{sp} of $Cu(OH)_2$ is 18.6 at $\mu = 0$, what is it at $\mu = 0.1$?
4. Is $SrSO_4$ ($K^0_{sp} = 2.5 \times 10^{-7}$) sufficiently insoluble to be useful for quantitative separations?
5. Using activity coefficients, calculate the solubility of $BaSO_4$ in (a) 0.1 *M* NaCl, (b) 0.01 *M* $BaCl_2$, and compare with results obtained by taking the activity coefficients to be unity.
6. The solubility of silver iodate in 0.5 *M* potassium nitrate is 2.67×10^{-4} *M* at 25°C. To 500 ml of the saturated solution is added 0.0005 mole of solid silver nitrate. Calculate the final silver and iodate concentrations in the solution, the amount of silver iodate precipitated, and the amount of silver iodate left in solution. *Ans.* $[Ag^+] = 1.067 \times 10^{-3}$; $[IO_3^-] = 6.7 \times 10^{-5}$; $AgIO_3$ precipitated = 0.0283 g; $AgIO_3$ left in solution = 0.0094 g.
7. A metal hydroxide $M(OH)_2$ which is a strong base has a solubility product equal to 4×10^{-15}. To 0.1 *M* solution of the metal salt, sodium hydroxide is added. Assuming that the volume does not change, calculate the pH after precipitation of (a) 1 per cent of the metal; (b) 50 per cent of the metal; (c) 99 per cent of the metal. *Ans.* (a) 7.30; (b) 7.45; (c) 8.30.
8. The following substances are to be precipitated: silver bromide, barium sulfate, calcium oxalate, barium iodate (iodic acid is a strong acid), zinc ammonium phosphate, magnesium hydroxide, calcium fluoride. In what cases is a quantitative precipitation likely to be obtained from acid medium?
9. Water is saturated with both silver iodate and silver hydroxide. Calculate the solubility of each constituent. $(K_{sp})_{AgIO_3} = 2.0 \times 10^{-8}$; $(K_{sp})_{AgOH} = 2 \times 10^{-8}$.
10. What is the intrinsic solubility of calcium sulfate and what fraction of the calcium in a saturated solution of calcium sulfate is present in the undissociated form?

$$\beta_1^0 = \frac{[CaSO_4]}{[Ca^{++}][SO_4^{--}]} = 200$$

$$K^0_{sp} = [Ca^{++}][SO_4^{--}] = 2.5 \times 10^{-5}$$

Ans. 0.005 *M*; 0.5.

11. Find (a) the solubility of magnesium hydroxide (inactive form) in water; (b) the pH of the saturated solution; (c) the solubility in 0.001 *M* NaOH; (d) the solubility in, and resulting pH of, a 0.001 *M* $MgCl_2$ solution. Take the activity coefficients to be unity. (Is this analytically justifiable?)

$$[Mg^{++}][OH^-]^2 = K_{sp} = 1.3 \times 10^{-9}$$

$$[MgOH^+][OH^-] = 10^{-8.3} \text{ (satd. soln.)}$$

12. At 25°C and 0.1 *M* ionic strength (0.1 *M* HCl) the dissociation constant and solubility product of tetraphenylarsonium chloroiridate (R_2IrCl_6) have the values

$$\frac{[R^+]^2[IrCl_6^{--}]}{[R_2IrCl_6]} = 4.4 \times 10^{-9}$$

$$[R^+]^2[IrCl_6^{--}] = 1.1 \times 10^{-14}$$

The dissociation constant of RCl ($\mu = 0.1$) is 8×10^{-2}. Find the solubility of tetraphenylarsonium chloroiridate in a 5×10^{-4} *M* solution of tetraphenylarsonium chloride in 0.1 *M* HCl.

13. Show that the solubility of BA_n, the salt of a weak acid, in (a) acid; (b) acid with excess A^-; and (c) acid with excess B^{+n} is given by

(a) $K_{sp}^{1/n+1}\left(\dfrac{K_a + [H^+]}{nK_a}\right)^{n/n+1}$

(b) $\dfrac{K_{sp}([H^+] + K_a)^n}{(K_aC_A)^n}$ (C_A = analytical concentration of A, including contribution from precipitate)

(c) $\left\{\frac{K_{sp}}{[B^+]}\right\}^{1/n}\left\{\frac{K_a + [H^+]}{nK_a}\right\}$

14. (a) Taking the activity coefficients to be unity, find the solubility of CaF_2 in a buffer of pH 4 that is 0.05 *M* in $CaCl_2$; (b) applying activity coefficients, calculate the solubility of $CaC_2O_4 \cdot H_2O$ in a buffer of pH 4.0 ($\mu = 0.1$).

$$K_{sp} = 2 \times 10^{-9} \qquad pK_{a1} = 1.27 \qquad pK_{a2} = 4.27$$

$[CaC_2O_4 \cdot H_2O]_{satd} = 1 \times 10^{-6}$ (intrinsic solubility)

15. (a) What is the molar solubility of $BaSO_4$ in 0.1 *M* HCl? (b) What is the molar solubility of $BaSO_4$ in 0.01 *M* $BaCl_2$ + 0.05 *M* HCl? Is $BaSO_4$ quantitatively insoluble?

$$K^0_{sp} = 1.4 \times 10^{-10}$$

$$K^0_2 = \frac{[H^+][SO_4^{--}]}{[HSO_4^-]} = 1.1 \times 10^{-2}$$

Apply activity coefficients.

16. (a) Find the solubility of $CaCO_3$ in water and the pH of the saturated solution:

$$K_{sp} = 5 \times 10^{-9} \qquad \frac{[H^+][CO_3^{--}]}{[HCO_3^-]} = 5 \times 10^{-11}$$

(b) What is the solubility in 0.05 *M* Na_2CO_3?

Ans. (a) $S = [Ca^{++}]$

$$= \frac{5 \times 10^{-9}}{3.9 \times 10^{-5}} = 1.3 \times 10^{-4}.$$

17. The solubility of copper(II) cupferrate has been reported as 1.3×10^{-4} *M* in 0.01 *M* acetic acid. Calculate the solubility product of $CuCf_2$. Does hydrolysis of Cu^{++} need to be considered?

$$(K_a)_{HAc} = 1.8 \times 10^{-5} \qquad (K_a)_{HCf} = 7 \times 10^{-5}$$

18. Find the solubility of AgBr in (a) 1.0 *M* NH_4OH; (b) 1.0 *M* ammonia saturated with AgCl.

19. (a) Find the solubility of lead iodide in 0.10 *M* KI solution. (b) What are the concentrations of Pb^{++}, PbI^+, PbI_2, PbI_3^-, PbI_4^{--}, and I^- in a saturated solution of lead iodide? What is the solubility of PbI_2 in water?

$$K_{sp} = [Pb^{++}][I^-]^2 = 3 \times 10^{-8}$$

Logs of formation constants of Pb–I complexes: β_1, 1.3; β_2, 2.8; β_3, 3.4; β_4, 3.9.

20. Tetragonal HgI_2 has a solubility of 0.061 g/liter at 25°C. What would be its solubility if its solution contained no Hg–I complexes (including undissociated HgI_2)? Logs of formation constants of Hg–I complexes: β_1, 12.9, β_2, 23.8; β_3, 27.6; β_4, 29.8.

21. Construct the logarithmic concentration–pH diagram of a lead(II) solution saturated with yellow PbO, using the following constants:

$$(K_{sp})_{\text{yellow PbO}} = [Pb^{++}][OH^-]^2 = 10^{-15.1}$$

$$\text{Solubility of yellow PbO} = [Pb(OH)_2] = 10^{-4.2}$$

$$\frac{[PbOH^+][H^+]}{[Pb^{++}]} = 3 \times 10^{-7}$$

$$\frac{[Pb(OH)_3^-][H^+]^3}{[Pb^{++}]} = 7 \times 10^{-29}$$

What are the relative amounts of Pb^{++}, $PbOH^+$, $Pb(OH)_2$, and $Pb(OH)_3^-$ in a solution at pH 7? pH 11?

22. (a) From the equilibrium constants

$$\frac{[Be^{++}][OH^-]}{[BeOH^+]} = 3 \times 10^{-8} \qquad (K_{sp})_{Be(OH)_2} = 2 \times 10^{-18}$$

$$\frac{[Be^{++}]^2[OH^-]}{[Be_2OH^{3+}]} = 3 \times 10^{-11}$$

find the pH at which the sum of concentrations of all ionic species of beryllium in solution is 1×10^{-6}, if the original concentration of Be^{++} is 0.02 *M*, and the volume is taken as constant. (b) At what calculated pH will the precipitation of $Be(OH)_2$ begin from a 0.02 *M* Be solution? If no precipitate appears at this pH what is the probable reason?

23. Is precipitation of aluminum hydroxide quantitative at pH 6.0?

24. Using the value 6×10^{-18} for $(K_{sp})_{Ni(OH)_2}$, 2×10^{-5} for K_b of NH_4OH, and the following values for nickel ammine dissociation constants,

$$K_n = \frac{[Ni^{++}][NH_3]^n}{[Ni(NH_3)_n^{++}]} \qquad ([NH_3] = [NH_4OH])$$

$$K_1 = 1.6 \times 10^{-3}$$

$$K_2 = 9 \times 10^{-6}$$

$$K_3 = 1.7 \times 10^{-7}$$

$$K_4 = 1.1 \times 10^{-8}$$

calculate the concentration of dissolved nickel species in a solution of pH 7 which is 0.5 *M* in NH_4Cl and 0.004 *M* in Ni. The concentration of $NiOH^+$ may be neglected.

25. Will aluminum be precipitated quantitatively if a 0.001 *M* Al solution which is 0.025 *M* in F^- is adjusted to pH 7? $K_{sp} = [Al][OH^-]^3 = 10^{-32}$. Al–F complexes:

$$\log \beta_1 = 6.1 \qquad \log \beta_4 = 17.7$$

$$\log \beta_2 = 11.15 \qquad \log \beta_5 = 19.4$$

$$\log \beta_3 = 15.0 \qquad \log \beta_6 = 19.7$$

26. Find the approximate pH at which precipitation of Hg(II) will begin from a 0.1 *M* sodium chloride solution which is 0.01 *M* in mercury when sodium hydroxide solution is gradually added. $[Hg^{++}][Cl^-]^4/[HgCl_4^{--}] = 1 \times 10^{-15}$; other complex mercury(II) chloride species may be neglected. The intrinsic solubility of HgO is 2.5×10^{-4} *M* and pK_{sp} of $Hg(OH)_2$ is 25.4.

27. Find the solubility of HgS in 1.0 *M* HCl solution which is 0.001 *M* in H_2S at 20°C.

$$Hg^{++} + 3Cl^- \rightleftharpoons HgCl_3^- \qquad \log \beta_3 = 14.1$$

$$Hg^{++} + 4Cl^- \rightleftharpoons HgCl_4^{--} \qquad \log \beta_4 = 15.1$$

Ans. Since $[HgCl_4^{--}] = 10^{-11.9}$ and [HgS] probably does not exceed 10^{-9}, the solubility is given by $[Hg(HS)_2] = 1 \times 10^{-9}$.

28. Find the solubility of silver sulfide in 0.1 *M* H^+ solution saturated with hydrogen sulfide at room temperature, given the following constants for $\mu = 0.1$:

$$(K_{sp})_{Ag_2S} = 6 \times 10^{-49}$$

$$\frac{[AgHS]}{[Ag^+][HS^-]} = 2 \times 10^{13}$$

$$\frac{[Ag(HS)_2^-]}{[Ag^+][HS^-]^2} = 5 \times 10^{17}$$

$$\frac{[AgS^-]}{[Ag^+][S^{--}]} = 6 \times 10^{16}$$

Assume that the solubility product of AgHS is not exceeded and that the intrinsic solubility of Ag_2S can be neglected.

29. A solution of manganese(II) in 100 ml of 0.1 *M* ammonia, which is also 0.1 *M* in ammonium chloride, is saturated with hydrogen sulfide. 2 ml of 14 *M* ammonia is added and the solution is allowed to stand overnight. Will manganese be precipitated quantitatively?

30. The solubility product of silver iodate in water is 3×10^{-8} and that of silver chloride 10^{-10} at room temperature. 100 ml of 0.1 *M* silver nitrate is added to 100 ml of a mixture 0.1 *M* in chloride and 0.1 *M* in iodate. (a) What is the sum of the concentrations of chloride and iodate after the addition of silver nitrate? (b) How large are the chloride and iodate concentrations, respectively, in the solution? (c) What is the chloride concentration in the solution at the moment when silver iodate separates with the chloride on addition of more silver nitrate? *Ans.* (a) $[Cl^-] + [IO_3^-] = 0.05$; (b) $[Cl^-] = 1.66 \times 10^{-4}$; $[IO_3^-] = 4.98 \times 10^{-2}$; (c) $[Cl^-] = 1.67 \times 10^{-4}$.

31. Find the solubility of AgCN in (a) a buffer of pH 9; (b) water in equilibrium with the air (pH 5.7); (c) 0.050 *M* $HClO_4$, neglecting activity coefficients.

$$[Ag^+][CN^-] = 1 \times 10^{-16}$$

$$\frac{[Ag^+][CN^-]^2}{[Ag(CN)_2^-]} = 1 \times 10^{-21}$$

$$\frac{[H^+][CN^-]}{[HCN]} = 5 \times 10^{-10}$$

Ans. (a) 6.3×10^{-6} *M*; (b) 6.3×10^{-6} *M*; (c) 1.0×10^{-4} *M*.

32. 0.244 g of $BaSO_4$ is heated with 100 ml of 0.10 *M* Na_2CO_3 until equilibrium is attained and the mixture is then cooled to room temperature. What percentage of the barium sulfate has been converted to $BaCO_3$? At room temperature and $\mu = 0.1$ the solubility-product constants of barium carbonate and barium sulfate are $10^{-7.5}$ and $10^{-9.2}$.

33. Above a pH of ~ 2.5, $[HSO_4^-]$ can be neglected compared to $[SO_4^{--}]$, so that the effective (conditional) solubility product of $BaSO_4$ can be written

$$(K'_{sp})_{BaSO_4} = [Ba'][SO_4^{--}] = \alpha_{Ba}[Ba^{++}][SO_4^{--}] = \alpha_{Ba}(K_{sp})_{BaSO_4}$$

where [Ba′] represents the analytical concentration of Ba in the solution (Ba^{++} and Ba complexes) and $\alpha_{Ba} = [Ba']/[Ba^{++}]$. At pH 8 in 0.1 *M* EDTA solution $\alpha_{Ba} = 10^{4.2}$. Find the solubility of $BaSO_4$ in this solution, taking $(K_{sp})_{BaSO_4}$ as $10^{-9.2}$. *Ans.* $10^{-2.5}$ *M*.

34. Calculate the effective (conditional) solubility product of CaF_2 at pH 2.

$$(K_{sp})_{CaF_2} = 10^{-9.8} \qquad (K_a)_{HF} = 10^{-3.05}$$

Ans. $10^{-9.8+1.09} = 10^{-8.7}$.

35. A metal chelate complex ML_2 has a molecular solubility of 1×10^{-6} *M* in water. What may be the lower limit of its dissociation constant

$$\frac{[M^{++}][L^-]^2}{[ML_2]} = K$$

in order that its solubility (molecular and ionic) shall not exceed 1×10^{-5} *M* when $[L^-] = 0.01$? *Ans.* 9×10^{-4}.

36. Suppose the molar solubility (S) of $M(OH)_3$ in pure water is to be calculated from constants available in the literature. What is wrong in the following approach?

$$[M^{+3}][OH^-]^3 = K_{sp} = 1 \times 10^{-32}$$

$$S = [M^{+3}]$$

$$S(3S)^3 = 1 \times 10^{-32}$$

$$\text{Ergo, } S = \left(\frac{1 \times 10^{-32}}{27}\right)^{1/4} = 4 \times 10^{-9}$$

37. A method for preparing sodium hydroxide solutions low in carbonate is based on the precipitation of the carbonate with calcium ion. A 1.0 *M* sodium hydroxide solution is shaken with a slight excess of calcium hydroxide (solid phase at equilibrium). Calculate the CO_3^{--} concentration of the solution (constants in Appendix). Estimate the probable accuracy of the result.

CHAPTER 8 *Electrode Potentials. Oxidation–Reduction Equilibria*

ELECTRODE POTENTIALS

Single electrode potential and electrode reactions

A metal may be looked upon as being composed of metal ions and electrons. Consider a metal that is placed in a solution containing ions of the metal, e.g., zinc in a solution of zinc sulfate. Both the metal and the solution contain zinc ions, but the metal–ion activity in the metal, in general, will be different from that in the solution and there will be a distribution of the zinc ions between metal and solution until a state of equilibrium is attained.

Consider the situation in which more zinc ions tend to leave the metal per unit time than are transferred from the solution to the metal. The metal will acquire a negative charge (excess of electrons) and the solution, a positive charge. The attractive coulombic forces between these negative and positive charges are very strong and hold the charges closely together. As a result, there is extremely little charge transfer and these charges will be located at the surface of the metal–solution phases, often called the *interface* (see Fig. 8-1). An electrical double layer is established with a corresponding potential difference between metal and solution;

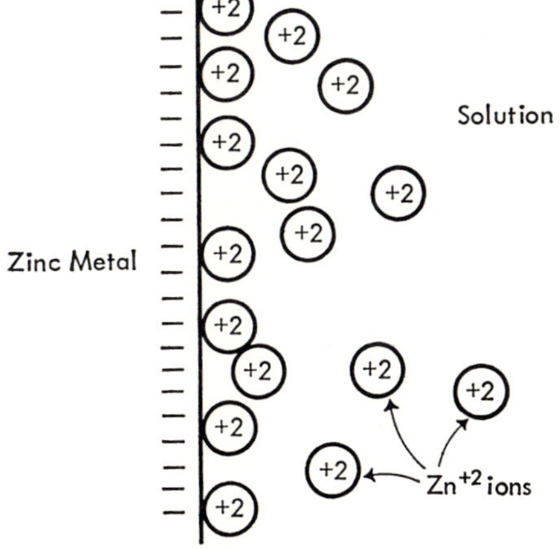

Fig. 8-1. Zinc metal-zinc ion solution interface.

the zinc metal with its excess negative charges has a negative potential, the potential difference between the bulk of the zinc phase and the bulk of the solution phase often being called the *electrode potential.* If more metal ions enter the metal than are transferred from the metal to the solution, the metal would acquire a positive potential. Such a situation probably exists with the more noble metals in solutions of their salts.

When we withdraw free negative charges (or electrons) from the zinc metal, for example, by making it part of an appropriate electrochemical cell (see below), there is a relative increase of zinc ions over electrons in the metal, and the above charge-transfer equilibrium is disturbed. More zinc ions must be transferred from the metal to the solution to "neutralize" or "equalize" the deficiency of negative charges. Formally, this reaction may be represented by the equation

$$\mathrm{Zn} \rightarrow \underset{\text{(to solution)}}{\mathrm{Zn}^{++}} + \underset{\text{(to other electrode in cell)}}{2e} \qquad (1)$$

in which the electrons formed are removed by the metal. On the other hand, if an excess of negative charges is imposed on the zinc (in an electrolysis cell), more zinc ions will be transferred from the solution to the metal, in order to restore the charge-transfer equilibrium. Formally the reaction is represented by

$$\mathrm{Zn}^{++} + 2e \rightarrow \mathrm{Zn} \qquad (2)$$

Under conditions of equilibrium, we have

$$\mathrm{Zn}^{++} + 2e \rightleftharpoons \mathrm{Zn} \qquad (3)$$

Reactions (1) and (2) are called *electrode reactions* or *electrolysis reactions.* Reaction (3) is reversible at the zinc electrode, which may be called a *reversible electrode.*

The absolute value of an electrode potential cannot be measured, but the *difference* in potential between two electrodes, in general, can be measured accurately.

The tendency of a metal to send ions into the solution, which is sometimes called the *electrolytic solution tension*, differs for different metals. It is the greatest for the base metals and the smallest for the noble metals.

Oxidation–reduction reactions and potentials

A reaction that involves the transfer of electrons is called an *oxidation–reduction*, or *redox, reaction.* A substance that combines with electrons is called an oxidizing agent or oxidant, whereas a substance that donates one or more electrons is called a reducing agent or reductant. Upon combining with electrons, an oxidant is reduced; while a reductant, upon donation of electrons, is oxidized. Thus Eq. (1) represents the oxidation of the reductant, zinc, to zinc ions, whereas Eq. (2) represents the reduction of zinc ions to zinc metal. In general, we represent an oxidation–reduction reaction by the equation

$$\mathrm{Ox} + ne \rightleftharpoons \mathrm{Red} \qquad (4)$$

From left to right the reaction is a reduction, and from right to left an oxidation.

As explained in the previous section, reactions (1), (2), and (4) cannot take place spontaneously. For example, to oxidize ferrous iron,

$$\mathrm{Fe}^{++} \rightleftharpoons \mathrm{Fe}^{3+} + e \qquad (5)$$

an oxidizing agent must be added (e.g., chlorine), which is the acceptor of electrons donated by the ferrous ions:

$$\tfrac{1}{2}Cl_2 + e \rightleftharpoons Cl^- \qquad (6)$$

The combination of the two half-reactions (4) and (5) yields the overall oxidation–reduction (redox) reaction:

$$Fe^{++} + \tfrac{1}{2}\,Cl_2 \rightleftharpoons Fe^{3+} + Cl^-$$

A reductant with its conjugate oxidized form represents a redox system, e.g., ferric–ferrous and chlorine–chloride systems. An oxidation or reduction always involves two redox systems. A metal placed in a solution containing the corresponding metal ions is a redox system. The greater the electrolytic solution tension or the more base the metal, the stronger a reductant the metal is. For example, zinc is a much stronger reducing agent than copper, or copper ions are a much stronger oxidizing species than zinc ions. This means that copper ions can oxidize zinc [compare Eqs. (4) and (5)]:

$$\begin{array}{r} Zn \rightleftharpoons Zn^{++} + 2e \\ Cu^{++} + 2e \rightleftharpoons Cu \qquad \\ \hline Cu^{++} + Zn \rightleftharpoons Zn^{++} + Cu \end{array} \qquad (7)$$

Redox equation (7) illustrates the well-known fact that a base metal precipitates a more noble metal from its solutions. In agreement with the above terminology, the potential of a metal in a solution of its ions (electrode potential) may be called either the oxidation or the reduction potential. As will be made clear below, these latter two terms have very specific meaning when numerical calculations are involved. The reduction potential of nonmetallic systems, like ferric–ferrous, ferricyanide–ferrocyanide, or uncharged bromine with its conjugate reduced form, bromide, can be measured with a platinum electrode or another noble metal (unattackable) electrode, which does not react chemically with the system. An electrical double layer is formed at the interface platinum–solution by transfer of electrons from the solution to the metal or vice versa. The noble metal is an *indicator electrode* in the sense that it indicates the tendency of the redox system to give off or take up electrons, and the potential is a quantitative measure of this tendency.

Electrode potential and emf of a cell

As already stated, a single potential or potential difference at the metal–solution interface cannot be measured with any degree of accuracy, and the absolute values of redox potentials are unknown. What can be measured accurately is the potential difference between two indicator electrodes in two half-cells. Two reversible half-cells in electrolytic contact with each other constitute an electrochemical, or *galvanic*, cell. Neglecting liquid-junction potentials (see below), the electromotive force (emf) of such a cell is equal to the difference in potential between the two electrodes. When one of the two half-cells is always the same, the difference in emf between two cells corresponds to the *difference* in potential between the two reduction potentials of the different half-cells. The half-cell that is always the same is called the *reference electrode*.

Example: Cell 1: reference electrode ‖ $AgNO_3(C_{Ag^+})$ | Ag
(half-cell) (half-cell)

In cell 1 the vertical line represents a phase boundary across which there exists a potential difference, and the double vertical line represents an electrolytic contact between the two half-cells which has zero potential difference across the phase boundary. In any real cell this contact may be the source of a (small) emf which is called the liquid-junction potential. By making contact with an appropriate electrolyte of suitable concentration, the liquid-junction potential, in general, can be kept small. When the double vertical line appears in a diagrammatic representation of an electrolytic cell, the liquid-junction potential is assumed to be zero. The use of a single vertical line would signify that an appreciable liquid-junction potential exists and should be taken into account. The liquid-junction potential is neglected in the present introduction to potentiometry but must be considered in any exact treatment.

Calling the reduction potential of the silver electrode $E_{Ag+/Ag}$ in the right-hand half-cell, and that of the reference electrode E_{RE}, we have

$$E_1 = E_{Ag+/Ag} - E_{RE} \tag{8}$$

where E_1 is the emf of cell 1.

Now consider cell 2:

Cell 2: reference electrode | $Fe^{3+}(C_{Fe^{3+}})$, $Fe^{++}(C_{Fe^{++}})$ | Pt

$$E_2 = E_{Fe^{3+}/Fe^{++}} - E_{RE} \tag{9}$$

where $E_{Fe^{3+}/Fe^{++}}$ represents the reduction potential of the solution containing ferric and ferrous ions. Evidently

$$E_1 - E_2 = E_{Ag^+/Ag} - E_{Fe^{3+}/Fe^{++}}$$

i.e., the difference between the emf of both cells is equal to the difference in potential between the silver and platinum electrodes.

By international agreement, all potentials are referred to the *normal hydrogen* electrode (N.H.E.) as the reference electrode. The potential of this electrode is *arbitrarily* taken equal to zero.[1]

Sign of electrode potential

According to the above, an electrode potential (hydrogen scale) is *numerically* equal to the emf of the cell:

Cell 3a: Pt, H_2 ($p_{H_2} = 1$ atm) | $H^+(a = 1)$ || $Ag^+(C_{Ag^+})$ | Ag^+

or

$$\text{N.H.E.} \parallel Ag^+(C_{Ag^+}) \mid Ag$$

Associated with the schematic representation of a complete electrochemical cell is a *cell reaction.* This cell reaction is found by *assuming* that oxidation occurs in

[1] This arbitrary zero potential is defined as the potential of the reversible hydrogen electrode in equilibrium with hydrogen gas at 1 atm partial pressure, and immersed in a solution containing hydrogen ions at unit activity:

$$\text{Pt, } H_2\ (p_{H_2} = 1\text{ atm}) \mid H^+(a = 1)$$

The scale of potentials based on this arbitrary zero point is called the *hydrogen scale.*

It is immaterial whether we refer a potential to the N.H.E. (as absolute zero) or any other reference electrode. The difference between electrode potentials is independent of the electrode used as reference.

the left half-cell and reduction occurs in the right half-cell. Thus for cell 3a, the cell reaction is given by

Right half-cell: $Ag^+ + e \rightleftharpoons Ag$

Left half-cell: $\frac{1}{2} H_2 \rightleftharpoons H^+ + e$

Cell reaction: $Ag^+ + \frac{1}{2} H_2 \rightleftharpoons Ag + H^+$

For cell 3b

Cell 3b: $Ag \mid Ag^+(C_{Ag^+}) \parallel$ N.H.E.

the cell reaction is given by

Right half-cell: $H^+ + e \rightleftharpoons \frac{1}{2} H_2$

Left half-cell: $Ag \rightleftharpoons Ag^+ + e$

Cell reaction: $Ag + H^+ \rightleftharpoons Ag^+ + \frac{1}{2} H_2$

The internationally accepted convention is to assign the cell emf a positive sign when the electrode on the left is negative and that on the right is positive. This is the case for cell 3a, in which the N.H.E. is the negative electrode and the silver electrode is the positive electrode. Since we conventionally adopt the potential of the N.H.E. as zero, the emf of cell 3a is equal to the potential of the silver electrode [Eq. (8)]:

$$E_{3a} = E_{Ag^+/Ag} - E_{NHE} = E_{Ag^+/Ag} \qquad (10)$$

Thus the silver electrode in a solution of silver ions has a *positive* potential. The electrode reaction assumed to occur at the silver electrode in cell 3a is a reduction, and this is the origin of the term *reduction potential* (or half-cell e.m.f.[2]) as we shall use it in this text. The subscript Ag^+/Ag appears in the same order as in the half-reaction for the right side of cell 3a.

For cell 3a, the right electrode is positive and the cell reaction represents the actual chemical reaction that would occur if the contents of both half-cells were placed in one beaker. Hence the cell reaction and the spontaneous reaction are the same. This is not the case for cell 3b, in which the right electrode is more negative than the left electrode and cell 3b is said to have a negative emf.

A base metal such as zinc (in a solution of zinc ions),

$$\text{N.H.E.} \parallel Zn^{++}(C_{Zn^{++}}) \mid Zn$$

has a negative potential, and the spontaneous chemical reaction on mixing the contents of the two half-cells

$$2\,H^+ + Zn \rightleftharpoons H_2 + Zn^{++}$$

does not correspond to the cell reaction.[3]

[2] J. J. Lingane, *Electroanalytical Chemistry,* Wiley-Interscience, New York, 2nd ed., 1958.

[3] It is evident from the above that the numerical value of the potential, as well as its sign, are entirely a matter of convention. The above sign convention is not universally adopted, and large groups assign the potential a sign opposite to that used above. This potential is frequently called the *oxidation* potential because it represents the emf of cell 3b, in which oxidation is assumed to be occurring at the silver electrode:

$$E_{3b} = E_{Ag/Ag^+} - E_{NHE} = E_{Ag/Ag^+}$$

The oxidation potential is the negative of the reduction potential:

$$E_{Ag^+/Ag} = -E_{Ag/Ag^+}$$

The sign convention of the potential was adopted in 1953, by the International Union of Pure and Applied Chemistry in Stockholm (referred to as the Stockholm convention). In this convention the sign of the *electrode potential* is the same as the *reduction potential.*

Electrochemical (galvanic) cells and electrolysis cells

Consider cell 4:

Cell 4: $-\mathrm{Zn} \mid \mathrm{ZnSO_4}(C_{\mathrm{Zn}^{++}}) \parallel \mathrm{CuSO_4}(C_{\mathrm{Cu}^{++}}) \mid \mathrm{Cu}+$

The copper electrode is the positive electrode, and the zinc electrode is the negative electrode. According to the above, the emf, E, of the cell is

$$E_4 = E_{\mathrm{Cu}^{++}/\mathrm{Cu}} - E_{\mathrm{Zn}^{++}/\mathrm{Zn}}$$

and the cell reaction is

$$\mathrm{Cu}^{++} + \mathrm{Zn} \rightleftharpoons \mathrm{Cu} + \mathrm{Zn}^{++}$$

Figure 8-2 presents a diagrammatic picture of the cell. ABC represents a glass tube filled with a solution of an electrolyte, like potassium chloride or nitrate, which serves to bring about electrolytic contact between the two half-cells. It is called a *salt bridge.*

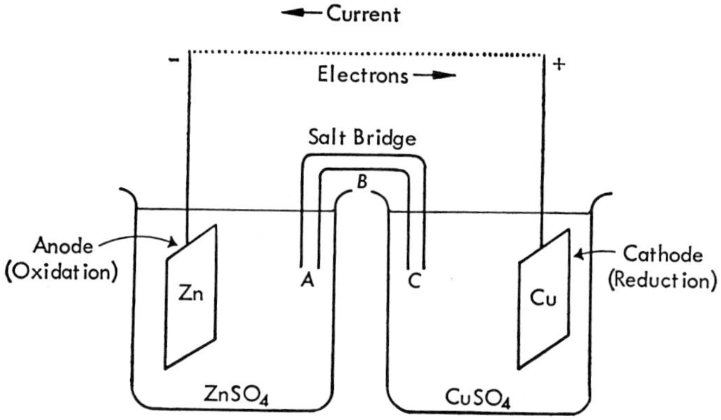

Fig. 8-2. Electrochemical cell.

No electrochemical reaction can occur in the cell until the copper and zinc electrodes are connected by a metal wire (dotted line in Fig. 8-2). When the circuit is closed, the cell operates *spontaneously.* The copper has a more positive potential than the zinc, and electrons flow from the zinc to the copper. The flow of electrons is called an *electric current.* Conventionally, the current is assumed to flow in a direction opposite to that of electrons, i.e., from a region of positive potential to a region of more negative potential.

With the flow of current, reactions occur at both electrodes. For the copper electrode to maintain its potential, the electrons reaching the copper electrode react with copper ions at the surface of the electrode:

$$\mathrm{Cu}^{++} + 2e \rightleftharpoons \mathrm{Cu}$$

In other words, copper ions are being reduced. To maintain the potential of the zinc electrode, electrons must be provided, which is done by the reaction

$$\mathrm{Zn} \rightleftharpoons \mathrm{Zn}^{++} + 2e$$

indicating that zinc is being oxidized. When the cell operates spontaneously, the reaction occurring in the cell is

$$Cu^{++} + Zn \rightleftharpoons Cu + Zn^{++} \tag{11}$$

Reaction (11) is exactly the reaction that would occur if the contents of both half-cells were mixed together. In this case the spontaneous cell reaction for cell 4 corresponds to the cell reaction. The electrode at which oxidation occurs (Zn electrode) is called the *anode* and the electrode at which reduction occurs is the *cathode*. In a galvanic cell, the negative electrode is the anode and the positive electrode is the cathode.

We shall now consider the question of whether it is possible to force cell reaction (11) to occur in the opposite direction, i.e., from right to left. This can be done by imposing an opposing external voltage. This is pictured in Fig. 8-3. *BA* represents a

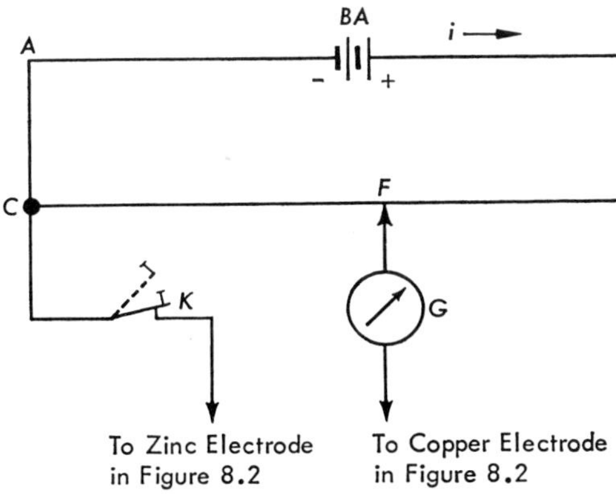

Fig. 8-3 Simple slidewire potentiometer.

battery, the emf of which is greater than that of the cell. Positive current flows through the circuit in the direction of the arrow. *AC* and *A'D* represent metallic wires of negligible resistance. Thus the potential difference between *C* and *D* is

$$E_C - E_D = E_{BA}$$

CD represents a resistance, for example, in the form of a thin, uniform wire. *F* represents a sliding contact that can be moved from point *C* to point *D*, *G* a galvanometer between the copper electrode and *F*, and *K* a tapping key (momentary switch), which is closed briefly during measurements. When both electrodes of the Zn–Cu cell are connected to *C*, the cell operates spontaneously and positive current flows through *G* from the positive (copper) electrode of the cell to *C*. When the sliding contact is now moved from *C* in the direction of *D*, an opposing voltage is applied to the cell which opposes the current flowing through the cell in the direction $G \rightarrow F \rightarrow C \rightarrow K \rightarrow \text{cell} \rightarrow G$. As long as $E_C - E_F$ is smaller than E_{cell}, the cell operates spontaneously and positive current goes from $G \rightarrow F \rightarrow C \rightarrow K$. When $E_C - E_F$ becomes equal to E_{cell}, no current flows through $C \rightarrow F \rightarrow G \rightarrow \text{cell} \rightarrow K \rightarrow C$ as indicated by the galvanometer *G*, a current-indicating instrument. When $E_C - E_F$ becomes greater than E_{cell}, positive current flows in the direction $C \rightarrow F \rightarrow G \rightarrow$ cell $\rightarrow K \rightarrow C$, which is opposite to that when the cell operates spontaneously. Electrons are now flowing through the metallic conductor from the copper electrode to the zinc electrode.

To maintain this flow of electrons, zinc ions are reduced and copper ions oxidized:

$$\begin{array}{ll} Zn^{++} + 2e \rightleftharpoons Zn & \\ \quad Cu \rightleftharpoons Cu^{++} + 2e & \\ \hline Zn^{++} + Cu \rightleftharpoons Zn + Cu^{++} & \text{(electrolysis reaction)} \end{array}$$

The zinc electrode has now become the cathode and the copper electrode, the anode. When the above reaction is forced to occur in the direction from left to right, the electrochemical cell has been transformed into an *electrolysis cell*.[4]

Measurement of emf. Potentiometry. Reference electrode

It would seem at first glance that the emf of a cell can be measured easily by means of a voltmeter. However, under these conditions current flows through the circuit and electrode reactions occur that affect the potential of the electrode. To prevent this, use is made of the principle described in the preceding section. We have seen that $E_C - E_F$ (Fig. 8-3) is equal to the emf of the cell, E_{cell}, when the galvanometer indicates that no current flows through $C \rightarrow K \rightarrow$ cell $\rightarrow G \rightarrow F$; hence no current is drawn from the cell. Under these conditions, when E_{cell} is compensated by an equal opposing emf:

$$E_{cell} = E_C - E_F \tag{12}$$

We can now apply Ohm's law, which states that $E = iR$, where i is the current and R the resistance. When the galvanometer indicates zero deflection and Eq. (12) is satisfied, we have

$$i = \frac{E_C - E_F}{R_{CF}} = \frac{E_C - E_D}{R_{CD}} = \frac{E_{BA}}{R_{CD}}$$

or

$$E_{cell} = E_C - E_F = \frac{R_{CF}}{R_{CD}} E_{BA}$$

The instrument that gives us R_{CF} and R_{CD} when the system is balanced is called a *potentiometer*. To determine E_{BA}, a cell with an exactly known emf is used instead of the unknown cell.[5] The cell with known emf is called a *standard cell*.

In analytical work the cell is usually composed of a reference electrode and the unknown half-cell, the electrode potential of which is to be determined (indicator

[4] This designation of electrolysis is not logical. Electrolysis reactions also occur in an electrochemical cell, but in the opposite direction to those in an "electrolysis" cell. It would be more logical to call the electrolysis in a galvanic cell "spontaneous (internal) electrolysis" and that in an electrolysis cell, "forced or imposed electrolysis." When the term electrolysis is used in this text, it refers to forced electrolysis unless stated otherwise.

[5] In an actual potentiometer, the sliding contact F in Fig. 8-3 is coupled to a scale that is calibrated in volts, and a variable resistance is inserted between points A and C. The instrument is calibrated to read directly in volts by setting F at the voltage of the standard cell and adjusting the variable resistance until the galvanometer indicates zero deflection. The standard cell is then replaced by the electrochemical cell, and the sliding contact, F, moved until no current passes through G. The e.m.f of the electrochemical cell is read directly from the scale in volts.

electrode). Conventionally, E_{cell} is taken equal to the potential of the electrode with reference to that of the reference electrode, i.e., the reference electrode is assumed to be on the left. The measurement of an electrode potential is called *potentiometry*.

In a previous section it was mentioned that, by international agreement, the potential of the normal hydrogen electrode is arbitrarily taken equal to zero. In experimental work it is not convenient to use this electrode as a reference electrode. Instead, an easily reproducible reference electrode is used. The commonest and most generally useful reference electrode is the *saturated calomel electrode* (S.C.E.), consisting of a mercury electrode in a solution saturated with potassium chloride and mercurous chloride. *The potential of this electrode at* 25°C *is* 0.242 V *with respect to the standard hydrogen electrode*, i.e., the emf of cell 5,

Cell 5: $$\text{N.H.E.} \left\| \begin{matrix} \text{KCl (sat.)} \\ \text{Hg}_2\text{Cl}_2 \text{ (sat.)} \end{matrix} \right| \text{Hg}$$

is

$$E_5 = E_{\text{SCE}} = 0.242 \text{ V}$$

For example, in measuring the potential of a silver electrode in a solution containing silver ions using a S.C.E. as reference, e.g., cell 6:

Cell 6: $$\text{S.C.E.} \parallel \text{AgNO}_3(C_{\text{Ag}^+}) \mid \text{Ag}$$
$$E_6 = E_{\text{Ag}^+/\text{Ag}} - E_{\text{SCE}} = E_{\text{Ag}^+/\text{Ag}} - 0.242$$
$$E_{\text{Ag}^+/\text{Ag}} = E_6 + 0.242 \text{ (versus N.H.E.)}$$

Reversible and irreversible electrode reactions

We have seen that when cell 4, the zinc–copper cell, operates spontaneously, the cell reaction $Cu^{++} + Zn \rightleftharpoons Cu + Zn^{++}$ occurs from left to right, while upon electrolysis, the reaction occurs from right to left. Considering each electrode separately, the reaction $Cu^{++} + 2e \rightleftharpoons Cu$ runs from left to right when the cell operates spontaneously, whereas in the electrolysis cell the reaction goes from right to left. *An electrode reaction is called reversible when the same reaction occurs, but in the opposite direction, when the flow of current is reversed.* When a different reaction occurs when the flow of current is reversed, the electrode reaction is *irreversible*. The copper and zinc electrodes in the above cell are reversible. The same is true of a platinum electrode placed in a solution containing iodine and iodide or ferric and ferrous iron. Mercury in the saturated calomel electrode and silver in a system containing silver ions (e.g., silver halides) are reversible electrodes:

$$Hg_2^{++} + 2e \rightleftharpoons 2\,Hg$$
$$Ag^+ + e \rightleftharpoons Ag$$

A platinum electrode in an air-free solution of lead or copper or thallium ions is *irreversible*. If this electrode is made the anode in an electrolysis cell and no base metal is on the electrode to be oxidized to metal ions, water or hydroxyl ions will be oxidized to oxygen:

$$H_2O \rightarrow \tfrac{1}{2}\,O_2 + 2\,H^+ + 2e$$

After the electrode has been made the cathode in the electrolysis cell and the platinum has been covered with a layer of the base metal, the electrode will become reversible. When it is now made the anode, the metal will be oxidized to metal ions.

Quite generally a platinum electrode in an aqueous solution of an acid, alkali, or an alkali salt is *irreversible*. When the electrode is made the cathode, the electrode reaction corresponds to the reduction of hydrogen ions or water:

$$2\,H^+ + 2e \rightleftharpoons H_2 \tag{14}$$

or

$$2\,H_2O + 2e \rightleftharpoons H_2 + 2\,OH^- \tag{14a}$$

On the other hand, when it is made the anode, reaction (15) or its equivalent (15a) occurs:

$$H_2O \rightleftharpoons \tfrac{1}{2}\,O_2 + 2H^+ + 2e \tag{15}$$

$$2\,OH^- \rightleftharpoons \tfrac{1}{2}\,O_2 + H_2O + 2e \tag{15a}$$

The Nernst equation

Consider the cell

$$\text{N.H.E.} \parallel M^{n+}\ (C_{M^{n+}}) \mid M$$

which, for simplicity, is written as the half-cell

$$M^{n+}\ (C_{M^{n+}}) \mid M$$

with the half-cell reaction

$$M^{n+} + ne \rightleftharpoons M \tag{16}$$

The reasons for the existence of a potential difference across the interface were discussed above, but no consideration was given to what effect the concentration of M^{n+} has on the electrode potential. It is easily seen that increasing the concentration of M^{n+} increases the rate at which M^{n+} ions enter the metal, and thereby makes the electrode potential more positive. Similar considerations hold when both the oxidized and reduced forms are soluble in the solution; increasing the concentration of the oxidized form increases the electrode potential, whereas increasing the concentration of the reduced form decreases the electrode potential.

The quantitative relationship between reduction potential and activities of the reduced and oxidized forms is known as the *Nernst equation*.

The Nernst equation[6] for the generalized half-reaction

$$aA + bB + cC + \cdots + ne \rightleftharpoons pP + qQ + rR + \cdots \tag{17}$$

is

$$E = E^0 + \frac{RT}{nF} \ln \frac{(A)^a(B)^b(C)^c \cdots}{(P)^p(Q)^q(R)^r \cdots} \tag{18}$$

where R is the gas constant, T the absolute temperature, F the Faraday, n the number of electrons taking part in the reaction, ln the natural logarithm, and E^0 the standard reduction potential. The value of E^0 is a function of temperature, and values at 25°C are given in Table A-12.

[6] We omit the derivation of this fundamental equation because it may be found developed in any elementary textbook of physical chemistry. As originally derived by Nernst (1889), concentrations rather than activities were used, because the concept of activity was then virtually unknown.

At 25°C, $T = 298$, Eq. (18) may be written in the form

$$E = E^0 + \frac{0.059}{n} \log \frac{(\mathrm{A})^a(\mathrm{B})^b(\mathrm{C})^c \cdots}{(\mathrm{P})^p(\mathrm{Q})^q(\mathrm{R})^r \cdots} \tag{19}$$

where log is the logarithm to the base 10. Note that in the most precise calculations, 0.05916, rather than 0.059, is used.

Identical conventions are used in Eq. (19) as are used for equilibrium constants (p. 36), and all solid phases involved in the half-reaction (17) are assigned unit activity. Similarly, in dilute solution the activity of water is assumed to be unity. It is conventional to express the activity of gaseous substances in terms of their fugacity. Usually, no significant error is made by substituting the partial pressure of a gas in atmospheres(p) for its fugacity, provided the pressure is not significantly in excess of one atmosphere.

In practice, the application of Eq. (19) may be complicated, because we usually know the concentration of a species, not its activity. If we substitute $f_\mathrm{A}[\mathrm{A}] = (\mathrm{A})$, etc., into Eq. (18) we obtain

$$E = E^0 + \frac{RT}{nF} \ln \frac{f_\mathrm{A}^a f_\mathrm{B}^b f_\mathrm{C}^c \cdots}{f_\mathrm{P}^p f_\mathrm{Q}^q f_\mathrm{R}^r \cdots} + \frac{RT}{nF} \ln \frac{[\mathrm{A}]^a[\mathrm{B}]^b[\mathrm{C}]^c \cdots}{[\mathrm{P}]^p[\mathrm{Q}]^q[\mathrm{R}]^r \cdots} \tag{20}$$

where[7]

$$E^0 + \frac{RT}{nF} \ln \frac{f_\mathrm{A}^a f_\mathrm{B}^b f_\mathrm{C}^c \cdots}{f_\mathrm{P}^p f_\mathrm{Q}^q f_\mathrm{R}^r \cdots} = E^{0'} \tag{21}$$

$E^{0'}$ is related to E^0 in much the same manner that K, the concentration constant, is related to K^0, the thermodynamic constant (p. 42).

Frequently, one or more of the species participating in the half-reaction reacts with water or a constituent in the solution and is present in more than one form. For example, if the half-reaction were $Fe^{3+} + e \rightleftharpoons Fe^{++}$, the pH of the solution about 3, and chloride ion were present, the equilibria

$$\begin{aligned} Fe^{3+} + H_2O &\rightleftharpoons FeOH^{++} + H^+ \\ Fe^{3+} + Cl^- &\rightleftharpoons FeCl^{++} \end{aligned} \tag{22}$$

would produce two additional ferric species.

Instead of calculating $[Fe^{3+}]$ from the equilibrium constants, it is convenient to write the Nernst equations in a form involving the analytical concentration of ferric ion. Analogous situations occur in numerous other cases. In the general case, given by Eqs. (17) and (18), we shall denote *analytical* concentrations by C_A, C_B, C_C, etc. Then the Nernst equation for the generalized half-reaction (17) is written

$$E = E^f + \frac{RT}{nF} \ln \frac{C_\mathrm{A}^a C_\mathrm{B}^b C_\mathrm{C}^c \cdots}{C_\mathrm{P}^p C_\mathrm{Q}^q C_\mathrm{R}^r \cdots} \tag{23}$$

where E^f is the *formal* reduction potential.

The formal reduction potential is defined as the potential of a half-cell in which the concentration term appearing in the logarithmic term of the Nernst equation is unity. Concentrations are expressed in molarity. E^f is usually measured in a

[7] If the ionic strength of the solution is less than 0.1, the activity coefficients of ionic species can be estimated using Table A-1.

solution containing a relatively large excess of some added electrolyte (supporting electrolyte) as compared to the concentration of the electroactive species. Species whose concentrations are fixed by the supporting electrolyte, e.g., hydrogen ion or complexing species, are not written in Eq. (23). E^f is related to E^0 in much the same manner that K', the conditional constant, is related to K^0, the thermodynamic constant (p. 123).

Tables A-12 and A-13 list values of E^0 and E^f, respectively, for redox systems of analytical interest.

Some examples of the Nernst equation, using E^0, $E^{0'}$, and E^f, for several half-reactions at 25°C are given below.

Redox system

(a) Ox + $ne \rightleftharpoons$ Red. Equivalent forms of the Nernst equation are

$$E = E^0_{\text{Ox/Red}} + \frac{0.059}{n} \log \frac{(\text{Ox})}{(\text{Red})} = E^{0'}_{\text{Ox/Red}} + \frac{0.059}{n} \log \frac{[\text{Ox}]}{[\text{Red}]}$$

$$= E^f_{\text{Ox/Red}} + \frac{0.059}{n} \log \frac{C_{\text{Ox}}}{C_{\text{Red}}} \tag{24}$$

The reduction potential of a redox system that contains only the ratio of the concentration of oxidized to reduced form in the logarithmic term of the Nernst equation is unaltered by dilution (with supporting electrolyte). The reduction potential is determined solely by the ratio $C_{\text{ox}}/C_{\text{red}}$. In Eq. (24), the reduction potential increases with an increase in the value of the ratio of $C_{\text{ox}}/C_{\text{red}}$. If $n = 1$, and the ratio is 10, $E = E^f + 0.059 \log 10 = E^f + 0.059$. If the ratio is 100, $E = E^f + 0.118$; if it is $\frac{1}{10}$, $E = E^f + 0.059 \log \frac{1}{10} = E^f - 0.059$, etc.

(b) Ox + $n\text{H}^+$ + $ne \rightarrow$ Red. In the presence of excess strong acid,

$$E = E^0_{\text{Ox/Red}} + \frac{0.059}{n} \log \frac{(\text{H}^+)^n(\text{Ox})}{(\text{Red})} = E^{0'}_{\text{Ox/Red}} + \frac{0.059}{n} \log \frac{[\text{H}^+]^n[\text{Ox}]}{[\text{Red}]}$$

$$= E^f_{\text{Ox/Red}} + \frac{0.059}{n} \log \frac{C_{\text{Ox}}}{C_{\text{Red}}} \tag{25}$$

Note that the expression involving E^f does not show a term involving C_{H^+}. By definition of the formal potential, hydrogen-ion concentration is omitted, if a specified concentration of hydrogen ion is present in the supporting electrolyte.

Formal potentials of some analytically important redox couples

Ferricyanide–ferrocyanide couple: $\text{Fe(CN)}_6^{3-} + e \rightleftharpoons \text{Fe(CN)}_6^{4-}$

E^f is a typical conditional constant and is especially useful in practical analytical calculations. For example, ferrocyanide reacts with hydrogen ions:

$$\text{Fe(CN)}_6^{4-} + \text{H}^+ \rightleftharpoons \text{HFe(CN)}_6^{3-}$$

The Nernst equation, using formal potentials, is

$$E = E^f_{\text{Fe(CN)}_6^{3-}/\text{Fe(CN)}_6^{4-}} + 0.059 \log \frac{C_{\text{Fe(CN)}_6^{3-}}}{C_{\text{Fe(CN)}_6^{4-}}} \tag{26}$$

E^f includes all ionic strength [Eq. (21)] and solution equilibrium effects of this system. The formal reduction potential of the ferro–ferricyanide couple becomes more positive as the hydrogen-ion concentration is increased. At infinite dilution, $E^0 = 0.365$ V, while $E^f = 0.56$ V in 0.1 *M* HCl and 0.71 V in 1 *M* HCl.

Ceric–cerous couple: $Ce(IV) + e \rightleftharpoons Ce(III)$. Ceric cerium reacts with water,

$$Ce^{4+} + H_2O \rightleftharpoons CeOH^{3+} + H^+$$
$$CeOH^{3+} + H_2O \rightleftharpoons Ce(OH)_2^{++} + 2\,H^+$$

and forms complex ions with chloride, sulfate, and nitrate ions. Thus $E^f_{Ce(IV)/Ce(III)}$ is a function of the concentration and kind of strong acid present, and is most positive in perchloric acid.

E^f has the following values in 1 *M* solutions of the indicated acids: 1.28 V, HCl; 1.44 V, H_2SO_4; 1.60 V, HNO_3; and I.70 V, $HClO_4$.

Ferric-ferrous couple: $Fe(III) + e \rightleftharpoons Fe(II)$. Protolytic equilibria affect the potential of the ferric-ferrous couple, through the formation of various hydroxy compounds. For example, pK = 2.2, at 25°C for the reaction

$$Fe(H_2O)_6^{3+} \rightleftharpoons Fe(H_2O)_5OH^{++} + H^+$$

Ferric ion forms complexes with many anions:

$$Fe^{3+} + X^- \rightleftharpoons FeX^{++}$$

and the formal reduction potential becomes less positive with increasing concentration of X^-. For example, fluoride ion reduces the reduction potential greatly, because Fe^{3+} is converted into FeF^{++}, FeF_2^+, FeF_3, and FeF_6^{3-}. Phosphoric acid has qualitatively the same effect upon the reduction potential. Thus, although ferric ions will ordinarily react with iodide,

$$Fe^{3+} + I^- \rightleftharpoons Fe^{++} + \tfrac{1}{2}\,I_2$$

no iodine is liberated in the presence of excess fluoride ion or phosphoric acid.

The quantitative effect of complexation on the reduction potential of a redox system, $Ox + ne \rightleftharpoons Red$, where both Ox and Red form stable complexes is illustrated in the case of the 1,10-phenanthroline complexes of ferrous and ferric iron. Ferrous phenanthroline (ferroin) is an important oxidation–reduction indicator used with ceric cerium as the oxidant. The phenanthroline (Ph) complex of ferrous iron is more stable than that of ferric iron.[8]

$$Fe^{3+} + 3\,Ph \rightleftharpoons Fe(Ph)_3^{3+} \qquad K_{III} = \frac{[Fe(Ph)_3^{3+}]}{C_{Fe(III)}[Ph]^3} \tag{27}$$

$$Fe^{++} + 3\,Ph \rightleftharpoons Fe(Ph)_3^{++} \qquad K_{II} = \frac{[Fe(Ph)_3^{++}]}{C_{Fe(II)}[Ph]^3} \tag{28}$$

In 1 *M* sulfuric acid $K_{II}/K_{III} = 2.8 \times 10^6$. The Nernst equation for $Fe^{3+} + e \rightleftharpoons Fe^{++}$ is

$$E = E^f_{Fe^{3+}/Fe^{++}} + 0.0591 \log \frac{C_{Fe(III)}}{C_{Fe(II)}} \tag{29}$$

[8] K_{III} and K_{II} (Eqs. (27) and (28)) are written as a mixture of equilibrium and analytical concentrations. $C_{Fe(III)}$ designate the total concentration of all ferric species not containing Ph. $C_{Fe(II)}$ denotes a similar quantity for the ferrous species. This device is convenient when we wish to indicate that the denoted species may exist in more than one form without making it necessary to discuss the nature of these species.

At equilibrium, we may substitute Eq. (30)

$$\frac{C_{Fe(III)}}{C_{Fe(II)}} = \frac{K_{II}}{K_{III}} \frac{[Fe(Ph)_3^{3+}]}{[Fe(Ph)_3^{++}]} \tag{30}$$

which was obtained by dividing Eq. (27) into Eq. (28) into Eq. (29) to get

$$E = E^f_{Fe^{3+}/Fe^{++}} + 0.059 \log \frac{K_{II}}{K_{III}} + 0.059 \log \frac{[Fe(Ph)_3^{3+}]}{[Fe(Ph)_3^{++}]} \tag{31}$$

Equation (31) is readily related to the Nernst equation for the half-reaction $Fe(Ph)_3^{3+} + e \rightleftharpoons Fe(Ph)_3^{++}$:

$$E = E^f_{Fe(Ph)_3^{3+}/Fe(Ph)_3^{++}} + 0.059 \log \frac{[Fe(Ph)_3^{3+}]}{[Fe(Ph)_3^{++}]} \tag{32}$$

Comparing Eqs. (31) and (32), we see

$$E^f_{Fe(Ph)_3^{3+}/Fe(Ph)_3^{++}} = E^f_{Fe^{3+}/Fe^{++}} + 0.059 \log \frac{K_{II}}{K_{III}} \tag{33}$$

In 1 M sulfuric acid, $E^f_{Fe^{3+}/Fe^{++}} = 0.68$ V; therefore,

$$E^f_{Fe(Ph)_3^{3+}/Fe(Ph)_3^{++}} = 0.68 + 0.059 \log 2.8 \times 10^6 = 1.06 \text{ V}$$

Ferric iron has become a considerably stronger oxidant as a result of the preferential stabilization of ferrous iron by *o*-phenanthroline.

As implied above, the converse situation exists for nearly all other complexing agents. Ethylenediaminetetraacetic acid is a quadraprotic acid, H_4Y, which forms a more stable ferric complex, FeY^-, than ferrous complex, FeY^{--}. By analogy with Eq. (32) we write

$$E^f_{FeY^-/FeY^{--}} = E^f_{Fe^{3+}/Fe^{++}} + 0.059 \log \frac{K_{II}}{K_{III}} \tag{34}$$

where K_{II} and K_{III} refer to the formation of FeY^{--} and FeY^-. At pH = 5, $E^f_{FeY^-/FeY^{--}} = 0.12$V. We estimate $E^f_{Fe^{3+}/Fe^{++}} = 0.74$V in noncomplexing media of ionic strength comparable to that of the buffer in which $E^f_{FeY^-/FeY^{--}}$ was determined. Using these formal potentials, we rearrange Eq. (34) and calculate K_{III}/K_{II}:

$$0.059 \log \frac{K_{III}}{K_{II}} = E^f_{Fe3+/Fe^{++}} - E^f_{FeY-/FeY^{--}} = 0.74 - 0.12 = 0.62$$

$$\frac{K_{III}}{K_{II}} = 10^{0.62/0.059} = 10^{10.5}$$

Cobaltic–cobaltous couple: Co(III) + $e \rightleftharpoons$ Co(II). EDTA also forms stable complexes with both cobaltic and cobaltous cobalt. In 4 M nitric acid the formal reduction potential for this system is reported to be about 1.85 V, sufficiently positive to oxidize water to oxygen. However, in the presence of EDTA, the formal potential is reduced to about 0.60 V (it varies with pH). Thus it is possible to oxidize Co(II) quantitatively under the proper conditions with a strong oxidizing agent such as dichromate or hydrogen peroxide in the presence of EDTA. The Mn(III)–Mn(II) behaves qualitatively like the cobalt system.

Vanadium and uranium couples: The strong effect of pH on the reduction potentials has been already noted above in the ferri–ferrocyanide system. In that system, the pH effect arises from the existence of protolytic solution equilibria. If the oxidant and reductant contain a different number of oxygen atoms, the Nernst equation predicts a pH dependence of the reduction potential.

The reduction of vanadate and uranyl ions is strictly reversible.

$$VO_3^- + 4\,H^+ + e \rightleftharpoons VO^{++} + 2\,H_2O$$

$$E = E^{0\prime} + 0.059 \log \frac{[VO_3^-][H^+]^4}{[VO^{++}]}$$

$$E = E^{0\prime} = 0.92 \text{ if } [H^+] = 1 \text{ and } [VO_3^-] = [VO^{++}]$$

Thus, at a pH of 1, the reduction potential is reduced to

$$E = 0.92 + 0.059 \log (10^{-1})^4 = 0.68 \text{ V} = E^f$$

In the VO_3^-–VO^{++} system it is obvious that

$$E^f = E^{0\prime} + 0.059 \log [H^+]^4$$

For the uranyl–uranic couple,

$$UO_2^{++} + 4\,H^+ + 2e \rightleftharpoons U^{4+} + 2\,H_2O$$

$$E = E^{0\prime} + \frac{0.059}{2} \log \frac{[UO_2^{++}][H^+]^4}{[U^{4+}]}$$

and it is apparent that

$$E^f = E^{0\prime} + 0.059 \log [H^+]^2$$

Permanganate and dichromate: Consider the two analytically important oxidants, permanganate and dichromate.

$$MnO_4^- + 8\,H^+ + 5e \rightleftharpoons Mn^{++} + 4\,H_2O$$

$$E = E^{0\prime} + \frac{0.059}{5} \log \frac{[H^+]^8[MnO_4^-]}{[Mn^{++}]}$$

$$Cr_2O_7^{--} + 14\,H^+ + 6e \rightleftharpoons 2\,Cr^{3+} + 7\,H_2O$$

$$E = E^{0\prime} + \frac{0.059}{6} \log \frac{[H^+]^{14}[Cr_2O_7^{--}]}{[Cr^{3+}]^2}$$

In these cases the equations do not hold rigorously. The reactions have been written as reversible reactions, which is not quite correct. Ions intermediate between permanganate and manganous manganese are formed during the reduction of the former, and the same is true in the reduction of dichromate to chromic chromium. Equilibrium between the various oxidation states is not established instantaneously. Nevertheless, it is true that the hydrogen ions have an enormous effect upon the oxidation potential of the above oxidizing agents. At a pH of 6, for example, it is found that the oxidation potential of permanganate is about 0.6 V lower than in 1 *M* acid solution (pH = 0). Use is made of this fact, for example, in the selective oxidation of the halides to free halogen. At a pH of 5 to 6 iodide is oxidized to iodine by permanganate, whereas bromide and chloride are not affected. At a pH of about 3 (acetic acid) bromide is oxidized, but chloride is still unoxidized. The latter is oxidized only at a much higher hydrogen-ion concentration.

A few examples of systems in which Ox or Red are insoluble are discussed below.

Metal ion–metal couple: In the general case $M^{n+} + ne \rightleftharpoons M$, three equivalent forms of the Nernst equation are

$$E = E^0_{M^{n+}/M} + \frac{0.059}{n}\log(M^{n+}) = E^{0'}_{M^{n+}/M} + \frac{0.059}{n}\log[M^{n+}]$$
$$= E^f_{M^{n+}/M} + \frac{0.059}{n}\log C_{M^{n+}} \quad (35)$$

The reduction potential of a metal ion–metal couple depends only upon the metal–ion concentration, since the activity of the metal is constant. In Eq. (35), E^0, $E^{0'}$, and E^f would be the reduction potential when the appropriate logarithmic term was zero. In actual practice, it is inconvenient, or sometimes impossible, to adjust the metal-ion concentration to such a high value, and E^0, $E^{0'}$, and E^f are calculated by extrapolating measured potentials obtained at lower concentrations to the concentration at which the logarithmic term is zero.

Consider the mercurous ion–mercury couple. The half-cell reaction is $Hg_2^{++} + 2e \rightleftharpoons 2\ Hg$ and the Nernst equation is

$$E = E^0_{Hg_2^{++}/Hg} + \frac{0.059}{2}\log(Hg_2^{++}) \quad (36)$$

or

$$E = E^f_{Hg_2^{++}/Hg} + \frac{0.059}{2}\log C_{Hg_2^{++}}$$

where $E^f_{Hg_2^{++}} = 0.776$ V in 1 *M* perchloric acid. Thus the reduction potential when $C_{Hg_2^{++}} = 0.01$ is

$$E = 0.776 + \frac{0.059}{2}\log 0.01 = 0.776 - 0.059 = 0.717\ \text{V}$$

Fairly noble metals with large, positive, reduction potentials can often be made fairly strong reducing agents by keeping $C_{M^{n+}}$ extremely small, either by adding an anion that forms an insoluble compound, or a stable complex.

Insoluble metal salt–metal couple: On adding chloride ion to a mercurous ion solution, the insoluble mercurous chloride (calomel) precipitates:

$$Hg_2^{++} + 2\ Cl^- \rightleftharpoons Hg_2Cl_2\downarrow \qquad K^0_{sp} = (Hg_2^{++})(Cl^-)^2 \quad (37)$$

We may substitute Eq. (37) into (36), obtaining

$$E = E^0_{Hg_2^{++}/Hg} + \frac{0.059}{2}\log K^0_{sp} - 0.059\log(Cl^-) \quad (38)$$

Equation (38) has the same form as the Nernst equation for the half-cell reaction $Hg_2Cl_2 + 2e \rightarrow 2\ Hg + 2\ Cl^-$:

$$E = E^0_{Hg_2Cl_2/Hg} - 0.059\log(Cl^-) \quad (39)$$

Therefore,

$$E^0_{Hg_2Cl_2/Hg} = E^0_{Hg_2^{++}/Hg} + \frac{0.059}{2}\log K^0_{sp} \quad (40)$$

The thermodynamic value of $K^0_{sp} = 10^{-17.88}$. Neglecting activity coefficients, we calculate in 1 M perchloric acid that

$$E^0_{Hg_2Cl_2/Hg} = 0.776 + \frac{0.059}{2} \log 10^{-17.88} = 0.248 \text{ V}$$

Complex metal ion–metal couple: Silver ion forms an EDTA complex,

$$Ag^+ + Y^{4-} \rightleftharpoons AgY^{3-} \qquad K = \frac{C_{AgY^{3-}}}{C_{Ag^+}C_{Y^{3-}}} \tag{41}$$

where $K = 10^{7.3}$ at an ionic strength of 0.1. The Nernst equation for this couple, $AgY^{3-} + e \rightleftharpoons Ag + Y^{4-}$ is

$$E = E^f_{AgY^{3-}/Ag} + 0.059 \log \frac{C_{AgY^{3-}}}{C_{Y^{4-}}} \tag{42}$$

while for the simple Ag^+–Ag couple it is

$$E = E^f_{Ag^+/Ag} + 0.059 \log C_{Ag^+} \tag{43}$$

At equilibrium, we may eliminate C_{Ag^+} between Eqs. (41) and (43) to obtain

$$E = E^f_{Ag^+/Ag} - 0.059 \log K + 0.059 \log \frac{C_{AgY^{3-}}}{C_{Y^{4-}}} \tag{44}$$

Comparing Eqs. (44) and (42), we see that

$$E^f_{AgY^{3-}/Ag} = E^f_{Ag^+/Ag} - 0.059 \log K \tag{45}$$

An experimental value for $E^f_{Ag^+/Ag}$ is not available at $\mu = 0.1$, but a satisfactory estimate can be made as follows. In noncomplexing media

$$E^f_{Ag^+/Ag} = E^{0\prime}_{Ag^+/Ag} = E^0_{Ag^+/Ag} + 0.059 \log f_{Ag^+} \tag{46}$$

From Table A-1, $-\log f_{Ag+} = 0.11$ and substituting into Eq. (46),

$$E^f_{Ag^+/Ag} = 0.7994 - 0.059(0.11) = 0.792 \text{ V}$$

Therefore, substituting into Eq. (45), at $\mu = 0.1$,

$$E^f_{AgY^{3-}/Ag} = 0.792 - 0.059 \log 10^{7.3} = 0.360 \text{ V}$$

Now, consider a metal, such as copper, which has two oxidation states. The formation of a complex ion can determine which oxidation state will be stable in contact with the metal. The various aquo copper couples are

$$Cu^+ + e \rightleftharpoons Cu \qquad E^0_{Cu^+/Cu} = 0.521 \text{ V}$$

$$Cu^{++} + 2e \rightleftharpoons Cu \qquad E^0_{Cu^{++}/Cu} = 0.337 \text{ V}$$

$$Cu^{++} + e \rightleftharpoons Cu^+ \qquad E^0_{Cu^{++}/Cu^+} = 0.153 \text{ V}$$

Since aquo cuprous ion is a stronger oxidant than aquo cupric ion, the oxidation of copper metal in noncomplexing media always produces cupric ion; i.e., the equilibrium constant for the disproportionation reaction

$$2\,Cu^+ \rightleftharpoons Cu^{++} + Cu \tag{47}$$

is very large ($10^{6.23}$).

In the presence of excess ammonia, the equilibria in solution can be represented fairly well by the overall reactions

$$Cu^+ + 2\,NH_3 \rightleftharpoons Cu(NH_3)_2^+ \qquad K_I = 10^{10.8} \tag{48}$$

$$Cu^{++} + 4\,NH_3 \rightleftharpoons Cu(NH_3)_3^{++} \qquad K_{II} = 10^{13.6} \tag{49}$$

We are now concerned with the half-reactions

$$Cu(NH_3)_2^+ + e \rightleftharpoons Cu + 2\,NH_3 \quad E^0_{Cu(NH_3)_2^+/Cu} = E^0_{Cu^+/Cu} - 0.059 \log K_I \tag{50}$$

$$Cu(NH_3)_4^{++} + 2e \rightleftharpoons Cu + 4\,NH_3\; E^0_{Cu(NH_3)/_4^{++}Cu} = E^0_{Cu^{++}/Cu} - \frac{0.059}{2} \log K_{II} \tag{51}$$

$$Cu(NH_3)_4^{++} + e \rightleftharpoons Cu(NH_3)_2^+ + 2\,NH_3 \qquad E^0_{Cu(NH_3)_4^{++}/Cu(NH_3)_2^+}$$

$$= E^0_{Cu^{++}/Cu^+} + 0.059 \log \frac{K_I}{K_{II}} \tag{52}$$

Equations (50) and (51) are derived in a manner similar to (45), and Eq. (52) similarly to (34). Substituting the values of K_{II} and K_I yields $E^0_{Cu(NH_3)_2^+/Cu} = -0.12$ V, $E^0_{Cu(NH_3)_4^{++}/Cu} = -0.065$ V, and $E^0_{Cu(NH_3)_4^{++}/Cu(NH_3)_2^+} = -0.01$ V. The disproportionation reaction involving the ammonia complexes of cuprous and cupric copper,

$$2\,Cu(NH_3)_2^+ \rightleftharpoons Cu(NH_3)_4^{++} + Cu \qquad K = 10^{-2.2}$$

is much smaller than for the aquo ions, and $Cu(NH_3)_4^{++}$ is a stronger oxidant than $Cu(NH_3)_2^+$.

Iodine (triodide)–iodide couple. Iodine is a valuable oxidant used in many analytical procedures. The Nernst equation for the reaction

$$I_2(aq) + 2e \rightleftharpoons 2I^-$$

is

$$E = E^0_{I_2(aq)/I^-} + \frac{0.059}{2} \log \frac{(I_2)}{(I^-)^2} \tag{53}$$

where $E^0_{I_2(aq)/I^-} = 0.6197$ V. Iodine is only slightly soluble in water, 0.00133 *M*, and another convenient form of the Nernst equation for this couple is based on the half-reaction

$$I_2(s) + 2e \rightleftharpoons 2I^-$$

$$E = E^0_{I_2(s)/I^-} + \frac{0.059}{2} \log \frac{1}{(I^-)^2} \tag{54}$$

The activity of iodine does not appear in Eq. (54), since the activity of solid iodine is constant. At equilibrium, we may equate Eqs. (53) and (54) to obtain

$$E^0_{I_2(s)/I^-} = E^0_{I_2(aq)/I^-} + \frac{0.059}{2} \log (I_2) \tag{55}$$

where the logarithmic term contains the activity of iodine in a saturated solution. It is a good approximation to substitute the molar solubility for (I_2) to obtain

$$E^0_{I_2(s)/I^-} = 0.6197 + \frac{0.059}{2} \log 0.00133 = 0.6197 - 0.0852$$

$$= 0.5345 \text{ V}$$

Iodine reacts with iodide ion to form the stable triodide ion

$$I_2 + I^- \rightleftharpoons I_3^- \qquad K^0 = \frac{(I_3^-)}{(I^-)(I_2)} = 700 \tag{56}$$

Consider a solution saturated with respect to solid iodine, containing an analytical concentration of iodide equal to 0.1 M. To calculate the reduction potential of this solution, reaction (56) must be taken into account. Since $f_{I^-} \simeq f_{I_3^-}$, and $f_{I_2} \simeq 1$, as a good approximation, $(I_3^-)/(I^-)(I_2) = [I_3^-]/[I^-][I_2]$. Since $0.1 = [I^-] + [I_3^-]$ and $[I_2] = 0.00133$, according to (56),

$$\frac{0.1 - [I^-]}{[I^-]0.00133} = 700$$

or $$[I^-] = 0.0516\ M \qquad \text{and} \qquad [I_3^-] = 0.0484\ M$$

According to (54)

$$E = 0.5345 + \frac{0.059}{2} \log \frac{1}{(0.0516)^2}$$
$$= 0.458\ \text{V}$$

It is also of interest to calculate the reduction potential of a solution containing analytical concentrations of 0.05 M I_2 and 0.5 M I^-. In this problem,

$$0.5 = [I^-] + [I_3^-] \tag{57}$$

$$0.05 = [I_2] + [I_3^-] \tag{58}$$

Subtracting (57) from (58), and rearranging yields

$$[I^-] = 0.45 + [I_2] \tag{59}$$

According to Eq. (56),

$$\frac{0.05 - [I_2]}{(0.45 + [I_2])([I_2])} = 700 \tag{60}$$

Examining Eq. (60) it is seen that as a good approximation, $[I_2]$ is small as compared to both 0.45 and 0.05. Therefore,

$$[I_2] = \frac{0.05}{0.45} \times \frac{1}{700} = 1.59 \times 10^{-4}$$

$[I^-] = 0.45$, and $[I_3^-] = 0.05$. Using Eq. (53),

$$E = 0.6197 + \frac{0.059}{2} \log \frac{1.59 \times 10^{-4}}{(0.5)^2}$$
$$= 0.523\ \text{V}$$

We round E to 0.001 V, since we have assumed activities equal to concentrations.

It is seen from the last problem that I_3^- formation is nearly quantitative in the presence of excess iodide. In such problems it is more convenient to use a Nernst equation involving the triodide–iodide couple.

$$I_3^- + 2e \rightleftharpoons 3\,I^- \qquad E = E^0_{I_3^-/I^-} + \frac{0.059}{2} \log \frac{(I_3^-)}{(I^-)^3} \tag{61}$$

The relationship between $E^0_{I_3^-/I^-}$ and $E^0_{I_2(aq)/I^-}$ is found by substituting Eq. (56) into (53) to obtain

$$E^0 = E^0_{I_2(aq)/I} - \frac{0.059}{2} \log K^0 + \frac{0.059}{2} \log \frac{(I_3^-)}{(I^-)^3} \tag{62}$$

Comparison of Eqs. (61) and (62) demonstrates that

$$E^0_{I_3^-/I^-} = E^0_{I_2(aq)/I^-} - \frac{0.059}{2} \log K^0 = 0.6197 - \frac{0.059}{2} \log 700$$

$$= 0.5355 \text{ V}$$

Gas electrodes

Some oxidants and reductants exist, at room temperature, as gases that are only slightly water soluble.

The hydrogen electrode is the best known example of a gaseous half-cell involving a reductant, and the chlorine and oxygen electrodes are two examples of half-cells involving gaseous oxidants. We will consider the hydrogen and oxygen electrodes. The potential of the hydrogen electrode is

$$\tfrac{1}{2} H_2 + e \rightleftharpoons H^+$$

$$E = 0.059 \log \frac{p_{H_2}^{1/2}}{(H^+)} = 0.059 \log p_{H_2}^{1/2} - 0.059 \text{ p}_a\text{H}$$

A platinized platinum electrode is usually used as a noble metal electrode in a hydrogen electrode half cell.

The potential of the oxygen electrode is

$$\tfrac{1}{2} O_2 + H_2O + 2e \rightleftharpoons 2\, OH^-$$

$$E = E^{0'}_{O_2/OH^-} + \frac{0.059}{2} \log \frac{p_{O_2}{}^{1/2}}{[OH^-]^2}$$

Since

$$[OH]^- = \frac{K_w}{[H^+]}$$

it is found that

$$E = E^{0'}_{O_2/OH^-} + \frac{0.059}{2} \log \frac{p_{O_2}{}^{1/2}}{K_w^2} + \frac{0.059}{2} \log [H^+]^2$$

Oxygen gas is very slightly soluble in water, and p_{O_2} is constant as a rule. Since K_w is constant at a given temperature, the expression $(0.059/2) \log (p_{O_2}{}^{1/2}/K_w^2)$ becomes constant. We find then that

$$E = E^f + \frac{0.059}{2} \log [H^+]^2 = E^f + 0.059 \log [H^+]$$

in which E^f again is a constant but has a value different from $E^{0'}_{O_2/OH^-}$ in the previous equation. Reversible oxygen electrode potentials are very difficult to measure.

Equilibrium constants and potentials of solutions

In the transformation of Ox to Red or *vice versa* we are always dealing with two redox couples, e.g., Ce(IV) + Fe(II) $\rightleftharpoons$ Ce (III) + Fe(III). The equilibrium constant for the oxidation-reduction reaction

$$\mathrm{Ox_1 + Red_2 \rightleftharpoons Red_1 + Ox_2} \tag{63}$$

is easily derived from the formal potential of both systems.

System 1: $\mathrm{Ox_1} + ne \rightleftharpoons \mathrm{Red_1}$:

$$E = E^f_{\mathrm{Ox_1/Red_1}} + \frac{0.059}{n}\log\frac{C_{\mathrm{Ox_1}}}{C_{\mathrm{Red_1}}} \tag{64}$$

System 2: $\mathrm{Ox_2} + ne \rightleftharpoons \mathrm{Red_2}$:

$$E = E^f_{\mathrm{Ox_2/Red_2}} + \frac{0.059}{n}\log\frac{C_{\mathrm{Ox_2}}}{C_{\mathrm{Red_2}}} \tag{65}$$

At equilibrium we may equate Eqs. (64) and (65), because the reduction potential of both systems must be equal.

$$E^f_{\mathrm{Ox_1/Red_1}} + \frac{0.059}{n}\log\frac{C_{\mathrm{Ox_1}}}{C_{\mathrm{Red_1}}} = E^f_{\mathrm{Ox_2/Red_2}} + \frac{0.059}{n}\log\frac{C_{\mathrm{Ox_2}}}{C_{\mathrm{Red_2}}} \tag{66}$$

or

$$\frac{0.059}{n}\log\frac{C_{\mathrm{Ox_2}}}{C_{\mathrm{Red_2}}}\frac{C_{\mathrm{Red_1}}}{C_{\mathrm{Ox_1}}} = E^f_{\mathrm{Ox_1/Red_1}} - E^f_{\mathrm{Ox_2/Red_2}}$$

The equilibrium constant for reaction (63) is $K = C_{\mathrm{Ox_2}}C_{\mathrm{Red_1}}/C_{\mathrm{Red_2}}C_{\mathrm{Ox_1}}$; thus

$$\frac{0.059}{n}\log K = E^f_{\mathrm{Ox_1/Red_1}} - E^f_{\mathrm{Ox_2/Red_2}} \tag{67}$$

or

$$K = \exp_{10}\{n(E^f_{\mathrm{Ox_1/Red_1}} - E^f_{\mathrm{Ox_2/Red_2}})/0.059\} \tag{68}$$

Note that $\exp_{10} E = 10^E$

The equilibrium constant for the general redox reaction

$$m\mathrm{Ox_1} + n\mathrm{Red_2} \rightleftharpoons m\mathrm{Red_1} + n\mathrm{Ox_2} \tag{69}$$

can also be readily derived from the formal potential of both systems.

System 1: $\mathrm{Ox_1} + ne \rightleftharpoons \mathrm{Red_1}$:

$$E_{\mathrm{Ox_1/Red_1}} = E^f_{\mathrm{Ox_1/Red_1}} + \frac{0.059}{n}\log\frac{C_{\mathrm{Ox_1}}}{C_{\mathrm{Red_1}}} \tag{70}$$

System 2: $\mathrm{Ox_2} + me \rightleftharpoons \mathrm{Red_2}$:

$$E_{\mathrm{Ox_2/Red_2}} = E^f_{\mathrm{Ox_2/Red_2}} + \frac{0.059}{m}\log\frac{C_{\mathrm{Ox_2}}}{C_{\mathrm{Red_2}}} \tag{71}$$

At equilibrium, $E_{\mathrm{Ox_1/Red_1}} = E_{\mathrm{Ox_2/Red_2}}$, and equating Eqs. (70) and (71) gives

$$\frac{0.059}{n\cdot m}\log\left\{\left(\frac{C_{\mathrm{Ox_2}}}{C_{\mathrm{Red_2}}}\right)^n\left(\frac{C_{\mathrm{Red_1}}}{C_{\mathrm{Ox_1}}}\right)^m\right\} = E^f_{\mathrm{Ox_1/Red_1}} - E^f_{\mathrm{Ox_2/Red_2}} \tag{72}$$

We see that the equilibrium constant for reaction (69)[9] is

$$K = \left(\frac{C_{\mathrm{Ox}_2}}{C_{\mathrm{Red}_2}}\right)^n \left(\frac{C_{\mathrm{Red}_1}}{C_{\mathrm{Ox}_2}}\right)^m$$

Thus

$$\frac{0.059}{n \cdot m} \log K = E^f_{\mathrm{Ox}_1/\mathrm{Red}_1} - E^f_{\mathrm{Ox}_2/\mathrm{Red}_2} \tag{73}$$

or

$$K = \exp_{10} \{n \cdot m(E^f_{\mathrm{Ox}_1/\mathrm{Red}_1} - E^f_{\mathrm{Ox}_2/\mathrm{Red}_2})/0.059\} \tag{74}$$

It is clear that the equilibrium constant for an oxidation–reduction reaction can be calculated from the reduction potential of the individual redox couples, or that the reduction potential of one of the couples can be calculated from the equilibrium constant and the other couple's reduction potential.

The reduction potential, E_{Eq}, of a mixture containing *equivalent* analytical concentrations of an oxidizing agent, Ox_1, and a reducing agent, Red_2, corresponds to the equivalence point in the titration of one species by the other. Assume that Eq. (69) describes the redox reaction and that $C_{\mathrm{Ox}_1} + C_{\mathrm{Red}_1} = mc$ and $C_{\mathrm{Ox}_2} + C_{\mathrm{Red}_2} = nc$. If

$$C_{\mathrm{Ox}_1} = my \qquad C_{\mathrm{Red}_1} = mc - my$$
$$C_{\mathrm{Red}_2} = ny \qquad C_{\mathrm{Ox}_2} = nc - ny$$

Then

$$\frac{C_{\mathrm{Ox}_1}}{C_{\mathrm{Red}_1}} = \frac{C_{\mathrm{Red}_2}}{C_{\mathrm{Ox}_2}} = \frac{y}{c - y} \tag{75}$$

Substituting Eq. (75) into (70) and (71) yields

$$E_{\mathrm{Eq}} = E^f_{\mathrm{Ox}_1/\mathrm{Red}_1} + \frac{0.059}{n} \log \frac{y}{c - y} \tag{76}$$

$$E_{\mathrm{Eq}} = E^f_{\mathrm{Ox}_2/\mathrm{Red}_2} + \frac{0.059}{m} \log \frac{c - y}{y} \tag{77}$$

Multiplying Eq. (76) by n, and (77) by m, and adding the two equations, we obtain

$$nE_{\mathrm{Eq}} + mE_{\mathrm{Eq}} = nE^f_{\mathrm{Ox}_1/\mathrm{Red}_1} + mE^f_{\mathrm{Ox}_2/\mathrm{Red}_2}$$

Therefore,

$$E_{\mathrm{Eq}} = \frac{nE^f_{\mathrm{Ox}_1/\mathrm{Red}_1} + mE^f_{\mathrm{Ox}_2/\mathrm{Red}_2}}{n + m} \tag{78}$$

Let us take as a specific example of an oxidation–reduction reaction the oxidation of Fe(II) with ceric perchlorate in 1 *M* perchloric acid. The redox reaction is

$$\mathrm{Ce(IV)} + \mathrm{Fe(II)} \rightleftharpoons \mathrm{Ce(III)} + \mathrm{Fe(III)} \tag{79}$$

[9] Note that the equilibrium constant for reaction (69), K_{69}, differs from that for reaction (63), K_{63}. When $n = m$, it is easily shown that $K_{69} = (K_{63})^n$.

with the half-reactions

$$\mathrm{Ce(IV)} + e \rightleftharpoons \mathrm{Ce(III)}$$

$$E = E^f_{\mathrm{Ce(IV)/Ce(III)}} + 0.059 \log \frac{C_{\mathrm{Ce(IV)}}}{C_{\mathrm{Ce(III)}}} \tag{80}$$

and

$$\mathrm{Fe(III)} + e \rightleftharpoons \mathrm{Fe(II)}$$

$$E = E^f_{\mathrm{Fe(III)/Fe(II)}} + 0.059 \log \frac{C_{\mathrm{Fe(III)}}}{C_{\mathrm{Fe(II)}}} \tag{81}$$

The formal potentials are $E^f_{\mathrm{Ce(IV)/Ce(III)}} = 1.70$ V and $E^f_{\mathrm{Fe(III)/Fe(II)}} = 0.735$ V. According to Eq. (68) the conditional equilibrium constant for Eq. (79) is

$$K = \exp_{10}\left\{\frac{[E^f_{\mathrm{Ce(IV)/Ce(III)}} - E^f_{\mathrm{Fe(III)/Fe(II)}}]}{0.059}\right\} = 10^{(1.70\,-\,0.735)/0.059} = 10^{16.36}$$

$$= 2._3 \times 10^{16}$$

and we see that the reaction between ceric cerium and ferrous ion will be quantitative. For example, in a mixture containing an analytical concentration of ferrous iron of 0.1 M and 0.05 M ceric cerium it is obvious that $C_{\mathrm{Ce(III)}} = 0.05$, $C_{\mathrm{Fe(III)}} = 0.05$, and $C_{\mathrm{Fe(II)}} = 0.05$. Since

$$K = 2._3 \times 10^{16} = \frac{C_{\mathrm{Ce(III)}]}\,C_{\mathrm{Fe(III)}]}}{C_{\mathrm{Ce(IV)}}\,C_{\mathrm{Fe(II)}}} = \frac{(0.05)\,(0.05)}{C_{\mathrm{Ce(IV)}}\,(0.05)}$$

and

$$C_{\mathrm{Ce(IV)}} = \frac{0.05}{2._3 \times 10^{16}} = 2._2 \times 10^{-18}$$

The reduction potential of this solution is calculated using Eq. (81):

$$E = 0.735 + 0.059 \log \frac{0.05}{0.05} = 0.735 \text{ V}$$

In a mixture containing an analytical concentration of ferrous iron of 0.1 M and 0.099 M ceric cerium, $C_{\mathrm{Ce(III)}} = 0.099$, $C_{\mathrm{Fe(III)}} = 0.099$, and $C_{\mathrm{Fe(II)}} = 0.001$. $C_{\mathrm{Ce(IV)}}$ is still negligibly small as compared to the other species, because

$$C_{\mathrm{Ce(IV)}} = \frac{C_{\mathrm{Ce(III)}}C_{\mathrm{Fe(III)}}}{C_{\mathrm{Fe(II)}}K} = \frac{0.099(0.099)}{0.001(2._3 \times 10^{16})} = 4._3 \times 10^{-16}$$

The reduction potential of this mixture is, according to Eq. (81),

$$E = 0.735 + 0.059 \log \frac{0.099}{0.001} = 0.735 + 0.059 \times 2.00$$

$$= 0.735 + 0.118 = 0.853 \text{ V}$$

Now consider a solution containing equivalent concentrations of Ce(IV) and Fe(II). If the analytical concentration of each couple is 0.1 M, and $C_{\mathrm{Ce(IV)}} = x$, then $C_{\mathrm{Ce(III)}} = 0.1 - x$, $C_{\mathrm{Fe(II)}} = x$, and $C_{\mathrm{Fe(III)}} = 0.1 - x$. Substituting these quantities into the expression for the constant for reaction (79), we obtain

$$2._3 \times 10^{16} = \frac{(0.1 - x)^2}{x^2}$$

Obviously $0.1 \gg x$, therefore,

$$x^2 = \frac{(0.1)^2}{2._3 \times 10^{16}}$$

$$x = 6._6 \times 10^{-10} = C_{\text{Fe(II)}} = C_{\text{Ce(IV)}}$$

Only a minute percentage (7×10^{-7} per cent) of the ferrous iron remains unreacted.

The reduction potential of this solution is calculated according to Eq. (78), where $n = m = 1$.

$$E_{\text{Eq}} = \frac{E^f_{\text{Ce(IV)/Ce(III)}} + E^f_{\text{Fe(III)/Fe(II)}}}{1 + 1} = \frac{1.70 + 0.735}{2}$$

$$= 1.218 \text{ V}$$

Thus the potential at the equivalence point in the titration of ferrous ion with ceric ion in 1 *M* perchloric acid is 1.218 V. For practical purposes, the analytical concentration of the reactants does not affect this potential, because the ratios $C_{\text{Ce(IV)}}/C_{\text{Ce(III)}}$ and $C_{\text{Fe(III)}}/C_{\text{Fe(II)}}$ at the equivalence point (and all other parts of the titration) are not altered by dilution with supporting electrolyte.

In a solution containing an analytical concentration ferrous ion of 0.1 *M* and 0.101 *M* ceric cerium, after reaction, $C_{\text{Fe(III)}} = 0.1$, $C_{\text{Ce(III)}} = 0.1$, and $C_{\text{Ce(IV)}} = 0.001$. We calculate that

$$C_{\text{Fe(II)}} = \frac{C_{\text{Ce(III)}} C_{\text{Fe(III)}}}{K C_{\text{Ce(IV)}}} = \frac{(0.1)(0.1)}{2._3 \times 10^{16}(0.001)} = 4._3 \times 10^{-16}$$

and the reduction potential, according to Eq. (80)[10] is

$$E_{\text{Eq}} = 1.70 + 0.059 \log \frac{0.001}{0.100} = 1.70 - 0.118$$

$$= 1.582 \text{ V}$$

Additional numerical examples are to be found in Chapter 37, particularly as applied to the calculation of reduction potentials during redox titrations.

The calculation of equilibrium constants from standard (or formal) potentials is of great practical significance. For example, the ability of silver metal to reduce various ions in various supporting electrolytes,

$$n\text{Ag} + \text{M}^{p+} \rightleftharpoons n\text{Ag(I)} + \text{M}^{(p-n)+} \tag{82}$$

can be predicted from the equilibrium constant for reaction (82); according to Eq. (74) it is

$$K = \exp_{10}\{(n(E^f_{\text{M}^{p+}/\text{M}^{(p-n)+}} - E^f_{\text{Ag(I)/Ag}})/0.059)\}$$

[10] It is possible to calculate E_{Eq} using $C_{\text{Fe(III)}} = 0.1$ and $C_{\text{Fe(II)}} = 4._3 \times 10^{-16}$ and Eq. (81). However, as a general rule, the equilibrium potential is calculated using the half-cell reaction which involves the redox couple whose equilibrium concentrations can be calculated most directly. For example, Eq. (81) was used to calculate E_{Eq} in the two solutions for which $C_{\text{Fe(III)}}$ and $C_{\text{Fe(II)}}$ could be obtained by assuming that the reaction between Ce(IV) and Fe(II) was quantitative. Since $C_{\text{Ce(III)}}$ must be known to use Eq. (80), such a procedure avoids calculating $C_{\text{Ce(III)}}$ from $C_{\text{Fe(III)}}$, $C_{\text{Fe(II)}}$, $C_{\text{Ce(IV)}}$ and the equilibrium constant.

In 1 M sulfuric acid $E^f_{Ag(I)/Ag} = 0.77$ V while $E^f_{Fe(III)/Fe(II)} = 0.68$ V. The constant for the redox reaction

$$Ag + Fe(III) \rightleftharpoons Ag(I) + Fe(II)$$

is

$$K = 10^{(0.68-0.77)1/0.059} = 10^{-(0.09/0.059)} = 0.03$$

and Fe(III) will be only partially reduced when the analytical concentration of iron is in the concentration range normally encountered. On the other hand, in 1 M HCl $E^f_{Ag(I)/Ag} = 0.47$ V, $E^f_{Fe(III)/Fe(II)} = \sim 0.68$ V, and $K = 3._5 \times 10^3$. Hence Fe(III) is quantitatively reduced. Silver chloride is quite soluble in 1 M HCl ($\sim 2 \times 10^{-4}$ M) because of the formation of anionic chloro complexes of silver.

The effect of complexation on the stability of various oxidation states is also readily estimated. Consider the disproportionation of aquo cuprous ions:

$$2\,Cu^+ \rightleftharpoons Cu^{++} + Cu \qquad K^0_{1,2} = \frac{(Cu^{++})}{(Cu^+)^2} \tag{83}$$

Reaction (83) is the difference between the half-reactions $Cu^+ + e \rightleftharpoons Cu$ and $Cu^{++} + e \rightleftharpoons Cu^+$. Therefore, according to Eq. (68), $K^0_{1,2}$ is given by

$$K^0_{1,2} = 10^{(E^f_{Cu^+/Cu} - E^f_{Cu^{++}/Cu^+})/0.059}$$

In the absence of supporting electrolyte, $K_{1,2} = K^0_{1,2}$, $E^f = E^{0\prime} = E^0$, $E^0_{Cu^{++}/Cu^+} = 0.153$ V, and $E^0_{Cu^+/Cu} = 0.521$ V. Therefore,

$$K^0_{1,2} = 10^{(0.521-0.153)/0.059} = 10^{6.2}$$

In the presence of ammonia, the equilibrium constant for the disproportionation reaction

$$2\,Cu(NH_3)_2^+ \rightleftharpoons Cu(NH_3)_4^{++} + Cu \qquad K_{I,II} = \frac{[Cu(NH_3)_4^{++}]}{[Cu(NH_3)_2^{++}][NH_3^+]^2} \tag{84}$$

is less than unity and is calculated from

$$K_{I,II} = \exp_{10}\{(E^{0\prime}_{Cu(NH_3)_2^+/Cu} - E^{0\prime}_{Cu(NH_3)_4^{++}/Cu(NH_3)_2^+})/0.059\} \tag{85}$$

Substituting for the $E^{0\prime}$-values

$$K_{I,II} = 10^{(-0.12-(0.01))/0.059} = 10^{-2.1}$$

Cu(I) is very much more stable toward disproportionation in ammonia solution than in water.

Relationships between the reduction potentials of an element with several oxidation states

Consider a metal that can give rise to ions with different valences; in addition to the half-reactions that can be written for the reduction of these ions to the metal, it is possible to write half-reactions for the reduction of one soluble oxidation state to another. These various half-reactions are related to each other, and it is possible to obtain the standard reduction potential of one half-reaction given two other standard reduction potentials. For example, consider the following half-reactions and Nernst equations:

$$M^{n+} + ne \rightleftharpoons M \qquad E = E^0_{M^{n+}/M} + \frac{0.059}{n} \log (M^{n+}) \tag{86}$$

$$M^{(n-a)+} + (n-a)e \rightleftharpoons M \qquad E = E^0_{M^{(n-a)+}/M} + \frac{0.059}{n-a} \log (M^{(n-a)+}) \tag{87}$$

If we multiply Eq. (86) by n and (87) by $(n - a)$ and subtract, we obtain

$$aE = nE^0_{M^{+n}/M} - (n-a)E^0_{M^{(n-a)+}/M} + 0.059 \log \frac{(M^{n+})}{(M^{(n-a)+})}$$

$$E = \frac{nE^0_{M^{+n}/M} - (n-a)E^0_{M^{(n-a)+}/M}}{a} + \frac{0.059}{a} \log \frac{(M^{n+})}{(M^{(n-a)+})} \tag{88}$$

Equation (88) is the Nernst equation for the half-reaction

$$M^{n+} + ae \rightleftharpoons M^{(n-a)+}$$

Hence

$$E^0_{M^{n+}/M^{(n-a)+}} = \frac{nE^0_{M^{n+}/M} - (n-a)E^0_{M^{(n-a)+}/M}}{a} \tag{89}$$

or

$$nE^0_{M^{n+}/M} = (n-a)E^0_{M^{(n-a)+}/M} + aE^0_{M^{n+}/M^{(n-a)+}} \tag{90}$$

Consider the cupric–copper and cuprous–copper couples. $E^0_{Cu^+/Cu} = 0.521$ V, $E^0_{Cu^{++}/Cu} = 0.337$ V; therefore, according to Eq. (89),

$$E^0_{Cu^{++}/Cu^+} = \frac{2(0.337) - 0.521}{1} = 0.153 \text{ V}$$

PROBLEMS

Tables A-12 and A-13 in the Appendix list values of E^0 and E^f.

1. Calculate the emf of the following cells. Assume $E^0 = E^{0\prime}$.
   (a) N.H.E. ‖ $Tl^+(10^{-3})$ | Tl $\qquad E^0_{Tl^+/Tl} = -0.336$ V
   (b) N.H.E. ‖ $Zn^{++}(10^{-3})$ | Zn $\qquad E^0_{Zn^{++}/Zn} = -0.763$ V
   (c) Tl | $Tl^+(10^{-3})$ ‖ $Zn^{++}(10^{-3})$ | Zn
   (d) N.H.E. ‖ pH = 4.00, $p_{H_2} = 1$ atm | Pt
   Ans. (a) −0.513 V; (b) −0.853 V; (c) −0.339 V; (d) −0.236 V.
2. Write the half-cell reaction for the right half-cells in Problem 1.
3. Devise cells whose cell reactions are (a) $\frac{1}{2}H_2 + Ag^+ \rightleftharpoons H^+ + Ag$; (b) $Zn + 2Tl^+ \rightleftharpoons Zn^{++} + 2Tl$; (c) $VO_2^+ + V^{3+} \rightleftharpoons 2\,VO^{++}$; (d) $Hg + Hg^{++} \rightleftharpoons Hg_2^{++}$; (e) $H^+ + OH^- \rightleftharpoons H_2O$; (f) $PbSO_4 \rightleftharpoons Pb^{++} + SO_4^{--}$.
4. Given the equilibrium constant for the reaction, $K = 15$,
$$Br_2(aq) + Br^- \rightleftharpoons Br_3^-$$
   and $E^0_{Br_2/Br^-} = 1.085$ V, calculate $E^0_{Br_3^-/Br^-}$.
   Ans. 1.05 V.
5. What is the solubility product of lead sulfate? $E^0_{PbSO_4/Pb} = -0.356$ V and $E^0_{Pb^{++}/Pb} = -0.126$ V.
   Ans. $10^{-7.8}$.
6. G. F. Smith and F. P. Richter [*Anal. Chem.* **16**, 580 (1944)] measured the reduction potentials for systems of the type
$$Fe(Phen)^{3+} + e \rightleftharpoons Fe(Phen)^{++}$$
   where Phen represents various substituted 1,10-phenanthrolines (ferroins).

Calculate the ratio of the stability constants for the ferrous and ferric complexes, given that in 1 M sulfuric acid $E^f_{Fe^{3+}/Fe^{++}} = 0.68$ V and $E^f_{Fe(Phen)_3^{3+}/Fe(Phen)_3^{++}} =$ 1.25 V (nitroferroin), 1.11 V (chloroferroin), and 0.97 V (2,2′-dipyridylferroin).

7. (a) What is the equilibrium constant for the reaction 2 Fe(III) + Sn(II $\rightleftharpoons$ 2 Fe(II) + Sn(IV) in 1 M HCl? $E^f_{Fe(III)/Fe(II)} = 0.69$ V, $E^f_{Sn(IV)/Sn(II)} = 0.14$ V. (b) What are the $C_{Fe(III)}$ and $C_{Sn(II)}$ in a solution containing 0.1 M Fe(II) and 0.05 M Sn(IV)? (c) What is the reduction potential, E_{eq}, in part (b)?

8. Silver ion forms a soluble complex with cyanide ion

$$Ag^+ + 2\,CN^- \rightleftharpoons Ag(CN)_2^- \qquad K = \frac{[Ag(CN)_2^-]}{[Ag^+][CN^-]^2} = 10^{18.7}$$

provided there is an excess of cyanide ion as compared to silver ion. (a) What is the standard reduction potential for the half-reaction $Ag(CN)_2^- + e \rightleftharpoons Ag + 2\,CN^-$? (b) What is the reduction potential if $[Ag(CN)_2^-] = 0.01\,M$ and $[CN^-] = 0.10\,M$? *Ans.* (a) −0.31 V; (b) −0.31 V.

9. What is the reduction potential of the Sn(IV)–Sn(II) couple in 1 M HCl when the ratio Sn(IV)/Sn(II) is (a) 10^{-2}; (b) 10^{-1}; (c) 1; (d) 10; (e) 100?
Ans. (a) 0.081 V; (b) 0.110 V; (c) 0.140 V; (d) 0.170 V; (e) 0.199 V.

10. Will platinum metal dissolve appreciably in aqueous chlorine solutions? $E^0_{PtCl_4^{--}/Pt} = 0.73$ V, $E^0_{Cl_2/Cl^-} = 1.359$ V.
Ans. Yes, because $K \gg 1$ for $Pt + Cl_2 + 2\,Cl^- \rightleftharpoons PtCl_4^{--}$.

11. Will silver metal dissolve appreciably in 1 M HCl (a) in the absence of oxygen; (b) in the presence of oxygen? $E^f_{Ag(I)/Ag} = 0.47$ V. Assume that the reaction $O_2 + 4\,H^+ + 4e \rightleftharpoons 2\,H_2O$ is reversible and $E^f \simeq 1.23$ V in part b.
Ans. (a) No; (b) yes.

12. Given $E^0_{Tl^{3+}/Tl^+} = 1.28$ V and $E^0_{Tl^+/Tl} = -0.336$ V, calculate $E^0_{Tl^{3+}/Tl}$.
Ans. 0.74 V.

CHAPTER 9 *Phase Equilibria. Stability of Solids. Adsorption on Solids*

This chapter discusses equilibria between one or more phases (solid, liquid, or gas) in systems of one or more components. Equilibria between slightly soluble substances and their saturated aqueous solutions have been treated in Chapter 7 with respect to the calculation of the solubility. In the present chapter the point of view is entirely different. The discussion is restricted to the general conditions for equilibria among phases without regard to the quantitative composition of the phases. The subject is of analytical significance in several respects, the most important of which are the stability of solid hydrates and the stability of certain solids upon ignition.

Definitions

A *system* is an isolated group of one or more substances in one or more phases. A *phase* is a homogeneous part of a system separated from other phases by physical boundaries. For example, substances may exist as solid, liquid, or vapor. Many solids, including water, may exist in more than one solid modification of different physical properties (density, refractive index, and so forth). Each such solid modification constitutes a phase. Gas phases always are continuous. Liquid or solid phases may be subdivided. Thus solid sugar in the form either of a single crystal or of many small crystals is a single phase. However, a mixture of solid sugar and salt contains two phases, regardless of the size of the individual crystals. Many pairs of solids form solid solutions over all or part of the composition range. A solid solution of any number of constituents constitutes a single phase. When more than one phase is present the system is called heterogeneous. The *components* of a system are those substances (either elements or compounds) the amounts (or concentrations) of which can undergo independent variations. Some examples will illustrate the definition. In a system containing only water, hydrogen and oxygen, which are the constituents of water, are not components, because the relative amounts of hydrogen and oxygen cannot be varied. Therefore, the only component is water. A system in which solid calcium oxide, solid calcium carbonate, and carbon dioxide are present contains three distinct chemical substances. The amounts of these substances are not independent because of the equilibrium

$$CaCO_3(s) \rightleftharpoons CaO(s) + CO_2 \quad (1)$$

Therefore, only two amounts can be varied independently and the system contains two components. Similarly, various mixtures of sodium carbonate and water may contain one or more of three different solids (the anhydrous material, monohydrate, or decahydrate) in addition to solution and vapor. Again the amounts of these substances or species are governed by equilibria similar to (1), and there are only two components, sodium carbonate and water.

Phase equilibria in one-component systems

The simplest example is provided by a substance that can exist in a single solid form in addition to liquid and vapor. The variation of pressure p with temperature T is shown in Fig. 9-1. In the entire discussion only *equilibrium* situations are considered; the rate of attaining equilibrium involving a solid may be small.

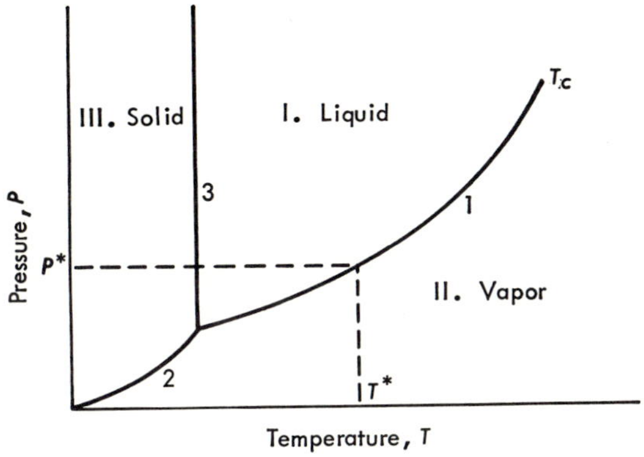

Fig. 9-1. *p-T* diagram for a one-component system.

The curved line (1) represents the p–T relation for a system containing liquid and vapor phases: thus p is the *vapor pressure* of the liquid. The line ends at the critical temperature T_c, above which there is no distinction between liquid and vapor. The amounts of the two phases are of no consequence.[1] Evidently if T is specified, p is completely determined; p can be changed only by a change in T. Therefore, a system of one component in two phases has *one degree of freedom*.[2]

Suppose the liquid–vapor mixture is contained in a cylinder with a movable piston. The temperature is kept constant at T^* and the volume is decreased by movement of the piston. The amount of vapour decreases but p remains constant at p^*. Upon continued compression all the vapor is condensed. Now the system contains only one phase and does not correspond any more to line 1; the system can have any pressure greater than p^*. Thus the one-component, one-phase system has two degrees of freedom, and in general can lie anywhere in area I. Similarly, a system containing only vapor has two degrees of freedom and lies in area II. Curves 2 and 3 represent, respectively, the vapor pressure of solid–vapor and pressure (note—not vapor pressure) of solid–liquid systems. The sign of the slope of line 3 depends upon whether the solid expands or contracts upon melting. Usually the former is the case and the slope is positive, as shown. The reverse is true for water. In all cases the slope of line 3 is very large and the solid–liquid line is nearly vertical.

Lines 1, 2, and 3 intersect at the *triple point*. This is an invariant point with no degree of freedom. If either p or T is changed, one of the phases must disappear. The triple point is not identical with the melting point, because the latter usually is determined at a pressure of 1 atm. However, the difference always is very small, amounting for water to about 0.01°C.

[1] When the amount of the liquid is very small, or when the liquid is present in very small drops, this statement is not exact because the vapor pressure of a finely divided liquid (or solid) is greater than that of the bulk liquid (or solid). This complication is irrelevant here but is mentioned in Chapter 10.

[2] Gravitational, electrical, and other effects are excluded.

Table 9-1 gives a few values of the vapor pressure of water.

TABLE 9-1. **Vapor Pressure of Water at Various Temperatures**

Temperature, °C	0	10	15	20	25	30	35	100
Pressure, mm Hg	4.6	9.21	12.79	17.53	23.76	31.82	41.6	760

The phase rule

The preceding paragraphs indicate a connection between the number of phases (P), number of degrees of freedom (F), and number of components (C). All the above can be summarized in the phase rule:

$$P + F = C + 2 \tag{2}$$

This simple rule shows that the maximum number of phases that can coexist at equilibrium is $C + 2$. In the system sodium carbonate–water, for example, the rule shows that the three solid forms cannot exist in equilibrium with solution and vapor at any temperature. Four-phase, two-component systems are permitted but it is not possible to predict whether or not such systems exist solely on the basis of Eq. (2).

Equilibria in two-component systems

An example of analytical significance is a system containing an anhydrous solid, a hydrated solid, and water vapor. In the sodium bromide–water system the equilibrium is represented by

$$NaBr \cdot 2\,H_2O(s) \rightleftharpoons NaBr(s) + 2\,H_2O \tag{3}$$

These two solids are present as separate phases; i.e., they do not form a solid solution. Therefore, $F = 1$; at a given temperature the vapor pressure is constant and *independent* of the amounts of the two solids.[3] It is important to note that there is no definite pressure where only one of the solids is present. Thus if only the anhydrous form is present, the vapor pressure can have any value from zero up to, but less than, the equilibrium pressure over the mixture of the two solids. The identical principle applies to the general equilibrium between two hydrates:

$$S \cdot mH_2O(s) \rightleftharpoons S \cdot nH_2O(s) + (m - n)H_2O \tag{4}$$

Therefore the vapor pressures of such *systems* depend only on temperature. Since vapor pressures generally increase exponentially with temperature, it is convenient to tabulate the *relative* vapor pressure (i.e., actual pressure ÷ vapor pressure of water at the same temperature), which does not vary much with temperature. Illustrative values are given in Table 9-2. When the relative vapor pressure is very small (e.g., $CaCl_2$–$CaCl_2 \cdot H_2O$) the mixture may serve as a desiccant (see Problem 10).

Another important two-component, three-phase system is a mixture of a solid (either anhydrous or hydrated, depending upon the substance), saturated solution, and vapor. In these systems also the vapor pressure depends only on temperature; illustrative values are in Table 9-3. The mixtures either of Table 9-2 or 9-3 may be used to maintain the relative vapor pressure (humidity) at a constant value in a closed container. Equilibrium is attained much more quickly in systems containing saturated solution and solid than in those containing two solids. If, for example, it were desired to maintain a constant humidity of about 60 per cent at 25°C, a saturated solution of ammonium nitrate would be much preferable to a mixture of the di- and heptahydrates of disodium hydrogen phosphate.

[3] The same conclusion is reached directly by writing the equilibrium constant K for reaction (3). The value of K depends only on temperature. Since the activities of the two solids are constant, the activity or vapor pressure of water also must be constant.

TABLE 9-2. **Relative Vapor Pressures of Mixtures of Anhydrous Solids and Their Hydrates or of Two Hydrates**

	Relative Vapor Pressures at Various Temperatures, °C				
Mixture	10	15	20	25	30
$KOH + KOH \cdot H_2O$					0.02
$CaCl_2 + CaCl_2 \cdot H_2O$			0.015		
$CaCl_2 \cdot H_2O + CaCl_2 \cdot 2\ H_2O$		0.05			
$NaOH + NaOH \cdot H_2O$			0.03		
$K_2CO_3 + K_2CO_3 \cdot 2\ H_2O$				0.05	
$NaI + NaI \cdot 2\ H_2O$		0.12	0.13	0.14	0.15
$Na_2CO_3 + Na_2CO_3 \cdot H_2O$				0.20	
$CaCl_2 \cdot 2\ H_2O + CaCl_2 \cdot 6\ H_2O$	0.21	0.21	0.22	0.22	0.23
$Na_2HPO_4 + Na_2HPO_4 \cdot 2\ H_2O$		0.27	0.28	0.29	
$NaBr + NaBr \cdot 2\ H_2O$	0.30	0.32	0.34	0.36	0.38
$Na_2HPO_4 \cdot 2\ H_2O + Na_2HPO_4 \cdot 7\ H_2O$	0.50	0.53	0.56	0.59	0.62
$Na_2CO_3 \cdot H_2O + Na_2CO_3 \cdot 10\ H_2O$	0.66	0.69	0.72	0.76	
$Na_2HPO_4 \cdot 7\ H_2O + Na_2HPO_4 \cdot 12\ H_2O$	0.65	0.70	0.75	0.81	
$Na_2SO_4 + Na_2SO_4 \cdot 10\ H_2O$	0.69	0.72	0.76	0.80	

TABLE 9-3. **Relative Vapor Pressures of Mixtures of Solids with Their Saturated Solutions in Water**

	Relative Vapor Pressures at Various Temperatures, °C				
Saturated Solution of	10	15	20	25	30
P_2O_5			0		
KOH			0.05		
$NaOH$			0.06		
$CaBr_2 \cdot 6\ H_2O$	0.23	0.21	0.19	0.17	0.15
$KC_2H_3O_2 \cdot 1.5\ H_2O$[a]				0.22	
$CaCl_2 \cdot 6\ H_2O$	0.38	0.35	0.32	0.29	0.26
$MgCl_2 \cdot 6\ H_2O$[a]				0.33	
$K_2CO_3 \cdot 2\ H_2O$			0.44	0.45	
$Ca(NO_3)_2 \cdot 4\ H_2O$	0.65	0.60	0.55	0.50	0.45
$Mg(NO_3)_2 \cdot 6\ H_2O$[a]				0.53	
$NaBr \cdot 2\ H_2O$	0.63	0.61	0.59	0.57	0.55
NH_4NO_3	0.69	0.66	0.63	0.60	0.57
$SrCl_2 \cdot 6\ H_2O$[a]				0.71	
$NaCl$	0.75	0.75	0.75	0.75	0.75
$NaNO_3$	0.77	0.76	0.75	0.74	0.73
KCl			0.86	0.85	0.85
$Na_2SO_4 \cdot 10\ H_2O$		0.95	0.90	0.85	0.78
$Na_2CO_3 \cdot 10\ H_2O$			0.91	0.89	0.87
$BaCl_2 \cdot 2\ H_2O$[a]				0.90	
KNO_3			0.95	0.94	0.94

[a] From R. H. Stokes and R. A. Robinson, *Ind. Eng. Chem.* **41**, 2013 (1949).

Vapor pressure of water–salt systems: Dependence on composition

In this and the following examples temperature is assumed constant. In the first example (Fig. 9-2), the solid (e.g., sodium nitrate) exists only in the anhydrous form. Suppose that increasing amounts of the salt are dissolved in water, the temperature being kept constant. The water vapor pressure decreases until the solution is saturated with salt. Further addition of salt (solid phase) does not change the composition of the saturated solution (liquid phase); a mixture of a saturated solution of a salt with the solid has a constant vapor pressure at a fixed temperature. Graphically the change of the vapor pressure on going from pure water to the pure salt is shown in Fig. 9-2. Point *a* corresponds to 100 per cent water with a vapor pressure corresponding to *ae*. Point *c* represents the pure anhydrous salt with an aqueous vapor pressure of zero. On dissolving increasing amounts of salt, the vapor pressure of the water decreases from *e* to *d*; at the latter point the solution is saturated with the salt. The composition

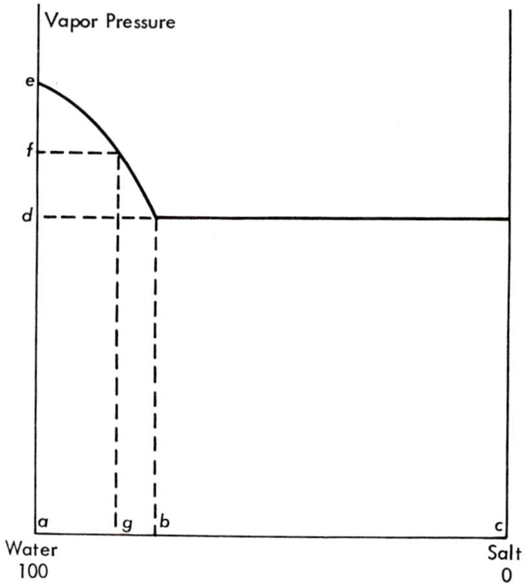

Fig. 9-2. Vapor pressure of water in the system water-anhydrous salt at constant temperature.

of the saturated solution is given by point *b* on the abscissa; the ratio of water to salt is equal to *bc* : *ba*. If the saturated salt solution is evaporated at constant temperature, the vapor pressure does not change until all the water has been removed from the solution.

The composition may be expressed on a weight or mole basis, as preferred. It is important to note that the composition given on the abscissa is that of the combined liquid and solid phases *exclusive* of the amount of water present in the vapor phase. The total amount of water in a system at a given vapor pressure depends, of course, upon the volume of the container. Practically the amount present as vapor usually is very small compared to the total amount.

If the solid may exist either as the anhydrous material or as a hydrate, the situation is as shown in Fig. 9-3. Point *a* corresponds to pure water and point *b* to the composition of the saturated solution (see Problem 8). As additional salt is added beyond point *b*, it forms a solid phase of the hydrated solid. The pressure remains constant at *d*. The formation of the hydrate uses up water, so the amount of the saturated solution decreases. Point f corresponds exactly to the composition of the solid hydrate (e.g., 66.7 mole per cent water for sodium bromide dihydrate). At point f, therefore, the only phases are solid and vapor. The system has two degrees of freedom, and at the specified temperature the pressure may have any value between d and g. Upon further addition of salt the system obviously contains a mixture of the hydrate and anhydrous form and the pressure again is constant at g. Finally, if only the anhydrous salt is present as a solid, the pressure may have any value between zero and g. The latter point is frequently misunderstood. When some water vapor is added to a dry container containing anhydrous salt, no hydrate is formed until the pressure reaches the value g. Thus all the added water is present as vapor. Upon further addition of water, the pressure does not increase until all the solid material has been converted to the solid hydrate.

If the salt can form various hydrates, the vapor pressure of the system water–salt at constant temperature is represented by a diagram such as that shown in Fig. 9-4.[4]

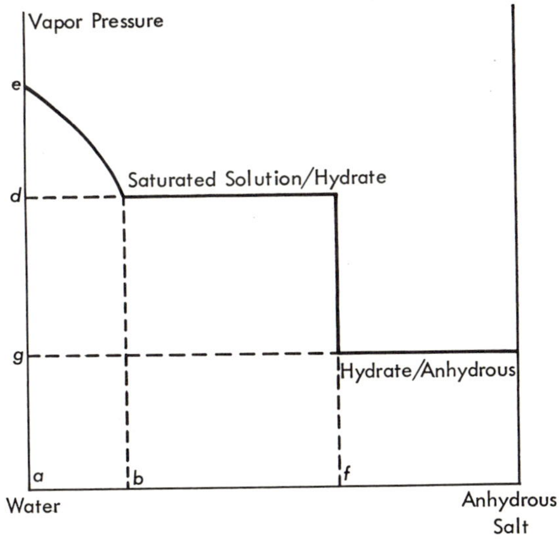

Fig. 9-3. Vapor pressure of water in the system water-salt hydrate-anhydrous salt at constant temperature.

This diagram represents the system water–sodium carbonate. Ordinary crystallized sodium carbonate usually consists of the decahydrate; in addition, there exist a monohydrate and, of course, the anhydrous form. At point b the solution is saturated with $Na_2CO_3 \cdot 10\ H_2O$. The vapor pressure remains constant up to c, corresponding to the pure decahydrate. The constant vapor pressure of mixtures of $Na_2CO_3 \cdot 10\ H_2O$ with $Na_2CO_3 \cdot H_2O$ is represented by de, and that of mixtures of the monohydrate with the anhydrous salt by fk. With the aid of diagrams such as the above we can easily predict what will happen when anhydrous salts or salt hydrates are exposed to the atmosphere. Atmospheric air contains varying amounts of water vapor, depending upon climatological and meteorological conditions. During rainy weather the air is nearly saturated with water vapor, and the aqueous pressure at various temperatures corresponds approximately to the data given in Table 9-1. As a rule the atmosphere is not saturated with water vapor. It is customary to express the *relative humidity* or the *degree of humidity* by the ratio of the prevailing water vapor pressure to that of air

[4] From N. Schoorl, *Commentaar op de Nederlandsche Pharmacopee*, A. Oosthock, Utrecht, Vyfde Uitgave, p. 37. 1927,

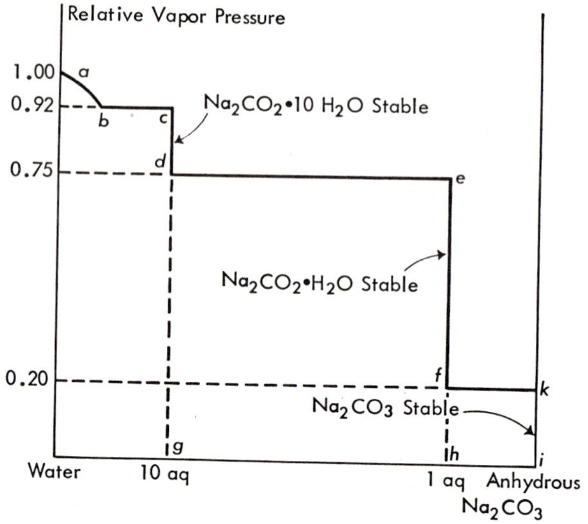

Fig. 9-4. Vapor pressure of water in the system water-sodium carbonate.

saturated with water at the same temperature. By multiplying this relative humidity by 100, the *percentage of relative humidity* is obtained. Suppose, for example, that the water vapor pressure of the atmosphere at 25°C is 11.88 mm Hg. Then since air saturated with water vapor at 25°C has a vapor pressure of 23.76 mm Hg, the relative humidity is equal to $11.88/23.76 = 0.50$, and the percentage humidity is 50. The humidity of the air is determined with a hygrometer or psychrometer. It should be realized that the humidity of the air in laboratories changes considerably with the season. In winter in cold climates, when the buildings are heated, the laboratory air is nearly always extremely dry, relative humidities as low as 0.10 to 0.20 often being observed. In the summer, however, the humidity of the laboratory air is close to that of the atmosphere and in damp weather can reach values close to saturation.

Let us consider what happens when various types of salts or salt hydrates are exposed to air of different humidities. Figure 9-2 represents the vapor pressure of the system water–sodium nitrate. Point d represents the vapor pressure; the ratio of the vapor pressure of the solution to that of pure water at the same temperature is 0.74 at 25°C. If the relative humidity of the air is larger than 0.74, say f, the salt will attract water and form a saturated solution. After all the salt has dissolved, it will continue to absorb water until the composition of the solution (g) is such that it has the same relative vapor pressure as the atmosphere. This phenomenon is called *deliquescence* and occurs only when the humidity of the air is greater than the vapor pressure of the saturated salt solution. If the relative humidity of the air is less than 0.74, solid sodium nitrate will remain unchanged. When we are dealing with a salt hydrate, the latter may, depending upon the conditions, remain unchanged or deliquesce or give off water. Sodium bromide, for example, can exist as dihydrate and as anhydrous salt. The saturated solution of the hydrate has a relative vapor pressure of 0.57 at 25°C. Therefore it will deliquesce when the relative humidity of the air is greater than 0.57. The dihydrate itself has a relative vapor pressure of 0.36 at 25°C and therefore will give off water when the humidity of the air becomes smaller than 0.36. This process is called *efflorescence*. Sodium bromide dihydrate is stable only when kept at relative humidities between 0.36 and 0.57 (at 25°C). From Fig. 9-4 it is seen that $Na_2CO_3 \cdot 10\,H_2O$ is stable at relative humidities between 0.92 and 0.75. Therefore it effloresces very easily with the formation of the monohydrate. The latter effloresces in turn when the relative humidity is less than 0.20. Or, starting from the other side, we may say that anhydrous sodium carbonate is not hygroscopic unless the humidity of the air is greater than 0.20. In the latter case it will be transformed into the monohydrate. No liquid phase is formed at humidities less than 92 per cent.

The phenomena described are of great analytical importance:

1. Analysis of salts which deliquesce very easily. A saturated solution of potassium hydroxide in water has a relative vapor pressure of 0.05 at 25°C. Therefore, the solid will deliquesce very rapidly when exposed to air. A sample for analysis should be weighed in a well-closed weighing bottle; if the weighing were made in a vessel open to the air, the sample would continuously increase in weight.

2. For volumetric analysis it is of great importance to have pure salts or salt hydrates available as primary standard substances and to keep them in an unchanged state. Borax, for example, having the composition $Na_2B_4O_7 \cdot 10\ H_2O$, is an excellent primary standard for the standardization of acids. A saturated solution of borax has a relative water vapor pressure of 0.99 at 20°C; therefore, deliquescence is not to be feared under any conditions. The relative vapor pressure of borax is 0.39 at 20°C, when it is in equilibrium with $Na_2B_4O_7 \cdot 5\ H_2O$. Therefore, if borax is exposed to air of relative humidity less than 0.39, it will effloresce to the pentahydrate. The latter is stable only at relative humidities between 0.39 and 0.25; if the relative humidity becomes less than 0.25, the pentahydrate will effloresce to the anhydrous salt. It has been the experience of the authors that substances kept in ordinary glass-stoppered bottles are usually not protected from atmospheric influences. With different types of bottles it was found that pure borax always effloresced during the winter season. However, it can be kept unchanged in bottles which are hermetically closed. It is much safer, however, to preserve a sample of the pure borax in an atmosphere in which it is stable, i.e., at a relative humidity between 0.99 and 0.39 (20°C). This can be done by placing the sample in a desiccator in which the humidity is kept constant by a suitable mixture. We have seen, for example, that a mixture of sodium bromide dihydrate and its aqueous solution has a constant humidity at a given temperature (0.59 at 20°C, 0.57 at 25°C). Therefore, if we place some anhydrous sodium bromide in a desiccator and add enough water to transform it into the dihydrate, and in addition a little more to assure the presence of a saturated solution of the latter, the atmosphere in this desiccator will have a relative humidity of 0.57 (25°C), and borax will be quite stable. It is still better to keep the borax at a relative humidity of 0.70. Such arrangements, by which the humidity in a certain space is kept constant, are called *hygrostats* by Schoorl. After nearly all the solid sodium bromide has dissolved by taking on water, the dish in which it has been placed in the hygrostat can be heated on a steam bath to recover the original product.

The use of hygrostats is also of great practical importance in the preparation of pure salt hydrates. The latter can be purified by recrystallization from water, but the final product always contains occluded mother liquor. The latter may be removed by Schoorl's method, which consists of efflorescing the salt to a lower hydrate or the anhydrous form by placing it over a dehydrating agent (see p. 470). By this process the crystals usually break down into an extremely fine powder, and consequently the mother liquor is removed. The product obtained is then placed in a hygrostat of the proper humidity to take up hydrate water and is not used until the salt has reached constant weight. Pure oxalic acid dihydrate, borax, and sodium thiosulfate are prepared and kept in the above way. As mentioned previously the use of a salt or a salt hydrate with its saturated solution is more advantageous than the use of a solid mixture of a salt with its lower hydrate or its anhydrous form. In the former case equilibrium is attained much more rapidly than in the latter.

Conversion of carbonates to oxides by ignition

MCO_3 represents a metal carbonate, M being calcium, lead, etc. In general upon ignition at a suitable temperature the carbonate may decompose according to

$$MCO_3(s) = MO(s) + CO_2 \qquad (5)$$

Evidently the equilibrium is governed by the same general principles discussed above. At any temperature the equilibrium vapor pressure of carbon dioxide above the mixture of the two solids is constant; this is called the *dissociation pressure*. Numerical values of dissociation pressure for the important system $CaCO_3$–CaO are given in Table 9-4.

TABLE **9-4. Dissociation Pressures of Calcium Carbonate**[a]

Temperature, °C	*Pressure of* CO_2 (mm Hg)
20	2.2×10^{-20}[b]
200	7.8×10^{-19}[b]
400	0.3×10^{-3}[b]
500	0.15
600	2.98
700	31.2
800	208
882	760
900	984

[a] From L. Andrussow, *Z. Physik. Chem. (Leipzig)* **116**, 95 (1925). The measured dissociation pressures of calcium carbonate depend somewhat upon particle size, the degree of perfection of the crystals, and the presence of traces of impurities. J. C. Southard and P. H. Royster, *J. Phys. Chem.* **40**, 435 (1936), found the dissociation pressure of calcite to equal 1 atm at 894.4°C.

[b] Extrapolated value.

From Table 9-4 it is seen that the dissociation pressure of calcium carbonate equals 1 atm at 882°C. Therefore, if the pressure of carbon dioxide is maintained at 1 atm and the temperature is less than 882°C, calcium carbonate is perfectly stable with respect to conversion into the oxide. Conversely, under the same conditions the oxide is entirely unstable and will be converted into the carbonate. It is important to examine the behavior in ordinary air, which contains about 0.03 per cent by volume of carbon dioxide.[5] When the atmospheric pressure is 760 mm Hg, the partial pressure of carbon dioxide therefore is 0.23 mm Hg. In atmospheric air, then, calcium carbonate is perfectly stable as long as its dissociation pressure does not exceed 0.23 mm Hg. Referring to Table 9-4 we find that this pressure is reached at slightly more than 500°C. Stated in other words, calcium carbonate can be maintained indefinitely at a temperature of 500°C in ordinary air with no possibility of formation of calcium oxide. All these considerations are of practical significance in the gravimetric determination of calcium by precipitation as calcium oxalate. Upon suitable ignition the oxalate may be converted either to the carbonate and oxide for weighing. The advantages and practical details, including the rate of attainment of equilibrium, are discussed in Chapter 28.

Similar considerations evidently apply to the decomposition of oxides. For example,

$$PbO(s) \rightleftharpoons Pb(s) + \tfrac{1}{2} O_2 \qquad (6)$$

cannot occur until the dissociation pressure reaches approximately 150 mm Hg.

Other analytical implications of the phase rule

Many methods of separation require a knowledge of the phase rule for their interpretation. For example, we may mention the preparation of pure compounds, such as glacial

[5] In a laboratory in which gas burners are being used the carbon dioxide content is higher than in normal air.

acetic acid or benzene, by freezing. A simple and important application of the rule states that no pure compound separates when the impurity forms a mixed crystal with the host crystal. Also, the conditions to be fulfilled in separations by fractional distillation are deduced from the phase rule.

Adsorbed and sorbed water

Solids exposed to the atmosphere quite generally absorb some water vapor on their surfaces. The phenomenon is different in nature from the equilibrium phenomena discussed above. Generally true equilibrium values are not attained, as will be evident from the examples given below. For a given solid with a surface of reproducible character, the relation between the amount absorbed and the water vapor pressure is given by the *adsorption isotherm*, an example of which is presented in Fig. 9-5. The adsorption is

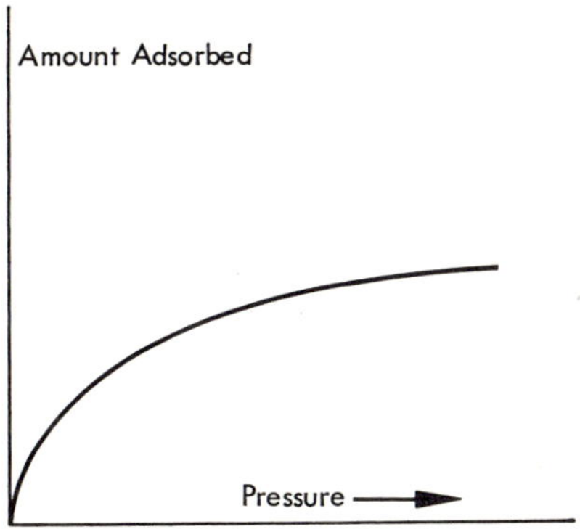

Fig. 9-5. Adsorption isotherm (gas-solid).

relatively large at small pressures and changes rapidly in this region with the vapor pressure. Quite generally the adsorption decreases with increasing temperature.

Ordinarily when considering the adsorption of water by solids, we speak of *surface condensation*, meaning by this that we are not dealing with a simple gas adsorption but with a condensation of absorbed vapor. To get such a liquid layer it is necessary that the surface be wetted by the absorbed substance.

In precise analyses (atomic-weight determinations) the adsorption taking place on the surface of the containers in which weighings are being made can become significant. Warburg and Ihmori[6] determined the condensation on various surfaces in air partially saturated with water vapor. Some results are given in Table 9-5. The data reported indicate the order of magnitude of the water condensation. In some cases the thickness of the layer may correspond to thousands of water molecules.

If glass is heated to temperatures above 150°C, it is difficultly wetted by water, and, if left in contact with water vapor, it is only slowly wetted again. From the above it is clear that slight deviations in analytical results may occur if weighings are made in large glass containers. If the latter are heated and then cooled in a desiccator, the weight will gradually increase when they are placed on the balance. The final equilibrium is reached much more rapidly from the wet side than from the dry. Therefore, it is recommendable to wipe a glass container (e.g., calcium chloride tube in water determinations) with a damp cloth and let it stand for about 15 min in the balance case before weighing. It should be realized that the weight of the glass container remains the same only when the humidity and temperature of the atmosphere remain unchanged.

[6] E. Warburg and T. Ihmori, *Wied. Ann.* **27**, 481 (1886); T. Ihmori, *ibid.* **31**, 1006 (1887); Cohnstaedt, *Physik. Z.* **10**, 643 (1909); Parks, *Phil. Mag.* [6]**5**, 517 (1903).

TABLE 9-5. **Surface Condensation of Water**[a,b]

Kind of Surface	*Thickness of Water Layer,* mm	*Weight of Adsorption Layer of Water/*dm^2, mg
Platinum	10^{-5}	0.1
Washed Jena glass	10^{-5}	0.1
Unwashed Jena glass	3.3×10^{-5}	0.33
Ordinary glass	10^{-4}–10^{-3}	1–10
Agate	10^{-3}	10

[a] T. Ihmori, *Wied. Ann.* **31**, 1006 (1887).
[b] R. Strömberg, *Avhandl. Akad. Stockholm* III[6], No. 2 (1929), gives data on the adsorption of water by fused silica, gold, and platinum at different humidities. As an example we cite his data for platinum (mg/dm^2): r.h. 20, 0.019; r.h. 40, 0.034; r.h. 60, 0.063; r.h. 80, 0.134; r.h. 100, 0.378.

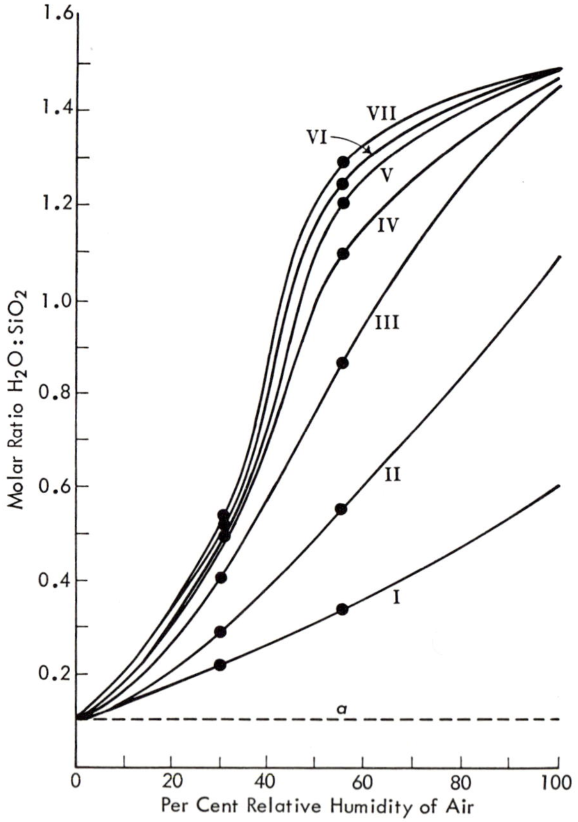

Fig. 9-6. Adsorption of water by silica gel. The dotted line represents the initial ratio of H_2O to SiO_2, and the curves I to VII give the ratio after 4, 10, 24, 48, 96, 192, and 750 hours exposure, respectively, of the gel to atmospheres of the relative humidity indicated.

Finely divided substances absorb water from the atmosphere and hold it on their surfaces. Equilibrium is usually attained with 5–10 min. The water absorbed is usually called *water of hygroscopicity*; its amount increases with increasing surface area and increasing humidity of the air. Therefore such substances, after cooling in a desiccator, should be weighed in well-closed weighing bottles, for otherwise they will absorb water very rapidly again.

Many substances of colloidal character have a huge internal surface development and can absorb very large amounts of water. In such cases equilibrium is reached very slowly; instead of adsorption we speak of a *sorption* process. The water taken up by the material is sometimes called *water of imbibition*. Silica gel, hydrous oxides, and various organic colloids such as cellulose, starch, agar, and proteins show such behavior.[7] In the air-dry state they may contain more than 20–30 per cent of water and still retain the appearance of perfectly dry powders. The water content of such substances changes considerably with varying humidities of the atmosphere. Potato starch, for example, may contain about 10 per cent of water at a relative humidity of 0.20, and 21 per cent of water at a relative humidity of 0.70. Thus, if such substances are kept in ordinary bottles which are not hermetically closed, their water content may change greatly with the seasons. The change in the moisture content with varying water vapor pressure of the atmosphere is usually given by an S-shaped curve. Figure 9-6, which represents the water content of silica gel at various humidities, shows this clearly.[8] The top curve, giving the molar ratios of sorbed water to SiO_2 after 750 hr of exposure to various vapor tensions, shows the S-shape plainly. It is also seen that equilibrium is reached very slowly; even after 10 days it was not attained. It is of interest to note that fused quartz scarcely absorbs water at all, in contrast to colloidal silica, which is a useful desiccant.

In Fig. 9-7 the water contents of gelatin, potato starch, and cellulose, respectively, are given for various relative humidities. As a rule it is not a very simple matter to determine the water content of such materials. On drying at temperatures of 100–115°C, the water is given off slowly; moreover, it is impossible to remove the last traces of water, because the surrounding air has a certain humidity, and the sorption at very low aqueous vapor pressures is still relatively large. Heating at much higher temperatures (180–200°C) involves the danger of decomposition, especially when dealing with organic materials. In such cases it is necessary to dry in a vacuum oven at 70–100°C and remove the water given off or to dry in a current of air that has been deprived of its water vapor. Great precaution must be taken in the final weighings, because the dried material is extremely hygroscopic and absorbs water very rapidly when exposed to the air. Therefore, the weighings should be made in well-closed containers.

The hygroscopic properties of cellulose are of analytical interest, because filter paper consists of this material. If a substance were collected on filter paper, dried at 100–120°C, and weighed after cooling, the dried paper would attract water very readily (see curve for cellulose in Fig. 9-7). Since the sintered glass, porous porcelain, or Gooch filter crucibles (see p. 464) do not show such behavior to any appreciable extent, they are always used for such purposes.

The states of water in solids

Water of hygroscopicity. In a previous section we saw that finely divided substances absorb water on their surfaces. The amount of hygroscopic water increases with increasing surface area. As a rule it is extremely small after ignition of a precipitate, at least if the latter does not combine chemically with the water. If the precipitate is not

[7] Pioneer work in this field has been done by J. M. van Bemmelen, *Z. Anorg. Allgem. Chem.* **5**, 466 (1894); **13**, 233 (1897); **18**, 14, 98 (1898); **30**, 265 (1902); J. M. van Bemmelen and E. A. Klobbie, *J. Prakt. Chem.* **46**, 497 (1892); *Rec. Trav. Chim.* **7**, 37 (1888); *Z. Anorg. Allgem. Chem.* **49**, 125 (1906).

[8] I. M. Kolthoff and V. A. Stenger, *J. Phys. Chem.* **38**, 250 (1934).

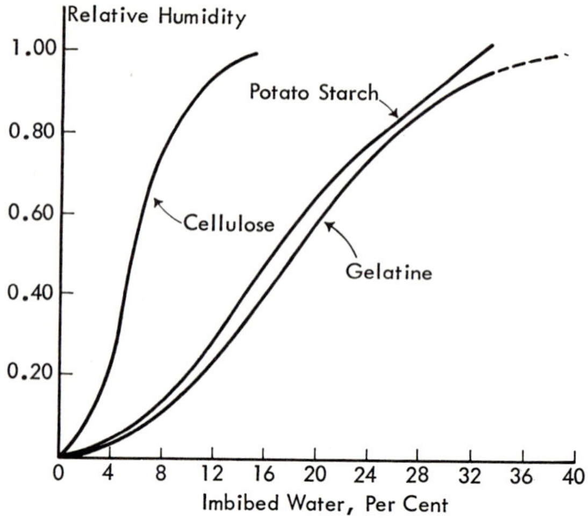

Fig. 9-7. Imbibition of water by colloidal substances.

ignited, the amount of hygroscopic water can be relatively large, since it is absorbed at both the external and internal surfaces (see p. 221). It is found, for example, that calcium oxalate monohydrate formed under analytical conditions is very hygroscopic after drying at 100–105°C and can take up 1 per cent of water. If such a dried precipitate is weighed unprotected in the air, its weight increases rapidly. It should be understood that absorbed water is not always quantitatively removed by heating at 100–130°C. The processes of adsorption and desorption of water are reversible.

Occluded water. Precipitates, and solids in general, contain occluded water in the interior of the material. Occluded water is not removed by heating at 100°C but disappears continuously at much higher temperatures. Most crystals separated from an aqueous medium contain occluded water. In most macroscopic crystals cavities filled with mother liquor may be seen with the aid of the microscope (Fig. 9-8). But even if

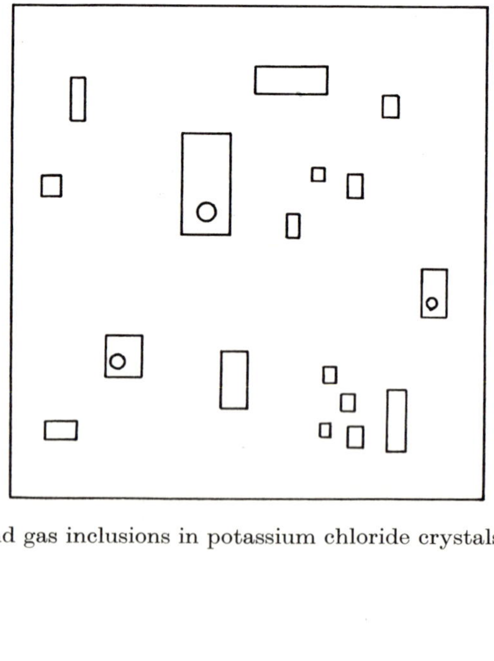

Fig. 9-8. Liquid and gas inclusions in potassium chloride crystals (× 100).

these cavities are not microscopically visible, water is usually present in amicroscopic cavities or in an adsorbed state at the internal isolated surface. On heating such crystals containing cavities, the former burst open at higher temperatures (*decrepitation*). In the coprecipitation of foreign ions the latter are absorbed during the growth of the precipitate in more or less hydrated state.

In addition, water may be distributed homogeneously throughout the solid. Such cases occur, for example, with natural glasses, which may contain several per cent water.

Summarizing then, occluded water may be present in cavities containing mother liquor, in the absorbed state at the internal surface, and finally in solid solution.

Water of imbibition.[9] In a previous section it was mentioned that sorbed water in organic lyophilic colloids is removed slowly and incompletely by heating at temperatures from 100 to 150°C. By working in a vacuum at higher temperatures, it usually escapes completely; in this way it is often possible to determine sorbed water and occluded water separately.

Hillebrand,[10] in a discussion of the condition of hydrogen in minerals, mineral aggregates, and mineral powders, gives the following classification of nonessential hydrogen:

"*Non-essential hydrogen*—its presence not necessary for characterization of the mineral. (Considered as water or other compounds.)

I. *In liquid solution*, specially in deliquescent powders, also in amorphous material, including undercooled liquids.
II. *Held by surface forces*
    a. Absorbed on surfaces in films of molecular thickness [usually monomolecular thickness].
        1. On walls of cavities within grains of aggregates.
        2. On exteriors of grains.
    b. Held by capillarity, in liquid form.
        1. In colloids with mesh structure.
        2. In definite openings within grains of aggregates.
        3. In spaces between separate grains.
III. *Included*
    a. As liquid droplets in cavities.
    b. As included grains of minerals belonging to one of the above listed classes."

Essential water

Water of crystallization. Various precipitates crystallize as hydrates that are stable in the air between certain limits of humidity. This, for example, is the case with $MgNH_4PO_4 \cdot 6\,H_2O$ and $CaC_2O_4 \cdot H_2O$, which can be weighed as such.

[9] Water of imbibition can be taken up in different ways:
1. The water is sorbed without changing the lattice of the original material.
    a. The original lattice may contain vacant spaces easily accessible to water and other vapors. Water may condense in these spaces. This occurs, for example, with substances of a zeolitic nature.
    b. The crystallites in lyophilic organic colloids absorb water at their surfaces, this process being accompanied by a swelling (starch, agar).
2. The sorbed water changes the lattice of the material.
    a. Water penetrates into the lattice, giving rise to a continuous increase of the distance between the planes of a layer lattice, and therefore to a swelling. The case occurs, for example, with graphitic acid. [J. R. Katz and J. C. Derksen, *Rec. Trav. Chim.* 53, 652 (1934).]
    b. A definite hydrate is formed.

[10] W. F. Hillebrand, *U.S. Geol. Surv. Bull.* 700, 1919, p. 64; W. F. Hillebrand and G. E. F. Lundell, *Applied Inorganic Analysis*, p. 676, Wiley, New York, 1929,

Water of crystallization in salts can be determined as a rule by indirect analysis, i.e., by determining the weight before and after heating. If the salt decomposes on heating, it is often possible to find a temperature range over which the water of crystallization is completely removed, without decomposition of the salt. If the water cannot be removed without decomposing the solid, a direct water determination must be resorted to.

Water of constitution. The water is not present as such but in the form of hydroxyl or hydrogen chemically bound in the compound. On heating, decomposition takes place with formation of water, which may be called *water of constitution.*

Examples. On heating Na_2HPO_4 or $KHSO_4$, water of constitution is expelled:

$$2\ Na_2HPO_4 \rightarrow Na_4P_2O_7 + H_2O$$

$$2\ KHSO_4 \rightarrow K_2S_2O_7 + H_2O$$

In these cases the hydrogen is removed as water. On heating of calcium and magnesium hydroxides, hydroxyl is also removed as water:

$$Ca(OH)_2 \rightarrow CaO + H_2O$$

Hillebrand (Ref. 10) gives the following classification of essential hydrogen in minerals:

"*Essential hydrogen*—present in regular atomic arrangement in molecular or crystal structure.

I. Present in definite, stoichiometric proportions:
   a. As hydrogen—called acidic by analogy to behavior in solutions.
   b. As hydroxyl—the hydrogen called basic, again by analogy.
   c. As water—especially water of crystallization.
II. Not present in stoichiometric proportions with reference to the substances as a whole, but present, in one of the above forms, in one or more of the constituents of mixed crystals, including isomorphous mixtures."

When dealing with substances such as minerals that contain various types of essential and nonessential water, an exact determination of each may be difficult. Nevertheless, considerable information may be obtained by appropriate analytical procedures and application of physicochemical methods. For a more detailed discussion the student is referred to the excellent treatment of Hillebrand (Ref. 10).

In connection with the above, a few words may be said with regard to sampling of solids for analysis. On grinding of mineral powders, the surface is increased and hence the water adsorption from the atmosphere. They may retain appreciable amounts of this absorbed water, even in excess of 1 per cent, after heating to temperatures far above 100°C. Moreover, it should be realized that the water content of the sample changes with the humidity of the air, if it is not kept in tightly closed containers.

Although more water is adsorbed by the ground material than the coarse powder, water of crystallization may be expelled from solids by long-continued grinding.

PROBLEMS

1. How may a sample of partially effloresced sodium thiosulfate pentahydrate be transformed entirely into the pentahydrate? The pentahydrate is stable between relative aqueous vapor pressures of 0.23 and 0.69 at 20°C.
2. Na_2HPO_4 is kept at the following relative vapor tensions: 0.25, 0.40, 0.65, 0.75, respectively, at 25°C. What is the composition of the solid after equilibrium has been reached in each case? How could the water of hydration and the water of constitution be determined separately?
3. What is the weight of water vapor in 1 liter of air at 20°C when its relative humidity is 0.40 and the barometric pressure is 750 mm Hg? When the relative humidity is 0.0005?

4. What phase or phases (in addition to water vapor) are present at equilibrium in a sodium carbonate–water system at the following relative humidities (25°C): 0.10, 0.20, 0.50, 0.76, 0.80, and 0.95?
5. 1 g of pure $NaBr \cdot 2\,H_2O$ is placed in a dry container of 10 liters volume at 25°C. At equilibrium, what weights of $NaBr(s)$ and $NaBr \cdot 2\,H_2O(s)$ are present?
6. Construct a pressure–composition diagram (at constant temperature) for the system calcium–carbon dioxide.
7. Is the vapor pressure of water constant in a desiccator in which concentrated sulfuric acid is used as desiccant?
8. The solubility of sodium bromide in water at 25°C is 48.6 g of NaBr per 100 g of solution. Calculate the mole per cent water in the saturated solution.
   Ans. 85.8 mole per cent.
9. Suppose you wish to keep the water content of an air stream constant at approximately 7 mm Hg at 25°C. How could this be done?
10. Calculate the weight of water vapor per liter of air in equilibrium with $CaCl_2(s)$ and $CaCl_2 \cdot H_2O(s)$ at 20°C.

CHAPTER 10 *Formation and Particle Size of Precipitates*

Crystal habit and supersaturation

The size and habit of the particles composing a precipitate depend upon the conditions under which the precipitate has been formed, the individual characteristics of the particular substance and its treatment (aging) after precipitation. Most precipitates, even those of colloidal dimensions, separate in a crystalline state.[1]

A precipitate separates from its supersaturated solutions. When dealing with precipitates composed of ions, the ions combine in the supersaturated solution to form ion associates or clusters. When these clusters have attained a certain size they form a separate (solid) phase in the solution and they grow to larger particles by deposition of ions from the supersaturated solution. The clusters have been called crystallites, germs, and, more generally, *nuclei*; they are much too small to be observed under a microscope. Depending upon the degree of supersaturation, a certain period of time elapses before visible particles separate; this is called the *induction period*. Quite generally, the particle size of a precipitate depends on the relation between the rate of formation of nuclei and the rate of growth of nuclei. When the former is very large compared to the latter, the precipitate will be composed of a large number of small particles (precipitation from concentrated highly supersaturated solutions). Under conditions of slight supersaturation the number of nuclei formed is small and, in general, the precipitate is composed of relatively few particles of microscopically well-defined crystals. Early in this century von Weimarn[2] described a large number of observations on the formation of precipitates and he recognized the importance of supersaturation with reference to the size of the particles. He introduced the concept of *relative supersaturation*

$$\frac{Q - S}{S}$$

where Q is the total concentration of the substance to be precipitated from solution and S is the solubility of crystals of microscopic size (greater than 1 μ). In equimolar solutions, e.g., of barium and sulfate, $Q = [Ba^{++}] = [SO_4^{--}]$. In nonequimolar solutions $Q = ([Ba^{++}][SO_4^{--}])^{1/2}$. In general, the particle size of a given precipitate becomes smaller with increasing relative supersaturation. For several reasons the von Weimarn expression is of little quantitative use. Even from a qualitative viewpoint it is of doubtful significance. According to the von Weimarn rule the particle size should increase continuously with decreasing supersaturation. Actually at a certain

[1] A body is said to be crystalline when the molecules, atoms, or ions that make it up are regularly arranged in a lattice as revealed by X-ray examination; solids that are not crystalline are called amorphous (see below).

[2] P. P. von Weimarn, *Zur Lehre von den Zuständen der Materie*, 1913; *Die Allgemeinheit des kolloiden Zustandes*, T. Steinkopff, Dresden, 1925; *Chem. Rev.* **2**, 217 (1926).

(*a*)

(*b*)

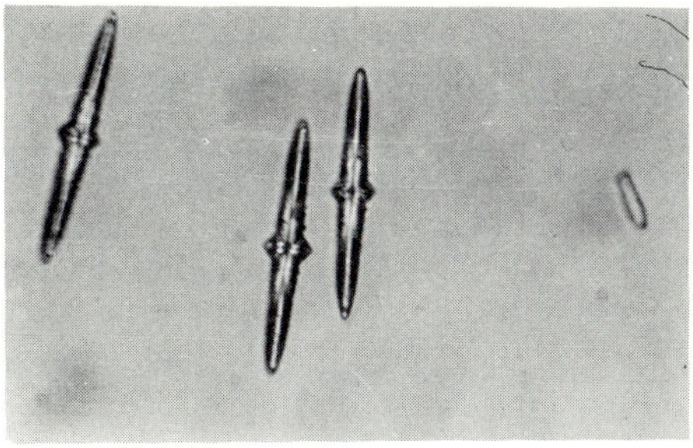

(*c*)

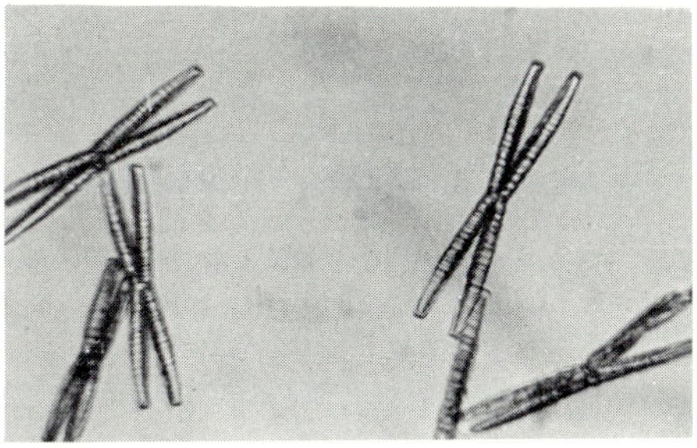

(*d*)

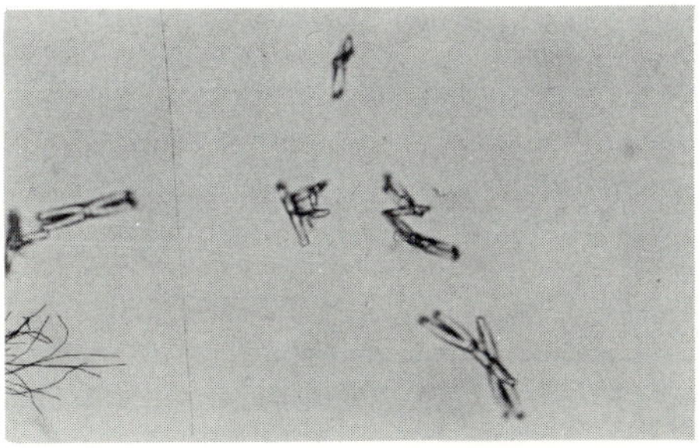

(*e*)

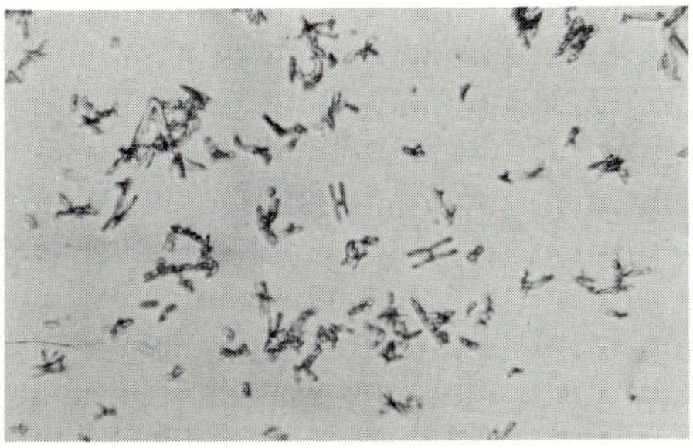

(*f*)

Fig. 10-1. Precipitates of lead sulfate from equimolar solutions of lead perchlorate and sodium sulfate. Magnifications 530×.

(*a*) $[Pb^{+2}] = [SO_4^{-2}] = 1.25 \times 10^{-3}$
(*b*) 2.5×10^{-3}
(*c*) 5×10^{-3}
(*d*) 6.4×10^{-3}
(*e*) 7×10^{-3}
(*f*) 1×10^{-2} *M*

relatively small supersaturation a maximum particle size is obtained; with further decrease of Q the particle size decreases continuously (see below). As will be discussed below, the rate of nucleation is an exponential function of Q. However, the rate of growth is an exponential function of $Q - S$, and when S is small with respect to Q it becomes an exponential function of Q. With reference to the formation of nuclei it should be realized that S should represent the solubility of a nucleus, which, in general, is much greater than that of a microscopic or macro crystal (see p. 210).

In studies of the particle size of a precipitate formed under different degrees of supersaturation, the dimensions observed immediately at the end of the crystallization period should be reported. As a result of aging (p. 220), the dimensions and shape of the particles may change considerably when the precipitate is allowed to stand in contact with the mother liquor. This is especially true when precipitates are formed in highly supersaturated solutions and the primary particles are very small and highly imperfect (p. 221). When dealing with microcrystalline precipitates such as barium and lead sulfate and calcium oxalate it is generally observed that at the smallest supersaturations the precipitates are composed of well-defined small crystals. With increasing supersaturation the particle size increases and the crystals remain well defined. With still more supersaturation the crystal size often increases, but the particles then have the appearance of less perfect crystals. At a given supersaturation a maximum particle size is observed, and the size then decreases rapidly with increasing supersaturation. This behavior is observed, for example, in the precipitation of lead sulfate at room temperature by mixing equimolar solutions of lead perchlorate and sodium or ammonium sulfate or sulfuric acid.[3] The results with sodium sulfate are reproduced in Fig. 10-1. In Fig. 10-2 the length of the crystals is plotted as a function of the concentration of lead or sulfate ions in the original supersaturated solution. The concentration at which maximum particle size is formed is sharply defined. In other studies,

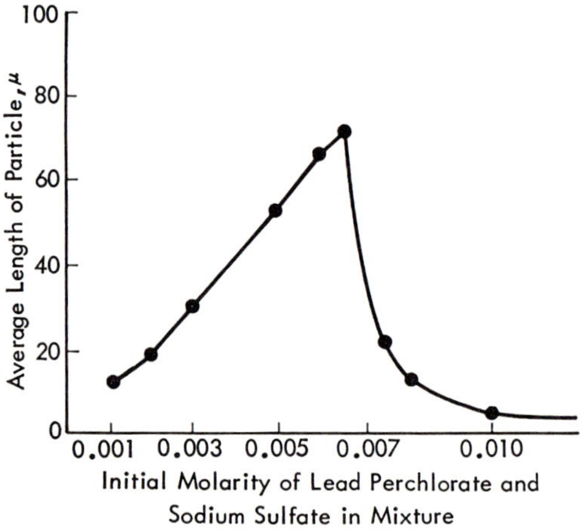

Fig. 10-2. Relation between particle size of lead sulfate and concentration in equimolar mixtures of solutions of lead perchlorate and sodium sulfate.

Van't Riet and Kolthoff[4] found that the maximum size is always found at the same value of the product of activities of lead and sulfate ions, immaterial whether equimolar solutions of lead and sulfate were used or whether lead or sulfate was in excess. This "critical value" of $(Pb^{++})(SO_4^{--})$ was found to be 4×10^{-6}, compared to a value of

[3] I. M. Kolthoff and B. van't Riet, *J. Phys. Chem.* **63**, 817 (1959).
[4] B. van't Riet and I. M. Kolthoff, *J. Phys. Chem.* **64**, 1045 (1960).

the solubility product of lead sulfate $K^0_{sp} = (Pb^{++})(SO_4^{--})$ of 1.65×10^{-8} at 25°C, corresponding to a ratio of 250 or a supersaturation of $(250)^{1/2} = 16$.

Many years ago, extensive studies of the particle size of barium, strontium, calcium and lead sulfate, barium carbonate, silver chloride, and antimony trisulfide were carried out by Odén.[5] As an example, his results obtained by rapid mixing at room temperature of equimolar solutions of barium thiocyanate and ammonium sulfate are presented in Table 10-1. From his experimental data he derived the following expression:

$$n = 1.6 \times 10^{16} m^2$$

in which n is the number of nuclei and m the molarity of barium sulfate in the precipitation mixture. He assumed that the number of nuclei formed is equal to the number of primary particles.

There has been an abundant literature on nucleation and growth of nuclei since Oden's papers appeared. Admittedly, the mechanism of the birth of a precipitate is still incompletely understood. However, studies that have appeared in the last few years have clarified the subject considerably. Considering the many conflicting statements in the literature we shall first discuss in more detail the phenomena of nucleation and crystal growth and then present an interpretation of the experimental results.

Nucleation

In the transformation of a molecular solution into a suspension of a precipitate, it is clear that ions composing a precipitate must unite to form a polymeric molecule that may serve as a nucleus or germ for further growth. These nuclei are extremely small and no measurable change in electrical conductivity is found in the nucleation period. According to a certain group of workers (see below) a nucleus contains no more than four to eight ions of which the precipitate is composed, or only from two to four ion pairs. On very close approach, two ions of opposite charge but of the same valence, e.g., Ba^{++} and SO_4^{--}, may form an ion pair $(Ba^{++}SO_4^{--})$ and may deposit as such on a growing particle. From the measurement of the *induction period*, I, which is the time that elapses between the mixing of reagent solutions and the first detectable appearance of precipitate, Christiansen and Nielsen[6] first concluded that a nucleus of barium sulfate is composed of from seven to eight ions,[7] of silver chromate of six and of calcium fluoride of from eight to nine ions. To find the kinetic order of the nucleation reaction they plotted the logarithm of the ion product, e.g., $[Ba^{++}][SO_4^{--}]$ versus the logarithm of the induction period and found a straight line that satisfies the relation

$$c^n I = \text{constant} \tag{1a}$$

The slope of the straight line is equal to n. [However, as we shall see later (see p. 205) the exponent thus found is considerably greater than the number of ions in the nucleus and changes when a certain supersaturation is attained.] For many years it had been known[8] that for many precipitates the induction period I is an exponential function of the initial concentration and that the following empirical relation holds:

$$I = kc^{-n} \tag{1b}$$

Christiansen and Nielsen interpret the relationship in the following way. For convenience we take as a concrete example the formation of barium sulfate, but the views

[5] S. Odén, *Arkiv Kemi* **1**, No. 26 (1920); **9**, Nos. 23 and 32 (1926).

[6] J. A. Christiansen and A. E. Nielsen, *Acta Chem. Scand.* **5**, 673, 674 (1951); see also A. E. Nielsen, *ibid.* **13**, 784 (1959).

[7] See V. K. LaMer, *Ind. Eng. Chem.* **44**, 1270 (1952).

[8] See, e.g., W. M. Fischer, *Z. Anorg. Allgem. Chem.* **145**, 311 (1925); A. Tovborg-Jensen, *Z. Physik. Chem.* (*Leipzig*) **180A**, 93 (1937); K. Kobayashi, *J. Chem. Soc. Japan*, **70**, 125 (1949); S. Odén and D. Werner, *Arkiv Kemi Mineral. Geol.* **9**, No. 32 (1926); A. van Hook, *J. Phys. Chem.* **44**, 751 (1940).

TABLE 10-1. **Number of Nuclei (Primary Particles) in Barium Sulfate Suspensions Prepared by Rapid Mixing of Equal Volumes of Equimolar $Ba(SCN)_2$ and $(NH_4)_2SO_4$ Solutions**

Molarity of Reacting Solutions	*Molar concn. m of* $BaSO_4$ *before Precipitation*	*Relative Supersaturation,* $U = m/S$	*Number of Nuclei n, Calculated from Equation*	$BaSO_4$, g/liter	*Average Weight of One Particle,* g	*Average Radius,* μ (calc.)	*Average Radius* (μ) *Found by Sedimentation Analysis*
0.8	0.4	40,000	2.56×10^{15}	93.4	3.65×10^{-14}	0.13	53% 0.2
0.2	0.1	10,000	1.6×10^{14} [a]	23.35	1.46×10^{-14}	0.20	Max. 0.24
0.02	0.01	1,000	1.6×10^{12}	2.335	1.46×10^{-12}	0.43	Max. 0.5
0.002	0.001	100	1.6×10^{10}	0.2335	1.46×10^{-11}	0.93	Max. about 1μ
0.0012	0.0006	62	5.8×10^{9}	0.140	2.43×10^{-11}	1.11	Obs. microsc. about 1μ

[a] Experimental value.

are the same for other ionic and nonionic precipitates. Barium sulfate is a strong electrolyte and its (supersaturated) solutions are practically speaking completely dissociated into barium and sulfate ions. After formation of a supersaturated solution association of ions takes place with the formation of clusters or polymeric forms:

(a) $Ba^{++} + SO_4^{--} \rightleftharpoons Ba^{++}SO_4^{--}$ (ion pair or dimer)
(b) $Ba^{++}SO_4^{--} + Ba^{++} \rightleftharpoons (Ba_2SO_4)^{++}$ (ion triplet or trimer)

or

(b′) $BaSO_4 + SO_4^{--} \rightleftharpoons \{Ba(SO_4)_2\}^{--}$ (ion triplet or trimer)
(c) $(BaSO_4)^{++} + SO_4^{--} \rightleftharpoons (BaSO_4)_2$ (ion quadruplet or tetramer)

or

(c′) $\{Ba(SO_4)_2\}^{--} + Ba^{++} \rightleftharpoons (BaSO_4)_2$
$(BaSO_4)_2 + Ba^{++} \rightleftharpoons \{Ba_3(SO_4)_2\}^{++}$ etc.

A nucleus is an ionic cluster of a certain size, which may correspond to that of a unit cell. It is considered to be no longer part of the liquid phase but the embryonic form of the crystalline phase. The nucleus grows rapidly by addition of ions, a process comparable to crystal growth. The rapidly growing nucleus is sometimes called a "*crystallite.*" We shall assume now that the size of the nucleus is that of the species $(BaSO_4)_2$. It is also assumed that equilibrium of reactions (a) and (b) is established rapidly during the induction period. The *rate* of formation of the nuclei according to equation (c) is then given by

$$\text{rate, nuclei formation} = \text{constant}\,[Ba_2(SO_4)^{++}][SO_4^{--}] \tag{2}$$

Application of the mass-action law yields

$$[BaSO_4] = \text{const.}\,[Ba^{++}][SO_4^{--}]\cdot[Ba_2SO_4^{++}] = \text{const.}'\,[Ba^{++}]^2[SO_4^{--}]$$

Thus we find that V_{nucl} = rate of nuclei formation

$$V_{nucl} = K[Ba^{++}]^2[SO_4^{--}]^2 \tag{2a}$$

When $[Ba^{++}] = [SO_4^{--}]$ in the supersaturated solution,

$$V_{nucl} = K[Ba^{++}]^4 = K[SO_4^{--}]^4 \tag{2b}$$

When $[Ba^{++}]$ is greater or smaller than SO_4^{--}, an average concentration C can be written, where $C = ([Ba^{++}][SO_4^{--}])^{1/2}$. Thus

$$V_{nucl} = KC^4 \tag{2c}$$

In all these and the following expressions *activities* should be written instead of concentrations. The value 4 is smaller than the value of the exponent n in Eq. (1), which is found from the plots of the logarithm of concentration versus the log of induction period. The reason is that the formation of visible nuclei also involves a growth factor.

The rate of formation of nuclei can also be expressed

$$\frac{dN}{dt} = KC^p \tag{2d}$$

where N is the number of nuclei formed. From careful experimental work Johnson and O'Rourke[9] calculated for barium sulfate a value of p equal to 4. Under their experimental conditions they obtained evidence that nucleation takes place at a *uniform* rate. At the end of the induction period the growth of the nuclei to crystallites and larger crystals is so rapid that supersaturation rapidly decreases, no new nuclei are formed after the induction period, and the sole precipitation reaction is the growth of the nuclei. Naturally the induction period becomes smaller with increasing concentration, but under the specified conditions the number of nuclei formed at the end of the

[9] R. A. Johnson and J. D. O'Rourke, *J. Am. Chem. Soc.* **76**, 2124 (1954).

induction period is constant and independent of the concentration C. This result is discussed further below. Since no more stable nuclei are formed after termination of the induction period the number of crystals (particles) formed is also constant and equal to the number of nuclei at the end of the induction period. Johnson and O'Rourke rapidly mixed equal volumes of *aged* equimolar solutions of barium chloride and sodium sulfate. The molar concentration C of barium and of sulfate in the mixture was varied between 2.5 and 20×10^{-4}. The number of particles formed was $1.1 \pm 0.1 \times 10^9$ per liter. (Nielsen observed 10^8 particles per liter.)

Referring to the rate of nucleation [Eq. (2c)] it is of interest to mention the effect of dilution of a supersaturated solution during the induction period. When the exponent in Eq. (2c) is 4, the rate of nucleation decreases 16 times when the supersaturated solution is diluted twofold immediately after its preparation, and consequently the induction period increases correspondingly. If the original supersaturated solution is allowed to stand for a period of time that is shorter than the induction period, I will decrease with increasing time elapsed before dilution. This is exactly what was observed by Jensen[10] in his experiments with supersaturated solution of calcium fluoride. He prepared such a solution which was 1.7×10^{-3} M in calcium fluoride. At 23°C this solution had an induction period of 9 min. At several intervals during the induction period he diluted this solution twofold and then observed the induction period of the diluted solution. A freshly prepared solution 8.5×10^{-4} M in calcium fluoride had an induction period greater than 14 days. The observations made upon dilution of the 1.7×10^{-3} M solutions are given in Table 10-2.

TABLE **10-2. Induction Period upon Twofold Dilution of** 1.7×10^{-3} M **CaF_2 Solution**

Diluted after time, min	2	4	6	7
Turbidity observed after	> 3 weeks	7 days	3 days	6 hr

Homogeneous and heterogeneous nucleation

Work in recent years has shown that the number of particles obtained at relatively small supersaturations greatly depends upon the pretreatment of the solutions containing the reacting ions. This has led to the conclusion that foreign suspended material (e.g., dust) in the reacting solutions and impurities on the glass walls of the containers are responsible for nucleation at relatively small supersaturations. This type of nucleation is called *heterogeneous* nucleation, in contrast to *homogeneous* nucleation at higher supersaturations. The experimental work leading to this conclusion is of direct analytical interest and is summarized below.

Using the same barium and sulfate solutions Nielsen[11] obtained 2000 particles|mm^3 when the glass containers had been cleaned by ordinary chemical methods. The number decreased to 100 per mm^3 or even less after cleaning by steaming. Apparently, ordinary glass walls hold tiny particles which can serve as growth centers (seeders) for the crystals and which are removed by steaming. Interesting also are the experiments of Schmid and Jetter[12] on the effect of exposure of their reagent mixtures to ultrasonics on the size and homogeneity of barium sulfate particles. They studied the precipitation from 3 N hydrochloric acid and found that not only did exposure to ultrasound waves during the induction period greatly affect the results, but even exposure of the reagent solutions before their mixing had a profound effect. For example, a sulfate solution was exposed for 1 min to ultrasound waves. The field was turned off and the solution mixed

[10] A. T. Jensen, *Z. Physik. Chem.* (*Leipzig*) **A180**, 93 (1937).
[11] A. E. Nielsen, *Acta Chem. Scand.* **11**, 1512 (1957).
[12] G. Schmid and A. Jetter, *Z. Elektrochem.* **56**, 760 (1952).

with the barium solution. The induction period was reduced by the treatment from 6 sec to 1 sec, while the precipitate obtained from the exposed solution was entirely homodisperse. Schmid assumes that by the exposure a seed is formed from the glass wall. The "glass nuclei" were visible under an electronmicroscope and had a size of 5–10 mμ. Upon repeated exposures, the glass could be "denucleated." When a suspension of a washed coarse barium sulfate precipitate was exposed in a "denucleated" glass vessel to ultrasound waves and the suspension filtered, a slightly turbid filtrate was obtained which was a most effective seed. The precipitate obtained with such seeds was homodisperse. Actually the number of particles formed was proportional to the volume of seed used, indicating that each particle of seed by growth produced a particle of precipitate and that no spontaneous (homogeneous) nucleation occurred. "Nuclei" of copper obtained by exposure of copper foil to ultrasonic waves also yielded an effective seed. The use of a seeder to obtain a homodisperse reaction product is not of recent date. It was first applied by Zsigmondy[13] to obtain homodisperse solutions of gold and other hydrophobic colloids.

Several authors[14] have reported that the size of barium sulfate particles depends on the age of the barium chloride solution. As an illustration, some results obtained by O'Rourke and Johnson[15] on the effect of the age of the barium chloride solution on the number of particles of barium sulfate formed on rapid mixing of equal volumes of 0.002 *M* solutions of barium chloride and sodium sulfate are reported in Table 10-3. Fischer[16] found in general that addition of a solid reagent to a solution of the ions to be precipitated causes a much smaller particle size than when a solution of the reagent is used. Apparently upon dissolution of the solid reagent tiny fragments of the reagent are formed temporarily which act as seeders. Also, addition of an inert solid electrolyte to the fresh mixture of solutions of two reactants has a similar effect.

TABLE **10-3. Number of Particles with Varying Age of Barium Chloride**

Age $BaCl_2$ solution, hr	0	3	4	10	22	72
Number of particles $\times 10^{-9}$ per liter	16.4	4.0	3.5	1.9	1.3	1.1

Most solids contain traces of insoluble material (dust) and the same is true of solvents such as water. Fischer[17] showed that much larger particles of barium sulfate were obtained from solutions prepared in freshly distilled water than when the water was kept for 24 hours or longer. Considering the effect of steaming of the container in which precipitation was carried out on the particle size (Christiansen and Nielsen, Ref. 11) one would expect that a previous steaming of a glass condenser used in the preparation of conductivity water may have an effect on particle size. To avoid these accidental impurities the use of noble metal (silver) containers and condenser would be preferable to the use of glass.

All present evidence indicates that in all studies on precipitate formation, heterogeneous *nucleation* is *being induced by extraneous material*, whether tiny particles on the wall of containers or (and) accidental impurities in the form of tiny insoluble particles. Even analytical-grade reagents contain impurities. A tolerance of 0.1 ppm of "insoluble material" is very reasonable. Assume that this insoluble material is composed of cubes with an edge of 5 mμ and a density of 2. If the number of particles of the impurity per gram of reagent is N we have $125 \times 10^{-21} \times 2 \times N = 10^{-7}$ and $N = 4 \times 10^{11}$

[13] R. Zsigmondy, *Z. Physik. Chem.* **56**, 65, 77 (1906); *Z. Anorg. Allgem. Chem.* **99**, 105 (1917).

[14] E. J. Bogan and H. V. Moyer, *Anal. Chem.* **28**, 473 (1956); R. B. Fischer and T. B. Rhinehammer, *ibid.* **25**, 1544 (1953), **26**, 244 (1954).

[15] J. D. O'Rourke and R. A. Johnson, *Anal. Chem.* **27**, 1699 (1955).

[16] R. B. Fischer, *Anal. Chim. Acta* **22**, 508 (1960).

[17] R. B. Fischer, *Anal. Chim. Acta* **22**, 501 (1960).

per gram. If these particles would all become sites for development of nuclei a large number of particles of precipitate could be anticipated. Fischer[18] considers these particles and particles on the wall of the container as the main if not the only sites of nucleation.

Under carefully controlled conditions, LaMer and Dinegar[19] determined the induction period in the precipitation of barium sulfate as a function of the supersaturation ratio $(K^0_{ss}/K^0_{sp})^{1/2}$ in which K^0_{ss} is the initial *ion-activity product* $(Ba^{++})(SO_4)^{--}$ and K^0_{sp} is the solubility product expressed in activities. To avoid effects of mode and speed of mixing of solutions they generated sulfate ions in the barium solution by making use of the reaction

$$S_2O_8^{--} + 2\,S_2O_3^{--} \rightarrow 2\,SO_4^{--} + S_4O_6^{--}$$

The rate of this reaction was determined and the sulfate concentration was known at the time of apearance of barium sulfate particles. For supersaturation ratios above 12 the induction period was sharply defined. The plot of $(K^0_{ss}/K^0_{sp})^{1/2}$ versus induction period extrapolated to zero time gave the ratio needed for instantaneous formation of particles which was called the critical (limiting) supersaturation ratio below which no spontaneous precipitation occurs and which was found equal to 21.5. In accordance with other experiments on induction periods it has been found that the critical supersaturation value, CSSV, is greatly dependent on presence of accidental suspended material which affects the induction period. For example, Collins and Leineweber[20] found the value of the CSSV to be reproducible under a given set of conditions, but quite dependent on accidental impurities (insoluble particles). Upon repeated recrystallization of the reagents a reproducible value of 32 of the CSSV instead of 21.5 was obtained.

Essentially heterogeneous nucleation leads to no new particles being formed, the number of particles being equal to the number of foreign nuclei accidentally present. It has been observed that at relatively large supersaturation (0.01 M barium sulfate) the number of particles produced suddenly increase (see curve I, Fig. 10-3). Nielsen,[21] who observed this behavior with barium sulfate, identified this supersaturation ($C/C_0 = 10^3$ with $BaSO_4$) with the minimum concentration required to induce *homogeneous nucleation*. The rate of homogeneous supersaturation increases so rapidly above this "*critical homogeneous*" supersaturation that the heterogeneous nucleation is entirely obscured. As an illustration we refer to a publication by Suito and Takiyama,[22] who determined by electron microscopy the particle size of barium sulfate formed at different values of C/C_0. Their results are given in Table 10-4. When the initial concentration C of barium sulfate became 0.01 M there was a sudden large increase in the

TABLE **10-4. Mean Size of Barium Sulfate Formed at Different Molar Concentrations**

Initial concentration of $BaSO_4$	0.0001	0.0003	0.001	0.01	0.05
Mean particle size, mμ	110	850	2030	1650	270

number of particles produced. This corresponds to a value of $C/C_0 = 1000$, which is identical with the value reported by Nielsen. Walton[23] determined the homogeneous supersaturation ratio for the following precipitates: $BaSO_4$, $CaC_2O_4 \cdot H_2O$, AgCl,

[18] R. B. Fischer, *Anal. Chim. Acta* **22**, 501 (1960).
[19] V. K. LaMer and H. Dinegar, *J. Am. Chem. Soc.* **73**, 380 (1951).
[20] F. C. Collins and J. P. Leineweber, *J. Phys. Chem.* **60**, 389 (1956).
[21] A. E. Nielsen, *Acta Chem. Scand*, **15**, 441 (1961).
[22] E. Suito and K. Takiyama, *Bull. Chem. Soc. Japan* **27**, 121 (1954).
[23] A. G. Walton, *Mikrochim. Acta* **3**, 422 (1963).

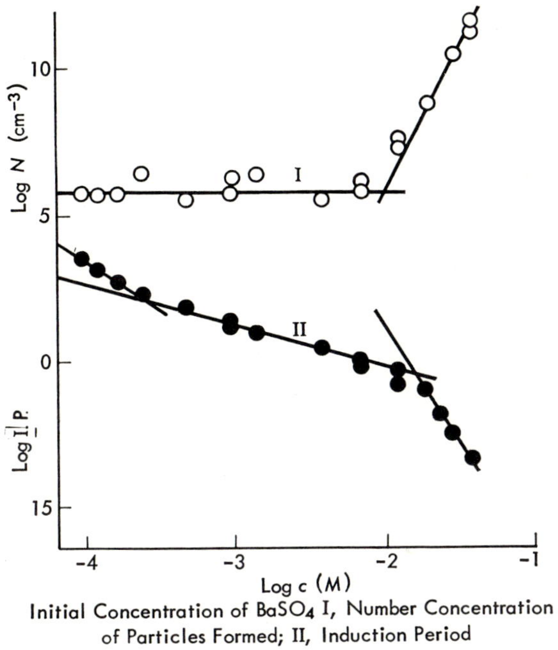

Fig. 10-3. I, logarithm of number of particles *vs.* log molarity of barium sulfate; II, log of induction period t *vs.* log $[BaSO_4]$.

$SrSO_4$, $PbSO_4$, $PbCO_3$, $SrCO_3$, and CaF_2. His results are presented in Table 10-7 and discussed in the section dealing with the relation between solubility and particle size. It may be expected that the exponent in the equation for the induction period [Eq. (1c)] will be different for heterogeneous than for homogeneous nucleation. This indeed was found by Nielsen (see curve II, Fig. 10-3). Below the first small break at 0.0004 M ($\log C < -3.4$) the crystals grow by a fourth-order surface reaction (see below); at $C > 0.0004\ M$ the rate-determining step is diffusion through the liquid. The second break occurs at 0.01 M $BaSO_4$, corresponding to the critical homogeneous supersaturation ratio. Kolthoff and Van't Riet (Ref. 3) had also observed two breaks (see Fig. 10-4) in the plot of the log of the molarity of lead sulfate versus log t (t is the induction period, I.P.). The slope of the first line corresponds to a value of n_1 of 1.7 in the expression

$$\text{I.P.} = k_1 c_i^{-n_1}$$

The slope n_2 of the second line is 2.8 and n_3 of the third line is approximately 8. Again, the point of intersection B of the second and third straight lines occurs approximately at the homogeneous critical supersaturation value, where the number of particles increased abruptly (see Fig. 10.4). This occurred at a molar lead sulfate concentration between 0.007 and 0.01 M. They also found that at or above 0.01 M lead sulfate the particles had a much smaller degree of perfection than particles formed below this critical value.

Growth of crystalline particles

The process of crystal growth is not simple. If the growth rate of particles were determined by diffusion of the lattice ions to the surface, the kinetics of the growth process would be first order with respect to the concentration c of the ions in the solution or to the supersaturation $c - c_s$, c_s referring to the solubility. If the concentration of lattice cations and anions is not the same, $c = ([B^{+n}][A^{-n}])^{1/2}$ when B^{+n} and A^{-n}

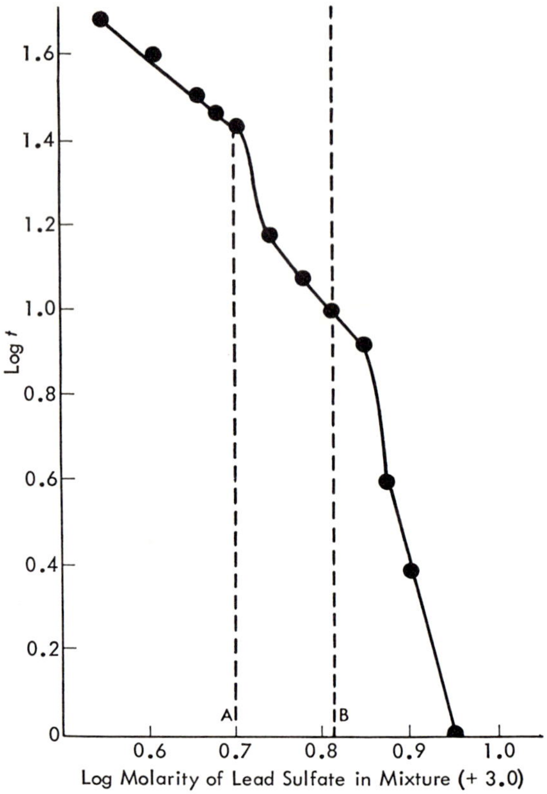

Fig. 10-4. Induction period in precipitation of lead sulfate.

have the same charge ($BaSO_4$, CaC_2O_4) or $c = ([B^{++}][A^-]^2)^{1/3}$ for a salt such as CaF_2. In a study of the precipitation of barium sulfate, Nielsen[24] concluded that at barium and sulfate concentrations higher than 4×10^{-3} *M*, the rate of growth became proportional to the concentration and the growth process became diffusion controlled. With initial concentrations between 0.5 and 1.5×10^{-3} *M*, the particles were found to be more or less distorted prisms, the corners having grown more than the middle of the faces. At concentrations greater than 1.5×10^{-3} *M*, the crystals had the appearance of stars. This appearance is to be expected when diffusion determines the growth rate, since the concentration then is greatest at the corners. The rectangular (orthorhombic) shape is obtained only when the concentration is approximately the same all over the surface, and this situation does not prevail when the consumption of the ions is fast compared with diffusion. Crystals of highly irregular shape are often formed from highly supersaturated solutions (see, e.g., Fig. 10-1). Most studies on the kinetics of growth of precipitates have been carried out at relatively small supersaturation. Nielsen (Ref. 24) found that at initial barium and sulfate concentrations less than 4×10^{-4} *M* the growth process is kinetically controlled by a surface reaction of the fourth order with respect to *c*. This result is in good agreement with the results of a careful study made by O'Rourke and Johnson,[25] who found the rate of growth of barium sulfate proportional to $c - c_s$.

[24] A. E. Nielsen, *Acta Chem. Scand.* **12**, 951 (1958).
[25] J. D. O'Rourke and R. A. Johnson, *Anal. Chem.* **27**, 1699 (1955).

Davies and Jones[26] postulate that a crystal in contact with an aqueous solution always becomes covered with a monolayer of hydrated lattice ions. For silver chloride they found the rate of crystal growth to be proportional to $(C - C_0)^2$. From this relation they concluded that crystallization of silver chloride occurs through the simultaneous dehydration of pairs of silver and chloride ions (ion pairs). There is general agreement that a surface reaction rate determines the process of crystal growth at low supersaturations. This surface reaction may be a dehydration or an adsorption[27] or both. When the supersaturation is so large that the rate is diffusion-controlled, the rate of deposition of ions [proportional to $(C - C_0)^m$] becomes so great that diffusion is the slowest, and therefore rate-determining, step. Adsorbed impurities can greatly affect the rate of crystallization (see the next section).

For a mathematical treatment of the kinetics of precipitation the student is referred to a monograph by Nielsen.[28]

Effect of adsorbable material on crystallization

It has been known for many years that foreign substances which are adsorbed by growing crystals may affect the length of the induction period, the rate of crystallization, and the crystal habit. More than a century and a half ago this effect was first described by the French mineralogist Haüy,[29] who observed that sodium chloride in the presence of urea crystallizes in the form of octahedra instead of cubes. Since that time the effect of dyes and other adsorbable substances on the crystal habit of freely soluble crystals has been investigated by many workers,[30] who found that the adsorbed substance is incorporated in the growing crystals. Few systematic studies have been made on the effect of the presence of adsorbed substances on the properties of slightly soluble precipitates. Fischer[31] noticed that lyophilic colloids, such as gelatin and agar, markedly increase the induction period in the precipitation of barium sulfate and also inhibit the rate of growth of the crystals. Jensen[32] showed the tremendous effect of starch on the induction period in the precipitation of calcium fluoride. A few of his results are given in Table 10-5. It is of interest to note that the induction period in Jensen's

TABLE **10-5. Effect of Starch on Induction Period of Calcium Fluoride at 23°C (Heterogeneous Nucleation)**

Absence of Starch			*In* 0.1% *Starch*		
$m^a \times 10^3$	$-\log [Ca^{++}][F^-]^2$	I.P., min	$m^a \times 10^3$	$-\log [Ca^{++}][F^-]^2$	I.P., min
2.1	7.43	2	2.29	7.32	(90)
2.00	7.49	4	2.17	7.39	50
1.90	7.56	4	2.07	7.45	38
1.80	7.63	10	2.00	7.50	78
1.70	7.71	11	1.84	7.61	36 hr
1.60	7.79	15	1.59	7.80	> 72 hr
1.50	7.87	27			

[a] Molarity of calcium fluoride.

[26] C. W. Davies and A. L. Jones, *Discussions Faraday Soc.* **5**, 103 (1949); *Trans. Faraday Soc.* **51**, 812 (1955).

[27] R. H. Doremus, *J. Phys. Chem.* **62**, 1068 (1958).

[28] A. E. Nielsen, *Kinetics of Precipitation*, Macmillan, New York, 1954.

[29] René Haüy, *Traite Mineral.* **2**, 356 (1800).

[30] For a review see H. E. Buckley, *Crystal Growth*, Wiley, New York, 1951, especially pp. 330–385.

[31] W. M. Fischer, *Z. Anorg. Allgem. Chem.* **145**, 311 (1925).

[32] A. T. Jensen, *Z. Physik. Chem.* (*Leipzig*) **180**, 94 (1937).

experiments refers to heterogeneous nucleation (p. 204). Hence it appears that starch is absorbed on the heterogeneous nuclei (accidental suspended impurities) and inactivates them. Note that the I.P. at $m = 1.60 \times 10^{-3}$ M is equal to 27 min, but becomes more than 72 hr in the presence of 0.1 per cent starch.

In their studies on the effect of potassium benzoate, sodium dodecyl sulfate, cetyl trimethyl ammonium nitrate, potassium naphthalene-2-sulfonate, and potassium eosinate, Davies and Nancollas[33] found that these substances retard the rate of crystallization of silver chloride. The effect of eosin (and probably similar compounds) is particularly striking because it can completely arrest crystal growth if present at sufficient concentrations. Even at a concentration of 1×10^{-6} M, it has a large retarding effect on the crystallization. In the presence of 4.6×10^{-5} M eosin no precipitation occurred from highly supersaturated solutions; apparently the dye hinders the growth of nuclei. The presence of adsorbable substances during the precipitation process not only affects the induction period and crystal habit, but also the filterability (and purity) of analytical precipitates. For example, Lindsly[34] reported that barium sulfate precipitated in the presence of picric acid is composed of markedly larger crystals than when formed in the absence of this acid. A large effect of some substances on the appearance of lead sulfate crystals separating from mixtures 0.01 M in lead sulfate was observed by Hähnert and Kleber.[35] Figure 10-5 shows the striking effect of picric acid, naphthol yellow, potassium permanganate, and ferric chloride. With the aid of electron micrography, Fischer[36] illustrated the effect of some foreign material upon the degree of perfection of barium sulfate crystals.

Even traces of substances, called poisons, may affect growth behavior. For example, a few parts per million of lead alter the growth process of potassium chloride from supersaturated solution.[37] For a discussion of such poisons reference is made to the literature.[38]

Solubility and particle size

Finely divided solids have a greater solubility than larger crystals. This was pointed out as early as 1813 by Wollaston[39] and in 1885 by Curie.[40] In the last quarter of the last century Gibbs[41] (1878) treated the relation between solubility and particle size from a thermodynamic viewpoint. Thomson[42] worked out the relationship, relating the vapor pressure of liquid droplets to the size of the droplets. Ostwald[43] applied the Gibbs–Thomson relationship to the relation between solubility and size of particles. Freundlich[44] perfected the expression known as the Ostwald–Freundlich equation:

$$\frac{RT}{M} \ln \frac{S_2}{S_1} = \frac{2\sigma}{\rho} \left(\frac{1}{r_2} - \frac{1}{r_1} \right) \tag{3}$$

in which S_2 and S_1 are the solubilities of spherical particles of radii r_2 and r_1, respectively. M is the molecular weight, ρ the density of the solid (M/ρ molecular volume), and σ is the surface tension at the solid–liquid interface. Later, Enüstün and Turkevich[45]

[33] C. W. Davies and G. H. Nancollas, *Trans. Faraday Soc.* **51**, 823 (1955).
[34] C. H. Lindsly, *Ind. Eng. Chem., Anal. Ed.* **8**, 176 (1936).
[35] M. Hähnert and W. Kleber, *Kolloid-Z.* **162**, 36 (1959).
[36] R. B. Fischer, *Anal. Chem.* **23**, 1667 (1951); **25**, 1544 (1953).
[37] G. W. Sears, *J. Chem. Phys.* **29**, 979 (1958).
[38] F. C. Frank, *Discussions Faraday Soc.* **243**, 299 (1951); G. W. Sears, *J. Chem. Phys.* **29**, 1045 (1958); A. R. Verma, *Crystal Growth and Dislocation*, Academic Press, New York, 1959.
[39] W. H. Wollaston, *Phil. Trans. Roy. Soc. London* **103**, 57 (1813).
[40] P. Curie, *Bull. Soc. Mineral.* **8**, 145 (1885).
[41] See J. W. Gibbs, *Collected Works*, vol. 1, Yale Univ. Press, New Haven, Conn., 1948.
[42] W. Thomson, *Phil. Mag.* [4]**42**, 448 (1881).
[43] W. Ostwald, *Z. Physik. Chem.* (*Leipzig*) **34**, 495 (1900).
[44] H. Freundlich, *Kapillarchemie*, Akademische Verlagsgesellschaft, Leipzig, 1909.
[45] B. V. Enüstün and J. Turkevich, *J. Am. Chem. Soc.* **82**, 4502 (1960).

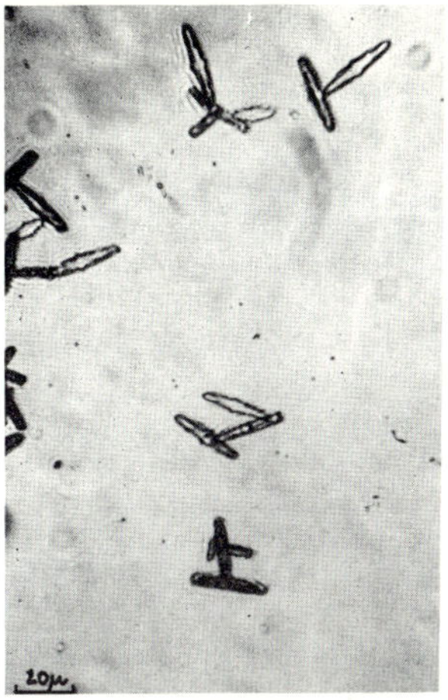

(a)

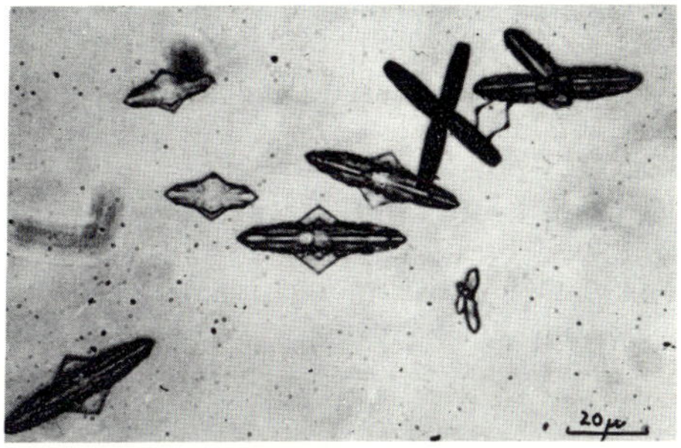

(b)

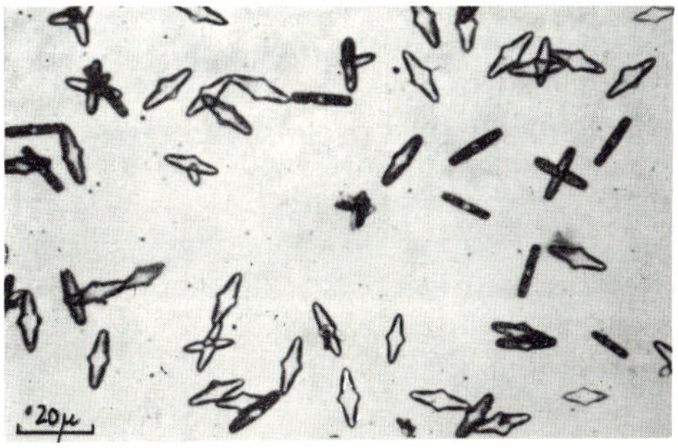

(c)

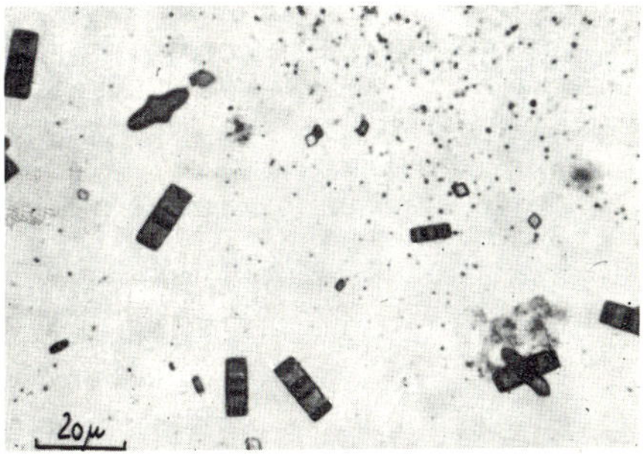

(d)

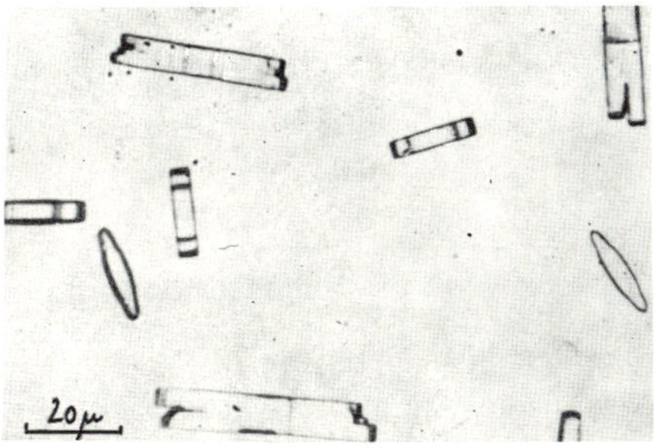

(e)

Fig. 10-5. (a) 0.01 M $PbSO_4$, no addition
(b) as (a), with 0.005 M picric acid, aged 24 hours
(c) as (a), with 0.005 M naphthol yellow
(d) as (a), with 0.005 M $KMnO_4$
(e) as (a), with 0.005 M $FeCl_3$

introduced a correction for particles that are cubes and not spheres. When we consider the relationship between very small particles with size r and that of macrocrystals with size r_1, $1/r_1$ becomes negligible and we obtain

$$\frac{RT}{M} \ln \frac{S_r}{S} = \frac{2\sigma}{\rho r} \tag{4}$$

in which S_r is the solubility of a particle of size r. Actually, instead of solubility the activity product should be written.[46]

Many authors have tried to verify Eq. (4) experimentally. Hulett[47] made the first experimental investigation to test the relation and used small particles of $CaSO_4 \cdot 2\,H_2O$ and $BaSO_4$.

Dundon and Mack[48] repeated Hulett's work and extended it to $SrSO_4$, PbF_2, PbI_2, Ag_2CrO_4, and CaF_2. The results of the classical work of Hulett and of Dundon and Mack, which are quoted in most texts appear to be unreliable for several reasons. They could not be confirmed with $BaSO_4$ by Cohen and Blekkingh[49] or by Balarev.[50] On the other hand, other authors were able to find a definite increase in solubility where particles become very small. May and Kolthoff[51] determined the solubility of freshly precipitated and aged lead chromate by analyzing both for lead and chromate in the saturated solutions. They found that the fresh precipitate had a particle radius of 0.086μ and a solubility 1.7 times greater than that of large particles. In this connection the work of Alexander,[52] who measured the solubility of amorphous silica is of special interest. He found a supersaturation of the order of 300 per cent for particles with a specific surface of 600 M^2/g and calculated 46 ergs/cm^2 for the silica–water interfacial tension. The most exact measurements of the relation between solubility and particle size have been made by Enüstün and Turkevich[46] with $SrSO_4$. Their results are summarized in Table 10-6. From these data they calculate a value for

TABLE **10-6. Relation between Particle Size x and Solubility C of Strontium Sulfate**

x (Å) ± 10 Å	C/C_0^a	$\log a/a_0$
96	1.50	0.154
130	1.41	0.130
155	1.32	0.106
168	1.33	0.109
252	1.18	0.063
378	1.11	0.040
500	1.08	0.029

[a] C_0 refers to solubility of macro crystals.

the interfacial tension $SrSO_4$/solution of 84 ± 8 ergs/cm^2 against a value of 1400 reported by Dundon and Mack (Ref. 48).

[46] B. V. Enüstün and J. Turkevich, *J. Am. Chem. Soc.* **82**, 4502 (1960).
[47] G. A. Hulett, *Z. Physik. Chem.* (*Leipzig*) **37**, 385 (1901); **47**, 357 (1904).
[48] M. L. Dundon and E. Mack, *J. Am. Chem. Soc.* **45**, 2479, 2658 (1923).
[49] E. Cohen and J. J. A. Blekkingh, *Z. Physik. Chem.* (*Leipzig*) µ**186**, 334 (1940).
[50] D. Balarev, *Kolloid-Z.* **96**, 19 (1941).
[51] D. R. May and I. M. Kolthoff, *J. Phys. Chem.* **52**, 836 (1948).
[52] G. B. Alexander, *J. Phys. Chem.* **61**, 1563 (1957).

In the section dealing with nucleation a distinction has been made between homogeneous and heterogeneous nucleation. The critical supersaturation, defined as the supersaturation below which spontaneous precipitation does not occur, does not provide us with a relation among solubility, size, and surface tension of particles, because precipitation usually occurs on heterogeneous nuclei. Nielsen[53] (see p. 204) made the first experiments that allowed the determination of the "homogeneous critical supersaturation" (of barium sulfate). Walton[54] determined this value for several precipitates (Table 10-7) and made use of the Volmer theory[55] to find the relation

TABLE 10-7. Surface Tension σ of Precipitates Calculated from Heterogeneous Supersaturation[a]

Salt	C'/C_0	σ, ergs/cm^2	r^*, (Å)
$BaSO_4$	1000	123	4.3
$CaC_2O_4 \cdot H_2O$	31	66	5.8
AgCl	5.5	77	5.4
$SrSO_4$	39	86	5.1
$PbSO_4$	28	79	5.3
$PbCO_3$	106	112	4.5
$SrCO_3$	30	92	5.0
CaF_2	21	114	4.3

[a] From A. G. Walton, *Mikrochim. Acta* **3**, 422 (1963).

between the homogeneous critical supersaturation C'/C_0, C_0 denoting the solubility of macrocrystals and the surface tension σ:

$$\ln Bn_0 = \frac{16\pi\sigma^3 v^2}{3k^3T^3(\ln C_1/C_0)^2}$$

where Bn_0 is equivalent to the binary collision frequency[56] and equal to 10^{24}, k the Boltzmann constant, and v the volume occupied by one ion. Also the much simpler Ostwald–Freundlich relation [Eq. (3)] can be applied, where S_r/S is now equal to the homogeneous critical supersaturation C'/C_0. Walton's results are summarized in Table 10-7, in which r^* is the radius of a critical nucleus. The value of σ for $SrSO_4$ is in gratifying agreement with that of 84 ± 8 calculated by Enüstün and Turkevitch (Ref. 45) from direct solubility determinations.

It is interesting to note that the radius of the critical nucleus of the various precipitates is very small and of the order of 5 Å. This small dimension conforms with conclusions drawn by Christiansen and Nielsen[57] from their kinetic experiments.

Van't Riet and Kolthoff[58] found that the activity product $a\text{Pb}^{++}a\text{SO}_4^{++}$ at the homogeneous critical supersaturation (see p. 213) of lead sulfate precipitated from mixtures of sodium sulfate and lead perchlorate at varying ionic strength was equal $4.5 \pm 0.5 \times 10^{-6}$ and constant between 0 and 50°C and unaffected by 10 per cent ethanol. The homogeneous critical supersaturation HCSS (activity ratio) is $(K_{ss}/K_{sp})^{1/2}$ (expressed in activities), K_{sp} being equal to 1.65×10^{-8} at 25°C, and is 15. Walton, who expressed the HCSS in terms of molar concentrations, found 28. When corrected for activity coefficients the two values become of the same order of magnitude.

[53] A. E. Nielsen, *Acta Chem. Scand.* **15**, 441 (1961).
[54] A. G. Walton, *Mikrochim. Acta* **3**, 422 (1963).
[55] M. Volmer, *Kinetik der Phasenbildung*, Edwards Bros., Ann Arbor, Mich., 1945.
[56] See A. A. Frost and R. G. Pearson, *Kinetics and Mechanism*, Wiley, New York, 1953, p. 20; D. Turnbull and J. C. Fisher, *J. Chem. Phys.* **17**, 71 (1949).
[57] J. A. Christiansen and A. E. Nielsen, *Acta Chem. Scand.* **5**, 673 (1951).
[58] B. van't Riet and I. M. Kolthoff, *J. Phys. Chem.* **64**, 1045 (1960).

Amorphous precipitates

Most precipitates have a crystalline structure. In some cases, however, *amorphous* precipitates which do not show the x-ray pattern of a crystalline substance may be formed. In order to understand this, we may, with Haber[59] consider two factors that determine the nature of the primary precipitate, the *aggregation velocity* and the *orientation velocity*. If the solubility limit is exceeded, the molecules, or aggregates of molecules, will have a tendency to unite and thus form larger aggregates. This aggregation velocity is a function of the degree of supersaturation; the larger the latter, the less regular the separated aggregates will be. Besides the supersaturation, the absolute concentration of the reacting ions is of significance. The primary aggregates formed, in which the molecules are arranged in a more or less chaotic manner, are not stable. By loss of energy they tend to reach a state of equilibrium in which the molecules become regularly arranged in a crystal lattice. The speed with which this process takes place is called the *orientation velocity*. It is evident that the form in which a precipitate separates depends upon the competition between the aggregation and orientation velocities. If the supersaturation is extremely large, the aggregation velocity may dominate, and the separated particles will not show an X-ray pattern; in other words, they are *amorphous*. On standing (aging), the amorphous precipitate may transform more or less rapidly into a crystalline modification. The orientation velocity is different for different substances. Strongly polar salts such as silver chloride and barium sulfate have a high orientation velocity and separate in a crystalline form.

In cases where we are dealing with substances such as the hydrous oxides, the orientation velocity decreases with the number of hydroxyl ions attached to the metal ion. The hydroxides of the divalent metals magnesium, cadmium, and zinc are obtained in a crystalline form; hydrous ferric oxide, on the other hand, if precipitated from cold solutions is amorphous but is transformed on aging, especially on heating, into a crystalline product. Hydrous oxides of the tetravalent metals (thorium, etc.) are usually obtained in the amorphous form and on aging change extremely slowly into crystalline modifications. Sometimes the aging (see p. 220) is accompanied by chemical interaction between the particles. This happens when the primary, amorphous particles are highly reactive and are of such a nature that they react with one another in a chemical way. An example of this kind is found in the aging of hydrous ferric oxide (*o*-ferric hydroxide). According to Krause,[60] in the early stages of aging two molecules of *o*-hydroxide react to form a polymerization product containing eight atoms of iron:

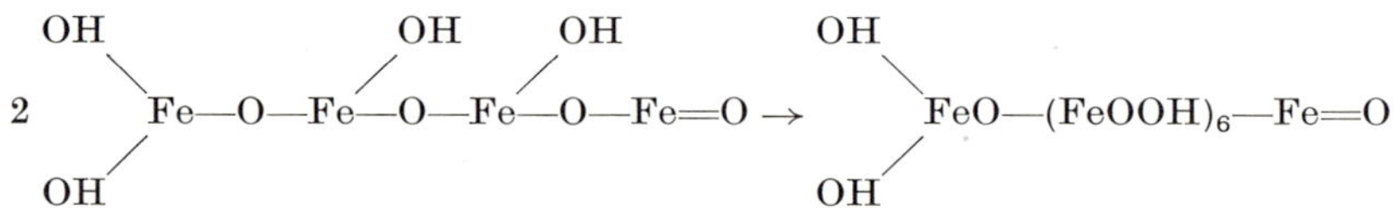

More molecules of the *o*-hydroxide can be added, and finally agglomerates with a chain structure containing 40 to 50 atoms of iron are formed. Although there is a definite ordering in these agglomerates, they are not yet crystalline. The polymerization is accompanied by a dehydration and a change of color. The aging may result in the formation of yellow amorphous ferric acid with a ring structure and finally in the formation of crystalline goethite. The speed and also the kind of aging are greatly dependent upon the composition of the aging medium.

Hydrous oxides formed under ordinary analytical conditions age rather slowly, and the precipitates retain their original gelatinous character for a long time. If possible, one should avoid the formation of such gelatinous precipitates in gravimetric procedures. This may sometimes be accomplished by choosing such conditions of precipitation that the supersaturation is not too great.

[59] F. Haber, *Ber.* **55**, 1717 (1922).

[60] See review by A. Krause, *Kolloid-Z.* **72**, 18 (1935).

Gelatinous precipitates are hard to filter and purify by a washing process. The primary particles are very much smaller than the pores of the filter paper, but they cannot pass through, because they are held together in the form of much larger secondary aggregates. Gelatinous precipitates have a marked tendency to clog the filter, and therefore a coarse-textured filter paper should always be used when dealing with such hydrated gels. But even then it sometimes happens that the jelly-like precipitate will form a nearly impervious layer over the surface of the filter paper. Filtration will then entirely cease, for although the capillary interstices between the primary particles are filled with liquid, it is impossible to drive this through even with the aid of suction below or pressure above the filter, for obviously if sufficient force is used to move the water between the primary particles, it would also disintegrate the secondary aggregates and force the primary particles through the filter. Doubtless an ultra filter, the pores of which are smaller than the primary particles, could be used, but the slowness of the process would make it of little value for ordinary analytical purposes. Membrane filters, made from esters of cellulose by evaporation from a suitable solvent, have been strongly recommended for the filtration of gelatinous precipitates. Macerated filter paper is very useful in filtering and washing gelatinous precipitates.

Finely dispersed precipitates can be coagulated by adding very small amounts of organic substances, which behave as lyophilic colloids, such as gelatin, agar, and gum arabic. Contrary to ordinary behavior the latter usually do not stabilize suspensoid sols if present in minute traces, but they bring about rapid and complete flocculations. This phenomenon is called *sensitization*. This sensitization is of great analytical importance, as shown by Caldwell and Moyer.[61] Especially in cases where we are dealing with finely dispersed precipitates which by their slimy character are hard to filter, the addition of a suitable sensitizer may improve the conditions considerably. Thus it was shown by Caldwell and Moyer[62] that a well-coagulated, easily filterable zinc sulfide is obtained if 0.5–2.0 mg of gelatin is added before or after the precipitation with hydrogen sulfide from a solution containing much ammonium sulfate. Gelatin has also been recommended[63] to obtain an easily filterable precipitate of calcium fluoride. It also seems useful in obtaining filterable precipitates of hydrous oxides, such as those of zirconium, titanium, ytterbium, tantalum, columbium, and nickel. Positively charged barium sulfate (supernatant liquid contains an excess of barium; see Fig. 10-6) is easily flocculated by a trace of agar–agar.[64] Use is made of this effect in the determination of sulfate. The filterability of negatively charged barium sulfate is not improved by addition of agar–agar.

Gelatin seems to be of special importance in the quantitative precipitation of silica, which has a strong tendency to stay dispersed as a hydrophylic colloid. Interesting results in this connection were obtained by Weiss and Sieger.[65] These authors claim that under the proper conditions a quantitative precipitation of the colloidal silica is obtained by the addition of gelatin. This procedure would obviate the usual one, which involves evaporation to dryness, heating of the residue, and evaporation of the filtrate.

Colloidal solutions

In gravimetric analysis we are usually dealing with slightly soluble substances which behave like strong electrolytes. The saturated solutions of most precipitates in water do not contain the salts in the molecular form but in the form of their ions. These ions are extremely small, their sizes being of the order of a few Angstrom units

[61] J. R. Caldwell and H. V. Moyer, *Ind. Eng. Chem., Anal. Ed.* **7**, 38 (1935); J. R. Caldwell, *J. Am. Chem. Soc.* **57**, 96 (1935); see also W. Wepritzkaja, *Z. Anal. Chem.* **111**, 287 (1937–1938).

[62] J. R. Caldwell and H. V. Moyer, *J. Am. Chem. Soc.* **57**, 2372 (1935).

[63] N. F. Michailowa, *Chem. Zentr.* [I]**110**, 4507 (1939).

[64] E. J. Bogan and H. V. Moyer, *Ind. Eng. Chem., Anal. Ed.* **14**, 849 (1942).

[65] L. Weiss and H. Sieger, *Z. Anal. Chem.* **119**, 245 (1940); see also S. J. H. Spronck, *Chem. Weekblad* **43**, 259 (1947).

(Å = 10^{-8} cm)[66] and of the same order of magnitude as those of the molecules of the solvent. Such true solutions, as well as other (molecular) solutions, are optically "empty." By this we mean that if part of the solution is illuminated by a powerful beam of light, no scattering of the light is observed when the solution is viewed at right angles to the incident light (Tyndall effect). If particles having a size larger than about 100 Å are suspended in the solvent, a distinct Tyndall effect will be observed. A true solution is perfectly clear under the ultramicroscope, and no particles can be observed. If particles with a size larger than 10 mμ are present in the solution, they will scatter light and will thus become visible in the ultramicroscope. With an ordinary microscope the limit of resolvability is of the order of 0.25 μ only.

The fact that a solution appears clear when examined with the naked eye or even with an ordinary microscope is no proof that the state of subdivision of the dissolved substance is of the molecular order. This can be easily demonstrated by passing a current of hydrogen sulfide gas through a solution of arsenious oxide. The resulting liquid has an orange color and appears clear if examined with the naked eye or with the microscope. Under the ultramicroscope, however, the picture is entirely different; myriads of bright particles in rapid, random motion are observed. Also, if the liquid is placed in a strong beam of light, a marked scattering is noticed.

The orange "solution" of arsenious sulfide has properties that are entirely different from those of "true" solutions. It will be recalled that arsenious sulfide is very slightly soluble in water, much less so than barium sulfate for example. Nevertheless, we can prepare "solutions" containing more than 10 g of As_2S_3 per liter. The solute in this solution is not present in the molecular or ionic state, but in the form of aggregates, each of which contains many molecules. In distinction from true molecular solutions, a solution of this type is called a *colloidal* solution. The colloidal state is characterized by particles whose size varies between about 100 and 1 mμ. Colloidal solutions have this in common with molecular solutions: As a rule they can be filtered unchanged through quantitative filter paper. The pores of ordinary filter paper are large enough to let the colloidal particles pass through. However, if filters with extremely small pores, called *ultrafilters*, are used, molecular solutions will pass unchanged, whereas colloidal particles are retained. The colligative properties of molecular solutions (osmotic pressure, freezing-point lowering, and boiling-point elevation) are relatively simple functions of the molecular concentration. In a colloidal solution, however, the finely divided particles have hardly any influence upon the osmotic properties of the solvent. Since the colloidal particles do not pass a collodion or parchment membrane, they can thus be separated from electrolytes by the process of *dialysis*.

We may distinguish between two types of colloidal solutions:

1. *Suspensoids, or hydrophobic* (*lyophobic*) *colloids*. The particles of these show no, or only a slight, attraction for water, as the name indicates. The solutions are only slightly viscous; if flocculated by the addition of a suitable electrolyte, the particles settle down in a filterable form. Such solutions are often called *sols*. Under suitable conditions, sols of most slightly soluble inorganic compounds may be prepared (metal sulfides, ammonium phosphomolybdate, silver iodide, etc.).

2. *Emulsoids, or hydrophylic colloids*. The particles of these have a strong affinity for water, and the colloidal solutions are highly viscous. Relatively large amounts of electrolytes are necessary to cause precipitation; the flocculate keeps water strongly absorbed and has a tendency to form jelly-like masses. Many of the organic compounds which are of great practical importance, such as the proteins and higher carbohydrates, form *gels*. A few of the inorganic precipitates are usually obtained in the hydrophylic form, for example, silicic acid and hydrous aluminum and stannic oxides. On addition of acid to a dilute sodium silicate solution, the latter remains perfectly clear. The silicic acid formed is in colloidal solution in a finely divided and highly hydrated state.

[66] Units and dimensions: 1 μ = 10^{-3} mm; 1 mμ = 10^{-6} mm; 1 angstrom unit = 1 Å = 10^{-7} mm = 0.1 mμ.

On evaporation to dryness, it separates but not in an entirely insoluble form. If the residue is treated with water, a small part of the silica goes into colloidal solution again. Evaporation with hydrochloric or perchloric acid is necessary to transform the silicic acid to a less hydrated and also less "soluble" state. Some gelatinous precipitates have properties intermediate between those of suspensoids and emulsoids.

Gels are very hygroscopic, and it is hard to remove all the adsorbed water, even on ignition. This case is met with in the determination of aluminum by precipitation as the hydrous oxide. The precipitate is ignited at a high temperature and must be heated a long time before all water is removed. The ignited residue is still hygroscopic.

From the analytical viewpoint sols are very important, and for this reason some of their properties will be discussed. In the first place the formation of colloidal solutions must be avoided because the particles pass through ordinary filters. Colloidal solutions must also be avoided as far as possible, because the flocculated particles have an enormous surface development and correspondingly great adsorptive powers. In a later chapter the adsorptive properties of precipitates will be dealt with in detail. It may simply be mentioned here that foreign constituents of the solution may be adsorbed at the surface of the flocculated particles and thus contaminate the precipitate.

Stability of sols. Flocculation. Peptization. Finely divided particles of a slightly soluble substance can be kept in colloidal solution if they are given an opportunity to acquire an electric charge. This is accomplished by the adsorption of anions or cations on the surface of the particles; in the former case they will have a negative charge and in the latter case a positive charge. Every precipitate has a tendency to adsorb its own ions. For example, it has been shown experimentally that metal sulfides have strong adsorptive properties toward sulfide ions, and silver halides for silver ions and halide ions, respectively. It should be realized that if adsorption of the sulfide ion by a metal sulfide occurs, an equivalent adsorption of some cation present in the solution must also take place, for otherwise electroneutrality of the solution would not be maintained. In the preparation of colloidal sulfide solutions the sulfide is usually absorbed in the form of hydrogen sulfide, if the latter is in excess in the solution. Sulfide ions are primarily adsorbed, and they are attracted so strongly that they are fixed on the surface of the particles. The other ions, which may be called the *counterions* (in this particular case hydrogen ions), cannot approach the surface so closely and will stay at a very small distance, although in the immediate neighborhood of the surface, since the electrostatic forces between the negative ions on the surface and the positive counterions are very large. According to this picture an *electrical double layer* is set up between the particles and the solution (see Fig. 10-6). A colloidal particle of arsenious

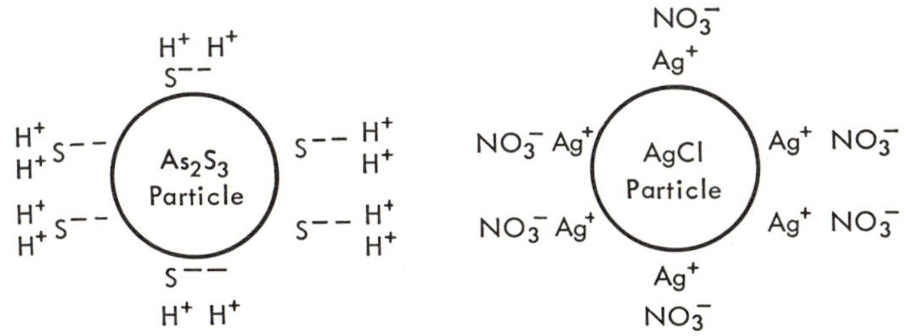

Colloidal particle of As_2S_3 with negative charge. S^{-} is inner or stabilizing ion (fixed on surface). H^+ is counter ion.

Colloidal particle of AgCl with positive charge. Ag^+ is inner or stabilizing ion (fixed on surface). NO_3^- is counter ion.

Fig. 10-6.

sulfide has a negatively charged surface, whereas the surrounding liquid is positive with respect to the particle. If an electric current is passed through the solution, the negative particles will move toward the anode, just as anions do, although the mobility of the colloidal particles is much smaller than that of ions. The existence of the electrical double layer is responsible for the stability of the colloid. All the particles have a charge of the same sign, and upon close approach they will not collide but will repel each other. With the ultramicroscope a rapid Brownian movement is observed. If by some means the electrical double layer is destroyed, the sol is no longer stable and the particles flocculate. This destruction of the double layer can be accomplished by the addition of a sufficient amount of electrolyte. The minimum amount of electrolyte necessary to flocculate the colloid is called the *flocculation value*. The latter depends primarily upon the kind of charge (+ or −) of the colloidal particles and the valence, and more or less upon the specific properties of the ions of opposite charge of the flocculating electrolyte. In experiments with arsenious sulfide, for example, Linder and Picton[67] found that the ratios of the "flocculation" values of electrolytes containing uni-, di-, and trivalent cations were of the order of 500 : 10 : 1. If, on the other hand, we are dealing with a positively charged colloid, the flocculation value depends primarily on the valence of the anion.

An extensive theoretical and experimental treatment of flocculation is given in *Colloid Science*.[68] Some data illustrating clearly the effect of the valence of a cation on the flocculation value F.V. of a negative sol and of that of the anion on a positive sol are given in Table 10-8. The ratio refers to the F.V. of a monovalent ion. The individual nature of the flocculating ion has some influence as well. This holds especially

TABLE 10-8. **Mean Flocculation Values (millimoles/liter) for Sols[a] of Negative and Positive Charges**

	Negative Sols					
Charge of cation[b]	As_2S_3 F.V.	Ratio	Au F.V.	Ratio	AgI F.V.	Ratio
1+	55	1	24	1	142	1
2+	0.69	0.013	0.38	0.016	2.43	0.017
3+	0.091	0.0017	0.006	0.0003	0.068	0.0005

	Positive Sols			
	Fe_2O_3 F.V.	Ratio	Al_2O_3 F.V.	Ratio
1−	11.8	1	52	1
2−	0.21	0.018	0.63	0.012
3−			0.080	0.0015
4−			0.053	0.0010

[a] See H. R. Kruyt (ed.), *Colloid Science*, vol. 1, Elsevier, Amsterdam, 1952, pp. 308, 309, Tables 2 and 3.

[b] The value for a tetravalent cation is omitted in this table because the cations used are highly hydrolyzed.

[67] S. E. Linder and H. Picton, *J. Chem. Soc.* **67**, 63 (1895).

[68] *Colloid Science*, (H. R. Kruyt, ed.), Elsevier, Amsterdam, 1952. This volume presents an authoritative treatment of colloidal systems: see also E. J. W. Verwey and J. T. G. Overbeek, *Theory of the Stability of Lyophobic Colloids*, Elsevier, Amsterdam, 1948.

for organic aromatic ions, the flocculation values of which as a rule are much smaller than those of inorganic ions of the same valence.

The ions that determined the original charge of the particles will remain on the surface during, and more or less after, the flocculation. This process involves an equivalent adsorption of the counterions by the flocculated precipitate. Experimentally it has been proved that strong flocculating ions replace the originally adsorbed counterions at the surface of the particles. If, for example, an arsenious or mercuric sulfide sol is flocculated with barium chloride, the precipitate will contain an adsorbed layer of barium sulfide or, if we use lanthanum chloride, of lanthanum sulfide. On the other hand, if the flocculation is carried out with an electrolyte, the cation of which is only slightly adsorbable, for example, with potassium or sodium chloride, the replacement is incomplete, and the precipitate will contain both adsorbed hydrogen and potassium sulfide. From the above it is evident that a flocculated colloid is always contaminated by surface adsorption. Part of the adsorbed electrolyte can be removed by thorough washing with water. In the washing one is often confronted with another difficulty. During the washing process the electrolyte concentration in the supernatant liquid will become smaller than the flocculation value, with the result that the particles may go into colloidal solution again. This phenomenon is called *peptization* and has to be considered in analytical procedures. In the precipitation of silver with an excess of halide or of phosphate as ammonium phosphomolybdate or of nickel as nickel sulfide, one often encounters the difficulty that, on washing, the precipitate "runs through the filter." This peptization can be prevented by washing with a suitable electrolyte solution that does not interfere in the subsequent steps of the determination. Silver halides can be washed with a dilute solution of a strong acid, and the phosphomolybdate or nickel sulfide with a solution of an ammonium salt. When the washed precipitate is ignited, the adsorbed electrolyte is removed by volatilization. If the flocculated colloid has a positive charge, as is the case in the precipitation of halides with an excess of silver, it is harder to make the adsorbed electrolyte harmless.

A fortunate circumstance, however, is that flocculated colloids usually undergo a rapid aging, especially on digestion, which decreases their total surface and the amount of absorbed electrolyte. This aging of precipitates will be discussed later.

An understanding of the simple properties of colloidal solutions is therefore of great analytical value. Sometimes one is faced with the difficulty that a precipitate shows a pronounced tendency to remain in colloidal solution. This occurs, for example, in the precipitation of lead with an excess of chromate. Part of the lead chromate is peptized by the excess of chromate and has a negative charge. Addition of a trace of a trivalent cation, such as aluminum in the form of its chloride, is very effective in obtaining complete flocculation and an easily filterable precipitate (the adsorbed aluminum does not interfere in the volumetric lead determination). In other cases one must beware of the adsorptive properties of the precipitate formed. Hydrous oxides precipitated from weakly acid medium are positively charged. The hydroxyl ions in the solid oxide have a great affinity for hydrogen ions and consequently tend to adsorb an equivalent amount of anions with the hydrogen ions. The higher the valence of the anion, the stronger it will be adsorbed. If, for example, hydrous ferric oxide is quantitatively precipitated at a pH of 4–5 and the solution contains sulfate, a great deal of the latter will be adsorbed, and it will not be removed by washing. If, on the other hand, the precipitation is carried out from ammoniacal medium, a primary adsorption of hydroxyl ions takes place, and the precipitate acquires a negative charge. It will no longer adsorb other anions but exerts a strong adsorptive effect on cations. Again the adsorbability of the latter increases with increasing valence. By precipitation of hydrous oxides from ammoniacal medium, there is always an adsorption of calcium, magnesium, nickel, copper, etc., when these are present in the solution. These cations cannot be removed by washing, and even a reprecipitation of the oxide is often not effective in affording a quantitative separation. If possible, therefore, one should endeavor to separate hydrous oxides from anions by precipitation in an alkaline medium, and from cations in an acid medium.

In certain cases the adsorptive properties of colloidal precipitates may advantageously be used for the removal of interfering ions from solutions. For example, hydrous stannic oxide in very dilute nitric acid has strong adsorptive properties for phosphoric and arsenic acids. The latter can be quantitatively removed with the aid of this adsorbent. Manganese dioxide, precipitated from weakly acid medium by addition of permanganate to a solution that contains a manganous salt, carries down quantitatively antimony, tin, and bismuth.[69] Other constituents such as copper, arsenic and others are also coprecipitated but not quantitatively.

Summarizing the facts important in analytical chemistry, we find:

1. That peptization of colloidal precipitates can be prevented by addition of, and washing with, a suitable electrolyte.
2. That colloidal precipitates contain impurities in the adsorbed state. By proper washing with dilute acid or ammonium salt solutions, it is often possible to replace the adsorbed electrolyte with a compound that is volatile at higher temperatures.
3. That in many cases slimy precipitates can be obtained in an easily filterable form by the addition of a trace of gelatin (see p. 215).
4. That use can be made of the adsorptive properties of finely divided precipitates in removing certain ions from solution or for concentrating traces of ions prior to their determination.

Experimental conditions of precipitation. Aging

Introduction. From previous sections it is clear that it is not possible to predict the size of a precipitate under given conditions of precipitation because, in addition to the relative supersaturation the individual characteristics of the precipitate are of major importance. Under ordinary conditions of precipitation the supersaturation is extremely large. When sulfate is added at room temperature to 0.1 M barium solution the mixture becomes supersaturated when the sulfate concentration is of the order of 2×10^{-9} M. By adding the precipitant to a hot solution, conditions become somewhat more favorable, partially because of greater solubility of the precipitate and, more important, because the rate of crystal growth increases with increasing temperature. For many years it has been common practice in gravimetric determinations to carry out precipitations from hot solutions by slow addition of a dilute solution of precipitant. Factors that increase supersaturation by increased solubility tend to increase the size of the precipitate. The precipitation of barium sulfate is carried out in relatively strong acid medium, in which a considerable fraction of the sulfate is present in the form of bisulfate ions. To keep supersaturation low during the precipitation, Hahn and Otto[70] recommended the following procedure. Some water or dilute acid is placed in a beaker. With efficient stirring a measured volume of the solution to be analyzed is delivered from one buret, and from another buret a solution of reagent. The speed of addition of both solutions is regulated in such a way that there is never a large excess of either the ion to be precipitated or of the precipitant, until the precipitation is complete. This method is now only of historical interest.

In general, supersaturation is kept at a minimum when the precipitant is (1) either generated slowly in the solution containing the ion to be precipitated or (2) the precipitate remains dissolved in an excess of acid and the hydrogen ions are removed slowly by addition of a suitable agent. The method has been called "*precipitation from homogeneous solution*,"[71] in which the rate of formation of the precipitant is kinetically controlled.

[69] H. Blumenthal, *Z. Anal. Chem.* **74**, 33 (1928); S. Kallmann and F. Pristera, *Ind. Eng. Chem., Anal. Ed.* **13**, 8 (1941).

[70] F. L. Hahn and R. Otto, *Z. Anorg. Allgem. Chem.* **126**, 257 (1923).

[71] L. Gordon, M. L. Salutsky, and H. H. Willard, *Precipitation from Homogeneous Solution*, Wiley, New York, 1959. For a more recent review see P. F. S. Cartwright, E. J. Newman, and D. W. Wilson, *Analyst*, **92**, 663 (1967).

Removal of hydrogen ions by some chemical reaction in precipitation processes was already suggested by Moser[72] in 1925. Moser and coworkers made use of the decomposition of nitrite, thiosulfate, and cyanate by hydrogen ions in the precipitation of hydrous oxides. These precipitates are still more or less gelatinous because they are formed too rapidly. Willard and Tang[73] used the combination of a buffer system with urea to produce dense precipitates of hydrous oxides which are easily filterable. They introduced the method of precipitation from homogeneous solution into analytical practice. Various modes of precipitation and the properties of our precipitates will now be discussed in greater detail.

Precipitation from relatively concentrated solutions at room temperature. Aging

Under these conditions the supersaturation, in general, is very large and the precipitate, which is formed immediately upon rapidly adding the reactants together, consists of very small and highly imperfect crystals, or agglomerates of imperfect crystals or of amorphous particles. Upon aging of such primary precipitates drastic changes occur, resulting in formation of larger crystals, which are considerably less contaminated than the primary precipitate is. (See p. 243.) Aging has been defined[74] to include all irreversible changes that occur in a precipitate after it has been formed. On digestion at 100°C, the primary imperfect crystals are subjected to successive recrystallizations. Recrystallization also occurs at room temperature. For example, precipitates of lead sulfate or barium sulfate formed at room temperature by mixing 0.1 *M* reactants undergo rapid recrystallizations on standing or shaking with the supernatant liquid at room temperature. This was shown by adding a radioactive isotope to the supernatant liquid and following the rate of incorporation of the isotope in the solid. When radioactive lead as thorium B or sulfate containing radioactive sulfur was added to the supernantant liquid of freshly formed lead sulfate from 0.1 *M* reactants, homogeneous distribution of the radioactive component was found after 10 min of shaking. When the radioactive constituent was added 10 min after the precipitation, homogeneous distribution was obtained after 1 hour of shaking, showing that the recrystallizations become less rapid with time of aging after the precipitation (for details see Kolthoff, Ref. 74). As a result of the recrystallizations the number of primary particles decreases and the resulting crystals have a much higher degree of perfection than the original particles. Impurities coprecipitated with the primary particles are expelled during the recrystallizations (see p. 244).

On the other hand, foreign constituents which form solid solutions (e.g.) in the precipitate can be incorporated during the aging (recrystallization) process. For example, potassium permanganate can form a solid solution in barium sulfate. Rapidly precipitated barium sulfate on shaking with a permanganate solution immediately after precipitation assumes a slightly pink color after a few minutes. The intensity of the latter, an indication of the amount of permanganate in the crystals, increases with time and approaches a maximum after a day or two. Such a precipitate is not decolorized by washing and retains its color if shaken with a solution of hydrogen peroxide or oxalic acid in 2 *N* hydrochloric acid. However, if the barium sulfate, after precipitation, is allowed to stand for some days before adding the permanganate, hardly any of the latter compound is taken up, showing that the primary precipitate has undergone a drastic perfection.

Recrystallization occurs because the primary particles are very small and have a greater solubility than the macrocrystals (p. 212), and the content of ionic and nonionic impurities incorporated in the lattice is large. These impurities enhance the tendency toward recrystallization. This is illustrated schematically in Fig. 10-7. The

[72] L. Moser and W. Maxymowicz, *Z. Anal. Chem.* **67**, 248 (1925).

[73] H. H. Willard and N. K. Tang, *J. Am. Chem. Soc.* **59**, 1190 (1937); *Ind. Eng. Chem., Anal. Ed.* **9**, 357 (1937).

[74] I. M. Kolthoff, *Analyst* **77**, 1000 (1952).

contaminating ions (*a* in Fig. 10-7) are freely soluble in the mother liquor and leave open spots in the surface of the lattice (*b* in Fig. 10-7). The remaining lattice ions are not subject to the electrostatic force of the missing neighboring ions and thus are held less tightly in the lattice than when the latter is perfect. These ions go into solution and deposit on more perfect crystals. By their disappearance a second imperfect layer is exposed, which for similar reasons will go into solution. When the primary particles are relatively pure (precipitates of silver halides), the recrystallization is mainly the result of the greater solubility of the very small particles as compared to macrocrystals, and the larger particles grow at the cost of the small ones. This process has been known for many years and has been called *Ostwald ripening.* It is observed in *colloidal* solutions of silver halides. Electromicrographs of silver bromide[75] in the colloidal state after various periods of aging are reproduced in Fig. 10-8 and typically illustrate the progress of Ostwald ripening during the aging process.

Fig. 10-7.

In addition to aging by recrystallization, via a liquid phase, we also may consider *thermal* aging as a result of thermal agitation of the lattice ions. Precipitates with a low melting point, such as the silver halides, exhibit thermal aging at room temperature. For example, the particles of freshly flocculated silver bromide are "fused" together as a result of the high thermal mobility of silver and bromide ions in a few layers below the surface.[76] Thus it is not possible to restore the original number of particles upon peptization of flocculated silver halides.

Tammann and Sworykin[77] postulated that the ions in a lattice acquire a high thermal mobility at about half the absolute melting point. This might be called the *relaxation temperature.* Barium sulfate melts at 1580°C or 1853°A; and its relaxation temperature should be about 925°A or 650°C. Actually, Kolthoff and MacNevin[78] observed pronounced thermal aging of small grain and highly imperfect barium sulfate at temperatures above 700°C. The product used was obtained by adding at room temperature with vigorous stirring an 0.8 *M* solution of barium chloride to an equal volume of 0.8 *M* sodium sulfate. The precipitate was collected and washed with water and alcohol and dried at room temperature. The product was very impure and contained 2 per cent occluded water, 0.22 per cent sodium chloride, and 1.4 per cent sodium sulfate. By heating to 500°C all the occluded water was removed but not the occluded sodium. Above 800°C the occluded sodium chloride was volatilized and the occluded sodium sulfate was expelled from the crystals and could be removed by washing with water. From specific surface measurement and microscopic observations it was concluded that at about 500°C a pronounced thermal aging, accompanied by a decrease of the specific surface, occurred as a result of sintering of the primary particles in the agglomerates. The thermal aging, as a result of the mobility of the lattice ions, became very pronounced at temperatures above 650°C, close to the relaxation temperature. As a result of this high thermal mobility, extensive internal recrystallization occurred with formation of large and well-developed crystals with a high degree of perfection. This recrystallization was accompanied by expulsion of foreign materials not fitting

[75] I. M. Kolthoff and R. C. Bowers, *J. Am. Chem. Soc.* **76**, 1503 (1954).
[76] I. M. Kolthoff and R. C. Bowers, *J. Am. Chem. Soc.* **76**, 1510 (1954).
[77] G. Tammann and A. Sworykin, *Z. Anorg. Allgem. Chem.* **176**, 46 (1928).
[78] I. M. Kolthoff and W. M. MacNevin, *J. Phys. Chem.* **44**, 921 (1940).

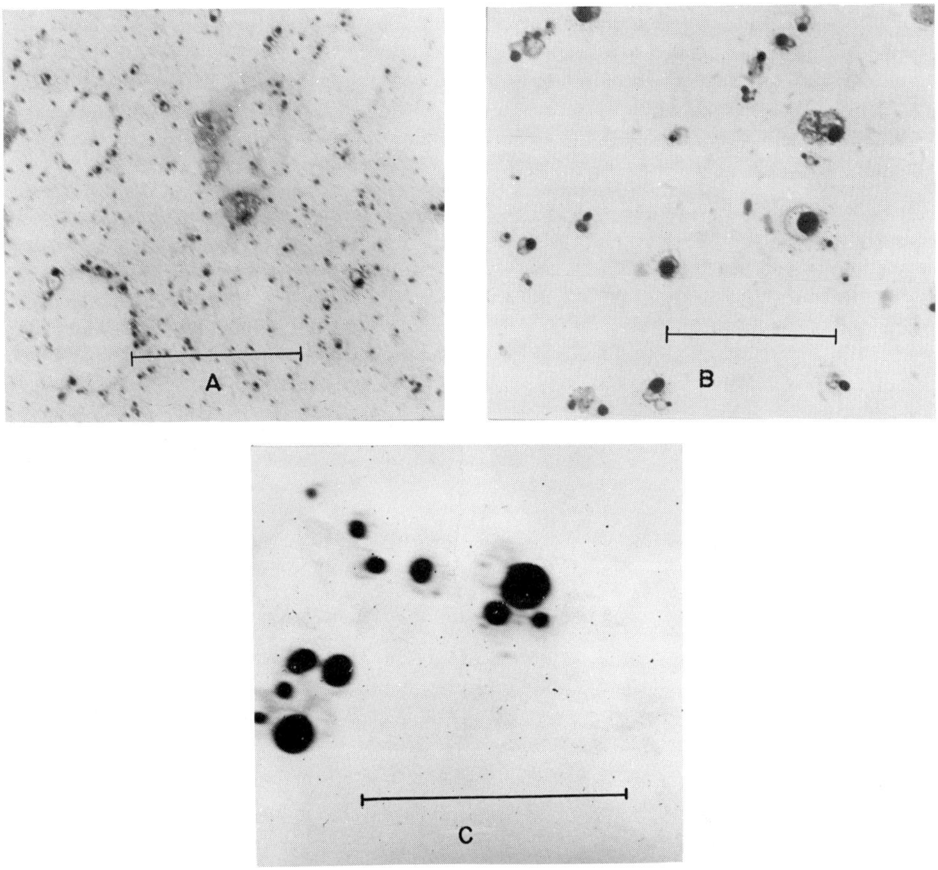

Fig. 10-8. (A) Fresh (5-minute old) silver bromide colloid. Particles have diameters of from less than 0.02 to 0.1 μ. Arrow 1 μ.
(B) Seventeen-hour old non-dialyzed colloid. Particles have diameters of from 0.05 to 0.3 μ. Arrow 1 μ.
(C) Six-week old dialyzed colloid. Particle diameters are from 0.06 to 0.3 μ. Arrow 1 μ.

into the lattice. The sintering and the crystal growth as a result of thermal mobility of the lattice ions is strikingly illustrated by some microphotographs in Fig. 10-9.

Another mode of aging, which is relatively rare, is the transformation of a metastable into a stable crystalline form. This case is encountered, for example, in the precipitation of calcium oxalate at room temperature from fairly concentrated solutions. The primary precipitate consists of higher hydrates, which on digestion are transformed into the stable monohydrate, with the result that a purer precipitate is usually obtained after digestion. This is readily understood when one realizes that the stable modification is formed very slowly and consists of much more perfect crystals than the primary precipitate.

Precipitation from relatively dilute, hot solutions

This is the classical and still the most common mode of precipitation in analytical chemistry. The crystals thus formed are much more perfect than those obtained from

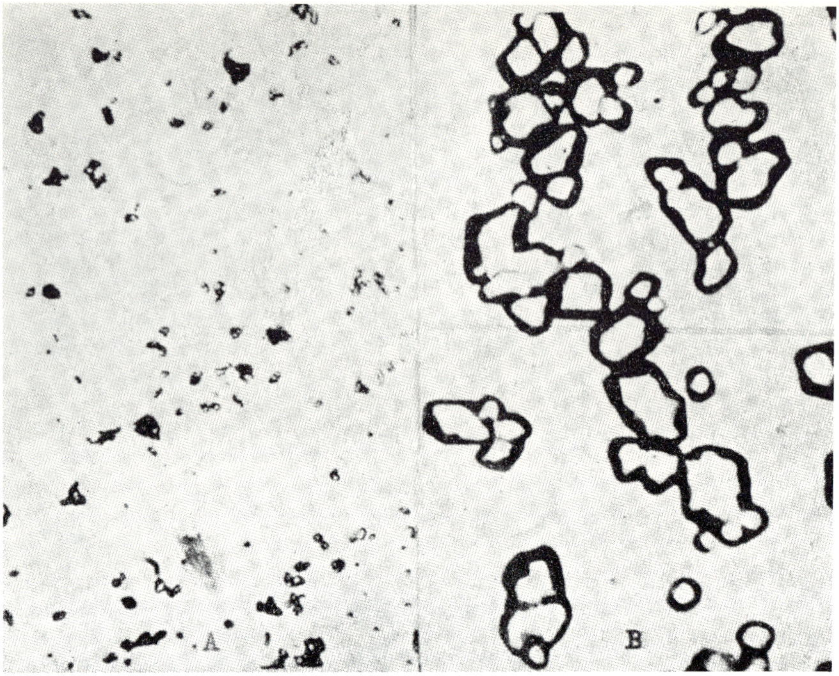

(a) (b)

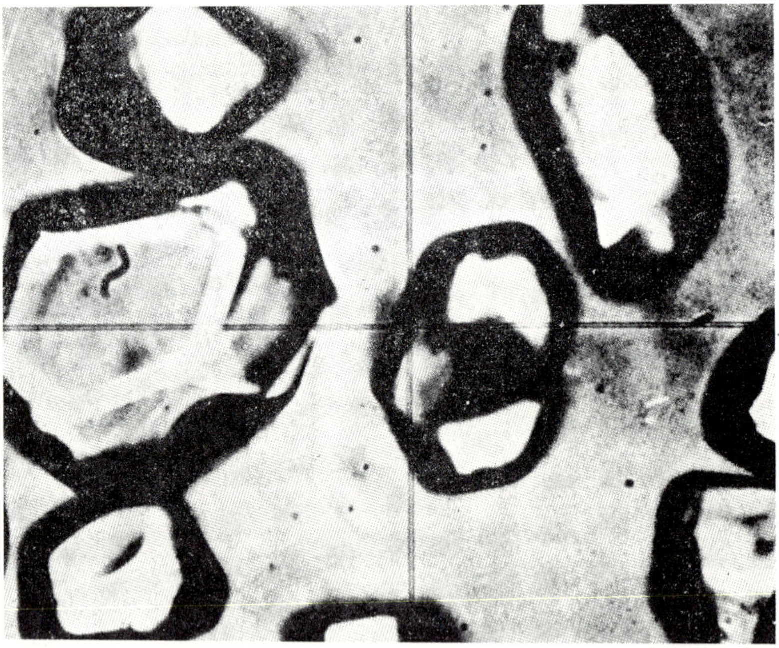

(c)

Fig. 10-9. Barium sulfate after heating for 1 hour at (a) 520°C, (b) 860°C, (c) 1020°C. Magnification × 1000.

relatively concentrated solutions at room temperature. Still the primary precipitate is formed rapidly during addition of precipitant to the solution of the constituent to be precipitated. Consequently, the primary precipitate often is composed of small particles, and is relatively impure. On aging on a steam bath, perfection by recrystallization and formation of a purer precipitate occur. This explains the common analytical practice of digesting a precipitate for several hours or more before filtration.

Precipitation from homogeneous solution

The rate of formation of the precipitant in the solution containing the constituent to be precipitated is kinetically controlled. By varying the experimental conditions this rate can be controlled at will and can be kept so small that the supersaturation is always small. The number of nuclei formed is then relatively small and they grow with continued formation of precipitant. Consequently, in general relatively large crystals are formed with a high degree of perfection. The precipitant is usually produced in either of the two following ways.

1. The precipitate is soluble in acid. Enough acid is added to the mixture of the ion to be precipitated and precipitant so that no precipitation occurs and a suitable reagent is added which slowly consumes hydrogen ions. A similar principle can be used when the precipitate is soluble in alkali (aluminum hydroxide) and a reagent (e.g., an ester) is added which slowly consumes hydroxyl ions.
2. The reagent is generated by slow decomposition of a suitable compound. When the reagent is an anion a suitable ester often can be used for this purpose. As mentioned previously, Willard and Tang (Ref. 73) used urea for the slow consumption of hydrogen ions in the precipitation of hydrous oxides.

$CO(NH_2)_2 + 2\,H_2O \rightarrow (NH_4)_2CO_3$. Urea can also be used in the homogeneous precipitation of calcium oxalate and barium chromate. Esters hydrolyze in acid and alkaline media.
$(COOR)_2 + H_2O \rightarrow 2\,ROH + (COOH)_2$. Thus dimethyl oxalate and diethyl oxalate have been used in the homogeneous precipitation of calcium and rare earth oxalates. Similarly, triethyl phosphate, trimethyl phosphate, and metaphosphoric acid have been used in the homogeneous precipitation of zirconium and hafnium phosphates and dimethyl sulfate in the slow generation of sulfate ions. In the precipitation of silver chloride from an ammoniacal medium Gordon et al.[79] used β-hydroxy-ethyl acetate for the removal of ammonia.

Other types of hydrolytic reactions include the formation of sulfate from sulfamic acid,

$$NH_2SO_3H + H_2O \rightarrow NH_4^+ + H^+ + SO_4^{--}$$

for the precipitation of the sulfates of barium, lead, strontium, and calcium, and the generation of hydrogen sulfide for the precipitation of a host of sulfides[80]:

$$CH_3CSNH_2 + H_2O \rightarrow CH_3CONH_2 + H_2S$$

Nonhydrolytic reactions have also been utilized in the production of a precipitant. For example, the thermal decomposition of peroxydisulfates with formation of sulfate ions has been recommended for the precipitation of barium sulfate. Iodate can be generated by interaction of periodate with a higher-valent alcohol such as glycol:

$$HOCH_2CH_2OH + IO_4^- \rightarrow 2\,HCOH + H_2O + IO_3^-$$

[79] L. Gordon, J. I. Peterson, and B. P. Burtt, *Anal. Chem.* **27**, 1770 (1955).

[80] For a review see E. H. Swift and F. C. Anson, *Advan. Anal. Chem. Instr.* **1**, 293 (1960), see also F. Burriel-Marti and A. M. Vidan, *Anal. Chim. Acta* **26**, 162 (1962), and P. F. S. Cartwright, E. J. Newman, and D. W. Wilson, *Analyst* **92**, 663 (1967).

This reaction is fast in acid medium. To obtain the desirable speed Stine and Gordon[81] used the monoethyl acetate ester, which does not reduce periodate and which hydrolyzes with formation of glycol:

$$HOCH_2CH_2OOCCH_3 + H_2O \rightarrow HOCH_2CH_2OH + CH_3COOH$$

This hydrolysis is the rate-determining step. The method has been used in the precipitation of thorium. In the precipitation of cerium(IV) iodate, the cerium is first reduced to the trivalent state and then slowly oxidized in an iodate-containing acid medium with peroxydisulfate. Several oxidation–reduction reactions can be used in the slow generation of precipitant. Arsenic trioxide is slowly oxidized by nitric acid to arsenic acid, which precipitates zirconium arsenate. Lead chromate can be precipitated by slowly oxidizing chromium(III) to chromium(VI) with bromate in an acetate buffer.

Finally, the solubility of a precipitate sometimes can be increased by adding a suitable complexing agent, which then is slowly displaced. For example, barium sulfate dissolves in an excess of ethylenediamine tetracetate at a pH above 9. Addition of nickel slowly displaces the barium from its complex and makes it precipitate as sulfate. A similar principle has been applied by Firsching[82] in the precipitation of barium chromate. Barium ions are released by slow addition of magnesium ions. For detailed information concerning precipitation from homogeneous solution, refer to the works by Gordon et al. (Ref. 79) and also Firsching.[83]

From the above review it is clear that very few methods are available for the slow generation of cations to be used in the precipitation of anions. No general use has been made of slow electrolytic generation of cations. Since the rate of electrolysis can be controlled within extremely wide limits, it is possible to generate metal ions with a controlled speed by anodic oxidation of mercury, silver, copper, bismuth, lead, etc., or of their amalgams. There is one example in the literature of the slow electrolytic generation of an anion: sulfate obtained by electrolytic oxidation of thiocyanate.[84]

PROBLEMS

1. In the precipitation of silver with an excess of iodide, part of the silver iodide remains in colloidal solution. How could the suspension be flocculated? How should the precipitate be washed to obtain a pure precipitate after ignition?
2. A solution containing ferric iron, magnesium, and sulfate is to be analyzed. Ferric iron interferes in the sulfate and magnesium determinations and must be removed by precipitation as hydrous oxide. The latter can be precipitated from a weakly acid or from an ammoniacal solution. Which condition should be selected?
3. Barium chromate precipitated under ordinary conditions consists of very fine particles. How could one proceed to get crystals of larger size?
4. Zinc sulfide can be precipitated from neutral or alkaline medium by the addition of an excess of alkali sulfide, or from weakly acid medium (pH 3–4) by hydrogen sulfide. Under what conditions is the coarser precipitate to be expected?
5. An excess of 0.5 *M* barium chloride is added to 0.1 *M* sulfate solution. A very fine precipitate of barium sulfate is obtained, the suspension runs through an ordinary filter. How can one flocculate the precipitate? (See the discussion on flocculation.)
6. Gelatin is positively charged at a pH of 2 or less and negatively charged at a pH of 8 or greater. Will an acidic or alkaline solution sensitize the precipitation of barium sulfate in the suspension described in Problem 5? *Ans.* Alkaline.

[81] C. R. Stine and L. Gordon, *Anal. Chem.* **25**, 1519 (1953).
[82] F. H. Firsçhing, *Talanta* **2**, 326 (1959).
[83] F. H. Firsching, *Advan. Anal. Chem. Instr.* **4**, 1 (1965).
[84] D. H. Klein and B. Fontal, *Talanta* **10**, 808 (1963).

7. Alkyl sulfates or alkyl acid sulfates do not yield ionic sulfate when dissolved in water. Upon heating of their aqueous solutions slow hydrolysis to inorganic sulfate occurs. Can use be made of this property in the preparation of a coarse precipitate of barium sulfate?

BIBLIOGRAPHY

For a general review of formation of precipitates see A. G. WALTON, *The Formation and Properties of Precipitates*, Wiley-Interscience, New York, N.Y., 1967.

CHAPTER 11 *Coprecipitation and Postprecipitation*

CLASSIFICATION

Precipitates separating from a solution quite generally are contaminated, and contain greater or lesser amounts of foreign material, including solvent. The incorporation of impurities into a precipitate by substances which under experimental conditions are usually soluble in the liquid phase is denoted by the general term *coprecipitation.* For example, barium sulfate obtained by adding barium chloride to a solution of potassium sulfate contains, in addition to coprecipitated water, potassium sulfate and bisulfate, although these salts are freely soluble in the liquid phase. Contamination of a precipitate can also occur as a result of *postprecipitation.* In this case the impurity precipitates after formation of the primary precipitate. The postprecipitated impurity is separated from its supersaturated solution. In the absence of the primary precipitate the postprecipitated constituent separates extremely slowly from its supersaturated solution, its precipitation being promoted (catalyzed) by the primary precipitate. Kolthoff[1] has proposed the following classification of coprecipitation phenomena:

1. *Adsorption of foreign ions at the surface of the precipitate.* The amount of coprecipitation increases with increasing surface area of the precipitate.
2. *Occlusion (or carrying down) by incorporation of foreign ions (and solvent) during the formation of a precipitate.*

Evidently coprecipitation is a source of error in precipitation analysis and should be reduced to a minimum. On the other hand, coprecipitation can be used to advantage in the separation of trace constituents from a solution. For example, a trace of lead in solution can be coprecipitated quantitatively with strontium sulfate. Much use of this kind of coprecipitation of a trace constituent into a precipitate is made in radiochemistry. The major precipitate is called the *carrier.*

The various characteristics of coprecipitation and postprecipitation are discussed in greater detail below.

COPRECIPITATION

Adsorption of salts having an ion in common with the precipitate

Figure 11-1 gives a schematic cross section of a crystal having the simple structure of the sodium chloride type, the + signs representing the positive ions, the − signs the negative ions. A positive ion *A* in the interior of the crystal is subjected to the attractive forces of four neighboring negative ions, *a*, *b*, *c*, and *d*, and also of course of one directly above it and one directly below. *A*, therefore, is acted upon by the electrical forces of six surrounding negative ions. An ion *B* in the surface experiences the attractive

[1] I. M. Kolthoff, *J. Phys. Chem.* **36**, 860 (1932).

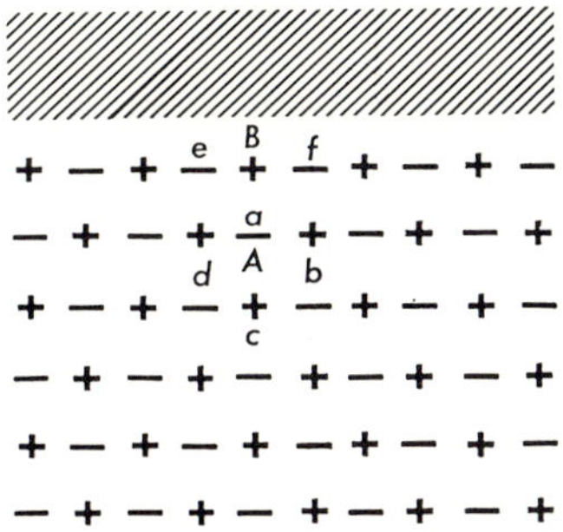

Fig. 11-1.

forces of five negative ions (*e*, *a*, *f*, and two in the neighboring planes). The positive ions in the surface therefore have free residual valences by virtue of which they can attract negative ions. Preferentially they will adsorb the negative lattice ions, since the forces acting between lattice cations and anions must be great as indicated by the slight solubility; for the same reason the negative ions in the surface will attract positive lattice ions from the solution. If lattice cations present in the solution are attracted and adsorbed by the surface, an equivalent number of foreign negative ions must also be attracted to the surface, since otherwise the solution would acquire a free negative charge and the surface a free positive charge, which is impossible. As the result of such attractive forces, then, we find that a salt containing a lattice ion will be adsorbed from the solution on the surface of the precipitate.

Let us now consider the adsorption of various salts all having the same lattice ion in common; for example, the adsorption of barium salts by barium sulfate. Experimentally it is found that the adsorption at the same final salt concentration in solution is not the same for the various barium salts but *increases more or less with decreasing solubility of the barium salt*. The relation is formulated by the *Paneth–Fajans–Hahn adsorption rule*[2]: Those ions whose compounds with the oppositely charged constituent of the lattice are slightly soluble in the solution in question are well adsorbed by the ionic lattice. At the same final concentration in the solution it is found, for example,[3] that silver acetate is much more strongly adsorbed than silver nitrate by silver iodide, the acetate being less soluble than the nitrate. In addition to the solubility of the adsorption compound, there are other factors which determine the adsorbability. Among these the following two are of great importance: the *electrical dissociability* of the adsorbed compound and *deformability* of the adsorbed ions. The adsorption increases with decreasing dissociation of the adsorption compound. For example, hydrogen sulfide, a weak electrolyte, is strongly adsorbed by metal sulfides.

The adsorbability also increases with increasing deformability of the adsorbed ions. Ions do not have a rigid structure, and it is possible by application of electrical forces to deform the shell of the outer electrons. Anions as a rule are more easily deformed than cations; the deformability of the former increases with increasing size of the ions. Anions of the large aromatic molecules (dyes) especially are distinguished by great deformability. Use is made of this fact in the application of adsorption indicators in volumetric analysis (see p. 721, where the mechanism of the action of adsorption indicators is discussed). *From the viewpoint of coprecipitation it is most important to remember that a precipitate absorbs salts containing the lattice cation or lattice anion and that the adsorbability as a rule increases with decreasing solubility and decreasing electrolytic dissociation of the adsorbed compound.*

[2] See O. Hahn, *Applied Radiochemistry*, Cornell Univ. Press, Ithaca, N.Y., 1936.
[3] J. S. Beekley and H. S. Taylor, *J. Phys. Chem.* **29**, 942 (1925).

Adsorption of salts having no ion in common with the precipitate. Exchange adsorption

It was explained above that the ions in the surface of an ionic lattice exhibit residual valence forces. The negative ions in the surface therefore attract positive ions from the solution, and the positive ions attract negative ions. Therefore, it might be expected that a salt having no ion in common with the precipitate would be adsorbed as such on the surface. Actually another process may take place as well. This equivalent adsorption of foreign cations and anions occurs, for example, in the adsorption of sodium bromate on the surface of barium sulfate.[4]

In a discussion of the effect of aging upon the structure of a precipitate, the process of kinetic exchange was mentioned (p. 220). Kinetic equilibrium between the precipitate and the surrounding solution means that the speed with which lattice ions leave the surface is equal to the speed with which ions from the solution deposit again on the surface. Let us consider what will happen when barium sulfate is shaken with a dilute solution of lead perchlorate. The lead ions are strongly attracted by the barium sulfate, since lead sulfate is slightly soluble and the lead sulfate fits into the lattice of barium sulfate. The perchlorate ions, on the other hand, have no pronounced tendency to be adsorbed, since the barium perchlorate is very easily soluble. Let us assume that *e* and *B* in Fig. 11-1 represent a sulfate and a barium ion, respectively, and that at a certain moment both leave the surface and go into the solution. There will be a pronounced tendency for ions to be deposited again at this vacant spot. Both barium and lead ions are strongly attracted by the lattice, and it may happen that a lead ion deposits at *B* instead of a barium ion. The sulfate ions, on the other hand, are much more strongly adsorbed than the perchlorate ions, and the former will be deposited at *e*. Therefore on shaking barium sulfate with a very dilute lead perchlorate solution, the following *surface* exchange reaction takes place:

$$\underset{(surface)}{BaSO_4} + \underset{(solution)}{Pb^{++} + 2\,ClO_4^-} \rightleftharpoons \underset{(surface)}{PbSO_4} + \underset{(solution)}{Ba^{++} + 2\,ClO_4^-}$$

As a result of this surface reaction, barium ions in the surface are partly replaced by lead ions, i.e., the latter are adsorbed, and an equivalent amount of barium ions enters the solution. This exchange adsorption takes place in spite of the fact that the solubility product of lead sulfate is not exceeded. Frequently, a nonequivalent adsorption occurs, even when the adsorbed ion does not fit in the crystal lattice. For example, on shaking calcium oxalate with potasium iodate or sulfate only the anion is adsorbed,[5] and an equivalent amount of oxalate ions enters the solution:

$$CaOx + 2\,IO_3^- \rightarrow \underset{(surface)}{Ca(IO_3)_2} + Ox^{--}$$

With solutions of manganese or sodium chloride only the cation is adsorbed:

$$CaOx + 2\,Na^+ \rightarrow \underset{(surface)}{Na_2Ox} + Ca^{++}$$

Obviously these adsorptions cannot be the result of an exchange between the lattice anion or cation with the adsorbed ions. It is impossible to replace one oxalate ion in the lattice with two iodate ions or one calcium ion with two sodium ions. The interpretation of these *apparent* exchange reactions is the following.

In a saturated solution of calcium oxalate there are present calcium and oxalate ions. Calcium iodate is fairly slightly soluble and therefore is markedly adsorbed on calcium oxalate (p. 229). Thus on addition of alkali iodate to the suspension an adsorption of calcium iodate occurs. By this process the solution becomes undersaturated.

[4] L. H. Reyerson, I. M. Kolthoff, and K. Coad, *J. Phys. Chem.* **51**, 371 (1947).

[5] I. M. Kolthoff and E. B. Sandell, *J. Am. Chem. Soc.* **55**, 2170 (1933); see also I. M. Kolthoff and C. Rosenblum, *ibid.* **55**, 2664 (1933).

To establish solubility equilibrium, oxalate ions must go into solution until the solubility product is attained. The resulting solution contains an excess of oxalate over calcium ions. A similar interpretation of the apparent cation-exchange adsorption can be given.

Exchange between foreign counterions adsorbed at the surface of impure precipitates and other ions in the solution

In a discussion of the properties of colloidal solutions (p. 214) we saw that the particles are kept in colloidal solution by adsorption of their own ions on the surface. Upon adding silver nitrate solution to 0.01 *N* potassium iodide (not enough to precipitate all the iodide), a stable colloidal solution of silver iodide is obtained. The particles are kept in colloidal solution by a primary adsorption of iodide ions, and an equivalent amount of potassium ions is present as "counterions." By addition of sufficient electrolyte, say potassium nitrate, the sol is flocculated. The fresh precipitate still contains absorbed potassium iodide, which is hard to remove by washing. The counterions, in this case potassium ions, can be replaced by other ions present in the solution. If the precipitate is washed with a dilute solution of sulfuric acid, for example, they are replaced by hydrogen ions:

$$\underset{(particle)}{AgII^-} \vdots K^+ + H^+ \rightleftharpoons AgII^- \vdots H^+ + K^+$$

($AgII^- \vdots$ denotes adsorbed iodide fixed on the surface of the silver iodide; K^+ is the counterion.)

Ammonium ions have, qualitatively, the same effect as hydrogen ions. If the precipitate has been thoroughly washed with a solution of an acid or an ammonium salt, it still may contain adsorbed iodide in the form of hydrogen or ammonium iodide. Since these two are easily volatilized, a pure precipitate results on heating. Not all countercations are easily exchanged by washing with a strong acid or an ammonium salt. In the section dealing with colloids a table has been presented (p. 218) which illustrates that a divalent cation flocculates a negatively charged sol at a concentration of the order of 80 times smaller than that of a monovalent ion, whereas the flocculation value of a tervalent ion is of the order of 500 times smaller than that of a monovalent ion. The higher-valent ions are now the counterions, they are held much more strongly in the compressed double layer than the monovalent ions, and, in general, are not easily removed by washing with hydrogen or ammonium ions. In the precipitation of halides with an excess of silver a silver salt will be adsorbed, which cannot be removed completely by washing with water or an electrolyte solution. Quite generally, adsorbed material on silver halides can be removed to a large extent when the precipitate is subjected to an aging process, especially by digestion at higher temperature. This aging results in a marked decrease of the specific surface and of the amount of foreign material adsorbed.

Hydrous oxides usually are obtained as amorphous or very small crystalline particles. These precipitates have a large surface area and can adsorb relatively large amounts of foreign electrolytes. When precipitated from an ammoniacal medium hydroxyl ions are primarily adsorbed with ammonium ions as counterions. In the presence of divalent ions such as copper, nickel, cobalt, and calcium, replacement of ammonium by divalent ions occurs and a bad coprecipitation is observed. When precipitated at room temperature the amount of foreign ion coprecipitated during the separation of the amorphous precipitate is of the same order of magnitude as the amount adsorbed when a solution of the same foreign ion is added to a suspension of the fresh precipitate, the composition of the supernatant liquids in both cases being the same (see second and third columns in Table 11-1). This, for example, is found with hydrous ferric hydroxide precipitated at room temperature by an excess of ammonia. The precipitate is amorphous and the amount of coprecipitated cations is only slightly greater than the amount adsorbed if added after the precipitation, all other conditions being the same. A few representative

TABLE 11-1. **Comparison of Adsorption on and Coprecipitation with Orthoferric Hydroxide**

Metal	*Adsorbed on Ppt. Formed at Room Temp., %*	*Coprecipitated at Room Temp., %*	*Adsorbed on Ppt. Formed at 98°C, %*	*Coprecipitated at 98°C, %*
Copper	19.0	27.3	—	—
Copper	13.5	16.5	3.6	17.0
Zinc	11.9	17.6	—	17.7
Nickel	20.3	28.3	—	—
Nickel	9.3	13.4	5.7	18.7
Cobalt	10.9	17.9	7.4	19.7
Magnesium	9.3	16.5	4.4	12.7
Calcium	12.2	12.9	—	7.7

figures taken from studies by Kolthoff and Moskovitz[6] and Overholser[7] are given in Table 11-1. The values given for a metal in a horizontal column are comparable, because the final composition of the supernatant liquids in ammonia, ammonium salt, and metal were identical.

When *o*-ferric hydroxide is precipitated at 98°C with an excess of ammonia it is still very finely divided but already shows properties of crystals as revealed by X-ray investigations. The specific surface of a product formed at 98°C is still extremely great but it is much less than that of a product precipitated at room temperature (see Table 11-1). On the other hand, the coprecipitation of the various cations investigated is almost identical when the precipitation is carried out at room temperature and at 98°C, as shown in Table 11-1. At 98°C about 75 per cent of the coprecipitated contaminant is incorporated within the precipitate. An increase in the concentration of ammonium salts decreases the adsorption of divalent ions. This is illustrated in Fig. 11-2. which refers to the precipitation of hydrous ferric hydroxide at room temperature. The effect of ammonium salt is twofold: (1) It competes with the divalent ions present as counterions, and (2) it decreases the hydroxyl-ion concentration. The hydroxyl ions are primarily adsorbed on the precipitate and their adsorption decreases with decreasing pH. The effect of pH is much greater than the competition effect between the counterions, as can be seen from Fig. 11-2. The adsorption of hydroxyl ions increases with increasing ammonia concentration (at constant concentration of ammonium salt). One would then expect that the adsorption of all divalent ions would increase with the ammonia concentrations. Actually, this was found to be true for calcium and magnesium (curves II and III in Fig. 11-3). On the other hand, the same figures show that the adsorption of cobalt, zinc, and nickel decreases markedly with increasing ammonia concentrations. Analytically it could be shown that these cations are adsorbed in the form of aquo ions and not as ammines.[8]

Summarizing, the following statements of analytical interest in the precipitation of hydrous ferric and some other oxides may be made. To decrease the coprecipitation or adsorption of all divalent metal ions, it is recommended to have a large amount of ammonium chloride in the solution. If the divalent metal forms ammine complexes a high ammonia concentration is desirable to minimize the coprecipitation. On the other hand, if the cation does not form an ammine at reasonable ammonia concentrations (up to 1 to 2 *M*), the concentration of ammonia should be kept very small.

[6] I. M. Kolthoff and B. Moskovitz, *J. Phys. Chem.* **41**, 629 (1937).
[7] I. M. Kolthoff and L. G. Overholser, *J. Phys. Chem.* **43**, 767, 909 (1939).
[8] See L. Levy, *Compt. Rend.* **189**, 426 (1929); I. M. Kolthoff and L. G. Overholser, Ref. 7.

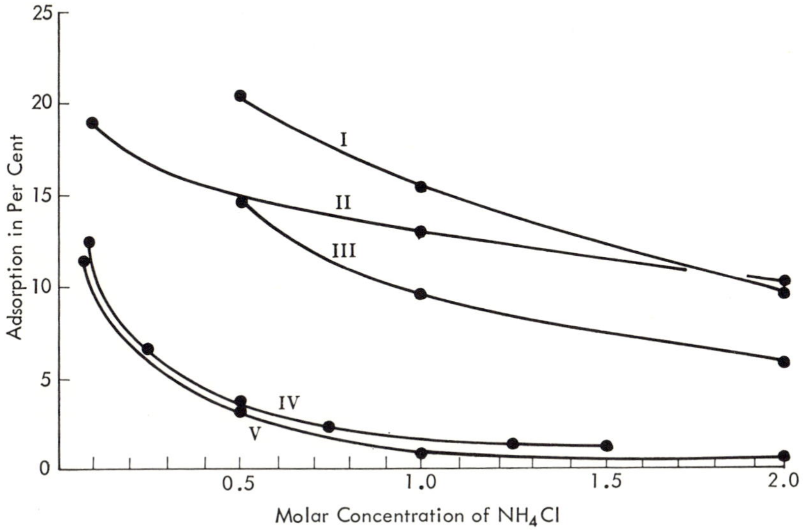

Fig. 11-2. Effect of ammonium chloride concentration on the adsorption of the various metals. I, Co; II, Ca; III, Mg; IV, Zn; V, Ni. Concentration of ammonia present: I, II, III, and V, 0.90 *M*; IV, 0.73 *M*.

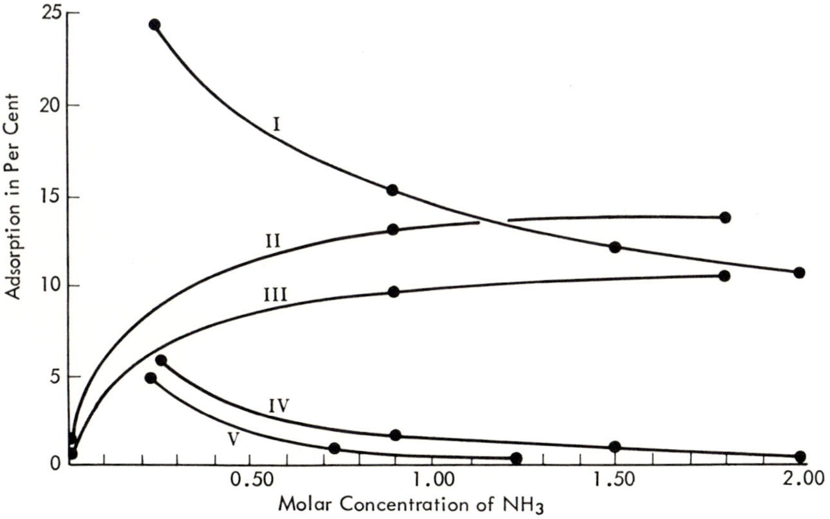

Fig. 11-3. Effect of concentration of ammonia on the adsorption of the various metals. I, Co; II, Ca; III, Mg; IV, Ni; V, Zn. The concentration of ammonium chloride was 1.00 *M* in all cases.

Adsorption of solvent (water) by precipitates. Hygroscopic water

Molecules that have a polar character (dipoles) are attracted by the ions in the surface of a crystal lattice and are held there very strongly. An air-dried precipitate, for example, always contains some adsorbed water, which may be called water of hygroscopicity, in addition to occluded water. The amount of water adsorbed increases with increasing surface development of the precipitate. Certain authors recommend weighing precipitates in the air-dry form by proceeding as follows. After the ordinary washing process, the precipitate is washed with alcohol, which is followed by ether. Air is then drawn through to remove the ether, and the precipitate is weighed. Under these conditions one is liable to find high results, owing to the adsorption of water vapor and also to occlusion of water. Thus it has been found[9] that calcium oxalate monohydrate precipitated under ordinary analytical conditions (dilute weakly acidic solutions, precipitation at 100°C) contains as much as 1 per cent of hygroscopic water in the air-dry state. This water is partially removed by heating to 105–110°C but is readily taken up again when the precipitate is exposed to the air. If the total surface of the precipitate is decreased by digestion at higher temperatures (recrystallization accompanied by perfection of crystals, pp. 220 and 223), the amount of hygroscopic water decreases proportionally. Barium sulfate loses adsorbed water readily at 100°C, but occluded water is removed slowly at temperatures below 500°C.[10]

Occlusion

Mixed-crystal[11] or solid-solution formation. Mixed crystals are easily formed by substitution of a cation or an anion in a crystal lattice by a cation or anion of the same charge and about the same size. Examples are solid solutions of lead (sulfate) in barium sulfate, the lead and barium ions having about the same size. Similarly, barium chromate can form a solid solution in barium sulfate, the chromate and sulfate ions having the same size. Solid solutions can also be formed if the solute has the same type of chemical formula and the ions have the same size as those of the host crystal but the ions have different charges. A typical example is the solid solution of $KMnO_4$ in $BaSO_4$, both compounds having the general formula ABO_4. On shaking a fresh and imperfect precipitate of barium sulfate with potassium permanganate the latter is incorporated in the form of a solid solution in the recrystallizing sulfate. The permanganate cannot be removed by washing with water.[12] Many examples of this type of solid solution are known, e.g., sodium nitrate in calcium carbonate (type ABO_3). Even where the type of chemical formula is not the same, solid solution may occur when the ions are of the same size. For example, lanthanum (and other lanthanides) fluoride, LaF_3, can form a solid solution in calcium fluoride, CaF_2, and tetravalent ions such as Th^{4+}, U^{4+}, and Ce^{4+} can form solid solutions in lanthanum fluoride.[13] Evidently, it is not possible to simply substitute one divalent Ca by one trivalent La ion in the lattice. In order not to violate electroneutrality, there should be created some vacant spots or holes where calcium ions were located. This type of solid solution with formation of lattice imperfections is quite common (semiconductors).

As a matter of fact, evidence is presented in the literature that, to a limited extent, many substances can be coprecipitated with the host crystal in the form of a solid solution, even though this would not be expected on the basis of similarity in size or

[9] E. B. Sandell and I. M. Kolthoff, *J. Phys. Chem.* **37**, 153 (1933).

[10] I. M. Kolthoff and W. M. MacNevin, *J. Phys. Chem.* **57**, 2573 (1935).

[11] The term *mixed crystal* is commonly applied to a crystal in which an atom or ion M is replaced by another atom or ion N. Such a crystal can also be described as a *solid solution* of M and N. The two terms are not quite synonymous, because solid solution can include other forms of incorporation of N in the lattice of M, e.g., interstitial solid solution, in which N does not occupy the normal atomic or ionic sites but occurs between the lattice particles.

[12] I. M. Kolthoff and G. E. Noponen, *J. Phys. Chem.* **42**, 237 (1938).

[13] K. Schlezter and L. G. Sillen, *Acta Chem. Scand.* **4**, 1323 (1950).

formula. For example, by exact X-ray measurements, Walden and Cohen[14] found that barium nitrate coprecipitated with barium sulfate is incorporated in the form of a solid solution with expansion of the barium sulfate lattice. From X-ray measurements Walton and Walden[15] concluded that coprecipitated ammonium, potassium, sodium, and lithium with barium sulfate are incorporated as solid solutions, partially as normal sulfate and partially as hydrogen sulfates. Even water coprecipitated with barium sulfate is present as a solid solution. The volume of the unit cell increases with the water content. It was concluded that three water molecules can replace one barium sulfate molecule in the lattice. Streuli et al.[16] concluded from X-ray measurements that ferric iron, nitrite, and chloride are coprecipitated with barium sulfate as solid solutions.

Distribution laws. When the coprecipitated contaminant is homogeneously distributed as a solid solution, the distribution law is the same as in the extraction of a constituent from one liquid phase into another liquid phase. Here the Berthelot–Nernst distribution law holds. For example, if lead is coprecipitated with barium sulfate, the lead being homogeneously distributed into the solid phase and the latter is in thermodynamic equilibrium with the liquid phase, we have

$$BaSO_4 + Pb^{++} \rightarrow PbSO_4 + Ba^{++}$$

$$\left(\frac{[Pb^{++}]}{[Ba^{++}]}\right)_{\text{solid}} = D\left(\frac{[Pb^{++}]}{[Ba^{++}]}\right)_{\text{solution}} \tag{1}$$

in which D is the *homogeneous distribution coefficient*. For convenience, concentrations instead of activities are written in Eq. (1). If x represents the quantity of the microconstituent (Pb^{++}) and y that of the macroconstituent (Ba^{++}) left in the solution, and if a and b are the corresponding amounts originally present in the solution, Eq. (1) can be written in the general form

$$\left\{\frac{(a - x)}{(b - y)}\right\}_{\text{solid}} = D\left(\frac{x}{y}\right)_{\text{solution}} \tag{1a}$$

which is similar to Eq. (1).

A second condition may be realized, in which during the precipitation kinetic equilibrium is attained between the ions in solution and the ions on the solid surface as each new layer is formed, barring recrystallization and solid diffusion effects. This condition results in an *inhomogeneous distribution* of the contaminant in the precipitate (unless $D = 1$; radioactive isotopes). With the formation of each new layer, Eq. (1) holds for the new surface layer. When D differs from 1 the ratio of Pb^{++} and Ba^{++} originally present in the solution changes continuously during the formation of the precipitate and we obtain the following differential equation:

$$\frac{d(a - x)}{d(b - y)} = \frac{dx}{dy} = \lambda\frac{x}{y} \tag{2}$$

where λ is a constant for a given system. Upon integration of Eq. (2) from the initial concentrations a and b to the final concentrations x and y, respectively, there results

$$\ln\frac{a}{x} = \lambda \ln\frac{b}{y} \tag{3}$$

or, in our example of lead sulfate,

$$\ln\frac{(\text{total } Pb^{++})}{[Pb^{++}]_{\text{solution}}} = \lambda \ln\frac{(\text{total } Ba^{++})}{[Ba^{++}]_{\text{solution}}}$$

[14] G. H. Walden, Jr., and M. A. Cohen, *J. Am. Chem. Soc.* **57**, 2591 (1935); **59**, 906 (1937).
[15] G. Walton and G. H. Walden, Jr., *J. Am. Chem. Soc.* **68**, 1742, 1750 (1946).
[16] C. A. Streuli, H. A. Scheraga, and M. L. Nichols, *Anal. Chem.* **25**, 306 (1953).

λ is called the *heterogeneous distribution* coefficient. This logarithmic distribution law was originally derived by Doerner and Hoskins.[17] Equation (3) is generally referred to as the *Doerner–Hoskins equation*. Under ideal conditions D and λ are numerically equal. If D or λ are greater than unity the microconstituent is enriched in the precipitate. The opposite is true where D or λ are smaller than unity. The enrichment is greater under conditions of heterogeneous than of homogeneous distribution (see Fig. 11-4). For example, when $D = \lambda = 25$ and 20 per cent of the macrocomponent is precipitated, the microcomponent is coprecipitated to the extent of 99 per cent under heterogeneous and 86 per cent under conditions of homogeneous distribution.

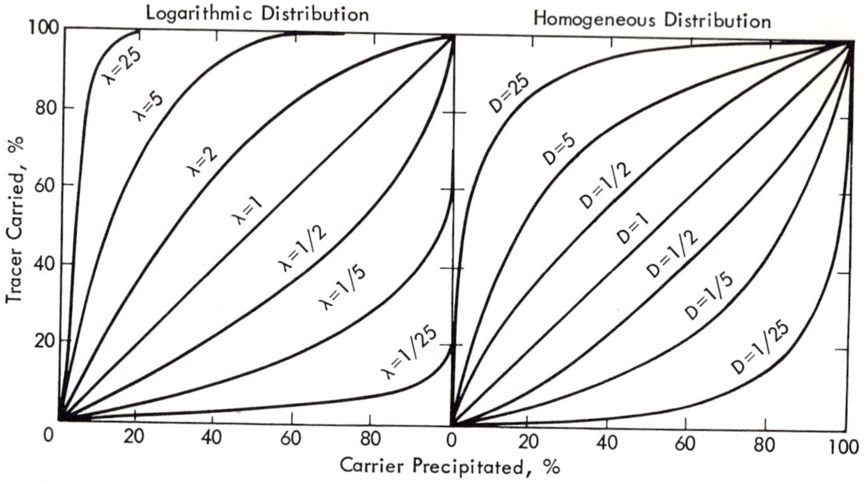

Fig. 11-4. Coprecipitation with varying values of λ and D.

Under ordinary conditions of precipitation it is relatively rare to get homogeneous distribution of the contaminant. However, it occurs where the thermal mobility of the ions with the host crystal is great. This, for example, is the case with the silver halides when formed under conditions such that the precipitate separates in colloidal solution. Silver chloride forms a solid solution in silver bromide; hence chloride is coprecipitated upon precipitation of silver bromide in the presence of chloride. The distribution coefficient for the chloride is very small, but the coprecipitation becomes appreciable near the equivalence point. When the precipitate is flocculated during the addition of silver, the particles fuse together and the thermal mobility in the agglomerates formed becomes negligibly small. Under such conditions heterogeneous distribution is approached. As an example, some results obtained by Kolthoff and Eggertsen[18] are presented in Table 11-2.

In the absence of aluminum there is clearly homogeneous distribution. In the presence of aluminum the distribution is heterogeneous. The higher values in the last column as compared to the values calculated from λ (third column) are due to rapid recrystallization as a result of aging. Homogeneous distribution resulted when the precipitate formed in the presence of aluminum was aged for 24 hr.

It has been found[19] in enriching systems with heterogeneous distribution that an increased rate of precipitation leads to decreased incorporation of the microcomponent

[17] H. Doerner and W. Hoskins, *J. Am. Chem. Soc.* **49**, 738 (1927).

[18] I. M. Kolthoff and F. T. Eggertsen, *J. Am. Chem. Soc.* **61**, 1036 (1939).

[19] J. N. Hermann, Ph.D. thesis, Los Alamos Scientific Laboratory, 1957; *Treatise on Analytical Chemistry* (I. M. Kolthoff and P. J. Elving, eds.), part I, vol. 3, Wiley-Interscience, New York, 1961, p. 1367.

TABLE **11-2. Coprecipitation of Chloride with Silver Bromide from Equimolar Mixture** $D = \lambda = 0.00254$.

Ag⁺ added in % of bromide	*Chloride Precipitated, %*			
	Calculated from		*Experimental*	
	D	λ	*Absence of* Al^{3+}	*With* Al^{3+}
99	4.33	0.99	4.30	—
100	4.80	1.13	4.81	2.18
101	5.33	1.39	5.26	2.52
102	5.90	2.03	5.87	2.19

and hence lower values of λ. The opposite is true in "depletion" systems. For example, Hermann determined the coprecipitation of americium with lanthanum oxalate from homogeneous solution, varying the rate of precipitation. His results are summarized in Table 11-3.

TABLE **11-3. Coprecipitation of** Am **with** $La_2(C_2O_4)_3 \cdot 11\ H_2O$. **Enrichment System. Precipitation from Homogeneous Solution**

λ	6.6	6.1	5.8	5.0	3.8	1.5
Average rate of ppt., % La per hour	0.01	0.07	1.2	1.9	4.1	5.3

When the rate of precipitation was kept constant, λ remained essentially constant during the precipitation. The variation of λ with rate of precipitation is attributed by Hermann to precipitation from supersaturated solution. Only when the precipitation occurs from saturated solution will λ be equal to D. In very fast precipitations λ approaches unity. This has been confirmed by Chang.[20] He studied the coprecipitation of lead with barium sulfate. The precipitation was made from homogeneous solutions generating the sulfate ion by hydrolysis of sulfamic acid at 99°C. Under a given set of conditions a constant value of λ was found during the precipitation, indicating Doerner and Hoskins distribution. With changes of stirring and of rate of precipitation λ was found to vary between 0.42 and 0.60. The average value of λ is some 20 times as large as the value of 0.027 of D at 95°C (or 0.015 by extrapolation at 99°C) reported for the homogeneous distribution of lead into barium sulfate[21] (for the method see p. 225). The difference is probably accounted for by a relatively large supersaturation during the separation of the mixed crystals. By adding to the solution a few grams per 100 ml of well-aged barium sulfate of large surface development before initiating precipitation, supersaturation during the precipitation was greatly decreased and the values of λ found under such conditions became much smaller than those observed without seeder and approached the value of D.

When the supersaturation is very large during the precipitation, as is realized by rapidly adding with constant stirring sulfate to the solution containing barium and lead, the distribution coefficients approached unity. Apparently, under these extreme conditions, the barium and sulfate are incorporated in the precipitate in the same ratio as they are present in the solution.

[20] C. Chang, Ph.D. thesis, Univ. of Minnesota, Minneapolis, 1958.
[21] I. M. Kolthoff and G. E. Noponen, *J. Am. Chem. Soc.* **60**, 197 (1938).

Distribution of strontium between 20 per cent methanol solution and barium sulfate at 83°C was studied by Gordon et al.[22] Until 50 per cent of barium was precipitated, a slight "negative" coprecipitation of strontium was found. Undoubtedly, this result must be attributed to experimental errors. The coprecipitation of strontium is very small and the authors analyzed the supernatant liquid for strontium. Above 50 per cent barium precipitated they found a λ value of 0.03 ± 0.004. Some systems which involve unexpected solid solution formation yield a Doerner–Hoskins distribution when precipitation is made from homogeneous solution. The coprecipitations of rare earth iodates with thorium iodate are examples of such a distribution. It must be assumed that the quadrivalent ion is replaced in the lattice by trivalent rare earth ions. Whether this replacement leads to formation of anion holes or to lattice distortions is not known. Quite generally a limited solubility of the microcomponent into the host crystal may be expected in the formation of abnormal mixed crystals. Shaver[23] investigated the coprecipitation of lanthanum, praseodymium, promethium, europium, yttrium, and scandium with thorium oxalate, the precipitation taking place from homogeneous solution. The systems are of analytical interest because the distribution coefficient of the rare earths into the solid is small. Therefore, very little coprecipitation can occur, until the precipitation of thorium is practically complete. This is illustrated in Table 11-4 for the coprecipitation of lanthanum. Note the constancy of λ during the precipitation.

TABLE 11-4. **Coprecipitation of La^{3+} with Thorium Oxalate**

Th *precipitated,* %	*Rare Earth Coppt.,* %	$1/\lambda \times 10^4$	$D \times 10^6$
88.2	0.15	7.0	200
96.2	0.24	7.3	95
99.0	0.33	7.2	33
99.800	0.32	5.2	6.4
99.880	0.38	5.6	4.6
99.920	0.46	6.4	3.7
99.967	0.51	6.4	1.6
99.987	0.60	6.7	0.8
		Av. 6.4	

A constant value of λ in the coprecipitation of the tervalent rare earths with calcium oxalate prepared by precipitation from homogeneous solution was obtained.[24] On the other hand, Hoste et al.,[25] who investigated the coprecipitation of molybdate and vanadate with cinchonine or β-naphthoquinoline tungstate, found neither homogeneous nor heterogeneous distribution, but found instead that the amount coprecipitated was directly proportional to the amount of precipitate:

$$\frac{x}{y} = K\frac{a}{b} \tag{4}$$

The precipitation was carried out from homogeneous medium by dissolving tungstic acid in a nitric acid solution containing hydrogen peroxide. The pertungstic acid

[22] L. Gordon, C. C. Reimer, and B. P. Burtt, *Anal. Chem.* **26**, 842 (1954).
[23] K. J. Shaver, *Anal. Chem.* **28**, 2015 (1956).
[24] B. C. Purkayastha and S. N. Bhattacharya, *J. Inorg. Nucl. Chem.* **10**, 103 (1959).
[25] J. Hoste, F. Adams, and R. Dams, *Chem. Weekblad* **60**, 35 (1964).

formed decomposes slowly at higher temperature. Some results are presented in Table 11-5.

TABLE **11-5. Coprecipitation of Mo with Tungstate**

% *Precipitated*	% *Coppt.*	*D*	λ	*K*
17.4	13.4	0.73	0.75	1.30
29.0	24.1	0.78	0.80	1.20
44.2	27.2	0.47	0.55	1.62
58.2	43.2	0.55	0.65	1.35
70.1	54.1	0.50	0.65	1.30
80.2	60.8	0.38	0.58	1.32
90.1	69.7	0.26	0.58	1.29

It is possible that the molybdate may diffuse rapidly into the solid. In the precipitation from homogeneous solution of cadmium in 2 *M* perchloric acid in the presence of zinc, using thioacetamide as the generating agent for hydrogen sulfide, Tanigawa et al.[26] found a value of λ which was constant during the precipitation of the cadmium. In equimolar (0.02 *M*) solutions of cadmium and zinc λ was found equal to 0.0041 when precipitated at 33°C and 0.0026 at 40°C. However, λ increased rather markedly with increasing concentration of thioacetamide, being 0.0041 when the initial thioamide concentration was 0.060 *M*, and 0.0060 when it was 0.120 *M*. A higher temperature and a small concentration of thioacetamide would appear to be favorable as far as the coprecipitation of zinc is concerned. The remarkable result was reported that, when the zinc concentration was made 40 times as large as that of cadmium, λ decreased 40 times. If substantiated by additional work, this would probably mean that zinc sulfide has only a limited solubility in cadmium sulfide which is attained when the original solutions of cadmium and zinc are 0.02 *M*. As a matter of fact, this limited solubility is well known in geochemistry.

Occlusion by entrapment of foreign ions and solvents. It has been mentioned that Walton and Walden (Ref. 15) and Streuli et al. (Ref. 16) found from X-ray analysis that alkali sulfates, barium salts, ferric, nitrite, and water can be incorporated into barium sulfate to a limited extent in the form of solid solutions. Contamination to a limited extent by solid-solution formation may also be expected to occur with other precipitates. Larger amounts of contaminants will cause lattice imperfections. Quite generally, the order of mixing the reactants will be of importance in determining the kind and extent of coprecipitation. Crystals adsorb salts that have an ion in common with the lattice.

If a precipitate such as barium sulfate is formed in the presence of an excess of a barium salt, say barium chloride, it will adsorb barium ions at its surface during the growth of the crystals. Owing to the electroneutrality of both phases, an equivalent amount of chloride ions must be dragged to the surface with the barium ions. If the precipitate is formed under such conditions that the crystals grow very slowly, the sulfate ions in the solution will replace most of the chloride, and the occlusion of barium chloride will be small. If, on the other hand, the crystals grow rapidly, as is usually the case in the precipitation of slightly soluble substances, there will be no time available for a complete exchange between the contaminating ions and the sulfate ions, and a larger coprecipitation will result. If the coprecipitation of different kind of anions is compared, one must realize that their adsorbability, and therefore their occlusion also, is roughly determined by the Paneth–Fajans–Hahn adsorption rule. In general, the less soluble the barium salt, the more it will be occluded by barium

[26] Y. Tanigawa, S. Hasegawa, and K. Takiyama, *Japan Analyst* **12**, 1300 (1962).

sulfate. However, it is emphasized that in addition to solubility other factors determine the adsorbability. Nitrate is more strongly occluded than chloride, because barium nitrate is less soluble than the chloride. If barium sulfate is precipitated from solutions containing an excess of sulfate, the foreign anions are given less opportunity to be adsorbed and occluded, since a primary adsorption of sulfate ions may be expected. Under these conditions the occlusion of foreign cations will predominate, the adsorbability and occlusion again being determined to a large extent by the Paneth–Fajans–Hahn adsorption rule. From the above it follows that if one wishes to cut down the coprecipitation of anions with barium sulfate, the precipitation should be carried out in a medium containing an excess of sulfate.

Ferric iron gives a very pronounced coprecipitation in the determination of sulfate as barium sulfate. The salts of ferric iron are strongly hydrolyzed in aqueous medium and contain many colloidal particles of positively charged ferric oxide. If the sulfate solution containing the iron is slowly precipitated with barium salt, the primary precipitate acquires a negative charge by adsorption of sulfate ions. The negative and positive particles attract and flocculate each other, and therefore a strong occlusion of iron occurs. The hydrolysis of the ferric iron salt is repressed by the addition of strong acid; most of the iron is then present in the form of aquoferric ions (or complexes). The amount of coprecipitation decreases in the presence of acid but is not completely eliminated, since the ferric ions are adsorbed by the growing particles. If the sulfate solution containing iron is added to an excess of acid barium chloride solution, the iron is not given much chance to be adsorbed, and a precipitate practically free of iron is formed.

The following rules guide us in predicting the kind and amount of occlusion:

1. *Occlusion of foreign cations may be expected to predominate when the solution contains an excess of lattice anions during the precipitation. The amount of occlusion increases with increasing concentration of lattice anions and decreases (sometimes to zero) if the precipitation is made from a solution containing an excess of lattice cations. The reverse rule holds for the coprecipitation of foreign anions.*

2. *Under identical conditions the coprecipitation of foreign ions is of the same order as their adsorbability on the surface of the precipitate.*

A few examples of the coprecipitation of foreign cations and anions with calcium oxalate monohydrate[27] which demonstrate the general validity of the above rules are given below. The order of the adsorbability of the alkali oxalates by calcium oxalate was experimentally found to be $Na^+ > K^+ = NH_4^+$. In agreement herewith, the coprecipitation decreases in the same order. In the precipitation of calcium oxalate at 100°C in the presence of the same number of equivalents of alkali chlorides, the following occlusion was found:

Excess of calcium during precipitation: Occluded alkali oxalate in millimoles per gram of calcium oxalate:

$$0.037\ Na^+; \qquad 0.0051\ K^+; \qquad 0.0064\ NH_4^+$$

Excess of oxalate during precipitation:

$$0.0064\ Na^+; \qquad 0.0072\ K^+; \qquad 0.0083\ NH_4^+$$

The coprecipitation of chloride, bromide, and iodide with calcium oxalate is extremely small under all conditions. This is to be expected, because the calcium halides are very soluble and therefore their adsorbabilities are very small. Hydroxide, sulfate, and iodate are more strongly occluded, as could be expected. In the precipitation of calcium oxalate at 100°C in the presence of 2 g of ammonium sulfate, an occlusion of 13.0 mg of calcium sulfate per gram of calcium oxalate was found when the precipitation was

[27] See I. M. Kolthoff and E. B. Sandell, *J. Phys. Chem.* **37**, 443, 459 (1933).

made in the presence of an excess of calcium, whereas only 5.3 mg of sulfate were occluded in the presence of an excess of oxalate.

The coprecipitation of calcium hydroxide with calcium oxalate is of practical importance. If calcium oxalate is precipitated from ammoniacal or even from neutral medium, an occlusion of calcium hydroxide occurs. If the precipitate is weighed in the form of oxide or sulfate, the hydroxide occlusion does not interfere with the calcium determination. However, if the precipitate is titrated with permanganate, low results will be found, because part of the calcium is present as hydroxide and not as oxalate. Precipitation from weakly acid medium (acetic acid and ammonium acetate) prevents the occlusion of hydroxide but may result in a coprecipitation of oxalic acid or bioxalate.

The effect of the kind of lattice ion present in excess during the precipitation of barium sulfate upon the occlusion of anions is clearly demonstrated by experiments of Weiser and Sherrick[28]:

Occlusion of: Cl^-	Ba^{++} in excess: 1.58 g eq./100 moles of $BaSO_4$
Occlusion of: Cl^-	SO_4^{--} in excess: 0.125 g eq./100 moles of $BaSO_4$
Occlusion of: ClO_3^-	Ba^{++} in excess: 5.84 g eq./100 moles of $BaSO_4$
Occlusion of: ClO_3^-	SO_4^{--} in excess: 2.27 g eq./100 moles of $BaSO_4$
Occlusion of: MnO_4^-	Ba^{++} in excess: 2.85 g eq./100 moles of $BaSO_4$
Occlusion of: MnO_4^-	SO_4^{--} in excess: 1.37 g eq./100 moles of $BaSO_4$

Schneider and Riemann[29] determined the coprecipitation of various ions with barium sulfate, which was precipitated at room temperature. Some results are summarized in Table 11-6. The effect of the kind of lattice ion in excess during the precipitation

TABLE **11-6. Coprecipitation of Ions with $BaSO_4$**

Amount of coprecipitated ion in millimoles per mole $BaSO_4$	NO_3^-	NO_2^-	Cl^-	Na^+
Excess of barium	146	102	27	41
Excess of sulfate	54	44	45	89

on the amount of coprecipitation of a given cation or anion is also observed when normal solid solution occurs, as is the case in the isomorphous system $BaSO_4$–$PbSO_4$. Kolthoff and Noponen[30] precipitated barium sulfate at room temperature from 3 *M* ammonium acetate solution, in which lead sulfate is fairly soluble. In one series of experiments barium nitrate was added from a buret to the sulfate–lead mixture and in another series sulfate was added to the mixture of barium and lead nitrates. In both sets of experiments the number of moles of sulfate was equal to that of the sum of barium and lead. With barium in excess during the precipitation the precipitate contained 4.8 per cent lead and 31.7 per cent with sulfate in excess.

Effect of conditions of precipitation on amount of coprecipitation

There are two extreme methods of precipitation, from homogeneous solution at higher temperature and by rapid mixing of relatively concentrated solutions of reactants at room temperature. Precipitation from homogeneous solution has been used mainly for the quantitative precipitation of cations. Unfortunately, no systematic studies have been made of the amount of coprecipitation when precipitation is made from homogeneous solution as compared to that when precipitation is made under normal analytical conditions, i.e., by adding precipitant slowly to a hot and well-stirred dilute solution of the ion to be precipitated. Under the usual analytical conditions the precipitate is formed relatively rapidly and much faster than by precipitation

[28] H. B. Weiser and J. L. Sherrick, *J. Phys. Chem.* **23**, 205 (1919).
[29] F. Schneider and W. Riemann, *J. Am. Chem. Soc.* **59**, 359 (1937).
[30] I. M. Kolthoff and G. E. Noponen, *J. Am. Chem. Soc.* **60**, 508 (1958).

from homogeneous solution. Still, under both conditions solid solution of contaminants usually occurs. For example, Chang (Ref. 20) found that barium sulfate precipitated at 100°C from homogeneous solution contained 2 to 3 per cent occluded water.

Precipitates formed by rapid mixing of relatively concentrated solutions of reactants are usually badly contaminated. The crystals are highly imperfect and on aging subject to repeated recrystallizations. This is easily illustrated by the results of a series of experiments in which homogeneous mixed crystals of barium and lead sulfate were obtained.[31]

To a mixture of 25 ml of 0.11 *M* barium nitrate and 10 ml of 0.1 *M* lead nitrate was added 25 ml of 0.1 *M* sodium sulfate. The fresh precipitate contained 68 per cent of the lead in the precipitate. This amount is much greater than corresponds to equilibrium between the solid and the supernatant liquid immediately after precipitation. Upon shaking of the suspension the lead content of the solid decreased continuously, until after a month of shaking it became equal to 13.7 per cent of the total amount of lead added. This final state corresponds to homogeneous distribution of the lead between the solid and the solution. This was proved by reaching the equilibrium from the other side. To 25 ml of 0.11 *M* barium nitrate was added 25 ml of 0.10 *M* sodium sulfate. Immediately after the precipitation 10 ml of 0.10 *M* lead nitrate was added and the mixture shaken for various periods of time. After 2½ months the lead content became practically equal to that of the precipitate, which originally contained coprecipitated lead. Some representative figures are given in Table 11-7.

TABLE 11-7. **Distribution of Lead between Barium Sulfate and Solution. Experiments Carried out at Room Temperature (25°C)**

Time of shaking	5 min	1 hr	4 hr	18 hr	24 hr	4 days	1 wk	1 month	2½ months
Total lead[a] in ppt., %	67.6	59.0	—	—	32.0	24.1	17.3	13.7	—
Total lead[b] in ppt., %	9.4	16.7	20.4	20.9	20.5	19.3	16.2	14.4	14.1

[a] Lead was coprecipitated.
[b] Lead was added after precipitation of barium sulfate.

Similarly, the same amount of potassium permanganate was found occluded in the form of a solid solution in barium sulfate when the permanganate was present during the precipitation or added after the precipitation, and the precipitates were aged for longer periods of time.[32]

1. To a mixture 26 ml of 0.3 *M* barium nitrate and 25 ml of 0.5 *M* potassium permanganate was added 25 ml of 0.3 *M* sodium sulfate at room temperature.
2. Same as 1, but the permanganate was added after the precipitation.

TABLE 11-8. **Distribution of $KMnO_4$ between Barium Sulfate and Solution**

Time of shaking	10 min	1 hr	1 day	7 days	30 days
$KMnO_4$ in mg per gram of $BaSO_4$					
1	19.3	17.2	11.5	9.2	—
2	—	9.6	11.3	8.1	8.3

[31] I. M. Kolthoff and G. E. Noponen, *J. Am. Chem. Soc.* **60**, 197 (1938).
[32] I. M. Kolthoff and G. E. Noponen, *J. Phys. Chem.* **42**, 237 (1938).

Quite generally, it is found that, as a result of repeated recrystallizations a pronounced purification occurs when a precipitate formed from concentrated solutions at room temperature is aged by digestion at higher temperatures. As an illustration, Table 11-9

TABLE **11-9. Coprecipitation of Iodate with Calcium Oxalate under Various Conditions**

Method of Precipitation and Treatment	*Average Size of Crystals, μ*	*Millimoles Iodate in 1 g of Calcium Oxalate Monohydrate*
1. Oxalate to excess calcium; room temp.	0.5	0.017
2. Calcium to excess oxalate; room temp.	0.5	0.011
3. Oxalate to excess calcium; 100°C	1	0.044
4. Calcium to excess oxalate; 100°C	1	0.028
5. Extreme dilution[a]; calcium in excess; room temp.	ca. 2	0.075
6. Extreme dilution; oxalate in excess; room temp.	2	0.09
7. Extreme dilution; calcium in excess; 100°C	5	0.06
8. Extreme dilution; oxalate in excess; 100°C	4–10	0.017
9. As in 1, but digested for 20 hr	ca. 2	0.0012
10. As in 2, but digested for 20 hr	ca. 2	0.001
11. As in 4, but digested for 20 hr	1–2	0.004
12. As in 7, but digested for 20 hr	ca. 5	0.03

[a] Precipitation at "extreme dilution" was carried out in the following way: The calcium chloride and ammonium oxalate solutions were added simultaneously from two burets to 60 ml of water containing the potassium iodate. Either the calcium or the oxalate was kept 1 ml in excess during the entire addition.

lists the amount of iodate coprecipitated with calcium oxalate precipitated under different conditions.[33]

As a matter of fact, Njegovan and Marjanovic[34] recommend for quantitative gravimetric determinations of sulfate or of barium to precipitate by mixing concentrated solutions of reactants at room temperature and then age at higher temperatures. From the analytical viewpoint the method has limited applicability, even though on aging a drastic purification occurs. The following examples show the gradual removal of adsorbed and occluded sodium sulfate from 2 g of barium sulfate when this was allowed to stand in the mother liquor at 20°C.[35]

Coprecipitated Na_2SO_4 mg	22.6	19.0	17.4	14.0	10.4	8.2	7.6
Time of standing	15 min	3 hr	18 hr	2 days	48 days	5 months	7 months

[33] I. M. Kolthoff and E. B. Sandell, *J. Phys. Chem.* **37**, 443, 459 (1933).

[34] V. Njegovan and V. Marjanovic, *Z. Anal. Chem.* **73**, 271 (1928); **74**, 191 (1928); **93**, 353 (1933); **108**, 33 (1937); V. Njegovan, *Anal. Chim. Acta* **5**, 55 (1951).

[35] J. W. Mellor, *A Treatise on Quantitative Inorganic Analysis*, Griffin, London, 1913, p. 614, quoted by T. B. Smith, *Analytical Processes*, E. Arnold, London, 2nd ed., 1940, p. 53.

The beneficial effect of aging upon the removal of nitrate coprecipitated with lead sulfate was shown by Kolthoff and Halversen.[36] Ten milliliters of 0.1 *M* sodium sulfate were added quickly to 10.5 ml of 0.1 *M* lead nitrate at room temperature. The fresh precipitate contained 0.64 per cent coprecipitated nitrate. Upon aging at 95°C in the supernatant liquid, which was made 0.01 *N* in nitric acid, the percentage of nitrate in the precipitate was 0.12 after 1 hr, 0.04 after 6 hr, and 0.01 after 24 hr.

The rule that aging under favorable conditions results in the formation of a much purer precipitate does not hold when:

1. The contaminant forms a solid solution in the host crystal (precipitate). In such a case the final precipitate obtained upon drastic aging contains the mixed-crystal-forming contaminant in distribution equilibrium with the solution. The mixed-crystal-forming contaminant could be more or less completely eliminated if the supernatant liquid were removed after the precipitation and the aging carried out in a medium free of excess precipitant.

2. The contaminant slowly forms a chemical compound with the original precipitate. This rare case was met with in the aging of ferric hydroxide in an ammoniacal medium in the presence of several divalent cations.[37] Apparently, these cations are able to form ferrites with ferric hydroxide even at relatively low temperatures. Some typical figures are given in Table 11-10, in which the precipitate was formed at room tempera-

TABLE **11-10. Change of Coprecipitation of Divalent Metals with Ferric Hydroxide Upon Aging at 90°C**[a]

Time of Aging at 98°C	*Zinc Coppt.*, %	*Nickel Coppt.*, %	*Cobalt Coppt.*, %	*Magnesium Coppt.*, %	*Calcium Coppt.*, %
Fresh	18.3	14.3	17.4	15.1	12.9
15 min	24.1				
30 min		37	33		
5 hr	35	51	58.5	14.0	4.0
2 days	52.4			12.1	

[a] (The final metal concentration was 0.02 *M* and the supernatant liquid 0.90 *M* in ammonia and 0.077 *M* in ammonium chloride.)

ture and aged at 98°C. It is seen that the amounts of coprecipitated zinc, nickel, and cobalt rapidly increase upon aging at 98°C; the amount of coprecipitated magnesium decreases very slightly, whereas the amount of coprecipitated calcium decreases considerably. Apparently, the calcium does not form a ferrite under the experimental conditions.

Reprecipitation

Reprecipitation is a powerful means of obtaining purer precipitates. In the first precipitation only a small fraction of the total amount of the contaminating constituent is carried down. Upon dissolving and reprecipitating, the concentration of the contaminant in the solution is much smaller than the original. Again a fraction of the amount still present is carried down upon precipitation, which naturally is much smaller than the amount coprecipitated in the first precipitation. In many cases the amount of impurity carried down in the second precipitation is so small that it may be neglected. However, it should not be inferred that reprecipitation always leads to precipitates of adequate purity. Generally, the percentage adsorption is much greater at small

[36] I. M. Kolthoff and R. A. Halversen, *J. Phys. Chem.* **43**, 605 (1939).
[37] I. M. Kolthoff and L. Overholser, *J. Phys. Chem.* **43**, 767, 909 (1939).

than at large concentrations of the contaminant. The change of the amount adsorbed with the concentration is determined by the so-called adsorption isotherm. Empirically it has been found that within a certain range of concentration frequently the following relation exists between the amount adsorbed (x) by a given weight of adsorbent and the final concentration of the adsorbate in the solution (c):

$$x = \alpha c^{1/n}$$

in which α and n are constants. The absorption isotherm in general has the shape given in Fig. 11-5.

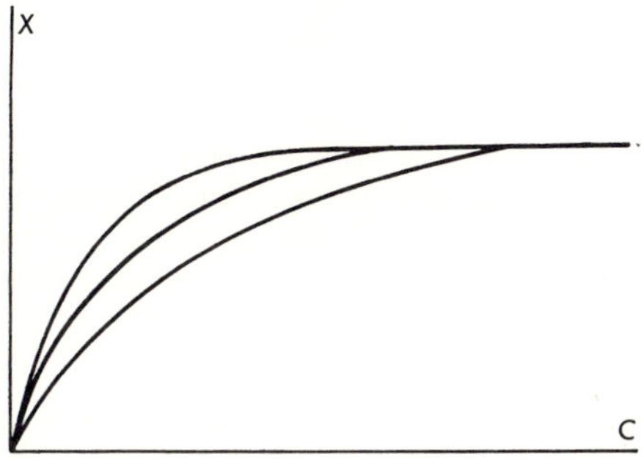

Fig. 11-5. Adsorption isotherms of solutions.

POSTPRECIPITATION

When a solution contains two ions, one of which is rapidly and the other slowly precipitated by the same reactant, the first precipitate is usually contaminated by the second one. This has been called *postprecipitation*.[38]

In the precipitation of calcium as oxalate in the presence of magnesium, a small amount of the latter is actually coprecipitated with the calcium oxalate. However, if the suspension is digested for a long time before filtration, some magnesium may precipitate as magnesium oxalate and contaminate the calcium oxalate precipitate.

The phenomena of postprecipitation are of extreme importance in the separation of metals from each other by means of hydrogen sulfide. In qualitative as well as quantitative analysis we make use of the fact that metals belonging to the "second group" (mercury, copper, etc.) form sulfides insoluble in dilute mineral acids, whereas those of the third group (zinc, manganese, etc.) are not precipitated from dilute acid medium by hydrogen sulfide. From the solubility product and the hydrogen-ion concentration, the amount of a given metal sulfide dissolved in a solution of a certain acidity saturated with hydrogen sulfide can be calculated. Thus we find that in a solution which is 0.1 N with respect to hydrogen ions the amount of dissolved zinc sulfide corresponds to about 5×10^{-5} M. Nevertheless, when hydrogen sulfide is passed through a solution that is 0.01 M in zinc and 0.1 N in hydrochloric acid, no precipitate appears until after a long time of standing. The original solution saturated with hydrogen sulfide is supersaturated with respect to zinc sulfide. The latter, however, separates very slowly, since the concentration of one of the reacting ions, the sulfide ion, is extremely small and of the order of 10^{-19} M. Finely divided substances in general will aid in overcoming the supersaturation and in this particular case will promote the precipitation of zinc sulfide. This actually happens, but the effect of most of them is extremely small. The metal sulfides of the second group have a very pronounced catalytic effect on

[38] I. M. Kolthoff and E. Pearson, *J. Phys. Chem.* **36**, 549 (1932). For a review of postprecipitation see I. M. Kolthoff and R. Moltzau, *Chem. Rev.* **17**, 293 (1935).

the speed of precipitation of zinc sulfide. If we have a mixture of copper or mercuric mercury and zinc in, let us say, 0.2 *N* sulfuric acid and pass in hydrogen sulfide, the copper or mercuric sulfide will precipitate first without carrying down any zinc (there is no coprecipitation). However, the precipitated sulfide promotes the separation of zinc sulfide and more of the latter is found in the precipitate the longer one waits before filtration. The strong promoting effect of the metal sulfides upon the precipitation of zinc sulfide is explained by the fact that they have strong adsorptive properties towards hydrogen sulfide or other sulfides.

The concentration of sulfide ions at the interface metal sulfide and solution is greater than in the bulk of the latter. Since the speed of precipitation is a function of the concentration of the reacting components, it may be expected that zinc sulfide will precipitate faster at the surface of the metal sulfide than in the bulk of the solution. Another way of explaining the promoting effect of the metal sulfides is the following: The particles have a pronounced tendency to adsorb sulfide ions, which are fixed on the surface. Hydrogen ions form the counterions in the double layer and may be replaced by other ions (p. 231). Zinc ions in the solution may replace the hydrogen ions in the double layer and give rise to the formation of an adsorbed layer of zinc sulfide. If the solubility product of the zinc sulfide has been exceeded, it will subsequently precipitate. It separates in an extremely finely divided form; each new layer of zinc sulfide formed promotes the precipitation of more material from the solution for the reasons mentioned.

It has been found that the promoting effect of mercuric sulfide upon the postprecipitation of zinc sulfide is extremely great; Table 11-11 clearly shows how the postprecipitation increases with the time of standing under hydrogen sulfide before filtration.

TABLE 11-11. **Postprecipitation of Zinc Sulfide with Mercuric Sulfide**[a]
(25 ml of 0.05 *M* mercuric chloride, 25 ml of 0.05 *M* zinc sulfate in 0.35 *N* sulfuric acid)

Time of Passing in Hydrogen Sulfide, min	*Time of Shaking under Hydrogen Sulfide Pressure before Filtration, min*	*Percentage of Total Zinc in Precipitate*
3	0	37.3
4	10	89.2
3	20	91.5
3	60	94.5
3	30	0.1 (without $HgCl_2$ at the same acidity)

[a] I. M. Kolthoff and R. Moltzau, *J. Phys. Chem.* **40**, 779 (1936); postprecipitation of zinc with bismuth sulfide, see I. M. Kolthoff and F. S. Griffith, *ibid.* **42**, 531 (1938); postprecipitation of nickel with sulfides of copper, mercuric mercury, and zinc, see I. M. Kolthoff and F. S. Griffith, *ibid.* **42**, 541 (1938); postprecipitation of ferrous sulfide with copper sulfide, see I. M. Kolthoff and F. S. Griffith, *J. Am. Chem. Soc.* **60**, 2036 (1938).

In making quantitative separations of various metals at different acidities with the aid of hydrogen sulfide, one must consider the possibility of the postprecipitation of a second sulfide promoted by the freshly precipitated primary sulfide.

PROBLEMS

1. Sulfate is to be determined in a solution containing calcium. Why would one expect an occlusion of calcium in the barium sulfate formed? Will this occlusion be greater if barium chloride is added slowly to the sulfate solution or if the latter is added to an excess of barium at the same rate?

2. Analyst A separates zinc from ferric iron by the addition of enough ammonia to keep all the former in solution as the zinc ammine ion. Analyst B prefers to use a large excess of ammonia and to add a few grams of ammonium chloride. Which is the better of the two procedures? Explain.
3. In the precipitation of barium in the presence of lead with an excess of sulfate, part of the lead is occluded by the barium sulfate in the form of mixed crystals. Lead sulfate is soluble in ammonium acetate. Explain why all the lead cannot be extracted from the precipitate with ammonium acetate.
4. An excess of standard barium hydroxide is added to a solution of pure ferric chloride, and the mixture is diluted to the mark in a volumetric flask. An aliquot part of the supernatant liquid is back-titrated with standard acid. Does this procedure give good results? Are they low or high?
5. Copper is separated from zinc in a medium of 0.5 N sulfuric acid by precipitation with hydrogen sulfide. What measures would you take to eliminate or minimize the post-precipitation of zinc? One analyst precipitates the copper by saturating the solution with hydrogen sulfide. Another analyst prefers to precipitate the copper from homogeneous solution, using thioacetamide as the hydrogen sulfide–generating agent. Under which conditions do you expect the least post-precipitation?
6. In a separation of copper from ferric iron and aluminum the trivalent metals are precipitated as hydrous oxides. Would you precipitate from acidic or alkaline medium? Explain.
7. In qualitative analysis it is often found that upon washing of nickel sulfide which has been precipitated with ammonium sulfide, the precipitate "runs through" the filter. What types of electrolyte would you add to the wash liquid to prevent peptization?
8. Upon drying of a barium sulfate precipitate at temperatures between 100 and 400°C, it was found to lose 0.3 per cent water. Upon ignition to 900°C, 0.6 per cent more water was given off. In what form was the water present in the original precipitate?
9. The value of D for CrO_4^{--} between water and barium sulfate has been determined by adding a known volume of 0.1 M sodium chromate to a mixture of 25 ml of 0.1 M barium nitrate and 25.5 ml of sodium sulfate. Describe how you would proceed further to find D.
10. Thorium B (radioactive lead) is coprecipitated with strontium sulfate. Calculate the percentage coprecipitation (a) when you have the ThB present in 0.1 M sodium sulfate solution and you precipitate 10 per cent of the sulfate with 0.1 M strontium chloride; (b) the same as (a) but here the thorium B is in the strontium solution and 10 per cent of the strontium is precipitated. Take $D = \lambda = 1$ and K_{sp} of $SrSO_4$ 10^{-8}.
11. A solution was 0.0244 M in barium perchlorate, 0.000244 M in lead perchlorate, and 0.00268 M in sulfamic acid. A volume of 41.0 ml was heated at 100°C and at various intervals the barium and lead in the precipitate were determined. Calculate λ and D from the following data of Chang (Ref. 20):

Ba ppt., %	19.2	33.7	36.8	37.0	51.1	56	63.2	72.6	83.4	89.6	91.2
Pb ppt., %	10.2	20.3	23.8	23.0	33.8	38.7	44.5	50.7	58.0	63.9	62.7
λ	0.53		0.59	0.56	0.58	0.60	0.59	0.55	0.48	0.45	0.41
D	0.48	0.50	0.55	0.51	0.48	0.47	0.45	0.40		0.24	0.16

CHAPTER 12 *Quantitative Separations. General Aspects; Precipitation Methods*

By separation is meant the process by which the constituents of a sample are isolated from each other. In many analyses, particularly those involving the complex products of nature and man, interference of other constituents prevents the direct determination of a given constituent. It then becomes necessary to separate the constituent in question before proceeding with the determination. Whenever possible, especially in simple chemical determinations, separations are avoided by utilizing differential complex formation or change in oxidation states to prevent interferences, but sooner or later such measures fail. Even in physical determination methods, e.g., in activation analysis, separations may be required.

A separation proper usually requires the formation of a new phase[1] and the subsequent separation of the new phase from the original one (usually liquid). In the separation of constituents A and B the new phase formed should ideally contain A to the exclusion of B, while the other phase should contain B to the exclusion of A. This initial step is the difficult and most essential one in a separation, and indeed is the only one in the whole process worthy of the name, for it is here that the real separation of A and B takes place. The separation can be brought about by a chemical reaction or a physical process or more often by a combination of both (e.g., the formation of a precipitate).

The first step in a separation (process I in the adjoining diagram) is physical in the sense that the analysis is almost always begun with a homogeneous phase (i.e., after the sample has been brought into solution) and a new phase is formed. Examples of an essentially physical process are the precipitation of a constituent by the addition of a solvent which is miscible with water and a distillation in which volatile A is separated from nonvolatile (or less volatile) B by varying the temperature or the pressure (or both) of the system.

The second and last step in a separation involves the segregation of the two phases of the heterogeneous system formed in the first step. Sometimes the two steps are virtually simultaneous. For example, in the electrolytic separation of a metal by deposition on an electrode the formation of the new phase and its segregation are concurrent; the formal segregation of the solid and liquid phase is accomplished by lifting the electrode out of the solution. At other times the second step becomes an operation in itself, as when the particles of a precipitate are segregated from the solution by filtration and washing. Generally speaking, the last step in a separation is a mechanical process, but it may be physicochemical or even chemical in nature.

The separation of phases without changing the state of aggregation is a mechanical process. In the separation of various solid phases use can be made of the difference in density (Chap. 22). The separation of a solid (precipitate) from a liquid is a very common

[1] The equivalent of forming a new phase is bringing a suitable phase into contact with the system and effecting a transfer of one of the constituents into this phase (extraction).

Stage I		Stage II		Stage III
A, B	process I → physical or chemical transformation (or both)	A + B	process II → mechanical, physical, rarely chemical	A/B
Homogeneous system (usually an aqueous solution in inorganic analysis)		Heterogeneous system		Phases containing A and B separated

operation in analytical chemistry and is discussed in detail in Chapter 20. The separation of immiscible liquids from each other is usually effected in a separatory funnel. In extraction procedures a solvent immiscible with water is added to extract a given constituent (process I in the diagram). Process II is then the simple phase separation in a separatory funnel.

In the separation of *constituents* we make use of suitable physical or chemical properties of the particular constituents present. For our purpose, the following simple classification of separation methods will suffice[2]:

1. Methods involving precipitation and (less often) coprecipitation.
2. Chromatographic methods, including ion exchange (Chapter 13).
3. Liquid–liquid extraction (Chapter 14).
4. Volatilization methods (Chapter 15).

Recovery factor and separation factor

Two factors are of fundamental importance in analytical chemical separations:

1. The completeness of recovery of the desired constituent in the sample which is carried through the separation procedure.

2. The extent of separation of the constituent from the other constituents of the sample.

The completeness of recovery of the constituent A in a given procedure is expressed by the recovery factor R_A:

$$R_A = \frac{Q_A}{(Q_A)_0} \tag{1}$$

where $(Q_A)_0$ = quantity of A in the sample taken and Q_A = quantity of A isolated. If desired, this quotient may be multiplied by 100 to give the percentage recovery.

The values of R_A required in quantitative analysis usually depend upon the relative amounts of A in the sample. When A is a major constituent, R_A should be $\geqq 0.999$; for a minor constituent (0.01–1 per cent), R_A should be > 0.99; for a trace constituent

[2] A number of methods which are quite specialized and which are not generally applied by inorganic analysts in day to day work are not treated in this book. These include ultrafiltration, dialysis, electrodialysis; thermal diffusion; gravitational, centrifugal, magnetic, and electrostatic methods; flotation, and others. See A. Weissberger (ed.), *Technique of Organic Chemistry*, especially part I, vol. 3, Wiley, New York, 1957, and also *Treatise on Analytical Chemistry* (I. M. Kolthoff and P. J. Elving with E. B. Sandell, eds.), Wiley-Interscience, New York, 1961.

(<0.01 per cent), a similar value of R_A is desirable, but for very low contents of A (e.g., 0.001–0.0001 per cent), $R_A \geqq 0.95$ may be acceptable.

The degree of separation of two constituents A and B is given by the *separation factor*. We shall define the separation factor for B, the undesired constituent, with respect to A, as the value which multiplied by the original ratio of B to A gives the final ratio:

$$\frac{Q_B}{Q_A} = S_{B/A} \frac{(Q_B)_0}{(Q_A)_0} \tag{2}$$

Therefore,

$$S_{B/A} = \frac{(Q_A)_0}{(Q_B)_0} \times \frac{Q_B}{Q_A} = \frac{R_B}{R_A} \tag{3}$$

Since in quantitative work $R_A \sim 1$,

$$S_{B/A} \cong R_B \tag{4}$$

The degree of separation required depends upon

1. The original ratio of B to A, $(Q_B)_0/(Q_A)_0$.
2. The permissible value of Q_B/Q_A, which is determined by the extent of interference of B in the determination of A and the accuracy desired.

Suppose that in the final determination—whether gravimetric, volumetric, photometric, or other—B exerts the same effect as an equal weight of A, so that 1 mg of B will be counted as 1 mg of A. Table 12-1 may then be set up, in which the last two

TABLE **12-1.**

$(Q_A)_0/(Q_B)_0$	Q_B/Q_A	R_B	*Relative Error in* A, %	*Absolute Error in* A, %
$0.1/99.9 = 0.001$	0.05	5×10^{-5}	5	0.005
$1/99 = 0.01$	0.02	2×10^{-4}	2	0.02
$10/90 \sim 0.1$	0.01	0.001	1	0.1
$50/50 = 1$	0.002	0.002	0.2	0.1
$90/10 \sim 10$	0.001	0.01	0.1	0.1
$99/1 \sim 100$	0.001	0.1	0.1	0.1

columns give the relative and absolute errors in A which should not be exceeded in accurate work, corresponding to various ratios of A to B in the sample. The third column contains the values of the separation factors for B required to provide the specified accuracy for various initial ratios of A to B. For constituents making up more than 10 per cent of the sample, the separation factor for the undesired constituent need not be much less than 0.001.

It might be thought that the modest requirements for the recovery and separation of major constituents could be met without great difficulty. They can indeed be met for most elements, but not always easily. In precipitation separations, R_A values of $\sim$0.999 can be reached, but not always easily exceeded. An initial concentration of 1 g of A per liter (0.01-0.03 M) is common in precipitation separations. If $R_A = 0.999$, not more than about 1 mg of A may remain in a liter of solution. Often the solubility of compounds of B under the conditions will be such that B will not be precipitated at like concentrations. This does not, however, mean that $R_B = 0.000$, because B may be coprecipitated with A. In liquid–liquid extraction, R_A can be given any value within reason, but, other factors remaining the same, an increase in R_A also increases R_B. These problems are considered later.

INORGANIC PRECIPITATION SEPARATIONS

The course of systematic separations

The separations required in the analysis of a given inorganic sample depend on a number of factors, of which the most important are the composition of the sample, the elements to be determined, and the type of the final determination, i.e., whether gravimetric, volumetric, spectrophotometric, or other. A wide and thorough acquaintance with the analytical reactions of the elements is required for the successful solution of complex separation problems. This is the most difficult field of chemical analysis and calls for the services of a professional.

A general scheme, the fruit of generations of analytical chemical labors, is available for the analysis of inorganic materials in which a large number of elements must be determined. It involves precipitation separations with inorganic reagents. In an elementary text such as this it would not be fitting to enter into details, but an abbreviated outline of the procedure may be given. The classic work of Hillebrand et al. supplies a full treatment.[3] After the sample has been decomposed, its hydrochloric (or other mineral acid) solution is evaporated to render the bulk of the silica insoluble. Niobium, tantalum, and tungsten will also be precipitated almost completely as the respective acids; various elements can contaminate the precipitate. The filtrate from these elements (the "acid group") is treated with hydrogen sulfide in approximately 0.3 N mineral acid solution to throw down more than a score of elements whose sulfides are insoluble in the acid medium (p. 258). The filtrate from the sulfides is treated with nitric acid to oxidize iron(II) and then with ammonia to precipitate trivalent and quadrivalent metals and also Be and U, as well as such amounts of Nb and Ta as may still be present (see p. 254). Precipitation with ammonium sulfide follows (Mn, Co, Ni, Zn, Tl). Finally, precipitations with ammonium oxalate and phosphate separate calcium (strontium) and magnesium. These separations are, for the most part, group separations and further separations must be made within the groups. Precipitation methods may again be applied, with more or less selective precipitants.[4] Newer separation methods, such as ion exchange and extraction, may also be utilized within the groups. Depending on the composition of the sample, abbreviations and modifications of the group separations can be introduced. Silicate rocks and many silicate minerals, as well as artificial materials of similar composition, are analyzed essentially according to the general scheme when the highest accuracy is required. Sulfide separations are of little importance for silicates because they are composed predominantly of elements of atomic number less than 27.

In the following pages we shall consider some aspects of sulfide and hydroxide separations (beginning with the latter for didactic reasons) to illustrate the underlying principles and to touch on some of the problems that arise.

Separation of metals by hydroxide precipitation

The solubilities of metal hydroxides (or hydrous metal oxides) in aqueous solution, as determined by the acidity, vary over such a wide range that useful separations of metals can be made by adjusting the hydrogen-ion concentration to a value that will allow quantitative precipitation of some while leaving others in solution. The pH at

[3] W. F. Hillebrand, G. E. F. Lundell, H. A. Bright, and J. I. Hoffman, *Applied Inorganic Analysis*, Wiley, New York, 2nd ed., 1953. The behavior of all elements, at least as far as they are known, with the common analytical reagents is given in G. E. F. Lundell and J. I. Hoffman, *Outlines of Methods of Chemical Analysis*, Wiley, New York, 1938.

[4] Organic precipitants of nonselective nature are occasionally used to separate a large number of elements as a group. For example, cupferron is useful for precipitating and separating relatively small amounts of Fe, Ti, Zr, V, Sn, Nb, and Ta from Al; 8-hydroxyquinoline precipitation in ammonical solution separates most metals from the alkali metals. Thioacetamide precipitates metal sulfides.

which various metals begin to precipitate as the acidity of a solution is reduced is indicated in Fig. 12-1. The values given are only approximate for a number of reasons. Obviously the pH of incipient precipitation will depend on the metal-ion concentration, being higher the lower the metal concentration. For example, zinc hydroxide begins to precipitate at pH 5.8 from 0.25 *M* zinc sulfate solution, at 6.17 from 0.05 *M* and at 6.5 from 0.01 *M* solution. The temperature and rate of attainment of equilibrium, as well as other precipitation variables, are also important factors. Moreover, the pH of precipitation depends upon the kind of anion present in solution. In general, if ammonium or alkali hydroxide is added to a metal solution, the hydroxide precipitated is not pure but is contaminated by anions as a result of the formation of basic salts[5] or adsorption of anions and is not the equivalent of the pure hydroxide. The higher the charge of the anion, the greater the tendency for coprecipitation. Sulfate is carried down to a larger extent than chloride, and chloride usually to a larger extent than perchlorate or nitrate.

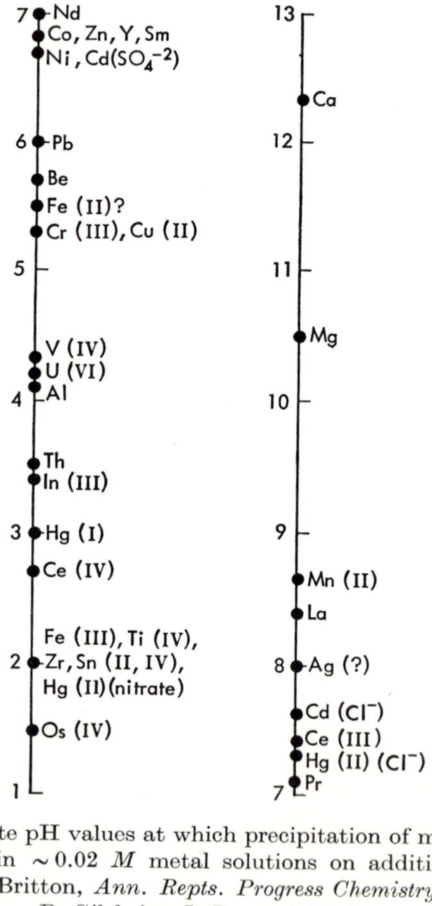

Fig. 12-1. Approximate pH values at which precipitation of metal hydroxides begins at room temperature in ~0.02 *M* metal solutions on addition of NaOH solution. Mostly from H. T. S. Britton, *Ann. Repts. Progress Chemistry* (Chem. Soc. London) **40**, 43 (1943). See, further, R. Gilchrist, *J. Research Natl. Bur. Standards* **30**, 89 (1943).

[5] With some metals the precipitate obtained with hydroxide is a basic salt approaching stoichiometric composition. Thus copper in sulfate solution gives a precipitate corresponding to $CuSO_4 \cdot 3\,Cu(OH)_2$. The formation of such basic salts invalidates any treatment of solubility based on the solubility product of the more or less pure hydroxide.

In spite of its approximate character, a compilation of pH precipitation values such as in Fig. 12-1 is not devoid of utility in making rough predictions of separability. Thus hydroxide separation of iron(III) from magnesium and calcium, but not from indium(III), appears feasible.

A closer examination of hydroxide separations requires a knowledge of solubility-product constants of the precipitates (Table A-10) and, when necessary, of constants of hydrolysis reactions resulting in the formation of charged hydroxo metal-ion species.

Consider the separation of two cations M^{m+} and N^{n+}, M forming the less soluble hydroxide. From the known values of the solubility-product constants of the hydroxides we calculate (1) the pH at which a specified fraction of M at a given initial concentration is precipitated, and (2) whether the solubility product constant of $N(OH)_n$ is exceeded at this pH. A quantitative recovery of M on the macro scale requires that its recovery factor, R_M, equal or exceed 0.999. To allow a margin for error in adjustment of the pH and in the value of $(K_{sp})_M$, and for other uncertainties, a more realistic R_M to aim for is 0.9999. If the predominant species of M in the solution at the end of the precipitation is M^{m+}:

$$[M]_{soln} \cong [M^{m+}] = C_M(1 - R_M) = 1 \times 10^{-4} C_M = \frac{(K_{sp})_M}{[OH^-]^m} \tag{5}$$

where C_M is the initial concentration of M in the solution. Suppose the minimum value of C_M in macroanalysis is 0.01 M. Then

$$[M]_{soln} \cong [M^{m+}] = 1 \times 10^{-6} = \frac{(K_{sp})_M}{[OH^-]^m}$$

and

$$[OH^-]_{M=10^{-6}} = \left\{\frac{(K_{sp})_M}{10^{-6}}\right\}^{1/m} = \{10^6(K_{sp})_M\}^{1/m} \tag{6}$$

Knowing the concentration of N ($=C_N$) we can determine whether any $N(OH)_n$ is precipitated at this hydroxyl-ion concentration. Precipitation of N will not occur if

$$C_N \cong [N^{n+}] < \frac{(K_{sp})_N}{[OH^-]^n_{M=10^{-6}}} < \frac{(K_{sp})_N}{\{10^6(K_{sp})_M\}^{n/m}} \tag{7}$$

If we stipulate that $C_N \cong [N^{n+}]$ shall not exceed 1 M, the condition for separability becomes

$$(K_{sp})_M < 10^{-6}(K_{sp})_N^{m/n} \tag{8}$$

An assumption made in the preceding discussion requires examination—that the concentration of metal in solution can be set equal to the simple cation concentration, $[M^{m+}]$ and $[N^{n+}]$. Actually, the simple cations can hydrolyze[6] and form charged species MOH^{m-1}, $M(OH)_2^{m-2}$, etc. Neglect of such species may give calculated solubilities smaller than the true values. The precipitation of ferric hydroxide may be taken as an example, The following equilibrium constants are involved:

$$[Fe^{3+}][OH^-]^3 = K_{sp} = 2 \times 10^{-39}$$

$$\frac{[FeOH^{++}][H^+]}{[Fe^{3+}]} = 9 \times 10^{-4}$$

$$\frac{[Fe(OH)_2^+][H^+]^2}{[Fe^{3+}]} = 5 \times 10^{-7}$$

$$\frac{[Fe_2(OH)_2^{4+}][H^+]^2}{[Fe^{3+}]^2} = 1.1 \times 10^{-3}$$

$$[Fe(OH)_3]_{satd} < 2 \times 10^{-9}$$

[6] Alternatively expressed, the H_2O ligands of the hydrated cation ionize to give OH^- bound to the metal.

The concentrations of iron species in equilibrium with ferric hydroxide at three different pH's of interest are (see Fig. 7-6, p. 140):

pH	$[Fe^{3+}]$	$[Fe(OH)^{++}]$	$[Fe(OH)_2^+]$	$[Fe_2(OH)_2^{4+}]$	$[Fe(OH)_3]$
3	2×10^{-6}	1.8×10^{-6}	1×10^{-6}	4×10^{-9}	$<2 \times 10^{-9}$
4	2×10^{-9}	2×10^{-8}	1×10^{-7}	4×10^{-13}	$<2 \times 10^{-9}$
7	2×10^{-18}	2×10^{-14}	1×10^{-10}	4×10^{-25}	$<2 \times 10^{-9}$

The assumption that the total concentration of iron in solution can be set equal to $[Fe^{3+}]$ is not badly in error at pH 3. Actually a little more than one half of the dissolved iron is present as $Fe(OH)^{++}$ and $Fe(OH)_2^+$, but this discrepancy is absorbed in the factor of safety we have adopted, whereby an R_M of 0.999 is assured by aiming for an R_M of 0.9999. At pH 3 the concentration of dissolved iron is $\sim 5 \times 10^{-6}$ M or about 0.3 mg of Fe per liter, if the constants used are reliable. At higher pH's, other iron species outweigh Fe^{3+}. Thus at pH 4 the concentration of $FeOH^{++}$ exceeds that of Fe^{3+} by a factor of ~ 10, and the concentration of $Fe(OH)_2^+$ exceeds that of Fe^{3+} by ~ 100. At pH 7, $[Fe^{3+}]$ is much smaller than $[Fe(OH)_2^+]$ and $[FeOH^{++}]$, but now the ionic species all have such low concentrations that they can be considered negligible from the analytical standpoint. Dissolved $Fe(OH)_3$ may be expected to be comparable in concentration to ionic iron species, chiefly $Fe(OH)_2^+$, at pH ~ 7. The dimer $Fe_2(OH)_2^{4+}$ can be neglected at all the pH's considered.

Results of calculations, such as the foregoing, based on solubility-product constants taken from the literature, should be looked upon as approximate only. The solubility of a metal hydroxide depends, as pointed out in an earlier chapter, on the age of the precipitate, the manner of its formation and other factors, so that recorded K_{sp} values (corrected to concentration constants if necessary) may not correspond at all closely to the K_{sp} of an analytical hydroxide precipitate. From Table A-10 one sees that the K_{sp} of a metal hydroxide can vary over a tenfold or even hundredfold range, depending on the state of the precipitate. The accuracy of older K_{sp} values is often questionable because the formation of metal–hydroxo ions was neglected. Moreover, it must be emphasized that fulfillment of the solubility requirements does not alone assure a successful separation. Coprecipitation of N with M may invalidate conclusions that separation is possible. Especially when the pH of a solution is being increased by addition of an external reagent, precipitation of the more soluble hydroxide may occur locally and serious entrainment of N in $M(OH)_m$ may occur. This danger is lessened by homogeneous precipitation (p. 256), but coprecipitation may still take place. Kinetic effects (rate of attainment of equilibrium) may also introduce uncertainties. Generally the greater the difference in K_{sp}'s, the more certain will be the separation. An estimate of the required precision in pH adjustment may be made from a graph such as that in Fig. 12-2. Calculations of the type considered here, serve as a useful guide to the analyst, particularly in establishing minimum requirements, but they always need experimental verification.

Ammonia precipitation. One of the commonest operations in classical analysis is precipitation of metal hydroxides with ammonium hydroxide. The ammonia–ammonium salt buffer system conveniently allows adjustment of the pH of a solution from an approximately neutral to weakly or moderately basic reaction. An obvious advantage of ammonia as a precipitant when metal hydroxides are to be filtered off, washed, and ignited for weighing as the oxides is the volatility of its salts at low temperatures. Ammonia serves as a group precipitant for metals prior to other separations. In applied analysis, precipitation at pH 6.5–7.0 in hot solution (boiling for a few minutes) separates Fe(III), Al, Cr(III), Ti, and Zr, as well as a number of other

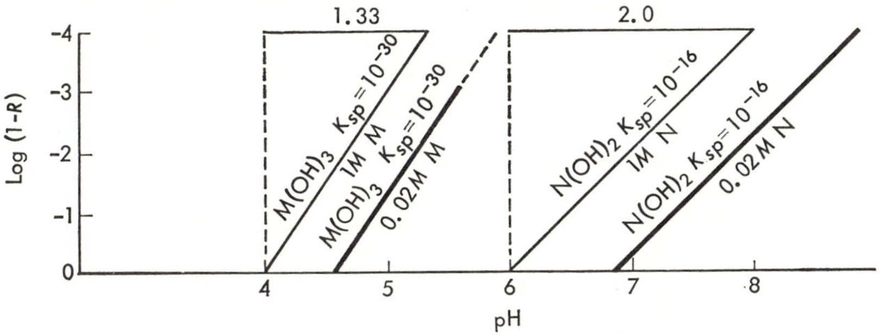

Fig. 12-2. Theoretical precipitation lines of $M(OH)_3$, $K_{sp} = 1 \times 10^{-30}$, and $N(OH)_2$, $K_{sp} = 1 \times 10^{-16}$, in which the log of the fraction of unprecipitated metal (1-R) is plotted against pH. For each metal, two lines are shown, corresponding to original concentrations of 1 and 0.02 M metal ion. All metal is assumed to be present in the solution as M^{3+} and N^{++}. Note that to span the range 0 to 0.9999 of metal precipitated, a pH increase of 1.33 is needed for M^{3+} and 2.0 for N^{++}.

metals, from Mg, Ca, Sr, and Mn.[7] If the trivalent or quadrivalent metals are present in excess, P, As, and V will be quantitatively precipitated as phosphate, arsenate, and vanadate. The alkaline earth metals, Mg and Mn, will be precipitated as phosphates if not enough of the trivalent and quadrivalent metals are present to precipitate all phosphate.

The separation of aluminum and iron from cobalt, zinc, and copper at pH ~7 is not satisfactory, as might be expected from the pH at which the latter elements begin to be precipitated (Fig. 12-1). The formation of ammine complexes increases the solubility of these metals under the precipitation conditions, but at best they remain borderline cases. The precipitation of nickel apparently does not occur and its coprecipitation is not serious. Table 12-2 illustrates the extent of carrying down of these divalent metals with aluminum hydroxide. A high concentration of ammonium ion improves the separation of Al and Fe from divalent metals (see Chap. 25). A large excess of ammonia further decreases the carrying down of Zn and Cu (ammine formation), but is permissible only when iron alone is present, because aluminum hydroxide dissolves when the pH is too high (p. 139).

The coprecipitation of the alkaline earths and magnesium with the trivalent metal hydroxides is very small if the solution contains sufficient ammonium salt, and a reprecipitation is not always necessary. The presence of gram amounts of alkali and alkaline earth metals makes a double precipitation necessary. Contamination of the precipitate with manganese is slight (Table 12-2), but the presence of P_2O_5 (Al and Fe in excess) seems to increase it. Manganese(II) shows little tendency to be oxidized to Mn(IV) at pH ~7, but at higher pH values oxidation occurs and manganese is precipitated.

The ammonium hydroxide used must be free of silica (taken up from glass bottles) and of carbonate (else the alkaline earths may be precipitated). Fluoride, tartrate, citrate, etc., must be absent from solutions in which hydroxide separations are made; they prevent or impede hydroxide precipitations by complexing heavy metal cations.

Other hydroxide or basic salt precipitations. Various methods have been worked out for the separation of metals at controlled pH in weakly acidic buffered solutions. Most of these call for the presence of an organic acid, whose anion is carried down by the precipitate, which is of indefinite composition but may consist to a large extent of a

[7] For procedures see W. F. Hillebrand, G. E. F. Lundell, H. A. Bright, and J. I. Hoffman, *Applied Inorganic Analysis*, Wiley, New York, 2nd ed., 1953.

TABLE 12-2. **Contamination of $Al(OH)_3$ by Divalent Metals**[a,b]

Foreign Element, g	*No. of Precipitations*	*Amount of Element in* $Al(OH)_3$, g
0.05 Ni	1	0.0006
0.05 Ni	2	None
0.05 Co	1	0.0041
0.05 Co	2	0.0012
0.05 Cu	1	0.0211
0.05 Cu	2	0.0077
0.05 Zn	1	0.0216
0.05 Zn	2	0.0108
1.0 Mn[c]	1	0.0017
1.0 Mn	2	0.00002

[a] From G. E. F. Lundell and H. B. Knowles, *J. Am. Chem. Soc.* **45**, 676 (1923).

[b] 200 ml of solution containing 0.10 g of Al and 5 g of NH_4Cl treated with ammonia at 100°C to give a pH of 6.5–7.5, boiled for 1–2 min before filtration.

[c] A single joint precipitation of 0.05 g each of Al, Fe, and Ti in the presence of 0.05 g of Mn gave a precipitate containing less than 0.00002 g of Mn. A double precipitation of 0.05 g each of Al, Ti, Fe, and P_2O_5 in the presence of 0.05 g of Mn gave a precipitate containing 0.00024 g of Mn.

basic salt. In the basic acetate method the precipitation is made in an acetic acid–acetate buffer of somewhat indefinite pH but usually at 5–5.5. Fe(III) is separated from Cu, Zn, Ni, Co, Mn, and most other divalent metals; the complete precipitation of aluminum is uncertain unless iron is present. Formic acid and succinic acid buffer systems have also been proposed. Precipitation at pH 3.8 in an ammonium benzoate–benzoic acid buffer provides quantitative separation of Al, Fe(III), and Cr(III) from most of the divalent metals; a double precipitation is required, especially in the presence of Zn and Co.[8]

A number of homogeneous precipitation methods are available for the separation of trivalent (and quadrivalent) metals from the divalent. By heating an acidic sample solution containing succinate, urea, and ammonium salt for an hour or two at the boiling point, the pH is gradually brought to a value of 4.2–4.6 and aluminum is precipitated in a dense, easily filterable form.[9] A single precipitation separates 0.1 g of aluminum satisfactorily from an equal weight of Ni and Co. A reprecipitation is required in the presence of zinc. Aluminum can be separated from iron(III) after reducing the latter with phenylhydrazine, and from copper(II) if it is reduced with hydroxylamine hydrochloride or sulfite.

Precipitation of hydroxides in approximately neutral solution can be brought about by utilizing reactions such as the following to remove hydrogen ions:

$$S_2O_3^{--} + 2\,H^+ \rightarrow SO_2\uparrow + S + H_2O$$

$$NO_2^- + H^+ \rightarrow HNO_2\uparrow \quad (\text{also } 3\,HNO_2 \rightarrow 2\,NO + NO_3^- + H^+ + H_2O)$$

$$IO_3^- + 5\,I^- + 6\,H^+ \rightarrow 3\,I_2 + 3\,H_2O$$

[8] I. M. Kolthoff, V. A. Stenger, and B. Moskovitz, *J. Am. Chem. Soc.* **56**, 812 (1934); A. A. Smales, *Analyst* **72**, 14 (1947). Note that aluminum is precipitated at a lower pH than when ammonia is the precipitant.

[9] H. H. Willard and N. K. Tang, *Ind. Eng. Chem., Anal. Ed.* **9**, 357 (1937).

An ion such as Al^{3+} will then hydrolyze in the hot, almost neutral, solution to give $Al(OH)_3$. The use of thiosulfate permits aluminum to be separated from iron. Ferric iron is reduced by SO_2 to ferrous. Even if the pH should rise to 7 (it is likely to remain a little below this), the concentration of Fe^{++} at the point of incipient precipitation of $Fe(OH)_2$ will be

$$[Fe^{++}] = \left(\frac{8 \times 10^{-16}}{10^{-14}}\right)^{1/2} = 0.3\ M$$

so that precipitation of iron is unlikely. Experimentally this conclusion is verified. The sulfur formed improves the filterability of the aluminum hydroxide. As ordinarily carried out, this method does not seem to give quite complete precipitation of aluminum.

Many different reagents have been suggested for precipitation of metal hydroxides at controlled pH in a basic medium. Weak organic bases such as aniline, pyridine, hexamethylenetetramine, and phenylhydrazine find more or less application.[10]

Slightly soluble metal oxides and carbonates such as ZnO, HgO, $CaCO_3$, and $BaCO_3$ are occasionally used to adjust the pH of a solution roughly in hydroxide separations. For example, if zinc oxide is added in excess to a solution (a) 0.02 N in HCl, and to one (b) 0.2 N in HCl, the approximate $[OH^-]$ values at equilibrium will be

$$\text{(a)} \qquad [OH^-] = \left(\frac{K_{sp}}{[Zn^{++}]}\right)^{1/2} = \left(\frac{8 \times 10^{-18}}{10^{-2}}\right)^{1/2} = 2.8 \times 10^{-8} \text{ or pH} = 6.45$$

$$\text{(b)} \qquad [OH^-] = \left(\frac{8 \times 10^{-18}}{0.1}\right)^{1/2} = 9 \times 10^{-9} \text{ or pH} = 5.95$$

Thus by adjusting the mineral acid concentration of the solution to approximate constancy (closer than above) at the beginning, a known and roughly constant pH is obtained so that metals of suitable K_{sp}'s will be separated.

Oxidation–reduction and complex formation in hydroxide precipitations. A change in the oxidation state of a metal can sometimes be put to good use in hydroxide separations. Separation of aluminum from iron by reduction of the latter to Fe(II) has already been mentioned. Chromium remains in solution in hydroxide precipitations if first oxidized to Cr(VI), except for such small amounts as may be coprecipitated. Differences in the precipitability of Mn(II) and Mn(IV), Tl(I) and Tl(III), Ce(III) and Ce(IV) also find analytical utilization.

Complexing as a means of keeping certain metals in solution is applied to some extent in hydroxide precipitations. 2,2′-Bipyridine complexes iron so firmly after reduction to the ferrous state that hydroxide separation of Al, Be, Ti, and even Mg becomes possible. α-Picolinic acid has been suggested as a similar complexing agent. A useful method for the separation of aluminum and other metals precipitating at about the same alkalinity is by ammonia precipitation at pH ~ 8 in the presence of mercaptoacetic (thioglycolic) acid.[11] Iron remains in solution as, apparently, the red-purple ion $Fe(OH)_2(SCH_2COO)_2^{--}$, containing iron in the ferric state. Ferrous iron gives the same product by air-oxidation. The coprecipitation of iron amounts to about 0.1 mg of Fe_2O_3 when 40–50 mg of Al_2O_3 is precipitated from 200 ml of solution containing an equal weight of Fe_2O_3. Reprecipitation provides satisfactory separation of aluminum from larger amounts of iron.

[10] See E. A. Ostroumov, *The Application of Organic Bases in Analytical Chemistry*, Pergamon, New York. An excellent survey of the use of these and other precipitants for hydroxides, with special reference to the separation of aluminum, may be found in the chapter by H. Fischer, A. von Unruh, and F. Kurz in R. Fresenius and G. Jander, *Handbuch der Analytischen Chemie*, part III, vol. III, Springer, Berlin, 1942, p. 82.

[11] C. Mayr and A. Gebauer, *Z. Anal. Chem.* **113**, 200 (1938); R. A. Hummel and E. B. Sandell, *Anal. Chim. Acta* **7**, 308 (1952).

At a sufficient concentration of EDTA, the only metals precipitated by ammonia are Be, Ti, Sb(III), Sb(V), Sn(IV), and U(VI), and various separations can thus be made, as, for example, Be from Al.[12] In the presence of hydrogen peroxide and EDTA, thorium and cerium can be precipitated with ammonia and separated from many other metals.

Separation of metals by sulfide precipitation

Sulfide separations date back to the earliest days of chemical analysis and are still important notwithstanding the introduction of organic precipitants, to say nothing of other types of separation than precipitation.[13]

Precipitation by hydrogen sulfide from the solubility standpoint has already been discussed (p. 141). The wide range of values of sulfide solubility-product constants suggests that separations of metals should be possible by regulation of the acidity and this is borne out by experiment. The following classification of metals can be made on the basis of the acidity allowing quantitative precipitation.

1. Precipitation from strongly acidic solutions, $[H^+] > 0.1$. Metals of this group, usually called the hydrogen sulfide group, include:

a. Cu, Ag, Hg, Pb, Bi, Cd, Ru, Rh, Pd, and Os (copper subgroup).

b. As, Sn, Sb, Mo, Se, Te, Ge, Au, Pt, and Ir (arsenic subgroup, complex sulfide formers).

These metals, and some not listed, can in general be separated from those precipitated only in solutions less acidic than 0.1 *M* in hydrochloric acid. Coprecipitation (by various mechanisms) and postprecipitation (p. 245) present troublesome problems not easily solved in some of these separations. Tl, In, and Ga are coprecipitated with some members of this group. The coprecipitation of W and V with members of the arsenic subgroup can be prevented by addition of tartrate. With some exceptions (precipitation of arsenic in strong hydrochloric acid solutions, separation of Sn(IV) from As(III) and Sb(III) in oxalic acid medium), sulfide separations of metals within the subgroups are usually not attempted.

2. Precipitation from weakly acidic solutions. Zinc is precipitated at pH 2–3, and Ni and Co at pH 5–6 (separation from manganese in the absence of iron). Tl and In can also be precipitated in nearly neutral solution.

3. Precipitation from basic solutions. In practice, precipitation from alkaline medium follows removal of elements precipitated in acid solution. A useful separation in basic medium is that of iron as FeS in the presence of tartrate to keep Al, Ti, Cr, U, and similar elements (whose sulfides are hydrolyzed in aqueous medium) in solution. Ni, Co, and Mn accompany iron. Various separations, as of the copper subgroup from the arsenic subgroup, can be made by the use of alkaline sulfide or polysulfide to form complex sulfo anions of As, Sb, Sn, etc. A consideration of such separations lies beyond the scope of this book.

The hydrogen-ion concentration at which a specified recovery factor, R_M, of a divalent metal is obtained from a solution of an initial metal concentration C_M is given by the expression

$$[H^+] = \left\{\frac{1 \times 10^{-21}[H_2S]C_M(1 - R_M)}{(K_{sp})_{MS}}\right\}^{1/2} \qquad (1 \times 10^{-21} \text{ holds for } \mu = 1, 20°C) \quad (9)$$

if M^{++} is the only metal species in solution that needs to be considered. This proviso frequently does not hold, inasmuch as complex sulfide or hydrosulfide metal anions are likely to be present (p. 142); the molecular solubility of metal sulfides seems to be

[12] R. Pribil, *Komplexone in der chemischen Analyse*, VEB Deutschen Verlag der Wissenschaften, Berlin, 1961. The use of EDTA in other quantitative precipitations is also discussed.

[13] For a detailed treatment see E. Brennecke, *Schwefelwasserstoff als Reagens in der quantitativen Analyse*, Enke, Stuttgart, 1939.

small enough to neglect for most analytical purposes. If the complex constants are known, the above equation is readily modified to take other metal species into account. It must be borne in mind that equilibrium is slowly attained in heavy metal sulfide systems, so that calculations of this kind should be looked upon as giving values serving as a rough guide, and useful particularly for arranging experimental conditions for testing. Even if the solubility relations of two metal sulfides meet the conditions for separability (see metal hydroxides, p. 253), coprecipitation phenomena may vitiate the possibility unless a reprecipitation is made.

Thioacetamide is often used as a source of sulfide (homogeneous precipitation), which it gives by hydrolysis in hot aqueous solutions.[14]

Electroseparations

In Chapter 46 precipitation on the surface of an electrode is discussed and explained. A cation that is deposited as a metal on the surface of an electrode combines with electrons: $Ag^+ + e \rightarrow Ag$. Thus, in electroreductions, electrons take the place of a chemical reducing agent, the intensity of the reduction being determined by the applied voltage (potential of the cathode). By working at controlled cathode potentials successive precipitation of various metals may be obtained (comparable to differential precipitation).

From the viewpoint of quantitative separations it is advantageous that in electroprecipitations, and in electroreductions and oxidations in general, no specific chemical precipitants are introduced.

Electrolysis with a mercury cathode (Chap. 51) serves to separate such metals as iron, chromium, nickel, cobalt, zinc, cadmium, gallium, copper, tin, molybdenum, bismuth, silver, gold, and some of the platinum metals from aluminum, titanium, zirconium, phosphorus, vanadium, and uranium, which are not deposited from dilute sulfuric acid.[15] This method is chiefly of value in separating large amounts of the first-named metals preparatory to the determination of the last named. The process may be looked upon as akin to extraction with an immiscible solvent.

ORGANIC REAGENTS AS PRECIPITANTS

Organic substances find important uses in quantitative inorganic analysis[16]:

1. As precipitants in gravimetric analysis and for separations.
2. To yield reaction products with inorganic cations or anions that can be extracted from aqueous solution with an immiscible organic solvent. Usually these reaction products are insoluble and may be those utilized in precipitations. (See Chapter 14.)
3. As photometric reagents, forming colored or fluorescent products, which are often extractable by organic solvents and may be the same as in 2 (see Chaps. 14 and 61).
4. As ion-exchange resins (see Chap. 13).
5. As titrants in volumetric determinations, forming soluble complexes, which are usually charged chelates. The outstanding example is ethylenediaminetetraacetic acid

[14] E. H. Swift and F. C. Anson, *Advan. Anal. Chem. Instr.* **1**, 293 (1960).

[15] For a review see J. A. Page, J. A. Maxwell, and R. P. Graham, *Analyst* **87**, 245 (1962).

[16] The following books, among others, treat the use of organic reagents in inorganic analysis: F. J. Welcher, *Organic Analytical Reagents*, Van Nostrand, Princeton, N.J., 1947–1948; J. F. Flagg, *Organic Reagents Used in Gravimetric and Volumetric Analysis*, Wiley-Interscience, New York, 1948; W. Prodinger, *Organische Fällungsmittel in der quantitativen Analyse*, Enke, Stuttgart, 1957; A. I. Busev and N. G. Polianskii, *The Use of Organic Reagents in Inorganic Analysis*, Pergamon, New York, 1960; W. C. Johnson, *Organic Reagents for Metals*, Hopkin and Williams, London, 1955; W. C. Johnson, ed., *Organic Reagents for Metals and Certain Radicals*, vol. 2, Hopkin and Williams, London, 5th ed., 1964. D. D. Perrin, *Organic Complexing Agents*, Wiley-Interscience, New York, 1964. H. A. Flaschka and A. J. Barnard (eds.), *Chelates in Analytical Chemistry*, Dekker, New York, 1967.

(Chap. 41). Titrations can also sometimes be based on the determination of the organic portion of a washed precipitate, as of 8-hydroxyquinoline in a metal hydroxyquinolate (p. 277).

6. As auxiliary complexing agents in conjunction with other reagents (for examples see p. 257).

7. As reducing and (less often) oxidizing agents. For example, ascorbic acid is a useful strong reducing agent.

8. As indicators.

9. As volumetric standards.

The discussion that follows is limited to the use of organic reagents as precipitants, whether for determination or separation. A considerable number of organic precipitants serve well for the gravimetric determination of inorganic constituents, particularly the metals. They may provide greater selectivity (p. 265) than inorganic precipitants. Often the precipitate is very slightly soluble and has a definite composition after drying at a low temperature ($\sim 100°C$). The chemical factor is likely to be favorable. These characteristics are not universal, however. Some organic precipitates cannot be dried without decomposition. The organic precipitant may itself be only slightly soluble in water, and, even if added as a solution in an organic solvent, may contaminate the precipitate of the inorganic constituent. Often the precipitate can be filtered and washed readily, although there is a tendency toward bulkiness, which can, however, be counteracted by proper precipitation conditions (homogeneous precipitation is often useful). If the precipitate of a metal cannot be weighed as such, it can always be transformed into an inorganic weighing form. Some organic precipitants have the disadvantage of being unstable (e.g., cupferron). At times, the high cost of some organic reagents may limit their use.

Precipitation reactions of organic reagents

Two requirements must evidently be fulfilled for the formation of an "insoluble" product in the reaction of an organic reagent with an inorganic ion (or substance):

1. The intrinsic (molecular) solubility of the product must be very low.
2. The dissociability of the product must not be too great.

Suppose a metal cation M^+ is precipitated as MR by an organic reagent anion R^-:

$$M^+ + R^- \rightleftharpoons MR$$

The concentration of the metal in the saturated solution in equilibrium with MR is

$$S = [MR]_{sat} + [M^+] \tag{10}$$

(where $[MR]_{sat}$ is the intrinsic solubility of MR) if other species of M are assumed to be negligible in concentration. The dissociation constant of MR in aqueous solution is

$$\frac{[M^+][R^-]}{[MR]} = \frac{K_{sp}}{[MR]_{sat}} = K_{diss} \tag{11}$$

Therefore,

$$S = [MR]_{sat} + \frac{[MR]_{sat}K_{diss}}{[R^-]} = [MR]_{sat} + \frac{K_{sp}}{[R^-]} \tag{12}$$

It is seen from Eq. (11) that K_{sp} is determined by the intrinsic solubility of MR and its dissociation constant.

Many, but by no means all, uncharged metal-organic reagent compounds have such a small intrinsic solubility in water that the ionic concentration of metal determines the solubility, if other metal species can be neglected:

$$S \sim \frac{K_{sp}}{[R^-]} \tag{14}$$

Most of the organic precipitants used in inorganic analysis form either chelate or ion-association compounds, and these two classes will now be considered.

Chelate complexes. The greater number of organic metal precipitants are of the type HL (or H_2L, less often $H_3L, \ldots$), where H represents an acidic hydrogen that can be replaced by an equivalent of metal. The resulting uncharged product tends to be slightly soluble in water, if the metal is a heavy metal, or (sometimes) an alkaline earth. Occasionally the product is an electrovalent or heteropolar salt (see arsonic acids, p. 269), but the great majority are chelate complexes.

The nature of chelate complexes has already been touched upon (p. 103). Chelating precipitants contain groups with replaceable hydrogen atoms, such as

—COOH	carboxyl
—OH	hydroxyl
—SH	mercapto
—SO_3H	sulfonic

together with functional groups of basic character such as

—NH_2	amino
—NH—	imino
=N—	amino(cyclic)
>CO	carbonyl
>CS	thioketo

with which the reacting metal is coordinated to form a five- or six-membered ring. The reagent molecule thus occupies more than one coordination position of the metal. For example, the acidic hydrogen of the – OH group of 8-hydroxyquinoline is replaced by an equivalent of Mg, and magnesium is coordinately bound to the basic =N— group, each 8-hydroxyquinoline radical (ligand) forming a five-membered ring structure with the metal atom:

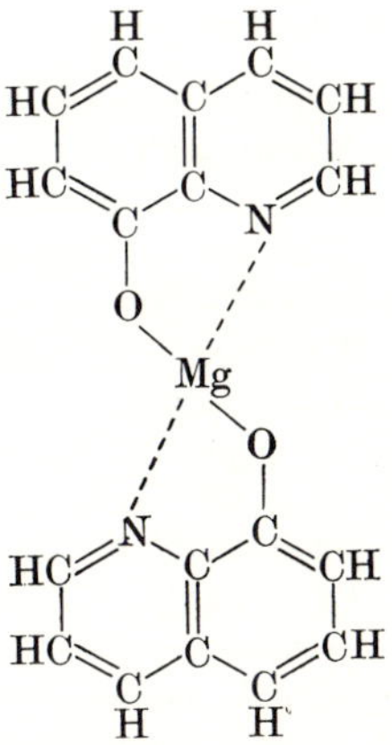

(--- or → is used to show a coordinate bond.) The formula as written is not quite correct, inasmuch as magnesium has a coordination number of 6, and only 4 positions are shown occupied. Actually, two molecules of water occupy the other two positions. The presence of water has an important bearing on the analytical properties of magnesium 8-hydroxyquinolate, both from the standpoint of gravimetric analysis (weighing form) and extraction separations.

The very low solubility of many organic compounds in water is due to the presence of hydrophobic groups such as —C_2H_5 and —C_6H_5, which are unable to form hydrogen bonds either as acceptors or donors and thus break down the water network resulting from hydrogen bonding. The effect of hydrophobic groups can be repressed by hydrophilic groups which lead to hydration of the molecule and make it more or less water-like and allow it to fit into the water structure more or less easily. The groups —OH and —COOH, even if not ionized, promote the solubility of an organic compound in water. Likewise —NH_2 is a hydrophilic group, especially when the H is replaced by alkyl groups to make the nitrogen more basic, thus enhancing hydration and solubility. Acidic groups (—COOH, —SH, often —OH) will of course lead to solubility of the organic compound in alkaline aqueous solutions. Sulfonic groups, —SO_3H, are very favorable for water solubility by ion formation (—SO_3^-).

Metal chelate formation involves incorporation of a metal ion in an organic matrix. If the product is uncharged or lacking in solubilizing groups, the combination is likely to be very slightly soluble in aqueous solution. Because of their electric charge, metal ions are surrounded by a hydration sphere of water molecules and are hydrophilic. In order that an insoluble product may be formed, the charge on the cation must be neutralized, its retinue of water molecules must be largely removed, and it must be combined with a species that is hydrophobic and only slightly soluble in water. Many chelating organic reagents are able to fulfill these functions. The formation of a metal 8-hydroxyquinolate (oxinate) furnishes a typical example:

$$M^{m+} + mOx^- \rightarrow MOx_m$$

When 8-hydroxyquinoline is present at sufficient concentrations in anionic form, the uncharged chelate is formed and precipitated. 8-Hydroxyquinoline itself is somewhat soluble in water because of the presence of the hydrophilic —OH group and possibly also, to a small extent, because of the nitrogen; the bulk of the molecule is hydrophobic. In the chelate the —OH group is no longer present and interaction of N with water is prevented by the bonding of the metal atom. The result is the formation of a very slightly soluble product. As we have seen, the hydroxyquinolate may contain coordinated water molecules but these do not seem to have an unfavorable effect on water insolubility. The intrinsic (molecular) solubility of some oxinates are listed in Table 12-3.

TABLE 12-3. Intrinsic (Molecular) Solubilities of Some Metal Oxinates in Water (25°C)[a]

	Solubility	
	Mole/liter	Mg of metal/liter
$CoOx_2 \cdot 2\,H_2O$	5×10^{-6}	0.3
$CuOx_2 \cdot 2\,H_2O$	3×10^{-7}	0.02
$ZnOx_2 \cdot 2\,H_2O$	3×10^{-8}	0.002
$FeOx_3$	1.6×10^{-7}	0.009

[a] Values for Co, Cu, and Zn oxinates (0.1 *M* ionic strength) from J. Fresco and H. Freiser, *Anal. Chem.* **36**, 372 (1964); for $FeOx_3$ (0.1 *M* ionic strength) from T. Turnquist and E. B. Sandell, *Anal. Chim. Acta* **42**, 239 (1968). See the latter paper for formation constants of Fe(III)-oxine complexes.

A matter of interest to the analyst is the effect of increasing the molecular weight of a chelating precipitant, as by increasing the length of a hydrocarbon chain or by replacing —CH_3 by —C_2H_5 or a phenyl group with a naphthyl group. In general, it is found that the solubility of an organic compound in water decreases as its hydrophobic groups increase in number or their molecular weight is increased (intermolecular attraction in the solid as well as solute–solvent interaction is involved). It may be expected that the intrinsic solubility of a metal chelate precipitate will be decreased also as the molecular weight of the reagent is increased in this way. This is generally found to be true. The effect is sometimes called *weighting*, although the increase in volume is likely a more important factor than the increase in weight per se. (Steric factors may lead to a dramatic reversal of expectations; see page 267.) Substitution of —Cl, —Br, or —I for —H tends to lower water solubility.

The total effect of weighting on the analytical solubility of a precipitate is complex. As we have seen (p. 260) the solubility of a metal-chelate precipitate depends on its intrinsic solubility, the solubility-product constant, and the anionic reagent concentration under the conditions of precipitation. The intrinsic solubility may not directly be of importance in an analytical precipitation, but it is important indirectly because the value of K_{sp} is equal to the product of the intrinsic solubility and the overall dissociation constant of the precipitate. As a rule, an increase in molecular weight brought about as described not only decreases the intrinsic solubility but also decreases the dissociation constant of the chelating reagent HL and of the metal chelate ML, so that the effect is doubly favorable. In comparing the effect of weighting on the solubility of the metal chelates, the solubilities must be determined at acidities where the concentrations of L^- of the two reagents are the same.

The requisite data for making a comparison of the effect of weighting on solubilities by evaluating all the factors involved are usually not available. The following empirical example, however, illustrates the analytical effect of an increase in the molecular weight —and volume—of a chelating precipitant. The sensitivities of the precipitation of Cu(II) in 0.2 *M* hydrochloric acid with thioglycolic acid, its anilide, and its α-naphthylamide are as follows[17]:

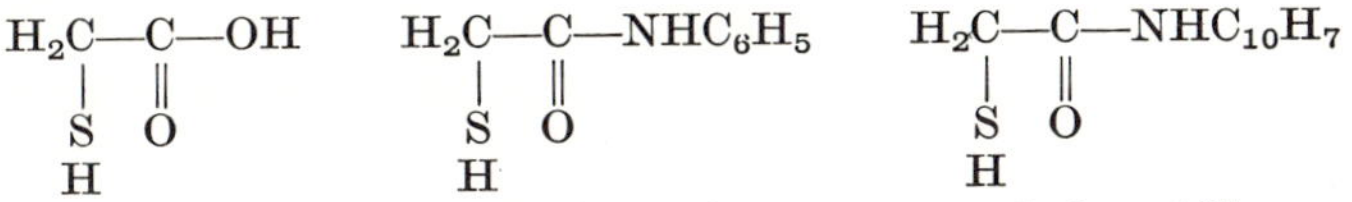

$1:2 \times 10^5$ $\quad$ $1:4 \times 10^6$ $\quad$ $1:1 \times 10^7$

The decrease in the solubility of a chelate is usually obtained at the cost of decreasing the solubility of the reagent, which may lead to difficulties in gravimetric analysis, because the reagent may contaminate the precipitate. Conversion of the precipitate to an inorganic weighing form is then required.

Chelating agents may combine with metal ions to give species that are charged, not electrically neutral. For example, 8-hydroxyquinoline can react with Fe(III) in acidic solutions to form $FeOx^{++}$ and $FeOx_2^+$ as well as $FeOx_3$; in basic solutions ions such as $FeOx_2(OH)_2^-$ can be formed. The possible formation of complex ions of this type needs to be taken into account in the treatment of the solubility of metal chelate precipitates.

In passing it may be noted that when a soluble metal chelate is desired (as in colorimetric methods), introduction of the sulfonic acid group, —SO_3H, into the chelating reagent is often effective, because this group ionizes extensively into —SO_3^- and H^+. Nevertheless, the solubility of the sulfonic acid may sometimes be quite low. Also, the solubility of some heavy metal sulfonates may be low. Thus the solubility of zinc anthracene 2-sulfonate is 1.3×10^{-4} *N* in water at room temperature.

Ion-association compounds. Another class of organic reagent consists of compounds ionizing extensively in aqueous solutions to form cations or anions of large size which

[17] R. Berg and W. Roebling, *Ber.* **68**, 403 (1935).

can combine with ions of opposite charge, also of large size, to give slightly soluble salts. The products are not chelates, but ion-association compounds. Large ions are only slightly hydrated and the neutral compounds they form do not fit well into the water structure and hence are slightly soluble. Most of these reagents are of the type R^+ or RH^+. For example, tetraphenylarsonium chloride in water ionizes to give tetraphenylarsonium cation, which reacts especially with certain bulky oxy or complex metal halide anions to form slightly soluble precipitates:

$$(C_6H_5)_4As^+ + MnO_4^- \rightleftharpoons (C_6H_5)_4AsMnO_4$$

$$2(C_6H_5)_4As^+ + HgCl_4^{--} \rightleftharpoons [(C_6H_5)_4As]_2HgCl_4$$

Rhodamine B in acid solution reacts as rhodaminium B cation to precipitate certain metal halide anions, preferentially of single negative charge:

$$R + H^+ \rightleftharpoons RH^+$$

$$RH^+ + SbCl_6^- \rightleftharpoons RHSbCl_6$$

The base benzidine forms benzidinium cation in acidic solutions, which precipitates sulfate (p. 271) and tungstate.

Tetraphenylboron anion (tetraphenylborate) is an excellent precipitant for potassium and other univalent metals of large ionic size:

$$B(C_6H_5)_4^- + K^+ \rightleftharpoons KB(C_6H_5)_4$$

It may be noted that the slight solubility of such salts as K_2PtCl_6, $KClO_4$, $K_2NaCo(NO_2)_6$, and $Cs_3Bi_2I_9$ in water is due to the large ionic sizes of cation and anion. Any large anion is a potential precipitant of K, Rb, Cs, and Tl(I).

The complex chromium thiocyanate anion $Cr(SCN)_6^{3-}$ precipitates the complex osmium(III) thiourea cation, in which neutral thiourea molecules are coordinated to osmium to form a complex of the ammine type:

$$Os(NH_2 \cdot CS \cdot NH_2)_6^{3+} + Cr(SCN)_6^{3-} \rightleftharpoons Os(NH_2 \cdot CS \cdot NH_2)_6Cr(SCN)_6$$

Many analogous examples could be given.

The solubilities of ion-association compounds vary over a wide range. The determination of sulfate by precipitation as benzidinium sulfate is not very satisfactory because the precipitate is too soluble. Similarly, the determination of tetrafluoroborate by precipitation with tetraphenylarsonium cation as $(C_6H_5)_4AsBF_4$ leaves something to be desired because the solubility of this compound in water at 25°C is $5 \times 10^{-4}\ M$. On the other hand, tetraphenylarsonium chloroiridate has both low molecular and ionic solubility:

$$[R^+]^2[IrCl_6^{--}] = K_{sp} = 1.1 \times 10^{-14} \text{ at } 25°C,\ 0.1\ M\ \mu\ (HCl) \qquad (R^+ = (C_6H_5)_4As^+)$$

$$[R_2IrCl_6]_{satd} = 2.6 \times 10^{-6}$$

Its precipitation is quantitative at relatively low tetraphenylarsonium concentrations ($10^{-3}\ M$). Tetraphenylarsonium perchlorate is sufficiently slightly soluble to be a useful precipitation form for ClO_4^-.

Specificity and selectivity

A quantitative reagent is said to be *specific* for an element if it allows the determination of that element in the presence of all other elements without separations. Adjustment of the acidity, use of differential complexing agents, and changes in oxidation states of elements are permitted under this definition. A given reagent could thus theoretically be specific for a number of elements depending on conditions. A few reagents come close to this ideal, but, except possibly for some spectrophotometric reagents, it is doubtful that any examples can be given.[18] A reagent may be called *selective* if it reacts with a limited number of elements, so that it can be used to determine at least one of these under suitable conditions, when the others are absent or have been removed. Specificity and selectivity are actually relative terms. Whether or not a foreign element is to be counted as an interfering element depends on its ratio to the element being determined and the accuracy required. With extremely few exceptions, procedures given for the determination of an element are usable only because some elements are not present in significant amounts and others do not exceed certain limits.

Precipitation reagents vary greatly in their selectivity. On the one hand, we have inorganic precipitants such as ammonium hydroxide which can precipitate half the elements in the periodic table, and, on the other, organic precipitants such as dimethylglyoxime, which is almost specific for nickel and palladium. As a class, organic reagents possibly tend to be more selective than inorganic reagents. At least, because of their greater variety, they offer more possibilities. However, a few inorganic precipitants are highly selective. Thus chloride is almost a specific precipitant for silver, if mercury and thallium are present as Hg(II) and Tl(III). Lead chloride is sufficiently soluble ($K_{sp} \sim 10^{-4}$) to allow ~ 0.5 g of Pb^{++} to remain dissolved in 100 ml of 0.05 *M* Cl^- solution (coprecipitation of $PbCl_2$ with AgCl is not considered in this connection). BiOCl is only slightly soluble, but reasonable amounts of bismuth can be kept in solution by increasing the hydrogen-ion concentration of the solution to repress the hydrolysis of Bi^{3+}. Antimony(III) behaves similarly to bismuth.

Reactivity of a chelating reagent of the type HL with various metals is associated with definite atomic groupings in the molecule.[19] Some metals will react more readily than others with a given complexing group, and there is thus a certain measure of selectivity. Reagents containing the —SH group are derivatives of H_2S and may be expected to react with metals bonding strongly to sulfur, corresponding roughly to metals forming slightly soluble sulfides in aqueous solution. Such a reagent may be expected to react with silver or copper but not aluminum or zirconium. On the other hand, a reagent containing an —OH group (8-hydroxyquinoline, for example) may be expected to react quite generally with most heavy metals, which bond to oxygen, forming slightly soluble hydroxides in aqueous solution. The coordinating atom is also of importance, as indeed is the whole chelating structure, in determining the extent of complex formation. In general, many metals will react with a particular complexing group but to different extents.

Reaction of a chelating reagent HL with a metal cation does not necessarily mean precipitation of the product. As we have seen, a necessary (but not sufficient) condition for the formation of a precipitate is that it have low molecular (intrinsic) solubility. The molecular solubility is a measure of the competition of the solvent and the lattice of the solid for the molecule. There is no necessary relation between the stability of a metal chelate and its solubility. For example, nickel dimethylglyoximate has a very low molecular solubility (1.2×10^{-6} *M*), whereas copper(II) dimethylglyoximate has a fairly high molecular solubility (8×10^{-3} *M*) in water. However, the copper complex is less dissociated than the nickel complex, the value of the constant

$$\frac{[M^{++}][HDx^-]^2}{[M(HDx)_2]}$$

being $10^{-19.24}$ for Cu and $10^{-17.24}$ for Ni.[20]

[18] It has been said half-seriously that in the future an analyst wishing to determine an element gravimetrically will reach for a bottle from the shelf of specific precipitants, selecting solution 20 for calcium, 29 for copper, and so on. At the present time this shelf seems to be empty.

[19] The doctrine of specific and selective groups in reagents has been expounded by F. Feigl in *Specific, Selective and Sensitive Reactions*, Academic Press, New York, 1949.

[20] D. Dyrssen and M. Hennichs, *Acta Chem. Scand.* **15**, 47 (1961).

Dimethylglyoxime and other oximes containing the 1,2-dioxime group

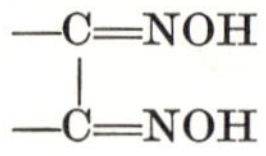

give slightly soluble precipitates with Pd, Pt(II), Ni, and Bi.[21] Dimethylglyoxime can be called a highly selective reagent for palladium, because Pt(IV) is not precipitated in cold solution and Ni and Bi will not be precipitated if the solution is acid ($\sim 0.5\ M$ HCl). The possibility of precipitating Pd in the presence of Ni by pH adjustment rests on the great difference in the K_{sp}'s of the two dimethylglyoximates, which in turn probably arises from the difference in the dissociation constants, because the intrinsic solubilities are not likely to be greatly different. Dimethylglyoxime misses being a specific reagent for Pd by a narrow margin. It is mainly disqualified by precipitating gold and selenium as the elements because of its reducing power.

Dimethylglyoxime (and other dioximes) is also a selective reagent for nickel, by virtue of the fortunate circumstance that although it forms quite stable complexes with copper, cobalt, zinc, and other metals, these complexes are soluble in water. The insolubility of nickel dimethylglyoximate in water has its origin in a structural factor. The molecule has a planar configuration:

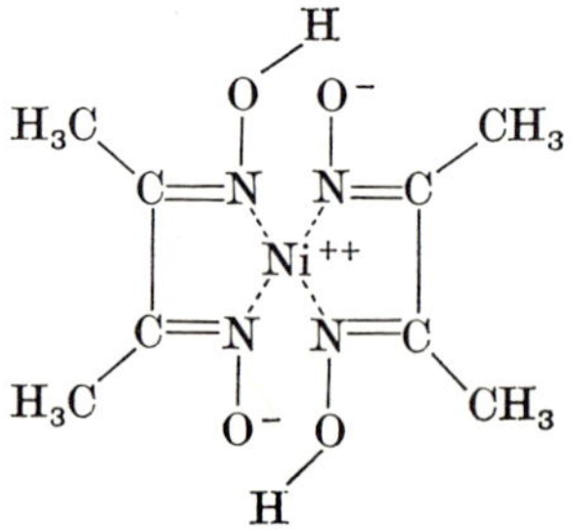

The groups —OH and —O$^-$ are strongly hydrogen-bonded internally and are not readily solvated, so that the water solubility is low. Moreover, nickel tends to form weak metal to metal bonds in the solid, which may contribute to the water insolubility of nickel dimethylglyoximate. Copper(II) and cobalt(II) dimethylglyoximates do not have the planar structure of the nickel complex, and since the chelate rings do not lie in the same plane, the —OH groups can be bonded to water molecules, and metal to metal bonding does not take place.[22]

Another type of structural factor can be of importance in promoting selectivity. If a sufficient number of chelating anions of an organic reagent cannot be arranged around a metal cation to neutralize its charge, the formation of an otherwise slightly soluble product is prevented. With larger cations the required number of anions for neutralization may be bonded, and precipitation selectivity is thus achieved. For

[21] The yellow bismuth precipitate, $(BiO)_2Dx$, is of a different type from the palladium and nickel dimethylglyoximates. It is formed at pH 8.5 or at pH 11–11.5 in the presence of EDTA and cyanide, which prevent the precipitation of nickel and various other metals. Tartrate or citrate inhibit or prevent the formation of the bismuth precipitate so that bismuth is not likely to interfere with the determination of nickel. See P. F. Lott and R. K. Vitek, *Anal. Chem.* **32**, 391 (1960).

[22] Dyrssen and Hennichs, Ref. 20, determined the solubilities of $Ni(HDx)_2$ and $Cu(HDx)_2$ in moist chloroform and found them to be much the same (4.75×10^{-4} and $1.05 \times 10^{-3}\ M$), indicating that the forces between the metal dimethylglyoximate molecules in the respective lattices are of the same order of magnitude. The difference in water solubilities of the two dimethylglyoximates is therefore attributed to solvation of $Cu(HDx)_2$, which is a consequence of the tendency of Cu^{++} to be five-coordinated, whereas Ni^{++} is only four-coordinated.

example, 8-hydroxyquinaldine (2-methyl-8-hydroxyquinoline) does not precipitate aluminum, but does precipitate Fe(III), Sc, Cu(II), Zn, and various other metals.[23] The failure of aluminum to be precipitated under any conditions is explained by the steric effect of the methyl group.[24] Its presence prevents the arrangement of three bulky 8-hydroxyquinaldinate radicals around the small aluminum ion to form an uncharged chelate complex (compare 8-hydroxyquinoline, p. 277).[25] Some other examples of steric hindrance in analysis are given in Chapter 14.

Reagents forming ion-association compounds also tend to show greater or less selectivity. If the reagent is cationic, a precipitate may not be formed at all if the size of the anion is too small. The charge of the anion is also of importance. If the cation (reagent) has a single charge, the tendency is for R^+A^- to be formed more easily than $(R_2)^{++}A^{--}$ or $(R_3)^{3+}A^{3-}$, although other factors enter in. For example, tetraphenylarsonium cation precipitates ReO_4^- but not MoO_4^{--}, and $IrCl_6^{--}$ but not $RhCl_6^{3-}$, enabling otherwise difficult separations of these metals to be made easily. Steric effects are probably involved here, but increasing hydration of an ion with increasing charge may also be a factor.

Solubility and separability

The extent of interference of element N in the precipitation of element M by a given reagent is primarily determined by the solubility relations of the two products. Can a specified recovery of M be obtained before N begins to precipitate? The answer to this question will be considered for a system in which the organic precipitant is of the type HL.

If an organic precipitant is a weak acid—and all useful chelating reagents are—the acidity of the solution will of course determine the fraction of metal precipitated with a given concentration of the reagent. Let

C_M = initial concentration of metal M
C_{HL} = initial concentration of reagent HL
R_M = recovery factor for metal (fraction precipitated)
K_{HL} = ionization constant of HL
K_{sp} = solubility-product constant of ML_m

On the assumption that the only metal species present in appreciable amounts in the solution is M^{m+}, its concentration at the end of the precipitation, if the volume remains constant, is

$$[M^{m+}] = C_M(1 - R_M) = \frac{K_{sp}[H^+]^m}{K_{HL}^m(C_{HL} - mR_MC_M - [L^-])^m} \tag{15}$$

If the acidity is such that the reagent is present predominantly as HL (upper limit of pH $\sim$ pK $-$ 1), the term $[L^-]^m$ may be disregarded, and the expression may be rearranged to give

$$[H^+]_{R_M} = \frac{K_{HL}(C_{HL} - mR_MC_M)C_M^{1/m}(1 - R_M)^{1/m}}{K_{sp}^{1/m}} \tag{16}$$

$[H^+]_{R_M}$ is the acidity at which R_M will have a specified value, if all the assumptions are valid. Theoretically, the pH range from incipient ($R_M = 0$) to 99.9 per cent precipitation is $3/m$, if excess HL is constant.

[23] L. L. Merritt and J. K. Walker, *Ind. Eng. Chem., Anal. Ed.* **16**, 387 (1944); J. P. Phillips and H. P. Price, *J. Am. Chem. Soc.* **73**, 4414 (1951); J. P. Phillips, J. F. Emery, and H. P. Price, *Anal. Chem.* **24**, 1033 (1952).

[24] H. Irving, E. J. Butler, and M. F. Ring, *J. Chem. Soc.* **1949**, 1489 (steric effects in the use of organic reagents).

[25] Actually it appears that some uncharged aluminum hydroxyquinaldinate is formed because, according to some reports, the chelate can be extracted to some extent into chloroform. In other words, aluminum hydroxyquinaldinate is highly dissociated.

In order that a specified recovery of M be obtained without precipitation of the more soluble precipitate NL_n, the K_{sp} values must differ at least by the factor

$$\frac{(K_{sp})_M^{1/m}}{(K_{sp})_N^{1/n}} = \frac{(1 - R_M)^{1/m} C_M^{1/m}}{C_N^{1/n}} \quad \text{(see hydroxide precipitation, p. 253)} \qquad (17)$$

or, if $m = n$ and $C_M = C_N$,

$$\frac{(K_{sp})_M}{(K_{sp})_N} \leqq 1 - R_M \qquad (18)$$

If $C_N = 10C_M$, $(K_{sp})_M/(K_{sp})_N$ must be $0.1(1 - R_M)$ or less, and so on.

As an illustration of these principles, consider the possibility of separating Cu(II), Cd, and Zn with quinaldic acid:

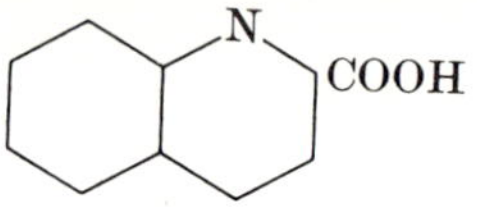

The pK_{sp} values of these quinaldates are[26]:

	pK_{sp}
Cu	16.8
Cd	12.3
Zn	13.8

$K_{\text{quinaldic acid}} = 1.2 \times 10^{-5}$. We see at once that the separation of Cd and Zn from each other is impossible. The separation of Cu and Cd should be possible (barring serious coprecipitation). The separation of Cu and Zn when present in equal concentrations appears dubious because, from Eq. (18),

$$1 - R_{Cu} = 0.001$$
$$R_{Cu} = 0.999$$

It would be difficult to adjust the pH so that 0.999 of the copper would be precipitated without precipitating zinc. If Zn/Cu = 0.1, the situation is more favorable, since $R_{Cu} = 0.9999$ when zinc begins to precipitate. Experiment has shown that the two elements can be separated when present in this ratio.

In practice, the state of affairs may be more complex than these simple considerations imply. Some chelating precipitants form intermediate, charged, complex species that may not be neglected; sometimes molecular solubilities of chelates are not negligible. The metals may be present to a greater or less extent in hydrolyzed forms, in addition to the simple ions. Foreign anions may complex the metals. Resolution of such systems requires a knowledge of additional equilibrium constants, which are not always available.

The possibility of coprecipitation of other elements is a disturbing factor, which must be evaluated experimentally. In general, coprecipitation will be greatest for those elements forming precipitates not much more soluble than that of the element being determined. Part of the coprecipitation may be due to entrainment of the more soluble precipitate.

[26] J. F. Flagg and F. T. McClure, *J. Am. Chem. Soc.* **65**, 2346 (1943); J. F. Flagg, *Organic Reagents Used in Gravimetric and Volumetric Analysis*, Wiley-Interscience, New York, 1948.

COMMON ORGANIC PRECIPITANTS

Some half-dozen organic precipitation reagents are considered in more or less detail in the following section to illustrate their use in inorganic analysis. Additional organic precipitants are listed in Table 12-4. The list is somewhat arbitrary and the references are not intended to be complete.

Arsonic acids

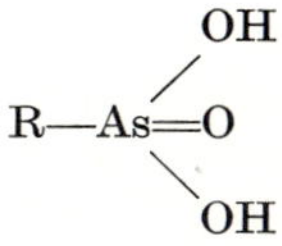

Like arsenic acid itself, these organic derivatives of H_3AsO_4 preferentially precipitate the quadrivalent metals in a dilute mineral acid medium. The precipitates are of the normal salt type and, when pure, contain 1 mole of quadrivalent metal to 2 moles of reagent. Thus phenylarsonic acid precipitates zirconyl phenylarsonate:

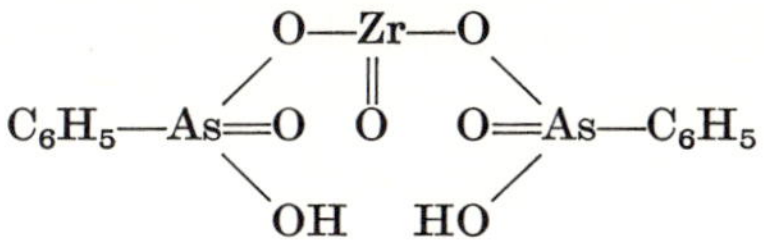

which is analogous to zirconyl arsenate:

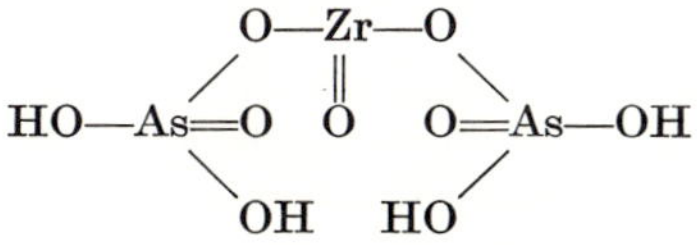

Especially at higher acidities, the normal zirconium salt can probably form also:

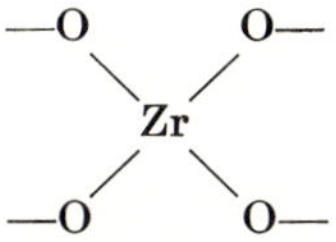

By proper choice of R the selectivity of the reagent can be increased. Thus while phenylarsonic acid precipitates both titanium and zirconium (and hafnium),[27] *n*-propylarsonic acid under the proper conditions precipitates only zirconium[28] and enables this element to be determined in the presence of aluminum, chromium, iron, copper, zinc, cadmium, nickel, cobalt, manganese, thorium, uranium, vanadium, and molybdenum, but not antimony, bismuth, or tin. *p*-Hydroxyphenylarsonic acid may be used for the determination of titanium and of zirconium if hydrogen peroxide is used to complex titanium.[29] The interference of iron is prevented by addition of ammonium thiocyanate. The arsonic acid precipitates do not have a definite composition after drying and they must be converted to the metal oxide for weighing after ignition.

[27] A. C. Rice, H. C. Fogg, and C. James, *J. Am. Chem. Soc.* **48**, 895 (1926). Th, Sn(IV), Nb, and Ta are also precipitated quantitatively. See, for Nb and Ta, I. P. Alimarin and B. I. Frid, *Mikrochemie* **23**, 17 (1937); K. Rankama, *Bull. Comm. Geol. Finlande*, No. 133 (1944).

[28] F. W. Arnold and G. C. Chandlee, *J. Am. Chem. Soc.* **57**, 8 (1935); G. C. Chandlee, *ibid.* **57**, 591 (1935). For determination in steel, see H. H. Geist and G. C. Chandlee, *Ind. Eng. Chem., Anal. Ed.* **9**, 169 (1937).

[29] C. T. Simpson and G. C. Chandlee, *Ind. Eng. Chem., Anal. Ed.* **10**, 642 (1938); A. Claassen, *Rec. Trav. Chim.* **61**, 299 (1942).

The discovery that mandelic acid is a selective precipitant for zirconium has made the arsonic acids less important as precipitants for this element.

α-Benzoinoxime

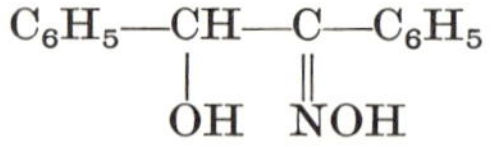

Mol. wt. 227.3

This compound was originally proposed as a specific reagent for copper[30] in ammoniacal solution, from which the green 1 : 1 complex is precipitated (tartrate keeps Fe(III), Al, and other metals in solution). However, it has found greater favor for the determination of molybdenum,[31] which is quantitatively precipitated in mineral acid medium as MoO_2Bx (H_2Bx = α-benzoinoxime). The precipitate is ignited to the oxide for weighing. Tungsten accompanies molybdenum, as do niobium and tantalum, but aside from these elements there are few interferences. A procedure for molybdenum in steel is given in Chapter 65.

Cupferron (ammonium salt of nitrosophenylhydroxylamine)

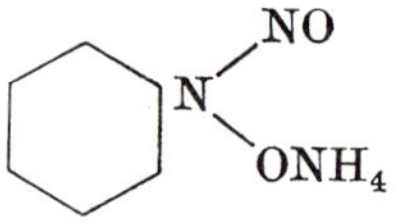

Mol. wt. 155.15

Nitrosophenylhydroxylamine is a weak acid ($K_a = 4 \times 10^{-5}$) which is only slightly soluble in water ($\sim 0.02\ M$, 25°). As a reagent, it is used in the form of its ammonium salt. The salt is freely soluble in water, but addition of a strong acid causes the free acid to separate as fine white crystals. The free acid is very unstable and decomposes on standing into nitrosobenzene and other products; the reagent is always used in cold solution, because heat promotes the decomposition of the acid. Cupferron is fairly stable in neutral or slightly basic solutions. In the dry state it can be kept for a long time if ammonium carbonate is placed in the bottle to maintain an alkaline environment.

Cupferron is one of the few organic precipitants used in strongly acidic solutions. In such a medium (5–10 per cent by volume of sulfuric or hydrochloric acid) it quantitatively precipitates iron(III),[32] titanium, zirconium, vanadium(V), uranium(IV), tin(IV), niobium, and tantalum and enables them to be separated from such elements

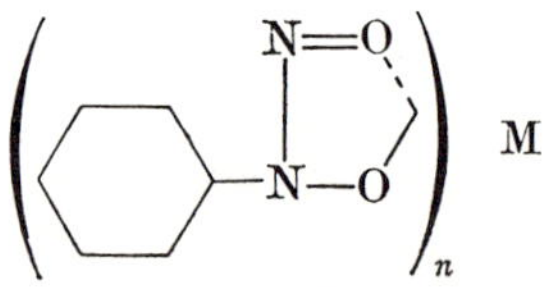

as uranium(VI), aluminum, chromium, manganese, nickel, cobalt, zinc, magnesium, and phosphorus.[33] In weakly acidic solution almost all the heavy metals are precipi-

[30] F. Feigl, *Ber.* **B56**, 2083 (1923).

[31] H. B. Knowles, *J. Res. Natl. Bur. Std.* **9**, 1 (1932).

[32] O. Baudisch, *Chem. Ztg.* **33**, 1298 (1909).

[33] W. F. Hillebrand and G. E. F. Lundell, *Applied Inorganic Analysis*, Wiley, New York, 1929, pp. 109–112; G. E. F. Lundell and H. B. Knowles, *J. Ind. Eng. Chem.* **12**, 344 (1920); G. F. Smith, *Cupferron and Neo-Cupferron*, G. Frederick Smith Chemical Co., Columbus, Ohio, 1938.

tated, but precipitation under these conditions is of little or no value. The metal cupferrates are never weighed as such because they are contaminated with nitrosophenylhydroxylamine and are unstable; they can be converted into the corresponding oxide by ignition. Cupferron is of great value in separations in applied analysis, either used alone or in conjunction with other separations, such as those with the mercury cathode (p. 259).

An important property of metal cupferrates is their solubility in organic solvents such as chloroform.[34] In general, metals forming slightly soluble cupferrates in acid solution can be separated by extraction from those that do not. Thus iron can be separated from aluminum in this way.

The solubility products of a number of cupferrates have been determined[35]:

$CuCf_2$	1×10^{-16}	$BiCf_3$	6×10^{-28}
$FeCf_3$	1×10^{-25}	$SnCf_4$	8×10^{-25}
$AlCf_3$	2×10^{-19}		

TABLE **12-4. Additional Organic Precipitants**

Reagent	*Remarks*	*Ref.*
Anthranilic acid 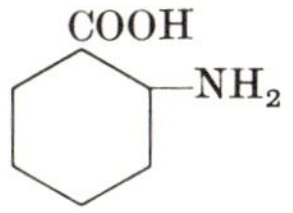 Mol. wt. 137.1	Metals such as Cd, Co, Cu, Fe(II, III), Pb, Mn, Hg, Ni, Ag, and Zn are precipitated in weakly acidic medium. Cd, Co, Cu, Pb, Ni, and Zn can be determined as anthranilates after drying at 105°C; a volumetric bromination method may also be applied. The reagent is of little value in separations.	H. Funk and M. Ditt, *Z. Anal. Chem.* **91**, 332 (1933); **93**, 241 (1933); **96**, 385 (1934). H. Funk and F. Römer, *ibid.* **101**, 85 (1935), H. Funk, *ibid.* **123**, 241 (1942). P. Wenger and E. Masset, *Helv. Chim. Acta* **23**, 34 (1940). For the use of 5-bromo-2-aminobenzoic acid as precipitant, see R. J. Shennan, *J. Soc. Chem. Ind.* **61**, 164 (1942).
Benzidine  Mol. wt. 184.2	Chiefly of importance as a precipitant for sulfate, forming $C_{12}H_{12}N_2 \cdot H_2SO_4$. The precipitate (which is not very insoluble) can be weighed or titrated with sodium hydroxide or iodometrically. The benzidine derivative 4-chloro-4′-aminodiphenyl  forms a less soluble sulfate than the parent compound.	See books in reference 16. R. Belcher, A. J. Nutten, and W. I. Stephen, *J. Chem. Soc.* **1953**, 1334.

[34] N. H. Furman, W. B. Mason, and J. S. Pekola, *Anal. Chem.* **21**, 1325 (1949); E. B. Sandell and P. Cummings, *ibid.* **21**, 1356 (1949).

[35] I. V. Pyatnitskii, *Zh. Analit. Khim.* **1**, 57 (1946).

TABLE 12-4. **Additional Organic Precipitants**—*Continued*

Reagent	*Remarks*	*Ref.*
Benzotriazole 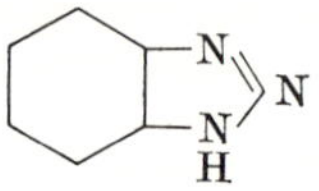 Mol. wt. 119.1	Precipitates Cu as $Cu(C_6H_4N_3)_2$ from tartrate–acetate medium at pH 7–8.5. Cd, Co, Fe(II), Ni, Ag, Zn are partially or completely precipitated also. Al, Sb(III, V), As, Cr, Fe(III), Mo, Se, Te are not precipitated. In the presence of EDTA in ammoniacal solution, only Ag precipitated. Os and Pd are precipitated from acetic acid–acetate solution.	J. A. Curtis, *Ind. Eng. Chem., Anal. Ed.*, **13**, 349 (1941); J. E. Fagel and G. W. Ewing, *J. Am. Chem. Soc.* **73**, 4360 (1951). For Ag: K. L. Cheng, *Anal. Chem.* **26**, 1038 (1954).
Bismuthiol II 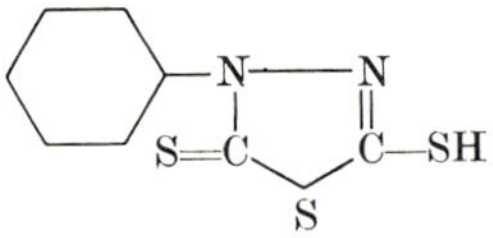	Precipitates numerous heavy metals in weakly acidic and neutral medium. Bi, As, Sb, but not Co, Ni, Mn, Fe(II) precipitated in 0.1 *M* HCl.	A. K. Majumdar and B. R. Singh, *Z. Anal. Chem.* **154**, 262, 413 (1957); **155**, 1, 81, 86, 166 (1957); **161**, 261 (1958).
Cinchonine	For quantitative precipitation of tungstic acid (Chap. 65).	
Isatin-β–oxime	Precipitates U(VI) from acetate-buffered or tartrate solution and separates from Mg, Mn, Zn, Cd, Ni, and Co. Precipitate ignited to U_3O_8.	V. Hovorka, V. Sykora, and J. Vorisek, *Chim. Anal.* **29**, 268 (1947).
Mandelic acid $C_6H_5 \cdot CHOH \cdot COOH$ Mol. wt. 152.1	Precipitates Zr (and Hf) from hydrochloric acid solution as the tetramandelate and separates it from Ti, Fe, V, Al, Cr, Th, Ce, Sn, Ba, Ca, Cu, Bi, Sb, and Cd. Derivatives, such as *p*-bromomandelic acid, can also be used. Mandelic acid has also been used as precipitant for Pu, Sc, and the rare earths.	C. A. Kumins, *Anal. Chem.* **19**, 376 (1947). R. E. Oesper and J. J. Klingenberg, *Anal. Chem.* **21**, 1579 (1949). Pu: E. Merz, *Z. Anal. Chem.* **166**, 417 (1959). Sc: I. P. Alimarin and K. Shen, *Zh. Analit. Khim.* **15**, 31 (1960). R.E. B. Weaver, *Anal. Chem.* **26**, 476 (1954).

TABLE 12-4. **Additional Organic Precipitants**—*Continued*

Reagent	*Remarks*	*Ref.*
α-Mercaptoacetanilide carbamate (formerly believed to be phenylthiohydantoic acid) 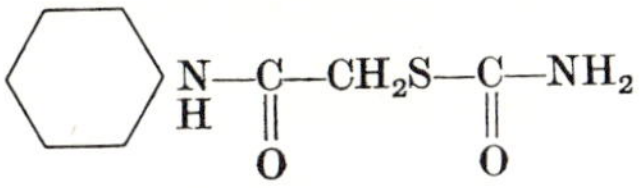	Co, Sb, and Cu are precipitated quantitatively in ammoniacal citrate medium. The reagent is used for separation of cobalt, but nickel and iron are partially precipitated. Cobalt precipitate is of indefinite composition.	H. H. Willard and D. Hall, *J. Am. Chem. Soc.* **44**, 2219, 2226, 2237, 2253 (1922).
Mercaptobenzothiazole 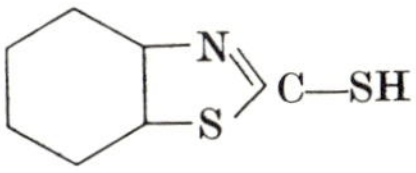 Mol. wt. 167.2	Precipitates numerous metals in weakly acidic or ammoniacal medium. Cu is precipitated in acid solution; precipitate must be ignited. Cd is precipitated in ammoniacal solution, following precipitation of copper.	G. Spacu and M. Kuras, *Z. Anal. Chem.* **102**, 24, 108 (1935); **104**, 88 (1936).
8-Mercaptoquinoline	Precipitant for Pd and Ni.	J. A. W. Dalziel and D. Kealey, *Analyst* **89**, 411 (1964).
5,6-Naphthoquinoline 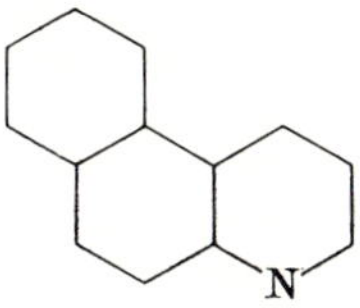	A reagent of the type RH^+ which precipitates halide and thiocyanate complex anions of Bi, Cd, Cu, Fe(III), Hg, U(VI), and Zn. Cd, but not Zn, precipitated as complex iodide. Has been used for separation of Mo(VI) and W(VI); former precipitated in weak acid, latter in strong.	R. Berg and O. Wurm, *Ber.* **60B**, 1664 (1927). A. Pass and A. M. Ward, *Analyst* **58**, 667 (1933). F. Hecht and R. Reissner, *Z. Anal. Chem.* **103**, 88 (1935). Mo and W: R. B. Golubtsova and F. M. Shemyakin, *Zh. Analit. Khim.* **3**, 118 (1948) **4**, 232 (1949).
Nitron (4,5-Dihydro-1,4-diphenyl-3,5-phenylimino-1,2,4 triazole) 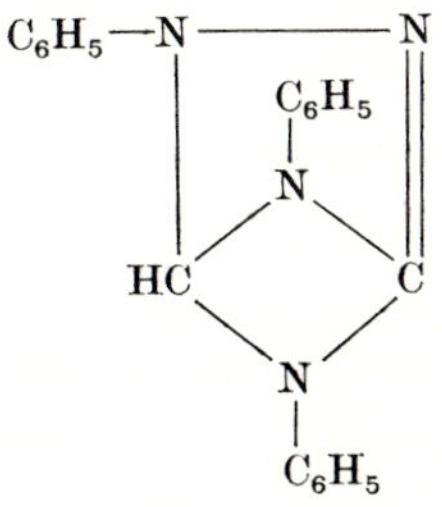	Precipitates nitrate, perchlorate, perrhenate, and tungstate as $C_{20}H_{16}N_4 \cdot HNO_3$, etc. Many anions ($Br^-$, I^-, CNS^-, etc.) interfere.	M. Busch, *Ber.* **38**, 861 (1905), A. Gutbier, *Z. Angew. Chem.* **18**, 494 (1905). For later work, see the books of Welcher and Flagg.

TABLE 12-4 **Additional Organic Precipitants**—*Continued*

Reagent	*Remarks*	*Ref.*
1-Nitroso-2-naphthol 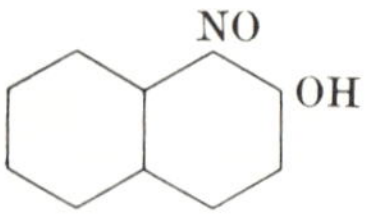 Mol. wt. 173.2	Co, Cu, Fe(III) and Pd are quantitatively precipitated from slightly acidic solutions; Cr(III), Sn, Ag, V, U, and Bi are partially precipitated; Hg, Pb, Cd, As, Sb, Al, Mn, Zn, Ni, etc., are not precipitated. The reagent is chiefly of value in separation of cobalt; the precipitate is of variable and uncertain composition. Iron(III) can be precipitated in the presence of zirconium if oxalic acid is present. The precipitates are soluble in chloroform.	G. Knorre, *Z. Angew. Chem.* **17**, 677 (1904); for separation of Pd from Pt, see W. Schmidt, *Z. Anorg. Allgem. Chem.* **80**, 335 (1913); F. Krauss and H. Dencke, *Z. Anal. Chem.* **67**, 86 (1925); C. Mayr, *Z. Anal. Chem.* **98**, 402 (1934).
Pyrogallol 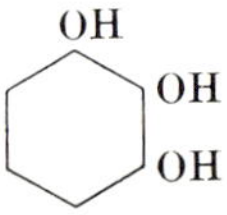 Mol. wt. 126.1	Gives crystalline precipitates with Sb(III) and Bi in mineral acid solution. *Gallic acid* has been used for separation of Bi from Pb, Cu, Fe, etc.; Sb, Sn, Hg, and Ag interfere.	F. Feigl, *Z. Anal. Chem.* **64**, 41 (1924); F. Feigl and H. Ordelt, *ibid.* **65**, 448 (1924); L. Kieft and G. C. Chandlee, *Ind. Eng. Chem., Anal. Ed.* **8**, 392 (1936).
Quinaldic acid 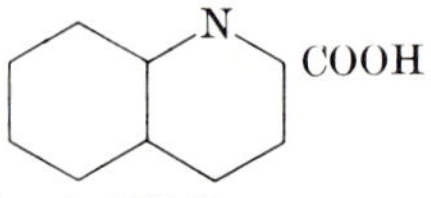 Mol. wt. 173.2	Precipitates copper, cadmium, and zinc in weakly acid solutions and is used for their gravimetric determination. Many other metals are also precipitated: Co, Fe(II, III), Pb, Hg, Mo, Ni, Pd, Ag, W, Al, Th, etc. The Cu salt $Cu(C_{10}H_6NO_2)_2 \cdot H_2O$ is one of the least soluble and can be precipitated in the presence of Cd, Ni, Co, and Pb by regulating pH. Zn can be precipitated in the presence of Cu(I), Hg, and Ag by complexing these with thiourea.	P. Ray and M. K. Bose, *Z. Anal. Chem.* **95**, 400 (1934); **100**, 324 (1935); P. Ray and N. K. Dutt, *ibid.* **115**, 265 (1939); *Mikrochemie* **17**, 11 (1935); P. Ray and J. Gupta, *ibid.* **17**, 14 (1935), **18**, 89 (1935). A. K. Majumdar, *Analyst* **68**, 242 (1943); C. F. Pritchard and R. C. Chirnside, *ibid.* **68**, 244 (1943). J. F. Flagg and F. T. McClure, *J. Am. Chem. Soc.* **65**, 2346 (1943); J. F. Flagg and D. W. Vanos, *Ind. Eng. Chem., Anal. Ed.* **18**, 436 (1946).
Quinoline-8-carboxylic acid 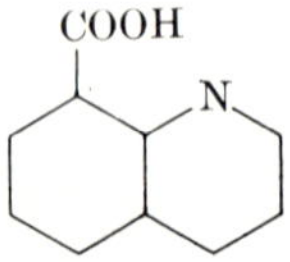 Mol. wt. 173.2	A gravimetric reagent for copper and other divalent metals.	J. R. Gilbreath and H. M. Haendler, *Ind. Eng. Chem., Anal. Ed.* **14**, 866 (1942); A. K. Majumdar, *J. Indian Chem. Soc.* **22**, 309 (1945).

TABLE **12-4** **Additional Organic Precipitants**—*Contiuued*

Reagent	*Remarks*	*Ref.*
Salicylaldoxime 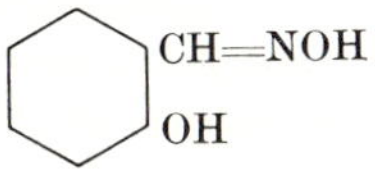 Mol. wt. 137.1	Precipitates copper at pH 2.5–3; such metals as Ag, Pb, Cd, Co, Zn, and Fe(II) do not precipitate at this acidity; with care a separation from Ni can be obtained. Fe(III) is coprecipitated. $Cu(C_7H_6O_2N)_2$ can be weighed after drying at 105°C.	F. Ephraim, *Ber.* **63**, 1928 (1930); **B64**, 1210, 1215, 2819 (1931); N. H. Furman and J. F. Flagg, *Ind. Eng. Chem., Anal. Ed.* **12**, 663, 738 (1940); L. P. Biefeld and D. E. Howe, *ibid.* **11**, 251 (1939); W. Reif, *Z. Anal. Chem.* **88**, 38 (1932).
Tannin (tannic acid)[a] $C_{14}H_{10}O_9$ Mol. wt. 322	Tannin is an organic precipitating reagent falling in a different category from others discussed here. It is a negatively charged colloid and appears to precipitate (flocculate) hydrous oxides of Nb, Ta, W, etc., by neutralizing their positive charge. Nb and Ta can be separated from each other and from Zn, Th, and Al; Ti from Zn; U from Nb, Ta, and Ti; Al from Be; Ga from Zn, Ni, Be, Th, etc.; Zr from U, V, and Th.	Separations of Nb and Ta: A. R. Powell and W. R. Schoeller, *Analyst* **50**, 485 (1925); **57**, 550 (1932) W. R. Schoeller, *ibid.* **57**, 750 (1932); W. R. Schoeller and C. Jahn, *ibid.* **59**, 465 (1934); W. R. Schoeller and E. F. Waterhouse, *Analyst* **53**, 515 (1928); W. R. Schoeller, *The Analytical Chemistry of Tantalum and Niobium*, Chapman & Hall London, 1937. Ti from Zr: A. R. Powell and W. R. Schoeller, *Analyst* **55**, 605 (1930); U from Ta, Nb, and Ti: W. R. Schoeller and H. W. Webb, *Analyst* **58**, 143 (1933); Zr from U, V, and Th: W. R. Schoeller, *Analyst* **69**, 272 (1944). Al and Ga: L. Moser and M. Niessner, *Monatsh.* **48**, 113 (1927); L. Moser and J. Singer, *ibid.* **48**, 673 (1927). L. Moser and A. Brukl, *Monatsh.* **50**, 657 (1928). M. L. Nichols and J. M. Schempf, *Ind. Eng. Chem., Anal. Ed.*, **11**, 278 (1939); G. W. Sears and H. Gung, *ibid.* **16**, 598 (1944).
Tetraphenylarsonium chloride $(C_6H_5)_4AsCl$ Mol. wt. 418.8	An anion precipitant: ClO_4^-, ReO_4^-, MoO_4^{--}, WO_4^{--}, $HgCl_4^{--}$, $SnCl_6^{--}$, $CdCl_4^{--}$, $ZnCl_4^{--}$, etc. The precipitates are of the type $[(C_6H_5)_4As]_2HgCl_4$.	H. H. Willard and G. M. Smith, *Ind. Eng. Chem., Anal. Ed.* **11**, 186, 269 (1939).

TABLE 12-4 **Additional Organic Precipitants**—*Continued*

Reagent	*Remarks*	*Ref.*
Tetraphenylboron, Na salt $B(C_6H_5)_4^- Na^+$	Precipitant for K, NH_4, Rb, Cs, Cu(I), Hg(I), Ag, and Tl.	See Chap. 32. A review is given by H. Flaschka and A. J. Barnard, Jr., *Advan. Anal. Chem. Instr.* **1**, 1 (1960).
Thiocyanate and organic bases	Pyridine, quinoline, isoquinoline, benzidine, ethylenediamine, and other organic bases together with an alkali metal thiocyanate precipitate slightly soluble compounds of zinc, cadmium, copper, and other divalent metals of the type $M(Py)_n(SCN)_2$. Thiocyanate alone is a selective precipitant for Cu(I).	See, e.g., for pyridine: G. Spacu and J. Dick, *Z. Anal. Chem.* **71**, 97, 442 (1927); **74**, 188 (1928); **76**, 273 (1929); for isoquinoline see A. E. Spakowski and H. Freiser, *Anal. Chem.* **21**, 986 (1949).

[a] Historical note: It is of interest that tannin is the earliest organic reagent of which there is record. Pliny the Elder (ca. 23–79 A.D.) states in his *Natural History* that adulteration of verdigris by vitriol (ferrous sulfate) was tested for in his time by the use of papyrus steeped in an extract of nut galls (tannin solution), whereby a dark color (ferrous tannate) was obtained in the presence of iron.

Neocupferron (the ammonium salt of nitrosonaphthylhydroxylamine) forms less soluble metal complexes than does cupferron. The K_a of the acid is 6×10^{-5}.

N-Benzoyl-N-phenylhydroxylamine. This compound is an analogue of cupferron

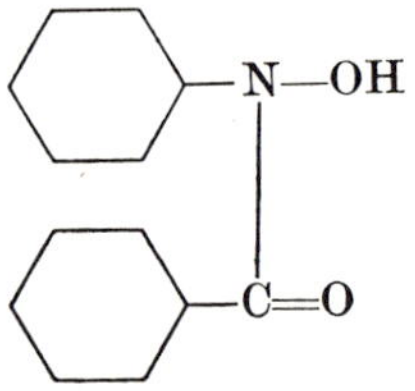

which is more stable.[36] Its precipitation reactions are similar to those of cupferron. The chelates of Al, Fe(III), Cu(II), Be(II),[37] Mo(VI),[38], U(VI),[39] La,[40] and other metals can be weighed as such after drying at 110°C. Various separations can be made by adjusting the pH and using complexing agents. The reagent has been used to separate Nb, Ta, and Ti.[41] Ti is kept in solution by the combined complexing action of EDTA and hydrogen peroxide.

This reagent is a weaker acid ($K_a = 3 \times 10^{-9}$) than nitrosophenylhydroxylamine and, in general, requires less acid solutions for quantitative precipitations of metals. However, Zr, Nb, Ta, and Sn(IV) can be precipitated from 0.5–1 N H^+.

[36] S. C. Shome, *Analyst* **75**, 27 (1950).
[37] J. Das and S. C. Shome, *Anal. Chim. Acta* **24**, 37 (1961).
[38] S. K. Sinha and S. C. Shome, *Anal. Chim. Acta* **24**, 33 (1961).
[39] J. Das and S. C. Shome, *Anal. Chim. Acta* **27**, 58 (1962).
[40] J. Das and S. C. Shome, *Anal. Chim. Acta* **32**, 52 (1965).
[41] F. J. Langmyhr and T. Hongslo, *Anal. Chim. Acta* **22**, 301 (1960); A. K. Majumdar and A. K. Mukherjee, *ibid.* **19**, 23 (1958); **21**, 245 (1959).

Dimethylglyoxime (diacetyldioxime) and other dioximes[42]

$$\begin{array}{l} CH_3—C{=}NOH \\ \quad\quad\;\; | \\ CH_3—C{=}NOH \end{array} \qquad \text{Mol. wt. 116.1}$$

First synthesized by Tschugaeff[43] and used by him for the detection of nickel and shortly thereafter by Brunck[44] for the determination of nickel in steel, this reagent and its analogues may be said to remain in a class by themselves. Dimethylglyoxime (and other dioximes) provides perhaps the closest approach to a specific organic precipitant. As already mentioned, palladium is quantitatively precipitated in dilute mineral acid medium. Nickel is precipitated in weakly acidic ($pH > 5$) or ammoniacal solution as a scarlet chelate complex $Ni(C_4H_7O_2N_2)_2$. Precipitation of iron and other metals as hydroxides is prevented by addition of tartrate. Iron should be present in the ferric state, since ferrous iron tends to be coprecipitated. Cobalt and a number of other metals form soluble complexes with the reagent and a greater excess must be added in their presence. The nickel and palladium precipitates can be weighed after drying at 105°C or higher temperature.

Dimethylglyoxime is soluble in alcohol (and in alkaline aqueous solutions in consequence of its acidic character) but only to a limited extent in water (0.063 g in 100 ml at 25°C). Therefore, too large an excess of the reagent must not be added for fear that it may crystallize out. Moreover, the nickel complex is soluble to some extent in alcoholic solutions. The error from this source is very small in cold solutions containing a small excess of reagent.[45]

Various other dioximes (e.g., diaminoglyoxime, methylbenzoylglyoxime, α-furildioxime, 1,2-cyclopentanedionedioxime, 1,2-cyclohexanedionedioxime, and 1,2-cycloheptanedionedioxime) have been recommended as superior to dimethylglyoxime because of their greater solubility in water.[46] However, there are objections to the use of all these reagents except the heptoxime for the determination of nickel in applied analysis. The heptoxime precipitates nickel quantitatively in slightly acidic solution ($pH > 3.5$), as shown by Voter et al. 1,2-Cyclohexanedionedioxime is reported to be preferable to dimethylglyoxime for the determination of palladium, chiefly because of the greater insolubility of the palladium complex.[47]

8-Hydroxyquinoline (oxine)[48]

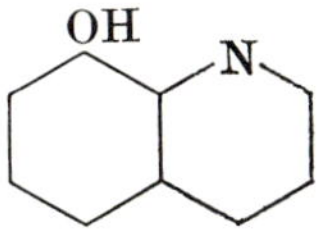

Mol. wt. 145.15

[42] A review with a bibliography of the applications of the dioximes to analytical chemistry is given by H. Diehl in a monograph published by the G. Frederick Smith Chemical Co., Columbus, Ohio, 1940. Stability constants and solubilities of metal dioximates are tabulated by D. Dyrssen, *Trans. Roy. Inst. Technol., Stockholm, Pure Appl. Chem.* **1964**, No. 220. The same author discusses their coordination chemistry in *Svensk Kem. Tidskr.* **75**, 618 (1963).

[43] L. Tschugaeff, *Ber.* **38**, 2520 (1905); *Z. Anorg. Chem.* **46**, 144 (1905).

[44] O. Brunck, *Z. Angew. Chem.* **20**, 834 (1907).

[45] P. Nuka, *Z. Anal. Chem.* **91**, 29 (1932).

[46] See review by R. C. Voter and C. V. Banks, *Anal. Chem.* **21**, 1320 (1949).

[47] R. C. Voter, C. V. Banks, and H. Diehl, *Anal. Chem.* **20**, 653 (1948). For the use of 4-methyl- and 4-isopropyl-1,2-cyclohexanedioxime for the determination of Ni and Pd, see C. V. Banks and D. T. Hooker, *Anal. Chem.* **28**, 79 (1956).

[48] F. L. Hahn, *Chem. Ztg.* **50**, 754 (1926); *Angew. Chem.* **39**, 1198 (1926); F. L. Hahn and K. Vieweg, *Z. Anal. Chem.* **71**, 122 (1927); R. Berg, *Z. Anal. Chem.* **70**, 341; **71**, 23, 171, 321, 369 (1927); **72**, 177 (1927); **76**, 191 (1929); R. Berg, *Die Chemische Analyse*, vol. XXXIV, Enke, Stuttgart, 2nd ed., 1938; R. G. W. Hollingshead, *Oxine and Its Derivatives*, Butterworth, London, 1954–1956.

Oxine (HOx) is an amphiprotic substance:

$$\left.\begin{array}{ll} HOx \cdot H^+ \rightleftharpoons HOx + H^+ & \dfrac{[HOx][H^+]}{[HOx \cdot H^+]} = 8 \times 10^{-6} \\ HOx \rightleftharpoons H^+ + Ox^- & \dfrac{[H^+][Ox^-]}{[HOx]} = 2 \times 10^{-10} \end{array}\right\} \mu = 0.1$$

Less than 5 per cent of the neutral species exists as the zwitter ion

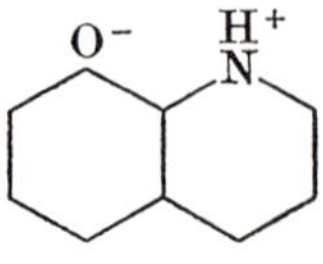

in aqueous solution.

The solubility is at a minimum at pH 7.4, which is the isoelectric point, where it amounts to about 0.5 g in 1 liter (0.0035 M) at room temperature. The restricted solubility should be borne in mind when the reagent is used in approximately neutral solutions.

This reagent can be made to give precipitates with almost all metals except the alkali metals, which serve for their determination and also are of use in separations. It forms five-membered chelate rings with metals and the solids are generally of the type MOx_n, in which the hydrogen of the —OH group is replaced by an equivalent of metal, but there are a number of exceptions, e.g., $ScOx_3 \cdot HOx$, $ThOx_4 \cdot HOx$, WO_2Ox_2, $UO_2Ox_2 \cdot HOx$ (actually HUO_2Ox_3), and $VOOHOx_2$. Cations that are extensively hydrolyzed, such as Ti(IV) and Th, may not give oxinates of stoichiometric composition. Bismuth gives a precipitate of the theoretical composition in the presence of tartrate. The divalent metal oxinates are likely to have two molecules of hydrate water (coordination number 6).

From the equations given above, it is evident that the oxinate ion (Ox^-) concentration and therewith the solubility of the metal oxinate depend very much on the acidity of the solution. By regulation of the hydrogen-ion concentration, some measure of precipitation selectivity can be obtained. The pH ranges of quantitative precipitation of many oxinates are shown in Fig. 12-3. The values should be looked upon as approximate because they depend to some extent on the conditions (e.g., excess of oxine), but they serve as a guide to possible separations. Metals whose pH of initial quantitative precipitation lie sufficiently far apart can be separated by pH adjustment (see p. 268), if coprecipitation is not serious.[49] Aluminum (Chap. 26) and iron(III), for example, can be determined in the presence of, and separated from, magnesium, the alkaline earth metals and beryllium by precipitation in an acetic acid–acetate buffer. Magnesium (Chap. 29) requires an ammoniacal solution for precipitation, and the excess of reagent must be controlled to avoid its precipitation. There appears to be some tendency for the reagent to be coprecipitated with magnesium oxinate so that the determination of magnesium is not as accurate as that of aluminum, which can be determined to within ~1 ppt in pure solution.

The selectivity of oxinate precipitations can be increased by the use of suitable metal-complexing agents. For example, in strongly basic tartrate solution, Cu, Cd, Zn, and Mg are precipitated, while Al, Fe(III), Cr(III), Pb, Sn(IV), and some other

[49] For coprecipitation in oxine separations, see H. V. Moyer and W. J. Remington, *Ind. Eng. Chem., Anal. Ed.*, **10**, 212 (1938). Homogeneous precipitation may enable separations to be made that are not successful by ordinary precipitation. Thus, homogeneous precipitation of zinc oxinate allows zinc to be separated from lead and magnesium, which are strongly coprecipitated in ordinary precipitation, as shown by J. P. Jones, O. E. Hileman and L. Gordon, *Talanta* **10**, 111 (1963). See further, F. H. Firsching, *Talanta* **10**, 1169 (1963), who reviews precipitation of metal chelates from homogeneous solution.

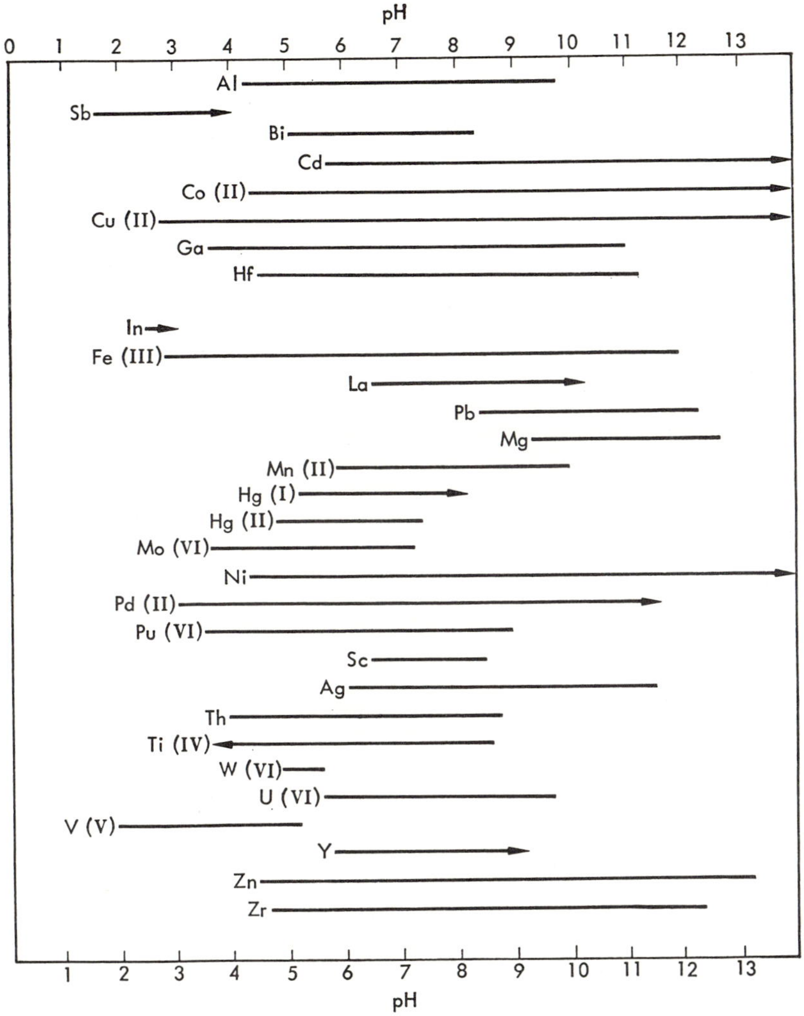

Fig. 12-3. pH ranges for quantitative precipitation of 8-hydroxyquinolates. Values for Bi, Ga, Hf, La, Hg, Pd, Ag, and Ti from R. Bock and F. Umland, *Angew. Chem.* **67**, 420 (1955); Co, Mg, Mo, Ni, W, U, and Zn from H. R. Fleck and A. M. Ward, *Analyst* **58**, 388 (1933); Al, Cu, Cd, Fe, Pb, and Mn from H. Goto, *J. Chem. Soc. Japan* **54**, 725 (1933), **56**, 314 (1935); others from various sources.

elements remain in solution. EDTA and other complexing agents also find application.

The oxinates can be prepared for weighing by drying at 105°C or higher. $MgOx_2 \cdot 2\,H_2O$ retains its hydrate water at 105°C, but like most other hydrated oxinates becomes anhydrous at ~135°C. Most anhydrous oxinates, including $AlOx_3$, can be heated to 150–200°C without decomposition. The higher drying temperature, if permissible, is preferred when the precipitate may be contaminated with small amounts of oxine, which is volatilized. When the oxinate does not have a stoichiometric composition, it can be ignited to the oxide after adding oxalic acid to prevent volatilization of the oxinate.

A volumetric method for the determination of metal oxinates is available (Chap. 45), as well as spectrophotometric methods based on chloroform extraction of certain oxinates.

Oxine also finds some use as a cationic precipitant, as in the determination of bismuth as $(HOx \cdot H)BiI_4$ and phosphorus as oxinium phosphomolybdate.

The reagent solution is prepared in ethyl alcohol or acetic acid. Some oxinates—those precipitated in anhydrous form (Al, Fe(III), and Ga, for example)—are appreciably soluble in dilute alcohol and must be precipitated with an acetic acid or a very dilute hydrochloric acid solution of the reagent.

Thionalide (thioglycolic-β-aminonaphthalide)

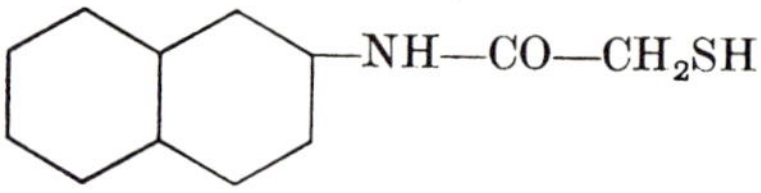

Mol. wt. 217.1

This reagent forms slightly soluble white or pale-colored chelates with most of the metals of the hydrogen sulfide group,[50] including antimony, arsenic, tin, bismuth, copper, mercury, silver, gold, and the platinum metals, in acid ($\sim 0.1\ N$) solution. The hydrogen of the —SH group is replaced by an equivalent of metal. In a general way, thionalide is similar to hydrogen sulfide in its reactions; lead and cadmium are not precipitated in mineral acid medium, however. The thionalates have a definite composition and most of them can be weighed after drying at 105°C. They can also be determined volumetrically by titration with a standard iodine solution:

$$2\,RSH + I_2 \rightarrow RS\text{—}SR + 2\,H^+ + 2\,I^-$$

Thionalide can be used for determining copper in acid solution in the presence of cadmium, cobalt, iron(II), lead, magnanese, nickel, zinc, and thallium; a disadvantage of the procedure is the interference of chloride, which may be present only in small amounts. In tartrate solution made basic with sodium carbonate, gold, copper, mercury, cadmium, and thallium are precipitated; in alkaline cyanide-tartrate solution, gold, thallium, tin, lead, antimony, and bismuth are precipitated; in sodium hydroxide–cyanide–tartrate solution only thallium is precipitated.

The reagent is susceptible to oxidation (as by ferric iron), forming RSSR, and hydroxylamine sulfate should be added when oxidizing agents are present. The reagent solution (in alcohol or glacial acetic acid) is stable for only a few hours.

SEPARATION OF TRACE CONSTITUENTS BY PRECIPITATION

The isolation of traces of substances is an important branch of modern analytical separations. Although newer separation methods (immiscible solvent extraction and ion exchange) are usually preferred when applicable, precipitation methods are still useful and play an important role in separations of radioactive elements.

Precipitation of a constituent at very low concentrations may fail because the precipitate is too soluble under the conditions or because it forms supersaturated solutions. Moreover, it may form colloidal solutions so that the insoluble material cannot be collected easily. Satisfactorily complete precipitation of a trace constituent may often be obtained by adding a small amount of another substance, also forming a slightly soluble product with the precipitant. The precipitate of the added substance can function as a *collector* (*carrier*) for the trace constituent in a number of ways (see Chapter 11). Sometimes its effect is more or less mechanical or physical, as when one

[50] R. Berg and W. Roebling, *Ber.* **68**, 403 (1935); *Z. Anal. Chem.* **109**, 305 (1937); **112**, 161 (1938); W. J. Rogers, F. E. Beamish, and D. S. Russel, *Ind. Eng. Chem., Anal. Ed.* **12**, 561 (1940).

precipitate drags down another. Adsorption effects are often predominant, as with hydroxide precipitates. Slightly soluble hydroxides frequently function as general collectors (scavengers) because of their adsorptive properties, which are enhanced by their enormous surface development.

Collectors are always used in sulfide precipitation of trace constituents. They aid in overcoming supersaturation and, if properly chosen, can reduce the solubility of the trace sulfide below the normal solubility (mixed crystal formation sometimes takes place, as with Sb_2S_5 and MoS_3). When CuS is used as a collector for HgS, as little as 0.02 μg of Hg can be recovered from a liter of solution, even though, in the absence of copper, the solubility of mercury under the conditions is greater than this. According to F. Haber, ammonium sulfide precipitation with lead sulfide as collector can be used to recover 1 μg of Au from 1000 liters (1 ton) of seawater.

Collectors are particularly effective when they incorporate the trace component by mixed crystal or solid solution formation[51] and the distribution coefficient is favorable (see Chapter 11). Two components M^+ and N^+ can form mixed crystals with X^- if the ionic radii are sufficiently similar. The empirical rule of V. Goldschmidt states that replacement of M^+ by N^+ in the lattice MX may be expected if the (ionic) radii do not differ by more than ±15 per cent.[52]

An example of the use of mixed-crystal formation in trace separations is provided by the isolation of lead by coprecipitation with strontium sulfate. Since the ionic radii of Pb and Sr are similar (1.32 and 1.27 Å according to V. Goldschmidt), it may be expected that $PbSO_4$ and $SrSO_4$ form mixed crystals. This is true. $SrSO_4$ should be a good collector for Pb, since it is a little more soluble than $PbSO_4$. Assuming that the homogeneous distribution expression holds approximately (see p. 235), we estimate a rough value for the distribution coefficient:

$$D = \frac{(K_{sp})_{SrSO_4}}{(K_{sp})_{PbSO_4}} = \frac{2.8 \times 10^{-7}}{1.06 \times 10^{-8}} \sim 25$$

This is a favorable distribution coefficient. Suppose 50 per cent of the strontium in solution is precipitated with sulfate. On the assumption of homogeneous distribution (less favorable than heterogeneous distribution for lead recovery):

$$\frac{[Pb]_{cryst}}{[Pb]_{soln}} = 25\frac{[Sr]_{cryst}}{[Sr]_{soln}} = 25 \times \frac{1}{1} \sim 25$$

This corresponds to the precipitation of ~95 percent of the lead. The order of magnitude of this value has been verified. Hahn showed that precipitation of 50 per cent of the Sr coprecipitated 99 per cent of the Pb (present as ThB).[53] If the precipitation of Sr is made more complete, more Pb is found in the precipitate. Thus if 99 per cent of the Sr is precipitated (a value that can be attained easily):

$$\frac{[Pb]_{cryst}}{[Pb]_{soln}} = 25 \times \frac{99}{1} \sim 2500$$

Even if D were only 1 (the value varies with the rate of precipitation), 99 per cent of the lead would be precipitated. The order of magnitude of this value permits the conclusion that quantitative isolation of lead can probably be obtained by coprecipitation with strontium sulfate. Experiment has verified this conclusion.[54]

[51] See page 234 for the meaning of these terms.

[52] Isomorphous substances, whose crystals have almost identical interfacial angles, do not necessarily form mixed crystals. They cannot if M^+ and N^+ differ too greatly in size. Contrarily, substances that are not isomorphous can form mixed crystals (e.g., CaF_2–YF_3), at least over a certain range, if the ionic sizes are similar. Differences in charge can be compensated by lattice vacancies or incorporation of another ion of suitable charge and size. Thus MnO_4^- can replace SO_4^{--} in $BaSO_4$ if Ba^{++} is simultaneously replaced by K^+.

[53] O. Hahn, *Applied Radiochemistry*, Cornell Univ. Press, Ithaca, N.Y., 1936, p. 105.

[54] I. T. Rosenqvist, *Am. J. Sci.* **240**, 356 (1942) (isolation of lead from rocks).

Compound formation can sometimes be put to good use in the isolation of traces of elements by coprecipitation. If a solution containing traces of Ag, Au, Hg, Pd, or Pt is treated with stannous chloride in the presence of Te(IV), the precipitated elemental tellurium carries down these easily reducible elements, no doubt in the form of tellurides.[55] Separation of the noble metals from large amounts of such metals as Fe, Zn, Ni, and Co is easily effected.

Coprecipitation of traces of metals with a solid organic precipitation agent has been applied, as in the isolation of such metals from seawater. For example, if an acetone solution of α-benzoinoxime is added to an acidic solution of Mo(VI), and the solution is boiled to vaporize the acetone, the benzoinoxime crystallizing out carries down the molybdenum[56]; under proper conditions satisfactory recoveries of microgram quantities of molybdenum in 1 liter of solution can be obtained. It is to be expected that solid α-benzoinoxime will strongly adsorb molybdenum benzoinoximate and incorporate it in the crystals, whatever the mechanism of the incorporation may be. Under some conditions, the distribution of coprecipitated trace metals roughly follows the Doerner–Hoskins expression. This type of collector is not likely to be very selective in its action, but it serves to concentrate traces from large volumes of solution.

Concentration of solutions by freezing

Traces of ions in very dilute aqueous solutions have been concentrated by freezing most of the solution while it is shaken. Salts such as NaH_2PO_4, NaF, NaCl, and $Mg(NO_3)_2$ have been collected in a small volume of the unfrozen sample with good recoveries.[57]

PROBLEMS

1. Pr is to be determined in a mixture with Nd, in which the weight ratio Pr/Nd is 0.10. In the final gravimetric determination of Pr, after separation from Nd, Nd will have the same effect as an equal weight of Pr. If the error in the determination of Pr, from the presence of Nd, is not to exceed 2 parts per 1000, what is the required separation factor $S_{Nd/Pr}$? Assume that the recovery of Pr is virtually complete.
2. On the basis of solubility considerations, decide which of the following separations may be possible by hydroxide precipitation: (a) Be, Mn; (b) Al, Ti; (c) Al, Be; (d) Ca, Mg; (e) Zr, Th; (f) Th, Ce(III).
3. (a) At what pH will the precipitation of Fe^{++} as $Fe(OH)_2$ from a 0.02 *M* solution be 99.99 per cent complete? At what pH will it begin? (b) If the ferrous iron solution is initially 1 *M* in HCl, can the pH adjustment conveniently be made with 4 *M* NH_4OH for complete precipitation?

$$\frac{[Fe^{++}][OH^-]}{[FeOH^+]} = 1 \times 10^{-4} \qquad (K_{sp})_{Fe(OH)_2} = 8 \times 10^{-16}$$

Ans. (a) pH = 9.34 (9.30 neglecting $Fe(OH)^+$); 7.3.

4. Ce(III) is to be separated from Mg and the alkaline earths by precipitation as $Ce(OH)_3$. At what pH will 99.99 per cent of the cerium be precipitated if its original concentration is 0.01 *M*? Is there danger of precipitation of Mg at this pH if its concentration is 1.0 *M*?

$$(K_{sp})_{Mg(OH)_2} = 10^{-9.2} \qquad (K_{sp})_{Ce(OH)_3} = 10^{-20.1}$$

[55] References to these methods may be found in Sandell, *Colorimetric Determination of Traces of Metals*, Wiley-Interscience, New York, 3rd ed., 1959.

[56] H. V. Weiss and M. G. Lai, *Talanta* **8**, 72 (1961). Co has been coprecipitated with 1-nitroso-2-naphthol, alkaline earth elements with potassium rhodizonate, etc. An earlier scheme proposed by V. I. Kuznetsov, *Zh. Analit. Khim.* **9**, 199 (1954), utilized one organic reagent to bind the trace element and another organic reagent to precipitate the first.

[57] G. H. Smith and M. P. Tasker, *Anal. Chim. Acta* **33**, 561 (1965); S. Kobayashi and G. F. Lee, *Anal. Chem.* **36**, 2197 (1964); J. Shapiro, *Science* **33**, 2063 (1961).

5. Show by calculations that separation of bismuth from 10 times its weight of Cu(II) by precipitation of bismuth hydroxide with excess ammonia appears feasible. See the tables in the Appendix for constants.
6. On the basis of solubility relations alone, which is the better approach to the separation of Al and Fe: (a) precipitation of $Al(OH)_3$ with iron left in solution as Fe^{++}; (b) precipitation of $Fe(OH)_3$ with Al^{3+} left in solution? Use the following $\log K_{sp}$ values: $Al(OH)_3$, -31.2; $Fe(OH)_3$, -38.7; $Fe(OH)_2$, -14.0. What other factors need to be taken into account in evaluating the two possible methods?
7. Derive Eq. (16), page 267.
8. Disregarding possible complications from coprecipitation and postprecipitation, should copper(II) be separable from an equal amount of zinc by hydrogen sulfide precipitation, if $R = 0.9999$ is accepted as a satisfactory value for both? What should be a suitable acidity for 0.02 *M* concentrations of each metal if the solution is saturated with H_2S at room temperature? Does this result agree with experiment?

$$(K_{sp})_{Cu} = 4 \times 10^{-36} \qquad (K_{sp})_{Zn} = 1 \times 10^{-24}$$

9. Does separation of cobalt from manganese by sulfide precipitation appear feasible on the basis of the solubility difference? Cadmium from zinc? Copper from cadmium?
10. It is proposed to determine barium gravimetrically as $BaSO_4$ in a sample of silver nitrate containing approximately 1 per cent $Ba(NO_3)_2$. Disregarding the possibility of coprecipitation of silver with $BaSO_4$, decide whether this proposal is sound.

$$K_{sp} \text{ of } BaSO_4 = 1 \times 10^{-10} \qquad K_{sp} \text{ of } Ag_2SO_4 = 2 \times 10^{-5}$$

11. Iron(III) in an amount ~ 0.05 g is to be separated by cupferron precipitation from 0.1 g of Al in 200 ml of solution. In cupferron precipitations it is customary to add the reagent in such excess that nitrosophenylhydroxylamine precipitates (solubility in water, 0.02 *M*). What is a suitable acidity for the separation?

$$K_{sp} \text{ of } FeCf_3 = 1 \times 10^{-25} \qquad K_{sp} \text{ of } AlCf_3 = 2 \times 10^{-19}$$

$$K_a \text{ of } HCf = 7 \times 10^{-5} \qquad (\mu = 0.1)$$

Ans. To be on the safe side, set the recovery of iron at 0.9999. The molar concentration of unprecipitated iron will then be (neglecting small volume changes)

$$0.0001 \times \frac{0.05}{56} \times \frac{1000}{200} = 5 \times 10^{-7}$$

Take this to consist chiefly of Fe^{3+}. We see that at this ferric ion concentration, all Al^{3+} will remain in solution because aluminum does not precipitate until

$$[Al^{3+}] > 5 \times 10^{-7} \times \frac{2 \times 10^{-19}}{10^{-25}} > 1$$

The cupferrate ion concentration required to bring the Fe^{3+} concentration to 5×10^{-7} is

$$[Cf^-] = \left(\frac{1 \times 10^{-25}}{5 \times 10^{-7}}\right)^{1/3} = 6 \times 10^{-7}$$

From the ionization constant of HCf

$$\frac{[H^+]}{[HCf]} = \frac{7 \times 10^{-5}}{6 \times 10^{-7}} \cong 100$$

Therefore,

$$[H^+] \sim 0.02 \times 100 \sim 2$$

Now it must be realized that if an acid such as HCl is used (which complexes Fe^{3+}) instead of $HClO_4$ to give an acidity of 2 *M*, the precipitation of iron will not be 0.9999 complete. A rough calculation, using the formation constants for $FeCl^{++}$–$FeCl_3$ at $\mu = 0.1$, gives a ratio $[Fe^{3+}]/\sum[Fe]$ of $\sim 1:40$. Because of the large difference in μ the ratio is uncertain. If it is approximately correct, the precipitation of iron will be $(1 - 40 \times 0.0001) = 0.996$ complete, which would mean that 0.2 mg of Fe remains in solution. Moreover, there is some uncertainty in the value of K_{sp} for ferric cupferrate. The ionic strength at which it was determined is not specified (certainly not $\mu = 0$, but is μ 0.1 or ~ 1?). The calculated value of the acidity can therefore be regarded as only rough. Actually it is confirmed by experiment. Quantitative precipitation of iron can be effected at this acidity in hydrochloric and even sulfuric acid solution.

12. Does separation of bismuth and tin(IV) from each other by cupferron precipitation appear feasible?
13. Can Al and Ca be separated from each other by phosphate precipitation (and if so specify the conditions)?

$$(K_{sp})_{AlPO_4} = 7 \times 10^{-19}$$
$$(K_{sp})_{Ca_3(PO_4)_2} = 1 \times 10^{-28}$$
$$(K_{sp})_{Ca(HPO_4)_2} = 3 \times 10^{-7}$$

14. From the solubility products (Appendix), decide whether it is likely that magnesium and copper(II) can be separated from each other by 8-hydroxyquinoline precipitation.
15. Specify the conditions under which barium can be separated from strontium by precipitation of the former as chromate.

$$(K_{sp})_{BaCrO_4} = 1 \times 10^{-10}$$
$$(K_{sp})_{SrCrO_4} = 4 \times 10^{-5}$$
$$\frac{[H^+][HCrO_4^-]}{[H_2CrO_4]} = 1.2$$
$$\frac{[H^+][CrO_4^{--}]}{[HCrO_4^-]} = 3.5 \times 10^{-7}$$
$$\frac{[HCrO_4^-]^2}{[Cr_2O_7^{--}]} = 3 \times 10^{-2}$$

16. Outline a method by which 1 mg of Cd can be separated from 100 mg of Ni, showing the supporting numerical considerations.
17. Utilizing the pertinent equilibrium constants, evaluate the possibilities in the separation of Ti from Fe(III) by precipitation of the former with ammonia in the presence of EDTA.
18. When P_2O_5 is in molar excess of $Al_2O_3 + Fe_2O_3$, Mn is carried down strongly by the $Al(OH)_3$–$Fe(OH)_3$ precipitate obtained by ammonia precipitation at pH ~ 7. Why? May a basic acetate precipitation be expected to improve the separation of Al and Fe from Mn in the presence of phosphate?
19. Propose methods (see the literature, or devise) for the separation of (a) ~ 50 mg of Pd and ~ 50 mg of Ag (determination of both); (b) 50 mg of Pd and 0.1 mg of Ag (determination of Ag); (c) 50 mg of Ag and 0.1 mg of Pd (determination of Pd).
20. A solution containing Ni, As(V), and Bi is analyzed for nickel by adding citrate, dimethylglyoxime, and ammonia in slight excess, and filtering off and weighing nickel dimethylglyoximate. Will the results for nickel be high or low? Suggest means for avoiding the error.
21. Give reasons for believing that the following quantitative schemes are or are not sound: (a) separation of Cu(II) from Ba and Fe(III) by precipitation of copper from weakly acidic solution with hydrogen sulfide; (b) Separation of Fe(III) from

Ca by addition of NH_4OH to a solution containing Fe(III), Ca^{++}, Na^+, Cl^-, and phosphate; (c) separation of Fe(III) from Mn(II) by addition of excess $BaCO_3$ to a slightly acidic solution of the two.

22. Can ~1 μg of Fe(III) in 100 ml of 1 *M* $CaCl_2$ be satisfactorily recovered by ammonia precipitation without a collector? Suggest a possible collector for $Fe(OH)_3$ if one appears needed.

CHAPTER 13 *Principles of Chromatographic Separations. Ion Exchange*

The neophyte is apt to conclude that this technique [*of ion-exchange chromatography*] *is at the same time the panacea for analytical separations and the death knell to the need for theoretical and descriptive chemistry. On the contrary! Delving more deeply into these marvelous applications, one soon realizes that these separations are based on a thorough knowledge of solution chemistry as well as imaginative versatility in modifying old and developing new experimental procedures.*

—RIEMAN AND BREYER (1961)

This chapter deals in an elementary way with the general principles of chromatography, especially as illustrated by the use of ion-exchange resins because of their importance in inorganic analysis, the branch of analysis chiefly treated in this text. Other types of chromatographic separations are considered only briefly.

CHROMATOGRAPHY: THE UNDERLYING PRINCIPLES

A powerful separation method of wide applicability is based on the flow of a fluid sample through a column of finely divided solid or of solid serving as a support for a liquid phase. In general, different solutes will partition to different extents between the stationary solid phase and the mobile fluid phase (liquid solution or gas) so that bands will be formed at different positions in the column. Depending on the nature of the stationary phase, the partition process may involve surface adsorption of the solutes, distribution of the solutes between the mobile phase and a liquid phase held on the solid support, and ion exchange. Separation of the bands without serious overlap requires the passage of a suitable fluid through the column after the sample has been added; this process is known as *elution*. If the separated substances are colored, a series of colored zones may be seen along the length of the column, whence the name *chromatography* for this type of separation.

The principles underlying column chromatography (and related multistage separations) are illustrated in Fig. 13-1.[1] The column at the far left of the diagram represents the stationary phase (e.g., a liquid phase fixed on an inert solid, or an ion-exchange resin) and the adjoining column represents the mobile phase (liquid solution or gas), which is slowly moving downward past the stationary phase. Consider the behavior of a single solute S originally present in a small volume of mobile phase at the top of

[1] The theory of column chromatography sketched in the following pages is only one of several. It is an equilibrium theory based on the concept of effective "plates," which has the merit of being simple and of giving a useful approximation of column operation.

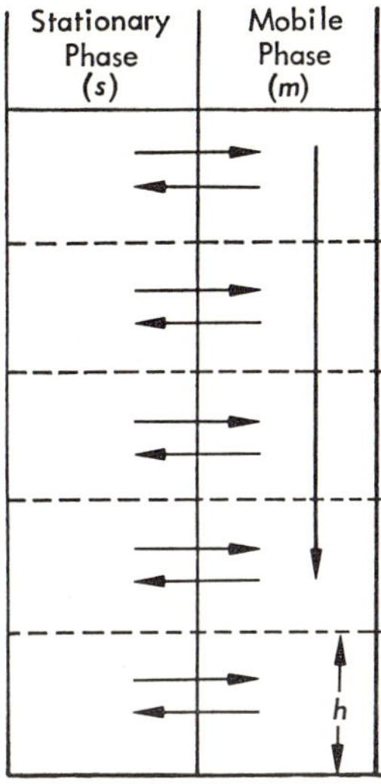

r	n=0	n=1	n=2	n=3	n=4
0	1.0	0.5	0.25	0.125	0.0625
1	0	0.5	0.5	0.375	0.25
2	0	0	0.25	0.375	0.375
3	0	0	0	0.125	0.25
4	0	0	0	0	0.0625

Fig. 13-1. Schematic representation of chromatographic and other multistage processes.

the column. The solute partitions between the stationary (s) and the mobile phase (m) in accordance with its distribution constant:

$$D = \frac{[\mathrm{S}]_s}{[\mathrm{S}]_m} \tag{1}$$

(Strictly D is the *partition* constant of S according to extraction terminology.) The column is imagined to be divided into cells designated r_0, r_1, r_2, . . ., indicated by the dashed lines in the figure. Equilibration is looked upon as taking place in discrete stages represented by the hypothetical cells. These cells are usually called *theoretical plates* from analogy with distillation theory.

Assume for simplicity of illustration that the cross-sectional areas of the stationary and mobile phases are equal and that $D = 1$. At the beginning, all of S is present in plate r_0, and at equilibrium S is divided evenly between the two phases, i.e., the fraction 0.5 is present in m and in s. On addition of a volume of pure solvent equal to the volume of mobile phase in a plate, the mobile phase in plate r_0 moves downward into the next plate r_1 and is replaced by the pure solvent. After this first transfer ($n = 1$), plate r_0 contains 0.5 of the original amount of S, which, on attainment of distribution equilibrium is equally divided between the mobile and stationary phases, so that $(r_0)_s$ contains 0.25 and $(r_0)_m$ contains 0.25 of the original amount of S. After the second transfer ($n = 2$), $r_0 = 0.25$, $r_1 = 0.25 + 0.25 = 0.5$, and $r_2 = 0.25$. The fractions of solute in successive plates after three and four transfers are shown in the tabular portion of the diagram. When $n = 4$, plate 2 contains the largest fraction of solute. With increasing n, the maximum moves down the column but lags behind the pure solvent front.

More generally, if the stationary and mobile phases are equal in volume, the fraction of S left in s after the first transfer ($r = 0$, $n = 1$) is

$$(\mathrm{F}_s)_{0,1} = \frac{[\mathrm{S}]_s}{[\mathrm{S}]_s + [\mathrm{S}]_m} = \frac{[\mathrm{S}]_s}{[\mathrm{S}]_s + [\mathrm{S}]_s/D} = \frac{D}{D+1} \tag{2}$$

and the fraction left in m is

$$(\mathrm{F}_m)_{0,1} = 1 - \frac{D}{D+1} = \frac{1}{D+1} \tag{3}$$

After n transfers, the fraction of S in each plate is given by the successive terms of the expansion

$$\left(\frac{D}{D+1} + \frac{1}{D+1}\right)^n \tag{4}$$

Thus, for any solute with the distribution coefficient D, we have the following distribution shown in Table 13-1 with respect to r and n.

TABLE **13-1**

				n	
r	0	1	2	3	4
0	1	$\frac{D}{D+1}$	$\frac{D^2}{(D+1)^2}$	$\frac{D^3}{(D+1)^3}$	$\frac{D^4}{(D+1)^4}$
1		$\frac{1}{D+1}$	$\frac{2D}{(D+1)^2}$	$\frac{3D^2}{(D+1)^3}$	$\frac{4D^3}{(D+1)^4}$
2			$\frac{1}{(D+1)^2}$	$\frac{3D}{(D+1)^3}$	$\frac{6D^2}{(D+1)^4}$
3				$\frac{1}{(D+1)^3}$	$\frac{4D}{(D+1)^4}$
4					$\frac{1}{(D+1)^4}$

The fraction of solute in plate r after n transfers, each involving the volume of the mobile phase of a plate, is given by the rth term of the binomial expansion, Eq. (4)[2]:

$$\begin{aligned} F_{r,n} &= \frac{\{n(n-1)\cdots(n-r+1)\}}{r!}\left(\frac{D}{D+1}\right)^{n-r}\left(\frac{1}{D+1}\right)^r \\ &= \frac{n!}{r!\,(n-r)!} \times D^{n-r} \times \left(\frac{1}{D+1}\right)^n \end{aligned} \tag{5}$$

It is seen that bands of substances of different D's move down the colum at different rates. If D is large, the rate is low, and conversely. Since D's of different components are generally independent of each other, separations can evidently be effected if a column consisting of a sufficient number of theoretical plates is used. Even if h (= HETP, height equivalent to a theoretical plate) is as large as 1 mm, a relatively short column will often suffice to separate substances of similar distribution coefficients, as will be demonstrated later. As we shall see (p. 312), h is a function of the particle size of the absorbent and the flow rate.

The foregoing principles may be extended as follows for a quantitative theory of column chromatography.[3] It is assumed that equilibrium is attained in each theoretical

[2] This equation, used here to illustrate column chromatography by discrete transfers from cell to cell (which is a simplification and indeed a fiction), represents exactly the distribution of a solute in the "countercurrent distribution" method of L. C. Craig, *J. Biol. Chem.* **155**, 519 (1944), when the immiscible phases are equal in volume, the distribution coefficient (or constant) being defined as the reciprocal of D given here if the system so requires. (See Chap. 14.)

[3] A. J. P. Martin and R. L. M. Synge, *Biochem. J.* **35**, 1358 (1941). The theory was developed by them to deal with partition chromatography, but the same treatment is generally applicable to other types of chromatography.

plate (within a theoretical plate the concentration is uniform both in the stationary and mobile phases), that diffusion from one plate to another is negligible compared to flow transfer, and that the distribution coefficient of the solute is independent of its concentration. The flow of solution through the column is taken to be continuous, not discontinuous as in the preceding introduction (Fig. 13-2). Let A_s = cross-sectional area of the stationary phase and A_m = cross-sectional area of the mobile phase. The symbols r, n, h, and D have the same significance as before.

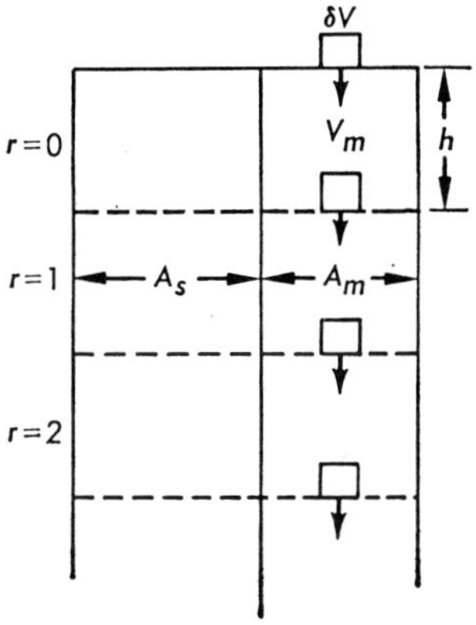

Fig. 13-2. Schematic representation of chromatographic column with continuous transfer of mobile phase.

Imagine a volume δV of solvent to be added at the top of the column so that the increment δV of the mobile phase is transferred from one plate to the next down the column. Since all the solute was initially present in plate r_0, with the fraction

$$(F_m)_{0,0} = \frac{[\mathrm{S}]_m A_m}{[\mathrm{S}]_m A_m + [\mathrm{S}]_s A_s} = \frac{A_m}{A_m + DA_s} \tag{6}$$

in the mobile phase (volume = V_m) at equilibrium, the fraction of solute transferred to plate r_1 is

$$F_{1,1} = \frac{\delta V}{V_m} \times \frac{A_m}{A_m + DA_s} = \frac{\delta V}{hA_m} \times \frac{A_m}{A_m + DA_s} = \frac{\delta V}{h(A_m + DA_s)} = \frac{\delta V}{V_h} \tag{7}$$

where $V_h = h(A_m + DA_s)$, the effective plate volume. The fraction of solute not transferred is

$$1 - \frac{\delta V}{V_h} \tag{8}$$

The fraction of solute in plate r after n transfers is expressed by the terms of the binomial expansion [see Eq. (5)]

$$\left\{\left(1 - \frac{\delta V}{V_h}\right) + \frac{\delta V}{V_h}\right\}^n$$

or

$$F_{r,n} = \frac{n!}{r!\,(n - r)!}\left(1 - \frac{\delta V}{V_h}\right)^{n-r}\left(\frac{\delta V}{V_h}\right)^r \tag{9}$$

Some approximations must now be introduced to make this expression manageable:

$$\frac{n!}{r!\,(n-r)!} \sim \frac{n^r}{r!} \qquad \text{when } n \gg r \gg 1\,^4 \tag{10a}$$

$$\left(1-\frac{\delta V}{V_h}\right)^{n-r} \sim \exp\left\{-(n-r)\frac{\delta V}{V_h}\right\} \sim \exp\left(-\frac{n\,\delta V}{V_h}\right) \qquad \text{when } n \gg r \tag{10b}$$

$$r! \sim e^{-r}r^r(2\pi r)^{1/2} \qquad \text{(Stirling's approximation)}\,^5 \tag{10c}$$

Introducing these approximations in Eq. (9) and noting that $n\delta V = V$, the total volume of solvent added, we obtain

$$F_{r,n} = \frac{1}{(2\pi r)^{1/2}}\left(\frac{V}{rV_h}\right)^r \exp\left(r - \frac{V}{V_h}\right) \tag{11}$$

This equation approaches the normal error function as r (and the number of transfers) becomes large. (See the result in Problem 1, p. 331.) The maximum value of $F_{r,n}$ is attained when $V/rV_h = 1$ and is then equal to $1/(2\pi r)^{1/2}$. Thus the concentration of solute at the band maximum decreases as the band moves down the column and is inversely proportional to the square root of the distance traveled from the top of the column. Another important relation can be derived—the rate of movement of the band maximum compared to the rate of movement of the solvent front. This ratio is the celebrated R_f of paper chromatography (p. 328):

$$R_f = \frac{\text{distance moved by band maximum}}{\text{distance moved by solvent front}}$$

$$= \frac{r_{\max}h}{V/A_m} = \frac{V}{A_m + DA_s} \times \frac{A_m}{V} = \frac{A_m}{A_m + DA_s} \tag{12}$$

[since $r_{\max}V_h = V$ and $V_h = h(A_m + DA_s)$]. Note that R_f is equal to the fraction of solute in the mobile phase at equilibrium at any level in the column.

It can be shown[6] that

$$\frac{C}{C_{\max}} = \exp\left\{-\frac{N(V_{\max} - V)^2}{2V_{\max}V}\right\} \tag{13}$$

where C = concentration of solute in effluent at volume V

$C_{\max}$ = concentration of solute at $V_{\max}$, i.e., when the peak (center) of the elution band is at the base of the column, and one half of the solute has been eluted and the other half remains in the column.

N = number of theoretical plates in the column

The solute is assumed to have been present originally in a narrow band a negligible distance from the top of the column.

Equation (13) allows the calculation of the number of theoretical plates in a column from the ratio of concentrations of solute at $V_{\max}$ and at any other known elution volume at which C is measurable if the elution curve is Gaussian. It is convenient to take C equal to $C_{\max}/e$ ($= 0.37C_{\max}$), so that $V = V_{\max/e}$ ($= V_e$). Equation (13) then becomes

$$\exp\left\{-\frac{N(V_{\max} - V_e)^2}{2V_{\max}V_e}\right\} = e^{-1} \tag{14}$$

[4] When $n = 10$ and $r = 5$, the approximation gives 8.3×10^2 instead of the correct value 2.5×10^2; when $n = 100$ and $r = 5$, 8.3×10^7 instead of 7.5×10^7; when $n = 100$ and $r = 50$, 3×10^{35} instead of 1×10^{29}.

[5] Stirling's formula gives good approximations even for small values of r, e.g., 1.92 for 2! and 118.0 for 5! (= 120).

[6] E. Glueckauf, *Trans. Faraday Soc.* **51**, 34 (1955).

and
$$N = \frac{2V_{\max}V_e}{(V_{\max} - V_e)^2} \tag{15}$$

or
$$N \cong \frac{2V_{\max}^2}{(V_{\max} - V_e)^2} \tag{16}$$

if the difference between $V_{\max}$ and V_e is small. If the elution band width is defined as $2(V_{\max} - V_e) = \beta$ (Fig. 13-3), the number of theoretical plates is given by

$$N \cong \frac{8V_{\max}^2}{\beta^2} \tag{17}$$

Further treatment of column theory will be found in the following section.

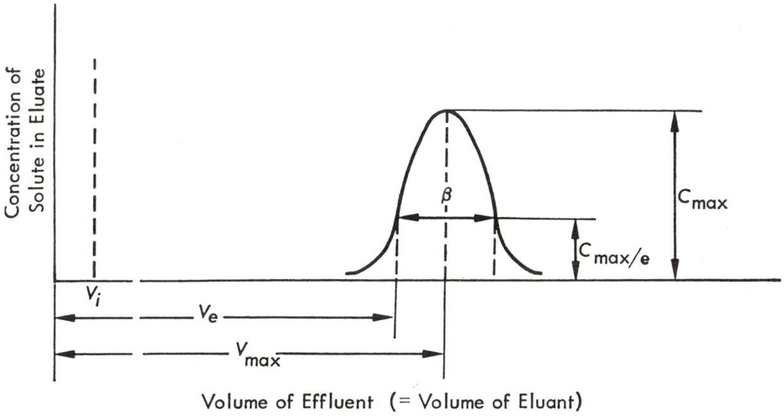

Fig. 13-3. Illustration of an elution curve. $V_{\max}$ is the volume of eluting solution required to elute the center of the solute band to the base of the column; V_e is the volume of eluate at which the concentration of the solute is $1/e$ of that at the maximum; V_i is the interstitial volume of the column of adsorbent (initially occupied by solvent). $C_{\max}$ is given by $(m/V_{\max})\,(N/2\pi)^{1/2}$, where m is the total amount of the eluted ion.

ION–EXCHANGE RESINS AND THEIR USE

The most widely used ion exchangers are the ion-exchange resins. The ion-exchange properties of certain synthetic resins were discovered by Adams and Holmes in 1934, but useful resins were not prepared until about 10 years later.

Nature and general behavior

Ion-exchange resins[7] consist of a cross-linked polymer network containing ionized or ionizable groups. These functional groups ionize into an ion fixed to the polymer chain and a counterion in the liquid phase which can be replaced by an ion of the same charge.

[7] A thorough treatment of ion exchangers may be found in F. Helfferich, *Ion Exchange*, McGraw-Hill, New York, 1962. Analytical applications are not discussed. For the latter see J. Inczedy, *Analytical Applications of Ion Exchangers*, Academic Press, New York, 1966 and Samuelson, Ref. 25.

Cation-exchange resins contain acidic groups:

$$\left.\begin{array}{l}-SO_3H \\ -COOH \\ -OH \text{ (phenolic)}\end{array}\right\} H^+ \text{ can be replaced by cations}$$

Anion-exchange resins contain weakly basic groups:

$$\begin{array}{l}-NH_2 \text{ (i.e., } NH_3^+) \\ -NH \cdot CH_3 \\ -N \cdot (CH_3)_2\end{array}$$

or strongly basic quaternary ammonium groups, e.g.,

$$-N(CH_3)_3OH$$

in which OH^- is replaceable by anions. The strong-acid ($-SO_3H$) and strong-base types of resin are of much greater importance in analytical practice than the weak-acid and weak-base types, because the former are effective over a wide range of pH.

The action of a cation-exchange resin is illustrated by the reaction occurring with a sodium chloride solution:

$$RH + Na^+ + Cl^- \rightleftharpoons RNa + H^+ + Cl^-$$

The reaction can be made to proceed quantitatively to the right by passing the solution through a column of the resin. Sodium chloride can thus be converted into an equivalent amount of hydrochloric acid. The resin can be regenerated (converted back to the acid form) by treatment with a strong acid. With a strong-base anion-exchange resin the reaction taking place with sodium chloride is

$$ROH + Na^+ + Cl^- \rightleftharpoons RCl + Na^+ + OH^-$$

Preparation

The most useful ion-exchange resins are based on cross-linked polystyrene. A copolymer of styrene and divinylbenzene is usually employed to give a three-dimensional polymer network in which polystyrene chains are linked with divinylbenzene units:

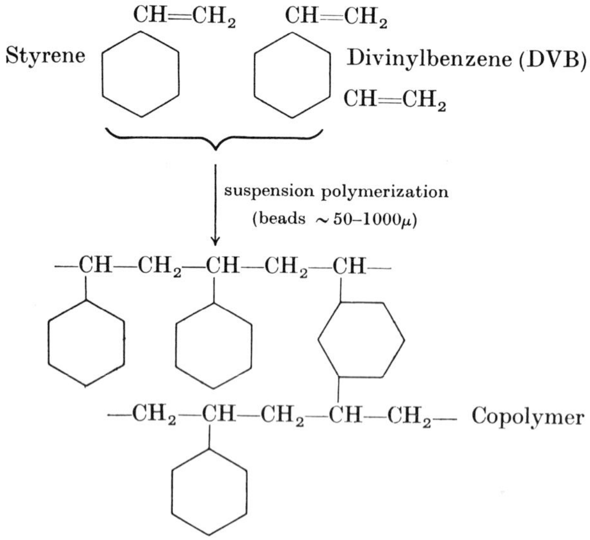

Sulfonation of cross-linked polystyrene gives strong-acid cation-exchange resins, which can be represented as

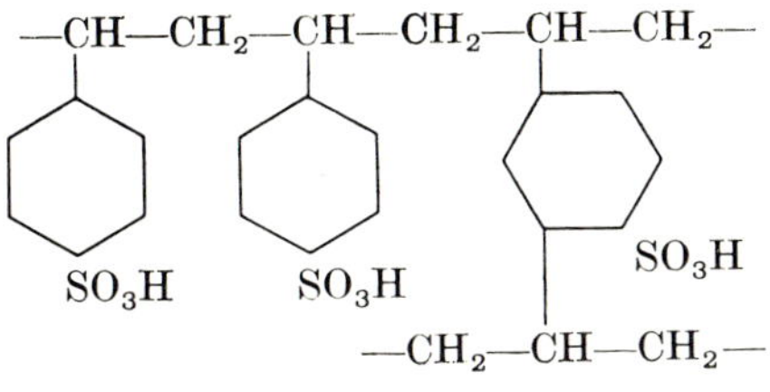

These resins are chemically very stable (even at 100°C) and are resistant to concentrated acids and alkalies and are not easily attacked by oxidizing and reducing agents.[8] Dowex 50[9] and Amberlite IR-120 are examples of resins of this type.

A strong-base anion-exchange resin is represented by the following:

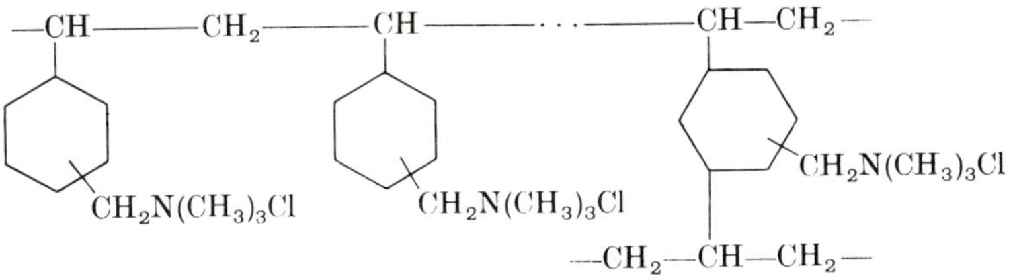

Dowex 1 and Amberlite IRA-400 are examples of this type of resin. Dowex 2 and Amberlite IRA-410 contain the ionic group $—NCH_2(CH_3)_2C_2H_4OH^+$. Anion-exchange resins are not as stable chemically and thermally as cation-exchange resins.[10]

The two-dimensional representation of the three-dimensional network in the resins serves in some measure to indicate the relatively wide channelways that traverse the structure and allow solutions to permeate, and ions to diffuse into, the grains of resin. The mesh width of the matrix may vary from a few to 100 Å or more, depending on the degree of cross-linking and swelling by solutions.

Properties of ion-exchange resins

Absorption of water: swelling. Particles of a dry resin placed in water or an aqueous solution take up water (largely by hydration of ions) and increase in volume. The degree of swelling is especially dependent on the degree of cross-linking. The greater

[8] The action of nitric acid is slight below 3 *M*. H_2O_2 attacks the resins, especially in the presence of Fe(III) and Cu(II), which act as catalysts. In acid solution, Cr(VI), Mo(VI), and V(V) are partially reduced. Permanganate reacts in both acid and alkaline solution. Reduction of Au(III), some Pt metals [Ir(IV)], and even Fe(III) can occur; pretreatment of the resin with chlorine largely prevents the reduction of these.

[9] The degree of cross-linking is indicated by the X number. Thus Dowex 50-X8 is a resin prepared from a styrene–DVB copolymer containing 8 per cent DVB. This resin is available in various particle sizes: 20–50, 50–100, 100–200, and 200–400 mesh. Dowex 50W is an improved form of Dowex 50.

[10] The free-base form of a strong-base anion-exchange resin is the least stable form, but even it can be used at room temperature. The chloride form shows only slight decomposition at 95°C. Low-molecular-weight amines and methanol are liberated on decomposition of the resin. It must be realized that all ion-exchange resins dissolve in water to a minute extent or are leached of soluble organic substances, so that separations should not be made (nor water purified) with ion-exchange resins when a trace of organic matter will subsequently interfere in a determination. Ion-exchange columns that have stood for any length of time (overnight) should be washed before use. Both cation- and anion-exchange resins are reported to be slowly attacked by mercury salts; mercury attaches to the aromatic rings.

the extent of cross-linking, the smaller the swelling. A sulfonated polystyrene resin (H^+ form) having cross-linking corresponding to about 5 per cent nominal DVB was found to absorb 1.5 g of water per gram of dry resin, compared to about 0.9 g of water for a resin with 10 per cent nominal DVB. Swelling is not as great in concentrated electrolyte solutions as in water or dilute solutions.

The phenomenon of swelling is of practical significance inasmuch as the ion-exchange capacity based on volume decreases with the extent of swelling, and moreover the packing of a resin bed is affected by swelling or contraction. Shrinkage leads to channeling. Dry resin packed into a glass tube may shatter it when moistened.

Uptake of electrolyte. A cation-exchange resin immersed in an electrolyte solution may take up appreciable amounts of the anion—in addition to the cation entering by exchange—as a result of the Donnan membrane equilibrium. Suppose a strongly acidic cation-exchange resin in the sodium form is brought into contact with a sodium chloride solution. At equilibrium the following relation holds (parentheses indicate activities):

$$(Na^+)_r(Cl^-)_r = (Na^+)_s(Cl^-)_s$$

and

$$\frac{(Na^+)_r}{(Na^+)_s} = \frac{(Cl^-)_s}{(Cl^-)_r} \quad (r \text{ refers to resin, } s \text{ to solution}) \tag{18}$$

For the sake of illustration, set concentrations equal to activities. If $[Na^+]_r = 4$ (a possible value) and the aqueous sodium chloride concentration is 1 M at equilibrium, the calculated chloride concentration in the resin is

$$[Cl^-]_r \sim \frac{1 \times 1}{4} = 0.25\ M$$

Resin in equilibrium with a 0.1 M sodium chloride solution will have only $\frac{1}{100}$ of this chloride concentration:

$$[Cl^-]_r = \frac{0.1 \times 0.1}{4} = 2.5 \times 10^{-3}\ M$$

Factors affecting extent of ion exchange. The relative adsorbabilities of ions by resins depend on a number of factors, among which the following are important.

Nature of the ions. Other factors being the same, the higher the charge of an ion, the more strongly it is adsorbed.[11] The trivalent rare earth ions are adsorbed more strongly than calcium by a cation-exchange resin, and calcium is adsorbed more strongly than the alkali metal ions. The size of the hydrated ion is also important. Adsorption increases with a decrease in the solvated ionic volume. In the alkali metal series the order of increasing adsorption is Li < Na < K < Rb < Cs. Sodium and potassium can be separated fairly easily by ion exchange, because potassium is adsorbed several-fold more strongly than sodium on some cation resins. In general, adsorption increases as the hydration of an ion decreases and its polarizability increases. Large organic ions are often found to be adsorbed strongly; however, if they are too large they may not be able to enter the resin structure. Strong-base anion exchangers show strong affinity for complex anions.

Character of the ion-exchange resin. The nature of the functional groups influences the extent of exchange of a particular ion. For example, hydrogen ion is preferentially exchanged by resins containing weakly acidic groups. Attempts have been made, with some success, to make resins more selective by introducing chelating groups that strongly bind certain metal ions.

The extent of cross-linking also affects the relative adsorbabilities of ions. With an increase in cross-linking, differences in adsorbability due to different ionic sizes are accentuated.

[11] The term *adsorption* is used in this chapter to cover the uptake of a solute by a solid irrespective of the mechanism, but usually by exchange. Some workers prefer the use of the general term *sorption*.

It appears that there is stronger adsorption of ions resembling the matrix of the exchanger. Thus benzylammonium cation is adsorbed more strongly than aliphatic amines of about the same molecular weight, which may in part be attributed to a greater "affinity" of the former anion for the matrix of the resin.[12]

Composition of the solution. The relative adsorbabilities of two ions may depend upon their concentrations in the solution, as will be demonstrated later. The concentration of substances combining with ions to form slightly dissociated uncharged molecules or complex ions in solution will have an enormous effect on adsorbability. Thus cobalt is not adsorbed at all from 1 *M* hydrochloric acid (in which it is present chiefly as Co^{++} and $CoCl^{+}$) by a strong-base anion-exchange resin, whereas it is quite strongly adsorbed from 8–9 *M* hydrochloric acid in which anionic cobalt chloro complexes are formed.

Capacity and loading. The number of equivalents of ion that can be adsorbed by exchange with a unit weight or volume of resin is the capacity of the resin. For resins in common use, capacities range from about 3 to 6 meq/g of dry resin. On the basis of resin *bed* volume the capacity is roughly 1–2 meq/ml, and depends on the swelling.[13]

By loading, L, is meant the fraction of the exchange resin converted:

$$L = \frac{a[A^{\pm a}]_r}{c} \tag{19}$$

where a is the charge of the ion, $[A^{\pm a}]_r$ is the concentration of adsorbed ion (replacing some other ion) in the resin, and c is the capacity. Especially with highly cross-linked resins, adsorbabilities and selectivities depend on the loading.

Rate of exchange. Consider the exchange reaction

$$RA + B^+ \rightleftharpoons RB + A^+$$

Three processes are involved which can determine the rate:

1. Diffusion of B^+ to the surface of the resin particle and simultaneous diffusion of A^+ from the surface into the solution.
2. Diffusion of B^+ into the resin to the exchange site and diffusion of A^+ out of the resin.
3. Replacement of A^+ by B^+ at the exchange site.

The third process is very rapid and never seems to be of importance (conceivably it could be the limiting factor in some chelating exchangers). The exchange rate is therefore diffusion-controlled, and, depending on conditions, either process 1 or 2 may be rate-determining.

The limiting factor in process 1 is the diffusion of the ions through the "Nernst film," the stagnant film existing at the surface of the particles even when the bulk of the solution is well mixed, but whose thickness diminishes with an increase in agitation. The rate of film diffusion increases with a rise in temperature. In step 2 (particle diffusion), the degree of cross-linking of the resin plays an important role. At low cross-linking, swelling is greater and porosity is increased so that the diffusion rate is increased. At high cross-linking, where the channels are more tortuous and make up a smaller fraction of the volume, diffusion coefficients in the resin may be several orders of magnitude less than at low cross-linking. The diffusion rate within the resin may approach that in aqueous solution at very low cross-linking.

[12] More or less anomalous adsorption of substances by exchange resins has been attributed to adsorption by the matrix. For example, weak acids, such as phosphoric, may be retained by a cation-exchange resin as a result of the action of van der Waals' forces. For an examination of resin selectivity in aqueous solutions, see R. M. Diamond and D. C. Whitney, Chapter 8 in J. A. Marinsky (ed.), *Ion Exchange*, vol. 1, Dekker, New York, 1966.

[13] Dowex 50-X8 is reported to have a capacity of 5.2 meq/g of dry H^+ resin and 1.8 meq/ml of fully swollen H^+ resin. The corresponding values for Dowex 1 or Amberlite IR-400 (6–8 per cent DVB) are approximately 3 meq/g and 1 meq/ml.

According to Helfferich,[14] the rate-determining step can be predicted from the value of the expression

$$\frac{Xd_r\delta}{Cdr}(5 + 2\alpha_{\mathrm{B}}^{\mathrm{A}}) \tag{20}$$

where X = concentration of fixed ionic groups in resin (or more generally in exchanger)
C = concentration of solution in equivalents
d_r = interdiffusion coefficient of ions A and B in resin
d = interdiffusion coefficient in film
r = bead radius
δ = film thickness ($\sim 10^{-2}$–10^{-3} cm, depending on agitation)
$\alpha_{\mathrm{B}}^{\mathrm{A}} = (C_{\mathrm{A}})_r C_{\mathrm{B}}/(C_{\mathrm{B}})_r C_{\mathrm{A}}$

If the value is $\ll 1$, the rate-determining step is particle diffusion; if the value is $\gg 1$, it is film diffusion.

The exchange rate may be determined by film diffusion when the exchanger has a high concentration of fixed ions, a low degree of cross-linking, small particle size, and when the solution is dilute and the agitation poor.

In column operation, particle diffusion is likely to be the rate-controlling process. The necessity for attaining an approach to equilibrium in a reasonable time period limits the degree of resin cross-linking that is practically useful. Decreasing the particle size is very beneficial in increasing the rate of exchange, but it cannot be carried too far, else the flow rate through the column becomes prohibitively slow (unless pressure is applied). Raising the temperature increases the exchange rate and flow rate of solution through the resin. Small ions exchange more rapidly than large.

Ion-exchange equilibria

The concentration equilibrium constant of the reversible reaction

$$\mathrm{R^-A^+ + B^+ \rightleftharpoons R^-B^+ + A^+}$$

with a given resin may be written

$$\frac{[\mathrm{B}]_r[\mathrm{A}^+]}{[\mathrm{A}]_r[\mathrm{B}^+]} = K \qquad \text{(constant at specified ionic strength and loading of resin)} \tag{21}$$

where concentrations without subscripts refer to the aqueous phase and concentrations with subscript r refer to the resin phase.

Introducing activity coefficients, f for the solution and g for the resin, we obtain the activity-exchange constant

$$\frac{[\mathrm{B}]_r g_{\mathrm{B}}[\mathrm{A}^+]f_{\mathrm{A}}}{[\mathrm{A}]_r g_{\mathrm{A}}[\mathrm{B}^+]f_{\mathrm{B}}} = K^0 \tag{22a}$$

or

$$\frac{[\mathrm{B}]_r[\mathrm{A}^+]}{[\mathrm{A}]_r[\mathrm{B}^+]} = K^0\left(\frac{g_{\mathrm{A}}f_{\mathrm{B}}}{g_{\mathrm{B}}f_{\mathrm{A}}}\right) = K \tag{22b}$$

and

$$\frac{[\mathrm{B}]_r}{[\mathrm{A}]_r} = K^0\left(\frac{g_{\mathrm{A}}f_{\mathrm{B}}}{g_{\mathrm{B}}f_{\mathrm{A}}}\right)\frac{[\mathrm{B}^+]}{[\mathrm{A}^+]} = K\frac{[\mathrm{B}^+]}{[\mathrm{A}^+]} \tag{22c}$$

The exchange constant K may be looked upon as the relative affinity coefficient of the two ions for the resin; it is sometimes called the selectivity coefficient for B^+ with respect to A^+. If $K > 1$, B has a greater affinity than A for the resin. The selectivity increases with cross-linking. Table 13-2 contains some illustrative values of K for cations on a sulfonated polystyrene resin at 25°C.

[14] Helfferich, Ref. 7, p. 255.

TABLE 13-2. **Some Values for Ion-Exchange Equilibrium Constants with Dowex 50 Resin, Based on $K_H = 1$[a]**

$$K_{nH}^{M} = \frac{[M]_r[H^+]^n}{[M^{+n}][H]_r^n}$$

	4% DVB	8% DVB
Li	0.8	0.8
H	1	1
Na	1.2	1.6
K	1.7	2.3
Rb	1.9	2.5
Cs	2.0	2.6
Ag	3.6	6.7
Mg	1.0	1.3
Zn	1.1	1.5
Cu	1.2	1.8
Ca	1.9	3.2
Sr	2.5	5.2
Ba	6.3	16.2
Ce, La	6.9	22

[a] Based on values given by A. Ringbom, *Complexation in Analytical Chemistry*, Wiley-Interscience, New York, 1963, p. 202, whose constants are derived from those of O. D. Bonner and coworkers. Concentrations in the resin are expressed as milliequivalents of exchanging ions per gram of dry H-form resin, in solution as milliequivalents per milliliter, i.e., as N. The constants in the table are to be looked upon as very approximate because of their variation with electrolyte concentration and loading of the resin. Besides, commercial resins are not uniform in composition, and a relatively small variation in the degree of cross-linking can have a relatively large effect on the value of exchange constant. Constants such as the ones given are useful for calculations at low loadings, especially when complexing of cations is involved and accurate exchange constants are not required. Exchange constants for anions on Dowex 1 and Dowex 2 are also tabulated by Ringbom. The range of the constants for anions tends to be greater than for cations.

Equation (22a) may be written

$$\frac{[B]_r}{[B^+]} = D_{B^+} = K^0\left(\frac{g_A f_B}{g_B f_A}\right)\frac{[A]_r}{[A^+]} = K\frac{[A]_r}{[A^+]} \tag{23}$$

D_{B^+} is the distribution coefficient (actually *constant*) of the ion B^+ between resin and aqueous solution and is equal to the distribution coefficient of B if B^+ is the only form of B in the solution. When other forms of B (hydrolyzed species and cationic, anionic, or neutral complexes) are present, the distribution coefficient is given by

$$\frac{[B]_r}{\sum [B]} = D_B$$

D_{B^+} will be approximately constant, independent of the concentration of B^+, if $[A]_r$ is large compared to $[B]_r$ and $[A^+]$ is large compared to possible values of $[B^+]$. This condition is fulfilled if B is a trace constituent and $[A^+]$ is not too small, so that it does not change much on exchange. If again B is a trace constituent, so that $[A]_r \sim$ constant, D_{B^+} varies inversely with $[A^+]$:

$$D_{B^+} \cong \frac{\text{constant}}{[A^+]} \tag{24}$$

and a plot of log D_B^+ versus log $[A^+]$ gives a straight line of slope -1.

If B has a charge of $+2$ (and A $+1$ as before), $D_{B^{++}}$ varies inversely with $[A^+]^2$ when $[A]_r$ remains approximately constant and a plot of log $D_{B^{++}}$ versus log $[A^+]$ gives a straight line of slope -2. If the ions A and B have charges a and b, the slope of the line in the log plot is $-b/a$. Figure 13-4 illustrates the increasing adsorption of Ca^{++} compared to that of Na^+ by a cation-exchange resin in the hydrogen form as the acidity of the solution is decreased. This effect—the favored adsorption of ions of higher charge at low concentrations—is sometimes called *electroselectivity*. It accounts for the success of water softening by ion exchangers.

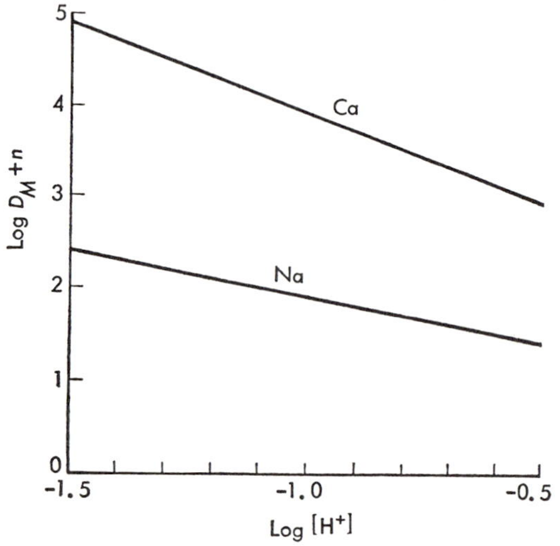

Fig. 13-4. Distribution coefficients of Na^+ and Ca^{++} adsorbed on Dowex 50–X8 resin in the H^+ form as a function of H^+ concentration in solution at equilibrium (calculated from exchange constants in Table 13-2). Capacity of resin taken as 5 meq per gram. Effect of variable ionic strength not considered.

The analyst is primarily interested in D_B, the distribution coefficient based on all forms of B in the aqueous solution. D_B is proportional to D_{B^+},

$$D_B = \frac{[B]_r}{\sum [B]} = F_B \frac{[B]_r}{[B^+]} = F_B D_{B^+} \tag{25}$$

when $[B^+]$ is a fixed fraction, F_B, of $\sum [B]$. This condition is fulfilled at a given pH and a given concentration of anionic complexing agent if the excess of the latter is such that it remains essentially constant when B varies, and if all the complex species of B contain only one atom of B per ion or molecule (*and these are not adsorbed*). For example, if B forms the chloro species BCl, BCl_2^-, BCl_3^{--}, . . . , the ratio $[B^+]/\sum [B]$ at a given chloride concentration will be essentially independent of the total concentration of B if the concentration of Cl^- is large compared to that of B. The analytical distribution coefficient can then be written

$$D_B = \frac{[B]_r}{\sum [B]} = \frac{F_B K [A]_r}{[A^+]} \quad \text{[from Eq. (21); at specified ionic strength and loading of resin]} \tag{26}$$

The value of D_B can be obtained experimentally as indicated on page 303, or by shaking a known weight (or volume) of resin with a solution of B having the specified composition and determining the amount of B adsorbed. D_B can be calculated approximately if the exchange constant K and the formation constants of the complex species are known and only B^+ is adsorbed.[15]

If the term $g_A f_B/g_B f_A$ in Eq. (22b), or more generally $g_A^b f_B^a/g_B^a f_A^b$, is not constant, the exchange constant K, based on concentrations, will vary. This quotient will not be constant if the resin imbibes electrolyte, or if the electrolyte content of the solution changes (except at low concentrations when A and B are univalent so that $f_A \sim f_B$). At low ionic strengths, when little electrolyte enters the resin, g_A/g_B depends essentially on the loading. Serious deviations from constancy may occur when $L > 0.01$.[16] However, useful approximate calculations of the distribution coefficient can often be made at higher loadings than this from tabulated values of K for a given resin.

Example 1: 2.0 g of Dowex 50-X8 resin in the H^+ form is shaken with (a) 100 ml of 0.1 N hydrochloric acid containing a trace of calcium, (b) with 100 ml of a solution 0.1 N in hydrochloric acid and 0.001 M in calcium ion. Find the fraction of Ca remaining in solution at equilibrium, given that

$$\frac{[\mathrm{Ca}]_r[\mathrm{H}^+]^2}{[\mathrm{Ca}^{++}][\mathrm{H}]_r^2} = 3.2$$ (concentrations in the resin are expressed in meq/g, in the solution in meq/ml)

and the capacity of the resin is 5 meq/g.

Ans. (a) $$\frac{[\mathrm{Ca}]_r}{[\mathrm{Ca}^{++}]} = 3.2 \times \frac{[\mathrm{H}]_r^2}{[\mathrm{H}^+]^2}$$

$[\mathrm{H}]_r$ remains practically constant, so that

$$\frac{[\mathrm{Ca}]_r}{[\mathrm{Ca}^{++}]} = 3.2 \times \frac{25}{0.01} = 8000$$

$$\text{Ca in solution/Ca in resin} = \frac{1 \times 100}{8000 \times 2} = \frac{1}{160}$$

Therefore,

$$\text{fraction of Ca in solution} = \frac{1}{161} \text{ (or 0.62 per cent)}$$

(b) Assume that most of the calcium enters the resin, so that

$$[\mathrm{H}]_r = 5 - 0.1 = 4.9 \text{ and } [\mathrm{H}^+] = 0.102$$

$$\frac{[\mathrm{Ca}]_r}{[\mathrm{Ca}^{++}]} = 3.2 \times \frac{4.9^2}{0.102^2} = 7400$$

$$\text{Ca in solution/Ca in resin} = \frac{1 \times 100}{7400 \times 2} = \frac{1}{148}$$

$$\text{Fraction of Ca remaining in solution} = \frac{1}{149} \text{ (or 0.67 per cent)}$$

The result shows that the assumption of almost complete exchange of calcium is justified.

Example 2: Find the distribution coefficient of calcium when Dowex 50-X8 resin in the sodium form is equilibrated with the following solutions containing trace amounts

[15] The tables given in A. Ringbom, *Complexation in Analytical Chemistry*, Wiley-Interscience New York, 1963, are useful in such calculations.

[16] For a fuller discussion, see K. A. Kraus in J. H. Yoe and W. J. Koch (eds.), *Trace Analysis*, Wiley, New York, 1957, p. 71.

of calcium: (a) 0.1 M Na^+, (b) 0.1 M Na^+ buffered at pH 6, 0.001 M EDTA (H_4Y).

$$\frac{[Na]_r[H^+]}{[Na^+][H]_r} = 1.56$$

$$\frac{[CaY^{--}]}{[Ca^{++}][Y^{4-}]} = 5 \times 10^{10}$$

$[Y^{4-}] = 1.5 \times 10^{-8}$ M in 0.001 M EDTA at pH 6
Capacity of resin is 5 meq/g

Ans. (a) $$\frac{[Ca]_r[Na^+]^2}{[Ca^{++}][Na]_r^2} = \frac{3.2}{1.56^2} = 1.32$$

$$D_{Ca} = \frac{[Ca]_r}{[Ca^{++}]} = 1.32\frac{[Na]_r^2}{[Na^+]^2} = \frac{1.32 \times 25}{0.01} = 3300$$

(b) $$\frac{[CaY^{--}]}{[Ca^{++}]} = 5 \times 10^{10} \times 1.5 \times 10^{-8} = 750$$

$$D_{Ca} = \frac{[Ca]_r}{[CaY^{--}]} = \frac{3300}{750} = 4.4$$

ION-EXCHANGE CHROMATOGRAPHY

With few exceptions, ion-exchange resins (or more generally, ion exchangers) are used in the form of a column and the principles of chromatography apply. The use of a column ensures quantitative exchange by bringing the sample solution into contact with fresh resin as it flows down the column. If the different ionic constituents of the sample, having the same sign, have different distribution coefficients, as they almost always will, separations can be effected by elution as will be shown.

Replacement of one ion by another

Consider first the simple case in which a solution containing only the salt BCl is percolated through a well-packed column of cation resin in the A^+ form. A replacement of A^+ in the resin by B^+ in the solution occurs:

$$RA + B^+ \rightleftharpoons RB + A^+$$

The upper part of the column will soon consist of exhausted resin, i.e., all in the RB form, and the exchange zone moves down the column as solution continues to be introduced (Fig. 13-5) and A^+ is carried off in the effluent. Other factors being constant, the rate at which the exchange zone moves down the column depends on the concentration of BCl in the sample solution and the capacity of the resin. When the front of the exchange zone reaches the bottom of the resin column, B^+ appears in the effluent, marking *breakthrough* (Fig. 13-6). The breakthrough capacity of the resin column is obtained by multiplying the volume added up to the breakthrough point by the concentration of BCl in the solution. It will be a little less than the column capacity and it varies with conditions (e.g., flow rate). The *column* capacity can be determined by determining A^+ in the effluent after complete replacement (excess of BCl) or by eluting B^+ with a suitable solution and determining it. Thus, if A^+ is H^+, the latter can be titrated with base after replacement with Na^+ added in excess.

For a given column and a given flow rate the sharpness of the boundary between B and A—the width of the exchange zone—depends on the relative affinities of B^+ and A^+ for the resin. If B^+ is more strongly adsorbed than A^+, so that its adsorption iso-

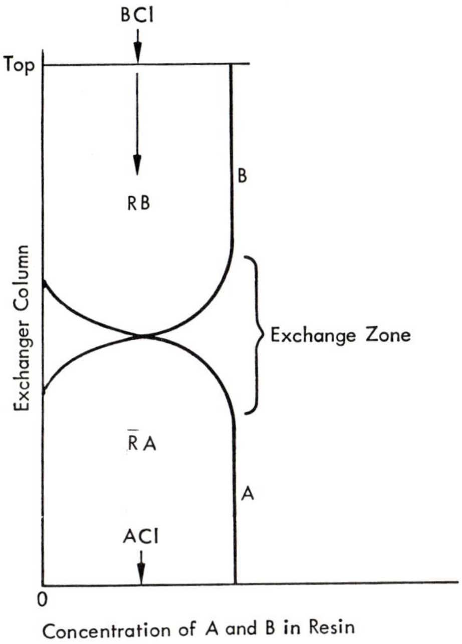

Fig. 13-5. Ideal concentration curves (isochrones) of B^+ and A^+ in column of exchanger RA on continuous introduction of B^+.

therm is represented by the upper curve in Fig. 13-7, the zone will be narrow and the effect of factors tending to broaden the zone will be diminished by the "self-sharpening" character of the boundary. The broadening factors include nonattainment of equilibrium, mixing of solution in adjacent levels during flow past the grains of resin, inhomogeneous flow because of irregularities in the packing, and, usually less important, longitudinal diffusion of ions. If B^+ is less strongly adsorbed than A^+, the boundary broadens as it moves down the column, and the extraneous factors promoting broadening are not counteracted. In the first case, B^+ adsorbed more strongly than A^+, zones of high concentration of B^+ tend to overtake zones of low concentration of B^+ as the exchange zone moves down the resin column.

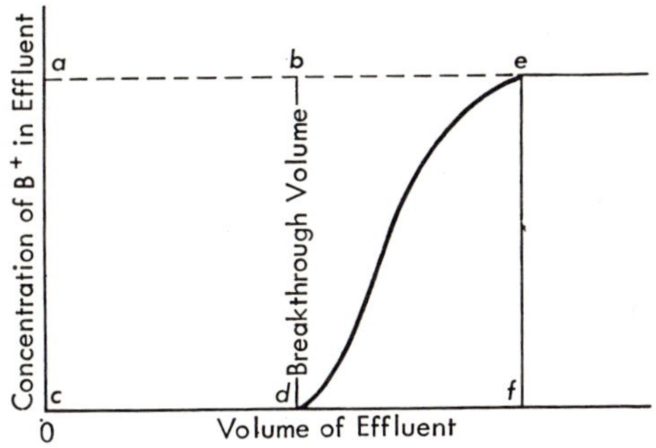

Fig. 13-6. Elution curve (isoplane) in replacement of A^+ in RA by B^+. B^+ is first detected in effluent at *d*. The breakthrough capacity is represented by the area *abdc*, the column capacity by the area *aefc*.

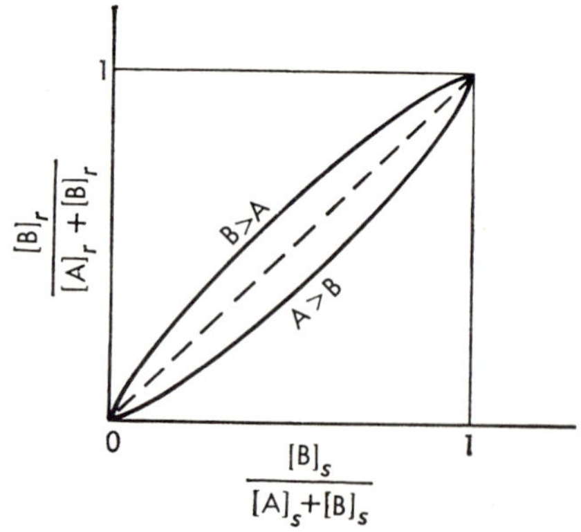

Fig. 13-7. Ion-exchange isotherms. B > A: ion B adsorbed more strongly than ion A; A > B, A adsorbed more strongly than B.

In practice, the resin column would be made sufficiently long so that when all BCl has been added, the center of the exchange zone will not have reached the midpoint of the column (and usually not this far). When water is now added to displace the BCl in the exhausted upper part of the column into the lower active portion, the zone will move a relatively short distance to complete the replacement of B^+ by A^+. Most of A^+ will by this time have passed out of the column as ACl, because the solvent front moves much more rapidly than the exchange zone (p. 290). The remainder of ACl can be washed out with a relatively small volume of water.

The exchange zone will be narrower the smaller the height of a theoretical plate of the column, which is especially a function of the particle size of the exchanger and the flow rate. The closer the approach to equilibrium, the narrower the zone. The smaller the particle size and the slower the flow rate, the narrower the zone will therefore be.

Elution

We now consider the separation of two (or more) constituents from each other by the use of resin (more generally, ion-exchanger) columns. The problem is simple if the two constituents can be transformed into ions of opposite charge or if one is present in uncharged form. But the solution of the general problem requires separation of ionic species of the same charge. This can be done by differential *elution*. After the ions (A, B, . . .) have been adsorbed at the top of the column, a suitable solution (whose counterions are *less* strongly adsorbed than those of the constituents) is passed through the column. The adsorption bands of A and B move down the column at different rates, depending on the distribution coefficients of A and B in the eluting solution. The distance between the two bands increases directly with the distance traveled, although the bands become broader. If the bands do not overlap significantly when near the bottom of the column, separation has been effected, and A and B can be collected in successive fractions of the eluate.[17]

This process requires closer examination. If the volume of sample solution added to the column is relatively small and the ion in question is moderately strongly adsorbed,

[17] The use of an eluting solution containing a counterion C which is more strongly adsorbed than either A or B results in development by *displacement*, in which the bands of A and B are driven down the column in front of C. If A is held more strongly than B, C displaces A and A displaces B, so that A and B move down the column in adjacent bands not separated by C. This type of chromatography is of value in preparative work (for example, in obtaining pure rare earth salts), because the bands are fairly uniform in concentration of A and B except in a relatively narrow region of overlap. It is not suitable for quantitative separations because of this overlap.

the adsorption band soon assumes, on elution, an approximation to the Gaussian curve (p. 290, Fig. 13-3) as it moves down the column. The distance, l, traversed by the band maximum is proportional to the elution constant $\mathscr{E}$ and to the volume of eluting solution (V), and inversely proportional to the cross-sectional area (a) of the column[18]:

$$l = \frac{\mathscr{E} V}{a} \tag{27}$$

The value of $\mathscr{E}$ for a particular ion and a solution of specified composition in a given resin column can be determined by obtaining the volume, V_{max}, of eluting solution required to bring the peak of the elution band to the bottom of the column ($l = l_{col}$):

$$\mathscr{E} = \frac{al}{V} = \frac{al_{col}}{V_{max}} \tag{28}$$

But $al_{col} = V_{col}$, the geometric column volume, so that

$$\mathscr{E} = \frac{V_{col}}{V_{max}} \tag{29}$$

If the band is colored or radioactive, $\mathscr{E}$ can be found by measuring the distance the center of the band has moved on addition of a known volume of eluant to a column of known cross-sectional area. Otherwise, the solute in question must be determined in successive small volumes of eluate near V_{max}.

We wish to relate $\mathscr{E}$ to D. If the fractional interstitial volume of the column, occupied by the mobile phase, is denoted by $i\,[= A_m/(A_m + A_s)$ (p. 289)], the eluting solvent front has advanced the distance V/ai while the band maximum has moved the distance $\mathscr{E}V/a$. Therefore,

$$R_f = \frac{\text{distance moved by band maximum}}{\text{distance moved by solvent front}} = \mathscr{E}i \tag{30a}$$

or

$$\mathscr{E} = \frac{R_f}{i} \tag{30b}$$

But

$$R_f = \frac{i}{i + D(1 - i)} \qquad \text{[Eq. (12), p. 290]} \tag{31a}$$

and

$$\mathscr{E} = \frac{1}{i + D(1 - i)} \tag{31b}$$

Now $D(1 - i)$ is the distribution coefficient based on a unit volume of resin *bed*. It may be denoted D_v. Therefore,

$$\mathscr{E} = \frac{1}{D_v + i} \qquad \text{and} \qquad D_v = \frac{1 - i\mathscr{E}}{\mathscr{E}} \tag{32}$$

Also,

$$V_{max} = V_{col}(D_v + i) = \frac{V_{col}}{\mathscr{E}} \tag{33}$$

The value of i can be found by determining $\mathscr{E}$ of a substance not adsorbed on the resin column. Another way is to fill the interstices of a resin column RH or RCl with 0.01 M HCl, wash this out with water, titrate with base, and apply a correction for the acid in the space below the disk and in the stopcock; the geometric column volume is found easily enough. In approximate calculations an average value of 0.4 may be taken for i (resin beads). When D_v is large, $\mathscr{E} \cong 1/D_v$.

[18] Kraus and coworkers, Ref. 16.

The extent of overlap of the elution bands of two ions A and B for a given ion-exchange resin column with a given eluting solution depends upon (1) the interval separating the peaks, which is determined by the distribution coefficients (or elution constants) of A and B, and the distance the adsorption bands have moved down the column, and (2) the widths of the bands, which depend upon the number of theoretical plates in the column (and consequently on the height of a theoretical plate) as well as on the distance the bands have moved down the column. If these parameters are known, the amounts of A and B recovered as fractions of the total amounts can be calculated for specified volumes of eluate, and hence the degree of separation can be obtained. Or, given the degree of recovery and separation required, the length of column needed can be calculated. Such calculations are based on theoretical Gaussian curves and linear adsorption isotherms are assumed. Elution curves are not always good approximations of symmetric Gaussian curves. The tail of an elution curve may be drawn out, so that a good deal more than the calculated volume of eluate may be needed to obtain a specified recovery of an ion. Tailing is especially likely to occur when the flow rate is not slow enough to allow approach to approximate equilibrium in the column. A well-packed column of uniform resin grains of small particle size, lightly loaded, is necessary for separation of substances having similar elution constants. Even under favorable conditions, the calculated results cannot be looked upon as highly accurate, but even rough values are of great aid in predicting the feasibility of separations and in lessening the labor of finding the best experimental conditions. The results of favorable calculations must always be checked experimentally. Unfavorable calculated results are not likely to be found better on test.

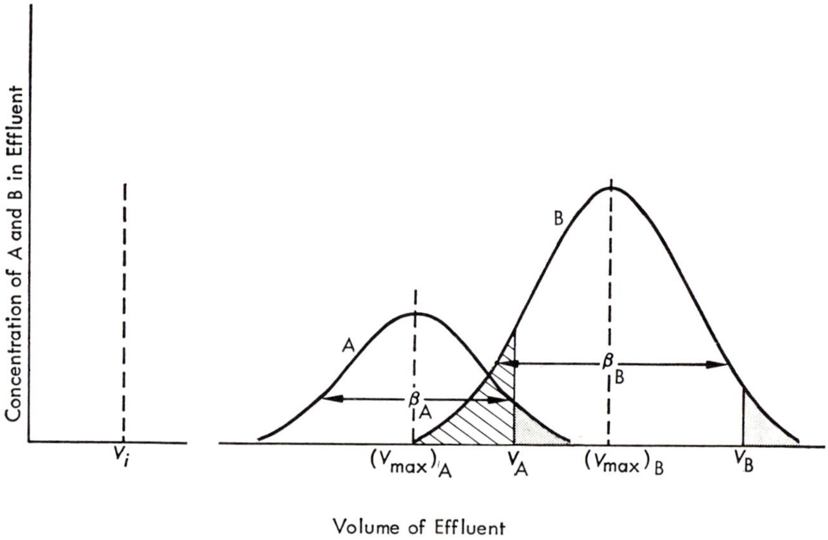

Fig. 13-8. Illustrating separation of A and B elution. The stippled area under the Gaussian curve A represents the fraction of A not eluted when the volume V_A of effluent is collected (this fraction is later collected with B) and the stippled area under curve B represents the fraction of B not eluted when a further volume of effluent $V_B - V_A$ is collected. In addition, the fraction of B represented by the cross-hatched area of B is collected with the major portion of A in the first portion V_A of effluent and thus escapes recovery. V_i represents the interstitial volume. The elution curves are shown as symmetrical, but this is not strictly correct since the band width increases as the band moves down the column. If β, in volume units, is small compared to V_{max}, the error is usually not important.

The procedure for finding the extent of separation can be illustrated with the aid of Fig. 13-8, which represents the elution curves of the two ions A and B. The volumes of eluting solvent or solution required to bring the band maxima from the top to the bottom of the resin column are designated as $(V_{max})_A$ and $(V_{max})_B$. The volume V_A of effluent (equal to the volume of the eluant added) is collected to contain most of A and a little of B, and the further volume $V_B - V_A$ is collected to contain most of B and a little of A. We wish to find the amounts of A and B in the two portions of effluent. The fraction of A in the portion V_A is given by the area under the elution curve of A from V_i (representing $-\infty$) to the vertical boundary at the abscissa V_A, the total area being taken as unity. The volume required to elute to a concentration $1/e$ (or other specified fraction) of the concentration at $(V_{max})_A$ must be known. This gives the elution band width,[19] which is a function of the number of theoretical plates (see p. 291). With $V_A - (V_{max})_A$ known in terms of $(V_e)_A - (V_{max})_A$ $(= \frac{1}{2}\beta_A)$, the required area is found with the aid of a table of the normal curve of error.[20] Similarly,

TABLE **13-3. Areas under Gaussian Curve Representing Adsorption Band in a Chromatographic Column Giving Fraction of Solute Eluted**[a]

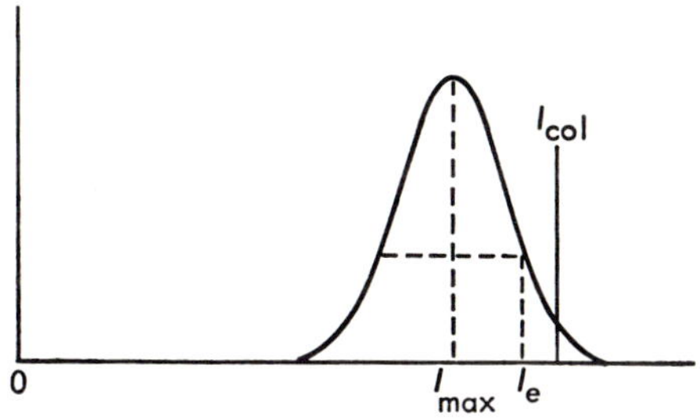

$\frac{l_{col} - l_{max}}{l_e - l_{max}} = \varphi$	*Fraction Not Eluted*[b]	*Fraction Eluted (= Recovery Factor)*	*Relative Concentration*
0	0.500	0.500	1.00
0.5	0.76	0.24	0.78
1	0.921	0.079	0.37
1.5	0.983	0.017	0.106
2	0.998	0.002	0.018
2.19	0.999	0.001	
2.5	0.9998	0.0002	0.002
2.64	0.9999	0.0001	
3	0.99999	0.00001	0.0001

[a] l_{col} = length of column representing point on adsorption band where cut is made; l_{max} = distance of adsorption band maximum from column top when cut is made; l_e ($> l_{max}$) = distance from column top to point on adsorption band at which solute concentration (C_e) is $1/e$ (= 0.37) of that at band maximum (C_{max}); $l_e - l_{max}$ = the half-band width $[= (2hl_{max})^{1/2}]$. The difference $l_{col} - l_{max}$ is expressed in terms of $l_e - l_{max}$. The volume of effluent collected at the point of cut is the volume of eluant required to bring the peak of the adsorption band to l_{max} $(= al_{max}/\mathscr{E})$.

[b] When $l_{max} > l_{col}$, this column gives fraction eluted.

[19] Some workers prefer to express the band width as the standard deviation, σ, of the Gaussian curve. The concentration corresponding to the standard deviation, C_σ, is $0.61C_{max}$.

[20] We designate $(V - V_{max})/(V_e - V_{max})$ as φ. Values of areas under the normal error curve corresponding to values of φ may be obtained from Table 13-3 and Fig. 13-9.

the fraction of the total amount of B in this portion of effluent is found as the ratio of the area under curve B from $V = V_i$ to V_A to the whole area. This area is shown cross-hatched in Fig. 13-8. The values so obtained are the recovery factors (p. 249), R_A and R_B, for A and B in the volume V_A of effluent. The ratio of B to A in this portion of effluent is given by

$$\frac{Q_B}{Q_A} = \frac{R_B(Q_B)_0}{R_A(Q_A)_0} \cong \frac{R_B(Q_B)_0}{(Q_A)_0} \cong S_{B/A}\frac{(Q_B)_0}{(Q_A)_0} \tag{34}$$

where $(Q_A)_0$ and $(Q_B)_0$ are the initial (total) amounts of A and B, and $S_{B/A}$ is the separation factor for B with respect to A under the specified conditions (see p. 250).

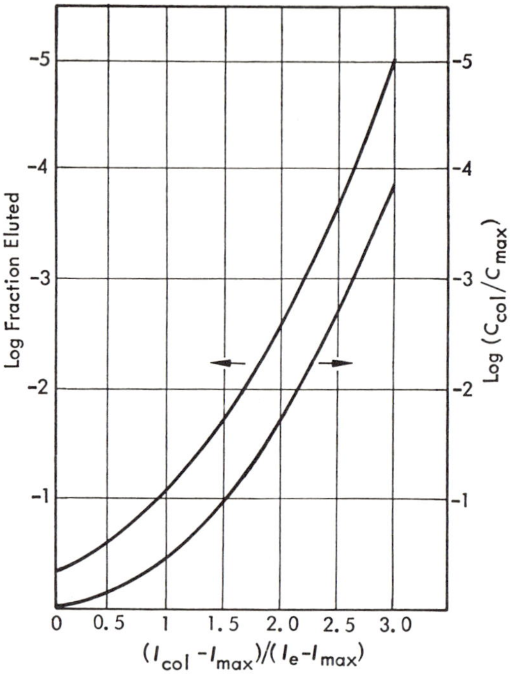

Fig. 13-9. Log of fraction of solute eluted and log (C_{col}/C_{max}) as functions of $(l_{col} - l_{max})/(l_e - l_{max})$. C_{col} is the concentration of the solute in the band at l_{col}.

The second portion of effluent collected, $V_B - V_A$, contains the fraction of B represented by the area under curve B between these limits. In addition, an amount of A corresponding to the area under curve A from V_A to V_B is present (shown stippled in the figure). These amounts can be found analogously to A and B in the first fraction of eluate V_A.

Example 3: With a certain cation-exchange column, $(V_{max})_{Na} = 205$ ml, $(V_{max})_{Li} = 165$ ml, and $(V_e)_{Na} = 194.5$ ml when 0.75 M HCl is used as eluant. 185 ml of effluent is collected to contain most of the lithium and an additional 40 ml to contain most of the sodium. Find the recovery factors for Li and Na in the two fractions. Is the separation acceptable?

Ans. The elution band width for Na, $2(205 - 194.5) = 21$ ml, is the band width corresponding to the passage of 194.5 ml of eluting solution through the column, for which the peak of the adsorption band lies within the column and not at the base of the column. The analogous elution band width for Li is not given, so it will be taken proportional to the Na band width, i.e.,

$$\frac{165}{205} \times 21 = 17 \text{ ml} \qquad \text{[Eq. (17), } h \text{ and } N \text{ assumed the same for Na and Li]}$$

and

$$(V_e)_{Li} = 165 - 8.5 = 156.5 \text{ ml}$$

Lithium. For an elution volume of 185 ml, at which the cut is made, the lithium elution half-band width is

$$\left(\frac{185}{156.5}\right)^{1/2} \times 8.5 = 9.2 \text{ ml}$$

Accordingly,

$$\varphi = \frac{V - V_{max}}{\beta/2} = \frac{185 - 165}{9.2} = 2.17$$

which (from Fig. 13-9) corresponds to a fractional recovery of 0.9989, which, of course, should be rounded off to 0.999.

Sodium. In the first fraction

$$\varphi = \frac{V_{max} - V}{\beta/2} = \frac{205 - 185}{(21/2)(185/194.5)^{1/2}} = 1.95$$

which corresponds to a fractional recovery of 0.003 (since $V_{max} > V$).

In the second fraction,

$$\frac{V - V_{max}}{\beta/2} = \frac{225 - 205}{(21/2)(225/194.5)^{1/2}} = 1.77$$

which corresponds to a recovery of 0.994. The net recovery of sodium is 0.994 − 0.003 = 0.991 of the total.

We may conclude from these results that the separation under the conditions specified is on the borderline of acceptability. The recovery of lithium (~99.9 per cent) is barely acceptable for macro amounts, and the separation from sodium is not sharp. If the sample ratio Na/Li = 1, the ratio Na/Li in the first fraction of effluent collected will be 0.003; if Na/Li = 10, the effluent ratio will be 0.03, etc. The calculations are based on ideal Gaussian curves, which may not be obtained in practice. An increase in column length is called for. Increasing the volume of the second fraction of eluate collected will increase the recovery of sodium, so that (with the same column length) the recovery factor for sodium can easily be increased to 1.000 − 0.003 = 0.997.

Required column length

In practice, the foregoing considerations are more likely to be applied in the reverse way to find the length of column of known characteristics required to effect a specified degree of separation when the $\mathscr{E}$'s (or D's) of A and B are known. In addition, the adsorption band widths of A and B must be known at any level in the column. If h is the same for A and B, the band widths of the two substances, measured in the column, are equal when the band maxima have traversed the same distance, i.e., when $(l_{max})_A = (l_{max})_B$. Actually h_A and h_B may differ appreciably and both band widths may need to be determined.

Let the quantity Q_A of A and Q_B of B be adsorbed at the very top of the column. As we have seen, on elution with a suitable solvent or solution, the distance between the peaks of the adsorption bands increases directly with the distance from the top of the column (the relative distances remain the same), but the widths of the peaks increase in proportion to the square root of the distance traversed. A net increase in separability thus results from an increase in the column length. The column length must be such that when the cut is made to recover the specified fraction R_A of A, the recovery of B, R_B, in that portion of eluate does not exceed a specified value, which is determined by the required separation factor (p. 250). The elution is continued with the same or another solvent or solution until the required recovery of remaining B has been obtained.

Consider the adsorption bands in the resin column (Fig. 13-10). The length of column required is indicated as l_{col}. The base of the column cuts the adsorption band of B so that the fraction $1 - R_B$ of B remains in the column, and it cuts the adsorption band of A (shown by the dashed curve, indicating its position *as if still in the resin*), so that the fraction R_A of A has been eluted from the column and $1 - R_A$ has not been eluted. The elution bands are assumed to be represented by the normal error curve. The following conditions must be satisfied:

$$l_{col} = (l_B)_{max} + \varphi_B (l_B)_{max}^{1/2} \left(\frac{w_B}{2}\right) \tag{35}$$

$$l_{col} = (l_A)_{max} - \varphi_A (l_A)_{max}^{1/2} \left(\frac{w_A}{2}\right) \tag{36}$$

where w is the band width in the column (measured at the concentration which is 0.37 of the maximum concentration in the band) when $l_{max} = 1$, and the φ's are the factors by which the respective half-band widths must be multiplied to give the stipulated recoveries. (See Table 13-3 or Fig. 13-9 for values of φ.)

The ratio $(l_A)_{max}/(l_B)_{max}$ is equal to the known ratio (ϵ) of the elution constants:

$$\frac{(l_A)_{max}}{(l_B)_{max}} = \frac{\mathscr{E}_A}{\mathscr{E}_B} = \epsilon \qquad \text{[from Eq. (27)]} \tag{37}$$

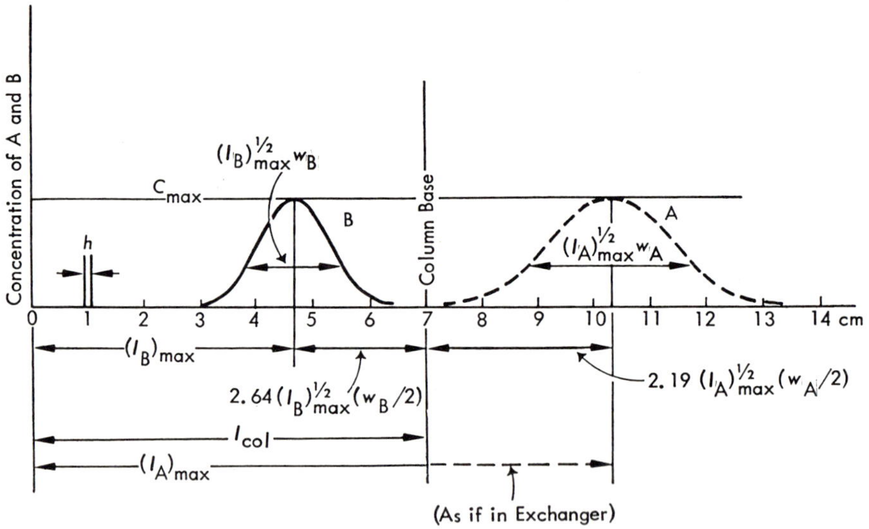

Fig. 13-10. Diagram illustrating required column length for chromatographic exchanger separations with special reference to Example 4. Concentrations of A and B are in terms of C_{max} for each (= 1). The adsorption band for A to the right of "column base" is hypothetical since it lies outside the column and is represented as if in the exchanger; the ratio of the area under curve A to the left of "column base" to the whole area as shown gives the fraction of A not eluted from the column ($= 1 \times 10^{-4}$). The height of a theoretical plate shown is the average ($h_A = 0.11$, $h_B = 0.08$ cm).

From Eqs. (35), (36), and (37) we find

$$(l_B)_{max} = \frac{(\varphi_B w_B + \varphi_A w_A \epsilon^{1/2})^2}{4(\epsilon - 1)^2} \tag{38}$$

and

$$l_{col} = \frac{(\varphi_B w_B + \varphi_A w_A \epsilon^{1/2})}{4(\epsilon - 1)} \left\{ \varphi_B w_B + \frac{\varphi_B w_B + \varphi_A w_A \epsilon^{1/2}}{\epsilon - 1} \right\} \tag{39}$$

The volume of added eluant (or collected effluent) at which the cut is to be made is given by

$$(V_{elu})_A = \frac{a(l_B)_{max}}{\mathscr{E}_B} = \frac{a(l_A)_{max}}{\mathscr{E}_A} \tag{40}$$

where a = cross-sectional area of column.

Similar considerations apply when a specified recovery of B is required with the ratio Q_B/Q_A given.

When the band widths of A and B are equal ($w_A = w_B = w$), we have

$$(l_B)_{max} = \frac{w^2(\varphi_B + \varphi_A \epsilon^{1/2})^2}{4(\epsilon - 1)^2} = \frac{2h(\varphi_B + \varphi_A \epsilon^{1/2})^2}{(\epsilon - 1)^2} \tag{41}$$

[since $w = 2(2h)^{1/2}$] and

$$l_{col} = w^2 \left\{ \frac{\varphi_B + \varphi_A \epsilon^{1/2}}{4(\epsilon - 1)} \right\} \left\{ \varphi_B + \frac{\varphi_B + \varphi_A \epsilon^{1/2}}{\epsilon - 1} \right\}$$

$$= 2h \left\{ \frac{\varphi_B + \varphi_A \epsilon^{1/2}}{\epsilon - 1} \right\} \left\{ \varphi_B + \frac{\varphi_B + \varphi_A \epsilon^{1/2}}{\epsilon - 1} \right\} \tag{42}$$

If the band peaks lie close together, h being small, so that $(l_B)_{max} \sim l_{col}$ when the cut is made, the approximate required column length may be found from Eq. (38) or (41).

It is possible to obtain an approximate value for $\mathscr{E}$ from the distribution coefficient D if this has been determined [batch equilibrium, conversion of D to D_v; Eq. (32)] or possibly calculated [Eq. (26), D converted to D_v]. The required diameter of the resin column will depend on the absolute amounts of A and B in the sample solution. If the amounts of the exchangeable ions exceed ~50 mg/cm² of resin column, the adsorption isotherm is not linear. Overloading of the column results in a drawn-out trailing edge of the adsorption band.

Example 4: Two ions A and B in the ratio $(Q_B)_0/(Q_A)_0 = 100$ are to be separated, with the requirements $R_A = 0.999$ and $Q_B/Q_A = 0.01$. With a 1 cm² × 10.0 cm column of the exchanger, $(V_{max})_A = 50$ ml, $(V_e)_A = 58$ ml, $(V_{max})_B = 110$, and $(V_e)_B = 125$ ml. What is the minimum length of exchanger column theoretically required for the separation?

Ans. $w_A = \dfrac{2\{(58/50 \times 10) - 10\}}{11.6^{1/2}} = 0.94$ cm

$w_B = 0.804$ cm

$\varphi_A = 2.19$

$\varphi_B = 2.64 \left(\text{corresponding to } R_B = \dfrac{0.01}{100} = 10^{-4}\right)$

$\epsilon = \dfrac{\mathscr{E}_A}{\mathscr{E}_B} = \dfrac{10/50}{10/110} = 2.2$

$(l_B)_{max} = 4.7$ cm [from Eq. (38)]

$l_{col} = 4.7 + 2.3 = 7.0$ cm [from Eq. (39) or (35)]

$(V_{elu})_A = 51.6$ ml [from Eq. (40)]

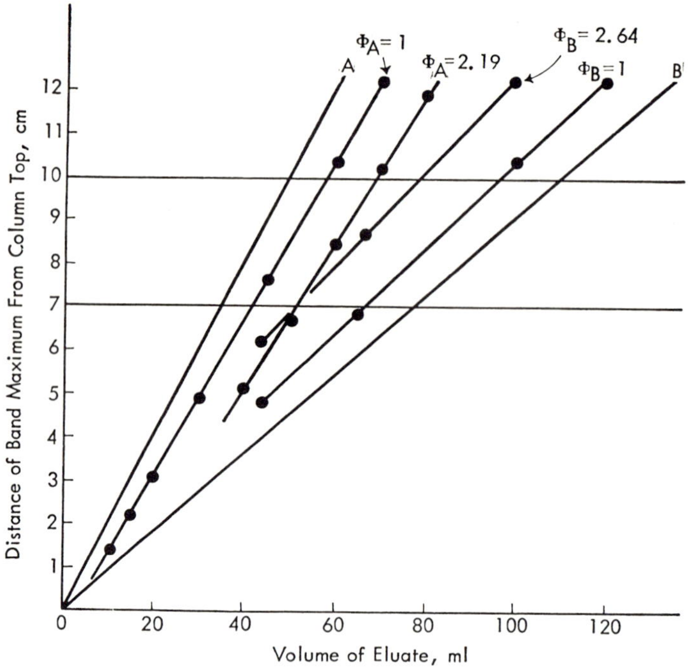

Fig. 13-11. Graphical determination of chromatographic column length for separations (Example 4). Lines A and B give distance of respective band maxima from top of column as function of eluate volume. Points defining curves for φ_A and φ_B are calculated from quantity $\varphi(l_{max})^{1/2}\, w/2$. Curves $\varphi_A = 2.19$ and $\varphi_B = 2.64$ intersect at 7.0 cm, which is the minimum column length for the specified separation; the corresponding eluate volume is 51.5 ml.

A graphical solution of this problem is shown in Fig. 13-11. Figure 13-10 represents the adsorption bands in the column.

Suppose B is also to be recovered. It seems reasonable to use an elution volume that theoretically will recover 99.999 per cent of B, thus introducing a margin of safety for departure of the actual elution curve from the theoretical curve. (Actually, since 0.01 per cent of B has been lost in the elution of A, the maximum recovery of B attainable is 99.99 per cent.) The factor corresponding to 99.999 per cent recovery is

$$\frac{l_{max} - l_{col}}{l_{max} - l_e} = 3$$

The band width at l_{max} is $(l_{max})^{1/2} \times 0.804$. Accordingly l_{max} is found by solving the quadratic equation

$$\frac{l_{max} - 7}{(0.804/2)(l_{max})^{1/2}} = 3$$

$$l_{max} = 10.98 \text{ cm, or } 11.0 \text{ cm}$$

The elution volume (total for B) is then $(11.0/10.0) \times 110 = 121$ ml (or an additional elution volume for B of $121 - 51.6 = 69$ ml).

Because of the uncertainties that have been mentioned, the analyst would choose a column length longer than the calculated minimum, e.g., 10 cm instead of 7 cm.

Other factors remaining the same, the more $\mathscr{E}_A/\mathscr{E}_B$ (or D_B/D_A) differs from unity, the easier will be the separation of A and B.[21] If A^{a+} and B^{b+} are cations eluted from a resin, which is mostly in the form RC, with an ion C^+, we have the relation

$$\frac{D_B}{D_A} = \frac{K_{AC}}{K_{BC}} \frac{[C]_r^{b-a}}{[C^+]^{b-a}} \tag{43}$$

where K_{AC} (and K_{BC} similarly) is given by

$$K_{AC} = \frac{[C]_r^a [A^{a+}]}{[A]_r [C^+]^a}$$

A and B are assumed to be present entirely as simple ions in the solution. If $a = b$, the C terms drop out. But this does not mean that the concentration of C^+ is not important. Even though a change in $[C^+]$ does not (ideally) alter the ratio D_B/D_A if $a = b$, it does change the absolute values of D_A and D_B. A suitable value of the *ratio* D_B/D_A (or $\mathscr{E}_A/\mathscr{E}_B$) is not the sole requirement in analytical separations. If the values of D are unduly large, an undesirably large volume of eluting solution will be needed. Thus, if $D_v = 100$ and $i = 0.4$, $\mathscr{E} \cong 0.01$, and 100 geometric column volumes ($V_{col}/0.01$) of solution will be needed to elute the band maximum to the base of the column. It may be advantageous to increase the concentration of C^+ to decrease the volume of solution required to elute A and B.

If A and B have different charges, variation in the concentration of C^+ in the eluting solution alters the separability. If $a = 1$ and $b = 2$, D_B/D_A increases as $[C^+]$ is decreased, $[C]_r$ being assumed approximately constant. An increase (or decrease) in D_B/D_A could be either favorable or unfavorable, depending on its original value. Suppose $D_B/D_A = 2$ at $[C^+] = 1$ ($a = 1, b = 2$). At $[C^+] = 2$, $D_B/D_A = 1$, at $[C^+] = 0.5$, $D_B/D_A = 4$.

With most elements, much greater changes in D_B/D_A can be obtained by differential complexing with a suitable complexing (or "masking") agent. If F_A and F_B are the fractions of the total amounts of A and B present as simple cations in the solution at a specified concentration of complexer and specified pH,

$$\frac{D_B}{D_A} = \frac{F_B}{F_A} \frac{K_{AC}}{K_{BC}} \frac{[C]_r^{b-a}}{[C^+]^{b-a}} \qquad \text{(see p. 298)}$$

F_B/F_A, and therewith D_B/D_A, can often be given a desired value by adjusting the concentration of the complexer, the pH, and possibly changing the oxidation state of one or both cations.[22] Anionic complexing agents find use both in separations of cations on cation exchangers and of complex metal anions on anion exchangers. With citrate, lactate, or EDTA as the complexing agent, even the rare earths can be separated from each other on a cation-exchange resin, although a long column with a large number of theoretical plates is necessary, because the distribution coefficients of these very similar metal ions are not greatly different under the most favorable conditions.

After one ion has been quantitatively eluted from a column, it is often advantageous to alter the composition or even the nature of the eluting solution to speed the elution of the second (and succeeding) ions.

Sometimes elutions are carried out with a continuous change in the concentration of the eluant (gradient elution).

[21] The ratio D_B/D_A is often called the *separation factor* in the literature. Actually, the extent of separation cannot be calculated from this ratio. The term separation factor should be restricted to R_B/R_A (p. 250).

[22] If the concentration of the complexing anion is made too high, the affinity of the complexed cation for the resin may become so low that elution development gives way to displacement development (p. 302).

Calculation of *h*. Glueckauf has proposed the following formula for approximate calculation of the height of a theoretical plate[23]:

$$h = 4r + \left[\frac{D_v}{(D_v + i)^2} \times \frac{0.142r^2u}{d_r}\right] + \left[\frac{D_v^2}{(D_v + i)^2} \times \frac{0.267r^2u}{d(1 + 70ru)}\right] \tag{44}$$

where r = particle radius (cm)
u = linear flow rate (ml/cm²/sec)
d_r = diffusion coefficient of solute in resin ($\sim 10^{-6}$ cm²/sec)
d = diffusion coefficient of solute in solution ($\sim 10^{-5}$ cm²/sec)

and D_v and i have their earlier meaning.

If $D_v > 5$ (so that $D_v > 10i$), i can be neglected for all practical purposes, so that

$$h \sim 4r + \frac{0.14r^2u}{d_rD_v} + \frac{0.27r^2u}{d(1 + 70ru)} \tag{45}$$

As u approaches 0, $h \sim 4r$.

The first term in the equation allows for particle size, irregular flow, and irregular packing of exchanger (channeling), the second term for the rate of attainment of equilibrium in the resin phase, and the third term for the rate of attainment of equilibrium across the liquid film around the particles.

Suppose the particle radius is 0.005 cm (corresponding to about 150 mesh), the flow rate is 0.6 cm/min (or 0.01 cm/sec) and $D_v = 10$. Using average values for diffusion coefficients, we find

$$h \sim 0.02 + \frac{0.14 \times 0.005^2 \times 0.01}{10^{-6} \times 10} + \frac{0.27 \times 0.005^2 \times 0.01}{10^{-5}(1 + 70 \times 0.005 \times 0.01)}$$
$$\sim 0.02 + 0.0035 + 0.007 \sim 0.03 \text{ cm}$$

It is seen that h is not constant for a given column; it varies with the flow rate (unless this term is small compared to the particle radius), the distribution coefficient, and the diffusion coefficients of the ions undergoing exchange. If average values are used for these quantities, curves can be constructed which give h as a function of particle radius and flow rate.[24] Such empirical curves are useful for obtaining a rough value of h from column variables. The empirical values agree with the experimental values within a factor of 3 or less, except for anion resins used in strongly acid solutions. For a 100-mesh resin ($r \sim 0.07$ mm), a rough average value of h is 0.5 mm for a flow rate of 0.1 ml/cm²/min (or 0.1 cm/min), 1 mm for 1 ml/cm²/min, and 3 mm for 5 ml/cm²/min. Even at the slow flow rate of 1 cm/min, h is more than $10r$.

APPLICATIONS OF ION EXCHANGE IN QUANTITATIVE ANALYSIS[25]

The following survey of the uses of ion exchange in analysis may be prefaced with some experimental details.

Ion-exchange resins for analytical work must have adequate purity. Some resins sold as "analyzed" or "reagent" are actually standard, commercial-grade resins and will require purification before use. Such purified ion-exchange resins may be purchased.

[23] E. Glueckauf, *Ion Exchange and Its Applications*, Society for Chemical Industry, London, 1955, p. 34.

[24] F. W. Cornish, *Analyst* **83**, 634 (1958); F. W. Cornish in A. A. Smales and L. R. Wager, *Methods in Geochemistry*, Wiley-Interscience, New York, 1960, p. 408.

[25] An extensive treatment (400 pages, general and applied inorganic) is offered by O. Samuelson, *Ion Exchange Separations in Analytical Chemistry*, Wiley, New York, 1963. Also see J. Inczedy, Ref. 7. Literature references are mostly omitted in the discussion that follows because abundant citations can be found in Samuelson's book and also in H. F. Walton, *Ion Exchange*, in L. Meites (ed.), *Handbook of Analytical Chemistry*, McGraw-Hill, New York, 1963.

TABLE **13-4**

	Maximum Content, ppm	
	Analytical-Grade (Purified) Dowex 1	*Typical Specification, Commercial Grade Dowex* 1 *or Dowex* 50
Fe	0.5	100
Cu	0.2	20
Ni	0.05	15
Pb	0.005	20
Al	5	800
Aliphatic amines	5	—
Minimum % in specified size range	95	50

Table 13-4 compares specifications for truly analytical-grade resins with commercial grade.[26]

The polystyrene strong-acid or strong-base type of resin finds by far the most use in inorganic analysis.

Cation-exchange resins can be purified by soaking in warm 4 *M* hydrochloric acid, washing with *pure* water, and repeating one or more times. Anion-exchange resins can be purified by treatment with 1 *M* hydrochloric acid, water, 0.5 *M* sodium hydroxide, and water and repeating the cycle. When traces of extracted organic matter may interfere, the resin should be washed with alcohol or acetone followed by hot water for cation resins and warm water (60°C) for anion resins (in Cl form). During this purification treatment, fines can be removed by decanting the supernatant liquid. In critical chromatographic separations, uniformity of grain size is important—a small percentage of coarse grains in a fine-grained resin may prevent successful separations. Specially sized resins are available commercially.

The ionic form of a resin will usually be H^+ or Cl^-.

An ion-exchange column is easily prepared. As tersely stated by Samuelson, "An ion exchange column consists of a glass tube." The dimensions of the tube may vary somewhat depending on whether a solution is simply to be percolated through the resin, as in the separation of cations from anions, or elution is to be carried out for a chromatographic separation, as of metals from each other. In the first case, a resin column 10 to 15 cm long will usually be ample, and this length may even suffice for some chromatographic separations, but it may need to be increased for this purpose. The flow rate will of course decrease as the column is lengthened. The diameter of the column is primarily determined by the amount of exchangeable material in the sample. Usually a tube of 1.0–2.0 cm internal diameter is used for a macro sample, the greater width for simple replacement reactions. A strong acid cation-exchange resin of 8 per cent cross-linking will hold roughly 2 meq of exchangeable ion per milliliter of bed; a strong-base anion exchanger (Dowex 50-X8) holds about 1 meq of ion/ml.

The glass tube should be provided with a stopcock and a reservoir (Fig. 13–12). A sintered glass disk or a small plug of glass wool near the stopcock supports the resin bed. In chromatographic separations a disk should always be used.

Resins of 100–200 mesh size are suitable for most inorganic analytical work. Coarser-grained resins (50–100 mesh) may be satisfactory in simple percolation procedures. In difficult chromatographic separations, when the plate height needs to be small, 200–400

[26] From catalog of Bio-Rad Laboratories, Richmond, Calif., May 1965.

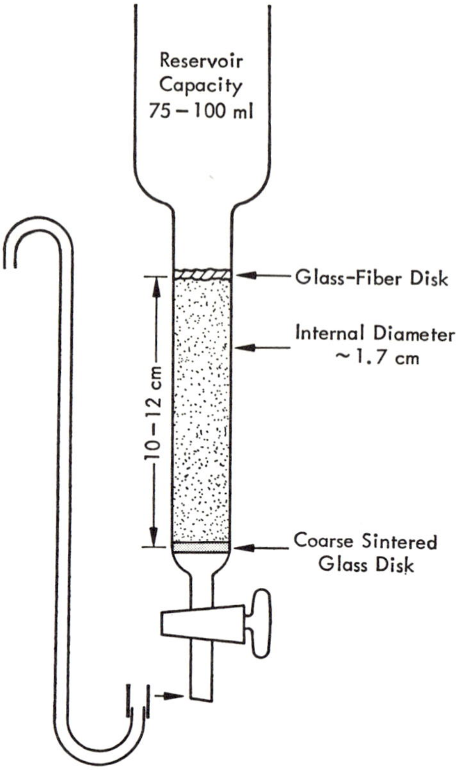

Fig. 13-12. Typical ion-exchange column for macro analysis. The dimensions are more or less arbitrary. The resin bed volume in the column shown is ~25 ml. The column is fairly short and is suitable for simple replacement of one ion by another. For chromatographic separations, especially the less easy ones, a longer column might be desirable. A plug of glass wool can be used in place of the sintered disk if chromatographic separations are not to be made. If desired, a siphon tube (as shown) can be attached to the column to prevent the liquid from draining below the top of the resin bed when not attended by the operator. If air enters the bed, the column is likely to be ruined, since on the subsequent addition of liquid, channels are likely to be formed in the resin. If the hydrostatic column length (top of resin to end of exit tube) does not exceed 15 cm, capillarity probably will hold solution in the resin (200-mesh or finer) when the meniscus reaches the top of the bed.

mesh resin may be required. The flow rate will then be low (but pressure can be applied). A resin of 8 per cent cross-linking is used in most inorganic analytical work.

Prepare the column as follows. Partially fill the tube with water (or eluant) and open the stopcock to displace air in and below the disk or plug. Add a slurry of the resin (1 volume resin : 2 volumes water) to the tube from a beaker; the resin should be free of air bubbles. Allow water to flow from the tube and build up the resin column to the required height. It is important that no air bubbles be trapped in the column. The water level should not be allowed to drop below the top of the resin column. Place a glass–fiber or filter-paper disk at the top of the resin bed to prevent its disturbance when solution is added. Wash the column with water or eluting solution as required. Drain the water or solution to the top of the bed before adding sample solution. It is advisable to wash the column before use if it has stood overnight.

In simple replacement of one ion by another, as in exchanging Na^+ in a solution for H^+ in a resin, a flow rate of 5 ml/cm²/min is reasonable, and a rate of 10 ml/cm²/min may be permissible. Of course, such rates are possible only with coarse resins.[27] In chromatographic separations the flow rate must be lower. Often a rate of 1 ml/cm²/min is satisfactory, but it may need to be lower.

Purification of water and some electrolyte solutions

See Chapter 17.

Determination of electrolyte content

Salts such as nitrates, sulfates, and perchlorates, whose anions are not easily determinable by titrimetric methods, can be determined rather simply by passing the solution through a strong-acid-type cation exchanger in the H^+ form and titrating the acid in the effluent. For example, Samuelson obtained the following results with nitrates.

	mmoles NO_3^- *Present*	mmoles NO_3^- *Found*
$LiNO_3$	2.566	2.569, 2.570, 2.567
$NaNO_3$	2.566	2.570, 2.568, 2.570
$Ni(NO_3)_2$	1.985	1.982, 1.986, 1.981

Approx. 50 ml of solution, sulfonated phenolic resin of 0.2–0.4 mm particle diameter, column 13 cm long and 1.2 cm in diameter, ~10 meq capacity; time for passage and washing with 75 ml of water = 10 min.

The same method can be used to determine the total salt concentration in a solution. For example, the sum of Ca, Mg, Na, and K in natural waters, expressed in equivalents, can readily be determined in this way. Carbon dioxide in the effluent is removed by boiling or bubbling air through the solution, and titration with standard base is then carried out. The electrolyte content of many other materials can be determined similarly (salts in blood serum, water-soluble salts in soil, salt in butter, etc.). The use of a strong-base anion exchanger in the —OH form, with titration of OH^- in the effluent, is usually much less satisfactory than the use of a cation exchanger.

For practical details, see Chapter 39.

Removal of ions interfering in analytical procedures

Usually the separation of a cation from interfering anions, or the converse, is involved. A few examples:

1. Removal of interfering anions such as sulfate and phosphate in the determination of the alkali metals. The solution is passed through a sulfonic acid-type resin (H^+ form) to adsorb the alkali metals, which are then eluted with hydrochloric acid.

2. Removal of iron and other heavy metals before determination of sulfate by precipitation as barium sulfate. A cation-exchange column (H^+ form of resin) is used. This separation simplifies the determination of sulfur in pyrite. With a 0.5-g sample, a column of 30-meq capacity is needed. About 1 hr is required for the separation (see Chap. 27). The elimination of Fe(III) and other metals reduces errors due to coprecipitation.

3. Removal of cations (Ca, Mg, Al, Fe) before determination of phosphorus in phosphate rock. Again a cation resin in the H^+ form is used. Phosphoric acid in the eluate

[27] 10 ml of water requires approximately 30 min to percolate through a 6.5-cm-long bed of 200–400 mesh Dowex 1-X8 resin in a tube of 1-cm² cross section.

can be determined gravimetrically or titrimetrically. If the metals are also to be determined, they can be eluted from the column with 4 *M* hydrochloric acid.

Many similar separations of cations and anions useful in applied analysis could be mentioned, not only for gravimetric and titrimetric analysis but also for spectrophotometric and other determination methods.

Separations of metals

Cation exchange. With the exception of separation of the alkali metals and sometimes of the alkaline earths, the separation of metals by a cation-exchanger rests on differential elution by one or more complexing agents forming anionic complexes of varying stabilities; occasionally an eluant altering the oxidation state of a cation is used. A great many procedures have been worked out. Complexing agents used include ethylenediamine, EDTA, oxalate, citrate, fluoride, and many others. The rare earth ions were separated from each other in the 1940s by elution with citrate solutions. Cation-exchange separations are also important in the separation of the transplutonium elements.

Separation of metals on cation exchangers by differential elution with anionic complexing agents has a drawback from the general analytical viewpoint: the complexing agent may interfere in the subsequent determination of the element by titrimetry, colorimetry, and other methods. The use of chloride as a complexing eluant in separations of heavy metals on a cation-exchange resin is not always successful. The anionic chloride complexes may be anomalously adsorbed (by the matrix) on the cation exchanger.[28]

The alkali metals can be separated by the use of a strong-acid type of cation-exchange resin of sufficiently high cross-linking (to accentuate selectivity) with hydrochloric acid as the eluant. Figure 13-13 shows how an excellent separation of sodium and potassium

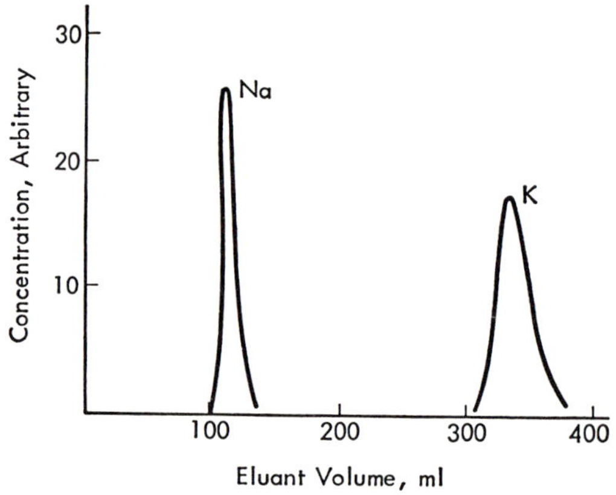

Fig. 13-13. Separation of Na (3 mg) and K (9 mg) on Dowex 50–X12 resin, 200 to 400 mesh. Column dimensions 9 × 200 mm. Flow rate 0.97 ml/cm^2/min, 0.6 *M* HCl as eluant at room temperature. From F. W. Cornish, *Analyst* **83**, 634 (1958).

can be obtained with a fine-grained resin, albeit at the cost of considerable time and inconveniently large elution volumes. Even better separations of Na, K, Rb, and Cs can be obtained with inorganic cation exchangers (Figs. 13-14 and 13-15).

The alkaline earths can be separated from each other (for example, Mg from Ca, Ca from Sr) and from the alkali metals, Fe(III) and Al by elution with hydrochloric

[28] K. A. Kraus, D. C. Michelson, and F. Nelson, *J. Am. Chem. Soc.* **81**, 3204 (1959).

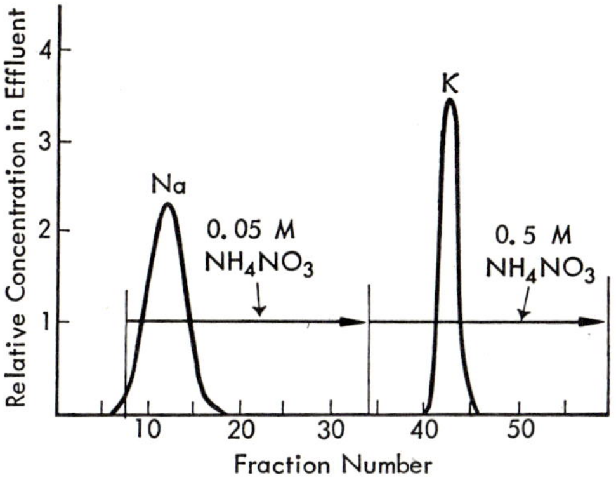

Fig. 13-14. Separation of 104 mg Na and 1.02 mg K on ammonium phosphomolybdate + asbestos (1.48 g + 0.74 g) column, 0.8 cm^2 × 6.8 cm, by elution with ammonium nitrate. After J. van R. Smit, W. Robb, and J. J. Jacobs, *J. Inorg. Nucl. Chem.* **12**, 104 (1959).

acid. However, complexing eluants such as citrate, lactate, ammonium α-hydroxyisobutyrate, EDTA, and others are used extensively in the separation of the alkaline earths from each other and from other elements.

Mixtures of an organic solvent such as ethyl alcohol or acetone with water are sometimes used to improve separations on cation-exchange (and also anion-exchange) resins. The general effect of the organic solvent is to decrease the dissociation of complexes. For example, a number of metals, including Bi, Ga, Zn, and Cd are only weakly adsorbed on a cation-exchange resin from a water–isopropanol–HCl solution because of the formation of stable anionic chloro complexes in this medium; others, such as U, Al, Co, Ni, and the alkaline earth metals are strongly held by the resin under the conditions.[29] A small amount of uranium can be separated from much bismuth in this way.

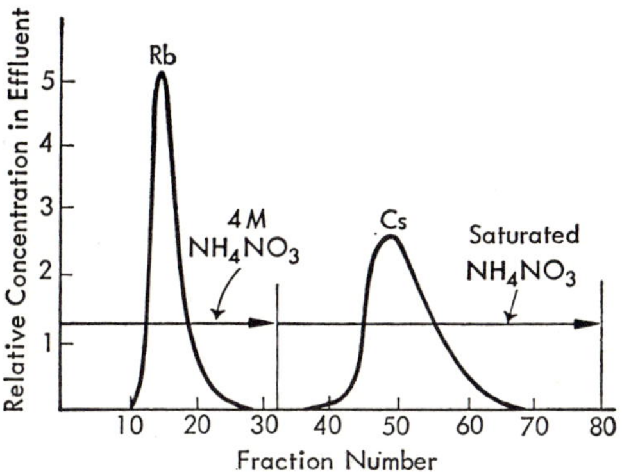

Fig. 13-15. Separation of 12.3 mg Rb and 13.5 mg Cs on 1:1 ammonium phosphomolybdate-asbestos column, 1.54 cm^2 × 8.9 cm. Each fraction of effluent ≡ 10 drops. Elution with ammonium nitrate. Flow rate ~ 1 cm/min. After J. van R. Smit, W. Robb, and J. J. Jacobs, *J. Inorg. Nucl. Chem.* **12**, 104 (1959).

[29] J. Korkisch and S. S. Ahluwalia, *Anal. Chem.* **37**, 1009 (1965). J. S. Fritz and J. E. Abbink, *Anal. Chem.* **37**, 1272 (1965), separate small amounts of Cu, Ni, Co, and Mn from much Fe(III), Zn, and Cd using Dowex 50 resin and HCl–acetone as eluant.

Fig. 13-16. Adsorption of elements from hydrochloric acid solutions by an anion exchange resin. No ads.: no adsorption from 0.1–12M HCl; sl. ads.: slight adsorption ($0.3 < D < 1$) from 12M HCL; str. ads.: strong adsorption ($D \gg 1$). From Kraus and Nelson, Reference **30**, (1955).

Anion exchange. The difference in stability of the chlorocomplexes of heavy metals has been made the basis of powerful methods for their separation by differential adsorption on strong base anion–exchange resins. Distribution coefficients have been determined for almost all adsorbable elements as a function of the hydrochloric acid concentration,[30] from which various separation possibilities can be deduced. Some elements, e.g., Be, Mg, Ca, Sr, Ba, Ni, Al, the rare earths, Y, and Th, are not adsorbed by strong-base anion exchangers no matter what the hydrochloric acid concentration. Others show strong differences in adsorbability as the hydrochloric acid concentration is varied (Fig. 13-16). The intrinsic adsorbability of the complex chloro anion or chloro acid is also a factor determining the distribution coefficient of a metal. It is interesting to note in Fig. 13-16 that those chloro complexes, which as acids are readily extracted into ether and other oxonium solvents—of Fe(III), Ga, Sb(V), Au(III), Tl(III)—have unusually high distribution coefficients ($\sim 10^5$–10^6) on quaternary amine resins. The resin matrix shows some characteristics of these immiscible solvents.

It was pointed out by Kraus that the distribution coefficient of a metal reaches its maximum value when the average charge of all the metal species in solution is zero. An equivalent statement is the following[31]: The distribution coefficient of a metal adsorbed on an anion exchanger saturated with a univalent ligand attains its maximum value when the average ligand number is equal to the charge of the free metal ion. This rule enables one to predict the ligand concentration at which maximum metal adsorption occurs, if the formation constants of the metal complexes are known for any ligand. Metals such as Au, Ag, and the platinum metals, which form slightly dissociated chloride complexes, are adsorbed strongly at very low chloride concentrations. Increasing the chloride concentration gives rise to increased competitive adsorption of Cl^- and the distribution coefficient of the metal falls.

With some metals (Au, for example) the distribution coefficient is not the same for HCl and alkali chloride solutions. This is not unexpected, because the chloro acid (e.g., $HAuCl_4$) may not be completely ionized.

The separation of Mn, Fe(II), Co, Ni, Cu, and Zn furnishes a good illustration of the utility of complex anion adsorption in chloride solution.[32] If the hydrochloric acid concentration at which noticeable adsorption ($\mathscr{E} = 1$) first occurs is taken as an index of complexing strength, the order of decreasing complexing is Zn $>$ Cu $>$ Co $>$ Fe(II) $>$ Mn(II) $>$ Ni. These acid concentrations probably indicate the points at which the elements form approximately equal (small) concentrations of MCl_3^-. The minima in the elution curves (adsorption maxima) probably indicate hydrochloric acid concentrations at which MCl_4^{--} become the principal species, whereas the extent of adsorption here corresponds to the concentration of the MCl_3^- complexes (Fig. 13-17). Beyond the elution minima, the adsorption of chloride ions begins to outweigh the adsorption of the anionic chloro metal complexes.

Since the elution constants of these metals differ considerably, it is to be expected that they can be separated on an anion-exchange resin by elution with hydrochloric acid solutions of the proper concentration. The results shown in Fig. 13-18 were obtained by placing 6 mg of each ion in 0.85 ml of concentrated hydrochloric acid on a 0.3 cm^2 $\times$ 26 cm Dowex 1 column ($-200 + 230$ mesh) and eluting with hydrochloric acid of different concentrations as indicated. The flow rate was 0.5 cm/min. Quite satisfactory separations of these metals from each other are shown. The ease of separation of zinc from nickel and cobalt is noteworthy. This separation is not easily effected by precipitation and extraction methods.

[30] K. A. Kraus and F. Nelson, *Proc. Intern. Conf. Geneva*, Aug. 1955, **7**, Session 9B.1, P/837, pp. 113, 131; K. A. Kraus and F. Nelson in W. J. Hamer (ed.), *The Structure of Electrolytic Solutions*, Wiley, New York, 1959, Chap. 23.

[31] S. Fronaeus, *Svensk Kem. Tidskr.* **65**, 1 (1953); *Acta Chem. Scand.* **8**, 1174 (1954).

[32] K. A. Kraus and G. E. Moore, *J. Am. Chem. Soc.* **75**, 1460 (1953).

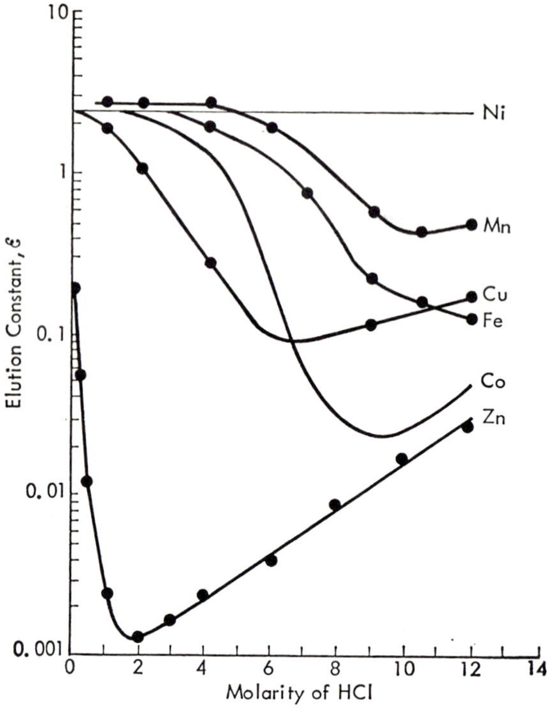

Fig. 13-17. Elution constants of divalent transition metals in HCl. After Kraus and Moore, *J. Am. Chem. Soc.* **75,** 1460 (1953).

One great advantage in the use of hydrochloric acid as an eluting agent is its general harmlessness in the subsequent determination of the separated metals. When metals are separated on a cation-exchange resin column with an anionic complexing agent as eluant (e.g., EDTA), the latter may interfere later, as already mentioned.

Differential complex formation in other halide solutions and in nitric and sulfuric acid also finds use in metal separations with anion-exchange resins. Hydrofluoric and hydrofluoric–hydrochloric solutions are useful in separations involving Zr, Hf, Ti, Nb, Ta, Sc, and Be. Th, Bi, and the rare earths are adsorbed on anion-exchange resins from nitric acid solutions. Uranium(VI) is strongly adsorbed from sulfuric acid solutions, chiefly as the ion $UO_2(SO_4)_2^{--}$, and this behavior enables it to be separated from most other metals. Separations of thorium are also possible in sulfate solutions.

Complex formation with acetate, ascorbate, citrate, oxalate, thiocyanate, and other anions has also been put to use in separations of metals on anion exchangers.

Separations of nonmetals

By the use of anion exchangers, simple anions can be separated from each other. For example, chloride, bromide, and iodide can be separated with sodium nitrate as an eluting agent. Orthophosphate, pyrophosphate, triphosphate, tetraphosphate, and polyphosphate can be separated by using potassium chloride for elution. It may be mentioned that perchlorate is held very strongly by anion-exchange resins and is not easy to elute; nitrate is more effective than chloride as eluant.

Isolation of traces

An analyst may need to concentrate one or more constituents in a large volume of solution into a small volume before they can be satisfactorily determined. Simultaneous separation from other constituents may or may not be required; it is usually desirable.

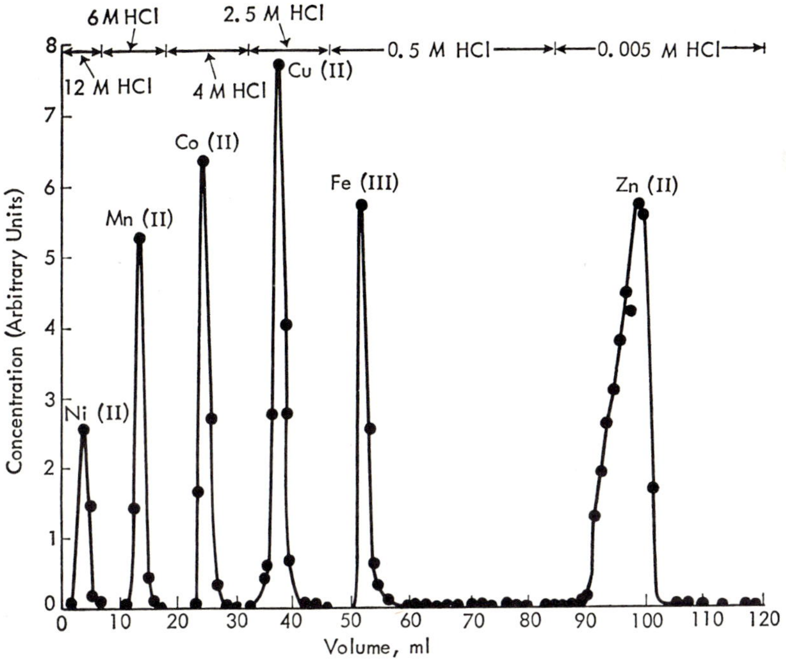

Fig. 13-18. Separation of transition metals Mn to Zn by echelon elution with hydrochloric acid of different concentrations. After Kraus and Moore, *J. Am. Chem. Soc.* **75**, 1460 (1953).

The isolation of traces of constituents by ion exchange is simple enough when the concentration of other electrolytes is not sufficiently high to seriously compete with the exchange of the trace constituents. After adsorption from a large volume, the trace constituents can be desorbed in a small volume of eluant.

Other factors remaining the same, the coefficient for the distribution of an ion between an exchanger and a solution remains constant down to extremely low concentrations. Ghiorso and coworkers established the existence of element 101 (Md) by isolating 17 atoms on a small ion-exchange column. Ion-exchange chromatography has been applied with great success in the isolation of transuranium elements and in their separation from each other and from large amounts of other elements; in fact, it finds general use in the concentration and separation of radioactive elements.

Some examples of more mundane applications of ion exchange in the isolation of traces follow: copper in milk, lead in maple sugar, and trace metals in wine by use of a cation (H^+-form) exchanger (separation from nonelectrolytes); concentration of Ca, Mg, Fe, Cu, Pb, and many other trace metals from natural and effluent waters by the use of an RH (strong-acid type) resin; concentration and separation of F^- in water with an anion-exchange resin.

When a trace element is accompanied by large amounts of other ions, it may be possible to effect a concentration and separation by complex formation. For example, in chloride solutions of the proper concentration, a considerable number of elements are adsorbed strongly on an anion exchanger (Fig. 13-16) and such elements can be isolated from brines and sea water. Gold and bismuth have been determined in sea water collected off Cape Town, South Africa, after adsorption in a small anion-exchange column (Amberlite IR-400).[33] The sea water (0.5 M Cl^-) was made 0.1 M in HCl and

[33] R. R. Brooks, *Analyst* **85**, 745 (1960).

250 liters was run through a 0.5 cm^2 × 13.2 cm resin column (100 days for passage). The isolated Au and Bi were determined spectrographically (0.009 mg of Au and 0.017 mg of Bi found in 1 ton of sea water).

Chelating resins are of value in isolating certain elements in the presence of large amounts of others (p. 323). Dowex A-1 has been used to concentrate heavy metals in water at pH ~5.

Separations of nonionic substances

Ligand exchange. A cation exchanger loaded with a metal forming complexes with anions or uncharged species, X, Y, . . ., can be used for the separation of such ligands by elution chromatography:

$$R^-M^+X + Y \rightleftharpoons R^-M^+Y + X$$

For example, amines adsorbed on R_2Cu or R_2Ni can be differentially eluted with ammonia and thus separated. Sugars adsorbed on anion exchangers loaded with borate ions can be separated by differential elution, and likewise aldehydes adsorbed on bisulfite-loaded anion-exchange resins. The applications are mostly, if not entirely, organic and will not be considered further here.

Ion exclusion, salting out, and solubilization chromatography;[34] gel filtration. Consider, as an example of ion exclusion, a solution of sodium chloride and a nonelectrolyte percolating through a bed of cation resin in the sodium form. The sodium chloride will largely remain in the interstitial solution because of the high sodium concentration in the solution within the resin grains (p. 294), but the nonelectrolyte will distribute between the solutions inside and outside the grains. The nonelectrolyte concentration within the grains may even be larger than outside the grains, because of adsorption of the nonelectrolyte by the resin matrix. Accordingly, sodium chloride moves more rapidly down the column than does the nonelectrolyte, and a separation can be effected by elution with water. In this way, sodium chloride or other sodium salt and acetic acid can be separated, as can sodium chloride and glycerin, etc. By the use of an electrolyte solution (e.g., ammonium sulfate) for elution, mixtures of nonelectrolytes of polar character can be separated (salting-out chromatography). Substances that are more easily salted into the resin phase (those less hydrophilic) are eluted less rapidly by the ammonium sulfate solution than those which are less easily salted out. By the use of ammonium sulfate solutions of different concentrations, the range of separable substances can be expanded. In solubilization chromatography, elution is carried out with increasing concentrations of such solvents as acetic acid and methyl alcohol in water, in which organic substances are more soluble than in water alone. These separations are a type of partition chromatography (p. 327), but since they are carried out with ion-exchange resins they are mentioned here.

Resins of cross-linked, nonionic copolymers of styrene and divinylbenzene have been introduced (1965) for the separation of organic substances from each other and from inorganic substances. Since these resins lack functional groups, the chromatographic separability of organic substances rests on their varying adsorbability by the hydrophobic polystyrene surface. The pore size of the resins also plays a role (molecular-sieve effect) in the separations. The latter effect is predominant in chromatography by "gel filtration" or "gel permeation," a recent type of column chromatography, which is of great importance in biochemistry for the separation of substances of high molecular weight. Sephadex is an example of a gel used in this way. It consists of a modified dextran, having a network of cross-linked molecules, which swells in water to form a structure with interstices whose size depends upon the extent of cross-linking. Molecules above a certain size will not be able to enter the interstices and will pass down the

[34] See W. Rieman and R. Sargent in W. G. Berl (ed.), *Physical Methods in Chemical Analysis*, vol. IV, Academic Press, New York 1961, for an account of these separations.

column on "filtration," whereas smaller molecules will enter the gel and their movement down the column will be retarded, the retardation being a function of their size. Other gels can be used in polar and non-polar organic solvents. Substances in the molecular weight range from $\sim 10^2$ to $> 10^6$ can be fractionally separated by selection of the proper gel.

Other ion exchangers

More or less special ion-exchange resins and ion-exchange materials other than resins are available and have their uses. They are considered briefly in this section.

Chelating resins.[35] The selectivity of ion-exchange resins can be increased by introducing complexing groups that strongly bind certain metals or groups of metals. For example, Dowex A-1 cation-exchange resin contains iminodiacetate chelating groups:

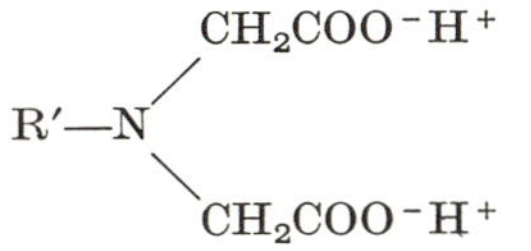

Cations that form chelate complexes with iminodiacetate anions are strongly adsorbed by this exchanger (Ca can be separated from much Li, as in determination of traces of Ca in Li salts). Moreover, since the metal complexes differ in stability, distribution coefficients can be quite different. Thus in 0.2 M KNO_3 solution, $\log(D_{\mathrm{Cu}}/D_{\mathrm{Zn}}) \sim 2$ (Ringbom). Because iminodiacetic acid is a weak acid, the distribution coefficients vary with pH. A chelating resin with mercapto groups has been prepared which is said to be specific for Au, Ag, and Hg.[36] It has been used to isolate 1.4 μg of Au from 100 liters of sea water.

Chelating resins have not yet been studied extensively, nor are they used extensively. They have some drawbacks. Equilibrium tends to be attained slowly. Because of the strong binding of the chelated cation the mobility of the cation is reduced. Chelation of the cation by the resin is sometimes slow. Selectivity is gained at the cost of a decrease in exchange rate. Chelating resins may be difficult to regenerate.

An anion exchanger saturated with a strongly complexing anion behaves like a selective cation exchanger. For example, an anion-exchange resin saturated with EDTA may be able to retain both cations and anions; thus

$$R_4Y + M^{++} + SO_4^{--} \rightleftharpoons R_2MY + R_2SO_4$$

Liquid ion exchangers.[37] This is the name given to organic electrolytes of high molecular weight which are soluble in organic solvents immiscible with water but insoluble in water, and which have a small, mobile ion that can be exchanged for an ion of the same sign in the aqueous phase. Long-chain aliphatic amines can function as liquid anion exchangers and dialkylphosphates as cation exchangers. Liquid ion exchangers are extractants and some substances of this type are referred to in Chapter 14.

Oxidation–reduction ion exchangers.[38] These are characterized by the presence of a reversible oxidation–reduction couple, such as Fe^{3+}–Fe^{++} or methylene blue–leucomethylene blue. They can be prepared from ordinary cation or anion-exchange resins by replacing H^+ by Fe^{3+} (Fe^{++}) or Cl^- by SO_3^{--}, for example, or by incorporating (by sorption) methylene blue or other suitable oxidant or reductant. An *electron*

[35] Review: G. Schmuckler, *Talanta* **12**, 281 (1965); J. R. Millar, *Chem. Ind.* (*London*) **1957**, 606.

[36] E. Bayer, *Angew. Chem.* **76**, 76 (1964).

[37] Review: H. Green, *Talanta* **11**, 1561 (1964). See also Chap. 14.

[38] H. G. Cassidy and K. A. Kun, *Oxidation-Reduction Polymers* (*Redox Polymers*), Wiley-Interscience, New York, 1965.

exchanger has no fixed ionic groups, but possesses an incorporated oxidation–reduction couple such as quinone–hydroquinone. The great advantage in the use of such a solid oxidant–reductant would be its ease of removal after the desired oxidation or reduction has been brought about in the liquid phase. Unfortunately, the available electron exchangers have serious limitations. They react slowly and lack chemical stability. They find limited use.

Ion-exchange celluloses. Cellulose fibers (e.g., cotton) can be treated to introduce ionic groups, which convert the cellulose into an ion exchanger.[39] Large molecules, such as proteins, nucleic acids, and enzymes, can be effectively separated on the ion-exchanging celluloses. The hydrophilic and open nature of the cellulose lattice, together with the low density of the weakly ionized groups, contribute to the effectiveness of chromatographic fractionation by such substances. Nonionic celluloses find use in partition chromatography (p. 327) of proteins, enzymes, hormones, and other large molecules.

Inorganic ion exchangers. Observations on ion exchange were made as early as 1845, when Thompson and Spence noted that water which had percolated through a mixture of soil and ammonium sulfate contained calcium sulfate. Way[40] investigated the phenomena in more detail and showed that the amount of calcium salt formed was equivalent to the amount of ammonium, sodium, and other salt added; he attributed the exchange to aluminosilicates in the soil and, in fact, found that aluminosilicates prepared in the laboratory behaved similarly. Gans, in 1905, demonstrated that synthetic and natural zeolites (aluminosilicates) could be used for water softening. Very little use was or is made of synthetic zeolites, manufactured for water softening, in chemical analysis. The principal reason for this lies in the narrow range of pH in which the silicate zeolites are stable; they dissolve in acid solutions and are peptized in basic solutions. Moreover, quantitative elution of a solute is usually difficult. The so-called molecular sieves are synthetic silicate zeolites of regular crystal structure having narrow, uniform pores, which are selective molecular adsorbents. They also show selectivity in the exchange of cations from aqueous solution but have the serious limitation of being stable only in the pH range 5–12.

Nonsilicate inorganic ion exchangers have greater utility. Zirconium phosphate and zirconium tungstate have found application in separations of the alkali and alkaline earth metals. A possible structure of such (nonstoichiometric) compounds of group IV oxides with acidic oxides of groups V and VI is (according to Helfferich) illustrated by

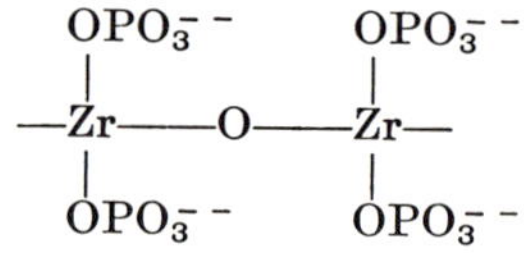

Inorganic ion exchangers of this type have high capacity (as great as 12 meq/g), show high rates of exchange, and are superior to organic exchangers in thermal and radiation stability.[41]

Ammonium phosphomolybdate, whose cation-exchange properties have been known for more than 50 years, provides a separation of the alkali metals; the sharp separation of rubidium and cesium is noteworthy (Figs. 13-14 and 13-15).

Aluminum oxide "for chromatographic adsorption according to Brockmann" functions as an anion exchanger in acidic solution. Strongly adsorbed anions include HSO_4^-, F^-, and $H_2PO_4^-$; $Cr_2O_7^{--}$ is moderately strongly adsorbed, while Cl^-, NO_3^-, and ClO_4^- are weakly adsorbed. This behavior has been put to good use in the deter-

[39] Diethylaminoethylcellulose is an anion exchanger of value in biochemical chromatography and cellulose phosphate is a strong-acid cation exchanger, which has been studied considerably. See I. Smith (ed.), *Chromatographic and Electrophoretic Techniques*, vol. I, Wiley-Interscience, New York, 1960, chap. 29.

[40] J. T. Way, *J. Roy. Agr. Soc.* **11**, 313 (1850); **13**, 123 (1852); **15**, 491 (1855).

[41] See C. B. Amphlett, *Inorganic Ion Exchangers*, Elsevier, Amsterdam, 1964.

mination of sulfur in steel.[42] Bisulfate ion is adsorbed from a hydrochloric–perchloric acid solution and thus separated from Fe^{3+}, which is a troublemaker in the determination of sulfur as $BaSO_4$:

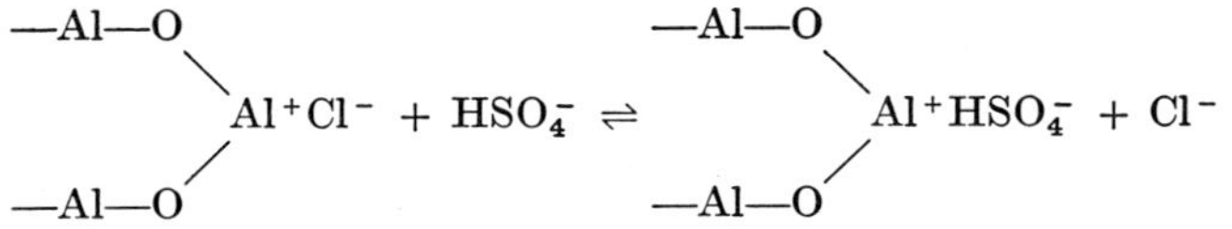

Sulfate is eluted from the column with ammonia (replacement of HSO_4^- by OH^-). The capacity of alumina at a pH less than 1 is about 0.17 meq/ml of column, much less than the capacity of an anion-exchange resin.

A procedure for the determination of cobalt in steel has been described in which the nitroso-R salt complex anion is adsorbed on alumina which has been washed with perchloric acid; the organic anion easily replaces the adsorbed perchlorate ion. Other metals, such as iron and nickel, are not adsorbed.

If washed with an alkali solution, aluminum oxide functions as a cation exchanger. Thus lithium can be separated from the other alkali metals by adsorption on aluminum oxide at pH 12.6.[43]

Ion exchange and immiscible solvent extraction compared

Both these methods of separation are powerful and of wide applicability. In any particular case, one or the other may be preferable. The prime criterion of choice is of course satisfactory recovery and separation of the constituents involved. When this criterion can be met by both methods, secondary factors may be decisive. When applicable, extraction methods often have an advantage in speed and in concentration of the constituent into a small volume of organic solvent free of complexing electrolytes used to make the separation more selective. Ion-exchange separations tend to be slow. Moreover, the constituent may finally be separated in an unpleasantly large volume of solution accompanied by complexing substances in the eluting solution, which may cause trouble in the final determination. The latter characteristic is by no means universal and there may also be problems of transferring an extracted substance from the organic solvent phase to an aqueous phase in extraction methods (on the other hand, extraction with a colored reagent may at the same time be the determination step).

Ion exchange is usually superior to extraction when the distribution coefficients of the substances to be separated lie close together, because of the greater ease of fractionation in the former method. Ion-exchange methods are clearly superior in the separation of the alkali, alkaline earth, and rare earth metals.

OTHER TYPES OF CHROMATOGRAPHY

Adsorption chromatography[44]

The use of columns of finely divided adsorbents for the separation of substances in solution was first clearly set forth by Tswett in 1903–1906.[45] He described his procedure for separating plant pigments as follows:

If a petroleum ether chlorophyll solution filters through a column of adsorbent (I use chiefly calcium carbonate which is tightly tamped into a glass tube), the pigments

[42] F. Nydahl, *Anal. Chem.* **26**, 580 (1954). In a later procedure for sulfur in inorganic and organic materials [W. J. Kirsten, K.-A. Hansson, and S. K. Nilsson, *Anal. Chim. Acta* **28**, 101 (1963)], a tandem column of aluminum oxide and cation-exchange resin is used to separate sulfate, which is then titrated with lead nitrate in alcoholic solution (dithizone as indicator).

[43] H. Ballczo and J. Sinabell, *Mikrochemie* **35**, 178 (1950).

[44] See, for this and other types of chromatography, E. Lederer and M. Lederer, *Chromatography*, Elsevier, Amsterdam, 1957; H. G. Cassidy, *Fundamentals of Chromatography* in A.

separate themselves from top to bottom in different colored zones according to the adsorption series, in which the more strongly adsorbed substances displace the more weakly held ones farther toward the bottom. This separation is practically complete if one, after the passage of the pigment solution, passes a stream of solvent through the column of adsorbent. Just like the rays of light in the spectrum, so are the different components of a pigment mixture separated in an orderly way in the calcium carbonate column and thus may be qualitatively and quantitatively determined in this way. Such a preparation I call a chromatogram and the corresponding method, the chromatographic method.

Needless to say, the described adsorption phenomena are suited not only to chlorophyll pigments. It is to be expected that all kinds of colored or colorless chemical compounds are subject to the same laws.[46]

This novel method of separation apparently did not attract the attention, or meet with the favor, of the organic chemists of the time. A quarter century of desuetude followed, and it was not until 1931 that the method was revived and again applied. In succeeding years it was widely used, and today is a standard method in preparative and analytical organic chemistry and biochemistry.

Only a few general remarks regarding this type of chromatography will be made here. Other factors being the same, the adsorbability of an organic compound increases with its polar character, i.e., with the kind and the number of polar groups in the molecule. The order of increasing adsorbability is usually: saturated hydrocarbons; unsaturated hydrocarbons; halogen compounds; hydroxy, amino, nitro, and thio compounds; acids and bases. Adsorbability usually increases with an increase in molecular weight and molecular volume. The geometric structure plays an important role—isomers are adsorbed to different extents. The adsorbability of a given compound depends greatly upon the polarity of the solvent, which is understandable from the competition between the molecules of the solvent and the substance to be adsorbed on the adsorbents usually employed. The more strongly polar the solvent, the less strongly the solute is adsorbed, so that the order of decreasing adsorption with solvent is petroleum ether, carbon tetrachloride, ether (anhydrous and free from alcohol), benzene, chloroform, alcohols, and water. The more strongly polar solvents serve for elution. The strength of adsorption normally increases in the following order with the (hydrophilic) adsorbent: sucrose and starch, talc, calcium carbonate, magnesium carbonate, activated silicic acid, and activated alumina and magnesia. The water content of the adsorbent may strongly affect its adsorptive properties.

Weissberger, *Technique of Organic Chemistry*, Vol. X, Wiley-Interscience, New York, 1957; J. C. Giddings, *Dynamics of Chromatography*, part I, Dekker, New York, 1966. A. H. Gordon and J. E. Eastoe, *Practical Chromatographic Techniques*, Newnes, London, 1964; E. Heftmann (ed.), *Chromatography*, Reinhold, New York, 1967; R. Stock and C. B. F. Rice, *Chromatographic Methods*, Reinhold, New York, 1963; I. Smith (ed.), *Chromatographic and Electrophoretic Techniques*, vol. I. *Chromatography*, Wiley-Interscience, New York, 1960; T. I. Williams, *The Elements of Chromatography*, Philosophical Library, New York, 1953; R. C. Brimley and F. C. Barrett, *Practical Chromatography*, Reinhold, New York, 1953; M. Lederer, *Chromatographic Reviews*, vol. 7, Elsevier, Amsterdam, 1965. J. C. Giddings and R. A. Keller (eds.), *Advances in Chromatography*, vol. I, Dekker, New York, 1966. More than 200 books on chromatography and related methods are listed in *Woelm Publication AL 12*, M. Woelm, Eschwege, Germany, 1966. The *Journal of Chromatography* appears monthly.

[45] Earlier, indefinitive use of adsorption columns is recorded in the literature, one of the earliest attempts being made by L. Reed, *J. Chem. Soc.*, Abstr. Proc. VIII–IX, 123 (1892–1893). He showed that, for example, ferric chloride and copper sulfate in aqueous solution gave bands at different places in a kaolin column. He also tried to separate organic substances on a column.

[46] M. Tswett, *Ber. Deut. Botan. Ges.* **24**, 235 (1906). The translated excerpt is in T. Robinson, *J. Chem. Educ.* **36**, 144 (1959).

Equilibrium is usually quickly established in adsorption chromatography so that faster flow rates can be used than in other types of chromatography. The adsorption isotherm is usually not linear and, as a consequence, asymmetrical bands tend to be produced.

The chief interest of the inorganic analyst in adsorption chromatography lies in the possibility of its use in connection with organic (chelating) reagents and their metal complexes. The purity of colored organic reagents can often be checked by running their solutions in organic solvents through columns of aluminum oxide and other absorbents mentioned, and seeing whether more than one colored band is obtained.[47] Different adsorbents and solvents should be tried. This can be done on a small scale, with tubes a few millimeters in diameter. Working on a larger scale, it may be possible at times to purify reagents chromatographically if large amounts are not required.[48]

Not much work has been done on the separation of metals by running an organic solvent solution of their chelate complexes through an adsorbing column. Aluminum oxide has been used as adsorbent for metal dithizonates in organic solvents, but nothing of particular analytical value has resulted. The hydroxyquinolates have also been tried, but apparently without noteworthy findings.

Thin-layer chromatography

In thin-layer chromatography,[49] the finely divided adsorbent is coated on a glass plate with the aid of a binder to form a layer ~ 0.25 mm thick. Silica gel is frequently used as the adsorbing layer. Thin-layer chromatography combines the use of the adsorbents of column chromatography with the technique of paper chromatography (p. 328). The resemblance to paper chromatography has been made closer by the introduction of a flexible sheet of solvent-resistant polyester coated with silica gel in a binder of poly(vinyl alcohol), which does not alter the adsorbing properties of silica. Another thin-layer form is a glass-fiber sheet impregnated with the adsorbent—silica gel, alumina gel, etc. Thin-layer chromatography provides rapid separation of complex organic substances on the ultramicro to milligram scale. It is more versatile than paper chromatography.

Partition chromatography (liquid–liquid)

In 1941 Martin and Synge separated acetylated amino acids by using silica gel as an inert support for a stationary aqueous phase and a mixture of butanol, chloroform, and water as the mobile phase. Since then many other applications of column liquid–liquid chromatography have been made. Although of special importance in the separation of organic substances, especially in biochemical work, it also finds use in inorganic separations for analytical purposes.

This type of chromatography is based on the difference in distribution of different solutes between two immiscible liquid phases, the general theory of which has already been considered. It shares with ionic-exchange adsorption the advantage of giving linear, or almost linear, adsorption isotherms over a considerable range of concentrations, so that symmetric adsorption bands tend to be formed. By the use of columns of suitable size, macro amounts of sample can be taken.

[47] It is easy to show that the 1-nitroso-2-naphthol put on the market by one U.S. laboratory supply house as a certified reagent is grossly impure.

[48] Activated alumina is of value in purification of solvents. Peroxides can be removed from ethers by percolating them through a column of alumina. Ethyl ether, benzene, chloroform, and other solvents can be dehydrated (H_2O content reduced to 0.01 per cent or less) in the same way. Ethyl alcohol, added as a stabilizer, can be removed from chloroform at the same time.

[49] E. Stahl (ed.), *Thin-Layer Chromatography*, Academic Press, New York, 1965; J. M. Bobbitt, *Thin-Layer Chromatography*, Chapman & Hall, London, 1963; K. Randerath, *Thin-Layer Chromatography*, Academic Press, New York, 1963. E. V. Truter, *Thin Film Chromatography*, Cleaver-Hume, London, 1963.

Most columnar inorganic partition chromatographic separations are carried out with cellulose (pulp or powder) as the column filling. Uranium and thorium can be separated from each other and from most other elements by the use of a cellulose column and an ether solution containing nitric acid as the mobile phase.[50] Uranium is eluted with 97 per cent ether–3 per cent concentrated nitric acid, and thorium with ether containing 12.5 per cent nitric acid. Uranyl and thorium nitrates are more soluble in the ether phase than in the aqueous phase and move down the column, whereas most of the other metal nitrates move very little. This type of separation is closely allied to paper chromatography and may in fact be considered one of its variants.

Another example is of interest from the standpoint of trace analysis. Lead, zinc, manganese, and other heavy metals reacting with dithizone can be isolated from natural waters by running the sample through a column of cellulose acetate holding a carbon tetrachloride solution of dithizone as the stationary phase.[51] The metals can be eluted with acid and ammonia.

Paper chromatography[52]

A popular type of partition chromatography is based on the use of adsorbent paper strips which furnish ready-made columns of a suitable medium. A drop of aqueous sample solution is applied at one end of a thick paper strip (2–3 cm wide, 25 or more cm long) or sheet and allowed to evaporate. The strip (or sheet) is then hung inside a closed glass vessel and the chromatogram is developed by allowing a suitable organic solvent or organic–aqeous solution to ascend the paper by capillarity or to descend by a combination of capillarity and gravity. The constituents being separated move at different rates to form a series of spots or bands, which can be made visible by spraying the strip with a reagent solution forming colored products with the constituents. If the position of the bands is known (as by the use of a comparison strip that has been sprayed) the strip can be cut into segments from which the constituents can be leached with a suitable solvent and determined colorimetrically or by some other method. Less exactly, the colored areas produced by spraying can be compared with standards or the color intensity can be measured photometrically in transmitted or reflected light. The paper strip must be long enough to provide separation of the zones of the constituents without serious overlap if determinations are to be made. Several hours or more may be required for the migration process. Chromatographic papers of different densities and flow rates are available commercially.

The general principles underlying separability by paper chromatography are the same as in other types of chromatography (p. 289). Since the elution solvent usually does not traverse the whole length of the strip it is convenient to refer the distance traveled by a particular solute band to the distance traveled by the solvent front. This ratio is the R_f (flow ratio) value of the solute under the conditions. R_f values obtained under standardized conditions are of aid in identification of the solutes in a chromatogram. The front edge of a band is usually sharper than the trailing edge, which may be quite diffuse because of the asymmetry of the band, and is therefore used in the measurement of R_f.

Paper chromatography is not simple partition chromatography in the sense that the solutes are partitioned between two immiscible liquid phases. In fact, the mobile solvent need not be, and often is not, immiscible with water. The cellulose fibers are not an inert support for a stationary liquid phase. Cellulose is a hydrophilic substance which binds water by hydrogen bridges or secondary valence forces, so that we might say

[50] N. F. Kember, *Analyst* **77**, 78 (1952).

[51] D. E. Carritt, *Anal. Chem.* **25**, 1927 (1953).

[52] R. J. Block, E. L. Durrum, and G. Zweig, *A Manual of Paper Chromatography and Paper Electrophoresis*, Academic Press, New York, 1958; F. Cramer, *Paper Chromatography* (2nd ed. trans. by L. Richards), Macmillan, New York, 1955; I. M. Hais and K. Macek, *Handbuch der Papierchromatographie*, Fischer, Jena, 1960.

it forms cellulose–water complexes. Even if the mobile phase consists of an organic solvent miscible with water or an organic solvent–water solution, an equilibrium can be set up such that the interior and exterior phases have quite different liquid compositions. Moreover, cellulose possesses adsorptive properties toward solutes, so paper chromatography may also involve adsorption chromatograpy. Separations with water alone as the mobile solvent have been reported. Cellulose contains carboxylic and other groups introduced in bleaching which can give rise to ion exchange. The mechanism of paper chromatography is accordingly complex and is not to be considered as restricted to partition between liquid phases, although this may be the predominant process in some separations, as of metal nitrates with organic solvents containing nitric acid as the mobile phase.

As an example of the usefulness of paper chromatography in inorganic analysis,[53] we may cite the separation of the platinum metals Pt, Pd, Rh, and Ir (Ru and Os are usually not considered in such separation schemes because they can be removed beforehand by volatilization).[54] Chromatographic paper strips, 3 × 45 cm, are employed with downward diffusion of solvent. A mixture of isobutyl methyl ketone, pentanol, and concentrated hydrochloric acid in the volume ratio 60 : 10 : 30 gives three well-defined equally spaced bands of Rh(III), Pd(II), and Pt(IV). Iridium(III) remains with Rh. It can be separated if oxidized to Ir(IV) with hydrogen peroxide, and *n*-butanol saturated with 3 *M* hydrochloric acid is used as the mobile phase. After development of the chromatogram, the strip is cut into sections containing the respective metal bands (with the aid of a comparison strip sprayed with stannous chloride to reveal the location of the bands) and the metals are washed out of the paper. Pt, Pd, and Rh are determined spectrophotometrically and Ir titrimetrically. The accuracy attainable is illustrated by Table 13-5.

TABLE **13-5**

	Minor Components	
Major Component, μg	*Present, μg*	*Found, μg*
545 Pt	2.1 Pd	1.9
	2.7 Rh	2.5
	1.8 Ir	1.5
3170 Pd	2.0 Pt	2.2
	3.0 Rh	3.2

The usual technique of paper chromatography limits the sample to a few milligrams and a very small volume. Larger amounts can be handled in a cellulose pulp column.

Many schemes have been worked out for the detection of cations and anions by chromatographic separation followed by spraying with color-producing reagents. A small number of strips in combination with a small number of color reagents suffice for the detection of a large number of ions.

Electrophoresis in stabilizing media such as filter paper and starch gel superficially resembles chromatography. In this process, different ions or charged particles migrate

[53] F. H. Pollard and J. F. W. McOmie, *Chromatographic Methods of Inorganic Analysis*, Butterworth, London, 1953; F. H. Pollard, *Chromatographic Methods of Inorganic Analysis with Special Reference to Paper Chromatography*, Academic Press, New York, 1953; O. C. Smith, *Inorganic Chromatography*, Van Nostrand, Princeton, N.J., 1953; E. Blasius, *Chromatographische Methoden in der analytischen und präparativen anorganischen Chemie unter besonderer Berücksichtingung der Ionenaustaucher*, Enke, Stuttgart, 1958.

[54] N. F. Kember and R. A. Wells, *Analyst* **80**, 735 (1955).

at different rates in an electric field and separations can thus be effected.[55] Electrophoresis can be combined with chromatography. Thus an electric field can be applied at right angles to the direction of flow of the solution in paper or in a thin layer of adsorbent and a double separation effect is obtained.

Gel filtration[56]

Brief reference has already been made to this type of chromatographic separation (p. 322). The limited scope of this chapter does not permit detailed treatment.

Gas chromatography[57]

This is one of the later branches of chromatography to be developed; the first analytical applications were made in 1952 by James and Martin.[58] It is divided into gas–solid chromatography and gas–liquid chromatography, the latter being far more important. The stationary phase in gas–liquid chromatography is a liquid held on a solid support such as diatomaceous earth; the gas or vapor partitions between the nonvolatile liquid phase and the inert carrier gas (N_2 or He, for example). As in other types of chromatography, the gaseous components form a series of bands in the column which are eluted by the stream of carrier gas and can be detected and determined in the issuing stream by their effect on the thermal conductivity, ionization, and other properties of the gas. These properties can be measured automatically and continuously as a function of the time at uniform gas flow (Fig. 13-19). The theory of gas–liquid partition chromatography is essentially the same as that of other types of chromatography if the compressibility of the gas (which produces a variation in velocity along the column) is taken into account. The area or height of a peak, compared to that of a standard, gives the amount of the component.

Gas chromatography is of great importance in the analysis of a wide variety of complex organic samples. By the use of heated columns it can be applied to substances that have vapor pressures as low as 10 mm Hg at 300°C or higher. Similar organic compounds can often be separated with ease. The sample need not exceed a few milligrams and the time required is usually not greater than $\frac{1}{2}$ hr.

The low viscosity of gases allows the use of very long columns and therefore better separability can be attained than in liquid–liquid partition chromatography. Since higher flow rates can be used than in the latter type of chromatography, gas chromatography is more rapid. Moreover, the concentration of the components in the carrier gas is easily determinable by a variety of devices. The accuracy is often within a few per cent.

Additional details concerning gas chromatography are given in Chapter 62.

[55] Review: D. Gross, *Analyst* **90**, 380 (1965); inorganic applications, p. 391.

[56] H. Determann, *Gelchromatographie*, Springer, Berlin, 1967.

[57] See, for example, A. I. M. Keulemans, *Gas Chromatography*, Reinhold, New York, 1959; D. Ambrose and B. Ambrose, *Gas Chromatography*, Newnes, London, 1961; S. Dal Nogare and L. W. Safranski, *Gas Chromatography*, in Mitchell et al. (eds.), *Organic Analysis*, vol. 4, Wiley-Interscience, 1960; D. H. Desty and C. L. A. Harbourn, *Vapor Phase Chromatography*, Academic Press, New York, 1958; R. L. Pecsock (ed.), *Principles and Practice of Gas Chromatography*, Wiley, New York, 1959; B. A. Rose, review in *Analyst* **84**, 574 (1959); V. J. Coates, H. J. Noebels, and L. S. Fagerson (eds.), *Gas Chromatography*, Academic Press, New York, 1958; J. H. Knox, *Gas Chromatography*, Wiley, New York, 1962; H. A. Szymanski, *Lectures on Gas Chromatography—1962*, Plenum Press, New York; H. A. Szymanski, *Progress in Industrial Gas Chromatography*, Plenum Press, New York, 1961; A. V. Signeur, *Guide to Gas Chromatography Literature*, Plenum Press, New York, 1964; H. Purnell, *Gas Chromatography*, Wiley, New York, 1962; E. Bayer, *Gas Chromatography*, Elsevier, New York, 1961; A. B. Littlewood, *Gas Chromatography*, Academic Press, New York, 1962; H. Kaiser (trans. by P. H. Scott), *Gas Phase Chromatography*, Butterworth, London; R. W. Moshier and R. E. Sievers, *Gas Chromatography of Metal Chelates*, Pergamon, New York, 1965. Also see *Gas Chromatography Abstracts*, Elsevier, New York.

[58] A. T. James and A. J. P. Martin, *Analyst*, **77**, 915 (1952).

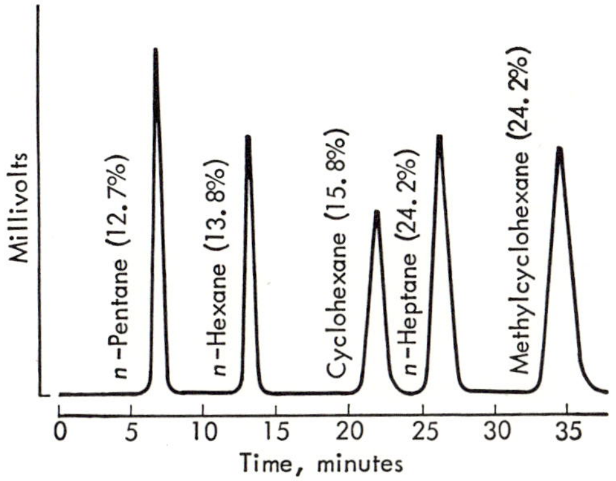

Fig. 13-19. Chromatogram of hydrocarbon mixture illustrating separation of similar compounds by gas chromatography (thermal conductivity detector). From M. Dimbat, P. E. Porter, and F. H. Stross, *Anal. Chem.* **28**, 291 (1956).

PROBLEMS

1. Consider a chromatographic column with equal volumes of mobile and stationary phases, in which all of the solute is originally present in the first theoretical plate ($r = 0$). If $D = 15$, what fraction of solute (F) will be found in $r = 0, 1, 2, \ldots, 12$ after 100 transfers, in each of which the volume of transferred mobile phase is equal to that in a plate? Plot F versus r and compare with the normal error curve having the same F_{max}. Also calculate $F_{6,100}$ and $F_{10,100}$ by Eq. (11) and compare with the values obtained on the basis of discontinuous transfer.
Ans.

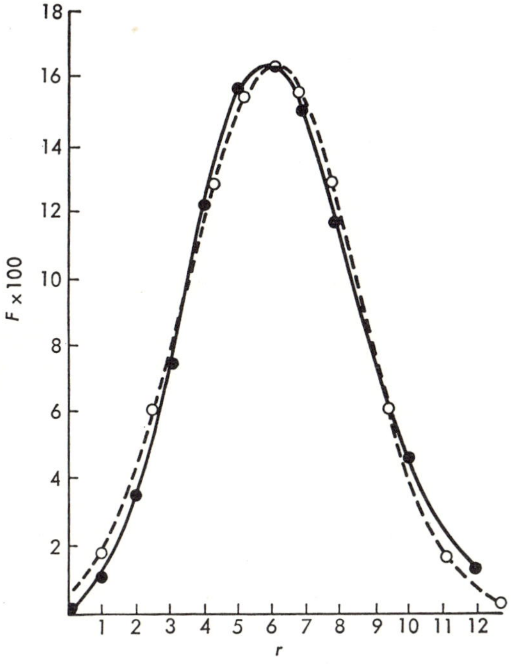

$F_{6,100}$ calculated on basis of continuous transfer = 0.162.

2. Given the following data: length of chromatographic column = 10 cm, cross-sectional area = 1 cm^2, $D = 25$, $h = 0.2$ cm, and $i = 0.4$, find the plate r containing the largest fraction of solute after 80 ml of solution (for which $D = 25$) has been passed through the column, and the value of this fraction. What is the band width in centimeters (defined as on p. 291)? Assume that all the solute was initially present in plate $r = 0$.
   Ans. $r = 26$, $F_{26} = 0.078$, adsorption band width $= 2(2hl)^{1/2} = 2.86$ cm.
3. If, for a certain solute and flow rate, $V_{max} = 60$ ml and $V_e = 30$ ml, what is the elution band width (i.e., the band width when the peak of the band is at the base of the column)? How many theoretical plates does the column contain?
   Ans. 85 ml band width.
4. A small amount of solute having a D_v (distribution coefficient based on resin *bed* volume) of 500 is originally present in the first ($r = 0$) theoretical plate of a column. If the length and cross-sectional area of the column are 3.0 cm and 1.0 cm^2, the height of a theoretical plate is 0.3 cm, and the fractional interstitial volume is 0.4, what fraction of the solute will be present in the last plate of the column when 12.0 and 120 ml of solvent have been passed through the resin column initially filled with the solvent? What volume of solvent must be passed through the column to bring F_{max} to the last plate of the column and what is then the value of F_{max}?
   Ans. (partial) $(V_{max})_{r=9} = 1351$ ml; $(F_{max})_{r=9} = 0.133$.
5. Derive the relation (see the legend of Fig. 13-3)

$$C_{max} = \frac{m}{V_{max}} \left(\frac{N}{2\pi}\right)^{1/2}$$

6. A small volume of 2.0 M HCl containing 1 mg of the metal M is added to an anion-exchange column so that essentially all of M is adsorbed in the first millimeter of the resin bed. The length of the column is 3.0 cm, its cross-sectional area is 1.0 cm^2, and the height equivalent to a theoretical plate is 0.3 cm. If D_v for M in 2.0 M HCl is 100, what distance down the column will the band maximum have moved after elution with 100 ml of this acid and what fraction of M will have been eluted? Take $i = 0.4$. What fraction of M will be eluted if $h = 0.6$ cm, other factors remaining the same?
7. Passage of 24.5 ml of eluting solution removes 90 per cent of A from a 1-cm^2 resin column of 8.0 cm length. What is the height of a theoretical plate if $\mathscr{E} = 0.35$? (Take the adsorption band of A to lie at the top of the resin column at the beginning.)
8. Collection of successive 0.1-ml portions of eluate near the elution maximum showed that the latter corresponded to 25.0 ml and that the first 23.0 ml of eluate contained 25.0 per cent of the total amount of solute. If the column length is 5.0 cm and the cross section is 1 cm^2, what are the values of h and N?
   Ans. $h = 0.070$ cm, $N = 71$.
9. Given $V_{max} = 35.0$ ml, $V_e = 30.0$ ml and $l_{col} = 15.0$ cm, (a) find N and h by the method outlined on page 290 and (b) compare with values obtained from Eqs. (15) and (16), page 291.
   Ans. $h = 0.171$ cm, $N = 88$.
10. Can 1 μg of Pd(II) as $PdCl_4^{--}$ in 10 liters of 1 M HCl be quantitatively retained (95 per cent) by running the solution slowly through a 1 cm^2 × 5 cm anion-exchange resin column if D_v for Pd(II) under the conditions is 1×10^3?
11. Evaluate the feasibility of separating Ca and Sr (weight ratio Ca/Sr = 100) by the use of Dowex 50-X8 resin (Table 13-2) with HCl as eluant.
12. Using the equilibrium constants in Table 13-2, specify the approximate column length and concentration of HCl that should be used to separate 0.1 mg of Ce(III) from 50 mg of Ca on Dowex 50-X8 resin. Take the exchange capacity of the resin bed to be 2 meq/ml.
13. Examine the feasibility of effecting a marked concentration of Cs (reducing the ratio K/Cs) by slowly running 100 ml of 0.1 M KCl which is 10^{-5} M in CsCl

through a relatively short column of Dowex 50-X8 resin in the K form. Essentially complete recovery of Cs is required. See Table 13-2 for exchange constants. Take the exchange capacity of the resin bed to be 1 meq/ml.

14. Sodium and potassium are to be separated by ion exchange on a Dowex 50 resin column 20 cm long and 0.9 cm in diameter. The eluting solution is 0.6 M HCl, in which the distribution coefficients are $D_{Na} = 15$ and $D_K = 47$ (these coefficients are based on the *weight* of resin). Take the density of the resin as 1.1, the fractional interstitial volume of the column as 0.45, and h as 0.35 mm. Assume sodium and potassium to be present in a narrow band at the top of the column at the start. (a) Find V_{max} for Na and K. (b) Calculate the width of the elution peaks at 0.37 of the peak height. (c) Calculate the theoretical volumes of 0.6 M HCl required to elute 0.99999 of each element. (d) Graph the elution curve (volume of elution solution versus concentration of Na and K). (e) Is the separation satisfactory for comparable amounts of Na and K?
15. Using the data of Problem 14, find the minimum length of column which, with 10 times as much Na as K in the sample, will allow recovery of 99.999 per cent K when contamination with 0.001 per cent Na is permissible. At what volumes of eluant shall the cuts be made? What factors may impair the accuracy of the calculated results?
16. Two ions A and B in the ratio $(Q_B)_0/(Q_A)_0 = 200$ are to be separated, with the requirements that $R_A = 0.999$ and $Q_B/Q_A = 0.001$. Find the length of column required if $h = 1$ mm, $\mathscr{E}_A = 0.3$, and $\mathscr{E}_B = 0.1$.
17. The elution constants of Co(II) and Fe(III) in hydrochloric acid of various concentrations are as follows:

	$\mathscr{E}$	
HCl, M	Co(II)	Fe(III)
1.0	2.5	1.3
2.0	2.2	0.4
3.0	1.8	0.04
4.0	1.3	0.015
5.0	0.7	
6.0	0.23	0.0022
7.0	0.06	
9.0	0.002	0.0002
10.0	0.023	

Outline a procedure for separating Co and Fe(III), specifying the concentration of hydrochloric acid, approximate elution volumes, and length of column.

18. From the following data on the elution of A from a resin column with 0.01 M hydrochloric acid: length of column = 6 cm, concentration of A = 0.0123 M at $V = V_{max} = 21.0$ ml, and concentration of A = 0.0067 M at $V = 28.5$ ml, find: (a) The volume of 0.01 M hydrochloric acid required to elute 99.9 per cent of A; (b) the number of theoretical plates in the column. (Assume A to be present in a narrow band at the top of the column at the start.)
19. If the equilibrium constant of the reaction

$$RA + B^+ \rightleftharpoons RB + A^+$$

is 10, find the fraction of B removed from 100 ml of solution originally containing 1 meq of B^+ by shaking with 1 g of RA, having a cation-exchange capacity of 5.0 meq/g, when (a) the solution contains no A^+ at the beginning; (b) the solution contains 10 meq of A^+ at the beginning.

20. If the equilibrium constant for the exchange reaction

$$2\,RA + B^{++} \rightleftharpoons R_2B + 2\,A^+$$

has the value 1.0 and the cation-exchange capacity of the resin is 5.0 meq/g, find the concentrations of A and B in the resin after shaking 1.0 g of the resin RA with 100 ml of solution having the initial concentrations (a) 0.005 *M* A^+ and 0.005 *M* B^{++}; (b) 0.05 *M* A^+ and 0.05 *M* B^{++}.

21. It has been reported that gold in fine flakes occurs to the value of 5 cents or less per ton in the upper part of the St. Peter sandstone of the Minneapolis–St. Paul district. Suppose that it is desired to check this value of gold content to the nearest 1 cent (~8 mg/ton). It is judged that 1 kg of sample will furnish more than enough gold for the analysis (8 μg of Au in 1 kg will correspond to about 1 cent Au/ton, and as little as 1–2 μg of Au can be determined spectrophotometrically within 10 per cent in the absence of interfering elements). The gold can be brought into solution by treatment of the crushed sample with aqua regia or hydrochloric acid containing Cl_2. Suppose then that the gold as $AuCl_4^-$, accompanied by 1.0 g of Fe(III) from iron compounds in the sample, is obtained in 1 liter of 2 *M* HCl. Outline an ion-exchange method whereby the gold can probably be isolated in a final solution volume of not more than 25 ml with a recovery of at least 90 per cent and accompanied by not more than 0.1 mg of Fe. [*Hint:* See Fig. 13-16. A "batch process" might be tried. Compare the rationale of immiscible solvent extraction (Chap. 14). Note that it is not easy to remove gold from an anion-exchange resin with hydrochloric acid.]

22. 1 g of a cation-exchange resin in the H^+ form is shaken with 100 ml of 0.01 *M* HCl containing a trace of silver. Find the distribution coefficient of silver and the recovery factor, if only Ag^+ exchanges.

$$\frac{[Ag]_r[H^+]}{[Ag^+][H]_r} = 6.7$$

$$\frac{[AgCl]}{[Ag^+][Cl^-]} = 8 \times 10^2$$

$$\frac{[AgCl_2^-]}{[Ag^+][Cl^-]^2} = 5 \times 10^4$$

$$\frac{[AgCl_3^{--}]}{[Ag^+][Cl^-]^3} = 1 \times 10^5$$

Capacity of resin = 5 meq/g.

23. Utilizing information in this chapter, devise a procedure for separating ~5 μg of Tl from 0.5 g of K^+ with a recovery of >90 per cent of the thallium, accompanied by not more than an equal weight of potassium. As far as possible, specify size of exchange column, volume, and composition of influent and eluant solutions, and other factors of importance in the separation.

CHAPTER 14 *Separations by Immiscible Solvent (Liquid–Liquid) Extraction*

A substance distributes between two contacting immiscible liquids—water and a suitable organic solvent, for example—roughly in the ratio of its solubility in each if it does not react with either and if it exists in the same form in both. If, at equilibrium, its concentration is much greater in the organic solvent phase than in the aqueous phase, the distribution behavior may be put to analytical use in concentrating the substance into a small volume of the organic liquid and, more importantly, in separating it from substances that do not distribute similarly. If the extracted substance is colored or fluorescent, the extract may serve for the direct determination of the substance. Or the extract can be used for determination of an element by flame photometry, X-ray fluorescence, optical spectrography, and possibly other methods.

Organic chemists have used organic solvents for extraction of organic substances from water solutions for more than 150 years. In inorganic analysis liquid–liquid extraction did not become widespread until relatively recent years, especially in connection with the use of organic reagents. Atomic energy programs have provided strong stimuli to investigations of immiscible solvent extractions. Extraction separations are particularly valuable for the isolation and separation of small amounts of substances, but, analytically, they also find use on the decigram and even gram scale.

The purpose of this chapter is to consider briefly the different types of liquid–liquid extraction systems of use in inorganic analysis, to give some general information on a few of the more important ones, and to show how, from a knowledge of the reaction equilibria in such systems, the analyst can formulate and predict the extractability of a substance and its separability from other substances.[1]

[1] For a 250-page treatment of the analytical aspects of extraction, see G. H. Morrison and H. Freiser, *Solvent Extraction in Analytical Chemistry*, Wiley, New York, 1957. Also see H. Irving and R. J. P. Williams in *Treatise on Analytical Chemistry* (I. M. Kolthoff and P. J. Elving with E. B. Sandell, eds.), part I, vol. 3, Wiley, New York, 1961, chap. 31, and F. A. von Metzsch in *Physical Methods in Chemical Analysis*, vol. IV, Academic Press, New York, 1961. Reviews: R. M. Diamond and D. G. Tuck, *Progr. Inorg. Chem.* **2**, 109 (1960); H. Irving, *Quart. Rev. (London)* **5**, 200 (1951); F. S. Martin and R. J. W. Holt, *Quart. Rev. (London)* **13**, 327 (1959); Y. Marcus, *Chem. Rev.* **63**, 139 (1963); L. I. Katzin in J. J. Lagowski (ed.), *Chemistry of Non-Aqueous Solvents*, vol. I, Academic Press, New York, 1966, chap. 5; D. F. Peppard, *Advan. Inorg. Chem. Radiochem.* **9**, 1 (1966). (This listing is not complete.) Also see the reviews in *Anal. Chem.* A considerable amount of information, general and specific, may be found in E. B. Sandell, *Colorimetric Determination of Traces of Metals*, Wiley-Interscience, New York, 3rd ed., 1959, and O. G. Koch and G. A. Koch-Dedic, *Handbuch der Spurenanalyse*, Springer, Berlin, 1964. D. D. Perrin in *Organic Complexing Agents*, Wiley-Interscience, New York, 1964, devotes a chapter to extraction into organic solvents. A detailed treatment of chelate extraction is given by J. Stary in *The Solvent Extraction of Metal Chelates*, Macmillan, New York, 1964.

With few exceptions, liquid–liquid extraction systems of interest in inorganic analysis have an aqueous solution as one phase and an organic solvent as the other.[2] A different type of extraction, less versatile, is liquid–solid extraction (leaching), in which one component of a solid mixture is dissolved by a suitable solvent, whereas the other (or others) is not. For example, sodium and potassium can be separated by converting to the chloroplatinates and leaching with 80 per cent ethyl alcohol. Sodium chloroplatinate dissolved readily, potassium chloroplatinate hardly at all.

Partition constant and extraction coefficient[3]

The partitioning of a substance A between an aqueous and an immiscible organic solvent phase:

$$\mathrm{A}_w \rightleftharpoons \mathrm{A}_o \qquad (o = \text{organic phase},\ w = \text{aqueous phase})$$

is governed by the equilibrium relation

$$\frac{(\mathrm{A})_o}{(\mathrm{A})_w} = \text{equilibrium constant} = P_{\mathrm{A}}^0 \qquad \text{(at constant temperature)} \tag{1}$$

in which the parentheses indicate the activity of the substance in solution. P_{A}^0 is the *activity partition constant* of A. It is assumed that a variation in the amount of A does not greatly affect the solubility of the two "immiscible" solvents in each other.

At a specified ionic strength in the aqueous phase and a specified concentration of A (and other solutes) in the organic phase

$$\frac{[\mathrm{A}]_o}{[\mathrm{A}]_w} = P_{\mathrm{A}} \tag{2}$$

[2] The fire assay for gold, silver, and the platinum metals—an old method which is still standard—is based on a high-temperature liquid–liquid extraction. The sample is fused with a mixture of flux, reducing agent, and lead oxide, whereby the noble metal is extracted by the molten lead, in which it is much more soluble than in the molten slag.

[3] *A note on extraction terminology and symbology.* The usage of terms applying to extraction is not uniform and some comments on our choice of terms and symbols may be desirable. *Distribution* is the general term referring to the transfer of substances between two (or more) phases whether liquid–liquid, solid–liquid, liquid–gas or gas–solid. Liquid–liquid extraction is distribution between two liquids (solutions) in contact—which therefore must be "immiscible." (A gaseous substance can distribute between two liquids not in contact—for example, ammonia in the Conway diffusion method for its determination; this is not extraction.) As shown above, it is essential to distinguish between the distribution of a single substance and that of a constituent existing in a number of forms (species) in one or both liquids. To avoid ambiguity we refer to the distribution of a single unchanged species as *partitioning* and use P as the symbol for the *partition constant* (or *partition coefficient*—the *coefficient de partage* of Berthelot and Jungfleisch). Sometimes K_D or other symbols are used by other authors and the quantity is called *distribution coefficient*. It seems to us that P is less ambiguous than K_D and has the advantage of avoiding a subscripted symbol to which additional subscripts must likely be added when equilibrium expressions are written. We use *extraction coefficient* (E) to denote the stoichiometric (analytical) concentration ratio of a constituent in the two phases (the organic phase concentration in the numerator). This quantity is often called the *distribution ratio* and represented by D, λ, or q. We believe there is little danger of confusing *partition constant* and *extraction coefficient*, in contrast to *distribution coefficient* and *distribution ratio*. (That E is used as a symbol for other quantities does not seem to be an important objection to its use here; there are more terms in chemistry than Roman and Greek letters to symbolize them, and short of resorting to the Coptic and Sanskrit alphabets some duplication is to be expected and doubtless borne.) Finally, it may be noted that the currently popular term *solvent extraction* is a dubious combination of words. Extraction necessarily involves a solvent, whether organic liquid, molten metal or fused salt. Most of the time, little ambiguity is likely to result from the use of the simple term *extraction*. In a chemical context it is unlikely to be mistaken for the extraction of roots (mathematical) or the drawing of teeth. *Liquid–liquid extraction* or *immiscible solvent extraction* are permissible or necessary terms as long as leaching is referred to as *liquid–solid extraction*.

where P_A is the partition constant of A based on concentrations. Since A is uncharged, P_A will often approximate P_A^0 within 10 per cent, if the electrolyte concentration of the aqueous phase does not exceed 0.2–0.3 *M*.

The foregoing relation—the partition law—was established experimentally by Berthelot and Jungfleisch in 1872, derived thermodynamically by E. Riecke in 1891, and further treated by Nernst in the same year, who took into account the ionization and association of the partitioning substance.

It is sometimes of interest to know how the partition constant of an un-ionized substance (nonelectrolyte) varies with the *indifferent* electrolyte concentration in the aqueous phase. Usually the solubility of a nonelectrolyte is smaller in an aqueous electrolyte (salt) solution than it is in water. This is the *salting-out* effect, which is ascribed principally to the withdrawal of water molecules by hydration of the electrolyte cations. Less often, electrolytes increase the solubility of a nonelectrolyte (*salting-in* effect); HCN is an example of such a nonelectrolyte.

The quantitative effect of electrolytes on the solubility of a nonelectrolyte is given by the following relation (Setchenow equation):

$$\log f = \log\left(\frac{\text{solubility of nonelectrolyte in water}}{\text{solubility of nonelectrolyte in electrolyte solution}}\right)$$

$$= k\mu$$

where f = activity coefficient of nonelectrolyte in electrolyte solution
k = a constant whose value depends on the nonelectrolyte and the electrolyte
μ = ionic strength of electrolyte in aqueous solution

This equation often holds very well up to surprisingly high electrolyte concentrations. For example, for dimethylglyoxime in aqueous sodium chloride solutions the value of k increases only 12 per cent as the sodium chloride concentration is increased from 0.05 to 5 *M*:[4]

NaCl, *M*	f	$\log f$	k
0.05	1.02	0.0086	0.17
0.5	1.24	0.0934	0.187
5.0	9.0	0.954	0.191

If the electrolyte exerts no effect on the solubility of the nonelectrolyte in the organic solvent, i.e., if it does not affect the mutual solubilities of the aqueous and organic phases or dissolve in the organic solvent, the partition constant of the nonelectrolyte will increase in proportion to the decrease of its solubility in the aqueous phase, so that

$$P_{\text{elect}} = fP_w$$

where P_{elect} is the partition constant for an electrolyte solution of specified concentration and P_w is the partition constant for water.

For dimethylglyoxime,

$$P_{5\,M\,\text{NaCl}} = 9 \times P_w = 9 \times 0.08_3 = 0.75$$

since the solubility of water in chloroform is low (0.072 g H_2O in 100 g $CHCl_3$ at 23°C), and it may be assumed that the slight change in water content of chloroform in equilibrium with 5 *M* sodium chloride solution compared to chloroform–water will not affect the dimethylglyoxime solubility in chloroform to any marked extent.

[4] H. Christopherson, M.S. thesis, Univ. of Minnesota, Minneapolis, 1954.

Occasionally the increase in partition constant produced by simple salting out (Setchenow type) can be of analytical importance (see also p. 346). For example, the extraction of tin(IV) iodide by benzene from aqueous solution becomes feasible if the electrolyte (perchlorate) concentration of the solution is made $\sim 5\ M$.[5] Salting out of the Setchenow type is one factor contributing to increased extraction when an electrolyte having an ion in common with the extracted species is added in high concentration, as when aluminum nitrate is used to augment the extraction of uranyl nitrate from nitric acid solution with ethyl acetate.[6]

In analytical work, the partition law can rarely be applied directly in the simple form of Eq. (2) to find the distribution of a constituent between two liquid phases. The constituent being extracted is likely to be present as more than one species in the aqueous phase and sometimes as more than one species in the organic phase, so that Eq. (2) does not give the desired information. The analyst usually wishes to know the ratio of the concentrations of the element A in all its forms in the two phases. He is interested in the value of the *extraction coefficient*, E, which gives this ratio:

$$E_{\text{A}} = \frac{\sum [\text{A}]_o}{\sum [\text{A}]_w} \tag{3a}$$

$\sum [\text{A}]$ signifies the *stoichiometric* concentration of A.

The extraction of iodine from an acidic aqueous iodide solution by carbon tetrachloride provides a simple illustration of P and E. The partition constant of iodine at 25°C is

$$P = \frac{[\text{I}_2]_{\text{CCl}_4}}{[\text{I}_2]_w} = 89.9 \quad \text{(at infinite dilution; } P \text{ varies slightly with the iodine concentration)} \tag{3b}$$

The extraction coefficient of iodine, considering all its forms (hydrolysis products neglected), is given by

$$E = \frac{2[\text{I}_2]_{\text{CCl}_4}}{2[\text{I}_2]_w + 3[\text{I}_3^-]_w + [\text{I}^-]_w} \tag{3c}$$

(Since E is based on analytical, i.e., stoichiometric, concentrations, the proper coefficients for the species concentrations must be introduced.) The value of E can be found from the known values of P and K,

$$K = \frac{[\text{I}_2]_w[\text{I}^-]_w}{[\text{I}_3^-]_w} = 0.00130 \tag{3d}$$

at a specified value of $[\text{I}^-]_w$ (or $[\text{I}_3^-]_w$). In practice, in isolation of iodine by extraction, a suitable oxidizing agent would be used to reduce $[\text{I}^-]_w$ to virtually zero, so that E would be given by Eq. (3b).

If we are interested only in the distribution of free iodine, I(0), between the two phases, we can write the expression for an extraction coefficient involving only the species $(\text{I}_2)_{\text{CCl}_4}$, $(\text{I}_2)_w$, and $(\text{I}_3^-)_w$:

$$\rho = \frac{[\text{I}_2]_{\text{CCl}_4}}{[\text{I}_2]_w + [\text{I}_3^-]_w} \tag{3e}$$

(I_5^- and higher polyiodide complexes may need to be included at high I_2 and I^- concentrations.) ρ can be found from P and K if $[\text{I}^-]_w$ is specified or known. Conversely, P and K can be found by obtaining values of ρ as a function of $[\text{I}^-]_w$. The determination of ρ is very easy because $[\text{I}_2]_{\text{CCl}_4}$ and $[\text{I}_2]_w + [\text{I}_3^-]_w$ are readily found by thiosulfate

[5] D. Gilbert and E. B. Sandell, *Microchem. J.* **4**, 491 (1960).

[6] For a discussion of salting effects, in the wider sense, in the extraction behavior of inorganic compounds, with special reference to indium chloride and oxygenated organic extractants, see R. M. Diamond, *J. Phys. Chem.* **63**, 659 (1959).

titration of I(0). The following relation can be derived [from Eqs. (3e) and (3d)]:

$$\frac{1}{\rho} = \frac{1}{P} + \frac{[I^-]_w}{PK} \tag{3f}$$

A plot of $1/\rho$ versus $[I^-]_w$ at constant ionic strength should give a straight line (for low I_2 concentrations), whose slope is $1/PK$ and whose intercept on the $1/\rho$ axis is $1/P$. The value of P is obtained easily enough from points at low $[I^-]_w$. K can then be found, an iterative calculation being made to obtain $[I^-]_w$ at equilibrium from its original concentration.

A question of importance in the application of extraction methods is whether the extraction coefficient remains constant as the concentration of an element or a substance varies in the aqueous phase, other factors remaining the same. In some systems E remains essentially constant; in others it does not. An example of variable E is furnished by a system in which association of one of the species with itself occurs in one phase or the other. Suppose A exists only as such in the aqueous phase and as A and A_n in the organic phase:

$$A_w \rightleftharpoons A_o$$
$$(nA)_o \rightleftharpoons (A_n)_o$$

The partition constant of the monomer is

$$P = \frac{[A]_o}{[A]_w} \tag{4a}$$

and the association (polymerization) constant of A in the organic phase is

$$K_{\text{assoc}} = \frac{[A_n]_o}{[A]_o^n} \tag{4b}$$

A_n is (by stipulation) the only associated species present, so that

$$E = \frac{\Sigma\,[A]_o}{\Sigma\,[A]_w} = \frac{[A]_o + n[A_n]_o}{[A]_w} = \frac{[A]_o + nK_{\text{assoc}}[A]_o^n}{[A]_w}$$
$$= P(1 + nK_{\text{assoc}}P^{n-1}[A]_w^{n-1}) \tag{4c}$$

When $n = 2$ (dimerization),

$$E = P + 2K_{\text{assoc}}P^2[A]_w \tag{4d}$$

A plot of E versus $[A]_w$ gives a straight line whose intercept on the E axis is equal to P and whose slope is $2K_{\text{assoc}}P^2$, from which K_{assoc} can be found.

Aliphatic acids commonly dimerize in solvents of low dielectric constant. Tetraphenylarsonium chloride polymerizes in chloroform (p. 359).

Recovery factor in extractions

If E_A is independent of the concentration of the extracted substance, the fraction of A in the combined organic extracts after n extractions, in each of which the volumes of the aqueous and organic solvent phases are equal (complete immiscibility assumed), is given by

$$(R_A)_n = \frac{(Q_A)_n}{(Q_A)_0} = 1 - \left(\frac{1}{E_A + 1}\right)^n = \frac{(E_A + 1)^n - 1}{(E_A + 1)^n} \tag{5a}$$

where $(R_A)_n$ is the *recovery factor* (p. 249). The percentage recovery is given by

$$(R_A)_{n,\%} = \frac{100\{(E_A + 1)^n - 1\}}{(E_A + 1)^n} \tag{5b}$$

If the aqueous phase volume is V_w and the volume of each portion of organic phase is V_o, and letting $r = V_o/V_w$,

$$(R_A)_n = 1 - \left(\frac{1}{rE_A + 1}\right)^n = \frac{(rE_A + 1)^n - 1}{(rE_A + 1)^n} = \frac{(E'_A + 1)^n - 1}{(E'_A + 1)^n} \tag{5c}$$

rE_A ($= E'_A$) is the effective extraction coefficient. Approximately (exactly if $n = 1$), when $E_A > \sim 10$ and n is small,

$$(R_A)_n = \frac{(E')^n}{(E')^n + 1} \tag{5d}$$

With a fixed total volume of organic solvent and an infinite number of extractions,

$$R_{max} = 1 - \exp(-r'E_A) \tag{5e}$$

where $r' = (\sum V_o)/V_w$ as $V_o \rightarrow 0$.

Intelligent application of an extraction reaction requires a knowledge of the value of E as a function of the variables in the extraction system. E can always be obtained experimentally but this may not be necessary. If the system is not too complex, it may be possible to calculate E from known values of the equilibrium constants of the reactions occurring. Before showing how this can be done for certain systems, we shall consider briefly a number of different types of extractions useful in inorganic analysis.

Classification of inorganic extraction systems

A simple classification of liquid–liquid extractions involving inorganic substances is difficult if not impossible because the different types cannot always be sharply differentiated. Some are a combination of types. For the purpose of an elementary treatment the following broad classification of analytically important extraction systems will be made.

1. Systems having simple inorganic species (molecules, rarely atoms) in both the organic solvent and aqueous phases. Examples are I_2 and other halogens (including At); OsO_4, RuO_4; covalent halides such as $AsCl_3$, $GeCl_4$, and SnI_4; Hg metal. These substances are all more or less volatile, evidence of their nonpolar nature. In such systems the organic solvent is usually largely inert in the sense that it simply dissolves the partitioning substance and does not react with it chemically. However, interaction may occur, as with iodine and benzene. A covalent molecule, having no particular affinity for water, may be accommodated more easily within the disordered structure of an inert organic solvent than within the ordered structure of water (hydrogen bonding of molecules), so that it will partition preferentially into the organic solvent. Nonoxygen solvents such as carbon tetrachloride or chloroform are preferred for the extraction of these substances, because they are less likely than oxygen solvents to extract other substances in solution. An ether, for example, would be a bad choice for extracting $GeCl_4$ or $AsCl_3$ from hydrochloric acid solution; it would also extract complex chloro acids from such a solution.

2. Systems in which uncharged chelate complexes of metals with organic reagents are extracted into the organic phase. 8-Hydroxyquinolates, dithizonates, and many other chelate complexes of metals can be extracted by a variety of organic solvents. The cation in such a complex is more or less surrounded by hydrophobic organic radicals and it is not surprising to find that the uncharged complex is, as a rule, only slightly soluble in water and much more soluble in organic solvents.

The metal cation in some chelate complexes is not coordinatively saturated by the anionic reagent radicals. For example, a divalent metal of coordination number 6 forming the chelate ML_2 with the reagent HL will have two coordination positions unoccupied, if each ligand anion L occupies two coordination positions (usually true). The remaining coordination sites can be filled by water molecules, organic solvent molecules (S), or uncharged reagent molecules, HL (Fig. 14-1). These addends can have a great effect on the extractability of the chelate complex. The presence of water molecules, e.g., $ML_2 \cdot 2\,H_2O$, may make the chelate very slightly soluble in organic solvents and thereby virtually nonextrable. Replacement of H_2O by HR (*adduct* formation with the chelating reagent) or by organic solvent molecules increases the solubility of the metal complex in the organic solvent and therewith its extractability if its water solubility is not increased significantly (the opposite effect is more likely). The extent of replacement of H_2O by other addends will vary from one system to another. The properties of the organic solvent (dipole moment, dielectric constant, etc.) will play an important role. When there are no good reasons to the contrary, non-oxygenated solvents are generally preferred to oxygen solvents in chelate extractions as being less likely to extract other substances. Chloroform often provides favorable extraction coefficients for chelates and is frequently used.

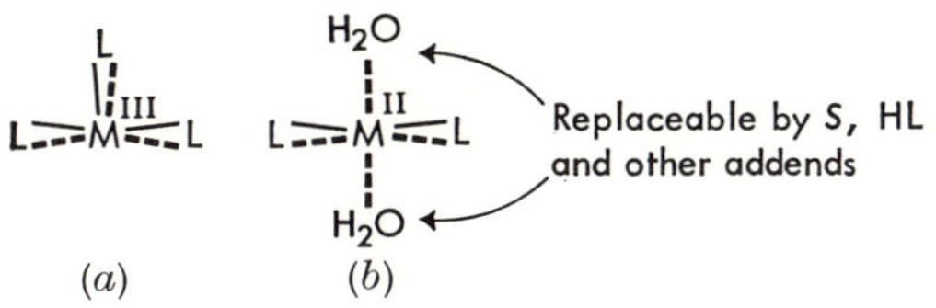

Fig. 14-1. Simplified representation of chelate complexes. (*a*) All the coordination sites of the cation M^{3+} (coordination number of six) occupied by chelate radicals L^-; (*b*) coordination sites of the cation M^{2+} (coordination number of six) partly occupied by chelate radicals. The addends are shown attached to the metal cation, but this is not the only possibility; they may be fixed to atoms of the anionic radical.

The organic addend need not be HL. It can be another organic molecule which, properly chosen, can have a very favorable effect on the extractability of the chelate complex. (See 8-Hydroxyquinoline, p. 358).

The extracted metal species in a chelate system may thus not be the simple chelate. It may resemble the species considered in classes (3) and (4) following.

3. Systems in which unsolvated ion-association compounds are formed with organic reagents. As pointed out in Chapter 12, large cations and large anions are but little solvated and the salts (ion-association compounds) they form with each other tend to have a low solubility in water, but frequently are more or less soluble in organic solvents if they have an organic component. There are hundreds of compounds of this type of demonstrated or potential usefulness in analytical separations. Some examples are:

Tetraphenylarsonium perrhenate, $(Ph_4As^+)(ReO_4^-)$.

Triphenylmethylarsonium tetrathiocyanatocobaltate, $(Ph_3MeAs^+)_2[Co(CNS)_4^{--}]$.

Rhodaminium B chloroantimonate, $(RH^+)(SbCl_6^-)$, where R = rhodamine B.

Cuprous biquinoline salt, $(CuBq_2^+)X^-$, where $X^- = Cl^-$, ClO_4^-, etc.

Benzyltrimethylammonium dithizonate, $(R^+)(HDz^-)$, where R^+ = benzyltrimethylammonium and HDz^- = primary dithizonate anion.

Tetraalkylammonium uranyloxinate, $(R_4N^+)(UO_2Ox_3^-)$, where R = alkyl and Ox^- = oxine (8-hydroxyquinoline) anion.

Cesium tetraphenylborate, $Cs^+(BPh_4^-)$.

Ion-association compounds composed of singly charged cations and anions are usually more easily extracted than others. Thus Ph_4As^+ forms a chloroform-extractable compound with ReO_4^- but not with MoO_4^{--}; $(Ph_4As)MnO_4$ is more readily extracted than $(Ph_4As)_2CrO_4$ by chloroform.

The extractability of these compounds varies from one organic solvent to another. Chloroform is often suitable and is preferred, when it can be used, as a nonpolar solvent not likely to extract other (inorganic) substances. $(CuBq_2)Cl$ can be extracted into amyl alcohol but not into chloroform. Extraction of $CsBPh_4$ requires a solvent of high dielectric constant (nitromethane or nitrobenzene, $\epsilon \sim 35$, or hexone–cyclohexane). The role of the organic solvent in the extractions of this type is largely passive in that it does not coordinate with the extracted species. However, it may promote extraction when its dielectric constant is high by favoring dissociation of the ion pairs.[7]

Adducts may be formed by ion-association compounds with the reagent in organic solvents.

Some useful ion-association reagents are discussed later in this chapter (p. 359).

4. Systems in which coordination complexes of inorganic species with the organic solvent molecules are formed. The complex species formed are of more or less different types. Thus the molecules of a suitable solvent can coordinate to the hydrogen ion of a complex metallo acid (or a mineral acid) or to the cation of a salt or ion-association compound. The strength of the bonding varies over a wide range. In some systems definite compounds are formed which can be isolated. In others there is an indefinite solvation of the extracted species. An important factor is always the displacement (not necessarily complete) of water molecules around the species to be extracted.

The extraction of iron(III) and a number of other trivalent metals from hydrochloric acid solution by ethyl ether and other ethers involves the formation of a coordination complex between the complex metal acid and the ether which, as might be expected, is ether-soluble. The hydrogen of $HMCl_4$ bonds to the basic oxygen atom of the ether. An ion pair is formed having a large cation consisting of donor solvent molecules and water associated with a proton. Other "oxygen solvents" of basic character, forming oxonium complexes (as below)

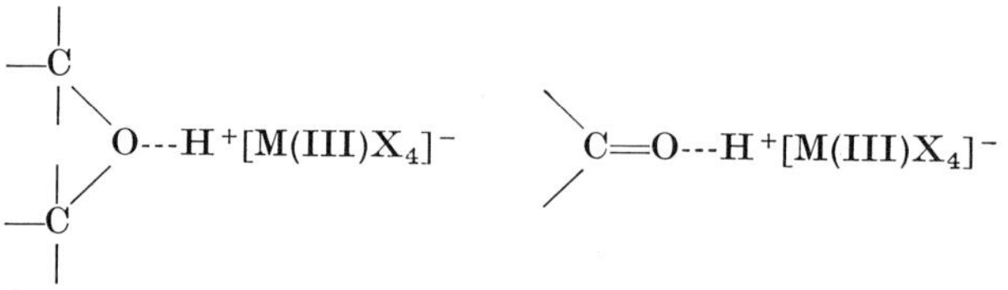

can be used instead of ether. Thus esters and ketones extract $HFeCl_4$ and other complex halo acids (and salts) very well. The immiscible organic liquid functions both as reagent and solvent. Nonoxygen solvents such as chloroform and benzene are useless for the extraction of complex metal acids of the type HMX_4.[8]

The solvation of the inorganic species by the immiscible solvent need not take place through the agency of an acidic hydrogen atom. Some solvents will coordinate with the metal. For example, tri-*n*-butyl phosphate (TBP) is a solvent of strong donor properties, which can bond directly to cations, often forming definite isolatable compounds with the extracted salt. It readily extracts uranium(VI) from nitric acid solutions (p. 346). The bonding ability of TBP is so great that it can be used as extractant in an inert solvent such as benzene. The strong solvating power of TBP and related compounds is explained by the basicity of the oxygen atom in the phosphoryl group

$$\equiv P{=}O \qquad \text{(more accurately } \equiv P \rightarrow 0\text{)}$$

and its ready steric availability. TBP can occupy a variable number of sites in the coordination sphere of a cation. For example, from strong hydrochloric acid solution the species $H[FeCl_4(TBP)_2]$ can be extracted, from dilute hydrochloric acid, $FeCl_3(TBP)_3$. The sum of the number of Cl anions and TBP molecules is 6, the coordination number of Fe^{3+}; TBP replaces water molecules coordinated to iron.

[7] See Diamond and Tuck (Ref. 1) for a discussion of the effect of various equilibria on the extraction coefficient in such systems.

[8] However, if H is replaced by a bulky organic cation such as tetraphenylarsonium cation, the ion pair $(R^+)(MX_4^-)$ will extract into chloroform.

EXTRACTION SYSTEMS USEFUL IN INORGANIC ANALYSIS

Extraction of chlorides

Skey in 1880 and Rothe and Henriot in 1892 showed that large amounts of ferric iron could be separated from many elements by extraction with ethyl ether from 6 *M* hydrochloric acid. This was the first serious use of organic solvent extraction in inorganic analysis and since that time this extraction, or variations of it, has been a standard method for removing large quantities of iron that would otherwise interfere in the determination of other elements. Ether extraction is also of occasional value in the extraction of other metals (Table 14-1).

TABLE 14–1. **Extraction of Metals as Chlorides from Hydrochloric Acid Solutions with Ethyl Ether**[a]

(100 ml of hydrochloric acid solution containing 1 g of metal as chloride extracted with an equal volume of ether saturated with water)

	Metal Extracted, %		
	10% HCl (2.9 *M*)	15% HCl (4.4 *M*)	20% HCl (6.0 *M*)
Au(III)	98		95
Fe(III)	8	92	99
Tl(III)		99	98
Ga			~97
Sb(III)	22	13	6
Sb(V)	6	22	81
As(III)	7	37	68
Ge			~50
Te(IV)	3	12	34
Sn(IV)	23	28	17
Hg(II)	0.4		0.2
Cu(II)	0.05		0.05
Zn	0.03		0.2
Ir(IV)	0.02		5

[a] Data mostly from F. Mylius and C. Hüttner, *Ber.* **44,** 1315 (1911). Other data for ~6 *M* HCl: As(V), extracted slightly; B, partially; Mo(VI), 80–90%; Sn(II), 15–30%; Tl(I), 0.3 mg/100 ml, ~10%; In, 3 mg/100 ml, ~5%; Se(IV), very slightly; V(V), slightly; V(IV), hardly at all. Not extracted appreciably: Al, Be, Bi, Cd, Cr, Co, Fe(II), Pb, Mg, Mn, Ni, Pd, Os, Pt, rare earths, Ag, Ti, Th, W, U, and Zr.

The ethyl ether extraction of iron (and other elements) has been much studied, but a quantitative interpretation of the extractability as a function of the hydrochloric acid and iron concentrations is not easy, and this system illustrates the difficulties that arise when aqueous and organic solvent phases are appreciably soluble in each other and the electrolyte concentration is high and variable.[9] The extractability of iron

[9] R. M. Diamond, *J. Phys. Chem.* **61**, 69 (1957), in a paper on general equations for extraction of inorganic compounds, considers the distribution of a metal halide between an aqueous hydrohalic acid and an organic solvent.

rises with the hydrochloric acid concentration up to 6.5 M HCl in consequence of increasing formation of $FeCl_4^-$ and decreasing dissociation of $HFeCl_4$, the extracted species.[10] Above 6.5 M HCl the extractability of iron decreases because of increasing solubility of ethyl ether in the aqueous phase as the acidity is increased (Fig. 14-2).

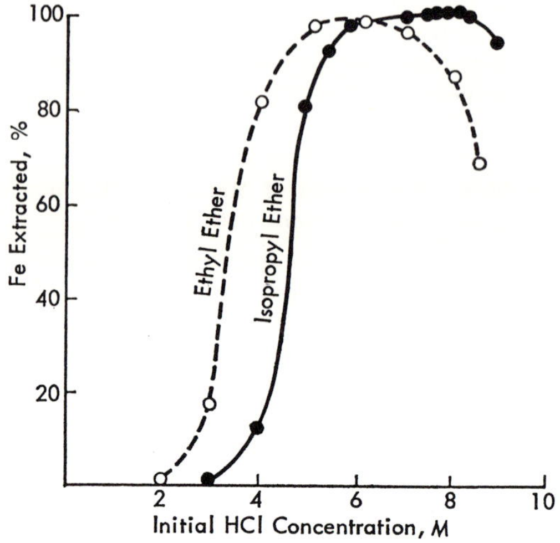

Fig. 14-2. Extraction of Fe(III) from hydrochloric acid solution by ethyl and isopropyl ether; 25 ml $FeCl_3$ − HCl solution (125 or 250 mg Fe with isopropyl ether, 250 mg Fe with ethyl ether) shaken with 25 ml of dry ether. Two isopropyl ether phases above 7.5 M HCl. After Dodson, Forney, and Swift, *J. Am. Chem. Soc.* **58**, 2573 (1936).

The two phases approach each other in composition and the extraction coefficients of iron must decrease. Iron(III), as well as gold(III) and gallium, is extracted to a larger extent at high than at low metal concentrations. This behavior is at least in part attributable to association of $HFeCl_4$ at higher iron concentrations in ethyl ether, a solvent of low dielectric constant ($\epsilon \sim 4$). Such association must increase E:

$$E = \frac{[FeCl_4^-]_o + [HFeCl_4]_o + 2[(HFeCl_4)_2]_o + \cdots}{[Fe^{3+}] + [FeCl^{++}] + \cdots + [FeCl_4^-]}$$

$[FeCl_4^-]_o$ will be small because there is little ionization in ethyl ether and this term can be neglected. Considering only dimerization of $HFeCl_4$ in ethyl ether,

$$2\ HFeCl_4 \rightleftharpoons (HFeCl_4)_2$$

$$\frac{[(HFeCl_4)_2]_o}{[HFeCl_4]_o^2} = K_{\text{dim}}$$

Therefore,

$$E \sim \frac{[HFeCl_4]_o + 2\ K_{\text{dim}}[HFeCl_4]_o^2}{\sum_{n=0}^{4} [FeCl_n]}$$

E for iron extraction with isopropyl ether also increases with the iron concentration, other variables being constant or nearly so.

[10] The extracted chloroferric acid is hydrated and carries water (and some HCl) into the ether phase with it. Of course, water and HCl also partition into ether in the absence of iron.

Isopropyl ether shows some advantages over ethyl ether for the extraction of metal chloro acids. The less soluble an ether is in hydrochloric acid, the higher will be the acid concentration at which the iron extraction begins to decline. With isopropyl ether the maximum lies at 7.75–8.0 *M* (Fig. 14-2) and the extractability is almost constant over the range 7.0–8.5 *M*.[11] With 250 mg of Fe(III), the extraction is 99.9 per cent complete at 7.75–8.0 *M* HCl when the phase volumes are each 25 ml. The extraction is less complete at lower iron concentrations, but if successive extractions are made, even 1 mg of Fe(III) can be satisfactorily recovered. The lesser volatility of isopropyl ether (boiling point 68°C) is another advantage over ethyl ether (b.p. 34.5°C).[12] Isopropyl ether extraction separates Fe(III) from Cu, Co, Mn, Ni, Zn, Al, Cr, Ti, and SO_4^{--}. Vanadium(V) is extracted to a considerable extent, V(IV) much less so; Mo(VI) and phosphoric acid accompany iron to a large extent into the ether phase.

Other oxonium solvents can also be used to extract $HFeCl_4$.[13] Various ketones are even more effective than ethers. The extraction coefficient of Fe(III) in the system methyl amyl ketone–8 *M* HCl is approximately 3300.[14] With equal volumes of phases, methyl isobutyl ketone extracts ~99.99 per cent of Fe(III) in a single extraction from 7 *M* HCl.[15] Tributyl phosphate and triamyl phosphate are even better iron extractants (~99.999 and ~99.9999 per cent from 8 *M* HCl). The iron extraction becomes rather less selective with the high extraction coefficient solvents. However, by carrying out the extraction from 7 *M* LiCl instead of HCl, with methyl isobutyl ketone as the extractant, the recovery of iron(III) is hardly affected (99.944 instead of 99.996 per cent), whereas the coextraction of As(V), V(V), Cr(VI), Mo(VI), Co, Ni, and some other metals is reduced to zero.[16]

Isopropyl or ethyl ether extraction is also useful in separations of Ga, Sb(V), Tl(III), and Au(III). By adding Ti(III) or ascorbic acid to reduce Fe(III), a virtually specific separation of gallium can be obtained. Ethyl acetate is a good extractant for chloroauric acid. It may be noted that all metals easily extracted by ether and other oxonium solvents from hydrochloric acid form chloro anions with a single negative charge. One reason seems to be that singly charged anions are much less hydrated than higher charged anions and thus form chloro acids partitioning more readily into the ether. Steric factors also play a role.

Germanium and arsenic (III) are extracted by nonoxygen solvents such as carbon tetrachloride from strong hydrochloric acid solutions. The extracted species are the covalent chlorides $GeCl_4$ and $AsCl_3$. These can also be extracted by oxonium solvents, but obviously such solvents should not be used because the selectivity of the extraction is then much decreased.

Extraction of other metal halides

Ethyl ether extracts ~99 per cent or more of Tl(I), Tl(III), In, and Au(III), as well as much Fe(III), from about an equal volume of 4 *M* HBr, when these metals are

[11] R. W. Dodson, G. J. Forney, and E. H. Swift, *J. Am. Chem. Soc.* **58**, 2573 (1936). No extraction maximum is shown by bis(2-chloroethyl)ether, which dissolves relatively slightly even in concentrated hydrochloric acid.

[12] Ethers, and especially isopropyl ether, tend to form peroxides by air oxidation on standing, which are highly explosive. Peroxide-containing ethers should not be distilled (unless the peroxide is first destroyed with a suitable agent) and should not be allowed to stand indefinitely in a laboratory.

[13] Extraction of many metals by 2-ethylhexanol from hydrochloric acid solution has been studied by K. A. Orlandini, M. A. Wahlgren, and J. Barclay, *Anal. Chem.* **37**, 1149 (1965).

[14] V. I. Kuznetsov, *J. Gen. Chem. USSR* (*Engl. Transl.*) **17**, 175 (1947).

[15] E. Bankmann and H. Specker, *Z. Anal. Chem.* **162**, 18 (1958). A. Claassen and L. Bastings, *ibid.* **160**, 403 (1958), recommend a 1:1 or 2:1 volume mixture of methyl isobutyl ketone and amyl acetate for extraction of Fe(III) from 7 *M* HCl; the extraction coefficient is ~4000. These mixtures are free of emulsification difficulties encountered with methyl isobutyl ketone alone.

[16] H. Specker, M. Cremer, and E. Jackwerth, *Angew. Chem.* **71**, 492 (1959).

present in ~100-mg amounts in 20 ml. Other elements extracted to a large extent include Sn(II, IV) and Sb(V). An interesting fact is the 99 per cent extraction of Tl(III) from 0.1–0.5 *M* HBr, a concentration at which very little Fe(III) is extracted.

Metals extracted to a large extent by ethyl ether from ~7 *M* HI include Sb(III), Hg(II), Cd, Au(III), and Sn(II); they are all extracted as iodo acids. Indium is extracted extensively from 0.5–2 *M* HI, apparently as InI_3 ($HInI_4$ is extracted at higher iodide concentrations).[17] Other oxonium solvents, such as isobutyl methyl ketone, have been proposed for iodide extractions. It is of interest that various esters can be used to extract bismuth mostly as BiI_3 from acid aqueous solutions of low iodide concentration (~0.005 *M* I^-).[18] Isoamyl acetate is a suitable extractant for BiI_3. It extracts $HBiI_4$ less well; isoamyl alcohol extracts both BiI_3 and $HBiI_4$ well. Unlike SbI_3, BiI_3 does not dissolve in such solvents as chloroform or benzene. The number of extracted elements is greatly reduced by using a nonoxygen solvent. Antimony is extracted as SbI_3 by benzene at low iodide concentrations, where the concentration of SbI_4^- is small compared to that of SbI_3. In 5 *M* sulfuric acid solution that is approximately 0.01 *M* in iodide, the extraction coefficient of Sb attains a maximum value of ~3000.

As already mentioned, tin(IV) can be successfully extracted as SnI_4 by benzene from a perchloric acid solution containing sodium perchlorate to reduce the aqueous solubility of SnI_4 by its salting-out effect. Sulfuric acid solutions can also be used. A high iodide concentration increases *E* in this system, since tin(IV) has little tendency to form negatively charged iodo complexes.

The ether extraction of fluorides has received some study. The extractability of the fluoro acids of Nb and Ta is of special analytical interest. Ether extraction of thiocyanates has also been studied and some important analytical applications have been made but these cannot be entered into here.

Extraction of nitrates

The need for extraction methods for uranium has stimulated the investigation of nitrate extractions.[19] From 8 *M* nitric acid, an equal volume of ethyl ether extracts 65 per cent U(VI), 35 per cent Th, 97 per cent, Ce(IV), 97 per cent Au(III), and smaller amounts of Fe(III) and other elements. The extraction of uranium can be improved by adding a salting-out agent such as lithium or aluminum nitrate, which increases the nitrate-ion concentration both directly (thus decreasing the dissociation of uranyl nitrate) and indirectly (by withdrawing water molecules in hydrating the salt cation), the latter effect being allied to the indifferent electrolyte effect considered earlier in this chapter.

Various other oxygen-type solvents have been studied for the extraction of uranium and congeneric elements and are now used in preference to ether. Ethyl acetate is very suitable for uranium extraction from nitric acid solution salted with aluminum nitrate. Tributyl phosphate extracts uranyl nitrate readily, as already noted. By using it as a reagent in an inert solvent, a separation of uranium from thorium and other elements can be obtained. Under suitable conditions, thorium can be extracted and separated from associated elements. Tri-*n*-octyl phosphine oxide (p. 362) is an even more effective extracting reagent for U(VI) in nitric acid solution. The literature on these separations is extensive.[20]

A different type of extraction from the above (which belong to type 4 in our classification) is the extraction of the ion association compound tetrapropylammonium uranyl nitrate into methyl isobutyl ketone (p. 362).

[17] H. Irving and F. Rossotti, *Analyst* **77**, 801 (1952).

[18] H. A. Mottola and E. B. Sandell, *Anal. Chim. Acta* **24**, 301 (1961).

[19] The use of organic solvents for the separation of uranium had its inception in the observation of E. Peligot in 1842 that solid uranyl nitrate is soluble in ethyl ether.

[20] For important applications see C. J. Rodden, *Analysis of Essential Nuclear Reactor Materials*, U.S. Atomic Energy Commission, 1964, chap. I.

Extraction of heteropolyacids

Alcohols, esters, ketones, ethers, and other oxonium solvents extract the heteropoly acids of Mo, As, P, W, V, and some other elements. Molybdenum blue, phosphomolybdic acid, and phosphotungstovanadic acid are some of the compounds of this type which are readily extractable. These extractions have not been much studied.

Extraction of metal chelates

As pointed out in Chapter 12, the greater number of organic reagents for metals are of the type HL, which react with metal cations (M^{n+}) by replacement of the acidic hydrogen to form a chelate complex ML_n. Because of its organic character, the uncharged complex is likely to be soluble in some organic solvent but only slightly soluble in water, so that the extraction coefficient will, as a rule, have a favorable value. The uncharged complex is not necessarily the only chelate complex formed. Charged complexes (ML_{n-1}^+ and analogous species) are also possible with some reagents, and their formation diminishes the extractability of a metal since they are not extractable as such. The formation of chelate adducts was noted earlier.

An expression can easily be obtained for the extractability of a metal cation as a function of the reagent concentration and the acidity when there are no complicating side reactions.[21] Assume that only the neutral complex ML_n is formed:

$$\underset{(w)}{M^{n+}} + \underset{(o)}{nHL} \rightleftharpoons \underset{(o)}{ML_n} + \underset{(w)}{nH^+}$$

The equilibrium constant for this two-phase reaction (w = aqueous, o = organic phase) can be written, just as any other equilibrium constant is written, as follows:

$$K_{\text{ext}} = \frac{[ML_n]_o[H^+]_w^n}{[M^{n+}]_w[HL]_o^n} \tag{6}$$

K_{ext} is the extraction equilibrium *constant* of the reaction at a fixed temperature and a specified ionic strength.

Equation (6) can be derived from a consideration of the individual equilibria involved, and a better understanding of the system is thus obtained:

$$(ML_n)_w \rightleftharpoons M^{n+} + nL^- \qquad K_m = \frac{[M^{n+}][L^-]^n}{[ML_n]_w} \tag{7a}$$

$$(ML_n)_w \rightleftharpoons (ML_n)_o \qquad P_m = \frac{[ML_n]_o}{[ML_n]_w} \tag{7b}$$

$$(HL)_w \rightleftharpoons H^+ + L^- \qquad K_l = \frac{[H^+][L^-]}{[HL]_w} \tag{7c}$$

$$(HL)_o \rightleftharpoons (HL)_w \qquad P_l = \frac{[HL]_o}{[HL]_w} \tag{7d}$$

From Eqs. (7a) and (7b):

$$\frac{[ML_n]_o}{[M^{n+}]_w} = \left(\frac{P_m}{K_m}\right)[L^-]_w^n \tag{7e}$$

[21] A general treatment of liquid–liquid extraction (chelate and other systems) is presented by H. Irving, F. J. C. Rossotti, and R. J. P. Williams, *J. Chem. Soc.* **1955**, 1906, and by J. Rydberg, *Arkiv Kemi* **8**, 101 (1955).

and from Eqs. (7c) and (7d):

$$[L^-]_w = \left(\frac{K_l}{P_l}\right)\left(\frac{[HL]_o}{[H^+]_w}\right) \tag{7f}$$

Substitution of Eq. (7f) into (7e) gives

$$K_{ext} = \frac{P_m K_l^n}{K_m P_l^n} = \frac{[ML_n]_o[H^+]_w^n}{[M^{n+}]_w[HL]_o^n} \quad \text{(for a specified ionic strength and temperature)} \tag{7g}$$

Comparing various chelate extractions, we see that the value of the extraction equilibrium constant increases with decreasing ionization of the chelate in the aqueous phase and its increasing partitioning into the organic phase, and with increasing ionization of the chelating agent (acid) and its increasing partitioning into the aqueous phase.

The ratio of the concentration of the metal complex in the organic phase to the concentration of metal cation in the aqueous phase is obtained by rewriting Eq. (6):

$$\frac{[ML_n]_o}{[M^{n+}]_w} = K_{ext}\frac{[HL]_o^n}{[H^+]_w^n} \tag{8}$$

If HL partitions predominantly into the organic phase, its concentration therein is found from the known amount added, corrected if necessary for the amount that has reacted with the metal, and the known volumes of the phases; otherwise its partition constant and ionization constant must be known. The same applies to ML_n.

Under certain conditions, $[ML_n]_o/[M^{n+}]_w$ can be set equal or approximately equal to the extraction coefficient of the metal,

$$E = \frac{\sum [M]_o}{\sum [M]_w} \cong \frac{[ML_n]_o}{[M^{+n}]_w} = K_{ext}\frac{[HL]_o^n}{[H^+]_w^n} \tag{9}$$

namely when:

1. The only metal species in the organic phase is ML_n. This means that no dissociation or association (polymerization) of ML_n occurs. This condition is fulfilled in nonoxygen solvents as a rule, and often in others. Moreover, adducts, such as $ML_n \cdot HL$, are assumed not to be formed nor products containing OH, etc.
2. The concentration of ML_n in the aqueous phase is small compared to the concentration of M^{n+}. This is often true.[22]
3. Charged metal-ligand complexes are absent in the aqueous phase. This condition may or may not be essentially fulfilled, depending on the reagent and the metal.
4. No metal species other than M^{n+} are present in the aqueous phase. This proviso frequently is not met. If the solution is not sufficiently acid, hydrolyzed metal species are likely to be present. Furthermore, anions in the solution will usually complex the metal to a greater or less extent, forming cationic, neutral, or anionic species. The heavy metals especially are more or less complexed by citrate, tartrate, cyanide, phosphate, chloride, and other halides—in fact, by almost all anions except perchlorate, or, sometimes, nitrate. The extent of complexing will often depend on the acidity of the solution. Last, polymerization or formation of polynuclear species is assumed not to take place.

An expression for E of the metal can be written if all the equilibrium constants for the hydrolysis and complexing reactions are known. As an example, consider the system involving a divalent metal, the chelating reagent HL, and a univalent complexing anion forming the complex species MX^+, MX_2, MX_3^-, ... The hydrolyzed metal

[22] It should be noted that E can never exceed P of the chelate. As the concentrations of M^{n+} and its hydrolyzed and complexed forms approaches 0, $E \rightarrow P$.

species MOH^+, $M(OH)_2$, $M(OH)_3^-$ (or HMO_2^-), and $M(OH)_4^{--}$ (or MO_2^{--}), as well as polynuclear species, may be present.

$$\sum [M]_w = [M^{++}] + [MOH^+] + [M(OH)_2] + \cdots + [MX^+] + [MX_2] + \cdots$$
$$= [M^{++}] + \frac{k_1[M^{++}]}{[H^+]} + \frac{k_2[M^{++}]}{[H^+]^2} + \cdots$$
$$+ \beta x_1[M^{++}][X^-] + \beta x_2[M^{++}][X^-]^2 + \cdots \tag{10a}$$

$$\sum [M]_w = [M^{++}]\left\{1 + \frac{k_1}{[H^+]} + \frac{k_2}{[H^+]^2} + \cdots \right.$$
$$\left. + \beta x_1[X^-] + \beta x_2[X^-]^2 + \cdots\right\} \tag{10b}$$

where the complex formation constants involving OH^- (H^+) and X^- are defined as follows:

$$k_1 = \frac{[MOH^+][H^+]}{[M^{++}]} \qquad \left(\text{whence } [MOH^+] = \frac{k_1[M^{++}]}{[H^+]}\right) \tag{11a}$$

$$k_2 = \frac{[M(OH)_2][H^+]^2}{[M^{++}]} \tag{11b}$$

and similarly for k_3, etc.

$$\beta x_1 = \frac{[MX^+]}{[M^{++}][X^-]} \qquad (\text{whence } [MX^+] = \beta x_1[M^{++}][X^-]) \tag{11c}$$

$$\beta x_2 = \frac{[MX_2]}{[M^{++}][X^-]^2} \tag{11d}$$

and similarly for βx_3, etc. HX and M—X species are assumed to be nonextractable. If HX is a weak acid, $[X^-]$ will of course depend on the pH. Therefore,

$$E = \frac{[ML_n]_o}{\sum [M]_w} = \frac{[ML_n]_o}{[M^{++}]\{1 + k_1/[H^+] + \cdots + \beta x_1[X^-] + \cdots\}} \tag{12}$$

Since $[ML_n]_o/[M^{++}] = K_{ext}[HL]_o^2/[H^+]^2$,

$$E = \frac{K_{ext}[HL]_o^2}{[H^+]^2\{1 + (k_1/[H^+]) + (k_2/[H^+]^2) + \cdots + \beta x_1[X^-] + \beta x_2[X^-]^2 + \cdots\}} \tag{13}$$

(The subscript w may be omitted where no confusion is likely to result.)

Additional terms will need to be introduced if appreciable concentrations of metal-ligand species occur in the aqueous phase or if more than one species is extracted into the organic phase. Likewise, formation of mixed species or of polymerized species will require additional terms. Undissociated ML_n will always be present in the aqueous phase and will need to be taken into account at high E's if an accurate value of E is required. If the only metal species in the water phase are M^{n+} and ML_n, the expression for the extraction coefficient is

$$E = \frac{K_{ext}[HL]_o^n}{[H^+]^n + (K_{ext}[HL]_o^n/P_m)} \tag{14}$$

The upper limit of E is P_m, as already noted.

Usually, at a given pH, a few species in the aqueous phase will predominate to the extent that the others can be neglected, so that in application the expression for E will be simpler than it appears.

It may be noted that E will be independent of the total metal concentration if no species of M in either phase contains more than one atom of M per charged or uncharged molecule, in other words, if there is no actual or virtual association of M. E would not be constant if the reacting anion L were divalent and the cation M triply charged, for example, since then the extracted product would be M_2L_3 (virtual association of M), and (neglecting hydrolyzed and complexed species):

$$\frac{[M_2L_3]_o}{[M^{3+}]^2} = \text{const.} \qquad \text{and} \qquad E = \text{const.} \times [M^{3+}]$$

Other factors remaining constant, E will decrease with a decrease in the amount of M in this system.

If the concentration of the complexing substance X is large compared to the concentration of M, so that its excess remains essentially constant as E varies, we may write (disregarding hydrolyzed forms of M):

$$E \simeq \frac{[ML_n]_o}{\sum [M]_w} = K'_{ext} \frac{[HL]_o^n}{[H^+]^n}$$

where K' is a constant whose value depends upon the concentration of X in the aqueous phase. Again it is assumed that real or virtual association of M does not occur in either phase. The use of these restricted (effective) constants is convenient in practical work.[23]

In examining chelate extraction equilibria it is often advantageous to plot log E versus pH for a constant excess of the extraction reagent in the system. Figure 14-3 illustrates an uncomplicated chelate extraction—of cupric dithizonate from an acid ($HClO_4$) solution with a carbon tetrachloride solution of dithizone. Log E versus pH curves are shown for three constant concentrations of dithizone in the organic solvent phase. (In order that, in practice, the concentration of dithizone in the carbon tetrachloride phase remain essentially constant, the amount of copper would need to be minute, so that varying copper extraction would not significantly affect the excess of reagent.) When, at equilibrium, $[Cu^{++}]$ outweighs $[CuL_2]$ in the aqueous phase, log E increases linearly with pH, and the slope of the line is 2, in accordance with

$$\log \frac{[CuL_2]_o}{[Cu^{++}]_w} = \log E = 2 \log[HL]_o + 2 \text{ pH [from Eq. (8)]}$$
$$= \text{const.} + 2 \text{ pH}$$

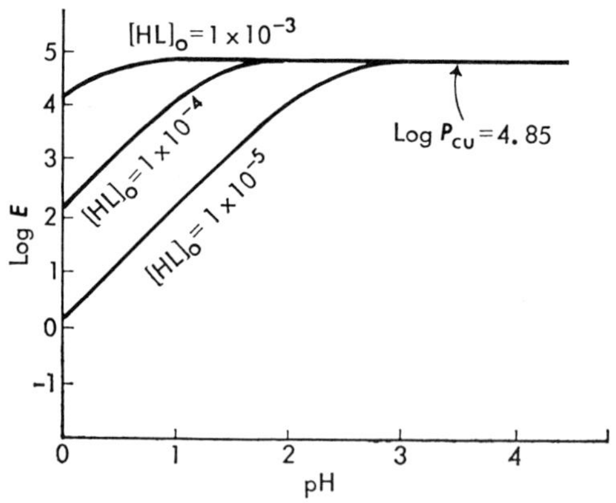

Fig. 14-3. Illustrating extraction of a chelate ML_2. Log E for Cu(II) in the system Cu^{++}-$H_2O(HClO_4)$-HL(dithizone)-CCl_4 is plotted as a function of pH and excess HL in the organic solvent phase. $K_{ext} = 1.5 \times 10^{10}$ ($\mu = 1$), $P_{Cu} = 7 \times 10^5$.

[23] See further A. Ringbom, *Complexation in Analytical Chemistry*, Wiley-Interscience, New York, 1963, chap. 7, in which the use of effective constants in extractions is systematically developed.

(If the extracted metal chelate is ML_3, the slope of the line would be 3, etc.) As $[Cu^{++}]$ decreases and $[CuL_2]_w$ increases with increasing pH, the extraction curve gradually bends toward the horizontal, parallel to the pH axis, in accordance with Eq. (14). Eventually $E = P_{Cu}$. In the pH range covered, the hydrolysis of Cu^{++} is not appreciable and can be ignored.

The very large value of K_{ext} in this system leads to large values of E at relatively low pH's and low chelating agent concentrations. Thus, even with $[HL]_o$ of only 1×10^{-4}, $E = 1000$ at pH ~ 0.5. Extractability of a substance from quite strongly acidic solutions such as this is always an analytical advantage.

It is interesting to compare the constants for the extraction systems Cu^{++}–H_2Dz (dithizone)–CCl_4 and Ni^{++}–H_2Dx (dimethylglyoxime)–$CHCl_3$ (Table 14-2). K_{ext} for the latter system has the small value of ~ 10, arising mostly from unfavorable values of chelate constants in the nickel system [P_{ML_2}/K_{ML_2} is 1×10^{20} for Ni, $\sim 7 \times 10^{28}$ for Cu; $(K_{HL}/P_{HL})^2$ is 9×10^{-20} for Ni, 7×10^{-18} for Cu]. Nevertheless, by carrying out the nickel extraction at pH 5, for example, a favorable partitioning of nickel into the chloroform phase can be obtained. At pH 5 with $[H_2Dx]_o = 1 \times 10^{-4}$, the calculated value of $[Ni(HDx)_2]_o/[Ni^{++}]$ is 1000. E cannot be this large because $P_{Ni(HDx)_2} = 410$ and nickel species other than Ni^{++} are present in a buffered solution at pH 5, but it will still have a favorable value.

TABLE **14-2. Comparison of Constants in the Extraction Systems Cu^{++}–Dithizone–CCl_4 and Ni^{++}–Dimethylglyoxime–Chloroform at 25°**[a]

	P_{HL}	K_{HL}	P_{ML_2}	K_{ML_2}	K_{ext}
Cu^{++}–H_2Dz–CCl_4	1.1×10^4	3×10^{-5}	7×10^5	1×10^{-23}	1.5×10^{10}
Ni^{++}–H_2Dx–$CHCl_3$	8.3×10^{-2}	2.6×10^{-11}	4.1×10^2	4×10^{-18}	10

[a] Constants for Ni^{++}–H_2Dx are for $\mu = 0.05$ and K_{ext} is calculated from its component constants. In the Cu^{++}–H_2Dz system, P_{HL} and K_{HL} are for $\mu = 0.1$, P_{ML_2} and K_{ML_2} for $\mu = 1$, and K_{ext} is the experimental value at $\mu = 1$.

Examples of organic extraction reagents

Diphenylthiocarbazone.[24] With some justice this reagent (Fig. 14-4), usually called dithizone, may be termed the heavy metal reagent par excellence. It reacts with more than a score of metals to form strongly colored chelate complexes which can be extracted into organic solvents such as carbon tetrachloride and chloroform (Table 14-3). Its chief use is for the colorimetric determination of traces of metals by extraction procedures (Chap. 61), but it also serves for the isolation of metals to be determined by other methods.

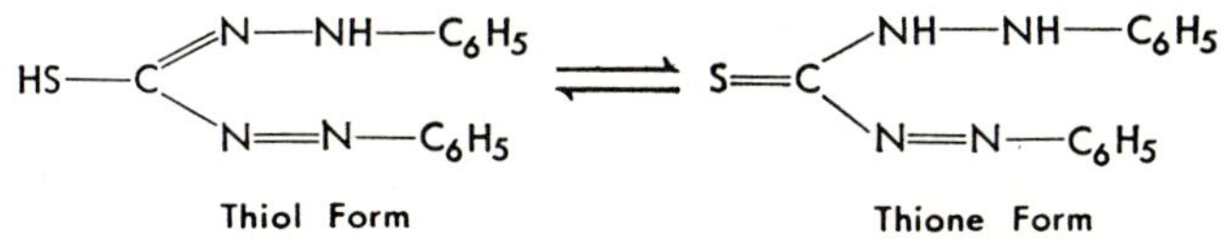

Fig. 14-4. Tautomeric forms of dithizone believed to exist in organic solvents. The primary metal dithizonates are probably derived from the thiol form by replacement of the H of the HS-group by an equivalent of metal. The two peaks in the absorbance-wavelength curve of dithizone in organic solvents (see Fig. 61–7) correspond to the two tautomers, whose relative amounts vary from one solvent to another.

[24] For a monographic treatment, see G. Iwantscheff, *Das Dithizon und seine Anwendung in der Mikro- und Spuren-analyse,* Verlag Chemie, Weinheim, 1958.

TABLE **14-3.** **Elements Reacting with Dithizone**[a]

H																
Li	Be											B	C	N	O	F
Na	Mg											Al	Si	P	S	Cl
K	Ca	Sc	Ti	V	Cr	Mn	Fe	Co	Ni	Cu	Zn	Ga	Ge	As	Se	Br
Rb	Sr	Y	Zr	Nb	Mo	Tc	Ru	Rh	Pd	Ag	Cd	In	Sn	Sb	Te	I
Cs	Ba	La	Hf	Ta	W	Re	Os	Ir	Pt	Au	Hg	Tl	Pb	Bi	Po	—
—	Ra	Ac	Th	Pa	U											

[a] The enclosed elements are extracted more or less completely from aqueous solution into a dithizone solution in CCl_4, $CHCl_3$, and other organic solvents. Ru, Rh, Os, and Ir may possibly react in lower oxidation states. Ge reacts in the absence of water.

Dithizone is a dibasic acid, but the secondary ionization ($K_2 < 10^{-15}$) is so small that it can usually be disregarded and only the primary ionization need be considered in extraction equilibria:

$$\underset{(w)}{H_2Dz} \rightleftharpoons \underset{(w,\ \text{yellow})}{HDz^-} + H^+$$

$$\frac{[H^+][HDz^-]}{[H_2Dz]_w} = K_{pd} = 1.5 \times 10^{-5} \qquad (\mu = 0) \tag{15a}$$

(The subscript *pd* signifies the *primary* ionization constant of dithizone.) The molecular solubility of dithizone in water is small ($2 \times 10^{-7}\ M$), but dithizone dissolves in alkaline aqueous solutions because of its acidic character. Its solubility in carbon tetrachloride amounts to 0.0021 M at room temperature (in chloroform $\sim 0.08\ M$). When a solution of dithizone in an organic solvent is shaken with an aqueous solution, dithizone is distributed between the two phases in accordance with the relation

$$E_d = \frac{[H_2Dz]_o}{[H_2Dz]_w + [HDz^-]} = \frac{[H^+]P_d}{K_{pd} + [H^+]} \tag{15b}$$

The value of E_d, the extraction coefficient of dithizone, depends on the ionic strength of the aqueous solution, which therefore should be specified. At low acidities, E_d will usually be expressed in terms of hydrogen-ion activity (H^+), instead of $[H^+]$.

For carbon tetrachloride–water, $P_p = 1.1 \times 10^4$ ($\mu = 0.1$, 25°C), so that in basic solutions, in which $[H^+]$ is small compared to K_{pd},

$$E_d = \frac{[H_2Dz]_o}{[HDz^-]} \cong 7 \times 10^8\ (H^+) \tag{15c}$$

(Henceforth in this chapter, the subscript w will be omitted when it is clear that the species concentration, as of an ion, is that in the aqueous phase.)

When a solution of dithizone in carbon tetrachloride (green color) or other immiscible solvent is shaken with an aqueous metal solution at an appropriate pH, the metal dithizonate is formed and dissolves in the carbon tetrachloride to which it imparts its characteristic color, usually some shade of red, violet, or orange. If the reaction takes place in acidic solution and dithizone is in excess, the carbon tetrachloride phase will show a mixed color resulting from the green of the remaining dithizone and the red or violet of the dithizonate. If the solution is sufficiently alkaline, the excess dithizone distributes mostly into the aqueous phase and the color of the carbon tetrachloride phase will be that of the dithizonate.

Dithizone shows little tendency to form intermediate, charged complexes, so that the simple expression (9) often adequately describes the extraction of a metal from acid solution. The extraction coefficient for a primary[25] dithizonate is then given by

$$E \sim \frac{[M(HDz)_n]_o}{[M^{n+}]} \cong K_{ext} \frac{[H_2Dz]_o^n}{[H^+]^n}$$

if $[M^{n+}] \gg [M(HDz)_n]_w$ at equilibrium and $\sum [M]_w \sim [M^{n+}]$ (substances complexing M absent).

Values of K_{ext} for primary cupric dithizonate, already used to illustrate chelate extraction as a function of $[H^+]$ and reagent concentration, are listed in Table 14-4 for $HClO_4$ and HCl solutions of different concentrations.

TABLE 14-4. **Extraction Equilibrium Constants (K_{ext}) for Primary Cupric Dithizonate**[a]

Medium	$K_{ext} \times 10^{-10}$	$K'_{ext} \times 10^{-10}$
$HClO_4$, 0.1 *M*	3	
$HClO_4$, 0.6 *M*	1.9	
$HClO_4$, 1.15 *M*	1.5	
HCl, 1.0 *M*		1.0
HCl, 2.0 *M*		0.7

[a] R. W. Geiger and E. B. Sandell, *Anal. Chim. Acta* 8, 197 (1953). K_{ext} for $HClO_4$ solutions is a concentration constant and its decrease with increasing $HClO_4$ concentration illustrates the effect of ionic strength. The constant for HCl solutions is an effective constant, as indicated by the prime mark, in that no distinction is made between Cu^{++} and $CuCl^+$ ($\sum[Cu]_w = [Cu^{++}] + [CuCl^+]$). The complexing of Cu^{++} with Cl^- is relatively weak in 1 *M* HCl, so that K'_{ext} for 1 *M* HCl does not differ much from K_{ext} for 1 *M* $HClO_4$.

Other metals extracted quantitatively from mineral acid solution with dithizone are Pd(II), Pt(II), Au(III), Hg(I, II), and Ag. These react in even more acid solutions than copper. Bismuth requires a less acid solution than copper for quantitative extraction. Most of the other dithizone-reacting metals require a very weakly acidic or slightly basic solution for satisfactory extraction (Fig. 14-5).

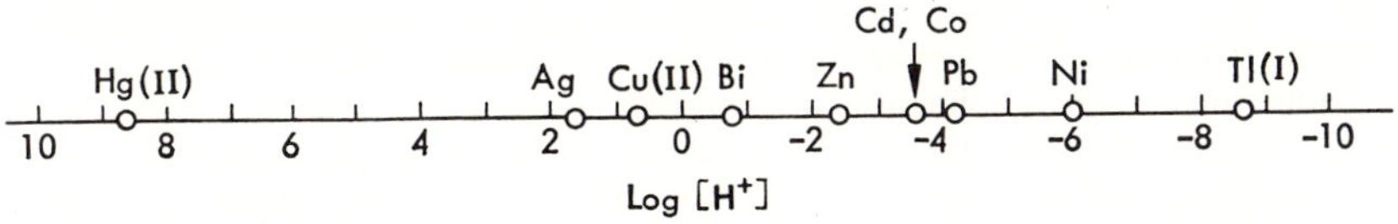

Fig. 14-5. Calculated hydrogen-ion concentrations at which $[M(HDz)_n]_o/[M^{+n}] = 99$ when an aqueous solution is shaken with an equal volume of dithizone in carbon tetrachloride whose concentration is 4×10^{-4} *M* (0.010% *w/v*) at equilibrium.

[25] Some metals (Cu, Hg, and Ag) can also form secondary dithizonates, in which both acidic hydrogens of dithizone are replaced by an equivalent of metal, e.g., Hg(II)Dz. The secondary dithizonates are preferentially formed at low hydrogen-ion concentrations and with a deficiency of dithizone. Analytically they are of lesser importance than the primary dithizonates.

Dithizone extraction of metals from a basic solution will usually require the presence of a metal-complexing reagent such as citrate or tartrate to prevent the precipitation of hydroxides. In spite of such complexing and formation of charged hydrolysis products, quantitative extraction of almost all reacting metals can be obtained in a suitable pH range. If no other reactions occurred, the extractability of a metal would reach an almost constant value at the pH where most of the dithizone in the organic solvent has been transferred to the aqueous phase as dithizonate. Actually, some metals show a sharp decrease in extractability at high pH's. This is true of lead and zinc, which form hydroxo complexes, i.e., plumbite and zincate, in strongly basic solutions.

Differential complexing (masking) of metals is of great importance in dithizone separations. A number of possibilities are indicated in Table 14-5. See the procedure for the spectrophotometric determination of lead (Chap. 61) for the use of cyanide to prevent the reaction of copper, zinc, nickel, cobalt, and most other heavy metals.

TABLE 14-5. **Complex-Forming Agents in Dithizone Reactions**

	Metals Reacting
Cyanide in basic solution	Pb, Sn(II), Bi, Tl(I), In
Cyanide in slightly acidic solution	Pd, Hg, Ag, Cu
Thiocyanate in dilute acid	Hg, Au, Cu
Bromide or iodide in dilute acid	Pd, Au, Cu
Cyanide + thiosulfate at pH $\sim$5 (CCl_4)	Zn, Sn(II)
EDTA, pH $\sim$4.5	Au, Ag, Hg

The extraction of dithizonates from basic solution is rapid, as might be expected from the high concentration of dithizonate ion in such a medium. In an acid medium, extraction may be quite slow, but the rate varies with the metal and the solvent. Carbon tetrachloride solutions extract more rapidly than chloroform because P_{H_2Dz} is smaller for the former, thus giving a higher concentration of dithizone in the aqueous phase. In general, the larger the value of K_{ext} the more rapid the extraction: Hg and Ag dithizonates are extracted rapidly from acid solutions, Cu, Bi, and Zn more slowly. The purity of the solvent is important. With carefully purified carbon tetrachloride, equilibrium in Cu and Bi extractions is reached in 1 or 2 min of vigorous shaking compared to as much as $\frac{1}{2}$ hr reported by some workers. The presence of strong complexing agents may also decrease the rate by diminishing the metal-ion concentration.[26] The rate of dithizone extraction in an acid medium should be tested unless definite information on the rate under the conditions is available.

When shaking is vigorous, the chemical reaction rates of HDz^- or H_2Dz with metal ions or their complexes may be more important than mass-transfer rates of reactants from one phase to the other in determining the rate of attainment of equilibrium in dithizone extractions.[27] However, in practice, the mass-transfer rates are often important and may even be the limiting factor. The transfer rate is a function of the diffusion rates of the reacting species and the products in the aqueous and organic solvent solutions and through the interface. The area of the interface is obviously

[26] For observations on this and other factors, see a general review of extraction kinetics by Y. A. Zolotov, I. P. Alimarin, and V. A. Bodnya, *J. Anal. Chem. USSR (Engl. Transl.)* **19**, 23 (1964).

[27] C. B. Honaker and H. Freiser, *J. Phys. Chem.* **66**, 127 (1962), found that the rate of extraction of zinc dithizonate (CCl_4 and $CHCl_3$ solutions of dithizone) is determined by the rate of formation of $ZnHDz^+$ by reaction of HDz^- with hydrated Zn^{++} when shaking is sufficiently vigorous. The extraction rate is first order with respect to zinc and inverse first order with respect to hydrogen-ion concentration.

important. Usually the organic solvent volume is smaller than the aqueous volume and on vigorous shaking the mixture consists predominantly of organic solvent droplets suspended in the aqueous phase. For a given volume of organic phase, the interfacial area is inversely proportional to the radius of the drops. Increasing the vigor of shaking reduces the drop size (at least to a certain point) and also tends to increase mixing within the phases and at their interface.

8-Hydroxyquinoline (oxine) and derivatives. If a metal cation M^{n+} is extracted as the oxinate MOx_n (oxine = HOx) into an immiscible organic solvent from an aqueous solution, the extraction coefficient of the metal is given by

$$E = \frac{[MOx_n]_o}{\sum [M]_w} \tag{16a}$$

Because oxine extractions are usually made from weakly acidic (or even basic) solutions, the approximation

$$E \sim \frac{[MOx_n]_o}{[M^{n+}]} = K_{ext} \frac{[HOx]_o^n}{[H^+]^n} \qquad \text{[Eq. (9), p. 348]} \tag{16b}$$

may not be admissible, since hydrolyzed metal species may be present in appreciable or even preponderant amounts. Moreover, intermediate, charged, metal-oxine complexes may be formed in the aqueous phase. For example, ferric iron forms $FeOx^{++}$ and $FeOx_2^+$ in acidic solutions and mixed oxine–hydroxide complexes in basic solutions.[28]

In any event, $[MOx_n]_o/[M^{n+}]$ can be calculated from Eq. (16b), if K_{ext} is known, as a function of the hydrogen-ion concentration.[29] If the constants for all the equilibria

$$M^{n+} \rightleftharpoons \text{M-containing species}$$

are available, $\sum [M]_w$ can be found, and therewith E.

Oxine is an ampholyte (p. 53) and the distribution of excess oxine between the organic solvent and the aqueous phase will be a function of the acidity. Chloroform is often used as the organic solvent in oxine extractions. The partition constant of HOx for water–chloroform has the value

$$P_{HOx} = \frac{[HOx]_{CHCl_3}}{[HOx]_w} = 3.4 \times 10^2 \qquad (25°C)$$

From P_{HOx} and the ionization constants given earlier (p. 278), the extraction coefficient of oxine can be calculated for any acidity:

$$E_{HOx} = \frac{[HOx]_{CHCl_3}}{[HOx]_w + [HOx \cdot H^+]_w + [Ox^-]_w} = \frac{5 \times 10^2}{1 + ([H^+]/K_1) + (K_2/[H^+])}$$

$$= \frac{5 \times 10^2}{1 + 1.25 \times 10^5[H^+] + (2 \times 10^{-10}/[H^+])}$$

In a chloroform–water–oxine system having equal volumes of the two phases, 95 per cent or more of the total oxine is present in the chloroform phase in the pH range 4–11. (The values of the constants K_1 and K_2 are actually based on the activity of hydrogen ion, not $[H^+]$.)

[28] $FeOx_2(OH)_2^-$ is formed in 0.01–0.07 M OH^- solutions.

[29] Values of K_{ext} for about 30 metals are listed by J. Stary, *Anal. Chim. Acta* **28**, 132 (1963). Stability constants of metal complexes with oxalate, tartrate, EDTA, and other ligands are also given.

An alternative expression for $[MOx_n]_o/[M^{n+}]$ is

$$\frac{[MOx_n]_o}{[M^{n+}]} = \frac{P_m[Ox^-]^n}{K_m} \qquad \text{[Eq. (7f), p. 348]} \tag{17a}$$

in which

$$P_m = \frac{[MOx_n]_o}{[MOx_n]_w} \tag{17b}$$

$$K_m = \frac{[M^{n+}][Ox^-]^n}{[MOx_n]_w} \tag{17c}$$

Accordingly,

$$\frac{P_m}{K_m} = \frac{[MOx_n]_o}{[M^{n+}][Ox^-]^n} = \text{const.} \tag{17d}$$

If the solubility of the metal oxinate in the organic solvent and its solubility product in water are known, P_m/K_m can be found. Solubility-product constants of oxinates have been reported, but not all such values are reliable, because the presence of intermediate oxinate complexes has not been taken into account in treating the water solubility. It is of course possible to evaluate P_m/K_m experimentally if $[M^{n+}]$ can be found experimentally or calculated, which may not be easy.

TABLE **14-6. Constants for Some Trivalent Metal Oxinates**[a]

	Solubility in $CHCl_3$, M	$K_{sp}([M^{3+}][Ox^-]^3)$	P_m/K_m
$AlOx_3$ (18°C)	0.045	$10^{-32.3}$	$10^{31.0}$
$GaOx_3$ (18°C)	0.092	$10^{-40.8}$	$10^{39.8}$
$InOx_3$ (18°C)	0.24	$10^{-33.9}$	$10^{33.3}$
$FeOx_3$ (25°C)	0.019	$10^{-43.5}$	$10^{41.8}$

[a] Data for oxinates of Al, Ga, and In from S. Lacroix, *Anal. Chim. Acta* **1**, 260 (1947), those for $FeOx_3$ from T. Turnquist, Thesis, Univ. of Minnesota, Minneapolis, 1965. The intrinsic solubility of $FeOx_3$ in water at 25°C is 1.6×10^{-7} M, so that $P_{FeOx_3} = 1.2 \times 10^5$.

Some illustrative values of P_m/K_m for trivalent metals with chloroform as extracting solvent are listed in Table 14-6. The magnitudes of the values are such that values of E = 10–100 are to be expected in feebly acidic solution with relatively dilute chloroform solutions of oxine, e.g., 1 per cent (0.07 M). Moreover, the values for $AlOx_3$ and $GaOx_3$ differ by a factor of $\sim 10^9$, so that by adjustment of pH it should be possible to extract gallium and leave aluminum in solution. Experimental work has shown that at pH 2.6, traces of gallium can be separated from $\sim 10^5$ times as much aluminum.[30] Iron(III) in small amounts should be separable from aluminum and from many divalent metals, most of which have much lower values of P_m/K_m. Table 14-7, based on experimental work, indicates various separation possibilities.

Complexing of otherwise extractable metals with ancillary reagents or by changes in their oxidation states can, of course, be applied to advantage in making oxine extractions more selective. Oxine extraction finds an important application in the isolation of aluminum, an element lacking outstanding separation reactions. For example, in the

[30] E. B. Sandell, *Ind. Eng. Chem., Anal. Ed.* **13**, 844 (1941). Gallium oxinate fluoresces in chloroform and very small amounts of gallium can be determined fluorimetrically.

TABLE 14-7. **pH Ranges for Quantitative Chloroform Extraction of Some Oxinates**[a]

Metal	pH *Range*
Al	4.5–10.7
Cr(III)	6.0–8.0
Co (as $CoOx_2 \cdot 2\,HOx$)	7.3–8.2
Cu	2.8–
Ga	3.0–11.7
Fe(III)	2.4–
Pb	8.2–11.0
Mn(II)	9.0–10.5
Mo(IV) (as MoO_2Ox_2)	2.0–5.6
Ni	5.5–8.8
Nb	2.8–10.5
Th	7.5–9.5
Ti (as $TiOOx_2$)	4.0–9.0
U(VI) (as $UO_2Ox_2 \cdot HOx$)	7.0–9.0
V(V) (as VO_2Ox)	3.2–5.1
W(VI) (as WO_2Ox_2)	3.0–4.3

[a] From K. Motojima and H. Hashitani, *Japan Analyst* **9,** 151 (1960). 50 ml of aqueous solution 0.06 per cent (0.004 *M*) in HOx extracted with 10 ml of $CHCl_3$ (1 min shaking). Metals, such as aluminum, which are precipitated as hydroxides, cannot be extracted at the pH where these form, unless oxine is added in aqueous solution to the acidic solution, the pH adjusted, and the extraction made with chloroform.

determination of aluminum (0.001–> 0.1 per cent) in plain carbon steel,[31] after the bulk of the iron(III) has been removed by extraction as the chloride with amyl acetate, Al, Ti, and Zn are extracted with oxine in chloroform at pH 5.5 in the presence of cyanide, 1,10-phenanthroline, and hydroxylammonium chloride. The residual Fe(II) and Cu, Ni, Co, and Mn are not extracted from this solution. Ti and Zr are stripped from the chloroform extract by shaking the latter with an aqueous solution of pH 9.2, leaving Al in the chloroform. The concentration of aluminum as oxinate in the extract can be found by measuring its absorbance at 390 mμ, a wavelength at which the excess oxine does not absorb appreciably. Very low concentrations of aluminum oxinate (and some other oxinates) in chloroform can be determined by measuring the fluorescence intensity of the extract.[32]

In the oxinates of ferric iron, aluminum, and other trivalent metals, all six coordination positions of the metal are occupied by oxinate groups and no water is coordinated. Such unhydrated oxinates extract smoothly into chloroform. Divalent metal oxinates may show a less simple behavior. Some divalent metals prefer octahedral coordination, so that the two coordination positions remaining after bonding of two oxinate groups $[6 - (2 \times 2) = 2]$ are occupied by two strongly bonded water molecules. Such hy-

[31] R. M. Dagnall, T. S. West, and P. Young, *Analyst* **90,** 13 (1965). EDTA was found to be unsatisfactory as a complexing agent. Contrary to other reports, 2-methyl-8-hydroxyquinoline in chloroform was shown by these authors to extract aluminum partially.

[32] See the procedure in Chapter 61.

drated oxinates are not appreciably soluble in chloroform.[33] Replacement of at least one, and preferably both water molecules, by organophilic groups is a requisite for successful extraction into an organic solvent. Oxine itself, as a neutral molecule, may serve if its concentration is sufficiently high. Thus strontium can be extracted as $SrOx_2 \cdot 2\ HOx$ with a 10 per cent chloroform solution of oxine.[34]

A study[35] of the distribution of zinc oxinate between aqueous solutions and various organic solvents has led to the conclusion that the following are the extracted zinc species:

4-Methyl-2-pentanone, 1-Butanol, 4-Methyl-2-pentanol	$ZnOx_2 \cdot 2\ S$	(S = solvent molecule)
Chloroform, Benzene	$ZnOx_2 \cdot HOx \cdot S$	
Butyl ether, Pentyl ether	$ZnOx_2 \cdot S \cdot H_2O$	
Carbon tetrachloride	$ZnOx_2 \cdot HOx \cdot H_2O$	

The extraction of zinc oxinate into hexane and cyclohexane is so poor (E = 0.01–0.0025 at pH 2–7, 30°C) that the extracted species is postulated to be $ZnOx_2 \cdot 2\ H_2O$. The replacement of only one water molecule by butyl ether, pentyl ether, and carbon tetrachloride is believed to be related to the small dielectric constants and dipole moments of these solvents. The organic solvent molecules are thought to occupy coordination positions of zinc ion when the ketone and the two alcohols are the immiscible solvents. The sites of the chloroform and benzene molecules are considered uncertain; attachment by hydrogen bonding to O or N of Ox or HOx is a possibility. As might be expected, the species containing one H_2O molecule do not extract as well as those that contain none.

4-Methyl-8-hydroxyquinoline, like 8-hydroxyquinoline, gives an adduct $ZnL_2 \cdot HL$ (solvent molecules not specified) when chloroform is the extracting solvent.[36] But the extracted species with 2-methyl-8-hydroxyquinoline is ZnL_2 without the addend HL. This finding is reasonably explained by the steric hindrance to hydrate and adduct formation imposed by the methyl group in the 2-position (see p. 267). When adducts are formed, the extraction expressions given earlier must of course be modified.

Copper(II), having a coordination number of 4 and forming square-planar complexes, gives anhydrous $CuOx_2$ and extracts facilely into chloroform.

The extraction of calcium and magnesium oxinates with chloroform can be carried out at a low oxine concentration if primary or secondary alkyl amines are present. Ion-association complexes $(RNH_3^+)(MOx_3^-)$ are reported to be formed and these are easily soluble in chloroform.[37] Similarly, the water of hydration of zinc and cadmium

[33] C. H. R. Gentry and L. G. Sherrington, *Analyst* **71**, 432 (1946), report that anhydrous Mg, Zn, and Cd oxinates are first extracted by a chloroform solution of oxine but that the extracts become cloudy on continued shaking because of the formation of the insoluble hydrated oxinates. Similar observations were made by Stary. The solubility of $ZnOx_2 \cdot 2\ H_2O$ in chloroform at room temperature is approximately 2×10^{-6} M.

[34] D. Dyrssen, *Svensk Kem. Tidskr.* **67**, 311 (1955).

[35] G. K. Schweitzer, R. B. Neel, and F. R. Clifford, *Anal. Chim. Acta* **33**, 514 (1965). The number of solvent molecules (CH_3Cl, butanol, etc.) in the extracted species were found by determining the extraction coefficient of zinc as a function of increasing amounts of $CHCl_3$, butanol, etc., added to the inert immiscible solvent cyclohexane.

[36] F. Chou, Q. Fernando, and H. Freiser, *Anal. Chem.* **37**, 361 (1965).

[37] F. Umland and W. Hoffmann, *Anal. Chim. Acta* **17**, 234 (1957); Umland and K.-U. Meckenstock, *Z. Anal. Chem.* **165**, 161 (1959). 2-Butoxyethanol allows extraction of calcium oxinate by chloroform at pH 12.6, as shown by C. L. Luke, *Anal. Chim. Acta* **32**, 221 (1965).

oxinates can be replaced partially or entirely by polar organic ligands—alcohols or primary and secondary alkyl amines—to form $MOx_2 \cdot 2\ ROH$, $MOx_2 \cdot H_2O \cdot RNH_2$, $MOx_2 \cdot H_2O \cdot R_2NH$, or $MOx_2 \cdot ROH \cdot RNH_2$, all extracted by chloroform.[38]

Introduction of a bulky group in the 7-position in the 8-hydroxyquinoline molecule gives reagents that show diminished reactivity with trivalent metals in consequence of steric hindrance. For example, 7-(α-[*o*-carbomethoxyanilino]benzyl)hydroxyquinoline in chloroform,

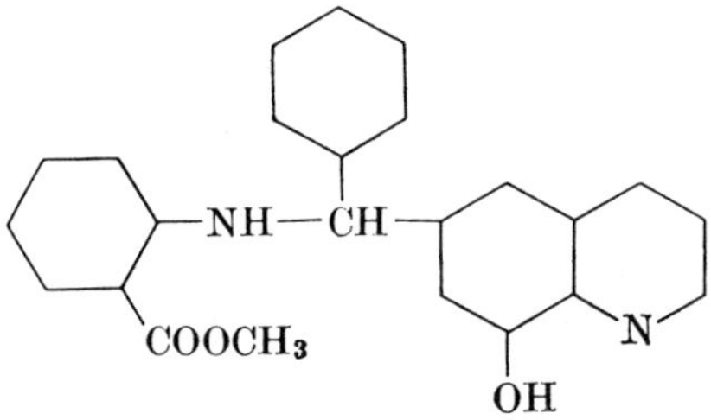

does not react with trivalent metals even on long shaking, whereas divalent metals such as Cu, Pb, Zn, Ni, and Mg are extracted.[39]

The rate of attainment of equilibrium in oxine extractions varies with the metal, the pH, and the presence of complexing substances. Most metals are extracted quite rapidly, equilibrium being reached in a few minutes of vigorous shaking, but Ni, Pd, Mo, and W require some hours of shaking.[29] Cr(III) hardly extracts at room temperature. The rate of extraction increases with pH as well as with oxine concentration. Tartrate, oxalate, and cyanide have little effect on the rate of extraction of most metals. In the presence of EDTA and other aminopolycarboxylates, which form very stable complexes and greatly reduce the cationic metal concentration, some hours of shaking may be needed to reach equilibrium with divalent metals, and even longer periods with tri- and quadrivalent metals. Conflicting reports on the oxine extractability of metals by various workers are probably, in part, attributable to differences in extraction times.

Chloro, bromo, and iodo derivatives (in the 5- and 7-positions) of oxine are more acidic and less basic than oxine itself, and their metal complexes can be extracted at higher acidities than the oxine complexes. These reagents are less soluble than oxine in water.

2-Methyl-8-hydroxyquinoline in chloroform extracts aluminum very poorly, but otherwise is similar to oxine in its extraction behavior; the extraction of most metals begins at a higher pH than when oxine is used.

Thiooxine (8-mercaptoquinoline) forms extractable chelates with many heavy metals.[40]

Tetraphenylarsonium chloride, $(C_6H_5)_4AsCl$, is a good example of a reagent whose cation forms unsolvated ion-association compounds with certain (large) anions, some of which can be extracted into organic solvents. Chloroform is the usual extracting solvent.

Tetraphenylarsonium chloride is incompletely ionized in water, especially in the presence of an ionic chloride, and polymerizes in chloroform, so that its extraction

[38] F. Umland, W. Hoffmann, and K.-U. Meckenstock, *Z. Anal. Chem.* **173**, 211 (1960). The formation of these adducts illustrates *synergism* in extractions. For an account of synergism in chelate extractions, see H. M. N. H. Irving in D. Dyrssen, J-O. Liljenzin, and J. Rydberg (eds.), *Solvent Extraction Chemistry*, Wiley, New York, 1967, p. 91. (Proceedings of International Conference at Gothenburg, Sweden, 1966).

[39] F. Umland and K.-U. Meckenstock, *Z. Anal. Chem.* **177**, 244 (1960); determination of Mg: F. Umland, B. K. Poddar, and K.-U. Meckenstock, *ibid.* **185**, 362 (1962).

[40] Survey and references in Stary, Ref. 1, p. 132.

coefficient depends both on the chloride concentration of the aqueous phase and the reagent concentration in the chloroform (or aqueous) phase. The values of the polymerization constants of tetraphenylarsonium chloride (RCl) in chloroform are[41]

$$\beta_2 = \frac{[(\text{RCl})_2]_{\text{CHCl}_3}}{[\text{RCl}]^2_{\text{CHCl}_3}} = 1 \times 10^2 \tag{18a}$$

$$\beta_4 = \frac{[(\text{RCl})_4]_{\text{CHCl}_3}}{[\text{RCl}]^4_{\text{CHCl}_3}} = 3.5 \times 10^5 \tag{18b}$$

From these constants and the partition constant and dissociation constant of the monomer in water ($\mu = 0$, 25°C)[41]

$$\frac{[\text{RCl}]_{\text{CHCl}_3}}{[\text{RCl}]_w} = 3.2 \tag{18c}$$

$$\frac{[\text{R}^+]_w[\text{Cl}^-]_w}{[\text{RCl}]_w} = 7.1 \times 10^{-2} \tag{18d}$$

the extraction coefficient of tetraphenylarsonium chloride ($CHCl_3/H_2O$) can be found (25°C):

$$E_R = \frac{3.2ff_+f_-[\text{Cl}^-]}{7.1 \times 10^{-2}f + f_+f_-[\text{Cl}^-]} \times (1 + 2 \times 10^2[\text{RCl}]_{\text{CHCl}_3} + 1.4 \times 10^6[\text{RCl}]^3_{\text{CHCl}_3}) \tag{18e}$$

(f, f_+, and f_- are the activity coefficients of RCl, R^+, and Cl^-). This expression may be used for RCl concentrations up to 0.2 M in chloroform, and for aqueous HCl concentrations up to 0.1 M. At higher hydrochloric acid concentrations, species such as RCl · HCl are formed, which are not taken into account in the foregoing expression.

The chloroform extraction of tetraphenylarsonium perrhenate has been reported[42] to follow the relation

$$\frac{[\text{RReO}_4]_{\text{CHCl}_3}}{[\text{ReO}_4^-]} = K_{\text{ext}} \frac{[\text{RCl}]_{\text{CHCl}_3}}{[\text{Cl}^-]} \qquad (K_{\text{ext}} = 1 \times 10^4)$$

corresponding to the overall extraction reaction

$$\text{ReO}_4^- + \underset{\text{CHCl}_3}{\text{RCl}} \rightleftharpoons \underset{\text{CHCl}_3}{\text{RReO}_4} + \text{Cl}^-$$

Presumably the extraction of MnO_4^- can be formulated similarly. Dissociation of the ions pairs does not occur in chloroform.

At least some of the tetraphenylarsonium complex chlorides form adducts in chloroform. Thus $IrCl_6^{--}$ is extracted as $R_2IrCl_6 \cdot 2\,RCl$ as well as R_2IrCl_6. From the equilibrium constant

$$\frac{[\text{R}_2\text{IrCl}_6 \cdot 2\ \text{RCl}]_{\text{CHCl}_3}}{[\text{R}_2\text{IrCl}_6]_{\text{CHCl}_3}[\text{RCl}]^2_{\text{CHCl}_3}} = 1.6 \times 10^5 \tag{19a}$$

we find that in a chloroform solution having a monomeric tetraphenylarsonium chloride concentration of 0.01 M, the ratio of the two iridium species is

$$\frac{[\text{R}_2\text{IrCl}_6 \cdot 2\ \text{RCl}]_{\text{CHCl}_3}}{[\text{R}_2\text{IrCl}_6]_{\text{CHCl}_3}} = 16 \tag{19b}$$

Additional constants needed for the formulation of the extraction coefficient of iridium (ionization constant of R_2IrCl_6 in water, and its partition coefficient) have been determined. The following expression can be written for the extraction coefficient of iridium

[41] J. Fok, Ph.D. thesis, Univ. of Minnesota, Minneapolis, 1963.
[42] S. Tribalat, *Anal. Chim. Acta* **3**, 113 (1949).

from 0.1 *M* HCl as a function of the monomeric tetraphenylarsonium chloride concentration in the chloroform phase and the total tetraphenylarsonium concentration (RCl and R^+) in the aqueous phase:

$$E_{Ir} = \frac{[R_2IrCl_6]_{CHCl_3} + [R_2IrCl_6 \cdot 2\,RCl]_{CHCl_3}}{[IrCl_6^{--}]_w + [R_2IrCl_6]_w}$$

$$= \frac{19.4(1 + 1.6 \times 10^5[RCl]^2_{CHCl_3})}{[1.36 \times 10^{-8}/(\sum [R]_w)^2] - 1} \quad (25°C) \qquad (19c)$$

Tetraphenylarsonium cation gives salts with I^-, SCN^-, ClO_3^-, ClO_4^-, MnO_4^-, and NO_3^- which are readily extracted into $CHCl_3$. Sulfate, F^-, IO_3^-, IO_4^-, phosphate, and borate are among the anions extracted only slightly.[43] Chromium(VI) is extracted from acidic solution but only partially from basic solution. Among extractable anionic complexes may be mentioned BF_4^-, BiI_4^-, $OsCl_6^{--}$, $AuCl_4^-$, $FeCl_4^-$, and thiocyanates of Co, Cu, Fe, and Mo; $Fe(CN)_6^{3-}$ is very slightly extracted, as would be expected from its high charge. Tetraphenylarsonium is thus far from a selective reagent. Nevertheless, it is valuable for some otherwise difficult separations, as of Re from much Mo, and Ir(IV) from much Rh(III). Some measure of selectivity can be obtained by varying the Cl^- or HCl concentration of the sample solution. Fe(III) is extracted only slightly from dilute hydrochloric acid solutions. Changes in oxidation state are sometimes helpful in achieving separations. For example, microgram amounts of Pt can be separated from 100 times as much Ir by reducing $IrCl_6^{--}$ to $IrCl_6^{3-}$ with ascorbic acid. The reduction of Pt(IV) as $PtCl_6^{--}$ is so slow that it can be extracted into chloroform, whereas $IrCl_3^{--}$ remains in the aqueous phase (P. Peng, University of Minnesota, 1965). Raising the charge on the iridium chloro anion from -2 to -3 by decreasing the oxidation state of iridium from $+4$ to $+3$ is possibly the reason for the success of this separation.

Tetraphenylphosphonium cation gives reactions similar to those of tetraphenylarsonium.

Tertiary high-molecular-weight amines, R_3N (tri-*n*-octylamine, for example), and their salts find use in the extraction of metals from acid solutions. Polar and nonpolar organic liquids can be used to dissolve these reagents; benzene is often chosen. High-molecular-weight amines and their salts are more soluble in organic solvents and less soluble in aqueous solutions than low-molecular-weight amines. An amine in an immiscible organic solvent can extract acids (H_2SO_4, HCl, HNO_3, H_3PO_4, $H_2Cr_2O_7$, molybdic acid, vanadic acid, etc.) from an aqueous phase[44]:

$$\underset{(o)}{R_3N} + \underset{(w)}{H^+} + \underset{(w)}{A^-} \rightleftharpoons \underset{(o)}{R_3NHA}$$

When an amine salt is the extracting agent, a reaction of the following type can occur:

$$\underset{(o)}{R_3NHA} + \underset{(w)}{B^-} \rightleftharpoons \underset{(o)}{R_3NHB} + \underset{(w)}{A^-}$$

B^- can be a complex anion such as $FeCl_4^-$ or $UO_2(SO_4)_2^{--}$. This type of reaction is similar to the one occurring with anion-exchange resins, and hence the amines are often called liquid anion exchangers.[45] By proper choice of acid (HCl, HNO_3, H_2SO_4, etc.) and its concentration, various useful separations can be carried out. For example,

[43] R. Bock and G. M. Beilstein, *Z. Anal. Chem.* **192**, 44 (1963).

[44] F. L. Moore, *Anal. Chem.* **29**, 1660 (1957).

[45] For extensive tabulations of liquid ion exchangers see C. F. Coleman, C. A. Blake, and K. B. Brown, *Talanta* **9**, 297 (1962); D. Dyrssen, *Svensk Kem. Tidskr.* **77**, 387 (1965); H. Green, *Talanta* **11**, 1561 (1964).

niobium can be separated from tantalum by extraction with tribenzylamine in chloroform from strong hydrochloric acid solution.[46] With 8 per cent tribenzylamine solution and 11 *M* HCl, $E_{Nb} = 74.4$ and $E_{Ta} = 0.002$. Niobium is thought to form the anion $NbOCl_4^-$ or $NbOCl_5^{--}$ in hydrochloric acid solution, whereas tantalum does not form analogous complex chloro anions. Microgram quantities of zinc can be separated from decigram quantities of nickel by extracting the former with triisooctylamine in xylene form 2 *M* hydrochloric acid solution.[47] Nickel is not extracted because it does not form complex chloro anions.

Quaternary amine cations form ion-association complexes with a considerable number of metals in NaOH, HCl, HNO_3, H_2SO_4, and HF systems, which can be extracted into methyl isobutyl ketone and other immiscible organic solvents. The extractability of 57 metallic elements from such media into methyl isobutyl ketone with tetrapropyl, tetrabutyl, and tetrahexyl amine cation as reagent has been studied.[48] The extracted metal anions all seem to have a single negative charge. Ta, but not Nb, Ti(III), or Zr, is extracted from HF solution as dilute as 1 *M*. This should serve as a separation of Ta from Nb and many other metals. Tetrapropyl ammonium-ion extracts U(VI) from an acid-deficient aluminum nitrate–salted aqueous phase as $(R_4N)^+ \cdot (UO_2(NO_3)_3)^-$ into methyl isobutyl ketone, and provides a good separation of uranium from almost all elements.[49]

Organophosphorus compounds. Derivatives of phosphoric acid, and related organic compounds, have been much investigated in recent years as reagents for the extraction of metals. Tri-*n*-butyl phosphate (TBP) was used as early as 1949 (by

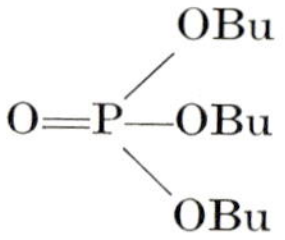

Warf) as an extracting reagent and solvent for U(VI), Th, and Ce(IV) in nitric acid solution. A coordinately solvated salt, such as $UO_2(NO_3)_2 \cdot 2$ TBP, is formed by bonding of the phosphoryl group to the cation. Analogous solvates are formed with the chlorides (p. 345). Other metals, such as Sc, Y, Zr, Nb, Mo, Fe(III), Sb, and Te, are also extracted to varying extents. TBP is often diluted with an inert hydrocarbon solvent to increase the selectivity of the extractions.

If BuO— is replaced by Bu—, or other alkyl group R, the polarity of the O—P bond in TBP (or other trialkyl phosphate) is increased and the extraction coefficient of metals rises with increasing solvating power in the order

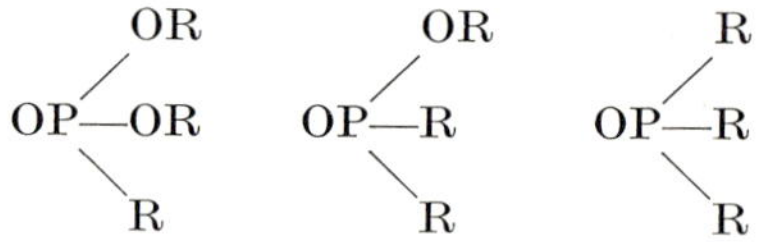

dialkyl alkylphosphonate | alkyl dialkyl phosphinate | trialkylphosphine oxide

(When an alkyl group is replaced by a phenyl group, C_6H_5—, the extraction coefficient

[46] J. Y. Ellenburg, G. W. Leddicotte, and F. L. Moore, *Anal. Chem.* **26**, 1045 (1954). For extraction of complex anions of metals with organic acids, including EDTA (e.g., $M(III)Y^-$), by high-molecular-weight amines in xylene and hexone, see F. L. Moore, *Anal. Chem.* **37**, 1235 (1965).

[47] W. L. Ott, H. R. MacMillan, and W. R. Hatch, *Anal. Chem.* **36**, 363 (1964); L. E. Scroggie and J. A. Dean, *Anal. Chim. Acta* **21**, 282 (1959).

[48] W. J. Maeck, G. L. Booman, M. E. Kussy, and J. E. Rein, *Anal. Chem.* **33**, 1775 (1961).

[49] W. J. Maeck, G. L. Booman, M. C. Elliott, and J. E. Rein, *Anal. Chem.* **30**, 1902 (1958); **31**, 1130 (1959).

decreases because the basicity of the OP≡ group decreases.) Tri-*n*-octylphosphine oxide,

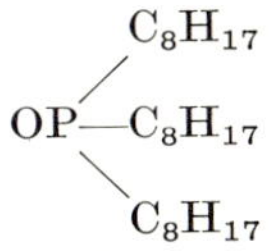

is an especially effective extraction reagent for such metals as U(VI), Ce(IV), Pu(IV), Th, and Zr from 1 *M* or stronger HCl and HNO_3 solutions.[50] It is often used in cyclohexane or cyclohexane–carbon tetrachloride solution.

Trialkyl thionophosphates[51] and dialkyl phosphordithioic acids[52]

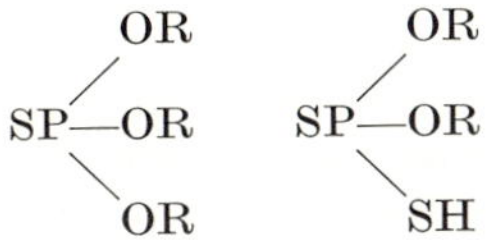

have been used as extraction reagents for metals that bind strongly to sulfur. The latter compounds are similar to sodium diethyldithiocarbamate in extraction behavior but extract in more strongly acid solutions.

Certain combinations of organophosphorus compounds can give much higher extraction coefficients than either alone (synergism). This phenomenon, arising from various causes, is shown by other extraction systems as well.

Other extraction reagents. The slightly soluble chelates formed by cupferron (p. 270) in mineral acid solution (~ 1 *N*) with Fe(III), Ga, Sb(III), Ti(IV), Sn(IV), Hf, Zr, V(V), U(IV), and Mo(VI) can be extracted with chloroform and other immiscible organic solvents.[53] In spite of the instability of the reagent, various useful separations of these elements can be made from those not precipitated (nor extracted) in this acidic medium [most divalent elements, As, Al, Sb(V), Cr(III), rare earths, Ge, Se, Te, U(VI), P, and some others]. The cupferrates are quite soluble in chloroform so that larger amounts of the extractable metals can be handled than in many other metal-organic reagent-immiscible solvent systems.

Other reagents reacting with a greater or less number of metals and finding use in immiscible solvent extractions include diethyldithiocarbamic acid (as the sodium salt),[54] acetylacetone and derivatives (thenoyltrifluoroacetone),[55] and rhodamine B and analogous compounds.

A fairly small number of extraction reagents are distinguished by almost specific behavior. For example, dimethylglyoxime in chloroform extracts nickel but hardly

[50] Review of use of phosphine oxides: J. C. White and W. J. Ross, *Natl. Acad. Sci. Nucl. Sci.* NAS-NS-3102 (1961); phosphorus-based extractants in general: Peppard, Ref. 1.

[51] T. H. Handley and J. A. Dean, *Anal. Chem.* **32**, 1878 (1960), who show that triisooctyl thiophosphate and tri-*n*-butyl thiophosphate are selective extractants for silver and mercury(II) in nitric acid medium.

[52] T. H. Handley and J. A. Dean, *Anal. Chem.* **34**, 1312 (1962); T. H. Handley, *ibid.* **37**, 311 (1965); H. Bode and W. Arnswald, *Z. Anal. Chem.* **185**, 179 (1962); A. I. Busev et al. (references in Handley and Dean).

[53] N. H. Furman, W. B. Mason, and J. S. Pekola, *Anal. Chem.* **21**, 1325 (1949); E. B. Sandell and P. F. Cummings, *ibid.* **21**, 1356 (1949); J. Stary and J. Smizanska, *Anal. Chim. Acta* **29**, 545 (1963); J. Ruzicka and J. Stary, *Talanta* **11**, 692 (1964).

[54] H. Bode, *Z. Anal. Chem.* **143**, 182 (1954); **144**, 165 (1955).

[55] Extraction constants (benzene) for chelates of acetylacetone, benzylacetone, and dibenzoymethane are listed by J. Stary and E. Hladky, *Anal. Chim. Acta* **28**, 227 (1963); see also H. Freiser and A. Krishen, *Anal. Chem.* **29**, 288 (1957).

any cobalt, and few other metals, from a slightly basic medium containing citrate or tartrate.[56] 1-Nitroso-2-naphthol (and also the 2-1 isomer) furnishes an almost specific extraction of cobalt.[57] α-Benzoinoxime in chloroform extracts molybdenum and tungsten from a mineral acid solution and separates them from nearly all other elements.[58]

The small amount of space devoted to the reagents mentioned in this section is not a reflection on their usefulness but rather an indication of the extensive treatments readily available on their application; a more extended discussion here would deal more with factual information than with new principles.

RATIONALE OF SEPARATIONS BY IMMISCIBLE SOLVENT EXTRACTION

Simple extraction methods

By these we mean methods that can be carried out in a small number of steps with simple apparatus. Separatory funnels serve very well for mixing and separating the phases. They can be used for sample sizes ranging from milligrams to grams. The great majority of extraction separations in inorganic analysis are made by simple methods and the process deserves examination.

Consider the separation of two elements A and B by immiscible solvent extraction with a given reagent, A being more extractable than B. The problem is to attain a specified value, close to 1, for the recovery factor (R_A) of A, while keeping R_B below a specified value close to 0. Another factor is involved which cannot be so explicitly formulated but which nevertheless is real and important: the economy factor O (oeconomia). O is inversely proportional to the work that must be expended or the time required to carry out the separation. It is an inverse function of the number of extractions needed to effect the separation. If no restriction is placed on O, R_A can theoretically be made to approach 1 and R_B to approach O as closely as desired. In practice there is always a limit to the amount of time or effort that can be expended. A separation problem may not be solvable if some limit is set on O, as by saying that a certain number of extractions must not be exceeded. Factors such as the number of extractions which increase R_A also tend to increase R_B, so that a good separation is attained only at the cost of decreasing R_A, and conversely, if O is restricted. Usually R_A is fixed, e.g., for a major constituent of a sample it should have a value of 0.999 or close thereto in a good analysis. We are then left with the problem of arranging the extraction steps so that as small a value as possible is obtained for R_B. Usually some permissiveness is associated with O.

R_A is determined by E_A, the relative volumes of the phases, and the number of successive extractions [Eq. (5c)], and similarly for R_B. The E's are a function of the reagent concentration, the hydrogen-ion concentration (usually), and the concentration of complexing substances. Obviously the greater the ratio E_A/E_B the easier or better will be the separation provided E_A (actually E'_A, p. 340) and E_B have suitable values,[59] i.e., $E_A > 1$, $E_B < 1$ for equal phase volumes. In some extraction systems a change in the acidity of the aqueous phase and in the reagent concentration alters E_A and E_B by the same factor, so that E_A/E_B remains constant or almost so. This

[56] E. B. Sandell and R. W. Perlich, *Ind. Eng. Chem., Anal. Ed.* **11**, 309 (1939); H. Christopherson and E. B. Sandell, *Anal. Chim. Acta* **10**, 1 (1954); A. Claassen and L. Bastings, *Rec. Trav. Chim.* **73**, 783 (1954).

[57] E. Boyland, *Analyst* **71**, 230 (1946); A. Claassen and A. Daamen, *Anal. Chim. Acta* **12**, 547 (1955); H. Baron, *Z. Anal. Chem.* **140**, 173 (1953).

[58] P. G. Jeffery, *Analyst* **81**, 104 (1956); G. B. Jones, *Anal. Chim. Acta* **10**, 584 (1954); V. Pfeifer and F. Hecht, *Z. Anal. Chem.* **177**, 175 (1960); P. Peng and E. B. Sandell, *Anal. Chim. Acta* **29**, 325 (1963).

[59] In the literature, E_A/E_B is sometimes called the separation factor. Actually, this *ratio* does not allow the extent of separation to be calculated (individual E values are needed) and the name is inappropriate. The separation factor is R_B/R_A.

may be true in the extraction of chelates of cations A and B having the same charge $+a$ with the reagent HL:

$$E_A = (K_{ext})_A \frac{[HL]^a}{[H^+]^a} \tag{20a}$$

(This is an approximation which is not always valid. See p. 348 et seq.)

$$E_B = (K_{ext})_B \frac{[HL]^a}{[H^+]^a} \tag{20b}$$

$$\frac{E_A}{E_B} = \frac{(K_{ext})_A}{(K_{ext})_B} \tag{21a}$$

If A and B have different charges, a and b:

$$\frac{E_A}{E_B} = \frac{(K_{ext})_A}{(K_{ext})_B}[H^+]^{b-a}[HL]^{a-b} \tag{21b}$$

and the ratio is a function of the acidity and the free-reagent concentration (Fig. 14-6).

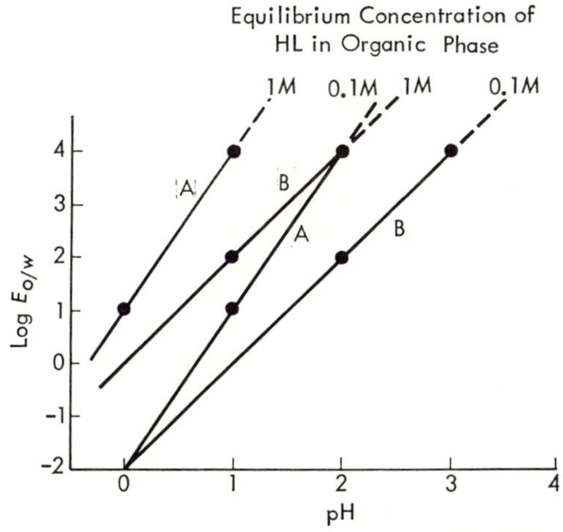

Fig. 14-6. Theoretical extraction coefficients of the cations A^{3+} and B^{++} in the system $(AL_3, BL_2)_o$-$(A^{3+}, B^{++})_w$, in which $(K_{ext})_A = 10$ and $(K_{ext})_B = 1$. When $[HL]_o = 0.1$ and pH $= 0$, $E_A = E_B$. The values of E_A and E_B are actually limited by the values of P_A and P_B. The latter are here assumed to be greater than 10^4.

In many extraction systems the factors controlling E_A (and E_B) can be adjusted so that E_A can be given the desired value—that which will provide the required R_A in one or two, or possibly three, extractions. E_A should not be made larger than actually necessary, the relative volumes of the phases being taken into account, in order to maintain E_B at a minimum. E_A being fixed, E_B is also fixed in the type of system we are considering, and we can calculate R_A and R_B and therewith the extent of separation, which is expressed by the separation factor

$$S_{B/A} = \frac{R_B}{R_A} \tag{22}$$

Consider a concrete example in which $E_A/E_B = 10{,}000$ and the values of E_A and E_B can be varied over a wide range. Let the phase volumes be equal. The E's are assumed independent of concentrations. The reagent concentration is taken to be large compared to the metal concentration, so that the former remains essentially constant. For different values of E_A and E_B, the recoveries of A and B after 1, 2, and 3 extractions are as follows:

		R_A			R_B		
E_A	E_B	$n = 1$	$n = 2$	$n = 3$	$n = 1$	$n = 2$	$n = 3$
10,000	1	0.9999			0.50	0.75	
1,000	0.1	0.999	0.999999		0.099	0.178	
100	0.01	0.99	0.9999		0.01	0.020	
10	0.001	0.91	0.992	0.999	0.001	0.002	0.003

The separation factor (R_B/R_A) improves (becomes smaller) as E_A and E_B are decreased. Increasing the number of extractions makes up for the decrease in E_A. If we accept 0.999 as a satisfactory value for R_A (roughly true for the requirements of ordinary determinations of macro constituents), the separation factor $S_{B/A} = 0.003$ after three successive extractions. This means that if equal amounts of A and B were present at the start, the ratio of B to A in the combined extracts is $\sim 1:300$. This may or may not be satisfactory, depending on the method of determination of A and the required accuracy. If the stipulation is made that $S_{B/A}$ must be 0.001 or less, with $R_A = 0.999$, the required separation cannot be achieved in three direct extractions.

The separation can be improved by introducing a *retrograde* extraction ("backwashing"). After two extractions have been made at $E_A = 50$ and $E_B = 0.005$, so that $R_A = 0.9996$ and $R_B = 0.010$, let the combined organic solvent phases be shaken with an equal volume of aqueous solution having a composition such that $E_A = 1000$ and $E_B = 0.1$. On attainment of equilibrium $R_A = 0.9986$ and $R_B = 0.0010$. The stipulated degree of separation has now been attained but R_A falls a very little short of that desired. Retrograde extraction is more efficacious at large values of E_A/E_B. It is especially effective in providing an improved separation if R_A need not be very close to 1 (as in trace analysis). Thus if an R_A of 0.99 is acceptable, an $S_{B/A}$ of 0.00006 can be obtained by two direct extractions at $E_A \sim 30$ and $E_B \sim 0.003$, followed by a retrograde extraction at $E_A = 100$ and $E_B = 0.01$. By a sacrifice of about 1 per cent in R_A, R_B has been decreased approximately 15 times, compared to the preceding example. At suitably high E_A and low E_B, repeated retrograde extraction decreases R_A in small constant decrements while decreasing R_B geometrically. Retrograde extraction can be applied only when the extraction is reversible (usually but not universally true) and equilibrium is attained in a reasonable period of time.[60]

[60] If the extraction is not sufficiently reversible, it may be possible to bring both A and B back into an aqueous solution by more or less drastic measures such as destroying the excess and combined reagent (for example, dithizone and dithizonates can be destroyed with permanganate, the excess of which can be removed with a reducing agent). The extraction with the organic phase can then be repeated under the same conditions as before, so that the recovery of A is quantitative, whereas the extraction of B is reduced by a large factor. Because of the extra work required, this method is not likely to be utilized often in applied analysis. Apart from separations, a substance may eventually need to be transferred to an aqueous phase for determination. When possible, the substance is stripped from the organic phase by shaking with an aqueous solution of proper pH. This treatment succeeds more often than not, but when it fails, more drastic approaches, as mentioned, are needed. As a last resort, the solvent is volatilized and the residue is wet-oxidized if organic matter is present. A few metals will be lost in such treatment. Occasionally addition of another organic solvent to the extract may facilitate the stripping of a substance into an aqueous solution. Thus addition of carbon tetrachloride to an ester solution of BiI_3 aids transfer of bismuth to an acid aqueous solution.

The foregoing example shows that when $E_A/E_B \leq 10^4$, the separation of A and B with an R_A of 0.999 is not particularly easily attained, since at least three extractions are required. In practice a greater number of extractions are almost certain to be needed, because a margin for error must be allowed in adjustment of the E's. Clearly, some general way of reducing E_B without greatly affecting E_A is needed. A powerful means of decreasing E_B without decreasing E_A seriously, if at all, is by the use of a suitable differential complexing agent to transform B into a nonreactive (more generally, into a less reactive) species. For example, lead forms a much less stable cyanide complex than zinc, so that the extraction of zinc (and many other metals) by dithizone can be reduced to inappreciable values without impeding the extraction of lead if cyanide is added in suitable amount. A change in the oxidation state of B may eliminate its extraction. Thus the extraction of iron as Fe(III) chelates can often be prevented by reduction to Fe(II) with ascorbic acid, titanous chloride, etc.

The proper choice of organic solvent, may, as we have seen, sometimes provide a favorable ratio for E_A/E_B.

Separation of the desired constituent A from B by extracting the latter and leaving A in the aqueous phase is generally less desirable than the reverse process, but it is sometimes of value. For example, large amounts of Fe(III) can be extracted from hydrochloric acid solution by one of the solvents mentioned on page 345 and separated from elements having very small extraction coefficients under the conditions. For the success of this method when $Q_B \gg Q_A$, a large value of E_B is desirable[61] and E_A must of course be much less than unity. Many chelate complexes of metals are not very soluble in organic solvents, and their extraction would require inconveniently large volumes of solvent. When retrograde extraction is applied, A may be left in an inconveniently large volume of aqueous phase. Moreover, the extraction of B instead of A may be less selective than the reverse; A may be left in the aqueous phase with many other elements.

Continuous extraction

When the extraction coefficient of the desired substance is low, it may be advantageous or necessary to use some type of continuous extraction system. A limited volume of the extracting organic solvent passes dropwise through the aqueous solution continuously, the cycle usually being mantained by vaporization and condensation of the solvent. Many different styles of apparatus are in use, and descriptions can be found in the more comprehensive treatments of liquid–liquid extraction.

Countercurrent extraction

The simple extraction methods that have been described evidently will not quantitatively separate two substances having extraction coefficients differing by a small factor. An intensive, systematic fractionation will then be required. The so-called *countercurrent extraction methods* provide such fractionation. The simplest method of this kind (sometimes termed the *fundamental method*) calls for the use of a series of separatory funnels or, more commonly, a series of tubes in an automatic apparatus that transfers one of the solvent phases from tube to tube in sequence after the two immiscible phases have been equilibrated. Let the tubes be numbered 0, 1, 2, ... Each contains a fixed volume of, say, the organic solvent, which is heavier than the aqueous phase. The aqueous solution containing the substances, A and B, to be separated is added to tube 0 and the liquids are shaken together. After the phases have separated, the aqueous phase of tube 0 is transferred to tube 1 and an equal volume of aqueous solution (not containing A and B) having a composition giving the desired extraction coefficients for A and B is added to tube 0. The contents of both tubes are mixed, and the aqueous phase in tube 1 is transferred to tube 2, that in tube 0 is transferred to tube 1, and fresh aqueous solution is added to tube 0. The process is repeated again

[61] Application of continuous extraction to remove B circumvents this requirement.

and again (see Fig. 13-1, p. 287). At the end of the whole process, aqueous phase will have reached the last tube in the series. A simple extension of the procedure leads to the single withdrawal modification in which addition of fresh aqueous solution is continued beyond the point where aqueous phase has reached the last tube. In other words, the number of portions of fresh aqueous solution added exceeds the number of tubes, and the additional portions are drawn off at the last tube.

The fraction of solute in any tube r after n transfers ($V_o = V_w$) is given by the expression (see p. 288)

$$F_{r,n} = \frac{n!}{r!\,(n-r)!}\,\frac{E^{n-r}}{(E+1)^n} \tag{23}$$

If the number of tubes and transfers are sufficiently large, the distribution of solute among the tubes will be approximately Gaussian, even if the effective extraction coefficient differs considerably from 1 (for example, compare the curves in the figure accompanying Problem 1, Chapter 13, p. 331). The relative sharpness of a peak in the plot of the fraction of solute against the serial numbers of the tubes increases with the number of tubes and the number of transfers. The distance between the peaks of A and B will be a function of E'_A/E'_B [$E' = (V_o/V_w)E$], becoming larger the more this quotient differs from unity. The extent of separation of A from B is given by the overlap of the curves of A and B (see Fig. 13-8). From considerations closely similar to those for the analogous case of column-chromatographic separations (p. 304), R_A and R_B corresponding to a cut at a certain tube can be found.[62]

If possible, conditions should be so adjusted that the product $E'_A E'_B$ approaches unity, in order to minimize the number of fractionation steps.

A double-withdrawal countercurrent extraction pattern is illustrated in Fig. 14-7.[63] The circles represent vessels in which aqueous solutions and organic solvent are mixed and allowed to separate after attainment of equilibrium. All phase volumes are taken equal. The organic phase from the first vessel C is transferred to R1 and the aqueous phase to L1, in the second row (first extraction). Organic solvent is added to L1 and water or aqueous solution of the proper composition to provide the desired extraction coefficients is added to R1. After equilibration, o (organic phase) in L1 is transferred

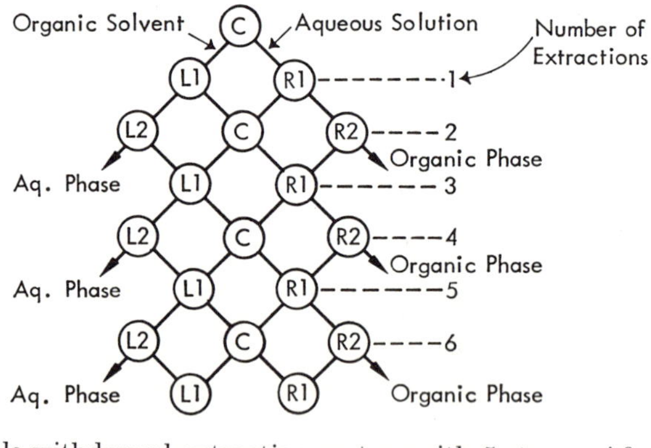

Fig. 14-7. Double withdrawal extraction system with 5 stages. After Hibbits, Reference 63.

[62] See also L. Rogers in *Treatise on Analytical Chemistry*, part I, chap. 22, vol. 2; E. Hecker, *Verteilungsverfahren in Laboratorium*, Verlag Chemie, Weinheim, 1955; H. M. Rauen and W. Stamm, *Gegenstromverteilung*, Springer, Berlin, 1953; von Metzsch, Ref. 1; Peppard, Ref. 1.

[63] J. O. Hibbits, *Anal. Chim. Acta* **24**, 113 (1961). For a discussion of old and new processes of multiplicative distribution (liquid–liquid extraction) refer to W. Fischer, K. Biesenberger, J. Heppner, and U. Neitzel, *Angew. Chem. Intern. Ed. Engl.* **3**, 791 (1964).

to C in the third row and w (aqueous phase) is transferred to L2; o in R1 is transferred to R2 and w to C. This constitutes the second extraction. Solvents o and w are added to L2 and R2, respectively, in the third row, and the contents of the three vessels are shaken together. o of R2 and w of L2 are both withdrawn from the system. o and w in C are transferred, respectively, to R1 and L1 in the fourth row; o in L2 is transferred to L1 and w in R2 to R1. This is the third set of extractions. The process is carried on, with addition of fresh solvents, for as many extractions as may be required. Note that o and w phases are withdrawn from the system after every other extraction step. The system shown in Fig. 14-7 has five stages ($N = 5$): L2, L1, C, R1, and R2. The first withdrawals are made after $(N + 1)/2$ extractions.

The mathematics of the double-withdrawal extraction scheme is not simple. Electronic computers provide solutions. The mathematics will not be entered into here, but the possibilities of the scheme will be illustrated by giving graphs showing recoveries in a 5-stage (Fig. 14-8) and 15-stage (Fig. 14-9) system. In a 15-stage system, with $E'_A = 4$ ($\alpha_A = 0.8$) and $E'_B = 0.43$ ($\alpha_B = 0.3$), $\sim$40 extraction steps give $R_A = 99.9$ per cent and $R_B \sim 0.1$ per cent. If the phase volumes are 10 ml at the beginning and are maintained the same throughout the procedure, 99.9 per cent of A will be contained in 70 ml of organic solvent, i.e., in (220–80)/2 ml. The relatively slow increase in volume with continued fractionation is a distinct advantage of the double-withdrawal countercurrent extraction method.

Countercurrent extraction has found little application in *inorganic* analysis (in biochemical separations it is important but often inferior to chromatographic methods). Even if the effective extraction coefficients differ by a factor of 10, and the values are most favorably disposed, i.e., $E'_A E'_B \cong 1$, the number of extractions required to attain $R_A \sim 99.9$ per cent and $R_B < \sim 0.1$ per cent is by no means small. As we have seen,

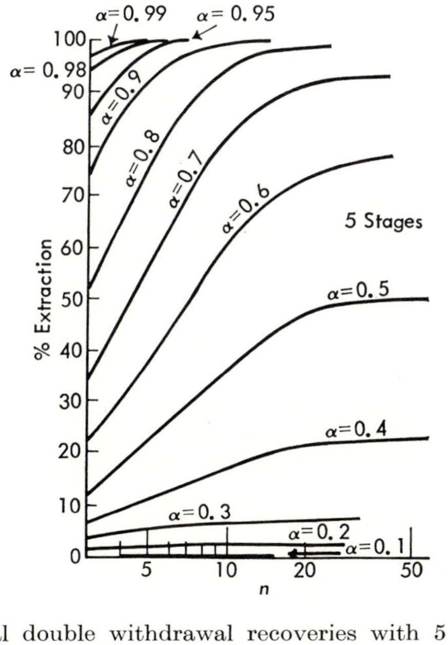

Fig. 14-8. Theoretical double withdrawal recoveries with 5 stages. α = fractional recovery in a single extraction (equal volumes of phases) $= E/(E + 1)$, or $E = \alpha/(1 - \alpha)$. Note that in order to obtain quantitative recovery ($\sim$99.9%), α must equal or exceed 0.9, that is, $E \geqq 9$. To obtain quantitative separation, α of the less easily extractable component may not exceed $\sim$0.1 ($E < \sim 0.1$). After Hibbits, Reference 63.

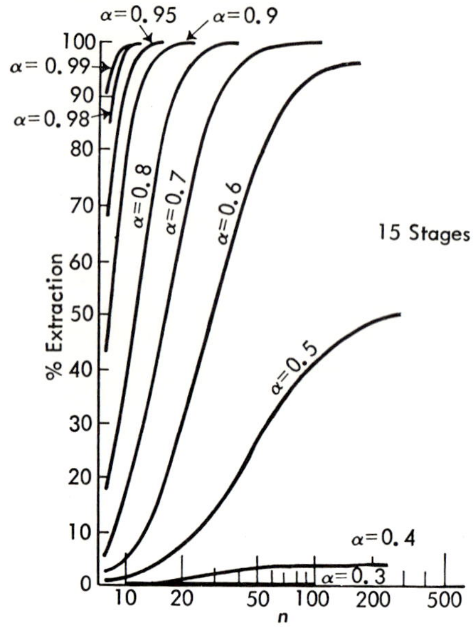

Fig. 14-9. Theoretical double withdrawal recoveries with 15 stages. For quantitative separations, E of the less easily extractable component should not exceed ~0.5 if E of the more easily extractable component ~10 (equal volumes of phases). After Hibbits, Reference 63.

approximately 40 extractions in a 15-stage system are needed in the double-withdrawal counter-current system. Even if the extractions can be carried out by an automatic apparatus,[64] one is likely to consider a different type of separation as an alternative to extraction. Frequently, ion exchange or other kind of column chromatography will provide better separations more easily than countercurrent extraction. Immiscible solvent extraction is at its best when E_A and E_B differ by a factor of 10^4 or more and the simple extraction methods can be applied. In inorganic analysis, especially when metals are being determined, the combination of the proper extraction reagent with a suitable differential complexing agent will often fulfill the desideratum for a number of elements at the same time.

PROBLEMS

1. A solution of OsO_4 in 50 ml of water is to be extracted by chloroform with a required recovery of 99.8 per cent of the OsO_4. How many extractions will be needed with (a) 5-ml portions of $CHCl_3$, (b) 10-ml portions of $CHCl_3$? $[OsO_4]_{CHCl_3}/[OsO_4]_w = 19.1$. *Ans.* (a) 6.
2. Suppose that the extraction of OsO_4 in Problem 1 is carried out by continuous extraction, in which $CHCl_3$ droplets of 0.010 ml volume fall through the aqueous solution and partition equilibrium is attained. What volume of $CHCl_3$ must pass through the aqueous solution (50 ml) to extract 99.8 per cent of the OsO_4? Assume that each droplet is removed from the aqueous phase when it has traversed the aqueous column. *Ans.* 16.3 ml.

[64] Slowness in attainment of extraction equilibrium may rule out its use in many metal-organic reagent systems. Partition chromatography (p. 327) is an alternative.

3. The extraction coefficient of the substance A in the system H_2O–$CHCl_3$ is $E = [A]_{CHCl_3}/[A]_{H_2O} = 10$. What is the recovery factor for A when 50 ml of aqueous phase is shaken with three 10-ml portions of chloroform, each of which is washed with 1 ml of water, the washings being discarded? Can this recovery be considered satisfactory for 0.1 mg of A? For 50 mg of A? How can the recovery be improved?
4. The partitition constant of the acid HA between water and an organic solvent is $[HA]_o/[HA]_w = 31$. The ionization constant of HA in water is 2×10^{-5}. What percentage of A will be extracted if 50 ml of aqueous solution is shaken with three 5.0-ml portions of the organic solvent at pH 1 and 5.0?
5. Plot the log of the extraction coefficient of *p*-nitrosodimethylaniline in benzene–water as a function of pH in the range pH 2–12.

$$P = 51 \text{ (benzene/water)} \qquad \frac{[R][H^+]}{[RH^+]} = 2 \times 10^{-10}$$

6. (a) HCN dimerizes in benzene but not in water. The following relation connects the extraction coefficient ($E = \sum [HCN]_{bz}/[HCN]_w$) with the concentration of HCN in the water phase:

$$E = 0.246 + 0.04[HCN]_w$$

Obtain the value of P_{HCN} and the dimerization constant of HCN. (b) Derive an expression for the extraction of the substance HA which is a weak electrolyte in water and dimerizes in both water and the immiscible organic solvent.

Ans. (a) $P = 0.246$, $K_{dim} = 0.33$.

7. The value of k in the Setchenow equation for the solubility of RuO_4 in sodium chloride solutions is 0.07. What is $[RuO_4]_{CCl_4}/[RuO_4]_w$ for a 4 M NaCl solution if the partition constant is 58 for water?
8. Calculate E for dithizone in the system H_2O–CCl_4 at pH 5 and 8.

Ans. 4400 at pH 5.

9. Oxine has two pH values at which $E_{oxine} = 1$ for an aqueous solution and a particular immiscible organic solvent. What are they when the solvent is $CHCl_3$?
10. The adjacent plot shows the relation between the equilibrium concentration of oxine in a certain immiscible organic solvent S and the extraction coefficient of zinc at pH 5 (data of Schweitzer, Neel, and Clifford, Ref. 35). What is the formula of the extracted zinc species?

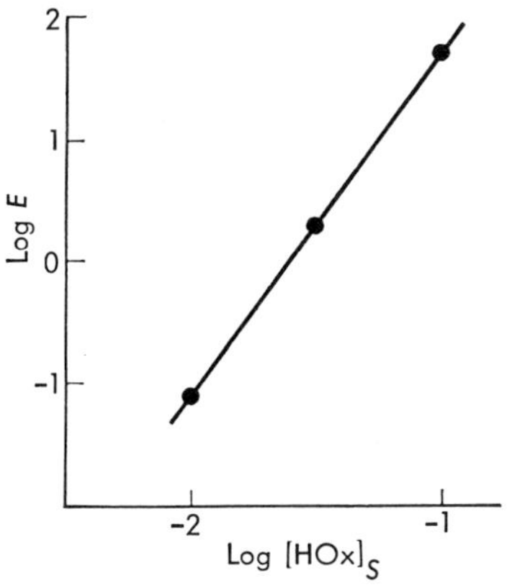

11. A 0.10 M $HClO_4$ solution of copper(II) containing 1.0 μg Cu/ml is shaken with an equal volume of 0.0100 per cent (wt./vol.) dithizone solution in carbon tetrachloride. What is the concentration of copper in the aqueous phase when equilibrium has been attained? $K_{ext} = 3 \times 10^{10}$, $P_{Cu(HDz)_2} = 7 \times 10^4$.

12. Find the percentage of nickel extracted from 100 ml of an aqueous solution containing the equivalent of 10.0 ml of 1.0 *M* HCl and 10.0 ml of 4.0 *M* sodium acetate by shaking with 10 ml of chloroform saturated with dimethylglyoxime (H_2Dx). Assume that the concentration of dimethylglyoxime does not change appreciably by reaction with the small amount of nickel present, and disregard the minute molecular solubility of $Ni(HDx)_2$ in water.

$$\frac{[Ni(HDx)_2]_o[H^+]^2}{[Ni^{++}][H_2Dx]_o^2} = 0.9$$

$$\frac{[NiAc^+]}{[Ni^{++}][Ac^-]} = 4.7$$

$$\frac{[NiAc_2]}{[NiAc^+][Ac^-]} = 18$$

$$\frac{[H_2Dx]_o}{[H_2Dx]} = 8.3 \times 10^{-2}$$

$$\frac{[H^+][HDx^-]}{[H_2Dx]} = 2.6 \times 10^{-11}$$

$$[H_2Dx]_{o,\text{satd}} = 4.5 \times 10^{-4}$$

(subscript *o* refers to chloroform phase).

13. Given the following data:
ionization constant of nitrosophenylhydroxylamine = 7×10^{-5}
solubility of nitrosophenylhydroxylamine in water = 0.025 *M*, in $CHCl_3$ = 5.5 *M*
solubility-product constant of aluminum cupferrate in water = 2×10^{-19}
solubility of aluminum cupferrate in chloroform = 1.4 *M*
find the value of the extraction constant ($CHCl_3$–H_2O) for aluminum cupferrate:

$$K_{ext} = \frac{[AlCp_3]_o[H^+]^3}{[Al^{3+}][HCp]_o^3}$$

What factors limit the accuracy of the calculated value?

14. Obtain the extraction coefficient of lead in the following aqueous–CCl_4 system:

H_2O	CCl_4
pH 12.0	$[H_2Dz] = 4 \times 10^{-5}$ (initial)
$[CN^-] = 0.10$	Volume = 25 ml
$[Cit^{3-}] = 0.10$	
0.1 μg of Pb (initial)	
Volume = 25 ml	

Use the following values for the equilibrium constants:

	K
$Pb^{++} + 2(H_2Dz)_{CCl_4} \rightleftharpoons \{Pb(HDz)_2\}_{CCl_4} + 2\,H^+$	5
$(H_2Dz)_{CCl_4} \rightleftharpoons H^+ + HDz^-$	1.4×10^{-9}
$PbOH^+ \rightleftharpoons Pb^{++} + OH^-$	1.5×10^{-8}
$Pb(OH)_2 \rightleftharpoons Pb^{++} + 2\,OH^-$	4×10^{-11}
$HPbO_2^- + H_2O \rightleftharpoons Pb^{++} + 3\,OH^-$	3×10^{-14}
$PbCit^- \rightleftharpoons Pb^{++} + Cit^{3-}$	2×10^{-5}
$Pb(CN)_4^{--} \rightleftharpoons Pb^{++} + 4\,CN^-$	5×10^{-11}

15. An aqueous solution (10 ml) of pH 4.0 containing 1 μg of Pb and 100 μg of Tl(I) is extracted with two portions of a 0.0102 per cent (w/v) carbon tetrachloride solution of dithizone, each having the same volume as the aqueous phase. The carbon tetrachloride phases are combined and shaken with 20 ml of pH 5.0 buffer. The excess dithizone remains quantitatively in the carbon tetrachloride. Calculate (a) the recovery factor for lead; (b) the separation factor for thallium from lead.

$$\left(S_{\text{Tl/Pb}} = \frac{(Q_{\text{Pb}})_o}{(Q_{\text{Tl}})_o} \times \frac{Q_{\text{Tl}}}{Q_{\text{Pb}}}\right)$$

$$\frac{[\text{M(HDz)}_n]_o[\text{H}^+]^n}{[\text{M}]_{\text{H}_2\text{O}}[\text{H}_2\text{Dz}]_o^n} = 5 \text{ for Pb, } 5 \times 10^{-4} \text{ for Tl}$$

16. Copper (0.0010 per cent) is to be separated from a 1-g sample of metallic zinc containing inappreciable amounts of other reacting metals. Outline a dithizone procedure by which this may be done with recovery of 99 per cent of the copper if the weight ratio Zn/Cu in the separated copper may not exceed 0.01. Take the volume of the aqueous phase to be 50 ml. For H_2O–CCl_4, $(K_{\text{ext}})_{\text{Cu}} = 1 \times 10^{10}$, $(K_{\text{ext}})_{\text{Zn}} = 100$.

Ans. It will be convenient to use a 0.01 per cent (4×10^{-4} *M*) solution of dithizone in carbon tetrachloride. The choice of the volume of carbon tetrachloride is arbitrary. 10 ml is a reasonable choice, because it combines the convenience of a small volume with an excess of dithizone sufficient to maintain the dithizone concentration at roughly its original level.

To be assured of $R_{\text{Cu}} = 0.99$, we will aim for $R_{\text{Cu}} = 0.999$, thus giving a tenfold factor of safety. Adopting a cautious approach, we shall attain this recovery factor for copper in two successive extractions (10 ml of CCl_4 each). We wish to find E_{Cu} and therefrom the acidity at which the extraction should be carried out.

From Eq. (5d),

$$E_{\text{Cu}} = \left\{\frac{(R_{\text{Cu}})_2}{1 - (R_{\text{Cu}})_2}\right\}^{1/2} \times \frac{1}{r} = \left(\frac{0.999}{0.001}\right)^{1/2} \times \frac{1}{0.2} = 160$$

The hydrogen-ion concentration giving this E_{Cu} is

$$[\text{H}^+] = \left\{\frac{(K_{\text{ex}})_{\text{Cu}}[\text{H}_2\text{Dz}]_o^2}{E_{\text{Cu}}}\right\}^{1/2}$$

$$= \left\{\frac{1 \times 10^{10} \times 16 \times 10^{-8}}{160}\right\}^{1/2} = 10^{1/2} \text{ or } \sim 3$$

(Note that the dithizone concentration has been assumed to be unchanged. Actually, in the first extraction, it is diminished by 10 per cent by reaction with copper. For simplicity this decrease will be neglected, since the error resulting is taken care of by the safety factor.)

$$E_{\text{Zn}} \text{ at } [\text{H}^+] = 3 \text{ is equal to } \frac{160}{10^{10}/10^2} = 1.6 \times 10^{-6}$$

The required separation factor for zinc is

$$S_{\text{Zn/Cu}} = R_{\text{Zn}} = \frac{1 \times 10^{-5} \times 10^{-2}}{1} = 1 \times 10^{-7}$$

The actual separation factor is

$$S_{\text{Zn/Cu}} = R_{\text{Zn}} = 2 \times 0.2 \times 1.6 \times 10^{-6} = 6 \times 10^{-7}$$

which means that the goal has not been reached. Accordingly, a retrograde extraction must be made. An acidity of 0.1 M provides a satisfactory extraction coefficient for removal of zinc:

$$E_{Zn} = 100 \times 16 \times 10^{-8}/0.01 = 1.6 \times 10^{-3}$$

After shaking 20 ml of the carbon tetrachloride extracts with 20 ml of 0.1 N mineral acid,

$$S_{Zn/Cu} = 1 \times 10^{-7} \times 1.6 \times 10^{-3} = \sim 2 \times 10^{-10}$$

which is a value small enough to furnish an ample factor of safety. The concentration of copper dithizonate in the carbon tetrachloride is not altered appreciably since E_{Cu} in the retrograde extraction is $1.6 \times 10^{-3}/10^{-8} \sim 1.6 \times 10^{5}$. Contamination of the carbon tetrachloride phase by entrained droplets of aqueous zinc solution and from incomplete separation of the two phases must be considered. To reduce this, it is advisable to carry out the retrograde extraction with two 10-ml portions of acid instead of one 20-ml portion.

The problem has now been solved. However, an analyst familiar with dithizone methods probably would alter the conditions slightly. The use of a 1 M acid (HCl) solution for the first extractions would be preferred to the ~ 3 M solution because equilibrium is reached more quickly on shaking. With the same volumes and dithizone concentration as before:

$$E_{Cu} = 1600$$

$$(R_{Cu})_2 = 0.9999$$

$$E_{Zn} = 1.6 \times 10^{-5}$$

$$(R_{Zn})_2 = S_{Zn/Cu} = 6 \times 10^{-6}$$

After one retrograde extraction as before at $[H^+] = 1$:

$$R_{Cu} = \sim 0.999$$

$$R_{Zn} = S_{Zn/Cu} = 6 \times 10^{-6} \times 1.6 \times 10^{-3} = 1 \times 10^{-8}$$

The calculated $S_{Zn/Cu}$ is 0.1 of the required value, and a reasonable safety factor is thus present.

17. A sample solution contains Fe(III) and U(VI) in the ratio $Fe/U = 1 \times 10^{5}$. It is desired to separate the two metals with a recovery of 95 per cent of the uranium and with a final ratio Fe/U not greater than 0.1 by extracting with one of the following solvents from 6 M HCl:

	E (*Extraction Coefficient*), 6 M HCl	
	Fe(III)	U(VI)
Ethyl ether	10^2	$<10^{-3}$
Methyl isobutyl ketone	10^3	0.12
Tributyl phosphate	$>10^4$	99

Values from Specker, Cremer, and Jackwerth, Ref. 16.

Assuming, as an approximation, that E is independent of the metal concentration, select the most suitable extractant, and specify the conditions (relative volumes, number of extractions, etc.) to effect the required separation.

18. Compare the number of steps required to obtain $R_A = 0.99$ and $R_B = 0.01$ by countercurrent extraction when (a) $E'_A = 10$ and $E'_B = 1$; (b) $E'_A = 3$ and $E'_B = 0.33$.
19. What is the extraction coefficient of silver with dithizone (CCl_4) from an aqueous solution 1 M in H^+ and 0.1 M in thiourea, if at equilibrium $[H_2Dz]_{CCl_4} = 1 \times 10^{-4}$?

$$K_{ext} \text{ of AgHDz(CCl}_4) = 1 \times 10^7$$
$$\beta_3 \text{ of Ag(thi)}_3^+ = 1 \times 10^{13}$$

Thiourea is such a weak base that it exists almost entirely as the neutral molecule in 1 M acid. *Ans.* 1×10^{-7}.
20. A metal is extracted into an organic solvent as $MOx_n \cdot aHOx$. Write the expression for K_{ext} [analogous to K_{ext} in Eq. (16b), p. 355]. Also write the expression analogous to Eq. (17a).
21. A chloroform solution is approximately 0.01 M in oxine and 0.001 M in dithizone. Devise a procedure based on liquid–liquid extraction which will separate the two compounds in purities of 99 per cent.
22. Suppose that $HMCl_4$ is extracted from aqueous HCl solution into an immiscible organic solvent of moderate dielectric constant in which it is partially dissociated into H^+ and MCl_4^-. Derive the expression for the extraction coefficient of M, representing the concentration of M in the aqueous phase as $\sum_{n=0}^{4} (MCl_n)$. (*Hint:* $[MCl_4^-]_o + [Cl^-]_o = [H^+]_o$ (electroneutrality condition), $\{[H^+]_o[Cl^-]_o\}/\{[H^+]_w[Cl^-]_w\} = \alpha_{HCl}$, and a similar expression for H^+ and MCl_4^-.)

CHAPTER 15 *Separations by Volatilization*

A considerable number of elements can be separated advantageously by volatilization methods.[1] This type of separation is of special value for the nonmetals and, in fact, is indispensable for some of them. Such methods often provide sharp separations and allow large samples to be taken for the determination of minor constituents. Volatilization methods may be divided into three classes, which grade into each other.

Volatilization at elevated temperatures, with or without chemical reaction

A familiar example of this class is the volatilization of "essential" water from inorganic materials by heating at suitable temperatures (p. 573).

Argon (and other inert gases) can be liberated from silicates by fusion with sodium hydroxide in vacuum. This separation is of great importance in the age determination of silicate rocks and minerals by measurement of the ratio $^{40}Ar/^{40}K$.

Combustion methods are included in this class, e.g., determination of carbon (Chap. 65) and sulfur in steel by burning the sample in a stream of oxygen to form CO_2 and SO_2, and likewise the determination of carbon and hydrogen in organic compounds by combustion in oxygen. (In such methods the separation is a step, an integral part, of the determination.) By heating an organic material (and some inorganic materials) with cupric oxide in a stream of carbon dioxide and leading the gases over cupric oxide and copper, any nitrogen it contains is converted into N_2, which can be determined by measuring its volume after absorbing the carbon dioxide in potassium hydroxide solution (Dumas method). Oxygen in organic compounds can be determined by volatilizing the sample in a stream of dry nitrogen or helium and passing over hot carbon (1100–1200°C) to convert oxygen to CO, which is led over I_2O_5 to form CO_2 and I_2, the latter being caught in sodium hydroxide solution and titrated (Unterzaucher method). Oxygen in metals is determined by fusing the sample with carbon in a vacuum and pumping off CO for measurement.

Mercury metal and mercury compounds can be volatilized at relatively low temperatures. Other metals such as zinc, cadmium, lead, and thallium can be vaporized at high temperatures ($\sim$1100°C), and small amounts are occasionally isolated in this way.[2] Lead has been isolated from stone and iron meteorites and silicate rocks by heating the sample (as much as 25 g) in a graphite crucible in an induction furnace at 1400°C while a stream of nitrogen is passed through the apparatus. Tin, even in the form of SnO_2, is transformed to volatile SnI_4 by heating a finely divided solid sample with ammonium iodide. Tellurium behaves similarly.

[1] For a short review see H. F. Walton in F. J. Welcher (ed.), *Standard Methods of Chemical Analysis*, vol. 2A, 6th ed., 1963, pp. 201–208.

[2] See, for example, W. Geilmann, *Z. Anal. Chem.* **160**, 410 (1958). For a critical survey of vapor pressures of elements see A. N. Nesmeyanov, *Vapor Pressure of the Elements*, Academic Press, New York, 1963. See J. R. DeVoe and W. W. Meinke, *Anal. Chem.* **35**, 2 (1963), for applicability of vacuum volatilization of radioactive metals.

In connection with separations by volatilization, it may be observed that the analyst occasionally purifies reagents or standard substances by such methods. Thus iodine, arsenious oxide, mercuric chloride, and benzoic acid can be obtained free of nonvolatile impurities by sublimation.

Evolution methods

In these, the sample is treated with a reagent that liberates the desired constituent as a gas. For example, CO_2 and H_2S are evolved by treatment of carbonates (Chap. 63) and sulfides (Chap. 65) with mineral acids, arsenic(III) can be transformed into AsH_3 by the action of zinc in an acid medium, SiF_4 is formed from fluorides in the presence of silica and sulfuric acid, etc. Usually the quantitative expulsion of the gaseous product is assured by boiling the aqueous solution so that these methods grade into those of the next section. Thus the high solubility of NH_3 in water requires that an aqueous solution be boiled for some time to drive it out of a solution of an ammonium salt that has been treated with a strong base (see Kjeldahl method, Chap. 39). Passage of an inert gas through the system to sweep out the volatile constituent is sometimes advantageous.

A system of analysis on the microgram scale developed by Conway[3] utilizes diffusion of the volatile constituent. A shallow sealed container is used, which consists of two chambers. In one, the volatile constituent is liberated by mixing the sample with a suitable reagent (e.g., NH_3 is released by alkali); in the other it is absorbed (NH_3 in a nonvolatile acid such as sulfuric). The temperature of the system is uniform and the concentration gradient is established chemically. The process is sometimes called *isothermal distillation*.

Distillation methods

A liquid or a solid substance that has a relatively high vapor pressure can be vaporized by boiling the solution in which it is dissolved. In inorganic analysis the solution will generally be an aqueous solution. The condensate will contain the substance dissolved (usually) in much water. The process is known as *distillation* of the substance, the codistillation of water being understood.

The semimetallic elements form volatile halogen compounds which serve for their separation from most elements. Distillation from hydrochloric acid solution is a standard method for the isolation of As(III), Sb(III), and Ge. The distillation of tin(IV) can be carried out more easily from a hydrochloric–hydrobromic acid solution ($SnBr_4$ is formed). Selenium can be distilled as $SeBr_4$. Chromyl chloride, CrO_2Cl_2, can be distilled by addition of chloride to a perchloric acid solution of Cr(VI). Some oxides are sufficiently volatile to permit easy separation by distillation. Osmium distills as OsO_4 from 5 *M* nitric acid. Ruthenium can be distilled as RuO_4 from an acid solution containing a strong oxidizing agent such as sodium bismuthate. Rhenium can be volatilized as Re_2O_7 from a sulfuric acid solution. These oxides are volatile because of the high proportion of oxygen that they contain. Fluosilicic acid can be distilled from sulfuric or perchloric acid–aqueous solution and fluorine is usually isolated in this way. A standard method for separation of boron is based on the volatility of methyl borate, $B(OCH_3)_3$. This is an example of one of the very few uses of an organic reagent—here methyl alcohol—in a volatilization method in inorganic analysis.

In passing it may be noted that purification of organic solvents (used in liquid–liquid or solid–liquid extractions) by distillation is often called for.

The theory of simple distillation (without fractionation) will be briefly considered here. A volatile substance partitions between a liquid (aqueous solution) and the vapor

[3] E. J. Conway, *Microdiffusion Analysis and Volumetric Error*, Crosby Lockwood, London, 4th ed., 1957. For an application of the diffusion method in the isolation of microgram quantities of fluorine, see H. W. Wharton, *Anal. Chem.* **34**, 1296 (1962).

phase in equilibrium with the former in accordance with its volatility, and the process of distillation may be looked upon as an extraction of the substance by water vapor formed on boiling and then condensed. The distribution coefficient of the substance A between the vapor (v) and liquid (l) phase (of specified composition) is defined as

$$D_A = \frac{[A]_v}{\sum [A]_l} \quad \text{(at a given temperature and pressure)} \tag{1}$$

$[A]_v$ is expressed as the concentration of A in the solution obtained by condensation of the vapor. Usually A will be present in more than one form in the liquid phase; only one species may be volatile. For example, in the distillation of $AsCl_3$, arsenic will be present in the hydrochloric acid solution in part as arsenious acid, possibly also as mixed chlorooxide species, as well as volatile $AsCl_3$. D is thus the counterpart of the extraction coefficient in liquid–liquid extraction. The value of D is found by condensing a portion of the vapor in equilibrium with the boiling liquid phase as a whole and determining the concentration of A in both phases. In practice this can often be done by distilling off a volume of solution which is small compared to the volume of the original solution, so that the concentration of A does not change significantly in the liquid distilled. The greater the volatility of A, the smaller must be the volume of distillate collected for this condition to hold.

When D is known, the volume of distillate that must be collected to obtain a specified recovery can be calculated. The expression for the recovery factor is readily derived. First consider the case in which the volume and composition of the solution distilled are maintained constant by addition of water. Let:

w = weight of A in the distillate
W = initial weight of A in the solution being distilled (distilland)
v = volume of distillate (mostly water)
V = volume of distilland (= constant)
D = distribution coefficient as already defined. It will be assumed to remain approximately constant during the course of the distillation, since, by stipulation, the composition of the solution remains constant except for a decrease in A
R = recovery factor for A

Then

$$dw = \frac{D(W - w)\,dv}{V} \tag{2a}$$

$$\int_0^w \frac{dw}{W - w} = \frac{D}{V}\int_0^v dv \tag{2b}$$

$$\ln \frac{W - w}{W} = -D\left(\frac{v}{V}\right) \tag{2c}$$

$$\log \frac{W - w}{W} = \log(1 - R) = -0.43D\left(\frac{v}{V}\right) \tag{2d}$$

The preceding equation may be written

$$R = 1 - \exp\left(-\frac{Dv}{V}\right) \tag{3}$$

If the volume of the solution distilled is not maintained constant, the following expression holds if D remains constant:

$$\log \frac{1}{1 - R} = D \log \frac{V}{V - v} \tag{4}$$

Actually D cannot be expected to stay constant if the volume of the solution changes appreciably. Even if no other factors are altered, the solution will become more concentrated with respect to the electrolytes usually present when its volume decreases and the boiling point will rise.

If the volatility of a substance is not sufficiently great at $\sim 100°C$ for successful isolation, it may be possible to volatilize it from a solution of perchloric or sulfuric acid at 200–300°C.

Substances of different volatilities can be separated by fractional distillation. Such separations are, or were, of importance in organic analysis (now largely replaced by gas chromatography); since they find little use in the inorganic field they will not be discussed here.[4]

The possibility of loss of constituents by volatilization in decomposition of samples by acid attack or by fusion always needs to be borne in mind.

PROBLEMS

1. The volatile substance A has a distribution coefficient of 10 (condensed vapor/aqueous solution). Find the percentage recovery of A when 10, 50, and 75 per cent of the original volume of the solution have been distilled, if (a) the volume of distilled solution is kept constant by addition of water or by steam distilling; (b) the water distilled is not replaced. *Ans.* (a) 63.2, 99.3, 99.94 per cent.
2. J. D. H. Strickland and G. Spicer, *Anal. Chim. Acta* **3**, 543 (1949), found that permanganic acid or Mn_2O_7 can be distilled from sulfuric acid solutions containing periodate, and obtained a value of 1.08 for D in 10.2 M sulfuric acid at 166°C. If the volume of the solution distilled is maintained constant at 100 ml, what volume of distillate must be collected to recover 99 per cent of the manganese in the distilling flask?
3. Nitrogen is slowly bubbled through 100 ml of water originally saturated with air (20.95 per cent O_2 by volume) at 25°C and 1.0 atm. Assuming equilibrium to be attained, what volume of nitrogen will be required to reduce the oxygen content of the water to 1×10^{-7} M O_2? The solubility of O_2 in water at 25°C and 1 atm is 3.16 ml/100 ml of H_2O.
4. In the Unterzaucher method for oxygen in organic compounds the following reactions are utilized: (a) conversion of organic oxygen to CO; (b) reduction of I_2O_5 by CO to I_2; (c) absorption of I_2 in NaOH; (d) treatment of solution c with an excess of an acetic acid solution of bromine; (e) treatment of solution d with an excess of formic acid to remove bromine; (f) addition of excess sulfuric acid and iodide to solution e; (g) titration of solution f with standard $Na_2S_2O_3$ solution (starch as indicator). Write balanced equations for all the reactions. To what weight of O is 1 ml of 0.01 N $Na_2S_2O_3$ equivalent? *Ans.* 6.67×10^{-5} g.

[4] See H. Laitinen, *Chemical Analysis*, McGraw-Hill, New York, 1960, p. 474, and E. W. Berg, *Physical and Chemical Methods of Separation*, McGraw-Hill, New York, 1963, chap. 2. The separation of $AsCl_3$ and $SbCl_3$ by fractional distillation is discussed by H. Biltz, W. Biltz, and W. Fischer, *Ausführung Quantitativer Analysen*, S. Hirzel, Stuttgart, 6th ed., 1953, p. 350.

CHAPTER 16 *Errors in Quantitative Analysis*

Suche die Methode der Analyse aufzufinden, bei welcher die Genauigkeit des Resultats am wenigsten von der Geschicklichkeit des operierenden Chemikers abhängt, und wenn diese ausgewählt ist, so sinne nach, welche unvermeidlichen Umstände vorhanden sind, die das Resultat unrichtig machen und ob es durch sie vergrössert oder verringert wird.

—J. J. BERZELIUS (1814)

The more numerous the number of observations and the less they vary among themselves the more their results approach the truth. We fulfill this last condition by the choice of the methods of observations, by the precision of the instruments, and by the care which we take to observe closely; then we determine by the theory of probabilities the most advantageous mean results or those which give the least value of the error. But that is not sufficient; it is further necessary to appreciate the probability that the errors of these results are comprised in the given limits; and without this we have only an imperfect knowledge of the degree of exactitude obtained. Formulae suitable to these matters are then true improvements of the method of sciences, and it is indeed important to add them to this method.

—P. S. LAPLACE (1814) (translated)

Any measurement involves an ensemble of three components: the system, some property of which is being measured; the instrument; and a human being who may be called the observer or operator. In these days of automated instrumentation, the human being may be reduced to the role of pushing buttons, reading a value printed by the instrument, or making pointer settings and estimating tenths of a scale division. This is still far from being generally true in chemical analysis, however. More often than not, the analyst is an operator rather than an observer. He must frequently establish the system, without which there would be nothing to measure. Even in more or less automated instrumental analysis, the sample must be selected and often suitably prepared before the analysis can be made.

Errors originate in all three components, but to variable extents, depending on the type of analysis, the composition of the sample, and other factors. A chemical analysis is usually more than a simple measurement. Instrumental errors per se may not be the most important. The steps that precede the determination proper may be quite complex in applied analysis and may introduce errors larger than those in the determination step.

The question of accuracy must, in general, receive attention both before and after an analysis. The choice of an analytical method may be of crucial importance. The errors to which an analysis is subject need to be considered and their effect on the final result evaluated as far as possible.

Theoretically, it is impossible to obtain the exact value of a physical quantity because it is incommensurable with the unit in which it is expressed. In practice the problem is to obtain a value that can be relied upon within certain limits, which are determined by the use of the result.

In this chapter we shall examine the factors affecting the reliability of chemical analyses and discuss methods of testing and increasing their accuracy. The statistical treatment of analytical data will be considered in an elementary way.

Accuracy and precision

At the outset it is necessary to define *error* and to distinguish between accuracy and precision.

The error in a measurement is the difference between the observed, i.e., measured, value and the true value of the quantity measured:

$$\epsilon = X - X_\tau$$

where ϵ = absolute error
X = measured value
X_τ = true value

The error of a measurement cannot be stated if the true value of the quantity is not known; since the true value is established by measurement, it might seem impossible to find the error associated with any measurement, and this is indeed true in the strict sense of the term. The difficulty is not insurmountable, however, as far as practical requirements are concerned, as will be shown later.

The error of a measurement is a measure of the inaccuracy of that measurement, and an inverse measure of its accuracy; i.e., the smaller the error, the greater the accuracy. Errors are usually not expressed in absolute values (as in the preceding formula), but in relative values, with respect to X_τ, because the value of ϵ without regard to the value of X_τ is not likely to be of practical importance. Usually we are interested in the value of ϵ/X_τ. This is the expression for *relative error*. It is convenient to express relative error in terms of percentage (parts per hundred), $100\epsilon/X_\tau$, or parts per thousand (ppt), $1000\epsilon/X_\tau$, preferably the latter in chemical analysis involving major constituents. The relative error is an (inverse) index of the accuracy of a chemical analysis.

The difference between a particular measured value and the arithmetical mean of a series of measurements of which it is a member is called its *deviation*:

$$d = X - \bar{X}$$

where d = absolute deviation
X = measured value
$\bar{X}$ = arithmetical mean of a series of measurements

The magnitude of the deviations of a series of measurements is a measure of the imprecision of the measurements or an inverse measure of their *precision*. Usually, we are more interested in the relative than in the absolute values of the deviations, and they are usually expressed in ppt (or percentage) with respect to $\bar{X}$.

Accuracy expresses the correctness of measurement and precision its reproducibility.[1]

[1] The concept of reproducibility (precision) is not as simple as it may appear at first sight. There are, in fact, different kinds of reproducibility. Assuming a well-defined system on which measurements are being carried out, the reproducibility can be based on replicates obtained (1) by a single observer at a given time (day and hour); (2) by the same observer at another time (say in winter instead of summer, in early morning instead of late afternoon); (3) by another observer in the same laboratory; (4) by still another observer in a different laboratory; and so on. The reproducibilities could vary to a greater or less extent from one situation to another. The reproducibility in one laboratory might be worse than in another because of unfavorable environmental factors such as fluctuating temperatures or high humidity or because of the use of inferior instruments. One observer may be more skillful or conscientious than another. The American Society for Testing and Materials, ASTM Committee E-15,

Accuracy without precision is difficult, but precision does not guarantee accuracy. The arithmetical mean of a series of measurements does not usually give the true value of the quantity measured, and there may be a considerable difference between the two. (See Fig. 16-1.)

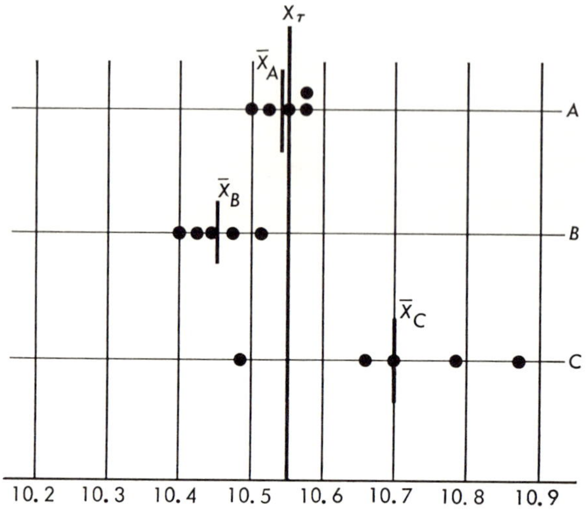

Fig. 16-1. Illustrating accuracy and precision in quantitative analysis.

The true value X_τ must itself be established experimentally and is known only within certain limits. When X_τ has been found by refined methods it may be considered fixed in such a narrow range that the error in a less accurate method can be found from it. For example, a chloride solution whose concentration is known within 1–2 parts per 10,000 can be prepared by weighing out purified, dried sodium chloride and used for testing the *accuracy* of chloride determinations. The atomic weights of sodium and chlorine are known with such high accuracy that they can be considered errorless from the standpoint of ordinary, careful chemical analysis. See also the discussion on page 414.

Errors can be divided into two classes: (1) determinate errors, and (2) indeterminate (random) errors.

Determinate errors

As the name implies, determinate errors are those whose magnitude can be determined (at least in principle) and the measurements thereby corrected. A determinate error may have the same value under a variety of conditions and may remain constant from one measurement to another; such a determinate error is called a *constant error*. The presence of an impurity in a substance used for the standarization of a volumetric solution would be a source of a constant error; inaccuracies due to uncalibrated weights fall in the same category (if the same pieces are used in successive weighings). Other

---

Method E 180-61T, uses the term *repeatability* to mean the precision of a method expressed as the agreement attainable between independent determinations run by a single analyst using the same apparatus and technique, and *reproducibility* to mean the precision of a method expressed as the agreement attainable between determinations run in different laboratories. Evaluation of interlaboratory tests is considered by K. Doerffel, *Z. Anal. Chem.* **184**, 81 (1961).

determinate errors, on the other hand, will vary in magnitude and even in sign with the conditions. The law of the change may be known, however, so that the value of the error can be determined under any circumstances. Thus the expansion and contraction of volumetric solutions with a change in temperature give rise to variable errors, but the magnitude of the change can be determined by noting the temperature, because the coefficient of expansion of the solution (and glass) is known. These variable determinate errors are sometimes called *systematic errors*. See Fig. 16-8. (In the quotations at the head of the chapter, Berzelius had in mind chiefly determinate errors, Laplace indeterminate errors.)

The determinate errors that must be considered in chemical analysis are rather numerous. No attempt will be made here to enumerate them all, but the following list is suggestive; the more important ones are included. Four general classes may be distinguished:

1. *Instrumental errors and those due to apparatus and reagents*
   a. Balance and weights. These include: unequal arm lengths (not always a source of error), insufficient sensitivity, uncalibrated weights, etc.
   b. Volumetric apparatus. Use of uncalibrated glassware.
   c. Vessels and utensils. Introduction of foreign material through attack of glassware, porcelain, etc. Loss in weight of platinum crucibles when strongly heated.
   d. Reagents. Presence of impurities, either the same as the substance sought or interfering substances.

2. *Operative errors.* These errors are mostly physical in nature and are associated with the manipulations of an analysis. They are independent, to a great extent, of the instruments and utensils employed, they are not related to the chemical properties of the system in hand, and their magnitude depends more upon the analyst himself than on any other factor. Operative errors may assume serious proportions if the analyst is inexperienced, careless, or thoughtless. These errors are reduced to insignificance by careful, skillful, and understanding work. The following operative errors may be cited as typical: leaving vessels uncovered and thus introducing dust and other foreign materials into solutions, spilling of liquids and solids, loss of material from solutions by effervescence and "bumping," failure to remove precipitates quantitatively from vessels, loss in transfer of precipitates, underwashing or overwashing of precipitates, use of vessels of unsuitable size, placing crucibles in the wrong position when heating so that the flame gases gain access to the interior, insufficient time of heating or the use of improper temperatures in igniting precipitates, weighing of crucibles before they are completely cold, failure to guard against absorption of moisture by the substance weighed, failure to apply weight corrections, failure to apply temperature corrections in volumetric analysis, errors in calculations, use of nonrepresentative sample, etc. When such errors are due to carelessness, they are inexcusable. At first the beginner will be lacking in skill and understanding, and he may make more or less serious operative errors, often being unaware of so doing. Soon, however, he gains the necessary dexterity and insight, and his operative errors will then assume negligible proportions. The errors inherent in certain manipulations can never be entirely eliminated, but they can be reduced to such a small magnitude by the proper procedure that they will hardly come into consideration.

3. *Personal errors*
   a. Personal equation. Somewhat different from the operative errors are the personal errors of the analyst; these originate in the constitutional inability of an individual to make certain observations accurately. For example, some persons are unable to judge color changes exactly in titrations and may, for example, always go a little past the end point. Color-blindness is a more serious form of such inability. These errors are likely to be quite constant in magnitude; they are included in the "personal equation" of the observer.

b. Prejudice. When it is a question what tenth of a division is to be taken in reading a scale, the operator is likely to choose the one that will make the result agree more closely with the preceding one; if there is some doubt as to the exact location of the end point, he is apt to stop the titration at the point which will give a result in agreement with the previous titration if he knows what the buret reading should be for agreement, and so on. These are errors due to prejudice. Even conscientious workers may be influenced unconsciously by prejudice.

4. *Errors of method.* These errors have their origin in the *chemical* or *physicochemical* properties of the analytical system. They are by far the most serious errors encountered in chemical analysis. The errors already mentioned can be eliminated or reduced to very small magnitude by the proper technique, because they are mostly physical in nature. It is otherwise with the *methodic errors*; these are inherent in the method, and no matter how skillfully and carefully the analyst works, their magnitude remains the same unless the conditions of the determination are altered.

Some sources of methodic errors are these:

1. Solubility of a precipitate in the solution in which it is precipitated, and in the wash liquid.
2. Failure of a reaction to proceed to quantitative completion.
3. Precipitation of another substance with the reagent used.
4. Coprecipitation and postprecipitation.
5. Decomposition or volatilization of a precipitate on heating; hygroscopicity of the weighing form.
6. Induced reactions and side reactions.

Of course, it is impossible to differentiate sharply between errors of operation and errors of method. Theoretically at least, methodic errors are operative errors in the sense that adoption of a better procedure would eliminate or diminish the error. Some methodic errors might be classed as operative, as, for example, the decomposition of a precipitate on heating, if the range of stability is wide, so that the analyst can easily regulate the temperature to prevent decomposition. The solubility of a precipitate in a wash liquid may lead to no sensible error if the washing is properly done, whereas a significant error may be made if the amount of liquid used is excessive. Errors due to coprecipitation may often be reduced to negligible proportions by dissolving the contaminated precipitate and reprecipitating it. Sometimes it is difficult or practically impossible to alter the conditions so that better results can be obtained, and then the errors may be called methodic in the true sense of the word. From the standpoint of the student it is well to distinguish between methodic and nonmethodic errors. For operative errors greater than an irreducible minimum, he must be held responsible; methodic errors are no fault of his.

The methodic errors of common determinations will be considered in the following chapters in the discussion of the respective procedures. Here it need only be emphasized that errors of method constitute the most serious source of inaccuracy in quantitative methods, because they often are very difficult to eliminate.

Additive and proportional determinate errors. An additive error is one which has the same absolute value no matter what the amount of the constituent being determined, while a proportional error is one whose absolute magnitude depends upon the quantity of the constituent. Benedetti-Pichler[2] gives an instructive example dealing with a method for aluminum involving precipitation of the hydroxide with ammonia and ignition of the washed precipitate to the oxide at a high temperature. In the first series

[2] A. A. Benedetti-Pichler, *Ind. Eng. Chem., Anal. Ed.* 8, 373 (1936).

of experiments the following amounts of aluminum oxide were obtained from the indicated amounts of potassium alum:

K *Alum*, g	Al_2O_3 *Found*, g	Al_2O_3 *Taken*, g	*Difference*, g
1.0000	0.1288	0.1077	+0.021
2.0000	0.2384	0.2154	+0.023
3.0000	0.3489	0.3231	+0.026
4.0000	0.4588	0.4308	+0.028
Blank	0.000	0.000	0.000

It will be seen that the errors (values in the last column) are nearly constant; they show only a slight increase as the amount of aluminum is increased 400 per cent. The major portion of the error was found to be due to silica in the ammonia, which was coprecipitated nearly completely with the aluminum hydroxide. Since practically the same volume of ammonia was used in each determination, it is evident why the error due to silica should be additive.

In the second series of experiments freshly distilled ammonia was used in the precipitation, and the following results were obtained:

K *Alum*, g	Al_2O_3 *Found*, g	Al_2O_3 *Taken*, g	*Difference*, g	ppt
1.0000	0.1087	0.1077	+0.0010	+9
2.0000	0.2178	0.2154	+0.0024	+11
3.0000	0.3258	0.3231	+0.0027	+8
4.0000	0.4352	0.4308	+0.0044	+10

The errors here are clearly proportional, amounting to about 10 ppt. The high results are due to the presence of water in the ignited residue; even after strong ignition aluminum oxide may retain water (p. 598). It is now apparent that the errors of the first series are actually a composite of additive and proportional errors, the former predominating (Fig. 16-2).

This example illustrates how valuable an analysis of errors may be in locating the source of inaccuracies in a method. Moreover, it shows that in testing a method one should always take various amounts of sample in order that the possible presence of additive and proportional errors may be revealed.

Indeterminate errors

The existence of another class of errors,[3] called *indeterminate* or *random*, is revealed by small fluctuations in the successive values of a measured quantity when the measurements are made by the same person with great care under conditions as nearly the same as possible. These errors arise from small chance variations in the instrument, the system, or even the operator. They have their origin in the limited ability of the operator to control all the variables that affect the measurements.

It is generally stated that random errors are beyond the power of the observer to prevent or allow for, but this statement refers to unchanging conditions and methods of measurement. By working under different conditions and by taking certain precautions,

[3] The word *error* is used somewhat loosely in the term *indeterminate error* (see p. 381). Strictly we should speak of *deviation* but no confusion is likely to arise from the use of *error* in a general sense in this connection.

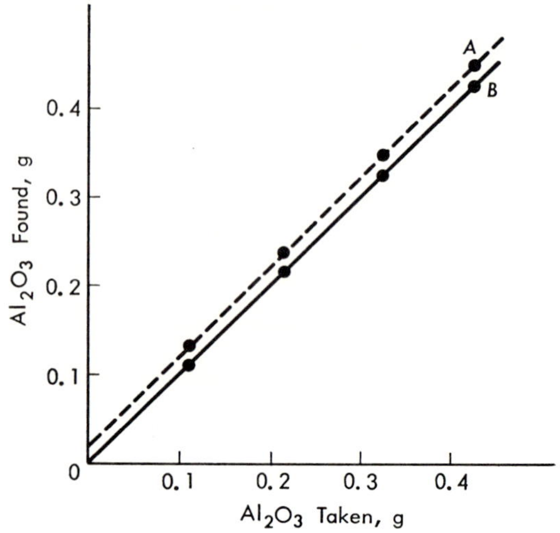

Fig. 16-2. Determination of aluminum by ammonia precipitation and ignition of $Al(OH)_3$ to Al_2O_3. *A*, precipitation with ammonia containing dissolved silica as an impurity. The amount of silica carried down by the aluminum hydroxide precipitate is almost independent of the weight of the latter and is approximately represented by the intercept on the *Y*-axis. *B*, precipitation with silica-free ammonia. The only significant error now is due to water in the ignited residue. The straight line passes through the origin, since the amount of water is proportional to the weight of Al_2O_3. The slopes of *A* and *B* are the same within experimental error.

it may be possible to reduce the variations in successive measurements. The weighing of a hygroscopic substance provides an illustration. Suppose an analyst is unaware of, or underestimates, the hygroscopic nature of a substance and weighs it after successive dryings in obtaining constant weight. He will find the observed weights to be variable, the exact weight in any weighing depending upon the time required to make the weighing (which determines the amount of moisture taken up by the substance) and perhaps upon the variable humidity of the atmosphere. To him the variable results may appear inexplicable, and he will attribute the variations to accidental errors. Actually these variable values are only the result of variable determinate errors. A practiced analyst would at once ascribe the variability in weight to the hygroscopic nature of the substance and would make the weighings under conditions such that the absorption of moisture would be impossible or difficult. The fluctuations would then disappear or assume a trifling magnitude.

Because of their nature, indeterminate errors can be treated by statistical methods. Given sufficient data, the analyst can estimate the uncertainty introduced by random errors and arrive at conclusions concerning the reliability of an analytical result as affected by them.[4]

[4] The following books may be mentioned as some dealing with the statistical evaluation of scientific, especially chemical, data: C. A. Bennett and N. L. Franklin, *Statistical Analysis in Chemistry and the Chemical Industry*, Wiley, New York, 1954; O. L. Davies et al., *Statistical Methods in Research and Production with Special Reference to the Chemical Industry*, Oliver & Boyd, Edinburgh, 3rd ed., 1957; W. L. Gore, *Statistical Methods for Chemical Experimentation*, Wiley-Interscience, New York, 1952; W. J. Youden, *Statistical Methods for Chemists*, Wiley, New York, 1951; G. Gottschalk, *Statistik in der quantitativen chemischen Analyse*, Enke, Stuttgart, 1962 (vol. 49 in the series *Die chemische Analyse*); V. V. Nalimov, *The Application of Mathematical Statistics to Chemical Analysis* (transl. by P. Basu), Pergamon, New York, 1963;

Normal law of indeterminate errors. In dealing with indeterminate errors (deviations), the assumption is usually made that their distribution obeys the normal law of error, which is represented by the normal frequency curve (of Gauss), whose equation is

$$Y = \frac{h}{\pi^{1/2}} \exp\{-h^2(X - \mu)^2\} \tag{1a}$$

where Y = frequency of occurrence of a deviation having a given magnitude
$X - \mu$ = magnitude of the deviation, X being the measured value and μ the arithmetical mean ($\mu = X_\mu$)
h = a constant whose value depends upon the character of the measurements

The value of h is given by $1/(\sigma\sqrt{2})$, in which σ is the standard deviation (p. 389), so that

$$Y = \frac{1}{\sigma(2\pi)^{1/2}} \exp\left\{-\frac{1}{2}\left(\frac{X - \mu}{\sigma}\right)^2\right\} \tag{1b}$$

The normal frequency curve is represented in Fig. 16-3 (and Fig. 16-4). It is of course based on a large number—we can say an infinite number—of measurements. In the language of statistics it represents the *universe* (infinite population) of data. The true mean, μ, of the entire population divides the curve into two symmetrical halves. Positive and negative deviations are equally probable. The form of the curve demonstrates that small deviations are much more probable than large ones, and that the arithmetic mean is the most probable value.

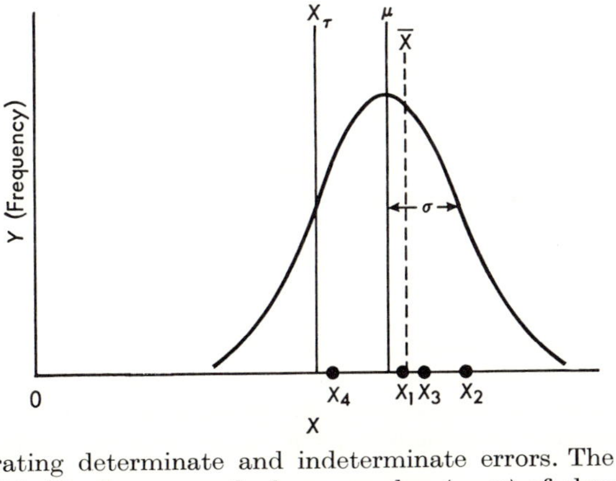

Fig. 16-3. Illustrating determinate and indeterminate errors. The true value of the quantity measured is X_τ, the mean of a large number ($\rightarrow\infty$) of observations is μ, and the frequency of the observed values is represented by the normal error curve shown. If a limited number of observations are made (X_1, X_2, X_3, X_4,) the mean is $\bar{X}$, which will not, except by chance, coincide with μ. The greater the number of measurements the smaller, in general, will be the difference between $\bar{X}$ and μ.

R. A. Fisher, *Statistical Methods for Research Workers*, Oliver and Boyd, Edinburgh, 12th ed., 1951; R. A. Fisher, *The Design of Experiments*, Oliver & Boyd, Edinburgh, 6th ed., 1951. J. Mandel, *The Statistical Analysis of Experimental Data*, Wiley-Interscience, New York, 1964; K. Eckschlager, *Fehler bei Chemischen Analysen*, Akad. Verlags., Leipzig, 1964. A chapter in H. Laitinen, *Chemical Analysis*, McGraw-Hill, New York, 1960, is devoted to statistics in quantitative analysis; likewise in C. L. Wilson and D. W. Wilson (eds.), *Comprehensive Analytical Chemistry*, vol. IA, Elsevier, Amsterdam (and Van Nostrand, Princeton, N.J.), 1959, and I. M. Kolthoff and P. J. Elving with E. B. Sandell (eds.), *Treatise on Analytical Chemistry*, part I, vol. 1, Wiley-Interscience, New York, 1959.

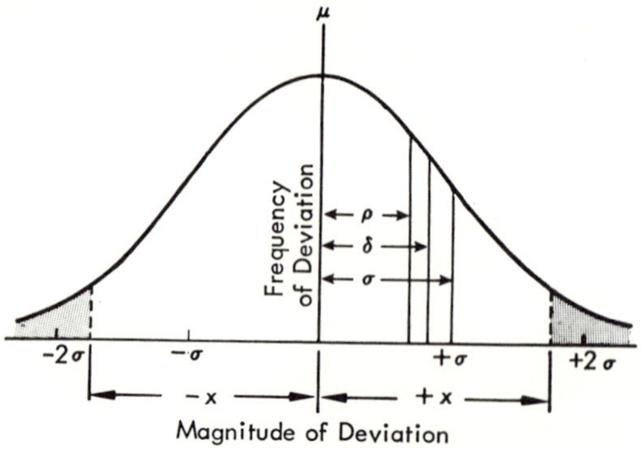

Fig. 16-4. The normal error curve, with δ, σ, and ρ indicated. The abscissa of the inflection point of the curve gives σ. The probable deviation is given by the abscissa of the line dividing the area under the curve on either side of μ into two equal parts. $x = (X - \mu)/\sigma$.

The mean of a finite number (N) of measurements, $\bar{X}$ ($= \sum X/N$), will not in general be the same as the mean, μ, obtained from an infinite (very large) number of measurements.[5] $\bar{X}$ is an *estimate* of μ, and the larger N is, the better will be the estimate. As already noted, μ need not coincide with X_τ, the true value, since determinate errors may be present. The difference between μ and X_τ represents the determinate error or sum of errors in the measurements.

The origin of indeterminate errors can be illustrated by the titration of successive samples of sodium carbonate, from the same lot of homogeneous material, with $\sim 0.1\ N$ standard acid and a visual indicator. Small deviations will occur in each of the several steps of the procedure. Slight irregularities in the action of the balance and possible errors in the estimation of the tenths of a scale division[6] will lead to small errors in the recorded weights of the samples. The hygroscopic character of sodium carbonate is likely to result in the absorption of a little water, the amount of which will fluctuate slightly from one sample to another depending on the time required for the weighing. The end point in any titration will not be exactly located; i.e., the analyst will stop a little short of, or go a little past, the exact point at which the indicator shows the hue corresponding to the equivalence point. Small errors will be made in reading the buret and volume errors may also result from drainage. Other sources of error could be mentioned. The final result will have an error which is the algebraic sum of the small component random errors. If the component errors arising from independent variables have approximately the same magnitude, it can be seen (at least in a general way)

[5] Values, such as μ, based on an infinite statistical sample are *parameters*, whereas those, such as $\bar{X}$, based on a finite sample are known as *statistics*. It is customary to represent the former by Greek letters.

[6] The magnitude of error in reading the scale of a balance or other instrument will depend on such factors as the experience of the operator, the size of the scale divisions and parallax. If the divisions of a scale are not too small ($\sim$1 mm or larger), the standard deviation of an experienced observer estimating to the nearest tenth of a division from a viewing distance of 25–30 cm will be approximately 0.05 division, or 0.05 mm, corresponding to $\sim$1 min of arc, roughly the resolving power of the human eye. If the scale divisions are too widely spaced, there will obviously be a loss in precision as far as absolute distances (and corresponding angles) are concerned, the relative error (fraction of a division) remaining much the same. If the divisions are spaced closer than 1 mm, the standard deviation in terms of a division will rise, but the absolute angular deviation should remain approximately the same if the lines are very fine. A magnifier will of course increase the precision of estimates when the scale is closely divided, if parallactic errors are avoided. Some instruments (balances, spectrophotometers) are so reproducible that the scale-reading error may be the major error.

that in a series of titrations they would combine to give totals whose distribution approaches the normal Gaussian. The random errors in weight and volume measurements in this example ($\sim$0.1 mg versus 0.01–0.02 ml) would be much the same in terms of sodium carbonate. The error in the end-point location might be similar or it might be considerably larger, depending on the indicator, the conditions of color observation, and the analyst. If the random error introduced by one factor is, on the average, more than twice as great as the sum of the others, the effect of the latter can usually be neglected (p. 401).[7] If a single factor controls the precision, approach to the Gaussian distribution becomes less certain.

It appears that random-error distributions in chemical analysis frequently are nonnormal.[8] Conclusions drawn from the application of simple statistical tests must accordingly be accepted with some caution. See also p. 393.

Precision of a single measurement. Consider a series of measurements of the same quantity made under constant conditions. As noted earlier, the extent of scatter (dispersion) of the individual values about the mean indicates (inversely) the precision of the measurements: i.e., the smaller the scatter, the greater the precision. Numerically, the precision is expressed by precision measures, defined as follows when N, the number of measurements, is large ($\rightarrow \infty$):

Average deviation (average of differences between measured values and arithmetical mean without regard to sign) =

$$\delta = \frac{\sum |X - \mu|}{N} \tag{2}$$

Standard deviation (deviation whose square is the mean of the squares of the differences between the measured values and the arithmetical mean) =

$$\sigma = \left\{\frac{\sum (X - \mu)^2}{N}\right\}^{1/2} \tag{3}$$

σ^2 is known as the *variance*.

Probable deviation (deviation such that in an infinite number of measurements, the number of deviations having values greater than it will be equal to the number having smaller values) = ρ = 0.67σ (infrequently used).

When N is large and the normal frequency distribution is followed,

$$\delta = 0.80\sigma = 1.18\rho \tag{4}$$

In practice the number of measurements in a series will usually be small (especially in complex, time-consuming analyses), so that only *estimates* of δ and σ can be obtained. These estimates are denoted by $\bar{d}$ and s. Thus the *sample* (i.e., *statistical* sample) average and standard deviations are given by

$$\bar{d} = \frac{\sum |X - \bar{X}|}{N} \tag{5}$$

$$s = \left\{\frac{\sum (X - \bar{X})^2}{N}\right\}^{1/2} \tag{6}$$

[7] It follows that when an analytical method is being studied for means of improving precision, special attention should be paid to locating the least precise factor and then reducing its effect.

[8] V. J. Clancey, *Nature* **159**, 339 (1947), examined 250 distributions based on 50,000 analyses and found only 10–15 per cent to be essentially normal. On the other hand, papers could be cited in which approximately normal distribution has been demonstrated. G. Wernimont, *Anal. Chem.* **21**, 116 (1949), has suggested that a reasonable explanation for some nonnormal distributions is assignable causes of variation affecting the results, which if removed should allow the distributions to approach the normal more closely. See also Nalimov (Reference 4, pp. 74–82), who points out that some error distributions are necessarily nonnormal.

As $N \to \infty$, $\bar{X} \to \mu$ and $\bar{d} \to \delta$ and $s \to \sigma$. Since $\bar{d}$ and $s = 0$ when $N = 1$, it is obvious that $\bar{d} < \delta$ and $s < \sigma$ for finite values of N.

The estimated standard deviation, s, is a better measure of precision than the estimated average deviation, $\bar{d}$, because it gives greater weight, as required, to large deviations. Especially when the number of measurements is small (say less than 5), a better estimate of the standard deviation is obtained by replacing N by $N - 1$ in the denominator of (6):

$$s = \left\{\frac{\sum (X - \bar{X})^2}{N - 1}\right\}^{1/2} \tag{7}$$

The use of $N - 1$ instead of N compensates for the difference between $\bar{X}$ and μ. The divisor $N - 1$ is known as the *number of degrees of freedom* available for calculating s; one degree of freedom has been used in obtaining the mean, leaving $N - 1$ independent deviations from the mean. s^2 is the *estimated variance.*

The relative standard deviation in parts per hundred, $100s/\bar{X}$ is known as the *coefficient of variation.* In analytical practice, however, the relative standard deviation is more usually expressed in parts per thousand.

The standard deviation can be estimated from related groups of measurements as well as from a single series. For example, the precision of a procedure, expressed as the standard deviation, can be obtained from the results of duplicate determinations of a constituent in different samples of about the same composition. To obtain the standard deviation from a total of N measurements in g groups, sum the squares of the deviations in each group, add the sums, divide by $N - g$, and extract the square root of the result:

$$s = \left\{\frac{1}{N - g}\sum_g \sum_i (X_i - \bar{X}_i)^2\right\}^{1/2} \tag{8}$$

The number of degrees of freedom is $N - g$, because one degree of freedom is lost in each group.

Precision of the mean. It can be shown that the arithmetical mean derived from N measurements is $N^{1/2}$ times as reliable as a single measurement. Accordingly:

$$\text{Estimated average deviation of the mean} = \bar{d}_{\bar{X}} = \frac{\bar{d}}{N^{1/2}} \tag{9a}$$

$$\text{Estimated standard deviation of the mean} = s_{\bar{X}} = \frac{s}{N^{1/2}} \tag{9b}$$

Example 1: Five determinations of iron in an iron ore by a volumetric method yielded the percentages 67.48, 67.37, 67.47, 67.43, and 67.40. Find the average deviation and standard deviation of a single determination and of the mean.

Percentage	$d = X - \bar{X}$	$(X - \bar{X})^2$
67.48	0.05	0.0025
67.37	0.06	0.0036
67.47	0.04	0.0016
67.43	0.00	0.0000
67.40	0.03	0.0009
Mean 67.43	$\sum$ 0.18	$\sum$ 0.0086

$$\bar{d} = \frac{0.18}{5} = 0.036 \text{ per cent absolute or 0.5 ppt}$$

$$s = \left(\frac{0.0086}{4}\right)^{1/2} = 0.047 \text{ per cent absolute or 0.7 ppt}$$

$$\bar{d}_{\bar{X}} = \frac{0.036}{(5)^{1/2}} = 0.016 \text{ per cent absolute or } 0.24 \text{ ppt}$$

$$s_{\bar{X}} = \frac{0.047}{(5)^{1/2}} = 0.021 \text{ per cent absolute or } 0.31 \text{ ppt}$$

(The deviations of a single measurement and of the mean expressed in parts per thousand should usually be rounded off to the first decimal unless the values are to be used in calculations.)

Precision may be expressed by giving the standard deviation of a single measurement. Thus in the above example the precision of the analyses is described by stating that the standard deviation of a determination is 0.7 ppt.

The accuracy of a result can be increased as far as random errors are concerned by increasing the number of measurements. However, a point is soon reached when the slight increase in precision thus attained does not justify the expenditure of the time and labor required (Fig. 16-5). To increase the precision threefold, almost 10 times as much work must be done on the average.

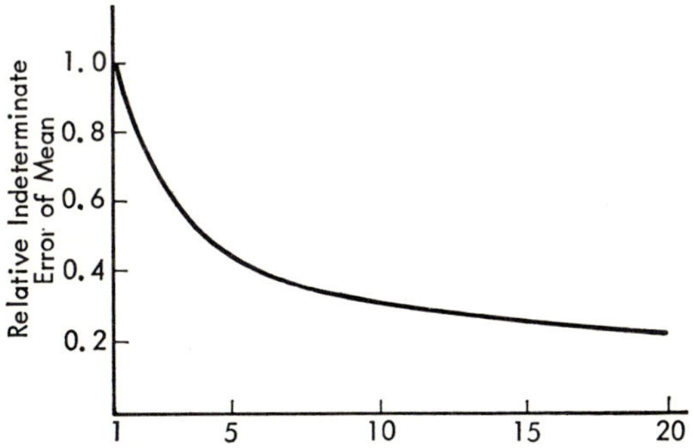

Fig. 16-5. Precision of the mean as a function of the number of measurements

Probability of deviations. If a sufficiently large number of measurements has been made, so that the population mean and standard deviation are known fairly exactly, the probability of the occurrence of a deviation lying outside the limits $\pm x = (X - \mu)/\sigma$ can be found by utilizing a table such as Table 16-1. A table of this kind can be computed with the aid of Eq. (1b) (p. 387). The probability of $\pm x$ being exceeded is given by the ratio of the area under the error curve from $-\infty$ to $-x$ and from $+x$ to $+\infty$ to the total area under the curve, which represents a probability of unity (= certainty) (Fig. 16-4). We see from Table 16-1 that the probability that the deviation of a single measurement will be greater than one half the standard deviation is 0.62; that it will be greater than the standard deviation, 0.32; that it will be greater than twice the standard deviation, 0.046; and so on. We may be 95 per cent confident, if the normal frequency distribution is followed and the system and conditions of measurement remain invariable, that the deviation of any single measurement will not exceed twice the standard deviation. In other words, if a single measurement X has been made, we may be 95 per cent confident that the population mean lies in the range

$$\mu = X \pm 2.0\sigma$$

and 99 per cent confident that it lies in the range

$$\mu = X \pm 2.58\sigma$$

If N measurements have been made,

$$\mu = \bar{X} \pm \frac{2.0\sigma}{N^{1/2}} \quad \text{for 95 per cent confidence} \tag{10}$$

TABLE 16-1. **Probability of Occurrence of Deviations in Terms of Standard Deviation, Based on Normal Frequency Distribution**

$x = (X - \mu)/\sigma$	*Probability of a Deviation Numerically* (±) *Greater than* x
0.00	1.00
0.10	0.92
0.20	0.84
0.30	0.76
0.40	0.69
0.50	0.62
0.60	0.55
0.70	0.48
0.80	0.42
0.90	0.37
1.0	0.32
1.5	0.13
2.0	0.046
2.5	0.012
3.0	0.0027
4.0	0.00006
5.0	0.0000006

In analytical practice, considerations of probability and confidence encounter the difficulty that the population mean and standard deviation usually are not known at all well because of the small number of measurements (determinations) that have been made. If $\bar{X}$ and s obtained from a small number of measurements are used as estimates of μ and σ, calculations of probability based on Table 16-1 will give inaccurate (too favorable) values. The probability of a deviation being greater than a specified value will be underestimated because the estimated value of σ will be too small. This difficulty may be remedied by taking into account the effect of the number of measurements. This can be done by the use of a quantity t, which is defined as

$$\pm t = (\bar{X} - \mu)\frac{N^{1/2}}{s} \tag{11}$$

(The quantity t is the analogue of x used for a large number of observations.) Tables are available in which values of t are tabulated for given probabilities (or confidence levels) at specified degrees of freedom. A condensed table of this kind is Table 16-2. The t test (examples below) takes into account both the variation of $\bar{X}$ from μ and s from σ.[9]

From Eq. (11) the confidence interval is

$$\mu = \bar{X} \pm \frac{ts}{N^{1/2}} \tag{12}$$

The normal frequency distribution is of course assumed to be followed. If $N = 1$, $\bar{X}$ becomes a single observation and the foregoing expression can be applied if s is

[9] The t test was devised by W. S. Gosset, writing under the pseudonym "Student" [*Biometrika* **6**, 1 (1908)]. For the derivation of the t distribution, see E. B. Wilson, *An Introduction to Scientific Research*, McGraw-Hill, New York, 1952, p. 274.

TABLE 16-2. **Values of *t* Corresponding to Given Probabilities**

Degrees of Freedom, n	*Probability of Numerical Deviation (±) Greater than t* 0.005 (99.5%)[a]	0.01 (99%)	0.05 (95%)	0.10 (90%)	0.50 (50%)	0.90 (10%)
1	127.3	63.66	12.71	6.31	1.00	0.16
2	14.1	9.93	4.30	2.92	0.82	0.14
3	7.45	5.84	3.18	2.35	0.77	0.14
4	5.60	4.60	2.78	2.13	0.74	0.13
5	4.77	4.03	2.57	2.02	0.73	0.13
6	4.32	3.71	2.45	1.94	0.72	0.13
7	4.03	3.50	2.37	1.90	0.71	0.13
8	3.83	3.36	2.31	1.86	0.71	0.13
9	3.69	3.25	2.26	1.83	0.70	0.13
10	3.58	3.17	2.23	1.81	0.70	0.13
15	3.25	2.95	2.13	1.75	0.69	0.13
20	3.15	2.85	2.09	1.73	0.69	0.13
25	3.08	2.79	2.06	1.71	0.68	0.13
∞	2.81	2.58	1.96	1.645	0.67	

[a] Confidence levels.

known from a series of other measurements of the same kind, as sometimes happens.[10] It would be risky to accept a confidence interval based on a single X, because the normal distribution may not be followed. If, however, the confidence interval is based on the mean of a number of X's, more faith may be placed on the result because the distribution of the means will tend to approach normality even though the individual values are not normally distributed.

The t test enables us to decide whether the difference between two means is statistically significant. The means to be compared might be the means of determinations of a constituent by two or more different methods, the means of determinations of a constituent by two analysts using the same method on the same sample, etc.

Another test, the F test, is used to decide whether two *variances* are significantly different. It serves to show whether the precision of two different methods is the same within specified probability limits. The ratio of the two variances in question ($s_1^2 > s_2^2$) is the variance ratio F:

$$F = \frac{s_1^2}{s_2^2} \tag{13}$$

If experimental F is smaller than the corresponding tabular value, no statistical difference between s_1 and s_2 is indicated; if experimental F is larger, s_1 is significantly greater than s_2. Tables are available that give values of F for one or more significance levels for various degrees of freedom n_1 and n_2. Table 16-3 is a brief F table (two-sided) for the 95 per cent confidence level. An example of its use is given later.

In attempting to decide whether the occurrence of a deviation of a given value is likely the result of chance, what shall be considered significant probability limits? Usually a probability of $\gtreqless 0.05$ is considered sufficiently small to be taken as significant;

[10] For example, an analyst determining iron volumetrically in ores of much the same composition may know the precision of the method in his hands. He might have available a series of duplicates, from which the standard deviation could be estimated (see Problem 11).

TABLE **16-3. Values of F at 95 Per Cent Confidence Level**[a]

n_2	n_1[b]									
	2	3	4	5	6	7	8	9	10	∞
2	19.00	19.16	19.25	19.30	19.33	19.36	19.37	19.38	19.39	19.50
3	9.55	9.28	9.12	9.01	8.94	8.88	8.84	8.81	8.78	8.53
4	6.94	6.59	6.39	6.26	6.16	6.09	6.04	6.00	5.96	5.63
5	5.79	5.41	5.19	5.05	4.95	4.88	4.82	4.78	4.74	4.36
6	5.14	4.76	4.53	4.39	4.28	4.21	4.15	4.10	4.06	3.67
7	4.74	4.35	4.12	3.97	3.87	3.79	3.73	3.68	3.63	3.23
8	4.46	4.07	3.84	3.69	3.58	3.50	3.44	3.39	3.34	2.93
9	4.26	3.86	3.63	3.48	3.37	3.29	3.23	3.18	3.13	2.71
10	4.10	3.71	3.48	3.33	3.22	3.14	3.07	3.02	2.97	2.54
∞	3.00	2.60	2.37	2.21	2.10	2.01	1.94	1.88	1.83	1.00

[a] Values have 0.05 probability of being exceeded.
[b] n_1, number of degrees of freedom for larger variance.

i.e., a deviation of this probability will normally, as a result of chance, occur less than 5 per cent of the time, corresponding to a confidence level of 95 per cent that it will not be exceeded. A probability of $\lesseqgtr 0.01$ is considered highly significant. The confidence level adopted in any particular case will depend to a large extent on the penalty that must be paid if the wrong conclusion is drawn and a wrong decision is made. If it is important that the result of an analysis be as free as reasonably possible of indeterminate error, a confidence level of 99.9 per cent might be made the basis of arranging the experimental conditions.

The following examples illustrate some applications of probability considerations in chemical analysis.[11]

Example 2: The following zero points (each determined in the same way) were obtained in testing the reproducibility of an analytical balance (sensitivity 3.55 divisions per milligram) at zero load:[12]

1.40	1.43	1.53	1.35
1.46	1.56	1.43	1.43
1.43	1.53	1.44	

(a) What is the probability that a single zero point will show a deviation as large as ± 0.05 mg if normal frequency distribution of errors holds?

(b) What is the probability that the mean of the zero points of the series lies within 0.02 mg of the true mean?

(c) How many zero points must be taken with this balance to enable one to say that their mean lies within 0.02 mg of the true mean at the 99 per cent confidence level? (As before, s has been determined from 11 rest points.)

[11] For an interesting discussion of pitfalls in applying statistical methods to chemical analyses, see W. G. Schlecht, *U.S. Geol. Surv. Bull.* **992**, 1953, p. 57.

[12] A series of readings such as these should be examined for trends or jumps, which would indicate that the system is not under control, i.e., that disturbing factors have entered and deviations are not random. There is no trend in this series (Fig. 16-6). If a consistent drift to higher or lower values were shown, a change in temperature, which is likely to affect the rest point of a balance, would be suspected. All the values would then be rejected and a new series obtained with the system under control, i.e., at constant temperature.

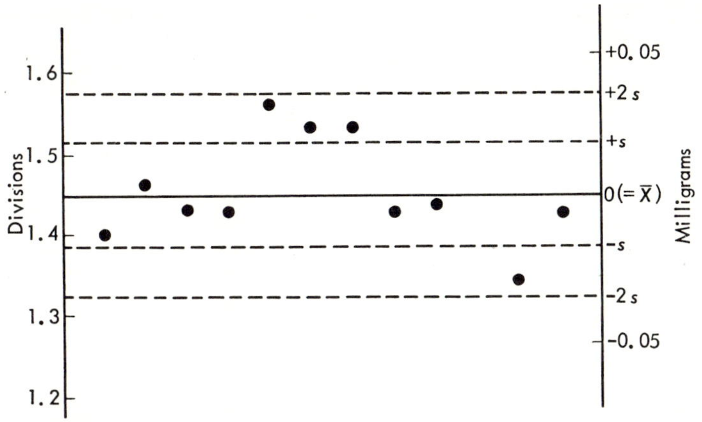

Fig. 16-6. Sequential plot of zero points of an analytical balance. s and $2s$ of a single zero point are indicated. If $\pm 2\sigma$, taken as control limits (p. 402), were indicated (instead of $\pm 2s$) the plot becomes a control chart.

Ans. (a) It will be instructive to find the probabilities by (1) taking s as an estimate of σ and utilizing Table 16-1, (2) using the t test:

(1) The estimated standard deviation of a zero point is $\sigma \sim s = (392 \times 10^{-4}/10)^{1/2} =$ 0.063 division or 0.018 mg. The ratio of the specified deviation to the estimated standard deviation of a zero point is

$$\pm x = \frac{X - \mu}{\sigma} \sim \frac{X - \overline{X}}{s} = \frac{0.05}{0.018} = 2.8$$

From Table 16-1,

$$\text{probability } (|x| > 2.8) < 0.01 > 0.003$$

(2) $$\pm t = \frac{(\overline{X} - \mu)N^{1/2}}{s} = \frac{0.05 \times 1}{0.018} = 2.8$$

From Table 16-2, opposite $n = 10$, we find that the probability of $t > \pm 2.8$ lies between 0.05 and 0.01. This probability is a little less favorable than that found in (1), but both results show that a deviation of 0.05 mg is unlikely as long as the environmental conditions of the weighing remain constant.

(b) Again using the two methods:

(1) $\pm x = 0.02(11)^{1/2}/0.018 = 3.7$, and, from Table 16-1,

$$\text{probability } (|x| > 3.7) < 0.003, \; > 6 \times 10^{-5}$$

(2) $\pm t = 3.7$.

From Table 16-2, opposite $n = 10$, the probability corresponding to $|t| > 3.7$ is a little less than 0.005 ($t = \pm 3.58$ at probability 0.005 for $n = 10$). Accordingly, it is very unlikely that the mean of 11 rest points deviates by more than 0.02 mg from the true mean. The result in (1) gives rise to too-optimistic expectations.

(c) $$N^{1/2} = \frac{ts}{\overline{X} - \mu} = \frac{0.018t}{0.02} = 0.9t$$

From Table 16-2, at $n = 10$, $t = 3.17$ for the 99 per cent confidence level. Therefore,

$$N^{1/2} = 0.9 \times 3.17 = 2.85$$
$$N = 8$$

Example 3: Is there a significant difference in the precision of the determination of total iron in a silicate rock by methods A and B? [In A, Fe(III) is reduced with H_2S, in B with metallic silver; Fe(II) is then titrated with dichromate in both.]

Method A: Fe_2O_3, %	*Method* B: Fe_2O_3, %
2.01	1.88
2.10	1.92
1.86	1.90
1.92	1.97
1.94	1.94
1.99	Av. 1.92
Av. 1.97	

.056
.002

Ans. Find the variances of the two methods and apply the F test.

$$s_A^2 = \frac{0.0344}{5} = 0.0069$$

$$s_B^2 = \frac{0.0049}{4} = 0.0012$$

$$F = \frac{s_A^2}{s_B^2} = \frac{0.0069}{0.0012} = 5.8$$

In Table 16-3, $F = 6.26$ at $n_A = 5$ and $n_B = 4$. Therefore, method A has not been shown to be less precise than method B at the 95 per cent significance level. However, the experimental F is so close to 6.26 that we may *suspect* method A to be actually less precise than method B; a greater number of determinations will be needed to test this belief. The 95 per cent significance level is, after all, a somewhat arbitrary dividing line. At the 90 per cent significance level, tabulated $F = 4.05$ (not in Table 16-3). On the basis of the present data, an analyst would be justified in selecting method B in preference to method A as more likely to give greater precision in future work.

Example 4: Is there a significant difference between the means of methods A and B in Example 3?

Ans. The question to be answered is whether the random deviations in the determination of total iron by the two methods are of such magnitude that they may account for the difference between the means; alternatively, the difference between the means might be due to the existence of determinate errors in either or both methods. If the precisions of the two methods are substantially the same, the probability of a specified difference between the means can be found in a simple way. We obtain the value of

$$t = \frac{\bar{X}_A - \bar{X}_B}{s_{\bar{X}}}$$ ($s_{\bar{X}}$ = estimated standard deviation of the difference between the means)

and compare with the appropriate value of t in Table 16-2 at, say, the 95 per cent significance level. If experimental t is the larger, we conclude that there is a significant difference between the means, on the basis that by chance an experimental t as large or larger will be obtained only once in 20 times.

We proceed as follows, first finding s, the standard deviation of a single determination in the two methods:

$$s = \left\{\frac{\sum (X_A - \bar{X}_A)^2 + \sum (X_B - \bar{X}_B)^2}{N_A + N_B - 2}\right\}^{1/2}$$ [from Eq. (8), p. 390]

$$= \left\{\frac{0.0344 + 0.0049}{9}\right\}^{1/2} = 0.00437^{1/2} = 0.066$$

The standard deviation of the difference of the means is given by

$$s_{\bar{X}} = s\left\{\frac{1}{N_A} + \frac{1}{N_B}\right\}^{1/2} = s\left\{\frac{N_A + N_B}{N_A N_B}\right\}^{1/2} \qquad \text{(see Table 16-4)}$$

$$= 0.066\left(\frac{11}{30}\right)^{1/2} = 0.066 \times 0.61 = 0.040$$

$$t = \frac{0.05}{0.04} = 1.25$$

In Table 16-2, at $n = 9$, we find that the probability ($|t| > 1.25$) lies between 0.10 and 0.5. We conclude that the difference between the means of the results obtained by the two methods is not greater than that attributable to chance at the 0.05 probability level.

As already pointed out, the foregoing treatment is based on the assumption that the precisions of the two methods are essentially the same. This assumption is barely supported by the F test at the 95 per cent confidence level. The conclusion that the means are not significantly different may be tested further by applying the t test with the smaller standard deviation, s_B:

$$t = \frac{0.05}{0.0012^{1/2}}\left(\frac{30}{11}\right)^{1/2} = 1.44 \times 1.65 = 2.4$$

$$\text{Probability } (|t_{n=9}| > 2.4) \sim 0.05$$

Even if method A is as precise as method B, there is a barely significant difference between the means at the 95 per cent confidence level. It must be concluded that the difference of 0.05 per cent between the means is not clearly statistically significant.

Rejection of a measurement.[13] Occasionally an apparently discrepant value will appear in a series of measurements. The observer may be unable to assign a cause for such a large deviation, but he may question it on the basis of experience or intuition. He may feel that it is unlikely to have occurred by chance alone, and will be inclined to attribute it to the action of some erratic or new factor on the system. For example, a mistake may have been made in reading a buret or recording a weight, a sample may not be entirely homogeneous, a transient voltage fluctuation may have occurred in a photometric absorbance measurement, etc. It may be granted that in a well-regulated analysis such mistakes and errors should seldom occur, but it must be recognized that they are always a possibility, particularly with less experienced analysts. Although experience is a valuable general guide, an apparently aberrant result cannot justifiably be omitted in the calculation of the average of a series unless it meets some more or less reasonable statistical criterion for rejection. The question of rejection is of special importance when the number of measurements is small, say 4 or 5, and the standard deviation as well as the theoretical mean must be estimated from this small number.

Consider a series of four values, one of which differs markedly from the rest. If its deviation is greater than approximately 4 (more precisely > 4.3) times the standard deviation calculated from the three replicates, the probability of its occurrence is < 0.05 if the normal frequency distribution is followed, since

$$X - \mu = (ts)_{n=2} \leq 4.3s_{n=2} \text{ at the 95 per cent confidence level}$$
[Eq. (11) and Table 16-2]

[13] See also W. J. Blaedel, V. W. Meloche, and J. A. Ramsay, *J. Chem. Educ.* **28**, 643 (1951). They prefer a rejection test proposed by R. B. Dean and W. J. Dixon, *Anal. Chem.* **23**, 636 (1951), in a paper on simplified statistics for small numbers of observations. To avoid deciding whether an apparently discrepant value should be rejected, it may be desirable, as pointed out by the latter authors, to use the median of a series in place of the average. The median is the middle value or the average of the two values nearest the middle of a series. It will be much less affected by a gross error than will the mean.

In the long run, 1 value in 20 will exceed the deviation of $4.3s$ as a result of chance, or one series in five, each consisting of four values, will have a value whose deviation is greater than this. The $4s$ criterion thus supplies a sufficiently liberal basis for the rejection of supposedly aberrant values. This test may be criticized for allowing the rejection of large valid deviations more often than justified, thus leading to an average of decreased accuracy. If aberrant values never occur, we err once in five times (one series in five) in rejecting a result by the $4s$ criterion, without obtaining any possible compensatory advantage. If aberrant results do in fact occasionally occur, the error incurred by rejecting a legitimate deviation is compensated to an uncertain extent by exclusion of a possibly aberrant value. The question is whether we are wise to run the risk of making a wrong decision 20 per cent of the time to exclude a possibly aberrant value. No definite answer can be given as long as the probability of aberrant values is unknown. The easiest solution is to deny the possibility of aberrant values and never reject any result, i.e., to assume that all deviations are always normally distributed. This may not be a bad general rule, but it is subject to possible exceptions. For example, if three replicates are run one day and a fourth on the following day and the deviation of the fourth value is greater than $4s$ of the first three, there might be good reason to suspect the fourth value. Or if four samples are titrated visually in, say, 30 min, and the first result has a deviation greater than $4s$, we may suspect that the first titration is less certain because of unfamiliarity of the analyst with the end point, or possibly for some other reason.

In a series of five values, one being suspect,

$$X - \mu = (ts)_{n=3} \leqq 3.2s_{n=3} \qquad \text{at the 95 per cent confidence level}$$

Now one group in four, each having five values, will have a value whose deviation is greater than $3.2s$. We may as well retain the $4s$ criterion for a five-value series (probability of a $4s$ deviation ~ 0.025, or one series in eight will normally contain a deviation greater than this).

The $4s$ criterion for four- and five-membered series (and analogous criteria can be easily derived for other series) may be useful to the analyst on a long-run basis but obviously does not carry much weight in any particular case. If it is important that the mean of a series be as accurate as possible, additional determinations should be run in a series in which a supposedly discrepant result occurs. The appearance of a large deviation gives rise to the suspicion that some new factor has entered or some old factor has assumed greater importance in determining random errors. Especially in analyses extending over hours or days, environmental factors may change and cause trends or jumps in a series of values. Before additional determinations are run, the conditions and circumstances of the analyses should be reviewed in an attempt to locate possible causes of discrepant results.

Example 5: Four determinations of chlorine in a sample of chloride yielded the values 30.34, 30.21, 30.42, and 30.38 per cent Cl. Should any of the results be rejected by the $4s$ criterion? Compare the averages with and without the suspected value with the median (see Ref. 13).

Ans. Evidently 30.21 is the doubtful value. The mean of the three most concordant determinations is 30.38, the standard deviation is 0.04, and the deviation of the suspected value is 0.17 per cent. Since $0.17/0.04 > 4$, omit 30.21 per cent in averaging, if the $4s$ test is accepted.

The median of the series is 30.36, compared to an average of 30.34 per cent of all four values. The median could well be taken to represent the average of the series; it represents a compromise between retaining and rejecting the questioned value.

Error in a calculated result. The final result of an analysis is almost always derived from the measurement of more than one quantity, e.g., the determination of the weight percentage of a constituent requires (1) weighing the sample and (2) obtaining the weight of the weighing form in gravimetric analysis, the volume of standard solution in volumetric analysis, etc. How will the errors in the component quantities combine to give the error in the calculated result? Table 16-4 supplies the answer to this important question. The results are self-evident if it is borne in mind that uncorrelated variances are additive.

TABLE **16-4. Error of a Calculated Result**[a]

R = calculated result
A, B, C = measured quantities from which R is derived
$\epsilon_R, \epsilon_A, \epsilon_B, \epsilon_C$ = absolute determinate errors of R, A, B, C
s_R, s_A, s_B, s_C = absolute standard errors (deviations) of R, A, B, C

	R calculated as sum or difference ($R = A + B - C$)	R calculated as product or quotient ($R = AB/C$)
Determinate errors	$\epsilon_R = \epsilon_A + \epsilon_B - \epsilon_C$	$\frac{\epsilon_R}{R} = \frac{\epsilon_A}{A} + \frac{\epsilon_B}{B} - \frac{\epsilon_C}{C}$
Indeterminate errors	$s_R = \pm \{s_A^2 + s_B^2 + s_C^2\}^{1/2}$	$\frac{s_R}{R} = \left\{\left(\frac{s_A}{A}\right)^2 + \left(\frac{s_B}{B}\right)^2 + \left(\frac{s_C}{C}\right)^2\right\}^{1/2}$

[a] For the general case $R = f(A, B, C, \ldots)$, we have for determinate errors,

$$\epsilon_R = \frac{\partial R}{\partial A}\epsilon_A + \frac{\partial R}{\partial B}\epsilon_B + \cdots$$

and for indeterminate errors

$$s_R = \left\{\left(\frac{\partial R}{\partial A}\right)^2 s_A^2 + \left(\frac{\partial R}{\partial B}\right)^2 s_B^2 + \cdots\right\}^{1/2}$$

These expressions are strictly valid only for linear functions of $A, B, C, \ldots$ but hold approximately otherwise if the errors are small compared to $A, B, C, \ldots$. $A, B, C, \ldots$ must be independent variables.

Example 6: What is the estimated standard deviation in a weighing with the balance of Example 2 if the load is such that the precision can be taken as essentially constant? Neglecting determinate errors, what is the probability that the result of a (single) weighing is not in error by more than 0.05 mg?

Ans. Two rest points (zero point and rest point with load) must be found, so that

$$s = (0.018^2 + 0.018^2)^{1/2} = 0.025 \text{ mg}$$

The probability that the indeterminate error does not exceed 0.05 mg lies between 0.9 and 0.95 ($t = 0.05/0.025 = 2$, $n = 10$).

Example 7: If the standard deviation of a rest point of a balance is 0.0001 g, what is the standard deviation (ppt) in the determination of chlorine in a 0.20-g sample of chloride yielding 0.25 g of silver chloride?

Ans. Two rest points must be obtained in weighing the sample, and four in weighing the silver chloride precipitate. Accordingly, the standard deviation in the chloride content, expressed in parts per thousand is

$$\left\{2\left(\frac{0.0001 \times 1000}{0.2}\right)^2 + 4\left(\frac{0.0001 \times 1000}{0.25}\right)^2\right\}^{1/2}$$

$$= \{2 \times 0.5^2 + 4 \times 0.4^2\}^{1/2} = 1.14^{1/2} = 1.1 \text{ ppt, or } 1 \text{ ppt}$$

If it is desired to express the standard deviation in terms of the absolute percentage of chlorine, we have

$$s = \frac{1.1}{1000} \times \frac{\text{Cl}}{\text{AgCl}} \times \frac{0.25}{0.20} \times 100 = \pm 0.03 \text{ per cent}$$

Correlated errors (deviations). In the preceding section the random deviations affecting the final result R were taken to be independent of each other, i.e., uncorrelated, so that if R is obtained as the sum of A and B, or their difference,

$$s_R^2 = s_A^2 + s_B^2$$

which is represented by (*a*) in Fig. 16-7. If the deviations are not independent of each other,[14] the following relation holds:

$$s_R^2 = s_{A \pm B}^2 = s_A^2 \pm 2r_{AB}s_As_B + s_B^2$$

in which r_{AB} is the correlation coefficient for the deviations of A and B. The + sign of the middle term means that s_A and s_B vary in the same direction, the − sign that they vary oppositely (one increases while the other decreases). The correlation coefficient is a measure of the linear relation between s_A and s_B, and its value can vary from $+1$ to -1. When $r = 0$, there is no correlation; when $r = \pm 1$ correlation is complete. Geometrically $r_{AB} = \pm\cos\alpha$ (Fig. 16-7*b*). When $r = \pm 1$ ($\alpha = 0$), the errors and not the variances become algebraically additive (transition to determinate errors).

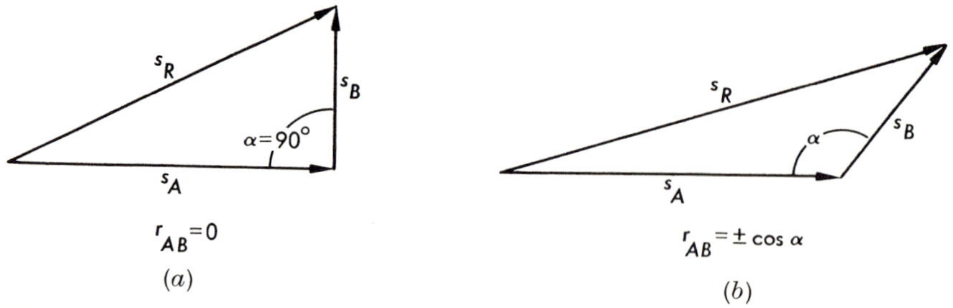

Fig. 16-7. Geometrical representation of additivity of random correlated errors (*a*), and uncorrelated errors (*b*). Slightly modified from Nalimov, *The Application of Mathematical Statistics to Chemical Analysis*, p. 31.

Sample size. A problem of some importance in analytical work concerns the size of sample that must be taken in order that the relative error in the final result arising from indeterminate errors of known magnitude shall not exceed specified limits at given probability levels. By increasing the size of the sample we decrease the relative errors due to indeterminate *absolute* errors in weighing the sample and the weighing form in gravimetric analysis. For reasons of convenience, expediency, or necessity (especially in microanalysis), the amount of sample taken should be no larger than really required, however.

First, let us find the weight of sample which will give a specified *standard deviation* in the percentage of a constituent. The percentage, P, of a constituent is given by the general expression

$$P = \frac{100FM}{U} \tag{14}$$

where F = chemical factor

M = *measure* of the constituent (weight of the weighing form in gravimetric analysis, volume of the standard solution in a titrimetric analysis, etc.)

U = weight of sample

Designate the absolute standard deviations of P, M, and U by s_P, s_M, and s_U and the standard deviation of a rest point of a balance (expressed as a weight) by w. The relative

[14] For example, deviations in buret readings are likely to be correlated (Problem 38).

standard deviation of P in terms of the deviations of the quantities from which it is derived is given by

$$\frac{s_P}{P} = \pm\left\{\left(\frac{s_M}{M}\right)^2 + \left(\frac{s_U}{U}\right)^2\right\}^{1/2}$$

$$\frac{1000s_P}{P} = \pm\left\{\left(\frac{1000s_M}{M}\right)^2 + \left(\frac{1000s_U}{U}\right)^2\right\}^{1/2} \quad \text{in ppt} \qquad (15)$$

(In the following equations we shall represent $1000s_P/P$ by p_r, signifying the relative standard deviation in P in parts per thousand for the sample weight U.) This expression is a general one that can be applied in gravimetric, titrimetric, photometric, and other types of analyses. In gravimetric precipitation analysis it becomes

$$p_r = \pm\left\{4\left(\frac{1000w}{M}\right)^2 + 2\left(\frac{1000w}{U}\right)^2\right\}^{1/2} \quad \text{in ppt} \qquad (16)$$

[In finding the weight of a precipitate, four rest points are determined: empty balance, crucible, empty balance, and crucible with precipitate, whence the term $4(1000w/M)^2$. Only two rest points are involved in weighing the sample.]

Substitution of $UP/100F$ for M in Eq. (16) gives

$$U = \frac{1.4 \times 10^3 w}{p_r}\left\{\left(\frac{1.4 \times 10^2 F}{P}\right)^2 + 1\right\}^{1/2} \qquad (17)$$

If $1.4 \times 10^2 F/P \lesseqgtr 0.5$, it may usually be disregarded in the preceding expression, and if $1.4 \times 10^2 F/P > 2$, the 1 in the square-root term may be disregarded.[15] U, then, is the weight of sample, which in the long run gives a specified relative standard deviation of $\pm 1000s_P/P$ (in ppt) in the percentage of constituent if the standard deviation of a rest point is w. Generally, however, the analyst desires to know the size of sample that must be taken in order that a specified deviation $\pm p_r$ shall not be exceeded at specified probability levels. If the size of sample is increased, the relative standard deviation of P resulting from weighing error decreases, since w remains the same. Suppose the weight of sample is doubled over that giving a specified relative standard deviation p_r, i.e., increased from U_{p_r} to $2(U_{p_r})$. The relative standard deviation now becomes $(p_r/2)$ ppt. The probability that this relative deviation will not be exceeded can be found with the aid of the expression

$$\pm t_9 = (\bar{X} - \mu)\frac{N^{1/2}}{s_9}$$ (For illustration, s is assumed based on 10 replicates of w)

or $$\pm t_9 = 2 \quad \text{since } \frac{X - \mu}{s_9} = 2 \quad (N = 1)$$

The probability that $\pm t_9 > 2$ is approximately 0.08; i.e., the probability that p_r will not be exceeded with a sample weighing $2U_{p_r}$ is about 0.92. Similarly, we find that to obtain the probability 0.99 that p_r will not be exceeded, a sample weighing $3.25U_{p_r}$ is required. This probability may be taken as furnishing reasonable assurance that the specified deviation p_r will not be exceeded in a single determination. The following approximate expression may then be written for the minimum weight of sample, $U_{0.99}$, to be taken in a gravimetric determination so that the relative deviation, in ppt, of the percentage of constituent shall not exceed a specified value p_r 99 times in 100, if the normal frequency distribution is followed:

$$U_{0.99} = \frac{5000w}{p_r}\left\{\left(\frac{140F}{P}\right)^2 + 1\right\}^{1/2} \quad \text{or} \quad \frac{5000w}{p_r}\left\{\frac{20{,}000F^2}{P^2} + 1\right\}^{1/2} \qquad (18)$$

[15] If R is derived from two measurements, A and B, and the relative indeterminate error of B is 0.5 or less that of A, one may as well forget about the error contributed by B, because the indeterminate error in R will at the worst be only 12 per cent greater because of the error in B: $(1 + 0.5^2)^{1/2} = 1.12$, compared to 1.

The figure 5000 has been obtained by rounding $3.25 \times 1400 = 4550$ (more exactly $3.25 \times 1414 = 4596$). The standard deviation of a rest point, w, is taken to be derived from 10 replicates.

Example 8: What minimum weight of $CaSO_4 \cdot 2\,H_2O$ (20.93 per cent H_2O) must be taken for the gravimetric determination of water by loss of weight on ignition to obtain a value lying in the range 20.93 ± 0.10 per cent with a probability of 0.99 if w (from 10 replicates) $= 0.010$ mg?

Ans. Since $p_r = (0.10 \times 1000)/21 = 5$ ppt and $F = 1$,

$$U_{0.99} = \frac{5000 \times 0.010}{5}\left\{\left(\frac{140}{21}\right)^2 + 1\right\}^{1/2} = 10 \times 7 = 70 \text{ mg}$$

It will be realized that random weighing errors may not be the only errors affecting the reproducibility of gravimetric results (or results of other methods). It is quite possible that other variables dependent on sample size or the quantity of the weighing form may obscure the effect of weighing error.

Control charts

An analytical chemical control chart is a plot of the measured values of a quantity (such as the percentage of a constituent or the concentration of a solution) made for the purpose of learning whether the deviations of the measurements are random and whether they fall within expected limits. Thus a control chart might consist of the results of a number of analysts analyzing the same sample, or the series of results obtained by a single analyst analyzing the same sample over a period of time. Suppose a constituent A is being determined in a laboratory day after day in samples of more or less the same composition. It would be wise to intersperse one or more standard samples (not marked as such) from time to time among the samples being analyzed for A. The analytical values obtained on the standard are then plotted in time sequence, the results of different analysts, if more than one, being differentiated. The content of A in the standard sample is indicated by a horizontal line. Usually the precision of the determination is known under the conditions, so that lines representing a deviation of 2σ, for example, can be drawn on both sides of the central line representing the true value and accepted as the control limits. The probability of a value falling outside the limit $\pm 2\sigma$ is 1 in 20, so that a tendency for a greater scatter would indicate that the precision is not under control. More important, a tendency for the values to be greater or less than the accepted value would indicate the presence of one or more determinate errors in the determination. Errors arising from deterioration of reagents, malfunctioning of instruments, environmental factors, etc., would not go undetected.

The plot in Fig. 16-6 could be converted into a simple control chart by drawing lines for $\pm 2\sigma$ ($\sim \pm 2s$).

Determination of the precision and accuracy of a method

Analytical methods differ considerably in accuracy and precision. The precision of a method is more easily determined than its accuracy. In a pure solution of a constituent both may be determined with little difficulty. The accuracy of a given method depends to a great extent upon the nature of the sample. In practical work the sample may be complex, separations may be required, and the accuracy of the final result may depend more upon the preliminary steps than on the final determination.

The precision and accuracy of a method are established by taking known amounts of the constituent in the form of a carefully purified element or compound of definite composition. Since atomic weights are known with greater accuracy than any constituent can be determined by ordinary analysis, the true amounts of the constituent present may be considered known. A series of determinations—from five to ten—are then run under specified conditions. The arithmetical mean of the results and the standard deviation are calculated. The standard deviation of a single determination may be used as a measure of the precision of the method. The difference between the mean and the theoretical value gives the determinate error of the method.

Example 9: Benedetti-Pichler[16] and coworkers investigated the gravimetric microdetermination of aluminum with 8-hydroxyquinoline by determining the element in from 1.7 to 6.5 mg of $KAl(SO_4)_2 \cdot 12\,H_2O$. The following percentages of Al_2O_3 were found in nine determinations:

10.74	10.77	10.77
10.77	10.81	10.82
10.73	10.86	10.81

$\bar{X} = 10.79$ per cent Al_2O_3 (from nine determinations)
$s = 0.042$ per cent absolute or 3.9 ppt[17]
$s_{\bar{X}} = 0.014$ per cent, or 1.3 ppt
$X_\tau = 10.77$ per cent Al_2O_3
$\bar{\epsilon} = \bar{X} - X_\tau = 10.79 - 10.77 = +0.02$ per cent Al_2O_3 absolute, or $+2$ ppt

The apparent determinate error of the method thus amounts to 0.02 per cent absolute, or about 2 ppt Al. Is this deviation of the mean from the theoretical value significant? Apply the t test:

$$t_8 = \frac{0.02}{0.014} = 1.4$$

The probability that $t_8 > 1.4$ (note that we are not concerned with the negative value) lies between 0.05 and 0.25. The existence of a determinate error in the procedure has not been demonstrated at the 95 per cent confidence level. The difference between the mean and the true value could be due to indeterminate errors in the procedure, although we may *suspect* there is a tendency for the procedure to give very slightly high results.

It may be expected that determinations made in the future by this procedure *under the same conditions* would rarely show a deviation as large as ± 12 ppt, i.e., the probability that $\pm t_8 > 12/3.9$ is a little greater than 0.01.

The precision and accuracy of an analytical method or procedure cannot be dissociated from the analyst unless the analyses are automatic.[18] It is often considered, however, that inaccuracy and imprecision (especially the former) arise mostly from deficiencies in the method or procedure, not in the analyst. Still, results obtained by different laboratories in the analysis of a standard sample by the same method usually vary to a greater extent than results obtained by the same analyst or by analysts in the same laboratory (p. 411). Local variations in environmental and other factors can account for some such differences. The ability of the analyst may be even more important.

Figure 16-8 is offered as an additional commentary on the determination of the accuracy of analytical methods.

Precision of determinations based on standard curves

On an earlier page it was pointed out that the relation between the quantity of a constituent and its *measure* can sometimes be represented by a straight line, which may or may not pass through the origin of the plot. Such curves are most frequently encountered in physical or physicochemical methods of analysis. For example, in

[16] A. A. Benedetti-Pichler, *Mikrochemie*, Pregl Festschrift, p. 6 (1929).

[17] The relative precision will depend on the sample size and the content of constituent. As we have seen, a variation in the amount of constituent affects the precision, other factors remaining constant, but this effect may be obscured by other, larger random variations as long as the amount lies in a reasonable range.

[18] *Automatic*: "Having self-acting or self-regulating devices that cause required acts to be performed at given points in an operation, without human intervention" (Commission on Nomenclature, Analytical Chemistry Division, IUPAC).

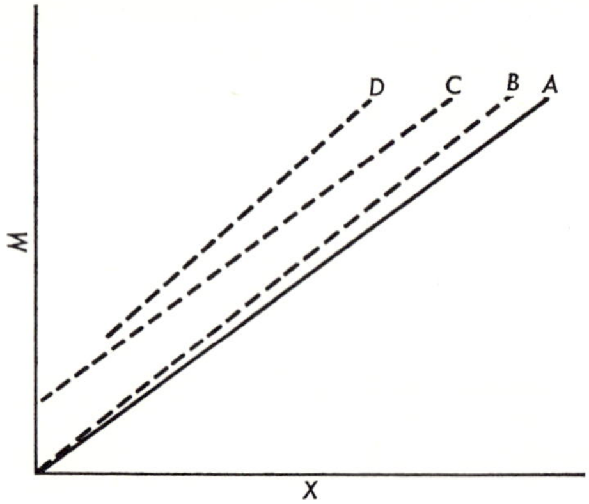

Fig. 16-8. General illustration of determinate errors in the determination of a constituent in a "pure" solution under specified conditions. The quantity X of constituent corresponds to a measured quantity M (weight of weighing form, volume of titrant, absorbance of a colored solution, and so forth). Ideally, M is directly proportional to X (line A). In errorless gravimetric, volumetric, and some physicochemical methods, M for a particular value of X is known, that is, M_τ/X = known constant. If X_1 gives M_1 instead of $(M_\tau)_1$, the absolute error is $\epsilon = M_1 - (M_\tau)_1$. In most physical and physicochemical methods, M_τ for a given X is not known or is nonexistent; only M as given by a standard curve (calibration curve) is known, that is, it has been experimentally determined. Accordingly, *under constant, specified conditions*, there is no determinate error in such a method, but in practice the italicized requirement is not easily fulfilled (for example, sample composition varies). In some, supposedly stoichiometric, analytical methods, ϵ may be independent (or almost so) of the quantity X. A plot of M versus X then gives a straight line (C) having the intercept ϵ on the M axis. The solubility of a precipitate in a gravimetric determination would give rise to a constant ϵ, which if fairly large might outweigh other errors. If ϵ is proportional to X, the slope of the line A is changed ($=B$). If both constant and proportional errors are present, the intercept and slope are both changed, as represented by D. In actuality, the various errors to which a determination is subject may not all be of the constant or proportional type, so that a plot of M versus X may give a curve; cf. Fig. 16–13. Statistical tests may be required in practice to eliminate the effect of random deviations and establish the existence of an intercept or a deviation in slope of a straight line from the theoretical slope at a given probability level.

spectrophotometric analysis a plot of absorbance ($\log I_0/I$) versus the concentration of constituent often gives a standard curve (calibration curve) which is a straight line but which may not pass through zero absorbance when the concentration of constituent is zero. The intercept on the absorbance axis represents the blank value (Fig. 16-9). A positive blank may result from the presence of a small amount of the constituent as an impurity in the reagents or simply from the absorbance of some substance in the reagents. The slope and intercept of the standard curve need to be established as exactly as possible. In this section we consider the procedure for finding the best straight line passing through the points $X_1, Y_1; X_2, Y_2; \ldots; X_n, Y_n$, and the precision of determinations based on such a line.

Let the X coordinates represent the quantity (or concentration) of a constituent and the Y coordinates the measure of the constituent. The X values are assumed to be known so precisely that their errors can be neglected; this assumption is usually valid because almost always the quantities $X_1, X_2, \ldots$ of the constituent can be measured out with high precision, or at least with much greater precision than the measures Y can be determined. It can be shown that the statistically best straight line through the

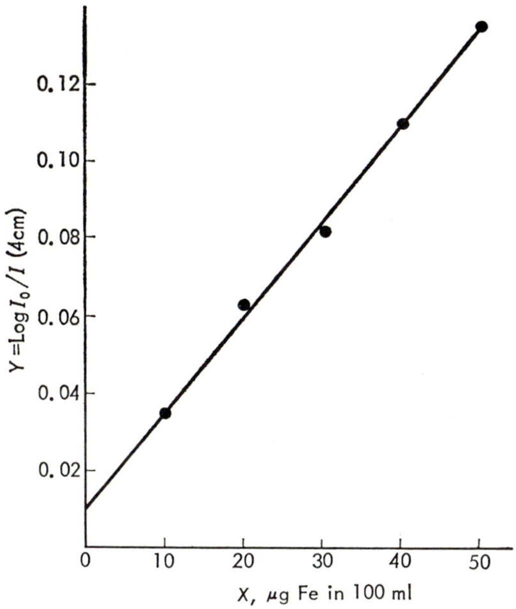

Fig. 16-9. Standard curve for the spectrophotometric determination of iron with thioglycolic acid.

series of experimental points is that for which the sum of the squares of the deviations (in the Y direction) is a minimum (the method of least squares). The values of the constants in the equation for this line,

$$Y_f = aX + b \qquad (Y_f \text{ are the fitted values}) \tag{19}$$

are found from the relations

$$a = \frac{\sum_{i=1}^{n} (Y_i - \bar{Y})(X_i - \bar{X})}{\sum_{i=1}^{n} (X_i - \bar{X})^2} \qquad (i = 1, 2, \ldots, n) \tag{20}$$

$$b = \bar{Y} - a\bar{X} \tag{21}$$

Example 10[19]*:* A standard curve for the spectrophotometric determination of iron with thioglycolic acid was established by measuring out various amounts of iron(II), treating with reagent and ammonia, diluting to volume, and determining the absorbance of the solution (100 ml) in a 4-cm cell with a 605-mμ filter:

X(μg of Fe/100 ml)	0	10	20	30	40	50
$Y(\log I_0/I)$	0.009	0.035	0.061	0.083	0.109	0.133

Find a and b of the best straight line representing these data.

Ans. $\bar{X} = 25$
$\bar{Y} = 0.072$
$\sum (X - \bar{X})(Y - \bar{Y}) = 4.32$
$\sum (X - \bar{X})^2 = 1750$
$\sum (Y - \bar{Y})^2 = 0.01067$, or 0.0107
$a = 4.32/1750 = 0.002469$, or 0.00247
$b = 0.072 - 0.00247 \times 25 = 0.010$

[19] J. Higgins, *Analyst* **89**, 211 (1964).

Therefore,

$$Y_f = 0.00247X + 0.010 \qquad \text{(Fig. 16-9)}$$

The value of X corresponding to a particular measured value Y_f is given by

$$X = \frac{Y_f - b}{a} = \frac{Y - 0.010}{0.00247} \qquad \text{[from Eq. (19)]} \tag{22}$$

When the relative error in $Y_f - b$ is large compared to that of a, the random error in $Y_f - b$ will chiefly determine the precision of X, and conversely. The relative error in $Y_f - b$ is likely to be large when this term is small, and the random errors in Y_f and b then determine the precision.

Example 11: Derive an expression for the indeterminate error of X in the preceding example.

Ans. The indeterminate error of X arises from the indeterminate error in the measurement of the absorbance ($\log I_0/I$) of the solutions (sample and standards). The random absorbance error may be largely attributable to the instrument, as in this case, but it could also be due to fluctuations in the measured absorbance of the solutions (variability due to lack of reproducibility of color development, fading of the color with time or by photochemical decomposition). We make the assumption here that the absolute deviation in the absorbance is essentially independent of the concentration of X in the range 0–50 μg of Fe in 100 ml. This approximation is justified for this range (see also Chapter 55). Any absorbance value is the difference between two absorbance readings, one of which may be a setting such as when the absorbance of a solution is referred to water (for the latter, $\log I_0/I = 0.000$ or transmission = 100.0 per cent). Suppose, as a reasonable assumption, that the reproducibility of a transmission reading is $\sim \pm 0.1$ per cent (0.1 division of a 100-division scale) near 100 per cent transmission. Take the reproducibility of an absorbance difference to correspond to ± 0.2 per cent transmission. Near zero absorbance, ± 0.2 per cent transmission corresponds to an absorbance of ± 0.0009; at 0.13 absorbance, ± 0.2 per cent (absolute) transmission corresponds to an absorbance of ± 0.0012. These values are not greatly different and as a first approximation they can be taken equal. Accordingly, take the absolute standard deviation of the absorbance Y to be constant over the range covered and represent its estimate by s ($= s_Y$). Its value will be found from the data. If s depends mostly on the spectrophotometer, we would expect it to be $\sim \pm 0.001$ in absolute absorbance. If noninstrumental factors also enter in, it could be appreciably larger.

A difficulty is immediately encountered. In Eq. (22) a and b are not statistically independent, and an exact treatment requires that their (negative) correlation be taken into account.[20] However, for simplicity and as a basis for discussion, an approximation will be made here. The estimated absolute standard deviations of Y_f and b will be taken equal to s, so that the standard deviation of $Y_f - b$ will be $\sqrt{2} \cdot s$. On these assumptions, the estimated relative standard deviation of X, expressed in per cent, is given by

$$\frac{100 s_X}{X} = 100\left\{\frac{2s^2}{(Y - b)^2} + \frac{s_a^2}{a^2}\right\}^{1/2} \tag{23}$$

where s_X and s_a are the estimated absolute standard deviations of X and the slope a.

[20] See J. Mandel, *The Statistical Analysis of Experimental Data*, Wiley, New York, 1964, p. 279, which gives a quadratic equation by which the confidence interval for the "true" value of X can be obtained. This equation is given on page 409. See also a paper on accuracy in chemical analysis using linear calibration curves, J. Mandel and F. J. Linnig, *Anal. Chem.* **29**, 743 (1957).

This equation follows from the rules for the combination of random errors in a quotient (Table 16-4). Substitute aX for $Y - b$:

$$\frac{100s_X}{X} = 100\left\{\frac{2s^2}{a^2X^2} + \frac{s_a^2}{a^2}\right\}^{1/2}$$

$$= \frac{100}{a}\left\{\frac{2s^2}{X^2} + s_a^2\right\}^{1/2} \tag{24}$$

The values of s and s_a are calculated as follows from the absorbance–concentration data of Example 11. For the computation of s, the values of $(Y - Y_f)^2$ are found as shown in Table 16-5. Then

$$s^2 = s_Y^2 = \frac{\sum (Y - Y_f)^2}{N - 2} = \frac{5.15 \times 10^{-6}}{N - 2} = \frac{5.15 \times 10^{-6}}{4} = 1.29 \times 10^{-6}$$

$$s = 1.14 \times 10^{-3}$$

TABLE **16-5**

X	Y	Y_f	$Y - Y_f$	$(Y - Y_f)^2$
0	0.009	0.0100	−0.0010	100×10^{-8}
10	0.035	0.0347	0.0003	9
20	0.061	0.0594	0.0016	256
30	0.083	0.0841	−0.0011	121
40	0.109	0.1088	0.0002	4
50	0.133	0.1335	−0.0005	25
				$\sum 515 \times 10^{-8}$

(The number of degrees of freedom is 2 less than the number of Y values because two parameters, a and b, have been calculated.)

It is seen that there is no pattern in the magnitudes or signs of $Y - Y_f$. This is a good indication that the relation between X and Y is linear. When the measurements of Y are of high precision, it may be difficult to tell by looking at the plot of the data whether the points tend to lie along a straight line or a slightly curved line. If, above, the values of $Y - Y_f$ had first been positive, then had become smaller and negative, and finally positive again, one would infer that the Y's lie along a curve concave upward. Also, no value of $Y - Y_f$ is abnormally large, so that a gross error in any Y is excluded.

The estimated variance of the slope a of the standard curve is given by

$$s_a^2 = \frac{s^2}{\sum (X_i - \bar{X})^2} = \frac{Ns^2}{N \sum X_i^2 - (\sum X_i)^2} = \frac{6 \times 1.29 \times 10^{-6}}{6 \times 5500 - (150)^2}$$

$$= \frac{7.74 \times 10^{-6}}{10{,}500} = 7.4 \times 10^{-10} \tag{25}$$

$$s_a = 2.7 \times 10^{-5}$$

The estimated standard deviation of X in per cent is accordingly given by

$$\frac{100s_X}{X} = \frac{100}{2.47 \times 10^{-3}}\left\{\frac{2 \times 1.29 \times 10^{-6}}{X^2} + 7.4 \times 10^{-10}\right\}^{1/2}$$

$$= 4.05 \times 10^4\left\{\frac{2.58 \times 10^{-6}}{X^2} + 7.4 \times 10^{-10}\right\}^{1/2} \tag{26}$$

Let us compare the estimated percentage standard deviations in the determination of 10 and 50 μg of iron under the specified conditions and the error contribution of the two terms under the square-root sign:

X	$2s^2/X^2$	s_a^2	$100s_X/X$
10	2.58×10^{-8}	7.4×10^{-10}	6.6
50	1.03×10^{-9}	7.4×10^{-10}	1.7

In the determination of 10 μg of Fe, the random error in the slope of the calibration curve has a negligible effect on the accuracy of the result. Most of the error originates in Y and b. If the assumption made, that $s_b = s = 0.00114$, is correct, the errors in Y and b contribute equally to the combined error. Now actually s_b will be somewhat less than s because the points near the blank point ($X = 0$) in conjunction with the least-squares slope aid in fixing b more exactly than if it were based on only the blank value. The calculated standard error, 6.6 per cent, in the determination of 10 μg of Fe is therefore a little larger than that obtained by using the least-squares intercept. However, from the practical standpoint, the difference is not great, and it is better, as a rule, to err with a bias to the pessimistic side. If there were no error at all in the standard curve, the standard error in the determination of 10 μg of Fe would be 4.6 per cent.

In the determination of 50 μg of Fe, the estimated standard error of the result is reduced to 0.9 per cent if the standard curve is errorless. If *duplicate* determinations of 10 and 50 μg of Fe are made, the calculated percentage standard errors of the results are:

10 μg of Fe		50 μg of Fe	
No Error in Standard Curve	*Error in Intercept*	*No Error in Standard Curve*	*Error in Standard Curve*
3.2%	5.6%	0.64%	1.56%

The range in which the "true" value of iron will lie at a given probability level can be obtained by using the appropriate t value. Suppose that the calculation of the iron content of 100 ml of solution by the use of Eq. (22) gives the value 10.0 μg of Fe. The standard error has already been estimated as 6.6 per cent for this amount of iron. We wish to know the range in which the true iron content (*true* in the sense of the result being affected by random errors) will lie with a probability of 95 per cent; i.e., we wish to find the 95 per cent confidence interval. The value of t corresponding to $n = N - 2 = 4$ and a confidence level of 95 per cent is 2.78 (Table 16-2). The required interval is therefore $\mu = 10 \pm 2.78 \times 0.066 \times 10 = 10 \pm 1.8$, or 8.2–11.8 μg f Fe.

When a standard curve is used from day to day, or simply at some time after it has been established, a and b may change. The intercept may change as a result of a slight contamination of the reagents (attack of glass vessels holding the solutions, introduction of foreign material from the atmosphere, etc.) or for other reasons. The slope of the standard curve may also change (deterioration of reagents, temperature change, etc.). The easiest and most effective way of checking the standard curve is by obtaining one or more absorbance readings of a standard against a blank solution.[21] The standard should of course be one giving an absorbance at or near the high end of the curve. If the slope thus obtained agrees within experimental error with the calibration slope,

[21] It may be asked why the determination of b was not omitted in setting up the standard curve. The answer is that it could often be omitted. However, it is of interest to know what the blank value is. Obviously a large blank is undesirable when a very small amount of constituent is being determined, and when such a blank is found, steps may need to be taken to reduce it. Measurement against the same blank in a series of determinations would be disadvantageous if the blank gave an appreciable absorbance changing fairly rapidly with time. This behavior is of rare occurrence and it could be countered by preparing a fresh blank for each sample.

that slope may be used, or the new value may be incorporated with the old to give what is presumably a more reliable average. But suppose that the new slope differs significantly from the old (statistical tests may be applied, if thought worthwhile, to establish a difference), so that the new value is accepted. The standard error (estimated) of this slope is, in per cent,

$$\frac{100s}{X_{\text{stand}}} \qquad (X_{\text{stand}} = \text{quantity or concentration of } X \text{ in the standard solution})$$

The value of s found in the calibration is used. The absorbance of the sample (Y_{samp}) is then measured against the blank, which is reset to show 100.0 per cent transmission. Then

$$X_{\text{samp}} = X_{\text{stand}} \times \frac{Y_{\text{samp}}}{Y_{\text{stand}}}$$

$$\frac{100s_X}{X_{\text{samp}}} = \left\{\left(\frac{100s}{Y_{\text{stand}}}\right)^2 + \left(\frac{100s}{Y_{\text{samp}}}\right)^2\right\}^{1/2}$$

The best precision is obtained when X_{stand} is at the upper limit of the standard curve and $X_{\text{samp}} \sim X_{\text{stand}}$. For 50 µg of Fe,

$$\frac{100s_X}{X_{\text{samp}}} = \frac{100 \times 0.00114 \times 1.4}{0.125}$$

$$= 1.3 \text{ per cent}$$

The standard error can, of course, be reduced by replicating Y_{stand} and Y_{samp}. If four measurements each of the absorbances of the standard and sample are made, $100s_X/X_{\text{samp}}$ under the conditions just specified becomes 0.65 per cent and the 95 per cent confidence limits are 49.1–50.9 µg of Fe. Further treatment of the accuracy of spectrophotometric determinations will be found in Chapter 55. Here we are only concerned with the effect of random errors on precision of determinations based on standard curves as illustrated by absorbance measurements.

If the photometer provides very stable absorbance readings, it would be possible to measure the absorbance of both standard and sample against the same blank, with resultant increase in precision. When $X_{\text{samp}} \sim X_{\text{stand}}$ the error in setting or reading the blank absorbance cancels out and the estimated standard deviations of Y_{samp} and Y_{stand} become $(100 \times 0.7s)/Y_{\text{samp}}$, so that $100s_X/X$ becomes $100s/Y_{\text{samp}}$, which is 0.9 per cent for 50 µg of Fe when one reading each of Y_{samp} and Y_{stand} is made and 0.46 per cent when four readings each of these are made. These last conditions are beginning to approach what might be called the ultimate, in that one would not often be willing to make more than four measurements of each quantity. The corresponding confidence interval at 95 per cent is now 49.4–50.6 µg of Fe. All through this explication we have based s on six calibration points (four degrees of freedom). The confidence interval can be narrowed by basing s on a larger number of replications. Suppose s is found from the standard deviations of a series of 11 replicate absorbance readings of the same solution. The value of t is now reduced from 2.78 to 2.23 (n = 10), and (if s is assumed to be found the same) the 95 per cent confidence interval becomes 49.5–50.5 µg of Fe. The assurance gain is small.

Finally, we give the equation referred to earlier, in which account is taken of the correlation between a and b in finding the confidence interval of X. The interval in which the statistical true value of X_i corresponding to a given Y_i lies at a specified confidence level is obtained by solving the following quadratic equation[20] for v:

$$\left\{a^2 - \frac{N(ts)^2}{N\sum X_i^2 - (\sum X_i)^2}\right\}v^2 - \{2a(Y - \bar{Y})\}v + \left\{(Y - \bar{Y})^2 - \frac{N+1}{N}(ts)^2\right\} = 0 \qquad (27)$$

provided $a^2\{N \sum X_i^2 - (\sum X_i)^2\} > N(ts)^2$. The value of the absolute standard deviation of Y ($= s$) is taken to be the same as in the calibration. If n replicate determinations of Y are made in the unknown sample, the quantity $(N + 1)/N$ is replaced by $(1/n) + (1/N)$ in the equation. The limits of the range in which the "true" value of X lies at the stipulated confidence level are obtained by adding the two values of v to the mean of X in the standardization runs:

$$X_1 = \bar{X} + v_1$$
$$X_2 = \bar{X} + v_2$$

Accuracy and precision sought and attained in inorganic quantitative analysis

The accuracy required in chemical analysis varies. Primarily, the accuracy sought in any analysis is determined by the use to which the result will be put. Analyses made to find the values of certain constants of nature (atomic weights, for example) or to establish laws of nature (Faraday's laws, for example) must obviously be highly accurate, and efforts are made to keep the error below 1 part in 10,000 or even in 100,000.

Next, in order of decreasing accuracy, come analyses of fairly high grade in which the error should not, in general, exceed the limits given in Table 16-6, depending on the relative amount of the constituent in the sample. It will be understood that the limits cannot be rigidly specified. The accuracy attainable will vary to some extent with the constituent and especially with the composition of the sample. In scientific analyses (as in rock analysis) and in some industrial analyses, accuracy corresponding to these values will be sought—and sometimes attained.

TABLE **16-6**

Percentage of Constituent	*Permissible Error, Parts per* 1000
~100	1–3
~50	3
~10	10
~1	20–50
~0.1	50–100
0.01–0.001	~100

Then there are other types of analyses in which high accuracy is not aimed at, because it is not required. Many determinations in food and physiological analysis, for instance, are entirely satisfactory if the results are within a few per cent of the truth. Obviously it is a waste of time and labor to determine a constituent with great accuracy if there is no particular demand for a very accurate result. Generally speaking, the accuracy that can be attained in any analysis is a function of the time that can be placed upon the determination. By making double and triple precipitations in separations, for example, the accuracy can usually be increased, but only at the expense of extra time and effort. If one precipitation gives results of fair accuracy, it would be foolish to make reprecipitations unless the higher accuracy actually is needed. In this book special attention is paid to procedures that are capable of giving rather high accuracy. The analyst who can perform accurate analyses and who has a grasp of the factors involved in any analysis will understand what legitimate changes can be made in a procedure when results of lesser accuracy are satisfactory.

In taking into account the factors that influence accuracy, special attention must be paid to the substances which accompany the constituent and to their relative amounts. Usually, pure solutions of substances offer no difficulty in analysis, and high accuracy is attainable if proper attention is paid to manipulative details. It is otherwise when the constituent in question is accompanied by others which may seriously interfere, for reasons indicated elsewhere in this chapter. Thus there is no particular difficulty in determining calcium accurately when it occurs as a pure salt, but the matter is different when it is accompanied by magnesium, as it usually is. Naturally, larger errors must be expected in the analyses of complex natural materials or artificial products than in academic analyses of pure solutions.

The accuracy and precision attainable at the present time in the analysis of a complex material entailing separations are illustrated by the analysis of a synthetic granite glass in 12 laboratoties of the United States, England, Finland, South Africa, and New Zealand (Table 16-7).[22] The sample was prepared with painstaking care from weighed amounts of pure compounds so that the composition was accurately known. The analyses were made by professional rock analysts, who were free to use methods of their own choosing. For the most part these were the classical methods of rock analysis. The agreement among laboratories leaves something to be desired for SiO_2 and Al_2O_3, whereas for MgO, CaO, Na_2O, and K_2O it is quite satisfactory on the whole. It can be seen at once, without the application of numerical statistical tests, that sources of determinate errors must be present in some of the methods. Thus 11 of 12 laboratories obtained high values in the determination of MgO, and 10 of 12 obtained high values for CaO, results that are unlikely to be due to chance. However, the magnitude of the error for each constituent is not serious. There seem to be no significant determinate errors in the determination of Na_2O and K_2O. The procedures followed by most of the

TABLE **16-7.** **Analysis of Synthetic Granite Glass**

Laboratory	SiO_2, %	Al_2O_3, %	MgO, %	CaO, %	Na_2O, %	K_2O, %	*Total,* %
1	72.30	16.20	0.88	2.05	3.82	5.40	100.65
2	72.40	15.85	0.87	1.84	3.35	5.72	100.03
3	72.01	16.78	0.75	1.92	3.15	5.76	100.37
4	72.37	16.39	0.90	1.67	3.25	5.57	100.15
5	72.42	16.12	0.82	1.89	3.08	5.70	100.03
6	72.75	16.12	0.81	1.85	3.13	5.65	100.31
7	72.25	16.27	0.82	1.99	3.07	5.76	100.16
8	72.24	16.34	0.88	2.00	3.11	5.62	100.19
9	72.11	15.91	0.89	2.35	3.27	5.75	100.28
10	72.35	15.98	0.86	1.90	3.27	5.79	100.15
11	71.38	16.14	0.86	1.90	3.14	5.84	99.26
12[a]	72.65	15.79	0.86	1.79	3.23	5.64	99.91
Av.	72.27	16.20	0.85	1.93	3.23	5.69	100.12
Standard composition	72.64	15.78	0.80	1.82	3.19	5.76	99.99
Error of av. %[b]	−0.37	0.42	0.05	0.11	0.04	−0.07	0.13
Error of av. ppt	−5	27	63	60	13	−12	13

[a] Average of three analysts.
[b] Absolute per cent.

[22] H. W. Fairbairn and J. F. Schairer, *Am. Mineralogist*, **37**, 744 (1952). For results obtained in routine rock analysis, see H. W. Fairbairn et al., *U.S. Geol. Surv. Bull.* **980** (1951), and R. E. Stevens et al., *U.S. Geol. Surv. Bull.* **1113** (1960). These compilations and discussions of analytical data show that it is much more difficult than generally realized to obtain accurate results in analyses of complex inorganic materials.

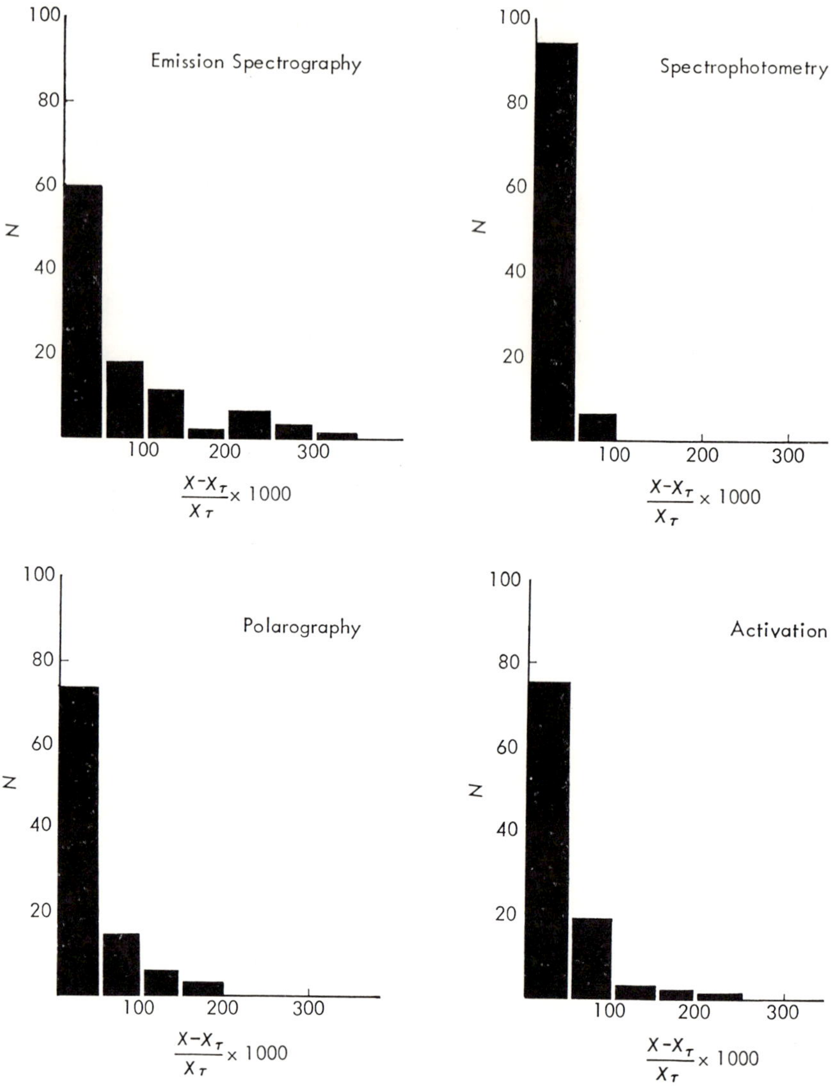

Fig. 16-10. Histograms showing accuracy of trace determinations of Cu, Mn, Cr, and Hg by different methods. Mean error in ppt, without regard to sign, is plotted against number of results, N, in each group. From Cook, Crespi, and Minczewski, Reference 24.

laboratories gave significantly low results for silica. The unprecipitated (and unrecovered) silica is later counted with Al_2O_3, so that alumina is reported too high (correlation of errors).

Information is available on the accuracy attainable in the determination of many elements in applied analysis.[23]

[23] W. F. Hillebrand and G. E. F. Lundell, *Applied Inorganic Analysis*, Wiley, New York, 1929, pp. 874–887. For data on accuracy of steel analysis, see G. E. F. Lundell, J. I. Hoffman, and H. A. Bright, *Chemical Analysis of Iron and Steel*, Wiley, New York, 1931.

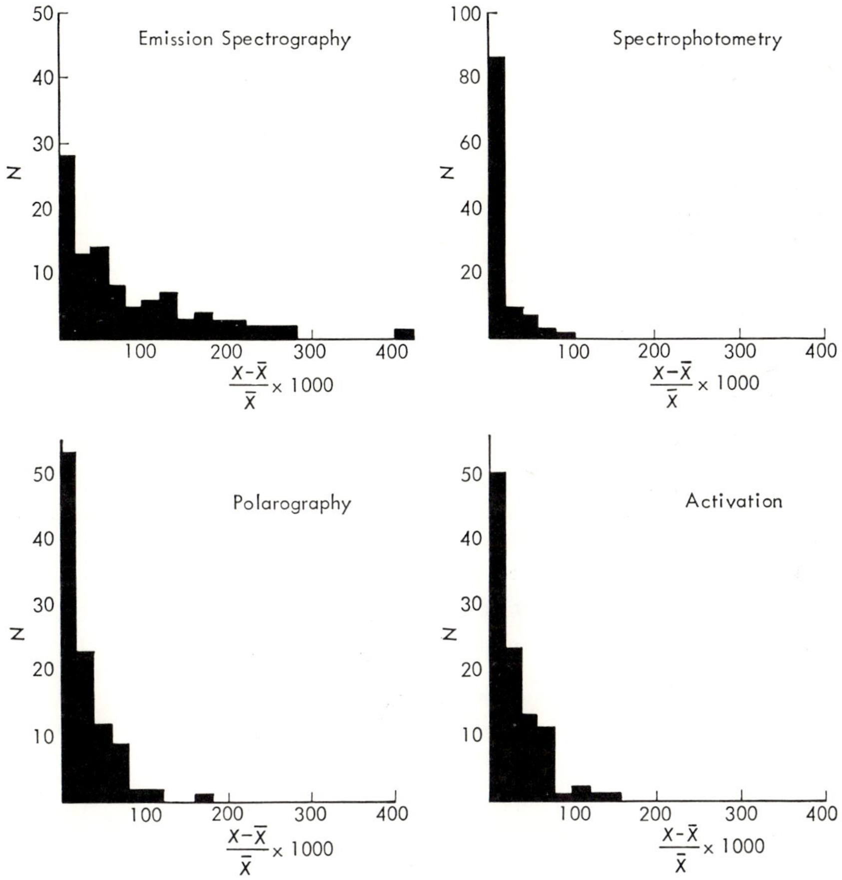

Fig. 16-11. Histograms showing precision of trace determinations of Cu, Mn, Cr, and Hg by different methods. Mean deviation in ppt plotted against number of results, *N*, in each group. From Cook, Crespi and Minczewski, Reference 24.

Trace analysis. "One of the most characteristic aspects of present day analytical chemistry is the growing importance of the determination of impurities at very low concentrations in materials of interest to modern industry. Trace analysis is today no longer a very specialized activity applied almost exclusively in the study of particular chemical problems but represents an important part of the daily routine work of many analytical laboratories."[24]

Trace determinations are almost always carried out by physicochemical or physical methods. The precision and accuracy attainable in routine trace analysis by four such methods—emission spectrography, absorption spectrophotometry, polarography and neutron activation—have been reported.[24] A solution containing approximately 5 ppm each of Cu and Mn and 15 ppm each of Cr and Hg was analyzed in nine laboratories. The results of this international comparison, embracing 551 determinations, are summarized in Figs. 16-10 and 16-11 and Table 16-8.[25] Since the four elements

[24] G. B. Cook, M. B. A. Crespi, and J. Minczewski, *Talanta* **10**, 917 (1963).

[25] The accuracy and precision for each element by each method in the nine laboratories individually are listed in the paper by Cook et al. (Ref. 24).

TABLE 16-8. **Summary of Accuracy and Precision of Trace Determinations of Cu, Mn, Cr, and Hg by Four Different Methods**[a]

Method	*No. of Determinations*	*Error of Average, %*[b]	*Deviation Having Probability* ≤ 0.05, *%*[c]
Emission spectrography	124	−0.2	37.8
Absorption spectrophotometry	172	+0.8	9.8
Polarography	116	+2.0	26.1
Activation	139	+2.5	20.4

[a] From G. B. Cook, M. B. A. Crespi, and J. Minczewski, *Talanta* **10**, 917 (1963).

[b] $100(\bar{X} - X_\tau)/X_\tau$.

[c] The deviation of a single determination, $100(X - \mu)/\mu$, having a probability of $\gtreqless 0.05$ of being exceeded; we may be 95 per cent confident that the deviation of a single determination will not exceed the figure given.

were present in much the same concentrations and the combination was such that serious interferences would not be expected (separations were apparently not made), the accuracy found is no doubt better than that which would be found in most applied trace analyses, where conditions are less favorable.

Methods of checking the accuracy of analyses

Parallel determinations. The results obtained in running parallel (replicate) determinations serve as a check on the result of a single determination and thus increase our confidence in the reliability of the value obtained (assuming that good concordance is secured) and also give an average value that is probably more nearly correct than a single value as far as random errors are concerned.

The values obtained in parallel determinations should agree reasonably well among themselves. In most analyses in which the percentage of the constituent present is not too small (major constituents), the differences should usually not be greater than 3 ppt. Obviously, if the results of duplicate or triplicate determinations show considerable differences, not much confidence can be placed on any result, and it is then necessary to run extra determinations until satisfactory concordance is obtained. The running of duplicates is of special value in detecting mistakes and blunders. If two or more results agree excellently, it is by no means justifiable to conclude that the value obtained is correct. Good concordance merely indicates that the accidental errors, or variations in the determinate errors, of all the determinations are the same or nearly the same. There may still be a large determinate error associated with the result (Fig. 16-1). Good agreement in parallel determinations signifies that the determinations have been made under closely similar conditions; it does not guarantee the accuracy of the results. Failure to obtain satisfactory agreement in parallel determinations is, however, evidence that operative errors or mistakes have been made.

The larger the number of measurements, the smaller is likely to be the deviation of the mean from the true value, as far as indeterminate errors are concerned (p. 390). The number of parallel determinations that should be run depends to a great extent upon the importance attached to the result. In an atomic-weight determination, for example, eight or ten or even a greater number of determinations are made. Similarly, in establishing the accuracy of a method or the suitability of a standard substance (p. 445), a large number of parallels is run. In ordinary analysis, duplicate determinations usually suffice. In noninstructional laboratories, single analyses are likely to be the rule in day to day work. The expense of duplicate analyses would often be prohibitive. In such laboratories, the analysts as a class are skilled and experienced, and random errors are reduced to a minimum. Often, but by no means always, the accuracy of the analytical results will depend more on the procedure than on the analyst. This generalization is more likely to be true when physicochemical ("instrumental") methods are used.

In standardizing a volumetric solution, it is desirable to make three parallel determinations. The error in the titer must be as small as practicable, since an error in the standardization is transmitted and added to the error made in the determination itself.

Summation. When all the constituents of a sample have been determined directly, a check on the general accuracy of the whole analysis can be obtained by adding up the percentages of the constituents. Because of unavoidable errors, the theoretical figure of 100.0 per cent is never obtained except by chance. The summation may yield a figure either higher or lower than 100 per cent. A good analysis will give a summation that is very close to 100 per cent. The permissible difference cannot be exactly stated, since it will depend upon such factors as the kind and the number of constituents in the sample. Some analyses are naturally more difficult to make than others. A rock analysis in which 20 or so constituents are determined may be considered satisfactory when the summation gives a value lying between 99.75 and 100.30 per cent (see Table 16-7). In a simpler analysis a better summation would usually be demanded. An analysis summing up to slightly more than 100 per cent is sometimes considered more satisfactory than one giving a sum slightly less, because small quantities of foreign materials may be introduced from the reagents and the utensils. A summation decidedly less than 100 per cent points to the omission of one or more constituents in the analysis.

A nearly perfect summation may be due, of course, to compensation of relatively large errors,[26] but usually such a summation may be taken as evidence of the substantial accuracy of the analysis.

The ionic balance. In inorganic analysis we often deal with samples composed of electrolytes. The latter are built up of cations and anions, which naturally must be present in equivalent amounts in order that electroneutrality may be preserved. This fact may be applied in testing the accuracy of analyses of electrolyte mixtures, either solids or solutions. The sum of the cations, expressed in equivalents or milliequivalents, is equal to the sum of the anions in an ideally perfect analysis. Illustration:

A solution containing calcium, magnesium, sodium, potassium, chloride, sulfate, and bicarbonate ions was found to have the following composition:

	g/100 ml		g/100 ml
Ca^{++}	0.4080	Cl^-	0.3763
Mg^{++}	0.1339	SO_4^{--}	0.9990
Na^+	0.2070	HCO_3^-	0.6322
K^+	0.0586		

The *ionic balance* is tested by adding up separately the milliequivalents of cations and anions:

Cations	*meq/liter*	*Anions*	*meq/liter*
Ca^{++}	203.6	Cl^-	106.1
Mg^{++}	110.1	SO_4^{--}	208.0
Na^+	90.0	HCO_3^-	103.6
K^+	15.0		
	418.7		417.7

[26] Compensation of related errors can occur in separations when an element not completely precipitated at one stage is precipitated with another element at a later stage, as already illustrated on page 412.

The agreement is satisfactory, because the difference is only 1 part in approximately 400, or 0.25 per cent. The analysis therefore is probably reliable. On the other hand, if the respective figures 420 and 412 had been obtained, the analysis could not be regarded as satisfactory, for this difference is equivalent to 2.0 per cent, which is too large. An error in the determination of one or more constituents, or the presence of other ions, would then be indicated.

The balance principle can be applied to analyses of solids composed of positive and negative components. For example, it can be used to test the accuracy of an analysis of a more or less complex mineral such as $(Pb, Ca)_5(PO_4, AsO_4)_3Cl$ (pyromorphite). See Problem 31.

Independent methods of analysis. Sometimes it is possible to establish the accuracy of a result by making the analysis in an entirely different way. For example, iron may first be determined gravimetrically by precipitation as ferric hydroxide after removal of interfering elements, followed by ignition of the precipitate to ferric oxide. Then it may be determined volumetrically by reduction to the ferrous state, followed by titration with a standard solution of an oxidizing agent such as potassium dichromate. If the results of two such radically different procedures have substantially the same value, they probably are correct within small limits of error, because the same methodic errors, and even the same operative errors, can hardly be associated with both methods. This is a very good way of verifying a result and should be used whenever possible when much importance is attached to the result. It is especially valuable for complex materials in which the effect of foreign constituents may be difficult to evaluate.

Methods of increasing the accuracy of analyses

Such self-evident means of reducing the errors of chemical analyses as calibration of weights and volumetric glassware, purification of reagents when necessary, use of resistant vessels, and adherence to the proper quantitative technique need only be mentioned. A sample of suitable size must be taken. Without attention to such details, no accuracy at all can be expected. Moreover, the use of instruments of adequate sensitivity and reproducibility is mandatory. In the following we shall consider chiefly the methods of minimizing the methodic errors, for these are the most serious and least easy to eliminate or reduce. Reduction of random errors by running parallel determinations has already been mentioned.

Application of corrections. Sometimes, when it is impossible to eliminate an error, it is possible to apply a correction for the effect it produces. Thus an impurity in a weighed precipitate may be determined and its weight deducted. The amount of a substance which remains unprecipitated may often be determined, and a correction thus found for solubility loss. The loss in weight which a platinum crucible may undergo on strong ignition can be determined by careful cleaning and reweighing. Frequently the labor involved in finding the correction values is so great that it is not worthwhile to make the corrections in ordinary analyses. Instrumental corrections (weights and buret especially) are often easily obtained and applied, however.

Running of blanks. Errors arising from the introduction of impurities through the reagents and vessels, and sometimes from other sources, can be allowed for by running what is called a blank determination. The sample is omitted; otherwise the details of the procedure are followed exactly as far as possible. The same amount of reagent is added, the same volume of solution is employed, the same temperature is used in the operations, the same period of time is allowed for each step, and so on. The precipitate finally obtained is weighed, if the determination is gravimetric, and the amount found is deducted from the weight of precipitate given by the sample. The blank is best carried along simultaneously with the sample. Although the errors arising from the sources mentioned are often greatly reduced by running a careful blank, they are not always entirely eliminated or even minimized, because the conditions in the determination itself cannot be exactly reproduced in the blank. For example, the large amount of precipitate formed in the precipitation proper may carry down much larger quantities of a foreign substance than the small precipitate of the blank will do. If no precipitate is formed in the blank, as may happen when the reagents are very pure, the loss of precipitate resulting from its solubility is not included. In general, a small blank correction is better than none at all. A large blank correction is undesirable, because the exact value of the correction may then become uncertain; purer reagents should then be procured, or whatever other steps are necessary should be taken to reduce the blank.

The necessity for running a blank is not the same in every determination. In some, a blank can safely be omitted, while in others it is unconditionally necessary. For example, a gravimetric chloride determination with silver nitrate as reagent hardly requires a blank if the distilled water and nitric acid used are tested for chloride and found to be free of the ion. On the other hand, a gravimetric determination of sulfur in a steel as barium sulfate would give erroneous results unless the correction obtained in the blank is applied. Small amounts of sulfur derived from the reagents and the gases of the flame will cause a large error in the result, because of the small percentage of sulfur in the sample.

Control determinations. The result of an analysis can be checked, and the value obtained can be corrected by taking a known amount of constituent and determining it by the procedure used for the "unknown" sample. Since the amount of constituent that should be found is known, the error of the *control* is obtained directly and this will also be the error of the sample determination, provided the control and sample are identical or almost so. This requirement may not be easily fulfilled in practice. More often than not, the composition of the sample is not sufficiently well known to allow the preparation of a synthetic sample equivalent to the unknown sample in all respects except for the absence of the constituent in question. Occasionally the composition of the sample may be known, as in the analysis of sea water, so that synthetic samples can readily by prepared. Sometimes it may be possible to remove the constituent in question from the solution of the sample by a selective method, so as to obtain a solution to which known amounts of the constituent can then be added; some trace element determinations can be studied in this way. Synthetic samples having varying composition may have to be prepared to test the effect of foreign constituents. A knowledge of the analytical reactions of all elements that may be present is helpful, if not indispensable, in planning and carrying through such tests.

Checking the accuracy of analyses, or finding correction values to increase the accuracy, by the method of controls is likely to require much time and labor and is therefore not often applied to a *determination*. It is, however, frequently used to investigate the accuracy of a *method*, and indeed is of fundamental importance for this purpose. Control determinations are carried out with varying amounts of constituent corresponding to the range that will need to be covered in the application of the procedure. The *measure* of the constituent (weight of the weighing form in gravimetric analysis, titration volume in titrimetric analysis, absorbance in spectrophotometry, etc.) is compared against the theoretical measure as calculated from the amount of constituent. The error (which becomes the correction value with reversal of sign) is conveniently plotted versus the amount of constituent. Nonlinear curves are frequently obtained (Fig. 16-12).

The known amounts of constituent for testing an inorganic analysis can usually be obtained without much difficulty by weighing out sufficiently pure compounds or elements. Many elements, particularly metals, are now available commercially in high purity. When necessary, compounds can be purified, as by recrystallization of soluble salts and by other methods (Chapter 17). Qualitative tests for purity can be made by sensitive color reactions, by emission spectrography, etc. The presence of water in the standard substance must be guarded against. Hydrated salts are best avoided as standards when high accuracy is required, because of the danger of variation in the water content and because of the difficulty in showing that the stoichiometric amount is present. Normally anhydrous substances are usually easy to test for foreign water.

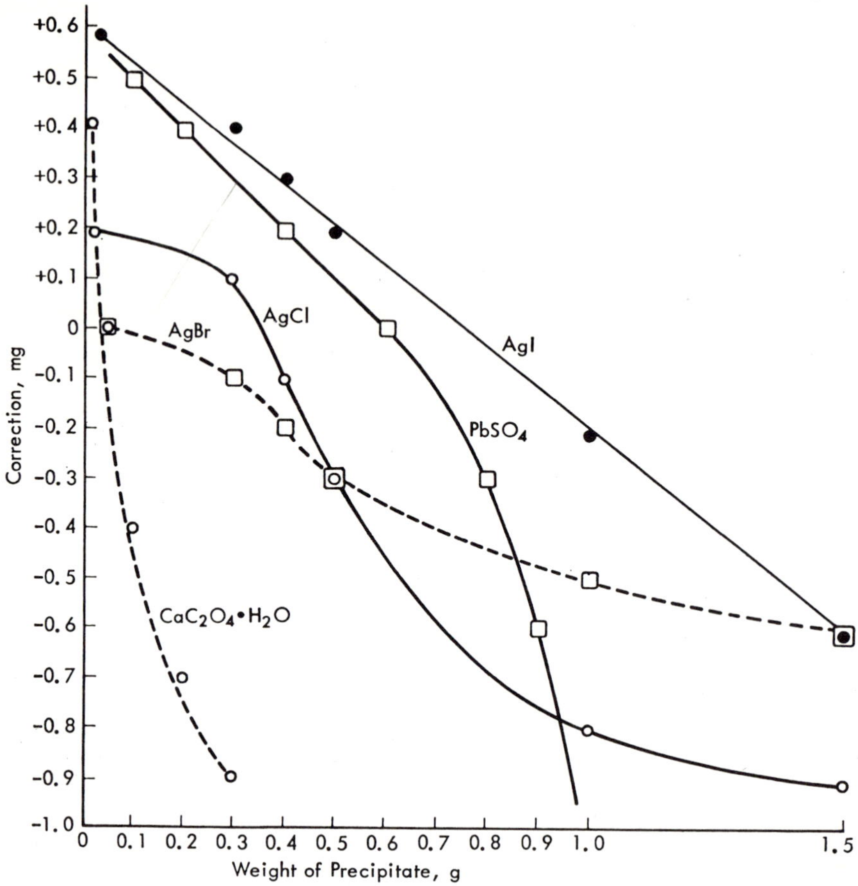

Fig. 16-12. Illustration of empirically determined correction values from control determinations. The corrections pertain to the gravimetric determination of Ca^{++}, Pb^{++}, Cl^-, Br^-, and I^- by the weighing forms indicated; drying temperatures: $CaC_2O_4 \cdot H_2O$ 20°, $PbSO_4$ 132°, AgCl 132°, AgBr 132°, AgI 132°. Data of L. W. Winkler, *Ausgewählte Untersuchungsverfahren für das chemische Laboratorium*, pp. 101, 104, 106, etc. (Vol. XXIX of the Margosches' series, *Die chemische Analyse*), Verlag Ferdinand Enke, Stuttgart, 1931. See also Winkler's papers in *Z. Angew. Chem.* When such correction values are used, the analysis must be made exactly as the procedure directs. Foreign substances can, of course, alter the corrections.

The substances used to supply the other elements of the synthetic samples must of course be free from significant amounts of the constituent whose determination is being tested.

In some physicochemical analyses (spectrographic and polarographic, for example) the method of *standard additions* may be applied as a means of increasing accuracy. M is obtained for the sample and for one or more samples (all of the same size) to which known amounts of the constituents being determined have been added.[27] The principle

[27] The statistical aspects of the standard addition method have been discussed by G. Ehrlich and R. Gerbatsch, *Z. Anal. Chem.* **209**, 35 (1965). In general, the amount of added constituent should, if possible, be larger than that originally present.

is illustrated in Fig. 16-13. Provided M is directly proportional to the quantity (q) of the constituent over the whole range ($M = 0$ when $q = 0$), the content of constituent, q_0, in the "unsalted" sample is given by the intercept of the extrapolated line $M_0M_1 \cdots$ on the q axis. The standard addition method especially provides a means of avoiding errors resulting from the presence of a foreign substance altering the response of a constituent by the same relative amount. Thus this method would be successful in eliminating the error resulting from a change in the standard curve from A to B in Fig. 16-8.

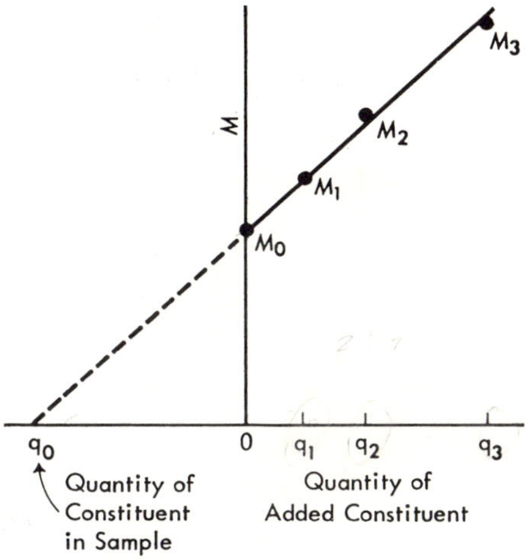

Fig. 16-13. Illustrating the principle of the standard addition method. Amounts q_1, q_2, and q_3 of constituent added to constant amounts of sample (containing the unknown amount q_0 of constituent) give responses M_1, M_2, and M_3, and q_0 gives M_0. Then $q_0 = M_0/a$ (a = slope of the line).

The recovery of an element in an analytical procedure can be determined by adding a radioactive isotope of the element and comparing the initial and final activities, allowance being made for radioactive decay. More accurately, the loss of an element at various stages of a separation can be found from the activity of precipitates or solutions not supposed to contain it. Conversely, in neutron-activation analysis, the recovery of the activated trace element is found by adding a relatively large known amount of inactive element after activation and determining its amount (usually gravimetrically) at the end of the separations.

Standard samples. Related to the method of controls is the use of standard samples. In dealing with more or less complex natural materials or artificial products, it is advantageous to prepare a sample of the material in question and to analyze it carefully by reliable methods, preferably by having a number of skilled analysts make the analysis independently. The averages of the results obtained in the most careful analyses by the cooperating analysts are taken as the most probable values. A sample thus analyzed is known as a standard sample. Standard samples serve the purpose of revealing operative and personal errors, as well as some methodic errors, made either in the analysis of the sample itself or of a sample having a similar composition. Standard samples are very useful in the investigation of the accuracy of a new method. Moreover, they may be used for standardizations.

The U.S. National Bureau of Standards prepares, analyzes, and sells standard samples of irons, steels, nonferrous alloys, ores, glasses, etc., as well as primary standards, e.g., sodium oxalate, benzoic acid, potassium biphthalate, and arsenious oxide.

For a listing see *Natl. Bur. Std. Misc. Publ. 260* (1965). A typical Bureau of Standards Certificate of Analyses is reproduced in Fig. 16-14. See *Anal. Chem.* **38**, 27A (1966), for an account of the NBS Standard Reference Materials Program.

Two standard rock samples (G-1, granite; W-1, diabase) were issued by the U.S. Geological Survey in about 1950. Extensive data on the major, minor, and trace constituents of these samples, representing an immense amount of work by scores of

U.S. DEPARTMENT OF COMMERCE
WASHINGTON, D.C. 20234

National Bureau of Standards
Certificate of Analyses

Standard Sample 19G
Acid Open-Hearth Steel, 0.2% Carbon

ANALYST	C	Mn	P		S		Si	Cu	Ni	Cr	V	Mo	Nb	Sn	Al	Ti	Co
	Direct combustion	Persulfate-Arsenite	Gravimetric (weighed as $Mg_2P_2O_7$ after removal of arsenic)		Gravimetric (direct oxidation and precipitation after reduction of iron)	Combustion Iodate titration	Perchloric acid dehydration		Weighed as nickel dimethylglyoxime	$FeSO_4$–$KMnO_4$ titration		Photometric			Total	H_2O_2 photometric	Photometric Nitroso-R-salt
1	0.226	[a]0.558	0.046	[b]0.049	0.032	[c]0.032	[d]0.186	[e]0.090	0.067	[f]0.377	[g]0.012	0.013	[h]0.027	[i]0.008	[j]0.027	[k]0.029	0.012
2	.225	.557	.045	[l].045	.033	.034	[d].184	[m].092 / [n].093	.062 / [o].064	[p].369 / [f].371	[q].010	.012	[r].028	[i].009	[s].031 / [t].033	[u].029 / .027	[v].011
3	.221	[w].552	-------	[l].045	-------	.034	.189	[x].100	[o].071	[p].371 / [f].370	[y].014	.011	[z].024	[a'].008	[b'].032	.028	-------
4	.221	[w].559	[c'].052	-------	-------	[w].033	[d',d].180	[m].089	.066	[f,w].376	[e'].014	.013	[z].025	-------	[f'].029	.024	-------
5	.219	[w].55	-------	[l].044	-------	[w].035	[d',d].189	[m].094	[o].066	.380	-------	.015	[g'].021	.009	.032	[h'].024	-------
6	.220	-------	-------	[b].046	-------	.031	.185	[e].097	[o].06	-------	-------	.015	-------	-------	-------	-------	-------
7	[i'].230	[w].550	-------	[l].046	-------	[w].034	[d].191	[n].091	.068	.376	[g].011	.013	[z].030	.008	[j'].031	.026	-------
Average	0.223	0.554	0.048	0.046	0.032	0.033	0.186	0.093	0.066	0.374	0.012	0.013	0.026	0.008	0.031	0.027	0.012
General average	0.223	0.554	0.046		0.033		0.186	0.093	0.066	0.374	0.012	0.013	[k']0.026	0.008	0.031	0.027	0.012

[a] Potentiometric titration.
[b] Molybdenum-blue photometric method. See J. Res. NBS **26**, 405 (1941) RP1386.
[c] 1-g sample burned in oxygen at 1,450 °C, and sulfur dioxide absorbed in starch-iodide solution. Iodine liberated from iodide by titration, during the combustion, with standard KIO_3 solution. Titer based on 93 percent of the theoretical factor.
[d] Double dehydration.
[e] Diethyldithiocarbamate photometric method. See J. Res. NBS **47**, 380 (1951) RP2265.
[f] Persulfate oxidation and potentiometric titration with ferrous ammonium sulfate.
[g] Mercury cathode. Vanadium oxidized with HNO_3 and titrated potentiometrically with ferrous ammonium sulfate.
[h] Ion-exchange. Hydroquinone photometric method. See J. Res. NBS **62**, 1 (1959) RP2923.
[i] Sulfide-iodine method. See BS J. Res. **8**, 309 (1932) RP415.
[j] Mercury cathode-cupferron-aluminon photometric method. See J. Res. NBS **64a**, No. 3, 235 (1960).
[k] Cupferron separation after solution of the sample in diluted HCl (1+2). Vanadium separated by treatment with NaOH.
[l] Alkali-molybdate method.
[m] Electrolytic method.
[n] Neocuproine photometric method.
[o] Photometric method.
[p] Perchloric acid oxidation.
[q] Cupferron—$FeSO_4$–$(NH_4)_2S_2O_8$–$KMnO_4$.
[r] Cupferron—ion-exchange—Nb_2O_5 gravimetric method.
[s] Ether—mercury cathode—8 hydroxyquinoline—Al_2O_3.
[t] Ether—cupferron—eriochrome cyanine R photometric method.
[u] Ether—cupferron—H_2S—cupferron—TiO_2.
[v] Chromium removed with ZnO.
[w] Titrating solution standardized by use of a standard steel.
[x] Diethyldithiocarbamate photometric method.
[y] $NaHCO_3$—$FeSO_4$–$(NH_4)_2S_2O_8$–$KMnO_4$.
[z] Niobium hydrolyzed with $HClO_4$ and H_2SO_3. ASTM method E30–56.
[a'] Tin preferentially precipitated with ammonium hydroxide in the presence of ferrous iron, solution of the precipitate is reduced with Stanreduce and titrated with standard iodate.
[b'] CrO_2Cl_2—ether—aluminon photometric method.
[c'] Weighed as ammonium phosphomolybdate.
[d'] Sulfuric acid dehydration.
[e'] Nitric acid oxidation and potentiometric titration with ferrous ammonium sulfate.
[f'] Weighed as $AlPO_4$.
[g'] H_2SO_3 hydrolysis—tannic acid—pyrogallol photometric method.
[h'] Vanadium separated with Na_2CO_3.
[i'] Differential gaseometric method.
[j'] Mercury cathode—cupferron—eriochrome cyanine R photometric method.
[k'] Values reported for niobium by analysts 3, 4, 5, and 7 include small amounts of tantalum present in the sample.

List of Analysts

1. Ferrous Laboratory, National Bureau of Standards. J. I. Shultz, in charge. Analysis by B. B. Bendigo and J. I. Shultz.
2. R. H. Rouse, Bethlehem Steel Co., Sparrows Point Plant, Sparrows Point, Md.
3. H. W. Huston, A. M. Byers Co., Ambridge, Pa.
4. C. A. Trathowen, Jones and Laughlin Steel Corp., Pittsburgh, Pa.
5. D. P. Robertson, Weirton Steel Co., Weirton, W. Va.
6. W. E. Walters, Pittsburgh Testing Laboratory, Pittsburgh, Pa.
7. W. F. Horscroft, Bethlehem Steel Co., Homer Research Laboratory, Bethlehem, Pa.

The steel for the preparation of this standard was furnished by the Bethlehem Steel Co., Bethlehem, Pa.

WASHINGTON, D.C.
September 30, 1964.

A. V. ASTIN, *Director.*

U.S. GOVERNMENT PRINTING OFFICE: 1964—O-748-843

Fig. 16-14. A U.S. National Bureau of Standards Certificate of Analyses.

analysts, have been published.[28] These samples are no longer generally available but other silicate rock standards have been issued by the Survey.[29]

Varying the conditions of analysis. A *method* of analysis can often be improved by varying the conditions in a systematic manner. In an analysis involving the formation of a precipitate, the most important variables are:

1. The chemical composition of the solution in which the precipitate is formed.
2. The concentration of the solution.
3. The temperature of precipitation.
4. The manner of precipitation: speed of addition of reagent, manner of mixing (reagent to solution or vice versa).
5. Treatment after precipitation: time of standing, digestion, wash solution used, etc.

The first factor, the composition of the solution, is a most important one. Constituents which would be precipitated with the substance sought must necessarily be removed or otherwise rendered incapable of precipitation or interference. Generally, if such interference is possible, a *separation* of the interfering substances from the constituent in question must be made, either by precipitating the undesired constituents or by precipitating the constituent sought. A reprecipitation is usually required. Sometimes physical and semiphysical methods of separation are applied in preference to precipitation. The accurate determination of a substance in a pure solution is nearly always an easy matter. The determination of the same substance when it is associated with others may be, and often is, one of great difficulty, as already emphasized. Methods of separation are often imperfect, and there is room for much improvement in this field.

Even when the foreign substances do not yield precipitates with the reagent added, the precipitate of the desired constituent often contains greater or lesser amounts of the former (coprecipitation, Chapter 11). Coprecipitation is a troublesome source of error in analyses involving precipitations and one cannot predict with certainty whether a given precipitate can be obtained sufficiently pure for quantitative purposes if other substances, normally unprecipitable, are present. Great caution must therefore be used in applying to a mixture a method which works well for a pure substance, even if the other substances do not precipitate of themselves with the reagent used. Thus sulfur in a sulfate solution cannot be determined with any accuracy by precipitation as barium sulfate if chlorate or nitrate is simultaneously present. In spite of the easy solubility of barium chlorate and nitrate under the conditions of the determination, relatively large amounts of these salts are present in the barium sulfate precipitate. Good results can be obtained only by destroying chlorate and nitrate before precipitating sulfate with barium.

Foreign substances may increase the solubility of a precipitate and even prevent its formation as the result of chemical interaction (see the effect of large amounts of magnesium on the precipitation of calcium oxalate, p. 625).

Sometimes the interference of foreign substances can be prevented in a simple manner by a slight alteration in the conditions of the determination. Thus, by changing the acidity of the solution or by adding a complex-forming reagent, the precipitation of a foreign substance can often be prevented or minimized.

The effect of the temperature of precipitation and the concentration of the solution at the time of precipitation are considered in detail elsewhere. Variations in these conditions lead to a change in the amount of coprecipitated substance and in the filterability of the precipitate and are therefore of great importance.

[28] See H. W. Fairbairn et al. and R. E. Stevens et al., cited in Ref. 22, and M. Fleischer and R. E. Stevens, *Geochim. Cosmochim. Acta* **26**, 525 (1962).

[29] F. J. Flanagan, *Geochim. Cosmochim. Acta* **31**, 289 (1967). For other sources of geochemical standards (rocks, ores, cement, refractories, glass, etc.), see F. J. Flanagan and M. E. Gwyn, *Geochim. Cosmochim. Acta* **31**, 1211 (1967). For analyses of some of these, see S. S. Goldich et al., *Canad. J. Earth Sci.* **4**, 747 (1967), and G. R. Webber, *Geochim. Cosmochim. Acta* **29, 229** (1965).

A method of analysis can be improved empirically by varying in turn the factors enumerated at the beginning of this section, while maintaining the others constant, and noting the effect of such changes upon the results.[30] Or from theoretical considerations it may be possible to decide immediately what conditions should be changed to attain the desired results. The last is becoming more and more the method of modern analytical chemistry. One of the purposes of this book is to show how the accuracy of analytical methods can be increased by the intelligent application of physicochemical principles.

Resort to other methods. If a given method of analysis is unsatisfactory, either because it gives results of insufficient accuracy or otherwise, it is sometimes possible to reject it in favor of another. An example is afforded by the methods for determining aluminum. The old procedure, which is still extensively used, involves precipitation of hydrous aluminum oxide with ammonium hydroxide and ignition of the precipitate to aluminum oxide. In the newer procedure the metal is precipitated as aluminum 8-hydroxyquinolate; the precipitate is dried at 130°C and weighed as the hydroxyquinolate. Aluminum oxide must be ignited at a very high temperature to render it anhydrous, and the ignited residue is likely to be hygroscopic; aluminum hydroxyquinolate, on the other hand, is easily dried at a low temperature, and is not hygroscopic. Moreover, the high molecular weight of the organic weighing form reduces the weighing error and enables even small amounts of aluminum to be determined with accuracy:

$$\frac{2\ \mathrm{Al}}{\mathrm{Al_2O_3}} = 0.5291 \qquad \frac{\mathrm{Al}}{\mathrm{Al(C_9H_6ON)_3}} = 0.05872$$

It should not be forgotten that a radical change in procedure may allow determinations to be made with accuracy that could otherwise not be made at all. Where gravimetric and volumetric methods fail, a colorimetric method, may, for example, give the desired results. Neutron activation now enables extremely low concentrations of elements to be determined, which earlier could not be determined at all. To be sure, a particular analyst or laboratory may not always have at his or its command the most advantageous method for a given analysis.

Accuracy in calculations

Care must be taken not to decrease the accuracy of the value reported in an analysis by unwarranted approximations in computations. On the other hand, the final value of an analysis should be expressed in such a way that the accuracy of the result is indicated; i.e., the number of figures in the reported result should be consistent with the probable precision of the analysis.

Significant figures. As the name implies, significant figures of a value are those that have some practical meaning. The number of significant figures in an experimentally determined value expresses the precision of its measurement. *The significant figures of a number include all the certain digits and the first (and first only) doubtful digit of that number.* Thus, if an object has been weighed to the nearest 0.1 mg on the analytical balance, its weight is recorded correctly by stating the value in grams to the fourth decimal place, e.g., 12.1230 g. There are six significant figures in this number. It would be incorrect to give the weight as 12.123 g, for this would imply that the weighing was made only to the nearest milligram and that the probable precision is ±1 mg. On the other hand, if the object were weighed on a rough balance sensitive to 0.01 g, it would be incorrect to record the value as 12.120 g; the weight should then be put down as 12.12 g. The final zeros of numbers must never be omitted when they are significant figures or included when they are not.

[30] A more efficient method, which has the advantage of showing possible interaction among variables, involves a statistical design in which the effect of several factors is found at the same time by the analysis of variance. Consideration of such factorial experiments lies beyond the scope of this book; reference should be made to statistical works (see Ref. 4).

When the zero comes before the decimal point and no decimals are given, it is impossible, of course, to tell whether it is a significant figure. Thus, in 100 ml, the zeros are needed to place the decimal point and they may or may not be significant. The use of exponential notation resolves the ambiguity. If the volume has been measured to within 1 ml, 100 has three significant figures and should be writttn 1.00×10^2.

Zeros lying immediately after the decimal point are not significant figures. Thus in the number 0.0019 there are only two significant figures.

Hurley[31] has made the following suggestion regarding the writing of the last significant figure:

Perhaps the most satisfactory way to treat the last significant figure is to give a rough indication of its doubtfulness by the use of subscripts. Using this convention, a measurement written as 21.23 would indicate a precision (average deviation of a single observation) of ±0.01 unit. If the average deviation is greater than this, the number would be written as 21.2_3. *The adoption of this convention reduces all rules for the use of significant figures to the following. In writing a number which represents a measured quantity, use the number of significant figures which comes nearest to expressing the real precision of the measurement, using a subscript if the last figure is less precise than ±1.*

Computations. In carrying through a series of computations, it is best to retain one digit beyond the last significant figure in order that the latter may not be altered in the arithmetical processes. Thus, in obtaining the average of 25.05, 25.07, and 25.07, the result should be written 25.063 if this number is to be used in calculations; otherwise the last significant figure alone is retained: 25.06. In rounding off values to the significant figure, the succeeding digit is dropped if it is less than 5; if it is 5 or greater, the significant figure is increased by 1. Thus the average of 0.1246, 0.1244, and 0.1247 is 0.1246 (0.12456).

In multiplications and divisions it must be remembered that the percentage of uncertainty, and not the absolute uncertainty, is transmitted to the result. Thus $(10.01 \pm 0.01) \times (10.01 \pm 0.01) = (100.2 \pm 0.2)$. The product 5×23.41 should be written 117.1; or if the value is to be used in further computations, one more figure is retained: 117.05. The use of logarithms is very advantageous when a number of multiplications and divisions are to be made. Not only is the time required for the computations much less than in the long-hand method, but the superfluous figures obtained in ordinary multiplication are automatically dropped and the last figure rounded off. Four-place logarithm tables are sufficiently precise for most analytical calculations, but the use of five-place tables is convenient to avoid interpolating. A slide rule may be used for checking results to detect errors in computation of the order of a tenth of 1 per cent, but it should not be used in the first place to obtain the results, since it usually is not sufficiently exact.

Statement of final results. The result of a calculation involving several experimentally determined quantities can be no more accurate than the least accurate value included. As an illustration consider an analysis for sodium by precipitation as sodium zinc uranyl acetate and weighing as such. Suppose that the weight of sample was 0.0496 g and the weight of precipitate obtained was 1.0220 g. The percentage of sodium present is

$$\frac{1.0220[Na/NaC_2H_3O_2 \cdot Zn(C_2H_3O_2)_2 \cdot 3\,UO_2(C_2H_3O_2)_2 \cdot 6\,H_2O]}{0.0496} \times 100$$

$$= \frac{1.0220 \times 0.01495 \times 100}{0.0496} = 30.8$$

[31] F. H. Hurley, *J. Chem. Educ.* **17**, 334 (1940).

The result is reported as 30.8 per cent, and not as 30.80 or 30.802 for several good reasons. First, since the error in weighing the sample may amount to about 0.2 mg, the result may be in error by 1 part in 250 (i.e., 2/496) from this source alone. The error in weighing the dried precipitate is only 1 part in 5000, but this high precision is more than cancelled by the error in weighing the sample. Then the probable accuracy of the *method* must be taken into account. The method probably is accurate to 1 part in 250 in the hands of a practiced analyst. Taking these facts into consideration, it is apparent that the percentage should be reported as 30.8, because the error of this determination is not likely to be less than 1 part in 300. The figure following the decimal point is the first doubtful digit. It would be incorrect to write the result as 30.80, because this would imply that the value is correct to the figure in the first decimal place, which is not true. Incidentally, it is a waste of time to weigh the dried precipitate to the nearest 0.1 mg; weighing to the nearest 1 mg is sufficiently precise.

In ordinary analysis a result can seldom be relied on to ± 0.1 per cent (relative error), so that more than four significant figures should never be used in reporting a result. Sometimes three significant figures will suffice. The agreement between duplicates should be taken into consideration when deciding how many figures to include in the average. If the results of a duplicate analysis give, say, the percentages 62.04 and 62.39, it would be useless and absurd to state the average as 62.22, because even the third figure is uncertain; the mean should be written 62.2 per cent. On the contrary, if the duplicates gave the values 62.03 and 62.09 per cent, should be given as 62.06 per cent, because when so written the figures indicate that the result is correct to the first decimal place but not to the second. More than one uncertain figure should not be included in a final result.

PROBLEMS

1. Classify the following errors as additive or proportional (or as both or neither): impurities in reagents, solubility of a precipitate, adsorption of moisture by an ignited residue and by a weighing vessel, electrification of object weighed, co-precipitation, volatilization of a constituent during decomposition of the sample, use of uncalibrated weights, use of wrong chemical factor, personal equation, buoyancy of air in weighing, balance arms of unequal length.
2. Give several examples of errors in quantitative analysis whose signs can be only positive or negative.
3. E. Mitscherlich (the discoverer of isomorphism) in 1829 determined the sodium chloride content of $NaCl \cdot 2\,H_2O$ by heating to expel the water and found 61.98 per cent NaCl compared to the true value 61.86 per cent. What were the absolute and relative errors of his determination of (a) NaCl; (b) H_2O?
4. The following values were obtained in parallel determinations of sulfur in a phosphor-bronze bearing metal: 0.097, 0.098, 0.100, 0.100, and 0.098 per cent. Obtain the average and standard absolute and relative deviations of a single determination and of the mean.
5. Compare the standard deviations of rest points of a balance obtained by (a) multiple swings, four readings on one side, three on the other; (b) single deflection, assuming that the standard deviation in reading the scale is 0.1 division (see p. 488 et seq.) and other random errors can be neglected.
6. From long experience an analyst knows that his standard deviation in the micro determination of chlorine in soluble chloride samples containing from 48 to 60 per cent Cl is 2.5 ppt and that the determinate error of the procedure is negligible. A single determination of chlorine in a particular sample gave him the value 51.14 per cent. What is the probability that the true chlorine content of the sample lies in the range 50.88–51.40 per cent? What assumptions are made in obtaining the probability? *Ans.* ~ 0.95.

7. Five determinations of Cl in a chloride sample gave a mean of 32.30 per cent and $s = 0.13$ per cent absolute. State the interval within which the population mean lies at the 95 per cent confidence level. *Ans.* 32.14–32.46 per cent.
8. Six gravimetric determinations of water in a 100-mg sample of a hydrate gave a mean of 15.10 per cent H_2O with a standard deviation of 3.3 ppt for a single determination. (a) What is the probability that the value 15.10 per cent lies within 3.3 ppt of the true mean? (b) What is the probability that in the future the average of duplicate determinations of water in the same sample under the same conditions will lie within 0.10 per cent absolute of the true mean?
9. In a certain determination of A the value 0.064 per cent (uncorrected for the blank) was obtained. The blank gave a value of 0.032 per cent A. The standard deviation of the determination of A in both sample and blank is 100 ppt, based on 10 runs. What is the range in which the percentage of A in the sample lies with a probability of 0.95, the usual assumptions being made? *Ans.* 0.016–0.048 per cent.
10. The extraction constant K of a metal (M^{3+}) extracted as MR_3 into an immiscible organic solvent (o) from aqueous solution (w) by the reagent HR,

$$K = \frac{[MR_3]_o[H^+]_w^3}{[M^{3+}]_w[HR]_o^3}$$

is determined by finding $[MR_3]_o$ and $[HR]_o$ at equilibrium. The random error in $[MR_3]_o$ and $[HR]_o$, expressed as standard deviation, is estimated as 50 ppt for each. $[M^{3+}]_w$ at equilibrium is found by difference from the known initial concentration of M^{3+} in the aqueous phase, the error of which may be considered negligible. Approximately one third of the metal is extracted into o, and the phase volumes are equal. $[H^+]$ is found from pH measurement (pH 2.00), which has a standard error of ± 0.02 pH unit. What is the standard error in K in ppt?
11. Duplicate gravimetric micro determinations of silver in similar copper–silver alloys gave the following values:

Ag, %	
17.77,	17.87
20.94,	21.11
18.54,	18.65
18.14,	18.18
18.91,	19.08

The sample weight was approximately 15 mg in each determination. Estimate the standard deviation of the procedure under the conditions. [*Hint*: First show that $s = \{(1/2n) \sum_n (X_2 - X_1)^2\}^{1/2}$, where n = number (i.e., sets) of duplicate determinations (and also the number of degrees of freedom), and $X_2 - X_1$ is the difference between paired values.

Ans. $s = 0.09$ per cent absolute, or 5 ppt. Because all the samples have much the same Ag content, s is assumed to have about the same absolute value for each.
12. (a) How many parallel determinations must an analyst run to be 95 per cent confident that $\bar{X} - \mu$ will not exceed the standard deviation of a single determination? (b) How many at the 99 per cent confidence level? *Ans.* (a) 6.
13. Referring to the tabulation on page 385, what is the average error (ppt) in the determination of from 0.1 to 0.4 g of Al_2O_3 by precipitation with ammonia, followed by ignition to aluminum oxide? State the range in which the error of a single determination will lie at a confidence level of 95 per cent. If a triplicate determination is run, in what range will the error of the mean lie at the same confidence level?

Ans. $\epsilon/X_\tau \times 1000 = (9.3 + 11.4 + 8.4 + 10.2)\frac{1}{4} = 10$ ppt average error; range of error at 95 per cent confidence level $= 10 \pm (1.3 \times 3.18) = 6$–14 ppt; range of error of mean of three determinations = 7.6–12.4 ppt.

14. What is the probability that the error of the mean of the four determinations of Al_2O_3 on page 385 (9.8 ppt) is due to indeterminate errors?
Ans. Much less than 0.005 since $t = 9.8(4)^{1/2}/1.3 = 15$, and the probability of $t > 5.8$ is 0.005 (i.e., 0.01/2) for $n = 3$ from Table 16-2. Moreover, the individual values are all higher than the true value. Besides, the high values can be explained. We can be certain that random errors cannot account for a deviation of the mean as large as 10 ppt.
15. In a study of the reliability of $CaC_2O_4 \cdot H_2O$ as a weighing form for calcium, solutions containing the equivalent of 0.3001 g of calcium carbonate (99.92 per cent $CaCO_3$, the remainder moisture) were precipitated under appropriate conditions, and the precipitate dried at 105°C and weighed with the following results: 0.4420, 0.4422, 0.4410, 0.4415, and 0.4409 g. (a) Find the standard deviation of a single determination and of the mean; (b) obtain the average *error* of the determinations. (c) May it be concluded that there is a determinate error in the method?
Ans. (a) $s = 1.3$ ppt; $s_{\bar{x}} = 0.6$ ppt; (b) +8.4 ppt.
16. In an investigation of the suitability of a certain method for the determination of sodium, the following results were obtained:

NaCl *Taken*, g	NaCl *Found*, g
0.01263	0.01257
0.01127	0.01128
0.01215	0.01207
0.01185	0.01186
0.01317	0.01312
0.01258	0.01252

(a) Find the standard relative error of the determinations. (b) May it be concluded that there is a determinate error in the method?
17. An electrolytic determination of Cu in a 20-mg sample of pure $CuSO_4 \cdot 5\,H_2O$ (25.46 per cent Cu) gave the value 25.32 per cent Cu. If the standard deviation of a rest point is 0.009 mg, based on 21 replicates, can the error reasonably be attributed to weighing error alone, assuming normal distribution of deviations?
Ans. $p_r = 3.5$ ppt Cu (standard deviation from weighing error)

$$t = -\frac{0.14 \times 1000}{25.5 \times 3.5} = -1.6$$

Probability ($t_{n=20} > -1.6$) $\sim$ 0.2.

The difference could thus be due to weighing instead of a determinate error, such as incomplete deposition of Cu.
18. Standardization of a solution of hydrochloric acid against sodium carbonate and borax gave the following results:

Na_2CO_3, *N*	*Borax*, *N*
0.18079	0.18081
0.18087	0.18080
0.18087	0.18081
0.18091	0.18085
0.18082	0.18085
0.18083	
0.18085	

(a) Is there a statistically significant difference in the precision of the two series of titrations at the 95 per cent confidence level? (b) Do the means of the two series differ significantly?

$$Ans. \text{ (a) } F = \frac{15.6 \times 10^{-10}}{6 \times 10^{-10}} = 2.6;$$

for $n_1 = 6$, $n_2 = 4$, $F = 6.16$ at the 95 per cent confidence level. There is no statistically significant difference in precision.

19. The results obtained by the cooperating analysts in the determination of molybdenum in Bureau of Standards steel No. 106 by two different methods are as follows:

Gravimetric Method: Mo, %	*Colorimetric Method:* Mo, %
0.164	0.161
0.166	0.165
0.164	0.156
0.159	0.163
0.165	0.156
0.160	0.171
0.174	0.160
0.159	
0.170	

(a) How do the two methods compare in precision? (b) Is the difference between the means of the methods statistically significant? (c) What can be said about the accuracy of the figure for molybdenum in the sample?

20. Six analysts determined Cu in brass by the same micro method and obtained the following results, each value of Cu found being the average of two determinations:

Cu *Present,* %	Cu *Found,* av., %
60.00	60.14
60.00	60.3
72.14	72.39
72.14	72.26
70.34	70.61
71.5	71.8

Is there a constant error in the method at the 95 per cent confidence level?

21. Two procedures for filling cells in absorption spectrophotometry were compared for precision. In procedure A the absorption cell was removed from the spectrophotometer for each filling with a potassium chromate solution. In procedure B the cell was emptied, washed, and filled (with a similar chromate solution) without being removed from the instrument. The same wavelength (373 mμ) was used for both series. Replicate absorbances had the following values for the two procedures

Procedure A	*Procedure* B
0.493	0.546
0.492	0.546
0.490	0.547
0.491	0.546
0.491	0.547

May it be concluded that the two procedures of cell filling give significantly different precisions (95 per cent confidence level)?

22. The partition coefficient (P) of 8-hydroxyquinoline between chloroform and water was determined after various times of shaking, with 8-hydroxyquinoline initially present in the aqueous phase:

Time, min	$P(CHCl_3/H_2O)$
2	334
4	337, 335
10	329, 341
20	339

Is equilibrium attained after 2 min of shaking?

23. G. P. Baxter and A. H. Hale, *J. Am. Chem. Soc.* **56**, 615 (1934), obtained the following values for the reacting ratio I_2O_5/Na_2CO_3:

3.14949
65
45
56
55
50
46
48
37
Av. 3.14950

Two of these values seem aberrant. Adopting the criterion that they may be rejected if their deviations from the mean of the other seven have a probability $\leqq$ 0.05 of occurring, should either, both, or neither be discarded in computing the average? Answer the same questions for the probability level $\leqq 0.01$. Which criterion seems preferable?

24. Lack of material limits the sample weight to 25 mg in a microgravimetric determination of water of hydration in $CaSO_4 \cdot 2\,H_2O$ by loss of weight on ignition. The standard deviation of a rest point of the microbalance is 0.003 mg, based on 10 replicates. (a) What is the standard deviation of the water content in ppt and in absolute per cent from weighing error? (b) What is the range in which the true water content will lie (if determinate errors are insignificant) at the 99 per cent confidence level? ($2\,H_2O/CaSO_4 \cdot 2\,H_2O = 0.209$.)

Ans. (a) ± 1.2 ppt, ± 0.025 per cent.

25. What is the minimum weight of sample that may be taken in the gravimetric determination of S in Na_2SO_4 as $BaSO_4$ for reasonable assurance (probability 0.99) that the relative deviation of the S percentage will not exceed 5 ppt if the standard deviation of a rest point is 0.05 mg? Assume that the standard deviation of a rest point has been estimated from 10 replicates. $S/BaSO_4 = 0.137$, $S/Na_2SO_4 = 0.226$.

Ans. 66 mg (more exactly, 61 mg). A sample of ~70 mg would be weighed.

26. What weight of sample should be taken to give reasonable assurance (probability of 0.99) in the gravimetric determination of silver as AgCl in an alloy containing 50 per cent silver that the weighing error will not result in an error greater than 5 ppt in the percentage of Ag, if the standard deviation of a rest point is 0.005 mg? Assume that the standard deviation of a rest point is based on 10 replicates. $Ag/AgCl = 0.753$.

27. Show that if P = percentage of constituent
N = normality of standard solution
E = equivalent weight of constituent in titration
w = standard deviation of a rest point in grams
v = standard deviation of buret reading in milliliters

the weight of sample U that must be taken in a volumetric determination to yield a standard deviation $\pm p_r$ in parts per thousand in the percentage of the constituent found is

$$U = \frac{140}{p_r}\left[100w^2 + \left(\frac{NEv}{P}\right)^2\right]^{1/2}$$

(a) What weight of sample must be taken to make the standard deviation of the percentage of the constituent equal to 1 part per 1000 in a titration with 0.1 N solution if the percentage of constituent is 100, its equivalent weight is 50, $w = 1 \times 10^{-4}$ g and $v = 0.01$ ml? (b) What weight of sample will provide a probability of 0.99 that the indeterminate error in P arising from weight and volume deviations will not exceed ± 1 ppt? Assume that w and v are based on 10 replicates.

28. Suppose that under the conditions of Problem 27 the uncertainty in locating the end point (based on six replications) is 0.04 ml. What weight of sample is required for an assurance of 0.95 that the indeterminate error in a titration will not exceed ± 2 ppt?

29. F. Haber in the 1920s determined the gold content of sea water by isolating gold as the sulfide and finally obtaining the metal in the form of a minute sphere whose mass was found by measuring its diameter under the microscope. From the following data, calculate the standard deviation of a determination in ppt:
Gold content of seawater: $\sim 1\ \mu g/m^3$
Sample size: 2 liters
Standard deviation of diameter measurement: $\pm 0.2\ \mu$
Density of Au: 19.3 *Ans.* ~ 90.

30. A mixture of anhydrous NaCl and KCl weighing 0.3000 g gave 0.6000 g of AgCl. If the standard deviation of a rest point is 0.1 mg, what is the standard deviation in absolute per cent of the NaCl and KCl contents?

31. Analysis of a specimen of pyromorphite gave the following percentages:

PbO	81.21	
CaO	0.45	
P_2O_5	15.77	
As_2O_5	Trace	
Cl	2.64	
Remainder	0.48	(Fe_2O_3, 0.02; Al_2O_3, 0.02; H_2O, 0.14; insoluble, 0.30)
	100.55	
Less O$\equiv$Cl	0.60	
	99.95	

The summation is very satisfactory. Is the charge balance satisfactory?
Ans. Yes. The positive and negative components balance to within 2 parts in ~ 740.

32. (a) Five determinations of ZrO_2 in a clay yielded the percentages 0.21, 0.23, 0.24, 0.22, and 0.23. How shall the average be reported? Why? (b) In a determination of water in a sample of $CaSO_4 \cdot \frac{1}{2} H_2O$, 0.2402 g of the material was ignited to expel the water and the loss in weight found. If the weighings were made to ± 0.1 mg, how many figures shall be used in expressing the percentage?
Ans. (a) 0.23 per cent.

33. (a) In obtaining the weight of CrO_3 equivalent to 0.1067 g of $K_2Cr_2O_7$, which of the following factors should be used: 0.67983, 0.6798, 0.68? (b) Which factor should be used in converting 0.4106 g of Fe_2O_3 to Fe: 0.69944, 0.6994, 0.699, 0.7?
Ans. (a) 0.68.

34. A procedure for the spectrophotometric determination of manganese in steel calls for a 1-g sample. If the Mn content of the sample is ~0.1 per cent, how accurately should the weighing be made?

35. Round the results of the following operations to the proper number of digits: (a) $106.4 + 2 \times 35.457$ (the molecular weight of $PdCl_2$); (b) $\frac{4}{3} \times \pi \times [(5.0/2) \times 10^{-2}]^3 \times 13.546$ (the mass in grams of a mercury globule of diameter 5.0×10^{-2} cm); (c) the number whose $-\log = 2.00$ ($[H^+]$ of a solution of pH 2.00); (d) $\log 100.0 - \log 95.5$ (absorbance of a solution of 95.5 per cent transmittance).

36. Vance and Borup (Chap. 64) determined zinc gravimetrically by precipitating with 8-hydroxyquinoline at 60°C and drying the washed precipitate at 125°C. When the theoretical factor, $Zn/Zn(C_9H_6ON)_2 = 0.1850$, was used, the errors in 12 determinations were the following:

Zn *Taken*, mg	*Error*, mg *of* Zn	Zn *Taken*, mg	*Error*, mg *of* Zn
77.0	−0.34	51.9	−0.33
78.9	−0.34	53.0	−0.30
78.8	−0.36	52.4	−0.35
76.8	−0.53	51.5	−0.23
77.8	−0.52	52.9	−0.30
78.0	−0.46	53.6	−0.36

The error due to solubility of the precipitate is not more than 0.01 mg of Zn under the conditions.

Plot the errors versus the amounts of zinc and show that it may reasonably be concluded that the error is proportional to the amount of zinc. Obtain the best value for the empirical factor. What is the probability that for ~50 mg of Zn, the error will not exceed ±0.15 mg or ±3 ppt when the empirical factor is used?

37. F. Nydahl (1954) determined sulfate in an HCl solution under constant conditions by precipitating and weighing as $BaSO_4$. Three sets of results, four determinations in each, gave:

S *Taken:* $BaSO_4$, mg	S *Found:* $BaSO_4$, mg	$\bar{d}_{\bar{X}}$, mg
1.22	1.13	0.025
12.24	12.34	0.027
36.58	37.19	0.034

On the assumption that the values of $BaSO_4$ found are subject to a constant solubility error and a proportional coprecipitation error, set up an equation of the form $y = ax + b$ which will give the amount of sulfur (as $BaSO_4$) present within a few hundredths of a milligram.

38. For a certain group of laboratory personnel the standard error of a single buret reading was 0.021 ml, whereas the standard error of the difference of two readings was 0.018 ml. Calculate the correlation coefficient r.

39. Apply Eq. (27), page 409, to find the interval in which the iron content (50 μg) lies at the 95 per cent confidence level in Example 11.

40. Show, using additivity of variances, that $s_{\bar{X}} = s/N^{1/2}$, in which s is the standard deviation of a single measurement in a series of N measurements.

41. Students in a beginning quantitative analysis class obtained the following values (average of duplicates) in the gravimetric determination of Cl (p. 584) in a sample of soluble chloride containing 51.12 ± 0.02% Cl:[32]

Cl (%)	Cl (%)
51.02	51.19
51.02	51.21
51.04	51.22
51.06	51.24
51.08	51.25
51.11	51.26
51.14	51.42
51.17	51.45
51.19	51.51
	52.11

(a) Plot N vs. percentage of Cl, using size classes 51.0–51.1, 51.1–51.2, ... As far as can be judged, is the distribution Gaussian? Indicate X_τ and $\bar{X}$ (obtained in *b*) in the plot. (b) Rejecting any aberrant values, calculate $\bar{X}$ and s of the average of duplicates and of the mean. (c) What are the 95 per cent and 99 per cent confidence intervals for the mean obtained by the class? (d) Is there a significant difference between the mean value obtained by the class and the true Cl percentage in the sample, i.e., does there seem to be a determinate error in the procedure as carried out by this class?

[32] We are indebted to Professor Peter J. Lingane for these data.

PART TWO

Practical Fundamentals

CHAPTER 17 *Reagents*

Solid reagents

In quantitative analysis only chemicals of the highest purity should be used as a rule; impure reagents may entirely vitiate the results of an analysis. The U.S.P. grade of chemical (conforming to the standards of the *United States Pharmacopoeia*) is usually less pure than the C.P. ("chemically pure") grade, and the latter less so than the "analyzed." The "analyzed," or reagent quality chemical as it is also called, usually comes with a label bearing a statement of the percentages of the important impurities present. Only reagent grade chemicals should, in general, be used in quantitative analysis.[1] Analyzed reagents should conform to the specifications of the American Chemical Society Committee on Analytical Reagents.[2] The statement of analysis on the bottle must not be accepted, however, as always indicating the true amounts of impurities present:

1. The analysis of the manufacturer may not always be correct, owing to the use of faulty methods or for other reasons. Often only the maximum limits are indicated.
2. Certain impurities may not have been tested for at all.
3. In packing the chemical, foreign material such as dust, bits of cork, and paper may have been introduced.
4. The reagent may have been contaminated after its receipt from the manufacturer if the bottle has been opened.

If there is any doubt as to the purity of a reagent, it should be tested by standard methods for the impurities that would cause errors in the determination.[3] There is no such thing as an entirely pure reagent; however, the amount and kind of foreign substance present in a particular case may be so small that it will not affect the results. If the purest reagent obtainable is still too impure for the purpose, it must be purified by an appropriate method; sometimes a blank can be run along with the determination (see p. 416) and the result corrected.

[1] Naturally, in certain analyses certain impurities in a reagent may be entirely unobjectionable. An impurity may be harmful or harmless depending upon circumstances. Thus ammonium hydroxide containing ammonium carbonate may be used for precipitating ferric hydroxide in pure solutions in a gravimetric determination of iron, but not in the presence of the alkaline earth metals, which would be precipitated partially as the carbonates. It is best to have available the purest reagents, because usually these will be used for a variety of purposes, and an impurity that does not interfere in one analysis may do so in another. The use of the purest reagents is especially important in analyses involving separations.

[2] Am. Chem. Soc. Committee on Analytical Reagents, *Ind. Eng. Chem.* **17**, 756 (1925); **18**, 636, 759 (1926); **19**, 645, 1369 (1927); **20**, 979 (1928). *Ind. Eng. Chem., Anal. Ed.* **1**, 171 (1929); **2**, 351 (1930); **3**, 221 (1931); **4**, 154, 347 (1932); **5**, 289 (1933); **12**, 631 (1940); **16**, 281 (1944); **19**, 210 (1947).

[3] Am. Chem. Soc. Committee on Analytical Reagents (Ref. 2); *Reagent Chemicals*, American Chemical Society, Washington, 1961; J. Rosin, *Reagent Chemicals and Standards*, Van Nostrand, Princeton, N.J., 5th ed., 1966.

Purification of solids

Recrystallization from solution.[4] The most common method of purifying soluble solids is by recrystallization from a suitable solvent, usually water. If the substance A containing a small amount of the soluble substance B as impurity is dissolved in the proper volume of hot water to give a saturated solution which is then allowed to cool, the crystals of A deposited will usually contain a much smaller amount of B than the original substance, because the cooled saturated solution of A will generally not be saturated with respect to B, which therefore remains in solution. The recrystallized salt will not be entirely free of the impurity even if carefully washed, because crystals separating from a solution always contain included mother liquor in the form of vacuoles (p. 194), the amount of liquid thus held often amounting to a few tenths of a per cent. It is of interest to note that the number of liquid inclusions in water-soluble crystals may depend upon the presence of small amounts of foreign substances in the solution as well as on the rate of crystallization. Very low concentrations of manganese result in the formation of numerous liquid inclusions in sodium chloride (Fig. 17-1). On the other hand, 250 ppm of Ni^{++} in solution allows ammonium dihydrogen phosphate to be crystallized from water without even microscopic inclusions and with only very slight contamination by nickel.

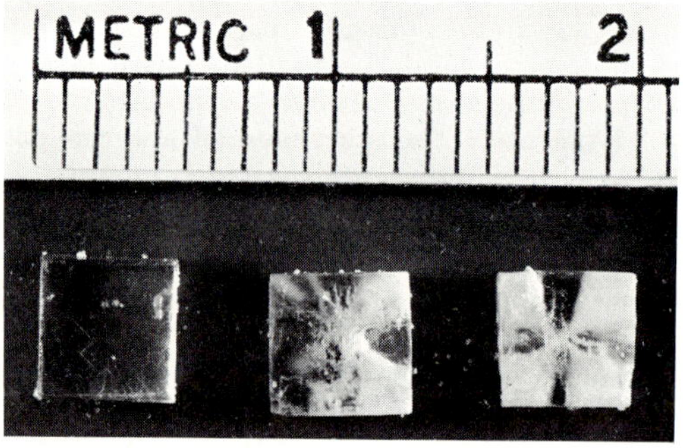

Fig. 17-1. Effect of small amounts of manganese on formation of liquid inclusions in sodium chloride. Crystals grown from brines containing respectively 0.00, 0.02, and 0.4 ppm of Mn(II); solution evaporated at ~50°C with mechanical stirring. From C. Surasiti and E. B. Sandell, *Nature* **203**, 60 (1964).

By a second recrystallization A can be obtained virtually free of B, provided mixed crystals are not formed (see p. 438).

In purifying a solid by recrystallization, a roughly weighed amount is dissolved in a sufficient volume of water to give a saturated or nearly saturated solution at the boiling point. The hot solution is filtered to remove insoluble material, which is always present in greater or less amount. Filter paper placed in a funnel with a broken-off stem (to hinder crystallization in the funnel) may be used for filtration, but a Büchner funnel with a sintered-glass plate (p. 468) is more suitable; a coarse sintered-glass filter crucible may be employed if the volume of liquid is small. Then the filtered solution is cooled with stirring. In the case of very soluble substances, it may be advisable to cool in ice to obtain a good yield. Finally, the solid is collected in a Büchner funnel with a sintered-glass disk (filter paper is often objectionable because fibers of cellulose are likely to be introduced), and as much as possible of the mother liquor is sucked off.[5] The solid may be pressed down in the funnel with the flat side of a glass stopper or the like, to aid in removing the liquid. To remove adherent solution, the crystals are washed with small portions of cold water, suction being applied after each addition. The recrystallized salt, if sufficiently purified by one recrystallization, is dried at or above

[4] For a discussion of purification by recrystallization, see T. S. Tipson, *Anal. Chem.* **22**, 628 (1950).

[5] Centrifugal drainage can be used to advantage in ridding the crystals of mother liquor. See T. W. Richards, *J. Am. Chem. Soc.* **27**, 104 (1905); N. F. Hall, *ibid.* **39**, 1148 (1917).

room temperature, depending upon the material. The solid thus dried will contain included (vacuole) water unless it has been heated to a high temperature. This water may be removed by reducing the crystals to a very fine powder by grinding in an agate mortar and again heating. The complete expulsion of water is necessary only when the recrystallized solid is to be used as a standard substance (p. 445). The pure dry salt thus obtained should be kept in a glass-stoppered bottle to protect it from atmospheric contamination (dust, fumes of acids and ammonia, etc.). Bottles with mouths ground on the outside (Fig. 17-2), instead of on the inside as usual, are very suitable for holding standard substances, especially hydrates which may lose water in air of low humidity. By lightly coating the ground-glass surface with vaseline, the bottle may be hermetically sealed, and there will be no danger of contaminating the solid with the sealing medium; furthermore, glass dust formed by rubbing the ground surfaces of stopper and bottle together cannot fall into the bottle and contaminate the solid.

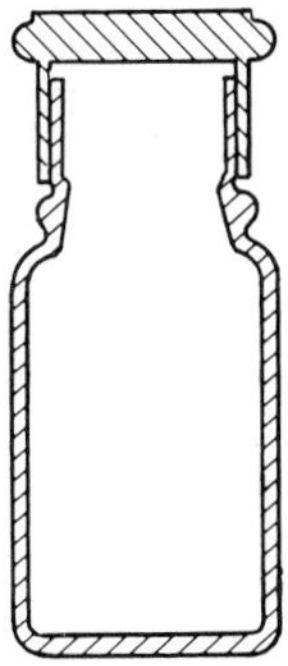

Fig. 17-2

Vacuole water can be removed from hydrated salts by allowing these to effloresce in an atmosphere of low humidity (in a desiccator over sulfuric acid, for example). The crystals then fall to a very fine powder, both foreign and hydrate water being lost. The substance can be rehydrated, if the hydrate is required, by exposure to an atmosphere of proper humidity (p. 189).

Table 17-1, showing the amount of water remaining in recrystallized sodium chloride after successive heatings at various temperatures, indicates how tenaciously water may be retained in salts.

TABLE **17-1. Moisture Content of Sodium Chloride Heated Successively at Various Temperatures**[a]

Temp., °C	*Time of Heating, hr*	*Water,* %
180–250	4	0.19
180–250	12	0.14
295–300	12	0.09
295–300	12	0.07
380	12	0.00

[a] From S. P. L. Sörensen, *Z. Anal. Chem.* **44**, 149 (1905).

If the substance to be purified is only slightly more soluble in hot water than in cold, the ordinary method of crystallization is unpractical. It is then necessary to resort to *isothermal crystallization*, in which the solvent is evaporated from the saturated solution at a constant temperature. For example, by boiling a saturated solution, the solid will be deposited little by little as the solvent evaporates. However, it is usually not advisable to evaporate the solvent at a high temperature because the containing vessel is likely to be attacked if made of glass or porcelain, and the substance crystallized out may be contaminated. Isothermal crystallization is best carried out at room temperature by evaporating the solution under diminished pressure in a vacuum desiccator.

Recrystallization does not always lead to purification of the substance recrystallized —a fact that must always be borne in mind. Recrystallization may fail to separate B from A when A and B form mixed crystals (p. 234). In such a case more or less of B will separate with A, the exact amount depending upon the percentage of B in the original crystals, the speed of crystallization, the time of contact between crystals and mother liquor before filtration, and upon the distribution coefficient of B between A and the solution; it may even happen that B will be concentrated in the recrystallized substance. The behavior of mixed crystals of A and B on recrystallization is represented by a curve such as that in Fig. 17-3, which represents the composition of crystals separating from a solution of mixed crystals ferrous ammonium sulfate hexahydrate and zinc ammonium sulfate hexahydrate. The ordinates give the composition of the solution (ranging from 100 per cent $ZnSO_4 \cdot (NH_4)_2SO_4 \cdot 6\ H_2O$ to 100 per cent $FeSO_4 \cdot (NH_4)_2\ SO_4 \cdot 6\ H_2O$). The composition of the first crystals separating from a solution of any composition can be read from the curve. Thus, if mixed crystals of these two sulfates containing 90 per cent of the ferrous salt are dissolved in hot water and the solution allowed to cool, the composition of the first crystals separating can be found by extending a horizontal line from the point on the *Y* axis corresponding to 90 per cent of iron salt in solution to the curve (which represents the equilibrium conditions in the system: isomorphous salts in solution—isomorphous crystals) and thence down to the *X* axis to the point *a*. The point *a* indicates the composition of the first crystals separating and corresponds to the composition 78 or 79 per cent $FeSO_4 \cdot (NH_4)_2SO_4 \cdot 6\ H_2O$—21 per cent $ZnSO_4 \cdot (NH_4)_2SO_4 \cdot 6\ H_2O$. It is seen that the crystals separating are enriched in zinc salt, and this is always true no matter what the ratio of iron to zinc salt may be in the original crystals. It will never be possible to free $FeSO_4 \cdot (NH_4)_2SO_4 \cdot 6\ H_2O$ from the isomorphous zinc salt by recrystallization; on the contrary, the zinc salt is

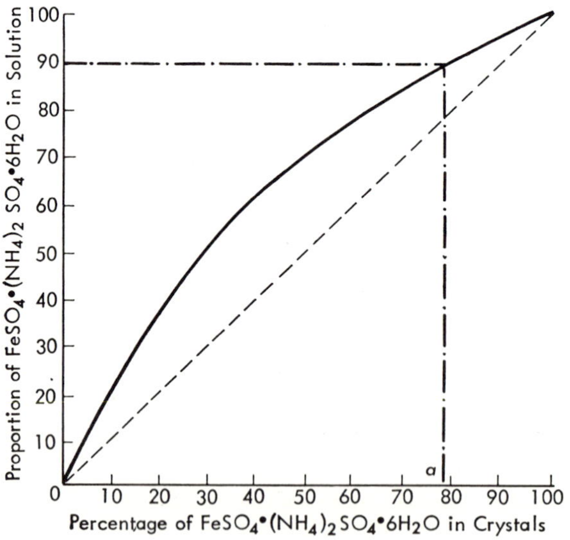

Fig. 17-3. The system $FeSO_4 \cdot (NH_4)_2SO_4 \cdot 6H_2O$–$ZnSO_4 \cdot (NH_4)_2SO_4 \cdot 6H_2O$. From T. V. Barker, *The Study of Crystals*, p. 124.

concentrated in the recrystallized material. $FeSO_4 \cdot (NH_4)_2SO_4 \cdot 6\ H_2O$ (Mohr's salt) has been recommended as a primary standard in oxidimetry (Chap. 43); it is evident that it cannot be prepared pure by recrystallization if zinc, and other metals behaving like zinc, are present, and this is usually the case. However, it will be seen from an inspection of Fig. 17-3 that *zinc ammonium sulfate* can be freed from the corresponding ferrous salt by a sufficient number of recrystallizations. If the zinc and iron salts were not isomorphous, a single recrystallization of zinc ammonium sulfate containing ferrous salt in small amounts would eliminate nearly all the iron. Since they are isomorphous, however, one recrystallization results in only a partial separation, the exact extent of separation depending upon how much the curve representing equilibrium conditions deviates from the diagonal line (drawn dashed in Fig. 17-3), which represents the case in which the crystals separating have the same composition as the solution from which they are deposited. When the equilibrium curve deviates but slightly from a straight line, it is difficult to separate the two components; a great number of recrystallizations will then be necessary, and the procedure (fractional crystallization) is not suitable for analytical use.

Figure 17-4 is a diagram of the system $CoSO_4 \cdot 7\ H_2O$–$FeSO_4 \cdot 7\ H_2O$. By starting with ferrous sulfate containing a small amount of cobalt sulfate, pure crystals of ferrous sulfate can be obtained by recrystallizing a sufficient number of times; but starting with cobalt sulfate containing a small amount of ferrous sulfate, the recrystallized cobalt sulfate will become richer and richer in iron. It would, therefore, be necessary to evaporate the mother liquor to obtain crystals of cobalt sulfate less rich in iron than the original and to repeat this process a great number of times—a procedure that would not be used in the purification of ordinary reagents. In such a case as this, another method of separation—not one involving recrystallization—would be used (see p. 444).

A third form of equilibrium curve for a binary isomorphous mixture is given in Fig. 17-5, which represents the system ammonium magnesium sulfate–ammonium magnesium chromate.

It is clear from the above that if the impurity forms mixed crystals with the substance to be purified, it will be difficult or even impossible to obtain the substance pure by simple recrystallization. Therefore, before recrystallization is adopted as a means of purification, the literaure should be consulted to determine, if possible, whether the impurities to be removed form isomorphous mixtures with the substance to be purified.

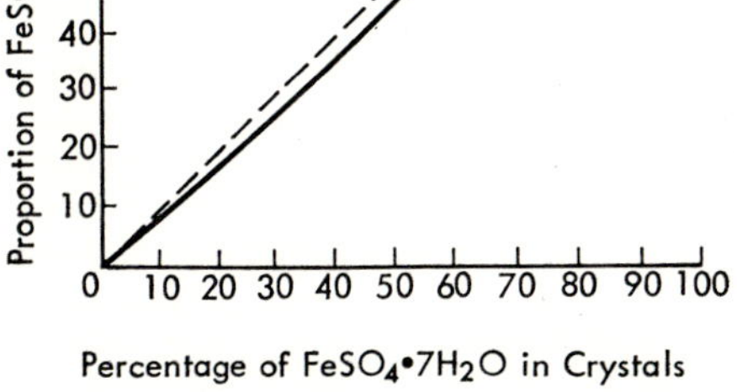

Fig. 17-4. The system $FeSO_4 \cdot 7H_2O$–$CoSO_4 \cdot 7H_2O$.

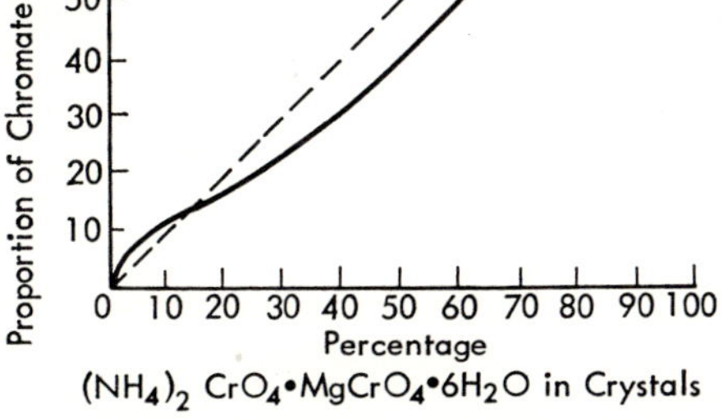

Fig. 17-5. The system $(NH_4)_2SO_4 \cdot MgSO_4 \cdot 6H_2O$–$(NH_4)_2CrO_4 \cdot MgCrO_4 \cdot 6H_2O$.

Even when B, the impurity, is not isomorphous with A, it may happen that B will be concentrated in A on recrystallization. This phenomenon is known as anomalous mixed-crystal formation. As an example of anomalous mixed-crystal formation, the system potassium chloride–lead chloride may be mentioned. Potassium chloride containing a small amount of lead chloride becomes continuously richer in lead as it is recrystallized.[6] Another example is ammonium chloride and ferric chloride. Although these two salts are not isomorphous, ammonium chloride recrystallized from a solution containing ferric chloride will always contain iron. Many other cases of anomalous mixed-crystal formation are known.

Although recrystallization is generally the best method of purifying an easily soluble solid, it is clear that it cannot always be relied upon. Merely because a substance has been recrystallized several times, it is not permissible to conclude that it is pure. In every case the purified substance should be tested by sensitive qualitative reactions for impurities possibly present.

Precipitation by change of solvent. Sometimes it is not advantageous, or even possible, to purify a solid by making a saturated solution in hot water and crystallizing it out by cooling. The substance may decompose or undergo some other change on heating, its solubility may vary only slightly with the temperature, or it may be so soluble in water that it cannot be crystallized satisfactorily. In such cases the substance can be crystallized or precipitated out of its concentrated solution by the addition of a liquid, miscible with water, in which it is much less soluble. Alcohol, in which most inorganic solids are sparingly soluble, is most commonly used. Potassium bicarbonate, which is partially decomposed into carbonate on heating above room temperature, can be purified very easily by adding alcohol to a saturated solution of the salt. The amount of alcohol, or other solvent added, must not, in any case, be so large that the impurities are also precipitated. Sodium chloride, which has a very small temperature coefficient of solubility, can be precipitated out of its saturated solution by the addition of hydrochloric acid or by passing in hydrogen chloride gas (common-ion effect).

When this method is used, the possible effect of the included solvent must always be considered.

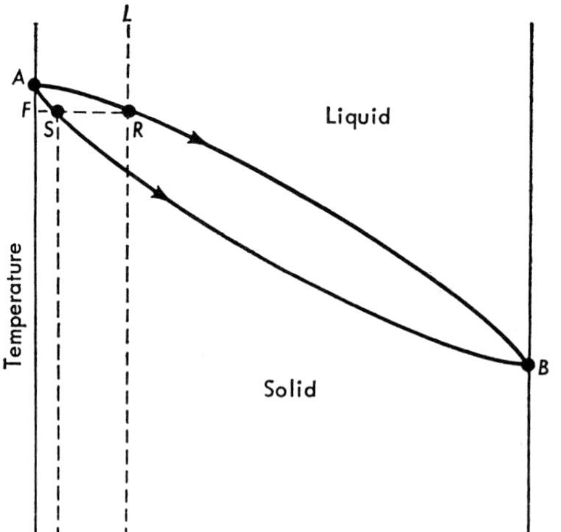

Fig. 17-6. Liquidus-solidus curves in the solid solution system A–B in which the minor component B lowers the freezing point of the major component A. The solid separating contains less B than the liquid with which it is in equilibrium.

[6] R. Mumbrauer, *Z. Physik. Chem. (Leipzig)* **A156**, 113 (1931); H. Käding, *ibid.* **162**, 174 (1932).

Crystallization from a melt. Fractional crystallization by freezing a liquid is frequently an effective means of purifying a substance. Suppose that the impurity B forms a continuous series of solid solutions (mixed crystals) with the major pure component A. If B lowers the melting point of A, Fig. 17-6 represents the liquidus–solidus curves of the system. When a liquid having the composition L is cooled, freezing occurs at the temperature F, and the solid first separating has the composition S. On continued freezing the composition of the solid changes in the direction shown by the arrow on the lower curve, and the solid becomes richer in B. It is evident that the first crystals separating contain less B than the liquid with which they are in equilibrium. Pure A is not obtained at any point.

If the distribution coefficient D_B $(= [B]_{solid}/[B]_{soln.})$ remains constant throughout the freezing, if the liquid is well mixed, and if diffusion is negligible in the solid, the concentration C of the impurity in the solid as a function of the fraction g of the original liquid frozen is given by

$$C = DC_0(1 - g)^{D-1} \tag{1}$$

where C_0 is the original concentration of B in the solution. C/C_0 is plotted as a function of g for a few values of D in Fig. 17-7. If $D > 1$, the impurity has a higher concentration in the solid than in the liquid in equilibrium with the solid. It can be seen from Fig. 17-7 that when $D < 1$, the concentration of B in A when, say, 20 per cent has been frozen, is not greatly different from the concentration in the first crystals formed. An appreciable fraction of A in purer form can therefore be collected. This fraction can then be melted and frozen again to obtain a like fraction of higher purity. A better yield can be obtained by increasing the fraction frozen each time and increasing the number of fractional freezings. In practice, D will depend on the conditions of freezing and is likely to be larger than the true value. With decreasing vigor of stirring, for example, the concentration of B at the solid–liquid interface builds up and more B will be incorporated in the solid than predicted from the equation given. With a slow

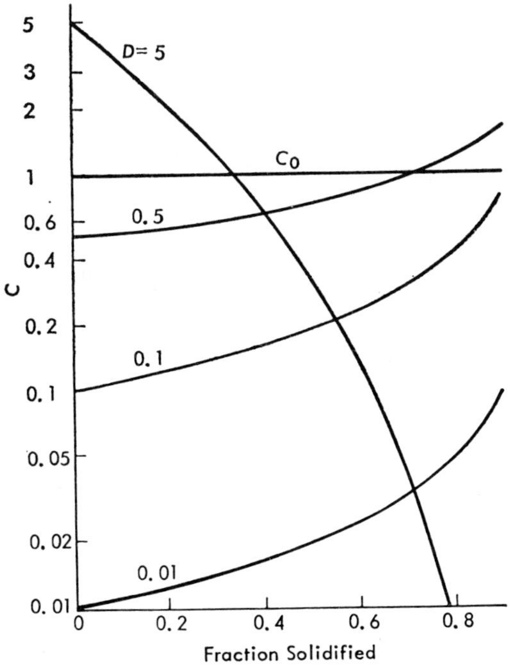

Fig. 17-7. Relative concentration, C, of impurity (original concentration $C_0 = 1$) in solid as a function of the fraction of the melt solidified in normal freezing for $D = 5$, 0.5, 0.1, and 0.01. Partial reproduction of figure in Pfann and Theuerer, *Anal. Chem.* **32**, 1574 (1960).

rate of freezing, the impurity has time to diffuse away from the interface and effective D becomes more favorable to separation.

Benzoic acid can be purified more effectively by freezing than by crystallization from water or benzene. Schwab and Wichers[7] purified benzoic acid by twice freezing very slowly a charge of 10 kg on the walls of a large spherical flask and rejecting 40 per cent of the melt each time. The acid had a purity of 99.986 per cent after the first fractionation and 99.998 per cent after the second, compared to an original purity of 99.91 per cent.

The fractional normal freezing process described is equivalent to melting all the substance in a tube and freezing it from one end, rejecting as much of the other end of the column as is necessary for the desired purity. (If the impurity gathers in the solid, with $D > 1$, the last frozen fractions would be saved.) There is another possible procedure, due to Pfann (reported to have occurred to its originator while napping at his desk after lunch). A narrow melted zone of the material produced by a heater encircling the tube is made to transverse repeatedly the material from one end of the tube to the other (Fig. 17-8). If D of the impurity is less than 1, the impurity is transferred

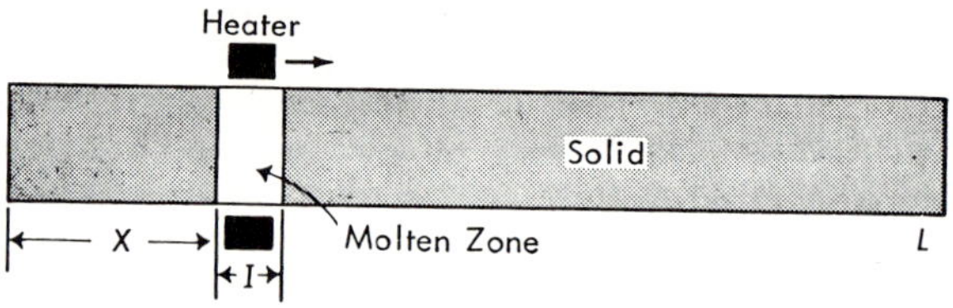

Fig. 17-8. Schematic representation of zone melting. The molten zone of length l, produced by the heater moves from left to right through the column of material of length L.

more or less rapidly to the end of the tube toward which the melted zone is moved. If l is the length of the melted zone of constant cross-sectional area, the concentration of the impurity at the distance X from the starting end of the column after one pass of the zone is ideally given by the following equation (due to Pfann):

$$C = C_0\{1 - (1 - D)e^{-DX/l}\} \tag{2}$$

This equation is not applicable to the last zone length, for which Eq. (1) usually applies (unless a eutectic separates). The calculation of C after a finite number of passes, n, is less simple but methods and results are available (see Pfann[8]). The variation of C/C_0 with X for some values of D and n is shown in Figs. 17-9 and 17-10. If $D > 1$, the impurity collects at the opposite end of the column.

This process of purification by passage of a melted zone is called *zone melting*, more accurately *zone refining* (see Ref. 8). The advantage of zone refining over normal freezing for purification lies in its greater efficiency, simplicity (automatic operation), and saving of time. It may be possible to obtain a substance with much lower contents of a number of impurities having D's greater and less than 1 by using a column longer than usual and recovering the central portion. The process of purification is rapid if D

[7] F. W. Schwab and E. Wichers, *J. Res. Natl. Bur. Std.* **32**, 253 (1944). They also describe freezing benzoic acid in a long glass tube, with cutting of the tube to obtain a 99.997 mole per cent pure product after three fractionations with 40 per cent yield.

[8] W. G. Pfann, *Zone Melting*, Wiley, New York, 2nd ed., 1966; H. Schildknecht, *Zonenschmelzen*, Verlag Chemie, Weinheim, 1964. A chapter in E. W. Berg, *Physical and Chemical Methods of Separation*, McGraw-Hill, New York, 1963, is devoted to zone melting. For reviews, with comments on analytical applicability, see E. F. G. Herington, *Analyst* **84**, 680 (1959), and W. G. Pfann and H. C. Theuerer, *Anal. Chem.* **32**, 1574 (1960).

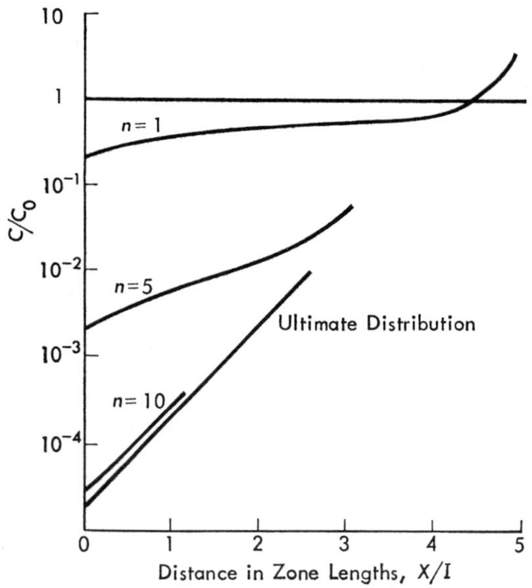

Fig. 17-9. Relative impurity concentration, C/C_0, as a function of distance in zone lengths, X/l, for $n = 1$, 5, and 10 passes when $D = 0.2$. Partial reproduction of figure in Pfann and Theuerer, *Anal. Chem.* **32**, 1574 (1960).

is not too close to unity. The rate of zone travel may be as slow as 1 cm/hr for organic substances and as great as 10 cm or more per hour for metals and semiconductors.

Zone refining has been applied especially for purification of metals (Al, Cu, Ga, Fe, Ni, Pb, Sn, Ag, and others), semiconductors, and organic compounds. For example, impurities in germanium have been reduced to 10^{-9} of their original content. If the melted substance attacks the container, it may be possible to use a rod clamped at both ends, with the molten zone being held in place by surface tension or an electromagnetic field. Chemical separation may be combined with zone refining, as in the

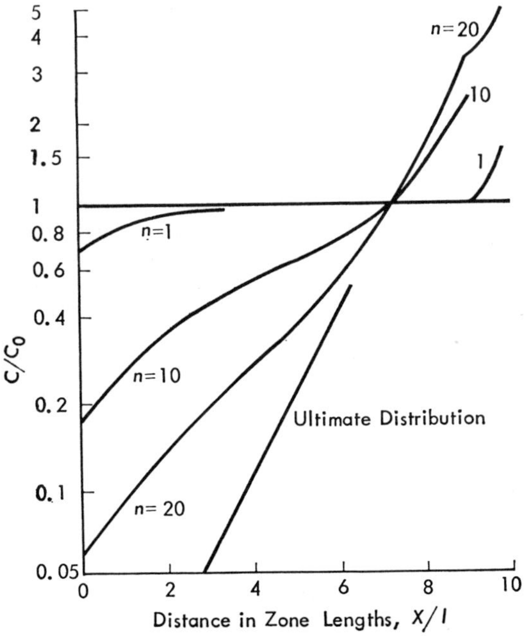

Fig. 17-10. Relative impurity concentration, C/C_0, as a function of distance in zone lengths, X/l, for $n = 1$, 10, and 20 passes when $D = 0.7$. Partial reproduction of figure in Pfann and Theuerer, *Anal. Chem.* **32**, 1574 (1960).

volatilization of boron from a bar of silicon by carrying out the zone melting in a stream of hydrogen containing water vapor. Organic substances to be purified by zone refining must be stable at the melting point.

Sublimation. Sublimation, which is crystallization from the vapor phase, can occasionally be made the basis of purification of inorganic solids. Ammonium chloride, arsenious oxide, and iodine can be thus purified. Organic solids can sometimes be purified in this way.

Direct removal of impurity by physical or chemical means. These methods can render valuable service. As an example of physical removal of impurity, the removal of water by heating may be mentioned. The temperature of heating must be such that the substance dried is not decomposed or otherwise changed. Similarly, volatile acids, ammonium salts, and volatile organic compounds may often be removed by heating the material to the proper temperature.

It is sometimes possible to dissolve out or extract a foreign substance. Ferric chloride is a common impurity in aluminum chloride. It is a simple matter to extract the iron from a hydrochloric acid solution of aluminum chloride with ether. Conversely, the substance desired can sometimes be dissolved out of a mechanical mixture with an impurity. Dyes, which are usually hard to recrystallize or to precipitate, can be separated from sodium sulfate or sodium chloride, which are common contaminants, by extraction with alcohol. The dye dissolves, leaving the insoluble sodium chloride or sulfate behind.

Sometimes chemical methods can be applied. For example, potassium chloride contaminated with lead (and other heavy metals) can be treated in solution with hydrogen sulfide. Lead will be precipitated as sulfide and can be filtered off. After expulsion of hydrogen sulfide, the potassium chloride solution can be evaporated down to crystallize out the salt. The potassium chloride would then have to be recrystallized to remove any sulfate, for example, formed by oxidation in the precipitation. Similarly, copper sulfate can be freed of ferrous iron[9] by adding hydrogen peroxide to the solution to oxidize the iron and then precipitating ferric hydroxide by the addition of a little sodium hydroxide.

Extraction with a nonselective organic reagent in an immiscible organic solvent may serve to remove a considerable number of trace metals from the solutions (neutral or slightly acidic) of certain salts not reacting with the reagent. Dithizone is sometimes used for this purpose.

Ion exchange can be useful in purification of reagents. For example, alkali chlorides, alkaline earth chlorides, nickel chloride, aluminum chloride, and a few other chlorides not forming complex chloro anions can be freed of iron(III), zinc, copper, and other heavy metals by running their solutions through a column of strong-base anion-exchange resin. The chloro anions of the heavy metals (such as $FeCl_4^-$) are adsorbed by the resin even at high electrolyte concentrations. Hydrochloric acid can be purified in the same way (see p. 319). Cation exchangers are not likely to be of use in separating cationic trace species because the exchange is too small at high electrolyte concentrations.

Carbonate can be removed from sodium hydroxide solutions by percolation through a strong-base anion exchanger, which preferentially adsorbs CO_3^{--}. A 50-ml column of Amberlite IRA-400 (Cl^- form) or equivalent resin will furnish 1 liter or more of 0.1 *M* carbonate-free sodium hydroxide. The first portion of effluent, containing chloride, is discarded. The column can be regenerated with hydrochloric acid.

Occasionally, it is necessary to prepare a pure reagent by synthesis. Thus calcium carbonate suitable for use in the J. Lawrence Smith method for the determination of the alkalies is obtained by precipitation, using pure ammonium carbonate and calcium chloride as reagents.

[9] Copper sulfate and ferrous sulfate cannot be readily separated by recrystallization because of the formation of isodimorphous mixtures.

Organic reagents

Notwithstanding the resources of the modern separations armamentarium, the purity of commercial organic analytical reagents is variable and some products sold are of very poor quality. Purification is often necessary. Recrystallization of solids from alcohol or other organic solvent as required is a common purification method. Sometimes sublimation is suitable. Organic liquids are usually purified by distillation, combined if necessary with chemical treatment to remove foreign substances. A knowledge of the impurities likely to be present is a prerequisite to the choice of purification method.

Organic reagents may be unstable, even in the solid state (e.g., cupferron). Solutions decompose more or less rapidly, and are best freshly prepared.

Standard Substances

These are substances of high purity whose content of the major component has been established within a small range of error by one means or other. The National Bureau of Standards of the United States issues a number of substances of this kind of certified content, including potassium hydrogen phthalate, benzoic acid, sodium oxalate, arsenic trioxide, and potassium dichromate. These are used as primary standards in titrimetry, combustion calorimetry (benzoic acid), and thermometer calibration (benzoic acid). Such substances must be essentially free of water as well as other impurities.[10]

Fractional freezing methods readily allow the preparation of benzoic acid having a purity of 99.997–99.999 and even 100.000 mole per cent. The purity can be established by measuring the slope of the freezing curve by thermometry sensitive to 0.0001°C. Crystallization from the melt eliminates the problem of contamination by solvent although not entirely by impurities dissolved in inclusions.

Such substances as potassium hydrogen phthalate, potassium dichromate, and sodium oxalate necessarily have to be purified by recrystallization from an aqueous solution, and solvent inclusions will be present even in crystals that have been grown very slowly. The purity of potassium hydrogen phthalate can be determined by comparison with pure benzoic acid (titration of both with a base). Such comparison was verified at the Bureau of Standards by growing large single crystals containing inclusions (microscopic examination), dissolving, and titrating the inclusion-free crystals. Single crystals of potassium hydrogen phthalate, free of aqueous inclusions, which had been obtained by recrystallization of NBS standard sample 84d (99.987 per cent pure), were found to be 99.999–100.001 per cent pure when compared against purest benzoic acid of 100.000 per cent purity. This agreement may be considered to establish the purity of both substances as 100.000 $\pm$ 0.001 per cent.

Liquid reagents

It is convenient to keep solutions of common reagents that are frequently used; yet since most liquid reagents soon become contaminated or altered in composition, only those solutions should be prepared which are rapidly used up.

There are several reasons for avoiding the use of liquid reagents that have stood for any length of time:

1. Glass bottles in which solutions are kept are attacked to a greater or less extent, depending upon the solution. Alkaline solutions especially act on glass, causing sodium, calcium, colloidal silica, and silicates, etc., to go into solution and some silica to separate. Sodium hydroxide and ammonium hydroxide solutions are preferably kept in plastic

[10] For an account of the preparation and evaluation of standard substances, with special reference to methods used at the National Bureau of Standards, see E. Wichers, *Anal. Chem.* **33**, 23A (1961).

containers. Sodium oxalate and ammonium phosphate—$(NH_4)_2HPO_4$—are other common reagents that act on glass. Ammonium molybdate, whether in acid or ammoniacal solution, should not be kept in glass for any length of time, because the silica introduced may cause errors in the determination of phosphorus (Chap. 30). Hydrofluoric acid is never kept in glass, of course, but in plastic bottles.

2. Solutions not especially protected from the atmosphere may undergo change by absorption of carbon dioxide, oxygen, ammonia and other fumes, etc. For example, ammonium hydroxide absorbs carbon dioxide to give ammonium carbonate, an objectionable constituent of ammonia for a number of analytical uses; neutral or acid solutions of iodides liberate iodine as a result of oxidation, a reaction promoted by light; ferrous iron solutions are also oxidized; and so on.

3. Solutions may decompose spontaneously (ammonium and potassium persulfates) or by photochemical action (ammonium or sodium oxalate gives the corresponding carbonate), or as the result of bacterial action (sodium thiosulfate) and the growth of molds.

These changes may sometimes be prevented or hindered by taking precautionary measures. Bottles made of borosilicate glass are more resistant to attack by solutions than those of ordinary glass and are more suitable for holding reagents. Polyethylene bottles may be required in trace analysis. Opaque or semiopaque bottles should be used for solutions sensitive to light.

It is a common observation that heavy metal ions especially (for example, lead and silver) are markedly adsorbed on glass from neutral or feebly acidic solutions. On standing, the concentration of dilute standard solutions of metals may decrease seriously as a result of such adsorption. Much of the adsorption arises from exchange of the metal ion with alkali or alkaline earth counterions at the glass surface.[11] The adsorption increases if the glass is appreciably attacked by water and the pH of the solution increases. In acid solutions (~ 0.1–$1\ M$), adsorption is usually small because of the replacing effect of H^+. According to this explanation, adsorption of metals should be very small on polyethylene and other plastics that have no fixed ionic groups, and this seems to be generally true.[12] However, molecular adsorption can occur on polyethylene. Iodine penetrates into polyethylene and its solutions should never be stored in plastic bottles. Traces of organic matter are extracted by aqueous solutions in polyethylene and other plastics and can have harmful effects, e.g., reducing OsO_4 to insoluble $OsO_2 \cdot xH_2O$, or absorbing in the UV range in spectrophotometric determinations.

Solutions, as well as solids, should be protected from external contamination. Stoppers should never be laid down upon the table top, unused portions of solution should not be returned to the bottle, and necks of bottles should be kept free of dust and deposits of ammonium salts by wiping with a *clean* damp cloth. Reagent bottles for solids as well as those for liquids are best provided with the flat-top form of stopper that projects over the lip of the bottle and to a great extent prevents the deposition of dust. Narrow-mouth (liquid) reagent bottles should be provided with a loosely fitting cap (Fig. 17-11) (or paper cup) to keep out dust and fumes. Rubber or cork stoppers should not be used for liquid reagent bottles.

[11] It appears that anion exchange can also take place, on $\equiv$Si—OH groups, as of $AuCl_4^-$ [P. Beneš, *Radiochim. Acta* **3**, 159 (1954)].

[12] The superiority of plastics to borosilicate glass for use with highly dilute solutions in trace work has been questioned by G. G. Eichholz, A. E. Nagel, and R. B. Hughes, *Anal. Chem.* **37**, 863 (1965), who studied the adsorption of radioactive isotopes of Cs, Sr, Y, Ce, Ba–La, Zr, I, and Ru–Rh on polypropylene. They concluded that "Contrary to popular belief, it appears that plastic containers are not necessarily superior to glass in tracer work; in fact, adsorption on glass [borosilicate] was less than on polypropylene in the majority of cases, the exception being notably ruthenium and zirconium, which are notorious for their tendency to form colloids in water over a wide range of pH values, and cesium, which showed higher adsorption on glass during initial exposures."

The concentration of a liquid reagent is best expressed in terms of moles, i.e., the number (or fraction) of gram molecular weights present in 1 liter of solution (M). Concentrations may be expressed in terms of normality, if ambiguity is not likely to arise (Chap. 33). It is convenient to express the strength of dilute acids and ammonium hydroxide with reference to the volume of water used to prepare them from the concentrated solutions. Thus 1:5 sulfuric acid is the solution obtained by mixing one volume of concentrated sulfuric acid (sp. gr. ca. 1.84) with five volumes of water. This diluted acid is ca. 6 N (see the Appendix).

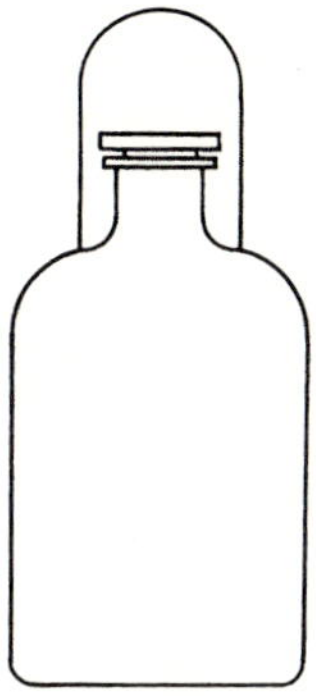

Fig. 17-11. Liquid reagent bottle with protecting cap.

The common acids, hydrochloric, nitric, sulfuric, phosphoric, and acetic, can be purchased very pure and do not need to be purified for ordinary careful analytical work. If purification should ever be necessary, it is accomplished by distillation (except phosphoric acid). Table 17-2 gives analyses of reagent-grade and electronic-grade hydrochloric acid which indicate the purity that can be expected.

TABLE **17-2**

	Reagent Grade, ppm	*Electronic Grade*, ppm, *Maximum*
SO_4	0.4	0.25
SO_3	<1	0.5
Free Cl	<0.5	
Heavy metals as Pb	0.3	
Hg		0.1
Cu	0.3	0.005
Ni	0.3	0.005
Co		0.005
Fe	0.04	0.2
As	0.02	0.005
Pb		0.001
Al		0.01
Zn		0.02
Na		0.05
K		0.05
Ca		0.05
NH_4	3	
Residue after ignition	1	

See Ref. 1, Chapter 18 (Robertson), for the content of Zn, Fe, Sb, Co, Cr, Sc, Cs, Ag, Cu, and Hf in acids, distilled water, organic solvents, and organic reagents.

Distillation is the method commonly used for purifying *liquids*. A nonvolatile impurity can be completely eliminated by distillation if precautions are taken to prevent mechanical carrying over of the boiling liquid. Volatile impurities can sometimes be converted into nonvolatile substances and thus removed; if this is impossible, fractional distillation will separate the two substances if their boiling points are different.

Fractional freezing (especially by zone melting) should be a powerful method for purifying some liquids. Probably sulfuric acid can be purified in this way (H_2SO_4 freezes at 10.5°C, $H_2SO_4 \cdot H_2O$ at 8.6°C, $H_2SO_4 \cdot 2\ H_2O$ at −38.9°C, $H_2SO_4 \cdot 4\ H_2O$ at −24.5°C). Acetic acid is commonly purified by freezing. Some organic solvents used as extractants in inorganic analysis, e.g., benzene, can be freeze-purified.

Pure ammonium hydroxide is prepared most easily by passing ammonia from a cylinder into water contained in a polyethylene bottle which must be cooled by running water. Alternatively, the ammonia gas can be obtained by heating strong ammonia water containing a little calcium hydroxide to retain any carbonate present.

Pure solutions of some gaseous or volatile reagents can be obtained in a simple way by *isothermal distillation* at room temperature. For example, hydrochloric acid of high purity can be prepared by placing a dish of concentrated hydrochloric acid and a smaller dish (polyethylene) of water in an empty desiccator and allowing the closed vessel to stand for a day or two until equilibrium has been attained. A fairly strong hydrochloric acid solution can thus be obtained, whose only impurities will be the few substances volatile under the conditions or dissolved out of the dish. Pure ammonia can be prepared in the same way. Isothermal distillation enables hydrofluoric acid to be purified in a simple way by using two polyethylene bowls, one inverted over the other, in place of a desiccator, together with polyethylene beakers.[13]

Distilled water

Special attention must be paid to the quality of the distilled (or deionized) water used for analytical purposes. Since large quantities of distilled water are used in every determination, a mere trace of an impurity may cause appreciable errors. The fact that water has been distilled is no guarantee that it is pure. It sometimes happens that, owing to the use of faultily constructed apparatus, water which has been distilled is more impure than the original. The foreign substances that may be present in distilled water include the following:

1. *Dissolved solids of the original water carried over as spray or by spurting.* Natural waters contain appreciable amounts of sodium, potassium, calcium, magnesium, iron, chloride, carbonate, and sulfate. By testing the distilled water with the appropriate reagents it is possible to tell whether these are present in significant amounts. The water may be tested directly for chloride with silver nitrate solution acidified with nitric acid, for sulfate with barium chloride acidified with acetic acid, for calcium with ammonium oxalate in ammoniacal solution, and for magnesium with titan yellow.[14] If all these reactions are negative (the solutions should be allowed to stand overnight in the precipitation tests), the distilled water may be accepted as satisfactory for analytical use, provided it contains no suspended matter.

2. *Dissolved gases.* The most objectionable is usually carbon dioxide. Distilled water often contains more carbon dioxide than the original water. It may be largely removed by bubbling purified air[15] through the water for a few hours. Water thus

[13] W. Kwestroo and J. Visser, *Analyst* **90**, 297 (1965).

[14] To 10–20 ml of water add 1 drop of 0.1 per cent titan yellow solution in water and then a few drops of 6 *N* sodium hydroxide. In the presence of magnesium a pink color is obtained, whereas the solution becomes orange if it is absent.

[15] The air is bubbled through dilute sulfuric acid to remove traces of ammonia, then through water to remove droplets of acid spray, and is then drawn through a fritted-glass disk to remove dust.

obtained is called equilibrium water (in equilibrium with the atmosphere as far as carbon dioxide is concerned) and is about 1.5×10^{-5} M with respect to carbon dioxide. Such water may be used for preparing practically carbonate-free alkali solutions (0.1 N and stronger) by dilution of strong sodium hydroxide (Chap. 39). Water may be freed of carbon dioxide, and other gases as well, by boiling for a short time. Such water quickly absorbs carbon dioxide from the air so that if water entirely free of carbon dioxide is required, it must be cooled and stored with protection from the air by attaching a soda-lime tube to the vessel in which it is kept. For most purposes for which "carbon dioxide–free" water is specified, equilibrium water will serve.

Distilled water always contains traces of ammonia. Chlorine from chlorinated tap water may be present.

3. *Dissolved material from pipes and containers.* The amount and kind of foreign material introduced in this way depends upon the equipment supplying the distilled water. Block tin and fused silica are considered the best materials for the construction of condensers.[16] Copper in very small amounts is often found in distilled water, and other metals may be present depending upon the metals with which the condensed water has come into contact.[17] The presence of copper, even in very small amounts, is often objectionable because of its catalytic properties. Thus sodium thiosulfate solutions made up with water containing minute amounts of copper decompose quite rapidly (Chap. 42). Distilled water that has stood for some time in glass (even in the most resistant varieties) will contain quite appreciable amounts of solids. Therefore, water that is as fresh as possible should be used for analytical work.

4. *Suspended matter. Dust.* The amount of insoluble matter present in some distilled water is large enough to be a source of error in gravimetric determinations. Organic matter (probably present to a large extent as insoluble material) is objectionable because of its reactivity with various reagents (silver nitrate, potassium permanganate, etc.).

To prepare good distilled water, the distillation should first be made from alkaline permanganate solution to oxidize organic matter. The first portions of the distillate are rejected and the middle fractions collected. Water so prepared will be practically ammonia-free, since ammonia is volatilized with the first portions of the distillate. A second distillation from a slightly acid solution of sulfuric acid will remove the last traces of ammonia. Care must be taken to prevent the carrying over of spray in the distillations. The distillate should be protected from the atmosphere to exclude acid and basic fumes of the laboratory.

[16] A laboratory water still with polyethylene condenser and receiver has been described by J. V. Jordan and G. D. Wyer, *Chemist-Analyst* **48**, 39 (1958). Such a still will supply water very low in metals. A trace of organic matter will be present. Also see H. Stegemann, *Z. Anal. Chem.* **154**, 267 (1957). The amount of silica introduced from a silica condenser is very small, approximately 2 μg of SiO_2 per liter. Water distilled in Jena-glass apparatus has been reported to contain 30 μg of SiO_2 per liter.

[17] Traces of heavy metals (copper, lead, and zinc) in distilled water may be detected with dithizone (see p. 351). To detect copper, add a drop of 1:1 hydrochloric acid to 10 or 15 ml of water in a 1.8×15 cm glass-stoppered flat-bottomed tube and shake for ½ min with 1 or 2 ml of 0.001 per cent dithizone solution in pure carbon tetrachloride. If copper is present, the originally green solution of carbon tetrachloride changes to blue-green, bluish, or violet depending upon the amount of metal present. To detect lead, add a few drops of 10 per cent sodium citrate and 1 or 2 ml of 10 per cent potassium cyanide solution to 10 ml of the water and shake for a few seconds with 1 or 2 ml of dithizone. If the carbon tetrachloride layer becomes pink, lead is present. Zinc may be detected by adjusting the pH of the sample to about 5 with an acetate buffer, adding 0.5 ml of 10 per cent sodium thiosulfate to prevent the reaction of any copper or lead present, and shaking for 1 min with 1 ml of dithizone. A change in color to bluish, violet, or red shows the presence of zinc. Less than 1 μg of each metal may be detected by these tests.

Deionized water

Water purified by ion exchange can often be used instead of distilled water in the analytical laboratory; it may even be superior to ordinary distilled water in some respects. Deionization of tap water is best carried out with strong-acid cation- and strong-base anion-exchange resins (Chap. 13); the latter remove silica. Mixed-bed resins are preferably used. Nonelectrolytes and colloidal substances are not removed, so deionized water is not always equivalent to good distilled water. Traces of organic substances are always taken up by the water from the resin, and these are sometimes objectionable.

Ion exchange is an effective means of removing traces of heavy metals from water (Table 17-3).

TABLE 17-3. **Removal of Heavy Metals from Distilled Water by Passage through a Column of Cation-Exchange Resin (Amberlite IR-100)**[a]

	Metal Present, ppm		
	Cu	Pb	Zn
Laboratory distilled water	0.2	0.055	0.02
Ion-exchange water, one passage	0.0035	0.0015	0.00
Ion-exchange water, five passages	0.0000	0.0010	0.00
Redistilled water (all-Pyrex still)	0.0016	0.0025	0.00

[a] From G. F. Liebig, Jr., A. P. Vanselow, and H. D. Chapman, *Soil Sci.* **55**, 371 (1943).

Gases

Oxygen, hydrogen, nitrogen, chlorine, carbon dioxide, ammonia, sulfur dioxide, and hydrogen sulfide are most conveniently obtained from cylinders of the compressed or liquefied gases. If necessary, the gas is purified by bubbling through appropriate solutions.[18]

Hydrogen sulfide can be prepared by action of dilute acid on ferrous sulfide in a Kipp generator, if no other supply is available. The gas may be purified by bubbling it successively through saturated sodium bicarbonate solution and water. Hydrogen sulfide prepared in this way will contain considerable amounts of hydrogen derived from metallic iron in the ferrous sulfide, but this will not matter in most analytical procedures.

REFERENCES ON PURIFICATION OF SOLIDS

E. H. Archibald, *The Preparation of Pure Inorganic Substances,* Wiley, New York, 1932. Deals especially with purification of substances for atomic-weight determinations, but the methods described have some application in purification of analytical reagents and standard substances.

I. M. Kolthoff and V. A. Stenger, *Volumetric Analysis,* vol. II, Wiley-Interscience, New York, 1947; I. M. Kolthoff and R. Belcher, *Volumetric Analysis,* vol. III, Wiley-Interscience, New York, 1957. Contains directions for preparing and testing standard substances.

D. D. Perrin, W. L. F. Armarego and D. R. Perrin, *Purification of Laboratory Chemicals,* Pergamon, Oxford, 1966.

[18] For purification of gases see L. Moser, *Die Reindarstellung von Gasen,* Enke, Stuttgart, 1920.

T. W. RICHARDS, *Stähler's Handbuch der Arbeitsmethoden in der anorganischen Chemie*, vol. II, De Gruyter, Berlin, 1919, p. 378. General discussion of purification of substances by crystallization.

L. VANINO, *Handbuch der präparativen Chemie*, Band I, Anorganischer Teil, Enke, Stuttgart, 1921. Gives directions for purifying and testing chemicals.

PROBLEMS

1. (*Note*: A knowledge of calculus is required to work this problem.) (a) Assuming that the relation

$$\frac{[Zn(NH_4)_2(SO_4)_2 \cdot 6\ H_2O]_{cryst}}{[Fe(NH_4)_2(SO_4)_2 \cdot 6\ H_2O]_{cryst}} = K\frac{[Zn(NH_4)_2(SO_4)_2 \cdot 6\ H_2O]_{soln}}{[Fe(NH_4)_2(SO_4)_2 \cdot 6\ H_2O]_{soln}}$$

holds for the mixed-crystal system ferrous ammonium sulfate–zinc ammonium sulfate described on page 438, calculate the value of K when the proportion of ferrous ammonium sulfate in solution is 30:70, 50:50, 70:30, and 90:10 (obtain data from Fig. 17-3). (b) Assuming that the expression given above holds during the course of the crystallization ($K = 2.5$, heterogeneous distribution), calculate the percentage of zinc ammonium sulfate in the crystals obtained when one half of the ferrous ammonium sulfate has been crystallized from a solution containing ferrous ammonium sulfate and zinc ammonium sulfate in the ratio 1000:1.

Ans. (a) $K = 2.6, 2.6, 2.4, 2.5$; (b) 0.164 per cent.

2. Potassium chloride and potassium bromide form mixed crystals, with the distribution coefficient (for amounts of Br < 1 per cent in KCl):

$$\frac{[Br^-]_{cryst}}{[Br^-]_{soln}} = 0.8$$ (concentrations expressed as weight per cent; room temperature to 60–70°C)

Devise a method (any method) to obtain potassium chloride containing not more than 1 ppm Br from potassium chloride containing 100 ppm Br.

3. A substance A containing 0.010 per cent of the impurity B is melted in a vertical tube to form a liquid column 10 cm high. The tube is slowly cooled in such a way that solidification begins at the bottom and proceeds to the top. If the distribution coefficient of B between solid and liquid A is

$$D = [B]_{solid}/[B]_{liquid} = 0.10$$

find the percentage of B in solid A in the following segments from the bottom: 0–1, 1–2, 4–5, 8–9, 9–10 cm. Assume that the liquid phase is mixed during cooling so that B is uniformly distributed through liquid A as A freezes.

4. From the data given on page 442 for the Schwab–Wichers purification of benzoic acid by freezing, calculate the recovery factor and the mean separation factor for the impurities (p. 249) after two freezings.

CHAPTER 18 *Apparatus*

This chapter deals with some facts concerning the material and construction of apparatus which have an important bearing on the proper performance of analytical work.

THE MATERIALS OF ANALYTICAL UTENSILS

Glass

In all quantitative work it is necessary to guard against the entrance of foreign substances that may directly or indirectly cause serious errors in the determinations. Special attention must be paid to the quality of the glassware used to hold solutions. Beakers and flasks are commonly made of borosilicate glass,[1] which has many

TABLE 18-1. **Composition of Pyrex and Vycor[a] and Alkali-Resistant Glass[b]**

	Pyrex,[c] %	*Vycor,*[d] %	*Alkali-Resistant Glass,* %
SiO_2	81.0	96.3	71
B_2O_3	13.0	2.9	<0.2
R_2O_3 (chiefly Al_2O_3)	2.2	0.4	16
ZnO[e]	N. d.	N.d.	
CaO	Negligible	Negligible	
BaO[e]	N. d.	N. d.	
MgO[e]	N. d.	N. d.	
Na_2O	3.6	<0.02	12 (Na_2O + K_2O)
K_2O	0.2	<0.02	
As_2O_3	0.002	0.005	
Sb_2O_3	N. d.	~ 0.1	
Undetermined	—	0.3	

[a] From E. Wichers, A. N. Finn, and W. S. Clabaugh, *Ind. Eng. Chem., Anal. Ed.* **13**, 419 (1941).
[b] Mostly from J. P. Williams, *Microchem. J.* **4**, 187 (1960).
[c] Chemical Pyrex glass (Corning 774, now designated 7740); annealing point 560°C, softening point 820°C.
[d] Corning Glass Works' 96 per cent silica glass 790 (now designated 7900).
[e] Negligible according to other reports.

[1] For example, Pyrex (Corning 7740) and Kimax (Kimble KG-33). See Table 18-1 for composition. For the content of certain trace elements (Zn, Fe, Sb, Co, Cr, Sc, Cs, Ag, Cu, and Hf) in glass as well as polyethylene, Teflon, etc., see D. E. Robertson, *Anal. Chem.* **40**, 1067 (1968).

favorable properties. The thermal coefficient of expansion of such glass is much lower than that of ordinary glass (Table 18-2), and therefore vessels of the former can be made thicker to prevent breakage due to mechanical shock and still resist cracking due to rapid heating or cooling. Borosilicate glass has a higher softening point than ordinary glass. In heating all articles of glass, care should be taken to avoid sudden temperature changes or uneven heating. Glassware should usually not be heated with a bare flame; beakers and flasks should be placed upon wire gauze.

TABLE 18-2. **Properties of Selected Glasses**[a]

Glass Type	Thermal Expansion Coefficient, per °C	Softening Point, °C	Service Temp. Limits, °C: Continuous	Service Temp. Limits, °C: Short-Time Maximum	Remarks
Soda-lime (Corning 0088, Kimble R6)	9×10^{-6}	650	400	450	"Ordinary" (flint) glass; for rod and tubing, sometimes graduated ware
Alkali-resistant (Corning 7280)	6.4×10^{-6}	870			Low in boron (<0.2% B_2O_3); heating and cooling should be gradual
Borosilicate (Corning 7740, Kimble KG-33)	3.2×10^{-6}	820	500	600	General laboratory ware
Borosilicate (red-colored) (Corning 7740, low actinic; Kimble Ray-Sorb)					Protection from light (<500 mμ); outer surface color-treated
Aluminosilicate (Corning 1720)	4.2×10^{-6}	915	675	850	High-temperature service (microcombustion tubes)
96% silica (Corning 7900)	0.8×10^{-6}	1500	900	1200	High-temperature service

[a] Mostly from J. P. Williams, *Microchem. J.* **4**, 187 (1960).

It should be realized that Pyrex and similar glasses, although much more resistant to the action of solutions than ordinary glass, are by no means exempt from attack. Water and solutions in general, especially when hot, act upon all glass; the amount of material brought into solution depends upon the composition of the glass, the nature of the solution, the time of contact, and so on. The data in Tables 18-3 and 18-4 indicate to what extent glass is attacked by some common solutions on heating. Alkaline solutions especially attack all kinds of glass. On the other hand, glass is generally acted upon only very slightly by acid solutions (except, of course, hydrofluoric acid).

Vycor. In 1940 a silica glass was put on the market under the trade name Vycor. It is produced by leaching a glass of special composition to leave a residue consisting essentially of silica, which is then fired at a high temperature to give a transparent vitreous product. Silica constitutes 96 per cent of this glass; the remainder is chiefly boric oxide with traces of aluminum, sodium, iron, antimony (0.1 per cent), and arsenic. Because of its silica content it is very resistant to the action of water and acids and may be used up to 900°C; above this temperature devitrification sets in. Beakers, crucibles, evaporating dishes, flasks, etc., constructed of this new glass are obtainable.

An alkali-resistant glass (Corning 7280), substantially boron-free, is also available.

TABLE 18-3. **Chemical Resistance of Pyrex and Vycor**[a,b]

Liquid	*Av. Loss in Weight per* 250 ml *Flask*	
	Pyrex, mg	*Vycor*, mg
Water	2	1
1 *N* H_2SO_4	4	1
6 *N* HCl	22	4.5
5% NaCl	7	—
5% NaCl in 0.001 *N* HCl	Negligible	—
0.05 *N* NaOH	90	40
0.5 *N* NaOH	285	—
100 ml of NH_4OH and 100 g of NH_4Cl in 1000 ml of aqueous solution	1.4	—
Concentrated ammonium hydroxide at room temperature for 8 and 28 days	1.9	—

[a] E. Wichers, A. N. Finn, and W. S. Clabaugh, *Ind. Eng. Chem., Anal. Ed.* **13**, 419 (1941). See also G. H. Wyatt, in C. L. Wilson and D. W. Wilson (eds.), *Comprehensive Analytical Chemistry*, vol. IA, chap. 2, Elsevier, Amsterdam, and Van Nostrand, Princeton, N.J., 1959.

[b] 200 ml of solution was boiled in a covered 250-ml Erlenmeyer flask for 6 hr, with replenishment of the solution with water midway in the boiling.

TABLE 18-4. **Additional Data on Chemical Resistance of Glasses**[a]

Liquid	*Temp.*, °C	*Time, hr*	*Plate Weight Loss*, mg/cm²		
			Borosilicate	96% *Silica*	*Alkali-Resistant*
H_2O	80	48	0.002	<0.001	0.002
5% HCl	100	24	0.005	0.0005	0.01
0.02 *N* Na_2CO_3	100	6	0.12	0.07	0.01
5% NaOH	100	6	1.4	0.9	0.04

[a] Mostly from J. P. Williams, *Microchem. J.* **4**, 187 (1960).

Fused silica (vitreous silica or fused quartz)

Fused silica is produced in two principal forms: the transparent and the translucent (from gas bubbles); the former variety is better, but usually the latter can be used in place of the more expensive transparent kind. The material has a number of desirable properties. It begins to soften only at very high temperatures (about 1700°C) and is not appreciably volatile below the softening point. However, it begins to devitrify at 1100–1200°C. Fused silica is very resistant to thermal shock, owing to its very low coefficient of expansion (5×10^{-7} for the clear variety), which is about 1/15 that of glass. It is not attacked by acids except hydrofluoric acid and, at high temperatures, phosphoric acid, which forms silicon phosphate. Alkaline solutions attack fused silica, especially at high temperatures (Table 18-5); with fused alkali hydroxides and carbonates the attack is very rapid. In general, silica is acted upon by metal oxides at high temperatures, silicates being formed.[2] Fusion with alkali pyrosulfates can be made in

[2] Owing to its acidic character, silica is also acted upon by nitrates, sulfates, and even chlorides at high temperatures.

TABLE 18-5. **Action of Reagents on Silica Glass**[a,b]

Reagent	*Temp.*, °C	*Time of Exposure*, days	*Loss in Weight of Flask*, mg
H_2O	18–100	Many	0.0
NH_4OH, 10%	18	2	0.8
NaOH, 10%	18	2	0.4
NaOH, 30%	18	2	0.0
KOH, 30%	18	4	1.2
NaOH, 2 *N*	100	3	48
Na_2CO_3, 2 *N*	100	3	12
KOH, 2 *N*	100	3	31
NaOH, 2 *N*	100	3	33

[a] From F. Mylius and A. Meusser, *Z. Anorg. Chem.* **44**, 221 (1905).

[b] In each test a 76-ml fused silica flask with 89 cm^2 of exposed surface was used. The tests were made with the same flask, the solutions being used in the order shown.

silica crucibles if the introduction of a small amount of silica is not objectionable. Fused silica beakers and dishes are especially valuable for evaporation of solutions that must be kept free of the alkali metals. Silica crucibles are easily electrified and slightly hygroscopic; they require a longer time for heating and cooling than do platinum crucibles, and it is more difficult to heat them to as high a temperature over a burner. Silica glass is permeable to hydrogen above 1000°C. It is more brittle than ordinary glass.

Porcelain

Laboratory articles called porcelain are actually glazed pottery. The resistance of porcelain to the action of solutions depends primarily upon the quality of the glaze. It seems to have been generally believed that porcelain is commonly less attacked than resistance glass by solutions, and porcelain has been preferred to glass in operations in which hot liquids remain in contact with the vessel for relatively long periods of time. Thus casseroles (porcelain dishes with handles) have been preferred to beakers for evaporations. The general superiority of porcelain over glass to attack by hot acidic solutions is questionable, but there is evidence that porcelain may be affected less than glass by alkaline solutions.[3] The principal constituents of the glaze are Si, Al, Ca, and Na (and K). In exact analysis, evaporations should be carried out in platinum (or in silica when the solutions are acidic).[4] Porcelain is mechanically stronger than glass. Its thermal expansion coefficient is $3\text{–}4 \times 10^{-6}$. Thick-walled vessels should not be subjected to sudden temperature changes.

Porcelain crucibles can be heated to high temperatures (1200°C, if the glaze is good) and are very frequently used for igniting precipitates, because of their cheapness; they do not change in weight to any appreciable extent on strong heating. For certain common operations porcelain crucibles cannot be used at all. Fusions with sodium carbonate and other alkaline substances cannot be made in them, and hydrofluoric acid,

[3] C. E. Waters, *Natl. Bur. Std. Tech. Papers* **105**, 1917, found that porcelain dishes lost about 2 mg in weight when 2.5 *M* solutions of sodium hydroxide were heated in them for 6 hr on the steam bath.

[4] Hot sulfuric acid attacks porcelain to form aluminum sulfate. L. A. Shirley, P. Shinners, and W. F. Wagner, *Chemist-Analyst* **54**, 11 (1965), found that porcelain filter crucibles which had been washed with dilute sulfuric acid and ignited at 700°C or below gained in weight because of incomplete conversion of $Al_2(SO_4)_3$ to Al_2O_3 at this temperature. At 800°C the conversion to Al_2O_3 is complete and no weight increase is shown.

of course, must be excluded from all porcelain ware. Pyrosulfate fusions should not be made in porcelain, although the extent of attack is not very great. In certain ignitions requiring a high temperature, porcelain crucibles are less satisfactory than platinum because the latter can be heated to a much higher temperature by the same flame. (This is not true if a furnace is used.) Thus it is not possible to ignite calcium oxalate satisfactorily to the oxide in porcelain by heating in a gas flame, whereas the conversion is fairly easy in platinum.

By replacing the quartz and a part of the kaolin in porcelain by sillimanite ($Al_2O_3 \cdot SiO_2$), a superior porcelain is obtained.[5] The thermal coefficient of expansion of sillimanite porcelain is less than that of ordinary porcelain, and sillimanite ware is therefore less likely to break when subjected to sudden temperature changes. The material is also stronger than porcelain and is a better conductor of heat. Sillimanite crucibles are to be preferred to those of porcelain.

Metal oxides and other basic substances may react with the glaze of porcelain crucibles at high temperatures. This will usually not affect the accuracy of the weighing after ignition but the crucible will become etched.

Platinum

Platinum is an indispensable metal to the analyst. Without it many analytical methods would be impossible.

Platinum has a very high melting point of 1774°C. It is not altered chemically when heated in air to the highest temperatures and is not attacked to any serious extent by most reagents (exceptions are noted below). The ability of the metal to withstand the action of molten alkali carbonates and hydrofluoric acid is especially important. Platinum is a very good conductor of heat and a good radiator. It does not absorb water vapor to any appreciable extent. For these reasons it is very suitable for the construction of crucibles for the ignition of precipitates.

Pure platinum is too soft for general use, and it is therefore hardened and stiffened with some alloying element such as iridium or rhodium, Iridium is the metal that has been most often used for this purpose.[6] Platinum thus alloyed, as well as the pure metal itself, is slightly volatile at elevated temperatures, a fact of which account must be taken when substances are to be weighed in platinum after strong ignition. The data in Table 18-6 will give an idea of the magnitude of the volatilization loss of platinum

TABLE 18-6. **Approximate Loss in Weight, mg/100 cm²/hr, at Temperatures Indicated for Platinum Nearly Free of Iron**[a]

		Platinum Containing:		
Temp., °C	*Pure* Pt	1% Ir	2.5% Ir	8% Rh
900 or less	0	0	0	0
1000	0.08	0.30	0.57	0.07
1200	0.81	1.2	2.5	0.54

[a] From G. K. Burgess and R. G. Waltenberg, *Natl. Bur. Std. Sci. Paper 280*, p. 372.

[5] The somewhat unfavorable properties exhibited by porcelain on strong heating are due to the presence of free quartz grains. Quartz at high temperatures is transformed into other crystal forms with considerable changes in volume. This change in volume causes strains in the porcelain that may result in cracking and weakening. Sillimanite does not show this behavior and therefore is used to replace the quartz in the porcelain body.

[6] Platinum–iridium alloys for chemical use ordinarily contain 0.3–2 per cent iridium. Crucible alloy usually contains 0.3–1 per cent Ir.

and its iridium and rhodium alloys. Below 900°C there is no loss, at 1000°C the loss is appreciable, and above this temperature it increases rapidly. Platinum alloyed with iridium loses weight more rapidly than pure platinum, owing to the greater volatility of iridium. A correction for volatility losses can be applied by running a blank with the vessel alone (p. 416). Rhodium-alloyed platinum is now commonly used for crucibles to avoid the volatility problem. A 2 per cent Rh alloy is suitable. Good platinum ware should contain no other element besides the alloying element. Small amounts of iron are frequently found in platinum; it may be present at the surface of the metal as a result of the use of iron tools in the shaping of the article, but it is also present in the interior. The presence of much iron is revealed by a dark color assumed by the utensil when heated to redness (formation of iron oxide). Iron is an objectionable impurity, first, because the vessel may gain in weight when heated and, second, because substances heated or treated in such platinum are likely to be contaminated. The superficial iron may be removed by treatment with hot concentrated hydrochloric acid or fused potassium pyrosulfate. Iron in the interior of the platinum will slowly diffuse out on continued heating and give ferric oxide at the surface, so that it is not easily eliminated. Other metals besides iron are sometimes found.[7]

At high temperatures platinum is permeable to various gases, especially hydrogen. This fact should be remembered when there is a possibility of reduction of the substance heated by the reducing gases of the flame.

Rules for the use of platinum ware

1. Platinum is a soft metal, even when alloyed, and therefore all platinum ware must be handled with care to prevent deformation and denting. Glass rods and the like should never be used to detach solids adhering to the walls, because they will scratch the metal. Neither should very hot vessels be suddenly cooled, as for example by plunging a red-hot crucible into water. The wooden or plastic forms commonly supplied with platinum crucibles and dishes are useful for ironing out any dents that the vessels may receive.

2. Platinum must not be heated in contact with any other metal than itself, because of the danger of forming alloys. Therefore, only triangles of platinum, clay, sillimanite, or silica may be used for supporting crucibles of platinum; dishes should be placed on asbestos, never directly on wire gauze. Crucible tongs should preferably be platinum-tipped. However, platinum utensils may be handled with nickel tongs if allowed to cool below red heat; only solid nickel or rust-resisting iron alloy tongs should be used, never nickel-plated or ordinary iron tongs.

3. Certain substances that directly attack platinum or will do so in the presence of another substance must be rigorously excluded from platinum. Common substances of this character are:

a. Carbon from smoky or luminous gas flames. Platinum vessels may be heated only by nonluminous flames, and, moreover, they must never be allowed to come into contact with the inner cone of a gas flame that contains reducing gases. If these precautions are not observed, platinum carbide may be formed, and this compound makes platinum very brittle and liable to crack. Filter paper may be ignited in platinum without harm if the heating is gradual, so that the paper does not burst into flame, and if the carbon is burned off at a low temperature with free access of air.

b. Compounds of the easily reducible metals and nonmetals, as well as the elements themselves. Included in this group are salts of silver, mercury, lead, bismuth, antimony,

[7] Regarding the quality and testing of platinum ware and causes of decrease in weight on heating see, G. K. Burgess and P. D. Sale, *J. Ind. Eng. Chem.* **6**, 452 (1914); **7**, 561 (1915); *Bur. Std. Sci. Paper 254*, 1915; G. K. Burgess and R. G. Waltenberg, *Natl. Bur. Std. Sci. Paper 280*, 1916; W. F. Hillebrand, P. H. Walker, and E. T. Allen, *J. Ind. Eng. Chem.* **3**, 686 (1911); R. W. Hall, *J. Am. Chem. Soc.* **22**, 494 (1900); G. A. Hulett and H. W. Berger, *ibid.* **26**, 1512 (1904).

and tin, sulfides, and compounds of arsenic and phosphorus.[8] These may be reduced by the carbon of filter paper or by reducing gases of the flame gaining access to the interior of the crucible.

c. Solid alkali oxides and hydroxides; barium oxide; nitrates, nitrites, and cyanides of the alkalies. All these substances attack platinum strongly when heated or fused in it. Sodium and potassium carbonates (but not lithium carbonate) may be fused in platinum with safety. Sodium borate may also be fused.

d. Solutions of the halogens and mixtures that furnish them, e.g., aqua regia, hydrochloric acid with oxidizing agents (permanganate, chromate, manganese dioxide, etc.). Ferric chloride in hydrochloric acid acts on platinum appreciably.

4. Substances of unknown nature should not be heated or treated in platinum.

5. All platinum ware should be kept clean and bright, inside and out. On long-continued heating, platinum becomes gray as a result of recrystallization. If the recrystallization, which begins at the surface, is allowed to continue, the vessel may eventually develop cracks; the deterioration can be stopped by burnishing the metal with moist sand composed of well-rounded grains (sea sand). In the case of a crucible this treatment may well be repeated after every five or six ignitions.

If a platinum utensil has become stained, it should be digested first in hydrochloric or nitric acid *singly*. If this treatment is without effect, potassium pyrosulfate should be fused in the vessel at a low temperature for 5 or 10 min. The fused mass is poured out onto a dry stone or metal slab before it solidifies. The vessel should then be digested in hydrochloric acid, and the treatment repeated if necessary. If fusion with pyrosulfate is without effect, sodium carbonate should be tried.

It should be borne in mind that many substances heated or fused in platinum attack the metal to a slight extent. The attack may be so slight that no sensible injury is done the vessel, but the amount of platinum brought into solution in this manner may be large enough to cause an appreciable error in the determination if it is not subsequently removed. Fused alkali pyrosulfate, for example, acts on platinum slightly, so that in an ordinary pyrosulfate fusion a milligram or more of platinum may be dissolved. Sodium carbonate attacks platinum to a lesser extent than does alkali pyrosulfate; usually only a few tenths of a milligram of metal are brought into solution in a sodium carbonate fusion.

In addition to crucibles, platinum is used for dishes in which acid or alkaline liquids can be evaporated without introduction of the material of the container.[9] Platinum is also used for electrodes (Chap. 51).

Gold

Gold is an indifferent metal chemically and would find wider use in the laboratory except for its relatively low melting point (ca. 1050°C). It is not attacked to any serious extent by molten alkali hydroxides and is therefore used instead of platinum for such fusions. Dishes of gold are very suitable for the evaporation of acid or basic solutions, being equal or superior to platinum for this purpose.

[8] Magnesium ammonium phosphate may be ignited to the pyrophosphate in platinum if care is taken to maintain oxidizing conditions.

[9] Surprisingly large amounts of platinum are brought into solution when hydrochloric acid–potassium chloride solutions are evaporated in platinum in air, as shown by H. von Wartenberg, *Z. Anorg. Allgem. Chem.* **273**, 91 (1953). Thus, evaporation (~80°C) to dryness of 100 ml of 10 per cent HCl containing 2 g of KCl in a platinum dish, dissolved platinum equivalent to 2.5 and 3.6 mg of $PtCl_6^{--}$ in two trials. With 0.2 g of KCl under the same conditions, 0.5 mg of $PtCl_6^{--}$ was dissolved. No attack resulted from evaporation of HCl–NaCl solutions. The attack in the presence of potassium is explained by the reaction

$$Pt + O_2 + 6\,Cl^- + 4\,H^+ \rightarrow PtCl_6^{--} + 2\,H_2O$$

being aided by the precipitation of K_2PtCl_6.

Silver

The melting point of silver is 960°C, and it is therefore not safe to heat articles of this metal over a bare flame. Silver oxide is unstable at high temperatures but has a certain stability at lower temperatures (in the neighborhood of 200°C), so that silver vessels may become covered with a superficial film of oxide when heated and are therefore unsuitable for weighing precipitates. Silver crucibles are sometimes used for fusions with sodium or potassium hydroxide. Silver is slightly attacked by molten hydroxide at the upper edge of the melt (combined action of alkali and oxygen).[10] It is less resistant than gold to the action of molten alkalies. Silver vessels serve very well for evaporation of alkaline solutions.

Nickel

Nickel oxidizes in air, and crucibles made of it are therefore useless for heating precipitates to be weighed. Nickel crucibles are advantageous in sodium peroxide fusions ($\sim$500°C) and are also suited for sodium hydroxide fusions. No metal is entirely unaffected by fused sodium peroxide. Nickel, however, is not seriously attacked, so that the same crucible can be used for a number of sodium peroxide fusions before it becomes perforated. The nickel introduced into the fusion mixture is usually unobjectionable as long as the metal does not contain the constituents being determined. Alloys of nickel and chromium (Nichrome, Chromel) are not easily oxidized when heated in air. A thin, firmly adherent coat of oxide is formed which protects the metal from further oxidation. Such articles as triangles and wire gauze, which must withstand a high temperature in an oxidizing environment, are made of this alloy in preference to iron, which is soon destroyed and which, moreover, rusts.

Nickel tends to become brittle and crack when heated with sulfur or sulfides. Fusions of sulfur compounds should be made in an oxidizing environment (excess of sodium peroxide). Silver, mercury, and vanadium compounds should not be heated in nickel. Borax should not be used as a flux.

Iron

Iron crucibles, although not as durable as those of nickel, may be substituted for the latter in sodium peroxide fusions; their cheapness makes up for their short life.

Crucibles for the determination of the alkalies by the J. Lawrence Smith method may be made of iron or nickel,[11] but platinum is preferable.

Zirconium

Crucibles of zirconium are commercially available and find use in sodium peroxide fusions. In a typical peroxide fusion at 550°C, approximately 2 mg of Zr is dissolved (less than in Ag, Fe, Ni, or Pt.)[12] A hundred fusions can be made in a crucible. If necessary, zirconium crucibles can be used above 550°C, although the attack will then be greater.

Fused alumina

Highly sintered articles of pure aluminum oxide (corundum) have been produced.[13] The refractory character (m.p. 2050°C) and great hardness of corundum make these utensils of value for certain purposes.

[10] W. Dittmar and D. Prentice, *J. Soc. Chem. Ind.* **12**, 248 (1893), found that a silver crucible in which 8.5 g of potassium hydroxide had been kept in the molten state for 10 min lost 10.8 mg in weight.

[11] A. W. Epperson and R. B. Rudy, *J. Ind. Eng. Chem.* **17**, 35 (1925).

[12] C. B. Belcher, *Talanta* **10**, 75 (1963).

[13] H. Gerdien, *Z. Elektrochem.* **39**, 13 (1933); H. Kohl, *Keram. Rundschau* **41**, 30 (1933).

Plastics

A great variety of laboratory ware (including filter funnels and crucibles, separatory funnels and pipets) is made from the thermoplastic polymers polyethylene and polypropylene, which are resistant to attack by most chemical reagents and to mechanical shock. Polyethylene is produced in two general types differing in density and melting point.[14] The lower-density (0.92) material melts at ~108°C and should not be heated above 70°C; the higher density (~0.95) material, melting at ~135°C, should not be heated above 100°C. The lower-density polyethylene is more flexible and is used for squeeze-type wash bottles, the higher-density type for bottles, flasks, etc., requiring greater rigidity. Polypropylene is harder and more rigid than polyethylene and can be used at higher temperatures (m.p. ~170°C, maximum service temperature ~130°C). Polyethylene and polypropylene are not affected at room temperature by concentrated HCl, HF, and H_3PO_4, or strong hydroxide solutions. They are slowly attacked by strong sulfuric acid (>60 per cent), concentrated nitric acid, bromine, and other strong oxidizing agents (including perchloric acid). H_2S and NH_3 are absorbed and their solutions are best not stored in plastic bottles. In general, organic solvents should not be stored in bottles of these materials. Softening, swelling, and cracking are likely to occur, especially with hydrocarbons and chlorinated hydrocarbons.

Because polyethylene is virtually unaffected by aqueous solutions, and because the low-density polymers are almost free of metals (<1 ppm Fe, Mg, Al, Pb, Cu, etc.), water, standard solutions, and reagent solutions are often stored in polyethylene (or polypropylene) bottles in preference to glass. If the stopper or cap fits well there seems to be no appreciable loss of water in a period of a few months. Most metals, even in very dilute solution, show little tendency to be adsorbed by polyethylene, and cation exchange cannot occur.[15]

Measuring cylinders made of glass-clear polycarbonate plastic are available.

Teflon is a tetrafluoroethylene polymer advertised as "the most heat-resistant organic material known to man." Its inertness to almost all reagents[16] combined with its ability to withstand temperatures up to about 300°C (~600°F) makes it very useful as a material for beakers and dishes in which strongly acid and alkaline solutions can be heated. Hydrofluoric acid decomposition of silicates, for example, can be carried out in Teflon dishes at temperatures above 100°C. Above approximately 250°C (~500°F) Teflon gives off tetrafluoroethylene in small amounts, so that at higher temperatures Teflon vessels should be used in a hood. Flames should not, of course, be used to heat Teflon vessels. Hot plates can be used, with temperatures not exceeding 300°C at any spot. Teflon is a soft material and should not be cleaned with abrasive powders. In addition to beakers and evaporating dishes, Teflon is used for plugs in burets and separatory funnels, stoppers, stirring rods, and watch glasses.

Rubber

Except for tubing and occasional use as stoppers, rubber is of little importance in the analytical laboratory. Rubber stoppers should not be allowed to come into contact with solutions and should be used as little as possible. Rubber contains sulfur and various other substances such as zinc oxide, antimony sulfide, and organic compounds, and these may be extracted by solutions. Polyethylene stoppers (solid) are less flexible than rubber, but when they can be used they are superior to rubber in resistance to

[14] The low- and medium-density polymers are formed by polymerization of ethylene at high pressures and temperatures of approximately 150–250°C without metal catalysts. The high-density (linear) polymers are formed at relatively low pressures and temperatures in the presence of metal catalysts. The linear polymers are 85–95 per cent crystalline, compared to 50–85 per cent for the lower-density, branched polymers.

[15] J. Murphy and J. P. Riley, *Anal. Chim. Acta* **14**, 318 (1956), state that phosphate is strongly adsorbed by polyethylene. Also see Ref. 12 in Chapter 17.

[16] It is attacked by molten alkali metals and liquid fluorine.

attack by aqueous solutions. Neoprene and silicone rubber stoppers have some advantages over ordinary rubber stoppers. Butyl rubber tubing is less permeable to air and carbon dioxide than ordinary rubber tubing. Silicone tubing is usable from -100 to 500°F. Tygon (halogenated vinyl plastic) and polyethylene tubing sometimes are substituted for rubber tubing because of greater chemical resistance but are not always suitable because of lesser flexibility.

Burners

Tirrill. A Tirrill burner (Fig. 18-1) is an improved type of Bunsen burner in which the gas and air supply can both be regulated. With this type of burner it is possible to attain a temperature of 1000°C or more in a covered platinum crucible; the maximum temperature obtainable in a porcelain crucible is considerably lower than this. Tirrill burners are generally used when a particularly hot flame is not required; many ignitions can be satisfactorily made with them.[17]

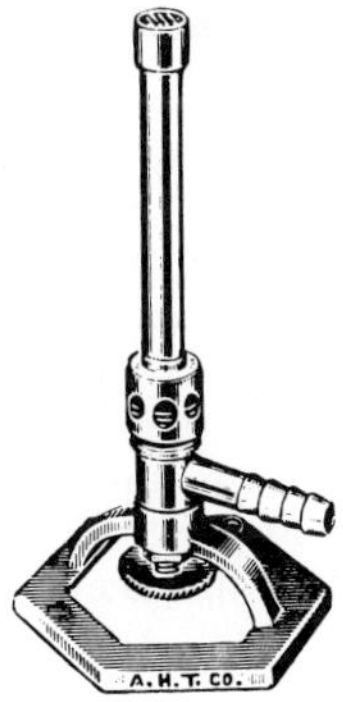

Fig. 18-1. Tirrill burner. Courtesy of A. H. Thomas Co., Philadelphia.

Meker. The Meker burner (Fig. 18-2) is so constructed that the gas can be mixed with sufficient air for complete combustion without having the flame "strike back." It furnishes a very hot oxidizing flame in which a platinum crucible can be heated to a temperature of approximately 1200°C. This burner is used for the ignition of precipitates that require a high temperature for a complete conversion into the weighing form, and for some fusions.

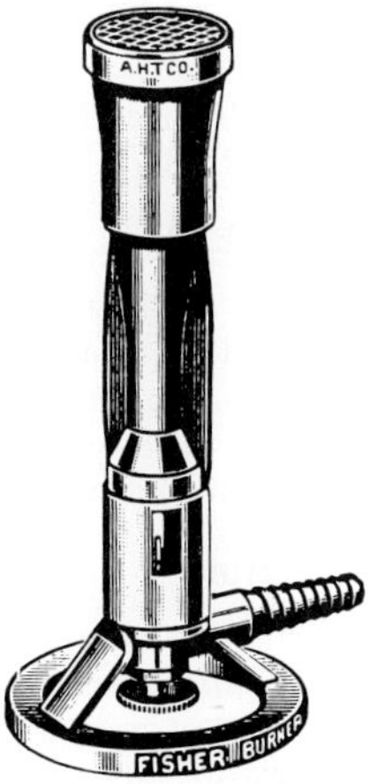

Fig. 18-2. Meker-type burner. Courtesy of A. H. Thomas Co., Philadelphia.

[17] The design of burners varies with the fuel used, i.e., whether manufactured or natural gas, and the proper burner must be available.

Blast lamps. These require compressed air for operation (Fig. 18-3). Temperatures obtainable with a blast lamp do not much exceed those given by a Meker burner, which as a rule can be used with equally satisfactory results.

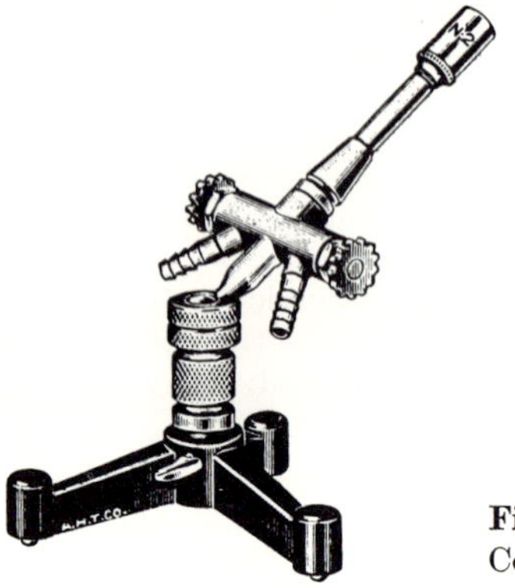

Fig. 18-3. Blast lamp. Courtesy of A. H. Thomas Co., Philadelphia.

Electric furnaces

Every analytical laboratory should be provided with an electric muffle furnace capable of being heated to at least 1000°C.[18] It is very desirable that such a furnace be equipped with a thermocouple and an indicating pyrometer, so that the temperature can be regulated to permit the ignition of precipitates which require heating in a narrow temperature interval.

For the determination of carbon in steel, organic combustions, etc., a combustion-tube furnace is required.

HEATING APPARATUS FOR PRODUCTION OF LOW TEMPERATURES

Constant-temperature electric oven

An electrically heated, thermostatically controlled drying oven having a temperature range extending from room temperature to 180°C, or still better to 250°C, is indispensable. It is used for drying precipitates, or solids in general, at low controlled temperatures. Liquids should not be evaporated, nor liquid mixtures digested, in drying ovens. Great care should be taken not to spill any material in ovens or furnaces.

Hot plate

An electrically heated hot plate capable of giving temperatures of 100–200°C should be available in the analytical laboratory. A gas-heated hot plate may be substituted.

Steam bath

For the slow evaporation of liquids, digestion of precipitates, etc., a source of heat with a temperature not rising above 100°C must be available. Such temperatures are most conveniently supplied by a steam bath. A bath connected to a steam line that can be run day and night is of great aid in expediting analytical work. In the absence of a steam bath a water bath of some kind, improvised if need be, must be used instead.

Radiator

This useful appliance (Fig. 18-4) is simply an air bath in which liquids can be evaporated rapidly without danger of loss. It consists of a large iron or, better, nickel crucible

[18] Furnaces with metallic resistance elements cannot in general be used above 1000°C for extended periods. Silicon carbide (Globar) elements allow attainment of temperatures of about 1450°C.

or special vessel fitted with a clay or silica triangle, in which the crucible containing the liquid is placed. By heating the external crucible moderately strongly with a burner, the liquid in the inner crucible can be made to evaporate rapidly without boiling. A radiator is especially valuable in the volatilization of sulfuric acid.

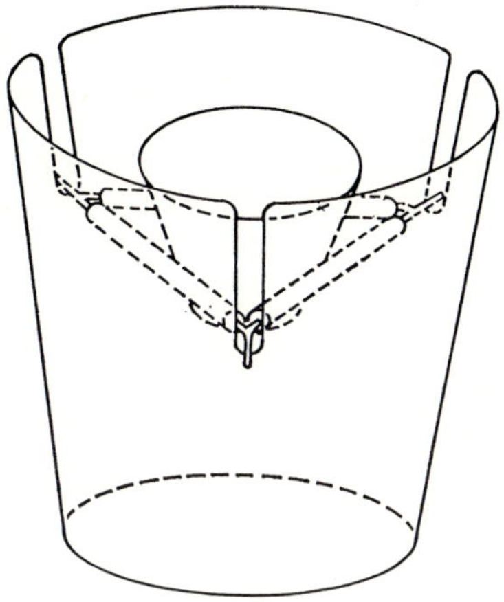

Fig. 18-4. Radiator.

Heat lamps

Infrared lamps are useful for evaporations, drying precipitates, and charring filter papers. Solutions that tend to creep can advantageously be evaporated in this way.

APPARATUS FOR FILTRATION

Funnels

Funnels should be carefully selected. The angle of the cone should be 60°. The stem should not be too wide (else it will be difficult to keep it filled with liquid) and should taper slightly to the bottom. Funnels are supported on a wooden filter stand.

Filter paper[19]

Filter paper (of cellulose fibers) for quantitative use must be practically ash-free. Such paper is prepared (commercially) by digestion with hydrochloric and hydrofluoric acids followed by washing with pure water. An 11-cm circle of quantitative (ash-free) filter paper yields approximately 0.0001 g of ash on ignition. Ordinary (qualitative) filter paper contains much ash and cannot be used for collecting precipitates that are later to be weighed after burning the paper.

Quantitative filter paper is made in various degrees of fineness. The choice of a particular grade of paper, as regards retentiveness, that should be used in any filtration is determined by the fineness of the precipitate to be filtered. Thus ferric hydroxide and other hydrous oxides which consist of large, slimy agglomerates are filtered through soft, loose-textured paper of relatively large pore size, which provides rapid filtration without letting any precipitate pass through. A precipitate of medium grain size, for example, calcium oxalate precipitated from a weakly acid solution, may be filtered through paper of moderately fine texture, whereas barium sulfate, which usually tends to give very fine crystals, not much agglomerated, requires a dense paper that has small pores. The denser (finer) the paper, the slower the filtration will be, and naturally a very fine paper is used only when actually needed.

[19] Regarding standards for analytical filter papers see B. W. Scribner and W. K. Wilson, *J. Res. Natl. Bur. Std.* **39**, 21 (1947). Recommendations include: α-cellulose content > 95 per cent, copper number < 0.5, pH > 6. Results of tests of commercial papers are described.

The 589 series of Schleicher and Schuell quantitative papers (available in 5.5, 7, 9, 11, 12.5 cm, and larger sizes) illustrates the range of retentivities:

No. 589, black ribbon, is a soft paper of very loose texture which filters very rapidly. It is suitable for gelatinous precipitates (hydrous oxides).

No. 589, white ribbon, is more retentive than the preceding and is suitable for many precipitates.[20]

No. 589, blue ribbon, is still more retentive and filters rather slowly. It is used for fine precipitates. The average pore diameter is about 2μ.

No. 589, red ribbon, is the most retentive of all and is intended for extremely fine precipitates. (No. 507 is a retentive hardened paper.)

The quantitative filter papers of two other manufacturers are designated as follows:

	Coarse	*Medium*	*Fine*
Munktell	OOR	OK OO	OOH
Whatman	41 41H (= 54) (hardened)	44 40 (lower ash)	42

Hardened filter paper is made by treating ash-free paper with nitric acid. It has a hard smooth surface, is quite strong when moist, and resists the action of strong acids and alkalies. It is used for filtration with vacuum, for liquids that would destroy unhardened paper, and in Büchner funnels for collecting recrystallized salts that must be kept free of filter-paper fibers.

Macerated filter paper (pulp) is used as an aid to rapid filtration of gelatinous precipitates (and to some extent other precipitates) that tend to clog the pores of a filter paper. It is prepared by moistening quantitative paper with concentrated hydrochloric acid, allowing the acid to remain in contact with the paper for 2 or 3 min (not longer), then adding water and stirring to disintegrate the paper, and finally filtering it off and washing it with water to remove all the acid. The fibers are then suspended in water and reserved for future use. Filter pulp may be purchased.

Filter paper made from glass fibers has advantages over cellulose filter paper in some analytical applications.[21]

Filter papers must be carefully protected from dust and laboratory fumes.

Filter crucibles

For the quantitative collection of precipitates to be weighed after heating, various kinds of crucibles with porous bottoms are employed.[22] They are all used with suction.

Gooch crucibles. The earliest filter crucible suitable for analytical purposes was devised by Gooch.[23] This kind of crucible has enjoyed great popularity and is still extensively used. A Gooch crucible is a tall-form porcelain (less often platinum) crucible having a bottom perforated with numerous small holes. The filtering medium is a mat of asbestos fibers obtained by pouring a suspension of asbestos into the crucible. On applying suction, the fibers are drawn to the bottom and form a compact mat which will retain even fine-grained precipitates.

[20] Also available: No. 589, green ribbon, approximately double the thickness of the other papers, and intermediate between black and white ribbons in retentivity.

[21] G. Giebler and T. Kempf, *Z. Anal. Chem.* **199**, 23 (1964).

[22] Filter crucibles of stainless steel with a fritted disk of the same material are useful for filtering strong-base solutions that would seriously attack glass and porcelain. Such crucibles can be used in filtering saturated sodium hydroxide solutions (preparation of carbonate-free hydroxide). They are not intended for weighing precipitates.

[23] F. A. Gooch, *Proc. Am. Acad.* **13**, 342 (1878).

Preparation of Gooch crucible. The asbestos used should be the long-fibered amphibole variety. To prepare the asbestos, a conveniently large amount of a good product[24] is cut to give fibers about 0.5 cm long, macerated with water in a mortar, and digested overnight with concentrated hydrochloric acid on the steam bath. The digested suspension is filtered off on a Büchner or sintered-glass funnel, and the fibers are thoroughly washed with hot distilled water. The washed asbestos is shaken vigorously with a large volume of water, and the coarser fibers allowed to settle. The supernatant liquid containing the fine fibers is poured off into another bottle and reserved, and the coarser fibers are resuspended in a sufficient volume of water to give a thin suspension. There should be no undisintegrated asbestos remaining in the material thus prepared.[25]

A clean Gooch crucible is placed in an adapter (Fig. 18-5) attached to a suction flask and is filled with a thin suspension of asbestos. Gentle suction is applied, and more asbestos poured in if necessary, so that a mat 1 to 2 mm in thickness is finally formed. It is possible to tell when the mat has acquired the correct thickness by holding the crucible up to the light and looking through it; the outline of the holes should barely be visible. If the precipitate to be filtered is very fine, it may be advisable to put a thin layer of very fine fibers upon the mat by adding some of the special suspension (see above). A perforated porcelain plate (Witt plate) is now placed upon the asbestos mat, and a little more suspension is poured in to barely cover the plate and hold it in place. The Witt plate protects the asbestos mat when liquid is poured into the crucible. Water must now be drawn through the crucible to wash out the loose fine fibers. The suction is increased, and water is passed through the crucible until fine fibers are no longer visible in the filtrate when examined in good light. Usually not less than 1 liter of water should be used for the washing. The crucible is then dried to constant weight at the same temperature to which the precipitate will be heated later.

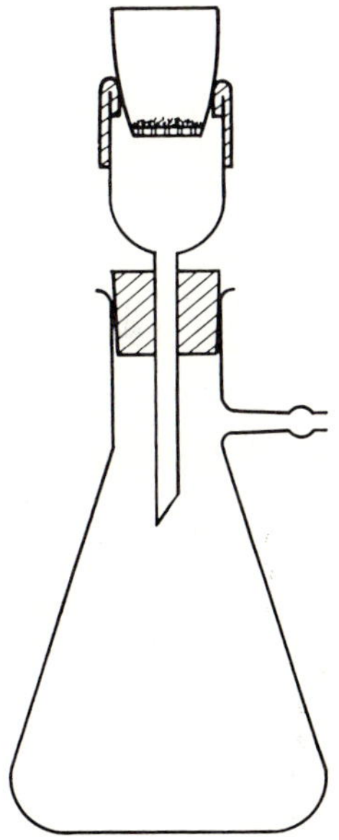

Fig. 18-5. Arrangement for suction filtration with Gooch and other filter crucibles. Adapters other than the type shown are often used; one consists of a rubber ring used with a short-stemmed funnel.

[24] The quality of asbestos on the market varies greatly; only the best available material should be used for chemical purposes.

[25] Depending upon the use to which it is to be put, the asbestos may require other additional treatment. For example, if it is to be used for filtering a permanganic acid solution, as in the bismuthate method for manganese, it should be digested with a nitric acid solution of potassium permanganate to destroy any substances that might later reduce permanganate.

Asbestos of the amphibole variety is very resistant to the action of dilute acids if it has been given the preliminary treatment described above. If dilute alkaline solutions are to be filtered through asbestos, the latter should be treated with hot dilute sodium hydroxide solution before being made into a mat. If there is a possibility of appreciable solvent action by the liquid filtered, it is best to run a blank using the same crucible as employed in the determination. It is rather difficult to obtain an asbestos mat that is entirely stable and does not furnish a small amount of very fine fibers on continued washing. In exact work it is recommendable to filter the liquid that has passed through a Gooch crucible through quantitative paper, to ignite the latter after washing, and to weigh the residue obtained, which will then represent the asbestos lost in the filtration of the precipitate.

Asbestos is somewhat hygroscopic; this fact should be borne in mind when using Gooch crucibles. Gooch crucibles can be heated to high temperatures. When heated over a flame, they should be placed in a porcelain capsule or ordinary crucible to prevent the entrance of gases through the asbestos.

Glass and silica mats. The asbestos mat of a Gooch crucible can, for some filtrations, be replaced by a mat of glass fibers. Such pads are available commercially in various sizes and need merely be fitted into the crucible and covered with a Witt plate (or two pads can be used without the plate). Obviously, glass-fiber mats cannot be used for ignition of precipitates at high temperatures ($>500°C$). A mat of fused silica fibers can be used in place of glass when the precipitate must be strongly ignited.[26]

Munroe crucibles. A Munroe crucible is a perforated platinum crucible having a mat of porous metallic platinum.[27] This mat is prepared by covering the bottom of the crucible to a depth of 0.25 to 0.5 cm with precipitated ammonium chloroplatinate moistened with alcohol and igniting cautiously at low redness. On heating, ammonium chloroplatinate decomposes, leaving a spongy layer of platinum which can be shaped into a coherent felt by working with a glass rod. The mat thus prepared is capable of retaining very fine precipitates.[28] Filtration is rapid. The chief advantage of this filter crucible lies in its ability to resist attack by most solutions and the possibility of heating to very high temperatures. Suitable solvents must be used to dissolve the precipitate after ignition, because the platinum mat once prepared must not be disturbed.[29]

Porous porcelain filter crucibles.[30] Porcelain crucibles having porous porcelain *bottoms* are superior to Gooch crucibles[31]. They can be heated to high temperatures[32] in a furnace, or they may be placed inside an ordinary porcelain crucible or a capsule and heated with a burner. Heating and cooling should be gradual to prevent cracking.

Water, dilute acids, and ammonia do not change the weight of these crucibles appreciably (at most 0.1 to 0.2 mg) when filtered through them, even when hot. They are, however, quite strongly attacked by alkali hydroxide solutions (Table 18-7). They are slightly hygroscopic. A 15-g porcelain crucible dried at 200°C gained 1.0 mg in

[26] W. W. Russell and J. H. A. Harley, Jr., *Ind. Eng. Chem., Anal. Ed.* **11**, 168 (1930).

[27] C. E. Munroe, *J. Anal. Appl. Chem.* **2**, 241 (1888); *Chem. News* **58**, 101 (1888). This type of crucible is also known as the Neubauer crucible, *Z. Anal. Chem.* **39**, 501 (1900). A filter crucible of porcelain with a filtering medium of porous platinum has been devised by O. Brunck, *Chem. Ztg.* **33**, 649 (1909).

[28] W. O. Snelling, *J. Am. Chem. Soc.* **31**, 456 (1909).

[29] O. D. Swett, *J. Am. Chem. Soc.* **31**, 928 (1909).

[30] For an extensive description and discussion of the use of porcelain, glass, and other filter crucibles see P. H. Prausnitz, *Glas- und Keramische Filter*, Akademische Verlagsgesellschaft, Leipzig, 1933.

[31] For results of tests of these crucibles, important from the analytical standpoint, see especially G. F. Hüttig and K. Schmitz, *Z. Anal. Chem.* **64**, 224 (1924), and L. Moser and W. Maxymowicz, *Chem. Ztg.* **50**, 326 (1926).

[32] The weight remains constant at high temperatures. Thus E. Eigenberger, *Z. Anal. Chem.* **68**, 220 (1926), ignited a porcelain filter crucible over a blast lamp at 7-min intervals and found the successive weights to be 21.9714, 21.9712, 21.9711, and 21.9711 g.

TABLE 18-7. **Change in Weight of Porous Porcelain and Jena Glass Filter Crucibles Treated with Sodium Hydroxide Solutions**[a,b]

Concentration of Sodium Hydroxide Solution, N	*Temp.,* °C	*Loss in Weight,* mg	
		Porcelain	*Jena Glass*
0.5	20	0.5	0.1
0.125	20	0.4	0.0
0.5	100	46.2	9.4
0.125	100	11.6	3.2
0.0156	100	0.7	0.0

[a] Partial data of L. Moser and W. Maxymowicz, *Chem. Ztg.* **50**, 326 (1926).
[b] In each case the solution was allowed to remain in contact with the crucible for 2 hr. The porcelain crucible weighed ca. 15 g and the glass crucible ca. 30 g.

weight after exposure to the atmosphere for $\frac{1}{2}$ hr (no increase in weight after this time).[33] The temperature at which the crucible has been dried may have an appreciable effect on its weight. Thus a 15-g crucible ignited strongly was found to weigh 0.7 to 0.9 mg less than when dried at 110°C to constant weight.[34] For general use in the analytical laboratory porcelain filter crucibles are to be preferred to the glass crucibles, because the former can be heated to much higher temperatures and are therefore more generally applicable.

Porcelain filter crucibles are available in various porosities. Thus Coors crucibles can be obtained in three grades of fineness, having average pore diameters of 15μ (for coarse precipitates), 5μ, and 1.2μ (for very fine precipitates).

Sintered-glass filter crucibles. These (Fig. 18-6) are made of borosilicate or other resistant glass and have a porous bottom of sintered ground glass fused on to the body of the crucible. They can be obtained in various degrees of fineness.[35]

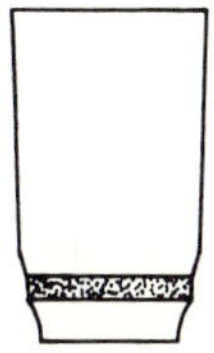

Fig. 18-6. Sintered-glass filter crucible.

[33] G. F. Hüttig and K. Schmitz, *Z. Anal. Chem.* **64**, 224 (1924). The humidity of the atmosphere in the experiment is not stated.

[34] L. Moser and W. Maxymowicz, *Chem. Ztg.* **50**, 326 (1926).

[35] Jena glass filtering apparatus is designated by certain numbers and letters. In the case of a filter crucible bearing the designation 1G4, the symbols have the following significance: the 1 indicates the size and form of the crucible, the G the kind of glass used ("Jenaer Geräteglas"), and the number 4 the grain size of the fritted-glass plate. The following are the average pore sizes of the different plates:

No.	*Pore Size,* μ
1	100–120
2	40–50
3	20–30
4	5–10

No. 3 is suitable for moderately fine precipitates and No. 4 for very fine precipitates such as barium sulfate.

These crucibles are used to collect precipitates that can be dried at low temperatures. The highest temperature to which a glass crucible can be safely heated is 500°C, and this only with very gradual heating and cooling. Sintered Pyrex and Jean glass crucibles are very resistant to the action of hot water and dilute acids, being the equal of, or even surpassing, the porcelain filter crucibles in this respect; but like the latter they are attacked by alkaline solutions, especially when hot, although to a smaller extent (Table 18-7). Concentrated ammonium hydroxide is without effect.[36] Glass-filter crucibles are but slightly hygroscopic (Table 18-8).

Silica filter crucibles with sintered silica bottoms are also manufactured. They can be heated to very high temperatures (see p. 454).

TABLE 18-8. **Hygroscopicity of Various Types of Filter Crucibles**[a,b]

	Gain in Weight, mg		
Time after Removal from Desiccator	*Glass Crucible* (G 5–6)	*Brunck Crucible*	*Gooch Crucible*
2 min	0.1	0.1	0.1
3 min	0.1	0.1	0.2
5 min	0.2	0.2	0.6
10 min	0.3	0.5	1.2
30 min	0.3	0.6	2.4
1 hr	0.3	0.6	3.4
2 hr	0.3	0.6	4.2
6 hr	0.3	0.6	4.9
24 hr	0.3	0.6	5.2

[a] From G. F. Hüttig, *Z. Angew. Chem.* **37**, 48 (1924).

[b] The crucibles were heated at 200°C for 1 hr, cooled in a desiccator, then exposed to the atmosphere and weighed at various intervals as shown. Each crucible weighed ca. 15 g.

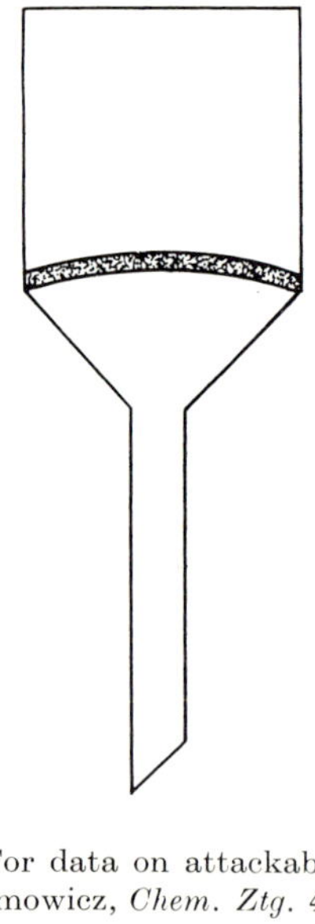

Fig. 18-7. Büchner funnel with sintered-glass disk.

[36] For data on attackability of sintered-glass crucibles by solutions, see L. Moser and W. Maxymowicz, *Chem. Ztg.* **48**, 693 (1924); A. Simon and W. Neth, *Chem. Fabrik*, **1**, 41 (1922).

Pyrex filter crucibles are available in a number of porosities: C, with a nominal maximum pore diameter or 40–60μ; M, 10–15μ; and F, 4–5.5μ.

After ignition or heating, a filter crucible containing a precipitate is cleaned by shaking out as much as possible of the solid and dissolving the remainder in a suitable solvent.

A Büchner funnel with a sintered-glass disk (Fig. 18-7) is useful for collecting salts after crystallization.

OTHER APPARATUS

Desiccators

A desiccator is a large glass vessel, provided with a tightly fitting cover, in which the atmosphere is kept free of water vapor by a drying agent (desiccant) placed in the lower compartment (Fig. 18-8). It is used chiefly for holding crucibles to protect them and their contents from moisture; these are supported above the desiccant in a porcelain plate provided with holes of suitable size for their reception. The desiccator plate is to be kept clean and free of desiccant; crucibles must not come in contact with the latter. The ground-glass rim of the desiccator and its cover are lightly coated with vaseline or a special grease[37] to make the vessel airtight. A desiccator with a stopcock is necessary when substances are to be dried in vacuo or when crucibles with very hygroscopic ignited precipitates need to be cooled under the same conditions. Desiccators are also used as hygrostats (p. 189) by filling the lower compartment with a solution, or mixture of solid and solution, that will give a definite water-vapor pressure.

Desiccants. Substances that have the property of absorbing water vapor strongly through chemical or physical action are used to charge desiccators, and also for drying gases and absorbing water in its determination (p. 576).

Granular or fused anhydrous calcium chloride is the drying agent most frequently used in desiccators. It has good capacity but is not a powerful desiccant, although sufficiently satisfactory in this respect to be usable in most work. Concentrated sulfuric acid is a more effective absorber of water than is calcium chloride; when it is used in a desiccator, the lower compartment should be partially filled with pumice, glass beads, or the like, to increase the surface of the acid exposed and to minimize splashing when the desiccator is carried about. On continued use the strong acid becomes darkened from organic matter (dust), and sulfur dioxide is liberated. Discolored acid should be replaced by fresh concentrated acid. Because of these disadvantages sulfuric acid is not

Fig. 18-8. Desiccator.

[37] A mixture of equal parts of vaseline and beeswax is very good; special mixtures are on the market. These are also used for lubricating stopcocks of burets.

much used as a desiccant. Ignition residues such as calcium oxide and aluminum oxide which are very hygroscopic must be cooled in a desiccator charged with a powerful dehydrator, such as magnesium perchlorate.

Anhydrous magnesium perchlorate[38] and magnesium perchlorate trihydrate have some favorable properties as dehydrating agents. The anhydrous salt rivals phosphorus pentoxide (probably the most powerful dehydrating agent known) in its drying power, is much more convenient to use than the latter, and can absorb approximately 50 per cent of its weight of water. Anhydrous calcium sulfate[39] is one of the more powerful drying agents. Other desiccants that deserve mention are barium oxide,[40] calcium oxide, aluminum oxide,[41] and silica gel.

The data in Table 18-9 indicate the comparative drying powers of some common desiccants.

TABLE **18-9**[a]

Material	*Residual Water per Liter of Air*, 30.5°C, mg
$CaCl_2$, granular (average composition = $CaCl_2 \cdot H_2O$)	1.5
$CaCl_2$, "technical anhydrous" (average composition $CaCl_2 \cdot \frac{1}{4} H_2O$)	1.25
$Ba(ClO_4)_2$, anhydrous	0.82
NaOH, sticks	0.80
$CaCl_2$, anhydrous	0.36
$Mg(ClO_4)_2 \cdot 3 H_2O$	0.03
Silica gel (functions by adsorption)	0.03
KOH, sticks	0.014
Al_2O_3 (functions by adsorption)	0.005
$CaSO_4$, anhydrous	0.005
CaO	0.003
$Mg(ClO_4)_2$, anhydrous	0.002
BaO	0.0007

[a] From J. H. Bower, *Natl. Bur. Std. J. Res.* **12**, 241 (1934).

Wash bottles

A wash bottle of the classical type is a Florence flask fitted up to deliver a fine stream of liquid for use in the transfer and washing of precipitates. The construction of an ordinary wash bottle will be evident from Fig. 18-9. Attention should be called to the following points: The rubber stopper should be of such size that it fits the flask tightly yet is not difficult to remove. The mouth tube is bent smoothly at an angle of 135°, and the delivery tube at an angle of 45°;[42] the latter should extend nearly to the bottom of the flask and should be bent slightly at its lower end to permit emptying the flask nearly to its last drop when inclined, as it is when used to wash precipitates. The dimensions of the mouth tube and the delivery tube should be such that the parts

[38] H. H. Willard and G. F. Smith, *J. Am. Chem. Soc.* **44**, 2255 (1922); G. F. Smith, *Dehydration Studies Using Anhydrous Magnesium Perchlorate*, G. F. Smith Chemical Co., Columbus, Ohio; F. Trusell and H. Diehl, *Anal. Chem.* **35**, 675 (1963). See the last paper for a comparison of the efficiency of various desiccants.

[39] W. A. Hammond and J. R. Withrow, *Ind. Eng. Chem.* **25**, 653, 1112 (1933).

[40] H. S. Booth and L. H. McIntyre, *Ind. Eng. Chem., Anal. Ed.* **2**, 12 (1930).

[41] F. M. G. Johnson, *J. Am. Chem. Soc.* **34**, 911 (1912); J. B. Barnitt, R. B. Derr, and E. W. Scripture, Jr., *Ind. Eng. Chem., Anal. Ed.* **2**, 355 (1930).

[42] It is almost unnecessary to say that the ends of all glass tubing must be rounded by holding in the flame after cutting.

above the stopper lie in the same straight line; both tubes should lie in the same plane. The top of the delivery tube may extend 2 or 3 cm above the stopper. The inclined portion of the tube should not be too long, say not more than 5 or 6 cm. The gently tapered nozzle, which should have a length of about 5 cm, is connected to the delivery tube with a piece of gray rubber tubing. The orifice of the jet should be approximately 0.5 mm in diameter so that it will deliver a rather fine stream of liquid. It is convenient to employ a 1000-ml flask for a wash bottle intended to hold water. One or two wash bottles of 250- or 500-ml capacity for use with liquids other than water should be at hand.

Only flasks of Pyrex or similar resistant glass should be used for wash bottles. A hot-water wash bottle should have some insulating material wrapped around its neck to protect the hand. Sheet cork held in place with copper wire or cord is very suitable, as is a wrapping of heavy cord. Asbestos paper should not be used because particles are liable to be detached from this material during use. Flasks with wicker-covered necks can be purchased. It is convenient to use a three-hole stopper in a hot-water flask, a short piece of glass tubing being inserted in the third hole. The back rush of steam into the mouth when the blowing is stopped is prevented by removing the finger from the glass tube, where it has been placed during the expulsion of the hot liquid. As soon as rubber stoppers show signs of disintegration, they should be replaced by new ones to avoid contamination by particles of rubber. All-glass wash bottles can be purchased.

Squeeze-type wash bottles of polyethylene (Fig. 18-10) are now in common use. They are suitable for wash liquids at room temperature.

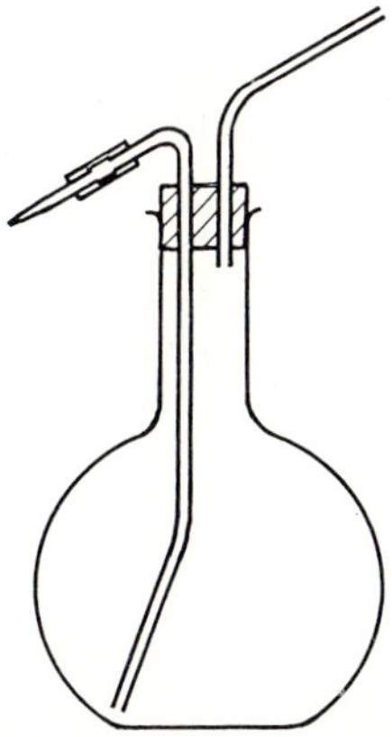

Fig. 18-9. Wash bottle.

Fig. 18-10. Polyethylene (squeeze-type) wash bottle.

Mortars

The two kinds of mortars finding common use in the analytical laboratory are the (1) agate (Fig. 18-11) and (2) Plattner (Fig. 18-12) mortars. Agate mortars and pestles are used to grind substances, already coarsely powdered, to a fine powder. Samples must often be ground before they can be brought into solution or satisfactorily fused. Naturally the sample must not be contaminated in the process, and therefore agate is used for the material of such mortars because of its great hardness and therewith its resistance to abrasion. It should be realized, however, that very hard substances ground in agate mortars may be appreciably contaminated by silica.[43] Substances

[43] This is especially true in silicate analysis, where many very hard materials are encountered. Fortunately, all samples of this nature contain high percentages of silica, so that if 0.05 per cent, let us say, of abraded silica were added to the sample in the grinding, no serious error would result.

should never be pounded in an agate mortar; a blow may break it. Liquids should not be allowed to stand in agate because agate is porous and may absorb them. Agate mortars should not be heated; such treatment may crack them.

Mullite ($3Al_2O_3 \cdot 2\ SiO_2$) has been used to some extent as a substitute for agate in analytical mortars and pestles. It has a hardness comparable to that of agate. Vitreous alumina is a superior substitute. It has a hardness of 9 on Mohs scale, compared to 7 for quartz or agate. For special purposes, mortars and pestles of boron carbide are available. This material is exceeded in hardness only by the diamond.

Porcelain mortars and pestles have no place in a quantitative laboratory. Their use is entirely inadmissible for grinding samples, which may contain hard material, and they should not be used for grinding salts on account of the danger of introducing porcelain dust.

For reducing a sample to a coarse powder that can be ground in agate, a specially constructed mortar (Plattner) made of hardened steel is used.

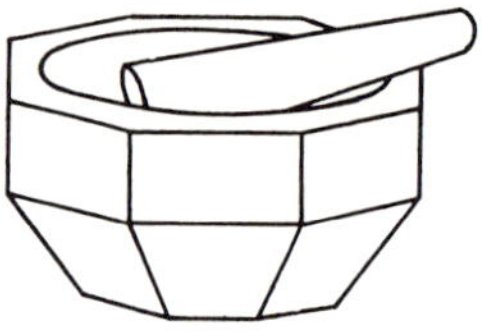

Fig. 18-11. Agate mortar and pestle.

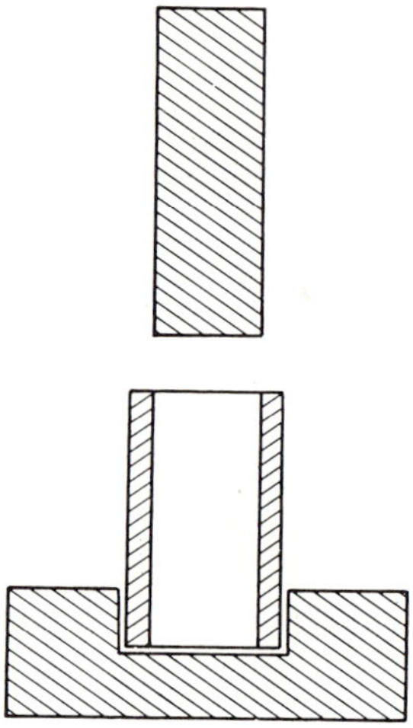

Fig. 18-12. Plattner mortar.

Balance and weights

The balance is the most important instrument used by the analyst, and its construction and use are considered in a special chapter (Chapter 19).

Volumetric glassware

See Chapter 21.

Electrolytic apparatus

See Chapter 51.

Some miscellaneous items

Stirring rods. Stirring rods are made by cutting a glass rod 3 or 4 mm in diameter (not thicker) into suitable lengths and rounding both ends in the flame to prevent scratching the vessels in which they are used. The glass should be borosilicate. One stirring rod should be provided with a short piece of rubber tubing sealed at one end ("policeman") for removing particles of precipitate adhering to walls of vessels. The rubber should fit the rod tightly.

Cover glasses. These should be borosilicate, best with ribs. Condensed steam can extract undesirable amounts of silica and other constituents from soda-lime cover glasses. Teflon cover glasses are available and are preferable to borosilicate for some purposes.

Wire gauze. Squares of wire gauze are used to support beakers, flasks, etc., when heated over a flame. They are commonly made of steel wire, but gauze of Nichrome or Chromel is to be preferred because it is nonrusting and lasts much longer. For heating platinum dishes, squares with asbestos-covered centers are used.

Triangles. These are used to support crucibles when heated with a burner. Because of their cheapness, pipestem clay triangles with steel wire are often used. Triangles of Nichrome or Chromel wire covered with sillimanite or, better, with fused silica, are very durable and are much to be preferred.

Tongs (double-bent form). As already mentioned, these (Fig. 18-13) should be of solid nickel or some rustless iron alloy. For handling platinum utensils it is desirable, but not strictly necessary, to have platinum-tipped tongs. If platinum vessels are allowed to cool below redness, they may be handled with pure nickel tongs or, better, with nickel tongs having chromel tips. Long-handled tongs are needed for use with a muffle furnace.

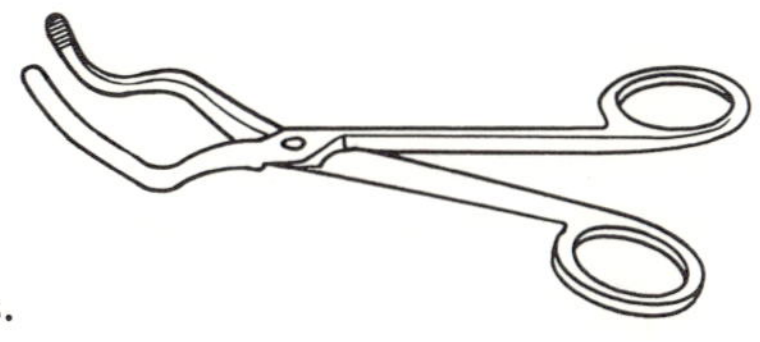

Fig. 18-13. Crucible tongs.

Desirable equipment for the analytical laboratory

Centrifuge. An electrically driven centrifuge is a valuable adjunct to any analytical laboratory. It is used in freeing recrystallized salts from mother liquor, for collecting difficultly filterable precipitates, and for approximate quantitative analysis by measurement of the volume of centrifuged precipitates.

Compound microscope. This instrument finds application in the analysis of heterogeneous mixtures, the examination of unknown samples, the determination of grain size of precipitates, etc.

A simple magnifier (10 × or 20 ×) is often useful for examining samples and for other purposes.

Spectrophotometer or filter photometer. See Chapter 55.

Apparatus for potentiometric and conductometric measurements. See Chapter 60.

Refractometer (Abbe)

Apparatus for determining densities of liquids (hydrometers, Westphal balances, and pycnometers).

PROBLEM

What kind of crucible (platinum, gold, silver, nickel, iron, silica, porcelain) may be used for fusions with the following fluxes: sodium carbonate, sodium hydroxide, potassium hydroxide, sodium peroxide, potassium pyrosulfate, sodium nitrate, a mixture of sodium carbonate and sodium nitrate, and a mixture of calcium carbonate and ammonium chloride?

CHAPTER 19 *Balance, Weights, and Weighing*

If the balance and weights are not accurate, and are not carefully taken care of, the labor and time expended on an analysis will largely go for naught. The balance and weights should therefore be regarded with a feeling akin to reverence, and the balance case looked upon, so to speak, as a sanctum sanctorum.

H. S. WASHINGTON

The result of a chemical analysis is usually expressed as a weight percentage. It is therefore necessary to determine the weight (strictly mass, as explained below) of the sample taken and the weight of the constituent sought. Every analysis involves, directly or indirectly, at least two sets of weighings. It is evident that the operation of weighing is of fundamental importance in quantitative analysis.

Principle of the method of weighing with the equal-arm balance

Before describing in detail the construction of the analytical balance and its use, it may be well to indicate the principle involved in weighing with the equal-arm balance.[1]

The equal-arm balance is essentially a lever of the first class[2] in which the arms are of equal length (schematically represented in Fig. 19-1). Suppose that a body having the mass M_1 is placed on the left-hand pan of the balance.[3] To obtain the "weight" of the body M_1, weights (bodies of known mass) are added to the right-hand pan until the pointer P returns to its original position.[4] Equilibrium is then restored, and by the principle of moments the following relation holds:

$$F_1 l_1 = F_2 l_2$$

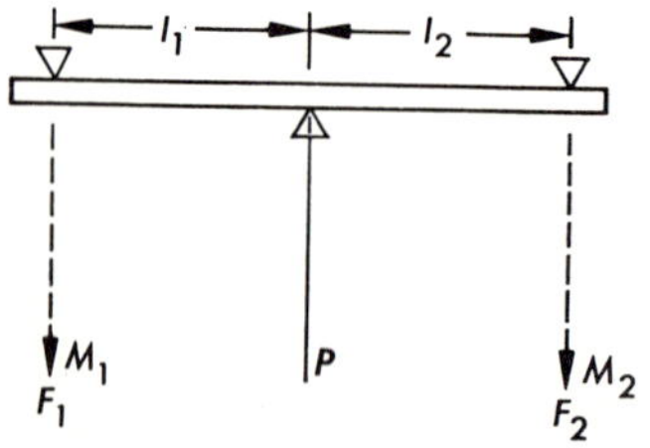

Fig. 19-1. Illustrating the principle of the equal-arm balance.

[1] Single-pan balances are replacing double-pan equal-arm balances to a large extent, but the principles of weighing remain essentially the same and it is advantageous to present them in terms of the equal-arm balance.

[2] A lever in which the fulcrum lies between the points of application of the forces.

[3] By the term *mass* is meant quantity of matter. It is defined and expressed numerically by the relation $f = ma$ (Newton's second law).

[4] To avoid complications due to buoyancy of the air, the operations are assumed to be carried out in vacuo.

where F_1 and F_2 are the forces acting on the left- and right-hand sides of the beam (lever) at the points of suspension of the balance pans, and l_1 and l_2 are the respective distances of the points of suspension of the left and right pans from the fulcrum. Since the balance is equal-armed, $l_1 = l_2$, and at equilibrium therefore $F_1 = F_2$. The origin of the forces F_1 and F_2 lies, of course, in the attraction of the earth (gravity) on the bodies on the left- and right-hand balance pans. F_1, then, is a measure of the force exerted by gravity on the body M_1, and this strictly is the *weight* of the body M_1. The numerical value of F_1 is given by the expression

$$F_1 = M_1 g$$

where g is the acceleration of gravity.[5] In quantitative analysis we are interested in the amount of matter in a body, i.e., its mass, and not in the force with which it is attracted by the earth, i.e., its weight. We wish to determine M_1. If M_2 is the mass of the weights on the right-hand pan, it is evident that $M_1 = M_2$, because

$$\frac{F_1}{F_2} = \frac{M_1 g}{M_2 g} = \frac{M_1}{M_2}$$

and F_1 is equal to F_2. Simple as this may appear, it is necessary to call attention to the following facts. In the first place, the equal-arm (analytical) balance is an instrument for determining mass, not weight. The value of g, the acceleration of gravity, varies from place to place on the earth's surface, depending chiefly upon the altitude and latitude of a particular locality, and therefore the weight of a body varies correspondingly. The mass of a body, however, is independent of its position on the earth. Since at any one locality the weights of bodies are proportional to their masses, g being constant, it has become customary to speak of the mass of a body as its weight; and the operation of comparing masses which is performed on the equal-arm balance has become known as weighing. It is permissible to refer to the mass of a body as its weight, provided it is understood that the latter term is used as a synonym for mass and not in its strict sense.

The construction of the analytical balance. An analytical balance is a balance capable of giving the weight of a body weighing as much as 200 g with a precision of 0.1 mg. As already indicated, it consists essentially of an equal-armed lever, supported at the center and free to swing in a vertical plane, from the ends of which are suspended two pans for the reception of the object weighed and the weights, respectively. Analytical balances differ in various constructional details, but all have the same fundamental features.

The following description applies specifically to a rider balance. The metal beam, B (Fig. 19-2), as rigid and as light in weight as possible, rests by means of a prism-form agate[6] knife-edge, K, set in its center, upon a highly polished agate plate, A, placed upon the central pillar or column, C. The two terminal agate knife-edges, K_1 and K_2, with edges turned upward, are mounted over the ends of the beam at equal distances from the central knife-edge. These knife-edges support the stirrups, St_1 and St_2, the contact being made against agate plates set into the tops of the latter. The pans, P_1 and P_2, are hung from the stirrups. Attached to the center of the beam and extending downward is a long pointer, N, which indicates deflections of the beam from the horizontal position on the scale, S, at the base of the column.

To protect the knife-edges from injury when objects are placed on, or removed from, the pans, and to prevent needless wear, the moving parts of the balance can be raised so that the knife-edges and planes are very slightly separated. The device by which the

[5] Again Newton's second law, i.e., a special case of $f = ma$.

[6] Corundum (synthetic sapphire) is also used for knives and planes.

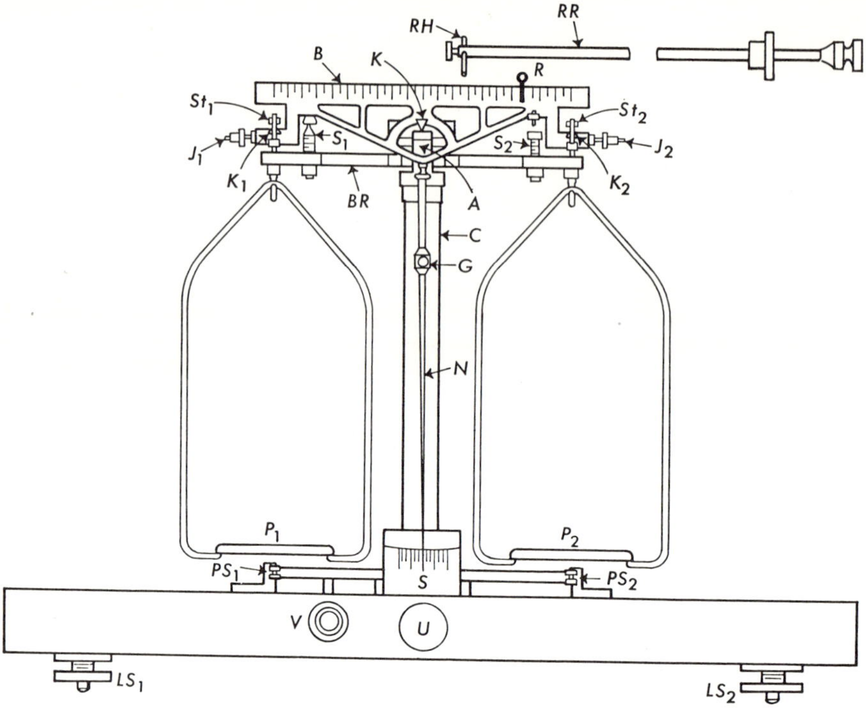

Fig. 19-2. Essential parts of the equal-arm analytical balance.

beam and the stirrups are raised is called the *beam rest* or the *beam arrest*. The beam-rest crossbar, BR, can be raised or lowered by turning the milled head, U, at the base of the balance. When the beam-rest bar is raised, the studs, S_1 and S_2, engage a cam and a slot on the under side of the beam, thereby raising the latter so that the central knife-edge and its agate plane are very slightly separated. At the same time the stirrups are lifted so that the terminal knife-edges and their planes are also separated. When the balance is not in use or when objects are being placed on, or removed from, the pans, the knife-edges and planes are thus separated by raising the beam to prevent injury or dulling of the knife-edges, or grooving of the planes. The *pan-arrest* mechanism serves the purpose of keeping the pans from swinging when the beam is lowered. The pan arrests are worked by the button, V, at the base of the balance, to the left side of the center. When this button is pushed in, the pan supports, PS_1 and PS_2, are lowered so that the pans are free to move up and down if the beam is released. To stop the swinging of the pans or to steady them, the button is allowed to move slowly outwards; the pan rests are then elevated and barely touch the bottom of the pans, thus lightly holding them in position. The beam of the balance is graduated (sometimes only on the right side) so that weight adjustments smaller than 10 mg (sometimes 5 mg) can be made by moving a light piece of wire along the beam. This piece of wire, the rider R, is bent so that it will hang on the beam. The rider is manipulated by means of the rider rod, RR, which can be moved above the beam, being handled by the end which extends through the balance case on the right side. At its left end the rider rod is provided with a hook, RH, by which the rider is lifted and transferred to any part of the graduated arm. When the rider is placed on the division directly over one of the terminal knife-edges, it produces the same effect as a weight of equal mass placed on the pan below. Therefore, if the division over the terminal knife-edge is numbered 10, the weight of the rider should be

10 mg;[7] if this division is numbered 12, the weight of the rider must be 12 mg in order that its effective weight in any position on the beam may correspond to the number of the graduation upon which it is placed. The right-hand arm of the balance is divided into 5, 10, or 12 large divisions, which are numbered. *The weight of the rider in milligrams should correspond to the number of divisions into which the arm is graduated from the central to the terminal knife-edge.* The large divisions are subdivided, usually into 10 parts, to indicate tenths of milligrams. In this manner the weight of an object can be obtained to 0.1 mg.

A movable mass, G, is attached to the pointer for the purpose of providing a means of altering the sensitivity of the balance. By moving G upward, the center of gravity of the moving parts is raised, and thereby the sensitivity of the instrument is increased, as will be explained later. Adjusting screws, J_1 and J_2, are placed at the ends of the beam. By moving the nuts in or out, either side of the balance can be made lighter or heavier than the other so that the beam can be made to rest horizontally when the pans are empty, if for any reason it does not do so originally.

Every balance is provided with some kind of device for indicating accurate leveling. This may consist of a plumb line suspended from the column, a pair of spirit levels, or a leveling plate. The instrument is leveled by means of the leveling screws, LS_1 and LS_2.

The balance proper is placed inside a glass case with a sliding door in front to protect it from air currents which would otherwise make accurate weighing impossible, and also to keep out dust and fumes as much as possible.

Chainomatic balances. The chainomatic balance (Fig. 19-3) differs from the rider balance in that weight adjustments below 50 or 100 mg are made with a fine gold chain attached to the right-hand arm of the beam. The other end of the chain is fastened to a block that can be moved up and down a graduated pillar by turning a crank at the base

Fig. 19-3. Chainomatic balance. Courtesy of A. H. Thomas Co., Philadelphia.

[7] Provided the beam is in equilibrium when the rider is over the central knife-edge. Some balances have beams graduated from 0 over the left terminal knife-edge to 10 over the right terminal knife-edge. In such a balance, the rider must be placed on 0 to maintain the beam horizontal, and the rider weighs 5 mg.

of the case on the right side. In this way more or less of the effective weight of the chain is transferred to the beam, the exact amount being indicated by the position of the chain on the pillar. By means of a vernier the weight on the right side of the balance can be read to 0.1 mg (Fig. 19-4). Some types of chainomatic balances have a beam notched at regular intervals so that by using a special rider it is possible to make weighings up to 1 g[8] without the use of loose weights. The advantage of chainomatic balances is that weighings can be made more rapidly than with the rider type.

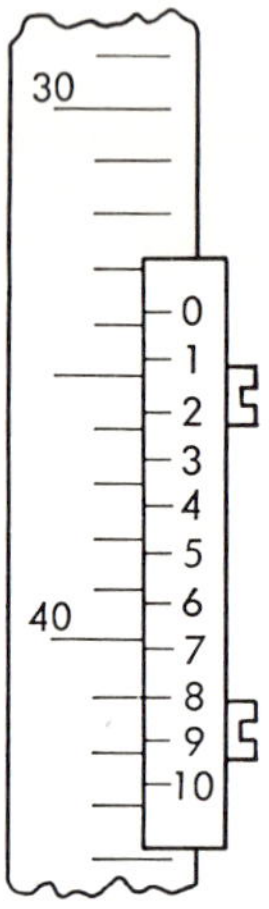

Fig. 19-4. The vernier reads 33.8 mg.

Air-damped and magnetic-damped balances

An air-damped balance has large pistons attached to the beam which move loosely in fixed cylinders. In this way the oscillations are so completely damped that if one side of the balance is given a slight overweight, the pointer will swing to one side when the pans are released and come to rest at the end of its excursion. In weighing, the point of rest is simply noted. Sensitivity is unimpaired. In some types of air-damped balances the scale is graduated to correspond to milligrams, and the point at which the pointer comes to rest is read with a microscope. In this manner weights less than 100 mg can be read off directly to 0.1 mg.

Balances can also be damped magnetically through eddy currents by hanging a counterpoised aluminum plate from one stirrup and permitting it to swing between the poles of a permanent magnet attached to the pillar.

Single-pan balances

Around the middle of the present century, a new type of analytical balance was introduced, in which one pan and a special set of weights (usually of stainless steel, density 7.76) on one arm are counterbalanced by a fixed weight on the other (Figs. 19-5 and 19-6). The arms are not of equal length and only two knife-edges are used. (Another type of one-pan balance is an equal-arm balance, in which the second pan is replaced by a carrier on which weights can be rapidly placed by rotation of knobs; it does not differ essentially from a two-pan balance.) A weighing is made by removing weights by means of a knob-operated mechanism to restore equilibrium. An optical scale allows 0.1-g weighings to be made to 0.1 mg. Air damping is used.

[8] In other types of balances a keyboard or knob-actuated mechanism is provided by which special weights (as large as 100 g) are quickly placed on, and removed from, a bar attached to the right stirrup.

One-pan balances are replacing the two-pan type. The advantages of the new type are speed, convenience, and sensitivity independent of load (the load is constant).[9] On the other hand, balances of this type may be more sensitive than equal-arm balances to temperature changes (as those produced by lamp illumination of the scale) and to slight flexing of the balance table.

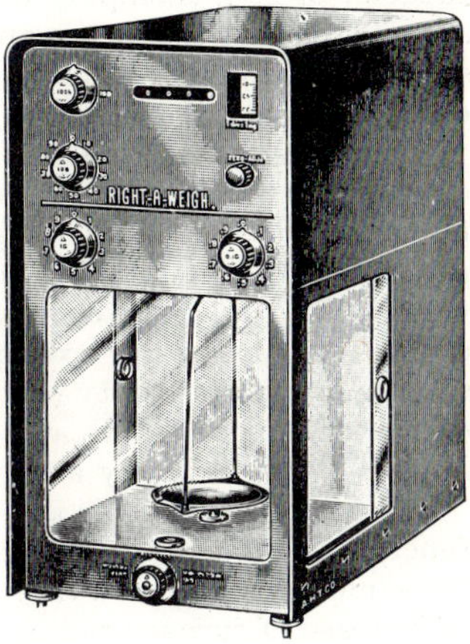

Fig. 19-5. A single-pan balance. Courtesy of A. H. Thomas Co., Philadelphia.

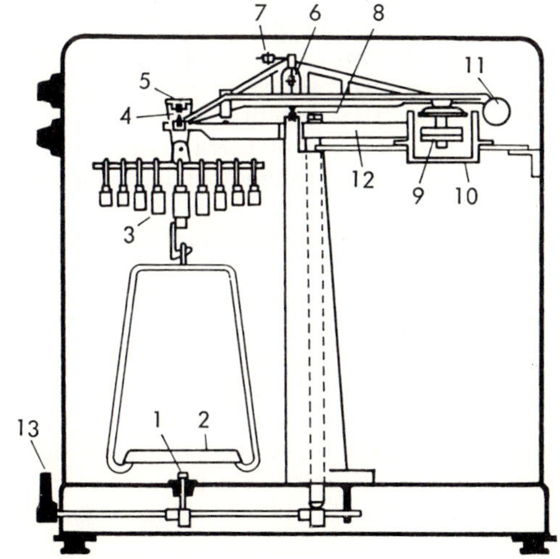

1 Pan Brake
2 Pan
3 Weights
4 Terminal Knife Edge (Sapphire)
5 Stirrup
6 Movable Weight for Adjustment of Sensitivity
7 Movable Weight for Initial Adjustment of Zero Point
8 Main Knife Edge (Sapphire)
9 Counterweight
10 Air Damper
11 Engraved Optical Scale
12 Lifting Device
13 Arrest Lever

Fig. 19-6. Essential parts of a single-pan balance (Mettler) of substitution type.

Null-type balances

In this type of balance, a mechanical or electromagnetic force is applied to restore the beam to its zero position, and this force, together with the weights used, gives the mass of the sample. Balances operating on the electromagnetic principle have a precision of approximately 1/10,000. The quartz fiber ultramicrobalance (see below) is a null-type balance. Torsion analytical balances are now (1966) coming on the market.

[9] For discussions of advantages of the "single-arm" balance, see, e.g., H. E. Almer et al., *J. Res. Natl. Bur. Std.* **68C**, 141 (1964); L. Bietry, *Chimia* (*Aarau*) **11**, 92; H. S. Peiser, *Natl. Bur. Std. Misc. Publ.* **235** (1960); I. Henins, *J. Res. Natl. Bur. Std.* **68A**, 529 (1964) (microchemical balance).

Balances for special analytical work

For weighing the small amounts of the precious metals obtained in assaying, an assay balance is used. This balance is smaller and lighter than the ordinary analytical balance and has a much smaller capacity but is sensitive to 0.005 or 0.01 mg.

In microchemical analysis a microbalance having a sensitivity of 0.001 mg and a capacity of 20 g is employed. The ordinary type of microbalance is constructed in the same manner as the macro analytical balance, the only essential difference being in the size and weight of the parts. A semimicrobalance has a sensitivity of 0.01 mg and a capacity of 100 g.

Weighings to less than 1 μg are made with an *ultramicrobalance*, in which a horizontal quartz fiber takes the place of the central knife-edge. An overweight on one side twists the fiber and an opposing torque is applied to the fiber to restore the beam to its original position. The amount of restoring twist, read on a calibrated dial, corresponds to the weight. The capacity (taring weight) of an ultramicrobalance is low, but the sensitivity can be very high. For example, one commercial ultramicrobalance, with a capacity of 2.5 g, has a sensitivity (and precision) of 0.1 μg, and weights up to 2 mg can be read on the optical scale. Greater sensitivities can be attained in this type of balance by decreasing the capacity and using quartz fibers of small diameters.

Thermobalance. This is a balance in which the weight of a sample is recorded as a function of the time as its temperature is gradually increased.[10] The container holding the sample is suspended from one arm of the balance inside a special furnace. The rate of increase of furnace temperature can be varied. A slow and uniform rate is usually required. As long as the substance heated remains unchanged, the weight–time (temperture) curve is a straight line parallel to the time (temperature) axis. If decomposition occurs, with a loss in weight, the curve trends downward, to reach another plateau when the decomposition is complete. Figure 28-1 (p. 622) shows some typical *pyrolysis* curves (thermograms). The thermobalance allows the stability ranges of weighing forms to be determined. Also it enables the composition of certain mixtures to be determined—those consisting of components undergoing stoichiometric decomposition at sufficiently well-separated temperatures.

A cautionary note needs to be sounded on the use of thermobalances: "Only after recognizing their limitations can the results of dynamic thermogravimetry be related to the ignition temperatures generally recommended. The appearance of a plateau for a compound on a dynamic thermogram does not necessarily imply that the compound is isothermally stable, either in a thermodynamic or practical sense, at all or any temperatures that lie on that plateau. The temperature limits between which it extends apply only to the particular conditions of that pyrolysis . . . . The temperature at the end of a plateau on a dynamic thermogram is also a function of the experimental conditions and therefore cannot be used to set an upper limit to the isothermal stability of the compound that it represents. Indeed, it is invariably higher than the maximum safe temperature for isothermal ignition."[11]

Other balances of the analytical laboratory

For the calibration of flasks (p. 552) and other work involving the weighing of heavy loads with a moderate degree of precision, there is required a large, sturdy equal-arm balance capable of taking a load of 2000 g and having a sensitivity of at least 0.01 g at full load.

[10] Originally devised by Chevenard. See C. Duval, *Thermogravimetric Analysis* (transl. by R. E. Oesper), Elsevier, Amsterdam, 2nd ed., 1963; W. W. Wendlandt, *Thermal Methods of Analysis*, Wiley-Interscience, New York, 1964; P. D. Garn, *Thermoanalytical Methods of Investigation*, Academic Press, New York, 1965. For a review of thermogravimetric analysis, see A. W. Coats and J. P. Redfern, *Analyst* **88**, 906 (1963).

[11] Remarks by A. E. Newkirk and E. L. Simons, *Anal. Chem.* **37**, 146 (1965), in connection with the thermogravimetry of tungsten trioxide.

A trip scale (Fig. 19-7) is used for weighing out rough amounts of solids or liquids in preparing reagent solutions of approximately known strength. Such a balance should have a capacity of 2 kg and a sensitivity of approximately 0.1 g.

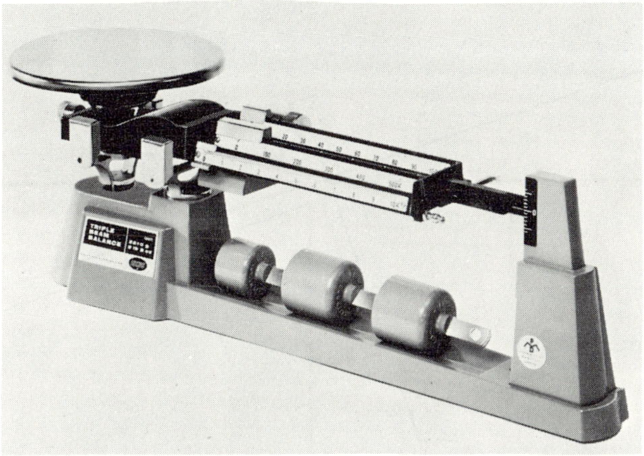

Fig. 19-7. Trip scale. Courtesy of A. H. Thomas Co., Philadelphia.

It is often convenient to have available a balance of 1-mg sensitivity and a capacity of 100 or 200 g for weighing out relatively large samples (5 g or more) and for use with a weight buret (p. 560).

Top-loading balances, introduced relatively recently, have a single pan on top; a parallelogram linkage provides stabilization. They are used for rougher weighings of the type mentioned above. In lower capacities they are quite sensitive and precise. Thus a model of 120-g capacity provides a precision of 0.5 mg, a model of 1200-g capacity a precision of 5 mg. Capacities up to 10 kg are available. Weighings can be made in seconds.

The requirements of an analytical balance[12]

1. *The balance must be accurate.* This condition is fulfilled if the beam retains its original equilibrium position when equal masses are placed on the pans, and if it gives consistent weighings. In order that a balance be accurate, the arms of the beam must be of equal length,[13] i.e., the distance of the central knife-edge from the right terminal knife-edge must be the same as that from the left, and furthermore the weights of each arm and pan must be the same (or differences compensated by movement of the adjusting screws at the ends of the beam). The consistency of a balance, denoted by the same results being obtained in successive weighings of the same object, depends upon the mechanical perfection of the knife-edges and planes and other factors (see p. 485).

2. *The balance must be stable.* When the beam is deflected from its horizontal position, it must return to its original position of equilibrium and not remain in any position in which it is put (neutral equilibrium) or topple over (unstable equilibrium). This condition is satisfied if the center of gravity of the moving parts lies below the point of suspension (central knife-edge).

[12] For testing of analytical balances see A. Craig, *Ind. Eng. Chem., Anal. Ed.* **11**, 581 (1939); L. C. Kreider, *ibid.* **13**, 117 (1941). W. M. MacNevin, *The Analytical Balance*, Handbook Publishers, Sandusky, Ohio, 1951.

[13] Since it is impossible to make the arms of exactly equal length, the excellence of a balance in this respect depends upon the extent to which the ideal is approached. The inequality in arm lengths of analytical balances usually falls in the range 5–20 ppm.

3. The balance must be sensitive. The sensitivity of a balance is expressed by the deflection of the beam or pointer produced by a small excess of weight on one of the pans. A general expression for sensitivity may be derived as follows, assuming friction to be negligible (Fig. 19-8).

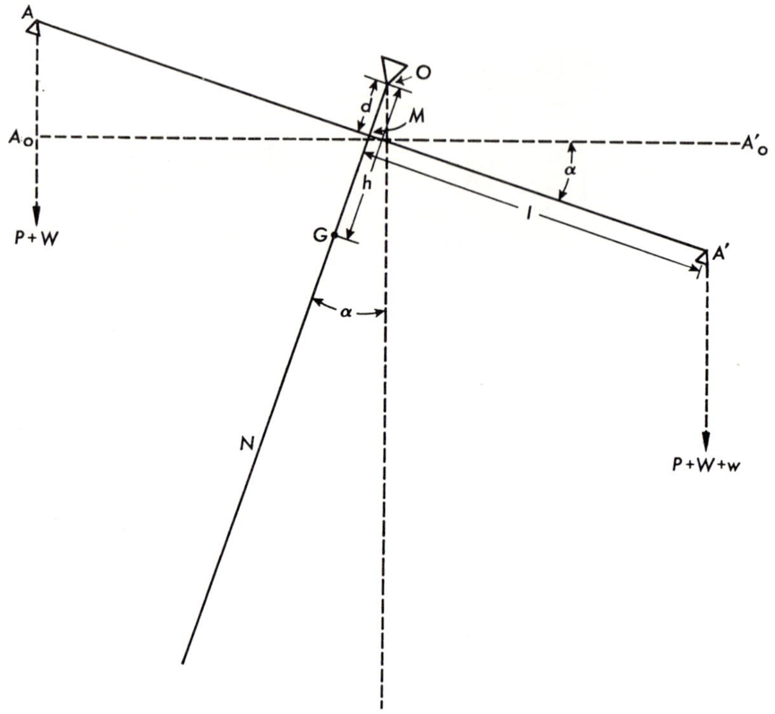

Fig. 19-8. Schematic diagram illustrating derivation of sensitivity of an equal-arm balance. The distances d and h are greatly exaggerated.

Let l = length of balance arm (beam length = $2l$)
h = distance of center of gravity G below central knife-edge fulcrum O
d = distance between O and M, the latter being the midpoint of the line joining the terminal knife-edges. This distance is taken as positive if O lies above M, otherwise negative.
w = excess weight on one of the pans producing an angular deflection α (measured in radians)
B = weight of the beam
P = weight of a pan and stirrup
W = load on each pan

In Fig. 19-8, the initial equilibrium position of the beam (with the load W on each pan) is $A_0A'_0$. With the small overweight w on the right-hand pan, the beam turns through the angle α and assumes the new equilibrium position AA'. The forces turning the beam counterclockwise are:

$2(P + W)$ acting at M through the distance $d \sin \alpha$
B acting at G through the distance $h \sin \alpha$

and clockwise:

w acting at A' through the distance $l \cos \alpha$

At equilibrium the clockwise and counterclockwise moments are equal, so that

$$wl \cos \alpha = 2d(P + W) \sin \alpha + hB \sin \alpha \tag{1}$$

$$wl = \frac{\sin \alpha}{\cos \alpha} \{2d(P + W) + hB\} = \tan \alpha \{2d(P + W) + hB\} \tag{2}$$

or

$$\tan \alpha = \frac{wl}{2d(P + W) + hB} \tag{3}$$

Since α is small, we may write

$$\alpha = \frac{wl}{2d(P + W) + hB} \tag{4}$$

If the length of the pointer is N cm and it is moved over the arc a (cm) by an overload of 1 mg ($w = 0.001$ g):

$$\alpha_{0.001} = \frac{a}{N} = \frac{0.001l}{2d(P + W) + hB} \tag{5}$$

The value of a, expressed in scale divisions, is usually taken to represent the sensitivity of a particular balance; i.e., the shift in the zero or rest point in scale divisions produced by a weight of 1 mg is the sensitivity of an analytical balance. Alternatively, the sensitivity may be defined as the weight corresponding to a change of one scale division in the zero point. This may be called the *reciprocal sensitivity*.

According to Eq. (4), the sensitivity will be independent of the load if $d = 0$ and h is constant. This ideal is represented in (a) of Fig. 19-9, in which the central and terminal knife-edges are coplanar. If the beam undergoes no flexure with increasing load the sensitivity does not change. If the central knife-edge lies above the plane of the terminal knife-edges, as in (b) of Fig. 19-9, the sensitivity decreases with the load (Fig. 19-10). This is likely to be the arrangement in most analytical balances. If the central knife-edge lies below the plane of the terminal knife-edges, as in (c), the sensitivity will increase with load unless counteracted by flexure of the beam. The balance may become unstable if d is approximately equal to h, i.e., when $2d(P + W) \geqq hB$, so that this arrangement is usually avoided by the manufacturer.

(A) All Knife-edges in Same Plane

(B) Plane of Terminal Knife-edges Below Central Knife-edge

(C) Plane of Terminal Knife-edges Above Central Knife-edge

Fig. 19-9.

The value of h can be found from the sensitivity at $P + W = 0$ (obtained by extrapolation), given the weight and length of the beam. The value(s) of d can then be found from the sensitivities at various loads.[14]

As already mentioned, the sensitivity of a balance can be varied by raising or lowering the center of gravity of the beam so that h increases or decreases. A balance adjusted to give a deflection of 2.5–5 scale divisions per milligram has a sensitivity satisfactory for quantitative analysis on the macro scale. If the sensitivity is too high, the period of oscillation becomes inconveniently long and the maximum overweight that can be determined by the method of swings (p. 488) is decreased.

[14] See Problem 1.

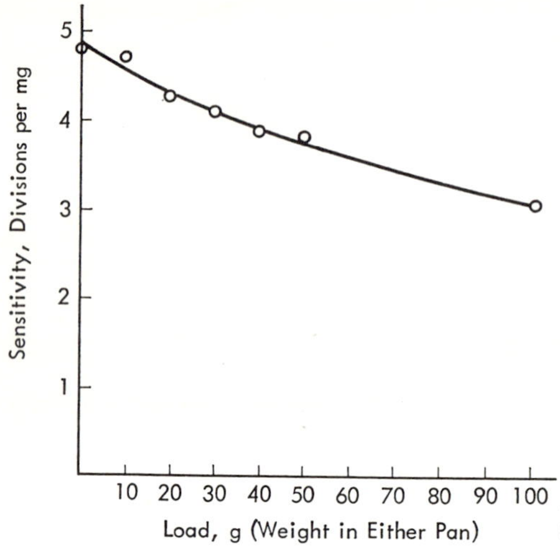

Fig. 19-10. Sensitivity curve of an analytical balance.

4. *The period of swing of the balance should not be too long.* This is a practical requirement. If the period of oscillation is too long, the process of weighing becomes tedious and requires too much time.

If I_B = moment of inertia of the beam (and pointer) (= Bk^2, where k is the radius of gyration of the beam, i.e., the distance from O at which the mass of the beam could be concentrated without changing the moment of inertia)

I_{P+W} = moment of inertia of the pans (+ stirrups) and the load

the period of oscillation of the balance, neglecting friction, is approximately

$$t = 2\pi \left\{ \frac{I_B + I_{P+W}}{g[B_h + 2d(P + W)]} \right\}^{1/2} \quad \text{(seconds)} \tag{6a}$$

$$t = 2\pi \left\{ \frac{Bk^2 + 2(P + W)l^2}{g[Bh + 2d(P + W)]} \right\}^{1/2} \tag{6b}$$

$$t = 2\pi \left\{ \frac{Bk^2/l + 2(P + W)l}{g} \times 1000\alpha_{0.001} \right\}^{1/2} \tag{6c}$$

[g is the acceleration of gravity ($\sim$980 cm/sec^2).]

For a given balance the period of oscillation increases as the square root of the sensitivity. Increasing the load increases the period. In general, the period of oscillation of an empty balance should not exceed 10–12 sec. Other factors remaining the same, short-arm balances swing more rapidly than long-arm balances.

Testing the precision of a balance

Although the reproducibility of an analytical balance is usually good enough for ordinary analytical work, it is advisable to test it. During the test, the balance should be free from the action of extraneous factors, such as temperature changes, vibration, etc., which may affect the reproducibility. The test is carried out by determining the rest point 5 or (preferably) 10 times in succession and calculating the standard deviation (p. 394) which is taken as the measure of precision (see p. 389). In order to reproduce the actual conditions of weighing, the object (weight) should be lifted off the pan and replaced each time, and the rider or chain should also be moved. The precision of the balance should be determined at loads of 0, 10 or 20, and 50 g.

The standard deviation obtained in this way includes the error in reading the scale (estimating tenths of a division). If the true reproducibility, which expresses the mechanical perfection of the balance, is desired, the reading error must be made small compared to the true reproducibility deviation. This can be done by substituting a finer scale and using a magnifier, or increasing the sensitivity (p. 483) so that the reading error becomes unimportant.[15] A good analytical balance may have an intrinsic reproducibility corresponding to a standard deviation as small as 0.005 mg. Random variations in successive rest points are the result of minute changes in arm lengths l_1 and l_2 such that l_1/l_2 is not constant. When the knife-edges and planes are separated by arrestment and brought together again, it is to be expected that because of irregularities in knife-edges and planes, the same points will not be in contact and very small changes in l_1 and l_2 will have occurred. The reproducibility will depend on the degree of perfection of the knife-edges and their planes and the arrestment mechanism. It is important that all the knife-edges be set as closely as possible at right angles to the beam. If a terminal knife-edge is not parallel to the central knife-edge, a minute shift in the front to back position of the terminal plane on the knife-edge can result in appreciable variations in arm length. A reproducibility of 0.01 mg, with, say, an effective weight of 20 g at the terminal knife-edges, means that l_1/l_2 changes on the average by 1 part in 2×10^6, corresponding to a change of $\sim 0.04\ \mu$ in a 7.5-cm arm ($1\ \mu$ = micron = 0.001 mm).

Typically, a new moderately priced analytical balance ($\sim$\$300 in 1962) with a sensitivity of 3.5 divisions per milligram shows an overall reproducibility (standard deviation of rest point) of ~ 0.02 mg at low loads.[16] (This does not include errors in setting the rider or the chain.) Single-pan balances can show similar reproducibility (see Ref. 42). A weighing requires the determination of two rest points, so that the standard deviation in a weight obtained by a balance of 0.02-mg reproducibility is $2^{1/2} \times 0.02 = 0.03$ mg (see Chapter 16). Even with balances of considerably poorer reproducibility, and including random errors from chain or rider, the error in weighing arising from the balance itself is not likely to exceed 0.1 mg. Exceptions can occur with old and worn balances.

WEIGHTS

The determination of the mass of an object requires the use of a series of reference masses called *weights.*

The system of weights and measures used in scientific work is the international metric system.[17] The unit of mass in the metric system is the *gram*, which is defined as the one-thousandth part of the mass of the international prototype kilogram at the International Bureau of Weights and Measures near Paris.

[15] R. A. Johnson and C. B. Graham, *Anal. Chim. Acta* **12**, 408 (1955), found that the reproducibility in terms of mass units was independent of the sensitivity in the high-sensitivity range (15 to 30–50 divisions per milligram) where reading error was unimportant.

[16] Much of the deviation is accounted for by reading error, which can be calculated roughly. The standard deviation in the reading of a scale (estimating tenths) by an experienced observer is not likely to be as large as 0.1 division if the divisions are not smaller than 1 mm and parallax errors are insignificant. A reasonable estimate is 0.05 division. If the zero or rest point is obtained by reading two swings (one reading on the left side, two on the right), the standard deviation of the zero point due to reading error is $\{0.05^2 + (0.05/2^{1/2})^2\}^{1/2} = 0.0038^{1/2} \sim 0.06$ division, or roughly the same as the standard deviation of a scale reading. This deviation corresponds to $(0.06/3.5) \times 1 = 0.01_7$ mg.

[17] For a short account of the international metric system of weights and measures, see *Natl. Bur. Std. Misc. Publ.* **135**, 1932.

Subdivisions and multiples of the gram are the following:

1 kilogram (kg)	= 1000 grams (g)
1 decigram (dg)	= 0.1 g
1 centigram (cg)	= 0.01 g
1 milligram (mg)	= 0.001 g
1 microgram (μg or γ)	= 10^{-6} g
1 nanogram (ng)	= 10^{-9} g
1 picogram (pg)	= 10^{-12} g

Two copies of the international prototype kilogram are deposited at the National Bureau of Standards, Washington, D.C. These are known as the United States prototype kilograms. Other nations possess similar copies of the international prototype kilogram. The mass of each copy has been carefully determined by direct comparison with the international prototype kilogram. The copies of the international kilogram are used for comparison of secondary standards employed in the calibration of weights for scientific work.

An ordinary set of analytical weights includes the following pieces[18]:

Grams: (100), 50, 30, 20, 10, 5, 3, 2, 1
Milligrams: 500, 300, 200, 100, 50, 30, 20, 10, 5, 3, 2, 1
Riders

Weights smaller than 10 mg (sometimes 5 mg) are not actually used in weighing, since weight adjustments smaller than this are made with a rider. A chainomatic balance does not require weights smaller than 1000, 200, 100, or 50 mg, depending upon its construction; and indeed some types of analytical balances do not require any external weights, as already indicated.

The larger, or integral, weights (1 g to 100 g) are made of (1) brass or bronze and plated with gold, platinum, rhodium, chromium, or nickel, or simply lacquered in the case of the cheaper varieties, or (2) an iron–chromium–nickel alloy. The latter material (stainless steel, 20 per cent Ni–25 per cent Cr) is resistant to abrasion, so such weights are subject to little change in mass as the result of wear. They are constant in weight in atmospheres free of acid and gases or vapors and are nonmagnetic. Electroplated weights are not entirely satisfactory, especially in humid atmospheres, which can lead to corrosion (entrance of moisture through defects in the plating). One-piece integral weights are superior to the two-piece type but are more expensive. Lacquered weights are not suitable for the most exact analytical work because of adsorption of moisture by the coating.

The fractional weights (those smaller than 1 g) are usually made of platinum, aluminum, or tantalum. The nominal mass is stamped on each weight, the numbers indicating milligrams or grams. Duplicate weights are distinguished by stamping with one and two dots (alternatively only one of the duplicates is stamped), and triplicates with one, two, and three dots. The weights are kept in a box containing compartments for each weight, the larger compartments being lined with velvet to prevent possible abrasion of the weights in removal and replacement. The weight box is provided with a cover to exclude dust when it is not being used. Weights are handled gently with bone-tipped forceps. All-metal forceps should not be used.

Before a set of weights can be safely used in quantitative analysis, it must be calibrated (p. 501). No reliance can be placed upon uncalibrated weights. Weights should be recalibrated at yearly or half-yearly intervals, oftener if used constantly.

[18] Other combinations are possible. Formerly the usual combination was 5-2-2-1 or 5-2-1-1-1. These combinations have the disadvantage of requiring differentiation of weights of the same denomination and the handling of a larger number of pieces on the average.

The U.S. National Bureau of Standards has set up specifications and tolerance limits (Table 19-1) for various classes of weights, of which the following are of analytical interest:

TABLE **19-1. Tolerances for Analytical Weights**[a]

	Class M		*Class* S		*Class* S-1		*Class* P
Denomination	*Individual Tolerance*[b] mg	*Group Tolerance*[b,c] mg	*Individual Tolerance* mg	*Group Tolerance* mg	*Individual Tolerance* mg	*Group Tolerance* mg	*Individual Tolerance* mg
100 g	0.50		0.25		1.0	Not more than ⅓ of the weights of a new set may be in error by more than ½ of the individual tolerances.	2.0
50 g	0.25		0.12		0.60		1.2
30 g	0.15		0.074		0.45		0.90
20 g	0.10		0.074	0.154	0.35		0.70
10 g	0.050		0.074		0.25		0.50
5 g	0.034		0.054		0.18		0.36
3 g	0.034	0.065	0.054	0.105	0.15		0.30
2 g	0.034		0.054		0.13		0.26
1 g	0.034		0.054		0.10		0.20
500 mg	0.0054		0.025		0.080		0.16
300 mg	0.0054	0.0105	0.025	0.055	0.070		0.14
200 mg	0.0054		0.025		0.060		0.12
100 mg	0.0054		0.025		0.050		0.10
50 mg	0.0054		0.014		0.042		0.085
30 mg	0.0054	0.0105	0.014	0.034	0.038		0.075
20 mg	0.0054		0.014		0.035		0.070
10 mg	0.0054		0.014		0.030		0.060
5 mg	0.0054		0.014		0.028		0.055
3 mg	0.0054	0.0105	0.014	0.034	0.026		0.052
2 mg	0.0054		0.014		0.025		0.050
1 mg	0.0054		0.014		0.025		0.050

[a] From *Natl. Bur. Std. Circ.* **547**, 1954.

[b] Acceptance tolerances for new weights; maintenance tolerances twice the acceptance tolerances are allowed for class M weights from 500 through 100 mg and for Class S weights above 100 mg.

[c] Any combination of bracketed weights must fall within group tolerance.

Class M. For use as reference standards (calibration) of high weight constancy and in highly accurate weighings; one-piece construction; nonmagnetic material meeting certain hardness requirements; integral weights of tarnishable materials must be coated with rhodium, platinum, gold, or other inert materials; fractional weights to be of material not requiring protection; surfaces to be smooth, polished, free of pits and configurative features conducive to collection of dust and abrasion.

Class S. For use as working standards and in accurate weighing; two-piece construction, with body cavity for weight adjustment; integral weights of stainless-steel alloy or brass (which may be lacquered); fractionals below 30 mg may be aluminum.

Class S-1. For use in routine analytical work; broader tolerances than for S.

Class P (formerly S-2). For routine analysis of moderate accuracy; requirements similar to S weights except tolerances are wider and all fractionals may be aluminum.

Class J. For microanalysis; 0.003-mg tolerance for readily compared weights.

METHODS OF WEIGHING

Direct weighing

The object is placed on the left-hand balance pan and weights are added to the other until the beam returns to its original position of equilibrium. By the principle of moments the mass of the object will then be equal to the mass of the weights, provided the balance arms are of equal length and the density of the object and the weights is the same. There are several methods of direct weighing, differing only in the method of determining the equilibrium position of the beam.

The method of swings. The first step in any method of weighing is the determination of the equilibrium position of the beam, as indicated by the position of the pointer on the scale, when the balance is empty. The equilibrium point is called the zero point or the rest point. We shall employ *zero point* to denote the equilibrium position of the balance when both pans are empty, and *rest point* to denote the equilibrium position of the balance when it is loaded.

Finding the zero point. The beam is first slowly lowered by turning the knurled head at the front base of the balance. The rider should be removed from the beam unless the beam is numbered from zero at the left end, in which case the rider must be placed at the zero mark to maintain equilibrium. The knife-edges will now rest against the agate planes, but the beam is not free to swing, because the pans are held in place by the pan arrests. The pan supports are next carefully lowered by pushing in the pan-arrest button.[19] Now the beam is free to swing and will usually do so. If it does not, or if the amplitude of the oscillations is too short, the rider may be momentarily placed upon the beam. It is convenient to have the pointer swing from two to five divisions on either side of the center mark of the scale. After the balance door has been closed, it is always best to let the beam make two or three swings before taking any readings, to let air currents subside. The farthest point reached by the pointer in each swing on both sides of the center mark of the scale is noted and recorded, the readings being made to a tenth of a division by estimation. An odd number of readings is taken on one side and an even number on the other (see Fig. 19-11 for justification of this procedure). Usually two swings are read on one side and three on the other. Readings to the right of the center mark are assigned positive values and those to the left negative values. Each set of readings is added separately and the average of each obtained. The algebraic

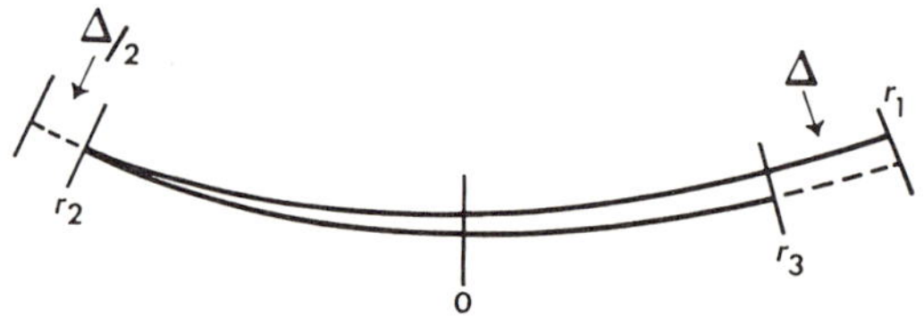

Fig. 19-11. Illustrating determination of equilibrium point of a balance from damped oscillations. In one oscillation (two swings), begun at r_1, the amount of damping is $\Delta(=r_3 - r_1)$. In half an oscillation the damping is approximately $\Delta/2$. The equilibrium point is the midpoint of an undamped swing,

$$\frac{r_1 + r_2 + \Delta/2}{2} = \frac{1}{2}\left(r_1 + r_2 + \frac{r_3 - r_1}{2}\right) = \frac{1}{2}\left(r_2 + \frac{r_1 + r_3}{2}\right).$$

Similar considerations hold for two readings on one side and three on the other, and so forth. See further, I. F. Stacy, *J. Chem. Educ.* **32**, 90 (1955).

[19] If the pans tend to oscillate, they should be arrested a few times with the pan supports while the beam is arrested. Such oscillations could cause erratic swings.

sum of the averages is divided by 2 to obtain the zero point. An illustration will make the procedure clear.

Readings on Left		Readings on Right	
(1)	-4.2	(1)	$+4.4$
(2)	-4.0	(2)	$+4.2$
(3)	-3.8		$+8.6$
	-12.0		

$$\text{Av. } \frac{-12.0}{3} = -4.0 \qquad \text{Av. } \frac{8.6}{2} = 4.3$$

$$\text{Mean } \frac{4.3 - 4.0}{2} = \frac{0.3}{2} = 0.15 \text{ or } 0.2\text{—the zero point}$$

Therefore, the zero point of the balance lies 0.2 of a division to the right of the center line (the nominal zero).

In observing the swings, the observer should be seated at a convenient height directly in front of the balance. The pointer should swing very close to the scale to prevent parallax errors as much as possible. In a well-adjusted balance the zero point will lie within one division of the zero of the scale. It is not necessary or even advisable that the two coincide. The zero point of a balance will change slightly from day to day, or even from hour to hour, owing to slight changes in temperature and other causes. If the zero point lies more than one division from the center line of the scale, one of the adjusting nuts attached to the ends of the beam should be moved in or out to make that side of the balance heavier or lighter as the case requires.[20] The balance should then be left to itself for half an hour to allow equalization of temperature.

Determination of sensitivity of balance at various loads. The sensitivity of a balance depends upon the weight on the pans (p. 482), usually decreasing with increasing load, and it must therefore be determined for various loads. The procedure is the following. After the zero point of the balance has been obtained, the pans are arrested, and the rider is placed exactly on the 1-mg division of the beam. The beam is then set swinging and the rest point of the balance obtained in exactly the same manner as described above. The difference in scale divisions between the zero point and the rest point expresses the sensitivity of the empty balance.[21] Thus, if the zero point of the balance is 0.2 and the rest point after adding a milligram is found to be -3.8, the sensitivity is 1 mg $\equiv$ 4.0 divisions. In like manner the sensitivity is determined with the following weights placed successively on both pans: 10, 20, 30, 40, 50, and 100 g. The rest point of the balance must be found for each set of weights before the excess milligram is added. The results should be graphed as shown in Fig. 19-10. From this curve the sensitivity at any load can be found. The number of the balance to which these data apply should be noted on the graph. In very exact work, the sensitivity for a particular load should be found at the time of the weighing because the sensitivity may vary slightly with temperature and for other reasons.

The weighing. After the zero point of the balance has been found, the object is ready to be weighed. The beam must be raised and the pans arrested before an object or a weight is placed on the pans. The object is placed in the center of the left pan.

An object is counterbalanced by adding weights in decreasing order of size. The weight estimated to be slightly heavier than the object is removed from the box with the forceps and placed on the right-hand pan, at the center. The beam is slowly

[20] The student should have the instructor make this adjustment.

[21] The sensitivity may also be expressed as the weight corresponding to a zero-point shift of one scale division (reciprocal sensitivity).

lowered,[22] and the movement of the pointer is observed. If the mass of the object is considerably different from that of the weight, the pointer will move one way or the other in spite of the fact that the pan supports are in place, for these are depressed by an excess weight of about 1 g on one of the pans. If the pointer moves to the left, the object is evidently lighter than the weight; the latter is replaced, after the beam has been raised, by the next smaller, and the beam lowered as before. If the weight is now too small, the next smaller weight is also added and the test for equilibrium made again. Weights are thus added and removed systematically until the pointer does not move when the beam is lowered. The excess of weight on one side or the other is now too small to depress the pan supports, and additional tests must be made by lowering the pan arrests so that the beam can move freely. Therefore the pan-arrest button is pushed in slightly, and the movement of the pointer is observed without letting it swing far from the center of the scale. The addition and removal of the smaller weights are continued as before, always systematically, until the object is balanced to within 5 or 10 mg, depending upon the weight of the rider. It is not necessary to raise the beam when adding *fractional* weights, but the pan supports should be up. When *any* weight is *removed*, the beam must be raised and the pans arrested to prevent injury to the knife-edges. The final counterbalancing is done with the rider. The latter should be placed systematically, first near the center of the arm, then one fourth of the distance from one end, etc. In precise weighings the rider is adjusted to the nearest milligram, and the tenths (or hundredths) of a milligram are then calculated from the difference between the zero point and the rest point, the sensitivity of the balance being known for the load on the pans.[23] An example will make the calculation clear.

Example

Zero point of balance = +0.5

Rest point of balance when weights on pan = 14.530 g and rider is placed on beam at 3.0 mg = +2.0 (weights and rider on right side of beam)

Sensitivity of balance at 14.5-g load = 4.0 divisions per milligram

Evidently the object is not quite counterbalanced. It is necessary to find the position of the rider at which the rest point would coincide with the zero point. Moving the rider from the 3-mg division of the beam to the 4-mg division will shift the rest point 4.0 scale divisions to the left at this load. By moving the rider

$$\frac{2.0 - 0.5}{4.0} = \frac{1.5}{4.0} = 0.38 \text{ large (milligram) division}$$

the original position of equilibrium of the beam will be restored. Since 0.38 large beam division ≡ 0.38 mg, the weight of the object is 14.53000 + 0.00300 + 0.00038 = 14.53338 g. (In ordinary analytical weighings this value would be rounded off to the fourth decimal place: 14.5334 g.)

Alternative method of obtaining the tenths of a milligram. Instead of *calculating* the point at which the rider should be placed to make the zero point at rest point coincide, the rider may be moved along the beam until a point is found at which the original position of the beam is restored. This is the method usually employed in quantitative work in which weighings need not be made with an accuracy greater than 0.1 mg. In this method the sensitivity of the balance need not be known. If, however, the sensitivity is approximately known (as it should be), it is possible to determine by mental calculation the point at which the rider should be placed to make the rest point coincide with

[22] The beam should not be lowered further than is necessary to indicate which side is heavier; i.e., the pointer should not be allowed to move more than a division or two from the center of the scale.

[23] The method of calculation is based on the assumption that the shift in the equilibrium point of the balance as indicated by a pointer playing over a scale which is a circular arc graduated in equidistant divisions is proportional to the increase in weight. For the small angles and small weights we are concerned with here, the assumption is valid.

the zero point. The rider is then placed at this point, and the correctness of the position is verified by observation of swings. If the two points agree within one fourth of a division, the weight of the object has been found to within approximately 0.05 mg if the sensitivity of the balance is 4 or more divisions per milligram. In ordinary quantitative work weighings are made to the nearest 0.1 mg only.

When it is not necessary to determine the zero point. If a substance is weighed by difference at one sitting, it is not necessary to find the zero point of the balance. The zero point may be taken to lie at the center of the scale, and the weight of the container + substance is found on this assumption. The substance, or a part of it, is poured out and the weight of the container is found by adding weights until the rest point lies at the center of the scale again. The difference between the zero point and the rest point in both weighings will be the same, provided the zero point has not shifted during the weighing, and therefore need not be known. It is not likely that the zero point will change a detectable amount during the weighing if the interval between the first and second weighings is small, i.e., a few minutes. If there is a possibility that the zero point will change between weighings, this method may not be used. For example, if the weight of an ignited residue is to be obtained, it is not permissible to weigh the empty crucible assuming the zero point to lie at the zero of the scale, then to place the precipitate in the crucible and to make the ignition, and finally to reweigh, adjusting the weights until the rest point lies at the center of the scale. During the time required for the ignition and cooling of the crucible, there is a possibility that the zero point of the instrument has changed.

The method of short single swing.[24] For ordinary analytical weighings in which it is not necessary to weigh with an accuracy greater than 0.1 mg, the method of short swings may be used. Indeed, this method can often be employed for making weighings to 0.02 mg. In this method the zero or rest point is found by adding the two successive readings of a short swing and dividing by 2 to obtain the mean value. This mean value—indicating the center of the swing—will give the equilibrium point of the balance if the swings are short. If the swings are short enough, the decrease in amplitude of each swing due to the resistance of the air and friction in the bearings will be so small that it will be inappreciable, and it is then permissible to take the mean of the end points of a swing as the zero point. Thus, in the example given in the preceding section, the zero point as determined by this method from the first two readings is $(-4.2 + 4.4)/2 = +0.1$. The deviation from the true zero point $(+0.15)$ is so small as to be entirely negligible in a balance having the usual sensitivity of three to four divisions per milligram. The difference could be made smaller by using swings shorter than four divisions. Thus, if a swing of two divisions were used, the decrease in amplitude with each swing would be approximately one half of what it is at 4 divisions, since the retardation is approximately proportional to the amplitude. Furthermore, any error made in determining the zero point in this manner would be compensated in the weighing if the amplitude were approximately the same in both and the swing were read in the same direction each time, e.g., from right to left. The length of the swings for use in this method may vary from one to four divisions.

It is recommended that this method of weighing be used in preference to the more elaborate one involving the determination of the zero or rest point from several swings. The latter method requires much more time, and the accuracy is probably no greater in the majority of cases. When a sample is weighed out by the method of short swings, the container and substance are counterbalanced with weights until the pointer swings equal short distances, say one or two divisions, on both sides of the center mark of the scale; the sample is then poured out and the weights adjusted until the pointer again swings equal distances on both sides.

For the calibration of weights, in which the results must be of the highest accuracy, the method of several swings should be used.

[24] H. L. Wells, *J. Am. Chem. Soc.* **42**, 411 (1920).

The single-deflection method.[25] One arm of the balance to be used for this method of weighing is made permanently heavier than the other, so that when the beam is lowered and the pans released the pointer will swing over a number of divisions to one side. The limit of the swing is taken as representing the zero point of the instrument. The weight of an object is obtained by adding weights until the pointer swings out the same number of divisions as before.

It is highly important that the pan supports be so adjusted that no impulse is delivered to the pans when they are released. The method cannot be used on balances that have a combined beam and pan-arrest mechanism. One side of the beam is made heavier than the other by turning the adjusting nut at the end of one of the arms. Usually the left-hand side is made heavier so that the pointer will swing to the right. If it is desired not to disturb the adjustment of the balance, a milligram weight or any other light object is placed on the left pan. The distance that the pointer swings out should preferably lie in the range three to six divisions. The pans should be released smoothly and not too slowly when using this method. If the pan-arrest button is pushed in too slowly, there is danger that the pointer will give swings of variable length. Before this method is used, the balance should be carefully tested to see whether consistent results can be obtained with it. When the pan supports are properly adjusted, a balance will give very constant deflections. A balance having a sensitivity of 1 mg $\equiv$ 4.0 divisions of scale was found to give deflections differing at the most by only 0.1 division.

It should not be forgotten when this method is used that the length of the deflection depends upon the sensitivity and, therefore, upon the load on the balance. When a large amount of material or a heavy object is to be weighed with great accuracy, the sensitivity, or the length of swing at different loads for the same overweight, must be determined. Usually this is not necessary, because most analytical weighings involve obtaining the difference between two weights, and this difference is small (ordinarily less than 1 g), so the change in sensitivity for this small change in load is inappreciable. On the other hand, if a large amount of substance is weighed, the *percentage* error due to change in sensitivity will be very small and usually can be neglected.

The single-deflection method of weighing can be used to advantage in the weighing of very hygroscopic substances. The weight of the ignited residue, for example, is first found approximately to a milligram; and then, after heating and cooling, the crucible is placed on the balance and the deflection read. From the length of the swing of the empty balance and the sensitivity at the load in question, the true weight of the crucible and its contents is obtained by a simple calculation in a manner analogous to that described under the method of swings. The weighing takes such a short time that no sensible increase in weight due to absorption of moisture occurs.

The microchemical method. It is customary in weighing with the microchemical balance to call the tenths of a pointer division *units* and to represent the zero or rest point by the *deflection sum.* For example, if the successive pointer readings are 1.8 (left of zero), 2.8 (right), and 1.6 (left), the zero point by this method would be $-(18 + 16)\frac{1}{2} + 28 = 11$. This method can of course be used in weighing with the ordinary analytical balance.

Inequality of length of balance arms

The direct method of weighing will give the true mass of the object only if the two arms of the beam have the same length (and the weighings are made *in vacuo*). If the right arm is slightly longer than the left and the object is placed on the left pan, it is evident that the direct weighing will assign a mass to the object which is a trifle less than the true value, the deviation depending upon the ratio of the arm lengths. In a good analytical balance the arms will seldom differ in length by more than 1 part in

[25] P. H. M.-P. Brinton, *J. Am. Chem. Soc.* **41**, 1151 (1919).

50,000. If an *absolute* weight must be determined with an accuracy greater than that corresponding to the difference in the length of the arms, it will be necessary to weigh by the method of transposition or substitution, to be described in the next section.

In the determination of the percentage of a constituent in a substance, it is not at all necessary to determine absolute weights. If the sample and the ignited or dried residue are both weighed on the same balance, and on the same pan, it is clear that the percentage of the constituent sought will not be affected by the difference in length of the balance arms, because the weights of sample and residue are altered in the same proportion:[26]

$$\text{Percentage of constituent} = \frac{\text{true weight of constituent}}{\text{true weight of sample}} \times 100$$

$$= \frac{\text{apparent weight of constituent}}{\text{apparent weight of sample}} \times \frac{L_r/L_l}{L_r/L_l} \times 100$$

$$= \frac{\text{apparent weight of constituent}}{\text{apparent weight of sample}} \times 100$$

where L_l = length of left-hand balance arm
L_r = length of right-hand balance arm
(Apparent weights refer to vacuo weights.)

Whenever the value of a *ratio* is to be obtained, apparent weights may be used in place of absolute or true weights. However, objects to be weighed must then always be placed on the same pan on the same balance.

It is sometimes necessary to find the true weight of an object. True weights, for example, are desirable or required in the calibration of volumetric apparatus (p. 547), and in other cases not so frequently encountered. In the calibration of weights the true mass of a weight may have to be determined with an accuracy of 1 part in 1 million, or even better. Since it is a mechanical impossibility in manufacture to set the knife-edges in the beam so that the difference in lengths of the arms will be this small, it is absolutely necessary to take the inequality into account. In the calibration of volumetric apparatus the difference in length of the arms can be neglected if the weighing can be made on an analytical balance. When a $\frac{1}{2}$-liter or 1-liter flask is to be calibrated by weighing out an absolute amount of water, an analytical balance cannot be used; a rougher balance of greater capacity must then be employed, and since in such a balance the arm lengths may be quite different, it is necessary to avoid this possible error by weighing by one of the methods described in the next section.

Weighing by transposition (double weighing; method of Gauss)

In this method the object is weighed first on one pan and then on the other. The mass of the object is the square root of the product of the weights required to counterbalance it in the two cases.

Let L_l = length of left arm
L_r = length of right arm
M_o = mass of object (i.e., true mass)
W_l = mass of weights on left pan required to counterbalance object = apparent mass of object placed on right pan
W_r = mass of weights on right pan required to counterbalance object = apparent mass of object placed on left pan

[26] This is based on the assumption that the ratio of the arm lengths always remains the same, which is approximately true (p. 494). The ratio is more variable in microchemical than in macro balances. See J. Lindner, *Mikrochemie* **34**, 67 (1948).

At equilibrium, when the object is on the left pan, it follows from the principle of moments that

$$M_o L_l = W_r L_r \tag{7}$$

and when the object is on the right pan that

$$M_o L_r = W_l L_l \tag{8}$$

Multiplying Eq. (7) by (8),

$$M_0^2 L_l L_r = W_l W_r L_l L_r$$

or

$$M_0^2 = W_l W_r$$

Therefore,

$$M_o = (W_l W_r)^{1/2} \tag{9}$$

Since W_l and W_r are very nearly equal, it is permissible to write[27]

$$M_o = \frac{W_l + W_r}{2} \tag{10}$$

It is assumed that no corrections need be applied for air buoyancy.

Ratio of lengths of balance arms. By dividing Eq. (8) by eq. (7) and solving for L_l/L_r, the following expression is obtained:

$$\frac{L_l}{L_r} = \left(\frac{W_r}{W_l}\right)^{1/2} \tag{11}$$

By means of this equation it is easy to calculate the ratio of the balance-arm lengths at the time the weighing was made with the load M_o on the pans. It should be realized that L_l/L_r is not a constant but a value that varies depending upon conditions. If one arm should be warmed slightly more than the other, the ratio would be changed. Even if both arms are heated equally, one may expand slightly more than the other, for although both arms are of the same material it is not necessarily true that each arm has the same coefficient of expansion.[28] Furthermore, the value of the ratio may vary with the load upon the pans and may depend upon other factors. Therefore, the ratio of arm lengths determined for one set of conditions may not be used for the calculation of the true mass of an object weighed under different conditions, except when an approximation is desired.

Weighing by substitution (method of Borda)

This is another method for finding the true mass of an object, irrespective of difference in arm lengths. The object is placed on the right-hand pan and counterbalanced with another body of about the same weight, called a *tare*. The rest point of the balance is determined, and the object is removed and replaced by weights that give the same rest point. The mass of the object is then equal to the mass of the weights (leaving out of consideration the buoyancy effect of the air), since both object and weights have been acting through the same arm length, L_r.

The precision attainable by this method is but half of that attainable by transposition, since the difference in rest points is only one half of that in the latter. Therefore the latter method should be used when the true mass of a body must be determined with utmost precision.

One-pan balances, as we have seen, operate on the substitution principle.

[27] If the difference between W_l and $W_r = d$, the percentage error introduced by this approximation is equal to $100(d^2/8M_0^2)$.

[28] The coefficient of thermal expansion of a solid depends somewhat upon its physical condition, and the two arms of a balance may not be exactly similar in physical properties.

ERRORS IN WEIGHING

On the basis of origin, weighing errors[29] may be classified as follows:

1. Due to balance and weights

a. Inherent. Included here are such sources of error as faulty arrestment mechanism and imperfections in knife-edges and planes that may lead to significant variations in zero points; inequality in arm lengths, already considered; and other mechanical errors (such as irregularity in spacing of subdivisions and notches in a beam). Inaccuracy of weights would be a serious source of error except for calibration. Riders and chains also require calibration, as do direct-reading scales.

b. Environmental. Temperature changes, especially those resulting in unequal heating of the beam, cause zero points to shift. The drift due to gradual room-temperature changes are not likely to cause trouble in macro weighings (they may be serious in micro weighing), where an error of 0.1 mg usually does no great harm. The zero point must, however, be determined every time a weighing is made. If a balance has not been used for sometime (as overnight) it is advisable to lower and raise the beam a few times to equalize any strains that may have been set up by temperature changes.

Vibration, above a minimum, can make exact weighing difficult. The gathering of dust on the balance bearings can give rise to erratic behavior.

2. Due to changes in the condition of the weighing vessel, the substance weighed, or the atmosphere. Errors arising from air buoyancy and adsorption of moisture are discussed in the following pages.

3. Due to the analyst. Some errors in weighing are definitely attributable to the analyst, and result from carelessness or inexperience, e.g., mistakes in reading scales and recording weights, failure to avoid parallax, poor estimation of tenths of a division, etc. Some of the errors mentioned in 1 and 2 are in part under the control of the analyst. For example, unfavorable environmental conditions can sometimes be improved by him.

Errors due to buoyancy of air

A body immersed in a fluid, i.e., a liquid or a gas, is buoyed up by a force equal to the weight of the fluid displaced. No error would be introduced from this cause in weighing if the weights used to counterbalance the object would lose exactly as much in weight as the object itself. But the density of the weights is usually different from that of the object, and consequently the volume of air displaced by the two will be different. If the density of the weights is greater than the density of the object weighed (the usual case in quantitative analysis), the latter will displace a greater volume of air than the former and will weigh a little less in air than in vacuum.

Suppose sodium chloride is being weighed with brass weights in air at 20°C, the atmospheric pressure being 760 mm Hg. The density of sodium chloride is 2.13, that of brass[30] is 8.4. One milliliter of dry air at 20°C and 760 mm Hg pressure weighs 0.00120 g. Therefore the apparent loss in weight of 1 g of sodium chloride will be $0.00120/2.13 = 0.00056$ g, and that of 1 g of brass will be $0.00120/8.4 = 0.00014$ g. Therefore, 1 g of sodium chloride will weigh $0.00056 - 0.00014 = 0.00042$ g less in air than in vacuo, a difference of -0.04 per cent. In the same manner the loss in weight of any substance weighed in air can be calculated if its density and the density of the air at the time of weighing are known. Conversely, the vacuum or true weight of a body can be calculated from its apparent weight in air if its density is known. The true weight of an amount of

[29] For a further discussion, especially from the statistical standpoint, see A. A. Benedetti-Pichler, *Mikrochemie Ver. Mikrochim. Acta* **34**, 153 (1949); *Mikrochim. Acta* **1956**, 565.

[30] The density of brass depends upon its composition. Brass used for the manufacture of weights may be taken to have the density 8.4. In very accurate work the density of the particular weights used must be determined.

sodium chloride weighing 1.0000 g in air at standard conditions is 1.0004 g. The following relation holds between the true weight of a body and its apparent weight in air[31]:

$$W = W_A + d_A\left(\frac{W_A}{d} - \frac{W_A}{d_W}\right)$$

where W = true (vacuum) weight of the body

W_A = apparent weight of the body in air = mass of the weights
d_A = density (g/ml) of air at time of weighing (depends upon pressure, temperature, and humidity of atmosphere)
d = density of the body
d_W = density of the weights

The buoyant effect of the air will not necessarily introduce an appreciable error into a determination. Most analytical results are stated in the form of a ratio:

$$\text{Percentage of constituent} = \frac{\text{weight of constituent}}{\text{weight of sample}} \times 100$$

Now, if the density of the weighing form of the constituent is equal, or nearly equal, to that of the sample, no appreciable error can result from using apparent weights in air for the calculation of percentage. Only when the densities of the two are quite different will it be necessary to correct the air weights to true weights. In most instances the densities of the two differ relatively little, and it is unnecessary in ordinary careful analytical work to reduce weights to vacuo. A rather unfavorable case may be taken for illustration. Soluble chlorides are determined by precipitating with silver nitrate to give silver chloride, which is weighed after drying. The density of silver chloride, 5.6, is very different from that of an alkali chloride, say sodium chloride (d = 2.1). In the following table are given the percentage corrections to be added (algebraically) to the air weights of sodium chloride and silver chloride (obtained with weights of various materials) to obtain the true weights in vacuum:

	Correction, %		
Substance	*Aluminum Weights* (d = 2.65)	*Brass Weights* (d = 8.4)	*Platinum Weights* (d = 21.5)
Silver chloride	−0.024	0.007	0.017
Sodium chloride	0.012	0.043	0.053
Difference	−0.036	−0.036	−0.036

The percentage of chloride in a sample is calculated from the formula

$$\frac{\text{weight of silver chloride} \times \text{factor} \times 100}{\text{weight of sample}}$$

in which the true weights of the substances are written. If air weights are used, it is evident that the true weight of chlorine can be found by applying the corrections above. Suppose that the percentage of chlorine in a sample of impure sodium chloride calculated

[31] Strictly $W = W_A + d_A[(W/d) - (W_A/d_W)]$, but since W is very nearly equal to W_A, it may be replaced by the latter in the case of solids and liquids.

from air weights is 60.02. The true percentage will be 60.02 × [(1 − 0.00024)/(1 + 0.00012)] if aluminum weights were used for weighing. This value is equivalent to 60.02 − (60.02 × 0.00036) = 60.02 − 0.022 = 60.00 per cent. The correction difference is, of course, independent of the material of the weights, provided both the sample and the precipitate are weighed with the same kind of weights. The difference between the true and apparent percentages is small in this case, amounting to 0.036 or 0.04 per cent (relative). In most determinations the difference would be even smaller. In ordinary careful analytical work it is usually justifiable to neglect the air corrections, since an accuracy better than 0.1 per cent, i.e., 1 part per 1000, is not sought or easily attainable. In more precise work, for example in testing the accuracy of a method and especially in atomic-weight determinations, it is necessary to apply the corrections. When standard substances are intercompared, weights must be corrected to vacuo. The difference in titer of an acid solution standardized against borax ($d = 1.7$) and sodium carbonate ($d = 2.5$) amounts to 0.02 per cent when vacuum corrections are not made.

When absolute weights are desired, it is important to apply vacuum corrections, e.g., in the calibration of volumetric glassware by weighing water. Since the density of water is accurately known at any temperature, it is easy to weigh out an amount of water that will indicate a certain desired volume (p. 549). It is necessary to correct for the buoyancy of the air in such work. If this correction should not be applied, the error will amount to

$$0.0012\left(1 - \frac{1}{8.4}\right) \times 100 \text{ per cent} = 0.106 \text{ per cent}$$

if the weighings are made with brass weights. This error corresponds to slightly more than 1 ml in the calibration of a liter volumetric flask and is too large to neglect.

Customarily, fractional weights in a set are adjusted so that their apparent masses in normal air ($d = 0.0012$ g/ml) are the same as those of the brass weights ($d = 8.4$) having true masses equal to their face values (corrected when necessary).[32] Fractional weights adjusted in this way will have a buoyancy correction equal to that for brass, no matter what their material, at and near normal air density.

For laboratories situated below an altitude of 1000 ft, the density of the air will rarely fall outside the limits 0.0011 and 0.0013 g/ml, and it is usually permissible to use the average value, 0.0012, in the calculations when vacuum corrections must be applied. For exact calculation see p. 549.

Reducing air buoyancy errors by use of rational atomic weights. A device for minimizing the errors resulting from air buoyancy in weighing is the use of the so-called rational atomic weights.[33] The principle involved may be brought out by an example. In calculating the percentage of chlorine in a chloride, using the customary gravimetric method of weighing precipitated silver chloride after drying, the formula employed is

$$\frac{\text{weight of silver chloride} \times (\text{Cl}/\text{AgCl}) \times 100}{\text{weight of sample}}$$

Since in this formula *weight* refers to vacuum weight, the air weight of both sample and silver chloride must be reduced to vacuo as described above; then the value of the ratio Cl/AgCl calculated from the international atomic weights can be introduced. The international atomic weights are, of course, based upon weighings in vacuo, and therefore the weight of chlorine thus found is the vacuum weight. Now it is possible to proceed in a converse manner to find the percentage of chlorine in the sample from air

[32] If the integral weights are of a material other than brass, i.e., stainless steel, they are usually adjusted with reference to brass ($d = 8.4$) in normal air. Since the density of stainless-steel weights used in single-pan balances is, on the average, 7.76 g/ml, there is only a difference of ~12 ppm compared to brass when the air density is 0.0012 g/ml.

[33] N. Schoorl, *Z. Anal. Chem.* **57**, 209 (1918).

weighings. Suppose we can find the air weight of the true amount of chlorine in the sample of given air weight. Then the true percentage of chlorine can be calculated. Actually it is possible to do this by using what may be called air-atomic weights, i.e., atomic weights that have been corrected for the buoyancy of the air. These are the *rational atomic weights* (Table 19-2). Rational atomic weights may be used in computing rational molecular weights, the atomic volumes being additive within limits. The rational atomic weights are not calculated from the atomic volumes of the elements themselves but from the average values which they have in various compounds. The calculations have been based on the assumption of the use of brass weights[34] (d = 8.4) and weighing in air under standard conditions (0°C, 760 mm Hg pressure).

TABLE **19-2.** **Rational Atomic Weights (1961)**

Aluminum	26.97	Mercury	200.59
Antimony	121.745	Molybdenum	95.94
Arsenic	74.91	Nickel	58.71
Barium	137.34	Nitrogen	14.00
Bismuth	208.99	Oxygen	15.995
Boron	10.80	Phosphorus	30.96
Bromine	79.89	Platinum	195.1
Cadmium	112.40	Potassium	39.08
Calcium	40.07	Selenium	78.96
Carbon	12.01	Silicon	28.09
Chlorine	35.44	Silver	107.87
Chromium	51.99	Sodium	22.98
Cobalt	58.93	Strontium	87.61
Copper	63.54	Sulfur	32.05
Fluorine	18.99	Tellurium	127.59
Gold	197.0	Thallium	204.37
Hydrogen	1.001	Tin	118.69
Iodine	126.88	Titanium	47.90
Iron	55.85	Tungsten	183.8
Lead	207.20	Uranium	238.03
Lithium	6.94	Vanadium	50.94
Magnesium	24.31	Zinc	65.37
Manganese	54.94		

Returning now to the example, it will be seen that by using the rational atomic weights in the factor Cl/AgCl, we obtain the weight of chlorine under the same conditions as the weight of the sample, and thereby the correct percentage of chlorine, provided the atomic volumes of the elements concerned are the same as those upon which the table of rational atomic weights has been constructed. This method of correcting for buoyancy has the advantage of rapidity and does not require a knowledge of the density of the weighing form and the sample itself. Unfortunately, great accuracy cannot be attained in this method, because the atomic volumes of the elements vary to a considerable extent in their various compounds[35] and the values of the rational atomic weights in the table are necessarily average values. As already mentioned, the use of the rational atomic weights is based upon weighing in air under standard conditions, and therefore change in the density of the air due to variation in temperature and

[34] The use of other weights introduces no error if they have been adjusted against brass in air.

[35] For example, the atomic volumes of calcium in various compounds are as follows: Ca (metal), 26; CaO, 11; $CaCl_2$, 20; $CaBr_2$, 18; CaI_2, 11; and $CaSO_4$, 6.

pressure is not taken into account. The error due to the latter deviations is ordinarily so small that it hardly enters into consideration in even refined analytical work (excluding, of course, atomic-weight determinations).

In spite of these objections, the use of the rational atomic weights is preferable to that of the international atomic weights in ordinary analytical work when air buoyancy introduces an appreciable error. The results obtained by using the rational atomic weights in the calculations will be at least as accurate as those obtained with the international; even when the error due to deviation of the atomic volumes from the average is pronounced, the results will be better, although perhaps not sufficiently accurate for the purpose. Even if they served no other purpose, the rational atomic weights would provide a rapid means of obtaining the magnitude of the error due to buoyancy, thus serving to indicate whether a correction is necessary.

It is emphasized that the international atomic weights are fundamental constants and that the rational atomic weights simply provide a means of minimizing the errors resulting from air buoyancy in practical analysis.

Changes in weight or condition of a substance or container during weighing

Absorption of moisture. Any material placed in a moist atmosphere will adsorb water at its surface, and its weight may thus be appreciably increased (p. 191). The amount of water adsorbed depends upon the properties of the material (extent of surface and especially the adsorptive power) and upon the relative humidity and temperature of the atmosphere in which it is placed. A material can be freed of surface water by heating to a high temperature. After cooling, it can be kept moisture-free by placing it in a desiccator containing a suitable drying agent. When the dried material is placed on the balance pan, it will adsorb or absorb more or less moisture from the air, and an increase of weight may be noticed during the weighing. Porcelain and platinum crucibles are so very slightly hygroscopic that the increase in weight of crucibles that have been ignited and cooled will usually be inappreciable in the few minutes required to make a weighing. On the other hand, large objects of glass may increase sensibly in weight during the weighing if originally moisture-free. Since the adsorption of moisture may be slow, it is recommended to wipe the container with a damp lintless cloth to saturate the surface with water, then with a dry cloth to remove the excess moisture, and finally to let it stand in the balance until the weight is constant. This procedure, which is used when the container cannot be heated to a high temperature, is based upon the fact that a surface containing an amount of water greater than the equilibrium concentration will come to equilibrium sooner than one which contains less than the required amount. The best method, however, for eliminating errors due to adsorption of water on surfaces of vessels is the use of a *tare*, a vessel which is as much like the vessel to be weighed as possible, both in weight and surface exposed. The tare is exposed to the same influences as the vessel itself during the determination and is used to counterbalance the latter before and after any treatment to which it may be subjected. In this way errors arising from adsorption of moisture by the container are canceled. Errors resulting from a change in humidity between the two weighings are eliminated.

The adsorption of moisture by a substance to be weighed is usually more serious than adsorption by a container. The substance may be very hygroscopic by nature, or on account of its great surface development it may adsorb large amounts of water when it would not do so in a compact form. Nearly all finely divided precipitates are appreciably hygroscopic, and precautions must be taken in weighing them. If the substance is only slightly hygroscopic, it may often be weighed with sufficient accuracy by carrying out the operation rapidly, the container being covered or closed as well as possible. Even markedly hygroscopic materials may be weighed with accuracy if the weights are on the pan, and the rider or chain is nearly adjusted as indicated by a preliminary approximate weighing. When the container and its contents are placed on the pan the second time after heating and cooling, the rest point can be found from two swings, or still better, from a single deflection, and the weight calculated from the zero point of the empty balance and the sensitivity. If a crucible is thus being weighed, it should be kept covered at all times. If a substance can be weighed in a weighing bottle, it is best to use this, for such a container can be tightly closed. An unusually hygroscopic substance contained in a crucible can be weighed by placing the crucible in a large closed weighing bottle and weighing the latter against a similar bottle used as a tare (Fig. 19-12). This procedure, however, rarely needs to be adopted.

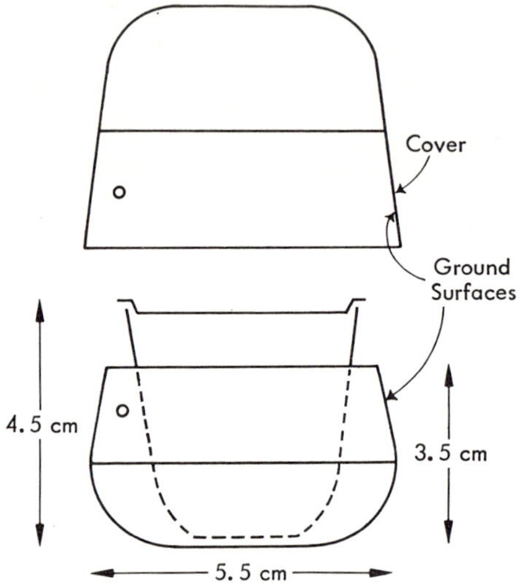

Fig. 19-12. Weighing bottle for crucibles. From Hoffman and Lundell, *J. Res. Natl. Bur. Std.* **18**, 1 (1937).

Electrification of containers. Electrification of glass containers may have to be guarded against. Dry glass vessels rubbed with a cloth may acquire a considerable charge of static electricity, and such objects placed on the balance will show an erroneous weight on account of the attraction of the charged container for the walls or floor of the balance. The effect is most noticeable when the atmosphere has low humidity (<40 per cent).[36] Pyrex is more easily electrified by friction than ordinary glass, and silica more easily than Pyrex. The charge can be dissipated by exposing the vessel to ultraviolet light[37a] or a high-frequency electrical discharge, or placing a radioactive[37b] substance in the balance case. Ionizing units containing ^{210}Po, an α emitter, are available commercially; their useful life is about 1 year.

Differences in temperature of container and balance. If the object weighed does not have the same temperature as the balance, serious errors may be made.[38] The apparent weight of an object warmer than the balance will be less than the true weight because the rising convection currents about the object will buoy it up. Moreover, if the warm body is hollow—especially if closed (covered crucible)—the air within will be less dense than the air in the balance case, and the buoyancy will result in a negative error. For example, a 25-ml covered crucible 1°C above the balance-case temperature will weigh 0.1 mg less than when at the same temperature [i.e., at 26°C, 25 ml of air weighs this much less than at 25°C: $25 \times 0.0012 \times (1/273 + 25) = 0.1$ mg]. Errors from a temperature difference can arise when crucibles are not cooled sufficiently long in a desiccator after ignition.

[36] D. F. Hayman, *Ind. Eng. Chem., Anal. Ed.* **8**, 343 (1936), states that electrification does not take place above a relative humidity of 45 per cent.

[37] (a) C. J. Rodden, *Ind. Eng. Chem., Anal. Ed.* **12**, 693 (1940); (b) F. W. Van Straten and W. F. Ehret, *Ind. Eng. Chem., Anal. Ed.* **9**, 443 (1937).

[38] E. Blade, *Ind. Eng. Chem., Anal. Ed.* **12**, 330 (1940); A. H. Corwin, *ibid.* **16**, 258 (1944).

The length of time a crucible should be cooled in a desiccator cannot be specified precisely, because it depends upon the temperature of the crucible, its size, and its material. Moreover, it depends upon the desiccator. If more than one crucible is cooled at the same time, a longer cooling time is required. Other factors being the same, porcelain crucibles need to be cooled longer than platinum crucibles, chiefly because the specific heat of porcelain is much greater than that of platinum.

It has been shown that the required cooling period of porcelain crucibles is longer than usually supposed.[39] For example, two crucibles (13 g each) allowed to cool to just below redness before being placed in a desiccator (17 cm in diameter, same height) required 60 min to reach a temperature 2°C above room temperature, and 76 min to reach a temperature 1°C above room temperature. If the same two crucibles were allowed to cool in air for 1 min before placing in the desiccator, a period of 50 min was needed to reach a temperature 2°C above room temperature. The error in the weight depends on the rapidity of the weighing. If the weight of a crucible was obtained after 2 min on the pan, a 1°C temperature difference corresponded to ~ 0.1 mg in weight.

In general, a porcelain crucible should be cooled for not less than 1 hour in a desiccator, if first allowed to cool for 1 min in the air after strong heating. The cooling period for a platinum crucible under the same conditions may be taken as 45 min.

When weighing out samples from weighing bottles, care should be taken not to heat these by excessive handling. If the container is held lightly with the finger tips only a short time, no serious error will result, although grasping the bottle with a folded strip of filter paper is to be preferred.

CALIBRATION OF WEIGHTS

Only calibrated weights may be used in quantitative analysis.[40] The simplest, but not necessarily the most accurate, method of weight calibration consists in comparing each weight against a corresponding weight in a standard set that has been reliably calibrated. The assumption must be made that the pieces of the master set have undergone no mass change since their calibration. The validity of this assumption is not always certain.

The standard weight set used in this method of calibration is likely, in the United States, to be a set calibrated at the National Bureau of Standards. The accuracy of weight calibration by the Bureau is the following:

Class M. 1 ppm for weights 10 g and above, 0.01 mg for weights 10 g to 1 g, 0.001 mg for weights below 1 g.

Class S. 3 ppm for weights 10 g and above, 0.03 mg for weights 10 g to 1 g, 0.003 mg for weights below 1 g.

Class J. ~ 0.0002 mg for microbalance weights in a standard series (up to 50 mg); ~ 0.0003 mg for microbalance weights not in a standard series.

The method of calibration to be described is that of Richards, in which the weights of a set are compared with each other.[41] The 10- (or 5-)mg weight is first arbitrarily assigned the value of 10.00 mg (or 5 mg), and the relative masses of all the other weights are determined and expressed in terms of the 10-mg weight chosen as standard. First, the 10-mg weight is compared with the 10-mg rider, then the 10-mg weight plus the rider with the 20-mg weight, and so on. The values for the larger weights obtained by this method will differ considerably from the nominal values because such a small

[39] J. Agterdenbos, *Anal. Chim. Acta* **15**, 429 (1956). The cooling rate of a crucible (or more accurately of the desiccator containing it) approximately follows a logarithmic relation (see Problem 11).

[40] For most analytical work, weights certified by the National Bureau of Standards as meeting the acceptance tolerances of Class S weights (Table 19-1) may be used without corrections.

[41] T. W. Richards, *J. Am. Chem. Soc.* **22**, 144 (1900).

standard is used. It is, therefore, customary to assume that one of the large weights, say the 10-g, weighs 10.00000 g and to express the weights of all the other pieces in terms of the 10-g weight. If the absolute value of the 10-g weight, or any other large weight, is determined by comparison with a weight of known mass (a standard weight) it is a simple matter to calculate the absolute values of all the other weights, since the ratios of the 10-g weight to the others have been determined.

Weighings in calibrations are made either by substitution or transposition. Weighing by substitution requires another set of weights, the pieces of which serve as tares; of course, these pieces need not be accurate. The accuracy obtainable by the method of transposition is twice as great as that by substitution. Directions for both methods are given below. We recommend the transposition method (p. 506).

Calibration of weights according to Richards[42]

Calibration of weights using the substitution method of weighing. First, the rider is compared with a weight of the same mass. Suppose that the balance takes a 10-mg rider. The 10-mg weight from the auxiliary set that is to serve as a tare is placed on the left balance pan, and the rider is placed *exactly* on the 10-mg mark on the right arm of the beam. The rest point of the balance is now found by the precise method of swings, the calculations being carried to the nearest 0.05 division.[43] The rider is removed from the beam, one of the 10-mg weights is placed on the right pan, and the rest point is found for this weight. From the known sensitivity of the balance, the difference in weight between the rider and the 10-mg weight (taken as 10.00 mg) is calculated. In the same manner the following comparisons of weights are made[44]:

0.01″ g	against	0.01′ g
0.02	against	0.01′ + 0.01″
0.05	against	∑0.05 (0.02 + 0.01′ + 0.01″ + rider at 10 mg)
0.1′	against	∑0.1
0.1″	against	0.1′
0.2	against	0.1′ + 0.1″
0.5	against	∑0.5
1′	against	∑1
1″	against	1′
1‴	against	1′
2	against	1′ + 1″
5	against	2 + 1′ + 1″ + 1‴
10′	against	5 + 2 + 1′ + 1″ + 1‴
10″	against	10′
20	against	10′ + 10″
50	against	20 + 10′ + 10″ + 5 + 2 + 1′ + 1″ + 1‴

The data are recorded systematically, as in Table 19-3. Calculations are made to the nearest 0.01 mg.

[42] For the rationale of Richards' method, see E. Blade, *Ind. Eng. Chem., Anal. Ed.* **11**, 499 (1939); F. H. Hurley, Jr., *ibid.* **9**, 239 (1937); F. C. Eaton, *J. Am. Chem. Soc.* **54**, 3261 (1932), and *Chemist-Analyst* **38**, 28, 52 (1949). A. A. Benedetti-Pichler, *Mikrochemie Ver. Mikrochim. Acta* **34**, 241, 298 (1949), gives practical details and discusses the precision of the method. For a detailed treatment of the testing of a one-pan, constant-load-type balance, see T. W. Lashof and L. B. Macurdy, *Anal. Chem.* **26**, 707 (1954). The calibration of the dial-operated weights and estimation of precision are described. The standard deviation of a weighing (two rest points determined) on a number of balances of the same manufacturer was found to be $\sim$0.02 mg. See also V. J. Norman and R. H. Schunke, *Australian J. Appl. Sci.* **14**, 180 (1963).

[43] The sensitivity of the balance should be at least 1 mg $\equiv$ three scale divisions, and preferably 1 mg $\equiv$ five divisions.

TABLE **19-3. Calibration of Weights by the Method of Substitution**

T, g	(W), g	R, div.	S, div./mg	$d = \frac{(R_1 - R_2)}{S}$, mg	$W_{0.01}$ (0.01′ = 0.01000 g), g	W_{10} (10′ = 10.00000 g), g	Corr., mg
0.01	0.01′	+0.4	5.0		Standard	0.00998	+0.02
0.01	(Rider at 10.0)	+0.4		0.00	0.01000	0.00998	+0.02
0.01	0.01″	+0.4		0.00	0.01000	0.00998	+0.02
0.02	{0.01′, 0.01″}	+0.5					
0.02	0.02	+0.55		−0.01	0.01999	0.01997	+0.02
0.05	{0.02, 0.01′, 0.01″, (Rider at 10.0)}	+1.0					
0.05	0.05	+1.3		−0.06	0.04993	0.04992	+0.01
0.1	0.05 etc.	+0.5					
0.1	0.1′	+1.0		−0.10	0.09982	0.09985	−0.03
0.1	0.1″	+1.0		0.00	0.09982	0.09985	−0.03
0.2	{0.1′, 0.1″}	−0.2					
0.2	0.2	0.0		−0.04	0.19960	0.19969	−0.09
0.5	0.2 etc.	+0.2					
0.5	0.5	−0.4		+0.12	0.49928	0.49923	+0.05
1	0.5 etc.	+1.0					
1	1′	+1.0	5.0	0.00	0.99844	0.99846	−0.02
1	1″	+0.75		+0.05	0.99849	0.99846	+0.03
1	1‴	+0.65		+0.02	0.99851	0.99846	+0.05
2	{1′, 1″}	0.0					
2	2	+0.2	5.0	−0.04	1.99689	1.99691	−0.02
5	{2, 1′, 1″, 1‴}	−1.0					
5	5	−0.5	4.9	−0.10	4.99223	4.99228	−0.05
10	{5, 2, 1′, 1″, 1‴}	+3.0					
10	10′	+3.0	4.8	0.00	*9.98456*	*9.98456*	0.00
10	10″	+3.2		−0.04	9.98452	9.98456	−0.04
20	{10′, 10″}	+2.4					
20	20	+2.1	4.4	+0.07	19.96915	19.96912	+0.03
50	{20 1′, 10′ 1″, 10″ 1‴, 5, 2}	−3.6					
50	50	−1.8	3.6	−0.50	49.92229	49.92280	−0.51

Explanation of Table 19-3 *and method of calculation of corrections.* The first column, headed T, gives the denomination of the tare placed on the left-hand pan. The second column contains the denomination of the weight being compared with the tare; (W) indicates the nominal weight of the piece in grams. The rest point obtained in each weighing is recorded in the next column, headed R. The following column, S, gives the sensitivity of the balance at various loads; the figures denote the number of scale divisions that the rest point of the balance is shifted by an excess weight of 1 mg.[45] In the fifth column, headed d, the difference in weight, expressed in milligrams, between the weight in question and the preceding weight or sum of weights is recorded. The difference is calculated from the expression $(R_1 - R_2)/S$, where R_1 is the rest point given by the preceding weight, R_2 is the rest point with the weight in question, the mass of which is to be expressed in terms of the preceding, and S is the sensitivity for the load on the pans. If R_1 is greater than R_2, the difference, d, will be positive, which means that the mass of (W) is greater than that of the preceding weight or sum of weights. In the column headed $W_{0.01}$ the mass of (W) is expressed in terms of the mass of the 0.01′-g weight; i.e., the assumption is made that the latter has a mass of 0.01000 g, and the mass of (W) is calculated from the preceding successive weights on this basis from the values of d. The values thus obtained will differ considerably from the face values of the weights, as already stated, because the standard chosen is so small. It is therefore convenient, after having obtained the relative masses of the weights, to express the masses in terms of a standard having a larger mass. There are two ways of proceeding:

1. It may be assumed that one of the larger weights, e.g., the 10′-g weight, has a mass exactly equal to its nominal value, i.e., 10.00000 g; then one can calculate the masses of the other weights on this basis. From the new series of values thus obtained, a set of corrections is derived, and these corrections are applied to all weighings made with the set in which a ratio of weights is determined. It should be emphasized again that in most analytical processes the ratio of two weights and not an absolute weight is required. As long as only the value of a ratio is to be determined, any weight in the set can be arbitrarily assumed to be correct, and the masses of the other weights can be derived on this basis from the experimentally determined relative weights, $W_{0.01}$.
2. Any *large* piece in the set may be compared with a standard weight of the same denomination, whose mass is known in terms of the international kilogram.[46] As soon as the true weight of any one piece in the set has been found, the true weights of all

[44] The chain of a chainomatic balance should be calibrated at 10-mg intervals. If the beam is notched, the effective weight of the rider should be found at each notch.

[45] The rider is assumed to have a mass which is close to 10 mg, or at least does not differ much from the 10-mg weight which is made the basis of the series of relative weight values. If this assumption does not hold, the values for the sensitivity, S, will be in error, i.e., not consistent with the weights based on the 5- or 10-mg weight. The result of the comparison between the rider and the 5- or 10-mg weight (the second observation in Table 19-3) tells us at once whether the assumption is justified. A difference of 0.2 or 0.3 mg with a 10-mg rider does not lead to any serious error in the value of S so far as an ordinary weight calibration is concerned. However, it is desirable that the rider and the 10-mg weight have the same mass within the experimental error of weighing so that it will not be necessary to apply a correction for different positions of the rider on the beam. All difficulties can be avoided by making the rider the standard and finding the masses of all the weights in terms of it (Hurley, Ref. 42).

One other assumption is involved in the procedure described above—that when the rider is placed, for example, on the 1-mg mark, it has an effective weight equal to one fifth or one tenth of its actual weight, or in other words that the beam is accurately graduated. This possible error need not cause much concern. It would be a bad analytical balance indeed that would have an error of 1 or 2 per cent in its beam markings.

[46] In the United States weights can be sent to the National Bureau of Standards in Washington for certification.

the others can be calculated from the relative values.[47] A weight set calibrated in this way can be used for all purposes, i.e., for the determination of absolute masses as well as for the determination of the values of ratios.

Suppose procedure 1 is adopted. The first step involves the calculation of the mass that each weight should have if one of the large weights—the 10′-g piece, let us say—is assigned the value indicated by its denomination, i.e., 10.00000 g. These values are found by proportion. For the sake of illustration we shall consider the calibration data given in Table 19-3. The relative weight of the 10′-g piece is 9.98456, the 0.01′-g weight being assumed to weigh 0.01000 g. If the mass of this weight is now assumed to be 10.00000 g, the mass of the 5-g piece on this basis will be (4.99223/9.98456) × 10.00000 = 4.99995. Therefore, this piece is too light on the assumption that the 10′-g piece is correct. Its correction value is 4.99995 − 5.00000 = −0.00005 or −0.05 mg. In a similar manner the corrections for all the other weights are found. To avoid the somewhat lengthy computations required in this method, it is permissible to proceed in a slightly different way. It is assumed that the 10′-g weight actually has the mass 9.98456. Therefore, it has no correction. The 5-g piece on this basis should weigh $\frac{5}{10}$ × 9.98456 or 4.99228 g. The values in the column headed W_{10} have been obtained in a similar manner by calculating the appropriate proportional parts of the standard value 9.98456. These values are the masses that the respective pieces should have if the 10′-g weight is assumed to weigh 9.98456 g. The difference between corresponding values of W_{10} and $W_{0.01}$ gives the correction to be applied in each case. If $W_{0.01} - W_{10}$ is positive, the weight is heavier than it should be on the assumption that the 10′-g piece weighs 9.98456, and the correction is therefore positive. In the case of the 5-g weight, $W_{0.01} - W_{10} = 4.99223 - 4.99228 = -0.00005$ g or −0.05 mg. In this way the corrections recorded in the last column of the table have been obtained for all the weights. The corrections are given in milligrams.

If procedure 2 is adopted, the true weight of one of the larger pieces (10 g or over) is determined by comparing it with a standard weight of the same denomination, the weighing being made, of course, by the method of substitution or transposition. Suppose that the true weight of the 10′-g piece is thus found to be 10.00002 g. Evidently the correction of the 10′-g weight is equal to +0.02 mg. The correction for the 5-g weight will be 4.99223 − (9.98456 − 0.00002)/2 = 4.99223 − 4.99227 = −0.00004 g or −0.04 mg.[48] In like manner the corrections are obtained for all the other weights.

[47] It should be realized that if the weight compared and the standard weight are not of the same density, i.e., of different materials, the calibrated weight will have the indicated value only when it is used in air of normal density (0.0012 g/ml). However, the usual changes in air density are not likely to affect appreciably the apparent mass difference between brass and other weights. Thus a pressure change from 760 to 684 mm Hg (−10 per cent change) would change the apparent weight of a Pt–Ir weight against brass by −8 ppm. It may be noted that at 3000 ft elevation, the nominal barometric pressure may be taken as 681 mm Hg, corresponding to an air density of 0.00108 g/ml at 20°C; at 5000 ft as 632 mm Hg and 0.00100 g/ml.

[48] The strictly correct expression is

$$\left[\left(4.99223 - \frac{9.98456}{2}\right)\frac{10.00002}{9.98456}\right] + \frac{0.00002}{2} = -0.040 \text{ mg}$$

but this exact form need be used only when the correction runs into milligrams or the factor corresponding to 10.00002/9.98456 (the ratio of the absolute weight to the relative weight) is considerably different from unity. If the absolute (cgs) corrections for the weights are desired, they may be obtained simply by adding the proper aliquots of the absolute correction to the relative corrections (the values in the last column of Table 19-3), the latter being multiplied if necessary by the ratio absolute weight/relative weight. These corrections are entered directly in a final column in the table. *Example*: What is the absolute correction for the 20-g weight in Table 19-3? *Ans*. [+0.03 × (10.00/9.98)] + (2 × 0.02) = +0.07 mg. See E. Blade, *Ind. Eng. Chem., Anal. Ed.* **11**, 499 (1939).

Calibration of weights using the transposition method of weighing. This method is chiefly valuable because the precision obtainable is twice as great as in weighing by substitution.[49] An auxiliary set of weights to serve as tares is not necessary. It does not require more time to calibrate by this method than by the other, since in either method two weighings are necessary to get the difference in weight between two weights.

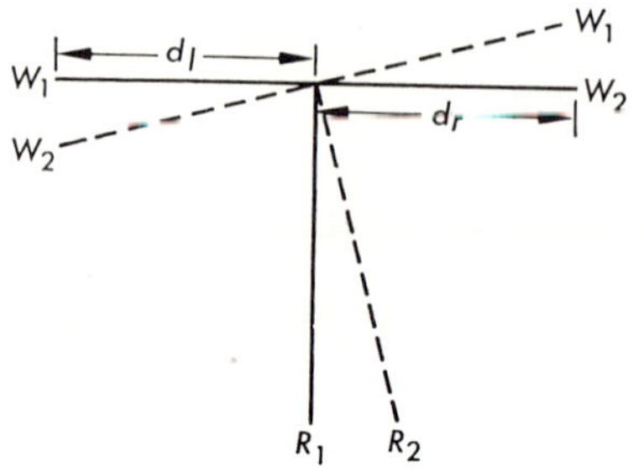

Fig. 19-13.

The method of calculating the difference in the mass of two weights which are compared is the following (Fig. 19-13). Let

W_1 = weight which is *first* placed on the left pan
W_2 = weight which is *first* placed on the right pan
R_1 = rest point when W_1 is on the left pan and W_2 on the right
R_2 = rest point when W_2 is on the left pan and W_1 on the right
w = weight to be added to, or taken from, the right pan to make R_2 equal to R_1
d_l = length of left balance arm
d_r = length of right balance arm

Then

$$W_1 d_l = W_2 d_r$$
$$W_2 d_l = (W_1 + w) d_r$$
$$(W_1 - W_2) d_l = (W_2 - W_1 - w) d_r$$

d_l may be considered equal to d_r. Then

$$W_1 - W_2 = W_2 - W_1 - w$$
$$W_1 = W_2 - \frac{w}{2}$$

w is calculated from R_1, R_2, and the sensitivity of the balance:

$$w = \frac{R_2 - R_1}{S}$$

where S = sensitivity. Therefore,

$$W_1 = W_2 - \frac{R_2 - R_1}{2S}$$

This expression gives the value of W_1 in terms of W_2. If the 0.01′-g piece is to be compared with the 0.01″-g and expressed in terms of the latter, the 0.01′-g piece is first placed on the left pan and R_1 determined. Then the weights are interchanged and R_2 found. Proceeding in this way, the value of any weight in terms of the preceding is found. To avoid confusion, the "unknown" weight is placed first on the left pan.

[49] P. F. Weatherill, *J. Am. Chem. Soc.* **52,** 1938 (1930).

TABLE 19-4 **Calibrations of Weights by Transposition**

(W_1), g	(W_2), g	R_1, div.	R_2, div.	S, div./mg	$\frac{R_2 - R_1}{2S}$, mg	W_2, g	W_1 $\left(= W_2 - \frac{R_2 - R_1}{2S}\right)$, g	$W_{1\,\text{ideal}}$ (10′ g = 10.00000 g), g	*Corr.*, mg
0.1							0.10000	0.10004	−0.04
0.1_T	0.1	+0.2	+0.4	4.0	+0.025	0.10000	0.09997		
0.2′	{0.1, 0.1_T}	+0.75	−0.05	4.0	−0.10	0.19997	0.20007	0.20008	−0.01
0.2″	0.2′	−0.3	+0.6	4.0	+0.11	0.20007	0.19996	0.20008	−0.12
0.5	{0.1, 0.2′, 0.2″}	+0.5	−0.3	4.0	−0.10	0.50003	0.50013	0.50020	−0.07
1′	{0.1, 0.5, 0.2″, 0.2″}	+0.8	−0.7	4.0	−0.19	1.00016	1.00035	1.00039	−0.04
1″	1′	0.0	0.0	4.0	0.00	1.00035	1.00035	1.00039	−0.04
1‴	1′	0.0	0.0	4.0	0.00	1.00035	1.00035	1.00039	−0.04
2	{1′, 1″}	−0.1	0.0	4.0	+0.01	2.00070	2.00069	2.00078	−0.09
5	{1′, 1″, 1‴, 2}	−0.15	−0.4	4.0	−0.03	5.00174	5.00177	5.00196	−0.19
10′	{1′, 1″, 1‴, 2, 5}	+1.0	−2.2	3.9	−0.41	10.00351	*10.00392*	*10.00392*	0.00
10″	10′	−0.3	−1.0	3.9	−0.09	10.00392	10.00401	10.00392	+0.09
20	{10′, 10″}	+0.1	−2.4	3.5	−0.36	20.00793	20.00829	20.00784	+0.45
50	Σ1′—20	+0.4	−1.6	2.7	−0.37	50.10973	50.02010	50.01960	+0.50

Explanation of Table 19-4. It is convenient to record the data obtained in the transposition method of weighing in a form such as that used in Table 19-4, which gives the results obtained in the calibration of a set of weights used with a chainomatic balance. (W_1) indicates the denomination of the weight first placed on the left pan and (W_2) that placed on the right pan. The first rest point obtained in comparing two weights is given in the column headed R_1; and the second rest point, i.e., the one obtained after the interchange of the weights, is given in column R_2. Column S gives the sensitivity at various loads. The difference in mass between the two weights compared is given in the next column. The difference is given by the expression $(R_2 - R_1)/2S$. W_1, the mass of (W_1) in terms of the mass of $(W_2)(= W_2)$, is obtained from the formula

$$W_1 = W_2 - \frac{R_2 - R_1}{2S}$$

the mass of the 0.1-g piece being assumed to be 0.10000 g. The relative masses of the pieces on the basis of 10′ g = 10.00000 g are recorded in the column headed $W_{1\,\text{ideal}}$. The calculations by which these values have been obtained are the same as those described in the previous section. The last column furnishes the corrections for each weight in milligrams.

Use of the corrections

To find the corrected weight of an object weighed with a particular set of calibrated weights, it is only necessary to find the algebraic sum of the corrections of the pieces used and to add this sum to the nominal weight. Thus, if the nominal weight of an object weighed with the set of Table 19-4 was found to be 7.5237 g, the corrected weight would be $7.5237 + (-0.00019 - 0.00009 - 0.00007) = 7.5233$ g (the chain of the chainomatic balance was tested and found to be correct). In adding up the corrections, the fifth decimal is retained until the sum has been obtained, when the value is rounded off to the fourth place, i.e., to the nearest 0.1 mg. The corrected value thus obtained will be the true weight in air if the corrections used are absolute, i.e., if the true weight of any piece in the set has been determined and the corrections derived therefrom as already indicated; otherwise, the value obtained will be merely a corrected value that can be used in obtaining ratios of masses.[50]

Instead of adding up the respective corrections for a given combination of weights each time, it is better to construct a table that will give immediately the correction to be applied to any combination of weights. The manner of setting up and using such a table will be evident from Table 19-5, the values in which refer to the integral weights of the set of Table 19-4, used with a chainomatic balance. In using a table such as this, it must be remembered that the smallest number of pieces possible is used in making a given weighing and that an unprimed weight is always used in preference to a primed, or a single-primed weight in preference to a double-primed, etc., because the table has been constructed on this basis.

TABLE **19-5.** **Cumulative Weight Corrections: Integral Weights**

g	0	1	2	3	4	5	6	7	8	9
0		−0.04	−0.09	−0.13	−0.17	−0.19	−0.23	−0.28	−0.32	−0.36
10	0.00	−0.04	−0.09	−0.13	−0.17	−0.19	−0.23	−0.28	−0.32	−0.36
20	+0.45	+0.41	+0.36	+0.32	+0.28	+0.26	+0.22	+0.17	+0.13	+0.09
30	+0.45	+0.41	+0.36	+0.32	+0.28	+0.26	+0.22	+0.17	+0.13	+0.09
40	+0.54	+0.50	+0.45	+0.41	+0.37	+0.35	+0.31	+0.26	+0.22	+0.18
50	+0.50	+0.46	+0.41	+0.37	+0.33	+0.31	+0.27	+0.22	+0.18	+0.14

The error in the corrections is a function of the standard deviation of weighings on the balance used for calibration. For a single weight, the error in the calibration value at the 95 per cent confidence level may be as large as twice the standard deviation of a single weighing on the calibrating balance; for a combination of weights it may be four times the standard deviation of a single weighing at the same confidence level.[51]

RULES FOR THE USE OF THE BALANCE

1. The balance should be placed in a suitable location, in a room separate from the laboratory proper to protect it from fumes. A room, with windows facing the north, which can be kept at reasonably constant temperature makes the best balance room. The balance should not stand near a window or a radiator. It should be placed upon a solid support to protect it from vibration as much as possible. The instrument should not be exposed to direct sunlight or drafts.
2. The balance should be level.
3. When not in use, the beam should be raised and the pan supports should be up to prevent injury to the agate knife-edges and planes. The door of the case should be kept closed. Weights and other objects should not be left standing on the pans.
4. The beam must be raised and the pans arrested before any object is placed on the pans. The zero point should be determined before each weighing unless the weighing is made by difference at one sitting. In testing for equilibrium, the beam is first slowly lowered and the pans then released. Before any object or weight is taken from the pans, the latter must be arrested by bringing up the pan supports when the pointer is near the center of the scale and then raising the beam. *Fractional* weights may be *added* to the pans when the beam is released, if the pans are supported. The pans should be arrested before the rider is moved.

[50] It must be remembered that the same set of weights must then be used for all the weighings made in one analysis.

[51] R. A. Chalmers and R. N. Curnow, *Analyst* **89**, 567 (1964).

5. All substances are to be weighed in containers to prevent injury to the pans. Suitable containers for weighing are: weighing bottles, watch glasses, crucibles, and, in special cases, small Erlenmeyer flasks (50 or 100 ml). Special care must be taken with corrosive liquids and solids that are volatile; these must be weighed in tightly closed vessels.

6. Objects to be weighed must have the temperature of the balance.

7. The balance must never be overloaded. Ascertain the maximum safe load for the balance used.

8. The balance must be kept clean. If it should happen that any substance is spilled on the pans or on the floor of the case—there is hardly any excuse for this—it must be brushed up immediately. The pans should be lightly brushed up with a camel's-hair brush now and then to remove dust that may have entered. Liquids must not be used for cleaning the pans.

9. Concerning weights: Weights are to be handled only with bone- or plastic-tipped forceps. In counterbalancing an object, the large weights are first tried, and then the smaller are added in systematic order. The smallest number of weights possible is always used; e.g., a 2-g weight is used in preference to 1 + 1 g. An unprimed weight is always used in preference to a primed. The weights should be arranged systematically on the balance pan; fractional weights should not be placed upon one another. Large weights should be placed in the center of the pan to avoid troublesome oscillations that would otherwise result. Great care should be taken to avoid dropping weights. In counting the weights, the empty compartments in the box are first checked (weights should be kept in their designated places in the box), then the weights on the pan are counted, and finally, to be sure, the weights are counted as they are returned to the box. The weight box is always to be kept covered when not in use. The result of a weighing should be recorded immediately in a notebook.

10. If anything appears to be wrong with the balance or if adjustments must be made, obtain the aid of the instructor.

GENERAL REFERENCES

W. Felgentraeger, *Feine Waagen, Wägungen und Gewichte*, Springer, Berlin, 1932.

Article "Balance" in Thorpe's *Dictionary of Applied Chemistry*, vol. I, Longmans, London, 4th ed., 1937, p. 587.

Department of Scientific and Industrial Research, *Balances, Weights and Precise Laboratory Weighing*, H.M. Stationery Office, London, 1954.

PROBLEMS

1. Find h and l for a balance whose sensitivity varies as follows with the weight on each pan:

P, g	*Sensitivity, div.*/mg
0	3.55
10	3.31
20	3.20
50	2.86
100	2.53

if the arm length is 7.5 cm, the beam weighs 90 g, a pan and stirrup weigh 24 g, the pointer is 20 cm long, and one scale division = 1.0 mm.

Ans. It will be advantageous to plot the reciprocal sensitivity instead of the sensitivity against $P + W$. [K. J. Mysels, *J. Chem. Educ.* **32**, 518 (1955).]

$$\frac{1}{\alpha} = \frac{2d(P + W) + hB}{wl}$$

Let w' = weight (g) corresponding to one scale division (s cm). Then

$$w' = \frac{s}{Nl}\{2d(P + W) + hB\}$$

w' is plotted versus $P + W$ (see the graph). If the beam does not bend appreciably with increasing load (d constant), a straight line should be obtained. If bending occurs, d becomes larger as W increases and a curve with its concave side upward will be obtained.

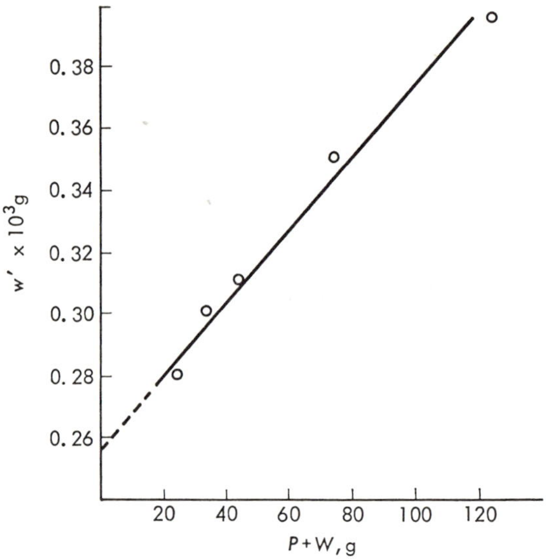

The plot is a straight line, within the experimental error, so that flexure of the beam is not significant up to a load of 100 g on each pan.
When $P + W = 0$,

$$w_0' = \frac{hBs}{Nl} = 2.57 \times 10^{-4} \text{ and } h = 4.3 \times 10^{-3} \text{ cm}$$

The slope of the line is 1.18×10^{-6}. Hence

$$d = \frac{Nl}{2s} \times 1.18 \times 10^{-6} = 8.9 \times 10^{-4} \text{ cm}$$

2. A balance with the constants

$P = 25$ g
$l = 7.5$ cm
$N = 24$ cm
One scale division = 0.10 cm
Sensitivity (divisions/mg) = 3.5 at 0 load, 2.6 at 100-g load

has an oscillation period of 11.5 sec with no load on the pans. Find the period with a load of 100 g (take $g = 980$ cm/sec²).

3. An object on the left pan of a balance has an apparent weight of 22.4367 g. After interchanging the object and the weights, the apparent weight is 22.4377 g. (a) What is the weight of the object, and what is the ratio of the lengths of the balance arms? (b) Supposing that an error of +0.1 mg is made in each weighing, what will be the error of the final result?
Ans. (a) True weight = 22.4372 g; ratio of arm lengths = 1.00000:1.00002.
4. What is the true weight of a brass weight (d = 8.4) having an apparent weight of 10.00000 g in dry air at 20°C and 760 mm Hg pressure? *Ans.* 10.00143 g.
5. At sea level a 0.5-g platinum weight (d = 21.5) is found to have the correction 0.01 mg when the 10-g brass weight (d = 8.4) of the set is taken to be correct. What will be correction of the 0.5 gram weight at Lhasa (elevation 11,800 ft) where the barometric pressure is 500 mm Hg?
6. In checking possible volatilization loss of a Pt–Ir crucible (d = 21.5) at high temperatures, an analyst found the weight of a 35-g crucible to be 0.15 mg more after ignition than before. If the barometric pressure and temperature were 730 mm Hg, 29°C, and 759 mm Hg, 24°C, in the first and second weighing, what was the true change in weight of the crucible?
7. If the maximum permissible error due to air buoyancy in weighing with a tare is 0.1 mg, what may be the difference in volumes of object and tare if the greatest pressure and temperature changes are, respectively, 25 mm Hg, and 5°C? Take the means to be 735 mm Hg and 25°C.
8. Find the weight of sodium chloride weighed in air (d = 1.2 mg/ml) required to prepare 1 liter of 1.0000 M solution, by (a) making exact buoyancy corrections; (b) using rational atomic weights. *Ans.* (a) 58.418 g; (b) 58.42 g.
9. What weight of hydrochloric acid of density 1.1006 and containing 20.04 per cent HCl must be taken to furnish 1.000 mole of HCl if the weighing is carried out in air at 20°C and 760 mm Hg pressure with brass weights?
10. Copper is determined in copper sulfate pentahydrate by electrolytic deposition of the metal and weighing as such. What will be the absolute error in the copper content found if the weighings are made in air under standard conditions (density of copper sulfate pentahydrate = 2.3, of copper = 8.9)?
11. Assume that for relatively small temperature differences the cooling rate of an object follows the relation

$$t = k \log \frac{T_1 - T_R}{T_2 - T_R}$$

where t = time required to cool the object from temperature T_1 to temperature T_2
T_R = room temperature
k = constant

If the cooling time of a desiccator from 35 to 26°C is 60 min when the room temperature is 25°C, find the cooling time from (a) 40–26°C; (b) 27–26°C; (c) 40–27°C.
Ans. (a) 70.6 min; (b) 18 min.
12. The uncorrected weight of an object obtained by use of the weights of Table 19-3 is 25.2003 g. What is the corrected weight?
13. If the true weight of the 10′-g weight of Table 19-4 is 9.99880 g, what are the absolute corrections for the 5-, 2-, and 1′-g weights?
Ans. −0.79 mg for 5-g weight.
14. A weight calibration gave the following relative values:

1′ g	1.00248	1‴	1.00256	5	5.01238
1″	1.00302	2	2.00391	10	10.02468

If the 10-g weight is assigned a mass of 10.00000 g, what are the respective weight corrections?
15. The integral weights of a set composed of a chromium–nickel–iron alloy have all been accurately intercompared among themselves and the "absolute" weights

found by comparison against a standard 50-g brass weight whose correction is based on the assumption of its use in air of density 1.2 mg/ml. If the chromium–nickel–iron weights are later used in very accurate work requiring buoyancy corrections, shall the corrections be calculated by using the density of brass or that of the chromium alloy?

16. The pointer of a balance swings in a plane 1 mm from the plane of the scale behind it, whose divisions are 1 mm apart. (a) If the eye of the observer, in a plane 25 cm from the scale, may, as a maximum, occupy a position 3 cm from the plane normal to the zero of the scale, what is the maximum parallax error in scale divisions? (b) How can the error be eliminated?
Ans. (b) Align an ink dot on the window of the balance case with the zero of the scale when observing swings.

17. Suppose the balance of Problem 1 is subjected to a temperature change such that one arm becomes 0.1°C warmer than the other. Taking the coefficient of expansion of the beam to be 2×10^{-5} per °C, what will be the shift in zero point in radians and in scale divisions?

18. A 10-g Al_2O_3 crucible ($d = 3.7$) is heated to 1000°C in the furnace of a thermo-balance. What is its apparent change in weight compared to the weight at room temperature?

CHAPTER 20 *Technique of Common Operations of Quantitative Analysis*

Though it may be true of some analysts that, like poets, they are born, not made, yet granted intelligence, some general chemical knowledge and a fair amount of dexterity and application, the necessary manipulative skill will come with practice, often in a surprisingly short time.

—H. S. WASHINGTON

A gravimetric determination includes most or all of the following operations: sampling, preparation of the sample, weighing the sample, solution or fusion, evaporation, precipitation, filtration and washing, ignition or drying of the precipitate, and the weighing of the residue. A volumetric determination may include most of these operations and, in addition, the exact measurement of volume. Leaving the measurement of volume until later, we shall consider in this chapter the operations mentioned. We shall be concerned here chiefly with the mechanical side of these operations and shall touch upon their theoretical aspects only when necessary.

Quantitative analysis is an art as well as a science. Some individuals are by nature more skillful than others, but anyone can acquire sufficient skill to make an ordinary quantitative analysis satisfactorily, if he will work carefully and give some consideration to the manner in which the operations should be performed. The beginner will probably make some mistakes at first, but he should avoid repeating such mistakes and should constantly strive to improve his technique.

The student should read this chapter carefully before beginning the laboratory work and should refer back to it in the course of his first analyses.

General remarks on the laboratory work

All quantitative work demands attention to details.

The desk and the apparatus must be kept scrupulously clean and in order. All glass vessels should be rinsed with distilled water before use. They should be free of grease internally, so that water will wet the surface uniformly.[1] Vessels should be wiped dry on the outside with a clean lintless cloth; the interiors, however, must never be touched

[1] Whenever possible, use a soap or detergent solution for cleaning glassware. Sometimes, as in removing grease from volumetric ware, a more powerful cleaning agent may be needed. It can be obtained by adding 15 g of powdered sodium or potassium dichromate to 500 ml of concentrated sulfuric acid. The vessel to be freed of grease is rinsed with the liquid, and the latter is allowed to moisten the walls for a few minutes, longer in stubborn cases. The excess is then poured back into the stock bottle, and the vessel is rinsed successively with tap and distilled water. The liquid is extremely corrosive and must not come in contact with the skin, clothes, or desk.

with a cloth or anything else; usually it makes no difference if the interior of a vessel is left moist. As soon as a piece of apparatus has been used, it should be cleaned and returned to its proper place.

During the course of the operations all vessels should be kept covered to prevent the entrance of dust, rust from ring stands, etc. Vessels containing solutions and solids should be numbered or labeled[2] to avoid mistakes or mix-ups. All observations should be recorded first-hand in a suitable notebook (see below).

The coating of ammonium salts that reagent bottles acquire on standing should be wiped off with a clean moist cloth, special attention being paid to the mouths of the bottles. It is advisable to protect the mouths of the liquid-reagent bottles by inverting small beakers over the necks if special protecting caps are not provided (p. 447).

A worker in a quantitative laboratory does not need to wear a laboratory coat or apron. "If the analyst is liable to drop acids on his clothes he is more than liable to spill some of the solutions he is analyzing" (H. S. Washington).

A few words may be said concerning precautionary measures to be taken in certain operations involving the element of danger to the worker. Operations with hydrogen sulfide, hydrogen cyanide, bromine, and other toxic gases and liquids must always be carried out in a well-ventilated hood. Likewise the volatilization of such substances as mercury and its compounds, arsenic compounds, and oxalic acid should be carried out in a hood. Hydrofluoric acid can produce very serious burns and must be handled with the greatest care; moreover hydrofluoric acid fumes are poisonous. In general, all strong acids should be volatilized in a hood. Strong perchloric acid must not be heated alone with organic matter or even easily oxidized inorganic substances for fear of violent explosions.[3] A pipet should not be used to withdraw portions of corrosive or poisonous liquids unless a trap of some kind is placed between it and the mouth. Pipet-filling attachments are available.

Organic solvents present hazards. Some are inflammable (ethers, alcohols, etc.), others are toxic (carbon tetrachloride and other chlorinated hydrocarbons, benzene, etc.). Ethers containing peroxides (formed on standing) should not be distilled (explosive).

Safety glasses are best worn in any operation in which spattering of strong acids, alkaline solutions, or molten fluxes can occur.

Before beginning a determination, the student should read through the directions and make sure that he understands them. The apparatus and reagents required should be obtained if not already at hand, and a mental plan of operations made. The work should be so arranged that it can be interrupted at a suitable point. For example, a filtration should not be begun unless the washing of the precipitate can be finished immediately afterward. Precipitations are advantageously made at the end of the laboratory period so that the precipitate can settle out, and digest if necessary, in the hours following. Lengthy evaporations should be allowed to take place overnight if possible. There should be no waste of time in an analytical laboratory. If one determination is unavoidably held up, a second should be started. For example, while crucibles are cooling in the desiccator, samples may be weighed out for another analysis; while an ignition is in progress, filtrations may be made, etc. Operations should not be hurried, for hurry may lead to loss of time through the necessity of having to repeat analyses. The aim should be to avoid repeating determinations. However, any analysis

[2] Beakers or flasks can be marked by writing with a pencil on the circular etched areas usually provided. Glassware in general can be marked with a wax pencil. Porcelain crucibles can be numbered with an iron-base ink on the unglazed bottoms before igniting and weighing.

[3] Perchloric acid or perchlorates must never be heated with alcoholic solutions. See E. Deisz, *Z. Anal. Chem.* **107**, 8 (1936). Condensation of fumed perchloric acid on the woodwork of hoods or in flues (especially if dusty or carrying organic vapors) can result in fires or even explosions. Anyone using perchloric acid in analytical work should read the precautions listed in *Analyst* **84**, 214 (1959). The use of a shield is always advisable in oxidations with perchloric acid.

in which a known error has been made should be immediately stopped (even if the effect of the error on the result may apparently be slight), and another one begun.

For some common operational errors, refer to page 383.

Laboratory notebook

All data should be recorded directly in a suitable laboratory notebook. The habit of making notes on loose sheets of paper must be avoided. The record of each determination should be well arranged, and no essential information should be omitted. The following items should be included: origin and/or nature of sample; date or dates of the analysis; constituents determined; method of analysis (reference to source of procedure is generally sufficient); all numerical data, such as weight of sample, weight of ignited or dried precipitate in gravimetric procedures, and titration volumes, normalities, and temperature in volumetric ones; calculations; and statement of results. All heatings to constant weight must be indicated. Any observations having a bearing on the results should be recorded.

The form of record illustrated below is satisfactory for use in instructional laboratories.

Problem ___*2*___ Date of report *Jan. 16, 1943*

Analysis of ___*Soluble chloride*___

Determination of ___*Chlorine*___

Weights of Samples

	Det. I	Det. II
App. wt. of vial + subst.	*10.7481* g	*10.5370* g
App. wt. of vial − subst.	*10.5370*	*10.3378*
App. wt. of subst.	*0.2111*	*0.1992*
Corr.	*−0.0001*	*−0.0002*
True wt. of subst.	*0.2110*	*0.1990*

Weights of Ignited Residues

	Det. I	Det. II
	(4)	(4)
	(3) *17.8453*	(3)
	(2) *17.8453*	(2) *17.4063*
App. wt. of xble. + res.	(1) *17.8456* g	(1) *17.4064* g
App. wt. of xble.	*17.4002*	*16.9862*
App. wt. of res.	*0.4451*	*0.4201*
Corr.	*0.0000*	*0.0000*
True wt. of ___*AgCl*___	*0.4451*	*0.4201*

Results: Det. I: *Cl = 52.19%* Det. II: *Cl = 52.23%*

Av. = *52.21% Cl*

Remarks: *Determination made according to directions on page 000 of the text. Sample air-dried. Analysis begun Jan. 15, finished Jan. 16.*

Calculations

$$\text{Chem. factor: } \frac{Cl}{AgCl} = 0.2474$$

Det. I:

$$\frac{0.2474 \times 0.4451 \times 100}{0.2110} = 52.19$$

$$\begin{aligned} \log 0.2474 &= 9.3934 - 10 \\ 0.4451 &= 9.6485 - 10 \\ 100 &= 2. \\ & \overline{1.0419} \\ 0.2110 &= 9.3243 - 10 \\ & \overline{1.7176} \ldots\ldots 52.19 \end{aligned}$$

Det. II: $\dfrac{0.2474 \times 0.4201 \times 100}{0.1990} = 52.23$

$$
\begin{array}{rl}
\log 0.2474 & = 9.3934 - 10 \\
0.4201 & = 9.6234 - 10 \\
100 & = 2. \\
\hline
 & \ 1.0168 \\
0.1990 & - 9.2989 - 10 \\
\hline
 & \ 1.7179 \ldots\ldots\ldots\ldots 52.23
\end{array}
$$

Sampling

All the careful work put into a good analysis will go for naught if the sample taken is not representative of the material whose composition is to be determined. Two radically different problems are encountered in obtaining a proper sample for analysis:

1. The composition of a particular substance or material is required when it may be mixed with extraneous substances. For example, the composition of a substance in the pure state may be required. It may be necessary to secure the substance in as pure a form as possible by recrystallization or other means. The analysis proper of a mineral may have to be preceded by its separation from other minerals mechanically associated with it. In the analysis of plant tissue, a sample free of adhering soil must be used. In other words, sampling in this category calls for the selection of a particular part of the material presented to the analyst.
2. The average composition of a mass of material is required. This, the most frequent case, is encountered in the analysis of natural materials, such as ores, and of artificial products as well, when the object is to determine their commercial value or to decide whether compositional specifications are met. Such materials may be very heterogeneous. An ore is a mixture of minerals, an alloy may vary in composition from point to point owing to segregation, etc. Formidable problems are likely to arise when hundreds of tons of bulk material are to be represented by a few hundred grams of laboratory sample.

It is not possible in this book to describe adequately the procedure followed in the sampling of heterogeneous materials, and it must suffice merely to indicate how the sample is obtained.[4] For bulk materials such as ore and coal, a *gross* or *composite* sample is collected by combining small portions taken by appropriate methods from all parts of the mass. Random sampling is most easily carried out on a stream of the material. The gross sample may weigh hundreds or even thousands of kilograms. The number and the size of the portions collected will depend on a number of factors, such as the degree of inhomogeneity, the particle size, and the accuracy required. The more careful the sampling, the greater the cost, so that a compromise is sought such

[4] The literature on sampling is extensive. For general surveys of principles and methods, see W. W. Walton and J. I. Hoffman in I. M. Kolthoff, P. J. Elving, and E. B. Sandell (eds.), *Treatise on Analytical Chemistry*, vol. 1, part 1, Wiley-Interscience, New York, 1959, chap. 4 (contains sources of information on sampling of specific materials); R. C. Tomlinson in C. L. Wilson and D. W. Wilson (eds.), *Comprehensive Analytical Chemistry*, sect. 3, vol. IA, Elsevier, Amsterdam (and Van Nostrand, Princeton, N.J.), 1959, chap. II. Statistical aspects of sampling are considered by H. A. Laitinen, *Chemical Analysis*, McGraw-Hill, New York, 1960, chap. 27, and A. A. Benedetti-Pichler in W. M. Berl, *Physical Methods in Chemical Analysis*, vol. 3, Academic Press, New York, 1956. A. D. Wilson, *Analyst* **89**, 18 (1964), has shown that in silicate rock analysis a 1-g sample crushed to 72 mesh (per inch) is sufficiently large for the determination of the major constituents.

that the error due to sampling does not exceed a specified limit imposed by the accuracy needed. Probability considerations apply. The gross sample is crushed, mixed, and subdivided, the process being repeated as many times as necessary until a laboratory sample weighing from, say, 100 g to 1 kg or more is obtained.

Up to this point the analyst may have had nothing to do with the sampling. His responsibility begins when he receives the laboratory sample. This will have to be mixed and sampled to give a representative subsample of perhaps a few grams from which the analytical samples are weighed out. Small portions of the laboratory sample will have to be collected from the whole for this purpose. Additional crushing or grinding, as in an agate mortar, with sieving may be necessary. All samples must be kept in a tightly closed container of suitable material.

Sampling of metals and alloys demands a different procedure because of the impossibility of pulverizing them by crushing. Drillings or chips representative of the cross section of the piece are obtained by suitable methods.[5]

Heterogeneity of alloys results from segregation and care must be taken to obtain portions from all parts of the mass. In sampling cast iron, loss of dust, which contains a higher proportion of graphite, must be avoided. Foreign substances, such as oil and surface scale, must be avoided or removed.

Crushing and grinding

A sample received by the analyst may require further reduction in particle size. If, initially, the particle size of the sample is just adequate to provide a composition having a specified standard deviation, the particle size must be reduced as the weight of sample is reduced so that the number of particles remains the same. Grinding is also necessary beyond the requirements of representative sampling to produce very fine particles that will dissolve in a reasonable period of time in solvents, including molten fluxes.

Coarse hard material is crushed in a hardened steel mortar (p. 472) to a size that can be ground in an agate mortar without having the particles fly out.

An agate mortar and pestle are ordinarily used for fine grinding in the analytical laboratory. A *small* amount of powdered material is transferred to the mortar and ground with a circular movement of the agate pestle. The material that accumulates on the upper part of the mortar should be pushed down occasionally with the pestle, and the grinding continued until it is evident that the powder is sufficiently fine. Substances must never be *pounded* in an agate mortar, for then chips of agate may be flaked off the pestle. When the first portion of the sample has been ground sufficiently, it is transferred from the mortar to a weighing bottle or vial, if only a small amount of sample is to be ground; or else, if a considerable quantity is to be pulverized, it is carefully poured onto a square of glazed paper and covered with an inverted dry beaker.[6] Then another small portion of powder is placed in the mortar and ground as before. Grinding should not be continued longer than really necessary, for the reasons given below. The combined ground portions should be thoroughly mixed. This is easily done when the powder has been put on glazed paper, by successively lifting the corners of the sheet and causing the powder to roll over itself many times. Care should be taken not to raise any dust. Finally, the powdered sample is transferred to a suitable well-stoppered vial or bottle.

Effect of crushing and grinding on composition of sample. Crushing of hard substances in a steel mortar may introduce significant amounts of metallic iron by abrasion. For example, crushing 9 g of quartz in $\sim$ 1-cm lumps in a Plattner mortar to a coarse powder of 0.7 mm maximum particle size introduced 280 ppm of iron.[7]

[5] For a full account of the sampling of ferrous materials see G. E. F. Lundell, J. I. Hoffman, and H. A. Bright, *Chemical Analysis of Iron and Steel*, Wiley, New York, 1931, p. 50.

[6] It may be advantageous to sift the powder at this point.

[7] *Anal. Chem.* **19**, 652 (1947).

Minute amounts of the minor elements of the steel will also be introduced, which might be of importance in trace analysis.

Fine grinding in agate may alter the composition of a sample appreciably:

1. If the substance ground is very hard, there may be considerable abrasion of the agate, and significant amounts of silica may be introduced into the sample.[8] Generally, however, this is not a source of serious error.

2. The water content of the sample may be altered. The finer the state of subdivision of a solid, the greater will be the amount of adsorbed (hygroscopic) moisture that it can retain (p. 191).[9] It is not possible to take account of the water thus introduced in fine grinding by drying the ground sample at 100–110°C, because the adsorbed water may not be entirely expelled at low temperatures.

A solid containing water of crystallization may lose such water to a greater or less extent on grinding, as a result of heat developed. Thus the partial dehydration of gypsum ($CaSO_4 \cdot 2\,H_2O$) on long-continued grinding is well known.[10] Furthermore, water held mechanically in the interior of crystals (vacuole water, p. 194) will be lost when the crystals are reduced to smaller dimensions, since the vacuoles are exposed and the liquid in them evaporates.

3. The chemical composition of the sample may be changed. Thus ferrous iron may be oxidized to ferric[11] with the result that erroneous results will be obtained, for instance, in the important determination of ferrous iron in rocks.[12] Calcite ($CaCO_3$) on fine grinding may lose small amounts of carbon dioxide.

It follows from all this that samples should never be ground finer than is really necessary. Just how fine a substance must be ground before analysis depends upon the sample, the method of decomposition, etc.; the analyst must use his judgment.

Drying the sample

Often the analyst is called upon to determine the composition of a sample as he receives it. Since the water content of a sample is subject to change, due regard must be paid to this matter, for a change in the percentage of water will naturally result in a change in the percentages of the other constituents as well. It may be necessary to reserve a portion of the sample as received and determine its moisture content by drying at 105–110°C; the determination of the other constituents may then be made on a sample suitably ground and dried at the same temperature; the percentage of the various constituents can then be reported on the original "moist basis." For possible errors, see p. 193.

Often ground air-dried samples are weighed out for analysis and the results reported on the air-dry basis.[13] This method is satisfactory for materials that are not appreciably hygroscopic or for those in which the constituents sought are present in such small amounts that a variation in water content within reasonable limits will not sensibly affect the results. For example, metals and alloys are always weighed in the air-dry state. The chief objection to the use of air-dried samples is, of course, the possibility of a change in the water content. Variation in the moisture content of a sample may be caused by a change in the humidity of the atmosphere (see the notorious case of starch, p. 193), as well as by a change in the degree of subdivision of a substance at constant humidity, as already noted. Consequently, it may happen that the results of

[8] W. Hempel, *Z. Angew. Chem.* **14**, 843 (1901).

[9] See W. F. Hillebrand, G. E. F. Lundell, H. A. Bright, and J. I. Hoffman, *Applied Inorganic Analysis*, where additional references will be found.

[10] W. F. Hillebrand, *J. Am. Chem. Soc.* **30**, 1120 (1908).

[11] R. Mauzelius, *Sveriges geol. Undersökning*, Årsbok **1**, 1907, No. 3; W. F. Hillebrand, U.S. *Geol. Surv. Bull.* **700**, 1919, pp. 60–64.

[12] Such oxidation can be minimized by grinding under alcohol (Hillebrand).

[13] Air-dried samples are to be used for procedures in this text unless otherwise stated.

different analysts in different localities, or of the same analyst at different times, for the same sample will differ on account of a change in the moisture content. For this reason it may be necessary to dry the sample at some arbitrary temperature and report the results on the "water-free basis." It is customary to dry samples of most substances at 105°–110°C. A sample so dried may not be entirely free of nonessential water (p. 195), but the procedure has the advantage of furnishing a sample of fairly constant composition with respect to water. If constant weight is only slowly attained at 105 to 110°C, it may be necessary to dry at some higher temperature; the temperature of drying should be reported. The possibility of change in the composition of the sample on heating must be considered (loss of water of crystallization, oxidation, etc.).

Even when the results are to be reported on the "dry basis" it may be advantageous to weigh out air-dried samples and determine moisture in a separate portion, thus enabling the analyses to be calculated to the water-free basis. This procedure is to be recommended when the dried sample is appreciably hygroscopic; it is hardly possible to weigh out successive portions of such a dried sample without having it take up moisture from the atmosphere. However, when only one or two portions are to be taken for the analysis and hygroscopic water is not to be determined, it is simpler to place the approximate weight of sample in a weighing bottle and to heat to constant weight (or sometimes simply overnight) at 105 to 110°C in a drying oven. The stopper should be laid across the mouth of the bottle, and this placed in a small beaker covered with a watch glass resting on glass hooks. After drying, the bottle is loosely stoppered, allowed to cool in a desiccator, and then weighed closed as described in the next section.

Weighing the sample

The general operation of weighing was discussed in Chapter 19.

Weighing by difference. This is the method that should usually be used in weighing samples for analysis. A suitable amount of sample (sufficient to give several portions) is placed in a glass-stoppered weighing bottle (Fig. 20-1), and the weight of the whole is obtained to the nearest tenth of a milligram on the analytical balance. The bottle is then removed from the balance with clean dry fingers[14] and unstoppered directly above

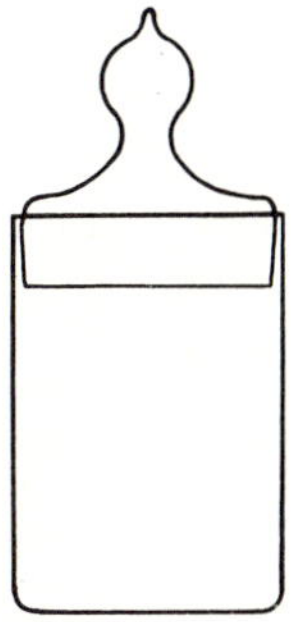

Fig. 20-1. Weighing bottle with internal stopper.

[14] It is hardly necessary when not striving for an accuracy greater than 0.1 mg to handle the sample bottle with a cloth or a strip of paper, provided the fingers are clean and dry and care is taken not to heat the bottle unduly by too long or too full contact with the skin. Thus it was found that a 15-ml weighing bottle lost in weight by 0.1 mg after being held between the thumb and two fingers for 20 sec (temperature of room 25°C, bottle loosely stoppered). In the most careful work, however, no object to be weighed should be handled directly with the fingers; oil and moisture from the skin may increase the apparent weight, and the rise in temperature may appreciably decrease that weight. The last effect may be the more serious.

the vessel into which the sample is to be poured. The stopper is held in the left hand and the bottle in the right. What is judged to be a proper weight of substance is then poured out carefully by tilting the weighing bottle slightly from the horizontal and rotating. Care should be taken to prevent raising dust. The bottle is kept over the vessel while being returned to the vertical position; it should be rotated slightly and tapped lightly when nearly vertical to cause any powder adhering to the lip to slide down into the tube; any particles on the very edge of the lip will fall into the receiving vessel. In reinserting the stopper care must be taken to avoid blowing out any particles of powder. The bottle is then reweighed. The difference in weight gives the weight of substance taken. The weighings should be entered immediately in the *laboratory notebook*.

The approximate amount of substance that should be poured out to give a certain weight is soon learned by experience. A beginner should pour out a portion estimated to be safely smaller than the weight desired, then reweigh, and pour out more if necessary. If too much sample is poured out, it is not permissible to return part of it to the weighing bottle; in such an event the receiving vessel should be washed out and another attempt made. This method of weighing is not suitable for weighing out an exactly predetermined amount of sample, but this rarely needs to be done. Thus, if a procedure calls for a 0.3-g sample, one weighing 0.25–0.35 g will usually serve equally well; but whatever the amount, the weighing must naturally be made to 0.1 mg. If it should be necessary to take an amount of substance quite close to a specified value, successive small portions may be transferred to a weighing bottle whose weight is approximately known until the required weight is present; then the whole is weighed exactly, the contents of the weighing bottle are poured out, and the bottle with any small amount of adhering powder is weighed.[15]

In weighing air-dried samples, no special precautions need to be taken to guard against errors due to hygroscopicity. Substances dried at 105°C or higher temperatures may be hygroscopic. It is generally not advisable to weigh out several portions of a hygroscopic substance from the same weighing bottle. Hygroscopic materials absorb moisture from the atmosphere most rapidly during the first few minutes of exposure. When dealing with such substances, it is safest to dry a single portion of sample as already described, to weigh the stoppered weighing bottle after cooling in a desiccator, to pour out the contents, and to reweigh the empty bottle.

Liquids should never be weighed in open vessels. They may be weighed by difference if a special weighing bottle (having a ground-in pipet with a stopcock) is used. Liquids may be conveniently weighed by the method of addition as described later, by introducing them into a weighed stoppered vessel. Great care must be taken in weighing volatile corrosive liquids to prevent the escape of fumes which might injure the balance.

Weighing by addition. Sometimes it is advantageous to weigh out a sample by addition. In this method the weight of the empty dry vessel to which the sample is to be transferred is first obtained, and the substance is then added, little by little if necessary, until the weight required is present[16]; the whole is then weighed carefully. The vessel into which the substance is weighed should be light, and should not present a large area of exposed surface in order that possible change in weight due to adsorption of moisture, electrification, heating, etc., during weighing may be kept at a minimum. It would be unthinkable to weigh a sample into a 250-ml Erlenmeyer flask, for example, if an accuracy of 0.1 mg were required.

This method of weighing is not suitable when the sample is to be treated directly in a beaker or flask; it may be used to weigh out a sample into a crucible prior to fusion. The method is also used when comparatively large quantities of substances must be

[15] It is advantageous to have a weighing bottle tared by a similar lighter one containing enough finely divided material (such as quartz sand or metal) to counterbalance the first within a few tenths of a milligram. With such a tared weighing bottle, weighing of samples is facilitated, and wear on the weights is reduced.

[16] A rough balance may be used first to obtain the approximate weight.

weighed out, as in the preparation of standard solutions in volumetric analysis (Chap. 33). A watch glass is employed as the weighing receptacle. It is first counterbalanced with weights or, better, with another watch glass of about the same weight; then it is removed from the pan and placed on a sheet of paper beside the balance case. The bulk of the substance is added, and the watch glass is replaced on the balance pan and tested against the weights that have been added, to see if it is heavy or light. The final adjustment in weight by adding or removing substance can be effected without removing the watch glass from the pan, if proper care is taken to avoid spilling substance in the balance. Thus, if the weight is a little too light, a spatula holding a little of the powdered substance at the end of the blade is poised over the center of the watch glass and gently tapped against a finger of the other hand to cause the powder to fall off *a little at a time*. The beam of the balance may be kept released and the pan arrests lowered all the while, if care is taken, so that the point of approximate equilibrium can be immediately noted. The excess powder on the spatula is discarded, and a small amount of the material on the watch glass is removed (after raising the beam) so as to cause the weight to become a little too light. The process is repeated with the powder now on the spatula, and in this manner a predetermined amount of substance is quickly weighed out, if necessary, to 0.1 mg. It hardly needs to be pointed out that substances that are hygroscopic, efflorescent, or volatile cannot be weighed accurately in this manner.

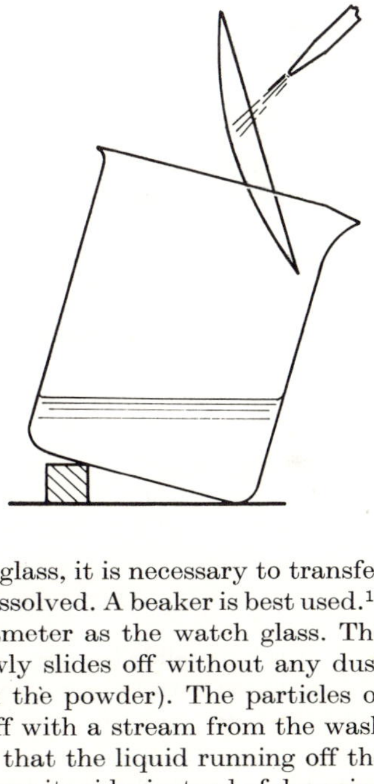

Fig. 20-2. Quantitative transfer of material from a watch glass to a beaker.

After the substance has been weighed out on a watch glass, it is necessary to transfer the solid without loss to some vessel in which it can be dissolved. A beaker is best used.[17] One should be chosen which has at least the same diameter as the watch glass. The latter is inclined over the beaker so that the solid slowly slides off without any dust being raised (sometimes it is advantageous to moisten the powder). The particles of material still adhering to the watch glass are washed off with a stream from the wash bottle. In this operation the beaker should be tilted so that the liquid running off the watch glass will strike the side of the beaker and flow down its side, instead of dropping directly into the solution below and splashing (Fig. 20-2). It is not advisable to use a brush to remove the adhering dust.

[17] If it is intended to transfer the solid to a volumetric flask, it is often advantageous to place a large dry funnel with a short wide stem in the neck of the latter to receive the material. If the solid is in the form of rather large crystals which will not pass through the stem of the funnel, or if it dissolves only slowly, the solid should be dissolved in a beaker and then transferred.

A counterbalanced weighing scoop (Fig. 20-3) is useful for weighing substances into narrow-mouthed vessels.

Fig. 20-3. Weighing scoop and counterweight.

Factor weights

By weighing out a certain calculated amount of sample—a factor weight—the weight of ignited residue in milligrams, or the titration volume in milliters, can be made to represent the percentage (or some simple multiple or fraction of the percentage) of the constituent desired. The analyst is thus saved the time and trouble of making the calculations; on the other hand, the time required to weigh out exactly a predetermined amount of sample may more than counterbalance the time saved in not making the calculation, and nothing is gained. The method is not applicable if the sample changes in weight rapidly when exposed to the air or if it is not in a finely divided form. The device of factor weights is frequently used by analysts in the industries. Circumstances determine whether or not it is recommendable to use factor weights; it is safe to say that nothing is gained by their use when weighings must be made with an accuracy of 0.1 mg.

Accuracy of weighings[18]

It is a waste of time to weigh with an accuracy much greater than that which can be attained in the subsequent steps of the analysis. This means that since we are concerned with the percentage accuracy of a result (p. 381), the actual weight of sample and the percentage of the constituent sought must both be taken into account when deciding the decimal place to which the weighing shall be made. The following expressions gives the magnitude of error in absolute per cent of the constituent sought resulting from a certain error in weighing a sample of a given weight:

error in percentage of constituent

$$= \frac{\text{error in weighing sample (g)}}{\text{weight of sample (g)}} \times \text{percentage of constituent}$$

For example, if a 1-g sample of steel containing 1.0 per cent manganese is weighed with an accuracy of 0.2 mg, the maximum error in the percentage of manganese resulting from the weighing error cannot exceed the value

$$\pm \frac{0.0002}{1.0} \times 1.0 = \pm 0.0002 \text{ absolute per cent}$$

Now, it is impossible to carry out the other steps of the procedure in the determination of manganese with an accuracy such that the final result will differ by not more than 0.0002 per cent from the true value. Therefore, weighing a 1.0 g sample to 0.1 mg is an unnecessary refinement. A sample of such size need not in this case be weighed closer than 0.5 mg, corresponding to a possible error of ± 1.0 mg in weighing, which in turn corresponds to a possible maximum error of 0.001 per cent of manganese—an accuracy reasonable to strive for, but not easily attained.

On the other hand, in weighing out a 0.5-g sample of iron ore containing, say, 90 per cent of Fe_2O_3, each weighing must necessarily be made to 0.1 mg to assure a combined weighing error less than 0.1 per cent, this being the approximate error associated with the other steps of the procedure in ordinary careful analysis.

[18] See Chap. 16, where a more exact treatment based on probability considerations is given.

The size of sample that can be taken is limited by such factors as the volume of the precipitate obtained in a gravimetric procedure, or the volume of titrating solution required in a volumetric method, and the volume for proper dilution of solutions in precipitations.

Decomposition of samples

All chemical methods of analysis, and even some physical methods, require bringing the sample into solution (almost always an aqueous solution for inorganic samples). Solution usually entails a breakdown in the composition of the material. The manner of decomposition depends on the nature of the sample, the constituent to be determined, and sometimes other factors. Usually the analyst knows the type of material and often the qualitative composition of the sample being analyzed, and more or less standard methods of decomposition based on experience apply.[19] Table 20-1 gives a survey of the common solvents and fluxes (fusion reagents). Refractory materials, which do not yield to acids, must be decomposed with fluxes. The effectiveness of fluxes is explained by the high temperature at which they are used and the high concentration of the

TABLE **20-1. Solvents and Fluxes**

Acids[a]	
HCl	For many salts of weak acids: carbonates, phosphates, some oxides (e.g., Fe_2O_3, Chap. 43), some sulfides; for metals above hydrogen in electromotive series. At high temperatures (250–300°C) in a sealed tube, HCl dissolves strongly ignited Al_2O_3, BeO, SnO_2, and some silicates; at 350–400°C with $HClO_4$ or other oxidizing agent dissolves Rh, Ir, Os, and Ru.
H_2SO_4	Used when its high fuming temperature (~300°C) is an advantage, as in expelling a volatile product or increasing reaction rate, e.g., for decomposing CaF_2 and monazite (phosphate of rare earths and thorium). Formation of sulfato complexes and dehydrating and oxidizing properties at high temperatures of advantage for some materials.
HNO_3	Provides oxidizing attack for metals not dissolved by HCl and other nonoxidizing acids; Au, Pt metals (except Pd), Nb, Ta, and Zr not dissolved; Al and Cr are passivized; Sn, Sb, and W give insoluble hydrous oxides. Dissolves most sulfides (not HgS). Usually not good for oxides (exceptions: UO_2 and U_3O_8).
$HCl–HNO_3$ (3:1 = aqua regia)	For sulfides, including HgS, and Au, Pd, and Pt.
$HClO_4$	At fuming temperatures (constant-boiling acid, 203°C = 72 per cent $HClO_4$) is a strong oxidizing agent and destroys organic matter. For this purpose, usually mixed with HNO_3 which will first oxidize easily attacked organic matter that might otherwise react violently with $HClO_4$. H_2SO_4 (dehydrating agent) increases oxidizing power. Good solvent for stainless steel, sulfides; with H_3PO_4 dissolves chromite ($FeO \cdot Cr_2O_3$). Does not dissolve Pt metals (except Pd), Nb, Ta, or Zr.

[19] See G. E. F. Lundell and J. I. Hoffman, *Outlines of Methods of Chemical Analysis*, chap. IV; H. H. Willard and C. L. Rulfs in *Treatise on Analytical Chemistry*, part I, vol. 2, chap. 24; and J. Dolezal, P. Povondra, Z. Sulcek (transl. by D. O. Hughes, P. A. Floyd and M. S. Barrett), *Decomposition Techniques in Inorganic Analysis*, Iliffe, London, 1968.

Table 20-1 (*continued*)

HF	For decomposition of many silicates in combination with H_2SO_4 or $HClO_4$ to eliminate fluorides (see p. 676). With HNO_3, dissolves Ti, W, Nb, and Zr (also their carbides, nitrides, and borides) in consequence of formation of complex fluorides. Certain refractory silicates and other minerals not decomposed (see listing under $LiF–H_3BO_3$ below).
Fluxes	
Na_2CO_3	For silicates (p. 651), oxides, phosphates, sulfates, at 1000°C or higher.[b] Sometimes used with oxidizing agents ($NaNO_3$, Na_2O_2).
$Na_2CO_3–S$	For Sn, As, and Sb sulfides and alloys; these elements give thio salts soluble in water, other metals insoluble sulfides. "Freiberger Aufschluss."
NaOH	Similar to Na_2CO_3 ($\sim$500°C) except not ordinarily used for a complete silicate analysis (p. 651).
$Na_2B_4O_7$	For zircon, chromite, and other refractory minerals. Because of introduction of boron, usually avoided. A mixture of $Na_2B_4O_7$ and Na_2CO_3 is sometimes used for TiO_2, Al_2O_3, and Fe_2O_3.
$LiBO_2$[c]	For silicates, with special reference to silica.
$CaCO_3–NH_4Cl$	J. Lawrence Smith decomposition, for alkali metals in silicates (p. 669).
Na_2O_2	A powerful alkaline oxidizing flux ($\sim$600°C) used when Na_2CO_3 is ineffective (e.g., for chromite). Attack of crucible is a problem (p. 459). Na_2O_2 + NaOH suitable for sulfides (p. 613).
$K_2S_2O_7$	For oxides (e.g., Fe_2O_3, Al_2O_3, TiO_2 in rock analysis) not easily attacked by acids (450–500°C).
HKF_2[d]	For silicates not easily decomposed by HF and for oxides of Nb, Ta, and Zr.
B_2O_3	For refractory oxides, silicates, and sulfates (1000–1100°C). Boron subsequently removed as methyl borate by evaporation with $CH_3OH–HCl$. Not commonly applied.
$LiF–H_3BO_3$[e]	Fusion at 800–850°C. Excess fluoride and boron volatilized by fuming with sulfuric acid. Decomposes zircon, spinel, corundum, rutile, kyanite, sillimanite, tourmaline, garnet, topaz, tantalite, chromite, graphite, and pyrite.

[a] Aqueous alkaline solutions (in contrast to alkaline fluxes) are infrequently used. NaOH or NH_4OH with Br_2 or H_2O_2 dissolves sulfides of As, Sb, Mo, etc. Al and a few other metals can be dissolved in NaOH; also acidic oxides such as As_2O_3. Na_2CO_3 is occasionally used for transposition of sulfates (e.g., $SrSO_4$) to carbonates, which, after washing, can be dissolved in HCl.

[b] Sintering of refractory silicates and oxides with sodium carbonate or a mixture of sodium carbonate and tetraborate sometimes is preferred to fusion. Less attack on the platinum crucible occurs. See A. N. Finn and J. F. Klekotka, *J. Res. Natl. Bur. Std.* **4**, 809 (1930); R. C. Chirnside, H. J. Cluley, R. J. Powell, and P. M. C. Proffitt, *Analyst* **88**, 851 (1963).

[c] C. O. Ingamells, *Talanta* **11**, 665 (1964).

[d] Fusion with KF (m.p. 880°C) followed by pyrosulfate fusion has been recommended for decomposition of refractory silicates and Al, Be, Th, and Zr oxides and phosphates [C. W. Sill, *Anal. Chem.* **33**, 1684 (1961)].

[e] V. S. Biskupsky, *Anal. Chim. Acta* **33**, 333 (1965). [Concerning decomposability of minerals of silicate rocks by HF see F. J. Langmyhr and S. Sveen, *Anal. Chim. Acta* **32**, 1 (1965); decomposition by HF at high temperatures and pressures is described by I. May and J. J. Rowe, *Anal. Chim. Acta* **33**, 648 (1965).]

active agent. Substances of acidic nature are decomposed with a basic flux (e.g., silicates with sodium carbonate), those of basic nature with an acid flux (e.g., Fe_2O_3 with potassium pyrosulfate, $K_2SO_4 \cdot SO_3$), amphoteric substances (such as Al_2O_3) with either. Oxidizable substances can be broken down with an oxidizing flux (chromite, $FeO \cdot Cr_2O_3$, with Na_2O_2). Fluxes may have the disadvantages of (1) introducing relatively large amounts of cations (occasionally anions) which can be undesirable in an analysis, and (2) attacking the crucible and so contaminating the sample with Pt, Fe, Ni, etc.

The possible loss of constituents by volatilization during decomposition must always be borne in mind. Acid attack at room temperature evolves CO_2, H_2S, hydrides of As and P (from arsenides and phosphides in metals), and other volatile compounds of the nonmetals. From basic solution NH_3 will be lost, also SiH_4 (as from Al + NaOH). From hot, especially boiling, hydrochloric acid solutions, a number of volatile chlorides, including $GeCl_4$, $AsCl_3$, $SbCl_3$, and $SnCl_4$, will escape. Boron as well as silicon volatilizes from hydrofluoric acid solutions. All compounds of mercury as well as the element itself are volatile above $\sim 100°C$, even to some extent at room temperature. Phosphorus (H_3PO_4) is lost from fuming sulfuric or perchloric acid. This enumeration is by no means complete.

A few words may be said concerning the common operation of dissolving a sample in water or acid. If the determination is gravimetric, the sample will usually be weighed into a beaker, unless certain preliminary operations, such as evaporation, must be performed prior to precipitation. The beaker should be immediately covered with a watch glass of suitable size, convex side downward. The solvent is then carefully added by pouring it down the side of a glass rod having its lower end resting against the wall of the breaker to prevent splashing, the cover glass being displaced meanwhile. If the liquid added causes evolution of gas (acids with carbonates, metals, etc.), the beaker must be kept covered during the addition. It is then best to add the reagent through a small funnel with a bent stem inserted beneath the watch glass at the lip (Fig. 20-4), or from a pipet; the stirring rod should not be placed in the beaker until later in order that as little open space as possible may be left between the cover glass and the vessel. These precautions are necessary because when bubbles of evolved gas break at the surface of the liquid a fine spray is produced, and minute droplets would be projected out of the vessel or carried away by the escaping gas if not intercepted by the cover glass. If there is a possibility of a vigorous reaction when the reagent is added, water should first be added to the sample to moderate the reaction. When gas evolution has ceased, the underside of the watch glass is rinsed off with a stream of water from the wash bottle, care being taken not to allow the water to fall directly into the solution. Solution of the sample may be hastened by stirring the liquid, or heating when necessary; boiling should be avoided except when really necessary (as in the expulsion of dissolved gases). Liquids are more safely boiled in Erlenmeyer flasks, with a buret funnel in the neck to prevent mechanical loss, than in covered beakers.

Care should be taken to see that all components of a heterogeneous sample are brought into solution. Grains remaining after an acid attack must be filtered off and treated further, by fusion if necessary.

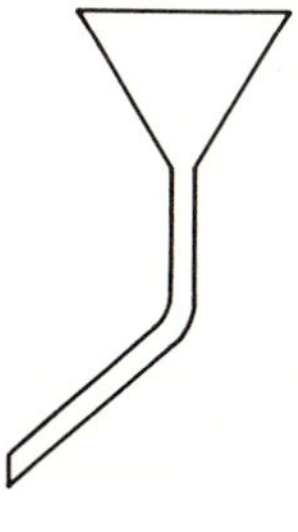

Fig. 20-4. Funnel with bent stem.

A few general statements regarding the performance of fusions may be made. Before a sample can be satisfactorily fused, it must be reduced to a very fine state of subdivision by grinding in an agate mortar, and then mixed very intimately with the flux. The mixing may usually be done in the crucible with the aid of a short stirring rod, as described on page 658. The kind of crucible to be used in the fusion is determined by the nature of the flux. Platinum crucibles are used in sodium carbonate and potassium pyrosulfate fusions.[20] For sodium hydroxide fusions nickel, silver, or gold crucibles are employed; with sodium peroxide, nickel, iron, or zirconium crucibles are used. The crucible should never be more than half filled with the fusion mixture and must be kept covered during the whole process. At the beginning, the heating should be very gradual to permit the quiet escape of moisture. The final temperature should not be higher than actually necessary, to minimize attack on the crucible by the flux (e.g., nickel with sodium hydroxide). When the fusion has been completed, the crucible is allowed to cool somewhat; then, while the contents are still fluid, it is grasped by the crucible tongs and tilted with a rotating motion so that the fused mass will be distributed up around the walls, giving a thin shell on cooling that permits easy detachment.

For detailed directions for making a sodium carbonate fusion, see page 659.

Evaporation

In the course of an analysis it is often necessary to reduce the volume of a solution or even to carry it to dryness. The vessels that are commonly used for the purpose are porcelain casseroles, platinum or silica dishes, and, under certain circumstances, beakers of resistant glass. The possibility of attack of the container by the hot liquid during the prolonged contact must always be borne in mind. It has already been pointed out (p. 454) that glass, even if of the resistant variety (borosilicate), is attacked by alkaline and neutral solutions and that porcelain is not immune to the action of these hot solutions. The nature of the analysis will determine whether the impurities thus introduced can give rise to errors. In work of the highest accuracy platinum dishes should usually be used for evaporations, since they are very resistant to attack by most solutions (p. 456). Solutions to be evaporated in porcelain or glass should be acidified before evaporation, if this is permissible, to lessen attack.

Evaporations must be conducted on a steam bath, or a low-temperature hot plate, in order that evaporation may take place without ebullition. Boiling may lead to mechanical loss even if the vessel is covered. Matters should be so arranged that large volumes of liquid can be left to evaporate on the steam bath overnight.

Now and then a solution must be boiled, as for example, when dissolved gases are to be expelled. Smooth boiling without *bumping* (the almost explosive conversion of superheated liquid to vapor) can be obtained by placing an antibump rod in the liquid. This is a rod having a short section of tubing ($\sim$ 1 cm) of the same diameter fused to the rod (Fig. 20-5). The air pocket thus formed effectively promotes smooth boiling.

Fig. 20-5. Anti-bump rod.

Such a rod should be replaced by an ordinary rod when precipitations are made.

During evaporation, liquids must be so covered that dust and other foreign material cannot fall into the vessel, and yet the vapor is given free opportunity to escape. This may be accomplished by supporting a watch glass of Pyrex having a diameter slightly greater than the container on a glass triangle or on small V-shaped glass hooks over the top of the vessel (Fig. 20-6). More conveniently, a special cover glass with three molded radial ribs on the bottom that hold it away from the vessel may be used. The

[20] High-temperature fusions in platinum over a flame, of a sample containing iron compounds, is likely to result in the reduction of some iron and its alloying with platinum. This can be prevented by using a muffle furnace.

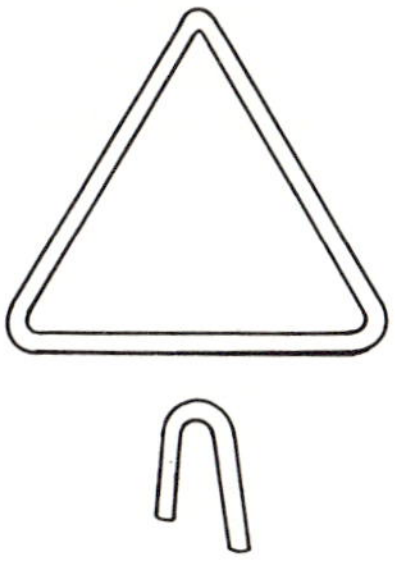

Fig. 20-6. Glass triangle and hook.

vessel should not be too full. If necessary the liquid can be added a little at a time until all has been evaporated.

For evaporating small volumes of liquid rapidly, a platinum crucible placed in a "radiator" serves admirably (see the next section).[21]

At the end of the evaporation, the sides of the vessel, the lower side of the cover glass, and the triangle or glass hooks should be rinsed off with distilled water.

Volatilization of sulfuric acid and expulsion of ammonium salts

Sulfuric acid may be expelled by careful heating on a high-temperature hot plate, but the volatilization is more expeditiously accomplished in an air bath. If the solution is contained in a crucible, a "radiator" such as that shown in Fig. 18-4 may be used very conveniently. The radiator consists simply of a large iron or nickel crucible fitted with a clay triangle or some other device for supporting the crucible; such a contrivance, especially made for the purpose, may also be purchased. The radiator is heated by a burner, which is adjusted so that the contents of the crucible evaporate rapidly without boiling. The liquid is heated uniformly in this arrangement, and there is no danger of loss by creeping or spattering. All evaporations of sulfuric acid, and other acids as well, should be conducted in a well-drawing hood.

Ammonium salts (particularly the chloride) are expelled by cautious heating, with a free flame, of the porcelain, silica, or platinum vessel containing the dry residue. The burner should be held in the hand and moved beneath and around the edges of the vessel. At the beginning of the heating the container should be covered with a watch glass or crucible cover, as required, to prevent loss of flying particles resulting from the expulsion of water of decrepitation; when decrepitation has ceased, any particles on the cover glass are transferred to the vessel, and further heating is continued with the latter uncovered. A crucible may conveniently be heated in a radiator. The expulsion of ammonium salts by volatilization can be accomplished quite easily when they are contained in a small platinum or porcelain vessel (crucible), but less easily when large quantities are present in a large porcelain vessel (casserole).

When large amounts of ammonium chloride must be removed, destruction by oxidation with nitric acid[22] is much to be preferred to volatilization (Chap. 63). The solid or concentrated solution is treated with an excess of nitric acid (at least 3 g of concentrated nitric acid for each gram of ammonium chloride) in a porcelain or glass vessel, warmed on the steam bath, and finally evaporated to dryness.

Precipitation

Gravimetric precipitations are made in beakers. It is often very important that the solution of the precipitant be added slowly and with efficient stirring of the solution, to avoid as much as possible local concentration of reagent, which generally leads to

[21] An infrared lamp is also useful here. Because it heats from above, rapid evaporation can be carried out without bumping or creeping.

[22] J. L. Smith, *Am. J. Sci.* **15**, 94 (1853).

increased contamination of the precipitate (p. 241). For this reason it is often advisable to add the reagent solution dropwise from a pipet or buret rather than from a graduate or beaker, since the latter does not permit such uniformly gradual addition. The addition must be made without splashing; i.e., the solution of the precipitant should be allowed to flow down the side of the beaker and should not be allowed to drop in from a height. Precipitations are usually made in hot solution, since the high temperature is conducive to the formation of an easily filterable precipitate composed of large crystals; moreover, the contamination of the precipitate may be less at higher temperatures. There is no necessity, however, for heating the solution to the boiling point; it is much better to add the precipitant to a solution slightly below the boiling point to avoid the possibility of loss through the sudden liberation of steam through release of superheating. In stirring, care should be taken not to bring the stirring rod into contact with the sides or bottom of the beaker for fear of scratching the glass.

A considerable excess of precipitant is always added in a quantitative precipitation; an unnecessarily large excess is, however, to be avoided as possibly leading to increasing solubility or contamination of the precipitate. Often it is possible to calculate roughly the amount of reagent required from the weight of sample taken, assuming the latter to be composed entirely of the compound of the constituent having the lowest equivalent weight. After the precipitate has settled, a few drops of reagent should always be added to see whether any more precipitate forms.

It is not advisable to filter off a precipitate immediately after it has been formed. Solubility equilibrium may not be reached at once, especially with substances tending to form supersaturated solutions. The precipitate may sometimes be obtained in a coarser, more easily filterable form, by letting it stand in contact with the hot mother liquor for some time (Chap. 10). Moreover, the amount of coprecipitated substance in the precipitate may decrease on standing, as a consequence of the growth and perfection of the crystalline particles of the precipitate. For these reasons it is frequently recommendable to *digest* a precipitate before filtration by letting it stand on the steam bath for some hours or overnight. Sometimes, however, so little is gained by this procedure that it is as well to filter off the precipitate an hour or so after precipitation; indeed, it is sometimes disadvantageous to let a precipitate stand for any length of time (precipitation of ferric hydroxide and other hydroxides or hydrous oxides with ammonia, especially in the presence of calcium; precipitation of calcium with oxalate in the presence of magnesium, etc.). Circumstances decide the procedure to be followed.

If the solubility of the precipitate is large, it may be necessary to let the solution come to room temperature before filtration. The liquid should then be stirred occasionally during the cooling.

Precipitation with hydrogen sulfide. It is usually best to make precipitations with hydrogen sulfide by leading the washed gas through a one-hole rubber stopper into an Erlenmeyer flask holding the solution. The entrance tube should not dip into the liquid but should be close to the surface (Fig. 20-7). After the air has been displaced from the flask, the stopper can be pressed into place and the absorption of hydrogen sulfide

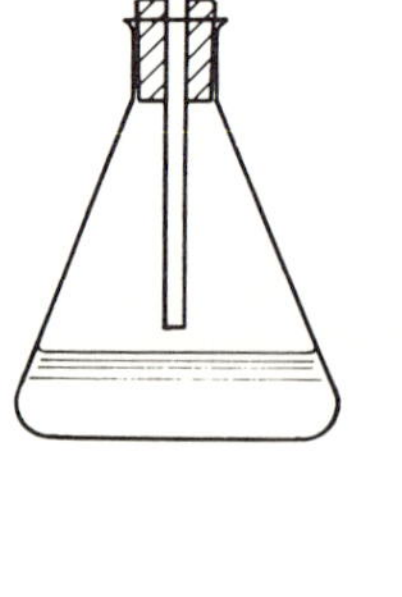

Fig. 20-7. Precipitation with hydrogen sulfide.

allowed to take place under the slight pressure of the gas, aided by swirling of the liquid. Sometimes the precipitation can be carried out in a beaker with slow bubbling of hydrogen sulfide. The solution should usually be hot.

Filtration and washing of precipitates

Filtration through paper. Filter paper is especially suitable for the collection of precipitates that are gelatinous[23] and those that must be heated to a very high temperature before weighing.

Owing to the hygroscopic nature of paper, precipitates cannot be weighed in a paper filter after washing and drying; it is necessary to ignite the filter and its contents in a crucible and weigh the residue, which corresponds to the precipitate and a negligible amount of ash. If in burning off the paper the precipitate tends to be reduced by the carbon or the gases formed, or otherwise affected, it is necessary to separate the bulk of the precipitate from the paper and ignite the latter, then to convert the small amount of adhering precipitate, which may have been reduced, to the original form by the use of proper reagents, and finally to add the main portion of the precipitate, which has been reserved in the meantime. Such a procedure is so troublesome and conducive to error that it may be stated as a general rule that if a precipitate will be changed chemically by the action of the paper during the ignition, it should not be collected on paper. The silver halides, for example, should not be caught on paper because they are easily reduced to free silver on ignition. On the other hand, barium sulfate may be collected in paper, for although it may undergo slight reduction to sulfide on ignition, it is easily oxidized back to sulfate on continued heating with free access of air.

Before beginning a filtration through paper, it should be decided what size and what fineness of *quantitative* filter paper is best suited for the particular filtration. In the first place, the size of the paper chosen should not be larger than is really necessary, and it should be remembered that the size is determined by the amount of precipitate to be collected rather than by the volume of liquid to be filtered. The precipitate should preferably fill not more than one third of the filter and never more than one half. Circles 9 cm in diameter find most frequent use. When the precipitate is to be weighed, quantitative ("ashless") paper (p. 463) must always be used. If the ash amounts to more than 0.1 mg, its weight should be deducted from the weight of the ignited residue.[24]

The method of preparing a filter for use is as follows. First, the dry paper is folded exactly in half and then once again through the center so that the two halves of the first crease do not quite coincide. When a paper so folded is opened, the apical angle of the cone will be slightly greater than 60°. Such a filter when placed in a 60° funnel will not quite fit the walls of the latter at the apex, but will fit at the top. A paper folded in this manner will filter more rapidly than one which adheres to the funnel over its whole area, for the liquid that has penetrated the former filter will run down unimpeded through the narrow space between paper and glass, and at the same time the top of the paper will fit the funnel so well that no air is likely to be drawn in between the two. The exact angle at which the second fold should be made depends upon the apical angle of the funnel used—which may not be exactly 60°. It may be necessary to refold the paper once or twice to get it to fit properly. The crease at the apex of the filter should not be flattened too much for fear of weakening the tip. The corner of the fold on the outside should be torn off (Fig. 20-8) in order that the paper may be fitted snugly to the funnel when it is later moistened.

The filter paper is now placed in a funnel of proper size; the top of the filter should fall 2 to 1 cm below the rim of the funnel. The paper is moistened with water from the

[23] The use of filter-paper pulp to aid the filtration of gelatinous precipitates has already been referred to (p. 464). The pulp should be added after the precipitate has been formed.

[24] The weight of ash of a single filter can be determined by igniting a number of papers together.

wash bottle, and its upper third gently pressed against the glass with the finger. When water is now poured into the filter, the stem of the funnel should fill. If it does not do so (as is likely to happen if the stem is too wide or the glass is greasy), the end of the funnel stem should be closed with the finger and the stem allowed to fill until the water rises into the apex and above. The air entrapped between the glass and paper is permitted to escape by pulling the paper gently aside at one point. The paper is then carefully refitted to the glass all around, and the finger removed; the stem should remain filled with liquid. The glass above the paper should not be moistened in these preliminaries. It is very important to have an unbroken column of liquid in the stem, because the weight of such a column produces a gentle suction that greatly hastens filtration.

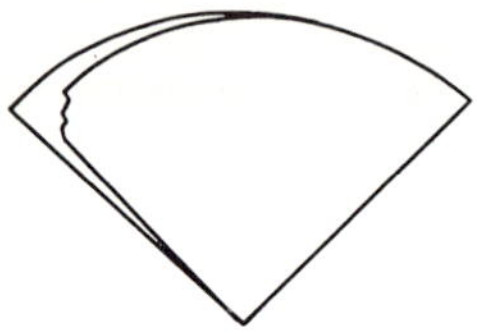

Fig. 20-8. Folding a filter paper.

When the filter paper has been properly fitted to the funnel, the latter is placed in a funnel stand, and a clean beaker is put into position beneath. The funnel stem should touch the side of the beaker to prevent splashing and should be well above the bottom of the vessel (Fig. 20-9); the stem of the funnel must never extend into the liquid in the receiving vessel. The beaker containing the liquid to be filtered is taken in the right hand and held slightly above and to the right of the funnel. The stirring rod is removed from the beaker, care being taken that no drops of liquid are lost, and held vertically over the funnel. The lower end of the stirring rod should be very close to, but should not touch, the filter paper on the side having three thicknesses of paper. The lip of the beaker is placed against the rod, and the vessel is slowly inclined until the liquid flows down the rod into the paper without splashing. It is inadmissible to transfer liquids in quantitative operations without the use of a stirring rod. The filter should never be

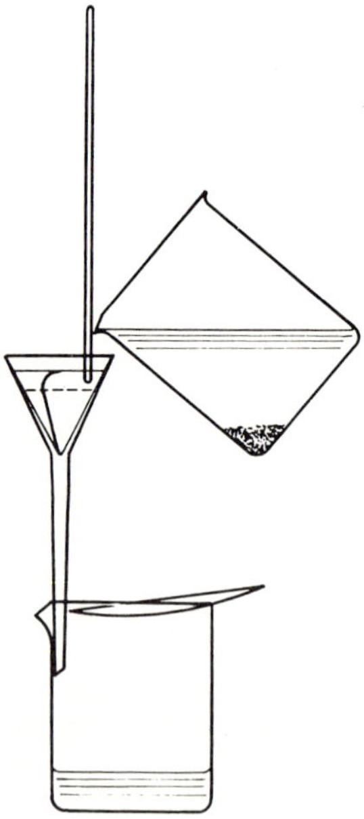

Fig. 20-9. Filtration through paper.

filled to the brim—the level of the liquid should not rise closer than to within 2 or 3 mm of the top of the paper. In transferring the liquid to the filter care should be taken not to disturb the precipitate in the bottom of the vessel. Even if the supernatant liquid appears to be entirely clear, it must nevertheless be passed through the filter.

As long as the liquid flows through the filter freely, the beaker is kept inclined, and a slow stream is poured into the funnel. When most of the supernatant liquid has been poured off, particles of precipitate will come over, and the speed of filtration will gradually decrease, owing to the filling of the pores of the paper with precipitate. When this occurs, the stirring rod is returned to the beaker, and the latter is set down while one waits for the filter to empty. It is advisable to keep the filter always fairly full of liquid to prevent the stem from emptying. The precipitate in the bottom of the beaker should be disturbed as little as possible. The stirring rod is kept out of the lip of the beaker, which will contain adhering precipitate. At this point the filtrate should be carefully examined for particles of precipitate that may have run through the paper. If such are found, the liquid in the receiving beaker must be passed through the filter, and the new filtrate caught in a fresh beaker. If the precipitate persists in running through, the filter paper used is too coarse or the precipitate is too fine for a quantitative filtration.

When as much as possible of the supernatant liquid has been poured through and a relatively small volume of liquid with precipitate is left, washing by *decantation* is begun. A stream of wash liquid from the wash bottle is directed against the sides of the beaker to wash down adhering particles, the precipitate is stirred up and allowed to settle. The amount of wash liquid to be added cannot be specified. It depends upon the amount of precipitate, its character, etc.; the analyst must use his own judgment. When the supernatant liquid is clear, it is poured into the filter, as much as possible of the precipitate being kept in the beaker. Another portion of wash liquid is added to the beaker, and the process is repeated. The filter is allowed to drain completely after each addition. The operation is repeated three or four times or as many times as may be thought advisable.

It is not always advantageous to wash by decantation. If the precipitate is coarse-grained and filters well, not much is gained by applying the decantation method. In such a case it is better to transfer the precipitate to the paper immediately after the supernatant precipitation liquid has been passed through the filter. Gelatinous precipitates that tend to clog the paper and make filtration slow can sometimes be washed free of most of the soluble salts by decantation, and the final washing made upon the paper. But even with such precipitates, not much is gained by washing by decantation if the precipitate is very voluminous and settles only slowly. Washing by decantation is objectionable if the volume of the filtrate must be kept as small as possible. The method may well be applied when the precipitate is to be dissolved and reprecipitated. All soluble salts need not then be removed; it is only necessary to make their concentration very small to prevent appreciable coprecipitation (p. 244), and this can be done by washing a number of times by decantation. To dissolve the precipitate for additional treatment, the beaker containing the bulk of the precipitate is put in the place of the one containing the filtrate, and the relatively small amount of precipitate on the filter is dissolved by filling the latter with the proper solvent and adding more as required. When all the precipitate has been dissolved, the paper is washed and the washings collected in the same beaker. All the precipitate in the latter is dissolved by stirring and bringing the solution into contact with all parts of the vessel and rod. The substance, now entirely in solution, is reprecipitated by adding the proper reagent and a little of the original precipitant to diminish the solubility. After standing, the precipitate is filtered off (possibly on the same filter if this has been washed with water to remove the solvent) and carefully washed as described below. The filtrate from the second precipitation is combined with that from the first, and the combined solutions are reserved if necessary.

Whether or not washing by decantation has been resorted to, the main bulk of the precipitate is transferred to the filter by mixing with wash solution and pouring off the suspension, the process being repeated until most of the solid is on the paper. The precipitate adhering to the sides and bottom of the precipitation beaker is removed in the following manner. The beaker is taken in the left hand, and the stirring rod is put across its top and held in place with the left index finger. The rod is made to rest against the lip of the beaker and should project 2 or 3 cm beyond. The beaker is then

Fig. 20-10. Transfer of precipitate from beaker to filter paper.

inclined over the funnel so that liquid can run out, down the rod into the paper without spattering. The wash bottle is grasped in the right hand, and a stream of wash solution is directed against the inside of the beaker so that the particles of precipitate are washed out into the filter (Fig. 20-10). After this treatment there will usually still be small amounts of precipitate adhering to the walls. These are removed by moistening the walls and rubbing gently with a "policeman" (Fig. 20-11). The whole interior of the beaker, as well as the surface of the stirring rod, is thus gone over. When all particles have been dislodged, the policeman is rinsed off and removed from the beaker.[25]

Fig. 20-11. Stirring rod with policeman.

[25] The policeman should not be allowed to stand in a solution for any length of time; it is not a substitute for a stirring rod and should not be used in transferring liquids.

The small amount of precipitate thus loosened is washed into the filter in the same manner as before. The beaker and stirring rod should finally be held up to the light and carefully examined for specks of precipitate or unrubbed surface, and the treatment repeated if necessary.[26]

As soon as all the precipitate has been transferred to the filter, washing should be begun. A precipitate should not be allowed to stand on the filter for any length of time unwashed. The mass will crack and cake together, and it will be virtually impossible to wash it free of soluble salts when it it is in this condition. The kind of wash solution to be used depends upon the precipitate (and to a lesser extent upon the substances in solution). Two factors especially must be considered: the solubility of the precipitate and its peptizability. If a precipitate is appreciably soluble in water, it cannot, of course, be washed with this liquid. This difficulty can be overcome, when a cation is being determined, by washing the precipitate with a dilute solution of an ammonium salt having an ion in common with the precipitate. The ammonium salt remaining with the precipitate after the washing is volatilized on ignition at a low temperature.[27] Thus in the gravimetric determination of calcium as oxalate, the precipitate is washed with dilute ammonium oxalate. Even if the precipitate is so slightly soluble in water that no error is to be feared from solubility loss, it is sometimes not advisable to wash with pure water, for peptization (p. 217) may then occur and the washings will run through the filter turbid.[28] This phenomenon can occur with coagulated colloidal precipitates but usually not with microcrystalline precipitates. To prevent peptization, a solution of some electrolyte is used for washing. The electrolyte should be one that can be removed easily (volatilized) in the preparation of the precipitate for weighing. For this reason dilute acids, ammonia, or ammonium salts are used. The silver halides, for example, are washed with very dilute nitric acid, in which they are not appreciably soluble. Sometimes washing with a solution of a salt having an ion in common with the precipitate is out of the question, as, for example, when calcium oxalate is to be titrated with permanganate after precipitation (Chap. 43). Then, water is of necessity used for washing, and care is taken to limit the amount to that actually necessary.[29] Whether a wash liquid should be used cold or hot depends on the solubility of the precipitate at higher temperatures. When permissible, hot wash liquids are employed because the speed of filtration is then greater, owing to the decrease in viscosity of liquids with an increase in temperature.

Hydrolyzable precipitates require the use of a wash liquid containing a substance repressing the hydrolysis (p. 633).

The wash solution running through the filter is never saturated with the precipitate, so that solubility losses are often less than would be expected from solubility data.

Washing is done more efficiently with many small portions of liquid than with few large portions, the total volume being the same. In other words, a smaller volume of liquid will suffice if small portions are added at a time. Let

v = volume (ml) of liquid remaining with the precipitate after draining
V = volume of wash liquid added each time

[26] Some precipitates (e.g., ammonium phosphomolybdate, nickel dimethylglyoximate) tend to form troublesome films on the side of the beaker and to creep on the filter paper. This film formation can be prevented by the addition of a commercial anticreep fluid which lowers the surface tension at the liquid–solid interface (wetting action).

[27] If the precipitate is to be dried at a low temperature, e.g., 110°C, the nonvolatile constituents of the wash solution can often be removed by finally washing with a water–alcohol solution, in which most inorganic precipitates are virtually insoluble.

[28] Moreover, hydrolysis of certain substances in solution may occur, leading to contamination of the precipitate. Thus ferric salts are hydrolyzed to basic salts by water. Using dilute hydrochloric acid as wash liquid prevents this (as in washing hydrous silica).

[29] Sometimes it is possible to use a saturated solution of the slightly soluble substance in water for washing. Mixtures of alcohol and water can be used at times.

C_0 = concentration (g/ml) of soluble substances in original; C_1 = the same concentration in the adhering solution after the first washing, C_2 = the concentration after the second washing, etc.

After the precipitate has been washed once and allowed to drain, the concentration of foreign substances in the adhering solution is evidently given by the following expression, if complete mixing is assumed:

$$C_1 = \frac{v}{V + v} C_0$$

After the second washing the concentration is

$$C_2 = \frac{v}{V + v} C_1 = \left(\frac{v}{V + v}\right)^2 C_0$$

and after n washings,

$$C_n = \left(\frac{v}{V + v}\right)^n C_0$$

The weight of foreign subtance, W_n, in grams remaining in the solution adhering to the precipitate after the nth washing will be

$$W_n = vC_n = \left(\frac{v}{V + v}\right)^n vC_0$$

It will be apparent from this expression that with a given volume of wash liquid the precipitate can be more completely washed if many small portions are used than if a few large portions are employed. By way of illustration, suppose that a precipitate is washed in one case with five portions of water of 10 ml each, and in the other with two portions of 25 ml. Assume that $v = 1$ and $C_0 = 1$. In the first case, after washing five times, the amount of foreign substance remaining with the precipitate will be

$$W_5 = \left(\frac{1}{1 + 10}\right)^5 \times 1 \times 1 = 7 \times 10^{-6} \text{ g}$$

whereas in the second, after two washings with the same total volume of liquid,

$$W_2 = \left(\frac{1}{1 + 25}\right)^2 \times 1 \times 1 = 1.5 \times 10^{-3} \text{ g}$$

Now, it must be emphasized that foreign substances are not so easily removed from the precipitate by washing as the above formulas would indicate. The preceding expressions have been derived on the assumption that the solution of the foreign substances is simply mechanically associated with the particles of the precipitate. This is far from being true. Actually parts of the foreign substances are more or less tenaciously retained at the surface of the particles as the result of *adsorption* (p. 228). Consequently, the concentration of the foreign substances in the solution adhering to the precipitate decreases more slowly on successive washings than is indicated above, and a greater number of portions of wash liquid must be used to reduce the contaminating substances to a permissible small amount than would be expected from the mechanical theory of washing.[30] Some of the solution is, moreover, more or less isolated in the interior of the mass of the precipitate, being held by capillarity between the grains, and is only slowly removed by the percolating wash liquid. However, it is still true that the washing will be more efficient, the smaller the portions used.

[30] For an expression taking adsorbability into account, see W. Ostwald, *The Scientific Foundations of Analytical Chemistry* (transl. by G. M'Gowan), Macmillan, New York, 1908, p. 20.

Washing does not remove to any appreciable extent contaminants occluded in the interior of the particles of the precipitate.

To return to the technique of washing. Before beginning the washing on the paper, it is advisable to substitute a clean empty beaker for the one containing the filtrate, so that if the precipitate should run through at the start of the washing, it will not be necessary to refilter a large volume of solution. The stream from the wash bottle is now directed around the top of the filter paper—which part will be free from precipitate if the previous directions have been followed—to remove soluble substances. It is important not to neglect this part of the filter. After it has been carefully washed all the way around with three or four small portions of solution, the thin layer of precipitate on the upper part of the paper is attended to. This precipitate is washed down into the filter with a gentle stream of liquid from the wash bottle. The jet must not be allowed to strike the precipitate suddenly, for then, more likely than not, some of it will be thrown out of the filter. When the filter is about one half full of liquid, the stream is stopped, and the paper is allowed to drain completely.[31] The washing down of the precipitate is continued until it has been collected in the apex of the filter (Fig. 20-12). When this has been accomplished, the jet from the wash bottle is again directed at the upper portions of the paper, beginning very near the top edge, and continued downward. Each time the filter is filled approximately half full—if the precipitate is compact and small in amount, less liquid need be used, whereas if it is very voluminous (hydrous oxides), a larger volume is required; in any event, the collected precipitate should be covered with liquid each time.

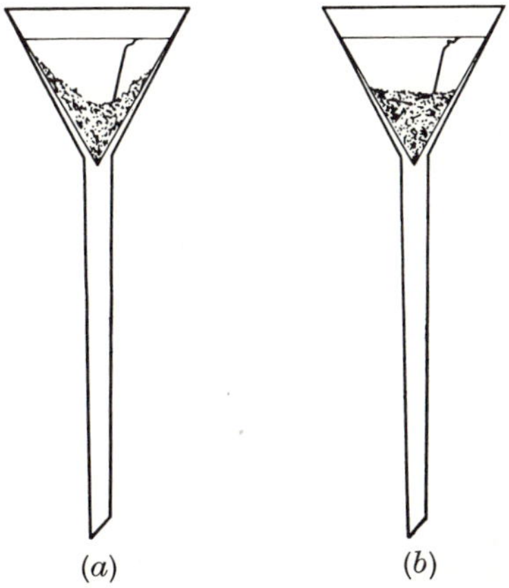

Fig. 20-12. Illustrating the washing of precipitates in paper: (*a*) before washing; (*b*) after washing.

After, say, 8 or 10 drainings, completeness of washing should be tested by adding to the last washings a reagent that will give a precipitate or coloration with a substance in the solution in which the precipitate was formed. For example, the presence of chloride in the washings, as revealed by the formation of a precipitate or opalescence when silver nitrate is added, may frequently be made the criterion of complete washing, since chloride ion is so often present in solutions. In testing for complete washing, the stem of the funnel is washed off with a few drops of water to remove any of the earlier solution possibly adhering on the outside tip (the stem should never be allowed to dip into the liquid in the receiving vessel), and 1 ml or so of the last washings is collected in a rinsed test tube. A drop of the appropriate reagent is added, and the solution is examined for turbidity (Fig. 20-13). If a precipitate is formed, the washing is repeated two or three times and the test again made; the process is repeated until the washings

[31] When some sulfides are washed, complete drainage is inadvisable, because of the tendency of the exposed precipitate to air-oxidize.

give no reaction, or at most only an exceedingly faint one. It is evident that the washings should not be tested too early in the process if the filtrate is to be analyzed for another constituent. The ignition of the precipitate is described on page 538.

Filtration of a solution containing a precipitate without altering the concentration of the dissolved substance is sometimes required. For example, it may be desired to take aliquot parts of an ore sample by diluting a solution containing undissolved gangue material to the mark in a volumetric flask and then measuring out definite amounts of the clear solution. The filtration is made through a dry filter paper placed in a dry funnel, and the filtrate is collected in dry vessel. The first portions of liquid running through the paper should be discarded, since they may contain less of the dissolved substance than the original solution because of adsorption of solute by the filter paper. Naturally care must be taken to prevent evaporation of solvent by keeping the vessels covered.

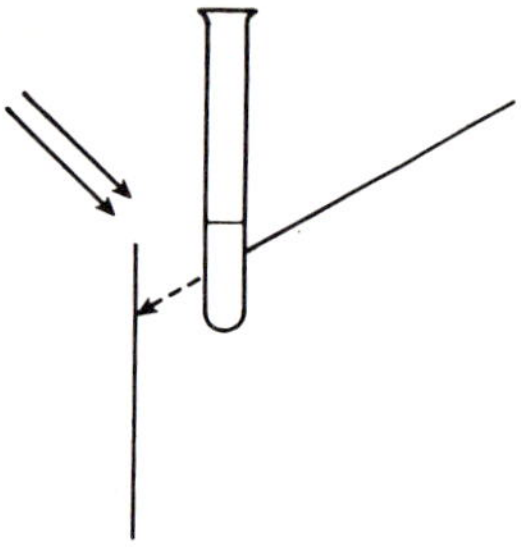

Fig. 20-13. Examination of a solution for turbidity against a dark background.

Filtration with filter crucibles.[32] Filter crucibles have the following advantages over paper in filtration:

1. Owing to the use of suction, filtration and washing require less time.
2. Washing is much easier and more efficient.
3. There is no danger of change in the composition of the precipitate by the action of reducing gases and carbon as with paper, or of decomposition or volatilization at the temperature required to burn away the carbon.
4. Drying or heating of the precipitate requires no special attention and is rapid.

Sometimes filter crucibles are not suitable:

1. Gelatinous precipitates cannot be filtered.
2. Precipitates that cannot easily be dissolved out of the crucible usually cannot be filtered advantageously.
3. The crucible may not resist the temperature required for the ignition (sintered-glass crucibles).
4. If the ignited residue requires treatment with a liquid reagent, a filter crucible may not be suitable.

Unless there is a special objection to its use, a filter crucible should be used in preference to filter paper.

On account of the labor involved in their preparation and the somewhat unstable character of their asbestos mats, Gooch crucibles are inferior to other kinds of filter crucibles. They are, however, cheap, can be heated to a high temperature, and are often satisfactory. The choice between sintered-glass and porous porcelain filter crucibles is determined mainly by the temperature to which the precipitate in question is to be heated. The former can be heated to a maximum temperature of 500°C, but only if the heating and cooling are very gradual. Porcelain filter crucibles can be heated to 1000°C with safety.

[32] See page 464 for a description of filter crucibles and preparation of the Gooch crucible.

Whatever the crucible used, it must first be washed with water or dilute acid to remove soluble substances and dried to constant weight at the same temperature to which the precipitate is later to be heated. It is then set in a Gooch funnel (p. 465), care being taken that the bottom of the crucible extends well into the latter so that there will be no chance of the filtrate coming in contact with the rubber. After gentle suction has been applied, the supernatant liquid in the precipitation vessel is carefully poured into the crucible with the aid of a stirring rod touching the side of the latter. The precipitate is transferred to the crucible in the manner already described. In washing, the stream from the wash bottle is first directed to the top of the crucible; the jet should not strike the precipitate in the bottom. The suction should not be so strong that the wash solution is immediately drawn through the crucible; it is essential that the wash liquid remain in contact with the precipitate for some moments. In testing for completeness of washing, a small test tube suspended within the suction flask by means of a cord may be used to collect small portions of the liquid. If the filtrate is to be used, care should be taken to see that no solution is carried off into the vacuum line; there is not much danger of this occurring if the stem of the funnel projects well into the flask and too strong suction is not used.

Filtration rate. The rate, expressed as the volume of liquid passing through a unit area of the filter medium in a unit time, is a function of the diameter and length of the (cylindrical) capillaries, their number per unit area, the viscosity of the liquid, and the differential pressure on the liquid. The volume (ml) of liquid of viscosity η flowing through a capillary of radius r (cm) and length l (cm) per second when the difference in pressure at the two ends is p (dynes/cm^2) is given approximately by Poiseuille's equation:

$$v = \frac{\pi r^4 p}{8\eta l} \tag{1}$$

η is expressed in dyne-sec/cm^2 (poises). For water, $\eta = 1.01$ at 20°C, 0.89 at 25°C, and 0.28 at 100°C.

If a sheet or plate of filter material has uniform capillaries of radius r at right angles to the sheet of thickness l, and these make up the fraction a of the surface (or the volume), the volume (ml) of liquid passing through 1 cm^2 of the sheet per second is approximately

$$v = \frac{ar^2 p}{8\eta l} \tag{2}$$

In any actual filter material, the capillaries will be interstices having varying effective radii and lengths greater than the thickness of the sheet or plate. The filtration rate will be largely determined by the predominant pore size if the size distribution of the pores shows a sharp peak. In any event, the filtration rate is strongly dependent on the pore radius. When a layer of precipitate has been deposited on a filter, the rate of passage of liquid may be determined by the particle size of the precipitate, not by the filter. Other factors remaining the same, the filtration rate might then be expected to vary approximately as the square of the particle size.

A sheet of filter material does not function wholly as a sieve. Fine particles of precipitate may be held more or less strongly on the walls of channels, and it may thus be possible to retain particles whose radii are smaller than the radii of the narrowest channels. Cellulose fibers swell when wetted and the pore size of filter paper then decreases.[33] It is often found that refiltration of a turbid filtrate gives a clear solution.

Adsorption by filter paper. Cellulose shows adsorptive properties. Partially, perhaps largely, these are attributable to ion exchange by cellulose, which contains fixed anionic (carboxyl) groups with cations as counterions. In neutral or weakly acidic solution,

[33] Filtration should therefore be made through an initially moistened paper. The operation of fitting the paper to the funnel (p. 529) assures this.

filter paper adsorbs lead, copper, and other cations (see Fig. 20-14). The adsorption is much decreased in fairly strongly acid solutions, as a consequence of replacement of metal cations by hydrogen ion. F. Nydahl mentions that barium is so strongly adsorbed by filter paper in filtration of barium sulfate in the determination of sulfate that it is not completely removed in the washing. The ash content of a 7-cm filter paper (Munktell No. 00) increased from 0.018 to 0.09 mg because of barium adsorption in a sulfate determination. In precise work, this type of error is eliminated by a blank run.

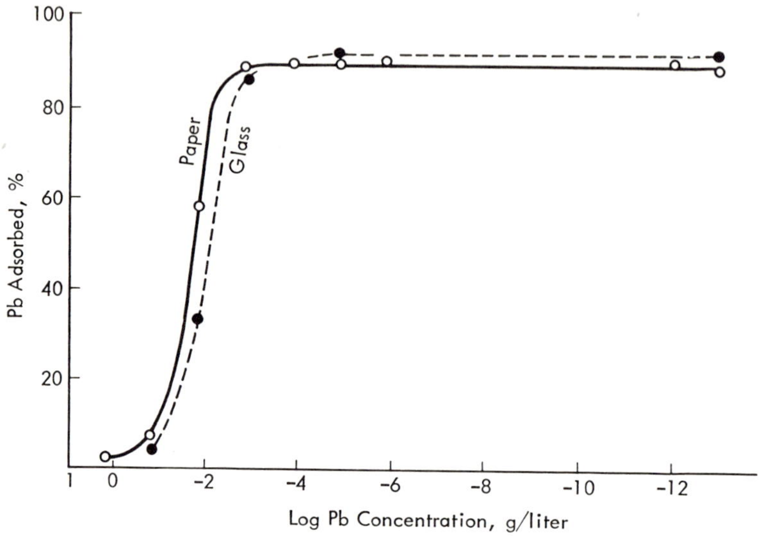

Fig. 20-14. Adsorption of Pb(II) by paper and glass at pH 6 according to T. Schönfeld and E. Broda, *Mikrochem. ver. Mikrochim. Acta* **36/37**, 537 (1951). Thirty ml of solution, 0.06 g of paper fibers (unsized, lignin-free) and 0.30 g of glass wool (250 cm^2 geometrical surface). The principal lead species in solution at this pH are Pb^{++} and $PbOH^+$.

Drying and ignition of precipitates

Ignition of precipitates in paper. The filter paper and its contents may first be dried, if desired, by covering the funnel with a circle of ordinary filter paper folded down around the rim, placing the whole in a drying oven, and heating at 100°C. This preliminary drying is by no means necessary, although advantageous for bulky gelatinous precipitates that hold much water. Usually the well-drained filter may be placed directly in the weighed crucible and dried therein before ignition; indeed, this is often to be preferred, because folding up a dry filter may lead to a slight loss of the contents.

The filter is carefully loosened from the funnel, lifted out, and the edges of the paper folded down and over the precipitate so as to enclose it completely. The packet is placed in a previously weighed crucible with the side having the three thicknesses of paper uppermost; if necessary, the whole is gently pressed down into the bottom of the crucible without breaking the paper. The crucible is set vertically in a triangle on a ring stand[34] and heated over a small Tirrill flame[35] (see Fig. 20-15). A flame about 2 cm high is suitable when the crucible is placed about 10 cm above the tip. The

[34] The triangle supporting the crucible should be placed on the ring of a ring stand and not on a tripod, in order that the crucible may be raised or lowered as necessary during the course of the drying and ignition.

[35] Alternatively, an infrared lamp can be used to dry and char the paper.

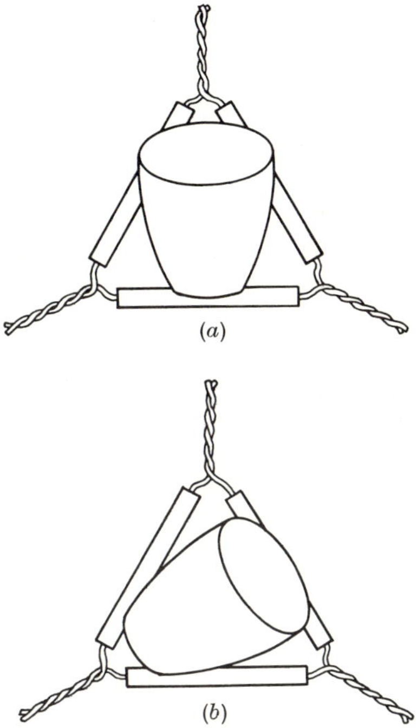

Fig. 20-15. Correct (*a*) and incorrect (*b*) position of inclined crucible on triangle.

crucible should be covered, but the cover is displaced a little to one side so that the steam can easily escape. During the heating the water must not boil or spatter. The heating must be slow, else the precipitate may be projected against the cover or even out of the crucible. When all the moisture has been expelled and there are no drops of water on the underside of the cover, the crucible is gradually lowered and the height of the flame increased if necessary, to slowly carbonize the paper. The crucible is kept covered during the carbonization, but the cover is lifted now and then to enable the progress of the charring to be followed. The paper must never burst into flame, nor should the escaping gases ignite. When the paper is completely charred, as revealed by its black color, and no more gases escape, the crucible is inclined on the triangle at an angle of about 45°, and the cover is placed on one side so that air can enter. It may even be advantageous to incline the crucible even more and lean the cover steeply against the mouth, as shown in Fig. 20-16. The crucible is heated to low redness at the bottom. The flame must not be allowed to approach the mouth of the crucible, for then currents of air set up might sweep out particles of light precipitate, and besides, the reducing gases finding admittance might reduce the precipitate and even injure the crucible if this is platinum. A flame larger than really necessary should be avoided.

After all the carbon has been burned away, the crucible is placed in the upright position and covered (any carbon on the cover may be burned away by holding the latter with the crucible tongs over the flame). The flame is adjusted to give the temperature desired, and the ignition is continued for 15 to 30 min in the usual case, or longer if required. The crucible must not be enveloped by the flame and must be kept out of

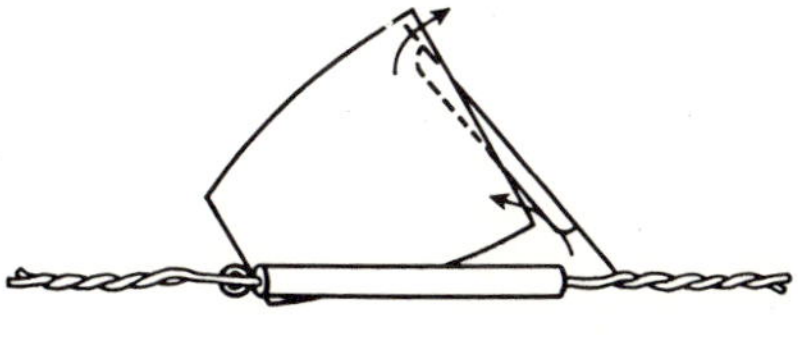

Fig. 20-16. Position of crucible (platinum) and cover in burning off carbon of filter paper.

the inner cone. If the temperature of ignition must be regulated over a narrow range to prevent decomposition or volatilization of the weighing form, an electric muffle furnace with a pyrometer is used in place of a burner. Such a source of heat is, moreover, to be preferred when a reducing atmosphere or contamination by gases of the flame (sulfur oxides) is to be avoided, or when porcelain crucibles must be heated to high temperatures.

If, in gas heating, a Tirrill burner will not supply a sufficiently high temperature, a Meker burner or a blast lamp is used. With the latter, the covered crucible should be set vertically in the triangle and the flame of the blast directed obliquely against the bottom and the lower part of the crucible; the flame must not strike the upper part of the crucible.

When the ignition is ended, the crucible is allowed to cool below redness in the air, and then while still hot it is placed covered[36] in the desiccator.[37] After the crucible has come to room temperature (p. 501), it is weighed, and ignited again for 10 to 15 min. The heating, cooling, and weighing are repeated until constant weight[38] is attained. In transferring the cooled crucible from the desiccator to the balance, the crucible tongs should always be used.

Platinum crucibles are to be preferred to those of porcelain, because the former heat and cool more quickly than the latter and can be raised to a higher temperature in the flame. In some ignitions platinum must be used, as for example in the ignition of silica when treatment with hydrofluoric acid is to follow. Owing to the high cost of platinum, porcelain crucibles are often substituted, especially in instructional laboratories. As already noted (p. 457), platinum crucibles must not be used in certain ignitions or fusions because of the danger of damage.

When platinum is heated at high temperatures—as with a Meker burner, a blast lamp, or in a muffle furnace—the slow loss in weight that it may undergo (p. 456) must be taken into account. The loss in weight may be allowed for by running a blank, i.e., heating the empty crucible at the same temperature and for the same period of time as in the ignition itself, and finding the change in weight. Or, if the platinum is not attacked in the slightest by the substance that has been ignited in it, the crucible may be cleaned after the ignition and reweighed; naturally, the solvent used to remove the adhering ignited residue must not affect the platinum at all, and mechanical cleaning (as with sand) is out of the question. In exact work the weight constancy of porcelain crucibles also should be checked by blanks.

Drying and ignition of precipitates in filter crucibles. It is sometimes possible to dry a precipitate that has been collected in a filter crucible by passing successive small portions of alcohol and then ether,[39] or acetone alone, through it after the washing with water or aqueous solution has been completed. The crucible is then allowed to stand in the air or in a desiccator for a short time to allow evaporation of the last traces of ether and water. This method of drying at room temperature gives good results with certain precipitates which are not finely divided and which do not absorb or occlude appreciable amounts of water. However, it is not generally applicable.

[36] After opening and closing, the humidity of the air in a desiccator may not be greatly different from that of the atmosphere. Since the rate of removal of water by the desiccant may be quite slow, the crucible should be well covered to minimize the uptake of moisture by the ignited residue during the cooling period. In work of the highest accuracy, ignited precipitates that are very hygroscopic should be cooled in a vacuum desiccator which is evacuated as soon as the hot crucible is inserted. When the crucible is cold and ready to be weighed, air is gradually let into the desiccator through a column of desiccant.

[37] When opening the desiccator after the crucible has cooled, the cover should be shoved aside slowly to avoid violent inrush of air.

[38] As noted elsewhere, constant weight is usually denoted by two successive weighings differing by not more than 0.2 mg.

[39] When the humidity is fairly high, there is danger of condensation of water vapor as a result of the rapid cooling.

When drying at low temperatures, the filter crucible is sucked as free of water as possible (often one or two rinsings with acetone are advantageous) and placed in a small beaker loosely covered with a watch glass to keep out dust. The beaker is placed in an electrically heated drying oven adjusted to the proper temperature. The first period of heating should be from 1 to $1\frac{1}{2}$ hr. Subsequent heatings to constant weight may be limited to half-hour periods.

For heating precipitates in filter crucibles at higher temperatures, an electric muffle furnace is preferably used. However, a flame may also be used. The covered filter crucible is then set in an ordinary porcelain crucible or capsule; filter crucibles must not be heated directly over the flame of a burner. If the precipitate is moist, heating must be gradual at first until the water is expelled.

PROBLEMS

1. Suggest means of bringing the following substances into solution, specifying the crucible to be used if a flux is employed: (a) AgBr; (b) Carborundum, CSi; (c) $Cu_2[Fe(CN)_6]$; (d) ferrotungsten; (e) As_2S_3; (f) cassiterite, SnO_2; (g) HgS; (h) BN; (i) Pb zirconate.
2. Compare the following two schemes for washing a precipitate BA whose K_{sp} is 1×10^{-8} and mol. wt. = 200. (a) The precipitate is washed with five portions of water, each 10 ml in volume. After each washing 1 ml of liquid remains with the precipitate and filter. Assume that after the precipitation the concentration of B^+ in the solution is 0.01 M. (b) The precipitate is washed with five portions of a 0.001 M solution of BCl (1 ml remains adhering to precipitate and filter each time). The washing is finished with three portions of water, each 2 ml. BCl is nonvolatile and remains unchanged on ignition of the precipitate; the molecular weight of BCl is 100. What is the error in (a) and (b)? Which washing scheme should be recommended?
3. Compare the filtration rates through the same medium with (a) a hydrostatic pressure of 15 cm of water (funnel stem filled with water); (b) a pressure difference of 0.5 atm (suction filtration), other factors remaining the same.
4. Calculate roughly the rate of flow of water at 20°C through a porous porcelain plate 1 mm thick having a capillary diameter of 20 μ and a = 0.2 [Eq. (2), p. 537] when the pressure differential is 0.75 atm.

CHAPTER 21 *Volumetric Apparatus and Its Use*

Common volumetric glassware

The commonly used articles of volumetric glassware[1] are flasks, pipets, and burets.

Flasks. A volumetric flask is a flat-bottomed pear-shaped vessel with a long narrow neck (Fig. 21-1). A thin line etched around the neck indicates (usually) a certain volume of liquid that is held at some definite temperature, and the flask is then said to be graduated to *contain*. A flask may also be marked to indicate a certain volume of liquid that will be delivered if the liquid is poured out under specified conditions, and then it is graduated to *deliver*. Flasks graduated to contain are used much more frequently than those graduated to deliver; the latter are not suitable for exact work because it is virtually impossible to deliver the same volume every time.

Since the graduation extends completely around the neck of the flask, it is easy, when making the liquid up to the mark, to avoid errors due to parallax[2] (Fig. 21-2) by

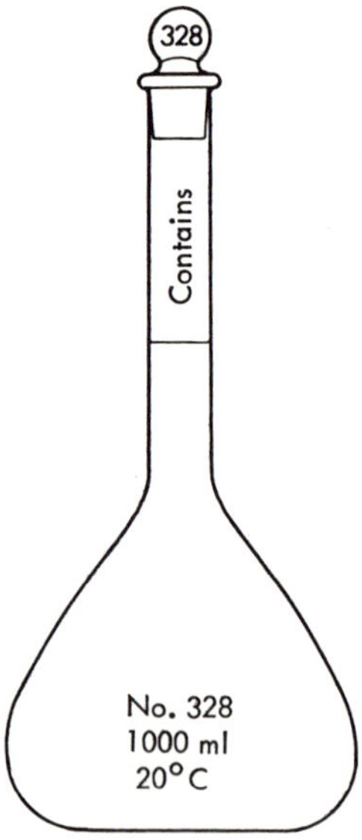

Fig. 21-1. Volumetric flask. Flasks of square cross-section are also available.

[1] For an extensive treatment see V. Stott, *Volumetric Glassware*, H. F. and G. Witherby, London, 1928.

[2] Parallax: The apparent displacement of an object as seen from two different points. It is evident that parallax will give rise to errors in reading the position of the meniscus of a liquid in a marked or graduated tube if the line of sight of the observer is not at right angles to the tube (Fig. 21-2).

aligning the eye and the mark so that the nearer and farther side of the ring are tangent to the lower edge of the meniscus. The neck of a volumetric flask is made comparatively narrow so that a small change in volume of liquid will cause a considerable difference in the height of the meniscus; the error made in adjusting the meniscus to the mark will consequently be very small. The distance from the mark to the stopper should be relatively great in order that there may be sufficient room for mixing after the volume has been made up to the mark. Volumetric flasks should be fitted with glass stoppers. The commonly used sizes are 50, 100, 250, 500, 1000, and 2000 ml.

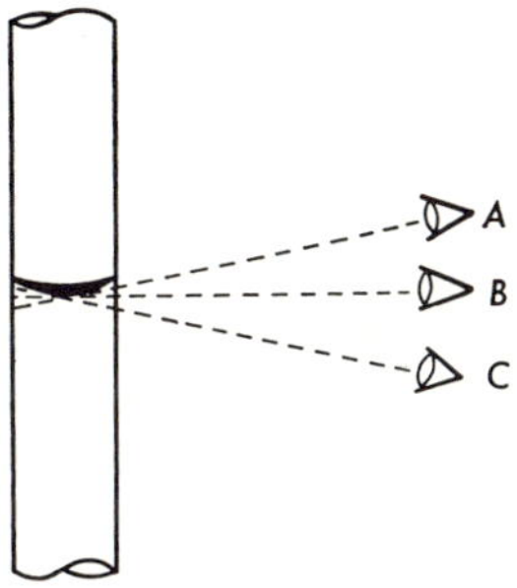

Fig. 21-2. Errors due to parallax in reading the level of a liquid in a graduated tube: A, meniscus reads too high; B, correct position of eye—no parallactic error; C, meniscus reads too low (volume too large).

Volumetric flasks are used for preparing solutions of known concentrations. The required amount of substance is weighed or measured out and transferred to a flask, water is added, and the volume is made up to the mark at a known temperature.

Pipets. Common pipets are cylinders drawn out at each end into narrow open tubes (Fig. 21-3). The upper limb of a pipet has a ring engraved upon it which fixes the volume of liquid that will be delivered if the pipet filled to this mark is allowed to deliver under specified conditions. The method of use is described later. A pipet used in this way to measure out one definite volume of liquid is known as a *transfer pipet.* Measuring pipets (Fig. 21-4) are narrow tubes subdivided into many divisions, which are employed for measuring out variable amounts of liquid. They do not find much application in exact work.

The orifice of a pipet must be of such size that the outflow of liquid does not take place too rapidly, for otherwise errors due to slight differences in drainage time will become too pronounced. Pipets of 10, 25, and 50 ml capacity are the most frequently used.[3]

Burets. Burets are long, graduated tubes of uniform bore provided with a stopcock (Fig. 21-5). They are used for delivering variable amounts of liquid and are therefore subdivided into many small divisions. Burets are used principally in titrations. The common size is 50 ml,[4] graduated to 0.1 ml. Glass stopcocks are attacked by alkaline solutions and tend to stick if such solutions are left in a buret. Teflon stopcock plugs are superior to glass in not being attacked by solutions, never "freezing," and requiring no lubrication. They are excellent for organic solvents.

Burets should be engraved with graduations extending halfway around the tube, so that parallactic errors can be easily avoided. To aid in reading the position of the meniscus, devices such as illustrated in Fig. 21-6 are used. These cause the bottom of the meniscus to be darkened and sharply outlined against the white background,

[3] Micropipets for accurate measurement of volumes less than 0.5 ml are available but will not be described here.

[4] It would be advantageous to have burets graduated slightly beyond 50 ml, say to 55 ml. Then it would be possible to take 50 ml of a standard solution with a pipet for titration without fear of having to refill the buret if the titrating solution were slightly weaker.

so that the level of the liquid can be read with great exactness.[5] Burets are always read to 0.01 ml by estimation, when graduated to 0.1 ml.

The glass stopcock of a buret must be lightly lubricated so that it turns easily (regarding the lubricant see p. 469). Too much grease must be avoided—it is likely to plug the stopcock or the tip. Burets are filled with the aid of a buret funnel, which should be removed after the filling.

For semimicro work, burets of 1 to 10 ml capacity graduated to 0.01 or 0.02 ml are employed. For volumes less than 1 ml, microburets based on the syringe principle are used. Volumes of 0.1 ml can be measured to 10^{-5} ml.

Fig. 21-3. Transfer pipet.

Fig. 21-4. Measuring pipet.

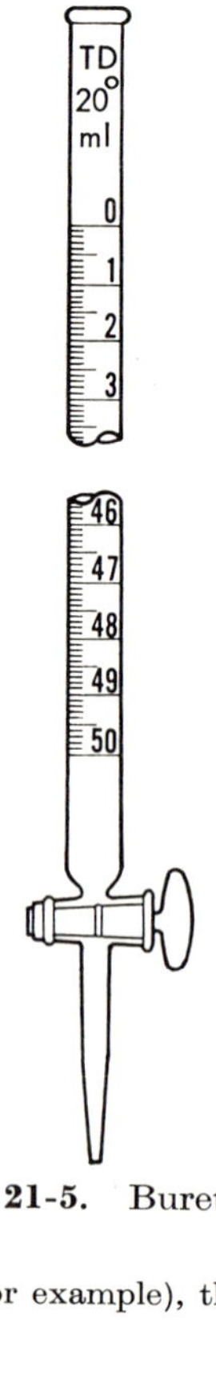

Fig. 21-5. Buret.

[5] When the liquid is strongly colored (potassium permanganate solution for example), the upper edge of the meniscus must be read.

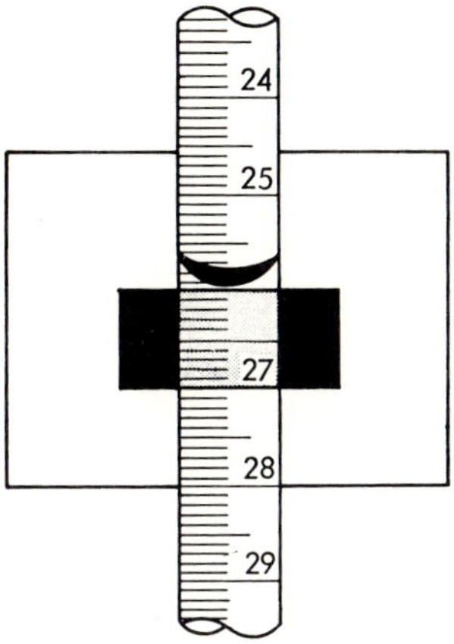

Fig. 21-6. Card for reading meniscus in a graduated tube (the upper edge of the dark portion should be placed ~ 1 mm below the meniscus).

Specifications for volumetric glassware

The following excerpts from circulars[6] of the National Bureau of Standards (U.S.) pertain to specifications and tolerances for volumetric ware. The outlet of any transfer pipet must be of such size that the free outflow shall last not more than 1 min and not less than the following times for the respective sizes:

Minimum Outflow Time for Pipets

Capacity, ml, up to and including	5	10	50	100	200
Outflow time, sec	15	20	30	40	50

The rate of outflow of burets and measuring pipets must be restricted by the size of the tip, and for any graduated interval the time of free outflow must not be more than 3 min or less than the following for the respective lengths:

Minimum Outflow Time for Burets

Length Graduated, cm	*Time of Outflow,* sec	*Length Graduated,* cm	*Time of Outflow,* sec
70	160	35	60
65	140	30	50
60	120	25	40
55	105	20	35
50	90	15	30
45	80		
40	70		

[6] *Natl. Bur. Std. Circ.* **434**, 1941; **602**, 1959. Volumetric glassware is available in two grades, often designated class A and class B. Class A is the precision grade, whose design and tolerances generally conform to the National Bureau of Standards requirements. See also National Physical Laboratory, Department of Scientific and Industrial Research, *Volumetric Glassware; Scientific Aspects of Design and Accuracy* (Notes on Applied Science 6) London, 1954.

Tolerances

(a) Flasks

Capacity, ml	*Limit of Error If to Contain*, ml
10	0.02
25	0.03
50	0.05
100	0.08
200	0.10
300	0.12
500	0.15
1000	0.30
2000	0.50

(b) Transfer Pipets

Capacity, ml *Less Than and Including*	*Limit of Error*, ml
2	0.006
5	0.01
10	0.02
30	0.03
50	0.05
100	0.08
200	0.10

(c) Burets and Measuring Pipets

Capacity (ml) *of Total Graduated Portion Less Than and Including*	*Limit of Error of Total or Partial Capacity*, ml	
	Burets	*Measuring Pipets*
2		0.01
5	0.01	0.02
10	0.02	0.03
30	0.03	0.05
50	0.05	0.08
100	0.10	0.15

Cleaning volumetric glassware

It is important that all volumetric apparatus be scrupulously clean and that the inside walls be entirely free of grease. Water will not uniformly wet a glass surface contaminated with any fatty substance; when water or an aqueous solution comes into contact with such a surface, droplets will adhere when the water drains away, or if a film is formed at first, it will soon break and leave patches of liquid. Moreover, the meniscus of a solution will be distorted if the neighboring walls are greasy. Upon a clean, grease-free glass surface water spreads, and leaves an invisible film when it is allowed to run off. No exact measurements of volume can be made if the walls of the apparatus are greasy even to the slightest extent. Therefore, before any volumetric work is attempted, the walls of the measuring vessel should be examined for greasiness, and the vessel cleaned if water does not uniformly wet the surface.

A number of cleaning agents can be used for removing grease films from glass surfaces. Common household detergents are sometimes satisfactory. A solution of sodium or potassium dichromate in concentrated sulfuric acid (p. 513) is effective in most cases. The article to be cleaned is filled with the cold cleaning solution and allowed to stand for a few hours, or overnight if the glass is very greasy. Sometimes a shorter period of contact is sufficient. Large flasks are cleaned by pouring in a small volume of the solution, inverting, and shaking at intervals so that the walls are kept continuously wetted. Pipets are kept filled with the cleaning mixture by attaching a piece of rubber tubing to the top of the suction tube and closing it with a pinch clamp. Great care must be taken when filling pipets not to get the corrosive liquid into the mouth. It is best to interpose a safety bottle between the pipet and the mouth, or to use the suction line. Burets are likewise allowed to stand filled with the cleaning mixture.

A solution of potassium hydroxide in alcohol is a very good degreasing agent; it is more rapid in its action than chromic–sulfuric acid and is used when the latter is ineffective. This alkaline liquid—and other strongly alkaline solutions as well—should not be allowed to remain in contact with glass for any great length of time because of the slow attack on the glass. An aqueous solution of sodium hydroxide containing potassium permanganate is also very effective. It may be left in the glass vessel for 5 to 10 min. Manganese dioxide will separate on the glass at the places where it is contaminated with grease or other organic substances; the coating of the dioxide is removed by rinsing with strong hydrochloric acid; the chlorine thus formed aids in removing the organic material. A dilute acid solution containing sulfite may also be used for removing the manganese dioxide.

Whatever cleaning solution is used, the vessel is finally rinsed out carefully, first with tap water and finally with distilled water, and allowed to drain. If the glass has been entirely freed of fatty substances, the walls will be uniformly wetted and a thin film will remain. It is usually not necessary, nor even advisable, to dry volumetric vessels before use.

Calibration of volumetric apparatus

Necessity for calibration. Ordinary volumetric glassware should always be tested for accuracy before use, because the errors of graduation may exceed the permissible errors of a determination. Indeed, no matter how accurate a piece of glassware is stated to be, the user should test it for himself and be assured beyond all doubt that it is sufficiently accurate for the work in hand.

Unit of volume. The unit of volume in the metric system is the liter. The *liter* is defined as the volume occupied by a *mass* of 1 kilogram (kg) of water at the temperature of maximum density (3.98°C) and under normal atmospheric pressure.[7] The one-thousandth part of the liter is called the *milliliter* (ml).

[7] *Travaux et mémoires du Bureau International des Poids et Mesures*, vol. XII, Paris (1902). Some confusion has resulted from a decision of the twelfth "Conférence Générale des Poids et Mesures" in 1964 to rescind the definition of the liter as 1.000028 dm^3. According to the new definition, *liter* is merely a special name for the cubic decimeter, but the resolution in which this action was taken points out that *liter* should not be used to express results of high precision. After 4 or 5 years there is little indication that analytical chemists have adopted the new definition. In this book, *liter* refers to the volume represented by one kilogram of water at its temperature of maximum density and under standard atmospheric pressure. Any qualmish analysts had best express their volume measurements in cubic decimeters or cubic centimeters by using the factor 1.00003 (more exactly 1.000028) to convert from liters or milliliters as defined in this book. Because the analyst will hardly ever, if ever, make volume measurements to 1 part in 36,000, this suggestion is equivalent to calling the old milliliter the cubic centimeter.

It was originally intended that the mass of the kilogram should be equal to that of one cubic decimeter of water at its temperature of maximum density. Because of experimental difficulties this simple relation was not quite realized, and it has been shown that 1 liter is equivalent to 1.000028 cubic decimeters or 1000.028 cubic centimeters[8] (cc). For most purposes the difference between a cubic centimeter and a milliliter is so small, amounting to only 1 part in 35,000 approximately, that no harm is done in calling a milliliter a cubic centimeter.

The composition of ordinary water is approximately:

	Mole per cent
$H_2{}^{16}O$	99.75
$H_2{}^{18}O$	0.20
$H_2{}^{17}O$	0.04
$HD^{16}O$	0.015

with minute amounts of $HD^{18}O$, $HD^{17}O$, $D_2{}^{16}O$, etc. The content of the heavier O and H isotopes in natural waters is variable over a moderate range (about 3 per cent for ^{18}O and 15 per cent for D) but the resultant variation in density is too small to be of significance in calibration of volumetric ware. Tap waters from different parts of the world have been found to show a range in density of less than 1 ppm.

General method of calibration. Volumetric glassware is generally calibrated, or tested for accuracy, by weighing the amount of pure water held or delivered at a given temperature and calculating the volume therefrom. Indirectly, the volume of a vessel can be determined by comparing its volume with that of a vessel that has been calibrated by the direct method. The first method is better.

Before entering upon the method of deriving the volume of a vessel from the weight of water held, it is important to realize that (1) the density of water varies with the temperature, (2) the volume of a glass vessel varies with the temperature, and (3) the water filling the vessel is weighed in air.

First, a temperature must be chosen at which the vessel will hold the designated volume. A flask holding exactly 1 liter at one temperature will no longer do so at any other temperature. Therefore a definite temperature must be decided upon as the *standard temperature* for volumetric glassware. The standard temperature chosen should be close to the average laboratory temperature at which the apparatus is used. In the United States the temperature of 20°C has been designated by the Bureau of Standards as the standard temperature for glassware. Therefore, in the United States (and most other countries),[9] the calibrated vessel will hold the nominal volume at 20°C; above 20°C it will hold more than the designated capacity, and below 20°C less.

Having decided upon the temperature at which the vessel shall hold the designated volume, it remains to derive the capacity at this temperature from the weight of water, determined in air with brass weights, held at any laboratory temperature. Usually the temperature of the laboratory will not be exactly 20°C, and the calibration must be carried out by weighing the water in the apparatus at some temperature other than 20°C. In order to calculate the volume of the vessel at the *temperature of the measurement*, the density of the water must be known at the temperature in question. By dividing the *mass* of the water by the density, the volume is directly obtained at the temperature of measurement. The mass of the water is obtained from its weight in air

[8] C. E. Guillaume, *La Création du Bureau International des Poids et Mesures et son œuvre*, Paris, 1927, p. 258.

[9] In some tropical countries, 27°C is taken as the standard temperature.

by reducing the weight to *vacuo*. For the latter calculation it is necessary to know, besides the density of the water, the density of the weights and the density of the air. The weights used in weighing large volumes of water are usually made of brass, and the density of brass may be taken equal to 8.4 (in very exact work the density must be determined). The density of the air can be calculated if the temperature, barometric pressure, and humidity are known. The following formula applies:

$$d_a = \frac{0.0012931p - 0.00049h}{760(1 + 0.00367t)}$$

where d_a = density of air, g/ml
p = atmospheric pressure, mm Hg (reduced to 0°C)
h = aqueous tension of the atmosphere, mm Hg[10]
t = temperature, °C

Usually it suffices to assume that the density of the air is 0.0012 g/ml and to calculate the air buoyancy on this assumption.

To make the calculation entirely clear, let us take a specific example. Suppose a liter flask is filled to the mark with water at 22°C (the laboratory temperature) and the weight of the water is found to be 997.00 g (weighing made by substitution). The mass, or vacuum weight, of the water weighing 997.00 g in air is

$$997.00 + 0.0012\left(\frac{997.0}{0.998} - \frac{997.0}{8.4}\right) = 997.00 + 1.06 = 998.06 \text{ g}$$

The density of water at 22°C is 0.99780 g/ml (Table 21-1). Therefore the volume of the water, and the volume of the flask at 22°C, is 998.06/0.99780 = 1000.26 ml.

TABLE **21-1. Calibration Table**

Temp., °C	*Vacuo Weight of* 1000 *ml of Water,* g	*Correction A,* g	*Correction B,* g	*Correction C,*[a] g	*Corrections A + B + C,* g	1000 − (*A + B + C*)	*Temp.,* °C
15	999.13	0.87	1.07	0.13	2.07	997.93	15
16	998.97	1.03	1.07	0.10	2.20	997.80	16
17	998.80	1.20	1.07	0.08	2.35	997.65	17
18	998.62	1.38	1.06	0.05	2.49	997.51	18
19	998.43	1.57	1.06	0.03	2.66	997.34	19
20	998.23	1.77	1.05	0.00	2.82	997.18	20
21	998.02	1.98	1.05	−0.03	3.00	997.00	21
22	997.80	2.20	1.05	−0.05	3.20	996.80	22
23	997.56	2.44	1.04	−0.08	3.40	996.60	23
24	997.32	2.68	1.04	−0.10	3.62	996.38	24
25	997.07	2.93	1.03	−0.13	3.83	996.17	25
26	996.81	3.19	1.03	−0.15	4.07	995.93	26
27	996.54	3.46	1.03	−0.18	4.31	995.69	27
28	996.26	3.74	1.02	−0.20	4.56	995.44	28
29	995.97	4.03	1.02	−0.23	4.82	995.18	29
30	995.67	4.33	1.01	−0.25	5.09	994.91	30
31	995.37	4.63	1.01	−0.28	5.36	994.64	31
32	995.05	4.95	1.01	−0.30	5.66	994.34	32
33	994.73	5.27	1.00	−0.33	5.94	994.06	33
34	994.40	5.60	1.00	−0.35	6.25	993.75	34
35	994.06	5.94	0.99	−0.38	6.55	993.45	35

[a] For soda-lime glass; for borosilicate glass the correction is 0.4 as large.

[10] h = (relative humidity) × (aqueous tension of air saturated with water vapor at temperature t).

After the volume has been found for any particular temperature, it remains to calculate the volume of the vessel at the standard temperature of 20°C. The coefficient of expansion of the material of which the vessel is made must be known. The coefficient of cubical expansion of soda-lime glass is 2.5×10^{-5} per °C., of borosilicate glass, 1×10^{-5}. The volume of a glass vessel at 20° is then readily obtained from its volume at any other temperature by using the following expression:

$$V_{20} = V_t + 2.5 \times 10^{-5} V_t(20 - t)$$

where V_{20} = volume of vessel at 20°C
V_t = volume of vessel at t°C

The flask of the example above would have a volume of 1000.26 − (0.000025 × 2 × 1000.3) = 1000.26 − 0.05 = 1000.21 ml at 20°C, if made of soda-lime glass.

To avoid the labor involved in making the calculations illustrated above, it is convenient to employ tables which give the weight of water that must be weighed out at any temperature in air to give the desired volume of the glass vessel at 20°C. Table 21-1 is an example of such tables.

Explanation of Table 21-1. The first column gives the laboratory temperature, i.e., the temperature of the water weighed and the glass vessels being calibrated. The column headed $1000 - (A + B + C)$ indicates the weight of water that must be weighed out with brass weights[11] at the temperature of calibration and at normal atmospheric pressure (760 mm Hg), to furnish a volume of water that will indicate a capacity of 1 liter if the vessel is brought to the standard temperature of 20°C; i.e., if this weight of water is contained in a flask and a mark is made at the meniscus, the flask so marked will hold 1 liter at the temperature of 20°C. The other columns of the table contain the corrections that must be applied to obtain the values in the last column. This table can be used for testing or calibrating vessels of any volume by taking proportional parts or multiples of the values for 1 liter.

How the table was constructed. To obtain the values in the next to the last column of Table 21-1, it is necessary to take into account the density of the water at the temperature of calibration, the effect of the buoyancy of the air on the water and the weights, and the difference in volume of the glass vessel at the calibration temperature and at 20°C. The corrections for these effects for various laboratory tempratures are given in the columns headed "Correction A," "Correction B," and "Correction C," respectively.

At any temperature above 4°C it is necessary to weigh out less than 1000 g of water to obtain a volume that will represent 1000 ml. The amount to be weighed out *in vacuo* at any temperature is evidently given by $1000d$, where d is the density of the water at the working temperature. Values for $1000d$ are recorded in the second column of the table. Values in the column "Correction A" indicate the weight in grams that must be subtracted from 1000 g to give the weight of water in vacuo that would occupy a volume of 1000 ml at the temperature indicated.

Since it is desired to make the weighings in air, we must determine from the vacuum weight the weight of water that must be weighed out *in air with brass weights* to give an amount of water that will have a volume of 1000 ml at the calibration temperature. $(1000 - A)$ g of water weighed in air will occupy a volume greater than 1000 ml, because the density of water is less than that of the weights, and therefore the water displaces more air than the equal weight of brass, and accordingly the apparent loss in weight of the water is greater than that of the weights. It is necessary to weigh out in air $(1000 - A - B)$ g of water to furnish 1 liter, where B is the correction for the differential buoyancy of the air. The value of B is easily calculated. It is equal to the weight of air displaced by 1 liter of water less the weight of the air displaced by

[11] The face value of a weight, corrections applied if necessary, must be its true mass, i.e., vacuum weight.

the weight of brass required to counterbalance the mass of water which represents a capacity of the glass vessel equal to 1 liter if the vessel is at the temperature of 20°C:

$$B = \left[1000 - \frac{1000 - (A + B + C)}{8.4}\right] d_a$$

In this expression, d_a is the density (g/ml) of the air at the temperature and the pressure of the weighing. C is the correction factor for the expansion of the glass; it is explained below. Solving this expression for B, it is found that

$$B = (7400 + A + C)\frac{d_a}{8.4 - d_a}$$

In exact work the value of d_a is calculated from the temperature, the pressure, and the humidity of the air in the balance case at the time of the weighing by applying the formula given on page 549. The value 8.4 for the density of the brass weights is an average value.

The values of B in Table 21-1 have been calculated on the assumption that the atmospheric pressure is 760 mm Hg, that the temperature of the air is the temperature of the water used in the calibration, and that the relative humidity of the air is 50 per cent. Generally it is not necessary to take into account variations in the value of B caused by departures of pressure and humidity from these assumed values. If very accurate results are required, the exact value of B can be found by substituting in the equation the value for the density of the air at the time of the weighing. More approximately the corrected value of B can be found as follows:

Let B_t = the buoyancy correction when p = 760 and the water and the air have the same temperature t (this is the value given in the table)
p = atmospheric pressure in millimeters of mercury
t_a = temperature of *air* when weighing is made
B'_t = buoyancy correction for atmospheric pressure p and temperature t_a.

Then

$$B'_t = B_t \times \frac{p}{760} + \frac{1 + 0.00367t}{1 + 0.00367t_a}$$

If the corrected value of B differs by less than 0.005 g from the value of B in the table, it is useless to apply the correction, because the values for A, B, and C in Table 21-1 have been computed only to the nearest 0.01 g., i.e., to 1 part in 100,000, which is 0.001 per cent.

If $1000 - (A + B)$ g of water are weighed out in air with brass weights, the volume of the water will be exactly 1000 ml and this will be the volume of the glass vessel at the temperature of the calibration. But the volume of the vessel at 20°C is desired. If the calibration temperature is above 20°C, the indicated volume will be too small and if below 20°C, it will be too large. To take account of the expansion (or contraction) of the glass vessel when it is brought to 20°C from the calibration temperature, it is necessary to introduce the correction C. The value of C evidently depends upon the coefficient of cubical expansion of the glass (2.5×10^{-5} for soda-lime, 1×10^{-5} for borosilicate, per °C). Therefore, a soda-lime glass vessel that holds exactly 1000 ml at t°C will hold $1000 + (20 - t)\ 1000 \times 2.5 \times 10^{-5}$ ml at 20°C, and it follows that $1000 - [A + B + (20 - t)\ 1000 \times 2.5 \times 10^{-5}]$ g of water must be taken at the temperature t to indicate a capacity of 1000 ml at 20°C. The value of $(20 - t)\ 1000 \times 2.5 \times 10^{-5}$ is the correction C, the values of which are given in the table for various temperatures. It is to be noted that for calibration temperatures above 20°C the value of C is negative.

The sixth column in the table contains the sum of the corrections A, B, and C, and *the seventh column gives the values of* $1000 - (A + B + C)$, *furnishing directly the weight of water which must be taken at the given temperature to indicate, in a glass vessel, a volume of 1000 ml at 20°C.*

Illustrations of the use of the table

1. a. What weight of water must be taken at 30°C to calibrate a liter flask? *Ans.* Reading opposite 30°C, the value 994.91 g is found in the next to the last column; this is the weight of water required.

(b) What amount of water must be weighed out at 20°C to calibrate a 250-ml flask? *Ans.* Value for 1000 ml at 20°C = 997.18 g. Therefore, for the volume 250 ml, (250/1000) × 997.18 = 249.29(5) g of water is required.

2. A 1-liter flask was filled exactly to the mark at 23°C and the weight of the water in the flask determined. It was found to be 996.70 g. What is the capacity of the flask at the standard temperature? *Ans.* Reading opposite 23°C, the total correction value, $A + B + C$, is found to be 3.40. The volume of the flask at 20°C is therefore 996.70 + 3.40 = 1000.10 ml.

3. A glass vessel holds 35.681 g of water at 25°C. What is the capacity of the vessel at 20°C? *Ans.* 35.681/0.99617 = 35.817 ml. The value 0.99617, the apparent density of water including the glass correction at 25°C, is found by dividing 996.17 in the next to last column by 1000.

4. A 1-liter flask is found to hold 996.39 g of water at 24°C. What is the capacity of the flask at 24°C? *Ans.* The correction C must not be included. Therefore the volume of the flask at 24°C is 996.39 + $(A + B)$ = 996.39 + 2.68 + 1.04 = 1000.11 ml.

5. A 500-ml Pyrex flask filled to the mark contains 497.46 g of water at 30°C. What is the capacity of the flask at the standard temperature? The coefficient of cubical expansion of Pyrex is 0.000010. *Ans.* $C = 500 \times (-10) \times 0.00001 = -0.050$. Therefore, the capacity of the flask at 20°C = 497.46 + (4.33/2) + (1.01/2) − 0.05 = 500.08 ml.

Directions for calibration

The apparatus to be calibrated must be scrupulously clean and free of grease. The water and the glassware should be allowed to stand for some time in the room in which the calibration is to be made before the operations are begun in order that they may assume the temperature of the air. The room temperature should be as constant as possible to preclude the possibility of volume changes during the calibration. Ordinary distilled water is used; it need not be air-free. A thermometer graduated in single degrees is suitable for obtaining the temperature of the water. Temperatures should be read to the nearest 0.5°C. In very accurate work the correctness of the thermometer should be checked; an error of 1°C will lead to an error of approximately 0.02 per cent in the volume of the vessel. Care should be taken not to warm the apparatus by excessive handling.

Flasks. For the calibration of flasks having a capacity greater than 100 or 200 ml, a sturdy balance that will take a load of 2 kg and which has a sensitivity of at least 0.01 g at the maximum load should be available. The flask to be calibrated must first be dried. It is not permissible to hasten the drying by heating, for then slow changes in volume may occur after cooling. A stream of air may be passed into the inverted flask to hasten the evaporation of adhering water. The air is filtered through cotton wool to remove dust and oil droplets that may be present in the compressed air (originating in the compressors). Flasks are quickly dried in this way if they are first rinsed once or twice with alcohol or, better, with acetone, to remove the water.

The water for the calibration of a flask must be weighed by substitution, because the rougher balances may have arms that are quite unequal in length. The procedure to be followed depends upon whether the flask is to be calibrated or simply tested for accurcy.

In the first case the flask is placed on the right-hand pan of the balance together with weights numerically equal to the nominal volume of the flask; e.g., for a liter flask 1000.0 g would be taken. The flask and the weights are now tared by adding to the left-hand pan other weights, lead shot, or any other finely divided dense material. The nominal weight on the right pan is next removed, and a weight equal to $(A + B + C)$ (or its proportional part or multiple, as required) is put in its place. Water is then poured into the flask until the original equilibrium is restored. In this operation care should be taken not to wet the neck of the flask. It is best to use a long-stemmed funnel. If drops adhere to the neck above the meniscus, they are removed with a roll of filter paper; such adhering drops will not form if the flask and the water are grease-free. The final adjustment of weight is conveniently made by adding or removing water with a long narrow glass tube or a pipet. When the adjustment has been made, the flask will contain a volume of water that indicates the capacity of the flask when the latter has a temperature of 20°C. The flask is now removed from the balance, and a narrow stip of gummed label is pasted all around the neck to indicate the position of the meniscus. The upper edge of the label should be tangent to the lowest part of the meniscus. The strip of paper should be coated with a thin layer of paraffin, or with varnish, to protect it from water. If it is desired to mark the flask permanently, the neck is covered for some distance on either side of the strip with a thin coating of paraffin, and a ring is made all around the neck by scratching the paraffin with a sharp-pointed instrument in such a way that the glass is laid bare. By exposing the unprotected part of the glass to hydrofluoric acid vapors, a mark will be etched into the glass. The student should use the first method of marking.

If it is desired to *test* a flask, i.e., to determine the capacity indicated by the mark already on the vessel, the same procedure as described above is followed in principle. The dry flask is tared with the nominal weight, the latter is replaced by $A + B + C$ or its proportional part, and the flask is then filled exactly to the mark with water. If the original equilibrium position of the balance is now restored, it is clear that the flask is exactly correct. If equilibrium is not established, weights are added or removed until the flask is exactly balanced. The weights added or removed correspond to the difference in volume from the nominal volume, and the error of the flask is calculated therefrom.

It is immaterial whether a flask is calibrated or tested. The latter method possesses the advantage that a new mark need not be made; but then, on the other hand, a correction must be applied to every determination in which it is used, if the flask is found to be in error. If the student merely checks the accuracy of a flask, he must record the capacity in his notebook immediately, indicating also the number of the flask tested, so that there may be no possibility of confusing it with another. It is advisable to affix to the flask a small paraffined label bearing the true capacity.

Pipets. The pipet to be *tested* is filled above the mark with distilled water by applying gentle suction with the mouth and closing the upper end of the tube with the dry index finger, the tip is wiped off to remove any adhering droplets, and the water is then allowed to flow out slowly, while the end of the pipet is held against the wall of a beaker until the meniscus is brought into exact coincidence with the mark. The tip of the pipet is then brought into contact with the wall of the vessel in which the water is to be weighed (this vessel must be weighed beforehand to 0.005 g). Small (50- or 100-ml) stoppered Erlenmeyer flasks or large glass-stoppered weighing bottles are best used as containers for weighing the water. The pipet is allowed to deliver freely while held vertically with the tip in contact with the inclined wall. The pipet is kept in contact with the wall for 15 sec after free outflow has ceased and is then removed; the water remaining in the tip must not be blown out or otherwise disturbed.[12] From the weight of the water delivered, the capacity of the pipet is calculated, most conveniently by making use of the data in Table 21-1. If it is not desired to make a new mark, the capacity of the pipet is recorded in the notebook, and a label affixed to indicate the capacity when filled to the mark and allowed to deliver under the conditions described. In all operations involving the use of the calibrated pipet, the corrected volume is used when the calculations are made. At least two checks of the accuracy should be made. If the results of the duplicates differ by more than 0.01 g, the test

[12] Bureau of Standards directions for delivery call for keeping the tip in contact with the wall for a count of 2 after free outflow has ceased. This method is followed by manufacturers in calibration.

should be repeated until this degree of concordance is obtained. Varying results point to greasiness of the glass or carelessness in the work.

If it seems desirable to have the pipet deliver exactly the nominal volume, it will be necessary to make a new mark. The internal diameter of the upper stem of the pipet can be measured and the volume corresponding to 1 mm calculated. Since the error of the pipet has been determined, it is a simple matter to find the position at which the new mark should be placed. A thin strip of gummed paper is used to indicate the new mark, and the pipet is tested as before. If the correct volume is not delivered, it will be necessary to change the position of the mark until the pipet has the designated capacity. Then the new mark can be made permanent by etching a ring into the glass with hydrofluoric acid vapors. For the student the testing of the pipet will suffice.

When pipets are emptied at different temperatures, slightly different amounts of water (or aqueous solution) are left in the pipet, because of the variation of viscosity and surface tension of liquids with temperature. The differences in the amount of water left in the pipet at different laboratory temperatures will, however, be very small and need not be considered if the same conditions of delivery are always fulfilled. When the pipet is used, it must be emptied in exactly the same way as in the calibration or testing.

Burets. A buret is tested at 5- or 10-ml intervals. It is fixed vertically in a suitable stand and filled slightly above the upper mark. The water should be poured in from a beaker, not a wash bottle. The level of the water is brought to the zero mark by opening the stopcock. The tip of the buret should be examined for air bubbles. If present, they must be removed by adding more water and opening the stopcock to its full extent, so that the air will be swept out. It is not necessary to bring the meniscus into exact coincidence with the first graduation; it may be allowed to fall between the 0.00- and the 0.10-ml line, and its position estimated to 0.01 ml. However, it is generally possible to control the stopcock so that the water will flow out very slowly, and then the lower part of the meniscus can be brought into exact coincidence with the first graduation. The accuracy of setting the meniscus to 0.00 is greater than that of estimating the position of the meniscus when it falls between two divisions. Parallactic errors are avoided by placing the eye so that the front and back of the graduation (next to the meniscus) extending halfway around the tube appear to coincide. The bottom of the meniscus should be made easily visible by the use of the device described on page 545. Before the height of the water is read, it is necessary to be sure that drainage from the upper part of the buret will not alter its level. Since the zero mark is situated a comparatively short distance from the top of the buret, it is sufficient to wait for 1 min after filling before reading, to avoid drainage errors.[13] After the reading has been recorded and the tip wiped off to remove adhering water, the stopcock is slowly turned so that the liquid flows out dropwise into a weighed stoppered 100-ml Erlenmeyer flask (which naturally need not be dry inside). A period of 30 sec should be required to deliver 5 ml. The level of the water is brought as closely as possible to the 5-ml mark, although it is not necessary to attain exact coincidence. The tip of the buret is touched to the inside wall of the flask to remove the adhering droplet of water, and the stoppered flask is then weighed to 0.005 g on the analytical balance. The level of the water is read $\frac{1}{2}$ min after the end of delivery. Readings are always made to 0.01 ml by estimation. From the weight of water delivered, the volume is calculated, and the result compared with the observed volume as obtained by subtracting the first reading from the last. The difference between the observed and calculated values gives the error of the buret at the 5-ml graduation.

The buret is then filled to the zero mark again and the water is allowed to flow out dropwise as before to the 10-ml mark at a rate not exceeding 10 ml/min. The water is received in the same flask, the last weight of which is taken as the initial weight for

[13] It is well to check the buret for leakage by seeing whether the reading remains constant for 15 min or so. Slight leakage can sometimes be overcome by regreasing the stopcock.

obtaining the weight of the next portion of water. The meniscus should be read after $\frac{1}{2}$ min. Continuing in this manner, one tests the buret at 5- or 10-ml intervals to 50 ml. Each point should be checked twice. Duplicate tests should agree to 0.01 ml, less preferably to 0.02 ml. The data are conveniently recorded as shown in Table 21-2. The corrections hold for a volume of liquid delivered from the zero mark of the buret.

TABLE **21-2. Buret Calibration**

(Temp. 28.5°C)

Interval, ml	*Weight of Water*, g		*True Volume*, ml	*Correction*, ml
0.00–10.00	63.89 53.92 9.97	73.86 63.89 9.97	10.02	+0.02
0.00–20.00	93.77 73.86 19.91	64.03 44.12 19.91	20.00	0.00
0.00–30.00	93.90 64.03 29.87	74.05 44.17 29.88	30.02	+0.02
0.00–40.00	83.96 44.12 39.84	84.07 44.22 39.85	40.03	+0.03
0.00–50.00	93.91 44.13 49.78	93.90 44.12 49.78	50.02	+0.02

Practical details regarding the use of volumetric apparatus

Flasks. In making up a solution of definite concentration by weighing out a pure solid and making up to volume, the substance is usually first transferred to a beaker after weighing and dissolved therein. The solution is transferred quantitatively to the flask, and water is added until the level of the solution has risen nearly to the junction of the flask and the neck, but not into the neck. The flask is now swirled so that the liquid will be mixed fairly well. If this precaution is not taken and water is poured in upon the concentrated solution until the level is nearly at the mark and the final adjustment then made without mixing, it sometimes happens that a contraction or expansion takes place when the concentrated solution is mixed with the water in the final shaking, and the meniscus will no longer fall at the mark. A partial mixing of the solution before making up to the mark will prevent this occurrence. The final adjustment to the mark is made by adding water dropwise from a pipet. Time must be allowed for water to drain down the sides of the neck if the latter has been wetted; and, of course, drops must not adhere to the inside of the neck above the liquid. If it is intended that the solution shall be made up to the mark at room temperature, it will be necessary to let the solution stand for some time before finally adjusting the meniscus, to equalize the temperature of the solution, which may not be at room temperature because of absorption or evolution of heat resulting from the solution of the substance. It is usually best, however, to make up to the mark without waiting, to insert the dry stopper carefully, and to mix the solution thoroughly by inverting the flask many times with a simultaneous rotary motion. When the solution is homogeneous, the temperature is taken with a clean dry thermometer read to the nearest 0.5°C. Solutions should not, as a rule, be stored in volumetric flasks. After the solution has been mixed, it is poured into a dry glass-stoppered borosilicate bottle.[14] If the stock bottle is not dry, it is rinsed with about three small portions of the solution and drained well after each rinsing. Even if the bottle appears dry, it is best to rinse it once with the solution. The stock bottle is labeled with the name of the reagent, its concentration, the date of preparation, the temperature at which the solution was made up to volume, and the initials of the person who prepared the solution, together with any other pertinent data. The label below is an illustration.

Potassium Dichromate
0.1000 *N* (0.01667 *M*)
(2.452 g of recrystallized $K_2Cr_2O_7$ dried at 180°C in 500 ml at 25°C)
12–16–33 J. J. L.

Before use the standard solution should be well shaken to mix any condensed water at the top of the bottle with the bulk of the liquid.

Pipets. Unless it is known to be clean, a pipet before use should be rinsed with distilled water, which is allowed to drain out as much as possible. Before the pipet is dipped into the solution, the drop of water remaining in the tip is blown out, and the tip is wiped with a piece of filter paper in order that the solution to be pipetted will not be diluted. The pipet is rinsed with two or three small portions of the solution, each portion being brought into contact with the whole inside surface of the pipet before being allowed to run out. The stem of the pipet should not be immersed unnecessarily far into the solution; however, care should be taken that the tip of the pipet does not become uncovered, for then the liquid may be drawn up into the mouth with unpleasant results if it is caustic or poisonous.[15] Strong suction should not be applied. The temperature at which an aliquot portion is removed should be noted. This may be assumed to be the room temperature if the solution has stood for some time in the room; otherwise a thermometer should be used.

The pipet must be used in the same way that it was calibrated; i.e., the manner of delivery must be the same, and the same period of drainage must be allowed. When liquids other than water are pipetted, the volumes of liquid delivered may be considerably different from the volume found in calibration, owing to differences in viscosity, density, and surface tension. Thus a pipet delivering 25.00 ml of water will not deliver 25.00 ml of, say, alcoholic potassium hydroxide solution. With dilute ($< 1\ M$) aqueous solutions of salts, acids, and bases, the differences are so small that they may be disregarded. The same considerations apply to the measurements with burets.

When a precision of 1 part in 1000 is required, a pipet smaller than 25 ml should not be employed, because an error of 0.2 per cent or more can then result.

After use, a pipet should be rinsed with distilled water and set in a rack to drain until it is used again.

Burets

Titrations. Before use, a buret is rinsed two or three times with about 5-ml portions of solution, which is allowed to drain out well each time.[16] The buret is then filled

[14] Alkaline solutions should be kept in polyethylene bottles with air-tight screw caps. These bottles are also suitable for other solutions, which must be kept free of traces of silica, boron, and other constituents derived from glass.

[15] The use of a rubber bulb for filling a pipet with hazardous solutions is recommended. Alternatively, the suction line can be used, with a safety trap interposed.

[16] In permanent installations burets can be connected directly to the stock bottles (see Fig. 39-1).

slightly above the mark, the stopcock opened to fill the tip and expel all air, and the level of the liquid brought as closely as possible to the first graduation. Note is made of the temperature. The solution should not be run out too rapidly from the buret during the titration. Apart from the danger of overstepping the end point, the drainage error may then be considerable. The solution should be allowed to flow out rapidly *dropwise* as in the calibration, generally at a rate not exceeding 10 ml/min. In making a titration, the stopcock is regulated with the left hand and the stirring rod or flask is held in the right.

The vessels used in titrations may be beakers or Erlenmeyer flasks. If a beaker is used, a stirring rod must always be employed to mix the liquid. The solution from the buret must not be allowed to accumulate locally in the solution titrated; each drop of reagent should be dispersed by continuous gentle stirring or swirling. The tip of the buret should not be placed so far above the beaker that splashing results when the solution strikes the liquid below.

Erlenmeyer flasks can be recommended as titration vessels. No stirring rod is necessary. Efficient mixing is achieved by swirling the flask. There is very little danger of loss from splashing, and neither will the evolution of a gas lead to loss. Flasks are very convenient when liquids must be boiled, and their use is demanded when the solution after boiling must be cooled with protection against the air. For example, if a solution is to be made carbon dioxide-free, and kept so, it is boiled in a flask with excess acid and then cooled after inserting a stopper carrying a soda-lime tube. If necessary, carbon dioxide can be excluded from the flask during the titration by passing a stream of carbon dioxide-free air into it (Fig. 21-7), and in a similar manner titrations can be made in the absence of oxygen by passing in an inert gas (N_2 or CO_2).

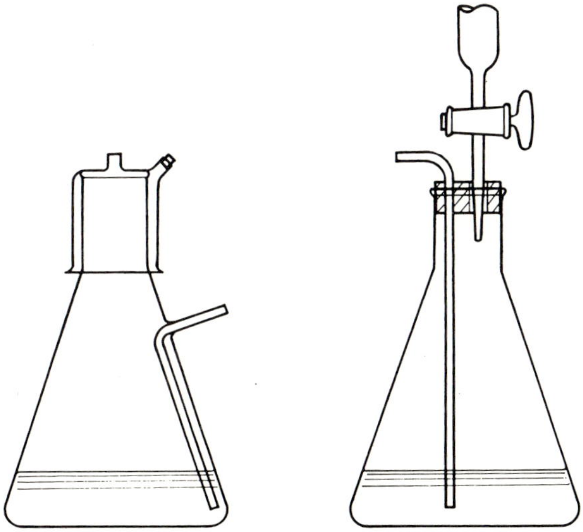

Fig. 21-7. Apparatus for titration in absence of carbon dioxide or oxygen.

It is a wise plan in any titration, especially if the amount of standard solution required is not known even approximately, to reserve a small portion of the solution to be titrated and to titrate without special care to the end point, perhaps running over the latter slightly. The reserved portion can then be added to the main portion and the titration continued carefully to the end point. In this way time can often be saved. The portion reserved should amount to only a few per cent of the whole. Thus, if a sample is dissolved in 50 ml of solution, about 1 ml may be removed with a piece of glass tubing and reserved in a small graduate or similar lipped vessel. When the end point has been attained or slightly overstepped, the reserved portion is transferred quantitatively to the titration vessel by rinsing out the graduate and the glass tubing. A special titrating flask (Fig. 21-8) may also be used.

Fig. 21-8. Titration flask. A portion of the solution to be titrated is reserved in the side tube.

Near the end point it is advisable to "split" the drops. The stopcock is slowly turned so that only a fraction of a drop flows out and remains adhering to the tip of the buret. This small volume of liquid may be removed with the stirring rod if a beaker is used, or by touching the tip to the walls if a flask is used. The flask must be tipped and the liquid carefully swirled so that the droplet will be mixed with the solution. It is best to wash down the sides of the flask with distilled water near the end point, paying special attention to the upper parts, upon which there may be (but should not be) adhering drops of reagent.

If it is uncertain that the end point has been reached, the buret should be read, and then another drop added to see whether any distinct change occurs in the solution.

Titrations should be carried out in good light. Direct sunlight is too bright and may cause undesirable chemical changes in the solutions as well (silver titrations with adsorption indicators, some iodometric titrations). Fairly strong indirect daylight is best. Artificial light is often objectionable; certain color changes are very difficult to see by it. "Daylight" or fluorescent lamps, which supply a greater amount of the short wavelengths of the spectrum than do the ordinary, may often be used to advantage.

The solution titrated should generally be viewed against a white background during the titration, and the vessels are therefore placed upon a large porcelain tile, milk-glass plate, or, what is just as effective, a piece of white paper. Color changes will be very easily visible against the light background. In certain titrations in which the end point is indicated by the appearance of a turbidity in the solution, a dark background, such as that supplied by black glazed paper, is used.

If the color change at the end point is gradual, it is advisable to use a *comparison solution*. The comparison solution should have approximately the same volume and the same composition as the solution titrated at the end point.[17] The amount of indicator added should also be the same in both cases. It will then be very easy to determine the point at which the color of the titrated solution matches that of the comparison solution.

When the titration has been finished, the solution remaining in the buret is allowed to run out (it should not be returned to the stock bottle) unless other titrations are to be made shortly. The buret is then rinsed out with distilled water and covered with an inverted test tube to keep out dust. Solutions should not be allowed to stand for long periods of time in burets, especially if they are of alkaline nature.

Temperature corrections. Attention must be paid to corrections for the change in volume of solutions with a change in temperature. Usually it suffices to assume that the coefficient of cubical expansion of water and dilute solutions (0.1 N and less) is the same and to use the average value, 0.00025 (0.025 per cent) per °C, in the interval

[17] Thus in standardizing hydrochloric acid by titrating against borax, $Na_2B_4O_7 \cdot 10\ H_2O$, using bromcresol green as indicator, the comparison solution is prepared by dissolving the correct amounts of sodium chloride and boric acid in water together with the indicator.

20–30°C when making corrections.[18] The expansion or contraction of the glassware itself rarely needs to be taken into account in practical work, since neglect of this factor introduces an error of only about 0.0025 per cent per degree difference of temperature (soda-lime glass).

Example: $\frac{1}{10}$ equivalent of silver nitrate is dissolved in water and diluted to 1 liter at 25°C in a flask which has been accurately calibrated for 20°C. What is the normality of the solution at 25°C and at 20°C?

Ans.

$$\frac{0.10000}{1.0000 + (5 \times 0.000025)} = 0.09999\ N \text{ at } 25°\text{C}$$

$$\frac{0.10000}{1.00013 - (5 \times 0.00025)} = 0.10012\ N \text{ at } 20°\text{C}$$

Use of special volume burets and weight burets. Measurements of volume of standard solutions delivered with a 50-ml buret usually cannot be relied upon to less than 0.03 ml or about 0.1 per cent when the volume is 30–40 ml. A titration made with a 50-ml volume buret can hardly be expected to be closer than 0.1 per cent to the theoretical value, assuming that the other operations have been carried out without any error at all. If an accuracy greater than this is desired in a volumetric procedure, it will be

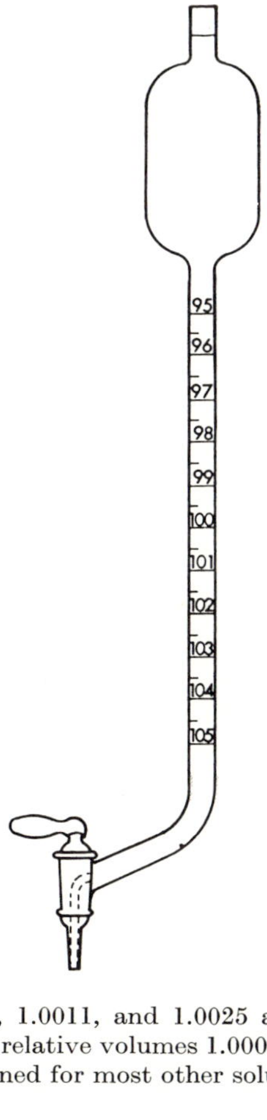

Fig. 21-9. Chamber buret.

[18] Thus water has the relative volumes 1.0000, 1.0011, and 1.0025 at 20, 25, and 30°C, respectively, while 0.1 M hydrochloric acid has the relative volumes 1.0000, 1.0012, and 1.0026 at the same temperatures; similar figures are obtained for most other solutions of comparable concentration.

necessary to make more refined measurements of the amount of the standard solution. There are two ways of proceeding. The first method involves the use of a special buret, sometimes called a chamber buret (Fig. 21-9), having a wide middle portion but rather narrow tubes above and below.[19] The upper tube bears a single graduation, which is the zero mark. The tube below the central enlarged portion is graduated like an ordinary buret into milliliters and fractions of a milliliter. The distance between the marks may be 0.01 ml; it will then be possible to estimate to 0.001 ml. The volume of the graduated portion of the buret may amount to only 5 or 10 ml, the total volume of the buret being perhaps 50–100 ml. Titrations can then be made with a precision approaching 0.01 per cent as far as the measurement of volume is concerned. It is, of course, necessary when using a buret of this kind to take a weight of sample such that when the titration is completed the meniscus will be in the graduated portion of the vessel. If the approximate content of the desired constituent in the sample is not known, a preliminary titration must be run using an ordinary buret. When using an instrument of this kind, great care must be taken to avoid errors of drainage and of change of volume due to temperature variations.

The second and better method of accurate determination of the amount of standard solution used in a titration is based on *weighing* the solution. The vessel used is called a weight buret (Fig. 21-10). It may have a volume of 50 to 100 ml and is roughly graduated. It is weighed before and after a titration, and the amount of reagent is calculated from the weight added, the solution having been standardized on the weight basis. The concentration of the solution is expressed in terms of the number of moles or equivalents in 1000 g of solution instead of 1000 ml. Since the weight of the solution delivered can be determined with great precision, the use of a weight buret leads to very precise results as far as the measurement of the amount of standard solution is concerned. It is impossible to measure 40 ml of solution from a buret with an accuracy

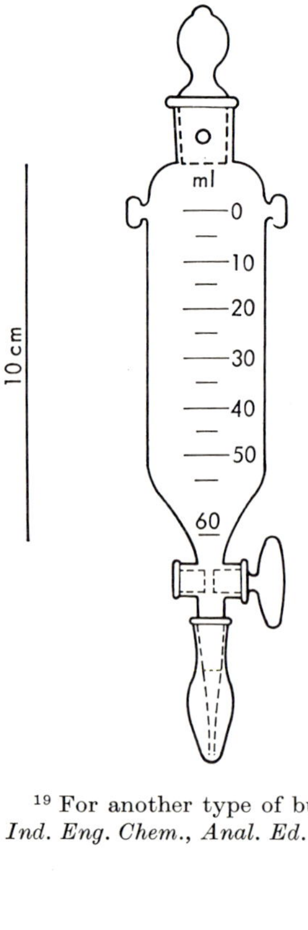

Fig. 21-10. Weight buret according to La Mer and Friedman, *Ind. Eng. Chem., Anal. Ed.* **2**, 54 (1930). For another design see H. N. Redman, *Analyst* **88**, 654 (1963).

[19] For another type of buret for precise measurement of volume see W. M. Thornton, Jr., *Ind. Eng. Chem., Anal. Ed.* **16**, 50 (1944).

of 0.01 ml, whereas it is not difficult at all to weigh the solution with an accuracy of 0.01 g. Generally it is more than sufficient to weigh the solution to the nearest 1 mg. This would correspond to a precision of 0.002 per cent in a weight of 50 g. Errors from drainage, change in temperature, etc., are entirely eliminated when weight burets are used. Buoyancy corrections must be applied.

The use of a weight buret or a special volume buret is of value only when the other steps of the procedure can be carried out with a precision that is comparable to that of the measurement of the amount of standard solution. In the first place, it must be possible to locate the position of the end point with great precision by a sharp change in color of the indicator or otherwise. Then it is necessary to be able to add very small increments of solution near the end point. If the end point were overstepped 0.01 ml by the inability to add a smaller amount than this volume, it would be illogical to weigh to 0.001 g. Since it is difficult to add portions of liquid smaller than 0.01 ml, it is often necessary to titrate in a special manner near the end point. The end point may be intentionally slightly overstepped and the excess of reagent then determined by titrating back with a very dilute solution—say 0.001 *N*—of an appropriate substance; or the addition of the titrating solution in the weight buret may be stopped shortly before the end point and the titration continued by adding a solution one tenth as strong. The measurement of volume of the dilute solution can be made with an ordinary buret.

PROBLEMS

1. A volumetric flask filled to the mark holds 997.00 g of water at 21.5°C. What is its volume at 20°C; at 21.5°C? *Ans.* 1000.10 ml at 20°C, 1000.14 ml at 21.5°C.
2. A pipet delivers 49.90 g of water at 23°C. What is its volume? *Ans.* 50.07 ml at 20°C.
3. A solution of hydrochloric acid is found to be 0.10055 *N* at 22°C. What will be its normality at 28°C? *Ans.* 0.10040 *N*.
4. What will be the capacity of a flask at 30°C which was found to hold 997.18 g of water at 20°C if the cubical coefficient of expansion of the glass is 0.000015 per °C?
5. What actual mass of water does a correctly marked liter flask hold at 20°C?
6. What weight of water must be taken at 25°C to indicate the mark representing 1 liter at the same temperature?
7. If a volumetric flask is to be calibrated at 20°C with an accuracy of 0.01 per cent, what is the greatest permissible error in obtaining the temperature of the water?
8. 5.8110 g of sodium chloride are transferred to a liter flask, dissolved in water, and the solution made up to the mark at 23°C. The mark on the flask has been found to correspond to a volume of 1000.48 ml at 20°C. What is the molarity of the solution? If 50.00 ml of the solution are pipetted out at 28°C, what weight of sodium chloride will be taken?
9. What weight of a liquid having a density of 0.8986 at 20°C and a cubical coefficient of expansion of 0.030 per cent/°C in the neighborhood of 20°C must be taken to indicate a volume of 500.0 ml in a glass vessel at 25°C?
10. A Pyrex buret (coefficient of cubical expansion 0.000010) is calibrated at 20°C. If it is later used at 35°C, will it be necessary to take into account the expansion of the glass?
11. A solution of potassium dichromate was standardized at 30°C. Later it was used at 25°C for titration of iron in an ore containing 60 per cent Fe. If temperature corrections are neglected, what absolute error in the percentage of iron will result?
12. A 1-liter flask was calibrated on the assumption that the air density was 0.0012 g/ml. Actually the barometric pressure was 735 mm Hg, the temperature 30°C, and the relative humidity 75 per cent. What error was made in the calibration (brass weights)?

13. Five 50.00-ml portions of water are delivered into a 250-ml volumetric flask and a mark is made at the meniscus. If the standard deviation in volume of a 50-ml portion is 0.02 ml, what is the standard deviation of the measured 250-ml volume? What is the probability that the mark lies within 0.10 ml of 250.00 ml?
14. It is desired to construct a 5-ml volumetric flask having a neck diameter such that the standard error in adjusting the level of the liquid to the mark will not exceed 0.005 ml. If the meniscus setting can be made to 0.2 mm (standard deviation), what may be the maximum neck diameter? *Ans.* 0.57 cm.
15. A series of weight-buret titrations are to be run in a laboratory at 1000 ft elevation over a period of several days, when the barometric pressure may vary by 4 cm Hg. Should weights of the standard solution ($\sim 0.1\ M$) be corrected to vacuum? Take the temperature as practically constant.

PART THREE

Gravimetric Analysis

CHAPTER 22 *Gravimetric Analysis: Introduction*

Classification of methods used in gravimetric analysis

In any direct gravimetric analysis, the constituent being determined is separated from the other constituents of the sample in the form of a pure phase which may be the constituent itself or a compound of known and definite composition; from the *weight* of the latter, the weight of the constituent sought is found. The separation, which must always precede the weighing of the constituent or a compound containing it, may be effected in a number of ways. The classification of gravimetric methods is based fundamentally on the methods of separation employed.

Precipitation methods. In a precipitation process the constituent being determined is precipitated as a very slightly soluble compound (or sometimes element), and the weight of the latter (or of the substance into which it may be advantageous to convert the precipitation form before weighing) is found.

For example, in determining chloride, a solution of the sample is treated with an excess of silver nitrate to precipitate silver chloride; the precipitate is filtered off, washed free of soluble substances, dried, and weighed as silver chloride. From the weight of the dried precipitate, the weight and percentage of chlorine as chloride are easily found (see below). Not infrequently the constituent in question is weighed in a form other than that in which it was precipitated. Thus calcium is precipitated as calcium oxalate monohydrate but is preferably weighed as calcium carbonate. Electrolytic determinations, in which metals are separated as such on a cathode or in which some anions are deposited as slightly soluble salts on a particular metal anode (halides as silver halides on a silver anode), belong in the class of precipitation processes.

Precipitation methods are of general applicability.

Volatilization or evolution methods. Here one or more of the constituents of the sample is volatile or can be transformed into a volatile substance.

Direct methods. The volatilized or evolved constituent is absorbed in a suitable medium, and the gain in weight is determined. The method is specific if no other volatile compounds which are absorbed are present. For example, water in a solid can be determined by heating the sample to the proper temperature and absorbing the volatilized water in a suitable desiccant such as anhydrous magnesium perchlorate. Even if a carbonate were present in the sample, the carbon dioxide evolved on heating would not interfere, since this gas is not retained by magnesium perchlorate. Analogously, carbon dioxide present as carbonate or bicarbonate can be determined by treating the sample with an excess of acid and absorbing the gas in a suitable substance, such as soda-lime, after the gas stream has been led through a desiccant to remove water vapor. Hence we are dealing here with a chemical separation of two volatile constituents; separations of volatile constituents are sometimes effected by physical methods such as fractional condensation. Evidently it is easy to arrange matters so that water and carbon dioxide can be determined simultaneously. Such a procedure finds an important application in the elementary analysis of organic compounds, in which the sample is burned in a stream of oxygen and the water and carbon dioxide formed from the oxidation of hydrogen and carbon respectively are absorbed separately in an appropriate absorption train.

Indirect (*difference*) *methods.* Here the weight of the residue remaining after the volatilization of a constituent is determined, and the amount of the constituent sought (which may either be volatilized or remain alone in the residue) is found from the change in weight. The method is very simple but not often applicable, because more than one volatile substance may be present or the sample may undergo chemical changes leading to change in weight of the residue (oxidation, for example). As a simple illustration, suppose it is desired to find the amount of a dissolved salt (or salts) in an aqueous solution. This can be done by evaporating the water, drying the residue at a suitable temperature, and weighing.

The method is of practical importance in the indirect determination of water (p. 573). If the sample does not decompose on heating, its water content can be determined by heating a weighed sample to a sufficiently high temperature to expel all the water and then weighing the residue. If the substance decomposes and gives off volatile products at the temperature required to expel the water, it is sometimes possible to add to the weighed sample a weighed amount of a nonvolatile solid that combines with and retains the volatile decomposition products.

Salts containing radicals of volatile acids can be treated with a volatile strong acid and heated to expel the former and the excess of the latter:

$$2\,NaCl + H_2SO_4 \rightarrow Na_2SO_4 + 2\,HCl$$

$$2\,NaNO_3 + H_2SO_4 \rightarrow Na_2SO_4 + 2\,HNO_3$$

In these examples, sodium chloride and nitrate are quantitatively transformed into sodium sulfate, so that it is possible to calculate the content of either alone in the sample from the increase (NaCl) or decrease ($NaNO_3$) in weight. Again the procedure is highly unspecific but nevertheless sometimes useful.

It is also possible to make use of chemical transformations in which certain constituents are converted into volatile compounds, e.g.:

$$CuO + H_2 \rightarrow Cu + H_2O$$

$$SiO_2 + 4\,HF \rightarrow SiF_4 + 2\,H_2O$$

The latter reaction is frequently applied in determining the true weight of silica in an impure ignited residue (p. 657).

For reducing purposes, it is possible to make use of the ammonium halides as reagents in "difference" methods.[1] On heating, these salts dissociate into ammonia and the halogen acid. The ammonia dissociates in turn into nitrogen and hydrogen, the degree of dissociation increasing with the temperature:

$$NH_4Cl \rightleftharpoons NH_3 + HCl$$

$$2\,NH_3 \rightleftharpoons N_2 + 3\,H_2$$

[1] L. Moser and S. Marian, *Ber.* **59**, 1335 (1926).

In addition, the halogen acid dissociates:

$$2\,HCl \rightleftharpoons H_2 + Cl_2$$

the degree of dissociation decreasing in the order HI > HBr > HCl at a given temperature. At 327°C the dissociation constant of the reaction is 1.5×10^{-6} for HCl, 3.5×10^{-3} for HBr, and 18.9 for HI. Potassium perchlorate can be quantitatively transformed into potassium bromide by heating to 400–500°C with an excess of ammonium bromide and repeating once or twice. After volatilization of the excess of the ammonium halide, the residue consists of pure potassium bromide. Similarly, it is possible to transform potassium and sodium sulfates into the corresponding bromides.

Miscellaneous methods. Sometimes methods other than those based on precipitation or volatilization are used in gravimetric analysis, and some of these will be mentioned here. When the components of a heterogeneous mixture are to be determined, simple methods of physical separation can often be applied. Thus, to determine sediment in a natural water, the sample is filtered, and the dried residue is weighed; to determine the relative amounts of different solid phases in a mechanical mixture, a liquid of suitable density, in which one (or more) of the components floats, can be used in effecting a separation, followed by weighing the different components. Special physical properties such as magnetic susceptibility can sometimes be used for such separations. Even differences in appearance (e.g., color) can be made the basis of separation by manual means. To be sure, such methods are rarely applied but are valuable now and then in the analysis of solid mixtures.

Differential solubility of the components of a mixture is also a property that finds application in gravimetric analysis. Methods based on this principle are frequently used in organic analysis but less often in inorganic analysis. The separation and determination of potassium in the presence of sodium by the extraction of the perchlorates of these metals by an organic solvent may be cited as an illustration. Potassium perchlorate is insoluble in the solvent used, whereas sodium perchlorate is easily soluble.

All these methods of separation can, of course, be used preparatory to the analysis of the sample by methods other than the gravimetric.

Precipitation analysis

We return now to precipitation processes, which deserve further discussion.

Precipitation forms. A precipitate must meet a number of requirements to a satisfactory degree if it is to be made the basis of an acceptable gravimetric method:

1. It must be so slightly soluble that the constituent in question is quantitatively precipitated.[2] The chief factor here is the solubility of the precipitate in the precipitation mixture, containing an excess of precipitant, and in the wash liquid, in which the presence of the precipitant may not be permissible. The temperature and time of standing before filtration may also be of importance.
2. It must be filterable. The primary or secondary particle size must be such that the precipitate is retained on the filter during filtration and subsequent washing. Most precipitates can be formed and treated so that this requirement is fulfilled without much difficulty, although precipitates differ in ease of filtration.
3. It must have a stoichiometric composition or be convertible into a weighing form of definite composition, preferably in a simple way.
4. It should be of high purity. As far as possible it should be free of coprecipitated substances, and it should not be contaminated with appreciable amounts of insoluble

[2] A macroprecipitation, or other macrooperation, is usually said to be *quantitative* when the loss of substance in the operation does not exceed ca. 0.1 mg, the ordinary limit of sensitivity of the analytical balance.

products of other constituents in the sample. Moreover, foreign substances should not hinder the precipitation. If weighed after drying at a low temperature it must not be contaminated by excess reagent, as sometimes happens with organic precipitants.

The last desideratum—purity—is, in practice, the most difficult one to fulfill. Precipitation reagents are quite commonly lacking in specificity or even selectivity. For this reason it is often necessary to separate the constituent whose amount is desired from the other constituents of the sample before the former can be precipitated in pure form. (The subject of separations is considered in Chapters 12–15.) If the separation is made by precipitation of the constituent sought, it is sometimes necessary to dissolve the *separation form* and reprecipitate the constituent in a form that is suitable for weighing or can be made so. Conversely, the interfering constituents can be precipitated by a suitable reagent and the desired constituent left in solution to be precipitated in another form. Interferences due to coprecipitation may require the removal of the disturbing foreign substances, or reprecipitations (p. 244). Sometimes regulation of the conditions of precipitation, addition of auxiliary complexing agents, etc., may permit the presence of the foreign substances during precipitation. Occasionally, when exact results are desired, corrections can be applied to the weight of the dried or ignited precipitate by determining the small amount of coprecipitated (or precipitated) foreign substance and deducting the latter.

Weighing forms. Frequently a constituent is not weighed in the form in which it is precipitated, and this for a number of reasons. For one, the precipitation form may be of indefinite composition. Thus ferric iron is often precipitated as hydrous ferric oxide, $Fe_2O_3 \cdot xH_2O$, but it is never weighed as such, because after drying at room temperature or 100°C the precipitate contains an indefinite amount of water, depending upon the exact conditions under which it has been precipitated and dried. It is, however, a simple matter to ignite[3] the hydrous oxide to convert it into the anhydrous form, Fe_2O_3, and so weigh it. The weighing of any precipitation form after drying at room temperature or at 100°C is a procedure to be looked upon with suspicion, because a precipitate so dried may still contain water held by adsorption or inclusion (p. 234). This is not to say that certain precipitation forms may not be weighed after drying in this manner, because much depends upon the precipitate in question, but the matter must always be put to critical test. Many substances, especially when in the finely divided condition, retain water tenaciously and can be rendered water-free only by heating at rather high temperatures—hence the frequency of ignition as a preparatory step to weighing. It is desirable, on the one hand, that the precipitation form be decomposed or transformed[4] into the weighing form at a relatively low temperature, and on the other, that the latter be stable (nondecomposable and nonvolatile) at high temperatures so that the conversion can easily be effected by the usual heating appliances of the laboratory without a critical control of temperature. Moreover, the ignited residue should not be appreciably hygroscopic. Other things being equal, a weighing form will be better for the purpose the greater its molecular

[3] As used in analytical chemistry, the term *ignition* means heating at a high or moderately high temperature.

[4] Occasionally the precipitation form is converted into a suitable weighing form by treatment with a suitable reagent, the excess of which is then volatilized. See the discussion of the weighing of calcium as the sulfate, page 624.

weight or, more exactly expressed, the larger the ratio of the weight of the substance weighed to that of the constituent being determined.

An advantage of gravimetric over volumetric analysis is that the ignited residue can be examined, after weighing, for foreign substances that might be present, and corrections applied if necessary.[5]

Calculations of gravimetric analysis

The computations of gravimetric analysis are very simple. Generally it is desired to find the *weight percentage* of a given constituent in a sample, and the following formula applies:

$$P = \frac{w}{W} \times 100$$

where P = percentage of the constituent sought
w = weight of the constituent
W = weight of the sample.

If the constituent is weighed in the form in which the percentage is to be expressed, it is only necessary to substitute the weight of the dried or ignited residue for w in the above equation. Usually, however, the constituent is weighed in the form of a compound which contains it in a known constant amount, this practice dating back to T. Bergman (1735–1784). Then it is necessary to find w from the weight of the dried or ignited precipitate, w', by making use of the formula

$$w = Fw'$$

in which F (variously called the chemical factor, conversion factor, gravimetric factor, and calculation factor) is a ratio that represents the weight of w that would be given by a unit weight of w' if the substance weighed were appropriately treated. Otherwise expressed, F is the numerical value of the ratio of a gram-equivalent of the constituent sought to a gram-equivalent of the compound weighed. Thus in determining chlorine as chloride by weighing the silver chloride obtained by treating the solution of the sample with an excess of silver nitrate, the weight of chlorine corresponding to the weight of dry silver chloride is obtained by using the factor

$$F_{Cl/AgCl} = \frac{Cl}{AgCl} = \frac{35.45}{143.32} = 0.2474$$

Examples: A 0.2501-g sample of a soluble chloride gave a precipitate of silver chloride weighing 0.5009 g. What is the percentage of chlorine in the sample?

Ans. $$P = \frac{F \times w' \times 100}{W} = \frac{0.2474 \times 0.5009 \times 100}{0.2501} = 49.55\% \text{ Cl}$$

In the ratio expressing the chemical factor, the formula of the substance sought occurs in the numerator and that of the substance weighed occurs in the denominator. Since the chemical factor expresses the ratio of two substances that

[5] When the amount of precipitate is very small, it is well to establish its identity qualitatively to be sure that the substance weighed is actually what it is supposed to be and not an entirely different substance.

are equivalent to one another, the same number of atoms of the constituent concerned must appear in numerator and denominator. Thus in calculating the weight of iron from the weight of an ignited ferric oxide residue, the factor is: $2\ Fe/Fe_2O_3$. The following examples should make the principle clear:

Substance Weighed	*Substance Sought*	*Factor*
$Mg_2P_2O_7$	MgO	$\frac{2\ MgO}{Mg_2P_2O_7}$
$Mg_2P_2O_7$	P_2O_5	$\frac{P_2O_5}{Mg_2P_2O_7}$
Fe_2O_3	Fe_3O_4	$\frac{2\ Fe_3O_4}{3\ Fe_2O_3}$
$BaSO_4$	$K_2SO_4 \cdot Al_2(SO_4)_3 \cdot 24\ H_2O$	$\frac{K_2SO_4 \cdot Al_2(SO_4)_3 \cdot 24\ H_2O}{4\ BaSO_4}$

The constituent sought need not be present in the substance weighed, as long as the stoichiometric relation between the two is known. Thus it is possible to determine the amount of potassium in a solution of pure potassium chloride by determining the weight of silver chloride yielded by the solution, because the weight relation between potassium and silver chloride is fixed. The factor is $K/AgCl$. No matter how many the intermediate compounds in a series of stoichiometric reactions, only the first and last need be considered in setting up the calculation factor.

Indirect analysis. There are various types of indirect analyses, but they all have the common feature that all the constituents being determined are not isolated in a pure form and weighed. Thus it is possible to obtain the weight of the sum of n constituents, then determine $(n - 1)$ of them by isolation (separation), and finally the remaining one by difference. Or all the constituents (usually limited to the case in which only two are present) may be determined indirectly. For example, the amounts of two substances may be determined from the change in weight that a known amount of the mixture, composed only of these, undergoes when transformed into two other substances that differ in molecular weight not only between themselves but from the original ones as well. It must be possible to set up as many independent simultaneous equations connecting the quantities of the substances involved as there are substances to be determined. The relations expressed in these equations are the stoichiometric ones. For example—to take an illustration that is often given—it is possible to determine the amount of sodium and potassium chlorides in a pure mixture of the two by obtaining the weight of the mixture and the weight of silver chloride which the sample yields.

Let a = weight of sodium and potassium chlorides
b = weight of silver chloride
x = weight of sodium chloride
y = weight of potassium chloride

Then

$$x + y = a$$

$$\frac{AgCl}{NaCl}x + \frac{AgCl}{KCl}y = b$$

$$2.452x + 1.923(a - x) = b$$

$$x = 1.890b - 3.635a$$

Since a and b are known, x (and y) can be calculated.

Indirect methods must be used with caution because they usually cannot yield accurate results. A small error in weighing is generally multiplied manyfold in the result. An inspection of the above equation shows that an error of 0.1 mg in obtaining the weight of the mixed chlorides produces an error of nearly 0.4 mg in the sodium chloride found.

PROBLEMS

1. Write the chemical factors for the following:

	Weighed	Sought
(a)	Al_2O_3	Al
(b)	CaF_2	AlF_3
(c)	$Mn_2P_2O_7$	Mn_2O_3
(d)	$CaCO_3$	$Ca(HCO_3)_2$
(e)	Pt	KCl ($KCl \rightarrow K_2PtCl_6 \rightarrow Pt$)
(f)	$PbMoO_4$	P_2O_5 ($P_2O_5 \rightarrow (NH_4)_3PO_4 \cdot 12\, MoO_3 \rightarrow PbMoO_4$)
(g)	AgCl	As ($Ag_3AsO_4 \rightarrow AgCl$)

2. If 0.5000 g of an iron-containing compound yielded 0.4990 g of Fe_2O_3, find the percentages of Fe_2O_3 and Fe. *Ans.* 99.80 Fe_2O_3, 69.80 Fe.
3. An oxalic acid sample of 1.000 g was precipitated with an excess of calcium chloride and the washed precipitate ignited to calcium oxide. The residue weighed 0.4402 g. Calculate the percentage of $H_2C_2O_4$ in the sample. *Ans.* 70.68 per cent.
4. A 0.7500-g sample of a soluble sulfate containing 0.50 per cent moisture yielded 1.044 g of $BaSO_4$. What is the percentage of S in the sample on (a) the moist basis, (b) the dry basis? *Ans.* (a) 19.12 per cent; (b) 19.22 per cent.
5. If 25.00 ml of a solution of barium chloride containing no other substance gave a dried precipitate of silver chloride weighing 0.3408 g, calculate: (a) the weight of $BaCl_2$ in 1 liter of the solution, (b) the molarity of the solution, (c) the normality of the solution as a precipitant for sulfate, (d) the weight of K_2SO_4 precipitable by 25.00 ml of the solution.
6. A mixture of anhydrous chlorides of sodium and potassium weighing 0.1000 g gave 0.0155 g of K_2PtCl_6. Calculate the percentages of K and Na in the mixture.
7. By appropriate treatment a sample of sodium thiosulfate was transformed into sodium sulfate and the latter was treated with an excess of barium chloride. If 0.2508 g of sodium thiosulfate is found to give 0.4710 g of $BaSO_4$, what is the percentage of $Na_2S_2O_3 \cdot 5\, H_2O$ in the sample? To what percentage of anhydrous sodium thiosulfate does this correspond?
8. A 0.4000-g sample of a pure anhydrous salt known to contain only sodium, boron, and oxygen yielded 0.4914 g of H_3BO_3 (by a volumetric method) and 0.2324 g of NaCl. What is the formula of the compound?

9. A mixture of NaCl and KCl weighing 0.5986 g yielded 0.7072 g of anhydrous Na_2SO_4 and K_2SO_4 when treated with sulfuric acid. Calculate the percentages of NaCl and KCl. If a weighing error of +0.2 mg were made in finding the weight of the mixed sulfates, what would be the resulting error in the percentages of NaCl and of KCl?
10. If 10.00 ml of sulfuric acid solution of density 1.010 yield a pure barium sulfate precipitate weighing 0.3780 g, what is the percentage by weight of H_2SO_4 in the solution, and the molarity and normality of the latter?

    Ans. 1.572 per cent; 0.1619 *M*; 0.3238 *N*.
11. What weight of barium chloride dihydrate will give a loss of 0.1000 g on heating to expel completely its water of hydration?
12. A sample containing silver gave 0.7468 g of silver chloride. What weight of precipitate would have been obtained if silver had been precipitated as bromide?
13. Iron, aluminum, and titanium in a sample were jointly precipitated with ammonia and ignited to Fe_2O_3, Al_2O_3, and TiO_2. The residue, weighing 0.4444 g, was brought into solution by potassium pyrosulfate fusion. A colorimetric determination of titanium in the solution gave 0.0123 g of Ti, and a titrimetric determination of iron after reduction gave 0.2022 g of FeO. Find the weight of Al_2O_3.

CHAPTER 23 *Determination of Water*

Water may be an essential constituent of a sample (p. 195) and will need to be determined when a complete analysis is made. Moreover, any more or less finely divided material which has been exposed to the atmosphere will contain adsorbed water, whose determination may be required, as in a complete analysis. The determination of water may, of course, be required even when a complete analysis is not being made.

A large number of methods are available for the determination of water (p. 578). An obvious method is to volatilize the water in a sample by heating to a suitable temperature and finding the weight of the expelled water in some way. If the material contains no other constituent which is volatile at the temperature required to drive off the water and the residue undergoes no weight change (as by oxidation), the decrease in the weight of the sample represents the vaporized water. When these conditions are not fulfilled, the expelled water must be collected (as by absorption in a desiccant) and its amount found, usually by weighing. This is the direct method, the former the indirect.

INDIRECT GRAVIMETRIC METHOD

Hygroscopic water or *moisture* is commonly determined from the loss in weight shown by a sample after heating to constant weight at some more or less arbitrary temperature, usually 105°C (p. 518). It should be borne in mind that essential water can also be lost at a low temperature from such samples as hydrated salts, whereas, on the other hand, all the nonessential water may not be removed at 105°C.

The indirect method can also be applied to the determination of essential or total (*essential* + *nonessential*) water in substances which give physically and chemically stable residues at the temperature required for the complete expulsion of the water. Examples of such substances are gypsum ($CaSO_4 \cdot 2\ H_2O$), barium chloride dihydrate, and calcium hydroxide. The temperature to which a substance must be heated to render it water-free depends, of course, upon the nature of the substance. Many hydrated salts readily lose their water at 105°C, although some are stable at the same temperature. Water chemically held as hydroxide requires a high temperature for complete removal. Aluminum oxide, for example, retains traces of water tenaciously even at 1000°C. The dehydrated products obtained in the indirect water determination are often very hygroscopic, and due care must be taken in weighing them.

Organic materials, which decompose at relatively low temperatures, may sometimes be deprived of their water by heating at as low a temperature as possible *in vacuo* (use of a vacuum drying oven), or by exposure to the dry atmosphere of a desiccator charged with a vigorous dehydrating agent such as magnesium perchlorate or, better, by passing desiccated air over the sample. By resorting to such procedures, it may also be possible to determine both the essential and nonessential water content.[1]

It is sometimes possible to determine water indirectly in substances that would otherwise decompose, by adding a suitable flux. For example, aluminum sulfate cannot be completely dehydrated without expelling sulfur trioxide. If, however, the sulfate is mixed with previously ignited sodium paratungstate ($5\ Na_2O \cdot 12\ WO_3$), the water can be completely expelled by heating the mixture to a high temperature, since the flux retains the sulfur trioxide, forming stable sodium sulfate, and the decrease in weight is due only to the water lost.[2]

Determination of water of hydration of barium chloride dihydrate. As an example of the indirect method, the determination of the water of hydration of barium chloride dihydrate may be taken as typical. The aqueous vapor pressure of barium chloride dihydrate at 20°C is 1.3 mm Hg and at 35°C it is 11.8 mm Hg; the hydrate is therefore stable in the atmosphere except under conditions of unusually low humidity. The salt readily loses all its water at 100°C, but higher temperatures can be used in the dehydration since anhydrous barium chloride is stable and nonvolatile at even fairly elevated temperatures. The dihydrate shows little tendency to occlude water in vacuoles, so that the loss in weight represents very closely the water of hydration. Since the salt is weighed out in the form of coarse crystals, the amount of adsorbed water is negligibly small.

Procedure. Coarsely powder analytical-grade barium chloride dihydrate in a mortar, and weigh out 1-g duplicate samples, proceeding as follows. Dry a 15-ml glass-stoppered weighing bottle in an oven at 105°C, cool in a desiccator, and obtain the weight of the empty stoppered bottle to 0.1 mg. Transfer the approximate weight of the sample to the bottle with the aid of a spatula, and reweigh to 0.1 mg. Then place the weighing bottle and its contents in an oven heated to 105°C or higher. Lay the stopper on its side over the mouth of the bottle to permit the free escape of water vapor.[3] Heat for 1 hr at 105°C (or $\frac{1}{2}$ hr at 150–200°C), then stopper the weighing bottle, and place it in a desiccator to cool to room temperature (p. 500). Before weighing, loosen the stopper to equalize the pressure. Keep the bottle stoppered during the weighing. Unstopper the bottle after weighing, return it to the oven, and heat as before for $\frac{1}{2}$–1 hr. Cool, weigh, and repeat the heating if necessary until the weight is constant to 0.2 mg. Calculate the percentage of water in the sample from the loss in weight found. Each sample should give a result agreeing within 0.05 absolute per cent with the calculated value, $(2\ H_2O \times 100)/BaCl_2 \cdot 2\ H_2O = 14.75$ per cent.

DIRECT GRAVIMETRIC METHODS

If the sample volatilizes or decomposes at the temperature required to expel water quantitatively, the determination must be made by the direct method, in

[1] The thermobalance (p. 480) may also be of aid in determining water in different states of combination.

[2] S. B. Kuzirian, *Am. J. Sci.* [4]**36**, 401 (1913); *Z. Anorg. Chem.* **85**, 127 (1914); K. Wiskont and I. Alimarin, *Z. Anal. Chem.* **79**, 271 (1929).

[3] If the sample decrepitates, the vessel in which the heating is done should be covered loosely at first to prevent loss of flying particles. This need not be done with barium chloride.

which the vaporized water is (1) absorbed in a suitable drying agent and the increase in weight found, or (2) the weight of the condensed water is obtained (Penfield method). Absorption procedures are more generally applicable and usually more accurate.

Absorption method

It may be necessary to mix the sample with an anhydrous flux to retain volatile substances and to aid in the expulsion of water from substances not easily deprived of water on heating. Thus, in the determination of water in ferric sulfate by the direct method, sulfur trioxide as well as water is expelled on heating. Erroneous results would be obtained if no provision were made for the retention of the sulfur trioxide, since the latter would otherwise be absorbed to a large extent in the drying agent and would count as water. The volatilization of sulfur trioxide can be prevented by mixing the ferric sulfate with anhydrous sodium paratungstate, for example; the latter holds the sulfur trioxide, forming sodium sulfate, which is not decomposed unless the temperature is raised unnecessarily high. Lead oxide–chromate is a good general flux (p. 577).

Alternatively, the vapors can be passed over a "retainer" before being led into the absorption tube. Thus lead oxide (a mixture of PbO and PbO_2 is generally used) heated above 250°C fixes sulfur trioxide and dioxide by forming lead sulfate. This retainer is also useful for catching other acidic substances such as hydrogen chloride and is used frequently in organic elementary analysis. With the aid of lead oxide it is possible to determine water in hydrolyzable salts such as magnesium chloride. The hydrates of such salts hydrolyze to the oxide or a basic salt when heated to expel water, e.g., $MgCl_2 \cdot H_2O \rightleftharpoons MgO + 2\,HCl$. Some of the hydrogen which should be evolved as water is set free as hydrogen chloride instead. However, if the gases are passed over hot lead oxide, the hydrogen chloride forms lead chloride and gives the equivalent amount of water

$$PbO + 2\,HCl \rightarrow PbCl_2 + H_2O$$

In these and similar cases the hydrolysis can be prevented by heating in a current of dry hydrogen chloride.

Apparatus. The setup shown in Fig. 23-1 is representative of assemblies used for the direct determination of water in solids.

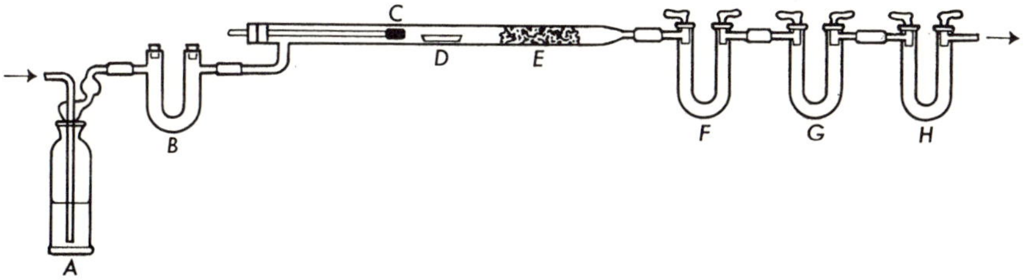

Fig. 23-1. Train for the direct determination of water.

The gas wash bottle *A* contains concentrated sulfuric acid and serves the double purpose of removing most of the water vapor from the entering air stream and indicating the speed with which air is being drawn or forced through the train. *B*, which may be a drying tower or U-tube, is filled with magnesium perchlorate or other efficient dryer having an aqueous vapor pressure less than that of sulfuric acid. The tube *C* may be made of borosilicate glass, porcelain, or fused silica, depending upon the temperature to which the sample must be heated. It may conveniently have a length of 40 to 50 cm and a diameter of 1.5 cm. Preferably it should be tapered at the exit end so that it can be connected directly to the absorption train. The sample is heated in the porcelain or platinum boat *D*. *E* is a mixture of lead oxide and lead dioxide intermixed with glass wool, for example, and placed between plugs of the same material. This retainer need only be included when the sample gives off sulfur oxides, hydrogen chloride, or other acidic compounds. The tube *C* may be heated with burners or, better, electrically in a tube furnace. The temperature of the hot zone of the tube corresponding to a given height and position of the burners, or to the setting of the rheostat of the electric furnace, should be determined prior to the determination by inserting a thermometer or thermocouple through the front end of the tube; or, better, either of these should be kept in place during the determination. The lead oxides *E* should be maintained at a temperature of 300–350°C.

The tube *C* is connected to the absorption train *F*, *G*, *H* by means of a thick piece of rubber tubing; the ends of the tubes of *C* and *F* should be in contact. As absorption vessels, U-tubes will serve.[4] Generally it is best to have two weighed absorption tubes (*F*, *G*) in the train. The absorbent in the last weighed vessel *G* should be the same as that in *B*, but that in *F* need not be so necessarily. Thus it may be advantageous to put the cheaper desiccant calcium chloride in the first absorber *F* and follow it with magnesium perchlorate in *G*. If carbon dioxide is evolved on heating the sample, calcium chloride should not be used in *F*, because this desiccant is likely to contain calcium hydroxide which will take up carbon dioxide; this impurity cannot always be completely removed by pretreating with carbon dioxide. These absorption tubes should be weighed against similar tubes used as tares. The last tube in the train, *H*, is simply a protecting tube for *G*; it should be filled with the same desiccant as *G*. *H* is connected to the vacuum line or an aspirator; as already mentioned, air can be forced instead of drawn through the system. Before use the assembly should be tested for air-tightness (see Chap. 63).

Determination of water of constitution of sodium bicarbonate.[5] As an example of the direct method for water, the determination of the elements of water in sodium bicarbonate may be taken. Sodium bicarbonate is quantitatively decomposed into sodium carbonate and water when heated to 270–300°C:

$$2\ NaHCO_3 \rightarrow Na_2CO_3 + H_2O + CO_2$$

Any hygroscopic or included water present is counted with the combined water when an air-dried sample is used.

Procedure. Set up a train similar to the one described in the previous section and depicted in Fig. 23-1. Place concentrated sulfuric acid in *A* and anhydrous magnesium perchlorate or calcium sulfate (drierite) in *B*, *F*, *G*, and *H*. Omit *E*.

Heat the central part of the tube *C* with the empty boat *D* in place (Fig. 23-1) to approximately 300°C; with *F*, *G*, and *H* disconnected, draw or force air through *A*, *B*, and *C* for about 10 min at the rate of two or three bubbles per second. Then connect in *F*, *G*, and *H*, and continue to pass air through the hot tube for 10 min. At the end of this time stop the air current, close *F*, *G*, and *H*, and place *F* and *G* in or near the balance case. After 15 min open the tubes momentarily, and weigh separately against a similar tare. Replace the tubes in the train, open them, and again pass air through the system for 10 min. Then weigh *F* and *G* as before. If their weights are constant to 0.3 mg, proceed with the determination as described in the next paragraph; otherwise replace them in the system and continue as before until constant weight is obtained, showing that all moisture has been swept out of the apparatus.[6]

[4] If the sample contains much water, it is advantageous to employ for *F* a U-tube with a bulb on the entrance side to hold the water expelled, thus saving the desiccant.

[5] Malachite (basic copper carbonate) or white lead may also be used as samples.

[6] In the most careful work the apparatus used and the procedure followed should be checked by running a control (p. 417) on a sample of known water content.

Weigh 0.5 g of air-dried reagent-grade sodium bicarbonate into the cold boat, and replace the latter in the central part of the cold tube *C*. With *F*, *G*, and *H* in the train, start a current of air through the system at the rate of two or three bubbles per second, and begin to heat the tube gently in the neighborhood of the boat. If any water condenses between *D* and the entrance end of the tube *C*, increase the rate of flow. Gradually increase the heat until the sample is at 300°C, and maintain it at this temperature for 15–30 min. If there is any condensed water on the walls of *C* between the boat and the exit end, cause it to be carried over into *F* by brushing the tube carefully with a flame. Continue the passage of air through the train until the tube *C* is cold and no water droplets are visible anywhere in the tube. Then close and disconnect *F*, *G*, and *H*. Allow *F* and *G* to stand in the balance case for 15 min and weigh as before against a tare. Replace the tubes in the train, and proceed as before until constant weight (0.3 mg) is attained. From the gain in weight obtained in duplicate analyses, calculate the percentage of water in the sample, and compare with the theoretical value.

Note: The carbon dioxide evolved can be determined at the same time as the water by inserting a tube filled with ascarite and desiccant (Chap. 63) between *G* and *H* and finding its increase in weight. A tube of ascarite must be included in the purification end of the train, between *A* and *B*, to remove carbon dioxide from the air stream. The loss in weight of the sample as determined from the weight of the residue should equal, within the experimental error, the sum of water and carbon dioxide found.

When so performed, the experiment simulates the determination of carbon and hydrogen in an organic compound by burning the latter in a stream of oxygen and obtaining the weight of carbon dioxide and water formed.

Penfield method

This method[7] is often suitable for the determination of small amounts of water (up to a few per cent) in solids. It is very simple and rapid. The sample, mixed with a flux, is strongly heated in a bulb at the end of a relatively long glass tube; the expelled water condenses in the cold part of the tube. The bulb is drawn off and the weight of water in the tube is found by weighing before and after the evaporation of the condensed water. The Penfield method is extensively used for the determination of total water in silicate rocks and minerals. Combined water can then be obtained by deducting hygroscopic water from the total. Directions follow.

Apparatus and reagent

Penfield tube. Prepare or obtain a borosilicate glass tube approximately 25 cm long with an internal diameter of 5–6 mm, having a bulb of 2–2.5 cm diameter at one end. A capillary tip, attached with a short length of rubber tubing, is used to minimize loss of water by evaporation during heating and standing (Fig. 23-2). If the sample contains much water, a slight enlargement midway in the tube may be advantageous.

Flux.[8] Heat 50 g of PbO (m.p. 888°C) and 25 g of $PbCrO_4$ (m.p. 844°C) separately at 800°C for 1 hr to remove water, mix well, and grind to pass a 20-mesh sieve. Store in a tightly closed bottle.

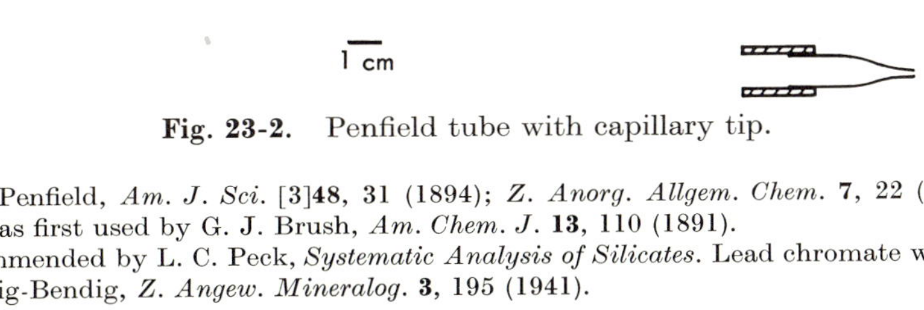

Fig. 23-2. Penfield tube with capillary tip.

[7] S. L. Penfield, *Am. J. Sci.* [3]**48**, 31 (1894); *Z. Anorg. Allgem. Chem.* **7**, 22 (1894). The method was first used by G. J. Brush, *Am. Chem. J.* **13**, 110 (1891).

[8] Recommended by L. C. Peck, *Systematic Analysis of Silicates.* Lead chromate was used by M. Hartwig-Bendig, *Z. Angew. Mineralog.* **3**, 195 (1941).

Procedure. Dry the Penfield tube by drawing air through it by means of a narrow tube reaching to the end while heating gently with a burner. Weigh out 1 g of powdered sample (0.5 g of micas) into a porcelain crucible and mix well into 3.8 g of flux, using a stirring rod. Transfer the mixture to the bulb of the Penfield tube, using a thistle tube. Tap the latter to dislodge any adhering particles and remove it without leaving any powder on the walls of the Penfield tube. Support the tube horizontally in a clamp or by placing it in a hole of a vertical asbestos board, 5 cm or so from the bulb. Wrap the remainder of the tube, to within a few centimeters of the open end, with a spiral strip of cloth. Keep this moist during the heating. Attach the capillary stopper.

Gradually heat the bulb with a burner[9] so that the softening temperature of the glass is reached in about 15 min. Continue to heat for 20 min at as high a temperature as the glass will stand, occasionally rotating the tube. Then, pull off the bulb with iron tongs by softening the junction of bulb and tube with a blast flame; round off the drawn-out tip.

Allow the tube to cool in a horizontal position, with the moist cloth in place. When room temperature is reached, remove the cloth, carefully wipe the tube dry with lintless cloth, and let it stand in the balance room for 15 min. Remove the capillary stopper and weigh the tube. Replace the tube in the holder, heat gently, and draw air through to evaporate all the water. Cool and weigh the tube. The decrease in weight represents the total water. To obtain the amount of combined water, deduct hygroscopic water.

Notes:

1. The use of a flux, whereby the sample is decomposed, enables water to be determined in minerals that otherwise would retain water tenaciously at 1000° C or higher.[10] Moreover, the flux retains sulfur, chlorine, and fluorine. Lead chromate, as an oxidizing agent, prevents the reduction of water to hydrogen by ferrous iron at high temperatures:

$$2\,FeO + H_2O \rightarrow Fe_2O_3 + H_2$$

2. The directions given above are intended for samples essentially free of carbonates (say < 0.1 per cent CO_2). Carbon dioxide causes error in two ways: (a) being heavier than air, that which remains in the tube with the condensed water gives high results, and (b) that which leaves the tube carries water vapor with it and leads to low results. The first error can be largely eliminated by connecting the tube, with condensed water, to a vessel whose atmosphere is saturated with water vapor and allowing CO_2 to diffuse out of the tube by standing for 2 hr.[11] If the carbonate content is known, a correction can be applied for the water lost with the carbon dioxide escaping from the tube. However, it is better to use the absorption method (p. 575) for water when carbonate is present in the sample.

3. The expelled water can be determined volumetrically.[12]

Other methods for water

A great many methods have been developed for the determination of water, of which only a few can be noted here.[13]

A useful method for water in materials of organic origin (plant materials, foodstuffs, oils and fats, etc.) consists of dissolving the sample in, or mixing with, an immiscible

[9] Peck, Ref. 8, describes a special furnace for the Penfield determination.

[10] Such minerals include micas, tourmaline, topaz, staurolite, epidote, and talc, containing —OH. See, for example, A. F. Smethurst, *Mineralog. Mag.* **24**, 173 (1935).

[11] Peck, Ref. 8. One of the authors (E. B. S), in a semimicro determination of water, used a glass rod fitting the tube to displace most of the air and carbon dioxide before weighing [*Mikrochem. Ver. Mikrochim. Acta* **38**, 487 (1951)].

[12] E. Dittler and H. Heuber, *Z. Anorg. Allgem. Chem.* **195**, 41 (1931); **199**, 17 (1931).

[13] For a survey see J. Mitchell, Jr., in *Treatise on Analytical Chemistry*, part II, vol. 1, 1961, and J. Mitchell and D. M. Smith, *Aquametry*, Wiley-Interscience, New York, 1948.

organic solvent having a higher boiling point than water and distilling.[14] Water vaporizes with the organic liquid and can be determined in the condensate by catching it in a calibrated trap and reading its volume. Toluene (b.p. 111°C) is often used as the liquid carrier for water in such procedures.

An important chemical method for water is based on titration with Karl Fischer reagent (Chap. 44).

Physical and physicochemical methods, e.g., infrared and ultraviolet spectrophotometry, nuclear magnetic resonance spectroscopy, mass spectrometry, neutron scattering, thermal conductivity, and dielectric measurements find use in the determination of water.

PROBLEMS

1. Devise methods (see the literature if necessary) for determining: (a) hygroscopic water in ammonium chloride; (b) total water in ferric chloride; (c) hydrate water in $Ba(OH)_2 \cdot 8\,H_2O$; (d) water content of strong sulfuric acid; (e) water in sulfur dioxide gas; (f) water in sodium hydroxide.
2. Suppose that the air in a Penfield tube, 15 cm long with an internal diameter of 0.5 cm, originally having zero humidity, becomes saturated with water at 25°C and 76 cm Hg atmospheric pressure. What weight of water (representing a negative error) is required for this?

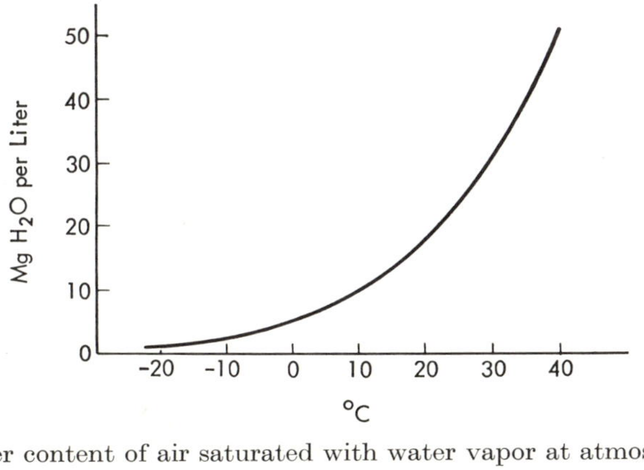

Fig. 23-3. Water content of air saturated with water vapor at atmospheric pressure.

3. Suppose that a 1-g sample containing 10 per cent calcium carbonate shows 1.00 per cent water by the Penfield method when no correction is made for the water vapor carried off by the carbon dioxide. If the carbonate is completely decomposed to oxide and all the carbon dioxide, saturated with water vapor at 25°C and 76 cm Hg, leaves the tube, what is the corrected value for water?
4. Calcium carbide (CaC_2) reacts with water to form acetylene (C_2H_2). A 1.00-g sample of moist material mixed with a weighed amount of excess CaC_2 was found to undergo a weight loss of 0.103 g. Calculate the percentage of moisture.
5. According to the International Critical Tables, a liter of air cooled to -190°C with liquid air or liquid nitrogen holds 1.6×10^{-23} mg of H_2O. How many molecules of H_2O does this weight represent?

[14] Water and the organic liquid distil in a fixed ratio (azeotropic mixture) at a constant boiling temperature, which is lower than that of the components.

CHAPTER 24 *Determination of Chlorine as Chloride*

Method

Chlorine in a soluble chloride is determined gravimetrically by precipitation as silver chloride by the addition of an excess of silver nitrate solution to the chloride solution containing nitric acid:

$$Cl^- + Ag^+ \rightarrow AgCl$$

The precipitate is collected in a filter crucible, washed with very dilute nitric acid to remove soluble salts, dried at 100–200°C, and finally weighed as AgCl. The presence of nitric acid prevents the precipitation of silver salts insoluble in neutral medium (silver carbonate, silver phosphate, etc.).

Solubility of silver chloride

One liter of water dissolves 1.6 mg of AgCl at 20°C, 1.8 mg at 25°C, and 21.1 mg at 100°C. The molecular (intrinsic) solubility of silver chloride is 2.4×10^{-7} *M* at 20°C and 4×10^{-7} *M* (corresponding to about 0.015 mg of Cl per liter) at 25°C. Short of changing the medium (as by adding alcohol) the chloride concentration can never be brought below 4×10^{-7} *M* by precipitation with silver ion at 25°C. A very slight excess of silver ion decreases the ionic solubility of silver chloride by the common-ion effect. Above $\sim$0.005 *M* Ag^+, the solubility of silver chloride increases because of the formation of the complex cations Ag_2Cl^+ and Ag_3Cl^{++}, whose activity dissociation constants at 20°C have the following values:[1]

$$\frac{(Ag^+)^2(Cl^-)}{(Ag_2Cl^+)} = 2.5 \times 10^{-5}$$

$$\frac{(Ag^+)^3(Cl^-)}{(Ag_3Cl^{++})} = 1.1 \times 10^{-5}$$

The solubility of silver chloride in silver nitrate solutions is given by the following expression, if concentrations and activities of ions are taken to be the same:

$$S = [Cl^-] + [AgCl] + [Ag_2Cl^+] + [Ag_3Cl^{++}]$$

$$= \frac{1.2 \times 10^{-10}}{[Ag^+]} + 2.4 \times 10^{-7} + 4.8 \times 10^{-6}[Ag^+] + 1.1 \times 10^{-5}[Ag^+]^2 \qquad (20°C)$$

[1] K. H. Lieser, *Z. Anorg. Allgem. Chem.* **304**, 296 (1960). The solubility was determined by the use of radioactive chlorine.

The concentration of Ag^+ at which the solubility of silver chloride is minimal is given by

$$\frac{dS}{d[Ag^+]} = -\frac{1.2 \times 10^{-10}}{[Ag^+]^2} + 4.8 \times 10^{-6} + 2.2 \times 10^{-5}[Ag^+] = 0$$

or $$[Ag^+] = 5 \times 10^{-3} \qquad (20°C)$$

At this silver ion concentration,

$$S = 2.4 \times 10^{-8} + 2.4 \times 10^{-7} + 2.4 \times 10^{-8} + 2.75 \times 10^{-10}$$
$$= 2.9 \times 10^{-7}\,M$$

which corresponds to 0.010 mg of Cl per liter at 20°C.

The solubility of silver chloride rises slowly with the excess silver concentration (Fig. 24-1). Even at 0.1 M $AgNO_3$ it corresponds to only 0.025 mg of Cl per liter at 20°C, so that the solubility loss in the precipitation is negligible except in the most exact work.

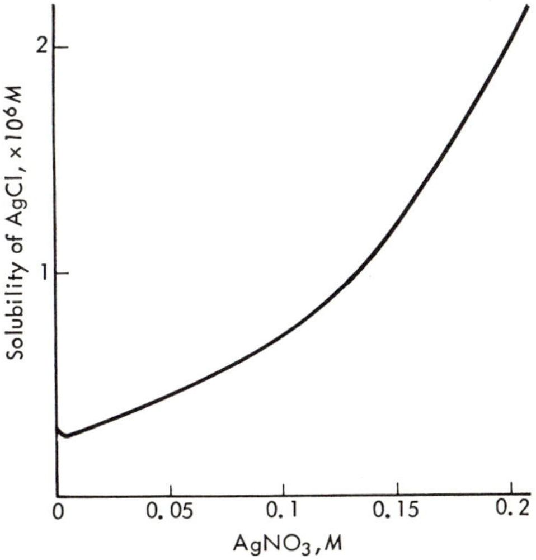

Fig. 24-1. Solubility of AgCl in $AgNO_3$, solutions at 20°C (experimental values of Lieser, Ref. 1).

Electrolytes not having an ion in common with a precipitate increase its ionic solubility to a slight extent (p. 129). In accordance with this fact the solubility of silver chloride is increased by the nitric acid added in the procedure, and such salts as the alkali nitrates. The increase in the solubility of the precipitate from this source is so small that it introduces no error in ordinary analytical work. In the determination of atomic weights and other highly exact analytical work involving the precipitation and weighing of silver chloride, the amount of silver chloride remaining in solution after precipitation is allowed for and that dissolved by the wash liquid is determined nephelometrically (Chap. 56).[2]

[2] In the gravimetric determination of silver as silver chloride by precipitation with hydrochloric acid, the solubility of the precipitate in the excess chloride becomes important. Several complex chloro anions are formed (p. 133). The minimum solubility of silver chloride lies between 0.001 and 0.01 M Cl^-. Analytically, its solubility is still negligible in 0.05 M Cl^-.

Coagulation and aging of the precipitate. Adsorptive properties

Silver chloride is a good example of a colloidal precipitate. When first formed by precipitation in cold dilute solutions, it consists of extremely minute, partially agglomerated particles. The coagulation and the concomitant agglomeration as well as aging of the precipitate are especially promoted by heating and stirring the suspension in the presence of nitric acid and silver nitrate. The precipitate settles in heavy curdy masses, leaving a nearly clear supernatant liquid.

Although the precipitation of silver chloride is quantitatively complete almost immediately after the addition of silver nitrate, the suspension is always heated and allowed to stand for a few hours to complete the coagulation and permit the aging of the precipitate. Because of its finely divided character, silver chloride might be expected to adsorb appreciable amounts of foreign chloride or silver salts. Actually the precipitate shows very little tendency to adsorb foreign substances, even if aged at room temperature for only a few minutes after precipitation. Precipitated under analytical conditions, silver chloride does not carry down alkali or alkaline earth metal salts that may be present, or silver nitrate, to any appreciable extent. The alkali content of a precipitate formed in a solution containing alkali salts is negligibly small.[3]

If silver chloride is washed with pure water, it may be peptized (p. 219). Some of the primary particles will then pass through the interstices of the filter and give a turbid filtrate. The peptization is easily prevented by using a dilute solution of nitric acid as the wash liquid. When the washed precipitate is dried, any nitric acid adhering to it evaporates readily. The strength of the nitric acid wash solution need not be greater than 0.01 *M*. Higher concentrations should not be used, for the solubility of silver chloride is greater in the stronger nitric acid solutions.

Completeness of washing of the precipitate is tested for by determining whether the excess of the precipitating agent, silver nitrate, has been entirely removed. Hydrochloric acid is a suitable reagent for this purpose. One or two drops (not more) of 0.1 *N* acid is added to 3 to 5 ml of the washings collected after the washing has been continued for some time. If the solution remains clear or shows only the slightest opalescence, all silver nitrate (and other soluble salts as well) has been washed out. It should be understood that a saturated solution of pure silver chloride will become opalescent when treated with a slight excess of silver nitrate or hydrochloric acid. Usually the wash liquid does not remain in contact with the precipitate long enough to become saturated with silver chloride, and therefore if the washings do not contain silver nitrate they may remain clear when tested with dilute hydrochloric acid in test tubes and examined in ordinary light.[4]

Effect of light on silver chloride

Silver chloride exposed to light is slowly decomposed into silver and chlorine, the silver remaining colloidally dispersed in the silver chloride and imparting a purplish blue color to it. The chlorine largely escapes from the crystals. This decomposition proceeds quite rapidly in strong light (direct sunlight), and even in

[3] G. F. Hüttig and E. Menzel, *Z. Anal. Chem.* **68**, 354 (1926).

[4] If the washings contain enough dissolved silver chloride to give an opalescence with dilute hydrochloric acid, completeness of washing is indicated by obtaining successive washings yielding equal turbidities.

diffuse light it is appreciable. Account must be taken of this phenomenon in analytical work, or serious errors may result.

In the determination of chloride, the chlorine liberated photochemically from the precipitate reacts with the excess of silver nitrate present in solution and precipitates more silver chloride:

$$3\,Cl_2 + 3\,H_2O + 5\,Ag^+ \rightarrow 5\,AgCl + ClO_3^- + 6\,H^+$$

The results would therefore be high under these conditions. If chloride is present in excess in the solution, as would be the case when *silver* is determined by precipitation with an excess of hydrochloric acid, the error would be negative, since then the chlorine simply escapes from the precipitate. Lundell and Hoffman,[5] who studied the effect of light on silver chloride from the analytical viewpoint, found that serious errors resulted if the solution containing the precipitate of silver chloride was exposed to strong light. Even when the precipitate was allowed to stand in reflected sunlight from a north window for 3 hr with occasional stirring, the results in a chloride determination were 0.2 per cent high. The same authors also found that dried silver chloride very slowly lost weight when exposed to light. It is evident therefore that the operations must be carried out in light as subdued as possible and under no circumstances in direct sunlight. When the solution containing the precipitate is set aside, it should be placed in the dark (locker of desk) and not allowed to stand on the desk. In the most accurate work, daylight should be avoided entirely, and the operations conducted in red light. In ordinary careful analysis in which an accuracy of 0.1 per cent is satisfactory, the precipitation of silver chloride may be carried out in subdued daylight; but, as already mentioned, the precipitate should not be allowed to stand unduly long in such light.

Drying the precipitate

The silver chloride precipitate is not collected on filter paper because the subsequent ashing of the paper would partially reduce silver chloride to silver and possibly volatilize some silver chloride. A filter crucible (porous porcelain, sintered glass, or Gooch) is used instead. The crucible and its contents are heated in an electric oven to dry the precipitate. Drying at 100°C or slightly above gives a residue sufficiently free of water to yield satisfactory results. Experiments by the authors have shown that silver chloride dried to constant weight at 110–120°C contains 0.03–0.04 per cent water. Silver chloride dried at 280°C retains 0.01 per cent water;[6] it must be heated to fusion (455°C) for removal of the last traces of water. The temperature must be carefully regulated in this operation, for the substance volatilizes rapidly above its melting point, and even at the melting point it is slowly volatile; the presence of organic matter causes the reduction of silver chloride. In ordinary analysis, fusion of the silver chloride preparatory to weighing is not necessary. The residue may be dried at 200–300°C or even at 100–200°C with satisfactory results. The dried residue is not appreciably hygroscopic. It has been proposed[7] to dry the precipitate without heating by washing

[5] G. E. F. Lundell and J. I. Hoffman, *Natl. Bur. Std. J. Res.* **4**, 109 (1930).

[6] W. F. Hillebrand, G. E. F. Lundell, H. A. Bright, and J. I. Hoffman, *Applied Inorganic Analysis*, Wiley, New York, 2nd ed., 1953, p. 207.

[7] J. Dick, *Z. Anal. Chem.* **77**, 352 (1929).

it successively with alcohol and ether and allowing the latter to evaporate at room temperature. All the water is not removed in this way, but the results are not high by more than 0.1 per cent, so the method is suitable for rapid work.

Interfering substances

Anions giving silver salts insoluble in dilute nitric acid must, of course, be absent. These include bromide, iodide, thiocyanate, cyanide, ferrocyanide, ferricyanide, sulfide, and thiosulfate. Other anions such as carbonate, phosphate, chromate, arsenate, and oxalate may be present, for the silver salts of these are soluble in the dilute nitric acid medium in which silver chloride is precipitated. In the presence of some of these anions (oxalate) it may be necessary to increase the nitric acid concentration in the solutions over that recommended in the procedure below. There is a tendency for sulfate to be coprecipitated, especially if the excess of silver is large or the period of digestion is short.[8]

Heavy metals may interfere, and if so it is necessary to remove them by appropriate methods before attempting to precipitate chloride. These metals may interfere in several ways:

1. They may hydrolyze in the dilute nitric acid solution and contaminate the silver chloride precipitate with their basic salts (antimony, bismuth, and tin, for example).
2. They may be coprecipitated with silver chloride (platinum and palladium, for instance), or even precipitated (lead chloride).
3. They may prevent complete precipitation of chloride by forming complex or slightly ionized chlorides (mercuric mercury, chromium).

Substances that reduce silver nitrate in acid medium must be oxidized before precipitating the chloride (hypophosphite is an example). Organic hydrophilic colloids such as gelatin and agar, which tend to keep the silver chloride in colloidal solution, must be destroyed. The procedure given below can, in general, be applied without modification only to the determination of chloride in the alkali and alkaline earth chlorides or hydrochloric acid solutions.

Procedure.[9] The following directions apply to the determination of chlorine in a soluble chloride containing no interfering substances.

Weigh out duplicate samples of 0.2–0.3 g (or an amount containing approximately 0.1 g of chlorine) into 250-ml beakers provided with stirring rods and covered with watch glasses. Proceed with each sample as follows. Add approximately 100 ml of water to the beaker, stir to dissolve the sample, and acidify with 1 ml of 1 : 1 nitric acid.[10] The distilled water and nitric acid used should be tested for the presence of chloride with silver nitrate solution.

Precipitate the chloride as silver chloride by slowly adding with stirring (p. 528) a slight excess of 0.1 N silver nitrate solution in the cold. If the chloride content of the sample is known approximately, calculate the volume of silver nitrate

[8] W. J. Blaedel, J. H. Jefferson, and H. T. Knight, *J. Chem. Educ.* **29**, 485 (1952).

[9] Before beginning this determination, the student should read carefully Chapter 20, dealing with the technique of precipitation, filtration, washing of precipitates, etc., and should refer to that chapter constantly during the determination for manipulative details.

[10] If the sample is alkaline, the solution must be neutralized with nitric acid (litmus paper as indicator) and a slight excess added as above.

solution required, and add an excess of 5–10 ml of 0.1 N silver nitrate over the calculated amount; otherwise assume that the sample is sodium chloride, and calculate the volume of precipitant required on this basis, again allowing an excess of 5 or 10 ml. *The precipitation and succeeding operations must be carried out in subdued daylight.*

After the silver nitrate solution has been added, heat the suspension nearly to boiling, and stir for a minute or two to aid the coagulation of the precipitate. Then remove the beaker from the heat, and allow the precipitate to settle. Test for complete precipitation of chloride by adding a few drops of silver nitrate solution to the supernatant liquid. If no more precipitate appears, set the beaker aside in the dark in a place free of HCl fumes and allow the solution to stand for at least 1 to 2 hr before filtration.

Collect the precipitate in a weighed filter crucible (Gooch, porcelain, or glass) which has been dried at the same temperature as is to be employed in heating the precipitate (ca. 100–200°C), following the directions given on pages 536–537 regarding the technique. Wash the silver chloride two or three times by decantation (p. 531) with cold 0.01 N nitric acid before transferring the precipitate to the crucible. Remove the last small particles of silver chloride from the beaker with a "policeman" (p. 532). Carefully examine the beaker to see that all the precipitate has been removed. Wash the precipitate in the filter crucible with 0.01 N nitric acid, added in small portions as described on page 537, until a few milliliters of the washings collected in a test tube give no turbidity, or at most only a faint opalescence, when tested with 1 drop of 0.1 N hydrochloric acid. Finally, wash the precipitate with one or two portions of water to remove most of the nitric acid.

Heat the crucible and its contents to constant weight ($\pm$ 0.2 mg) at 110–200°C, following the directions on page 541. Calculate the percentage of chlorine in the sample.

Other determinations based on the same principle

Bromide, iodide, and thiocyanate can also be determined gravimetrically by precipitation with silver nitrate and weighing of the corresponding silver precipitate by methods closely resembling those for the determination of chloride. Moreover, the oxygen acids of the halogens (chlorate, bromate, etc.) can be determined by reducing them to the halides with a suitable reducing agent and precipitating with silver nitrate.

By using hydrochloric acid as precipitant, silver can be determined with great accuracy as the chloride. Mercury is sometimes estimated by precipitation as mercurous chloride, but the results are somewhat low, owing to the appreciable solubility of the precipitate in an excess of chloride and volatility of the precipitate at relatively low temperatures.

Determination of chlorine in insoluble chlorides

If the chloride is insoluble, it is necessary to boil it with sodium carbonate solution or, better, to fuse it with sodium carbonate and extract the melt with water. In either case chloride goes into solution, and the filtrate is analyzed in the usual way after acidification with nitric acid.

Determination of chlorine in organic compounds

The sample is decomposed by an oxidizing attack and nonionic chlorine is converted to chloride. In the Carius method, the sample is heated with fuming nitric acid in the presence of silver nitrate at 300°C in a sealed tube of heavy-walled glass. Silver chloride is formed and is filtered off after opening the tube. Bromine and iodine in organic compounds are determined similarly.

In the Parr bomb method, the sample is heated with sodium peroxide (and potassium nitrate–sucrose as primer) in a sealed metal vessel. The melt is treated with water and the solution is boiled to decompose the excess peroxide. After acidification, chloride is determined gravimetrically or volumetrically with silver.

The organic compound may be burned in a stream of oxygen and the gaseous products absorbed in hydrogen peroxide. Or, the sample, wrapped in filter paper, can be burned in oxygen in a closed flask (micro method of Schöniger).[11] After combustion, the flask is shaken to absorb the reaction products in sodium hydroxide–hydrogen peroxide solution (added to the flask at the beginning) and chloride is determined volumetrically.

PROBLEMS

1. What volume of 0.10 N silver nitrate is required to precipitate the chloride in 0.500 g of NaCl? *Ans.* 85.5 ml.
2. What volume of silver nitrate solution containing 10.8 g of silver nitrate per liter is required to precipitate the chloride in 0.5 g of calcium chloride hexahydrate?
3. A 1.00-g sample of alloy yielded 0.5721 g of silver chloride. What is the percentage of silver in the alloy?
4. What weight of sample must be taken in the chloride determination in order that 1 mg of AgCl shall represent 0.10 per cent Cl? *Ans.* 0.247 g.
5. What weight of silver is contained in the silver chloride precipitate furnished by 0.2443 g of $BaCl_2 \cdot 2\,H_2O$?
6. A sample consisting of potassium sulfate and potassium nitrate was repeatedly evaporated with hydrochloric acid to convert the nitrate into chloride and then precipitated with silver nitrate. If a 0.4107-g sample yielded 0.2417 g of silver chloride, what was the percentage of potassium nitrate in the sample?
7. If 0.2500 g of an anhydrous mixture of sodium and potassium chlorides gives a dried silver chloride precipitate weighing 0.5248 g, calculate the percentages of NaCl and KCl in the mixture. *Ans.* 33.4 per cent NaCl, 66.6 per cent KCl.
8. A solution of potassium bromate was reduced to bromide and the latter precipitated with an excess of silver nitrate. If 0.4681 g of AgBr was obtained, what was the weight of the $KBrO_3$ present? *Ans.* 0.4160 g.
9. Assuming theoretical conversion, what volume of 0.100 N silver nitrate solution can be obtained from 10.00 g of pure silver chloride residues? *Ans.* 698 ml.
10. What volume of silver nitrate solution containing the equivalent of 2.000 g of silver per liter is required to precipitate the chloride and bromide in 0.5000 g of a mixture containing 60 per cent NaCl and 40 per cent NaBr?
11. A silver chloride precipitate containing free silver as a result of exposure to strong light gave an apparent chloride content of 40.00 per cent in a salt. If the weighed residue contained 0.75 per cent Ag, what is the absolute error resulting?
12. Exactly 50 ml of a pure cobalt chloride solution yielded 0.5395 g of AgCl. Find the weight of Co in a liter of solution. If it is assumed that the maximum error in weighing

[11] For a review of the determination of various elements in organic compounds by the oxygen flask method, see A. M. G. Macdonald, *Analyst*, **86**, 3 (1961).

the silver chloride precipitate is 0.2 mg and there are no other errors, what is the corresponding error in the concentration of the cobalt solution?

13. Find the solubility of AgCl in the following solutions at 25°C taking f_{Ag} and f_{Cl} as 0.96 at $\mu = 0.001$, 0.90 at $\mu = 0.01$, and 0.75 at $\mu = 0.1$: (a) 0.01 *M* HNO_3; (b) 0.001 *M* $AgNO_3$; (c) 0.01 *M* $AgNO_3$; (d) 0.01 *M* $AgNO_3$ + 0.09 *M* $NaNO_3$.
14. Estimate the maximum solubility loss, in terms of Cl, in the determination of chloride in 0.25 g of NaCl by the procedure on page 584, taking the final volume of solution as 150 ml and the volume of wash solution as 100 ml. What maximum error, in relative per cent, can result from the solubility loss?
15. Calculate the solubility of silver chloride in 0.05 *M* sodium chloride solution at 18°C.
16. Ten ml of 0.1 *M* silver nitrate is added to 90 ml of solution containing 0.50 g of $H_2C_2O_4 \cdot 2\,H_2O$ and 10 ml of 0.5 *M* nitric acid. Will silver oxalate be precipitated? Complexing of silver with oxalate may be neglected.

$$[Ag^+]^2[C_2O_4^{--}] = 3.5 \times 10^{-11}$$
$$\text{Oxalic acid: } pK_1 = 1.14, \; pK_2 = 3.85$$

17. 0.143 g of AgCl is digested with 100 ml of 1.0 *M* NaOH until equilibrium is attained. What fraction of the silver chloride will theoretically have been converted to Ag_2O? Disregard formation of AgO_2^-.

$$\tfrac{1}{2}\,Ag_2O + \tfrac{1}{2}\,H_2O \rightleftharpoons Ag^+ + OH^-$$
$$[Ag^+][OH^-] = K_{sp} = 3 \times 10^{-8}$$

18. Does digestion of $PbCl_2$ with sodium carbonate appear to be a feasible way of obtaining the chloride in solution without an appreciable amount of lead? (Consider the formation of $HPbO_2^-$.)
19. Devise a method for determining chloride in a mixture of silver chloride and lead chloride.
20. Determination of the solubility of silver chloride in silver nitrate solutions requires that Cl in low concentrations be determined in the presence of a thousandfold molar excess of silver. Outline two different methods by which this can be done, referring to the literature if necessary.

CHAPTER 25 *Determination of Iron*

Method

The hot solution containing iron in the ferric[1] state is treated with a slight excess of ammonia to precipitate ferric hydroxide (hydrous ferric oxide, $Fe_2O_3 \cdot xH_2O$).[2] The precipitate is filtered off on paper, washed, ignited to ferric oxide, and weighed as such.

Properties of ferric hydroxide

The precipitate formed by rapid addition of ammonia or other basic solution to an iron(III) solution at room temperature consists of amorphous ferric hydroxide having an indefinite composition (p. 214). On aging it may become more or less (submicroscopically) crystalline, depending on conditions. α-FeOOH (goethite) and α-Fe_2O_3 (hematite) are preferentially formed. Under special aging conditions, β-FeOOH and γ-FeOOH (lepidocrocite) are formed. Precipitation from hot solutions gives a precipitate showing an X-ray diffraction pattern of α-Fe_2O_3.[3] A freshly formed amorphous ferric hydroxide precipitate may have a K_{sp} value as large as 10^{-38} at room temperature. On aging, the solubility of the precipitate decreases with an increase in particle size so that the K_{sp} of the (amorphous) precipitate may fall to 10^{-39}. K_{sp} of α-Fe_2O_3 is $10^{-42.7}$.

The solubility of ferric hydroxide as a function of the hydrogen-ion concentration has already been discussed (p. 140). Even at as low a pH as 4, the solubility does not amount to more than about 0.02 mg of Fe per liter. The intrinsic solubility of ferric hydroxide is extremely small, not greater than 2×10^{-9} *M*.[4] Evidently

[1] Ferrous iron is not quantitatively precipitated by ammonia in the presence of ammonium salts.

[2] The terms *hydroxide* and *hydrous oxide* are often used more or less synonymously. Strictly speaking, a metal hydroxide has a definite composition corresponding to the chemical formula, such as $Fe(OH)_3$ and $Al(OH)_3$. Precipitates of quadrivalent, trivalent, and many divalent metal hydroxides are never obtained as such, but in the form of highly voluminous hydrous oxides. These precipitates have no definite stoichiometric composition but consist of the metal oxide or hydroxide with variable amounts of water. The latter may be bound partially chemically and partially physically (adsorption), the total amount of water held by the precipitate depending upon the manner of precipitation and of aging before filtration. See H. B. Weiser, *Inorganic Colloid Chemistry*, vol. II, Wiley, New York.

[3] H. B. Weiser and W. O. Milligan, *Chem. Rev.* **25**, 1 (1939).

[4] H. Lengweiler, W. Buser, and W. Feitknecht, *Helv. Chim. Acta* **44**, 805 (1961). Higher values reported earlier apparently were in error because of the presence of colloidal ferric hydroxide in solution. Ferric hydroxide is extremely weakly amphoteric. K for the reaction $Fe(OH)_3(s) + OH^- \rightleftharpoons Fe(OH)_4^-$ has a value of 10^{-4}–10^{-5}.

errors due to incomplete precipitation and solubility loss of iron in washing will be negligible.

Because the solubility of ferric hydroxide is so small, the relative supersaturation during precipitation will be very great and the precipitate will have a tremendous surface development. Even on aging, the size of the particles increases but slowly. Ferric hydroxide is a typical example of a flocculated colloid. The primary particles agglomerate to large aggregates that retain much adsorbed water. The temperature of precipitation is a most important factor in obtaining an easily filterable precipitate. The coagulation of a colloidal precipitate, and especially the agglomeration of the primary particles, is aided enormously by raising the temperature of the solution. Therefore the precipitation is made at or near the boiling point, and the liquid is kept at 100°C for a short time after precipitation. The liquid should not be allowed to boil for any length of time after precipitation, for then the precipitate is liable to become slimy and difficult to filter, apparently as the result of partial breaking up of the aggregates. More compact and more easily filterable precipitates can be obtained by homogeneous precipitation.

Filter paper is always used for the filtration of gelatinous precipitates such as hydrous ferric oxide. Suction filtration with filter crucibles results in clogging of the crucible by the submicroscopic particles of these precipitates being forced into the pores of the filtering medium. The use of filter paper pulp (p. 464) aids filtration and washing.

Because of their bulky nature the hydrous oxide precipitates are advantageously washed by decantation (p. 531). Hot water is a suitable wash liquid for hydrous ferric oxide, because the precipitate is not easily peptized. If the washing is likely to be long continued, a hot 1 per cent solution of ammonium chloride is recommended as a wash liquid and will do no harm in any case. Under no circumstances may the precipitate be allowed to stand in the filter paper before complete washing, because it shrinks rapidly as it partially dries. As a result, channels are formed in the precipitate, the wash liquid runs off through these, and thorough washing then becomes impossible. By virtue of its large surface development, hydrous ferric oxide (and other hydrous oxides) exhibits pronounced adsorptive properties.

From solutions having an acid reaction there will be primarily an anion adsorption. The surface of the hydrous oxide attracts the protons (hydrogen ions), and consequently an equivalent amount of anion is adsorbed (see p. 215):

$$\underset{(surface)}{Fe^{3+}\begin{matrix}OH^-\\OH^-\\OH^-\end{matrix}} + \underset{(acid\ in\ solution)}{H^+ + A^-} \rightarrow \underset{(surface)}{Fe^{3+}\begin{matrix}OH^-\\OH^-\\OH_2\end{matrix}}\Bigg| A^-$$

The adsorbability of the anions increases with increasing charge. For example, sulfate, chromate, and oxalate are much more strongly adsorbed than chloride, nitrate, etc. In addition, the adsorption increases with increasing hydrogen-ion concentration. Therefore a coprecipitation of anions can be expected if the precipitate is formed in acid medium. This is actually the case. On very slow addition of ammonium hydroxide to a solution of a ferric salt, a coprecipitation of anions occurs, because the solution remains acid as a result of the strong hydrolysis of the ferric ions as long as some of these remain in the solution. The coprecipitated anions are more or less rapidly given off again in the presence of an excess of ammonia.

In ammoniacal, or quite generally in alkaline, media, hydrous ferric oxide tends to adsorb hydroxyl ions, with the result that an equivalent amount of cation is carried down:

$$\begin{array}{c} OH^- \\ Fe^{3+}OH^- \\ OH^- \end{array} + OH^- + B^+ \rightarrow \left.\begin{array}{c} OH^- \\ Fe^{3+}OH^- \\ OH^- \\ OH^- \end{array}\right| B^+$$

(surface) *(base in solution)* *(surface)*

Therefore, in precipitation from ammoniacal medium there may be a pronounced coprecipitation of cations. The adsorbability of the cations increases with decreasing solubility of their hydroxides and with their charge. Thus calcium, magnesium, copper [$Cu(NH_3)_4^{++}$], zinc [$Zn(NH_3)_4^{++}$], etc., are much more strongly adsorbed and coprecipitated than sodium, potassium, and ammonium. A large excess of ammonium ion decreases the adsorption and coprecipitation of other cations. There is competition between the ammonium and other cations to be adsorbed; increase of the ammonium concentration therefore favors the adsorption of the ammonium ions at the cost of the other cations. Much more effective is the combination of a small excess of ammonia with large amounts of ammonium salts. Under these conditions the replacing effect of the ammonium ions is so great that the adsorption and coprecipitation of divalent cations can be reduced to extremely small values in many instances.

Properties of ferric oxide

When hydrous ferric oxide is strongly heated (800–1000°C), it is converted into ferric oxide:

$$Fe_2O_3 \cdot xH_2O \rightarrow Fe_2O_3 + xH_2O$$

The ignition of the precipitate should be made under good oxidizing conditions, especially during the burning of the paper filter, for otherwise ferric oxide may be partially reduced by carbon or reducing gases to the magnetic oxide, Fe_3O_4, or even to the metal. These reduction products can be converted back to ferric oxide by continued heating of the residue with free access of air, although the reoxidation may be slow if much magnetic oxide has been formed. Care should be taken to exclude reducing gases from the flame during the ignition (p. 539). When a platinum crucible is heated with a flame, there is possibility of transfusion of hot reducing gases through the walls.

On strong heating, ferric oxide is decomposed into the magnetic oxide and oxygen:

$$6\ Fe_2O_3 \rightleftharpoons 4\ Fe_3O_4 + O_2$$

Fortunately a very high temperature must be attained before this decomposition becomes analytically significant. Baxter and Hoover[5] found that a sample of pure ferric oxide heated in air to constant weight at approximately 1100°C lost only 0.004 per cent in weight, corresponding to the formation of 0.12 per cent Fe_3O_4. The magnetic oxide resulting forms a solid solution with the ferric oxide,[6] and

[5] G. P. Baxter and C. R. Hoover, *J. Am. Chem. Soc.* **34**, 1657 (1912).

[6] R. B. Sosman and J. C. Hostetter, *J. Am. Chem. Soc.* **38**, 807, 1188 (1916).

consequently the oxygen pressure of the system at equilibrium depends upon the ratio of ferrous iron to ferric in the solid phase (contrast this system with the $CaCO_3$–CaO–CO_2 system, p. 190). At higher temperatures the extent of the decomposition is greater. This solid solution formation also explains why the reoxidation of the magnetic oxide formed during the ignition of the hydrous oxide is a slow process.

Ferric oxide is rendered water-free by ignition at 1000°C. The precipitate can be satisfactorily ignited in a covered porcelain or platinum crucible over a Tirrill burner. The residue thus obtained is not appreciably hygroscopic, in contrast to aluminum oxide, which requires ignition at the minimum temperature of 1200°C to render it water-free and nonhygroscopic (p. 598).

Often it is necessary to obtain the combined weight of ferric oxide and aluminum oxide in a sample. Thus, in determining aluminum and iron in the presence of each other, the metals are precipitated as the hydrous oxides and ignited to the oxides; the weight of the ignited mixed oxides is obtained, the iron in the residue is determined by a volumetric method, and the aluminum is then found by difference. In such cases the choice of a proper ignition temperature for the mixed oxides of aluminum and iron is an important matter. If the temperature is too low, the aluminum oxide is not rendered water-free, because it tenaciously retains water, as already stated, up to temperatures of 1200°C or slightly higher. On the other hand, if the ignition temperature is too high, there is danger of converting some of the ferric oxide into the magnetic oxide of iron; it appears[7] that this decomposition can be analytically important at 1200°C.

It is sometimes stated that a precipitate of hydrous ferric oxide containing chloride (as from a wash solution of ammonium chloride) loses iron on ignition as the result of the volatility of ferric chloride at high temperatures. Such losses do not take place if only small amounts of chloride are present.[8] It is safe to ignite a precipitate of iron which has been washed with a 1 per cent solution of ammonium chloride without first washing out most of the chloride.

Interfering substances. Separations

Other elements that are precipitated by ammonium hydroxide, such as aluminum, trivalent chromium, titanium, zirconium, and many of the less common elements, must be absent in a solution in which iron is to be determined gravimetrically. Anions such as phosphate, arsenate, vanadate, and silicate, which yield insoluble iron salts in weakly basic media, must likewise be absent. Silica is a very common contaminant of the hydrous ferric oxide precipitate; it may be derived from the ammonium hydroxide used in the precipitation, or it may be taken up from the glass vessels during the precipitation.

Certain substances form complexes with ferric iron, from which it cannot be precipitated with ammonium hydroxide. Organic hydroxy compounds such as hydroxy acids (tartaric, citric, etc.) and sugars complex iron in this way. These can be destroyed by igniting the sample or oxidizing with an appropriate reagent in solution. Pyrophosphate and fluoride hinder the precipitation.

[7] K. Zimmermann, *Chem. Weekblad* **31**, 317 (1934), claims that ferric oxide is decomposed to a very appreciable extent even at 1050°C; L. C. Peck, *U.S. Geol. Surv. Bull.* **1170**, 1964, recommends an ignition temperature of 1100°C for the mixed oxides of iron and aluminum.

[8] H. W. Daudt, *J. Ind. Eng. Chem.* **7**, 847 (1915).

Precautions must be taken to prevent the coprecipitation, or even precipitation, of divalent metal hydroxides when precipitating ferric iron with ammonium hydroxide. With some of these metals a satisfactory separation cannot be made. Generally a high concentration of ammonium salt in the solution has a favorable effect. The decrease in the coprecipitation of such metals as magnesium and calcium under these conditions is to be attributed largely to the replacing effect of the ammonium ion (p. 590). No particular difficulty is encountered in separating ferric iron by ammonia precipitation from the alkali and alkaline earth metals[9] as well as magnesium if a sufficient amount of ammonium salt is present; a double precipitation gives a hydrous ferric oxide practically free of these metals (in the presence of much magnesium a triple precipitation may be required). Likewise a satisfactory separation of iron from nickel and manganese can be made if care is taken to limit to a minimum the excess of ammonium hydroxide added.[10] The separation from copper, zinc, and cobalt by the addition of a large excess of ammonium hydroxide is not satisfactory.

Precipitation of iron by the basic acetate method (p. 256) serves to separate the metal satisfactorily from zinc, nickel, cobalt, and manganese under the proper conditions.[11] By double precipitation with ammonium benzoate in hot, weak acetic acid medium, good separations of iron from zinc, manganese, nickel, cobalt, calcium, and magnesium can be effected.[12]

Other precipitation forms of iron

Cupferron precipitation of iron(III) in mineral acid medium serves as a separation from aluminum and some other elements (p. 270). 8-Hydroxyquinoline and derivatives are infrequently used to precipitate iron(III).

2-Mercaptopyridine-N-oxide ("thione") yields a black chelate, $Fe(C_5H_4NOS)_3$, with ferric (or ferrous) iron at pH 2–6, which has been proposed for its gravimetric determination.[13] This reagent is used as the bromide of its thiouronium

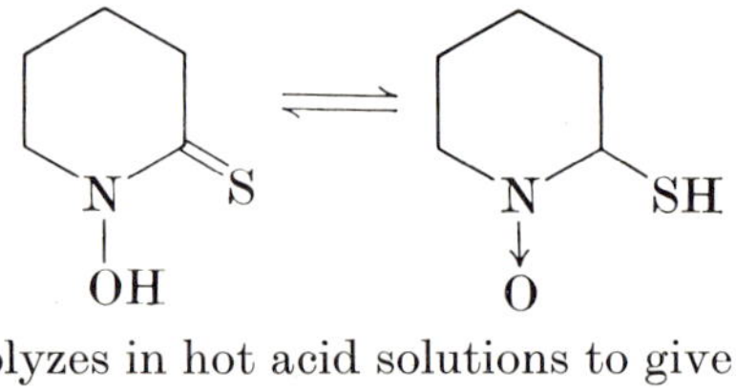

derivative, which hydrolyzes in hot acid solutions to give thione. The iron chelate formed by homogeneous precipitation is crystalline and can be weighed after drying at 160°. Ca, Mg, Al, Ti, and Th do not interfere, but, Cu, Co, Ni, Zn, and Cr(III) do; a limited amount of Mn may be present.

The separation of iron as sulfide from other elements of an ammonia precipitate is based on the nonprecipitability of these by ammonia in the presence of tartrate,

[9] When calcium or other alkaline earth metals are present, it is important to use carbonate-free ammonium hydroxide for the precipitation of iron (and other metals precipitable by ammonia), else calcium carbonate may be precipitated.

[10] G. E. F. Lundell and H. B. Knowles, *J. Am. Chem. Soc.* **45**, 676 (1923).

[11] H. H. Willard and J. L. Sheldon, *Anal. Chem.* **22**, 1162 (1950), show that iron can be separated satisfactorily from manganese, cobalt, nickel, copper, zinc, cadmium, and magnesium as basic ferric formate precipitated by slow hydrolysis of urea.

[12] I. M. Kolthoff, V. A. Stenger, and B. Moskovitz, *J. Am. Chem. Soc.* **56**, 812 (1934).

[13] J. A. W. Dalziel and M. Thompson, *Analyst* **89**, 707 (1964).

owing to the formation of tartrate complexes. The complex of iron is, however, decomposed by sulfide because of the very slight solubility of FeS, in ammoniacal medium. Such elements as Al, Ti, Zr, Cr, P, V, Be, Nb, and Ta are not precipitable as sulfides and a separation from iron is readily obtained.

Determination of iron in ferrous ammonium sulfate or other salt

The gravimetric determination of iron as Fe_2O_3 is rarely applied in practice because of the presence of other elements precipitated by ammonia. The volumetric determination (oxidative titration of ferrous iron) is much more important (Chap. 43). However, ammonia precipitation of iron, aluminum, titanium, and other elements is important in the systematic analysis of silicates. The gravimetric determination of iron is included here as an illustration of ammonia precipitation.

Procedure. Weigh out two 1.0-g samples of ferrous ammonium sulfate hexahydrate (or a sufficient amount of other sample to contain 0.15–0.2 g of iron) into 400-ml Pyrex beakers provided with stirring rods and cover glasses. Dissolve each portion in 40 or 50 ml of water and 10 ml of 1 : 1 hydrochloric acid. Heat the solutions to boiling, and treat each as follows: Measure out 1 ml of concentrated (16 N) nitric acid[14] in a 5- or 10-ml graduate, and add it dropwise with stirring to the solution, which should be boiling only gently. The solution will first become dark, owing to the formation of $FeSO_4 \cdot NO$; this soon decomposes, leaving a yellow solution. Continue to boil the solution gently until it is clear yellow. Usually boiling for 3–5 min is sufficient. At this point it is well to test the solution for the complete oxidation of the iron. Transfer a drop of the solution to a test tube by means of a stirring rod, and dilute with about 1 ml of water. Place a small drop of the mixed diluted solution on a spot plate, and add a drop of freshly prepared 0.01–0.02 per cent potassium ferricyanide solution. If a blue color appears, ferrous iron is still present in the solution, and more nitric acid must then be added to the solution in the beaker and the boiling continued to complete the oxidation. Rinse the solution remaining in the test tube into the beaker.

If all the iron is found to be in the ferric state, dilute the solution to 200 ml, heat nearly to boiling, and add pure 1 : 1 ammonium hydroxide, in a slow stream from a small beaker, stirring constantly, until a slight excess of reagent is present, as shown by the odor of the liquid after blowing away the fumes.[15] Remove the beaker from the heat, and allow the precipitate to settle. Then decant the supernatant liquid through a 9-cm ashless filter paper of loose texture (e.g., S and S 589, black band) well fitted to the funnel, being careful to leave as much as possible of the precipitate in the beaker. Wash the precipitate three or four times by decantation (p. 531) with hot water, using 75–100 ml each time. Then transfer the hydrous ferric oxide to the filter (p. 532) with the aid of a policeman (p. 472). Dislodge all particles of precipitate from the walls of the beaker, and add them

[14] Instead of nitric acid, bromine water or pure hydrogen peroxide may be used to oxidize the ferrous iron. Hydrogen peroxide is conveniently used in the form of 30 per cent solution (perhydrol); 1 ml of this solution is sufficient. The excess should be destroyed by boiling the solution.

[15] At this point it is advantageous, but not necessary, to add a little ashless filter-paper pulp to aid the filtration and washing of the precipitate (p. 464). The addition of paper pulp is of greater importance when the ignited residue must subsequently be brought into solution, as in rock analysis.

to the main portion. Finally, examine the beaker carefully for any remaining precipitate. Wash the precipitate on the filter (p. 535) with hot water until the washings give no reaction, or at most only a slight one, with silver nitrate acidified with nitric acid.[16]

If convenient, set the filter and its contents aside at this point to drain and partially dry overnight. This air drying is by no means necessary, but it saves time in the subsequent slow heating of the precipitate in the crucible to expel water. Fold over the edge of the filter paper, and place it in a weighed porcelain crucible. Proceeding as described on page 538, heat the crucible with the low flame of a Tirrill burner to expel the water, taking great care to avoid sudden expulsion of steam, which would be almost certain to lead to losses of precipitate. When the precipitate is dry, char the paper below red heat, and burn off the carbon at as low a temperature as possible, following the directions given on page 539. Increase the heat of the burner to full when all the carbon has been burned away, and maintain the crucible at this temperature for 15 min. Take care to exclude the gases of the flame from the interior of the crucible. Allow the crucible to cool below redness, and place it covered in a desiccator. After 60 min weigh the crucible and its contents, and repeat the ignition to constant weight (0.2 mg), employing 10- to 15-min periods of heating.

Calculate the percentage of iron in the sample from the weight of Fe_2O_3 obtained.

Other metals determined by precipitation as hydrous oxides

Of the more common metals the following can be quantitatively precipitated by ammonium hydroxide: aluminum, chromium(III), titanium(IV), and zirconium.[17] The precipitates can be ignited to the corresponding oxides Al_2O_3, Cr_2O_3, TiO_2, and ZrO_2. In the precipitation of aluminum and chromium the pH of the solution must be adjusted rather carefully because the hydroxides of these metals are somewhat soluble in ammonium hydroxide. The precipitation of hydrous aluminum oxide is considered in Chapter 26. The gravimetric determination of chromium possesses the disadvantage that the precipitate must be ignited in hydrogen when results of the highest accuracy are desired. Chromic oxide when strongly heated in air gives small amounts of $Cr_2(CrO_4)_3$ with consequent high results. It is much better to determine chromium volumetrically by oxidizing to chromic acid and titrating the latter (see, for example, Chap. 65).

PROBLEMS

1. A 1.000-g sample of partially effloresced ferrous ammonium sulfate yielded a residue of Fe_2O_3 weighing 0.2140 g after strong ignition. What was the percentage of water in the sample? *Ans.* 23.86 per cent.

[16] Naturally this thorough washing is not called for in the analysis of pure ferrous ammonium sulfate or other samples not containing appreciable amounts of nonvolatile matter.

[17] In the presence of an oxidizing agent such as bromine or ammonium persulfate, divalent manganese can be quantitatively precipitated as hydrous manganese dioxide by ammonia. On ignition the precipitate can be made to give a residue approximating the composition Mn_3O_4. Since the composition of the ignition residue is variable even under rather carefully regulated conditions, this method of determining manganese is not often applied. Besides, titrimetric methods are far superior to gravimetric for manganese.

2. A determination of iron in a sample, based on the weighing of Fe_2O_3, gave a value of 10.11 per cent Fe. If the ignited precipitate contained 3.00 per cent Fe_3O_4, what is the true percentage of Fe in the sample? *Ans.* 10.12 per cent.
3. (a) What volume of bromine water (3.5 g of bromine in 100 ml of solution) is theoretically required to oxidize the iron in 1.000 g of $FeCl_2 \cdot 4\,H_2O$?
   (b) What volume of 16 *M* nitric acid is required to oxidize the iron in 1.000 g of $FeSO_4 \cdot 7\,H_2O$? Assume reduction of nitric acid to NO. *Ans.* (a) 11.5 ml.
4. An oxide of iron contains 24.12 per cent iron in the ferrous state and 48.24 per cent in the ferric state. What is its formula?
5. How many milliliters of ammonium hydroxide solution having a density of 0.95 and containing 12.7 per cent by weight NH_3 are required theoretically to precipitate the iron in 0.50 g of ferric bromide hexahydrate?
6. Will the addition of 5 ml of 16 *M* ammonium hydroxide to 195 ml of 0.1 *M* ferrous ammonium sulfate solution (oxygen-free) produce a precipitate of ferrous hydroxide if the solubility product constant of the latter is 2×10^{-14}?
7. A 1 *M* solution of manganous chloride is about 10^{-6} *M* in ferric iron. Taking the various factors involved into account, discuss the feasibility of quantitatively separating the two metals from each other by hydroxide precipitation of iron. See the Appendix for constants.

CHAPTER 26 *Determination of Aluminum*

Like almost all other elements, aluminum can be determined without great difficulty if present alone in solution. However, it lacks distinctive separation reactions and its determination reactions are far from selective, so that the accurate determination of aluminum in the presence of associated elements is not likely to be either simple or easy. This chapter deals mostly with the determination of aluminum in the absence of interfering elements.

Gravimetric methods are still important in the determination of aluminum. The EDTA titration methods for aluminum leave something to be desired.[1] The two important gravimetric methods are (1) precipitation as aluminum hydroxide, followed by ignition to Al_2O_3 and weighing as such, and (2) precipitation as aluminum 8-hydroxyquinolate, which can be weighed as such after drying.

Determination as oxide

Even in a solution free of other elements, the determination of aluminum as Al_2O_3 is subject to important sources of error. Aluminum hydroxide is an amphoteric substance and the use of too large an excess of ammonia in its precipitation can result in appreciable solubility loss by aluminate formation. The precipitate must be ignited strongly to form anhydrous Al_2O_3 and the residue is hygroscopic unless it has been heated to a very high temperature. Aluminum hydroxide, or basic aluminum salt, is of course a very unselective precipitation form, because many other metals will be precipitated in the neutral or faintly acidic medium in which the precipitation is carried out. Besides, elements which when present alone are not precipitated, may be coprecipitated with aluminum hydroxide.

Composition of the precipitate. The system Al^{3+}–OH^-–H_2O is not a simple one. Successive replacement of water molecules in the hydrated aluminum ion $Al(H_2O)_6^{3+}$ by OH^- eventually leads to precipitation of aluminum hydroxide or aluminum hydroxide-oxide. The reaction is complex and will not be examined here. In addition to amorphous aluminum hydroxide, which can be represented as $Al(OH)_3 \cdot xH_2O$ or $Al_2O_3 \cdot xH_2O$, the following crystalline ortho- or metahydroxides of aluminum exist:

Gibbsite, $Al(OH)_3$	(also called hydrargillite)
Bayerite, $Al(OH)_3$	(metastable)
Boehmite, $AlO(OH)$	
Diaspore, $HAlO_2$	(not encountered in analytical aluminum hydroxide precipitates)

[1] For determination of aluminum in silicates by EDTA titration following cupferron–chloroform extraction, see M. Weibel, *Z. Anal. Chem.* **184**, 322 (1961); cf. V. Skrivanek and P. Klein, *ibid.* **184**, 360 (1961). E. Kiss, *Anal. Chim. Acta* **39**, 223 (1967), separates iron by MIBK extraction from hydrochloric acid solution and titanium by cupferron–chloroform extraction, and then titrates aluminum with DCTA; also see D. T. Pritchard, *ibid.* **32**, 184 (1965) and R. Pribil and V. Vesely, *Talanta* **9**, 23 (1962). This listing is incomplete.

The order of increasing stability of these forms seems to be amorphous aluminum hydroxide < boehmite (< bayerite) < gibbsite, but attainment of equilibrium is slow.

The conditions of precipitation and aging of aluminum hydroxide determine the composition of an analytical precipitate, which is usually a mixture of components. The phases present can be identified by X-ray diffraction and differential thermal analysis. Ammonia precipitation, as usually carried out, gives a largely amorphous precipitate, with some boehmite. Very slow addition of ammonia to a hot solution can yield a mixture of amorphous aluminum hydroxide and gibbsite. Slow passage of CO_2 into a hot aluminate solution precipitates mostly gibbsite, fast passage into a cold solution precipitates much boehmite, both with amorphous hydroxide.[2] Bayerite has also been reported to be formed by leading CO_2 into alkaline solutions of aluminum.

The change taking place in an aluminum hydroxide precipitate formed by ammonia precipitation on digestion under analytical conditions may be described as relatively slight. Coagulation of the precipitate is promoted, but transformation of amorphous aluminum hydroxide into crystalline modifications proceeds very slowly and little change may occur in a day at 100°C. The conditions existing during precipitation are much more important than the aging effect. Denser, more easily filterable, aluminum hydroxide precipitates (or precipitates of basic salts of organic acids) can be obtained by slow precipitation brought about by gradual decrease in the hydrogen-ion concentration of the solution as a result of reaction of H^+ with a suitable added substance at or near 100°C. See page 256 for brief observations on such methods.

It can be seen that the manner of precipitation and the aging of the aluminum hydroxide precipitate will affect its solubility. The K_{sp} of gibbsite has been reported as 10^3–10^4 times smaller than the K_{sp} of amorphous aluminum hydroxide. The composition of the precipitate would also be expected to affect coprecipitation of cations and anions.

Solubility of the precipitate. The solubility of aluminum hydroxide as a function of the pH has already been touched upon (p. 139). Because of the dependence of K_{sp} upon the manner of precipitation and incomplete knowledge of the equilibria at and near the isoelectric point (the pH of minimum solubility of aluminum hydroxide), the analytical solubility in this region cannot be satisfactorily calculated from equilibrium constants drawn from tables. However, a half-century ago, Blum made a careful study of the determination of aluminum as Al_2O_3 after ammonia precipitation, and obtained approximate values for the solubility of aluminum in the critical region.[3] He found that the precipitation is quantitative at or slightly below pH 7 and resolution of the precipitate is already significant at pH 9:

pH	Mg *of* Al_2O_3/*liter*
6.0	6
6.5	~0.5
7.6	1
9.0	2

The figures in the last column represent Al_2O_3 in the hot filtrate (after adjustment of the pH the solution was boiled for a few minutes before filtration).

Phenol red is a suitable indicator for adjusting the pH in the ammonia precipitation.[4] Its pH range is 6.4 (yellow) to 8.0 (red). A comparatively large amount of ammonium

[2] See L. Erdey, *Theorie und Praxis der Gravimetrischen Analyse*, Akad. Kiado, Budapest, 1964, vol. II, p. 293 et seq., for differential thermograms of such precipitates.

[3] W. Blum, *J. Am. Chem. Soc.* **38**, 1282 (1916).

[4] J. N. Frers, *Z. Anal. Chem.* **95**, 119 (1933).

chloride should be present in the solution during the precipitation. On the one hand, the buffering action of the ammonium salt enables the adjustment of the pH to the optimum value to be made more easily, and, on the other, the salt aids in the coagulation of the precipitate. Furthermore, a sufficient quantity of ammonium salt minimizes the coprecipitation of divalent metals such as calcium and magnesium, because the ammonium ion has a replacing effect upon the other cations. To obtain an easily filterable precipitate, the precipitation is made in hot solution. The precipitate must not be washed with water, for then the hydrous aluminum oxide is peptized and will run through the filter. A dilute solution of ammonium chloride made basic to phenol red with ammonium hydroxide is a suitable wash liquid.

Ignition of the precipitate. A high ignition temperature is needed to convert aluminum hydroxide to anhydrous Al_2O_3. Gibbsite begins to be transformed to γ-Al_2O_3 at about 300°C; boehmite is also formed at lower temperatures and converts to γ-Al_2O_3 at about 450°C. This form of alumina consists of very fine crystals and retains water tenaciously. After ignition of an ammonia precipitate at 950°C, 1–2 per cent water may be held in the Al_2O_3 residue. At about 1100°C, γ-Al_2O_3 is transformed into α-Al_2O_3 (corundum), the stable form, and the last of the water is expelled.[5] However, the residue is hygroscopic. If the ignition temperature is raised to 1200°C or higher, the oxide becomes much less hygroscopic and can be weighed without special care.[6] The ignition behavior of aluminum hydroxide or of the aluminum precipitates obtained in the presence of organic anions (which are in part basic salts) varies somewhat, depending, for example, on the particle size, but a high temperature is always required for satisfactory conversion to nonhygroscopic Al_2O_3.

The statement sometimes made that hydrous aluminum oxide containing chloride (as from the wash solution) loses aluminum chloride on ignition has no basis in fact. Precipitates formed in solutions containing sulfate are contaminated with basic aluminum sulfate, and this substance must be strongly heated to decompose it into aluminum oxide and sulfur trioxide. Whereas hydrous aluminum oxide containing no sulfate can be heated to constant weight at 1200°C in approximately 10 min, a precipitate containing moderate amounts of sulfate will require a $\frac{1}{2}$-hr ignition at the same temperature to effect the same result.

Coprecipitation with hydrous aluminum oxide. Aside from the fact that many other elements are quantitatively precipitated by ammonium hydroxide, it is necessary to bear in mind when precipitating hydrous aluminum oxide that a number of metal hydroxides or hydrous oxides are carried down by the precipitate. Cobalt, copper, zinc, and some other hydroxides are partially precipitated at the pH required to precipitate aluminum quantitatively. A reprecipitation does not give a pure precipitate when appreciable amounts of these metals are present (see p. 256).

Under the conditions specified in the procedure below, satisfactory separations of

[5] W. Blum, Ref. 3, found that a precipitate ignited in platinum over a blast lamp at 1100–1150°C for 5–10 min did not lose weight when heated further at 1450°C for three 10-min intervals, correction being made for the loss in weight of the platinum crucible, and great care being taken to prevent adsorption of moisture during cooling and weighing.

[6] J. N. Frers, *Z. Anal. Chem.* **95**, 131 (1933). H. Wiele, *Anal. Chim. Acta* **15**, 326 (1956), states that an ignition temperature of ~1200°C is required to dehydrate Al_2O_3, and a temperature of 1500°C to make it nonhygroscopic. For other observations on ignition of Al_2O_3, see O. I. Milner and L. Gordon, *Talanta* **4**, 115 (1960). According to Frers the hygroscopicity of a *porcelain* crucible must be taken into account. He found that porcelain crucibles which had been ignited strongly were appreciably hygroscopic; therefore, an error was likely to be made if they were weighed immediately after removal from the desiccator, because the change in weight due to adsorption of water is greatest when they are first exposed to the air. Frers, therefore, recommends cooling the crucible for 1 hr in the desiccator and then letting it stand for 10 min in the balance before weighing. Evidently this procedure can be followed only when the ignited residue is entirely nonhygroscopic.

aluminum from alkalies,[7] calcium, barium, strontium, and magnesium can be made, especially if the precipitate is dissolved and reprecipitated. Magnesium is the most strongly coprecipitated of these. Blum[8] found that when precipitated under conditions similar to those recommended below, 0.1 g of aluminum oxide contained less than 1.0 mg of magnesium oxide, when the equivalent of 0.2 g of magnesium oxide was present. A second precipitation reduces the amount of magnesium to negligible proportions. It is especially important to have sufficient ammonium chloride in solution when magnesium is present.

In the presence of cobalt, zinc, and nickel, it is best to precipitate aluminum as the basic benzoate in slightly acid solution (p. 256). For homogeneous precipitation of aluminum, see page 256.

Aluminum can be separated from iron by ammonia precipitation in the presence of thioglycolic acid (p. 257).

Procedure. To 200 ml of solution containing a suitable weight of sample (0.2–0.3 g of Al_2O_3), add 4–5 g of pure ammonium chloride and 0.5 ml of 0.1 per cent solution of phenol red. Heat the solution nearly to boiling, and add pure dilute (1 : 1) ammonium hydroxide dropwise until the color changes from yellow to orange. Keep the liquid at the boiling point for 2 or 3 min, and then filter through a loose-textured paper. Wash the precipitate with a hot 2 per cent solution of ammonium chloride made neutral with ammonia against phenol red. If necessary dissolve the precipitate in hot 1 : 1 hydrochloric acid and reprecipitate. Place the paper with the precipitate in a weighed platinum or porcelain crucible, dry, char the paper and burn off the carbon, and ignite the residue in a furnace at 1200°C or, less preferably, over a blast lamp (see Ref. 8). Heat for 10-min periods until constant weight is attained. Although the oxide should not be sensibly hygroscopic if the temperature of the ignition has been 1200°C or above, it is wise to observe the usual precautions called for in the weighing of hygroscopic substances. Cool the crucible in a desiccator charged with magnesium perchlorate or other effective desiccant, and keep it covered at all times.

Determination by precipitation with 8-hydroxyquinoline

8-Hydroxyquinoline quantitatively precipitates aluminum from an acetate-buffered solution (pH 5–6) as $Al(C_9H_6ON)_3$. The precipitate can be weighed as such after drying at 120°–150°C or it can be titrated bromometrically (Chap. 45). Precipitation of aluminum as the hydroxyquinolate has advantages over precipitation as the hydroxide. The hydroxyquinolate is crystalline and can be filtered more easily than the hydroxide. It can be dried at a low temperature, it is not hygroscopic, and the gravimetric factor is small ($Al/Al(C_9H_6ON)_3 = 0.05873$).

Under the proper conditions, excellent accuracy and precision are obtainable, as shown by the following data for the gravimetric method (procedure below) in pure solution[9]:

Al mg	Mean difference from theoretical value % (relative)	Standard deviation of a single determination % (relative)
2	−0.65	0.25
10	0.0	0.07
20	+0.02	0.06

[7] With small amounts of alkalies present a single precipitation will usually suffice, but when large amounts are in solution, as, for example, after a potassium pyrosulfate fusion, double precipitations must be made.

[8] If the ignition is made over a blast lamp, a platinum crucible should be used.

[9] A. Claassen and L. Bastings, *Analyst* **92**, 614 (1967). They also give directions for the volumetric method, not included here. For another study of the gravimetric oxine method for aluminum, see R. A. Chalmers and M. A. Basit, *Analyst* **92**, 680 (1967).

A number of important separations of aluminum from other elements can be effected by 8-hydroxyquinoline. Precipitation in faintly acidic solution separates aluminum from the alkalies, magnesium, calcium, and beryllium[10]; in ammoniacal solutions from phosphorus, arsenic, fluorine, and boron; in ammoniacal solutions containing hydrogen peroxide from tantalum, niobium, titanium, and molybdenum; and in ammonium carbonate solution from uranium.[11] On the other hand, in solutions containing tartrate and sodium hydroxide, copper, cadmium, zinc and magnesium are precipitated by 8-hydroxyquinoline, whereas aluminum and iron(III) remain in solution.[12] Precipitation of aluminum in ammoniacal cyanide-EDTA solution, with a titrimetric finish, allows determination of aluminum in the presence of many elements including the alkaline earths, magnesium, cadmium, cobalt, nickel, iron, copper, rare earths, manganese, tin, and zinc; interfering elements include beryllium, antimony, bismuth, gallium, indium, niobium, tantalum, thorium, uranium, and zirconium.[13]

Reagents

8-Hydroxyquinoline, 2.5 per cent. Dissolve 25 g of the pure reagent in 29 ml of 6 *M* hydrochloric acid, dilute with water, filter if necessary and make up to one liter.

Ammonium acetate, 20 per cent w/v. Dilute this stock solution with 9 times its volume of water before use. The pH of the diluted solution should be 6.5–7.0.

Wash solution. Dilute 8 ml of the 2.5 per cent 8-hydroxyquinoline solution to 500 ml with water, add 3 drops of bromocresol purple indicator solution and dilute ammonia until the color just changes to purple (pH 6), and dilute to one liter.

Procedure

Add dilute ammonia dropwise to the cold sample solution ($\sim$100 ml, 2–30 mg Al) until a slight precipitation of aluminum hydroxide occurs. Then add 2 *M* hydrochloric acid dropwise until the precipitate dissolves, followed by an excess of 0.2–0.5 ml. Add 0.7 ml of 8-hydroxyquinoline solution for each milligram of aluminum present plus an excess of 10 ml. Heat the solution to 70°C and add slowly, with stirring, 20 ml of ammonium acetate solution. Heat just to boiling and allow the beaker to stand on the steam bath for one-half hour. Cool to about 50°C (or allow to cool to room temperature overnight).

Collect the precipitate in a weighed filter crucible (glass or porcelain). Transfer and wash the precipitate with warm (50°–60°C) wash solution, using not more than 100 ml. Finally wash twice with 5 ml of cold water. Avoid sucking the precipitate dry during washing to prevent the formation of a compact mass which hinders washing. Dry to constant weight at 140°–150°C. $Al_2O_3/2\ Al(C_9H_6ON)_3 = 0.1110$.

Notes. 1. The use of a hydrochloric acid solution of 8-hydroxyquinoline instead of an acetic acid solution (hitherto commonly used) makes it easier to attain the desired pH (>5.2)—less acetate solution need be added.

2. At 50°C, with a 10 ml excess of 2.5 per cent 8-hydroxyquinoline solution, the amount of unprecipitated aluminum in 150 ml of acetate-buffered solution varies as follows with the pH:

pH	4.0	4.5	4.8	5.0	5.3 to 7
Al, μg	75	41	30	26	22–20

The solubility loss is much the same at 20° and 70° as at 50°C. The minimum solubility has thus nearly been reached at pH 5.0, and $\sim$20 μg seems to represent the molecular

[10] I. M. Kolthoff and E. B. Sandell, *J. Am. Chem. Soc.* **50**, 1900 (1928).

[11] Separations in ammoniacal solution are described by G. E. F. Lundell and H. B. Knowles, *Natl. Bur. Std. J. Res.* **3**, 91 (1929).

[12] R. Berg, *Z. Anal. Chem.* **71**, 370 (1927).

[13] A. Claassen, L. Bastings and J. Visser, *Analyst* **92**, 618 (1967).

solubility of $Al(C_9H_6ON)_3$ under the conditions. The solubility loss of aluminum on washing with the specified wash solution is ~5 μg per 100 ml at 20°–100°C. Washing with hot water (as in some earlier procedures) results in partial decomposition (hydrolysis) of the precipitate and the loss of 8-hydroxyquinoline gives rise to low results in the volumetric determination of aluminum.

3. The accuracy of the volumetric method equals that of the gravimetric (Ref. 13).

4. The above procedure may be used for the determination of aluminum in the presence of the alkalies, alkaline earths and magnesium, possibly with a reprecipitation when relatively large amounts of the alkaline earths and magnesium are present.

5. More than 1 mg of fluoride causes low results. Citrate and oxalate also interfere, but tartrate apparently may be present.

PROBLEMS

1. A 0.5000-g sample containing aluminum in the form of gibbsite, $Al(OH)_3$, yielded 0.2499 g of Al_2O_3. Obtain the percentage of gibbsite present.
2. A 0.9988-g sample gave a 0.5047-g mixture of ferric oxide and aluminum oxide. The oxides were brought into solution and iron was determined volumetrically, 0.0946 g of Fe being found. Find the percentages of Fe_2O_3 and Al_2O_3 in the sample.
   Ans. 13.69 per cent Fe_2O_3, 36.98 per cent Al_2O_3.
3. What weight of potassium alum must be taken to prepare 1 liter of solution of such strength that 1 ml will contain the equivalent of 1.00 mg of Al?
4. A 1.000-g sample containing 10 per cent Fe(III), 5 per cent Al, and 2 per cent Ti was brought into solution, the metals precipitated as hydroxides, and the washed precipitate ignited to the oxides. What was the weight of the residue?
5. What weight of sample must be taken in order that each milligram of aluminum hydroxyquinolate shall represent 0.10 per cent Al_2O_3?
6. Find the volume of 5.0 per cent 8-hydroxyquinoline solution required to precipitate the aluminum in a 0.5-g sample containing 10.0 per cent Al.
7. Demonstrate that separation of aluminum from iron by ammonia precipitation of aluminum hydroxide after reduction of iron to the ferrous state (by the use of a suitable reductant) is likely to be successful. Assume the solubility-product constant of ferrous hydroxide to be 2×10^{-14}.
8. Will oxalate (0.1 *M*) prevent quantitative precipitation of aluminum with ammonia?

$$[Al^{3+}][C_2O_4^{--}]^2/[Al(C_2O_4)_2^-] = 1 \times 10^{-13}$$
$$[Al^{3+}][C_2O_4^{--}]^3/[Al(C_2O_4)_3^{3-}] = 2 \times 10^{-17}$$

9. Suppose that in the Chancel (thiosulfate) method for the precipitation of aluminum hydroxide (p. 257), 1.0 mg of SO_2 remains in 200 ml of solution after boiling. Will the precipitation of aluminum be quantitative?

CHAPTER 27 *Determination of Sulfur as Barium Sulfate*

Method

A dilute solution of barium chloride is added to the solution containing sulfur as sulfate to precipitate barium sulfate:

$$SO_4^{--} + Ba^{++} \rightarrow BaSO_4$$

The precipitation is made in a solution slightly acidified with hydrochloric acid and near the boiling temperature. The precipitate is filtered off, washed with water, ignited at redness, and weighed as barium sulfate.

Solubility of barium sulfate. Functions of hydrochloric acid in the procedure

The solubility of barium sulfate in water is very small, amounting to 0.25 mg at 25°C and ~0.4 mg at 100°C per 100 ml of solution. Because the solubility is not much greater in hot than in cold solution, hot water may be used for washing the precipitate. Mineral acids increase the solubility very appreciably (p. 616). Some values for the solubility of barium sulfate in hydrochloric acid solutions of different concentrations are given in Table 27-1.

TABLE **27-1. Solubility of Barium Sulfate in Hydrochloric Acid Solutions at Room Temperature**[a]

Normality of Hydrochloric Acid	mg *of* $BaSO_4$ *in* 100 ml *of Saturated Solution*
0.00	0.4
0.10	1.0
0.30	2.9
0.50	4.7
1.00	8.7

[a] From Z. Karaoglanow, *Z. Anal. Chem.* **56**, 241 (1917). The temperature of saturation is not specified.

The presence of a small amount of hydrochloric acid in the solution in which barium sulfate is precipitated prevents the possible precipitation of the barium salts of such anions as carbonate and phosphate which are insoluble in neutral or basic solutions; moreover, the coprecipitation of barium hydroxide is prevented by carrying out the precipitation in a slightly acid medium. Hydrochloric acid has the additional important effect of promoting the formation of a dense precipitate that settles rapidly and is easily filterable. Barium sulfate precipitated in hot solution forms larger and more perfect crystals than when precipitated at room temperature. The concentration of hydrochloric acid that may be used is limited by the solubility of barium sulfate in the acid medium; a concentration of approximately 0.05 N is suitable. In the presence of an excess of barium chloride, the solubility of the precipitate is very small at this acidity. Except in the most exact work, the loss of barium sulfate by solubility in the wash water may be neglected.

Since barium sulfate is not freely soluble in any acid except hot concentrated sulfuric acid, it is not feasible to purify the precipitate by reprecipitation.

Coprecipitation with barium sulfate

Barium sulfate provides a notorious example of a precipitate subject to contamination through coprecipitation. Berzelius, in 1820, confessed his ignorance of the state of impurities in barium sulfate, a matter that remains obscure 150 years later. Since the determination of sulfate is of great practical importance, much attention has been paid to the problem of obtaining pure barium sulfate, and hundreds of papers on the subject are to be found in the literature. In spite of the immense amount of work that has been done, the determination of sulfate remains one of the less accurate determinations of quantitative analysis.

In this section we shall confine ourselves to a brief survey of coprecipitation with barium sulfate and reserve a more detailed examination, especially with reference to the minimization of the associated errors, to the next section.

Precipitated under analytical conditions, barium sulfate yields a microcrystalline precipitate. The contaminating substances are largely present in the interior of the crystals and not at the external surface. We are dealing here with *occlusion* of foreign substances (p. 228), which is the result of the incorporation in the crystals of ions or molecules absorbed during the growth of the precipitate. From the analytical point of view the coprecipitation of the following are important:

1. Foreign anions: chloride, nitrate, and chlorate. These are coprecipitated as the corresponding barium salts.
2. Foreign cations: hydrogen, the alkali metals, calcium, and ferric iron. These are coprecipitated as the sulfates or bisulfates.
3. Water: water is occluded (partially in solid solution) and adsorbed by barium sulfate. On ignition all the water is driven off, provided the temperature of heating is high enough. Precipitates dried at 105–120°C may contain several tenths of 1 per cent water.

In general, the phenomena of coprecipitation observed with barium sulfate are in accord with the predictions of the theory of coprecipitation (p. 239). In studying the contamination of barium sulfate, the influence of such factors as the specific adsorbability of the ions in question, the concentration and temperature of the solutions, the manner of addition of the precipitant, and the treatment of the precipitate after its formation must all be considered.

Barium sulfate adsorbs both soluble barium salts and foreign sulfates so strongly that often both are found in the same precipitate, the relative amounts of each depending upon the manner of precipitation. The higher the concentration of the foreign substance, the greater is the amount of coprecipitation, so that in the ordinary method of precipitation, solutions as dilute as are practically possible are used in the analysis to minimize contamination. Since the precipitation must usually be made in hot solution to obtain a sufficiently coarse precipitate to be readily filterable, the possible effect of varying the temperature is not of much importance in studying the conditions of precipitation, and not many data have been gathered on this point.

In agreement with the general rules of coprecipitation (p. 240), the carrying down of foreign anions is greater when the sulfate solution is added to the barium solution than when the precipitation is made in the reverse way; the opposite holds true for cation coprecipitation. As already indicated, both foreign anions and cations are likely to be found in the same precipitate; but, again in accord with the coprecipitation theory, cation coprecipitation predominates over anion coprecipitation when barium solution is added to sulfate solution and conversely, especially if the solutions are dilute and the mixing is slow and thorough, but exceptions can occur when the specific adsorbabilities of the foreign cation and anion are quite different. The simultaneous addition of a barium salt and a sulfate solution to a dilute hydrochloric acid solution (the method of Hahn)[1] yields a precipitate that is quite impure,[2] and this method of precipitation is not to be recommended. The precipitate thus obtained consists of very large crystals; but this is of no special advantage, because the precipitate obtained by the ordinary method is sufficiently coarse to be readily filterable.

The alkali sulfates are strongly coprecipitated with barium sulfate, potassium sulfate more strongly than sodium or lithium, although the difference is not very marked. In acid solutions, sulfuric acid or bisulfates are coprecipitated. Increasing the hydrogen-ion concentration tends to result in the displacement of adsorbed cations but increases the sulfuric acid coprecipitation at the same time. Because of its slight solubility, calcium sulfate is coprecipitated more strongly than the alkali sulfates. The divalent metals are all coprecipitated to a greater or less extent (Table 27-2). Ferric iron is more strongly carried down than ferrous.

The extent of coprecipitation of various anions is indicated in Table 27-4. The occlusion of chloride is troublesome but less serious than that of nitrate and chlorate.

TABLE **27-2.** **Coprecipitation of Metal Sulfates with Barium Sulfate**[a,b]

	mg *of Foreign Sulfate in* 1 g *of* $BaSO_4$	
Metal Sulfate	*No Chloride Present*	5 g *of Metal Chloride Present*
Na	4.1	8.7
Li	3.7	6.4
K	5.6	6.7
Cu	6.6	8.8
Cd	16.2	20.8
Zn	6.4[c]	6.1[c]
Ni	4.8	7.7
Mn	9.6	14.8
Fe(II)	5.7[d]	12.5[d]
Mg	1.5	
Al	5.8	8.5

[a] Based on data given by J. Johnston and L. H. Adams, *J. Am. Chem. Soc.* **33**, 829 (1911).

[b] Conditions of precipitation: Volume of solution 350 ml, 2 ml of 2 per cent HCl present, time of precipitation, 4 min. $BaCl_2$ added to sulfate solution, time of standing before filtration, 18 hr.

[c] 5 ml of 20 per cent HCl.

[d] 6 ml of 2 per cent HCl.

[1] F. L. Hahn and R. Otto, *Z. Anorg. Allgem. Chem.* **126**, 257 (1923).

[2] I. M. Kolthoff and M. J. van Cittert, *Z. Anal. Chem.* **63**, 392 (1923); but see L. Erdey, F. Paulik, É. Buzágh, and L. Pólos, *Acta Chim. Hung.* **41**, 109 (1964). The latter paper contains electron micrographs of various barium sulfate precipitates.

After ignition of a contaminated barium sulfate precipitate, part of the occluded substance can be washed out with water or dilute hydrochloric or other acid. Analytically this behavior is not very important because the contaminant cannot be quantitatively removed in this way.

Attention has been called in a previous chapter to the favorable effect that standing or digestion before filtration has on reducing the amount of coprecipitated substances in barium sulfate (p. 243).

Errors in the ordinary method of determination of sulfate as barium sulfate and their minimization

By the *ordinary method* we mean the customary one used in precipitating barium sulfate, in which fairly dilute solutions of barium and sulfate are employed and one solution is added directly to the other at or near 100°C. In discussing the errors that may occur, it is advantageous to consider the problem under three heads, depending upon the composition of the solutions.

Precipitation of barium sulfate by the addition of barium chloride to sulfuric acid in the absence of other cations and anions, and the reverse precipitation. This is the simplest case. When a dilute solution of barium chloride is added slowly with stirring to a hot dilute solution of pure sulfuric acid, the precipitate contains appreciable amounts of chloride as barium chloride. When the precipitation is made in the reverse way, the amount of chloride coprecipitated is greater than before. This is clearly shown in Table 27-3, which contains some of the data obtained by Richards and Parker[3]

TABLE **27-3. Coprecipitation of Chloride with Barium Sulfate**[a]

Conditions of Precipitation	Cl *in* 1 g *of* $BaSO_4$, mg
1. $BaCl_2$ (vol. ca. 20 ml) added slowly in slight excess to H_2SO_4 (vol. ca. 50 ml) at boiling temperature; $\frac{1}{4}$ ml strong HCl present	1.2, 1.3
2. As in 1 but with double the required amount of $BaCl_2$	1.2 (av.)
3. Reverse precipitation: H_2SO_4 to $BaCl_2$	4.6 (av.)
4. As in 1 except with 10 ml of strong HCl present	4.7, 4.2
5. As in 1 except with 20 ml of strong HCl present	9.1, 7.6
6. As in 1 except precipitation made in a solution four times as dilute	0.7, 0.7
7. As in 1 except $BaCl_2$ added very rapidly	1.8, 1.7

[a] From T. W. Richards and H. G. Parker, *Z. Anorg. Chem.* **8**, 419–421 (1895).

in a study of the coprecipitation of barium chloride with barium sulfate. It is also found that precipitates formed by either manner of precipitation contain sulfuric acid, the precipitate obtained by addition of barium solution to sulfuric acid containing more than the one formed in the reverse precipitation. The adsorption of sulfuric acid does not preclude the adsorption of barium chloride or conversely. The adsorption of the latter is the result of the adsorption of barium ions. When barium chloride is added to sulfuric acid drop by drop with vigorous stirring, there will be present in the neighborhood of the barium sulfate crystals already formed, ions of hydrogen, sulfate, barium, and chloride. The crystals will naturally adsorb barium and sulfate ions and build up their lattices; and since, if the mixing has been well done, there are more sulfate than barium ions in the region of the surface, there will be a greater chance for hydrogen ions than for chloride ions to be dragged to the surface. The ratio of the amounts of sulfuric acid and barium chloride coprecipitated will depend

[3] T. W. Richards and H. G. Parker, *Z. Anorg. Chem.* **8**, 413 (1895).

not only upon the relative concentrations of the two substances at the surfaces of the growing crystals but also upon their relative adsorbabilities. At any rate, the adsorption of both substances is to be anticipated, but the one that is in excess may be expected to be adsorbed and coprecipitated to a larger extent. If the mixing is not thorough, quite the opposite effect may be found.

The absolute amounts of barium chloride and sulfuric acid carried down depend upon the concentrations of the reacting solutions and the rate of mixing, as well as the temperature. Slow addition of reagent and thorough stirring during precipitation aid in decreasing the amount of occlusion.

When a precipitate of barium sulfate containing both sulfuric acid and barium chloride is heated, the two contaminants react to give barium sulfate, and hydrogen chloride escapes. If equivalent amounts of the two substances were originally present, no error would result; but usually the amounts are quite different, and after the heating some barium chloride will remain if it preponderated, or else sulfuric acid will escape, if it was in excess, leaving behind a pure residue of barium sulfate. In the first case high results will be obtained if sulfate is being determined, because barium chloride remains largely unchanged after moderate ignition, whereas in the second the results will be low, because sulfate is lost. If barium is being determined, the results will be low in the first case because $BaCl_2$ has a lower molecular weight than $BaSO_4$, whereas in the second case there will be no error at all.

In determining sulfuric acid in pure solution, barium chloride should be added to the former solution because the specific adsorbability of barium chloride is greater than that of sulfuric acid. When the precipitation is so made, rather good results are obtained through a compensation of errors. The small amount of barium chloride left in the ignited precipitate approximately compensates the solubility loss. Thus Richards and Parker (Ref. 3) found that two precipitates of barium sulfate precipitated from pure dilute sulfuric acid with barium chloride, each weighing ca. 0.8 g, contained 0.5 and 1.7 mg of barium chloride, respectively, after ignition, whereas the solubility losses were 1.1 and 1.7 mg, respectively. If results of the highest accuracy are required, it will be necessary to correct the properly ignited barium sulfate for the chloride it contains and determine the loss due to the solubility of barium sulfate. The chloride in the precipitate can be determined by fusing the barium sulfate with sodium carbonate, extracting the melt with water, and determining chloride in the extracts after acidification with nitric acid (Ref. 3), or by drawing a stream of air through a hot sulfuric acid solution of the barium sulfate into a solution of silver nitrate to absorb the hydrogen chloride evolved, finally determining the excess of silver by titration or weighing the silver chloride.[4] The chloride found is calculated to barium chloride and deducted from the weight of the precipitate. To determine the solubility loss, the filtrate and the washings are simply evaporated to dryness, the residue is taken up in a small volume of water acidified with a little hydrochloric acid, and the precipitate is collected by filtration, washed, and ignited. It will rarely be necessary to make these corrections.

It is much easier to determine barium than sulfate accurately in pure solution, because the precipitate can be moistened with sulfuric acid after ignition and reignited after fuming off the excess acid. Any coprecipitated barium chloride is transformed into sulfate.

Precipitation of barium sulfate in the presence of cations (as chloride or sulfates). Here the barium sulfate can contain barium chloride, foreign metal sulfate or bisulfate, and sulfuric acid, the amounts of these depending on the composition of the original solution, the method of precipitation, and the length of time between precipitation and filtration. On ignition such a precipitate may lose hydrochloric acid and sulfuric acid, and the residue may contain a foreign sulfate or its decomposition products, and barium chloride as well, if the latter was originally in excess over sulfuric acid or bisulfate.

[4] G. A. Hulett and L. H. Duschak, *Z. Anorg. Chem.* **40**, 196 (1904).

In the presence of the alkali metals. The determination of sulfate in alkali sulfates or sulfuric acid in the presence of alkali chlorides is important analytically. Allen and Johnston[5] and Johnston and Adams[6] carefully studied the effects of various factors on the sulfate determination in the presence of the alkalies. As a result of an investigation of the precipitation of sodium sulfate with barium chloride, the first authors arrived at the following conclusions regarding the occlusion of sodium sulfate by barium sulfate:

1. A decrease in the concentration of sodium sulfate decreases the coprecipitation slightly.
2. Varying the rate of addition of barium chloride has a rather small effect on the amount of coprecipitation, the tendency of rapid addition being to decrease the quantity of sodium sulfate carried down.
3. Adding the sodium sulfate to the barium chloride solution does not decrease the occlusion to any great extent.
4. Increasing the hydrochloric acid concentration slowly decreases the amount of sodium sulfate coprecipitated.
5. Allowing the precipitated barium sulfate to stand in the supernatant liquid for some time before filtration reduces the amount of occluded sodium sulfate considerably. Digestion of the precipitate with water, barium chloride, or sodium sulfate solution has no marked effect.
6. Sodium chloride increases markedly the amount of coprecipitated alkali sulfate. Sodium chloride itself is not coprecipitated.
7. The ignited precipitate contains only traces of chloride, even when the precipitation is made in the presence of as much as 30 g of sodium chloride, but before ignition considerable amounts of chloride are present.

The same authors also investigated the volatilization of free sulfuric acid from the precipitates (formed in the presence of sodium sulfate in hydrochloric acid solution) on heating and found that:

1. All precipitates formed in acid solution lose sulfuric acid when heated above 500°C.
2. The sulfuric acid loss increases with the concentration of acid present during the precipitation up to 0.1 N and then decreases.
3. In the reverse precipitation (sulfate to barium) a precipitate is obtained which loses much less sulfuric acid.
4. The loss of sulfuric acid increases with the concentration of sodium chloride present.
5. The loss of acid is decreased by increasing the rate of addition of barium chloride.
6. Allowing the precipitate to stand in the supernatant liquid decreases the amount of sulfuric acid volatilized on ignition.

In studying the precipitation of barium sulfate from potassium sulfate solutions, Allen and Johnston obtained similar results; the amount of potassium sulfate coprecipitated was found to be about the same as with sodium sulfate, although the volatilization losses were greater in the presence of potassium chloride (see p. 608 for effect of pH).

The determination of sulfate in alkali sulfates by slow precipitation with barium chloride in weakly acid solutions gives low results when the precipitate is ignited, because the residue weighed contains alkali sulfate and no appreciable amount of barium chloride to compensate the resulting deficiency in weight ($Na_2SO_4 < BaSO_4$, $K_2SO_4 < BaSO_4$), to say nothing of the sulfate lost as sulfuric acid by volatilization. With ammonium

[5] E. T. Allen and J. Johnston, *J. Am. Chem. Soc.* **32**, 588 (1910).
[6] J. Johnston and L. H. Adams, *J. Am. Chem. Soc.* **33**, 829 (1911).

sulfate the error is usually more serious than with the other alkali sulfates because a precipitate of barium sulfate containing ammonium sulfate loses it on ignition.

The error due to occlusion of the alkali sulfates and sulfuric acid can be reduced by:

1. Making the precipitation in the reverse way, i.e., adding the alkali sulfate solution to the acidified chloride solution. The amounts of alkali sulfate and sulfuric acid occluded are then decreased, while the amount of barium chloride carried down is increased; thereby the two errors, opposite in sign, more nearly compensate each other. As a matter of fact, the weight of precipitate thus obtained is usually greater than the true value because of the barium chloride in the ignited residue. When the precipitation is made by addition of barium chloride to sulfate, the amount of alkali sulfate coprecipitated far outweighs the barium chloride occluded. The method of reverse precipitation has been recommended by Popoff and Neuman.[7]

2. Adding the barium chloride solution *rapidly* to the sulfate solution.[8] Rieman and Hagen[9] compared various methods of precipitation of barium sulfate in use in 1940 and found that the most accurate results for sulfur in sodium sulfate, especially in the presence of sodium chloride, were obtained when the barium chloride solution was added rapidly to the sulfate solution according to the directions of Hintz and Weber. Although the weight of the ignited precipitate obtained in the Hintz and Weber method is close to the theoretical value, the good results are due to a compensation of errors. A study of the actual amounts of occluded substances in barium sulfate precipitated by various procedures would be desirable. Rapid precipitation is used in the procedure given on page 611 for the determination of sulfur in a soluble sulfate.

3. Making the precipitation at pH 5–6 when alkali chlorides are present in high concentrations.[10] The barium chloride solution is added rapidly in slight excess (10–20 per cent) to the sulfate solution, both near boiling. In 1.2 M KCl (100 times the K_2SO_4 concentration) the error in the sulfate determination is essentially zero (1 to -1 ppt) at pH 6; in the absence of KCl at about the same pH (5) the error is -1 ppt. In 1.2 M NaCl, the error is also essentially zero at pH 6; in the absence of NaCl, precipitation at both pH 1 and 5 gives almost correct results. The favorable effect produced by raising the pH is plausibly explained by very slight adsorption of HSO_4^- (because of its very low concentration) at this pH. Coprecipitation of K_2SO_4 (or Na_2SO_4) by solid-solution formation, i.e., replacement of Ba^{++} by K^+, requires simultaneous introduction of HSO_4^- to maintain electroneutrality. If HSO_4^- is not available, K^+ does not enter the lattice of $BaSO_4$ to a significant extent.

If a very accurate determination of sulfate in the presence of the alkali metals is required, it can be had by applying corrections for solubility, chloride and alkali sulfate in the precipitate, and loss of sulfuric acid on heating—all as described by Allen and Johnston.[11]

In the presence of the divalent metals (see Table 27-2). Magnesium is very slightly occluded by barium sulfate. Calcium causes serious errors, as already mentioned, because calcium sulfate is strongly coprecipitated, as expected from its slight solubility.[12]

[7] S. Popoff and E. W. Neuman, *Ind. Eng. Chem., Anal. Ed.* **2**, 45 (1930).

[8] E. Hintz and H. Weber, *Z. Anal. Chem.* **45**, 31 (1906).

[9] W. Rieman, III, and G. Hagen, *Ind. Eng. Chem., Anal. Ed.* **14**, 150 (1942).

[10] R. B. Fischer and T. B. Rhinehammer, *Anal. Chem.* **25**, 1544 (1953). See this paper also for observations on the size and morphology of barium sulfate crystals formed under various conditions.

[11] E. T. Allen and J. Johnston, *J. Am. Chem. Soc.* **32**, 588 (1910); see J. Johnston and L. H. Adams, *ibid.* **33**, 843 (1911).

[12] For a method that gives better results than the usual one in the determination of sulfate in the presence of calcium, see J. Johnston and L. H. Adams, *J. Am. Chem. Soc.* **33**, 829 (1911). Calcium can be separated from sulfate by cation exchange (p. 614).

Often in practice it is admissible to make the precipitation of barium sulfate in the presence of moderate amounts of ferrous iron or zinc, if the solution is quite strongly diluted.

In the presence of trivalent metals. The important trivalent metals encountered are ferric iron, aluminum, and chromium. Ferric iron has a great tendency to be carried down, in large part because of the hydrolysis of the iron with the formation of a positively charged colloid of basic ferric salt, which is strongly adsorbed by the negatively charged barium sulfate (p. 228). The coprecipitation can be decreased by making the precipitation in the reverse manner and in the presence of a fairly high concentration of hydrochloric acid (but only at the cost of solubility losses of barium sulfate in the latter case). The error can be eliminated by removing the trivalent iron by double precipitation with an excess of ammonium hydroxide followed by the removal of ammonium salts from the solution. A better way to remove iron(III) is by the use of a cation-exchange resin in the hydrogen-ion form (procedure on p. 614). Other undesirable cations, such as calcium, are removed at the same time. Often quite satisfactory results can be obtained by reducing iron to the ferrous state with hydroxylamine, metallic zinc, or aluminum, since divalent iron is much less strongly occluded than trivalent iron. Still another method of minimizing the error due to to iron(III) is by complexing with EDTA added in regulated amount.[13]

Aluminum is not particularly strongly carried down by barium sulfate (much less so than iron), but a high concentration of the metal is deleterious. Chromic chromium is an undesirable constituent in a solution from which barium sulfate is to be precipitated. On the one hand, it is coprecipitated, and, on the other, it may prevent the complete precipitation of sulfate (complex chromic sulfates). This complex can be transformed into a chromium acetate complex by the addition of an excess of alkali acetate, thus making the sulfate completely precipitable.

Precipitation of barium sulfate in the presence of foreign anions. The coprecipitation of barium chloride has already been discussed. The occlusion of barium nitrate, chlorate, and phosphate is much greater than that of chloride and can lead to serious analytical errors (cf. Table 27-4). Nitrate and chlorate must be destroyed before the precipitation

TABLE **27-4. Coprecipitation of Anions with Barium Sulfate**[a,b]

Anion	*Mole Anion Coprecipitated per Mole* $BaSO_4$
NO_3^-	0.0848
NO_2^-	0.0747
ClO_3^-	0.0584
$Fe(CN)_6^{4-}$	0.0330
MnO_4^-	0.0285
Cl^-	0.0176
$Fe(CN)_6^{3-}$	0.0090
Br^-	0.0083
CN^-	0.0031
CNS^-	0.0022
I^-	0.0006

[a] From H. B. Weiser and J. L. Sherrick, *J. Phys. Chem.* **23**, 205 (1919).

[b] Conditions of precipitation: 40 ml of 0.1 *M* sodium sulfate solution was added rapidly to 50 ml of 0.1 *M* solution of the barium salt diluted with 160 ml of water. Both solutions were near the boiling point when mixed. The precipitate was filtered off after 1 min and analyzed *directly* for the anion in question.

[13] Ref. 10; M. Nishimura, M. Saito, and Y. Uzumasa, *Japan Analyst* **13**, 544 (1964).

of barium sulfate is attempted. This can be done by evaporating the sample to dryness two or three times with concentrated hydrochloric acid:

$$NO_3^- + 3\,Cl^- + 4\,H^+ \rightarrow Cl_2 + NOCl + 2\,H_2O$$

$$ClO_3^- + 5\,Cl^- + 6\,H^+ \rightarrow 3\,Cl_2 + 3\,H_2O$$

Other anions are less frequently encountered. Barium sulfate precipitated in acid solution in the presence of chromate is contaminated by barium chromate (barium sulfate and barium chromate form mixed crystals). Determination of sulfate in the presence of molybdate is unsatisfactory.[14]

Precipitation of barium sulfate from homogeneous solution

An attempt to determine sulfate by adding magnesium salt to a sulfate solution containing barium complexed with EDTA (Mg forms the more stable EDTA complex), so as to release Ba^{++} slowly, was not successful.[15] Homogeneous precipitation of barium with excess sulfate is easier. It can be brought about by hydrolysis of sulfamic acid or dimethyl sulfate in hot acidic solution to yield sulfate ions.

Empirical corrections in the determination of sulfate

Since it is difficult to obtain very accurate results in the determination of sulfate by precipitation as barium sulfate, Winkler[16] (and others) proposed to apply empirical corrections to the result obtained. The corrections are found by precipitating various amounts of sulfate under the same conditions as in the determination and recording the deviations from the theoretical value in a table or constructing a graph.[17]

The principal weakness of this method is that the kind and amount of foreign substances present must be the same in the two cases, and the method can therefore be applied conveniently only when a large number of determinations of sulfate are being made on samples of nearly the same composition.

The ignition of barium sulfate

Barium sulfate when pure is not decomposed when heated in air until a temperature of about 1400°C is reached[18]:

$$BaSO_4 \rightarrow BaO + SO_3$$

The precipitate is, however, rather easily reduced to sulfide by carbon formed by the charring of the filter paper. This reduction can be avoided by burning off the filter slowly at a low temperature with free access of air. When necessary the precipitate can be moistened with sulfuric acid to convert sulfide to sulfate. The final ignition of the barium sulfate need not be made at a temperature higher than 800 or 900°C (a Tirrill burner is suitable)[19]; higher temperatures are best avoided as possibly leading to a partial decomposition when traces of impurities, such as iron, are present.

[14] D. A. Lambie and W. R. Schoeller, *Analyst* **65**, 281 (1940).

[15] F. H. Firsching, *Anal. Chem.* **33**, 1947 (1961).

[16] See, for example, L. W. Winkler, *Z. Angew. Chem.* **33**, I, 160 (1920).

[17] C. E. Waters, *Natl. Bur. Std. Tech. Paper* **177**, 25.

[18] W. Mosttowitsch, *Metallurgie* **6**, 450 (1909); *Chem. Abst.* **5**, 841 (1911).

[19] H. A. Fales and W. S. Thompson, *Ind. Eng. Chem., Anal. Ed.* **15**, 206 (1939), recommend drying barium sulfate at 110–120°C instead of igniting. Their method involves compensation of errors.

Other substances determinable as insoluble sulfates

Sulfur in forms other than sulfate, e.g., sulfide, sulfite, and thiosulfate, can be converted to sulfate by appropriate methods and determined by precipitation with barium chloride. As already indicated, barium can be determined by precipitation with a slight excess of sulfuric acid. Strontium can be precipitated quantitatively as the sulfate if alcohol is added to reduce the solubility of the precipitate. Rarely, calcium is precipitated and determined as sulfate in the presence of alcohol. Lead can be separated and determined as the sulfate with the proper excess of sulfuric acid (p. 1112).

Sulfur in organic compounds can be determined by oxidation to sulfate (see the discussion of methods for Cl in organic compounds, p. 586) followed by precipitation as barium sulfate, but a volumetric finish is more usual.

Determination of sulfur in a soluble sulfate

The following procedure is intended for the determination of sulfur in the sulfates of the alkali metals. The precipitation is made by rapidly adding barium chloride in slight excess to the hot sulfate solution.[20] Under these conditions the coprecipitation of barium chloride outweighs that of the alkali sulfate, and the ignited precipitate is usually heavier than the theoretical weight by a few parts per thousand (Table 27-5).[21] The special advantage of this method of precipitation is the relatively slight effect of a large amount of alkali chloride which may be introduced in the preparation of the sample.

TABLE **27-5.** **Error in Sulfate Determination by Rapid Addition of Barium Chloride**[a]

Molar ratio, $NaCl/Na_2SO_4$	0	1	2	5	10	20	40	110	190
Error, ppt	+2	+2	+1	+4	+4	+3	+3	+4	+4

[a] From W. Rieman, III, and G. Hagen, *Ind. Eng. Chem., Anal. Ed.* **14,** 150 (1942).

Small amounts of ferrous iron, magnesium, zinc, and aluminum may be present. Calcium and ferric iron should not be present in appreciable amounts. Strongly coprecipitable anions (nitrate, chlorate, etc.) must be absent.

As already mentioned (p. 606), good results can be obtained in the determination of sulfate in sulfuric acid solutions containing no alkali metals by adding the barium chloride solution slowly. This method of precipitation may be applied in the determination of sulfur in pyrite (p. 613).

Procedure. Weigh out a sufficient amount of sample to contain approximately 0.1 g of sulfur into a 600-ml beaker provided with a stirring rod, add 25–50 ml of

[20] See Refs. 8 and 9. If the sample contains only alkali metal sulfates and chlorides (no heavy metals) the procedure of Fischer and Rhinehammer (Ref. 10) is also suitable and may give more accurate results with much alkali chloride present.

[21] If the precipitation is made by adding the barium chloride solution slowly to the sulfate solution, the results are usually low to the extent of a few parts per thousand, owing to the occlusion of alkali metal sulfate or sulfuric acid, but this error becomes larger if alkali metal chlorides are present.

water, and stir to dissolve.[22] Acidify with 2 ml of concentrated hydrochloric acid and dilute to 350 or 400 ml with water. Dilute 20 ml of 5 per cent barium chloride (5 g of $BaCl_2 \cdot 2\,H_2O$ in 100 ml of solution—ca. 0.2 *M*) to 75 or 100 ml in a small beaker, heat nearly to boiling, and add the solution rapidly (over a period of about 5 sec) with stirring to the sulfate solution, which has been heated to near the boiling point. Allow the precipitate to settle, and test the supernatant liquid for complete precipitation by adding a little barium chloride solution. If a precipitate should form, add more barium chloride, allow the precipitate to settle as before, and test again. When an excess of precipitant has been added, set the covered beaker aside for an hour or two on the steam bath (or over a low-temperature hot plate or a low flame, so adjusted that the solution does not boil). If convenient, allow the solution to stand on the steam bath overnight.

After digestion, decant the clear supernatant liquid through a fine-grained ashless filter paper (p. 530), receiving the filtrate in a clean beaker. For safety, test the first portions of the filtrate for complete precipitation by adding a few drops of barium chloride solution. When most of the supernatant liquid has been filtered, and while nearly all the precipitate still remains in the precipitation beaker, empty the beaker containing the filtrate, and replace it beneath the funnel. Now transfer the precipitate to the paper with a jet of water from the wash bottle (p. 532). If any of the precipitate runs through the paper (it should not if a fine paper is used), pour the filtrate through the paper, and repeat if necessary until the solution runs through perfectly clear. Use a policeman to remove any precipitate adhering to the walls of the beaker or to the stirring rod, finally examining the vessel carefully to see that all barium sulfate has been removed. Wash the precipitate and paper 8 or 10 times with small portions of warm water, proceeding as directed on page 535. Test the last few milliliters of washings with a drop or two of silver nitrate solution for the presence of chloride, this ion being indicative of incomplete washing. If a positive test is obtained, continue the washing until no test, or at the most only a very faint opalescence, is given.

Fold the moist paper around the precipitate and place in a weighed porcelain crucible ignited to redness and cooled in a desiccator. Dry the paper by heating the loosely covered crucible placed in a triangle some centimeters above a small flame (p. 538). Then gradually increase the heat until the paper chars and volatile matter is expelled. Do not heat to redness at this stage, and do not let the paper burst into flame. When the charring is complete, raise the temperature of the crucible to low redness, and burn off the carbon with free access of air (crucible slightly inclined and with cover displaced). When the carbon is gone, ignite for 10–15 min at moderate redness over a Tirrill burner. Then allow the crucible to cool somewhat in the air, and place it in the desiccator to cool to room temperature. Weigh, and repeat the ignition with 10-min periods of heating until constant weight (± 0.2 mg) is attained.

Report the percentage of S, SO_3, or SO_4 in the sample as required.

Note: If desired, a Gooch or porous porcelain filter crucible may be used to collect the precipitate. The ignition (also of the empty crucible) is conveniently made in a muffle at 900°C or as described on page 541.

[22] If the sample is insoluble in water, add hydrochloric acid dropwise with warming until it is dissolved. Then add 2 ml of the concentrated acid in excess. If sulfuric acid is analyzed, omit the hydrochloric acid.

Determination of sulfur in a sulfide

The determination of sulfur in the naturally occurring sulfides of iron, copper, arsenic, etc., is generally based on the oxidation of sulfide to sulfate and the precipitation of the latter as barium sulfate.[23] Two general methods are in common use for the decomposition of insoluble sulfides:

1. Oxidation in the wet way by such reagents as aqua regia, fuming nitric acid, and bromine with nitric or hydrochloric acid. For example, the action of bromine and nitric acid on pyrite ("iron pyrites") may be represented as follows:

$$FeS_2 + 15\,Br + 3\,HNO_3 + 8\,H_2O \rightarrow Fe(NO_3)_3 + 2\,H_2SO_4 + 15\,HBr$$

2. Fusion with an alkaline oxidizing agent such as sodium peroxide[24] or a mixture of sodium carbonate and potassium nitrate:

$$2\,FeS_2 + 15\,Na_2O_2 \rightarrow 4\,Na_2SO_4 + Fe_2O_3 + 11\,Na_2O$$

Each method has its advantages and disadvantages. Fusion with sodium peroxide is a very effective method of decomposing refractory sulfides, but there are a number of objections to its use in accurate work, one of which is the introduction of a large amount of sodium. Oxidation in the wet way is free from this objection, but the action is slower, and greater care must be taken to obtain complete decomposition; moreover, the nitric acid introduced must be destroyed by evaporation with hydrochloric acid, and the iron brought into solution must be removed or reduced to the ferrous condition to prevent serious contamination of the barium sulfate. When the wet attack is used, sulfur present as an insoluble sulfate (e.g., barium or lead sulfate) largely escapes determination. For siliceous materials (rocks) containing very small amounts of sulfur (as sulfide, sulfate, or both), fusion with a 10 : 1 mixture of sodium carbonate and potassium nitrate is suitable. This fusion may be made in a platinum crucible since it is not seriously attacked by this flux. The introduction of alkali salts is of relatively small importance here, because the amount of sulfur present is very small; i.e., the coprecipitation of even 1 per cent of alkali sulfate is hardly of moment when the sulfur content of the sample is, let us say, 0.1 per cent. Sulfides soluble in nonoxidizing acids may be analyzed by collecting the hydrogen sulfide evolved on treatment with acid in a suitable medium and determining the sulfur therein gravimetrically or volumetrically (see the determination of sulfur in steel).

Attention must be called to the danger of contamination in the decomposition of sulfides when gas flames are used for heating. The products of combustion of gas contain the oxides of sulfur, and these are taken up avidly by alkaline fluxes and also by solutions. The error from this source will generally be small if care is taken to exclude the flame gases. Electrical heating is to be preferred. In the determination of small amounts of sulfur it is of the greatest importance to run blanks simultaneously with the determination itself to correct for sulfur introduced from the atmosphere and in the reagents. It should also be mentioned that in preparing a sulfide for analysis, the sample should not be ground unduly fine on account of danger of oxidation of sulfide to sulfur dioxide.[25]

An example of the determination of sulfur in a sulfide is provided by the analysis of pyrite, a material of great commercial importance. The interference due to iron is

[23] A rapid method consists of burning the sulfide in a stream of oxygen at 1200°C, absorbing the SO_2 and SO_3 in hydrogen peroxide solution, and titrating the sulfuric acid formed with base.

[24] A 1:1 mixture of sodium peroxide and sodium hydroxide has been recommended by L. S. Theobald, *Analyst* **67**, 288 (1942), for the decomposition of sulfides such as pyrite, chalcocite, and bornite. An alcohol lamp may be used for the fusion, and the danger of introducing sulfur from a gas flame is thus avoided.

[25] E. T. Allen and J. Johnston, *J. Ind. Eng. Chem.* **2**, 196 (1910).

minimized by reducing it to the ferrous state and making the precipitation in a cold dilute solution. Alternatively, iron (and other metals) can be removed by passing the solution of the sample through a column of cation-exchange resin.[26] Directions for both procedures follow.

Procedure.[27] Weigh 0.4 g of pyrite,[28] ground to pass an 80-mesh sieve, into a dry 250-ml covered beaker, and add 3–4 ml of a mixture of 2 parts by volume of bromine to 3 parts of carbon tetrachloride (hood!). Allow the mixture to stand for 15 min at room temperature with occasional shaking; then add 10 ml of concentrated nitric acid and let stand again for 15 min with occasional agitation. Next, heat below 100°C by placing the beaker on a thick asbestos board over the steam bath until all action has ceased and most of the bromine has been expelled. Raise the cover glass and evaporate to dryness by heating directly on the steam bath. Add 10 ml of concentrated hydrochloric acid, mix, evaporate to dryness, and bake the residue for 30–60 min at 100°C to dehydrate the silica. To the residue add 1–2 ml of hydrochloric acid and then, after a few minutes, 50 ml of water. Rinse down the sides of the beaker as well as the cover glass. Heat to boiling and keep at this temperature for about 5 min. Allow to cool for 5 min and then add 0.2 g of pure powdered aluminum to reduce the iron. Stir until the solution becomes colorless, allow to cool somewhat, and wash down the sides of the beaker and the cover glass. Filter off insoluble matter on a small filter, and catch the filtrate in a 600-ml beaker. Wash the residue and the paper thoroughly with hot water. Dilute the filtrate and washings to 400 ml, and add 2 ml of concentrated hydrochloric acid. Precipitate the sulfate in the cold by adding 50 ml of 5 per cent barium chloride solution dropwise, with stirring, at a rate not exceeding 5 ml/min. Allow the precipitate to settle for 2 hr or, better, overnight. Filter, using paper or a Gooch crucible, and wash the precipitate as described on page 612. Ignite the precipitate to constant weight (p. 612). Report the percentage of sulfur found. In the most accurate work, run a blank through all the steps of the procedure.

Alternative procedure: Separation of iron by ion exchange. Prepare a column of Dowex 50 or Amberlite IR-100 resin (50–100 mesh, H^+ form) by pouring a slurry into a tube of ~2 cm diameter to give a column length of ~10 cm (see p. 314). Pass 50 ml of 6 *M* hydrochloric acid through the column at a rate of about 15 ml/min (the effluent should be colorless), followed by 100 ml of water. Test the last 10–20 ml of effluent with barium chloride to be sure no sulfate is present.

Decompose the sample as described above. After dehydrating silica and taking up the residue in 1 ml of concentrated hydrochloric acid and 50 ml of hot water, filter through a small paper and wash with 25 ml of 1 : 50 hydrochloric acid in small portions. Run the combined filtrate and washings through the resin column at a rate of 15–20 ml/min, collecting the effluent in a 600-ml beaker. Wash the column with 200 ml of water. The column may be used for a second run. (The resin is regenerated with 150 ml of 6 *M* hydrochloric acid, or more as required to give a colorless effluent, and washed with water.)

Heat the sulfate solution (effluent and washings) to near boiling and add 50 ml of 5 per cent barium chloride solution dropwise with stirring. Allow the precipitate to

[26] O. Samuelson, *Svensk Kem. Tidskr.* **54**, 124 (1942) and E. H. Swift, *Introductory Quantitative Analysis*, Prentice-Hall, Englewood Cliffs, N.J., 1950, p. 368.

[27] W. S. Allen and H. B. Bishop, *8th Intern. Congr. Appl. Chem.* **1–2** (I–II), 48 (1912); *J. Ind. Eng. Chem.* **11**, 46 (1919). In an alternative decomposition procedure (Swift, *Introductory Quantitative Analysis*) the sample is treated with a mixture of 10 ml of concentrated nitric acid, 3 ml of concentrated hydrochloric acid, and 5 drops of liquid bromine. After 10–15 min at room temperature, the mixture is heated slowly on the steam bath until brown fumes are no longer evolved; if particles or globules of sulfur appear, a few drops of bromine are added. The solution is evaporated almost to dryness, 2 ml of concentrated hydrochloric acid is added, and evaporation is continued to dryness on a steam bath to dehydrate silica.

[28] If the percentage of sulfur is to be reported on the dry basis (p. 519), determine hygroscopic water in a separate 1-g portion of sample.

settle for 2 hr or, better, overnight. Filter, using paper or a filter crucible and wash the precipitate as described on page 612. Ignite the precipitate to constant weight. Report the percentage of S. In the most accurate work, run a blank through the whole procedure.

PROBLEMS

1. If 0.5500 g of a soluble sulfate yielded 0.6460 g of barium sulfate, what was the percentage of sulfur in the sample?
2. Calculate the volume of 0.2 *M* barium chloride solution required to precipitate the sulfate in 0.5 g of sodium sulfate. What weight of barium sulfate will be obtained? *Ans.* 17.60 ml; 0.8216 g.
3. Find the volume of 0.15 *M* barium chloride solution needed to precipitate the sulfate in 1.00 g of $Al_2(SO_4)_3$.
4. It is desired to prepare a solution of barium chloride of such strength that 1 ml shall represent 1.00 mg of SO_4. What weight of $BaCl_2 \cdot 2\,H_2O$ must a liter of such a solution contain? *Ans.* 2.543 g.
5. An impure sample of Na_2SO_4 weighing 1.500 g gave 2.390 g of $BaSO_4$. What is (a) the percentage of S in the sample, and (b) the percentage purity of the salt?
6. An analyst reports 16.70 per cent S in a sample of sulfate. If he made an error of +0.5 mg in weighing the barium sulfate precipitate from 1.0 g of sample, what is the true percentage of S?
7. From the following data obtain the formula of a sulfide mineral: 1.000 g yielded 0.3451 g of copper; 0.5000 g contained 0.1525 g of iron; and 0.3000 g yielded 0.7640 g of barium sulfate.
8. It is desired that each milligram of ignited barium sulfate residue shall represent 0.10 per cent sulfur. What weight of sample must be taken? *Ans.* 0.137 g.
9. Calculate the weight of an unignited barium sulfate precipitate obtained from 0.1000 g of SO_4 when the precipitate contains (a) 1 per cent Na_2SO_4; (b) 1 per cent K_2SO_4; (c) 1 per cent $(NH_4)_2SO_4$; (d) 1 per cent $BaCl_2$; (e) 1 per cent $Fe_2(SO_4)_3$; (f) 1 per cent $KHSO_4$; (g) 1 per cent $KHSO_4$ + 1 per cent $BaCl_2$. What will be the weights of these ignited precipitates if the ignition is made at 600°C? What are the relative errors in percentage of SO_4 in each case as calculated from the weights of the ignited residues?
10. An ignited barium sulfate precipitate weighing 0.5016 g which is suspected to contain barium sulfide is moistened with sulfuric acid, the excess of the latter is evaporated, and the residue is finally heated at redness. If the weight of the residue is 0.5024 g, what percentage of barium sulfide was present?
11. If 0.77 g (0.0033 mole) of barium sulfate is boiled with 100 ml of 0.10 *M* potassium carbonate solution until equilibrium is attained, what fraction of the precipitate is converted into carbonate if the equilibrium constant of the reaction

$$BaSO_4(s) + CO_3^{--} \rightleftharpoons BaCO_3(s) + SO_4^{--}$$

is 0.5 under the conditions? *Ans.* 100 per cent.
12. A solution of barium and lead is treated with sufficient ammonium acetate to keep lead sulfate in solution, and then an excess of ammonium sulfate is added. Is it to be expected that the precipitated barium sulfate will be free of lead?
13. The sulfate content of a ferric sulfate solution is determined by adding an excess of carbonate-free ammonium hydroxide followed by an excess of barium chloride solution; the mixture is acidified and the barium sulfate filtered off. What is the function of the ammonium hydroxide? What substance or substances are likely to be carried down by barium sulfate precipitated in ammoniacal solution? Will the results be high or low?
14. Devise a method for determining the sulfur content of celestite (strontium sulfate).

15. Show that the solubility of a slightly soluble sulfate MSO_4 in an acidic solution is given by

$$[M^{++}] = \left\{\frac{K_{sp}(K_2 + [H^+])}{K_2}\right\}^{1/2}$$

where K_{sp} is the solubility product of MSO_4 and K_2 is the second ionization constant of sulfuric acid.

16. Calculate the solubility of $BaSO_4$ in 0.10 *M* HCl, with and without the use of activity coefficients. Take K_2 of H_2SO_4 to be 2×10^{-2}.

17. Find the approximate weight of sulfur remaining in solution under the conditions of the procedure on page 611, applying activity coefficients.

CHAPTER 28 *Determination of Calcium*

Method

Calcium is precipitated as calcium oxalate by treating a hot hydrochloric acid solution with ammonium oxalate or oxalic acid and neutralizing with ammonium hydroxide:

$$Ca^{++} + C_2O_4^{2-} + H_2O \rightarrow CaC_2O_4 \cdot H_2O$$

The precipitate is washed with dilute ammonium oxalate solution, ignited to calcium carbonate or oxide by heating to the proper temperature or converted into calcium sulfate by treatment with sulfuric acid, and weighed in any of these forms.[1] Cations other than the alkalies and magnesium must be absent.

Solubility of calcium oxalate

One liter of water dissolves 7.2 mg of calcium oxalate monohydrate[2] at 25°C and 14.0 mg at 95°C. In neutral solutions containing ammonium oxalate the solubility is, of course, less and a dilute solution of this salt is therefore used to wash the precipitate in the gravimetric determination of calcium. The solubility in oxalate solutions is greater than calculated from the solubility product, taking the ionic strength effect into account (Table 28-1), which indicates some formation of complex calcium oxalate anion. Calcium oxalate has a minimum solubility (0.15 mg of Ca per liter at 25°C) in 0.01 *M* ammonium oxalate solution, but the solubility is almost constant in the range 0.001–0.1 *M* ammonium oxalate. The value of the constant

$$\frac{[CaC_2O_4]}{[Ca^{++}][C_2O_4^{--}]}$$

has been reported as approximately 1×10^3. In a saturated solution then, the concentration of undissociated calcium oxalate is $K_{sp} \times 1 \times 10^3 = (2 \times 10^{-9}) \times (1 \times 10^3) = 2 \times 10^{-6}\ M$, which is the intrinsic solubility if the value 1×10^3 is correct.

[1] Calcium oxalate monohydrate is also a weighing form for calcium under some conditions.

[2] Three hydrates of calcium oxalate are known: $CaC_2O_4 \cdot H_2O$, $CaC_2O_4 \cdot 2\,H_2O$, and $CaC_2O_4 \cdot 3\,H_2O$. The monohydrate is the stable form under ordinary conditions and is always obtained when the precipitation is made in hot solution. The di- and trihydrate, which are formed by precipitation in cold solution under special conditions, are transformed into the monohydrate on standing, especially rapidly at higher temperatures.

TABLE 28-1. **Solubility of $CaC_2O_4 \cdot H_2O$ in Ammonium Oxalate Solutions (25°C)**[a]

Concentration of $(NH_4)_2C_2O_4$, *M*	*Solubility, M* × 10^6 *Found*	*Calculated*
0	49.0	
0.001	5.9	3.4
0.003	4.2	1.5
0.01	3.8	0.76
0.03	4.3	0.44
0.1	5.3	0.35
0.3	8.8	0.26

[a] From F. Nydahl, *Acta Chem. Scand.* **5**, 669 (1951).

The solubility of calcium oxalate increases with increasing hydrogen-ion concentration, as a consequence of the removal of oxalate ions from the system $CaC_2O_4(s) \rightleftharpoons Ca^{++} + C_2O_4^{--}$. Oxalate ions are transformed into bioxalate ions and at higher acidities into oxalic acid. From the equilibria:

$$HC_2O_4^- \rightleftharpoons C_2O_4^{--} + H^+ \qquad K_2 = 6 \times 10^{-5} \tag{1}$$

$$H_2C_2O_4 \rightleftharpoons HC_2O_4^- + H^+ \qquad K_1 = 6.5 \times 10^{-2} \tag{2}$$

it is seen that as a monobasic acid oxalic acid is fairly strong, but as a dibasic acid it is a typically weak acid.

Quantitative precipitation of calcium oxalate is obtainable in weakly acidic solutions, a precipitation medium that is advantageous at times (p. 621). Let us find the concentration of calcium in a saturated calcium oxalate solution of pH 4 in which the total oxalate ($C_2O_4^{--}$, $HC_2O_4^-$, and $H_2C_2O_4$) concentration is 0.10 *M*. From Eqs. (1) and (2) we see that at this pH,

$$[H_2C_2O_4] = \frac{10^{-4}[HC_2O_4^-]}{6.5 \times 10^{-2}} = 1.5 \times 10^{-3}\,[HC_2O_4^-]$$

and

$$[C_2O_4^{--}] = \frac{6 \times 10^{-5}\,[HC_2O_4^-]}{10^{-4}} = 0.6\,[HC_2O_4^-]$$

Therefore,

$$[C_2O_4^{--}] = \frac{0.6}{1.5 \times 10^{-3} + 1 + 0.6} \times 0.10 = 0.0375$$

If the (weak) complexing of calcium oxalate with excess oxalate, and the ionic strength effect, could be neglected, the solubility would be given by

$$\frac{K_{sp}}{0.0375} = \frac{2 \times 10^{-9}}{3.75 \times 10^{-2}} = 5 \times 10^{-8}$$

But the complexing cannot be neglected. If it is assumed that the formation of soluble calcium oxalate complexes comes about by addition of $C_2O_4^{--}$, and not

$HC_2O_4^-$, to Ca^{++} we can make use of Table 28-1 to estimate the solubility. We see that at $[C_2O_4^{--}] = 0.0375$, the experimentally determined solubility is close to 4×10^{-5} *M*. The ion activities are lower, however, in the 0.1 *M* oxalate solution than in the 0.04 *M*. The resultant increase in solubility should not exceed ~50 per cent, so that the maximum value of the solubility is $\sim 6 \times 10^{-6}$ *M*, corresponding to ~0.05 mg of Ca in a working volume of 200–250 ml.

The presence of magnesium increases the solubility of calcium oxalate appreciably by formation of complex magnesium oxalate which withdraws oxalate ions. When magnesium is present in solution with calcium, care must be taken to add enough oxalate to transform magnesium into the complex oxalate, or serious solubility losses of calcium may ensue.

Precipitation of calcium oxalate

Precipitation of calcium from cold neutral or ammoniacal solutions gives an extremely finely divided precipitate that can be filtered and washed only with difficulty. Therefore, the precipitate is never formed in cold solution (except as noted below). Formerly it was the practice to make the precipitation by adding oxalate to a hot neutral or ammoniacal calcium solution. The precipitate thus obtained is coarse-grained enough to be filtered without any particular difficulty. It is much better, however, to precipitate calcium oxalate by adding ammonium oxalate to a hot *acid* solution of the calcium salt and finally neutralizing with ammonium hydroxide. Depending upon the acidity, more or less of the calcium oxalate will be precipitated on the addition of the oxalate. This precipitate will be coarse on account of the low supersaturation of the solution with respect to calcium oxalate. The remainder of the calcium is also thrown down in the form of relatively large crystals if the acid is neutralized by moderately slow addition of ammonium hydroxide to the hot solution. The precipitate thus obtained can be filtered and washed with ease.[3]

The precipitate formed by the neutralization of an acid solution may be contaminated to a slight extent with coprecipitated oxalic acid or acid oxalate, but this does not matter if the precipitate is ignited to oxide or carbonate, or converted into sulfate and weighed as such. This method of precipitation gives satisfactory, although slightly high, results when calcium is determined volumetrically with potassium permanganate (Chap. 43). In this method the oxalate precipitate is dissolved in hot sulfuric acid, and the liberated oxalic acid is titrated with standard permanganate solution. The volumetric method can give accurate results only when the ratio of $C_2O_4^{--}$ to Ca^{++} in the precipitate is 1 : 1. The precipitate formed in acid solutions in the way mentioned has a composition that closely approaches this ratio. However, a precipitate obtained by adding ammonium oxalate to a

[3] H. H. Willard and F. L. Chan (Willard and Furman, *Elementary Quantitative Analysis*, Van Nostrand, New York) neutralize a strongly acid solution of calcium, containing excess oxalate, by adding much urea and heating just below the boiling point. The urea is thus slowly hydrolyzed:

$$(NH_2)_2CO + H_2O \rightarrow CO_2 + 2\,NH_3$$

and extremely coarse crystals of calcium oxalate are formed as a result of the slow neutralization. The neutralization requires from 20 to 60 min for completion. See R. S. Ingols and P. E. Murray, *Anal. Chem.* **21**, 525 (1949). Directions for a procedure tested by the authors (footnote 8) are given on page 623.

neutral or ammoniacal solution is contaminated with basic calcium oxalate, and decidedly low results are then obtained in the volumetric method.

Interfering substances

The determination of calcium by precipitation as oxalate, in the manner described, requires the absence of elements that form slightly soluble oxalates or which are otherwise precipitated under the conditions of the determination. Because of this requirement, calcium must first be separated from virtually all metals except the alkalies and magnesium before the oxalate precipitation can be made (but see the modified oxalate precipitation, p. 621).[4] When *small* amounts (a few milligrams) of barium are present, double or triple precipitations of calcium oxalate suffice to separate the two elements.

In the general case it is necessary to precipitate calcium oxalate from a solution containing magnesium and the alkali metals. The coprecipitation of these cations as well as certain anions must therefore be taken into account. The contamination of the calcium oxalate precipitate by magnesium is discussed in a later section in this chapter.

Calcium oxalate tends to occlude the alkali metals, which are carried down as the oxalates. In the usual gravimetric method in which the precipitate is finally weighed as calcium oxide, carbonate, or sulfate, the possible presence of sodium and potassium must be borne in mind, and in the volumetric method (titration with permanganate) that of ammonium as well. It will be seen from Table 28-2 that sodium is much more strongly coprecipitated than potassium. Unless inordinately large amounts of potassium are present, the quantity of this cation carried down by the precipitate is so small that a double precipitation is not required in ordinary analytical work. The presence of sodium calls for a double precipitation of calcium oxalate. Precipitation from ammoniacal solution yields a precipitate containing less sodium (and potassium) than one formed by slowly neutralizing an acid solution with ammonium hydroxide. It is to be noted that calcium oxalate precipitated with ammonium oxalate contains appreciable amounts of ammonium oxalate.

The "digestion method" of precipitation described on page 243 can be applied with advantage in obtaining a fairly pure precipitate of calcium oxalate in the presence of sodium. In this method ammonium oxalate is added rapidly to the neutral calcium solution at room temperature. The mixture is then diluted, acidified with acetic acid, and heated on the steam bath from 12 to 24 hr. The digestion results in the conversion of the higher hydrates of calcium oxalate, which have been precipitated in the cold solution, into the monohydrate; the slow recrystallization thus effected eliminates the greater part of the coprecipitated sodium or ammonium. Under the proper conditions the amount of sodium in the precipitate is reduced to such an extent that a double precipitation is often not necessary.

Sulfate is quite strongly coprecipitated with calcium oxalate, owing to the low solubility of calcium sulfate. The presence of sulfate in the precipitate leads to high results when calcium carbonate is used as the weighing form and possibly also when the precipitate is ignited to the oxide, although in the latter case more or less of the calcium sulfate is decomposed into the oxide at the high temperature of the ignition. Of course, there is no error if the precipitate is converted into calcium sulfate and weighed as such. In the permanganate volumetric method the results for calcium are low in the presence of sulfates, unless there are compensating errors. Phosphate and

[4] Strontium when present is precipitated as the oxalate with calcium and is separated later by dissolving the ignited precipitate in nitric acid and extracting the anhydrous nitrates with 78 per cent nitric acid, in which strontium nitrate is insoluble.

TABLE 28-2. **Coprecipitation of Alkalies with Calcium Oxalate**[a]
(0.1 g of calcium as chloride used in each determination)

Conditions of Precipitation	*Percentage of Alkali Oxalate in* $CaC_2O_4 \cdot H_2O$
Sodium (1.7 g of sodium chloride present in each)	
1. 50 ml of 2% ammonium oxalate solution added to 250 ml of calcium solution containing 5 ml of concentrated hydrochloric acid; neutralization made with dilute ammonia added dropwise to hot solution. Precipitate filtered after several hours	1.3
2. 50 ml of 2% ammonium oxalate added slowly to ammoniacal (1 ml of concentrated ammonia) solution having volume of 250 ml, at boiling temperature	0.8
3. 25 ml of 4% ammonium oxalate added at one stroke to cold neutral calcium solution having volume of 250 ml. Precipitate digested in mother liquor at 100°C for 2 days	0.2
Potassium (1.5 g of potassium chloride present in each)	
1. 50 ml of 2% ammonium oxalate solution added to 200 ml of hot calcium solution containing 2 ml of concentrated hydrochloric acid; dilute ammonia then added slowly to neutralization	0.1–0.2
2. 50 ml of 2% ammonium oxalate solution added to 200 ml of hot ammoniacal (1 ml of concentrated ammonia) calcium solution	0.03
Ammonium	
1. 50 ml of 2% ammonium oxalate added to 200 ml of hot calcium solution containing 2 ml of concentrated hydrochloric acid (no added ammonium chloride); then neutralized with dilute ammonia	0.2–0.3
2. 50 ml of 2% ammonium oxalate added to 200 ml of hot ammoniacal (1 ml of concentrated ammonia) calcium solution	0.2–0.3
3. As in 1 except 1 g of ammonium chloride present	0.3–0.4

[a] From I. M. Kolthoff and E. B. Sandell, *J. Phys. Chem.* **37**, 448 (1932).

arsenate are not coprecipitated markedly if the precipitation of calcium is made in acid solution, i.e., if the acid solution containing excess oxalate is slowly neutralized with ammonium hydroxide.[5] Other anions less frequently encountered, such as iodate and chromate, are carried down by calcium oxalate, and double precipitations are required when they are present. The halides are coprecipitated in extremely small amounts, usually not more than a few hundredths of 1 per cent. The coprecipitation of the anions is greater in hot solutions than in cold, the reverse being true for the cations.

Modified oxalate method. Instead of finally making the solution ammoniacal in the precipitation of calcium oxalate, it is sometimes advantageous to keep it slightly acidic. When this is done, calcium can be precipitated in the presence of iron, titanium,

[5] However, if magnesium is simultaneously present, magnesium ammonium phosphate or arsenate is precipitated to a greater or less extent depending upon the final pH of the solution.

and other metals. In this method[6] the hot hydrochloric acid solution containing at least 1 per cent of ammonium oxalate monohydrate is treated with ammonium hydroxide until the pH is 3–4 (light green color with bromphenol blue). The mixture is digested at 100°C for 1 or 2 hr, and the precipitate is washed with a solution containing ammonium oxalate and oxalic acid (2 g and 1 g, respectively, per liter). Double precipitations should be made. The method may be used for the determination of calcium in phosphate rock.

Weighing forms of calcium

The following weighing forms of calcium have been proposed: $CaC_2O_4 \cdot H_2O$, CaC_2O_4, $CaCO_3$, CaO, $CaSO_4$, and CaF_2. The first four forms are obtained by heating the precipitate of calcium oxalate at suitable temperatures (Fig. 28-1), and the last two by treating the precipitate, or its ignition products, with sulfuric and hydrofluoric acids, respectively. Of these, the carbonate, oxide, and sulfate are well-established weighing forms that have long been used.

Calcium oxalate monohydrate. A number of workers[7] have proposed weighing calcium as $CaC_2O_4 \cdot H_2O$ after drying at 100–110°C. As ordinarily precipitated, calcium oxalate monohydrate is not an entirely satisfactory weighing form. After drying at 110°C it tends to contain foreign water and also to be hygroscopic.

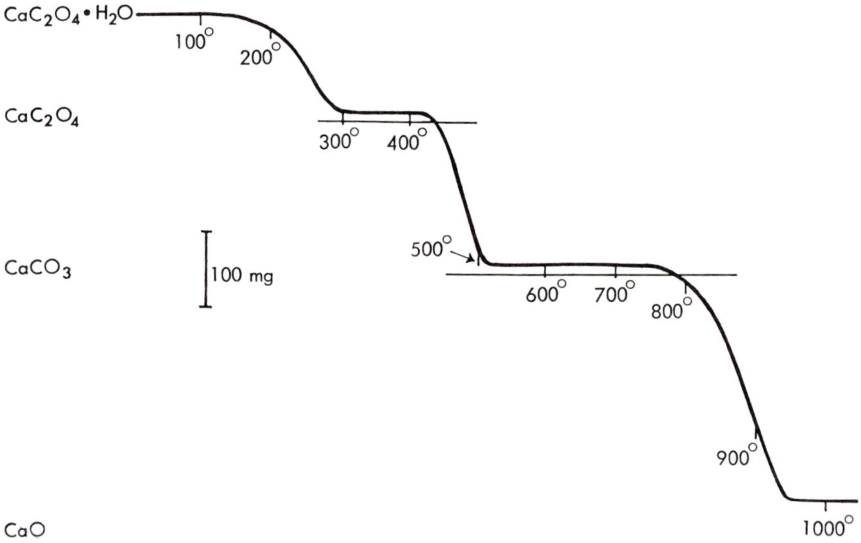

Fig. 28-1. Thermogram for calcium oxalate monohydrate (1.019 g, porcelain crucible, ambient air, heating rate 300°C/hr). From E. L. Simons and A. E. Newkirk, *Talanta* **11**, 549 (1964).

[6] J. I. Hoffman and G. E. F. Lundell, *J. Res. Natl. Bur. Std.* **20**, 607 (1938); H. D. Chapman, *Soil Sci.* **26**, 479 (1928); W. H. McComas, Jr., and W. Rieman, III, *Ind. Eng. Chem., Anal. Ed.* **14**, 929 (1942).

[7] See, for example, S. Goy, *Chem. Ztg.* **37**, 1337 (1913); L. W. Winkler, *Z. Angew. Chem.* **31**, 187 (1918); O. Brunck, *Z. Anal. Chem.* **94**, 81 (1933); H. Sibelius, *Suomen Kemistilehti* **A8**, 25 (1935); J. Dick, *Z. Anal. Chem.* **77**, 358 (1929), recommended washing the precipitate successively with water, alcohol, and ether, and then drying briefly in the air or in a vacuum desiccator at room temperature. L. Moser and L. von Zombory, *Z. Anal. Chem.* **81**, 95 (1930), raised objections to this procedure.

Coprecipitated ammonium oxalate remains in the dried precipitate. The error from these sources varies, depending on the manner of precipitation, but values high by 5 ppt or more are not unusual. In dry air, the monohydrate may lose water, being partially converted to the anhydrous oxalate, on heating at 110°C; this decomposition is less likely with precipitates formed in acidic than in ammoniacal or neutral solutions.

The coarse crystals of $CaC_2O_4 \cdot H_2O$ obtained by heating a mineral acid–oxalate solution of calcium with urea can be washed with acetone and weighed after drying at room temperature.[8] With 0.1–0.2 g of Ca, the results are within 1–2 ppt, with 0.05 g of Ca within ~5 ppt when the precipitation is made slowly. The procedure is given on page 629. Especially in microgravimetric determinations of calcium, $CaC_2O_4 \cdot H_2O$ is a desirable weighing form because of the favorable chemical factor.

At temperatures in the neighborhood of 200°C calcium oxalate monohydrate loses its water of crystallization, yielding the anhydrous oxalate, which has also been recommended as a weighing form. Anhydrous calcium oxalate is so hygroscopic, however, that it is not suitable for the purpose; besides, complete expulsion of water seems to be difficult.

Calcium carbonate. Calcium oxalate begins to decompose into the carbonate above 350°C.[9] Willard and Boldyreff[10] studied the suitability of calcium carbonate as a weighing form for calcium. They found that for the complete decomposition of calcium oxalate into carbonate in a reasonable period of time the temperature must be at least 450°C. These authors also found that above 550°C calcium carbonate loses carbon dioxide. When calcium is to be weighed as the carbonate, the oxalate precipitate should be ignited in the temperature range 475–525°C. Since the temperature must be controlled within rather narrow limits, an electric furnace provided with a pyrometer or a suitable thermometer must be used. If such equipment is available, it is recommended that calcium be weighed as the carbonate for the most accurate results. The precipitate must then be collected in a Gooch or porcelain filter crucible and not in paper, because the latter gives an ignited residue containing carbon.

In the old method of determining calcium as carbonate, the oxalate precipitate was ignited at low redness by means of a burner; the residue was moistened with ammonium carbonate solution to convert any oxide formed into carbonate, briefly ignited, and then weighed. The procedure was repeated until constant weight was attained. This method is not to be recommended.

If the precipitate is heated in a steam of carbon dioxide, the temperature may be raised to 850°C without fear of decomposing the carbonate. This method gives good results[11] but is not convenient.

Calcium oxide. At approximately 880°C the dissociation pressure of calcium carbonate is 760 mm Hg. This means that at temperatures very slightly above this value, calcium carbonate can be converted completely into the oxide even when surrounded by an atmosphere of carbon dioxide at atmospheric pressure. Therefore

[8] E. B. Sandell and I. M. Kolthoff, *Ind. Eng. Chem., Anal. Ed.* **11**, 90 (1939).
[9] See page 189 for a discussion of the heterogeneous equilibria in the system $CaCO_3$–CaO.
[10] H. H. Willard and A. W. Boldyreff, *J. Am. Chem. Soc.* **52**, 1888 (1930).
[11] H. W. Foote and W. M. Bradley, *J. Am. Chem. Soc.* **48**, 676 (1926).

by heating calcium carbonate in a covered crucible to this temperature, it may be expected that the carbonate will be converted quantitatively into the oxide in spite of the fact that the carbon dioxide is not allowed to escape freely. However, on cooling there would be a recombination of calcium oxide and carbon dioxide, with the result that some calcium carbonate would be formed. Actually, the ignition is made in a loosely covered crucible so that the carbon dioxide escapes by diffusion and convection, being replaced by air. In this manner the partial pressure of carbon dioxide is kept far below the atmospheric pressure, and there is not enough of the gas left in the crucible at the end to form any significant amount of calcium carbonate. It is found, however, that the rate of decomposition of calcium carbonate is so slow in the neighborhood of 900°C that it is advantageous practically to make the ignition at a higher temperature, say in the range 1100–1200°C. Moderate amounts of calcium carbonate can be quantitatively converted into the oxide in a relatively short time by ignition in a platinum crucible over a blast lamp or a Meker burner. The ignition cannot be successfully made in a porcelain crucible under these conditions, and an electrically heated muffle furnace must then be used.

Due care must be taken in cooling and weighing calcium oxide, because it avidly takes on water and, moreover, slowly reacts with the carbon dioxide of the atmosphere. The crucible should be kept *tightly covered* in the desiccator during cooling to prevent the taking up of water from the atmosphere of the desiccator. It must be remembered that after opening the desiccator the humidity of the air therein may not be very different from that of the external atmosphere. Pure concentrated sulfuric acid, freshly ignited calcium oxide, or an equally vigorous desiccant should be used as drying agent; calcium chloride is not suitable. In any case, the crucible and its contents should not be allowed to remain in the desiccator longer than is necessary. The crucible should likewise be kept covered during the weighing; this operation should be performed rapidly, the final weighing being made with all the weights on the pan, and the rider adjusted.

Although satisfactory results can be obtained by weighing calcium as the oxide, it is preferable to weigh as the carbonate if the proper facilities are available, because the latter weighing form is not appreciably hygroscopic and has a much higher molecular weight.

The reader will have noted in the discussion of calcium carbonate and oxide as weighing forms that the temperatures of decomposition do not agree well with the thermogram in Fig. 28-1. The latter is only a rough guide because the temperature is raised relatively rapidly in the thermobalance and equilibrium is not attained.

Calcium sulfate. By treating calcium oxalate, carbonate, or oxide with an excess of sulfuric acid, evaporating the latter, and igniting the residue to the proper temperature, anhydrous calcium sulfate is obtained. Calcium sulfate is not easily decomposed by heating; the temperature of beginning decomposition,

$$CaSO_4 \rightarrow CaO + SO_3$$

is given as 1200°C, but the decomposition may set in at a lower temperature if foreign substances are present. Calcium sulfate may safely be heated over a Tirrill flame to moderate redness. If the residue is not heated strongly enough, traces of water remain, and the results are slightly high.

Calcium sulfate would be a good weighing form for calcium except that the conversion of the precipitate into the sulfate is attended with some difficulties. The precipitate must be collected on paper, and the latter must be burned before the conversion is attempted. If the precipitate is not separated from the filter paper, carbonate is formed during the ignition; there will consequently be an evolution of carbon dioxide when sulfuric acid is subsequently added, and this is likely to lead to loss even if the crucible is covered. Therefore, it is better to ignite strongly to convert most of the carbonate to the oxide and to add water rapidly to the cold residue to slake the oxide. Sulfuric acid is then added. Or the dried precipitate is separated from the filter paper and reserved, while the paper, with a small amount of adhering precipitate, is ignited to burn away all the carbon. The main portion of the precipitate is then added, and the whole is treated with dilute sulfuric acid. In any case, a rather large excess of dilute sulfuric acid is added to ensure the conversion of the residue into sulfate, the conversion being hindered by the tendency of the slightly soluble calcium sulfate to form as a coating over the particles of the residue.[12] The excess sulfuric acid is removed by volatilization in a radiator (p. 463), and the final ignition is made over a medium Tirrill flame. A porcelain crucible may be used in this procedure.

Calcium fluoride. By treating calcium oxalate, carbonate, or oxide with excess hydrofluoric acid as described above for calcium sulfate, evaporating and igniting the residue, anhydrous calcium fluoride is obtained. Calcium fluoride may be heated to redness without fear of decomposition, and it is not hygroscopic. A platinum crucible is required for this conversion. This weighing form is rarely used.

Separation of calcium from magnesium by the oxalate method

It will be seen from Table 28-3 that the solubility of magnesium oxalate in aqueous solutions is relatively small, and hence it may be expected that a sharp separation of calcium from magnesium by precipitation with oxalate is not easily accomplished. There is, first, the danger that magnesium oxalate will actually be precipitated, owing to its limited solubility, unless proper care is taken. There is also the possibility that magnesium oxalate will be appreciably occluded by calcium oxalate in accordance with the coprecipitation rule (p. 229), which states that the tendency of a compound having an ion in common with the lattice of the precipitate to be adsorbed by the growing precipitate increases with decreasing solubility of the compound. Last, the solubility of calcium oxalate in the presence of a high concentration of magnesium ion is a source of error that must be taken into consideration when the ratio of magnesium to calcium is large.

TABLE 28-3. **Solubility of Magnesium Oxalate in Water and Ammonium Oxalate Solutions at Room Temperature**

Concentration of $(NH_4)_2C_2O_4 \cdot H_2O$, g/100 ml *of Solution*	*Solubility of* $MgC_2O_4 \cdot 2\,H_2O$ *at* 22 ± 1°C, g/100 ml *of Solution*
0	0.036
1	0.059
2	0.095
3	0.133
4	0.187

[12] Dissolving the slaked lime in a little hydrochloric acid before the addition of sulfuric acid is therefore recommended.

The danger of separation of magnesium oxalate as such is diminished by the tendency of this salt to form supersaturated solutions which are quite stable under certain conditions.[13]

The greater the degree of supersaturation, the greater is the danger of precipitation of magnesium oxalate. The supersaturation is broken down more readily when the solutions are boiled; and, in general, equilibrium is more rapidly established the higher the temperature. Solutions of magnesium oxalate having a high degree of supersaturation sooner or later deposit magnesium oxalate on standing at room temperature, although it may take some time for crystallization to begin and, once started, it will usually be quite slow unless the temperature is raised. Therefore when calcium oxalate is to be precipitated in the presence of much magnesium, the temperature of the solution should not be too high, and boiling must especially be avoided. It has been shown experimentally that by using a considerable excess of ammonium oxalate in the precipitation as directed in the procedure following, as much as 150 mg of magnesium may be present in 200–250 ml of solution without danger of precipitation of magnesium oxalate. If much more magnesium than this is present, there is the possibility of a slow deposition of magnesium oxalate on standing, especially if the solution is kept warm. The postprecipitation of magnesium oxalate is to be avoided, because when it occurs there is great danger that a double, or even a triple, precipitation of calcium oxalate will not yield a precipitate free of magnesium.

Even when the conditions are such that magnesium cannot precipitate as the oxalate, it is found that the calcium precipitate is contaminated to a greater or less extent by occluded magnesium oxalate. The amount of magnesium that is coprecipitated depends upon the conditions of precipitation, as the results in Table 28-4 indicate. The calcium precipitate obtained by neutralizing a strongly acid calcium–magnesium solution containing a relatively large amount of ammonium oxalate with ammonium hydroxide contains very little magnesium. The decrease in the magnesium coprecipitation under these conditions is to be ascribed to the excess of ammonium oxalate present during the whole course of the precipitation, whereby the solubility of magnesium oxalate is increased (complex magnesium oxalate ions are formed). It follows, then, that the precipitation of calcium in the presence of magnesium should be made by neutralizing an acid solution containing a considerable excess of ammonium oxalate and not by adding oxalate to an ammoniacal solution of calcium–magnesium, as has frequently been recommended. There is one objection to the addition of *very* large amounts of ammonium oxalate, and that is the interference of ammonium salts, especially of the oxalate, in the determination of magnesium (p. 638). It is true that ammonium oxalate may be destroyed prior to the determination of magnesium, but this may not be convenient. In any case, however, it is necessary to add a fairly large excess of ammonium oxalate to precipitate calcium oxalate quantitatively, because, as already mentioned, magnesium ions withdraw oxalate ions to form a complex oxalate, thus increasing the solubility of calcium oxalate. Calcium does not show much tendency to form a complex with oxalate, and the solubility of calcium oxalate remains negligibly small even in a saturated solution (ca. 5 per cent) of ammonium oxalate at room temperature[14] (cf. Table 28-1).

[13] W. Fischer, *Z. Anorg. Allgem. Chem.* **153**, 62 (1926); T. Holth, *Anal. Chem.* **21**, 1221 (1949).

[14] In strong ammonium oxalate solutions at 95°C, the solubility of calcium oxalate becomes appreciable, as shown by E. R. Wright and R. H. Delaune, *Anal. Chem.* **18**, 426 (1946). At this temperature, 100 ml of water containing 35 g of ammonium oxalate monohydrate dissolves 4 mg of Ca, and the same volume containing 15 g dissolves approximately 1 mg of Ca.

TABLE 28-4. **Coprecipitation of Magnesium Oxalate with Calcium Oxalate under Various Conditions**[a]

Conditions of Precipitation	*Magnesium in Calcium Oxalate Precipitate* (100 mg *of* Ca), mg
1. 11 ml of 0.5 *N* (3.5%) ammonium oxalate solution added dropwise over a period of 3 min to the cold (20–25°C) neutral calcium–magnesium solution. An excess of 5 g of ammonium oxalate in 50 ml of water then added	0.33
2. Conditions as in 1 except that the 0.5 *N* ammonium oxalate solution was added dropwise to the calcium–magnesium solution *heated to* 70°C	0.70
3. Precipitated as in 2 in hot (70°C) solution, but calcium–magnesium solution was made ammoniacal before precipitation by adding 2 ml of concentrated ammonium hydroxide and 1 g of ammonium chloride	0.64
4. Precipitated as in 2 in hot (70°C) solution, but calcium–magnesium solution was made acid before precipitation by adding 5 ml of 6 *N* acetic acid	1.06
5. 50 ml of hot 10% ammonium oxalate solution added very rapidly (5 sec) to neutral calcium–magnesium solution heated to 70°C	1.26
6. Calcium–magnesium solution acidified with 5 ml of concentrated hydrochloric acid, heated to 70°C, and treated with 11 ml of 0.5 *N* ammonium oxalate. No precipitate formed. Then made ammoniacal by adding dilute ammonium hydroxide over a period of 2 min	0.66
7. 50 ml of hot 10% ammonium oxalate solution added rapidly to hot (70°C) calcium–magnesium solution containing 5 ml of concentrated hydrochloric acid. After 5 min the solution was neutralized with dilute ammonium hydroxide	0.16
8. Conditions as in 7 except that precipitation was made at room temperature	0.15
9. 100 mg of calcium and 100 mg of magnesium as chlorides dissolved in 50 ml of water, and the solution added slowly (5 min) to 200 ml of warm (70°C) ammonium oxalate solution containing 5 g of the salt	0.72

[a] In each experiment 100 mg of calcium and 100 mg of magnesium as chlorides were present in 200 ml of solution. The solution was allowed to stand 1–2 hr after precipitation, the precipitate filtered off, washed, dissolved in hydrochloric acid, and reprecipitated by addition of ammonium hydroxide; the magnesium in the filtrate from the reprecipitated calcium oxalate was precipitated with 8-hydroxyquinoline and the precipitate titrated.

Usually a double precipitation of calcium oxalate must be made to secure a precipitate free of magnesium, even when the precipitation is made under favorable conditions. Thus in Experiment 7, Table 28-4, a single precipitation of calcium from an acid solution containing 2.5 per cent ammonium oxalate gave a precipitate containing nearly 0.2 mg of magnesium to 100 mg of calcium, equal amounts of the two metals being originally present. This amount of coprecipitated magnesium would be of no consequence in routine analysis but has to be taken into account in more accurate work.

As ordinarily carried out, the oxalate method for the separation of calcium from magnesium fails when the ratio of magnesium to calcium is very large. However, by using a large excess of ammonium oxalate to keep magnesium in solution, it is possible to separate as little as one part of calcium from 100 parts of magnesium.[15] If a large excess of ammonium oxalate is not added, calcium is precipitated incompletely or not at all. Such unprecipitated calcium is subsequently quantitatively precipitated as the phosphate with magnesium ammonium phosphate in the determination of magnesium and can be recovered by the method of Hillebrand.[16] In this procedure the ignited residue of magnesium pyrophosphate (p. 639) is dissolved in a small excess of dilute sulfuric acid, and the solution is treated with a large excess of 75 per cent ethyl alcohol. Calcium sulfate is insoluble in the alcoholic solution, whereas magnesium sulfate is quite soluble and is not precipitated. The precipitate of calcium sulfate is filtered off, washed with 75 per cent alcohol, dissolved in hydrochloric acid, and calcium is then precipitated as the oxalate and weighed as the oxide. The true weight of magnesium pyrophosphate is found by deducting the weight of $Ca_3(PO_4)_2$ corresponding to the weight of calcium found from the weight of the impure magnesium pyrophosphate.

Procedure. The following directions provide for the presence of magnesium and the alkali metals but no other cations.

Prepare in water or hydrochloric acid a solution of the sample containing not more than 200 mg of calcium or 100–150 mg of magnesium, dilute to 200 ml, and add a few drops of methyl red indicator. Add 5 ml of concentrated hydrochloric acid in excess and then 50 ml of a warm ammonium oxalate solution[17] containing 3 g[18] of $(NH_4)_2C_2O_4 \cdot H_2O$. Heat the solution to 70–80°C, and add 1 : 1 ammonium hydroxide dropwise with stirring until the color changes from red to yellow. Allow the solution to stand without further heating for 3 hr (not longer if magnesium is present),[19] and filter through paper. Wash the precipitate four or five times with cold 0.1 per cent ammonium oxalate solution, and then dissolve it in 50 ml of hot 1:4 hydrochloric acid. Wash the paper with hot 1 : 100 hydrochloric acid, and dilute the solution to 200 ml. Add 1 g of ammonium oxalate dissolved in a few milliliters of hot water, heat the solution nearly to boiling, and precipitate calcium oxalate again by neutralizing the acid solution with ammonium hydroxide as already described. Allow the solution to stand for at least 4 hr (standing overnight does no harm). Collect the precipitate in a weighed porcelain filter or Gooch crucible if calcium carbonate is to be used as the weighing form, or in filter paper if the precipitate is to be ignited to the oxide. Wash the precipitate thoroughly

[15] M. Bobtelsky and Frau Malkowa-Janowski, *Z. Angew. Chem.* **40**, 1436 (1927); E. R. Wright and Delaune, *Anal. Chem.* **18**, 426 (1946). The latter authors apply an empirical correction for solubility loss in the hot concentrated oxalate solution and obtain good results.

[16] W. F. Hillebrand, *U.S. Geol. Surv. Bull.* **700**, 1919, p. 152.

[17] This solution should be filtered hot if not entirely clear.

[18] If magnesium is not to be determined, add 5–6 g of ammonium oxalate. A like amount may be added if the oxalate is later destroyed by nitric acid (Chap. 63) prior to the determination of magnesium. If calcium alone is present, add only 1 g of ammonium oxalate in excess, and do not reprecipitate unless sodium is present. In the absence of magnesium make the precipitation at 100°C.

[19] It is advisable to stir occasionally to disperse the precipitate through the solution and so aid the precipitation of the last traces of calcium which tend to remain in supersaturated solution, especially if the amount of precipitate is small. In the presence of magnesium, better results are obtained when only one precipitation is made if the time of standing is only 1 hr. In this period of time the precipitation of calcium is not quite complete and the negative error partially compensates the positive error due to coprecipitated magnesium.

with cold 0.1 per cent ammonium oxalate solution. Reserve the filtrates and washings from the first and second precipitations if magnesium is to be determined (Chap. 63).

If the precipitate is to be ignited to the carbonate, dry the crucible first at a low temperature, and then place it in an electric furnace regulated to 500°C. The temperature should not vary by more than ± 25°C from this value. After 1 or 2 hr remove the crucible, cool, and weigh. Repeat the heating to constant weight, using 30-min periods of heating. Finally, to be sure that no decomposition of calcium carbonate has occurred, moisten the residue with 2 or 3 drops of saturated ammonium carbonate solution, allow to stand at room temperature for 15 min, dry at 110–120°C for $\frac{1}{2}$ hr, and weigh. There should be no change in weight. If an increase in weight is found, repeat the moistening with ammonium carbonate and heating until the weight is constant.

When the oxide is to be used as the weighing form, place the paper with the precipitate in a platinum crucible, char the filter at a low temperature, and then heat strongly for 15–30 min over a Meker burner or blast lamp. Keep the crucible covered except after the start of the strong heating, when the cover should be displaced to one side for a short time to allow the escape of carbon dioxide. Do not allow the flame gases to gain access to the interior of the crucible. Cool the covered crucible in a desiccator charged with magnesium perchlorate or similar effective drying agent. Do not permit the crucible to stand unnecessarily long in the desiccator. Weigh as rapidly as possible, and of course keep the crucible covered. Reheat strongly for 10 or 15 min and weigh again, first placing the weights on the pan so that only the rider or chain need be adjusted after the crucible is placed on the balance. Repeat the ignition until constant weight is obtained.

If a porcelain crucible is used for igniting the precipitate to the oxide, employ an electric muffle furnace, and heat to 1000–1100°C.

Calculate the percentage of calcium or calcium oxide.

Regarding the recovery of any calcium that may have escaped precipitation, see page 628.

Alternative procedure: homogeneous precipitation with urea. Add 5 ml of concentrated hydrochloric acid to an approximately neutral sample solution containing 75–200 mg of Ca and dilute to 175–200 ml. Heat almost to boiling and add 1.0 g of ammonium oxalate monohydrate dissolved in 20 ml of hot water (no precipitate should form), followed by 5.0 g of reagent-quality urea. Mix and heat the covered beaker at 80–90°C overnight. The solution should then be basic to methyl orange.

Cool the solution to room temperature and collect the precipitate in a porcelain or glass-filter crucible whose weight has been obtained after washing with water and acetone and drying in the air. Wash the precipitate with small portions of cold water and then with three or four 2- to 3-ml portions of reagent-quality acetone. Draw air through the crucible for 5 min and weigh the precipitate as $CaC_2O_4 \cdot H_2O$. It is well to let the crucible stand on the balance pan for 10–15 min and then reweigh to be certain that the weight is constant.

If appreciable amounts of sodium or magnesium (say >20 mg of Mg) are present, make a double precipitation. The first precipitation need not be made by the urea method.

PROBLEMS

1. 250 ml of tap water gave an ignited calcium oxide residue weighing 0.0476 g. Calculate the calcium content of the water in ppm (milligrams per liter) of calcium carbonate. *Ans.* 338.
2. What volume of 0.050 *M* calcium chloride solution must be taken to furnish 1.00 g of calcium carbonate? *Ans.* 200 ml.
3. What weight of pure calcium carbonate must be taken to prepare 1 liter of solution containing 1.000 mg of Ca per milliliter? What volume of 6.00 *M* hydrochloric acid is required to dissolve the carbonate?
4. Calculate the solubility of $CaC_2O_4 \cdot H_2O$ in a solution of pH 5.0 which is 0.03 *M* in oxalate; in a solution containing 2 g of ammonium oxalate and 1 g of oxalic acid in 1 liter.
5. How much more soluble is $CaC_2O_4 \cdot H_2O$ at pH 3 than at pH 6?
6. Find the solubility of calcium oxalate monohydrate in 0.02 *M* acetic acid.
7. An ignited residue of calcium carbonate contains 1.0 per cent CaC_2O_4. What error (in ppt of Ca) is made if it is assumed that the weighed residue consists entirely of calcium carbonate? *Ans.* +2.2 ppt.
8. What volume of sulfuric acid of sp. gr. 1.24 containing 32.3 per cent H_2SO_4 is theoretically required to convert 0.75 g of $CaC_2O_4 \cdot H_2O$ into $CaSO_4$? *Ans.* 1.26 ml.
9. A calorimetric determination of magnesium in 0.1267 g of calcium carbonate ignited at 500°C showed 0.0003 g of Mg to be present. What is the true weight of calcium carbonate? (*Query*: In what form is magnesium present in the ignited calcium carbonate?)
10. Find the solubility of lead oxalate in 0.1 *M* potassium oxalate solution disregarding the ionic strength effect.

$$[Pb^{++}][C_2O_4^{--}] = 4.8 \times 10^{-10}$$

$$\frac{[Pb^{++}][C_2O_4^{--}]^2}{[Pb(C_2O_4)_2^{--}]} = 2.9 \times 10^{-7}$$

11. Outline a method, consulting the literature if necessary, for determining calcium in a lead–calcium alloy containing ~1 per cent Ca.

CHAPTER 29 *Determination of Magnesium*

Two gravimetric methods for the determination of magnesium are considered in this chapter. The first is an old one that has been employed since the early nineteenth century and is still a standard method. The second method, a more recent one (1927), furnishes a good example of the use of an organic precipitant in quantitative analysis.

Titrimetric determination of Mg with EDTA is much more rapid than the gravimetric determination and is more commonly applied in practice.

Determination of magnesium by precipitation as magnesium ammonium phosphate and ignition of the precipitate to magnesium pyrophosphate

Method. An acid solution of the sample is treated with an excess of ammonium phosphate, and an excess of strong ammonium hydroxide is then added at room temperature to precipitate magnesium ammonium phosphate:

$$Mg^{++} + NH_4^+ + PO_4^{3-} + 6\,H_2O \rightarrow MgNH_4PO_4 \cdot 6\,H_2O$$

The precipitate thus obtained does not have quite the theoretical composition, and it is therefore dissolved in dilute hydrochloric acid after washing with a dilute solution of ammonia and is reprecipitated by the addition of excess ammonium hydroxide. A small amount of ammonium phosphate is added in the second precipitation to reduce the solubility of the precipitate. The pure precipitate thus obtained is washed with dilute ammonium hydroxide, ignited to magnesium pyrophosphate at a high temperature (1100°C), and weighed as such:

$$2\,MgNH_4PO_4 \rightarrow Mg_2P_2O_7 + 2\,NH_3 + H_2O$$

Solubility of magnesium ammonium phosphate. Magnesium ammonium phosphate is one of the most soluble precipitation forms used in quantitative analysis. The equilibria in solutions of magnesium ammonium phosphate are complicated. In any solution saturated with magnesium ammonium phosphate the following relation holds:

$$[Mg^{++}][NH_4^+][PO_4^{3-}] = K_{sp}$$

In a saturated solution of the salt in water, part of the NH_4^+ ions will be transformed into NH_4OH. This equilibrium is governed by the hydroxyl-, and hence by the hydrogen-ion concentration:

$$NH_4^+ + OH^- \rightleftharpoons NH_4OH$$

and

$$\frac{[NH_4^+]}{[NH_4OH]} = \frac{K_{NH_3}}{[OH^-]} = \frac{K_{NH_3}}{K_w} \cdot [H^+]$$

in which K_{NH_3} is the ionization constant of ammonium hydroxide and K_w the ionization product of water. The trivalent phosphate ions are strongly hydrolyzed, the amount transformed into HPO_4^{--} and $H_2PO_4^-$ being dependent upon the hydrogen-ion concentration of the solution and the third and second ionization constants of phosphoric acid:

$$PO_4^{3-} + H^+ \rightleftharpoons HPO_4^{--} \qquad \text{(third ionization)}$$

$$HPO_4^{--} + H^+ \rightleftharpoons H_2PO_4^- \qquad \text{(second ionization)}$$

A solution of magnesium ammonium phosphate in water therefore will contain the following components: Mg^{++}; NH_4^+, NH_4OH; PO_4^{3-}, HPO_4^{--}, $H_2PO_4^-$.

If the molar solubility of the salt is equal to c, then

$$[Mg^{++}] = c$$

$$[NH_4^+] + [NH_4OH] = c$$

$$[PO_4^{3-}] + [HPO_4^{--}] + [H_2PO_4^-] = c$$

With known c, known hydrogen-ion concentration, and the known values of the various ionization constants, it is possible to calculate $[NH_4^+]$ and $[PO_4^{3-}]$ in the saturated solution and therefore also the solubility product. Unfortunately no exact data are available. However, it is doubtful whether the figures would have an exact significance, because the solid phase probably is not stable in contact with water. Evidently the solubility of magnesium ammonium phosphate will be smaller in a mixture of ammonia and ammonium chloride than in water. The excess of ammonium ions represses the solubility according to the mass-action law:

$$[Mg^{++}][NH_4^+][PO_4^{3-}] = K_{sp}$$

and besides, the excess of ammonia lowers the hydrogen-ion concentration, thereby decreasing the hydrolysis of the PO_4^{3-} ions.

In the determination of magnesium an excess of ammonium phosphate is present in the solution containing ammonia and ammonium ions. Since the ammonium ions repress the dissociation of ammonium hydroxide, $[NH_3]$ and $[NH_4^+]$ may be taken equal to the analytical concentrations of ammonia and ammonium salts, respectively. On the other hand, the hydrolysis of the phosphate ions is appreciable.

$$PO_4^{3-} + H_2O \rightleftharpoons HPO_4^{--} + OH^-$$

$$HPO_4^{--} + H_2O \rightleftharpoons H_2PO_4^- + OH^-$$

As a result of this hydrolysis, $[PO_4^{3-}]$ is much smaller than the analytical concentration of ammonium phosphate.

Assume that the concentration of free ammonia in the supernatant liquid is 1 M and that of ammonium ion is 0.18 M. From the ionization constant of ammonia it is found that

$$[OH^-] = \frac{[NH_4OH]}{[NH_4^+]} K_{NH_3} = \frac{1}{0.18} \times 1.8 \times 10^{-5} = 10^{-4}$$

and

$$[H^+] = \frac{K_w}{[OH^-]} = \frac{10^{-14}}{10^{-4}} = 10^{-10} \qquad (25°C)$$

Further assume that the excess of ammonium phosphate corresponds to a concentration of 0.02 M:

$$[PO_4^{3-}] + [HPO_4^{2-}] + [H_2PO_4^-] = 0.02$$

The third ionization constant of phosphoric acid is approximately 5×10^{-13}:

$$\frac{[H^+][PO_4^{3-}]}{[HPO_4^{--}]} = 5 \times 10^{-13}$$

and since $[H^+]$ has been taken equal to 10^{-10},

$$\frac{[PO_4^{3-}]}{[HPO_4^{--}]} = 5 \times 10^{-3} \qquad (1)$$

The second ionization constant of phosphoric acid is 7.5×10^{-8}. Therefore,

$$\frac{[HPO_4^{--}]}{[H_2PO_4^-]} = \frac{7.5 \times 10^{-8}}{10^{-10}} = 750 \qquad (2)$$

From Eqs. (1) and (2) it follows that virtually all the phosphate is present as HPO_4^{--}. Since the total phosphate concentration was taken equal to 0.02 M, it is found from Eq. (1) that

$$[PO_4^{3-}] = 5 \times 10^{-3} \times 0.02 = 10^{-4}$$

The amount of magnesium unprecipitated is determined by the solubility product of magnesium ammonium phosphate and the ammonium- and phosphate-ion concentrations:

$$[Mg^{++}] = \frac{K_{sp}}{[NH_4^+][PO_4^{3-}]} \qquad (3)$$

It is now readily seen that with the same excess of ammonium and total phosphate in the solution, the amount of magnesium left in solution within a certain pH range increases linearly with the hydrogen-ion concentration.

If by varying the concentration of free ammonia the hydrogen-ion concentration were made equal to 10^{-9}, all other concentrations being kept the same as before, $[PO_4^{3-}]$ would be equal to 0.0005 instead of 0.005 M, and the amount of magnesium left in the solution would be 10 times greater [Eq. (3)].

Evidently the solubility of magnesium ammonium phosphate in dilute ammonia is much less than in water. Use is made of this fact in washing the precipitate.

In the precipitation of phosphate as magnesium ammonium phosphate (Chapter 30), the supernatant liquid contains an excess of magnesium, ammonium ions, and ammonia. The amount of trivalent phosphate ion left in solution is given by the expression

$$[PO_4^{3-}] = \frac{K_{sp}}{[NH_4^+][Mg^{++}]}$$

From this, one might infer that the addition of a large excess of ammonium ions would be desirable in order to decrease the solubility as much as possible. This, however, is not true, since the trivalent phosphate ions are hydrolyzed:

$$PO_4^{3-} + H_2O \rightleftharpoons HPO_4^{--} + OH^-$$

This hydrolysis—and therefore also the solubility and the amount of phosphate not precipitated—increases with increasing ammonium-ion concentration. Given the excess of ammonia and magnesium ions, the ammonium-ion concentration at which the solubility of the precipitate is a minimum can be calculated.

To avoid losses due to solubility, the precipitation of magnesium ammonium phosphate is made in relatively small volumes of solution, and care is taken to have the solution cold at the time of filtration. Since the precipitate tends to form supersaturated solutions, the precipitation mixture is allowed to stand for some time, preferably overnight, before filtration. For the reason already stated, the precipitate is washed with cold dilute ammonium hydroxide. A 1:20 solution of ammonium hydroxide is a suitable wash liquid.[1] Under the conditions recommended in the procedure below, less than 0.1 mg of magnesium should be lost in double precipitations as a result of solubility.[2]

Precipitation of magnesium ammonium phosphate

Possible contaminants of the precipitate. In order that the ignited magnesium ammonium phosphate precipitate may consist entirely of magnesium pyrophosphate it is necessary that the precipitate be composed entirely of $MgNH_4PO_4$. As will be shown later, a precipitate contaminated with other phosphates will not give pure magnesium pyrophosphate on ignition, and erroneous results will then necessarily be obtained. It is by no means an easy matter to obtain a pure precipitate of magnesium ammonium phosphate, as the extensive literature on the subject indicates.[3] The difficulty arises from the coprecipitation of ammonium phosphates and magnesium phosphates.

Magnesium forms the slightly soluble phosphates, $MgHPO_4 \cdot aq.$, and $Mg_3(PO_4)_2 \cdot aq.$ When an excess of ammonium hydroxide is added to an acid solution of a magnesium salt containing ammonium phosphate (the recommended method of making the precipitation), the possible precipitation of the preceding phosphates must be considered. As long as there is a sufficient concentration of ammonium salt in the solution, the precipitation of magnesium hydroxide is excluded. The substance or substances precipitated will depend upon the solubility of the various possible compounds, and the concentrations of PO_4^{3-}, HPO_4^{2-}, $H_2PO_4^-$, and NH_4^+ in a particular solution. In a solution of a phosphate the following equilibria exist:

$$H_3PO_4 \rightleftharpoons H_2PO_4^- \rightleftharpoons HPO_4^{--} \rightleftharpoons PO_4^{3-}$$
$$+ H^+ \quad + H^+ \quad + H^+$$
$$\xleftarrow{} \qquad\qquad \xrightarrow{}$$

(acid solutions) *(alkaline solutions)*

In the determination of magnesium there is no danger of the precipitation of $Mg(H_2PO_4)_2$ because the solubility of the acid phosphate is relatively great. As the addition of ammonium hydroxide to the acid magnesium solution is continued, the danger of precipitation of $MgHPO_4$ and $Mg_3(PO_4)_2$ increases, because the concentrations of HPO_4^{--} and PO_4^{3-} are increased. If in the meantime the solubility product of $MgNH_4PO_4$ has been exceeded, this phosphate will be precipitated, and there is then less danger of the precipitation of the other phosphates. It is rather easy to arrange conditions during the precipitation so that only $MgNH_4PO_4$ will be precipitated upon the addition of ammonia to the acid solution. It is difficult, however, to prevent the *coprecipitation* of other phosphates during the precipitation.

The possibility of contamination of magnesium ammonium phosphate through coprecipitation is especially great because of the numerous substances that may be carried down: H_3PO_4, $(NH_4)H_2PO_4$, $(NH_4)_2HPO_4$, $(NH_4)_3PO_4$, $Mg(H_2PO_4)_2$, MgH-PO_4, $Mg_3(PO_4)_2$, basic magnesium phosphates, $Mg(OH)_2$, $MgCl_2$, and NH_4Cl. Of

[1] According to R. F. Uncles and G. B. L. Smith, *Ind. Eng. Chem., Anal. Ed.* **18**, 699 (1946), the solubility of magnesium ammonium phosphate in 1 *N* ammonium hydroxide corresponds to 0.55 mg of Mg per liter. See this paper for other solubility values.

[2] J. I. Hoffman and G. E. F. Lundell, *Natl. Bur. Std. J. Res.* **5**, 288 (1930).

[3] Following are some references to the precipitation of magnesium ammonium phosphate W. Gibbs, *Am. J. Sci.* [3]**5**, 114 (1873). H. Neubauer, *Z. Anal. Chem.* **43**, 14 (1904); *Z. Angew. Chem.* **9**, 435 (1896). F. A. Gooch and M. Austin, *Am. J. Sci.* [4]**7**, 187 (1899); *Z. Anorg. Chem.* **20**, 121 (1899); *Chem. News* **79**, 233, 244, 255 (1899). K. K. Järvinen, *Z. Anal. Chem.* **44**, 335 (1905). B. Schmitz, *ibid.* **45**, 512 (1906); **65**, 46 (1924). K. Bube, *ibid.* **49**, 587 (1910). Z. Karaoglanow and P. Dimitrow, *ibid.* **57**, 353 (1918). D. Balarew, *Z. Anorg. Chem.* **101**, 225 (1917). G. Jörgensen, *Analyst* **51**, 61 (1926). A. Epperson, *J. Am. Chem. Soc.* **50**, 321 (1928). J. I. Hoffman and G. E. F. Lundell, *Natl. Bur. Std. J. Res.* **5**, 279 (1930).

these it is necessary to consider only the ammonium and magnesium phosphates, which have a special tendency to coprecipitate, owing to the presence of ammonium, magnesium, and phosphate ions in the lattice of the precipitate. The effect of the presence of these contaminants on the determination of magnesium by the pyrophosphate method is indicated in Table 29-1, which gives the ignition product of the various

TABLE **29-1. Ignition Products of Possible Contaminants of Magnesium Ammonium Phosphate**

Contaminant	*Ignition Product*	*Probable Direction of Error in Determination of Magnesium*
H_3PO_4	$HPO_3 + P_2O_5$[a]	Positive
$NH_4H_2PO_4$	$HPO_3 + P_2O_5$	Positive
$(NH_4)_2HPO_4$	$HPO_3 + P_2O_5$	Positive
$(NH_4)_3PO_4$	$HPO_3 + P_2O_5$	Positive
$Mg(H_2PO_4)_2$	$Mg(PO_3)_2$	Positive
$MgHPO_4$	$Mg_2P_2O_7$	None
$Mg_3(PO_4)_2$	$Mg_3(PO_4)_2$	Negative
Basic magnesium phosphates	$Mg_3(PO_4)_2 + MgO$	Negative
$Mg(OH)_2$	MgO	Negative
$MgCl_2$	Chiefly MgO	Negative
NH_4Cl	(Volatile)	None

[a] Volatilizes on long-continued ignition at elevated temperatures (1000–1100°C).

contaminants. Coprecipitation of the ammonium phosphates and $Mg(H_2PO_4)_2$ leads to high results for magnesium, and the coprecipitation of $Mg_3(PO_4)_2$, basic phosphates, and $Mg(OH)_2$ to low results; the presence of $MgHPO_4$ causes no error, for it gives the pyrophosphate on ignition.

Effect of conditions of precipitation upon nature of coprecipitated substances. It is interesting to consider the effect of the conditions of precipitation of magnesium ammonium phosphate upon the kind of coprecipitation. For the sake of illustration three cases will be considered, which have their counterparts in methods of precipitation that have been recommended in the literature:

1. The phosphate solution is added to a strongly ammoniacal solution of magnesium containing a relatively low concentration of ammonium salts. After ignition of the precipitate it is found that the residue weighs less than it should. This result is usually explained as being due to the *precipitation* of $Mg_3(PO_4)_2$. Actually there is a coprecipitation of this compound, or perhaps of a basic magnesium phosphate. It is evident that the conditions of precipitation are favorable for the adsorption and occlusion of these substances, owing to the high pH of the solution. The probability of coprecipitation of ammonium phosphate, $(NH_4)_3PO_4$, is small because it is not in excess during the precipitation.

2. The phosphate solution is added to a neutral or slightly ammoniacal solution of magnesium salt containing considerable amounts of ammonium salts. The results for magnesium are generally found to be high after the precipitate has been ignited, as a result of the coprecipitation of ammonium phosphates such as $(NH_4)_2HPO_4$ and $NH_4H_2PO_4$. Upon ignition these yield metaphosphoric acid and phosphorus pentoxide, which are only difficultly volatile at the temperature of ignition.

3. An acid solution of a magnesium salt containing excess ammonium phosphate is slowly neutralized with ammonium hydroxide. After ignition the precipitate weighs more than it should. Under the conditions given there is a possibility of the coprecipitation of $Mg(H_2PO_4)_2$, $MgHPO_4$, and ammonium phosphates. The presence of Mg-

$(H_2PO_4)_2$ is likely to lead to high values for magnesium, because it decomposes into $Mg(PO_3)_2$ on ignition and this phosphate loses P_2O_5 only slowly at high temperatures, finally, however, giving $Mg_2P_2O_7$. The presence of $MgHPO_4$ leads to no error, but again ammonium phosphates in the precipitate cause high results. The higher the concentration of ammonium salts, especially ammonium phosphate, the greater will be the tendency for ammonium phosphate to be coprecipitated, and the higher the results will be as long as the solubility of the precipitate is not unduly increased.[4]

TABLE 29-2. **Effect of Conditions on the Determination of Magnesium by the Pyrophosphate Method**[a,b]

Variation in Standard Procedure	*Number of Precipitations*	$Mg_2P_2O_7$ (*as* $MgCl_2$) *Taken*, g	$Mg_2P_2O_7$ *found*, g
Excess NH_4Cl, g			
1	1	0.1156	0.1179
5	1	0.1156	0.1180
10	1	0.1156	0.1189
16	2	0.1367	0.1367
0	2	0.1156	0.1161
Excess NH_4OH, ml			
25	1	0.1367	0.1395
25	2	0.1367	0.1367
Excess $(NH_4)_2HPO_4$, 150 ml of saturated solution	2	0.1367	0.1370

[a] From A. W. Epperson, *J. Am. Chem. Soc.* **50**, 321 (1928).
[b] The determinations were made substantially as described in the procedure on page 638, except as indicated in the first column.

Rationale of the double precipitation of magnesium ammonium phosphate. It is virtually impossible to determine magnesium accurately by a single precipitation of magnesium ammonium phosphate, unless the amount of magnesium present is known beforehand so that too large an excess of ammonium phosphate can be avoided, and large amounts of ammonium salts are not present. Usually these conditions cannot be met, especially in systematic analyses where separations, resulting in accumulation of ammonium salts in the solution, have preceded the precipitation of magnesium. However, if magnesium can be precipitated quantitatively the first time, it is relatively easy to obtain an accurate determination by dissolving the precipitate in dilute hydrochloric acid and reprecipitating with ammonium hydroxide under the proper conditions. In the reprecipitation the amount of ammonium salt can be controlled, and the excess of ammonium phosphate can be limited to the small amount required to reduce the solubility of the precipitate. The conditions are such that no appreciable amount of ammonium phosphate is coprecipitated, and the main source of error is thereby eliminated.

In the reprecipitation, the precipitate is dissolved in dilute hydrochloric acid, too large an excess being avoided, a very small amount of ammonium phosphate is

[4] Confirmed by the experimental results of A. W. Epperson, *J. Am. Chem. Soc.* **50**, 321 (1928). See Table 29-2.

added, and an excess of ammonium hydroxide is then slowly added at room temperature. Care must be taken to limit the ammonium-ion, and especially the phosphate-ion, concentration to relatively low values, or the second precipitate will also be contaminated with ammonium phosphate. The procedure for the determination of magnesium, as here briefly described and given in detail on page 638, has been shown to give accurate results.[5]

Ignition of the precipitate. Magnesium ammonium phosphate is usually ignited to magnesium pyrophosphate at a high temperature and weighed as such. It has been proposed to weigh the precipitate in the form of the hexahydrate after washing with alcohol and ether and allowing the latter to evaporate at room temperature.[6] In the latter method it is always necessary to precipitate at room temperature to ensure the presence of magnesium ammonium phosphate entirely in the form of the hexahydrate. At higher temperatures there is danger that the monohydrate will be formed. It is stated that the transition temperature for the transformation of the hexahydrate into the monohydrate lies at 40–60°C.[7] This method of weighing the precipitate has not been generally adopted for accurate work. There is always great danger in weighing a substance after drying at room temperature, owing to the possible presence of nonessential water. It is safer to ignite the precipitate to the pyrophosphate.

The ignition of magnesium ammonium phosphate to magnesium pyrophosphate is best made at 1100°C in an electric furnace; at this temperature the residue can be heated for prolonged periods without decomposition. At 1200°C the pyrophosphate slowly loses in weight, and at 1000°C the precipitate comes to constant weight only slowly.[8] The ignition may be made in porcelain or platinum crucibles.

Burning off the carbon of the filter paper requires great care. There is danger of reduction of the phosphate precipitate if the heating is strong while carbon still remains, with the possibility of injury to the platinum crucible, if such is used, by the reduced phosphorus; at high temperatures there is some loss of phosphorus when carbon is heated with pyrophosphate.[9] Moreover, if the heating is rapid at first, the residue is blackened by the carbon resulting from the decomposition of the carbonaceous products of the paper, and this carbon is so intimately associated with the particles of the residue that it remains even after prolonged heating at high temperatures; sometimes an apparently white residue will contain a core of very dark material.[10] The formation of a dark residue can be prevented by charring the paper and burning off the carbon at as low a temperature as possible. The

[5] A. W. Epperson, *J. Am. Chem. Soc.* **50**, 321 (1928); J. I. Hoffman and G. E. F. Lundell, *Natl. Bur. Std. J. Res.* **5**, 279 (1930).

[6] For an example of such a procedure see L. W. Winkler, *Z. Angew. Chem.* **31**, 211 (1918).

[7] The exact temperature depends upon the composition of the solution. See K. Bube, *Z. Anal. Chem.* **49**, 587 (1910).

[8] J. I. Hoffman and G. E. F. Lundell, *Natl. Bur. Std. J. Res.* **5**, 289 (1930).

[9] K. D. Jacob and D. S. Reynolds, *J. Assoc. Offic. Agr. Chemists* **11**, 128 (1928).

[10] S. J. Kiehl and H. B. Hardt, *J. Am. Chem. Soc.* **55**, 3555 (1933), made some observations upon the ignition of magnesium ammonium phosphate. With reference to the blackening of the precipitate, they state: "The magnesium ammonium phosphate hexahydrate during the process of losing water of hydration in the early stages of the ignition possesses the property of attaching firmly within its complex structure certain gaseous organic compounds, if present, which later break down and leave graphitic carbon." They have shown that the luminescence which magnesium ammonium phosphate shows on ignition at 550–600°C is caused by the heat liberated in the conversion of amorphous magnesium pyrophosphate into a crystalline form. This

temperature must be raised very gradually in the process. If a burner is used, the reducing gases of the flame must not be allowed to gain access to the contents of the crucible. The ignited residue is usually somewhat gray, not pure white.

A porcelain filter crucible may be used to collect the precipitate. The crucible should be heated in an electric muffle furnace at 1100°C. Gooch crucibles are less suitable because asbestos may be attacked by ammoniacal phosphate solutions. They may be employed if a blank shows that they do not undergo appreciable weight changes under the conditions of the determination.

Interfering substances. Before magnesium can be determined by precipitation as magnesium ammonium phosphate, all other metals possibly present (except the alkali metals) must be removed, because they are precipitated in ammoniacal phosphate solution. In the regular course of analysis, magnesium will remain with calcium and the alkalies after the other metals have been separated with hydrogen sulfide and ammonium hydroxide. Calcium is separated from magnesium by precipitation as calcium oxalate (Chapter 28).[11] Magnesium is thus finally left with the alkali metals, together with large amounts of ammonium salts that have accumulated in the course of the separations. The alkali metals are all coprecipitated to a greater or less extent with magnesium ammonium phosphate. Lithium forms a sparingly soluble phosphate. The amount of sodium coprecipitated is not large, and a precipitate virtually free of sodium can be obtained by a reprecipitation provided that excessive amounts are not present.[12] Potassium shows a much greater tendency to contaminate the precipitate, owing to the formation of mixed crystals of $Mg(NH_4, K)PO_4$. Triple precipitations may be required to yield a pure precipitate when large amounts of potassium are present. The unfavorable effect of large quantities of ammonium salts has already been noted; a double precipitation of magnesium ammonium phosphate under the proper conditions as already described gives a pure precipitate in their presence.

High concentrations of oxalate render the precipitation of magnesium ammonium phosphate incomplete, owing to the formation of complex magnesium-oxalate ions. Quantities of ammonium oxalate of the order of 1 g in 100–200 ml do not interfere sensibly. Sulfate in moderate concentrations and small amounts of citrate are without effect.

Determination of magnesium by the pyrophosphate method. The following directions apply to the determination of magnesium in a solution free from the interfering substances mentioned in the preceding section.

Reagent: Diammonium phosphate solution. Dissolve 25 g of $(NH_4)_2HPO_4$ in 100 ml of water.

Procedure. Prepare duplicate neutral or slightly acidic solutions of sample

---

crystallization may explain the great difficulty or impossibility of oxidizing the carbon by long heating at high temperatures. It seems highly probable that the crystallization of the pyrophosphate would result in the inclusion of the particles of carbon in the crystals, which then would be protected from the action of oxygen.

[11] A very small amount of calcium remains in solution after precipitation with oxalate and is later precipitated with magnesium. It can be recovered from the weighed magnesium pyrophosphate as described on page 628.

[12] It is best not to use sodium phosphate or sodium ammonium hydrogen phosphate as a precipitant, although if only a slight excess of either is used in the second precipitation of magnesium ammonium phosphate, correct results are usually obtained.

containing not more than 0.1 g of magnesium oxide. Proceed with each solution as follows. Add 5 ml of concentrated hydrochloric acid, and dilute to a total volume of approximately 150 ml. Add 10 ml of freshly prepared ammonium phosphate reagent and a few drops of methyl red indicator to the solution, which should be at room temperature or preferably lower (cool in ice). Now add pure concentrated ammonium hydroxide slowly with constant stirring until the indicator turns pure yellow. Avoid scratching the sides of the beaker with the stirring rod. Continue to stir the solution for a few minutes, adding ammonia dropwise to keep the solution yellow, then add 5 ml of concentrated ammonium hydroxide in excess, and stir again for a minute or two. Set the solution aside in a cool place for at least 4 hr, preferably overnight.

Filter through a retentive paper, and wash the precipitate moderately with cold 1:20 ammonium hydroxide solution. Dissolve the precipitate in approximately 50 ml of warm 1:10 hydrochloric acid and wash the paper well with hot 1:100 hydrochloric acid solution. Dilute the solution to 150 ml, add 0.5 ml of ammonium phosphate reagent, and cool to room temperature or, better, in ice. Then again precipitate magnesium by the addition of ammonium hydroxide as described above, and allow the solution to stand for at least 4 hr or, better, overnight. Filter through paper or a porcelain filter crucible, and wash the precipitate with cold 1:20 ammonium hydroxide until the washings, acidified with nitric acid, give no test for chloride ion with silver nitrate.

If a filter paper was used to collect the precipitate, fold it up, and place it in a weighed porcelain or platinum crucible. Char the paper, burn off the carbon at as low a temperature as possible with free access of air, and then ignite to constant weight at 1100°C in an electric muffle furnace or, less preferably, over a Tirrill burner if a platinum crucible is used or over a Meker burner if a porcelain crucible is employed.[13] If a porcelain filter crucible is used, dry it first in an oven or over a low flame, and then heat gradually to 1100°C in an electric muffle; if a furnace is not available, place the filter crucible inside an ordinary porcelain crucible, and heat gradually to the full heat of a Meker burner. Heat for $\frac{1}{2}$-hr periods until constant weight is obtained. Calculate the percentage of magnesium or magnesium oxide as required.

Other applications. Manganese, cadmium, zinc, cobalt, and other divalent metals as well, form double ammonium phosphates analogous to magnesium ammonium phosphate and can be determined by precipitation in this form under the proper conditions. Manganese ammonium phosphate is insoluble in excess ammonia, but the double phosphates of cadmium, zinc, and cobalt are soluble in ammoniacal solutions as consequence of the formation of metal ammine complexes. Therefore the latter metal ammonium phosphates must be precipitated in approximately neutral solutions and the precipitates washed with water. Fortunately, the double phosphates of the metals named are less soluble than magnesium ammonium phosphate, so washing with water does not lead to any appreciable solubility loss. For zinc ammonium phosphate the optimum pH of precipitation is 6.6[14] (compare Fig. 29-1).

[13] If a black residue should be obtained, the oxidation of carbon will be aided if the crucible is cooled at intervals during the ignition to permit the entrance of air.

[14] T. R. Ball and M. S. Agruss, *J. Am. Chem. Soc.* **52**, 120 (1930); J. E. Vance and R. E. Borup, *Anal. Chem.* **25**, 610 (1953).

Phosphorus in the form of orthophosphate can be precipitated as magnesium ammonium phosphate with an excess of magnesium chloride in solutions containing ammonia and ammonium salts, and weighed as magnesium pyrophosphate (Chapter 30).

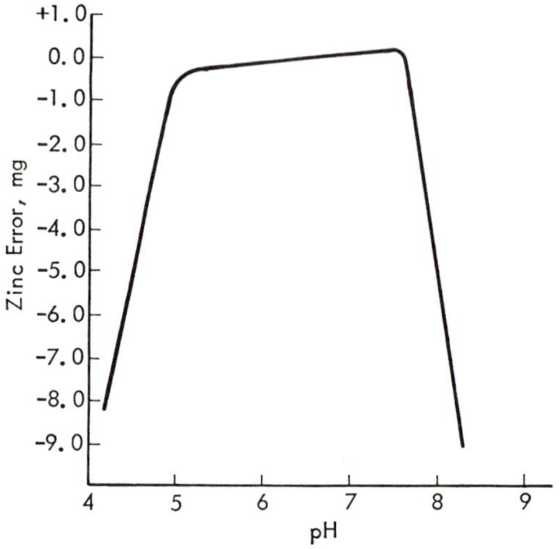

Fig. 29-1. Effect of pH on solubility loss in precipitation of $ZnNH_4PO_4$. From Ball and Agruss, Reference 14.

Determination of magnesium by precipitation as magnesium hydroxyquinolate

Magnesium can be precipitated quantitatively with 8-hydroxyquinoline (oxine)[15] in the pH range 9.5 to about 13.[16] The precipitation is best made by adding the oxine solution to the ammoniacal magnesium solution containing ammonium salt.[17] There is a tendency for oxine to be carried down with the precipitate, and any large excess of reagent must be avoided. The precipitate has the composition $Mg(C_9H_6ON)_2 \cdot 2\,H_2O$ after drying at 105°C. It rapidly becomes anhydrous when heated at 150–160°C, but a slight decomposition takes place at this temperature (Ref. 16). The precipitate is preferably dried at 105°C, although if it is contaminated with oxine better results are obtained by drying at 150°C because most of the oxine is volatilized at this temperature.

The precipitation of magnesium as the quinolate requires the absence of all other metals save the alkalies, for nearly all these yield slightly soluble hydroxyquinolates in ammoniacal solution. Small amounts of calcium are permissible, but usually when it is present double precipitations must be made, or, better, calcium should first be precipitated as the oxalate. A large excess of oxalate should be avoided in the calcium separation, for otherwise the precipitation of magnesium as quinolate may not be quantitative. In the presence of large amounts of the alkali metals, a double precipitation should be made.

The hydroxyquinolate method has a number of advantages over the classical phosphate method for magnesium. The advantage in drying the precipitate to

[15] R. Berg, *Z. Anal. Chem.* **71**, 23 (1927); F. L. Hahn and K. Vieweg, *ibid.* **71**, 122 (1927); J. C. Redmond and H. A. Bright, *J. Res. Natl. Bur Std.* **6**, 113 (1931).

[16] H. R. Fleck and A. M. Ward, *Analyst* **58**, 388 (1933).

[17] C. C. Miller and I. C. McLennan, *J. Chem. Soc.* **1940**, 656. See also R. C. Chirnside, C. F. Pritchard, and H. P. Rooksby, *Analyst* **66**, 399 (1941).

constant weight at a low temperature, instead of igniting at a high temperature, and the rapidity of the new method are especially noteworthy. By the use of 8-hydroxyquinoline magnesium can be separated from the alkali metals without the introduction of a nonvolatile reagent, which is a great advantage when the alkalies are to be determined.

The following directions are essentially those of Miller and McLennan.

Procedure. To 100 ml of solution containing the equivalent of 0.01–0.05 g of magnesium oxide, add 2 g of ammonium chloride, 0.5 ml of 0.02 per cent *o*-cresolphthalein in alcohol, and 6 *N* ammonium hydroxide in 2–3 ml excess of that required to give a violet color (pH 9.5). Heat the solution to 70 or 80°C, and add dropwise with stirring a 5 per cent solution of 8-hydroxyquinoline in 2 *N* acetic acid until a small excess is present as shown by a yellow color in the supernatant liquid. Digest for about $\frac{1}{2}$ hr on the steam bath, and filter the hot solution through a weighed filter crucible. A trace of sodium tauroglycocholate in the solution reduces the tendency of the precipitate to stick to the beaker. Wash with about 50 ml of warm water, and dry the precipitate at 105–110°C, first for 1 hr and then for $\frac{1}{2}$-hr periods until constant weight is attained. The dried precipitate contains two molecules of water of hydration.

If too large an excess of oxine (100 per cent) was added, dry to constant weight at 150°C and weigh as the anhydrous hydroxyquinolate.

PROBLEMS

1. What weight of sample must be taken in order that each milligram of $Mg_2P_2O_7$ residue shall represent 0.05 per cent MgO? *Ans.* 0.724 g.
2. If the solubility of $MgNH_4PO_4 \cdot 6\,H_2O$ in water is 0.11 g/liter at 23°C, how many milligrams of Mg and P_2O_5 does the saturated solution contain per liter? *Ans.* 10.9 mg of Mg; 31.8 mg of P_2O_5.
3. What volume of oxine solution containing 5 g of oxine per 100 ml is required to precipitate 0.05 g of MgO?
4. A precipitate of magnesium ammonium phosphate has 1 per cent of the ammonium replaced by the equivalent amount of potassium. What will be the error when it is assumed that the precipitate is pure magnesium ammonium phosphate hexahydrate and (a) it is weighed as the hexahydrate after drying at room temperature; (b) it is ignited to 1100°C and the residue weighed? *Ans.* (a) 0.09 per cent; (b) 0.42 per cent.
5. 25.00 ml of a magnesium chloride solution gave 0.5628 g of $Mg_2P_2O_7$. What is the molarity of the solution?
6. A residue of magnesium pyrophosphate weighing 0.2048 g was examined for calcium by the sulfate–alcohol method described on page 628 and the equivalent of 0.0005 g of CaO was found. Obtain the true weight of magnesium pyrophosphate.
7. Find the hardness of a water of which 250 ml yields 0.0565 g of $CaCO_3$ and 0.0386 g of $Mg_2P_2O_7$. (Hardness is expressed in ppm, i.e., milligrams per liter, of $CaCO_3$, magnesium being counted as the equivalent quantity of calcium carbonate.)
8. Assuming that the solubility product of magnesium ammonium phosphate is 2×10^{-13}, what will be the concentration of magnesium ion in milligrams per liter in a solution that is 0.5 *M* in ammonium ion, 1 *M* in ammonia, and 0.1 *M* in ammonium phosphate?
9. Find the solubility of magnesium ammonium phosphate in 0.1 *M* ammonium hydroxide.

CHAPTER 30 *Determination of Phosphorus*

Method

In the absence of interfering elements, phosphorus (in the form of orthophosphate) is precipitated as magnesium ammonium phosphate hexahydrate by magnesium chloride and ammonium chloride in ammoniacal solution:

$$PO_4^{3-} + NH_4^+ + Mg^{++} + 6\ H_2O \rightarrow MgNH_4PO_4 \cdot 6\ H_2O$$

The precipitate is ignited to magnesium pyrophosphate and weighed as such:

$$2\ (MgNH_4PO_4 \cdot 6\ H_2O) \rightarrow Mg_2P_2O_7 + 2\ NH_3 + 13\ H_2O$$

When interfering substances such as calcium and iron, which would otherwise be precipitated, are present, two courses are open:

1. The precipitation of magnesium ammonium phosphate may be made in the presence of much ammonium citrate, which prevents the precipitation of calcium and other phosphates. This procedure may be applied in the analysis of samples high in phosphorus (phosphate rock).
2. Phosphorus may first be separated from the interfering elements by precipitation as ammonium phosphomolybdate in acid medium. This precipitate may then be dissolved in ammonia, and magnesium ammonium phosphate precipitated. This procedure is used for samples that are low in phosphorus.

Solution of sample

Material attacked by mineral acids is decomposed with nitric acid if separation of phosphorus as ammonium phosphomolybdate is to follow. Other mineral acids are avoided, because they will interfere in the precipitation of ammonium phosphomolybdate; if their use is required in the decomposition of the sample, as small an amount as possible is employed, and the excess is removed by evaporation.[1]

Acid-insoluble material is treated with hydrofluoric acid or fused with sodium carbonate (leaching of the melt with water extracts the phosphate). Silicic acid is removed by dehydration in the usual manner (p. 653) before proceeding with the analysis. If titanium, tin, or other metals forming phosphates slightly soluble in acids are present, the corresponding phosphates will be left with the silica if the sample is decomposed with acids and steps must be taken to recover such phosphorus.

[1] Sulfuric acid cannot be removed by fuming, because phosphorus is lost.

Metallurgical products contain phosphorus mostly in the form of phosphide, and an oxidizing attack is required to convert the latter into orthophosphate and prevent loss of phosphorus as phosphine.

Precipitation of ammonium phosphomolybdate

When a nitric acid solution of ammonium molybdate is mixed with a solution of orthophosphate, a yellow precipitate of ammonium phosphomolybdate forms, having a composition approaching

$$(NH_4)_3PO_4 \cdot 12\ MoO_3 \cdot 12\ H_2O \qquad \text{or} \qquad (NH_4)_3(PO_4 \cdot 12\ MoO_3) \cdot 12\ H_2O$$

This compound is more correctly called ammonium molybdophosphate and formulated

$$(NH_4)_3\{P(Mo_{12}O_{40})\} \cdot 12\ H_2O$$

Recent work[2] indicates that the composition of the precipitated substance corresponds to the diammonium rather than the triammonium salt:

$$(NH_4)_2HPO_4 \cdot 12\ MoO_3 \cdot aq$$

When washed with ammonium nitrate solution, the precipitate is converted into the triammonium salt.

The exact composition of the yellow precipitate depends on such factors as the composition of the solution (concentration of molybdate, nitric acid, ammonium nitrate, etc.) and the temperature of precipitation.[3] When the precipitation is made according to the directions on page 1112, the weight of the precipitate after drying at $\sim 110°C$ corresponds quite closely to the composition $(NH_4)_3PO_4 \cdot 12\ MoO_3$. When phosphorus is precipitated as ammonium phosphomolybdate as a separation from other elements before final precipitation as magnesium ammonium phosphate, the composition of the yellow precipitate is of no importance as long as the precipitation of phosphorus is quantitative. When the yellow precipitate is weighed after drying or after conversion to $P_2O_5 \cdot 24\ MoO_3$, or titrated (Chap. 65), it is important that the composition of the precipitate correspond as closely as possible to the formula, if a theoretical factor is to be used. It is especially important that the ratio $MoO_3 : PO_4$ be 12. Methods based on weighing or titrating the yellow precipitate can hardly be relied on closer than 1–2 per cent, so such methods are suitable only for small amounts of phosphorus.

The precipitation of ammonium phosphomolybdate is carried out in nitric acid (1–1.5 M) solution, hydrochloric and sulfuric acids being avoided as far as possible, owing to the greater solubility of the precipitate in their presence. It is important to have present a rather large amount (5–15 per cent) of ammonium nitrate. This salt hastens the separation of the precipitate and decreases its solubility as a result of the common-ion effect. At concentrations above approximately 15 per cent of ammonium nitrate, the solubility of the yellow precipitate is increased. A large excess of ammonium molybdate reagent is required for successful precipitation. Precipitation is much more rapid in warm than in cold solutions, and precipitates formed at higher temperatures are easier to filter. However, precipitates formed above 50°C contain more molybdic oxide than corresponds to the ideal formula given above. Molybdic oxide may itself separate and contaminate the precipitate if the temperature is too high and the time of standing is too long. In warm solutions all the phosphorus is probably precipitated within $\frac{1}{2}$ hr if the solution is stirred and substances retarding precipitation are absent, but it is best to allow the precipitation mixture to stand overnight (at room temperature). Precipitation at room temperature is complete within 10 hr. Unless substances are present which contaminate the precipitate when formed in warm

[2] D. Stockdale, *Analyst* **83**, 24 (1958).

[3] Recent studies have been made by P. Cannon, *Talanta* **3**, 219 (1960), as well as Stockdale, Ref. 2.

solutions, the precipitation is always made warm. The precipitate must be washed with a solution of an electrolyte; pure water peptizes the precipitate and causes it to run through the filter. A dilute solution of ammonium nitrate is a suitable wash liquid if phosphorus is to be precipitated later as magnesium ammonium phosphate. If the precipitate is to be titrated (Chap. 65), a solution of potassium nitrate is used for the washing.

Common substances interfering in the precipitation of ammonium phosphomolybdate and methods of eliminating or minimizing the interferences are given in Table 30-1. Molybdate reagent may take up silica from a glass container. Such a reagent gives an ammonium phosphomolybdate precipitate contaminated with ammonium silicomolybdate. The impure precipitate has a Mo/P ratio greater than 12 and will therefore yield high results in the volumetric method.[4]

TABLE **30-1.** **Substances Interfering in the Precipitation of Phosphorus as Ammonium Phosphomolybdate**[a]

Substance	*Effect*	*Prevention of Interference*
Sulfuric and hydrochloric acids, and salts of these acids	Retard precipitation and increase solubility of precipitate when present in large amounts. Too much nitric acid also increases solubility of precipitate and favors precipitation of MoO_3	Avoid introduction in preparation of sample. Interference not very serious with moderate amounts and can be overcome by adding a larger excess of ammonium molybdate and increasing period of standing. As much as 10% by volume of hydrochloric acid and 5% sulfuric acid permissible in routine determinations
Hydrofluoric acid and fluorides	Retard precipitation and form a more soluble precipitate	If required in decomposition of sample, volatilize later by evaporation with nitric acid. For small amounts add boric acid to form harmless fluoboric acid
Arsenic	Forms insoluble arsenomolybdate and retards precipitation of phosphomolybdate	May be precipitated with hydrogen sulfide in strong hydrochloric acid solution or volatilized as arsenic trichloride by boiling strong hydrochloric acid solution containing bromide. Small amounts not precipitated if phosphomolybdate is thrown down in the cold
Silicon	Forms insoluble silicomolybdate; may retard precipitation	Remove by dehydrating and filtering off silica at the start
Tungsten	Precipitated	Remove at beginning by digesting with hydrochloric and nitric acids and filtering off tungstic acid; evaporate filtrate with nitric acid to remove chlorides

[4] N. Birnbaum and G. H. Walden, Jr., *J. Am. Chem. Soc.* **60,** 66 (1938).

TABLE 30-1. Substances Interfering in the Precipitation of Phosphorus as Ammonium Phosphomolybdate[a]

Substance	*Effect*	*Prevention of Interference*
Titanium, zirconium, and tin	Retard precipitation and may lead to loss of phosphorus through formation of insoluble phosphate	Remove by appropriate means before precipitation of phosphomolybdate. If only small amounts are present, interference can be prevented by addition of small amounts of citrate to ammonium hydroxide used to dissolve yellow precipitate
Vanadium	V(V) precipitated as vanadomolybdate. V(IV) precipitated only in warm solution. Both V(V) and V(IV) retard precipitation of phosphomolybdate	Prevent carrying down of vanadium by reducing to quadrivalent state and making precipitation at room temperature. Add larger excess of ammonium molybdate and allow to stand longer to overcome retarding of precipitation of phosphomolybdate
Selenium and tellurium (IV, VI)	Retard precipitation and contaminate precipitate	Allow at least 4 hr for precipitation and finish determination by precipitating $MgNH_4PO \cdot 6\ H_2O$ in the presence of citrate
Organic substances	May interfere with formation of phosphomolybdate precipitate by forming fairly stable complexes with molybdate (oxalates, tartrates, etc.), thus reducing concentration of active reagent	Destroy by appropriate means, or in some cases add larger excess of molybdate

[a] Based mostly upon statements of G. E. F. Lundell and J. I. Hoffman, *Ind. Eng. Chem.* **15**, 44 (1923). D. Stockdale (Ref. 2) found that precipitates formed in the presence of substantial concentrations of sulfate and Fe(III) contained extra MoO_3.

Precipitation of magnesium ammonium phosphate in the phosphorus determination

The errors associated with the precipitation of magnesium as magnesium ammonium phosphate have already been discussed (p. 634). Similar errors are encountered in the precipitation of phosphorus, but it must be remembered that here the conditions are different, the precipitation being made in the reverse way. If the magnesia reagent is added very slowly to the ammoniacal phosphate solution, the results for phosphorus are low, apparently owing to the formation of a precipitate contaminated with coprecipitated ammonium phosphate. If the reagent is added rapidly, $Mg_3(PO_4)_2$ is coprecipitated, and the results may be high or nearly correct as the result of compensation of errors. If an acid solution of phosphate containing an unregulated excess of magnesia reagent is treated with an excess of ammonium hydroxide, erroneous results are likewise obtained, owing to the presence of too much magnesium at the moment of precipitation. Since the amount of phosphorus present is not generally known, it is usually impossible to limit the excess of magnesia mixture to the smallest possible amount. The difficulty can be overcome by dissolving the precipitate in acid and reprecipitating by adding ammonium hydroxide to the acid solution containing very little magnesia reagent to reduce the solubility of the magnesium ammonium phosphate. Under these conditions the excess of magnesium ion present at the moment of precipitation is so small that a precipitate having the ideal composition is formed. This procedure, as that for magnesium, furnishes a striking example of the utility of double precipitations in the formation of pure precipitates. It seems vain to attempt to obtain a pure precipitate of magnesium ammonium phosphate by a single precipitation, and the methods of determination of phosphate calling for one precipitation with magnesia reagent are not applicable in accurate work.

The precipitations of magnesium ammonium phosphate are best made in the cold by neutralizing the hydrochloric acid solution slowly with ammonium hydroxide and finally adding a large excess. The loss of phosphorus on reprecipitation, from solubility of magnesium ammonium phosphate, is negligibly small when the precipitations are made under the proper conditions.

If phosphorus has first been separated as ammonium phosphomolybdate, account must be taken of the possible presence of such elements as titanium and iron in the yellow precipitate. If proper steps to eliminate their interference are not taken, these contaminants will lead to loss of phosphorus when the yellow precipitate is dissolved in ammonium hydroxide, because the phosphates of titanium, iron, and similar elements are insoluble in ammonia and will be left on the filter paper (some is likely to go through the paper and produce a turbid filtrate). By using an ammonium hydroxide solution containing a low concentration of ammonium citrate to dissolve the ammonium phosphomolybdate, this error is avoided. Owing to complex formation between citrate and iron and titanium, the phosphates of these metals dissolve in this solution. Washing the filter paper with dilute hydrochloric acid after passing ammonium hydroxide–ammonium citrate through it assures the complete solution of the phosphorus. The presence of citrate in the solution from which magnesium ammonium phosphate is precipitated prevents the carrying down of iron, titanium, etc.

Determination of phosphoric anhydride in phosphate rock

Phosphate rock is an important commercial material used chiefly for the production of phosphatic fertilizers. Its value depends upon its phosphorus content, and the determination of this element is therefore important. Phosphate rock is composed primarily of fluoriferous calcium phosphate; the minor constituents include carbonate, chloride, silicates, iron, aluminum, and magnesium. The sample is decomposed by a mixture of nitric and hydrochloric acids, or nitric acid alone.

The procedure described below for the determination of phosphorus in phosphate rock involves precipitation of magnesium ammonium phosphate in the presence of much citrate to keep calcium and other metals in solution.[5] This procedure is

[5] J. I. Hoffman and G. E. F. Lundell, *J. Res. Natl. Bur. Std.* **19**, 59 (1937).

more rapid than the one in which phosphorus is first precipitated as ammonium phosphomolybdate,[6] and gives equally accurate results.[7]

Procedure. Weigh duplicate 0.3–0.35-g samples of ground material dried at 105°C into 250-ml. Erlenmeyer flasks, and treat each as follows. Add 10 ml of concentrated hydrochloric acid and 2 ml of concentrated nitric acid, place a small watch glass over the mouth of the flask, and boil gently for ½ hr. Rinse off the cover glass and add 20 g of ammonium citrate, 5 ml of hydrochloric acid, and 60 ml of magnesia reagent.[8] Neutralize with concentrated ammonium hydroxide, using litmus paper as indicator, and add an excess of 2 ml. Dilute the solution to 150 ml with water, add a few glass beads, stopper with a snugly fitting rubber stopper, and shake at intervals for about ½ hr. Then allow to stand overnight.

Filter through a close-textured paper, and wash the flask and paper once with a small volume of 1:20 ammonium hydroxide. Discard the filtrate. Dissolve any precipitate in the flask in approximately 40 ml of 1:4 hydrochloric acid, and pour the solution through the paper. Wash the flask and paper with more of the same acid. Add 0.3 g of citric acid and 1 ml of magnesia mixture to the solution, which should have a volume of 75–100 ml. While stirring, add ammonium hydroxide until the solution is alkaline to litmus and then an excess of 5 ml. Stir occasonally during ½ hr, and allow to stand for 4 hr, or overnight.

Collect the precipitate on a 9-cm close-textured filter paper,[9] and wash with cold 1:20 ammonium hydroxide. Transfer the paper to a weighed porcelain or platinum crucible. Char the paper at a low temperature, burn off the carbon below 900°C, and ignite to constant weight at 1050–1100°C. Calculate the percentage of P_2O_5.

Note: The time of standing in the first precipitation can be reduced to 3 hr. and in the second to 2 hr if the solution is chilled before shaking or stirring and then allowed to stand in ice water or in a refrigerator at 5 or 10°C.

Other weighing forms of phosphorus

It was pointed out earlier in this chapter that the composition of ammonium phosphomolybdate is slightly variable. However, for small amounts of phosphorus the error introduced by this variability is not serious and is counterbalanced by the high molecular weights and resultant small chemical factors of weighing forms based on the yellow precipitate. The best of such weighing forms is phosphomolybdic anhydride, which is obtained by heating ammonium phosphomolybdate at 400–500°C:

$$2\,(NH_4)_3PO_4 \cdot 12\,MoO_3 \rightarrow P_2O_5 \cdot 24\,MoO_3 + 6\,NH_3 + 3\,H_2O$$

[6] G. E. F. Lundell and J. I. Hoffman, *J. Assoc. Offic. Agr. Chemists* **8**, 184 (1924).

[7] A titrimetric method is also available [A. J. Gondie and W. Rieman, *Anal. Chem.* **24**, 1067 (1952)], in which the sample solution is run through a strong-acid cation-exchange resin to remove cations, and the effluent containing HCl and H_3PO_4 is titrated with standard NaOH; the volume difference corresponding to end points at pH 4.6 ($H_2PO_4^-$) and 9.2 (HPO_4^{--}) represents H_3PO_4. This method is not generally as accurate as the gravimetric method but may be used when errors of ±10 ppt can be tolerated. A. Gee and V. R. Deitz, *Anal. Chem.* **25**, 1320 (1953), have reported that phosphorus in phosphate rock can be determined by differential spectrophotometry (as molybdiphosphovanadic acid) with an accuracy comparable to the gravimetric method (~ ±1 ppt error).

[8] Prepare as follows: Dissolve 400 g of magnesium chloride hexahydrate and 300 g of ammonium chloride in 1500 ml of warm water. Add ammonium hydroxide until the solution is alkaline to litmus. After 1 hr filter the solution and add hydrochloric acid to the filtrate until acid to litmus.

[9] A fine-porosity porcelain filter crucible may also be used (less likelihood of dark $Mg_2P_2O_7$). The ignition should then be carried out in a muffle furnace.

The effect of variable ammonium and water content is eliminated by the use of this weighing form. It allows the use of ammonium nitrate as a wash solution, in which the yellow precipitate is less soluble than in dilute nitric acid. The precipitate is collected in a porcelain filter crucible (or a Gooch crucible) and converted to $P_2O_5 \cdot 24\,MoO_3$ in a muffle furnace at 450°C, less preferably by heating at dull redness over a flame. Phosphomolybdic anhydride is bluish black; if heated at too high a temperature it becomes whitish and the weight is too low because of partial volatilization. See Chapter 65 for an application of this weighing form for the determination of phosphorus in steel.

When the yellow precipitate is dried at 110–140°C, its composition corresponds quite closely to $(NH_4)_3PO_4 \cdot 12\,MoO_3$. It may be noted that the molecular weight of this compound is virtually the same as that of $(NH_4)_2HPO_4 \cdot 12\,MoO_3 \cdot H_2O$, the composition reported by Stockdale (Ref. 2) for the yellow precipitate dried at 140°C. The value of the factor may vary by a few parts in a hundred, depending on the exact conditions of precipitating and drying the yellow precipitate. It appears that the empirical factor 1.63 per cent P is better than the theoretical factor 1.65 per cent P when the precipitate is dried at 110°C. The residue is hygroscopic.

Whether one or the other of these weighing forms is used, care must be taken in the precipitation of ammonium phosphomolybdate to prevent the precipitation of molybdic acid by keeping the temperature of the solution below 50°C.

Slightly soluble phosphomolybdates are formed by large cations other than ammonium. Gravimetric and titrimetric methods based on the formation of quinolinium phosphomolybdate, $(QH)_3PO_4 \cdot 12\,MoO_3$, and 8-hydroxyquinolinium phosphomolybdate, $(H_2Ox)_3PO_4 \cdot 12\,MoO_3$, are available.

Nongravimetric methods for phosphorus

This chapter has dealt almost wholly with the gravimetric determination of orthophosphate. In addition to gravimetric and titrimetric methods (Chap. 65), spectrophotometric methods are of importance in the determination of orthophosphate.[10] The complex acids of phosphorus with molybdic acid and vanadomolybdic acid have strong yellow colors and serve very well for the photometric determination of the element. In some procedures the absorbance of the aqueous solution of the heteropoly acid is measured; in others the heteropoly acid is extracted into an immiscible oxonium solvent or a mixture of an oxonium solvent with a non-oxonium solvent (butanol + chloroform, for example). In a second type of method, the heteropoly phosphomolybdic acid is reduced to heteropoly molybdenum blue, a strongly colored product of ill-defined composition in which the average oxidation state of molybdenum lies between 6 and 5 (see Chap. 65 for a procedure). With a suitable reducing agent, under proper conditions, only the molybdenum combined with phosphorus is reduced; the excess, free molybdic acid is unaffected. The P–Mo blue can be extracted (into butanol, for example) when high sensitivity is required. Si, Ge, and As(V) form colored heteropoly complexes similar to the P complexes, but by regulation of conditions the interference can often be reduced to a large extent. When necessary, Ge and As (which are more likely than Si to give trouble) can be separated by volatilization of the chlorides or bromides.

Another aspect of phosphorus analysis involves separation and determination of ortho- pyro-, meta-, and polyphosphates in mixtures. Besides precipitation and titration methods, paper chromatography, ion-exchange chromatography, X-ray

[10] For a survey of many methods see K. Kodama, *Methods of Quantitative Inorganic Analysis*, Wiley-Interscience, New York, 1963. A number of procedures are given in O. G. Koch and G. A. Koch-Dedich, *Handbuch der Spurenanalyse*, Springer, Berlin, 1945, and D. F. Boltz (ed.), *Colorimetric Determination of Nonmetals*, Wiley-Interscience, New York, 1958.

diffraction, infrared spectrophotometry, and nuclear magnetic resonance find valuable application in the resolution of such mixtures. Phosphorus in lower oxidation states (e.g., hypophosphorous, phosphorous, and hypophosphoric acids) is usually determined by oxidative titration.

Methods for such analyses may be found in standard analytical reference works.[11]

PROBLEMS

1. A 0.5204-g sample of phosphate rock containing 0.71 per cent of moisture yielded an ignition residue of $Mg_2P_2O_7$ weighing 0.3188 g. Calculate the percentage of P_2O_5 in the sample on the wet and the dry basis.
Ans. 39.12 per cent, wet basis; 39.40 per cent, dry basis.
2. A sample of granite weighing 1.15 g gave 0.0203 g of $P_2O_5 \cdot 24\ MoO_3$. Calculate the percentage of P_2O_5 and P in the rock. *Ans.* 0.068 per cent P_2O_5; 0.030 per cent P.
3. In the determination of P_2O_5 in 85 per cent phosphoric acid (density 1.69), what is the maximum weight of sample that may be taken without having the weight of $Mg_2P_2O_7$ exceed 0.50 g?
4. A sample containing phosphorus yielded 0.0769 g of $PbMoO_4$ after the following transformations: $P_2O_5 \rightarrow (NH_4)_3PO_4 \cdot 12\ MoO_3 \rightarrow PbMoO_4$. What weight of P_2O_5 was present in the sample?
5. Would an investigation of the determination of phosphorus as cesium phosphomolybdate (very slightly soluble) seem to be worthwhile?

[11] See chapters on phosphorus by W. Rieman and J. Beukenkamp in *Treatise on Analytical Chemistry* (I. M. Kolthoff and P. J. Elving with E. B. Sandell, eds), part II, vol. 5, Wiley-Interscience, New York, 1961, and S. Greenfield in *Comprehensive Analytical Chemistry* (C. L. Wilson and D. W. Wilson, eds.), vol. IC, Elsevier, Amsterdam (and Van Nostrand, Princeton, N.J.).

CHAPTER 31 *Determination of Silica*

Historical note. During the 18th century the composition of minerals was a leading subject of scientific study, which provided a strong stimulus to the development of chemical analysis. The groundwork for the analysis of silicates was laid at that time. In the middle 1700s, von Cronstedt used fluxes such as sodium carbonate and borax to decompose refractory silicates. Bergman first described[1] the determination of silica by fusion of the sample with sodium carbonate, and evaporation to dryness of the solution of the melt in hydrochloric acid to separate silica. Berzelius knew that the recovery of silica in this method—the standard method to this day—was not complete and he showed that the dissolved silica could be coprecipitated with the hydrous metal oxides precipitated with ammonia. The investigation of the gravimetric silica determination has continued down to the present time and we now know how an accurate determination can be made when accurate results are required, as in geochemical applications. Jolles and Neurath first proposed, in 1898, a colorimetric method for silica based on molybdate as reagent, and these methods have been developed to the point where they provide results almost as accurate as those of the gravimetric method. The newer instrumental methods, such as X-ray fluorescence, are not able to match the accuracy of the gravimetric method. Moreover, the latter is an absolute method, for which there is always likely to be a use no matter what new instrumental methods are devised or what improvements are made in the old ones.

Gravimetric method

If completely decomposable by acids, the sample is treated with hydrochloric acid; otherwise (and this is the usual case) it must be fused with sodium carbonate, and the melt, which contains the silicate in acid-decomposable form, is then treated with hydrochloric acid (occasionally other acids). The acid solution from either treatment is evaporated to dryness to separate the silica as insoluble $SiO_2 \cdot xH_2O$. The residue is extracted with hot dilute hydrochloric acid to dissolve salts of iron, aluminum, the alkali metals, etc., which may be present. The greater portion of the silica remains undissolved and is filtered off. The filtrate is evaporated to dryness, the residue is treated with dilute hydrochloric acid as before, and the second portion of silica is filtered off on a fresh filter. The two washed precipitates, containing practically all the silica in the sample, are ignited in a platinum crucible to silicon dioxide, and weighed. The ignited residue is usually not entirely pure but will contain small amounts of the oxides of titanium, aluminum, iron, etc. To obtain the weight of silicon dioxide present, the impure silica is treated with hydrofluoric and sulfuric acids to volatilize silica as silicon tetrafluoride. The residue is strongly heated to convert the sulfates of aluminum, iron, etc., into the

[1] T. Bergman, *Kgl. Vetenskapsakad. Handl.* **40,** 224 (1779).

corresponding oxides, in which form these metals were originally present, and the crucible is weighed. The loss in weight represents the amount of silica.

Decomposition of silicates

The solubility of a silicate in strong acids depends largely upon the proportion of silica in the silicate, and upon the solubility of the metal oxide or oxides present, which in turn is a function of the basicity of the metal. Other things being equal, the greater the ratio of silica to bases in a silicate, the less easily will the silicate be attacked by acids. The more basic the metals present, the more easily will the silicate be decomposed by acids. Thus the sodium silicates are soluble in water, calcium silicate ($CaSiO_3$) is insoluble in water but readily decomposed by dilute acids, whereas a silicate such as Al_2SiO_5 is hardly acted upon by acids. A silicate containing much essential water is likely to be acid-decomposable. Portland cement, blast-furnace slag, and the zeolites can be decomposed by heating with mineral acids.

The great majority of silicates found in nature, as well as many artificial ones (glass, porcelain), cannot be completely decomposed by acid treatment. However, all can be brought into a form completely decomposable with acids by fusion, or strong heating, with an alkali. This treatment results in an increase in the proportion of bases in the silicate, thereby rendering it acid-decomposable. The alkali commonly used is sodium carbonate.[2] A part of the sodium combines with the silica, and an equivalent amount of carbon dioxide is liberated. The fusion is always made with a large excess of sodium carbonate to ensure complete acid decomposability of the product. The silicate must be ground exceedingly fine, and a high temperature must be employed in the fusion, or else decomposition is likely to be incomplete. The fusion is made in platinum which is not attacked to any significant extent by molten sodium carbonate.

Although sodium carbonate is commonly the preferred reagent for the decomposition of refractory silicates, other alkaline substances find considerable application. One of these is sodium hydroxide, which acts much more vigorously upon silicates than does sodium carbonate. It can be obtained in a rather high degree of purity, but less easily so than sodium carbonate, and moreover is less easily kept in a pure state. Molten sodium hydroxide attacks platinum, and such crucibles cannot be used for the fusion. Nickel crucibles are suitable, although the melt is contaminated by nickel to a greater or less extent, depending upon the fusion temperature; if the temperature is not allowed to rise above 500°C (and this is sufficiently high for rapid decomposition), the amount of nickel dissolved is very small, provided bright crucibles are used (p. 459). Sodium hydroxide must first be fused alone in the crucible to expel the water it contains, which would otherwise cause spattering with losses; the sample is then added. Decomposition of silicates, and other materials as well, with sodium hydroxide finds valuable applications, especially in the analysis of technical materials in which only one or a few

[2] Sodium carbonate is used in preference to potassium carbonate, especially because the latter is deliquescent. Formerly a mixture of the two carbonates was employed, because it has a lower melting point than either component. In these days when it is easy to melt sodium carbonate with the burners found in every laboratory, it alone should be used because the decomposition is aided by the high temperature of the melt.

constituents are to be determined.[3] This flux is less often used when a complete analysis of a silicate is to be made (e.g., rock analysis).

Sodium peroxide is a powerful flux, sometimes employed in the analysis of siliceous materials when an oxidizing flux is required.[4] It is used in iron or nickel crucibles.

For the determination of the alkalies in silicates by the J. Lawrence Smith method, the flux employed is a mixture of calcium carbonate and ammonium chloride, which on heating yields calcium oxide and calcium chloride. It is the basic mixture of the latter two substances which is the active reagent (see p. 672).

When silica is not to be determined, silicates may be decomposed by heating with hydrofluoric and sulfuric acids. Silicon is volatilized as silicon tetrafluoride, and the metals are left as sulfates. A few silicates are not easily decomposed by hydrofluoric acid.

Analytically important properties of hydrous silica. Silicic acid, H_4SiO_4, is a very weak acid, with a primary ionization constant of 6×10^{-10}. Acidification of a (basic) solution of a silicate liberates the acid, a portion of which remains in true solution as monosilicic acid, the remainder forming polymers (polysilicic acids). The aggregation of polysilicic acid molecules forms colloidal hydrous silica, $SiO_2 \cdot xH_2O$, as sols or gels. The separation (flocculation) of silica from its colloidal solutions is aided by the presence of electrolytes and by heating. The polymerization and colloidal behavior of silica are complex; its sols have properties intermediate between the hydrophilic and the hydrophobic. Acidification of a sodium silicate solution followed by evaporation gives a gel of silicic acid. Similar treatment of a sodium carbonate melt of a silicate gives a flocculent precipitate of hydrous silica. The precipitates are amorphous (they give no X-ray pattern).

The intrinsic solubility of amorphous silica is by no means small. It amounts to approximately 140 ppm (mg/liter) of SiO_2 in water or dilute acidic solutions.[5] Accurate values are difficult to obtain because of extreme slowness in attainment of equilibrium. The solubility remains essentially constant in the pH range 1–8 or 9. Most or all of the dissolved silica is present as monosilicic acid. At 0°C the solubility is 60–80 ppm; at 90°C, 300–380 ppm (Krauskopf), 432 ppm (Lenher and Merrill).

How is a gravimetric determination of silica possible with such a high intrinsic solubility of amorphous silica? It is possible only because solubility equilibrium is reached slowly. Bergman devised a better method than he knew when he

[3] See, for example, C. J. van Nieuwenburg and H. H. Dingemans, *Chem. Weekblad* **25**, 266 (1928), and especially O. Brunck and R. Holtje, *Angew. Chem.* **45**, 331 (1932).

[4] W. F. Muehlberg, *Ind. Eng. Chem.* **17**, 690 (1925). For a sintering procedure in platinum vessels see T. A. Rafter, *Analyst* **75**, 485 (1950).

[5] G. B. Alexander, W. M. Heston, and R. K. Iler, *J. Phys. Chem.* **58**, 453 (1954); K. B. Krauskopf, *Geochem. Cosmochim. Acta* **10**, 1 (1956). Other workers have reported similar values. Silica gel and silica sols show essentially the same solubility. Early work on the solubility of silica was done by V. Lenher and H. V. Merrill, *J. Am. Chem. Soc.* **39**, 2630 (1917). The study by T. H. Elmer and M. E. Nordberg, *J. Am. Ceram. Soc.* **41**, 517 (1958), of the solubility of porous silica glass (96 per cent SiO_2) and silica gel in nitric acid is of analytical interest. They found that the solubility decreases with increasing nitric acid concentration, but the rate of solution increases with the acid concentration. At 65°C, the solubility of silica was found to be 200 ppm in 0.8 M HNO_3, 80 ppm in 3 M, 24 ppm in 6 M, and 10 ppm in 8.4 M. The solubility (S) follows the relation $-\log S = a[HNO_3] + b$. This is the equation for the salting out of a nonelectrolyte by electrolytes.

evaporated the acidified sodium carbonate melt of a silicate to dryness. The evaporation is a most effective way of coagulating colloidal silicic acid and reducing its solubility to zero, albeit temporarily. When dilute hydrochloric acid is added to the residue to bring soluble substances into solution, the hydrous silica begins to dissolve (see the next section), but the amount dissolved in the length of time required to leach the residue and to filter is far below the equilibrium amount. Nevertheless the loss may amount to a few per cent of the total silica under the usual conditions. Much of this is recovered by evaporating the filtrate and collecting the second portion of silica in a fresh paper. It was formerly believed that the loss of silica on taking up the residue in dilute acid was due largely to reversion of silica to a colloidal state. Actually, most, if not practically all, of the loss is due to true solution of silica as monosilicic acid.[6]

The procedure will now be examined in more detail.

Separation of silica with hydrochloric acid

This is the acid commonly used to decompose acid-decomposable silicates and the sodium carbonate melt of a refractory silicate. The evaporation to dryness is carried out at 100°C in a platinum dish or a porcelain casserole.[7] It appears that little if anything is gained by baking the residue (Table 31-1). The heating of the residue is often called dehydration of silica, but this term is not strictly correct. The silica residue, dried at about 100°C, has the approximate composition $SiO_2 \cdot \frac{1}{2} H_2O$, but is not to be regarded as consisting of a definite hydrate. Although

TABLE **31-1. Effect of Baking First Silica Precipitate on Recovery of Silica**[a,b]

Temp., °C	*Time of Heating, hr*	*Silica Lost, %*
100	1	0.76
(steam bath)	2	0.90
	4	0.90
105	1	1.05
	2	1.13
	4	1.12
125	1	1.11
	2	1.16
	4	1.13

[a] From L. H. Andersson, *Arkiv Kemi* **19**, 249 (1962).

[b] $\frac{1}{2}$-g samples of K-feldspar (SiO_2 69.3, Al_2O_3 17.2, CaO 0.3, K_2O 10.2, and Na_2O 2.2 per cent) analyzed essentially according to procedure on page 658. Residue leached for ~15 min with 80 ml of hot 1:15 HCl.

[6] For investigations of silica losses in analytical procedures, see W. F. Hillebrand, *J. Am. Chem. Soc.* **24**, 362 (1902); V. Lenher and E. Truog, *ibid.* **38**, 1050 (1916). A few procedures have been described in which gelatin is added with the intent of coagulating and precipitating silica without evaporation to dryness. In view of the large intrinsic solubility of silica, the accuracy of such procedures is questionable.

[7] A platinum vessel is generally to be preferred to one of porcelain, but in instructional laboratories the extensive use of platinum is usually out of the question. A small amount of silica tends to adhere tenaciously to porcelain. Samples containing much iron had best be evaporated in porcelain (Pt is attacked by $FeCl_3$). MnO_4^{--} (manganate) can be reduced with alcohol.

prolonged heating of the residue is valueless (it may even do some harm, especially if the temperature is raised above ~110°C, by increasing the contamination of silica), the evaporation should be carried to complete dryness. If this is not done, the silica may be difficult to filter.

The evaporation residue is taken up in dilute hydrochloric acid to bring soluble salts into solution. A 10 per cent solution of hydrochloric acid appears to dissolve less silica than stronger acid.[8] In the presence of appreciable amounts of hydrolyzable metal ions (titanium, ferric iron, aluminum) it is advantageous to moisten the residue with concentrated hydrochloric acid and then dilute with water after a short period of contact. The leaching out of soluble substances is aided by heating near 100°C. The time of contact of the acid solution with the residue should not in general exceed 15–20 min, lest the amount of silica brought into solution be unduly increased (Fig. 31-1).[9] The percentage loss decreases with increasing total amount of silica. The silica precipitate is washed with hot dilute hydrochloric acid to remove titanium, iron, and aluminum as completely as possible.

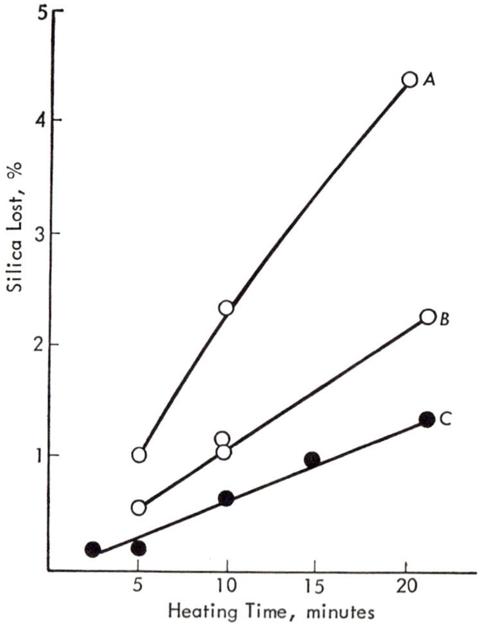

Fig. 31-1. Loss of silica on digestion: *A*, 64 mg SiO_2; *B*, 200 mg; *C*, 640 mg. Temperature ~100°C. From Andersson, Reference 9.

The filtrate and washings, ordinarily containing 1–3 per cent of the total silicon of the sample, are evaporated and the residue is treated much the same as before. A cold 1 : 100 hydrochloric acid solution is used for washing. A third evaporation

[8] Thus H. T. Bucherer and F. W. Meier, *Z. Anal. Chem.* **82**, 41 (1930), found that under otherwise constant conditions, 3.2 mg of silica went into solution with 10 per cent acid, 5.4 mg with 15 per cent, and 9.2 mg with 20 per cent acid.

[9] In any practical analysis, various metals will be present and it is of interest to know whether they affect the loss of silica. L. H. Andersson [*Arkiv Kemi* **19**, 249 (1962)] found that Al up to a molar ratio of 4:1 (with respect to Si) had no significant effect on the recovery of silica, but Mg increased the loss almost twofold in a 2–4:1 molar ratio.

is not carried out, for if the previous recoveries have been carefully made, very little, if any more, silica will be obtained. Notwithstanding that no more silica can be recovered by evaporation, a small quantity remains in the filtrate from the second evaporation.[10] The resulting error is usually small, but it cannot be ignored in accurate analysis. In a systematic analysis of a silicate, the remaining silica will be carried down, not necessarily quantitatively, by the Al, Fe(III), etc., hydroxide precipitate formed with ammonia. Silica therein can be determined by fusing the ignited oxides with potassium pyrosulfate, dissolving in dilute sulfuric acid, evaporating to fuming, and filtering off silica after diluting. Table 31-2 shows the distribution of silica at various stages in an analysis of a feldspar.

TABLE **31-2.** **Determination of Silica in Feldspar (U.S. Bureau of Standards Sample No. 99)**[a,b]

Per Cent SiO_2 *Found in* [c]							
A	B1	B2	C	D1	D2	E	*Total* SiO_2
68.18	0.50						68.68
67.94		0.60	0.02	0.03		0.10	68.69
		0.76	0.01		0.09	0.01	

[a] From L. H. Andersson, *Arkiv Kemi* **19**, 249 (1962).

[b] SiO_2 68.66, Al_2O_3 19.06, Fe_2O_3 0.07, MgO 0.053, CaO 0.36, Na_2O 10.73, K_2O 0.41 per cent. Separation of silica carried out in much the same way as in the procedure on page 658. The duplicate determinations of the original have been averaged above.

[c] A, first silica precipitate; B1, first filtrate and washings (spectrophotometrically); B2, second silica precipitate; C, hydrous oxide precipitate from filtrate from B2 reprecipitated twice, SiO_2 determined spectrophotometrically in filtrates; D1, hydrous oxide precipitate fused with $K_2S_2O_7$, melt dissolved in dilute sulfuric acid and silica collected; D2, sulfuric acid solution of D1 evaporated to fumes, diluted with water and silica collected; E, filtrate from D1 or D2.

Aluminum hydroxide is more effective than ferric hydroxide in the coprecipitation of silica. Andersson found that to recover ~99 per cent of the silica (2.4–6 mg) at pH ~8.2 in hot solution, mole ratios of Al/Si ~4 and Fe/Si ~25 are required.[11]

If an ammonia precipitation is not made, silica can be determined colorimetrically in the combined filtrates and washings from the silica precipitations.[12]

The procedure on page 658 omits the determination of the silica remaining in solution after two evaporations with hydrochloric acid and accordingly is not intended for the most accurate analyses. If porcelain is used for the evaporations and a blank is not run, the introduced silica may more than compensate for the unprecipitated silica.

[10] See the table in W. F. Hillebrand, G. E. F. Lundell, H. A. Bright, and J. I. Hoffman, *Applied Inorganic Analysis*, Wiley, New York, 2nd ed., 1953, p. 677, for the amounts of silica left in solution after double evaporations in the analysis of various materials.

[11] L. H. Andersson, *Arkiv Kemi* **19**, 249 (1962).

[12] P. G. Jeffery and A. D. Wilson, *Analyst*, **85**, 478 (1960), make a single silica precipitation and then apply a spectrophotometric determination of silica in the filtrate.

If fluoride is present in the sample, more or less silicon may be lost by volatilization on evaporation with acid. The presence of Al decreases the loss of silicon by forming aluminum fluoride complexes, and it appears that as much as 10 per cent F in the sample produces a negligible error if sufficient aluminum chloride is added before evaporation with hydrochloric acid. The addition of aluminum is, of course, not permissible when the filtrate from the silicate precipitate is to be analyzed for aluminum, iron, etc.

Contamination of silica. Amorphous silica possesses rather strong adsorptive power and shows a tendency to carry down heavy metals (e.g., iron, titanium, aluminum) from hydrochloric acid (and other acid) solutions. It is difficult to remove such metals completely by washing. Moreover, the treatment with hydrochloric acid may result in the precipitation of various slightly soluble compounds. Niobium, tantalum, tungsten, and silver will be precipitated quantitatively, or almost so, and certain combinations of elements in the sample can give rise to insoluble substances in the silica precipitate. Boron is carried down by silica; it can be removed by volatilization as methyl borate. Errors from most of these contaminants can be avoided by hydrofluorizing the weighed silica, as described later.

Use of acids other than hydrochloric in the separation of silica. Evaporation to fumes of a sulfuric or perchloric acid solution precipitates most of the silica in an acid-decomposable silicate. Sulfuric acid is not very often used because it gives anhydrous sulfates with a number of metals (e.g., Al, Fe, Cr, Ni, Co), which dissolve only slowly in the diluted acid mixture, and, moreover, barium, lead, and calcium sulfates are insoluble or only slightly soluble and therefore contaminate the silica. Perchloric acid does not have these disadvantages and sometimes is used instead of hydrochloric acid in the precipitation of silica.[13] With the exception of potassium, rubidium and cesium, the perchlorates are soluble in dilute perchloric acid or water. Perchloric acid would be substituted for hydrochloric acid in the presence of metals forming slightly soluble chlorides (silver, lead). The use of perchloric acid always calls for precautions to prevent explosions (organic matter in samples and in hoods). One evaporation to fuming with perchloric acid does not recover all the silica in a sample.[14]

Nitric acid has been used in a special way in the separation of silica. Fairly complete recoveries of silica can be obtained in a single treatment of acid-decomposable silicates (e.g., Portland cement), if the sample is mixed with ammonium nitrate and heated with concentrated nitric acid.[15] The separated silica is filtered off and washed with 5 per cent nitric acid containing a few drops of hydrogen peroxide. The silica precipitated in this way is said to be purer than that obtained by evaporating with hydrochloric or perchloric acid, and the separation is so complete that usually the filtrate need not be evaporated to recover the remaining

[13] H. H. Willard and W. E. Cake, *J. Am. Chem. Soc.* **42**, 2208 (1920).

[14] A procedure for the determination of silica in acid-soluble silicates with a mixture of perchloric and hydrochloric acids has been given by F. W. Meier and O. Fleischmann, *Z. Anal. Chem.* **88**, 88 (1932). H. R. Shell prefers to carry out the first "dehydration" of silica with hydrochloric acid, the second with perchloric acid. [I. M. Kolthoff and P. J. Elving with E. B. Sandell (eds.), *Treatise on Analytical Chemistry*, part II, vol. 2, Wiley-Interscience, New York, 1962, p. 191.] This procedure is said to give more complete recovery of silica than a double evaporation with hydrochloric acid.

[15] W. van Tongeren, *Chem. Weekblad* **34**, 774 (1937).

silica. The rationale of this method is clear: The high electrolyte concentration salts out silica.[16]

Obtaining the true weight of silica in the ignition residue

The washed precipitates of silica are heated to a high temperature[17] (Meker burner or blast lamp) in a platinum crucible to obtain anhydrous silica (SiO_2). Care must be taken in this operation to prevent mechanical loss of the light, fluffy residue. The ignited silica is hygroscopic and must therefore be cooled and weighed with due precautions.

As already mentioned, the weight of the ignited residue does not accurately represent the weight of silica present, because the silica is always contaminated with greater or less amounts of nonvolatile contaminants. Small amounts of titanium, aluminum, and iron are tenaciously retained by the hydrous silica and cannot be entirely removed by washing with dilute hydrochloric acid.[18] Furthermore, substances insoluble per se may be present in the separated silica. Thus it is common to find both titanium and phosphorus in silicates. These form titanium phosphate, which is only slightly soluble in dilute acids and is therefore left to a large extent with the dehydrated silica. If large amounts of calcium and sulfur should be simultaneously present (natural waters), insufficient washing of the silica precipitate may leave calcium sulfate associated with it.

Fortunately silicon forms a gaseous fluoride, and the silica in the impure ignited residue may therefore be obtained from the loss in weight resulting from treatment of the residue with hydrofluoric acid, provided the contaminants are in the same form, i.e., possess the same weight, before and after the hydrofluoric acid treatment and are not volatilized in the operation. The fulfillment of these requirements is assured by conducting the hydrofluorization in the presence of sulfuric acid as explained in the next paragraph. The reaction taking place between silica and hydrofluoric acid is represented by the following equations:

$$SiO_2 + 6\,HF \rightleftharpoons H_2SiF_6 + 2\,H_2O$$

$$H_2SiF_6 \rightleftharpoons SiF_4 + 2\,HF$$

The last reaction proceeds to the right on heating. It may be noted that the presence of sulfuric acid displaces the equilibrium in the first reaction to the right on evaporation, owing to the dehydrating properties of this acid.

[16] Concentrated hydrochloric acid can be used in place of nitric acid: N. van der Weel, *Chem. Weekblad* **47**, 845 (1951). The addition of ammonium chloride is not necessary, although it was used in some earlier procedures of this type for silica in Portland cement: M. I. Zaitzev, *Zavodskaya Lab.* **3**, 492 (1934); E. E. Maczkowske, *J. Res. Natl. Bur. Std.* **16**, 549 (1936). Hydrochloric acid may be expected to be less desirable than nitric acid in the presence of much sodium, because of the sparing solubility of sodium chloride in hydrochloric acid.

[17] W. Miehr, P. Koch, and J. Kratzert, *Z. Angew. Chem.* **43**, 250 (1930), state that after heating to the indicated temperatures for 1 hr and cooling for $\frac{1}{2}$ hr in a desiccator, the ignited silica contains the following percentages of water: 900°C, 0.9; 1000°C, 0.5; 1100°C, 0.2; 1200°C, 0.1. After ignition at 1200°C, for $\frac{1}{2}$ hr, SiO_2 shows an incipient X-ray pattern of cristobalite.

[18] The alkali and alkaline earth metals are usually not found in significant amounts in the silica precipitate, chiefly because the salts of these metals do not hydrolyze to form slightly soluble basic salts. K. Rankama, *Compt. Rend. Soc. Geol. Finlande* **14**, (1939), studied the composition of the residue from silica obtained in rock analysis.

After the silica precipitate has been strongly ignited, the principal contaminants will be present as the oxides (TiO_2, Al_2O_3, Fe_2O_3). Now, if sulfuric acid were omitted in the evaporation of the silica with hydrofluoric acid in the expulsion of silicon as the tetrafluoride, the contaminating metals would be left as the fluorides after the ignition of the residue, with the exception of titanium, which would be lost as volatile TiF_4. Consequently, erroneous results would be obtained in calculating the amount of silica present from the loss in weight. The difficulty is avoided by adding sulfuric acid with the hydrofluoric acid. After the evaporation of the latter the contaminating metals are left as sulfates (titanium is not volatilized). During the subsequent ignition the sulfates of titanium, aluminum, and iron are decomposed into the oxides; e.g.,

$$Fe_2(SO_4)_3 \rightarrow Fe_2O_3 + 3\ SO_3$$

The decomposition of the sulfates of these metals is complete far below 1000°C. The final ignition should be conducted over a Meker burner or blast lamp but need not be continued long because of the small amount of residue.[19]

Determination of silica in an acid-insoluble silicate

Procedure. Reduce 2 or 3 g of coarsely powdered sample to a very fine powder in an agate mortar, grinding only small portions at a time.[20] Weigh out duplicate samples of 0.8 g into platinum crucibles of 20- to 30-ml capacity. Next, weigh out on a watch glass 4 (± 0.2) g of purest anhydrous sodium carbonate. Place the crucible on a sheet of paper, and by means of a spatula transfer approximately three fourths of the sodium carbonate to the crucible. Intimately mix the silicate powder and the flux by stirring gently with a platinum spatula or a short glass rod with rounded ends; be careful not to scratch the crucible or to raise dust. The mixture should finally appear homogeneous; care should be taken to see that no unmixed sample remains at the bottom of the crucible. Tap the spatula or glass rod to cause any adhering particles to fall into the crucible, and rub the spatula on the sodium carbonate on the watch glass to remove any powder possibly sticking to it. Transfer all the remaining sodium carbonate to the crucible, covering the mixture with a uniform layer of flux.

[19] Mention may be made of the effect of other contaminants, less usually present in appreciable amounts, upon the indirect determination of silica by hydrofluorization. The sulfates of magnesium and manganese are decomposed completely into the oxides at temperatures slightly below 1000°C. In the case of manganese the oxide is of somewhat variable composition. Calcium sulfate when pure gives the oxide when heated to approximately 1200°C, but a lower temperature in the presence of silica. The presence of this sulfate in the impure silica gives rise to some uncertainty in the determination of silica by difference.

The presence of alkali chlorides is very objectionable. These should not be present in the silica, but if the precipitate has not been carefully washed (especially after a sodium carbonate fusion of the sample), they may be found in it. The alkali remaining unvolatilized will be weighed as silicate or possibly as chloride before the hydrofluoric acid treatment and as the sulfate after. The result for silica will then be low. The error can be reduced by moistening the precipitate with sulfuric acid before the first ignition but not entirely eliminated because an alkali silicate may be formed during the ignition (Hillebrand, Lundell, Bright, and Hoffman, *Applied Inorganic Analysis*, p. 681). When phosphate contaminates the silica, it is necessary to fume off the sulfuric acid at as low a temperature as possible to minimize the volatilization of phosphoric acid.

[20] Sifting through a 100-mesh silk bolting cloth, with return of the oversize fraction to the mortar, is advantageous.

Place the covered crucible vertically upon a clay triangle, and begin the heating with a very low Tirrill flame; it is safest to support the crucible about 10 cm above the flame. After 5–10 min lower the crucible and gradually increase the height of the flame until the bottom of the crucible is faint red. After another 10 min raise the temperature slowly until the mass is in quiet fusion (lift the cover slightly with the crucible tongs). Keep the contents at this temperature for about 30 min; there should be but little bubbling of the molten mass during this period; the crucible must be kept covered. Finally, apply the full heat of the Tirrill burner (or Meker if necessary) for 10–15 min. Remove the burner, lay aside the cover of the crucible, grasp the crucible with the tongs, and give it a slow rotary motion so that the molten contents will spread over the sides and solidify as a thin shell over the interior. This procedure will later facilitate removal of the contents. Set the crucible on a clean stone surface to cool, or return it to the triangle, replacing the cover.

When the crucible is thoroughly cold, fill it approximately one third full with water. After a few minutes heat the crucible cautiously with a low flame; do not boil the liquid. Touch the cake with a 6- to 7-cm glass rod to see if it has been loosened; if so, transfer it to the casserole. If it has not, pour the liquid into a 300-ml porcelain[21] casserole, being careful to wash off the drop that remains adhering to the outside of the crucible. Again add water to the crucible, and heat. If, after repeating this procedure once or twice, the cake does not loosen, rinse off the outside of the crucible, place it on its side in the casserole, and fill the latter about one third with water. Warm, and stir until the cake disintegrates and can be easily removed. By means of a 15-cm stirring rod lift the crucible out of the liquid and, while holding it a few centimeters above the liquid, wash its interior and exterior thoroughly with water. Hold the crucible cover over the crucible, pour a few drops of dilute hydrochloric acid upon the underside of the former to dissolve any sodium carbonate that may have spattered on to it during the fusion, and rinse the crucible, finally adding the solution to the casserole. When the cake is well disintegrated, pour very slowly 20 ml of concentrated hydrochloric acid into the cold *covered* casserole.[22] When the evolution of carbon dioxide has ceased, rinse off the cover glass and the sides of the dish above the liquid. Some flocculent silicic acid will usually be visible at this point. There should be no gritty particles present; if such are found, the decomposition is most likely incomplete, and a new sample must be fused with sodium carbonate, greater care being taken to ensure complete decomposition.

Place the dish on the steam bath, and evaporate the contents to dryness. Heat the residue for 1 hr on the steam bath, and then add 5 ml of concentrated hydrochloric acid. With the aid of a stirring rod bring the acid in contact with all the solid in the dish. Add 75 ml of water, rinse down the sides, and leave the stirring rod in the vesssel. Heat for 10–15 min on the steam bath or over a low flame, stirring to aid solution of soluble salts, until only silica remains undissolved. Filter through a 9-cm filter paper of medium texture, and receive the filtrate in a 250-ml beaker. Transfer as much as possible of the silica to the paper, rinse the casserole with a warm solution of 1 : 20 hydrochloric acid, and wash the precipitate thoroughly with the same wash liquid and finally once with water. Pour the filtrate that has

[21] A platinum dish is better.

[22] It is best to add the acid through a small funnel inserted beneath the cover glass (p. 525).

been collected in the beaker into the original casserole, rinse out the beaker, and evaporate the solution to dryness. Heat the residue at 100–105°C for $\frac{1}{2}$ hr. Moisten the separated salts with 1–2 ml of concentrated hydrochloric acid, add 25 ml of water, heat for 5 min to extract soluble salts, and filter through a fresh small (7 cm) filter paper. Rinse out the dish, and carefully wash the precipitate with cold 1 : 100 hydrochloric acid. To remove any silica adhering to the casserole, rub its interior with one or two small pieces of moist quantitative filter paper, using the stirring rod. Place this paper in the filter. Discard the filtrate (except when other constituents are to be determined in the solution).

Fold up the moist filters, and place them in a platinum crucible. Dry the filter papers by heating the covered crucible, placed in the vertical position, with a very low Tirrill flame. When all water has been expelled, lower the crucible and char the paper. When the volatile matter has been driven off, slide the cover slightly to one side, and increase the flame to burn off the carbon, still keeping the crucible vertical. Then, when all the carbon has been burned off, cover the crucible tightly, replace the Tirrill burner with a Meker, and apply its full heat for $\frac{1}{2}$ hr.[23] Cool the crucible in a desiccator, weigh, and reheat to constant weight, using 15 min periods of heating. Remember that ignited silica is hygroscopic.

Moisten the residue with 1–2 ml of water added from a pipet inserted below the cover. Add 4 or 5 drops of 1 : 2 sulfuric acid and then, carefully, 5 ml of hydrofluoric acid. Place the crucible in a radiator (less preferably on a hot plate), and evaporate off the hydrofluoric acid in a hood without boiling. When only sulfuric acid remains, increase the heat and fume it off. Then heat the crucible strongly for 10–15 min with the direct full flame of a Meker burner. Cool and weigh the crucible. From the loss in weight calculate the percentage of SiO_2 in the sample.

Regarding the running of a blank, see page 416. It is important that the hydrofluoric and sulfuric acids used be free from nonvolatile impurities. It is safest to test them by evaporating the same volumes as used in the procedure in a platinum crucible and obtaining the weight of the ignited residue.

Other methods for silicon

A number of gravimetric methods have been described in which silicon is converted into silicomolybdic acid and precipitated as the silicomolybdate of quinolinium, oxinium, and other organic cations of large volume. Titrimetric methods can be based on the formation of such salts (solution of washed precipitate in standard base and back-titration with acid).

Silicon can be determined by conversion to SiF_6^{--} with HF, precipitation of fluosilicate as K_2SiF_6 in the presence of a high concentration of potassium chloride and titration of the washed precipitate with standard base to form F^- and H_4SiO_4. Results of respectable accuracy have been obtained in some procedures based on this titration.[24]

The spectrophotometric determination of silica is of some importance. It is based either on the yellow color of α-silicomolybdic acid, formed by molybdate in acid solution, or the blue color (heteropoly molybdenum blue) resulting from

[23] If a suitable furnace is available, complete the ignition by heating for $\frac{1}{2}$ hr at 1200°C.

[24] R. J. W. McLaughlin and V. S. Biskupsi, *Anal. Chim. Acta* **32**, 165 (1965); A. G. C. Morris, *Analyst* **90**, 325 (1965).

the reduction of silicomolybdic acid with stannous chloride and other reducing agents.[25] Silica must be present in solution as monosilicic acid to react rapidly with molybdate.

Silicate analysis

This chapter has treated the determination of the characteristic element of silicates. The complete analysis of a silicate, more especially that of a mixture of silicates (silicate rock, glass, ceramic products), requires the determination of a considerable number of elements. In a standard rock analysis at least 14 major and minor elements are determined: SiO_2, Al_2O_3, Fe_2O_3, FeO, MgO, CaO, Na_2O, K_2O, H_2O^+, H_2O^-, TiO_2, P_2O_5, MnO, and CO_2. In addition, the determination of BaO, S, SO_3, F, Cl, Cr_2O_3, ZrO_2, and NiO may be required. Still other constituents are determined in special studies. When the most accurate results are needed, most of the major constituents of a silicate rock or other silicate material are determined by gravimetric and volumetric methods according to well-established procedures.[26] Physicochemical methods (spectrophotometric, emission spectrographic, and, less often routinely, neutron activation) are used for trace constituents and some of the minor constituents. A chemical rock analysis requires a high degree of skill and may be said to represent the apogee of classical analysis.

A rapid scheme of rock analysis providing results that are sufficiently accurate for some purposes is based on the spectrophotometric and flame-photometric determination of most of the major constituents.[27]

X-ray fluorescence methods also find use in the rapid determination of major and minor elements in silicate materials.

PROBLEMS

1. What is the minimum volume of hydrofluoric acid (sp. gr. 1.15, 46 per cent HF) and of sulfuric acid (sp. gr. 1.84, 96 per cent) required to decompose 1.00 g of a silicate of the following percentage composition and convert the metals into the normal sulfates?

SiO_2	60	Fe_2O_3	4	MgO	4	Na_2O	4
Al_2O_3	15	FeO	5	CaO	5	K_2O	3

[25] J. D. Strickland, *J. Am. Chem. Soc.* **74**, 862 (1952); J. B. Mullin and J. P. Riley, *Anal. Chim. Acta* **12**, 162 (1955); A. Ringbom, P. E. Ahlers, and S. Siitonen, *Anal. Chim. Acta* **20**, 78 (1959); L. H. Andersson, *Arkiv Kemi* **19**, 223, 235 (1962).

[26] Hillebrand, Lundell, Bright, and Hoffman, *Applied Inorganic Analysis*, pp. 793–957. This treatment, not entirely up to date, is based on Hillebrand, *U.S. Geol. Surv. Bull.* **700**, 1919. H. S. Washington, *The Chemical Analysis of Rocks*, Wiley, New York, 4th ed., 1930. A. W. Groves, *Silicate Analysis*, Allen and Unwin, London, 2nd ed., 1951. L. C. Peck, *U.S. Geol. Surv. Bull.* **1170**, 1964. R. C. Chirnside, *Trans. Soc. Glass Technol.* **43**, 5T (1959). J. A. Maxwell, *The Chemical Analysis of Rocks and Minerals*, Wiley-Interscience, New York (forthcoming). For a survey of physical and physicochemical methods in the analysis of geomaterials, see A. A. Smales and L. R. Wager (eds.), *Methods in Geochemistry*, Wiley-Interscience, New York, 1960. For the accuracy attained in the determination of silica in rock analysis see page 411.

[27] L. Shapiro and W. M. Brannock, *U.S. Geol. Surv. Circ.* **165**, 1952; see also E. L. P. Mercy, *Geochim. Cosmochim. Acta* **9**, 161 (1956). R. Hedin, *Colorimetric Methods for Rapid Analysis of Silicate Materials*, Swedish Cement and Concrete Institute, Royal Institute of Technology, Stockholm, 1947.

2. From the following analytical data obtain the formula of a hydrated magnesium silicate: (a) a 1.120-g sample yielded 0.1445 g of essential water; (b) a 0.3621-g sample gave 0.4362 g of $Mg_2P_2O_7$; (c) on treatment with hydrofluoric and sulfuric acids the ignited silica residue from a 0.9870-g sample showed a loss of 0.4278 g.
3. An ignited silica residue weighing 0.5050 g and containing 1.0 per cent sodium chloride is treated with hydrofluoric and sulfuric acids and silica determined by loss in the usual way. How great is the resulting error? *Ans.* −2.2 ppt.
4. If the sodium in Problem 3 is present as Na_2SiO_3, what is the error?
5. Silica is to be determined in a sample containing 10 per cent CaF_2. Outline a procedure.
6. Suppose that 1 per cent of a sample composed of $KAlSi_3O_8$ is not decomposed by sodium carbonate fusion but is decomposed in the hydrofluoric–sulfuric acid treatment of the ignited silica. What will be the error in SiO_2 in parts per thousand?
7. Suggest methods for decomposing carborundum and ethyl silicate prior to determination of silicon.
8. From the data in footnote 5, page 652, find by extrapolation the solubility of silica in water at 65°C.
9. The solubility of quartz in water has been reported as 1.8×10^{-4} *M* at 25°C. Find the solubility of quartz in 0.010 *M* Na_2CO_3 if $[H^+][H_3SiO_4^-]/[H_4SiO_4] = 10^{-9.2}$.
10. What weight of SiO_2 does 1 ml of 1 *N* NaOH represent in the titration of K_2SiF_6? What is a suitable indicator? *Ans.* 0.0150 g.

CHAPTER 32 *Determination of Sodium and Potassium*

Two different types of methods are of primary importance at present in the determination of sodium, potassium, and the other alkali metals: (1) gravimetric, volumetric, or colorimetric after precipitation as a slightly soluble salt, and (2) flame photometric (or atomic absorption). Neutron activation and isotope dilution are special methods of determination, not likely to be used routinely; X-ray fluorescence methods are sometimes used for potassium, rubidium, and cesium. Flame photometry is at its best in the determination of the alkali metals, and sodium and potassium are now routinely determined more frequently in this way than by classical methods, because of the speed, simplicity, and sensitivity of the former method. When the necessary conditions are fulfilled, it appears that the precision and accuracy of flame photometry matches or nearly matches that of the best classical methods. Flame photometry is especially advantageous when many analyses are to be made of samples having much the same composition. It has great value in the determination of traces of the alkali metals, and of rubidium, cesium, and potassium in mixtures, an analysis difficult or impossible to carry out satisfactorily by classical methods.

In this chapter we shall discuss in fair detail several gravimetric methods for sodium and potassium and consider more briefly their flame-photometric determination.[1] Applications will be illustrated by the determination of sodium and potassium in silicate rocks, an analysis recommended to any serious student. Lithium,[2] rubidium, and cesium will not be treated.

CHEMICAL METHODS

Frequently, the determination of both sodium and potassium is required in a sample. The most accurate procedure then consists of the determination of each separately, but it is also possible to determine one of the elements and obtain the other by difference knowing the sum NaCl + KCl (or other pure salt mixture). The latter alternative requires the separation of all other elements (see p. 669). Even if only sodium or potassium is to be determined, some separations from other elements may still be necessary.[3]

[1] An exhaustive treatment (404 pages) of the determination of the alkali metals by methods available up to about 1940 may be found in R. Fresenius and G. Jander (eds.), *Handbuch der analytischen Chemie*, vol. IA, part III, Springer, Berlin, 1940. A more recent survey is presented by S. Kallmann in *Treatise on Analytical Chemistry* (I. M. Kolthoff and P. J. Elving with E. B. Sandell, eds.) vol. I, part II, Wiley-Interscience, New York, 1961, and by H. Tollert, *Analytik des Kaliums*, Enke, Stuttgart, 1962.

[2] For the flame-photometric determination of traces of lithium, sensitive to 5 ppm, in silicate rocks and like materials, see R. B. Ellestad and E. L. Horstman, *Anal. Chem.* **27**, 1229 (1955).

[3] The limited scope of this chapter does not allow a general treatment of separations. Ion exchange is sometimes of value in separating the alkali metals from other elements and from each other.

Potassium

Two methods, the chloroplatinate and the tetraphenylboron, for the determination of potassium in the presence of sodium and other elements will be described here. Both reagents form slightly soluble precipitates with potassium by virtue of their large anionic size and single charge (p. 264). Chloroplatinic acid has been used for determining potassium for almost 150 years. It provides an accurate determination of potassium but requires the use of an alcoholic medium. Tetraphenylboron is a relatively new reagent. It forms a potassium salt less soluble than potassium chloroplatinate, thus allowing a simpler and more rapid procedure, and its use is subject to few interferences. Time may show tetraphenylboron to be the superior reagent.

Another method for potassium is based on the solubility of sodium perchlorate and the insolubility of potassium perchlorate in organic liquids such as *n*-butyl alcohol or a mixture of this with ethyl acetate.[4] Potassium forms a very sparingly soluble cobaltinitrite, which, when sodium cobaltinitrite is the precipitant, has the approximate composition $K_2NaCo(NO_2)_6$. The ratio K : Na is variable and the method does not have good accuracy. Precipitation of potassium in this way is sometimes used to separate it from the bulk of sodium in a high-sodium sample.

The chloroplatinate method. This method is based on the virtual insolubility of potassium chloroplatinate and the solubility of sodium chloroplatinate in 80 per cent ethyl alcohol.[5] The sample in the form of the chlorides of the alkali metals, and free of other substances, is treated with an excess of chloroplatinic acid and the solution is evaporated almost to dryness. The chlorides of sodium and potassium are thus converted into the corresponding chloroplatinates with expulsion of hydrogen chloride:

$$2\,KCl + H_2PtCl_6 \rightarrow K_2PtCl_6 + 2\,HCl$$

It is important to add enough reagent to convert sodium into the chloroplatinate, because sodium chloride is insoluble in 80 per cent alcohol. The evaporation residue is then treated with 80 per cent alcohol, which dissolves sodium chloroplatinate and the excess of chloroplatinic acid, leaving the potassium chloroplatinate undissolved. The latter is filtered off, washed with alcohol, dried at 135°C, and weighed as K_2PtCl_6.[6]

Alcoholic solutions stronger than 80 per cent by volume are avoided in the extraction of the mixed chloroplatinates as possibly leading to the decomposition of sodium chloroplatinate:

$$Na_2PtCl_6 \rightarrow 2\,NaCl + PtCl_4$$

The solubility of potassium chloroplatinate in 80 per cent alcohol at 20°C is 8.6 mg/100 ml. The solubility loss of potassium in the procedure is negligibly small if care is taken

[4] G. F. Smith and J. F. Ross, *J. Am. Chem. Soc.* **47**, 1020 (1925).

[5] Potassium and sodium chloroplatinates do not form mixed crystals or double salts. K_2PtCl_6 (anhydrous) crystallizes in octahedra (isometric system) and $Na_2PtCl_6 \cdot 6\,H_2O$ in triclinic plates, at opposite ends of the crystallographic spectrum, so to speak.

[6] Claims have been made that the potassium salt does not have quite the theoretical composition. H. Fresenius and P. H. M. Brinton, *Z. Anal. Chem.* **50**, 21 (1911), obtained highly accurate results by using the factor 0.3056 in place of the theoretical factor 0.3067 for $2\,KCl/K_2PtCl_6$. Although slight deviation from the stoichiometric composition in a salt such as K_2PtCl_6 is a plausible possibility, it appears that the theoretical composition can be accepted for all practical purposes; in some of the earlier work the issue was confused by a small error in the atomic weight of platinum. L. C. Peck (*U.S. Geol. Surv. Bull.* **1170**, 1964) states "Careful experiments showed that the best average values are obtained by using theoretical factors, although results tend to be slightly low for small amounts of potassium and slightly high for large amounts." In the most careful work the value of the factor for the particular reagent used and procedure followed should be obtained by finding the weights of potassium chloroplatinate yielded by known amounts of pure potassium chloride. Such accuracy is rarely required. The procedure on page 665 should give values within a few parts per thousand of the true values for potassium in the presence of comparable amounts of sodium.

to restrict the alcohol solution to the minimum volume required; the presence of an excess of the reagent as well as of the sodium chloroplatinate and the nonattainment of solubility equilibrium in washing lead to a smaller loss by solubility than indicated by the figure given.[7]

Ammonium, rubidium, and cesium all yield chloroplatinates insoluble in alcohol and are counted with potassium if present. Ammonium salts can be removed easily by ignition of the mixed chlorides at a low temperature. Lithium chloroplatinate is soluble in alcohol and accompanies sodium into the filtrate. Sulfate must not be present because sodium sulfate is insoluble in the alcoholic solution. In general, substances other than the alkali chlorides should be absent; however, small amounts of calcium, magnesium, and strontium chlorides are usually permissible.[8]

It is often advantageous in practical analysis to reduce the platinum in the washed potassium chloroplatinate precipitate to the metal and obtain the weight of the potassium present from the ratio 2 K : Pt. When the platinum in the precipitate, and not the precipitate itself, is weighed, it is possible to determine potassium in the presence of sulfates, phosphates, calcium, magnesium, and indeed most other metals.[9]

The following directions are intended for the determination of potassium in a mixture of potassium and sodium (lithium) chlorides containing no other elements, except possibly small amounts of the alkaline earth chlorides.

Procedure. Dissolve a weighed sample of suitable size (generally not over 0.3 g) in 5–10 ml of hot water in a small (35-ml) porcelain dish. Assume the sample is all sodium chloride and add enough pure chloroplatinic acid containing 0.10 g of Pt per milliliter to convert the sodium into the chloroplatinate (1 g of NaCl requires 16.7 ml of chloroplatinic acid of this strength), plus an additional 1 ml.

Place the dish on the steam bath and evaporate nearly, but not quite, to dryness; the syrupy liquid should solidify on cooling. Add 10 ml of 80 per cent (by volume) ethyl alcohol[10] to the cold dish, and break up the residue with a short stirring rod having a flattened end. After thorough disintegration, decant the supernatant solution into a fine-porosity sintered glass or porcelain filter crucible of 15-ml size, which has been weighed after drying at 130°C, leaving as much as possible of the precipitate in the dish. Continue to extract the residue in the dish, gently crushing the crystals, using 1- to 2-ml portions of 80 per cent alcohol each time and decanting into the crucible under suction until the alcohol extract is colorless. The salt remaining in the dish should be golden yellow in color, with no orange-red of sodium chloroplatinate. All these operations should be carried out in a place free of ammonia fumes. Transfer quantitatively the precipitate in the dish to the filter crucible with a minimum volume of 80 per cent alcohol. By this time all soluble substances should have been washed out of the crucible, but give it a final rinse, all around the interior, with a few milliliters of alcohol.

Dry the crucible at about 70°C for 30 min to remove alcohol. Then dry to constant weight at 130–140°C, keeping the crucible loosely covered to prevent possible loss of crystals by decrepitation. Calculate the weight or percentage of KCl (or K_2O or K) as required, using the theoretical factors.

[7] P. L. Hibbard, *J. Ind. Eng. Chem.* **9**, 504 (1917), washed 0.33 g of K_2PtCl_6 in a Gooch crucible with six portions of 80 per cent ethanol, 10 ml each, and found a loss of 1.5 mg (0.3 mg of K_2O); with 95 per cent ethanol the loss was 0.4 mg.

[8] According to Fresenius, 0.5 g of $CaCl_2$ or $MgCl_2$ leads to errors of +3 to +5 ppt in the determination of potassium in 0.5-g mixtures of KCl + NaCl; 0.5 g of $BaCl_2 \cdot 2\,H_2O$ gives an error of +6 ppt.

[9] The reduction of the platinum is conveniently effected by metallic magnesium. See W. B. Hicks, *J. Ind. Eng. Chem.* **5**, 650 (1913); T. E. Keitt and H. E. Shiver, *ibid.* **10**, 219, 994 (1918); **11**, 1049 (1919); R. C. Wells, R. K. Bailey, and J. G. Fairchild, *ibid.* **16**, 935 (1924).

[10] Eighty per cent alcohol by volume is prepared by mixing 5 volumes of 95 per cent alcohol with 1 volume of water. The specific gravity of the mixture should be 0.86. The alcohol must be free of aldehydes.

Save the weighed precipitate and filtrate and washings for the recovery of platinum.[11]

The tetraphenylboron method.[12] In 1949, Wittig and coworkers showed that potassium tetraphenylboron, $K[B(C_6H_5)_4]$, is a slightly soluble salt that can be used as a determination form for potassium. The sodium salt is very soluble in water (0.9 *M*, 25°C) and is used as the precipitant. The tetraphenylboron salts of NH_4, Rb, Cs, Tl(I), and Ag(I) are also very slightly soluble, as are those of many large organic cations of the type R^+ or RH^+. The combination of slight solubility, stoichiometric composition, favorable chemical factor, and satisfactory filterability has made tetraphenylboron a popular reagent for potassium.

Sodium tetraphenylboron is not indefinitely stable in the solid state. However, it may be kept for many months, especially if stored in a cold place, protected from light. In solution it decomposes more rapidly. Depending on the purity of the solid reagent, the solution may develop a turbidity and phenolic odor in days or weeks. The stability is greater in basic than in acidic solutions. At pH 9–10, a solution may show no appreciable decrease in titer (< 1 per cent) in several months at room temperature. The reagent now available will usually give a basic solution in water and ordinarily no special treatment is required to prepare a solution sufficiently stable for a gravimetric determination. It is best stored in a refrigerator; it may be used as long as it shows no turbidity.

The solubility of $KB(C_6H_5)_4$ is 1.8×10^{-4} *M* in water at 25°C, corresponding to 7 mg of K per liter. With an excess of $B(C_6H_5)_4^-$ the solubility is of course much less. The precipitate is composed of larger crystals and is more easily filterable if formed in a mineral acid medium. Such a medium is also more desirable than a neutral or slightly acidic medium from the standpoint of lesser interference by other cations, which can give trouble by hydrolysis and other reactions.[13] However, since the instability of $B(C_6H_5)_4^-$ increases with the acidity, this must be controlled. At room temperature, a mineral acid concentration as high as 0.1 *M* is permissible in gravimetric determinations. If the solutions are cooled to 0°C, an acidity as high as 0.6 *M* is permissible. The precipitation of potassium is rapid, and long standing is not necessary or desirable. Washing can be carried out with dilute reagent solution, followed by a small volume of water, or with a saturated solution of potassium tetraphenylboron. The precipitate is usually dried at 105°C or so, but a higher temperature can be used, because it is stable up to about 250°C.

Except for the metals giving slightly soluble precipitates with tetraphenylboron ion, very few interfere in reasonable amounts (say 10–20 times as much as the amount of K). Mercury(II) decomposes the reagent, forming C_6H_5HgCl and other products. None of the common anions interferes with the potassium determination in dilute acid medium.

Volumetric methods can be used to determine potassium indirectly in the washed precipitate. Thus $B(C_6H_5)_4^-$ can be titrated with silver. More than a score of volumetric procedures have been proposed, but at best they offer only marginal advantages over the simple gravimetric method. Likewise, the indirect spectrophotometric determination of potassium based on absorption by $B(C_6H_5)_4^-$ in acetonitrile–water solution is of little interest for macro amounts.

Since ammonium tetraphenylboron is approximately as insoluble as potassium tetraphenylboron, ammonium salts must not be present. Gentle ignition, carefully carried out, volatilizes ammonium salts from a solid sample (p. 527). Alternatively, formaldehyde can be added to an alkaline sample solution to form hexamethylenetetramine by reaction with ammonia. The subsequent precipitation of ammonia must then be carried out in alkaline, not acid, solution.

[11] See, for example, M. C. Swisher and F. F. Hummel, *Ind. Eng. Chem., Anal. Ed.* **11**, 162 (1939).

[12] For a review of the analytical applications of tetraphenylboron see H. Flaschka and A. J. Barnard, Jr., *Advan. Anal. Chem. and Instr.* **1**, 1 (1960).

[13] EDTA can be used as complexing agent for foreign cations in acetic acid or basic solutions.

Procedure. Add enough hydrochloric or nitric acid to 50 ml of sample solution (not above 20°C), preferably containing 5–12 mg of K, to make the acidity 0.1 *M*. While stirring, add slowly, from a pipet, 5 ml of 3 per cent sodium tetraphenylboron solution.[14] Allow to stand for 10–15 min (not more than $\frac{1}{2}$ hr). Collect the precipitate in a sintered glass or porcelain filter crucible of fine porosity. Draw off the solution as completely as possible, and finish the transfer and wash the precipitate with three portions of 0.1 per cent solution of sodium tetraphenylboron of about 3 ml each. Finally, wash with two 2-ml portions of water, drawing off each portion sharply. Dry the crucible to constant weight at 105°C. $K/KB(C_6H_5)_4 = 0.1091$, $K_2O/2\ KB(C_6H_5)_4 = 0.1314$.

Sodium

The direct determination of sodium in the presence of the other alkali metals (except lithium) is based upon its precipitation as a slightly soluble triple acetate with uranyl ion having the general formula

$$Na(C_2H_3O_2) \cdot M(C_2H_3O_2)_2 \cdot 3\ UO_2(C_2H_3O_2)_2 \cdot 6\ H_2O$$

in which M may be magnesium, zinc, nickel, or certain other divalent metals. The triple acetates with zinc and magnesium[15] are the most satisfactory precipitation forms for sodium.

Determination of sodium as sodium uranyl zinc acetate.[16] The solubility of sodium uranyl zinc acetate hexahydrate in water is 5.85 g/100 ml of solution at 21°C, so that a

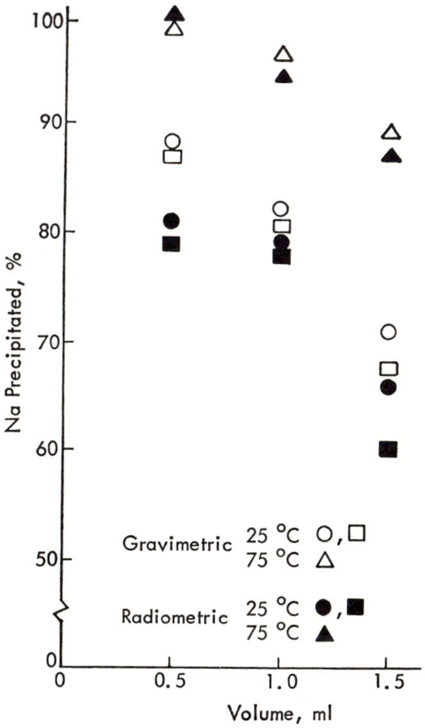

Fig. 32-1. Effect of volume of sample solution and temperature of precipitation on recovery of sodium in the triple acetate method. (From Ross, Reference 17.) 1 mg Na, 10 ml reagent, 2.5 hours standing. On longer standing, recoveries are improved: 89 per cent after 5 hours and 97 per cent after 16 hours (1 mg Na, room temperature precipitation). Recovery increases with the amount of sodium: 73, 87 and 91 per cent precipitated after 3 hours with 0.5, 1.0, and 2 mg Na (room temperature precipitation). Note coprecipitation of reagent when precipitation is made at room temperature.

[14] The reagent (potassium-free) solution should be clear.

[15] E. R. Caley and C. W. Foulk, *J. Am. Chem. Soc.* **51**, 1664 (1929); E. R. Caley, *ibid.* **52**, 1349, 4247 (1930).

[16] H. H. Barber and I. M. Kolthoff, *J. Am. Chem. Soc.* **50**, 1625 (1928).

special precipitation procedure must be followed to separate sodium quantitatively. Sodium can be precipitated completely by adding a comparatively large volume of a strong solution of the reagent, uranyl zinc acetate, saturated with the sodium salt, to a very small volume of the alkali metal solution. Especially with small amounts of sodium (1–2 mg) the precipitation is far from complete after several hours at room temperature (Fig. 32-1). Much better results are obtained by heating the reagent to 75°C before adding it to the sample solution (1 ml or less) and allowing it to cool to room temperature.[17] The precipitate is collected in a filter crucible and washed with successive small portions of the reagent solution saturated with the triple salt. The adhering solution of the reagent is next washed out with 95 per cent alcohol saturated with sodium uranyl zinc acetate. Finally, the precipitate is washed with acetone or ether to remove the alcohol, and a stream of air is drawn through the crucible. The precipitate is weighed in the air-dry state as

$$NaC_2H_3O_2 \cdot Zn(C_2H_3O_2)_2 \cdot 3\,UO_2(C_2H_3O_2)_2 \cdot 6\,H_2O$$

The very small calculation factor is of great advantage when a small amount of sodium must be determined but, on the other hand, it limits the weight of sample that can conveniently be taken when potassium and sodium are present in similar amounts, so that an aliquot part may then have to be taken.

The water content of the dried triple acetate is slightly greater than corresponds to the hexahydrate. The amount of foreign water ("zeolitic water," according to Schoorl[18]) depends upon the relative humidity of the atmosphere in which the precipitate is dried. Schoorl found the following water contents after drying at room temperature at various humidities:

Relative Humidity, %	*Molecules of* H_2O *in Precipitate*
0	6.17
0.34	6.45
0.68	6.53
0.94	6.64
0.98	6.80

The slightly high results due to foreign water are compensated by small solubility losses in the procedure. The hydrate water of the triple acetate is so firmly bound that it is not expelled when the precipitate is heated at 110°C.

Lithium interferes in the determination since it also forms a slightly soluble triple acetate[19] with the uranyl zinc acetate reagent, and it must therefore be removed by appropriate methods before the precipitation of the sodium. Potassium does not interfere except when present in relatively large amounts. If more than 50 mg of potassium chloride is present per milliliter of solution to be precipitated, the solution must be diluted until the concentration does not exceed this value. Since the reagent must be added in the ratio of at least 10 ml to 1 ml of the alkali solution to precipitate sodium quantitatively, such dilution is unpractical when very much potassium is present, and it is then better to remove most of the potassium by precipitation as the

[17] W. J. Ross, *Anal. Chem.* **37**, 168 (1965).

[18] N. Schoorl, *Rec. Trav. Chim.* **59**, 305, 729 (1940).

[19] By using a uranyl zinc acetate reagent of special composition, it is possible to determine lithium as $LiZn(UO_2)_3(C_2H_3O_2)_9 \cdot 6\,H_2O$ after separation from sodium by the extraction of the alkali chlorides with a suitable organic solvent. See C. C. Miller and F. Traves, *J. Chem. Soc.* **1936**, 1395.

perchlorate in a 70 per cent alcohol solution by the addition of ammonium perchlorate.[20] Ammonium salts, magnesium, calcium, and barium may be present in reasonable amounts; strontium leads to high results.[21] Organic acids such as oxalic and tartaric interfere, as do anions, such as phosphate,[22] which give precipitates with the reagent. Sulfate must be absent when potassium is simultaneously present, since potassium sulfate is only slightly soluble in the reagent. Sulfates can be converted into chlorides by treatment with barium chloride.

Sodium probably can be determined to within ~10 ppt by the triple acetate method if more than 1–2 mg is present.

Reagents: *Uranyl zinc acetate.* Prepare by mixing equal volumes of solutions *a* and *b*, allowing to stand 24 hr, and then filtering off the precipitate of sodium uranyl zinc acetate usually formed from traces of sodium in the reagents.[23] Preserve the solution in a Pyrex bottle.

Solution a. Mix 10 g of $UO_2(C_2H_3O_2)_2 \cdot 2\,H_2O$, 6 ml of 30 per cent acetic acid, and 50 ml of water, and warm to dissolve.

Solution b. Mix 30 g of $Zn(C_2H_3O_2)_2 \cdot 2\,H_2O$, 3 ml of 30 per cent acetic acid, and 50 ml of water, and warm to dissolve.

Ethyl alcohol saturated with sodium uranyl zinc acetate. Shake 95 per cent alcohol with pure sodium uranyl zinc acetate at room temperature until saturated.

Ether or acetone.

Procedure. Evaporate the sample solution (1–8 mg of Na, not more than 25 mg of K) to as small a volume as possible without separation or salts (1 ml or less). A 50-ml borosilicate beaker is a suitable precipitation vessel. Heat the zinc uranyl acetate reagent to 75°C and add 10 ml to the sample solution. Stir well and allow the mixture to stand covered for 1 hr to cool to room temperature. Stirring at intervals aids quantitative precipitation.

Filter off the precipitate in a filter crucible (sintered glass or porous porcelain), suck dry, and wash the precipitation beaker, crucible, and precipitate with five to eight 2-ml. portions of the reagent; drain the crucible well each time. Then wash five times with 2-ml portions of the alcohol wash solution and finally with a few small portions of ether or acetone. Draw air through the crucible for a few minutes to evaporate the ether or acetone. Place the crucible in the balance case, and weigh after 10 or 15 min. It is advisable to let the crucible stand on the balance pan for 10 min or so and then weigh again to be sure that the weight is constant. To find the weight of sodium multiply the weight of the precipitate by 0.01495.

When the highest accuracy is required, the recovery of sodium should be found by the use of the radioactive tracer ^{22}Na or ^{24}Na.

Sodium and potassium after J. Lawrence Smith decomposition

In the method to be described, the refractory material (silicate rocks and minerals, glass, ceramics, and the like) is decomposed and the nonalkali metals are separated from the alkali metals at the same time. Sodium and potassium (and the other alkali metals if present) are obtained in the form of the chlorides. After the weight of NaCl + KCl has been found, either K or Na (usually K) can be determined, and the other

[20] H. H. Barber and I. M. Kolthoff, *J. Am. Chem. Soc.* **51**, 3233 (1929).

[21] In the presence of much magnesium, calcium, etc., a double precipitation may be necessary. See C. C. Miller and F. Traves, *J. Chem. Soc.* **1936**, 1390. Consult this paper for a method of determining sodium directly in a silicate after decomposition by the J. L. Smith procedure.

[22] Phosphate can be removed by treating the solution with zinc carbonate. See O. R. Overman and O. F. Garrett, *Ind. Eng. Chem., Anal. Ed.* **9**, 72 (1937).

[23] If no precipitate is formed, add a small amount of sodium chloride to saturate the precipitant with the triple acetate. The solution should be filtered and used at approximately the same temperature (preferably 20°C) to avoid errors due to solubility changes.

found by difference; alternatively, both K and Na can be determined directly. When the alkali metals are to be determined by flame photometry a simpler method of decomposition—with hydrofluoric acid—may be advantageous (see p. 676).

In the J. Lawrence Smith method of decomposition,[24] sodium and potassium are brought into water-soluble form by heating the very finely divided sample with a mixture of calcium carbonate and ammonium chloride. The sintered mass is leached with water to extract the alkalies, and filtered. Silica, aluminum, magnesium, iron, etc., all remain undissolved, whereas much of the calcium goes into solution as hydroxide and chloride. Calcium is removed from the filtrate by precipitation with ammonium carbonate, sulfate is precipitated with barium chloride, and ammonium salts are removed by volatilization.

Procedure. The sample must be very finely ground, else it is certain that the decomposition will be incomplete.[25] Weigh out roughly 0.5 g or slightly more of 100-mesh silicate powder, and grind it exceedingly fine in an agate mortar. Transfer the powder to a weighing bottle, and weigh out 0.5 g of sample into a 100-ml silica dish or porcelain casserole. Mix 0.5 g of pure ammonium chloride very thoroughly with the powder, using a small agate pestle. Weigh out on a watch glass, to the nearest decigram, 4 g of calcium carbonate low in alkalies[26]; add three fourths of this, a little at a time, to the vessel containing the sample, mixing with extreme thoroughness.[27] Do not apply heavy pressure on the pestle, to avoid scratching the vessel. When mixing is complete, lay the pestle down in the watch glass containing the remainder of the calcium carbonate, and transfer the mixture with the aid of a spatula to a J. Lawrence Smith crucible[28] (Fig. 32-2) or an ordinary 30-ml platinum crucible. It is advisable to place a little calcium carbonate in the bottom of either crucible to prevent adhesion of sintered material at the end of the ignition. If a Smith crucible is used, first transfer the powder

Fig. 32-2. J. L. Smith crucible.

[24] J. L. Smith, *Am. J. Sci.* [2]**50**, 269 (1871).

[25] According to M. O. Lamar, W. M. Hazel, and W. J. O'Leary, *Ind. Eng. Chem., Anal. Ed.* **7**, 429 (1935), refractory samples can be decomposed in one treatment if reduced to 200-mesh powder; 100-mesh powder may not be completely decomposed.

[26] This is best prepared by precipitating a dilute solution of calcium chloride (from purest calcium carbonate and hydrochloric acid) at boiling temperature with excess ammonia and ammonium carbonate. The precipitate should be well washed with hot water and reserved especially for alkali determinations in silicates.

[27] Alternatively, the sample may be mixed with a previously prepared, finely ground mixture of calcium carbonate and ammonium chloride (obtained, for example, by grinding the two together in a ball mill). Some analysts prefer to use an agate mortar for mixing the sample with ammonium chloride and calcium carbonate.

[28] A Smith crucible is to be preferred for the decomposition because it allows the sample to be heated to a high temperature without danger of loss of alkalies by volatilization. However, an ordinary crucible will give satisfactory results if care is exercised in heating.

to a creased piece of glazed paper, and thus avoid the possibility of loss in filling the crucible. Pour the carbonate remaining on the watch glass into the dish, and rub it over the inside with the pestle to remove any small amount of adhering powder, and finally transfer the whole to the crucible.

If a Smith crucible is being used, insert it in a nearly vertical asbestos or transite board with a hole of suitable size to fit the crucible tightly about one fourth of the distance from the capped end. If an ordinary crucible is employed, place it in a hole in a square of asbestos. About one third of the crucible should extend below the asbestos. These asbestos shields must be used to prevent the possible loss of alkali chlorides by volatilization in the subsequent strong heating. Cap or cover the crucible, and place a 50-ml beaker containing water on the top of the cover of the ordinary crucible. Heat the crucible not above redness with a low flame for 10–15 min or until ammonia ceases to come off. Reduce the heat if ammonium chloride fumes appear. Then increase the heat until the lower third of the crucible is bright red. The finger-form crucible may be heated with the full flame of two Tirrill burners. Continue the heating for about $\frac{1}{2}$ hr. Allow the crucible and its contents to cool, add enough water to the residue to cover it, allow to stand for a few minutes, rinse off the cover, and then wash the residue into a silica dish or small porcelain casserole. Or, in the case of the Smith crucible, tap the inverted crucible to make the sintered mass fall into the dish. If the mass is not dislodged, add water to the crucible and allow to stand until cold; then transfer to dish and rinse the crucible. Add enough water to the residue in the dish to give a volume of 50–75 ml, and grind up the mass with a small agate pestle until no lumps remain. Then heat the liquid to boiling and keep at the boiling point for a few minutes. Decant the solution through a fast-filtering paper, leaving as much as possible of the residue in the dish; receive the filtrate in a 400-ml Pyrex beaker or, better, a large silica dish. Add 50 ml of hot water to the dish or casserole, heat to boiling, rub the insoluble mass with the pestle for some minutes, and again decant. Repeat the last operation three more times. Wash the paper with 50 ml of hot water or, preferably, 50 ml of hot saturated calcium hydroxide solution. Test the residue in the dish for complete decomposition by treating a portion with dilute hydrochloric acid. All should dissolve except possibly a few specks of iron ore minerals.

Add a few milliliters of concentrated ammonium hydroxide to the solution in the beaker, and then precipitate calcium by adding an excess of ammonium carbonate (2 g in 25 ml of cold water) to the warm solution. Heat the solution to boiling and allow the precipitate to settle. Filter the solution through paper into a large porcelain casserole or silica dish, wash the precipitate carefully with hot water,[29] and evaporate the filtrate to dryness on the steam bath.

Carefully heat the dried residue to expel ammonium salts, taking care not to volatilize the alkali chlorides by too strong heating.[30] The residue will be dark-colored, but this does not matter. Dissolve it in 5 ml of water, add one drop of 10 per cent barium chloride solution, and allow to stand for a few minutes. Then add a few drops of ammonium hydroxide and 1 ml of cold saturated ammonium carbonate solution, and evaporate to dryness on the steam bath. Take up the residue in a few milliliters of cold water, and filter through a small filter into a small platinum dish. Wash the dish and paper carefully with small portions (2–3 ml) of water containing a drop each of ammonium hydroxide and saturated ammonium carbonate solution in 20 ml; 20–25 ml of water will be sufficient for washing. Evaporate the filtrate and washings to complete dryness on the steam bath. Heat the dish very carefully (just below redness) to expel

[29] In the most exact work the calcium carbonate should be dissolved in hydrochloric acid and reprecipitated with ammonium carbonate to recover any small amounts of alkali chlorides coprecipitated.

[30] It will be easier to expel ammonium salts if the operation is carried out in a small, preferably platinum, dish. Therefore when the solution has been reduced to a small volume, transfer it to a platinum dish and complete the evaporation in it.

ammonium salts, keeping it covered at first to prevent possible loss of solid by decrepitation. Cool and weigh. Reheat to incipient fusion of the chlorides and weigh again, repeating the process, if necessary, until constant weight is obtained. Then dissolve the salts in 5–10 ml of hot water, filter to remove any insoluble material, and wash the paper, ignite it in the crucible, and weigh to obtain the weight of the insoluble material.

Determine potassium in the solution by the chloroplatinate or tetraphenylboron method. Find sodium by difference.[31] Apply the blank corrections. In running the blank, weigh out the same amounts of ammonium chloride and calcium carbonate as used in the determination itself, and follow all the steps of the determination as closely as possible.

Notes:

1. The active reagents in the J. L. Smith method of silicate decomposition are calcium oxide and calcium chloride. The ammonium chloride in the decomposition mixture furnishes calcium chloride by interaction with the calcium carbonate:

$$CaCO_3 + 2\,NH_4Cl \rightarrow CaCl_2 + 2\,NH_3 + CO_2 + H_2O$$

Calcium oxide (from the excess calcium carbonate) and chloride, acting jointly on the silicate at the high temperature, displace the alkalies from the silicate, forming alkali chlorides. The calcium-rich silicate formed is insoluble in water, so that when the sintered mass is leached with water only the alkali chlorides and the excess calcium chloride and hydroxide (from the calcium oxide) dissolve. Magnesium would be likely to dissolve to some extent were it not for the presence of calcium hydroxide in the solution, which precipitates magnesium hydroxide. Boron, molybdenum, and some fluorine accompany the alkali metals; when present in appreciable amounts, they must be removed by appropriate treatment.[32] Lithium is partially retained in the leached residue, and some of it is lost in the precipitation of calcium carbonate. Since the reaction mixture does not fuse, it is highly important to have the reactants present in a very fine form and intimately intermixed, else decomposition may be incomplete. When much iron is present in the sample, the mixture may partially fuse, and the mass will then be hard to remove from the crucible. This difficulty can be avoided by increasing the proportion of calcium carbonate in the reaction mixture.

Calcium chloride alone can be used to decompose silicates.[33]

It appears to be advantageous to use a mixture of calcium carbonate and calcium chloride directly in the silicate decomposition as proposed by van Tongeren,[34] who gives the following directions. The sample is weighed into an agate mortar, a solution of calcium chloride prepared by nearly neutralizing 0.5 g of pure calcium carbonate with hydrochloric acid is added, and the sample is ground for some minutes under this solution. Then there is added, with intervening mixing, most of a portion of 4 g of calcium carbonate. Alcohol is added gradually to give the mass a doughlike consistency. The mixture is then transferred by means of a spatula to the crucible (either the ordinary or J. L. Smith form), the bottom of which has been covered with a thin layer of calcium carbonate, and the mortar is rinsed with the remainder of the calcium carbonate. The crucible is placed in a drying oven; when the contents are dry, the

[31] Alternatively, sodium can be determined as described on page 667 and potassium found by difference, or the solution of the mixed chlorides can be divided into two parts and both sodium and potassium determined directly. Another procedure involves removal of platinum from the filtrate of the potassium chloroplatinate precipitate by reduction with alcohol and determination of sodium by the triple acetate method. See L. Lund, *J. Geol.* **56**, 490 (1948). The use of tetraphenylboron in the determination of potassium in silicates is treated in some detail by H. J. Cluley, *Analyst* **80**, 354 (1955).

[32] See G. E. F. Lundell and J. I. Hoffman, *Outlines of Methods of Chemical Analysis*, Wiley, New York, 1938, p. 76.

[33] E. Mäkinen, *Z. Anorg. Chem.* **74**, 74 (1912).

[34] W. van Tongeren, *Chem. Weekblad* **32**, 224 (1935).

crucible is heated directly to the decomposition temperature. The purpose of weighing the sample before grinding is to avoid possible weight changes due to uptake of water from the atmosphere and possible oxidation of ferrous iron. By grinding under liquid in the manner described, there is no danger of loss of powder, and much less time is required than when the sample is ground in the dry state.

Stevens[35] has proposed the use of barium chloride in place of ammonium chloride in the J. L. Smith method. The mixture may then be heated more rapidly, the alkali metal chlorides are obtained free of sulfate, and samples high in sulfate (alunite) do not yield a fused mass as in the usual method. The accuracy of the new procedure is equal to that of the old.

2. If the sample contains Rb and Cs, these metals will be included in the potassium chloroplatinate (or tetraphenylboron) precipitate, and can be determined therein by flame photometry. A corrected value for potassium is thus obtained.

If sodium is determined by difference and an appreciable amount of lithium is present in the mixed chlorides, the value for sodium will, of course, be high. Usually only traces of lithium will be present, but errors can be avoided by applying flame photometry (p. 663). Even if sodium is determined by the triple-acetate method, lithium will give rise to a positive error, as already pointed out (p. 668).

FLAME-PHOTOMETRIC METHODS

Under good working conditions, the precision of the flame-photometric determination of sodium and potassium may attain 5 ppt (standard deviation). The accuracy may be much poorer. It depends on the composition of the sample and on the composition of the solution sprayed into the flame. The effect of foreign substances varies with such factors as the flame temperature and the optical system (whether the instrument is a filter photometer or spectrophotometer, and if the latter, the resolving power). A relatively low-temperature flame suffices for the determination of the alkali metals, and is indeed preferable to avoid the excitation of other elements.

Free acids, introduced in decomposing the sample, usually depress the radiation of the alkali elements by anion interference, forming stable combinations that reduce the concentration of free ions in the flame. Foreign cations (e.g., NH_4^+, Al, Mg) can either repress or enhance the radiation of the alkali elements. The magnitude of the effect depends on the analytical conditions. If the composition of the sample is known, the error can be compensated by using calibrating solutions of the proper composition. This approach is not always feasible, and interfering cations may have to be removed.

The alkali metals exert radiation interference on each other. Thus sodium enhances potassium radiation (and potassium enhances sodium, but to a lesser extent). Again the magnitude of the effect varies with the analytical conditions. The use of suitable calibration curves is one way, albeit an inelegant one, of sometimes avoiding this difficulty. Another is by the use of a radiation buffer; e.g., when potassium is being determined, the sample solution is treated with so much sodium that the effect of the variable sodium content of the sample becomes insignificant. The use of an internal standard (p. 676) helps to reduce or eliminate errors due to radiation interference, as well as errors arising from flame variability due to changes in air or gas pressure and the functioning of the atomizer. Lithium is used as an internal standard for sodium and potassium, if the sample does not contain it in significant amounts.

Potassium

An example of an important application of the flame-photometric determination of an alkali metal is provided by the determination of potassium in rocks and minerals for geological dating. The potassium–argon method for the age determination of a geological

[35] R. E. Stevens, *Ind. Eng. Chem., Anal. Ed.* **12**, 413 (1940).

material is based on the determination of the $^{40}Ar/^{40}K$ ratio. Since the isotopic composition of natural potassium is essentially constant, the content of ^{40}K can be found from the potassium content of the sample. When the sample is rich in potassium (say 5 per cent K_2O), the chemical (gravimetric) determination after a J. Lawrence Smith decomposition probably gives results that are more accurate and precise than a flame-photometric determination. However, with suitable instrumentation and a sound procedure, flame photometry can give comparable accuracy and precision for such potassium contents, and superior accuracy and precision when the potassium content is less than ~1 per cent K_2O. Moreover, when a series of samples is to be analyzed, flame photometry is much more rapid than the chemical method. For these reasons, considerable work has been done to develop an accurate flame-photometric determination of potassium.

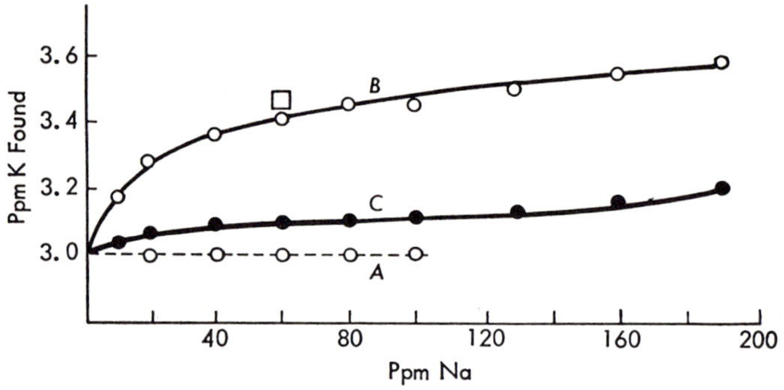

Fig. 32-3. Effect of Na on K emission (ammonium sulfate-carbonate solution, 3.00 ppm K in solution). From Cooper, Reference 36.

A. Beckman DU with flame-photometer attachment, oxyhydrogen flame
B. Perkin Elmer (model 146), single channel, propane-air flame
C. Same as *B* except double channel, lithium as internal standard
□ = Atomic absorption

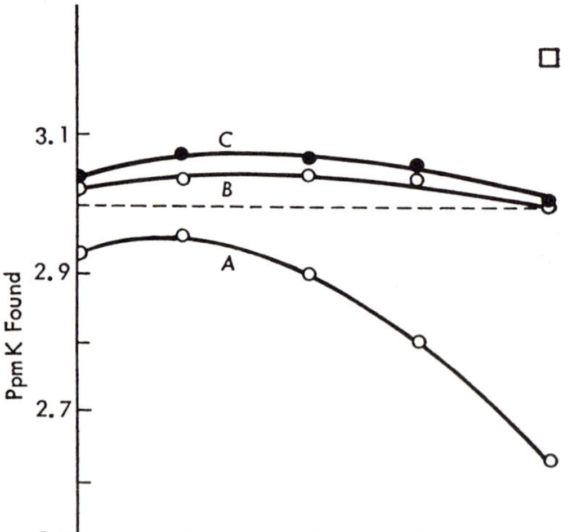

Fig. 32-4. Effect of Mg on K emission, 3.00 ppm K (ammonium sulfate-carbonate solution). (See Fig. 32-3 for significance of *A*, *B*, *C*, and □.)

Figures 32-3 to 32-7, from a recent study of the flame-spectrophotometric determination of potassium in geological materials, illustrate the effect of some foreign substances under different conditions of determination.[36] An examination of these figures makes it clear that the accuracy attainable is strongly dependent upon the instrument used and the procedure applied. The practice of flame photometry is not merely a matter of spraying a solution into a flame and reading a meter.

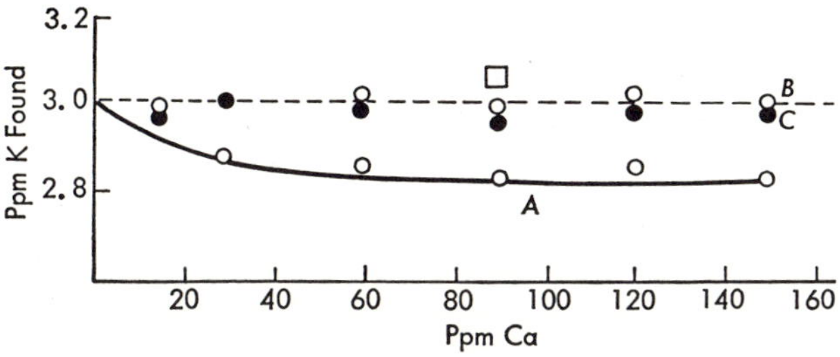

Fig. 32-5. Effect of Ca on K emission in H_2SO_4 solution, 3.00 ppm K. (See Fig. 32-3 for significance of *A*, *B*, *C*, and □.)

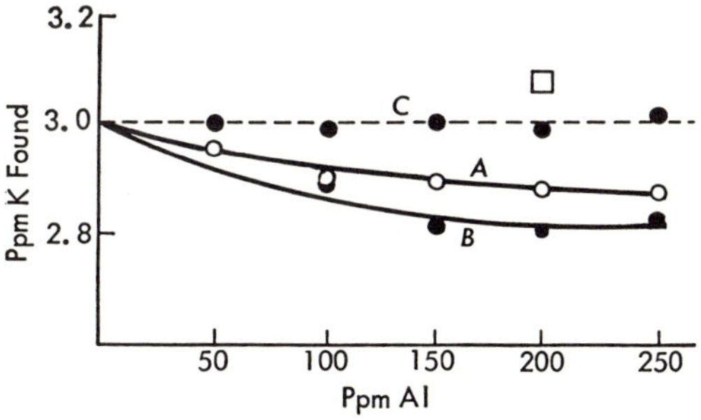

Fig. 32-6. Effect of Al on K emission in H_2SO_4 solution, 3.00 ppm K. (See Fig. 32-3 for significance of *A*, *B*, *C*, and □.)

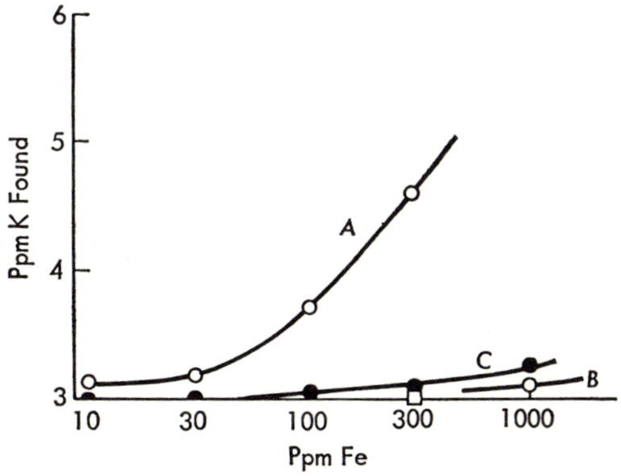

Fig. 32-7. Effect of Fe (III) on K emission in H_2SO_4 solution, 3.00 ppm K. (See Fig. 32-3 for significance of *A*, *B*, *C*, and □.)

[36] J. A. Cooper, *Geochim. Cosmochim. Acta* **27**, 525 (1963). Only a portion of the experimental results are presented above. For effect of Rb and Fe on K determination, see J. A. Cooper, I. D. Martin, and M. J. Vernon, *ibid.* **30**, 197 (1966). The mutual interference effects of the alkali elements and H^+ on the flame-photometric determination of the alkalies have been studied by E. L. Grove, C. W. Scott, and F. Jones, *Talanta* **12**, 327 (1965).

The following procedure[37] very nearly meets all the requirements of a potassium determination for K–Ar dating without separations. Lithium is used as an internal standard and sodium as a buffer to eliminate the enhancement of potassium radiation by sodium. The added sodium also practically eliminates the enhancement due to magnesium, but not that due to much iron. The presence of lithium, rubidium, and casium requires modification of the procedure.

PROCEDURE

Weigh a suitable amount of finely ground sample (biotite and muscovite 0.16 g, plagicolase and hornblende 0.30 g, K-feldspar 0.12 g) into a 70-ml platinum dish and add 6.0 ml of 50 per cent sulfuric acid and 25 ml of concentrated hydrofluoric acid. Evaporate down slowly on a water bath, add 10 ml more of hydrofluoric acid, and evaporate down again. Add 2 ml of concentrated nitric acid, mix, and leave on the bath for 5 min. Place the dish on a hot plate and heat to the first appearance of sulfuric acid fumes. Since the sulfuric acid concentration must remain constant for accuracy, do not prolong the fuming. Cool, add 50 ml of water, and evaporate down on a water bath. Add 50 ml of water, warm to bring all soluble matter into solution, and transfer quantitatively to a 250-ml beaker. Bring the solution to boiling and allow to cool.

Add to the beaker 100 ml of a solution containing 1250 ppm Na and 500 ppm Li. Transfer the solution quantitatively to a 250-ml volumetric flask and make up to volume. Prepare standard potassium solutions at convenient concentration intervals, e.g., 8, 10, . . . , 40, 50 ppm K, containing the same concentrations of sulfuric acid, sodium, and lithium as the diluted sample solution.

Carry out the photometry in a Perkin-Elmer or equivalent double-channel flame photometer with a propane–air flame at $1\frac{1}{4}$ and 10 lb/in.^{2} pressure, respectively. Adjust the wavelength selector to give maximum response for the potassium doublet 766.5 and 769.9 mμ and set the instrument to give almost full-scale deflection for the higher of the two standards bracketing the sample. Obtain, for each solution, the intensity of the potassium radiation compared to that of lithium. Make at least five runs in alternate directions.

After each sample run, extinguish the flame, and remove and rinse the jet to wash away salt encrustations. Remove and clean the burner daily.

PROBLEMS

1. What is the minimum volume of chloroplatinic acid solution containing 0.10 g of platinum per milliliter that should be taken to assure the conversion of a 0.200-g mixture of sodium and potassium chlorides into the chloroplatinates? *Ans.* 3.34 ml.
2. A determination of potassium in a 0.500-g sample by the chloroplatinate method yielded the result 20.27 per cent K_2O. If the solubility loss in the procedure was 1.0 mg of K_2PtCl_6, find the corrected K_2O value. *Ans.* 20.31 per cent.
3. The potassium chloroplatinate precipitate from a 0.7504 g sample was reduced to metallic platinum by magnesium. If 0.1122 g of platinum was obtained, what was the percentage of potassium in the sample?
4. A 0.2046-g mixture of anhydrous sodium and potassium chlorides was dissolved in water and a one-tenth aliquot taken for the determination of sodium by the triple-acetate method. If the sodium zinc uranyl acetate precipitate obtained weighed 0.2117 g, find the percentage of sodium and potassium in the mixture. *Ans.* 15.47 per cent Na, 31.82 per cent K.

[37] Cooper, Ref. 36. The decomposition of the sample essentially follows J. F. Evernden and J. R. Richards, *J. Geol. Soc. Australia* **9**, 1 (1962).

5. A sample consisting of the mixed chlorides of sodium and potassium weighing 0.3575 g yielded 0.1162 g of $KClO_4$. What is the percentage of Na_2O and K_2O in the sample?
6. If a weighed precipitate of sodium zinc uranyl acetate contained seven molecules of water of crystallization instead of the assumed six molecules, what would be the error in the sodium determination? *Ans.* +12 ppt.
7. In the determination of potassium by the chloroplatinate method, the use of 80 per cent alcohol by volume is called for. How many milliliters of 95 per cent (by volume) alcohol are required for the preparation of 1 liter of this alcohol solution?
8. Calculate the maximum absolute and relative solubility loss of potassium (with 12 mg of K) in the tetraphenylboron procedure on page 667.
9. An analyst is given an aqeuous solution of unspecified composition with a request for the determination of its potassium content to within ~ 10 ppt if the K concentration is > 100 mg/liter and to within ~ 100 ppt if the concentration is < 100 mg/liter. Outline a reasonable procedure.
10. With the aid of information obtained from the chemical literature when necessary, outline the essential steps of a suitable procedure for the determination of (a) K in seawater; (b) Na in rainwater; (c) K in a glass containing boron; (d) Na in reagent-grade $CaCO_3$; (e) Li in spodumene; (f) Cs in pollucite; (g) Rb and Cs in lepidolite.

PART FOUR

Theory of Volumetric Analysis

CHAPTER 33 *General Discussion of Titrimetry*

In *volumetric* or *titrimetric*[1] analysis, a substance is determined by *titration*, which is the process of adding a reagent solution (*titrant*) of known concentration or reacting strength (*titer*) until, ideally, all the substance has reacted, i.e., until an equivalent amount of the titrant has been added. The titrant solution is usually called the standard solution; it is added from a buret. In visual *end-point determination* (see below) titrant is added until a suitable *indicator* shows a change in color or fluorescence or light scattering. In physicochemical end-point determination (p. 769) titrant is added in increments and some physical property of the solution is measured after each addition. The line or curve representing the change of the physical property has a different slope before and after the end point and the latter is found graphically as the intersection of the two lines. In same methods the value of the property measured changes abruptly at the end point, e.g., the potential of an indicator electrode in a potentiometric titration.

The end point as a rule does not coincide with the *equivalence point*. The difference between the amount of titrant corresponding to the end point and that corresponding to the equivalence point represents the *titration* error. When the characteristics of the system titrated and the indicator are known, it is often possible to calculate the titration error. Also it can often be determined experimentally, by preparing a solution of the same composition as the titrated solution would have at the equivalence point and adding titrant (or sometimes a solution of known concentration of substance titrated) until the color of the indicator is the same as at the end point in the titration. For example, in the titration of ammonia with hydrochloric acid, methyl red or bromcresol green can be used as indicators (see p. 789). At the equivalence point the composition of the solution is that of pure ammonium chloride. This salt is easily obtainable in the pure state and the titration error can be determined with the aid of its solution.

Ordinarily, a proper amount of the indicator is added to the system to be titrated, and the color change is observed in the solution. Such indicators are called *internal* or *inside indicators*. In certain cases an interaction between the indicator and the system titrated takes place before the end point is reached. Under these conditions the end point appears too early, and an *external* or *outside* indicator is used. During the titration a drop of the reaction mixture is tested on a plot plate or on filter paper with a suitable reagent. The use of external indicators is not recommended.

[1] The time-honored term is volumetric analysis, but the more accurate, and increasingly favored, term is titrimetric, since this type of analysis is characterized by titration. Although the amount of reagent equivalent to the titrated substance is usually found from the volume of added solution (hence the name *volumetric*), it can be obtained (and is, in the most precise analyses) by weighing the solution.

CLASSIFICATION OF VOLUMETRIC METHODS

Methods based on combination of ions or molecules

Acid–base titrations

$$H^+ + OH^- \rightleftharpoons H_2O$$
$$H^+ + A^- \rightleftharpoons HA$$
$$B^+ + OH^- \rightleftharpoons BOH$$

The above reactions are used in the volumetric determination of substances behaving like acids or bases. Acids and salts of very weak bases can be titrated with a standard base; bases and salts of very weak acids can be titrated with a standard acid. In the latter case the term *acidimetry*, in the former *alkalimetry* are sometimes used.

Precipitation titrations

$$Ag^+ + Cl^- \rightleftharpoons AgCl$$
$$3\,Zn^{++} + 2\,K_4Fe(CN)_6 \rightleftharpoons K_2Zn_3[Fe(CN)_6]_2 + 6\,K^+$$

One of the most common reagents is silver nitrate. Titrations with silver nitrate are often called *argentimetric methods*.

Complexation titration

$$H_2Y^{--} + Cu^{++} \rightleftharpoons CuY^{--} + 2\,H^+$$

(Y^{4-} refers to ethylenediamine tetraacetate, pp. 744 and 803.) This type of titration is often referred to as *chelatometric*. Nonchelatometric complexing titrations are also applied:

$$2\,Cl^- + Hg^{++} \rightleftharpoons HgCl_2$$
$$2\,CN^- + Ag^+ \rightleftharpoons Ag(CN)_2^-$$

Methods based on combination of molecules. Such titrations are practically limited to titrations of organic compounds.

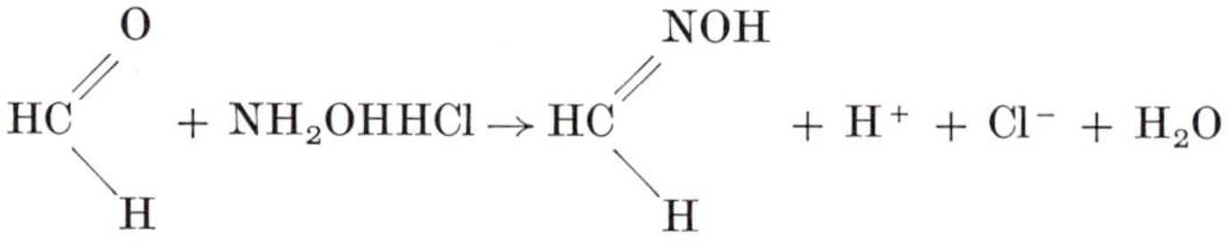

formaldehyde oxime

Strictly speaking the reagent is not a neutral molecule but the cation $NH_2OH_2^+$. Similarly, the reactions between a carbonyl group and bisulfite and of an ester with hydroxyl ions are those between a neutral molecule and an ion:

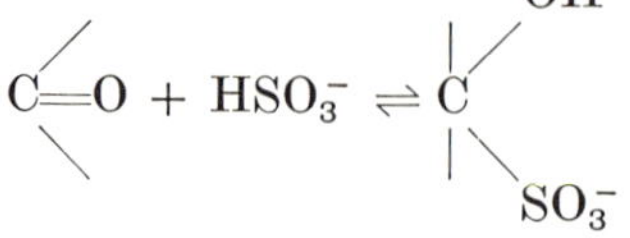

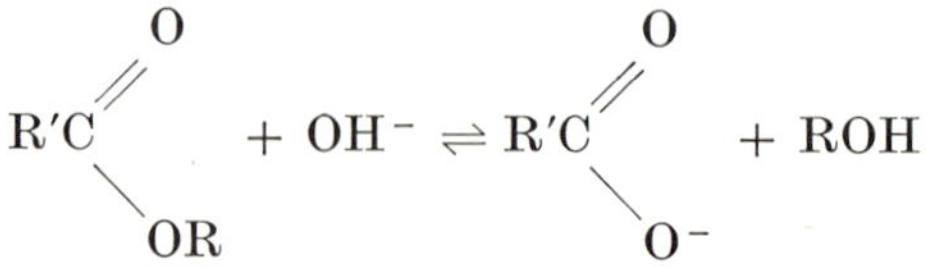

Reactions of this kind usually are slow. As a rule an excess of reagent is added and after completion of the reaction the excess of reagent is found by titration, as in the saponification of an ester (*indirect* titration). Sometimes a product formed in the reaction is titrated, for example, the acid formed in the reaction between a carbonyl group and hydroxyl ammonium chloride. In such instances the concentration of the added reagent need not be known.

Methods based on transfer of electrons

Oxidation–reduction titrations

$$Ce^{4+} + Fe^{++} \rightleftharpoons Ce^{3+} + Fe^{3+}$$

$$I_2 + 2\,S_2O_3^{--} \rightarrow S_4O_6^{--} + 2\,I^-$$

Limitations of volumetric methods

Not every chemical reaction can serve as the basis of a direct titration. The following conditions at least must be fulfilled:

1. The reaction between the substance to be determined and the reagent should occur with great speed. This condition is satisfied by most reactions used in acidimetry and alkalimetry and usually also those in which slightly dissociated compounds and complex compounds are formed. In precipitation reactions the precipitate does not always separate instantaneously (see Chapter 10). The silver halides are formed almost instantaneously; microcrystalline precipitates such as barium sulfate and lead sulfate separate more slowly, especially from dilute solutions. In such cases addition of alcohol may have a favorable effect, because as a rule it decreases the solubility of slightly soluble inorganic salts and consequently increases the speed of precipitation.

Various oxidation–reduction reactions do not take place instantaneously. The addition of a suitable *catalyst* (see p. 822) increases the speed of reaction. When the reaction velocity is slow or when the end point cannot be detected in a simple way, a known excess of the reagent is often added, and the excess is back-titrated with a suitable standard solution after the first reaction is complete.

2. The substance to be determined should react stoichiometrically with the reagent, and no side reactions should occur.

Sometimes it is possible to develop an *empirical* method when side reactions occur. The conditions under which the procedure is carried out must then be exactly defined and the directions followed very closely. Such a method is used, for example, in the determination of reducing sugars with an excess of an alkaline copper solution. Quite generally, however, such empirical methods are not to be recommended.

3. Other substances present in the solution should not react or interfere with the main reaction. A reducing agent frequently reacts slowly with atmospheric oxygen so that its solution is stable for a comparatively short time only. However, in the titration of reducing agents it is often found that the main reaction taking place promotes the reaction between the reducing substance and oxygen. We then speak of an *induced* reaction. A solution of sulfite or bisulfite, for example, is oxidized by air, this oxidation being induced by the reaction between sulfite or bisulfite and iodine.

4. An indicator should be available for the detection of the end point. If no suitable indicator is available, the latter very often can be found by the application of physico-chemical methods. As such, the change of the *potential of a certain electrode* during the course of a titration (*potentiometric titration*), the change of the *electrical conductivity* of the solution during the titration (*conductometric titration*), the change of the *current upon electrolysis* of the solution being titrated (*amperometric titration*), the change of light absorption of an indicator (*spectrophotometric titration*), or the change in temperature (*thermometric titration*) will serve for location of the equivalence point (p. 769).

Equivalent weight. Normality

The *equivalent weight*, the *gram equivalent*, or the *equivalent* of a substance is the weight in grams *which in its reaction* corresponds to a gram-atom of hydrogen, or of hydroxyl, or half a gram-atom of oxygen, or a gram-atom of a univalent ion. The equivalent weight is not a constant as the molecular weight is but depends upon the reaction in which the standard solution is used. This will be clear from the following examples.

In acid–base titrations it is always very simple to derive the equivalent weight. In the titration of monoprotic bases and acids an equivalent is equal to a mole:

$$H^+ + Cl^- + Na^+ + OH^- \rightarrow Na^+ + Cl^- + H_2O$$

With polyprotic acids and bases the equivalent is a variable. In aqueous solution sulfuric and oxalic acids, for example, are always titrated as diprotic acids, and therefore the equivalent of these is one half of the molecular weight. Phosphoric acid, however, can be titrated as a mono- or a diprotic acid, and in the presence of much calcium chloride as a tribasic acid:

$$H_3PO_4 + OH^- \rightleftharpoons H_2PO_4^- + H_2O \quad (1)$$

$$H_3PO_4 + 2\,OH^- \rightleftharpoons HPO_4^{--} + 2\,H_2O \quad (2)$$

$$H_3PO_4 + 3\,OH^- \rightleftharpoons PO_4^{3-} + 3\,H_2O \quad (3)$$

In the titration according to Eq. (1) the equivalent is equal to 1 mole, in (2) to $\frac{1}{2}$ mole, and in (3) to $\frac{1}{3}$ mole.

In precipitation and complex-formation analysis also, we may read directly from the reaction equation the relation between the molar and the equivalent weight. If cyanide is titrated with silver according to the Mohr method, the equivalent of the former is equal to 1 mole:

$$Ag^+ + CN^- \rightleftharpoons AgCN$$

If the cyanide is titrated according to the method of Liebig, the end point is reached when the following reaction is complete:

$$Ag^+ + 2\,CN^- \rightleftharpoons Ag(CN)_2^-$$

In this case 1 mole of silver corresponds to 2 equivalents of cyanide.

In oxidation–reduction reactions the equivalent of a substance is that part of 1 mole which in its reaction corresponds to the removal of one half of a gram-atom of oxygen or the combination with 1 gram-atom of hydrogen or any other univalent element. Quite generally the easiest way to find the equivalent is to determine the change in the *state of oxidation* in the reaction.

Example: In the titration of ferrous iron to ferric iron with any oxidizing agent, the state of oxidation of the iron changes from 2 to 3:

$$Fe^{++} \rightarrow Fe^{3+}$$

The equivalent of ferrous iron in this case is equal to 1 mole. If metallic iron, however, is determined by oxidation to the ferric state:

$$Fe^0 \rightarrow Fe^{3+}$$

the equivalent is equal to $\frac{1}{3}$ mole.

Copper can be titrated iodometrically by making use of the following reactions:

$$2\,Cu^{++} + 4\,I^- \rightarrow 2\,CuI + I_2 \qquad (4)$$

$$2\,S_2O_3^{--} + I_2 \rightarrow S_4O_6^{--} + 2\,I^- \qquad (5)$$

Considering that one copper ion reacts with two iodide ions [Eq. (4)], one might at first infer that the equivalent of copper is $\frac{1}{2}$ mole. However, in this titration use is made of the fact that the liberated iodine is measured (with sodium thiosulfate), and accordingly it is seen that 1 mole of copper corresponds to 1 equivalent of iodine and that the equivalent of copper therefore is 1 mole. Equation (5) shows us that the equivalent weight of sodium thiosulfate in iodometric titrations is the molecular weight.

The same considerations hold for oxidizing agents containing oxygen:

$$MnO_4^- + 8\,H^+ + 5e \rightarrow Mn^{++} + 4\,H_2O \qquad \text{Eq. wt.} = \frac{MnO_4}{5}$$

$$MnO_4^+ + 4\,H^+ + 3e \rightarrow MnO_2 + 2\,H_2O \qquad \text{Eq. wt.} = \frac{MnO_4}{3}$$

$$Cr_2O_7^{--} + 14\,H^+ + 6e \rightarrow 2\,Cr^{3+} + 7\,H_2O \qquad \text{Eq. wt.} = \frac{Cr_2O_7}{6}$$

$$VO_4^{3-} + 6\,H^+ + e \rightarrow VO^{++} + 3\,H_2O \qquad \text{Eq. wt.} = \frac{VO_4}{1}$$

The equivalent of a substance is not a fixed quantity but depends upon the reaction in which it takes part. One cannot answer offhand, for example, the following question: What is the equivalent weight of potassium iodate? If potassium iodate is determined by precipitation as silver iodate:

$$IO_3^- + Ag^+ \rightarrow AgIO_3$$

The equivalent weight $= KIO_3/1$. If it is reduced to iodine and this is titrated with thiosulfate:

$$IO_3^- + 5\,I^- + 6\,H^+ \rightarrow 3\,I_2 + 3\,H_2O$$

the equivalent weight $= KIO_3/6$. If it is used as an oxidizing agent in 6 N hydrochloric acid, the iodate is reduced to ICl:

$$IO_3^- + 2\,I^- + 3\,Cl^- + 6\,H^+ \rightarrow 3\,ICl + 3\,H_2O$$

the equivalent weight $= KIO_3/4$. In the last case the equivalent weight of iodide is I/2, since it is oxidized according to the electrochemical equation:

$$I^- \rightarrow I^+ + 2e$$

$$(ICl \rightarrow I^+ + Cl^-)$$

The number of equivalents of a substance dissolved in 1 liter of solution determines the *normality* (N) of the solution. If 1 equivalent is present in 1 liter, the solution is 1 N; if 0.05 equivalent is present, the solution is 0.05 N or $\frac{1}{20}$ N, etc.

The expression of the concentration of a solution in terms of normality may be ambiguous. If we are dealing with normal solutions of hydrochloric, sulfuric, and oxalic acids, of silver nitrate, iodine, and arsenic trioxide, the normality refers to the most common reactions in which these substances take part.

However, if the normality of a solution of ferric chloride, potassium ferrocyanide, or iodic acid is given, it is impossible to state what the molecular concentration is, unless the reaction is given to which the normality refers. A 1 N ferric chloride solution may be 1 M:

$$Fe^{3+} \rightarrow Fe^{++} + e$$

or $\frac{1}{3}$ M:

$$Fe^{3+} + 3\,OH^- \rightarrow Fe(OH)_3$$

A 1 N potassium ferrocyanide solution may be 1 M:

$$Fe(CN)_6^{4-} \rightarrow Fe(CN)_6^{3-} + e$$

or $\frac{1}{4}$ M:

$$Fe(CN)_6^{4-} + 2\,Pb^{++} \rightarrow Pb_2Fe(CN)_6$$

In such cases it is best to indicate on the label of the bottle the molarity of the solution with the normality, for example, in the following way:

$$1\,N\ K_4Fe(CN)_6\left(\frac{M}{1}\right) \qquad \text{or} \qquad 1\,N\ K_4Fe(CN)_6\left(\frac{M}{4}\right)$$

Sometimes the reacting strength of a standard solution is expressed as the weight of the titrated substance equivalent to 1 ml of the standard solution. This is its *titer*.

Calculations in volumetric analysis

The calculations of volumetric analysis are always very simple, because at the equivalence point the number of equivalents of standard solution used is equal to the number of equivalents of the substance titrated.

Examples: A weighed amount, a grams, of the substance to be determined is dissolved and titrated. V_1 milliliters of standard solution of normality N are used to reach the end point ($\sim$ equivalence point). What is the percentage of the substance titrated in the unknown?

One liter of the standard solution contains N equivalents. Therefore in V_1 milliliters there are $V_1(N/1000)$ equivalents. If the equivalent weight of the substance titrated is E, a grams of the unknown contain $V_1NE/1000$ g of the substance titrated, and the percentage is

$$\frac{V_1NE}{1000}\,\frac{100}{a} = \frac{V_1NE}{10a}$$

Often a weighed amount, a grams, of the substance to be titrated is dissolved and made up to a certain volume in a volumetric flask, and an aliquot part of this solution is then pipetted out. If, for example, 500 ml of solution is prepared and 25 ml is titrated, the above expression becomes

$$\text{percentage} = \frac{V_1NE}{10a}\,\frac{500}{25} = \frac{V_1NE}{10a}\cdot 20$$

In the standardization of a solution (see the next paragraph) an exactly weighed amount say b grams, of the standard substance with the equivalent weight E_s is dissolved and titrated with the solution. Y_1 milliliters of the latter is used to reach the end point. What is the normality, N? b grams of the standard substance corresponds to b/E_s equivalents. Therefore,

$$\frac{V_1N}{1000} = \frac{b}{E_s}$$

$$N = \frac{1000b}{V_1E_s}$$

If the normality, N_1, of a solution is determined by titrating a certain volume, V_1 milliliters, with a solution of a known normality, N_2, and V_2 milliliters of the latter is used to reach the end point, we have the relation

$$V_2N_2 = V_1N_1$$

$$N_1 = N_2\frac{V_2}{V_1}$$

The student should not make the calculations more involved than is necessary. If, for example, 25 ml of a ferrous iron solution requires 20 ml of 0.125 N potassium permanganate for titration, there is present in the former $20 \times 0.125 \times (55.84/1000)$ g of ferrous iron. The authors have found that a student sometimes calculates the number of grams of permanganate used in the titration, then states that 1 mole of $KMnO_4$ corresponds to 5 moles of Fe^{++}, etc. This complicated way of computation, although leading to the same result, is entirely unnecessary and is to be considered wrong. The concept of normality has been introduced because it allows the calculation to be made with equivalents and not with grams and moles.

Standard solutions

Primary standard solutions have an accurately known concentration of active substance (reagent). They are prepared by weighing out a known fraction of the equivalent weight of a substance of high degree of purity, called a *primary standard substance*. Fortunately, quite a few primary standard substances are available, although few are suitable for the preparation of a primary standard solution. Examples of primary standard solutions used as titrants are silver nitrate, potassium dichromate, and arsenic trioxide. A *secondary standard solution* is a solution whose concentration is obtained by standardization, preferably using a primary standard substance or a highly accurate gravimetric method. For example, a standard sodium hydroxide solution is of the secondary type; it is standardized with primary standards such as potassium biphthalate or benzoic acid. A standard solution of hydrochloric acid, not containing any other acid, usually is also of the secondary type. It is standardized by using the primary standard sodium carbonate or by determining chloride gravimetrically as silver chloride. As a matter of fact, the concentration of a more concentrated acid solution can be determined in this way; this solution can then be diluted to give a secondary standard solution of known concentration. *Secondary standard substances* are of a lower degree of purity than the primary ones. The concentration of the active agent of a secondary standard substance is found by comparison against a primary standard or by the use of a suitable analytical method of known accuracy. Secondary substances are frequently used in practice; they must be kept under conditions such that their composition does not change (hygroscopicity, air oxidation). Preferably they should have a high equivalent weight, so that weighing errors are negligibly small.

Standard solutions must be stored in such a way that they do not change their concentration on standing (see practical part).

Standardization. The standardization of a solution in general requires a higher degree of accuracy than an ordinary volumetric determination. An error in the standardization recurs in all the analyses that are made with the solution and therefore will add itself to the other errors.

For ordinary purposes one may be content with an accuracy of 0.1 per cent in the standardization, and it should be realized that this accuracy is not always easy to attain.

The following conditions must be fulfilled in a standardization:

1. A suitable primary standard substance must be available. Such a substance should fulfill the following requirements: It must be easy to obtain, to purify and dry, and to preserve in a pure state. The substance must not be so hygroscopic that it takes up water during weighing. It must be capable of being tested for impurities by qualitative tests of known sensitivity. In general, the total amount of impurities in a primary standard should not be greater than 0.01–0.02 per cent. Each newly prepared primary standard must be subjected to severe tests of purity.

It is desirable that standard substances have a high equivalent weight in order that the weighing error may be insignificant (see below). The reaction with the standard solution must be stoichiometric. The error in the determination of the end point (titration error or indicator blank) must be negligible or easy to determine exactly experimentally.

The most important primary standard substances are considered in the practical part of this book. In certain cases it is permissible to use a secondary standard substance if the amount of the active material is exactly known.

Various attempts have been made to find *universal* primary standard substances that can be used for the standardization of different solutions employed in the various groups of volumetric analysis. Potassium acid iodate, $KH(IO_3)_2$, approaches this ideal more or less. It can be used for the determination of the titer of standard base:

$$KH(IO_3)_2 + NaOH \rightarrow KIO_3 + NaIO_3 + H_2O$$

and of sodium thiosulfate,

$$KH(IO_3)_2 + 10\,KI + 11\,HCl \rightarrow 6\,I_2 + 6\,H_2O + 11\,KCl$$

It can be reduced with pure sulfurous acid to iodide; if the excess of the reducing acid is boiled off, a solution of iodide of exactly known concentration is obtained which may be used for the standardization of silver nitrate and permanganate solutions (p. 828).

Oxalic acid, $(COOH)_2 \cdot 2\,H_2O$, can be used as a primary standard for the standardization of bases and permanganate. Sodium oxalate serves as a primary standard for the standardization of acids and permanganate. For the standardization of acids it is transformed into sodium carbonate by heating:

$$Na_2C_2O_4 \rightarrow Na_2CO_3 + CO$$

Arsenic trioxide is used for the standardization of iodine, potassium, permanganate, and ceric sulfate solutions.

2. The weight of the standard substance used should not be too small. Two weighings are involved in every determination. If the error of an ordinary analytical weighing is taken as 0.1 mg, then for every standardization at least 200 mg of standard substance must be taken, if the error of weighing is not to exceed 0.1 per cent. For this reason standard substances with a high equivalent weight should be selected, if possible.

Potassium iodate, as a primary standard, has the disadvantage of a small equivalent weight (35.66). If, for example, 0.1 N thiosulfate is to be standardized with it and if 40 ml of the latter is used, the corresponding weight of potassium iodate is only about

0.14 g. In such instances it is more advantageous to work with a solution of exactly known content. In the preparation of a liter of 0.1 *N* potassium iodate, 3.566 g is used. An error in weighing amounting to 0.2 mg corresponds to a relative error of 0.006 per cent and hence does not come into consideration. The allowable error of a liter volumetric flask is about 0.2 ml, corresponding to 0.02 per cent, and that of a 50-ml pipet amounts to 0.07 per cent. Therefore, if we use 50 ml, the instrumental errors will be smaller than 0.1 per cent. In the standardization of very dilute solutions (0.025 *N* or weaker), an accuracy of 0.1 per cent is no longer obtainable by weighing out the primary standard directly, and here an exactly measured volume of a solution of exactly known content of the primary standard should be used.

3. The volume of standard solution used in the standardization should not be too small, if the result is not to be too greatly affected by the errors of reading and drainage. If each reading of the buret is subject to an error of 0.01 ml, and if there is a drainage error of 0.02 ml in addition, then the error that is possible in every determination amounts to 0.04 ml. If 40 ml of standard solution is used, this corresponds to an accuracy of 0.1 per cent. The student therefore should never use a volume of, say, 10 ml in a standardization.

4. In general, the titration must be made directly to the end point. A back-titration with another standardized solution increases the possibility of error.

5. In general, the standardization of a solution against another standardized liquid should be avoided, because in every single standardization errors are inherent that may become additive. Thus, sodium hydroxide, for example, should not be standardized with a standardized solution of hydrochloric acid but with a primary standard substance. However, it is desirable to test independently standardized solutions against each other as, for example, 0.1 *N* hydrochloric acid against 0.1 *N* alkali.

6. Every standardization should be based on at least three parallel determinations. The results should agree within 0.1–0.2 per cent.

Any titration, as a rule, involves a titration error (see p. 681), which can be corrected for by the determination of the indicator blank. As a rule, this correction is not applied. If the solution is standardized under the same conditions (amount of reagent used and final volume) as those which obtain in the actual determination, the titration error is practically eliminated. Considered from this point of view, it is preferable to standardize the solution by comparing with a known weight of the substance being determined at the same final volume and under the same experimental conditions.

A note on -imetry and -ometry. In this chapter and other chapters of this text we meet with a number of terms having one or the other of these postfixes. Usually they are meant to express the same thing and are only variant spellings, but this is not always true and confusion may sometimes arise. A few observations seem called for. The combining form *-metry* (Greek *metria*) means the art, process, or science of measuring something that is specified, and is joined to the first element of a word by *-i-* or *-o-*. The suffix *-i-* is used chiefly in terms taken from, or imitative of, Latin, the suffix *-o-* chiefly in terms taken from, or imitative of, Greek. They have the meaning, when combined with -metry, of *of* or *by*, and this ambiguity may give rise to difficulties. The meaning of such terms are colorimetry, nephelometry, and compleximetry (or complexometry) is clear enough. (Nephelometry is preferred to nephelimetry, because *nephelos* is Greek; compleximetry is preferred to complexometry, because *complexus* is Latin). But what does *acidimetry* mean? Is it measurement *of* or *by* an acid? In Webster's *New International Dictionary* (2nd ed.) we find the definition "measurement of the strength of acids, or of the quantity of free acid present in a solution." However, the majority of chemists, especially analytical chemists, in English-speaking countries (not in France) use acidimetry in the opposite sense as titration *with* an acid. In this book the common (English) chemical usage has been followed. To be sure, one can always say titration with an acid or of an acid, as required, and this had best be done when there is a possibility of ambiguity or misinterpretation. A somewhat awkward situation has arisen with respect to *iodimetry* and *iodometry*. Some persons (including the authors of the third edition of this book) used iodimetry to mean titration with iodine, iodometry to mean titration of iodine. This practice has not been widely followed, however. If acidimetry is taken to mean titration with an acid, why should not iodimetry mean titration with an iodine solution? The analogy is not complete; acidometry is not used for titration of an acid. It seems best to abandon the distinction between iodimetry and iodometry. Iodimetry (from the neo-Latin *iodum* through the Greek *iodes*) may be used to indicate both types of titration. We shall use the term *iodine titrations* in these senses.

PROBLEMS

1. Balance the following equations, and derive the equivalent weight (fractions of a mole) of the substance first mentioned:
   $H_2C_2O_4 + 2\ NaOH \rightarrow$
   $Na_2CO_3 + HCl \rightarrow$
   $Na_2CO_3 + 2\ HCl \rightarrow$
   $K_2CrO_4 + 2\ AgNO_3 \rightarrow$
   $K_2CrO_4 + \cdots + KI + \cdots + HCl \rightarrow$
   $H_3PO_4 + NaOH \rightarrow$
   $H_3PO_4 + 2\ NaOH \rightarrow$
   $HgCl_2 + 2\ HCN \rightarrow Hg(CN)_2 +$
   $2\ AsH_3 + 12\ AgNO_3 + 3\ H_2O \rightarrow As_2O_3 + 12\ HNO_3 + 12\ Ag$
   $\cdots K_4Fe(CN)_6 + \cdots KMnO_4 + \cdots H_2SO_4$
   $KCNS + AgNO_3 \rightarrow$
   $\cdots HCNS + 3\ O + H_2O \rightarrow HCN + H_2SO_4$
   $3\ Mn^{++} + 2\ MnO_4^- + 2\ H_2O \rightarrow 5\ MnO_2 + 4\ H^+$
   $As_2O_3 + 2\ I_2 + 2\ H_2O \rightarrow$
   $MnO_2 + 4\ H^+ + 2\ I^- \rightarrow$
   $\begin{cases} HCN + Br_2 \rightarrow BrCN + H^+ + Br^- \\ (BrCN + 2\ I^- \rightarrow I_2 + CN^- + Br^-) \end{cases}$
   $VO_4^{3-} + Fe^{++} + 6\ H^+ \rightarrow VO^{++} + \cdots$
   $\cdots K_2Cr_2O_7 + \cdots + KI + \cdots + HCl \rightarrow \cdots Cr^{3+} + \cdots$
2. A solution of hydrochloric acid is standardized with pure sodium carbonate: $Na_2CO_3 + 2\ HCl \rightarrow 2\ NaCl + H_2O + CO_2$. 0.2000 g of Na_2CO_3 requires 30.00 ml of the acid. What is the normality of the hydrochloric acid? *Ans.* 0.1258 *N*.
3. To 25.00 ml of 0.1000 *N* $KMnO_4$ an excess of potassium iodide and acid is added. The liberated iodine is titrated with sodium thiosulfate. Of the latter 26.00 ml is required. What is the normality of the thiosulfate? *Ans.* 0.0962 *N*.
4. If 50.00 ml of a sulfuric acid solution yields 0.3000 g of barium sulfate, what is the normality of the acid? *Ans.* 0.0514 *N*.
5. If 0.500 g of impure calcium carbonate is dissolved in 25.00 ml of 0.5100 *N* hydrochloric acid and the excess of acid (CO_2 does not interfere) is titrated back with 6.50 ml of 0.4900 *N* NaOH, what is the percentage of calcium carbonate in the sample? *Ans.* 95.75 per cent.
6. What should be the normality of a silver nitrate solution in order that each milliliter may correspond to 1 mg of Cl? *Ans.* 0.0282 *N*.
7. How many milliliters of 0.2000 *N* silver nitrate are required to precipitate the chloride and the bromide quantitatively in a mixture of 0.2000 g of KCl and 0.2000 g of KBr? *Ans.* 21.81 ml.
8. How much water must be added to 1 liter of 0.110 *N* hydrochloric acid to obtain 0.100 *N* acid, assuming no expansion or contraction on mixing? *Ans.* 100 ml.

CHAPTER 34 *Theory of Acid-Base Titrations*

For a general discussion of the calculations of pH in solutions of acids, bases, salts, and mixtures, the student is referred to Chapter 3. Applications of these derivations are made in this chapter.

ACID–BASE INDICATORS

The indicators used in acidimetry and alkalimetry are either weak organic acids (indicator acids) or weak organic bases (indicator bases), the acid form of which has a color different from that of the conjugate base form. Using the Brønsted concept of acids and bases we denote the acid form of an indicator by HI, which can have a charge of zero, or one or more positive or negative charges. The conjugate base, denoted by I, has a charge one less positive than that of HI. At a given pH the equilibrium between HI and I is given by

$$\mathrm{HI} \rightleftharpoons \mathrm{H}^+ + \mathrm{I} \tag{1}$$

and

$$\frac{[\mathrm{H}^+][\mathrm{I}]}{[\mathrm{HI}]} = K_{\mathrm{HI}} \tag{2}$$

in which K_{HI} is the indicator constant.

$$\frac{[\mathrm{I}]}{[\mathrm{HI}]} = \frac{[\text{form with alkaline color}]}{[\text{form with acid color}]} = \frac{K_{\mathrm{HI}}}{[\mathrm{H}^+]} \tag{3}$$

[1]

When $[\mathrm{H}^+]$ has the same numerical value as K_{HI} the ratio of [I] to [HI] becomes equal to unity, meaning that 50 per cent of the indicator is present in the acid form and 50 per cent in the alkaline form. From Eq. (3) it is seen that at any hydrogen-ion concentration both the forms I and HI are present. The human eye has a limited ability to detect either of the two colors when one of the two predominates. If the acid form of the indicator is red, for example, and the alkaline form is yellow and we compare a solution consisting entirely of the red form with one consisting of a mixture of 99 per cent red and 1 per cent yellow, the eye will not be able to detect any difference in color between the two solutions. Generally, this will still be true when the ratio of [HI] to [I] is *approximately* equal to 10. In any solution in which the ratio [I] : [HI] is smaller than 0.1, we obtain by visual

[1] The equilibrium should be expressed in terms of activities instead of concentrations. In general, we can neglect activity coefficients in the application of indicators in acid–base titrations. In the colorimetric pH determination, activity coefficients must be considered.

observation the impression that the indicator is present completely in the acid form. Similarly, we get the impression that the indicator is present completely in the alkaline form when the ratio [I] to [HI] is greater than 10. (These limits of $\frac{1}{10}$ and 10 are somewhat arbitrary, depending on the color sensitivity of the eye and the particular indicator, but are quite close to the truth.)

From the above, then, it is clear that we see only the acid color of the indicator when

$$\frac{[\mathrm{I}]}{[\mathrm{HI}]} < 0.1$$

and only the basic color when

$$\frac{[\mathrm{I}]}{[\mathrm{HI}]} > 10$$

Between the ratios 0.1 and 10 we observe an intermediate color. An indicator, therefore, does not change its color abruptly from that of the acid form to that of the alkaline form at a definite hydrogen-ion concentration but changes gradually over a certain range of $[\mathrm{H}^+]$. If we assume that the two limits are determined by the ratio $[\mathrm{I}]/[\mathrm{HI}] > 0.1 < 10$, it is found [Eq. (3)] that

$$\frac{[\mathrm{I}]}{[\mathrm{HI}]} = \frac{1}{10} = \frac{K_{\mathrm{HI}}}{[\mathrm{H}^+]} \qquad \text{or} \qquad [\mathrm{H}^+] = 10K_{\mathrm{HI}} \tag{4}$$

$$\frac{[\mathrm{I}]}{[\mathrm{HI}]} = 10 = \frac{K_{\mathrm{HI}}}{[\mathrm{H}^+]} \qquad \text{or} \qquad [\mathrm{H}^+] = \tfrac{1}{10}K_{\mathrm{HI}} \tag{5}$$

At $[\mathrm{H}^+] = 10K_{\mathrm{HI}}$ the indicator is present in the acid form, and at $[\mathrm{H}^+] = \frac{1}{10}K_{\mathrm{HI}}$ it is present in the alkaline form. The indicator changes color between

$$[\mathrm{H}^+] = \underset{(\textit{acid color})}{10K_{\mathrm{HI}}} \quad \text{and} \quad \underset{(\textit{alkaline color})}{\tfrac{1}{10}K_{\mathrm{HI}}}$$

or, expressed in pH, between

$$\mathrm{pH} = \mathrm{p}K_{\mathrm{HI}} \pm 1$$

in which $\mathrm{p}K_{\mathrm{HI}}$ is $-\log K_{\mathrm{HI}}$.

This range of pH over which the color change of the indicator takes place is called the *color-change interval* of the indicator. The color-change interval is determined experimentally with the aid of buffer solutions; its exact limits will depend more or less upon the subjective judgment of the observer. Table 34-1 indicates that the color-change interval of methyl orange is located between pH 3.1 (red) and 4.4 (yellow-orange). This means then that the pronounced color change takes place between these two pH values. However, another observer might fix these limits, for example, between pH 2.9 and 4.3.

An enormous number of compounds with indicator properties are found in nature and among the products of the laboratory.[2] A selected list of indicators for volumetric purposes (and also for the colorimetric determination of pH) is given in Table 34-1.

[2] Compare I. M. Kolthoff, *Acid-Base Indicators* (trans. by C. Rosenblum), Macmillan, New York, 1937; pp. 103 ff.

TABLE 34-1. **Color-Change Interval of Selected Indicators and Values of K_{HI} at Room Temperature**

Scientific Name	*Trade Name*	*Solvent*	*Nature of Uncharged Form*[a]	*Acid Color*	*Base Color*	*Interval in* pH	K^0_{HI}
Diphenylamino *p*-benzene sodium sulfonate	Tropeoline OO	Water	b	Red	Yellow	1.3– 3.0	2.0
Thymolsulfonphthalein	Thymol blue	Water	a	Red	Yellow	1.2– 2.8	1.65
Dimethylaminoazobenzene	Methyl yellow	90% alcohol	b	Red	Yellow	2.9– 4.0	3.2
Dimethylaminoazobenzene sodium sulfonate	Methyl orange	Water	b	Red	Yellow-orange	3.1– 4.4	3.4
Tetrabromphenolsulfonphthalein	Bromphenol blue	Water	a	Yellow	Purple	3.0– 4.6	4.1
Tetrabrom-*m*-cresolsulfonphthalein	Bromcresol green	Water	a	Yellow	Blue	3.8– 5.4	4.9
Dimethylaminoazobenzene sodium carboxylate	Methyl red (sodium)	Water	b	Red	Yellow	4.2– 6.2	5.0
Dichlorosulfonphthalein	Chlorophenol red	Water	a	Yellow	Red	4.8– 6.4	6.25
Dibromthymolsulfonphthalein	Bromthymol blue	Water	a	Yellow	Blue	6.0– 7.6	7.30
Phenosulfonphthalein	Phenol red	Water	a	Yellow	Red	6.4– 8.0	8.0
Dimethyldiaminophenazin chloride	Neutral red	70% alcohol	b	Red	Yellow-brown	6.8– 8.0	7.4
m-Cresolsulfonphthalein	Cresol purple	Water	a	Yellow	Purple	7.4– 9.0	—
Thymolsulfonphthalein	Thymol blue	Water	a	Yellow	Blue	8.0– 9.6	—
Phenolphthalein		70% alcohol	a	Colorless	Red-violet	8.0– 9.8	—
Thymolphthalein		90% alcohol	a	Colorless	Blue	9.3–10.5	—
p-Nitroaniline azosodium salicylate	Alizarine yellow	Water	a	Yellow	Violet	10.1–12.0	—

[a] a, indicator acid; b, indicator base.

In Fig. 34-1 the color-change intervals of various indicators are represented graphically. It is seen that the above set of indicators covers the pH range between 1 and 12. The color-change intervals in Table 34-1 and the K^0_{HI} values refer to a temperature of 25°C. When HI is uncharged or has a negative charge K^0_{HI} hardly changes with the temperature. Thus the color-change interval of the sulfophthaleins and the phthaleins hardly changes between 0 and 100°C. On the other hand, when HI has a positive charge, as with methyl orange, methyl yellow, and methyl red, the color-change interval changes to lower pH values; for example, for methyl orange from 3.1 to 4.4 at 25°C to 2.5–3.7 at 100°C; for methyl yellow from 2.9 to 4.0 at 25°C to 2.4–3.5 at 100°C, and for methyl red from 4.2 to 6.3 at 25°C to 4.0–6.0 at 100°C.

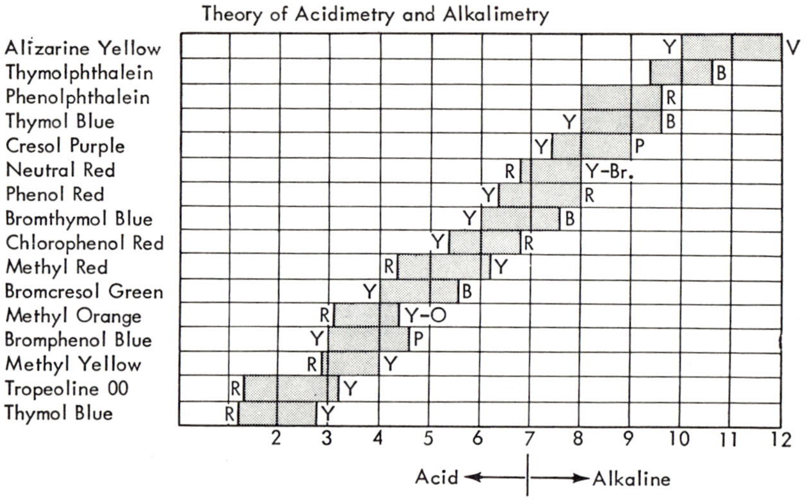

Fig. 34-1. Color-change intervals of some common indicators.

Preparation of indicator solutions

The stock solutions of the indicators contain 0.5–1 g of indicator per liter of solvent. Tropeoline OO, methyl orange, methyl red sodium salt, and alizarine yellow can be dissolved in water. Methyl yellow, neutral red, phenolphthalein, and thymolphthalein are dissolved in 70–90 per cent alcohol.

The sulfonphthaleins are soluble in water if enough sodium hydroxide is added to neutralize the sulfonic group; 100 mg of these indicators are ground in an agate mortar with the specified volume of 0.1 N sodium hydroxide solution and then diluted with water to 100 ml. The following volumes of 0.1 N sodium hydroxide are required for 100 mg of the sulfonphthaleins: thymol blue, 2.15 ml; bromphenol blue, 1.5 ml; bromcresol green, 1.45 ml; chlorophenol red, 2.35 ml; bromthymol blue, 1.6 ml; phenol red, 2.85 ml; cresol purple, 2.65 ml.

Color change and structural change of indicators

The color change of an indicator is accompanied by a structural change. Below the structural changes of some of the more representative indicators are given (for further details see the literature).[3]

[3] See Ref. 2.

Phenol red:

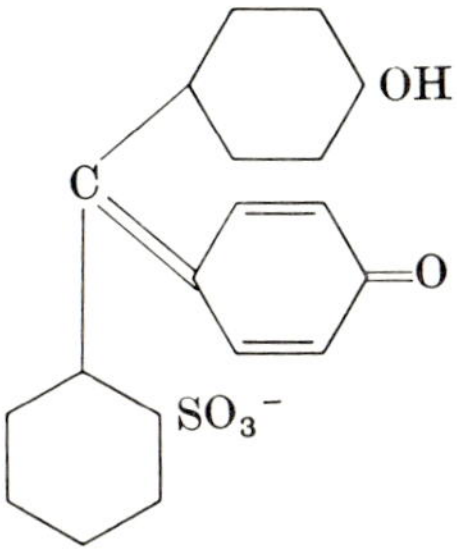

yellow
(*acid form*)

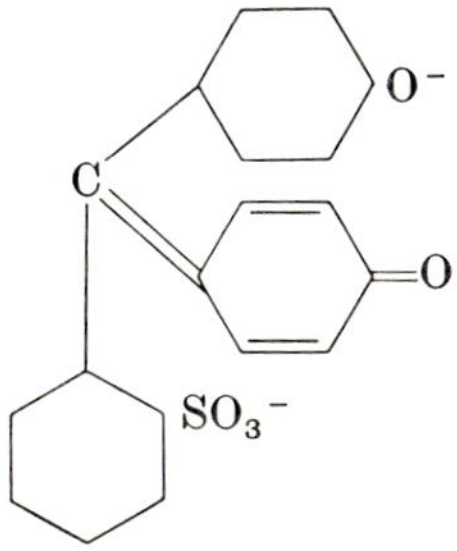

red
(*alkaline form*)

Phenolphthalein:

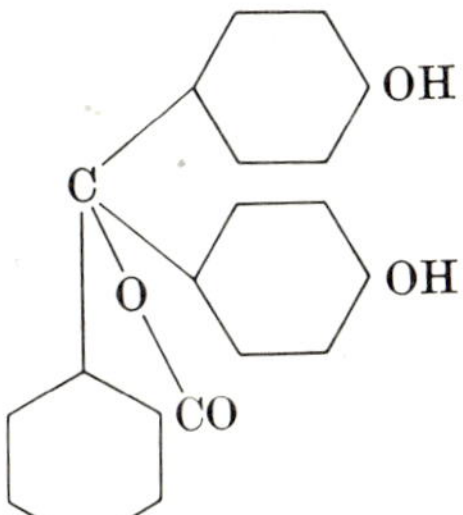

colorless
(*acid form*)

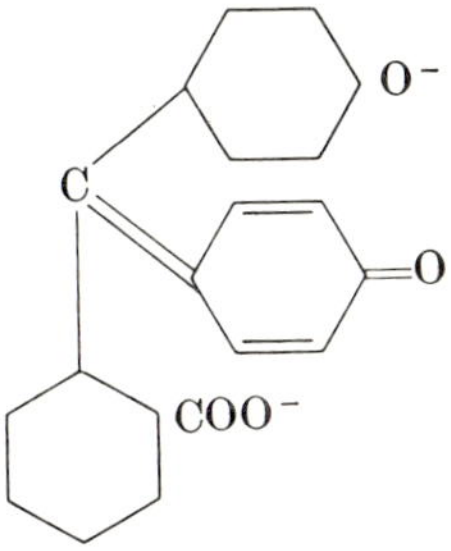

red-violet
(*alkaline form*)

Methyl orange:

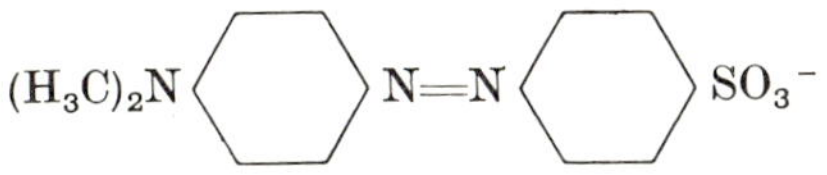

yellow-orange
(*alkaline form*)

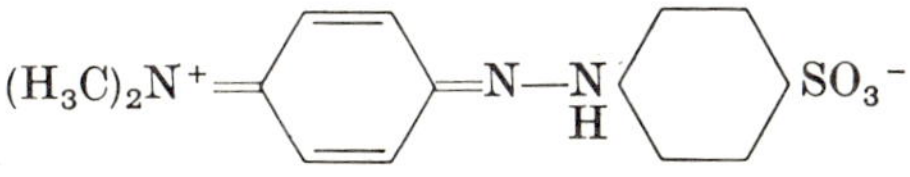

red
(*acid form*)

Mixed indicators

When the color change at the end point is not sharp, a mixture of two indicators or of an indicator and a dye can often be used to advantage. If a proper selection is made and the indicator and the dye are mixed in the proper ratio, the mixed indicator will give a sharp and pronounced color change at a definite pH and may be used in cases in which the color change is not easily seen with an ordinary indicator. A few of these mixed indicators are given below.

Methyl orange–indigo carmine: A solution containing 1 g of methyl orange and 2.5 g of indigo carmine in 1 liter of water. The alkaline color is green. The mixture exhibits a grayish neutral shade at pH 4.0 and changes toward violet in more acid solutions. The stock solution should be kept in brown bottles, best in the dark.

(a) *Bromcresol green* (0.1 per cent)–(b) *methyl red* (0.1 per cent): 3 parts of (a) are mixed with 2 parts of (b). Pronounced color change at pH 5.1; acid color, red; alkaline color, green.

(a) *Phenolphthalein* (0.1 per cent)–(b) *methylene green* (0.1 per cent): 1 part of (a) mixed with 2 parts of (b). Color in acid medium, green; at pH 8.8, pale blue; at pH > 9.0, violet.

(a) *Cresol red* (0.1 per cent)–(b) *thymol blue* (0.1 per cent): 1 part of (a) mixed with 3 parts of (b). Acid color, yellow; alkaline color, violet; pink at pH 8.2; violet at pH 8.4.

(a) *Cresol red* (0.1 per cent)–(b) *Methyl red* (0.1 per cent)–(c) *Methylene blue* (0.2 per cent): 16 ml of (a) + 20 ml of (b) + 4 ml of (c). At pH 8.8, violet; pH 8.6, gray-violet; pH 8.35, gray-green; pH 8.25, green.[3]

Neutralization of strong acids with strong bases. Neutralization curves

Since strong acids and bases are completely ionized, it is simple to calculate the change in the hydrogen-ion concentration and the pH during the course of the neutralization of a strong acid with a strong base, or vice versa.

If, for example, we start with 100 ml of 1 *N* hydrochloric acid and add 90 ml of 1 *N* sodium hydroxide solution, there will be 10 ml of 1 *N* hydrochloric acid left in a volume of 190 ml. Therefore,

$$[H^+] = \frac{10}{190} \times 1 = 5.3 \times 10^{-2}$$

and pH = 1.3. When 100 ml of 1 *N* sodium hydroxide has been added, the equivalence point is reached; and the pH should be 7 if no impurities are present. With 101 ml of 1 *N* sodium hydroxide there will be 1 ml of 1 *N* sodium hydroxide in excess in a volume of 201 ml; therefore, $[OH^-] = 5 \times 10^{-3}$, $[H^+] = 2 \times 10^{-12}$, and pH 11.7, etc.

In Table 34-2 the change in the pH during the neutralization of 100 ml of 1 *N*, 0.1 *N*, and 0.01 *N* solutions of a strong acid, respectively, after successive additions of sodium hydroxide of corresponding strength is given. The data are plotted in

TABLE 34-2. **Titration of 100 ml of HCl with NaOH at Room Temperature**

NaOH *Added*, ml (= % *Acid Neutralized*)	*N Solution*		0.1 *N Solution*	0.01 *N Solution*
	$[H^+]$	pH	pH	pH
0	1	0	1.0	2.0
50	3.3×10^{-1}	0.5	1.5	2.5
90	5×10^{-2}	1.3	2.3	3.3
98	1×10^{-2}	2.0	3.0	4.0
99	5×10^{-3}	2.3	3.3	4.3
99.8	1×10^{-3}	3.0	4.0	5.0
99.9	5×10^{-4}	4.3	4.3	5.3
E. P.[a] *100.0*	10^{-7}	*7.0*	*7.0*	*7.0*
100.1	2×10^{-11}	10.7	9.7	8.7
100.2	1×10^{-11}	11.0	10.0	9.0
101	2×10^{-12}	11.7	10.7	9.7
102	1×10^{-12}	12.0	11.0	10.0
110	2×10^{-13}	12.7	11.7	10.7

[a] *E. P.*, equivalence point.

[3] S. A. Jimeno and A. A. Chéoux, *Chim. Anal.* (*Paris*) **46**, 196 (1964).

Fig. 34-2; the curve obtained is called the *neutralization curve.* From the quantitative viewpoint we are especially interested in the changes in the pH near the equivalence point. For this reason, the part of the curve of Fig. 34-2, 2 per cent before and 2 per cent after the equivalence point, is reproduced on an enlarged scale in Fig. 34-3.

The color-change intervals of some of the ordinary indicators are also indicated on this graph. Inspection of Fig. 34-3 or an analysis of the data in Table 34-2 reveals: (1) the best titration conditions and (2) the error that is made if an indicator is used which does not change color in the neighborhood of the equivalence point. Let us consider methyl orange (or bromphenol blue) and phenolphthalein (or thymol blue) as indicators covering the most extreme changes of pH.

Methyl orange changes color between pH 3.0 (red) and 4.2 (yellow-orange), and phenolphthalein between pH 8.0 (colorless) and 9.8 (violet-red).

Titration of 1 *N* acid with 1 *N* sodium hydroxide. After the addition of 99.8 ml of 1 *N* sodium hydroxide, the pH is 3.0. Methyl orange is then still present in the red form. With 99.9 ml of sodium hydroxide the pH is 3.3, and therefore most of the indicator is still in the acid form. Upon addition of the last 0.1 ml of sodium hydroxide, the pH changes abruptly from 3.3 to 7. Therefore a very sharp color change is obtained with methyl orange. If sodium hydroxide is added until the alkaline color just appears, there is no indicator blank. At the equivalence point phenolphthalein is still present in the acid form; addition of a very small amount of 1 *N* sodium hydroxide will change the pH abruptly to the alkaline side. Thus the addition of 0.1 ml of 1 *N* base changes the pH from 7 to 10.7; at the latter pH phenolphthalein or thymol blue is present completely in the alkaline form. Summarizing then, it may be said that in the titration of 1 *N* hydrochloric acid with 1 *N* sodium hydroxide any indicator having a color-change interval lying between that of methyl orange and phenolphthalein (and even thymolphthalein) can be used. The indicator blanks are negligibly small,and the color changes are very pronounced.

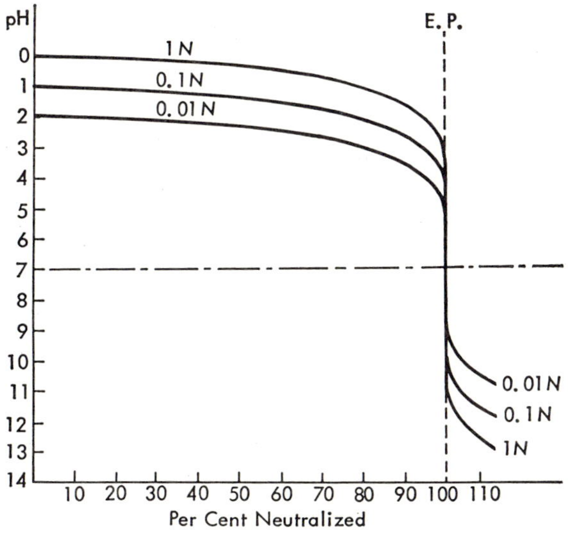

Fig. 34-2. Neutralization curves of 1 *N*, 0.1 *N*, and 0.01 *N* hydrochloric acid with sodium hydroxide.

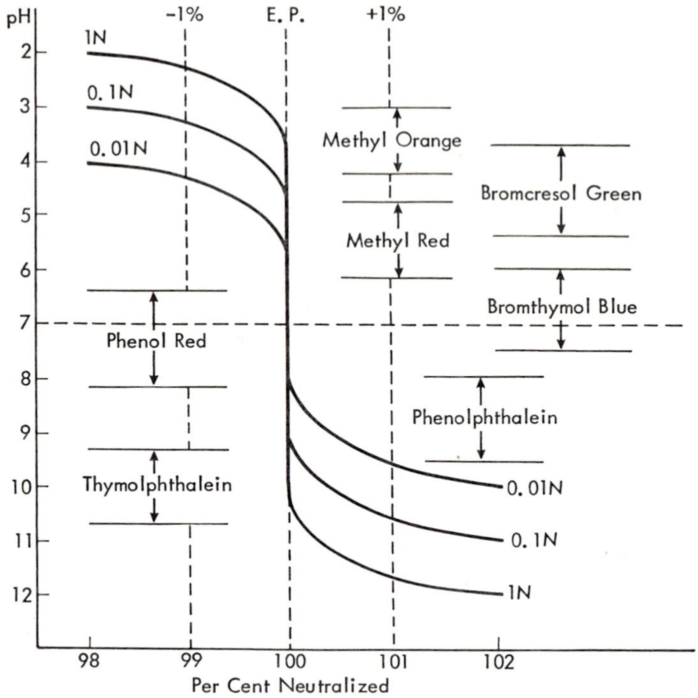

Fig. 34-3. Neutralization curves of 1 *N*, 0.1 *N*, and 0.01 *N* hydrochloric acid with 1 *N*, 0.1 *N*, and 0.01 *N* sodium hydroxide. From 2% before to 2% after the equivalence point.

Titration of 0.1 *N* hydrochloric acid with 0.1 *N* sodium hydroxide. With more dilute acids and bases the pH changes near the equivalence point become less abrupt. After the addition of 98 ml of 0.1 *N* sodium hydroxide to 100 ml of 0.1 *N* hydrochloric acid, the pH has become equal to 3. On further addition of sodium hydroxide, methyl orange begins to change color from red to the intermediate color. After 99.8 ml of sodium hydroxide has been added, the pH is 4; and most of the methyl orange is present in the alkaline form. If the titration is stopped here, we are still 0.2 ml from the equivalence point; the error then amounts to 0.2 per cent, and the indicator blank is 0.2 ml. (In this case the indicator blank must be added to the titration figure.) In the titration of 0.1 *N* hydrochloric acid with 0.1 *N* sodium hydroxide, it is advisable to add sodium hydroxide until the indicator is present completely in the alkaline form. For most practical purposes the error is then negligibly small. In the reverse titration of 0.1 *N* sodium hydroxide with 0.1 *N* hydrochloric acid with methyl orange as indicator, an excess of acid must be added to produce a change from yellow to pink (the first change in color should be considered as the end point). If this color change takes place at pH 4.0, the error again is 0.2 per cent, and the indicator blank is 0.2 ml (which in this case is to be subtracted from the amount of acid used). This indicator blank is experimentally determined in the following way.

The solution is titrated in the ordinary manner until the color change of methyl orange to the first pink is noticed. Then in another vessel the same amount of indicator as used in the titration is added to 200 ml of 0.05 *N* sodium chloride solution, and standard hydrochloric acid is added from a buret until the color of the solution matches that of the titrated solution at the end point.

It is evident that methyl red will give a much sharper color change than methyl orange in the titration of 0.1 *N* solutions. After the addition of 99.9 ml of 0.1 *N* sodium hydroxide the pH is 4.3, and methyl red is still present in the acid form. Upon the addition of the last 0.1 ml of sodium hydroxide, the pH changes abruptly from 4.3 to 7, the color of the indicator changes from red to yellow. The end-point error is also negligibly small with phenolphthalein as indicator. With 0.1 ml of 0.1 *N* sodium hydroxide in excess, the pH is 9.7, and most of the phenolphthalein is present in the alkaline form.

The above interpretation is correct only when we work with solutions which do not contain carbon dioxide. Under practical conditions this is never the case, since standard sodium hydroxide as a rule contains a small amount of carbonate. Carbon dioxide is such a weak acid (see p. 787) that it has no effect upon the color of methyl orange; however, it is titrated as a univalent acid with phenolphthalein as indicator. In working with carbon dioxide-free solutions, the difference in the amounts of sodium hydroxide used with methyl orange and with phenolphthalein as indicators is not larger than 0.15–0.2 ml of 0.1 *N* sodium hydroxide, when 100 ml of 0.1 *N* hydrochloric acid are titrated. A larger difference will be due to carbon dioxide present in the solution after the end point to methyl orange has been reached.

Titration of 0.01 *N* hydrochloric acid with 0.01 *N* sodium hydroxide. It is now a simple matter to specify what indicators should be used in the neutralization of 0.01 *N* hydrochloric acid with 0.01 *N* sodium hydroxide. Methyl orange does not give a sharp color change because the indicator is present in the alkaline form before the equivalence point is reached. After the addition of 90 ml of 0.01 *N* sodium hydroxide the pH is 3.3; upon further addition of base the color of the indicator changes gradually to yellow-orange. After the addition of 98–99 ml of 0.01 *N* sodium hydroxide, the indicator is practically completely present in the alkaline form. The end-point error then amounts to 1 or 2 per cent. In this case it is much better to use methyl red and to titrate to the yellow color. Other indicators which change color between a pH of 5 and 9, such as bromthymol blue, phenol red, or phenolphthalein, can be used as well. Quite generally, when very dilute solutions of a strong acid must be titrated with a strong base (or vice versa), it is preferable to carry out the titration at boiling temperature, using phenolphthalein, phenol red, bromthymol blue, or methyl red as indicator. Any carbon dioxide present is expelled by boiling the slightly acid solution. The latter should be cooled before the final color adjustment.

Neutralization of weak acids with strong bases or weak bases with strong acids

In the following discussion the titration of 0.1 *N* solutions of weak acids with strong bases or of 0.1 *N* solutions of weak bases with strong acids only will be considered. It is a simple matter to make similar derivations for other concentrations. Whereas in the titration of 0.1 *N* solutions of a strong acid with a strong base we can select any indicator changing color between a pH of 4 and 10, we are much more limited as regards choice of indicator in the titration of weak acids (with strong bases) or of weak bases (with strong acids).

This is evident from a study of the neutralization curves. In Chapter 3 equations for the calculation of the pH of a solution of a weak acid (p. 44), of mixtures of the weak acid and its salt (p. 51), and of solutions of the salt of the acid with a strong base (p. 48) were derived. These equations are used in the calculation of the neutralization curves.

In Table 34-3 the neutralization curves of 0.1 *N* acetic acid ($K_a = 1.8 \times 10^{-5}$), and of a weaker acid with an ionization constant of 10^{-7} with 0.1 *N* sodium hydroxide, are given.

TABLE **34-3. Neutralization Curves of 100 ml 0.1 *N* Acetic Acid ($K_a = 1.8 \times 10^{-5}$) and of 100 ml 0.1 *N* HA ($K_a = 10^{-7}$) with 0.1 *N* Sodium Hydroxide at Room Temperature ($K_w = 10^{-14}$)**

0.1 *N* NaOH *Added*, ml (= % *Acid Neutralized*)	0.1 *N Acetic Acid*, pH	0.1 *N Acid*, $K_a = 10^{-7}$, pH
0	2.87	4.0
10	3.80	6.0
50	4.75	7.0
90	5.70	8.0
99.0	6.75	9.0
99.8	7.45	9.57
99.9	7.75	9.70
E. P. 100.0	*8.72*	*9.85*
100.1	9.7	10.00
100.2	10.0	10.19
101	10.7	10.7

These data are plotted in Fig. 34-4; for comparison the neutralization curve of 0.1 *N* hydrochloric acid is added. In addition, the color-change intervals of methyl orange (M.O.), methyl red (M.R.), and phenolphthalein (Phpht.) are indicated.

An inspection of the neutralization curve of acetic acid shows that methyl orange cannot be used as indicator. It changes from red to yellow-orange between a pH of 3.0 and 4.2. At the latter pH only 30 per cent of the acid is neutralized, and the titration error would amount to 70 per cent. Methyl red is not a good indicator either because it begins to change color at a pH of about 4.6, i.e., before 50 per cent of the acid has been neutralized, and is present in the yellow form at a pH of 6.0 when about 93 per cent of the acid has been neutralized. In order to get a good result it is necessary to use an indicator which changes color in the neighborhood of the pH prevailing at the equivalence point. Since the solution has a slightly alkaline reaction at the equivalence point, an indicator changing color in alkaline medium must be used. It is evident that phenolphthalein and thymol blue are the best indicators. *Quite generally weak acids should be titrated with phenolphthalein, thymol blue, or thymolphthalein as indicators.*

The changes in pH near the equivalence point become less pronounced with a decrease in the ionization constant of the acid, because hydrolysis then begins to play a more important role. In the titration of 0.1 *N* solutions of acids having an ionization constant of the order of 10^{-7}, it is seen from the table that even thymol

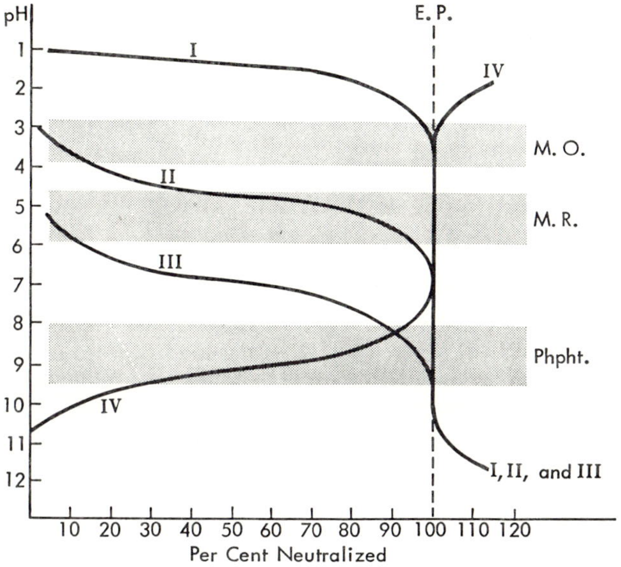

Fig. 34-4. Neutralization curves of 0.1 *N* acid and of 0.1 *N* ammonia (I) 0.1 *N* HCl; (II) 0.1 *N* acetic acid, $K_a = 1.8 \times 10^{-5}$; (III) 0.1 *N* HA, $K_a = 10^{-7}$; (IV) 0.1 *N* NH_3, $K_a = 1.8 \times 10^{-5}$.

blue and phenolphthalein do not yield good end points. After 90 per cent of the acid has been neutralized, the pH = 8.0; and upon further addition of sodium hydroxide the two indicators will only gradually change their colors to the alkaline side, and no sharp end point is obtained. Thymolphthalein is still useful in this case, since it begins to change color at a pH of about 9.6, 0.2 per cent before the end point. Acids having an ionization constant smaller than 10^{-7} cannot be satisfactorily titrated in 0.1 *N* solutions.

In Table 34-4 the neutralization curve of 0.1 *N* ammonium hydroxide with 0.1 *N* hydrochloric acid is given (see also Fig. 34-4).

TABLE **34-4. Neutralization Curve of 100 ml of 0.1** *N* **Ammonium Hydroxide** ($K_b = 1.8 \times 10^{-5}$) **with 0.1** *N* **Hydrochloric Acid at Room Temperature** ($K_w = 10^{-14}$)

0.1 *N* HCl *Added*, ml (= % *Base Neutralized*)	pH
0	11.13
10	10.20
50	9.25
90	8.30
99.0	7.25
99.8	6.55
99.9	6.25
E. P. 100.0	*5.25*
100.1	4.3
100.2	4.0
101.0	3.3

From the data it is evident that none of the indicators changing color in alkaline medium will yield a good end point. Phenolphthalein gives a very gradual color change and becomes colorless when about 94 per cent of the ammonia has been neutralized. Methyl red, methyl orange, and other indicators changing color at pH values between approximately 6 and 3 should be used. *Quite generally weak bases are titrated with methyl red, chlorphenol red, bromcresol green, bromphenol blue, or methyl orange as indicators, using a comparison solution for the end-point determination.*

In the titration of weak acids with weak bases (e.g., acetic acid with ammonia), the pH changes near the equivalence point are always very gradual, and no sharp end point can be found with any indicator. Useful results can be obtained by using an indicator which assumes an intermediate color at the end point and employing buffer solutions with the same pH as the solution at the equivalence point, with the same amount of indicator, for comparison.[4] If possible, however, one should always titrate a weak acid with a strong base and a weak base with a strong acid.

In the above examples the weak acid or weak base to be titrated was uncharged. The calculations remain the same when the acid has a positive charge or the base has a negative charge. For example, the ammonium ion is an acid with $K_{HA} = 5.5 \times 10^{-10}$. No sharp end point can be obtained in the titration of 0.1 N ammonium chloride with sodium hydroxide. On the other hand, the cations of aromatic amines, such as anilinium and pyridinium, are much stronger acids with a K_{HA} of the order of 10^{-5}–10^{-4}. Therefore, salts of such weak bases can be titrated with sodium hydroxide, using thymol blue or phenolphthalein as indicator.

Aquocations of the heavy metals are acids in the Bronsted sense. In addition, the corresponding hydroxides or hydrous oxides are slightly soluble. Under suitable conditions such cations can be titrated with sodium hydroxide. Examples include

$$Al^{3+} + 3\ OH^- \rightarrow Al(OH)_3$$

$$Cu^{2+} + 2\ OH^- \rightarrow Cu(OH)_2$$

Quite generally, the direct titration of these cations with a strong base yields an end point considerably before the equivalence point because of the precipitation of basic salts. Satisfactory results are usually obtained by the addition of an excess of base and back-titration with acid. The method can even be applied to the determination of magnesium in its salts. Although magnesium hydroxide is a strong base, it is slightly soluble ($K_{sp} = 10^{-11}$). An excess of standard base is added to a solution of the magnesium salt and the excess of base in the *clear* supernatant liquid or in the filtrate is back-titrated with standard acid.

Anions in salts of weak acids are bases and can be titrated with standard hydrochloric acid, provided the liberated acid is weak enough to give a distinct end point. The basic strength of anions such as cyanide and metaborate is of the same order of magnitude as that of ammonia (see Appendix) and they can be titrated with a strong acid. Borax is a salt, which in solution can be considered as half-neutralized boric acid:

$$Na_2B_4O_7 + 5\ H_2O \rightarrow 2\ H_3BO_3 + 2\ H_2BO_3^- + 2\ Na^+$$

$$2\ H_2BO_3^- + 2\ H^+ \rightarrow 2\ H_3BO_3$$

The end point is so sharp that borax can be used as a standard substance for hydrochloric acid.

[4] For further details see I. M. Kolthoff and V. A. Stenger, *Volumetric Analysis*, 2nd English ed., vol. I (1942), vol. II (1947), Wiley-Interscience, New York.

Polyprotic acids and their salts. Mixtures of two acids

Consider a diprotic acid with dissociation constants K_1 and K_2. For all practical purposes we can use for the calculation of $[H^+]$ at the first equivalence point of the equation (see p. 54),

$$[H^+] = (K_1K_2)^{1/2} \tag{6}$$

At the second equivalence point the same equation can be used as for a monoprotic acid. In an equivalent mixture of two weak acids with dissociation constants K_1 and K_2, Eq. (6) holds at the first equivalence point. In Table 34-5 is presented

TABLE 34-5. **Titration with Standard Sodium Hydroxide of a Mixture 0.1 *N* in Acetic Acid and 0.1 *N* in Ammonium Chloride**
($K_1 = 1.8 \times 10^{-5}$, $K_2 = 5.5 \times 10^{-10}$)

NaOH *Added Expressed in Equivalent % of Acetic Acid*	$[H^+]$	pH	$\frac{\Delta pH}{\Delta C_{NaOH}}$
50	1.8×10^{-5}	4.75	
90	2.0×10^{-6}	5.70	
98	3.7×10^{-7}	6.43	
99	2.2×10^{-7}	6.65	0.22
99.6	1.4×10^{-7}	6.85	0.33
100	1.0×10^{-7}	7.00	0.38
100.4	7.1×10^{-8}	7.15	0.38
101	4.5×10^{-8}	7.35	0.33

the change of pH in the titration of a mixture of 0.1 *N* in acetic acid and 0.1 *N* in ammonium (chloride) or 0.1 *N* in boric acid with standard sodium hydroxide.

It is seen that no large change in pH occurs at the first equivalence point; this means that no sharp color change of any indicator can be observed at the end point. Between 0.4 per cent before and 0.4 per cent after the equivalence point the change in pH is only 0.3. The naked eye can detect such a difference with the aid of a suitable indicator when a comparison solution is used which has the same pH and which contains the same amount of indicator as the titrated solution at the equivalence point. In order to eliminate ionic strength effects, which have been neglected in our derivation, it is best to use a comparison solution which is closely similar to the titrated solution at the first equivalence point. In the example of Table 34-5 a 0.1 *N* solution of pure ammonium acetate can be used. Mixed indicators (p. 695) are particularly useful in titrations in which the change in pH at the equivalence point is small.

The larger the ratio K_1/K_2, the more sharply the end point can be detected. When this ratio is smaller than 10^4, the accuracy of the end-point determination becomes less than 1 per cent.

Phosphoric acid can be titrated as a mono- or diprotic acid. The pH at the first equivalence point is $\frac{1}{2}(pK_1 + pK_2) = 4.66$. Bromocresol green or methyl yellow can be used as indicators. It is advisable to use a solution of pure NaH_2PO_4 for comparison in the detection of the end point. The pH at the second equivalence point is $\frac{1}{2}(pK_2 + pK_3) = 9.7$. At this pH phenolphthalein and thymol blue are present in the alkaline form and a color change of these indicators is observed before the equivalence point. Thymolphthalein is a better indicator for this titration, as it begins to change color at pH 9.6.

The third ionization constant of phosphoric acid is of the order of 5×10^{-13}. Therefore, a solution of Na_3PO_4 reacts strongly alkaline. A titration to the third equivalence point is not possible, unless the trivalent phosphate ion is removed. This is done by addition of a large excess of calcium chloride after reaching the second equivalence point:

$$2\ Na_2HPO_4 + 3\ CaCl_2 \rightarrow Ca_3(PO_4)_2 + 4\ NaCl + 2\ HCl$$

Carbonic acid (hydrated carbon dioxide) is a dibasic acid. A solution of alkali bicarbonate has a pH of $\frac{1}{2}(pK_1 + pK_2) = 8.40$.

Carbon dioxide can be titrated as a monobasic acid, using phenolphthalein or thymol blue (or, better, a mixed indicator, see pp. 695 and 787) as indicator. The color change is not sharp, and it is necessary to use a solution of freshly prepared pure sodium bicarbonate with the same amount of indicator for comparison.

The second ionization constant of carbonic acid is so small that it cannot be determined as a dibasic acid by a direct titration. However, the carbonate ion can be removed by precipitation. A measured excess of barium hydroxide is added to the solution of carbon dioxide or bicarbonate:

$$H_2CO_3 + Ba(OH)_2 \rightarrow BaCO_3 + 2\ H_2O$$

$$NaHCO_3 + Ba(OH)_2 \rightarrow BaCO_3 + NaOH + H_2O$$

and the excess is titrated back with standard acid, using phenolphthalein or thymol blue as indicator, without filtering off the barium carbonate.

In Table 34-6 data for the neutralization curve of carbon dioxide to $NaHCO_3$ and Na_2CO_3 are given (see also Fig. 34-5). It is clear that a solution of sodium carbonate can be titrated as a mono- or diprotic base.

TABLE 34-6. **Data for Neutralization Curve of Carbon Dioxide** (100 ml of 0.05 *M* carbon dioxide titrated with 0.1 *N* NaOH; $K_1 = 3 \times 10^{-7}$, $K_2 = 6 \times 10^{-11}$)

0.1 *N* NaOH *Added*, ml	pH
0	3.9
0.5	4.5
5	5.5
25	6.5
45	7.5
49	8.08
49.5	8.21
50 $NaHCO_3$	*8.35* [Phpht.; thymol blue (mixed indicator)]
50.5	8.54
51	8.67
55	9.27
75	10.2
95	11.2
100 Na_2CO_3	*11.4* (no good indicator here)
105	11.75

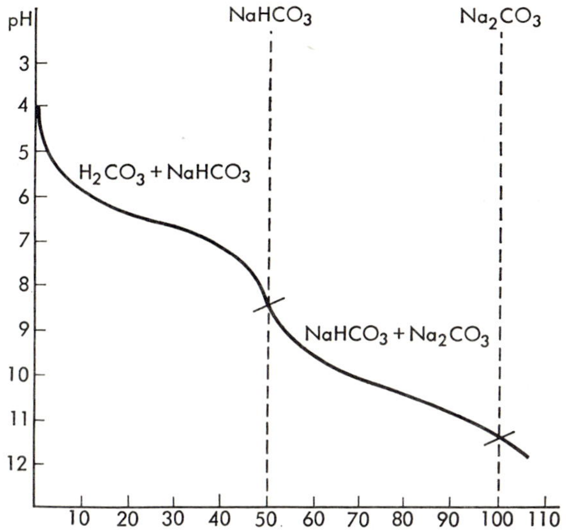

Fig. 34-5. Titration of 100 ml 0.05 M H_2CO_3 with 0.1 N sodium hydroxide.

Hydrolytic precipitation and complex formation reactions

A solution of potassium cyanide is alkaline.

$$CN^- + H_2O \rightleftharpoons OH^- + HCN$$

If cyanide ions are removed by the formation of a slightly soluble or slightly dissociated compound, the solution will not become alkaline until cyanide is in excess. Mercuric cyanide is a substance which is much less dissociated than mercuric chloride. Therefore, the following reaction runs quantitatively to the right:

$$HgCl_2 + 2\,CN^- \rightarrow Hg(CN)_2 + 2\,Cl^-$$

Upon titration of an alkali cyanide solution with mercuric chloride, the reaction remains alkaline until the end point. A solution of mercuric chloride reacts acid toward methyl red. A sharp color change from yellow to red is observed at the end point using methyl red as indicator.

A solution of an alkali cyanide can be titrated with standard silver nitrate, using an acid–base indicator such as phenolphthalein or phenol red for the detection of the end point. The cyanide is transformed into the complex $Ag(CN)_2^-$ ion:

$$2\,CN^- + Ag^+ \rightarrow Ag(CN)_2^-$$

which is the anion of a strong acid. The solution of its alkali salt accordingly has a neutral reaction. Therefore, after the removal of the last trace of cyanide with silver ions, the reaction changes from alkaline to neutral.

An application of a hydrolytic precipitation react is made in the determination of the hardness of water (sum of calcium and magnesium). Potassium palmitate solutions (soap) are strongly hydrolyzed and have an alkaline reaction. They form insoluble palmitates with calcium, magnesium, barium, strontium, and some other cations:

$$2\,KPalm + Ca^{++} \rightarrow Ca(Palm)_2 + 2\,K^+$$

For the determination of the sum of calcium and magnesium in water, the latter is neutralized with acid to methyl orange (titration of the bicarbonate). After boiling off the carbon dioxide, the solution will have a neutral reaction after cooling. It is then titrated with the soap solution, using phenolphthalein as indicator.

One might expect that sodium carbonate would be a useful reagent in the determination of the alkaline earths. The carbonates of calcium, strontium, and barium are very slightly soluble, and their saturated solutions have a pH of about 9. If an excess of the alkaline earth ion is present, the pH is less than 9; and with an excess of sodium carbonate it is greater. Actually the precipitation of the alkaline earth carbonates during titration with sodium carbonate is rather slow; direct titration, therefore, is not very useful. Addition of an excess of the standard carbonate solution and back-titration of the filtrate, however, yields good results.

Various other reagents are used in hydrolytic precipitation and complex formation reactions:

Potassium chromate for barium: $CrO_4^{--} + Ba^{++} \rightleftharpoons BaCrO_4$

$CrO_4^{--} + H_2O \rightleftharpoons HCrO_4^- + OH^-$

Aluminum chloride for fluoride: $Al^{3+} + 6\,F^- \rightleftharpoons AlF_6^{3-}$

$Al^{3+} + H_2O \rightleftharpoons AlOH^{++} + H^+$

Sodium sulfide for zinc: $Zn^{++} + S^{--} \rightarrow ZnS$

$S^{--} + H_2O \rightarrow HS^- + OH^-$

Acid salts of chelating agents such as disodium ethylenediaminetetraacetate (Na_2H_2Y) react with many cations with liberation of hydrogen ions:

$$H_2Y^{--} + Zn^{++} \rightleftharpoons ZnY^{--} + 2\,H^+$$

End-point error. This corresponds to the difference between the amount of standard reagent used between the equivalence point and the end point. The error —often called *titration error*—is expressed in per cent or ppt of the number of equivalents of substance titrated. The end-point error in the titration of a strong acid with a strong base has been discussed. When 0.01 N strong acid is titrated with a strong base, the end-point error (neglecting volume change) is only ± 0.1 per cent when pH $= 5$ or 9 at the end point, and ± 0.01 per cent when pH is 6 or 8 at the end point. The end-point error is also small in the titration of a carboxylic acid ($K_{HA} \geqq 10^{-5}$) with a strong base. The analytical concentration of the acid is denoted by C_a and that corresponding to the amount of standard base added by C_b. For convenience it is assumed that no volume change occurs during the titration. At the equivalence point, $C_a = C_b$. At other points, $C_a - C_b$, which can be positive or negative, gives the end-point error. At any point during the titration

$$C_a = [HA] + [A^-]$$

and

$$C_b = [A^-] + [OH^-] - [H^+]$$

In the titration under consideration the solution is weakly alkaline at or near the end point and $[H^+]$ can then be neglected. At the equivalence point

$$C_a = C_b = [A^-] + [HA] = [A^-] + [OH^-]$$

or

$$[HA] = [OH^-]$$

This result, of course, is simply arrived at when considering that at the equivalence point the solution is composed of pure NaA:

$$A^- + H_2O \rightleftharpoons HA + OH^-$$

Consider the titration of 0.1 M acetic acid ($K_{HA} = 1.8 \times 10^{-5}$) and, assuming no volume change, that the pH at the end point deviates ± 1 from that at the equivalence point. At the equivalence point $[HA] = [OH^-] = 7.5 \times 10^{-6}$ and pH $=$

8.88 ($K_{H_2O} = 10^{-14}$). When pH at the end point is 9.88 (9.9) [HA] $= 7 \times 10^{-7}$ and $[OH^-] = 7.6 \times 10^{-5}$. The end-point error in per cent is then

$$C_b - C_a = \left\{\frac{[OH^-] - [HA]}{C_a}\right\} \times 100 = 0.07 \text{ per cent}$$

Evidently the titration error is negligibly small when the change in pH at the equivalence point is pronounced and the proper indicator is used. When the break in the pH at the equivalence point is gradual, no sharp end point can be observed with any indicator. Under such conditions a comparison solution of the same pH, and preferably of the same composition as the titrated solution at the equivalence point, must be used with an indicator which is partially in the acid and partially in the basic form at this point. A suitable mixed indicator is highly recommended for this purpose. Assume that a color difference between the solution titrated and the comparison solution can be detected corresponding to a difference of ± 0.1 pH. Consider now the titration of 0.1 M solution with $K_{HA} = 10^{-9}$. Neglecting volume change we calculate in the same way as above that the end-point error is ± 0.5 per cent.

In the titration with standard base of polyprotic acids, such as phosphoric or carbonic, or of their salts to an acid salt with acid, no sharp color change at the equivalence point is observed. To get the best and most precise results, a comparison solution must be used. Consider the titration of sodium carbonate of molar concentration C_S to bicarbonate with standard acid. In the following A^{--} denotes carbonate and HA^- bicarbonate ion. $K_1 = 3.5 \times 10^{-7}$, $K_2 = 4.5 \times 10^{-11}$, $K_{H_2O} = 10^{-14}$, and $C_S = 0.1$. At the equivalence point we have a solution of NaHA and

$$[H^+] = (K_1K_2)^{1/2} = 4 \times 10^{-9}, \quad \text{pH} = 8.40$$

It is easily seen that at the equivalence point, $[H^+]$ and $[OH^-]$ are negligibly small as compared to $[H_2A]$ and $[A^{--}]$:

$$[H_2A] = \frac{[H^+][HA^-]}{K_1} = 1.1 \times 10^{-3}$$

$$[A^{--}] = \frac{[HA^-]}{[H^+]} K_2 = 1.1 \times 10^{-3}$$

At any point in the titration

$$C_S = [A^{--}] + [HA^-] + [H_2A]$$

and

$$C_a = 2[H_2A] + [HA^-] + [H^+] - [OH^-]$$
$$\cong 2[H_2A] + [HA^-]$$

Neglecting $[H^+]$ and $[OH^-]$, it follows from the last two equations that at the equivalence point when $C_a = C_S$, $[H_2A] = [A^{--}]$.

Assume that the end point can be detected within ± 0.1 pH of the pH at the equivalence point, e.g., at pH 8.30. At this pH,

$$[A^{--}] = 9 \times 10^{-4} \quad \text{and} \quad [H_2A] = 1.4 \times 10^{-3}$$
$$C_a - C_S = [H_2A] - [A^{--}] = 0.5 \times 10^{-3}$$

and the end-point error is $[(0.5 \times 10^{-3})/10^{-1}] \times 100 = 0.5$ per cent. For the titration of 0.01 M sodium carbonate the same end-point error is calculated as for 0.1 M solution. When the dilution becomes so large that approximation that $[H^+]$ and $[OH^-]$ are negligibly small as compared to $[H_2CO_3]$ and $[CO_3^{--}]$ no longer holds, and the equations become very involved.

Acid–base titrations in organic analysis

Below are presented some reactions that are applied in the determination of some important organic functional groups and which involve acid–base titrations. Details and procedures may be found in monographs on organic analysis and other references.[5]

Acids and bases. The titration curves are similar to those of inorganic acids and bases. Many acids and bases which are too weak to be titrated in aqueous medium can be titrated in a nonaqueous solvent. This is discussed in the next section.

Esters. These compounds are transformed into the corresponding alkali salts by heating with alkali hydroxide, e.g.,

$$CH_3COOC_2H_5 + Na^+ + OH^- \rightarrow CH_3COO^- + Na^+ + C_2H_5OH$$

The reaction is called *saponification*. Most esters are slightly soluble in water and, for quantitative purposes, they are usually saponified by heating with an excess of standard alcoholic alkali. The excess alkali is back-titrated, using phenolphthalein or thymol blue as an indicator. A blank is run with the reagent.

In the analysis of a mixture of two esters use can often be made of the difference in rates of saponification of the two esters. For example, the rate constant of saponification of ethyl acetate is considerably greater than that of isopropyl acetate.[6] The sum of the two esters is found from the amount of alkali consumed in the saponification of both esters. From a *kinetic analysis* of the rate of saponification the amounts of each ester can be found.[7]

Acid anhydrides. In general, these compounds react slowly with water to yield the corresponding acids:

$$(RCO)_2O + H_2O \rightarrow 2\ RCOOH$$

The reaction is catalyzed by a base. To a weighed amount of the substance in a flask is added an excess of standard barium hydroxide or carbonate-free sodium hydroxide. The flask is attached to a reflux condenser which has an efficient carbon dioxide trap, and the liquid is refluxed until all the anhydride is hydrolyzed. The excess of base is titrated with standard hydrochloric acid, using phenolphthalein or thymol blue as indicator.

Carbonyl compounds

Oximation reaction. Reaction with hydroxylamine hydrochloride (oximation) is the basis of the most widely used method for the determination of carbonyl compounds:

$$\underset{\displaystyle H}{R{-}C} = O + NH_2OH \cdot HCl \rightleftharpoons RCH = NOH + H_2O + HCl$$

$$\begin{matrix} R \\ & C = O + NH_2OH \cdot HCl \rightleftharpoons RR'C = NOH + H_2O + HCl \\ R' \end{matrix}$$

[5] J. F. Fritz and G. F. Hammond, *Quantitative Organic Analysis*, Wiley, New York, 1957. S. Siggia, *Quantitative Organic Analysis*, Wiley, New York, 3rd ed., 1963. J. Mitchell, Jr., I. M. Kolthoff, E. S. Proskauer, and A. Weissberger (eds.), *Organic Analysis*, vols. 1, 2, 3, and 4, Wiley-Interscience, New York, 1953–1960. I. M. Kolthoff and P. J. Elving (eds.), *Treatise on Analytical Chemistry*, part II, vols. 12 and 13, Wiley-Interscience, New York, 1965–1966.

[6] T. S. Lee and I. M. Kolthoff, *Ann. N.Y. Acad. Sci.* **53**, 1093 (1951).

[7] See T. S. Lee in *Organic Analysis*, vol. 2, p. 237.

The released hydrochloric acid is titrated with standard alkali to the proper pH. (The hydroxylamine salt has an acid reaction, pH ~ 4.)

Reaction with sodium sulfite. The carbonyl compound reacts with an excess of sodium sulfite with the liberation of free alkali:

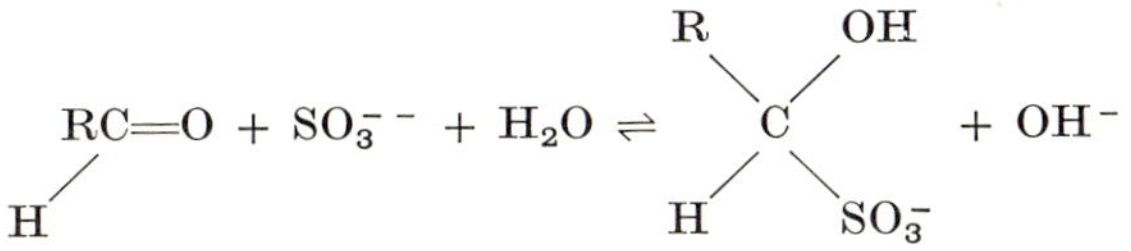

After completion of the reaction the alkali formed is titrated with standard acid. Since sodium sulfite has an alkaline reaction toward phenolphthalein, it is recommended to use thymolphthalein as indicator.

α-**Epoxy group.** α-Epoxides are a group of cyclic ethers in which the oxygen forms a three-membered ring with two adjacent carbon atoms:

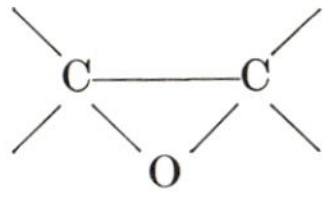

These epoxides react with hydrochloric acid, forming a chlorohydrin:

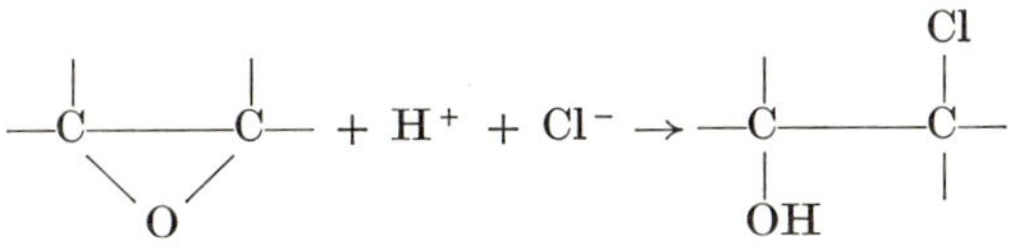

The difference between the amount of acid added and the amount unconsumed, as determined by titration with standard base, is a measure of the epoxide.

Hydroxyl groups. Compounds containing hydroxyl groups react with an excess of acetyl chloride to form an acetate ester and hydrochloric acid:

$$ROH + CH_3COCl \rightarrow CH_3COOR + HCl$$

The excess of acetyl chloride is hydrolyzed to yield two equivalents of acid:

$$CH_3COCl + H_2O \rightarrow CH_3COOH + HCl$$

In the actual determination an exactly measured amount of acetyl chloride dissolved in dry toluene is placed in a glass-stoppered flask; an excess of pyridine, which catalyzes the reaction, is added, followed by the (water-free) sample. After completion of the reaction at 60°C, water is added to decompose the excess of reagent and the acid is titrated with standard alkali using phenolphthalein as indicator. A blank which is run in a similar way yields two equivalents of acid, while 1 mole of alcohol liberates only one equivalent of acid. The hydroxyl content is calculated from the difference in the amounts of alkali used in the blank and the procedure. For details see Smith and Bryant.[8] Another procedure makes use of the amount of acid formed in the acetylation with acetic anhydride:

$$(CH_3CO)_2O + ROH \rightarrow CH_3COOR + CH_3COOH$$

Acid–base titrations in nonaqueous media

We distinguish between *amphiprotic* solvents, which have distinct acid and basic properties, and *aprotic* solvents, which are much weaker acids and usually much weaker bases than water (see Chapter 5). Among the amphiprotic solvents we distinguish

[8] D. M. Smith and W. M. D. Bryant, *J. Am. Chem. Soc.* **57**, 61 (1935).

among (1) hydroxylic solvents (water, alcohols), (2) protogenic solvents (acetic, sulfuric acid), and (3) protophylic solvents (ethylenediamine, dimethylformamide, ammonia). Hydroxylic solvents have acid and basic character comparable to that of water. However, they have a much smaller dielectric constant, DC, than water (DC of water is 78.5, of methanol 32.6, of ethanol 24.3, and of butanol 11.5). The dielectric constant of the solvent HS has a large effect on the dissociation of uncharged acids and bases:

$$HA + HS \rightleftharpoons H_2S^+ + A^-$$

$$B + HS \rightleftharpoons BH^+ + S^-$$

in which H_2S^+ is the solvated proton (lyonium ion) and S^- the solvent minus a proton (lyate ion). The electrostatic attraction between ions increases very much with decreasing dielectric constant. As a consequence K_{HA} and K_B decrease with decreasing dissociation constant (see Table 5-2).

Similarly, the autoprotolysis constant $K_{HS} = [H_2S^+][S^-]$ decreases with decreasing dissociation constant if acid and basic properties of the solvent remain about unchanged. For example, the dissociation constant of uncharged acids and bases in ethanol is about 10^6 times smaller than that in water, and so is the autoprotolysis constant of ethanol ($K_{HS} = 8 \times 10^{-20}$) (see Table 5-2). This means that the changes in pH at the end point in the titration of weak uncharged acids with a strong base and of weak uncharged bases with a strong acid are the same as in water. On the other hand, if we consider the protolysis of a cation acid or an anion base:

$$BH^+ + HS \rightleftharpoons B + H_2S^+$$

$$A^- + HS \rightleftharpoons HA + S^-$$

there is no dissociation into ions and K_{BH^+} and K_{A^-} remain about the same as in water. This is actually found to be the situation in ethanol. We have seen that in the titration of a weak cation acid, such as ammonium ion in water, no sharp break in pH occurs at the end point. Similarly, in the titration of an anion base, such as acetate ion, no sharp break at the end point in the titration with a strong acid occurs. By calculations similar to those made in aqueous medium it is easily shown that a sharp break in pH occurs in the titration in ethanol of ammonium with sodium ethylate and of acetate with ethanolium chloride (HCl) in ethanol.

Extensive investigations on acid–base titrations in tertiary butanol have been made by Marple and Fritz.[9] This solvent has a dielectric constant of only 11.5. At this low dissociation constant all electrolytes, even those which are completely dissociated in water, are incompletely dissociated. In such solvents positive and negative ions combine to form ion pairs, which makes calculation of acid–base equilibria more involved than in water.

Among the protogenic solvents glacial acetic acid is a very popular solvent for the titration of weak bases. All bases that have a K_b in water greater than 10^{-9} become strong bases in acetic acid (leveling effect by acid character of solvent) and yield about the same break in pH at the end point in titrations with perchloric acid.

The titrated base at the end point is present in the form of $BH^+ClO_4^-$, which is partially dissociated into its ions. The more the reaction

$$BH^+ + HS \rightleftharpoons H_2S^+ + B$$

proceeds to the right the less sharply the end point can be detected. Acetic acid is a much weaker base than water, and the above reaction proceeds much less to the right in acetic acid than in water. This accounts for the fact that bases, which are too weak to be titrated in water, can be titrated successfully in acetic acid. Similarly, it is easily

[9] L. Marple and J. S. Fritz, *Anal. Chem.* **35**, 1223, 1431 (1963).

seen that acids which are too weak to be titrated in water can be titrated in solvents which are weaker acids than water. In such a solvent the reaction

$$HA + S^- \rightleftharpoons A^- + HS$$

goes much less to the right than in water.

Acetic acid has a dielectric constant of only 6.1 and ion-pair formation is very pronounced. For example, the dissociation constant of perchloric acid is about 10^{-5}, of sulfuric acid K_1 is 6×10^{-8}, of alkali and amine acetates of the order of 10^{-6}–10^{-7}, and of sodium perchlorate of 3×10^{-6}. The autoprotolysis constant of acetic acid is 3×10^{-15}. From the equations presented in Chapter 5 it is simple to calculate the changes of pH on the entire neutralization curves of bases with perchloric acid.

Protophilic solvents such as dimethylformamide (DC 27) and ethylenediamine (DC 12.9) have gained popularity for the titration of weak acids such as phenols with tetraalkylammonium hydroxide. Pyridine (DC 12.5) (an aprotic protophylic solvent—it has no acid character) is particularly suitable for such titrations.[10]

Aprotic protophobic solvents are much weaker bases than water and are useful for the titration of very weak bases with perchloric acid. However, many strong electrolytes are insoluble in inert solvents and often soluble in acetic acid or protophylic solvents. For example, sodium salts of carboxylic acids are soluble in glacial acetic acid and insoluble in protophobic aprotic solvents. For this reason mixtures of such a solvent with acetic acid are often used to advantage when the protophobic solvents alone are less suitable because of solubility relations.

Aprotic protophobic solvents that have been used in titrations of weak bases are acetone (dielectric constant 20.7), acetonitrile (36), methyl isobutyl ketone (13.1), acetic anhydride (20.7), and nitromethane (35.9). Acetonitrile has no leveling effect on bases. It is highly recommended as a solvent for titration of amines and carboxylates with standard perchloric acid in acetic acid. A very sharp end point is observed with *p*-naphtholbenzein from yellow to green. Salts that are insoluble in acetonitrile are dissolved in acetic acid.[11] The color change and the break in pH at the equivalence point in acetonitrile are much sharper than in glacial acetic acid. Hydrocarbons, including benzene, toluene, and chloroform are less suitable; they have a very low dielectric constant, and most electrolytes are insoluble in such solvents.

The shape of neutralization curves of most acids is quite different in aprotic protophobic than in amphiprotic solvents and in aprotic protophylic solvents. Amphiprotic solvents are acids and stabilize anions of an acid by hydrogen bonding:

(a) $$HA + HS \rightleftharpoons H_2S^+ + A^-$$

(b) $$A^- + HS \rightleftharpoons A^-\text{—}HS$$

overall dissociation: $HA + 2\,HS \rightleftharpoons H_2S^+ + A^-\text{—}HS$. (For convenience only one HS is written in the partial equations; many molecules of HS may participate.) Reaction (b) cannot occur in aprotic protophobic solvents because they have no hydrogen-bonding donating properties. When no hydrogen-bond donors other than HA are present in the solution, the following reaction, called homoconjugation, occurs:

(c) $$A^- + HA \rightleftharpoons HA_2^-$$

with a homoconjugation constant

$$K^f_{HA_2^-} = \frac{[HA_2^-]}{[A^-][HA]}$$

[10] R. H. Cundiff and P. C. Markunas, *Anal. Chem.* **28**, 792 (1956).

[11] For details see I. M. Kolthoff, M. K. Chantooni, Jr., and S. Bhowmik, *Anal. Chem.* **39**, 1627 (1967).

The overall dissociation of the acid is then the sum of reactions (a) and (c):

(d) $$2\,HA \rightleftharpoons H^+ + HA_2^-$$

with an overall dissociation constant

(e) $$K_{2HA} = \frac{[H^+][HA_2^-]}{[HA]^2}$$

whereas the hydrogen-ion concentration of a weak acid in hydroxylic solvent changes in proportion to $[HA]^{1/2}$. We find that in an aprotic protophobic solvent it changes in proportion to [HA]. Thus, on tenfold dilution the hydrogen-ion concentration decreases 10 times, just as that of a strong acid in a hydroxylic solvent. The neutralization curve is also greatly affected by the homoconjugation. When $K^f_{HA_2^-}$ and HA are large enough, $[H^+]$ in the *beginning* of the neutralization curve is given by

$$[H^+] = \frac{[HA]^2}{[HA_2^-]} K_{2HA}$$

or, when C_a represents the initial acid concentration and C_b the concentration of strong base added

(f) $$[H^+] = \frac{[C_a - C_b]^2}{C_b} K_{2HA}$$

and

(g) $$K_{2HA} = K^f_{HA_2^-} K_{HA}$$

Again we notice a difference between the behavior in a hydroxylic and an aprotic protophobic solvent. In a hydroxylic solvent

$$[H^+] = \frac{C_a - C_b}{C_b} K_{HA}$$

and dilution of the mixture of a weak acid and its salt slightly affects $[H^+]$ because of a slight change of the activity coefficients. In the aprotic protophobic solvent, $[H^+]$ decreases with dilution, in proportion to the change of $C_a - C_b$. When the acid is 50 per cent neutralized (often called in the literature the half-neutralization potential, HNP, with reference to the glass electrode),

(h) $$[H^+]_{1/2} = K_{HA}$$

It is immaterial what the value of $K^f_{HA_2^-}$ is. As a result of the homoconjugation, neutralization curves of weak acids with a strong base in inert solvents have quite a different shape from those in which no homoconjugation occurs. This can easily be demonstrated by a simple example.

Take $K^f_{HA_2^-} = 10^4$; and $C_a - C_b = 0.01$ and $C_b = 0.001$. Then $[H^+] = 10^{-1} \times 10^4 \times K_{HA} = 10^3 K_{HA}$. Without homoconjugation $[H^+] = 10K_{HA}$. With or without homoconjugation $[H^+]_{1/2} = K_{HA}$. When C_b is greater than $\frac{1}{2}C_a$ (after 50 per cent neutralization), there is a marked concentration of unconjugated A^-. This will react with HA to give HA_2^- and

$$[HA] = \frac{[HA_2^-]}{[A^-]} \times \frac{1}{K^f_{HA_2^-}}$$

When $C_a - C_b = 0.001$, and $C_b = 0.01$, practically all HA is present in the form of its homoconjugate:

$$[HA] = \frac{0.001}{0.009} \times \frac{1}{10^4} = 1.1 \times 10^{-5}$$

and $$[A^-] = 9 \times 10^{-3}$$

Hence

$$[\mathrm{H}^+] = \frac{1.1 \times 10^{-5}}{9 \times 10^{-3}} K_{\mathrm{HA}} = 1.2 \times 10^{-3} K_{\mathrm{HA}}$$

as compared to

$$[\mathrm{H}^+] = \frac{0.001}{0.01} K_{\mathrm{HA}} = 10^{-1} K_{\mathrm{HA}}$$

in water. Without homoconjugation pH changes between 9 and 91 per cent neutralization from $\mathrm{p}K_{\mathrm{HA}} - 1$ to $\mathrm{p}K_{\mathrm{HA}} + 1$. In our example ($C_a = 0.011$, $K^f_{\mathrm{HA}_2^-} = 10^4$) pH changes from $\mathrm{p}K_{\mathrm{HA}} - 3$ to $\mathrm{p}K_{\mathrm{HA}} + 3$.

Figure 34-6 gives the shape of some neutralization curves of weak acids with a strong base in acetonitrile. Curve V refers to an acid the anion of which does not homoconjugate, e.g., the picrate ion. The other curves refer to the neutralization of weak acids, keeping the volune unchanged; i.e., $C_a + C_S$ remains constant, C_S denoting the salt concentration. The salt has been considered to be completely dissociated. If curve 1 represents the shape of the neutralization curve of a 0.1 M acid with $K_{\mathrm{HA}_2^-} = 10^4$, then curve 2 is that of 0.01 M, curve 3 of 0.001 M, and curve M of 0.0001 M acid. A more extensive and more exact discussion is found in the papers by Kolthoff and Chantooni.[12]

In aprotic solvents the ratio of K_1 to K_2 of diprotic acids is much greater than that in water. In water most dibasic acids do not give a break in pH after neutralization of the first proton, only one break being observed after complete neutralization of the acid. In aprotic solvents distinct breaks in pH are observed after neutralization of the first

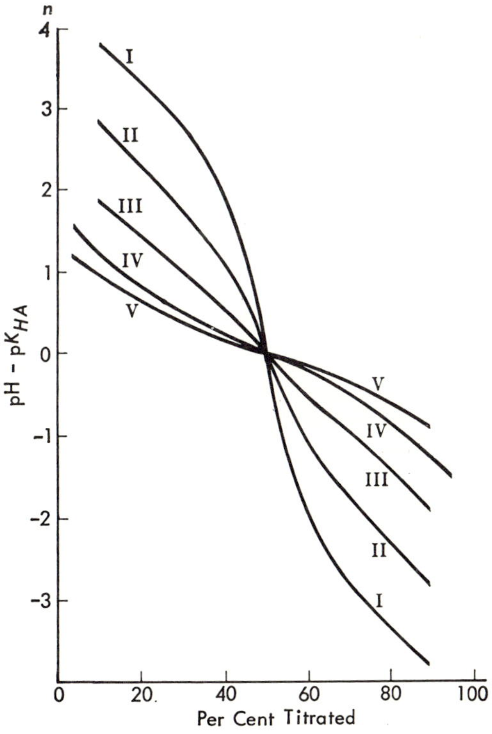

Fig. 34-6. The neutralization curves of some weak acids with a strong base in acetonitrile.

[12] I. M. Kolthoff and M. Chantooni, Jr., *J. Am. Chem. Soc.* **87**, 4428 (1965). In this paper are found references to other papers dealing with homoconjugation.

and second protons. A typical example is the titration in acetone[13] of succinic acid with tetraalkylammonium hydroxide. The neutralization curve is presented in Fig. 34-7. Sulfuric acid also gives a pronounced break in pH after neutralization of the first proton. pK_2 of sulfuric acid in water is 2.0, but in acetonitrile 25.9 and in dimethylsulfoxide 14.5. Several examples can be given of differentiating titrations in aprotic solvents, i.e., titration of a mixture of two acids, which only exhibit one break in pH in aqueous solution after neutralization of both acids, while in an inert solvent a break in pH is also observed after neutralization of the stronger acid. In an aprotic protophylic solvent such as dimethyl sulfoxide or pyridine, homoconjugation is very much less pronounced than in aprotic protophobic solvents. The reason is that the protophylic solvents hydrogen bond the acid HA:

$$HA + S \rightleftharpoons AH\text{---}S$$

Hence the anion A^- competes with the solvent in hydrogen-bonding HA:

$$AH\text{---}S + A^- \rightleftharpoons AH\text{---}A^- + S$$

Because of small homoconjugation, aprotic protophobic solvents are recommended as media for the titration of acids or mixtures of acids. On the other hand aprotic protophobic solvents (nitroalkanes, ketones, acetonitrile) are recommended for titrations of weak bases.

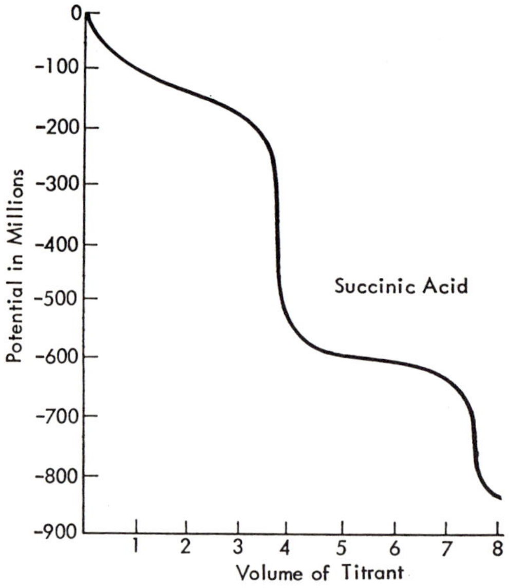

Fig. 34-7. Titration of succinic acid in acetone.

REFERENCES

There is an abundant literature on acid–base titrations and equilibria in nonaqueous solvents. Below is given a select list of references to theoretical and experimental treatment of the subject.

Theoretical

I. M. Kolthoff and S. Bruckenstein, in *Treatise of Analytical Chemistry* (I. M. Kolthoff and P. J. Elving, eds.), part I, vol. 1, p. 475, Wiley-Interscience, New York, 1959.

[13] J. S. Fritz and S. S. Yamamura, *Anal. Chem.* **29**, 1079 (1957).

R. P. Bell, *The Proton in Chemistry*, Cornell Univ. Press, Ithaca, N.Y., 1959.

Chemistry in Nonaqueous Solvents, vol. 4 (G. Zander, H. Spandau, and C. C. Addison, eds.), a multivolume treatise, Wiley-Interscience, New York, 1963.

I. Gyenes, *Titration in Nonaqueous Media*, Van Nostrand Co., New York, 1967.

The Chemistry of Nonaqueous Solvents, Vol. I, Principles and Techniques, J. J. Lagowski, ed., Austin, Academic Press, 1966.

W. Huber, *Titrations in Nonaqueous Solvents*, Academic Press, New York, 1967.

Experimental

Biannual reviews of *Analytical Chemistry.*

J. S. Fritz, *Acid Base Titrations in Nonaqueous Solvents*, G. F. Smith Chemical Co., Columbus, Ohio, 1952.

S. R. Palet, M. N. Das, and G. R. Somayulu, *Nonaqueous Titrations*, Indian Association for Cultivation of Science, Calcutta, India, 1959.

T. Kucharsky and L. Safarik, *Titrations in Non-Aqueous Solvents*, Elsevier, Amsterdam, 1965.

PROBLEMS

1. Derive the neutralization curve of 0.2 N hydrochloric acid with 0.2 N sodium hydroxide at room temperature. What is the titration error when methyl orange is used as indicator (pH at end point = 4.0), methyl red (pH at end point = 6.0), phenolphthalein (pH at end point = 9.0), and thymolphthalein (pH at end point = 10.0)?
2. Derive the neutralization curve of 0.1 N formic acid with 0.1 N sodium hydroxide at room temperature. What is the titration error with methyl red (pH at end point = 6.0), phenol red (pH at end point = 7.0), and phenolphthalein (pH at end point = 9.0)? $K_a = 10^{-4}$, $K_w = 10^{-14}$.
3. Derive the neutralization curve of 1 N ammonium hydroxide with 1 N hydrochloric acid at room temperature. What is the titration error with methyl orange (pH at end point = 4.0), methyl red (pH at end point = 6.0), phenol red (pH at end point = 7.0), and phenolphthalein (pH at end point = 9)? $K_{NH_3} = 1.8 \times 10^{-5}$, $K_w = 10^{-14}$.
4. Sulfurous acid is a dibasic acid. $K_1 = 1 \times 10^{-2}$, $K_2 = 1.0 \times 10^{-7}$. Can it be titrated as a monobasic acid? What is the pH of a solution of $NaHSO_3$? What indicators could be used? Can it be titrated as a dibasic acid? What is the pH of a 0.1 M solution of Na_2SO_3? ($K_w = 10^{-14}$.) What indicator should be used?
5. 0.1 N borax solution is to be titrated with 0.1 N hydrochloric acid. What will be the pH at the end point? What indicator should be used? $K_a = 10^{-9}$.
6. To 50.00 ml of 0.1012 N magnesium sulfate solution 25.00 ml of 0.2514 N sodium hydroxide solution is added, and the mixture is diluted to 100 ml in a volumetric flask. With 0.1046 N hydrochloric acid 50.00 ml of the filtrate is back-titrated. How much acid will be required? What indicator should be used? How much magnesium remains in solution after the addition of the sodium hydroxide? $[Mg^{++}][OH^-]^2 = 10^{-11}$. *Ans.* 5.85 ml of acid; $[Mg^{++}] = 6.9 \times 10^{-8}$ M.
7. A solution is 0.1 N in hydrochloric acid and 0.2 N (M) in boric acid. To 50 ml of this solution is added 50 ml of 0.1 N sodium hydroxide. What is the pH of the mixture after addition of the sodium hydroxide? ($K_{H_3BO_3} = 10^{-9}$.) What indicator should be used in the titration of hydrochloric acid in the presence of boric acid? *Ans.* pH = 5.0.
8. A solution of 0.1 N hydrochloric acid is titrated with 0.1 N sodium hydroxide. The pH at the end point is 4.0. Calculate the titration error. *Ans.* −0.2 per cent.
9. Pure water exposed to carbon dioxide was found to have a pH of 4.52. (a) Calculate the carbon dioxide concentration of the water. ($K_{1H_2CO_3} = 3 \times 10^{-7}$.) (b) How much 0.01 N barium hydroxide is necessary to precipitate all the carbon dioxide in 100 ml of the water? *Ans.* (a) 3×10^{-3} M.

CHAPTER 35 *Theory of Precipitation Analysis*

Argentimetry

The most important volumetric precipitation reactions are carried out with silver nitrate as the reagent. These methods, involving the use of a standard solution of silver nitrate, are called *argentimetric methods* and will be discussed in detail in this chapter. The general principles, however, also hold for any other precipitation method.

Titration curves

The change in the silver-ion concentration and that of the anion precipitated during the course of a titration can be calculated from the solubility product of the silver salt formed and the concentrations of the solution titrated and the silver nitrate solution. Let us assume that 50 ml of 0.2 N sodium chloride are titrated with 0.2 N silver nitrate. After the addition of 40 ml of 0.2 N silver solution, there are left 10 ml of 0.2 N sodium chloride in a volume of 90 ml (compare p. 128). Therefore,

$$[Cl^-] = \frac{10}{90} \times 0.2 = 0.022 \qquad \text{and} \qquad pCl^- = 1.66$$

Since the solubility product of silver chloride is approximately 10^{-10}, we have

$$[Ag^+][Cl^-] = 10^{-10} = K_{sp}$$

or

$$pAg^+ + pCl^- = 10 = pK_{sp}$$

and in this particular case

$$pAg^+ = 10 - 1.66 = 8.34$$

After the addition of 50 ml of 0.2 N silver nitrate, the equivalence point is reached, and a saturated solution of silver chloride with no excess of chloride or silver is obtained. At this point

$$[Ag^+] = [Cl^-] = \sqrt{K_{sp}} = (10^{-10})^{1/2}$$

$$pAg^+ = pCl^- = -\tfrac{1}{2}\log K_{sp} = 5$$

With 0.05 ml of silver nitrate in excess, $[Ag^+] = 10^{-4}$; hence

$$pAg^+ = 4 \qquad \text{and} \qquad pCl^- = 6$$

Quite near the equivalence point, the values of $[Ag^+]$ or $[Cl^+]$ are larger than correspond to the excess of either ion added to the solution, since the dissolved silver chloride contributes some silver and chloride ions to the solution. If the excess of Ag^+ corresponds to the concentration A and the solubility of silver chloride in this excess is x, then

$$[Cl^-] = x, \text{ and } [Ag^+] = A + x, \text{ and } x(A + x) = (K_{sp})_{AgCl} = 10^{-10}$$

If A is greater than approximately $10(K_{sp})^{1/2}$,$(A + x)$ becomes virtually equal to A, and the simple calculation applied above may be used.

In Table 35-1, values of pCl^- and pAg^+ for the titration of 100 ml of 0.1 N sodium chloride with 0.1 N silver nitrate are given (solubility product of silver chloride $= 10^{-10}$). In addition, the corresponding figures for pI^- and pAg^+ are given for the titration of 100 ml of 0.1 N alkali iodide with 0.1 N silver nitrate.

TABLE 35-1. **Titration of 100 ml of 0.1 N NaCl and 100 ml of 0.1 N KI, respectively, with 0.1 N $AgNO_3$ $[(K_{sp})_{AgCl} = 10^{-10}, (K_{sp})_{AgI} = 10^{-16})]$**

0.1 N $AgNO_3$ *Added*, ml	*Titration of Chloride* pCl^-	pAg^+	*Titration of Iodide* pI^-	pAg^+
0	1		1	
90	2.3	7.7	2.3	13.7
98	3.0	7.0	3.0	13.0
99	3.3	6.7	3.3	12.7
99.8	4.0	6.0	4.0	12.0
99.9	4.3	5.7	4.3	11.7
E. P. 100	*5.0*	*5.0*	*8.0*	*8.0*
100.1	5.7	4.3	11.7	4.3
100.2	6.0	4.0	12.0	4.0
101	6.7	3.3	12.7	3.3
110	7.7	2.3	13.7	2.3

In Fig. 35-1 are represented the changes of pAg^+ in the region between 10 per cent before and 10 per cent after the equivalence point in the titration of 0.1 N chloride and 0.1 N iodide, respectively, with 0.1 N silver nitrate. The corresponding changes of pCl^- in this region in the titration of chloride are read from the right-hand ordinates.

The changes of the ion exponents near the equivalence point are much more pronounced in the case of silver iodide than of silver chloride, since the solubility product of the former is about 10^6 times smaller than that of silver chloride. The values derived in Table 35-1 and plotted in Fig. 35-1 are actually found in potentiometric titrations, where pAg^+ is measured with the aid of a silver electrode.

Determination of the end point

Equal turbidity method. No indicator is used in the determination of the end point in this method. Use is made of the fact that in the titration of sodium chloride with silver nitrate the solution is saturated with silver chloride at the equivalence point and then contains equal concentrations of silver and chloride ions. If a small amount of chloride or silver is added to a portion of the clear supernatant liquid, the latter will become turbid, owing to the formation of silver chloride, since the

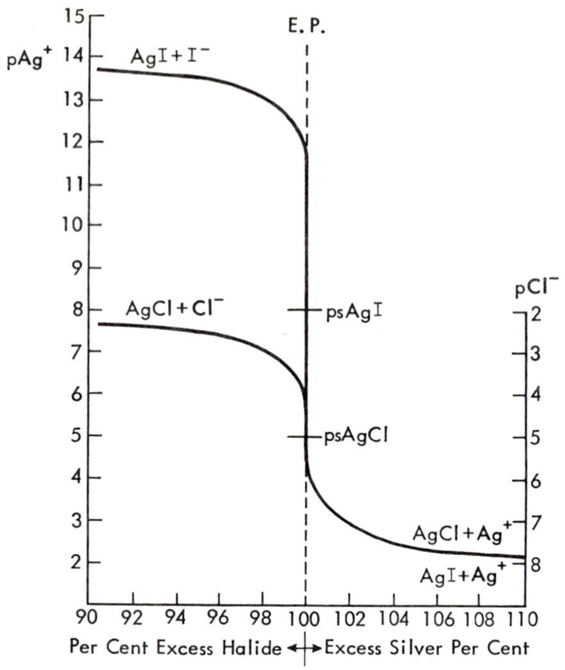

Fig. 35-1. Titration curves of 100 ml 0.1 *N* sodium chloride and 100 ml 0.1 *N* potassium iodide, respectively, with 0.1 *N* silver nitrate.

solubility of the silver chloride is decreased by an excess of either of the two ions. At the equivalence point the turbidity obtained with a slight excess of silver in the sample of the supernatant liquid is equal to that produced by an equal excess of chloride. If chloride were still in excess, the turbidity produced in a sample of the supernatant liquid by silver would be more intense than that produced by an excess of chloride, and the reverse effect would be noticed if silver nitrate had been added in excess during the titration. Therefore in applying this method, two equal portions of the clear supernatant liquid are withdrawn near the equivalence point and tested with silver nitrate and with chloride. If equally strong turbidities are not found, the titration is continued with chloride or silver nitrate to the point of equal turbidities. The *equal turbidity method* was introduced more than one hundred years ago by Gay-Lussac[1] for the determination of silver with chloride. The method, under the proper conditions, is one of the most exact of all titrimetric methods. It is of lasting significance for the atomic-weight determinations of silver, chlorine, and *metals analyzed in the form of pure metal chlorides*,[2] and is still used for the determination of silver in the mints of certain governments.

If the silver salt formed is very slightly soluble, as is the case with silver bromide, thiocyanate, and iodide, the saturated solution at the equivalence point contains so little of the dissolved salt that no turbidity is noticed with an excess of either of the two common ions. In such cases the end point can be found in a simpler way than

[1] J. L. Gay-Lussac, *Instruction sur l'essai des matières d'argent par la voie humide*, Paris, 1832.

[2] See, e.g., T. W. Richards and R. C. Wells, *J. Am. Chem. Soc.* **27**, 459 (1905); C. R. Johnson, *J. Phys. Chem.* **35**, 540, 830, 2237 (1931).

described above. In the titration of bromide, for example, with silver nitrate, small portions of the clear supernatant liquid are withdrawn near the equivalence point and tested with silver for the presence of an excess of bromide. A turbidity will be produced as long as there is an excess of bromide present. The titration is carefully continued until no turbidity is produced by the addition of silver to the supernatant liquid.

Titration to the clear point. Various slightly soluble salts remain in colloidal solution during titration. This is especially the case in the determination of an alkali iodide with silver nitrate. Upon addition of the latter to the iodide solution, the silver iodide formed remains in colloidal solution, the particles being negatively charged by adsorption of iodide ions at their surface (see p. 215). Near the equivalence point most of the iodide in the solution has been transformed into silver iodide, and upon further addition of silver nitrate the reagent will react with the adsorbed iodide. The stability of the suspension then decreases, owing to the removal of the peptizing iodide ions, and flocculation of the sol begins just before the equivalence point is reached. Still the supernatant liquid is turbid as long as some iodide remains in the absorbed state. After all the adsorbed iodide has reacted, the supernatant liquid becomes perfectly clear, and the end point is then reached. This "clear point" determination is very sharp and can be made with great precision. The method is even applicable if a high concentration of electrolyte is present.

Use of potassium chromate as indicator for the detection of an excess of silver. The Mohr titration. In the above methods no indicator is required for the determination of the end point. In the titration of chloride with silver nitrate, potassium chromate is frequently used as an indicator. This is the well-known Mohr[3] method for the determination of chloride and bromide. Chromate ions combine with the silver to form slightly soluble red silver chromate at the end point. The conditions during the titration must be such that the chloride is precipitated quantitatively as silver chloride before the precipitation of the reddish silver chromate is perceptible. On the other hand, the indicator should be sufficiently sensitive to give a sharp color change with a slight excess of silver.

We are dealing here with a case of differential precipitation, the general theory of which has been discussed on page 148.

When both solid phases, silver chloride and silver chromate, are in equilibrium with the solution, we have

$$[Ag^+][Cl^-] = (K_{sp})_{AgCl} = 10^{-10}[Ag^+] = \frac{(K_{sp})_{AgCl}}{[Cl^-]}$$

$$[Ag^+]^2[CrO_4^{--}] = (K_{sp})_{Ag_2CrO_4} = 2 \times 10^{-12}[Ag^+] = \left(\frac{(K_{sp})_{Ag_2CrO_4}}{[CrO_4^{--}]}\right)^{1/2}$$

At the point where silver chromate starts to precipitate and the indicator shows the end point, we have therefore

$$[Ag^+] = \frac{(K_{sp})_{AgCl}}{[Cl^-]} = \left(\frac{(K_{sp})_{Ag_2CrO_4}}{[CrO_4^{--}]}\right)^{1/2}$$

$$\frac{[Cl^-]}{([CrO_4^{--}])^{1/2}} = \frac{(K_{sp})_{AgCl}}{\sqrt{(K_{sp})_{Ag_2CrO_4}}} = \frac{10^{-10}}{(2 \times 10^{-12})^{1/2}} = 7 \times 10^{-5}$$

[3] F. Mohr, *Ann.* **97**, 335 (1856).

At the equivalence point $[Cl^-] = 10^{-5}$. If silver chromate starts to precipitate at this chloride-ion concentration,

$$[CrO_4^{--}] = \left(\frac{[Cl^-]}{7 \times 10^{-5}}\right)^2 = 0.02$$

Theoretically, therefore, the indicator will start to give silver chromate at the equivalence point when the chromate concentration of the solution is 0.02 *M*. However, it should be realized that a certain excess of silver must be added before the red silver chromate is visible.

Practically, an indicator concentration at the end point corresponding to about 5×10^{-3} *M* chromate has proved to be satisfactory. Theoretically the indicator should change color when $[Cl^-] = 7 \times 10^{-5}[CrO_4^{--}]^{1/2} = 7 \times 10^{-5} \times 7 \times 10^{-2} = 0.5 \times 10^{-5}$, which is slightly beyond the theoretical end point, and the silver-ion concentration would then be

$$[Ag^+] = \left(\frac{2 \times 10^{-12}}{5 \times 10^{-3}}\right)^{1/2} = 2.0 \times 10^{-5}$$

Actually the sensitivity of the indicator for silver ions is smaller, since a certain amount of silver chromate must be formed before it is visible. Experimentally this sensitivity is determined by adding 0.01 *N* silver nitrate from a microburet to 100 ml of 0.01 *M* potassium chromate until a distinct difference in color is noticed when compared with a blank. At 25°C the sensitivity corresponds to a silver-ion concentration of from 3.5 to 4×10^{-5} *M*.

At 25°C the sensitivity of 4×10^{-5} *N* silver is equal to the indicator blank; the sensitivity is unaffected by the presence of the white silver chloride. In the titration of 100 ml of 0.1 *N* chloride with 0.1 *N* silver nitrate, the color change is not observed until the silver-ion concentration is 4×10^{-5} (in a volume of 200 ml). This corresponds to an excess of 0.08 ml of 0.1 *N* silver nitrate, or to a titration error of 0.08 per cent, which is negligibly small. The titration error, however, increases with increasing dilution of the chloride. If the titration is carried out with 0.01 *N* solution, the titration error amounts to 0.8 per cent. In the titration of dilute solutions the indicator blank should be determined separately and subtracted from the amount of standard solution used.

The Mohr method is generally used for the determination of chloride and bromide. It cannot be applied to the accurate determination of iodide and thiocyanate, for the end point is obscured as a result of adsorption phenomena.

The titration must be carried out in neutral or weakly alkaline solution. The pH should be greater than 7 but not larger than 10.5, as in acid medium the sensitivity of the indicator decreases very strongly with the hydrogen-ion concentration. The second ionization constant of chromic acid is small, and therefore the chromate ions react with hydrogen ions:

$$CrO_4^{--} + H^+ \rightleftharpoons HCrO_4^-$$

and the decrease of the sensitivity is thus explained. On the other hand, the solution should not be too alkaline, since silver hydroxide might then precipitate before silver chromate. Sodium bicarbonate or borax is very useful for the adjustment of the acidity in the Mohr titration.

End-point detection with adsorption indicators. In a discussion of the properties of colloidal systems it was mentioned (p. 215) that suspensions of silver chloride or silver iodide in the presence of an excess of the halide ion acquire a negative charge, owing to the adsorption of the latter, whereas the charge becomes positive in the presence of an excess of silver ions:

$$[AgClCl^-] \vdots K^+ \qquad \text{(negatively charged)}$$
$$[AgClAg^+] \vdots NO_3^- \qquad \text{(positively charged)}$$

A precipitate has a tendency to adsorb its own ions. Therefore, silver chloride suspended in a medium containing chloride ions and other anions will in general tend to adsorb the chloride ions. As soon as the equivalence point is reached in an argentimetric chloride determination, part of the excess silver nitrate added will be adsorbed by the silver chloride particles. The nitrate ions are not strongly held by the particles; and if other anions are present which are more strongly adsorbed, they will take the position of the adsorbed nitrate ions.

The above explains the principle of the application of *adsorption indicators. These indicators do not give a color change in the solution but on the surface of the precipitate.* When a few drops of a dilute fluorescein solution are added to a 0.05 N silver nitrate solution, the color of the indicator remains the same as in water. The fluorescein, therefore, does not behave as an indicator for dissolved silver ions. If a little chloride is added to the silver solution, silver chloride is formed; and the color turns red or pinkish, since the silver chloride adsorbs silver ions and the latter attract the fluorescein ions to the surface with the formation of the red silver fluoresceinate at the *surface* of the precipitate. If more chloride is added, the suspension remains red until chloride is in excess. At this point the silver originally adsorbed at the surface is transformed into silver chloride, and the particles will adsorb chloride ions. The suspension will then have the white color of silver chloride; and the fluorescein ions, adsorbed in the red form in the presence of an excess of silver, are sent into the solution, in which they have a yellowish green color.[4]

Lecture experiment: Add to 50 ml of approximately 0.02 N silver nitrate a few drops of a 0.1 per cent fluorescein solution in alcohol. Then add a few drops of 0.1 N sodium chloride. The color turns red and remains red upon further addition of chloride until an excess has been added.

Fluorescein and dichlorofluorescein are very useful indicators in the titration of halides with silver nitrate. The following phenomena are usually observed during the titration: The silver halide remains partially in colloidal solution until the equivalence point is approached. About 1 per cent before the end point, flocculation occurs. The titration should then be continued carefully with vigorous agitation, until the precipitate appears reddish. Such a titration should not be carried out in direct sunlight, since silver halides with adsorbed indicator are extremely light-sensitive. The red color changes rapidly to gray and black in strong light.

[4] See K. Fajans and O. Hassel, *Z. Elektrochem.* **29**, 495 (1923); K. Fajans and W. Steiner, *Z. Physik. Chem.* **125**, 309 (1927); K. Fajans and H. Wolff, *Z. Anorg. Allgem. Chem.* **137**, 221 (1924); O. Hassel, *Kolloid-Z.* **34**, 304 (1924); for review see I. M. Kolthoff, *Chem. Rev.* **16**, 87 (1935).

Since the color change takes place on the surface of the particles, it is of importance to have as large a surface development of the precipitate as possible. Therefore, the method works best under those conditions in which the precipitate separates as a flocculated colloid. Large amounts of neutral salts obscure the end point somewhat, because they exert their flocculating effect before the end point is reached. With very dilute halide solutions the amount of precipitate formed is so small that the color change is no longer pronounced. For this reason chloride solutions more dilute than 0.005 N cannot be titrated with fluorescein or dichlorofluorescein as indicator.

Fluorescein itself is a very weak acid with an ionization constant of the order of 10^{-8}. The color change is to be attributed to an adsorption of fluorescein ions; and therefore it may be expected that hydrogen ions will interfere, because they remove the indicator ions from the solution. This is actually true: titrations with fluorescein as indicator must be carried out in neutral or slightly alkaline medium (pH between 7 and 10). Dichlorofluorescein[5] is a stronger acid than fluorescein, and for this reason it can be used in slightly acid solutions. It is a very useful indicator in the titration of chlorides.

Bromides, iodides, and thiocyanates are better titrated with eosin (tetrabromfluorescein) as indicator. The color change to red is extremely sharp; the method gives excellent results, even in highly diluted solutions. Eosin is a much stronger acid than fluorescein, and therefore titrations can be made in solutions with a pH of 2 or greater. Unfortunately, eosin cannot be used in the chloride titration because the adsorbability of the eosinate ions by the silver halides is much greater than that of the fluoresceinate. We have seen that a silver chloride suspension in a solution containing an excess of chloride ions adsorbs the latter on the surface of the particles. If eosin is added to such a suspension, there will be a competition between the chloride and eosin ions to be adsorbed, and the result is that part of the adsorbed chloride is replaced by eosin. This causes a color change toward red. Therefore, in the titration of chloride with silver nitrate, eosin cannot be used as an adsorption indicator, because the color will turn reddish long before the equivalence point is reached.[6]

Titration of silver with thiocyanate. The Volhard method for the determination of halides

Silver can be determined very accurately in acid medium with potassium or ammonium thiocyanate as the precipitant. Silver thiocyanate is a very slightly soluble salt, having a solubility product of 10^{-12}. An excess of thiocyanate can be detected very sharply with a ferric salt, the latter forming the intensely colored red complex ferric thiocyanate in acid medium. The indicator is very sensitive toward an excess of thiocyanate. One to two milliliters of a saturated solution of ferric ammonium sulfate (about 40 per cent) added to 100 ml of 0.2–0.5 N nitric acid produces a pale orange color with 0.01 ml of 0.1 N thiocyanate and changes to red-orange with a larger excess. The indicator blank therefore is negligibly small.

[5] I. M. Kolthoff, W. M. Lauer, and C. J. Sunde, *J. Am. Chem. Soc.* **51**, 3273 (1929).

[6] For further details concerning the use and application of various other dyes as adsorption indicators, the student is referred to I. M. Kolthoff and V. A. Stenger, *Volumetric Analysis*, vol. II, Wiley-Interscience, New York, 2nd ed., 1947; on the mechanism of the color change with adsorption indicators, see I. M. Kolthoff, *Kolloid-Z.* **68**, 190 (1934).

In the titration of silver with thiocyanate in acid medium, using ferric iron as an indicator, the color change appears 0.7 to 1 per cent before the equivalence point. The silver thiocyanate formed adsorbs silver ions, which are removed from the solution. Therefore, after the first end point has been reached, the titration must be continued with vigorous agitation until the orange color does not disappear, even after strong shaking. Then all the adsorbed silver is transformed into the thiocyanate, and highly accurate results are obtained.

An important application of this silver titration is made in the determination of chloride, and also of bromide and iodide in acid medium. The halides cannot be titrated in strongly acid medium by Mohr's method or by the use of adsorption indicators. Therefore, an excess of standard silver nitrate is added to the solution and the excess is back-titrated with standard thiocyanate. This is the well-known *Volhard* method.[7]

In the chloride determination according to Volhard, a difficulty is encountered in the back-titration of the excess of silver:

$$Cl^- + Ag^+(\text{excess}) \rightleftharpoons AgCl$$
$$Ag^+ + SCN^- \rightleftharpoons AgSCN$$

We have here two slightly soluble salts, AgCl and AgSCN, in equilibrium with the solution. Hence the following relation holds:

$$\frac{[Cl^-]}{[SCN^-]} = \frac{(K_{sp})_{AgCl}}{(K_{sp})_{AgCNS}} = 100$$

After all the silver has been titrated back, an excess of thiocyanate may react with the silver chloride, since silver thiocyanate is less soluble than silver chloride:

$$AgCl + SCN^- \rightleftharpoons AgSCN + Cl^-$$

Equilibrium is established when $[Cl^-]/[SCN^-]$ has become equal to 100, or, in other words, after practically all the excess thiocyanate has reacted. The end point, therefore, is not very sharp.

Let us take, by way of illustration, a case in which 100 ml of a silver chloride suspension is treated with 0.1 ml of 0.1 N thiocyanate, corresponding to a concentration of 10^{-4}. After equilibrium is established:

$$[Cl^-] \cong \frac{99}{100} \times 10^{-4}$$

and
$$[SCN^-] = \frac{1}{100} \times 10^{-4}$$

The sensitivity of the ferric indicator towards thiocyanate corresponds to a concentration of 10^{-5} N of the latter. Therefore, according to this calculation, it will be necessary to add an excess of 1.0 ml of 0.1 N thiocyanate before a permanent end point is obtained.

This reaction between silver chloride and the excess of thiocyanate is relatively slow. Rothmund and Burgstaller[8] have recommended that the aqueous suspenion

[7] J. Volhard, *J. Prakt. Chem.* **117**, 217 (1874).
[8] V. Rothmund and A. Burgstaller, *Z. Anorg. Chem.* **63**, 330 (1909).

be covered with a layer of ether, benzene, ligroin, or any other organic solvent immiscible with water. The silver chloride agglomerates at the water-organic solvent interface and is withdrawn from the action of the thiocyanate solution. This method gives fair results. Excellent results, however, are obtained if about 1 ml of nitrobenzene is added to the suspension before back-titrating the excess of silver. A sharp and permanent end point was found by Caldwell[9] under these conditions, and the authors were able to confirm the results. Another alternative is to add an excess of silver nitrate to the acid chloride solution, make up to the mark in a volumetric flask, and titrate an aliquot part of the clear supernatant liquid or the filtrate with thiocyanate. The advantage of the filtration method is that a sharp and permanent end point is obtained.

In the above we have made the *approximation* that all the dissolved thiocyanate is present as SCN^-. This is not correct, because more or less of the thiocyanate combines with the indicator, ferric ion, to form $FeSCN^{++}$. In a solution which is saturated with silver chloride and thiocyanate and which does not contain excess of any of the lattice ions, the following relation holds:

$$[Ag^+] = [Cl^-] + [SCN^-] \tag{1}$$

Moreover,

$$\frac{[Cl^-]}{[SCN^-]} \simeq 100 \tag{2}$$

In the presence of ferric iron part of the thiocyanate ions forms red $FeSCN^{++}$. Therefore, in a suspension composed of AgCl and AgSCN and ferric iron we have

$$[Ag^+] = [Cl^-] + [SCN^-] + [FeSCN^{++}] \tag{3}$$

In 100 ml of a solution which contains nitric acid and ferric iron the red color is detectable when $[FeSCN^{++}]$ is 6×10^{-6} *M*. Thus we have at the equivalence point in the Volhard titration

$$[Ag^+] = [Cl^-] + 0.01[Cl^+] + 6 \times 10^{-6} \tag{3a}$$

Since $[Ag^+][Cl^-] = 10^{-10}$, we find at the equivalence point that $[Cl^-] = 7 \times 10^{-6}$. From Eq. (2) it follows that $[SCN^-] = 7 \times 10^{-8}$.

The equilibrium constant[10]

$$K = \frac{[Fe^{3+}][SCN^-]}{[FeSCN^{++}]} = 7.2 \times 10^{-3}$$

From the above relations it follows that at the equivalence point the ferric iron concentration $[Fe^{3+}]$ should be 0.6. Actually, at the end point the ferric concentration must be somewhat smaller, because at this high concentration the iron color would interfere with the detection of the end point. Swift et al.,[11] who used the above theoretical considerations in the development of simplified Volhard procedures, found a sharp end point at a concentration of 0.2 *M* ferric iron. Swift and coworkers develped two procedures. In one procedure ferric thiocyanate is used as a direct indicator and no back-titration is required. The second procedure resembles the original Volhard method, but it requires two back-titrations. The end point in both titrations is stable (see p. 798).

[9] J. R. Caldwell, *Ind. Eng. Chem., Anal. Ed.* **7**, 38 (1935).
[10] H. S. Frank and R. L. Oswalt, *J. Am. Chem. Soc.* **69**, 1321 (1947).
[11] E. H. Swift, G. M. Arcand, R. Lutwack, and D. J. Meier, *Anal. Chem.* **22**, 306 (1950).

In the titration of bromide according to Volhard, no difficulties are encountered in the determination of the end point because silver bromide is slightly less soluble than silver thiocyanate. Iodide also can be determined very accurately by the Volhard method. In this case one should not add the iron indicator until there is an excess of silver, because dissolved iodide reacts with ferric iron with the liberation of iodine:

$$Fe^{3+} + I^- \rightleftharpoons Fe^{++} + \tfrac{1}{2} I_2$$

PROBLEMS

1. Calculate the bromide- and silver-ion concentrations in the titration of 100 ml of 0.1 N potassium bromide with 0.1 N silver nitrate after addition of 50, 90, 98, 99, 99.8, 99.9, 100.00, 100.1, 100.2, 101, and 110 ml of the reagent. [$(K_{sp})_{AgBr} = 4 \times 10^{-13}$.]
2. What is the bromide-ion concentration in a 0.01 M potassium chromate solution in equilibrium with solid AgBr and Ag_2CrO_4? [$(K_{sp})_{AgBr} = 4 \times 10^{-13}$; $(K_{sp})_{Ag_2CrO_4} = 2 \times 10^{-12}$.] *Ans.* 3×10^{-8} M.
3. What is the solubility of silver oxalate, $Ag_2C_2O_4$, in a solution which is 0.01 N with respect to silver ions? [$(K_{sp})_{Ag_2C_2O_4} = 5 \times 10^{-12}$.] *Ans.* 5×10^{-8} M.
4. What is the ratio of $[Br^-]$ to $[SCN^-]$ in a solution in equilibrium with solid AgBr and AgSCN? [$(K_{sp})_{AgBr} = 4 \times 10^{-13}$; $(K_{sp})_{AgSCN} = 7 \times 10^{-13}$.]
Ans. $[Br^-]:[SCN^-] = 0.57$.
5. Describe the difference in the mechanism of the indicator action of the following indicators: fluorescein (adsorption indicator), chromate (in the Mohr titration), ferric sulfate in the Volhard titration, and sodium nitroprusside in the titration of chloride with mercuric nitrate.
6. According to Volhard 25.00 ml of a sodium chloride solution is titrated by adding 50.00 ml of 0.1100 N silver nitrate solution and diluting the mixture to 100 ml; 50 ml of the supernatant liquid require 5.23 ml of 0.0980 N thiocyanate. How much sodium chloride does the solution contain? Make the same calculation, but correct for the amount of silver nitrate adsorbed by the silver chloride, assuming that this adsorption amounts to 0.1 ml of 0.1100 N silver nitrate.
Ans. 0.2615 g of NaCl, uncorrected; 0.2609 g of NaCl, corrected.
7. The silver salt AgA of an organic acid has a solubility in water of 10^{-3} M. To 25 ml of 0.1 N solution of the sodium salt of this organic acid are added 25 ml of 0.12 N silver nitrate. (a) What is the solubility of AgA after the addition of silver nitrate? (b) What is the percentage of organic ion not precipitated?
Ans. (a) 10^{-4} M; (b) 0.2 per cent.
8. A 0.1 M solution of alkali iodate is titrated with 0.1 N silver nitrate, chromate being used as an indicator. The chromate concentration at the end point is 2×10^{-3} M. Solubility product of silver iodate is 2×10^{-8}, of silver chromate 2×10^{-12}. Calculate (a) the concentration of iodate at the end point; (b) the titration error.
Ans. (a) $[IO_3^-] = 6.3 \times 10^{-4}$; (b) -1.3 per cent.

CHAPTER 36 *Theory of Compleximetric Titrations*

The principles of complexation equilibria were presented in Chapter 6. The application of these principles to the quantitative interpretation of compleximetric titrations is the subject of the present chapter.

Calculation of titration curves

In a preceding chapter it was shown how to calculate $[H^+]$ or pH in acid–base titrations. The quantity of interest in compleximetric titrations is [M] or pM, where [M] is the molar concentration of a metal ion such as Ca^{++} or Al^{3+}.

Reaction with one ligand. Consider the simplest complexation reaction:

$$M + L \rightleftharpoons ML \tag{1}$$

The equilibrium constant, or formation constant, is

$$K = \frac{[ML]}{[M][L]} \tag{2}$$

In analytical applications concentrations rather than activities usually are wanted and a concentration constant is used. When the value of K is relatively large, reaction (1) goes practically to completion. For example, suppose 10 ml of 0.1 M L is added to 100 ml of 0.1 M M. Assuming complete reaction of the added L, there is formed 10 ml of 0.1 M ML and there remains 90 ml of 0.1 M M. Therefore, [ML] and [M] are, respectively, $(10 \times 0.1)/110 = 0.0091$ M and $(90 \times 0.1)/110 = 0.082$ M; pM $= 1.10$. The calculation is valid only if [L] is negligibly small compared to [ML]. From Eq. (2),

$$[L] = \frac{[ML]}{[M] \cdot K} \tag{3}$$

which is exact. If $K = 10^{10}$, $[L] = 1.1 \times 10^{-11}$ M, confirming that reaction (1) is complete. With much smaller values of K the simple calculation is not correct (see below; see also Problem 1). Assuming that K is large, [M] is calculated exactly as above up to the equivalence point in the titration. Thus after addition of 99 ml, [M] and [ML] are equal, respectively, to 5.03×10^{-4} M and 0.0497 M. The student should verify that [L] is still negligibly small if $K = 10^{10}$. After addition of the equivalent amount of L, [M] and [L] are exactly equal, regardless of the value of K. Hence $[ML] = 0.05 - [M]$ and Eq. (2) becomes

$$\frac{0.05 - [M]}{[M]^2} = K \tag{4}$$

For large values of K Eq. (4) becomes simply

$$[\mathrm{M}] = \left(\frac{0.05}{K}\right)^{1/2} \tag{5}$$

For $K = 10^{10}$, $[\mathrm{M}] = 2.24 \times 10^{-6}$ M; pM = 5.65. Beyond the equivalence point there is an excess of L, and the method of calculation is identical with that used before the equivalence point. The calculations needed for construction of the titration curve are collected in Table 36-1, and the curve (pM versus volume L added)

TABLE **36-1. Titration of 100 ml of 0.1 M M with 0.1 M L** ($K = 10^{10}$)

Volume Added, ml	*Molar Concentrations* [M]	[ML]	[L]	pM
0	0.100	0	0	1.00
10	$\frac{90}{110} \times 0.1$	$\frac{10}{110} \times 0.1$	$\frac{10}{90} \times 10^{-10}$	1.10
20	$\frac{80}{120} \times 0.1$	$\frac{20}{120} \times 0.1$	$\frac{20}{80} \times 10^{-10}$	1.18
30	$\frac{70}{130} \times 0.1$	$\frac{30}{130} \times 0.1$	$\frac{30}{70} \times 10^{-10}$	1.27
50	$\frac{50}{150} \times 0.1$	$\frac{50}{150} \times 0.1$	10^{-10}	1.48
80	$\frac{20}{180} \times 0.1$	$\frac{80}{180} \times 0.1$	$\frac{80}{20} \times 10^{-10}$	1.95
90	$\frac{10}{190} \times 0.1$	$\frac{90}{190} \times 0.1$	$\frac{90}{10} = 10^{-10}$	2.28
99	$\frac{1}{199} \times 0.1$	$\frac{99}{199} \times 0.1$	99×10^{-10}	3.30
99.9	$\frac{0.1}{199.9} \times 0.1$	$\frac{99.9}{199.9} \times 0.1$	$\frac{99.9}{0.1} \times 10^{-10}$	4.30
100	2.24×10^{-6}	$\frac{100}{200} \times 0.1$	2.24×10^{-6}	5.65
100.1	$\frac{100}{0.1} \times 10^{-10}$	$\frac{100}{200.1} \times 0.1$	$\frac{0.1}{200.1} \times 0.1$	7.00
101	$\frac{100}{1} \times 10^{-10}$	$\frac{100}{201} \times 0.1$	$\frac{1}{201} \times 0.1$	8.00
110	$\frac{100}{10} \times 10^{-10}$	$\frac{100}{210} \times 0.1$	$\frac{10}{210} \times 0.1$	9.00
120	$\frac{100}{20} \times 10^{-10}$	$\frac{100}{220} \times 0.1$	$\frac{20}{220} \times 0.1$	9.30
150	$\frac{100}{50} \times 10^{-10}$	$\frac{100}{250} \times 0.1$	$\frac{50}{250} \times 0.1$	9.70

is shown in Fig. 36-1. The calculation for large K can be summarized simply, for the addition of v ml of cM L to 100 ml of cM M:

	[M]	[ML]	[L]	
$v < 100$	$\frac{100 - v}{100 + v} \cdot c$	$\frac{v}{100 + v} \cdot c$	$\frac{v}{100 - v} \cdot \frac{1}{K}$	(6)
$v > 100$	$\frac{100}{v - 100} \cdot \frac{1}{K}$	$\frac{100}{v + 100} \cdot c$	$\frac{v - 100}{v + 100} \cdot c$	

It is left as an exercise for the student to generalize these expressions to unequal concentrations of reactants.

Figure 36-1 also contains titration curves for values of log K equal to 14, 12, 8, 6, 4, and 2. For large values of K the curves are coincident before the equivalence point and parallel beyond the equivalence point. From the expression given above for large K, $v > 100$, there is obtained

$$pM_{(v>100)} = \log(v - 100) + \log K - 2 \tag{7}$$

Thus a given increase in log K causes an equal vertical displacement in the pM curve.

The method of calculation outlined above is incorrect for small values of K. Although the general method is described in Chapter 6, a few numerical examples here may be helpful. Consider the addition of 99 ml of 0.1 M L to 100 ml of 0.1 M M. When K is small, reaction (1) is not complete, and the exact expressions should be used. The procedure is as follows:

$$[M] + [ML] = \frac{100 \times 0.1}{199}$$

$$[L] + [ML] = \frac{99 \times 0.1}{199}$$

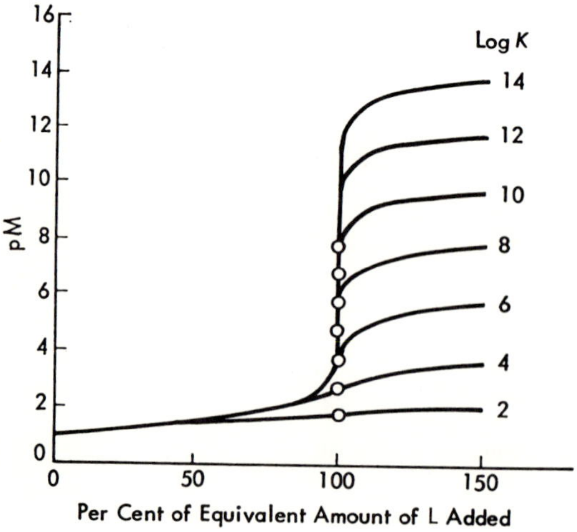

Fig. 36-1. Titration of 0.1 M M with 0.1 M L; log K = 14, 12, . . . 2. Circles indicate pM at equivalence points.

Eliminating [ML],

$$[M] = [L] + \frac{0.1}{199}$$

Eliminating [ML] and [L] from Eq. (2) the result is

$$[M]^2 + \left(\frac{1}{K} - \frac{0.1}{199}\right)[M] - \frac{10}{199K} = 0 \qquad (8)$$

The student should derive the corresponding equation applicable to any concentrations and ratios of volume. The solutions of Eq. (8) for various values of K are:

K	$[M]$
10^2	0.01817
10^4	0.00245
10^6	0.000587
10^8	0.000504

Evidently for very large K, [M] approaches the limit $0.1/199 = 0.000503\ M$.

For values of K larger than 10^4 it may be preferable to use a method of successive approximations. First approximations to [M] and [L] are calculated on the basis of complete reaction. A first approximation to [L] is obtained from Eq. (3). The second approximations to [M] and [ML] are obtained by addition and subtraction, respectively, of [L] to the first approximations. The process is continued until a constant value is reached. In suitable situations this method of calculation is much faster than writing out and solving the exact quadratic equation. The method is illustrated for $K = 10^6$, $v = 99$ ml. The numerical work is:

Approximation	[M]	[ML]	$\frac{[ML]}{[M]}$	[L]
First	0.000503	0.0497	99	9.9×10^{-5}
Second	0.000602	0.0496	82.4	8.2×10^{-5}
Third	0.000585	0.0496	84.8	8.5×10^{-5}
Fourth	0.000588	—	—	—

Change of pM around equivalence point

Table 36-2 contains values of pM within 1 per cent of the equivalence point for $\log K = 14, 12, \ldots, 2$, in the titration of equimolar solutions of concentration 0.1, 0.01, and 0.001 M. The values are accurate to three decimal places, although such accuracy usually is not attainable in practical measurements. By comparing the exact values with the results of approximate calculations [see Table 36-1, Eq. (7)], the student may learn the conditions under which the approximations are reliable to any given accuracy.

TABLE 36-2. **pM Close to Equivalence Point Titration of c*M* M with c*M* L**

	pM						
% of Equiv. Amount	log K = 14	12	10	8	6	4	2
$c = 0.1$							
99	3.299	3.299	3.299	3.298	3.231	2.610	1.741
99.8	4.000	4.000	3.999	3.979	3.555	2.650	1.746
99.9	4.301	4.301	4.300	4.232	3.603	2.655	1.746
100.0	7.651	6.651	5.651	4.651	3.651	2.660	1.747
100.1	11.000	9.000	7.001	5.069	3.700	2.665	1.747
100.2	11.302	9.301	7.301	5.321	3.748	2.670	1.748
101	12.005	10.000	8.000	6.001	4.069	2.710	1.753
$c = 0.01$							
99	4.299	4.299	4.299	4.291	4.001	3.165	2.434
99.8	5.000	5.000	4.997	4.864	4.123	3.178	2.436
99.9	5.301	5.301	5.292	5.000	4.138	3.180	2.436
100.0	8.151	7.151	6.151	5.151	4.154	3.181	2.436
100.1	11.000	9.000	7.008	5.301	4.169	3.183	2.437
100.2	11.301	9.301	7.303	5.437	4.184	3.184	2.437
101	12.000	10.000	8.000	6.009	4.305	3.198	2.439
$c = 0.001$							
99	5.299	5.299	5.298	5.231	4.610	3.741	3.319
99.8	6.000	5.999	5.979	5.555	4.650	3.746	3.321
99.9	6.301	6.300	6.232	5.603	4.655	3.746	3.321
100	8.651	7.651	6.651	5.651	4.660	3.747	3.321
100.1	11.000	9.001	7.069	5.700	4.665	3.747	3.322
100.2	11.301	9.301	7.321	5.748	4.670	3.748	3.322
101	12.000	10.000	8.001	6.069	4.710	3.753	3.324

Figure 36-2 is a plot of pM versus log K for addition of 99, 99.8, . . ., 100.2 per cent of the equivalent amount of 0.1 *M* L to 0.1 *M* M. The student may use Table 36-2 to prepare similar plots for 0.01 *M* and 0.001 *M* solutions. The change in pM in the neighborhood of the equivalence point may be obtained from such figures. For example, in the titration of 0.1 *M* solutions, pM changes by about two units within 0.1 per cent of the equivalence for log K = 9.2, and within 1 per cent of the equivalence point for log K = 7.0.

The paragraphs above contain everything that is needed to calculate the titration curve in any compleximetric titration in which a 1 : 1 complex is formed beween metal ion and chelating agent. If complications arise due to reaction of L with hydrogen ion, or reaction of M with hydroxide ion or with other substances, these complications may be absorbed into the value of K, as explained in Chapter 6 (pp. 123–124) by the use of a conditional equilibrium constant. Some numerical examples are given on pages 739–741.

Addition of two ligands. The reactions to be considered are

$$M + L \rightleftharpoons ML \tag{9}$$

$$ML + L \rightleftharpoons ML_2 \tag{10}$$

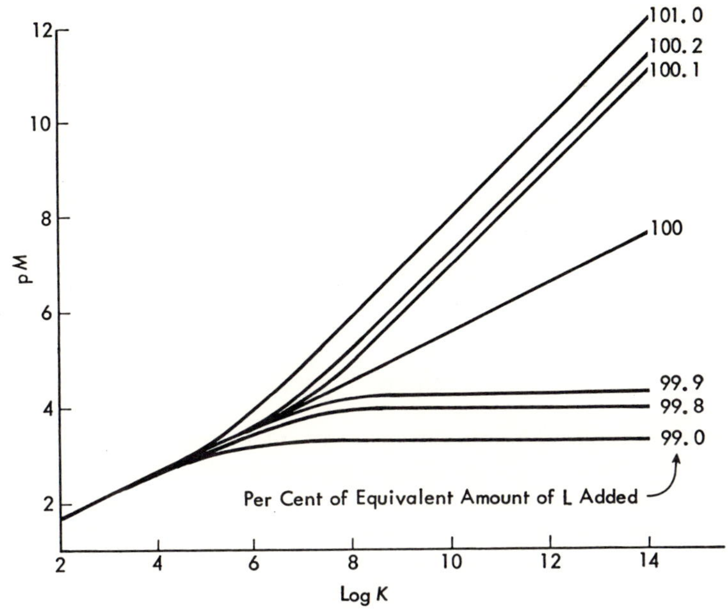

Fig. 36-2. Variation of pM with log K in titration of 0.1 M M with 0.1 M L.

The corresponding concentration equilibrium constants are

$$K_1 = \frac{[\mathrm{ML}]}{[\mathrm{M}][\mathrm{L}]} \tag{11}$$

$$K_2 = \frac{[\mathrm{ML_2}]}{[\mathrm{ML}][\mathrm{L}]} \tag{12}$$

Practical examples include the reactions of Ag^+ or Cu^+ with NH_3, although in these cases hydrolytic metal species have to be considered for an exact treatment.

There are two ways to calculate the titration curve. One way is to make at the outset some simplifying assumptions; if they are not valid, successive approximations must be made. The other way involves no assumption, and the volume of the titrant is calculated exactly as a function of pM. One or the other method may be preferable in practice, depending upon the values of K_1 and K_2, and both methods should be mastered. Both are illustrated for the case $K_1 = 10^6$, $K_2 = 10^4$ (approximately the values for Cu^+–NH_3).

Consider the addition of 10 ml of 0.1 M L to 100 ml of 0.1 M M. The approximate method is as follows. There are three M-containing species, and it is obvious that M is the major species. The question is whether ML or ML_2, or both, have to be considered. As a trial assume that reaction (9) goes to practical completion and reaction (10) is negligible. Then $[\mathrm{M}] = 0.082$ and $[\mathrm{ML}] = 0.0091\ M$, whence from Eq. (11), $[\mathrm{L}] = 1/9K_1 = 1.1 \times 10^{-7}\ M$ and from Eq. (12), $[\mathrm{ML_2}] = 10^4 \times 0.0091 \times 1.1 \times 10^{-7} = 10^{-5}\ M$. Evidently both assumptions are correct. The same is true after addition of 50 ml. After addition of 90 ml, however, the two assumptions lead to the following concentrations: $[\mathrm{M}] = 5.27 \times 10^{-3}$,

$[ML] = 0.0473$, $[L] = 9 \times 10^{-6}$, and $[ML_2] = 4.3 \times 10^{-3}$ *M*. Since $[ML_2]$ is not negligibly small, the calculation must be improved.

It is instructive to consider the problem from a different point of view. The various mixtures of M and L are identical with appropriate mixtures of ML and M. Thus a solution made by mixing 90 ml of 0.1 *M* L and 100 ml of 0.1 *M* M is identical with one made by mixing 90 ml of 0.1 *M* ML, 10 ml of 0.1 *M* M, and diluting to 190 ml. In a solution containing ML, reaction (13) may or may not have to be considered:

$$2\,ML \rightleftharpoons M + ML_2 \tag{13}$$

The equilibrium constant is

$$\frac{[M][ML_2]}{[ML]^2} = \frac{K_2}{K_1} \tag{14}$$

In the present example the equilibrium constant of reaction (13) is only 10^{-2}. When a sufficient amount of M (or ML_2) is present together with ML, reaction (13) is practically suppressed. This is the reason the approximate calculation works well in the first two cases (addition of 10 or of 50 ml of L). After addition of 90 ml, however, [M] is so small that (13) occurs to an appreciable extent, and this must be corrected for. The calculation is as follows. If (13) did not occur, [ML] and [M] would be 0.0473 and 0.00527 *M*, respectively. As a result of reaction (13),

$$[ML] = 0.0473 - [ML_2]$$
$$[M] = 0.00527 + [ML_2]$$

Rewriting so as to solve for [M],

$$[ML] = 0.0526 - [M]$$
$$[ML_2] = [M] - 0.00527$$

Substituting these values in Eq. (14), $[M] = 0.00778$ *M*. The same method of calculation can be used for values of L larger than 100 ml.

The above example shows that the approximate method is more suitable the larger the ratio K_1/K_2. When the latter quantity is as small as 10, as it is in many complexation systems, the method becomes tedious and an exact calculation is preferable.

This is carried out as follows. Equations (15)–(18), derived in Chapter 6, are used:

$$f_0 = \frac{1}{1 + K_1[L] + K_1K_2[L]^2} \tag{15}$$

$$\bar{n} = \frac{K_1[L] + 2K_1K_2[L]^2}{1 + K_1[L] + K_1K_2[L]^2} \tag{16}$$

$$[M] = f_0 C_M \tag{17}$$

$$C_L = \bar{n} C_M + [L] \tag{18}$$

A value of [L] is chosen, and f_0 and $\bar{n}$ are calculated using Eqs. (15) and (16). Equation (17) then yields [M] and (18) yields C_L/C_M. The calculation is repeated for as many values of [L] as are needed to construct the titration curve. This is all that is required if the total volume is practically constant. If the change in volume during titration must be corrected for, one additional arithmetical step is required. If v ml of C_L^0 M L is added to V ml of C_L^0 M M, we have

$$C_M = \frac{V}{V+v} C_M^0 \qquad C_L = \frac{v}{V+v} C_L^0$$

Then after $\bar{n}$ has been calculated as above, from Eq. (18)

$$\frac{v}{V+v} C_L^0 = \bar{n} \frac{V}{V+v} C_M^0 + [L] \tag{19}$$

whence v/V is found for any [L]. The numerical work for titration at constant volume is illustrated in Table 36-3. Practically, it is uneconomical of effort to calculate for evenly spaced values of [L]. The best procedure is to calculate C_L/C_M for a few widely spaced values of [L], and then decide which additional calculations are needed.

TABLE 36-3. **Titration of M with L, Two Complexes Formed**
($K_1 = 10^6$, $K_2 = 10^4$; $C_M = 0.1$ M, constant volume)

[L]	$1 + K_1[L] + K_1K_2[L]^2$	$K_1[L] + 2K_1K_2[L]^2$	$\bar{n}$	$\frac{[M]}{C_M}$	$\frac{C_L}{C_M}$	pM
10^{-8}	1.010	0.0100	0.0099	0.990	0.0099	1.004
10^{-7}	1.100	0.1002	0.0912	0.909	0.0912	1.040
10^{-6}	2.01	1.02	0.507	0.497	0.507	1.295
10^{-5}	12	12	1.000	0.0833	1.000	2.079
10^{-4}	201	300	1.492	0.00497	1.493	3.303
(At this stage it is evident from the big increase in pM that more values of [L] around 10^{-5} are needed)						
5×10^{-6}	6.25	5.50	0.880	0.1600	0.880	1.796
8×10^{-6}	9.64	9.28	0.963	0.1038	0.963	1.984
9×10^{-6}	10.81	10.62	0.982	0.0925	0.982	2.034
1.1×10^{-5}	13.21	13.42	1.016	0.0757	1.016	2.121
1.2×10^{-5}	14.44	14.88	1.031	0.0692	1.031	2.159
2×10^{-5}	25.00	28.00	1.120	0.0400	1.120	2.398
(The above calculations give sufficient points around the first equivalence point)						
10^{-3}	1.100×10^4	2.10×10^4	1.910	9.09×10^{-5}	1.920	5.041
10^{-2}	1.01×10^6	2.01×10^6	1.989	9.90×10^{-7}	2.089	7.004
10^{-1}	1.001×10^8	2.001×10^8	1.999	1.00×10^{-8}	2.999	9.000
(To define the second equivalence point, a few more values of [L] between 10^{-3} and 10^{-1} are needed)						
2×10^{-3}	4.20×10^4	8.20×10^4	1.955	2.38×10^{-5}	1.975	5.624
5×10^{-3}	2.55×10^5	5.05×10^5	1.980	3.92×10^{-6}	2.030	6.407
8×10^{-3}	6.48×10^5	1.288×10^6	1.985	1.543×10^{-6}	2.065	6.812
5×10^{-2}	2.505×10^7	5.005×10^7	2.00	4.0×10^{-8}	2.50	8.398

Figure 36-3 shows that no pronounced change in pM occurs at either equivalence point, and therefore these points cannot be located by the use of indicators. This is true despite the fact that ML_2 is relatively stable compared to M and L: $[ML_2] = 10^{10}[M][L]^2$. If, on the other hand, a bidentate ligand of comparable overall stability is used, i.e., $K = 10^{10}$ in Eq. (2), a pronounced change in pM occurs at the equivalence point and a precise titration can be carried out. The example shows why bi- or polydentate ligands are much superior for titrations to unidentate ligands such as ammonia or fluoride.

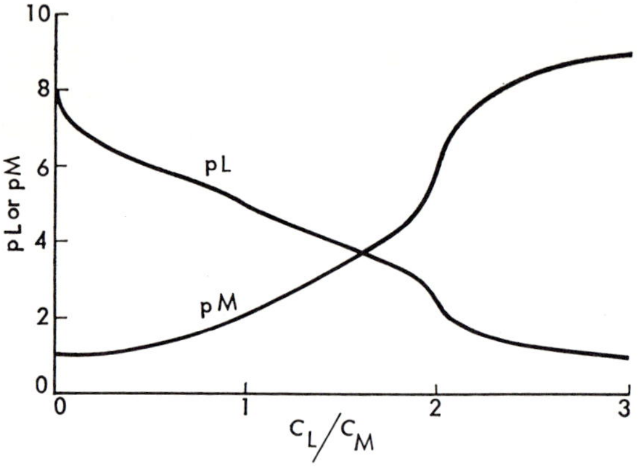

Fig. 36-3. Titration of 0.1 *M* M with L at constant volume; $\log K_1 = 6.0$, $\log K_2 = 4.0$

Of course, if K_1 and K_2 are very large, and their ratio also is large, a titration is possible. In this case, [ML] is the main species only at extremely small values of [L], and the only important reaction is

$$M + 2\,L \rightleftharpoons ML_2$$

This is the case, for example, in the titration of cyanide with silver (see pp. 741–742).

Addition of more than two ligands

The treatment is an obvious extension of that given above. The hypothetical case $Me^{3+} + F^-$, in which complexes $MeF^{2+}, \ldots, MeF_6^{-3}$ are formed, has been discussed in Chapter 6. Making use of the same values of K_n used before ($\log K_1, \ldots, \log K_6 = 6.2, 5.0, 3.9, 2.7, 1.6$, and 0.5), the exact titration curve at constant volume has been calculated [equations analogous to (15)–(18)] with the result shown in Fig. 36-4.

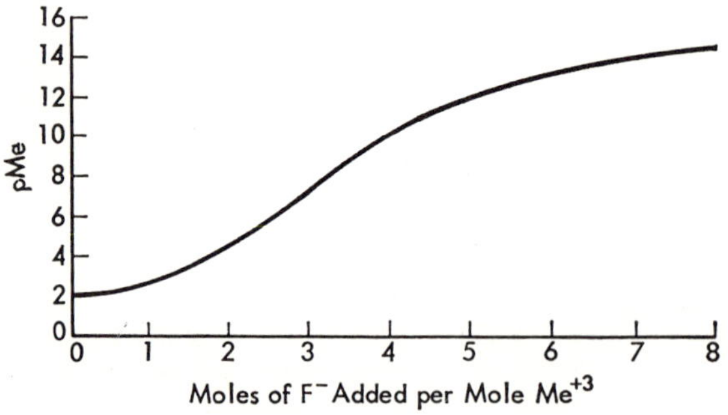

Fig. 36-4. Titration of 0.01 *M* Me^{+3} with F^- at constant volume. $\log K_1, \ldots, \log K_6 = 6.2, 5.0, 3.9, 2.7, 1.6, 0.5$.

Despite the magnitude of the overall formation constant, $\beta_6 = 7.3 \times 10^{19}$, the curve of pMe versus added F^- rises smoothly and no equivalence point can be detected. Again this curve should be compared with that corresponding to reaction with one hexadentate ligand.

Reaction of chelating agent with more than one metal ion: Consecutive Titrations

When a chelating agent L is added to a mixture of two metal ions, M_1 and M_2, with each of which it reacts in a 1 : 1 mole ratio, the metal ions compete for L:

$$M_1 + L \rightleftharpoons M_1L \tag{20}$$

$$M_2 + L \rightleftharpoons M_2L \tag{21}$$

If the equilibrium constants are, respectively, K_{M_1} and K_{M_2},

$$\frac{[M_1L]}{[M_1]} = \frac{K_{M_1}}{K_{M_2}} \frac{[M_2L]}{[M_2]} \tag{22}$$

If K_{M_1} is much larger than K_{M_2}, L reacts almost quantitatively with M_1 before any appreciable reaction with M_2 occurs. An example will illustrate for $K_{M_1} = 10^{15}$, $K_{M_2} = 10^{10}$.[1] Assume a solution 0.01 *M* in M_1 and in M_2 to which there is added the fraction f of the amount of L equivalent to M_1 (volume constant). The exact values of the various concentrations are found from the relations

$$[M_1] + [M_1L] = 0.01 \tag{23}$$

$$[M_2] + [M_2L] = 0.01 \tag{24}$$

$$[L] + [M_1L] + [M_2L] = 0.01f \tag{25}$$

plus the two equilibrium constants. Exact values are found by assuming various values of [L] (see Table 36-3). Values of $[M_1]$, pM, and $[M_2]$ thus calculated are given as a function of amount of L added in Table 36-4. pM_1 changes from

TABLE 36-4. **Addition of L to Mixture 0.01 *M* in M_1 and in M_2** ($K_{M_1} = 10^{15}$, $K_{M_2} = 10^{10}$; constant volume)

% L[a]	$[M_2]$	pM_1
99.0	0.009990	3.96
99.8	0.009977	4.37
99.9	0.009974	4.43
100.0	0.009969	4.50
100.1	0.009962	4.57
100.2	0.009957	4.64
101.0	0.009892	5.05

[a] Per cent of amount equivalent to M_1.

[1] These are the approximate values of the conditional constants for $M_1 = Fe^{3+}$, $M_2 = Ni^{++}$, L = EDTA, at pH 4.

approximately 4.0 to 5.0 within 1 per cent of the equivalence point corresponding to M_1. The corresponding change in pL is from approximately 13.0 to 12.0. At the second equivalence point, corresponding to M_2, the changes in pM_2 and pL are, of course, very pronounced (see Fig. 36-1).

The situation is entirely analogous to the titration of mixtures of weak acids (see Chapter 35). In mixtures in which successive formation constants differ by factors of 10^5 or more, titration of both M_1 and M_2 is possible, provided both K_{M_1} and K_{M_2} are sufficiently large.

Indicators for metal ions

These indicators can be divided into two broad classes: (1) colorless (or relatively weakly colored) substances which react with a metal ion to form an intensely colored or a slightly soluble product, and (2) dyes which react with a metal ion to form a dye of a different color. Several members of the first class have been used for analytical purposes for a long time. For example, colorless thiocyanate ion reacts with ferric ion to form intensely colored red complex ions; the color is detectable by the eye at concentrations of about 10^{-5} *M*. Iodide reacts with silver ion to form slightly soluble silver iodide, the presence of a very small amount of which can be recognized by turbidity. Quantitative discussion of the uses of these indicators is given elsewhere (pp. 723, 800) and further discussion of metal indicators is confined to members of the second class.

These substances are often called "metallochromic" indicators. They are chelating agents which contain a chromophoric ("color-forming") group. There are several requirements that an indicator should satisfy. A degree of selectivity is highly desirable. The stability of the dye-metal complex must be in the proper range; if the stability is relatively small, there is no possibility of a sharp color change, but obviously the stability must be less than that of the metal complex with the titrant. Finally, the reaction with metal ion should be rapid and reversible (many indicators are deficient in the last respect). The azo and phthalein groups are among the most common chromophores. Eriochrome Black T and PAN[2] are examples of indicators containing the azo group:

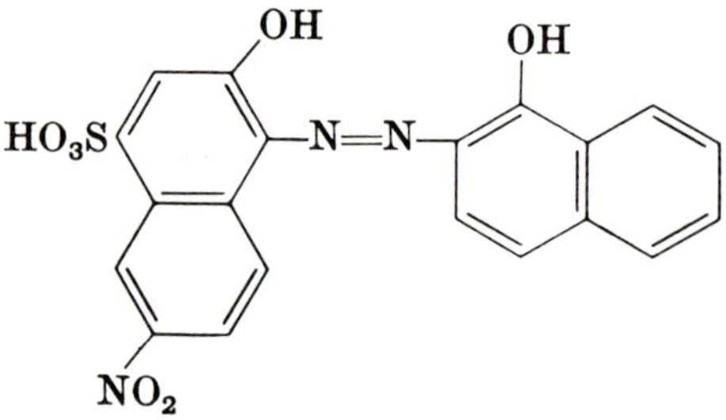

Eriochrome Black T

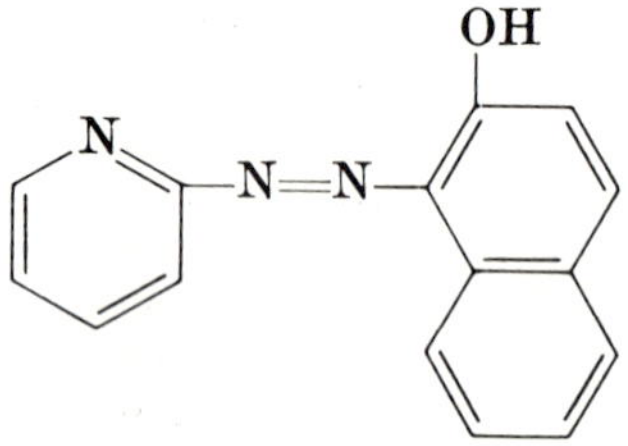

PAN

Dithizone (see p. 351) is another azo-containing indicator. Phthalein complexone

[2] PAN is the abbreviation for 1-2-(pyridylazo)-2-naphthol.

is an example of an EDTA-type of chelating agent containing the phthalein chromophore:

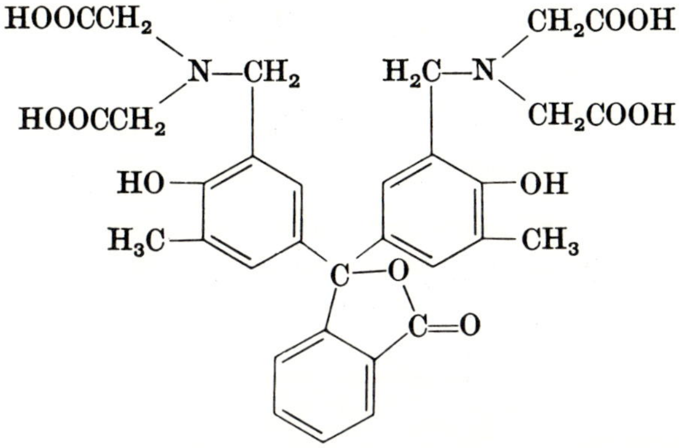

(compare the structure of the acid form of phenolphthalein, p. 695).

The three substances whose structures are given above, and all other metallochromic indicators, are acid–base indicators as well as chelating agents. For example, Eriochrome Black T has three replaceable hydrogens, and the substance therefore may be represented as H_3In. Depending upon the pH of the solution, the indicator, in the absence of a metal ion, may be present as H_3In, H_2In^-, HIn^{--}, or In^{3-}. The respective pK values are 1.6 (ionization from sulfonic group), 6.3, and 11.6. Therefore unless the solution is strongly acid, which is not usually the case in compleximetric titrations, only the species H_2In^-, HIn^{--}, and In^{3-} need be considered; their colors are red, blue, and orange, respectively. Evidently the acid–base transition intervals of the indicator are centered approximately at pH 6.3 and pH 11.6, and the indicator is blue in the range 7–11. The indicator reacts with many metal ions to form a metal complex; all the complexes are red and generally similar in color to H_2In^-. With a divalent ion Me^{++} (e.g., Mg^{++}, Ca^{++}, Ba^{++}, Zn^{++}) the complex has the formula $MeIn^-$.[3] When the metal complex is formed in the pH range 7–11 the reaction is

$$Me^{++} + HIn^{--} \rightleftharpoons MeIn^- + H^+ \tag{26}$$

and a color change occurs from blue to red.

Sensitivity of metal indicators

The equilibrium constant[4] of reaction (26) is

$$\frac{[MeIn^-][H^+]}{[Me^{++}][HIn^{--}]} = K \tag{27}$$

Equation (27) may be written

$$\frac{[MeIn^-]}{[HIn^{--}]} = \frac{[red]}{[blue]} = \frac{K \cdot [Me^{++}]}{[H^+]} \tag{28}$$

[3] Eriochrome Black T is a bidentate ligand and $MeIn_2^{4-}$ may be formed at high concentrations of indicator. This species is ignored here because it is of no consequence in titrations.

[4] Experimentally K is often determined using a pH meter to obtain hydrogen-ion activity and a spectrophotometer to obtain concentrations of the other species. Such a measurement yields a mixed activity-concentration equilibrium constant.

Therefore, half of the indicator is present in the red (metal complex) form, and half in the blue (HIn^{--}) form, when

$$[Me^{++}] = \frac{[H^+]}{K} \tag{29}$$

This value of $[Me^{++}]$ evidently is the transition value:

$$pMe_{trans} = pH + \log K \tag{30}$$

Metallochromic indicators accordingly have a transition interval of about two pMe units, exactly analogous to the pH interval of acid–base indicators.

By using the third dissociation constant of H_3In,

$$\frac{[H^+][In^{3-}]}{[HIn^{--}]} = K_{In^{3-}} \tag{31}$$

Eq. (27) may be written

$$\frac{[MeIn^-]}{[Me^{++}][In^{3-}]} = \frac{K}{K_{In^{3-}}} = K_{Me-In} \tag{32}$$

Values of K_{Me-In} are known for many systems; a few important values are given in Table A-8. By way of illustration the reaction between Mg^{++} and Eriochrome Black T is considered; $K_{Me-In} = 10^{7.0}$ (at ionic strength 0.08).

At high pH, reactions between Mg^{++} and OH^- must be considered. The formation constant for the reaction

$$Mg^{++} + OH^- \rightleftharpoons MgOH^+ \tag{33}$$

is $10^{2.6}$, and the solubility product is

$$[Mg^{++}][OH^-]^2 = (K_{sp})_{Mg(OH)_2} = 10^{-10.4} \tag{34}$$

both values at ionic strength 0.1. Equations (33) and (34) yield the following values for solutions saturated with magnesium hydroxide:

pH	13	12	11	10	9
$[Mg^{++}]$, M	$10^{-8.4}$	$10^{-6.4}$	$10^{-4.4}$	$10^{-2.4}$	$10^{-0.4}$
$[MgOH^+]$, M	$10^{-6.8}$	$10^{-5.8}$	$10^{-4.8}$	$10^{-3.8}$	$10^{-2.8}$

Consider the system at pH 10. A precipitate will form if $[Mg^{++}]$ exceeds $10^{-2.4}$ M $= 0.004$ M. At pH 10 almost all the indicator not bound to magnesium is present as HIn^{--}. To find the condition that half of the indicator is present as $MgIn^-$, use Eq. (27):

$$\frac{[MgIn^-]}{[HIn^{--}]} = \frac{K \times [Mg^{++}]}{[H^+]} = \frac{K_{In^{3-}} \times K_{Mg-In} \times [Mg^{++}]}{[H^+]}$$

$$= \frac{10^{-11.6} \times 10^{7.0} \times [Mg^{++}]}{10^{-10}} = 10^{5.4}[Mg^{++}]$$

Hence one half of the indicator is present as $MgIn^-$ when $[Mg^{++}] = 10^{-5.4}\ M$; i.e., $pMg = 5.4$. Stated in other words, $pMg_{trans} = 5.4$ at pH 10.

At pH 12 the color change upon reaction with magnesium is from orange[5] to red. The concentration of magnesium must be less than $10^{-4.4}\ M$ to avoid precipitation. The calculation of pMg_{trans} is a little more complicated than at pH 10 because appreciable amounts of indicator are present as In^{3-} and HIn^{--}; using $K_{In^{3-}}$, $[In^{3-}] = 2.5[HIn^{--}]$. Therefore, the total concentration of indicator, C_{In}, is given by

$$\begin{aligned} C_{In} &= [In^{3-}] + [HIn^{--}] + [MgIn^-] \\ &= 3.5[In^{3-}] + [MgIn^-] \end{aligned}$$

We wish to find the value of $[Mg^{++}]$ at which $[MgIn^-] = \frac{1}{2}C_{In}$. Substitution of these expressions into Eq. (32), with $K_{Mg-In} = 10^{7.0}$, yields

$$[Mg^{++}] = 3.5 \times 10^{-7}\ M \qquad pMg_{trans} = 6.5$$

Proceeding in this way, the sensitivity of the indicator for magnesium can be calculated at a given pH. The calculation really amounts to a calculation of the conditional constant (pp. 123–124). If for simplicity and generality electrical charges are omitted and M represents metal and In indicator, the conditional constant is

$$\frac{[MIn]'}{[M]'[In]'} = K'_{M-In} \tag{35}$$

in which $[M']$ represents the total concentration of all M-containing species *except* those containing In, $[In]'$ represents total concentration of all In-containing species *except* those containing M, and $[MIn]'$ is the total concentration of all species containing M and In.[6] Thus the value of $[MIn]'$ includes the concentration of species such as MHIn and MInOH, if such should be formed.

The solid line in Fig. 36-5 represents the conditional stability constant for the system magnesium–Eriochrome Black T. Evidently $pMg_{trans} = \log K'_{Mg-In}$. The figure, which can be interpreted as a plot of pMg_{trans} versus pH, also shows

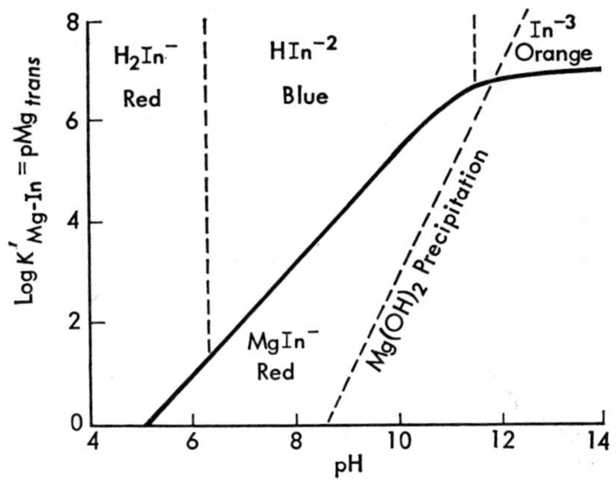

Fig. 36-5. Eriochrome Black T as indicator for magnesium.

[5] More exactly, bluish-orange.
[6] The discussion is restricted to 1:1 reaction products of M and In.

the colors of various regions and the conditions for precipitation of magnesium hydroxide.[7]

If the metal forms complexes other than hydroxy complexes, as, for example, zinc ion in an ammonia buffer, these must be taken into account in the calculation of K'_{M-In}. An example will illustrate: M = Zn^{++}, In = Eriochrome Black T; pH = 9; total concentration of ammonia = 0.1 M. The following equilibrium data are required: $K_{Zn-In} = 10^{12.9}$; solubility product of zinc hydroxide = $10^{-13.3}$; overall formation constants of hydroxy complexes, $\beta_1 = 10^{4.4}$, $\beta_3 = 10^{14.4}$, $\beta_4 = 10^{15.5}$; overall formation constants of ammonia complexes, $\beta_1 = 10^{2.27}$, $\beta_2 = 10^{4.61}$, $\beta_3 = 10^{7.01}$, $\beta_4 = 10^{9.06}$.[8] Using the ionization constant of ammonia, and assuming that the concentration of ammonia present in the zinc complexes is negligibly small compared to that of ammonia, the concentration of the latter is 0.036 $M = 10^{-1.44}$ M. Therefore when $[OH^-] = 10^{-5}$ M, we find

$$[Zn^{++}]' = [Zn^{++}](1 + 10^{4.4-5} + 10^{14.4-15} + 10^{15.5-20} + 10^{2.27-1.44} + 10^{4.61-2.88} + 10^{7.01-4.32} + 10^{9.06-5.76}) = 10^{3.41}[Zn^{++}]$$

No precipitation of zinc hydroxide occurs when $[Zn^{++}]$ is less than $10^{-5.3}$ M. Similarly, using the dissociation constants of the indicator, we find when $[H^+] = 10^{-9}$ M that

$$[In^{3-}] = 10^{-2.6}[HIn^{--}] \qquad \text{and} \qquad [HIn^{--}] = 10^{2.7}[H_2In^-]$$

Evidently practically all the indicator (not bound to zinc) is present as HIn^{--}, so $[In]' = 10^{2.6}[In^{3-}]$. Since $[ZnIn]'$ is assumed equal to $[ZnIn^-]$, we obtain

$$K'_{Zn-In} = \frac{[ZnIn^-]'}{[Zn^{++}]'[In]'} = \frac{K_{Zn-In}}{10^{3.4} \times 10^{2.6}} = \frac{10^{12.9}}{10^{6.0}} = 10^{6.9}$$

Accordingly, under these conditions $pZn_{trans} = 6.9$, corresponding to a high sensitivity for zinc.

Figure 36-6, due to Ringbom, gives a plot of K'_{Zn-In} calculated as above as a function of pH, in the absence of ammonia and in the presence of 0.1 M or 1 M total ammonia.[9] In the absence of ammonia, at pH around 9, the sensitivity of Eriochrome Black T for zinc is extremely high. In 1 M ammonia, on the other hand, the sensitivity is very low.

For a more detailed discussion, which is beyond the scope of this text, reference is made to the book by Ringbom.

Blocking of metal indicators

The name refers to a more or less irreversible reaction between metal ion and indicator. A conspicuous and troublesome example is the reaction between copper and Eriochrome Black T, because the $CuIn^-$ complex is formed rapidly and

[7] This figure is the same as one given by A. Ringbom, *Complexation in Analytical Chemistry*, Wiley-Interscience, New York, 1963, p. 85, except that $(K_{sp})_{Mg(OH)_2} = 10^{-10.4}$, instead of the value $10^{-10.0}$ used by Ringbom.

[8] These apply at ionic strength about 0.1, except the values for the hydroxy complexes, which refer to ionic strength close to zero.

[9] Ref. 7, p. 88.

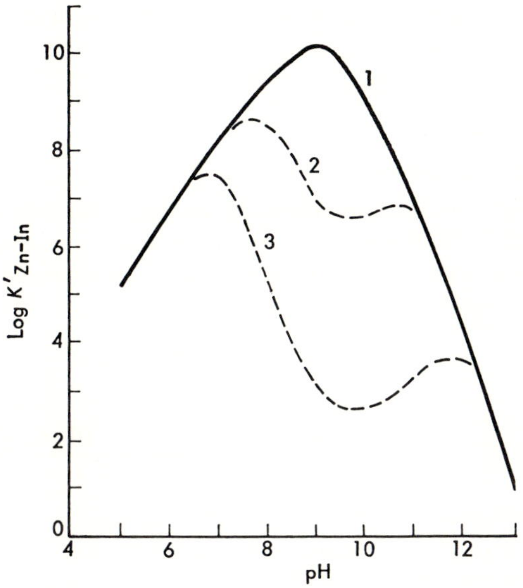

Fig. 36-6. Eriochrome Black T as indicator for zinc, in the absence and presence of ammonia. In curves 1, 2, and 3, $[NH_3] + [NH_4^+] = 0$, 0.1, and 1.0 M, respectively.

dissociates very slowly. Similar reactions occur with cobalt, nickel, and iron. Copper in particular is often present in water, and even a trace of copper spoils an EBT end point. Fortunately blocking by copper can be prevented by adding a little potassium cyanide, since the copper(I) cyanide complex does not react with EBT.

Specific compleximetric titrations

The following examples are of important compleximetric titrations. Given the various equilibrium constants, all the calculations are carried out as described above.

Titration of cyanide with silver nitrate. When silver ions are added to a solution of alkali cyanide, a very stable complex cyanide is formed:

$$Ag^+ + 2\,CN^- \rightleftharpoons Ag(CN)_2^- \tag{36}$$

the alkali salt of which is soluble. Although complexes containing three and four cyanides can be formed, their concentration is negligible at values of $[CN^-]$ of practical interest. After reaction (36) is complete, further addition of silver nitrate results in the formation of a precipitate of silver cyanide:

$$Ag^+ + Ag(CN)_2^- \rightarrow 2\,AgCN \tag{37}$$

Therefore the end point is marked by the appearance of a turbidity. The equilibrium constant is

$$\beta_2 = \frac{[Ag(CN)_2^-]}{[Ag^+][CN^-]^2} = 1.3 \times 10^{21} \tag{38}$$

at ionic strength around 0.1. Let us assume that at the end point $[Ag(CN)_2^-]$ is 0.05 M. Then, since at this point $[Ag^+] = \frac{1}{2}[CN^-]$, we find that

$$4[Ag^+]^3 = \frac{0.05}{1.3 \times 10^{21}} \tag{39}$$

or $[Ag^+] = 2.1 \times 10^{-8}\ M$, $[CN^-] = 4.2 \times 10^{-8}\ M$. The product $[Ag^+][CN^-]$ therefore would be 9×10^{-16}. However, the solubility product of silver cyanide is

$$K_{sp} = [Ag^+][CN^-] = 10^{-16} \tag{40}$$

Evidently the solution becomes saturated to silver cyanide just before the equivalence point is reached. If, in the titration of 100 ml of 0.2 M potassium cyanide, 99.9 ml of 0.1 M silver nitrate has been added, the amount of cyanide remaining corresponds to 0.1 ml of 0.2 M solution, so $[CN^-] = 1.00 \times 10^{-4}\ M$. Therefore, according to Eq. (38),

$$[Ag^+] = \frac{[Ag(CN)_2^-]}{[CN^-]^2} \frac{1}{1.3 \times 10^{21}}$$

$$= \frac{5 \times 10^{-2}}{10^{-8} \times 1.3 \times 10^{21}} = 3.8 \times 10^{-15}\ M$$

whereas precipitation of silver cyanide cannot begin until $[Ag^+]$ exceeds $10^{-12}\ M$ [Eq. (40)]. Therefore, the titration error is smaller than 0.1 per cent, because after the addition of 99.9 ml of silver nitrate the solution is still clear, and it becomes saturated to silver cyanide between this point and the equivalence point.

This method was proposed by Liebig in 1851. It yields good results if the reagent is added very carefully near the end point. Rapid addition may cause premature precipitation of silver cyanide, which dissolves slowly. Deniges (1893) improved the titration by adding ammonia and a little potassium iodide. At suitable concentrations of these reagents no silver cyanide can precipitate but silver iodide precipitates at the equivalence point. The amounts of ammonia and iodide added must be regulated carefully, for otherwise the end point may appear too early or too late.[10,11] With the proper concentrations the end point can be determined very precisely and is practically identical with the equivalence point (for procedure see pp. 811–812, and also Problem 3).

Formation of slightly dissociated mercury(II) compounds; determination of chloride, bromide, iodide, and thiocyanate. The logarithms of the overall formation constants for the system $Hg^{++} - X^-$, where X^- represents Cl^-, Br^-, I^-, or CNS^-, are as follows:

	Cl^-	Br^-	I^-	SCN^-
$\log \beta_1$	6.7	9.1	12.9	—
$\log \beta_2$	13.2	17.3	23.8	16.1
$\log \beta_3$	14.1	19.7	27.6	19.0
$\log \beta_4$	15.1	21.0	29.8	20.9

From these numbers it is evident that HgX_2 is only slightly dissociated (mercuric iodide is slightly soluble; see below). Therefore, when a fully ionized

[10] I. M. Kolthoff and V. A. Stenger, *Volumetric Analysis*, vol. I, Wiley-Interscience, New York, 2nd ed., 1942, p. 156.

[11] I. M. Kolthoff and V. A. Stenger, *Volumetric Analysis*, vol. II, Wiley-Interscience, New York, 2nd ed., 1947, p. 282.

mercury(II) salt, such as the nitrate or perchlorate, is added to a solution containing chloride or bromide, mercuric ions disappear by formation of the very slightly dissociated halides. A substance which reacts visibly with Hg^{++} is used as an indicator. Sodium nitroprusside, $Na_2Fe(CN)_5NO$, is a suitable indicator; it forms a slightly soluble compound, $HgFe(CN)_5NO$, with Hg^{++} but does not react with either mercuric chloride or bromide. The end point therefore is marked by the appearance of turbidity (see Problem 4). In the titration of chloride, which forms a less stable complex than bromide, there is a small indicator correction which can be determined experimentally. The correction increases slightly with the concentration of the chloride being titrated. The method is of great practical importance because it allows direct titration of chloride in acid medium. Evidently nitroprusside cannot be used as indicator for the determination of iodide, because mercuric iodide is slightly soluble and has an intense red color.

The equilibrium constants given above show that HgI_4^{--} is much more stable than the corresponding chloride and bromide complexes, and use is made of this fact for the determination of iodide. When mercuric nitrate is added to an acid solution containing iodide the main reaction is

$$Hg^{++} + 4\,I^- \rightleftharpoons HgI_4^- \qquad (41)$$

The end point is indicated by the appearance of a permanent red turbidity of mercuric iodide:

$$Hg^{++} + HgI_4^{--} \rightarrow 2\,HgI_2 \qquad (42)$$

The end point appears a little before the equivalence point. The error can be computed from the appropriate equilibrium constants[12] and can also be determined experimentally.

In the determination of chloride and bromide, and of very dilute iodide (less than 3×10^{-4} *M*), it is possible to use diphenylcarbazide and diphenylcarbazone as indicators. These substances give intense violet colors with mercuric ions at a suitable acidity.[13]

The formation of the slightly dissociated thiocyanate is made use of in the titration of a mercuric salt such as the nitrate or perchlorate with thiocyanate. Ferric iron is used as indicator. When reaction (43),

$$Hg^{++} + 2\,SCN^- \rightleftharpoons Hg(SCN)_2 \qquad (43)$$

is complete, the excess of thiocyanate reacts with ferric iron to form the red complex.

If the concentration of mercury is relatively large, a precipitate of mercuric thiocyanate is formed just before the end point. The precipitate does not interfere with the color change, however. The color change is very sharp and the indicator blank is negligibly small when the temperature is below 20°C.

The method affords a simple and accurate determination of mercury(II) in acid solution. It can be applied also to the titration of thiocyanate with a standard mercury solution. In this case mercuric nitrate is added to the acid thiocyanate

[12] See Ref. 10, p. 154.
[13] See Ref. 10, p. 344.

solution until the color of the ferric thiocyanate disappears. The method can be applied also to the determination of cyanides, because mercuric cyanide is much less dissociated than mercuric thiocyanate. Therefore an excess of standard mercuric nitrate may be added to the cyanide solution and the excess back-titrated with thiocyanate.

In the above titrations chloride and bromide obviously interfere.

Titrations with EDTA and related compounds. Since EDTA reacts with metal ions in a 1 : 1 mole ratio,[14] i.e., according to reaction (1), the titration curve can be calculated very simply if the conditional stability constant is available. Tables A3–7 contain the necessary data for a few important examples,[15] and their use is illustrated below by the construction of titration curves for zinc–EDTA at pH 2, 5, 9, and 11. From Table A-5

$$[ZnY^{--}] = 10^{16.5}[Zn^{++}][Y^{4-}] \tag{44}$$

The values of α_{Zn} (i.e., α_M) and α_Y (i.e., α_L) are calculated as in Chapter 6, Eq. (59):

$$\alpha_Y = 1 + 10^{10.22}[H^+] + 10^{16.38}[H^+]^2 + 10^{19.05}[H^+]^3 + 10^{21.04}[H^+]^4$$
$$\alpha_{Zn} = 1 + 10^{4.4}[OH^-] + 10^{14.4}[OH^-]^3 + 10^{15.5}[OH^-]^4$$

Whence we find:

pH	2	5	9	11
log α_{Zn}	0.0	0.0	0.2	5.4
log α_Y	13.4	6.4	1.2	0.1
log K'_{Zn-Y}	3.1	10.1	15.1	11.0

If the titration at pH 9 is carried out in an ammonia–ammonium buffer, for example, 0.05 *M* ammonium hydroxide, 0.09 *M* ammonium chloride, $[NH_3] = 10^{-1.30}$ *M*, and the value of α_{Zn} is different:

$$\alpha_{Zn} = \alpha \text{ (due to hydroxy complexes)} + (10^{2.27-1.30} + 10^{4.61-2.60} + 10^{7.01-3.90} + 10^{9.06-5.20})$$
$$\log \alpha_{Zn} = 3.9 \qquad \log K' = 11.4$$

Thus at pH 9 the conditional stability constant is smaller by $10^{3.7}$ in the presence of 0.05 *M* ammonia than in its absence. However, the formation constant is still large enough (compare Fig. 36-1 and Table 36-2) to permit a satisfactory titration even in dilute solution.

Figure 36-7 is a plot of K'_{M-L} versus pH for the system Fe^{3+}–EDTA.[16] The dashed curve is calculated exactly as outlined above, i.e., by considering the formation of hydroxy complexes of iron and the acid forms of EDTA. However, FeY^- is not the only complex formed; at low and high pH, respectively, complexes FeHY

[14] Except in a very few exceptional cases; e.g., the complex of UO_2^{++} and EDTA has the mole ratio 2:1.

[15] See Ringbom, Ref. 7, for extensive tables.

[16] From Ref. 7, p. 50.

and $FeYOH^{--}$ are formed. The conditional stability constant therefore is larger than it would be if only FeY^{-} were formed; the solid curve of Fig. 36-7 is the correct value calculated by taking these substances into account. A similar situation exists in the Al^{3+}–EDTA system. It may be noted that at high pH, the stability of the EDTA complexes is not large enough to prevent precipitation of ferric hydroxide.

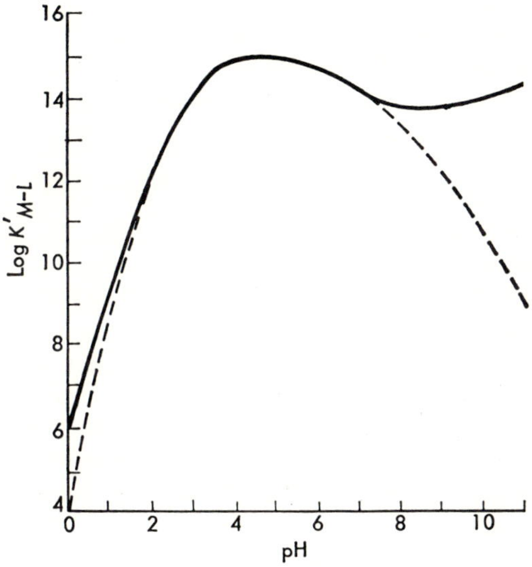

Fig. 36-7. Conditional stability constant for the system Fe(III)–EDTA.

Titrations with EDTA (and with other titrants) may be carried out in various ways. The simplest is *direct titration*, in which the pH of the solution is adjusted to a suitable value, and the titrant is added to reach the equivalence point which may be recognized by some color change or by an instrumental measurement. In some cases this procedure may be unsuitable. Naturally it cannot be used if the complexation reaction is slow. (Slightly sluggish reactions for example between magnesium, nickel, or iron with EDTA, are best carried out at elevated temperature.) The metal ion may precipitate as the hydroxide if the pH required for the titration is too high, and the hydroxide may dissolve slowly upon addition of titrant. Impurities may cause trouble by reacting in a practically irreversible manner with the metal indicator (blocking), or there may be no suitable indicator for the direct titration. The difficulties generally may be taken care of in one way or another as discussed below.

Precipitation of metal as hydroxide often may be prevented by addition of an auxiliary (relatively weaker) complexing agent. A practical example is the use of an ammonium–ammonia buffer which keeps ions such as Zn^{2+}, Cd^{2+}, Cu^{2+}, and Ni^{2+} in solution as ammonia complexes at pH 10.

If the complexation reaction is too slow to permit direct titration, a *back-titration* is used. An excess of titrant is added, sufficient time for complete reaction is allowed, the pH is adjusted, and the excess of titrant is determined by titration with a rapidly reacting metal ion. (This procedure usually also will eliminate difficulties arising from precipitation of a hydroxide or from blocking of the indicator.) The conspicuous example, already mentioned in Chapter 6, is the reaction of Cr(III) with EDTA. Even at 100°C the reaction is not fast enough for a direct titration. If an excess of EDTA is added and the solution is boiled for 2–15 min (depending upon the anion present in the chromium solution) the reaction is complete. It is remarkable that the CrY^- complex, once formed, is so stable that the back-titration may be carried out at pH as low as 4, using iron, nickel, copper, or manganese for the back-titration.

If no suitable indicator is available for the titration of M with Y, a *replacement titration* may be useful, if a second metal can be found with two properties: There must be a suitable indicator for N, and the complex NY must be less stable than the complex MY. Upon addition of NY to M a replacement reaction occurs:

$$NY + M \rightleftharpoons N + MY \tag{45}$$

and the liberated N is titrated with Y. This principle is made use of in many titrations; perhaps the most important example is the titration of calcium with EDTA using Eriochrome Black T (EBT) as indicator, at pH 10. At this pH, the value of K'_{CaY} is $10^{10.2}$. Therefore in the titration of 0.01 *M* Ca^{++} with 0.1 *M* EDTA, pCa at the equivalence point is 6.1. The value of K_{CaIn} is only $10^{5.4}$. Calculating in exactly the same way as for magnesium (see p. 739), the value of pCa_{trans} at pH 10 is found to be 3.8. Thus the sensitivity of EBT for calcium is relatively poor. Even in the titration of 0.01 *M* solutions, the indicator would change color more than 1 per cent before the equivalence point and the change would not be sharp. For more dilute solutions, EBT obviously would be a useless indicator. The difficulty is overcome if some magnesium is added. The sensitivity of EBT for Mg^{++} is high (see p. 739), and the conditional formation constant of the MgY^{--} complex is only $10^{8.2}$ at pH 10.

In practice either of two procedures may be followed. A suitable small volume of a mixture of equivalent amounts of Mg^{++} and EDTA may be added to the calcium solution before titration with EDTA. Alternatively, a small amount of magnesium chloride may be added to the EDTA solution. In this case, the EDTA solution must be standardized against calcium free of magnesium.

The end point in this titration is discussed further on page 748.

Many substances that do not react with complexing agents may be determined by an *indirect titration*. A simple example is provided by the determination of sulfate. A known amount of barium may be added; after removal of the precipitate the excess of barium may be titrated with EDTA. Similarly, phosphate may be determined indirectly by making use of the precipitation of magnesium ammonium phosphate, followed by compleximetric titration of the magnesium.

Reductions with liquid amalgams provide the basis for many indirect titrations that are especially applicable to mixtures. Consider, for example, the determination of Pb^{++} and Mn^{++} in a mixture. The sum of the two may be obtained by titration of one aliquot portion. If a second aliquot is shaken with a liquid zinc amalgam (absence of air), lead is replaced by an equivalent amount of zinc. Since zinc is easily masked by cyanide, titration of the second aliquot after addition of potassiun cyanide gives the manganese content, and hence the lead content by difference. As another example, consider the determination of Pb^{++} and Bi^{3+} in a mixture.

Titration of one aliquot portion with EDTA gives the sum of the number of gram atoms. If another aliquot is shaken with a lead amalgam, Bi^{3+} is replaced by Pb^{++}. Now each gram atom of Bi^{3+} originally present is replaced by 1.5 gram atom of Pb^{++}, and from a second titration with EDTA obviously the contents of Bi^{3+} and Pb^{++} can be calculated (see Problem 5).

Titrations with polyamines. Among the analytically important polyamines there may be mentioned triaminotriethylamine, $N(CH_2CH_2NH_2)_3$, known as "tren," and triethylenetetramine,

$$NH_2—CH_2—CH_2—NH—CH_2—CH_2—NH—CH_2—CH_2—NH_2,$$

known as "trien." As was pointed out in Chapter 6, these complexing agents offer a greater degree of selectivity than the polyaminocarboxylic acids such as EDTA, because the polyamines do not form complexes with the alkaline earth ions.

A comparison of the logarithms of stability constants of some metal ions with EDTA and tren is of interest:

	Cu^{++}	Ni^{++}	Fe^{++}	Mn^{++}	Ca^{++}
EDTA	18.8	18.6	14.3	14.0	10.7
Tren	18.8	14.8	8.8	5.8	—

The differences between the constants are much larger with tren than with EDTA, and the differences with Cu^{++}–Ni^{++}–Fe^{++} are large enough that successive determinations of the first three metals in a mixture are possible with tren.

The dependence of the stability constants upon pH is calculated exactly as described in the preceding sections.

Titration error

The discussion is limited to reactions in which a 1:1 complex ML is formed [reaction (1)]. At the equivalence point [M]′ is exactly equal to [L]′. If $\alpha_M = \alpha_L = 1$, [M] = [L], but this is rarely the case in practical compleximetric titrations. The reaction between calcium ion and EDTA at pH approximately 11 is, as already mentioned, one of the few examples.

A numerical example will illustrate. The value of K [Eq. (2)] is assumed to be 10^{13} and the reaction is carried out in a medium in which $[M] = 10^{-1}[M]'$, $[L] = 10^{-2}[L]'$, and $[ML] = [ML]'$. Therefore the conditional constant K' is 10^{10}. To 1 liter there is added exactly 0.01 mole of M and 0.01 mole of L. The molar concentrations are $[ML] = 10^{-2}$, $[M]' = [L]' = 10^{-6}$, $[M] = 10^{-7}$, and $[L] = 10^{-8}$.

The calculation of the titration error in compleximetric titrations is even simpler than in acid–base titrations. If M is titrated with L, the relative error at the end point is given by

$$\text{relative error} = \frac{[L]' - [M]'}{C_M} \tag{46}$$

in which C_M is the total concentration of M at the end point. The percent error is 100 × the relative error.

Equation (46) may be written in a form convenient for use when the difference in pM′ between the equivalence point and the end point is given. Denoting pM′ at the equivalence point as pM'_{EQ}, we have

$$\Delta pM' = pM' - pM'_{EQ} \tag{47}$$

whence

$$[M]' = [M]'_{EQ} \times 10^{-\Delta pM'} \tag{48}$$

In exactly the same way it is found that

$$[L]' = [L]'_{EQ} \times 10^{-\Delta pL'} \tag{49}$$

Substituting Eqs. (48) and (49) in (46), and noting that $[M]'_{EQ} = [L]'_{EQ}$, the general result is

$$\text{relative error} = \frac{[M]'_{EQ}}{C_M}(10^{-\Delta pL'} - 10^{-\Delta pM'}) \tag{50}$$

Equation (50) may be simplified in the majority of cases. If the reaction is practically quantitative (large K'):

$$[M]'_{EQ} = (C_M/K')^{1/2} \tag{51}$$

Also, if the error is not too large, the change in volume between equivalence point and end point is negligibly small, which means that

$$\Delta pL' = -\Delta pM' \tag{52}$$

Making use of Eqs. (51) and (52), the approximate result is

$$\text{relative error} = \frac{10^{\Delta pM'} - 10^{-\Delta pM'}}{(C_M K')^{1/2}} \tag{53}$$

The student may derive a still simpler version of Eq. (53) that applies when $\Delta pM'$ is small enough to permit expansion of $10^{\Delta pM'}$ in series (see Problem 6).

The calculation of titration error is illustrated by the following example. A solution 0.01 M in calcium and 10^{-4} M in magnesium–EDTA is titrated with 0.01 M EDTA at pH 10, using EBT. At the end point $C_{Ca} = 5 \times 10^{-3}$ M. Using the data given earlier in this chapter, the following values apply at the end point:

$$pMg = 5.4$$

$$[MgIn^-] = 10^{-5.4}$$

$$C_{Mg} = 5 \times 10^{-5}$$

$$[MgY^{--}] = 5 \times 10^{-5} - 2 \times 10^{-5.4} = 4.2 \times 10^{-5} = 10^{-4.38}$$

$$[Y^{4-}] = \frac{[MgY^{--}]}{10^{8.2} \times [Mg^{++}]} = 10^{-7.18}$$

$$[CaY^{--}] = 5 \times 10^{-3} = 10^{-2.30}$$

$$[Ca^{++}] = \frac{[CaY^{--}]}{10^{10.2} \times [Y^{4-}]} = 10^{-5.32}$$

Using Eq. (46),

$$\text{relative error} = \frac{10^{-7.18} - 10^{-5.32}}{10^{-2.30}} = -0.001, \text{ or } -0.1 \text{ per cent}$$

Alternatively, using Eq. (53),

$$\Delta pM = \Delta pM' = 5.32 - \tfrac{1}{2}(2.30 + 10.20) = -0.93$$

$$\Delta pL' = 0.93 \qquad (C_M K')^{1/2} = \tfrac{1}{2}(10^{-2.30+10.20}) = 10^{3.95}$$

$$\text{relative error} = \frac{10^{-0.93} - 10^{0.93}}{10^{3.95}} = -\frac{10^{0.93}}{10^{3.95}} = -10^{-3} \text{ or } -0.1 \text{ per cent}$$

The error may be read directly from a logarithmic diagram (Fig. 36-8; compare Fig. 6-6). The end point corresponds to point (a); the error is given by the (logarithmic) difference of points (a) and (b) as -0.001 or -0.1 per cent (Fig. 36-8 does not include the concentration of $MgIn^-$).

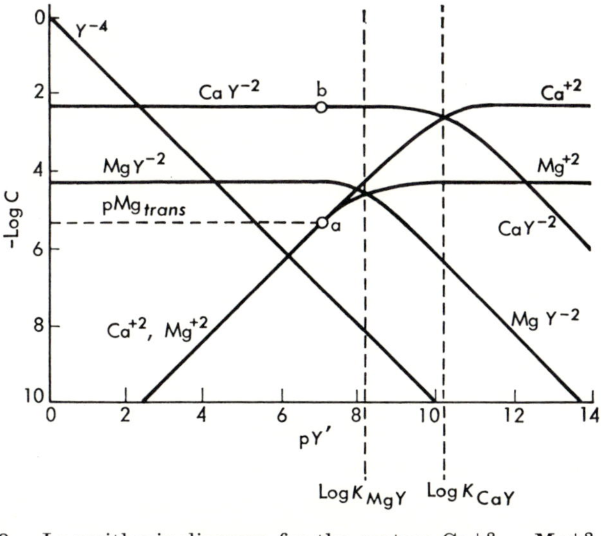

Fig. 36-8. Logarithmic diagram for the system $Ca^{+2} - Mg^{+2}$–EDTA.

End-point detection without metal indicators

Although metal indicators are used in the great majority of titrations with polydentate complexing agents, end-point detection is possible without the use of such indicators. In fact, in the early applications of EDTA, acid–base indicators were used.[17] The reaction of metal ion Me^{n+} with EDTA, added as the disodium salt, is

$$Me^{n+} + H_2Y^{--} \rightleftharpoons MeY^{n-4} + 2\,H^+ \qquad (54)$$

Therefore 2 moles of hydrogen ion are liberated per mole of Me^{n+} titrated. After addition of chelating agent the solution is titrated with standard base to determine the amount of hydrogen ion liberated.

[17] G. Schwarzenbach and W. Biedermann, *Helv. Chim. Acta* **31**, 459 (1948).

There are two major practical difficulties. If any complex other than MeY^{n-4} is formed, for example $MeHY^{n-3}$ or $MeYOH^{n-5}$, the method obviously is inapplicable. Even if this difficulty does not exist, the success of the method depends upon a pronounced difference in the pK_a's of the titrant. Unless the appropriate pK_a's are widely different, no sharp end point can be obtained. Actually such indicators are now mainly of historical interest for titrations with EDTA and related substances.

Acid–base indicators are more valuable in compleximetric titration involving mercury(II) or cyanide. Upon mixing solutions of mercuric chloride and hydrocyanic acid (caution! poison!) reaction (55) occurs:

$$HgCl_2 + 2\,HCN \rightleftharpoons Hg(CN)_2 + 2\,H^+ + 2\,Cl^- \tag{55}$$

This reaction may be applied very simply for the determination of mercuric chloride. To the solution, containing no excess of free acid, there is added slightly more than the equivalent quantity of hydrocyanic acid. The liberated strong acid is titrated with strong base using methyl yellow as indicator. Several mercuric salts, such as the perchlorate and nitrate, are highly dissociated but also are highly hydrolyzed. The free acid is simply determined by adding a sufficient excess of chloride so that the mercuric salt is transformed into $HgCl_4^{--}$. The free acid can then be titrated with methyl red. At the end point an excess of hydrocyanic acid is added, and the liberated strong acid is titrated as above.

Oxidation reduction indicators also may be used in compleximetric titrations. In Chapter 8 it was shown that the potential of a solution containing an oxidized (Ox) and reduced (Red) form of a metal is given by

$$E = E_0 + \frac{0.06}{n} \log \frac{[Ox]}{[Red]} \tag{56}$$

where n is the number of electrons in the *equilibrium*:

$$Ox + ne \rightleftharpoons Red \tag{57}$$

Assume both Ox and Red form 1 : 1 complexes with the complexing agent L, and the formation constants are

$$K_O = \frac{[OxL]}{[Ox][L]} \tag{58}$$

$$K_R = \frac{[RedL]}{[Red][L]} \tag{59}$$

Combining Eqs. (56), (58), and (59) there is obtained[18]

$$E = E_0 + \frac{0.06}{n} \log \frac{[OxL]}{[RedL]} \frac{K_R}{K_O} \tag{60}$$

Assuming that K_O is much larger than K_R, as it is in practically all cases (for example, in the systems Fe^{3+}–Fe^{++}–EDTA and V^{3+}–V^{++}–EDTA, the ratio K_O/K_R is about 10^{11} and 10^{13}, respectively), upon addition of L to a mixture of Ox and Red the former will react quantitatively with L before the latter does. After addition of L equivalent to Ox, there is a pronounced change in E which may be recognized with a suitable indicator. Because the use of oxidation–reduction indicators is subject to various interferences, they are rarely used anymore, and the use of the indicator Variamine Blue B as an indicator for the titration of iron(III) with EDTA is mentioned merely for completeness. This indicator has the structure

[18] Activity corrections are not included in Eq. (60).

This colorless substance can be oxidized (in the pH range 2–6) by Fe(III) to a blue-violet oxidation product. Therefore, upon addition of the indicator to a solution of Fe(III), the violet form and an equivalent amount of Fe(II) are formed. Upon addition of EDTA, Fe(III) is transformed into a complex so stable that it does not oxidize the indicator. At the equivalence point, when practically all the Fe(III) has been complexed, the indicator reverts sharply to the colorless form. The titration must be performed rapidly because Fe(II) in the presence of EDTA is easily air-oxidized.

Numerous physical or instrumental methods of end-point detection may be applied. These are discussed in Chapter 38 but are mentioned here for completeness. The most widely used instrumental method at present is the potentiometric titration, especially with a mercury electrode. Photometric, conductometric, high-frequency, thermometric, amperometric, and chronopotentiometric titrations have been used as well.

PROBLEMS

1. 10 ml of 0.1 *M* L is added to 100 ml of 0.1 *M* Me. The reaction is $Me + L \rightleftharpoons MeL$. What is the smallest value of K such that assumption of complete reaction of the added L does not introduce an error of more than 0.1 per cent in the calculated value of [MeL]/[Me]? *Ans.* $K = 1.2 \times 10^4$.
2. In a Liebig–Denigès titration, 50.00 ml of a potassium cyanide solution requires 20.00 ml of a 0.1100 *M* silver nitrate solution. How much cyanide does the solution contain? Can the titration be carried out in the presence of chloride?
3. Calculate the titration error in a Liebig–Denigès titration, assuming that 0.1 *M* KCN is titrated with 0.1 *M* $AgNO_3$, and that at the end point the concentrations of ammonia and potassium iodide are 0.1 *M* and 0.005 *M*. (See Tables A-3 and A-9 for the necessary equilibrium constants.)
4. 100 ml of 0.01 *M* NaCl is titrated with 0.01 *M* $Hg(NO_3)_2$ in acid solution. At the end point the concentration of sodium nitroprusside is 0.004 *M*. If the solubility product

$$[Hg^{++}][Fe(CN)_5NO^{--}] = 2.5 \times 10^{-9}$$

calculate what volume of mercuric nitrate must be added before the solution is saturated to mercuric nitroprusside.

5. A solution contains Bi^{3+} and Pb^{++}. A 50-ml aliquot is titrated with 0.01 *M* EDTA to determine the sum of the two metals, 25.00 ml being required. A separate 50-ml aliquot is shaken with a lead amalgam. After replacement of Bi^{3+} is complete, the aliquot is titrated with 0.01 *M* EDTA, 30.00 ml being required. Calculate the molar concentrations of bismuth and lead in the original solution.
6. Show that for small values of $\Delta pM'$, Eq. (53) reduces to

$$\%\ \text{error} = 460 \frac{\Delta pM'}{(C_M K')^{1/2}}$$

How small must $\Delta pM'$ be for this expression to be valid within 10 per cent relative?

CHAPTER 37 *Oxidation–Reduction Titrations*

Potentiometric titration curves

A potentiometric titration curve portrays the change of reduction potential in the course of a titration of a reductant with an oxidant or vice versa. In Chapter 8 the general equations have been presented which give the relation between reduction potential and the concentrations of the constituents which determine it. Most redox titrations are carried out at relatively large ionic strength. Therefore expressions for formal potentials will be used. (See appendix table for values of E^f.)

$$\text{Ox} + ne \rightleftharpoons \text{Red} \qquad E = E^f + \frac{0.059}{n}\log\frac{[\text{Ox}]}{[\text{Red}]} \qquad (25°\text{C})$$

$$\text{Ox} + m\text{H}^+ + ne \rightleftharpoons \text{Red} \qquad E = E^{0\prime} + \frac{0.059m}{n}\log[\text{H}^+] \tag{8.4}$$

$$+ \frac{0.059}{n}\log\frac{[\text{Ox}]}{[\text{Red}]}$$

$$= E^f + \frac{0.059}{n}\log\frac{[\text{Ox}]}{[\text{Red}]}$$

Titration of ferrous iron with ceric cerium in 1 *N* H_2SO_4

$$\text{Fe}^{++} + \text{Ce(IV)} \rightleftharpoons \text{Fe(III)} + \text{Ce}^{3+} \tag{1}$$

We are dealing with the partial systems

$$\text{Fe(III)} + e \rightarrow \text{Fe(II)} \qquad E = E^f_{\text{Fe(III)/Fe(II)}} + 0.059\log\frac{[\text{Fe(III)}]}{[\text{Fe(II)}]}$$

$$= 0.68 + 0.059\log\frac{[\text{Fe(III)}]}{[\text{Fe(II)}]}$$

$$\text{Ce(IV)} + e \rightarrow \text{Ce(III)} \qquad E = 1.44 + 0.059\log\frac{[\text{Ce(IV)}]}{[\text{Ce(III)}]}$$

The Roman numerals indicate the valence state of the various ions, and the bracketed valence states refer to the analytical concentration of the species and not to the single ion concentration. The equilibrium constant K of reaction (1) is given by (see Appendix)

$$\log K = \log \frac{[\mathrm{Fe(III)}][\mathrm{Ce(III)}]}{[\mathrm{Fe(II)}][\mathrm{Ce(IV)}]} = \frac{1.44 - 0.68}{0.059} = 12.9 \qquad K = 8 \times 10^{12}$$

$$E_{\mathrm{EP}} = \log \frac{E^f_{\mathrm{Ce(IV)/Ce(III)}} + E^f_{\mathrm{Fe(III)/Fe(II)}}}{2} = \frac{1.44 + 0.68}{2} = 1.06$$

in which E_{EP} is the potential (versus NHE) at the equivalence point. At this point

$$\frac{[\mathrm{Fe(III)}]}{[\mathrm{Fe(II)}]} = \frac{[\mathrm{Ce(III)}]}{[\mathrm{Ce(IV)}]} = (8 \times 10^{12})^{1/2} = 2.8 \times 10^{6}$$

and the reaction is quantitative.

It is a simple matter now to calculate the entire potentiometric titration curve. When 9 per cent of the iron(II) has been oxidized $[\mathrm{Fe(III)}]/[\mathrm{Fe(II)}] = 0.1$ and $E = 0.68 - 0.059 = 0.62$. After addition of 99.9 per cent of the equivalent amount of cerium $E = 0.68 + 0.18 = 0.86$. With 0.1 per cent cerium in excess $E = 1.44 - 0.18 = 1.26$. Thus the potential changes from 0.86 to 1.26 between 0.1 per cent before and 0.1 per cent after the end point. Any oxidation–reduction indicator with E^f of the order of 1.05 ± 0.1 will indicate a sharp end point. The change in reduction potential in the titration of ferrous iron with ceric sulfate is presented in Table 37-1 and graphically in Fig. 37-1.

TABLE **37-1.** **Titration of Iron(II) with Cerium(IV) in 1 *N* Sulfuric Acid**

Cerium(IV) Added in % of Iron(II)	Ratio $\frac{\mathrm{Fe(III)}}{\mathrm{Fe(II)}}$	E
9	0.1	0.62
50	1	0.68
91	10	0.74
99	100	0.80
99.9	1000	0.86
100	2.5×10^{6}	*1.06* (E_{EP})
	Ratio $\frac{\mathrm{Ce(IV)}}{\mathrm{Ce(III)}}$	
100.1	0.001	1.26
101	0.01	1.32
110	0.1	1.38

With a given initial concentration of iron(II) and taking the volume change into account [Fe(II)], [Fe(III)], [Ce(III)], and [Ce(IV)] are easily calculated at any point during the titration.

The equilibrium constant of the reaction

$$2\ \mathrm{Ox}_1 + \mathrm{Red}_2 \rightleftharpoons 2\ \mathrm{Red}_1 + \mathrm{Ox}_2$$

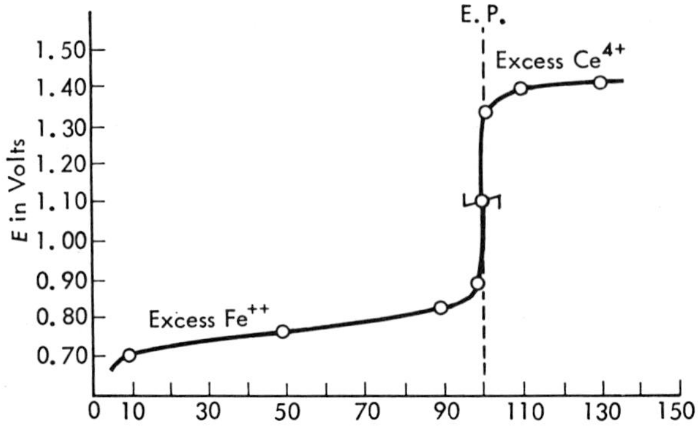

Fig. 37-1. Cerium (IV) added in per cent of Fe (II).

with the partial reactions

$$Ox_1 + e \rightarrow Red_1$$
$$Ox_2 + 2e \rightarrow Red_2$$

has been derived in Chapter 8 (p. 165). It has also been shown there that in a mixture containing equivalent amounts of Ox_1 and Red_2 (corresponding to the equivalence point in a titration) that

$$E = E_{EP} = \frac{E^f_{Ox_1/Red_1} + 2\,E^f_{Ox_1/Red_1}}{3}$$

As an example we consider *the titration of tin(II) with iron(III) in* 1 *N hydrochloric acid*:

$$E^f_{Fe(III)/Fe(II)} = 0.68 \qquad E^f_{Sn(IV)/Sn(II)} = 0.14$$

$$E = 0.68 + 0.059 \log \frac{[Fe(III)]}{[Fe(II)]} = 0.14 + \frac{0.059}{2} \log \frac{[Sn(IV)]}{[Sn(II)]}$$

The values of E in the above titration are readily calculated. The data are presented in Table 37-2. It is of interest to note that in the titration of iron(II) with cerium(IV)

TABLE **37-2. Titration of Tin(II) with Iron(III) in 1** *N* **Hydrochloric Acid**

Iron(III) Added in % of Sn(II)	Ratio $\frac{Sn(IV)}{Sn(II)}$	E
9	0.1	0.11
50	1	0.14
91	10	0.17
99	100	0.20
99.9	1000	0.23
100		*0.32* (E_{EP})
	Ratio $\frac{[Fe(III)]}{[Fe(II)]}$	
100.1	0.001	0.50
101	0.01	0.56
110	0.1	0.62

the change in potential between 0.1 per cent before the equivalence point and the equivalence point is equal to that between the equivalence point and 0.1 per cent after the equivalence point. The curve is symmetrical at both sides of the equivalence point. This is no longer true in the titration of tin(II) with iron(III). Here the change in potential between 0.1 per cent before and at the equivalence point is 0.09 V, but 0.18 V between that at the equivalence point and 0.1 per cent after the equivalence point. In the present chapter we are concerned with the visual detection of the end point. In Chapter 52 potentiometric titrations are discussed in which E is measured during the titration.

It is a simple matter to calculate titration curves of a mixture of two reducing agents with an oxidant or of a mixture of two or more oxidants with a reductant. Consider, for example, the titration of a mixture of tin(II) and iron(II) with cerium(IV). Until the first equivalence point the values of E are the same as in Table 37-2. After the first equivalence point iron(II) is oxidized and the potentials are the same as in Table 37-1 until the second equivalence point. At and after the second equivalence point the concentration of Ce(III) formed during the titration of tin(II) must be taken into consideration. This has only a small effect on E_{EP} and potentials thereafter.

Oxidation–reduction (Redox) indicators

An oxidation–reduction indicator is a substance the oxidized form of which has a color different from the reduced form. The oxidation and reduction of the indicator should be reversible.

$$I_{Ox} + ne \rightleftharpoons I_{Red}$$

I_{Ox} is the oxidized form of the indicator, which has a color different from that of the reduced form, I_{Red}. A dilute iodine or iodide solution, for example, may be considered as an indicator for oxidation–reduction reactions. The iodide solution is oxidized by substances that have a more positive reduction potential than iodine, whereas substances with a more negative reduction potential reduce iodine to iodide. If a strong reducing agent is titrated with a strong oxidizing agent, iodine will be liberated at the end point. *In general, the action of an oxidation–reduction indicator does not depend upon the specific nature of the oxidant or reductant titrated but upon the reduction potentials of the indicator and of the system titrated.*

The properties of some oxidation–reduction indicators which are very useful in titrations with strong oxidizing agents such as ceric sulfate and potasisum dichromate are discussed below.

The properties of oxidation–reduction indicators which change color at more negative reduction potentials have been studied in great detail by Clark[1] and Michaelis.[2] They are rarely used in volumetric work and are not discussed in detail in this text.

That a redox indicator does not react by combination with one of the constituents in the solution, but by an oxidation–reduction reaction, can be easily demonstrated

[1] W. M. Clark, *The Determination of Hydrogen Ions*, 1928; see also Bull. 151 of the U.S. Hygienic Laboratory, *Studies in Oxidation and Reduction;* see also I. M. Kolthoff and V. A. Stenger, *Volumetric Analysis*, vol. I, Wiley-Interscience, New York, 2nd ed., 1942.

[2] L. Michaelis, *Oxydations-Reductions Potentiale*, Springer, Berlin, 2nd ed., 1933.

by the following experiment. Add some hydrochloric acid and a few drops of diphenylamine or its sulfonic acid to a mixture of 0.01–0.02 M potassium ferricyanide and ferrocyanide. The color of the yellow solution is not changed by the addition of the indicator. In another vessel add some hydrochloric acid and indicator to a similar volume of about 0.1 M zinc sulfate. The solution remains colorless. However, when the zinc solution is added to the ferro-ferricyanide mixture, the beautiful violet color of the oxidation product of the indicator appears. Since zinc ferrocyanide is so much less soluble than zinc ferricyanide, the reduction potential of the original mixture is greatly increased by the addition of the zinc and becomes large enough to permit the oxidation of the indicator.

Diphenylamine and diphenylbenzidine. Diphenylamine and diphenylbenzidine are very slightly soluble in water. Stock solutions are therefore prepared in sulfuric acid. Knop[3] was the first to propose the use of diphenylamine as an indicator in the titration of ferrous iron with potassium dichromate. Upon oxidation, diphenylamine and diphenylbenzidine give a beautiful violet oxidation product called diphenylbenzidine violet.

The mechanism of this oxidation can be represented by the following reactions[4]:

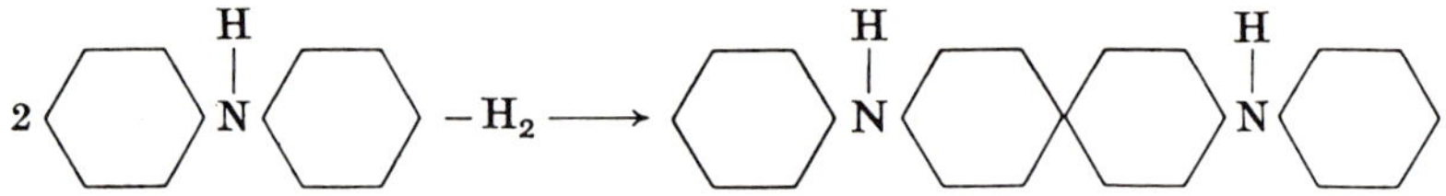

diphenylamine diphenylbenzidine (A)

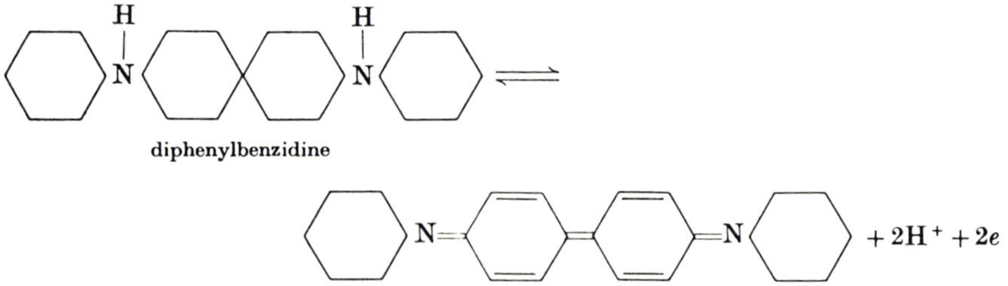

diphenylbenzidine violet (B)

First, the diphenylamine is oxidized to diphenylbenzidine [reaction (A)]; this reaction is irreversible. The oxidation of diphenylbenzidine to diphenylbenzidine violet is reversible. At a potential above 0.76 V (referred to the normal hydrogen electrode) the color of the system is violet; below 0.06 V it is colorless.

The velocity of oxidation of diphenylamine or diphenylbenzidine by dichromate in acid medium is relatively small; however, it is induced by the reaction between ferrous iron and the oxidizing agent. In the application of these indicators irregular results are sometimes found. This can be caused by the very small solubility of diphenylbenzidine in water or in dilute acids. Upon oxidation of the diphenylamine, the diphenylbenzidine may precipitate, and then the latter reacts very slowly with excess potassium dichromate. If too much indicator is used, the color at the end point is sometimes found to be green; the green product formed is an extremely slightly soluble addition compound of the benzidine with the oxidized benzidine. Neither diphenylamine nor diphenylbenzidine can be used as indicators in the presence of tungstate; they form insoluble tungstates, and no color change occurs.

[3] J. Knop, *J. Am. Chem. Soc.* **46**, 263 (1924).
[4] I. M. Kolthoff and L. A. Sarver, *J. Am. Chem. Soc.* **52**, 4179 (1930).

These indicators can be used in the titration of oxidizing agents such as permanganate, ceric cerium, dichromate, and vanadate with ferrous sulfate. Under these conditions the indicator shows a color change from violet to colorless. It should be realized, however, that the diphenylbenzidine violet is not stable in solutions of oxidizing agents; it is slowly and irreversibly oxidized to other organic compounds.

The slight solubilities of diphenylamine and diphenylbenzidine and their inapplicability in the presence of tungstate are a serious handicap to their general use. For this reason Sarver and Kolthoff[5] prepared a sulfonic acid of diphenylamine which is much more soluble in water and which can be used in the presence of tungstates.

Diphenylamine sulfonic acid.[6] The barium and sodium salts of this acid are commercially available. The stock solution is made up to contain 0.2–0.5 per cent of the salt in water. Upon oxidation the indicator gives a reddish violet color resembling that of a dilute permanganate solution. The color change is brilliant and very sharp, the mechanism of the oxidation being similar to that of diphenylamine. The color change takes place at a potential of 0.83 V, which is about 0.1 V higher than that of diphenylamine.

This indicator can be used for the direct titration of oxidizing agents with ferrous sulfate, but again it should be remembered that the reddish violet oxidation product is slowly decomposed on standing.

A few drops of the 0.2 per cent indicator solution (barium salt) in a volume of 50–100 ml will give sharp color changes. The indicator consumes a little oxidizing agent for its own oxidation. In titrations with 0.1 N reagents the indicator correction is so small that it can be neglected. In the titration of 0.01 N solutions the correction should be considered[7]; it corresponds to approximately 0.12 ml of 0.01 N oxidizing agent for each 0.1 ml of the 0.2 per cent indicator solution used. The correction can be decreased to about half of this value by using the indicator in a partially oxidized state. The red-violet oxidation product combines with unoxidized diphenylamine sulfonic acid to form a slightly soluble green addition compound of the violet and diphenyl benzidine sulfonic acid.

Willard and Young[8] recommend the following procedure for the preparation of a stock solution of the partially oxidized indicator.

One hundred milliliters of 0.01 M diphenylamine sodium sulfonate and 5 ml of concentrated sulfuric acid are diluted to about 300 ml. To this liquid 25 ml of 0.1 N potassium dichromate are added slowly, followed by 8 ml of 0.1 N ferrous sulfate. This green solution is allowed to stand for 3–4 days until a portion of the liquid gives no color when added to 100 ml of water containing 2 ml of 0.1 N dichromate and 5 ml of concentrated sulfuric acid. The supernatant liquid is then siphoned off slowly, care being taken not to disturb the green precipitate. Three hundred milliliters of water and 15 ml of concentrated sulfuric acid are added to the precipitate; the latter is allowed to settle again, and the supernatant liquid is siphoned off as before. The green precipitate is then shaken up with 100 ml of water. In a titration 0.1 ml of the suspension gives the same color intensity as 0.2 ml of 0.2 per cent diphenylamine barium sulfonate.

Variamine blue, N-(p-methoxyphenyl)-p-phenylene diamine hydrochloride. Erdy and coworkers[9] found that this colorless substance is easily oxidized to a blue product. Its aqueous solution is stable for weeks. Between pH 1.5 and 6.3 it behaves like a reversible indicator. In this pH range $E = 0.712 - 0.059$ pH (25°C). It has been found useful in the titration of iron(III) at pH 2–3 with a suitable reductant, of zinc with potassium ferrocyanide (see p. 800), and in several iodimetric titrations. In the pH

[5] L. A. Sarver and I. M. Kolthoff, *J. Am. Chem. Soc.* **53**, 2906 (1931).

[6] For other derivatives of diphenylamine, see H. H. Willard and G. D. Manalo, *Anal. Chem.* **19**, 167 (1947).

[7] L. A. Sarver and I. M. Kolthoff, *J. Am. Chem. Soc.* **53**, 2906 (1931).

[8] H. H. Willard and P. Young, *Ind. Eng. Chem., Anal. Ed.* **5**, 154 (1933).

[9] For a review see L. Erdy, *Chemist-Analyst* **48**, 106 (1959).

range 1.5–6.3 its sensitivity to iodine is the same as that of starch. For details and literature, see Ref. 9.

2-Hydroxy variamine blue, 2-hydroxy-4-amino-4′-methoxydiphenylamine. This indicator also exhibits a color change from colorless to blue at an acid pH. According to Erdy and Kasa[10] the reduction potential changes with pH in the following way:

$$E_0 = 0.68\text{–}0.059\ \text{pH} \qquad (\text{pH } 2\text{–}5)$$
$$E_0 = 0.66\text{–}0.059\ \text{pH} \qquad (\text{pH } 6\text{–}12)$$

Contrary to variamine blue, which can be used only in an acid medium, it can be used up to pH 12, at which pH the oxidized form is red. It is a very suitable indicator for the titration of iodine or ferricyanide with ascorbic acid in weakly alkaline medium, e.g., in bicarbonate solution at pH 8.6. Its reduction potential is then 0.24 V.

N,N-Dimethyl-4-aminodiphenyl. This compound has been recommended[11] as an indicator for titrations with ceric cerium and dichromate. The colorless indicator is oxidized to a blue product. It cannot be used for titrations with permanganate, which destroys the blue color. No additional details are presented here, because ferroin and its derivatives are much more suitable indicators.

Tris–1, 10–phenanthroline-ferrous ion, referred to as ferroin, and related indicators. The base *o*-phenanthroline, $C_{12}H_8N_2$, dissolves easily in solutions of ferrous salts, three molecules combining with one ferrous ion. The complex ferrous ions thus formed are intensely red in color; with strong oxidizing agents the ferric complex is formed, which has a blue color, the intensity of which is much less than that of the ferrous complex. Therefore in working with small amounts of indicator, the latter behaves virtually as a one-color indicator. Its properties have been studied by Walden, Hammet, and Chapman.[12] The complex ferrous ions are slowly decomposed by strong acids, or by salts of other metals (Co^{++}, Cu^{++}, Ni^{++}, Zn^{++}, Cd^{++}) which form stable complexes with the base. The ferrous iron complex does not form, or forms only very slowly, when ferrous ion is added to a strongly acid solution of the base. The blue oxidized form is remarkably resistant to further action of strong oxidizing agents. The color change red to blue is reversible:

$$Fe(C_{12}H_8N_2)_5^{++} \rightleftharpoons Fe(C_{12}H_8N_2)_3^{3+} + e$$

The point of visible color change (red to colorless) corresponds to 90 per cent oxidation of the indicator when one drop of a solution of the latter of the concentration given below is present. This color change occurs at a potential of 1.11 V,[13] and hence at a much higher potential than that of the diphenylbenzidine indicators. The indicator is very useful in titrations with ceric sulfate (for example, of ferrous iron, ferrocyanide, and vanadyl ions). In titrations with potassium dichromate the diphenylamine sulfonic acid, however, is much to be preferred.

A 0.025 *M* solution of phenanthroline–ferrous sulfate is a convenient one for indicator purposes. It is prepared by dissolving the stoichiometric amount of phenanthroline monohydrate (which is commercially available) in a 0.025 *M* solution of ferrous sulfate in water. One drop of this solution is used in a volume up to 300 ml. One drop is equivalent to less than 0.01 ml of 0.1 *N* oxidizing agent; the

[10] L. Erdy, I. Kasa, and T. Meisel, *Talanta* **10**, 471 (1963); L. Erdy and I. Kasa, *Acta Chim. Acad. Sci. Hung.* **41**, 59 (1964).

[11] M. Matrka and M. Syrová, *Collection Czech. Chem. Commun.* **28**, 3446 (1963).

[12] G. H. Walden, L. P. Hammett, and R. P. Chapman, *J. Am. Chem. Soc.* **55**, 2649 (1933).

[13] D. N. Hume and I. M. Kolthoff, *J. Am. Chem. Soc.* **65**, 1895 (1943).

indicator blank is therefore negligibly small with standard solutions of this concentration and needs to be determined only when more dilute solutions are used. The stock solution can be kept unchanged for at least 1 year.

Reduction potentials of a large number of iron(II) complexes with polysubstituted 1,10-phenanthroline and 2,2′-bipyridine have been determined by Brandt and Smith.[14] Schilt[15] prepared a number of iron, ruthenium, and osmium 2,2′-bipyridine complexes and studied their properties as redox indicators. Structurally these indicators are related to ferroin. Some useful indicators are incorporated in Table 37-3.

Tris(2,2′-bipyridine) ruthenium(II) (RuBP) and tris(4,4′-dimethyl-2,2′-bipyridine) ruthenium(II) (RuDMBP) have been recommended as *fluorescent* indicators.[16] The

TABLE **37-3. Selected List of Oxidation–Reduction Indicators**[a]

	Color Change		
Indicator	(Ox)	(Red)	E^f *at* pH = 0, V
Vanadium(II)-1,10-phenanthroline	Colorless	Blue-violet	0.14[b,c]
Indigo monosulfonate	Blue	Colorless	0.26
Phenosafranine	Red	Colorless	0.28
Indigo tetrasulfonate	Blue	Colorless	0.36
Methylene blue	Green-blue	Colorless	0.36
1-Naphthol-2-sulfonic acid indophenol	Red	Colorless	0.54
2,7-Diaminophenothiazonium chloride	Blue	Colorless	0.56
Variamine blue[d]	Colorless	Blue	0.59[d] (pH 2)
2-Hydroxyvariamine blue[d]	Colorless	Blue	0.56[d] (pH 2) 0.06 (pH 12)
Diphenylamine (diphenylbenzidine)	Violet	Colorless	0.76 ± 0.1[b]
Diphenylamine sulfonic acid	Red-violet	Colorless	0.80 ± 0.1[b]
$[Fe(bipy)_2](CN)_2$[e]	Pale violet	Orange	0.78
$[Os(bipy)_3]SO_4$[e]	Pale pink	Green	0.80
$[Os(bipy)_2](CN)_2$[e]	Pale violet	Orange	0.81
Erioglaucine	Red	Green	1.0[b]
Fe 5,6-Dimethylphenanthroline	Pale blue	Red	0.975
p-Ethoxychrysoidine	Pale yellow	Red	1.0
$[Fe(bipyr)_3]SO_4$	Red	Pale blue	1.03
p-Nitrodiphenylamine	Violet	Colorless	1.06[b]
$[Fe(II)(o\text{-phenanthroline})_3](ClO_4)_2$ (ferroin)	Pale blue	Red	1.11
$[Ru(bipyr)_3]SO_4$[e]	Green	Orange	1.24
$[Fe(II)(nitro\text{-}o\text{-phenanthroline})_3]$ $(ClO_4)_3$(nitrophenanthroline)	Pale blue	Violet-red	1.25

[a] A description of the preparation of some indicators with a reduction potential more positive than 0.7 V is given by T. H. Whitehead and C. C. Wills, Jr., *Chem. Rev.* **29**, 69 (1941).
[b] Transition potential.
[c] From W. P. Schaefer, *Anal. Chem.* **35**, 1746 (1963).
[d] See the text.
[e] bipy, bipyridyl.

[14] W. W. Brandt and G. F. Smith, *Anal. Chem.* **21**, 1313 (1949).
[15] A. A. Schilt, *Anal. Chem.* **35**, 1599 (1963).
[16] B. Kratochvil and D. A. Zatko, *Anal. Chem.* **36**, 527 (1964).

former compound had been used already as a color redox indicator,[17] but its color change from yellow to faint blue is not very satisfactory. Fluorescent properties of ruthenium(II) complexes of 1,10-phenanthroline, 2,2′-bipyridine, and several methyl derivatives had been studied by Veening and Brandt,[18] but Kratochvil and Zatko (Ref. 16) were the first to recommend the two ruthenium compounds as fluorescence indicators in titrations with ceric cerium and permanganate. They exhibit orange-red fluorescence when irradiated with ultraviolet light. The fluorescence disappears upon addition of Ce(IV) or Mn(VII). The reduction potential is not reported but seems to be of the same order of magnitude as that of ferroin. Insoluble compounds, iodine, ferricyanide, and to a lesser extent iron(III) and uranium(VI) interfere by quenching the fluorescence.

Geyer and Steinmetzer[19] had studied several 3,6-dihydroxyphthalic acids as fluorescence indicators. These were found to have reduction potentials varying between 0.85 and 0.95 V and they should be useful in titrations in which diphenylamine (benzidine) are used.

Erioglaucine A and Eriogreen B.[20] Erioglaucine is used in the form of a 0.1 per cent solution in water, in which it has a blue color. In strongly acid medium it is yellow, and in 0.3–0.4 *N* acid it has a green color. Upon the addition of a trace of permanganate or ceric sulfate, the color changes from greenish to bluish red. Eriogreen (0.1 per cent solution in water) in 0.3–0.5 *N* acid solution has a bluish green color, which is changed to orange-yellow by a trace of permanganate or ceric sulfate. The color change occurs at a potential of 0.71 V for erioglaucine and of 0.72 V for eriogreen. Rao and Rao[21] report a transition potential for both indicators at 1.01 V. The indicator blank is negligibly small. A few drops of the indicator in a volume of 200–400 ml give a sharp color change with 0.1 ml of 0.01 *N* permanganate or ceric sulfate. A disadvantage of these indicators is that the oxidized forms are rather unstable, especially in the presence of an excess of oxidizing agent; moreover, the color changes become less pronounced in strongly acid media. Therefore the ferrous-phenanthroline complex is, in general, much to be preferred.

Rao and Rao[22] have reported that copper phthalocyanine tetrasulfonic acid and nickel phthalocyanine sulfonic acid[23] can be used as indicators in titrations with ceric sulfate. The copper indicator gives a color change from violet blue to pink and the nickel indicator from blue to violet. More recently cobalt–phthalocyaninesulfonic acid was reported[24] to be more suitable than the copper or nickel compounds. However, the oxidized forms are extremely unstable and hence the color change is irreversible. Therefore ferroin is much to be preferred.

Several dyes with low reduction potentials can be reduced to the colorless leuco form, e.g., methylene blue and indigo sulfonates (see Table 37-3). Ruzicka and Adámek[25] found that several derivatives of phenothiazine are useful in titrations with titanium(III) chloride. Thionol(7-hydroxyphenolthiazone) gives a color change from red to colorless. Its reduction potential is 0.47 V at pH 1.4. Thionoline(7-ammophenothiazine-2) changes color from red-violet to colorless; its reduction potential is 0.54 at pH 0.66. Thionine (2,7-diaminophenothiazonium chloride) changes from blue to colorless; its reduction potential is 0.56 V at pH 0–5. Titrations were carried out at 70°C in 4 *N* hydrochloric acid.

A select list of indicators is presented in Table 37-3.

[17] J. Steigman, N. Birnbaum, and S. M. Edmonds, *Anal. Chem.* **14**, 30 (1942).
[18] H. Veening and H. Brandt, *Anal. Chem.* **32**, 1426 (1960).
[19] R. Geyer and H. Steinmetzer, *Angew. Chem.* **72**, 634 (1960).
[20] J. Knop, *Z. anal. Chem.* **77**, 111 (1929).
[21] G. G. Rao and V. P. Rao, *Z. Anal Chem.* **168**, 172 (1959).
[22] G. P. Rao and V. Rao, *Z. Anal. Chem.* **160**, 109 (1958); **163**, 1, 263, 266 (1958).
[23] G. P. Rao and V. Rao, *Z. Anal. Chem.* **190**, 213 (1962); **196**, 102 (1963).
[24] H. S. Gowda and R. Shakunthala, *Chemist-Analyst* **55**, 30 (1963).
[25] E. Ruzicka and J. Adámek, *Z. Anal. Chem.* **195**, 411 (1963).

It is now a simple matter to interpret the behavior of a redox reduction indicator, for example in the titration of ferrous sulfate with a strong oxidizing agent. Ferrous-phenanthroline changes from red to almost colorless at a potential of 1.11 V. From Table 37-1 we see that in the titration of ferrous iron with ceric sulfate the potential at the equivalence point is 1.06 V; the indicator therefore is still present in the red form. After the addition of a 0.1 per cent excess of ceric sulfate the potential changes to 1.26 V, and the indicator is oxidized. Evidently the titration error is negligibly small.

Diphenylamine sulfonic acid is oxidized to the violet form at a potential of 0.83 V, which is noticeably before the end point. To get good results with this indicator (and with diphenylamine and diphenylbenzidine), it is necessary to reduce the reduction potential of the iron system, which can be done as we have seen by transforming the ferric iron into a complex by the addition of phosphoric acid or fluoride. Under those conditions excellent results are obtained in the titration of ferrous iron with dichromate or ceric sulfate, using the diphenylamine or diphenylbenzidine as indicator for the detection of the end point.

The reagent may serve as its own indicator. If the reagent is strongly colored and is decolorized by the substance that is being determined—or if it is converted into slightly colored compounds—the end point is marked by the color of a slight excess of reagent. Thus, for example, in titrations with permanganate it is, as a rule, not necessary to add a special indicator, because a slight excess of permanganate can be easily detected by its pink color, even in the presence of slightly colored ions such as ferric ion or vanadate. If necessary, the excess of permanganate required for the detection of the end point can be determined in a blank and subtracted from the volume used in the titration.

Titrations with, or of, iodine can be carried out without an indicator. The end point is marked by the appearance, or the disappearance, of the intense yellow color of the iodine. Usually, however, a *specific indicator* for iodine is added near the end of the titration. Starch has the property of forming an adsorption compound with iodine in the presence of iodide which has an intense blue color. Solutions as dilute as 2 to 5 $\times$ 10^{-5} N in iodine still give a blue color with starch. Sometimes organic solvents are used for marking the appearance or disappearance of a trace of iodine at the end point. Iodine has an intense violet color in carbon tetrachloride and chloroform and is extracted by these solvents from aqueous medium. This test is even more delicate than the starch iodine reaction.

Ceric cerium has an intense yellow color, and the end point in titrations with it may be detected in colorless solutions without the addition of a special indicator. However, a much sharper end point is obtained if a suitable oxidation–reduction indicator is used.

Titrations with potassium dichromate without an indicator do not yield a sharp end point, because the greenish chromic ions formed obscure the yellow-orange color of the dichromate. Therefore, an indicator must be used.

Specific indicators for the system titrated are seldom used. An example of their application is found in the titration of ferric iron with a strong reducing agent such as titanous chloride or chromous chloride. Thiocyanate gives a reddish color with ferric ions in acid medium; when all the iron has been reduced, the solution turns colorless. The color change is sharp.

Before oxidation–reduction indicators were used, external indicators enjoyed some popularity. In the titration of ferrous iron with dichromate, for example, potassium ferricyanide or diphenylcarbazide was used as external indicator. The former reacts with ferrous iron to form Turnbull's blue, and the latter with dichromate to form an intensely red-colored compound. A drop of the solution being titrated is treated on a spot plate with a drop of ferricyanide or diphenylcarbazide. When the drops fails to give a coloration with the former or gives a red color with the latter, the end point is reached.

General procedures for oxidation or reduction of substances prior to titration

Various substances which can exist in different states of oxidation can be transformed by suitable means into a higher state of oxidation and titrated with a reducing agent, or into a lower state of oxidation and then titrated with an oxidizing agent. In all these cases the substance used for the oxidation or reduction must generally be absent when the titration is made.

Sodium bismuthate is a powerful oxidizing agent; it oxidizes manganous manganese to permanganate at room temperature in nitric acid solution. After the oxidation is complete the excess is easily removed by filtration, because the bismuthate is insoluble.

In solutions buffered with pyrophosphate to pH 2–6, lead(IV) oxide oxidizes manganese(II) to manganese(III), cerium(III) to cerium(IV), vanadium(IV) to V(V) and chromium(III) to Cr(VI).[26] The excess lead dioxide is removed by filtration.

Potassium persulfate, $K_2S_2O_8$, is often used as an oxidizing agent, usually in combination with a little silver salt which acts as a catalyst. Manganese, for example, is quantitatively oxidized to permanganate if the oxidation is carried out in the presence of phosphoric or hydrofluoric acid; trivalent chromium is oxidized to dichromate, and vanadyl to vanadate. The excess of persulfate can be decomposed by boiling:

$$K_2S_2O_8 + H_2O \rightarrow 2\,KHSO_4 + \tfrac{1}{2}\,O_2$$

Hydrogen peroxide is often used as an oxidizing agent in alkaline medium: it is decomposed by boiling the alkaline solution. This decomposition is catalyzed by the addition of a little nickel salt or more strongly by iodide.

Perchloric acid in hot concentrated (72 per cent) solution is a powerful oxidant. It oxidizes chromium(III) and vanadium(IV) to the pentavalent states but does not oxidize manganese(II).[27] After dilution it has no oxidizing power. Smith[28] recommends perchloric acid in combination with sulfuric and nitric acids for the destruction of organic material. Proper care is essential because of danger of explosion when concentrated perchloric acid is heated with various kinds of organic material.

Silver(II) oxide has been used as an oxidant for the oxidation of manganese(II), chromium(III) and cerium(III).[29] Based on this oxidation, Lingane and Davis[30] developed simple and rapid titrimetric methods for the determination of manganese and chromium in aluminum and ferrous alloys.

Willard and Merritt[31] introduced ozone as an oxidizing agent. The reduction potential of the system

$$O_3(g) + 2\,H^+ + 2e \rightarrow O_2(g) + H_2O$$

[26] See I. M. Kolthoff and R. Belcher, *Volumetric Analysis*, vol. III, Wiley-Interscience, New York, 2nd ed., 1957.

[27] H. H. Willard and W. E. Cake, *Ind. Eng. Chem.* **11**, 480 (1919); H. H. Willard and R. C. Gikon, *Ind. Eng. Chem., Anal. Ed.* **3**, 88 (1931).

[28] G. F. Smith, *Anal. Chim. Acta* **8**, 397 (1953).

[29] M. Tanaka, *Bull. Chem. Soc. Japan* **27**, 10 (1954).

[30] J. J. Lingane and D. G. Davis, *Anal. Chim. Acta* **15**, 201 (1956); see also C. P. Lloyd and W. F. Pickering, *Talanta* **11**, 1409 (1964).

[31] H. H. Willard and L. J. Merritt, Jr., *Ind. Eng. Chem., Anal. Ed.* **14**, 486 (1942).

is 2.07 V. Hence ozone is a powerful oxidizing agent. The excess of ozone and the oxygen formed are readily removed after the oxidation by passing an inert gas through the solution. Manganese in 1–2 *M* perchloric acid solution and in the presence of silver (nitrate) as a catalyst is quantitatively oxidized to permanganate. The latter can be titrated by standard procedures. Trivalent cerium is quantitatively oxidized to the quadrivalent form in a solution containing phosphoric and sulfuric acid. Ceric phosphate separates in the form of a white gel which is dissolved by sulfuric acid. Vanadyl ion is quantitatively oxidized to vanadate. Chromic ion in acid medium is scarcely oxidized unless silver nitrate is present as a catalyst and even then only in small amounts. Selenite and tellurite are quantitatively oxidized to selenate and tellurate in alkaline solution, but there is only slight oxidation in acid medium. Trivalent arsenic and antimony are completely oxidized in acid or alkaline solution. Mercurous perchlorate is readily oxidized to the mercuric form. Cobaltous and nickelous hydroxides form the corresponding trivalent hydroxides. Nitrite is oxidized to nitrate; hypophosphite and phosphite are oxidized to phosphate. Iodide is quantitatively oxidized to periodate in 1–10 per cent sodium hydroxide solution. Chloride can be present in any amount, but bromide interferes.

For reducing purposes, metals, metal amalgams, and stannous chloride are often used, more rarely hydrogen sulfide and sulfur dioxide. The excess of the latter two can be easily removed by boiling, but they are usually slow in their reducing action. Tin(II) chloride in 1 *N* hydrochloric acid is a powerful reducing agent ($E^f_{\text{Sn(IV)/Sn(II)}} = 0.14$ V). It is often used for the quantitative reduction of iron(III) to iron(II). The excess of tin(II) is oxidized with mercuric chloride:

$$\text{Sn(II)} + 2\,\text{HgCl}_2 \rightarrow \text{Sn(IV)} + \text{Hg}_2\text{Cl}_2 + 2\,\text{Cl}^-$$

For details see page 832.

Hydrazine in the form of hydrochloride, sulfate, or the free base is an attractive reductant, because in many instances it is oxidized to inert nitrogen. It finds useful application in the reduction of arsenic(V) to arsenic(III) in acid medium. The excess is destroyed by heating with concentrated sulfuric acid. The arsenic(III) is titrated with bromate[32] (p. 861).

Hydriodic acid is fairly concentrated solution in the presence of hypophosphorous acid finds special application in the quantitative reduction of sulfate to hydrogen sulfide, which is distilled and determined iodimetrically.[33]

Several metals are powerful reducing agents (see Table A-0). The relation between a metal electrode potential and the metal-ion concentration (activity) in the solution is given by $E = E^0 + (0.059/n)\log[\text{M}^{n+}]$. When a metal is used as a reducing agent, the metal-ion concentration increases during the reaction with an oxidant and therefore the reduction potential of the metal increases. By adding to the solution an anion which forms a slightly soluble salt (or complex) with the metal ions the potential of the metal can be kept practically constant and a value which is much more negative than the formal potential. In this manner it is possible to make fairly noble metals behave as powerful reductants.

Consider, for example, the potential of silver in 1 *N* hydrochloric acid. In this medium $[\text{Ag}^+] = (\text{K}_{\text{sp}})_{\text{AgCl}}/[\text{Cl}^-] \cong 10^{-10}$. The formal potential of silver in 1 *N* perchloric acid is 0.79 V. Hence in 1 *N* hydrochloric acid the reduction potential is

$$E = 0.79 + 0.059 \log 10^{-10} = 0.20 \text{ V} \qquad (25°\text{C})$$

Thus ferric iron is quantitatively reduced by silver in a medium of 1 *N* hydrochloric acid.[34]

[32] V. E. Schulek and P. von Villecz, *Z. Anal. Chem.* **76**, 81 (1929).
[33] L. Luke, *Ind. Eng. Chem., Anal. Ed.* **15**, 602 (1943).
[34] G. H. Walden, Jr., L. P. Hammett, and S. M. Edmonds, *J. Am. Chem. Soc.* **56**, 350 (1934).

Problems: The student may solve the following problems:

(a) A solution of ferric perchlorate in 1 *N* perchloric acid is shaken with silver until equilibrium is attained. Calculate the ratio ferric to ferrous.

(b) A solution of ferric iron in 1 *N* hydrochloric acid is reduced with silver. Calculate the equilibrium ratio of ferric to ferrous.

(c) As (b), but in 1 *N* hydrobromic instead of hydrochloric acid (take $(K_{sp})_{AgBr} = 10^{-12}$).[35]

Mercury in 1 *N* hydrochloric acid is about as strong a reductor as silver.

Problem: Calculate the potential of mercury in 1 *N* hydrochloric acid saturated with calomel.

Metals that have standard potentials more negative than hydrogen, such as zinc or cadmium, discharge hydrogen ions in acid medium:

$$Zn + 2\,H^+ \rightarrow Zn^{++} + H_2$$

Therefore their application to reductions is limited by the hydrogen potential. However, in Chapter 8 we saw that hydrogen has a very high overvoltage on mercury; therefore, if the metals are used in the form of amalgams, it is possible to prevent hydrogen formation. Such metal amalgams are powerful reducing agents and can be used for the reduction of vanadate, V(V), to V(IV), V(III), and V(II); of molybdate, Mo(VI), to Mo(III); of tungstate, W(VI), to W(V) or to W(III); of uranyl, U(VI), to U(IV) and U(III); of Ti(IV) to Ti(III); of Cr(III) to Cr(II), etc. The reduction potentials of amalgams saturated with the following metals are zinc, -0.76; cadmium, -0.40; lead, -0.13; and bismuth, 0.32 V. A simple apparatus for work with saturated amalgams is described by Smith and Kurtz.[36]

In general it is advisable to remove oxygen from the solution in reductions with metal and amalgam reductors, because oxygen may be reduced to hydrogen peroxide. By the use of different metal and amalgam reductors differential reductions can be made. For example, with the silver reductor in 1 *N* hydrochloric acid, the following reductions occur quantitatively: ferric iron, molybdate [Mo(VI) to Mo(V) in 2 *M* HCl], vanadate [V(V) to V(IV)], uranyl [U(VI) to U(IV)], copper [Cu(II) to Cu(I) in 2 *M* HCl]. Tetravalent titanium and chromic chromium are not reduced. On the other hand, these elements are quantitatively reduced by zinc amalgam. Moreover, zinc amalgam reduces Mo(VI) to Mo(III), V(V) to V(II), and Cu(II) to Cu(0).

A popular amalgam is that of zinc, which is used in the form of a filtering column (Jones reductor) or an amalgamated zinc spiral (see the procedure on p. 829).

The mercury cathode, especially when used at constant potential, has been applied for many years in quantitative separations of metal ions and also in the quantitative reduction of a metal from a higher to a lower oxidation state. For example iron(III) is reduced to iron(II), or to iron(0) amalgam at very negative potentials. Other reductions of titrimetric interest are Cr(III) to Cr(II), Cu(II) to Cu(I), Eu(III) to Eu(II), Ir(IV) to Ir(III), Pu(IV) to Pu(III), Re(VII) to Re(IV), Sn(IV) to Sn(II), U(VI) to U(IV), Yb(III) to Yb(II).[37]

Induced reactions

Various reducing agents, such as trivalent arsenic and antimony, hydrogen peroxide, or stannous tin can be titrated accurately with permanganate in the presence of hydrochloric acid. This is to be expected, because the substance to be titrated is much more easily oxidized than the chloride. Therefore oxidation of the latter, which moreover takes place slowly, does not occur until the substance titrated has been quantitatively

[35] For the silver–silver bromide reductor see F. Pantani, *Anal. Chim. Acta* **31**, 121 (1964).

[36] G. F. Smith and L. T. Kurtz, *Ind. Eng. Chem., Anal. Ed.* **14**, 854 (1942).

[37] For an extensive review of the literature on the use of the mercury cathode in analytical chemistry see J. A. Page, J. A. Maxwell, and R. P. Graham, *Analyst* **87**, 245 (1962).

oxidized. This behavior, however, is not observed in the titration of ferrous iron with permanganate. A small part of the chloride is oxidized to chlorine or hypochlorous acid, HClO, during the titration of the ferrous iron; and since the chlorine reacts only slowly with the ferrous iron, too much oxidizing agent will be used up. The reaction between ferrous iron and permanganate *induces* the reaction between hydrochloric acid and permanganate, and we encounter here a so-called *induced reaction.*

In general, it can be stated that if an oxidizing agent reacts with more than one electron in its reduction to the final state, there will be formed intermediate unstable valence forms of the oxidant in its reaction with a one-electron donor (reductant). For example, in the reduction of permanganate Mn(VII) with ferrous iron there will be formed intermediately Mn(VI), Mn(V), Mn(IV), and Mn(III). All these valence forms react with ferrous iron, but several of them also react with divalent manganese. If another reducing substance is present in the solution, one or more of the metastable valence forms might react to some extent with this reducing substance. It appears possible that Mn(V) is a strong oxidizing agent which reacts not only with ferrous iron and divalent manganese, but also with a comparable rate with hydrochloric acid. In the beginning of the titration of ferrous iron the divalent manganese concentration is very small and hydrochloric acid may react with the manganese of intermediate valence state. However, when much divalent manganese is added to the solution, the intermediate valence form will react with divalent manganese and practically no reaction with hydrochloric acid occurs. The manganous salt prevents the induced reaction. In a titration of ferrous iron in the presence of chloride, manganese is added as Zimmermann–Reinhardt solution, which is a solution of manganous sulfate in dilute sulfuric and phosphoric acids (see p. 828). In the reduction of ceric to cerous cerium, no intermediate valence form can be formed. Consequently, no induced oxidation of chloride is to be expected in the titration of ferrous iron with ceric sulfate, and this is indeed the case.

Conclusive evidence of the formation of intermediate valence forms is derived from the induced oxidation of divalent manganese by the reaction between chromate and arsenious oxide in acid medium. An acid chromate solution reacts quantitatively with As(III) but does not oxidize divalent manganese. However, on addition of arsenious oxide to an acid mixture of dichromate and divalent manganese a copious precipitate of manganese dioxide separates. Similarly, cerium (III) is not oxidized by chromium(VI) in 1 to 2 *N* sulfuric acid. However, this oxidation is induced by the reaction between chromium(VI) and arsenic(III). The mechanism of this reaction is much more involved than indicated by the overall reaction. From reaction kinetics it is known that the rate-determining step is first order to chromium(VI) and arsenic(III):

$$\mathrm{Cr(VI)} + \mathrm{As(III)} = \mathrm{Cr(IV)} + \mathrm{As(V)} \qquad (1)$$

Cr(IV) is a strong oxidizing agent but it also has weak reducing properties. It is amphielectronic, like V(III), Mo(V), etc. If there is no reducing agent in the solution other than arsenic(III) the Cr(IV) reacts with Cr(VI):

$$\mathrm{Cr(IV)} + \mathrm{Cr(VI)} = 2\,\mathrm{Cr(V)} \qquad (2)$$

The Cr(V) reacts rapidly with As(III):

$$\mathrm{Cr(V)} + \mathrm{As(III)} = \mathrm{Cr(III)} + \mathrm{As(V)} \qquad (3)$$

The sum of (1), (2), and (3) yields the overall reaction

$$2\,\mathrm{Cr(VI)} + 3\,\mathrm{As(III)} = 2\,\mathrm{Cr(III)} + 3\mathrm{As(V)}$$

In the presence of manganese(II) or cerium(III) there is competition between Cr(VI) and Mn(II) or Ce(III) to react with Cr(IV):

$$\mathrm{Cr(IV)} + \mathrm{Mn(II)} = \mathrm{Cr(III)} + \mathrm{Mn(III)} \qquad (4a)$$

[Mn(III) disproportionates into Mn(II) and MnO_2]:

$$\mathrm{Cr(IV)} + \mathrm{Ce(III)} = \mathrm{Cr(III)} + \mathrm{Ce(IV)} \qquad (4b)$$

Under suitable conditions reaction (4) may replace Eq. (2). The overall reaction then is the sum of (1) and (4). This means that under such conditions one Cr(VI) oxidizes one As(III) to As(V) and one Mn(II) to Mn(III). The ratio of the number of equivalents involved in the induced reaction (4) to that in the inducing reaction (1) is called the *induction factor*. Under favorable conditions it is actually found to be one half in the above system. This corresponds to the sequence of reactions (1) and (4).

In very dilute sulfuric acid chromic acid slowly oxidizes diphenylamine and diphenylbenzidine indicators. However, on addition of a trace of iron(II) the oxidation is induced and the violet or blue forms appear immediately. In 1 *N* sulfuric acid ferrous phenanthroline is not oxidized by Cr(VI); however, on addition of some arsenic(III) the oxidation is induced instantantaneously. The unstable chromium(IV) produced in reaction (1) rapidly oxidizes the indicator.

Of analytical interest are reactions induced by the reaction between hydrogen peroxide and ferrous iron. Hydrogen peroxide is a two-electron acceptor and iron(II) a one-electron donor. The reaction is first order to hydrogen peroxide and ferrous iron and the rate-determining step is given by

$$H_2O_2 + Fe^{++} = Fe(III) + OH^- + OH\cdot \qquad (5)$$

The hydroxyl free radical is very unstable and is a very strong oxidant. It reacts with hydrogen peroxide to form another unstable free-radical HO_2, called perhydroxyl:

$$H_2O_2 + OH\cdot = H_2O + HO_2\cdot \qquad (6)$$

The perhydroxyl free radical is a mild reducing agent and is oxidized by iron(III):

$$HO_2 + Fe(III) = Fe^{++} + H^+ + O_2 \qquad (7)$$

Thus iron(II) is reformed in reaction (7) and it reacts again with the peroxide according to reaction (5). Such a sequence of reactions is called a *chain reaction*. It would cause a quantitative decomposition of the peroxide to oxygen if it were not for the fact that the $OH\cdot$ free radical [reaction (2)] can also react with iron(II):

$$OH\cdot + Fe^{++} = OH^- + Fe(III) \qquad (8)$$

This reaction terminates the chain. The sum of (1), (5), and (8) corresponds to the stoichiometric reaction

$$H_2O_2 + 2\,Fe^{++} = 2\,OH^- + 2\,Fe(III) \qquad (9)$$

Actually, upon *slow* addition of ferrous iron to hydrogen peroxide there is a pronounced induced decomposition of peroxide to oxygen and the reaction deviates considerably from the stoichiometric one. However, when the iron(II) is added rapidly (e.g., from a pipet) and the mixture is well shaken the stoichiometric reaction (9) occurs. Under these conditions the concentration of iron(II) in the mixture is so large that reaction (8) occurs in favor of (6). However, even under these conditions the reaction is greatly interfered with by organic compounds even though these compounds react extremely slowly with the peroxide. The deviations from stoichiometry are particularly great in very dilute solutions of the reactants. When carried out in a nitrogen atmosphere much less iron is consumed than corresponds to reaction (9), but in the presence of oxygen there is a large overconsumption of iron(II). We will first consider the reaction in the presence of some ethanol in the absence of oxygen. Table 37-4 presents some experimental results[38] in the presence of ethanol and in the absence and in the presence of oxygen. Evidently the reaction between hydrogen peroxide and ferrous iron is not suitable for quantitative purposes, because even in the presence of traces of organic compounds the molar reaction ratio of iron(II) to peroxide is much smaller than 2 in the absence of oxygen and much greater than 2 in the presence of oxygen (induced oxidation by oxygen). In the experiments in Table 37-4 the iron(II) was added to the

[38] I. M. Kolthoff and A. I. Medalia, *J. Am. Chem. Soc.* **71**, 3777, 3784, 3789 (1949).

TABLE **37-4.** **Reaction between Fe^{++} and H_2O_2 in 1.5 N H_2SO_4**
$[Fe^{++}]_{initial} = 5 \times 10^{-4}\,M$, $[H_2O_2]_{initial} = 1.25 \times 10^{-4}\,M$
(instantaneous mixing)

Exp. No.[a]	*Concentration of Ethanol, M*	*Gas Atmosphere*	*Molar Reaction Ratio* Fe^{++}/H_2O_2
1	0	Nitrogen	1.99
2	10^{-4}	Nitrogen	1.64
3	10^{-3}	Nitrogen	0.54
4	10^{-2}	Nitrogen	*0.15*
5	0	Oxygen	2.01
6	10^{-4}	Oxygen	2.74
7	10^{-2}	Oxygen	*10.4*

[a] Note the difference in reaction ratio in expts. 4 and 7.

acid hydrogen peroxide solution by instantaneous mixing (see Ref. 38). If the iron were to have been added dropwise the deviation from stoichiometry would have been considerably greater. The large deviations from stoichiometry indicate that chain reactions are involved. The following mechanisms account for the results.

Absence of oxygen:

$$H_2O_2 + Fe^{++} = Fe(III) + OH^- + OH\cdot \qquad \text{(rate-determining)} \tag{5}$$

$$OH\cdot + C_2H_5OH = H_2O + CH_3\dot{C}HOH \tag{10}$$

$$CH_3\dot{C}HOH + Fe(III) = CH_3COH + H^+ + Fe^{++} \tag{11}$$

$$OH\cdot + Fe^{++} = OH^- + Fe(III) \qquad \text{(termination)} \tag{12}$$

Presence of oxygen: As an example the induced oxidation of acetone is considered:

$$OH\cdot + CH_3COCH_3 = H_2O + CH_3CO\dot{C}H_2 \tag{13}$$

$$CH_3CO\dot{C}H_2 + O_2 = CH_3CO\dot{C}H_2OO \qquad \text{(very fast)} \tag{14}$$

$$CH_3CO\dot{C}H_2OO + Fe^{++} + H^+ = CH_3COCH_2OOH + Fe(III) \tag{15}$$

$$CH_3COCH_2OOH + Fe^{++} = CH_3CO\dot{C}H_2O + OH^- + Fe(III) \tag{16}$$

$$CH_3CO\dot{C}H_2O + Fe^{++} + H^+ = CH_3COCH_2OH + Fe(III) \qquad \text{(termination)} \tag{17}$$

$$CH_3CO\dot{C}H_2O + CH_3COCH_3 = CH_3COCH_2OH + CH_3CO\dot{C}H_2 \tag{18}$$

Reaction (18) produces the same free radical as reaction (13), accounting for the chain mechanism.

The possibility of induced air oxidations must always be considered in oxidation–reduction titrations. Under ordinary conditions titrations are carried out in the presence of air. Although various reducing agents are only slowly oxidized by air, this reaction may be induced during the titration of the reducing agent with the oxidizing substance. Iodide, for example, is slowly oxidized by air in acid medium; various reactions, however, may induce the air oxidation of iodide. This, for example, takes place in the iodometric determination of vanadate. The acid solution of vanadate is treated with an excess of iodide, and the liberated iodine is titrated with sodium thiosulfate. The reaction between vanadate and iodide induces the air oxidation of the latter, and high results are found. The vanadate reacts with iodide, forming first V(III) which readily reacts with oxygen to form a "peroxide".[39] In such cases good

[39] G. Edgar, *J. Am. Chem. Soc.* **38**, 2369 (1916); W. C. Bray and J. B. Ramsey, *ibid.* **55**, 2279 (1933); J. B. Ramsay, E. L. Colithman, and L. C. Pack, *ibid.* **68**, 1695 (1946).

results can be obtained by making the titration in an atmosphere of carbon dioxide or other inert gas.

PROBLEMS

1. Calculate the reduction potential of a mixture of potassium ferrocyanide and ferricyanide in which the ratio of the two is 100 : 1, 10 : 1, 1 : 1, 1 : 10, and 1 : 100. Do the same for a mixture of stannous tin with stannic tin. The formal potential of the ferro-ferricyanide system is 0.44 V, that of the stannous-stannic tin, 0.14 V.
2. The formal potential of the vanadate–vanadyl system,

$$VO_4^{3-} + 6\,H^+ + e \rightleftharpoons VO^{++} + 3\,H_2O$$

at $[H^+] = 1$ is equal to 1.2 V. What is the reduction potential at 30°C and at pH = 1, and $[VO_4^{3-}] = [VO^{++}]$? *Ans.* 0.84 V.
3. What is the equivalent weight of vanadate in the titration with ferrous iron? Balance the equation and write the two partial reactions.
4. The formal potential of the $VO_4^{3-} \rightarrow VO^{++}$ system is 1.2; that of $VO^{++} \rightarrow V^{3+}$ is 0.4. Is it possible to reduce VO_4^{3-} to VO^{++} and V^{3+}, respectively, with ferrous sulfate or with titanous chloride in 1 *N* hydrochloric acid? [Formal potential Fe(III) → Fe(II) is 0.68. Formal potential Ti(IV) → Ti(III) is −0.04.]
5. What is the equivalent weight of stannous chloride in the titration with ferric sulfate? Is this reaction quantitative at the equivalence point? Can phenanthroline-ferrous complex be used as an indicator? [Formal potential Sn(IV) → Sn(II) is 0.14; Fe(III) → Fe(II) is 0.68.] In 1 *N* hydrochloric acid the indicator changes from red to almost colorless at a potential of 1.1 V. What should be the reduction potential of a useful indicator in this titration?

CHAPTER 38 *End-Point Determination and Physicochemical Titration Methods*

In the preceding discussion of titrimetry the end point of a titration was detected visually using an indicator that changes color, or more rarely produces a turbidity at the end point. There are many reactions that could be made the basis of a titration, but for which visible end-point determination is not suitable, either because there is no satisfactory indicator or because the reaction is too incomplete at the equivalence point to give a sharp color change. In the latter case the reaction often becomes quantitative with a relatively small excess of either one of the reactants. When visual end-point detection is not feasible, it is often possible to locate the end point using some physicochemical or physical property of the species during the titration. Such a property is usually directly proportional to the concentration or, more rarely, to the logarithm of the concentration, of the species present in solution. Values of the physicochemical property measured are plotted in an appropriate manner and the end point located graphically. For example, in amperometric titrations, the current measured is proportional to the concentration of one or more of the reactants or products, in spectrophotometric titrations the absorbance is proportional to the concentration of one or more of the reactants or products, and the same is true in radiometric titrations, where radioactivity is measured during the titration. The graphical location of the end point is the same in all these cases. To avoid repetition it is appropriate to discuss the graphical location of the end point in the present chapter.

The graphical plotting methods fall into two classes. In the first, data obtained relatively distant from the end point are plotted so that two straight lines are obtained. One line represents data obtained well before, and the other represents data obtained well after, the end point. Their point of intersection corresponds to the end point. This class includes spectrophotometric, amperometric, conductometric, radiometric, thermometric (enthalpometric), and cryoscopic titrations. The second class of plotting method uses data obtained in the immediate vicinity of the equivalence point and is employed mainly in the application of the potentiometric method, although it has been used for several spectrophotometric titrations.

Linear extrapolation methods

Consider the reaction

$$X + T \rightleftharpoons P + Q \qquad (1)$$

where X is the unknown constituent being determined, T is the titrating species, and P and Q are the products of the reaction.

First, we assume that the solution property which is being measured is directly proportional to the equilibrium concentration of a solute species present in the titration mixture. If Z represents the measured property, then

$$Z = a_x[\mathrm{X}] + a_t[\mathrm{T}] + a_p[\mathrm{P}] + a_q[\mathrm{Q}] \tag{2}$$

The coefficients a_x, a_t, a_p, and a_q represent constants that determine the amount [X], [T], [P], and [Q] contribute to the magnitude of Z. Titration curves of various shapes can be observed depending upon the relative magnitudes of a_x, a_t, a_p, and a_q. The solid line in Fig. 38-1 represents the shape of the titration curve obtained for reaction (1) when $a_t = a_p = a_q = 0$, and is derived from Eq. (3),

$$Z = a_x[\mathrm{X}] \tag{3}$$

assuming quantitative reaction. Let c_x equal the initial concentration of X, f equal the fraction titrated, and R equal the ratio of the volume of T added to the initial volume of X. Then (3) may be written

$$Z(1 + R) = Z' = a_x \cdot c_x - a_x \cdot c_x \cdot f$$

which is the equation of a straight line of negative slope ($= -a_x \cdot c_x$). This line intercepts the fraction-titrated axis at the equivalence point ($f = 1$). We plot Z' rather than Z to take into account dilution effects during the titration. As a practical matter, it is common to titrate with as concentrated a reagent as possible in order to make R as small as possible, i.e., so that $Z = Z'$.

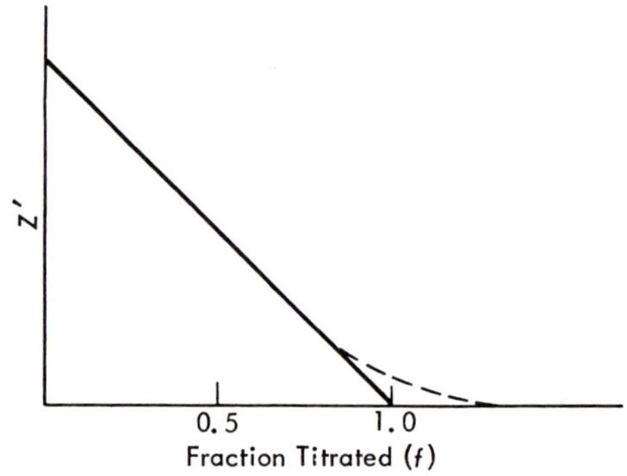

Fig. 38-1. Linear Titration Line. ——— Quantitative reaction; - - - - Nonquantitative reaction.

The solid titration line shown in Fig. 38-1 illustrates the advantages of the graphical extrapolation methods. Data need not be obtained in the vicinity of the point, and no problems are caused by overrunning the end point.

If reaction (1) is nonquantitative, the titration line starts to deviate from a straight line near the end point. Depending upon the completeness of reaction, more or less rounding occurs, as is illustrated by the dashed line in Fig. 38-1. This rounding is most pronounced at the equivalence point, and becomes less at points more distant from the equivalence point because the excess of X over T or *vice versa* forces reaction (1) to completion. Provided this rounding is not excessive, data obtained in the linear regions may be extrapolated and the end point located.

Various titration lines for reaction (1) are shown in Fig. 38-2. Line 1 corresponds to $a_t = a_x = a_q = 0$, and $a_p =$ some nonzero value, line 2 to $a_t = a_x = 0$, and

$a_q = a_p =$ some nonzero value. Note that the location of the equivalence point is sharper for line 2 than for line 1, because both the products formed in the reaction contribute equally to Z' in line 2, as compared to line 1, in which only P contributes to Z'. Line 3 corresponds to line 1, except that $a_t = 0.333a_p$. No break in the titration line would be observed if $a_t = a_p$.

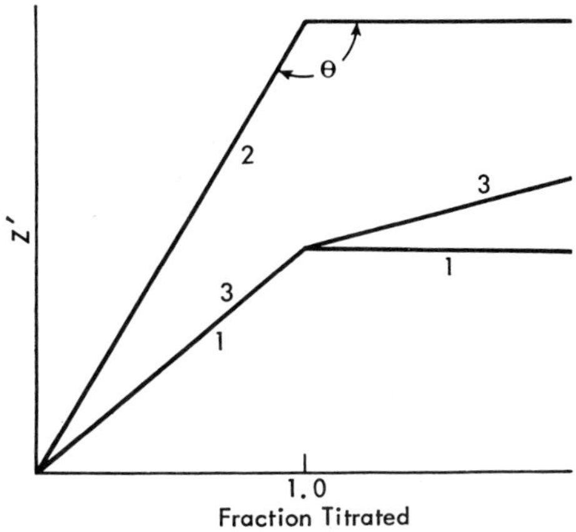

Fig. 38-2. Shapes of linear titration lines.

Figure 38-3 illustrates several other commonly observed linear titration lines. V-shaped line 1 is obtained if $a_p = a_q = 0$, $a_x =$ some nonzero value, and $a_t = 0.5a_x$. V-shaped line 2 is obtained instead of line 1 if $a_t = a_x$. Line 3 results if, of all the species in reaction (1), only T contributes to Z. Line 4 will be obtained if $a_x = a_p = a_t =$ some nonzero value, and $a_q = 0$.

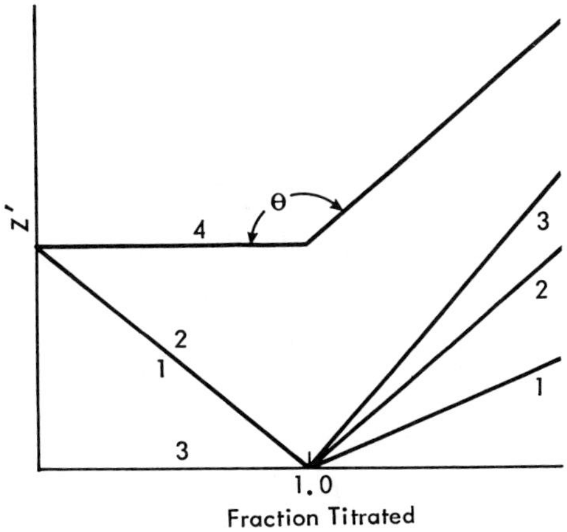

Fig. 38-3. Shapes of linear titration lines.

As is apparent from Figs. 38-2 and 38-3, the smaller the angle of intersection of the two straight-line portions, θ, the more accurate the location of the equivalence point. V-shaped lines such as line 2 of Fig. 38-3 are preferred. Line 4 of Fig. 38-3 represents an undesirable experimental situation, because it is necessary to measure changes in Z′ after the equivalence point which are small compared to the absolute value of Z′.

The accuracy in the location of the end point decreases with decreasing completeness of reaction. When the reaction is incomplete during the entire course of the titration, no straight line or lines are obtained, only a curve or curves.

Second derivative method

For all practical purposes, the potentiometric titration method is the only type in which the measured property, the potential, is a logarithmic function of the concentrations of the species being titrated or being used as titrant (see Chapter 8). Figure 38-4 represents a typical S-shaped curve obtained during a potentiometric

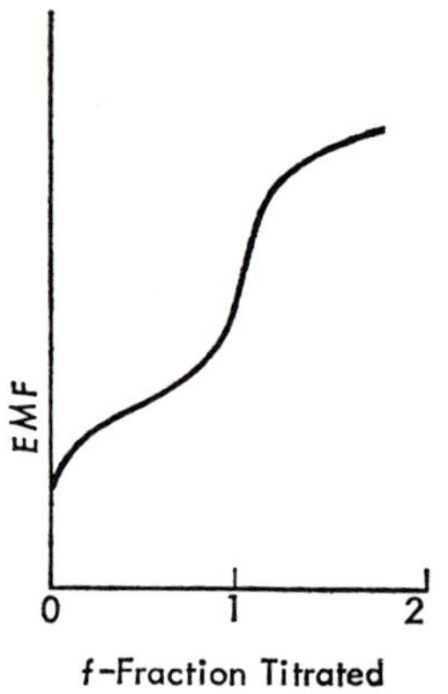

Fig. 38-4. Illustrative potentiometric titration curve.

titration. The inflection point in this curve corresponds to the end point, except principally in precipitation titrations, when the precipitate involves a cation and anion of different absolute charge and the precipitate has a fairly large solubility. Figures 38-5 and 38-6 represent the first and second derivatives of the curve in Fig. 38-4 in the immediate vicinity of the inflection point. The first derivative curve is a maximum, and the second derivative curve is zero, at the inflection point. Provided experimental data are obtained at closely spaced intervals in the immediate vicinity of the inflection point, it is possible to calculate the second derivative curve to a good approximation. Under such conditions, the change in potential for a change in fraction titrated is very nearly the first derivative of potential with

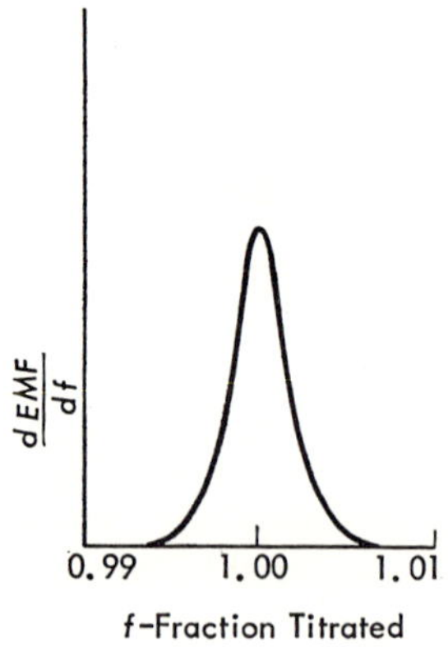

Fig. 38-5. First derivative curve for Figure 38-4.

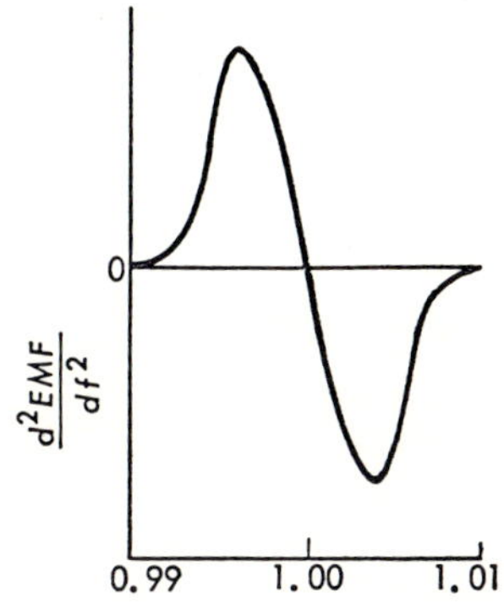

Fig. 38-6. Second derivative curve for Figure 38-4.

respect to fraction titrated. Similarly, the change in the first derivative with respect to a change in the fraction titrated approaches the value of the second derivative.

If experimental data are obtained at equally spaced volume increments,

$$\Delta V = V_{i+1} - V_i \tag{5}$$

and the corresponding potential values are E_{i+1} and E_i, we have

$$\frac{dE}{dV} \simeq \left(\frac{\Delta E}{\Delta V}\right)_{V_1+\frac{1}{2}\Delta V} = \frac{E_{i+1} - E_i}{\Delta V} \tag{6}$$

at the volume $V_i + \frac{1}{2}\Delta V$. By definition

$$\frac{\Delta^2 E}{\Delta V^2} = \frac{(\Delta E/\Delta V)_{V_{i+1}+\frac{1}{2}\Delta V} - (\Delta E/\Delta V)_{V_i+\frac{1}{2}\Delta V}}{\Delta V}$$

and

$$\frac{d^2E}{dV^2} \simeq \left(\frac{\Delta^2 E}{\Delta V^2}\right)_{V_{i+1}} = \frac{E_{i+2} - 2E_{i+1} + E_i}{\Delta V^2} \tag{7}$$

at the volume V_{i+1}.

An example of the calculation of second difference curve according to Eq. (7) is given in Table 52.1 on p. 996.

The accurate estimation of the first and second derivative curves using Eqs. (6) and (7) requires that the experimental data be closely spaced, and that several data points be obtained in the volume interval between the maximum and minimum in Fig. 38-6. For example, if Fig. 38-4 represents *any* aqueous potentiometric acid–base titration, the shape of the titration curve in the immediate vicinity of the inflection point is described by Eq. (145) in Chapter 4. Differentiating this equation three times, and setting $d^3pH/df^3 = 0$, leads to the general result that the maximum and minimum occur at exactly 0.31 pH unit on either side of the inflection point. If we arbitrarily require that three experimental data points be obtained between the maximum and minimum of Fig. 38-6, no more than a 0.10-pH-unit change should occur for the volume increment chosen in the vicinity of the inflection point. From Eq. (146) of Chapter 4 we obtain the result

$$\Delta V = \frac{0.1V}{\eta_{\max}} \tag{8}$$

where V is the volume required to reach the inflection point, ΔV the size of the volume increment which produces a pH change of 0.1 unit, and η_{max} the sharpness index at the inflection point. [For example, in the titration of 100 ml of 0.1 M weak acid ($K_{HB} = 10^{-5}$) with 0.1 M sodium hydroxide, the sharpness index at the inflection point is 1100, and $\Delta V = 0.01$ ml.]

PROBLEMS

1. Plot pH versus percent titrated and 10^{-pH} versus percent titrated for the titration of 0.01 M HCl with 1 M NaOH. Discuss the relative advantages of these two plotting methods.
2. Suppose that 0.01 M NaOH had been used as titrant in Problem 1. What change in plotting would be needed to obtain a straight-line extrapolation using the second plotting technique?
3. Radioactive chloride ion is titrated with nonradioactive silver ion. Sketch the shape of the plot of radioactivity in the solution versus fraction titrated.
4. If radioactive silver ion were substituted in Problem 3, what would the shape of the plot of radioactivity in solution versus fraction titrated be?
5. A 0.1 M solution of a weak base ($K_B = 10^{-5}$) is titrated with 0.1 M HCl. Calculate $[H^+]$ at 99, 100, 101, 102, 103, 104, and 105 per cent titrated. Suggest a linear plotting method based upon experimental data obtained from pH measurements.
6. A 0.001 M solution of chloride is titrated with $AgNO_3$ and $[Ag^+]$ measured during the titration. Neglecting volume change, plot $[Ag^+]$ after 25, 50, 75, 90, 95, 99, 100, 101, 125, and 150 per cent of the equivalent amount of silver. $(K_{sp})_{AgCl}$ at the titration temperature is 10^{-10}.
7. (a) Repeat Problem 6, for 0.0001 M chloride. (b) The solubility of silver chloride decreases markedly with decreasing temperature and in the presence of considerable quantities of organic solvents, such as methanol or acetone. How can you increase the accuracy of the titration of 0.0001 M chloride?
8. Arsenic trioxide is titrated with bromate in acid media in the presence of bromide (see p. 820). $[Br_2]$ is measured during the titration. Plot the titration line.

PART FIVE

Volumetric Determinations

CHAPTER 39 *Acidimetry and Alkalimetry*

Standard hydrochloric acid

The strong inorganic acids are commercially available in the form of concentrated solutions. The specific gravities of solutions of acids and bases of various concentrations can be found in handbooks.[1] Therefore, if the commercial product is pure, its strength can be found by the careful determination of the specific gravity, and a standard solution can be prepared by proper dilution. It is necessary, however, not to rely entirely upon the empirical tables giving the relation between concentration and specific gravity but to determine the normality of the acid after dilution.

Solutions of hydrochloric acid are entirely satisfactory as standard solutions for acidimetric titrations.

Hulett and Bonner[2] have recommended a *constant-boiling* mixture of hydrochloric acid and water for the preparation of standard solutions. The vapor of the constant-boiling acid has the same composition as the liquid. The concentration of this constant-boiling mixture depends upon the pressure under which it is distilled. To prepare the constant-boiling acid, Foulk and Hollingsworth [*J. Am. Chem. Soc.* **45**, 1223 (1943)] start with a hydrochloric acid solution of specific gravity 1.18 and distill with a speed of 3–4 ml/min. After about three-fourths of the original liquid has been distilled and rejected, the constant-boiling mixture is collected in a dry flask. It is best to begin with a liter or more of acid. The distillation is discontinued when a residue of only 50–60 ml remains in the distilling flask. The barometric pressure during the distillation should be read to 1 mm. The "equivalent weight" of the acid is given in Table 39-1. The constant-boiling mixture can be kept in well-closed bottles without change.

TABLE **39-1. Composition of Constant-Boiling Hydrochloric Acid**

Atmospheric Pressure during Distillation, mm Hg	*Concentration of Hydrochloric Acid, of* HCl/100 g (*Referred to Vacuum*)	*Weight of Distillate Containing* 1 *mole of* HCl, *When Weighed in Air*
780	20.173	180.621
770	20.197	180.407
760	20.221	180.193
750	20.245	179.979
740	20.269	179.766
730	20.293	179.551

[1] Compare I. M. Kolthoff and V. A. Stenger, *Volumetric Analysis*, vol. II, Wiley-Interscience, New York, 1947.

[2] G. A. Hulett and W. D. Bonner, *J. Am. Chem. Soc.* **31**, 390 (1909). See also W. D. Bonner and A. C. Titus, *ibid.* **52**, 633 (1930); A. C. Titus and D. E. Smith, *ibid.* **63**, 3266 (1941).

Preparation of 1 *N* hydrochloric acid

1. Weigh a dry, stoppered 250-ml Erlenmeyer flask to 0.01 g. Introduce the approximate amount of *constant-boiling acid* required for ½ liter of 1 *N* solution. Care must be taken to avoid wetting the neck of the flask. Reweigh the flask to 0.01 g, after replacing the stopper. Then add 100 ml of water (which is in equilibrium with the air), and transfer the contents quantitatively to a 500-ml volumetric flask. Dilute to the mark after the solution has attained room temperature, stopper, and mix. Transfer the acid to a dry glass-stoppered bottle, and label with the date of preparation and the normality.

2. If no constant-boiling mixture is available, a reagent-grade product of approximately known specific gravity can be taken. Measure with a graduate as carefully as possible the proper volume of the concentrated acid (12 *N*), and dilute as described above.

Preparation of 0.1 *N* hydrochloric acid. 0.1 *N* hydrochloric acid may be prepared by diluting the constant-boiling mixture or the concentrated acid. For the present purpose it may be obtained from the 1 *N* acid by taking two 50-ml portions of the latter with a pipet and diluting in a volumetric flask to 1 liter. Transfer the diluted solution to a glass stoppered liter bottle. Standardize as described below, and label.

Preparation of 0.01 *N* hydrochloric acid. Dilute two 50-ml portions of the 0.1 *N* acid in a volumetric flask to 1 liter.

Preparation of primary-standard sodium carbonate[3]

Procedure. Add gradually and with frequent stirring 769 g of analytical reagent-grade sodium hydrogen carbonate to 3 liters of distilled water at 86 ± 1°C in a hard-glass 4-liter beaker; maintain the solution at this temperature until all the solid has been added and has dissolved. Remove the beaker from the source of heat and stir the solution until the temperature falls to 75°C and a small crop of crystals is precipitated. Pass the solution, without further cooling, under suction through a close-grained filter-paper supported on a previously warmed large Buchner funnel. Cool the filtrate quickly and with stirring until the temperature falls to 18–20°C and the main crop of crystals is separated. Collect the crystals on a Whatman No. 541 filter paper supported on a Buchner funnel, and remove as much of the mother liquor as possible by suction. Wash the crystals on the filter paper once with a small amount of ice-cold distilled water, and dry them as completely as possible under suction. Transfer the damp crystals to a porcelain basin, and dry them by heating at 100°C; stir the mass from time to time during the drying. Lightly grind the dried crystals to a powder, mix the powder thoroughly, and convert it to sodium carbonate by heating to constant weight at 270 ± 10°C in a platinum basin.[4]

Procedure for standardization. Weigh out 0.2–0.25 g of sodium carbonate accurately, transfer to a 250-ml Pyrex flask, and dissolve in 25 ml of water. Add one drop of phenolphthalein solution (p. 694), and titrate with the 0.1 *N* hydrochloric acid to be standardized until the solution is colorless. At this point the sodium carbonate will have been transformed into bicarbonate with a little acid present in excess. Therefore, somewhat less than the volume of acid used up to this point must be added to transform all the carbonate into carbon dioxide (see Fig. 34-5 and pp. 705 and 786):

$$Na_2CO_3 + HCl \rightarrow NaHCO_3 + NaCl$$

$$NaHCO_3 + HCl \rightarrow H_2O + CO_2 + NaCl$$

Add a few drops of bromcresol green, and continue the titration until the color of the indicator begins to change from blue to green. At this point the solution still contains

[3] Procedure recommended by the Analytical Standards subcommittee of the Analytical Methods Committee of the Society for Analytical Chemistry, *Analyst* **90**, 251 (1965).

[4] Very pure sodium bicarbonate can be prepared from reagent-grade sodium carbonate by saturating a strong solution of the latter with carbon dioxide to precipitate the bicarbonate.

a trace of sodium bicarbonate and a large amount of carbon dioxide. Remove the latter by boiling the solution for 1–2 min, during which time the color changes to blue and finally to violet (mixed color of blue of bromcresol green and red of phenolphthalein). Cool the solution to room temperature, and continue the titration until the color changes to green. Repeat the procedure with two other portions of sodium carbonate. Calculate the normality of the acid.

Notes:

1. Methyl yellow, methyl orange, the mixed indicator methyl orange with indigo carmine (see p. 695), or bromphenol blue may be used as indicators instead of bromcresol green. With the first three indicators the solution should be cooled to room temperature after boiling, before continuing the titration to the end point.

2. Just before the equivalence point is reached, the solution is supersaturated with carbon dioxide. This explains why the first color change on titrating at room temperature is reached a trifle too early. If, after reaching the first color change at room temperature, the solution is boiled, the excess of carbon dioxide is expelled; and a sharp end point is obtained thereafter. The indicator blank can be determined by taking an equal volume of water containing approximately the same amounts of sodium chloride and indicator as are present in the solution titrated at the end point and adding hydrochloric acid until the color of the blank is equal to that of the solution at the end point.

3. No indicator correction is found if phenol red, bromthymol blue, or methyl red is used as indicator. The titration must then be carried out at boiling temperature, and hydrochloric acid added until the acid color of the indicator does not turn alkaline upon continued boiling. This procedure is recommended for the standardization of 0.01 N or more dilute acid. It is advisable to determine in a preliminary trial the approximate amount of acid required for reaching the end point, then to add in the standardization a little less than this amount, to boil 5 min, and finally to continue the titration in the manner described above. It is still simpler to reserve a small volume of the solution titrated, which is then added to the main portion when the end point in the latter has been reached.

4. Any of the indicators having a color-change interval located between those of phenolphthalein and methyl orange can be used if a slight excess of hydrochloric acid is added, the solution is boiled for 5 min or heated on a steam bath for $\frac{1}{2}$ hr to remove all carbon dioxide, and the excess of acid is finally back-titrated with 0.01 N sodium hydroxide.

5. In the standardization of 0.5 N or stronger acids with methyl orange, methyl yellow, or bromphenol blue as indicator, the titration can be carried out at room temperature directly to the end point; the end point is reached when the color of the solution becomes orange-red or yellow-purple, respectively. It is recommended that the solution be shaken vigorously just before the end point is reached, to remove that part of the carbon dioxide which is present in supersaturated solution.

6. Hydrochloric acid can be standardized gravimetrically by precipitation as silver chloride. This is one of the most exact methods for standardizing the acid, if the weight of the silver chloride is large enough to make the weighing errors negligibly small.

Other standard substances. Various standard substances can be used in place of sodium carbonate for the standardization of strong acids. A description of the preparation and purity test of these standards is given by Kolthoff and Stenger in *Volumetric Analysis*, vol. II. For details the student is referred to this text.

Borax, $Na_2B_4O_7 \cdot 10\,H_2O$. Sodium tetraborate decahydrate is obtained in a pure condition by two recrystallizations of a good product from water, followed by drying to constant weight in a desiccator over a solution saturated with respect to sucrose and sodium chloride.[5] Hurley[6] showed that the recrystallized salt can be obtained quickly

[5] H. Menzel, *Z. Anorg. Allgem. Chem.* **224**, 10 (1935).

[6] F. H. Hurley, *Ind. Eng. Chem., Anal. Ed.* **8**, 220 (1936); **9**, 237 (1937).

in the dry state by washing it with alcohol and ether and sucking it air dry. It should be kept in well-closed bottles, or else it may lose water and effloresce to the pentahydrate. Borax has the advantage of a large equivalent weight (190.71; that of sodium carbonate is 52.98), and the standardization can be carried out at room temperature with methyl red as indicator. The indicator blank is negligibly small. Its use as a standard substance is highly recommended.

Sodium oxalate, $Na_2C_2O_4$. This substance was originally recommended by Sörensen[7] as an ideal primary standard. It is easily obtained in a pure state and can be converted by heating at the proper temperature into sodium carbonate, in which form it is used for standardizing acids:

$$Na_2C_2O_4 \rightarrow Na_2CO_3 + CO$$

Sodium oxalate is also very useful in the standardization of potassium permanganate (p. 825).

Potassium bicarbonate, $KHCO_3$. This substance can be obtained in a pure state by recrystallization from aqueous solutions in the presence of carbon dioxide or by precipitation with alcohol. Its equivalent weight is almost twice that of sodium carbonate.

Potassium bitartrate. By ignition this substance is converted into potassium carbonate, which can be used for the standardization of acids. The bitartrate can also be used for the standardization of strong bases (p. 783).

Potassium iodate. Potassium iodate may be used for the direct standardization of acids by making use of the reaction

$$IO_3^- + 5\,I^- + 6\,H^+ \rightarrow 3\,I_2 + 3\,H_2O$$

By adding an excess of neutral alkali iodide and sodium thiosulfate to a solution of potassium iodate, hydrogen ions are removed upon addition of hydrochloric acid, until all the iodate has been converted into iodine. The iodine is removed by the excess sodium thiosulfate:

$$2\,Na_2S_2O_3 + I_2 \rightarrow Na_2S_4O_6 + 2\,NaI$$

Therefore, the liquid remains neutral and colorless during the titration with hydrochloric acid until the end point is reached. As soon as all the iodate has been used up, the solution becomes acid, as indicated by the color change of methyl yellow, methyl orange, bromcresol green, etc.

Mercuric oxide, HgO. Mercuric oxide has a high equivalent weight and can be kept unchanged. It contains no water of crystallization, and it is not hygroscopic. Acid solutions can be standardized against this substance by using phenolphthalein or methyl yellow as indicator. In the standardization use is made of the fact that mercuric oxide dissolves in potassium iodide with the formation of potassium mercuric iodide and an equivalent amount of potassium hydroxide:

$$HgO + 4\,KI + H_2O \rightarrow K_2HgI_4 + 2\,KOH$$

(or $$HgO + 4\,I^- + H_2O \rightarrow HgI_4^{--} + 2\,OH^-$$)

The colorless solution is titrated with acid, using an indicator with a color-change interval between pH 8 and 4. Instead of potassium iodide, much potassium bromide or sodium thiosulfate may be used.

Preparation of standard alkali. The hydroxides of sodium, potassium, and barium are used in the preparation of solutions of standard base; of these, sodium hydroxide is the most generally used. Most analytical grades of the latter reagent contain small amounts of sulfate, chloride, and silica, besides smaller or larger amounts of water and carbonate. Therefore a standard solution cannot be prepared by dissolving an exactly weighed amount in a certain volume of water. Moreover,

[7] S. P. L. Sörensen, *Z. Anal. Chem.* **36**, 639 (1897); **42**, 33, 512 (1903); **44**, 156 (1905); **45**, 217 (1906).

it is desirable in most cases to have carbonate-free standard base available. If barium hydroxide is used as the base, carbonate need not be removed from the solution by the addition of any special reagent, because barium carbonate is insoluble in barium hydroxide solution. Therefore, in the preparation of a solution of this base the precipitate is allowed to settle, and the supernatant liquid is siphoned off with protection against the atmosphere to exclude carbon dioxide.

Various methods are used for the preparation of carbonate-free sodium hydroxide. If for certain purposes (for example, in physiochemical research) it is desirable to have the base solution free of all impurities, it can be prepared by converting metallic sodium into sodium hydroxide (*care! explosion!*), or, better, sodium amalgam. Sodium and potassium hydroxides purified by treatment with alcohol are commercially available in the form of pellets which do not contain an appreciable amount of impurities, except some water and a little sodium carbonate. Solutions prepared from these purified products can be used for many analytical purposes.

Carbonate can be removed from sodium hydroxide solutions by the addition of an excess of barium or strontium chloride. The resulting solution of the base then contains barium or strontium ions; therefore, instead of applying this purification process, one might as well, in many cases, substitute barium hydroxide for sodium hydroxide. In general, however, the use of sodium hydroxide is to be preferred, because barium may give precipitates which are objectionable because of coprecipitation of hydroxide (in the titration of sulfuric acid, for example).

In the preparation of carbonate-free sodium hydroxide, use is also made of the fact that sodium carbonate is insoluble in a concentrated solution of sodium hydroxide.

Procedure for the preparation of concentrated carbonate-free sodium hydroxide. Dissolve 50 g of reagent-grade sodium hydroxide in 50 ml of water in a Pyrex flask. The sodium carbonate remains in suspension after the base has dissolved and settles very slowly. Therefore, filter the strong lye through a Gooch or sintered-glass crucible with exclusion of air to prevent contamination by atmospheric carbon dioxide. If a centrifuge is available, a clear carbonate-free solution may be obtained by centrifuging for a short time. The strong carbonate-free sodium hydroxide solution can be kept in paraffin-lined or ceresine-lined glass bottles.

The above method cannot be applied to the preparation of carbonate-free potassium hydroxide, because potassium carbonate is appreciably soluble in the concentrated base solution. In this case the lime method[8] is recommended for the preparation of a practically carbonate-free solution.

Preparation of standard sodium hydroxide solutions

Water for diluting. The water used for the dilutions should not contain an excess of carbon dioxide. Ordinary distilled water contains carbon dioxide. This can be removed either by boiling the water for a few minutes and cooling, or by drawing air, washed by alkali acid, and water, through it for some hours. Water in equilibrium with the atmosphere is 1.5×10^{-5} M with respect to carbon dioxide. If 0.01 N sodium hydroxide is prepared by diluting carbonate-free 0.1 N sodium hydroxide with this equilibrium water, only 0.3 per cent carbonate is introduced. Water can be tested in the following way for the presence of an excess of carbon dioxide: 500 ml of the water treated with 5 drops of 1 per cent phenolphthalein solution should not require more than 0.3 ml of 0.1 N alkali to give a permanent red color.

[8] I. M. Kolthoff, *Z. Anal. Chem.* **61**, 48 (1922).

Storing of standard alkali. Solutions of strong bases absorb carbon dioxide from the air; if alkali hydroxides are exposed to the atmosphere, they rapidly become contaminated with carbonate; barium hydroxide solutions become turbid, owing to the formation of insoluble barium carbonate. Short exposure of an alkali hydroxide solution to the atmosphere is not very harmful, at least when the temperature of the solution does not differ much from that of the surrounding air. When the standard solutions are quickly transferred to a buret, the contamination by carbon dioxide can be disregarded. However, the solution in the buret should be protected from the atmosphere by a soda-lime tube.

In general it is recommended that the burets be connected with the stock bottles in such a way that contamination by carbon dioxide is eliminated. An arrangement such as that shown in Fig. 39-1 serves the purpose quite well and is recommended for more permanent installations.

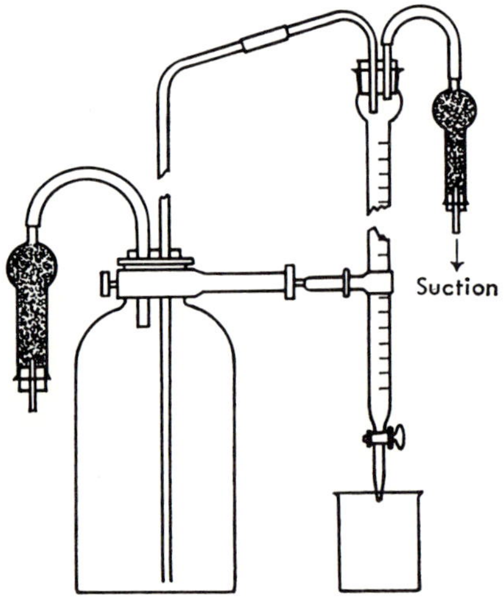

Fig. 39-1. Arrangement for protection of standard base against carbon dioxide.

Sodium hydroxide is tested for carbonate by filling a test tube to within 2 cm of the top with the solution and adding 1 ml of 0.5 *N* barium chloride. The tube is immediately stoppered and the contents mixed. No turbidity should appear. It is desirable to make this test at various times after the standard solution has been prepared.

When strongly alkaline solutions are kept in ordinary glass vessels, they become contaminated with silicates, owing to the attack of the glass. Solutions of barium hydroxide will form a white deposit, because barium silicate is insoluble. Therefore, it is quite generally recommended that the standard alkali be kept in a polyethylene bottle.[9]

Preparation of approximately 1 *N* sodium hydroxide. Weigh out approximately 40 g of the concentrated carbonate-free sodium hydroxide solution into a 100-ml Erlenmeyer flask, transfer to a volumetric flask of 500 ml, make up to the mark

[9] B. A. Soule, *Ind. Eng. Chem., Anal. Ed.* **1**, 109 (1929), showed that a coat of Thermoprene, a rubber paint made by the B. F. Goodrich Company, Akron, Ohio, is very good for preventing alkaline solutions from attacking glass.

with equilibrium water, and mix well. Transfer to a clean rubber-stoppered bottle, and label. Do not expose the solution to the air longer than is absolutely necessary.

Preparation of approximately 0.1 *N* sodium hydroxide. Pipet two 50-ml portions of the approximately 1 *N* sodium hydroxide solution into a 1-liter volumetric flask, and dilute with equilibrium water to the mark. Transfer the solution, after thorough mixing, to a dry bottle and keep tightly stoppered. Standardize as described below.

Preparation of approximately 0.1 *N* sodium hydroxide. Pipet 50 ml of the 0.1 *N* solution into a 500-ml volumetric flask, and dilute to the mark with equilibrium water.

Standardization of 0.1 *N* sodium hydroxide. Potassium biphthalate as standard substance. $C_6H_4(COOH)COOK$. Potassium biphthalate is obtained as a 99.97 per cent pure product from the Bureau of Standards. A reagent-grade product can be purified by recrystallization from water after the addition of enough potassium carbonate to ensure the absence of free phthalic acid, followed by recrystallization from pure water. The crystals should separate at a temperature above 30°C. They are collected, washed, and dried at 125°C. The salt is not hygroscopic and can be kept unchanged.

Procedure for the standardization. Weigh out three 0.8- to 0.9-g portions of the salt into 250-ml Pyrex flasks and dissolve in 25 ml of equilibrium water. Titrate each solution with the base to be standardized, using phenolphthalein or thymol blue as indicator. If the alkali solution is carbonate-free, a sharp color change is obtained which is permanent for at least 5 min. Calculate the normality.

Other standard substances

Benzoic acid, C_6H_4COOH. Pure benzoic acid can be obtained from the Bureau of Standards. The substance is very voluminous and contains hygroscopic water. It is therefore melted by heating to a temperature of 130–140°C and poured into a platinum dish; after cooling, the cake is ground up in an agate mortar. The acid is only slightly soluble in water but is very soluble in alcohol. Weighed portions of the acid are dissolved in 20 ml of alcohol which has been previously neutralized to phenolphthalein with alkali. The solution is then titrated with the solution to be standardized, using the same indicator.

Crystallized oxalic acid, $(COOH)_2 \cdot 2\,H_2O$. The reagent-grade acid is recrystallized by dissolving 500 g in an equal quantity of boiling 3–4 *N* hydrochloric acid, filtering, and cooling with stirring. Calcium is removed in this way. The crystal meal is transferred to a Büchner funnel with a sintered-glass bottom; and after removal of the mother liquor by suction, the crystals are washed several times with dilute hydrochloric acid. The product is then recrystallized from water until chloride-free. The final crystals are quite pure but can contain as much as 0.2 per cent occluded water. To remove this water the crystals are dehydrated in a desiccator over sulfuric acid at 60°C. The oxalic acid loses its water of crystallization, and the crystals break down to a very fine powder; therefore, the vacuole water is lost. The fine powder is then rehydrated by exposing it to the air or by keeping it in a desiccator over deliquescent sodium bromide until the weight is constant. The final product does not contain more than 0.01 per cent hygroscopic water and is stable at relative humidities between 5 and 95 per cent. About 0.25 g of the acid is dissolved in equilibrium water and titrated with the standard base, phenolphthalein or thymol blue being used as indicator.

Potassium biiodate, $KH(IO_3)_2$. Potassium biiodate possesses many advantages as a primary standard. Since iodic acid is a strong acid, the standardization can be made with any indicator showing a color change between pH 4 and 10. The salt has a high equivalent weight. It is anhydrous, nonhygroscopic, and can be kept unaltered.

Potassium bitartrate, $COOH \cdot CHOH \cdot CHOH \cdot COOK$. This salt is only slightly soluble in cold water but easily soluble at higher temperatures. A weighed portion of the pure salt is dissolved in warm water and titrated with standard base (phenolphthalein as indicator).

Hydrazine sulfate, $N_2H_4 \cdot H_2SO_4$. This substance is anhydrous and can be obtained in a pure state by recrystallization from water and drying at 150°C. In the standardization methyl red is used as indicator:

$$2(N_2H_4 \cdot H_2SO_4) + 2\ NaOH \rightarrow (N_2H_4)_2 \cdot H_2SO_4 + Na_2SO_4 + 2\ H_2O$$

Exercises. Titrate 25 ml of standard 1 *N*, 0.1 *N*, and 0.01 *N* hydrochloric acid with standard 1 *N*, 0.1 *N*, and 0.01 *N* sodium hydroxide, respectively, using methyl orange (or methyl yellow, or bromphenol blue), methyl red, phenol red (or bromthymol blue), and phenolphthalein (or thymol blue) as indicators. Make the same titrations by adding the acid to the base. Report the results as follows:

Titration of Acid with Standard Base

Indicator	*Color at End Point*	*ml of* 1 *N Alkali Required*	*Calculated ml*	*ml of* 0.1 *N Alkali Required*	*Calculated ml*	*ml of* 0.01 *N Alkali Required*	*Calculated ml*
Bromcresol green							
Methyl red							
Phenol red							
Phenolphthalein							

Titration of Base with Standard Acid

Indicator	*Color at End Point*	*ml of* 1 *N Acid Required*	*Calculated ml*	*ml of* 0.1 *N Acid Required*	*Calculated ml*	*ml of* 0.01 *N Acid Required*	*Calculated ml*
Bromcresol green							
Methyl red							
Phenol red							
Phenolphthalein							

1. The acid and the base were both standardized independently. Does the amount of base required in the titration of 25 ml of acid correspond to the calculated amount?

2. Interpret the results in those cases in which the amounts found and calculated do not agree. Consider the excess of acid or base present at the end point, and calculate the correction for the indicator blank (see Table 34-1, p. 698).

If the alkali solution is titrated with acid, the former is exposed to the air for a longer or shorter period and will absorb some carbon dioxide. Explain what influence a possible carbonate content will have on the results obtained with the various indicators.

Titration of a weak acid

Determination of the strength of glacial acetic acid. Weigh a dry, stoppered 50-ml flask, introduce 2–2.5 g of the acid, and reweigh, taking care that the acid does not touch the neck of the flask. After the weight has been determined, add 20–30 ml of water and transfer the solution quantitatively to a 500-ml volumetric flask. Make up to the mark with equilibrium water. After thorough mixing, titrate 25-ml portions of the acid with 0.1 N standard base, using the following indicators: methyl orange, methyl red, phenol red, and thymol blue.

1. Which indicator changes color sharply at the end point?
2. Explain why the unsuitable indicators give vague color changes, and indicate how large the errors may be with different indicators.
3. Calculate the acetic acid content of the sample titrated.
4. How many equivalents of acid does the sample contain per gram, and how many per milliliter, if the density = 1.05?

Determination of sodium bicarbonate and sodium carbonate in a mixture: Method 1. This method is especially recommended for those cases in which the sample contains small amounts of bicarbonate and relatively large amounts of carbonate. The total alkalinity of the sample is first determined by titration with standard acid, with methyl orange, methyl yellow, bromphenol blue, or bromcresol green as indicator (see the standardization of acid with sodium carbonate). The bicarbonate content is then determined by adding standard base to a solution of the sample in excess of the amount required to transform the bicarbonate into carbonate. All the carbonate is removed from the solution by adding an excess of barium chloride, and the excess alkali is finally titrated back without filtering, phenophthalein or thymol blue being used as indicator.

$$NaHCO_3 + NaOH \rightarrow Na_2CO_3 + H_2O$$

$$Na_2CO_3 + BaCl_2 \rightarrow BaCO_3 + 2\ NaCl$$

Since barium carbonate is the salt of a strong base and a very weak acid, it is strongly hydrolyzed in solution. Its solubility, however, is very small and is, moreover, diminished by the excess of barium salt present. A suspension of barium carbonate in a solution containing barium chloride is neutral toward phenolphthalein, since the carbonate-ion concentration, and therefore the concentration of the hydroxyl ions formed by hydrolysis, is negligibly small. The color change from red to colorless in the white suspension can be observed very sharply.

From the amounts of acid used in the determination of the total alkalinity and the amount of sodium hydroxide required to transform the bicarbonate into carbonate, it is possible to calculate the carbonate and bicarbonate content of the sample.

Procedure. Weigh out 1.5–2 g of sample and transfer to a 250-ml volumetric flask. Make the solution up to the mark with equilibrium water and mix thoroughly.

Total alkalinity. Titrate 25.00 ml of the solution with 0.1 N standard acid, using methyl yellow or any other indicator changing color near a pH of 4. The end point is reached when the indicator assumes an orange-red color. A sharper color change is obtained with the mixed indicator methyl orange–indigo carmine (p. 695); the end point is then reached when the indicator changes from neutral gray to violet. For a more exact determination of the end point see page 779.

Sodium bicarbonate content. Add 25.00 ml of 0.1 N sodium hydroxide solution to 25 ml of the sample and immediately thereafter 10 ml of 10 per cent barium chloride solution and 2 drops of phenolphthalein (or thymol blue). Back-titrate the suspension immediately with standard 0.1 N hydrochloric acid. The end point is reached when the color changes from pink to colorless (or from bluish to yellow if thymol blue is used). Run a blank under exactly the same conditions with 25.00 ml of standard base, 25 ml of water, and 10 ml of barium chloride solution. The difference between the amounts of acid used in the blank and in the determination itself corresponds to the sodium bicarbonate content of the solution.

Notes:

1. The method described is also very useful in the determination of the carbonate content of commercial alkali hydroxides. In this case the sum of carbonate and hydroxide is determined by titration with acid (methyl yellow as indicator), and then the hydroxide alone after removing the carbonate by precipitation with barium chloride. The method is also used for testing reagent-grade sodium carbonate for the presence of sodium hydroxide or sodium bicarbonate (p. 780).

2. The method is often applied to the determination of bicarbonate and carbonate in boiler water. In these cases too much bicarbonate will be found, since sulfate, which is always present in boiler water, interferes. If barium sulfate is precipitated from a solution containing an excess of hydroxyl ions, part of the latter is coprecipitated and, consequently, high results are obtained. If the sulfate content of the sample is known, a blank can be run with the same amount of sulfate, following the same procedure that is used in the determination.

3. Carbon dioxide in the free state, or bound as bicarbonate or carbonate, can be determined by evolution from an acid medium and absorption in an excess of standard barium hydroxide, or of standard sodium hydroxide containing an excess of barium chloride. It should be stated that the absorption of the carbon dioxide by the alkaline solution is relatively slow, and part of the gas may escape if it is bubbled through only one bottle containing the standard base. Therefore, it is necessary to pass the gas through a number of bottles in succession with known volumes of standard base or, better, to work in a closed system and circulate the gas a few times through the one bottle containing the standard base and barium salt.

4. The alkaline earth metals yield slightly soluble carbonates. Therefore, they can be determined by treating their neutral solutions with an excess of standard sodium carbonate and then titrating the excess in an aliquot part of the filtrate with standard acid, methyl yellow being used as indicator. It is also possible to make the determination by precipitating calcium, strontium, or barium with an excess of a sodium carbonate solution, filtering and washing the precipitate with 50 per cent alcohol, and finally dissolving it in an excess of standard acid. The excess acid is back-titrated with standard base.

Determination of sodium bicarbonate and sodium carbonate in a mixture: Method 2. In the discussion of the neutralization curve of carbonic acid (see Table 34-6, p. 704, and Fig. 34-5, p. 705) it was shown that a solution of sodium bicarbonate has a pH of 8.35 and that near this point the pH changes relatively slowly upon addition of a slight excess of sodium carbonate or carbon dioxide. Therefore if sodium carbonate is titrated to bicarbonate with standard acid,

$$Na_2CO_3 + HCl \rightarrow NaHCO_3 + H_2O$$

there is no indicator which will give a sharp color change.

To obtain good results the titration must be continued until the pH has reached a value of 8.35, which is the pH at the theoretical end point. This can be done by employing phenolphthalein or thymol blue as indicator and using for comparison a solution of sodium bicarbonate which has the same volume and concentration as the unknown at the end point and which contains the same amount of indicator. The end point is reached when the solution titrated has the same color as the comparison solution. It is still simpler to use a mixed indicator. Simpson[10] has recommended the use of a mixture

[10] S. G. Simpson, *Ind. Eng. Chem.* **16**, 709 (1924).

of 6 parts thymol blue and 1 part cresol red as a mixed indicator. The mixture has a purple color in the carbonate solution; in the neighborhood of the equivalence point it changes to blue. In making the titration, the acid is added slowly, until the solution assumes a rose color. This is the end point; upon further addition of acid the color changes to orange-yellow. The end point can be found with an accuracy of 0.5 per cent without a comparison solution. This method is not as accurate as the first one, if the sample contains a large excess of sodium bicarbonate.

Procedure. Use the same solutions as in method 1, and determine the total alkalinity (the sum of carbonate and bicarbonate) in the same way as there described.

Sodium carbonate content. Add a few drops of the mixed indicator (see above) to 25.00 ml of the solution, and titrate with 0.1 N standard acid until the color changes from blue to pink.

Notes:

1. Free carbon dioxide which occurs in most natural waters and also in distilled water can be titrated as a monobasic acid by using the mixed indicator, phenolphthalein, or thymol blue for the detection of the end point. In such titrations it is always noticed that the color of the indicator becomes alkaline before the end point is reached. The reason is that most of the carbon dioxide is present in the free state, and a small part only (less than 1 per cent) as carbonic acid:

$$CO_2 + H_2O \rightleftharpoons H_2CO_3$$

The carbonic acid reacts instantaneously with the base. The equilibrium is then disturbed, and the carbon dioxide, not being an acid, must first form carbonic acid before reacting with alkali. Therefore, if a carbon dioxide solution is titrated with standard base, with phenolphthalein as an indicator, the color will turn pink very early in the titration, although the greater part of the carbon dioxide is still in the solution in the free state. After $\frac{1}{2}$–1 min, the liquid will turn colorless. This phenomenon is repeated until all the carbon dioxide has been transformed into bicarbonate. The end point is reached when the basic color of the indicator persists for at least 5 min.

More or less carbon dioxide will escape from the solution during such a titration. Therefore, it is necessary first to make a preliminary titration to the permanent end point. A second titration is then made, and nearly the required amount of alkali as found in the first titration is added all at once, and from then on the addition is continued carefully to the end point.

2. Carbon dioxide can be determined in the above way in the presence of bicarbonate, and the latter can be determined with standard acid with methyl yellow as an indicator. It is also possible to determine the sum of carbon dioxide and bicarbonate by adding an excess of standard base and barium chloride, or of standard barium hydroxide, and then titrating back with standard acid, using phenolphthalein as indicator (method 1).

Boric acid and borax. Boric acid is such a weak acid that it cannot be titrated accurately with 0.1 N standard alkali. (The ionization constant of boric acid is 5×10^{-10}; see the discussion on page 701 and Fig. 34-4.) However, it can be transformed into a relatively strong acid by adding organic hydroxyl compounds containing more than one hydroxyl group per molecule, such as glycerol, mannitol, dextrose, or invert sugar. These polyhydroxy alcohols form complex acids with boric acid which are much stronger than the boric acid alone. Mannitol is more effective than glycerol; 0.5–0.7 g of the former per 10 ml of solution is sufficient. It is much more economical, however, to use a solution of invert sugar for the purpose. This is prepared in the following way: About 3 kg of sucrose are dissolved in 1 liter of distilled water, and the liquid is boiled several minutes until clear, best in a large tinned vessel. Then 25 ml of 3 N sulfuric acid is added quickly; and, after stirring for $\frac{1}{2}$ min, the solution is treated with $1\frac{1}{2}$

liters of water to which 25 ml of 3 N carbonate-free sodium hydroxide have been added, stirred again, and cooled. The resulting solution should be neutral and colorless; it contains about 55 per cent invert sugar.

For every 10 ml of 0.1 N boric acid, 3–4 ml of invert sugar solution is added, and the solution is then titrated to a faint pink with phenolphthalein as indicator. Then 5 ml more of the invert sugar solution is added. If the color of the indicator vanishes, more alkali is added until the pink color appears, and then more invert sugar is added. This procedure is repeated until the addition of invert sugar no longer causes a fading of the pink color. The invert sugar solution should be very carefully neutralized to phenolphthalein just before its use.

Sodium tetraborate ($Na_2B_4O_7 \cdot 10\ H_2O$ = borax). Borax in solution can be considered as boric acid which is 50 per cent neutralized:

$$Na_2B_4O_7 + 5\ H_2O \rightleftharpoons 2\ NaH_2BO_3 + 2\ H_3BO_3$$

The NaH_2BO_3 can be titrated with standard acid (methyl red, bromcresol green, methyl yellow, methyl orange, or bromphenol blue as indicator):

$$NaH_2BO_3 + HCl \rightarrow H_3BO_3 + NaCl$$

or

$$Na_2B_4O_7 + 2\ HCl + 5\ H_2O \rightarrow 4\ H_3BO_3 + 2\ NaCl$$

Therefore, this titration gives the amount of alkali bound to the boric acid.

Another portion of the sample can be titrated with standard base in the presence of invert sugar:

$$[H_3BO_3 \text{ complex}] + NaOH \rightarrow [NaH_2BO_3 \text{ complex}] + H_2O$$

$$Na_2B_4O_7 + 2\ NaOH + 3\ H_2O \rightarrow 4\ [NaH_2BO_3 \text{ complex}]$$

If the borax is pure, the amount of standard acid required in the titration with methyl red as indicator is equal to the amount of standard base used in the titration with phenolphthalein in the presence of invert sugar. In the following procedure, commercial borax, which usually is partially dehydrated, may be used as the sample.

Procedure. Weigh out 3–5 g of sample, transfer to a 250-ml volumetric flask, and make up to the mark.

Titration of the alkali. Pipet 25 ml of the well-mixed solution in to a flask, add 3 drops of methyl red, and titrate with 0.1 N standard acid until the color changes to orange-red.

Titration of the boric acid. Pipet 25 ml of the solution into a flask; add 10 ml of the neutral invert sugar solution and 3 drops of phenolphthalein. Titrate with 0.1 N standard alkali to the color change of the indicator. Add 5 ml more of the invert sugar solution, and continue the titration if the liquid becomes colorless. Repeat the test by adding more invert sugar.

Calculation. Calculate the number of equivalents of alkali and boric acid in the sample. If the salt has the composition $Na_2B_4O_7 \cdot xH_2O$, the number of equivalents of alkali required is equal to that of boric acid.

Calculate the $Na_2B_4O_7$ and water content of the sample in per cent.

Phosphoric acid H_3PO_4:

$$H_3PO_4 \rightleftharpoons H_2PO_4^- + H^+ \qquad K_1 = 7 \times 10^{-3};\ pK_1 = 2.16$$

$$H_2PO_4^- \rightleftharpoons H^+ + HPO_4^{--} \qquad K_2 = 7 \times 10^{-8};\ pK_2 = 7.16$$

$$HPO_4^{--} \rightleftharpoons H^+ + PO_4^{3-} \qquad K_3 = 5 \times 10^{-13};\ pK_3 = 12.3$$

Phosphoric acid as a monobasic acid

$$H_3PO_4 + NaOH \rightarrow NaH_2PO_4 + H_2O$$

The pH of a solution of monosodium phosphate (which is not too dilute) is found from the equation (see p. 54) to be

$$\text{pH} = \tfrac{1}{2}(\text{p}K_1 + \text{p}K_2) = 4.66$$

Experimentally it is found that a 0.05–0.01 M solution of monopotassium phosphate has a pH of 4.4–4.45. The change of the pH near the equivalence point is not very pronounced. If methyl orange or bromcresol green is used as indicator, it is desirable to employ a solution of monopotassium phosphate with the same amount of indicator for comparison. Calcium and magnesium in small amounts do not affect the determination of the end point.

Phosphoric acid as a dibasic acid. The pH of a 0.05 M disodium phosphate solution is 9.6. At this pH, phenolphthalein and thymol blue exist in their alkaline forms. Therefore, if phosphoric acid is titrated as a dibasic acid with phenolphthalein or thymol blue as indicator, the color change will occur too early. However, if at the point at which the indicator shows a color change to the alkaline side, the solution be half-saturated with sodium chloride, the hydrolysis is repressed and the final color change occurs at the theoretical end point.

Thymolphthalein begins to change color at a pH of 9.7 and therefore will just show a color change to bluish at the end point. In this case no sodium chloride should be added, and the standard base used must be carbonate-free.

It is possible first to titrate phosphoric acid as a monobasic acid with methyl orange or methyl yellow as indicator. After the first end point is reached, a few drops of thymolphthalein can be added and the titration continued to the second end point.

Note: Calcium and magnesium interfere in the titration of phosphoric acid as a dibasic acid, since in the presence of these metals precipitates of indefinite composition are formed. The interference can be prevented by the addition of an excess of neutral 10 per cent potassium oxalate after the first end point (methyl orange) has been reached. The precipitate of calcium oxalate does not render the color change of thymolphthalein or phenolphthalein less sharp. Barium, if present, can be made harmless by the addition of sulfate after the first end point has been reached.

Ammonia in an ammonium salt. Ammonia is a weak base that is volatile. Therefore, if an excess of a strong base is added to a solution of an ammonium salt and the mixture is distilled, the ammonia will be quantitatively expelled. Since ammonia volatilizes from its aqueous solutions at room temperature, it is necessary to absorb it in an excess of standard acid. The excess of acid can be back-titrated in the presence of methyl red (or methyl orange, bromcresol green, methyl yellow, or bromphenol blue) as indicator. (Why not phenolphthalein?)

A convenient form of distillation apparatus for this determination is shown in Fig. 39-2. It consists of a 500-ml long-necked Pyrex flask (Kjeldahl flask), to the top of which a spray trap is attached by means of a rubber stopper as shown. The purpose of the trap is to prevent droplets of the strongly alkaline liquid from being carried over into the condenser. The lower limb of the trap, which projects down into the neck of the flask, should have a beveled end to facilitate the return flow of liquid condensed in the

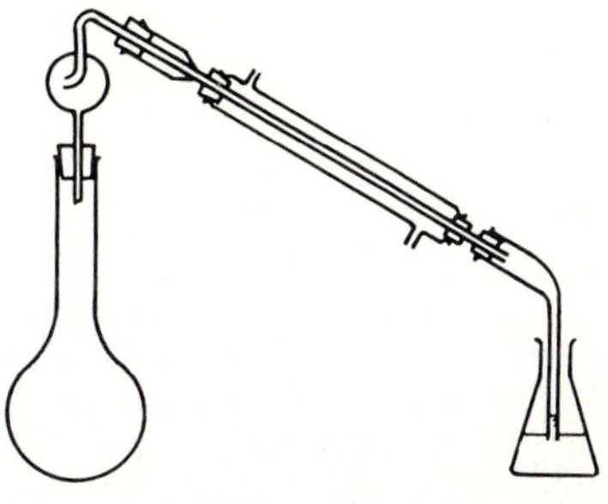

Fig. 39-2. Apparatus for ammonia distillation.

trap. The upper limb of the trap is connected to an ordinary glass condenser by means of a rubber stopper; the delivery tube should extend well into the cooled part of the inner tube of the condenser. The lower end of the condenser is provided with an adapter tube attached by means of a rubber stopper. The end of the adapter dips into the standard acid contained in the 250-ml Erlenmeyer flask which serves as a receiver.

Procedure. Weigh out a 1.5- to 2.0-g sample of crude ammonium sulfate into a 250-ml volumetric flask; dissolve the salt in water, dilute to the mark, mix thoroughly. Pipet 25 ml of the solution into the distillation flask and dilute with 100 ml of water.

Pipet 50 ml of 0.1 N standard hydrochloric acid into the receiver, and adjust the flask so that the end of the adapter tube barely dips beneath the surface of the acid.

Add about 1 g of granulated zinc to the solution in the distillation flask.[11] When all is in readiness, with the spray trap connected to the condenser and the receiver in place, quickly pour 50 ml of cold 50 per cent sodium hydroxide solution into the flask, and connect the latter immediately to the trap and condenser. Gently shake the contents of the flask with a rotary motion, and place a burner under the flask at once. It is very important to perform these operations as quickly as possible to prevent loss of ammonia.

Heat the contents of the flask rapidly to boiling, and then adjust the flame so that the boiling is continuous but not too rapid. The receiver should be carefully attended to during the distillation, because there is a tendency for some of the receiving acid to be drawn up into the condenser. This danger is minimized by bringing the liquid in the flask to boiling as quickly as possible and by adjusting the receiver so that the end of the adapter dips only slightly into the solution. When the distillation is correctly performed, there should be a gentle oscillation of a short column of the acid in the lower end of the adapter, but there should be no escape of bubbles from its end after the air has been expelled.

Continue the distillation until about two thirds of the liquid in the flask has distilled over. Then lower the receiver so that the end of the adapter is out of the acid, and remove the flame from the flask. Do not remove the flame until the receiver has been lowered or part of the acid may be sucked back. Disconnect the adapter from the condenser, and rinse it out into the flask with a small amount of water from the wash bottle. Add a few drops of methyl red indicator, and titrate the excess acid in the solution with 0.1 N sodium hydroxide (color change from red to yellow). Calculate the percentage of nitrogen and also of ammonia in the sample. In the most accurate work run a blank.

Notes:

1. Instead of absorbing the ammonia in an excess of standard acid, it is possible to reduce the volatility of the ammonia in water by adding a very weak acid. For this purpose there may be used 50 ml of a solution of 4 per cent boric acid in equilibrium water. The boric acid solution should be kept cool during the distillation. In the titration of the ammonia, 50 ml of the solution of the boric acid, mixed with enough water to make the volume equal to that of the solution at the end point and containing the same amount of indicator (methyl red or, better, bromcresol green), is used for comparison. The titration is continued slowly until the colors of the two liquids are identical.

2. Nitrate can be reduced quantitatively to ammonia with the aid of certain alloys. Arnd's alloy,[12] consisting of 60 per cent copper and 40 per cent magnesium, is very useful.

[11] The purpose of the zinc is to prevent "bumping" during subsequent distillation of the strongly alkaline liquid. The zinc reacts slowly with the sodium hydroxide solution with the evolution of hydrogen, which comes off in a fine stream of bubbles from each zinc particle. The fine gas bubbles prevent superheating of the liquid, which is the cause of "bumping." If nitrate is present, zinc cannot be used, because nitrate is reduced to ammonia by zinc in alkaline solution (see note 3 on p. 831); in such a case small pieces of porous porcelain plate should be employed.

[12] T. Arnd, *Z. Angew. Chem.* **30**, 169 (1917); **33**, 296 (1920).

About 0.25 g of the nitrate dissolved in 200 ml of water is transferred to the distillation flask, 10 ml of a 20 per cent magnesium chloride ($MgCl_2 \cdot 6\ H_2O$) solution and 5 g of the finely pulverized alloy are added, and the whole is boiled until about two thirds of the liquid has distilled over. The distillate is received in 50 ml of 0.1 N standard acid or in a boric acid solution.

Determination of crude protein in flour or wheat (Kjeldahl[13] nitrogen determination). The proteins comprise a group of complex nitrogen-containing compounds which constitute one of the three classes of important nutrient materials present in foods (carbohydrates, fats, and proteins). When protein-containing material is heated with concentrated sulfuric acid in the presence of mercury[14] as a catalyst to speed up the reaction, the acid is gradually reduced to sulfur dioxide and water, while the carbon and hydrogen of the organic material are oxidized to carbon dioxide and water, and the nitrogen is transformed into ammonium sulfate. The latter is determined by adding an excess of alkali to the solution after the digestion with acid, and distilling in the manner described under the ammonium determination. The percentage of protein is found by multiplying the percentage of nitrogen by a factor which varies with the nature of the protein present in the sample. For the protein in wheat and flour the factor commonly used is 5.7. The results obtained are referred to as *crude protein.*

Procedure. Weigh out a 1.5- to 2.0-g sample of flour or ground wheat,[15] and transfer to a *dry* 800-ml Kjeldahl flask of Pyrex glass. Weigh out roughly 10 g of anhydrous sodium sulfate (or potassium sulfate) and 0.6–0.8 g of mercuric oxide, and add the mixture to the flask. Pour 25 ml of concentrated sulfuric acid into the flask in such a way that any solid adhering to the neck is washed down.

Support the flask in an inclined position on a ring stand (or on the special rack provided for this purpose) in a well-ventilated hood, and heat the contents gently with a small flame. The mixture will froth and turn black at first. When the frothing ceases to be vigorous, increase the flame until the mixture boils gently. As the digestion proceeds, the original black color gradually fades to brown, and finally the solution becomes perfectly colorless. Continue the heating for 15 min after the solution has become colorless; then remove the flame and allow the flask to cool. The liquid will solidify to a mass of crystals or a solid cake on cooling.

When the flask has cooled to at least 50°C, carefully add 200 ml of water, and swirl the flask until the solid material has dissolved completely. Cool the solution to room temperature under the tap, and add 10 ml of 5 per cent potassium sulfide solution to precipitate the mercury. Then add granulated zinc and sodium hydroxide, and distil off the ammonia in exactly the same manner as described under the determination of ammonia.

Calculate the percentage of nitrogen in the sample, and multiply the value obtained by 5.7 to obtain the percentage of "crude protein."

In the most accurate work run a blank on the reagents through all the steps of the procedure, and correct the result obtained in the analysis of the sample.

Titrimetric determination of equivalent weight of salt by cation exchange. A simple method of determining the equivalent weight of a salt in solution consists in exchanging the cation with hydrogen ion by the use of a cation exchanger and titrating the acid with standard base. This method is especially advantageous when the anion of the salt cannot easily be determined (nitrate, perchlorate, and sulfate, for example). In a mixture of salts, the sum of the cations (as equivalents) can be found in this way.

[13] J. Kjeldahl, *Z. Anal. Chem.* **22**, 366 (1883).

[14] Copper sulfate and selenium can also be used as catalysts in the digestion. Alkali sulfates are often added also to raise the boiling point of the mixture and hasten the reaction.

[15] Samples should be kept in well-closed bottles, as the water content changes markedly with varying humidity of the atmosphere.

Prepare two resin colums and run the determination in duplicate. Allow at least 40 min to wash the columns and pass unknown through them.

Resin column. Drop a ball of glass wool (sideshelf), about the size of a marble, into the ion-exchange tube and push it to the bottom of the column with a stirring rod. Add a little water and run it through the stopcock to displace any air bubbles. A few milliliters of water should remain in the tube above the glass wool plug. Add a slurry of Dowex 50W-X8 resin (50–100 mesh) in 1 *M* hydrochloric acid, so that a column 10–12 cm long is obtained. The column must be uniformly packed with resin and free of air bubbles. Wash out the acid with 200 ml of water at a rate *not exceeding* 10 ml/min (200 drops/min). Never allow the level of water to fall below the top of the resin bed; otherwise channels form in the bed. Test the last 25-ml portion of effluent with a drop of bromcresol green to see whether all the acid has been washed out. The test portion should be green to blue if washing is complete. If not, wash the resin some more.

Procedure. On the analytical balance weigh out 2.0–2.5 g of sodium sulfate sample to 0.1 mg, dissolve in water, and dilute to 250 ml in a volumetric flask and mix thoroughly. Do the following in duplicate: Pipet a 25-ml portion of sample solution into the reservoir of the ion-exchange tube. Adjust the rate of flow to about 3 ml/min (60 drops/min) and collect the effluent in a 250-ml Erlenmeyer flask. Use the same flask to collect all portions of wash solution. Wash the column with three 25-ml portions of water. The first portion should be passed through the colume at a rate of 3 ml/min, the others more rapidly, 5–10 ml/min. After the last wash, run a few milliliters of solution from the column into a test tube, and test for completeness of washing with bromcresol green. If the solution is still acid, wash the column with another 25-ml portion of water, collecting the effluent in the 250-ml Erlenmeyer flask. Test again for completeness of washing as above. Titrate the combined effluent plus washings with standard 0.1 *N* sodium hydroxide using phenolphthalein as the indicator.

Calculate the equivalent weight of the unknown. Assuming that the only electrolyte present is sodium sulfate, report %S in the unknown.

Notes:

1. Chlorides, nitrates, perchlorates, sulfates, etc., of the alkali and alkaline earth metals (except sulfates of the latter) may be determined according to these directions. Heavy metal salts that do not hydrolyze extensively can also be analyzed in essentially the same way. Iron(III) solutions containing sulfate are likely to give trouble by forming basic ferric sulfate so that some sulfate remains in the column, especially if the flow rate is low. With phosphates and acetates slower flow rates and larger volumes of wash water are usually required, because of the tendency of phosphoric acid and acetic acid to be adsorbed by the resin. See the book by Samuelson[16] for specific details for various salts. Anions forming complexes with metals should, in general, be absent.

2. Salt solutions containing free acid (at low concentration) may be analyzed if the acid is titrated in an aliquot and deducted from the amount required for the solution passed through the resin column.

3. Anion exchangers of strongly basic type in the OH^- form may be used to determine anions by titration of the effluent with standard acid, but except in special cases their use is not as advantageous as that of cation exchangers because washing and regeneration are more time consuming. They can be used advantageously for phosphates, carbonates, sulfites, etc. They cannot be used for heavy metal salts forming insoluble hydroxides.

[16] O. Samuelson, *Ion Exchangers in Analytical Chemistry*, John Wiley and Sons, New York, 1953, p. 6.

Titration of a carboxylate in acetic acid as a solvent. Solvent: Reagent-grade acetic acid that does not contain more than 0.1 per cent water.

Reagent: 0.1 *N* perchloric acid in acetic acid. Weigh *P* g of 70 per cent perchloric acid into a 1-liter volumetric flask,

$$P = \frac{1000}{p}$$

in which p is the wt./wt. percentage of perchloric acid, as determined by titration. Add about 100 ml of anhydrous acetic acid and Q g of acetic anhydride in successive portions, stirring and cooling the mixture during the addition. Q is calculated from the reaction equation

$$(CH_3CO)_2O + H_2O = 2\ CH_3COOH,$$

in which H_2O is the number of grams of water in P g of perchloric acid used. Stopper the flask, let stand for 24 hr, and fill up to the mark with solvent acetic acid.

Indicator: 0.2 per cent crystal violet solution in acetic acid.

Standardization: Accurately weigh about 0.5 g of potassium biphthalate, previously dried, and dissolve in 30 ml of acetic acid applying gentle heat. Cool, add 2 drops of indicator, and titrate with 0.1 *N* perchloric acid until the blue color turns green. A precipitate of potassium perchlorate is formed during the titration. Run a blank with the same volume of acetic acid as the volume at the end point and add acid until the color in the blank is the same as in the titrated solution. Subtract the blank from the volume of titrant at the end point.

Assay of sodium acetate trihydrate. Prepare 0.05 *N* pure sodium perchlorate solution, add 0.2-ml indicator to 50 ml. Accurately weigh about 0.2 g of the salt, dissolve in 25 ml of acetic acid, add 0.2-ml indicator, and titrate with 0.1 *N* acid until the color of the indicator is the same as in the solution for comparison.

Note: Acetic acid has a large coefficient of expansion. It is desirable to carry out titration at the same temperature as that in the standardization. This is simpler than calculating the correction from the expansion coefficient α ($\alpha = 0.001$).

PROBLEMS

1. If 0.4212 g of pure potassium tetraoxalate, $KHC_2O_4 \cdot H_2C_2O_4 \cdot 2\ H_2O$, requires 48.90 ml of sodium hydroxide solution for neutralization with phenolphthalein as indicator, calculate the normality of the sodium hydroxide. *Ans.* 0.1017 *N*.
2. A precipitate of calcium oxalate obtained from a 0.5004-g sample of limestone is ignited to give a mixture of calcium carbonate and oxide which is then dissolved in 50.00 ml of 0.2000 *N* hydrochloric acid. If 4.50 ml of 0.1060 *N* sodium hydroxide are required for the back-titration of the excess acid (bromcresol green as indicator), what is the percentage of calcium carbonate in the limestone? *Ans.* 95.24 per cent.
3. 100 ml of a natural water require 41.0 ml of 0.0200 *N* acid for neutralization against methyl orange, and 2.0 ml of the same acid against phenolphthalein. Calculate the HCO_3^- and CO_3^{--} contents of the water in milligrams per liter. *Ans.* 452 mg of HCO_3^- and 24 mg of CO_3^{--} per liter.
4. It is found that 4.1 ml of 0.0100 *N* sodium hydroxide are needed for the titration of 200 ml of distilled water when the indicator used is phenolphthalein. What is the CO_2 contents of the water in milligrams per liter? *Ans.* 9.02 mg of CO_2 per liter.
5. How can sulfuric and boric acids be determined with standard base in a mixture of the two in water?
6. The sum of calcium and magnesium in a boiler water is determined after neutralizing the sample to methyl orange by precipitation of the calcium and magnesium with a standard mixture of sodium hydroxide and sodium carbonate, the excess being titrated in an aliquot part of the filtrate with standard acid. What is the composition of the precipitate? What indicator should be used in the back-titration with

standard acid? Fifty milliliters of the standard alkali mixture required 70.54 ml of 0.0910 N acid. In the actual determination 100 ml of the alkaline mixture were added to 100 ml of the solution and the mixture diluted to 250 ml. One hundred milliliters of the filtrate required 12.00 ml of 0.0910 N acid. What is the sum of the number of equivalents of calcium and magnesium in 1 liter of the water?

Ans. 0.1011 equivalent per liter.

7. An unweighed sample of calcium phosphate is dissolved in 50.00 ml of 0.0500 N hydrochloric acid. 10.12 ml of 0.0490 N base is required to back-titrate the excess acid, methyl orange being used as indicator. The pH at the end point is 4.5. After reaching the first end point, an excess of neutral potassium oxalate is added and the titration continued to the color change of thymolphthalein. If 30.15 ml of 0.0490 N alkali is used between the first and second end point, what is the ratio of CaO to P_2O_5 in the sample of calcium phosphate? *Ans.* 1 CaO : 0.423 P_2O_5.
8. A commercial product of sodium phosphate was found to consist of a mixture of Na_3PO_4, Na_2HPO_4, and water of crystallization. How can the composition of the sample be determined by titration with acid using two different indicators?
9. Devise a method for determining sodium carbonate and borax in a mixture consisting only of these two substances and water, by simple acidimetric and alkalimetric titrations.
10. A solution of hydrochloric acid is standardized by adding an excess of ammonia to 100.00 ml, evaporating to dryness, and weighing the residue, which was found to have a weight of 0.5350 g. Calculate the normality of the acid.
11. A solution of pure lead nitrate is saturated with hydrogen sulfide. The precipitate is filtered off and washed, and the filtrate is titrated with sodium hydroxide, methyl orange being used as an indicator; 24.50 ml of 0.5000 N sodium hydroxide is required. Calculate the amount of lead in the sample solution.

CHAPTER 40 *Precipitation Titrations. Argentimetry*

Preparation of standard solutions.

Silver nitrate. Metallic silver can be obtained in a very pure state.[1] However, it is not very suitable for the preparation of standard silver solutions, because it must be dissolved in nitric acid, and the excess of acid must then be removed in most cases. Therefore it is preferable to prepare standard silver solutions by dissolving exactly weighed amounts of pure silver nitrate in water.

Very pure preparations of silver nitrate can be purchased. Impure products can be purified by recrystallization from dilute nitric acid. The crystals are collected in a glass filter crucible or a Büchner funnel with a sintered-glass disk to prevent contamination by shreds of filter paper or traces of organic substances, which exert a reducing action on the salt. The salt is dried at 100°C and then, protected from dust, is heated 15 min at 220–250°C (melting point is 208°C). Fusion of the salt is required to remove the occluded water in the crystals, which is not expelled by heating at 110°C. Large unmelted crystals which are otherwise pure may contain 0.1–0.2 per cent water. The heating should not be continued for too long a period, or at a temperature higher than 250°C because silver nitrate will then slowly decompose with the formation of silver nitrite. The dried salt can be kept unchanged in a glass-stoppered bottle; it is not hygroscopic.

Prepare 500 ml of 0.1000 N silver nitrate solution by weighing out the proper amount of pure dried silver nitrate,[2] dissolving in water, and diluting to volume. Protect the solution from the light.

Sodium chloride. Although it is not necessary to standardize the silver nitrate solution, if it has been prepared from a pure product, it is desirable to have available a standard chloride solution to check the titer of the silver solution after some lapse of time.

Pure sodium chloride is prepared by precipitation of a filtered saturated solution of a reagent-grade product with concentrated hydrochloric acid or hydrogen chloride (p. 444). The precipitated salt is filtered off with suction, washed with water, dried, ground to a powder, and heated in an electric furnace at 500–600°C to constant weight. One may heat over a gas flame if the crucible is placed in a hole in an inclined asbestos board to prevent the entrance of sulfur-containing gases from the burner that might contaminate the salt.[3]

[1] T. W. Richards and R. C. Wells, *J. Am. Chem. Soc.* **27**, 459 (1905).

[2] For student purposes it is sufficient to use a pulverized analytical reagent dried at 150°C.

[3] For purity tests compare I. M. Kolthoff and V. A. Stenger, *Volumetric Analysis*, vol. II, Wiley-Interscience, New York, 2nd ed., 1947.

Prepare 500 ml of 0.1000 N sodium chloride solution by weighing out the calculated amount of the pure salt, dissolving in water, and diluting to volume.

Potassium thiocyanate. Potassium thiocyanate of reagent quality has been found to be 99.9 per cent pure.[4]

It can be further purified by two recrystallizations from water. The product is first dried in a desiccator at room temperature and then in an oven at 120–150°C. When thus dried, the salt contains only a few hundredths of 1 per cent occluded water. The last small trace of water can be removed by fusing the salt for 2–3 min at 190–200°C. The pure dry salt is not appreciably hygroscopic in an atmosphere of less than 45 per cent relative humidity. It should be kept protected from light and acid fumes.

Prepare 500 ml of 0.1000 N solution by weighing out 4.858 g of the pure dry salt, dissolving in water, and diluting to volume. Solutions of pure potassium thiocyanide in pure water are quite stable.

Exercises

Checking the titer of the standard sodium chloride solution against the standard silver nitrate solution. The Mohr titration. The theory of the Mohr titration has been discussed on page 719.

Indicator solution: A 5 per cent solution of potassium chromate in water. One milliliter of indicator is used for volume of 50–100 ml.

Procedure. Pipet 25 ml of 0.1000 N sodium chloride solution in a 250-ml Erlenmeyer flask, add 1 ml of indicator solution, and slowly titrate with 0.1 N silver nitrate solution, swirling the liquid constantly, until the first permanent color deviation from the pure yellow of the suspension is obtained.

Determine the indicator blank by adding 1 ml of the indicator to 50 ml of water or, better, to 50 ml of a suspension of a little chloride-free calcium carbonate (a few hundred milligrams), and then silver nitrate until the color of the blank matches that of the solution titrated. The indicator blank, which should not amount to more than 0.03–0.05 ml of 0.1 N silver nitrate, is deducted from the volume of silver nitrate used in the titration.

Calculate the normality of the silver nitrate solution.

Determination of chloride in a soluble chloride according to Mohr. Weigh out 2–2.5 g of an alkali chloride sample, dissolve in water, and dilute to 250 ml in a volumetric flask. Pipet out 25 or 50 ml (requiring between 25 and 40 ml silver nitrate), and titrate according to the procedure described above. Report the percentage of chlorine. The remainder of the solution is to be reserved for the titration with an adsorption indicator, and according to Volhard.

Notes:

1. Bromides can be titrated in the same way as chlorides. The Mohr method does not give satisfactory results in the titration of iodides and thiocyanates because of adsorption phenomena.

2. The titration must be carried out in neutral or weakly alkaline solution (p. 720); the pH should lie between 6.5 and 10.5 at room temperature. Acid solutions are conveniently neutralized with chloride-free borax, sodium bicarbonate, or calcium carbonate.

[4] I. M. Kolthoff and J. J. Lingane, *J. Am. Chem. Soc.* **57**, 2126 (1935).

3. Pure ammonium chloride can be titrated according to the standard procedure. If the solution is acid, the pH should be adjusted to a value between 6.5 and 7.2. At a higher pH too much free ammonia is formed, which increases the solubility of silver chloride and silver chromate; and the color change then appears too late.

4. Chlorides of cations which are hydrolyzed and give solutions of distinctly acid reaction (aluminum, iron, bismuth, tin, zinc, etc.) cannot be titrated according to the method of Mohr. Basic salts containing chloride precipitate upon the adjustment of the pH. Neither can the method be applied to the titration of solutions containing copper, nickel, and cobalt salts, the cations of which are colored. Lead and barium yield slightly soluble chromates with the indicator and interfere in the titration; barium can be made harmless by the addition of an excess of sodium sulfate.

5. The titration cannot be made in the presence of anions yielding slightly soluble silver salts, such as phosphate, arsenate, pyrophosphate, sulfite, sulfide, carbonate, and oxalate. Sulfite can be made harmless by oxidation to sulfate, and hydrogen sulfide can be removed by boiling. Anions which do not form slightly soluble or complex silver salts do not interfere and may be present in large quantities. If, however, the solution contains large amounts of noninterfering ions, the silver chromate formed by a local excess of reagent exhibits a marked tendency to agglomerate. It then reacts slowly with the chloride ions which are still present, so that the end point may apparently be reached too early. Under these conditions it is necessary to shake vigorously near the end point and to add silver nitrate until the reddish brown color no longer disappears after shaking and allowing the solution to stand.

6. The reverse titration of silver with chloride using chromate as indicator does not give good results, because the flocculated silver chromate reacts too slowly with chloride.

Chloride titration with an adsorption indicator. Before making this titration the student should read pages 215 and 721.

Indicator: (1) 0.1 per cent fluorescein in 70 per cent alcohol or 0.1 per cent sodium fluoresceinate in water, or (2) 0.1 per cent dichlorofluorescein in 70 per cent alcohol, or 0.1 per cent sodium dichlorofluoresceinate in water.

Procedure. Pipet 25 or 50 ml of the chloride solution (see p. 795) into a 250-ml Erlenmeyer flask. Add 5–10 drops of either indicator, and titrate the solution with silver nitrate in diffuse light with constant swirling. The silver chloride sol flocculates approximately 1 per cent before the equivalence point. Continue the titration dropwise with vigorous agitation until the precipitate suddenly becomes reddish. The indicator blank cannot be determined directly, owing to the great sensitiveness of the precipitate toward light; the correction fortunately is negligibly small with 0.1 N silver nitrate.

Notes:

1. The determination give accurate results in the titration of chlorides, bromides, iodides, and thiocyanates.

2. The method cannot be applied to the titration of solutions more dilute than 0.005 N.

3. The pH should be the same as that in the Mohr titration when fluorescein is used as indicator. Dichlorofluorescein, being a stronger acid than fluorescein, gives good results when the pH is greater than 4. For this reason the chlorides of copper, nickel, manganese, zinc, and aluminum, which cannot be titrated according to Mohr, can be determined by a direct titration when dichlorofluorescein is used as indicator.

4. The interference of anions is the same as in the Mohr titration. Large amounts of electrolytes which do not yield insoluble silver salts may obscure the end point. The coagulation of the silver halide then begins early in the titration, and the end point becomes less distinct.

Determination of bromide in alkali bromide with eosin as indicator. Indicator: Eosin, 0.1 per cent solution in 70 per cent alcohol, or 0.1 per cent solution of the sodium salt in water.

Procedure. Add 10 drops of indicator and a few milliliters of 6 N acetic acid to 50–100 ml of the neutral bromide solution, which should not be more concentrated than 0.05 N. Titrate with silver nitrate, with constant agitation, in diffuse light. Approximately 1 per cent before the equivalence point the silver bromide flocculates. Continue the titration dropwise with vigorous stirring or shaking until the precipitate suddenly becomes colored intensely red. The indicator blank is negligibly small.

Notes:

1. Iodide and thiocyanate can be determined in the same way. With chloride the color change occurs much too early (for interpretation see p. 723).
2. The titration yields good results at pH values between 3 and 10.
3. Even in the titration of very dilute solutions (0.001 N) a sharp color change is obtained.

Titration of iodide in an alkali iodide to the "clear point." Dilute the solution of the iodide to a concentration of less than 0.04 N. Carefully titrate with silver nitrate with constant shaking. Approximately 1 per cent before the equivalence point, the silver iodide begins to flocculate. Continue the titration carefully, with vigorous shaking after each addition of reagent, until the supernatant liquid is perfectly clear. The end point can be determined with great precision in solutions of pure iodide.

Titration of iodide in the presence of chloride with diiodofluorescein as adsorption indicator. Indicator: 0.5 per cent diiodofluorescein in 70 per cent alcohol (dimethyl diiodofluorescein and Bengal red are also suitable indicators).

Procedure. Add 1–2 ml of the indicator to 50–100 ml of the solution, which should not be stronger than 0.02 N in iodide. Titrate with 0.1 N silver nitrate until the color changes from orange-red to a bluish red. The color change is not very sharp; it goes through a brownish red and then a carmine-red before the end point. For exact determinations it is indispensable to use at the end point a reference solution of the same composition as the system titrated. The color change in the presence of chloride comes a little late; 0.8 per cent of the amount of silver nitrate required should be subtracted as a correction.

Note: Iodide ions are much more strongly adsorbed on silver iodide than are chloride ions; the indicator ions are adsorbed to a less extent than the iodide and much more strongly than the chloride ions. Therefore, after all the iodide has been transformed into silver iodide, the indicator will be adsorbed, and the color change then occurs. The precipitation of iodide is quantitative before the chloride begins to precipitate.

Titration of silver according to the Volhard method (see p. 724). Indicator: Saturated solution of ferric ammonium sulfate in water (approximately 40 per cent).

Procedure. Add 1 ml of indicator solution to 25–40 ml of standard silver nitrate solution, acidify with 5 ml of 6 N nitric acid, and titrate with thiocyanate. The first perceptible color change to *orange-red* occurs about 1 per cent before the equivalence point, owing to the fact that silver ions are present on the surface of the precipitate in the adsorbed state. After the first color change has occurred, continue the titration carefully with vigorous agitation to the appearance of a brownish tinge which is permanent on strong shaking. The indicator blank amounts to 0.01 ml of 0.1 N silver nitrate. Calculate the normality of the thiocyanate solution.

Determination of chloride in a suitable sample according to Volhard's method. (for the theory see p. 724).

Procedure 1. Add 5 ml of 6 *N* nitric acid to 25–50 ml of the chloride solution (see p. 796), and then add a measured excess of standard silver nitrate solution. Filter, and wash the precipitate thoroughly with dilute nitric acid. Add to the combined filtrate and washings 1–2 ml of ferric alum solution, and titrate the excess silver as described in the preceding exercise. Write the equations for the reactions involved, and calculate the chloride content of the solution.

Note: Instead of filtering off and washing the silver chloride, the following procedure may be applied. Add nitric acid and an excess of silver nitrate to a measured volume of the chloride solution in a volumetric flask of 100- or 250-ml capacity. Make up to the mark, mix thoroughly, and filter into a dry flask, discarding the first 5 or 10 ml of the filtrate. Pipet an aliquot part, e.g., 50 ml, of the filtrate into a clean flask, and titrate the excess of silver with 0.1 *N* thiocyanate.

Procedure 2.[5] Pipet a 25- or 50-ml sample of the chloride solution into a 250-ml Erlenmeyer flask, and add 5 ml of 6 *N* nitric acid. Add standard 0.1 *N* silver nitrate solution from a buret until an excess of 2–5 ml is present. Then add 3 ml of pure nitrobenzene and 1 ml of ferric alum indicator and shake vigorously to coagulate the precipitate. Titrate the excess silver with standard 0.1 *N* potassium thiocyanate solution until a permanent reddish-brown color appears that does not fade after 5 min.

Procedure 3 (Swift[6] et al., see p. 724). To approximately 30 ml of solution, which contains the equivalent of 20–40 ml of 0.1 *N* chloride, add 15 ml of 6 *N* nitric acid, 10 ml of 2 *N* ferric nitrate, and 1.00 ml of standard 0.01 *N* potassium thiocyanate solution (using a 1-ml pipet). Titrate with silver nitrate while vigorously swirling the mixture, to the disappearance of the color of the ferric thiocyanate complex. Compare the solution with one which has been similarly titrated, but to which an excess of 1–2 drops silver nitrate has been added. The final volume should be approximately 100 ml. Subtract 0.1 ml of 0.1 *N* silver from the volume read at the end point.

Notes:

1. In the determination of bromide and iodide it is not necessary to filter off the silver halide before the back-titration of the silver (see p. 798). In the iodide determination the indicator should be added after the addition of excess silver nitrate, for otherwise the ferric iron will liberate iodine from the iodide.
2. The great advantage of the Volhard method is that it can be used for the volumetric halide determination in strongly acid media.
3. The procedure can be applied to the determination of anions whose silver salts are slightly soluble in water but which are soluble in acid, such as oxalate, phosphate, arsenate, chromate, and sulfide; these may be precipitated in neutral solution with an excess of silver nitrate. The silver is back-titrated after filtration, or the washed precipitate is dissolved in dilute nitric acid and the silver titrated with thiocyanate.

Most slightly soluble sulfides, such as those of lead, bismuth, zinc, manganese, nickel, and cobalt, react with silver nitrate to form silver sulfide quantitatively (see the list of solubility products in the Appendix):

$$PbS + 2\,AgNO_3 \rightarrow Ag_2S + Pb(NO_3)_2$$

The excess of silver can be back-titrated; or the silver sulfide can be washed, dissolved in nitric acid, and boiled until all the nitrogen oxides are expelled, and then titrated with thiocyanate.

[5] J. R. Caldwell and H. V. Moyer, *Ind. Eng. Chem., Anal. Ed.* **7**, 38 (1935).

[6] E. H. Swift, G. M. Arcand, R. Lutwack, and D. J. Meier, *Anal. Chem.* **22**, 306 (1950).

Determination of zinc by precipitation with standard potassium ferrocyanide solution. Zinc ions in neutral or acid medium react with potassium ferrocyanide to form extremely slightly soluble potassium zinc ferrocyanide:

$$2\ K_4Fe(CN)_6 + 3\ Zn^{++} \rightarrow K_2Zn_3[Fe(CN)_6]_2 + 6\ K^+$$

This reaction has been used for more than 60 years for the determination of zinc. Until recently the end point was detected by means of external indicators such as uranyl nitrate, ammonium molybdate, or ferric chloride.

The proper use of an external indicator requires some skill, and it is therefore preferable to use an internal indicator whenever possible. As such, diphenylamine, diphenylbenzidine, or diphenylamine sodium sulfonate can be used (p. 756). We have seen that these substances are oxidation–reduction indicators. The reduction potential of a ferro-ferricyanide solution is given by (see p. 752)

$$E = E_0 + 0.060 \log \frac{[Fe(CN)_6^{3-}]}{[Fe(CN)_6^{4-}]}\ (30°C)$$

An acid solution of a mixture of ferro- and ferricyanide has a reduction potential much lower than that required to oxidize the indicators to their intensely colored oxidation products. However, if zinc is added to the mixture, the ferrocyanide is precipitated as potassium zinc ferrocyanide. Since ferrocyanide is removed from the solution, the reduction potential increases. After all the ferrocyanide has reacted, there is a sharp rise in the reduction potential; and the brilliant blue color of the oxidized form of the indicator develops with the slightest excess of zinc. The mathematical formulation of the process is somewhat complicated by the fact that zinc ferricyanide is also slightly soluble. Its solubility, however, is much greater than that of the ferrocyanide, so the formation of zinc ferricyanide does not interfere with the color change.

It is also possible to titrate in the reverse manner. When diphenylamine is added to an acid solution of zinc containing a little ferricyanide, the blue-violet oxidation product is formed. This coloration will persist upon titration with ferrocyanide until all the zinc has been transformed into potassium zinc ferrocyanide. The first excess of ferrocyanide reduces the violet form of the indicator to the colorless form. Use is made of this fact in the determination of zinc with ferrocyanide.

0.05 *M* potassium ferrocyanide. A reagent-grade product of $K_4Fe(CN)_6 \cdot 3\ H_2O$ is recrystallized from water and dried over a saturated solution of sodium bromide dihydrate to constant weight; it then consists of the trihydrate. The sample can also be dehydrated by heating at 100°C to constant weight. A 0.05 *M* solution is prepared and stabilized by the addition of 0.2 g of sodium carbonate per liter. Standardize the solution against the standard zinc solution according to the procedure given below.

0.1 *M* standard zinc solution. An exactly weighed amount of pure zinc is dissolved in dilute sulfuric acid and diluted to the mark with water in a volumetric flask. The solution should be approximately 0.1 *M*. From the amount of zinc taken, the exact concentration is calculated.

Reagent: potassium ferricyanide solution. 1 per cent solution of C.P. potassium ferricyanide in equilibrium water. It is kept in a dark bottle.

Indicator: 1 per cent solution of diphenylamine in concentrated sulfuric acid. Diphenylbenzidine and diphenylamine sodium sulfonate can be used, but a better end point is obtained with diphenylamine.

Procedure (for standardization and zinc determination in an unknown solution). To 25 ml of zinc sulfate solution in a 250-ml Erlenmeyer flask, add 25–75 ml of water, 2 g of ammonium sulfate, 10–30 ml of 6 *N* H_2SO_4, 2 or 3 drops of the potassium ferricyanide solution, and 2 or 3 drops of diphenylamine indicator solution. Wait until the blue color develops, and titrate past the end point with 0.05 *M* potassium ferrocyanide solution until the blue color changes to light, creamy yellowish green. Diminish the speed of titration as the end point is approached, and do not add more than 1–2 ml of ferrocyanide in excess. Back-titrate slowly with standard zinc solution to the appearance of the purple color. A sharp color change is produced by the addition of 1 drop of reagent. If the end point is overstepped, the excess of zinc can be back-titrated with the standard ferrocyanide solution. An indicator blank of 0.05 ml of 0.05 *M* ferrocyanide for the 3 drops of indicator should be added to the volume used.

Notes:

1. Iron and other metals which form insoluble ferrocyanides must be absent.
2. The appearance of the initial blue color is somewhat facilitated if a few milliliters of ferrocyanide solution are added before the addition of the ferricyanide and indicator.
3. In order to obtain accurate results (within 0.3 per cent), the conditions under which the titration is carried out must be rigidly controlled and should be exactly the same as in the standardization. The speed of titration near the end point and the excess of ferrocyanide added before the back-titration with zinc must be kept uniform. The inaccuracy increases with the amount of ferrocyanide added in excess.

PROBLEMS

1. Indicate how chloride can be determined volumetrically in each of the following salts: calcium chloride, manganous chloride, barium chloride, ferric chloride, sodium chloride + sodium phosphate, sodium cloride + sodium sulfite, sodium chloride + sodium sulfate.
2. A 0.2000-g sample of zinc sulfide was treated with 50 ml of 0.1000 *N* silver nitrate. After complete transformation, the mixture was filtered, washed, and the excess of silver back-titrated with 15.50 ml of 0.1 *N* thiocyanate. Calculate the percentage of zinc sulfide in the sample. *Ans.* 84.04 per cent.
3. When silver nitrate is added to a solution of Na_2HPO_4, a yellow precipitate of Ag_3PO_4 is formed. The precipitation can be made quantitative by adding enough sodium hydroxide (using phenolphthalein as indicator) to neutralize the acid formed. Write the equations. Show how this reaction may be made the basis of an argentimetric method for phosphate.
4. If 0.3074 g of a mixture of pure potassium chloride and pure potassium bromide required 30.98 ml of 0.1007 *N* silver nitrate solution for titration, calculate the percentage of potassium chloride and of potassium bromide in the mixture.
Ans. 34.85 per cent KCl; 65.15 per cent KBr.
5. Devise a method for determining silver volumetrically in an alloy consisting of silver and nickel.
6. A 0.3236-g mixture of anhydrous silver halides was dissolved in 50.00 ml of 0.1810 *M* potassium cyanide solution. The excess of cyanide required 28.14 ml of 0.1010 *N* silver nitrate solution for back-titration. Calculate the silver content of the mixture.
Ans. 56.13 per cent Ag.
7. How many milliliters of 0.0510 *M* ferrocyanide solution is required to titrate 25.00 ml of 0.1000 *M* zinc sulfate?

$$2\ K_4Fe(CN)_6 + 3\ Zn^{++} \rightarrow K_2Zn_3[Fe(CN)_6]_2 + 6\ K^+$$

Ans. 32.7 ml.
8. A potassium chlorate solution is reduced in acid medium with sulfur dioxide to chloride (write reaction equation). After removal of the excess sulfur dioxide, 50.00 ml of 0.1001 *N* silver nitrate is added, the precipitate is filtered off and washed, and the excess of silver is back-titrated with 20.35 ml of 0.09982 *N* ammonium thiocyanate solution. (What indicator is used?) (a) Calculate the amount of potassium chlorate in the solution. (b) Would it be possible to titrate the chloride formed after reduction and removal of the excess of sulfur dioxide according to Mohr? (Chloride-free reagents may be added to the solution to reduce the acidity.)

9. A sample consists of a mixture of alkali metal chloride and chromate. (a) How can one determine the chloride in this mixture by a direct titration? (b) Is it possible to determine the sum of chloride and chromate by an argentimetric titration?
10. To 100 ml of a fresh solution of hydrogen sulfide in water is added 50.00 ml of 0.1000 *N* silver nitrate. The precipitate is filtered off and washed. For the titration of the acid in the filtrate 38.50 ml of 0.1015 *N* sodium hydroxide is required. (a) What indicator is used? (b) How many milliliters of 0.0990 *N* thiocyanate would be required to titrate the excess of silver in the filtrate? (c) Calculate the molarity of the hydrogen sulfide solution.

CHAPTER 41 *Compleximetric Determinations*

This chapter contains practical details and procedures for some important determinations. The theory is given in Chapters 6 and 36.

Titrations with EDTA

Preparation of standard solutions of EDTA. The most important compound for analytical purposes is the disodium salt, formula $Na_2H_2Y \cdot 2\ H_2O$, molecular weight 372.24. Preparations commercially available at the present time are not sufficiently pure to be used as primary standards. Therefore unless a purification procedure is used (see below) the solution must be standardized against a suitable primary standard. If the solution is to be used within a short time, it may be stored in an ordinary glass bottle. Upon prolonged standing metal ions are leached from the glass and the effective concentration of EDTA therefore is decreased. Goetz et al.[1] found that the concentration decreases about 1 per cent after storage in glass for 1 month. Therefore, it is recommended to store the standard solution in a polyethylene or other plastic container. After 5 months of storage in polyethylene, there is less than 0.05 per cent change in titre of a 0.01 *M* solution.[2]

According to Blaedel the reagent-grade material may be purified easily for use as a primary standard. A nearly saturated aqueous solution (about 10 g/100 ml of water) is prepared, and ethyl alcohol is added slowly until a permanent precipitate is formed. The precipitate is removed by filtration and is discarded. An equal volume of alcohol is added to the filtrate. The resulting precipitate is filtered, washed with acetone, with ether, and is air dried; yield, about 75 per cent. The product thus obtained may contain as much as 0.5 per cent water, which practically cannot be removed at room temperature. After drying at 80°C for 4 days (50 per cent relative humidity) constant weight is obtained with an assay corresponding to 100.04 ± 0.01 per cent $Na_2H_2Y \cdot 2\ H_2O$. It is not recommended to dry at temperatures above 80°C because dehydration may occur (see below). The dihydrate is stable at room temperature over a wide range of humidities. Thus it can be weighed in the open air. It can be stored in a closed container or over calcium chloride or sulfuric acid without change in weight. (It should not be stored over phosphorus pentoxide.) Further information on stability is given by Duval[3] and Wendlandt.[4]

[1] C. A. Goetz, T. C. Loomis, and H. Diehl, *Anal. Chem.* **22**, 798 (1950).
[2] W. J. Blaedel and H. T. Knight, *Anal. Chem.* **26**, 741 (1954).
[3] C. Duval, *Anal. Chim. Acta* **16**, 545 (1957).
[4] W. W. Wendlandt, *Anal. Chem.* **32**, 848 (1960).

The anhydrous form may be obtained by drying overnight at 120°C in a vacuum oven. While the product has the composition corresponding to Na_2H_2Y, it is not suitable for use as a primary standard because it is hygroscopic under ordinary atmospheric conditions.

The acid form H_4Y, molecular weight 292.25, rarely is used as a primary standard. The solubility in water is about 0.2 g/liter. According to Schwarzenbach and Flaschka[5] it deserves wider use, because it is easily obtained in a pure state by addition of hydrochloric acid to a solution of the disodium salt. The precipitate is washed well and dried at 110°C. For the preparation of 0.10 *M* solution 14.61 g of the free acid and 4.5 g of sodium hydroxide are dissolved in 400 ml of water. After cooling to room temperature the solution is diluted to the mark in a 500-ml volumetric flask.

Procedure. The following procedure may be used for the preparation and standardization of 0.02 *M* EDTA solution containing approximately 5×10^{-4} *M* magnesium chloride. This solution may be used for the direct titration of calcium, or of calcium plus magnesium, with Eriochrome Black T as indicator.

Solution: Dissolve 7.4 g of $Na_2H_2Y \cdot 2\,H_2O$ and 0.1 g of $MgCl_2 \cdot 6\,H_2O$ in 1 liter of deionized water.

Buffer: Mix 6.8 g of ammonium chloride and 57 ml of concentrated ammonium hydroxide (in hood). Dilute with water to approximately 100 ml.

Indicator: Dissolve approximately 0.5 g of Eriochrome Black T in 100 ml ethanol (also see Note 2 below).

Standardization: Weigh 0.35–0.4 g of reagent-grade calcium carbonate (low in magnesium) to 0.1 mg. Dissolve in about 20 ml of 0.5 *M* HCl. Avoid using a large excess of acid. After the solid is completely dissolved, add water and transfer *quantitatively* to a 250-ml volumetric flask. Dilute to the mark and mix thoroughly. Carry out the following titration at least in duplicate. Pipet 50 ml of the calcium solution into a 250-ml volumetric flask. Add at least 20 ml of the buffer solution and 4 drops of the indicator solution. Titrate with the EDTA solution. At the end point the color changes sharply from red to pure blue with no reddish tint (see Note 3). Calculate the concentration of the EDTA solution.

Notes:

1. If magnesium chloride is not added to the EDTA solution and EBT is used as indicator, it is necessary to prepare separately a solution containing equimolar amounts of magnesium and EDTA, and to add a suitable amount of this solution to the calcium solution before titration.

Several other indicators are available (instead of EBT) which at the proper pH are sensitive to calcium and not to magnesium. Murexide (ammonium purpurate)

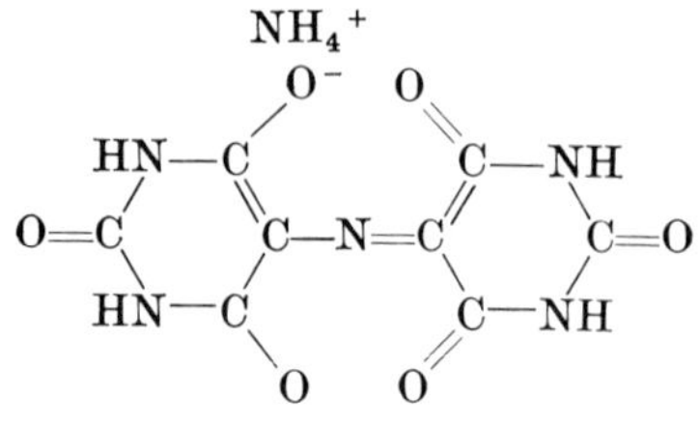

murexide

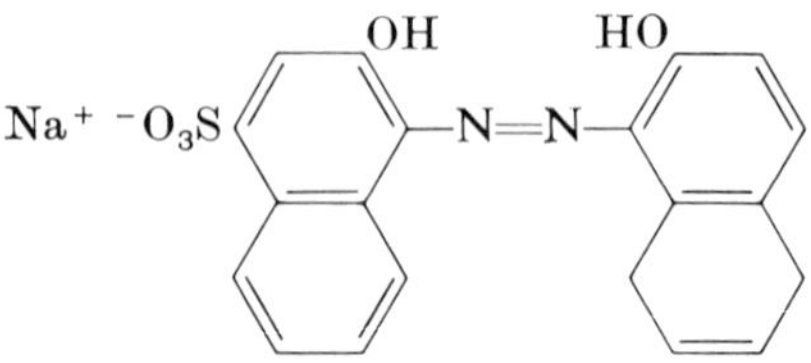

calcon

[5] G. Schwarzenbach and H. Flaschka, *Die Komplexometrische Titration*, Enke, Stuttgart, 1965.

was the first metallochromic indicator used.[6] At pH 12 this indicator gives a pink color with calcium which changes to blue-violet at the end point. At this pH magnesium is precipitated as the hydroxide and does not interfere. (Coprecipitation of calcium with magnesium may be minimized if the EDTA, in an amount about equivalent to calcium, is added to a neutral or acid solution and the solution is then made alkaline.) The color change with murexide is not as sharp as with certain other indicators. A much better indicator is *o*-cresol complexone, commonly known as metalphthalein (the formula is shown on p. 737). At pH 11 this yields a red color with calcium (also with barium and strontium; the color with magnesium is very weak). The change in the titration is from red to colorless, and is sharp. Calcon also may be used as indicator for calcium at pH 12.5–13 without interference from magnesium. Other indicators are listed by Schwarzenbach and Flaschka[7] and Perrin.[8] A convenient tabulation is given by Meites.[9]

2. Solutions of EBT deteriorate due to air oxidation and the indicator solution should be prepared fresh as needed. Many workers prefer to add a mixture of the solid indicator diluted with a several hundredfold excess of pure sodium chloride or other inert substance. Air oxidation of the indicator in the solution being titrated is catalyzed by several metal ions, including copper(II) and manganese(II). If this difficulty is experienced, ca. 1 g of ascorbic acid may be added before titration to eliminate interference.

3. If trouble is experienced due to a sluggish color change, the solution may be warmed to ca. 60°C.

4. Traces of copper in the water will interfere seriously by blocking the indicator. The water may be tested as follows. To 50–100 ml of water add a few milliliters of buffer. Upon addition of EBT the solution should be pure blue with no violet or red tint. If the solution is not pure blue, add a little EDTA. If the solution becomes blue, the water contains calcium or magnesium, which should be removed or corrected for. If the solution does not become blue, the water probably contains copper or iron, and a little potassium cyanide should be added *before* the indicator.

5. Provided the solution of EDTA is standardized as above and is used for titrations at the same pH as in the standardization, the water used for its preparation need not be free of metal impurities, because these are compensated for in the standardization. However, if the standard solution is used for titrations at various pH values, the effect of impurities (iron, copper, calcium) will be different at different pH. It goes without saying that if EDTA is purified and used as a primary standard, the water must be free of metal impurities.

6. Useful summaries of practical aspects should be consulted for further details[10,11] (see also Refs. 5 and 9).

Determination of calcium, or of calcium plus magnesium. Hardness of water. In the absence of interfering substances the determination is carried out in exactly the same way as the standardization.

The hardness of water is due to the presence of calcium and magnesium, usually in the form of bicarbonates. The total amount of calcium plus magnesium is expressed in terms of the equivalent amount of calcium carbonate in parts per million (milligrams per liter).[12] Thus a calcium carbonate hardness of 1000 ppm

[6] G. Schwarzenbach, W. Biedermann, and F. Bargerter, *Helv. Chim. Acta* **29**, 811 (1946).

[7] See Ref. 5, p. 136f.

[8] D. D. Perrin, *Organic Complexing Agents*, Wiley-Interscience, New York, 1964.

[9] L. Meites (ed.), *Handbook of Analytical Chemistry*, McGraw-Hill, New York, 1963.

[10] H. N. Flaschka in C. L. Wilson and D. W. Wilson (eds.), *Comprehensive Analytical Chemistry*, vol. 1B, Elsevier, Amsterdam, 1960, pp. 288–385.

[11] C. N. Reilley, R. W. Schmid, and F. S. Sadek, *J. Chem. Educ.* **36**, 619 (1959).

[12] The German degree of water hardness corresponds to 10 ppm calcium oxide, and therefore is equal to 18 × the hardness expressed as ppm calcium carbonate.

corresponds to a total concentration of calcium plus magnesium of 0.01 *M*. Water may contain heavy metal ions as impurities, including particularly copper, iron, aluminum, and manganese. Each of these substances [manganese when present as manganese(II)] interferes with the titration by blocking the indicator. The addition of sodium cyanide and of a reducing agent (ascorbic acid, or hydroxylamine) will prevent interference by all common metals except aluminum. Interference by aluminum is prevented by the addition of triethanolamine.

Unless the hardness is very low, the solution becomes supersaturated to calcium carbonate upon addition of the pH 10 buffer. Fortunately, the precipitation usually is slow, and no trouble is experienced unless the titration is carried out too slowly. If a precipitate of calcium carbonate is formed, it redissolves upon the addition of titrant; however, the presence of a precipitate during titration may cause a drifting and unsatisfactory end point. If such difficulty should be experienced, it can be prevented most simply as follows: A determination of the alkalinity (Chapter 39) is carried out on a separate portion of the sample. The appropriate amount of acid is added to the sample to be used for determination of hardness and the solution is stirred for a few minutes to expel carbon dioxide.

Procedure. Select a suitable size of sample (note 1). For each 50-ml sample, add about 5 ml of ammonium–ammonia buffer, add a few tenths of a gram of ascorbic acid, a few crystals of potassium or sodium cyanide (caution! poison!), and a few drops of 20% triethanolamine. Titrate with 0.02 *M* EDTA (containing magnesium) (note 2) using EBT as indicator.

Notes:

1. The proper size of sample depends upon the hardness. A preliminary titration using 50 ml of the water sample should be carried out to determine the proper-size sample for the main titration. If the hardness is low, the standard solution of EDTA may be diluted.
2. The magnesium content of the water may be too low to permit proper functioning of EBT. Therefore, an EDTA solution containing magnesium should be used.

The combination of ion exchange, acidimetric titration, and compleximetric titration without and with a masking agent to determine the composition of a four-component unknown is illustrated in the following procedure.

Analysis of a four-component unknown. The unknown contains H^+, Na^+, Zn^{++}, and Mg^{++}. H^+ is determined by titration with standard sodium hydroxide. In a separate portion all the cations are converted into the equivalent amount of H^+ by an ion-exchange column and determined with standard base. Titration of two other portions with standard EDTA, with and without potassium cyanide, determines Mg^{++} and the sum of Mg^{++} and Zn^{++}.

The preparation of and standardization of sodium hydroxide solution and the preparation of the ion-exchange column are described on pages 781 and 792.

Procedure. Weigh an appropriate amount of sample, dissolve, and dilute to 250 ml in a volumetric flask. Mix thoroughly and take 25-ml aliquots for the following titrations, doing each in duplicate.

Determination of H^+. Titrate a 25-ml aliquot with standard sodium hydroxide, using bromcresol green as indicator. Calculate millimoles of H^+ in the sample.

Determination of total cations. Pipet a 25-ml aliquot into the reservoir of the ion-exchange column. Adjust the flow rate to about 3 ml/min and collect the effluent in a 250-ml Erlenmeyer flask. Wash with three 25-ml portions of water; the first portion should be passed through the column at a rate of 3 ml/min, the others more rapidly, at 5–10 ml/min.

Titrate the solution with standard 0.1 *N* sodium hydroxide, using bromcresol green as indicator. Test for completeness of washing by running 25 ml of water through the column and seeing whether this consumes any base.

Calculate the milliequivalents of cations in the sample.

The resin can be regenerated by allowing 50 ml of 2–3 *M* hydrochloric acid to flow through at a rate of 2–3 ml/min.

Determination of ($Zn^{++} + Mg^{++}$). Pipet a 25-ml aliquot into a 250-ml Erlenmeyer flask, add approximately enough 0.1 *N* sodium hydroxide to neutralize the acid present, and add approximately 25 ml of buffer solution (see p. 804) and 4 drops of Eriochrome Black T indicator solution. Titrate with standard EDTA until the solution turns from pink to a purple to a pure blue. Calculate total mmoles Zn^{++} and Mg^{++}.

Determination of Mg^{++}. Pipet a 25-ml aliquot into a 250-ml Erlenmeyer flask, neutralize with the proper amount of 0.1 *N* sodium hydroxide, add 25 ml of buffer, approximately 2 g of potassium cyanide (caution! poison!) indicator, and 50 ml of water. Titrate with standard EDTA.

Calculate the number of mmoles of H^+, Na^+, Zn^{++}, and Mg^{++} in the sample.

Determination of aluminum, nickel, cobalt, and iron(III) in a mixture by EDTA titration after ion-exchange separation.[13] The separation of metals by differential adsorption of their chloro complexes on strong-base-type anion-exchange resins was touched upon in Chapter 13. The resolution of the system Al–Ni–Co–Fe(III) with the aid of ion exchange and EDTA titration is offered here as an example of such anion-exchange separations. Aluminum and nickel do not yield anionic chlorocomplexes at any hydrochloric acid concentration and therefore are not adsorbed by an anion-exchange resin under any conditions. Cobalt(II) and iron(III) are strongly adsorbed from 9 *M* hydrochloric acid. Since cobalt is only weakly adsorbed from 4 *M* hydrochloric acid, it can be eluted from the column with acid of this concentration, which does not elute Fe(III). Iron can be desorbed with 0.5 *M* hydrochloric acid.

Various EDTA procedures for the titration of aluminum and nickel (in the presence of each other), cobalt and iron are available. In the following procedure, EDTA standard solution is added in excess and a back titration is made with copper(II).

Ion-exchange column. Prepare a resin column (Dowex 1 X4, 100–150 mesh) 1.0–1.2 cm in diameter and 30 cm in length. Pass an excess (100 ml) of 9 *M* hydrochloric acid through the column. (See the comments in Chapter 39 regarding preparation of columns.)

Special solutions. EDTA, 0.02 *M*. Prepare and standardize as described on p. 804, but without magnesium.

Cu(II) solution, 0.03 *M*. Dissolve about 1.9 g of copper sulfate pentahydrate in 250 ml of water. Standardize against EDTA as directed below.

[13] The procedure is based on one given by D. H. Wilkins and L. E. Hibbs, *Anal. Chim. Acta* **18**, 372 (1958), for the determination of Al, Ni, Co, Cu, and Fe in the magnetic alloy Alnico. These authors first separate (and determine) copper by electrolytic deposition.

Acetate buffer, pH 5. Dissolve 27.3 g of sodium acetate trihydrate in water, add 60 ml of 1.0 *M* hydrochloric acid, and dilute to 1 liter with water.

PAN indicator (see p. 811). Dissolve 50 mg of the compound in 100 ml of methanol or ethanol.

Standardization of Cu(II) solution. Pipet 25 ml of the copper solution into a beaker, add 10 ml of acetate buffer, add 4–5 drops of PAN indicator, heat to boiling, and titrate with standard EDTA. The color change is from blue-violet to yellowish green.

Procedure. Separations. The Al–Ni–Co–Fe solution, containing 100–200 mg of each ion, should have a volume of 5–10 ml and should be 9.0 *M* in hydrochloric acid. Add the solution to the ion-exchange column and allow it to percolate through the resin (the rate is not critical). Elute the aluminum and nickel with 50 ml (or more if required) of 9 *M* hydrochloric acid, catching the effluent in a 250-ml beaker. Now begin the elution of cobalt with 4 *M* hydrochloric acid. When the blue cobalt band is within 5 cm of the base of the resin, remove the beaker containing the nickel and replace it with another. Pass approximately 75 ml of 4 *M* hydrochloric acid (or more if needed, as shown by the color) to elute cobalt completely. Remove the beaker containing the cobalt and elute iron with 75 ml (or more if required) of 0.5 *M* hydrochloric acid into a fresh beaker.

Determination of aluminum and nickel. Evaporate (hood) the solution containing aluminum and nickel to a volume of 10 ml, transfer the solution to a 250-ml volumetric flask, and dilute to volume. Transfer a 25-ml aliquot to a 250-ml beaker or flask and add 10 ml of acetate buffer and a sufficient volume of standard EDTA solution to complex both metals and provide a slight excess. Heat to boiling, add 4–5 drops of PAN indicator, and back-titrate the excess of EDTA with standard copper solution (color change from yellow to purple). Calculate the sum of Al and Ni.

Add 1 g of ammonium fluoride to the hot solution and back-titrate the EDTA (liberated from the Al complex) with standard copper solution. Calculate the amount of Al and find Ni by difference.

Determination of cobalt and iron. Treat the two solutions of these metals identically as follows. Evaporate to 10 ml, transfer to a 250-ml volumetric flask, and dilute to the mark with water. Transfer a 25-ml aliquot to a 250-ml beaker or flask and add standard EDTA solution in 5–10 ml excess, followed by 10 ml of acetate buffer. Heat to boiling and back-titrate the excess of EDTA with standard copper solution, using PAN as indicator. Calculate the amounts of Co and Fe.

Other Useful Chelating Titrants

Nitrilotriacetic acid, $N(CH_2COOH)_3$, abbreviated NTA. This substance is fairly widely used, in the form of its trisodium salt. The relative stabilities of metal complexes follow the same general pattern as those with EDTA, as shown in Table 41-1, in which *K* refers to the formation constant of the complex. In these examples (and several others not listed above) log *K* for NTA is $\frac{3}{5}$ to $\frac{2}{3}$ of the corresponding log *K* for EDTA. The stabilities of the alkaline earths are not great enough to permit satisfactory titrations.

1,2-Diaminocyclohexane-N,N,N′,N′-tetraacetic acid, abbreviated DCYTA. This substance forms complexes generally even more stable than those of EDTA (Table 41-1). So far as titrations (ordinary concentrations) in alkaline medium are concerned, the increase in stability is not of much practical value. However, the greater stability permits useful titrations to be carried out at greater dilution, or at lower pH, than with

EDTA. It should be mentioned, however, that the reactions in general are slower than those involving EDTA.

TABLE 41-1. **Comparison of Formation Constants of Some Metal Chelates**

	log K		
Metal Ion	NTA	EDTA	DCYTA
Ba^{++}	4.8	7.8	8.0
Ca^{++}	6.4	10.7	12.5
Fe^{3+}	15.9	25.1	29.3
La^{3+}	10.4	15.4	16.3
Mg^{++}	5.4	8.7	10.3
Zn^{++}	10.5	16.5	18.7

Ethyleneglycol bis(β-aminoethylether)-N,N,N′,N′-tetraacetic acid, abbreviated EGTA. One of the most useful applications of this titrant is suggested by the following pair of formation constants: $\log K = 11.0$ and 5.2 for Ca^{++} and Mg^{++}, respectively. The difference in the stability constants is so great that calcium can be titrated in the presence of magnesium; this is not possible using EDTA unless the magnesium hydroxide is precipitated, e.g., at pH 12 or more.

Polyamines. Several examples are mentioned in Chapter 6. The substances of greatest analytical significance are *trien* (triethylenetetramine) and *tetren* (tetraethylenepentamine). The great practical advantage of the polyamines is that they form water-soluble complexes of high stability with metal ions such as Cu^{++}, Hg^{++}, Ni^{++}, Cd^{++}, and Zn^{++}, but not with the alkaline earth ions. For the metal ions given, the respective values of log K with tetren are 24.3, 27.7, 17.6, 14.0, and 15.4. Thus it is possible for example to titrate Cu^{++} and Cd^{++} or Zn^{++}, stepwise, in the presence of the alkaline earths. Unfortunately the pure polyamines are not available commercially; the impurities invariably present in the commercial preparations spoil the end point.

Finally the compound *uramildiacetic acid*,

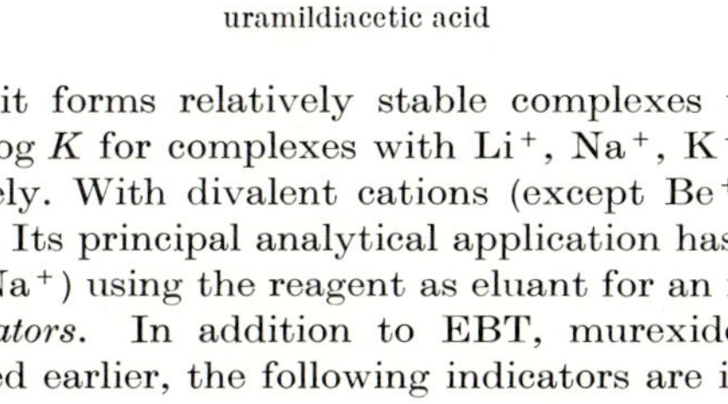

uramildiacetic acid

is of interest because it forms relatively stable complexes with some monovalent cations; the values of log K for complexes with Li^{+}, Na^{+}, K^{+}, and Tl^{+} are 5.6, 3.3, 1.9, and 6.7, respectively. With divalent cations (except Be^{++}) it forms complexes less stable than EDTA. Its principal analytical application has been the separation of the alkalies (e.g., K^{+}, Na^{+}) using the reagent as eluant for an ion-exchange resin.

Metallochromic indicators. In addition to EBT, murexide, and metalphthalein, which we have described earlier, the following indicators are important enough to be given special attention.

Xylenol orange is the trivial name for 3,3′-bis-N,N-di(carboxymethyl)aminoethyl-*o*-cresolsulfonphthalein.

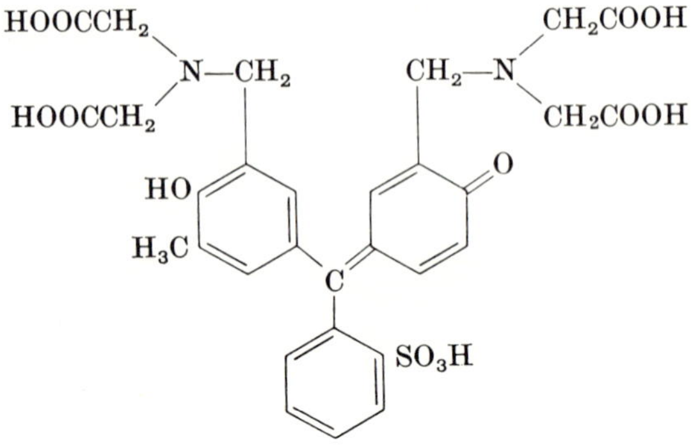

xylenol orange

This substance is an acid–base as well as a metal indicator. In acidic medium it is lemon yellow, and at pH 5.4–7.4 it is red. Since the complexes of xylenol orange with most metals are red, its use as a metal indicator is restricted to acid medium (below pH 5.4). It is usable even in 0.2 *M* nitric acid. Metal ions including Bi^{3+}, Th^{4+}, Sc^{3+}, La^{3+}, Zn^{++}, Pb^{++}, Hg^{++}, and several others can be titrated directly with EDTA in acid medium. Several other metal ions which react slowly with EDTA in acid medium, most importantly Fe^{3+}, Al^{3+}, Cu^{++}, and Cr^{3+}, may be determined by back titration using standard solutions of bismuth, thorium, or zinc nitrate.

Pyrocatechol violet (also known as catechol violet) has the structure

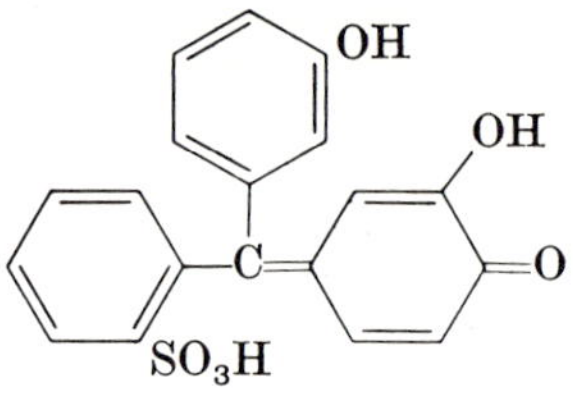

pyrocatechol violet

The neutral molecule is red. In the pH range 2–6 the hydrogen from the sulfonic group is dissociated and the resulting anion is yellow. At higher pH's the substance is violet (pH 7–10) or red purple (pH >10). Most of the metal complexes with this indicator are blue. Therefore, in a compleximetric titration in the pH range 2–6 a good color change from blue to yellow occurs. The indicator is particularly useful for the direct titration of Bi^{3+} with EDTA at pH 2–3. The only common metals that interfere under these conditions are thorium and zirconium. Interference by iron(III) and mercury(II) can be prevented by reduction of these ions, e.g., by ascorbic acid.

PAN is the common abbreviation for 1-(2-pyridylazo)-2-naphthol:

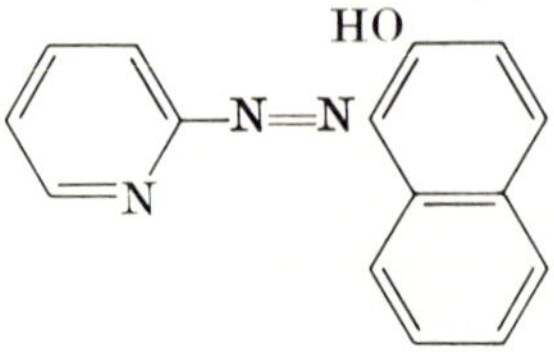

PAN

This red substance is insoluble in water but readily soluble in alcohols and other organic solvents. As compared to EBT it has the advantage that the solution in alcohol is stable indefinitely. PAN forms complexes (usually red) with many polyvalent metals, but does not form colored complexes with the alkaline earths or aluminum. The indicator is yellow at pH below 8. The metal complexes are insoluble in water and therefore react slowly with EDTA. For this reason titrations in which PAN is used as indicator are performed either in the presence of an organic solvent (e.g., 20–50 per cent ethanol) or in a boiling solution. Under these conditions direct titrations with EDTA of copper, zinc, cadmium, and scandium are possible. Many back titrations in hot acidic medium are possible, using a standard solution of copper nitrate as titrant; these include titrations of lead, iron(III), and nickel.

Calcon is the sodium salt of Eriochrome Blue Black R (the anion sometimes called Pontachrome Blue Black R):

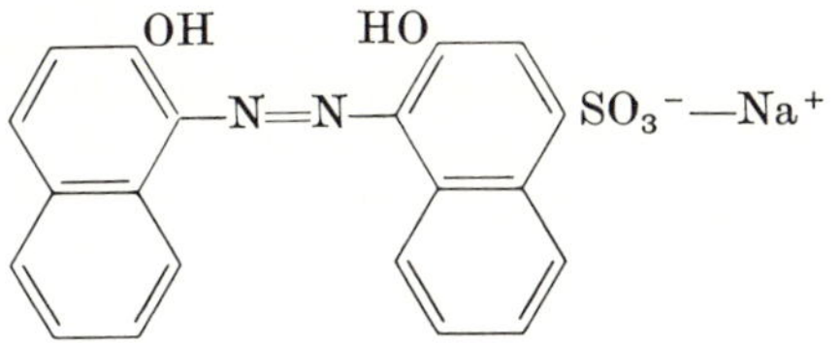

calcon

The neutral molecule, H_3In, which exists only in strongly acid medium, is green; color changes to red, blue, and orange occur at pH 7, 10, and 13.5, respectively. The important difference from EBT is that the blue color of HIn^{--} is maintained up to pH 13. Since the complexes with metals, including calcium, are red, the color change in titration at pH 13 is from red to pure blue.

Calmagite is the sodium salt of 1-(1-hydroxy-4-methyl-2-phenylazo)-2-naphthol-4-sulfonic acid:

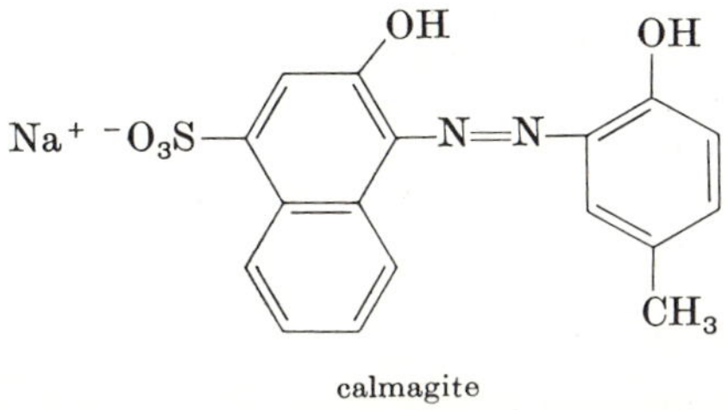

calmagite

The colors of the anions H_2In^-, HIn^{--}, and In^{3-} are red, blue, and orange, respectively. These are the same as of the anions of EBT, but the changes occur at higher pH (8.1 and 12.4) in calmagite than in EBT (6.3 and 11.6). Since the metal complexes are red, calmagite is a useful indicator in alkaline solution up to pH about 12. Generally speaking this indicator is a superior replacement for EBT for two reasons: the color change is sharper than the EBT, and solutions of the indicator are stable practically indefinitely.

Indicator constants and acid–base properties of metallochromic indicators are given in the Appendix.

Argentimetric determinations

Determination of cyanide in potassium cyanide. Weigh out 0.35–0.4 g of potassium cyanide (use extreme care in working with this salt; it is a deadly poison), dissolve in water, add 5–6 ml of 6 *N* ammonia and 0.2 g of potassium iodide, dilute to 100 ml, and titrate with 0.1 *N* silver nitrate solution until a permanent turbidity of silver iodide is obtained. The turbidity is best seen by viewing the liquid against a dark background. The reactions involved are

$$Ag^+ + 2\,CN^- \rightleftharpoons Ag(CN)_2^- \qquad \text{(quantitative at end point)}$$
$$Ag(CN)_2^- + Ag^+ \rightleftharpoons 2\,AgCN$$
$$2\,AgCN + 2\,NH_3 \rightleftharpoons Ag(NH_3)_2^+ + Ag(CN)_2^-$$
$$Ag(NH_3)_2^+ + I^- \rightleftharpoons AgI + 2\,NH_3$$

Calculate the cyanide content of the sample.

Notes:

1. For the preparation of standard silver nitrate solution, see page 795.

2. The method given is used for the determination of cyanide and silver. Quite generally it is preferable to titrate silver with thiocyanate. However, the latter method cannot be directly applied to the determination of silver in silver halides, for example in photographic films. The silver in these insoluble salts may be determined by dissolving the sample in an excess of standard cyanide solution and back-titrating with standard silver nitrate according to the method given above.

3. The titration of cyanide with silver nitrate can be applied to the determination of cations which form complex cyanides. Zinc and nickel, for example, react with cyanide to form slightly soluble cyanides:

$$Ni^{++} + 2\,CN^- \rightleftharpoons Ni(CN)_2$$

This cyanide dissolves in an excess of alkali cyanide, forming a stable complex:

$$Ni(CN)_2 + 2\,CN^- \rightleftharpoons Ni(CN)_4^{--}$$

The titration is carried out by using silver iodide as an indicator and adding standard potassium cyanide solution to the ammoniacal nickel solution until the opalescence due to silver iodide disappears. The end point is not reached exactly at the equivalence point. Therefore, the cyanide solution must be standardized against a nickel solution under approximately the same conditions as in the actual determination.[14] The method is used for the routine determination of nickel in steel; ferric iron can be made harmless by transforming it into a complex citrate.

Mercurimetric determinations

Preparation of 0.1 *N* mercuric nitrate solution. Weigh out approximately the calculated amount (4.3 g) of analytical reagent mercuric nitrate, and transfer to a 250-ml volumetric flask. Add about 5 ml of 6 *N* nitric acid and make up to the mark with water.

Standardization of mercuric nitrate solution against standard thiocyanate

Procedure. Add 1 ml of ferric alum indicator (saturated solution of ferric ammonium sulfate in water) and 5 ml of 6 *N* nitric acid to 25–40 ml of 0.1 *N* mercuric nitrate solution, and titrate at a temperature not higher than 20°C with standard 0.1 *N* thiocyanate solution until the color changes to brownish. The indicator blank is negligibly small. Calculate the normality of the mercury solution. For preparation of standard thiocyanate solution, see page 796.

[14] For a modified procedure see I. M. Kolthoff and F. S. Griffith, *J. Phys. Chem.* **42**, 542 (1938).

Notes:

1. The mercuric thiocyanate formed during the titration is slightly soluble but remains in solution by complex formation with mercuric ions:

$$Hg^{++} + 2\,SCN^- \rightleftharpoons Hg(SCN)_2$$
$$Hg(SCN)_2 + Hg^{++} \rightleftharpoons 2\,HgSCN^+$$

It also dissolves in an excess of thiocyanate solution with the formation of complex ions:

$$Hg(SCN)_2 + SCN^- \rightleftharpoons Hg(SCN)_3^-$$
$$Hg(SCN)_2 + 2\,SCN^- \rightleftharpoons Hg(SCN)_4^{--}$$

The solution remains clear during the titration until near the equivalence point. Then there are not enough mercury ions left to keep the mercuric thiocyanate in solution, and the latter separates in the form of colorless needles. The precipitate does not interfere with the sharpness of the color change.

2. The reverse titration can be applied to the determination of thiocyanate. In this case standard mercuric nitrate solution is added to the disappearance of the red-brown coloration. The color begins to fade before the end point is reached; the gradual fading indicates the approach of the latter. The last change from the brownish tinge to colorless can be recognized quite distinctly.

3. The procedure described is a very accurate one for the determination of mercuric compounds. However, chloride and bromide interfere because their mercuric salts are also slightly dissociated. Therefore mercury cannot be titrated directly in mercuric chloride. In this case it is necessary to add a solution of the mercury salt to an excess of sodium hydroxide, filter off and wash the mercuric oxide, and dissolve it in nitric acid.

4. Mercuric cyanide is much less dissociated than mercuric thiocyanate. The presence of mercuric cyanide does not interfere in the mercury titration with thiocyanate. Use can be made of this fact in the determination of cyanide. A solution of the latter is added to a measured excess of standard mercuric solution:

$$Hg^{++} + 2\,CN^- \rightleftharpoons Hg(CN)_2$$

and the excess of mercuric ions is titrated with thiocyanate.

5. A solution of mercuric thiocyanate in alkali thiocyanate gives a slightly soluble crystalline precipitate with zinc:

$$Hg(SCN)_2 + 2\,SCN^- \rightleftharpoons Hg(SCN)_4^{--}$$
$$Hg(SCN)_4^{--} + Zn^{++} \rightleftharpoons ZnHg(SCN)_4$$

Use is made of this reaction in the volumetric determination of zinc. The precipitate is allowed to settle, and the excess of thiocyanate is titrated in an aliquot part of the supernatant liquid with standard mercury solution. Aluminum and ferric iron do not interfere; copper, bismuth, nickel, cobalt, manganese, ferrous iron, chromium, and chloride must be removed.[15]

Titration of chloride with standard mercuric nitrate. The halides, as we have seen (Chapter 36), react with mercuric ions to give slightly dissociated mercuric halides:

$$Hg^{++} + 2\,Cl^- \rightleftharpoons HgCl_2$$

Therefore, if mercuric ions are added to a chloride solution, the former are removed

[15] For a more detailed discussion see I. M. Kolthoff and V. A. Stenger, *Volumetric Analysis*, vol. II, p. 340.

by the formation of slightly dissociated mercuric chloride. When the equivalence point has been reached, there is a pronounced increase in the mercuric-ion concentration. Sodium nitroprusside, diphenyl carbazide, or diphenyl carbazone can be used as indicators. The former reacts with Hg^{++} to yield a white insoluble salt, the latter to give an intense violet color with Hg^{++} at the proper acidity. We shall first consider nitroprusside. Since mercuric nitroprusside is only very slightly soluble, the solution becomes turbid as soon as an excess of mercury has been added.

It is necessary to determine the indicator blank at the end of the titration. This may be done as follows: From the amount of standard mercury solution used, the mercuric chloride concentration at the end point is calculated roughly (within 10 per cent). A mercuric chloride solution having the same concentration and volume as the solution titrated at the end point, and containing the same amount of indicator, is prepared. Standard mercuric nitrate solution is then carefully added to the synthetic solution until a turbidity appears. The blank is deducted from the volume of standard reagent used in the titration. The blank varies somewhat with the concentration of mercuric chloride at the end point, because part of the mercuric ions added in excess reacts with mercuric chloride, forming complex ions:

$$HgCl_2 + Hg^{++} \rightleftharpoons 2\, HgCl^+$$

The higher the concentration of mercuric chloride at the end point, the greater will be the amount of mercuric ions removed by the complex formation, and consequently the larger the indicator blank will be. For a solution of 50 ml of 0.05 N mercuric chloride, the correction amounts to 0.17 ml of 0.1 N mercuric nitrate.

This titration method for chloride is of great practical importance, since it allows the direct determination of this anion *in an acid medium even at great dilutions*. It deserves a much wider application than it enjoys at present.

Indicator solution: 10 per cent sodium nitroprusside. The solution should be kept in a dark bottle. For each 10 ml of solution titrated 0.1 ml of indicator is used.

Procedure. Add the indicator to a measured volume of chloride solution, and titrate with standard mercuric nitrate until a permanent turbidity is obtained. Standardize the mercuric nitrate solution against pure sodium chloride under the same conditions as obtain in the titration. Determine the indicator blank as described above.

Notes:

1. The presence of mineral acid has virtually no effect upon the results.
2. Bromides and thiocyanates may also be titrated in this way.
3. Of the metallic salts, those of copper, nickel, cobalt, and cadmium interfere because they form slightly soluble nitroprussides. Zinc, ferric iron, bismuth, manganese, aluminum, barium, calcium, strontium, and magnesium do not interfere in dilute solution.
4. The use of diphenylcarbazide or diphenylcarbazone as indicators yields good results only if a pH between 1.5 and 2.0 is maintained in the solution being titrated. The end point appears too soon when insufficient acid is present, too late when the acidity is higher.

Indicator: Either a saturated solution of diphenylcarbazide in 95 per cent ethanol or a 2 per cent solution of diphenylcarbazone in the same solvent.

Procedure for chloride. For titration with 0.1 *N* mercuric nitrate, adjust conditions so that the volume at the end point will be 80–100 ml. Free acid should be neutralized with sodium hydroxide, to an orange color with 5 drops diphenylcarbazide indicator. Add 4 ml of 0.2 *N* nitric acid and titrate with mercuric nitrate until the color changes first to pink-violet and finally to a deep blue-violet. Deduct a blank carried out with the same quantities of acid, water, and indicator.

Notes:

1. Interference by cadmium, cobalt, ferrous iron, lead, and copper is prevented by the addition of the specified amount of nitric acid.
2. The procedure can be used for determination of mercuric salts by addition of an excess of standard chloride solution and back titration.

PROBLEMS

1. A solution contains calcium and magnesium. A 50-ml aliquot was titrated at pH 12.5 with 0.02000 *M* EDTA (what indicator?), 20.00 ml being required. Another 50-ml aliquot was titrated at pH 10.0 (EBT), 32.50 ml being required. Give the concentrations of calcium and magnesium in the solution.
2. The hardness of a sample of water was determined by titration with EDTA; a 100.0-ml sample of water required 15.75 ml of 0.01500 *M* EDTA. Calculate the water hardness in terms of ppm of calcium carbonate.
3. Before the EDTA titration was developed, water hardness commonly was determined by titration with a standard solution of a soap, e.g., sodium stearate. Explain one or more ways in which the end point of such a titration could be found.
4. A 0.3236-g mixture of anhydrous silver halides was dissolved in 50.00 ml of 0.1810 *M* potassium cyanide solution. The excess of cyanide required 28.14 ml of 0.1010 *N* silver nitrate for back titration. Calculate the silver content of the mixture.
   Ans. 56.13 per cent Ag.
5. Devise a procedure for determining ca. 0.001 *M* Hg^{++} in the presence of ca. 0.1 *M* Ag^{+}.
6. Devise compleximetric titration procedures for the determination of each constituent of the following mixtures (specify pH, indicator, and masking agents, where needed): (a) Ca^{++}, Mg^{++}, and Fe^{3+}; (b) Ca^{++}, Cu^{++}; (c) Al^{+3}, Hg^{++}; (d) Pb^{++}, Ca^{++}, Bi^{3+}; (e) Zn^{++}, Cu^{++}, Ca^{++}.

CHAPTER 42 *Reagents Used in Oxidation–Reduction Titrations*

Oxidizing agents

Potassium permanganate. This oxidizing agent was first used by Margueritte[1] for the well-known titration of ferrous iron and soon thereafter became a common reagent in the analytical laboratory. It is a strong oxidizing agent, its oxidation potential in 1 *N* sulfuric acid being of the order of 1.5 V. In an acid medium the reduction of permanganate can be represented by

$$MnO_4^- + 8\,H^+ + 5e \rightleftharpoons Mn^{++} + 4\,H_2O$$

In feebly acid, neutral, or alkaline solutions the septavalent manganese is reduced to the quadrivalent state, thus causing the precipitation of hydrated manganese dioxide during the titration:

$$MnO_4^- + 2\,H_2O + 3e \rightleftharpoons MnO_2 + 4\,OH^-$$

In titrations of colorless or slightly colored solutions with permanganate it is not necessary to use an indicator, since an excess of the reagent is indicated by its color. One-tenth milliliter of 0.01 *N* permanganate imparts an extremely pale pink color to 100 ml of water, and 0.2 ml imparts a pale pink color; temperature and acid concentration are without influence upon the limit of visibility. Since the intensity of the color of oxidized diphenylamine, diphenylbenzidine, diphenylamine sulfonic acid (see p. 757), and the ferrous-phenanthroline complex is much greater than that of permanganate, the use of these indicators is recommended in titrations with very dilute permanganate, at least if the indicator is not attacked before the end point is reached.

In acid medium potassium permanganate reacts very rapidly with many reducing substances according to definite stoichiometric equations. There are, however, some practical disadvantages involved in the use of permanganate. In the first place, it is difficult to obtain potassium permanganate in an entirely pure state; it is generally contaminated by traces of manganese dioxide. Therefore a solution prepared by weighing out an exactly known amount of a reagent-grade product must be standardized. Moreover, ordinary distilled water is likely to contain reducing substances which will react with the permanganate to form manganese dioxide. Before using the solution, it is necessary to filter it through a nonreducing filter to remove the manganese dioxide. The presence of the latter is very objectionable because it catalyzes the autodecomposition of permanganate on standing. If proper care is taken, manganese dioxide-free solutions of permanganate can be kept unchanged for a long time; acid solutions cannot be kept, because they decompose faster than neutral solutions, the dioxide formed again accelerating further decomposition.

Boiling permanganate solutions, especially if acid, are subject to autodecomposition with the evolution of oxygen. Again this process is catalyzed by manganese dioxide.

[1] F. Margueritte, *Compt. Rend.* **22**, 587 (1846).

This fact must be considered in those cases in which a substance is oxidized in hot solution with an excess of permanganate (hypophosphite, phosphite, various organic substances), the latter being titrated back with a suitable reducing agent. It was found, for example, that on heating a 0.05 *N* potassium permanganate solution in 0.4 *N* sulfuric acid for 1 hr at 100°C the loss by decomposition amounted to 0.6 per cent; in 0.8 *N* sulfuric acid, to 1.7 per cent; in 2 *N* acid, to 58 per cent. The decomposition is also appreciable in hot alkaline solution. In the back titration of an excess of permanganate, an excess of a standard reducing agent (e.g., ferrous sulfate, oxalate) is usually added and the latter is determined with the standard permanganate. The direct titration of permanganate with a reducing agent often gives rise to difficulties, because compounds of manganese intermediate between divalent manganese and permanganate are formed, imparting a brownish color to the solution. Complete decoloration to the manganous state then often occurs slowly.

Many titrations with permanganate, such as those of arsenic trioxide, trivalent antimony, hydrogen peroxide, and ferrocyanide, can be carried out in the presence of hydrochloric acid. This is not the case with the titration of ferrous iron. The reaction between ferrous iron and permanganate *induces* (see p. 764) the oxidation of hydrochloric acid by the permanganate, and hypochlorous acid or chlorine is formed. These oxidation products react incompletely with ferrous iron, and the result is the consumption of too much permanganate in the titration of iron in the presence of hydrochloric acid. Addition of a manganous salt is quite effective in preventing the induced oxidation of hydrochloric acid.

Ceric sulfate. Ceric sulfate was first used as an oxidizing agent by Lange[2] and later applied by various workers. It was not until after the systematic investigations of Willard and Young at the University of Michigan and of Furman and coworkers at Princeton University that ceric sulfate became a common reagent in the analytical laboratory. It can now be obtained in a fairly pure state in the form of the double salt, ceric ammonium sulfate. Ceric sulfate is a powerful oxidizing agent and is used in acid media only.

$$Ce^{4+} + e \rightleftharpoons Ce^{3+}$$

The solution has an intense yellow color, and in solutions not too dilute the end point may be detected without the use of an indicator. In general, however, it is preferable to add a suitable indicator.

Ceric sulfate solutions have the advantage over permanganate that they are less subject to decomposition on standing and even on boiling. If oxidations must be carried out in acid solutions with an excess of oxidizing agent at boiling temperatures, it is preferable to use ceric sulfate instead of permanganate. Naturally the presence of chlorides interferes, because chlorides reduce the hot ceric cerium solution. Another advantage is that in the reduction of ceric cerium no intermediate products are formed. An advantage of ceric sulfate over permanganate is that the former can be used in the titration of ferrous iron in the presence of chloride and of many organic substances.

Erroneous results are obtained in the titration with permanganate of iron(II) in the presence of reducing substances, like phosphate and hypophosphate, even though these reductants react very slowly with permanganate at room temperature. The reason is that in the reduction of permanganate to manganese(II) by iron(II) various unstable intermediate valence forms of manganese are formed which can oxidize these reducing constituents. This is a typical example of an induced reaction (p. 764). The reduction of cerium(IV) to (III) is a one-electron transfer reaction and no intermediate valence states of cerium are formed. Thus the above type of induced reaction cannot occur. Hence iron(II) can be titrated with cerium(IV) in the presence of reducing substances which do not react or react very slowly with cerium(IV). This statement also holds true in the titration of iron(II) in the presence of organic substances. For

[2] L. T. Lange, *J. Prakt. Chem.* **82**, 129 (1861).

example, the titration of iron(II) in the presence of alcohols, glycerol, and sugar gives high results with permanganate as reagent but good results with cerium(IV) as oxidant.

The reduction potential of the cerium solution depends upon the concentration and kind of acid present. Smith and Getz[3] give the following reduction potentials of cerium solutions in 1–8 *N* solutions of various acids:

Perchloric acid	1.70–1.87 V
Nitric acid	1.61–1.56 V
Sulfuric acid	1.44–1.42 V
Hydrochloric acid	1.28 V (decomposes in 1 *N* HCl)

Analytical use is made of the fact that solutions of cerium in perchloric or nitric acid are much stronger oxidizing agents than ceric sulfate solution in sulfuric acid.

Smith et al.[4] described the preparation of pure $(NH_4)_2Ce(NO_3)_6$, which was called by them ammonium hexanitratocerate. This salt can now be purchased in pure form and is recommended by Smith[5] as a primary standard after drying at 85°C. Its equivalent weight is 548.26.

Potassium dichromate. This salt can be obtained in a pure state by recrystallization from water and drying at 150–180°C. Therefore, standard solutions of exactly known content can be prepared by weighing out the pure dry salt and diluting to the proper volume. Potassium dichromate solutions do not change their titer on standing. The reduction potential of dichromate in acid medium is less than that of permanganate or ceric cerium. In the reduction green chromic ions are formed, and therefore an indicator must be used for the detection of the end point:

$$Cr_2O_7^{--} + 14\,H^+ + 6e \rightleftharpoons 2\,Cr^{3+} + 7\,H_2O$$

Diphenylamine sulfonic acid, diphenylbenzidine, and diphenylamine are most useful; the ferrous-phenanthroline complex is less suitable because its oxidation potential is so high. A practical advantage in the use of dichromate lies in the fact that ferrous iron can be titrated in the presence of hydrochloric acid; the latter is not oxidized provided that the acid concentration is less than 1 or 2 *N*.

Iodine. The reduction potential of iodine is much lower than that of the oxidizing agents mentioned above (see Tables in Appendix); it is of the same order as that of the ferrous-ferric iron system. Iodine can be obtained in a pure state by sublimation, and standard solutions can be prepared by weighing out known amounts of the pure product. Iodine is only slightly soluble in water but dissolves readily in a potassium iodide solution with the formation of triiodide ions:

$$I_2 + I^- \rightleftharpoons I_3^-$$

Iodine is volatile even in potassium iodide solutions; an excess of iodide decreases the volatility because it displaces the equilibrium to the right in the above equation. In using standard iodine solutions, one should be careful in withdrawing the liquid not to expose the solution to the atmosphere for a longer time than is really necessary. Acid iodine solutions cannot be kept in storage because hydrogen iodide is easily oxidized by the air, resulting in an increase of the titer of the iodine solution.

Triiodide solution has an intense yellow color, and therefore titrations of 0.1 *N* or strong solutions can be made without an indicator. As a rule, however, it is preferable to use starch, which forms a blue adsorption compound with iodine of great color intensity, as an indicator. In certain titrations use is made of the solvents chloroform and carbon tetrachloride, in which the iodine has an intense purple color, for the detection of the end point.

[3] G. F. Smith and C. A. Getz, *Ind. Eng. Chem., Anal. Ed.* **10**, 191 (1938).

[4] G. F. Smith, V. R. Sullivan, and G. Frank, *Ind. Eng. Chem., Anal. Ed.* **8**, 449 (1936).

[5] G. F. Smith, *Anal. Chem.* **21**, 1233 (1949).

In determinations in which an excess of iodine is used, the excess being back-titrated after a certain time of standing, closed flasks must be employed to prevent losses of iodine by volatilization. Such determinations are made, as a rule, in glass-stoppered flasks called "iodine flasks."

Potassium iodate. By recrystallization from water and drying at 180°C, potassium iodate is easily obtained pure. It is very useful in the extemporaneous preparation of an acid iodine solution of known concentration:

$$IO_3^- + 5\,I^- + 6\,H^+ \rightarrow 3\,I_2 + 3\,H_2O$$

By adding an excess of iodide and acid to a standard iodate solution, an equivalent amount of iodine is obtained. In strong hydrochloric acid solution iodate is quantitatively reduced to iodine monochloride, according to Andrews.[6] If the concentration of hydrochloric acid at the end point is at least 3 *N*, iodate reacts quantitatively with the following reducing agents according to the equations:

$$2\,KI + KIO_3 + 6\,HCl \rightarrow 3\,KCl + 3\,ICl + 3\,H_2O$$

$$2\,I_2 + KIO_3 + 6\,HCl \rightarrow KCl + 5\,ICl + 3\,H_2O$$

$$As_2O_3 + KIO_3 + 2\,HCl \rightarrow As_2O_5 + ICl + KCl + H_2O$$

$$N_2H_4 + KIO_3 + 2\,HCl \rightarrow KCl + ICl + N_2 + 3\,H_2O$$

$$2\,SO_2 + KIO_3 + 2\,HCl + H_2O \rightarrow 2\,H_2SO_4 + ICl + KCl$$

Instead of using high concentrations of hydrochloric acid, it is possible to work in 0.5 *N* sulfuric acid in the presence of hydrogen cyanide. R. Lang[7] showed that iodine cyanide is then formed:

$$IO_3^- + 2\,I^- + 3\,HCN + 3\,H^+ \rightarrow 3\,ICN + 3\,H_2O$$

Although the "iodine cyanide" method yields excellent results, great care should be taken in its application since hydrogen cyanide is a dangerous poison.

Potassium bromate. This is obtained in a pure state by recrystallization from water and drying at 180°C. Bromate is a strong oxidizing agent in acid media and is reduced to bromide, which finally reacts with the first excess of bromate to form free bromine:

$$BrO_3^- + 5\,Br^- + 6\,H^+ \rightarrow 3\,Br_2 + 3\,H_2O$$

Bromine may be detected by its yellow color; but its presence is shown much more sensitively by azo indicators, such as methyl red (sodium salt in water) and methyl orange. These indicators have a red color in acid medium but are destroyed by free bromine, thus giving a color change from red to pale yellow. The color change is not reversible, because the indicators are decomposed by the bromine. The reaction between bromine and the indicator proceeds rapidly, especially in strongly acid medium and at higher temperatures; because the color change is irreversible, there is a possibility that the bromate present in local excess may react with the indicator before the end point is reached. A fading of the color is often noticed before the end point is reached.

Smith and Bliss[8] tested a great number of dyes with regard to their suitability as indicators in bromate titrations. The following are very useful: Bordeaux (Color Index[9] number 88), chrysoidine R (number 21), naphthol blue black (number 246). They are

[6] L. W. Andrews, *J. Am. Chem. Soc.* **25**, 756 (1903); see especially G. S. Jamieson, *Volumetric Iodate Methods*, Chemical Catalog Co., New York, 1926.

[7] R. Lang, *Z. Anorg. Allgem. Chem.* **122**, 332 (1922); **142**, 229, 279 (1925); **144**, 75 (1925); see also I. M. Kolthoff and V. A. Stenger, *Volumetric Analysis*, vol. II, Wiley-Interscience, New York, 2nd ed., 1947.

[8] G. F. Smith and H. H. Bliss, *J. Am. Chem. Soc.* **53**, 2091 (1931).

[9] Society of Dyers and Colourists, *Colour Index*.

used as 0.1 per cent solution in water. Indigo sulfonic acids may also be used as indicators; the blue color is destroyed by bromine.

In recent years reversible indicators have been found for the detection of the end point. Of special importance is *p*-ethoxychrysoidin (4′-ethoxy-2,4-diaminoazobenzene), an oxidation–reduction indicator with a reduction potential of 0.76 V (from red to faint yellow with ceric sulfate or permanganate), which has been introduced by Schulek et al.[10] In a titration with bromate the red color of the indicator becomes more and more intense as a result of bromination and at the end point the color disappears suddenly as a result of a reversible oxidation.

α-Naphthoflavone has also been recommended as a reversible indicator in titrations with bromate (bromine).[11] It gives a very sharp end point changing from a very pale straw color to a deep brownish orange. Sinn[12] recommends quinoline yellow, which gives a reversible color change from yellow-green to colorless.

Direct titration with bromate is of special advantage in the determination of trivalent arsenic and antimony, of thallous thallium, and of hydrazine in acid medium:

$$BrO_3^- + 3\,As(III) + 6\,H^+ \rightarrow Br^- + 3\,As(V) + 3\,H_2O$$

$$3\,N_2H_4 + 2\,BrO_3^- \rightarrow 2\,Br^- + 3\,N_2 + 6\,H_2O$$

Various substances cannot be oxidized directly with bromate but react quantitatively with an excess of bromine. Acid solutions of bromine of exactly known concentration are obtained from a standard bromate solution by adding an excess of bromide and acid:

$$BrO_3^- + 5\,Br^- + 6\,H^+ \rightarrow 3\,Br_2 + 3\,H_2O$$

Bromine is very volatile, and therefore such operations with an excess of bromine must be carried out in iodine flasks or in tightly stoppered glass bottles at as low a temperature as possible. The excess of bromine is determined iodometrically (addition of potassium iodide and titration with thiosulfate) or by the quick addition of an excess of standard arsenious oxide, which is then back-titrated with standard bromate.

An important use is made of bromate in organic analysis. Substances with a phenol group react with bromine to form brominated substitution products. Ordinary phenol, for example, reacts with an excess of bromine with the formation of tribromophenol bromide, which upon back titration of the excess of bromine (iodometrically or with arsenious oxide) is transformed into tribromophenol:

$$C_6H_5OH + 3\,Br_2 \rightarrow C_6H_2Br_3OH + 3\,H^+ + 3\,Br^-$$

The well-known method of Koppeschaar[13] for the titration of phenols is based on this reaction.

An important use of this method is made in the determination of various metals which form insoluble compounds of definite composition with 8-hydroxyquinoline. 8-Hydroxyquinoline (frequently called *oxine*) yields a dibromo substitution product with bromine[14]:

$$C_9H_7ON + 2\,Br_2 \rightarrow C_9H_5Br_2ON + 2\,H^+ + 2\,Br^-$$

The application of the method for the determination of magnesium is described on page 862.

[10] E. Schulek and P. Rozsa, *Z. Anal. Chem.* **115**, 185 (1939); E. Schulek and Z. Somogyi, *ibid.* **128**, 398 (1948).

[11] F. L. Uzel, *Casopis Cesk. Lékár.* **15**, 143 (1935); E. Schulek, *Z. Anal. Chem.* **102**, 111 (1935); R. Belcher, *Anal. Chim. Acta* **3**, 578 (1949).

[12] V. Sinn, *Analyst* **73**, 236 (1948); R. Belcher, *Anal. Chim. Acta* **5**, 26 (1951).

[13] W. F. Koppeschaar, *Z. Anal. Chem.* **15**, 233 (1876).

[14] R. Berg, *Pharm. Ztg.* **71**, 1542 (1926).

Calcium hypochlorite. One-tenth normal solutions of calcium hypochlorite are quite stable. Hypochlorite is a powerful oxidizing agent and is of special advantage in oxidations carried out in neutral or alkaline medium. It reacts fairly rapidly with various substances which are not at all or only slowly oxidized in acid medium by other common oxidizing agents such as permanganate and ceric sulfate.

In many cases hypobromite reacts faster than hypochlorite; a hypobromite solution of known concentration can be prepared extemporaneously by adding an excess of alkali bromide to a standard hypochlorite solution:

$$OCl^- + \underset{(excess)}{Br^-} \rightarrow OBr^- + Cl^-$$

Usually an excess of hypochlorite (or hypobromite) is used in oxidations. The excess can be determined iodometrically by adding an excess of potassium iodide and acid and titrating the liberated iodine with sodium thiosulfate:

$$OCl^- + 2\,I^- + 2\,H^+ \rightarrow I_2 + Cl^- + H_2O$$

$$2\,S_2O_3^{--} + I_2 \rightarrow S_4O_6^{--} + 2\,I^-$$

For the back titration one can also add an excess of standard arsenious oxide solution and titrate the excess with standard hypochlorite or bromate, using Bordeaux as indicator (see p. 861):

$$2\,OCl^- + As_2O_3 \rightarrow As_2O_5 + 2\,Cl^-$$

Belcher[15] found that tartrazine gives a reversible color change from yellow to colorless with hypobromite; with hypochlorite the color change is not reversible. Tartrazine is a good indicator for the titration of arsenious trioxide with hypobromite. Quinoline yellow gives a reversible color change with both hypochlorite and hypobromite.[16]

In bicarbonate medium ammonia is quantitatively oxidized by hypobromite according to the equation

$$2\,NH_3 + 3\,OBr^- \rightarrow N_2 + 3\,Br^- + 3\,H_2O$$

This reaction is made the basis of an exact method for the determination of ammonia. Urea is oxidized in bicarbonate medium to nitrogen, carbon dioxide, and water:

$$\begin{array}{l} \quad NH_2 \\ \ / \\ CO + 3\,OBr^- \rightarrow CO_2 + N_2 + 2\,H_2O + 3\,Br^- \\ \ \backslash \\ \quad NH_2 \end{array}$$

In alkaline medium sulfide, thiosulfate, and tetrathionate are quantitatively oxidized to sulfate:

$$S^{--} + 4\,OCl^- \rightarrow SO_4^{--} + 4\,Cl^-$$

$$S_2O_3^{--} + 4\,OCl^- + H_2O \rightarrow 2\,SO_4^{--} + 4\,Cl^- + 2\,H^+$$

Nitrite in weakly alkaline medium is oxidized to nitrate:

$$NO_2^- + OCl^- \rightarrow NO_3^- + Cl^-$$

This reaction furnishes one of the best methods for the determination of nitrites.

Finally, it may be mentioned that hydrogen peroxide in weakly alkaline medium behaves as a reducing agent toward hypochlorite or hypobromite. The reaction proceeds quantitatively according to the equation

$$OCl^- + H_2O_2 \rightarrow H_2O + O_2 + Cl^-$$

[15] R. Belcher, *Anal. Chim. Acta* **4**, 468 (1950).

[16] V. Sinn, *Analyst* **73**, 236 (1948); R. Belcher, *Anal. Chim. Acta* **5**, 26 (1951).

Reducing agents

Standard solutions of reducing agents are used in the titration of oxidizing agents. Most of these reducing agents are not air-stable, because they are slowly oxidized by the oxygen of the air. Standard solutions of strong reducing agents such as titanous chloride or sulfate, or chromous chloride, must be protected from the oxygen of the air; and titrations must be made in an inert atmosphere of oxygen-free hydrogen, nitrogen, or carbon dioxide. In contact with the air they are rapidly oxidized, especially during a titration. Titanous chloride and chromous chloride are valuable reagents. On account of the great care required in their use, they are not employed as ordinary laboratory standard solutions but find application in routine analyses.

Ferrous salts. Solutions of ferrous salts are slowly oxidized by air. This air oxidation is relatively rapid in neutral medium but is inhibited by an excess of acid. A maximum stability is found in solutions containing 0.5–1 *N* sulfuric acid. But even such solutions must be standardized daily.

Standard ferrous iron solutions are often used in the determination of oxidizing substances. In many cases the reaction is sufficiently rapid to enable the titration to be made directly to the end point, using one of the suitable oxidation–reduction indicators (p. 759). Acid solutions of ceric sulfate, dichromate, and vanadate can be titrated directly with a ferrous iron solution. In the titration of permanganate a slight excess of ferrous iron is added and the latter back-titrated, because the direct titration does not proceed smoothly. Various oxidizing substances react slowly with ferrous iron; this, for example, is the case with lead peroxide, minium, pyrolusite (MnO_2), and nitric acid, the latter in strongly acid medium:

$$MnO_2 + 2\,Fe^{++} + 4\,H^+ \rightarrow Mn^{++} + 2\,Fe^{3+} + 2\,H_2O$$

$$NO_3^- + 3\,Fe^{++} + 4\,H^+ \rightarrow NO + 3\,Fe^{3+} + 2\,H_2O$$

The latter reaction is catalyzed by molybdate.[17]

In such cases the oxidizing substance to be determined is heated with an excess of standard ferrous iron solution until the reaction is complete. However, solutions of ferrous iron are air-sensitive, especially at higher temperatures; and therefore it is necessary in the above cases to protect the mixtures containing excess ferrous iron from air oxidation. This can be accomplished by displacing the air in the reaction vessel by an inert gas, such as carbon dioxide or oxygen-free nitrogen, and passing the gas through the solution during the entire procedure. As a rule it is possible to get satisfactory results in a much simpler way by attaching a *Contat–Göckel trap* to a flask containing the solution (Fig. 42-1), or by use of the arrangement depicted in Fig. 42-2.

The *Contat–Göckel*[18] *trap* is filled with sufficient water or sodium bicarbonate solution to cover the end of the siphon tube. At the end of the boiling the trap is filled about half full of saturated sodium bicarbonate or 20 per cent potassium bicarbonate solution. As the flask cools, the solution is automatically drawn in until the pressure of the carbon dioxide in the flask is equal to the external pressure; the bicarbonate in the trap protects the contents of the flask from the oxygen of the air. The apparatus of Fig. 42-2 is used in a similar way. The apparatus is also useful in other cases when solutions must be protected from air oxidation after complete reduction [Sn(IV) → Sn(II); Ti(IV) → Ti(III), etc.].

Sodium thiosulfate. Sodium thiosulfate is the most important reagent used in iodometry. Solutions prepared from a pure salt in pure water (conductivity water)

[17] For a volumetric nitrate determination based on the reduction of an excess of ferrous iron, see I. M. Kolthoff, E. B. Sandell, and B. Moskovitz, *J. Am. Chem. Soc.* **55**, 1454 (1933); W. Leithe, *Anal. Chem.* **20**, 1082 (1948).

[18] A. Contat, *Chem. Ztg.* **22**, 298 (1898); H. Göckel, *Z. Angew. Chem.* **13**, 620 (1899).

are quite stable. However, if the water contains an excess of carbon dioxide, a slow decomposition takes place:

$$S_2O_3^{--} + H^+ \rightarrow HSO_3^- + S$$

with the formation of some sulfur.

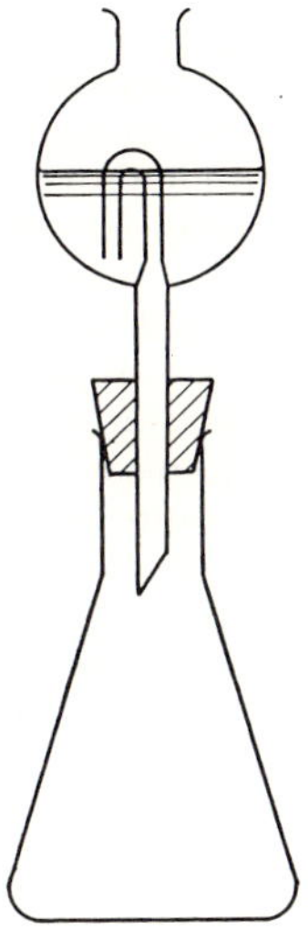

Fig. 42-1. Flask with Contat-Göckel trap.

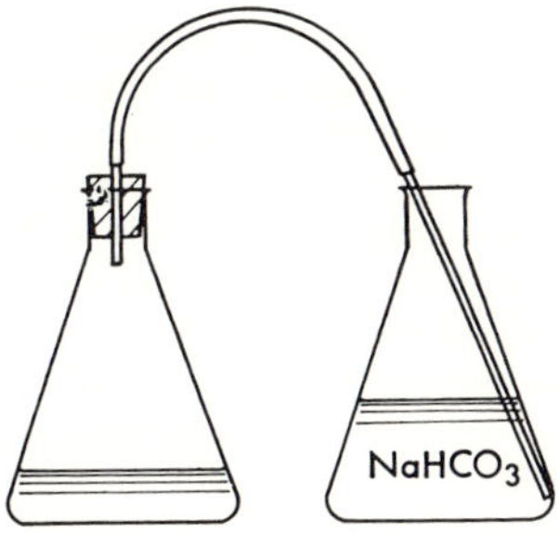

Fig. 42-2. Arrangement for excluding air from an acid solution.

Thiosulfate is slowly oxidized by oxygen of the air; sulfite is slowly formed and is then rapidly oxidized to sulfate:

$$S_2O_3^{--} \rightarrow SO_3^{--} + S \qquad \text{(very slow)}$$

$$SO_3^{--} + \tfrac{1}{2}\,O_2 \rightarrow SO_4^{--} \qquad \text{(measurable)}$$

$$\overline{S_2O_3^{--} + \tfrac{1}{2}\,O_2 \rightarrow SO_4^{--} + S}$$

There is also a possibility of the intermediate formation of tetrathionate. This is especially the case in the presence of copper, which strongly catalyzes the decomposition of thiosulfate:

$$2\,Cu^{++} + 2\,S_2O_3^{--} \rightarrow 2\,Cu^+ + S_4O_6^{--} \qquad \text{(instantaneous)}$$

$$2\,Cu^+ + \tfrac{1}{2}\,O_2 + 2\,H^+ \rightarrow 2\,Cu^{++} + H_2O \qquad \text{(measurable)}$$

Ordinary distilled water often contains traces of copper derived from the storage system. Since the copper has a detrimental effect upon the stability of thiosulfate, it is desirable to use water free of this metal for preparing the solution. In addition, it is best to add 0.01 per cent of sodium carbonate as a preservative to the freshly prepared thiosulfate solution. The sodium carbonate has two functions: It makes harmless any carbon dioxide present in the water (acid decomposition of thiosulfate, see above), and it removes copper ions, if present.

Even with the abovementioned precautions it is necessary to restandardize a thiosulfate solution after it has stood for some time. The entire mechanism of its decomposition is extremely complicated. In addition to these causes of slow decomposition, a deterioration may occur as a result of the growth of microorganisms (thiobacteria). Organisms that consume sulfur apparently exist in the atmosphere; and these can withdraw sulfur from thiosulfate, converting it into sulfite, which in turn is easily oxidized to sulfate by air. If a decomposition by microorganisms occurs, a strong turbidity is usually noticed on long standing, and the titer of the solution decreases fairly rapidly. In such cases it is best to discard the solution and prepare a new one, which is kept in a bottle cleaned with cleaning mixture.

In some texts the stabilization of thiosulfate solutions by the addition of sodium hydroxide is recommended. This procedure should not be followed; under such conditions an increase in the titer of the thiosulfate due to the formation of sulfite and sulfide may occur.

Arsenic trioxide. This substance can be obtained relatively easily in a pure form and is valuable in the standardization of iodine solutions in bicarbonate medium and of ceric sulfate and permanganate in acid medium (p. 827). It also finds use in the titration of bromate or hypochlorite. The trioxide dissolves very slowly in water but readily in sodium hydroxide. A standard solution is usually prepared by dissolving a weighed amount of the pure substance in an excess of sodium hydroxide and adding hydrochloric acid until the reaction is neutral or slightly acid to litmus paper. An alkaline solution of the arsenic trioxide should not be kept, because arsenite is slowly oxidized to arsenate by air under these conditions, thus causing a decrease in the normality. Neutral or weakly acid solutions, on the other hand, are quite stable and can be kept unchanged for long periods of time.

Sodium oxalate and oxalic acid, $H_2C_2O_4 \cdot 2\,H_2O$. Both substances can be obtained in a pure state and are used in the standardization of permanganate and ceric sulfate. In addition, sodium oxalate is an excellent primary standard for the standardization of acids (see p. 780), and oxalic acid for bases. Solutions of oxalates and oxalic acid decrease in strength on standing, especially when exposed to light. A slow decomposition takes place:

$$H_2C_2O_4 \rightarrow CO_2 + CO + H_2O$$

Standard solutions of sodium oxalate and oxalic acid should be kept in the dark, preferably in resistant glass bottles. Solutions of the acid can be kept unchanged under these conditions for more than 1 year; solutions of the sodium salt, however, are much less stable.

Potassium ferrocyanide, $K_4Fe(CN)_6 \cdot 3\,H_2O$. This salt is easily obtained in a pure state by recrystallization from water. It can be kept as the trihydrate over deliquescent sodium bromide, or it can be converted into the anhydrous form by drying at 100°C. Its aqueous solutions are not very stable; an air oxidation takes place and, in addition, a decomposition catalyzed by light. Acids (carbon dioxide in the water) catalyze the decomposition very strongly. It is therefore best to add 0.2 per cent of sodium carbonate to the freshly prepared solution. If kept in a dark bottle, it is stable for more than a month, although restandardization at frequent intervals is recommended. Potassium ferrocyanide is a valuable standard for ceric sulfate. It is also of practical importance in the titration of zinc (for procedure see p. 800).

CHAPTER 43 *Permanganate, Cerium (IV), and Dichromate Methods*

PERMANGANATE

Preparation of 0.1 *N* potassium permanganate

Since distilled water usually contains traces of reducing substances which react slowly with permanganate to form hydrous manganese dioxide, it is necessary to remove these oxidizable substances to prepare a stable potassium permanganate solution, because finely divided manganese dioxide promotes the autodecomposition of permanganate (p. 816). This is most simply accomplished by boiling the solution for some time, cooling, and filtering off the manganese dioxide.

Dissolve about 3.2 g of reagent-grade potassium permanganate in 1 liter of water, heat the solution to boiling, and keep slightly below the boiling point for 1 hr. Alternatively, allow the solution to stand at room temperature for 2 or 3 days. Filter the liquid through pure asbestos (free of organic matter) or through a sintered-glass filter crucible. Transfer the filtrate to a clean glass-stoppered bottle freed from grease with cleaning mixture. Protect the solution from evaporation, dust, and reducing vapors, and keep it in the dark or in diffuse light. If in time manganese dioxide settles out, refilter the solution and restandardize it.

If 0.01 *N* permanganate is required, it is prepared by dilution of 0.1 *N* solution with a good grade of water. The dilute solution should be prepared just before use and should not be stored for any length of time, because of the relatively rapid rate of deterioration.

Standardization. Standard substances

Sodium oxalate, $Na_2C_2O_4$. **Crystallized oxalic acid,** $H_2C_2O_4 \cdot 2\,H_2O$. In acid medium oxalic acid is oxidized by permanganate to carbon dioxide and water:

$$5\,C_2O_4^{--} + 2\,MnO_4^- + 16\,H^+ \rightarrow 10\,CO_2 + 2\,Mn^{++} + 8\,H_2O$$

For the purity of these standard substances compare pages 780 and 783.

Procedure. Weigh out 0.25–0.30 g of pure sodium oxalate, dried at 105–110°C, into a 250-ml Erlenmeyer flask, dissolve in 60 ml of water, add 15 ml of 1:8 sulfuric acid (free from reducing substances), heat to 80–90°C, and titrate slowly with permanganate to the first permanent pink tinge, swirling the solution constantly and being careful to wait until the solution is colorless before adding further amounts of permanganate. The temperature of the solution should be above 60°C at the end point. Calculate the normality of the permanganate.

Notes:

1. The mechanism of the reactions between oxalate and permanganate ions is extremely complicated.[1] The first few drops of permanganate react slowly with the oxalic acid; but after a small amount of manganous salt has been formed, the reaction occurs almost instantaneously in hot solution. It appears that the manganous ions catalyze the reaction between permanganate and oxalic acid. That this is true can be shown by adding a small amount of manganous salt at the very start of the titration; the permanganate is then reduced rapidly from the start.

In the titration of hydrogen peroxide with permanganate, a similar phenomenon is observed. The first few drops of permanganate are decolorized slowly, but again the reaction is catalyzed by adding a trace of a manganous salt at the very beginning of the titration. Permanganate and manganese(II) react to give intermediate valence forms of manganese which react more rapidly with oxalic acid or peroxide than permanganate does.

2. During the titration of oxalic acid in the presence of oxygen an induced oxidation of oxalic acid by oxygen occurs. The same is observed in the titration of oxalic acid with cerium(IV) sulfate or in its reaction with dichromate. The induced reaction is promoted by manganese(II), which forms upon addition of permanganate a complex manganese (III) oxalate. This complex is not stable and decomposes by the following mechanism:

$$\mathrm{Mn(III)} + C_2O_4^{--} \rightarrow Mn^{++} + CO_2 + CO_2^- \quad (1)$$

$$\mathrm{Mn(III)} + CO_2^- \rightarrow Mn^{++} + CO_2 \quad (2)$$

$$2\,\mathrm{Mn(III)} + C_2O_4^{--} \rightarrow 2\,Mn^{++} + 2\,CO_2 \quad (3)$$

Equation (3) represents the overall reaction. Manganese(III) is a one-electron acceptor and oxalate is a two-electron donor. They react with the intermediate formation of a free-radical CO_2^- (or possibly first $C_2O_4^-$, which disproportionates into CO_2 and CO_2^-). The CO_2^- free radical is very unstable and quickly reacts with manganese(III) according to Eq. (2). The CO_2^- free radical is strongly reducing and rapidly reacts with oxygen. Here we are dealing with a one-electron donor (CO_2^-) and a two-electron acceptor. They react with the intermediate formation of an unstable free-radical HO_2, called *perhydroxyl*:

$$CO_2^- + O_2 + H^+ \rightarrow HO_2 + CO_2 \quad (4)$$

HO_2 is unstable; it can oxidize oxalate:

$$HO_2 + C_2O_4^{--} + H^+ \rightarrow H_2O_2 + CO_2 + CO_2^- \quad (5)$$

with formation of the CO_2^- free radical. Equations (4) and (5) represent a *chain reaction*; the CO_2^- formed in reaction (5) is being oxidized in reaction (4). The sum of reactions (4) and (5) yields the overall induced reaction

$$C_2O_4^{--} + O_2 + 2\,H^+ \rightarrow H_2O_2 + 2\,CO_2 \quad (6)$$

H_2O_2 consumes the same number of equivalents of oxidant as oxalate does. Hence the formation of peroxide does not interfere in the titration; it is partially oxidized (and reformed) during the titration and completely oxidized at the end point.

Alternatively, the formation of H_2O_2 can be accounted for by disproportionation of HO_2 formed in reaction (5):

$$2\,HO_2 \rightarrow H_2O_2 + O_2$$

3. Hot solutions of oxalic acid decompose slowly:

$$H_2C_2O_4 \rightarrow CO_2 + CO + H_2O$$

[1] For details see I. M. Kolthoff and V. A. Stenger, *Volumetric Analysis*, vol. III, Wiley-Interscience, New York, 2nd ed., 1957.

This reaction is catalyzed by manganous salts, but it occurs only to a negligibly small extent if the titration is carried out according to the standard procedure. Too much permanganate, on the other hand, may be used as a result of the autodecomposition of permanganate. In hot acid solution the latter is slowly decomposed with evolution of oxygen, and therefore too rapid titration and insufficient stirring of the liquid may lead to error.[2]

4. In the presence of hydrochloric acid, too much permanganate may be used, owing to an induced oxidation of chloride by the reaction between oxalic acid and permanganate. The error is eliminated by the addition of manganous salt. If the titration is made at temperatures above 70°C, hydrochloric acid does not interfere[3] even in the absence of manganous salt.

5. The correction for the excess of permanganate required to give a visible color in the titrated solution may be obtained as follows. Cool the titrated solution to room temperature, add 0.1 g of solid potassium iodide and 1–2 ml of 0.5 per cent starch solution, and titrate slowly with 0.01 N sodium thiosulfate solution to the disappearance of the blue color. Obtain the ratio between the permanganate and thiosulfate solutions by running 2–3 ml of 0.1 N permanganate solution into 50 ml of 1 N sulfuric acid containing 0.1 g of potassium iodide and titrating the liberated iodine with the 0.01 N thiosulfate solution, using starch as indicator.

6. To avoid errors caused by decomposition at higher temperatures, Fowler and Bright[4] recommended the following procedure for the standardization at lower temperatures. They found results which were in close agreement with the theoretical value.

Procedure according to Fowler and Bright. Transfer 0.3 g of sodium oxalate (dried at 105°C) to a 600-ml beaker. Add 250 ml of dilute sulfuric acid (5 H_2SO_4 to 95 water) previously boiled for 10–15 min and then cooled to 27 ± 3°C. Stir until the oxalate has dissolved. Add 39–40 ml of 0.1 N permanganate (theoretical for 0.1 N is 44.73 ml) at a rate of 25–35 ml/min while stirring slowly. Let stand until the pink color disappears. Heat to 55–60°C and complete the titration by adding the permanganate until a faint color persists for 30 sec. Add the last 0.5–1 ml dropwise with particular care to allow each drop to become decolorized before the next is introduced.

If necessary determine the excess of permanganate required to impart a pink color to the solution by comparing with a blank containing the same amount of sulfuric acid at the same temperature as the solution titrated at the end point.

Other standard substances

Arsenious oxide, As_2O_3 (see p. 824)

$$5\,As_2O_3 + 4\,MnO_4^- + 12\,H^+ \rightarrow 5\,As_2O_5 + 4\,Mn^{++} + 6\,H_2O$$

Arsenious oxide is easily obtainable in a pure state (Bureau of Standards product is 99.97 per cent pure). Under the proper conditions the reaction between arsenious oxide and permanganate proceeds smoothly without side reactions. Therefore, *the authors prefer the use of arsenious oxide as a standard substance to that of sodium oxalate.* The titration of arsenious oxide with permanganate in sulfuric acid medium proceeds smoothly at the beginning, but a yellow-green or brown coloration soon appears (complex manganic arsenates) and disappears only slowly, even at high temperatures. If made in hot solution in the presence of hydrochloric acid, the titration may be carried out directly. It is better, however,

[2] See especially the extensive studies of R. S. McBride, *J. Am. Chem. Soc.* **34**, 393 (1912).
[3] G. P. Baxter and J. E. Zanetti, *Am. Chem. J.* **33**, 500 (1905).
[4] R. M. Fowler and H. A. Bright, *J. Res. Natl. Bur. Std.* **15**, 493 (1935).

to add a catalyst to accelerate the reaction between arsenious oxide and permanganate. A trace of iodate or iodide is a suitable catalyst.

Procedure. Weigh out 0.2–0.25 g of pure arsenious oxide, dissolve in 10 ml of 3 *N* sodium hydroxide (free of reducing substances), add 15 ml of 1 : 1 hydrochloric acid, dilute with 50 ml of water, add 1 drop of 0.002 *M* potassium iodate, and titrate with permanganate to the first rose color.

Note: This titration is accurate to within 0.02 per cent. If ferrous phenanthroline is used as indicator, the end point coincides with the equivalence point.[5]

Mohr's salt. $FeSO_4 \cdot (NH_4)_2SO_4 \cdot 6H_2O$. Although Mohr's salt has a high reaction weight and the iron titration with permanganate is accurate and simple, the salt has the disadvantage of usually being contaminated with manganese, zinc, and magnesium. These metals can occur in the form of double ammonium sulfates isomorphous with Mohr's salt, and they cannot then be removed by recrystallization of the salt from water or precipitation from an aqueous solution with alcohol.

Mohr's salt, therefore, cannot be relied upon as a primary standard substance. Considering, however, its high equivalent weight and the simplicity of the procedure of standardization, it is sometimes advantageous to have available a large sample of Mohr's salt of *known activity* toward permanganate and to keep it in well-closed bottles to prevent efflorescence and oxidation of the ferrous iron. The salt should have a uniform green color.

Procedure. Weigh out 1.2 g of the salt, dissolve in 50 ml of 1 *N* sulfuric acid, add 5 ml of 85 per cent phosphoric acid, and titrate to the first permanent pink.

Potassium iodide, KI. This substance can be prepared in a pure state and has been found very useful as a primary standard for silver nitrate,[6] potassium permanganate, and ceric sulfate.[7] The standardization of permanganate is done potentiometrically or according to Andrews[8] in 4 *N* hydrochloric acid (see p. 819) or by the iodocyanide method of Lang[9] (see p. 819).

Titrations with permanganate

Iron. The titration of ferrous iron with permanganate is a simple matter. The solution is acidified with sulfuric acid and titrated directly to the end point. The color change is somewhat sharper if some 85 per cent phosphoric acid is added. The latter forms a colorless complex with ferric iron, and the color change therefore takes place from colorless to pink, whereas the change is from yellow to pinkish in the absence of phosphoric acid. Hydrofluoric acid has the same effect as phosphoric acid; however, it etches the wall of the flask and is therefore not used unless its presence is required. Also, the titration should be made in the absence of air, since hydrofluoric acid accelerates the oxidation of ferrous iron by oxygen. Chloride interferes in the iron titration. It has already been mentioned (p. 765) that the reaction between ferrous iron and permanganate induces the oxidation of hydrochloric acid to hypochlorous acid or chlorine. Therefore, high results are found in the presence of chloride, and the end point is not permanent. Chloride is made practically harmless by the presence of a manganous salt, which is added in the form of the so-called Zimmermann–Reinhardt solution (p. 833), and by slow titration. However, under these conditions the titration is not highly accurate in the presence of *much* hydrochloric acid. The end point is fleeting since an excess of permanganate oxidizes the chloride slowly. Even with the use of the Zimmermann–Reinhardt mixture, it is often advisable to remove most of the hydrochloric acid by evaporation.

[5] I. M. Kolthoff, H. A. Laitinen, and J. J. Lingane, *J. Am. Chem. Soc.* **59**, 429 (1937).
[6] Compare I. M. Kolthoff and J. J. Lingane, *J. Am. Chem. Soc.* **58**, 1524 (1936).
[7] I. M. Kolthoff, H. A. Laitinen, and J. J. Lingane, *J. Am. Chem. Soc.* **69**, 429 (1937).
[8] L. W. Andrews, *J. Am. Chem. Soc.* **25**, 756 (1903).
[9] R. Lang, *Z. Anorg. Allgem. Chem.* **122**, 332 (1922).

The titration of ferrous iron is often required in the indirect determination of strong reducing agents. Solutions of stannous tin, titanous titanium, and cuprous copper, for example, can be titrated *directly* with permanganate in a current of carbon dioxide or nitrogen. These strong reducing substances are easily oxidized by air, and therefore precautions must be taken to protect them from air oxidation. This air oxidation may be prevented by adding an excess of ferric iron to the solution immediately after the reduction of these metals to their lower valence states.

$$2\,Fe^{3+} + Sn^{++} \rightarrow 2\,Fe^{++} + Sn^{4+}$$
$$Fe^{3+} + Cu^{+} \rightarrow Fe^{++} + Cu^{++}$$
$$Fe^{3+} + Ti^{3+} \rightarrow Fe^{++} + Ti^{4+}$$

An equivalent amount of ferrous iron is formed, which may be titrated with permanganate or any other suitable oxidizing agent (potassium dichromate, ceric sulfate) in the presence of air.

The reduction of ferric iron is also involved in the determination of hydroxylamine. In boiling acid solution the latter reduces ferric iron according to the equation[10]

$$2\,NH_2OH + 4\,Fe^{3+} \rightarrow N_2O + 4\,Fe^{++} + H_2O + 4\,H^+$$

Reduction of ferric to ferrous iron. Iron very often occurs in the ferric form or as a mixture of ferric and ferrous iron. If all the iron is to be determined oxidimetrically, the ferric iron must first be quantitatively reduced to the ferrous form. The excess of reducing agent must then be removed, or made harmless before the titration. A number of substances are suitable for the reduction of ferric iron.

1. Hydrogen sulfide and sulfurous acid, the excess of which can be removed by boiling, with protection of the ferrous iron solution against air oxidation (p. 823).
2. Hydrogen with platinum black or platinized platinum gauze as catalyst. This method is hardly ever applied.
3. Metals, such as zinc, cadmium, aluminum, silver (p. 764), and others or their amalgams which quantitatively reduce ferric iron to the ferrous state:

$$2\,Fe^{3+} + Zn \rightarrow 2\,Fe^{++} + Zn^{++}$$

4. Stannous chloride solution, which reduces ferric iron rapidly and the excess of which can be destroyed by oxidation to the stannic form by mercuric chloride:

$$2\,Fe^{3+} + Sn^{++} \rightarrow 2\,Fe^{++} + Sn^{4+}$$
$$Sn^{++} + 2\,HgCl_2 \rightarrow Hg_2Cl_2 + Sn^{4+} + 2\,Cl^-$$

Of the methods mentioned, those involving the use of metals or stannous chloride are generally employed.

Reduction with metals. The Jones reductor. Zinc is the metal most frequently used as reducing agent. It can be purchased in a pure state but should always be tested for the presence of iron. This can be done by dissolving about 10 g in 100 ml of 1 : 5 sulfuric acid and adding 1–2 drops of 0.1 *N* permanganate. The solution should become distinctly colored. Instead of zinc, cadmium and aluminum are often used. All three of these can be used in the form of wire wound in a spiral or a chain which can be easily removed from the solution after the reduction is complete. Cadmium is obtainable in an iron-free state; aluminum as a rule contains a trace of iron and is therefore less suitable. It is possible of course to run a blank on the latter metal, but this is not very convenient. Then one must determine the amount of aluminum dissolved in order to apply the proper correction.

Very pure zinc and cadmium react rather slowly with acids. The presence of traces of other metals such as lead and copper accelerates the reaction markedly. However,

[10] W. C. Bray, M. E. Simpson, and A. A. MacKenzie, *J. Am. Chem. Soc.* **41**, 1363 (1919).

these promote the solution of the metals in acid as well, and large quantities are lost by this simultaneous dissolution process. This can be prevented by the use of amalgamated metals. By treating zinc or cadmium with a dilute solution of a mercuric salt, the metal is covered with a layer of mercury amalgam, since the latter metal is lower in the electromotive series than zinc and cadmium:

$$Hg^{++} + Zn \rightarrow Zn^{++} + Hg$$

Hydrogen has a high overvoltage at a mercury surface which explains the fact that during the reduction of the ferric iron in acid solution little hydrogen is formed.

Reduction with amalgamated zinc is usually carried out in the Jones reductor,[11] which consists of a column of amalgamated granulated zinc contained in a glass tube provided with a stopcock, through which the solution to be reduced may be drawn. Because of the large surface exposed, such a zinc column is much more efficient than pieces of zinc placed in the solution.

A convenient form of the reductor is shown in Fig. 43-1. The tube should be of sufficient length to hold a column of zinc 25–35 cm in height and should have an internal diameter of approximately 2 cm. The zinc column rests on a pad of glass wool which is supported by a perforated porcelain plate; the top of the column should extend to within a short distance of the bottom of the upper reservoir. The outlet tube is connected to a 500-ml suction flask. (The use of an Erlenmeyer flask is not safe because it may collapse under vacuum.)

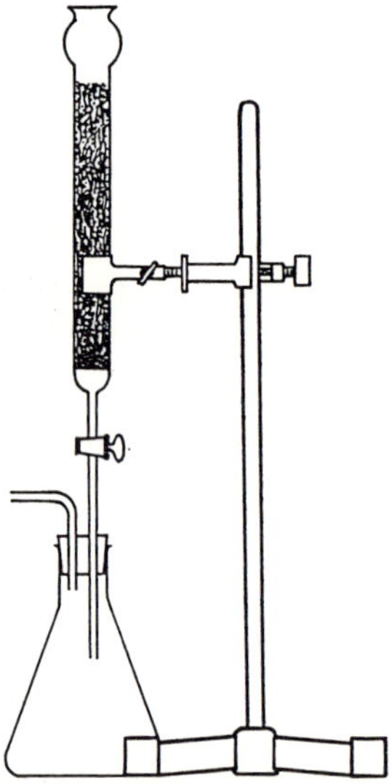

Fig. 43-1. Jones reductor.

Amalgamated zinc for use in the reductor is prepared as follows. Add 300 ml of 2 per cent mercuric nitrate (or chloride) solution and 1–2 ml of concentrated nitric acid to 300 g of *pure* 20- to 30-mesh zinc in a beaker. Stir the mixture thoroughly for 5–10 min, then decant the solution from the zinc, and wash two or three times by decantation. The amalgamated zinc should have a bright silvery luster. Fill the reductor tube with water, then add the zinc slowly until the column is completely packed. Wash with 500 ml of distilled water, using gentle suction; leave the reductor full of water after the washing to prevent the formation of basic salts, which tend to clog the zinc column.

[11] C. Jones, *Trans. AIME*. **17**, 411 (1888–1889); for a discussion of significant factors in efficiency of amalgamated zinc, see H. W. Stone and D. N. Hume, *Ind. Eng. Chem., Anal. Ed.* **11**, 598 (1939).

Procedure for reduction of ferric iron with the Jones reductor. The ferric sulfate solution to be reduced may have a volume of 100–150 ml; it should be about 2 *N* in sulfuric acid and should contain not more than 0.25 g of iron. Wash out the reductor first with 150–200 ml of 1:20 sulfuric acid, added in 30-ml portions; attach a clean flask to the reductor, and, using gentle suction, pass the cold ferric iron solution through the column at a rate not exceeding 75 ml/min. Do not allow the level of liquid in the reductor to fall below the top of the zinc column during the reduction.

When all the iron solution has been drawn into the reductor, wash out the container with three 25-ml portions of 1:20 sulfuric acid, add the washings to the reductor, and draw each portion through before adding the next. Finally, wash out the reductor with three 25- to 35-ml portions of water. The final washing with water is necessary to remove all acid and prevent excessive consumption of zinc. A test for complete reduction of the iron may be made by adding one drop of the reduced solution to a drop of potassium thiocyanate on a spot plate. No red color will be obtained if reduction is complete.

Disconnect the flask from the reductor, wash off the end of the delivery tube, add 25 ml of Zimmermann–Reinhardt solution, and titrate immediately with 0.1 *N* potassium permanganate solution in exactly the same manner as described on page 833.

Run a blank determination by passing 200 ml of 1:20 sulfuric acid through the reductor in exactly the same manner as described above, and titrate the solution with 0.1 *N* permanganate. Subtract the blank thus found from the volume of permanganate used in the actual determination. If the zinc is pure, the correction should be less than 0.05 ml of 0.1 *N* permanganate. It is best to run the blank before passing the solution to be reduced through the reductor to be sure that the reductor is in good working order.

Notes:

1. As regards acidity, solutions 0.5–5 *N* in hydrochloric or sulfuric acid can be used. Sulfuric acid is preferable to hydrochloric acid because the latter may interfere in the subsequent titration (p. 765).

2. Solutions containing compounds of copper, tin, arsenic, antimony, and other reducible metals must not be used. If these substances are present they must be removed before the reduction. Platinum, derived in small amounts from platinum vessels, is quickly reduced to metal in the upper part of the zinc column and hence does not interfere.

3. Nitric acid or nitrates must not be present in the solutions to be reduced, since a more or less complete reduction to hydroxylamine and other compounds takes place. If nitric acid is present, evaporate the solution to dryness, add 3 ml of water and 2 ml of concentrated sulfuric acid, and evaporate to fumes of sulfuric acid. Repeat the evaporation with sulfuric acid a second time to ensure complete removal of the nitric acid, then dilute to 100 ml with water, add 4–5 ml of concentrated sulfuric acid, and proceed with the reduction.

4. Organic compounds, which may consume permanganate, must be absent. These may be destroyed by heating with concentrated nitric and sulfuric acids, finally evaporating several times with sulfuric acid to assure the expulsion of all nitric acid.

5. Other elements which are reduced to a definite stage in the reductor are molybdenum, chromium, titanium, and vanadium. Uranium is reduced mostly to the quadrivalent state but partially to the trivalent; since trivalent uranium is very unstable in air, the overreduction may be easily corrected for by passing a current of air through the solution after reduction. In this way any trivalent uranium is quickly oxidized to the quadrivalent state, in which it is quite air-stable.[12]

[12] G. E. F. Lundell and H. B. Knowles, *J. Am. Chem. Soc.* **47**, 2637 (1925).

The quadrivalent uranium may then be titrated with permanganate in the same manner as ferrous iron.

6. With the exception of iron and uranium (UO^{++}), the reduced solutions are unstable in air, and hence they must be protected from air oxidation after reduction. This is easily done by arranging the receiver so that the end of the outlet tube of the reductor dips below a solution of ferric sulfate or ferric alum contained in the receiver. When the reduced solution flows into the ferric solution, an equivalent amount of ferric iron is reduced to the ferrous state, and the ferrous iron thus formed may be titrated with a suitable oxidizing agent.

7. It has been claimed that hydrogen peroxide may be formed in the column when air is present. However, Lundell and Knowles[13] found that hydrogen peroxide is completely reduced when passed through a Jones reductor. On the other hand, with a less strongly reducing metal than zinc, for example, silver reductor, hydrogen peroxide is formed in the presence of oxygen. This can be avoided by saturating the solution with carbon dioxide before passing it through the column. The washing should also be done with air-free dilute acid.

Reduction with stannous chloride. In hot solution ferric ions are rapidly reduced by stannous ions (added in the form of stannous chloride). The excess of stannous tin can be destroyed by adding mercuric chloride, which is reduced to mercurous chloride. Mercurous chloride is not acted upon appreciably by the oxidizing agents used to titrate ferrous iron and may be left in the solution after the reduction. This method of reducing iron is very often used in the analysis of iron ores.

Great care must be taken not to add more than a very slight excess of stannous chloride in reducing the iron, or else metallic mercury may be formed on the addition of mercuric chloride:

$$Sn^{++} + HgCl_2 \rightarrow Sn^{4+} + Hg + 2\ Cl^-$$

The formation of free mercury (indicated by a gray or black precipitate) ruins the determination, because the metal in a finely divided form is oxidized by permanganate (and ceric cerium and dichromate). The reaction between stannous chloride and mercuric chloride is always allowed to take place in a cold solution, and the mercuric chloride is added as rapidly as possible. The presence of much mercurous chloride in the solution is objectionable, because it may be oxidized somewhat by potassium permanganate or ferric iron; this is another reason for limiting the excess of stannous chloride to the minimum quantity necessary. Detailed directions for the reduction of ferric iron with stannous chloride are given below under the determination of iron in an ore.

Determination of iron in an iron ore. The principal ores of iron are composed of oxides or hydrated oxides of iron:

Fe_2O_3 (hematite)

Fe_3O_4 (magnetite)

$Fe_2O_3 \cdot H_2O$ (goethite); $[Fe_2O_3 \cdot H_2O] \cdot xH_2O$ (limonite)

The carbonate, $FeCO_3$ (siderite), is a minor ore.

Ores consisting of the anhydrous oxides, Fe_2O_3 and Fe_3O_4, dissolve only slowly in acids, the others rather easily. Concentrated hydrochloric acid is the best acid to use in effecting solution. Sulfuric acid and oxidizing acids such as nitric acid are not suitable.

[13] G. E. F. Lundell and H. B. Knowles, *Ind. Eng. Chem.* **16**, 724 (1924).

The addition of stannous chloride greatly aids in the solution of hematite and magnetite ores. It is very important to employ a finely ground sample to hasten solution. A completely decomposed ore will give a residue of silica that is white or nearly so. If any dark material remains after the acid treatment, it is necessary to fuse the washed residue with sodium carbonate and treat the fusion product with hydrochloric acid to recover all the iron.

The minor constituents of iron ores include: silica, aluminum, calcium, magnesium, manganese, the alkalies, phosphorus, sulfur, titanium, vanadium, chromium, and other elements as well. Organic matter is sometimes present. Of these, it is necessary in a volumetric determination of iron to consider the possible effect of titanium, vanadium, chromium, and organic matter. Titanium does not interfere if the iron is reduced with stannous chloride but does do so if the reduction is effected by metals, e.g., the Jones reductor. The use of the latter is generally inadmissible in the analysis of iron ores because titanium is usually present, sometimes in quite large amounts. Vanadium is reduced by stannous chloride to vanadyl ion, and the latter is not oxidized by mercuric chloride and therefore remains to be oxidized by potassium permanganate. If the titration is made with potassium dichromate, using diphenylamine sulfonic acid as an internal indicator instead of potassium permanganate solution, there is no interference from vanadium at the proper acidity. Chromium is not often met with in appreciable quantities in iron ores; it is not reduced by stannous chloride and therefore does not interfere. Organic matter is easily destroyed by treating the acid solution with potassium chlorate or permanganate before adding stannous chloride or by igniting the sample before dissolving.

Procedure. Special solutions required:

Stannous chloride. 150 g of iron-free $SnCl_2 \cdot 2\,H_2O$ in 1 liter of 1:2 hydrochloric acid. This solution should be freshly prepared.

Mercuric chloride. 5 per cent solution of $HgCl_2$ in water.

Zimmermann–Reinhardt solution. Dissolve 70 g of $MnSO_4 \cdot 4\,H_2O$ in 500 ml of water, add with stirring 125 ml of concentrated sulfuric acid and 125 ml of 85 per cent phosphoric acid, and dilute to approximately 1 liter.

Transfer 0.3–0.35 g of finely ground air-dried ore to a 250-ml beaker[14]; add 10 ml of concentrated hydrochloric acid and 3 ml of stannous chloride solution, and heat on the steam bath until the iron has been dissolved as indicated by a *white residue.* If any dark particles remain, add a few drops of concentrated nitric acid and continue the digestion, adding more nitric acid a little at a time until the residue is entirely white or it is seen that nitric acid is without effect. If the addition of nitric acid was necessary, evaporate the liquid to dryness, and heat for a short time at 110°C to ensure the expulsion of the acids. Take up the residue in 10 ml of 1:1 hydrochloric acid. The presence of any dark particles at this point calls for fusion of the residue with sodium carbonate.[15] Proceed as follows. Filter off the insoluble material, using a small filter paper, and wash the residue with dilute hydrochloric acid. Place the paper and its contents in a porcelain crucible,[16] burn off the paper, ignite the residue, and fuse it with 0.5 g of sodium carbonate. Take up the melt in a small excess of hydrochloric acid and add the solution, with any suspended material, to the main iron solution.

Dilute or concentrate the iron solution, as the case requires, to 15 or 20 ml. Not much more than the equivalent of 5 ml of concentrated hydrochloric acid should be

[14] If it is known, or there is reason to suspect, that the ore contains organic matter, ignite it, after weighing, at low redness in an uncovered porcelain crucible and then transfer quantitatively to a beaker.

[15] In the most accurate analyses the residue should be fused with sodium carbonate even if it does not appear to contain any iron.

[16] Platinum crucibles are not to be used in the sodium carbonate fusion in this case because the small amount of platinum thus introduced would interfere subsequently in the reduction of the ferric iron with stannous chloride.

present. Add dropwise a 2 or 3 per cent solution of potassium permanganate until a yellow color is produced. Heat the solution nearly to boiling, and add the stannous chloride solution carefully from a pipet drop by drop with stirring, until the yellow color of the ferric ions has disappeared, and then *1 or 2* drops more. Cool the solution to room temperature (not over 25°C), and add 10 ml of mercuric chloride solution all at once. The precipitate obtained should be white and silky, and *small in amount*[17]; if it is gray, indicating the presence of finely divided elemental mercury, the solution must be rejected. Wait 2 or 3 min, not longer, and then pour the solution into a 600-ml beaker or a large casserole containing 25 ml of Zimmermann–Reinhardt solution and 300 ml of water. Titrate slowly with 0.1 *N* potassium permanganate solution, not adding the standard solution more rapidly than the drops can be counted. The end point is indicated by a faint permanent pink coloration. Report the percentage of iron in the sample.

In the most accurate work a blank determination should be carried through all the steps of the procedure.

Further titrations with permanganate

Nitrite. Nitrite does not react with permanganate in neutral or alkaline medium; in warm acid medium it is oxidized quantitatively to nitrate:

$$5\,NO_2^- + 2\,MnO_4^- + 6\,H^+ \rightarrow 5\,NO_3^- + 2\,Mn^{++} + 3\,H_2O$$

If an acidified nitrite solution is slowly titrated with permanganate, poor results are obtained because nitrous acid is volatile and unstable; and in addition an air oxidation takes place giving rise to the formation of nitrate. It is better therefore to add the nitrite solution from a buret to a measured volume of acidified standard permanganate solution until the latter is decolorized. This is the classical procedure of Lunge[18]; it is not very accurate, since nitrite does not react instantaneously with the permanganate and losses may occur by volatilization or decomposition of the nitrous acid before it has reacted with the oxidizing agent.

It is better to use the second procedure described below, in which the nitrite solution is mixed with an excess of permanganate, the mixture acidified, and the excess of permanganate titrated back iodometrically or according to any of the other methods mentioned for the titration of permanganate (see p. 828).

Procedure 1 (of Lunge). Prepare an approximately 0.1 *N* solution of the nitrite. Pipet 25 ml of 0.1 *N* potassium permanganate solution into a 500-ml flask, add 25 ml of 1 : 5 sulfuric acid and 250 ml of water, heat to 40°C, and titrate with the nitrite solution. Add the nitrite very slowly with constant stirring toward the end of the titration when the solution has a pale violet color. The accuracy of the titration is usually about 0.5–1 per cent.

Procedure 2. Pipet 50 ml of 0.1 *N* potassium permanganate solution into a glass-stoppered flask, add 5 ml of 1 : 5 sulfuric acid and 25 ml of approximately 0.1 *N* nitrite solution. Stopper the flask, shake frequently, and after 15 min standing add 2 g of potassium iodide and titrate with standard thiosulfate solution (p. 846).

Hydrogen peroxide. In acid solution, hydrogen peroxide reduces permanganate, oxygen being formed:

$$5\,H_2O_2 + 2\,MnO_4^- + 6\,H^+ \rightarrow 5\,O_2 + 2\,Mn^{++} + 8\,H_2O$$

The first drops of permanganate are decolorized slowly, but afterwards the reaction runs smoothly to the end point (see p. 825). If a little manganous salt is added at the beginning, the first few drops of permanganate are reduced at once.

[17] If no precipitate appears, sufficient stannous chloride was not added in the reduction of the iron, and the sample must be discarded.

[18] G. Lunge, *Z. Angew. Chem.* **4**, 629 (1891); *Chem. Ztg.* **26**, 501 (1904).

Procedure. Weigh out 3 g of commercial hydrogen peroxide solution (about 3 per cent H_2O_2). Transfer the sample to a 250-ml volumetric flask, and dilute with water to the mark. Pipet out 25 ml, add 10 ml of 1 : 5 sulfuric acid, and titrate with 0.1 *N* permanganate solution.

Notes:

1. Hydrogen peroxide solutions are not stable over a long period of time. Small amounts of organic substances, such as acetanilide, urea, and uric acid, are often added to commercial products to render them more stable. Since these organic substances, with the exception of urea, consume permanganate, the above method does not give accurate results in their presence. Under such circumstances it is better to carry out the titration with ceric sulfate or to resort to the iodometric method (p. 854).

2. Alkali and alkaline earth peroxides, percarbonates, and perborates can be titrated in the same way as hydrogen peroxide, if they are dissolved in dilute acid.

Determination of available oxygen in higher oxides of manganese and lead. The higher oxides of lead, such as lead dioxide, PbO_2, minium, Pb_3O_4, and the higher manganese oxides (e.g., pyrolusite, MnO_2) may be determined by reducing with an excess of standard arsenious oxide or oxalic acid solution, and after solution back-titrating the excess of the latter with standard permanganate solution or any other suitable oxidizing agent. The use of arsenious oxide is to be preferred to that of oxalic acid, because the latter slowly decomposes at higher temperatures to carbon dioxide and carbon monoxide (p. 826). This decomposition is catalyzed by manganous salts, which are formed during the reduction of the pyrolusite. In technical laboratories an acid ferrous sulfate solution is often used as the reducing agent. Good results are obtained if the hot solution is protected from the air during the reaction:

$$2\,MnO_2 + As_2O_3 + 4\,H^+ \rightarrow 2\,Mn^{++} + As_2O_5 + 2\,H_2O$$

$$MnO_2 + H_2C_2O_4 + 2\,H^+ \rightarrow Mn^{++} + 2\,H_2O + 2\,CO_2$$

$$MnO_2 + 2\,Fe^{++} + 4\,H^+ \rightarrow Mn^{++} + 2\,Fe^{3+} + 2\,H_2O$$

Procedure for determination of available oxygen in pyrolusite. Boil 0.2 g of the very finely powdered sample of pyrolusite in an Erlenmeyer flask with 50 ml of 0.1 *N* arsenious oxide solution (p. 848) and 10 ml of concentrated sulfuric acid to complete decomposition as indicated by the disappearance of brown or black particles. Cool the solution, add 100 ml of water, 5 ml of hydrochloric acid, and 1 drop of 0.002 *M* potassium iodate solution (catalyst, see p. 838), and titrate with permanganate. Standardize the arsenious oxide solution by running a blank, following the steps of the procedure. In the analysis of PbO_2 or Pb_3O_4, 0.5-g samples are taken, and 10 ml of concentrated hydrochloric acid is used instead of sulfuric acid. A condenser should be used during the heating to prevent volatilization of $AsCl_3$.

Determination of calcium after precipitation as calcium oxalate. Metals which give slightly soluble oxalates may be determined permanganimetrically by dissolving the washed precipitate in hot dilute sulfuric acid and titrating with standard potassium permanganate solution. This method is very often used for the determination of calcium.

The precipitation of calcium as oxalate has already been considered (p. 619). It is only necessary here to call attention to the fact that if the precipitate is formed by adding ammonium oxalate to a neutral or ammoniacal solution of calcium, the calcium oxalate obtained is contaminated with basic oxalate or calcium hydroxide, and the results of the permanganate titration will be low, sometimes to the extent of 1 per cent. If the precipitation is made as described under the gravimetric determination of calcium by the addition of oxalate to a dilute hydrochloric acid solution of the element, followed by the neutralization of the acid with dilute ammonia, there is obtained a precipitate of nearly the theoretical composition. A small amount of coprecipitated ammonium oxalate and probably oxalic acid is present in precipitates so formed, so that the results are high by approximately 0.2 per cent, if the volume of wash water used is restricted to the smallest necessary amount, to avoid solubility losses.[19] The accuracy of the volumetric method compares favorably with that of the gravimetric, and the former is more rapid when many samples are to be analyzed.

The precipitation of calcium may be carried out as described in Chapter 28, a single or double precipitation being made as the nature of the sample and the required accuracy demand. In addition, we give below an alternative procedure, with special reference to the determination of calcium in limestone or dolomite.[20] A single precipitation of calcium oxalate is made from acidic medium. As much as 100 mg of Fe_2O_3, 100 mg of Al_2O_3, and 500 mg of P_2O_5 do not appreciably affect the determination of about 100 mg of CaO. Manganese produces high results and not more than 1–2 mg may be present. Titanium also causes high results and its amount should not exceed 5 mg. The good results obtainable by this simple method[21] are in part the result of compensation of errors (solubility loss versus coprecipitation of magnesium and ammonium oxalates).

Procedure (for interfering substances, see p. 817). Weigh out duplicate samples containing 0.07–0.09 g of calcium, and precipitate calcium oxalate as described on page 619. Allow the precipitate to stand for 1 hr, and then filter it off in a glass or porcelain filter crucible. Wash crucible and precipitate with small portions of cold water until the washings are chloride-free, avoiding the use of more water than is really necessary. Dissolve the precipitate by pouring 1 : 8 sulfuric acid[22] into the crucible, allowing the acid to remain in the crucible for a minute or two before drawing it into the suction flask.[23] Solution of the calcium oxalate is hastened by stirring it up in the liquid with a short stirring rod kept in the crucible. Continue drawing hot dilute sulfuric acid through the crucible until all the precipitate is dissolved. Not more than 50 ml should be required. Examine the filter crucible carefully at the end, to be sure that no precipitate remains undissolved. Dilute the solution with water until the sulfuric acid concentration is 1–1.5 N, heat to 80°C, and titrate slowly with standard potassium permanganate to the first pink tinge (for correction for excess see p. 827).

Determination of calcium in limestone or dolomite. To 0.3–0.4 g of finely ground sample[24] (0.12 g of CaO) in a covered 250-ml beaker, add 5 ml of water and 10 ml of concentrated hydrochloric acid. Warm until decomposition is complete.

[19] H. Bassett, *J. Chem. Soc.* **1934**, 1273, recommends the use of saturated calcium oxalate solution for washing.

[20] J. J. Lingane, *Ind. Eng. Chem., Anal. Ed.* **17**, 39 (1945).

[21] Bureau of Standards dolomite 88, containing 30.49 per cent CaO, gave 30.53 per cent, and Bureau of Standards limestone 1a containing 41.32 per cent CaO gave an average of 41.24 per cent.

[22] The sulfuric acid solution used should be tested for reducing substances with permanganate.

[23] It is convenient to receive the acid solution directly in a 250-ml flask so that it will not be necessary to transfer the solution before the titration.

[24] To destroy organic matter and to render more easily soluble in acid the calcium in siliceous material the sample should be ignited strongly in a platinum crucible for 5–10 min if its composition calls for this special treatment.

Dilute to 50 ml, heat almost to boiling, and add 100 ml of hot, filtered 5 per cent ammonium oxalate solution. Add a few drops of methyl orange indicator solution and precipitate calcium oxalate (at an initial temperature of about 80°C) by dropwise addition of 1 : 1 ammonia over a period of 5–10 min with continuous stirring. Stop the addition of ammonia at a pH of 4.0, corresponding to a yellowish pink color (a pH between 4.4 and 4.5 is permissible). Allow the solution to stand for 30 minutes.

Collect the precipitate in a filter crucible (a Gooch crucible is satisfactory) and wash the beaker, crucible, and precipitate with 8 or 10 small portions of cold (preferably ice-cold) water. Dissolve the precipitate in dilute sulfuric acid and titrate with permanganate as described in the previous procedure.

CERIC SULFATE

Use of ceric sulfate

The advantages of ceric sulfate over potassium permanganate have already been mentioned (p. 817). Ceric sulfate may be used in most titrations in which permanganate is employed.

Sulfuric acid solutions of pure ceric sulfate are stable even at boiling temperatures. Hydrochloric acid solutions of the salt are unstable, owing to reduction of the ceric to cerous cerium by the hydrochloric acid, with simultaneous liberation of chlorine:

$$2\,Ce^{4+} + 2\,Cl^- \rightarrow 2\,Ce^{3+} + Cl_2$$

This reaction takes place quite rapidly at boiling temperature, and hence hydrochloric acid cannot be used in oxidations which necessitate boiling with an excess of ceric sulfate in acid solution. In such oxidations sulfuric acid must be used. However, direct titrations with ceric sulfate in hydrochloric acid medium may be successfully performed, even in hot solution. For example, oxalic acid may be titrated with ceric sulfate in hot hydrochloric acid solution, and hydrochloric acid solutions of ferrous iron may be accurately titrated at room temperature. In this respect, ceric sulfate distinguishes itself very favorably from permanganate.

It should be mentioned that the presence of hydrofluoric acid is very harmful in ceric sulfate titrations. Fluoride forms a stable complex with ceric ions and decolorizes the yellow solution; with much fluoride at low acidities a precipitate of ceric fluoride is formed. Cerous salts also form complexes with fluoride, but these are less stable than the ceric complexes. Consequently, fluoride decreases the reduction potential of the ceric–cerous system very strongly. A dilute sulfuric acid solution of ceric sulfate in the presence of much fluoride does not oxidize iodide for this reason.

Since ceric phosphate is only slightly soluble, a precipitate is obtained in dilute sulfuric acid solutions containing phosphate; but the precipitate is soluble in stronger acid. Ceric sulfate is always used in acid solution, since ceric hydroxide is a weak base, insoluble in water, and precipitates from neutral or weakly acid solutions as a result of hydrolysis.

The ferrous–phenanthroline complex is an excellent indicator in many titrations with ceric sulfate; diphenylamine sulfonate may also be used in certain titrations, for example, that of ferrous iron. In titrations of colorless solutions with 0.1 *N* ceric sulfate, the end point can be detected by the yellow color of the reagent itself (ceric ions are yellow, cerous colorless) without the use of a special indicator; the detection of the end point in this way is more satisfactory in hot than in cold solutions because of an increase in the intensity of the yellow color of ceric ions with a rise in temperature. An appreciable blank correction must be applied when the reagent serves as its own indicator, and for this reason a special indicator should be used whenever possible.

Preparation of 0.1 *N* standard ceric ammonium sulfate

Solutions of ceric sulfate are most conveniently prepared from ceric ammonium sulfate, a salt that is commercially available in a fairly pure state. The salt has the composition $Ce(SO_4)_2 \cdot 2(NH_4)_2SO_4 \cdot 2\,H_2O$. An approximately 0.1 *N* solution is prepared as follows. Slowly add 30 ml of concentrated sulfuric acid to 500 ml of water with constant stirring, then add 63 g of ceric ammonium sulfate dihydrate, and stir until solution is complete. Cool to room temperature, filter the solution if turbid, and dilute to 1 liter.

The commercial salt often contains traces of phosphate, which give rise to a slow precipitation of a small amount of ceric phosphate on standing. The rate of precipitation of ceric phosphate from the strongly acid solution is very slow, so that no errors result if the solution is used soon after standardization. However, if the solution is to be kept for some time, it should be allowed to stand for 2 weeks or longer and then filtered and standardized. A solution prepared in this way is stable for a long time.

Solutions of ceric sulfate may be standardized against sodium oxalate in hot hydrochloric or sulfuric acid solution, but the end point must then be determined without a special indicator or by the potentiometric method, because the oxidation–reduction indicators available at present are all oxidized before the equivalence point is reached. For this reason the method cannot be recommended for highly accurate work because of the appreciable blank correction that must be applied when the reagent serves as its own indicator. It is better, therefore, to standardize against arsenious oxide, pure electrolytic iron, or potassium ferrocyanide.

If arsenious oxide is used, it is necessary to add a trace of osmium tetroxide or, less preferably, iodine monochloride as catalyst because otherwise the reaction between arsenious oxide and ceric sulfate is extremely slow. When osmium tetroxide is used as catalyst, the reaction is rapid at room temperature, and an excellent color change is obtained at the equivalence point with the ferrous-phenanthroline complex as indicator. With iodine monochloride as catalyst the reaction is somewhat slower than with osmium tetroxide, but accurate results are obtained if the titration is finished at 50°C in the presence of sufficient hydrochloric acid.[25] It should be mentioned that when osmium is used as catalyst, the titration must be made in dilute sulfuric acid because hydrochloric acid interferes seriously.

Standardization against arsenious oxide[26]

Weigh out 0.15–0.22 g of pure dry arsenious oxide into a 250-ml Erlenmeyer flask, and add 15 ml of 2 *N* sodium hydroxide; warm gently to hasten solution. When the sample has completely dissolved, cool to room temperature, and add 25 ml of 1 : 5 sulfuric acid. Dilute to 100 ml; add 3 drops of 0.01 *M* osmium tetroxide (0.25 g of OsO_4 in 100 ml of 0.1 *N* sulfuric acid) as catalyst and 1 drop of ferrous-phenanthroline indicator. Titrate with 0.1 *N* ceric ammonium sulfate solution until the reddish-orange color changes sharply to colorless or very pale blue.

Note: If iodine monochloride is used as catalyst, proceed as follows. Prepare the iodine monochloride solution by dissolving 0.0575 g of potassium iodate and 0.0890 g of potassium iodide in 125 ml of water and adding all at once 125 ml of concentrated hydrochloric acid. Cool to room temperature and adjust the solution as follows. Add 5–10 ml of chloroform and titrate very carefully with either dilute potassium iodate or dilute potassium iodide, as the case may require, until the chloroform layer shows only a very faint iodine color after vigorous shaking.

Dissolve the sample of arsenious oxide as described above, and add 20 ml of concentrated hydrochloric acid. Dilute to 100 ml; add 2.5 ml of the iodine monochloride

[25] H. H. Willard and P. Young, *J. Am. Chem. Soc.* **55**, 3260 (1933).
[26] K. Gleu, *Z. Anal. Chem.* **95**, 305 (1933).

solution and 1 drop of ferrous-phenanthroline indicator. Titrate with the ceric sulfate solution until the reddish-orange color begins to fade and returns only slowly after the addition of each drop of oxidizing agent. Heat the solution to 50°C (use a thermometer), and continue the dropwise addition of ceric sulfate until the color is discharged and there is no return of the orange color of the indicator within 1 minute.

Determination of the ferrocyanide content of commercial potassium ferrocyanide

Weigh out a 1.0- to 1.2-g sample of commercial potassium ferrocyanide and dissolve in 80 ml of water. Add 20 ml of 1 : 5 sulfuric acid and 1 drop of ferrous-phenanthroline indicator. Titrate with 0.1 N ceric ammonium sulfate solution until the color changes sharply from orange to pure yellow.

Calculate and report the percentage of $K_4Fe(CN)_6$ in the sample.

Notes:

1. Instead of using the ferrous-phenanthroline complex as indicator, 2–3 drops of a 2 per cent solution of ferric chloride may be added to the solution.[27] The greenish coloration due to ferric ferrocyanide disappears sharply at the equivalence point in consequence of the oxidation of ferro- to ferricyanide, and the color of the solution then changes to pure yellow.

2. The above method cannot be applied if the impure potassium ferrocyanide contains organic material, thiocyanate, or other substances oxidizable by ceric sulfate.

POTASSIUM DICHROMATE

Use of potassium dichromate

Potassium dichromate is obtained pure by twice recrystallizing a reagent-grade product from water, followed by drying of the pulverized crystals at 150–200°C in an electric oven.

Solutions of potassium dichromate in water are stable indefinitely; even when acidified with dilute sulfuric, perchloric, or hydrochloric acids, there is no reduction on boiling. On boiling with more concentrated hydrochloric acid, however, a reduction takes place with the simultaneous evolution of chlorine. At room temperature, titrations can be made with dichromate in the presence of hydrochloric acid if the concentration of the acid is less than about 1–2 N. This fact is of great importance in the titration of ferrous iron, because hydrochloric acid interferes in this case when permanganate is used. Dichromate solutions are also less easily reduced by organic matter than is permanganate.

The best indicator for the detection of the end point is sodium diphenylamine sulfonate, although diphenylamine and diphenylbenzidine can also be used. The ferrous-phenanthroline complex is less suitable, as has already been mentioned (p. 758).

Preparation of 0.1000 N standard potassium dichromate solution

Weigh out 4.903 g of pure dry potassium dichromate, dissolve in water, and dilute to exactly 1 liter.

Titration of ferrous iron

If ferrous iron is titrated in the presence of hydrochloric or sulfuric acid, any of the diphenylamine compounds being used as indicator, the color change is not sharp and occurs too early, because the indicator is partially oxidized before the end point is reached (p. 757). However, a sharp color change is produced at the equivalence point if phosphoric acid is added to the solution. Phosphoric acid lowers the reduction potential of the ferric-ferrous iron system by forming a complex with the ferric ions, which explains its beneficial effect when using the above indicators. According to Smith

[27] N. H. Furman and O. M. Evans, *J. Am. Chem. Soc.* **51**, 1128 (1929).

and Brandt[28] a sharp end point is also obtained with 5,6-dimethyl-1,10-phenanthroline (see Table in Appendix) in 2 *M* hydrochloric or sulfuric acid.

The indicator blanks are negligibly small in titrations with 0.1 *N* solutions.

Determination of iron in an iron ore

Indicator solution: A 0.2 per cent solution of sodium diphenylamine sulfonate in water.

Procedure. Bring the sample into solution, and reduce the iron as described under the permanganate method (p. 829). After the reduction with stannous chloride and addition of mercuric chloride, add 200 ml of water, 10 ml of 1:5 sulfuric acid, 5 ml of 85 per cent phosphoric acid, and 6–8 drops of indicator. Titrate slowly with 0.1 *N* standard dichromate solution, stirring constantly, until the pure green color changes toward a gray-green. Then add the dichromate very slowly, one drop at a time, until the first tinge of purple or violet-blue appears. Run a blank determination with the reagents and make the proper correction, if necessary. Calculate the percentage or iron.

Notes:

1. The color change of the indicator is reversible. Therefore, if the solution is accidentally overtitrated, the excess oxidizing agent can be back-titrated with a standard ferrous iron solution. The oxidized form of the indicator is slowly destroyed by an excess of oxidizing agent, and the back titration should therefore be made quickly.

2. The reverse titration of dichromate with ferrous sulfate also gives good results; use is made of it in the determination of chromium in chromium steels.

3. Vanadic acid can be titrated in the same way as chromic acid:[29]

$$VO^{3-} + Fe^{++} + 6\,H^+ \rightarrow VO^{++} + Fe^{3+} + 3\,H_2O$$

Titration of uranium and its application to the volumetric determination of sodium

It has already been mentioned (p. 762) that sexivalent uranium (uranyl) is reduced to a mixture of tri- and quadrivalent uranium in the Jones reductor. After aeration to oxidize trivalent uranium to the quadrivalent state, the solution may be titrated with potassium dichromate, using ferric iron as a catalyst[30] and either diphenylamine, diphenylbenzidine, or diphenylamine sulfonate as indicator.

$$3\,UO^{++} + Cr_2O_7^{--} + 8\,H^+ \rightarrow 3\,UO_2^{++} + 2\,Cr^{3+} + 4\,H_2O$$

Potassium permanganate or ceric sulfate may also be used for titrating reduced uranium solutions.

The titration of uranium may be used for a volumetric sodium determination by precipitating the sodium as $NaZn(UO_2)_3(CH_3COO)_9 \cdot 6\,H_2O$, as in the gravimetric determination of sodium (p. 667). The precipitate is dissolved in dilute sulfuric acid, the solution is reduced in the Jones reductor, and the reduced solution is titrated with standard dichromate solution. Since the ratio of uranium to sodium in the precipitate is constant, the result of the uranium titration can be used to calculate the amount of sodium present. This volumetric method is of practical importance in the determination of very small quantities of sodium (less than 1 mg) when the amount of precipitate is too small to allow accurate weighing.

[28] G. F. Smith and W. W. Brandt, *Anal. Chem.* **21**, 948 (1949).

[29] N. H. Furman, *Ind. Eng. Chem.* **17**, 314 (1925).

[30] In the absence of iron the end point is not good, evidently because the reaction between uranous ions and dichromate is relatively slow. Ferric iron, however, reacts very quickly with uranous ions,

$$UO^{++} + 2\,Fe^{3+} + H_2O \rightarrow UO_2^{++} + 2\,Fe^{++} + 2\,H^+$$

and thus accelerates the titration.

Procedure. Precipitate, filter, and wash the sodium uranyl zinc acetate from a suitable weight of sample, as described in the gravimetric procedure (p. 667). The weighed precipitate from the gravimetric procedure may be used.

Place the filter crucible containing the triple salt in a Gooch funnel attached to a 250-ml suction flask. Pour 50–75 ml of 2 *N* sulfuric acid slowly through the crucible in several small portions, using gentle suction; then wash out the crucible and funnel with water. Dilute the solution to 100 ml, if not already at this volume, and add sufficient 6 *N* sulfuric acid to make the total acid concentration approximately 2 *N*. Reduce the solution in a Jones reductor in the same manner as described under the determination of iron (p. 839). The reduced solution should have the dark olive-green color characteristic of trivalent uranium. Place a long glass tube in the solution, and bubble a rapid stream of air through the liquid for 5–10 min; this treatment will oxidize all trivalent uranium to the quadrivalent state, and the solution will assume the clear green color of quadrivalent uranium. Wash off the tube with a small amount of water from a wash bottle; add 40 ml of 2 per cent ferric chloride solution, 15 ml of 85 per cent phosphoric acid, and 6–8 drops of diphenylamine sulfonate indicator. Titrate immediately with standard 0.1 *N* dichromate in the same way as described under the titration of ferrous iron (p. 839).

PROBLEMS

1. A 0.2250-g sample of a mixture consisting only of iron and ferric oxide (Fe_2O_3) was brought into solution, and the iron was reduced to the divalent state and titrated with potassium permanganate. If 37.50 ml of 0.0991 *N* permanganate were required, calculate the percentages of Fe and Fe_2O_3 in the sample.
Ans. 74.1 per cent Fe; 25.9 per cent Fe_2O_3.
2. 25.00 ml of a solution containing cobalt and nickel was treated with an excess of potassium nitrite in the presence of acetic acid to precipitate cobalt. After washing, the precipitate was dissolved in 50 ml of 0.0990 *N* permanganate. Then 20.00 ml of 0.1010 *N* ferrous iron solution was added, and the excess ferrous iron was titrated with 0.0990 *N* potassium permanganate, 12.37 ml of the latter being required. Calculate the weight of the cobalt in the solution. Given equations.
Ans. 0.0222 g of Co.
3. Outline methods for the determination of ferrous, ferric, and total iron in a mixture of divalent and trivalent iron.
4. Manganous manganese is quantitatively oxidized by permanganate in hot neutral or weakly acid medium according to the equation

$$3\,Mn^{++} + 2\,MnO_4^- + 2\,H_2O \rightarrow 5\,MnO_2 + 4\,H^+$$

If zinc sulfate or a mixture of the latter with zinc oxide is added to prevent adsorption of manganous ions by the manganese dioxide, the end point is indicated by the appearance of a pink color in the supernatant liquid after the precipitate has settled. In such a titration it was found that 35.70 ml of standard permanganate were required. The latter had been standardized against sodium oxalate according to the usual procedure and found to be 0.1000 *N*. What is the manganese content of the solution? *Ans.* 0.0588 g of Mn.

Ferric iron if present does not interfere since it is precipitated in the form of hydrous ferric oxide by the zinc oxide added. How could manganese be determined by the above method in a solution containing ferrous iron?
5. A solution contains ferric chloride and hydrogen peroxide. Give procedures by which one can determine the peroxide and the ferric iron by titration with permanganate.
6. A solution contains ferrous sulfate and potassium bromide. How may the bromide (argentimetrically) and the ferrous iron be determined? (Bromide interferes with the titration of ferrous iron with permanganate.)

CHAPTER 44 *Iodimetry*

Iodimetric methods

The reaction

$$I_2 + 2e \rightleftharpoons 2\,I^-$$

is reversible. Substances with a reduction potential much lower than that of the iodine–iodide system are oxidized by iodine and can be titrated with a standard iodine solution. Examples are

$$SO_3^{--} + I_2 + H_2O \rightarrow SO_4^{--} + 2\,I^- + 2\,H^+$$

$$S^{--} + I_2 \rightarrow S + 2\,I^-$$

$$2\,S_2O_3^{--} + I_2 \rightarrow S_4O_6^{--} + 2\,I^-$$

$$Sn^{++} + I_2 \rightarrow Sn^{4+} + 2\,I^-$$

On the other hand, iodide exerts a reducing action upon strongly oxidizing systems, with the formation of an equivalent amount of iodine. The liberated iodine is titrated with standard thiosulfate solution. Examples are

$$2\,Ce^{4+} + 2\,I^- \rightarrow 2\,Ce^{3+} + I_2$$

$$BrO_3^- + 6\,I^- + 6\,H^+ \rightarrow Br^- + 3\,I_2 + 3\,H_2O$$

$$2\,MnO_4^- + 10\,I^- + 16\,H^+ \rightarrow 2\,Mn^{++} + 5\,I_2 + 8\,H_2O$$

The reduction potential of the iodine–iodide system is virtually independent of the pH of the solution as long as the latter is less than about 8; at higher alkalinities iodine reacts with hydroxyl ions to form hypoiodite and iodide. The hypoiodite is extremely unstable and is rapidly transformed into iodate:

$$I_2 + 2\,OH^- \rightarrow IO^- + I^- + H_2O$$

$$3\,IO^- \rightarrow 2\,I^- + IO_3^-$$

Substances that do not possess oxidizing or reducing properties hardly affect the reduction potential of the iodine–iodide system. This is not true of many other oxidation–reduction systems. On page 169 it was mentioned that the reduction potential of substances which upon reduction give up oxygen with the formation of water increases very much with increasing hydrogen-ion concentration of the solution. This, for example, is the case with systems involving permanganate, dichromate, vanadate, bromate, etc. A neutral chromate solution, for example, does not react with iodide, whereas in acid medium chromate is quantitatively reduced to trivalent chromium. By the proper adjustment of the pH it is sometimes possible to titrate the reduced form of a substance with iodine, and its oxidized form with thiosulfate after addition of iodide. This, for example, is the case with the arsenious oxide–arsenate system.

$$As_2O_3 + 2\,I_2 + 5\,H_2O \rightleftharpoons 2\,H_3AsO_4 + 4\,I^- + 4\,H^+$$

This reaction is completely reversible; at pH values between 9 and about 4, arsenious oxide can be titrated with iodine solution. In strongly acid medium, on the other hand, arsenate is reduced to arsenious oxide, the reaction then going from right to left as written above. Upon titration with thiosulfate the iodine is removed, and the displacement of the equilibrium from right to left is thus aided.

The reduction potential of various systems can be changed by the addition of a substance that forms a complex, a slightly dissociated, or a slightly soluble compound with one of the two components (or with both, but then of different stability or solubility). The reaction between ferric iron and iodide,

$$2\,Fe^{3+} + 2\,I^- \rightarrow 2\,Fe^{++} + I_2$$

can be made the basis of an iodimetric iron titration. However, if fluoride or phosphoric acid is added, the ferric iron is transformed into a complex which no longer reacts with iodide. Pyrophosphate and EDTA have a similar effect. Use is often made of this behavior in preventing the reduction of ferric iron by iodide in the iodimetric determination of other substances.

The reduction potential of the cupric–cuprous system is lower than that of iodine–iodide. In spite of this the reaction

$$2\,Cu^{++} + 4\,I^- \rightleftharpoons 2\,CuI + I_2$$

runs quantitatively to the right. The cuprous ions formed disappear in the form of the extremely slightly soluble cuprous iodide, and consequently the reduction potential of the copper system is raised. If, on the other hand, the cupric ions are transformed into a complex as, for example, by the addition of tartrate or citrate, the above reaction runs quantitatively from right to left.

From the above it is evident that the conditions prevailing in a system being titrated may have a great effect upon the equilibrium conditions. If, in an iodimetric titration, equilibrium is such that a measurable amount of the substance is still present in the higher state of oxidation, the complete reduction may be aided by increase of the hydrogen-ion concentration, increase of the iodide concentration, slow titration with thiosulfate by which the iodine is removed, addition of a suitable catalyst, or removal of the iodine by extraction with organic solvents (chloroform or carbon tetrachloride) or by distillation. The distillation method is applied with slight modifications to the determination of available oxygen in the higher oxides of manganese and lead. In the evaluation of pyrolusite the sample may be boiled with hydrochloric or hydrobromic acid and the liberated chlorine or bromine absorbed in potassium iodide solution:

$$MnO_2 + 4\,H^+ + 4\,Cl^- \rightarrow Mn^{++} + 2\,H_2O + Cl_2 + 2\,Cl^-$$

$$Cl_2 + 2\,I^- \rightarrow 2\,Cl^- + I_2$$

It is not advisable to boil the sample with an acid iodide solution, since the latter is a much stronger reducing agent than chloride and will react with ferric iron, which is a common impurity in pyrolusite:

$$2\,Fe^{3+} + 2\,I^- \rightleftharpoons 2\,Fe^{++} + I_2$$

An important application of iodometric methods is made in the determination of the acid content of solutions. The reaction

$$IO_3^- + 5\,I^- + 6\,H^+ \rightarrow 3\,I_2 + 3\,H_2O$$

proceeds very rapidly and is quantitative. If an excess of a neutral solution of potassium iodate and iodide is added to a dilute solution of hydrochloric acid or any other strong acid, the hydrogen ions yield an equivalent amount of iodine, which can be titrated with thiosulfate. This method is especially advantageous in the titration of very dilute solutions of strong acids, because a very sharp color change is obtained at the end point.

The iodimetric titration of weak acids is not so simple. Upon the removal of the hydrogen ions by reaction with iodate and iodide, a buffer system of the weak acid and its salt is formed. At the resulting low hydrogen-ion concentration becomes very small. In such a case good results are obtained by adding an excess of standard thiosulfate after the addition of the iodate–iodide solution. The iodine formed in the reaction is thus removed, and the reaction velocity is thereby greatly increased. The excess of thiosulfate is finally back-titrated with a standard iodine solution.

Sources of error in iodimetry

Two important sources of error in iodimetric titrations are the air oxidation of an acid iodide solution and the loss of iodine by volatilization.

Iodide in acid medium is slowly oxidized by oxygen:

$$4\,I^- + 4\,H^+ + O_2 \rightarrow 2\,I_2 + 2\,H_2O$$

This reaction is extremely slow in neutral medium, but the velocity increases with increasing hydrogen-ion concentration and is greatly accelerated by direct sunlight, by the reaction between iodide and the substance to be reduced (an induced reaction, very pronounced in the iodometric vanadate titration in strongly acid medium), and by the presence of certain substances which exert a catalytic effect. Cuprous ion, for example, has an appreciable effect, and it is easily oxidized by air; the cupric copper thus formed reacts in turn with more iodide, etc. Nitrite and nitric oxide (NO), even if present only in the smallest traces, interfere markedly. Nitrite in acid medium reacts with iodide according to the reaction

$$2\,NO_2^- + 2\,I^- + 4\,H^+ \rightarrow 2\,NO + I_2 + 2\,H_2O$$

Nitric oxide forms higher oxides of nitrogen by reaction with the oxygen of the air, and these oxides then react with iodide to form more iodine and nitric oxide, the cycle repeating itself indefinitely. If a trace of nitric oxide is present in the solution titrated, no permanent end point is found in the titration with thiosulfate; the starch–iodine color returns very quickly after the removal of the iodine.

In the iodimetric titration of oxidizing agents in acid medium with an excess of iodide, the solution must not be allowed to stand any longer than is necessary before the titration of the iodine is made. If an appreciable period is required for the reaction to go to completion, the air in the vessel should be replaced by carbon dioxide. This can often be done in the following simple way: Acid is added to the solution to be titrated in an iodine flask; the air is then removed by successive additions of three or four portions of a few hundred milligrams of sodium bicarbonate. The potassium iodide is added in solid form and the flask stoppered immediately thereafter. Dry ice is often found suitable for the removal of oxygen.

The other source of error lies in the comparative ease with which iodine is volatilized from its solutions. This volatility is materially decreased by the presence of a large excess of iodide, which reacts with iodine to form triiodide ions[1]:

$$I^- + I_2 \rightleftharpoons I_3^-$$

The excess of iodide displaces the equilibrium from left to right. Iodine solutions should never be exposed for any length of time to the atmosphere. In titrations at room temperature ($<25°C$) losses are negligibly small if the solution contains at least 4 per cent of potassium iodide.

Detection of the end point. Starch as indicator

Iodine in aqueous iodide solutions has an intense yellow to brown color, and its presence is thus made evident even at high dilutions. A drop of 0.1 *N* iodine solution imparts a pale yellow color to 100 ml of water. In otherwise colorless solutions iodine can serve as its own indicator. A more pronounced color change at the end point is obtained if starch is used as an indicator.

Starch is made up of at least two different components, called amylose and amylopectin; the former gives a deep blue color with iodine, the latter a weak violet color. The relative amounts of these vary with the source of the starch. Thus potato starch consists of about 20 per cent amylose and 80 per cent amylopectin, whereas cornstarch is almost all amylopectin. Potato starch is, therefore, much better for the preparation of an indicator solution for iodimetry and iodometry. The use of the sodium salt of starch glycolic acid has been recommended in place of starch.[2] Starch reacts with iodine in the *presence of iodide* to form an intensely blue-colored adsorption complex, visible even at very low iodine concentration. One hundred milliliters of water to which 50–100 mg of potassium iodide and a few milliliters of 0.2 per cent starch solution have been added should become distinctly blue (not violet) upon the addition of 1 drop of 0.1 *N* iodine solution. The sensitivity of the starch reaction corresponds to an iodine concentration of 1 to 2×10^{-5} *N* at room temperature in the presence of a small amount of iodide (0.001 *N* or more); in the absence of iodide the reaction is less sensitive.

The sensitivity of the reaction decreases with increasing temperature of the solution. At 50°C it is about 10 times less sensitive than at 25°C. Ethyl and methyl alcohols materially decrease the sensitivity of the reaction. In solutions containing more than 50 per cent alcohol the blue color does not appear at all. On the other hand, small amounts of alcohol (less than 5 per cent) have hardly any effect.

The starch solution should be fresh or else should be properly preserved. It often happens that an old starch solution does not impart a blue color to a dilute iodide–iodine solution, but a violet or reddish one. Such a solution should not be used, because the reddish-violet adsorption compound is only slowly decolorized by thiosulfate and no sharp end point is obtained. A stable starch solution can be obtained by the following procedure.

Preparation of starch solution

Triturate 2 g of soluble starch and 10 mg of mercuric iodide (preservative) with a little water, and add the suspension slowly to 1 liter of boiling water. Continue boiling until the solution is clear; cool, and transfer to a glass-stoppered bottle.

Use 5 ml of this starch solution for each 100 ml of solution to be titrated. In the titration of iodine, starch should not be added until just before the end point is reached, because the fading of the color of the iodine is a good indication of the approach of the end point.

[1] Standard iodine solutions are always prepared by dissolving iodine in potassium iodide solutions. Pure water dissolves very little iodine.

[2] S. Peat, E. J. Bourne, and R. D. Thrower, *Nature* **159**, 810 (1947).

The indicator blank is negligibly small in iodimetric or iodometric titrations of 0.1 *N* solutions. With more dilute solutions it must be determined in a liquid having about the same composition that the solution titrated has at the end point.

Organic liquids for the detection of the end point

The limit of visibility of the reddish-violet color of iodine in carbon tetrachloride and chloroform exceeds that of the iodine–starch compound. On shaking 50 ml of 4×10^{-5} *N* iodine solution with 10 ml of chloroform or carbon tetrachloride, the lower layer of organic liquid becomes distinctly violet. The use of these organic liquids for the detection of the end point offers advantages in the titration of extremely dilute solutions, in cases in which the reaction between iodine and the substance to be titrated is relatively slow, and in titrations in extremely strongly acid medium (titrations with iodate according to Andrews; see p. 819). The titrations must be made in bottles or flasks with ground-glass stoppers; after each addition of reagent the vessel must be well shaken, and some time must be allowed for the separation of the liquid into two layers.

Reactions between thiosulfate and iodine and arsenious oxide and iodine

The most important reactions applied in iodometric titrations are those between sodium thiosulfate and iodine and between arsenious oxide and iodine.

Thiosulfate is oxidized to tetrathionate by iodine:

$$2\,S_2O_3^{--} + I_2 \rightarrow S_4O_6^{--} + 2\,I^- \tag{1}$$

Strong oxidizing agents, such as bromine and hypochlorite, oxidize thiosulfate quantitatively to sulfate. Other oxidizing agents, such as potassium permanganate, potassium dichromate, and ceric sulfate, give an incomplete oxidation to sulfate. For this reason an excess of iodide is always added in the determination of these oxidizing substances before titration with thiosulfate. Hypoiodite oxidizes thiosulfate to sulfate; a weakly alkaline iodine solution (for example, in sodium carbonate) gives a partial oxidation to tetrathionate and a partial oxidation to sulfate:

$$S_2O_3^{--} + 4\,I_2 + 10\,OH^- \rightarrow 2\,SO_4^{--} + 8\,I^- + 5\,H_2O \tag{2}$$

According to the normal reaction (oxidation to tetrathionate), 1 mole of thiosulfate corresponds to 1 *equivalent* of iodine ($\frac{1}{2}$ I_2), whereas in the oxidation to sulfate 1 mole of thiosulfate consumes 8 equivalents of iodine. Therefore, one will find less thiosulfate required for the titration of iodine in weakly alkaline medium than in neutral or acid medium [Eq. (1)]. More of the thiosulfate will be oxidized to sulfate [Eq. (2)] the higher the hydroxyl-ion concentration. The iodine and iodide concentrations, the temperature, and the presence of foreign substances also affect the extent to which reaction (2) takes place.

In certain cases it is necessary to neutralize an acid iodine solution before titration with thiosulfate. This neutralization should not be made with an excess of sodium bicarbonate, because erroneous results [Eq. (2)] will then be obtained. Thus in the titration of 25 ml of 0.1 *N* iodine solution in the presence of 0.5 g of sodium bicarbonate, 24.00 ml of 0.1 *N* sodium thiosulfate were used, whereas with 2 g of sodium bicarbonate only 22.5 ml of 0.1 *N* thiosulfate were required to reach the end point. Neutralization with an excess of ammonium carbonate, disodium phosphate, or borax may cause errors as well. In the titration of 0.1 *N* iodine solutions the pH should be less than 7.6; of 0.01 *N* solution, less than 6.5; and of 0.001 *N* solutions, less than 5. Under these conditions the reaction between iodine and thiosulfate takes place quantitatively according to Eq. (1).

If acid iodine solutions must be neutralized before titration, a buffer solution of the proper pH should be used. At the end of the titration the approximate value of the pH should be determined to be sure that the alkalinity was not too high during the titration. Ordinarily iodine solutions can be titrated in quite strongly acid medium if care is taken to stir or shake well. If it is necessary to make the titration of iodine in a weakly alkaline medium, standard arsenious oxide solution should be used instead of sodium thiosulfate for the purpose. In such a case an acid solution can be neutralized with an excess of sodium or potassium bicarbonate. However, since iodine vapor may be carried along with the escaping carbon dioxide, it is better to use instead a sufficient amount of borax or disodium phosphate.

At a pH of 8 the reaction between arsenious oxide and iodine is represented by the equation

$$As_2O_3 + 2\,I_2 + 5\,H_2O \rightleftharpoons 2\,HAsO_4^{--} + 4\,I^- + 8\,H^+$$

This reaction is reversible, and it is seen that the hydrogen-ion concentration plays an important role in determining the state of equilibrium. With a change in pH the following equilibria must be considered as well:

$$HAsO_4^{--} + H^+ \rightleftharpoons H_2AsO_4^-$$
$$H_2AsO_4^- + H^+ \rightleftharpoons H_3AsO_4$$

Washburn[3] and Roebuck[4] found that the equilibrium constant at room temperature is

$$K = \frac{[H_3AsO_4][H^+]^2[I^-]^2}{[H_3AsO_3][I_2]} = 5.5 \times 10^{-2}$$

In the titration of arsenious oxide with iodine, the pH of the solution may have any value between 4 and 9. Results accurate to 0.1 per cent are obtained in the pH region between 5 and 9. At pH values less than 5 the reaction becomes rather slow, and the end point is reached only very gradually.

In the reverse titration—iodine with arsenious oxide—the pH should not exceed 9; at higher pH values some iodine reacts with the hydroxyl ions and escapes titration by formation of iodate. Trivalent antimony can be titrated in the same way as trivalent arsenic. To prevent the precipitation of basic antimony salts, the antimony is kept in solution as a complex by the addition of tartrate in the form of potassium sodium tartrate (Rochelle salt).

Preparation of standard solutions

0.1 *N* sodium thiosulfate solution. Dissolve in 1 liter of freshly boiled, cooled water 25 g of $Na_2S_2O_3 \cdot 5\,H_2O$, and add 0.1 g of sodium carbonate to the solution. Allow the solution to stand for a day before standardizing.

[3] E. W. Washburn, *J. Am. Chem. Soc.* **30**, 31 (1908); **35**, 681 (1913).
[4] J. R. Roebuck, *J. Phys. Chem.* **6**, 365 (1902); **9**, 727 (1905).

It is not possible to prepare sodium thiosulfate solutions of definite strength by weighing out the pure salt *as purchased*, because sodium thiosulfate pentahydrate effloresces very easily and such a product therefore is not likely to contain the proper amount of water. If desired, sodium thiosulfate can be recrystallized and dried in a hygrostat (p. 189) over a saturated solution of calcium chloride hexahydrate, and the pentahydrate thus obtained can be weighed out accurately to give a solution of definite strength. However, sodium thiosulfate solutions prepared with the greatest care may change in titer slowly with the lapse of time, and it is therefore best to standardize the solution at the time it is used. Thiosulfate solutions from which sulfur has separated should be discarded.

0.1 *N* iodine solution. Place 12.7 g of reagent-grade iodine (weighed on a trip scale) in a 250-ml beaker; add 40 g of iodate-free potassium iodide[5] and 25 ml of water. Stir at intervals to hasten solution of the iodine, and when all has dissolved dilute to approximately 1 liter. Keep the solution in a *glass-stoppered* bottle, and store in a cool place. Protect the solution from the light as much as possible, and keep the stock bottle well stoppered.

It is possible to prepare standard iodine solutions by weighing out specially purified (resublimed) dry iodine, dissolving in potassium iodide solution, and diluting to a definite volume.

Purification of iodine. Commercial iodine is contaminated by chlorine, bromine, and water. For the purification 10–12 g are triturated with 1 g of potassium iodide and 2 g of ignited lime to bind the water. This mixture is placed in a dry beaker, which is covered with a round-bottomed flask containing cold water. The beaker is carefully heated over a very small flame and the iodine sublimed against the cold flask. The coarse crystals are collected, ground in an agate mortar, and dried in a calcium chloride desiccator with an ungreased cover.

Preparation of 0.1 *N iodine by exact weighing of pure iodine.* Place approximately 20 g of potassium iodide in a large weighing bottle and dissolve in 5 ml of water. Allow to stand until the bottle and its contents have come to room temperature; then stopper the bottle and weigh to 0.1 mg. Without splashing, introduce 6–6.5 g of dry resublimed iodine (previously weighed on a rough balance), stopper, and again weigh to 0.1 mg. Transfer the solution quantitatively to a 500-ml flask without undue delay and make up to the mark with water. Pour the well-mixed solution into a *glass-stoppered*, preferably brown, bottle and store in a cool dark place.

When the solution is taken from the bottle, the latter is opened for as short a time as possible to avoid possible loss of iodine by volatilization.

Upon storage of iodine solutions in ordinary glass bottles, the glass may give off slight amounts of alkali which will transform iodine into iodate. Upon titration with arsenious oxide only the free iodine left reacts. On the other hand, when the iodine solution is allowed to react in acid medium, all the iodate formed is recovered again as iodine and the titer corresponds to that of the original iodine solution. It is best to store standard iodine solutions in alkali resistant glass. Iodine solutions should not be titrated in beakers but in Erlenmeyer flasks.

Standard 0.1 *N* arsenious oxide solution. Pure arsenious oxide can be obtained from the Bureau of Standards[6] and also from various manufacturers. The substance may be purified by recrystallization from hot 20 per cent hydrochloric acid. The crystal

[5] Test potassium iodide for iodate as follows: Dissolve 1 g in 25 ml of water; add 2 ml of starch solution and 1 ml of 6 *N* sulfuric acid. There should be no immediate appearance of a blue color.

[6] Prepared according to directions of R. M. Chapin, *J. Am. Chem. Soc.* **41**, 351 (1919).

powder thus obtained is dried in a desiccator over sulfuric acid. For highly accurate work it is further purified by sublimation.

Preparation of 0.1 N solution. Weigh out 2.4725 g of pure dry arsenious oxide and dissolve in 20 ml of 1 *N* sodium hydroxide solution. Add 1 *N* sulfuric or hydrochloric acid to the solution until the reaction is neutral or only slightly acid to litmus paper. Transfer the solution to a 500-ml volumetric flask and make up to the mark.

Standardization of 0.1 *N* thiosulfate solution

Various standard substances are available. Potassium iodate, potassium bromate and potassium dichromate are the most useful. Electrolytic copper, iodine, oxalic acid dihydrate, and potassium ferricyanide may also be used.[7]

Standardization against potassium iodate. Potassium iodate is obtained in a pure state by recrystallization from water and drying at 180°C. Even with only a slight excess of hydrogen ions the following reaction rapidly goes to completion:

$$IO_3^- + 5\,I^- + 6\,H^+ \rightarrow 3\,I_2 + 3\,H_2O$$

Procedure. Weigh out 0.14–0.15 g of pure potassium iodate, dissolve in 25 ml of water, and add 2 g of potassium iodide (iodate-free) and 10 ml of 1 *N* sulfuric or hydrochloric acid. Titrate with standard thiosulfate solution with constant stirring. Add a few milliliters of starch solution when the color of the liquid has become a pale yellow, and continue the titration till the color changes from blue to colorless.

Notes:

1. The only disadvantage involved in the use of potassium iodate is its small equivalent weight. If 0.150 g is weighed out and the weighing error amounts to 0.2 mg, an error of 0.13 per cent is thus introduced. Therefore it is more accurate to prepare a standard solution of potassium iodate of exactly known normality (about 0.09 *N*); 50 ml of this solution are pipetted out and treated according to the above procedure.

2. Potassium bromate may also be used for standardizing thiosulfate solutions. It is obtained in a pure state by recrystallizations from water and drying at 180°C. With iodide and acid it reacts quantitatively according to the equation

$$BrO_3^- + 6\,I^- + 6\,H^+ \rightarrow 3\,I_2 + Br^- + 3\,H_2O$$

This reaction, however, does not proceed as rapidly as that with iodate, and a higher acidity is required than in the case of the latter. At least 5 ml of 4 *N* hydrochloric acid or 10 ml of 4 *N* sulfuric acid should be added to 25 ml of the bromate–iodide mixture if the titration is to be made immediately after adding the acid. Molybdate catalyzes the reaction very strongly, and it is advantageous to add a few drops of a 3 per cent ammonium molybdate solution in an iodometric bromate titration.

Standardization against potassium dichromate. In acid medium dichromate reacts with iodide according to the equation

$$Cr_2O_7^{--} + 14\,H^+ + 6\,I^- \rightarrow 3\,I_2 + 2\,Cr^{3+} + 7\,H_2O$$

The reaction does not take place instantaneously, but the speed increases very strongly with increasing hydrogen-ion concentration of the solution. For this reason the standardization should be made in a strongly acidified solution. However, the high acidity may introduce an error through the air oxidation of hydrogen iodide. If the standardization

[7] For a more detailed discussion see I. M. Kolthoff and R. Belcher, *Volumetric Analysis*, vol. III, Wiley-Interscience, New York, 2nd ed., 1957.

is carried out according to the procedure given below, the error resulting from the air oxidation is not greater than 0.05 per cent and therefore negligibly small for most cases.

In work of the highest accuracy it is best[8] to work at a lower acidity and to let the mixture stand in a closed flask from which all air has been displaced by carbon dioxide until the reaction is complete.

Procedure. Weigh out 0.2 g of pure potassium dichromate (p. 839) and dissolve in 50 ml of water. Add 2 g of potassium iodide and 8 ml of concentrated hydrochloric acid. Mix thoroughly and titrate with thiosulfate, swirling the liquid constantly until the brown color has changed to a yellowish green. (What would happen if there were a local excess of thiosulfate in the acid solution, and how would this affect the results?) Add a few milliliters of starch solution and continue the titration until the color changes sharply from blue to light green.

Note: The iodimetric chromate titration finds application in the volumetric determination of barium and lead. Both of these ions yield very slightly soluble chromates with an excess of potassium chromate in neutral medium or in a medium buffered with a mixture of acetic acid and sodium acetate:

$$Ba^{++} + CrO_4^{--} \rightarrow BaCrO_4$$

$$Pb^{++} + CrO_4^{--} \rightarrow PbCrO_4$$

The precipitate is washed with water, dissolved in dilute hydrochloric acid, and titrated according to the above directions. It is also possible to use an excess of standard chromate solution (standardized against thiosulfate) to precipitate the barium or lead, and back-titrate an aliquot part of the filtrate.

Standardization against potassium ferricyanide. Potassium ferricyanide is obtained in a pure state by recrystallization from water and drying at 100°C. It has the advantage of a high equivalent weight. Ferricyanide reacts fairly slowly with iodide in acid medium, but the reaction is accelerated by the addition of zinc sulfate:

$$2\ Fe(CN)_6^{3-} + 2\ I^- \rightarrow 2\ Fe(CN)_6^{4-} + I_2$$

Zinc ions yield precipitates with ferrocyanide and ferricyanide; since the precipitate obtained with the former is much less soluble than that with the latter, the addition of zinc sulfate to a ferricyanide solution increases its reduction potential (p. 175). The zinc ferricyanide precipitate reacts very quickly with iodide to form potassium zinc ferrocyanide, and an equivalent amount of iodine is liberated.

Procedure. Weigh out 1.3 g of pure potassium ferricyanide and dissolve in 40 ml of water contained in a glass-stoppered flask. Add 2 g of potassium iodide, 2 ml of 4 *N* hydrochloric acid, and stopper the flask. After 1 min, add 10 ml of a 30 per cent zinc sulfate solution, and titrate the mixture containing the gelatinous precipitate with thiosulfate. Add 5 ml of starch solution after the color has faded to a pale yellow, and continue the titration to the disappearance of the blue color.

Standardization of 0.1 *N* iodine solution against arsenious oxide

Procedure. If a 0.1 *N* solution of arsenious oxide of accurately known content is available (p. 848), pipet out 25 ml into a 250-ml Erlenmeyer flask, add 25–50 ml of water, 1 g of sodium bicarbonate, and a few milliliters of starch solution, and titrate with the iodine solution to the first blue color.

Otherwise, weigh out 0.2 g of pure dry arsenious oxide, dissolve in 10 ml of 1 *N* sodium hydroxide solution, and add 15 ml (an excess) of 1 *N* sulfuric acid. Then carefully add 2 g of sodium bicarbonate dissolved in 50 ml of water. Titrate with the iodine solution to the first permanent blue tinge, using starch as indicator.

[8] C. R. McCroskey, Ph.D. Dissertation, Ohio State Univ., Columbus, 1918; W. C. Vosburgh, *J. Am. Chem. Soc.* **44**, 2120 (1922); W. C. Bray and H. E. Miller, *ibid.* **46**, 2204 (1924); S. Popoff and J. L. Whitman, *ibid.* **47**, 2259 (1925); F. L. Hahn, *ibid.* **57**, 614 (1935).

Checking the titer of a thiosulfate solution against standard potassium permanganate or ceric sulfate. In Chapter 33 it was emphasized that a standardization involves the use of a primary standard substance, and that standardized solutions should not be used in obtaining the normality of another reagent, since errors may accumulate. However, it is good practice to check one standardized solution against another. Thus the normality of a thiosulfate solution can be compared with that of a potassium permanganate or ceric sulfate solution:

$$2\ MnO_4^- + 10\ I^- + 16\ H^+ \rightarrow 2\ Mn^{++} + 5\ I_2 + 8\ H_2O$$

Procedure. Measure out 40 ml of 0.1 *N* standard permanganate in an Erlenmeyer flask, and add 10 ml of 2 *N* hydrochloric or sulfuric acid and 3 g of potassium iodide dissolved in 10 ml of water. After mixing thoroughly, titrate with thiosulfate, adding starch toward the end of the titration. Calculate the normality of the thiosulfate and compare with the figure found by direct standardization. The normality of a thiosulfate solution can be also checked against a standard iodine solution, and vice versa, as indicated below.

Exercise 1:

(a) Titrate 25.00 ml of 0.1 *N* standard iodine solution with 0.1 *N* standard thiosulfate solution.
(b) As in (a), but add 0.5 g of sodium bicarbonate to the iodine solution.
(c) As in (a), but add 0.5 g of sodium carbonate to the iodine solution.
(d) As in (a), but add 5 ml of 4 *N* hydrochloric acid to the iodine solution.

Report the results and explain any deviations from the calculated value.

Exercise 2:

(e) Titrate 25 ml of 0.1 *N* sodium thiosulfate solution with 0.1 *N* standard iodine solution.
(f) As in (a), but add 5 ml of 4 *N* hydrochloric acid to the thiosulfate solution.

Compare the results with the calculated value and explain any deviations.

Exercise 3:

(g) Titrate 25 ml of 0.1 *N* standard iodine solution with 0.1 *N* standard arsenious oxide solution.
(h) As in (g), but add 2 g of sodium bicarbonate to the iodine solution.
(i) As in (g), but add 3 g of borax to the iodine solution.

Report the results, compare them with the calculated values, and explain any deviations found.

Determination of available chlorine in bleaching powder

Iodometric method. The active constituent of bleaching powder is hypochlorite, which is responsible for the bleaching action of the material. In addition, calcium chloride, calcium chlorate, and calcium oxide are usually present. There is some evidence that the calcium is present in the form CaCl(OCl). On treating bleaching powder with acid, chlorine is liberated:

$$CaCl(OCl) + 2\ H^+ \rightarrow Ca^{++} + Cl_2 + H_2O$$

In the evaluation of bleaching powder the result is usually expressed in terms of available chlorine, which is the chlorine liberated in acidifying the solution.

Available chlorine can be determined iodometrically by adding potassium iodide and acid to the suspension of the material:

$$OCl^- + 2\ I^- + 2\ H^+ \rightarrow Cl^- + I_2 + H_2O$$

The amount of iodine liberated is equivalent to the available chlorine. It is often recommended that the solution of the sample be acidified with acetic acid. The above reaction is completed quickly at this relatively low acidity. However, it often happens that moist samples of hypochlorite partly decompose on storage, with the formation of chlorite and chlorate. Chlorite in acid medium is reduced by iodide:

$$ClO_2^- + 4\,I^- + 4\,H^+ \rightarrow 2\,I_2 + Cl^- + 2\,H_2O$$

In acetic acid media this reaction proceeds very slowly. Therefore, if this acid is used for acidifying the hypochlorite–iodide mixture and chlorite is present, no sharp end point is found. The iodine color returns, owing to the slow reaction between chlorite and iodide. For this reason it is better to acidify with sulfuric acid, because in the presence of this acid the chlorite reacts rapidly with the iodide, whereas chlorate does not interfere.

Procedure. Weigh out a 5-g sample of bleaching powder and triturate it in a mortar with successive portions of water, until it is distintegrated and no lumps remain. Transfer the mixture to a 500-ml volumetric flask, make up to the mark with water, and shake until a homogeneous suspension is obtained. Pipet out 50 ml of suspension and transfer to a 250-ml Erlenmeyer flask. Add 2 g of potassium iodide and 15 ml of 4 *N* sulfuric acid, and titrate with standard thiosulfate. Report the percentage of available chlorine in the sample.

Arsenite method. Hypochlorite reacts very quickly with arsenious oxide according to the equation

$$2\,OCl^- + As_2O_3 \rightarrow As_2O_5 + 2\,Cl^-$$

No internal indicator is available for the detection of the end point, because hypochlorite is such a powerful oxidizing agent that it destroys all organic dyes. The end point may be found with iodide–starch paper. As long as there is an excess of hypochlorite present, a drop of the solution being titrated turns the iodide–starch paper blue; the end point is reached when a drop of the liquid does not color the indicator paper.

Instead of using an external indicator (the use of which requires some practice), one can add an excess of standard arsenious oxide solution and titrate back with standard iodine solution using starch as indicator. Or, if a standard hypochlorite solution is available, the excess of arsenious oxide can be back-titrated with this solution with Bordeaux as an inside indicator (see p. 861).

Indicator paper. Dip strips of filter paper in 50 ml of starch solution containing 1 g of potassium iodide. Drain the strips and spread them on a watch glass to dry. Keep the dried strips in a dark bottle.

Procedure. Transfer 50 ml of the suspension of bleaching powder (see iodometric procedure) to a 250-ml beaker. Titrate with standard arsenious oxide solution until no blue color develops when the iodide–starch paper is thoroughly wetted by a drop of the liquid.

Report the percentage of available chlorine in the sample.

Iodide

Various iodometric methods are available for the determination of iodides. Two simple procedures of great practical importance are described below.

Oxidation of iodide to iodate and iodometric titration of the latter. Iodide in neutral or weakly acid medium is quantitatively oxidized to iodate by an excess of chlorine or bromine. If much iodide is present, the chlorine or bromine first liberates free iodine, which precipitates but dissolves in an excess of reagent:

$$2\,I^- + 6\,Cl_2 + 6\,H_2O \rightarrow 2\,IO_3^- + 12\,Cl^- + 12\,H^+$$

The excess of chlorine or bromine may be removed by boiling or by adding an excess of a phenol or aniline solution. A phenol solution is often used for the removal of free chlorine or bromine from a solution. Depending upon the molecular ratio of the phenol and the halogen, mono-, di-, or trichloro (or bromo) phenol and an equivalent amount of halogen acid are formed:

$$C_6H_5OH + Br_2 \rightarrow C_6H_4BrOH + H^+ + Br^-$$
$$C_6H_5OH + 2\,Br_2 \rightarrow C_6H_3Br_2OH + 2\,H^+ + 2\,Br^-$$
$$C_6H_5OH + 3\,Br_2 \rightarrow C_6H_2Br_3OH + 3\,H^+ + 3\,Br^-$$

After the excess of chlorine or bromine has been removed, the iodate formed is titrated iodometrically:

$$IO_3^- + 5\,I^- + 6\,H^+ \rightarrow 3\,I_2 + 3\,H_2O$$

Since 1 mole of iodide yields 1 mole of iodate and the latter corresponds to 6 equivalents of iodine, it is evident that the equivalent weight of the iodide in this particular titration is one sixth of the molecular weight.

The method has the practical advantage that small amounts of iodide yield relatively large amounts of iodine.

Determination of iodide in iodized salt. Iodized table salt usually contains approximately 0.02 per cent potassium iodide. Since the amount of iodide present is so small, it is necessary to take a large sample and to use 0.005 *N* sodium thiosulfate to titrate the iodine formed by following the method described above.

Procedure.[9] Dissolve a 5-g sample of the iodized salt in 100 ml of water in a 250-ml Erlenmeyer flask, add a drop of methyl orange indicator, acidify carefully with 2 *N* sulfuric acid, and add 2 ml in excess. Introduce bromine vapor just above the surface of the solution with gentle swirling until the solution retains a permanent yellow color (or add freshly prepared bromine water). Add a few glass beads to prevents bumping and boil the solution until the yellow color due to bromine disappears and then for 2 min longer. Cool, wash down the sides of the flask, and add 0.2 g of potassium iodide and 2–3 ml of starch solution. Mix the contents of the flask, and titrate the liberated iodine immediately with 0.005 *N* thiosulfate.

Calculate the percentage of potassium iodide in the salt.

Notes:

1. Bromine vapor is introduced into the flask by blowing a slow current of dried air through a small amount of pure bromine[10] contained in a small gas-washing bottle. The delivery tube from the bromine bottle should end just above the surface of the solution in the flask and should be drawn out to a small tip to prevent the introduction of too large quantities of bromine.

2. Small amounts of bromide (less than 20 times the amount of iodide) do not interfere in the determination. Too little iodine is found when excessive quantities of bromide are present. Quantities of ferric iron in excess of 0.1 mg interfere, but the interference may be eliminated by the addition of phosphoric acid. Even small amounts (1 mg) of nitrites interfere, but the interference may be eliminated by the azide method of Reith.[11]

Oxidation of iodide to iodine with sodium nitrite solution. Nitrite oxidizes iodide quantitatively to iodine in an acid medium:

$$2\,NO_2^- + 2\,I^- + 4\,H^+ \rightarrow 2\,NO + I_2 + 2\,H_2O$$

[9] J. F. Sadusk, Jr., and E. G. Ball, *Ind. Eng. Chem., Anal. Ed.* **5**, 386 (1933).
[10] G. M. Karns and H. C. Donaldson, *J. Am. Chem. Soc.* **54**, 442 (1932).
[11] J. F. Reith, *Rec. Trav. Chim.* **48**, 386 (1929).

In this case the equivalent weight of the iodide is equal to the atomic weight. The liberated iodine can be titrated with standard thiosulfate after removal of the excess of nitrous acid and nitric oxide, both of which would otherwise interfere (p. 844). Both the nitrous acid and the nitric oxide can be destroyed by the addition of an excess of urea, which is oxidized to nitrogen, carbon dioxide, and water; nitrous acid is reduced to nitrogen and water. The reaction between nitrite and iodide in acid medium is rapid, whereas the reaction between the nitrous acid (and also the nitric oxide) and urea is fairly slow. The urea can therefore be added to the acidified iodide solution before the nitrite; the latter reacts first with the iodide; the excess nitrous acid and nitric oxide formed are made harmless by subsequent shaking with the urea. Chlorides do not interfere in this determination, but high results are found if more than an equivalent amount of bromide (with respect to iodide) is present.

Procedure. Prepare 250 ml of a solution 0.1–0.15 *M* with respect to iodide. Transfer 25 ml to a 250-ml glass-stoppered flask (iodine flask), and add 1 g of urea, 8 ml of 0.5 *M* sodium nitrite solution, and 5 ml of 4 *N* sulfuric acid. Stopper the flask and allow to stand with frequent shaking for about 10 min. Cool in an ice bath, add 2 g of potassium iodide to dissolve the iodine, and titrate with standard thiosulfate solution. Shake the contents well to dissolve iodine in the gas phase. Report the percentage of iodine in the sample.

Hydrogen peroxide

Hydrogen peroxide reacts with iodide in acid medium according to the equation

$$H_2O_2 + 2\,I^- + 2\,H^+ \rightarrow I_2 + 2\,H_2O$$

The reaction is fairly slow, but it is catalyzed by molybdate.

Procedure. Dilute a commercial sample of hydrogen peroxide according to the directions given under the permanganate procedure (p. 834). Pipet a 25-ml portion into a 250-ml Erlenmeyer flask; add 10 ml of 4 *N* sulfuric acid, 1 g of potassium iodide, and 3 drops of a neutral 3 per cent ammonium molybdate solution; and titrate with standard thiosulfate, using starch as indicator.

Notes:

1. Other peroxides soluble in acid, such as sodium, alkaline earth, and zinc peroxides, may be determined by the above procedure; percarbonates and perborates may likewise be determined in this way.

2. Higher oxides of manganese (pyrolusite) and of lead which are insoluble in acid react very slowly with iodide. Therefore if the determination is made according to the above procedure, the mixture must be allowed to stand a long time before all the higher oxide is dissolved. This will introduce a material error because of air oxidation of the hydrogen iodide unless special precautions are taken (see p. 844). Therefore, it is better to make use of the fact that the higher oxides of manganese and lead oxidize hydrochloric acid to chlorine and hydrobromic acid to bromine. The sample is heated in a distillation apparatus with hydrochloric acid, and the chlorine evolved is absorbed in a potassium iodide solution. The liberated iodine is titrated with thiosulfate. Since the procedure described on page 835 is simpler and more accurate, no detailed directions are given here.

Iodometric determination of copper

In neutral or weakly acid solutions cupric copper reacts with iodide to give insoluble cuprous iodide and iodine:

$$2\,Cu^{++} + 4\,I^- \rightleftharpoons 2\,CuI + I_2$$

Under proper conditions the reaction runs quantitatively to the right, and copper can be determined iodometrically in this manner by titrating the liberated iodine with thiosulfate. The iodometric method for copper rivals the electrolytic method in accuracy and is more rapid and less subject to interference from other elements.

The reaction between cupric ions and iodide is reversible. Any conditions that lead to an increase in the solubility of cuprous iodide or to a decrease in the concentration of cupric ion (formation of complex or slightly ionized cupric salts, etc.) may prevent the quantitative reduction of cupric copper. A reasonable excess of iodide is always added to ensure the complete reduction of copper. The acidity of the solution must be adjusted within certain limits. If the pH of the medium is too high, the reaction does not proceed quantitatively to the right (equation above), and the reaction is slow. Under these conditions a fleeting end point is obtained, the blue color of starch–iodine returning soon after the decolorization by thiosulfate as the equilibrium is shifted to the right. However, for reasons that will be mentioned immediately, it is desirable to maintain the pH of the solution at as high a value as possible consistent with quantitative reduction of copper. To maintain the acidity of the solution at the proper value, a buffer is added. The upper pH limit is determined by the kind of buffer system employed, e.g., the reduction of cupric ions is complete at a higher pH in a phthalate buffer solution than in a phosphate mixture. These differences are due to the different activities of cupric ion in various solutions. The acidity of the reaction mixture must not be too high, or else the results will be slightly high, owing to the air oxidation of iodide; copper catalyzes this reaction. Moreover, in too strongly acid solutions there is interference from such elements as arsenic and antimony.

The most common elements found associated with copper in nature which interfere in the method, and which must be rendered inactive, are iron, arsenic, and antimony. Ferric iron is reduced by iodide:

$$2\,Fe^{3+} + 2\,I^- \rightleftharpoons 2\,Fe^{++} + I_2$$

The interference of iron can be prevented by transforming it into a complex that will not react with iodide in slightly acid solutions. Alkali pyrophosphates and fluoride are commonly used for this purpose. The latter, which is preferable, gives the slightly dissociated ion FeF_6^{3-}. When added in the form of ammonium bifluoride, the iron is made harmless, and moveover the pH is adjusted to the correct value (hydrogen fluoride mixed with fluorides gives a buffer system). Arsenic and antimony must not be present in the trivalent form, for these consume iodine. These elements will generally be present in the quinquevalent form as a consequence of the oxidizing attack used to bring the sample into solution. Arsenic and antimony in the quinquevalent form will not oxidize iodide in a medium having a pH greater than 3.5. By buffering the solution with ammonium bifluoride as described below, it is possible to prevent the interference of these elements and still maintain the pH of the solution at a value at which the reduction of cupric ion is complete. Oxidizing agents such as the oxides of nitrogen and free bromine, possibly remaining after bringing the sample into solution, must be expelled. Elements such as silver, lead, bismuth, and zinc do not interfere, at least if present in relatively small amounts, as is usually the case.

In neutral solution cupric copper does not react with iodide if it is transformed into a tartrate complex by the addition of sodium potassium tartrate. Use is made of this behavior when trivalent arsenic must be titrated with iodine in the presence of copper (for example, in the analysis of Paris green). The bright blue solution containing an excess of tartrate is titrated with standard iodine solution. At the end point the color changes sharply to greenish blue (starch indicator need not be added). It is possible to titrate the copper in the same solution after the arsenic titration, after adding sulfuric acid to give the proper acidity.

The iodometric copper titration is extensively applied in the determination of reducing sugars (dextrose, levulose, etc.). These are oxidized at boiling temperature by an alkaline solution of cupric copper containing tartrate or citrate to hold the metal in solution (complex formation), the copper being reduced to cuprous oxide, which is precipitated. The precipitate may be filtered off and weighed, or more conveniently (1) it may be dissolved and oxidized, and the copper determined iodometrically, or (2) the excess of copper remaining in solution may be determined iodometrically after acidification. The reaction between the reducing sugar and the copper is not stoichiometric, the amount of cuprous oxide precipitated by a given sugar depending upon such factors as the concentration of the sugar solution, the composition of the reagent, and the time of boiling. Therefore the amount of sugar corresponding to a given weight of cuprous oxide must be found by reference to a table which has been constructed empirically. The procedure used in the construction of the table must be carefully followed in the determination.

Determination of copper in ores. The following rapid method for copper in ores, mattes, slags, etc., is due to Park.[12] The presence of iron, arsenic, and antimony is provided for. The sample is brought into solution by treatment with hot nitric acid, which oxidizes these elements, as well as any cuprous copper present, to their highest valences. The interference of iron is prevented by transforming it into complex ferric fluoride. By addition of the proper amount of ammonium bifluoride the pH is adjusted to a value of 3.3–4.0.[13] At this acidity there is no interference from quinquevalent arsenic or antimony.

Procedure. Weigh out 1 g[14] of finely ground material into a 150-ml beaker, add 20 ml of concentrated nitric acid, and warm until it is certain that all the copper has been brought into solution. Evaporate the solution to a volume of 5 ml, add 25 ml of water, and boil to bring all soluble material into solution and to expel the oxides of nitrogen. Filter off the residue on a small filter paper, receiving the filtrate in a 250-ml Erlenmeyer flask (if the residue is small in amount and light-colored, the filtration may be omitted). Wash the residue thoroughly with hot 1 per cent nitric acid. Concentrate the filtrate to 25 ml, cool, and add dropwise dilute ammonium hydroxide to precipitate all the iron, being careful to avoid an excess; the liquid should smell very faintly of ammonia after blowing away the fumes. Add 2.0 g of ammonium bifluoride and shake to dissolve the ferric hydroxide, and then add 3 g of potassium iodide. Titrate the liberated iodine at once with 0.1 N sodium thiosulfate (note 2), adding 3 ml of 0.2 per cent starch solution near the end point. The color change is from blue to yellowish white. The blue color should not return within 10 or 15 min. A quick return of the color may be caused by the addition of too much ammonium hydroxide or too little fluoride, as well as by the presence of oxides of nitrogen. Report the percentage of copper.

Notes:

1. Most copper ores can be decomposed by heating with concentrated nitric acid. If such treatment fails to bring all the copper into solution, heating with hydrochloric and nitric acid followed by evaporation to fumes with sulfuric acid, dilution, and boiling with bromine water to reoxidize any arsenic or antimony that may have been reduced by free sulfur will usually bring all the copper into solution. All bromine must, of course, be expelled by boiling before adding iodide. Some silicate ores of copper and slags may not be completely decomposed by acid treatment. In such a case the acid-insoluble material must be fused with sodium carbonate after extraction with ammonium acetate and ammonium hydroxide solutions to remove any lead sulfate and silver chloride, respectively, that may be present, which would otherwise attack the platinum crucible.

[12] B. Park, *Ind. Eng. Chem., Anal. Ed.* **3**, 77 (1931).

[13] Park also added potassium biphthalate for the adjustment of pH. According to W. R. Crowell, T. E. Hillis, S. P. Rittenberg, and R. F. Svenson, *Ind. Eng. Chem., Anal. Ed.* **8**, 9 (1936), the addition of the biphthalate can be omitted since the bifluoride has the proper buffering action. See also W. R. Crowell, S. H. Silver, and A. T. Spiher, *ibid.* **10**, 80 (1938).

[14] This is the proper weight of sample for material containing 10–30 per cent of copper; for lower contents take a larger amount.

2. Although not strictly necessary, it is preferable to standardize the thiosulfate solution against pure copper. Dissolve 0.20–0.25 g of pure electrolytic copper foil in 10 ml of 6 *N* nitric acid, and evaporate the solution to the point at which it will solidify on cooling, but not further, or basic salts of copper may be formed that will be hard to dissolve. Dissolve the copper nitrate in 25 ml of water, and add dropwise a 1 *N* solution of sodium hydroxide until a slight permanent precipitate is obtained. Add 7 ml of 6 *N* acetic acid to the liquid and dilute to 50 ml with cold water. To the cool solution add 3 g of potassium iodide, stir or shake to dissolve, and titrate with the 0.1 *N* sodium thiosulfate solution, adding starch when the color of the iodine has become pale yellow.

3. Approximately 1 g of ammonium bifluoride should be added for each 0.1 g of iron present. Aluminum forms a slightly soluble complex fluoride, so that if it is present in any quantity, more bifluoride must be added to take the place of the fluoride bound by this metal. Manganese when present alone does not interfere in the method, but if iron is present with manganese high results are obtained. A few milligrams of manganese with 0.1 to 0.2 g of iron lead to no serious error; when larger amounts are present (which rarely happens), iron must be removed before the titration.

Other iodimetric methods

Inorganic. It is beyond the scope of this text to consider all iodometric and iodimatic procedures.[15] However, the principal reactions on which a number of determinations are based are given briefly below:

Hydrogen sulfide. Hydrogen sulfide in acid medium is oxidized to sulfur by iodine:

$$H_2S + I_2 \rightarrow S + 2\,H^+ + 2\,I^-$$

The solution must not be alkaline (alkali sulfides), for part of the sulfide is then oxidized to sulfate:

$$S^{--} + 4\,I_2 + 8\,OH^- \rightarrow SO_4^{--} + 8\,I^- + 4\,H_2O$$

The hydrogen sulfide titration is often used to determine sulfur in iron and steel and the dissolved gas in mineral waters.

Sulfurous acid and its salts

$$HSO_3^- + I_2 + H_2O \rightarrow SO_4^{--} + 2\,I^- + 3\,H^+$$
$$SO_3^{--} + I_2 + H_2O \rightarrow SO_4^{--} + 2\,I^- + 2\,H^+$$

The method is not accurate, because low results are obtained as a result of air oxidation of the sulfurous acid or its salts during the titration. Good results are obtained when sulfurous acid or its salts are added to an excess of iodine solution.

Tin. Quadrivalent tin is reduced by metallic iron, lead, or zinc to the bivalent form

$$Sn^{4+} + Fe \rightleftharpoons Fe^{++} + Sn^{++}$$

and the latter may be titrated in acid medium with iodine. Stannous tin is easily oxidized by air, especially during the titration. Therefore, air oxidation must be prevented by working in an atmosphere of carbon dioxide.

Ferric iron

$$2\,Fe^{3+} + 2\,I^- \rightarrow 2\,Fe^{++} + I_2$$

The equilibrium in acid medium is displaced to the right by using a very large excess of potassium iodide. The method is not of much practical significance for the determination of iron, because titration with permanganate, dichromate, or ceric sulfate, after reduction of the ferric iron to the ferrous state, is much more accurate.

[15] Cf. I. M. Kolthoff and R. Belcher, *Volumetric Analysis*, vol. III.

Water with Karl Fischer reagent. Some 30 years ago Fischer[16] developed a method for the iodimetric determination of water. This method has gained great importance and is generally used for the determination of water in inorganic and organic materials, especially in nonaqueous solvents. Water in a sample dispersed in an anhydrous solvent such as methanol is titrated directly with a reagent containing iodine, sulfur dioxide, pyridine, and methanol. The reactions involved have been clarified by Smith et al.,[17] who found that 1 mole of iodine reacts with 1 mole of water in the presence of pyridine and methanol:

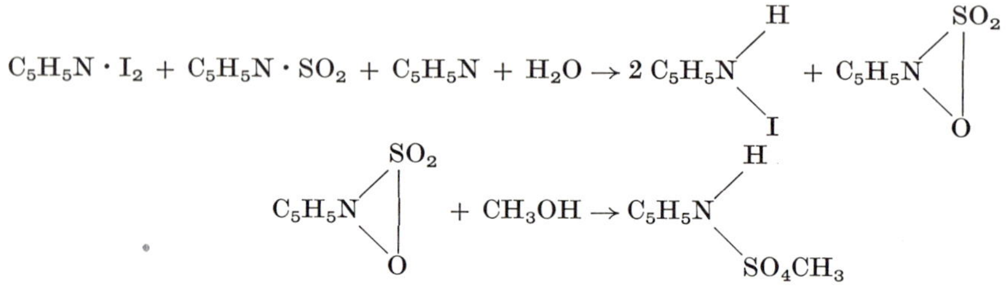

The overall reaction is the sum of these two partial reactions, from which it is clear that each mole of water requires 1 mole of iodine, 1 mole of sulfur dioxide, 3 moles of pyridine, and 1 mole of methanol.

Preparation and stability of Karl Fischer reagent. Transfer 84.7 g (0.33 mole) of resublimed iodine to a dry 1-liter glass-stoppered bottle and add 269 ml (3.3 moles) of C.P. pyridine. Shake the mixture until all the iodine is dissolved and add 667 ml of anhydrous methanol. This stock solution is stable and can be kept indefinitely. Several days before the reagent is required cool the stock solution at ice temperature. Collect 64 g (1 mole) of anhydrous sulfur dioxide in a cold trap immersed in a solid carbon dioxide bath and protected from atmospheric moisture. Carefully add with shaking the sulfur dioxide to the chilled solution. Stopper the mixture, allow it to return to room temperature, and keep it several days before using. The resulting reagent should be equivalent to about 3.5 mg of water per milliliter.

A two-reagent modification has been proposed by Johannson[18] and perfected by Seaman et al.[19]

Karl Fischer reagent is a very active desiccant; the apparatus used to store and dispense the reagent and the titrations flask assembly must be designed to protect the reagent from moisture in the air. Several titration assemblies are commercially available. The end point in a titration can be detected visually or potentiometrically or amperometrically. The reagent has a dark-brown color which is changed to a chromate yellow color by reaction with water.

Standardization: The reagent must be standardized at least once each day by using a solution of methanol containing a known amount of water. A blank must be run with the ethanol.

Procedure. Weigh 2–15 g of water into a dry 1-liter glass-stoppered volumetric flask and dilute to the mark with methanol containing less than 0.1 per cent water. Titrate several 10-ml portions of the standard water solution and of the anhydrous methanol in a 125-ml dry Erlenmeyer flask (kept in a desiccator) with the reagent.

[16] K. Fischer, *Angew. Chem.* **48**, 394 (1935).

[17] D. M. Smith, W. M. D. Bryant, and J. Mitchell, Jr., *J. Am. Chem. Soc.* **61**, 2407 (1939); for details on the reaction mechanism and procedures see J. Mitchell, Jr., and D. M. Smith, *Aquametry*, Wiley-Interscience, New York, 1948; see also I. M. Kolthoff and R. Belcher, *Volumetric Analysis*, vol. III, p. 409.

[18] A. Johannson, *Acta Chem. Scand.* **3**, 1058 (1949).

[19] W. Seaman, W. H. McComas, and G. A. Allan, *Anal. Chem.* **21**, 510 (1949).

As an aid in the detection of the end point, prepare one solution having the color of the reagent just before the end point and another having the color of the iodine end point.

Note: The reagent can be used for the determination of water in a host of organic compounds. Substances which react with iodine or which oxidize iodide to iodine or which react with a constituent in the reagent to yield water interfere.

Water in inorganic compounds (crystal hydrates) can be determined with the reagent. Oxidizing and reducing agents, oxides, and hydroxides interfere. Salts of weak oxyacids and borates interfere.

$$MSO_3 + C_5H_5N = MO + C_5H_5NSO_2$$
$$MO + 2\,HI = MI_2 + H_2O$$
$$H_3BO_3 + 3\,CH_3OH = B(OCH_3)_3 + 3\,H_2O$$

Organic compounds. Acetone. Iodine reacts with alkali to give hypoiodite, which is unstable and disproportionates to iodate and iodide. Acetone reacts with hypoiodite according to the equation

$$CH_3COCH_3 + 3\,KOI = CHI_3 + CH_3COOK + 2\,KOH$$

Procedure. To 10 ml of less than 0.03 M acetone solution add 5 ml of 4 N or at least 25 ml of N sodium hydroxide and 25 ml 0.1 N iodine. Shake thoroughly, let stand for 10 min and acidify with hydrochloric acid. Titrate the excess of iodine with 0.1 N thiosulfate.

Notes:

1. Very dilute solutions of acetone can be determined in the above way by using more dilute iodine solution.
2. Acetaldehyde and ethanol interfere, but methanol does not.

Carbon disulfide. This compound is quantitatively converted to xanthogenate by alcoholic potassium hydroxide. This reaction product reacts quantitatively with iodine in neutral solution to form dixanthogen:

$$CS_2 + KOH + C_2H_5OH = SC(OC_2H_5)_2SK + H_2O$$
$$2\,CS(OC_2H_5)_2SK + I_2 = 2\,KI + (SCOC_2H_5S)_2$$

Procedure. Dissolve a weighed sample of the solution in 1–2 ml of freshly prepared alcoholic potassium hydroxide solution and carefully add 1 N acetic acid until phenolphthalein is just bleached but is still alkaline to litmus. Add starch and titrate with 0.01 N iodine.

Mercaptans. Compounds with the sulfhydryl group generally react with iodine to form disulfide:

$$2\,RSH + I_2 = RSSR + 2\,HI$$

Mercaptans are easily oxidized by air, especially at pH greater than 7.

Thioglycolic acid. Dissolve the sample in air-free water and add this solution from a buret to a measured volume of standard iodine solution until all iodine has been consumed. Accuracy is about 0.5 per cent.

Notes:

1. The iodimetric titration can be used for the determination of cysteine, glutathione, and water-insoluble mercaptans. Thioglycolic-*p*-aminonaphthalide (thionalide), $C_{10}H_{17}NH{-}CO{-}CH_2SH$, precipitates with Cu, Ag, Au, Hg, Sn, As, Sb, BO, Pt, Pd (see p. 280). The precipitate can be collected, treated with an excess of iodine, and the latter back-titrated with thiosulfate. Detils of procedures are given in *Volumetric Analysis*.[20]

[20] I. M. Kolthoff and R. Belcher, *Volumetric Analysis*, vol. III, pp. 386–394.

2. Disulfides can be reduced with zinc to the corresponding mercaptans, which can be titrated iodimetrically.

PROBLEMS

1. A sample of calcium hypochlorite was found to contain 50 per cent available chlorine. How many grams of a 3 per cent solution of hydrogen peroxide are required to reduce all the hypochlorite in 10 g of the sample if the reaction takes place according to the equation

$$OCl^- + H_2O_2 \rightarrow Cl^- + H_2O + O_2?$$

What is the volume of the oxygen evolved at 25°C and 760 mm Hg pressure?

Ans. 79.8 g; 1.725 liters.

2. How could trivalent arsenic be determined iodimetrically in a solution containing potassium chromate?
3. The lead in 25 ml of a solution is precipitated with an excess of potassium chromate; the precipitate of lead chromate is filtered off, washed, dissolved in acid, and titrated iodometrically. (Give equations.) If 31.50 ml of 0.1010 N sodium thiosulfate are required to titrate the liberated iodine, what is the molarity of the solution with respect to lead? *Ans.* 0.0424 M.
4. If 0.2000 g of sodium oxalate requires 31.00 ml of potassium permanganate solution for complete oxidation in acid solution, how many milliters of 0.1000 N sodium thiosulfate will be required in the iodometric titration of 25 ml of this permanganate solution? Give equations. *Ans.* 24.07 ml.
5. A solution of chromic acid containing iron is to be titrated iodometrically. How can the iron be made harmless?
6. 25.00 ml of a potassium iodide solution are treated with dilute hydrochloric acid and 10.00 ml of 0.0500 M potassium iodate solution (an excess). The iodine liberated is volatilized by boiling the solution, and after cooling an excess of potassium iodide is added to react with the iodate remaining. 21.14 ml of 0.1008 N sodium thiosulfate solution are required to titrate the liberated iodine. Calculate the molarity of the potassium iodide solution. *Ans.* 0.0288.
7. To 25.00 ml of a potassium iodide solution an excess of potassium iodate and 1 g of a weak acid are added. The iodine formed (give reaction equation) is titrated with thiosulfate. How may the acidity be adjusted? Under the proper conditions of titration 30.00 ml of 0.1048 N thiosulfate are used. Calculate the iodide content of the solution. *Ans.* 0.1048 M.
8. A solution contains cupric copper, trivalent arsenic, and ferric iron. How can one determine iodometrically the arsenic and the copper in the mixture? (The oxalate complex of cupric copper does not react with iodide in neutral medium.)
9. A solution is 0.1 N in iodine, 0.005 M in iodate, and 0.2 M in iodide. An aliquot of the solution is titrated with arsenious trioxide in the presence of excess bicarbonate. Another aliquot is acidified and titrated with thiosulfate. What is the difference in "iodine titer" found in the two titrations?

CHAPTER 45 *Potassium Bromate and Calcium Hypochlorite as Titrimetric Reagents*

Potassium bromate as an oxidizing agent

For a general discussion see page 819.

Standard potassium bromate solution is an excellent reagent for the determination of trivalent arsenic and antimony, thallous thallium, and hydrazine in acid medium. Cuprous copper and stannous tin may also be determined if air is excluded.

$$BrO_3^- + 3\,As(III) + 6\,H^+ \rightarrow Br^- + 3\,As(V) + 3\,H_2O$$

$$BrO_3^- + 3\,Tl^+ + 6\,H^+ \rightarrow Br^- + 3\,Tl(III) + 3\,H_2O$$

$$BrO_3^- + 6\,Cu^+ + 6\,H^+ \rightarrow Br^- + 6\,Cu^{++} + 3\,H_2O$$

$$2\,BrO_3^- + 3\,N_2H_4 \rightarrow 2\,Br^- + 3\,N_2 + 6\,H_2O$$

The equivalent weight of potassium bromate is one sixth the molecular weight.

It has already been mentioned that the indicators used, such as methyl orange, the sodium salt of methyl red (solution in *water*), Bordeaux, and the indigo sulfonic acids, are irreversibly oxidized by an excess of bromate. Therefore, if an excess of bromate is accidentally added, it cannot be back-titrated directly using these indicators. In such a case an excess of standard arsenious oxide is added, which is then titrated with the bromate.

p-Ethoxychrysoidine hydrochloride is a suitable reversible indicator. It is used as a 0.1 per cent solution in water. The color change is from red to yellow (p. 820). According to Belcher[1] methylthymol blue, which changes color from red to yellow, behaves like a reversible indicator, provided no bromide is added to the sample that is being titrated.

In the titration of arsenic and antimony the solution should be at least 1.5 *N* with respect to hydrochloric acid at the end point.

Preparation of 0.1 *N* standard potassium bromate solution. Weigh out 2.783 g of pure potassium bromate which has been dried at 150°C, and dilute with water to 1 liter.

Titration of standard arsenious oxide solution against standard bromate

Procedure. Pipet 25 ml of standard 0.1 *N* arsenious oxide solution into an Erlenmeyer flask and add 25 ml of water, 15 ml of concentrated hydrochloric

[1] R. Belcher, *Talanta* **12**, 775 (1965).

acid, 0.5 g of potassium bromide, and 1–2 drops of 0.1 per cent solution of methyl orange or methyl red. Titrate slowly with the standard potassium bromate solution with constant swirling. Near the end point add the bromate dropwise with intervals of a few seconds between drops until the color changes sharply from red to colorless or very pale yellow. Add another drop of indicator to make sure that the end point has been reached. If the color of the indicator fades before the end point is reached, owing to too rapid titration and insufficient mixing, add another drop of methyl orange so that the solution assumes a distinct pink color; then continue the titration. Compare the volume of bromate used with the calculated volume.

Use of potassium bromate in bromine substitution reactions (see page 820)

Determination of metals yielding slightly soluble compounds with 8-hydroxyquinoline. Various metals such as aluminum, iron, zinc, nickel, cobalt, copper, manganese, and magnesium under specified conditions of acidity or alkalinity yield well-defined crystalline precipitates with 8-hydroxyquinoline (pp. 277 and 820). Use is made of these precipitations in gravimetric determinations (p. 640). The 8-hydroxyquinoline in the precipitates can be determined by bromometric titration. The volumetric method is of special advantage when small amounts of the metals are to be determined. The precipitates obtained with 8-hydroxyquinoline have the general formula $(C_9H_6ON)_nMe$, in which n is the valence of the metal. In the bromometric determination of hydroxyquinoline in acid medium, 1 mole of the latter requires 4 equivalents of bromine:

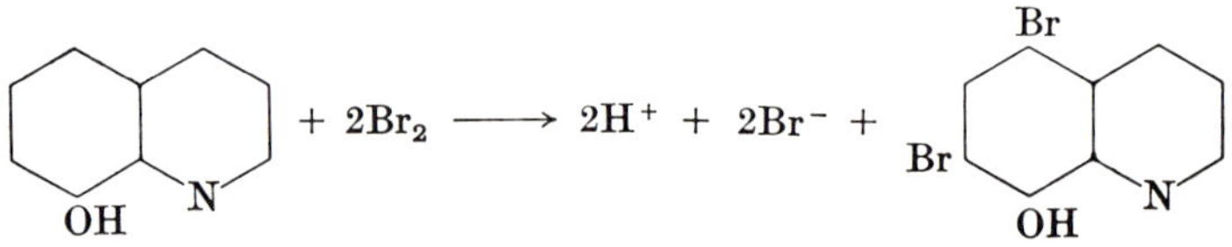

Therefore 1 mmole of a divalent metal in the oxine precipitate requires 8 milliequivalents of bromine; and 1 mmole of a trivalent metal, 12 meq.

Determination of magnesium

Procedure. Precipitate magnesium with 8-hydroxyquinoline from a solution containing 8–12 mg of magnesium, and wash the precipitate, proceeding as described on page 640. Dissolve the precipitate on the filter in 2–2.5 N hydrochloric acid, collecting the filtrate in a glass-stoppered flask; and wash the filter with the same acid until the washings are colorless. Adjust the volume of the filtrate with the same acid to 50–75 ml, add 1 g of potassium bromide and 2–4 drops of a 0.1 per cent solution of the sodium salt of methyl red. Titrate slowly with standard bromate solution, swirling constantly, until the color changes from red to yellow. A slight but distinct excess of bromine is then present. Stopper the flask, allow to stand for 2 min, add 1 g of potassium iodide, and titrate the liberated iodine with standard 0.1 N sodium thiosulfate, using starch as indicator.

The direct titration of the dissolved precipitate with bromate does not give accurate results, since it is hard to detect the end point. However, the addition of the indicator is recommended because it permits the detection of a slight excess of bromine. The indicator blank is negligibly small.

Upon the addition of potassium iodide to the solution a chocolate brown precipitate of an iodine addition compound of the dibromohydroxyquinoline is formed, which, however, decomposes rapidly on titration with thiosulfate.

Instead of back-titrating iodometrically, an excess of standard arsenious oxide may be added and the excess of the latter titrated directly with standard bromate.

Phenol. Phenol reacts with an excess of bromine to form tribromophenolbromide, $C_6H_2Br_3OBr$, which reacts with an excess of iodide to yield iodine and tribromophenol. The overall reaction after addition of iodide is

$$C_6H_5OH + 3\,Br_2 = 3\,HBr + C_6H_2Br_3OH$$

Procedure. In an iodine flask add to 25 ml of 0.01 *M* phenol solution, 25 ml of 0.1 *N* bromate, 0.5 g of potassium bromide and 5–10 ml of 4 *N* hydrochloric or sulfuric acid. Close the flask and after 10 min, add 1 g of potassium iodide, shake well, and titrate the iodine with 0.1 *N* thiosulfate.

Calcium hypochlorite or sodium hypobromite as an oxidizing agent

(See page 821.)

One-tenth *N* calcium hypochlorite solution has been proposed by Kolthoff and Stenger[2] as a standard oxidizing agent, because it has been found that the solution prepared from a commercial product[3] can be kept in dark glass bottles for a long time without appreciable change in titer.

Preparation of approximately 0.1 *N* solution. Six to ten grams of calcium hypochlorite, depending upon the amount of available chlorine in the commercial product, is added to 250 ml of distilled water, the mixture is shaken and the solution filtered to remove iron oxide, excess of calcium hydroxide, and any other insoluble material present in the commercial product. The filtrate is diluted to 1 liter. The formation of a turbidity on standing as the result of the precipitation of calcium carbonate is of no consequence. The standard solution should be kept in a glass-stoppered bottle and protected from the light (the bottle may be painted black on the outside).

Composition of the standard solution. An analysis of an approximately 0.1 *N* hypochlorite solution (prepared from the above product) gave the following composition: hypochlorite, 0.1068 *N*; calcium, 0.0743 *N*; sodium, 0.0306 *N*; free hydroxide, 0.0037 *N*; chloride, 0.0478 *N*. This corresponds to $Ca(OCl)_2$, 0.0267 *M*; $Ca(OH)_2$, 0.0019 *M*; $CaCl_2$, 0.0086 *M*; and NaCl, 0.0306 *M*.

Stability of the standard solution. When kept in a dark bottle and in a dark compartment, it was found that the normality of various solutions decreased 0.0–0.3 per cent after 3 months, 0.3–0.5 per cent after 6 months, and 0.3–1 per cent after 1 year. More dilute solutions (0.01 *N*) are less stable.

In many instances bromide is added to the solution to be titrated before the addition of hypochlorite. Thus the active agent then is hypobromite. Many workers prefer to use as reagent a standard hypobromite instead of hypochlorite solution.

Preparation of 0.1 *N* sodium hypobromite. Dissolve 24 g of pure sodium hydroxide in 1 liter of distilled water and cool to −4°C in a mixture of ice and salt. Add 8 g of pure bromine. Shake thoroughly and leave the solution in the cooling mixture for 2 hr. The solution should be standardized before use.

Standardization of the hypochlorite solution against arsenious oxide. Indicator solution: 0.2 per cent Bordeaux (British Colour Index No. 88). Color change from pink to very faint yellow-green. An indicator correction of 0.03 ml of 0.1 *N* hypochlorite should be subtracted per 0.1 ml of indicator solution in a final volume of 50–75 ml. For the use of tartrazin and of quinoline yellow as indicators see page 820.

Cerano[4] observed that bromthymol blue acts like a reversible indicator for hypochlorite. It changes color from deep blue to greenish yellow. Belcher (Ref. 1) found bromcresol purple, cresol red, xylenol orange, thymol blue, and methyl thymol blue suitable indicators. He prefers bromcresol purple, which changes color from purple to yellow.

[2] I. M. Kolthoff and V. A. Stenger, *Ind. Eng. Chem., Anal. Ed.* **7**, 79 (1935).
[3] Manufactured by the Mathieson Alkali Co., Niagara Falls, N.Y.
[4] L. A. Cerano, *Rev. Fac. Ing. Quim. Univ. Nacl. Litorial, Santa Fe, Arg.* **29**, 105 (1960).

Procedure. Pipet 25 ml of standard 0.1 N arsenious oxide solution into an Erlenmeyer flask; add 1 g of potassium bromide and 0.5 g of sodium or potassium bicarbonate. Titrate with 0.1 N hypochlorite solution, added moderately rapidly until within a few milliliters of the expected end point. Add 1 drop of indicator, and then titrate dropwise with constant swirling until the color of the indicator fades. Add 1 more drop of indicator; if the color does not fade, continue the titration until 1 drop of hypochlorite causes the solution to flash from pink to colorless or light yellow-green. Subtract the indicator correction, and calculate the normality of the hypochlorite solution.

Notes:

1. Since the indicator reaction is not reversible, one must be sure that the end point is actually reached when the color has faded; this can be tested by adding more indicator. It is advantageous, however, to use as little indicator as possible. When the approximate end point is known, one can perceive the end point with only 1 drop of Bordeaux solution.

2. The solution can also be standardized iodometrically. For 25 ml of hypochlorite solution add 1–1.5 g of potassium iodide, 5 ml of 6 N sulfuric acid, and titrate with standard sodium thiosulfate.

Determination of ammonia in ammonium sulfate. In the presence of sodium bicarbonate and bromide, ammonia is oxidized quantitatively to nitrogen and water by calcium hypochlorite.

$$2\,NH_3 + 3\,OBr^- \rightarrow N_2 + 3\,Br^- + 3\,H_2O$$

It is necessary to add an excess of hypochlorite since the reaction is fairly slow. Then a known excess of arsenious oxide solution is added, and the latter is back-titrated as described above.

Procedure. Weigh out a 1-g sample of ammonium sulfate, dissolve in water, and make up to the mark in a 250-ml volumetric flask. Pipet out 25 ml of this solution, add 1 g of potassium bromide and 0.5 g of sodium or potassium bicarbonate, and titrate with 0.1 N hypochlorite solution until a permanent light yellow color, indicating a slight excess of hypobromite, appears.

Allow to stand 5 min, and then add from a pipet 10 ml of 0.01 N arsenious oxide (obtained by proper dilution of the standard 0.1 N solution). Add 1 drop of Bordeaux indicator, which should impart a pinkish color to the solution. (If the color fades, there is not an excess of arsenious oxide present.) Continue the titration with hypochlorite as in the standardization of the latter against arsenious oxide.

Calculate the amount of hypochlorite corresponding to 10 ml of 0.01 N arsenious oxide solution, and deduct this volume from the total volume of hypochlorite used. The difference is equivalent to the amount of ammonia present. Calculate the percentage of NH_3 and of N.

Note: The excess of hypochlorite (or actually hypobromite) may also be determined iodometrically by adding first 0.5 g of potassium iodide, then 10 ml of 6 N H_2SO_4 (slowly to avoid vigorous effervescence), and finally titrating with standard thiosulfate. The advantage in the use of arsenious oxide in bicarbonate medium is that interference of foreign oxidizing agents such as ferric iron, chromate, or arsenate is eliminated. On the other hand, the presence of iodide will cause an error in the arsenious oxide procedure but not in the iodometric method.

PART SIX

Theory of Electroanalytical Chemistry

CHAPTER # 46 *Methods of Electroanalysis*

In recent years a wide variety of electroanalytical methods have been devised. These new methods, and the more classical techniques, commonly involve at least one of the three electrical quantities—voltage, current, and resistance. In the various electrochemical titration techniques, titrant is added to a solution and the following electrical quantity determined: (1) the potential of an indicator electrode (potentiometric titrations), (2) the resistance of the solution (conductometric titrations), (3) the current at an indicator electrode held at fixed potential (amperometric titrations), or (4) the potential of an indicator electrode with a constant impressed current (constant-current potentiometric titration).

In the nontitration techniques, one of the electrical quantities is controlled during the experiment and one or both of the others determined. In some cases the measured electrical quantity changes with time, and this time dependence is also recorded. Sometimes the product produced by the electrode reaction is quantitatively determined by using a physical property, e.g., weight, volume of gas evolved, etc., or some chemical procedure is used to determine the amount of product.

In this introductory chapter, the more common electrochemical methods are classified and discussed briefly. These methods are considered in more detail in the following chapters. However, it is necessary, initially, to review the definition of the various electrical units and the relationships that exist among them.

LAWS AND UNITS

Electrical units

The unit of current is called the *ampere* in honor of the French physicist and chemist A. M. Ampère (1775–1836). One ampere is that current which deposits 1.11800 mg of silver (or 0.3296 mg of copper, etc.) from a solution in 1 second.[1] Current is measured by means of an instrument called an *ammeter*.

The unit *quantity* of electricity is called the *coulomb* in honor of the French physicist C. A. Coulomb (1736–1806). The coulomb is defined as the quantity of electricity which flows through a given cross section of a conductor in 1 second when the current is 1 ampere. One coulomb, therefore, will deposit 1.11800 mg of silver.

[1] The above definition is that of the "international ampere." The "absolute" ampere is defined as follows: If, in a conductor, 6.25×10^{18} electrons flow through a certain cross section in 1 second, the current is equal to 1 ampere. The quantity of electricity associated with the electron is 4.774×10^{-10} esu (electrostatic units). The "international ampere" and the "absolute ampere" are practically identical.

The unit of electrical resistance is the *ohm*, so called after the German physicist G. S. Ohm (1787–1854), who discovered the relation between current, potential, and resistance. One ohm is equal to the resistance at 0°C of a column of mercury 106.300 cm in length of uniform cross section (very nearly 1 mm^2) and weighing 14.4521 g.[2]

The unit of electromotive force (and potential difference) is called the *volt* after the Italian physicist Count A. Volta (1745–1827). One volt is the electromotive force required to maintain a current of 1 ampere through a conductor having a resistance of 1 ohm. Electromotive force (emf) is measured by means of a *voltmeter*.

The unit of capacitance is named in honor of Michael Faraday (1791–1867). One *farad* is equal to the capacitance of a capacitor (condenser) which has 1 coulomb of charge stored when a 1-volt difference exists between the two plates. Most capacitors have capacitances which are very much smaller than 1 farad, and the units used are the micro (10^{-6}), nano (10^{-9}), and pico (10^{-12}) farad.

The unit of inductance is named for Joseph Henry (1797–1878), the American physicist who contributed much to the early research on electromagnetism. One *henry* is equal to the self-inductance of a coil with a 1-volt potential difference across its ends through which current flows with a constant rate of change of 1 ampere per second.

The unit of frequency is named for Heinrich Rudolph Hertz (1857–1894), who is known for his contributions to the study of electromagnetic radiation. One *hertz* represents a frequency of 1 cycle per second.

Laws of electrolysis

The brilliant research of the English scientist Michael Faraday (1791–1867) resulted in the discovery of the relation which exists between the quantity of electricity which passes through a solution and the quantity of matter which separates at the electrodes.

The first law of Faraday states: *The quantities of substances set free at the electrodes are directly proportional to the quantity of electricity which passes through the solution.* Thus the quantity of a metal liberated at the cathode during the passage of 10 coulombs of electricity through a solution of the metal salt is five times that which is deposited when 2 coulombs pass through the solution.

The second law states: *The same quantity of electricity sets free the same number of equivalents of any substance at the electrodes.* This law implies that the passage of the same quantity of electricity through solutions of copper chloride, silver sulfate, and hydrobromic acid sets free quantities of copper, chlorine, silver, oxygen, hydrogen, and bromine that are proportional to the equivalent weights, or in a ratio of:

$$\frac{63.54}{2} : \frac{35.453}{1} : \frac{107.870}{1} : \frac{15.9994}{2} : \frac{1.00797}{1} : \frac{79.909}{1}$$

If in a certain case two ions are discharged simultaneously by a quantity of electricity sufficient to discharge 1 gram equivalent of a substance, then the sum

[2] The above definition is that of the "international ohm." The "absolute ohm" would correspond to a column 106.247 cm in length.

of the quantities of the two substances discharged simultaneously will be equal to 1 gram equivalent.

The number of grams of an element set free by the passage of 1 coulomb of electricity (or 1 ampere-second) is called the *electrochemical equivalent* of the element. According to Faraday's laws, electrochemical equivalents are proportional to chemical equivalents. The electrochemical equivalent of silver is, as we have seen, 0.001118 g. It is of interest to calculate the amount of electricity required to liberate 1 equivalent of an element. From Faraday's first law we learn that this amount is found by dividing the gram-equivalent weight by the electrochemical equivalent. Thus in the case of silver: 107.88/0.001118 = 96,494, or approximately 96,500 coulombs. This quantity of electricity is called the *faraday* (F). In accordance with the second law of electrolysis, the passage of 1 faraday of electricity will liberate 1 gram equivalent of some substance at each electrode.

Faraday's laws of electrolysis are among the most exact laws of physical science; they hold under all conditions, independent of the concentration of the electrolyte, of the temperature, and of the solvent used.

Relations between electrical quantities

The relation between the three fundamental quantities, current, resistance, and electromotive force, is expressed by *Ohm's law*, which states: *The current, i, is directly proportional to the electromotive force, E, and inversely proportional to the resistance, R:*

$$i = \frac{E}{R} \qquad \text{or} \qquad E = iR \tag{1}$$

The charge, Q, on a capacitance, is directly proportional to the potential difference, E, and the capacitance, C:

$$Q = CE \tag{2}$$

The potential difference, E, across an inductance is proportional to the time rate of change of current, di/dt, through it, and its inductance, L,

$$E = L\frac{di}{dt} \tag{3}$$

When alternating current flows in an electrical circuit, the current and potential vary with time in a sinusoidal fashion. If only resistive elements are present in a circuit, Ohm's law describes the relationship between current and emf at any instant in time. If capacitances or inductances are present, a phase difference exists between the sinusoidal current function and emf function. The behavior of such circuits is described in terms of the *impedance*, Z, which is defined as

$$Z = \frac{E_{\text{max}}}{I_{\text{max}}} = \frac{E_{\text{rms}}}{I_{\text{rms}}} = \frac{E_{\text{ave}}}{I_{\text{ave}}} \tag{4}$$

where the subscripts signify the maximum, root mean square, and average values, respectively. The impedance of a circuit is related to the resistance, the capacitance, and the inductance of the circuit. The contribution of the latter two terms appears in the reactance term, X, and when all circuit elements are in series

$$Z = (R^2 + X^2)^{1/2} \tag{5}$$

where

$$X = 2\pi f L - \frac{1}{2\pi f C}$$

and f is the frequency of the alternating current.

If the resistance, capacitance, and inductance are all in parallel, the admittance, $Y(=1/Z)$, is related to the conductance, $G(=1/R)$, and the susceptance, B, of the circuit by

$$Y = (G^2 + B^2)^{1/2} \tag{6}$$

where

$$B = 2\pi f C - \frac{1}{2\pi f L}$$

CLASSIFICATION

It is most convenient to classify electrochemical methods of analysis in terms of the changes which are produced on the sample being analyzed. Depending upon the exact experimental conditions (stirring, size of the electrode, volume of solution, and time of electrolysis) the electrochemical procedure may convert all, some, or none of the reactant in the cell to products. We denote *bulk electrolysis* methods as those in which all the electroactive reactant in the cell is quantitatively converted to products. *Diffusion-layer* electrolysis methods refer to those techniques in which only a thin layer of solution immediately adjacent to the electrode, the diffusion layer, undergoes electrolysis. In these methods, concentration is related to the measured electrical variable, and ordinarily the time of electrolysis is so short that only a negligible fraction of the reactant is electrolyzed. *Zero-current* methods involve experimental procedures which should, theoretically, not alter the concentration of any species in solution.

Bulk-electrolysis methods

Quantitative electrodeposition. The constituent to be determined is separated on a cathode or an anode (bulk electrolysis). The reaction product is determined by weighing (hence the name electrogravimetry), or by other means. The product of the electrolysis may be soluble, as in the case when Fe(III) is reduced electrolytically to Fe(II), which can be determined in various ways, *v. i.*

Coulometry. Instead of determining the amount of constituent plated out quantitatively, the amount of electricity required for the quantitative electrolytic separation is measured, and the amount of the constituent is calculated by making use of Faraday's classical laws of electrolysis. The amount of electricity is measured in *coulombs*, and this type of measurement is called *coulometry*.

Coulometric titration. Many substances cannot be conveniently electrolyzed directly to a higher or lower oxidation state. Frequently, in such cases, a species capable of reaction with the original substance can be generated electrochemically. The equivalence point for the reaction between the original substance and the electrochemically generated species is then determined using a visual indicator or some physiochemical technique. By determining the number of coulombs required to reach the equivalence point, the quantity of the initial substance can be calculated.

Diffusion-layer electrolysis methods

Voltammetry. In voltammetry the potential of a suitable electrode is held constant and the current through the electrode measured under such conditions that only the surface layer at electrode is depleted of electroactive material. This surface layer is called the *diffusion layer* when the solution is unstirred, because material is transported to the electrode surface across the surface film by diffusion. If the solution is stirred, the effective thickness of the surface layer is reduced, but it is conventional to still use the term diffusion layer to describe the surface film across which electroactive material is transported.

To quantitatively describe the shape of the current–potential ($i - E$) curve obtained by varying the voltage applied to the two electrodes making up an electrochemical cell, it is necessary to consider how the potential of each electrode varies as current passes through the cell.

First consider the cell shown in Fig. 46-1, in which a rotating platinum-wire electrode of area ~0.2 cm^2 is one electrode, and a large-area (~10 cm^2) silver electrode covered with silver chloride is the other electrode. The solution contains an indifferent electrolyte, 0.1 M potassium chloride, and is very dilute in thallous chloride, e.g., 10^{-4} M. This indifferent electrolyte, more commonly called the *supporting electrolyte*, is added to prevent electrical migration from contributing to mass transport (see p. 893).

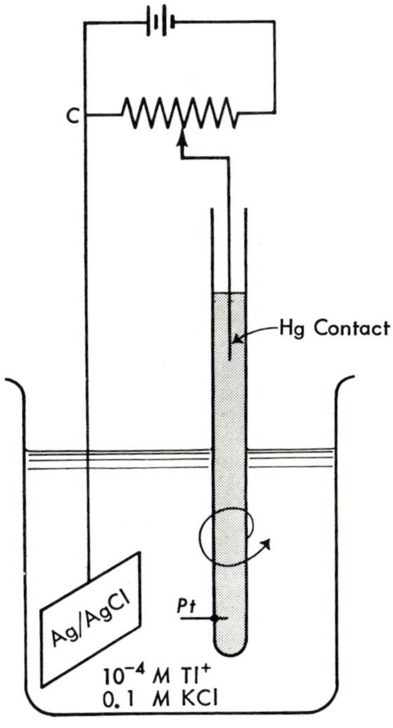

Fig. 46-1. Representation of the voltammetric cell: Ag/AgCl, KCl(0.1 M), $Tl^+(10^{-4}$ $M)/Pt$. Rotating platinum wire electrode used.

The silver–silver chloride electrode is a reversible electrode; i.e., if current passes in one direction the reaction

$$AgCl + e \rightarrow Ag + Cl^- \tag{7}$$

occurs, while reversal of the current causes the reaction

$$Ag + Cl^- \rightarrow AgCl + e \tag{8}$$

to occur. If no current passes through the electrode, its potential is given by the Nernst equation:

$$E = E^0_{AgCl/Ag} - \frac{RT}{F} \ln (Cl^-) = E^{0\prime}_{AgCl/Ag} - 0.059 \log [Cl^-] \text{ at } 25°C \tag{9}$$

where $[Cl^-] = 0.1$. For simplicity $E^{0\prime}_{AgCl/Ag}$ and $[Cl^-]$ are written instead of $E^0_{AgCl/Ag}$ and $(Cl^-]$ (see p. 165). This does not affect this derivation or further ones because f_{Cl^-} is constant. If current passes through the electrode, the potential can be calculated from

$$E = E^{0\prime}_{AgCl/Ag} - 0.059 \log [Cl^-]^0 \tag{10}$$

where $[Cl^-]^0$ is the concentration of chloride ion at the electrode surface when current is flowing. The notation [ ] and $[\ ]^0$ will be used below to represent the bulk and electrode surface concentrations of a species. If reaction (7) occurs, $[Cl^-]^0 > [Cl^-]$, while $[Cl^-]^0 < [Cl^-]$ if reaction (8) takes place. Provided the area of the silver electrode is sufficiently large, or the current through the electrode sufficiently small, $[Cl^-]^0$ will not differ significantly from $[Cl^-]$. Hence the potential of the electrode will not differ appreciably from the zero-current value of 0.28 V (versus N.H.E.). Such an electrode is called a *depolarized electrode*. All reversible electrodes can function as depolarized electrodes provided the current density (current flowing through 1 cm^2 of electrode surface) is sufficiently small.

According to the definition given on page 163, the platinum electrode is a typically irreversible electrode in this solution. When it is made the anode in an electrolysis, oxygen evolution occurs (see p. 163), and when it is made the cathode, the electrode reaction is

$$Tl^+ + e \rightleftharpoons Tl \tag{11}$$

Upon short-circuiting the cell (point C in Fig. 46-1), no current flows. The platinum electrode adopts the potential of the silver–silver chloride electrode. The potential versus the N.H.E. of the Tl^+/Tl electrode in 10^{-4} M thallous chloride solution is about -0.57 V N.H.E. or -0.85 V versus the Ag/AgCl electrode. It is necessary to impose this potential, which may be called the *decomposition potential*, upon the platinum electrode to reduce thallium ions to thallium on the platinum. The cell has now become an *electrolysis* cell (see p. 162).

We shall consider now the current–potential curve (Fig. 46-2), which gives the current that flows at various applied potentials. For reasons to be stated on page 886, the platinum electrode is rotated rapidly at a constant rate during the experiment. In this experiment we shall vary the potential of the platinum electrode from positive potentials to negative potentials and then back positively. Until point A, the current is extremely small and close to zero. This small current is called the *residual current*. At point A the *decomposition potential* is attained

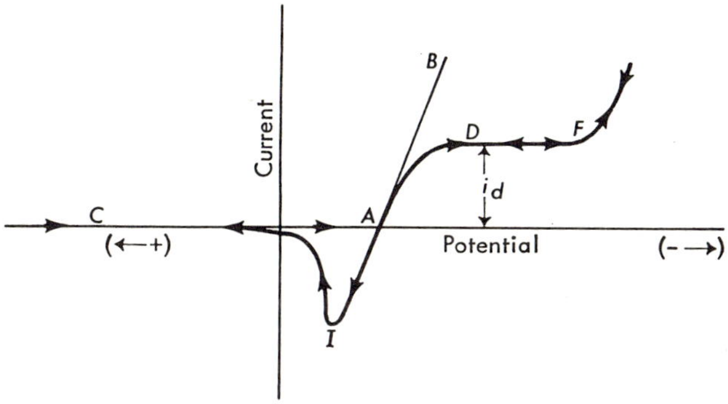

Fig. 46-2. Current–potential curve obtained with the cell in Fig. 46-1. Arrowheads indicate direction of potential scan.

and reduction of thallium ions starts. If the concentration of the thallium ions around the electrode would remain the same as in the solution and if the electrode reaction is ideally reversible, then the current would increase with increasing applied potential according to the straight line AB, the slope of which is inversely proportional to the resistance, R. Under the specified conditions, Ohm's law would hold: $E = iR$. However, during the electrolysis, the concentration of the thallium ions at the electrode must be smaller than in the bulk of the solution.

This concentration can be calculated from the Nernst equation:

$$E = E^{0\prime}_{Tl^+/Tl} + 0.59 \log [Tl^+]^0 \ (25°C) \tag{12}$$

At constant electrode potential, $[Tl^+]^0$ is seen to be constant. Since thallium ions are being plated out continuously on the electrode, it follows that thallium ions are being transferred from the bulk of the solution to the electrode surface in order to hold $[Tl^+]^0$ constant. This transfer results from the concentration difference, between the electrode surface and the bulk of the solution, and occurs by diffusion and convection caused by rotating the electrode. The rate of transfer by convective–diffusion exactly equals the rate at which thallium ions are plated out.

When the current is very small, $[Tl^+]^0$ does not differ very much from $[Tl^+]$. However, with increasing current, the difference between $[Tl^+]$ and $[Tl^+]^0$ increases. Consequently, the actual current–potential curve deviates to the right of the line AB (Fig. 46-2), which was calculated on the basis that $[Tl^+] = [Tl^+]^0$. The difference in potential between the theoretical and experimental lines is often called *concentration overpotential*. The concentration difference, $[Tl^+] - [Tl^+]^0$, extends only over a very thin layer in the immediate neighborhood of the electrode, the diffusion layer (see p. 901). At a rotating electrode a diffusion layer of constant thickness is established almost instantaneously.

At any point on the current–potential curve, the current is proportional to the number of equivalents of thallium plated out per second. Since the current is constant at a given potential, the amount plated out must be equal to the amount transferred per second to the surface because of the concentration difference $([Tl^+] - [Tl^+]^0)$. At any point the rate of supply (by convective–diffusion) is equal to the rate of disappearance (current), or

$$i = k([Tl^+] - [Tl^+]^0) \tag{13}$$

The current cannot be increased indefinitely, because $[Tl^+]^0$ becomes smaller and smaller and soon approaches zero. When this is the case, $([Tl^+] - [Tl^+]^0)$ in Eq. (13) becomes equal to $[Tl^+]$. Hence the current is proportional to $[Tl^+]$ (point D in Fig. 46-2):

$$i_d = k[Tl^+] \tag{14}$$

The current, i_d, is determined by the rate of mass transfer (convection plus diffusion), but for convenience i_d is called the *diffusion current* rather than the convective-diffusion current.

The constant k in Eq. (14) is called the *mass-transfer coefficient* and depends upon the form of the electrode and mass-transport conditions. The factors that affect k are discussed in Chapter 47 for several different kinds of electrodes.

Equation (14) is the basis of the quantitative application of voltammetry. In voltammetry, an electrode at which mass transfer limits the current is called the indicator electrode, since it indicates the concentration of electroactive substance. From the analytical point of view it is important to emphasize that loss of electroactive material by electrolysis is negligibly small when a microindicator electrode is used.

Upon continuous increase of the negative potential, the current will suddenly increase at a certain potential, when another constituent is being discharged. At the platinum electrode in a neutral potassium chloride solution, the current at point F will start to increase due to the reduction of water:

$$H_2O + e \rightarrow \tfrac{1}{2} H_2 + OH^-$$

In an acid medium, point F is displaced to less negative potentials, corresponding to the discharge potential of hydrogen ions at the particular hydrogen-ion concentration. This hydrogen discharge usually occurs at a more negative potential than corresponds to the thermodynamically reversible hydrogen ion–hydrogen potential. This *overpotential* (see p. 932) is small on bright platinum and still smaller at platinized platinum. The greatest hydrogen overpotential is found on mercury, where it is of the order of 0.8 V (see Table 51.1, p. 933). This accounts for the use of mercury electrodes in quantitative voltammetry and in separations at a mercury cathode (see p. 936). It is even possible to separate alkali metals as amalgams at a mercury electrode. The deposition of alkali metals is impossible at a platinum electrode.

Returning to Fig. 46-2, we continue the experiment from the negative potential F and make the applied potential less and less negative. If the electrode reaction is perfectly reversible, we will follow the line $FDAIC$ because after the previous electrolysis there is formed a thallium deposit on the electrode, which dissolves when the potential is made less negative than A (path AI). The cell operates now as an electrochemical (galvanic) cell (see p. 160).

Curves of the sort shown in Fig. 46-2 are usually obtained using an instrument known as a *polarograph* which continuously plots the current as the voltage is varied automatically.

The potential at A corresponds to that of thallium in a solution of thallous chloride at a concentration, $[Tl^+]$:

$$E = E^0_{Tl^+/Tl} + \frac{RT}{F} \ln (Tl^+) = E^{0'}_{Tl^+/Tl} + 0.059 \log [Tl^+] \text{ (25°C)} \tag{15}$$

It is a simple matter to derive a relation between current and potential on part *AD* (often called the *wave*), on the voltammogram, provided the electrode reaction is perfectly reversible, i.e., that the potential at the electrode corresponds to the surface concentration $[Tl^+]^0$. We have seen already that the following relations hold:

$$i_d = k[Tl^+] \quad (14)$$

$$i = k([Tl^+] - [Tl^+]^0) \quad (13)$$

or

$$i = i_d - k[Tl^+]^0$$

and

$$[Tl^+]^0 = \frac{i_d - i}{k} \quad (16)$$

in which k is a constant and, in general, equal to the rate of transfer of the electroactive ions under the experimental conditions (rate of stirring, temperature).

Introducing the relation in Eq. (16) into Eq. (12), we find that at any point on the "wave"

$$E = E^{0\prime}_{Tl^+/Tl} + 0.059 \log [Tl^+]^0 = E^{0\prime}_{Tl^+/Tl} + 0.059 \log \frac{(i_d - i)}{k}$$

$$= E^* + 0.059 \log (i_d - i) \quad (17)$$

In practical voltammetry, it is customary to separate the reference electrode from the indicator electrode by a suitable salt bridge or diaphragm. Thus the cell

reference electrode ‖ $Ag^+(10^{-4}\ M)$, $KNO_3(0.1\ M)$ | Pt

would be used to obtain the experimental current–voltage curve for Ag^+ at a rotating platinum electrode. The saturated calomel electrode, S.C.E., is the most used reference electrode, and in this case a potassium nitrate salt bridge would be employed in place of a potassium chloride salt bridge to interconnect the S.C.E. and the silver-ion half-cell. *It is almost universal practice to report current–potential curves with respect to the S.C.E., whatever reference electrode is used.*

A suitable reference electrode for voltammetry must have a large surface area, and a high concentration of soluble, electroactive species which takes part in the reference electrode half-reaction. Such an electrode cannot develop a significant concentration difference in electroactive constituents between the electrode surface and the solution and will have a low internal resistance. A low internal resistance is highly desirable, because all the current flows through the reference electrode, producing a voltage drop.

Similarly, the salt bridge connecting the two half-cells should have a wide cross section, should be short, and should contain a concentrated electrolyte solution to provide a highly conductive path.

The potential used in the wave equation (17), E, is calculated from

$$E = E_{applied} - iR$$

where $E_{applied}$ is the voltage across the cell, i the current through the cell, and R the sum of the cell resistance, the resistance of the voltage source, and of the resistance of the microammeter which is usually inserted in series with the sliding contact and the rotating platinum electrode in Fig. 46-2.

Polarography. Of the various types of indicator electrodes, the dropping mercury electrode (D.M.E.) introduced by Heyrovsky[3] has gained the most acceptance in inorganic and organic analysis. The term polarography is applied to voltammetry using the D.M.E.

The D.M.E. consists of a glass capillary usually 10–15 μ in radius and 5–10 cm long, from which mercury falls as a series of drops. The capillary is connected by flexible tubing to a leveling bulb filled with mercury. The difference in height between the mercury in the leveling bulb and the tip of the capillary determines the rate at which mercury flows through the capillary. A drop falls when its weight exceeds the surface tension force holding it to the end of the capillary. Drop times of 3–7 sec are considered to include the optimum values.

During electrolysis the current at a D.M.E. increases continuously with time as the drop grows and decreases abruptly to zero when the drop falls. This cycle repeats itself with each new drop, and currents through successive drops are extremely reproducible.

Diffusion currents proportional to the bulk concentration, C, of electroactive substances are observed at the D.M.E., i.e.,

$$i_d = k'C$$

where k' depends upon the diffusion coefficient of the electroactive substance, the dimensions of the capillary, and other experimental conditions. For convenience we shall discuss the current when the drop falls, although the current at any other time in the drop life could be used.

A description of the factors that affect the diffusion current at a D.M.E. is given in Chapter 47.

The current–potential curve for the reduction

$$\mathrm{Hg} + \mathrm{Tl}^+ + e \rightarrow \mathrm{Tl}_{\mathrm{amalgam}} \tag{18}$$

in 0.1 M KNO_3 has the shape shown in Fig. 46-3. Such curves are called polarograms when the D.M.E. is the indicator electrode and voltammograms for all other indicator electrodes. The equation for this wave is obtained by evaluating the surface concentrations of Tl^+ in the aqueous phase, and of Tl in the mercury phase, and substituting into the Nernst equation for Eq. (18):

$$E = E^{0'}_{\mathrm{Tl}^+/\mathrm{Tl}_{\mathrm{amal}}} + 0.059 \log \frac{[\mathrm{Tl}^+]^0}{[\mathrm{Tl}_{\mathrm{amal}}]^0} \tag{19}$$

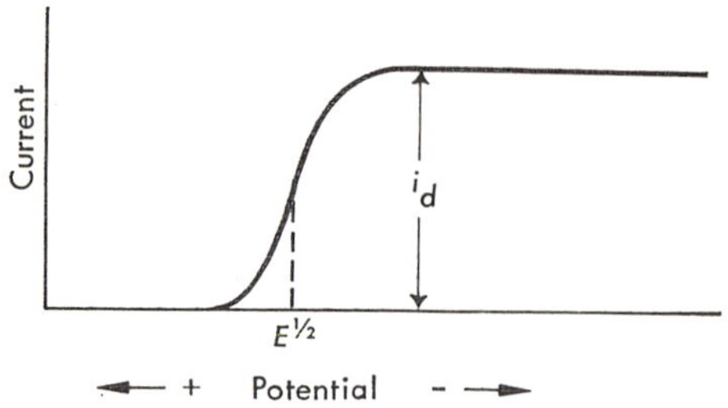

Fig. 46-3. Polarogram of 10^{-3} M Tl^+ in 0.1 M KNO_3.

[3] J. Heyrovsky, *Chem. Listy* **16**, 256 (1922); *Phil. Mag.* **45**, 303 (1923). Heyrovsky received the Nobel Prize in Chemistry in 1959 for his development of the polarographic method of analysis.

At any point on the polarographic wave, the concentration of thallous ion at the D.M.E. surface is given by an equation of the same form as (16):

$$[\mathrm{Tl}^+]^0 = \frac{i_d - i}{k_s} \tag{20}$$

while the surface concentration of thallium in the amalgam is directly proportional to the current

$$[\mathrm{Tl}_{\mathrm{amal}}]^0 = \frac{i}{k_a} \tag{21}$$

k_s and k_a are the mass-transfer coefficients for thallium in the solution and the amalgam, respectively. Substituting Eqs. (20) and (21) into (19) yields

$$E = E^{0\prime}_{\mathrm{Tl}^+/\mathrm{Tl}_{\mathrm{amal}}} + 0.059 \log \frac{k_a}{k_s} + 0.059 \log \frac{i_d - i}{i} \tag{22}$$

When $i = i_d/2$,

$$(E)_{i_d/2} = E^{0\prime}_{\mathrm{Tl}^+/\mathrm{Tl}_{\mathrm{amal}}} + 0.0591 \log \frac{k_a}{k_s} + 0.059 \log \frac{i_d - (i_d/2)}{i_d/2}$$

The value of $\log [i_d - (i_d/2)]/i_d/2$ is zero. Hence

$$(E)_{i_d/2} = E^{0\prime}_{\mathrm{Tl}^+/\mathrm{Tl}_{\mathrm{amal}}} + 0.0591 \log \frac{k_a}{k_s} = E^{1/2}_{\mathrm{Tl}^+/\mathrm{Tl}_{\mathrm{amal}}} \tag{23}$$

The potential at which $i = i_d/2$ is known as the *half-wave potential*, $E^{1/2}$. Combining Eqs. (22) and (23) yields

$$E = E^{1/2}_{\mathrm{Tl}^+/\mathrm{Tl}_{\mathrm{amal}}} + 0.059 \log \frac{i_d - i}{i} \tag{24}$$

The half-wave potential for reaction (18) is independent of the concentration of Tl^+ present. In this case, the half-wave potential is almost equal to $E^{0\prime}_{\mathrm{Tl}^+/\mathrm{Tl}_{\mathrm{amal}}}$ since the ratio k_a/k_s is approximately unity. (See page 904 for the relationship between k_a and k_s.) Thus the half-wave potential is characteristic of Tl^+, and the D.M.E. supplies qualitative as well as quantitative information.

Multicomponent analysis is possible. As an example, consider a mixture of Cu^{++}, Cd^{++}, and Zn^{++}, all about 10^{-3} M, in the presence of a supporting electrolyte, 0.1 M $\mathrm{NaClO_4}$. As the potential of the D.M.E. is made progressively more negative, starting from a potential positive with respect to the reduction potential for Cu^{++} and going to a potential more negative than that required to reduce Zn^{++}, we obtain the current–potential curve shown in Fig. 46-4.

Starting at potential A, in Fig. 46-4, as the D.M.E. is made more negative, the reaction

$$\mathrm{Cu}^{++} + 2e + \mathrm{Hg} \rightarrow \mathrm{Cu}_{\mathrm{amal}} \tag{25}$$

starts to occur, and at potential C, the current has become constant because the rate of supply of Cu^{++} to the D.M.E. surface has become completely diffusion-controlled:

$$(i_d)_{\mathrm{Cu}^{++}} = k_{\mathrm{Cu}^{++}}[\mathrm{Cu}^{++}] \tag{26}$$

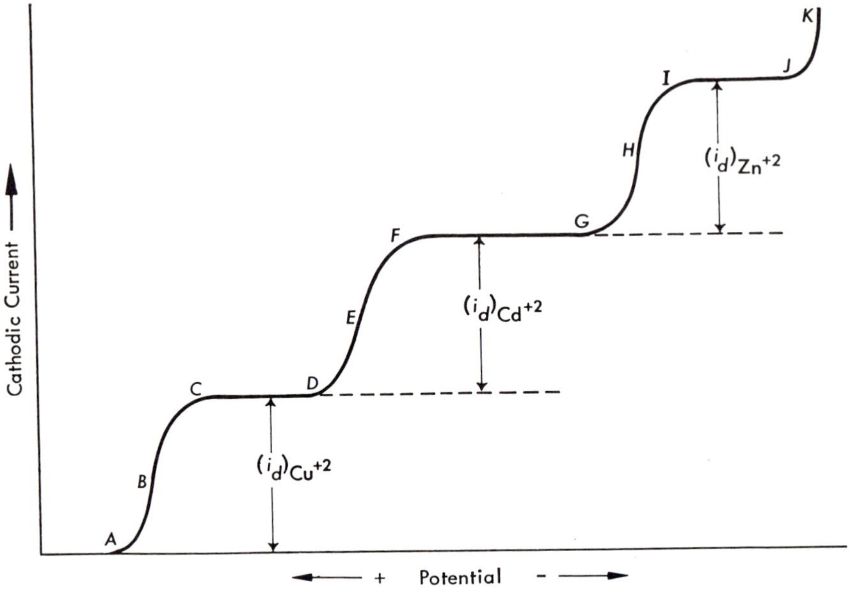

Fig. 46-4. Polarogram of a three-component mixture containing 10^{-3} *M* Cu^{++}, Cd^{++}, and Zn^{++} in 0.1 *M* $NaClO_4$.

where $k_{Cu^{++}}$ is a constant characteristic of the D.M.E. and cupric copper in 0.1 *M* sodium perchlorate (see p. 888). The potential at $i = (i_d)_{Cu^{++}}/2$, point *B*, is the half-wave potential for reaction (25). $E^{1/2}_{Cu^{+}/Cu_{amal}}$ is characteristic of the Cu/Cu_{amal} couple in 0.1 *M* $NaClO_4$.

Until potential *D*, the current at the D.M.E. remains constant, but at more negative potentials it increases, because the reaction

$$Cd^{++} + 2e + Hg \rightarrow Cd_{amal} \qquad (27)$$

occurs in addition to Eq. (25). In the potential range *FG*, the current in excess of $(i_d)_{Cu^{++}}$ is controlled solely by the diffusion of Cd^{++} to the surface of the D.M.E.; hence this *extra* current is the cadmium-ion diffusion current,

$$(i_d)_{Cd^{++}} = k_{Cd^{++}}[Cd^{++}] \qquad (28)$$

The *total* current in the potential range *FG* is the sum of the currents arising from the electrochemical reactions (25) and (27). In general, when more than one electrochemical reaction occurs simultaneously at an indicator electrode, the total current is the sum of the individual currents for each reaction. The quantitative determination of mixtures of electroactive species is based on this principle.

The half-wave potential for reaction (27), $E^{1/2}_{Cd^{++}/Cd_{amal}}$, occurs at potential *E*, when the current is $(i_d)_{Cu^{++}} + \frac{1}{2}(i_d)_{Cd^{++}}$.

The increase in current in the potential range *GHI* corresponds to the reaction

$$Zn^{++} + 2e + Hg \rightarrow Zn_{amal} \qquad (29)$$

where the diffusion current for Zn^{++} is that current in excess of the sum of the diffusion currents of Cu^{++} and Cd^{++}. Again

$$(i_d)_{Zn^{++}} = k_{Zn^{++}}[Zn^{++}] \tag{30}$$

Potential H is the half-wave potential characteristic of Zn^{++} in 0.1 M sodium perchlorate, and occurs at a current $(i_d)_{Cu^{++}} + (i_d)_{Cd^{++}} + \frac{1}{2}(i_d)_{Zn^{++}}$.

Amperometric titrations. Use of the relation

$$i_d = kC \tag{31}$$

is made in *amperometric titrations* with one indicator electrode. Consider the titration of Cl^- with Ag^+, using a rotating platinum electrode, at an applied potential where Eq. (31) holds for Ag^+.

Neglecting the solubility of silver chloride, the current remains zero (or equal to a very small value, the residual current) until the equivalence point. When excess silver ion is present, the current is proportional to the excess concentration of silver ion. The current variation during the titrations is shown in Fig. 46-5; the equivalence point corresponds to the intersection of the two straight-line segments for the reasons given in Chapter 38. Amperometric titrations are considered in more detail in Chapter 49 and extensively discussed by Stock in his recent book on the subject.[4]

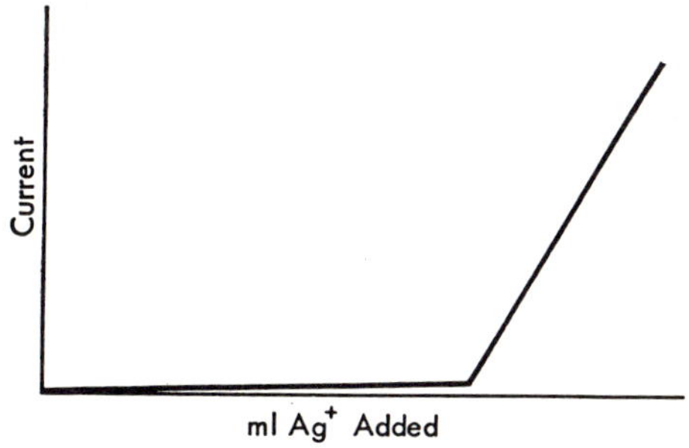

Fig. 46-5. Amperometric titration of Cl^- with Ag^+.

Linear-scan voltammetry. Specialized forms of voltammetry exist in which the potential of the indicator electrode is varied in a controlled way with time and the time-dependent current observed. Some of these techniques are finding considerable acceptance in electroanalytical chemistry as research tools, and one, linear-scan voltammetry, is of interest in trace analysis. Consider a cell in which the conditions for semiinfinite linear diffusion are met, i.e., a cell in which only diffusion normal to the indicator electrode need be considered (see p. 886). The solution is unstirred and contains supporting electrolyte and a reducible species, e.g., Ag^+. The potential of the indicator electrode is now caused to vary linearly with time from a potential positive with respect to the decomposition potential of silver to one that is considerably more negative. At some time during the voltage scan, silver will start to deposit on the electrode, and the current would rise exponentially with voltage (time) were it not for the fact that the diffusion-layer thickness also increases. This increase in diffusion-layer thickness tends to decrease the current

[4] J. T. Stock, *Amperometric Titrations*, Wiley-Interscience, New York, 1965.

inversely as the square root of time. These two opposing effects produce a curve with a maximum, as shown in Fig. 46-6.

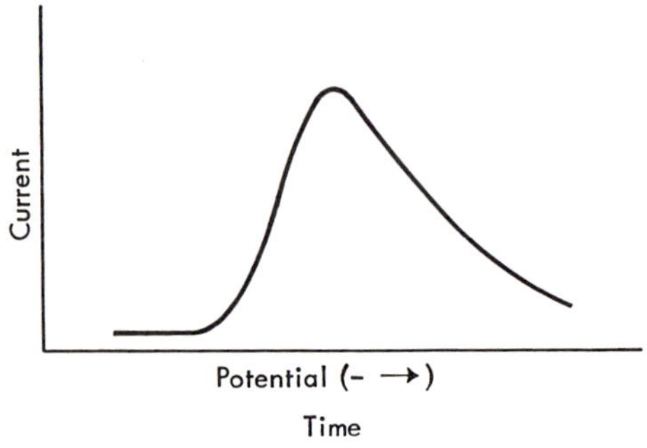

Fig. 46-6. Theoretical current–potential curve for the reduction of Ag^+ using linear-scan voltammetry.

The peak current density, i_p^0 (μA/cm²), is given by

$$i_p^0 = kn^{3/2}v^{1/2}D^{1/2}C \qquad (32)$$

where n = number of faradays consumed or given up at the electrode surface per mole of diffusing substance
D = diffusion coefficient, cm²/sec
v = rate of voltage scan, V/sec
C = concentration of electroactive solute, mmoles/liter

and k has different values, depending upon whether or not the electrode process is reversible, and also whether or not the product produced by the electrochemical reaction deposits as a solid on the electrode or diffuses away from the electrode–solution interface. For example, in the case of the reversible reduction of silver ion to silver metal at a solid electrode, $k = 3.67 \times 10^5$, whereas for a reversible process such as the reduction of thallous ion at a mercury electrode, $Tl^+ + Hg + e \rightarrow Tl_{amal}$, where the reduced form is soluble and diffuses away from the electrode surface, $k = 2.27 \times 10^5$.

Linear-scan voltammetry is readily used with concentrations of 10^{-5} M, and Streuli and Cooke,[5] using a mercury electrode, report the method is applicable to the determination of concentrations as low as 5×10^{-7} M.

The great analytical sensitivity of linear-scan voltammetry as compared to polarography is *not* related to the absolute magnitude of the currents obtained using the two methods. As in all voltammetric techniques, the analytical sensitivity is determined by the ratio of the current density resulting from the electrolysis of the diffusing species as compared to the current density required to charge the electrical double layer (see p. 892). This ratio is larger in the case of linear-scan voltammetry than in polarography.

Linear-scan voltammetry is finding application with the dropping mercury electrode. Commercial instruments exist that scan the entire potential range (0.4 to −2.0 V versus S.C.E.) in 1 or 2 sec. This scan is caused to occur late in the

[5] C. A. Streuli and W. D. Cooke, *Anal. Chem.* **25**, 1691 (1953); *J. Phys. Chem.* **57**, 824 (1953); also see P. Delahay, chap. 6 of Ref. C2 at the end of this chapter, for a comprehensive discussion of linear-scan voltammetry.

life of the mercury drop, when the rate of change of area with time is small, so that the conditions for semiinfinite linear diffusion are approximately fulfilled. Such instruments will undoubtedly find wide acceptance in the analysis of solutions at the micromolar level.

A discussion of cyclic (or triangular wave) voltammetry, alternating current polarography, square-wave polarography, and other voltage scanning techniques are beyond the scope of this text, and the student is referred to specialized texts for a more complete discussion. (See references under C, p. 884.)

Chronopotentiometry. A constant current is passed through an electrode submerged in an unstirred solution containing electroactive material and indifferent electrolyte. The electrolysis of the diffusion layer produces a change in the concentration of the potential-determining system at the electrode surface and hence a change in the electrode potential. The changes of electrode potential with time are recorded, yielding a plot such as that shown in Fig. 46-7, which is a representation of the reduction of Ag^+ at a platinum electrode. The time difference between the sharply changing potential region when the current is turned on and the potential which occurs when the surface concentration of electroactive material approaches zero is known as the transition time, τ. τ is readily related to the concentration, C, of the diffusing species through the Sand equation,[6] Eq. (33), when conditions for semiinfinite linear diffusion are met:

$$i^0\tau^{1/2} = \frac{nF}{2}(\pi D)^{1/2}C = 8.55 \times 10^4 nD^{1/2}C \tag{33}$$

where i^0 is the current density in $\mu A/cm^2$, F is the Faraday constant, and the other quantities have their customary significance.

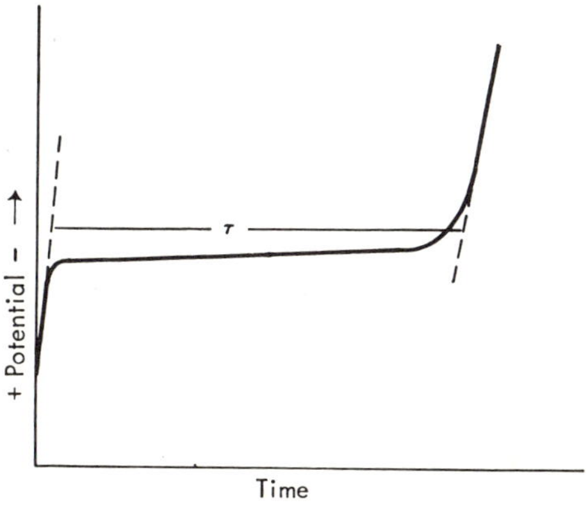

Fig. 46-7. Chronopotentiogram for the reduction of Ag^+ on platinum. Charging currents assumed to be almost negligible.

Chronopotentiometry has about the same analytical sensitivity as polarography. Its principal application until now has been in the fundamental studies of electrochemical reactions rather than in analysis. Further details can be found in the references C.1, C.2, and C.3 at the end of the chapter.

Methods related to voltammetry. A number of electrochemical techniques combine more than one of the abovementioned procedures and cannot be simply

[6] H. J. S. Sand, *Phil. Mag.* **1**, 45 (1901).

classified. Usually they involve some kind of voltammetry, along with some other technique. Stripping analysis is an excellent example of such a method, and exists in a number of variations.

In stripping analysis, the desired constituent is deposited upon an electrode, and the amount of the plated material determined by electrooxidizing or electroreducing the deposit. The quantity of electricity required to *remove* the deposit is then determined and Faraday's law is used to calculate the quantity of plated material. Ordinarily, the deposition time is much longer than the time required to strip the electrode of the deposit. Hence the stripping currents are much larger than the plating currents. If the deposit is a metal, an anodic current must be used to oxidize the deposit—hence the name *anodic stripping*. If an insoluble salt, i.e., silver iodide, is deposited by oxidizing a silver electrode in the presence of iodide ion, a cathodic current is required to reduce the silver iodide to silver metal and soluble iodide ion (*cathodic stripping*).

If the initial (plating) electrolysis involves the quantitative electrolysis of the bulk of the solution, obviously the amounts of material initially present in solution and stripped from the electrode are the same. This procedure is advantageous if the amount of material is too small to be weighed and ordinary coulometry fails because the current during the plating cycle is not an accurate measure of the amount of deposited material.

In other stripping-analysis methods, carefully controlled stirring conditions are used during the plating cycle and only a small, but reproducible, portion of the total amount of the unknown constituent is deposited. Using solutions of known concentration, calibration curves are constructed by plotting coulombs stripped versus concentration. Unknown solutions are analyzed under identical conditions to those used for the known solutions and their concentrations found from the calibration graph.

Multicomponent anodic stripping analysis is possible using a liquid mercury electrode. The plating cycle may either involve quantitative electroreduction of the bulk solution or only partial electrolysis of selected metal ions under conditions of reproducible stirring. The latter case is more usual. Diffusion of the metals deposited on mercury occurs away from the solution–mercury interface into the mercury electrode. This diffusion phenomenon presents a complication that is minimized by oxidizing the various metals from the mercury in a time short as compared to the plating time. Often the preferred electroxidation method is linear-scan voltammetry, and peak currents are observed near potentials characteristic of the metal being oxidized. These peak currents are proportional to the concentration of metal ion originally present in the solution.

A thorough discussion of anodic stripping techniques has been given by Shain.[7]

Zero-current methods

Potentiometry and potentiometric titrations. Theoretically, the potential of an electrode can be determined without passing any current, and thus without any electrolysis. In practice, this ideal situation can be approached quite closely using modern electronic devices.

[7] I. Shain, chap. 50 of Ref. C3, listed on p. 884.

The quantitative relation between the potential of a metal electrode and the concentration of the corresponding metal ions in solution and also between the potential of a noble metal and the ratio of the concentrations of the oxidized and reduced forms of an oxidation–reduction system were given in Chapter 8. Quite generally we can write

$$E = E^0 + \frac{0.059}{n} \log \frac{(\mathrm{Ox})}{(\mathrm{Red})} \tag{34}$$

When dealing with a metal in a solution of its ions, (Red) is constant and we get

$$E_{\mathrm{M}} = E^0 + \frac{0.059}{n} \log (\mathrm{M}^{n+}) \tag{35}$$

Thus it would seem that the activity of a metal ion in a solution can be determined by measuring the potential of the corresponding metal in the solution and the concentration then calculated. For two reasons, the value of the potential does not allow an exact calculation of the metal-ion concentration:

1. As stated on page 421, the following relation exists between the activity, (M^{n+}), and the concentration $[\mathrm{M}^{n+}]$ of the metal ions:

$$[\mathrm{M}^{n+}] = \frac{(\mathrm{M}^{n+})}{f}$$

in which f denotes the activity coefficient. Its value may be known approximately, but not exactly.

2. The measurement of the potential involves the measurement of the emf of a cell (p. 162). The two half-cells are connected by electrolytic contact, which is the seat of a small (and variable) emf called the *liquid-junction potential.*

Although in specific instances there are ways to reduce the uncertainties caused by the above sources to small dimensions, no general use can be made of Eq. (35) for the *exact* determination of ion concentrations in solutions. (See p. 940.)

Analytically, an important use of Eqs. (34) and (35) is made in *potentiometric titrations.* In earlier chapters (34, 35, and 37), we have calculated the change of pH, pAg, and the ratio of the concentrations of oxidized to reduced forms of a system in acid–base, precipitation, and redox titrations. When the equilibrium constant of the reaction is favorable, a large break in pH, pAg, or log (Ox)/(Red) occurs at the equivalence point. As can be seen from Eqs. (34) and (35), corresponding large changes in potential must also occur at the equivalence point, which can then be located precisely using the second derivative method (p. 946). During a titration, in the vicinity of the equivalence point, the activity coefficient of the electroactive constituents remains approximately constant; hence the change in potential indicates the change in concentrations. The liquid-junction potential may also be considered constant during a titration near the equivalence point.

Under certain conditions it is advantageous to pass a small, constant current through the indicator electrode when measuring the potential. This technique is known as *constant-current potentiometry* and is discussed on page 948.

Conductometry and conductometric titrations. Solutions of electrolytes conduct an electric current. If the conductivity of the solution is determined using an alternating-current method, under ideal conditions, no net electrochemical reaction occurs. Therefore we classify conductometry as a zero (on the average)-current method.

The electrical conductivity of a solution depends both upon the kind and concentration of the ions present. For example, the conductivity of a solution containing four ions, A^+, B^+, C^-, and D^-, is given by

$$\text{conductance} = \text{const.}\,[N_{A^+}\lambda_{A^+} + N_{B^+}\lambda_{B^+} + N_{C^-}\lambda_{C^-} + N_{D^-}\lambda_{D^-}] \quad (36)$$

where N represents normality of the ion and λ the *equivalent conductance* of the ion (p. 952). The constant is determined by the geometry of the conductance cell (p. 952).

Consider the conductances of 0.001 N solutions of hydrochloric acid, sodium chloride, and sodium hydroxide. Expressed in a certain unit, they are approximately 426, 126, and 248, respectively. Assume that we titrate 0.001 N hydrochloric acid with a concentrated solution of sodium hydroxide and that the change in volume during the titration is negligible. At the beginning of the titration, the conductance is 426, at the equivalence point 126, and when the excess of sodium hydroxide corresponds to 0.001 N, it is $248 + 126 = 374$. The conductance changes linearly during the titration. Thus a plot of conductance versus volume of sodium hydroxide added has the appearance of curve 1 of Fig. 38-3 (p. 771). In this figure $Z' =$ conductance $-$ 126. The equivalence point corresponds to the intersection point of the two straight lines. Conductance titrations represent one of the earliest physiochemical techniques in which the equivalence point was found graphically from observations made before and after the equivalence point.

REFERENCES

The following list of references is intended to introduce the student to recent and readily available older sources which deal with the fundamentals or some specific aspect of electrochemical methods of analysis. It is not intended to be complete, but only to serve as a starting point for students interested in a more detailed discussion than is possible in this book.

A. Electrochemical data tabulations

1. W. M. LATIMER, *Oxidation States of the Elements and Their Potentials in Aqueous Solution*, Prentice-Hall, Englewood Cliffs, N.J., 2nd ed. This book is the classic reference source for oxidation potentials.

2. B. E. CONWAY, *Electrochemical Data*, Elsevier, Amsterdam, 1952.

B. General electrochemistry

1. D. A. MACINNESS, *Principles of Electrochemistry*, Reinhold, New York, 1939.

2. G. KORTUM and J. O'M. BOCKRIS, *Textbook of Electrochemistry*, Elsevier, Amsterdam, 1951.

These books deal with the physical chemical basis of electrochemistry.

C. Electroanalytical chemistry

1. J. J. LINGANE, *Electroanalytical Chemistry*, Wiley-Interscience, New York, 2nd ed., 1958. A text suitable for the student who wishes to obtain a more comprehensive treatment of many of the topics discussed in this and later chapters.

2. P. DELAHAY, *New Instrumental Methods in Electrochemistry*, Wiley-Interscience, New York, 1954. A rather mathematical approach to newer electrochemical techniques, principally those involving diffusion-layer electrolysis methods.

3. Section D-2, *Electrical Methods of Analysis* in *Treatise on Analytical Chemistry* (I. M. KOLTHOFF and P. J. ELVING, eds.), part I, vol. 4, Wiley-Interscience, New York, 1964. Contains individual chapters by various authors covering the electrochemical material in this text in a thorough manner.

4. Reviews of various electroanalytical topics appear biannually (even years) in *Analytical Chemistry*.

D. Potentiometry and potentiometric titrations

1. I. M. Kolthoff and N. H. Furman, *Potentiometric Titrations*, Wiley, New York, 2nd ed., 1932.
2. I. M. Kolthoff and H. A. Laitinen, *Determination of pH: Electrometric Titrations*, Wiley, New York, 2nd ed., 1941.
3. H. A. Laitinen in *Physical Methods of Chemical Analysis* (W. G. Berl, ed.), Academic Press, New York, 1951.
4. N. H. Furman, *Potentiometry*, Ref. C3, chap. 45.

E. Polarography, voltammetry, and amperometric titrations

1. I. M. Kolthoff and J. J. Lingane, *Polarography*, Wiley-Interscience, New York, 2nd ed., 1952.
2. G. W. C. Milner, *Polarography*, Longmans, London, 1956.
3. L. Meites, *Polarographic Techniques*, Wiley-Interscience, New York, 2nd ed., 1965.
4. J. Heyrovsky and J. Kůta, *Principles of Polarography*, Publishing House of the Czechoslovak Academy of Sciences, Prague, 1965.

Ref. D2.

5. J. T. Stock, *Amperometric Titrations*, Wiley-Interscience, New York, 1965.
6. R. N. Adams, *Voltammetry at Electrodes with Fixed Surfaces*, Ref. C3, chap. 47.
7. R. N. Adams, *Electrochemistry at Solid Electrodes*. Marcel Dekker, New York, 1969.
8. *Progress in Polarography* (P. Zuman, ed., with collaboration of I. M. Kolthoff), vols. I and II, Wiley-Interscience, New York, 1962.

F. Conductometry and conductometric titrations

1. I. M. Kolthoff, *Konductometrische Titrationen*, Theodore Steinkopff, Dresden, 1923.
2. G. Jansler and O. Pfunelt, *Die visuelle Leitfähigkeitstitration*, Enke, Stuttgart, 2nd ed., 1934.
3. H. T. S. Britton, *Conductometric Analysis*, Van Nostrand, Princeton, N.J., 1934.
4. Ref. D3.
5. J. W. Loveland, *Conductometry and Oscillometry*, Ref. C3, chap. 51.
6. C. N. Reilley, *High Frequency Methods*, Ref. C1, chap. 15.
7. G. G. Blake, *Conductometric Analysis at Radio Frequency*. Chapman & Hall, London, 1950.
8. K. Cruse and R. Huber, *Hochfrequenztitration*, Verlag Chemie, Weinheim, 1957.

CHAPTER 47 *The Dropping Mercury Electrode and Other Indicator Electrodes*

A suitable voltammetric indicator electrode must meet two requirements. First, current at the electrode must be reproducible and insensitive to vibration conditions found in an ordinary laboratory environment. Second, the electrode metal must not react with the supporting electrolyte, or oxidize or reduce the solvent, in the potential range of interest. A third attribute of considerable significance in trace analysis is desirable: The rate of mass transfer to the electrode surface should be high in order that the current density at the electrode be large.

The requirement of current reproducibility precludes, in general, electrochemical experiments longer than 30–60 sec in the laboratory, when diffusion is the only mode of mass transport. Consider a glass tube with an indicator electrode closing off one end. The tube is filled with supporting electrolyte and electroactive material, and a reference electrode inserted into the other end of the tube. Assume that the voltage applied across the indicator electrode and the S.C.E. causes the indicator electrode surface concentration of the electroactive material to become zero instantaneously. Then, a very large current which decreases rapidly flows through the cell; this current ultimately approaches zero if there is no vibration or natural convection. Material is constantly electrolyzed at the electrode surface, and, with time, the layer of electrolytically depleted solution gets thicker; i.e., the distance from the surface where the electroactive species concentration is zero to a point in the solution approaching the bulk concentration increases. This distance is the diffusion-layer thickness (δ) and increases with time. Since the current is inversely proportional to δ (see p. 902), the current decreases with time. If vibration occurs, the diffusion layer is disrupted, and irregular currents result.

This experiment corresponds to the conditions for semiinfinite linear diffusion. The current–time dependence is given by

$$i^0 = \frac{nFD^{1/2}C}{(\pi t)^{1/2}}$$

where the terms have their customary significance[1] (p. 881).

At very long times, on the order of several minutes, relatively steady currents can be observed at wire or plate electrodes immersed in a supporting electrolyte containing electroactive material. These currents are the result of fluid motion induced by natural convection. A density difference is established between the electrode surface and the bulk of the solution because of the concentration gradient across the diffusion layer, and a fairly reproducible motion of the fluid ensues.

Specialized techniques making use of time-dependent currents have been discussed briefly (p. 880) but are not yet used routinely.

[1] For details see P. Delahay, *Newer Instrumental Methods of Analysis*, Wiley-Interscience, New York, 1954, p. 46ff.

We shall consider now two types of indicator electrodes which meet the first two criteria listed above in radically different ways and have found wide acceptance. The first indicator electrode to be discussed is the dropping mercury electrode (D.M.E.), which satisfies the above criteria by keeping the time of electrolysis short. The second kind includes the various types of rotated electrodes, which function by creating reproducible fluid flow adjacent to the electrode surface, thereby swamping out the effects of natural convection.

THE DROPPING MERCURY ELECTRODE (D.M.E.)

Figure 47-1 represents one form of dropping mercury electrode as described earlier on page 876. When a drop falls, virtually all the electrolyzed solution is carried along with the detached drop. Every new drop forms, and electrolysis begins, in an environment independent of the history of the previous drop.[2] Since the electrolysis time is so short, ordinary levels of vibration in the laboratory do not disturb the relatively small diffusion layer.

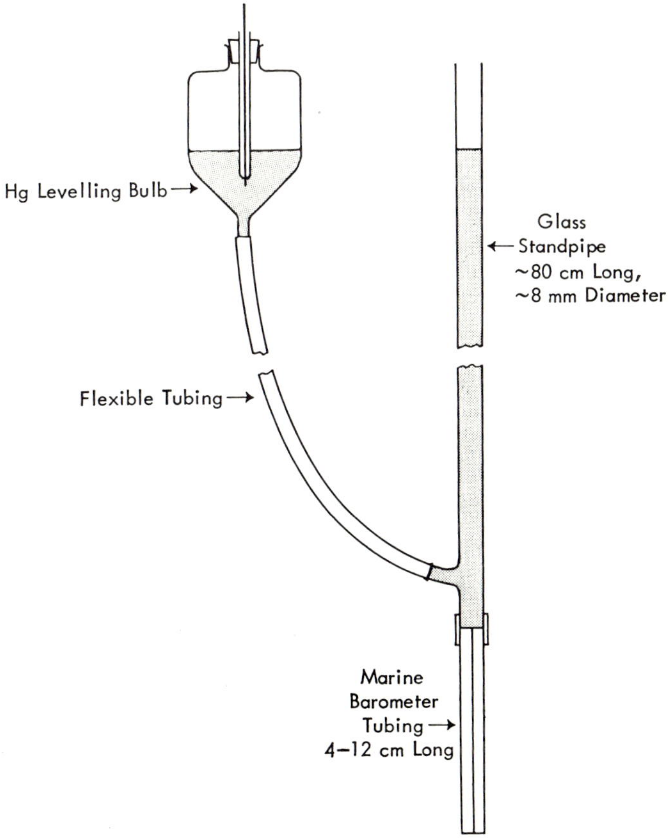

Fig. 47-1. Dropping mercury electrode

[2] There is some experimental evidence indicating that the solution surrounding the first mercury drop at which electrolysis is begun is slightly richer in electroactive material than subsequent drops. See H. Strehlow and M. von Stackelberg, *Z. Electrochem.* **54**, 51 (1950); also L. Meites, *Polarographic Techniques*, Wiley-Interscience, New York, 2nd ed., 1965, p. 144 ff.

The diffusion current

Using some simplifying assumptions Ilkovic[3] derived an equation

$$i_t = 708nm^{2/3}t^{1/6}D^{1/2}C \tag{1}$$

giving the relation between the diffusion current i_t (microamperes) at any time t (seconds) during the life of a drop, and the characteristics of the capillary m (milligrams per sec)—the rate of flow of mercury, the concentration of electroactive species C (millimoles per liter), the diffusion coefficient D (centimeters squared per second) of the electroactive species, and the number (n) of electrons transferred to or from the diffusing species at the electrode. Equation (1) is known as the *Ilkovic equation*. Experimentally it is usual to measure the average diffusion current, i_d, during the growth of the drop,

$$i_d = 607nm^{2/3}t_m^{1/6}D^{1/2}C = kC \tag{2}$$

in which t_m is the drop time in seconds. Alternatively, the maximum current,

$$(i_d)_M = 708nm^{2/3}t_m^{1/6}D^{1/2}C = k'C \tag{3}$$

can be measured using a galvanometer with a short period (of about 1 sec or a strip-chart recorder with a pen response of at least 1 sec full scale).

The constants k and k' of Eqs. (2) and (3) are the mass-transfer coefficients for the D.M.E.[4]

Figure 47-2 is a graph of the variation of the diffusion current with the age of the mercury drop according to Eq. (1). Obviously, an ordinary laboratory galvanometer cannot follow such rapid current changes but under appropriate conditions can accurately measure the average diffusion current.

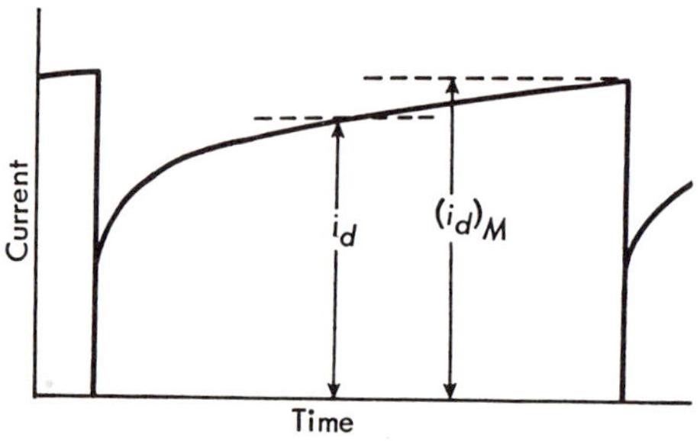

Fig. 47-2. Instantaneous current–time curve during the lifetime of a drop. i_d (average diffusion current) = 6/7 $(i_d)_M$ (maximum diffusion current).

[3] D. Ilkovic, *Collection Czech. Chem. Commun.* **6**, 498 (1934); *J. Chim. Phys.* **35**, 129 (1938); I. M. Kolthoff and J. J. Lingane, *Polarography*, Wiley-Interscience, New York, 1952, chap. 4; L. Meites, *op. cit.*, chap. 3.

[4] Ilkovic made a number of simplifying assumptions in obtaining Eq. (1). Various authors have proposed more exact equations, the most elaborate, proposed by Koutecky, has the form

$$i_d = 607nD^{1/2}m^{2/3}t^{1/6}[1 + 34.7D^{1/2}t^{1/6}m^{-1/3} + 100(D^{1/2}t^{1/6}m^{-1/3})^2]C$$

This equation agrees with Eq. (2) to within 2–3 per cent for a D.M.E. with the usual values of m and t. There is still an unexplained discrepancy between experimental current–time curves and the various equations that have been proposed. Agreement between theory, Eq. (2), and experiment for the average current is quite good and seems to involve compensation of errors. See J. Koutecky and M. von Stackelberg, *Progress in Polarography*, vol 1 (P. Zuman and I. M. Kolthoff, eds.), Wiley-Interscience, New York, 1962, p. 21; J. J. Lingane and B. A. Loveridge, *J. Am. Chem. Soc.* **72**, 438 (1950); H. Strehlow and M. von Stackelberg, *Z. Electrochem.* 54, 51 (1950).

To make possible comparison of polarographic data obtained with different D.M.E.'s or the same D.M.E. under different conditions, it is customary to report m and t, and the diffusion current constant, I, defined as

$$I = \frac{i_d}{m^{2/3}t_m^{\ 1/6}C} = 607nD^{1/2} \tag{4}$$

Factors affecting the diffusion current

Electrode potential. The electrocapillary curve. In the absence of oxygen or any other electroactive species, the surface tension of the mercury–solution interface, σ_{Hg}, varies as a function of the D.M.E. potential, E, as shown in Fig. 47-3. As mentioned earlier, all potentials are given with respect to the S.C.E. reference electrode. Such a plot is called an *electrocapillary curve*. The maximum surface tension occurs at a potential where there is no net excess of anions or cations at the mercury–solution interface. On either side of the electrocapillary maximum (E.C.M.) either cations (− side) or anions (+ side) will be in excess at the electrode surface, thereby decreasing the surface tension.

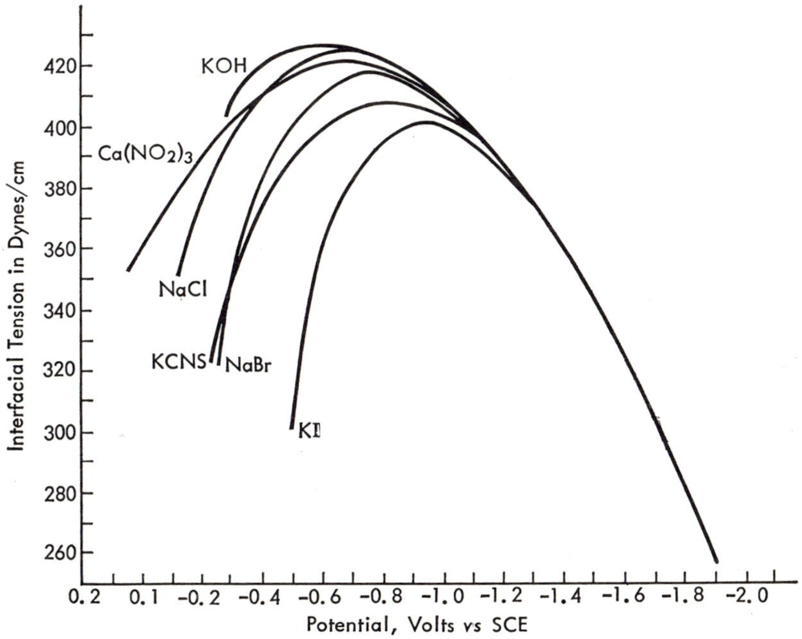

Fig. 47-3. Electrocapillary curves at 18°C according to D. C. Grahame, *Chem. Rev.* **41**, 441 (1947).

The potential dependence of σ_{Hg} is important, because the drop weight and drop time are determined by σ_{Hg} and the capillary radius a. If d_{Hg} and d_s represent the density of mercury and the solution, respectively, from elementary physics we see that the drop will fall when the force supporting it ($= 2\pi a\sigma_{Hg}$) is exceeded by the gravitational force acting to dislodge it ($mtg[1 - d_s/d_{Hg}]$), where g is the gravitational constant. Hence the drop weight, mt, is given by

$$mt = \frac{2\pi a\sigma_{Hg}}{(1 - d_s/d_{Hg})g} = k^* \tag{5}$$

where k^* is a constant only at constant electrode potential. As shown in Table 47-1, m is virtually unaffected by changes of potential in the range 0 to -1.90 V, with only a change of about 2 per cent in m occurring. This experimental result is confirmed by Eq. (13) on page 891, which states that m is constant at constant applied pressure of mercury. Combining this result with Eq. (5), it is seen that

$$t = \sigma_{\mathrm{Hg}} \left\{ \frac{2\pi a}{m(1 - d_s/d_{\mathrm{Hg}})g} \right\} = \text{const.} \times \sigma_{\mathrm{Hg}} \tag{6}$$

The drop time is proportional to the surface tension of the mercury–solution interface. Thus Fig. 47-3 represents both the variation of σ_{Hg} and t with E.[5]

The diffusion current depends upon the product $m^{2/3}t^{1/6}$ [Eq. (2)], and from Table 47-1 it is seen that this product changes much less than t over the potential range 0 to -1.90 V, only a 10 per cent decrease in $m^{2/3}t^{1/6}$ accompanying a 50 per cent decrease in t.

TABLE **47-1. Effect of Potential upon Flow Rate and Drop Time of a D.M.E. in 0.1 *M* KCl**[a,b]

E vs S.C.E.	m, mg/sec	t, sec	$m^{2/3}t^{1/6}$
0.0	2.651	2.75	2.27
−0.5	2.621	3.08	2.29
−0.80	2.626	2.88	2.28
−1.10	2.625	2.66	2.24
−1.50	2.672	2.18	2.18
−1.90	2.681	1.56	2.08

[a] From J. J. Lingane and I. M. Kolthoff, *J. Am. Chem. Soc.* **61**, 825 (1939).

[b] Mercury height, 21.8 ± 0.2 cm, 0.04% sodium methyl red added.

This change in $m^{2/3}t^{1/6}$ is significant in quantitative polarography. For example, consider an approximately 10^{-3} *M* Pb(II) solution in 0.1 *M* KCl. A sketch of the polarogram for the reduction of lead ($Pb^{++} + 2e + Hg \rightarrow Pb_{\mathrm{amal}}$) is given in Fig. 47-4. $E^{1/2}_{\mathrm{Pb^{++}/Pb(amal)}} = -0.386$ V versus S.C.E., and a potential of -0.80 V (E_1 in Fig.

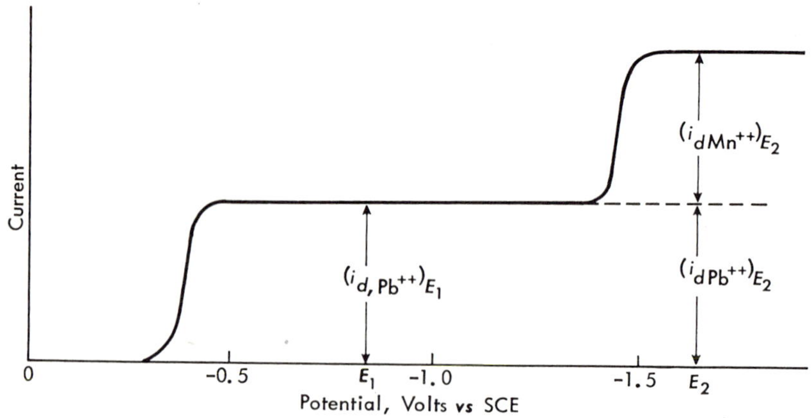

Fig. 47-4. Polarogram of Pb^{+2} and Mn^{+2} in 0.1 *M* KCl supporting electrode (theoretical).

[5] It is usual to construct a plot of t versus E for each supporting electrolyte in quantitative polarography (see p. 1035).

47-4) is convenient for measuring the diffusion current for lead. At this potential, $m^{2/3}t^{1/6}$ is nearly a maximum, and thus the diffusion current is a maximum according to Eq. (2). At a potential of -1.7 V, $m^{2/3}t^{1/6}$ is considerably smaller, as is the diffusion current, as shown by the dashed line in Fig. 47-4. From Eq. (2), the diffusion currents of lead ion at two different potentials are related to each other by

$$(i_{d,\mathrm{Pb}})_{E_2} = (i_{d,\mathrm{Pb}})_{E_1}\left\{\frac{(m^{2/3}t^{1/6})_{E_2}}{(m^{2/3}t^{1/6})_{E_1}}\right\} \tag{7}$$

where the subscripts E_1 and E_2 indicate the potential at which the various quantities are measured. Since m is relatively insensitive to potential variation (Table 47-1), it is permissible for many applications to use the simpler expression

$$(i_d)_{E_2} = (i_d)_{E_1}\left(\frac{t_{E_2}}{t_{E_1}}\right)^{1/6} \tag{8}$$

Calibration plots of i_d versus C must be corrected for the change in $m^{2/3}t^{2/6}$ if used at a potential different from that used in constructing the plot.

This correction must be made in the quantitative determination of a mixture of two or more substances at the D.M.E. The curve in Fig. 47-4 represents the polarogram that would be obtained in a mixture containing about 10^{-3} *M* Pb(II) and Mn(II). The second diffusion current results from the two electron reduction of Mn(II), with a half-wave potential of -1.48 V versus S.C.E. The total diffusion current i_{total} is conveniently measured at -1.70 V (E_2) and the diffusion current caused by the manganese reduction is given by

$$(i_{d,\mathrm{Mn}})_{E_2} = i_{\mathrm{total}} - (i_{d,\mathrm{Pb}})_{E_2} \tag{9}$$

Eliminating between Eqs. (9) and (7) yields

$$(i_{d,\mathrm{Mn}})_{E_2} = i_{\mathrm{total}} - (i_{d,\mathrm{Pb}})_{E_1}\left\{\frac{(m^{2/3}t^{1/6})_{E_2}}{(m^{2/3}t^{1/6})_{E_1}}\right\} \tag{10}$$

Problem 1 is a numerical exercise based upon Table 47-1 and Fig. 47-4.

Dependence of *m* upon mercury height. The rate of flow of mercury through the capillary is given by Poiseuille's equation:

$$m' = \frac{\pi a^4 P}{8\eta l} \tag{11}$$

where m' is the volume rate of flow of mercury (cm^3/sec), a is the capillary radius (cm), η is viscosity of mercury (dyne-sec/cm^2 [poise]), l is the capillary length (cm), and P is the pressure forcing the mercury through the capillary (dyne/cm^2). At 25°C, Eq. (11) reduces to

$$m = 4.64 \times 10^9 \frac{a^4}{l}\left\{h_{\mathrm{Hg}} - \frac{3.1}{(mt)^{1/3}} - \frac{h_{\mathrm{soln}}d_{\mathrm{soln}}}{13.5}\right\} \tag{12}$$

where m is the mass rate of mercury flow in milligrams per second, h_{Hg} is the height of the mercury column above the capillary tip, and h_{soln} is the depth of immersion in the solution with a density d_{soln}.

The term $3.1/(mt)^{1/3}$ arises from the pressure required to create a sphere of mercury in the solution, and ordinarily has a value equivalent to 1 or 2 cm Hg. The term $h_{\mathrm{soln}}d_{\mathrm{soln}}/13.5$ represents the hydrostatic back-pressure, and is rarely larger than 0.2 cm. It can usually be neglected. We write

$$m = 4.64 \times 10^9 \frac{a^4}{l} h' \tag{13}$$

where

$$h' = h_{\mathrm{Hg}} - \frac{3.1}{(mt)^{2/3}} - \frac{h_{\mathrm{soln}}d_{\mathrm{soln}}}{13.5}$$

is called the *effective height* of mercury. For any particular D.M.E., Eq. (13) can be written

$$m = k''h' \tag{14}$$

where k'' is the proportionality constant determined by the capillary length and radius. From Eq. (13) it is seen that the rate of flow of mercury varies inversely with the length of the capillary and from Eq. (14) that it is proportional to the effective mercury height.

The effect of variation of mercury height upon a diffusion current is obtained by substituting Eqs. (14) and (5) into the Ilkovic equation, (2), to give

$$i_d = 607nC(k^*)^{1/6}(k'')^{1/2}h'^{1/2}$$

i.e., i_d is proportional to $h'^{1/2}$.

Effect of temperature. The principal effect of temperature upon the diffusion current occurs as a result of the variation of the diffusion coefficient with temperature.[6] The relative change in i_d, with temperature T, $(di_d/dT)/i_d$, is found by differentiating Eq. (2) with respect to temperature T, and dividing the result by Eq. (2):

$$\frac{di_d/dT}{i_d} = \frac{1}{2}\frac{dD/dT}{D} = \frac{1}{2}\frac{d \ln D}{dT} \tag{15}$$

The value of $(d/dT) \ln D$ ranges from 0.02 to 0.04 for most species at 25°C, yielding a change of 1–2 per cent per degree in the diffusion current. Temperature control on the order of 0.1–0.2°C is satisfactory for most polarographic studies, considering the other possible sources of error.

Limiting current i_l and residual current i_r. The "apparent" diffusion current i_l measured in polarography is the sum of the true diffusion current i_d, a non-Faradaic current i_c, and usually a small Faradaic current i_F caused by reduction of impurities (usually oxygen in the solution), which can be kept very small. The sum, $i_c + i_F$, is called the *residual current* i_r; hence

$$i_d = i_l - i_c - i_F = i_l - i_r \tag{16}$$

In polarographic measurements i_r must always be determined.

Even though i_F can be made negligibly small, i_c cannot, because it depends upon factors that cannot be altered appreciably. As mentioned in Chapter 8, a double layer exists at the interface of an electrode and a solution and may be represented approximately by a plane of negative charges at a definite distance from a plane of positive charges. (This representation corresponds to the compact or Helmholtz double layer.) There is an excess of cations at the electrode surface at potentials on the negative side of the electrocapillary maximum and an excess of anions on the positive side of the electrocapillary maximum. At the electrocapillary maximum, $E_{\max}$, there is zero net charge, hence the term "point (or potential) of zero charge."[7]

This charge distribution corresponds to a charged condenser, and the net charge is given by

$$Q = k_0(E - E_{\max})A \tag{17}$$

where k_0 is the capacitance per unit area (μF/cm²) and A is the area of the electrode (cm²), when Q is given in microcoulombs.

[6] See I. M. Kolthoff and J. J. Lingane, *Polarography*, Wiley-Interscience, New York, 1952, p. 90.

[7] See I. M. Kolthoff and J. J. Lingane, Ref. 6, p. 134, and J. Heyrovsky and J. Kûta, *Principles of Polarography*, Publishing House of the Czechoslovak Academy of Sciences, Prague, 1965, pp. 17–24, 229–232, for further details.

At a D.M.E. the electrode area changes with time; hence according to Eq. (17), Q must change with time, i.e., current (dQ/dt) must flow, even though the electrode potential is kept constant. Differentiating Eq. (17) with respect to time yields

$$i_c = \frac{dQ}{dt} = k_0(E - E_{\text{max}})\frac{dA}{dt} \tag{18}$$

The area of the D.M.E. is readily expressed in terms of the density of mercury, d_{Hg}, m, and t, to give

$$\bar{i}_c = -\alpha k_0(E - E_{\text{max}})t^{-1/3} \tag{19}$$

where

$$\alpha = 4\pi m^{2/3}(\tfrac{3}{4}\pi d_{\text{Hg}})^{2/3} \tag{20}$$

and $\bar{i}_c$ is the average current due to double-layer charging in a drop lifetime. At 25°C (Eq. 19) becomes

$$\bar{i}_c = 8.53 \times 10^{-3}k_0 m^{2/3}t^{-1/3}(E_{\text{max}} - E) \tag{21}$$

It is of considerable analytical importance to compare the magnitude of the average charging current to average diffusion current by dividing Eq. (21) by (2):

$$\frac{\bar{i}_c}{\bar{i}_d} = \frac{1.4 \times 10^{-5}k_0'(E_{\text{max}} - E)}{nCD^{1/2}t^{1/2}} \tag{22}$$

Let us calculate when $\bar{i}_c = \bar{i}_d$, assuming $t = 4$ sec, $k_0 = 30$ μF/cm²,[8] $D = 9 \times 10^{-6}$ cm²/sec, and $E_{\text{max}} - E = 1$ V; i.e.,

$$1 = \frac{1.4 \times 10^{-5} \times 30 \times 1}{nC(9 \times 10^{-6})^{1/2}(4)^{2/3}}$$

which yields

$$nC = 0.07 \tag{23}$$

From Eq. (23) we see that if $n = 1$, at a concentration of 7×10^{-5} M, the average residual and diffusion currents are equal, whereas if $n = 2$, the concentration for the condition corresponding to Eq. (23) is 3.5×10^{-5} M. Current–potential curves for such low concentrations of diffusing species are extremely distorted, and these concentrations are approaching the lower limit of the polarographic method. Obviously, if the limiting current is measured at a value of E close to E_{max}, the effect of charging current is very much reduced. Techniques for compensating for the capacity current have been devised, but they are generally difficult to use since k_0 is a complicated function of electrode potential.

Composition and kind of supporting electrolyte. It is necessary to add an excess of an indifferent (supporting) electrolyte to a solution that is to be studied polarographically. Ions migrate in an electric field; hence in the absence of supporting electrolyte, an electroactive ion will be transported to the D.M.E. both by electrical migration and by diffusion. As supporting electrolyte is added, the electrical migration of the electroactive ion decreases, since the ion's transference number[9] decreases. Electrical migration becomes negligible when a 50- to 100-fold excess of supporting electrolyte is present. Under such conditions, a true diffusion current can be measured, and the many complications caused by the contribution of electrical migration avoided.

At large concentrations of supporting electrolyte the diffusion coefficient of the electroactive species may decrease markedly. In those cases where size of the diffusing

[8] $k_0 = \sim 40\mu F/cm^2$ if E_2 is positive compared to E_{max}, and ~20 μF/cm² if E is negative compared to E_{max} in dilute alkali chloride solutions.

[9] The transference number of an ion is that fraction of the total conductance of a solution which can be ascribed to the ion (see p. 952).

species is not changing, the effect can be quantitatively interpreted in terms of the increasing viscosity of the supporting electrolyte.[10]

Complexation produced by altering the supporting electrolyte can affect the diffusion coefficient of the electroactive species if there is a marked change in its size. For example, the diffusion coefficient of Pb(II) in 1 *M* sodium hydroxide is 17 per cent less than that in 1 *M* nitric acid. Evidently, $HPbO_2^-$ is a larger species than hydrated lead ion (or lead nitrato complex, if it is the species present). Similarly, the lead complex present in a supporting electrolyte containing 0.25 *M* disodium tartrate and 0.1 *M* sodium hydroxide has a diffusion coefficient 2.4 times smaller than in 1 *M* nitric acid, because the tartrato complex is much larger than the other two lead species.

The diffusion coefficient in some cases is sensitive to the removal of a proton from a species. Using the data of Kolthoff and Auerbach,[11] the diffusion coefficient of FeY^{-1} (ferric complex of ethylenediaminetetraacetate ion) has a diffusion coefficient of 6.1×10^{-6} cm²/sec, whereas the apparent diffusion coefficient of a mixture of hydroxy complexes, $FeYOH^{--}$ and $FeY(OH)^{3-}$, is $5.5_5 \times 10^{-6}$ cm²/sec. The diffusion coefficient of the ferrous complex, FeY^{--}, is $5.6_3 \times 10^{-6}$ cm²/sec.

Potential range of the D.M.E.

The composition and concentration of supporting electrolyte determine the maximum cathodic and anodic potentials which can be applied to the D.M.E. without producing a significant Faradaic current from oxidation of the mercury or reduction or oxidation of the supporting electrolyte.

If the anion of the supporting electrolyte does not form a stable complex or a precipitate with Hg_2^{++} or Hg^{++}, (ClO_4^-, NO_3^-, acetate), the limiting anodic process is the oxidation of the D.M.E. to produce Hg_2^{++}. The practical potential limit depends upon the magnitude of the diffusion current to be measured, as compared to the anodic (dissolution) current, and is approximately 0.4 V versus S.C.E. If complex or precipitate formation occurs, the anodic potential limit before oxidation of the D.M.E. occurs is shifted in a negative direction. In 0.1 *M* solution of supporting electrolyte the limits are approximately 0.1 V for Cl^-, -0.2 V for OH^-, and -0.8 V for S^{--} or CN^-. The concentration dependence of the anodic potential limit can be calculated for reversible processes according to the results given in Chapter 48.

The cathodic potential limit depends upon the cation of the supporting electrolyte. In 0.1 *M* acid, the overvoltage for H_2 evolution is sufficiently large on the D.M.E. to attain a potential of ~ -1.5 V in practice. For 0.1 *M* Na^+ or K^+, a limit of ~ -1.9 V is usually given before the cathodic current due to alkali amalgam formation becomes appreciable. Quaternary ammonium cations, such as tetra(*n*-butyl)ammonium ion, are usable at potentials as negative as -2.6 V. The limiting cathodic process involves the reduction of the quaternary cation by a number of paths determined by the nature of the organic group.[12] Quaternary ammonium salts are used as supporting electrolytes in the determination of alkali ions in aqueous solutions. They are also extensively used in nonaqueous polarography because of favorable solubility considerations and the large negative potentials which can be applied.

Interfering electrode reactions

It has been stated (p. 878) that the total current at an indicator electrode is the sum of the currents arising from the simultaneously occurring electrochemical reactions. Practical polarography depends upon the fact that there is no interference of one reaction with another when both occur simultaneously. This is the usual situation, but examples exist where this is not the case.

[10] D. M. Brasher and F. R. Jones, *Trans. Faraday Soc.* **42**, 775 (1946); H. A. McKenzie, *J. Council Scien. and Ind. Res. Australia* **21**, 210 (1948).

[11] I. M. Kolthoff and C. A. Auerbach, *J. Am. Chem. Soc.* **74**, 1452 (1952).

[12] B. C. Southworth, R. Osteryoung, K. D. Fleischer, and F. C. Nachod, *Anal. Chem.* **33**, 208 (1961).

Oxygen is irreversibly reduced in two steps in 0.1 *M* KCl supporting electrolyte. The first wave occurs at $E^{1/2} \sim -0.05$ V according to

$$O_2 + 2\,H_2O + 2e \rightarrow H_2O_2 + 2\,OH^-$$

while the second step occurs at $E^{1/2} \sim -0.8$ V in neutral or unbuffered media according to

$$O_2 + 2\,H_2O + 4e \rightarrow 4\,OH^-$$

or in the presence of excess hydrogen ions according to

$$O_2 + 4\,H^+ + 4e \rightarrow 2\,H_2O \tag{24}$$

Hydrogen ions are irreversibly reduced according to

$$2\,H^+ + 2e \rightarrow H_2 \tag{25}$$

yielding a diffusion current region between -1.7 and -1.8 V. In the presence of oxygen, the OH^- ions formed by the reduction of oxygen neutralize hydrogen ions diffusing toward the D.M.E., and the diffusion current of hydrogen decreases and becomes zero when sufficient oxygen is present.

Depending upon the relative concentrations of O_2 and H^+, the current at -1.7 V may be due entirely to Eq. (24), or to Eq. (24) and partially to (25).[13]

When a constituent is reduced with the formation of OH^-, it may interfere in the polarography of mixture of the constituent and a metal ion. For example, when iodate is reduced,

$$IO_3^- + 3H_2O + 6e \rightarrow I^- + 6\,OH^-$$

OH^- can diffuse away from the D.M.E. and react with a metal ion. In the presence of Cd^{++}, iodate does not interfere with the observed limiting current for Cd^{++}, because Cd^{++} is reduced first. However, the apparent diffusion current of IO_3^- is much less than the true value because of the precipitation of $Cd(OH)_2$. On the other hand, in a mixture of iodate and manganous ion, the correct diffusion current for iodate ion is observed, because iodate is reduced before manganous ion.

Maxima. Polarograms are often distorted by the occurrence of maxima, such as those shown in Fig. 47-5. These curves represent the sharp and rounded maxima seen with *maxima of the first kind*. Maxima are always accompanied by rapid motion of the solution adjacent to the mercury drop and are observed for both irreversible and reversible processes. This motion of solution causes increased transport of electroactive material to the D.M.E. and an increase in the current above the normal diffusion current. Maxima of the first kind are not observed if the half-wave potential of the electroactive substance occurs close to or at the electrocapillary maximum. *Maxima of*

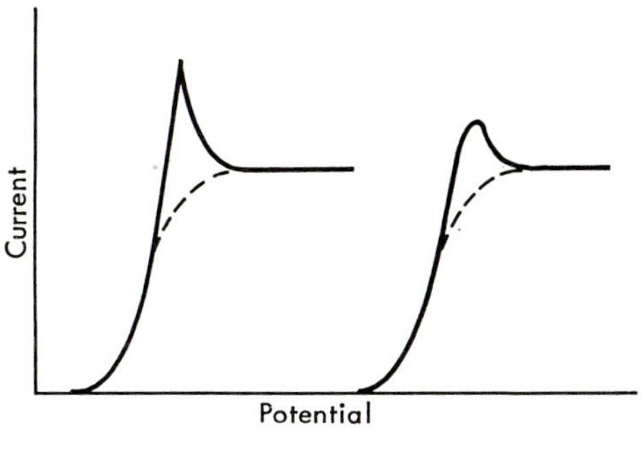

Fig. 47-5. Two examples of maxima at the D.M.E. dotted curve in the presence of surface active material.

[13] See I. M. Kolthoff and J. J. Lingane, Ref. 6, chap. 6, for a quantitative discussion of this phenomenon.

the second kind occur chiefly in solutions of high supporting electrolyte concentration and are markedly dependent upon mercury flow rate. They are most pronounced at the electrocapillary maximum, in contradistinction to maxima of the first kind. Maxima of the first kind can be prevented by the addition of a suitable *maximum suppressor* in very low concentration ($\sim 10^{-2}$ wt. per cent).

Maximum suppressors are surface active substances (surfactants, denoted by SAS), much as gelatine, Triton X-100, polyacrylamide (PAM), and various other organic substances. They are adsorbed at the surface of the mercury drop and may seriously affect the location and shape of the current–potential curves and the value of i_d. Hence the effect of a particular SAS at a given concentration should always be tested in a search for a SAS that yields a normal current–potential curve.

The interpretation of polarographic maxima is a very complicated subject and the student is referred to Levich,[14] Kolthoff and Lingane,[15] and Heyrovsky and Kůta[16] for more details.

ROTATING ELECTRODES

In a solution that is stirred by a mechanical device, it is generally impossible to obtain steady and reproducible currents at a stationary electrode. Usually the currents observed fluctuate greatly about an average value because of the shifting and variable liquid flow patterns caused by the stirrer in the vicinity of the electrode. Although the flow pattern within several millimeters of a stirrer is relatively stable and insensitive to cell geometry or positioning of the stirrer in the cell, fluid motion at more distant points is quite sensitive to small changes in the latter two variables and is not particularly stable for any length of time. However, this difficulty is avoided by incorporating the electrode into the stirrer, i.e., rotating the electrode. The diffusion-layer thickness is much less than 1 mm at a rotating electrode, and a relatively reproducible fluid-flow pattern exists for some distance outside this region. Since cell geometry and electrode positioning are no longer of primary importance, only modest care need be exercised in reproducing the physical location of a rotating electrode within a cell.

In the presence of supporting electrolyte, the maximum current that can flow at a rotating electrode is limited by convective and diffusive transport of electroactive substance. As stated on page 874, this current is properly referred to as a convective-diffusion current; however, for convenience we shall use the term diffusion current, because there is little opportunity for confusion when the nature of the indicator electrode is known.

A number of generalizations are possible about rotating electrodes. First, they are more sensitive than the D.M.E., because current flows by convective transport in addition to diffusive transport. (Sensitivity is used in this context in the sense of being able to detect a lower concentration of electroactive material.) Second, the electrode material is not restricted to mercury and may be platinum, gold, graphite, or any other suitable conductor, generally a solid. The choice of electrode material depends upon the electroactive species to be studied. Third, the behavior of rotating electrode, with one exception (see Ref. 27, p. 897), is influenced by the previous electrochemical history of the electrode surface, since it is not periodically renewed as is the case with the D.M.E. This property causes most of the problems experienced with a solid electrode.[17]

If the electrode is made of a metal which can be readily sealed into glass, such as platinum, the rotated wire electrode made popular by Laitinen and Kolthoff[18] is useful.

[14] V. G. Levich, *Physicochemical Hydrodynamics*, Prentice-Hall, Englewood Cliffs, N.J., 2nd ed., 1962, p. 561ff. Translated from the Russian.

[15] I. M. Kolthoff and J. J. Lingane, Ref. 6, chap. 10.

[16] J. Heyrovsky and J. Kůta, Ref. 7, chap. 19.

[17] See L. Meites, Ref. 2, p. 432ff., for a discussion of this point.

[18] H. A. Laitinen and I. M. Kolthoff, *J. Phys. Chem.* **45**, 1079 (1941).

In Fig. 47-6 the platinum wire is sealed into a soft glass tube, and mercury is placed inside the tube to make electrical contact. The electrode is then mounted in a hollow spindle that is driven by a constant-speed (synchronous) motor through a gear or pulley system. A wire dips into the mercury through the open end of the glass tube comprising the electrode and completes electrical contact to platinum wire.[19]

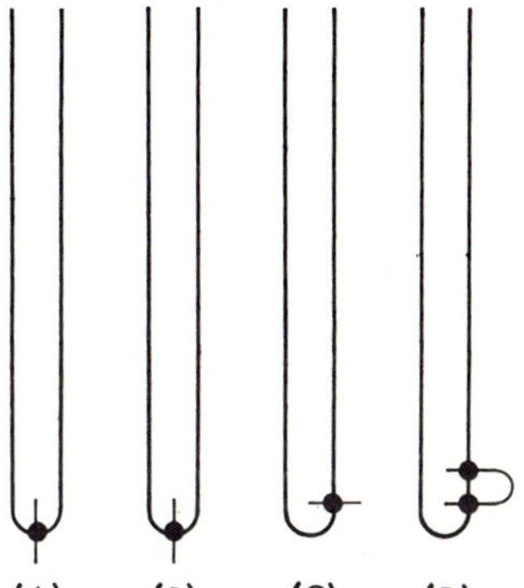

Fig. 47-6. Forms of some rotating platinum microelectrodes. 16 or 18 gauge wire, ~1 cm total length. Typical surface area of form D electrode is 0.2 to 0.3 cm^2.

The diffusion current at a rotating electrode depends markedly upon its shape. With few, rare exceptions, it is not possible to derive an expression relating current to the various relevant parameters. Among these parameters are the electrode area A, the angular velocity of rotation ω, the diffusion coefficient D, and the concentration C of the electroactive species. It has been found experimentally for a number of different electrode configurations that

$$i_d = \text{const.} \times nAD^{\alpha}\omega^{\beta}C \tag{26}$$

where α ranges from 0.5 to 1 and β from 0.4 to 1.

Wire electrodes suffer from the problem of rapid, irregular fluctuations in i_d resulting from turbulent eddies in the solution adjacent to the electrode. These fluctuations increase the uncertainty in i_d and reduce the effective analytical sensitivity to about 10–20 times that of a D.M.E. Such electrodes have a day-to-day reproducibility of 1–3 per cent in i_d, although substantially better results are obtainable over short time periods. Form D is slightly more rigid than the other electrode forms in Fig. 47-6 and has an advantage when maximum reproducibility is needed.

By choosing the shape of the electrode so that laminar fluid flow occurs at the electrode surface, a net increase in sensitivity is obtained. This is the case for a rotating disk electrode, which is shown in Fig. 47-7. While smaller diffusion-current densities (i_d/A) are obtained, fluctuations in i_d are tremendously reduced because of the absence of turbulent regions near the electrode, and the accuracy in the measurement of i_d is substantially increased as compared to a rotating wire electrode.

Levich[20] solved the complex hydrodynamic problem at a rotating disk electrode to obtain the expression for the diffusion current,

$$i_d = 0.62nFA\gamma^{-1/6}\omega^{1/2}D^{2/3}C \tag{27}$$

[19] A commercial synchronous rotator is available from the E. H. Sargent Co., Chicago, as are some forms of wire electrodes.

[20] V. G. Levich, *Zh. Fiz. Khim.* **18**, 335 (1944); *Acta Physiochem. URSS* **17**, 257 (1942); **19**, 117 (1944); *Zh. Fiz. Khim.* **23**, 575 (1948).

where γ is the kinematic viscosity of the solution (viscosity/density of the solution) and the other quantities have the customary significance.

From the analytical point of view it is important to note that i_d is proportional to C in both Eqs. (26) and (27). The rotating disk electrode is the preferred form of a rotating electrode for three reasons:

1. The dependence of i_d upon fundamental quantities is well understood.
2. Sensitivity is superior to that at a wire electrode.
3. Electrode materials other than platinum are readily fabricated into the form of a disk electrode. Construction details are available in the literature.[21,22]

This electrode will undoubtedly find more general use in analytical chemistry than it has in the past.

The first application of a rotating platinum electrode was reported by Nernst and Merriam,[23] who obtained current–potential curves for the reduction of iodine, bromine, chlorine, silver ion, and permanganate ion. The student is referred to the citations below[24,25,26] for further details and particulars concerning other electroreductions and oxidations at rotating solid electrodes.

To date, the principal analytical application of rotating electrodes has been as an indicator electrode in amperometric titrations, as discussed on page 915, although a number of procedures, such as the determination of dissolved oxygen, by electroreduction at a rotating platinum electrode are used.

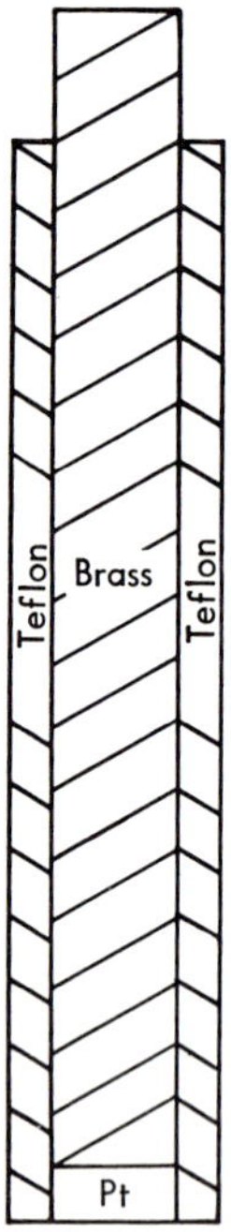

Fig. 47-7. Rotating platinum disk electrode. The platinum is soldered to the brass shaft and the brass rod is press-fitted into an insulating Teflon tube. The electrode is rotated about its long axis, and electrical contact is made to the platinum disk by means of a brush which rests on the rotating brass shaft.

[21] S. Azim and C. A. Riddiford, *Anal. Chem.* **34**, 1023 (1962).
[22] W. J. Albery and S. Bruckenstein, *Trans. Faraday Soc.* **62**, 1920 (1966).
[23] W. Nernst and E. S. Merriam, *Z. Physik. Chem.* (*Leipzig*) **52**, 235 (1905).
[24] I. M. Kolthoff and J. J. Lingane, Ref. 6, p. 412ff.
[25] L. Meites, Ref. 2, p. 420ff.
[26] D. T. Napp and S. Bruckenstein, *Anal. Chem.* **39**, 481 (1967); D. C. Johnson, D. T. Napp, and S. Bruckenstein, *Ibid.*, **40**, 482 (1968); P. Beran and S. Bruckenstein, *Ibid.*, **40**, 1044 (1968); G. W. Tindall and S. Bruckenstein, *Ibid.*, **40**, 1051, 1402, and 1637 (1968).

The potential limits available for studying electrochemical process depend upon the supporting electrolyte and electrode material. Table 47-2 summarizes the limits that can be obtained using mercury, platinum, and gold as the electrode metal. As was the case with the D.M.E., the presence of complexes will make the anodic potential limit less positive. Neither gold nor platinum can be used at as negative potentials as mercury can, gold being slightly superior to platinum in this regard. These cathodic limits are determined by the overpotential for hydrogen evolution on the metal. The larger anodic potential range accessible on platinum and gold make these metals especially useful in the study of oxidants such as Ce(IV), MnO_4^-, Br_2, and I_2.

TABLE 47-2. **Potential Limits at Platinum, Gold, and Mercury Electrodes in the Absence of Complex Forming Agents as a Function of the pH of the Supporting Electrolyte**

Metal	*Cathodic Limit, Volt* *versus* S.C.E.	*Anodic Limit, Volt* *versus* S.C.E.
Platinum	−0.06 pH	1.2–0.06 pH[a]
Gold	−0.4–0.06 pH[b]	1.45–0.06 pH[c]
Mercury	−1.0 to −2.6[d]	0.4

[a] Platinum oxide forms at (0.80–0.06 pH) V and inhibits some electrochemical reactions.

[b] pH > 2. In 1 *M* H^+, the cathodic limit is only −0.2 V more negative than Pt.

[c] Gold oxide forms at (0.84–0.06 pH) V and inhibits some electrochemical reactions.

[d] −1.0 V in strong acid solution, −1.9 V in alkali metal salt supporting electrolyte, and −2.6 V in long-chain quaternary ammonium salts.

Mercury electrodes are well suited for investigating ions such as Zn^{++} and Mn^{++} which are reduced at potentials more negative than those attainable at platinum or gold.

The sensitivity of the D.M.E. can be increased about tenfold by using a rotated D.M.E. For details see Okinaka and Kolthoff.[27]

PROBLEMS

1. A solution contains three species, A, B, and C, which reduce at the D.M.E. Three diffusion currents of 10.0, 19.6, and 28.0 μA are observed at 0.0, −0.5, and −1.10 V versus S.C.E., corresponding respectively to the diffusion current of A, of A + B, and of A + B + C. Calculate the diffusion current for (a) A at −0.50 and −1.1 V; (b) B at −0.50 and −1.1 V; (c) C at −1.1 V. Use the data in Table 47-1. *Ans.* (a) 10.1 and 9.87 μA; (b) 9.5 and 9.3 μA; (c) 8.8 μA.
2. The diffusion current for an unknown concentration of Cd^{++} in air-free 1.0 *M* KNO_3 is 10.0 μA at −0.8 V versus S.C.E. The volume of solution in the polarographic cell is 100 ml. 10.0 ml of 0.01 *M* $Cd(NO_3)_2$ are added to the cell and the diffusion current becomes 17.0 μA. What is the initial concentration of Cd^{++}? *Ans.* 1.15×10^{-3} *M*.
3. The drop time of a capillary 10.50 cm long is 12.0 sec at the electrocapillary maximum when the effective height of mercury is 40 cm. A drop time of 6.0 sec is desired at an effective mercury height of 50 cm. How much should be cut from the capillary? *Ans.* 3.94 cm.

[27] Y. Okinaka and I. M. Kolthoff, *J. Am. Chem. Soc.* **79**, 3326 (1957).

4. In 0.5 *M* potassium chloride, the diffusion coefficient of Cu(II) is 6.8×10^{-6} cm^2/sec at 25°C. (a) What is the convective diffusion current density on the limiting current region for the reactions

$$Cu(II) + 2e \rightarrow Cu$$

at a rotating platinum disk electrode when the angular velocity of rotation is 600 rad/sec? Assume the kinematic viscosity of the supporting electrolyte is 8×10^{-3} centistoke. $C_{Cu(II)} = 1.00 \times 10^{-3}$ *M*. (b) What is the value of the diffusion current density when the drop falls for the reaction

$$Cu(II) + Hg + 2e \rightarrow Cu_{amal}$$

at the D.M.E. described in Table 47-1, at −0.5 V versus S.C.E. (in the diffusion-limited region)? $C_{Cu(II)} = 1.00 \times 10^{-3}$ *M*.

Ans. (a) 2.36×10^3 $\mu A/cm^2$; (b) 2.12×10^2 $\mu A/cm^2$

CHAPTER 48 *Current–Potential Equations*

In Chapter 46 the wave (current–potential) equations have been given for two frequently encountered cases, the reduction of a metal ion to a solid metal, and the reduction of one soluble species to another which diffuses away from the electrode surface. Several other situations exist which are of considerable importance in the analytical application of voltammetry. Before these examples are discussed, it is advantageous to consider briefly the region across which the concentration gradient arises at an indicator electrode during electrolysis, i.e., the diffusion layer. It will be assumed that the reaction

$$Tl^+ + e \rightarrow Tl$$

occurs at a rotating platinum electrode, although in general it is not necessary to specify the reaction or the electrode used.

A schematic representation of the diffusion layer adjacent to the platinum electrode is shown in Fig. 48-1, where y represents the distance from the electrode surface. Two concentration profiles are shown for different electrode surface concentrations of thallium ion, C^0 and $C^{0'}$. The exact equation for the concentration of thallium at a point y in the diffusion layer C^y is extremely complicated, and can be calculated only for a very special shape of rotating electrodes. Nernst[1] recognized that it would be a good approximation to assume that C^y changes linearly in going from the electrode surface to the bulk of the solution and suggested that a tangent be drawn to the $C - y$ curve at $y = 0$ (dashed lines in Fig. 48-1), thereby defining the diffusion-layer thickness δ as the distance from the electrode surface to the intersection of the tangent with the line $C_{Tl^+} = C$ (point A). The concentration profile is then given by simple geometry as

$$C^y - C^0 = (C - C^0)\frac{y}{\delta} \qquad (1)$$

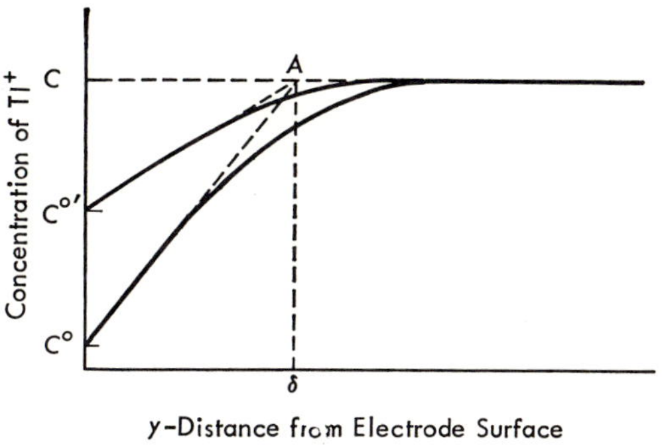

Fig. 48-1. Concentration profiles for two different surface concentrations of Tl^+ adjacent to an electrode surface.

[1] W. Nernst, *Z. Physik. Chem.* (*Leipzig*) **47**, 52 (1904); also see A. A. Noyes and W. R. Whitney, *ibid.* **23**, 689 (1897).

The slope of the concentration profile, the *concentration gradient*, at the electrode surface is

$$\left(\frac{dC^y}{dy}\right)_{y=0} = \frac{C - C^0}{\delta} \tag{2}$$

If N_{Tl} represents the number of moles of thallium metal plated out on the electrode in time t, the rate of reduction of thallium ions is $-dN_{Tl}/dt$ ($-$ sign because thallium ions are disappearing). According to the Fick's first law of diffusion, at the electrode surface the rate of supply of Tl^+ to the electrode surface, dN_{Tl^+}/dt, is

$$\frac{dN_{Tl^+}}{dt} = -\frac{dN_{Tl}}{dt} = AD\left(\frac{dC^y}{dy}\right)_{y=0} = \frac{dN}{dt} \tag{3}$$

where D is the diffusion coefficient of thallium ion and A is the electrode area. Combining Eqs. (2) and (3) gives

$$\frac{dN}{dt} = \frac{AD}{\delta}(C - C^0) \tag{4}$$

According to Faraday's law, the rate of reduction is proportional to the rate at which charge passes (current); hence

$$\frac{dN}{dt} = \frac{i}{nF} \tag{5}$$

where $n = 1$ for thallium ion and F is the Faraday constant. Combining Eqs. (4) and (5) yields

$$i = \frac{nFAD}{\delta}(C - C^0) \tag{6}$$

where $nFAD/\delta = k$, a constant for a given electrode, temperature, and rotation speed but different for different ions. Hence

$$i = k(C - C^0) \tag{7}$$

As mentioned earlier, k is called the mass-transfer coefficient.

In the material that follows, the nature of the indicator electrode is not specified unless this is required by the electrode reaction. Substitutions of the appropriate mass-transfer coefficient into the general equations will yield the particular result required.

Composite waves. Oxidized and reduced forms both soluble and present simultaneously

Consider a solution containing a soluble oxidant, Ox, and a soluble reductant, Red, which are partners in the reversible electrode reaction

$$\text{Ox} + ne \rightleftharpoons \text{Red}$$

Their concentrations in the bulk of the solution are [Ox] and [Red]. The current–potential curve for this system is shown in Fig. 48-2. Curve FGA represents the wave obtained when only Ox is present ([Red] $= 0$), and only reduction can occur at the indicator electrode surface. A cathodic diffusion current, $(i_c)_d$, occurs at potential A, when $[\text{Ox}]^0 = 0$ because of electroreduction. Curve CDE represents the curve obtained when only Red is present in the bulk of the solution and only oxidation can occur at the electrode surface. An anodic diffusion current, $(i_a)_d$, occurs at a potential F when the rate of electrooxidation becomes large enough to reduce the surface concentration of Red to zero. When both Ox and Red are present in the bulk of the solution, curve ABC is obtained, and both oxidation and reduction can occur at the electrode surface, depending upon the applied potential. We observe both cathodic and anodic diffusion currents. Curves FGA, CDE, and ABC are called the *cathodic*, *anodic*, and *composite* waves of the redox system.

As stated on page 165, the potential of this redox system is conveniently written

$$E = E^{0'}_{\mathrm{Ox/Red}} + \frac{0.059}{n} \log \frac{[\mathrm{Ox}]}{[\mathrm{Red}]} \tag{8}$$

at 25°C. In Fig. 48-2, the potential at point B on the composite wave ABC corresponds to the reduction potential of the system since $i = 0$. (More accurately the reduction potential is observed when i equals the residual current. See page 944 for a discussion of the consequences of this statement in terms of the limitations placed upon potentiometric titrations at low concentrations.)

The wave equation for this system is obtained by substituting the surface concentrations of the oxidized, $[\mathrm{Ox}]^0$, and reduced, $[\mathrm{Red}]^0$, forms into Eq. (8). Since no assumption concerning the nature of Red was made in obtaining Eq. (7), we may write for the reduction of Ox

$$i = k_{\mathrm{Ox}}\{[\mathrm{Ox}] - [\mathrm{Ox}]^0\} \tag{9}$$

and

$$[\mathrm{Ox}]^0 = \frac{(i_c)_d - i}{k_{\mathrm{Ox}}} \tag{10}$$

where $(i_c)_d$ is the cathodic diffusion current for the reduction process and is given by

$$(i_c)_d = k_{\mathrm{Ox}}[\mathrm{Ox}] \tag{11}$$

The current i is proportional to the rate of disappearance of Ox at the electrode surface and therefore is proportional to the rate of appearance of Red at the electrode surface. $[\mathrm{Red}]^0$ depends upon two factors: (1) the current, which tends to increase it, and (2) diffusion from the electrode surface (either back into solution or into the mercury if an amalgam forms), which tends to decrease it.

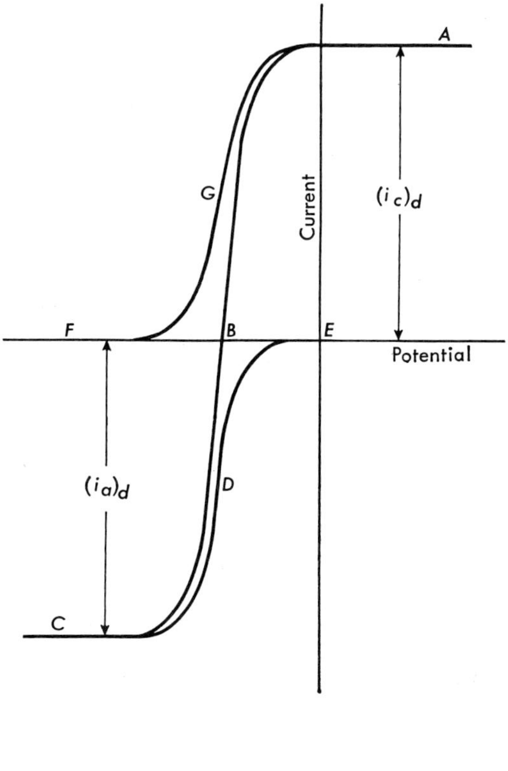

Fig. 48-2. Current-potential curve for the redox system Ox + $ne \rightleftharpoons$ Red. Curve FGA—cathodic wave. Curve CDE—Anodic wave. Curve ABC—composite wave.

After a very short time, the rate of production of Red by electrolysis and the rate of diffusion of Red away from the electrode surface become equal and a steady-state surface concentration, $[\text{Red}]^0$, is produced. The relationship between current and concentration is

$$i = -k_{\text{Red}}([\text{Red}] - [\text{Red}]^0) \tag{12}$$

which has the same form as Eq. (9), *except for the minus sign, which allows for the fact that* $[\text{Red}]^0 > [\text{Red}]$ *when Ox is being converted to Red at the electrode surface.*

In terms of the Nernst diffusion-layer picture,

$$k_{\text{Red}} = \frac{nFAD_{\text{Red}}}{\delta_{\text{Red}}}$$

where D_{Red} is the diffusion coefficient of Red, and δ_{Red} is the diffusion-layer thickness for Red. δ_{Red} is generally a function of D_{Red}, and therefore differs for δ_{Ox}.

For the rotating disk electrode, it can be shown from Eq. (27) (p. 897) that δ varies as $D^{1/3}$; hence $k_{\text{Ox}}/k_{\text{Red}} = (D_{\text{Ox}}/D_{\text{Red}})^{2/3}$, while for the DME we see from Eq. (3) (p. 888) that δ varies as $D^{1/2}$; hence $k_{\text{Ox}}/k_{\text{Red}} = (D_{\text{Ox}}/D_{\text{Red}})^{1/2}$.

The surface concentration of Red is found by calculating the value of the anodic diffusion current. At a potential corresponding to F in Fig. 48-2, $[\text{Red}]^0$ is zero, and the maximum anodic current passes through the electrode,

$$i = (i_a)_d = -k_{\text{Red}}[\text{Red}] \tag{13}$$

Conventionally anodic currents have a negative sign, whereas cathodic currents have a positive sign. Combining Eqs. (12) and (13) yields

$$[\text{Red}]^0 = \frac{-(i_a)_d + i}{k_{\text{Red}}} \tag{14}$$

Substituting Eqs. (10) and (14) into the Nernst equation (8) yields

$$E = E^{0'}_{\text{Ox/Red}} + \frac{0.059}{n} \log \left\{ \frac{(i_c)_d - i}{i - (i_a)_d} \frac{k_{\text{Red}}}{k_{\text{Ox}}} \right\} \tag{15}$$

or

$$E = E^{*}_{\text{Ox/Red}} + \frac{0.059}{n} \log \frac{(i_c)_d - i}{i - (i_a)_d} \tag{16}$$

as the wave equation for the composite wave, where

$$E^{*}_{\text{Ox/Red}} = E^{0'}_{\text{Ox/Red}} + \frac{0.059}{n} \log \frac{k_{\text{Red}}}{k_{\text{Ox}}} \tag{17}$$

If $[\text{Red}]^0 = 0$, $(i_a)_d = 0$, and the cathodic wave equation is obtained (curve FGA, Fig. 48-2):

$$E = E^{*}_{\text{Ox/Red}} + \frac{0.059}{n} \log \frac{(i_c)_d - i}{i} \tag{18}$$

while, if $[\text{Ox}] = 0$, $(i_c)_d = 0$, and the anodic wave equation (curve CDE, Fig. 48-2) results:

$$E = E^{*}_{\text{Ox/Red}} - \frac{0.059}{n} \log \frac{(i_a)_d - i}{i} \tag{19}$$

The log term in Eq. (16) is zero when[2]

$$i = \tfrac{1}{2}\{(i_a)_d + (i_c)_d\}$$

[2] Note that $(i_a)_d$ is a negative quantity, since it represents an anodic current.

and the potential corresponding to this current, $E^{1/2}$, is called the *half-wave potential*, i.e.,

$$E^{1/2}_{\mathrm{Ox/Red}} = E^* = E^{0'}_{\mathrm{Ox/Red}} + \frac{0.059}{n} \log \frac{k_{\mathrm{Red}}}{k_{\mathrm{Ox}}} \qquad (20)$$

This potential is independent of concentration for systems of the type Ox + $ne \rightleftharpoons$ Red. If [Red] = 0, $E^{1/2}_{\mathrm{Ox/Red}}$ occurs at $i = (i_c)_d/2$, while if [Ox] = 0, $E^{1/2}_{\mathrm{Ox/Red}}$ occurs at $i = (i_a)_d/2$.

A short table of $E^{1/2}$ values for processes at the D.M.E. is given in Table 48-1 for three different supporting electrolytes. A more comprehensive table can be found elsewhere.[3]

TABLE 48-1. **Half-Wave Potentials at the D.M.E.**[a]

Species and Oxidation-State Changes	*Supporting Electrolyte*		
	0.1 *M* Cl^- [b]	1 *M* HCl	1 *M* NH_3 + 1 *M* NH_4Cl
Bi(III → 0)		−0.09	
Cd(II → 0)	−0.60	−0.642	−0.81
Cr(II → III)	−0.34[c]		−0.85[d]
Cr(III → II)	−0.61[c]		−1.43 (II → O, −1.71)[e]
Cu(II → I)		0.04	−0.24
Cu(I → 0)		−0.22	−0.51
Fe(II → 0)	−1.3[c]		−1.49[c] (II → III, −0.34)
Ga(III → 0)	−1.1[c]		−1.6
In(III → 0)	−0.56	−0.56[f]	
Mn(II → 0)	−1.48		−1.65
Ni(II → 0)	−1.1[c]		−1.10[c]
Pb(II → 0)	−0.40	−0.44	
Sb(III → 0)		−0.15	
Sn(II → IV)		−0.1	
Sn(II → 0)		−0.47	
Tl(I → 0)	−0.46		−0.48
U(VI → V)		−0.18[f]	−0.8 (V → III, −1.4)
U(VI → IV)		−0.21	
U(IV ⇌ III)		−0.89	
		−0.94	
V(III → II)		−0.51	
V(V → IV)		> 0[f,g]	
V(IV → II)			−1.28 (IV → V, −0.94)
Zn(II → 0)	−1.00[h]		−1.35

[a] Taken mainly from *Handbook of Analytical Chemistry* (L. Meites, ed.), McGraw-Hill, New York, 1963, Table 5-23.
[b] 0.1 *M* KCl, LiCl, NH_4Cl, or $(Me)_4$ NCl.
[c] Reported as irreversible.
[d] 0.1 *M* NH_3, 5 *M* NH_4Cl.
[e] Data in parentheses refer to multiple waves, seen in addition to the wave listed on the left.
[f] 0.1 *M* HCl.
> 0 signifies that the diffusion current occurs at all potentials more negative than the anodic dissolution potential for the D.M.E.
[g] 1 *M* KCl.

[3] *Handbook of Analytical Chemistry* (L. Meites, ed.), McGraw-Hill, New York, 1963, Table 5-23.

For a redox system in which one form is present at constant activity, e.g., the reduction of a metal ion onto a platinum electrode,

$$M^{n+} + ne \rightarrow M$$

the wave equation is (see p. 895)

$$E = E^{0'}_{M^{n+}/M} + \frac{0.059}{n} \log \frac{i_d - i}{k_{M^{n+}}} \tag{21}$$

and the half-wave potential is defined as the potential where $i = i_d/2$; i.e.,

$$E^{1/2}_{M^{n+}/M} = E^{0'}_{M^{n+}/M} + \frac{0.059}{n} \log \frac{i_d}{2k_{M^{n+}}} = E^{0'}_{M^{n+}} + \frac{0.059}{n} \log \frac{[M^{n+}]}{2} \tag{22}$$

$E^{1/2}_{M^{n+}/M}$ is a function of the bulk concentration of M^{n+}, $[M^{n+}]$.

The half-wave potential is independent of concentration only if Ox and Red are both soluble, and the reduction of one Ox species produces exactly one Red species. Thus the electroreduction of an organic disulfide compound (RSSR) to a sulfhydryl compound (RSSH), according to

$$RSSR + 2\,H^+ + 2e \rightleftharpoons 2\,RSH$$

does not yield a concentration-independent half-wave potential.

For several reasons, the potential upon passage of current in several systems may not correspond exactly to the ideally reversible oxidation–reduction system; i.e., the Nernst relation between surface concentration and the potential may not hold exactly. This phenomenon is called *polarization*. A number of different situations give rise to polarization. For example, the rate of the electron transfer reaction (Ox + $ne \rightarrow$ Red or Red $\rightarrow$ Ox + ne) may be slow compared to the maximum rate of mass transfer to the electrode surface. Alternatively, either Ox or Red may be involved in a slow equilibrium with other species present in solution, and the rate of these chemical reactions must be taken into account in calculating the rate of transfer of Ox and Red from the bulk of the solution to the electrode surface. These and other complications are discussed in references C2, C3, E1, E2, E3, and E4 on pages 884 and 885.

The reversibility of a reaction may be tested experimentally using the wave equation for the electrode reaction. For example, the wave equation for the reduction of 10^{-4} *M* $Fe(CN)_6^{3-}$

$$Fe(CN)_6^{3-} + e \rightarrow Fe(CN)_6^{4-}$$

in 0.1 *M* KCl at a rotating platinum disk electrode is, according to Eq. (18),

$$E = E^{1/2}_{Fe(CN)_6^{3-}/Fe(CN)_6^{4-}} + 0.059 \log \frac{i_d - i}{i}$$

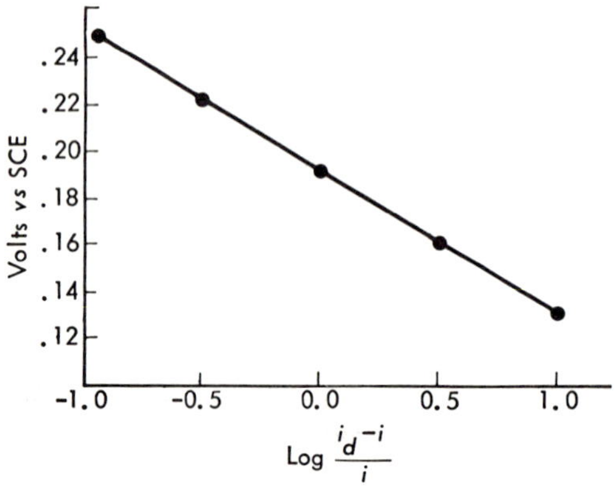

Fig. 48-3. Test of reversibility of $Fe(CN)_6^{-3} + e \rightleftharpoons Fe(CN)_6^{-4}$ at a rotating platinum disk electrode in 0.1 *M* KCl. $[Fe(CN)_6^{-3}] = 1 \times 10^{-4}$ *M*. Angular velocity of rotation = 11.1 radians/sec.

A plot of E versus $\log (i_d - i)/i$ is shown in Fig. 48-3. The experimental slope of 0.060 agrees with the theoretical slope and we conclude that this reduction is reversible under these experimental conditions.

A similar analysis for the reduction of mercurous ion in 0.1 M nitric acid at a dropping mercury electrode

$$Hg_2^{++} + 2e \rightleftharpoons 2\,Hg$$

yields the wave equation

$$E = E^{0'}_{Hg_2^{++}/Hg} + \frac{0.059}{2} \log \frac{i_d - i}{k_{Hg_2^{++}}}$$

according to Eq. (21). However, a plot of E versus $\log (i_d - i)$ yields a slope of 0.045, considerably different from the theoretical value of 0.0296 at 25°C.

EFFECT OF COMPLEXATION ON $E^{1/2}$

Complexation of one or more of the species diffusing to an indicator electrode can alter the half-wave potential because the formal potential of a redox couple can be shifted by complex formation. We shall consider some of the more important situations below, avoiding, as much as possible, references to a particular indicator electrode.

The shift in $E^{1/2}$ which occurs on complexation is of considerable analytical significance. First, it provides a way of separating waves of two redox couples whose formal potentials lie so close together in noncomplexing supporting electrolytes that two well-resolved waves cannot be obtained. An example of this sort is the separation of the waves of Tl^+ and Pb^{++} at a D.M.E. At the D.M.E. in noncomplexing media, 1 M perchloric acid, $E^{1/2}_{Tl^+/Tl_{amal}} = -0.48$ V and $E^{1/2}_{Pb^{++}/Pb_{amal}} = -0.38$ V; only one wave is observed experimentally in solution containing both ions.

In 1 M NaOH, $E^{1/2}_{Tl^+/Tl_{amal}} = -0.48$ V while $E_{Pb(II)/Pb_{amal}} = -0.755$ V; the half-wave potential of lead ion has been shifted because of the formation of the complex ion, $HPbO_2^-$, and it is possible to determine both lead and thallium simultaneously from the limiting currents of the separate waves.

Second, the shift in $E^{1/2}$ provides a convenient way to determine the stability constants of complex ions. The principles underlying the voltammetric determination of stability constants are given below. These results are quite general, and it is only necessary to specify the mass-transfer coefficients [i.e., the value of k in equations such as Eqs. (10) and (14)] for a given electrode to obtain the particular result. In all cases supporting electrolyte is assumed to be present, and a temperature of 25°C is used.

Complex metal ion–metal couple

Consider a metal ion that forms a complex, e.g., lead ion with hydroxide ion

$$Pb^{++} + 3\,OH^- \rightleftharpoons HPbO_2^- + H_2O \qquad K = \frac{[HPbO_2^-]}{[Pb^{++}][OH^-]^3} \tag{23}$$

and is reduced at a D.M.E. reversibly, forming a soluble metal amalgam according to

$$HPbO_2^- + 2e + 2\,H_2O + Hg \rightleftharpoons Pb_{amal} + 3\,OH^- \tag{24}$$

The potential of the D.M.E. is given by

$$E = E^{0'}_{HPbO_2^-/Pb_{amal}} + \frac{0.059}{2} \log \frac{[HPbO_2^-]^0}{[Pb_{amal}]^0[OH^-]^3} \tag{25}$$

In Eq. (25), as pointed out above, the symbol $[\ ]^0$ denotes the surface concentration of the specified species, not the concentration raised to the zeroth power.

If hydroxyl ion is present in large excess as compared to lead ion in the bulk of the solution, the surface concentration of OH^- is constant and unaffected by the electrode process (24), and by writing Eq. (25) as

$$E = E^{0'}_{HPbO_2^-/Pb_{amal}} - \frac{0.059 \times 3}{2} \log [OH^-] + \frac{0.059}{2} \log \frac{[HPbO_2^-]^0}{[Pb_{amal}]^0} \tag{26}$$

it follows that

$$E = \text{const.} + \frac{0.059}{2} \log \frac{[HPbO_2^-]^0}{[Pb_{amal}]^0} \tag{27}$$

Thus we see that Eqs. (27) and (8) are of identical form; the mass-transfer problem does not differ in principle from the reduction of lead ion (Ox) in 0.1 M potassium nitrate at the D.M.E. to form lead amalgam (Red). Hence, from Eq. (9),

$$i = k_{HPbO_2^-}([HPbO_2^-] - [HPbO_2^-]^0) \tag{28}$$

When the potential becomes sufficiently negative, $[HPbO_2^-]^0 = 0$, and

$$i_d = k_{HPbO_2^-}[HPbO_2^-] \tag{29}$$

or, from Eqs. (28) and (29),

$$[HPbO_2^-]^0 = \frac{i_d - i}{k_{HPbO_2^-}} \tag{30}$$

All currents are cathodic currents. From Eq. (12), the current is related to the surface concentration of lead amalgam,

$$i = k_{Pb_{amal}}[Pb_{amal}]^0 \tag{31}$$

Substituting Eqs. (30) and (31) into (26), we find

$$E = E^{0'}_{HPbO_2^-/Pb_{amal}} - \frac{0.059}{2} \log \frac{k_{HPbO_2^-}}{k_{Pb_{amal}}} - \frac{0.059 \times 3}{2} \log [OH^-] + \frac{0.059}{2} \log \frac{i_d - i}{i} \tag{32}$$

When $i = i_d/2$

$$E = E^{1/2}_{HPbO_2^-/Pb_{amal}} = E^{0'}_{HPbO_2^-/Pb_{amal}} - \frac{0.059}{2} \log \frac{k_{HPbO_2^-}}{k_{Pb_{amal}}} - \frac{0.059 \times 3}{2} \log [OH^-] \tag{33}$$

which demonstrates that the half-wave potential is independent of biplumbite-ion concentration.

Studies of dropping amalgam electrodes indicate that the Ilkovic equation (p. 888) applies to the oxidation of a metal amalgam. Hence $k_{HPbO_2^-}/k_{Pb_{amal}} = (D_{HPbO_2^-}/D_{Pb_{amal}})^{1/2}$.

In the absence of OH^-, the half-wave potential for the reduction of lead at a D.M.E. is given, in analogy with Eq. (20), by

$$E^{1/2}_{Pb^{++}/Pb_{amal}} = E^{0'}_{Pb^{++}/Pb_{amal}} - \frac{0.059}{2} \log \frac{k_{Pb^{++}}}{k_{Pb_{amal}}} \tag{34}$$

The shift in half-wave potential by adding excess OH^- to a lead-ion solution is

$$\Delta E^{1/2} = E^{1/2}_{HPbO_2^-/Pb_{amal}} - E^{1/2}_{Pb^{++}/Pb_{amal}} = E^{0'}_{HPbO_2^-/Pb_{amal}} - E^{0'}_{Pb^{++}/Pb_{amal}} - \frac{0.059 \times 3}{2} \log [OH^-] - \frac{0.059}{2} \log \frac{k_{HPbO_2^-}}{k_{Pb^{++}}} \tag{33}$$

In Chapter 8, an expression was derived for the difference between the formal potentials of an aquo and a complex ion in terms of the complex-ion-formation constant. This equation [Eq. (4)] holds when amalgam electrodes are employed, i.e.,

$$E^{0'}_{HPbO_2^-/Pb_{amal}} - E^{0'}_{Pb^{++}/Pb_{amal}} = -\frac{0.059}{2} \log K \tag{34}$$

Combining Eqs. (34) and (33) yields

$$\Delta E^{1/2} = -\frac{0.059}{2}\log K - \frac{0.059 \times 3}{2}\log [OH^-] - \frac{0.059}{2}\log \frac{k_{HPbO_2^-}}{k_{Pb^{++}}} \tag{35}$$

Thus the half-wave potential should shift 89 mV more negative for a tenfold increase in $[OH^-]$. Lingane[4] reports a shift 83 mV; the difference between theory and experiment undoubtedly arises from the assumption that $E^{0'}_{HPbO_2^-}$ remains constant while $[OH^-]$ changes.

In 0.1 *M* potassium nitrate, $E^{1/2}_{Pb^{++}/Pb_{amal}} = -0.388$ V versus S.C.E., whereas in 0.1 *M* sodium hydroxide, $E^{1/2}_{HPbO_2^-/Pb_{amal}} = -0.681$ V versus S.C.E. Hence $\Delta E = -0.293$ V. Also, the ratio $k_{HPbO_2^-}/k_{Pb^{++}}$ was found to be 1.05 by measurement of the diffusion currents of equal concentrations of Pb^{++} and $HPbO_2^-$. Thus, according to Eq. (35), we estimate the stability constant for reaction (23):

$$-0.293 = -\frac{0.059}{2}\log K - \frac{0.059 \times 3}{2}\log 10^{-1} - \frac{0.059}{2}\log 1.05$$

or

$$K = 7.8 \times 10^{12}$$

at an ionic strength of 0.1.

Complexation of both forms of a redox couple

A number of metals form stable ions in two different valence states, both of which can form complexes with the same ligand. As was seen in Chapter 8, both ferric and ferrous iron form EDTA complexes, and these give a perfectly reversible wave at a dropping mercury electrode[5] according to

$$FeY^- + e \rightleftharpoons FeY^{--} \tag{36}$$

$$E = E^{0'}_{FeY^-/FeY^{--}} + 0.059 \log \frac{[FeY^-]}{[FeY^{--}]} \tag{37}$$

Consider a solution that contains only FeY^-. Both FeY^- and FeY^{--} are diffusing species, and, by analogy with Eq. (18), we see that the wave equation is given by

$$E = E^{0'}_{FeY^-/FeY^{--}} + 0.059 \log \frac{k_{FeY^{--}}}{k_{FeY^-}} + 0.059 \log \frac{i_d - i}{i} \tag{38}$$

with a half-wave potential

$$E^{1/2}_{FeY^-/FeY^{--}} = E^{0'}_{FeY^-/FeY^{--}} + 0.059 \log \frac{k_{FeY^{--}}}{k_{FeY^-}} \tag{39}$$

According to Eq. (34), Chapter 8,

$$E^{0'}_{FeY^-/FeY^{--}} = E^{0'}_{Fe^{3+}/Fe^{++}} + 0.059 \log \frac{K_{II}}{K_{III}} \tag{40}$$

where

$$K_{II} = \frac{[FeY^{--}]}{[Fe^{++}][Y^{4-}]} \quad \text{and} \quad K_{III} = \frac{[FeY^-]}{[Fe^{3+}][Y^{4-}]}$$

Therefore, the half-wave potential is given by

$$E^{1/2}_{FeY^-/FeY^{--}} = E^{0'}_{Fe^{3+}/Fe^{++}} + 0.059 \log \frac{K_{II}}{K_{III}} + 0.059 \log \frac{k_{FeY^{--}}}{k_{FeY^-}} \tag{41}$$

[4] J. J. Lingane, *Chem. Rev.*, **29**, 1 (1941).
[5] I. M. Kolthoff and C. Auerbach, *J. Am. Chem. Soc.* **74**, 1452 (1952).

At the D.M.E., in the pH range 3.5–6.5, and $\mu = 0.1$, the iron–EDTA system behaves reversibly,[5] and $E^{1/2}_{FeY^-/FeY^{--}} = -0.13$ V versus S.C.E. or 0.12 V versus N.H.E. Using the value of $E^{0'}_{Fe^{3+}/Fe^{++}} = +0.75$ V versus N.H.E. at $\mu = 0.1$, the ratio K_{II}/K_{III} is calculated to be $10^{10.8}$, in excellent agreement with the potentiometric value of Schwarzenbach.[6]

The student is referred to the references E1 to E5, on page 885, for derivations of a number of other current–potential curves such as the reduction of $Fe(C_2O_4)_3^{3-}$ to $Fe(C_2O_4)_2^{--}$ or $Fe(C_2O_4)_3^{4-}$, the oxidation of mercury in the presence of ions that form a soluble complex, and the oxidation of mercury in the presence of ions to form an insoluble precipitate, among others. General expressions are given, and methods for determining the formula of a complex (or precipitate) using the current–potential and $E^{1/2}$ data are described. The relative ease, as compared to potentiometric methods, for determining stability constants (or their ratios), has made the polarographic method quite popular, and the method is widely used.

PROBLEMS

1. Consider the cell illustrated in Fig. 46-1, in which $C_{Tl^+} = 10^{-4}\,M$, $C_{Cl^-} = 10^{-2}\,M$, and potassium nitrate has been added until it is 1.0 M. The area of the silver–silver chloride electrode is 10 cm^2 and that of the rotating platinum wire electrode is 0.2 cm^2. Assume $D_{Cl^-} = D_{Tl^+}$. The cathodic diffusion current is 10 μA. At the diffusion current, what is (a) the current density through the platinum electrode? (b) The current density through the Ag/AgCl electrode? (c) The increase in chloride concentration at the Ag/AgCl electrode surface? (Neglect any differences in the diffusion-layer thickness at the Ag/AgCl electrode and the platinum electrode.) (d) The change in potential of the Ag/AgCl electrode as i goes from 0 to i_d? (e) Repeat part (c) assuming that the diffusion-layer thickness at the Ag/AgCl electrode is 10 times that at the rotating platinum electrode. (f) Repeat part (d), for the conditions specified in (e).

   Ans. (a) 50 μA/cm^2; (b) 1 μA/cm^2; (c) $2 \times 10^{-6}\,M$; (d) 0.03 mV.

2. A chemist determines the polarogram of a solution thought to contain *either* Pb^{++} *or* Tl^+, and then discards the solution. The electrode reactions are either

$$Tl^+ + e + Hg \rightarrow Tl_{amal}$$

   or

$$Pb^{++} + 2e + Hg \rightarrow Pb_{amal}$$

   After determining the current–potential curve, he realizes that he forgot to note the potential when he began to record the current–potential curve. The difference in potential when $i = \frac{3}{4}i_d(E^{3/4})$ and $i = \frac{1}{4}i_d(E^{1/4})$ is

$$E^{3/4} - E^{1/4} = -0.028 \text{ V}$$

   Which ion was in the solution? *Ans.* Pb^{++}.

3. Derive the wave equation at the D.M.E. for the electrode reaction

$$MX_p^{n-pb} + ne + Hg \rightarrow M_{amal} + pX^{b-}$$

   and express the result in terms of the stability constant for the formation of MX_p^{n-pb} i.e.,

$$M^{n+} + pX^{b-} \rightarrow MX_p^{n-pb}$$

   and the half-wave potential for the reaction

$$M^{n+} + ne + Hg \rightarrow M_{amal}$$

[6] G. Schwarzenbach et al., *Helv. Chim. Acta* **34**, 576 (1951).

4. Derive the wave equation for the formation of an insoluble mercurous salt, i.e., the oxidation

$$2\ Hg + 2X^- \rightarrow Hg_2X_2 + 2e$$

in terms of $E^0_{Hg_2^{++}}$ and the solubility product of Hg_2X_2. In practice, except at very low concentrations, poor diffusion currents are obtained because the film of insoluble Hg_2X_2 interferes with diffusion.

5. (a) In Problem 4, what fraction of the bulk concentration of X^- exists at the D.M.E. surface when $i = i_d/4$? (b) What is the shift in $E^{1/2}$ for this process for a tenfold increase in C_{X^-}? *Ans.* (a) $\frac{3}{4}$; (b) -0.059 V.

CHAPTER 49 *Amperometric Titrations*

Amperometric titrations with one indicator electrode

The principle underlying the classical type of amperometric titration was given in Chapter 38 in the section dealing with linear extrapolation methods and was also mentioned on page 879. Since i_d is proportional to C for all indicator electrodes, it is not necessary to specify the indicator electrode in discussing the details of this technique.

Consider the reactions of T, a titrant, with X, a species to be determined, yielding Y. Y may be a soluble species or a precipitate. As a first example we choose a reaction that is not quantitative at the end point, e.g., a precipitation reaction in which the precipitate has a significant solubility:

$$\mathrm{X} + \mathrm{T} \rightarrow \mathrm{Y} \tag{1}$$

Assume that a potential can be chosen such that a cathodic diffusion current for X is observed and that T and Y are electroinactive at this potential. As T is added to X, the diffusion current decreases and becomes constant in the presence of excess T.

In Fig. 49-1 the ordinate, i', corresponds to the diffusion current corrected for dilution (V = initial volume of X and v = the volume of T added). The equivalence point corresponds to C, the intersection of the extrapolated lines AB and EF. The current region BDE indicates that reaction (1) is not quantitative in the vicinity of the equivalence point and excess T is required to force the reaction to completion. Since only a short extrapolation is required to point C, satisfactory results are obtained. If the reaction is quantitative (precipitate completely insoluble), the curved region is not observed and the titration curve consists of two straight lines, ABC and CEF.

The measurable current in the presence of enough excess titrant to completely suppress the solubility corresponds to the residual current of the solution. It would be incorrect to extend line AB to the volume axis ($i' = 0$) and take that volume as the equivalence point.

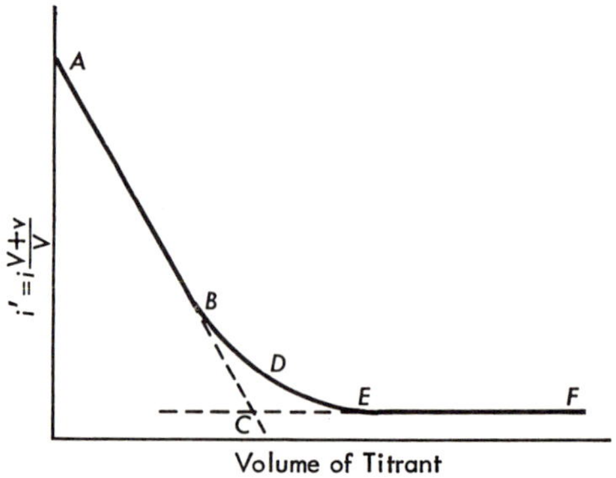

Fig. 49-1. Amperometric titration of an electroactive species with a nonelectroactive titrant. The product is nonelectroactive at the applied potential, or insoluble.

When the substance titrated is not reduced at the applied emf, but the reagent is, the titration lines are just the reverse of those given in the previous case (Fig. 49-2). Thus it has been found possible[1] to titrate sulfate solutions with lead nitrate with a high degree of accuracy at an applied emf at which lead yields a diffusion current.

When both the substance titrated and the reagent yield a diffusion current at the applied emf, the titration lines have the shape given in curve 2, Fig. 38-3, p. 771. Such titration lines are obtained, for example, in the titration of dilute lead solutions with chromate[2] or in the titration of nickel with dimethylglyoxime.[3]

Fractional precipitation titrations can also be carried out amperometrically. Suppose that there are in the solution two ions A and B which yield insoluble salts with the reagent and that B does not start to precipitate until practically all of A is precipitated. We shall further assume that A and the reagent yield a diffusion current at the applied emf, while B is not reduced at this potential. It is easily seen, then, that the titration line will have the shape given in Fig. 49.3. Point A corresponds to complete precipitation of A and point B to that of B. Provided A does not coprecipitate with B, the fractional-precipitation titration should yield useful analytical results.

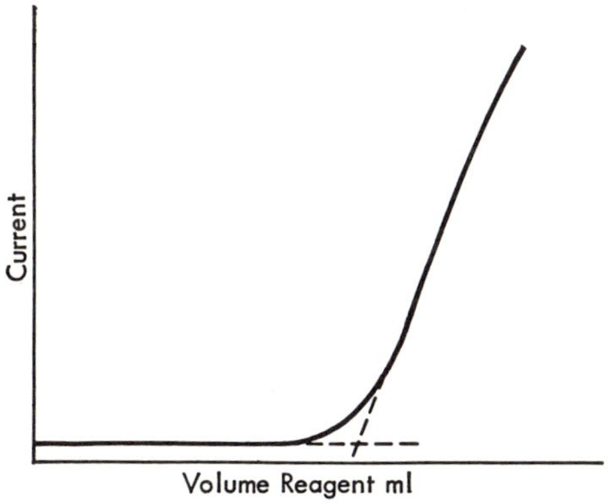

Fig. 49-2. Amperometric titration of a nonelectroactive species with an electroactive titrant.

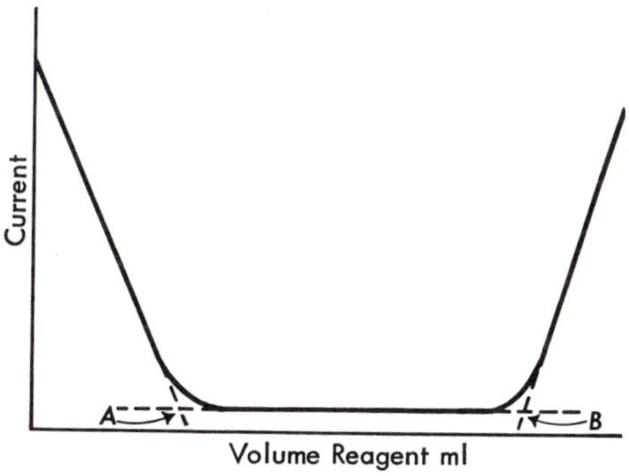

Fig. 49-3. Amperometric titration curve obtained in the course of a fractional precipitation titration using an electroactive titrant. The most insoluble reactant is electroactive.

Amperometric titrations are not limited to precipitation reactions but can also be applied to neutralization, complexation, and oxidation–reduction reactions.

Redox titrations corresponding to

$$\mathrm{Ox_1 + Red_2 \rightarrow Red_1 + Ox_2} \tag{2}$$

where Ox_1 is the titrant and Red_2 is the species being determined, yield more complicated amperometric titration curves than precipitation reactions, because there are four species capable of yielding a diffusion current. First, assume that both the Ox_1/Red_1 and Ox_2/Red_2 couples are reversible. Current–potential curves for the two redox systems are given in Fig. 49-4. Illustrated are cathodic, anodic, and composite waves for both couples assuming that the mass-transfer coefficients for all species are the same. Current–potential curves at various stages in the titration according to reaction (2) are shown in Fig. 49-5.

[1] I. M. Kolthoff and Y. D. Pan, *J. Am. Chem. Soc.* **62**, 3332 (1940).
[2] I. M. Kolthoff and Y. D. Pan, *J. Am. Chem. Soc.* **61**, 3402 (1939).
[3] I. M. Kolthoff and A. Langer, *J. Am. Chem. Soc.* **62**, 211 (1940).

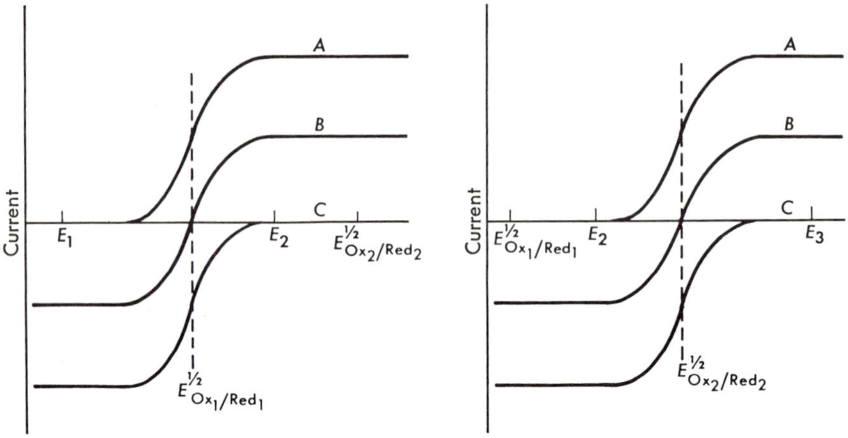

Ox_1/Red_1 System (A) Ox_2/Red_2 System (B)

Fig. 49-4. Current-potential curves for titrant and unknown redox couple. Curve *A*, pure-oxidized form; Curve *B*, $C_{ox}/C_{red} = 1$; Curve *C*, pure reduced form.

Amperometric titrations obtained at potentials E_1, E_2, and E_3 are given in Fig. 49-6. Straight lines are found in all cases. The most useful curve is E_2, and the end point is taken to be where the current is zero. If the residual current is not zero and is unknown, a titration error of unknown magnitude occurs. In practice, differences exist between the mass-transfer coefficients of the four species, and small breaks occur in the titration lines at the equivalence point. Usually these breaks are very poorly defined, and again curve E_2 yields the most useful titration curve.

If the titrant redox couple is reversible, while the couple of the species being determined is inert, e.g., $S_2O_3^{--}$ being titrated with I_2 at a rotating platinum electrode, better amperometric titration lines are obtained than in the completely reversible

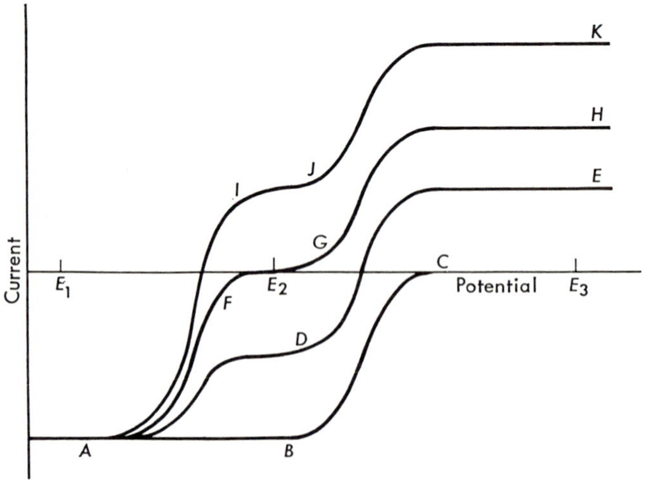

Fig. 49-5. Current-potential curves during a redox titration. 0%—*ABC*, 50%—*ADE*, 100%—*AFGH*, and 150%—*AIJK*.

situation. For example, if the applied potential corresponds to the limiting current region for I_2 ($I_2 + 2e \rightarrow 2I^-$) a "reverse-L"-shaped curve similar to that shown in Fig. 49-2 is obtained.

Obviously it is highly advantageous to use an irreversible (electrochemically inert) titrant to titrate a species that is part of a reversible redox system.

Amperometric titrations have certain limitations. Suppose that the three reducible substances A_1, A_2, and A_3 are present in the solution (Fig. 49-7); A_1 is reduced at the potential E_1, giving the diffusion current i_1, proportional to the concentration of A_1; A_2 is reduced at the potential E_2, giving the diffusion current i_2; and A_3 is reduced at the potential E_3, giving the diffusion current i_3. Now suppose that a titration of (or with) A_1 is to be carried out in the presence of any amount of A_2 and A_3. If the applied e.m.f. is kept between E_1 and E_2, only A_1 is reduced, and the current is proportional to the concentration of A_1 alone, A_2 and A_3 being without influence. However, if A_2 is titrated in the presence of A_1, the current is the sum of the diffusion currents of A_1 (i_1) and the diffusion current of A_2 (i_2), the latter decreasing continuously during the titration. With increasing ratio of i_1/i_2 the accuracy in the measurement of i_2 becomes smaller. If A_3 is titrated, analogous considerations apply. For example, it is not readily possible to carry out the amperometric titration of sulfate with barium (or the reverse) in the presence of large amounts of metal ions more noble (more easily discharged) than barium, e.g., zinc, cadmium, nickel, cobalt, etc. In such a case, E_3 of Fig. 49-7 would correspond to the discharge potential of barium and E_1 and E_2 to those of cadmium and zinc, for instance. In this case it would be better to use a lead solution instead of barium, because the discharge potential of lead is found at a potential more positive than E_1.

Many organic reagents yield precipitates with metals that are useful in amperometric titrations. For example, nickel can be titrated with dimethylglyoxime; cobalt, copper, and palladium with α-nitroso-β-naphthol; and copper with benzoinoxime, using the D.M.E.

The rotating platinum electrode is also useful in amperometric titration involving metal ions. For example, it has been used for the titration of silver with chloride (gelatine should be present) and copper in ammoniacal solution with ethylenediaminetetraacetic acid. It is especially valuable in amperometric titrations involving oxidants. For example, the titration of arsenic trioxide with bromate in an acid bromide medium can be made with high precision and accuracy even when the concentration of arsenic

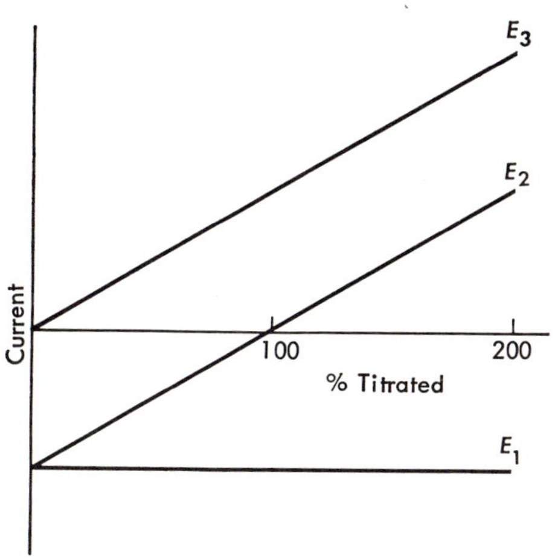

Fig. 49-6. Amperometric redox titrations obtained at potentials E_1, E_2, and E_3.

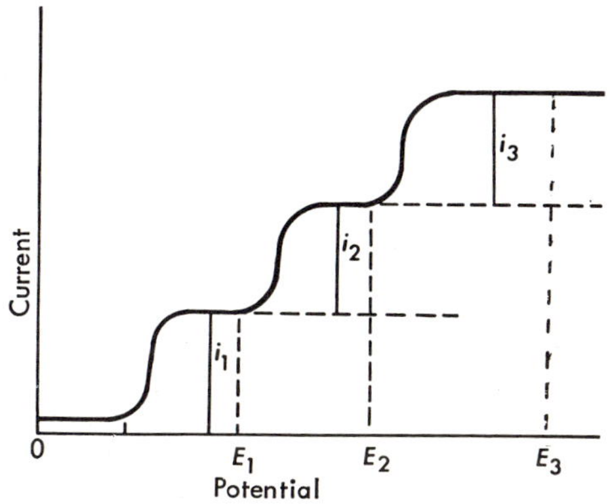

Fig. 49-7. Current-potential curves of three reducible substances.

is as small as 10^{-5} *N*. The same is true of the titration of arsenic trioxide with iodine or the reverse titration.

Amperometric titrations with two indicator electrodes: Biamperometric titrations

Biamperometric titrations are performed by immersing two indicator electrodes into the titration flask, imposing an emf across the electrodes, and observing the current changes during the titration. Depending upon the species present, the current at the equivalence point may go to zero, go through an acute minimum, or increase from zero to a measurable value.

Ordinarily, no attempt is made to use electrodes of identical area, nor are the electrodes rotated. The abrupt variations of current in the vicinity of the equivalence point are sufficiently obvious to locate the equivalence point, and it is not necessary to have highly reproducible currents.

For convenience in discussing the principle of this technique, assume that two electrodes of identical area are used, and both are rotated, so that reproducible mass transfer occurs, thereby making it possible to use current–potential curves to predict currents at various points in the titration. The titration to be discussed is the reverse of reaction (2). Assume that the material to be determined, Ox_2, exhibits the current–potential curves shown in Fig. 49-4B of the Ox_2/Red_2 system, while the titrant redox couple, Ox_1/Red_1, is electrochemically inert. At the start of the titration only Ox_2 is present, and the titrant, Red_1, is added; e.g., I_2 is titrated with $S_2O_3^{--}$. Hence Fig. 49.4B gives the current–potential curves at 0, 50, and 100 per cent titrated.

Figure 49-8 is an expanded view of Fig. 49-4B near the zero-current axis and includes a curve corresponding to 99 per cent titrated.

If a small potential, ΔE of 10–20 mV, is applied across both indicator electrodes, the anodic current through one indicator electrode must equal the cathodic current through

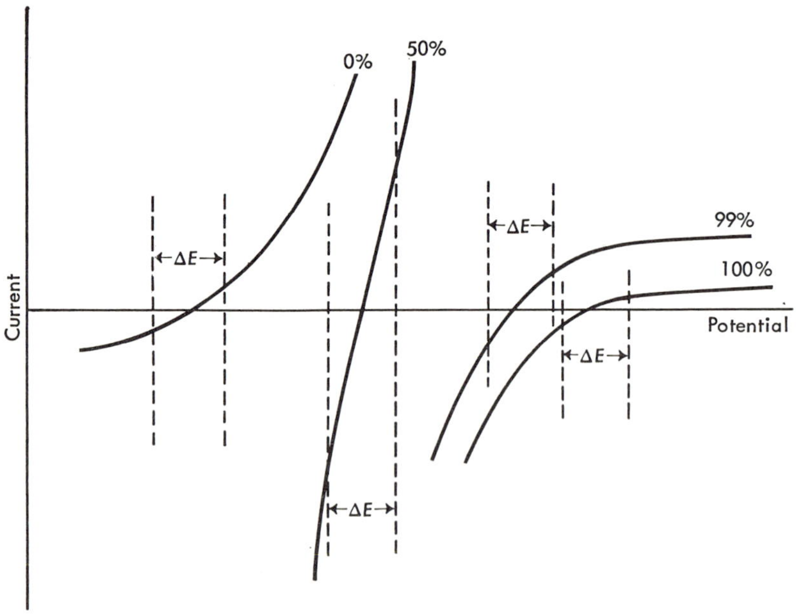

Fig. 49-8. Current-potential curves near the zero current axis at various stages in titration of Ox_2 with nonelectroactive reductant. The Ox_2/Red_2 couple is assumed to be reversible.

the other indicator electrode, with the potential of both electrodes adjusting until their difference is ΔE and the current condition is satisfied. This is shown graphically in Fig. 49-8. Initially, at 0 per cent titrated, the current is very small. It rises to a maximum at 50 per cent titrated and decreases to a negligibly small value at 100 per cent or more titrated, as shown in Fig. 49-9. Such a titration has been called the "dead stop" titration by Foulk and Bawden,[4] who applied the technique to the titration of iodine with thiosulfate. This technique is readily applicable to solutions as dilute as 10^{-5} *M*.[5]

The reverse of this titration is frequently used to detect the equivalence point in the Karl Fischer titration for water[6] (see p. 858).

The shape of the titration curve is sensitive to the applied voltage and depends upon the electrochemical properties of the species participating in the titration.

It is left as an exercise for the student to demonstrate that when both the unknown redox couple and titrant redox couple are reversible a sharp minimum in the current through the indicator electrodes occurs at the equivalence point (curve *ABCE*, Fig. 49-9). If the applied potential difference is small compared to the difference in $E^{1/2}$ values for the two redox couples, this minimum current approaches negligibly small values. In this particular case (two reversible systems), the biamperometric titration is superior to the conventional amperometric titration, which shows little or no break in the titration curve at the equivalence point. (Compare Figs. 49-9 and 49-6.)

The biamperometric method can be used for precipitation titrations, especially those involving silver ion. In the latter case, two silver electrodes are used. For example, Cl^- [7] and Br^- [8] have been determined using the biamperometric technique. Further details about biamperometric and amperometric titrations can be found elsewhere.[9]

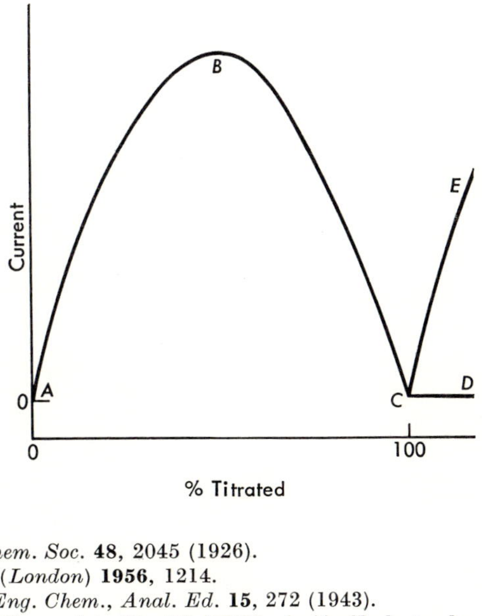

Fig. 49-9. Biamperometric redox titration curves. Curve *ABCD*-unknown is electroactive, titrant is non-electroactive; Curve *ABCE*—titrant and unknown are both electroactive.

[4] C. W. Foulk and A. T. Bawden, *J. Am. Chem. Soc.* **48**, 2045 (1926).

[5] E. C. Potter and J. F. White, *Chem. Ind.* (*London*) **1956**, 1214.

[6] G. Wernimont and F. J. Hopkinson, *Ind. Eng. Chem.*, *Anal. Ed.* **15**, 272 (1943).

[7] S. Samson, *Nature* **172**, 1043 (1952); *Anal. Chem. Acta* **13**, 473 (1955). Ag/AgCl electrodes were used.

[8] G. Marinenko and J. K. Taylor, *J. Res. Natl. Bur. Std.* **67A**, 31 (1963).

[9] J. T. Stock, *Amperometric Titrations*, Wiley-Interscience, New York, 1965; L. Meites, (ed.), McGraw-Hill, New York, 1963, p. 502ff.; P. Delahay, *Newer Instrumental Methods of Analysis*, Wiley-Interscience, New York, 1954, p. 258ff.; G. Charlot, J. Badoz-Lambling, and B. Trémillon, *Electrochemical Reactions*, Elsevier, Amsterdam, 1962, p. 204ff.

PROBLEMS

1. Lead may be titrated at the D.M.E. with dichromate ion in a neutral medium according to

$$Cr_2O_7^{--} + 2\,Pb^{++} + H_2O \rightarrow 2\,PbCrO_4\downarrow + 2\,H^+$$

Construct the amperometric titration curves (a) at a potential where only $Cr_2O_7^{--}$ is reduced, Cr(VI) + 3*e* → Cr(III); (b) at a potential where Pb^{++} and $Cr_2O_7^{--}$ are both reduced.

2. Lead chromate is more insoluble than barium chromate, and Ba^{++} is reduced only at extremely negative potentials at a D.M.E. Construct the amperometric titration curve for an equimolar mixture of Pb^{++} and Ba^{++} with CrO_4^{--} at a potential where Pb^{++} and CrO_4^{--}, but not Ba^{++}, are reduced.

3. The As(V)/As(III) couple is inert at a rotating platinum electrode in 1 *M* sulfuric acid, where the reaction

$$As(III) + Br_2 \rightarrow As(V) + 2\,Br^-$$

proceeds quantitatively. The diffusion current for $Br_2 + 2e \rightarrow 2\,Br^-$ is observed over at 0 V versus the S.C.E. (a) Construct the amperometric titration curve for this reaction; (b) using the biamperometric technique, how would the current change near the end point? Assume that the Br_2^-/Br^- couple is reversible.

4. 0.01 *M* ferrous iron is titrated with ceric cerium in 1 *M* sulfuric acid. Use the formal potentials in Table A-13 and assume that both systems are reversible (actually there is considerable deviation). Sketch the (a) amperometric titration curve using a rotating platinum electrode. At which potential (versus S.C.E.) would you carry out the titration? (b) Sketch the biamperometric titration curve using two platinum electrodes.

CHAPTER 50 *Coulometry and Coulometric Titrations*

The branch of electrochemistry known as coulometry encompasses all techniques that determine the number of coulombs required to convert, by electrolysis, one species quantitatively into another. Suppose that A is the species that is to be converted into B. Transfer of electrons may occur strictly from (to) the electrode to (from) A in solution, producing B (electron transfer via a *primary process*) or a second species may be the electrochemically active compound producing an electrode intermediate that reacts chemically with A to produce B (electron transfer via a *secondary process*).

Examples of primary processes are the conversion of Cu(II) to Cu(I) in 1 *M* potassium chloride, or the deposition of Ag(I) as silver metal on a platinum electrode. The technique of controlled-potential coulometry (p. 920) can be used in such cases.

The technique of constant-current electrolysis is usually used when secondary processes occur. An example of a secondary process is the coulometric determination of As(III) by oxidation to As(V) in the presence of 0.1 *M* potassium bromide in 1 *M* sulfuric acid at a platinum anode. The reactions are

Electrochemical:	$2\,Br^- \rightarrow Br_2 + 2e$	(1)
Chemical	$Br_2 + As(III) \rightarrow 2Br^- + As(V)$	(2)
Net reaction:	$As(III) \rightarrow As(V) + 2e$	(3)

and no detectable concentration of free bromine can exist until all the As(III) has been converted to As(V). In the above reaction sequence, bromide ion serves as catalyst in the electrochemical oxidation of As(III).

If an olefinic compound is substituted for As(III), bromination of the organic compound occurs via a secondary chemical reaction

$$R_2C{=}CR_2 + Br_2 \rightarrow R_2CBrCBrR_2 \tag{4}$$

but bromide ion is not regenerated in this instance.

In the examples given above involving As(III) and the olefin, bromine was generated in the cell containing the species to be determined. The procedure is known as *internal generation of reagent.*

Current efficiency

In using Faraday's law to calculate the amount of material electrolyzed, it is assumed that only one electrochemical reaction occurs. This is not always the case, and the term *current efficiency* is defined as the percentage of the charge passed through a coulometric cell which produces the desired electrochemical reaction. The electrode at which the desired electrochemical reaction occurs when current passes is known as the *working electrode* and the other electrode required to complete the electrical circuit is called the *auxiliary electrode.*

Dissolved oxygen is reducible, and must be removed in those cases where it might interfere in order to maintain 100 per cent current efficiency. (Bubbling the solution with nitrogen gas is a convenient method of removing oxygen.)

Water may be reduced to hydrogen unless the working electrode potential E_{work} is more positive than the right side of

$$E_{\text{work}} = 0.059 \log \frac{(\text{H}^+)}{\text{p}_{\text{H}_2}^{\frac{1}{2}}} \tag{5}$$

Also water will be oxidized to oxygen unless

$$E_{\text{work}} \leq 1.23 + 0.059 \log (\text{H}^+)\, \text{p}_{\text{O}_2}^{\frac{1}{4}} \tag{6}$$

holds. Hence the region of thermodynamic stability of water has a positive potential limit given by Eq. (6) and a negative potential limit given by Eq. (5). Both the oxidation and reduction of water are irreversible and depend markedly upon the electrode metal and its previous history (overvoltage phenomena). Thus it is not practical to calculate how much current will actually be used for the oxidation or reduction of water at a particular potential. It is customary to perform a blank experiment by omitting the unknown from the coulometric cell, and determine the quantity of charge which flows in the time required to electrolyze the unknown at constant potential, or the time required to reach the end point when using a constant-current coulometer. The charge consumed in the blank is then subtracted from the charge consumed in analyzing the unknown.

In coulometry of dilute solutions, the coulombs required to change the electrical double layer may become an appreciable fraction of the total coulombs required for the desired electrochemical process. Correction for this effect is included in the blank correction described above, to a first approximation.

Current efficiencies apparently less than 100 per cent may result if precautions are not taken to prevent electroactive species produced at one electrode from diffusing to the other electrode in the coulometric cell.

It is necessary to separate the working and auxiliary electrodes by a suitable diaphragm, such as porous porcelain or sintered glass, which serves to prevent species produced at one electrode from diffusing to the other one. This diffusion barrier may be omitted when the electrode reaction products of both electrodes are insoluble, e.g., the electrolysis of Cu^{++} in sulfuric acid produces copper metal at the cathode and oxygen gas at the anode.

Under appropriate conditions it is possible to design an electrochemical cell which apparently has 0 per cent current efficiency—e.g., a cell consisting of two platinum electrodes dipping into a well-stirred solution containing I_3^- and I^- will pass current without producing any change in the total amount of I_3^- or I^-. ($I_3^- + 2e \rightleftharpoons 3\,I^-$.)

When the species to be determined exhibits a well-developed, diffusion-limited current region, the technique of controlled potential coulometry is feasible. If no potential region exists where the current is mass-transfer limited by the species to be analyzed, and some other process occurs simultaneously, less than 100 per cent current efficiency results. An example is the oxidation of Ce(III) to Ce(IV) in not too acid solutions at a platinum anode, where water is being simultaneously oxidized to oxygen.

When 100 per cent current efficiency cannot be obtained by a primary electrode process, it is necessary to add a species that can be electrolyzed with 100 per cent current efficiency to produce an electrode intermediate which reacts rapidly with the species to be determined. In this case, either constant-potential or constant-current coulometry may be employed. As is shown on page 925, when speed is important, constant-current coulometry should be used.

Controlled-potential coulometry

The advantages of maintaining the working electrode potential at a constant value are readily made clear using Fig. 50-1, which represents the current–potential curve for the reductions

$$\text{A} + ne \rightleftharpoons \text{B} \tag{7}$$

and

$$\text{C} + me \rightleftharpoons \text{D} \tag{8}$$

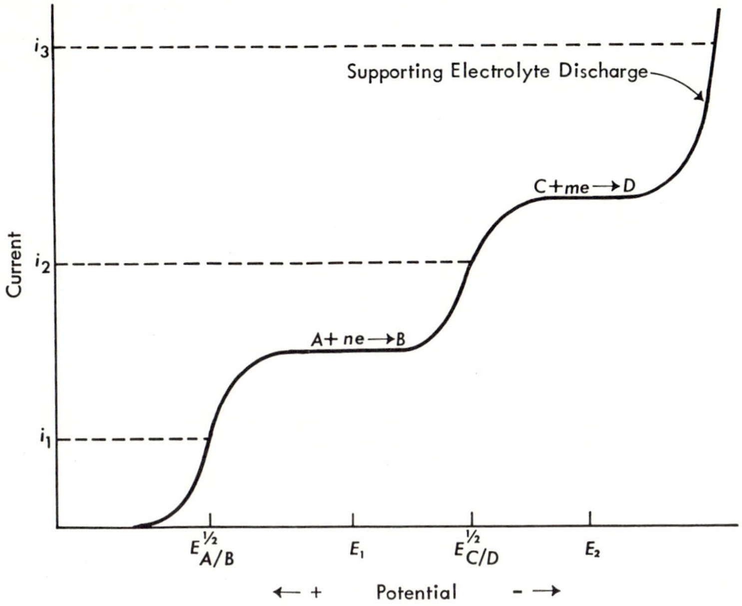

Fig. 50-1. Current-potential curves for the reaction $A + ne \rightarrow B$ and $C + me \rightarrow D$. E_1 is the controlled potential at which reaction (7) occurs and E_2 is the controlled potential at which reaction (8) occurs. Identical concentrations of A and C are present initially.

where all species are soluble in water. As long as the working electrode potential is held on the first diffusion current region, e.g., potential E_1, only Eq. (7) can occur to a significant extent.

The current due to Eq. (7) will decrease as the concentration of A decreases throughout the solution, and will ultimately decay to zero. From the number of coulombs that flowed, the quantity of A initially present can be calculated.

Figure 50-2 represents, in a schematic manner, the components necessary to construct a simple controlled-potential coulometer. All parts are labeled and defined in the legend to Fig. 50-2. Two separate electrical circuits exist in the cell; one measures the potential of the working electrode with respect to the S.C.E., using the potentiometric method (p. 161) so that no current flows in that circuit. Current, i_1, flows in the other circuit through the auxiliary electrode, the variable resistance, the storage battery, the coulometer, the working electrode, and the solution between the working and auxiliary electrodes.

By adjusting the variable resistance, E_{work} can be adjusted to the desired potential and maintained at this value throughout the experiment.

The potential distribution in the auxiliary-working electrode circuit is described by

$$V = i_1R + i_1R_{\text{cell}} + E_{\text{work}} + E_{\text{aux}} \tag{9}$$

where R_{cell} is the cell resistance between the working and auxiliary electrodes, and E_{work} and E_{aux} are the potentials of the auxiliary and working electrodes. It is assumed that the resistance of the coulometer, Q, is zero. Rearranging Eq. (9) yields

$$E_{\text{work}} = V - (i_1R_{\text{cell}} + i_1R + E_{\text{aux}}) \tag{10}$$

As i_1 decreases with electrolysis time, R_{cell} will change in a way governed by the details of the electrochemical processes at the auxiliary and working electrodes. Also, E_{aux} will change, since the surface concentration of species determining E_{aux} changes with i_1. By adjusting R, the numerical value of the sum of the terms in the parentheses of Eq. (10) can be maintained constant, and hence the value of E_{work} held constant at the value set on the potentiometer in the other circuit. In Fig. 50-2 the salt bridge has been pulled to a sharp tip and placed close to the working electrode on the side nearer to the auxiliary electrode. This shape of capillary is known as a Luggin capillary, and, when positioned as described, helps prevent most of the ohmic potential drop ($i_1 R_{cell}$) within the cell from interfering with the potentiometric measuring circuit. Harror and Shain have examined this problem in a number of commonly used electrolysis cells.[1]

Figure 50-2 represents a simple manual *potentiostat* incorporating a current integrator. Electromechanical and electronic potentiostats have been developed which automatically control E_{work} to within a few millivolts and are in common use.

Returning to a consideration of Fig. 50-1, assume that a potential E_1 is established at the working electrode using a potentiostat, and electrolysis is allowed to proceed for

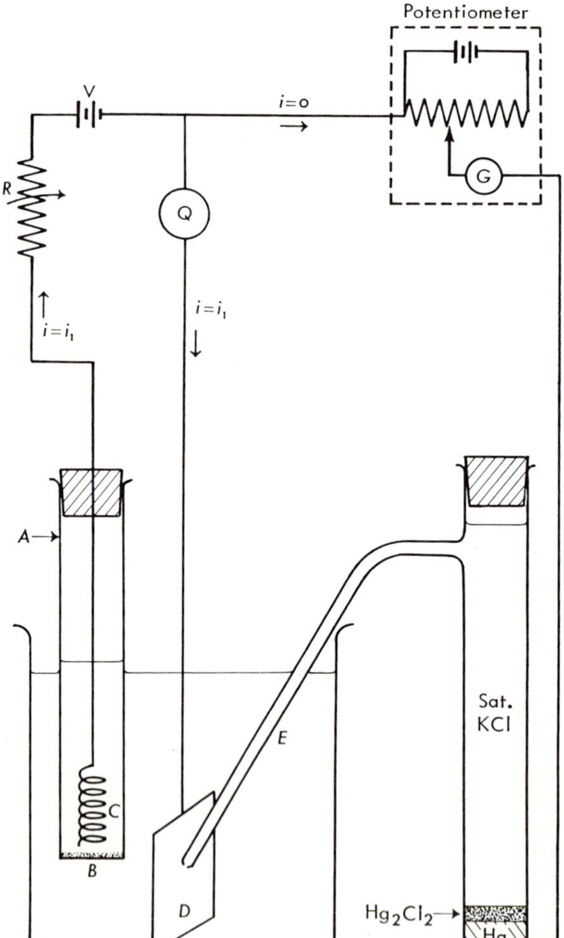

Fig. 50-2. Schematic representation of a controlled potential coulometer. *A*—auxiliary electrode compartment; *B*—fritted (porous) glass disk; *C*—auxiliary electrode; *D*—working electrode; *E*—salt bridge in form of a Luggin capillary, *F*—S.C.E.; *G*—galvanometer; *R*—variable resistance; *V*—storage battery; *Q*—coulometer, Stirrer not shown, but solution is well stirred.

[1] J. E. Harror and I. Shain, *Anal. Chem.* **38,** 1148 (1966).

a sufficiently long time (see below) such that the Nernst expression

$$E_1 = E_{\text{work}} = E^f_{A/B} + \frac{0.059}{n} \log \frac{C_A}{C_B} \tag{11}$$

holds at every point in the solution. The ratio C_A/C_B is

$$\frac{C_A}{C_B} = 10^{(E_1 - E^f_{A/B})/0.059/n} \tag{12}$$

and thus, for a one-electron process, the difference $E_1 - E^f_{A/B}$ must be -0.177 V in order that 99.9 per cent of A be converted into B, and -0.089 V when $n = 2$. More complete conversion of A into B is possible if E_1 is made still more negative, but ultimately this gain is offset by the start of the reduction of C, which results in a decrease in current efficiency. Hence we are interested in establishing what the minimum difference in formal potentials for reactions (7) and (8) may be, while still maintaining a specified current efficiency and extent of conversion of A into B.

At potential E_1, the Nernst equation for Eq. (8) is

$$E_1 = E^f_{C/D} + \frac{0.059}{m} \log \frac{C_C}{C_D} \tag{13}$$

Subtracting Eq. (11) from (13) and rearranging terms yields

$$E^f_{C/D} - E^f_{A/B} = 0.059 \log \left[\left(\frac{C_D}{C_C}\right)^{1/m} \left(\frac{C_A}{C_B}\right)^{1/n} \right] \tag{14}$$

As an example of the application of Eq. (14), suppose that $m = n = 1$, that A is to be converted 99.9 per cent into B, and that a current efficiency of 99.9 per cent is desired when the starting concentrations of A and C are equal. In terms of concentration ratios these conditions require that

$$C_A/C_B = 10^{-3} \qquad \text{and} \qquad C_D/C_C = 10^{-3}$$

Substitution in Eq. (14) gives

$$E^f_{C/D} - E_{A/B} = 0.059 \log (10^{-3} \times 10^{-3}) = -0.354 \text{ V}$$

If $n = m = 2$, a difference in formal potentials of only -0.177 V would be sufficient to satisfy the requirements. If the initial concentrations of A and C are different, the current efficiency calculations must be modified to allow for this difference.

Equations (12) and (14) apply to coulometry at a mercury pool electrode, in which the reduced forms B and D are soluble metal amalgams. It is necessary to allow for the relative volumes of the solution and mercury phases in calculating current efficiency and the degree of completion of the electrochemical reaction (see Problem 3).

If either B or D is insoluble (i.e., a metal), Eqs. (12) and (14) do not apply. The treatment in such cases is discussed on page 930.

Referring again to Fig. 50-1, it has been stated that the current for reaction (7) decreases continuously with time; the law governing this current decay is given by

$$i = k \cdot C_A e^{-\alpha t} \tag{15}$$

where α is a constant for a particular electrode, cell, and stirring conditions, t the time of electrolysis, C_A the bulk concentration when $t = 0$, and k the mass-transfer coefficient for the electroactive species under the particular experimental conditions (see Chapter 47). α is related to k by the expression

$$k = nFv\alpha \tag{16}$$

where v is the volume of solution in the cell.

From Eq. (15) it is seen that current decays exponentially with time (first-order kinetics). If $T_{1/2}$ represents the time required to halve the concentration of A throughout the solution, *the half-time*, the electrolysis is 99.9 per cent complete when

$$t = 9.9T_{1/2} \tag{17}$$

Half-times of 5 min are readily obtained under ordinary experimental conditions, and quantitative electrolysis is possible in 50 min.

After A has been electrolyzed at potential E_1 (Fig. 50-1), C may be determined by electrolysis at potential E_2, E_2 being chosen on the basis of the considerations discussed above. It is evident that controlled-potential coulometry is an especially attractive method of analysis in multicomponent mixtures, since modern instrumentation makes possible completely automatic control of the working electrode potential and also very precise (± 0.01 per cent) determination of the number of coulombs that flow.

Controlled potential coulometry has had numerous applications. Lingane and Page[2] successfully analyzed mixtures of nickel and cobalt at a mercury cathode by electrolysis from an aqueous pyridine supporting electrolyte. Meites and Meites[3] demonstrated that trichloroacetate may be determined according to

$$Cl_3COO^- + H^+ + 2e \rightarrow Cl_2HCOO^- + Cl^-$$

at controlled potential in the presence of di- and monochloroacetate ions. Lingane and Small[4] used controlled-potential coulometry at a silver working electrode to determine Cl^-, Br^-, and I^- separately, and in mixtures. The electrode reaction was $Ag + X^- \rightarrow AgX + e$. More details are to be found elsewhere.[5]

Constant-current coulometry. Coulometric titrations

In well-stirred solutions containing two species A and C, electrolysis may be carried out with a constant current flowing through the working and indicator electrodes. Under the appropriate conditions, 100 per cent current efficiency for reaction (7) may be obtained. These conditions may be deduced by examining Fig. 50-1.

If a constant current i_1 is forced to flow through the cell, only reaction (7) occurs *initially*. Similarly, if a current i_2 is imposed, *initially* reactions (7) and (8) occur simultaneously at the electrode surface, and, as the product of reaction (8) diffuses back into solution, the reaction

$$D + A \rightarrow B + C \tag{18}$$

can occur if reaction (18) is rapid. The net result of applying current i_2 is the conversion of A into C by direct electrolysis, (7), and by chemical reaction, (18); i.e., some C has been produced via a secondary process involving the electrode intermediate D. Hence 100 per cent current efficiency for production of B is maintained.

If a current i_3 is imposed, supporting electrolyte will be reduced, e.g., hydrogen gas is evolved, and the current efficiency for the production of B is less than 100 per cent. Hence as long as the impressed constant current does not exceed the sum of the diffusion currents for reactions (7) and (8) the current efficiency for the production of B will be 100 per cent.

Assume that a current i_2 is applied until all A is converted to B. As the concentration of A decreases, the diffusion current for A decreases, and when 50 per cent of A has been converted into B, i_2 will exceed the sum of the two diffusion currents for reactions (7) and (8). From this time on, hydrogen will be evolved, and the current efficiency will no longer be 100 per cent.

[2] J. J. Lingane and J. A. Page, *Anal. Chem. Acta* **13**, 281 (1955).

[3] T. Meites and L. Meites, *Anal. Chem.* **27**, 1530 (1955).

[4] J. J. Lingane and L. A. Small, *Anal. Chem.* **21**, 1119 (1949).

[5] G. A. Rechnitz, *Controlled Potential Analysis*, Macmillan, New York, 1963; Y. Israel and L. Meites in *Handbook of Analytical Chemistry* (L. Meites, ed.), McGraw-Hill, New York, 1963, pp. 5–195ff.

This difficulty could be prevented by restricting the constant current to a value less than that of the diffusion current for reaction (8). As a practical matter, to speed up the analysis it is customary to add a relatively large excess of C as compared to A, ensuring that all the current not consumed by (7) will be used in (8) and thus permitting the use of larger constant currents. Current-potential curves for this situation, at 0, 50, and 100 per cent electrolyzed, are shown in Fig. 50-3. The diffusion current region for reaction (7) is shown, while the rapidly increasing current at potential E_2 represents the rising portion of the wave for reaction (8). In this solution, a constant current corresponding to i^* could be used without decreasing the current efficiency below 100 per cent, since i^* is very much less than the diffusion current for reaction (8).

A method of detecting the time at which all A has been converted to B is needed in constant-current coulometry, because current flows whether or not A is present in solution. Any classical or physiochemical technique suitable for detecting the end point of a titration may be employed, and the coulombs Q required for the quantitative conversion of A to B are calculated from

$$Q = i \times t \tag{19}$$

where i is the constant current used and t is the time to reach the end point.

From Fig. 50-3 and Eq. (19) it is apparent that it is not necessary that the species A be electroactive, as long as reaction (18) is rapid.

In order not to overrun the end point, thorough and rapid stirring of the solution is necessary, the rate of mixing becoming more important, the higher the constant currents employed.

The constant-current electrolysis of A using current i^* (Fig. 50-3) will be completed considerably faster than the controlled-potential electrolysis at potential E_1. Assuming that the current used throughout a constant-current electrolysis is equal to the current at the start of a controlled-potential electrolysis, and all other conditions are held constant, the constant-current electrolysis will be 100 per cent complete in 1.42 half-times versus the 9.96 half-times [equation (17)] required for 99.9 per cent completion at controlled potential. This is a sevenfold saving in time (see Problems 6–10).

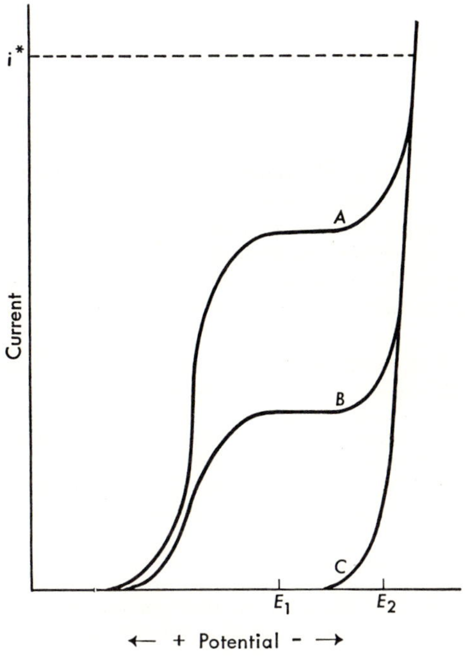

Fig. 50-3. Current-potential curves for the constant-current coulometric titration: $A + ne \rightarrow B$ in the presence of a large excess of C; Curve A—0% titrated; curve B—50% titrated; curve C—100% titrated. i^* is constant titration current.

Constant-current coulometry has been widely applied in analysis, and a bibliography is given elsewhere.[6]

In discussing a few typical coulometric titrations below, it is convenient to classify methods according to the electrode intermediate which is generated when the mass-transfer controlled current for the species to be determined is exceeded, i.e., in terms of D in Eq. (17). D is known as the *generated titrant.*

Hydroxide ion may be generated by the reduction of water, $H_2O + e \rightarrow OH^- + \frac{1}{2} H_2$, and hydrogen ion by the oxidation of water, $\frac{1}{2} H_2O \rightarrow H^+ + \frac{1}{4} O_2 + e$, at platinum electrodes; the sample should not contain electroactive substances which are reduced more easily than water in the titration acids, or oxidized more easily than water in the titration of bases. The auxiliary electrode should be separated from the working electrode by a suitable diaphragm. If the solution contains halide ion, the use of a silver auxiliary electrode when OH^- is generated at the working electrode avoids the use of a diaphragm.[7] The silver half-reaction is $Ag + X^- \rightarrow AgX + e$. In this case, no hydronium ions are produced, as would be the case at a platinum auxiliary electrode.

Iodine may be generated from potassium iodide solutions by oxidation at a platinum electrode. As(III) is readily titrated in the presence of phosphate buffer (pH 8).[8]

Depending upon the As(III) concentration, starch may be used to locate the end point, or amperometric, potentiometric, or photometric end-point detection used.

Bromine is generated from potassium bromide in 0.1 *M* sulfuric acid using a platinum electrode. As(III) can be oxidized to As(V)[9]; the end point is detected by a conventional amperometric method, or a biamperometric method with 0.200 V across the indicator electrode. At high concentration (greater than 10^{-3} *M*) the potentiometric detection of the end point is satisfactory.

Ce(IV) can be generated at 99.8 per cent current efficiency from 0.1 *M* Ce(III) solutions in sulfuric acid of concentration greater than 3 *M*, provided the current density at the working electrode is in the range 1–10 mA/cm^2:[10]

$$\text{Ce(III)} \rightarrow \text{Ce(IV)} + e$$

Before these optimum conditions were known, titrations of Fe(II),[11] $Fe(CN)_6^{4-}$,[12]

$$\text{Ce(IV)} + \text{Fe(II)} \rightarrow \text{Ce(III)} + \text{Fe(III)}$$

and Ti(III),[13]

$$\text{Ce(IV)} + \text{Ti(III)} \rightarrow \text{Ce(III)} + \text{Ti(IV)}$$

were reported.

Even under the less than ideal generation conditions used, satisfactory analytical results were obtained. The good analytical results stem from the fact that considerable direct electrooxidation of the substances being determined also was occurring, in addition to the electrogeneration of Ce(IV).

Silver ion has been generated by direct oxidation of silver metal and used to determine the various halide ions.[14] Thiocyanate may also be determined, as well as ions that form stable soluble complexes with silver ion.

[6] D. D. Deford in *Treatise on Analytical Chemistry*, part I, vol. 4 (I. M. Kolthoff and P. J. Elving, eds.), Wiley-Interscience, New York, 1963, p. 2516ff.; see also chaps. 19, 20, and 21, Ref. C1, on p. 884.

[7] L. Szebelledy and Z. Somogyi, *Z. Anal. Chem.* **395** (1935); J. J. Lingane and L. A. Small, *Anal. Chem.* **21**, 1119 (1949).

[8] W. J. Ramsey, P. S. Farrington, and E. H. Swift, *Anal. Chem.* **22**, 332 (1950).

[9] R. J. Meyers and E. H. Swift, *J. Am. Chem. Soc.* **70**, 1047 (1948).

[10] J. J. Lingane, C. H. Langford, and F. C. Anson, *Anal. Chim. Acta* **16**, 165 (1957).

[11] N. H. Furman, W. D. Cooke, and C. N. Reilley, *Anal. Chem.* **27**, 945 (1951).

[12] R. U. Dilts and N. H. Furman, *Anal. Chem.* **27**, 1275 (1955).

[13] R. U. Dilts and N. H. Furman, *ibid.* **27**, 1596 (1955).

[14] J. J. Lingane, *Anal. Chem.* **26**, 622 (1954).

Ethylenediaminetetraacetate ion has been generated from its mercury complex at a mercury electrode[15]:

$$HgY^{--} + 2e \rightarrow Hg + Y^{4-}$$

and metals ions such as calcium, zinc, and lead may be determined. Other ligands can be generated in a similar fashion.

In all the examples given above, internal generation of reagent was used. Techniques have been devised for generating the reagent in a flowing stream of electrolyte, just before the stream enters a vessel containing the species to be determined.[16] This procedure is advantageous when suitable in situ methods for generation of a particular reagent do not exist, or cannot be used because of the presence of substances that would interfere with the electrode reaction generating the titrant.

PROBLEMS

1. What is the potential range in which water is thermodynamically stable with respect to oxidation to oxygen or reduction to hydrogen at (a) pH = 0; (b) pH = 7; (c) pH = 14? Assume p_{H_2} = 1 atm and p_{O_2} = 1 atm.
   Ans. (a) 1.23 to 0.00 V; (b) 0.82 to −0.41 V; (c) 0.40 to −0.83 V.
2. Consider Fig. 51-1 and reactions (7) and (8). It is desired to reduce 99.99 per cent of A to B at a platinum working electrode, with a current efficiency of 99.99 per cent. $n = 2$ in reaction (7) and $m = 1$ in (8). (a) What is the minimum required potential difference between $E^f_{C/D}$ and $E^f_{A/B}$? (b) How much more negative than $E^f_{A/B}$ should the controlled electrolysis potential E_1 be?
   Ans. (a) 0.354 V; (b) 0.118 V.
3. Repeat Problem 2, assuming that the working electrode is a mercury pool, that the reduced forms of A and D are metals which form a soluble amalgam, that (a) the volume of the aqueous phase and the mercury pool are equal, and that (b) the volume of the aqueous phase is 10 times that of the mercury pool.
   Ans. (1a) 0.354 V, (1b) 0.118 V; (2a) 0.325 V, (2b) 0.148 V.
4. Derive Eq. (15) starting with Faraday's law: $i = nAF(dN/dt)$, where dN/dt is the instantaneous rate of disappearance of the species, A, which undergoes electrolysis; the Nernst diffusion expression: $i = nAFDC_A/\delta$, where C_A is the initial bulk concentration of A; and the relation N/solution volume = C_A. (This derivation may be found on p. 227 of Reference C1, p. 884.)
5. How may the time required for a controlled-potential electrolysis be decreased?
6. Rewrite Eq. (15) in terms of the initial current at the start of electrolysis i^0 and the half-time of electrolysis $T_{1/2}$. *Ans.* $i = i^0 e^{-(\ln 2)t/T_{1/2}} = i^0/2^{t/T_{1/2}}$.
7. Calculate the number of coulombs that have passed through the working electrode after a time t during a controlled-potential electrolysis, using the result from Problem 6.
   Ans. $Q_{cp} = i^0 T_{1/2}\{[1 - (\frac{1}{2})^n]/\ln 2\}$, where n is the time measured in half-times, i.e., $n = t/T_{1/2}$.
8. After infinite time, how many coulombs will flow through the working electrode in Problem 7? *Ans.* $Q_{cp} = i^0 T_{1/2}/\ln 2$.
9. If a constant current equal to that which flows initially through the working electrode in Problem 7 is forced through the cell, how much time will be required to pass the same number of coulombs? *Ans.* $T_{1/2}/\ln 2$ sec.
10. What is the time saving in performing a constant-current coulometric titration (Problem 9) as compared to using the controlled-potential technique?
    Ans. Constant-current technique is seven times faster reaching 99.9% titrated.
11. What effect does doubling the constant current in Problem 9 have on the answer to Problem 10?

[15] C. N. Reilley and W. W. Porterfield, *Anal. Chem.* **28**, 443 (1956).
[16] D. D. Deford, Ref. 6, p. 2510ff.

CHAPTER 51 *Electrogravimetry and Electrolytic Separations*

Wolcott Gibbs'[1] analysis of copper solutions by weighing copper metal that had been electrodeposited on a platinum electrode was the first application of an electrochemical method to quantitative analysis. Within the next 30 years, electrogravimetry was employed regularly for the determination of many metals. In general, electrogravimetry is the term applied to a bulk electrolysis (p. 870) in which the product of an electrochemical reaction is deposited quantitatively upon an electrode in an adherent form suitable for weighing. This deposit may be a metal or an insoluble material of definite stoichiometry such as silver chloride or lead dioxide, while the electrode may be any suitable metal. Most commonly, platinum is used for the electrode material because of its chemical inertness, but other metals have been employed.[2]

In this classical technique, the simple circuit shown in Fig. 51-1 is quite satisfactory in solutions containing one electrodepositable species, e.g., Cu^{++} in sulfuric acid–

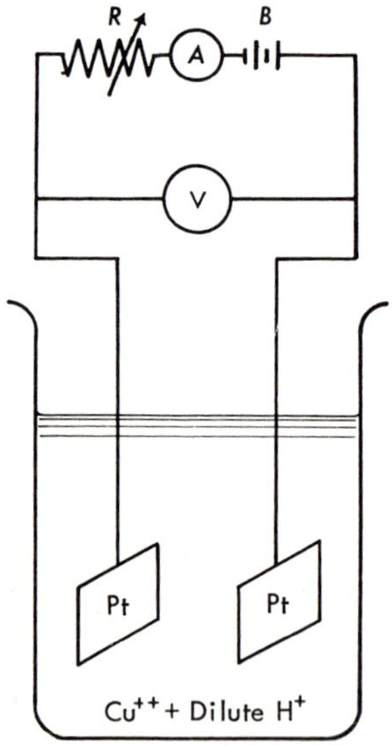

Fig. 51-1. Schematic electrodeposition apparatus. *B*—storage battery; *R*—variable resistance; *A*—ammeter; *V*—voltmeter; Cathode Reaction—$Cu^{+2} + 2e \rightarrow Cu$; Anode Reaction—$2H_2O \rightarrow 4H^+ + O_2 + 4e$.

[1] W. Gibbs, *Z. Anal. Chem.* **3**, 327 (1864).

[2] W. Böttger, *Elektroanalyse* in *Physikalische Methoden der analytischen Chemie*, Part II, Akademische Verlagsgesellschaft, Leipzig, 1936, p. 129; A. Schleicher, *Elektroanalytische Schnellmethoden*, Enke, Stuttgart, 1947, p. 42; J. J. Lingane, *Electroanalytical Chemistry*, Wiley-Interscience, New York, 2nd ed., 1958, p. 351ff.

supporting electrolyte containing some nitric acid (p. 1114). In the latter case copper deposits at the platinum cathode and oxygen evolves at the platinum anode. The working electrode in this case is the cathode and the anode is the auxiliary electrode. Frequently, a substance known as a depolarizer is added to prevent an undesirable reaction from occurring. For example, if the supporting electrolyte in the above example had been hydrochloric acid, chlorine would be evolved at the anode and the platinum anode would corrode. If hydrazine is added to the hydrochloric acid as an anodic depolarizer, chlorine will not be evolved, because the reaction

$$N_2H_5^+ \rightarrow N_2 + 5\,H^+ + 4e$$

occurs more readily than the evolution of chlorine.

In nitric acid–supporting electrolyte, reduction of nitric acid to the lower oxides of nitrogen can occur partially, instead of complete reduction to NH_4^+ (see p. 937). The addition of urea, which reacts with nitrous acid, avoids the problems caused by the lower oxides.

It is possible to maintain an approximately constant current through, or approximately constant voltage across, the cell by manually adjusting the resistance in Fig. 51-1 when necessary. The constant-voltage situation is discussed on page 934. As was described in Chapter 50, a constant-current electrolysis does not proceed with 100 per cent current efficiency once the impressed current exceeds the diffusion current at the electrode. The current efficiency is not of special consequence in electrogravimetry because only the weight of deposit is determined. However, constant-current electrolysis is not feasible if two species that can be electrodeposited are present in the solution. At some stage of the electrolysis, the diffusion current of the more electroactive species, which is plated out at more positive potentials, will fall below the value of the impressed constant current and the second species will deposit. In such cases, control of the electrode potential is extremely valuable and this technique is known as *controlled-potential electrogravimetry*. The principle of this method is the same as that given for controlled-potential coulometry on page 923, with the minor difference that the quantity of electroactive material deposited is determined directly by weighing, rather than by calculating from Faraday's law. Usually a suitable depolarizer is used, and the diaphragm between the anode and cathode compartments shown in Fig. 50-2 is omitted.

The method of controlled-potential electrogravimetry was first employed in 1907 by Sand,[3] and independently by Fischer.[4] It was generally recognized as being a potentially powerful tool for the analysis of mixtures, but was unpopular until automatic potentiostats made the laborious manual adjustment of potential unnecessary. Early contributors to the development of potentiostats were Hickling,[5] Caldwell et al.,[6] and Lingane.[7] In the past 15 years, numerous designs for potentiostats have appeared in the literature and a number of commercial instruments are available (p. 1041).

The possibility of using controlled-potential electrolysis as a method of quantitatively converting a species to a specific oxidation state for determination by a nonelectrochemical method should be mentioned, e.g., quantitative reduction of iron(III) to iron(II). The principle has been discussed earlier in the discussion of controlled-potential coulometry (p. 923).

When it is desired to remove all electrodepositable metals from solution, constant-current electrolysis is quite useful. In this way nondepositable species are rapidly and conveniently separated from depositable metals. For example, the separation of Zn^{++}

[3] H. J. S. Sand, *J. Chem. Soc.* **91**, 401 (1907).

[4] A. Z. Fischer, *Angew. Chem.* **20**, 134 (1907).

[5] A. Hickling, *Trans. Faraday Soc.* **38**, 27 (1942).

[6] C. W. Caldwell, R. C. Parker, and H. Diehl, *Ind. Eng. Chem., Anal. Ed.* **16**, 532 (1944); H. Diehl, *Electrochemical Analysis with Graded Cathode Potential Control*, G. F. Smith Chemical Co., Columbus, Ohio, 1948.

[7] J. J. Lingane, *Ind. Eng. Chem., Anal. Ed.* **17**, 332 (1945); J. J. Lingane and S. L. Jones, *Anal. Chem.* **22**, 1169 (1956).

from Ag^+ and Cu^{++} in a dilute nitric acid medium is readily accomplished using a platinum cathode. The reduction of Ag^+ and Cu^{++} to the metal is possible in this medium, while Zn^{++} cannot be reduced.

Electrodeposition is often applied in situations where the electroseparated constituent is soluble in the electrode material. Of great analytical importance is the *mercury cathode*, which can be used to separate a large number of elements from other non-depositable species (see p. 936). Generally, the determination of the quantity of metal deposited by weighing has not been popular. The main objections to electrogravimetry at a mercury electrode are (1) the high vapor pressure of mercury and (2) the relatively large weight (~50 g) of mercury required for an electrode of useful area. When quantitative information is desired about the material deposited in the mercury, current practice favors the use of (1) controlled-potential coulometry during the plating process, (2) electrolytic stripping of the amalgam at controlled potential or constant current, (3) chemical decomposition followed by analysis of the aqueous phase, or (4) voltammetry using the mercury electrode as an anode and determining the diffusion currents corresponding to the oxidation of the metals present in the amalgam.

Electrolysis under conditions of reversible electrode behavior at controlled potential

In this section it is assumed that reversible electrode behavior is exhibited by the working electrode in this discussion of constant-potential electrogravimetry. Further, it will be assumed that the solution is stirred and that reproducible mass-transfer conditions exist. Hence current–potential curves at the electrodes can be described by the same equations as those given for microelectrodes (p. 902), and the effect of adding complexing agents upon half-wave potentials is also the same (Chapter 48).

Completeness of deposition of a metal. To use the method of controlled-potential electrogravimetry, it is necessary to establish that quantitative electrodeposition of a constituent is possible when other depositable species are present. We consider the situation at a solid electrode, e.g., platinum, below. The situation at a mercury electrode has already been treated in the chapter on coulometry (p. 923). Figure 51-2 represents a current–potential curve for a solution that contains two metal ions, M^{m+} and N^{n+}, both of which are reduced to the metals according to

$$M^{m+} + me \rightarrow M \qquad E_{M^{m+}/M} = E^f_{M^{m+}/M} + \frac{0.059}{m} \log C_{M^{m+}} \tag{1}$$

$$N^{n+} + ne \rightarrow N \qquad E_{N^{n+}/N} = E^f_{N^{n+}/N} + \frac{0.059}{n} \log C_{N^{n+}} \tag{2}$$

at the working electrode.

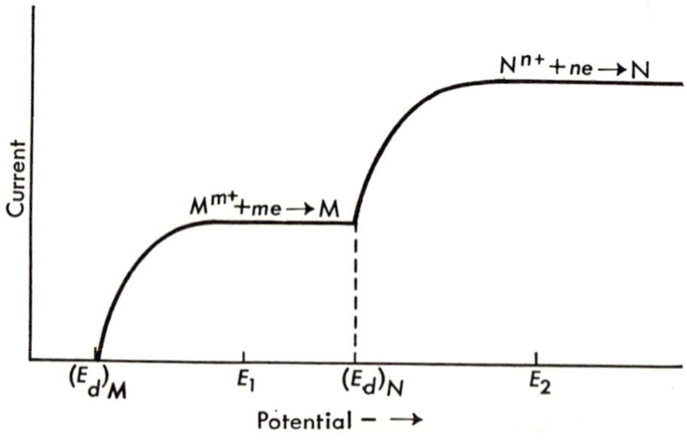

Fig. 51-2. Current–potential curve for a solution containing two different ions, M^{m+} and N^{n+}, which deposit as metals on a solid electrode.

The potential at which M just starts to plate on the electrode, the decomposition potential, $(E_d)_M$, is calculated directly from Eq. (1), using the initial concentration of M^{m+}, C_i:

$$(E_d)_M = E^f_{M^{+m}/M} + \frac{0.059}{m} \log C_i \tag{3}$$

while the concentration of $C_{M^{m+}}$ remaining in the solution after completion of the electrolysis, C_f, at potential E_1 is

$$E_1 = E^f_{M^{+n}/M} + \frac{0.059}{m} \log C_f \tag{4}$$

Subtracting Eq. (3) from (4) yields

$$E_1 - (E_d)_M = \frac{0.059}{m} \log \frac{C_f}{C_i} \tag{5}$$

Equation (5) may be used to calculate the potential at which the working electrode should be controlled in order to reduce the solution concentration of M^{m+} to a desired value. In order that the deposition of M be quantitative, C_f/C_i should be decreased to a value of at least 10^{-3}. Substituting the latter value into Eq. (5) yields

$$E_1 - (E_d)_M = \frac{-0.177}{m} \tag{6}$$

From Eq. (6) it is seen that the potential of the working electrode should be controlled at least -0.177 V more negative than the decomposition potential for reaction (1), $(E_d)_M$, when $m = 1$, and at potentials of -0.089 and -0.059 V when $n = 2$ and 3, respectively. No interference from reaction (2) is experienced as long as the controlled potential required for quantitative deposition of M is less than the decomposition potential of N [$(E_d)_N$ in Fig. 51-2].

After weighing the deposited M on the working electrode, N can be deposited by selecting another suitable working electrode potential, E_2.

The time dependence of the current during a controlled-potential electrolysis has been discussed on page 923 and decays exponentially with time. A few typical controlled-potential electrogravimetric analyses are discussed below.

In ammoniacal medium, silver may be separated from copper[8] by controlled-potential electrolysis at -0.24 V versus S.C.E. (All potentials will be referred to the S.C.E. below, as this is the usual practice.) Cadmium may be separated from zinc using a cathode potential of -0.80 V in acetate buffer of pH about 4.[9] Nickel is deposited at -1.10 V from an ammoniacal tartrate medium containing sodium sulfite in the presence of zinc, aluminum, and iron.[10]

Lingane[11] has shown that copper can be deposited without interference from antimony arsenic, lead, nickel, zinc, and practically all other common metals at a controlled-cathode potential of -0.36 V in a tartrate medium of pH about 4.5. Hydrazine dihydrochloride is added as anodic depolarizer. In the presence of bismuth at a pH of above 5.2, a cathode potential of -0.30 V is required to effect the separation of copper. This procedure is satisfactory for determination of copper in simple brasses and bronzes.

Lingane and Jones[12] developed a procedure for the successive deposition of copper, bismuth, lead, and tin based on successively plating the first three from tartrate

[8] H. Diehl, Ref. 6.

[9] H. J. S. Sand, *Electrochemistry and Electrochemical Analysis*, vol. II, Blackie, Glasgow, 1940.

[10] S. Torrance, *Analyst* **63**, 488 (1938).

[11] J. J. Lingane, *Ind. Eng. Chem., Anal. Ed.* **17**, 640 (1945).

[12] J. J. Lingane and S. L. Jones, *Anal. Chem.* **23**, 1798 (1951).

medium of pH 5.9–5.2 at cathode potentials of −0.30, −0.40, and −0.60 V, respectively. Tin is deposited after acidifying with hydrochloric acid at −0.60 V, since the stannic tartrate complex is so stable above a pH of 2 that tin cannot be deposited.

Further examples of controlled-potential electrogravimetry are given elsewhere.[13]

Overpotential

In the prior discussion of current–potential curves (Chapters 46 and 48), it was assumed that redox couple behaved in a reversible manner; i.e., the rate of electron transfer for the reaction

$$\text{Ox} + ne \rightleftharpoons \text{Red} \qquad (7)$$

was infinitely faster than mass transfer or any other step in the overall electrode reaction. Strictly speaking, no reaction is completely reversible, because the rate of electron transfer always occurs at a finite rate.

The expression for the current at an electrode in terms of the current density i^0 $(= i/A)$ and the difference between the electrode potential and formal electrode potential, ΔE, is

$$i^0 = nFk_s^0\{C_{\text{Ox}}^0 e^{-\alpha nF\Delta E/RT} - C_{\text{Red}}^0 e^{(1-\alpha)nF\Delta E/RT}\} \qquad (8)$$

where k_s^0 is the rate constant of the forward and backward reaction at $E_{\text{Ox/Red}}^f$, α (the transmission coefficient) is a constant for the reaction, and the other terms have their usual significance.

An extensive discussion of this topic is beyond the scope of this chapter, and the student is referred elsewhere for details.[14] However, one particular situation occurs frequently and will be discussed as it applies to electrodeposition.

Suppose that the rate of electron transfer is the *slowest* step in reaction (7) and the current density at the electrode is small compared to the diffusion current for the reaction; then

$$\eta = \gamma - \beta \ln i^0 \qquad (9)$$

where η is the overpotential and γ and β are constants characteristic of the electrode and solution. *η is positive for anodic processes and negative for cathodic processes,* when, as in the case of this text, the Stockholm sign convention (p. 1591) is used. The overpotential, η, is the difference between the reversible electrode potential and the actual electrode potential. Values of γ and β are tabulated elsewhere.[15]

Table 51-1 gives the overpotential for hydrogen evolution for various electrode materials. The overpotential for platinized platinum is small compared to smooth platinum, as a result of the large surface area resulting from the presence of finely divided platinum on the surface, and, from Eq. (8), it can be shown that the overpotential is zero when the current density is zero (see Problem 13).

The overpotential of hydrogen on mercury is extremely high and accounts for the ability to deposit metals on mercury, whose electrode potentials are considerably more negative than that of the H^+/H_2 couple (see p. 936).

The anodic overpotential for oxygen evolution in 1 N sulfuric acid on smooth platinum and gold is given in Table 51-2, according to Hickley and Hill.[16]

[13] J. J. Lingane, *Electroanalytical Chemistry*, Wiley-Interscience, New York, 2nd ed., 1958, chap. 15; H. Diehl, Ref. 6; J. A. Page in *Handbook of Analytical Chemistry* (L. Meites, ed.), McGraw-Hill, New York, 1963, pp. 5–172ff.

[14] J. O'M Bockris in *Electrical Phenomenon at Interfaces* (J. A. V. Butler, ed.), Methuen, London, 1951; P. Delahay, *Newer Instrumental Methods of Analysis*, Wiley-Interscience, New York, 1952, chap. 2.

[15] J. O'M. Bockris. Ref. 14, pp. 164–165.

[16] A. Hickling and S. Hill, *Trans. Faraday Soc.* **46**, 550 (1950). Data in phosphate buffer, (pH 6.84) and in 1 N KOH are also given. For both platinum and gold, the change of pH from 0 to 14 causes very little change in overpotential at the lowest current densities. At the highest current densities the overpotential is increased by ~0.4 V on platinum and ~0.6 V on gold in 0.1 N KOH.

TABLE 51-1. **Overpotential (volts) of H_2 from 1 M H_2SO_4 on Different Electrodes**[a]

	Current Density, mA/cm^2			
Electrode Material	1	10	100	1000
Ag	0.475	0.762	0.875	1.089
Au	0.241	0.390	0.588	0.798
Bi	0.78	1.05	1.14	1.23
Cu	0.479	0.584	0.801	1.254
Graphite	0.600	0.779	0.977	1.220
Ni	0.563	0.747	1.048	1.241
Pb	0.52	1.090	1.179	1.262
Platinized Pt	0.015	0.030	0.041	0.048
Smooth Pt	0.024	0.068	0.288	0.676
Hg[b]	0.879	1.042	1.070	1.115

[a] From J. A. Page in *Handbook of Analytical Chemistry* (L. Meites, ed.), McGraw-Hill, New York, 1963, pp. 5–185.

[b] Interpolated from data at other current densities assuming that the Tafel equation is obeyed over the range of interpolation.

TABLE 51-2. **Overpotential (volts) of O_2 from 1 N H_2SO_4 on Platinum and Gold**[a]

	Current Density, mA/Cm^2					
Electrode Material	0.01	0.1	1	10	100	1000
Platinum	0.52	0.67	0.80	0.92	1.07	1.15
Gold	0.73	0.82	0.86	0.93	1.00	1.08

[a] From A. Hickling and S. Hill, *Trans. Faraday Soc.* **46**, 550 (1950).

Classical Electrogravimetry

To explain the behavior of an electrolysis cell such as that shown in Fig. 51-1, overpotentials for the anodic, η_a and cathodic, η_c, processes must be included. If V is the applied voltage to the cell, i the current in the circuit, and R_T the sum of the cell resistance and ohmic resistances external to the cell,

$$V = E_a + \eta_a - E_c + \eta_c + iR_T \tag{10}$$

where E_a and E_c are the reversible electrode potentials of the anode and cathode, which may be calculated from the Nernst equation. (See Problems 2, 5, 10, and 11.)

A plot of i versus V of Eq. (10) has the shape shown in Fig. 51-3, with a sharply rising portion that appears to be linear. Extrapolation of this linear portion of the curve back to zero-current yields E_d, the *decomposition voltage* or *back emf for the cell reaction.* In the case where η_a and η_c are zero, E_d is calculated from the Nernst equation for the cell reaction (p. 161) $E_d = E_a - E_c$. If this is not the case, experimental determination of E_d is required.

In Fig. 51-4 the cathode potentials (versus N.H.E.) of various metal sulfates, as calculated from standard potentials, are plotted as a function of the negative logarithm of the metal-ion concentration in solution. The solid horizontal line at a voltage of -0.07 V represents the decomposition potential of hydrogen in 1 M sulfuric acid ($\eta = -0.07$ V at a current density of 10 mA/cm^2); that at -0.6 V, the same on copper ($\eta = -0.6$ V). The dashed lines give the change of cathode potential in sulfuric acid on smooth platinum and copper at various pH values, given on the scale at the top of the graph. A study of this idealized figure teaches us many things of practical importance.

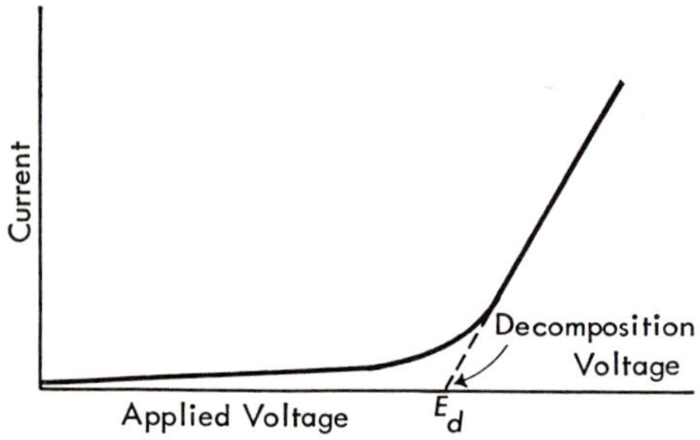

Fig. 51-3. Current-voltage curve for an electrolysis cell.

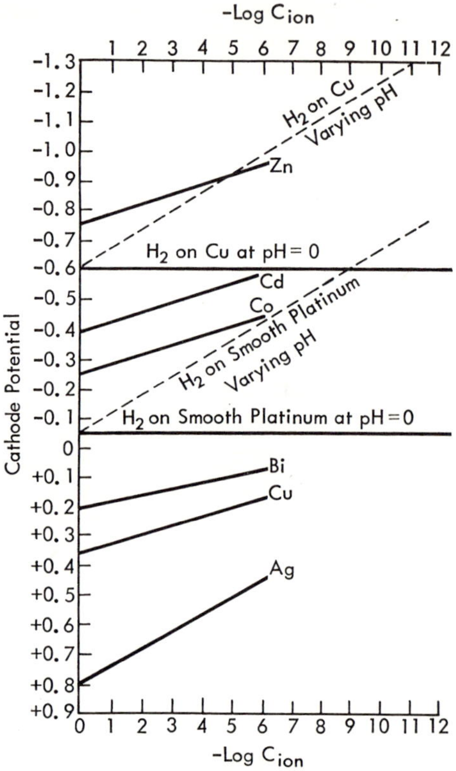

Fig. 51-4. Decomposition voltage of metal sulfates in 1 M H_2SO_4.

Completeness of deposition of a metal. In the classical electrogravimetric technique (Fig. 51-1), the cathode potential is not controlled, but either we control the cell current in a very approximate manner or apply an approximately constant cell voltage V. When the anode potential remains constant during electrolysis, the constant applied cell-voltage technique approaches the controlled-potential method in the theory of the electrolysis of a single species.

Suppose that we electrolyze a solution of 1 M copper in 1 M sulfuric acid. The anode potential (oxygen, with overpotential) is 1.70 V and the copper potential 0.35 V; hence the electrolysis starts at an applied emf of $1.70 - 0.35 = 1.35$ V (decomposition voltage, Fig. 51-3). Suppose that during the electrolysis we keep the applied voltage reasonably constant at 2.0 ± 0.1 V and that the anode potential does not vary. The equilibrium concentration of copper at a voltage of 2.0 is calculated with the aid of the Nernst equation:

$$1.35 - 2.0 = -0.65 = \frac{0.059}{2} \log [\mathrm{Cu}^{++}]$$

$$-\log [\mathrm{Cu}^{++}] = 22.0$$

$$[\mathrm{Cu}^{++}] = 1 \times 10^{-22}$$

The same value, of course, is calculated from the cathode potential. An emf of 2.0 V corresponds to a cathode potential, E, of $1.7 - 2.0 = -0.30$ V:

$$E = E^f + \frac{0.059}{2} \log [\mathrm{Cu}^{++}] = 0.35 + \frac{0.059}{2} \log [\mathrm{Cu}^{++}] = -0.30$$

and the same result is found.

The hydrogen overpotential on copper is approximately -0.6 V (current density 0.01 A/cm^2); therefore, hydrogen is not evolved until the voltage is raised from 1.7 to about 2.3 V. Therefore, a copper solution can be electroanalyzed in acid medium without any interference from hydrogen evolution. In addition, it is seen that when the copper concentration has reached the value of approximately 10^{-22} M the electrolysis automatically comes to a stop, since Eq. (10) may be written

$$V = E_d + iR_T$$

When V becomes equal to E_d, i becomes equal to zero. Thus in the electrolyanalysis of a copper sulfate solution one could use a lead storage cell having an emf of about 2.1 V as a source of current and let the apparatus stand overnight at room temperature. The next day all the copper would be separated, and practically no current is wasted. What has been said for copper naturally holds qualitatively for metals such as silver and bismuth.

If a zinc sulfate solution is electrolyzed under analogous conditions in a fairly strong acid medium, hydrogen will be evolved before the zinc is deposited (see Fig. 52-4), even if a copper-plated platinum cathode is used. (Zinc should never be plated on pure platinum, because it forms an alloy with this metal which is not completely soluble in acid, and the platinum is eventually spoiled.)

To electroanalyze a zinc solution one can resort to the following modifications:

1. Decrease the hydrogen-ion concentration. Suppose that we are working in an acetic acid–acetate buffer with a pH of 6. The hydrogen electrode potential is calculated from the Nernst equation:

$$E = 0 + 0.059 \log 10^{-6} = -0.354$$

(see the dashed line on smooth Pt, Fig. 51-4). If a copper-plated electrode is used, the hydrogen potential becomes $-0.6 - 0.354 = -0.954$ V (Fig. 52-4). The calculated zinc-ion concentration at a zinc electrode with a potential of -0.954 V is

$$-0.954 = -0.76 + \frac{0.059}{2} \log [\mathrm{Zn}^{++}]$$

$$\log [\mathrm{Zn}^{++}] = -6.57$$

$$[\mathrm{Zn}^{++}] = 10^{-6.57} = 2.7 \times 10^{-7}$$

Hence under these conditions the zinc is quantitatively deposited. If the hydrogen-ion concentration is further decreased by addition of sodium hydroxide, zinc hydroxide will precipitate and then dissolve as zincate-ion in an excess of the base. If ammonia is used as the base, the zinc hydroxide will dissolve as an ammino complex. In both cases most of the zinc is transformed into complex ions; consequently a large decrease in the zinc-ion concentration results, and therefore the zinc electrode becomes more negative. The hydrogen potential also becomes more negative, because of the decrease of the hydrogen-ion concentration and the increase of the overpotential in an alkaline medium.

If all data (complex constant of the zinc complex formed, and the overpotential of hydrogen) are known, it is possible to calculate whether the electroanalysis can be carried out without interference from hydrogen liberation. Empirically, it has been found that various metals, such as zinc and nickel, can be advantageously electroanalyzed in an ammoniacal medium or in other complexing solutions (cyanide).

2. Use a cathode at which the overpotential of hydrogen is extremely large. An inspection of Table 51-1 shows that mercury is distinguished by its very high hydrogen overpotential. Böttger[17] has made extensive investigations on the use of the mercury cathode in electroanalysis and has shown that it is an important tool in the quantitative electroanalysis of various metals that cannot be deposited on platinum because of hydrogen evolution. Another advantage of mercury is that it forms amalgams with the separated metals. The amalgams may be considered as dilute solutions of metals in mercury. When dissolved in mercury, metals have less tendency to go into solution than when present in their pure state (they behave more like noble metals and have more positive standard potentials). Therefore by using a mercury cathode in a solution of low hydrogen-ion concentration, it is even possible to determine strongly electropositive alkali metals by electroanalysis.

In dilute sulfuric acid solution it is possible to deposit iron, chromium, zinc, cadmium, gallium, germanium, copper, tin, molybdenum, lead, bismuth, silver, gold, the platinum metals (except ruthenium and osmium), arsenic, selenium, tellurium, thallium, rhenium, and mercury at a mercury cathode. Manganese is only partially removed. Remaining in solution are aluminum, zirconium, titanium, phosphorous, vanadium, uranium, magnesium, and the alkaline and rare earth metals.[18]

Using the constant-current electrolysis method, the potential of the cathode becomes increasingly more negative as the electrolysis proceeds (see Fig. 50-1, p. 921). In a solution containing only one metal ion, the final cathode potential is determined by the reduction of the supporting electrolyte. Given this potential value, the completeness of metal deposition is calculated as above. The time required for a constant-current electrodeposition can never be less than that required using a constant cell voltage or controlled-potential electrodeposition. Any current in excess of the diffusion current does not lead to electrodeposition of the desired constituent, and if it produces a gas, may have an unwanted effect on the character of the deposit (see below).

Influence of stirring and temperature. Character of the deposit. Present-day practice is to stir the solution vigorously, thereby reducing the thickness of the Nernst diffusion layer (p. 901) and increasing the rate of electrolysis. In addition, stirring minimizes the magnitude of concentration gradients *parallel* to the electrode surface, thus preventing evolution of hydrogen from parts of the electrode which are locally depleted with respect to the reducible species.

Various stirring methods are used in practical work; rotating the cathode or the anode, or agitation by means of an independent stirrer, can all give good results. Increase of temperature increases the rate of mass-transfer in solution although stirring is more efficient than heating. Still, working at higher temperatures has some other advantages; it decreases the resistance of the electrolyte, increasing the current and

[17] W. Böttger, *Ber.* **42**, 1824 (1909); T. P. McCutcheon, *J. Am. Chem. Soc.* **29**, 1445 (1907); H. S. Lukens and E. F. Smith, *ibid.* **29**, 1455 (1907); T. P. McCutcheon and E. F. Smith, *ibid.* **29**, 1460 (1907); especially W. Böttger, in *Physikalische Methoden der analytischen Chemie*, vol. I, Akademische Verlagsgesllschaft, Leipzig, 1936, p. 137. For a review of the use of the mercury cathode in chemical analysis see J. A. Maxwell and R. P. Graham, *Chem. Rev.* **46**, 471 (1950), and J. A. Page, J. A. Maxwell, and R. P. Graham, *Analyst* **87**, 245 (1962).

[18] W. F. Hillebrand and G. E. F. Lundell, *Applied Inorganic Analysis*, Wiley, New York, 1929, p. 105; G. E. F. Lundell and J. I. Hoffman, *Outlines of Methods of Chemical Analysis*, Wiley, New York, 1931, p. 94; H. O. Johnson and L. Lykken, *Anal. Chem.* **19**, 481 (1947); T. D. Parks, H. O. Johnson, and L. Lykken, *ibid.* **20**, 148 (1943); see also E. B. Sandell, *Colorimetric Determination of Traces of Metals*, Wiley-Interscience, New York, 3rd ed., 1959, pp. 48, 225, and 274, for further details.

giving a denser and more adherent deposit. Therefore, for many rapid determinations the temperature is maintained between 60 and 90°C, and the solution is well stirred.

As far as the nature of the deposit is concerned, it is absolutely necessary that it shall adhere firmly to the cathode in order that the solution may be rinsed off at the end without loss of metal. For analytical purposes the deposit must be coherent, dense, and smooth. Such a deposit is finely crystalline, with a metallic luster. Flaky, spongy, and powdery deposits are less dense and less pure; they adhere only loosely to the cathode, and for these reasons they should be avoided. As a rule more satisfactory deposits are obtained when the metal is deposited from a solution in which it is present as a complex rather than as a simple ion. Increase of current density leads to a decrease in the grain size of the deposit; and if it is raised sufficiently, evolution of hydrogen may occur, owing to depletion of metal ions near the cathode. If an appreciable evolution of hydrogen occurs, the deposit will become broken up and irregular. In certain cases the addition of nitric acid to the electrolyzed solution is recommended. Nitric acid prevents the formation of hydrogen at the cathode, functioning as a depolarizer

$$NO_3^- + 10\,H^+ + 8e \rightarrow NH_4^+ + 3\,H_2O \qquad (10)$$

since reaction (10) occurs more easily than hydrogen evolution.

For analytical purposes the current densities that are employed are seldom in excess of 0.1 A/cm^2 and usually are about 0.01 A/cm^2 (stirring). No adjustment of current density is possible using the method of controlled-potential electrolysis (except by choosing a potential that yields a current less than the diffusion current).

It should be emphasized that at the end of the electrolysis the current should not be shut off as long as the electrodes are in the solution. If the current is broken, the metallic deposit may slowly go back into solution.

Electrolysis without applied emf. Internal electrolysis. Electroanalysis by the spontaneous discharge of a galvanic cell in which the solution to be analyzed is the catholyte has been named "*internal electrolysis*."[19] To make the principle clear, let us consider the cell

$$Pt \mid Cu^{++} \parallel Zn^{++} \mid Zn$$

Although the potential of the platinum electrode in the copper-ion solution is ill-defined, it will always be considerably more positive than that of the zinc electrode. Hence, when the cell is short-circuited, electrons flow from the zinc to the platinum electrode, giving rise to a reduction of the copper ions:

$$Cu^{++} + 2e \rightarrow Cu$$

The platinum becomes coated with copper and behaves like a copper electrode which has a well-defined potential in the copper-ion solution. Simultaneously with the reduction of copper ions, an oxidation of zinc occurs at the zinc electrode, the net cell reaction being

$$Cu^{++} + Zn \rightarrow Cu + Zn^{++}$$

Because of the great difference in potential between the zinc and copper electrodes, the separation of copper ions will be quantitative. This would be true even if the zinc–zinc sulfate were replaced, say, by cadmium–cadmium sulfate. The advantages, limitations, and specific applications of internal electrolysis have been discussed by Clarke et al.[20] in a paper in which they describe the apparatus and the application of the method to the determination of copper and bismuth in an alloy, using a lead anode and a platinum-gauze cathode. For the determination of small quantities of mercury in the presence of copper and zinc by internal electrolysis, see Fife.[21]

[19] H. J. S. Sand, *Analyst* **55**, 309 (1930); A. Schleicher, *Z. Anal. Chem.* **83**, 127 (1931).

[20] B. L. Clarke, L. A. Wooten, and C. L. Luke, *Ind. Eng. Chem., Anal. Ed.* **8**, 411 (1936); see also J. J. Lurie and L. B. Ginsburg, *ibid.* **10**, 201 (1938).

[21] J. G. Fife, *Analyst* **63**, 650 (1938).

PROBLEMS

1. During an electrolysis the applied emf as read on a voltmeter is equal to 3 V. The back emf is equal to 1.5 V, and the resistance of the system is 10Ω. What is the current expressed in amperes? What is the current density at the cathode, if the latter has a surface area of 20 cm^2? What is the current when the applied emf is reduced to 1.5 V? *Ans.* 0.15; 0.0075 A/cm^2; zero.
2. Calculate the decomposition voltage of 0.01 *N* silver sulfate in 1 *N* sulfuric acid. The potential of the silver electrode in 0.01 *N* silver nitrate is 0.69 V, that of the oxygen electrode 1.3 V. The oxygen overpotential is 0.4 V. The silver is transformed into a complex and the silver-ion concentration reduced to 10^{-10} *N*. What is the decomposition voltage now? Assuming that the hydrogen-ion concentration remains 1 *N* and that the overpotential of hydrogen on silver is equal to -0.5 V (hence the potential of the cathode when hydrogen is discharged is -0.5 V), would there be any danger of hydrogen evolution before all the silver is deposited? *Ans.* 1.01 V; 1.49 V; no; a total voltage of 2.20 would be required for H_2 evolution.
3. How could the current be increased during an electrolysis if the difference between the applied emf and back emf, and the temperature, are kept constant?
4. The back emf in the electrolysis of 0.1 *N* silver nitrate at the start is 0.76 V. What is the value after the silver-ion concentration has been reduced to 0.001 *N*? *Ans.* 0.88 V.
5. The potential of a dilute magnesium amalgam in a 0.001 *M* magnesium sulfate solution having a hydrogen-ion concentration of 10^{-10} is equal to -1.46 V. The overpotential of hydrogen on mercury is -1 V; the potential of the oxygen anode is 1.5 V. What is the magnesium-ion concentration when hydrogen begins to be liberated? *Ans.* 2.3×10^{-8} *M*.
6. A bromide solution is electrolyzed with a silver anode. What is the reaction occurring at the anode? How much of the anode reaction product is formed by a current of 0.1000 A in 1 min? *Ans.* 0.01168 g.
7. The potential of Cu | 0.1 *M* Cu^{++} is 0.315 V, that of Cd | 0.1 *M* Cd^{++} is 0.43 V. In 0.4 *M* potassium cyanide solution which is 0.1 *M* in copper, the potential of copper is -0.97 V; in 0.4 *M* cyanide which is 0.1 *M* in cadmium, the cadmium potential is -0.87 V. (a) Is it possible in the absence of cyanide to plate out the copper quantitatively in the presence of cadmium? (b) Which metal plates out first in 0.4 *M* cyanide? (c) Calculate the cadmium concentration in the 0.4 *M* cyanide solution when the cathode potential has become -0.96 V. Assume that all the cadmium is present as $Cd(CN)_4^-$ and that the cyanide concentration remains unchanged during the electrolysis. *Ans.* (c) 10^{-4} *M*.
8. A solution that is 0.1 *M* in ferric and 0.1 *M* in ferrous sulfate is electrolyzed between two platinum electrodes, the solution being stirred vigorously. (a) What is the decomposition voltage? (b) What reactions occur at the cathode and at the anode? (c) How does the composition of the solution change during the electrolysis?
9. The hydrogen overpotential on mercury is taken equal to -1 V. The formal potential of zinc is equal to -0.76 V. A solution of zinc perchlorate in 1 *M* perchloric acid is electrolyzed at a mercury cathode, the potential of the latter being kept at -0.97 V. (a) What is the final zinc concentration in the solution? (b) Does hydrogen evolution interfere with the electrolysis of the zinc? *Ans.* (a) 1.2×10^{-7} *M*; (b) no.
10. The overpotential of lead on platinum is zero; the anodic overpotential of oxygen is 0.47 V. The equilibrium potential (not including overpotential) of Pt | O_2, 1 *M* $HClO_4$ is equal to -1.23 V; the formal potential of lead is -0.13 V. A solution of 1 *M* lead perchlorate in 1 *M* perchloric acid is electrolyzed between two platinum electrodes. (a) Calculate the decomposition voltage. (b) Calculate the decomposition voltage of the 1 *M* lead perchlorate solution in a buffer solution with a pH of 4. *Ans.* (a) 1.83 V; (b) 1.59 V.

11. The decomposition voltage of 1 N perchloric acid at smooth platinum electrodes is 1.70 V. (a) Calculate the overpotential of the oxygen electrode, assuming that there is no hydrogen overpotential. (b) Calculate the decomposition voltage when the cathode consists of copper and the hydrogen overpotential on copper is -0.6 V. (c) Calculate the decomposition voltage of 0.01 N perchloric acid solution at smooth platinum electrodes, assuming that the conditions of overpotential are the same as (a).

$$E^0_{H^+/H_2} = 0.0 \text{ V} \qquad E_{O_2/H_2O} = 1.23 \text{ V}$$

Ans. (a) 0.47 V; (b) 2.3 V; (c) 1.70 V.

12. A solution contains 10^{-1} M Ag^+ and 1 M Cu^{++}. The formal potential for silver is 0.80 and for copper 0.341 V. (a) What is the most negative potential that can be applied without depositing copper metal? (b) How much silver would be left in solution at this potential? (c) What is the least negative potential required to deposit 99.9 per cent of the silver?

Ans. (a) 0.34 V; (b) 1.7×10^{-8} M; (c) 0.62 V.

13. Obtain the Nernst equation for reaction (7) from Eq. (8). (Hint: What is i° during the potentiometric measurement of potential?)

CHAPTER 52 *Potentiometry and Potentiometric Titrations*

Ideally, the emf of the cell

$$\text{reference electrode} \parallel (M^{n+}) \mid M$$

should give the value of the activity of M^{n+}. The liquid-junction potential always introduces an uncertainty in the interpretation of the cell emf, but under appropriate conditions this problem can be minimized. For analytical purposes, until recently, only the extremely important measurement of p_aH (with the H_2 or glass electrode) made use of direct potentiometry. In such measurements, it is possible to approach the true value of p_aH, but we never get the exact value.[1]

Recently, specific ion-sensitive membrane electrodes have been introduced which hold considerable promise for the measurement of sodium, calcium, and fluoride ion activities (p. 943).

For ions other than H^+, cell emf measurements have rarely been used for determination of ion concentrations. In addition to the liquid-junction potential, ionic activity coefficients are not known exactly. In rare cases it is possible to construct a concentration cell in which both these effects are minimized, e.g.,

$$\text{Ag} \left| \begin{matrix} C_{Ag^+} = x \\ 1\text{–}2\ M\ KNO_3 \end{matrix} \right\| \begin{matrix} C_{Ag^+} = a \\ 1\text{–}2\ M\ KNO_3 \end{matrix} \right| \text{Ag}$$

In this cell the same large excess of inert electrolyte (KNO_3) is present on both sides, and the emf is given by

$$E = 0.059 \log \frac{a}{x}$$

where a is the known silver concentration and x the unknown silver-ion concentration.

By far the most common application of potentiometry is in potentiometric titrations, and a short discussion of the classical potentiometric (zero current) and constant-current potentiometric titrations is given below (p. 943).

CLASSIFICATION OF INDICATOR ELECTRODES

Electrodes of the first kind

Metal ion–metal electrodes

$$M^{n+} + ne \rightleftharpoons M$$

are known as electrodes of the first kind. Silver, mercury, and, under certain conditions, copper and bismuth behave reversibly. Ordinarily base metals do not behave reversibly.

[1] R. G. Bates, *Determination of pH*, Wiley, New York, 1964.

More generally, we include half-cell reactions of the form

$$\mathrm{Ox} + ne \rightleftharpoons \mathrm{Red}$$

in this class, where a "noble" metal electrode is used to indicate the potential. Any metal that is not oxidized at the potential of the system can be used to indicate the electrode potential. Most frequently, platinum or gold are used, although sometimes mercury (the D.M.E.) is convenient.

Electrodes of the second kind

These electrodes have half-cell reactions of the form

$$\text{insoluble salt} + ne \rightleftharpoons \text{metal} + \text{anion}$$

and include systems such as AgX/Ag (X = Cl^-, Br^-, or I^-), $PbSO_4$/Pb, Hg_2Cl_2/Hg, etc. The potential of the electrode varies with the logarithm of the activity of the anion (p. 170).

Electrodes of the third kind

These electrodes may be represented by a half-reaction of the form

$$\mathrm{M}_x\mathrm{X}_n + x\mathrm{N}^{n+} + xne \rightleftharpoons x\mathrm{M} + \mathrm{N}_x\mathrm{X}_n$$

or by the half-cell

$$\mathrm{N}^{n+},\ \mathrm{N}_x\mathrm{X}_n,\ \mathrm{M}_x\mathrm{X}_n \mid \mathrm{M}$$

where M and N represent different metals which form the insoluble salts or stable complexes M_xX_n and N_xX_n. In the case where the latter two species are precipitates, the electrode potential is given by

$$E = E^0 + \frac{0.059}{n}\log(\mathrm{N}^{n+})$$

LeBlanc and Harnapp[2] have studied a number of systems of this sort and have considered in detail the various factors that influence the electrode potentials. Some of the electrode systems studied were Tl^+, TlI, AgI/Ag; Cu^{++}, CuS, HgS/Hg; Ca^{++}, $CaSO_4$, $PbSO_4$/Pb; Ca^{++}, $CaWO_4$, Hg_2WO_4/Hg; and Ca^{++}, CaC_2O_4, MC_2O_4/M, where M = Zn, Cu, Pb, and Hg(mercurous). All such systems are relatively slow in reaching solubility equilibrium (and a stable potential) and can be used only over a relatively narrow range of N^{n+} concentration.

A more practical example, discussed by Reilley and Schmid,[3] is based on the electrode system where both M_xX_n and N_xX_n are soluble, e.g., ethelenediaminetetraacetic acid, H_4Y, complexes. An example is

$$\mathrm{HgY}^{--} + \mathrm{Cd}^{++} + 2e \rightleftharpoons \mathrm{Hg} + \mathrm{CdY}^{--}$$

where the electrode potential is given by

$$E = E^0 + \frac{0.059}{2}\log\frac{(\mathrm{Cd}^{++})(\mathrm{HgY}^{--})}{(\mathrm{CdY}^{--})}$$

under suitable conditions. In this case it is usual to add a fixed amount of HgY^{--} to the solution, making the potential dependent upon the ratio $(Cd^{++})/(CdY^{--})$. Such electrodes may be used in the potentiometric titration of a large number of cations, including Ca^{++}, Sn^{++}, Ba^{++}, Mg^{++}, Zn^{++}, Cd^{++}, Cu^{++}, Pb^{++}, Mn^{++}, Co^{++}, Ni^{++}, VO^{++}, Al^{3+}, Ga^{3+}, In^{3+}, Tl^{3+}, Cr^{3+}, Sc^{3+}, Y^{3+}, La^{3+}, Ce^{3+}, Pr^{3+}, Nd^{3+}, Sm^{3+}, Zr^{4+}, Hf^{4+}, and Bi^{3+}.[4]

[2] M. LeBlanc and D. Harnapp, *Z. Physik. Chem.* (*Leipzig*) **166**, 321 (1933).
[3] C. N. Reilley and R. W. Schmid, *Anal. Chem.* **30**, 947 (1958).
[4] C. N. Reilley, R. W. Schmid, and D. W. Lamson, *Anal. Chem.* **30**, 953 (1958).

The species M_xX_n must be considerably more stable (or insoluble) than N_xX_n; otherwise the reaction

$$M_xX_n + xN^{n+} \rightarrow xM^{n+} + N_xX_n$$

will occur and significantly alter the concentration of N^{n+} present on the addition of M_xX_n.

Electrodes of the third kind are used principally in the potentiometric titration of metal ions for which no convenient electrode of the first kind exists.

Membrane electrodes

Membrane electrodes differ from the four kinds of electrodes discussed above in three ways: (1) ion transfer, rather than electron transfer, occurs at the electrode surface; (2) they do not respond to the redox potential of the solution; and (3) frequently they are extremely high resistance and require special instrumentation (an electrometer) to measure the cell emf.

pH-sensitive glass electrode. The glass electrode is the most common membrane electrode in use today and is used to measure the pH of solution. Its emf response is described in terms of the phase boundaries represented by

$$\text{solution, } (H^+)_1 \mid \text{glass} \mid (H^+)_2\text{, solution}$$

The emf developed across the glass membrane at 25°C is related to the hydrogen-ion activities, $(H^+)_1$ and $(H^+)_2$ on both sides of the membrane according to

$$E_g = 0.059 \log \frac{(H^+)_1}{(H^+)_2}$$

Hence, if the pH on one side of the glass membrane is held constant using a buffer, the potential across the membrane will depend upon the pH of the solution on the other side. To measure this membrane potential, reference electrodes can be inserted in the solutions on both sides of the glass membrane. A practical glass electrode system may be represented schematically by

$$\underbrace{\text{Ag} \mid \text{AgCl, HCl} \mid \text{glass}}_{\text{glass electrode}} \mid \text{solution of unknown pH} \mid \text{S.C.E.}$$

The portion between the dashed lines represents the glass electrode, a glass tube with a bulb of pH-sensitive glass, about 0.5–1.0 cm in diameter and a few hundredths of a centimeter thick, closing one end. In one version the tube contains a hydrochloric acid solution, and a silver–silver chloride electrode dips into the tube, passing through a liquid-tight seal which closes off the other end of the tube (Fig. 52-1a).

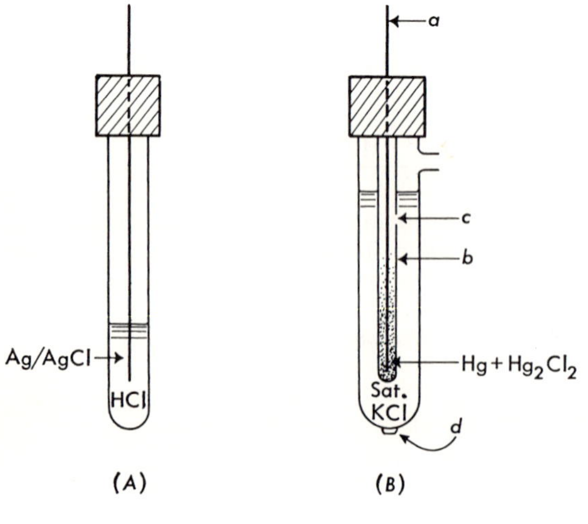

Fig. 52-1. *A*. Glass electrode. *B*. Calomel electrode; *a*—electrical lead terminating in a platinum wire which contacts Hg; *b*—glass capillary with hole, *c*—blown in side; *d*—small hole or porous diaphragm.

This unit can be dipped directly into the unknown solution and the emf determined with respect to a calomel electrode. A commercial calomel electrode is usually constructed to have a cylindrical form resembling that of the glass electrode and includes the salt bridge at the bottom so that it can also be dipped directly into the unknown solution (Fig. 52-1b).

The emf of the glass–calomel electrode assembly is given by

$$E = E_{\text{S.C.E.}} - E^0_{\text{AgCl/Ag}} + 0.059 \log(\text{H}^+)(\text{Cl}^-) - 0.0591 \log(\text{H}^+)_{\text{X}} \qquad (1)$$

where (H^+) and (Cl^-) represent the activity of hydrogen and chloride ions in the standard buffer and $(\text{H}^+)_{\text{X}}$ is the unknown hydrogen-ion activity.

The first three terms on the right side of Eq. (1) are constant; hence

$$E = \text{constant} - 0.059 \log(\text{H}^+)_{\text{X}} \qquad (2)$$

It is most accurate to calibrate the electrode assembly by dipping it into two buffers of known p_aH and to adjust the controls so that the pH meter scale is correct at these two known p_aH values. The known solutions are chosen so that their p_aH values bracket the p_aH of the unknown solution. In this way p_aH values accurate to ± 0.02 unit can be determined, provided the composition of the unknown is not too different from that of the standard buffers.

Specific ion-sensitive electrodes. Recently glass electrodes showing a high specificity toward sodium ion have become available commercially.[5] Such electrodes obey the relationship

$$E = \text{const.} - 0.059 \log(\text{Na}^+)$$

when used in a cell analogous to that described above for the pH-sensitive glass electrode. The range of application is from $1M$ to $5 \times 10^{-7}\ M$ Na^+ in the pH range 7–10. A nonspecific glass electrode, sensitive to Ag^+, K^+, NH_4^+, Rb^+, Na^+, Cs^+, and Li^+, in decreasing order of sensitivity, is also available.

Calcium ion-sensitive electrodes[6] have also been announced and are described as containing a liquid ion exchanger separated by a membrane from the unknown solution. Claims are made that these electrodes are usable in solutions as dilute as $10^{-5}\ M$ Ca^{++} in the pH range 3–11. A perchlorate ion-sensitive electrode based on this principle is also available.[7] Membrane electrodes sensitive to Cl^-, Br^-, I^-, SO_4^{--}, and phosphate are also available. These involve a silicone membrane containing a precipitate of the ion being determined.

Lanthanum fluoride single crystals doped suitably to increase the conductivity of the crystal respond to fluoride ion quite selectively. A commercial fluoride ion electrode is available[7] and is being used extensively in determining fluoride at the ppm level in water. Such electrodes are solid state ionic conductors. Bromide and iodide ion-sensitive electrodes based on solid state ionic conductors are also available.[7] Rechnitz has recently published an excellent review of ion-selective electrodes.[8]

POTENTIOMETRIC TITRATIONS

Beherend[9] performed the first potentiometric titration in 1893, titrating chloride, bromide, and iodide with mercurous nitrate. Since that time thousands of publications have appeared which employ this method, whose accuracy is limited mainly by the titrametric technique rather than by the interpretation of emf values. Before discussing some methods of end-point detection it is important to consider certain experimental problems which ultimately place a lower limit upon the minimum concentration of unknown which can be titrated using this method.

[5] Beckman Instruments, Inc., Fullerton, Calif.

[6] *Chem. Eng. News*, **44** (22), **50** (1966).

[7] Orion Research Inc., 11 Blackstone St., Cambridge, Mass.

[8] G. A. Rechnitz, *Chem. Eng. News*, **45** (25), 146 (1967).

[9] R. Beherend, *Z. Physik. Chem.* (*Leipzig*) **11**, 466 (1893).

Limitations on potentiometric titrations. Mixed potentials

In Chapter 8, relationships were derived for the potential at various stages of a redox titration:

$$\text{Red}_1 + \text{Ox}_2 \rightleftharpoons \text{Ox}_1 + \text{Red}_2 \tag{3}$$

The reduction potential of such a system is not a function of the analytical concentration of the redox couple and depends only upon the ratio of [Ox]/[Red]. If reaction (3) is quantitative, plots of electrode potential, E, versus per cent titrated (%T) should be independent of the analytical concentration of the species being titrated. This is normally the case at concentrations larger than 10^{-5} N of the species involved in the half-cell reaction which determines the electrode potential. At lower analytical concentrations, the exact value depending upon the particular experimental situation, E versus %T plots begin to differ significantly from the theory presented in Chapter 8.

This phenomenon occurs because the experimentally measured value of the electrode potential is a *mixed potential* and may differ appreciably from the true electrode potential. The difference between the true and experimental value of the electrode potential arises because of the presence of a residual current at the indicator electrode.

As was stated in Chapter 47, the total current at an indicator electrode is the sum of faradaic current arising from the reaction being investigated and any residual current. The theoretical Nernst potential corresponds to a current in the external circuit equal to the residual current, while the experimental value of the electrode potential corresponds to a zero value for total current at the electrode. Hence theory and experiment can agree only if the residual current is exactly zero.

The residual current arises from the presence of electroactive species other than the redox couple being titrated, e.g., trace impurities that are electroactive at potentials close to the Nernst potential. Oxygen is irreversibly reduced at a platinum electrode, and the current due to its reduction is markedly dependent upon the history of the electrode surface. Only platinum is discussed, although analogous behavior is displayed by other indicator electrodes. At oxidizing potentials, in noncomplexing media, the electrode itself is oxidized to form a monolayer of PtO or PtO_2. This reaction is relatively slow and the varying current arising from it can cause drifting potentials for a considerable time.[10] At sufficiently oxidizing potentials, water can be oxidized to produce O_2 (or the oxidized form of the supporting electrolyte is produced). At negative potentials evolution of hydrogen can occur at the electrode surface.

The concentration of the redox couple, its reversibility, and the magnitude of residual current determines the difference between the theoretical and experimentally measured electrode potential. This point is illustrated in Fig. 52-2, where current–potential curves have been drawn for three redox systems.

The solid lines represent the current–potential curves that would be obtained in the absence of residual current, while the dashed lines include the residual current. Curve ABQ is the residual current, while curve CDE is the current–potential curve for a solution containing significant amounts of both forms of a reversible redox system. D, the point of intersection of ABQ with CDE, is the theoretical Nernst potential. The intersection of $C'D'E'$ with the zero-current axis D' represents the potential that is measured with a potentiometer and occurs at a potential significantly more positive than D.

Curve FG represents the same ratio of [Ox]/[Red] as in CDE but at four times higher an analytical concentration. In this case, the difference between the current–potential curve corrected for residual current, and including residual current ($F'G'$), is much smaller in the vicinity of the zero-current axis; therefore, the error in potential measurement is smaller than in the first example.

[10] J. N. Ross and I. Shain, *Anal. Chem.* **28**, 548 (1956).

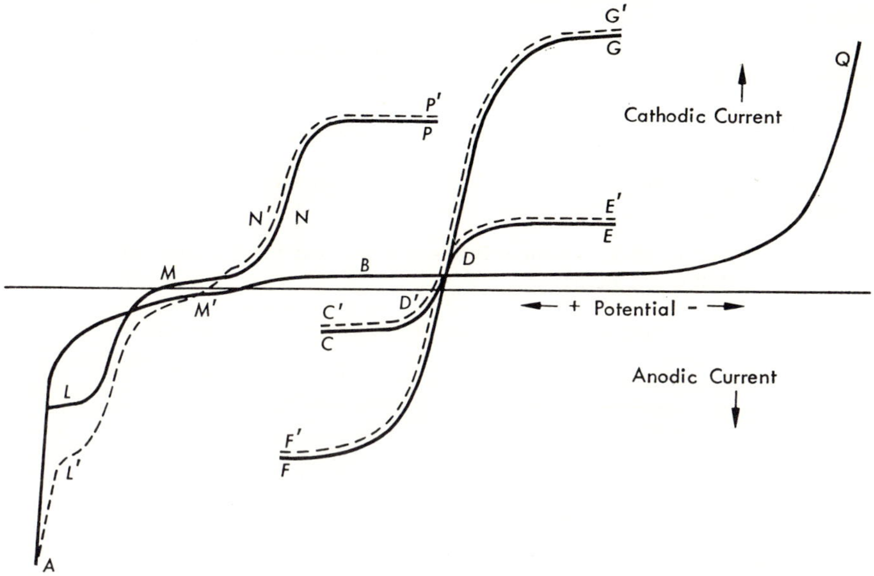

Fig. 52-2. Effect of mixed potentials on the measurement of redox potentials.

The current–potential curve of an irreversible system with a positive formal potential is represented by curve *LMNP*. In this case the reversible electrode potential corresponds to *M*. Curve *AL′N′P′* represents the experimental current–potential curve which includes the residual current. The measured potential occurs at *M′* and is considerably more negative than the theoretical electrode potential.

Clearly the difference between the true and experimental value of the redox potential increases the more dilute or the more irreversible the redox couple is and the closer the formal potential of the system approaches the decomposition potential of the supporting electrolyte. Thus it is impossible to state a lower concentration limit for potentiometric titrations in general, the exact limit depending upon the characteristics of the redox couples participating in the titration reaction.

The student should not infer from the above that if a redox indicator electrode indicates 10^{-10} *M* of an ion, this number always is in error because of a mixed potential. The solution may still be well-poised; i.e., an 0.01 *M* ammonia–ammonium chloride buffer is well poised as far as a hydrogen electrode is concerned, because the electrode reaction can be written

$$2NH_4^+ + 2e \rightleftharpoons 2NH_3 + H_2$$

Similarly, while the concentration of Ag^+ in a 0.1 *M* chloride ion solution saturated with silver chloride is about 10^{-9} *M*, an Ag/AgCl electrode would be well-poised in this solution because the electrode reaction can be written

$$AgCl + e \rightleftharpoons Ag + Cl^-$$

EQUIVALENCE-POINT LOCATION

A number of techniques have been proposed for the location of the equivalence point in a potentiometric titration and are discussed below briefly. Additional details are given elsewhere.[11]

[11] I. M. Kolthoff and N. H. Furman, *Potentiometric Titrations*, Wiley, New York, 2nd ed., 1949, chap. 6; I. M. Kolthoff and H. A. Laitinen, *pH and Electro-Titrations*, Wiley, New York,

Second derivative method

This technique is derived from the work of Hostetter and Roberts,[12] and involves locating the point in the titration curve where the second derivative of potential with respect to volume of titrant becomes zero (see p. 772).

An example of the method of calculation is illustrated in Table 52-1. The pH was *calculated* in the vicinity of the equivalence point for a 0.2 M solution (100 ml) of a weak acid, $pK_{HB} = 7.000$, with 0.2 M sodium hydroxide as titrant. Equation (145) of Chapter 4 was used and equal volume increments assumed.

TABLE 52-1. **Second Difference Method for Location of Inflection Point**[a]

ml, NaOH		$\frac{\Delta pH}{\Delta V}$	$\frac{\Delta^2 pH}{\Delta V^2}$
V	pH		
99.52	9.295		
		0.93	
99.62	9.388		2.0
		1.13	
99.72	9.501		3.1
		1.44	
99.82	9.645		3.9
		1.83	
99.92	9.828		3.3
		2.16	
100.02	10.044		−1.0
		2.06	
100.12	10.250		−3.9
		1.67	
100.22	10.417		−3.7
		1.30	
100.32	10.547		−2.6
		1.04	
100.42	10.651		

The way in which the various differences are formed is clear from the table format; e.g., the first entry in the column headed $\Delta pH/\Delta V$ is obtained from

$$\frac{\Delta pH}{\Delta V} = \frac{(pH)_{V_2} - (pH)_{V_1}}{V_2 - V_1} = \frac{9.388 - 9.295}{99.62 - 99.52} = 0.93$$

and is the estimated value of dpH/dV at $(V_2 + V_1)/2$ (99.57 ml).

Similarly, the estimated value of $\Delta pH/\Delta V$ at 99.67 ml is calculated to be 1.13.

The first entry in the last column is obtained from

$$\frac{\Delta^2 pH}{\Delta V^2} = \frac{\left(\frac{\Delta pH}{\Delta V}\right)_{99.67\text{ ml}} - \left(\frac{\Delta pH}{\Delta V}\right)_{99.57\text{ ml}}}{99.67 - 99.57}$$

$$= \frac{1.13 - 0.93}{99.67 - 99.57} = 2.0$$

2nd ed., 1941, chap. 7; H. A. Laitinen, *Potentiometric Analysis in Physical Methods in Chemical Analysis*, Vol. II (W. G. Berl, ed.), Academic Press, New York, 1951; E. Müller, *Die electrometrische Massanalyse*, Steinkopff, Dresden, 6th ed., 1942; see P. Delahey, *Instrumental Methods of Analysis*, Macmillan, New York, 1957, p. 60, for a more complete bibliography.

[12] J. C. Hostetter and H. S. Roberts, *J. Am. Chem. Soc.* **41**, 1337 (1919).

and is the estimated value of d^2pH/dV^2 at the midpoint of the volume interval, (99.67–99.57)/2, 99.62 ml.

The inflection point lies between 99.92 and 100.02 ml. It is not necessary to plot these data, because linear interpolation is satisfactory. The volume at inflection point, V_{IP}, is

$$V_{IP} = 99.92 + 0.1\left(\frac{3.3}{3.3 + 1.0}\right) = 99.92 + 0.08 = 100.00 \text{ ml}$$

Because of the approximation involved in assuming $\Delta^2pH/\Delta V^2 = d^2pH/dV^2$, it is not correct to use more than one significant figure for the result obtained by interpolating within the volume increment over which the second difference changes sign.

The pH values listed in Table 52-1 are given to ± 0.001 unit. This precision is just at the limit of the best commercially available pH meters, while an error 10–20 times this value is usually obtained with the usual pH meter found in the laboratory. It is apparent that if the pH were uncertain to ± 0.01 unit in the titration of HB, the inflection point could not be located with the accuracy shown in Table 52-1, and that a substantially larger volume increment would be required to obtain measurable pH differences.

In ordinary laboratory work, equal volume increments are added by allowing successive drops of titrant to fall from the buret, emf readings being taken after each drop. The volume of a drop is then determined by noting the change in volume for 20–50 drops. If greater precision is required, any of the modern piston-type microburets capable of delivering 1 ml of titrant with an accuracy of ± 0.001 ml (or 0.1 ml $\pm$ 0.0001 ml) should be used. Alternatively, weight burets used with a modern single-pan balance with projection scale corresponding to 1 g (error ± 0.1 mg) can be used if the highest accuracy is required, without any great increase in the time required for titration.

An automatic titrator, which is commercially available, has been designed which stops the titration when the second derivative is zero.[13] When the titration curve is not symmetrical on both sides of the equivalence point, differences may exist between the equivalence point and the point at which $d^2E/dV^2 = 0$. In most acid–base titrations, this difference is negligibly small but has been calculated; e.g., Ricci[14] and Kolthoff and Furman[15] consider this problem in connection with redox titrations, as have Kolthoff and Stenger in connection with precipitation titrations.[16] In general, if the equilibrium constant for the titration reaction is large, the difference between the equivalence point and end point is of no practical importance. Fortuin has compared this method with a more elaborate calculation technique.[17]

Titration to a fixed potential

Müller[18] proposed titration to a predetermined potential, a technique that is relatively simple to apply with modern high-input impedance electronic devices which indicate potential on a meter (pH meter).

Pinkhof[19] substituted for the calomel electrode a reference electrode whose potential exactly corresponded to the potential at the equivalence point. No potentiometer is

[13] H. V. Malmstadt, *Record Chem. Progr.* (*Kresge-Hooker Sci. Lib.*) **17**, 1 (1956), summarizes the results obtained with this instrument. Its principal limitation is rate of attainment of electrode equilibrium.

[14] J. E. Ricci, *Hydrogen Ion Concentration*, Princeton Univ. Press, Princeton, N.J., 1952, pp. 164, 218.

[15] I. M. Kolthoff and N. H. Furman, Ref. 10, chap. 3.

[16] I. M. Kolthoff and V. Stenger, *Volumetric Analysis*, vol. I, Wiley-Interscience, 1942, p. 143, and Ref. 14, p. 21.

[17] J. M. H. Fortuin, *Anal. Chim. Acta* **24**, 175 (1961).

[18] E. Müller, Ref. 11, p. 78.

[19] J. Pinkhof, *Über die Toepannung der Electrometrische Titraties*, Dissertation, Amsterdam 1918.

needed with this method, only a galvanometer and tapping key to determine when the two electrodes have the same potential.

Gran's method

Gran[20] proposed that plots of $\Delta\text{ml}/\Delta E$ or $\Delta\text{ml}/\Delta\text{pH}$ versus ml be made to locate the end point of a potentiometric titration. Such plots are linear just before and after the end point, having a V-shape for all potentiometric titrations. An example of such a plot is shown in Fig. 52-3, for the titration of 7.5×10^{-4} *M* benzoic acid with 0.02 *M* sodium hydroxide. In some titrations, the lines become curved at points more distant from the end point. The reader is referred to the original reference for more details.

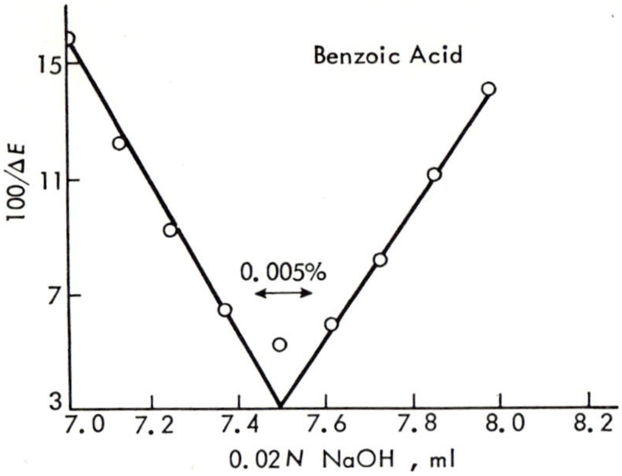

Fig. 52-3. Graphical determination of the inflection point using Gran's method, according to R. G. Bates and E. Wichers, *Journal of Research of the National Bureau of Standards*, **59**, 9 (1957).

CONSTANT-CURRENT POTENTIOMETRY

One indicator electrode

Classical potentiometry at zero current is difficult to apply when the electrode potential is established slowly. Slow equilibrium generally occurs when (1) no reversible redox couple is present, and (2) when the solution is poorly poised; i.e., one of the species participating in the electrode reaction is present at extremely low concentration. Under such conditions, the method of constant-current potentiometry is quite useful. In this technique, a small constant current is passed through the indicator electrode and its potential is measured throughout the titration.

This technique was first applied by Willard and Fenwick,[21] and 29 years later by Reilley et al.,[22] who applied it to titrations with ceric sulfate and interpreted the results using current–potential curves. Gauguin[23] and Gauguin and coworkers[24] obtained mathematical expressions for various titrations, while Kolthoff reviewed the relationships between amperometric titrations and potentiometry.[25]

Consider the titration of Red_1 with Ox_2 according to

$$\text{Red}_1 + \text{Ox}_2 \rightleftharpoons \text{Ox}_1 + \text{Red}_2 \tag{3}$$

For simplicity, it is assumed that the diffusion coefficients of all species are the same,

[20] G. Gran, *Acta Chim. Scand.* **4**, 559 (1950); see also *Analyst*, **77**, 661 (1952), for another method proposed by Gran.

[21] H. H. Willard and F. Fenwick, *J. Am. Chem. Soc.* **44**, 2504 (1922).

[22] C. N. Reilley, W. D. Cooke, and N. H. Furman, *Anal. Chem.* **23**, 1223 (1951); *ibid.* **23** 1226 (1951).

[23] R. Gauguin, *Anal. Chim. Acta* **7**, 172 (1952).

[24] K. Gauguin, G. Charlot, and J. Coursier, *Anal. Chim. Acta* **7**, 184 (1952).

[25] I. M. Kolthoff, *Anal. Chem.* **26**, 1685 (1954).

and that the residual current is negligibly small, although this need not be done. The current–potential curves for this system are shown at various stages in the titration – 0, 50, 99, 100, 101, and 150 per cent titrated in Fig. 52-4. E_1 is the formal potential for the electrode reaction

$$Ox_1 + ne \rightleftharpoons Red_1$$

and E_2^f is the formal potential for the electrode reaction

$$Ox_2 + ne \rightleftharpoons Red_2$$

In Fig. 52-4 the two dashed horizontal lines represent a constant anodic current, i_a, and a constant cathodic current, i_c. The potentiometric titration curves at currents of i_c, 0, and i_a are constructed by plotting the intersection of the appropriate constant-current line with the various current–potential curves versus per cent titrated. The results are shown in Fig. 52-5. It is apparent that the direction of current flow shifts the location of the end point, and it occurs slightly late when a cathodic current is used. The titration error is small provided i_a (or i_c) is small compared to the diffusion current for Ox_1 near the equivalence point.

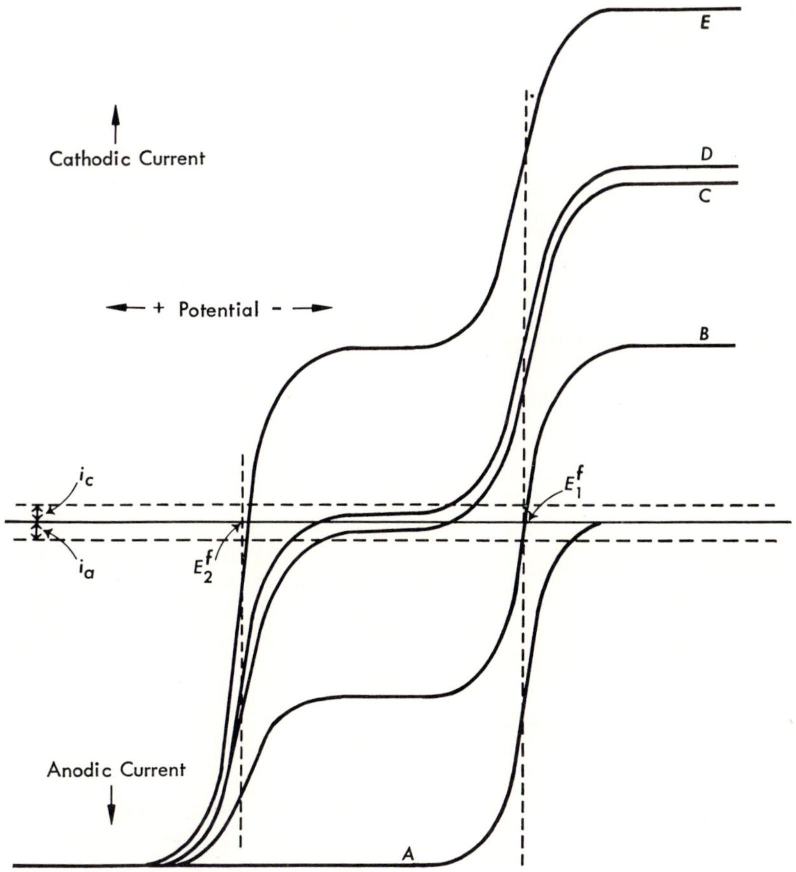

Fig. 52-4. Current–potential curves during the Redox titration $Red_1 + Ox_2 \rightleftarrows Ox_1 + Red_1$. Per cent titrated: Curve *A*—0%; Curve *B*—50%; Curve *C*—99%; Curve *D*—101%; Curve *E*—150%.

Two indicator electrodes

If two identical indicator electrodes are used, and a constant current passed through them, one serves as an anode and the other as a cathode, and the potential difference across them corresponds to the vertical difference between curves *A* and *C* in Fig. 52-5. (*Note:* No reference electrode is used.) This difference is plotted versus per cent titrated in Fig. 52-6 and produces a curve similar in appearance to the first derivative of the potentiometric titration curve at zero current. The maximum occurs at the equivalence point and becomes sharper the smaller the impressed current.

A more detailed discussion and a detailed bibliography on the subject of constant-current potentiometry is given by Charlot et al.[26]

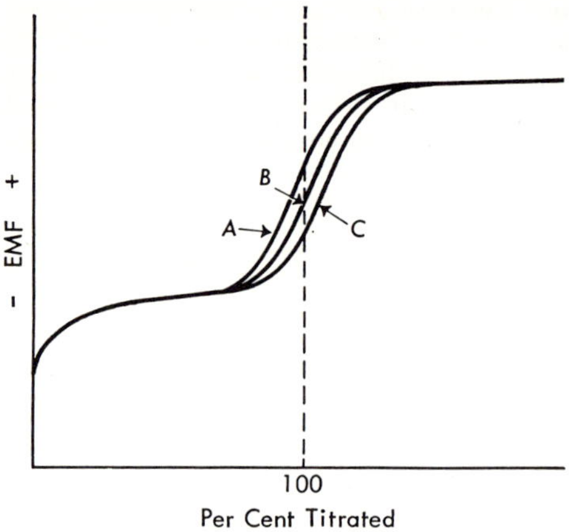

Fig. 52-5. Constant-current potentiometric titration. Curve *A*—small anodic current, Curve *B*—zero current, and Curve *C*—small cathodic current through the indicator electrode.

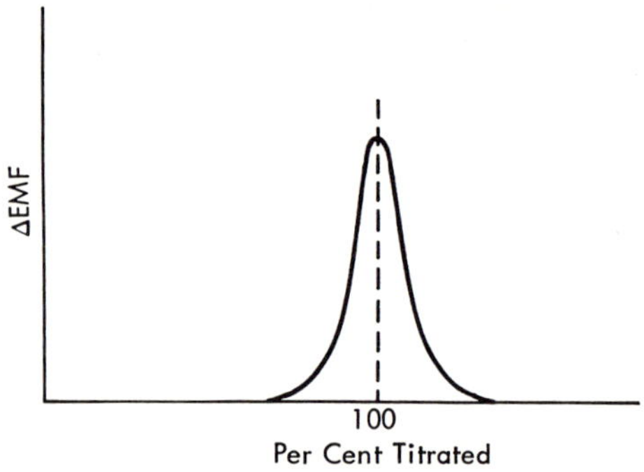

Fig. 52-6. Constant-current potentiometric titration with two indicator electrodes.

[26] G. Charlot, J. Badoz-Lambling, and B. Trémillon, *Electrochemical Reactions*, Elsevier, Amsterdam, 1962, p. 173ff.

CHAPTER 53 *Conductometry and Conductometric Titrations*

Since 1800 chemists have been interested in the resistance of aqueous and nonaqueous solutions. The present distinction between weak, strong, and nonelectrolytes is based upon conductivity experiments dating back to the nineteenth century, and the earliest estimates of the dissociation constants of weak electrolytes are based on conductometric measurements. Many distinguished chemists have made experimental and theoretical contributions to conductometry and associated solution chemistry, laying the foundations of much of present-day solution physical chemistry. This group includes Hittorf, Nernst, Kohlrausch, Ostwald, Walden, Debye, Onsager, Hückel, Falkenhagen, Jones, Shedlovsky, Kraus, and Fuoss. This chapter deals with the elements of conductometry and conductometric titrations, techniques that are sometimes overlooked in modern times of more sophisticated instrumentation.

Conductance laws in solution

The electrical conductance of a solution depends upon the kind and concentration of ions present, and is not a specific property of a particular ionic species. The *specific conductance* κ may be defined as the conductance across the two opposite faces of a cube of solution, 1 cm on an edge. If 1 gram mole of a completely dissociated electrolyte is dissolved in the 1-cm^3 volume of the cube, the conductance corresponds to the *molar conductance* μ, while if 1 gram equivalent is dissolved, the conductance is called the *equivalent conductance*, Λ, of the electrolyte. When the volume of solute containing 1 mole of electrolyte is V cm^3, we have

$$\kappa = \frac{\mu}{V} \tag{1}$$

If C is the molar concentration of the electrolyte (we neglect the difference between ml and cm^3),

$$\kappa = \frac{\mu C}{1000} \tag{2}$$

If the electrolyte has the empirical formula $M_x^{m+}X_m^{x-}$, and we define the normality N of the solute as the total concentration of positive (or negative) charge produced on dissociation, we see that

$$N = mxC \tag{3}$$

Substituting Eq. (3) into (2),

$$\kappa = \frac{\mu N}{1000mx} = \frac{\Lambda N}{1000} \tag{4}$$

where $\Lambda = \mu/mx$.

Since electrical current is carried both by cations and anions in solution, the equivalent conductance of the salt $M_x^{m+}X_m^{x-}$ is the sum of the equivalent ionic conductances of the cation, $\lambda_{M^{m+}}$, and the anion, $\lambda_{X^{x-}}$,

$$\Lambda_{M_x^{m+}X_m^{x-}} = \lambda_{M^{m+}} + \lambda_{X^{x-}} \tag{5}$$

and, substituting Eq. (5) into (4), yields

$$\kappa = \frac{N}{1000}(\lambda_{M^{m+}} + \lambda_{X^{x-}}) \tag{6}$$

Laboratory conductance cells used do not have the cubical geometry discussed above, because this geometry is inconvenient to build (see Fig. 60-7, p. 1045). It is conventional to calibrate experimental conductance cells using solutions of known specific conductance and calculate correction factors by which the observed conductance in a particular cell must be multiplied to obtain the specific conductance. By definition, the *cell constant* θ is given by

$$\theta = \frac{\kappa}{1/R} \tag{7}$$

where κ is the specific conductance of the standard substance (as determined in a cell of ideal geometry) and R is the experimentally observed resistance. The specific conductance of some potassium chloride solutions suitable for calibrating conductometric cells is given in Table 53-1.

TABLE **53-1. Specific Conductance of Potassium Chloride Solutions at 25°C**[a]

Grams KCl/1000 g *of solution*	*Specific Conductivity,* Ω^{-1}/cm
0.74526	0.0012205
7.4191	0.012856
71.135	0.11134

[a] From G. Jones and B. C. Bradshaw, *J. Am. Chem. Soc.* **55**, 1780 (1933). Data at 0 and 18°C are also given.

In a solution containing a mixture of different electrolytes, the specific conductance is proportional to the sum over all ions, of the equivalent conductance of an ion multiplied by its normality, i.e.,

$$\kappa = \sum \frac{N_i \lambda_i}{1000} \tag{8}$$

Comparing Eqs. (7) and (8) we see that the conductance of a solution in any conductance cell is proportional to $\sum N_i\lambda_i$.

The transference number of an ion t_i is defined by

$$t_i = \frac{\lambda_i N_i}{\sum \lambda_i N_i}$$

and represents the fraction of the total current passing through the solution which is carried by a particular ion. Adding a large excess of one electrolyte to a solution containing a small amount of a second electrolyte decreases the transference number of the ions of the second. This is the rationale for the use of a supporting electrolyte in voltammetry.

The equivalent conductance of an electrolyte depends upon the ionic strength of a solution in a predictable manner, and tends toward the *equivalent conductance at infinite dilution* Λ^0 (or λ_+^0 or λ_-^0 for ions) as the solution is diluted more and more.[1] Table A-14 lists some equivalent ionic conductances at infinite dilution. λ^0 values in water for most ions lie in the range 30–80 ohm^{-1}/cm equivalent. Two notable exceptions are the solvated proton, H_3O^+, and hydroxyl ion, OH^-, for which $\lambda_{H^+} = 350$ and $\lambda_{OH^-} = 198$. The exceptionally high mobilities H^+ and OH^- are related to the special mechanism of transport available to these ions in water and involve the shift of hydrogen bonds in the water structure to produce an H_3O^+ or OH^- ion distant from the original ion, without the necessity of having to move the oxygen atom associated with the ion.

CLASSICAL CONDUCTOMETRIC TITRATIONS

Because of the effect of ionic strength upon conductance, and its nonspecific nature, conductometry is only rarely capable of being used to measure concentration of a species in solution. However, much use has been made of conductometry in *conductometric titrations*.[2] The principle is identical to that discussed in Chapter 38 in the section dealing with equivalence-point location using linear extrapolation methods. As an example, consider the titration of 10^{-3} *M* hydrochloric acid with concentrated sodium hydroxide. We will assume $\lambda = \lambda^0$ and use $\lambda_{H^+} = 350$, $\lambda_{Cl^-} = 76$, $\lambda_{Na^+} = 50$, and $\lambda_{OH^-} = 198$. According to Eq. (8), at the start of the titration when $N_{H^+} = N_{Cl^-} = 10^{-3}$,

$$\kappa = \frac{10^{-3} \times 350 + 10^{-3} \times 76}{1000} = 426 \times 10^{-6}$$

At the equivalence point, the solution consists of 10^{-3} *M* sodium chloride and

$$\kappa = \frac{10^{-3} \times 50 + 10^{-3} \times 76}{1000} = 126 \times 10^{-6}$$

After a 100 per cent excess of sodium hydroxide has been added,

$$\kappa = \frac{N_{Na^+}\lambda_{Na^+} + N_{Cl^-}\lambda_{Cl^-} + N_{OH^-}\lambda_{OH^-}}{1000}$$

$$= \frac{(2 \times 10^{-3})(50) + (10^{-3})(76) + (10^{-3})(198)}{1000}$$

$$= 374 \times 10^{-6}$$

and the titration curve has the shape shown in Fig. 53-1.

Conductometric titration curves for a precipitation reaction and a weak acid–strong base titration are shown in Figs. 53-2 and 53-3. It is left as an exercise for the student to derive the shape of these curves (see Problems 2 and 3).

A summary of conductometric titrations which have been used to determine various anions and cations is given by Kolthoff and Laitinen.[3]

[1] According to the Onsager equation, $\Lambda = \Lambda^\circ - B\mu^{1/2}$ (μ = ionic strength), where B is a constant depending upon the solvent, electrolyte, and temperature. See Refs. B1 and B2, page 884, for further details. Strictly, Eq. (5) is valid only at infinite dilution and in the form $\Lambda^\circ = \lambda_+^0 + \lambda_-^0$ is an exact statement of the *law of independent migration* of ions originally formulated by Kohlrausch in 1876.

[2] F. W. Küster and M. Grüter, *Z. Anorg. Chem.* **35**, 454 (1903), were the first to use conductometry to locate the equivalence point of a titration.

[3] I. M. Kolthoff and H. A. Laitinen, *pH and Electrotitrations*, Wiley, New York, 1941, pp. 134–135.

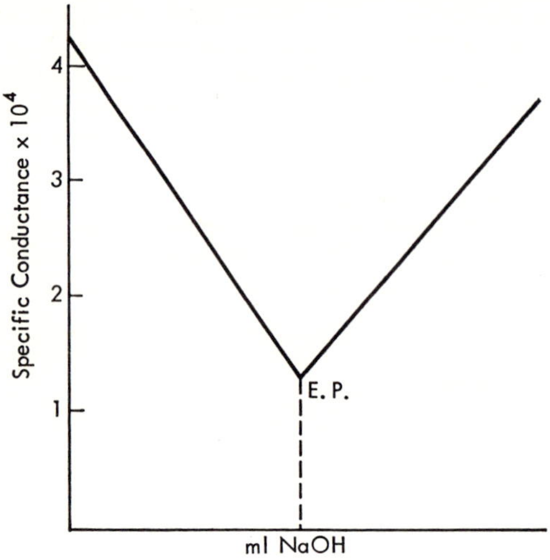

Fig. 53-1. Conductometric titration of 10^{-3} *M* hydrochloric acid with sodium hydroxide (no volume change).

The conductance of electrolyte solutions is a marked function of the solution temperature, decreasing approximately 2 per cent per °C. For most conductometric titrations it is sufficient to maintain the temperature constant to within 1 degree.

Conductometric titrations become less accurate as more and more foreign electrolyte is added to the solution. The accuracy of a titration is closely related to the relative change of conductance which occurs during the titration. By using more sensitive and precise methods of measuring conductance, the method can be applied in the presence of relatively large amounts of indifferent electrolytes. This is accomplished at the cost of simplicity in the method, and other methods, such as amperometric titrations or potentiometric titrations, should be given careful consideration under such circumstances.

Additional details concerning conductometry and conductometric titrations can be found in the references listed on page 885.

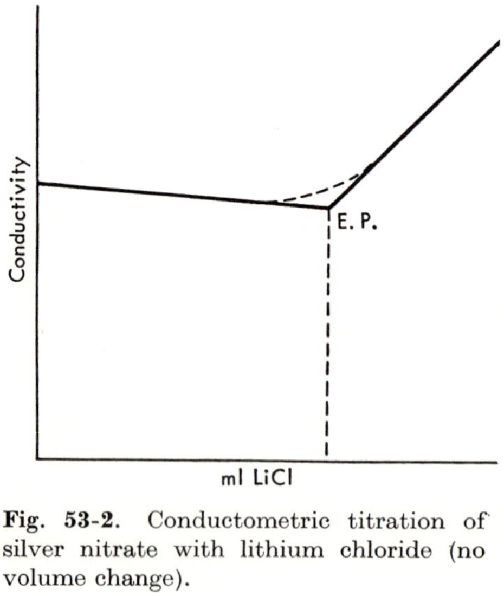

Fig. 53-2. Conductometric titration of silver nitrate with lithium chloride (no volume change).

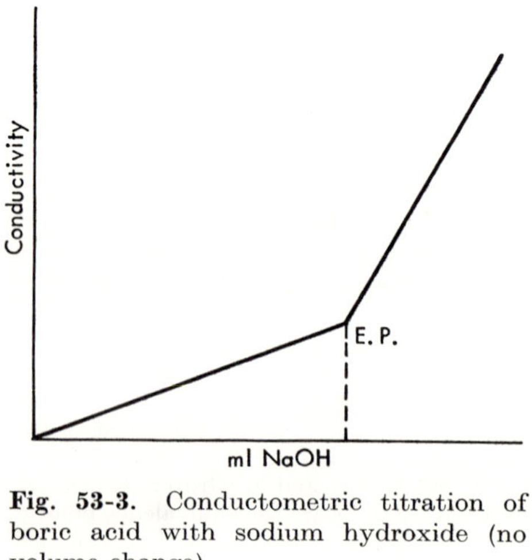

Fig. 53-3. Conductometric titration of boric acid with sodium hydroxide (no volume change).

HIGH-FREQUENCY METHODS

Classical conductometry makes use of relatively low frequency alternating current (10^2–10^4 Hz) in determining the conductivity of an electrolyte solution. Any real conductometric cell behaves as if there are capacitive elements associated with it, and the experimental measurement of conductance takes this into account (p. 1043). The current flowing through the cell, because of its properties as a capacitor, is governed by a *susceptance* term, while the current flowing because of the mobility of ions in an electric field is described by the conductance.

Electrodeless conductance measurements take advantage of the fact that at high frequencies (> 1 MHz), current will flow through a cell even if the electrodes are separated from the solution by a glass wall. The glass wall acts as a capacitance, which couples the solution to the source of alternating current. An explanation of the principle of "conductance" measurement using an electrodeless cell is beyond the scope of this book (see Ref. F6, p. 885). The quantity measured is a function of the capacitance arising from the container and the solution, and the low-frequency conductance of the solution. It is not surprising, in view of the complexity of the situation, that the shapes of high-frequency titration curves obtained in electrodeless cells differ markedly from low-frequency curves and depend upon the frequency used. An example of the titration of chloride ion with silver is shown in Fig. 53-4.

High-frequency conductometry is especially valuable in those situations where the introduction of metallic electrodes is impossible or difficult.

Further information is to be found in References F6, F7, and F8, page 885.

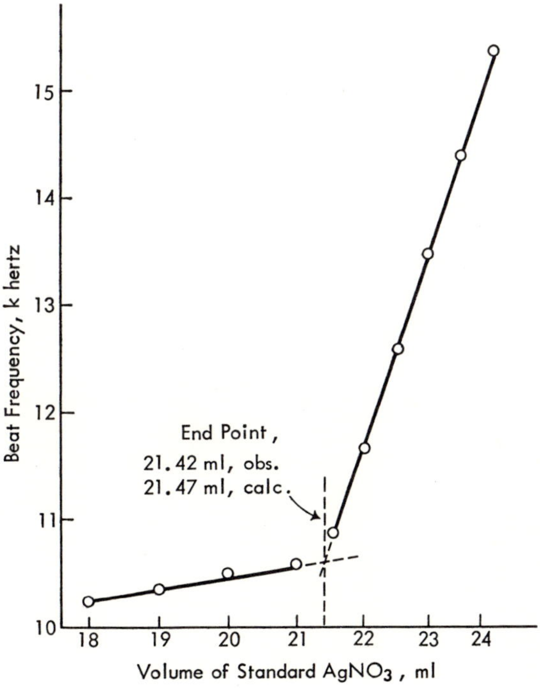

Fig. 53-4. High-frequency titration of sodium chloride with silver nitrate, using 5 mc titrimeter. 25 ml of 0.01 *N* sodium chloride titrated in initial volume of 125 ml with 0.01163 *N* silver nitrate. [From W. J. Blaedel and H. V. Malmstadt, *Anal. Chem.* **22**, 734 (1950).]

PROBLEMS

Neglect volume change in these problems. Use the data listed in Table A-14 and assume $\lambda = \lambda^0$, and assume constant volume during titration.

1. Assuming that all reactions are quantitative, plot the conductometric titration curves for the titration of (a) 10^{-3} *N* HCl with (1) sodium hydroxide and (2) potassium hydroxide; (b) 10^{-3} *N* calcium nitrate with (1) potassium oxalate and (2) lithium oxalate.
2. Calculate and plot the conductometric titration curve of silver nitrate with lithium chloride, taking into account the solubility of AgCl, $K_{sp} = 10^{-10}$, when (a) $C_{Ag^+} = 10^{-2}$ *N*; (b) $C_{Ag^+} = 10^{-3}$ *N*, at 0, 20, 40, 60, 80, 100, 120, 140, 160, 180, and 200 per cent of the equivalent amount of chloride.
3. Calculate and plot the conductometric titration curve of 0.1 *N* weak acid, $K_{HB} = 5.9 \times 10^{-10}$, with sodium hydroxide at 0, 20, 40, 60, 80, 100, 120, 140, 160, 180, and 200 per cent of the equivalent amount of hydroxide. $\lambda_{B^-} = 50$.
4. Repeat Problem 3 for 0.1 *N* weak acid, $K_{HA} = 10^{-5}$, $\lambda_{A^-} = 40$.
5. Repeat Problem 3 for 10^{-4} *N* weak acid, $K_{HA} = 10^{-5}$, $\lambda_{A^-} = 40$.
6. (a) A solution is 0.001 *N* in KIO_3 and 0.1 *N* in KI. Construct the titration curve with HCl. $\lambda^0 = 40.5$ for IO_3^-.

$$IO_3^- + 5\,I^- + 6\,H^+ \rightarrow 3\,I_2 + 3\,H_2O$$

   (b) A solution is 0.002 *N* in HCl and 0.01 *N* in KI. Construct the titration curve with KIO_3.
7. Compare the conductometric and potentiometric titration curves of 0.01 *M* solution of a weak acid, HA, $K_{HA} = 10^{-7}$, with sodium hydroxide. $\lambda_{A^-} = 40$. Plot equivalent conductance (or pH) after addition of 10, 50, 70, 80, 90, 95, 99, 100, 101, 105, 110, 120, and 150 per cent of an equivalent amount of added base.

PART SEVEN

Theory of Optical Methods of Analysis

CHAPTER 54 *Introduction to Optical Methods of Analysis*

Chapters 54–58 present the principles of various optical methods of analysis. The list of methods discussed is not exhaustive, but it includes those of the greatest analytical significance. The present chapter serves as a general introduction. Practical procedures based upon these principles are given in Chapter 61.

Optical methods may be divided into two broad groups: methods in which an external source of radiation is used, which is incident upon the sample, and methods in which the sample itself emits or is caused to emit radiation. In various processes of the first group the incident radiation may be absorbed (spectrophotometric analysis) or may be scattered with or without a change in wavelength (Raman scattering, fluorescence, nephelometry, turbidimetry). If neither absorption nor scattering occurs, there may be a change in the direction of the radiation due to refraction or a change in some property of the radiation such as its state of polarization. Measurement of any of these effects in suitable cases may permit a quantitative determination of one or more constituents, as explained below. In processes of the second group the wavelength or intensity of the emitted radiation may be measured to provide a qualitative or quantitative determination of one or more constituents.

Electromagnetic radiation

Since all optical methods involve electromagnetic radiation, a brief description of such radiation is given first. Light, or electromagnetic radiation in general, may be described as a wave motion characterized by electric and magnetic displacements at right angles to the direction of propagation. As far as the optical phenomena to be described are concerned, only the electric displacement needs to be considered. We consider ideal monochromatic radiation first. This corresponds to a wave in which the displacement varies with time according to a simple sine or cosine dependence; the radiation corresponds to propagation of a wave of a single frequency. The frequency ν is directly proportional to energy E:

$$E = h\nu \tag{1}$$

where h is Planck's constant, 6.625×10^{-27} erg-sec/molecule. Frequency, therefore, is independent of the medium in which the radiation is propagated. The wavelength λ, velocity v, and ν are related:

$$v = \lambda\nu \tag{2}$$

When the radiation is propagated in vacuo, λ and v are denoted as λ_v and c, respectively; $c = 2.997930 \times 10^{10}$ cm/sec. Both λ and v depend upon the medium in which the radiation is propagated. If n denotes the index of refraction of the medium with respect to vacuum,

$$\begin{cases} v = \dfrac{c}{n} \\ \lambda = \dfrac{\lambda_v}{n} \end{cases} \tag{3}$$

For all transparent substances n is greater than unity and depends upon the wavelength. By international convention wavelength values are those in standard dry air at 15°C. The values are smaller than λ_v by about 0.029 per cent, because n of air (in the visible region) at 1 atm pressure and 15°C is 1.00029. Thus the wavelength of a certain sodium line which is 5895.92 Å in air corresponds to 5894.21 Å in vacuo.

Energy may be expressed also in terms of calories per mole or electron volts. These two quantities are equal to $9.5376 \times 10^{-11}\nu$ and $4.1354 \times 10^{-15}\nu$, respectively.

Energy E must not be confused with intensity, which is proportional to the square of the amplitude of vibration. When radiation traverses a hypothetical perfectly transparent medium, there is no decrease in intensity; in all real substances intensity decreases with increasing distance as a result of absorption or scattering or both.

Monochromatic radiation is characterized also by its state of polarization. In natural (unpolarized) radiation the direction of the electric vector around the axis of propagation is completely random. Plane-polarized radiation consists of waves in which the vibrations are confined to a single plane. Ordinary unpolarized radiation can be resolved into two mutually perpendicular plane polarized beams of equal intensity.

The values of λ_v and ν cover an enormous range. Figure 54-1 represents the portion from $\lambda_v = 1$ to 10^{-8} cm, which includes the region associated with "optical" methods, from microwaves to X rays. Radiation of still larger and smaller λ_v also exists. "Light," which is the visible portion of the spectrum, covers only one octave, the range of λ_v being about 3.8×10^{-5} cm (violet) to 7.6×10^{-5} cm (red). These limits are not sharply defined. The corresponding range of ν is from 7.9×10^{14} sec^{-1} to 3.95×10^{14} sec^{-1}. In the visible region λ is usually expressed in angstrom units (Å), 10^{-8} cm, or in millimicrons (mμ), 10^{-7} cm.[1] Thus the wavelength of green light, $\lambda = 5 \times 10^{-5}$ cm, is 5000 Å or 500 mμ. In the infrared region the common unit is the micron (μ), 10^{-4} cm.

Monochromatic radiation does not exist. All sources emit radiation that is more or less heterochromatic. Some sources, for example those used in atomic absorption spectroscopy (p. 1002), emit a few spectral lines that are close to being monochromatic. Sources such as tungsten lamps, hydrogen discharge tubes, and carbon arcs emit almost continuous radiation over relatively wide ranges of frequencies. By the use of filters, more or less narrow spectral regions may be isolated from a continuous spectrum. In most instances the radiation thus obtained is far from

[1] The symbol nm for nanometer, 10^{-9} meter, is becoming increasingly used in place of mμ.

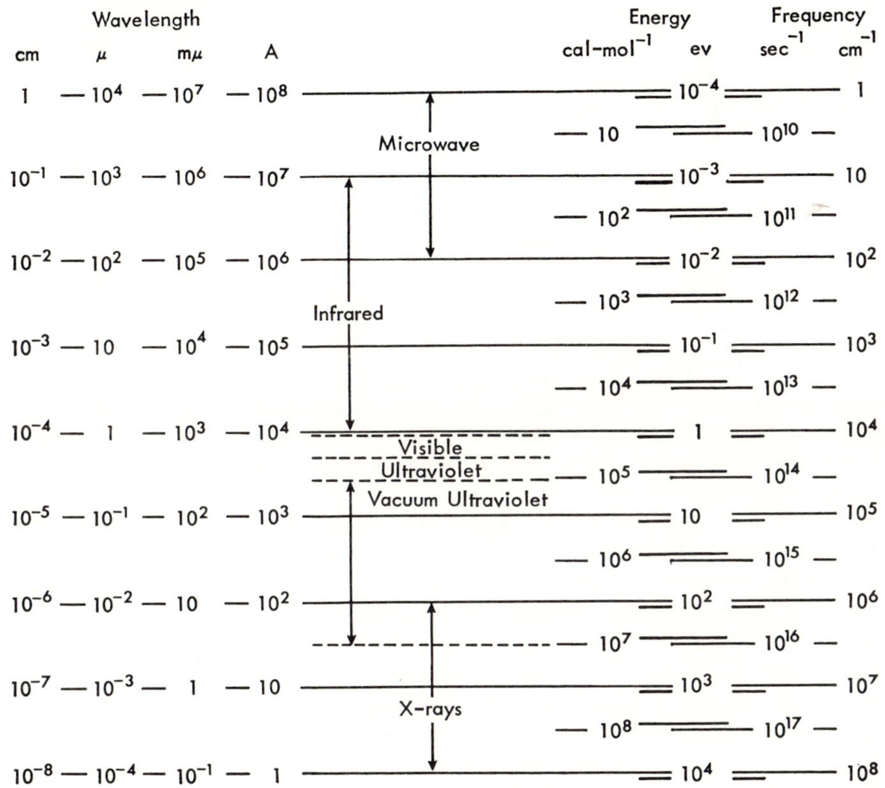

Fig. 54-1. Wavelength, energy, and frequency in the optical part of the electromagnetic spectrum.

monochromatic. For example, a standard "projection blue" filter, Wratten number 47, has a maximum transmission of 51 per cent at 440 mμ; the transmission drops to 10 per cent at 406 and 500 mμ. Interference filters may have much narrower transmission bands. Filters are useful in simple colorimeters or "filter photometers" (p. 975). However, in the majority of practical applications, approximately monochromatic radiation of the desired frequency is obtained by the use of a continuous source and a monochromator.

Monochromators. The essential elements are an entrance slit for the heterochromatic radiation from the source, a lens or mirror to make the radiation parallel, a dispersing element (one or more prisms or gratings or both), a lens or mirror to focus the dispersed radiation, and an exit slit through which a more or less narrow spectral band passes. Two or more of these functions may be performed by one element. Figure 54-2 is a schematic diagram of a monochromator using a prism and a mirror; the latter serves both to collimate (i.e., make parallel) the incident radiation and to focus the dispersed radiation. Mirrors are preferred to lenses because of the freedom of the former from chromatic aberrations. The wavelength passing through the exit slit is changed by rotation of the prism.

Spectrophotometers. A spectrophotometer is the combination of a mono-

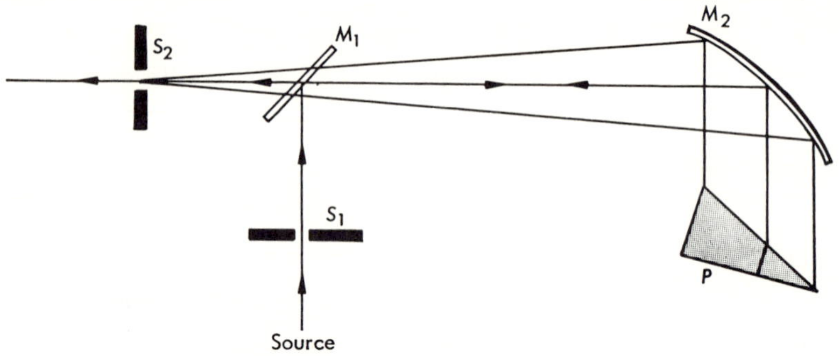

Fig. 54-2. Schematic diagram of a single-beam prism monochromator. S_1 and S_2, entrance and exit slits; M_1, plane mirror; M_2, parabolic mirror; P_1, prism with reflecting surface. The path of dispersed radiation is above or below M_1. Only one dispersed ray is shown. Approximately monochromatic radiation passes through S_2.

chromator with a photometer, the latter being used to measure the intensity of the monochromatic radiation. It is placed to the left of the exit slit in Fig. 54-2. In all spectrophotometric measurements the intensity transmitted by the sample is compared with the intensity transmitted by a reference; hence relative rather than absolute measurements are made. In a single-beam instrument (Fig. 54-2) the reference and the sample are placed successively in the monochromatic beam. Denoting the intensity transmitted by reference and sample as I_0 and I, respectively, the transmittance T is I/I_0. In such instruments it is essential that the intensity of the incident beam remain constant. In a double-beam instrument the monochromatic beam is divided by optical means into two beams of equal intensity. One beam passes through the reference and the other through the sample, and the value of T is measured directly. In such instruments minor fluctuations in the incident intensity are of no consequence.

It is beyond the scope of this text to describe even a small fraction of the optical and instrumental problems that are involved in optical methods of analysis, and reference is made to specialized works for details. The following will be found useful. A classical book on the problems of energy measurement is that of Forsythe.[2] The books by Baly,[3] Harrison et al,[4] and Sawyer[5] are especially useful in regard to the construction and use of spectroscopic instruments. Thorough treatments of absorption spectrophotometry are given by Bauman,[6] Kortüm,[7] Lothian,[8] and Mellon.[9] A discussion of the general principles[10] and practical

[2] W. E. Forsythe (ed.), *Measurement of Radiant Energy*, McGraw-Hill, New York, 1937.

[3] E. C. C. Baly, *Spectroscopy*, vol. I, Longmans, London, 1924.

[4] G. R. Harrison, R. C. Lord, and J. R. Loofbourow, *Practical Spectroscopy*, Prentice-Hall, Englewood Cliffs, N.J., 1948.

[5] R. A. Sawyer, *Experimental Spectroscopy*, Prentice-Hall, Englewood Cliffs, N.J., 1944.

[6] R. P. Bauman, *Absorption Spectroscopy*, Wiley, New York, 1962.

[7] G. Kortüm, *Kolorimetrie, Photometrie und Spektrometrie*, Springer, Berlin, 1962.

[8] G. F. Lothian, *Absorption Spectrophotometry*, Hilger & Watts, London, 1958.

[9] M. G. Mellon (ed.), *Analytical Absorption Spectroscopy*, Wiley, New York, 1950.

[10] E. J. Meehan, *Optical Methods of Analysis*, Wiley-Interscience, New York, 1964; E. J. Meehan in *Treatise on Analytical Chemistry* (I. M. Kolthoff and P. J. Elving, eds.), part I, vol. 5, Wiley-Interscience, New York, 1964, chaps. 53–55.

details[11] are found in a comprehensive treatise. Finally, reference is made to a useful survey of commercially available apparatus.[12]

PROBLEM

Express the mercury radiation, wavelength 4358.35 Å (in air), as the following: microns, millimicrons, sec^{-1}, cm^{-1}, ergs, $molecule^{-1}$, electron volts, and calories per mole. Use $n_{air} = 1.00029$.

[11] A. A. Schilt and B. Jaselkis in *Treatise on Analytical Chemistry*, part I, vol. 5, chap. 58.
[12] S. Z. Lewin, *J. Chem. Educ.* **37**, A197, A271, A341, A401, A455, A507, A797 (1960).

CHAPTER 55 *Absorption Spectrophotometry*

This method of analysis is based upon the fact that different substances absorb radiation of a given frequency to different extents, or, stated in other words, that each substance has a characteristic *absorption spectrum*. The existence of a color is due to absorption in the visible region of the spectrum. Solutions of potassium permanganate, for example, have two *absorption bands* in the visible region, with pronounced maxima at 526 and 545 mμ. The transmittance of a solution at these wavelengths, or at any other wavelength where absorption occurs, is related to the concentration of permanganate (see the next section). Potassium dichromate has a much different absorption spectrum. By measuring transmittance at two suitable wavelengths, it is possible to determine permanganate and dichromate in a mixture of the two substances. (For theory, see p. 976; for procedure, see p. 1052.)

Spectrophotometric analysis is not restricted to colored substances such as permanganate and dichromate. Almost all substances (with the exceptions mentioned in the next paragraph) absorb somewhere in the ultraviolet, and modern ultraviolet spectrophotometers are used routinely for measurements down to 200 mμ or somewhat lower wavelengths. Benzene and toluene, for example, are easily distinguished by their UV spectra. A similar remark applies to nitrite and nitrate.[1]

The relatively few liquids that do not absorb in the UV may serve as solvents. Water is transparent to below 200 mμ and is used for most inorganic substances. Saturated aliphatic compounds such as heptane and isooctane also are transparent to below 200 mμ, as are alcohols such as methanol and ethanol. Each of these solvents must be especially pure, because even small amounts of most unsaturates or aromatic compounds make them useless for measurements at the shorter wavelengths. Carbon tetrachloride, benzene, and acetone, among other common solvents, are useful down to about 270, 290, and 340 mμ, respectively.

Measurements in the infrared region from 2 to 15 μ, particularly above about 7 μ, are extremely useful for identification of organic compounds. This is because the absorption spectra in this region are so complex and full of detail that the spectrum serves as a remarkably characteristic property ("fingerprint") of a pure organic compound. A few examples are given in Figs. 55-1 to 55-4. In fact, comparison of the IR spectrum of a substance with that of an authentic pure sample provides one of the easiest tests of its purity, although the sensitivity for impurities is not great. A discussion of the use in qualitative analysis in general, and a specific

[1] A useful summary of UV absorption spectra of many inorganic compounds is given by R. P. Buck, S. Singhadeja, and L. B. Rogers, *Anal. Chem.* **26**, 1240 (1954).

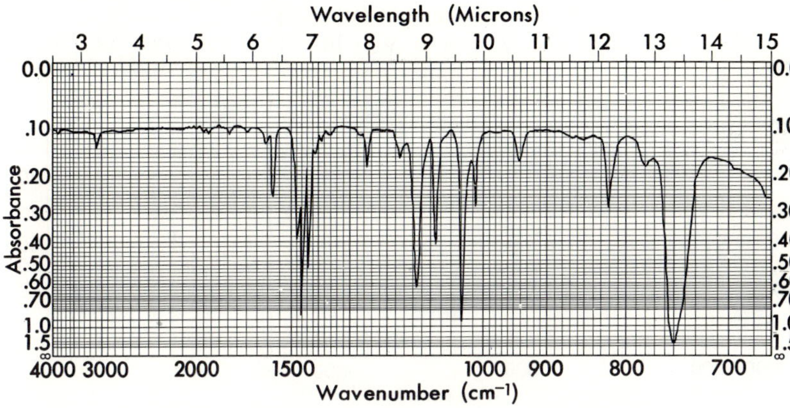

Fig. 55-1. Infrared spectrum of o-dichlorobenzene.

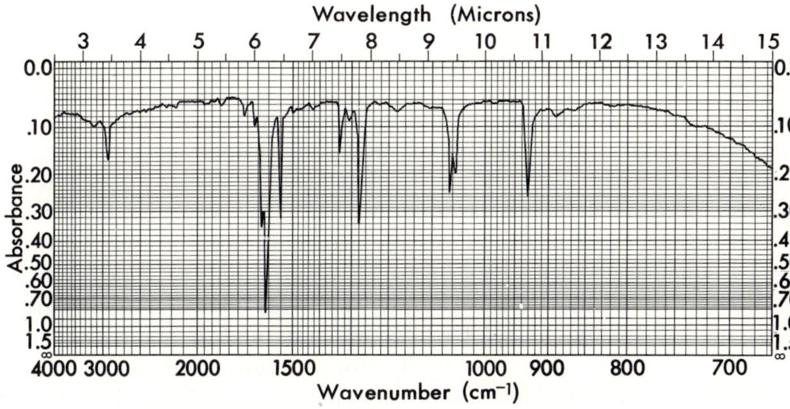

Fig. 55-2. Infrared spectrum of benzoquinone (melt).

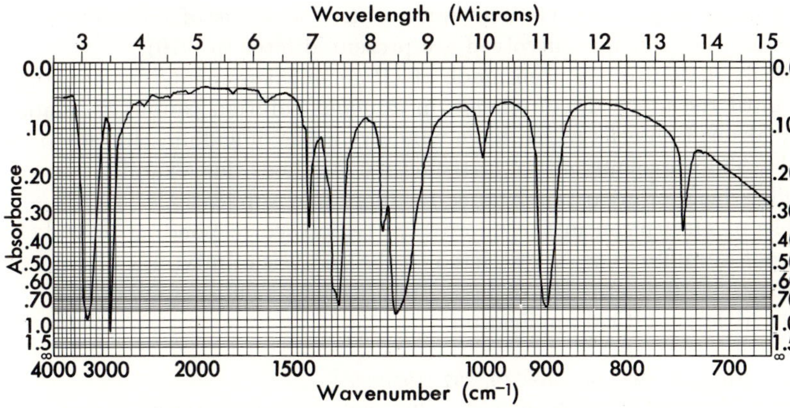

Fig. 55-3. Infrared spectrum of t-butyl alcohol.

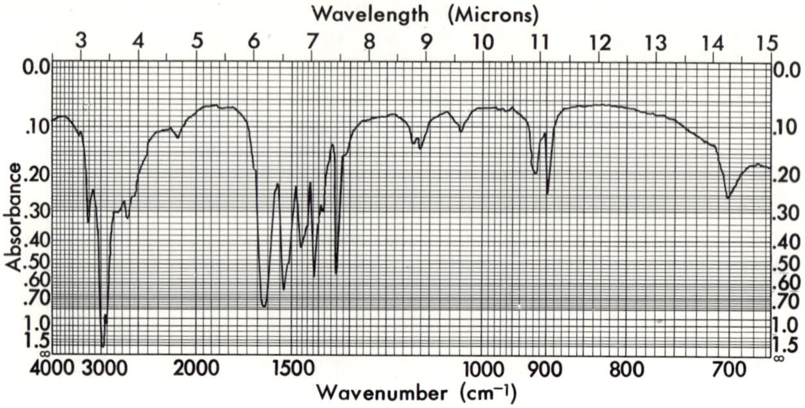

Fig. 55-4. Infrared spectrum of glycine.

example of the use of IR to solve a simple problem in organic qualitative analysis, are given in Chapter 61.

Aside from proof of identity of specific compounds IR is indispensable in two other respects.

1. In a mixture of compounds, it may or may not be possible to identify the individual components. (The use of vapor-phase chromatography prior to IR measurements is a very great aid in solving such problems.) Even when the individual components may not be identified, it is usually possible to identify functional groups, such as carbonyl, amine, etc. This is due to the fact that the vibrational frequencies which give rise to IR absorption are approximately the same in different molecules. For example, a C—H vibration occurs at about 2840–2970 cm^{-1} (3.4–3.5 μ) in aliphatic compounds and at about 3030 cm^{-1} in aromatic compounds. The C═C and C≡C groups in aliphatic compounds absorb at about 1590–1690 and at about 2100–2270 cm^{-1}, respectively. Based upon the observation of the spectra of thousands of compounds, many correlation charts for absorption frequency have been prepared. One of the best known, which is reproduced in virtually all texts on IR analysis, is that of Colthup.[2]

2. The other important use is in quantitative analysis of mixtures of organic compounds. The principles involved are presented later (p. 976).

Relatively little use has been made of IR for measurements on inorganic compounds until recently. Measurements in aqueous solution cannot be made using ordinary absorption cells which have windows made of sodium chloride or similar materials. The development of satisfactory methods for examination of solid samples has been of great importance for inorganic applications.[3]

The use of absorption in the microwave region, corresponding to wavelengths of the order of millimeters, is restricted to the analysis of mixtures of dipolar gases at pressures of 10^{-3} mm Hg or less. For this reason microwave absorption has relatively little analytical application.

[2] N. B. Colthup, *J. Opt. Soc. Am.* **40,** 397 (1950).

[3] A survey of spectra for a large number of inorganic substances is given by F. A. Miller and C. H. Wilkens, *Anal. Chem.* **23,** 1253 (1952).

Absorption in the X-ray region is due to transitions between K, L, . . . electrons, and the valence electrons are not involved (except for elements of low atomic number). The process therefore is substantially independent of physical and chemical effects such as the nature of the sample (solid or liquid) or chemical bonding. The X-ray absorption thus depends on the elementary composition. Since the absorption coefficient increases rapidly with atomic number, it is especially useful for problems such as the determination of lead in gasoline.

Quantitative aspects of spectrophotometry. The Beer–Lambert law

A solution containing a single invariable absorbing species, the concentration of which is c, is placed in an absorption cell having optically flat and parallel windows for entrance and exit of radiation. The internal path length in the cell is b cm. As is explained below, c is measured in various units, but it is universal practice to measure b in centimeters. An identical absorption cell contains solvent. The following discussion applies both to single-beam and double-beam instruments, in any region of the spectrum.

Consider first the ideal case in which each cell is illuminated with a parallel beam of monochromatic radiation of intensity I'_0 (Fig. 55-5). Upon incidence at point P_1 on the cell containing the solvent, a fraction of the radiation is reflected directly backward. The extent of this reflection depends upon the index of refraction n of the cell window; if n is 1.5 or 1.6, the fraction reflected is 0.04 or 0.14, respectively.[4] Since the windows are optically flat, there is no diffuse reflection. Depending on the transparency of the cell window, there may be some absorption on the path from P_1 to P_2. At the interface between window and solution, P_2, another reflection occurs, the extent of which depends upon the relative values of n of cell window and solution. In almost all practical cases the extent of reflection at P_2 is small; when the solvent is water ($n = 1.33$) and the window is glass ($n = 1.50$), the fraction reflected amounts to about 0.0036. Some absorption by solvent may occur between P_2 and P_3. The path from P_3 through P_4 is exactly equivalent to the path through P_1 to P_2. If f represents the fraction transmitted at or between each of the points $P_1, \ldots, P_4$:

$$\begin{aligned} I_0 &= I'_0 f_1 f_{12} f_2 f_{23} f_3 f_{34} f_4 \\ &= I'_0 f_1^2 f_{12}^2 f_2^2 f_{23} \end{aligned} \tag{1}$$

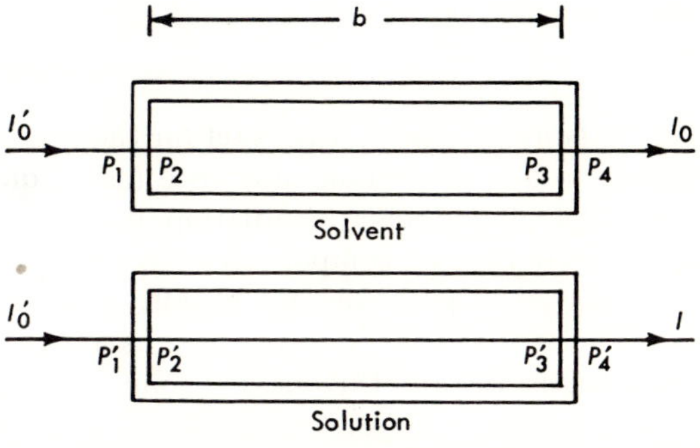

Fig. 55-5. Transmission through reference cell and sample cell.

[4] In general upon perpendicular incidence from a medium of refractive index n_1 to one of index n_2, the fraction reflected is $\{(n_1 - n_2)/(n_1 + n_2)\}^2$; n for air is practically unity.

In the same way,

$$I = I'_0 f'^2_1 f'^2_{12} f'^2_2 f'_{23} \qquad (2)$$

The values of f_1 and f'_1, and of f_{12} and f'_{12}, are exactly equal. If the values of n for solvent and solution are the same, f_2 is exactly equal to f'_2; this situation exists in practically all cases in visible-UV spectrophotometry. (Note, however, that $f_2 \neq f'_2$ if the solution is very concentrated.) When $f_2 = f'_2$, division of Eq. (2) by Eq. (1) yields

$$T = \frac{I}{I_0} = \frac{f'_{23}}{f_{23}} \qquad (3)$$

Therefore, the measured transmittance is the ratio of the transmissions of solution and of solvent, regardless of the magnitude of reflections and of cell transmissions. (In this treatment the effect of multiple reflections has been ignored; see p. 975.)

The quantitative relationship of T to b and to c is derived as follows. The process of absorption (no reemission) can be regarded as an irreversible reaction between the absorbing species and a photon, in which the energy of the photon is converted into thermal energy. Therefore, the "reaction rate," which determines the rate at which photons are being removed from the beam, is proportional to the product of the concentration of absorbing species and of photons; the concentration of the latter evidently is proportional to intensity.

We consider first the relation between T and b. When the parallel monochromatic beam of intensity I traverses the thickness db of absorber, the change in intensity is given by

$$dI = -k'I\,db \qquad (4)$$

since the number of absorbers is exactly proportional to the thickness db but is independent of I. The proportionality constant k' depends upon wavelength for a given absorbing species and upon solvent and concentration in the case of a solution; generally k' depends also upon temperature. For a given substance or a given solution, k' depends only upon wavelength; therefore Eq. (4) can be integrated exactly to yield

$$I = I_0 \exp(-k'b) \qquad (5)$$

which can also be written

$$T = \exp(-k'b) \qquad (6)$$

This is *Lambert's law*.[5] It is exact for the conditions specified in the derivation and applies to any absorbing substance (gas, liquid, solution, or homogeneous solid) in parallel monochromatic radiation (it being understood, of course, that reflections are corrected for by comparison with a suitable reference, as discussed above).

In analytical problems one usually deals with solutions; the relation of T to c may be discussed in the same way as above. Consider a parallel monochromatic beam traversing any thickness of solution of concentration c. As before, c refers to

[5] J. H. Lambert, *Photometria sive de mensura et gradibus luminus, colorum et umbrae*, 1760. The Lambert law is also known as the Bougeur law: P. Bougeur, *Essai d'optique sur la gradation de la lumiere*, 1729.

the concentration of a single absorbing species and not necessarily to the analytical concentration. If c is increased to $c + dc$, the change in transmitted intensity is

$$dI = -k''I\,dc \tag{7}$$

where k'' is proportional to b. This differential equation is exact. If it can be assumed that the number of absorbers is exactly proportional to c, and that the probability of absorption is independent of c, the equation can be integrated to yield

$$I = I_0 \exp(-k''c) \tag{8}$$

or

$$T = \exp(-k''c) \tag{9}$$

in which, for a given solute–solvent, temperature, and path length, k'' depends only on wavelength. This is the *Beer law*.[6] It involves two essential assumptions about the constancy of k''; no such assumptions are involved in the Lambert law.

If both b and c are variable, the combined Beer–Lambert law may be written

$$I = I_0 \exp(-k'''bc) \tag{10}$$

The forms usually used are

$$I = I_0 \times 10^{-abc} \tag{11}$$

$$T = 10^{-abc} \tag{12}$$

where $k''' = 2.303a$. Defining

$$A = \log (I_0/I) \tag{13}$$

we obtain

$$A = abc \tag{14}$$

$$A = \log T^{-1} = -\log T \tag{15}$$

The quantity A is the *absorbance* (formerly called optical density, or extinction, or sometimes absorbancy). The quantity a is the *absorptivity*, when c is measured in grams/liter; when c is measured in moles/liter, the Beer–Lambert law is usually written

$$A = \epsilon bc \tag{16}$$

and ϵ is called the *molar absorptivity*. Evidently ϵ is equal to a times the molecular weight of the absorbing substance. Both a and ϵ were formerly known under a variety of names, including specific or molar extinction coefficient, absorption coefficient, and absorbancy index.

Since the product abc is an exponent, it must be dimensionless. Therefore, the dimensions of a must be the same as the dimensions of $b^{-1}c^{-1}$, equal to cm/1000 cm^3/g, or 1000 cm^2/g. The physical meaning therefore is absorption cross

[6] A. Beer, *Ann. Physik* **86**, 78 (1852); also see F. Bernard, *Ann. Chim. Phys.* [3] **35**, 385 (1852).

section per unit weight. Similarly, the value of ϵ corresponds numerically to 1000 cm^2/mole. In microchemical work this mode of expression or the equivalent often is chosen (see Problem 12).

The values of ϵ cover a wide range. ϵ is practically zero in regions of transparency (i.e., for pure water or octane throughout the visible and ultraviolet). The maximum values of ϵ occur at the peaks of absorption bands, and values larger than 10^3 are common. As an example of the calculation of ϵ from experimental data, consider a solution of potassium chromate in 0.05 N potassium hydroxide. A solution containing 0.0400 g of potassium hydroxide per liter, measured in a cell having $b = 1.000$ cm, at 25°C, has the transmittances given in Table 55-1.[7] The solution serves as a standard for calibration of spectrophotometers.

TABLE 55-1. **Absorption of Solution of Potassium Chromate, 0.0400 g liter, in 0.05 N Potassium Hydroxide, $b = 1.000$ cm, 25°C**

λ, mμ	T	A[a]	a	ϵ[b]
230	0.674	0.1713 ± 0.0006	4.28	8.32×10^2
275	0.175	0.757 ± 0.0025	18.93	3.68×10^3
290	0.373	0.428 ± 0.0012	10.70	2.08×10^3
315	0.900	0.0458 ± 0.0005	1.15	2.22×10^3
370	0.103	0.987 ± 0.004	24.68	4.79×10^3
450	0.927	0.0329 ± 0.0005	0.82	1.60×10^2

[a] The uncertainties in A are calculated assuming a constant uncertainty of ±0.001 in T; see page 978.
[b] $\epsilon = 194.2a$; $a = A/0.04$.

Maximum values of ϵ between 10^3 and 10^4 are typical of many inorganic substances. Some organic substances, particularly dyes, and metal organic complexes have higher values. For example, the maximum ϵ for copper dithizonate is 3.56×10^4.

Assuming for illustration a value of ϵ of 10^4, the product bc must be $0.5\text{–}1 \times 10^{-4}$ to give a value of A in the range 0.5–1. Therefore, for $b = 1$ cm (the most common value), c must be $0.5\text{–}1 \times 10^{-4}$ M. The precision that can be attained in such a measurement is discussed on page 978. The *limit of detection* is very low; assuming that a value of $A = 0.001$ (i.e., $T = 0.998$) can be detected with certainty, the minimum detectable c is $10^{-7}/b$ M. The minimum *amount* detectable depends of course both upon b and upon the cross section of the cell, which determines the volume of solution required. In most spectrophotometers using 1-cm cells, 2–3 ml is required. Many microcells have been devised to use very small volumes of solution. A practical example is illustrated in Problem 13.

[7] G. W. Haupt. *J. Res. Natl. Bur. Std.* **48**, 414 (1952); *J. Opt. Soc. Am.* **42**, 441 (1952). Values at many other wavelengths are given. The preparation and stability of the solution are described in detail. Earlier values are given by K. S. Gibson in *Analytical Absorption Spectroscopy* (G. Mellón, ed.), Wiley, New York, 1950, p. 261. Other substances for which standardization data are given by Gibson are solutions of copper sulfate and of cobalt ammonium sulfate.

Graphical presentation of data

The abscissa usually is either wavelength or frequency. All visible–UV spectrophotometers are graduated either in millimicrons or in angstrom units, and a wavelength scale usually is chosen for presentation of visible–UV data. However, this is not always the case; especially in publications dealing with the interpretation of absorption spectra, rather than with their application, a frequency scale is generally preferred. In chemical applications the frequency unit is usually cm^{-1}; in many physics journals the electron volt is chosen. (*Example*: $\lambda_v = 5000$ Å $= 20{,}000\ cm^{-1} = 2.4795$ eV.) Infrared spectrophotometers may have scales marked in wavelength (μ), in frequency (cm^{-1}), or in both, and both units are commonly used for graphical presentation (see Figs. 55-1 to 55-4). As the ordinate, one of the following choices may be made: transmittance T or per cent transmission, $100T$; absorbance A; or log A. Sometimes plots of T^{-1} or of $1 - T$ are seen. The latter evidently is a mirror image of a plot of T.

Figure 55-6 shows plots of T, A, and log A versus wavelength for three concentrations of a substance which follows Beer's law. A plot of T versus λ generally is most useful for comparing the spectral transmission of filters; evidently such a plot does not show the effect of changes in c on T at either very high or very low T. A plot of A for various values of c emphasizes the effect of c at low values of T. The principal value of plotting log A versus λ is that it shows at a glance whether or not the substance follows Beer's law at all wavelengths; if it does, the log A–λ curves for different values of c are parallel.

Sources of error in spectrophotometric measurements

Several errors may affect the precision. Naturally the precision with which the transmittance or absorbance scale may be read is a factor. In single-beam instruments the stability of the light source obviously is important. (In double-beam instruments, as already mentioned, this is not the case.) The absorption cells must be properly cleaned and must be matched in transmission over the wavelength range in use. Failure to check this is a common source of error. Deposits invisible to the naked eye may affect the cell transmission. As a routine practice, the relative transmission of both cells, both filled with solvent, should be measured before every experiment. Any difference in transmission may be corrected for.

Example. At a certain wavelength, cell 1 transmits 90.6 per cent as much as cell 2, when both are filled with solvent. This corresponds to an absorbance difference of 0.043. Suppose a measurement is made with solvent in cell 1 and solution in cell 2. If the solution has an apparent absorbance of 0.615, the correct absorbance is $0.615 + 0.043 = 0.658$. If the solvent had been placed in cell 2 and solution in cell 1, the apparent absorbance would have been $0.658 + 0.043 = 0.701$.

Such differences in transmissions of cells are by no means uncommon at shorter wavelengths in the UV, and they should always be checked. Also, the values of b should either be the same or should be known accurately. (See page 1060 for a method of determining small differences in b.) Another factor that affects precision is the placement of the cells in the beam in a correct, reproducible manner, exactly at right angles to the beam (Fig. 55-7a). If the beam is not perpendicular to the cell window, the path length within the cell is longer than the correct value (Fig. 55-7b). Also the effect of reflections depends upon this placement (see below). Naturally solution and reference must be at the same temperature.

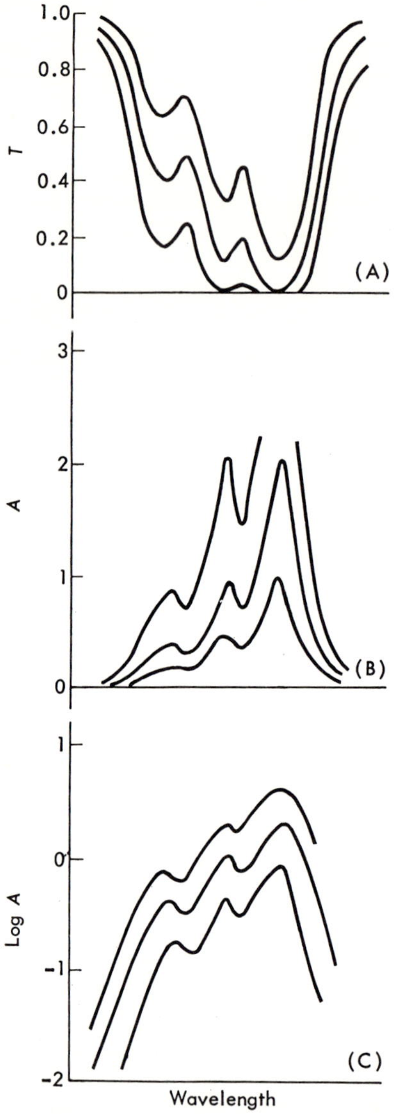

Fig. 55-6. Three ways of plotting spectrophotometric data.

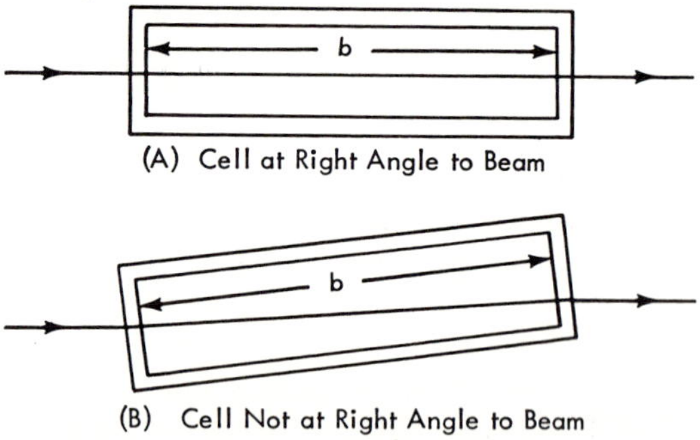

Fig. 55-7. If cell is not at right angle to beam, path length is greater than correct value b.

Various other factors affect the accuracy. First, and usually most important, is the fact that the radiation may not be sufficiently near monochromatic. The magnitude of the effect depends upon the substance being investigated. Suppose two substances X and Y with absorption bands as shown in Fig. 55-8 are measured in the same spectrophotometer. In any spectrophotometer the width of the spectral band depends mainly upon the product of two quantities. One is the dispersion of the instrument at the exit slit (compare Fig. 54-2). This is commonly expressed as wavelength spread per millimeter at the exit slit.

Example. An instrument in which the dispersing element is a diffraction grating has a practically constant dispersion. If the visible spectrum (380–760 mμ) is spread at uniform dispersion over 38 mm, the dispersion is 380/38 = 10 mμ/mm or 100 Å/mm. (Note that a *large* dispersion corresponds to a *small* numerical value.)

The dispersion of prism instruments depends very markedly upon the wavelength; for a quartz prism, the dispersion is about 40 times as large at 200 mμ as it is at 589 mμ. The other quantity determining spectral band width is the width of the exit slit (S_2 in Fig. 54-2).[8] If the width of the exit slit with the grating referred to above is 0.5 mm, the emergent beam has a band width of 5 mμ or 50 Å.

In an internationally used prism spectrophotometer, the dispersion at the exit slit varies from about 1 mμ/mm at 200 mμ, 10 mμ/mm at 400 mμ, up to 50 mμ/mm at 700 mμ. The actual slit widths required in measurement depend on the transparency of the cells and solvents, and thus may vary widely at the low end of the UV. Through the visible region it is usually possible to use slit widths of 0.01 mm (or even less); a width of 0.01 mm corresponds to a spectral band width of 0.1 mμ (or 1 Å) at 400 mμ. A popular grating spectrophotometer operates at a fixed band width of 0.5 mμ (5 Å) over the entire spectral range. These values illustrate the order of magnitudes of band widths used in present instruments. Thus from a knowledge of the dispersion and the slit width, the actual band spread of the radiation can be calculated. The crosshatched area in Fig. 55-8 represents the

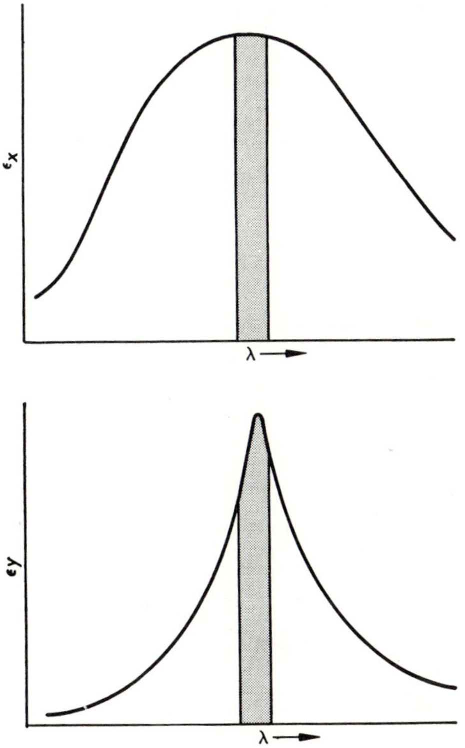

Fig. 55-8. Comparison of spectral band width (crosshatched area) to absorption band.

[8] Other factors are involved as well, which are omitted here for simplicity. See Ref. 2.

spectral slit width calculated in this way. It is evident that so far as measurements on X (with a broad absorption band) are concerned, the radiation is effectively monochromatic. In the case of Y (with a narrow absorption band) the value of ϵ varies markedly within the spectral band used for the measurement. The important result is this: In the situation illustrated in Fig. 55-8, *X follows Beer's law, Y does not.* It is emphasized that this is not a real deviation from the law but only an *apparent* one due to the use of heterochromatic radiation. The A–c curves shown in Fig. 55-9 illustrate the effect, which is quite general. As c_x is increased, A_x increases in proportion. As c_y is increased, the A–c relation is not linear; the curve always bends over to the horizontal axis as shown.[9]

Figure 55-8 also shows that to obtain precision (entirely aside from accuracy) it is essential that measurements be made at the same value of exit slit width, when the width of the slit is a variable, as it is in most instruments. Many workers ignore this effect.

Exactly the same effect upon A–c curves is shown when instruments with appreciable amounts of *stray radiation* are used. This is radiation of other than the desired wavelength passing through the exit slit. It arises in general from reflections within the monochromator, and the amount is critically dependent upon the optical design. In modern double monochromators (two dispersing elements in series) the amount of stray radiation is very small. Several instruments now available have less than 0.01 per cent stray radiation at any wavelength within their range (see Problem 15).

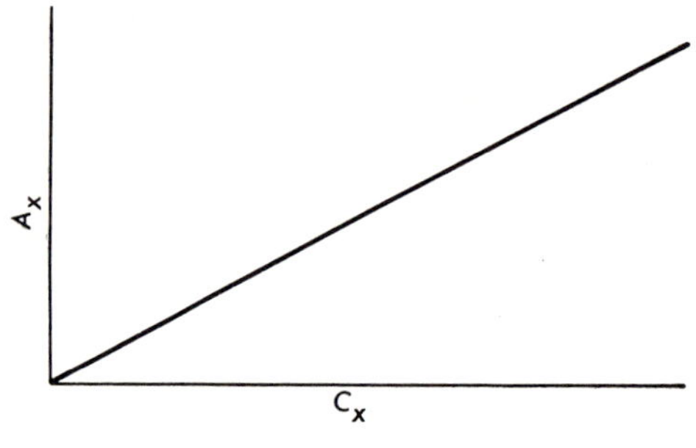

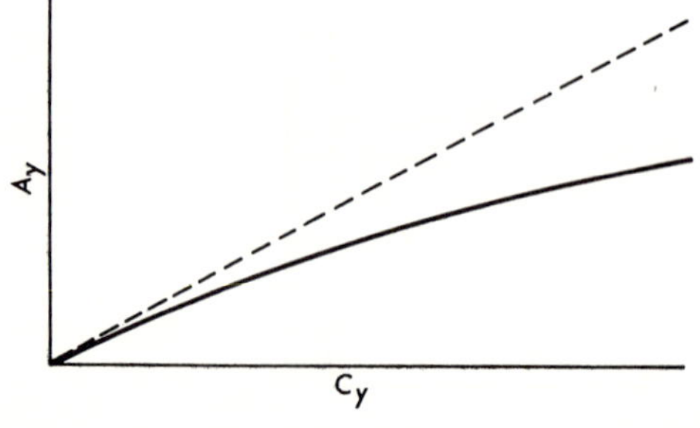

Fig. 55-9. Effect of spectral band width (compare Fig. 55-8) on A-C relation.

[9] For additional details see Ref. 10, chap. 54.

The use of nonparallel radiation also causes apparent deviation from Beer's law, but the effects are relatively minor; they become most pronounced at high absorbances (two or more).

Although the majority of measurements are made using absorption cells with flat entrance and exit windows, it is sometimes necessary or expedient to use cylindrical cuvettes (test tubes). Beer's law *cannot be followed* in such cells. The deviation is of the same form as shown in the A_y–c_y curve in Fig. 55-9. However, if the cells are matched properly and always are reproducibly positioned (this is a much more critical operation than it is with ordinary cells), satisfactory precision can be attained with their use.

Finally, a word may be said about multiple reflections. It was mentioned previously that about 4 per cent reflection occurs upon perpendicular incidence from air onto a surface with $n = 1.5$ (points P_1 and P_4 in Fig. 55-5). Evidently radiation reflected from P_4 may be reflected again at P_1. This radiation thus passes three times through the cell, instead of once. This effect causes a deviation from Beer's law that is most important at low transmissions. It becomes less important at higher absorbances and is not discussed further.[10]

Filter photometers

In the introduction to Chapter 54 it was mentioned that filters may be used to isolate a portion of the spectrum emitted by a *continuous* source. Examples have been given of the efficiency of a typical filter (p. 961) as compared to a monochromator (p. 973) in this respect. From these numbers it is evident that all such filter photometers operate with much wider spectral band widths than do spectrophotometers. Figure 55-8, shows, therefore, that one may not expect adherence to Beer's law in a filter photometer, when the transmittance of the solution is measured with respect to a solvent, unless the measurements are made on substances having unusually broad absorption bands. The form of the deviation always will be of the form shown for Y in Fig. 55-9. A special situation exists if a source such as a mercury arc is used. This source emits relatively few lines at wavelengths 546, 436, and 405 mμ, etc. When the arc is run at low pressure, the lines are very sharp. Filters are available for the effective transmission of only one selected line of the mercury spectrum. With a proper source and filter it is possible to approach more closely to monochromatic radiation than it is with any spectrophotometer using a continuous source. However, only a few wavelengths are available.

A different situation exists in a *colorimeter*. The term colorimeter refers to an instrument in which a filter may be used, as in a filter photometer, but in which the transmittance of two *solutions* is *matched*, usually by visual means. The match is accomplished by varying the thickness of solutions viewed. Suppose the concentration of one of the solutions is known to be c_1, and the intensity transmitted by thickness b_1 of this solution is the same as that transmitted by thickness b_x of unknown. Then the concentration c_x of the unknown is simply given by $c_1(b_1/b_x)$. It is very important to note that in this kind of measurement the mass per unit cross section of absorbing substance in the two beams is the same, and heterochromaticity cannot cause apparent deviation from the Beer law.

[10] A thorough treatment is given by L. S. Goldring, R. C. Hawes, G. H. Hare, A. O. Beckman, and M. E. Stickney, *Anal. Chem.* **25**, 869 (1953).

Spectrophotometric analysis of mixtures

The principle is very simple and is illustrated in Fig. 55-10. The absorption cells are assumed to have the same length, and reflections are assumed to be corrected for. Evidently in (a) and (b), respectively, $I_x = I_0 \times 10^{-A_x}$, and $I_y = I_0 \times 10^{-A_y}$. In (c), the radiation incident upon Y is equal to $I_0 \times 10^{-A_x}$; hence $I_y = I_0 \times 10^{-A_x} \times 10^{-A_y} = I_0 \times 10^{-(A_x+A_y)}$. Example (d) is identical to (c). If no chemical reaction or any interaction occurs between X and Y upon mixing, the result in (e) is identical to (c) and (d). *The absorbance of the mixture is the sum of the individual absorbances.*

Example. The solution of X transmits 50 per cent; $A_x = 0.301$. The solution of Y transmits 40 per cent; $A_y = 0.398$. The solution containing X and Y transmits 50 per cent of 40 per cent, or 20 per cent; $A_{x+y} = 0.699$.

The practical analysis of mixtures is carried out as follows. Assuming the mixture contains two absorbing solution X and Y, it is necessary first to measure the absorption spectrum of solutions of X and of Y of known concentration. In this way ϵ_x and ϵ_y are obtained as a function of wavelength. (It is entirely proper to use a_x and a_y if concentrations in grams/liter are desired.)

Two appropriate wavelengths are selected (Fig. 55-11) and the absorbance of the unknown mixture is measured at each wavelength. Evidently, if the absorbances of the mixture are A_1 and A_2, respectively, we have

$$\begin{aligned} A_1 &= \epsilon_{x1}C_x + \epsilon_{y1}C_y \\ A_2 &= \epsilon_{x2}C_x + \epsilon_{y2}C_y \end{aligned} \tag{17}$$

where $\epsilon_{x1}, \ldots, \epsilon_{y2}$ are the molar absorptivities of X and Y at λ_1 and λ_2. All the quantities are known except C_x and C_y. The pair of simultaneous equations in two unknowns is solved:

$$\begin{aligned} C_x &= \frac{\epsilon_{y2}A_1 - \epsilon_{y1}A_2}{\epsilon_{y2}\epsilon_{x1} - \epsilon_{y1}\epsilon_{x2}} \\ C_y &= \frac{\epsilon_{x1}A_2 - \epsilon_{x2}A_1}{\epsilon_{x1}\epsilon_{y2} - \epsilon_{x2}\epsilon_{y1}} \end{aligned} \tag{18}$$

The practical success of the method depends upon choosing λ_1 and λ_2 such that X is the main absorbing species at one wavelength and Y is the main absorbing species at the other. In the limiting case in which $\epsilon_{y1} = 0$ and $\epsilon_{x2} = 0$ (or $\epsilon_{y2} = 0$, $\epsilon_{x1} = 0$) the problem reduces to two separate, independent spectrophotometric determinations. In any practical case in which $\epsilon_{y1} \neq 0$, $\epsilon_{x2} \neq 0$, the precision necessarily is less than in the determination of single constituents. In the limiting case $\epsilon_{x1}/\epsilon_{x2} = \epsilon_{y1}/\epsilon_{y2}$, the problem becomes indeterminate and neither C_x nor C_y can be found by spectrophotometric measurements (see Problem 16).

An important special case of Eq. (17) arises when

$$\epsilon_{x1} = \epsilon_{y1} \qquad \text{or} \qquad \epsilon_{x2} = \epsilon_{y2} \tag{19}$$

Assuming the former relation,

$$A_1 = \epsilon_{x1}(C_x + C_y) \tag{20}$$

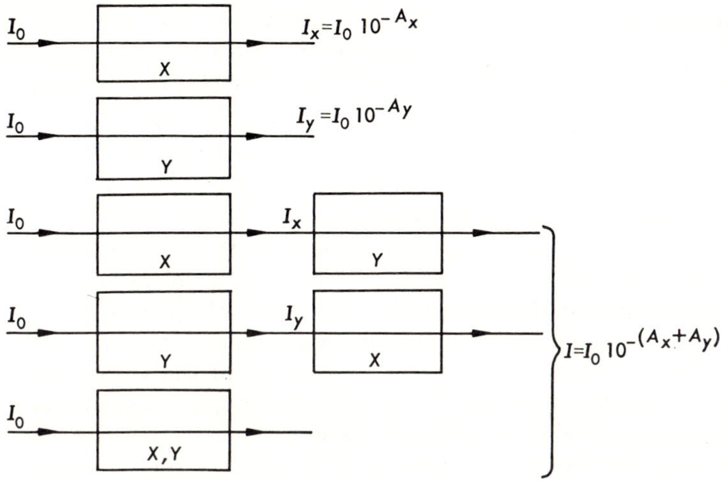

Fig. 55-10. Transmission through solutions of X, of Y, and of $X + Y$.

and the absorbance is proportional to the *total* concentration. Such a point is called an "isosbestic" point. At least one such point must exist at some wavelength in the case of any two substances having dissimilar spectra.

An example will illustrate the calculations for analysis of mixtures. Cells of the same length are used for measurements on X, Y, and the mixtures, with the following results:

	λ_1		λ_2	
Solution	T	A	T	A
5×10^{-5} *M* X	0.800	0.098	0.040	1.398
6×10^{-5} *M* Y	0.060	1.222	0.750	0.188
Mixture	0.500	0.301	0.400	0.398

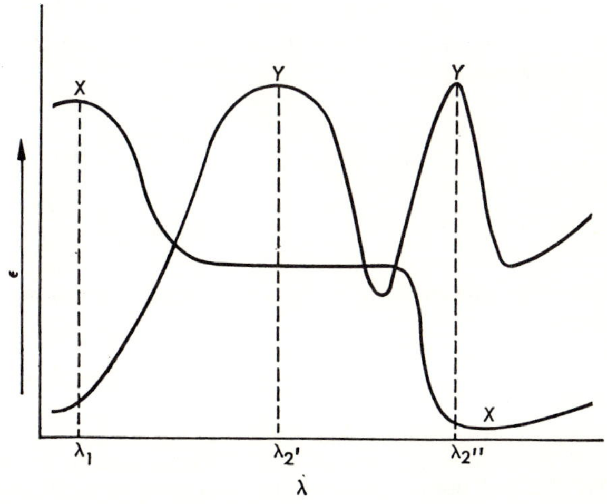

Fig. 55-11. Choice of wavelengths for analysis of mixtures.

$b\epsilon_{x1}$ and $b\epsilon_{x2}$ are found to be 1.98×10^3 and 2.80×10^4, respectively; $b\epsilon_{y1}$ and $b\epsilon_{y2}$ are 2.04×10^4 and 3.13×10^3, respectively. Since the value of b is the same in all the measurements, b may be omitted. Equations (17) thus assume the form

$$0.301 = 1.98 \times 10^3 C_x + 2.04 \times 10^4 C_y$$
$$0.398 = 2.80 \times 10^4 C_x + 3.13 \times 10^3 C_y$$

Obtaining C_x and C_y is left as an exercise.

The choice of the proper wavelengths is a matter of crucial importance. A hypothetical situation is shown in Fig. 55-11. λ_1 is an obvious choice, since it is the only wavelength available at which ϵ_x is appreciably greater than ϵ_y. The choice between λ_2' and λ_2'' is harder. λ_2'' offers a large difference between ϵ_x and ϵ_y, which is advantageous; however, owing to the narrowness of the absorption band of Y, it is possible (see Fig. 55-8) that Y may show an apparent deviation from Beer's law at λ_2''. If so, it would be essential to choose λ_2'. Thus the practical choice depends upon the band width of the spectrophotometer that is to be used.

Precision of spectrophotometric determinations

Referring to Table 55-1, it is noted that measurements of a given constant precision (± 0.001 in T) do not yield equally precise values of A. Thus at $T = 0.900 \pm 0.001$, A is 0.0458 ± 0.0005; the precision in A is about 1.2 per cent. At $T = 0.103 \pm 0.001$, A is 0.987 ± 0.004 (± 0.43 per cent). At $T = 0.373 \pm 0.001$, A is 0.428 ± 0.0012 (± 0.28 per cent). The value of A obtained at $T = 0.373$ is known more precisely than either of the other two values of A. This simple calculation illustrates a very important principle, which is explained in detail below.

When the maximum precision in the determination of C is required, this must correspond to the maximum precision in A (assuming, as we shall, that Beer's law is followed). This corresponds to the minimum value of $\delta A/A$, where δA is the experimental uncertainty in A. The uncertainty in A arises in general from the combined uncertainty in *three* measurements: namely of I, of I_0, and of the experimental setting for $T = 0$, prior to measuring either I or I_0. An exact treatment must take all these uncertainties into account.[11] For our purposes it is sufficient to consider the uncertainty only in I, which is the most important one for most measurements.[12] (However, the result of the more elaborate treatment is given later.) Since

$$A = \log \frac{I_0}{I} \tag{21}$$

$$\delta A = \frac{1}{2.303} \frac{\delta I}{I} \tag{22}$$

since the uncertainty in I_0 is neglected. Therefore,

$$\frac{\delta A}{A} = \frac{1}{2.303} \frac{\delta I}{A \cdot I} \tag{23}$$

[11] N. T. Gridgeman, *Anal. Chem.* **24,** 445 (1952).

[12] The treatment is due to H. Von Halban and L. Ebert, *Z. Physik. Chem.* (*Leipzig*) **A112,** 321 (1924). A useful discussion has been given by A. Ringbom, *Z. Anal. Chem.* **115,** 332 (1939).

The arithmetic is somewhat simpler if A is replaced by $E = 2.303A = \ln (I_0/I)$:

$$\frac{\delta A}{A} = \frac{\delta E}{E} = \frac{\delta I}{E \cdot I} \tag{24}$$

The problem is to find the minimum value of Eq. (24). There are two important practical cases.

1. In most nonrecording (manually operated) spectrophotometers the measurement is made on a scale that is linear in T, i.e., one that is graduated in equal divisions from $T = 0$ to $T = 1$. In such instruments it appears that the uncertainty in measuring T is constant, independent of T (analogous to the constant uncertainty ± 0.001 in T in Table 55-1). Therefore, $\delta I/I_0$ is constant, and Eq. (24) may be written

$$\frac{\delta E}{E} = \frac{\delta I}{I_0 \cdot E \cdot e^{-E}} \tag{25}$$

The minimum value of Eq. (25) occurs at the maximum value of Ee^{-E}, that is, at $E = 1$; therefore, $A = 0.434$, $T = 0.368$. At this value,

$$\left(\frac{\delta E}{E}\right)_{\min} = \left(\frac{\delta A}{A}\right)_{\min} = e\,\frac{\delta I}{I_0} = 2.72\,\frac{\delta I}{I_0} \tag{26}$$

Accordingly, when T is measured with a constant uncertainty of ± 0.001, the maximum precision in A or in C is $e \times 0.001$, or 0.27 per cent. If T is measured to ± 0.5 per cent, the maximum precision in A, or in C, is 1.4 per cent. Figure 55-12 shows how the relative precision depends on T, for this case (T measured to 0.5 per cent). If T is 0.20 or 0.60, the precision in C is slightly poorer, 1.6 per cent. Using the same spectrophotometer, a determination accurate to 2.0 per cent could be carried out within the range $T = 0.13$–0.70 (13–70 per cent transmission). Beyond those limits the error increases very rapidly.

As a practical matter, the use of a high rather than a low absorbance is preferred, because it reduces the relative importance of errors due to mismatched cells.

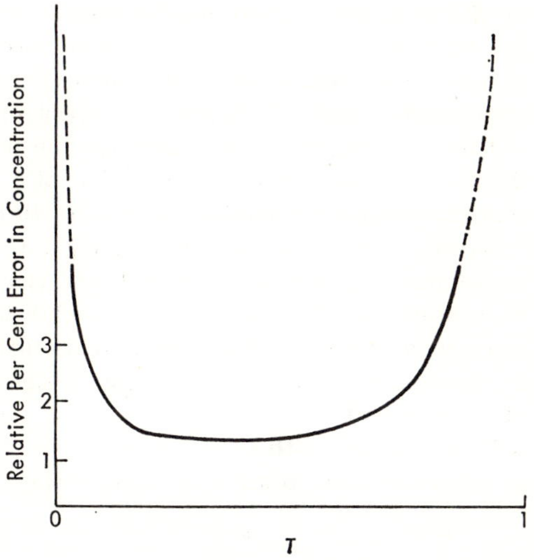

Fig. 55-12. Relative photometric error as a function of T. (Measurements made to ± 0.005 in T)

The curve shown in Fig. 55-12 is not exact because it does not include the uncertainties in setting $T = 0$ and $T = 1$. When both these are taken into account, and both are assumed to be the same as the uncertainty in measuring T, it is found (see Ref. 11) that the maximum precision in A occurs at $T = 0.388$ (instead of at $T = 0.368$), but the minimum value of $\delta A/A$ is $3.38(\delta I/I_0)$ [instead of $2.72(\delta I/I_0)$].

2. The other important practical case applies to recording (double-beam) spectrophotometers in which a scale *linear in absorbance* is used for the measurement. It appears in such instruments that the uncertainty in reading the absorbance (not the transmittance) is constant, at least up to absorbances of 1 or somewhat more. (At high A, the intensity of the beam transmitted through the solution is small, and noise in the amplifying circuit limits the precision with which A can be measured.) A few instruments permit precise measurements up to $A = 4$ (0.01 per cent transmission). Therefore $\delta I/I$ is constant and Eq. (23) shows directly that the maximum precision in A is obtained, not at $A = 0.43$, as in the preceding case, but at the maximum A. It appears in practice with most recording spectrophotometers now in use that the most precise values are obtained at an A around unity.

To summarize briefly: With careful work and with the best instruments, determinations of concentrations precise to about 0.1 per cent are possible under optimum conditions. Additional improvement may be achieved using the principles outlined in the next section.

Differential spectrophotometry

A determination of concentration precise to 0.1 per cent requires a determination of percentage transmission accurate to 0.03 per cent. A substantial increase in the precision of the determination of concentration may be attained in a relatively simple way. This is done by using a conventional spectrophotometer to compare the intensities transmitted by the solution of unknown concentration and by a comparison solution whose concentration is known exactly and the transmission of which is approximately the same as that of the solution of unknown concentration. This is the method of "differential spectrophotometry"; as will be seen from the example below, a large gain in precision can often be realized using solutions of relatively low transmission, whence come the names "high-absorbance spectrophotometry" and "high-precision spectrophotometry."

The following example illustrates the principle. Consider a solution of a solute that follows Beer's law. Suppose that at a given (constant) path length and at the concentration c the solution transmits 40 per cent of the incident monochromatic radiation.[13] The absorbance A is $\log(100/40)$ and $E = \ln(100/40) = 0.916$. If the transmission is measured to 0.1 per cent absolute, the corresponding absolute uncertainty in E is $\delta E = \delta I/I$, or 0.0025. The relative uncertainty in E, and consequently in c, is 0.27 per cent. Solutions of concentrations $4c$ and $5c$ would transmit 2.56 and 1.02 per cent under the same conditions. If the intensity of the incident radiation is increased by the factor 100/2.56, a full-scale reading will be obtained with the solution of concentration $4c$ and a reading of 40 per cent of full-scale with solution $5c$. Under these circumstances, the concentration $5c$ is

[13] This is close to the optimum value and is chosen for convenience in the further comparisons.

known relative to $4c$ with the same precision that c is known relative to zero concentration (solvent). If the concentration $4c$ is known relatively exactly, the concentration $5c$ is determined with a precision of 0.05 per cent, corresponding to a fivefold increase in precision.

This example indicates that an unlimited increase in precision could be attained by using standard and unknown solutions of sufficiently low transmission. In practice the increase is limited by two factors. First, full-scale readings can be obtained with standards of low transmission only by using relatively wide spectral bands, i.e., wide exit slits on the monochromator. Therefore the radiation is no longer monochromatic. With such radiation all solutes deviate from Beer's law (Fig. 55-9); the direction of the deviation is such as to decrease the concentrational sensitivity. The importance of this factor depends on the spectrophotometer used and on the width of the absorption band of the solute (Fig. 55-8). Second, the presence of stray radiation due to scattering and reflection from the optical surfaces makes measurements with solutions of too low transmission essentially meaningless. This factor must be evaluated for each spectrophotometer at each wavelength. Some commercial instruments permit measurements at transmissions less than 0.1 per cent, but this is by no means the case in all instruments now in use. The question must be settled experimentally for the solvents and instrument being used.

The use of comparison solution instead of solvent was made years ago by Kortüm[14] for concentration measurements accurate to 1 part in 1000. The method has been investigated further by Ayres, Bastian et al., and Hiskey et al.[15] In many cases concentration determinations accurate to better than 0.1 per cent have been made spectrophotometrically.

In the practical application of the method, it is essential that reference and unknown be at the same temperature. (In fact, differential measurements provide a sensitive method for the determination of differences of temperature.)

An example of the use of this method for the precise determination of relatively large amounts of copper is given on page 1057. Further details may be found elsewhere.[16]

Reference also is made to a simple and ingenious differential method developed by Ringbom and Osterholm.[17] A two-cell instrument, which may be a simple filter photometer, is used. The change of transmission is measured when an accurately known volume of solvent is added to an accurately known volume of solution. With the aid of appropriate calibration curves, results of very high precision can be obtained.

Chemical apparent deviations from the Beer law

In the preceding pages various instrumental reasons for apparent deviations from Beer's law have been discussed. In some cases deviations may occur for other reasons, discussed below.

[14] G. Kortüm, *Angew. Chem.* **50**, 193 (1937). See also Ref. 7, chap. 54.

[15] G. H. Ayres, *Anal. Chem.* **21**, 652 (1949); R. Bastian, *ibid.* **21**, 972 (1949); R. Bastian, R. Weberling, and F. Palilla, *ibid.* **22**, 160 (1950); R. Bastian, *ibid.* **23**, 580 (1951); C. F. Hiskey, *ibid.* **21**, 1440 (1949); C. F. Hiskey, J. Rabinowitz, and I. G. Young, *ibid.* **22**, 1464 (1950); I. G. Young and C. F. Hiskey, *ibid.* **23**, 506 (1951).

[16] See Ref. 10, chap. 54.

[17] A. Ringbom and K. Osterholm, *Anal. Chem.* **25**, 1798 (1953).

In the derivation of Eq. (8) from (7) it was assumed that k'' is independent of concentration. This assumption may be incorrect for either of two reasons: (1) the absorptive properties of the absorbing species (molecule or ion) may be affected by the presence of other molecules, or (2) the concentration of the *actual* absorbing species is not proportional to the measured (analytical) concentration of the *solute*. This first effect is of very minor importance at the concentrations of solutes normally used in visible-UV spectrophotometry; that is, the solutes always are so dilute that they behave essentially as ideal solutes. In a very precise investigation Kortüm[18] measured the molar absorptivities of potassium ferricyanide in aqueous solution, with the results shown in Table 55-2.

TABLE 55-2. Molar Absorptivity of Potassium Ferricyanide at 436 mμ

C, moles/liter	ϵ
8.680×10^{-5}	742.3
3.103×10^{-4}	742.4
8.175×10^{-4}	742.5
1.6715×10^{-3}	742.6
2.929×10^{-3}	742.7
1.5813×10^{-2}	742.7

Over an approximately 200-fold range in concentration, the total measured variation in ϵ is only about 0.05 per cent.

If concentrated solutions should be examined, however, deviations must be anticipated.

The second cause is of quite frequent occurrence; it may be illustrated by a few simple examples.

Example: In a solution of a weak acid HA, the acid is present to some extent as A^-. If the two molar absorptivities are ϵ_{HA} and ϵ_{A^-}, the observed absorbance is

$$A = \epsilon_{HA}C_{HA} + \epsilon_{A^-}C_{A^-} \tag{27}$$

Since the extent of ionization depends upon the total concentration, it is evident that the solution cannot in general follow Beer's law; i.e., A/C is not constant in general. A/C may increase or decrease with c, depending upon whether the measurement is made at a wavelength at which $\epsilon_{HA} > \epsilon_{A^-}$, or vice versa. However, at any wavelength at which $\epsilon_{HA} = \epsilon_A$ (isosbestic point, p. 977), A/C is constant; therefore, measurement at such a wavelength is suitable for the determination of the analytical concentration of HA, regardless of its extent of ionization.

At any other wavelength, the system may be made to follow Beer's law if the paH and ionic strength are kept constant. Then the extent of ionization is independent of concentration. The observed ratio A/bc depends upon paH and lies between the two extremes ϵ_{HA} and ϵ_{A^-}.

Example: In an aqueous solution of potassium chromate, three species are present to a greater or lesser extent: CrO_4^{--}, $HCrO_4^-$, and $Cr_2O_7^{--}$. These substances

[18] See p. 261 of Ref. 7, chap. 54.

have different absorptivities and solutions of chromate accordingly deviate from Beer's law. Even if the pH is kept constant at some value at which all three species are present, deviation from the law occurs because the equilibrium involving $Cr_2O_7^{--}$ is concentration-dependent. In fact, careful measurements of absorbance as a function of concentration yield the equilibrium constant of the reaction[19]

$$Cr_2O_7^{--} + H_2O = 2\,HCrO_4^-$$

(Also see page 1052.) If the pH is high enough, all the chromium is present as CrO_4^{--} and Beer's law is followed exactly. It has been mentioned already that solutions of potassium chromate in 0.05 *N* potassium hydroxide serve as photometric standards (see Ref. 7).

Example: In systems such as iron(III)–thiocyanate, more than one complex may be formed depending upon $[SCN^-]$. Again, since the complexes $FeSCN^{++}, \ldots, Fe(SCN)_6^{3-}$ have different absorptivities, an iron(III)–thiocyanate system does not follow Beer's law upon dilution. If mononuclear complexes are formed, the fraction of metal present as any complex is determined by the concentration of free ligand and is independent of total concentration of metal (Chapter 6). Therefore, systems containing relatively small amounts of iron (so that only mononuclear complexes are formed) together with a large, constant concentration of thiocyanate, follow the Beer law exactly (see p. 1049). Precisely the same remarks apply to copper(II)–ammonia and similar systems.

Collected procedures: References

For many practical procedures involving visible and ultraviolet measurements the student is referred to comprehensive works on spectrophotometric determination of metals[20] and of nonmetals.[21] A collection of colorimetric procedures by Allport et al. is useful.[22,23] Finally, reference is made to a comprehensive collection of spectrophotometric data that is valuable for many analytical applications.[24]

Theory and practice of infrared measurements are discussed extensively by Herzberg,[25] Smith,[26] and Brugel[27]; Nakamoto[28] gives a survey of inorganic applications.

[19] W. G. Davies and J. E. Prue, *Trans. Faraday Soc.* **51**, 1045 (1955).

[20] E. B. Sandell, *Colorimetric Determination of Traces of Metals*, Wiley-Interscience, New York, 1959.

[21] D. F. Boltz (ed.), *Colorimetric Determinations of Non-Metals*, Wiley-Interscience, New York, 1958.

[22] N. L. Allport and J. W. Keyser, *Colorimetric Analysis*, vol. I, Chapman & Hall, London, 1957.

[23] N. L. Allport and J. E. Brocksopp, *Colorimetric Analysis*, vol. II, Chapman & Hall, London, 1963.

[24] I.U.P.A.C. Commission on Spectrochemical and Other Optical Properties for Analysis, *Spectrophotometric Data for Colorimetric Analysis*, Butterworth, London, 1963.

[25] G. Herzberg, *Infrared and Raman Spectra of Polyatomic Molecules*, Van Nostrand, Princeton, N.J., 1945.

[26] L. A. Smith in *Treatise on Analytical Chemistry* (I. M. Kolthoff and P. J. Elving, eds.), part I, vol. 6, Wiley-Interscience, New York, 1965, chap. 66.

[27] W. Brugel, *An Introduction to Infrared Spectroscopy*, Methuen, London, 1962.

[28] K. Nakamoto, *Infrared Spectra of Inorganic and Coordination Compounds*, Wiley, New York, 1963.

PROBLEMS

1. From the following data obtained in a filter photometer determine whether an acid solution of ferric iron with an excess of thiocyanate follows Beer's law:

Transmittancy (1 cm)

% Fe	%
0.0001	80.0
0.0002	64.5
0.0004	44.0
0.0008	22.0

Can you suggest an explanation for the results?

2. A 1-cm layer of a 0.0100 per cent solution of a compound having a molecular weight of 194 transmits 9.0 per cent light of wavelength 490 mμ. What is the molar absorptivity for this wavelength?
3. If the molar absorptivity of potassium permanganate is 2235 at 520 mμ, what will be the transmittance of a 0.0010 per cent solution in a 2-cm cell at this wavelength?
4. A solution of a colored substance has an absorption band in the range between 420 and 470 mμ with a maximum between 445 and 450 mμ. (a) What is the color of the solution? (b) What condition has to be fulfilled by other colored substances present in the solution in order not to have interference if the absorbance of the unknown is determined with a spectrophotometer at a wavelength region between 445 and 450 mμ?
5. What color of light filter would you expect to be useful in the determination of copper in ammoniacal solution?
6. The molar absorptivity of a colored substance with a molecular weight of 100 is equal to 10,000 in a certain solvent. Calculate the value of k in the equation

$$\frac{I_t}{I_0} = 10^{-klc'}$$

in which c' is the concentration of the substance in grams per 100 ml.

7. A solution of dithizone (mol. wt. 256) in chloroform, containing 2.0 mg in 1 liter, transmits 14 per cent of the incident light of wavelength 610 mμ in a 2.5-cm cell. Find the molar absorptivity of dithizone in chloroform at this wavelength.

Ans. 43,500.

8. The colorless ions A^+ and B^- combine to form the colored compound AB:

$$A^+ + B^- \rightleftharpoons AB$$

When 0.0025 mole of A^+ is mixed with 0.004 mole of B^- and diluted to 1 liter, the absorbance of the resulting solution is 0.80 of that obtained by mixing 0.0025 mole of A^+ with 0.020 mole of B^- and diluting to the same volume. What is the approximate value of the equilibrium constant $K = [AB]/[A^+][B^-]$?

Ans. $K = 2 \times 10^3$ (exact value, 1.65×10^3).

9. Bendig and Hirschmüller [*Z. Anal. Chem.* **120**, 385 (1940)] found that the equilibrium constant for the reaction between titanium and hydrogen peroxide in acid solution,

$$\text{Ti(IV)} + H_2O_2 \rightleftharpoons \text{Ti(IV)} \cdot H_2O_2 \text{ (yellow complex)}$$

$$\frac{[\text{Ti(IV)}][H_2O_2]}{[\text{Ti(IV)} \cdot H_2O_2]} = K$$

has a value of 10^{-4}. What volume of 0.9 *M* (ca. 3 per cent) hydrogen peroxide must be added to a solution containing 0.01 mole of Ti(IV) in order that after dilution to 1 liter, 99 per cent of the titanium shall be present in the form of the yellow complex? *Ans.* $[H_2O_2]_{total} = 1.98 \times 10^{-2}$; volume = 22 ml.

10. An equilibrium of the type

$$\underset{(colored)}{AB} \rightleftharpoons \underset{(colored)}{A} + \underset{(colorless)}{B}$$

exists in a mixture of AB, A, and B. The transmission curves of two mixtures in which the sum of concentration of the components is constant but the ratio AB/A is different are found to intersect at a point corresponding to the wavelength λ. Prove that the transmission curve of any such equilibrium mixture must pass through the same point (isosbestic point). What relation exists between the extinction coefficients of AB and A at the isosbestic point?

11. Show that the ionization constant of a weak acid (HA and A^- colored) can be obtained spectrophotometrically by measuring at a suitable wavelength the absorbance, $\log(I_0/I)$, of a mixture of constant total concentration (not necessarily known) at three different acidities, of which one must be such that only A^- exists in solution, and applying the relation

$$K_a = \frac{[H^+]_1[H^+]_2(A_2 - A_1)}{[H^+]_2(A_1 - A_0) - [H^+]_1(A_2 - A_0)}$$

where A_1 and A_2 are the absorbances of mixtures of HA and A^- at acidities $[H^+]_1$ and $[H^+]_2$ and A_0 is the absorbance when the acid has been practically completely transformed into its salt A^-. From the standpoint of the precision of determining K_a, what can be said about the choice of $[H^+]_1$ and $[H^+]_2$?

12. At a certain wavelength the value of k in the Lambert–Beer law expression for Cr as chromate is 0.006 cm²/µg. (a) What is the weight of sample containing 10 per cent Cr that will provide maximum precision in the spectrophotometric determination of Cr in a final volume of 5 ml in a 1-cm cell? (b) What is the absorptivity? *Ans.* (a) 3.6 mg.

13. An absorption cell is constructed from capillary tubing with an internal diameter of 0.030 cm; the path length is 10 cm. Calculate the volume required to fill this cell, and the amount of solute having $\epsilon = 10^4$ required to yield an absorbance of 0.01.

14. Calculate transmittance of solutions of potassium chromate of concentration 0.08 and 0.02 g/liter, in 0.05 *N* potassium hydroxide, at the wavelengths given in Table 55-1. Beer's law is obeyed.

15. A given solution is known to follow Beer's law. Transmissions are measured at a wavelength at which a 1×10^{-4} *M* solution transmits exactly 10.0 per cent. Suppose the transmission of a 1×10^{-4} *M*, 2×10^{-4} *M*, and 3×10^{-4} *M* solutions are measured in a spectrophotometer with 1 per cent stray radiation which is not absorbed at all by the solution. What would the apparent absorbances be? Repeat, using 0.01 per cent stray light. [*Hint:* If the true transmission is 50 per cent, the apparent transmission with 1 per cent stray radiation is

$$\frac{(0.50 \times 99) + 1}{100} \times 100 \Big]$$

16. In a certain binary mixture $a_{x1}/a_{x2} = a_{y1}/a_{y2}$. Can any information whatever be obtained by spectrophotometric measurement on such a mixture?

17. The sensitivity of a spectrophotometric determination of copper with dithizone is specified as follows: 0.0022 µg of Cu per cm² corresponds to $\log(I_c/I) = 0.001$ (533 mµ). (a) Calculate the molar absorptivity. (b) How much copper must be present in a 10-ml solution to yield an absorbance of 0.010 when measured in a 10-cm cell?

CHAPTER 56 *Fluorescence and Scattering*

Both fluorescence and scattering are examples of photoluminescent processes. By photoluminescence is understood a process in which the incidence of radiation upon the sample causes emission of radiation by the sample; the emitted frequency may or may not be the same as that of the incident radiation. It is contrasted in this way with the process of pure absorption, discussed in Chapter 55, in which all the absorbed energy is converted into other forms of energy. The photoluminescent processes of greatest analytical significance are fluorescence, induced either by visible–UV radiation or by X-rays or electrons, scattering without change of wavelength, and Raman scattering with change of wavelength. The term *fluorescence* used without qualification refers to a process caused by visible or UV radiation.

Fluorescence

More than 100 years ago it was observed that the path of a beam of sunlight through a solution of chlorophyl was marked by a cone of red light. Soon afterward it was discovered that a colorless solution of quinine sulfate held before a strong light appeared blue on the surface of the solution. This kind of phenomenon was named fluorescence by Stokes in 1852. Stokes showed that it is due to absorption of illuminating light with emission of light of a different color. Fluorescence is distinguished from Rayleigh or Mie scattering (see below) in that in the former case absorption occurs and the emitted light is of a wavelength different from the absorbed light. It is distinguished from Raman scattering, in which a change of wavelength occurs, in that Raman scattering does not involve absorption. The distinction between fluorescence and phosphorescence is not so simple and will not be discussed.[1]

The simplest example of fluorescence is provided by a monatomic vapor (e.g., sodium vapor) illuminated with monochromatic radiation of energy that corresponds exactly to the energy required to raise the valence electron from the lowest to the first excited energy level. In the case of sodium this corresponds to the yellow light emitted by sodium, of wavelength 5890 and 5896 Å[2]. Such radiation is absorbed strongly by sodium atoms and is reemitted without change of wavelength, the intensity being equal in all directions (Fig. 56-1). This is known

[1] For a complete discussion see P. Pringsheim, *Fluorescence and Phosphorescence*, Wiley-Interscience, New York, 1949, chap. 4.

[2] The first excited energy level in sodium is a close doublet. See also chap. 57.

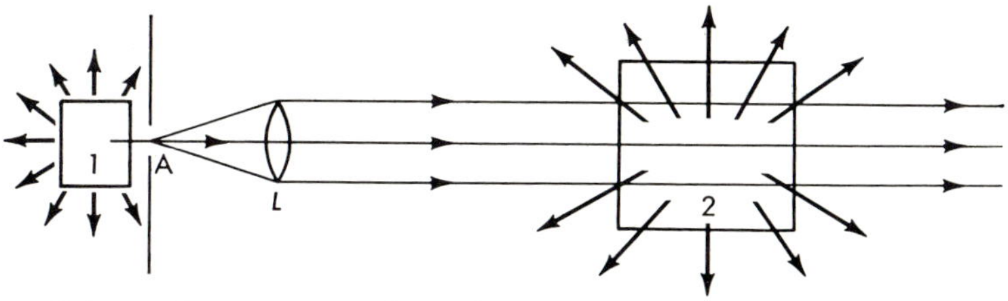

Fig. 56-1. Schematic representation of resonance radiation. 1, Sodium vapor at relatively high temperature; *A*, aperture; *L*, lens; 2, sodium vapor at a lower temperature than 1; emission occurs in all directions.

as *resonance radiation.*[3] If the absorbed energy is so large that the valence electron is raised to a state higher than the first excited state (wavelength 3303 Å in the case of sodium), the possibility exists for the electron to return by steps to the ground state, and in this way two or more light quanta may be emitted. In such a case, the emitted radiations are of longer wavelength than the absorbed radiation.

Reemission solely of the exciting radiation is observed only in monatomic vapors under the conditions of resonance radiation described above. When molecules absorb energy sufficient to excite an electron, changes also occur in vibrational and rotational energy. For this reason an electronic transition in a molecule consists of many closely spaced lines, whereas it is a single line (or a very few lines) in an atom. With increasing molecular weight the vibrational and rotational levels become more closely spaced. If the lines are too close to be resolved, a *band* is observed. Therefore, when a polyatomic molecule absorbs a single frequency (e.g., the 3663-Å line emitted by a mercury arc), it may emit a band of radiation. In this case the process is called *fluorescence*. If the emitted band is of longer or shorter wavelength than the absorbed radiation, it is called Stokes or anti-Stokes radiation, respectively. The former is always more intense. Figure 56-2 shows in highly schematic form how the two kinds of emission may occur.

Fluorescence occurs with many substances in different states of aggregation. The behavior of dissolved substances is of greatest interest in analytical applications. In cases of analytical importance a visible fluorescence occurs upon illumination by a mercury arc when the arc is used with a filter that transmits only UV radiation, or violet-blue light. For qualitative purposes the substance thus illuminated is viewed in a dark room. For quantitative work, the intensity of fluorescence is measured and compared with that emitted from solutions of known concentration. A practical procedure is described on page 1075.

The fluorescence of some organic compounds can be detected at extremely low concentrations. Thus disodium fluoresceinate can be detected at concentrations as low as 10^{-14} g/ml[4]; many other compounds can be detected below 10^{-10} g/ml.[5]

Comparatively few inorganic compounds are fluorescent in the dissolved state. Relatively strong fluorescence in solution is shown by uranyl compounds

[3] Striking and beautiful experiments with sodium vapor are described by R. W. Wood, *Physical Optics*, Macmillan, New York, 1934.

[4] See Ref. 1, p. 348.

[5] A list of limiting detectable concentrations is given by M. Haitinger, *Mikrochemie* **10**, 321 (1935). With modern methods of detection the limits may be smaller than those given by Haitinger.

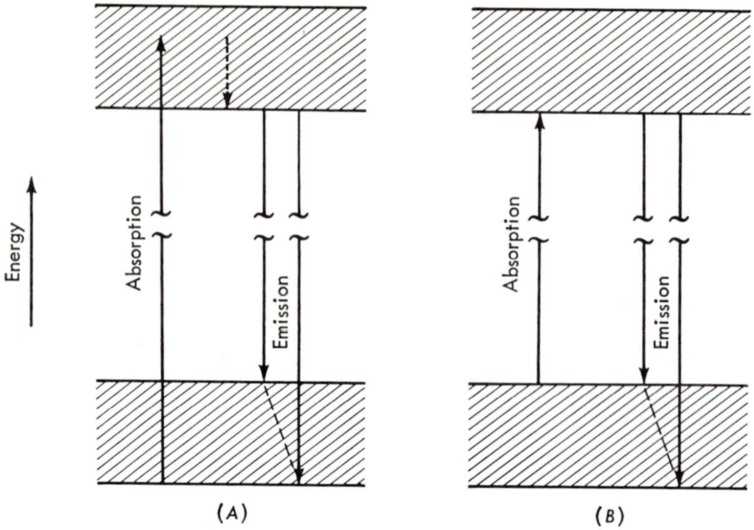

Fig. 56-2. Schematic representation of Stokes (*A*) and anti-Stokes (*B*) fluorescence by a polyatomic molecule. The upper and lower electronic levels are represented as vibrational-rotational continua. In (*A*), the excess of vibrational energy is lost by conversion to thermal energy (dotted line).

(containing the cation UO_2^{++}) but not by uranous or other uranium compounds; however, the intensity is far smaller than that shown by fluorescein or other dyes at the same concentration. U(VI) also shows fluorescence in sodium fluoride or sodium fluoride–carbonate solid solutions, and this behavior forms the basis of a widely used method for determining traces of uranium. Some thallous compounds and some of the rare earth (atomic number 57–71) compounds also are fluorescent in solution. Thus samarium, europium, and terbium can be identified from the fluorescence of their salts in aqueous solutions at concentrations of 10^{-3}–10^{-4} *M* or less.

The reaction products of several inorganic cations with suitable organic reagents have fluorescence properties, use of which is made in the determination of these metal ions. Sensitive tests for beryllium and aluminum are based upon the fluorescence of their compounds with morin, which is a benzopyran compound. The test for beryllium is sensitive to 10^{-3} ppm and can be made nearly specific for beryllium. The determination of aluminum is described on page 1075.

It may be shown easily (Problem 2) that the fluorescence intensity of a substance that obeys Beer's law is proportional to concentration at small concentrations. With increase of concentration one would expect a limiting intensity to be reached corresponding to essentially complete absorption of the exciting light. Often, however, at a certain concentration the intensity goes through a flat maximum and decreases with further increase of concentration. Thus the same intensity could correspond to two different values of concentration. This phenomenon is due to "self-quenching." Analytically it is easily detected and corrected for by measurement of the sample and of a diluted sample.

Various substances have specific quenching effects. For example, traces of ethanol or iodide cause a marked reduction in the fluorescence of an aqueous solution of uranyl sulfate. The quenching of fluorescence by constituents present in the final solution must always be considered in quantitative applications, or large errors may result. In the quantitative determination, a calibration curve of fluorescence intensity versus concentration is established. The amount of constituent in the sample is obtained from the calibration curve. To be sure that no error is caused by quenching by certain constituents in the sample, a known amount of the constituent being determined is added after measurement of the unknown. The increase in fluorescence should correspond to that expected on the basis of the calibration curve. If it does not, interfering substances are present.

References. Pringsheim's classical work (*cf.* Ref 1) is a comprehensive summary of fact and theory up to 1949. The books by Danckwortt and Eisenbrand[6] and Radley and Grant[7] are useful for analytical applications. Many procedures are summarized by Conrad,[8] and a concise summary of modern developments is given by Wotherspoon and Oster.[9] A book edited by Hercules[10] describes the principles with some discussion of analytical applications.

X-ray fluorescence

Emission of x-rays may occur if a sample is exposed to electrons or to x-rays of sufficiently high energy. The process has been known for a long time and was used extensively for analysis in the 1920s.[11] For various reasons, excitation by electrons was less convenient than by x-rays until comparatively recently. Practical methods exist now to confine the section of the sample illuminated by an electron beam to a microscopic area, with diameter of the order of 10^{-4} cm, and to measure the intensity and wavelength of the x-rays thus produced. The apparatus is relatively costly; however, the technique, known as *electron-probe microanalysis*, is one of the most powerful modern methods for analysis of microscopic sections of samples.[12]

Excitation by x-rays also provides a very powerful method for analysis of metals. The minimum area of sample is perhaps 10^4 times as large as in electron-probe microanalysis. The equipment is less costly than that required for electron probe, and the technique at present is more widely practiced. The method also is called x-ray emission spectrography[13] and x-ray spectrochemical analysis.[14]

The principle is exactly the same as in ordinary fluorescence. By the absorption of an x-ray of suitably high energy, an electron is removed from an atom, and the vacancy thus created is filled by an electron from a higher level. The difference in energy between upper and lower level appears as a fluorescent x-ray. In ordinary

[6] P. W. Danckwortt and J. Eisenbrand, *Lumineszenz-Analyse im filtrierten ultravioletten Licht*, Akademie-Verlag, Guest and Partig, Leipzig, 6th ed., 1956.

[7] J. A. Radley and J. Grant, *Fluorescence Analysis in Ultraviolet Light*, Van Nostrand, Princeton, N.J., 4th ed., 1954.

[8] A. Conrad, in *Treatise on Analytical Chemistry* (I. M. Kolthoff and P. J. Elving, eds.), part I, vol. 5, Wiley-Interscience, New York, 1964, chap. 59.

[9] N. Wotherspoon and G. Oster in *Technique of Organic Chemistry* (A. Weissberger, ed.), vol. I, part III, Wiley-Interscience, New York, 1960, chap. 31.

[10] D. M. Hercules (ed.), *Fluorescence and Phosphorescence Analysis*, Wiley, New York, 1966.

[11] Hafnium was discovered in 1923 by Coster and Hevesy by means of x-ray fluorescence excited by electrons.

[12] L. S. Birks, *Electron Probe Microanalysis*, Wiley-Interscience, New York, 1963.

[13] H. A. Liebhafsky, H. G. Pfeiffer, E. H. Winslow, and P. D. Zemany, *X-Ray Absorption and Emission in Analytical Chemistry*, Wiley, New York, 1960.

[14] L. S. Birks, *X-Ray Spectrochemical Analysis*, Wiley-Interscience, New York, 1959.

fluorescence valence electrons are involved, whereas in x-ray fluorescence the $K, L, M, \ldots$ electron shells are involved. Therefore, as explained in Chapter 54, x-ray fluorescence is an atomic process and the physical or chemical state of an element is of no consequence (except for the very light elements). Figure 56-3 illustrates the principle of the measurement. The sample S may be a solid or solution. The emitted x-rays have wavelengths that are characteristic of the elements in the sample. After being made parallel by the collimator, the x-rays strike the analyzing crystal, which may be quartz, sodium chloride, lithium fluoride, etc. Depending on the angle θ between the crystal and the beam, a single wavelength will be diffracted according to the Bragg relation

$$m\lambda = 2d \sin \theta$$

where m is the order of diffraction (almost always, $m = 1$), λ the wavelength, and d the interplanar spacing in the analyzing crystal. To measure the emitted spectrum, the crystal is rotated from $\theta = 0°$ to $\theta = 90°$ and the detector D is rotated at twice the angular speed of the crystal. Thus the wavelength (characteristic of the element) and the intensity (related to amount) are measured. In practice, corrections are required for absorption of both the incident and emitted radiation dependent upon the thickness of the sample and the presence of other elements. It is not possible to carry out exact quantitative work without taking such effects into account. With suitable calibration (see Refs. 12 and 13) the method is rapid and much more precise than ordinary spectrochemical analysis (Chapter 57). The following results, due to Brissey,[15] show the precision that can be obtained in the analysis of a complex material (results in percent by weight):

	% Mo	% Ni	% Co	% Fe	% Cr
Mean	9.60	53.64	10.32	2.58	18.17
Standard deviation[a]	0.09	0.73	0.06	0.03	0.16

[a] Of a single determination.

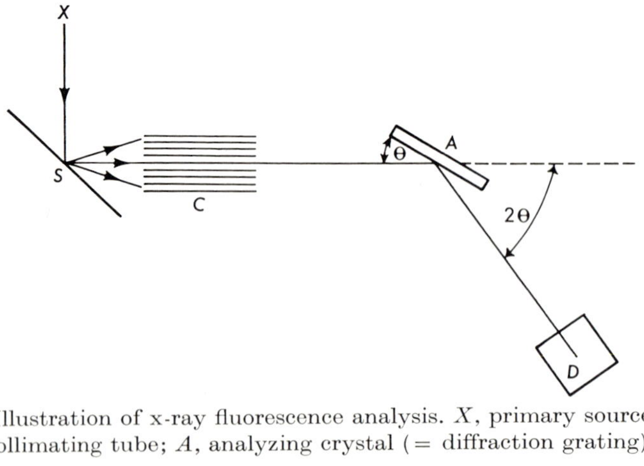

Fig. 56-3. Illustration of x-ray fluorescence analysis. X, primary source of x-rays; S, sample; C, collimating tube; A, analyzing crystal (= diffraction grating); D, detector.

[15] R. M. Brissey, *Anal. Chem.* **25**, 190 (1953); R. M. Brissey, H. A. Liebhafsky, and H. G. Pfeiffer, *Am. Soc. Testing Mater. Spec. Tech. Publ.* **157**, 1954, p. 43.

Rayleigh and Mie scattering

The discussion is limited to scattering by transparent (nonabsorbing) systems. Rayleigh and Mie scattering are distinguished, as is explained in the next paragraph, by the ratio of the size of the scattering unit (molecule, ion, colloidal particle, etc.) to the wavelength of the incident radiation. In either process, the scattering unit is transparent (for example, a sugar molecule in water or a droplet of benzene in water) and no absorption occurs; the system is not raised to an excited energy level by absorption of radiation. Instead, the incident radiation induces an oscillating electric dipole, which acts as a new source emitting radiation of the same frequency as the incident radiation. In additional contrast to fluorescence, the angular distribution of scattered intensity is not uniform but in general varies markedly with the angle.

Rayleigh scattering[16] occurs when the radius (or other linear dimension) of the scatterer is $\frac{1}{20}$ or less of the wavelength of the incident radiation. The intensity of scattering is proportional to the fourth power of the frequency, i.e., inversely to the fourth power of the wavelength. Therefore blue light (wavelength about 450 mμ) is scattered about 5.86 times as strongly as red light (700 mμ).[17] This accounts for the bluish tinge observed in the scattering of white light by suspensions of *small* particles. The most familiar example is the blue color of the sky. Also, the intensity scattered increases as the square of the volume, i.e., as the sixth power of the radius. The intensity is independent of the angle, provided incident radiation is used which is polarized perpendicular to the plane of observation. This kind of scattering of visible radiation is exhibited by gases, liquids, and dissolved substances except those of very high molecular weight. The amount of scattering is so small that it is not generally measurable from the reduction in transmitted intensity but is measured at an angle to the incident radiation.

When the radius or other linear dimension is not so small compared to the wavelength, the situation is entirely different. A fourth-power dependence on frequency is not observed, and the scattered intensity becomes more and more concentrated in the forward direction. Specifically it is emphasized that (contrary to an erroneous statement given in many textbooks) the fourth-power dependence is not observed in typical colloidal systems. The variation of scattered intensity with frequency becomes smaller with increasing particle size; scattering by particles of radius only slightly larger than the wavelength is practically independent of frequency. Typical colloidal suspensions of colorless substances such as barium sulfate are therefore white.

A special case exists if the scattering substance (e.g., a polystyrene molecule) is dissolved in a solvent (e.g., benzene) of about the same refractive index as the scatterer. This situation gives rise to Rayleigh–Gans scattering. The experimental and theoretical aspects were developed considerably by Debye. Measurements in such systems now are used extensively to determine molecular weight and certain other properties of high-molecular-weight solutes.[18-22]

[16] Named after Lord Rayleigh, who gave a quantitative treatment in 1871.

[17] The refractive index is assumed to be the same at both wavelengths.

[18] P. Debye, *J. Phys. Colloid Chem.* **51**, 18 (1947).

[19] G. Oster, *Chem. Rev.* **43**, 319 (1948).

[20] K. A. Stacey, *Light Scattering in Physical Chemistry*, Butterworth, London, 1956.

In all applications of inorganic interest the refractive indices of scatterer and solvent (or suspending medium) are quite different, and Rayleigh–Gans theory does not apply. Typical examples are suspensions of silver chloride ($n = 2.07$) or of barium sulfate ($n = 1.65$) in water ($n = 1.33$). In such systems (unless the particles are extremely small, which is never the case in practical examples) the full Mie theory[23] must be used to understand and interpret the scattering behavior. Because of the complexities of such calculations, quantitative interpretation has become feasible only recently; most studies of inorganic colloid systems up to now have been entirely empirical.

Despite the empirical nature, many procedures of practical value have been developed with which the student should be acquainted.[24-26] Two kinds of measurement may be made. If the amount of scattering is large enough to cause an appreciable reduction in the intensity of the transmitted beam, the transmittance may be measured. This operation, which is exactly analogous experimentally to a spectrophotometric measurement, is called *turbidimetric measurement.* Such measurements are possible with the majority of colloidal suspensions. If the suspension is very dilute, or the particles are relatively small, the value of transmittance may be practically unity and the measurement is of no value. In this case the intensity at an angle to the incident beam must be measured. In some applications (see Refs. 18–22) the intensity is measured as a function of angle. In most analytical applications, the light scattered at (about) 90° is measured; this is called *nephelometric measurement.*[27] Such measurements were made visually a century ago. T. W. Richards introduced the word nephelometry (from Greek nephelos, cloud) in 1894. He used such measurements to estimate the amount of silver chloride remaining in colloidal dispersion after precipitation. He compared the amount of light scattered by the unknown suspension with that scattered by a suspension containing a known amount of dispersed silver chloride. Exactly the same principle is used in nephelometric analysis today. Very simple apparatus can serve for visual measurements. Richards used simply a source of light, two test tubes inclined at an angle so that both could be viewed from above at the same time, and two tubes of black paper which served to vary the extent of tube illuminated. Visual colorimeters of the duBoscq type are easily adapted to nephelometry. Since most measurements now are made objectively rather than visually, the practice of visual nephelometry will not be discussed further. (References 24 and 25 contain a comprehensive discussion.)

A brief description of a nephelometric microdetermination of sulfur will illustrate

---

[21] D. McIntyre and F. Gornick (eds.), *Light Scattering from Dilute Polymer Solutions,* Gordon & Breach, New York, 1964.

[22] F. W. Billmeyer in *Treatise on Analytical Chemistry* (I. M. Kolthoff and P. J. Elving, eds.), part I, vol. 5, *Wiley-Interscience,* New York, 1964, chap. 56.

[23] Named after the German physicist Gustav Mie, who solved the problem of scattering by a sphere (either transparent or absorbing) of any size in 1908.

[24] J. H. Yoe and H. Kleinman, *Photometric Chemical Analysis,* vol. II, *Nephelometry,* Wiley, New York, 1929.

[25] F. D. Snell and C. T. Snell, *Colorimetric Methods of Analysis,* Van Nostrand, Princeton, N.J., 3rd ed., 1948–1954.

[26] F. P. Hochgesgang in *Treatise on Analytical Chemistry* (I. M. Kolthoff and P. J. Elving, eds.), part I, vol. 5, Wiley-Interscience, New York, 1964, chap. 63.

[27] In nephelometers the angle viewed is at least 70–110°.

the typical procedure. By suitable treatment of the sample (combustion, wet oxidation, or other treatment as required), the sulfur is converted into sulfate ion and brought into aqueous solution. The solution is acidified with hydrochloric acid, and ethyl alcohol and gelatin are added, followed by addition of barium chloride. (The alcohol is added to decrease the solubility of barium sulfate and thus increase the sensitivity for sulfur. The gelatin is a protective colloid and increases the stability of the suspension.) The suspension is illuminated with white light and the relative intensity at approximately right angles (see Ref. 27) is measured. A calibration curve is constructed using known amounts of sulfur.

Hundreds of such methods for sulfur have been published, with variations in the acidity, the amount or kind of alcohol, the nature of the protective colloid, and the mode of addition of barium chloride. There is no difficulty in measuring sulfur at the level of parts per million; since the volume of suspension need be only of the order of 10 ml, the absolute sensitivity is high. With a rigorously controlled method of formation, and in the absence of interfering substances (see next paragraph), the reproducibility may be as good as a few per cent (but not at the 1-ppm level).

Nephelometric and analogous turbidimetric methods have advantages of simplicity, speed, and fairly high sensitivity. The difficulties arise *not* in the optical measurement, which is simple, but in the preparation of the suspension. The amount of scattering in colloidal systems is a sensitive function of the particle size.[28] Consider a suspension of silver bromide in water illuminated with green light (wavelength 546 mμ). From exact Mie theory calculations[29] it may be shown that three suspensions of the same weight concentration, containing particles with radii equal to 0.066, 0.112, and 0.132 μ, have scatterings the relative values of which are 0.94, 1.76, and 1.59, respectively. Thus in the latter two examples, an *increase* of radius of 15 per cent causes a *decrease* of 10 per cent in the amount of scattering.

Thus the conditions used must be such that a given weight of the substance to be measured will always produce a suspension containing particles of the same size. In an early investigation Hibbard[30] showed that simply by varying the concentration and manner of mixing the reagents, it is possible to produce a suspension of barium sulfate having three times the opacity of another containing the same amount of sulfate.

The following factors are of critical importance:

1. The concentration of the two ions which combine to form the suspension.
2. The ratio of concentrations in the solutions mixed.
3. The manner of mixing.
4. The rate of mixing.
5. The time before measurement.
6. The stability of the dispersion.
7. The temperature.
8. The presence of other electrolytes.
9. The presence of nonelectrolytes. The solubility of most inorganic precipitates is decreased by the addition of organic solvents miscible with water. Since the supersaturation thus is affected (Chapter 10), the number and size of particles usually are changed. Also, the change in refractive index invariably changes the amount of scattering.

[28] The sensitivity to size is still greater in Rayleigh scattering, because the scattering per particle increases as the sixth power of the radius.

[29] E. J. Meehan and Z. Hugus, *J. Opt. Soc. Am.* **51**, 260 (1961).

[30] P. L. Hibbard, *Ind. Eng. Chem.* **16**, 804 (1924).

10. If the suspensions settle too rapidly, a protective colloid may be added as noted above. The particle size is usually entirely different when the suspension is prepared in the absence and presence of protective colloids.

Turbidimetric or nephelometric measurements sometimes are used to locate the equivalence point in a titration, for example, of sulfate with barium or of chloride with silver. Experimental conditions may be found under which the amount of scattering increases during the titration, reaching a maximum at or near the equivalence point. There is an immense amount of empirical literature.[31] The student should be aware of the fact that such methods are completely empirical. Actually, it is entirely possible for the amount of scattering to pass through a maximum, or even several maxima, *before* the equivalence point, depending upon the particle size and refractive index of the colloidal particles of the slightly soluble reaction product.[32] Since the particle size is an extremely sensitive function of experimental conditions, the observed variation of turbidity during titration depends on many factors other than the amount of substance being titrated; not all of these factors can be controlled by the experimenter.

Raman scattering

When a beam of monochromatic radiation is incident upon a transparent sample, some radiation is scattered, as was discussed in detail above. In some cases there is a pattern of much weaker lines at longer wavelength, and possibly a pattern of still weaker lines at shorter wavelength. The phenomenon is called the *Raman effect*.[33] The difference in frequency between the Raman lines and the exciting line is the *Raman shift*; this is characteristic of the scattering substance and independent of the exciting frequency. The energy abstracted from the incident beam raises the molecule to a higher vibrational level.[34] Therefore, measurement of the shifts serves for identification and measurement of intensity serves for determination. Suppose, for example, that a substance (e.g., pentene-2) has Raman shifts of 1675 and 2930 cm^{-1}. If the sample is illuminated with mercury wavelength 4358 Å (22,938 cm^{-1}),[35] Raman lines are observed at (22,938–1675) cm^{-1} or 4701 Å, and at (22,938–2930) cm^{-1} or 4996 Å. Qualitatively, Raman spectra resemble infrared spectra (Chapter 55); however, they are *not* identical, because some frequencies appear only in the infrared and some appear only in the Raman spectra. For example, the 1675-cm^{-1} line of pentene-2 is strong in the Raman spectrum and barely detectable in the infrared. The spectra generally are sufficiently characteristic to allow easy identification. One big advantage, as compared to infrared analysis, is that water may be used as a solvent.

There are practical limitations to the analytical value of Raman measurements. The intensity is relatively low, which means that convenient measurements are

[31] See Refs. 24 and 25; also M. Bobtelsky, *Heterometry*, Van Nostrand, Princeton, N. J., 1960.

[32] E. J. Meehan and Grace Chiu, *Anal. Chem.* **36**, 536 (1964).

[33] The effect was discovered in 1928 by Raman and Krishnan in India and independently by Landsberg and Mandelstam in Russia.

[34] Shifts to shorter wavelength arise by interaction of photons with molecules initially in an excited vitrational level. Because of their relatively very low intensity they are of no analytical value and will not be discussed further.

[35] The frequency refers to vacuo, whereas the wavelength refers to air and 15°C and 1 atm.

possible only with liquids and solutions. The sensitivity for minor constituents is only of the order of per cent. Moreover, relatively large samples are required. The most severe practical limitations are due to the fact that the sample should be nonabsorbing and should be free of traces of substances that cause either turbidity or fluorescence.

References. Comprehensive early works are those of Hibben,[36] Kohlrausch,[37] and Herzberg.[38] A comprehensive general survey is given by Jones and Sandorfy,[39] and a brief summary with emphasis on analytical applications is given by Rosenbaum.[40]

PROBLEMS

1. For Rayleigh particles the *scattering coefficient* is

$$K = \frac{8}{3}\alpha^4\left(\frac{m^2 - 1}{m^2 + 2}\right)^2$$

and the transmission through a thickness l is

$$\frac{I}{I_0} = e^{-n\pi r^2 Kl}$$

(r = radius, $\alpha = 2\pi r/\lambda$, λ = wavelength, m = refractive index ratio, and n = number of particles per milliliter.) An aerosol is composed of spherical particles all 2×10^{-6} cm in radius; the particles have density 2 g/ml, and $m = 1.50$. Calculate transmission of 500 mμ through 100 cm of such an aerosol containing 0.01 g/liter.

Ans. 84 per cent.

2. Assuming that a constant fraction a of absorbed radiation is reemitted, derive the relation between fluorescent intensity and concentration for a substance that follows Beer's law. What relation is observed at very small concentrations?
3. Calculate the wavelengths scattered by a substance with Raman shifts of 1000 and 2000 cm^{-1} when illuminated by (a) 4358 Å; (b) 4047 Å.

[36] J. H. Hibben, *The Raman Effect and Its Chemical Applications*, Reinhold, New York, 1939.

[37] K. W. F. Kohlrausch, *Ramanspektren, Hand- und Jahrbuch der Chemischen Physik*, vol. 9 (A. Eucken and K. L. Wolf, eds.), Becker and Erles, Leipzig, 1943, and Edwards Bros., Ann Arbor, Mich., 1945.

[38] G. Herzberg, *Infrared* and *Raman Spectra of Polyatomic Molecules*, Van Nostrand, Princeton, N.J., 1945.

[39] R. N. Jones and C. Sandorfy, *Technique of Organic Chemistry* (A. Weissberger, ed.), vol. IX, Wiley-Interscience, New York, 1956, chap. IV.

[40] E. J. Rosenbaum in *Treatise on Analytical Chemistry* (I. M. Kolthoff and P. J. Elving, eds.), part I, vol. 6, Wiley-Interscience, New York, 1965, chap. 67.

CHAPTER 57 *Spectrochemical Analysis and Atomic Absorption Analysis*

In these analytical methods the sample is heated to a relatively high temperature. Chemical compounds present in the sample are dissociated more or less completely into atoms. In *spectrochemical analysis* the radiation *emitted* by the atoms is measured. Measurement of wavelength provides identification, and measurement of intensity emitted at a particular wavelength gives a quantitative determination. In *atomic absorption analysis* a separate source is used, which emits radiation characteristic of the element to be determined. For example, if sodium is to be determined, the source contains sodium. This radiation, which can be absorbed only by sodium atoms, is passed through the sample, and the amount of *absorption* is measured. This method is used only for quantitative determinations.

Since in both methods any compounds present in the sample are decomposed, an elementary analysis is obtained.[1] Furthermore, as will be explained, the analysis usually is for metallic constituents. Thus if samples of potassium chloride, potassium chlorate, or potassium perchlorate were analyzed by ordinary spectrochemical means, only potassium would be detected. By using special procedures, chlorine and oxygen could be detected but the sensitivity would be far smaller than for potassium.

Qualitative analysis is based upon the fact that each element has a characteristic emission spectrum. These emission spectra range from simple (lithium, sodium) to very complex (iron, uranium). More than one million atomic and ionic spectral lines have been measured and assigned to the various elements. In contrast to the absorption spectra of dissolved substances, practically all these lines are extremely well defined, and their wavelengths can be expressed with great accuracy. For example, the wavelength of a certain orange line of krypton-86, which is the international standard of length, is 6056.1252 Å in dry air. Many thousands of lines have wavelengths known with an accuracy of 0.001 Å. In principle, therefore, a sufficiently precise measurement of the wavelength of a line serves to identify the element. In practice one must take account of the possibility of confusion with lines of nearly the same wavelength. A widely used compilation lists 109,000 of the principal spectral lines in the wavelength region 10,000–2000 Å.[2] The average

[1] In a few exceptional cases the emission by a molecule may be measured. An example is CaF (stable at high temperature).

[2] G. R. Harrison, Massachusetts Institute of Technology, *Wavelength Tables*, Wiley, New York, 1939. This table contains some misidentifications, especially among the rare earths (lanthanides), atomic number 57–71; see R. N. Kniseley, V. A. Fassel, and C. F. Lentz, *U.S. At. Energy Comm. Rept. IS-50*, Aug. 1, 1959.

spacing between the lines in this table is 0.07 Å. Practical spectrochemical analysis in general does not depend upon a single extremely precise determination of wavelength; in doubtful cases confirmatory lines are looked for.

Most spectrochemical analysis is carried out using a spectrograph. A spectrograph includes an entrance slit, collimating lens or mirror, dispersing element [prism(s) or grating(s)] and focusing lens or mirror, as shown in Fig. 54-2, but a photographic plate (or film) is placed at the focus of the dispersed spectrum instead of an exit slit. In this way the emission spectrum is photographed. Usually two or more spectra are photographed side by side on the same plate. If one of these is a spectrum with many lines of known wavelength, for example, the spectrum of iron, this serves as a comparison spectrum; wavelengths of lines in the sample are obtained by measuring their positions relative to the positions of the known lines in the comparison spectrum. Generally instruments used for spectrochemical analysis are required to have higher dispersion (p. 973) and higher resolving power[3] than spectrophotometers. This is because absorption spectra of dissolved substances never contain the fine details that atomic emission spectra do. (For further information on construction and use of spectrographs and related equipment, the references cited in Chapter 54, p. 962, should be consulted.) However, in some cases emission spectra may be examined successfully with an ordinary spectrophotometer. This is the case if the sample contains only elements with relatively simple spectra (e.g., the alkalies) or, in general, if the temperature of the source is not high enough to excite elements having complicated spectra. This is the situation when flames are used (see below), and spectrochemical analyses using flame sources commonly are carried out with spectrophotometers rather than with spectrographs.

Intensity of emission of spectral lines

The intensity of a spectral line is directly proportional to the number of atoms (or ions) possessing sufficient electronic energy to emit the line. As an example, consider the emission of yellow light by the sodium atom. The wavelengths (in air) of 5889.95 and 5895.92 Å correspond to energies of 16,983.0 and 16,965.8 cm^{-1}, respectively. Both lines arise in electronic transition to the lowest or ground-energy level. Therefore, a sodium atom must possess energy corresponding to the above values to emit the line. (The intensity also depends upon the transition probability. This is temperature-independent and need not be considered for the present purpose.) The relative number of atoms possessing energy E above the ground state, or *excitation energy*, is given by the Boltzmann expression

$$\frac{N'}{N} = \frac{p'}{p} \exp\left(-\frac{E}{kT}\right)$$

Here N' and N are numbers of atoms in the excited and ground state, respectively, T is absolute temperature, and k is the Boltzmann constant, the value of which depends on the units of E. When E is expressed as cm^{-1}, k is 1.4388. The factor

[3] Resolving power is $\lambda/\Delta\lambda$, where $\Delta\lambda$ is the minimum difference in wavelength λ that must exist between two lines if they are to be recognized as separate; otherwise they merge into an apparent single line. For example, sodium emits two lines of wavelength 5890 and 5896 Å. A resolving power of about 1000 is required to recognize the presence of two lines.

p'/p is the ratio of the statistical weights of the two levels.[4] In the present case p'/p is equal to 2 and 1 for the energy levels corresponding to 5890 and 5896 Å respectively. The Boltzmann expression yields the values shown in Table 57-1 for the relative numbers of sodium atoms capable of emitting the spectral lines.

TABLE **57-1**

	Fraction capable of emitting	
Temp., °K	5890 Å	5896 Å
1000	5×10^{-11}	2.5×10^{-11}
2000	10^{-5}	5×10^{-6}
6000	0.034	0.017

In a system in which the total concentration of sodium atoms in the vapor state is constant, the intensity of emission therefore is increased 200,000-fold when the temperature is raised from 1000 to 2000°K. The intensity of a line originating from a still higher energy level is even more strongly dependent on temperature. For example, sodium vapor also emits a line in the near UV at 3303 Å, which corresponds to a transition to the ground level from a level 30,263 cm^{-1} above the ground level. Application of the Boltzmann equation shows that between 1000 and 2000°K the intensity of this line increases by a factor of 2.9×10^9. Also, assuming the transition probabilities for 3303 Å and the yellow lines to be the same,[5] the intensity at 1000°K of the UV line would be only about 5×10^{-9} that of the yellow line. With increase of temperature, *all* the intensities increase, but the intensities of lines of high excitation energy increase much more rapidly than those of low E.

As an additional example, we may compare the intensity of emission from zinc and from sodium vapor. The value of E for the strongest line in the spectrum of zinc (2139 Å) is about 2.75 times as great as E for the sodium lines 5890–5896 Å. In a sample containing equal concentrations of sodium and zinc atoms, the number of zinc atoms capable of emission is of the order of 10^{-9} and $10^{-3.5}$ that of the number of sodium atoms, at 2000 and 5000°K, respectively. Therefore, the sensitivity for zinc is orders of magnitude smaller than for sodium.

The numerical examples above demonstrate the great effect of temperature upon emitted intensity, and explain the great difference both in sensitivity of detection and in number of elements detectable by emission in ordinary flames and in electric sources.

In order to produce the complete emission spectrum of a neutral atom, energy equal to the ionization potential must be absorbed by the atom. The ionization potentials of the alkali metals, in the gas state, range from 3.9 V for Cs to 5.4 V for Li; for the alkaline earths the values range from 5.2 V for Ba to 7.6 V for Mg and 9.3 V for Be. Most of the metals have ionization potentials smaller than 10 V,

[4] A discussion of this factor is beyond the scope of this text. See standard texts on atomic spectra, e.g., G. Herzberg, *Atomic Spectra and Atomic Structure*, Prentice-Hall, Englewood Cliffs, N.J., 1937; H. E. White, *Introduction to Atomic Spectra*, McGraw-Hill, New York, 1937.

[5] This is not exactly correct, but the error introduced is negligible for the present purpose.

whereas the halogens and rare gases have larger potentials; as examples, the values for F, Cl, I, and He are 17.3, 13.0, 10.6, and 24.4 V, respectively.

Although it is usually not possible to state exact values of temperatures in electric arcs and sparks, the energies available in most such sources are equivalent to temperatures of 4000 to 8000°K or more. These equivalent temperatures are much greater than those attained in flames. The temperatures of ordinary air–gas and oxygen–hydrogen flames are around 2100 and 2700°K, respectively. This explains why it is possible using a simple direct-current arc to detect more than 70 elements, and to detect more by using a high-energy spark; whereas in an acetylene flame Lundegårdh[6] found fewer than 40 elements detectable. (With modern photoelectric methods of detection, 57 elements are detectable; see Chapter 61.)

Applications of spectrochemical analysis

The spectroscopic method has been in use for a long time in the sensitive detection of many elements (Bunsen, Kirchhoff). Spectrum analysis for quantitative purposes was introduced long ago by Lockyer[7] and Hartley.[8] As explained above, the intensities of various lines of an element are different, and in general the intensity of a given line is proportional to the concentration of the element. On dilution, the weakest lines will disappear first, and finally a concentration is reached at which only the most persistent lines (raies ultimes of de Gramont)[9] remain. By photographing the spectra at various dilutions of sample and observing the lines which are left, it is possible to estimate the concentration of the element. One of the first quantitative methods, used in the last century, depended upon counting the number of lines detected under specified conditions. In more accurate methods, the intensity of a certain line is determined by a photometric method and compared with a nearby line of another element present at a constant concentration. This "internal standard method" was developed by Gerlach[10] and Schweitzer.[11] A concise summary of modern procedures is given by Scribner and Margoshes.[12]

The main advantages of the spectroscopic method are the following: The method is specific for the element to be determined, although as mentioned earlier the possibility of overlapping of spectral lines must be borne in mind. The method is time-saving; it allows the quantitative determination of traces of elements in a sample, especially in an alloy, without any preliminary treatment and, in particular, without any previous knowledge as to whether or not the element may be present. Thus specific chemical tests are unnecessary.[13] A permanent record may be obtained on the photographic plate. By spectrographic analysis the following impurities in

[6] H. Lundegårdh, *Die quantitative Spektralanalyse der Elemente*, G. Fischer, Jena, 2nd ed., 1934.

[7] N. Lockyer, *Phil. Trans. Roy. Soc. London* **164**, 749 (1874).

[8] W. N. Hartley, *Phil. Trans. Roy. Soc. London* [II] **175**, 49, 325 (1884).

[9] A. de Gramont, *Ann. Chim. Phys.* [8] **17**, 437 (1909); *Ann. Chim.* **3**, 269 (1915).

[10] W. Gerlach, *Z. Anorg. Allgem. Chem.* **142**, 383 (1925).

[11] E. Schweitzer, *Z. Anorg. Allgem. Chem.* **164**, 127 (1927).

[12] B. F. Scribner and M. Margoshes, *Treatise on Analytical Chemistry* (I. M. Kolthoff and P. J. Elving, eds.), part I, vol. 6, Wiley-Interscience, New York, 1965, chap. 64.

[13] In many cases, however, preliminary separations or concentrations may be used to great advantage.

tin could be determined directly: Cu, 0.1; Pb, 0.1; Fe, 0.005; Zn, 0.001; Ni, 0.001 per cent.[14]

In the first applications of spectrochemical analysis, the accuracy was far less than that of gravimetric, volumetric, and most colorimetric procedures. An accuracy of 10 per cent was considered satisfactory. Actually such accuracy was relatively good, considering that the method was applied mainly to the determination of elements present in amounts so small that the application of conventional procedures would have been impractical or would have yielded results of even less accuracy than those of the spectrochemical method. Within recent years considerable work has been done on the excitation of arc and spark spectra in a reproducible manner. Substantial progress has been made toward the development of stable and reproducible sources and also in the instrumental problem of the accurate measurements of the intensity of spectral lines. As a result, spectrochemical analysis can be made in some cases with an accuracy of about 1 per cent. However, to attain such accuracy, as contrasted with precision, it is essential to know the approximate composition of the sample. The reason that the semiquantitative composition must be known is that in many instances the intensities of lines of a given element in a given source are increased or decreased by the addition of other elements. The direction and magnitude of the effect in some instances can be correlated in a qualitative way with the relative excitation potentials and volatilities of the elements or their compounds, but the effects are not understood quantitatively.

Spectrochemical analysis is particularly suited for work of a repetitive nature, for example, the situation encountered in large industrial plants in which hundreds of similar determinations may be required daily. Direct-reading instruments have been developed for such purposes.[15] In these instruments the photographic part of the process, which is time-consuming, is eliminated. Instead, narrow mirrors are placed in the focal plane of the spectrograph so as to direct radiation of selected wavelength to sensitive electron-multiplier phototubes. Each such receiver is illuminated by, and measures the intensity of, a selected spectrum line of an element to be determined. With such devices analytical data are available for 12 or more elements in less than 1 min. Although the cost of such instruments is considerable, the cost per determination may be reduced to a small value compared to that of conventional methods when large numbers of determinations are made.

Flame photometry

Flame tests for sodium and potassium have been used for many years in qualitative analysis. Practically no other general use was made of flame spectra for either qualitative or quantitative analysis until recently, with the exception of the pioneer work of Lundegårdh.[6] On page 999 it was explained why the sensitivity of a flame source is many orders of magnitude smaller than of an electric arc or spark, because of the effect of temperature.

Therefore, the flame is less useful for qualitative analysis than either the arc or spark. On the other hand, the temperature of a flame can be kept more nearly

[14] See, e.g., W. F. Meggers, C. C. Kiess, and F. J. Stimson, *Natl. Bur. Std. Sci. Paper 444*, 1922.

[15] See, e.g., M. F. Hasler and H. W. Dietert, *J. Opt. Soc. Am.* **34**, 751 (1944).

constant than is the case for any electric source. This makes flame spectroscopy particularly attractive for the quantitative determination of those elements for which its sensitivity is adequate. In practice, quantitative work usually is carried out with aqueous solutions. The solution is introduced into the flame as a fine spray or mist by means of an atomizer operated at constant air pressure. The temperature of the flame is constant over long periods of time when the rates of consumption of the flame gases and the rate of entry of spray are kept constant.

The intensities of the emitted spectrum lines may be determined in the photographed spectrum or may be measured directly by isolating the desired spectrum line by means of a spectroscope equipped with a suitably narrow exit slit and a photometer. Both recording and nonrecording spectrophotometers are being widely used for this purpose. By measuring the intensity of radiation at suitable characteristic wavelengths, it is possible to detect (for example) approximately 0.01 ppm of Na, 0.02 ppm of Li, 1 ppm of Ba, 10 ppm of Fe, and 5000 ppm of Se. The total volume of solution consumed need not exceed 1 ml, and the accuracy in the most favorable cases may be better than 1 per cent. A practical procedure is given in Chapter 61, page 1071. A comprehensive discussion of the principles and applications of flame photometry should be consulted for further details.[16]

A particularly simple procedure has been developed for the estimation of sodium and potassium.[17] It is based on the principle that at relatively low temperatures such as correspond to the ordinary gas flame, an appreciable amount of radiation is emitted only by those elements of low excitation potential, specifically the alkalies. Since the alkali metals have simple spectra with comparatively few lines, it is not essential to use a spectroscope to isolate radiation of the desired wavelength; radiation characteristic of sodium may be isolated from that characteristic of potassium, and vice versa, by filters of appropriate colors, and the intensity of the filtered transmitted radiation is measured to obtain a quantitative estimate of the amount of each element present. For exact work it is necessary to correct for imperfect spectral resolution and for the effects of other easily excited elements if these are present.[18] However, the method offers considerable advantages for the determination of sodium and potassium with greater ease and, under favorable conditions, of comparable accuracy as the gravimetric method. Especially in flame photometry the kind and amount of anions in the sample may have a great effect upon the intensity of the emitted light. For a given cation the degree of volatility and the heat of dissociation depend upon the anion in the salt. In general, it is best to transform alkali salts into the chlorides. Phosphates and sulfates are much less volatile than the chlorides.

Atomic absorption analysis

The previous discussion has pointed out two inherent weaknesses of spectrochemical analysis: the intensity (upon which quantitative measurements depend)

[16] R. Herrmann and C. T. J. Alkemade, *Chemical Analysis by Flame Photometry*, Vol. 14 (transl. by P. T. Gilbert), Wiley-Interscience, New York, 2nd ed., 1963.

[17] R. B. Barnes, D. Richardson, J. W. Berry, and R. L. Hood, *Ind. Eng. Chem., Anal. Ed.* **17**, 605 (1945). For determination in rocks see G. H. Osborn and H. Johns, *Analyst* **76**, 410 (1951).

[18] There is a large amount of contradictory literature regarding so-called "interelement effects" in flame photometry. Usually the observed effects arise from imperfect spectral resolution.

is extremely sensitive to small fluctuations in temperature, and the sensitivity for elements of high E (such as zinc) is much smaller than for elements of low E (such as sodium). Atomic absorption analysis, introduced by Walsh in 1955,[19] practically eliminates both these disadvantages. The principle is simple and may be illustrated by the analysis of a mixture of sodium and zinc.

In a mixture containing equal numbers of sodium atoms and zinc atoms at 2000°K the number of zinc atoms capable of *emitting* 2139 Å is only 10^{-9} times the number of sodium atoms capable of emitting 5890 Å. But the number of zinc atoms capable of *absorbing* 2139 Å is exactly the same as the number of sodium atoms capable of absorbing 5890 Å, because in each case practically all the atoms are in the lowest electronic level. This equality holds good no matter what the temperature of the sample, unless it is extremely high. Therefore, if a source is available that emits 2139 Å at constant intensity, the absorption is determined by the concentration of zinc atoms, because sodium atoms do not absorb this frequency. An analogous remark applies to the use of a source that emits sodium radiation.

By the use of the characteristic sources, interferences are eliminated in the ideal case. For reasons that are the same as those involved in spectrophotometry, and the failure of Beer's law in heterochromatic radiation (p. 972), it is essential to use a source that emits lines sharper than the absorbed lines. The width of an absorption or emission line is determined mainly by the temperature of the vapor. The production of sharp emission lines is achieved in practice by using hollow-cathode sources.

From the above discussion it is evident that the temperature of the sample would be of no consequence if it were high enough to decompose all compounds *completely* into atoms, but not so high as to cause appreciable ionization. In fact, not all compounds are completely dissociated in any source that is now used. Moreover, the extent of dissociation naturally depends upon the anion. For this reason it is still necessary, even in atomic absorption analysis, to maintain a reasonable control over the temperature and the composition of the sample.

As a practical matter, it is also necessary to eliminate the effect of *emission* by the sample. This is accomplished by interrupting the radiation from the source, at a specified frequency, and measuring the transmission of the radiation with a receiver sensitive to this frequency but not sensitive to uninterrupted (d-c) radiation.

[19] A. Walsh, *Spectrochim. Acta* **7**, 108 (1955).

CHAPTER 58 *Polarimetry, Refractometry, and Microscopic Methods*

Polarimetry

In Chapter 54 it was mentioned that natural (unpolarized) light can be resolved into two components of equal intensity which are plane-polarized at right angles to each other; that is, in each component the vibration of the electric vector is confined to a single plane. Various optical devices permit one to obtain one or the other of these components.[1] One of the simplest methods is the use of a sheet of Polaroid film. For precise optical measurements it is preferable to use a polarizing crystal made of calcite (e.g., a Nicol prism or Glan–Thompson prism) which produces complete polarization over a wide spectral range.

The analytical application of polarimetry depends upon the following fact. Many substances *rotate* the plane of vibration, and the amount of rotation by a given substance depends upon its concentration. The phenomenon is known as *optical rotation*; substances that cause such rotation are *optically active*. The effect was discovered by Biot in 1815. It was recognized not long afterward (Pasteur, van't Hoff, Lebel) that optical activity is associated with the presence in a molecule of an asymmetric carbon atom, i.e., an atom with four different substituents. Such a molecule can exist in two forms which are mirror images. Many asymmetric atoms other than carbon (tetravalent tin, sulfur, selenium, pentavalent nitrogen, to name a few) also can give rise to optical activity.

The physical picture of optical rotation is the following. A beam of plane-polarized light is exactly equivalent to two beams of circularly polarized light of equal intensity but rotating with equal frequencies in opposite directions. Upon passage through an optically inactive medium the two components travel at the same speed. In an optically active medium, however, one of the components is retarded,[2] and the student may easily show that the net result is a *rotation* of the plane of the resultant plane-polarized light. The angle of rotation evidently is proportional to the thickness traversed. If the direction of rotation by the sample is clockwise or counterclockwise as the observer views the light source, the sample is called dextrorotatory (*d*) or levorotatory (*l*), respectively. A mixture of equal amounts of *d* and *l* forms has no rotation.

[1] A very comprehensive discussion is given by W. Heller in *Technique of Organic Chemistry* (A. Weissberger, ed.), part III, vol. I, Wiley-Interscience, New York, 3rd ed., 1960, chap. 33.

[2] In other words, the refractive index is different for right- and left-hand polarization.

A polarimeter is a device for the measurement of optical rotation.[3] The measurement (for analytical applications) is carried out on a solution of concentration c, contained in a polarimeter tube (cylindrical tube with flat windows) of length l at a given temperature and using a given wavelength. It is a general convention to express c in grams per ml and l in decimeters. If the measured angle of rotation in degrees of the yellow light of sodium (D lines) at 20°C is α, the *specific rotation* is[4]

$$[\alpha]_{\mathrm{D}}^{20} = \frac{\alpha}{cl} \tag{1}$$

In dilute solution $[\alpha]$ is practically independent of concentration. For example, $[\alpha]_{\mathrm{D}}^{20}$ of sucrose in aqueous solution is 66.456 at infinite dilution, 66.464 at $c = 0.01$, and 66.519 at $c = 0.1$. However, $[\alpha]$ is very much dependent upon the solvent and in a given solvent generally is affected by the presence of optically inactive substances. Therefore, for quantitative applications it is necessary to control not only temperature and wavelength[5] but also to know the approximate composition of the solvent. The most frequent application of polarimetric measurement is the determination of sugars. It provides an easy and accurate way to follow the conversion of sucrose into fructose and glucose, because each of these sugars has a different $[\alpha]$. In some instances (notably steroids) it may be used for the approximate analysis of a mixture of two optically active substances. Polarimetry finds rare application in inorganic analysis. However, various inorganic salts can be determined polarimetrically by virtue of the fact that they form complexes with optically active molecules in which the rotation of the latter is altered markedly (i.e., molybdate with tartaric acid and tartrates).

For further information the student may consult footnotes 6–10.

Refractometry

The refractive index n can be regarded as a relatively unspecific property. It is true that every substance has a unique value of n. However, the values of n of hundreds of thousands of organic liquids cover the relatively small range from 1.26 to 1.8.[11] Since the ordinary accuracy of measurement does not exceed a few units in the fourth decimal, it is evident that measurement of n alone cannot serve

[3] Ref. 1 gives a summary of visual, photoelectric, and photographic methods of measurement.

[4] Some workers prefer to express c in grams per 100 ml, in which case

$$[\alpha]_{\mathrm{D}}^{20} = 100\,\frac{\alpha}{cl}$$

[5] Dependence of $[\alpha]$ upon wavelength is known as *optical rotatory dispersion*. It is not at present of direct analytical application, but it provides one of the most powerful tools for structure determination in organic chemistry. See, e.g., C. Djerassi, *Optical Rotatory Dispersion. Applications to Organic Chemistry*, McGraw-Hill, New York, 1960.

[6] F. J. Bates, et al., *Natl. Bur. Std. Circ. 440*, 1942.

[7] C. A. Browne and F. W. Zerban, *Physical and Chemical Methods of Sugar Analysis*, Wiley, New York, 3rd ed., 1941.

[8] W. A. Struck and E. C. Olson in *Treatise on Analytical Chemistry* (I. M. Kolthoff and P. J. Elving, eds.), part I, vol. 6, Wiley-Interscience, New York, 1965, chap. 71.

[9] T. M. Lowry, *Optical Rotatory Power*, Longmans, London, 1935.

[10] H. Landolt, *Optical Rotation of Organic Substances*, Chemical Publishing Co., New York, 2nd ed., 1902.

[11] Fluorinated hydrocarbons have lower n than any other organic liquid.

in general to prove the identity of a substance, in the way that (e.g.) the infrared spectrum can serve. Nevertheless, when used together with other nonspecific properties such as density and boiling point, it is valuable in confirming the identity of a substance.

It should be realized that many values quoted in the literature for supposedly pure substances are not reliable beyond the third decimal. For example, Bauer and Lewin[12] quote five modern values of n_D^{20} (i.e., n measured at 20°C using sodium D light) of *o*-xylene. Three of the values are given to five decimals. The average value is 1.5058 and the average deviation is 0.0010.

In the following paragraphs a brief discussion is given of the uses of refractive index measurements for the quantitative determination of the composition of solutions; this is the principal analytical application.

Measurement of refractive index

When a ray passes from isotropic medium 1 to isotropic medium 2, the angles of incidence I and refraction R are connected by Snell's law:

$$\frac{\sin I}{\sin R} = \frac{n_2}{n_1} \tag{2}$$

where n_1 and n_2 are the refractive indices of mediums 1 and 2, respectively, with respect to vacuum. The refractive index of vacuum is unity. In practice, determinations usually are made with medium 1 being air at about 1 atm and 20°C. Under such conditions n of air differs from unity by less than 0.03 per cent, and Eq. (2) usually is written

$$\frac{\sin I}{\sin R} = n \tag{3}$$

it being understood that n refers to the denser medium. Most instruments for the measurement of n, i.e., *refractometers*, are based directly or indirectly upon measurement of the critical angle.[13] The critical angle is defined by $I = 90°$ (grazing incidence), $\sin I = 1$, $\sin R = 1/n$; for glass of $n = 1.5$, $R = 41°\,49'$. Any ray in the denser medium approaching the surface at an angle greater than this is totally reflected internally. Figure 58-1 illustrates the principle. The critical angle is defined by the boundary of light and dark fields. In practice, of course, the boundary is not observed within the material but after the rays emerge again into air, and allowance must be made for the additional refraction. (See Ref. 12 for complete details.)

The most widely used refractometer for chemical purposes is the Abbe. A few drops of liquid are placed on the surface of a prism of accurately known n, and another prism is placed in contact to form a continuous film of uniform thickness, ca. 0.1 mm. The critical angle depends upon n of the liquid and the prisms, and by interchanging prisms different ranges of n may be covered. Illumination is by ordinary white light, but by the use of a compensator for dispersion, the value

[12] N. Bauer and S. Z. Lewin in *Technique of Organic Chemistry* (A. Weissberger, ed.), part II, vol. 1, Wiley-Interscience, New York, 1960, chap. 18.

[13] Measurement of intensity of reflection also may provide a sensitive measurement of n. An exhaustive and critical treatment of methods is given in Ref. 12. The same authors give a discussion of the analytical applications in *Treatise on Analytical Chemistry*, part I, vol. 6 (I. M. Kolthoff and P. J. Elving, eds.), Wiley-Interscience, New York, 1965, chap. 70.

obtained corresponds to yellow D light of sodium, i.e., n_D. Table 58-1 lists the precisions attainable with various commercial refractometers. The measurements require only a few minutes and can generally be made with 0.2 ml of sample, or less. Many refractometers also allow measurements on solids (transparent or opaque).

TABLE 58-1. **Absolute Accuracy, Relative Accuracy, and Range of Critical-Angle Refractometers**

Refractometer	*Absolute Accuracy, n*	*Accuracy, Δn^a*	*Range*
Abbe	1×10^{-4}	2×10^{-4}	1.30–1.70; 1.45–1.84
Precision Abbe	2×10^{-5}	2×10^{-5}	1.33–1.64; 1.36–1.50; 1.40–1.70
Pulfrich	1×10^{-4}	1×10^{-5}	1.33–1.61; 1.41–1.74; 1.64–1.86
Dipping (immersion)	7×10^{-5}	7×10^{-5}	1.32–1.54

[a] Accuracy in comparing n of two samples on the same refractometer.

Several differential refractometers are available for liquids which permit measurements of Δn to 3×10^{-6} or better. As the name indicates, the difference in n between sample and a reference liquid is measured directly. No critical-angle refractometer can give six-decimal accuracy in n; interferometers must be used.[14]

The value of n always decreases with increasing temperature, and close control of temperature is essential for accuracy. For water and dilute aqueous solutions

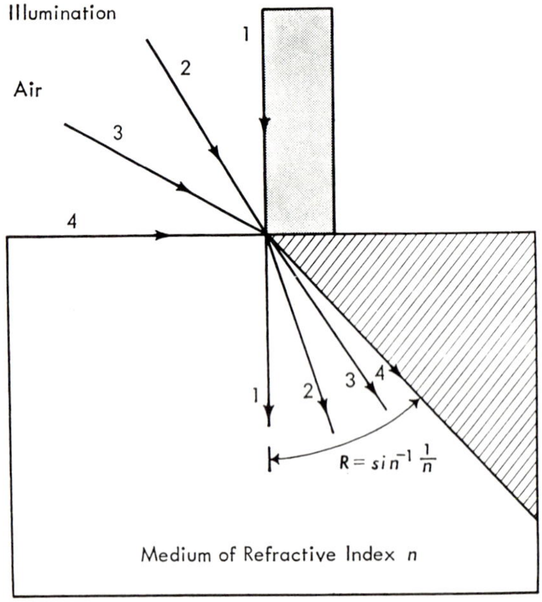

Fig. 58-1. When illuminated as shown, the cross-hatched area, for which sin $R > 1/n$, is dark.

[14] Values of n of aqueous sucrose solutions have been measured interferometrically, and these solutions are often used as references in differential refractometers.

dn/dT is about -1×10^{-4} per °C. Most organic liquids have temperature coefficients from three to five times this value; an average value of -4.5×10^{-4} per °C is commonly used for approximate temperature corrections. Evidently, therefore, temperature must be controlled within 0.2°C, even for four-decimal accuracy.

Binary mixtures and solutions

No general relation exists between the composition and refractive index of a binary mixture of components A and B. In an ideal solution n varies linearly with the volume fraction V_A:

$$n = n_A V_A + n_B(1 - V_A) \tag{4}$$

This kind of relation is observed, for example, in mixtures of pentane and mesitylene but is relatively uncommon. Solutions of naphthalene in benzene exhibit a linear relation between n and weight fraction w_A:

$$n = n_A w_A + n_B(1 - w_A) \tag{5}$$

Sometimes a relation linear in mole fraction f_A is observed:

$$n = n_A f_A + n_B(1 - f_A) \tag{6}$$

Mixtures of carbon tetrachloride and acetone follow Eq. (6). It is emphasized that there is no theoretical basis for any such equations [except Eq. (4), which is of limited occurrence], and an experimental determination must be made in every case. Binary systems including water very often show great deviations from any additive relationship. For example, aqueous solutions of ethanol and of acetone have maxima in n-composition curves. Given an empirically established curve, however, refractometry provides a quick method of analysis. The precision depends of course upon the slope of the n-composition curve (Problems 3–5). In practice, as mentioned above, a complete calibration curve must be established, under the same conditions as are used for analysis.

In many aqueous solutions n varies approximately linearly with weight concentration. In solutions of proteins, for example, dn/dc is practically constant up

TABLE 58-2. **Index of Refraction of Aqueous Solutions of Sucrose**

Wt. % *Sucrose*	n_D^{20}	Wt. % *Sucrose*	n_D^{20}
0	1.33299	10	1.34783
1	1.33443	11	1.34937
2	1.33588	12	1.35093
3	1.33733	13	1.35250
4	1.33880	14	1.35408
5	1.34027	15	1.35567
6	1.34176	16	1.35728
7	1.34326	17	1.35890
8	1.34477	18	1.36053
9	1.34629	19	1.36218
		20	1.36384

to 10 per cent by weight.[15] Table 58-2 gives the values of n of aqueous solutions of sucrose.[16] Use of this table is illustrated in Problem 5.

Microscopical methods

Microscopical methods are those in which a microscope is used. Such methods can sometimes be applied when other methods fail; they are particularly valuable in the quantitative analysis of complex heterogeneous mixtures which are not readily amenable to ordinary chemical methods.[17] To determine the relative amounts of the components of a powdered mixture, the number of particles of each may, for example, be counted under the microscope. The numerical percentages thus obtained can be converted into weight percentages if the density of each component is known; if necessary, account is taken of the difference in particle size of the various components. To be sure, such a method does not give results of high accuracy; but not infrequently an analysis can be made in no other way, and, moreover, the time required is usually not great. In an analogous manner the percentages of the components of a compact material such as a rock, an ore, or an alloy can be found by measuring microscopically the linear or areal intercepts of the components in a prepared section of the material.[18]

The microscope also serves for the indirect measurement of very small masses beyond the sensitivity limits of the microbalance. Thus the weights of small spheres of gold, platinum, mercury, etc., obtained by suitable procedures can be found by determining their diameters with a microscope micrometer.[19] Even the analysis of a minute amount of gas in the form of a spherical bubble can be based on this principle.[20]

The polarizing microscope in skillful hands is a powerful tool for identification of substances; its proper use requires a higher degree of skill and practice than the use of most instruments. An excellent survey of what can be accomplished is given by McCrone and McCrone.[21] As only one example, there may be mentioned the application of fusion methods. In such methods the sample is first melted and then allowed to solidify under the microscope. It was found that, with experience, it was possible to identify any one of 60-odd high-explosive compounds or mixtures in a few seconds on samples weighing less than 1 mg.

Reference should be made also to the use of phase-contrast microscopy, which is of fairly recent development.[22] The principle is the following. Ordinarily objects

[15] The quotient dn/dc is known as the refractive index increment (c expressed as g/ml). It is measured routinely in connection with light-scattering determination of molecular weights.

[16] Table 58-2 is abridged from an extensive table given in *Official Methods of Analysis of the Association of Official Agricultural Chemists*, p. 828, 10th ed., 1965. This table gives values at intervals of 0.2 wt. per cent up to 8.5 per cent.

[17] See E. M. Chamot and C. W. Mason, *Handbook of Chemical Microscopy*, vol. I, Wiley, New York, 2nd ed., 1938, p. 431.

[18] See the work of Chamot and Mason cited in Ref. 17, and also A. Johannsen, *Manual of Petrographic Methods*, McGraw-Hill, New York, 2nd ed., 1918, p. 291.

[19] For an application of this method see G. Lunde, *Mikrochemie* **5**, 16, 102 (1927).

[20] A. Krogh, *Abderhaldens Handbuch der biochemischen Arbeitsmethoden*, vol. 8, p. 495.

[21] W. C. McCrone and L. B. McCrone in *Treatise on Analytical Chemistry* (I. M. Kolthoff and P. J. Elving, eds.), part I, vol. 6, Wiley-Interscience, New York, 1965, chap 72.

[22] F. Zernicke, *Z. Tech. Physik* **16**, 454 (1935); A. H. Bennett, H. Osterberg, H. Jupnik, and O. W. Richards, *Phase Microscopy*, Wiley, New York, 1951. A clear review is given by O. W. Richards, *Science* **120**, 631 (1954).

are discerned (with or without the aid of a microscope) by virtue of difference in transmission (i.e., a colored object on a colorless background), or difference in refractive index (an oil droplet in water). When there is no difference in color and the difference in refractive index is small, the border is indistinct. Zernicke showed that the contrast could be exaggerated (or could be reversed) by the use of absorbing or phase-changing plates in the back focal plane of the microscope objective. These are known as zone plates. As a single example, it is possible by the use of phase-contrast microscopy to detect dispersed particles of butadiene–styrene rubber (n_D 1.54) in polystyrene (n_D 1.59); this would not be possible by ordinary microscopic methods.

PROBLEMS

1. The specific rotation $[\alpha]_D^{20}$ of sucrose is 66.5. Calculate the angle of rotation of plane-polarized light upon passing through 10 cm of solution of concentration 0.1 g/ml.
2. $[\alpha]_D^{20}$ for fructose and glucose is equal to -92.4 and $+52.7$, respectively. When the sucrose of Problem 1 is completely converted to fructose and glucose, what is the rotation?
3. The values of n_D^{20} of benzene and ethanol are 1.5012 and 1.3610, respectively, and the densities are 0.879 and 0.789 g/ml. The refractive index of a solution is accurately proportional to the volume fraction [Eq. (4)]. (a) Calculate the composition of a solution having n_D^{20} 1.3900. (b) Assuming refractometers are available with an accuracy of 10^{-4} and 10^{-5}, what is the smallest detectable concentration of benzene in ethanol?
4. Look up (International Critical Tables, or elsewhere) the refractive index of water–ethanol solutions. Assuming the use of an Abbe with an accuracy of $\pm 10^{-4}$, how accurately can the composition of solutions ca. 10, 65, and 90 wt. per cent ethanol be determined?
5. What is n_D of a solution containing 9.6 wt. per cent sucrose? How accurately must n be measured to give a concentration reliable to ± 0.5 per cent?

PART EIGHT

Other Physical Methods

CHAPTER 59 *Other Physical Methods*

Physical methods have become increasingly important in analysis. They are briefly discussed in this chapter with their analytical applications, especially in titrations.

Generally, any measurable property that is a function of the amount or the concentration of a constituent can be made the basis of a method for the determination of that constituent. Properties vary in the extent to which they are characteristic of a constituent. Colligative properties, e.g., osmotic properties, are especially uncharacteristic. Freezing-point depression, boiling-point elevation, and osmotic pressure of ideal solutions are proportional to the molar concentration of a substance. On the other hand, an emission spectrum is characteristic of a given constituent and is used for the identification and determination of the constituent in the presence of others. In polarography the half-wave potential of a dissolved constituent is more or less characteristic of the electroactive substance and can be made more characteristic by masking effects. Thus the combination of the determination of the half-wave potential and the diffusion current allows the identification and determination of a constituent.

Considerable use of noncharacteristic properties is made in quantitative analysis. We have already seen how electrical conductance is utilized in conductometric titrations. Heat of reaction between two reacting constituents is the basis of thermometric titrations (see p. 1016). Use of nonspecific properties is also made in the analysis of binary mixtures of known constituents. The composition of a mixture of water and a solute can be found from the density or refractive index. If the solute is an electrolyte, the concentration can be found by measuring the electrical conductance. Tables are sometimes available which give the relation between composition and the property measured. It should be realized that in a binary mixture there is a possibility that two mixtures of different composition may have numerically the same value of the measured property. In Fig. 59-1 point a gives the value of some property for pure compound A and b that for pure compound B.

On going from 100 per cent A to 100 per cent B, the numerical value of the property as a rule does not change linearly or, in other words, not according to the straight line 1. If the latter would give the numerical value of a property of a mixture of A and B, the composition of any mixture could be easily calculated from the numerical value of property of the mixture and the values of the same property of pure A and B. Suppose, for example, that the mixture contains x per cent B; then it contains $(100 - x)$ per cent A. The following relation holds:

$$100 \cdot m = (100 - x)a + xb$$

where m denotes the numerical value of the property of the mixture. As a rule, however, there is no linear relation as represented by line 1; curves of types such as 2 or 3 in Fig. 59-1 represent the more general case. Even then it is possible to derive the composition of a binary mixture from a single measurement of an unspecific property with the aid of suitable tables.

The matter becomes more involved if the curve giving the numerical values of the property of mixtures of A and B shows a definite maximum (curve 4) or minimum

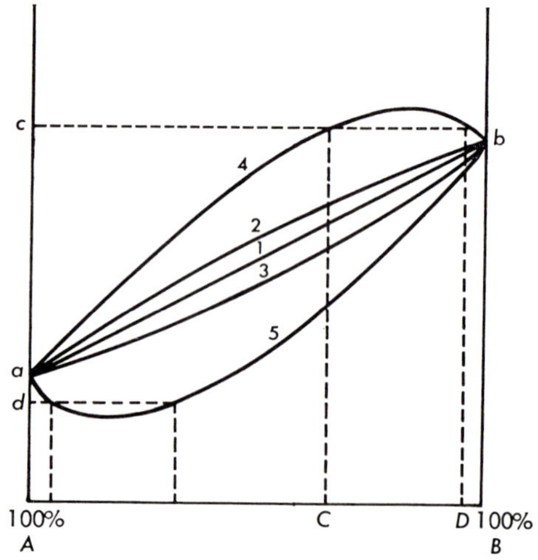

Fig. 59-1.

(curve 5). If we are dealing with the case represented in curve 4, it is easily seen that any measured value greater than b corresponds to two different compositions of the mixture. Let the value measured be equal to c. The composition may then correspond either to C (ratio A : B = BC : AC) or to D (ratio A : B = BD : AD). Usually there are simple methods for deciding which of the two is being dealt with.

If we are dealing with a ternary mixture, the measurement of one property does not suffice to find the composition. Determination of the values of two different unspecific properties often enables us to solve the problem in a graphical way, as illustrated in Fig. 59-2.

BWA is a right-angled triangle in which $BW = WA$. Let W represent pure water, B the pure solute B, and A the pure solute A. Points on the line WB indicate mixtures of water and B, whereas points on WA correspond to mixtures of water and A.

Any point in the triangle represents a mixture of W, A, and B, the composition of which may be found graphically. If P be such a point, it is seen that the corresponding mixture contains $MB = NA$ per cent water and $(100 - NA)$ per cent of B plus A,

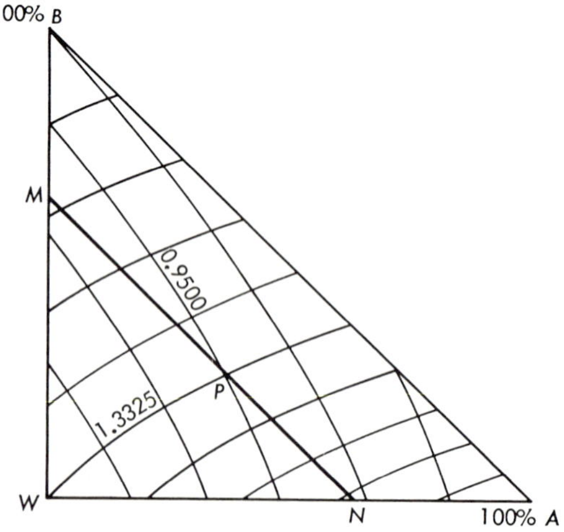

Fig. 59-2.

the latter in a ratio of *PN* of *B* to *PM* of *A*. Empirically the numerical values of two unspecific physical properties, say the density and the refractive index, are determined for a great number of mixtures of varying composition. The lines in the figure represent, respectively, the compositions of mixtures having the same density and the compositions of mixtures having the same refractive index. The density and the refractive index of the unknown are determined. Suppose that the former value is 0.9500 at 25°C; then the composition of the unknown must lie on some point of the line marked 0.9500. If the refractive index is 1.3325, the composition must be on the line marked 1.3325. Evidently the point of intersection, *P*, of these two lines gives the composition of the unknown. Actually, as a rule, one will plot not the entire diagram *BWA* but only a small part of it on a large scale in order to make more accurate interpolation, if necessary.

Theoretically it would be possible to make a similar physicochemical analysis if three solutes were present in the solution. In this case three properties would have to be measured. However, the solution becomes less simple, and moreover small errors in the determination of the physical properties would affect the results very seriously. For a more detailed discussion see Kurz in *Treatise on Analytical Chemistry*, part I, Wiley-Interscience, New York, 1967.

For numerical values of properties of substances and mixtures, the student is referred to extensive tables to be found in the following:

National Research Council, *International Critical Tables of Numerical Data*, McGraw-Hill, New York, 1926–1933.

Landolt-Bornstein-Roth, *Physikalisch chemische Tabellen* (W. A. Roth and K. Scheel, eds.), and supplementary volumes, Springer, Berlin, 1951–1965.

Tables annuelles de constantes er données numeriques, Gauthiers-Villars, Paris and McGraw-Hill, New York.

L. Meites (ed.), *Handbook of Analytical Chemistry*, McGraw-Hill, New York, 1963.

Handbook of Chemistry and Physics, Chemical Rubber Publishing, Cleveland, Ohio, 1965.

N. A. Lange (ed.), *Handbook of Chemistry*, McGraw-Hill, New York, 10th ed., 1961.

In addition to density and refractive index, the following nonspecific properties are used in analysis.

Viscosity

When a body is moved through a liquid, it meets with a resistance that tends to stop the motion. This resistance is due to the "viscosity" of the liquid. A liquid, such as ether, which is very mobile, has a relatively small viscosity, whereas a liquid such as glycerol or a concentrated sugar solution, which is syrupy in character, shows a relatively high viscosity. Although the viscosity of water is changed by a solute, hardly any use is made of viscosity determination in quantitative inorganic analysis. Such determinations are of more importance in organic analysis, especially in the evaluation of mineral oils. Measurements of viscosity are as a rule made by the determination of the rate of flow of a liquid through a capillary tube under definite conditions. The apparatus is called a *viscometer*.

Surface tension

Dissolved substances change the surface tension of the solvent. Most inorganic substances increase the surface tension of water, but the effect is relatively small and of no analytical importance. Various organic substances decrease the surface tension of water materially, especially in relatively dilute solutions; such substances are said to be capillary-active. The concentration of such a substance may be found by determining the surface tension of the solution. Use of an abrupt change of the surface tension of a solution at the end point of a titration can be made in *tensiometric titrations*.

For example, salts of alkaloids, which are weak bases, hardly affect the surface tension of water. On the other hand, the free alkaloids, which are weak bases, greatly lower the surface tension. Similarly, salts of the lower fatty acids (C_6–C_8) have little effect on the surface tension of water, but the free acids greatly lower the surface tension. Thus alkaloid salts or salts of the lower fatty acids can be used as surface-tension indicators in acid–base titrations. The method has little practical significance.

Miscibility-phase titrations

When two liquids are only partially soluble in each other, they may separate into two layers after mixing. Increase of temperature as a rule increases the mutual solubilities, and often a temperature is found above which the two liquids are completely miscible in any ratio. This temperature is called the *critical solution temperature.*

Two liquids that are miscible in any ratio at room temperature may separate into two phases on addition of a third substance. Alcohol and ligroin, for example, are completely miscible. However, on addition of water the solution becomes turbid, and two layers are formed. Water can be added from a buret until upon thorough shaking the liquid phase remains turbid (*phase titration*). Another method is to add an excess of water and to heat the mixture in a closed tube somewhat above the temperature at which the two layers become completely miscible. Upon slow cooling with stirring the temperature is read at the point where a permanent turbidity appears. The determination of this "critical solution temperature" is often used in the determination of small concentrations of water in alcohols and some other organic solvents.

Thermal methods of analysis

The determination of the critical solution temperature might be called a *thermometric method.* By an exact determination of the freezing- or melting-point curve of a compound it is often possible to determine with a very high accuracy its degree of purity. The method does not give information regarding the kind of impurity.

In recent years *thermometric titrations* have been gaining popularity for titrations in aqueous and nonaqueous media. Here the heat of reaction is measured upon titration of a constituent with a suitable reagent. Use is made of the enthalpy change during the titration in the location of the end point. The change in temperature during the titration is plotted versus the volume of standard reagent added. Two lines are obtained which intersect at the end point.[1]

Another type of thermometric titration is the *cryoscopic* one. In recent years good use has been made of the advances in the technique of freezing-point measurements in cryoscopic titrations. The end point is found graphically from a plot of freezing-point lowering versus volume of reagent added.[2]

Thermal conductivity. The rate of heat transfer from a heated solid in contact with a gas is related to the thermal conductivity of the gas and the temperature difference between the solid and the gas. If the solid is a temperature-sensitive resistance, e.g., a thermistor or a fine metal wire, and the resistance is heated by the passage of current, its final temperature will be determined by the thermal conductivity of the gas. This resistance can be measured extremely accurately by using a Wheatstone bridge in which the current through the bridge serves to heat the temperature-sensitive resistance. In a mixture of two gases, the thermal conductivity of the mixture, k, is related to the thermal conductivity of the two components k_1 and k_2 by

$$\frac{1}{k} = \frac{1 - x}{k_1} + \frac{x}{k_2} \tag{1}$$

where x represents the mole fraction of the second component. Equation (1) applies

[1] For a detailed discussion see J. Jordan in *Treatise on Analytical Chemistry*, part I, vol. 8, Wiley-Interscience, New York, 1967, p. 5,175.

[2] S. Bruckenstein and N. E. Vanderbourgh, *Anal. Chem.* **38** 687 (1966).

when $x < 0.1$.[3,4] Values of the thermal conductivity of some common gases are given in Table 59-1.

TABLE 59-1. **Thermal Conductivities of Gases and Organic Vapors**[a]

Gas or Organic Vapor	$k \times 10^5$ *at* 0°C
Hydrogen	41.6
Helium	34.8
Methane	7.2
Oxygen	5.9
Nitrogen	5.8
Air	5.8
Nitric oxide	5.7
Carbon monoxide	5.6
Ammonia	5.2
Ethane	4.3
Carbon dioxide	3.7
Propane	3.6
Methanol	3.4
Hydrogen sulfide	3.1
Acetone	2.4
Benzene	2.2
Sulfur dioxide	2.0

[a] From S. Dal Nogare and R. S. Juvet, Jr., *Gas-Liquid Chromatography*, Wiley-Interscience, New York, 1962, p. 192.

The thermal-conductivity technique is capable of considerable accuracy and has high sensitivity if a sufficient difference exists between the thermal conductivities of the two components. For example, the Perkin-Elmer Corporation has marketed an instrument for the quantitative microanalysis of organic compounds which determines the carbon, hydrogen, and nitrogen content by a series of thermal-conductivity measurements after combustion of the organic compound. Using helium ($k = 34.8 \times 10^{-5}$) as the carrier gas, the results quoted rival the classical gravimetric and gasometric techniques.

Differential thermal conductivity cells have been widely used as detectors in gas chromatographs. The greatest sensitivity is obtained using helium as the carrier gas, but cost factors make the use of nitrogen ($k = 5.8 \times 10^{-5}$) common. Further data are to be found in the monograph by Dal Nogare and Juvet.[5]

Other thermal methods. A number of other thermal methods of analysis have been developed in recent years. Among these are:

1. Thermogravimetry, which involves the determination of the weight of a substance as a function of temperature.
2. Differential thermal analysis (DTA), which involves determining the temperature difference between the substance being studied and a thermally stable material, such as γ-Al_2O_3, when both are heated under identical conditions. These temperature differences are related to chemical reactions or phase changes that may occur.

[3] R. S. Brokaw, *Ind. Eng. Chem.* **47**, 2398 (1955).

[4] H. Lehman, *Technik* (*Berlin*) **9**, 530 (1957).

[5] S. Dal Nogare and R. S. Juvet, Jr., *Gas-Liquid Chromatography*, Wiley-Interscience, New York, 1962, chaps. X and XI.

3. Pyrolytic techniques involving all methods in which the sample is thermally decomposed, and the products of decomposition analyzed by some chemical or physical method.

Further details concerning thermal methods of analysis are to be found in Wendlandt's monograph.[6]

Comprehensive review of physical properties

In the last 25 years a host of new nuclear, subatomic, atomic, and molecular properties have been discovered on which new analytical techniques are based. These advances in techniques are being made possible, not only by new discoveries and theories, but also by the phenomenal progress of instrumentation. This progress has its impact both on the development of techniques based on new discoveries, and also on the development of techniques based on the measurement of a host of classical properties. It is beyond the scope of this book to discuss in any detail the host of modern techniques; references are made to texts dealing with such techniques.[7–9] Techniques based on electrical and optical properties have been presented at length in the sections dealing with Electroanalysis and Optical Methods of Analysis.

Magnetic susceptibility determinations find use in the analysis of gases. For example, the magnetic susceptibility of molecular oxygen is much greater than that of any other gas. Based on this difference, Pauling et al.[10] developed the "oxygen meter."

Magnetic titrations. Coryell et al.[11] introduced this type of titration, which is based on a change in orbital moment that occurs when a diamagnetic reagent is added to a paramagnetic reagent or vice versa. The change of the magnetic susceptibility is measured during the titration and the end point is found graphically. The bulk susceptibility of a solution is measured by Gouy's method.[12] Heit and Ryan[13] simplified the apparatus for a magnetic titration. They give the titration curves obtained in the reaction between nickel and cyanide, and of iron(II) with 1,10-phenanthroline or bipyridyl. It is not expected that the method will find general analytical application.

An introduction to the theory underlying magnetic methods of analysis is given by Mulluy.[14]

Nuclear magnetic resonance. The work of Purcell and coworkers[15] and of Bloch and coworkers[16] in 1946 led to the development of one of the most important techniques presently used for the elucidation of the structure of organic molecules—nuclear magnetic resonance (NMR) spectroscopy. NMR techniques have had an impact rivaled by few other discoveries made in this century. From the analytical viewpoint, the application of NMR spectroscopy lies principally in the area of determining the detailed structure and configuration of molecules rather than in determining the concentration of a particular species. A short introduction to the principles underlying NMR spectroscopy is given below, along with references to more detailed works on the subject.

[6] W. W. Wendlandt, *Thermal Methods of Analysis*, Wiley-Interscience, New York, 1964.

[7] W. G. Berl (ed.), *Physical Methods in Chemical Analysis*, Academic Press, New York, 1950–1956.

[8] A. Weissberger (ed.), *Technique of Organic Chemistry*, Wiley-Interscience, New York, 1960.

[9] I. M. Kolthoff and P. J. Elving (eds.), *Treatise on Analytical Chemistry*, part I, vols. III to X, Wiley-Interscience, New York, 1950–1968.

[10] L. Pauling, R. E. Wood, and J. H. Sturdevant, *J. Am. Chem. Soc.* **68**, 795 (1946).

[11] C. D. Coryell, L. Pauling, and M. W. Dodson, *J. Phys. Chem.* **43**, 825 (1939).

[12] L. G. Gouy, *Compt. Rend.* **109**, 935 (1889).

[13] M. L. Heit and D. E. Ryan, *Anal. Chim. Acta* **29**, 524 (1963).

[14] L. N. Mulluy, *Magnetic Methods*, Wiley-Interscience, New York, 1963.

[15] E. M. Purcell, H. C. Torrey, and R. V. Pound, *Phys. Rev.* **69**, 37 (1946).

[16] F. Bloch, W. W. Hansen, and M. E. Packard, *Phys. Rev.* **69**, 127 (1946).

Nuclei that have an odd number of neutrons *or* protons act as if their nuclear charges are spinning about an axis, and thus behave as a small magnet. Such nuclei possess an *angular* momentum, which is described by the *spin quantum number*, I, and a *nuclear magnetic moment*, μ. Depending upon the particular nucleus, the value of I varies between $\frac{1}{2}$ and $\frac{9}{2}$, taking on half-integral or integral values. If the mass number of the nucleus is odd, I is half integral, while if the mass number is even and the nuclear charge number is odd, I is integral. Nuclei that contain an even number of protons *and* neutrons have no net angular momentum ($I = 0$). The total angular momentum of a nucleus equals $(h/2\pi)\,[I(I + 1)]^{1/2}$, where h is Planck's constant.

The spin quantum numbers of some nuclei are given in Table 59-2.

TABLE 59-2. **Nuclear Spin Quantum Numbers of Some Nuclei**

Nucleus	*Spin Quantum Number* (I *in units of* $h/2\pi$)
^{1}H	$\frac{1}{2}$
^{2}H	1
^{6}Li	1
^{7}Li	$\frac{3}{2}$
^{10}B	3
^{11}B	$\frac{3}{2}$
^{12}C	0
^{13}C	$\frac{1}{2}$
^{14}N	1
^{15}N	$\frac{1}{2}$
^{16}O	0
^{17}O	$\frac{5}{2}$
^{19}F	$\frac{1}{2}$
^{29}Si	$\frac{1}{2}$
^{31}P	$\frac{1}{2}$
^{32}S	0
^{33}S	$\frac{3}{2}$
^{35}Cl	$\frac{3}{2}$
^{45}Sc	$\frac{7}{2}$
^{73}Ge	$\frac{9}{2}$

Pople et al.[17] give a very complete listing of the magnetic properties of nuclei.

In the presence of a uniform external magnetic field, H, a spinning nucleus with a magnetic moment may take up only a certain set of discrete orientations with respect to the magnetic field. The number of allowed orientations is $2I + 1$. Associated with each orientation is a *magnetic quantum number*, m, which takes on values of I, $(I - 1)$, $(I - 2)$, . . . , $(-I + 1)$, and $-I$. For the proton, $I = \frac{1}{2}$; hence there are two possible magnetic quantum numbers (and orientations) which have the values $\frac{1}{2}$ and $-\frac{1}{2}$. For deuterium, ^{2}H, there are the three values 1, 0, and -1.

The two possible orientations for the proton correspond to one in which there is a component of the magnetic moment in the direction of the magnetic field (low-energy state) and the other in which the component is opposite to the external magnetic field

[17] J. A. Pople, W. G. Schneider, and H. J. Bernstein in *High-Resolution Nuclear Magnetic Resonance*, McGraw-Hill, New York, 1959, p. 480ff.

direction (high-energy state). These components are illustrated in Fig. 59-3. In this figure, the lines that terminate with an arrowhead and are labeled μ are vectors that represent the orientation of the axis of the spinning nucleus with respect to the external magnetic field. Their projections along the direction of the magnetic field are the vectors labeled μ_H.

The energy associated with such states as compared to their energy in the absence of a magnetic field is $m\mu H/I$. Thus, for the proton, the two states have energies of $-\mu H$ and μH, with an energy difference of $2\mu H$.[18]

If electromagnetic radiation (emr), properly oriented with respect to the external magnetic field, of energy exactly equal to $2\mu H$ is incident upon the system containing ^1H nuclei, some nuclei will change their spin orientation, going from the lower-energy state to the higher-energy state. The separation between these two levels is very small. For the proton, $\mu = 1.42 \times 10^{-22}$ erg/gauss; hence in a magnetic field of 10 kG, as is common in a commercial NMR spectrometer, the energy difference between these levels is $2(1.42 \times 10^{-22})\ (10^3) = 2.8 \times 10^{-19}$ erg. As a result of this small energy separation, only 6×10^{-4} per cent more protons are found in the lower-energy level as compared to the higher-energy level at room temperature. Despite this small difference in population of the two spin states, it is possible to detect the absorption of emr of the appropriate frequency.

The energy associated with a quantum of emr of frequency ν is $h\nu$. Hence the condition for absorption of radiation by a proton going from one spin state to another is (2)

$$h\nu = 2\mu H \tag{2}$$

or the "resonance frequency" is

$$\nu = \frac{2\mu H}{h} \tag{3}$$

In the presence of a magnetic field, the axis of rotation of the proton may be thought of as precessing about the direction of the magnetic field just as the axis of rotation of a spinning top precesses about the earth's gravitational field if the top is not exactly vertical. The angular velocity of precision of the proton, ω, is related to the resonance frequency, ν, by

$$\omega = 2\pi\nu$$

Hence from Eq. (3)

$$\omega = \frac{4\pi\mu H}{h} \tag{4}$$

The magnitude of the angular momentum vector for the proton is determined by the spin quantum number, I, and equals $h/4\pi$; hence Eq. (4) becomes

$$\omega = \frac{\mu H}{I} = \gamma H \tag{5}$$

where γ is known as the gyromagnetic (or magnetogyric) ratio.

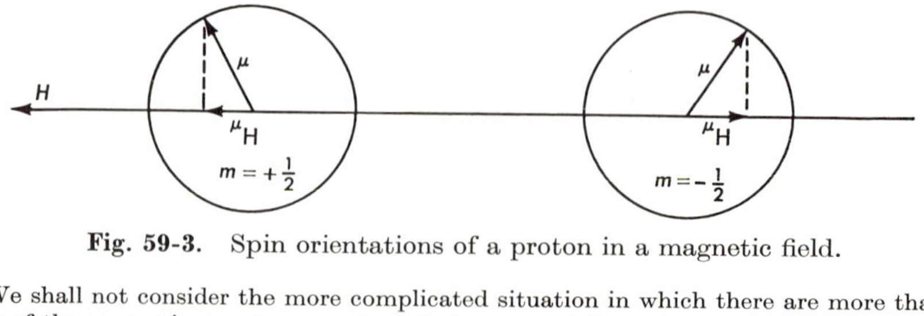

Fig. 59-3. Spin orientations of a proton in a magnetic field.

[18] We shall not consider the more complicated situation in which there are more than two values of the magnetic quantum number. It is apparent that if more than two energy levels exist more than one transition may occur.

In an NMR spectrometer, the sample containing the nuclei being studied is placed in a uniform magnetic field. A coil leading to a radiofrequency (rf) oscillator is positioned adjacent to the sample such that the coil axis is perpendicular to the direction of the magnetic field while the plane of the coil is parallel to the direction of the magnetic field. This arrangement produces the equivalent of a rotating electromagnetic field whose magnetic component moves in a plane perpendicular to the direction of the applied magnetic field. Absorption of energy, resonance, occurs when the condition given in Eq. (2) is satisfied.

Commercial NMR spectrometers irradiate the sample with a constant rf frequency, and sweep the magnetic field over a small range of field strengths, producing a spectrum that has the appearance of Fig. 59-4. In the Varian Model A-60 NMR instrument, the rf oscillator operates at 60 Mhz, and the proton resonance is observed when the field is 14,092 G. The absorption of energy can be observed by measurements made at the terminals of the rf coil, or a second detector coil can be used. This detector coil is positioned so that its axis is mutually perpendicular to the axis of the rf coil and the direction of the magnetic field. Any signal arising in the detector coil originates in the coupling produced between it and the rf coil by changes in spin orientation within the sample.

The application of NMR spectroscopy to the study of the structure of molecules depends upon the fact that the local magnetic field in the vicinity of a nucleus differs from the external magnetic field. This difference is related to the local electronic environment of a specific nucleus, and very detailed conclusions can be drawn from a high-resolution NMR presented on the next page in Fig. 59-5.

Since all charged particles are influenced by a magnetic field, in the presence of an applied magnetic field electronic circulations about a nucleus will occur so as to induce a magnetic field that opposes the external one. This type of magnetic shielding is known as *diamagnetic shielding*, and is proportional to the applied magnetic field. Hence the field at a nucleus can be represented by

$$H_{\text{eff}} = H(1 - \sigma) \tag{6}$$

where σ is the screening (or shielding) constant. Its value for protons is about 10^{-5}. In Eq. (2) above, H_{eff} rather than H should have been used, and the more exact expression for resonance is

$$\nu = 2\mu H(1 - \sigma)/h \tag{7}$$

The relative difference in the resonance field strength of a proton in a sample as compared to the resonance field strength in a standard is called the *chemical shift*, δ.

$$\delta = (\mathrm{H_{sample}} - \mathrm{H_{ref}})/\mathrm{H_{ref}}$$

δ is also given by

$$\delta = \sigma_{\text{ref}} - \sigma_{\text{sample}}$$

and expressed in units of parts per million (ppm).

NMR spectrometer magnetic-field sweep ranges are frequently calibrated by the *side-band technique* using a low-frequency oscillator to modulate the magnetic field. Side-bands appear on either side of the resonance signals in the spectrum, displaced by a frequency which is an integral number times the modulating frequency.

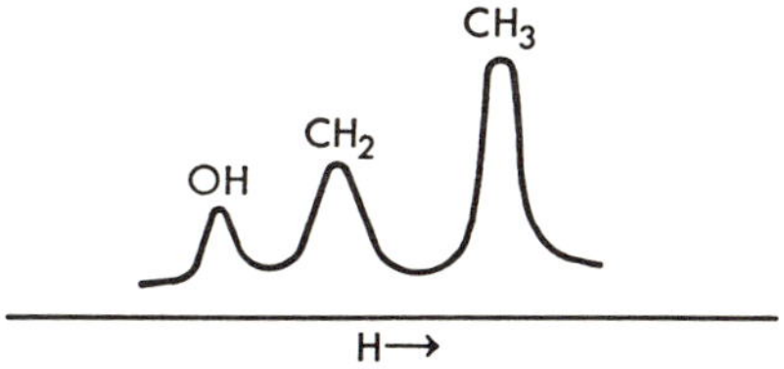

Fig. 59-4. Low-resolution proton NMR spectrum of liquid ethanol.

Provided the amplitude of the modulating frequency is small enough, only the side-band corresponding to the frequency of the low-frequency oscillator will have appreciable intensity as compared to the original resonance line. The difference between two different resonance lines can be interpolated from the side-band markers introduced by the low-frequency oscillator. The chemical shift is then given by

$$\delta = (\nu_{\text{ref}} - \nu_{\text{sample}})/\nu$$

where ν is the frequency of the rf oscillator, since ν is very nearly equal to ν_{ref}.

δ is ordinarily expressed in units of parts per million (ppm). Hence, if the rf oscillator frequency is 60 Mhz

$$\delta = \frac{\nu_{\text{ref}} - \nu_{\text{sample}}}{60}$$

The usual reference substance in organic solvents is tetramethyl silane. Tetramethyl silane, $(CH_3)_4Si$, has 12 identical protons in an electronic environment which produces high diamagnetic shielding as compared to protons in the usual organic compound. For this reason, Tiers[19] suggested another unit of chemical shift, defined as

$$\tau = 10.0 - \delta$$

so that negative values of chemical shift would not be encountered in most studies.

Considering a specific example, it is customary to denote covalently bound hydrogen ions as protons in the particular organic group; for example, the OH proton in ethanol denotes the hydrogen of the hydroxyl, the methylene protons refer to the two hydrogens in the CH_2 group, etc.

The spectrum of an acidified sample of ethanol as obtained on a high-resolution NMR spectrometer is given in Fig. 59-5 and should be compared to the low-resolution spectrum shown in Fig. 59-4.

In both these figures the absorption due to the OH proton appears as a single peak, while the absorption due to the methyl protons results in a triplet and that due to the methylene protons results in a quartet in the high-resolution spectrum, as compared to the single peaks observed in Fig. 59-4.

These *splittings* are related to changes in the magnetic field resulting from different spin orientations of the protons on adjacent groups. Ramsey[20] has proposed a mechanism that accounts for the principal cause of this phenomenon, known as *spin-spin splitting*. The two methylene protons have spins of either $+\frac{1}{2}$ or $-\frac{1}{2}$, which we represent by vertical arrows $\uparrow$ or $\downarrow$, respectively. The various permutations of spin states of the two protons are

$$\uparrow\uparrow;\ \uparrow\downarrow,\ \downarrow\uparrow;\ \downarrow\downarrow$$

and have equal probability. The two states with spins $+\frac{1}{2}$ and $-\frac{1}{2}$, and $-\frac{1}{2}$ and $+\frac{1}{2}$, are equivalent. These orientations produce variations in the magnetic field adjacent

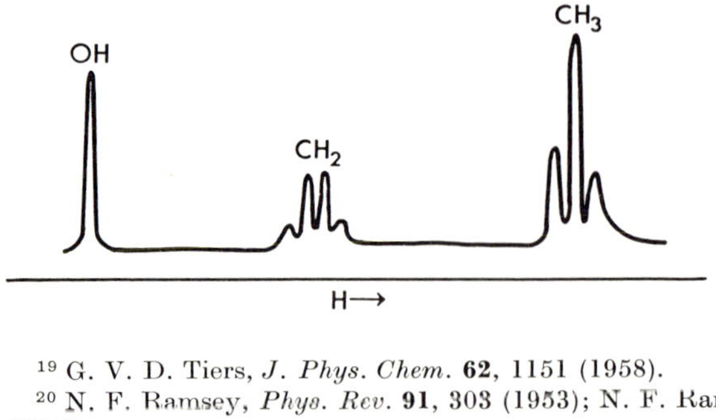

Fig. 59-5. High-resolution proton NMR spectrum of liquid ethanol.

[19] G. V. D. Tiers, *J. Phys. Chem.* **62**, 1151 (1958).

[20] N. F. Ramsey, *Phys. Rev.* **91**, 303 (1953); N. F. Ramsey and E. M. Purcell, *ibid.* **85**, 143 (1952).

to the methyl protons by means of an indirect coupling mechanism via the electrons in the molecule and split the methyl proton resonance into three peaks with areas in the ratio of 1 : 2 : 1. No effect is seen on the OH proton since it exchanges very rapidly with protons on other ethanol molecules in acidified solution. This exchange is rapid compared to the time of the experimental measurements and results in an averaging of the various magnetic environments seen by the OH proton.

In highly purified ethanol the OH proton exhibits a triplet in the ratio 1 : 2 : 1, since proton exchange between ethanol molecules is much slower in the absence of acid (or base).

The various permutations of spin orientations for the three methyl protons are

$$\uparrow\uparrow\uparrow;\ \uparrow\uparrow\downarrow,\ \uparrow\downarrow\uparrow,\ \downarrow\uparrow\uparrow;\ \uparrow\downarrow\downarrow,\ \downarrow\uparrow\downarrow,\ \downarrow\downarrow\uparrow;\ \downarrow\downarrow\downarrow$$

There are four nonequivalent combinations of the methyl proton spin states and these split the methylene proton resonance into four peaks, with areas in the ratio 1 : 3 : 3 : 1.

The frequency difference (in hertz) between successive peaks in a multiplet is equal to the *spin-spin splitting constant*, J, in a limiting case. This limiting case is known as the zero-order approximation and corresponds to the situation where J is much smaller than δ, when both have the same units. $J = 7$ Hz for protons in the methylene and methyl groups in ethanol as compared to a chemical shift of 149 Hz when the rf oscillator operates at 60 MHz. J is independent of the external magnetic field, hence spin-spin splitting peaks can be differentiated from chemical shift peaks by altering the magnetic field of the NMR instrument. Alternatively one may vary the solvent, since J is independent of solvent while δ is not. If the difference in the chemical shift for the proton on two groups approaches within an order of magnitude of the spin-spin splitting constant, the spectra are considerably more complicated than would be inferred from the above treatment. Many spin-spin splitting situations fall into this category.

It is apparent that, if the NMR spectrum of a compound can be related to previously determined chemical shifts and spin-spin splitting constants, the detailed structure of a molecule can be determined. Applications of this sort number in the thousands, and typical examples are to be found in a number of introductory and advanced monographs.[21–25]

The actual application of the NMR method involves details that cannot be discussed here and depends upon such factors as the spatial arrangement and detailed chemical environment of protons. Applications include the determination of the conformation of rings in steroid molecules and the rotational barriers in molecules, in addition to the exact position of various functional groups in newly prepared organic molecules.

Varian Associates has published catalogs containing the NMR spectra of over 1000 organic molecules.[26]

The application of NMR spectroscopy to quantitative analysis depends upon the fact that the signal strength is proportional to the number of magnetic nuclei, as discussed by Pople et al. (Ref. 17). The NMR method has been applied to the determination

[21] J. D. Roberts, *Nuclear Magnetic Resonance*, McGraw-Hill, New York, 1959.

[22] L. M. Jackman, *Applications of Nuclear Magnetic Resonance Spectroscopy in Organic Chemistry*, Pergamon, New York, 1959.

[23] K. Wiberg and B. J. Nist, *The Interpretation of NMR Spectra*, Benjamin, New York, 1962.

[24] J. D. Roberts, *An Introduction to the Analysis of Spin–Spin Splitting in High-Resolution Nuclear Magnetic Resonance Spectra*, Benjamin, New York, 1961.

[25] N. S. Bhaca and D. H. Williams, *Applications of NMR Spectroscopy in Organic Chemistry*, Holden-Day, San Francisco, 1964.

[26] N. S. Bhaca, L. F. Johnson, and J. N. Shoolery, *NMR Spectra Catalog*, Varian Associates, 1962, and N. S. Bhaca, D. P. Hollis, L. F. Johnson, and E. A. Pier, *NMR Spectra Catalog*, Varian Associates, 1963.

of moisture in agricultural materials, food products, and other biological and organic materials. Details are to be found in the literature.[27]

The nondestructive NMR method has been used to follow the concentration of species in reaction mixtures.[28] It has been used to determine the relative concentration of various anions of the oxy acids of phosphorus,[29] using the P resonance.

Applications to quantitative analysis are becoming more frequent but are still limited to situations in which a nondestructive technique is required, or to problems for which no other satisfactory technique presently exists.

Mass spectrometry

In a mass spectrometer a gas or a vapor derived from a liquid or a solid is bombarded by a beam of electrons in an evacuated chamber called the *ionization chamber*. The bombardment causes both ionization and rupture of chemical bonds. The positive ions thus formed are accelerated in an electric field and pass through a narrow slit into the analyzer chamber. In this chamber the action of a magnetic field at right angles to the electric field causes an ion of a given m/e ratio (m = molecular weight, e = charge, usually equal to unity) to travel in a circular path, the radius of which is proportional to the square root of m/e. Thus the heterogeneous beam of ions entering the analyzer is spread out into a fan of ion beams of various values of m/e. This spread-out fan may be photographed, as in the early instrument of Aston (1919), but in the majority of present applications an electric method of measurement is used. The radius of the circular path is proportional to the square root of the accelerating voltage and inversely proportional to the square root of the magnetic field. Therefore, by varying either the voltage or the magnetic field, ions of different m/e ratios travel in the same path and pass through a narrow slit into the detector where the corresponding small electric current is amplified and is either measured or recorded.

The *kind* and *amount* of the fragments produced in the ionization chamber, for a given energy of bombarding beam, are characteristic of the molecule, so that every chemical compound has a unique cracking pattern or *mass spectrum*. For example, a simple molecule such as ethane, C_2H_6, yields the fragments listed in Table 59-3.

TABLE **59-3**[a]

m/e	*Ion*	m/e	*Ion*
1	H	25	C_2H
2	H_2	26	C_2H_2
12	C	27	C_2H_3
13	CH	28	C_2H_4
14	CH_2	29	C_2H_5
15	CH_3	30	C_2H_6
24	C_2		

[a] Fragments containing ^{13}C and deuterium are observed, also, in amounts corresponding to their isotopic abundance.

[27] T. M. Shaw and R. H. Elsken, *J. Chem. Phys.* **18**, 1113 (1950); *J. Appl. Phys.* **26**, 313 (1955); T. M. Shaw, R. H. Elsken, and C. H. Kunsman, *J. Assoc. Offic. Agr. Chemists* **36**, 1070 (1953); W. L. Rollwitz, *Proc. Natl. Electron. Conf.*, **12**, 113 (1957); T. F. Coway, R. F. Cohee, and R. J. Smith, *Food Eng.*, June 1957.

[28] A. G. Whittaker, D. W. Moore, J. N. Shoolery, and R. Jones, *J. Chem. Phys.* **25**, 336 (1956); N. Muller and J. Goldenson, *J. Am. Chem. Soc.* **78**, 5182 (1956).

[29] C. F. Callis, J. R. Van Wazer, and J. N. Shoolery, *Anal. Chem.* **28**, 269 (1956).

The relative amounts of these fragments depend upon the energy of the bombarding beam. Normal butane and isobutane have different mass spectra, because the relative proportions of the fragments are different. By establishment of the cracking pattern of a number of pure compounds, the observed pattern permits identification and analysis of very complex mixtures. This analytical technique was introduced on a practical basis by Hoover and Washburn in 1940 and was immediately adopted very widely in the petroleum industry.

All substances without exception yield a mass spectrum, the only requirement being that the vapor pressure be sufficiently high. In practice, substances with a boiling point below 450°C can be measured easily. One milliliter of gas at atmospheric pressure is ample, and much smaller amounts may be used if necessary. Because of the unique cracking patterns mentioned above, determination can be carried out in mixtures of 20 or more components. Also, the detection of traces of homologous impurities is possible, for example, a few ppm of toluene in benzene. Aside from analysis per se, one of the major uses of mass spectrometry has been in the determination of isotopic traces, particularly ^{13}C, ^{15}N, and ^{18}O, for the elucidation of reaction mechanisms.

Within the past few years the use of the mass spectrometer for analyses of complicated mixtures has to a certain degree been both replaced and supplemented by the use of the gas chromatograph.[30] At the same time new developments have given entirely new fields of application for mass spectrometry. In most commercial instruments in use until recently the resolution of m/e peaks was not high. A *high-resolution* instrument provides a unique kind of chemical analysis. Suppose, for example, a substance may contain C, H, N, and O. A peak at mass number 28 could correspond to CO^+, N_2^+, CH_2N^+, and $C_2H_4^+$.[31] But the *accurate* masses of these ions are different, amounting to 27.994914, 28.006147, 28.018723, and 28.031299 mass units, respectively, and by use of a high-resolution instrument these peaks may be separated and measured. Several commercial instruments are available now by means of which peaks as illustrated above due to the four nominal mass numbers 28 are easily resolved. Assuming that the peak of largest mass number corresponds to the parent molecule, the molecular weight thus is measured directly to the third decimal place or better. This gives a *unique* determination of the empirical formula which is found by reference to a suitable table.[32] Furthermore, by combination of the empirical formula and the major fragmentation peaks, it has been possible even to establish structural formulas of very complex organic substances solely on the basis of mass-spectrometric examination of very limited amounts of material.[33]

General references

F. W. Ashton, *Mass Spectra and Isotopes*, E. Arnold, London, 2nd ed., 1948.

J. H. Beynon, *Mass Spectrometry and Its Applications to Organic Chemistry*, Elsevier, Amsterdam, 1960.

American Petroleum Institute, *Catalogue of Mass Spectral Data*, ADI Research Project 44, Carnegie Institute of Technology, Pittsburgh.

D. W. Stewart, in *Technique of Organic Chemistry*, part IV, vol. I (A. Weissberger, ed.), Wiley-Interscience, New York, 1960, chap. LI.

Dielectric constant

Measurements of the dielectric constant may be of analytical importance in a study of a mixture of two liquids which are nonelectrolytes. Various interesting examples are

30 The combination of gas chromatograph for preliminary separation, with the effluent being led directly to a mass spectrometer for identification, is now in wide use.

31 ^{13}C and other isotopes are ignored here.

32 J. H. Beynon and A. E. Williams, *Mass and Abundance Tables for Use in Mass Spectrometry*, Elsevier, Amsterdam, 1963.

33 J. H. Beynon, *Proc. X Colloquium Spectroscopicum Internationale*, p. 743, Spartan Press, New York, 1963.

mentioned by Ebert,[34] who also applies the method to the determination of water in strongly hygroscopic solids.

Radioactive methods

Radioactive decay produces ionization when the high-energy particles or radiation that are emitted pass through the medium surrounding the sample. The amount of ionization is a function of the type of nucleus undergoing decay, the surrounding medium, and, most important, the quantity of the radioactive species present in the sample. The measurement of this ionization (by means of electroscopes, Geiger-Müller counters, ionization chambers, and other instruments) has been utilized as a sensitive method of analysis of such naturally occurring radioactive elements as radium, protactinium, and polonium, because quantities as small as 10^{-10}–10^{-12} g of these elements can readily be determined. For radioactive nuclides of short half-life, radon, ThB (Pb), and UX_I(Th), for example, which are in radioactive equilibrium with their parent nuclides, analysis is only possible by measurement of their radioactivity. Uranium and potassium have been determined radiometrically, but chemical methods are more precise and much more sensitive.

The advent of methods to prepare artificial radioactive nuclides of all the elements has extended the range of applications of radioactivity in the field of analytical chemistry.

A radioactive isotope is characterized by the type and energy of the radiation, alpha, beta, and gamma, the half-life and the decay scheme. Use of some of these characteristics is made in the following radioactive methods. The sensitivity inherent in these methods is extremely great; as few as several thousand molecules can be detected.

Below some types of radioactive methods are briefly discussed.

Tracing an element throughout chemical procedures. If a radioactive isotope of an element is added to a system, the behavior of that element in chemical operations is readily established by measuring the radioactivity of all fractions. Methods of separating gold and platinum[35] have been developed by examing classical procedures, with a radioactive gold isotope to follow losses of gold in the various steps. Radioactive ruthenium has been used[36] to evaluate the accepted method of analysis of ruthenium in ores. This involved fusion of the sample, collection of silver, gold, and the platinum metals in a lead button, cupellation of the lead button, separation of ruthenium by distillation, and finally gravimetric determination. For samples containing 6–8 mg of ruthenium, high losses were found in the fusion and cupellation steps, with no loss in the separation and gravimetric operations.

Coprecipitation has long been studied[37] using natural radioactive nuclides, and this work has been extended with artificial radioactive nuclides. Detailed studies of the coprecipitation of transuranic elements with various materials have been made to develop processes for concentration and isolation of these new elements. For example, approximate oxidation potentials have been obtained for neptunium and plutonium by measuring the coprecipitation of these elements with lanthanum fluoride after treatment with various oxidizing agents.[38] Studies of postprecipitation of a radioactive nuclide which is isotopic with one of the constituents of a precipitate and which is added to the supernatant liquid after the precipitate has been formed have thrown much light upon the mechanism of recrystallization of precipitates upon aging.[39]

[34] L. Ebert, *Angew. Chem.* **47**, 305 (1934); see also P. Cohen Henriquez, *Rec. Trav. Chim.* **54**, 327 (1935).

[35] O. Erbacher and K. Philipp, *Angew. Chem.* **48**, 409 (1935).

[36] R. Thiers, W. Graydon, and F. E. Beamish, *Anal. Chem.* **20**, 831 (1948).

[37] O. Hahn, *Applied Radiochemistry*, Cornell Univ. Press, Ithaca, N.Y., 1936.

[38] G. T. Seaborg and A. C. Wahl, in *The Transuranium Elements, National Nuclear Energy Series*, Paper 1.6, Div. IV, Vol. 14B, McGraw-Hill, New York, 1949.

[39] For a review see A. C. Wahl and N. A. Bonner, *Radioactivity Applied to Chemistry*, Wiley, New York, 1951.

Methods of separating by ion exchange such similar elements as zirconium–hafnium,[40] barium–radium,[41] and all the rare earth metals[42] have been developed using artificially radioactive isotopes of these elements to furnish an easy means of analysis.

Other applications of radioactivity include measurement of partition coefficients of elements and compounds between two immiscible liquids, volatility, and adsorption of small quantities on various surfaces.

Isotopic dilution. If one is dealing with a complex mixture, it is sometimes difficult to obtain a quantitative analysis of a constituent because of the inefficiency of separation procedures. The method of isotopic dilution can often be utilized to carry out the analysis. To a mixture containing x grams of an element or compound, a known weight, a grams, of this element or compound containing a known quantity of a radioactive isotope is added. If the specific activity, S_1, in units of radioactivity per gram of this added material is A, the specific activity in the mixture, S_2, is $aA/(a + x)$. The element or compound is isolated in pure form without regard to the per cent recovery, and S_2 is determined. The weight x can then be calculated from S_2 or, as is more customary, from the ratio $S_1/S_2 = R$ and $x = a(R - 1)$. In most cases R is large in comparison with unity and $x = aR$. The method of isotopic dilution finds important application in such fields as biochemistry, biology, and physiology, where the types of systems dealt with are so complex that usual analytical procedures are difficult. Reverse isotopic dilution is a common analytical method in nuclear chemistry for estimation of yields in nuclear reactions. Here a known weight of nonradioactive material is added to a sample containing various radioactive nuclides and the per cent recovery of the inactive material is used to estimate the original amount of radioactive species present in the sample.

General references

G. D. Chase and J. R. Rabinowitz, *Principles of Radioisotope Methodology*, Burgess, Minneapolis, 1962.

R. A. Faires and B. H. Parks, *Radioisotope Laboratory Techniques*, G. Allen, London, 2nd ed., 1960.

R. T. Overman and H. M. Clark, *Radioisotope Techniques*, McGraw-Hill, New York, 1960.

Activation analysis. Activation analysis is a nuclear method of quantitative elemental analysis. It determines elements regardless of their chemical or physical state. When a sample is exposed to a flux of bombarding particles such as the high-energy helium ions from a cyclotron or the neutrons in a chain reacting pile, nuclear reactions occur and artificially radioactive nuclides are formed. The number of nuclei formed is governed by the following equation:

$$\frac{dN}{dt} = \text{rate of formation in bombardment} - \text{rate of radioactive decay}$$

$$\frac{dN}{dt} = N_0\sigma f - \lambda N$$

$$N = \frac{N_0\sigma f}{\lambda}(1 - e^{-\lambda t})$$

where N = number of nuclei formed
N_0 = number of target nuclei
σ = probability of a particular nuclear reaction occurring
f = flux of bombarding particles
λ = radioactive decay constant of species N
t = length of bombardment time

[40] K. Street, Jr., and G. T. Seaborg, *J. Am. Chem. Soc.* **70**, 4268 (1948).

[41] E. R. Tompkins, *J. Am. Chem. Soc.* **70**, 3520 (1948).

[42] See, e.g., E. R. Tompkins, J. X. Khym, and W. E. Cohn, *J. Am. Chem. Soc.* **69**, 2769 (1947).

If σ, f, and λ are known and t and N (proportional to the radioactivity produced) are measured experimentally, it is possible to determine N_0, the amount of some element in a sample. This is the basis for activation analysis.

In general, a standard sample and an unknown are subjected to the same flux of bombarding particles for the same length of time. The samples are dissolved and a known weight of the particular element is added and then isolated by standard chemical procedures. The yield of the chemical steps is determined by reverse isotopic dilution and a comparison of the radioactivity in standard and unknown is carried out. Since σ, f, λ, and t are the same for each sample, the value of N_0 in the unknown can be calculated directly from

$$(N_0)_u = (N_0)_s \times \frac{N_u}{N_s} = (N_0)_s \times \frac{(\text{radioactivity})_u}{(\text{radioactivity})_s}$$

For activation with pile neutrons, the size of the sample is not particularly critical. The weight of an element that can be determined by this method of analysis varies with the element. While, for example, as little as 0.00003 μg of europium and dysprosium and of the order of 0.005 μg of sodium, manganese, copper, bromine, and gold can be detected, only 9 μg of iron can be detected. Accuracy is probably good to ± 10 per cent. Difficulties are sometimes encountered when the sample contains other materials with high σ, since the flux then is no longer the same in standard and unknown samples.

For a typical element, detectable to 0.001 μg, the concentration limit of detection of that element is 0.1 per cent in a 1-μg sample or 1 ppm in a 1-mg sample. The weight of an element that can be determined varies with the element. For example, at a thermal-neutron flux of 10^{13} (1 hr irradiation), 0.0004 μg of sodium, 0.04 μg of magnesium, 0.02 μg of aluminum, 0.009 μg of silicium, 0.02 μg of phosphorus, 2 μg of sulfur, 0.002 μg of chlorine, 5×10^{-5} μg of bromide and of potassium, 0.1 μg of calcium and 0.0004 μg of vanadium, 4×10^{-6} μg of manganese, 0.00002 cobalt, 2×10^{-4} μg of copper, 2×10^{-3} μg of arsenic, 5×10^{-6} μg of indium, and 10^{-6} μg of europeum can be detected by β counting.

General References

H. J. M. Bowen and D. Gibbons, *Radioactivation Analysis*, Oxford, New York, 1963.

J. M. A. Lenihan and S. J. Thomson (eds.), *Activation Analysis: Principles and Applications*, Academic Press, New York, 1965.

D. Taylor, *Neutron Irradiation and Activation Analysis*, Van Nostrand, Princeton, N.J., 1964.

W. B. Werthmann, *Activation Analysis Bibliography*, Gmelin Institute, Frankfurt, 1964.

A comprehensive and up to date review of the subject is given by V. P. Guinn in *Treatise on Analytical Chemistry* (I. M. Kolthoff and P. J. Elving, eds.), part I, vol. 8, Wiley-Interscience, New York, 1967.

Radioactive titrations. The principle of this type of titration is simple, and end-point determination, in principle, is similar to that used in photometric or amperometric titrations, except that a phase separation between reaction product and solution must be made. For example, chloride can be titrated with a radioactive silver solution, and radioactivity of the solution is plotted versus volume of silver added (corrected for change of volume by addition of reagent).[43]

Automation in analytical chemistry

In the past decade considerable progress has been made in developing instruments capable of performing a sequence of chemical steps and physical manipulations without

[43] For a review see T. Braun and J. Tölgyessy, *Radiometric Titrations*, Pergamon, New York, 1967; see also *Talanta* **11**, 1278 (1964).

human intervention. Ingenious automatic devices have been developed which can withdraw known amounts of sample and reagent solutions and mix them. If necessary the sample can be brought to the desired temperature and separations by extraction or other means can be done automatically before further automatic treatment. Devices have been developed that will then introduce the treated sample into spectrophotometers, coulometers, gas chromatographs, and a number of other existing instruments used in physicochemical and physical methods of analysis. The signal is then recorded.

Blaedel and Laessig give a table summarizing a number of applications of automation to analytical chemistry.[44] Applications to industrial pharmaceutical process and quality control are reported elsewhere,[45] as are other reviews.[46,47]

Note on the meaning of automatic analysis. "Automation" is a popular word today but what does it really mean? Sometimes it is used in the sense of *mechanization* or *instrumentation*, when these terms would be preferable. The following extract from the summary of a lecture by T. R. F. W. Fennel[48] illustrates the meaning that should be attached to these terms:

[Consider] the performance of an acid–base titration. The operations involved are addition of titrant, mixing, observation of the effect on the indicator and noting of the volume of standard solution at the end-point. These human operations might be replaced by—

(a) Use of magnetic stirrer (mechanization).
(b) Instrumental (photometer or pH meter) detection of the effect of the addition of titrant to the solution (instrumentation).
(c) Use of a motor-driven syringe buret (mechanization).
(d) Use of a chart recorder to monitor the response of the indicator system (more advanced instrumentation).

The titration could now proceed, after the apparatus had been switched on, without human attention. The process had been mechanized and instrumented, but . . . was it automatic? There was no control mechanism. The apparatus would continue to feed titrant and record the response of the indicator system until the buret jammed and the recorder ran out of paper. It was true that, with a knowledge of the rate of addition of titrant and of chart speed, the result of this unattended titration could be calculated. Before considering how to classify this stage in development of the titration, . . . further developments can be envisioned—

(e) Use of a safety device, a stop mechanism, to switch off the buret drive motor and recorder when the buret had delivered its maximum volume of titrant.
(f) Use of a device that switched off the buret drive motor when the indicator system showed that a pre-set optical density or pH had been reached in the solution.
(g) Digital recording of the volume of titrant added.

Further developments could be visualized after this stage, e.g., washout and refilling systems for the titration vessel, a refilling system for the buret and a print-out system for the volumes of titrant added in each titration.

At stage (d), a titration could be performed without attention, but, apart from having constant-speed motors fitted to the buret and recorder, without control. Stage (e) provided a form of control system to the apparatus, but it was "blind" control, merely

[44] W. J. Blaedel and R. H. Laessig, *Advan. Anal. Chem. and Instr.* **5**, 69 (1966).

[45] Automation in Industrial Pharmaceutical Process and Quality Control, Conference, *Ann. N.Y. Acad. Sci.* **130**, 483 (1965).

[46] H. Malissa, *Z. Anal. Chem.* **222**, 100 (1966).

[47] R. H. Muller, *Anal. Chem.* **38**, 101A (1966).

[48] *Proc. Soc. Anal. Chem.* **1**, 130 (1964). A few words have been changed for the present purpose.

a safety device to stop damage to the apparatus. This device did perform a desired act. It switched the apparatus off at a predetermined point in an operation, the point when the buret was empty, but was this enough to classify the apparatus as automatic? At stage (f), a new principle had been introduced. Indicator response had been linked to the motor drive and this was an example of feedback. The predetermined point was now the end-point of the titration and the volume of titrant might be read directly from the buret. Stage (g) was a slight advance on this in that it made the volume easier to read. There was no doubt that the apparatus could be called "automatic" from stage (f) onwards . . ., but there was some doubt about the earlier stages.

In order to differentiate between these stages. . .the nomenclature given below is proposed—

(i) Up to stage (d), the development of the process should be classified as "mechanization."
(ii) At stage (e), the process should still preferably be classed as "mechanized," but could be classed "semi-automatic."
(iii) From stage (f) onwards, the process should be classed "automatic."

These proposals require that the term "automation" should not be used when no control mechanism was incorporated in the system, and that it should not be used unqualified until some form of feedback was incorporated.

. . .There is the possibility of introducing the terms *automatize* and *automatization*, which have not yet been used to describe analytical operations, for the final development of automation, i.e., the complete replacement of the analyst, when sampling, analysis, and reporting of the results was done without human attention. It was a sobering thought to end with, that by the time this stage was reached the earlier terms discussed would have become as obsolete as the analytical chemist.

PART NINE

Physicochemical Methods of Analysis: Practical Part

CHAPTER 60 *Electrochemical Methods of Analysis*

In this chapter experimental procedures are given that illustrate some typical applications of electrochemical methods of analysis. The fields of polarography, constant-current coulometric titrations, constant-potential electrogravimetry, and conductance titrations are each represented by one example. The application of the amperometric method of end-point detection is included as an alternative in the coulometric titration procedure. One example of potentiometric titration using the glass electrode is included—the nonaqueous titration of a weak base in glacial acetic acid as solvent.

In some of the discussion material preceding the experimental procedures, preliminary experiments are given which should be performed by the novice to develop a familiarity with the equipment and the principles underlying the method.

Polarography. Analysis of a mixture of thallium, lead, cadmium, zinc, and manganese

This experiment consists of 10 parts. They are the determination of:

1. Rate of flow of mercury at constant applied pressure through the capillary making up the dropping mercury electrode assembly.
2. Critical height.
3. Electrocapillary curve in airfree 0.1 *M* KCl.
4. Pressure dependence of drop time for an open-circuited D.M.E.
5. Current–voltage curve for air-saturated 0.1 *M* KCl.
6. Current–voltage curve for N_2-saturated 0.1 *M* KCl (the residual current curve).
7. Current–voltage curve of air-free 0.5×10^{-3} *M* Cd^{++} in 0.1 *M* KCl.
8. Concentration dependence of i_d upon Cd^{++} concentration.
9. Effect of applied pressure upon i_d.
10. Qualitative and quantitative analysis of an unknown containing Pb^{++}, Tl^{+}, Cd^{++}, Zn^{++}, and/or Mn^{++}.

The dropping mercury electrode apparatus may be purchased or constructed using Fig. 47-1 as a guide. Any commercial polarograph may be used to perform the experiments described below.

A simple polarographic cell that can be constructed from readily available materials is shown in Fig. 60-1. This cell or its commercial equivalent should be used in all subsequent experiments.

Care of the D.M.E. Always raise the mercury reservoir enough to start the flow of mercury through the capillary before submerging the capillary in the supporting electrolyte. Always wash the tip of the capillary thoroughly with distilled water and blot the tip dry with lintless paper before lowering the reservoir.[1]

[1] *Caution:* Mercury vapor is very poisonous. Store mercury in a sealed container and clean up all spills immediately. Notify the instructor in the event of a large mercury spill.

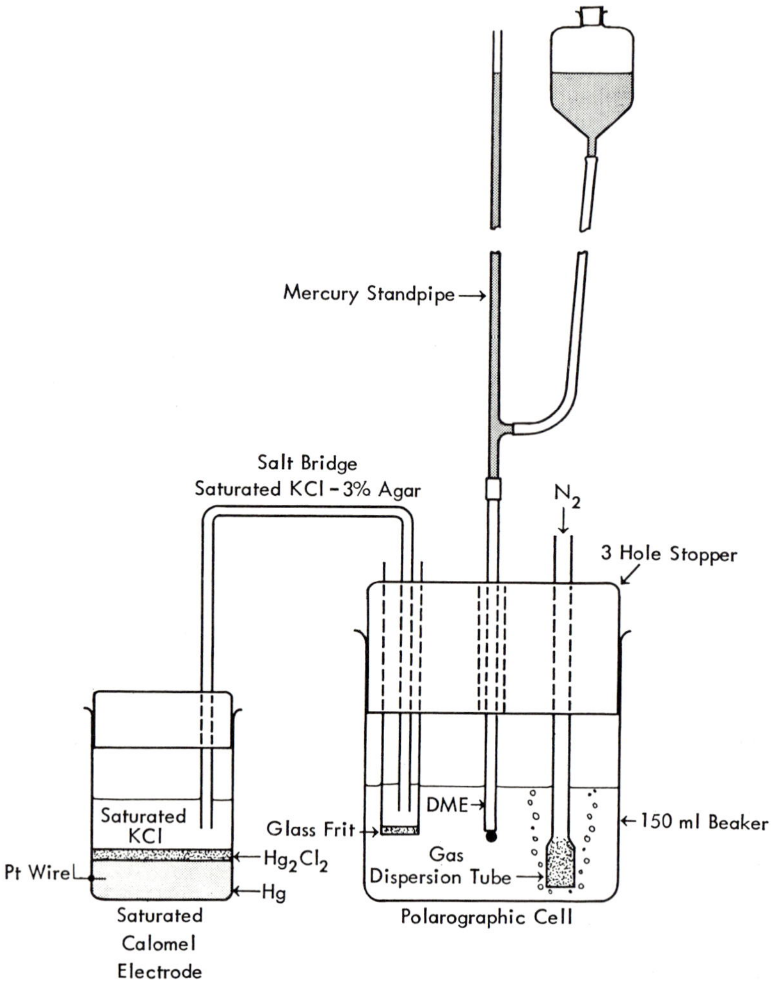

Fig. 60-1. Polarographic cell. The gas dispersion tube is in the position required to deoxygenate the solution. It is raised above the solution level and the flow of nitrogen continued after all oxygen has been removed. The fritted glass compartment is filled with supporting electrolyte to the level of the solution in the polarographic cell after introducing the salt bridge.

Experimental procedure

1. Determination of *m*, the rate of flow of mercury. Adjust the mercury level to successive markings on the standpipe until a droptime between 4 and 6 sec is obtained when the capillary is submerged in 0.1 *M* KCl and the D.M.E. is open-circuited. Using a previously weighed weighing bottle as the container, record the time required for 50 drops to fall with a stopwatch. Remove the 0.1 *M* KCl solution with an eyedropper, wash the mercury drops collected and the inside walls of the weighing bottle with distilled water several times to remove all traces of the KCl solution, and finally wash twice with acetone. Allow the acetone to evaporate and weigh. Calculate *m*, the rate of flow of mercury in milligrams per second, and the drop weight.

From Poiseuille's equation, Eq. (11) of Chapter 47, m' may be written in terms of m by introducing the density of mercury, d_{Hg}

$$m = \frac{\pi a^4 d_{Hg} P}{8\eta l}$$

This equation reduces at 25°C to

$$m = 4.64 \times 10^9 \frac{a^4}{l}\left[h - \frac{3.1}{(mt)^{1/3}}\right] \approx 4.64 \times 10^9 \frac{a^4 h}{l}$$

where m has the units of mg/sec, h is the height of mercury in cm, a is the capillary radius in cm, and l is the capillary length in cm.

In exact work the term $3.1/(mt)^{1/3}$, which arises because of the back pressure (see p. 891), must be taken into account, because it usually is of the order of 1 cm.

Using the above expression, calculate a and the back pressure in cm.

The drop weight is determined by the radius of the capillary and the surface tension of the solution. Setting the force supporting the drop (which arises from surface tension) equal to the gravitational force acting to dislodge the drop, it follows that the drop falls when

$$mtg\left(1 - \frac{d_{H_2O}}{d_{Hg}}\right) = 2\pi a\sigma$$

Calculate a from the drop weight assuming σ = 400 dynes/cm. How does this value compare with the value of a calculated from m?

2. Determination of critical height. If the mercury reservoir is slowly lowered while the capillary tip is submerged in the supporting electrolyte, a height will be reached at which the mercury ceases to flow. This height is called the *critical height* and is related to the radius of the capillary at its tip by the equation

$$h_{crit} = \frac{2\sigma}{agd_{Hg}}$$

Determine the critical height and calculate the capillary radius. How does this compare with the radius calculated from the drop-weight method and from the rate of flow of mercury?

3. Determination of the electrocapillary curve. The surface tension of mercury in contact with a potassium chloride solution depends upon the potential difference across the mercury–solution interface.

The plot of t (or σ) versus E is called the *electrocapillary curve*. The shape of the curve and the values of t obtained are the same for all supporting electrolytes consisting of capillary inactive ions, but vary when capillary active ions are present (see p. 889).

Determine the drop time (at a fixed mercury pressure) at 0.1-V intervals in deoxygenated 0.1 M KCl, starting at 0.1 V versus the S.C.E. and stopping at −1.8 V. Sufficiently accurate results are obtained by timing 5 drops with a stop watch. Plot the electrocapillary curve and determine E when $t = t_{max}$. How does this value compare with the reported value?

In this experiment, and all subsequent experiments, raise the mercury reservoir to a height such that a drop time between 4 to 6 sec is obtained with an open-circuited D.M.E.

4. Pressure dependence of drop time. Poiseuille's equation leads to the conclusion that m is proportional to h. Since the product mt is determined only by the surface tension for a given capillary, what variation of t with h should be expected at constant E?

Determine t as a function of h for h = 20, 30, 40, 50, and 60 cm, using an open-circuited D.M.E. in 0.1 M air-free KCl and compare with the predicted result.

5. O_2 reduction. Oxygen is reduced at the D.M.E., yielding two waves in 0.1 *M* KCl corresponding to the reactions

$$O_2 + 2\,H_2O + 2e \rightarrow H_2O_2 + 2\,OH^-$$
$$H_2O_2 + 2e \rightarrow 2\,OH^-$$

Both waves are irreversible. While these waves can be used for the determination of oxygen, their presence interferes with the polarographic determination of other species. The object of this experiment is to indicate the magnitude of the currents to be expected with air-saturated 0.1 *M* KCl.

Determine the current–voltage curve of air-saturated 0.1 *M* KCl between +0.2 V and −1.8 V using the 100-μA full-scale range of the polarograph.

6. Residual current of 0.1 *M* KCl. The current–voltage curve of air-free 0.1 *M* KCl shows that a measurable current flows through the D.M.E. even when no electrochemical processes are possible. This current, the so-called charging current, arises because the mercury–solution interface acts like a capacitor. See page 893 for further details.

Besides the charging current, any reducible or oxidizable impurities will contribute to the measured current. The sum of the charging and faradiac (impurity) currents is called the *residual current.* This residual current–voltage curve must be subtracted from all other current–voltage curves to obtain the actual current–voltage curve of the electroactive species.

In 0.1 *M* KCl the positive potential limit arises because of the oxidation

$$2\,Hg + 2\,Cl^- \rightarrow Hg_2Cl_2 + 2e$$

while the negative potential limit arises from the reaction

$$K^+ + e + Hg \rightarrow K_{amal}$$

Determine the current–voltage curve of air-free 0.1 *M* KCl from +0.2 V to −1.8 V using the 10-μA current range of the polarograph.

7. Current–voltage curve of an air-free solution of 0.5×10^{-3} *M* Cd^{++} in 0.1 *M* KCl. When Cd^{++} is reduced at a D.M.E. in 0.1 *M* KCl as supporting electrolyte, the reaction occurring is

$$Cd^{++} + 2e + Hg \rightarrow Cd_{amal}$$

Assuming the Nernst equation can be applied using surface concentrations, and, substituting from the Ilkovic equation, the following result is obtained:

$$E = E^{1/2} + \frac{RT}{2F} \ln \frac{i_d - i}{i}$$

where i is the current observed at applied potential E, i_d is the limiting diffusion current and $E^{1/2}$ is the applied potential when $i = \frac{1}{2}i_d$ (see Chapter 48).

Determine the current–voltage curve of an air-free solution of 0.5×10^{-3} *M* Cd^{++} in 0.1 *M* KCl. Plot E versus $\log_{10}\{(i_d - i)/i\}$. Report $E^{1/2}$ and the slope. In calculating E for this plot, remember that $E = E_{appl} + iR$, where i is the measured current (+ for cathodic and − for anodic) and R is the resistance of the entire circuit external to the voltage source. This resistance stems from two principal sources: (1) the electrolyte resistance of the D.M.E.–S.C.E. assembly, and (2) the voltage drop across the current measuring resistance in a potentiometric-type polarograph. The electrolyte resistance must be measured with an a-c Wheatstone-bridge instrument or some other type of a-c ohmmeter. D-c instruments do not give meaningful results. The second correction depends upon the sensitivity of the potentiometric recorder.

The potential of the D.M.E. is calculated from the expression

$$E = E_{appl} + i(R_{cell} + R_{recorder})$$

As an example, suppose the cell resistance, R_{cell}, is 1000 Ω, the applied potential is −0.590 V, and the current is 40 μA. The potentiometric recorder has a full-scale sensitivity of 1.000 mV and the current range switch is set at 50 μA. What is the potential of the D.M.E.?

In order that the potentiometric-recorder full-scale deflection correspond to 50 μA, a resistance, R_{recorder}, whose value is $0.001/(50 \times 10^{-6}) = 20\ \Omega$ must be inserted into the polarographic circuit by the current-range switch of the polarograph. Hence

$$E = -0.590 + 40 \times 10^{-6}(1000 + 20) = -0.549 \text{ V}$$

8. Concentration dependence of ***id*** **upon C.** The Ilkovic equation (p. 888) states that the maximum current is given by

$$(i_d)_m = 708nm^{2/3}t_m^{1/6}D^{1/2}C$$

If only C is varied, $(i_d)_m$ should be proportional to C. Note that it is better to record maximum currents if the polarograph response time is fast enough. Most recording instruments are equipped with various "damping positions" which reduce the fluctuations of the current. A little experimentation will reveal, in general, that none of the damped or undamped positions is exactly $\frac{6}{7}$ of the maximum current as measured with an oscilloscope. However, the maximum current in the undamped position often agrees with the oscilloscopic measurement. Generally, a potentiometric recorder with a response time of one second (full scale) is capable of determining the maximum current. One should consult the manufacturer's literature or experimentally establish the polarograph's ability to record maximum currents if there are any doubts.

Determine the current–potential curves for 0.5, 1.0, 2.0, 3.0, and 4.0×10^{-3} *M* Cd^{++} in air-free 0.1 *M* KCl, scanning from −0.2 V to −1.0 V. Plot $(i_d)_m$ versus C. Calculate D and the diffusion current constant, I, where

$$I = \frac{\frac{6}{7}(i_d)_m}{m^{2/3}t_m^{1/6}C} = 607nD^{1/2}$$

9. Effect of ***h*** **on** $(i_d)_m$. According to the Ilkovic equation, $(i_d)_m$ is proportional to $m^{2/3}t^{1/6}$. Also m and t depend upon h (part 4). Derive the relationship between $(i_d)_m$ and h and verify it using a 2.0×10^{-3} *M* Cd^{++} solution. Measure the diffusion current at −0.9 V versus S.C.E., varying h between 20 and 60 cm, at 10-cm intervals.

This relationship is a test for a diffusion-controlled process and is positive proof that the current is limited by diffusion at the applied potential.

10. *Qualitative and quantitative analysis of an unknown.* The procedure given below is suitable for the analysis of a mixture containing Pb^{++}, Tl^{+}, Cd^{++}, Zn^{++}, or Mn^{++}.

From Table 48-1 it is apparent that in 0.1 *M* KCl as supporting electrolyte, no interferences will occur in a solution that contains either Pb^{++} or Tl^{++} and any of the other ions. However, it is not possible to determine whether both Pb^{++} and Tl^{+} are present using this supporting electrolyte, and it is necessary to employ 1 *M* NaOH as supporting electrolyte in which the half-wave potential for Pb^{++} is −0.755 V as compared to −0.48 V for Tl^{+} (see p. 905).

Procedure. The unknown sample should be treated to yield a solution containing these ions in dilute nitric acid (0.01 *M*). Two aliquots should be taken. The first should be used in 0.1 *M* KCl containing 0.01 per cent gelatine as supporting electrolyte.[2] Prior to determining the current–potential curve, the pH should be adjusted to lie between 6 and 7 using 0.1 *M* NaOH. If the polarogram does not rule out a mixture of Pb^{++} and Tl^{+}, the second aliquot should be used in 1.0 *M* NaOH as supporting electrolyte. The volume of the aliquot taken should be chosen to yield concentrations of the various ions of about 0.001–0.005 *M*.

The ions present should be identified from the half-wave potentials of the polarographic waves observed in O_2-free solutions. In each of the two supporting electrolytes,

[2] The usual quantitative volumetric technique should be used in preparing solutions using the supporting electrolyte. If the volume of the aliquot is less than 10 per cent of the volume of supporting electrolyte, the dilution of the supporting electrolyte may be ignored. If this is not the case, an appropriately more concentrated supporting electrolyte should be used.

prepare a standard containing 0.001 M of each species present in the unknown. Obtain the polarogram of the standard in each supporting electrolyte. Determine the residual current in each supporting electrolyte. Calculate the concentration of each ion present in the unknown. Be sure to take drop-time effects into account, when subtracting the diffusion current for a more easily reduced species from that of a less easily reduced species.

Constant-current coulometry. Coulometric titration with electrogenerated iodine

The principles of constant-current coulometry involving an internally generated reagent were discussed in Chapter 50. In this section a procedure is given for the coulometric titration of As(III) with electrogenerated iodine in sodium bicarbonate medium. The reactions are

Electrochemical:	$3\,I^- \rightleftharpoons I_3^- + 2e$
Chemical:	$I_3^- + As(III) \rightleftharpoons 3\,I^- + As(V)$
Net reaction:	$As(III) \rightleftharpoons As(V) + 2e$

The end point may be detected visually using starch as indicator, or amperometrically using a rotating platinum electrode held at a potential of zero volts versus S.C.E. The latter potential is in the limiting-current region for the reduction of I_3^-.

The electrochemical end-point-detection technique is applicable to the analysis of solutions of As(III) which are 50–100 times more dilute than can be analyzed using starch as indicator. The visual end-point-detection technique using starch as indicator is capable of detecting an excess of iodine equivalent to 1 or 2 × 10^{-5} M.

The electrical circuits and apparatus used for the electrogeneration of iodine are shown schematically in Fig. 60-2. Also shown is the amperometric end-point-detection circuit. The resistor R and the potentiometric strip-chart recorder are optional and replace

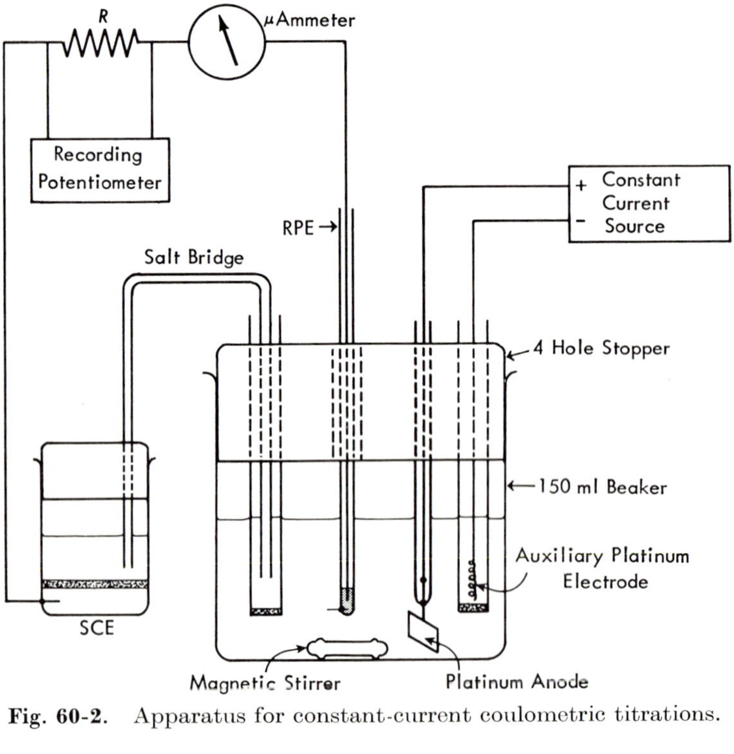

Fig. 60-2. Apparatus for constant-current coulometric titrations.

the microammeter when it is desired to automatically record the amperometric titration curve.[3]

The saturated calomel electrode and salt bridge should be of the low-resistance type used in conventional two-electrode voltammetry.

The constant current should be chosen so that between 100 and 200 sec are required to reach the end point, and the timer should be accurate to 0.1 sec. Using starch as indicator, a 0.02 M As(III) solution can readily be titrated with an accuracy of 0.1–0.2 per cent.

Procedure

Coulometric titration of As(III)

1. Fill both sintered glass tubes in the electrolysis cell of Fig. 60-2 three quarters full with 1 M Na_2SO_4. Place a platinum-wire electrode ($\sim$0.5 cm^2) in one of these tubes. The other tube is not needed unless the end point is detected amperometrically. Prepare 1 liter of supporting electrolyte containing 350 g of potassium iodide and 70 g of sodium bicarbonate per liter.

2. In the beaker that comprises the rest of the cell shown in Fig. 60-2, place 80 ml of supporting electrolyte. This volume should be sufficient to cover the platinum anode (Area $\sim$1 cm^2). If the level of supporting electrolyte in the cell is not the same as the level of the Na_2SO_4 in the fritted glass compartments, 1 M Na_2SO_4 should be added or removed as required.

3. Insert a Teflon-coated magnetic stirring bar, and stir at a rate that produces a small vortex.

Visual end-point detection with starch. This procedure is suitable for determining approximately 200 μeq of As(III), and will require about 200 sec to reach the end point using a current of $\sim$100 mA.

The supporting electrolyte may contain some I_3^- formed by the air oxidation of iodide ion. Add 5 ml of 0.2 per cent starch solution (p. 845) to the cell. If a blue color forms, add 0.05 M As(III) solution dropwise until the color disappears.

Select a current range on the constant-current source corresponding to $\sim$1 μeq/sec (96.5 mA). Turn on the electrolysis current and turn it off when the first sign of a blue color develops in the solution.[4] Any excess As(III) added in the previous step is now titrated. Add $\sim$200 μeq of As(III) unknown to the cell, and reset the timer on the constant-current source to zero. Start the electrolysis and turn off the current at the first faint tinge of blue color in the cell. Record the time required for the electrolysis and calculate the micromoles of As(III) present in the unknown from

$$\text{micromoles} = \frac{\text{const. current } (\mu\text{A}) \times \text{time (sec)}}{2 \times 96{,}493}$$

Amperometric end-point determination. The procedure given below is suitable for the determination of $\sim$20 μeq of As(III). Quantities of As(III) as much as 50 times smaller can be determined by using a proportionately more sensitive current-measuring device than specified below.

Repeat steps 1 and 2 above. Insert the salt bridge from the calomel electrode in the amperometric end-point-detection circuit into the unused fritted glass disk compartment, insert the platinum-wire electrode, and start the rotator after completing the electrical wiring shown in Fig. 60-2. Use a 30-μA full-scale microammeter to measure the current due to excess electrogenerated iodine. Repeat step 3 above.

[3] The E. H. Sargent Company Model SR recorder, and their constant-current source and timer combination, the Coulometric Current Source, may be readily interconnected so that starting the chart drive motor starts the constant-current source and timer simultaneously. Platinum-wire electrodes and a device for rotating the electrodes at constant speed (600 rpm) are also available from this manufacturer.

[4] The end point will be easier to detect if a piece of white paper is placed on the magnetic stirrer under the cell, positioned in such a manner as to act as white background for the platinum anode at which I_3^- is produced.

If a current is observed on the microammeter, iodine is present in the supporting electrolyte. Add 0.05 *M* As(III) dropwise, until the current drops to a constant value within several tenths of a microampere of zero current through the rotating platinum electrode. This residual current is ordinarily very small. Determine the excess As(III) by turning the constant-current source on,[5] and then off when a small current is seen on the microammeter. Record the current reading and the timer reading. Obtain three or four more sets of current and timer readings, stopping when 15 μA is passing through the rotating platinum electrode. Plot these data and locate the end point for the excess As(III). Add ~20 μeq of As(III) to the titration cell, and turn the constant-current source on. Determine the residual current by measuring the current in the rotating platinum electrode circuit after 50, 100, and 150 sec of titration. Stop the electrolysis current at the first sign of appreciable current flow through the rotating platinum electrode, and record the timer reading. Obtain three or four more pairs of current and timer readings just after the end point. The time required to titrate the As(III) unknown is the difference between the two end points obtained for the unknown As(III) sample and the blank As(III) titration.

Figure 60-3 is a reproduction of the automatically recorded amperometric titration curve for 20 μeq of As(III) using the equipment described in Reference 3. The value of *R* in Fig. 60-2 was 100 Ω, and recorder sensitivity was 1.25 mV.

Controlled-potential electrogravimetry. Separation and determination of copper

As was pointed out in Chapter 51, controlled-potential electrolysis is of considerable advantage in the analysis of metal-ion mixtures. A procedure is given in this section which quantitatively deposits copper from solutions containing antimony, arsenic, bismuth, tin, nickel, and zinc, without any interference from the latter ions. No interference is caused by almost all other commonly encountered metals.

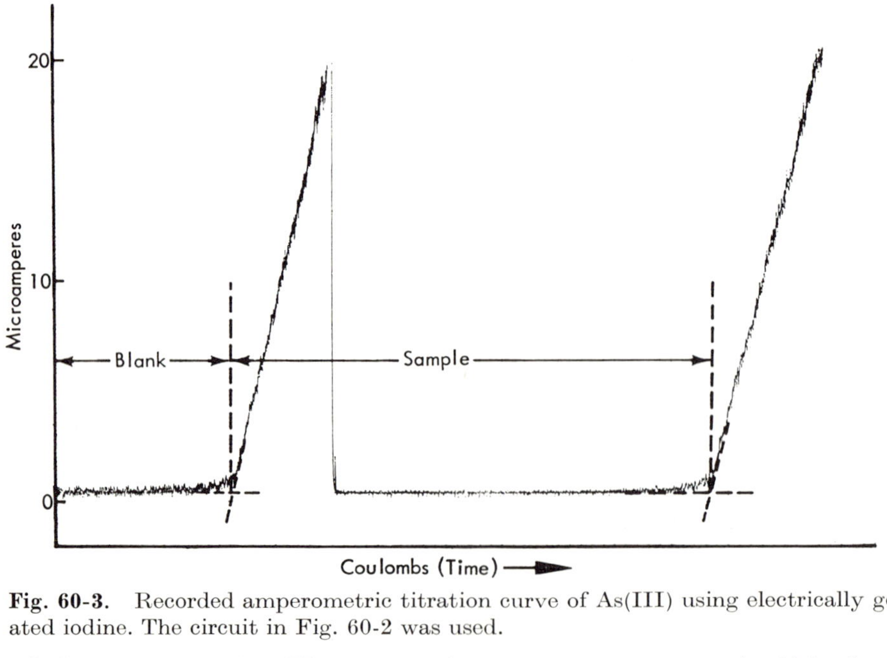

Fig. 60-3. Recorded amperometric titration curve of As(III) using electrically generated iodine. The circuit in Fig. 60-2 was used.

[5] The current range in milliamperes on the constant-current source should be chosen to equal one half the number of μeq of As(III) to be titrated. In this case, mA = 0.5 × 20 = 10.

Schoch and Brown[6] found that copper could be deposited from hydrochloric acid media at a potential of -0.4 V versus S.C.E. without any interference from lead, tin, and other metals with deposition potentials more negative than -0.4 V. Torrance[7] and Diehl and Brouns[8] developed procedures in $\sim 1\ M$ hydrochloric acid medium for the determination of copper in alloys that did not contain bismuth, antimony, or arsenic. Hydrazine hydrochloride, or hydroxylamine served as anodic depolarizer, making it unnecessary to separate the auxiliary electrode (anode) and working electrode (cathode) from each other by a porous barrier.

In a hydrochloric acid medium, antimony, bismuth, and arsenic are deposited simultaneously with copper. Another difficulty which occurs in 1 *M* hydrochloric acid was pointed out by Diehl and Brouns. If the cathode potential lies between $+0.27$ and -0.06 V, the chlorocuprous complex, $CuCl_3^{--}$ will be produced at the cathode, and, on migration to the anode, will be reoxidized back to the cupric state. Hence no copper metal will deposit. Lingane[9] suggests the use of hydrochloric acid concentrations less than 0.5 *M* to avoid problems associated with the formation of chlorocuprous ions.

Tartrate medium in the pH range 4–6 is considered by Lingane[10] to be far superior to hydrochloric acid solution for the deposition of copper. Besides yielding mirror-bright copper deposits with excellent properties, the separation of copper from antimony, arsenic, and bismuth is possible in tartrate medium by appropriate control of the pH.

Reference electrodes. Commercial reference electrodes as supplied with the potentiostat may be used or a suitable electrode fabricated in the laboratory. A very compact silver/silver chloride electrode can be constructed by anodizing a pure silver wire in 1 *M* potassium chloride solution. Then an agar gel prepared by warming 7.5 g of potassium chloride, 3 g of agar, and 100 ml of water is poured into a micro gas dispersion tube, just filling the fritted glass portion. After the gel has solidified, the rest of the tube is filled with 1 *M* potassium chloride, and the end of the tube sealed with a rubber policeman. The rubber policeman is pierced with a pin and the anodized silver wire inserted. The completed electrode is illustrated in Fig. 60-4.

Potentiostat. The necessity of using a three-electrode system to control the working electrode potential was discussed in Chapter 50. Various commercial devices are available. For example, an electromechanical device, equipped with a rotating counter (auxiliary) electrode, is manufactured by the Fisher Scientific Company, and an excellent electronic potentiostat, the Wenking design, is distributed by the Brinkman Company.

Recent developments in solid-state electronic devices have made completely electronic devices practical and economical, and it is not difficult to build a potentiostat using *operational amplifiers*. Operational amplifiers are electronic devices that can be made to perform a variety of control functions and mathematical operations such as addition, subtraction, multiplication, differentiation, and integration. The theory and application of such devices is beyond the scope of this text and can be found elsewhere.[11]

The procedure given below is suitable for the determination of copper in the presence of antimony, arsenic, lead, nickel, and zinc.

[6] E. P. Schoch and D. G. Brown, *Trans. Am. Electrochem. Soc.* **22**, 265 (1913).

[7] S. Torrance, *Analyst* **62**, 719 (1937); **63**, 488 (1938).

[8] H. Diehl and R. Brouns, *Iowa State Coll. J. Sci.* **20**, 155 (1945); H. Diehl, *Electrochemical Analysis with Graded Cathode Potential*, G. F. Smith Chemical Co., Columbus, Ohio, 1948, p. 35ff.

[9] J. J. Lingane, *Ind. Eng. Chem., Anal. Ed.* **18**, 430 (1946).

[10] J. J. Lingane, *Ind. Eng. Chem., Anal. Ed.* **17**, 640 (1945).

[11] George A. Philbrick Researches, Inc., *Applications Manual for Computing Amplifiers for Modelling, Measuring, Manipulating and Much Else*, Dedham Mass., 1966; Burr-Brown Research Corporation, *Handbook of Operational Amplifier Applications*, Tucson, Ariz., 1963; McKee-Pederson Instruments, *MP-System 1000, Operation and Applications*, Danville, Calif., 1966, Operational Amplifiers Symposium, *Anal. Chem.* **35**, 1770 (1963), and C. F. Morrison, Jr., *Generalized Instrumentation for Research and Teaching*, Washington State Univ., Pullman, 1964.

Procedure. Clean the platinum cathode by dipping in concentrated nitric acid, rinse in distilled water, and dip in reagent-grade acetone. Shake the excess acetone off the electrode, place in an oven at 110°C for approximately 2 min, and then allow to cool to room temperature in air for about 5 min. Weigh the cathode to the nearest 0.1 mg. Clean the platinum anode using concentrated nitric acid and wash with water.

Prepare a silver–silver chloride reference electrode as described above, if a saturated calomel electrode is not supplied with the potentiostat. It will be assumed that the sample is present in 50 ml of an acid solution containing approximately 200 mg of copper and that nitrate and/or chloride ion may be present, as is the case if an alloy were dissolved in nitric acid or aqua regia. Transfer the liquid sample quantitatively into a 250-ml beaker. The solution must be boiled very gently for a minute or two to remove most of the nitrogen oxides and chlorine. Add 100 ml of a solution containing 23 g (0.10 mole) of sodium tartrate dihydrate and 1 g of urea (to remove the last traces of oxides of nitrogen and chlorine). Dilute the solution to approximately 200 ml and adjust to pH 5.00 with 5.0 *M* sodium hydroxide. Use a pH meter for the latter step. Finally, add 1–2 g of hydrazine dihydrochloride as an anodic depolarizer.

Mount the platinum cathode and the platinum anode in the potentiostat, place the beaker containing the sample solution in position, and position the reference electrode next to the cathode. Stir the solution vigorously using the method provided with the potentiostat. If a nonuniform copper deposit appears on the cathode, it may be necessary to provide auxiliary stirring, e.g., by the use of a magnetic stirrer.

The potential of the platinum cathode is set at −0.34 V versus the Ag–AgCl reference electrode, or −0.36 V versus S.C.E. The initial current may be as large as 1 or 2 A and it usually decreases to 0.01 A or less after 45 min. Usually all the copper is deposited when the current has remained constant for 10 min. Verify this by raising the beaker slightly (or by adding a little more water) and electrolyzing an extra 10 min. There should be no visible deposition of copper on the freshly submerged platinum.

The correct procedure to stop the electrolysis is to lower the electrolysis beaker, without turning off the potentiostat, and wash the electrodes immediately with a fine stream of distilled water from the wash bottle. The washings (first 10 ml) are collected in the electrolysis beaker if the other constituents are to be determined. The cathode is washed with reagent-grade acetone, placed in an oven at 110°C for 2 or 3 min,

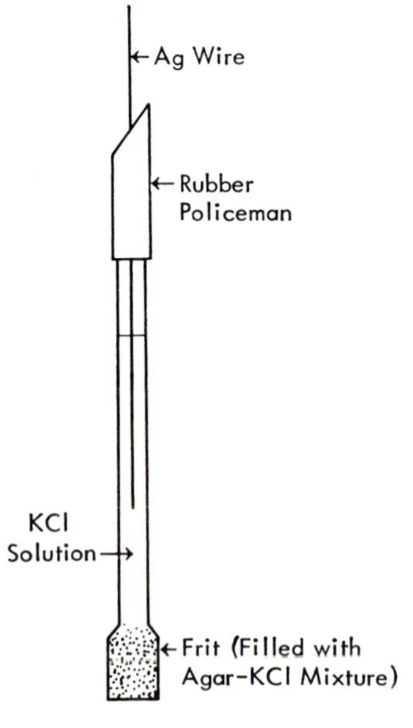

Fig. 60-4. Ag/AgCl reference electrode.

and then allowed to cool for about 5 min before weighing. Calculate the weight of copper deposited from the increase in weight of the platinum cathode. Remove the copper from the cathode by dipping in nitric acid.

Conductometric titration of a phenol

The principle of conductometric titrations was discussed in Chapter 53, and a procedure is given below that is suitable for the titration of phenol and substituted phenols with sodium hydroxide. It may also be used for the titration of other weak acids. However, before giving the analytical method the principle of the Wheatstone bridge will be presented.

Measurement of conductance using a Wheatstone bridge. The resistance of an electronic conductor can be determined using the direct-current Wheatstone bridge illustrated in Fig. 60-5.

R_x is the unknown resistance, R_1 and R_2 are standard resistors, and R_3 is an adjustable standard resistor. E is a battery, K a tapping key, and G a galvanometer. The bridge is balanced by adjusting R_3 until the galvanometer does not deflect when the tapping key is held down momentarily. Under these conditions, the potential at points B and D must be the same.

Current flows from point A to point C via two paths; i_1 through R_x and R_1, and i_2 through R_2 and R_3. The total current through the bridge is the sum of i_1 and i_2. Since the potential at points B and D is the same, the ohmic voltage drop through the resistors R_x and R_3 must be the same. Hence, from Ohm's law,

$$i_1R_x = i_2R_3 \tag{1}$$

Similarly,

$$i_1R_1 = i_2R_2 \tag{2}$$

Dividing Eq. (1) by (2) yields

$$\frac{R_x}{R_1} = \frac{R_3}{R_2}$$

and

$$R_x = \frac{R_1R_3}{R_2} \tag{3}$$

Since R_1, R_2, and R_3 are all known, R_x can be calculated.

The same principle can be used to measure the conductance of solutions, but certain modifications must be made.

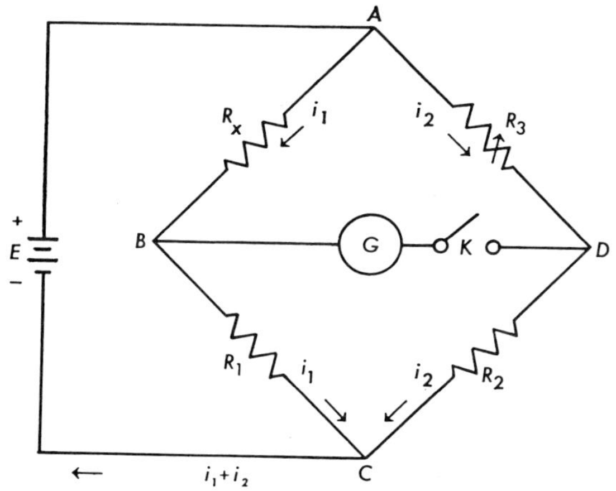

Fig. 60-5. Direct-current Wheatstone bridge. E = d-c voltage source (several volts): R_x = unknown resistor, R_1 and R_2 = standard resistors, R_3 = adjustable resistor, G = galvanometer, K = tapping key.

An alternating-current (a-c) voltage must be used to energize the bridge, rather than a direct-current (d-c) voltage, since the d-c voltage would polarize the electrodes in the conductance cell by electrolyzing the solution. Ordinarily, a-c voltages between 3 and 6 V with a frequency 1–10 kHz are used across points A and C of Fig. 60-6. In this figure, R_x represents the resistance of the conductivity cell, which is located between points A and B. However, the cell also acts like a small capacitor, which is represented by C_x. Capacitor C_x has a capacitive reactance which permits a-c current to flow through it, in addition to the current that flows through R_x, between point A and point B. Thus a variable capacitor, C_B, must be inserted into the bridge to permit balancing the capacitive reactance of the conductance cell. Both R_3 and C_B must be adjusted to make the magnitude and phase of the a-c voltages at points B and D identical. The value of C_B required to balance or *null* the bridge depends upon the values of C_x, R_1, R_2, and R_3. If R_3 is altered, it is necessary to readjust C_B to obtain a sharp null.[12]

The cell conductance is the reciprocal of R_x, and at the bridge balance may be calculated from

$$\text{conductance} = \frac{R_2}{R_1 R_3}$$

A d-c galvanometer cannot be used to detect the a-c null, but various other devices can be used. If the frequency of the a-c voltage is in the audio range, earphones are satisfactory as a null device. The various adjustments are made until a minimum in the sound level is obtained.

In several commercial conductance bridges, the voltage across points B and D is amplified and applied to an electron-ray tube, the so-called "magic-eye" tube. The top of this device emits a green fluorescence as a result of electrons striking a phosphor coating inside the glass tube. A null is indicated by the appearance of a pie-shaped wedge in the circular pattern. For extremely precise conductance measurements a cathode-ray oscilloscope is used as the null detector.

Direct-reading conductance meters are commercially available. The conductance in reciprocal ohms, mhos, is presented on a panel-mounted meter.[13]

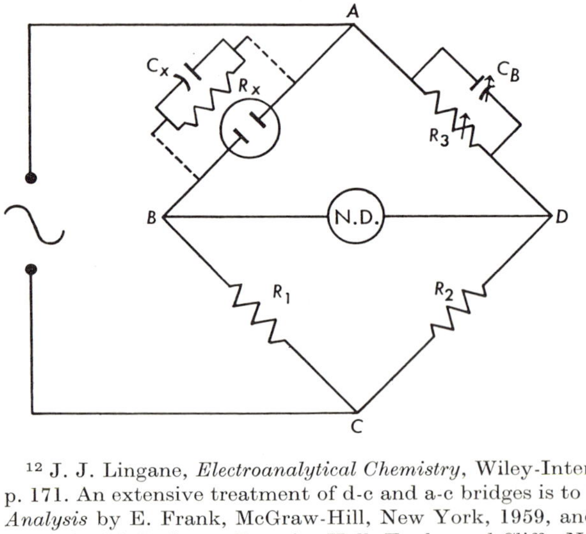

Fig. 60-6. Alternating-current Wheatstone bridge. N.D. = A-C null detector, C_B = variable capacitance, R_x = cell resistance, C_x = cell capacitance, ~ = a-c voltage source.

[12] J. J. Lingane, *Electroanalytical Chemistry*, Wiley-Interscience, New York, 2nd ed., 1958, p. 171. An extensive treatment of d-c and a-c bridges is to be found in *Electrical Measurement Analysis* by E. Frank, McGraw-Hill, New York, 1959, and also in *Basic Electrical Measurements*, by M. B. Stout, Prentice-Hall, Englewood Cliffs, N.J., 2nd ed., 1960. The latter reference is particularly valuable as a source of information concerning commonly used electrical devices.

[13] Such instruments are manufactured by the Leeds and Northrup Company, and by Barnstead. The latter device is available from the E. H. Sargent Co.

Conductance cells. Critical discussions of the design of conductance cells have been given by Parker,[14] Jones and Bollinger,[15] and by Shedlovsky.[16] Four types of conductance cells are shown in Fig. 60-7. In *A* a dip-type cell is represented. It may be placed directly in the beaker containing the solution to be titrated. It is convenient for titration purposes but yields conductometric data of limited accuracy, 0.5 per cent. *B* illustrates another cell which may be used for titrations; the electrodes are held in a fixed position by the standard-taper glass joints.

The Kohlrausch cell, which was classically used to measure conductance of solutions, is shown in *C*. It suffers from a relatively high cell capacitance caused by the close spacing of the leads to the electrodes. These problems are overcome to a considerable extent by using a cell such as shown in *D*.

Determination of phenol and substituted phenols. The procedure given below is based upon the work of Kolthoff[17] and is suitable for the titration of a number of substituted phenols.

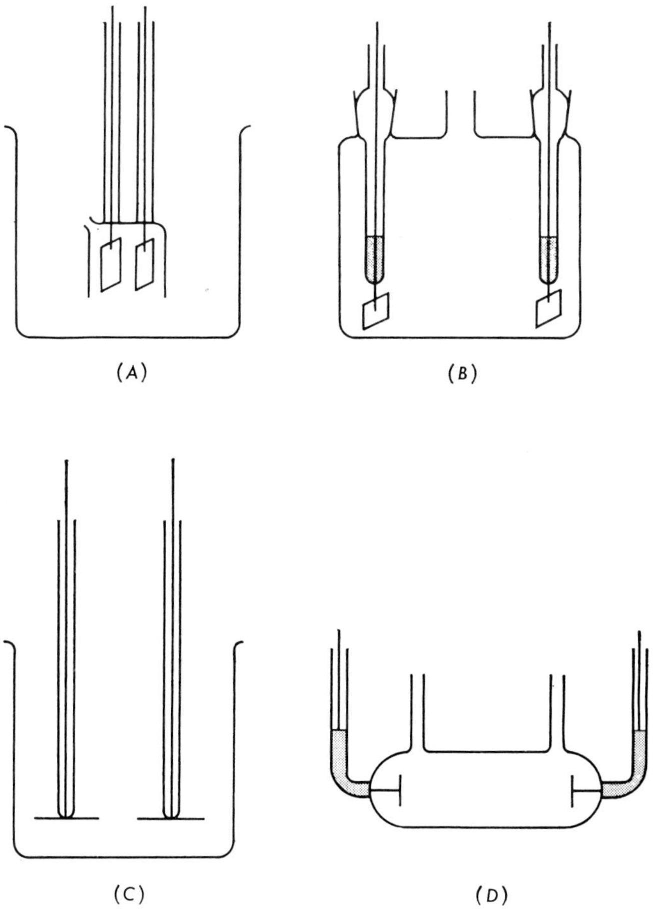

Fig. 60-7. Conductance cells.

[14] H. C. Parker, *J. Am. Chem. Soc.* **45**, 1366 (1923); H. C. Parker, *ibid.* **45**, 2017 (1923).
[15] G. Jones and V. M. Bollinger, *J. Am. Chem. Soc.* **53**, 411 (1931).
[16] T. Shedlovsky, *J. Am. Chem. Soc.* **52**, 1793 (1930).
[17] I. M. Kolthoff, *Z. Anorg. Chem.* **112**, 187 (1920).

Procedure for conductometric titration of phenols. Weigh to 0.1 mg approximately 2.5 meq of the unknown sample and transfer to a 100-ml beaker. Dissolve in 50 ml of a water–ethanol mixture (40 : 60 ratio), place in a room-temperature thermostat (± 0.25), introduce a dip-type conductivity electrode, and wait 10 min for thermal equilibrium. Determine the conductance, add 0.2 ml of 1.0 *M* sodium hydroxide from a 5-ml microburet, mix with a stirring rod, and determine the conductance. Continue adding 0.2-ml increments of sodium hydroxide until the end point has been exceeded by at least 50 per cent. Plot the conductance data and extrapolate the linear portions of the titration curve to their point of intersection to locate the end point.

Potentiometric titration of a base in glacial acetic acid

A discussion of nonaqueous acid–base equilibria and titrations has been given in Chapters 5 and 34. Considerable emphasis was placed on glacial acetic acid as a convenient solvent for the titration of weak bases, and tabulations of compounds which have been titrated in this solvent and the end-point-detection techniques used are given by Kolthoff and Bruckenstein.[18] Information concerning titrations in other solvents can be found in the work of Beckett and Tinely[19] and of Kucharsky and Safarik.[20]

Perchloric acid is the universally accepted titrant, and it is standardized against potassium acid phthalate using crystal violet as indicator.[21] The titration reaction is

$$HClO_4 + KOOC{-}C_6H_4COOH \rightarrow KClO_4\downarrow + (HOOC)_2C_6H_4$$

The indicator changes from blue, to bluish green, to yellow in the vicinity of the end point. Quinaldine red and *p*-naphtholbenzein are recommended by Higuchi[22] as being superior to crystal violet.

Sodium carbonate and triphenylguanidine may be used as primary standards in place of potassium acid phthalate.[23] Other primary standards that have been suggested are diphenylguanidine[24] and sodium salicylate.[25]

Preparation of 0.1 *M* perchloric acid in glacial acetic acid. Add 8.4 ml of concentrated perchloric acid (70–72 per cent) to 500 ml of glacial acetic acid in a 1-liter volumetric flask. Stir slowly, and slowly add, while stirring, 20 ml of acetic anhydride in several portions. If the mixture becomes excessively warm, cool before adding all the acetic anhydride. Dilute to 1 liter with glacial acetic acid. Allow the solution to stand overnight to ensure that the reaction between water and acetic anhydride is complete.[26]

Care should be taken not to add any more than the specified quantity of acetic anhydride, because excess acetic anhydride will react with primary and secondary amines during their titration by perchloric acid. If a very large excess of acetic anhydride is added, an explosive mixture may be produced. When prepared as described above, the perchloric acid titrant is stable and shows no change in normality over considerable periods of time (at least 1 year).

[18] I. M. Kolthoff and S. Bruckenstein, *Treatise on Analytical Chemistry* (I. M. Kolthoff and P. J. Elving, eds.), part I, vol. 1, Wiley-Interscience, New York, 1959, p. 535.

[19] A. H. Beckett and E. H. Tinely, *Titrations in Nonaqueous Solvents*, British Drug House, Poole, Eng., 2nd ed., 1960.

[20] J. Kucharsky and L. Safarik, *Titrations in Nonaqueous Solvents*, Elsevier, Amsterdam, 1965.

[21] W. Seaman and E. Allen, *Anal. Chem.* **23**, 592 (1951).

[22] T. C. Higuchi and J. I. Bodin in *Pharmaceutical Analysis* (T. C. Higuchi and E. Brockmann-Hanssen, eds.), Wiley-Interscience, New York, 1961, p. 393.

[23] T. C. Higuchi, C. Rehm, and C. Barnstein, *Anal. Chem.* **28**, 1506 (1956).

[24] J. K. Carlton and W. C. Bradbury, *Anal. Chem.* **27**, 67 (1955).

[25] M. E. Auerbach, *Drug Std.* **19**, 127 (1951).

[26] S. Bruckenstein, *Anal. Chem.* **31**, 1757 (1959).

Standardization of perchloric acid. Weigh, accurately, 0.8 g of potassium acid phthalate into a 250-ml Erlenmeyer flask containing 100 ml of glacial acetic acid. Solution can be hastened by gentle warming. Add 4–5 drops of 0.5 per cent crystal violet indicator (prepared in AcOH) and titrate to a blue-green color. Perform a blank titration to determine the volume of perchloric acid required to titrate 100 ml of glacial acetic acid.

The coefficient of thermal expansion for acetic acid is 1.1×10^{-3} per °C, about five times greater than that for water. The normality of the perchloric acid titrant should be corrected if the temperature at which it is used is 1 degree or more different from its standardization temperature by using

$$N_{T_1} = \frac{N_{T_0}}{1 + 0.0011(T_1 - T_0)} \tag{4}$$

T_0 is the temperature at the time of standardization and T_1 that at the time of analysis. N represents the normality at the temperature indicated by the subscript.

Determination of bases. Unknown base solutions are best titrated potentiometrically using a glass electrode as the hydrogen-ion indicator electrode. Commercial aqueous calomel reference electrodes may be used as the reference electrode, provided the saturated potassium chloride solution is replaced with an aqueous lithium chloride solution (approximately 0.1 M). Unstable emf readings are obtained during the titration if the potassium chloride solution is not replaced, because potassium perchlorate will precipitate in the region where the aqueous and acetous solutions meet. This precipitate can greatly increase the resistance of the reference electrode.

The following procedure is suitable for the titration of bases whose dissociation constants in water are greater than 10^{-11}.

Procedure for potentiometric titration of an unknown base. Weigh, accurately, 2.5–4.0 meq of the unknown base in a 100-ml beaker. Dissolve in 50 ml of glacial acetic acid, heating to aid solution if necessary.[27] Introduce the glass and calomel electrodes, set the pH meter to the emf scale, and stir the solution using a magnetic stirrer (turn off the magnetic stirrer when making emf readings). Determine the shape of the titration curve and the approximate end point by making emf readings at 2-ml volume increments of perchloric acid. Titrate a second sample to within 2 ml of the end point. Measure the emf. Add 1 drop of perchloric acid, stir, and read the emf after 30 sec.[28] Repeat this procedure until the end point is exceeded. Determine the volume of perchloric acid required to reach the end point using the second derivative method (p. 946).[29]

Uncharged bases and salts of acids (anionic bases) whose Brønsted base dissocation constants in water are greater than 10^{-9} can be titrated with perchloric acid using crystal violet as indicator according to the procedure given above for the standardization of perchloric acid with potassium acid phthalate.

Mercuric acetate is an extremely weak base in glacial acetic acid. In the presence of compounds containing ionic halide, e.g. Cl^-, reaction (5) occurs:

$$2\,BH^+Cl^- + Hg(OAc)_2 \rightarrow 2\,BH^+OAc^- + HgCl_2 \tag{5}$$

[27] If the base proves to be insoluble, it will be necessary to add an excess of perchloric acid and back-titrate with 0.1 M sodium acetate solution. The latter solution is prepared by dissolving 8.2 g of anhydrous sodium acetate in 1 liter of glacial acetic acid, and it is standardized against the perchloric acid titrant using crystal violet as indicator.

[28] Do not attempt to read the volume after adding each drop. Merely record the volume when dropwise addition is started and the volume when it is discontinued. The volume per drop can then be calculated from the number of drops and the difference in buret readings.

[29] If the points used to plot the second difference curve are too scattered, recalculate the curve using emf data for every other drop of titrant. Usually, omitting every other data point will smooth the second difference curve sufficiently in those cases where problems are encountered.

producing the strong base, BH^+OAc^-, because mercuric chloride is extremely slightly dissociated. Pifer and Wollisch[30] demonstrated that the BH^+Ac^- could be titrated in the presence of mercuric acetate using perchloric acid as titrant.

The method of Pifer and Wollisch may be used to titrate hydrohalide salts of amines, using either potentiometric or visual end-point detection. To apply their procedure, add 15 ml of 5 per cent $Hg(OAc)_2$ solution in glacial acetic acid to a 100-ml beaker (or 250-ml Erlenmeyer flask) containing 2.5–4 meq of the hydrohalide and ~50 ml of glacial acetic acid. Titrate using 0.1 *M* perchloric acid.

[30] C. W. Pifer and E. G. Wollisch, *Anal. Chem.* **24**, 300 (1952).

CHAPTER 61 *Optical Methods of Analysis. Procedures*

This chapter contains eight experimental procedures that are representative of optical methods of analysis. Six of the procedures involve spectrophotometric measurements and chemical operations of various degrees of difficulty. The first procedure (determination of iron with thiocyanate) is typical of the simplest kind of spectrophotometric method in which a single constituent is determined in a sample that is free of interfering constituents. The second and third procedures illustrate the analysis of a binary mixture and the method of differential (high-precision) spectrophotometry. One example of a spectrophotometric titration (copper with EDTA) is given. A procedure for the determination of a trace of lead as the dithizonate after extraction into an organic solvent is described, followed by an example illustrating the use of infrared spectroscopy to solve a simple problem in organic analysis. The last two procedures are for the qualitative and quantitative determination of metals by flame photometry, and for determination of a trace of aluminum by fluorescence measurement.

No experimental procedures are given for atomic absorption analysis, emission spectrochemical analysis, or light scattering.

Naturally suitable apparatus must be provided. Since instruments of many different manufacturers will be used, no specific operating instructions are given in this book, except for the flame attachment for a spectrophotometer.

Determination of ferric iron

Traces of iron are often determined colorimetrically or spectrophotometrically with thiocyanate as reagent. Ferric iron reacts with thiocyanate to give an intensely red-colored compound which remains in true solution, whereas ferrous iron does not react. The mechanism of the reaction between ferric iron and thiocyanate is complex. A great number of factors affect the color shades or the color intensity of the red compound formed, and the method is therefore far from ideal from an analytical viewpoint. However, with the proper procedure and with due consideration of possible interferences, satisfactory results are obtained.

More than one product can be formed when thiocyanate reacts with ferric iron (see Chapter 6). Depending upon the thiocyanate concentration a series of complexes represented by $Fe(SCN)_n^{+3-n}$, where $n = 1, \ldots, 6$, can be obtained. At low thiocyanate concentration the predominant[1] colored species is $FeSCN^{++}$. At

[1] H. E. Bent and C. L. French, *J. Am. Chem. Soc.* **63**, 568 (1941); S. M. Edmonds and N. Birnbaum, *ibid.* **63**, 1741 (1941); R. K. Gould and W. C. Vosburgh, *ibid.* **64**, 1631 (1942).

0.1 *M* thiocyanate concentration the predominant species[2] is $Fe(SCN)_2^+$, while at higher thiocyanate concentrations negatively charged complex ions predominate. The colors of all these complexes are red. However, the absorption spectra are not identical. At low concentrations of thiocyanate the absorption maximum is at about 460 mμ, while at larger concentrations (0.1–0.5 *M*) the maximum is at 480 mμ. Babko[3] has estimated the dissociation constants of the various complexes.

Effect of concentration of thiocyanate upon the color. From the above discussion it is evident that both the color and the extent of absorption depend upon the thiocyanate concentration. Therefore, in the determination of iron it is desirable to use a very large excess of reagent. Under these conditions Beer's law holds[4] over a fairly wide range of iron concentrations. In addition, the color obtained is more stable than with small amounts of thiocyanate.

Influence of acid upon the color. Owing to extensive hydrolysis, solutions of ferric salts in water contain relatively few ferric ions. Addition of a strong acid suppresses the hydrolysis and therefore intensifies the color obtained with thiocyanate. (Thiocyanic acid is a strong acid.) The effect of the acid depends upon its type and its concentration. Karsten[5] found that with a large excess of thiocyanate the color is practically independent of the concentration of nitric or hydrochloric acid, if the acid normality in the mixture is between 0.05 and 0.5 *N*. Karsten prefers nitric acid to hydrochloric acid, because the color obtained with the former is more stable. Sulfuric acid should not be used because sulfate ions form complexes with ferric ions, and consequently above a certain acidity the color intensity decreases with increasing sulfuric acid concentration.

A source of error that must be avoided is the fading of the color after the reactants are mixed. This is due to reduction of iron(III) by thiocyanate. This reaction is accelerated greatly when the solution is exposed to daylight. In diffuse daylight, the absorbance of a solution contining 4 ppm of iron in 0.5 *M* hydrochloric acid and 0.3 *M* thiocyanate decreases about 10 per cent in the first 10 min after mixing. A similar initial decrease occurs also in 0.2 *M* nitric acid, but the initial decrease is followed by an increase due to formation of a colored reaction product of nitric acid and thiocyanate. The fading can be practically prevented by the addition of a suitable amount of persulfate or of hydrogen peroxide.[6] However, such addition is unnecessary if the absorbance is measured *immediately* after mixing the reactants.

Interfering substances

Cations. Cupric copper, if present in quantities greater than 2 mg/liter (Karsten, Ref. 5), interferes by formation of a light-green coloration with thiocyanate. Mercuric mercury forms slightly dissociated $Hg(SCN)_2$ and complex $Hg(SCN)_4^{--}$ ions with an excess of thiocyanate. It therefore removes some of the reagent and gives a lighter color. By using a large excess of thiocyanate, the interference is

[2] See H. S. Frank and R. L. Oswalt, *J. Am. Chem. Soc.* **69**, 1321 (1947).

[3] A. K. Babko, *J. Gen. Chem. USSR* **16**, 1549 (1946); *C.A.* **41**, 4732 (1947).

[4] J. T. Woods and M. G. Mellon, *Ind. Eng. Chem., Anal. Ed.* **13**, 551 (1941). These authors used a spectrophotometer. If a filter photometer is used, the system deviates from Beer's law (Figs. 55-8 and 58-9).

[5] P. Karsten, Bÿdrage tot de Colorimetrie, thesis, Groningen, 1934. Also see T. C. J. Ovenston and C. A. Parker, *Anal. Chim. Acta* **3**, 277 (1949); W. Hacker, A. Zimmermann, and H. Rechmann, *Z. Anal. Chem.* **129**, 104 (1949).

[6] For details see E. B. Sandell, *Colorimetric Determination of Traces of Metals*, Wiley-Interscience, New York, 3rd ed., 1959, p. 529.

hardly noticeable with relatively small amounts of mercury (less than 1 g/liter). Bismuth in quantities greater than 100 mg/liter may give rise to the formation of complex bismuth thiocyanate ions with a yellow color. Zinc and cadmium in larger quantities interfere. Nickel and cobalt and large amounts of manganese give colored complexes with thiocyanate.

Anions. Phosphates, fluorides, and oxalates interfere even in small quantities, because they form fairly stable complexes with ferric ions. The interference by sulfates is less strong but still very pronounced. Nitrates and chlorides have a slight effect when present in large amounts, nitrates less than chlorides.

From the above it is evident that one must be careful in the application of the spectrophotometric or colorimetric determination of iron with thiocyanate in solutions containing unknown amounts of salts. The following procedure can be used without any special precautions in the determination of iron in drinking water, which as a rule contains very little if any salt. In other cases, when unknown amounts of salts are present, it is advisable to precipitate the iron with a slight excess of ammonia, dissolve the washed hydrous iron oxide in dilute nitric acid, and determine the iron in the solution so obtained.

Before starting the experiment, learn the operation of the spectrophotometer. Check the absorption cells as follows. Clean a pair of cells thoroughly. The exterior surfaces must be dry. It is unnecessary to dry the interiors, because the cells may be filled and rinsed with the solution to be contained. Reserve one cell for the reference solution and one for the solution containing iron. Measure the transmittance or absorbance[7] of the sample cell at 480 mμ with respect to the reference cell when both are filled with water or with reference solution (see below). If the absorbance is not zero, correct the subsequent readings, as explained on page 971.

If the number of spectrophotometers is limited, students may work in pairs to construct the calibration curve. Each student should analyze his own "unknown." The following standard solutions are sufficient for two students. The procedure assumes the use of 1-cm absorption cells and is suitable for iron contents up to about 1 mg.

Standard solutions: Potassium thiocyanate: dissolve 60 g of reagent-grade material in 1 liter of water (note 1). Nitric acid, 0.2 *M*: dilute 33 ml of 6 *M* reagent to 1 liter.

Standard iron solution (note 2): Prepare a solution of ferric nitrate or ferric ammonium sulfate containing about 0.5 g of iron and 10 ml of concentrated nitric acid per liter. Determine the iron content accurately (p. 828 f.).

Procedure. Transfer 5 ml (pipet) of the standard iron solution, containing 2.50 mg of iron, to a 100-ml volumetric flask, and dilute to the mark with 0.2 *M* nitric acid. Mix thoroughly. By means of a buret deliver 3, 6, 12, and 18 ml of the solution to each of four 100-ml volumetric flasks. Read the volume delivered to 0.01 ml. To a fifth 100-ml flask add 50 ml of thiocyanate solution and dilute to the mark with 0.2 *M* nitric acid. This is the reference solution ("solvent") which is placed in the reference cell to adjust the spectrophotometer to 100 per cent

[7] If the instrument is provided both with transmittance and absorbance scales, read the latter. If only a transmittance scale is provided, calculate the absorbance from the transmittance. Do not keep superfluous (insignificant) figures. For example, if the transmittance is 0.900 ± 0.002, the absorbance is 0.046 ± 0.001. As a practical matter, the third decimal in absorbance represents the limit of accuracy of most spectrophotometers.

transmission for all the following measurements. Treat each solution containing iron separately as follows. Do *not* add thiocyanate to any solution *until just before* the spectrophotometric measurement is to be made. Add 50 ml of thiocyanate (pipet) and dilute to the mark with 0.2 *M* nitric acid; mix thoroughly. Transfer a suitable amount to the absorption cell and measure absorbance at 480 mμ immediately. Plot absorbance versus concentration of iron (note 3).

Analysis of unknown. Transfer the unknown sample quantitatively to a 250-ml volumetric flask, using 0.2 *M* nitric acid to rinse the container. Add 125 ml of thiocyanate and dilute to the mark with 0.2 *M* nitric acid. Measure absorbance at 480 mμ (notes 4 and 5). Calculate the concentration by reference to the calibration curve.

Notes:

1. The thiocyanate must be free of iron. It may be tested by adding a few ml. of acetone to 1–2 g of salt; the solution should be colorless. If purification is necessary, add a few milligrams of an aluminum salt to the aqueous solution, precipitate aluminum hydroxide by adding a few drops of ammonia, and filter off the precipitate, which will contain all the iron.

2. For instructional purposes time may be saved if the instructor issues a solution containing an accurately known amount of iron together with the unknown solution.

3. If the spectrophotometer provides sufficiently monochromatic radiation a straight line will be obtained. Otherwise the curve will resemble Fig. 55-9. A similarly shaped curve will be obtained if round cuvettes are used instead of absorption cells with flat windows.

4. It is desirable to have an absorbance reading in the range 0.5–1. If the amount of iron is too large or too small, the dilution should be adjusted suitably. Because of the instability of the color system, solutions that are too concentrated should not be diluted. It is wise, when working with a sample of completely unknown concentration, to carry out a preliminary measurement with a portion of the sample to determine the proper dilution for accurate measurement.

5. The iron must be present as iron(III). If iron is present in the ferrous state, it is easily oxidized by adding a little bromine water to the solution. The excess of bromine need not be removed, as it is reduced immediately by the thiocyanate.

6. *o*-Phenanthroline reacts with iron(II) to give an intensely colored complex (p. 109). The reaction is simple and straightforward and is preferred for the accurate determination of iron. For procedure, see Ref. 6, p. 537 ff.

Determination of permanganate and dichromate in a mixture

The absorption spectra of 10^{-3} *M* solutions of potassium permanganate and dichromate in an acid medium (1.0 *M* sulfuric acid, 0.7 *M* phosphoric acid) are shown in Fig. 61-1.[8] From the pronounced difference in the spectra, it appears that the mixture should be well suited to spectrophotometric analysis. This is the case, despite some complications that are discussed below.

Solutions of potassium permanganate follow Beer's law in monochromatic radiation. The absorption peak at 545 mμ is chosen as one of the two analytical wavelengths in preference to the peak at 520 mμ because the absorption by

[8] J. J. Lingane and J. W. Collat, *Anal. Chem.* **22**, 166 (1950).

chromium(VI) is smaller at the former wavelength. Because the peak is narrow, the measurements must be made at the smallest possible slit width (p. 973). When the spectral band width is increased from 0.8 to 5.3 mμ,[9] the observed absorbance decreases by 2.4 per cent.

The complications arise from the equilibria involving chromium(VI). The most important equilibria are the following:

$$Cr_2O_7^{--} + H_2O \rightleftharpoons 2\,HCrO_4^- \quad (1)$$

$$HCrO_4^- \rightleftharpoons H^+ + CrO_4^{--} \quad (2)$$

Equilibria involving $Cr_3O_{10}^{--}$ and $Cr_4O_{13}^{--}$ are unimportant at the concentrations involved in spectrophotometric measurements. The equilibrium constant of reaction (2) is 3×10^{-7}. Therefore, in an acid medium all the chromium is present as $Cr_2O_7^{--}$ and $HCrO_4^-$, but the relative amounts of the two species depend upon the concentration. Contrary to a statement in a current text on advanced analytical chemistry (in a discussion of this method), the concentration of $HCrO_4^-$ is never negligible; actually in dilute solutions $HCrO_4^-$ is the main species. Assuming that reaction (1) represents the only important equilibrium, and taking the equilibrium constant equal to 0.030, which applies at an ionic strength of unity,[10] the values in Table 61-1 are calculated for the composition of solutions made by dissolving *c*

TABLE **61-1. Calculated Composition of** *cM* **Potassium Dichromate Solution (pH ca. 3, ionic strength 1)**

c	$[HCrO_4^-]$, *M*	$[Cr_2O_7^{--}]$, *M*	% $HCrO_4^-$
10^{-2}	1.138×10^{-2}	4.31×10^{-3}	56.9
10^{-3}	1.787×10^{-3}	1.064×10^{-4}	89.4
10^{-4}	1.974×10^{-4}	1.30×10^{-6}	98.7
10^{-5}	1.9973×10^{-5}	1.33×10^{-8}	99.87

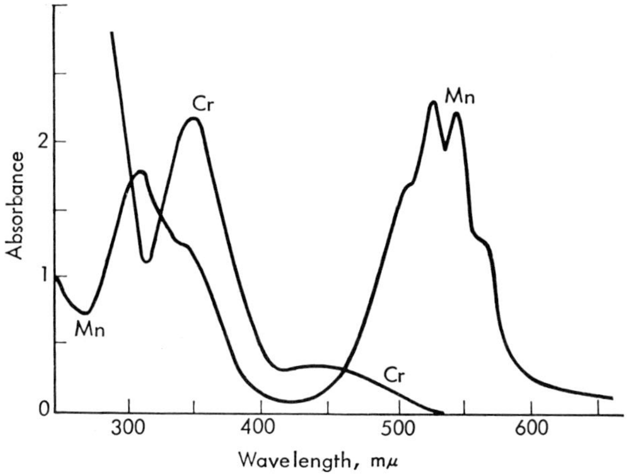

Fig. 61-1. Absorption spectra of 10^{-3} *M* potassium dichromate and potassium permanganate in 1 *M* sulfuric acid, 0.7 *M* phosphoric acid (1-cm cell).

[9] Corresponding to slit widths of 0.01–1.2 mm on a Beckman DU spectrophotometer.
[10] W. G. Davies and J. E. Prue, *Trans. Faraday Soc.* **51**, 1045 (1955).

moles of potassium dichromate in 1 liter of acid solution. The system is further complicated by the existence of $HCr_2O_7^-$ and H_2CrO_4, which are present in significant amounts at hydrogen-ion concentrations around 1 *M*;[11] also, mixed complex ions containing chromium and sulfur may be formed in sulfuric acid solutions.

Therefore, solutions of potassium dichromate much above 10^{-4} *M* cannot follow Beer's law, no matter what the acidity is, unless the measurements are made at an isosbestic point. In this system, ignoring species such as $HCr_2O_7^-$, etc., such a point corresponds to the relation $\epsilon_{HCrO_4^-} = 2\epsilon_{Cr_2O_7^{--}}$. Deviations from Beer's law were established by Hantzsch in 1910 and were studied very precisely by Kortüm.[12] In fact, the deviations from Beer's law in the region 366–404 mμ are used to evaluate the equilibrium constant of reaction (1) (see Refs. 10 and 11). It is conclusively established that in 0.001 *M* and 1 *M* perchloric acid no isosbestic point exists in the region 250–500 mμ; the closest approach in 1 *M* acid occurs at about 460 mμ (Ref. 11). Throughout this region, $\epsilon_{Cr_2O_7^{-2}}$ is more than $2\epsilon_{HCrO_4^-}$; therefore the apparent molar absorptivity decreases with decreasing concentration and approaches a limit at low concentration, at which all the chromium is present as $HCrO_4^-$. Table 61-2 shows some of Kortüm's data, representing the observed (average) molar absorptivity. (The wavelengths correspond to lines in the mercury spectrum.)

The variation of $\bar{\epsilon}$ with *c* at 366 mμ is quite marked. Therefore the peak at 350 mμ, which at first glance (Fig. 61-1) appears to be suitable for the analysis of the mixture, would in fact be unsuitable because of pronounced chemical deviation from Beer's law. At this wavelength an empirical calibration curve would have to be established.

According to Körtum's precise measurement, $\bar{\epsilon}$ at 436 mμ varies less with concentration than at 366 mμ. Therefore, the equivalent absorptivities of $HCrO_4^-$ and $Cr_2O_7^{--}$ are not very different at this wavelength. However, this is *not* an isosbestic point. The chemical deviations from the Beer law become greater with increasing concentration, exactly as indicated in Table 61-1.

TABLE **61-2. Apparent Molar Absorption of Potassium Dichromate**[a] **(0.005 *N* sulfuric acid, 20°C)**

436 mμ			366 mμ		
c, M	$\bar{\epsilon}$	*Ratio*[b]	*c, M*	$\bar{\epsilon}$	*Ratio*[b]
3.063×10^{-5}	449.0	1	2.3765×10^{-5}	2581.6	1
1.323×10^{-4}	449.7	1.0016	8.496×10^{-5}	2596.2	1.0057
4.723×10^{-4}	454.6	1.0125	2.2508×10^{-4}	2629.9	1.0187
1.3145×10^{-3}	462.8	1.0180	4.587×10^{-4}	2687.1	1.0409
2.5546×10^{-3}	472.0	1.0512	8.037×10^{-4}	2751.2	1.0657
4.713×10^{-3}	482.4	1.0744	4.346×10^{-3}	3172.4	1.2289
2.5556×10^{-2}	510.8	1.1376			

[a] From G. Kortüm, *Z. Physik. Chem.* (*Leipzig*) **B33**, 243 (1936).
[b] Ratio of absorptivity to that of the most dilute solution.

[11] J. Y. Tong and E. L. King, *J. Am. Chem. Soc.* **75**, 6180 (1953).
[12] G. Kortüm, *Z. Physik. Chem.* (*Leipzig*) **B33**, 243 (1936).

From the above it is clear that the determination of chromium(VI) and manganese(VII) in a mixture is much simpler, theoretically and practically, in an alkaline medium in which both substances follow Beer's law accurately over wide ranges of concentration. However, the determination in an acid medium that is given below is of great practical value, because it is applicable to the determination in steel and other iron-containing alloys; iron and several other constituents would be precipitated in an alkaline medium.

Lingane and Collat (Ref. 8), using the medium 1.0 *M* sulfuric and 0.7 *M* phosphoric acid, chose the wavelengths 440 and 545 mμ for the determination of chromium and manganese in steel. In this medium the Beer law was found to be obeyed at 440 mμ over the concentration range 1.7×10^{-4} to 4.2×10^{-3} *M* dichromate.[13] Also, in 1 *M* sulfuric acid the Beer law is obeyed within experimental error (of ordinary spectrophotometric measurement) at least up to 1.5×10^{-3} *M* dichromate, which corresponds to an absorbance of about 0.55 in a 1-cm absorption cell.

Equipment: Photoelectric spectrophotometer (visible region) and 1-cm glass absorption cells. See page 971 with regard to checking cells.

Standard solutions: Potassium permanganate, 0.003 *M*; prepare an approximately 0.15 *N* (0.03 *M*) solution and standardize as directed on page 825; dilute tenfold. Potassium dichromate, 0.015 *M*: weigh out 2.206 g of pure dry potassium dichromate and dilute to exactly 500 ml (note 1). Prepare 3.75 *M* sulfuric acid by diluting the 6 *M* reagent (use graduated cylinder).

Procedure. Pipet 1-, 2-, 3-, 4-, and 5-ml portions of 0.003 *M* potassium permanganate into individual 50-ml volumetric flasks. Add 20 ml (pipet) of 3.75 *M* sulfuric acid to each and also to another empty 50-ml flask. Dilute to the mark with water. Measure the absorbance of each solution at 440 and 545 mμ, using the 1.5 *M* sulfuric acid as reference (note 2). At each wavelength use the smallest possible slit width; make a note of the slit widths and use the same values for the subsequent measurements. Plot absorbance versus concentration and calculate molar absorptivity of Mn(VII) at each wavelength.

Repeat the above procedure exactly, using 0.015 *M* potassium dichromate instead of potassium permanganate. Calculate the molar absorptivity of chromium(VI) at each wavelength (note 3).

Analysis of unknown. Transfer the unknown solution quantitatively to a 50-ml volumetric flask, add 20 ml (pipet) of 3.75 *M* sulfuric acid, and dilute to the mark with water. Measure absorbances at 440 and 545 mμ against 1.5 *M* sulfuric acid, using the same slits as were used with the standard solutions. Calculate the concentration of chromium and manganese using Eq. (18) of Chapter 60 (note 4).

Procedure for determination in iron and steel

To a 1-g sample in a 300-ml Kjeldahl flask add 30 ml of water and 10 ml of concentrated sulfuric acid, and also 10 ml of 85 per cent phosphoric acid if tungsten is present. Boil gently until decomposition is complete or the reaction subsides.

[13] From the exact measurements of Kortüm at 436 mμ, one would expect an increase of absorptivity of a few per cent at the upper end of this range. It is possible that the observed value at the highest concentration ($A = 1.54$) is a few per cent too low for instrumental reasons.

Then add 5 ml of concentrated nitric acid in several small portions. If much carbonaceous residue persists, add 5 ml more of nitric acid, and boil down to copious fumes of sulfuric acid. Dilute to about 100 ml and boil until all salts have dissolved. Cool, transfer to a 250-ml volumetric flask, and dilute to the mark.

Pipet a 25- or 50-ml aliquot of the sample solution (centrifuged or filtered if not perfectly clear) into a 250-ml Erlenmeyer flask. Add 5 ml of concentrated sulfuric acid, 5 ml of 85 per cent phosphoric acid, and 1–2 ml of 0.1 *M* silver nitrate, and dilute to about 80 ml. Add 5 g of pure potassium persulfate, swirl the flask until most of the salt has dissolved, and heat to boiling. Keep at the boiling point for 5–7 min. Cool slightly and add 0.5 g of pure potassium periodate (note 5). Again heat to boiling and keep at the boiling point for about 5 min. Cool, transfer to a 100-ml volumetric flask, and dilute to the mark. Transfer a portion of the solution to a 1-cm absorption cell and measure absorbance at 440 and 545 mμ.

Because cobalt, vanadium, nickel, and iron absorb to some extent, correction for these elements is required. Calculate the first approximation to the contents of chromium and manganese from Eq. (18) of Chapter 55. The apparent percentages are corrected by the values given in Table 61-3.

TABLE **61-3. Corrections for Interfering Substances**[a]

Substance	*Correction to* Cr, 440 mμ[b]	*Correction to* Mn, 545 mμ[b]
Cr[c]	—	0.0025
Mn[c]	0.490	—
V	0.0266	0.0000
Co	0.0072	0.0011
Ni	0.0039	0.0001
Fe	0.0005	0.0000

[a] From J. J. Lingane and J. W. Collat, *Anal. Chem.* **22**, 166 (1950).

[b] Equivalent chromium or manganese percentages to be deducted from apparent percentages, for each per cent of interfering substance.

[c] These "corrections" are to be used when the content of either chromium or manganese is known by an independent method and the spectrophotometric method is used to determine the other. For example, in the determination of chromium in a steel known to contain 1 per cent manganese, the apparent chromium content found by measurement at 440 mμ must be decreased by 0.490 per cent.

Notes:

1. For instructional purposes time may be saved if the instructor provides the student with these standard solutions.

2. For work of the highest accuracy the reference solution for the spectrophotometric measurement must be 1.5 *M* sulfuric acid. If pure, this solution is completely transparent in the spectral range used and may be replaced by water as reference. The difference in refractive index causes a difference in transmittance which is negligibly small.

3. Specific absorptivities a (based on grams per liter) may be used as well. The values of a for Mn and Cr are equal to 2.877 and 2.829, respectively, times the molar absorptivities of permanganate and dichromate.

4. Because of differences among spectrophotometers even of the same model, absorbance data must not be taken from the literature or from other laboratories. The measurements on the unknowns must be made on the same instrument that is used for the standard solutions.

5. Both oxidants are needed. Persulfate (with silver as catalyst) satisfactorily oxidizes chromium but not manganese; periodate satisfactorily oxidizes manganese but not chromium.

Determination of large amounts of copper by differential spectrophotometry

Innumerable methods have been proposed for the colorimetric and spectrophotometric determination of copper. In the absence of all interfering substances, a determination of relatively low sensitivity may be based upon the color of aqueous solutions. However, the nature and amount of anion and the acidity must be known, because of the formation of copper complexes of different colors.[14] The sensitivity and accuracy are much improved in the presence of ammonia; the concentration of ammonia must be kept constant for the reasons explained on page 983. Some amines give a more intense color than ammonia does; the absorbance of a given amount of copper in the presence of tetraethylenepentamine is 3.5 times that in the presence of ammonia.[15] Generally speaking, the ammonia and amine methods are suitable for amounts in the order of 200 ppm. In either case cobalt(II) and nickel(II) obviously interfere by the formation of colored complexes. Another serious objection is the fact that many ions [aluminum, antimony, beryllium, chromium(III), iron(II), iron(III), to name a few] are precipitated in ammoniacal medium, and a separation of copper is required before the determination.

Several methods are suitable for the determination of micro amounts of copper, particularly methods using dithizone or diethyldithiocarbamate.[16] Such methods can be made relatively free of interference by other metals. For the determination of macro amounts of copper, however, the use of an ultrasensitive method has the disadvantage of requiring great dilutions before the spectrophotometric measurement.

A method of only moderate sensitivity, but much more nearly free from interference than the ammonia or amine method, is based upon absorption in relatively strong acid solution. The existence of yellow-green chloride complexes in concentrated chloride solution has been known for a long time. In 1 : 1 hydrochloric acid, the absorption maximum is at 970 mμ (near-infrared).[17] Of all the common metals only nickel absorbs appreciably, and only 1/170 as much as copper does. Bastian[18] found that 10 per cent perchloric acid was a suitable medium for the precise determination of large amounts of copper (for example, in the analysis of copper

[14] Compare the colors of solutions of copper sulfate and copper chloride.

[15] T. B. Crumpler, *Anal. Chem.* **19**, 325 (1947).

[16] E. B. Sandell, *Colorimetric Determination of Traces of Metals*, Wiley-Interscience, New York, 3rd ed., 1959, p. 442ff.

[17] D. G. Davies and H. M. Hershenson, *Anal. Chim. Acta* **13**, 150 (1955).

[18] R. Bastian, *Anal. Chem.* **21**, 972 (1949).

alloys) by means of differential spectrophotometry. Figure 61-2 shows the absorption spectra of copper and nickel in 10 per cent perchloric acid. Cupric perchlorate has a broad maximum at about 800 mμ. Nickel is the only common constituent of copper alloys that absorbs appreciably in this region. (Note that the concentration of nickel in Fig. 61-2 is four times that of copper.) The absorbance (1 cm thickness) of 1 g of cobalt, chromium, or iron per 100 ml is less than 0.03. In solutions known to contain only copper, the wavelength 800 mμ may be used, but in the presence of nickel the wavelength 870 mμ is preferable. Because the curve of absorptivity versus wavelength is not steep, Beer's law is followed at 870 mμ if a spectrophotometer working at the usual narrow slit widths is used.

From Fig. 61-2, A is about 0.41 (1-cm cell) at a copper content of 0.25 g/100 ml. A solution containing 1.5 g of Cu per 100 ml (ca. 0.25 M) therefore has an A of about 2.5, which is immeasurable on most (not all) spectrophotometers. Bastian (Ref. 18) found that a solution of this concentration could be used as the reference solution for differential measurements with a Beckman DU spectrophotometer. The slit width was 0.34 mm, corresponding to a spectral band width of 26 mμ.[19] Under these circumstances, solutions containing up to 1.8 g of Cu per 100 ml measured against the reference solution gave an A–c relation that was linear within experimental error. (With a reference of twice this concentration and a correspondingly wider slit width the A–c relation was not linear.) A solution containing 1.75 g of Cu per 100 ml measured in this way had an absorbance of 0.39, compared to the value 0.41 quoted above for a solution of 0.25 g/100 ml measured against water. The difference in the net absorbances (0.39 versus 0.41) is relatively small and in the direction to be expected for the reasons given on page 981. Summarizing, it may be anticipated from all the above that a copper content in the range 1.6–1.8 g/100 ml, measured against an accurately known reference, ought to be obtained about six times as precisely as that of a solution containing 0.1–0.3 g of Cu per 100 ml measured in the usual way. This is borne out by the experimental findings of a precision and accuracy of 1–2 ppt.

Nickel, chromium, iron, and cobalt, in amounts up to 4 per cent each, do not interfere in the determination. These percentages are above those normally contained in many copper-base alloys. Tin will be largely precipitated as metastannic acid during the dissolution process. If this is removed by filtration, loss of copper with the precipitate must be guarded against when large amounts of tin are present.

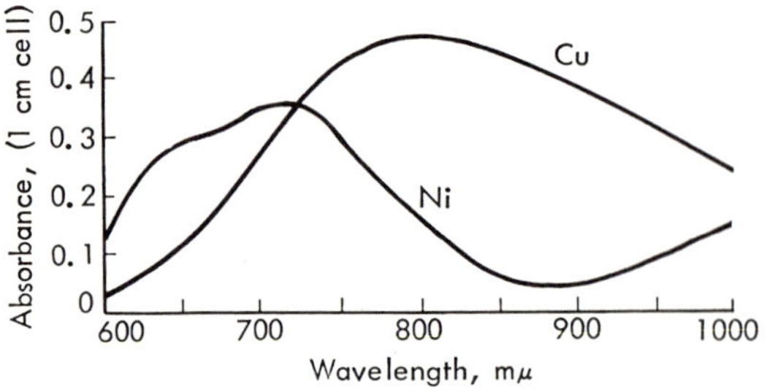

Fig. 61-2. Absorption spectra in 10 per cent perchloric acid. Concentration of copper, 0.25 g per 100 ml; of nickel, 1.0 g per 100 ml.

[19] No general significance should be attached to quoted values of slit widths, since they depend upon the intensity of the source (i.e., upon its age), the optical alignment, the quality of mirrors in the instrument, and particularly upon the sensitivity of the phototube or photomultiplier. Many instruments operate now at narrower slit widths than was the case in 1949.

In the procedure which is given below for samples containing much tin, the metastannic acid is filtered off, treated with perchloric acid and water, and most of the tin is then removed by volatilization after addition of hydrobromic acid.

Equipment. A spectrophotometer capable of measurement at the wavelength 870 mμ and with adjustable slit width; 1-cm absorption cell. It is essential that one cell be used throughout for the reference and another for the solutions. If for any reason[20] different cells are used, each must be marked for identification and its exact length (in relation to the reference cell) must be determined as explained in note 1. Also it is essential that reference and sample solution be at nearly the same temperature: A difference of 5°C will cause an error in concentration of more than 0.1 per cent.

Preparation of reference solution. Accurately weigh 1.5000 g of oxygen-free, highest purity copper into a 250-ml beaker. Add 20 ml of 1 : 1 nitric acid. After the copper has dissolved, add 10 ml of 60 per cent perchloric acid and take the sample to heavy fumes. Cool, dilute with 50 ml of water, and boil for 2 min to expel chlorine. Cool to room temperature, transfer quantitatively to a 100-ml volumetric flask, and dilute to the mark (note 2).

Preparation of standard curve. Prepare solutions containing 1.6, 1.7, and 1.8 g of copper per 100 ml exactly as described above. Measure the absorbance of these and also of the reference solution at 870 mμ against the reference solution, using the narrowest slit width possible. Depending upon the instrument used, the A–c curve will more or less approximate a straight line.

Analysis of unknown solution. For a student exercise the instructor may issue a solution of accurately known concentration, in the range 1.5–1.8 g of Cu per 100 ml, in 10 per cent perchloric acid. The student may determine the concentration by measuring the absorbance at 870 mμ and referring to the A–c graph prepared above.

Procedure for analysis of copper-base alloys. In a typical brass containing mainly copper, zinc, lead, and only a trace of tin, the sample may be treated exactly as described for the preparation of the reference solution (note 3). If the copper content is (e.g.) 60 per cent, the weight of sample should be 2.5–2.7 g.

If much tin is present, e.g., in bronzes, the following procedure may be followed (see Ref. 18). The weight of sample should be such that the weight of copper is in the range 1.5–1.8 g.

Accurately weigh a suitably sized sample into a 250-ml beaker. Dissolve in 20 ml of 1 : 1 nitric acid. After decomposition is complete, add 50 ml of hot water. Filter the solution and wash the residue twice with hot water. Place the filter paper in the original beaker and add 20 ml of nitric acid and 10 ml of 60 per cent perchloric acid. Heat to fumes to destroy the paper, cool the solution somewhat, and add a little water and 20 ml of 48 per cent hydrobromic acid. Heat to fumes again to remove tin (and antimony). Repeat twice. If any residue of tin and antimony remains, remove it by filtration. Wash the precipitate and combine the filtrate and washings with the original filtrate. Boil to fumes and dilute to 100 ml as described under preparation of the standard reference solution. Measure absorbance at 870 mμ against the reference solution.

[20] Measurements on many samples may be expedited by using one reference cell and three cells for measurement, which is possible in some instruments.

Notes:

1. The following procedure is used to determine the ratio of cell lengths of absorption cells. Let the cells be (1) and (2) with cell lengths b_1 and b_2. Fill each cell with a solvent known to be transparent at the wavelength in question. Using (1) as reference, measure absorbance A_1 of (2). Replace solvent in (2) by an absorbing solution and measure A_2. Finally, replace solvent in (1) by the same absorbing solution and measure A_3.

The measurements therefore are:

Cell 1	*Cell 2*	*Absorbance of (2) versus (1)*
Solvent	Solvent	A_1
Solvent	Solution	A_2
Solution	Solution	A_3

The value of A_1 corrects for unequal cell transmittances. The absorbance of the solution per unit path length therefore is $(A_2 - A_1)/b_2$; the difference in path lengths is equal to $A_3 - A_1$ divided by the absorbance per unit path length. Hence

$$b_2 - b_1 = b_2 \frac{A_3 - A_1}{A_2 - A_1}$$

Rearranging there is obtained

$$\frac{b_1}{b_2} = 1 - \frac{A_3 - A_1}{A_2 - A_1}$$

Evidently the precision of determining the ratio is the greater the greater A_2 is. The calculation assumes that the refractive indices of solution and solvent are equal; this introduces a negligible error in practically all cases.

2. The specific gravity of 60 per cent perchloric acid is ca. 1.54. Upon dilution of 10 ml to 100 ml with water, the concentration becomes a little less than 10 per cent.

3. If enough tin is present to cause a perceptible turbidity due to metastannic acid the solution should be filtered; the filtrate should be washed thoroughly and the washings added to the filtrate before being diluted to the mark.

Determination of copper by spectrophotometric titration with EDTA

In a spectrophotometric titration, the absorbance A of a solution is measured at a suitable wavelength after addition of known increments of titrant. Depending on the relative absorptivities of the substance being titrated, of the titrant, and of the reaction product, the relation between A and volume of titrant may have one of the shapes described in Chapter 38. If the reaction is quantitative, straight lines are obtained of different slopes before and after the end point. It is possible also to add a substance (not necessarily one of the usual indicators) to serve as an indicator. For example, both bismuth and copper form complexes with EDTA, but BiY^- is far more stable than CuY^{--}. Therefore, when EDTA is added to a mixture of the two substances, BiY^- is formed quantitatively before any CuY^{--} is formed. (The principle is illustrated in Table 36-4.) CuY^{--} has an absorption maximum at 745 mμ at which EDTA and BiY^- do not absorb. In a measurement at 745 mμ, during titration with EDTA, an abrupt increase in A signals the formation of CuY^{--}; thus copper serves as an indicator for the titration of bismuth. By spectrophotometric measurement it is possible to extend considerably the scope of compleximetric titrations, and in particular to carry them out at much lower concentrations than is possible with the use of metallochromic indicators. The procedure given below gives excellent results in titration of copper solutions in the range 10^{-4} M.

A *photometric* titration is the same in principle as a spectrophotometric titration except that monochromatic radiation is not used. The performance of such a titration is very simple. A beaker is used as the titration vessel. A light source is arranged to send a beam of light vertically through the beaker and through a suitable color filter to a photronic cell which is connected to a galvanometer. Titrant is added in the usual way and the transmittance of the reaction mixture is measured.

For the reasons discussed in Chapter 55, the use of monochromatic radiation offers increased selectivity and sensitivity, and therefore spectrophotometric titrations are to be preferred in general.

In a *colorimetric* titration, the end point is observed visually by using two flat-bottomed cylinders, one of which serves for comparison. Such titrations are now obsolete.

The construction of spectrophotometers is such that the arrangement of beaker plus vertical light path mentioned above is not practical. Many vessels have been improvised or constructed for the performance of spectrophotometric titrations.[21] Satisfactory vessels can be made by attaching a section of 1-cm precision square tubing to the side of a square bottle (Fig. 61-3). The bottle is closed with a perforated metal cap containing a self-sealing gasket. Titrant is added by a hypodermic syringe through the gasket. A light-tight cover is constructed for the cell compartment of the spectrophotometer which permits the bottle to rest on top of the cover

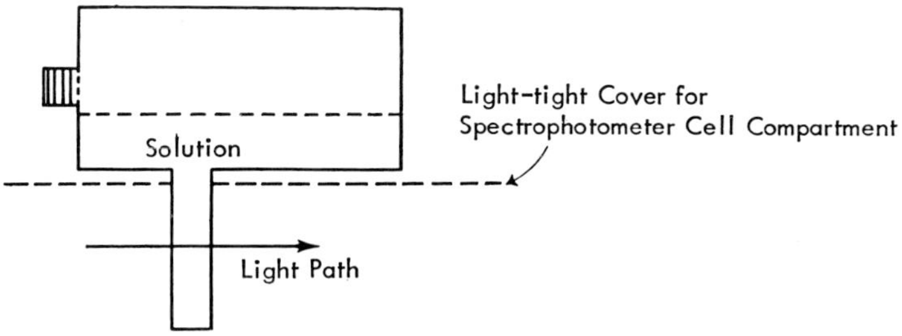

Fig. 61-3. Cell for spectrophotometric titration. A piece of precision square tubing, attached to the square bottle, serves as the absorption cell.

[21] Typical early examples are described by: (a) P. B. Sweetser and C. E. Bricker, *Anal. Chem.* **24**, 409 (1952): (b) H. V. Malmstadt and E. C. Gohbandt, *ibid.* **26**, 442 (1954); for a review, see (c) R. F. Goddu and D. N. Hume, *ibid.* **22**, 1314 (1950): also (d) J. B. Headridge, *Photometric Titrations*, Pergamon, New York, 1961.

while the square tubing projects through the cover in the path of the monochromatic beam.

Five- and 10-cm cylindrical absorption cells, which are widely available and extensively used, are very satisfactory as titration vessels when an ultramicroburet is available for addition of the titrant. These cells are shaped as in Fig. 61-4. A typical cell with a 5- or 10-cm light path has a total volume of about 15 and 30 ml, respectively. The free volume in the neck, which is used for filling and emptying the cells, amounts to several tenths of a milliliter and is sufficient to allow the addition of titrant in the form of a relatively concentrated solution.

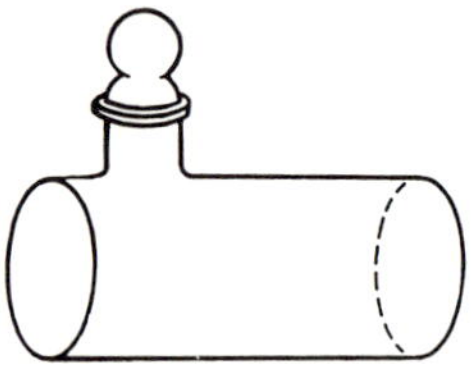

Fig. 61-4. Cylindrical absorption cell, suitable as vessel for spectrophotometric titration.

Titration of copper. As mentioned above, CuY^{--} absorbs at 745 mμ. It also has strong absorption in the near ultraviolet, and use is made of this band in the procedure described. Figure 61-5 (due to Malmstadt and Gohbandt, Ref. 21b) gives the spectra of the copper(II) EDTA complex, copper nitrate, and EDTA. In the region 340 to about 260 mμ, the complex absorbs much more strongly than either copper nitrate or EDTA (note the 20-fold difference in concentrations in Fig. 61-5). The wavelength 320 mμ is suitable for the use of a 5-cm cell as titration vessel and copper concentration up to around 5×10^{-4} $\dot{M}$. Although this wavelength does not correspond to an absorption maximum, the absorptivity curve is not steep and satisfactory agreement with the Beer law is observed in practice with the Beckman DU spectrophotometer or the equivalent. For other concentrations or path lengths a different wavelength may be chosen. By carrying out the titration

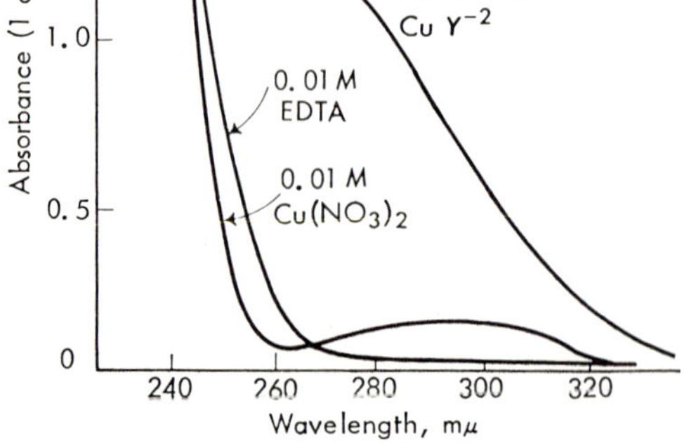

Fig. 61-5. Absorption spectra of copper nitrate, copper-EDTA complex, and EDTA. pH = 4.2.

in a buffer of pH 4.2, interference by calcium, barium, and magnesium is eliminated. These metals, of course, would interfere in titration at pH 10 or thereabouts. If other metals may be present, a still lower pH and a different wavelength may be preferable for the titration. At a pH of 2.4–2.8 and wavelength 745 mμ, according to Sweetser and Bricker,[22] considerable amounts of zinc, cadmium, manganese(II), tin(IV), and chromium(III), and smaller amounts of aluminum, cause little or no interference. Interference by small amounts of iron(III) and tin(II) can be prevented by adding sodium fluoride and potassium tartrate, respectively.

The procedure given below is designed to acquaint the student with the method of carrying out the titration and with the principles involved in selection of concentration and wavelength. For this reason, the necessary standard solution of copper and buffer are furnished by the instructor. A solution of EDTA of approximately known concentration, which the student standardizes against the known solution of copper, is also furnished.

Equipment: Beckman Model DU spectrophotometer with UV attachment or the equivalent and a 10-cm cell compartment; four 1.000-cm silica cells and two 5.00-cm cylindrical Pyrex cells; 0.1–1 ml ultramicroburet or the equivalent. Examine the microburet carefully to be sure there is no air bubble between the mercury column and the solution. See the instructor if the buret needs attention.

Standard solutions furnished: Standard cupric nitrate of exactly known concentration, approximately 0.0125 *M*. Disodium dihydrogen ethylenediamine tetraacetate, approximately 0.1 *M*, and pH 4.2 acetate buffer, total acetate = 0.2 *M*.

Procedure. Prepare the following solutions:

a. 0.0005 *M* cupric nitrate in acetate buffer. Pipet 2.00 ml of standard cupric nitrate into a 50-ml volumetric flask and dilute to the mark with the buffer.

b. 0.002 *M* disodium dihydrogen ethylenediamine tetraacetate. Using a Mohr pipet deliver 1 ml of 0.1 *M* versene solution to a 50-ml volumetric flask and dilute to the mark with buffer.

c. 0.0005 *M* copper–EDTA complex. Pipet 2.00 ml of standard cupric nitrate and 1 ml of 0.1 *M* versene into a 50-ml volumetric flask and dilute to the mark with buffer.

Standardize the 0.1 *M* EDTA solution against the standard copper solution as follows: Pipet 10 ml of solution (a) into a 5.00-cm Pyrex cell, add sufficient distilled water to raise the solution level just to the neck of the cell, insert the Teflon stopper, and invert four or five times, shaking gently. Determine the absorbance at 320 mμ versus a reference 5.00-cm cell containing acetate buffer. To the cell containing copper add 0.01 ml of 0.1 *M* EDTA from a 0.1-ml ultramicroburet and determine the absorbance. Always submerge the tip of the microburet when delivering titrant, and be sure to mix the solution thoroughly before measuring the absorbance. Add successive increments of 0.01 ml of versene solution until beyond the point at which the absorbance at 320 mμ becomes constant. Plot the data to determine the equivalence point. (Is it necessary to make a volume correction in treating the data?) Calculate the concentration of the EDTA solution.

Analysis of unknown. Transfer quantitatively the contents of the bottle containing the unknown to a 250-ml volumetric flask and dilute to volume with distilled water (see the note). Pipet 2.00 ml of this solution into a 50-ml volumetric

[22] P. B. Sweetser and C. E. Bricker, *Anal. Chem.* **25**, 253 (1953).

flask and dilute to volume with pH 4.2 acetate buffer. Determine the copper content of the unknown by transferring 10 ml of the solution to a 5-cm cell, titrating exactly as described in the preceding paragraph. Report the weight of copper present in the unknown.

Note: The directions for the dilution of the unknown correspond to a copper content of the order of 150 mg. For larger or smaller amounts the dilution may be modified suitably.

Spectrophotometric determination of lead with dithizone

Extraction of heavy metals from an aqueous solution with an organic solvent solution of dithizone was discussed in Chapter 14. Lead is extracted to the extent of 99 per cent or more from a cyanide–citrate solution in the pH range 7.5–11.5 with a slight excess of dithizone in carbon tetrachloride (see Fig. 61-6). The minimum pH for the extraction of 99 per cent of the excess dithizone into an equal volume of the aqueous phase from carbon tetrachloride is 10.85. Therefore, by maintaining a pH of ~11.0 in the aqueous phase, lead dithizonate can be extracted virtually completely into the carbon tetrachloride, and dithizone can be transferred almost completely to the basic aqueous phase. Actually, the complete removal of dithizone from the carbon tetrachloride phase is not essential, because absorbance due to it is taken into account in the standard curve, but greater care is then needed to maintain the pH constant to ensure constant partitioning of dithizone between the two phases. From the absorbance curves of dithizone and lead dithizonate

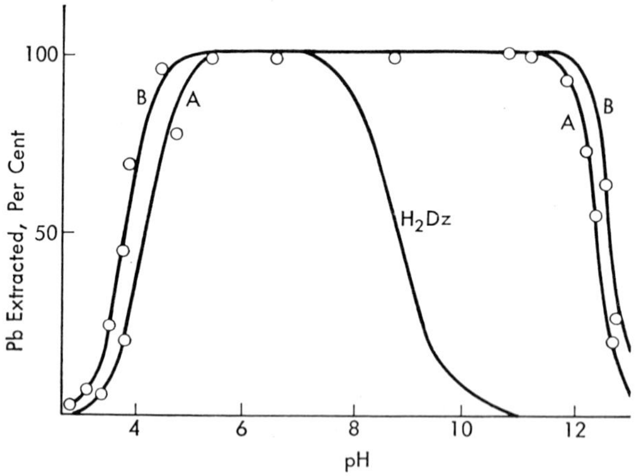

Fig. 61-6. Extraction of lead dithizonate and dithizone from sodium perchlorate solution. The solid curves are the calculated curves for the extraction of lead (0.0625 mg in 25 ml aqueous solution, $\mu = 0.1$) with 25 ml of 0.001 per cent (w/v) dithizone in carbon tetrachloride (A) and with 25 ml of 0.002 per cent (w/v) dithizone (B). The circles represent experimental values. The extraction curve for dithizone (H_2Dz) is calculated. In solutions containing citrate (0.01 *M*) and cyanide (0.02 *M*), the extraction of lead decreases below pH ca 7.5 instead of below pH ca 5.5 in their absence as shown above. From O. B. Mathre and E. B. Sandell, Talanta *11*, 295 (1964).

(Fig. 61-7), it can be seen that the absorption maximum of lead dithizonate (520 mμ) falls approximately at the absorption minimum of dithizone (510 mμ).[23]

Citrate is used to prevent precipitation of other heavy metal hydroxides, and cyanide to prevent reaction of such metals as zinc and copper with dithizone. The use of cyanide as a metal-complexing agent makes the dithizone determination of lead fairly selective. Under the recommended conditions, the only other metals reacting with dithizone are Sn(II), Bi, Tl, and possibly In. Tin(IV) does not react, so that the interference of tin(II) is easily avoided by a change in oxidation state. Bi and Tl can be separated from Pb by dithizone extraction at regulated acidity, but these methods will not be described here. Dithizone tends to be converted into a yellow oxidation product by oxygen and other oxidizing agents, and sulfite is used to prevent or minimize this.

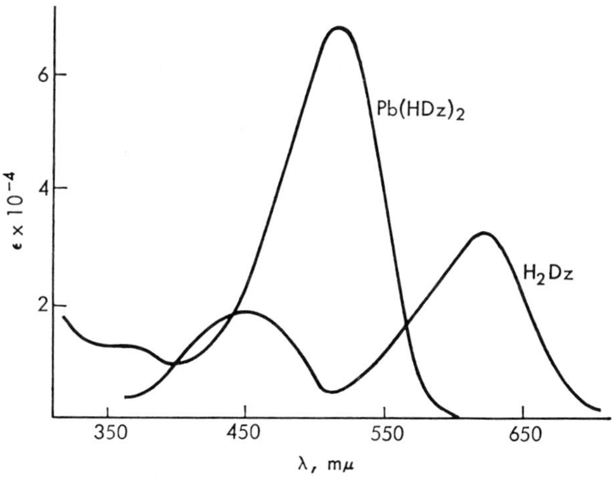

Fig. 61-7. Molar absorptivity curves of dithizone and lead dithizonate in carbon tetrachloride.

Another spectrophotometric method for the determination of lead and other heavy metals with dithizone is based on reversion.[24] The absorbance of the organic solvent extract of the metal dithizonate, containing more or less excess of dithizone, is first measured at a suitable wavelength at which dithizone absorbs. The extract is then shaken with an acid aqueous solution containing a complexing agent (hydrogen ion alone is sufficient for some metals, including lead) which converts the dithizonate into an equivalent amount of dithizone. The resulting increase in absorbance enables the concentration of liberated dithizone to be calculated, and therefrom the concentration of the initially combined metal. A calibration curve is used in practice.

The accuracy of the determination of lead with dithizone by the usual method is indicated in Fig. 61-8. By reducing the volume of dithizone–CCl_4 to 2 ml, 1–2 μg of Pb should be determinable to ~2 per cent.

[23] Similar correspondence holds for other red- or orange-colored dithizonates, so that metals whose dithizonates are extracted from an acid medium can be determined in the presence of the excess of dithizone, now remaining wholly in the carbon tetrachloride, by absorption measurements around 500 mμ. The standard curve will show a fairly large intercept on the *A* axis due to absorption by excess of dithizone.

[24] H. Irving and E. Butler, *Analyst* **78,** 571 (1953).

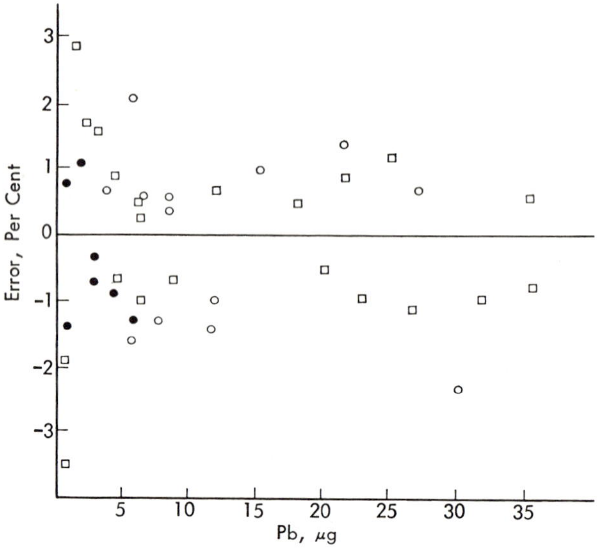

Fig. 61-8. Accuracy of determination of lead by mono-color dithizone method: ○ = 10 ml of aqueous phase, pH 10.8; 10 ml of 0.001 per cent dithizone in CCl_4; 1-cm cell, 520 mμ;
● = 10 ml of aqueous phase, pH 10.8; 25 ml of 0.001 per cent dithizone in CCl_4; 10-cm cell, 520 mμ;
□ = Reversion method: 10 ml of aqueous phase, pH 10.8; 25 ml of 0.001 per cent dithizone; reverted with 10 ml of 0.16 *M* HNO_3; 1-cm cell (10-cm cell below 8 μg of Pb), 620 mμ.

The following directions presuppose the absence of metals that react with dithizone in an alkaline cyanide medium either by forming dithizonates [Bi, Tl, Sn(II), In] or by oxidizing it [much Fe(III), Cu]. In applied analysis, a preliminary extraction of lead would be made from basic cyanide–citrate solution to separate it from such metals as Fe and Cu; by shaking the CCl_4 extract with 1:100 nitric acid, lead is transferred from the carbon tetrachloride to the aqueous phase. A constant specified acidity is also obtained in this way. This preliminary extraction is not included in the following procedure.

Reagents

Water, distilled in borosilicate glass or purified by ion exchange. It should give no test for lead.

Carbon tetrachloride, analytic reagent, preferably redistilled. Avoid breathing carbon tetrachloride.

Dithizone, solution A. 50 mg of good-quality dithizone in 1 liter of carbon tetrachloride. Keep cool and in the dark.

Dithizone, solution B. 10 mg of dithizone in 1 liter of carbon tetrachloride. Prepare by diluting the 50 mg /liter solution shortly before use.

Ammonium hydroxide. Pass ammonia from a cylinder into chilled redistilled water to give a saturated solution.

Ammonia–cyanide–sulfite solution. Dilute 725 ml of 8.0 *M* ammonia and 30 ml of 10 per cent potassium cyanide solution (low in lead) to 1 liter with water and dissolve 1.5 g of sodium sulfite in the solution.

Standard lead solution, 0.00100 per cent Pb in 1 : 100 nitric acid. This solution is preferably prepared shortly before use by diluting a 0.100 per cent Pb solution (0.160 g of dried lead nitrate in 100 ml of 1 : 100 nitric acid).

Procedure. Transfer 10 ml of a 1 : 100 (0.16 *M*) nitric acid solution containing 1–20 μg of Pb to a short-stemmed separatory funnel and add 10.0 ml of ammonia–cyanide–sulfite solution and 10.0 ml of dithizone solution B. Shake well for 30 sec and allow the phases to separate. If the carbon tetrachloride solution appears clear, run it directly into a 1- or 2-cm absorption cell, first allowing a few drops to run out of the stem of the funnel to displace any water or aqueous solution in the stopcock base. Aqueous droplets in the stem of the funnel can be removed with a thin roll of filter paper. If required, the carbon tetrachloride layer can be filtered through a small plug of glass wool in a very small funnel or, perhaps better, in the stem of the separatory funnel (discard the first few drops). Obtain the absorbance at 520 mμ against carbon tetrachloride in the reference cell without undue delay. Avoid exposure to strong light.

Establish the standard curve by taking 0, 1, 2, 5, 10, 15, and 20 μg of Pb, diluting to 10.0 ml with 1 : 100 nitric acid solution, adding 10.0 ml of ammonia–cyanide–sulfite solution and shaking with 10.0 ml of 0.001 per cent dithizone solution.

Analysis by infrared spectroscopy. Detection of impurities in carbon tetrachloride

Infrared measurements at the present time are carried out mainly for analytical purposes, and it is fair to state that an infrared spectrophotometer is an indispensable analytical instrument. This experiment is designed to acquaint the student with the practical use of infrared spectroscopy to solve a typical analytical problem. The analysis by infrared is simple; the analysis by any other method except gas chromatography would be very time-consuming.

The use of infrared measurements for quantitative determinations is based on exactly the same principles as the use of visible and ultraviolet measurements. The use for qualitative purposes (aside from detection of functional groups, p. 966) is either (1) to verify the identity of a sample or to check the purity of a sample, or (1) to identify the constituents of or impurities in a sample. Problem (1) is easily solved if an authentic pure sample is available; it is necessary only to record and compare the spectra of the unknown and the known materials. Problem (2) is much harder in general. As was pointed out on page 964, the characteristic spectrum serves as a fingerprint for a given compound. Just as in criminology, it is sometimes much harder to identify the fingerprint than to obtain it. The practical solution to problem (2) in general depends upon the availability of an extensive library of standard spectra plus an efficient, rapid method of searching the library.

There are various more or less convenient methods now in use for locating or assigning a spectrum. The situation as of 1962 is well summarized by Brügel.[25] A very extensive commercial collection which is widely used is the Sadtler series, issued by the Sadtler Research Laboratories, Philadelphia, Pa. This consists of reproductions of spectra over the range 2–15 μ; the instrument and operating conditions are given. The student should note that many different infrared spectro-

[25] W. Brügel, *An Introduction to Infrared Spectroscopy*, Methuen, London, 1962, p. 265ff.

photometers are in use, and these differ considerably in dispersion, resolving power, and stray radiation. As a result, the relative intensities of bands may sometimes appear to be different. [The positions (wavelengths), however, should always agree.] The differences are much more pronounced than in the visible–UV region. As of 1969 there were 29,000 spectra in the Sadtler series. The following procedure (given here in a much abbreviated version) is used to identify a spectrum. The wavelength region is divided in 13 steps of 1 μ each, i.e., 2–3, 3–4, . . . , 14–15. In each of these steps the *strongest* band is noted and its wavelength is recorded to 0.1 μ. Even if several strong bands may be present, only one is used for each step. Also the strongest band in the entire spectrum (2–15 μ) is noted similarly. The spectra are classified in a "Spec-Finder" according to the wavelength (to the nearest 0.1 μ) of the strongest band. For the reasons mentioned earlier, one or another band may seem to be the strongest, depending upon experimental conditions, so the same compound may be listed more than once in the Spec-Finder. Turning to the spectra listed under the strongest band, the positions of the bands in the 13 intervals are then compared. Generally several compounds thus coded will be found, and one finally compares the spectra of these possibilities with that of the unknown. The procedure is systematic and satisfactory, although the beginner might be discouraged to note, in his search for a compound with (e.g.) its strongest band at 6.2 μ, that 2168 possibilities are listed.

A useful compilation of an entirely different sort, and for a different purpose, is published in a standard format in the journal *Analytical Chemistry* under the heading "Infrared Quantitative Analysis Data."[26] Typical specimens are shown in Tables 61-4, 5, and 6.[27] The three tables represent recommended methods to solve specific problems that are evident from the headings. Referring to the analysis of carbon tetrachloride for six trace impurities, note that six wavelengths are given, at each of which only one of the six impurities absorbs. Thus a quantitative analysis of six impurities is possible if no other substances are present. As emphasized before, it is essential for quantitative work to determine the absorptivities on the particular instrument used for analysis. The meaning of most of the items in the tables is self-evident, but a few explanatory notes may be needed. Referring to the analysis of carbon tetrachloride, in the column headed "λ or ν, B.L. Pts" for carbon

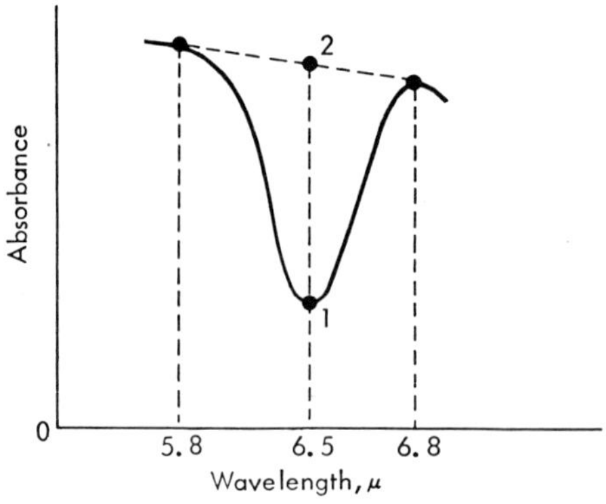

Fig. 61-9. Base line method of correction for background.

[26] The purpose and format are explained in *Anal. Chem.*, **29**, 1551 (1957).
[27] Modified from *Anal. Chem.* **31**, 157 (1959).

TABLE 61-4. **Determination of Geranyl Acetate in Geranyl Formate**

(A. J. FENTON, The Procter and Gamble Co., Cincinnati, Ohio CS-66)

Component			*Range*	*Accuracy*	*λ*	*Slit*	*Concn.*
No.	*Name*	*Formula*	*%*	*%*	*B.L. Points*	*(mm) Δλ*	*g/l Length mm*
1	Geranyl acetate	$C_{12}H_{20}O_2$	0–20	±0.2	8.1 7.93–8.22	300 0.040μ	200 0.05

Instrument: Perkin-Elmer Model 321, NaCl prism
Sample Phase: Solution in carbon disulfide
Cell Windows: NaCl
Absorbance Measurement: Base line
Calculation: Graphical
Relative Absorbance–Analytical Matrix:

Component/λ	8.1 μ
1	1.67

Material Purity: Reference compounds 99+% pure
Comments: The absorbance is expressed as the slope of the Beer's law curve in absorbance per 100% of constituent. Calibration curve passes through −0.033 absorbance at 0% acetate.

disulfide there is found "6.5 μ, 5.8–6.8." The abbreviation B.L. refers to "base line" and the meaning is shown in Fig. 61-9. The full curve represents the transmittance of a very dilute solution of carbon disulfide in carbon tetrachloride run against a reference plate of sodium chloride (the same material as the cell window). The absorption maximum of carbon disulfide is at 6.5 μ. The background absorption is

TABLE 61-5. **Determination of Citronellyl Acetate in Citronellyl Formate**

(A. J. FENTON, The Procter and Gamble Co., Cincinnati, Ohio CS-67)

Component			*Range*	*Accuracy*	*λ*	*Slit*	*Concn.*
No.	*Name*	*Formula*	*%*	*%*	*B.L. Points*	*(mm) Δλ*	*g/l Length mm*
1	Citronellyl acetate	$C_{12}H_{22}O_2$	0–20	±0.3	8.1 7.75	0.300 0.040 μ	200 0.05

Instrument: Perkin-Elmer Model 321, NaCl prism
Sample Phase: Solution in carbon disulfide
Cell Windows: NaCl
Absorbance Measurement: Base line
Calculation: Graphical
Relative Absorbance–Analytical Matrix:

Component/λ	8.1 μ
1	1.68

Material Purity: Reference compounds 99+% pure
Comments: Absorbance expressed as slope of Beer's law curve in absorbance per 100% of constituent. Calibration curve passes through 0.075 absorbance at 0% acetate.

TABLE **61-6. Analysis of Carbon Tetrachloride for Trace Impurities**[a]

(L. R. KILEY and R. T. SCHEDDEL, The Dow Chemical Co., Midland, Mich. CS-68)

Component			*Range*	*Accuracy*	*λ or ν*	*Slit (mm)*	*Concn.*
No.	*Name*	*Formula*	*%*	*%*	*B.L. Pts.*	*Δλ or Δν*	*% Length mm*
1	1,2-Dichloro-ethane	$C_2H_4Cl_2$	0–0.15	±0.02	3.35 μ 3.1–3.9	0.130 0.027 μ	100 3.5
2	2-Propanone	C_3H_6O	0–0.02	±0.001	5.75 μ 5.6–5.9	0.286 0.037 μ	100 3.0
3	Carbon di-sulfide	CS_2	0–0.10	±0.003	6.5 μ 5.8–6.8	0.342 0.039 μ	100 0.1
4	Trichloro-methane	$CHCl_3$	0–0.15	±0.05	8.2 μ 7.5–8.6	0.491 0.044 μ	100 0.1
5	Trichloro-ethylene	C_2HCl_3	0–0.15	±0.005	10.7 μ 10.6–10.9	0.782 0.054 μ	100 0.1
6	Perchloro-ethylene	C_2Cl_4	0–0.15	±0.005	10.95 μ 10.8–11.2	0.821 0.056 μ	100 0.1

Instrument: Baird Associates Model A, NaCl prism
Sample Phase: Full strength
Cell Windows: NaCl
Absorbance Measurement: Base line——— Point———
Calculation: Inverse matrix——— Successive approx.———
Graphical———

Relative Absorbances[b]*—Analytical Matrix:*

Component/λ	3.35	5.75	6.5	8.2	10.7	10.95
1	0.068	0.000	0.000	0.000	0.000	0.000
2	0.000	2.740	0.000	0.000	0.000	0.000
3	0.000	0.000	0.458	0.000	0.000	0.000
4	0.000	0.000	0.000	0.032	0.000	0.000
5	0.000	0.000	0.000	0.000	0.047	0.000
6	0.000	0.000	0.000	0.000	0.000	0.093

Material Purity: Reference compounds 99+% pure

[a] These data represent standard publication and submission is open to anyone in accordance with regulations of ANALYTICAL CHEMISTRY. The Coblentz Society is acting only as an aid to the journal.

To standardize procedures, ANALYTICAL CHEMISTRY requests that material be sent in quintuplicate to the chairman of the review committee: Robert C. Wilkerson, Celanese Corp. of America, Post Office Box 8, Clarkwood, Tex.

[b] Relative absorbances are given as the slope of the Beer's law concentration curves used expressed in terms of absorbances per 0.15% of constituent.

corrected for by the base-line technique, introduced by Wright.[28] The absorbance in the absence of carbon disulfide is assumed to be at point (2), which is defined by the base-line points at 5.8 and 6.8 μ and the absorption maximum at 6.5. Thus the net absorbance is (1) – (2). Many workers have shown that absorbances corrected in this way for background follow the Beer law.[29]

Equipment: Perkin-Elmer Infracord Model 137 (or equivalent) and 0.1-mm NaCl cell; one small desiccator containing silica gel; one 0.10-ml ultramicroburet and five 10-ml volumetric flasks. The prism in the instrument and all windows are *rock salt*. They must be protected from moisture at all times. If the cell windows are touched with the fingers, they will be etched; therefore, extreme care must be used when handling them. The cells are to be filled in the hood and kept in a desiccator when not in the spectrophotometer.

The cells furnished with the Infracord are filled with a hypodermic syringe. It is necessary to flush the cell several times with solution to be analyzed to ensure complete removal of previous solutions. At the conclusion of the experiment flush the cell with carbon tetrachloride, drain dry, and replace it in the desiccator. Do not dry the cell by flushing air through it with the syringe. This will cause atmospheric moisture to condense inside the cell.

Solutions: Carbon tetrachloride (reagent grade) and all other necessary solutions are furnished by the laboratory instructor.

Special instructions: Before attempting any work on this instrument, the student should read the instruction sheet for it. This should be done before the assigned laboratory period. The operation of this instrument will be demonstrated by the laboratory instructor, if necessary, during the assigned period.

Examine the 0.1-ml microburet carefully and be sure there is no air bubble present. Submerge the tip of the buret in solution before delivery. Handle the microburet with extreme care.

Procedure. Place a plate of sodium chloride in the reference beam and fill the 0.1-mm sample cell with reagent-grade carbon tetrachloride. Scan the spectrum from 2.5 to 15 μ. Refill the cell with your unknown, and determine its spectrum from 2.5 to 15 μ. (Record all your spectra on the same sheet of graph paper.) Refer to page 1070 and identify the impurities in your unknown. Prepare standard solutions (on a vol./vol. per cent basis) of suitable concentrations using a 0.1-ml ultramicroburet and a 10-ml volumetric flask. Scan a solution containing an intermediate concentration of impurity(ies) to verify the qualitative identification. Determine the spectrum of the other standards only in the vicinity of the region of low transmittance. From the corrected absorbances of the standard solutions, calculate the concentration of impurities in the unknown.

Qualitative and quantitative analysis by flame spectrophotometry

The procedure given below is designed for the identification and determination of five elements (calcium, copper, lithium, manganese, thallium) in an aqueous solution. The presence of any of these elements is detected by emission of radiation at a characteristic wavelength. The concentration then is determined by comparison of the intensity emitted with that emitted by solutions of known concentration

[28] N. Wright, *Ind. Eng. Chem., Anal. Ed.* **13**, 1 (1941); also see J. J. Heigl, M. F. Bell, and J. U. White, *Anal. Chem.* **19**, 293 (1947).

[29] However, for some criticisms, see G. Pirlot, *Bull. Soc. Chim. Belges* **58**, 28 (1949).

under identical experimental conditions. The experiment is designed to introduce the student to the practice of flame spectrophotometry and therefore includes a study of the effect of gas pressure upon intensity. From the results of these measurements the student may judge the degree of control of pressure that is required for a given precision in analysis. Also to save the time of the student, the number of possible constituents is limited to five, and standard solutions of each are furnished for the preparation of the calibration curve. Naturally the identical principles and measurements would be used in general if the number of constituents were not thus limited. It may be noted that the use of a recording spectrophotometer, which yields a record of intensity versus wavelength, is a great advantage in terms of time required for qualitative identification in general. For the determination of a given element, however, the nonrecording instrument is entirely suitable.

The experiment can be carried out with any commercially available flame spectrophotometer. The directions and detailed procedure have been written specifically for the use of a Beckman DU spectrophotometer, with flame attachment, and equipped with a photomultiplier. Older versions of this instrument, which were equipped with photoelectric cells rather than photomultipliers, can be used, but a lower sensitivity will be found.

For discussion of details beyond the scope of this text reference is made to the literature.[30]

Equipment: Beckman Model DU Spectrophotometer with photomultiplier and flame attachment, or the equivalent, six 50-ml volumetric flasks, and seven 5-ml beaker cells. At the start of the laboratory period, turn the DU selector switch to "check" and allow 15 min to warm up. The DU power supply is preset and is *left on*. Ask the laboratory instructor to light the burner.

Warning! Flame temperature is about 3000°C. *Keep your hands out of the burner housing and away from burner stack.*

To light burner: oxygen is turned on first (*and off last*):

1. Set oxygen tank regulator at 25 psi.
2. Set hydrogen tank regulator at 10 psi
3 Set oxygen DU regulator at 12 psi
4. Set hydrogen DU regulator at 3 psi and immediately insert lighted match into burner housing toward burner tip up from the base of burner; burner will light with a "pop."

To shut down temporarily:

1. Set hydrogen DU regulator to zero pressure.
2. Set oxygen DU regulator to zero pressure.

[30] A comprehensive treatment is given by R. Hermann and C. T. J. Alkemade, *Flame Photometry*, Wiley-Interscience, New York, 1963. This book contains, among many other practical details, an atlas of 95 flame spectrograms of 57 elements under various conditions. Also see F. Burriel-Marti and J. Ramirez-Muñoz, *Flame Photometry*, Elsevier, Amsterdam, 1957; B. L. Vallee and R. E. Thiers in *Treatise on Analytical Chemistry* (I. M. Kolthoff and P. J. Elving, eds.), part I, vol. 6, Wiley-Interscience, New York, 1965, chap. 65; J. A. Dean, *Flame Photometry*, McGraw-Hill, New York, 1960; the classical early work is H. Lundegårdh, *The Quantitative Spectral Analysis of the Elements*, G. Fischer, Jena, vol. I, 1929, vol. II, 1934. A comprehensive bibliography is given by R. Mavrodeneau, *Bibliography of Analytical Flame Spectroscopy*, Central Library of Spectroscopy, Ellicott City, Md., vol. I, 1962, vol. II, 1963.

To shut down overnight:

1. Shut hydrogen off at tank.
2. Shut oxygen off at tank.
3. Release all four pressure regulators *immediately*. (Gas regulator screw must be loose, full counterclockwise, when shut down, or damage to the diaphragm will result.)

Standard solutions: In automatic burets; aqueous solutions containing respectively 5 ppm Li, 20 ppm Ca, 25 ppm Mn, 30 ppm Cu, and 80 ppm Tl.

Qualitative analysis. Transfer the unknown quantitatively to a 50-ml volumetric flask and dilute to the mark. Mix thoroughly. First aspirate solvent (water), and then the unknown, and scan slowly the wavelengths selected from the metal-ion emission spectra (see Fig. 61-10), using the following technique:

1. DU selector switch at 0.1, DU sensitivity turned fully counterclockwise, slit setting 0.1, and $\%T$ at 100.
2. Position the cell containing the solvent under the capillary, open the shutter, and balance the null meter with the dark-current control at the chosen wavelength.
3. Substitute the unknown solution and *slowly* scan this wavelength over the range ± 15 mμ.

Samples may be conserved by spraying only when actually making an adjustment or taking a reading; *speed is essential*. Deflection of the needle to the left indicates an increased light intensity at detector, suggesting the presence of emitting species. If an ion is present, there should be a *large* deflection of the needle. Follow this procedure and report the presence or absence of each of the five ions. After identifying the constituents in your unknown *immediately* ask the instructor to shut off the burner.

Determination. Prepare standard solutions of the ions identified (to be used in analysis) by delivering 1.00-, 2.00-, 3.00-, 4.00-, and 5.00-ml portions of each of the standard solutions (of the identified species) into individual 50-ml volumetric flasks and diluting to the mark. Vary the sequence with different species so that the maximum concentration of each species occurs in different volumetric flasks. For example:

Flask	*A*	*B*	*C*	*D*	*E*
Li	1.00	2.00	3.00	4.00	5.00
Mn	5.00	2.00	3.00	4.00	1.00
Tl	1.00	5.00	3.00	4.00	2.00

Set the wavelength dial to the position for the determination of one of the identified species. Relight the burner and fill a cell with the most concentrated solution of the ion under analysis, position the cell, and open the shutter. Using the dark current to center the null meter, adjust the wavelength dial to give a maximum deflection (slit 0.1, selector 0.1, per cent response 100). After obtaining this wavelength setting *do not touch* wavelength knob again until all seven solutions have been analyzed for this particular species. The standard working curve for an individual ion obtained by adjusting the percent response dial to 100 with the most concentrated solution of the ion, using the dark current to balance the meter (slit

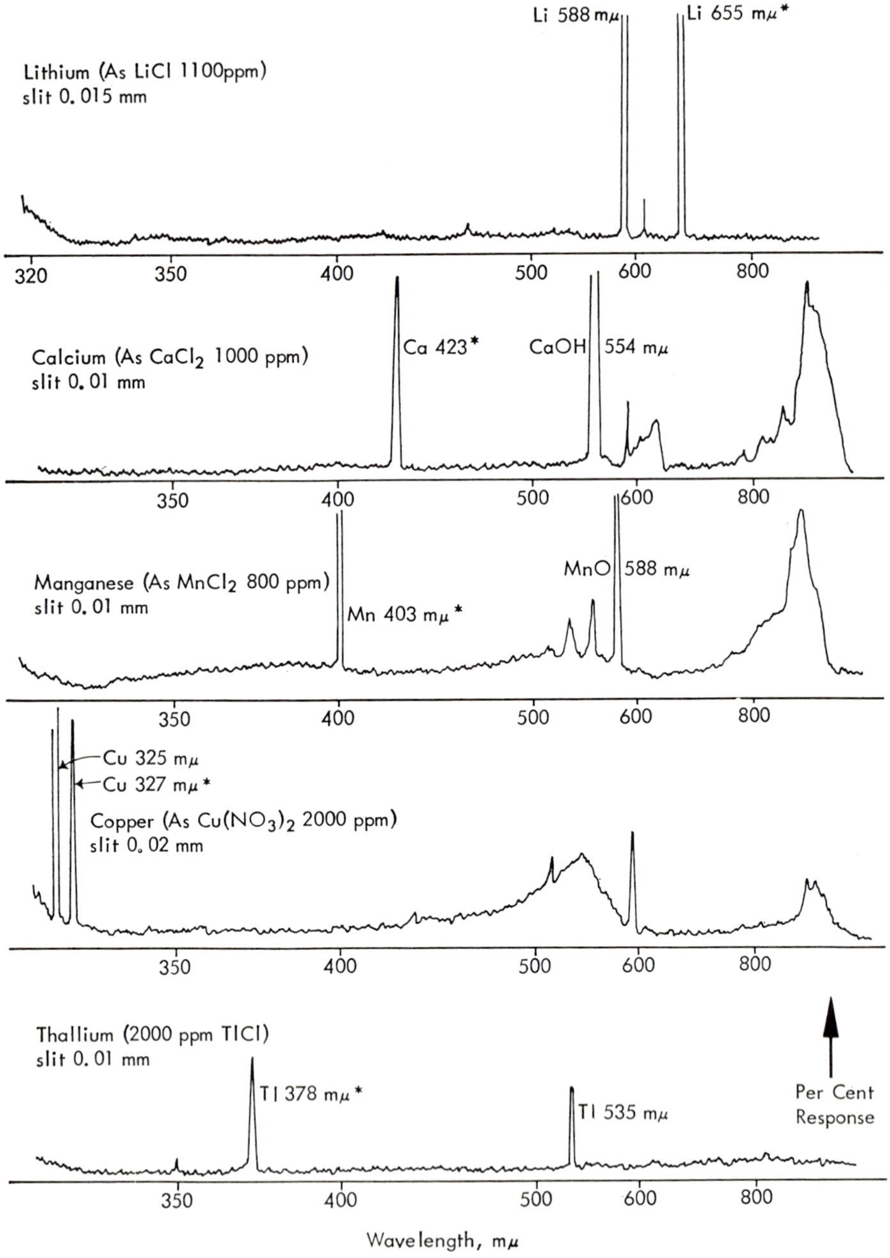

Fig. 61-10. Flame emission spectra of five elements. The asterisk indicates the wavelength used in analysis.

setting 0.01 for Ca, Mn, Tl; 0.015 for Li; 0.02 for Cu) and then adjusting the DU sensitivity (usually 0–5 turns from full counterclockwise) so that the solvent will balance the meter when the response dial reads about 10 per cent. This gives a scale spread of 10–100 per cent for 0 ppm to the most concentrated solution. Check ion concentration and solvent alternately until per cent response dial readings are 100 $\pm$ 1.5 and 10 $\pm$ 1.5; then, using only the shutter and the per cent response dial, measure the intensity in the less-concentrated standard solutions and the unknown. After each solution, run pure solvent and plot Δ versus concentration, where Δ is per cent response dial (solution)—per cent response dial (solvent). Follow the same procedure for each ion found in the unknown. From the calibration curve(s) thus obtained, report the concentration of metal ion(s) present.

Gas-pressure dependence. Using the DU as set for the last ion, determine the per cent response dial for the second most concentrated solution at varying hydrogen pressure (2.5, 3.0, 3.5, 4.0, 4.5, 5.0 psi), keeping oxygen constant at 12 psi. Next, with hydrogen at 3.0 psi, vary oxygen (10, 11, 12, 13, 14, 15 psi); do not let oxygen pressure decrease below 8 psi or the burner will be damaged.

Fluorimetric determination of aluminum. The analytical applications of 8-hydroxyquinoline (oxine) are discussed on page 355. Almost all metals except the alkali metals can be precipitated by this reagent: the completeness of precipitation depends markedly upon the pH. Most hydroxyquinolates are soluble in chloroform, yielding yellow solutions (ferric iron and vanadium yield green-black and magenta-black solutions, respectively). A photometric determination upon the chloroform solution, therefore, may be used to detect small amounts of many metals.[31] However, the similarity of colors noted above obviously indicates the need for separation before the spectrophotometric measurement.

Although the hydroxyquinolates are colored, relatively few fluoresce; strong fluorescence is exhibited by aluminum, gallium, and indium. Measurement of the fluorescence of the chloroform extract therefore has two great advantages over a spectrophotometric measurement; the determination is much less subject to interference and (for the reasons explained in Chapter 56) it is much more sensitive. The procedure for aluminum given below is suitable for amounts of the order of micrograms. As in any method, optical or other, for the determination of a trace of material, scrupulous care must be taken to avoid accidental contamination of sample or of reagents. The procedure is based upon those given by Tuelo et al.[32] and Goon et al.[33] These authors have shown that no detectable fluorescence is obtained after extraction from solutions (50 μg/100 ml) of beryllium (if free of aluminum), bismuth, calcium, cerium, chromium, cobalt, copper, iron (but see below), lead, magnesium, manganese, mercury, molybdenum, nickel, silver, thallium, thorium, tin, tungsten, uranium, and vanadium. Cadmium and zinc give fluorescence only about $\frac{1}{25}$ as intense as an equal amount of aluminum. Ions such as ferric iron and titanium, which form strongly colored hydroxyquinolates,

[31] For a general review on the behavior of 8-hydroxyquinoline as a precipitating and extracting agent see E. B. Sandell, *Colorimetric Determination of Traces of Metals*, Wiley-Interscience, New York, 3rd ed., 1959, pp. 179–188.

[32] J. W. Tuelo, W. J. Stringer, and G. A. F. Harrison, *Analyst* **74**, 296 (1949) (procedure for trace of aluminum in beer).

[33] E. Goon, J. E. Petley, W. H. McMullen, and S. S. Wiberley, *Anal. Chem.* **25**, 608 (1953) (procedure for aluminum in steel or bronze).

cause interference apparently by absorption either of the exciting, or of the fluorescent radiation, or both. For example, the apparent recoveries of a trace of aluminum in the absence of iron, and in the presence of iron in the weight ratio 1 : 1 and 6 : 1, were 100, 83, and 65 per cent, respectively. Relatively large amounts of citrate or tartrate (500 mg/10 μg of aluminum) lowered the recovery of aluminum from 100 to 25 or to 82 per cent, respectively. Other anions tested, which included acetate, chloride, nitrate, perchlorate, and sulfate, had virtually no effect (see Ref. 33).

Equipment: A Coleman photoelectric photofluorimeter or the equivalent is employed. (Study the operating instructions for the instrument before starting the experiment.) The illumination should be by means of a mercury arc equipped with a filter transmitting the mercury line at 366 mμ, but practically no visible radiation. The fluorescence should be measured using a green filter transmitting only above 450 mμ. Other equipment: six 50-ml separatory funnels; six 25-ml volumetric flasks; two 1-ml pipets; three fluorescence cells with stoppers. Preferably rectangular cells are used, but cylindrical cells will serve.

Reagents: Standard aluminum solution containing 2 μg of aluminum per milliliter of 2 per cent (wt./vol.) 8-hydroxyquinoline in 1 *M* acetic acid solution (note 1). 1 *M* ammonia–ammonium chloride buffer, pH 9.4 (note 2). Anhydrous sodium sulfate (note 3). Reagent-grade chloroform. Solutions must be prepared using "conductivity" water and stored in polyethylene containers. All equipment must be cleaned thoroughly at the start and end of the experiment.

Procedure. Add 25 ml of aluminum solution to each of five separatory funnels. Add 0, 2, 4, 6, or 8 ml of standard solution, respectively, to the five funnels. Add 1.0 ml of 8-hydroxyquinoline solution and 1.0 ml of buffer solution to each solution and extract twice with 10 ml of chloroform, shaking vigorously for 1 min each time; collect the chloroform in a 25-ml volumetric flask. Dilute the extracts to the mark with chloroform, add 1 g of anhydrous sodium sulfate, shake, and allow the sodium sulfate to settle. Decant the chloroform extracts into the fluorimeter cells.

Determine the fluorimeter reading for each of the five solutions without undue delay (note 4). Draw a calibration curve, plotting scale reading (proportional to intensity of fluorescence) versus concentration of aluminum (note 5).

Determine the aluminum content of an "unknown" sample by the above procedure. A considerable saving of time is made if the unknown sample is carried along with the standards when the calibration curve is prepared (note 6).

Notes:

1. 8-Hydroxyquinoline is soluble both in ethanol and in acetic acid. Since many metal hydroxyquinolates including aluminum are appreciably soluble in dilute aqueous ethanol, the use of acetic acid is to be preferred.

2. The pH is not critical in this experiment. Goon et al. (Ref. 33) found the extraction of aluminum was complete at pH 6.5–10.0 and recommended the value 8.0 ± 1.5.

3. Shaking with anhydrous sodium sulfate will clarify the chloroform solutions, which may be turbid, owing to a trace of suspended water.

4. The fluorescence intensity is developed practically immediately and does not change during (at least) 24 hr if the solution is stored in the dark or in diffuse light. If the chloroform solution is allowed to stand exposed to ultraviolet radiation in

the fluorimeter, there is a steady decrease in intensity due apparently to photochemical decomposition. In one experiment (see Ref. 33) the intensity decreased approximately 27 per cent in 1 hr of continuous UV illumination. The effect is entirely negligible if the measurement is made within a minute or two while in the fluorimeter.

5. A linear relation is observed between scale reading (approximately proportional to intensity) and concentration of aluminum up to concentrations of about 10–12 μg per 50 ml of chloroform. At higher concentrations a deviation is observed (see Chapter 56, Problem 2). The sensitivity of the measurement depends upon the intensity and steadiness of the exciting source and the accuracy with which the relative fluorescent intensity may be measured. With typical commercial fluorimeters a sensitivity of 0.1 μg per 50 ml should be attained.

6. When many samples are to be determined over an extended period of time, some workers prepare the calibration curve with solutions containing known amounts of aluminum, as above, only once. At the same time that the calibration curve is prepared, the relative fluorescent intensity of a solution such as quinine sulfate is measured. For later determinations the fluorimeter is adjusted so that a solution of quinine sulfate of the same concentration gives the same reading as was obtained when the calibration curve was prepared. This procedure saves time but introduces the danger that the reagents used on a later occasion may not yield the same fluorescence reading with a given amount of aluminum as those used to prepare the calibration curve.

CHAPTER 62 *Gas-Liquid Chromatography*

The purpose of this chapter is to familiarize the student with the practical aspects and the elementary calculations of gas chromatography, and to illustrate their application using experimental data obtained from a simple mixture of normal alkanes.

Apparatus

A block diagram representing the apparatus required to perform gas–liquid chromatography, GLC, is shown in Fig. 62-1. A carrier gas, such as helium, nitrogen, or argon, is allowed to escape from a cylinder of the compressed carrier gas through pressure-reducing and regulating valves, through the reference section of the detector, which is located in a controlled-temperature block, and then through the sample-injection system, through the column, through the other side of the detector and finally through a flow meter and out into the atmosphere.

The column is contained within a controlled-temperature oven and may be constructed from approximately 6 ft of $\frac{1}{4}$-in. diameter brass, stainless steel, or glass tubing. The actual chromatographic separations occur on the packed material within the column. The packing material consists of a solid support of large surface area, e.g., of specially treated diatomaceous earth. The solid support is impregnated with several per cent of a liquid which serves as a *stationary phase*. Under ideal conditions the chromatographic process involves the continuous distribution of the injected material between the carrier gas and the liquid phase. A large number of solid supports and liquid stationary phases have been used. The solid support should be completely inert so as to minimize adsorption of the material being chromatographed, and thereby prevent the occurrence of distorted chromatographic peaks. The liquid phase should be nonvolatile and chemically stable under the conditions existing in the column. Naturally, the liquid phase must be suitable for the separations required.

Depending on the concentration of the constituent in the sample and the nature of the detector being used, from 0.1 to 10 μl of sample is injected through a rubber septum into the heated sample-injection port. Samples larger than 10 μl may be injected, but the quality of the separation may suffer. Commercially available microliter syringes are used to introduce the sample, and accuracy of several per cent is possible. The sample is transported by the carrier gas through the column and, under optimum conditions, is separated into its various components by the chromatographic process. These individual components emerge at different times from the end of the column and pass through the detector, where an electrical

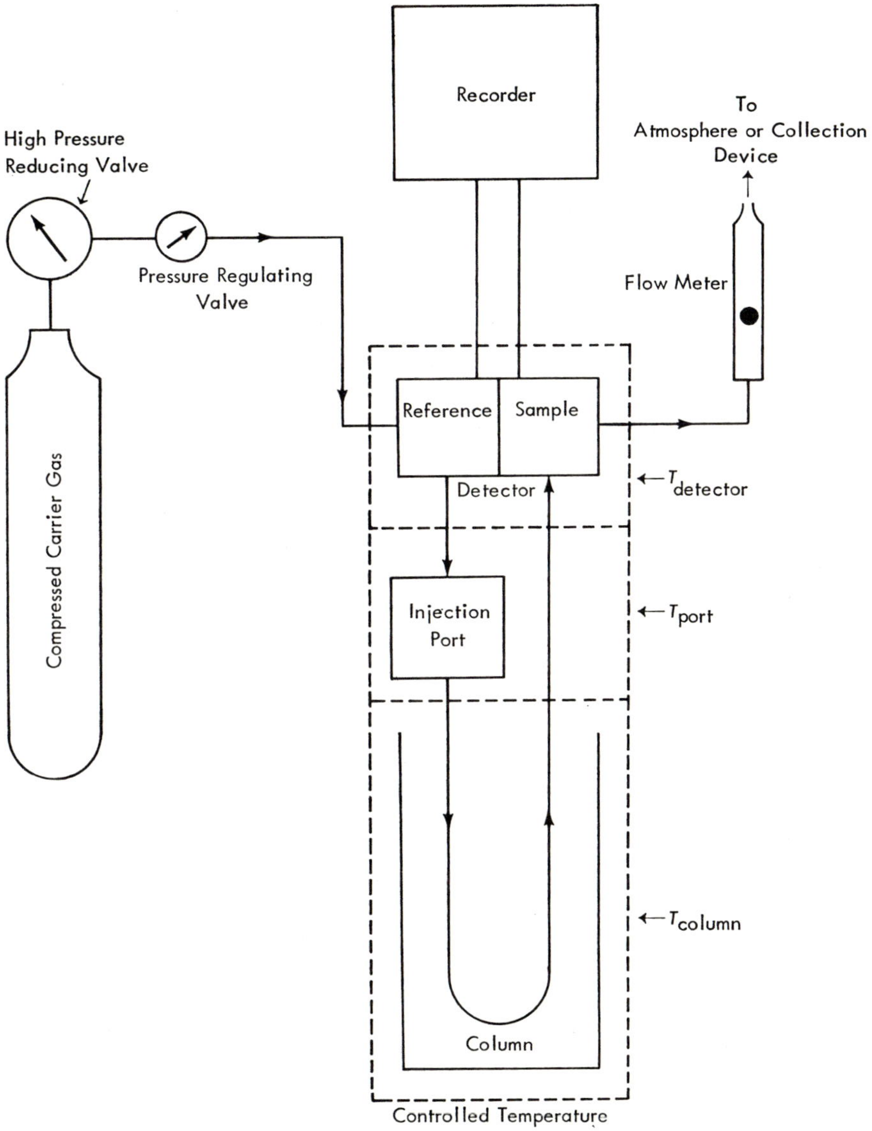

Fig. 62-1. Block diagram of a gas chromatograph.

signal related to the concentration of the constituent in the carrier gas is produced. This electrical signal is usually recorded on a potentiometric strip-chart recorder, producing a gas chromatogram such as is shown in Fig. 13.7.

Distribution theory

Plate theory is extremely useful in discussing the factors that determine the shape of the gas chromatographic peak. Using the principles of plate theory presented in Chapter 13, it is possible to obtain an expression for the concentration (or fraction) of the species partitioning (solute) between the gas phase and the stationary phase at any point in a gas chromatographic column [see Eq. (11) of Chapter 13]. Of immediate interest is the concentration C of the solute as it leaves a column of N theoretical plates and enters the detector. If w is the initial weight of the solute injected into the column, V^0 the volume of gas that has passed, and V_R^0 the volume of gas passed when the maximum concentration of solute leaves the column (the retention volume), C is given by

$$C = \frac{w}{V_R^0}\left(\frac{N}{2\pi}\right)^{1/2} \exp\left[-\frac{N}{2}\left(1 - \frac{V^0}{V_R^0}\right)^2\right] \tag{1}$$

where

$$V_R^0 = N(v_M + Dv_L) \tag{2}$$

v_M and v are the volumes of the mobile (gas) and stationary (liquid phases) in a single plate, and D is the distribution coefficient between the mobile and the stationary phases [Eq. (1), Chapter 13]. Figure 62-2 represents the hypothetical gas chromatogram calculated for a three-component mixture having species with retention volumes of 100, 200, and 300 (arbitrary volume units), assuming unit concentration for each of the three solute species and $N = 100$. The dashed curves represent the concentration associated with the peaks for $V_R^0 = 200$ and 300. From Eq. (1) it is seen that when $V^0 = V_R^0$, the solute concentration is a maximum, C_{max}, and is given by

$$C_{\text{max}} = \frac{w}{V_R^0}\left(\frac{N}{2\pi}\right)^{1/2} \tag{3}$$

For constant retention volume, the peak height increases as the square root of the number of theoretical plates. Commercially packed columns with $N = 400$–600 plates per foot are available.

As seen in Fig. 62-2, the peak width, ΔV^0, increases with V_R^0. One measure of peak width commonly used is the distance between the inflection points on the rising and falling portions of the peak. From Eq. (1) the inflection point can be shown to occur at $V^0 = V_R^0(1 \pm N^{-1/2})$ and the peak width is

$$\Delta V^0 = 2V_R^0/N^{1/2} \tag{4}$$

Hence the peak width of a substance is proportional to its retention volume and varies inversely as the square root of the number of theoretical plates.

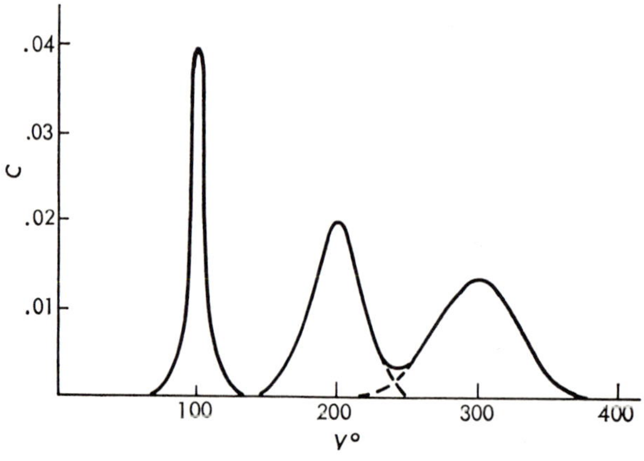

Fig. 62-2. Gas chromatogram. Calculated from Eq. (1). Three components present with $V_R^0 = 100$, 200, and 300. $N = 100$ and $w = 1$.

The relations given above are based on a discrete, multistaged-type extraction process (binomial distribution). A Poisson distribution theory is a better model for a continuously flowing system, such as in gas–liquid chromatography. However, both models approach a Gaussian distribution for columns with a large number of theoretical plates and Eqs. (3) and (4) are unaffected by the mathematical approach used.

The number of theoretical plates in a column divided into the column length (in centimeters) is known as HETP (the average height equivalent to a theoretical plate). It may be calculated from Eq. (4) by measuring the peak width at $C = C_{\max}/e^{1/2} = 0.607C_{\max}$; and $N = 4[V_R^0/\Delta V_R^0]^2$. Alternatively, if the peak width is measured at $C = C_{\max}/e = 0.368C_{\max}$, $N = 8[V_R^0/\Delta V_R^0]$.

Column efficiency

The number of theoretical plates is a measure of the chromatographic column efficiency. The relationship given above [Eq. (4)] was obtained on the assumption that the entire quantity of solute was introduced instantaneously into the first plate of the column. Despite the fact that experimental conditions do not quite coincide with this assumption, Van Deemter et al.[1] have shown that Eq. (4) holds provided the feed volume is less than $0.5V_R^0/N^{1/2}$. Under such conditions the required minimum number of theoretical plates to separate two species chromatographically when they are present in equal concentration is

$$N_{\min} \approx 4\left(\frac{D_1 + D_2}{D_1 - D_2}\right)^2 \tag{5}$$

where D_1 and D_2 are the distribution coefficients of the two solutes.

Van Deemter et al. derived an expression for HETP that combined rate theory and plate theory. Their result is of the form

$$\text{HETP} = A + \frac{B}{u} + Cu \tag{6}$$

where u is the linear gas velocity, A depends upon particle diameter and irregularity of packing of the solid support, B depends upon the tortuosity of the channels in the column and the first power of the diffusion coefficient of solute molecules in the gas phase, and C depends upon the distribution coefficient of the solute, the ratio of the stationary-phase volume and gas-phase volume, the effective film thickness and inversely with the diffusion coefficient of the solute in the stationary phase.

Figure 62-3 is a plot of Eq. (6) for arbitrary values of A, B, and C. There is an optimum value of gas velocity which results in a minimum value of HEPT for any given column and solute, $u_{\text{opt}} = (B/C)^{1/2}$. From the shape of the curve in Fig. 62-3 it is apparent that a gas velocity slightly higher than the optimum value is to be preferred over lower gas velocities. The reader is referred to the original article by Van Deemter et al. (Ref. 1) and Dal Nogare and Juvet[2] for a discussion of the significance of the various terms in Eq. 6.

[1] J. J. Van Deemter, F. J. Zuiderweg, and A. Klinkenberg, *Chem. Eng. Sci.* **5**, 271 (1956).

[2] S. Dal Nogare and R. S. Juvet, Jr., *Gas-Liquid Chromatography*, Wiley-Interscience, New York, 1962.

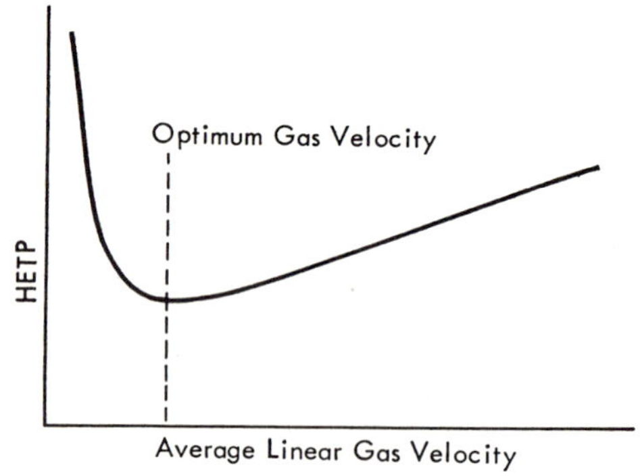

Fig. 62-3. Plot of the Van Deemter equation.

Integration of gas chromatographic peaks

The area under a gas chromatographic peak is proportional to the amount of material that was originally injected into the gas chromatograph. This area can be determined in a number of ways, and the most commonly used methods are listed in order of decreasing precision.

Electronic integration. The electrical signal from the detector is sent to an electronic integrator, most commonly a voltage to frequency convertor, whose output frequency is proportional to input voltage. The output count of the integrator is totaled using a high-speed electronic counter, and is proportional to the area of the peak. Rather sophisticated accessories must be included in the electronic integrator to allow for base-line drift, and, in multiple-peak chromatograms, to detect the beginning and end of a peak. The area of successive peaks is automatically printed.

Such devices are quite expensive but are becoming widely used in situations where integration accuracies of ~ 0.5 per cent are required or a very large number of chromatograms must be integrated each day.

The ball and disk integrator. A simple mechanical device known as a *ball and disk integrator* may be attached to the potentiometric strip-chart recorder used to trace the chromatogram. This device automatically integrates the area under the curve as it is being recorded. A second pen is coupled to the integrator, and the results presented in the form of sawtooth-like pen excursions on the margin of the recorder paper. The total area between two points on the recording is directly related to the number of excursions. The accuracy obtained with such a device is $\sim 1\frac{1}{2}$ per cent.

Cut and weigh. One of the classical methods for determining the area under a curve is to carefully cut out the curve and weigh it on an analytical balance. The area can then be calculated using as a standard the weight of a geometric shape of known area, such as a square, which has been cut from the same paper. The accuracy of this method is limited by the reproducibility in cutting out the chromatogram and the uniform density of the recording paper.

This method suffers from the defect that it is relatively slow and destroys the experimental curve. The latter problem can be alleviated by copying the chromatogram, using one of the readily available office copiers. Electrostatic type copiers are most suitable for this application, since the reproduction paper can be chosen for maximum uniformity. In the hands of a skilled operator, this technique yields a relative precision better than 2 per cent.

Height × width. An idealized chromatogram is shown in Fig. 62-4. *H* represents the height of the chromatographic peak and *W* represents the width of the chromatographic trace one half the distance between the base line and the peak height; the product *H* × *W* approximates the area of the curve. The results obtained in this manner are proportional to the true area but differ by several per cent. This deviation is not important, since the area calculated by this method is proportional to sample size. The principal difficulties encountered with the peak height × half-width method is the difficulty in measuring the half-width on narrow chromatographic peaks. If the peaks are symmetrical, precision close to 3 per cent can be obtained. Poor results are obtained in the case of asymmetric peaks.

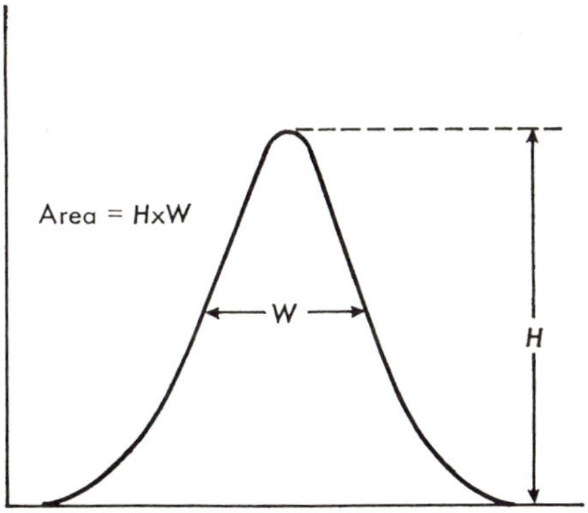

Fig. 62-4. Height × width method for area calculation.

Planimetry. The planimeter is a mechanical device whose principle of operation is discussed in many elementary calculus books. In operation the base of the instrument is held stationary and a sharp point attached to a movable arm is used to trace the perimeter of the curve whose area is being determined. The area traced out is presented on a dial. Precision is improved by tracing out the curve several times, but this can become very time-consuming. Precision of this technique is about 4 per cent.

Triangulation. A triangle is constructed by drawing two straight lines tangent to the rising and falling portions of the chromatographic peak at the inflection points, as illustrated in Fig. 62-5. The height of the triangle, *H*, is taken as the distance between the base line and the intersection of these two straight lines, and the area is calculated from $\frac{1}{2}$ H × *B*. If the peak is of Gaussian shape, the area calculated by this method is 97 per cent of the true area. The precision of this technique depends upon a number of factors: markedly upon the skill of the operator in drawing the correct tangents, the shape of the peak, and even the sharpness of the pencil used in drawing the lines. Under ideal conditions, the precision of this method is about 4 per cent.

Mobile phase

A pressure differential exists across the inlet and outlet of a gas chromatographic column. Hence the density of the carrier gas is a function of the position within the column. James and Martin[3] were the first to obtain the gas compressibility correction term, j,

$$j = \frac{3}{2}\left[\frac{(p_i/p_o)^2 - 1}{(p_i/p_o)^3 - 1}\right]$$

where p_i and the p_o are the inlet and outlet pressures of the column. Thus, the average flow rate through the column is jF_c, where F_c is the temperature corrected flow rate.

The observed flow rate F must be corrected to obtain F_c. If the column temperature is T_c(°K), and the flow-meter temperature T_f:

$$F_c = \frac{FT_c}{T_f} \tag{7}$$

Since R_f (p. 290) may be defined as the ratio of the rate of movement of solute to that of the mobile phase, it is possible to obtain from Eq. (7) an expression for the experimental retention volume V_R: This expression is

$$V_R = \frac{2al}{3R_f}\left[\frac{(p_i/p_o)^3 - 1}{(p_i/p_o)^2 - 1}\right] = \frac{2}{3}V_R^0\left[\frac{(p_i/p_o)^3 - 1}{(p_i/p_o)^2 - 1}\right] \tag{8}$$

where a is the gas-phase cross section, l the length, and p_i the inlet pressure of the column. V_R^0 is known as the *limiting* retention volume; i.e., V_R^0 is obtained from the experimental retention volume by correcting for the compressibility of the gas. It is dependent upon the characteristics of a particular column as shown by Eq. (2) and can be expressed in terms of the total volume of mobile and stationary phases in the column, V_L and V_M, respectively:

$$V_R^0 = V_M + DV_L \tag{9}$$

In order that chromatographic data obtained in one laboratory be of general use, a uniform method of presenting chromatographic data is highly desirable.[4]

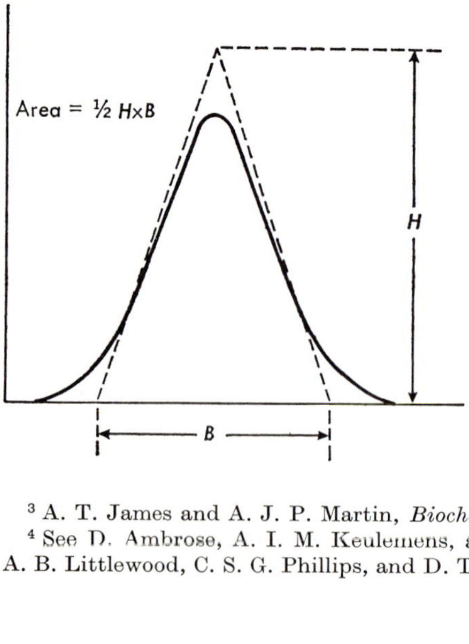

Fig. 62-5. Triangulation method for area calculation.

[3] A. T. James and A. J. P. Martin, *Biochem. J.* **50**, 679 (1951).

[4] See D. Ambrose, A. I. M. Keulemens, and J. H. Purnell, *Anal. Chem.* **30**, 1582 (1958); A. B. Littlewood, C. S. G. Phillips, and D. T. Price, *J. Chem. Soc.* **1955**, 1480.

Note that V_M can be determined by injecting, along with the unknown solute, some material such as air, nitrogen, or a rare gas for which $D \approx 0$. The adjusted retention volume V'_R is measured with respect to the peak maximum of one of these gases; hence

$$V'_R = V^0_R - V_M = DV_L = \frac{Dw_L}{\rho_L} \tag{10}$$

where w_L is the weight and ρ_L is the density of the liquid phase in the column. V'_R is readily related to the *net* retention volume V_N through the volumetric flow rate F_c and the corrected retention time t'_R, by taking into account the compressibility of the gas. The expression is

$$V_N = \frac{3}{2} V'_R \left[\frac{(p_1/p_0)^2 - 1}{(p_1/p_0)^3 - 1}\right]$$

$$= \frac{3}{2} t'_R F_c \left[\frac{(p_1/p_0)^2 - 1}{(p_1/p_0)^3 - 1}\right] \tag{11}$$

t'_R is the elution time difference between the air (or inert gas) peak and the solute peak.

To obtain V_N it is necessary to determine the column inlet and outlet pressures, volumetric flow rate, the column temperature and the flow-meter temperature, and the retention time of the solute with respect to air or N_2, etc.

Since V'_R, and therefore V_N, depend upon the weight of liquid phase in the column, the quantity V_g, the *specific* retention volume, is defined as

$$V_g = \frac{V_N}{w_L}\frac{273}{T_c} = \frac{V_N}{V_L \rho_L}\frac{273}{T_c} \tag{12}$$

and was first recommended by Littlewood et al. (Ref. 4). The correction term $273/T_c$, where T_c is the absolute temperature of the column, does not account for the temperature coefficient of V_g, but merely normalizes all data to a common temperature, 0°C.

In terms of experimental quantities the specific retention volume is given by

$$V_g = \frac{3}{2}\frac{t'_R F}{w_L}\left[\frac{(p_1/p_0)^2 - 1}{(p_1/p_0)^3 - 1}\right]\frac{273}{T_c} \tag{13}$$

The *relative* retention volume, α, is defined as the ratio of the specific retention volume of one solute to that of a standard solute, as obtained under the identical conditions. From the various relationships derived above, it is seen that

$$\alpha = \frac{V_g}{(V_g)_s} = \frac{V_N}{(V_N)_s} = \frac{t'_R}{(t'_R)_s}$$

where the subscript s denotes that the quantity refers to the standard solute.

The *Kovats* retention index (see Reference 19, below) is less temperature sensitive than α.

Solid supports

The ideal solid support in gas–liquid chromatography would be of uniform size, completely inert, and act only to distribute the liquid phase in a thin, uniform film.

A large number of supports have been suggested, and at present a number of specially processed diatomaceous earths, which are available commercially, are very popular. Depending upon the nature of the stationary liquid phase and the separations to be performed, these supports may be treated with alkali and/or acid, and in the latter case then with silanizing agents such as dimethyldichlorosilane or hexamethyldisilane if an especially inert support is needed. Silanizing the support markedly reduces adsorption of the compounds being chromatographed and minimizes asymmetrical chromatographic peaks, "tailing." Tailing is generally characterized by an abrupt chromatographic peak followed by a long, drawn-out region, as compared to the symmetrical Gaussian-shaped peaks obtained under ideal conditions. Tailing of this sort arises because of strong interactions between active sites on the solid support and the substance being chromatographed. Chemical interaction between the solid support and the substance being chromatographed may cause dehydration, isomerization, polymerization, etc. For this reason highly inactive supports must be used for the gas chromatography of sensitive compounds such as steroids, carbohydrates, pesticides, and others. Solid supports may be obtained from a number of commercial sources.[5-10]

Mesh size. The particle-size distribution of the support affects the quality of the separations that can be obtained. As mentioned in the discussion of the Van Deemter equation, the best separations are obtained with solid supports that have a uniform, small size. In practice, solid supports are screened and sold in terms of specified mesh sizes. For example, Chromosorb W, a commercially available diatomaceous earth calcined with sodium carbonate, is available in a variety of size distributions ranging from 30/60 to 120/140. The first number designates the screen size that just passes the particles of the solid support, and the second number specifies the screen size that stops the support.

The finer the mesh size of the solid support, the greater is the column efficiency. However, the pressure drop across the chromatographic column increases rapidly with decreasing mesh size. For the highest column efficiency, consistent with practical pressure requirements, 100/120 mesh is popular, while in less critical situations either 40/60 or 60/80 mesh may be used.

It is not possible to recommend any single support material for all applications. However, a calcined, acid-washed diatomaceous earth which has been silanized will be generally useful when coated with an appropriate stationary phase.

Size-graded fluorocarbon polymers such as Teflon T-6 are especially inert solid supports, and, when coated with an appropriate stationary phase, can be used to separate very polar compounds such as ammonia, amines, aliphatic acids, and water without appreciable tailing.

Glass beads have also been used as solid support and are particularly advantageous in the separation of high-molecular-weight polar compounds when lightly coated with an appropriate liquid phase (0.25 per cent).

[5] Beckman Instruments, Inc., Scientific and Process Instruments Division, Fullerton, Calif. 92634.

[6] Varian Aerograph, 2700 Mitchell Drive, Walnut Creek, Calif. 94598.

[7] Johns-Manville Corporation, Celite Division, 22 East 40th Street, New York, New York 10016.

[8] F. & M. Scientific Division of Hewlett-Packard, Avondale, 19311.

[9] Packard Instrument Company, Inc., P.O. Box 53917, Thomas Grove, Ill.

[10] Analabs, Inc., Hamden, Conn. 06518.

Solid adsorbents (gas–solid chromatography). In certain situations, such as when it is desired to separate mixtures of gases chromatographically, appropriate uncoated solid phases may be used. These supports function as solid adsorbents.

One of the substances used is silica gel, which separates carbon dioxide from oxygen and nitrogen and is occasionally used for separation of low-molecular-weight hydrocarbons.

Charcoal has analogous uses to silica gel, and unsaturated hydrocarbons elute before saturated hydrocarbons containing the same number of carbons.

Alumina is used in the chromatography of hydrocarbons containing less than five carbons, and the order of elution of unsaturated and saturated hydrocarbons is the reverse of that obtained with charcoal.

Molecular sieves are frequently used in gas chromatography. Mixtures containing helium, neon, argon, krypton, xenon, hydrogen, oxygen, nitrogen, methane, carbon monoxide, nitrous oxide, and silicon hexafluoride may be chromatographed on grade 5A molecular sieves.

Porous styrene–divinylbenzene copolymers having surface areas in the range 300–400 m^2/g, as compared to 1–10 m^2/g for the conventional solid supports, have recently become available. These resins are usually used without a liquid phase. Because of their high surface areas they can be used to separate the permanent gases and gaseous and low-molecular-weight organic compounds such as hydrocarbons, alcohols, glycols, ketones, chloromethanes, and acetate esters.

Stationary phases

Selection of an appropriate liquid stationary phase is critical to the success of gas–liquid chromatography. In most practical problems, no precise theoretical guides exist as to the choice of the optimum liquid phase and it is often necessary to refer to the literature to determine which stationary phase is most suitable for the problem at hand.[11–14] A summary of stationary phases used is given in Table 62-1.

Littlewood has proposed a classification scale for liquid phases based upon the polarity of the liquid phase.

Nonpolar stationary phases. The nonpolar class consists of high-boiling liquids or greases such as silicone and hydrocarbon-type compounds. Commercially available examples are Squalane and SE-3.

Polar stationary phases. Commercially available examples of this type are Carbowax and dimethyl sulfolane. Polar stationary phases have molecular structures that incorporate a large fraction of polar organic groupings.

Intermediate stationary phases. These stationary phases have properties intermediate to those of polar and nonpolar stationary phases. These materials usually have a relatively long nonpolar framework with polarizable or polar

[11] A. B. Littlewood, *Gas Chromatography*, Academic Press, New York, 1962.

[12] J. S. Lewis, *ASTM Spec. Tech. Publ. 343* (Compilation of Gas Chromatographic Data), 1963.

[13] Gas Chromatography Abstracting Service, Preston Technical Abstract Company, Evanston, Ill. (This is a source of a continually up-dated compilation of gas-chromatographic data.)

[14] Gas Chromatograph Abstracts, Gas Chromatograph Discussion Group of the Institute of Petroleum Hydrocarbon Research, Butterworth, London.

substituents. Commercially available members of this class are benzyldiphenyl, diisodecyl phthalate, and SE-52.

Hydrogen-bonding stationary phases. This class is really a subclass of polar stationary phases and consists principally of those materials that have a large number of hydrogen atoms available for hydrogen bonding. Examples are tetrahydroxythelendiamine and mannitol.

Specific stationary phases. Silver nitrate can be coated upon appropriate solid supports and is useful in the separation of olefins because a specific chemical reaction occurs between silver nitrate and olefinic compounds. Stationary phases that rely upon the principle of chemical interaction with the substance to be chromatographed are included in this group.

Table 62-1 lists various classes of organic compounds that have been separated on the indicated stationary phases. It is customary to designate most stationary phases by a trivial name or collection of capital letters and numbers. Space does not permit chemical identification of these compounds.

The column headed maximum temperature in Table 62-1 gives the maximum column temperature that can be applied to solid supports coated with the indicated stationary phase. Since it is conventional practice to precondition columns by heating them at least 20°C above the highest expected operating column temperature, 20° should be subtracted from the numbers listed to obtain the useful working temperature.

The purpose of preconditioning is to minimize "bleeding" of the stationary phase, a phenomenon that results in high and drifting base lines and ultimately contamination of the detector. Five to 10 ml/min of carrier gas is passed through a column while it is held under appropriate temperature. Special commercial devices are available for preconditioning columns. Quite satisfactory results are obtained using the gas chromatograph, provided the detector is disconnected so that the carrier gas exiting from the column does not flow through it.

Preconditioning drives off any volatile impurities present in the chromatographic column.

If an extremely sensitive detector, such as the electron capture detector, is used, longer preconditioning times and higher preconditioning temperatures must be used.

Detectors

Thermal conductivity detector. As mentioned in Chapter 59, the thermal conductivity of a two-component gas mixture is related to the composition of this mixture. A detector based upon this principle is readily constructed by replacing two opposite arms of a Wheatstone bridge with two identical temperature-sensitive resistors. The other two arms are variable resistors used to balance the bridge. Sufficient voltage is applied to the Wheatstone bridge to heat the temperature-sensitive resistors considerably above the temperature of the detector cell walls. The gas exiting from the pressure regulator (Fig. 62-1) passes over one of the temperature-sensitive resistors, through the column, and then passes over the other temperature-sensitive resistor. The voltage difference that arises across the bridge is proportional to the concentration of the material being eluted from the column.

TABLE 62-1. **Stationary Phases for Use in Gas Chromatography**[a]

Type of Compound	*Stationary Phase*	*Max. Temp., °C*	*Classification*[b]
Acids			
C_1–C_9 (free)	Porapak T; 6% FFAP		
C_1–C_{18} (free)	FFAP	275	S, P
Bile and urinary	SE-30	375	N
Fatty acid–methyl esters	DEGS	190	P
	FFAP	275	S, P
	Apiezon L	300	N
	TCEPE	180	P
	EGSS-X	200	
	BDS		
Alcohols			
C_1–C_5	Porapak Q		
	Hallcomid M-18 OL	150	I
	Carbowax 600 or 1540	125/200	P
C_1–C_{18}	FFAP	275	S, P
	Carbowax 20M	250	P
Di-poly	FFAP	275	S, P
	QF-1	250	I
Aldehydes			
C_1–C_5	DC 550	275	I
	b,b′-Oxydipropionitrile	100	P
	Ethofat	140	I
	Porapak Q		
C_5–C_{18}	Carbowax 20M	250	S, P
Alkaloids	QF-1	250	I
	SE-30	375	N
Amines			
See Nitrogen compounds			
Amino Acid Derivatives			
Butyl-N-trifluoroacetyl esters	DEGS/EGSS-X	190/200	P
Boranes	Apiezon L	300	N
	Nujol	200	N
Esters			
Mixed	Dinonyl phthalate	175	I
	Porapak Q		
Triglycerides	JXR	300	
Ethers	Carbowax 20M	250	S, P
	b,b′-Oxydipropionitrile	100	P
Glycols	Porapak Q		
Halogen compounds	Diisodecyl phthalate	175	I
	DC 550	275	I
	QF-1 (FS-1265)	250	I
	FFAP	275	S, P
Freons	Dibutyl tetrachlorophthalate	150	I, P
	UCON Polar 2000	200	
Fluorides (C_5–C_9)	Hexadecane	50	N

TABLE 62-1. **Stationary Phases for Use in Gas Chromatography**[a] *(continued)*

Type of Compound	*Stationary Phase*	*Max. Temp., °C*	*Classification*[b]
Hydrocarbons			
Aliphatic, C_1–C_5	Diisodecyl phthalate	175	I
	Propylene carbonate	60	P
	Carbowax 400	125	P
	Tributyl phosphate	50	I
C_5–C_{10}	Squalane	100	I, N
	SE-30	375	N
Aromatic	Tetracyanoethylated pentaerythritol	180	P
	Dibutyl tetrachlorophthalate	150	I, P
	Tricresyl phosphate	125	I
	Carbowax		
Hydroxy olefins	2,4-Xylenyl phosphate	175	I
C_1–C_8	$AgNO_3$/Benzyl cyanide		S
	Dimethyl sulfolane	50	P
	Propylene carbonate	60	P
C_6–up	D.C. 550	275	I
	D.C. 703	225	I
Naphthalenes	Diisodecyl phthalate	175	I
Polynuclear	SE-30 on DMCS-treated support	375	N
	FFAP on DMCS-treated support	275	S, P
	PMPE, 5 or 6 ring	250/300	I
Ketones	Lexan	300	P
	FFAP	275	P
	D.C. 550	250	I
	Porapak Q		
Nitrogen compounds			
Amides	Versamid 900	250	P
Amines	Dowfax 9N9/KOH	225	S
	Polyethyleneamine (PEI)	250	P
	Porapak R		
Ammonia	Ethofat or Carbowax 600 on Chromosorb T	140/125	I/P
	Porapak Q: 6% polyethylene-amine		
Nitriles	Tetracyanoethylated pentaerythritol	180	P
	FFAP	275	S, P
	XF-1150	200	I
Oils, essential			
General	Carbowax 20M	250	P
Organometallic	FFAP	275	S, P
	SE-30	375	N
Pesticides	Dow 200	250	N
	QF-1	250	I
	SE-30	375	N

TABLE 62-1. **Stationary Phases for Use in Gas Chromatography**[a] (*continued*)

Type of Compound	*Stationary Phase*	*Max. Temp., °C*	*Classification*[b]
Phenols	Dinonylphthalate	175	I
Phosphorus	SE-30	375	N
	STAP	255	P
Silanes	SF-96	300	N
	FFAP	275	S, P
Steroids	STAP	255	P
	XE-60	275	I
	QF-1	250	I
	SE-30	275	N
Sugar derivatives			
Trimethylsilyl ethers	QF-1	250	I
	SE-52	300	I
Sulfur compounds	Carbowax 20M	250	P
	FFAP	275	S, P
	Dinoylphthalate	175	I
	Porapak Q		
Water	Any Porapak		
Gases			
Ar and O_2	6-ft activated molecular sieve 5A at −72°C		
H_2, O_2, He, N_2, CO, CH_4	20-ft molecular sieve 13X		
CO_2, H_2S, CS_2, COS	Silica gel or Porapak Q		
H_2 isotopes	6-ft activated molecular sieve 13X		
CO_2, N_2, O_2, CH_4, CO	$2\frac{1}{2}$-ft silica gel internal column		
	20-ft molecular sieve 13X external column		
N_2O, CO_2, NO	Porapak Q		

[a] From *Gas Chromatograph Accessory Catalog*, Varian Aerograph, Walnut Creek, Calif., 1967, p. 5.

[b] A more quantitative classification of stationary phases has been proposed by L. Rohrschneider [*J. of Chromat.* **22**; 6 (1966) and *Advances in Chromatography*, Vol. IV, Marcel Dekker, 1967] which appears to be capable of predicting the R.I. of a compound (see Ref. 19, below) from previously determined properties of the stationary phase.

Thermal-conductivity detectors are one of the most inexpensive types of gas chromatographic detectors and are relatively easy to maintain and use. For this reason, even though they are not the most sensitive type of detector, they find wide application.

Electrochemical detector. A very sensitive detector for gas chromatography is available commercially[15] and is applicable to the quantitative analysis of molecules containing halogens. The principle can be illustrated in the case of a chlorine-containing organic molecule, e.g., the pesticide, Aldrin. After passing through the

[15] Dohrmann Instruments Company, San Carlos, Calif. 94071.

column of a gas chromatograph, the carrier gas containing the Aldrin is mixed with oxygen and passed through a furnace at 800°C. The exit gas from the furnace is bubbled through a solution that contains 80 per cent acetic acid. The potential of a silver electrode immersed in the well-stirred acetic acid solution is measured with respect to a silver–silver acetate reference electrode. In order to poise the indicator electrode, a small amount of silver ion is generated by oxidation of another silver electrode immersed in the electrolyte. Any chloride ion introduced in the solution by the gas from the furnace tends to reduce the concentration of the silver ion at the surface of the indicator electrode and changes its potential. This change in electrode potential is sensed electronically and an anodic current is passed through the silver generator electrode until the appropriate potential is reestablished at the silver indicator electrode. The current passing through the generator electrode is recorded as a function of time, using the conventional strip-chart recorder. The shape of the chromatograph resembles that obtained with a thermal-conductivity detector; the area under the peak corresponding to the halogen-containing compound is readily related by Faraday's law to the total quantity of material that was injected into the chromatograph.

In a recent innovation, at least 10 times more sensitivity is achieved by having the entering gas stream impinge directly upon the indicator electrode rather than dissolving the gas completely in the cell. In this mode of operation, the surface concentration of silver ion at the indicator electrode is decreased far more than is the case when the gas is uniformly dissolved in the entire volume of the electrolyte. Fifty nanograms of Aldrin are said to give a chromatographic peak corresponding to half-scale deflection on a 10-in. strip-chart recorder.

This principle is applicable to the determination of sulfur in organic compounds. SO_2 exiting from the furnace is passed into an electrolyte containing iodide and iodine, and platinum electrodes are used both as generator and indicator electrodes for iodine.

Ionization detectors. A number of very sensitive gas chromatographic detectors are based on the principle that a gas mixture is almost a perfect insulator at ordinary temperatures unless electrons, or electrically charged atoms or molecules, are present. The presence of these charged particles may be caused by an external source of radiation, for example, beta rays, by temperature as in a flame, or by radiofrequency energy. The number of charged particles present can be measured by applying a suitable voltage across the electrodes and measuring the current. Depending upon the nature of the particular detector, the current may increase or decrease when a particular species is eluted from the GLC column and enters the detector.

Flame ionization detector. In a flame ionization detector, the effluent gas from a GLC column is mixed with hydrogen and oxygen, passed through a hollow stainless-steel needle, and burned at the tip of the needle. A platinum-wire loop is placed just above the tip of the flame. An appropriate potential is applied across the stainless-steel needle and the platinum-wire loop, and the current that flows is determined using a suitable electrometer. The current that flows is proportional to the concentration of the material producing the ionization for currents in the range 10^{-12}–10^{-5} A. This detector will respond to all organic compounds with the exception of formic acid. It is insensitive to inorganic compounds that are not ionized in such a flame; i.e., a response is obtained from the elements in groups I and II. This detector is insensitive to water and permanent gases.

Cross-section ionization detectors. Electrons produced by a beta emitter such as ^{90}Sr may be used to ionize a gas mixture. In general, the higher the molecular weight of a gas, the more likely is the ionization by beta radiation. With this class of detector a carrier gas such as hydrogen or helium is used, and the effluent gas is passed into an ionization chamber where it is subjected to the beta radiation. The electrons produced in the ionization of the organic molecules present in the carrier gas are collected using an applied potential of 300–1000 V. The concentration of a particular molecule may be calculated from its cross section for the ionization process, which is equal to the sum of the cross sections of the atoms composing the molecule. This detector is not very sensitive.

Argon ionization detector. In this detector, argon atoms are raised to a metastable electronic state. Argon ions are produced by exposing argon gas to radiation and accelerating the ions across a field of about 1000 V. The metastable argon atoms are produced by the collision between these argon ions and argon atoms. When organic molecules present in the effluent of the GCL column collide with the metastable argon atoms, a transfer of energy takes place, and ionization and dissociation of the organic species occurs. It is a very sensitive detector but does not respond to water, oxygen, nitrogen, carbon monoxide, carbon dioxide, methane, ethane, or fluorocarbons. Argon is used as the carrier gas.

Electron-capture detector. Nitrogen serves as a carrier gas when the electron-capture detector is used. As the nitrogen flows through the detector, beta particles from a tritium source interact with the nitrogen molecules and form *slow* electrons. These electrons are collected at an anode whose potential is set between 10 and 100 V. On the introduction of an organic molecule, absorption of the electron occurs, and fewer electrons are collected at the anode. The mathematics of the absorption processes are equivalent to that in spectrophotometry; i.e., the logarithm of the ratio of the electrons collected in the presence of an absorbing species to the electrons collected in the absence of an absorbing species is proportional to the concentration of the absorbing species. The electron affinity of halogen-containing organic and inorganic compounds is extremely high, and great sensitivity can be obtained using this detector. Oxygen, ozone, organometallic compounds, carbonyl compounds such as anhydrides and 1,2-diketones, and organic peroxides are also detected with high sensitivity. A degree of selectivity toward classes of compounds is possible, because detector response ceases at a clearly defined potential for different molecules.

Experimental

This experiment is concerned with the gas chromatography of C_5–C_8 normal alkanes. Their gas chromatography is simple, and experiments are described which yield the data required for the various elementary calculations conventionally made in gas chromatography.

A relatively simple gas chromatograph may be used, and a thermal-conductivity detector will provide adequate sensitivity. A 4- to 6-ft-long $\frac{1}{4}$ in. stainless-steel column using an acid-washed, silanized diatomaceous earth for the support and Apiezon L (10 per cent loading) or Squalane (10 per cent loading) will prove quite satisfactory. Other stationary phases can be selected with the aid of Table 62-1.

In the experiments below, the following temperatures should be used: inlet (injection) block, ~200°C;[16] column, 50°C; detector, ~300°C.

Consult the manufacturer's literature for the particular chromatograph which is to be used, and adjust it according to the specific directions.

Determination of the optimum average linear carrier gas velocity. Adjust the inlet pressure to 10 lb/in.^{2}. The chromatograph should come to equilibrium rapidly, as indicated by a constant base line. Rapidly inject 0.5 μL of pure octane.[17] Mark the instant of injection on the chart paper and obtain the chromatogram. The average linear gas velocity u is equal t_a/l, where t_a is the time for the air peak to appear after the injection of the sample and l is the column length. Calculate the number of theoretical plates from the octane peak height and peak width (at 0.607 h_{max}) from Eq. (4). Since the length of recorder chart is proportional to time, and hence to the volume, the number of theoretical plates can be calculated from

$$N = \frac{4x^2}{\Delta x^2}$$

where Δx is the peak width and x the distance from the point of injection to the peak height h_{max} in recorder chart divisions. Calculate HETP from the number of theoretical plates and the column length (HETP $= l/N$).

Adjust the inlet pressure to 20 psi, allow the chromatograph to equilibrate and readjust the recorder-chart base line as required. Inject 0.5 μl of octane and continue as above. Repeat the entire procedure for inlet pressures of 30, 40, 50, 60, and 80 psi. Plot HETP versus u and determine the optimum average linear carrier gas velocity u_{opt}. Adjust the inlet pressure to yield a value of u slightly higher than u_{opt} and perform all experiments at this value of u. Note that u_{opt} depends upon the solute properties [Eq. (6)] and that in critical separations u_{opt} is determined for the most difficult-to-separate species.

Proportionality of peak area to concentration. Obtain the gas chromatograms for 0.5-, 1-, 2-, 5-, and 10-μl portions of pure octane. Calculate the area under the peaks using the peak height × half-width and the triangulation methods. Plot peak area versus concentration of octane for both integration techniques.

Chromatography of a mixture of alkanes. Prepare a mixture that contains equal weights of pentane, hexane, heptane, and octane. Inject 0.5 μl of this mixture into the gas chromatograph. Record the column inlet and outlet pressure, column temperature, volumetric carrier gas-flow rate, temperature of the flow meter, time to reach the air peak, and the solute peak times. Calculate the specific retention volume V_g for all components of the mixture, using Eq. (3).

[16] It is necessary that the temperature of the carrier gas in the injection block be slightly higher than the boiling point of the highest boiling compound to be injected. Unless thermal degradation occurs, there is little effect of higher temperatures. See C. E. Bennet, S. Dal Nogare, and L. W. Safranski, in *Treatise on Chemistry*, chap. 37, part I, vol. 3, Wiley-Interscience, New York, 1961 for a discussion of inlet and column temperatures.

[17] After filling the microliter syringe to the desired volume, a little air is usually drawn into the syringe so that it will be injected with the sample. This air provides the "air peak" from which the correct retention volume V_r', the net retention volume V_N, and the specific retention volume V_g can be calculated, using the experimental retention volumes.

Plot the log of the specific retention volume versus the number of carbon atoms in the alkane solute. A straight line should be obtained. See References 3, 18, and 19 for the significance of log retention volume plots in qualitative analysis.

Calculate the area under each of the chromatographic peaks and compare the response of the thermal-conductivity detector for each of these species. See Reference 2 for a discussion of the effect of the thermal conductivity of a solute upon the response of a thermal conductivity detector.

If all the components of the solution injected into the gas chromatograph yield separate gas chromatographic peaks, the percentage of each component in the mixture may be estimated by dividing the area under the peak for a particular component by the total area of all peaks. Calculate this fraction for each of the alkanes in your mixture, and compare the results with the fraction actually present.

Another technique that may be used with Gaussian chromatographic peaks involves calculation of $V_R^0 \times C_{max}$. As seen from Eq. (3), this product is proportional to the weight of the solute originally injected. Hence the relative amounts of materials responsible for different chromatographic peaks can be readily estimated by calculating $V_R^0 \times C_{max}$ for each solute species.

[18] N. H. Ray, *J. Appl. Chem.* (*London*) **4**, 21 (1954). See also Ref. 29, p. 242ff.

[19] E. Kovats, *Helv. Chim. Acta*, **41**, 1915 (1958) has proposed the term *retention index*, R.I., which may be determined as follows. Plot log (retention time) for a number of n-alkanes versus $100 \times n$, where n is the number of carbon atoms. By definition, R.I. $= 100 \times n$. Determine the retention time for the unknown and read its R.I. value from the logarithmic plot.

PART TEN

Analysis of Complex Materials

The limited scope of this book does not allow an extensive treatment of the analysis of inorganic materials having a more or less complex composition in which a number of constituents must be determined. The subject is so important, however, that it cannot be ignored, and some examples are given in the following chapters. There are two general approaches, or combinations of them, to the chemical analysis of complex materials. In the first, a number of constituents are determined successively, with intervening separations, in the same portion of sample. The separations that are required can often be utilized as steps in the determination of a series of elements. Thus, in classical silicate-rock analysis most of the major constituents are determined by successive precipitations in solutions derived from a single weighed portion. In the second approach, a separate sample (or aliquot) is taken for each constituent, which is then determined with minimal separations (often none) by a sufficiently selective procedure. This general method can be applied in steel analysis, in which minor or trace elements are determined. This method, when applicable, has the advantages of speed and simplicity and sometimes increased accuracy. To a large extent, the first method is used more by necessity than by choice, but it may be indispensable when the amount of sample is limited.

The analysis of limestone[1] (Chapter 63) has been selected as an example of an analysis in which most of the constituents can be determined, after successive separations, in one portion of sample. A complete scheme of analysis is not presented, but the procedure illustrates the general approach. The analysis of brass (Chapter 64) illustrates the seriate determination of the chief constituents in an alloy with intervening separations. Some alternative procedures utilize separate samples. Finally, the determination of a considerable number of constituents in steel (Chapter 65) serves as an example of analysis with individual samples, one for each constituent.

[1] A more satisfactory example of the analysis of a complex material would be that of a silicate rock, requiring the determination of 15 or more constituents. Such an analysis is extremely instructive, even in this day of analytical instrumentation, but the time allotted to analysis in most chemistry curricula is so limited that few students could carry out such a long analysis.

CHAPTER 63 *Analysis of Limestone*

Limestone is a rock consisting chiefly of calcium carbonate. A magnesian or dolomitic limestone is one containing considerable amounts of magnesium carbonate. *Dolomite* is strictly the equimolecular compound of calcium and magnesium carbonates, $CaCO_3 \cdot MgCO_3$, but the name is commonly applied to all carbonate rocks that contain sufficient magnesium carbonate to approach this composition. The procedure given below applies to the analysis of any of these varieties.

TABLE **63-1. Composite Analysis of Limestones**[a]

	A, %	*B*, %
SiO_2	5.19	14.09
TiO_2	0.06	0.08
Al_2O_3	0.81	1.75
Fe_2O_3, FeO	0.54	0.77
MnO	0.05	0.03
CaO	42.61	40.60
MgO	7.90	4.49
K_2O	0.33	0.58
Na_2O	0.05	0.62
Li_2O	Trace	Trace
$H_2O -$ (<110°C)	0.21	0.30
$H_2O +$ (>110°C)	0.56[b]	0.88[b]
P_2O_5	0.04	0.42
CO_2	41.58	35.58
S	0.09	0.07
SO_3	0.05	0.07
Cl	0.02	0.01
Organic		
	100.09	100.34

[a] From F. W. Clarke, *U.S. Geol. Surv. Bull. 770*, 5th ed., 1924, p. 564. Column A, composite analyses of 345 limestones by H. N. Stokes; column B, composite analyses by H. N. Stokes of 498 limestones used for building purposes.
[b] Includes organic matter.

The minor constituents of limestone include the carbonates of ferrous iron and manganese, as well as calcium sulfate, silicates, silica, pyrite, carbonaceous matter, and, of course, water. The composite analyses of a large number of limestones given in Table 63-1 serve to indicate the amounts of the subsidiary constituents.

Limestone and dolomite have many uses and are of great commercial importance. For judging the suitability of a limestone for a given technical purpose, a complete analysis is never required.[1] The following determinations are usually made in a shorter technical analysis:

1. Loss on ignition
2. Silica
3. Combined oxides, "R_2O_3." This is the residue obtained by igniting the precipitate formed by ammonium hydroxide. It consists of Al_2O_3, Fe_2O_3, TiO_2, Mn_3O_4, and P_2O_5, the latter combined as aluminum, iron, or titanium phosphate
4. Calcium oxide
5. Magnesium oxide
6. Carbon dioxide

The determination of these constituents is described under method I, which may be called classical analysis of limestone.

When only acid-soluble calcium and magnesium are to be determined, method II (p. 1107) may be used, in which calcium is determined by titration with EDTA. Magnesium is found by difference after the sum of calcium and magnesium has been determined by another EDTA titration. These determinations are very rapid.

METHOD 1

The outline below diagrams the procedure followed in a "proximate" analysis of limestone for the determination of the constituents mentioned above.

Loss on ignition

When heated at moderately elevated temperatures (1000–1100°C), a limestone loses all its carbon dioxide and moisture, and any carbonaceous matter present is oxidized. At the same time, reactions may occur which lead to an increase in weight, as, for example, oxidation of sulfides to sulfates and of ferrous iron and manganese. At a higher temperature (>1200°C) sulfur will be lost as the result of decomposition of sulfates, and the alkalies will be volatilized. The ignition loss represents approximately the amount of carbon dioxide (and water) in the sample.

Procedure. Weigh 0.5 g of ground air-dry sample into a covered crucible (porcelain will do, but platinum is better) and dry at 100°C for 1 hr. Cool and weigh. The loss in weight represents hygroscopic moisture. Then ignite the crucible and its contents strongly, preferably in an electric furnace regulated to 1000–1100°C, or over a Meker burner. The heating should be slow at first to prevent mechanical loss through sudden expulsion of carbon dioxide. Heat for $\frac{1}{2}$–1 hr, cool, and weigh. Repeat the ignition to constant weight, using $\frac{1}{2}$-hr periods of heating. Remember that the ignited residue is hygroscopic. Calculate the loss on ignition in per cent.

2. Silica

After ignition, step 1, the sample is in the proper condition for the determination of silica. Any acid-insoluble silicates originally present will have been converted into acid-decomposable calcium magnesium silicates by the intense heating with calcium oxide. The ignition residue is treated with excess hydrochloric acid, and the solution is evaporated to dryness to dehydrate the silica and render it insoluble, as described in

[1] For refined methods of carbonate rock analysis, such as are needed for scientific purposes, see W. F. Hillebrand, *U.S. Geol. Surv. Bull. 700*, 1919, p. 246; W. F. Hillebrand, G. E. F. Lundell, H. A. Bright, and J. I. Hoffman, *Applied Inorganic Analysis*, Wiley, New York, 2nd ed., 1953, p. 964.

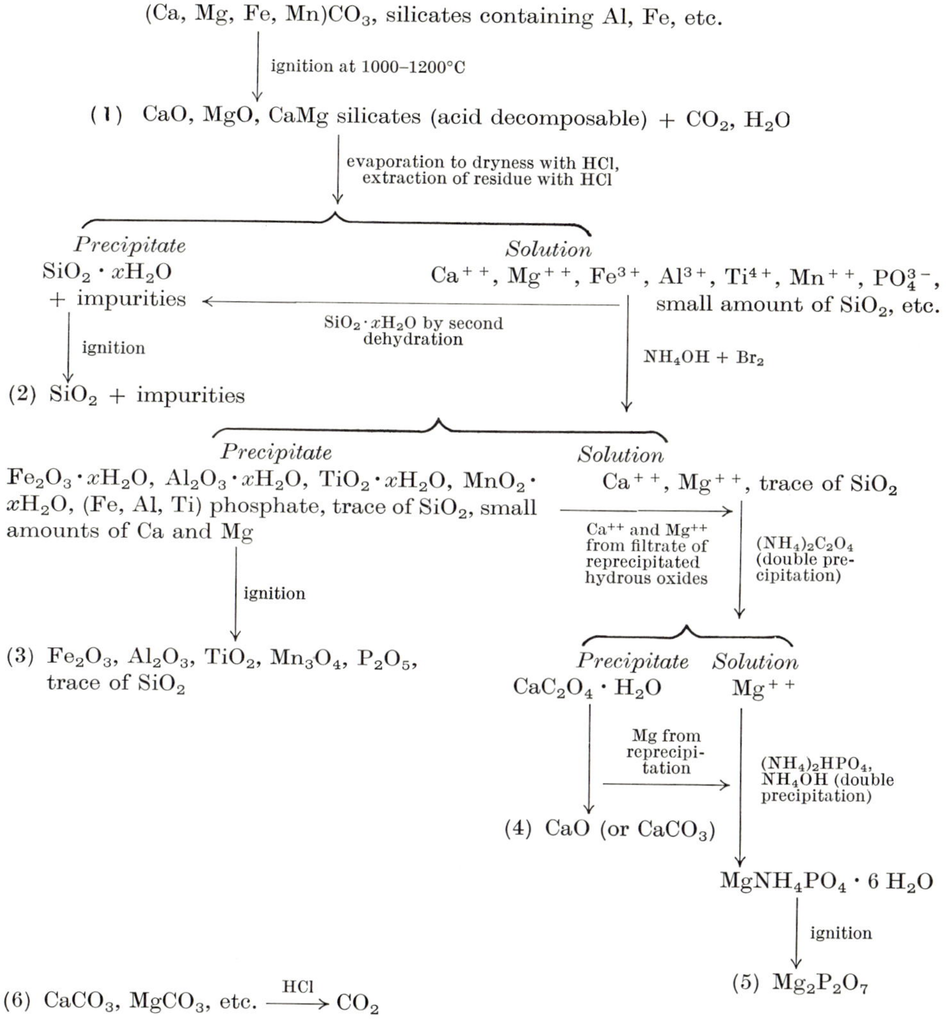

Chapter 31. Unless the amount of silica present is small, two dehydrations of silica with intervening filtration should be made to separate as much of it as possible. As stated in Chapter 31, a small amount of silica escapes separation even after two evaporations, but no account is taken of this in the technical analysis of limestone. When the silica that escapes separation after the second dehydration is not recovered, as here, the ignited silica should not be corrected for the impurities present by hydrofluorizing (p. 657), because the two errors will compensate each other to a certain extent.[2]

Procedure. Transfer the residue of the ignition to a small covered casserole, add 5 ml of water, and clean the crucible with a few drops of dilute hydrochloric acid, adding the

[2] The following data given by Hillebrand, Lundell, Bright, and Hoffman, *Applied Inorganic Analysis*, p. 982, are of interest in this connection. The percentage of silica in an argillaceous limestone containing 18.08 per cent SiO_2 as obtained by a *single* evaporation was 17.80 per cent when no corrections were made, 17.71 per cent when the residue was hydrofluorized, and 17.90 per cent when the preceding result was corrected for the silica in the ammonium hydroxide precipitate.

solution to the casserole. Add 5 ml of concentrated hydrochloric acid and hasten solution of the residue by disintegration with a glass rod. Evaporate the solution to dryness on the steam bath, and heat the residue for $\frac{1}{2}$–1 hr at 100–110°C. Pour 5 ml of concentrated hydrochloric acid over the residue, and allow the mixture to stand for 2–3 min. Then add 25 ml of water and heat on the steam bath for 5–10 min to bring all soluble material into solution, using a stirring rod to break up any lumps present. Filter off the silica on quantitative paper; wash the precipitate thoroughly with 1 : 100 hydrochloric acid and finally once or twice with water to remove most of the hydrochloric acid. Evaporate the filtrate and washings to dryness in the original casserole, heat the residue at 100–110°C, and take it up in hydrochloric acid and digest as before, using, however, smaller volumes of acid and water the second time. Filter off the second portion of silica on a fresh small filter paper, and wash as before with 1 : 100 hydrochloric acid and finally with one or two portions of water. Reserve the combined filtrates and washings for step 3. Place the two silica precipitates in a weighed crucible and ignite to SiO_2 over a Meker burner (p. 660). Report the percentage of SiO_2 present.

Notes:

1. If the limestone is very impure, i.e., contains much silicate, it is best to ignite strongly the finely powdered sample in a platinum crucible with half its weight of anhydrous sodium carbonate and then to treat the product with an excess of hydrochloric acid as above.

2. Perchloric acid can be used to advantage in the dehydration of silica in this determination.[3]

3. Combined oxides (Fe_2O_3, Al_2O_3, TiO_2, Mn_3O_4, and P_2O_5)

Addition of a slight excess of ammonium hydroxide to the filtrate from the silica gives a precipitate of the hydrous oxides of ferric iron, aluminum, titanium, and of phosphorus (as the phosphates of these metals). To precipitate manganese quantitatively it is necessary to add an oxidizing agent such as bromine, ammonium persulfate, or hydrogen peroxide to oxidize it to the quadrivalent state, in which form it is precipitated by ammonium hydroxide as hydrous manganese dioxide. The use of ammonium persulfate is open to the objection that sulfate is introduced, which later may give rise to the coprecipitation of calcium sulfate with calcium oxalate. Bromine water is free of these objections, but it is a little difficult to precipitate manganese quantitatively by its use. The oxidation of manganese is carried out in a weakly ammoniacal solution. Bromine reacts with ammonium ion in weakly acidic or basic solutions,

$$2\,NH_4^+ + 3\,Br_2 \rightarrow 8\,H^+ + N_2 + 6\,Br^-$$

and the excess added is thus quite rapidly destroyed. As a consequence, the oxidation and precipitation of manganese may be incomplete. However, it seems that for the small quantities of manganese usually present in limestones fairly satisfactory results are obtained by the use of bromine.[4]

Any large excess of ammonium hydroxide must be avoided in the precipitation, or else aluminum may be incompletely precipitated (p. 597). Magnesium and calcium hydroxides are coprecipitated to a small extent with the hydrous oxides (p. 255). Moreover, calcium carbonate may possibly be precipitated to a slight extent by the ammonium carbonate formed when the ammoniacal solution is exposed to the atmosphere or by that originally present in the ammonium hydroxide. For these reasons the ammonia precipitate is dissolved in hydrochloric acid and again precipitated. The first and second filtrates are combined and reserved for the determination of calcium and magnesium.

[3] F. H. Fish and F. M. Taylor, *J. Chem. Educ.* **10**, 246 (1933).

[4] E. V. Holt and H. F. Harwood, *Mineralog. Mag.* **21**, 318 (1928), investigated persulfate, hydrogen peroxide, chlorine, and bromine for the oxidation of manganese in rock analysis and found bromine the most satisfactory. G. J. Austin, *Analyst* **68**, 274 (1943), recommends periodate (pH 4–5). See also W. Klatt and C. Dozinel, *Z. Anal. Chem.* **126**, 97 (1943).

Any silica in solution at the time of the ammonia precipitation will be partially precipitated with the hydrous oxides. Small amounts of iron, titanium, etc., are retained in the main silica precipitate as impurities and thus escape determination in the procedure given; the error is very small.

Procedure. Heat the filtrate from step 2 (having a volume of 150–200 ml) nearly to boiling, add pure concentrated ammonium hydroxide slowly until a slight precipitate appears, and then add 5 ml of fresh saturated bromine water. Add ammonium hydroxide dropwise with stirring until a slight excess is present, as indicated by litmus paper. Allow the liquid to stand for 5 or 10 min near the boiling point, and then filter through a small rapid quantitative paper. Wash the precipitate three or four times with hot 1 per cent ammonium chloride solution. Test the filtrate with ammonium hydroxide for complete precipitation. Dissolve the precipitate in 10 ml or so of hot 1 : 1 hydrochloric acid, and receive the solution in the original beaker. Wash the filter paper thoroughly with hot water. Dilute the solution to 75 ml, and precipitate the hydrous oxides as before, again using 5 ml of bromine water. Filter, catching the filtrate in the beaker containing the filtrate from the first precipitation. Wash the precipitate thoroughly with 1 per cent ammonium chloride solution. Place the damp paper in a weighed crucible, char, burn off the carbon at a low temperature, and then ignite strongly (p. 594). Report the percentage of mixed oxides ("R_2O_3").

4. Calcium

The filtrate from the hydrous oxide precipitation will contain all the calcium and magnesium, as well as the small amounts of sodium and potassium originally present in the sample. For the accurate determination of calcium in limestones, especially the dolomitic varieties, a double precipitation of calcium oxalate should be made (p. 625). However, results of fair accuracy, sometimes satisfactory for the purpose, can be obtained by making only one precipitation under the conditions recommended in Chapter 28.

Procedure. Concentrate the filtrate and washings from step 3 to 200 or 250 ml after acidification with hydrochloric acid, and then proceed as directed. Report the percentage of calcium oxide present. (For the volumetric determination of calcium, see p. 836.)

5. Magnesium

The filtrate from the calcium precipitation will contain such large amounts of ammonium salts, and of the especially harmful oxalates, that it is advisable to destroy these in accurate work before precipitating magnesium as magnesium ammonium phosphate. This can be done most conveniently by oxidation with nitric acid in hot solution in the presence of chloride. Oxalate is thus oxidized to carbon dioxide and water, and ammonium ion to water, nitrous oxide, and nitrogen.

Procedure. Add 75 ml of concentrated nitric acid to the combined filtrates and washings from step 4, and evaporate to dryness on the steam bath. Only a small residue (of magnesium salts) should remain. Add 2 ml of concentrated hydrochloric acid and 25 ml of water, warm for a few minutes, and filter through a small paper if a slight residue of silica remains. Wash the paper carefully, and then determine magnesium in the filtrate as directed on page 638. Report the percentage of magnesium oxide present.

6. Carbon dioxide

The carbon dioxide content of a carbonate is determined by treating the sample with dilute hydrochloric or perchloric acid in a suitable apparatus and (1) finding the loss in weight due to the carbon dioxide liberated (indirect method) or (2) absorbing the carbon dioxide evolved in a suitable medium and ascertaining the gain in weight of the absorbent. The first method gives satisfactory results for samples containing relatively large amounts of carbon dioxide and is often used because of its simplicity and rapidity. The second method generally gives more accurate results but requires more elaborate apparatus.

Indirect method. For the indirect determination of carbon dioxide a compact apparatus called an *alkalimeter* (Fig. 63-1) is used. The alkalimeter is weighed after charging with sample and acid, the acid is allowed to react with the carbonate, and the apparatus is then weighed again after all the carbon dioxide has been expelled. Naturally the gas escaping from the reaction flask is saturated with water vapor, so it must be passed over a drying agent, which removes and retains the moisture. To remove the carbon dioxide quantitatively from the solution and the free space of the apparatus, a current of dry air is drawn through the warm liquid.

Procedure. First, see that the alkalimeter is clean and dry, inside and out. Place a wad of cotton in the bottom of B (Fig. 63-1), and fill this tube three-quarters full with one of the perchlorate driers (anhydrous magnesium perchlorate, magnesium perchlorate trihydrate, or barium perchlorate), anhydrous calcium sulfate, or, less preferably, granulated calcium chloride.[5] Place another plug of of cotton at the top of the column of drier, and be sure that no particles of desiccant adhere to the tube above the cotton. Insert the top part of the tube, and close it for the time being with a short piece of rubber tubing closed at one end with a short section of glass rod (t_2). Also fill the two drying tubes T_1 and T_2, provided with tightly fitting one-hole rubber stoppers, with the same drying agent as used in B, placing small plugs of cotton in the ends of the tubes. Keep these tubes closed with short sections of rubber tubing stopped with glass rod when not in use. Next fill the tube A about two-thirds full[6] with 4 N perchloric acid or, less preferably, with 4 N hydrochloric acid; be sure that the stopcock is closed so that no acid will run into the flask F, and be careful to keep the ground-glass surfaces dry.

When all is in readiness, weigh 1.5 g of powdered sample into the flask, and add a few milliliters of water to moisten the powder. Replace the top of the flask, and stopper B and A with the tubes t_2 and t_1. Place the whole on the balance and weigh, preferably using another similar alkalimeter as a tare (p. 499). Be sure that the alkalimeter is at room temperature (do not warm it by injudicious handling) and that it is clean and dry

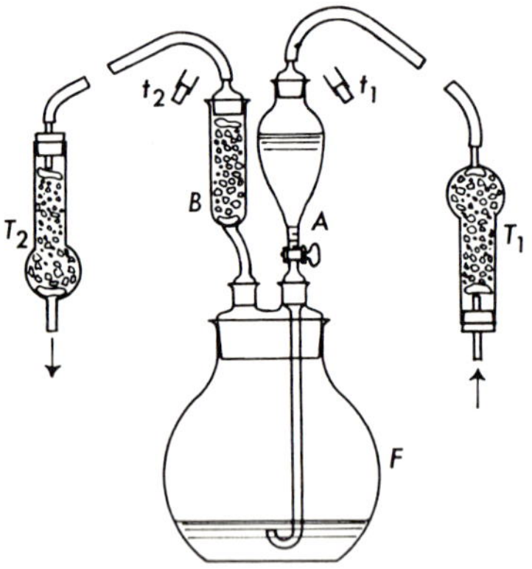

Fig. 63-1. Apparatus for indirect determination of carbon dioxide.

[5] Anhydrous calcium chloride as purchased always contains small amounts of calcium hydroxide. The presence of hydroxide is, of course, very objectionable when calcium chloride is used to dry carbon dioxide as in this determination. Therefore, the calcium chloride to be used for this purpose should be saturated with carbon dioxide previous to use by being allowed to stand overnight in an atmosphere of the gas, and the latter should then be washed out with dry air. Such treatment, however, does not guarantee the conversion of all hydroxide present into carbonate, so on absorption of water during subsequent use some of the unattacked hydroxide may become exposed. For this reason especially, the perchlorate driers are to be preferred.

[6] It is best to calculate the volume of acid required and to use an excess of, say, 50 per cent.

on the outside. It is best to allow the apparatus to stand near the balance for $\frac{1}{2}$ hr before weighing. The stopper t_2 should be removed momentarily before weighing to assure the same pressure inside the apparatus as outside.

Remove the apparatus from the balance, attach the drying tubes T_1 and T_2 with short lengths of rubber tubing, and allow the acid in A to run slowly into the flask at such a rate that there is a brisk but not violent evolution of carbon dioxide. Allow the apparatus to stand at room temperature for 15 min or until little or no carbon dioxide is being evolved. Then draw a slow current of air through the apparatus in the direction indicated in Fig. 63-1, and heat to 80 or 90°C but not to boiling. Remove the flame, and continue to draw air through the solution at the rate of two or three bubbles per second until it is cold. Disconnect the drying tubes, replace the plugs t_1 and t_2, place the alkalimeter (wiped off with a cloth if necessary) near the balance, and allow it to stand for $\frac{1}{2}$ hr. Then weigh as before against the tare, momentarily removing the plug t_2 to equalize the pressure within the flask. From the decrease in weight calculate the percentage of carbon dioxide in the sample.

Direct method. In the direct method for carbon dioxide, matters must be so arranged that the gain in weight of the carbon dioxide absorbent actually represents the amount of carbon dioxide present in the sample; i.e., all the carbon dioxide must be liberated and absorbed, no other gases may be absorbed, and the absorbing medium must neither gain nor lose water. These requirements are met in the apparatus described below.

Apparatus. The form of apparatus illustrated in Fig. 63-2 is suitable for use in an instructional laboratory.[7] Minor modifications can be made in the train as far as form of the separate pieces is concerned.

The decomposition vessel A (Fig. 63-2) is a 200- or 250-ml Erlenmeyer flask fitted with a two-hole rubber stopper through which pass a small condenser C and a dropping funnel B. The upper end of the dropping funnel is closed with a one-hole rubber stopper carrying a drying tube D filled with ascarite (sodium hydroxide–asbestos) resting on a small plug of cotton; the upper end of the drying tube is closed with a plug of cotton.

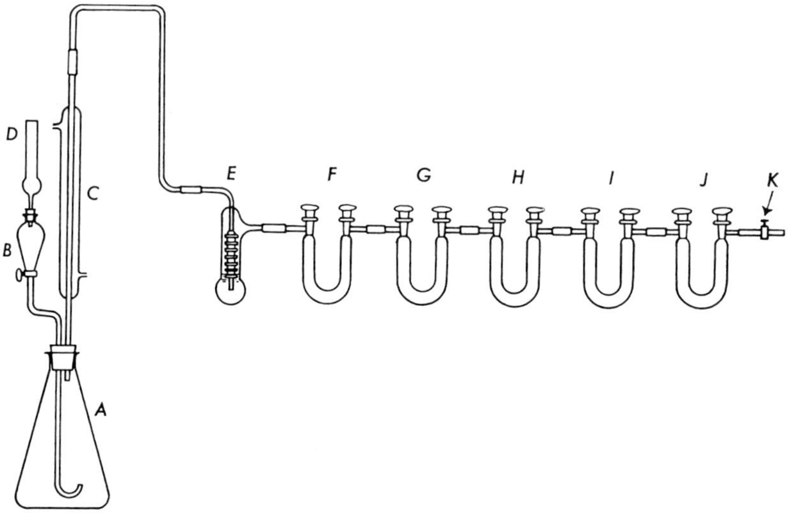

Fig. 63-2. Apparatus for direct determination of carbon dioxide.

[7] For a slightly better form, in which connections are made by mercury seals instead of rubber tubing, and Wesson tubes are used in place of U-tubes as absorption vessels, see Hillebrand, Lundell, Bright, and Hoffman, *Applied Inorganic Analysis*, p. 769.

The stem of the dropping funnel reaches nearly to the bottom of the flask and is bent upwards at its lower end to prevent the entrance of carbon dioxide. The upper end of the condenser is connected with the purification train by means of a short section of fresh gum tubing. In this, and other glass to glass connections, the two ends of the glass tubing should be in contact.

The bottle *E* is partially filled with concentrated sulfuric acid, the function of which is to remove most of the water vapor from the entering gas. The U-tube *F* is filled with dehydrated copper sulfate in pumice,[8] which removes hydrogen sulfide and hydrochloric acid from the gas stream. The second U-tube *G* contains anhydrous magnesium perchlorate or magnesium perchlorate trihydrate,[9] which serves to remove the water vapor still remaining in the gas. The U-tubes *H* and *I* are the absorption vessels and are each filled two-thirds full of ascarite[10] and one-third full of magnesium perchlorate or magnesium perchlorate trihydrate, as shown in Fig. 63-3 (the same desiccant must be used as is contained in *G*). The ascarite absorbs the carbon dioxide and, with the drying agent following, takes up the water formed in the reaction

$$2\,NaOH + CO_2 \rightarrow Na_2CO_3 + H_2O$$

Ascarite is not a particularly good drying agent, and it is therefore followed by a good desiccant to take up any water that may pass through. The use of two absorption tubes assures the complete absorption of carbon dioxide and gives an indication when the first tube must be recharged. The last tube *J*, containing magnesium perchlorate (anhydrous or trihydrate) together with ascarite, protects the absorption tubes. From the last U-tube a rubber tube provided with a screw clamp *K* to regulate the flow leads to the suction line or an aspirator. All the U-tubes are hung by copper or brass wire from a horizontal iron rod, and the arrangement is made as compact as possible.

Procedure. First test the apparatus for gas-tightness as follows. Pour enough water into flask *A* to cover the lower end of the dropping funnel, replace the stopper, and close the top of tube *D* tightly with a well-fitting stopper. Apply gentle suction, gradually open clamp *K* until one bubble per second passes through *E*, and keep it open until the bubbles cease (which will occur if the apparatus is free from leaks). Close *K* tightly, and allow the apparatus to stand under the slight vacuum for 5 min. Then close the stopcock of *B*, remove the stopper, gradually open the stopcock of funnel *B*, and note whether there is an inrush of air as shown by bubbles in *A*. If the vacuum has been maintained, it may be assumed that the apparatus is free of significant leaks. If the

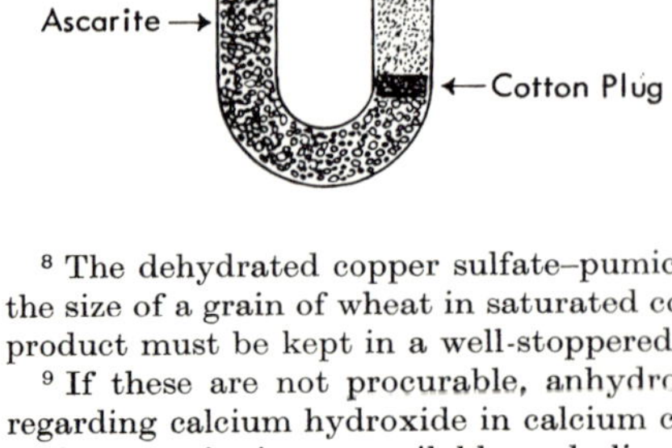

Fig. 63-3. Absorption tube for carbon dioxide.

[8] The dehydrated copper sulfate–pumice is prepared by soaking pulverized pumice having the size of a grain of wheat in saturated copper sulfate solution and drying at 150–180°C. The product must be kept in a well-stoppered bottle.

[9] If these are not procurable, anhydrous calcium chloride may be used. See footnote 5 regarding calcium hydroxide in calcium chloride.

[10] If ascarite is not available, soda-lime may be used in *H*, *I*, and *D*.

apparatus is not gas-tight, examine all the connections carefully, replace any that appear to be defective, wire the rubber tubing to the glass, and again test for leaks.[11]

When the train is gas-tight, draw a current of air through it at the rate of about two bubbles per second for 15 min. Then close the stopcocks of the absorption tubes *H* and *I*, disconnect the tubes from the train, and weigh them separately against a similar U-tube used as a tare. Allow the tubes to stand in or near the balance case for $\frac{1}{2}$ hr before weighing. Open a stopcock of each absorption tube and of the tare for an instant before weighing. Replace the absorption tubes in the train, and repeat the operation of drawing air through the system until the weight of each tube is constant to 0.5 mg.

When constant weight has been attained, weigh 1–1.5 g of sample into flask *A*, add 50 ml of water or enough to cover the lower end of the dropping funnel, disconnect the absorption tubes and the protecting tube *J* from the train, attach the suction line connection to *G*, and draw a slow stream of air through the rest of the apparatus for 5 min to remove any carbon dioxide introduced in opening flask *A* when weighing the sample. Then connect in the absorption tubes, leaving *K* open and the vacuum source disconnected. Pour 50 ml of 1 : 1 hydrochloric acid into funnel *B*, replace ascarite tube *D*, and allow the acid to flow slowly into the flask so that there is a slow evolution of carbon dioxide, corresponding to the passage of about three bubbles per second through *E*. When all the acid has been added and only a slow evolution of carbon dioxide is taking place, close the stopcock of *B*, start the water circulating in condenser *C*, and gradually heat the contents of the flask to boiling. Connect the train to the suction line, open the stopcock of *B*, and regulate *K* so that approximately two bubbles of gas per second pass through the train. Continue boiling gently for a few minutes, and then gradually reduce the heat, finally taking away the source altogether. In the meantime regulate the suction so that a constant slow stream of carbon dioxide-free air is being drawn through the system. Continue to draw air through the apparatus for 20–30 min after the cessation of heating. Reweigh the absorption tubes against the tare, taking the same precautions as in the first weighing. Repeat the determination until satisfactory checks are obtained. Report the percentage of carbon dioxide found.

METHOD II: CaO, MgO

The sample is dissolved in hydrochloric acid. Calcium and magnesium present in acid-insoluble silicates are therefore not included in the determination. The EDTA titration of calcium and magnesium is carried out according to well-known procedures.[12] The interference of subsidiary elements is prevented by adding triethanolamine, cyanide, and ascorbic acid (p. 806).

Procedure. Weigh accurately 0.5 g of finely ground sample into a 250-ml beaker. Add 10 ml of water and 8 ml of concentrated hydrochloric acid and boil (covered beaker) until all carbonate has been brought into solution. Transfer the solution quantitatively to a 500-ml volumetric flask, dilute to the mark with water, and mix.

Determination of CaO. Transfer 5 ml of 5 *M* potassium hydroxide, 2 ml of 1 per cent potassium cyanide (care!), 5 ml of 20 per cent triethanolamine, and 40.00 ml of standard 0.005 *M* EDTA to a 250-ml Erlenmeyer flask, and add a large drop (0.06 ml) of 0.02 per cent Calcein indicator. Pipet 25 ml of the sample solution into the flask.

[11] In careful work it is desirable to test the apparatus by determining the carbon dioxide content of a sample containing a known amount of carbonate (anhydrous sodium carbonate for standardization—see p. 778—may be used).

[12] The directions following are essentially those given in undated sheets issued by the G. Frederick Smith Chemical Co., Columbus, Ohio. Other indicators can be substituted for those mentioned (see Chapter 41). The standard deviations for the determination of CaO + MgO and CaO are approximately ±1 ppt. The accuracy of the methods is less well established but apparently is satisfactory. NBS Dolomite (No. 88), containing 60.37 per cent CaO and 21.48 per cent MgO, gave 60.29 per cent CaO and 21.44 per cent MgO.

If the solution is fluorescent, titrate with 0.005 M EDTA solution to the disappearance of fluorescence. If the solution is not fluorescent add a sufficient volume, accurately measured, of standard 0.010 M Ca solution to restore the fluorescence, and then titrate with EDTA to the disappearance of the fluorescence. Carry out the titration in the dark with an ultraviolet source and view the solution through a sheet of yellow cellophane. Calculate the percentage of CaO in the sample.

Standardize the EDTA solution against 25 ml of standard calcium solution according to the above procedure. Prepare the standard calcium solution by dissolving 1 g of $CaCO_3$ of verified Ca content in hydrochloric acid (as above) and diluting to 1 liter.

Determination of CaO + MgO and MgO by difference. Pipet a 25-ml aliquot of sample solution into a 250-ml flask and add 10–20 mg of ascorbic acid, 2 ml of 5 per cent potassium cyanide, 5 ml of ammonia–ammonium chloride buffer of pH 10 (570 ml of concentrated ammonia and 70 g of NH_4Cl in 1 liter), 5 ml of 20 per cent triethanolamine, and 2 drops of 0.05 per cent Calmagite indicator. Titrate with standard 0.01 M EDTA solution containing magnesium (see p. 804), stirring well, until the last of the pink color of the indicator disappears. Calculate, as percentages, the sum of CaO and MgO as CaO, and deduct from this the percentage of CaO already found. Convert the difference (MgO as CaO) to the percentage of MgO by multiplying by MgO/CaO = 0.7190.

Standardize the magnesium-containing EDTA solution against 25 ml of standard calcium solution (1 g of $CaCO_3$ dissolved in hydrochloric acid and diluted to 1 liter) according to the preceding directions.

PROBLEMS

1. Assuming a sample of limestone to consist entirely of $CaCO_3$, calculate (a) the volume of 4 N hydrochloric acid required to liberate all the carbon dioxide from a 1-g sample; (b) the volume of carbon dioxide obtained in (a) measured at 21°C and 745 mm Hg; (c) the volume of ammonium oxalate solution containing 4.0 g of $(NH_4)_2C_2O_4 \cdot H_2O$ in 100 ml of solution required to precipitate the calcium in a 1-g sample. *Ans.* (a) 5.00 ml; (b) 246.0 ml; (c) 35.5 ml.
2. Find the volume of 6 N hydrochloric acid needed to dissolve 1 g of an equimolar mixture of calcium and magnesium carbonates. How much CaO and $Mg_2P_2O_7$ will be furnished by the same weight?
3. A sample consisting only of calcium and magnesium carbonates is found to contain 20.0 per cent calcium. Find the percentage of magnesium in the material. *Ans.* 14.41 per cent.
4. An EDTA titration of Ca + Mg in a dolomite gave 60.37 per cent calculated as CaO, and a titration of calcium gave 30.50 per cent CaO. What is the percentage of MgO?
5. A 0.6033-g sample of air-dried limestone containing negligible amounts of silica, iron, aluminum, etc., gave a loss of 0.0022 g on heating to 100°C, and a further loss of 0.2710 g after strong ignition. Calculate the approximate molar ratio of magnesium to calcium in the limestone. Explain why the calculated result can be only approximate at best. *Ans.* 0.20.
6. A mixture consisting only of calcium carbonate, magnesium carbonate, and calcium sulfate is found to contain 25.0 per cent carbon dioxide and 27.0 per cent calcium. Calculate the percentages of calcium and magnesium carbonates and calcium sulfate. *Ans.* 32.94 per cent $CaCO_3$; 20.16 per cent $MgCO_3$; 46.90 per cent $CaSO_4$.
7. Deduce the formula of a mineral having the following percentage composition, counting the iron with magnesium (isomorphous replacement): CaO, 29.31; MgO, 15.78; FeO, 9.30; CO_2, 45.75.
8. Show that magnesium hydroxide will not be precipitated by ammonia in the separation of the hydrous oxides in dolomite under the conditions on page 1103.

CHAPTER 64 *Analysis of Brass*

Alloys, by time-honored classification, are divided into ferrous and nonferrous. The latter class includes dozens of alloy families, for example, copper-base, nickel–copper, zinc-base, lead–tin–antimony, aluminum, titanium, gold, silver, and platinum. In this chapter we consider the analysis of brass, a copper-base alloy in which zinc is an important constituent, and subsidiarily, the analysis of bronze, which is a copper alloy having tin or some other element (such as P or Al) as an essential constituent. The constituents of typical brasses and ordinary bronzes usually fall in the following percentage ranges:

Cu	Zn	Sn	Pb	Fe
60–90	15–40	0–6(10)	0–2	0–3

Iron and lead, and tin in brass, are usually adventitious constituents. Small amounts of various other elements will also be present.

In the following pages, procedures will be given for Cu, Zn, Sn, Pb, and Fe. Other elements are present as essential constituents in some bronzes, special brasses, and related alloys, e.g., in Mn–bronze, P–bronze, Al–brass, Si–bronze, and German silver (10–20 per cent Ni + Cu + Zn). These extra elements will not be considered in this chapter.[1] The procedures described are gravimetric and titrimetric. In industrial routine analysis, when speed is important and the highest accuracy is not demanded, physical methods (emission spectrography, X-ray fluorescence spectrography) may be applied. Colorimetric (photometric) methods are important for minor and trace constituents.

Outline of procedure

One portion of sample is ordinarily used for the entire analysis (determination of Cu, Zn, Pb, and Fe). Treatment of the alloy with nitric acid brings Cu, Zn, Pb, and Fe into solution and precipitates tin as hydrous stannic oxide (metastannic acid), which is filtered off, ignited, and weighed as SnO_2. The filtrate from the tin precipitate is

[1] For the determination of these elements in copper alloys and for the analysis of other nonferrous alloys, see the references in Chapter 1 (p. 19). L. Silverman in L. Meites (ed.), *Handbook of Analytical Chemistry*, McGraw-Hill, New York, 1963, outlines the determination of 20 elements in copper alloys. Reviews of nonferrous alloy analysis appear in *Anal. Chem.* See, for example, C. J. Leffault, Jr., and M. L. Moss, *Anal. Chem.* **33**, 77R (1961); F. Will, III, and M. L. Moss, *ibid.* **35**, 96R (1963); F. Will, III, *ibid.* **37**, 92R (1965) (light metals); R. Z. Bachman and C. V. Banks, *ibid.* **37**, 103R (1965) (Zr, Hf, V, Nb, Ta, Cr, Mo, W). As an indication of the types of analytical methods being studied, the following breakdown of methods found by Will to have been used in papers published in the interval September 1962 through August 1964 on analysis of light metal (Al, Be, Mg, Ti) alloys is of interest: colorimetric, 27 per cent; spectrochemical, 22; classical, 15 (including 4 per cent chelometric); activation analysis, 11; gas determinations, 6; polarographic, 5; ion exchange, 3; X-ray, 3; flame photometric, 3; atomic absorption, 2; and miscellaneous, 3.

evaporated to fumes with sulfuric acid to expel nitric acid and then diluted with water to precipitate $PbSO_4$, which is filtered off and weighed after gentle ignition. (Alternatively lead can be electrodeposited as PbO_2.) The filtrate from the lead precipitation is electrolyzed to deposit Cu. The electrolyzed solution, containing Fe(II) and Zn, is treated with bromine water to oxidize ferrous iron. A double precipitation with ammonia separates iron as ferric hydroxide from zinc, and the precipitate is weighed as Fe_2O_3 after ignition. Zinc is precipitated as $ZnNH_4PO_4 \cdot 6\,H_2O$, which is ignited to $Zn_2P_2O_7$ for weighing.

Alternatively, copper and zinc can be determined volumetrically, the former iodimetrically [thiosulfate titration of I_2 liberated by reaction of Cu(II) with I^-], the latter with EDTA.

Tin

In the absence of other acids and of much iron, the separation of tin as metastannic acid (hydrous stannic oxide) by digesting with nitric acid,

$$3\,Sn + 4\,HNO_3 + H_2O \rightarrow 3\,H_2SnO_3 + 4\,NO$$

is essentially quantitative; not more than 0.1–0.2 mg of Sn should remain in solution under the conditions of the following procedure. If more than about 0.5 per cent Fe is present in the alloy, precipitation of tin may not be quantitative. Evaporation of the solution to dryness (see silica, p. 653) is not necessary for maximal recovery of tin, and should, as a matter of fact, be avoided as increasing the contamination of metastannic acid. Hydrous tin oxide shows strong adsorptive properties, and carries down P, As, Sb, Si, and some other elements completely or almost so (if they are present in small amounts); Fe(III) to an appreciable extent; and Cu, Zn, and Pb to a slight extent. In the routine analysis of brass in which the tin and iron contents are low, the contamination of the weighed SnO_2 by Fe, Cu, Zn, and Pb may be neglected. Exact analyses (especially of bronze) require the separation of the contaminants from the tin residue and the recovery of these. Volatilization of tin as SnI_4 provides one means of separation (see the note on p. 1111).

Table 64-1 indicates the results that may be expected in the routine method, with no correction for the contaminants of SnO_2. The values for Cu, Sn, and Pb were obtained by procedures similar to those that follow.[2]

Procedure. Since the borings or turnings of brass are likely to be oily, wash the sample with ether, benzene, or similar organic solvent on filter paper, and dry by exposing to the air on a watch glass for a short time.

Weigh out 1-g portions into 100-ml beakers, add 10 ml of water and 15 ml of concentrated nitric acid (keep beaker covered), and evaporate at a temperature of 80–100°C to a volume of 5–10 ml but not to dryness. This digestion should be extended over a period of at least 1 hr to ensure the quantitative separation of the tin.[3] Then dilute with water to 50 ml, heat to 90–100°C, and keep at this temperature for 10–15 min. If the precipitate appears large in amount, stir in a little filter-paper pulp (p. 464). Filter hot through a dense filter paper, receiving the filtrate in a small beaker. If the filtrate is not clear, refilter it; add more paper pulp if the precipitate persists in running through. Reserve the filtrate and the (later) washings.

Wash the precipitate at least 10 times with hot 1 : 20 nitric acid. Place the paper and its contents in a weighed porcelain crucible, dry, char and burn off the carbon at as low a temperature as possible, and then ignite strongly for $\frac{1}{2}$ hr over a Meker burner under good oxidizing conditions (to prevent reduction of SnO_2 to Sn) or in a muffle

[2] Partial data from D. I. Walter and O. R. Gates in *Standard Methods of Chemical Analysis* (F. J. Welcher, ed.), vol. Two, part A, Van Nostrand, Princeton, N.J., 1963, chap. 26.

[3] If no precipitate has appeared at the end of this period, no appreciable amounts of tin are present.

TABLE 64-1. Comparison of Results of Brass and Bronze Analysis by Routine Methods with U.S. National Bureau of Standards Certificate Values

NBS *Sample*	Cu, %	Zn, %	Sn, %	Pb,[a] %	Ni,[b] %	Fe, %
37b: Brass						
NBS certificate value	70.36	27.09	0.99	0.90	0.45	0.21
Routine, U.S. Naval Research Lab.	70.37		1.01	0.92	0.46	0.18[c]
52b: Bronze, cast						
NBS certificate value	88.25	2.96	8.00	0.011	0.72	0.032
Routine, U.S. Naval Research Lab.	88.17		8.11		0.72	0.01
62b: Bronze, Mn (contains 1.29% Mn)						
NBS certificate value	57.39	37.97	0.96	0.28	0.27	0.82
Routine, U.S. Naval Research Lab.	57.41		1.00	0.27	0.29[d]	0.77[c]
63b: Bearing metal, phosphor bronze (contains 0.44% P, 0.12% Si, 0.54% Sb)						
NBS certificate value	77.94	0.71	9.78	9.36	0.33	0.47
Routine, U.S. Naval Research Lab.	77.76		11.51	9.42	0.32	0.29[e]

[a] Electrolytically as PbO_2.
[b] Photometrically with dimethylglyoxime–bromine after electrolytic separation of Cu and Pb.
[c] Photometrically with *o*-phenanthroline after separation from Zn and Ni by coprecipitation with MnO_2.
[d] First PbO_2 deposit dissolved and redeposited in presence of Cu to remove codeposited Mn.
[e] 0.0018 g of Fe was recovered from SnO_2 (1-g sample of 63b).

furnace at 1100°C. The flame must not be allowed to envelop the crucible; it should be placed in a slanting position, preferably in an asbestos shield, and left uncovered. Repeat the ignition to constant weight. The ignited SnO_2 may be brownish, owing to the presence of ferric oxide; it should not be dark gray, for this points to the presence of appreciable amounts of cupric oxide as the result of incomplete washing.

Calculate the percentage of tin in the sample.

Note: The results obtained for tin by the above procedure are likely to be slightly high, owing to the presence, as already mentioned, of small amounts of iron, copper, zinc, and possibly lead in the ignited residue. The weight of the foreign oxides present can be determined and a correction applied in the following way.[4] Mix the residue with about 15 times its weight of ammonium iodide, and heat at 425–475°C until fumes cease to come off (about 15 min). Tin is thus volatilized as stannic iodide, but the other metals remain. Cool and add 2 ml of nitric acid to the residue. Evaporate to dryness and ignite at low redness. Subtract the weight of the residue from the weight of the impure stannic oxide to obtain the true weight of the oxide. If antimony is present, it is volatilized with the tin. In accurate analyses, the residue from the tin volatilization should be dissolved in a milliliter or two of hot hydrochloric acid and this solution added to the filtrate and washings containing Zn, Pb, and Fe. The subsequent procedure should entail determination of lead as sulfate (HCl is then volatilized).

Lead

Either of the following two methods may be used for the determination of lead in brass.

[4] E. R. Caley and G. Burford, *Ind. Eng. Chem., Anal. Ed.* **8**, 114 (1936).

Determination of lead as sulfate. The solubility of lead sulfate in sulfuric acid of proper concentration[5] is sufficiently small to permit the quantitative separation of lead as sulfate from many other metals whose sulfates are soluble, including copper and zinc. Other acids, such as hydrochloric and nitric, must be absent, for they increase the solubility of lead sulfate; they are easily volatilized by heating the solution to fumes of sulfur trioxide after adding sulfuric acid.

The precipitate of lead sulfate is washed with dilute sulfuric acid to minimize solubility losses. The washed precipitate is ignited in a filter crucible at 500–600°C. The precipitate is not collected in paper, because the latter would partially reduce the lead sulfate during the ignition. Above 700°C lead sulfate begins to lose sulfur trioxide.

Procedure. Add 4 ml of concentrated sulfuric acid to the filtrate and washings from the tin precipitate and evaporate on the hot plate to fumes of sulfuric acid. Cool, wash down the sides of the beaker, and again evaporate to fumes to ensure the complete expulsion of nitric acid; do not carry the evaporation far past the fuming point. Cool, dilute with 25 ml of water, heat nearly to boiling, and stir until everything except lead sulfate has gone into solution. Then add 50 ml of water and allow to stand for at least 1 hr, stirring occasionally. The liquid must be at room temperature at the end.

Filter off the precipitate in a weighed Gooch or porcelain filter crucible and wash thoroughly with cold 1 : 20 sulfuric acid. Wash out the sulfuric acid with a few milliliters of alcohol. Reserve the filtrate and washings. Ignite the precipitate to constant weight at 500–600°C in a muffle or over a flame. In the latter case the filter crucible should be placed in an ordinary porcelain crucible and the latter heated to a barely visible red at the bottom; care should be taken to exclude the reducing flame gases from the crucible. Calculate the percentage of lead.

Electrolytic determination of lead as dioxide. By electrolysis in nitric acid solution lead can be separated as lead dioxide on a Pt-gauze anode. The solution must be quite strongly acid with nitric acid, else some of the lead may be deposited on the cathode; a concentration of 15 ml of concentrated nitric acid in 100 ml is sufficient. Among the elements that interfere by contaminating the deposit are manganese (above 1 per cent), silver, bismuth, antimony, and tin. Arsenic, mercury, and phosphate prevent or hinder complete deposition; chloride, of course, is very objectionable. Sulfate is undesirable; it promotes cathodic deposition of lead and may precipitate lead as sulfate if much lead is present.

[5] H. D. Crockford and D. J. Brawley, *J. Am. Chem. Soc.* **56**, 2600 (1934), give the following values for the solubility of lead sulfate in sulfuric acid solutions of various concentrations at 25°C (only a few of these authors' figures are given here):

Concn. of Sulfuric acid, wt. %	mg *of* $PbSO_4$ *in* 100 ml
1	0.22
5	0.20
10	0.16
20	0.12
30	0.12
40	0.12
50	0.12
60	0.12

Above a concentration of 60% of sulfuric acid, the solubility of lead sulfate increases. At 20°C, 100 ml of water dissolves 4.6 mg of lead sulfate.

Electrodeposited lead dioxide does not have quite the theoretical composition: It weighs more than it should.[6] An empirical factor for converting the weight of PbO_2 to Pb satisfactorily corrects for the departure from the stoichiometric ratio when small amounts of lead are involved. According to Cohen,[7] complete deposition of PbO_2 cannot be achieved. In 140 ml of electrolyzed solution, 0.4 mg of Pb remains dissolved. Cohen corrects the weight of PbO_2, dried at 120°C, by using the empirical factor $Pb/PbO_2 = 0.860$ (instead of the theoretical factor 0.8662) and adding 0.4 mg of Pb (for a 140-ml volume). This correction procedure is adopted here.

Procedure.[8] To the filtrate from the metastannic acid (which should not have a volume greater than 100 ml) add nitric acid in the ratio of 1.5 ml of the concentrated acid to 10 ml of solution. Connect a clean weighed platinum-gauze electrode (preferably sand-blasted) to the positive terminal of the source of current. A stout platinum wire placed within the gauze serves satisfactorily as the cathode. Bring the 150-ml beaker containing the solution beneath the electrodes, and adjust its height so that the electrodes extend very nearly to the bottom and the anode is at least two-thirds immersed. Turn on the current, and adjust the resistance to give 1.5–2 A at 2 V or more. Electrolyze for 1 or $1\frac{1}{2}$ hr. Test for completeness of deposition by adding a little water and noting whether any deposit is formed on the freshly immersed gauze or stem of the anode when the electrolysis is continued for 15 min. When deposition is complete, lower the beaker slowly, or raise the electrodes while rinsing off the latter thoroughly with the aid of the wash bottle, taking care that there is no loss of solution. When the electrodes are out of the solution, turn off the current. Dip the gauze anode into water in a beaker, and then rinse with acetone to remove water. Dry the anode with its deposit of lead dioxide in an oven at 120°C for $\frac{1}{2}$ hr, cool, and weigh. From the weight of lead dioxide calculate the percentage of lead, using the value 0.860 for the factor Pb/PbO_2 and adding 0.1 mg of Pb for each 40 ml of electrolyzed solution. Clean the anode by placing in dilute nitric acid containing hydrogen peroxide.

The cathode may have a deposit of copper. If so, place it in the electrolyzed solution, and let it stand therein until all the copper has been dissolved off. Then remove the cathode, rinse it with water, and reserve the solution in the electrolytic beaker for the determination of copper.

Note: The solution should not contain more than 0.1 g of lead. With large amounts of lead it often happens that some of the lead dioxide flakes off the electrode. It is better to electrolyze overnight with an amperage of 0.5, or slightly less, than to use a greater current over a shorter period of time.

Copper

The analyst has a choice between gravimetric and volumetric methods for the determination of copper. Two standard procedures are offered below, the first electrolytic, the second iodimetric. Volumetric EDTA procedures are also available. Physicochemical methods can also be applied. For the determination of copper by differential spectrophotometry, see Chapter 61.

[6] See L. Hertelendi, *Z. Anal. Chem.* **122**, 30 (1941), for values of the empirical factor as a function of the amount of PbO_2.

[7] A. Cohen, *Rationelle Metallanalyse*, Birkhaüser, Basel, 1948.

[8] In an alternative procedure given by Walter and Gates for the analysis of copper-base alloys, lead as PbO_2 and copper are electrodeposited simultaneously from the filtrate and washings from the tin precipitate, whose nitric acid concentration is 1–1.5 *M*. One drop of 0.1 *M* HCl is added to the solution (150 ml) to give a better copper plate, and the electrolysis is carried out overnight at a current density of 0.5 A/dm^2 or for $2\frac{1}{2}$ hr at a current density of 4 A/dm^2 (the latter procedure requires a gauze electrode for PbO_2 as well as Cu). The value 0.866 is used for the factor Pb/PbO_2 (PbO_2 dried at 110–120°C). The values for Cu and Pb in Table 64-1, obtained by this procedure, show excellent agreement with the Bureau of Standards certificate values.

Electrolytic determination of copper. The electrolytic deposition of copper has already been discussed (p. 928). The method is an excellent one for the separation of copper from zinc and lead.

Procedure.[8] If lead was separated by the sulfate method, concentrate the solution to 100 ml if necessary, add 5 ml of concentrated ammonium hydroxide to reduce the acidity and 2 ml of nitric acid to guarantee the formation of a good copper deposit. Dry and weigh a clean platinum gauze cathode, and attach, together with the anode, to the electrolysis apparatus (Fig. 51-1), having the source of current disconnected. Be sure that the cathode is connected to the negative terminal. If a rotating anode is used, see that it can rotate without coming into contact with the cathode at any point. Raise the beaker into position, and adjust its height so that the electrodes extend nearly to the bottom of the beaker and the cathode is nearly but not entirely covered by the solution. The beaker should be placed on a wooden pedestal or tier of blocks, so that it can be easily lowered away from the electrodes at the end of the electrolysis. Cover the beaker with a split watch glass; and with all the resistance in, so that only a small current will flow, close the circuit, and proceed as in (1) or (2), depending on whether or not stirring is used.

(1) *Electrolysis without stirring.* Decrease the resistance until a current of approximately 0.5 ampere flows, as indicated by an ammeter in the circuit, and electrolyze overnight, using an applied potential of 2 volts (a single lead accumulator may be used) or more.

Alternatively, use a current of about 1.5 amperes, and electrolyze for three hours.

(2) *Electrolysis with stirring.* Start the motor driving the anode or glass propeller, and adjust its speed so that the solution is vigorously stirred without there being any danger of mechanical loss of liquid. Regulate the resistance to give a current of 3 to 4 amperes; the voltage across the electrode terminals should be 2 to 4 volts (not above the latter figure in the presence of nickel). Continue the electrolysis for about forty-five minutes, or until the blue color of the copper has entirely disappeared.

Regardless of whether procedure (1) or (2) is used, rinse off the cover glass into the beaker when it is judged that the electrolysis is complete, and add more water, if necessary, to bring the level of the solution above the top of the cathode. Continue the electrolysis for one half-hour if procedure (1) is used, or for a quarter of an hour without stirring if procedure (2) is followed (in the latter case reduce the amperage to 0.5 to 1). If at the end of this time no copper deposit has appeared on the freshly immersed surface of the cathode, it may be assumed that the deposition is complete. If a coat of copper does form, continue the electrolysis as long as may be judged necessary, and again test as before by adding a little more water and seeing whether a further deposit is formed on the stem of the cathode.

When the electrolysis is complete, slowly lower the beaker (or raise the electrodes) while directing a stream of water from the wash bottle against the exposed part of the cathode. Do not break the current until the cathode is entirely out of the solution and has been thoroughly washed. Any acid solution allowed to remain on the deposit will partially dissolve it. Finally dip the cathode into a beaker of water to insure the removal of any soluble substances, and then rinse it with successive portions of alcohol or acetone to remove water. Place the cathode in an oven heated to 100°, and remove after two to three minutes. Cool to room temperature and weigh. Calculate the percentage of copper.

If lead was determined electrolytically, first remove the large amount of nitric acid present. Add 3 ml of concentrated sulfuric acid to the solution from which lead has been deposited electrolytically and evaporate until fumes of sulfur trioxide appear. Cool, dilute with water to 75 or 100 ml, add 2 ml of concentrated nitric acid, and electrolyze out the copper as described above.

Iodimetric determination of copper. An aliquot of the filtrate from the lead sulfate precipitation, corresponding to ~0.25 g of sample, can be used for the determination of copper by the usual iodimetric method. Or copper can be determined in a separate sample according to the following directions. The interference of iron is prevented by pyrophosphate (much more effective than orthophosphate). Such minor components of copper alloys as lead, bismuth, tin, antimony, and nickel do not interfere. Thiocyanate is added in the following procedure to diminish the required amount of potassium iodide.

Procedure.[9] Transfer 0.3–0.4 g of sample (not more than 0.25 g of Cu) to a 200-ml Erlenmeyer flask and dissolve, with warming, in 10 ml of acid mixture (500 ml of sulfuric acid and 150 ml of nitric acid diluted to 1 liter with water). Evaporate to incipient crystallization, add 10 ml of water, and warm, if necessary, to bring all copper salts into solution. Cool and add 2 g of solid sodium pyrophosphate (for < 2 per cent Fe) and 8 ml of 2 per cent starch solution, followed by 25 ml of iodide–thiocyanate solution (5 g of potassium iodide and 50 g of potassium thiocyanate in 1 liter of water). Immediately titrate with sodium thiosulfate (25 g/liter), allowing the titrant to flow rapidly from the buret until near the end point, which is marked by a color change from bluish to gray. Save the titrated solution if zinc is to be determined (p. 1116).

Standardize the thiosulfate solution against copper solutions obtained by weighing out 0.25 g of pure copper and treating in the same way as the sample.

Note: Since Cu(II) is slowly reduced to Cu(I) by thiocyanate, the titration must be made rapidly, and the standardization of the thiosulfate solution should be made against copper. When results of the highest accuracy are desired, it is best to add iodide alone at the beginning, then to titrate with thiosulfate, and to add thiocyanate shortly before the end point.

Iron

The separation of iron from zinc in brass is usually effected by precipitation as ferric hydroxide with ammonia after oxidation to the ferric state. Zinc is coprecipitated with ferric hydroxide. The use of an excess of ammonia and ammonium chloride improves the separation. The amount of iron present in brass is usually so small that not more than one reprecipitation of ferric hydroxide need be made. Any aluminum present will be partially precipitated with iron and weighed with ferric oxide, and the remainder will be precipitated with zinc ammonium phosphate in the determination of zinc.

Procedure. Heat the solution from the copper electrolysis to boiling, and add 2 ml of saturated bromine water and concentrated ammonia until a permanent precipitate forms, followed by an excess of 10 ml. Allow the precipitate to settle, filter it off on a small paper, and wash three or four times with 1 : 100 ammonium hydroxide. Replace the beaker containing the filtrate and washings with the precipitation beaker, and pour 20 ml of hot 1 : 3 hydrochloric acid through the filter to dissolve the precipitate. Wash the paper with hot 1 : 100 hydrochloric acid and then with water to remove acid. Precipitate the iron as before, using a 5-ml excess of ammonium hydroxide. Collect the precipitate in the same filter as before, and wash it thoroughly with the ammonium hydroxide wash solution, receiving the filtrate and washings in the beaker containing the first filtrate. Reserve the combined filtrates for the determination of zinc. Ignite the precipitate to ferric oxide (p. 594) and weigh. Calculate the percentage of iron.

Zinc

A number of methods can be used to determine zinc in the solution remaining from the electrolytic deposition of copper or the precipitation of ferric hydroxide. One of these, in which zinc is precipitated as $ZnNH_4PO_4 \cdot aq$, has been in use since 1900 and is usually considered a standard method.[10] It lacks selectivity, but this drawback is not important for the present purpose, because precipitable elements (except for some minor

[9] A. Cohen, Ref. 7, p. 269.

[10] For a critical study, see J. E. Vance and R. E. Borup, *Anal. Chem.* **25**, 610 (1953).

constituents) will have been removed at this stage. Zinc ammonium phosphate is soluble in both acidic and basic solutions, and the pH must be adjusted quite carefully for quantitative precipitation. The pH range of minimum solubility is 6.1–7.6 (see p. 640) in the presence of ammonium salts. Diammonium hydrogen phosphate is used as precipitant. Hydrogen ions liberated in the precipitation react with excess diammonium hydrogen phosphate to give dihydrogen phosphate and the pH is kept within permissible limits. The precipitate first formed is actually zinc phosphate; this flocculent precipitate is rapidly transformed into the crystalline double phosphate on digestion. A 1 per cent solution of $(NH_4)_2HPO_4$ adjusted to pH 6.8 is the best wash liquid, but cold water can be used; a 50 per cent solution of alcohol is used to remove residual ammonium phosphate. As much as 5 g of $(NH_4)_2SO_4$ or NH_4Cl in 150 ml of sample solution does not diminish the accuracy, but as little as 0.1 g of NH_4NO_3 (0.01 M) leads to low results. Zinc ammonium phosphate is obtained in anhydrous form by heating the precipitate at 110°C and is a good weighing form. It is readily transformed to $Zn_2P_2O_7$ by ignition at 500–600°C, which also is a good weighing form. The solubility loss in 150 ml of solution is less than 0.1 mg of Zn. The absolute error in the determination usually does not exceed ± 0.1 to ± 0.15 mg of Zn. Nickel is reported not to be precipitated.

Zinc can also be precipitated and weighed as the mercurithiocyanate, $ZnHg(CNS)_4$. A high electrolyte concentration should be avoided, but the adjustment of the acidity is not critical. Small amounts of iron, cobalt (<0.2 per cent in the sample), and nickel (<0.7 per cent) seem to be without much effect, although they are coprecipitated to some extent (mixed-crystal formation). Copper must be absent. The results for zinc in yellow-metal alloys by this method do not seem to differ greatly from those by the phosphate method.[11] This method is given below as an alternative to the phosphate method.

Sulfide precipitation of zinc is more selective than the previous methods. Hydrogen sulfide precipitation in cold sulfate–bisulfate buffered solutions at pH 2 provides good separations of zinc from Fe, Al, Mn, and Ni but not from Co. Zinc sulfide is ignited in air to ZnO for weighing.

A number of organic precipitants can be used for the gravimetric determination of zinc. 8-Hydroxyquinoline gives complete precipitation of zinc (within 0.01 mg), but the factor 0.1861 instead of the theoretical factor 0.1850 must be used for converting the weight of $Zn(C_9H_6ON)_2$, precipitated at pH 4.8–4.9 and dried at 125°C, to Zn.[10] Anthranilic acid also gives low results when the precipitation is made in hot solution (high in cold solution). These reagents are unselective for zinc.

The loss of zinc in the separations preceding its determination is small. In analyses of an alloy (1 g) containing 30 per cent Sn, 15 per cent Pb, 11 per cent Cu, 13 per cent Fe, 6 per cent Ni, and 25 per cent Zn, the SnO_2 residue contained 0.14 mg of Zn (average); $PbSO_4$, <0.01 per cent Zn; and electroplated Cu <0.01 mg of Zn.

Titrimetic determination of zinc. EDTA titration of zinc provides a rapid determination and one that is often sufficiently accurate. The procedure given below can be applied to a separate portion of sample if Sn and Pb do not exceed 0.1 per cent according to the originators of the procedure.[12] Copper is precipitated as iodide or thiocyanate in the presence of ascorbic acid, and thus is not titrated.

Procedure for phosphate precipitation. Evaporate the filtrate and washings from the ferric hydroxide precipitations to about 5 ml, add 10 ml of concentrated hydrochloric acid and evaporate to dryness or fumes of sulfuric acid to remove nitric acid. Add 1 ml of hydrochloric acid, dilute to 50 ml with water, and filter off any insoluble silica that may be present.

[11] C. L. Rulfs and L. J. Kirby, *Anal. Chem.* **27**, 1498 (1955).

[12] E. Koros and Z. Remport-Horvath, *Chemist-Analyst* **46**, 91 (1957). The method has not been investigated extensively. Brass samples containing 29.4 and 39.0 per cent Zn gave the values 29.3 and 39.0 per cent, a bronze containing 17.6 per cent Zn gave 17.7 per cent. A number of other EDTA procedures for zinc in alloys have been described.

Adjust the volume of the zinc solution to 150–200 ml; add a few drops of methyl red indicator and then pure ammonia solution carefully until the solution becomes yellow. Heat the solution almost to boiling and slowly add 25 ml of freshly prepared 10 per cent solution of diammonium hydrogen phosphate.[13] Keep the solution hot for $\frac{1}{2}$ hr; at the end of this time the precipitate should be crystalline and should settle readily when stirred up. Remove the beaker from the heat and allow to stand at room temperature for at least 4 hr. Collect the precipitate in a porcelain filter crucible and wash with small portions of cold water, or better with 1 per cent $(NH_4)_2HPO_4$ solution adjusted to pH 6.8, until the washings are chloride-free. Finally wash with small portions of neutral 50 per cent ethyl alcohol to remove ammonium phosphate.

Dry the precipitate to constant weight at 110°C and weigh as $ZnNH_4PO_4$ or ignite to $Zn_2P_2O_7$ at 500–700°C and weigh as such. The conversion to pyrophosphate is preferably carried out in a muffle furnace; if a burner is used for heating, place the filter crucible in an ordinary crucible and take care to prevent the entrance of flame gases that could reduce the precipitate.

Calculate the percentage of zinc. Sum up the percentages of tin, lead, copper, iron, and zinc.

Procedure for mercurithiocyanate precipitation. Transfer the solution from the electrolysis of copper to a 250-ml volumetric flask and make up to volume with water.[14] Dilute a 50-ml aliquot of this solution with 25 ml of water and add 25 ml (or an excess) of ammonium mercurithiocyanate precipitant (32 g of NH_4CNS and 27 g of $HgCl_2$ in 1 liter, filtered if necessary). Allow to stand with occasional stirring for 1 hr at room temperature. Collect the precipitate in a glass or porcelain filter crucible and wash with 50 ml of wash solution (10 ml of precipitant diluted to 1 liter with water), which has been cooled to 5–10°C. Dry the crucible and precipitate for 1 hr at 105–110°C and weigh as $ZnHg(CNS)_4$ [$Zn/ZnHg(CNS)_4 = 0.1312$]. Bear in mind that mercury salts are poisonous.

Procedure for EDTA titration. If the tin and lead contents of the brass sample are small enough to ignore, dissolve 0.1 g of brass in 5 ml of 6 *M* nitric acid and boil out oxides of nitrogen. Cool, transfer to a 100-ml volumetric flask, and dilute to the mark with water. Dilute a 20-ml aliquot to 80 ml with water and add 1–2 g of potassium iodide and enough ascorbic acid to reduce the liberated iodine. Add 6–8 drops of Xylenol Orange (0.1 g of indicator in 100 ml of water), followed by enough 20 per cent hexamethylenetetramine solution to change the color to red-violet. Titrate with 0.005 *M* EDTA solution (1.86 g of the disodium salt in 1 liter; 1 ml $\sim$0.3 mg of Zn) to a yellow color. Standardize the titrant against a zinc solution prepared from pure zinc metal.

If copper has been determined iodimetrically, the titrated solution can be diluted to volume and an aliquot taken for titration with EDTA. More generally, an aliquot of the solution from which copper has been electrolyzed and iron precipitated, and from which tin and lead were removed earlier, can be used for the zinc titration after adjustment of the acidity.

PROBLEMS

1. With recourse to the literature, outline a method for the accurate analysis of phosphor bronze bearing metal having the approximate percentage composition Cu 78, Sn 10, Pb 9.6, P 0.6, Sb 0.6, Zn 0.5, Fe 0.3, As 0.2, S 0.05, Al 0.05, and Ni 0.01. All the constituents listed are to be determined.

[13] This solution should give a very faint pink color with phenolphthalein; if it does not, add ammonia until a barely perceptible color appears.

[14] It may be possible to carry out the zinc precipitation in the filtrate from the iron precipitation, after acidification with nitric or sulfuric acid, but data are lacking.

2. May it be concluded that there is a systematic difference as large as 0.1 per cent absolute in the determination of zinc by the gravimetric pyrophosphate and mercurithiocyanate methods in the range of the plot below? A difference as large as 0.05 per cent absolute?

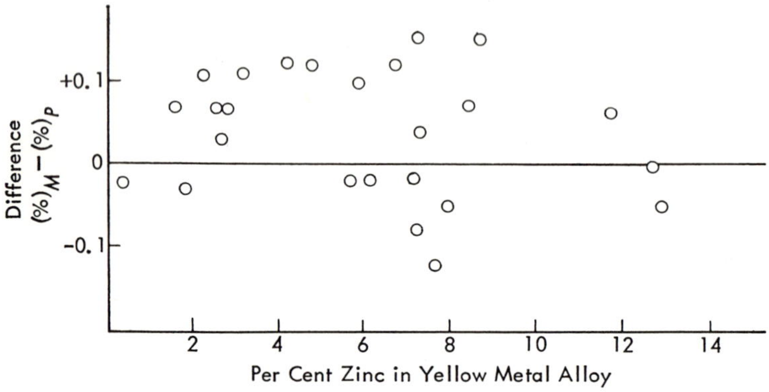

(Data of Ruffs and Kirby, Ref. 11.)

CHAPTER 65 *Analysis of Steel*

The analysis of steel and iron is one of the most important types of applied analysis. The physical and chemical properties of a ferrous material are determined by the kinds and amounts of the minor or subsidiary elements that are present, and the determination of these constitutes steel analysis; iron itself is rarely if ever determined.

The elements found in iron and steel may be divided into three groups from the analytical standpoint. The first group includes certain elements that are always present, such as carbon, manganese, phosphorus, sulfur, and silicon, which are often determined because of their important effect on the properties of the ferrous material. The limits of one or more of these are specified in the purchase and sale of the material. Also included in this group are elements that are added in small quantities to improve the properties of the product (boron and lead, for example). The second group comprises alloying elements whose determination is often required: copper, nickel, chromium, vanadium, molybdenum, tungsten, niobium, titanium, aluminum, and some others (thousands of alloys have been described). The third group of elements includes those whose presence in their usual amounts is not of much importance and which are infrequently determined. They may be introduced in the raw materials (arsenic) or in scrap (tin, zinc). We may also count as members of this last group elements that are added to steel for experimental purposes and whose determination is then called for.

The scope of this chapter is limited, as befits a general textbook.[1] The procedures given (for 10 elements) are gravimetric, titrimetric, and colorimetric. These methods can, at least in part, be replaced, and are replaced in production laboratories, by physical methods which are much more rapid than the classical methods and which may still be able to furnish the requisite accuracy. Emission and X-ray spectrography provide rapid determinations for a considerable number of elements in steel analysis. Atomic absorption spectrography is useful for some elements. Gas chromatography has been utilized in the determination of carbon. Even neutron activation has been applied, but not routinely. As in other sectors of the analytical field, classical methods in ferrous analysis serve as umpiring methods when results of high accuracy are required. They are also applied in the analysis of steel standards that are needed in physical methods.

In the directions that follow, it is assumed that a representative sample is available for analysis. Directions for sampling ferrous materials will not be given here—too much space would be required—and if sampling must be done by the analyst he should refer to the books listed on page 18.

The following procedures for the determination of manganese, phosphorus, sulfur, and silicon apply to carbon steels but they can often be used for alloy steels also. Iron meteorites are analyzable as an iron–nickel alloy (average composition approximately 90 per cent Fe, 9 per cent Ni, and 0.6 per cent Co).

[1] See page 18 for a list of books on analysis of ferrous materials. For a review of recent methods see H. F. Beeghly and L. C. Pasztor, *Anal. Chem.* **37**, 87R (1965); **33**, 70R (1961) and earlier years; B. Bagshaw, *Analyst* **84**, 475 (1959).

Carbon

Carbon may exist combined or free in ferrous products. In most steels it occurs combined as carbide; in cast iron it exists uncombined as graphite. The sum of combined and uncombined carbon is known as total carbon, and its determination is of great technical importance.

Determination of total carbon. The sample of ferrous material is burned in a stream of pure oxygen at a temperature of 1000°C or above, and the carbon dioxide formed is caught in an absorbent such as ascarite (asbestos impregnated with sodium hydroxide). From the gain in weight of the absorbent, the amount of carbon present is calculated.[2]

Apparatus. A simple assembly for the determination of carbon by combustion is shown in Fig. 65-1. The train represented is typical of those commonly used.

In Fig. 65-1, A is a tank of commercial oxygen (purity greater than 99.5 per cent and free of organic matter) with a reducing valve B (preferably with a pressure gauge), and M is a mercury pressure gauge and safety valve. C is a tower filled with ascarite, which absorbs any carbon dioxide in the oxygen. The wash bottle D contains concentrated sulfuric acid and is connected to the combustion tube G through a breech connector E. A one-hole rubber stopper can be substituted for the breech connector; the stopper should be protected on the inside from radiant heat by a perforated asbestos disk or plug.

The combustion tube may be of porcelain (commonly used), sillimanite, vitrified clay, or silica. The length of this tube may conveniently be 50 or 60 cm. The inside diameter should be approximately 2.5 cm. The exit end of the combustion tube should preferably be tapered to permit it to be connected directly to the train following. The electric furnace F for heating the combustion tube should be approximately 30 cm in length and should be capable of giving a temperature of 1000–1100°C. The rheostat H permits the temperature to be regulated. A thermocouple should be used to find the setting of the rheostat which gives the required temperature for combustion (usually 1000°C).

The boat in which the sample is burned may be of nickel (low in carbon), vitrified clay, alundum, or porcelain. The boat should be provided with a special slotted cover to prevent damage to the combustion tube by spattering molten iron oxide. Before a nickel boat is used, it should be heated to 1000°C in oxygen. The boat should be protected from the molten oxide by lining with an inert infusible material such as alundum (aluminum oxide).

The exit end of the combustion tube is connected to a tower or U-tube I packed with asbestos, the function of which is to screen out finely divided iron oxide and sulfur

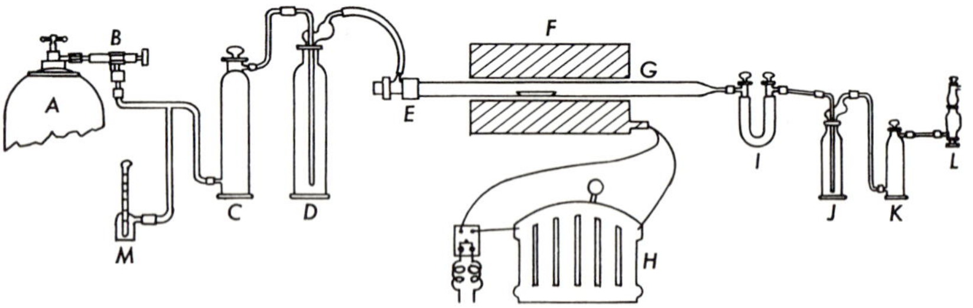

Fig. 65-1. Train for determination of carbon in steel by combustion.

[2] More rapid determinations of carbon can be carried out by nongravimetric methods. In an automated commercial apparatus, the change in volume occurring when the gaseous combustion products (mixture of CO_2 and O_2) are treated with alkali hydroxide is measured. Another commercial apparatus for carbon is conductivity: The carbon dioxide is absorbed in barium hydroxide solution and the decrease in conductance ($BaCO_3$ is precipitated) is observed.

trioxide. The wash bottle *J* contains concentrated sulfuric acid saturated with chromic acid. Here the sulfur dioxide in the gas stream is oxidized to sulfur trioxide, which is retained in the liquid and in the next tube. Tube *K* is filled with a good desiccant, preferably anhydrous magnesium perchlorate. Calcium chloride should not be used.

L is the carbon dioxide absorption bulb. This may have a variety of forms, as shown in Fig. 65-2. The best absorber for the carbon dioxide is sodium hydroxide–asbestos (ascarite), which can be purchased. The carbon dioxide absorbent is covered on the exit side with a layer of desiccant which should be the same as the one used in tube *K*, in order that there may be neither loss nor gain of moisture. Water is formed when carbon dioxide reacts with sodium hydroxide, and the ascarite cannot be relied upon to retain this moisture completely. The absorption bulb should be weighed against a like tube used as counterpoise, and it must naturally be filled with oxygen. Generally, it is not necessary to use a protecting tube for *L*.

Procedure (for plain carbon steels containing less than 0.1 per cent of sulfur).[3] Assemble the apparatus as described in the preceding section and see that the train is gas-tight.

A blank and a standard sample should be run before carbon is determined in the unknown. To run a blank, proceed as follows. Place the covered boat with its lining of alkali-free alundum (90 mesh) in the hot zone of the combustion tube and heat the furnace to 1000°C. Pass a stream of oxygen through the tube at the rate of about 100 ml/min for 15 min while the train is disconnected at *L* (Fig. 65-1). Then connect in the rest of the train and pass oxygen through at the same rate for 5 min. Close the absorption vessel *L*, disconnect it, and allow it to stand near the balance for 10 min. Then weigh (opening the vessel momentarily to equalize the pressure), using a similar tube as tare. Replace *L* in the train and pass oxygen through the system (combustion tube at 1000°C) at the rate of about 200 ml/min for 10 min. Turn off the oxygen, close the absorption tube, disconnect it, and weigh after it has stood in or near the balance case for 10 min. Make another blank determination as before and compare the two results. The increase in weight in either blank should not exceed 1 mg in the usual case, and the agreement should be within 0.5 mg. If satisfactory agreement is not obtained, repeat the blank determination until the gain in weight is constant.

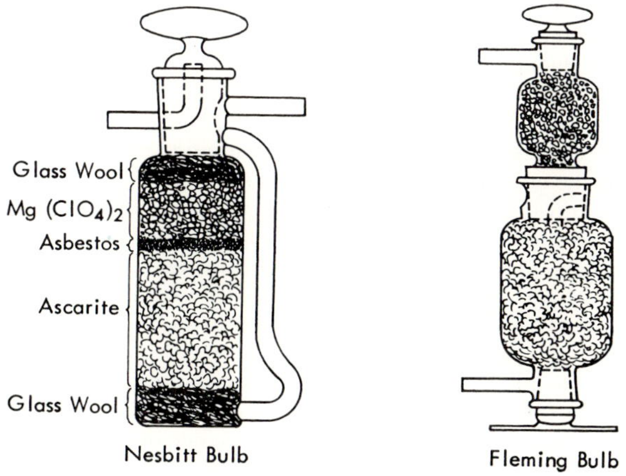

Fig. 65-2. Absorption vessels for carbon dioxide.

[3] The conditions specified in this procedure are essentially those recommended by G. E. F. Lundell, J. I. Hoffman, and H. A. Bright, *Chemical Analysis of Iron and Steel*, Wiley, New York, 1931, p. 170. The train in Fig. 65-1 is similar to one shown by them as in use at the Bureau of Standards.

When a satisfactory blank has been obtained, weigh 2–3 g of a standard sample (i.e., one with a known content of carbon) into the boat, placing the steel in the furrow made in the alundum lining. The sample should be in the form of small shavings or chips, and these should be closely packed in the groove of alundum. Disconnect the train at the entrance end of *G*, and push the covered boat into the center of the combustion tube, which should be at 1000°C. Close the tube, open the absorption vessel *L*, and after $\frac{1}{2}$–1 min pass in oxygen rapidly at the rate of 300–400 ml/min while the sample is burning. When the combustion is complete (after 2 or 3 min), decrease the oxygen current to approximately 200 ml/min, and continue it at this rate for 5 min to sweep out all the carbon dioxide. Then turn off the oxygen, close the absorption tube, remove it from the train, and place it near the balance. After 15 min (when it should be at room temperature), open the tube momentarily and weigh it against the tare. Examine the residue in the boat; if there is any evidence of incomplete combustion, reject the result. Otherwise, calculate the percentage of carbon in the standard sample, subtracting the blank correction. If a satisfactory check is obtained with the percentage of carbon known to be present, run the unknown sample in exactly the same way as described above. Report the percentage of carbon.

Manganese

For the amounts of manganese usually present in steel (<1 or 2 per cent), colorimetric determination as MnO_4^- after periodate oxidation is generally the best method. It is rapid and the error should usually not exceed ± 0.01 per cent absolute for percentages of Mn up to 1 per cent or somewhat more. Determination of higher percentages probably could be made with accuracy and precision comparable to those of volumetric methods by applying differential spectrophotometry.

Mostly for didactic reasons, we also give directions for the volumetric bismuthate method. It is a useful exercise for a serious student of inorganic quantitative analysis to determine a constituent by two different methods and compare both the results and the methods themselves. The spectrophotometric periodate and the volumetric bismuthate methods provide an opportunity for such comparison. It is true that both methods are based on oxidation of Mn(II) to Mn(VII), but the oxidation is carried out with different reagents.

Colorimetric method. A number of strong oxidizing agents will convert Mn(II) to Mn(VII) for colorimetric (spectrophotometric) determination. Periodate (as potassium or sodium periodate) is the preferred reagent[4]:

$$2\,Mn^{++} + 5\,IO_4^- + 3\,H_2O \rightarrow 2\,MnO_4^- + 5\,IO_3^- + 6\,H^+$$

The reaction is quite rapid in a nitric–phosphoric acid solution at 100°C. Phosphoric acid contributes to the smoothness of manganese oxidation, prevents the precipitation of ferric periodate, and decolorizes iron(III) by complex formation.

The slight effect of such colored elements as copper, cobalt, and nickel on the absorbance measurement can be corrected for by obtaining the absorbance of a sample blank which has not been treated with periodate. Chromium is the only element likely to cause difficulty. Cr(III) is oxidized (not quantitatively) to Cr(VI) by periodate and the absorbance curves of MnO_4^- and $HCr_2O_7^-$ overlap to some extent. The error due to Cr(VI) can be reduced by measuring the absorbance of MnO_4^- at its secondary absorption maximum, 545 mμ, without significant loss of sensitivity; the absorbance of Cr(VI) is still less at 575 mμ, but here the absorbance of Mn(VII) is changing rapidly with the wavelength and is little more than one half that at 545 mμ. At 545 mμ, with a $\sim$1 mμ band width, the absorbance of Cr(VI) is about 0.0025 that of an equal weight of Mn(VII), so that as much as 0.5–1 per cent Cr could be present in the sample without producing important error in the manganese determination. When more than about 1 per cent Cr is present a correction can be obtained by oxidizing Cr(III) to Cr(VI) by

[4] H. H. Willard and L. H. Greathouse, *J. Am. Chem. Soc.* **39**, 2366 (1917).

fuming with perchloric acid, oxidizing Mn(II) with periodate, measuring the absorbance, adding NO_2^- to reduce MnO_4^- (but not $HCr_2O_7^-$), and again measuring the absorbance. Directions will not be given here; they can be found in *ASTM Methods for Chemical Analysis of Metals*.[5]

Standard manganese solution. Weigh out (to 0.1 mg) 0.10 g of pure manganese metal ($>$ 99.9 per cent Mn) and dissolve in 20 ml of 1 : 1 nitric acid. Boil the solution to remove oxides of nitrogen, transfer to a 1-liter flask, and dilute to the mark with water.

Procedure. Transfer 0.5 g of steel (0.2–1.5 per cent Mn) to a beaker, add 50 ml of 1 : 3 nitric acid, and warm to dissolve. Boil for a few minutes to expel oxides of nitrogen. Cautiously add 1 g of ammonium persulfate (peroxydisulfate) and boil gently for 10–15 min to oxidize carbon compounds and destroy the excess of persulfate. If a color of permanganate is visible or dark manganese oxides have separated, add enough powdered sodium sulfite to give a clear solution. Boil 5 min to expel sulfur dioxide. Cool the solution, transfer to a 100-ml volumetric flask, and make up to the mark with water.

Pipet two 25-ml aliquots of the mixed solution into small beakers. Add 5 ml of 85 per cent phosphoric acid to each. Transfer one of the solutions to a 100-ml volumetric flask and dilute to the mark with water. Set this solution aside for use as the blank (reference) solution in the absorbance measurement.

Add 0.25 g of KIO_4 to the second beaker and boil, or keep very slightly below the boiling point (95°C), for 10 min. Cool, transfer the solution to a 100-ml volumetric flask, and dilute to the mark with water. Measure the absorbance of the mixed solution in a 1-cm cell at 545 mμ against the blank solution in the reference cell. In the absence of chromium, a filter photometer with a filter having maximum transmission at $\sim$550 mμ may be used for the absorbance measurement. If chromium is present in appreciable amounts, a spectrophotomer giving a narrow band width at 545 mμ must be used. Its use is also to be preferred when colored ions (Cu, Ni, Co) are present in the solution.

Obtain the concentration of manganese(VII) from the standard curve and report the percentage of Mn in the sample.

Standard curve. Pipet 5, 10, and 15 ml of standard manganese solution into beakers, and add 5 ml of phosphoric acid and enough nitric acid to make the solution 2 *M* in HNO_3 after dilution to about 30 ml. Add 0.25 g of KIO_4 and heat as described above. Finally, dilute to 100 ml in a volumetric flask and determine the absorbance at 545 mμ in a 1-cm cell against a blank solution having the same concentrations of nitric and phosphoric acids and KIO_4 as the standards. Plot log I_0/I versus manganese concentration. A straight line passing through the origin should be obtained. If desired, more points than corresponding to the above aliquots can be determined, and this should be done if a good straight line is not obtained with three points (four including the origin).

Permanganate solutions containing periodate are stable indefinitely.

Bismuthate method.[6] The steel is dissolved in nitric acid, and manganese is oxidized to permanganate in cold solution with sodium bismuthate:

$$2\,Mn^{++} + 5\,NaBiO_3 + 14\,H^+ \rightarrow 2\,MnO_4^- + 5\,Na^+ + 5\,Bi^{3+} + 7\,H_2O$$

The excess bismuthate is filtered off, and a measured excess of ferrous sulfate is added to the filtrate to reduce permanganate to manganous salt:

$$MnO_4^- + 5\,Fe^{++} + 8\,H^+ \rightarrow Mn^{++} + 5\,Fe^{3+} + 4\,H_2O$$

The excess of ferrous iron is then determined by titration with standard potassium permanganate solution (p. 828).

[5] Also see J. J. Lingane and J. W. Collat, *Anal. Chem.* **22**, 166 (1950), for simultaneous determination of Cr(VI) and Mn(VII) by absorbance measurements at 440 and 545 mμ. This method is discussed in chapter 61.

[6] W. Blum, *J. Am. Chem. Soc.* **34**, 1395 (1912); G. E. F. Lundell, *ibid.* **45**, 2600 (1923); T. R. Cunningham and R. W. Coltman, *Ind. Eng. Chem.* **16**, 58 (1924).

The oxidation of manganese by sodium bismuthate can be carried out in nitric or sulfuric acid solution. The acidity must be regulated, the concentration of manganese must not be too high, and a sufficient excess of bismuthate must be added to obtain correct results. Hydrochloric and hydrofluoric acids must be absent; moderate amounts of phosphoric acid do not interfere. The nitric acid used in the procedure must be free of nitrous acid. The method is not applicable to samples containing cobalt and cerium. The presence of chromium and vanadium is undesirable; but, by adopting certain precautions, satisfactory results can be obtained when not too large amounts of these are present. Sodium bismuthate will oxidize chromium to some extent; moreover, the permanganate formed will oxidize chromium slightly. The speed of both these reactions is decreased by cooling the solution, so that by working without undue slowness in cold solution (10°C) satisfactory results can be obtained when 1 per cent or so of chromium is present in the sample. Vanadium interferes slightly, because it is oxidized rather slowly by potassium permanganate after reduction by ferrous sulfate, thus causing a fleeting end point.

Reagents and standard solutions

Sodium bismuthate. Available oxygen should correspond to at least 75 per cent $NaBiO_3$. Chloride and manganese should be absent.

1:30 Nitric acid free of nitrous acid. Boil concentrated nitric acid (hood) for 1 or 2 min, cool, and dilute with water.

Potassium permanganate, 0.03 N. Dissolve 1 g of potassium permanganate in 1 liter of water, allow to stand for several days, and filter through asbestos. Keep the permanganate solution in a glass-stoppered bottle. Standardize the solution by weighing out 0.1 g of pure dried sodium oxalate, dissolving in 100 ml of hot water containing 15 ml of 6 *N* (1 : 5) sulfuric acid, and titrating slowly with continuous stirring (p. 825).

Ferrous ammonium sulfate, 0.03 N. Dissolve 12 g of $FeSO_4 \cdot (NH_4)_2SO_4 \cdot 6\,H_2O$ min 1 liter of 1 : 20 sulfuric acid.

Procedure. Weigh a 1-g sample (Mn = 0.3–1.0 per cent) into a 250-ml beaker, and dissolve in 50 ml of 1 : 3 nitric acid. Boil for a few minutes, remove the beaker from the heat, add 0.5 g of sodium bismuthate, and again boil for a few minutes. A precipitate of manganese dioxide or a pink color due to permanganate should appear; if neither a precipitate nor a color is produced, add more bismuthate. Then add dropwise a strong solution of sodium sulfite (or sulfurous acid) until the precipitate or permanganate coloration disappears. Boil for 4 or 5 minutes. Cool the solution to approximately 15°C, and add 0.5–1 g of sodium bismuthate; some of the reagent should remain undissolved. Stir for 1–2 min, add 50 ml of cold 1 : 30 nitric acid, and filter through a Gooch crucible (or a sintered-glass crucible) with the aid of suction. Of course, bismuthate must not run through the filter. Wash the residue well with dilute nitric acid. Add 3 ml of 85 per cent phosphoric acid to the filtrate, and slowly run in ferrous ammonium sulfate solution from a buret or pipet until a slight excess is present as indicated by the disappearance of the color due to permanganate. Then titrate the solution with standard permanganate to the first pink tinge (p. 834).

Standardize the ferrous ammonium sulfate solution by running a blank through all the steps of the procedure, taking the same volume of ferrous ammonium sulfate solution as used in the determination itself. Report the percentage of manganese.

Notes:

1. The solution of the sample is subjected to a preliminary oxidation with sodium bismuthate to destroy any reducing substances (carbonaceous matter) present that might later react slowly with permanganate.

2. The concentration of nitric acid during the oxidation should be at least 10 per cent and preferably 20 per cent by volume. As much as 50 mg of manganese can be present in 100 ml if the oxidized solution is filtered without undue delay. Approximately 0.25 g of 80 per cent sodium bismuthate is needed to oxidize 10 mg of manganese quantitatively.

3. The oxidation can be made at room temperature (up to 25°C) in the absence of chromium, but it is best to cool the solution to 10–15°C. The oxidation is complete, if the solution is stirred, in 1 min, provided the conditions mentioned above are fulfilled. Slightly longer standing does no harm.

4. The residue of sodium bismuthate should be thoroughly washed with dilute nitric acid, because it shows a tendency to retain permanganate, especially if a high concentration is present.

5. The filtrate should not be allowed to stand for any length of time before titration, because the permanganic acid may decompose.

6. The bismuthate method can be applied for the determination of manganese in spiegeleisen (16–21 per cent Mn) and ferromanganese ($\sim$80–85 per cent Mn).

Other methods. With silver ion as catalyst, manganous ion can be oxidized to permanganate by alkali persulfates in hot acid solution:

$$2\,Mn^{++} + 5\,S_2O_8^{--} + 8\,H_2O \rightarrow 2\,MnO_4^- + 10\,SO_4^{--} + 16\,H^+$$

In the presence of phosphoric acid, as much as 100 mg of manganese can be smoothly oxidized in this way.[7] The permanganate thus formed can be titrated with sodium arsenite solution. The reaction between permanganate and arsenite does not proceed stoichiometrically, and the titer of the latter solution must therefore be found by titrating a known amount of permanganate under the same conditions as the unknown solution. Heptavalent manganese is reduced by arsenite to a valence state approximately represented by $Mn^{3.3+}$, which has a yellow to brown color in solution. The end point of the titration is not very sharp, especially when comparatively large amounts of manganese are present, but the method has the advantage of being very rapid. The end point can be found more precisely potentiometrically or photometrically. It is also possible to use a solution containing sodium arsenite and sodium nitrite in equivalent amounts for the titration.[8] In this case the permanganate is reduced nearly stoichiometrically to manganous ion (if silver is first precipitated with chloride), and the color change is from pink to colorless at the end point, so that the latter can be found more exactly than when sodium arsenite alone is used. The mixed reducing solution must be standardized against a known amount of manganese oxidized under the same conditions as the sample.

In neutral pyrophosphate solution, Mn(II) can be stoichiometrically oxidized to Mn(III) by Mn(VII):

$$4\,Mn^{++} + MnO_4^- + 8\,H^+ + 15\,H_2P_2O_7^{--} \rightarrow 5\,Mn(H_2P_2O_7)_3^{3-} + 4\,H_2O$$

This reaction has been made the basis of a volumetric method for manganese over a wide range of concentrations.[9] It is remarkably free of interference. The end point must be found potentiometrically because of the red-violet color of $Mn(H_2P_2O_7)_3^{3-}$.

Phosphorus

The magnesia method (see below), in which phosphorus is precipitated as ammonium phosphomolybdate and finally weighed as magnesium pyrophosphate, is usually considered the most accurate and most generally applicable method for phosphorus in steels and ferro alloys. However, it is a lengthy method, so that it is not often used. Less accurate but more rapid methods based on the precipitation of ammonium phosphomolybdate are given on page 1126.

[7] H. A. Bright and C. P. Larabee, *Natl. Bur. Std. J. Res.* **3**, 573 (1929). Another meritorious oxidant (+2Mn $\rightarrow$ +7Mn) is argentic oxide, AgO, the excess of which is destroyed by heating the solution to 50°C for a few minutes, as shown by J. J. Lingane and D. Davis, *Anal. Chim. Acta* **15**, 201 (1956).

[8] E. B. Sandell, I. M. Kolthoff, and J. J. Lingane, *Ind. Eng. Chem., Anal. Ed.* **7**, 256 (1935).

[9] J. J. Lingane and R. Karplus, *Ind. Eng. Chem., Anal. Ed.* **18**, 191 (1946).

A colorimetric method (such as that described below) is more rapid than any of these and is likely to be as accurate (if not more so) for the amounts of phosphorus (<0.1 per cent) usually encountered in ferrous materials.

Phosphorus occurs in steel as phosphide and an oxidizing acid (usually nitric) is used to dissolve the sample, thus preventing loss of phosphorus as phosphine and converting it to orthophosphate, the form required in all the methods.

Especially when small amounts of phosphorus are being determined, contamination of glassware by detergents should be guarded against. Many detergents contain phosphates. These should not be used in cleaning glassware used in phosphorus determinations. Glass vessels that may have been cleaned with such detergents should be boiled in 1:1 hydrochloric acid and rinsed carefully.

Colorimetric method. All the colorimetric methods mentioned on page 648 have been applied for the determination of phosphorus in steel. The following procedure (the choice is somewhat arbitrary) is of the Mo–P blue type.[10] For carbon steels, the average error is about 0.002 per cent in the range 0.005–0.1 per cent P. As much as 35 per cent Ni, 15 per cent Cu, 5 per cent V, and 2 per cent Cr may be present in the sample without causing an error greater than 0.002 per cent P.

Special solutions

Sodium sulfite, 100 g of the anhydrous salt in 1 liter of solution. Filter if not clear.

Ammonium molybdate, 20 g of $(NH_4)_6Mo_7O_{24} \cdot 4\ H_2O$ in 1 liter of 10 *N* sulfuric acid.

Hydrazine sulfate, 1.5 g of $N_2H_4 \cdot H_2SO_4$ in 1 liter of water.

Ammonium molybdate–hydrazine sulfate. Dilute 25 ml of the ammonium molybdate solution to 80 ml with water, add 10 ml of hydrazine sulfate solution, and dilute to 100 ml. This solution is unstable and should be prepared immediately before use.

Procedure. Dissolve 0.5 g of steel (0.005–0.1 per cent P) in 50 ml of 1 : 2 nitric acid with heating. Cool the solution, transfer to a 100-ml volumetric flask, dilute to the mark with water, and mix. Transfer two 10-ml aliquots of this solution to 125-ml conical flasks, add 3 ml of perchloric acid, evaporate to fuming, and continue gentle fuming for 3–5 min to expel all nitric acid.[11] Add 10 ml of water and 15 ml of sulfite solution to the cooled solution. Heat to boiling and boil for ½ min. Set one of the aliquots aside (for preparing reference solution). To the other add 20 ml of ammonium molybdate–hydrazine sulfate reagent solution. Heat to about 90°C and maintain at this temperature for 5 min. Then heat the solution just to the boiling point. Cool rapidly to room temperature, transfer the solution to a 50-ml volumetric flask, and dilute to the mark with water. Measure the absorbance at 830 mμ in a 1-cm cell against the reference solution obtained by diluting the reserved treated aliquot to 50 ml in a volumetric flask after adding 5 ml of 10 *N* sulfuric acid. The color is stable for 12 hr or more.

Establish the standard curve by taking steel samples of known P content through the procedure. If necessary, two samples of known P content (one having low P) can be used to establish the curve (straight line), since absorbance is proportional to the phosphorus concentration.

Methods based on the precipitation of ammonium phosphomolybdate. The precipitate $(NH_4)_3PO_4 \cdot 12\ MoO_3 \cdot aq$ can be treated in a number of ways to determine phosphorus, the more important of which are:

1. Solution in ammonium hydroxide, and precipitation of magnesium ammonium phosphate from this solution by the addition of magnesia mixture, followed by ignition to magnesium pyrophosphate (p. 645).

2. (a) Drying at 110°C and weighing as $(NH_4)_3PO_4 \cdot 12\ MoO_3$ (p. 647), or (b) heating to 450°C and weighing as $P_2O_5 \cdot 24\ MoO_3$ (p. 647).

3. Dissolving in a measured volume of standard base and titrating the excess of the latter with standard acid.

[10] J. L. Hague and H. A. Bright, *J. Res. Natl. Bur. Std.* **26**, 405 (1941). Also see N. Lounamaa and W. Fugmann, *Z. Anal. Chem.* **199**, 352 (1965).

[11] If As exceeds 0.05 per cent, remove it at this point by adding 5 ml of 1:4 hydrobromic acid to the cooled solution, evaporating to fumes, and fuming gently for ~5 min.

Procedures for these methods follow.

Magnesia method. This method for phosphorus has already been considered at length in Chapter 30.

Reagent solutions

Ammonium molybdate reagent. Make up two solutions:

1. Dissolve 100 g of ammonium molybdate in a mixture of 400 ml of water and 80 ml of concentrated ammonium hydroxide.

2. Dilute 400 ml of concentrated nitric acid with 600 ml of water. Immediately before use, slowly pour a suitable volume of solution 1 into twice its volume of solution 2.

Magnesia mixture. Dissolve 50 g of magnesium chloride hexahydrate and 100 g of ammonium chloride in 500 ml of water. Make ammoniacal, let stand overnight, filter off any precipitate formed, make the filtrate very slightly acidic with hydrochloric acid, and dilute to 1 liter.

Procedure. Weigh out 3 g of steel into a 250-ml Erlenmeyer flask, add 75 ml of 6 N (3 : 5) nitric acid, and warm gently to dissolve. Then add 10 ml of 2 per cent potassium permanganate solution and boil for a few minutes to precipitate manganese dioxide; add more permanganate if no precipitate appears. Dissolve the precipitate by adding dropwise a solution of sodium sulfite. Boil again for a few minutes to expel oxides of nitrogen and sulfur dioxide. Cool the solution to approximately 50°C; add 35 ml of 6 N (3 : 4) ammonium hydroxide[12] and 75 ml of molybdate reagent. Stopper the flask and shake for 5 or 10 min. Wash off the stopper and set the flask aside for at least 2 hr, or, better, overnight.

Filter off the precipitate on an S and S No. 589 Blue Band (or equivalent) filter paper, wash with 1 : 100 nitric acid two or three times, and then five or six times with 5 per cent ammonium nitrate solution. The filtrate and washings should be mixed and reserved to see if further precipitation of ammonium phosphomolybdate occurs. Dissolve the precipitate in 20 ml of 6 N ammonium hydroxide containing 2 g of citric acid, catching the solution in a 250-ml beaker. Wash the filter paper three or four times with 1 : 20 ammonium hydroxide, then an equal number of times with hot water, and finally a few times with 1 : 20 hydrochloric acid. (If the ammoniacal solution of the precipitate is not clear at this point, see note 3.) Add a few drops of methyl red, and acidify the solution with hydrochloric acid. Next add 20 ml of magnesia reagent, cool the solution in ice water, and make alkaline with ammonium hydroxide. Stir for 5 min; or if no precipitate appears, set the solution aside for 1 hr or so. Finally add 10 ml of concentrated ammonium hydroxide, and allow the solution to stand overnight in a cool place.

Filter on paper, and wash the precipitate three times with 1 : 20 ammonium hydroxide. Place the original beaker under the funnel, and dissolve the precipitate in about 20 ml of 6 N hydrochloric acid. Carefully wash the filter paper with hot 6 N hydrochloric acid. Add 0.5–1 g of ammonium bromide to the combined filtrate and washings, and boil down the the solution to a volume of 5–10 ml but not further. Dilute to 50–75 ml; add 0.1 g of citric acid and 2–3 ml of magnesia reagent. After cooling the solution in ice water, make ammoniacal, stir, and allow to stand overnight before filtering.

Collect the precipitate on paper. Wash with 1 : 20 ammonium hydroxide, transfer the paper to a crucible, char, and burn off the carbon at the lowest possible temperature. Then ignite to constant weight at 1050–1100°C. Test the ignited residue for purity by adding 5 ml of 1 : 1 nitric acid and 20 ml of water. If there is a residue, filter it off, wash it with hot water, ignite in a platinum crucible, and weigh. Pour a few drops of hydrofluoric acid into the crucible, evaporate to dryness, ignite, and again weigh. Subtract the loss in weight, if any is found, from the weight of the impure magnesium pyrophosphate. Report the percentage of phosphorus.

[12] On the addition of the ammonia, the solution should turn dark reddish brown, but no permanent precipitate of hydrous ferric oxide should form (too much ammonia). Add the ammonium hydroxide slowly with shaking.

Notes:

1. The above procedure is applicable to plain carbon steels and many alloy steels. The directions must be modified for the analysis of cast iron, chrome–vanadium steels, tungsten steels, etc. (see p. 644 for interfering substances in the phosphomolybdate precipitation).

2. Citric acid is added to the ammoniacal solution of ammonium phosphomolybdate to prevent the contamination of the magnesium ammonium phosphate precipitate by such elements as iron, vanadium, titanium, and zirconium that may be present in the yellow precipitate.

3. If appreciable amounts of titanium and zirconium are present, the phosphates of these elements will remain on the filter paper after treatment of the yellow precipitate with ammonium hydroxide, some running through to give a turbid filtrate. If the solution is not clear at this point, it must be refiltered after heating to boiling, the filter paper washed with hot water and ignited in a platinum crucible, the residue fused with a little sodium carbonate, the melt extracted with hot water, and the extract filtered and added to the ammoniacal solution of the ammonium phosphomolybdate. In this way the phosphorus in the insoluble phosphate is converted into sodium phosphate.

4. If arsenic is present, as it usually is in small amounts in steel, it is precipitated by the magnesia reagent and will be weighed with the magnesium pyrophosphate if it is not removed prior to the final precipitation. Boiling the magnesium ammonium phosphate solution in hydrochloric acid after addition of ammonium bromide reduces arsenate to arsenite and volatilizes arsenic as arsenious chloride.

5. Treatment of the ignited residue of magnesium pyrophosphate with acid followed by hydrofluorization of any insoluble matter serves to correct for any silica present in the weighed magnesium pyrophosphate.

6. It is to be noted that variations in the composition of the ammonium phosphomolybdate precipitate do not affect the result for phosphorus in this method.

Weighing phosphorus as $P_2O_5 \cdot 24\,MoO_3$. As stated above, ammonium phosphomolybdate can be dried at 110°C to $(NH_4)_3PO_4 \cdot 12\,MoO_3$, or ignited to $P_2O_5 \cdot 24\,MoO_3$ at 450°C; the latter weighing form is the better. It is convenient to use this method in place of the alkalimetric (see below) when only a few determinations are occasionally made. The small factor in this method is of distinct advantage. Concerning the accuracy of this method, see page 648.

Procedure. Precipitate ammonium phosphomolybdate as described in the magnesia method,[13] and collect the precipitate in a weighed porcelain filter crucible or a Gooch crucible that has been heated to 400–450°C. Wash the precipitate thoroughly with 5 per cent ammonium nitrate solution. Heat the precipitate in an electric furnace at 400–450°C for 1 hr. Repeat the heating to constant weight. If a furnace is not available place the filter crucible inside a porcelain crucible, and heat the latter to low redness over a burner. The residue should finally have a bluish black color, with no admixture of yellow (undecomposed ammonium phosphomolybdate) or white (indicating too high a temperature of ignition). The residue is hygroscopic.

Alkalimetric method. The ammonium phosphomolybdate is dissolved in a known volume of standard sodium hydroxide solution, and the excess of the latter is back-titrated with standard acid, phenolphthalein being used as indicator. From the volume of sodium hydroxide solution required to react with the precipitate, the amount of phosphorus is calculated on the basis of the equation

$$(NH_4)_3PO_4 \cdot 12\,MoO_3 + 23\,NaOH \rightarrow 11\,Na_2MoO_4 + (NH_4)_2MoO_4 + NaNH_4HPO_4 + 11\,H_2O$$

It will be seen that the calculation is based upon a precipitate of ideal composition, i.e., one in which the ratio of phosphorus to molybdic oxide is 1 : 12 and which contains no nitric acid.

[13] Any insoluble material in the solution must be filtered off before the molybdate reagent is added.

Standard solutions

0.1 N sodium hydroxide. Dilute concentrated carbonate-free sodium hydroxide with equilibrium water to give an approximately 0.1 N solution (p. 781). Standardize against potassium biphthalate, using phenolphthalein as indicator (p. 783).

0.1 N nitric acid. Dilute 7 ml of concentrated nitric acid to 1 liter. Standardize the solution by running a ratio against the standard sodium hydroxide solution, using phenolphthalein as indicator.

Procedure. Weigh into a 250-ml Erlenmeyer flask a 1- to 3-g sample of steel, depending upon the phosphorus content, dissolve in nitric acid and treat with potassium permanganate and sulfite as described under the magnesia method. After expulsion of sulfur dioxide, cool the solution to 30°C, add 35 ml of 6 N ammonium hydroxide[14] and 40 ml of molybdate reagent. Stopper the flask, and shake the contents for 5–10 min. Allow the precipitate to settle, and decant the supernatant liquid through an S and S No. 589 Blue Band (or equivalent) filter paper. Wash the flask, precipitate, and paper twice with 5-ml portions of 1 : 100 nitric acid and then thoroughly with a 1 per cent solution of potassium nitrate until the washings give no acid reaction with methyl orange. Place the paper in the original flask, add approximately 20 ml of 0.1 N sodium hydroxide, and shake to dissolve the precipitate. If the precipitate does not entirely dissolve, add more sodium hydroxide solution and shake again until all is in solution. Then add 50 ml of equilibrium water and 5 drops of 0.1 per cent phenolphthalein solution, and titrate with 0.1 N nitric acid, adding a few drops in excess. Finish the titration by adding sodium hydroxide to the first faint pink color.

A blank should be run if the stoichiometric titer of the sodium hydroxide is to be used. The phosphorus titer of the sodium hydroxide solution can also be found empirically by standardizing against a steel (similar to the sample) of known phosphorus content.

Notes:

1. The above directions apply to plain carbon steels and some alloy steels. Alloy steels mentioned in note 1, under magnesia method, require slightly different treatment.

2. It is highly important that the ammonium phosphomolybdate precipitate obtained in the alkalimetric procedure have the proper ratio of phosphorus to molybdic oxide. For this reason, particular attention must be paid to the temperature of precipitation; if the temperature is too high, the precipitate will be enriched in molybdic oxide, and high results will be obtained for phosphorus. It is best to keep the temperature below 45°C during the precipitation. Moreover, at higher temperatures the contamination of the precipitate by vanadium and arsenic is greater.

In the presence of the interfering elements mentioned under the magnesia method, a longer period of precipitation must be used and more molybdate added.

3. The end point in the titration with phenolphthalein as indicator is not very sharp. At a pH of about 8 the molybdate is all present in the form of MoO_4^{--}; upon further addition of acid the pH decreases slowly with the formation of complex acid molybdates. For this reason phenolphthalein must be used as indicator. At the end point most of the phosphate is present in the form of HPO_4^{--}. The presence of the ammonium ion causes the end point to be vague. To neutralize all the ammonia to ammonium ion, methyl red should be used a indicator; however, under these conditions a large portion of the molybdate would be transformed into the acid complexes. Since the amount of ammonia is small compared with concentration of molybdate (see the equation), only a small error is introduced by using phenolphthalein as indicator.

Sulfur

The gravimetric determination of sulfur as barium sulfate is probably, on the whole, the most accurate method for sulfur in steel, but is so laborious that it is avoided when possible.

[14] A permanent precipitate of hydrous ferric oxide must not form.

In a rapid routine method, the procedure for which will not be given here, the sample is heated to 1425–1525°C (induction furnace) in oxygen to form SO_2 which is absorbed in $KI–I_2$ solution. In an automated commercial form of the apparatus, standard KIO_3 solution is automatically added at a rate to keep pace with the consumption of iodine in the solution as SO_2 is absorbed during the combustion. Starch is used as indicator and a photoelectric cell regulates the flow of standard KIO_3 solution: as the blue color fades as a result of the reaction of SO_2 with I_2, the cell actuates a mechanism releasing KIO_3 solution until the end point is reached. The buret is graduated to read the percentage of S. The cost of a unit is moderate (~$800 in 1962) so that when sulfur determinations are being run steadily, the saving in labor costs soon pays for the instrument. For some alloy steels the recovery of sulfur is not complete, and empirical calibration with known samples is required.

Another rapid titrimetric method, not always applicable, is based on evolution of H_2S (see below).

Gravimetric method: weighing as barium sulfate. The sample is dissolved in concentrated nitric acid; sulfur, which exists in ferrous materials as sulfide, is thereby oxidized to sulfate:

$$3\,FeS + 12\,HNO_3 \rightarrow Fe(NO_3)_3 + Fe_2(SO_4)_3 + 9\,NO + 6\,H_2O$$

Since nitrates and ferric iron are coprecipitated to a large extent by barium sulfate (p. 603), it is necessary to remove or destroy these ions. Nitrate is removed by evaporating the solution with hydrochloric acid. Ferric iron is reduced to ferrous by treatment with zinc in acid solution, and the excess of zinc, together with the silica, is filtered off. In the filtrate, sulfate is precipitated by adding barium chloride; the barium sulfate is filtered off, washed, ignited, and finally weighed.

Procedure. Dissolve 5 g of sample in 50 ml of concentrated nitric acid in a covered 250-ml Erlenmeyer flask. If solution is slow, 5 ml of hydrochloric acid may be added, a few drops at a time. When the sample has dissolved, add 0.5 g of analytical-grade sodium chloride, evaporate to dryness, and bake on the hot plate for 10 or 15 min. Cool, add 30 ml of concentrated hydrochloric acid, evaporate to dryness, and bake as before. Again add 30 ml of hydrochloric acid, and evaporate the solution to a syrupy consistency. Now add 5 ml of hydrochloric acid, 25 ml of water, and 5 g of 20- to 30-mesh sulfur-free zinc. Place the flask on the steam bath and warm until all ferric iron has been reduced and there is hardly any evolution of hydrogen. Filter the solution through a small filter, catching the filtrate in a 250-ml beaker. Wash with 50–75 ml of 1:50 hydrochloric acid. Warm the solution to 70°C, and precipitate sulfate by adding 10 ml of 10 per cent barium chloride solution. Allow the solution to stand at least 18 hr. Filter off the precipitate on a dense filter paper and wash first with a solution containing 10 ml of hydrochloric acid and 10 ml of 10 per cent barium chloride in 1 liter of water to remove iron, and then with cold water to remove chlorides (test with silver nitrate). Not more than 50–75 ml of water should be used for the last washing. Examine the filtrate and washings carefully for barium sulfate that might have run through the paper. Ignite the precipitate to constant weight in a weighed crucible over a Tirrill burner. For the best results use a platinum crucible, and treat the ignited residue with 1 drop of 12 N sulfuric acid and 0.5–1 ml of hydrofluoric acid, evaporate to dryness, ignite again, and then weigh.

It is very important to run a blank determination, carrying it through all the steps of the determination itself. Report the percentage of sulfur found after applying the correction obtained in the blank.

Notes:

1. The above procedure can be used for most alloy steels as well as for cast iron; tungsten steels require special treatment.

2. Ferrous iron and zinc are not carried down to any serious extent by barium sulfate. The precipitate of barium sulfate obtained in this procedure is never entirely

pure, containing, for example, chloride and other substances as well. Certain constituents of alloy steels may contaminate the precipitate (chromium, tungsten, etc.); when these are present, special steps must be taken to prevent their interference. Contamination of the precipitate by common contaminants (chloride) is not serious because of the relatively small amounts of sulfur present.

3. The precipitate must first be washed with a dilute solution of hydrochloric acid to prevent the hydrolysis of the heavy-metal salts present. Since barium sulfate is appreciably soluble in dilute acid solutions, even when barium chloride is present, as it is in the first wash solution, only the smallest volume required should be used. In very accurate work the barium sulfate dissolved by the wash solutions is recovered by evaporating the latter to dryness, taking up the residue in a small volume of water containing a little barium chloride and hydrochloric acid, filtering, washing, and igniting. According to Lundell et al. the barium sulfate thus recovered is equivalent to 0.003 per cent or less of sulfur on 4.57 g (factor weight) of sample.

4. The use of gas for heating should be avoided in the procedure (except in the ignition of the barium sulfate) to prevent the introduction of sulfur.

5. See N. Lounamaa and W. Fugmann, *Z. Anal. Chem. 199*, 98 (1965), for a method in which sulfate is adsorbed on alumina, eluted with ammonia, and determined by addition of an excess of barium which is back-titrated with EGTA.

H_2S evolution method. This method involves the evolution of hydrogen sulfide by solution of the sample in hydrochloric acid, the absorption of the hydrogen sulfide in an ammoniacal solution of zinc sulfate or cadmium chloride with the formation of zinc or cadmium sulfide, and the titration of the sulfide in an acid solution with a standard solution of iodine. A solution of potassium iodate containing an excess of potassium iodide is advantageously used as the standard iodine solution; in acid solution iodate and iodide react to give iodine, which oxidizes the sulfide to sulfur:

$$IO_3^- + 5\,I^- + 6\,H^+ \rightarrow 3\,I_2 + 3\,H_2O$$

$$H_2S + I_2 \rightarrow S + 2\,I^- + 2\,H^+$$

Starch is used to indicate the end point.

It will be seen that this method of determining sulfur is based upon the assumption that the sulfur exists in the sample entirely in the form of sulfide and that all the sulfur is evolved as hydrogen sulfide on treatment with acid. Since these conditions may not be fulfilled, the method is not as accurate as the gravimetric method; but because of its rapidity it is very often used, especially in control laboratories where the nature of the sample is known. The accuracy of the method can be increased by standardizing the iodine solution against a steel of known sulfur content as much as possible like the sample analyzed in its characteristics.

Apparatus: A 250-ml Florence flask with a two-hole rubber stopper carrying (1) a thistle tube extending nearly to the bottom of the flask and (2) a spray trap to which is connected a delivery tube dipping nearly to the bottom of a 250-ml Erlenmeyer flask.

Solutions: Ammoniacal zinc sulfate. Dissolve 20 g of zinc sulfate heptahydrate in 100 ml of water, and add 100 ml of concentrated ammonium hydroxide. Filter off any precipitate formed on standing.

Potassium iodate–potassium iodide. Dissolve 1.1 g of pure dry potassium iodate, weighed to the nearest milligram, and 12 g of potassium iodide in exactly 1 liter of water. Calculate the sulfur titer of the solution from the amount of potassium iodate weighed out.

Starch, 0.2 per cent (see p. 845). The solution must give a sharp color change with iodine.

Procedure. Place 5 g of steel in the flask, replace the stopper, and dip the delivery tube into the Erlenmeyer containing 100 ml of water and 10 ml of ammoniacal zinc sulfate solution. Pour 80 ml of 1 : 1 hydrochloric acid into the flask through the thistle tube, and heat with a flame so that there is a rapid steady evolution of gas. Boil for 1 min after the sample has dissolved, and then disconnect the delivery tube, leaving it in the beaker for a stirrer. Cool the solution to room temperature if necessary, add 5 ml of starch solution and 40 ml of 1 : 1 hydrochloric acid, and titrate immediately with the potassium iodate–iodide solution to the first permanent blue color.

Notes:

1. Since the method is more or less empirical, the directions must be followed exactly. Thus, it is not permissible to take a sample that weighs much more or less than 5 g, for then the percentage of sulfur found will be appreciably different.

2. The following table contains some results for the determination of sulfur in Bureau of Standards irons and steels by the oxidation and the evolution method as given in the Bureau's certificates:

	Percentage of Sulfur	
Material	*Weighing as* $BaSO_4$	*Evolution as* H_2S
Cast iron No. 4d	0.075	0.072
Bessemer steel No. 8d	0.083	0.084
B.O.H. steel No. 11d	0.041	0.041
Cr–V steel No. 30c	0.014	0.013
Ni steel No. 33b	0.032	0.031

Most cast irons and some alloy steels give erroneous (low) results by the evolution method. It fails for steels containing Ti, Zr, and Se.

Silicon

The steel or iron is dissolved in sulfuric, perchloric, or hydrochloric acid (sometimes with the aid of an oxidizing agent such as nitric acid), and the solution is evaporated to fumes (sulfuric acid or perchloric acid) or to dryness (hydrochloric acid) to dehydrate the silica and render it insoluble. Water or acid is then added to dissolve soluble salts, and the silica is filtered off. Single dehydrations never yield all the silica; a second dehydration recovers nearly all the remainder. However, in all but the most accurate work a single evaporation will suffice, provided the percentage of silicon in the material is not greater than 0.5. The silica thus obtained is not pure. It is ignited at a high temperature and weighed as anhydrous silica plus contaminants, then treated with hydrofluoric acid in the usual way to volatilize silica, and the residue weighed. The loss in weight gives the amount of silica (p. 657).

Procedure (sulfuric acid method). Transfer a 5-g sample to a 300-ml casserole and dissolve in 125 ml of 1 : 5 sulfuric acid. Solution is aided by gentle warming. (The determination may be made more rapidly by using dilute hydrochloric acid to dissolve the sample and then adding sulfuric acid.) Evaporate the solution to fumes of sulfuric acid, taking precautions against spattering. When dense fumes have been given off for about 3 min, remove the casserole from the heat and allow to cool somewhat. Then add all at once 100 ml of water warmed to 40°C. Stir to bring salts into solution, heating somewhat if necessary but not to boiling. Without delay filter off the silica on a rapid paper (S and S No. 589 Black Band or equivalent). With a policeman dislodge the silica adhering to the walls of the casserole. Wash the precipitate with 1 : 20 hydrochloric acid until the iron salts have been washed out (about six times will suffice), and then a few times with hot water. Place the paper and precipitate in a platinum crucible, heat slowly to char the paper and burn off the carbon, and then apply the full flame of a Meker burner. If an electric muffle is used, regulate the temperature to 1050–1150°C. When constant weight has been attained, add a few drops of 1 : 1 sulfuric acid to the residue and then approximately 5 ml of hydrofluoric acid. Evaporate the hydrofluoric acid, and fume off the sulfuric acid in a radiator. Ignite the residue strongly and weigh again. In very accurate work a blank should be run.

Report the percentage of silicon.

Notes:

1. A 5-g sample is suitable for steels containing less than 0.5 per cent silicon; for higher silicon contents a smaller sample should be taken. For cast iron a 1-g sample should be used. Alloy steels can be analyzed by the above procedure if they dissolve in sulfuric or hydrochloric acid, else nitric acid with or without acid must be used (see Tungsten).

2. Silicon exists as iron silicide in steels and in cast iron. There is no danger of losing silicon as the hydride by dissolving the sample in a nonoxidizing acid.

3. The following table indicates the results obtained for silicon in steels by dehydration with various acids.[15]

Silicon Obtained in Single Evaporations

Steel	*Percentage of Silicon Present*	*Percentage of Silicon by Several Methods*			
		H_2SO_4	HNO_3	$HClO_4$	HCl
Silicon steel	4.70	4.66	4.60	4.67	4.63
Acid electric steel	0.450	0.435	0.431	0.436	0.428
A.O.H. steel	0.112	0.105	0.105	0.106	0.103

The chief advantage in the use of perchloric acid lies in the easy solubility of the anhydrous perchlorates. Perchloric acid is profitably used for dehydration when appreciable amounts of chromium or nickel are present in the sample; these metals yield difficultly soluble sulfates when sulfuric acid is used for dehydration.

4. The ignited silica always contains greater or lesser amounts of foreign substances, depending upon the constituents of the steel and the method used. Among the possible contaminants the following may be mentioned: aluminum oxide, titanium oxide, titanium phosphate, and iron oxide. The weight of the silica must be obtained by hydrofluorizing by the usual procedure. In the presence of phosphorus the sulfuric acid must be expelled at as low a temperature as possible after evaporation of the hydrofluoric acid to avoid loss of phosphorus. When tungsten is present in the hydrofluorized residue, the final ignition must be made below 800°C to prevent volatilization of tungstic oxide.

5. Photometric methods for silicon are available. See, for example, R. T. Andrews and C. H. R. Gentry, *Analyst*, **81**, 339 (1956).

Nickel

Nickel can be determined gravimetrically or colorimetrically with dimethylglyoxime (p. 277). Only the gravimetric determination will be considered here. An acid solution containing iron in the ferric state is treated with tartaric acid and an alcoholic solution of dimethylglyoxime. The addition of a slight excess of ammonia precipitates nickel quantitatively as nickel dimethylglyoximate, which is washed and weighed as such after drying at 100–120°C. The presence of tartrate is required to prevent the precipitation of ferric iron in ammoniacal solutions; ferric iron forms a complex with tartrate from which iron cannot be precipitated as hydroxide by bases. Unnecessarily large amounts of the reagent should be avoided, because dimethylglyoxime is not very

[15] R. M. Fowler, *Ind. Eng. Chem., Anal. Ed.* **4**, 382 (1932).

soluble in water (or dilute alcohol) and may separate; if very much reagent is added, the solubility of the precipitate may become perceptible as a consequence of the alcohol introduced.

The following directions apply to simple nickel steels. For the determination of nickel in steel containing cobalt and more than approximately 0.3 per cent of copper, see the careful study of A. Claassen and L. Bastings, *Analyst* **91**, 725 (1966).

Procedure. Weigh out a 1-g sample (for ~ 1 per cent Ni) into a 400-ml beaker and dissolve in 50 ml of 1 : 1 hydrochloric acid. Heat, if necessary, to hasten solution. Cautiously add 10 ml of 1 : 1 nitric acid and boil for a few minutes to expel oxides of nitrogen. Dilute to 200 ml and add 7 g of tartaric acid. When the tartaric acid has dissolved, neutralize with concentrated ammonia and add approximately 1 ml in excess. If any insoluble material is visible in the solution, filter it off on paper, and wash the latter with hot water. Make the solution slightly acid with hydrochloric acid, heat to about 80°C, and add 20 ml of a 1 per cent alcoholic solution of dimethylglyoxime. Add ammonia until the liquid is slightly ammoniacal, and set the solution aside for 1 hr (or overnight if the amount of nickel is low, e.g., 0.1 per cent). Filter through a weighed filter crucible, and wash the precipitate with hot water until the washings give no reaction for chloride when tested with nitric acid and silver nitrate. Test the filtrate for complete precipitation by adding a little dimethylglyoxime solution. Dry the precipitate by heating in an oven at 100–120°C for 1 hr. Repeat the heating to constant weight. The dried precipitate contains 20.32 per cent nickel. Report the percentage of nickel in the steel.

Chromium

Nearly all methods for the determination of chromium in steel are based upon the oxidation of Cr(III) to Cr(VI) and the subsequent determination of the latter by volumetric methods (or colorimetrically if the amount is small). Volumetrically, Cr(VI) is determined by adding a measured amount of a reducing agent, usually ferrous sulfate, and back-titrating the excess of the reducing solution with permanganate, or by direct potentiometric titration with ferrous sulfate; in either case Cr(VI) is reduced to Cr(III).

Cr(III) can be oxidized to Cr(VI) in hot acid solution (the medium in which the oxidation is usually effected) by a number of oxidizing agents, of which the following may be mentioned: alkali persulfates in the presence of silver nitrate, perchloric acid (complete oxidation difficult), potassium permanganate, and potassium bromate. The excess of persulfate can be destroyed by boiling, and the excess of perchloric acid is rendered nonoxidizing by simple dilution with water. Vanadium, if present, is oxidized to vanadate, VO_4^{3-}. Vanadate is reduced to the vanadyl state, VO^{++}, by ferrous sulfate, but vanadyl ion is slowly reoxidized to vanadate by potassium permanganate in the back titration, so that this element does not interfere seriously in the procedure if the titration is properly made. The end point, however, is not very satisfactory. Substitution of standard arsenite solution for ferrous sulfate is advantageous.

Since vanadium is often found with chromium in steels, it is convenient to combine the determination of these two elements.

Vanadium

The volumetric determination of vanadium involves either (1) reduction of vanadate to the vanadyl stage, or (2) oxidation of vanadyl salt to vanadate, with proper provision for the effect of interfering elements that may be present. Since the determination of vanadium is most often required in chrome–vanadium steels, we shall give a procedure by which both of these elements can be determined at the same time. The following procedure[16] is one of several that can be used for the simultaneous determination.

[16] I. M. Kolthoff and E. B. Sandell, *Ind. Eng. Chem., Anal. Ed.* **2**, 140 (1930). Another method is given by H. H. Willard and P. Young, *ibid.* **6**, 48 (1934).

It involves the oxidation of chromium and vanadium in acid solution with potassium bromate,[17] and the destruction of the excess bromate together with oxidized manganese by boiling with ammonium sulfate in the presence of a low concentration of hydrochloric acid. After cooling, a measured excess of standard arsenious acid is added; chromic acid is thereby reduced to chromic salt, whereas vanadate is unaffected (see note 1). The excess of arsenious acid is titrated with permanganate in the presence of a trace of iodide as catalyst (note 2), and the chromium thus determined. Vanadium is then determined in the same solution by titration with ferrous ammonium sulfate, diphenylbenzidine being used as indicator ($VO_4^{3-} \rightarrow VO^{++}$).

Solutions: *Arsenious acid, 0.1000 N.* Weigh out 4.947 g of pure arsenious oxide, dissolve in 50 ml of 1 *N* sodium hydroxide, dilute nearly to 1 liter, make neutral or slightly acid with hydrochloric acid, and then dilute to exactly 1 liter. If pure arsenious oxide is not available, make up an approximately 0.1 *N* solution, and standardize against standard permanganate solution.

Potassium permanganate, 0.025 N. Standardize against the arsenious acid solution (make the latter solution 1 *N* in hydrochloric acid, add 1 drop of 0.002 *N* potassium iodide solution, and titrate with permanganate to the first pink tinge), or standardize against sodium oxalate.

Ferrous ammonium sulfate, 0.025 N, in 0.5 *N* sulfuric acid.

Potassium iodide solution (catalyst), 0.002 N.

Diphenylbenzidine solution (indicator), 0.1 per cent in concentrated sulfuric acid.

Procedure. Weigh a 2- to 3-g sample (chromium content 1 per cent) into a 250-ml Erlenmeyer flask, add 25 ml of water, 1.5 ml of concentrated sulfuric acid for each gram of steel, allowing a 1- to 2-ml excess, and 3–4 ml of 85 per cent phosphoric acid. Heat to hasten solution. When the sample has dissolved, add dropwise 1 ml of concentrated nitric acid for each gram of steel. Boil to expel the oxides of nitrogen, and dilute the solution to 100 ml. Now add 1.5 g of pure potassium bromate and boil for 5 min. Then add cautiously 5 g of ammonium sulfate, and boil until most of the bromine has been expelled. Next add 10 ml of 1 *N* hydrochloric acid, and boil until starch–iodide paper held at the mouth of the flask is no longer colored blue, indicating complete expulsion of bromine. The volume of the solution should be kept at 75–100 ml during the boiling by the addition of water, if necessary.

Cool the solution to room temperature; add 6–8 ml of 85 per cent phosphoric acid and 1 or 2 drops of 0.002 *N* potassium iodide. Run in standard arsenious acid solution from a buret until the color of the solution in the flask becomes green (indicating complete reduction of chromic acid); allow a few milliliters of arsenious acid in excess. After 3–5 min back-titrate the arsenious acid with potassium permanganate; the end point is indicated by a sudden deepening in the color of the solution to bluish green.

Add 1 drop of arsenious acid to reduce any permanganate present, and then sufficient sodium acetate to react with the acid present (allow approximately 5 g of sodium acetate trihydrate for each milliliter of concentrated sulfuric present, including also the acid in the indicator to be added, and 1.5 g for each milliliter of 12 *N* hydrochloric acid); no permanent precipitate should form. Add 0.5–1.0 ml of 0.1 per cent diphenylbenzidine indicator solution, allow to stand for approximately 5 min, and then titrate slowly with 0.025 *N* ferrous sulfate. The color change is from blue to green or bluish green. Correct for the indicator blank by adding 0.03 ml to the volume of the 0.025 *N* ferrous sulfate solution used for each 0.1 ml of 0.1 per cent diphenylbenzidine solution added.

Notes:

1. A slight reduction of vanadate by arsenious acid would lead to no error, because in the back-titration with potassium permanganate vanadyl ion would be reoxidized to vanadate ion.

[17] Other oxidizing agents such as perchloric acid and potassium persulfate may, of course, be used instead of potassium bromate.

2. Unless a trace of iodide (or iodate) is present as a catalyst in the titration of arsenite ion with permanganate, a quantitative reduction of the latter to manganous ion cannot be effected, a brown solution containing trivalent or quadrivalent manganese being obtained. In the presence of a trace of potassium iodide and a low concentration of chloride, the reaction proceeds smoothly and rapidly, and an excellent end point is obtained.

3. In very accurate work the titration error should be experimentally determined, both for the chromic acid and the vanadic acid titration.

4. Regarding the use of oxidized diphenylamine sulfonic acid and diphenylamine as indicators in the vanadium titration, see H. H. Willard and P. Young, *Ind. Eng. Chem., Anal. Ed.* **5**, 154, 158 (1933); if the diphenylamine sulfonic acid is used as indicator, vanadium can be titrated in the presence of tungsten.

Molybdenum

A selective determination of molybdenum in steel is provided by precipitation with α-benzoinoxime (H_2Bx) as MoO_2Bx, which can be ignited to MoO_3 for weighing.[18] The only other elements precipitated in mineral acid solution by the reagent are W(VI), Cr(VI), V(V), Pd, and Ta. Chromium and vanadium do not interfere when they are reduced to Cr(III) and V(IV). Tungsten in the weighed MoO_3 is corrected for as described in the procedure below.

Molybdenum can also readily be determined colorimetrically. The favorite reagent is thiocyanate, which gives strongly colored amber molybdenum(V) complex thiocyanates. Stannous chloride is the usual reducing agent; iron(III) is reduced to Fe(II) and gives no color. Directions will not be given here.

Reagent: 2 per cent solution of *α-benzoinoxime* in 95 per cent alcohol.

Wash solution: 50 ml of *reagent solution* and 10 ml of *concentrated sulfuric acid* diluted to 1 liter.

Procedure. Dissolve a 1-g sample of steel in 50 ml of 1 : 5 sulfuric acid, and treat the hot solution with the minimum amount of nitric acid required to oxidize the molybdenum and decompose carbides (1 ml will suffice for a 1-g sample). Boil the solution to expel oxides of nitrogen, and filter if the liquid is not perfectly clear. Dilute to 100 ml, cool, and add enough ferrous ammonium sulfate solution to reduce vanadic and chromic acids. Cool the solution to 5–10°C, add 10 ml of α-benzoinoxime reagent, then bromine water to impart a distinct color, and finally a few more milliliters of reagent. Allow the mixture to stand for 10 min, and then filter through an S and S No. 589 Black Band paper. Wash the precipitate with 150 ml of cold fresh wash solution. Transfer the paper and precipitate to a weighed crucible, and char the paper at a low temperature. Then place the crucible in an electric muffle and ignite to constant weight at 500–525°C. Treat the residue with a few milliliters of warm 1 : 3 ammonium hydroxide, and filter through a small filter if any insoluble material remains. Wash the filter thoroughly with warm water containing a little ammonium hydroxide, and reserve the filtrate and washings. Ignite the paper containing the insoluble residue, and weigh. Deduct this weight from the weight of the molybdic oxide. Examine the filtrate and washings for tungsten by adding excess hydrochloric acid and a little cinchonine solution (see Tungsten), and digesting overnight. If a precipitate of tungsten is formed, filter it off, wash, and ignite to the oxide at 525°C. Deduct the weight of tungstic oxide from the weight of molybdic oxide. Report the percentage of molybdenum.

Notes:

1. Molybdenum is precipitated in cold solutions in the presence of bromine to prevent possible reduction to lower valence states by α-benzoinoxime before complete precipitation.

From two to five times the theoretical amount of α-benzoinoxime should be used for the precipitation; 5 ml of reagent suffices to precipitate 0.01 g of molybdenum.

[18] H. B. Knowles, *Natl. Bur. Std. J. Res.* **9**, 1 (1932).

The solution should not be allowed to stand long before filtration. Low results are obtained if the precipitate stands ½ hr before filtration is begun. Precipitation is complete within a few minutes after the addition of reagent.

Molybdenum α-benzoinoxime cannot be dried at a low temperature and used as a weighing form.

2. The temperature of ignition of molybdic oxide must be carefully controlled. The oxide begins to volatilize at 550°C, but the rate of loss is very slow below 600°C. The most suitable ignition temperature is 500–525°C.

3. The treatment of the ignited molybdic oxide with ammonium hydroxide corrects for the presence of substances such as ferric oxide and silica that have been carried down by the precipitate. These are insoluble in ammonia, whereas molybdic oxide (and tungstic oxide if present) is readily soluble.

4. Tungsten, if present, will be precipitated by the reagent and must then be precipitated from the ammoniacal solution of the ignited residue by addition of hydrochloric acid and cinchonine, ignited to the oxide, and its weight deducted from that of the impure molybdic oxide.

5. The following results obtained by Knowles (Ref. 18) in the analysis of Bureau of Standards steels for molybdenum indicate the accuracy of the method:

	Percentage of Molybdenum	
Steel	*Added or Present*	*Found*
Chrome-molybdenum steel No. 72	0.149	0.166 0.160
B.O.H. steel No. 11d	0.51	0.52
	5.13	5.15

Tungsten

The steel is treated with hydrochloric or sulfuric acid, and tungsten is oxidized to tungstic acid, H_2WO_4, by the addition of nitric acid. Tungstic acid, sparingly soluble in acid solutions, separates nearly completely on digestion. By adding the alkaloid cinchonine, the precipitation of tungsten can be made quantitative, the last small amounts being precipitated as cinchonine tungstate. The tungstic acid thus separated is never pure but contains greater or less amounts of silica, iron oxide, and such elements as chromium, vanadium, and molybdenum, if these are present in the steel. The precipitate of tungstic acid is filtered off, washed, ignited at a low temperature, and treated with hydrofluoric acid and sulfuric acid to volatilize silica as silicon tetrafluoride.

Tungsten is not volatilized. The residue is ignited at not too high a temperature and weighed. The ignited residue is impure tungstic oxide containing such contaminants as mentioned above. To obtain the weight of tungstic oxide, it is necessary to determine the weight of the impurities and deduct from the weight of the crude oxide. For this purpose the residue is fused with sodium carbonate; tungstic oxide is thus converted into water-soluble sodium tungstate, whereas iron remains as insoluble ferric oxide, which can be filtered off, weighed, and its weight deducted from that of the impure tungstic oxide. If elements such as chromium, vanadium, and molybdenum are present in the impure oxide, they are converted into soluble sodium chromate, vanadate, and molybdate, respectively, and must be determined in the filtrate by colorimetric methods.

Procedure. Weigh a 2-g sample into a 400-ml beaker, add 50 ml of 1 : 1 hydrochloric acid, and heat. When the decomposition is complete, remove the beaker from the heat and add carefully 10 ml of 1 : 1 nitric acid. Digest the mixture at 90–100°C until the separated tungstic acid is bright yellow; no dark particles should be visible. Dilute to 100 ml, add 5 ml of cinchonine solution (10 per cent solution in 1 : 1 hydrochloric acid), and digest slightly below the boiling point for 30–45 min. Filter off the precipitate on a retentive quantitative filter paper and wash thoroughly with cinchonine wash solution (25 ml of the cinchonine reagent solution diluted to 1 liter). Fold up the filter paper and place it in a weighed platinum crucible. Dry the paper and contents by heating over a low flame, char the paper, and burn off the carbon at the lowest possible temperature. Cool the crucible, and add 1 or 2 drops of 1 : 1 sulfuric acid and about 2 ml of hydrofluoric acid. Evaporate off the hydrofluoric acid and expel the sulfuric acid (preferably in a radiator). Ignite the residue at 750–850°C, best in a muffle furnace. If no furnace is available, heat the covered crucible to low redness over a Tirrill burner. Cool and weigh to obtain the weight of $WO_3 + Fe_2O_3$, etc. Fuse the weighed residue with 5 g of purest anhydrous sodium carbonate. Extract the cold melt with 75–100 ml of water, and filter off any insoluble material on a quantitative filter paper; rinse out the crucible carefully, and wash the residue and paper thoroughly with hot water. Place the filter in the crucible, char the paper, burn off the carbon, and fuse the residue with 0.5–1 g of sodium carbonate. Take up the cooled melt in 25 ml of water, filter, and wash the crucible and paper with hot water. The second filtrate is to be combined with the first and reserved if chromium, vanadium, and molybdenum are to be determined (note 1). Place the filter in the crucible and ignite to obtain the weight of ferric oxide. Deduct the weight of ferric oxide from the weight of impure tungstic oxide to obtain the weight of tungstic oxide. Calculate the percentage of tungsten.

Notes:

1. If the steel contains chromium, vanadium, and molybdenum in appreciable amounts, these elements should be determined colorimetrically in the combined filtrates from the sodium carbonate fusion, and the ignited residue of tungstic oxide corrected for the amounts found. For the details of this procedure the student is referred to Lundell et al. (Ref. 3, p. 330).

2. In accurate work a blank must be run on the sodium carbonate to correct for the insoluble impurities it may contain.

3. Concerning coprecipitation with tungstic acid, see W. Gebauhr, *Z. Anal. Chem.* **197**, 212 (1963).

Simultaneous determination of silicon and tungsten

It is sometimes convenient to determine both silicon and tungsten in the same sample. This can be done by precipitating tungstic acid and silica, and obtaining the weight of the combined ignited residue of SiO_2 and WO_3, treating with hydrofluoric and sulfuric acids to volatilize silica, and reweighing. The loss in weight gives the amount of SiO_2 present in the mixture, and the weight of the hydrofluorized residue gives the amount of WO_3 (Fe_2O_3, etc.).

Procedure. Take a 2-g sample of steel and precipitate tungstic acid, as described above, with cinchonine. Filter off the tungstic acid, wash it, and reserve the precipitate. Evaporate the filtrate and combined washings to dryness, and heat the residue for an hour at 105–110°C. Add 10 ml of concentrated hydrochloric acid, warm for a few minutes, dilute to 100 ml with water, and warm until all salts have dissolved. Filter off the silica, and wash it with 1 : 20 hydrochloric acid and finally one or two times with 1 : 50 sulfuric acid.

Silicon. Place the filter papers containing tungstic acid and silica in a weighed platinum crucible, dry and char the papers, and burn off the carbon at a low temperature. Ignite at 750–850°C in an electric furnace or, less preferably, over a Tirrill burner. After weighing the crucible and its contents, add one or two drops of 1 : 1 sulfuric acid and 3 ml of hydrofluoric acid. Evaporate the hydrofluoric acid without boiling, and fume off the sulfuric acid. Ignite the residue at 750–850°C. The loss in weight represents the amount of silica. Report the percentage of silicon.

Tungsten. The residue of impure tungstic oxide left in the crucible is to be corrected for ferric oxide as described above under tungsten. Calculate the percentage of tungsten.

Note: If silica is to be determined very accurately in a tungsten steel, its determination should not be combined with that of tungsten. A temperature of 1000–1100°C, or even higher, is needed to remove the last traces of water from silica, and at these temperatures tungstic oxide volatilizes.

PROBLEMS

1. What weight of steel should be taken for analysis in order that each milliliter (20°C, 1 atm) of CO_2 formed by combustion shall represent 0.01 per cent C?
2. List the methods that can be used for the separation of much iron preparatory to the determination of the minor constituents of steel. (Consult other sections of this book and the literature.)
3. Why is the presence of hydrochloric and hydrofluoric acids objectionable in the bismuthate method for manganese?
4. Devise a method for determining the amount of active oxygen in sodium bismuthate.
5. What volume of 4 *N* nitric acid will be required to dissolve 1 g of iron if the acid is reduced to NO?
6. The permanganic acid formed in the oxidation of a 1.010-g sample of steel by the bismuthate method was treated with 50.00 ml of 0.0320 *N* ferrous ammonium sulfate solution. If 10.18 ml of 0.0203 *N* potassium permanganate solution was required for the back-titration of the ferrous solution, what is the percentage of manganese in the sample?
7. What weight of 80 per cent $NaBiO_3$ is theoretically required to oxidize the manganese in 0.50 g of steel containing 0.50 per cent Mn?
8. The $(NH_4)_3PO_4 \cdot 12\ MoO_3$ precipitate obtained from a 1.000-g sample of steel was dissolved in 10.00 ml of 0.1022 *N* sodium hydroxide solution, and the excess of base was back-titrated with standard nitric acid, phenolphthalein being used as indicator. If 6.05 ml of 0.1010 *N* acid was required, what is the percentage of phosphorus in the sample?
9. What weight of sample must be taken in the determination of phosphorus in order that each milligram of ignited magnesium pyrophosphate shall represent 0.01 per cent phosphorus?
10. A 5.00-sample of steel run for sulfur by the evolution method required 5.2 ml of potassium iodate–iodide solution (1.000 g of KIO_3 and 10.00 g of KI in 1 liter) for the titration of the zinc sulfide in acid solution. Calculate the percentage of sulfur.
11. Why should iron be oxidized to the ferric condition before the precipitation of nickel with dimethylglyoxime?
12. Write equations for all the important reactions occurring in the determination of chromium and vanadium by the procedure given.
13. (a) In the determination of vanadium by titration with ferrous iron, what weight of sample must be taken in order that each milliliter of 0.020 *N* ferrous ammonium sulfate shall correspond to 0.100 per cent vanadium? (b) What volume of sodium arsenite solution containing the equivalent of 5.000 g of As_2O_3 per liter is required to reduce the dichromate yielded by a 2.00-g sample of steel containing 1.00 per cent chromium?

Appendix

List of Tables in Appendix

TABLE **A-1. Average Values of Activity Coefficients of Ions**

	Activity Coefficient at an Ionic Strength of				
Ionic Charge[a]	0.005	0.01	0.025	0.05	0.1
1 ($\mathring{a} = 4$)	0.93	0.90	0.86	0.82	0.77
2 ($\mathring{a} = 5$)	0.74	0.67	0.56	0.47	0.37
3 ($\mathring{a} = 5$)	0.51	0.40	0.27	0.18	0.12
4 ($\mathring{a} = 6$)	0.32	0.21	0.105	0.055	0.027

[a] The values of $\mathring{a}$ correspond to an average value for the commonly encountered ions of the specified charge and were used to evaluate f_i using Eq. (5) of Chapter 3. Note that for hydrogen ion, a value of $\mathring{a} = 9$ should be used if the most precise results are required.

TABLE A-2. **Thermodynamic Dissociation Constants of Some Analytically Important Acids at 25°C**[a]

Acid	pK^0_{HB}	*Remarks*
Inorganic acids		
Ammonium ion	9.24	
Arsenic acid,		
pK_1	2.22	
pK_2	6.98	
pK_3	11.53	18°C, ionic strength varied
Arsenious acid	9.22	
Boric acid	9.23	
Carbonic acid,		
pK_1	6.35	
pK_2	10.33	
Chromic acid,		
pK_2	6.50	pK 1.64 for $2\ HCrO_4^- \rightarrow Cr_2O_7^{--} + H_2O$
Cyanic acid	3.66	Room temp., ionic strength varied
Ferrocyanic acid,		
pK_3	3.00	16–18°C
pK_4	4.30	K_1 and K_2 very large
Hydrazinium ion, $N_2H_5^+$	7.99	
Hydrazoic acid	4.72	
Hydrocyanic acid	9.2	
Hydrofluoric acid,		
pK	3.17	$HF \rightleftarrows H^+ + F^-$
pK	0.59	$HF + F^- \rightleftarrows HF_2^-$
Hydrogen selenide,		
pK_1	3.89	
pK_2	11.00	
Hydrogen sulfide,		
pK_1	7.00	
pK_2	12.92	

Acid	pK_{HB}	*Remarks*
Organic acids		
Acetic acid	4.76	
Adipic acid,		
pK_1	4.42	
pK_2	5.41	
Anilinium ion	4.62	
Benzoic acid	4.20	
Bromoacetic acid	2.80	Acid concn. = 0.031 *N*
n-Butyric acid	4.82	
Citric acid,		
pK_1	3.13	
pK_2	4.76	
pK_3	6.40	
Ethylenediammonium ion,		
pK_1	7.52	30°C
pK_2	10.65	30°C
Ethylenediaminetetra-acetic acid,		
pK_1	1.99	20°C, $\mu = 0.1$
pK_2	2.67	20°C, $\mu = 0.1$
pK_3	6.16	20°C, $\mu = 0.1$
pK_4	10.22	20°C, $\mu = 0.1$
Fumaric acid,		
pK_1	3.02	
pK_2	4.39	
Formic acid	3.77	
Glycine,	2.35	
pK_2	9.78	
Glycolic acid	3.88	
Lactic acid	3.86	

Hydroxylammonium ion	5.98	
Hypochlorous acid	7.53	
Nitrous acid	3.29	$\mu \cong 0.001$
Phosphoric acid,		
pK_1	2.23	
pK_2	7.21	
pK_3	12.32	
Phosphorous acid,		
pK_1	2.00	18°C, ionic strength varied
pK_2	6.58	18°C, ionic strength varied
Pyrophosphoric acid,		
pK_1	1.52	
pK_2	2.36	
pK_3	6.00	
pK_4	9.25	
Selenous acid,		
pK_1	2.62	
pK_2	8.32	
Sulfuric acid,		
pK_2	1.99	
Sulfurous acid,		
pK_1	1.76	
pK_2	7.21	
Monoethanolammonium ion	9.50	
Nitrilotriacetic acid,		
pK_1	3.03	20°C
pK_2	3.07	20°C
pK_3	10.70	20°C
Maleic acid,		
pK_1	1.92	
pK_2	6.22	
Malonic acid,		
pK_1	2.85	
pK_2	5.66	
Oxalic acid,		
pK_1	1.25	
pK_2	4.28	
1,10-Phenanthrolinium ion	4.96	
Phenol	9.95	
o-Phthalic acid,		
pK_1	2.95	
pK_2	5.41	
Propionic acid	4.88	
Pyridinium ion	5.17	
Salicylic acid	2.98	
Succinic acid,		
pK_1	4.19	
pK_2	5.48	
d-Tartaric acid,		
pK_1	3.04	
pK_2	4.37	

[a] From *Treatise on Analytical Chemistry* (I. M. Kolthoff and P. J. Elving with E. B. Sandell, eds.), part I, vol. 1, Wiley-Interscience, New York, 1959, p. 432.

TABLE A-3. Stepwise Formation Constants of Metal Complexes[a]

This table contains stepwise formation constants of complexes of various metal ions M and monodentate complexing agents A, plus ethylenediamine. K_n corresponds to the reaction

$$MA_{n-1} + A \rightleftharpoons MA_n$$

and therefore

$$K_n = \frac{[MA_n]}{[MA_{n-1}][A]}$$

In a few instances in which stepwise constants are not available, the table contains the overall constant β_n, which corresponds to

$$M + nA \rightleftharpoons MA_n$$

$$\beta_n = \frac{[MA_n]}{[M][A]^n}$$

Such values are in boldface type.

The values apply at 20–25°C and at an ionic strength of about 0.1, unless the latter is specified.

Metal Ion	*n*	log K_n
Ammonia complexes		
Ag^{+}	1, 2	3.4, 4.0[b]
Cd^{++}	1, . . . , 6	2.6, 2.1, 1.4, 0.9, −0.3, −1.7
Co^{++}	1, . . . , 6	2.1, 1.6, 1.0, 0.7, 0.1, −0.7
Cu^{++}	1, . . . , 5	4.1, 3.5, 2.9, 2.1, −0.5
Ni^{++}	1, . . . , 6	2.8, 2.2, 1.7, 1.2, 0.7, 0.0
Zn^{++}	1, . . . , 4	2.3, 2.3, 2.4,[c] 2.1
Chloride complexes		
Ag^{+} ($\mu = 0.2$)[d]	1, . . . , 3	2.7, 1.8, 0.3
Cd^{++}	1, . . . , 3	1.6, 0.5, −0.6
Fe^{3+} ($\mu = 1$)	1, . . . , 3	0.6, 0.1, −1.0
Hg^{++} ($\mu = 0.5$)	1, . . . , 4	6.7, 6.5, 0.9, 1.0[e]
Cyanide complexes:		
Ag^{+} [d]	1, . . . , 4	**21.1**,[f] 0.7, −1.1
Cd^{++} ($\mu = 3$)	1, . . . , 4	5.5, 5.1, 4.7, 3.6
Fe^{3+}	1, . . . , 6	**31**[g]
Hg^{++}	1, . . . , 4	18.0, 16.7, 3.8, 3.0
Ni^{++}	1, . . . , 4	**31.3**[h]
Zn^{++}	1, . . . , 4	**16.7**[h]
Fluoride complexes:		
Al^{3+} ($\mu = 0.5$)	1, . . . , 6	6.1, 5.1, 3.9, 2.7, 1.7, 0.3
Fe^{3+} ($\mu = 0.5$)	1, . . . , 3	5.2, 4.0, 2.7
Th^{4+} ($\mu = 0.5$)	1, . . . , 3	7.7, 5.8, 4.5
UO_2^{++} ($\mu = 1$)	1, . . . , 4	4.5, 3.4, 2.6, 1.3
Zr^{4+} ($\mu = 2$)	1, . . . , 3	8.8, 7.3, 5.8
Hydroxo complexes		
Ag^{+} ($\mu \cong 0$)	1, . . . , 3	2.3, 1.3, 1.2
Al^{3+}	1, . . . , 4	8.9, **33.3**[h]
Cr^{3+}	1, 2	10.2, 8.1
Cu^{++} ($\mu \cong 0$)	1	6.0
$Fe^{+,+}$ ($\mu = 1$)	1	4.5

TABLE A-3. Stepwise Formation Constants of Metal Compounds[a] *(continued)*

Metal Ion	n	$\log K_n$
Fe^{3+} ($\mu = 3$)	1	11.0
Hg^{++} ($\mu = 0.5$)	1, 2	10.3, 11.4[b]
Mg^{++} ($\mu \cong 0$)	1	2.6
Pb^{++} ($\mu \cong 0.3$)	1, . . ., 3	6.2, 4.1, 3.0
Zn^{++} ($\mu \cong 0$)	1	4.4
Iodide complexes		
Ag^{+} ($\mu = 1.6$)[c]	1, 2	13.9, −0.2
Cd^{++}	1, . . ., 4	2.4, 1.0, 1.6,[c] 1.2
Hg^{++} ($\mu = 0.5$)[c]	1, . . ., 4	12.9, 10.9, 9.3, 3.8, 2.2
Ethylenediamine complexes		
Ag^{+}	1, 2	4.7, 3.0
Cd^{++}	1, . . ., 3	5.5, 4.6, 2.1
Co^{++}	1, . . ., 3	5.9, 4.8, 3.1
Cu^{++}	1, 2	10.6, 9.1
Ni^{++}	1, . . ., 3	7.7, 6.4, 4.5
Zn^{++}	1, . . ., 3	5.7, 4.7, 1.7

[a] Mainly from A. Ringbom, *Complexation in Analytical Chemistry*, Wiley-Interscience, New York, 1963. A more extensive tabulation is given by L. G. Sillén and A. E. Martell, *Stability Constants of Metal-Ion Complexes*, The Chemical Society, London, 1964. Extensive tabulations are given also by K. B. Yatsimirskii and V. P. Vasil'ev, *Instability Constants of Complex Compounds*, Pergamon, New York, 1960, and by L. Meites (ed.), *Handbook of Analytical Chemistry*, McGraw-Hill, New York, 1963.

[b] $K_2 > K_1$.

[c] $K_3 > K_2$.

[d] Also see solubility product, Table A-9.

[e] $K_4 > K_3$.

[f] $\log \beta_2$.

[g] $\log \beta_6$.

[h] $\log \beta_4$.

TABLE A-4. **Formation Constants of Polyamine Complexes**[a,b]

	log *K*			
Metal Ion	*Penten*[c]	*Tetren*	*Trien*	*Tren*
Ag^+	—	—	7.7	7.8
Cd^{++}	16.8	14.0	10.8[d]	12.3
Co^{++}	15.8	15.1	11.0	12.8
Cu^{++}	22.4	24.3	20.4	18.8
Fe^{++}	11.2	11.4	7.8	8.8
Fe^{3+}	—	—	21.9	—
Hg^{++}	29.6	27.7	25.3	22.8
Mn^{++}	9.4	7.6	4.9	5.8
Ni^{++}	19.3	17.6	14.0[e]	14.8
Pb^{++}	—	10.5	10.4	—
Zn^{++}	16.2	15.4	12.1	14.7

[a] Negligible complex formation with Al^{3+}, Ba^{++}, Ca^{++}, Mg^{++}, and Sr^{++}.

[b] Penten, pentaethylenehexamine; Tetren, tetraethylenepentamine; Trien, triethylenetetramine; Tren, triethanolamine.

[c] Complexes with one, two, or three hydrogens are known; all these are much less stable than MA.

[d] $\log K_2 = 3.1$.

[e] $\log K_2 = 5.4$.

TABLE A-5. Formation Constants of Complexes with EDTA and Related Compounds[a,b]

	log *K*					
					NTA	
Metal Ion	*DCYTA*	*DTPA*	*EDTA*[c]	*EGTA*	$n = 1$	$n = 2$
Ag^{+}	—	—	7.3	—	—	
Al^{3+}	17.6	—	16.1	—	—	
Ba^{++}	8.0	8.8	7.8	8.4	4.8	
Bi^{3+}	24.1	—	22.8	—	—	
Be^{++}	—	—	9.3	—	—	
Ca^{++}	12.5	10.6	10.7	11.0	6.4	
Cd^{++}	19.2	19.0	16.5	15.6	10.1	4.4
Co^{++}	18.9	19.0	16.3	12.3	10.6	
Cr^{3+}	—	—	23	—	—	
Cu^{++}	21.3	20.5	18.8	17.0	12.7	3.6
Fe^{++}	18.2	16.0	14.4	—	8.8	
Fe^{3+}	29.3	27.5	25.1	—	15.9	8.4
Hg^{++}	24.3	27.0	21.8	23.2	12.7	
Mg^{++}	10.3	9.3	8.7	5.2	5.4	
Mn^{++}	16.8	15.5	14.0	11.5	7.4	
Ni^{++}	19.4	20.0	18.6	12.0	11.3	4.5
Pb^{++}	19.7	18.9	18.0	13.0	11.8	
Lanthanides	17–22	19	16–20	—	10–12	
Sn^{++}	—	—	22.1	—	—	
Sr^{++}	10.0	9.7	8.6	8.5	5.0	
Th^{4+}	23.2	27	23.2	—	10.5	
Ti^{3+}	—	—	21.3	—	—	
V^{++}	—	—	12.7	—	—	
V^{3+}	—	—	25.9	—	—	
VO^{++}	19.4	—	18.8	—	—	
Zn^{++}	18.7	18.0	16.5	12.8	10.5	

[a] Complexes containing H and OH are formed also, which are less stable than MA.

[b] DCYTA, 1,2-diaminocyclohexane-N,N,N′,N′-tetraacetic acid; DTPA, diethylenetriaminepentaacetic acid; EDTA, ethylenediaminetetraacetic acid; EGTA, ethyleneglycol-bis(β-aminoethyl ether)-N,N′-tetraacetic acid; NTA, nitrilotriacetic acid.

[c] See also Table A-7.

TABLE A-6. **Values of log α for EDTA and Related Compounds**[a]

pH	log α *DCYTA*	*DTPA*	*EDTA*	*EGTA*	*NTA*
13	0.0	0.0	0.0	0.0	0.0
11	0.8	0.1	0.1	0.0	0.0
9	2.7	1.6	1.3	0.8	0.9
7	4.8	5.1	3.3	4.3	2.8
5	7.9	9.2	6.4	8.3	4.8
3	12.5	14.6	10.6	12.5	7.0
1	19.8	23.1	17.1	19.0	11.4
	Acid dissociation constants used above:				
pK_1	2.43	1.89	1.99	2.00	1.97
pK_2	3.54	2.79	2.67	2.65	2.57
pK_3	6.10	4.29	6.16	8.85	9.81
pK_4	11.70	8.61	10.22	9.46	
pK_5		10.48			

[a] See (58)–(60), Chapter 6, for the definition of α.

TABLE A-7. **Conditional Stability Constants of EDTA Complexes**[a]

Metal Ion	log *K at* pH 1	3	5	7	9	11	13
Al^{3+}		5.5	9.4	7.8	4.2		
Ba^{++}				4.4	6.4	7.7	7.8
Ca^{++}			4.2	7.3	9.4	10.6	10.7
Cd^{++}		6.0	9.9	13.1	14.8	12.9	8.8
Co^{++}		5.9	9.8	12.9	14.9	13.0	8.6
Cu^{++}	3.6	8.4	12.3	15.4	16.4	15.7	13.8
Fe^{++}		3.8	7.8	11.0	12.9	12.7	10.8
Fe^{3+}	8.4	14.0	14.9	14.2	13.7		
Hg^{++}	6.6	11.4	11.6	10.7	9.0	7.9	6.0
Mg^{++}			2.2	5.3	7.4	8.6	8.7
Mn^{++}		3.7	7.5	10.7	12.7	13.9	14.0
Ni^{++}	3.6	8.3	12.1	15.3	16.9	14.1	8.6
Pb^{++}	2.6	7.6	11.5	13.8	13.9	12.9	7.8
Sr^{++}			2.1	5.3	7.3	8.6	8.6
Zn^{++}		6.1	10.0	13.1	15.0	11.0	4.0

[a] Based on α values of EDTA and stability of hydroxo complexes.

TABLE A-8. **Metallochromic Indicators**[a,b]

	log *K*								
Metal Ion	*CAL*	*CMG*	*EBT*	*MPT*	*MUR*	*NIN*	*PAN*	*PYV*	*XYO*
Al^{3+}								19.1	
Ba^{++}				6.2					
Bi^{3+}								27.5	5.5
Ca^{++}	5.6	6.1	5.4	7.8	5.0				
Co^{++}							>12	9.0	
Cu^{++}	21				17.9		16	16.5	
Fe^{3+}									5.7
Mg^{++}	7.6	8.1	7.0	8.9		6.2	8.5	4.4	
Ni^{++}					11.3	11.5	12.7	9.4	
Pb^{++}								13.3	
Zn^{++}	12.5		13.5	15.1		9.5	11.2	10.4	6.2
	Log formation constant of acid forms of In:								
HIn	13.5	12.4	11.6	12.0	10.5	11.6[c]	12.3	12.5	12.3
H_2In	7.4	8.1	6.3	11.4	9.2		<2	9.8	10.5
H_3In			1.6	7.8	0			7.8	6.4
H_4In				7.0				0.3	3.2
H_5In				2.9					2.6
H_6In				2.2					

[a] Log of formation constant for reaction Me + In $\rightleftharpoons$ MeIn; ionic strength about 0.1, except as noted.

[b] CAL, calcon; CMG, calmagite; EBT, Eriochrome Black T; MPT, metal phthalein; MUR, murexide; NIN, 1-nitroso-2-naphthol; PAN, pyridylazo naphthol; PYV, Pyrocatechol Violet; XYO, Xylenol Orange.

[c] In 75 per cent dioxane.

TABLE **A-9. Solubility Products (as pK_{sp}) of Salts**[a,b]

	Ag		
Ag acetate	2.7	AgI	16.08
Ag_3AsO_4 (dil.)	19.95	$AgIO_3$	7.51
Ag benzoate	4.0	Ag_2MoO_4	11.55
AgBr	12.31	$AgNO_2$	3.2
$AgBrO_3$	4.28	Ag_3PO_4	15.84
AgCN	15.9	Ag quinaldate	17.9
AgCNO	6.64	Ag reineckeate	9.26
Ag_2CO_3	11.1	Ag salicylate	4.85
$Ag_2C_2O_4$	11.0	AgSCN	11.97
AgCl	9.75	Ag_2SO_4	4.8
$Ag_3Co(CN)_6$	20.1	AgSeCN	15.4
Ag_2CrO_4	11.95	Ag_2SeO_3	15.0
$Ag_2Cr_2O_7$ (dil.)	6.7	Ag_2SeO_4	7.25
$Ag_4Fe(CN)_6$ (dil.)	40.8	$AgVO_3$	6.3
$Ag_2Fe(CN)_5NO$	12.1	Ag_2WO_4	11.26
	Al		
$AlAsO_4$ (var.)	15.8	$AlPO_4$ (var.)	18.2
Al oxinate	32.3		
	Ba		
$Ba_3(AsO_4)_2$ (var.)	50.1	$BaMnO_4$ (dil.)	9.6
$BaCO_3$	9.26	$Ba(HPO_4)$ (var.)	7.0
BaC_2O_4	6.8	$Ba_3(PO_4)_2$ (var.)	22.5
$BaCrO_4$	9.93	Ba oxinate	8.3
BaF_2	6.0	$BaSO_4$	9.97
$Ba(IO_3)_2$	8.82		
	Bi		
$BiAsO_4$ (var.)	9.36	$BiPO_4$ (var.)	22.9
BiI_3 ($\mu = 2$)	18.1	Bi reineckeate	12.85
	Ca		
$Ca_3(AsO_4)_2$ (var.)	18.2	$Ca_3(PO_4)_2$	26
$CaCO_3$	8.32	$CaHPO_4$	6.6
CaC_2O_4	8.64	$CaSO_3$	6.51
CaF_2	10.57	$CaSO_4$	4.62
$Ca(IO_3)_2$	6.15	$CaSeO_3$	5.53
$CaMoO_4$	7.38	Ca tartrate	6.11
Ca oxinate	11.0	$CaWO_4$	8.06
	Cd		
Cd anthranilate	8.4	$CdHg(SCN)_4$	5.42
$Cd_3(AsO_4)_2$ (var.)	32.7	$Cd_3(PO_4)_2$	32.6
$CdCO_3$	11.3	Cd quinaldate	12.3
CdC_2O_4	7.8	$CdSeO_3$	7.8
$Cd_2[Fe(CN)_6]$ (dil.)	15.2		

TABLE A-9. **Solubility Products (as pK_{sp}) of Salts**[a,b] *(continued)*

		Ce(III)	
$Ce_2(C_2O_4)_3$	25.4	$Ce_2(SeO_3)_3$	24.4
CeF_3 (dil.)	15.1	$CePO_4$ (var.)	22.8–23.7
$Ce(IO_3)_3$	9.5	Ce(III) tartrate	19.0
		Co	
Co anthranilate	11.0	Co oxinate	29.7
$Co_3(AsO_4)_2$ (var.)	28.1	$Co(HPO_4)$ (dil.)	6.7
$CoCO_3$	12.8	$Co_3(PO_4)_2$	34.7
$Co_2[Fe(CN)_6]$ (var.)	14.7	Co quinaldate	10.8
$Co[Hg(SCN)_4]$	5.8	$CoSeO_3$	6.8
		Cr	
$CrAsO_4$ (var.)	20.1	$CrPO_4$ (violet) (var.)	17
$CrPO_4$ (green) (var.)	22.6		
		Cu(I, II)	
Cu(II) anthranilate	14.2	$Cu_2[Fe(CN)_6]$ (var.)	15.9
$Cu_3(AsO_4)_2$ (var.)	35.1	$CuHg(SCN)_4$ (dil.)	7.5
CuBr	8.28	CuI	11.96
$CuCO_3$	9.63	$Cu(IO_3)_2$	7.14
CuC_2O_4	7.54	Cu(II) oxinate	29.7
CuCl	6.73	$Cu_3(PO_4)_2$ (dil.)	36.9
$CuCrO_4$	5.44	$Cu_2P_2O_7$ (var.)	15.1
CuCN	19.5	CuSCN (var.)	12.7
Cu(II) dimethylglyoximate (μ = 0.1)	21.34	$CuSeO_3$	7.7
		Eu	
$Eu(IO_3)_3$	11.3		
		Fe(II, III)	
$FeAsO_4$ (var.)	20.2	$FePO_4$ (var.)	21.9
$FeCO_3$	10.5	$Fe_4(P_2O_7)_3$ (μ = 0.1)	22.6
FeC_2O_4	6.7	Fe(III) quinaldate	16.9
$Fe_4[Fe(CN)_6]_3$ (var.)	40.5	$Fe_2(SeO_3)_3$	30.7
Fe(III) oxinate	43.5		
		Ga	
$Ga_4[Fe(CN)_6]_3$ (var.)	33.8	Ga oxinate	32.06, 40.8
		Hg(I, as Hg_2^{++}, and II)	
Hg(I) acetate	14.7	Hg_2I_2	28.35
Hg_2Br_2	22.24	$Hg_2(IO_3)_2$	13.7
$Hg_2(CN)_2$	39.3	Hg(I) quinaldate	17.9
Hg_2CO_3	16.05	Hg(II) quinaldate	16.8
$Hg_2C_2O_4$	13	$Hg_2(SCN)_2$	19.52
$Hg_2(CN)_2$	39.3	Hg_2SO_4	6.13

TABLE A-9. **Solubility Products (as pK_{sp}) of Salts**[a,b] (*continued*)

Hg_2Cl_2	17.88	HgSe	59.8
Hg_2CrO_4	8.7	$HgSeO_3$	13.7
$(Hg_2)_3[Fe(CN)_6]_2$	20.1	Hg_2SeO_3	14.2
$(Hg_2)_2[Fe(CN)_6]$	11.95	Hg(I) tartrate	10
$Hg_2(HPO_4)$	12.4	Hg_2WO_4 (var.)	17.0
	In		
$In_4[Fe(CN)_6]_3$ (var.)	43.7	In oxinate	31.34, 33.9
	K		
K bitartrate	3.5	K_2PtCl_6	4.96
	La		
$La_2(C_2O_4)_3$	27.7	$LaPO_4$ ($\mu = 0.5$)	22.4
$La(IO_3)_3$	11.2	La tartrate	18.7
	Mg		
$Mg_3(AsO_4)_2$ (var.)	19.7	$MgNH_4PO_4$ (dil.)	12.6
$MgCO_3$	5.0	Mg oxinate	15.4
$Mg_2C_2O_4$ (dil.)	4.1	$Mg_3(PO_4)_2$	27.2
MgF_2	8.15	$MgSeO_3$	4.9
	Mn		
$Mn_3(AsO_4)_2$ (var.)	28.7	$Mn_2[Fe(CN)_6]$ (var.)	12.1
$MnCO_3$	9.3	Mn oxinate	21.7
	Ni		
Ni anthranilate	11.7	$Ni_2[Fe(CN)_6]$ (var.)	14.9
$Ni_3(AsO_4)_2$ (var.)	25.5	Ni oxinate	26.1
$NiCO_3$ (var.)	8.2	$Ni_3(PO_4)_2$ (dil.)	30.3
Ni dimethylglyoximate ($\mu = 0.05$)	23.4	Ni quinaldate	10.1
		$NiSeO_3$	5.0
	Pb		
Pb anthranilate	8.4	PbI_2	8.19
$Pb_3(AsO_4)_2$ (var.)	35.4	$Pb(IO_3)_2$	12.6
$PbBr_2$	4.41	$Pb(HPO_4)$	9.85
$Pb(BrO_3)_2$	5.1	$Pb_3(PO_4)_2$	42.1
$PbCO_3$	13.2	$PbMoO_4$	9.7
PbC_2O_4	10.5	Pb quinaldate	10.6
$PbCl_2$	4.79	$Pb(SCN)_2$ (var.)	4.7
$PbCrO_4$	13.75	$PbSO_4$	7.78
PbF_2	7.57	PbS_2O_3 (var.)	6.6
PbFCl (dil.)	8.62	$PbSeO_3$	11.5
$Pb_2[Fe(CN)_6]$	6.9	$PbSeO_4$	6.85
PbI(OH)	15.2		
	Ra		
$RaSO_4$	10.4		

TABLE A-9. **Solubility Products (as pK_{sp}) of Salts**[a,b] *(continued)*

		Sr	
$Sr_3(AsO_4)_2$ (dil.)	17.8	Sr oxinate	8.7
$SrCO_3$	9.96	$SrHPO_4$ (var.)	6.24
SrC_2O_4	7.25	$Sr_3(PO_4)_2$ (var.)	27.4
$SrCrO_4$ (var.)	4.65	$SrSO_4$	6.6
SrF_2	8.61	$SrSeO_3$	5.74
$Sr(IO_3)_2$	6.48		
		Th	
$Th(C_2O_4)_2$	22	$Th(HPO_4)_2$ (var.)	21
ThF_4 (var.)	25.3	$Th_3(PO_4)_4$ (var.)	78.6(?)
		Tl(I)	
TlBr	5.47	$Tl_4Fe(CN)_6$	10.2
$TlBrO_3$ (dil.)	4.1	TlI	7.19
TlCl	3.73	$TlIO_3$	5.5
Tl_2CrO_4	12.0	TlSCN	3.77
		Tl(III)	
Tl oxinate	32.4		
		U	
UF_4 (var.)	21.24	$U(HPO_4)_2$	27.5
UO_2HAsO_4	10.5	UO_2HPO_4 (var.)	10.7
UO_2KAsO_4	22.6	UO_2KPO_4 (var.)	23.1
UO_2LiAsO_4	18.8	$UO_2NH_4PO_4$ (var.)	26.36
$UO_2NH_4AsO_4$	23.8	$(UO_2)_3(PO_4)_2$ (dil.)	49.1
UO_2NaAsO_4	21.9	UO_2SO_3 (var.)	8.6
$(UO_2)_2Fe(CN)_6$ (var.)	13.2		
		V	
$(VO)_3(PO_4)_2$	24.13		
		Y	
$Y_2(C_2O_4)_3$ (var.)	28.3		
		Zn	
Zn anthranilate	9.75	$ZnHg(SCN)_4$ (dil.)	6.66
$Zn_3(AsO_4)_2$ (μ = 0.1)	27.8	$Zn(IO_3)_2$	5.4
$Zn(BO_2)_2$	10.2	Zn oxinate	24.3
$Zn(CN)_2$	12.6	$Zn_3(PO_4)_2$ (var.)	32
$ZnCO_3$	10.7	Zn quinaldate	13.8
ZnC_2O_4	8.9	$ZnSeO_3$	6.6
$Zn_2Fe(CN)_6$ (var.)	15.4		

[a] Mostly from J. Bjerrum, G. Schwarzenbach, and L. G. Sillén, *Stability Constants*, Chemical Society, London 1957–1958. Possible hydration of solid phases not indicated.

[b] Room temperature, usually 25 or 20°C; μ = 0 unless stated otherwise; var., variable salt concentration; dil., dilute.

TABLE A-10. Solubility-Product and Related Constants of Metal Hydroxides or Oxides[a]

Ks_n constants refer to equilibria of solid hydroxide (or oxide) with OH^-, n indicating number of OH in dissolved species, e.g.:

$$M(OH)_2 \rightleftharpoons MOH^+ + OH^- \qquad Ks_1 = [MOH^+][OH^-]$$

$$M(OH)_2 \rightleftharpoons M(OH)_2 \qquad Ks_2 = [M(OH)_2]$$

$$M(OH)_2 + OH^- \rightleftharpoons M(OH)_3^- \qquad Ks_3 = \frac{[M(OH)_3^-]}{[OH^-]}$$

A two-digit subscript refers to polynuclear hydroxyl complexes, e.g.:

$$3\,M(OH)_2 \rightleftharpoons M_3(OH)_3^{3+} + 3\,OH^- \qquad Ks_{33} = [M_3(OH)_3^{3+}][OH^-]^3$$

The negative logarithms of the constants are given below. Temperature is 25°C unless otherwise indicated. Abbreviations: amorph., amorphous; act., active; inact., inactive; dil., dilute; var., variable; R.T., room temperature.

Solid Phase	μ	pK_{sp}	pKs_1	pKs_2	pKs_3	pKs_4	*Other*
Ag_2O	0	**7.71**	**5.7**	**3.72**			
AgO	0	31		5.7	3.7		
$Al(OH)_3$, amorph.	0	31.2				−1.3	
γ-AlOOH (boehmite)	0	31.7				−0.8	
γ-$Al(OH)_3$ (gibbsite)	0	33.0				0.5	
α-$Al(OH)_3$ (bayerite)	0	33.8				1.3	
$Au(OH)_3$	var.	46 (22°C)					
	0					3.3	2.6
$Be(OH)_2$, amorph.	0	**20.8**			**2.2**	**2.4**	pKs_{33} **29.3**
α-$Be(OH)_2$	0	**21.1**			**2.5**	**2.7**	pKs_{33} **30.3**
β-$Be(OH)_2$	0	**21.5**			**2.9**	**3.1**	pKs_{33} **31.4**
Bi_2O_3	0.1					5.3	
$Bi(OH)_2Cl$	0	30.75					
$Ca(OH)_2$	0	**5.4_3**	**4.0_3**				
β-$Cd(OH)_2$ (act.)	0	**13.7**					
β-$Cd(OH)_2$ (inact.)	0	**14.4**			**4.7**	**5.1**	
CdOHCl	0	**10.5**					
$Ce(OH)_3$ (act.)	var.	20.1					
$CeO_2 \cdot xH_2O$	var.	24.4	($K_{sp} = [CeO^{++}][OH^-]^2$)				
$Co(OH)_2$, blue, R.T.	0	**14.2** (R.T.)			**3.6**		
$Co(OH)_2$, pink (act.)	0	**14.8** (R.T.)			**4.2**		
$Co(OH)_2$, pink (inact.)	0	**15.7**			**5.1**		
$Co(OH)_3$	0	44.5					
$Cr(OH)_2$	0	17					
$Cr(OH)_3$, amorph. (act.)	var.	30				0.4 (20°C)	
Cu_2O	0	**14.7**					
$Cu(OH)_2$ (inact.)	0	**18.8**			**3.6**	**2.7**	pKs_{22} **20.6**
CuO (act.)	0	**19.7**			**4.5**	**3.6**	pKs_{22} **22.2**
CuO (inact.)	0	**20.5**			**5.3**	**4.4**	pKs_{22} **23.8**
$Dy(OH)_3$ (act.)	var.	23.1					
$Er(OH)_3$ (act.)	var.	23.3					

TABLE **A-10**. **Solubility-Product and Related Constants of Metal Hydroxides or Oxides**[a] *(continued)*

Solid Phase	μ	pK_{sp}	pKs_1	pKs_2	pKs_3	pKs_4	*Other*
$Eu(OH)_3$ (act.)	var.	22.5					
$Fe(OH)_2$ (act.)	dil.	**14**			**4**	**4.5**	
$Fe(OH)_2$ (inact.)	0	**15.1**			**5.1**	**5.5**	
$Fe(OH)_3$, amorph. (act.)	3	**38.7**	**27.5**	**16.6**		4.5	pKs_{22} **51.9**
$Fe(OH)_3$, amorph. (inact.)	3	**39.1**	**27.9**	**17.0**		**4.9**	pKs_{33} 52.7
α-Fe_2O_3	0	42.7					
$Ga(OH)_3$, amorph.	0	**37**				**−2.4**	
GaOOH	0	**39.1**				**−0.3**	
$HfO_2 \cdot xH_2O$	var.	25.4	($K_{sp} = [HfO^{++}][OH^-]^2$)				
$Hg_2(OH)_2$	0	23.7					
HgO (red)	0	**25.4**	**15.2**	**3.6**	**4.5**		
$Ho(OH)_3$ (act.)	var.	23.2					
$In(OH)_3$	0	**36.9**					
In_2O_3	0	**35.9**					
$La(OH)_3$ (act.)	var.	18.9					
$La(OH)_3$ (inact.) (18°C)	var.	21.0					
$Lu(OH)_3$ (act.)	var.	23.7					
$Mg(OH)_2$ (act.)	0	9.2					
$Mg(OH)_2$ (inact.)	0	10.9	8.3				
$Mn(OH)_2$	0	**12.8**			**5.0**		
$Nd(OH)_3$ (act.)	var.	21.5					
$Nd(OH)_3$ (inact.)	0	**23.1**					
$Ni(OH)_2$ (act.), R.T.	0	**14.7**			**1.7**		
$Ni(OH)_2$ (inact.)	0	**17.2**			**4.2**		
$Pb_2O(OH)_2$	0	**14.9**		**4.0**	**1.0**		
PbO, yellow	0	**15.1**		**4.2**	**1.2**		
PbO, red	0	**15.3**		**4.4**	**1.4**		
β-PbO_2	0	**64**					pKs_6 **4.5**
$Pd(OH)_2$		24?					
$Pm(OH)_3$ (act.)	var.	21.8					
$Pr(OH)_3$ (act.)	var.	21.1					
$Ru(OH)_3$	dil.	36					
$Sb(OH)_3$	var.	41.4			4.7		
$Sc(OH)_3$ (act.)	var.	33					
$Sm(OH)_3$ (act.)	var.	22.1					
SnO	0	**26.2**		**5.3**	**0.9**		
SnO_2, amorph. (20°C)	var.	58					
SnO_2 (act.) (20°C)	var.	61					
SnO_2 (inact.)	0	64.5					
$Tb(OH)_3$ (act.)	var.	22.9					
$ThO_2 \cdot xH_2O$, amorph.	var.	23.3	($K_{sp} = [ThO^{++}][OH^-]^2$)				
$TiO_2 \cdot xH_2O$, amorph.	0	29	($K_{sp} = [TiO^{++}][OH^-]^2$)				
Tl_2O_3	3	**45.0**	**31.9**	**19.2**			
$Tm(OH)_3$ (act.)	var.	23.5					
$UO_2(OH)_2$	0	22					
$Y(OH)_3$ (act.)	var.	22.8					
$Yb(OH)_3$ (act.)	var.	23.6					
$Zn(OH)_2$, amorph.	0	**15.50**			**1.4**	**0.2**	
β_1-$Zn(OH)_3$	0	**16.24**			**2.14**	**0.9**	

TABLE A-10. **Solubility-Product and Related Constants of Metal Hydroxides or Oxides**[a] *(continued)*

Solid Phase	μ	pK_{sp}	pKs_1	pKs_2	pKs_3	pKs_4	*Other*
β_2-$Zn(OH)_2$	0	**16.20**					
γ-$Zn(OH)_2$	0	**16.26**					
δ-$Zn(OH)_2$	0	**16.15**					
ϵ-$Zn(OH)_2$	0	**16.47**					
ZnO (act.)		**16.66**			**2.56**	**1.4**	
ZnO (inact.)		**16.83**			**2.73**	**1.5**	
$ZrO_2 \cdot xH_2O$, amorph.	var.	25.5	($K_{sp} = [ZrO^{++}][OH^-]^2$)				

[a] Most of these constants are taken from the IUPAC report by W. Feitknecht and P. Schindler in *Pure Appl. Chem.* **6**, 130 (1963). Values recommended by them are printed in boldface. Constants for basic salts given in the report are not included here.

TABLE **A-11. Solubility Products of Sulfides**[a]
(μ = 0, 25°C)

	K_{sp}	pK_{sp}
Ag_2S	6×10^{-50}	49.2
Bi_2S_3	1×10^{-97}	97
CdS	8×10^{-27}	26.1
CoS, α	4×10^{-21}	20.4
CoS, β	2×10^{-25}	24.7
Cu_2S	2.5×10^{-48}	47.6
CuS	6×10^{-36}	35.2
FeS	6×10^{-18}	17.2
HgS, black	1.6×10^{-52}	51.8
HgS, red	4×10^{-53}	52.4
In_2S_3	6×10^{-74}	73.2
MnS, pink (amorphous)	3×10^{-10}	9.6
MnS, green (crystalline)	3×10^{-13}	12.6
NiS, α	3×10^{-19}	18.5
NiS, β	1×10^{-24}	24.0
NiS, γ	2×10^{-26}	25.7
PbS	2.5×10^{-27}	26.6
PoS	5.5×10^{-29}	28.3
Sb_2S_3	1.5×10^{-93}	92.8
SnS	1×10^{-25}	25
SnS_2	2.5×10^{-27}	26.6
Tl_2S	5×10^{-21}	20.3
ZnS, sphalerite	1.6×10^{-24}	23.8
ZnS, wurtzite	5×10^{-25}	24.3

[a] Almost all from A. Ringbom, *Solubilities of Sulfides*, Report to Analytical Section, IUPAC, 1953. Values based on $[H^+]^2[S^{--}]/[H_2S] = 1.1 \times 10^{-20}$.

TABLE **A-12. Standard Reduction Potentials**[a]

Electrode Reaction	E^0, V
$F_2 + 2e \rightarrow 2\,F^-$	2.87
$H_2O_2 + 2\,H^+ + 2e \rightarrow 2\,H_2O$	1.77
$MnO_4^- + 4\,H^+ + 3e \rightarrow MnO_2 + 2\,H_2O$	1.69
$2\,BrO_3^- + 12\,H^+ + 10e \rightarrow Br_2 + 6\,H_2O$	1.52
$MnO_4^- + 8\,H^+ + 5e \rightarrow Mn^{++} + 4\,H_2O$	1.51
$Au^{3+} + 3e \rightarrow Au$	1.5
$Cl_2 + 2e \rightarrow 2\,Cl^-$	1.359
$Cr_2O_7^{--} + 14\,H^+ + 6e \rightarrow 2\,Cr^{3+} + 7\,H_2O$	1.33
$Tl^{3+} + 2e \rightarrow Tl^+$	1.28
$MnO_2 + 4\,H^+ + 2e \rightarrow Mn^{++} + 2\,H_2O$	1.23
$O_2 + 4\,H^+ + 4e \rightarrow 2\,H_2O$	1.229
$2\,IO_3^- + 12\,H^+ + 10e \rightarrow I_2 + 6\,H_2O$	1.19
$Br_2 + 2e \rightarrow 2\,Br^-$	1.087
$VO_2^+ + 2\,H^+ + e \rightarrow VO^{++} + H_2O$	0.9994
$Pb^{++} + 2e \rightarrow Pb$	0.987
$2\,Hg^{++} + 2e \rightarrow Hg_2^{++}$	0.907
$Ag^+ + e \rightarrow Ag$	0.7994
$Hg_2^{++} + 2e \rightarrow 2\,Hg$	0.792
$Fe^{3+} + e \rightarrow Fe^{++}$	0.771
$MnO_4^- + e \rightarrow MnO_4^{--}$	0.56
$H_3AsO_4 + 2\,H^+ + 2e \rightarrow HAsO_2 + 2\,H_2O$	0.559
$I_2(s) + 2e \rightarrow 2\,I^-$	0.5345
$I_3^- + 2e \rightarrow 3\,I^-$	0.5355
$Fe(CN)_6^{3-} + e \rightarrow Fe(CN)_6^{4-}$	0.356
$BiO^+ + 2\,H^+ + 3e \rightarrow Bi + H_2O$	0.32
$PbO_2 + H_2O + 2e \rightarrow PbO + 2\,OH^-$	0.28
$Hg_2Cl_2 + 2e \rightarrow 2\,Hg + 2\,Cl^-$	0.2680
$HAsO_2 + 3\,H^+ + 3e \rightarrow As + 2\,H_2O$	0.248
$AgCl + e \rightarrow Ag + Cl^-$	0.2224
$SbO^+ + 2\,H^+ + 3e \rightarrow Sb + H_2O$	0.212
$BiOCl + 2\,H^+ + 3e \rightarrow Bi + H_2O + Cl^-$	0.16
$Hg_2Br_2 + 2e \rightarrow 2\,Hg + 2\,Br^-$	0.1392
$AgBr + e \rightarrow Ag + Br^-$	0.071
$Hg_2I_2 + 2e \rightarrow 2\,Hg + 2\,I^-$	−0.040
$Pb^{++} + 2e \rightarrow Pb$	−0.126
$AgI + e \rightarrow Ag + I^-$	−0.152
$Sn^{++} + 2e \rightarrow Sn$	−0.140
$Ni^{++} + 2e \rightarrow Ni$	−0.23
$V^{3+} + e \rightarrow V^{++}$	−0.255
$Co^{++} + 2e \rightarrow Co$	−0.28
$Tl^+ + e \rightarrow Tl$	−0.336
$PbSO_4 + 2e \rightarrow Pb + SO_4^{--}$	−0.3563
$Ti^{3+} + e \rightarrow Ti^{++}$	−0.37
$Cd^{++} + 2e \rightarrow Cd$	−0.402
$Cr^{3+} + e \rightarrow Cr^{++}$	−0.41
$Fe^{++} + 2e \rightarrow Fe$	−0.440
$S + 2e \rightarrow S^{--}$	−0.48
$Cr^{++} + 2e \rightarrow Cr$	−0.56
$Zn^{++} + 2e \rightarrow Zn$	−0.7628
$Mg^{++} + 2e \rightarrow Mg$	−2.37

TABLE **A-12. Standard Reduction Potentials**[a] *(continued)*

Electrode Reaction	E^0, V
$Na^+ + e \rightarrow Na$	−2.713
$Ca^{++} + 2e \rightarrow Ca$	−2.87
$Ba^{++} + 2e \rightarrow Ba$	−2.90
$K^+ + e \rightarrow K$	−2.925
$Li^+ + e \rightarrow Li$	−3.02

[a] Largely from *Handbook of Analytical Chemistry* (L. Meites, ed.), McGraw-Hill, New York, 1963, Table 5.1

TABLE **A-13. Some Formal Reduction Potentials**[a]

Electrode Reaction	E^f	*Medium*
$Ag(II) + e \rightarrow Ag^+$	1.93	4 N HNO_3
$Ce(IV) + e \rightarrow Ce(III)$	1.70	1 N $HClO_4$
	1.61	1 N HNO_3
	1.44	1 N H_2SO_4
	1.28	1 N HCl
$Co^{3+} + e \rightarrow Co^{++}$	1.85	4 N HNO_3
$Co(en)_3^{3+} + e \rightarrow Co(en)_3^{++}$	−0.2	0.1 N KNO_3 + 0.1 M en
$Cr(III) + e \rightarrow Cr(II)$	−0.40	5 N HCl
$Cr_2O_7^{--} + 14\,H^+ + 6e \rightarrow 2\,(Cr^{3+}) + 7\,H_2O$	1.00	1 N HCl
	1.11	4 N H_2SO_4
$Fe(III) + e \rightarrow Fe(II)$	0.73	1 N $HClO_4$
	0.71	0.5 N HCl
	0.68	1 M H_2SO_4
	0.46	2 M H_3PO_4
$Fe(CN)_6^{3-} + e \rightarrow Fe(CN)_4^{4-}$	0.56	0.1 N HCl
$I_3^- + 2e \rightarrow 3\,I^-$	0.545	1 N H^+
$MnO_4^- + 8\,H^+ + 5e \rightarrow Mn^{++} + 4\,H_2O$	1.45	1 N $HClO_4$
$Os(VIII) + 4e \rightarrow Os(IV)$	0.79	5 N HCl
$Sn(IV) + 2e \rightarrow Sn(II)$	0.14	1 N HCl
$Sb(V) + 2e \rightarrow Sb(III)$	0.75	3.5 N HCl
$Ti(IV) + e \rightarrow Ti(III)$	−0.01	0.4 N H_2SO_4
	0.12	4 N H_2SO_4
	−0.04	1 N HCl
	−0.05	1 M H_3PO_4
$UO_2^{++} + 4\,H^+ + 2e \rightarrow U(IV) + 2\,H_2O$	0.41	1 N H_2SO_4

[a] Largely from *Handbook of Analytical Chemistry* (L. Meites, ed.), McGraw-Hill, New York, 1963, Table 5.1.

TABLE A-14. Ionic Equivalent Conductances at Infinite Dilution at 25°C[a]

Cations	λ^0_+	*Anions*	λ^0_-
H^+	349.8	OH^-	198.6
Li^+	38.7	F^-	55.4
Na^+	50.10	Cl^-	76.35
K^+	73.50	Br^-	78.14
Rb^+	77.8	I^-	76.8
Cs^+	77.3	N_3^-	69
Ag^+	61.9	NO_3^-	71.46
Tl^+	74.7	ClO_3^-	64.6
NH_4^+	73.6	BrO_3^-	55.7
Ca^{++}	59.50	ClO_4^-	67.4
Sr^{++}	59.5	IO_4^-	54.6
Ba^{++}	63.6	HCO_3^-	44.5
Cu^{++}	56.6	Acetate	40.9
Zn^{++}	52.8	SO_4^{--}	80.0
Co^{++}	55	$C_2O_4^{--}$	74.1
La^{3+}	69.7	CO_3^{--}	69.3
Ce^{3+}	69.8	$Fe(CN)_6^{3-}$	100.9
		$Fe(CN)_6^{4-}$	110.1

[a] From R. A. Robinson and R. H. Stokes, *Electrolyte Solutions*, Butterworth, London, 1955, app. 6.1.

TABLE **A-15. pH of Clark and Lubs Buffer Solutions at 25°C[a]**

p_aH	x
25 ml of 0.2 *M* KCl + x ml of 0.2 *M* HCl, diluted to 100 ml	
1.00	67.0
1.20	42.5
1.40	26.6
1.60	16.2
1.80	10.2
2.00	6.5
2.20	3.9
50 ml of 0.1 *M* $KHC_8H_4O_4$ (potassium hydrogen phthalate) + x ml of 0.1 *M* HCl, diluted to 100 ml	
2.20	49.5
2.40	42.2
2.60	35.4
2.80	28.9
3.00	22.3
3.20	15.7
3.40	10.4
3.60	6.3
3.80	2.9
4.00	0.1

TABLE A-15. **pH of Clark and Lubs Buffer Solutions at 25°C**[a] (*continued*)

p_aH	x
50 ml of 0.1 *M* $KHC_8H_4O_4$ + x ml of 0.1 *M* NaOH, diluted to 100 ml	
4.20	3.0
4.40	6.6
4.60	11.1
4.80	16.5
5.00	22.6
5.20	28.8
5.40	34.1
5.60	38.8
5.80	42.3
50 ml of 0.1 *M* KH_2PO_4 + x ml of 0.1 *M* NaOH, diluted to 100 ml	
5.80	3.6
6.00	5.6
6.20	8.1
6.40	11.6
6.60	16.4
6.80	22.4
7.00	29.1
7.20	34.7
7.40	39.1
7.60	42.8
7.80	45.3
8.00	46.7
50 ml of soln. 0.1 *M* with respect to both H_3BO_3 and KCl + x ml of 0.1 *M* NaOH, diluted to 100 ml	
8.00	3.9
8.20	6.0
8.40	8.6
8.60	11.8
8.80	15.8
9.00	20.8
9.20	26.4
9.40	32.1
9.60	36.9
9.80	40.6
10.00	43.7
10.20	46.2

[a] From V. E. Bower and R. G. Bates, *J. Res. Natl. Bur. Std.* **55**, 197 (1955). Values of x for pH 7.60, 7.80 and 8.00 buffers revised September 1965.

TABLE **A-16. pH, Composition, and Ionic Strength of Alkaline Buffer Solutions at 25°C**[a]

p_aH	*Composition*	*Ionic Strength*
50 ml of 0.1 *M* tris(hydroxymethyl)aminomethane + *x* ml of 0.1 *M* HCl, diluted to 100 ml		
7.00	46.6	0.047
7.20	44.7	0.045
7.40	42.0	0.042
7.60	38.5	0.039
7.80	34.5	0.035
8.00	29.2	0.029
8.20	22.9	0.023
8.40	17.2	0.017
8.60	12.4	0.012
8.80	8.5	0.009
9.00	5.7	0.006
50 ml of 0.025 *M* $Na_2B_4O_7$ + *x* ml of 0.1 *M* HCl diluted to 100 ml		
8.00	20.5	~0.025
8.20	18.8	~0.025
8.40	16.6	~0.025
8.60	13.5	~0.025
8.80	9.4	~0.025
9.00	4.6	~0.025
50 ml of 0.025 *M* $Na_2B_4O_7$ + *x* ml of 0.1 *M* NaOH, diluted to 100 ml		
9.20	0.9	0.026
9.40	6.2	0.031
9.60	11.1	0.036
9.80	15.0	0.040
10.00	18.3	0.043
10.20	20.5	0.046
10.40	22.1	0.047
10.60	23.3	0.048
10.80	24.25	0.049
50 ml of 0.05 *M* $NaHCO_3$ + *x* ml of 0.1 *M* NaOH, diluted to 100 ml		
9.60	5.0	0.035
9.80	7.6	0.040
10.00	10.7	0.046
10.20	13.8	0.053
10.40	16.5	0.058
10.60	19.1	0.063
10.80	21.2	0.067
11.00	22.7	0.070
50 ml of 0.05 *M* Na_2HPO_4 + *x* ml of 0.1 *M* NaOH, diluted to 100 ml		
11.00	4.1	0.085
11.20	6.3	0.090
11.40	9.1	0.095
11.60	13.5	0.104
11.80	19.4	0.116
12.00	26.9	0.131

TABLE **A-16. pH, Composition, and Ionic Strength of Alkaline Buffer Solutions at 25°C**[a] *(continued)*

p_aH	*Composition*	*Ionic Strength*
25 ml of 0.2 *M* KCl + *x* ml of 0.2 *M* NaOH, diluted to 100 ml		
12.00	6.0	0.062
12.20	10.2	0.070
12.40	16.2	0.082
12.60	25.6	0.101
12.80	41.2	0.132
13.00	66.0	0.182

[a] From R. G. Bates and V. E. Bower, *Anal. Chem.* **28**, 1322 (1956).

TABLE **A-17. Density of Water**[a]
(Under standard pressure of 760 mm Hg)

Temp., °C	*Density*, g/ml	*Temp.* °C	*Density*, g/ml
15	0.99913	26	0.99682
16	0.99897	27	0.99655
17	0.99880	28	0.99627
18	0.99862	29	0.99598
19	0.99843	30	0.99568
20	0.99823	31	0.99537
21	0.99802	32	0.99506
22	0.99780	33	0.99473
23	0.99757	34	0.99440
24	0.99733	35	0.99406
25	0.99708		

[a] From P. Chappuis, *Travaux et mémoires du Bureau International des Poids et Mesures*, Paris, vol. XIII, 1907. To obtain the density expressed as grams per cubic centimeter (absolute density), subtract 0.00003 from the density values in the table.

TABLE **A-18. Approximate Densities and Molarities of Concentrated Acids and Ammonia**

	wt. %	*Density*, g/ml	*Molarity*
Acetic acid	99.5	1.05	17.4
Ammonium hydroxide	27 (NH_3)	0.90	14.3
Hydrochloric acid	37	1.18	12.0
Hydrofluoric acid	48	1.15	27.6
Nitric acid	70	1.41	15.7
Perchloric acid	70	1.66	11.6
Phosphoric acid	85	1.69	14.7
Sulfuric acid	95.5	1.83	17.8

TABLE **A-19. Densities of Acids and Ammonia at 20°C**

	Density, g/ml					
wt. %	CH_3COOH	HCl	HNO_3	H_3PO_4	H_2SO_4	NH_3
5	1.0055	1.023	1.026	1.025	1.032	0.977
10	1.0125	1.047	1.054	1.053	1.066	0.958
15	1.0195	1.073	1.084	1.082	1.102	0.940
20	1.026	1.098	1.115	1.113	1.139	0.923
25	1.033	1.124	1.147	1.146	1.178	0.907
30	1.038	1.149	1.180	1.181	1.219	0.892
35	1.044	1.174	1.214	1.216	1.260	
40	1.049	1.198	1.246	1.254	1.303	
45	1.053		1.278	1.293	1.348	
50	1.058		1.310	1.335	1.395	
55	1.061		1.339	1.379	1.445	
60	1.064		1.367	1.426	1.498	
65	1.067		1.391	1.475	1.553	
70	1.069		1.413	1.526	1.611	
75	1.070		1.434	1.579	1.669	
80	1.070		1.452	1.633	1.727	
85	1.069		1.469	1.689	1.779	
90	1.066		1.483	1.746	1.814	
95	1.061		1.493	1.807	1.834	
100	1.050		1.513	1.870	1.831	

TABLE **A-20. Recommended Unit Prefixes**[a]

Multiples and Submultiples	*Prefixes*	*Symbols*
10^{12}	tera	T
10^{9}	giga	G
10^{6}	mega	M
10^{3}	kilo	k
10^{2}	hecto	h
10	deka	da
10^{-1}	deci	d
10^{-2}	centi	c
10^{-3}	milli	m
10^{-6}	micro	μ
10^{-9}	nano	n
10^{-12}	pico	p
10^{-15}	femto	f
10^{-18}	atto	a

[a] As adopted by International Committee on Weights and Measures. Reprinted from *Natl. Bur. Std. Tech. News Bull.* Feb. 1963. Also see *Natl. Bur. Std. Misc. Publ. 232,* 1961.

Index

FOUR-PLACE LOGARITHMS TO BASE 10

No.	0	1	2	3	4	5	6	7	8	9	1	2	3	4	5	6	7	8	9
10	0000	0043	0086	0128	0170	0212	0253	0294	0334	0374	4	8	12	17	21	25	29	33	37
11	0414	0453	0492	0531	0569	0607	0645	0682	0719	0755	4	8	11	15	19	23	26	30	34
12	0792	0828	0864	0899	0934	0969	1004	1038	1072	1106	3	7	10	14	17	21	24	28	31
13	1139	1173	1206	1239	1271	1303	1335	1367	1399	1430	3	6	10	13	16	19	23	26	29
14	1461	1492	1523	1553	1584	1614	1644	1673	1703	1732	3	6	9	12	15	18	21	24	27
15	1761	1790	1818	1847	1875	1903	1931	1959	1987	2014	3	6	8	11	14	17	20	22	25
16	2041	2068	2095	2122	2148	2175	2201	2227	2253	2279	3	5	8	11	13	16	18	21	24
17	2304	2330	2355	2380	2405	2430	2455	2480	2504	2529	2	5	7	10	12	15	17	20	22
18	2553	2577	2601	2625	2648	2672	2695	2718	2742	2765	2	5	7	9	12	14	16	19	21
19	2788	2810	2833	2856	2878	2900	2923	2945	2967	2989	2	4	7	9	11	13	16	18	20
20	3010	3032	3054	3075	3096	3118	3139	3160	3181	3201	2	4	6	8	10	13	15	17	19
21	3222	3243	3263	3284	3304	3324	3345	3365	3385	3404	2	4	6	8	11	12	14	16	18
22	3424	3444	3464	3483	3502	3522	3541	3560	3579	3598	2	4	6	8	10	12	14	15	17
23	3617	3636	3655	3674	3692	3711	3729	3747	3766	3784	2	4	6	7	9	11	13	15	17
24	3802	3820	3838	3856	3874	3892	3909	3927	3945	3962	2	4	5	7	9	11	12	14	16
25	3979	3997	4014	4031	4048	4065	4082	4099	4116	4133	2	3	5	7	9	10	12	14	15
26	4150	4166	4183	4200	4216	4232	4249	4265	4281	4298	2	3	5	7	8	10	11	13	15
27	4314	4330	4346	4362	4378	4393	4409	4425	4440	4456	2	3	5	6	8	9	11	13	14
28	4472	4487	4502	4518	4533	4548	4564	4579	4594	4609	2	3	5	6	8	9	11	12	14
29	4624	4639	4654	4669	4683	4698	4713	4728	4742	4757	1	3	4	6	7	9	10	12	13
30	4771	4786	4800	4814	4829	4843	4857	4871	4886	4900	1	3	4	6	7	9	10	11	13
31	4914	4928	4942	4955	4969	4983	4997	5011	5024	5038	1	3	4	6	7	8	10	11	12
32	5051	5065	5079	5092	5105	5119	5132	5145	5159	5172	1	3	4	5	7	8	9	11	12
33	5185	5198	5211	5224	5237	5250	5263	5276	5289	5302	1	3	4	5	6	8	9	10	12
34	5315	5328	5340	5353	5366	5378	5391	5403	5416	5428	1	3	4	5	6	8	9	10	11
35	5441	5453	5465	5478	5490	5502	5514	5527	5539	5551	1	2	4	5	6	7	9	10	11
36	5563	5575	5587	5599	5611	5623	5635	5647	5658	5670	1	2	4	5	6	7	8	10	11
37	5682	5694	5705	5717	5729	5740	5752	5763	5775	5786	1	2	3	5	6	7	8	9	10
38	5798	5809	5821	5832	5843	5855	5866	5877	5888	5899	1	2	3	5	6	7	8	9	10
39	5911	5922	5933	5944	5955	5966	5977	5988	5999	6010	1	2	3	4	5	7	8	9	10
40	6021	6031	6042	6053	6064	6075	6085	6096	6107	6117	1	2	3	4	5	6	8	9	10
41	6128	6138	6149	6160	6170	6180	6191	6201	6212	6222	1	2	3	4	5	6	7	8	9
42	6232	6243	6253	6263	6274	6284	6294	6304	6314	6325	1	2	3	4	5	6	7	8	9
43	6335	6345	6355	6365	6375	6386	6395	6405	6415	6425	1	2	3	4	5	6	7	8	9
44	6435	6444	6454	6464	6474	6484	6493	6503	6513	6522	1	2	3	4	5	6	7	8	9
45	6532	6542	6551	6561	6571	6580	6590	6599	6609	6618	1	2	3	4	5	6	7	8	9
46	6628	6637	6646	6656	6665	6675	6684	6693	6702	6712	1	2	3	4	5	6	7	7	8
47	6721	6730	6739	6749	6758	6767	6776	6785	6794	6803	1	2	3	4	5	5	6	7	8
48	6812	6821	6830	6839	6848	6857	6866	6875	6884	6893	1	2	3	4	4	5	6	7	8
49	6902	6911	6920	6928	6937	6946	6955	6964	6972	6981	1	2	3	4	4	5	6	7	8
50	6990	6998	7007	7016	7024	7033	7042	7050	7059	7067	1	2	3	3	4	5	6	7	8
51	7076	7084	7093	7101	7110	7118	7126	7135	7143	7152	1	2	3	3	4	5	6	7	8
52	7160	7168	7177	7185	7193	7202	7210	7218	7226	7235	1	2	2	3	4	5	6	7	7
53	7243	7251	7259	7267	7275	7284	7292	7300	7308	7316	1	2	2	3	4	5	6	6	7
54	7324	7332	7340	7348	7356	7364	7372	7380	7388	7396	1	2	2	3	4	5	6	6	7
	0	**1**	**2**	**3**	**4**	**5**	**6**	**7**	**8**	**9**	**1**	**2**	**3**	**4**	**5**	**6**	**7**	**8**	**9**